International Structural Engineering and Construction Conference (ISEC-10)

Interdependence of Structural Engineering and Construction Management

Chicago, Illinois, USA
20 - 25 May 2019

Volume 1 of 2

ISBN: 978-1-5108-8867-8

Printed from e-media with permission by:

Curran Associates, Inc.
57 Morehouse Lane
Red Hook, NY 12571

Some format issues inherent in the e-media version may also appear in this print version.

Printed by Curran Associates, Inc. (2019)

ISEC PRESS
PO Box 1334
Fargo, North Dakota 58107
USA

isec.sec@gmail.com

Additional copies of this publication are available from:

Curran Associates, Inc.
57 Morehouse Lane
Red Hook, NY 12571 USA
Phone: 845-758-0400
Fax: 845-758-2633
Email: curran@proceedings.com
Web: www.proceedings.com

TABLE OF CONTENTS

VOLUME 1

CHAPTER 1: BIM IN CONSTRUCTION

CHAPTER 2: GREEN INFRASTRUCTURE DESIGN

CHAPTER 3: INNOVATIVE HOUSING TECHNOLOGIES

CHAPTER 4: MANUFACTURING SYSTEMS FOR CONSTRUCTION

CHAPTER 5: DECISION MAKING AND NEURAL NETWORK MODELS IN CONSTRUCTION

CHAPTER 6: REHABILITATION AND RETROFITTING METHODS – PART 1

CHAPTER 7: REHABILITATION AND RETROFITTING METHODS – PART II

CHAPTER 8: RETROFIT OF COMPOSITE AND METAL STRUCTURES

CHAPTER 9: MODELING IN CONSTRUCTION MANAGEMENT

CHAPTER 10: PROJECT MANAGEMENT

CHAPTER 11: LAW AND DISPUTE RESOLUTION

CHAPTER 12: COST AND DURATION OPTIMIZATION IN PROJECT MANAGEMENT

CHAPTER 13: ALTERNATIVE MATERIALS FOR CONCRETE

VOLUME 2

CHAPTER 14: CONCRETE DESIGN

CHAPTER 15: GEOPOLYMER CONCRETE

CHAPTER 16: DISASTER PREVENTION STRATEGIES

CHAPTER 17: EDUCATIONAL COMPONENT AND PROFESSIONAL ETHICS IN CIVIL ENGINEERING

CHAPTER 18: BRIDGES

CHAPTER 19: REINFORCED CONCRETE STRUCTURES

CHAPTER 20: STEEL STRUCTURES PART I

CHAPTER 21: STEEL STRUCTURES PART II

CHAPTER 22: DYNAMIC RESPONSE OF STRUCTURES

CHAPTER 23: COMPOSITE SYSTEMS

CHAPTER 24: STRUCTURAL MEASUREMENTS AND COMPUTATIONAL METHODS

CHAPTER 25: FOUNDATION AND PAVEMENT DESIGN

Interdependence between Structural Engineering and Construction Management
Edited by Ozevin, D., Ataei, H., Modares, M., Gurgun, A., Yazdani, S., and Singh, A.

ISBN: 978-0-9960437-6-2

BUILDING INFORMATION MODELING AND OPERATION

ANICA MEINS-BECKER, AGNES KELM, MATTHIAS KAUFHOLD, MELANIE QUESSEL, and MANFRED HELMUS

Construction Management and Economics, University of Wuppertal, Wuppertal, Germany

The research project "BIM-based operation" contributes standardization and uniform understanding of the BIM process and the resulting changes in the construction and real estate sectors. The project develops an ideal-type target process chain, based on an actual process analysis, for using the BIM method in real estate operations. A standardized life-cycle process is intent to show required information for operation actives in concrete terms. Based on this, information and communication interfaces can be identified to analyze changing performance requirements of the parties involved, and open questions such as legal issues can be further elaborated or dealt with. Based on literature research and numerous expert interviews, it analyzes and determines how individual process steps are carried out and which parties need to deliver what kind of information to which participant at what time, to fulfill build owners' requirements to a successfully constructed and managed building. The focus is thus on the exchange of information between those involved in the operation phase. Existing standards are consciously integrated into the development. Companies, institutions, and other stakeholders are thus given an overview of the situation in relation to the important exchange of information. In particular, the real estate operating phase is attributed to a high optimization potential through digitization.

Keywords: Process management, Digitalization real estate life-cycle, Information delivery, Process-orientated method, ISO 19650, Implementation guideline.

1 INTRODUCTION

The digitization of the economy has been progressing steadily for several years, also in Germany. With regard to the building sector, the main focus is on the method of Building Information Modeling. Against the background of optimizing the effectiveness and efficiency and strengthening the competitiveness of the associated construction and real estate industry of Germany - even in international comparison - the implementation of the BIM method is also essential during the entire real estate life-cycle (Bauindustrieverband NRW e.V 2017). More and more information is being generated and stored digitally to improve the processes in the respective work steps. However, they are often not designed for further use in downstream phases. In practice, companies have to put the information into their systems and adapt it to their needs. The development of standards for the BIM-based processing of projects is necessary to ensure that the methodology can also be used across the board (ARGE BIM-Leitfaden 2013).

The research project "BIM and operation" is part of a longer-term overall concept to be set up. The aim is to promote the standardization efforts of building data models with regard to processes and the associated flow of information between the project participants in all lifecycle phases with the focus on the processes of the operation lifecycle phase of a building. At the same time, this will help to create transparency with regard to the BIM method for those involved in construction and real estate management (ISO 29481 2016). To compile relevant information, the method of business process management is used. The Focus of the project is on the use of data during the lifecycle phase operation of a building. It covers a total duration of 24 months with a start in September 2017. This paper presents the BUW's process information management approach and the present status of the project BIM-bases operations.

2 METHOD

2.1 Scope and Process Model

According to the definition of the University of Wuppertal (BUW), the lifecycle of a real estate consists of five phases: development, design, realization, operation and demolition. Within these phases, new information are generated from various roles from various information sources.

The extensive consideration of all these phases is not feasible in a single research project. For this reason, the BUW has decided to subdivide the lifecycle of a real estate in a structured manner and to develop the respective sub-areas in independent research projects, which are always interlinked. Through the close networking, synergies can be used as best as possible and a broad knowledge foundation can be established. The use of business process modelling allows to interlink information in a holistic system. This enables the chair of Construction Management & Economics at the BUW to publish the gained findings in a targeted way in the general public, thereby continuing promoting the digitization of the real estate value chain.

In the following, the above-described linkage of the research projects is presented and explained for the currently running projects in each case. The division takes place on the one hand by separating the sequence of phases or activities (horizontal axis). On the other hand, by means of the classification of the detailing process levels - subdivision of a process in sub processes (vertical axis).

- BIM - Processes - Lifecycle covers the overlapping frame over all phases of the lifecycle. The processes are viewed in coarser detail levels primarily from the point of view of the overall role of the building owner / operator. (Status: completed)
- BIM - Processes - Realization considers the phase of the realization and the resulting requirements for the upstream and downstream phases of the real estate lifecycle. The processes are viewed in deeper detail from the perspective of the execution companies. (Status: completed)
- BIM - Processes - Work Planning considers the step of the work planning of small and medium-sized enterprises (SMEs) in the implementation phase. (Status: ongoing)
- BIM - Processes - Health and Safety considers the planning, realization and operation phases from the specific point of view of occupational health and safety. The processes are viewed over all process levels and integrated into the overall landscape. (Status: completed)
- BIM - based operation -considers the operation phases from the specific point of view of occupational parties. (Status: ongoing)
- BIM - Processes - Demolition considers the phase of demolition and the resulting requirements in terms of the material cycle. The processes are viewed in deeper detail from the point of view of the demolition and recycling companies. (Status: ongoing)

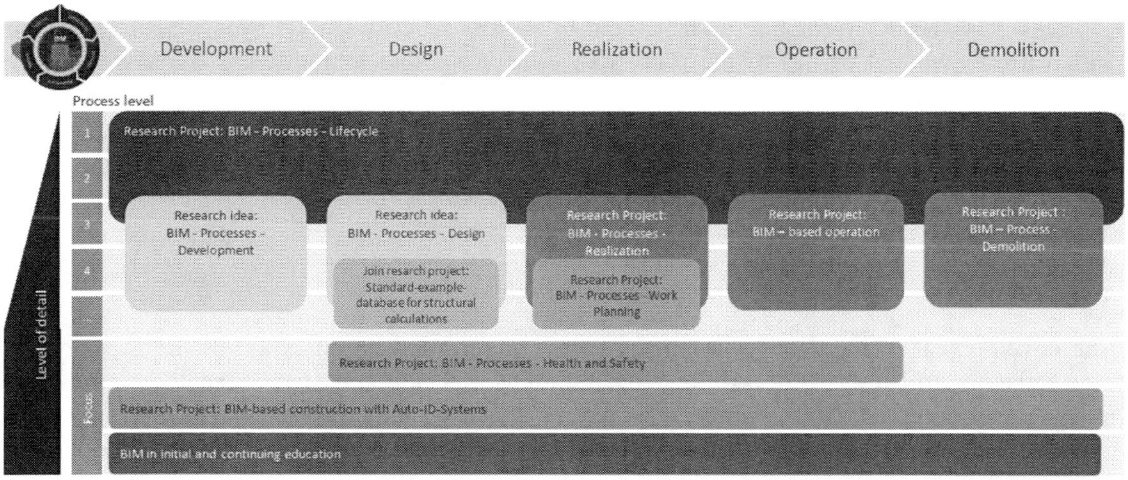

Figure 1. Integration of the BUW research projects in the real estate lifecycle.

Based on initial literature research and subsequent expert interviews, the respective research projects have brought together information from which basic principles have been developed. Based on this, the project teams use the information in a database-based BPM software. It enables a central collection of the entire knowledge and thus ensures that the described overall concept can be implemented. Finally, the gained and implemented findings are verified by the practice partners.

2.2 Ideal-Type BIM Process with the Focus on the Operation Phase

On the basis of the defined real estate lifecycle phases, processes from specific points of view were standardized and displayed in the common process model. This was used to investigate the extent to which the processes change by implementing the BIM method. Even with the actual processes, a consistent, continuous flow of information can be generated. This allows the required information to be linked to an associated process as input and the condition that each process produces exactly one output. The processes represent a human process flow (independent of auxiliaries), which will not be changed in principle for the introduction of BIM. Through the digitization, only parts of it are supported or carried out by help of the IT, thus supplemented by a technical process and requirements. The IT-supported or semi-automated handling of processes requires are more stringent regulations on storing information etc. This is because the automatisms are based on pre-programmed queries, which can in principle only cope with the situations that were taken into account during programming. So, the focus is thus on the exchange of information between those involved in the life cycle to answer the question "who needs which information from whom when and for whom", the so called information process.

Based on this the answer to the question for the ideal-type BIM (reference-) process sequence was created. In terms of construction, this is described by the so-called information management process. It describes "who has to do something that information can be generated and made loss-free available". This development is carried out by analysis of BIM guides, guidelines, standards, participation in various committees, the accompaniment of practical projects and support for the implementation of BIM projects of specialized companies. Subsequently, the information management process is assigned to the information process. The combination of both processes results in the specialized process flow of the ideal-type BIM target process chain.

The BUW BIM information management process is validity gained by different workshops with various national parties of the real estate lifecycle. The frame of the BUW information management process is also in accordance with the still not published international BIM information management process ISO 19650 to support existing standardization and regulation.

Especially for the operating phase of the life cycle process it is very interesting to define the relevant information for the processes of technical and infrastructural building management and to consider the interface of commercial building management aspects.

3 RESULTS

The development of an ideal typic target process chain for the application of the method BIM investigates the business process along the building lifecycle and environment of information management and production to achieve beneficial business outcomes to asset owners, operators and the project management. The range of business outcomes reach from e.g. better cost management or clearly defined responsibilities to improve the project management to visualizations and 3-D models for better public relations work. To clarify the questions in the information management process, the project participants must initially set up their requirements for both the information as well as the information management process. The collaborative process landscape of the BUW considers this with BIM targets, BIM uses and BIM requirements, which together result in the so-called BIM application container. The information to be entered there from the relevant viewpoints is used for the corresponding research projects. Based on this, lists can be derived as a basis for the establishment of the information management process and various other information for the most diverse tasks.

At present, the BUW process model contains information focusing on technical and infrastructural building management. For further classification, the operating phase comprises the areas allocation of operation, preparation for commissioning, commissioning, ongoing operation and change of operator/owner. Within these areas are business processes for carrying out functional technical- and infrastructural building management processes. In case of technical building management, it consists e.g. Maintenance services, inspections for special services, warranties, expert inspections, statutory inspections, invoicing and payment and warranty processes for all shown areas. Supplementary processes are further updated and validated in the exchange of technical experts. The process record is scheduled to be completed at the end of 2018.

As a result of the BUW Process map, decision points required information can be easier generated by interfaces through other lifecycle parties, e.g., planning and construction companies. Individually required information can be defined and imported into the process model as well. Within finishing the research of information requirements, exemplary business outcomes will be assigned into the categories of BIM targets, BIM uses and BIM requirements, the so-called BIM application container. The defined application containers, within in their information requirements, will be part of the tender documents, that the information can be most efficiently produced down supply chains. The guideline for action contains a step by step descriptions to give building owners, operators and project managers an understanding of the practical implementation and potential business outcomes.

4 DISCUSSION

The application of reference processes is discussed in many ways, especially in the context of individual construction projects. Against this background, it becomes even more evident how necessary the implementation and application of processes and consistent process models in the real estate life cycle are, to obtain a large number of example processes to develop further reference processes.

5 CONCLUSIONS

Over the next few years BIM will appear to be well established. In doing so, the processes as a basis for the planning of digitally supported or fully automated processes will inevitably gain in importance. To this end, standardized process models, such as the process landscape developed by the BUW, will serve as a basis for the specific situations in companies and projects to be displayed more quickly. The process landscape of the BUW is suited for this task. In addition, due to the structured structure, you can flexibly adapt and expand.

The essential basis for the implementation of BIM is to implement the process landscape as well as the will of the industry and the acceptance of the employees. An additional essential factor for a comprehensive use of the method BIM depends on the collective action of the lifecycle parties and the interests of the building owners, operators and project managers. The guideline for actions will be a part to bring the method BIM more into the focus of this important group.

Acknowledgments

This research projects are funded by the Federal Institute for Building, Construction and Spatial Development (BBSR) by the German Federal Ministry for Building, Building and Regional Planning (BBR), funded by the Forschungsinitiative ZukunftBAU of the Federal Ministry for the Environment, Nature Conservation, Construction and Nuclear Safety (BMUB)

References

ARGE BIM-Leitfaden, AEC3 & OPB 2013, BIM-Leitfaden für Deutschland, retrieved from www.bmvi.de, 2013. Retrieved from https://www.bmvi.de/SharedDocs/DE/Anlage/Digitales/bim-leitfaden-deu.pdf?__blob=publicationFile on March 1, 2017.

Bauindustrieverband NRW e.V, Positionspapier Digitalisierung der Wertschöpfungskette Bau. Retrieved from www.bauindustrie-nrw.de/politik/ positionspapiere/?cid=2285&did=813& sechash=71524b04 on March 1, 2017.

ISO 29481, *Building Information Models - Information Delivery Manual,* International Organization for Standardization, www.iso.org, 2016.

Interdependence between Structural Engineering and Construction Management
Edited by Ozevin, D., Ataei, H., Modares, M., Gurgun, A., Yazdani, S., and Singh, A.

ISBN: 978-0-9960437-6-2

DATA AND PROCESS MODEL FOR ADVANCED INTEGRATION OF MEP INTO BIM PROJECTS

RAINER PARTL[1], STEFAN HAUER[2], and MICHAEL MONSBERGER[1]

[1]*Institute of Construction Management and Economics, Graz University of Technology, Graz, Austria*

[2]*Sustainable Thermal Energy Systems, AIT Austrian Institute of Technology GmbH, Vienna, Austria*

The coordination of mechanical, electrical, plumbing (MEP) and architecture is a most important task in BIM projects. Open data exchange based on BIM formats such as the Industry Foundation Classes (IFC) is an important basis for advanced collaboration in complex project environments. It facilitates performance-based design concepts as multidisciplinary teams can more easily combine task-specific design and simulation tools. An analysis of the IFC4 standard pointed out, that MEP components are not sufficiently specified yet in order to provide such functionality. Moreover, MEP specific guidelines and process models describing the various interactions between architectural and MEP design in a BIM project are still lacking. This paper thus presents a comprehensive MEP data model that complements the current state of IFC. In addition, an associated BIM process model using Business Process Model and Notation (BPMN) that outlines the application of the data model in different design stages of a project is introduced. For this purpose, we take an air-handling unit as example and apply the developed model to an air-handling unit of a campus building at Graz University of Technology. The presented results support architects, MEP engineers, and BIM managers in the project-specific implementation of air-handling unit models and the set-up of a suitable process for the integration into the project.

Keywords: MEP coordination, Air-handling unit, LOG, LOI, IFC, BPMN.

1 INTRODUCTION

The coordination of mechanical, electrical and plumbing (MEP) systems is a key success factor in complex building projects. The knowledge required is usually tacit and seldom documented (Wang and Leite 2016). In complex buildings, (e.g., research laboratories, hospitals, and commercial buildings) the density of MEP systems is typically high resulting in substantial investment costs. The collaboration between different project participants is usually a challenging task in such projects (Riley *et al.* 2005). The changing of architectural design and input requirements from other stakeholders (owner, structural engineer, etc.) are usually a time-consuming factor for the project participants during the project lifecycle (Park and Lee 2017).

Boktor *et al.* (2014) and Park and Lee (2017) reported, that MEP engineering usually dominates the modeling and coordination process in both, traditional 2D-design as well as building information modeling (BIM) projects, due to high rates of information dissemination. Specifically, the interaction and exchange of information between architecture and MEP engineers is a very close process, compared to that of the other project participants (Park and Lee

2017). In future, the parameters of 3D objects in a BIM model will provide a higher quality of the needed information to improve the performance of collaboration (Eastman *et al.* 2011).

The objective of this paper is to introduce five levels of geometry (LOG) of an air-handling unit and to present a parametric data model that complements the current state of industry foundation classes (IFC). In addition, we introduce an associated BIM process model using business process model and notation (BPMN) that outlines the application of the data model in different design stages of a project.

2 METHODOLOGY

In this paper, an air-handling unit is considered as an example. Air-handling units are usually cost-intensive and spatially demanding in MEP equipment rooms. We refer to the planning data of the building PTC 2 (Production Engineering Centre) on the campus of Graz University of Technology. At the time when the building was designed (around 2010), the planning was done with traditional 2D-drawings. For this paper we selected an air-handling unit of this building, which provides the seminar-rooms with supply and exhaust air in the first floor. It is an air-handling unit with a heating function, equipped with a rotary wheel for heat recovery. Thermodynamic air treatment functions such as cooling, dehumidifying and humidifying are not part of this particular air-handling unit.

The geometric representation follows the recommendations of the BIM Forum (2018). Moreover, the LOG design of the air-handling unit was set up in compliance with the existing 2D-drawings of the building's ventilation system. Relevant parametric data for various design stages are mapped with the IFC-standard. In order to investigate the interaction between the MEP engineer and other project participants, we explore the information flow associated with such a unit during the planning phase by using BPMN as a modelling language.

3 RESULTS

The results are grouped into three parts and presented in the following. First, five levels of a geometric model of the air-handling unit are represented. In a next step, the relevant information of the air-handling unit for MEP planning are listed in tabular form as a parametric data model and compared with the IFC data schema. Finally, the collaboration and information data exchange during the various design stages are illustrated in a BPMN-model.

3.1 LOG Layout for an Air-Handling Unit

The relevant parts of the considered air-handling unit are represented as 3D model. In addition, peripheral equipment such as routes for supply and return systems as well as space for installation and maintenance are considered. The first two levels of geometry (LOG 100 and LOG 200) are shown in Figure 1. Subsequently, the progression to LOG 300, LOG 350 and LOG 400 is illustrated in Figure 2. The typical characteristics of every LOG stage are explained in the following:

LOG 100: In an early design stage, the air-handling unit is ashlar-formed. In the context of structural requirements, the approximate size of the plinth is illustrated for static load calculation. The space requirements for duct connection, maintenance and installation of supply lines are indicated in yellow coloring. These space requirements would usually be depicted as volumes, but for the sake of better illustration they are reduced to areas in the figures.

LOG 200: For LOG 200 the conceptual design of the air-handling unit is refined. Based on functional requirements set by the client, the basic subcomponents of the unit (e.g. heat exchangers, fans, filters etc.) are defined. Compared to the previous LOG, level 200 thus shows the individual components of the unit and indicates the function of each with an appropriate symbol. The preliminary size of the single components is for example important to determine the total length of the air-handling unit. The rough size of the components and their location are relevant to specify wall openings and transportation routes required to bring the assembled unit into the MEP equipment room. Furthermore, the duct connections of the air distribution systems are shown in rudimentary form.

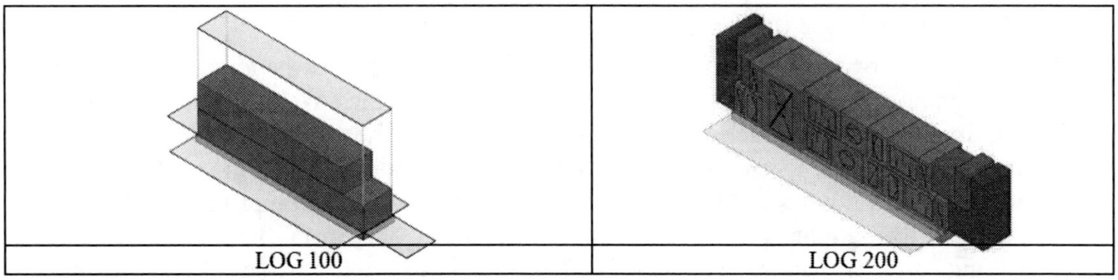

Figure 1. Geometric models for LOG 100 and LOG 200 of an air-handling unit.

LOG 300: Level 300 corresponds with the design development phase in the overall planning process. It is important to provide all relevant information for the bidding process and the award of contract and to provide other disciplines with required information. For this purpose it is necessary to prepare a detailed geometric design including information about associated electrical equipment such as frequency converters (FC), motors (M), pressure difference sensors (Δp) etc. required by the electrical engineering and building automation. In contrast to the previous representation, the duct connections are shown in detail now. Additionally, the ports of the waterside connection of the air heater are introduced.

LOG 350: This LOG refers to the construction phase. The unit is thus modeled with real dimensions obtained from product data by manufactures. It shows the specified geometry of a certain product, which differs from the design representation of LOG 300. In Figure 2 (middle), the heat recovery system of the chosen product is for example split up into the rotary wheel and non-specific (empty) sections for maintenance, which are connected by a damper. Moreover, the number of individual components differs. For example, two components (e.g. ventilator and silencer) are combined in a single unit.

LOG 400: Level 400 represents the status of an as-build model. The illustration of the components is basically equal to that of the previous level. The main difference is that components requiring maintenance are highlighted in yellow. In addition, the revision doors are also shown.

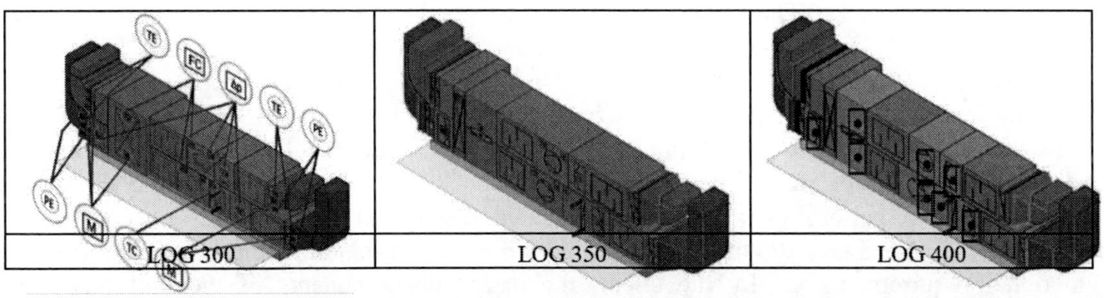

Figure 2. Geometric models for LOG 300, LOG 350 and LOG 400 of an air-handling unit.

3.2 Parametric Data Model of an Air-Handling Unit

In the following, the design stages from the conceptual design until the bidding and award of contract are considered. The most important parameters of the first three levels of information (LOI) are explained and compared with the IFC4 Add 2 standard. IFC is a neutral and software independent data format providing interoperability among project participants working with different software tools. In order to provide this basic functionality and additional features such as quality assurance based on model checking, it is essential that IFC data models including MEP component models are as complete as possible. For this reason, relevant parameters of the considered air-handling unit were identified and compared with the IFC data schema in order to evaluate the completeness of the IFC4 standard. The investigated parameters are grouped into parameters specifying the entire air-handling unit (Table 1) and into parameters describing specific sub-components of the unit (Table 2).

Table 1. Mapping of parameters of an air-handling unit for different LOIs and compared with IFC.

Parameters of the entire air-handling unit	Specific parameter	LOI 100	LOI 200	LOI 300	IFC data schema	
					Exist	Missing
Air volume flow		x	x	x	x	
Space requirements		x			x	
Material			x	x	x	
Geometry	Length, wide, height	x	x	x	x	
Weight		x	x	x	x	
Building automation control system	Number of switch cabinets	x				x
	Measurement equipment (sensor)		x	x	x	
	Digital information - Analog in- and output - Digital in- and output			x		x
Electrical parameters	Electric load estimate	x			x	
	Electric power (current, voltage)		x	x	x	
Internal pressure loss			x	x	x	

Table 2. Mapping of parameters of sub-components of an air-handling unit and compared with IFC.

Sub-components of the air-handling unit	Specific parameter	LOI 100	LOI 200	LOI 300	IFC data schema	
					Exist	Missing
Filter	Filter type		x	x	x	
	Filter grade			x		x
Heat recovery system	Type specification (e.g. rotating heat recovery system)		x	x	x	
	Exhaust and outlet temperature		x	x	x	
	Heat and humidity efficiency		x	x	x	
Air heater	Outlet temperature of the heat recovery system and supply air temperature		x	x	x	
	Thermal power		x	x	x	
	Proportion glycol / water		x	x	x	
	Return and flow temperature		x	x	x	
	Dimension of waterside connection			x		x
Duct silencer	Insertion loss (octave band)		x	x	x	
	Number of scenery			x		x
Fan	Sound power level (octave band)		x	x	x	
	External pressure loss (total pressure)		x	x	x	

The parameters in Table 1 are considered to be particularly relevant for the entire planning team whereas the parameters in Table 2 are considered to be mainly relevant for MEP planners.

As an example, the space requirements are essential for the early architectural design stages (LOI 100). In addition, load estimates for electrical engineers and building automation (BA) engineers are required. This information is relevant in the following design stages (LOI 200, LOI 300). Parameters such as geometry, weight, and the air volume flow are important in all design stages. In the building automation domain of IFC, the switch cabinet is not defined. Furthermore, detailed information on sensors and other parameters (filter grade, dimension of water side connections, number of scenery) are still lacking. The detailed specification of the sub-components takes mainly place in the full conceptual design phase (cf. Figure 3). The corresponding parameters are thus primarily relevant for LOI 200 and LOI 300 as shown in Table 2. The identified parameters are well covered by the IFC standard.

3.3 BPMN Model of an Air-Handling Unit for the Design Stages

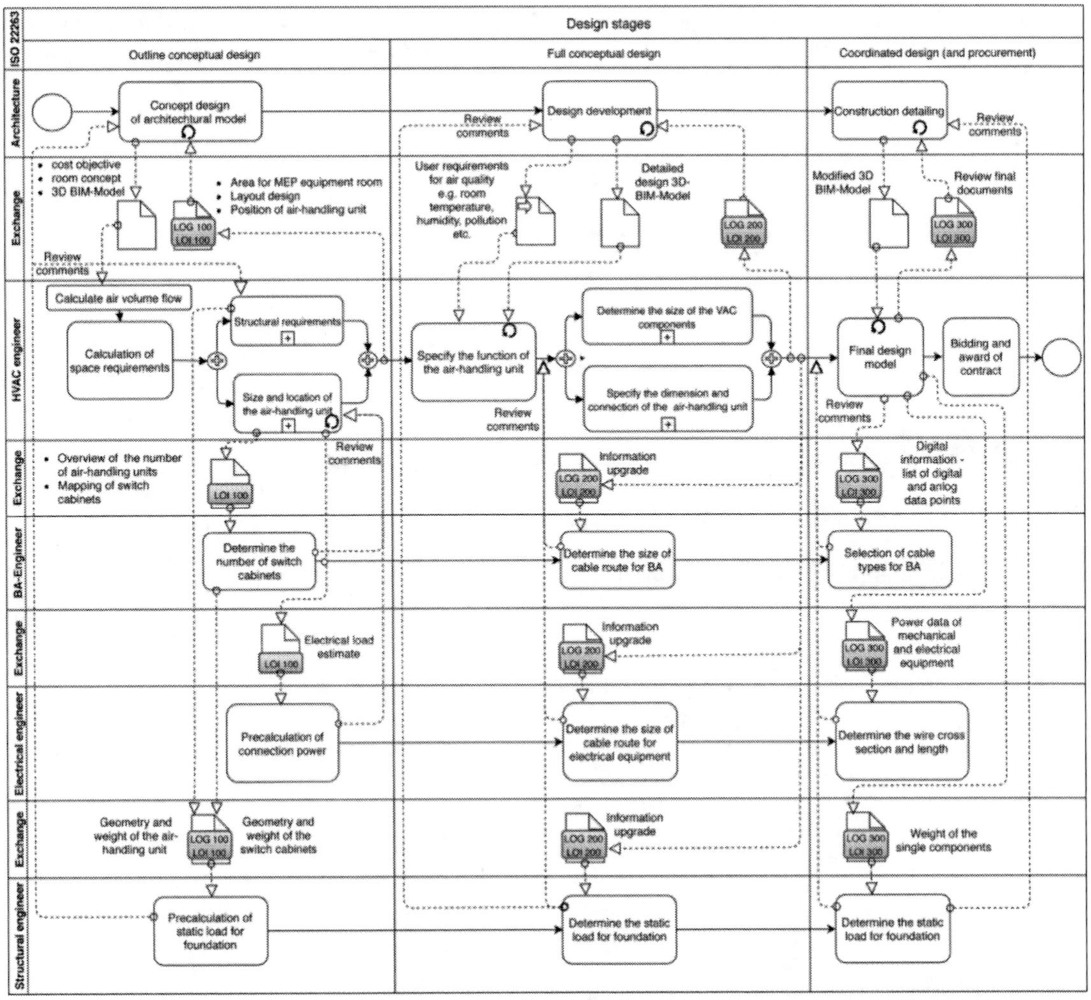

Figure 3. BPMN model of the interaction and workflow between the project participants for different design stages of an air-handling unit.

The information flow for the air-handling unit during the planning phase is explored in Figure 3 using the BPMN methodology in order to investigate the interaction between MEP engineering and the other project participants.

The structuring of the design stages is done according to the ISO 22263 (2008) standard. Figure 3 shows that even the design of a single component such as an air-handling unit requires the interaction of many different project participants. It illustrates the different disciplines and the required information exchange between individual project tasks. The iterative information exchange leads to a higher information level and more detailed geometric model. The requirements for the LOI and the LOG can be different for project participants. As an example, the building automation (BA) engineer and the electrical engineer primarily need alphanumeric information from the heating ventilation and air conditioning (HVAC) engineer in early design stages. The figure shows that the rate of information exchange between the HVAC engineer and the architect is high.

4 CONCLUSION

The paper discusses a generic BIM model of an air-handling unit. For this purpose, different levels of geometry and relevant parameters for designing an air-handling unit are explored. The LOG illustrations consider the requirements of MEP engineers as well as other project participants in different project phases. An evaluation of IFC4 Add 2 shows that most of the relevant parameters are covered by this standard. However, parameters in the building automation domain and some parameters required for the bidding and award of contract are lacking in the IFC data schema. The geometric and the parametric model are an integral part of the information exchange flow in the BPMN model. The interdisciplinary work flow during the design stages demonstrates that many project participants are involved in the planning process of an air-handling unit. The described approach could be applied to further components in order to map complete systems.

References

BIM Forum, Level of Development Specification Guide, Draft for Public Comment, October 03, 2017. Retrieved from http://bimforum.org/wp-content/uploads/2017/11/LOD-Spec-2017-Part-I-2017-11-07.pdf on July 15, 2018.

Boktor, J., Hanna, A., and Menassa, C., State of Practice of Building Information Modeling in the Mechanical Construction Industry, *Journal of Management in Engineering*, 30(1), 78-85, January, 2014.

Eastman, C. M., Teicholz, P., Sacks, R., and Liston, K., *BIM Handbook. A Guide to Building Information Modeling for Owners, Managers, Designers, Engineers, And Contractors*, 2nd ed., Wiley, New Jersey, 2011.

ISO 22263, *Organization of Information about Construction Works, Framework for Management of Project Information*, Beuth, January, 2008.

Park, J. H., and Lee, G., Design Coordination Strategies in a 2D and BIM Mixed-Project Environment: Social Dynamics and Productivity, *Journal of Building Research and Information*, Taylor and Francis Group, 5(6), 631-648, February, 2017.

Riley, D. R., Varadan, P., James, J. S., and Thomas, R., Benefit-Cost Metrics for Design Coordination of Mechanical, Electrical, and Plumbing Systems in Multistory Buildings, *Journal of Construction Engineering and Management*, 131(8), 877-889, August, 2005.

Wang, L., and Leite, F., Formalized Knowledge Representation for Spatial Conflict Coordination of Mechanical, Electrical and Plumbing (MEP) Systems in New Building Projects, *Journal of Automation in Construction*, Elsevier, 64(0), 20-26, April, 2016.

Interdependence between Structural Engineering and Construction Management
Edited by Ozevin, D., Ataei, H., Modares, M., Gurgun, A., Yazdani, S., and Singh, A.

ISBN: 978-0-9960437-6-2

KEY KNOWLEDGE ENABLER FACTORS FOR EFFECTIVE BIM IMPLEMENTATION IN CONSTRUCTION ORGANIZATIONS

VEERASAK LIKHITRUANGSILP and TRAN TUAN KIET

Dept of Civil Engineering, Faculty of Engineering, Chulalongkorn University, Bangkok, Thailand

Building information modeling (BIM) is an innovative approach that is widely used to overcome various challenges in the construction industry. Current BIM implementation mainly focuses on information management, but puts very limited effort on knowledge. That is, the current BIM practice is information-centered, but it is not mature enough to generate and capture experiential knowledge. In addition, knowledge management (KM) is limitedly adopted for BIM implementation. Based on this paper's rigorous literature review as well as a survey questionnaire, nine knowledge enablers and 52 knowledge enabler factors for BIM implementation (KEFBIs) are identified. In this paper, we deploy ISM fuzzy MICMAC to identify and classify the key KEFBIs that influence BIM implementation by contractors. The results can guide contractors to avoid difficulties while adopting BIM so that they can plan and allocate their resources optimally.

Keywords: Knowledge management, Delphi, Mean score, Interpretive structural modeling, ISM, Fuzzy MICMAC.

1 INTRODUCTION

Construction is a major sector of every nation's economy. It is vital to the creation of material and fixed assets for many countries (Flanagan and Norman 1993). However, construction projects always encounter many problems in both maintaining their position and meeting the needs of customers. This is because there have been limited innovations in this industry. Recently, BIM has been a new, innovative technology that can address various challenges in construction. BIM changes the execution of construction projects and is useful for various issues in construction management (Bryde *et al.* 2013). It can support the processes of design, scheduling, and budgeting of built assets (Gu and London 2010) because of the visualization of BIM models (Succar and Mohamad 2015).

2 BIM IMPLEMENTATION IN CONSTRUCTION ORGANIZATIONS

BIM is both a technological and organizational innovation (Succar and Mohamad 2015). As an organizational innovation, BIM implementation involves the changing of a firm's values, culture, structure, and infrastructure (Dossick and Neff 2010). BIM implementation encounters many difficulties because it encompasses change management and the resistance of individuals. Thus, it is very challenging to achieve full benefits of BIM in mainstream construction projects and companies. (Gu and London 2010).

As construction companies begin implementing BIM in their organizations, they must comprehend the factors that influence knowledge for BIM implementation, which are known as *knowledge enablers for BIM implementation* (KEBIs). Because enablers are the driving forces in carrying out BIM implementation, they generate knowledge in organizations and motivate group members to share their knowledge and experiences among one another. They allow organizational knowledge to grow concurrently and systematically (Stonehouse and Pemberton 1999). This paper identifies key KEFBIs for effective BIM implementation of contractors. The results can help contractors appreciate KEFBIs that are vital for the BIM implementation process.

3 LITERATURE REVIEW

Knowledge enablers are referred to as the key factors that determine the effectiveness of knowledge management (KM) implementation in an organization, and are the driving forces that solidify KM (Yeh *et al.* 2006). In other words, knowledge enablers can be viewed as critical success factors of KM (Hung *et al.* 2005). According to Dang *et al.* (2018), knowledge enabler factors (KEFs) can be used to measure the manageability of knowledge in organizations as well as the appropriateness of organizational mechanisms for intentionally and consistently fostering knowledge. Knowledge enablers, or enablers of BIM, have been investigated in some previous studies (e.g., Lee and Choi 2003, Teeragetgul and Charoenngam 2006, Dossick and Neff 2010, Arayici *et al.* 2011, Dang *et al.* 2018). However, so far, KEFBIs have not been investigated at the organization level of construction companies. Thus, this paper focuses on identifying KEFBIs (i.e., the items under KEBIs) at the organization level of construction companies.

4 RESEARCH METHODOLOGY

Figure 1 displays the research framework adopted for this paper. The two main stages are: the identification of KEFBIs in contractors and the identification of the contextual relationship among KEFBIs in contractors.

4.1 Delphi Method

In this paper, the Delphi method is used to gather responses through a consensus among a group of selected experts. The six experts have doctoral degrees in construction management and have more than ten years of work experience in construction, with at least five-year experience in BIM or KM. They were invited to participate in a series of discussions on the preliminary KEBIs and lists of KEFBIs. The experts were requested to review the appropriateness and adequacy of the potential KEBIs and KEFBIs in the context of contractors.

4.2 Questionnaire Survey

The questionnaire survey consisted of two steps. First, a preliminary questionnaire was sent to the experts. They were asked to rate the performance of each KEFBI on the Likert scale from 1 (very low) to 5 (very high). After the questionnaire characteristics were achieved among most experts, the final questionnaire was conducted and used to collect data. The survey focused only on the contractors in Vietnam. The respondents evaluated the performance of 52 KEFBIs in their firms. After two months, 63 responses were received. We eliminated some incomplete and invalid responses such as unidentified respondents and the respondents having less than one year in BIM fields. Finally, 59 responses were considered valid and used for further analyses.

4.3 Data Analysis

Mean score, which is a widely applied method, was used to calculate the mean values of KEFBIs. The ranking of KEFBIs helps contractors understand the importance of KEFBIs. Different sizes of organizations have different organizational and managerial structures, as well as different means of knowledge management (Dang *et al.* 2018). The mean values and ranking of KEFBIs are determined for two groups: small and medium contractors (SMCs) and large contractors (LCs). Per our rigorous literature review, nine KEBIs and 52 KEFBIs are listed in Table 1.

Table 1. Definitions of KEBIs and KEFBIs.

Code	Definition	References
SL	**Strategy and leadership**	1,2,3,4,5
SL1 to SL11	Clear or written policy on BIM implementation; Culture to promote BIM knowledge sharing; Strategic alliance to acquire BIM knowledge; Programs to improve BIM staff; Facilitating to gain BIM knowledge through experiential learning; Criteria for assessing BIM staff performance; Supportive supervisor management; Good leaders; Company leaders readiness for change; Developing knowledge management system; Defining company vision and mission clearly.	5,6,7,8,9
F	**Formalization**	5,6
F1 to F5	BIM implementation activities being covered by some formal procedures; Formal or planning contacts and communication; Ignoring regulations and reaching informal agreements; Making own regulations on the BIM implementation process; Setting benchmarking metrics for BIM implementation.	5,6,7
D	**Decentralization**	5,6,10
D1 to D5	Taking action without supervisor; Encouraged to make own decisions; Making decisions without refer others' experience; No need to ask supervisor before action; Making decisions without their supervisor's approval.	5,6,10
C	**Collaboration**	5,7
C1 to C6	Satisfied with level of collaboration; Willingness to help, support; Willingness to collaborate across organizational units; Willingness to admit responsibility for mistakes or failure; Open communication; Information sharing	5,6,7,8, 10
T	**Trust**	5,6,7
T1 to T6	BIM staff are generally trustworthy; Having reciprocal faith in other' intentions and behavior; Having reciprocal faith in others' ability; Having reciprocal faith in others' behaviors to work toward organizational goals; Having reciprocal faith in others' decision toward organizational interests than individual interests; Having relationships based on reciprocal faith.	5,6,7
L	**Learning**	5,6,12
L1 to L5	Provide various formal training programs on BIM; Provide opportunities for informal individual development other than formal training; Encourage people to attend BIM seminars, symposia; Provide various BIM programs such as clubs and community gatherings; Satisfied by the contents of BIM training or self-development programs.	5,6,7
I	**Incentive**	5,7,13
I1 to I4	Good teamwork is recognized and awarded; Provide opportunities for professional development; Provide monetary incentives; Provide non-monetary incentives (promotion).	5,7
IT	**Information technology**	5,6,7,14,15
IT1 to IT5	Provide IT support for collaboration among members; Provide IT support for communication among members; Provide IT support for searching and accessing information needs; Provide IT support for simulation and prediction; Provide IT support for systematic storing.	5,6,7
IS	**Individual skills**	5,6,7
IS1to IS5	Understanding others' tasks; Suggestions about others' tasks; Communication ability with various groups; Specialists on BIM fields; Performing their own tasks effectively without regard to environmental changes	5,6,7

References:(1) Pan and Scarbrough 1998,(2) Yeh *et al.* 2006,(3) Ho 2009,(4) Sin *et al.* 2009,(5) Dang *et al.* 2018,(6) Lee and Choi 2003,(7) Teeragetgul and Charoenngam 2006,(8) Dossick and Neff 2010 (9) Arayici *et al.* 2011,(10) Lee *et al.* 2012,(11) Homayouni *et al.* 2010,(12) Bixler 2002,(13) Wu 2012,(14) Lee and Kim 2001,(15) Hung *et al.* 2005.

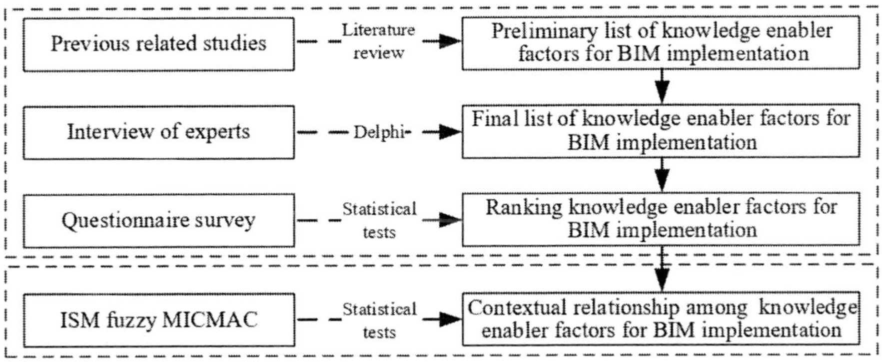

Figure 1. Research framework.

In addition, to examine the agreement within an individual group on the ranking of KEFBIs, whether the ranking orders of the two aforementioned groups are related, and whether the mean values of each item rated by the two aforementioned groups are different, Kendall's coefficient of concordance, Spearman's coefficient of rank correlation, and T-test were applied, respectively.

Subsequently ISM, which is a well-established methodology for identifying relationships among specific items (Attri et al 2013), was used to investigate the interrelationships among top ten knowledge enabler factors for BIM implementation of contractors. However, KM for BIM implementation of contractors is very complex; the relations among these factors are quite diverse. To consider the strength of relations and increase the sensitivity of MICMAC analysis, a fuzzy additional input of possibility of interaction among the KEFBIs was adopted.

5 RESULTS

5.1 Ranking of Knowledge Enabler Factors for BIM Implementation

Table 2 presents the ranking of the top ten KEFBIs, as well as the results of Kendall's coefficient (K-value) and T-test. The results of the ranking identify the KEFBIs in the two aforementioned groups. The company leaders' readiness for change (SL9) and the willingness to help and support (C2) are ranked first and second by both SMCs and LCs. It indicates that SL9 and C2 are evaluated as the most important KEFBIs for contractors. As a result, strategy and leadership (SL) and collaboration (C) are KEBIs that contractors should pay attention when implementing BIM.

The K-values for the ranking of the two groups are 0.091 and 0.197. It can be concluded that the response consensus within each group is achieved. The computed Spearman rank correlation coefficient between two groups is 0.622. It can be inferred that there are strong agreements among groups on ranking 52 KEFBIs. The results of T-test show that, at the significance level of 5 percent, there is no significant difference in the ratings of the two groups.

5.2 ISM Fuzzy MICMAC Analysis for Top Ten KEFBIs

Figure 2 presents the ISM models and four groups according to these dependent and driving powers of the top ten KEFBIs. From the ISM model, it has been observed that SL9 is at the first level of the ISM model. This KEFBI can be used to define the company vision and mission clearly (SL11) and strategic alliance to acquire BIM knowledge (SL3). SL11 and SL3 will provide programs to improve BIM staff (SL4), and provide opportunities for professional development (I2). SL11 and SL3 are at level two of ISM model. Level three in the ISM model contains SL4 and I2. SL4 increases the willingness to admit responsibility for mistakes or failure

(C4), which forms level four of the ISM model. If all the four levels of KEFBIs are implemented in the organization, then it leads to achieve the willingness to help, support (C2), having reciprocal faith in others' decision toward organizational interests than individual interests (T5), the willingness to collaborate across organizational units (C3), and the culture to promote BIM knowledge sharing (SL2), which are at level five.

Table 2. Top ten and results of Kendall's coefficient, T-test for KEFBIs.

Rank	SMCs (n=30)		LCs (n=29)		Overall (n=59)		T-test	
	KEFBI	Mean	KEFBI	Mean	KEFBI	Mean	F-value	p-value
1	SL9	4.00	SL9	4.07	SL9	4.03	1.012	0.319
2	C2	3.93	C2	4.07	C2	4.00	1.477	0.229
3	IT1	3.83	C3	4.07	C3	3.92	0.580	0.449
4	C5	3.83	SL2	4.03	SL11	3.88	2.109	0.152
5	SL11	3.80	SL3	4.03	C5	3.88	0.028	0.868
6	C3	3.77	SL4	4.03	C4	3.88	0.291	0.592
7	C1	3.77	C4	4.03	SL4	3.85	0.264	0.610
8	C4	3.73	T5	3.97	IT5	3.80	1.592	0.212
9	SL8	3.73	SL11	3.97	SL3	3.80	2.963	0.091
10	C6	3.73	I2	3.97	T5	3.80	3.845	0.055
K-value	0.091		0.197		0.119			
Chi-square	139.683		260.756		339.491			
p-value	0.000		0.000		0.000			

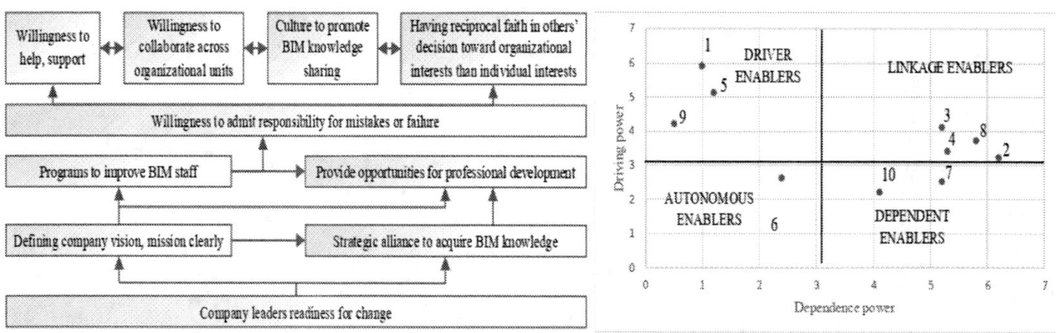

Figure 2. ISM model formulation and cluster of KEFBIs in large contractors.

From the driving-dependence power diagram, only I2 is identified autonomous, which are weak drivers, weak dependent, and relatively disconnected from the system. I2 and C4 are weak drivers, but are strongly dependent on the others. It indicates that two KEFBIs require all the others during BIM implementation to reduce the effect of these KEFBIs. C2, T5, C3, and SL2 have strong drive power as well as strong dependence power. Thus, if these KEFBIs have any change, it will affect others and themselves. The fourth cluster in the diagram includes SL9, SL11, and SL3, having strong driving power but weak dependence. These KEFBIs help achieve others, which appears at the top of ISM. Thus, management should facilitate these KEFBIs for successful BIM implementation in their organizations.

6 CONCLUSIONS

This paper presents 52 KEFBIs, the factor means and rankings of which are in conformity with different sizes of contractors. ISM fuzzy MICMAC analysis was used to provide accurate analysis, regarding the driving and dependence power of KEFBIs. The results can help contractors implement BIM more efficiently. Moreover, they can plan and allocate their resources optimally. This research only consists of country-related findings; the KEFBIs are identified for contractors in Vietnam. Without additional data collection, it cannot automatically be used in other countries and for other kinds of enterprises. Further studies should re-examine the results of this study in other countries and for other kinds of enterprises.

References

Arayici, Y., Coates, P., Koskela, L., Mike K., Usher C., and O'Reilly K., Building Information Modeling adoption and implementation for architectural practices, *Structural Survey*, Emerald, 29(1), 7-25, 2011.

Attri, R., Dev, N., and Sharma, V., Interpretive Structural Modelling (ISM) approach: An Overview, *Research Journal of Management Sciences*, 2(2), 3–8, 2013.

Bixler, C. H., Applying the Four Pillars of Knowledge Management, *KM World*, 11(1), 10-20, 2002.

Bryde, D., Broquetas, M., and Volm, J. M., The Project Benefits of Building Information Modeling (BIM), *International Journal of Project Management*, 31(7), 971–980, 2013.

Dang, N. C., Long L. H., and Kim, S. Y., Impact of Knowledge Enabling Factors on Organizational Effectiveness in Construction Companies, *JKM*, 22(4), 759-780, 2018.

Dossick, S., and Neff, G., Organizational Divisions in BIM-Enabled Commercial Construction, *Journal of Construction Engineering and Management*, 136(4), 459–467, 2010.

Flanagan, R., and Norman, G., *Risk Management And Construction*, Blackwell Science, Inc, 1993.

Gu, N., and London K., Understanding and facilitating BIM adoption in the AEC industry, *Automation in Construction*, 19 (8), 988-999, 2010.

Ho, C. T., The Relationship Between Knowledge Management Enablers and Performance, *Industrial Management and Data Systems*, 109(1), 98-117, 2009.

Homayouni, H., Neff, G., and Dossick, C., Theoretical Categories of Successful Collaboration and BIM Implementation within The AEC Industry, *Construction Research Congress*, 778-788, 2010.

Hung, Y., Huang, S., Lin, Q., and Tsai, M., Critical Factors in Adopting a Knowledge Management System for The Pharmaceutical Industry, *Industrial Management and Data Systems*, 105(2), 164-183, 2005.

Lee, H., and Choi, B., Knowledge Management Enablers, Processes, and Organizational Performance: An Integrative View and Empirical Examination, *JPIS*, 20(1), 179-228, 2003.

Lee, J. H., and Kim, Y. G., A Stage Model of Organizational Knowledge Management: A Latent Content Analysis, *Expert Systems with Applications*, 20, 299-311, 2001.

Lee, S., Kim, B. G., and Kim, H., An Integrated View of Knowledge Management for Performance, *Journal of Knowledge Management*, 16(2), 183-203, 2012.

Pan, S., and Scarbrough, H., A Socio-Technical View of Knowledge-Sharing at Buckman Laboratories, *Journal of Knowledge Management*, 2(1), 55-66, 1998.

Sin, C. H., Goh, G. G. G., and Eze, U. C., Knowledge Management Enablers Toward Successful New Product Development: A Case Study in A Semiconductor Manufacturing Firm, *JKMP*, 10(4), 16-32, 2009.

Stonehouse, G. H., and Pemberton, J. D., Learning and Knowledge Management in The Intelligent Organization, *Participation and Empowerment: An International Journal*, 7(5), 131-144, 1999.

Succar, B., and Mohamad, K., Macro-BIM Adoption: Conceptual Structures, *Automation in Construction*, 57, 64-79, 2015.

Teeragetgul, W., and Charoenngam, C., Factors Inducing Knowledge Creation: Empirical Evidence from Thai Construction Projects, *ECAM*, 13(6), 584-599, 2006.

Yeh, Y. J., Lai, S. Q., and Ho, C. T., Knowledge Management Enablers: A Case Study, *Industrial Management and Data Systems*, 106(6), 793-810, 2006.

Interdependence between Structural Engineering and Construction Management
Edited by Ozevin, D., Ataei, H., Modares, M., Gurgun, A., Yazdani, S., and Singh, A.

ISBN: 978-0-9960437-6-2

A MEASUREMENT INSTRUMENT FOR THE TECHNOLOGY EXTENT AND PROCESS EFFICIENCY IN BIM ENABLED CONSTRUCTION PROJECTS

GOZDE BASAK OZTURK

Dept of Civil Engineering, Aydin Adnan Menderes University, Aydin, Turkey

Technology use in all fields can play an important role as the booster in creating lean and efficient processes. Technology use may result in reduced duplications and delays in workflows while helping to speed up task realization. Building Information Modeling (BIM) enhances knowledge share, use and reuse for better collaboration, communication, coordination, and monitoring as a knowledge base platform. BIM enhanced construction projects may positively affect the process efficiency. The aim of this paper is to define a measurement instrument of BIM usage as technology enhancer tool and of process efficiency of construction projects. Research leans on the prominent literature to concrete the measurement instrument. A survey is established to construction project professionals to understand the technology efficiency and process efficiency in BIM enabled construction projects. The research based on the related data collected from 92 professionals experienced both in traditional project delivery methods and BIM enabled construction projects. The respondents are the construction industry experts comprised of construction project managers, BIM managers, and BIM implementation experts (architects, civil engineers, mechanical engineers, electrical engineers etc.). Data is analyzed and tested with structural equation modeling software to verify the proposed measurement instrument. The technology efficiency and process efficiency factors for BIM enabled construction projects are tested and refined. Research findings present the measurement instrument for both technology and process efficiency in BIM enabled construction projects.

Keywords: Building information modeling, Structural equation modeling, Construction firms.

1 RESEARCH BACKGROUND

Significant quantities of resources are wasted each year as a result of inefficient or non-existent quality management procedures (Arditi and Gunaydin 1997). Therefore, project management focuses on efficiency of process during project life cycle in order to minimize cost, to clarify routines and to clear division of resources (Farsi and Fillippini 2004). Efficiency, which is the status of organizations' utilized resources and achieved goals with less percentage of normal used resources, focuses on the resource investment versus output (Jacobs and Chase 2010). Therefore, it is clear that resource (money, time, etc.) saving is possible through efficiency improvement efforts (Reed *et al.* 1996). However, many researchers suggest that because of its easy to mimic techniques, efficiency is important but not enough for organizational performance (Frumkin and Kim 2001). On the other hand, the lack of process efficiency improvement intention may cause impotence of competitiveness among other companies those keep up with new approaches for more efficient processes. Technology can enhance process efficiency by improving the methods for task implementation. Construction companies need technological solutions for more efficiency. BIM, as a relatively new technological platform for improving knowledge share, use and reuse for better collaboration, communication, coordination, and monitoring, establishes a base for efficient processes in construction projects. This paper aims to develop a measurement instrument for both technology and process efficiency in BIM enabled construction projects. The following sections present a short description of methodology used in this research, and the research findings related to the development of measurement instrument.

2 TECHNOLOGY AND PROCESS EFFICIENCY IN BIM ENABLED PROJECTS

The AEC industry has long been focused on increasing quality, while decreasing time, cost, and change with an aim of fulfilling the scope by using construction management techniques. Integrated project delivery triggers efficiency among process enhancers (Azhar *et al.* 2008). Forging collaboration through extensive information exchange increases process efficiency (Glick and Guggemos 2009). Therefore, compared to traditional processes, BIM improves efficiency by enhancing integration and allowing more accurate and efficient collaboration in construction project processes (Carmona and Irwin 2007). BIM serves as a virtual model, which accurately simulates the real world building with fully encompassed information within and significantly changes the workflow and project delivery processes (Hardin 2009), enhances project collaboration and control among stakeholders, improves productivity (less re-work, conflicts and changes), increases quality and performance, accelerates project delivery, reduces wastages, reduces construction time and cost (Azhar 2011), triggers new revenue and business opportunities (Qian 2012, Barlish and Sullivan 2012, Succar 2009). Hence, the extent of technology efficiency may increase process efficiency in BIM enabled construction projects. In the light of the related literature, the measurement instrument is developed for the extent of technology efficiency in BIM enabled construction projects. The hypothesized factors are represented in Figure 1.

The ability to achieve efficient processes is vital for modern companies (Smith and Fingar 2003). Construction processes are complex and changes occur throughout the project. Process improvement in construction aims better product for less cost within specified time frame. The American General Contractors claims that "Building Information Model is a data-rich, object-oriented, intelligent and parametric digital representation of the facility. Views and data appropriate to various users' needs can be extracted and analyzed from the model to generate information that can be used to make decisions and improve the process of delivering the facility" (AGC 2006). BIM as a platform improves collaboration, communication, coordination, monitoring, and control for integrated project delivery may improve process efficiency in construction projects. According to related literature, the measurement instrument is developed for the extent of process efficiency in BIM enabled construction projects. The hypothesized factors are represented in Figure 1 (Lee *et al.* 2011, Jacobs and Chase 2010, PMI 2007).

3 METHODOLOGY

A methodology that contains a structured questionnaire was adopted in this study. A structured questionnaire is applied to 92 construction industry professionals experienced in both traditional method and BIM enabled construction projects to verify the measurement instrument for the extent of technology and process efficiency in BIM enabled construction projects. 92 respondents out of 140 randomly selected professionals, representing a response rate of 67%.

The respondents are the construction industry experts comprised of construction project managers, BIM managers, and BIM experts (architects, civil engineers, mechanical engineers, electrical engineers etc.). The collected data analyzed with structural equation modeling method.

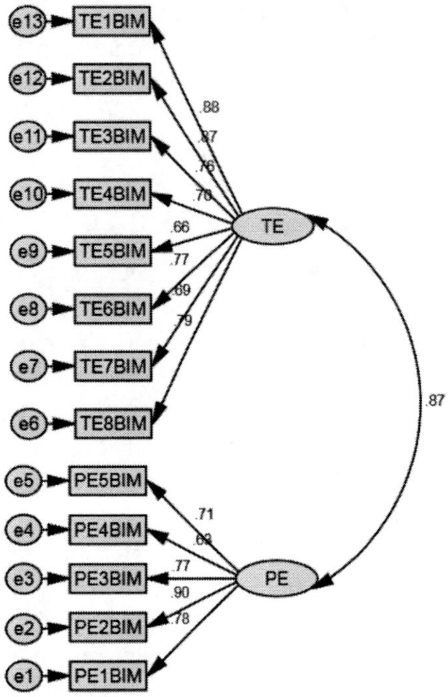

Factor Code	*The Extent of Technology Efficiency Factors*
TE1BIM	Minimized change
TE2BIM	Increased collaboration
TE3BIM	Robust communication
TE4BIM	Fulfilled operational requirements and specifications
TE5BIM	Fulfilled technical specifications
TE6BIM	Well maintained logistics
TE7BIM	High performance target cost projection
TE8BIM	Minimized life cycle cost

Factor Code	*The Extent of Process Efficiency Factors*
PE1BIM	Meeting short-term targets of deliverables
PE2BIM	Meeting procurement goals
PE3BIM	Meeting performance targets
PE4BIM	Knowledge capture
PE5BIM	Safety index

Figure 1. The conceptual measurement instrument of the extent of technology and process efficiency (including 8 factors for technology efficiency and 5 factors for process efficiency) (left) and the factor list with codes and names of technology efficiency and process efficiency (right).

4 FINDINGS

The goal is to identify factors that affect the extent of technology and process efficiency in BIM enabled construction projects. An intensive literature review was conducted to accomplish this task. Comprehensive item generation process is needed for construct validity (Nunnally 1978). A conceptual measurement model was developed in accordance with the theoretical background. 8 factors for the extent of technology efficiency and 5 factors for the extent of process efficiency were identified (Figure 1). The factors are formulated into statements to conduct a proper question to respondents. Five-point Likert Scale is used for collecting data from respondents, where 5 represents strongly agree, 4 indicates agree, 3 indicates neutral, 2 indicates disagree, and 1 indicates strongly disagree. The data that is collected from professionals, who are both experienced in traditional and BIM delivery methods, about the extent of technology and process efficiency in BIM enabled construction projects, is tested using AMOS Version 22.0 to validate measurement model. Data and analysis results related to measurement model are represented in observed variables (Figure 1), latent variables (Figure 1) and goodness-of-fit comparisons (Table 1). The data collected via questionnaire survey was tested for content validity, convergent validity, reliability, and multicollinearity. Goodness-of-fit of the models were assessed (Bollen and Long 1993). Content validity is assured with an extensive literature review in the research field for obtaining correlation between theoretical do006Dain and specific measurement variables

(Carmines and Zeller 1991). Convergent validity was tested with α = 0.001 level of significance for all factors loadings (Churchill, 1979).

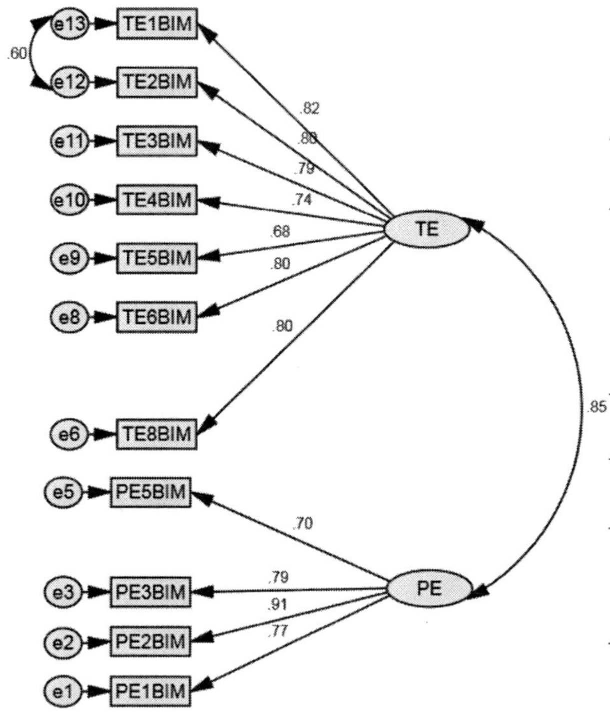

Factor Code	*Estimate (Conceptual Model)*	*Estimate (Refined Model)*	*P-Value*
TE1BIM	.877	.822	0.001
TE2BIM	.868	.799	0.001
TE3BIM	.755	.795	0.001
TE4BIM	.704	.743	0.001
TE5BIM	.661	.677	0.001
TE6BIM	.770	.795	0.001
TE7BIM	.691	-----	0.001
TE8BIM	.791	.708	0.001

Factor Code	*Estimate (Conceptual Model)*	*Estimate (Refined Model)*	*P-Value*
PE1BIM	*.782*	*.774*	*0.001*
PE2BIM	*.903*	*.907*	*0.001*
PE3BIM	*.772*	*.790*	*0.001*
PE4BIM	*.632*	-----	*0.001*
PE5BIM	*.710*	*.700*	*0.001*

Figure 2. The refined measurement instrument of the extent of technology and process efficiency (left) and factor loadings for conceptual and refined models with respective significance level value (including 7 factors for technology efficiency and 4 factors for process efficiency) (right).

The accuracy of the measurement instrument is assured with internal consistency of the constructs, which is a measure for reliability based on Cronbach's alpha value (Lin *et al.* 2005). According to Churchill (1979) the threshold value for reliability is 0.6 for exploratory studies, whereas it is .70 and higher for Nunnally (1978) and Flynn *et al.* (1990). The numbers on the arrows stands for the strength of relationship between observed variables (rectangular boxes, Figure 1,2) and latent variables (oval boxes, Figure 1, 2) (Joreskog and Sorbom 1993). Factors related to the extent of technology efficiency and process efficiency for BIM enabled construction projects (Figure 1) were refined over several iterations by removing the variable with low factor loading to improve model (Figure 2). The refined measurement model includes a total of 11 observed variables as opposed to 13 variables for a better model fit of measurement instrument. High performance target cost projection (TE7BIM) factor was eliminated from the measurement instrument because there is another cost related factor called minimized life cycle cost (TE8BIM) which is a better measure with a higher factor loading. Knowledge capture (PE4BIM) factor was eliminated because of it has the lowest factor loading. All factor loadings in the refined measurement model were statistically significant at .001 level. The refined measurement model indicates that the observed variables establish a valid measure for their respective latent variables. Model is also refined with using the modification indices to reduce $\chi 2$ levels for each possible path (Hair *et al.* 1998, Arbuckle 2007, Hoyle and Panter 1995). The refined measurement model is statistically significant at .022 levels which confirms the < .05 level of significance rule (Table

1). The conceptual and refined measurement models were compared using goodness-of-fit indices (Table 1). Multicollinearity is tested with Pearson correlation analysis conducted on the observed variables, which were found to be below 0.90 (Hair *et al.* 1998). The results show that there is no multicollinearity.

Table 1. Goodness-of-fit Indices of conceptual and refined measurement models.

Model	**P-Value**	**X^2 / df**	**p-value**	**GFI**	**TLI**	**CFI**	**RMSEA**
Goodness-of-fit Threshold (acceptable)	**< *0.05***	**< 3.00**	**< 0.05**	**> 0.76**	**> 0.70**	**> 0.73**	**> 0.80**
Conceptual Model	.000	1.974	.000	.833	.905	.922	.103
Refined Model	.022	1.485	.022	.896	.962	.971	.073

The fit of measurement instrument to data was tested via goodness-of-fit indices. One of the determinants of goodness-of fit is relative chi-square ($\chi 2$=df), which should be smaller than 3 (Jaspara 2003). The other determinants those used for model comparisons are the goodness-of-fit index (GFI> .76, Arrindell *et al.* 1998), the comparative fit index (CFI > .73), the Tucker-Levis fit index (TLI > .70), and the root-mean square error of approximation (RMSEA < .080-acceptable) (Klein 1998). Table 1 presents the results of the goodness-of-fit tests for measurement models. The refined measurement model is statistically significant and model fit is good. The statistical results show that the refined measurement instrument is approved. There is a high correlation with a value of .85 between the extent of technology and process efficiency in BIM enabled construction projects, which indicates a strong positive relationship between those variables. The relationship must be assessed in future researches.

5 CONCLUSION

An attempt was made in this study to assess the validity of measurement instrument for the extent of technology and process efficiency in BIM enabled construction projects. The results are satisfactory and the refined measurement instrument is accepted with a good. In conclusion, this research fulfilled its aim to validate a reliable measurement instrument for technology and process efficiency in BIM enabled construction projects, and evidenced a high correlation between two latent variables. In future studies, the relationship between the technology efficiency, process efficiency, and other variables in BIM enabled construction projects will be tested.

References

AGC, *The Contractors' Guide to BIM,* Associated General Contractors (AGC) of America, 2006. Retrieved from www.agc.org in April, 2008.

Arbuckle, J. L., *AMOS 16.0 User's Guide,* AMOS Development Corporation, Chicago, 2007.

Arditi, D., and Gunaydin, H. M., Total Quality Management in the Construction Process, *International Journal of Project Management*, 15(4), 235–43, 1997.

Arrindell, W. A., Sanavio, E., Aguilar, G., Sica, C., Hatzichristou, C., and Eisemann, M., The Development of a Short Form of the EMBU: Its Appraisals with Students in Greece, Guatemala, Hungary and Italy, *Personality Individual Differences,* 27(4), 613–628, 1999.

Azhar, S., Building Information Modeling: Trends, Benefits, Risks, and Challenges for the AEC Industry, *Leadership Manage. Eng.*, 11(3), 241-252, 2011.

Azhar, S., Hein, M., and Sketo, B., *Building Information Modeling: Benefits, Risks and Challenges,* Proc., 44th Associated Schools of Construction National Conference, Auburn, AL, 2008.

Barlish, K., and Sullivan, K., How to Measure the Benefits of BIM — A Case Study Approach, *Automation in Construction*, 24 149–159, 2012.

Bollen, K. A., and Long, J. S., *Testing Structural Equation Models*, Sage, Newbury Park, CA., 1993.
Carmines, E. G., and Zeller, R. A., *Reliability and Validity Assessment*, Sage, Newbury Park, CA, 1991.
Churchill, G. A., A Paradigm for Developing Better Measures for Marketing Constructs, *J. Marketing Res.*, 16(1), 64–73, 1979.
Farsi, M., and Filippini, M., An Empirical Analysis of Cost Efficiency in Nonprofit and Public Nursing Homes, *Annals of Public and Cooperative Economics*, 75(3), 339–365, 2004.
Frumkin, P., and Kim, M.T., Strategic Positioning and the Financing of Nonprofit Organizations: Is Efficiency Rewarded in the Contributions Marketplace?, *Public Administration Review*, 61(3), 2001.
Glick, S., and Guggemos, A., *IPD and BIM: Benefits and Opportunities for Regulatory Agencies*, Proc., 45th Associated Schools of Construction National Conference, Gainesville, FL, 2009.
Hair, J. F, Anderson, R. E., Tatham, R. L., and Black, W. C., *Multivariate Data Analysis*, 5th Ed., Prentice-Hall, Upper Saddle River, NJ, 88–92, 1998.
Hardin, B., *BIM and Construction Management*, Wiley, Indianapolis, IN, 2009.
Hoyle, R. H., and Panter, A. T., *Writing about Structural Equation Models, Structural Equation Modeling: Concepts, Issues, and Applications*, R. H. Hoyle, ed., Sage, Thousand Oaks, CA, 158–176, 1995.
Jacobs, F., Chase, R., *Operations and Supply Chain Management*, 13th. McGraw-Hill/Irwin, New York, NY, 2010.
Jaspara, A., Cognition, Culture and Competition: An Empirical Test of the Learning Organization, *Learn. Organization Int. J.*, 10(1), 31–50, 2003.
Joreskog, K. G., and Sorbom, D., *LISREL 8: Structural Equation Modeling with the SIMPLIS Command Language*, Scientific Software International, Chicago, 1993.
Klein, R. B., *Principles and Practice of Structural Equation Modeling*, Guilford Press, New York, 1998.
Lee, S., Lee, D., Schniederjans, M., Supply Chain Innovation and Organizational Performance in The Health Care Industry, *Int. J. Oper. Prod. Manag.*, 31(11), 1193–1214, 2011.
Lin, C., Chow, W., Madu, C., Kuei, C., Yu, P., A Structural Equation Model of Supply Chain Quality Management and Organizational Performance, *Int. J. Prod. Econ.*, 96(3), 355–365, 2005.
Nunnally, J. C., *Psychometric Theory*, 2nd Ed., McGraw-Hill, New York, 1978.
Reed, R., Lemak, D. J., and Montgomery, J. C., Beyond Process: TQM Content and Firm Performance, *The Academy of Management Review*, 21(1), 173-202, 1996.
PMI, *Construction Extension to the PMBOK® Guide, 3rd edition. Project Management Institute*, Newton Square, Pennsylvania, USA, 2007.
Qian, A., Benefits and ROI of BIM for Multi-Disciplinary Project Management. National University of Singapore, 2012. Retrieved from http://www.icoste.org/wp-content/uploads/2011/08/Benefits-and-ROI-of-BIM-for-Multi-Disciplinary-Project-Management.pdf on August 19, 2013
Smith, H., and Fingar, P., *Business Process Management—The Third Wave*, Florida: Meghan-Kiffer, 2003.
Succar, B., The Five Components of BIM Performance Measurement, *Part of Proceedings: W096 — Special Track 18th CIB World Building Congress (combined with W104)*, May 2010, Salford, UK, University of Newcastle, NSW Australia, 2010.

Interdependence between Structural Engineering and Construction Management
Edited by Ozevin, D., Ataei, H., Modares, M., Gurgun, A., Yazdani, S., and Singh, A.

ISBN: 978-0-9960437-6-2

IMPROVING OCCUPATIONAL HEALTH AND SAFETY BY USING ADVANCED TECHNOLOGIES AND BIM

AGNES KELM, ANICA MEINS-BECKER, and MANFRED HELMUS

Construction Management and Economics, University of Wuppertal, Wuppertal, Germany

Despite a slight decrease in the number of accidents at work on German construction sites, a large number of accidents on construction sites occurred as a result of the collision of different trades. An important potential is in the optimization of the safety and health by making available predefined information about work safety for all involved trades. The aim of the project is the application of the BIM-method and the use of RFID-technology and Augmented Reality to improve work safety in the real estate and construction industry. It has to identify, define, and standardize information relevant to work safety so that they can be used for prevention-measures with Building Information Modeling (BIM). To do so, a process map was developed to map all relevant occupational health and safety processes during the lifecycle phases of a building. Therefore, all work safety-relevant processes have to be identified and defined. Furthermore, a demonstrator was developed in order to show possibilities like the usage of Augmented Reality (AR) and a commercially available tablet with a web app, in combination with Bluetooth beacons and RFID technology, based on a common data base. This will lead to the possibility of cross-company planning regarding work safety-relevant processes with BIM as well as the optimization of work safety in additional use of AutoID-Technologies.

Keywords: Digitalization, Modern technologies, RFID, Auto-ID, Personal protective equipment (PPE), Augmented Reality (AR), Safety at work.

1 INTRODUCTION

The digitization of the economy has been progressing steadily for several years, also in Germany. With regard to the construction sector, the focus is on the method of building data modeling. Compared to other countries, the method has not yet been applied uniformly in Germany. Against the background of optimizing the effectiveness and efficiency and strengthening the competitiveness of the associated construction and real estate industry of Germany - even in international comparison - the implementation of the BIM method is also essential for health and safety during construction work and maintenance.

The research project "Application of the BIM-method and the use of RFID-technology to improve work safety in the real estate and construction industry" is part of a longer-term overall concept to be set up. The aim is to promote the standardization efforts of building data models with regard to processes and the associated flow of information between the project participants in all lifecycle phases (Bodtländer *et al.* 2017). At the same time, this will help to create transparency with regard to the BIM method for those involved in construction.

The BIM method provides detailed information about all aspects of the construction and adds value for everyone involved in the construction process. This includes, for example, the availability of a 3D representation, the description of the whole construction workflow and data about changes during the whole life cycle of a building. The goal is to show that using this compiled data when applying the BIM method, current, and future software solutions can be implemented to increase workplace safety fast and efficiently.

Emerging technologies such as Augmented Reality and the Internet of Things allow new ways to access, visualize and annotate the stream of information surrounding us. Using these tools, we are able to implement new safety and accident prevention measurements based on different data sources. The BIM method is an important keystone in this because it provides all the required data for the implement. Key benefits to improve safety are the capability to directly getting informed about dangers as well as showing important safety information when needed. Based on our process map, a demonstrator was developed that shows how a process can be implemented.

2 RESEARCH APPROACH

The goal of the research project is, to connect the preventive design of work with the BIM-method within the construction industry. For this connection, a prevention strategy BIM with new tools for bigger support of occupational health and safety will be created with prevention partners from the construction industry. The pilot-application shows exemplary application possibilities and describes new forms of the cooperation between prevention partners and constructing partners in an alliance of prevention. As a result, basics and examples for further systematic usage and working on digitization of preventive occupational health and safety during the stages of construction, usage, maintenance, and reconstruction of real estate are developed (Bodtländer *et al.* 2018).

3 RESEARCH METHOD

For BIM to be successfully used in the area of occupational health and safety, the processes must first be looked at in detail. According to the definition of the University of Wuppertal (BUW), the lifecycle of a real estate consists of five phases: development, planning, realization, operation & maintenance, and demolition. Within these phases, new information is generated from various roles from various information sources. The extensive consideration of all these phases is not feasible in a single research project. For this reason, the BUW has decided to subdivide the lifecycle of a real estate in a structured manner and to develop the respective sub-areas in independent research projects, which are always interlinked. Through the close networking, synergies can be used as best as possible and a broad knowledge foundation can be established. This enables the chair of Construction Management & Economics at the BUW to publish the gained findings in a targeted way in the general public, thereby continuing promoting the digitization of the real estate value chain.

Existing safety-specific software products accompanied were worked out by interviews and a survey among German software developers for safety-specific applications to evaluate the current impact of BIM for health and safety in Germany.

To show that the defined processes can be implemented using existing state-of-the-art hard- and software a prototype for the Microsoft Mixed Reality Headset HoloLens as well as an of-the-shelf tablet pc was developed. For the development, modern development tools were used that allowed automatic data distribution among all devices. Custom software libraries to import IFC files were developed so they can be used in the demonstrator. Because we not only had the 3D model but also the logical, representation we could use it for annotation and to connect devices inside the building with specific tasks.

4 RESULTS

Within the framework of the research project, a process map was developed to map all relevant occupational health and safety processes including the associated tasks and responsibilities during the lifecycle phases of the planning, realization, and operation of a building. The required information (information input), as well as the generated information (information output) per process, were determined, assigned and combined into a continuous information process. Through close collaboration with experts from the field, the developed process map could be verified in the sense of the information protection of relevant information provision. Evaluations can then be generated from the developed information process, for example, required information for each responsibility and process in the corresponding lifecycle phase. Further evaluation variants are possible. These evaluations are ultimately intended as an aid to provide project-specific requirements (BIM requirements) to digital applications (BIM application), e.g. a web app for individual trades. In order to provide detailed information in a targeted and digital manner, a project-specific information management is necessary, which is the responsibility of the client. This should be defined at the beginning of the project in order to make use of preventive possibilities. In this case, the project sharing roles (for example, BIM author or BIM users), amount of information (e.g. a 3D model or concrete attributes), delivery times (e.g. after confirmation of the order) and digital tools (e.g. Tablet or AR- Application) are defined. The previous points form the basis for efficient building information modeling in terms of occupational health and safety. Only after an extensive adjustment of the digital requirements, a contractor can meet a client's qualifications, equipment, and requirements. An individual view of the organizational possibilities of the contractors is crucial since BIM has not yet reached the protection of the construction and real estate sector.

The demonstrator shows that BIM-based data can easily be integrated into current state-of-the-art software. With the help of modern frameworks and technologies, the data provided by the BIM procedure can be combined with information from external sources such as the user position and additional information such as the escape and rescue plan. Changes in the data are forwarded to all users immediately. An important part of the demonstrator is the direct communication channel between the different users of the system. This not only increases the safety it also shortens the response time for questions and helps resolve problems faster.

Microsoft Augmented Reality Glasses HoloLens is perfectly suited to show the user information directly at his workplace. Because this device is still under development and not ready for the mass market, we also use a current of-the-shelf tablet to show what can be used today. Because both solutions are connected with each other, they allow their users to create a dialog with each other. Because text input on is not very accessible, especially on the HoloLens it takes some time to type even a word, we provide a boilerplate of common sentences for the input.

First data from Autodesk Revit were exported to an IFC file to display the 3D model of the building on mobile devices. The data was then imported using a custom script-based IFC open-shell that also swapped some of the objects within the model with a version with a lower number of polygons so that mobile devices could draw them. Another problem that had to be solved was the file size of the IFC file for use on the web and mobile devices. Instead of loading the entire building, the Web application only displays a single floor. To show the user information without loading much in advance, everything that was not part of the visual representation was exported to the JSON format, which was only a fraction of the original IFC. This file can be quickly loaded and processed on mobile devices to decide which floor or room to display and keep the file size low without affecting the functionality users need.

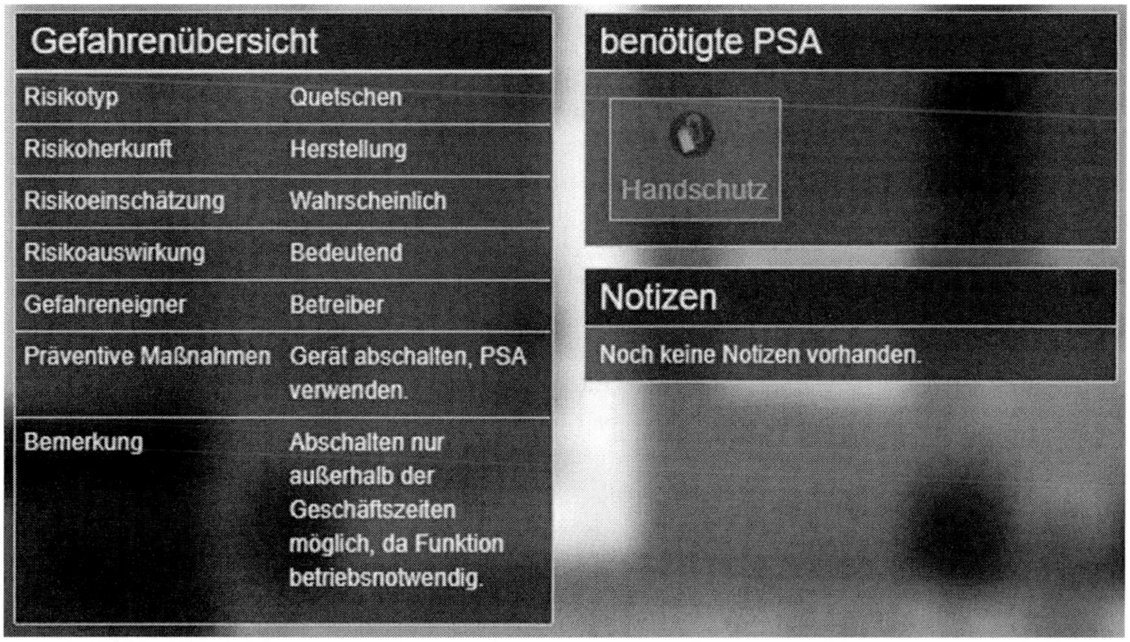

Figure 1. A detailed table with safety information as well as the required PPA for a task inside the building in the web and Android application.

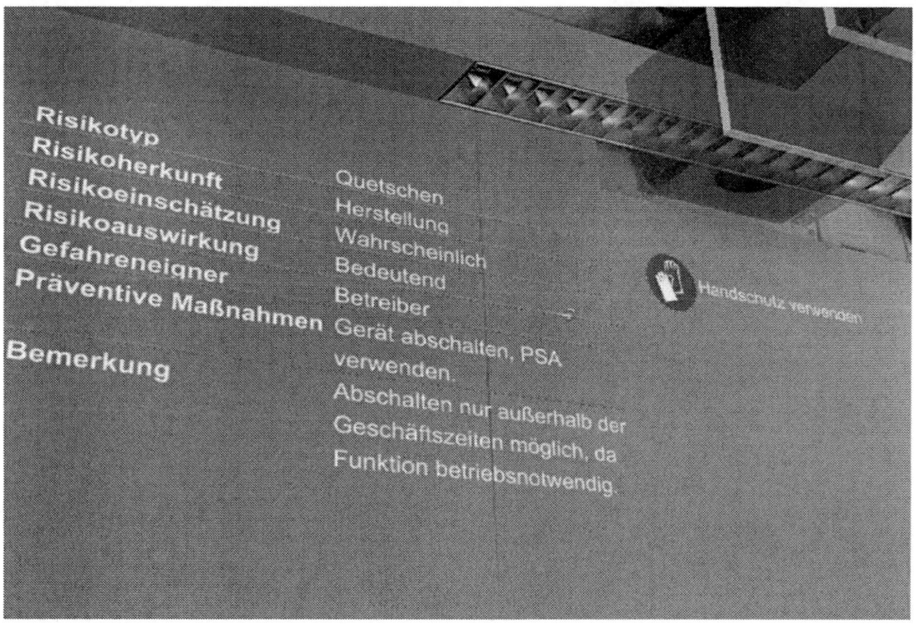

Figure 2. The same table and required PPA from figure 1 in AR on Microsoft HoloLens directly besides the work item (shown as a hologram in the top-right).

5 CONCLUSIONS

The application of reference occupational health and safety process is discussed in many ways, especially in the context of individual construction projects. Against this background, it becomes even more evident how necessary the implementation and application of processes and consistent process models in the real estate life cycle are in order to obtain a large number of examples processes and develop further reference processes.

The feedback for the developed demonstrator has been very positive, especially the Augmented Reality application shows that it can provide additional value for safety without interrupting the workflow because of the hands-free nature of the hardware. It needs to be seen how fast this technology will be established and how the final release version of HoloLens will establish itself in the construction industry. The user only benefits from these devices with a strong user base and the availability of these devices.

The implementations for Microsoft HoloLens and the Android Tablet show that it is possible to implement safety and health protection measurements that can be embedded into existing workflows quite easily. Important for the future is that those experiments will be adapted for existing solutions to really improve work safety for the single worker.

Acknowledgments

This research project is funded by the German Social Accident Insurance (DGUV). In order to ensure a direct reference to the applicability in practice, the following practice partners support the research project: BfGA - Beratungsgesellschaft für Arbeits- und Gesundheitsschutz mbH, ISSA construction, Ingenieurbüro Dudek, Office-4 Baubetreuung GmbH, BG BAU – Berufsgenossenschaft der Bauwirtschaft and Chichon & Stolberg Elektroanlagenbau GmbH.

References

Bodtländer, C., H., M., Kelm, A., Meins-Becker, A., *BIM - Processes - Health & Safety*, ISEC-9, Valencia, Spain. doi: 10.14455/ISEC.res.2017.27, 2017.

Bodtländer, C., Helmus, M., Kelm, A., Meins-Becker, A., Arbeitsschutz – Building Information Modeling - Forschungsbericht zum Projekt Anwendung der Methode Building Information Modeling und Einsatz der RFID-Technik zur Verbesserung des Arbeitsschutzes in der Bau- und Immobilienwirtschaf, 2018.

Interdependence between Structural Engineering and Construction Management
Edited by Ozevin, D., Ataei, H., Modares, M., Gurgun, A., Yazdani, S., and Singh, A.

ISBN: 978-0-9960437-6-2

INTEGRATING BUILDING INFORMATION MODELING AND VISUAL PROGRAMMING FOR BUILDING LIFE-CYCLE COST ANALYSIS

VEERASAK LIKHITRUANGSILP[1], HANG T. T. LE[1], NOBUYOSHI YABUKI[2], and PHOTIOS G. IOANNOU[3]

[1]*Dept of Civil Engineering, Faculty of Engineering, Chulalongkorn University, Pathumwan, Thailand*

[2]*Division of Sustainable Energy and Environmental Engineering, Osaka University, Osaka, Japan*

[3]*Dept of Civil and Environmental Engineering, University of Michigan, Ann Arbor, USA*

In recent years, the fierce competition in worldwide real-estate market has pushed the stakeholders towards the sustainability for buildings. Life-cycle cost (LCC) is an effective economic evaluation tool that provides a detailed account for all costs related to constructing, operating, maintaining, and disposing a construction project over a defined period of time. Awareness of better value of money throughout the LCC is beyond the initial price. Governments and Contracting authorities add the LCC as a key provision in the context of National Codes and Council Directives to promote the growth of sustainability concept. Current LCC analytical methods are costly, laborious, and time-consuming due to the difficulties of obtaining information and implementing many single LCC analyses for all building elements, which may be attributed to the inaccuracy of results. Building information modeling (BIM) is a modern technology that can potentially overcome the asperities that obstruct practical LCC implementation. This paper develops a new automated system for performing LCC analyses for new building projects by integrating BIM authoring programming with visual programming. The proposed system consists of two main modules. The BIM module is designed to retrieve 3D geometric and physical parameters of building element types. The life-cycle cost calculation module can perform automatic estimating and report results. This system provides an economic evaluation tool for the owner to manage the total life-cycle budget of their projects.

Keywords: Sustainability, Economic evaluation, BIM authoring programming, Contractor selection.

1. INTRODUCTION

Buildings serve many vital functions in human life. They provide space for work, education, living, and other public and private uses. In many developing countries, building construction is one of the most important sectors of the construction industry. It is also a sector that consumes natural resources the most and accounts for a high percentage of the greenhouse gas emission, which has an adverse impact on the environment and occupants' health. According to the U.S. Green Building Council (2015), buildings in the U.S. are responsible for 41 percent of total U.S. energy consumption. In addition, they account for 38 percent of total U.S. CO_2 emissions, 40 percent of total raw material use globally, and create 250 million tons of total U.S. solid waste output. The global goals for sustainable development have pushed investors toward sustainable buildings, which are driven with the low running costs in the construction industry. The life-cycle cost (LCC) is an economic evaluation tool that provides a detailed account for all costs related to the design, construction, operation, maintenance, and disposal of a construction project over a defined period of time. The LCC implementation is challenging as cost estimators must cope with various problems concerning collecting and storing relevant information, especially operation and maintenance costs. Assaf *et al.* (2002) pointed out that the difference and lack of data requirements were main barriers for the application of LCC analysis in the construction industry. The LCC calculation will be accurate if the collected data are reliable (Emblemsvåg 2003). In practice, capturing data for LCC analysis is challenging due to numerous data documented in paper forms, the limited existence of data sources, and errors in inputting and transferring data manually to implement many calculation processes.

Building information modeling (BIM) is a modern technology that can improve the productivity of the construction industry. Likhitruangsilp *et al.* (2016) presented thirty BIM uses in various aspects of construction management. Recently, the use of BIM has increased significantly. Integrating BIM platforms in various construction management aspects is another trend in academic research. This will transform the traditional construction industry to the digital industry. Likhitruangsilp *et al.* (2018a) also proposed a BIM-enabled system for calculating the time and cost impacts for construction change orders. This system provides the 3D visual outputs and can save a significant amount of time as compared with traditional paperwork methods. BIM adoption in the construction industry was only at 28 percent in 2007, but the level of adoption became 71 percent in 2012 and kept increasing. In addition, 93 percent of BIM users believed that there was potential that we will gain more value from BIM in the future. Two-thirds of BIM users said that they observed positive ROI on their investment in BIM, and 87 percent of experts were experiencing positive ROI with BIM (Koppinen and Morrin 2012).

Even though there have been several publications related to the LCC analysis, most of them did not consider all cost categories for their computations, especially the energy cost. This contributes to the limited use of LCC as a sustainability-based tool in the construction industry (Langdon 2007). Flanagan *et al.* (1989) proposed a framework for collecting data that are necessary for determining project life-cycle costs. Ashworth (1996) proposed a method for estimating the life expectancies of different building components by using a combination of objective analysis and subjective judgment. Alqahtani and Whyte (2016) used the multiple regression models and artificial neural networks to predict more accurate running costs of building projects. Kehily and Underwood (2016) presented a methodology to embed LCC within an existing 5D BIM technology (CostX) to produce a 5D BIM based LCC artifact. Although this research work provided a solution to develop a LCC tool, it is dependent on a commercial 5D BIM estimating software, which is a major cost for building designers. To alleviate the aforementioned difficulties, this paper proposes an automatic system for building LCC analysis using open-source visual programming, which functions with BIM via Dynamo script for Autodesk Revit. It allows users to extend BIM capacity by using the graphical algorithm editor and Python programming language. Some research works integrated BIM platforms to develop the automatic systems in various construction management areas. For example, Likhitruangsilp *et al.* (2018b) integrated Autodesk Revit Architecture, Excel spreadsheet, Visual Basic for Application, and Revit Dynamo to develop a BIM-enabled change detection system for assessing impacts of construction change orders. Dynamo programming language is easy to understand for cost managers who are usually nonprogrammers. The proposed system is a comprehensive, but user-friendly, tool that can replace the conventional LCC analysis. In addition, we also investigate the extension of current BIM capacities to the LCC calculation for building structure types concerning the element properties of BIM element models.

2 LIFE-CYCLE COST ANALYSIS FOR SUSTAINABLE BUILDINGS

A construction project can be divided into a number of phases: planning design, construction, operation, maintenance, and disposal. Most of them last from one to three years, except for the operation and maintenance phase that may last 30 to 50 years. This is why 30 to 40 percent of the total LCC for a facility incurs in these first three phases, and 60 to 70 percent incurs in the later phases (GCR 2004). For the traditional competitive tendering approach, contractors are often chosen based on their lowest construction costs, which are a small part compared to the running costs of the building life-cycle cost. This is an irrational method because the higher cost may have been created by building owners due to inadequate operation later on. In this paper, we focus on four main cost categories for the LCC analysis. The total life-cycle cost can be calculated as shown in Eq. (1):

$$C_{Life\_cycle} = C_{Construction} + \frac{(1+i)^n - 1}{i(1+i)^n} C_{Maintenance} + \frac{(1+i)^n - 1}{i(1+i)^n} C_{Operation} - \frac{1}{(1+i)^n} C_{Salvage} \tag{1}$$

where C_{Life_cycle}, $C_{Construction}$, $C_{Maintenance}$, $C_{Operation}$, and $C_{Salvage}$ are the total life-cycle cost, construction cost, maintenance cost, operation cost, and salvage cost of a building structure type, respectively. i is the discount rate of money, and n is the period of analysis.

The construction costs related to material, labor, and equipment to construct a building element are obvious compared with other cost categories in the LCC analysis. The maintenance costs are related to the costs of regular care, repair, and replacement of a building element or system. The data sources usually contain the average annual maintenance per structure unit. The expense for maintenance depends on the type of the building structure. This cost may be part or total of initial construction costs to replace such element. The operation costs encompass service, security, and insurance for operating all the building structures under the condition considered. The energy cost usually takes up the greatest share of the operation cost. Most of the building operation cost incur in its operation period. The amount of energy consumption for a building depends heavily on the use hours of building system operations, weather conditions, the performance level required by building owners, as well as the building design and insulation provisions. Thus, project managers should not overlook the energy cost of building elements. Each building element type has a unique expected service life. Once it reaches such expected life, the building element will lose its technical capacity, reliability, and quality due to natural ageing and use. The salvage cost is the value of a building structure at the end of the service life. It is calculated as the difference between the resale and the original values of the building structure. While construction costs are clear, operation and maintenance costs are not obvious in the design phase. Thus, the LCC analysis plays an important role in making a funding decision.

3 RESEARCH METHODOLOGY

To develop the BIM-enabled system for building life-cycle cost analysis, we first did comprehensive literature review on the life-cycle cost analysis, applications of BIM to cost management, and other relevant issues. We then defined the system requirements in accordance with the research objectives. The conceptual framework for proposed system was then established. The next step was to choose the potential platforms for the relevant functions of the proposed system. Once the system has entirely been developed, it was applied to an actual building project as a case study for illustrating and verifying the applicability and practicality of the proposed system.

4 RESEARCH SYSTEM ARCHITECTURE

The proposed BIM-enabled LCC analysis system is a computerized system, which is developed based on the synthesis of LCC analysis practice, the capabilities of BIM tools, and Python programming language. The system can address two main problems of the conventional LCC analysis. The first problematic issue is data management. The LCC calculation heavily relies on various information. In the design phase, owners must usually deal with the issue of information retrieval, especially energy costs. The proposed system can provide the coefficients of heat transfer for all materials, which are essential building element properties required for estimating energy consumption. In addition, the database management system is used to organize, store, and query data for BIM models.

The second challenge of the conventional LCC implementation is time consumption resulting from multiple computations such as performing quantity takeoff as well as estimating construction costs, maintenance costs, energy costs, and salvage costs for all the building structure types. All these calculations often require many inputs (e.g., coefficients), and each input requires many relevant parameters. By using visual programming, the proposed system can implement all the LCC computation in a single simulation process. The system architecture consists of two main modules: the BIM module and the LCC calculation module, as shown in Figure 1.

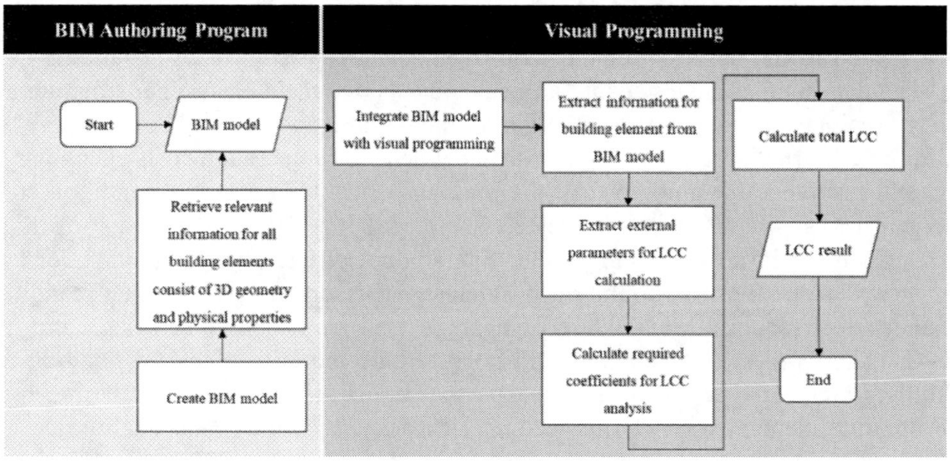

Figure 1. Overview of the architecture of the proposed system.

The BIM module has two significant roles. First, it represents the visual models with geometry parameters for all the building elements. Second, it stores other parameters that are exported from external database into the BIM internal database. The required data of the BIM module are collected from three data sources. The first source is historical data (i.e., project database). The second source is the RSMeans data 2018 book (Phelan 2018). The data consist of material cost, labor cost, and equipment cost for constructing the building element type. The third source is life-cycle costing for design professional books. The retrieved data consist of the expected service life, the annual service unit price, and the percent replaced for each building element type. The LCC analysis module is designed to implement the estimating function and report the results. The information requirements for cost estimating are extracted from the element properties and the spreadsheet LCC template. This module is based on Dynamo script with Python programming language.

5 DATA ACQUISITION AND INTEGRATION

An academic facility is used to illustrate the application of the proposed system. The department of civil engineering building is a five-story building located at Chulalongkorn University, Bangkok, Thailand. After creating and assigning all the element properties of the BIM model, the integrated model was analyzed to automatically extract information required for the LCC coefficient calculations. Figure 2 shows an example for applying the proposed system to a typical external wall type of the case study building.

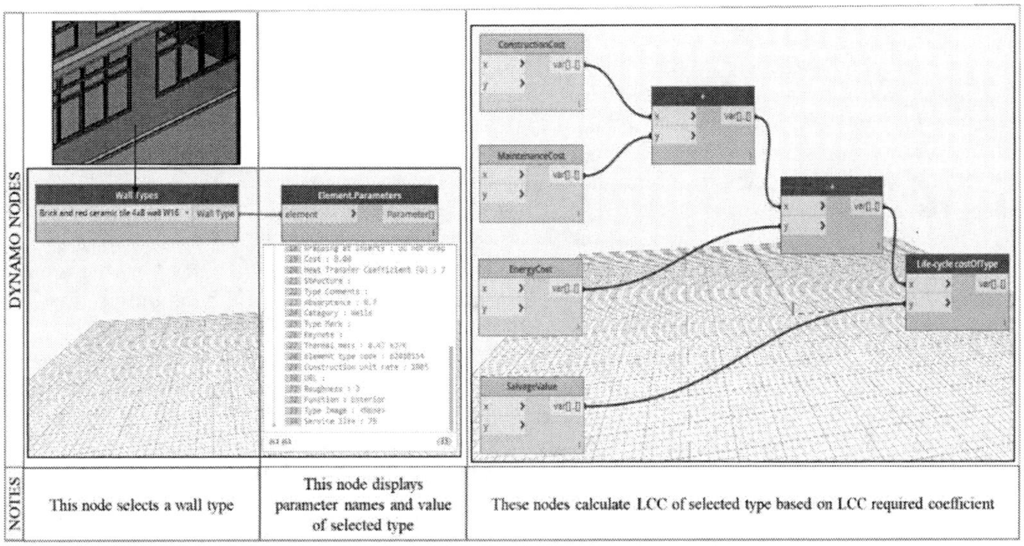

Figure 2. An example for applying the proposed system to a typical external wall type.

6 CONCLUSIONS

This paper presents a new BIM-based system for building life-cycle cost analysis. We propose a methodology for extending the capabilities of current BIM tools for addressing the limitations of conventional LCC implementation, including data management and lengthy estimating processes. The system is then applied to an academic building to verify its capacity and practicality in the construction industry. For future research, green procurement is a key to green growth in order to provide a sustainable approach for the building industry, we intend to investigate a method to integrate our proposed system with green procurement.

Acknowledgments

The authors are very grateful to the AUN/SEED-Net and Department of Civil Engineering, Faculty of Engineering, Chulalongkorn University for funding this research project.

References

Alqahtani, A., and Whyte, A., Estimation of Life-Cycle Costs of Buildings: Regression vs Artificial Neural Network, *Built Environment Project and Asset Management*, *6*(1), 30-4, 2016.

Ashworth, A., Estimating the Life Expectancies of Building Components in Life- Cycle Costing Calculations, *Structural Survey, 14*(2), 4-8. doi:10.1108/02630809610122730, 1996.

Assaf, S., A., Al-Hammad, A., Jannadi, O., A., and Saad, S., A., Assessment of the Problems of Application of Life Cycle Costing in Construction Projects, *Cost Engineering, 44*(2), 17-17, 2002.

Emblemsvåg, J., *Life-Cycle Costing: Using Activity-Based Costing and Monte Carlo Methods to Manage Future Costs and Risks*, John Wiley & Sons, 2003.

Flanagan, R., Norman, G., and Meadows, J., *Life Cycle Costing: Theory and Practice*, BSP Professional Books, 1989.

GCR, N., Cost Analysis of Inadequate Interoperability in The US Capital Facilities Industry, 2004. Retrieved from https://buildingsmart.no/sites/buildingsmart.no/files/b04022.pdf on March 2019.

Green Building Council, Feb 23, 2015. Retrieved from https://stg.usgbc.org/articles/green-building-facts on March 2019.

Kehily, D., and Underwood, J., Embedding Life Cycle Costing in 5D BIM, *Journal of Information Technology in Construction*, 22, 145-167, 2017.

Koppinen, T., and Morrin, N., Green BIM Innovation at Skanka, 2012. Retrieved from https://www.group.skanka.com/4a2286/globalassets/sustainability/green/our-journey-to-deep-green/how-we-define-green/greenbiminnovation.pdf.

Langdon, D., *Life Cycle Costing (LCC) as a Contribution to Sustainable Construction: A Common Methodology - Final Report*. 451-464, 2007.

Likhitruangsilp, V., Malvar, M. J. S., and Handayani, T. N., *Implementing BIM Uses for Managing Risk in Design-Build Projects*, Proceeding of 16th International Conference on Computing in Civil and Building Engineering, Osaka, Japan, 887-894, 2016.

Likhitruangsilp, V., Handayani, T., and Yabuki, N., *A BIM-Enabled Change Detection System for Assessing Impacts of Construction Change Orders,* Proceedings of the 17th International Conference on Computing in Civil and Building Engineering (ICCCBE 2018), Tampere, Finland, 2018a.

Likhitruangsilp, V., Handayani, T., Yabuki, N., and Ioannou, P. G., *A BIM-Enabled System for Evaluating Impacts of Construction Change Orders,* Proceedings of the Construction Research Congress 2018, New Orleans, Louisiana, USA, 2018b.

Phelan, M. (2018). *Assemblies Costs with RSMeans Data 2018 43rd annual edition*. Gordian RSMeans data Construction Publishers and Consultants, United States of America, ISSN 1524-2153.

Interdependence between Structural Engineering and Construction Management
Edited by Ozevin, D., Ataei, H., Modares, M., Gurgun, A., Yazdani, S., and Singh, A.

ISBN: 978-0-9960437-6-2

AN ENGINEERING RELIABILITY ANALYSIS METHOD BASED ON ADAPTIVE METAMODELS

QIAN WANG[1], ERICA JAROSCH[1], and HONGBING FANG[2]

[1]*Dept of Civil and Environmental Engineering, Manhattan College, Riverdale, USA*
[2]*Dept of Mechanical Engineering and Engineering Science, The University of North Carolina at Charlotte, Charlotte, USA*

In practical engineering problems, numerical analyses using the finite element (FE) method or other methods are generally required to evaluate system responses including stresses and deformations. For problems involving expensive FE analyses, it is not efficient or straightforward to directly apply conventional sampling-based or gradient-based reliability analysis approaches. To reduce computational efforts, it is useful to develop efficient and accurate metamodeling techniques to replace the original FE analyses. In this work, an adaptive metamodeling technique and a First-Order Reliability Method (FORM) were integrated. In each adaptive iteration, a compactly supported radial basis function (RBF) was adopted and a metamodel was created to explicitly express a performance function. An alternate FORM was implemented to calculate reliability index of the current iteration. Based on the design point, additional samples were generated and added to the existing sample points to re-generate the metamodel. The accuracy of the RBF metamodel could be improved in the neighborhood of the design point at each iteration. This procedure continued until the convergence of the reliability analysis results was achieved. A numerical example was studied. The proposed adaptive approach worked well and reliability analysis results were found with a reasonable number of iterations.

Keywords: Finite element (FE), First-order reliability method (FORM), Radial basis functions (RBFs).

1 INTRODUCTION

Reliability analysis of civil engineering problems has attracted considerable attention in the last few decades (Hohenbichler *et al.* 1987, Ang and Tang 1975, Li and Low 2010, Au and Wang 2014). Commonly used reliability analysis methods include first-order and second-order reliability methods (FORM/SORM) (Hohenbichler *et al.* 1987, Low and Tang 2007). In addition, various sampling methods are also available, e.g., Monte Carlo simulations (MCS) (Rubinstein 1981) and subset simulations (Au and Wang 2014). For engineering practice, a numerical analysis method such as the FE analysis is routinely used to evaluate system responses. Therefore, the numerical analysis codes needs to be integrated with MCS or FORM/SORM for practical applications.

For engineering applications that require expensive response analyses, it is useful to develop approximate models to replace the implicit performance functions. This is referred to as a metamodel and the system responses are written explicitly in terms of input variables (Bai *et al.* 2012). A single quadratic function is commonly used to replace an implicit function in a metamodel (Faravelli 1989). This is called response surface method (RSM), which can be used to solve a variety of engineering problems, including design optimization, reliability analysis, as well as reliability-based design optimization (Youn and Choi 2004, Lü and Low 2011, Lü *et al.* 2011). To improve the model accuracy for highly nonlinear functions, different techniques have been studied, including support vector machines (Zhao *et al.* 2014), radial basis functions (Wu 1995, Fang and Horstemeyer 2006, Yin *et al.* 2016), and high dimensional model representation (Tunga and Demiralp 2005). FORM/SORM and MCS can be employed to calculate the reliability index, once an explicit metamodel function becomes available.

It is useful to investigate accurate and efficient approximation models that are applicable to complex engineering problems. A new reliability analysis method is presented in this paper that integrates FORM and an adaptive metamodeling technique. The basic concept of an adaptive metamodel is that the model accuracy can be improved in each reliability analysis iteration, when additional sample points are introduced. The metamodels are based on augmented RBFs, and the reliability index is found based on an alternate FORM in literature. In the next section, the proposed method is first introduced. Finally, a numerical example and some concluding remarks are presented.

2 THE RELIABILITY ANALYSIS METHOD

The proposed method based on combined FORM and an adaptive RBF technique is introduced in this section.

2.1 Reliability Analysis

The failure probability, P_F, is based on the following integration (Madsen *et al.* 1986):

$$P_F \equiv P(g(x) \leq 0) = \int_{g(x)\leq 0} p_X(x)dx \tag{1}$$

where **x** represents all random variables in a vector form and $g(\mathbf{x})$ is a performance function. In Eq. (1), the failure probability P_F varies between 0.0 and 1.0.

2.2 An RBF-Metamodeling Technique

A RBF metamodel can be created using n sample points in Eq. (2) (Fang and Horstemeyer 2006):

$$\tilde{g}(\mathbf{x}) = \sum_{i=1}^{n} \lambda_i \phi(\|\mathbf{x} - \mathbf{x}_i\|) \tag{2}$$

where ϕ is a basis function and λ_i is the weighted coefficient. In this work, an augmented RBF metamodel combining the compactly supported basis function $\phi_{2,0}$ (Wu 1995) and a polynomial function was used (Fang and Wang 2008).

2.3 FORM

For correlated normal random variables, the reliability index is defined as (Hasofer and Lind 1974):

$$\beta = \min_{\mathbf{x}\in F} \sqrt{(\mathbf{x}\text{-}\boldsymbol{\mu})^T \mathbf{C}^{-1} (\mathbf{x}\text{-}\boldsymbol{\mu})} \tag{3}$$

An alternative interpretation of Eq. (3) was presented by Low and Tang (2007). Using an expanding ellipsoid concept in the original space of random variables, β is written as in Eq. (4):

$$\beta = \min_{\mathbf{x} \in F} \sqrt{\mathbf{n}^{\mathrm{T}} \mathbf{R}^{-1} \mathbf{n}} \tag{4}$$

where $\mathbf{n}$ is a column vector of $n_l = \frac{x_l - \mu_l^N}{\sigma_l^N} = \Phi^{-1}[F(x_l)]$. Φ^{-1} is the inverse cumulative distribution function (CDF) of a standard normal variable (Low and Tang 2007). A nonlinear optimization algorithm can be used to find β, by treating the vector $\mathbf{n}$ as design variables in optimization (Low and Tang 2007, Zhao *et al.* 2014).

2.4 An Adaptive Approach

After an initial metamodel is generated, an adaptive technique is applied. This will locally improve the RBF metamodeling accuracy. This local refinement can continue until a certain stop criterion is met. Note that the computational cost is mainly determined by the metamodel sample size, i.e., the number of FE simulations.

2.5 Overall Procedure

(i) Determine an initial sample size and generate initial sample points.
(ii) Calculate performance function values at the sample points. This step involves FE analyses or other numerical analysis methods to evaluate system responses.
(iii) Create an RBF metamodel of the performance function using all sample points.
(iv) Perform reliability analysis and find the reliability index β using the alternate FORM.
(v) Check convergence of the reliability index β. Stop the overall procedure if a convergence is achieved. Otherwise, go to the next step.
(vi) Include the current design point as an additional sample point. An additional FE analysis is required to evaluate the system responses. Go to step (iii).

3 A NUMERICAL EXAMPLE

Figure 1 shows a circular tunnel surrounded by homogeneous and isotropic rock mass (Hoek 1998, Lv *et al.* 2011, Zhao *et al.* 2014). The tunnel is subjected to a hydrostatic far field stress p_0. To support the tunnel, a uniform internal pressure p_i is applied. To be consistent with the studies found in literature, two performance functions used in this work are shown in Eq. (5) and (6):

$$g_1(\mathbf{x}) = 3 - \frac{r_p}{r_0} \tag{5}$$

$$g_2(\mathbf{x}) = 0.01 - \frac{u_{ip}}{r_0} \tag{6}$$

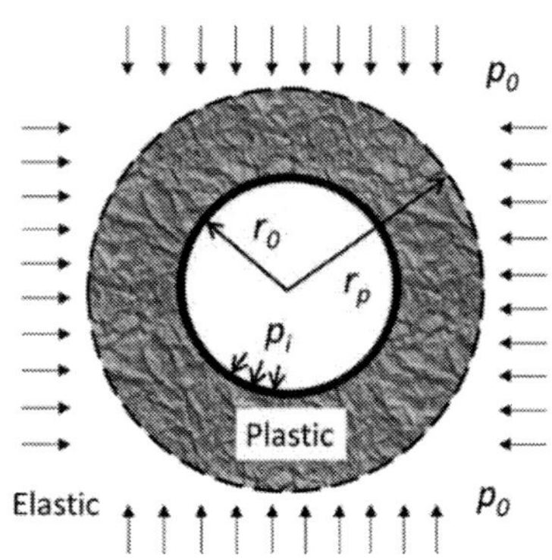

Figure 1. A circular tunnel example.

Table 1. Random variables of the tunnel example.

Variable	Distribution	Mean	Standard deviation
E	Normal	373 (Mpa)	48 (Mpa)
c	Normal	0.23 (Mpa)	0.068 (Mpa)
ϕ	Normal	22.85 (Degrees)	1.31 (Degrees)

The first performance function $g_1(\mathbf{x})$ is a criterion for tunnel plastic zone size, and the allowable size is 3. The second performance function $g_2(\mathbf{x})$ defines the inward displacement requirement. Table 1 lists random variables of the tunnel problem, i.e., E, c, and ϕ. All other variables are deterministic variables. Closed-form analytical performance functions are available and details of the functions are given in literature (Li and Low 2010). The correlation coefficient between cohesion c and friction angle ϕ was –0.5. For the first performance function, far field stresses p_O= 2.5 MPa and support pressures p_i= 0.0 MPa. For the second performance function, far field stresses p_O= 2.5 MPa and support pressures p_i= 0.868 MPa (Li and Low 2010). Twenty-one initial samples were generated based on the Latin hypercube sampling method. One additional sample point was generated in each adaptive RBF iteration.

For both performance functions, four adaptive iterations were required, and a total of 21+3=24 sample points were needed. The reliability analysis was also performed using FORM and closed-form analytical functions. The reliability indices and design points were calculated and the analysis results of both performance functions are listed in Table 2. For the first performance function, both methods led to β = 0.697. The reliability index β calculated using both methods was 2.504 for the second performance function. To further compare results, a global RBF metamodeling method was also used such that FORM could be applied. To obtain the global RBF metamodels, a total of 51 sample points were created based on the Latin hypercube sampling method. The global RBF method resulted in a reliability index of 0.692 and 2.510 for the two performance functions, respectively. These results are not as accurate as those obtained using the adaptive RBF approach. From this example it is seen that the adaptive RBF approach worked well and the accurate results of the reliability analysis were obtained within a few adaptive iterations.

Table 2. Reliability analysis results of the tunnel example.

		$g_1(\mathbf{x})$			$g_2(\mathbf{x})$		
		Analytical function	Adaptive RBF	Global RBF	Analytical function	Adaptive RBF	Global RBF
Far field stress	p_o (Mpa)		2.5			2.5	
Support pressure	p_i (Mpa)		0.0			0.868	
Reliability index	β	0.697	0.697	0.692	2.504	2.504	2.510
	E (Mpa)	373	373	372	259	260	260
Design point	c (Mpa)	0.185	0.185	0.186	0.189	0.187	0.186
	ϕ (Degrees)	23.019	23.017	23.005	22.653	22.609	22.630

4 SUMMARY AND CONCLUSIONS

This work studied an effective engineering reliability analysis method, which integrated an alternate FORM and an adaptive RBF approach. The proposed method was implemented in an iterative manner. The RBF model accuracy was locally improved, due to the additional sample points identified in each analysis iteration. The new method was applied to an example problem and accurate results were obtained. A few iterations and a relatively small number of sample points were required to achieve convergence of the iterations. The proposed approach is useful for the analysis and design of complex problems that require expensive response simulations.

References

Ang, A. H. S., and Tang, W. H., *Probability Concepts in Engineering Planning and Design, Vol. 1 Basic Principles*, Wiley, New York, 1975.

Au, S. K., and Wang, Y., *Engineering Risk Assessment with Subset Simulation*, John Wiley & Sons, 2014.

Bai, Y. C., Han, X., Jiang, C., and Liu, J., Comparative Study of Metamodeling Techniques for Reliability Analysis Using Evidence Theory, *Advances in Engineering Software*, Elsevier, 53, 61-71, 2012.

Fang, H., and Horstemeyer, M. F., Global Response Approximation with Radial Basis Functions, *Engineering Optimization,* Taylor & Francis, 38(04), 407-424, 2006.

Fang, H., and Wang, Q., On The Effectiveness of Assessing Model Accuracy at Design Points for Radial Basis Functions, *Communications in Numerical Methods in Engineering*, Wiley, 24(3), 219-235, 2008.

Faravelli, L., Response Surface Approach for Reliability Analysis, *Journal Engineering Mechanics*, 115(12), 2763-2781, 1989.

Hasofer, A. M., and Lind, N. C., Exact and Invariant Second Moment Code Format, *Journal Engineering Mechanics*, 100(1), 111-121, 1974.

Hoek, E., Reliability of Hoek-Brown Estimates of Rock Mass Properties and Their Impact on Design, *International Journal of Rock Mechanics and Mining Science*, Elsevier, 35(1), 63–68, 1998.

Hohenbichler, M., Gollwitzer, S., Kruse, W., and Rackwitz, R., New Light on First and Second-Order Reliability Methods, *Structural Safety*, Elsevier, 4, 267-284, 1987.

Li, H.-Z., and Low, B. K., Reliability Analysis of Circular Tunnel under Hydrostatic Stress Field, *Computers and Geotechnics*, Elsevier, 37, 50–58, 2010.

Low, B. K., and Tang, W. H., Efficient Spreadsheet Algorithm for First-Order Reliability Method, *Journal Engineering Mechanics.*, ASCE, 133(12), 1378–1387, 2007.

Lü, Q., Sun, H. Y., and Low, B. K., Reliability Analysis of Ground-Support Interaction in Circular Tunnels using Response Surface Method, *International Journal of Rock Mechanics and Mining Sciences*, Elsevier, 48(8), 1329–1343, 2011.

Lü, Q., and Low, B. K., Probabilistic Analysis of Underground Rock Excavations Using Response Surface Method and SORM, *Computers and Geotechnics*, Elsevier, 38, 1008–1021, 2011.

Madsen, H. O., Krenk, S., and Lind, N. C., *Methods of Structural Safety*, Prentice-Hall, Englewood Cliffs, NJ, 1986.

Rubinstein, R. Y., *Simulation and the Monte Carlo Method*, Wiley, New York, 1981.

Tunga, M. A., and Demiralp, M. A., Factorized High Dimensional Model Representation on The Nodes of a Finite Hyperprismatic Regular Grid, *Applied Mathematics and Computation*, Elsevier, 164(3), 865-883, 2005.

Wu, Z., Compactly Supported Positive Definite Radial Function, *Advances in Computational Mathematics*, Springer, 4, 283-292, 1995.

Yin, H., Fang, H., Wen, G., Wang, Q., and Xiao, Y., Adaptive RBF-Based Multi-Objective Crashworthiness Optimization of Functionally Graded Multi-Cell Tube, *Structural and Multidisciplinary Optimization*, Springer, 53(1), 129-144, 2016.

Youn, B. D., and Choi, K. K., A New Response Surface Methodology for Reliability-Based Design Optimization, *Computers and Structures*, Elsevier, 82, 241-256, 2004.

Zhao, H., Ru, Z., Chang, X., Yin, S., and Li, S., Reliability Analysis of Tunnel Using Least Square Support Vector Machine, *Tunnelling and Underground Space Technology*, Elsevier, 41, 14-23, 2014.

Interdependence between Structural Engineering and Construction Management
Edited by Ozevin, D., Ataei, H., Modares, M., Gurgun, A., Yazdani, S., and Singh, A.

ISBN: 978-0-9960437-6-2

ASSESSMENT OF BUILDING INFORMATION MODELING UPTAKE IN THE MANAGEMENT OF CONSTRUCTION PROJECTS

RAPHEAL ABIODUN OJELABI, IGNATIUS OWOICHO OMUH, ADEDEJI OLUSHOLA AFOLABI, and PATIENCE FIKIYEMO TUNJI-OLAYENI

Dept of Building Technology, Covenant University, Ota, Nigeria

The need for digitization of industries' production process has been given high attention in the 21st century. This is due to the upsurge in technology in the present age. Industries which include manufacturing, telecommunication and aviation among others have set the pace for others to follow in the use of digital tools in their production process. Therefore, this study is aimed at examining the awareness and level of uptake of a digital tool like Building Information Modeling (BIM) in the management of construction project and the challenges limiting its use within the built environment. The study adopted a survey method by randomly selecting 120 construction practitioners in construction firms operating in Lagos state Nigeria. Likewise, 120 structured questionnaires were distributed to the targeted respondents, however, 100 questionnaires were retrieved. The data collected from the survey were processed using Mean item score, percentage, and relative importance index. Findings garnered from the study revealed that construction organization awareness of the BIM tool is very high, however, its uptake in the construction industry is not encouraging. It further revealed that the driving factor in encouraging the BIM adoption cut across the government, the construction industry and the clients. In conclusion, the study recommends that the trio should provide an adequate support structure and push for BIM implementation in construction project delivery.

Keywords: Construction organization, Built environment, Digitization, BIM.

1 INTRODUCTION

The construction industry is one of the key sectors across the globe due to its capacity to cause physical transformation of the urban environment (Ojelabi *et al.* 2017). The industry relevance is evidenced in its deliverables, which are visible in nations across the globe. In a bid to sustain the leading role of the construction industry globally, stakeholders in the industry geared their effort in developing a sustainable system that can enhance the performance of the professional players within the construction industry. Such efforts have led to the development of a digital tool known as Building Information Modeling System (BIM). Takim *et al.* (2013) asserted that the introduction of BIM in the construction industry has brought about a positive shift in the built environment. The emergence of Building Information Modeling System in the built environment has changed the face of the construction industry. Bui *et al.* (2016) affirmed that BIM is a catalyst for innovation and accelerated productivity in the built environment.

BIM is a system that adopts computer technology in generating a model capable of causing collaboration as to populate information and stimulate the planning, design, construction, and operation of a facility (Construction Industry Development Board 2013, Amusan *et al.* 2018). The capacity of the system as revealed in the definition, which includes planning, design, construction, and operation provides a powerful synergy that makes it unequal to other software used in the built environment. BIM also has the capacity for collaboration of all stakeholders in the built environment. Prior to the initiation of the BIM system, Lindblad (2013) opined that the construction industry is one of the largest industry with the highest stakeholders and as well the most fragmented across the globe. The solo effort of the different stakeholders in the construction industry is unequal to any other industry in the past. However, the introduction of BIM system has helped foster a high level of collaboration among construction players, which have culminated into prompt and effective project delivery. Roberti and Buckett (2012) affirmed that BIM is an information and technology software initiated to encourage interdisciplinary collaboration and communication among construction professionals in the production of construction facilities.

It is evident that the benefits of BIM in the construction industry are enormous. However, much impact has not been experienced in developing nations built environment compared to the developed nations. This has warranted the study to assess the level of adoption of BIM in the developing nations of the world using Nigeria as a case study and also investigate the driving factors that can encourage BIM implementation.

2 LITERATURE REVIEW

The world is referred to as a global village with the advent of information technology (IT). There is no sector across the globe that do not tap into the benefits of information technology. The introduction of information technology in the construction industry is very instrumental to the cutting edge recorded in the built environment. Fischer and Kunz (2006) affirmed that blended history of the construction industry and information technology span between 3 decades and such have accounted for the positive shift experienced in the built environment. The integration of information technology (IT) in the construction industry has led to the development of different software's and BIM falls in the latest and trending software in the industry. Adoption of BIM in the construction process in the built environment has undoubtedly enhanced the performance of the construction professionals in achieving a better and faster delivery of infrastructures. Associated General Contractors of America (2005) affirmed that models generated from BIM system are highly rich in data for use of the individual professional player in the construction industry for analysis to aid their decision making in the delivery of infrastructures. Rokooei (2015) revealed further how instrumental BIM have been in the construction industry due to its capacity to create a common language among construction professionals by making them an integrated team in the process of project delivery. The synergy platform provided by BIM for all players in the construction industry has never been like what is in existence now. Such platforms have however culminated into a geometric increase in the production of infrastructures delivery.

However, despite the introduction of BIM across the globe, its deliverables are unequal probably due to the issue of adoption and implementation among different countries. Developed nations of the world undoubtedly have benefited immensely from the deliverables of BIM due to the visible result in their built environment. Smith (2014)'s study on "global strategies on BIM implementation" revealed that advanced nations of the world gives high credence to the adoption and use of BIM and such have contributed to the massive improvements in their built environment. Wong *et al.* (2009) identified United State as the leader in the development and implementation of BIM system. Azhar *et al.* (2008) revealed that 75% of construction companies in the United States are considering the adoption of BIM and will only employ those with BIM skills. Smith (2014) further revealed the stand of General Service Administration (GAS), a body responsible for the management of federal government's facilities as it mandates the use of BIM for all public works in the United State of America. This accounted for the jump in the adoption of the tool from 28% in 2007 to 71% in 2012 within the territory of North America (McGraw 2014). It is not surprising why they are referred to as one of the construction leaders in the world. The great transformation in their built environment can be traced to their effort in the adoption and use of sustainable tools like BIM. In the same vein, the United Kingdom is another country that has a significant record in BIM adoption. The country has something in common with the United State in their push for BIM adoption and usage among industry players. HM Government (2012) revealed that the UK government developed a strategy for implementation of BIM in all public projects as they considered it more cost effective and better management tool for construction-related projects. Smith (2014) revealed further that the BIM strategy plan has a significant impact on the stakeholders in the built environment in both public and private sector as they move towards developing manpower technically in order to meet the requirements. Likewise, Shaikh *et al.* (2016) asserted that countries in the Scandinavian region like Finland, Denmark, and Norway among others also shows remarkable push in the adoption and use of BIM.

Bui *et al.* (2016) asserted that developed and developing nations are not altogether equal in the adoption and implementation of BIM. Nigeria, for example, is a developing country with a fast-growing population equal to its level of infrastructural development; Nigeria ranked 132nd out of 138th infrastructural strength among nations across the globe (World Economic Forum 2016). It is evident that the construction industry productivity in Nigeria is not encouraging. This can be traced to the poor rate of infrastructures production due to the report of construction projects time and cost overruns, inadequate planning, and budgetary provisions, inefficient and poor service delivery among others. However, Lindblad (2013) revealed that one of the key issues affecting construction industry productivity can be attributed to the fragmentation of the professionals in the industry. In a bid to address the professional fragmentation issue, information and communication technology (ICT) tools are conceived to deal with the shortcoming (Johnson and Laepple 2003). The construction industry in Nigeria has gone through different technological phases, however, the adoption of the latest technological BIM tool is still subject to questioning.

3 RESEARCH METHODOLOGY

The study was aimed at assessing the level of adoption and implementation of building information modeling in Nigeria for construction related works in Lagos state. In a bid to achieve the aim, the study set the following objectives to examine the level of awareness and use of BIM and identify the factors that can encourage BIM adoption among the local professional's using the relative importance index (RII) to determine the level of importance of the factors. Data was sourced from both primary and secondary sources. Primary data was obtained with the aid of structured questionnaires administered to the construction professionals in the study area. The sample was selected using purposive sampling technique, one of the non-probabilistic techniques. Construction professionals' views in Lagos were sorted due to their engagement in high numbers of construction works ongoing in the state. A total of 120 questionnaires were distributed.

4 RESULT AND ANALYSIS

4.1 Characteristics of Respondents

In this section, the personal information of the respondents used for the study was analyzed using percentage. The results obtained are presented in Table 1.

Table 1. Respondents' characteristics.

Characteristics	Frequency	Percentage
Academic qualification		
HND	2	2.0
B.Sc/BTech	42	42.0
M.sc/MPM	56	56.0
Total	**100**	**100**
Profession		
Architect	30	30.0
Quantity surveyor	16	16.0
Builder	38	38.0
Structural engineer	11	11.0
Electrical engineer	5	5.0
Total	**100**	**100**
Years of industrial experience		
1-5 years	37	37
6-10 years	54	54
11-15 years	4	4
16-20 years	4	4
Above 21 years	1	1
Total	**100**	**100**
Professional qualification		
NIA	37	37
NIOB	43	43
NIQS	8	8
NSE	12	12
Total	**100**	**100**

Table 1 shows the summary of the demographic characteristics of the respondents. It was also discovered that 42% of the respondents have B.Sc/B.Tech as their academic qualification. Also, 56% of the respondents have MSc/MPM and only 2% possesses HND as their highest academic qualification. The analysis shows that the respondents are fairly represented academic wise. The findings further revealed the construction professionals were well represented in the study. Builders have the highest respondents of 38% followed by Architects with a percentage score of 30%. Quantity surveyors and structural engineers are also fairly represented with the percentage score of 16% and 11% respectively while the least professional respondents are the electrical engineers with a percentage score of 5%. The demographic results also showed that the respondents' industrial experience is sufficient for the study as 54% of the respondents' years of industrial experience falls within the range of 6-10 years. It was followed by respondents with 1-5 years of experience with 37%. The professional affiliation of the respondents was also captured and it was revealed that 37% of the respondents were NIA members, 43% are NIOB members, 12% are NSE members and 8% are NIQS members respectively.

4.2 Awareness and Use of BIM

The study revealed the level of awareness and use of BIM among the local professionals. The findings revealed that the level of awareness is not bad, but the usage level of BIM is very low. It is evidenced from the result in Table 2 that only 12% of the respondents are not aware of BIM, however, 83% are aware of the software but are not using it while the percentage of respondents using it are 5% as revealed from the study.

Table 2. Level of BIM awareness and use.

BIM awareness and use	Frequency	Percent
Not aware	12	12.0
Aware and using BIM	5	5.0
Aware but not using BIM	83	83.0
Total	**100**	**100.0**

4.3 Factors Encouraging BIM Adoption and Implementation

This section reveals the varying factors that can encourage BIM adoption and implementation. The factors are ranked in accordance to their level of importance as presented in Table 3.

Table 3. BIM drivers.

	RII	Rank
Affordability of BIM software.	0.98	1
Government enforcement support in the adoption and implementation of BIM.	0.95	2
Increasing investment in BIM software.	0.91	3
Competitive pressure.	0.87	4
Client pressure.	0.86	5
Availability of BIM training programs.	0.86	5
Attaching benefits to BIM competency.	0.85	6
Mandate construction staff to be BIM competent.	0.82	7
Creating an organizational structure that supports BIM.	0.82	7
Grant provision scheme for BIM training.	0.78	8

It is evidenced from the results that all the identified factors are significantly important as their relative importance values are all above 0.70. However, three of the factors have remarkably high importance index value above 0.90 and they are affordability of BIM software, government enforcement support in the adoption and implementation of BIM and increasing investment in BIM software with 0.98, 0.95, and 0.91 importance index respectively.

5 CONCLUSION AND RECOMMENDATION

The findings from the study revealed that the level of awareness of BIM among the construction experts is very high; however, its implementation in the construction industry is very poor. To enhance the tool adoption, factors that can contribute to the BIM implantation were assessed and the findings revealed that the drivers cut across the construction industry, clients, and the government. Therefore, the study recommends that it is essential for the three parties to collaborate in a bid to enhance the implementation of the tool within the built environment.

Acknowledgments

The researchers appreciate the publication support fund given by Covenant University towards the publication of this article.

References

Amusan, L., Ezewinwe, O., Osawaru, F., Joshua, O., Afolabi, A., and Ojelabi, R., Creating Sustainable Construction: Building Informatics Modelling and Lean Construction Approach, *Journal of Theoretical and Applied Information Technology*, 96(10), 3025-3035, 2018.

Associated General Contractors of America, *The Contractor's Guide to BIM*, 1st Ed., AGC Research Foundation, Las Vegas, NV, 2005.

Azhar, S., Hein, M., and Sketo, B., *Building Information Modelling (BIM): Risks and Challenges*, Proceedings of the 144th ASC Annual Conference, Aurburn, Alabama, 2008.

Bui, N., Merschbrock, C., and Munkvold, B. E., A Review of Building Information Modelling for Construction in Developing Countries, *Procedia Engineering*, 164, 487-494, 2016.

Construction Industry Development Board, *Workshop of BIM Portal and Collaboration Platform for Affordable BIM*, Holiday Inn Glenmarie, Shah Alam, Selangor, Kuala Lumpur, January 9-11, 2013.

Fischer, M., and Kunz, J., The Scope and Role of Information Technology in Construction, 2006. Retrieved from http:cife.stanford.edu/online.publications/TR156.pdf on June 15, 2018.

HM Government, *Building Information Modelling, Industrial Strategy- Government and Industry in Partnership*, Government Report, London, 2012.

Johnson, R. E., and Laepple, E. S., *Digital Innovation and Organizational Change in Design Practice*, CRS Center, Texas A & M University, 2003.

Lindblad, H., Study of the Implementation Process of BIM in Construction Projects, 2013. Retrieved from http://kth.diva-portal.org/smash/get/diva2:633132/FULLTEXTD1 on July 22, 2018.

McGraw, H., *The Business Value of BIM for Construction in Major Global Markets: How Contractors Around the World Are Driving Innovation with Building Information Modelling*, Smart Market Reports, 2014.

Ojelabi, R. A., Oyeyipo, O., and Afolabi, A. O., Built Environment Professionals' Perception of the Effectiveness of Building Control Measures in Lagos State, *Journal of Construction in Developing Countries*, 22(1), 41-54, 2017.

Roberti, J., and Buckett, N., *Analysis of the BIM Issues as a Basis for a BRANZ Research Program*, Memorandum, BRANZ, Wellington, New-Zealand, 2012.

Rokooei, S., Building Information Modelling in Project Management: Necessities, Challenges and Outcomes, *Procedia-Social and Behavioral Sciences*, 210, 87-95, 2015.

Shaikh, A. A., Raju, R., Malim, N. L., and Jayaraj, G. K., Global Status of Building Information Modelling (BIM) - A Review. *International Journal on Recent and Innovation Trends in Computing and Communication*, 4, (3), 300-303, 2016.

Smith, P., BIM Implementation – Global Strategies, *Procedia Engineering*, 85, 482-492, 2014.

Takim, R., Harris, M., and Nawawi, A. H., Building Information Modelling (BIM): A New Paradigm for Quality of Life within Architectural Engineering and Construction (AEC) Industry, *Procedia-Social and Behavioral Sciences*, 101, 23-32, 2013.

Wong, A., Wong, F., and Nadeem, A., *Comparative Roles of Major Stakeholders for the Implementation of BIM in Various Countries*, Hong Kong Polytechnic University, Hong Kong, 2009.

World Economic Forum, The Global Competitiveness Report, 2016. Retrieved from www.weforum.org/gcr on July 3, 2018.

Interdependence between Structural Engineering and Construction Management
Edited by Ozevin, D., Ataei, H., Modares, M., Gurgun, A., Yazdani, S., and Singh, A.

ISBN: 978-0-9960437-6-2

GREEN ROOFS TO IMPROVE THE ENERGY RENOVATION RESILIENCE OF HISTORIC BUILDINGS

GIOVANNI SANTI[1] and SARA BATTINI[2]

[1]*Dept of Energy, Systems, Territory, and Constructions Engineering (D.E.S.T.e.C.), University of Pisa, Pisa, Italy*

[2]*Dept of Civil and Industrial Engineering (DICI), University of Pisa, Pisa, Italy*

The introduction of vegetation in urban areas, through both green roofs and green walls, is a sustainable strategy for improving the environment and the quality of life, as well as crucial for urban biodiversity since the moment it is able to create new habitats for plant and animal species. The design and realization of green roof systems abroad is promoted and stimulated while in Italy, this subject, is still an innovation not supported by many real implementations. The application of this technological green system has a great importance for the redevelopment of existing building heritage, especially for historic buildings, to improve their energy-performance qualities, with respect for their architectural value. The aim of this study is to identify the technical issues for the realization of green roofs in urbanized contexts by focusing on the implementation of a green roof on a building of Leghorn following intervention guidelines developed. This research shows that not only does this system allow higher energy saving, but it also brings a decrease of load bearing on the structure.

Keywords: Greenery integrated system, Retrofits, Benefits, Energy saving, Costs, Components, Structural assessment.

1 INTRODUCTION

The application of green roofs in architecture and in the green technology field can be found more often in new constructions, but sometimes, they are considered as retrofit instruments for existing buildings. Green roofs, in particular the extensive ones, are a technology that was originally promoted in Europe several decades ago and are increasingly realized in cities, because they are an important strategy that addresses some key urban environmental issues, allowing the achievement of different benefits (Tabares-Velasco *et al.* 2012, European Commission 2013). Since the first studies in this field, it has been possible to appreciate the positive effects of green roofing at urban scales and on single building, such as the storm water management, the retention of polluting substances, the energy conservation, the mitigation of the urban heat island effect (UHI), the increased longevity of rooftop membranes, the absorption of the particulates and the noxious substance in the air, the oxygen production and the relative absorption of CO_2, in addition to the protection against the erosion due to atmospheric agents and climatic phenomena (Berardi *et al.* 2014, Galbrun and Scerri 2017). Green environments improve air quality and have proven economic health benefits such as stress reduction, lower blood pressure and muscle tension, and increased positive feelings (Coutts and Hahn 2015). In this context green roofs, in cities, can magnify these effects for business people and city residents, and extensive green roofs are a modern modification of the roof-garden concept because they have shallower substrates, require less maintenance, and are more strictly functional in purpose than intensive living roofs or roof gardens. Several study cases show the environmental, thermal and acoustic benefits of horizontal green roof systems, particularly in the case of retrofitting existing buildings. Nevertheless, it is also important to understand the relationships between these systems in architectural terms, therefore linked to both the urban context and the single building. Currently, commercially tested systems, assessed by various studios, are limited to find better technology solutions that optimize stratigraphy to be deployed or technology solutions with detail points to significant parts, but which can enter into crisis, for example, in building cases with particular architectural and historical features. Since we are taking into consideration an intervention on a pre-existent building, it is necessary to check that:

1. The increase of loads can be supported by the existing structure of the building.
2. Enough space can be found for the increased thickness of roofs and facades.
3. The new envelope may assure the expected performance.
4. Construction elements which can obstruct the implementation of the system be removed or mitigated.
5. Implementation in the pre-existent building of the irrigation plant of the greenery system can be possible.
6. Plant species should integrate with the context, so that they must be chosen carefully.

There are several studies which analyze the energetic performance of these systems according to the localization, the plant species implemented, planting substrates and ecological relevance (Francis and Lorimer 2011) but only few studies analyze the relation between the identity of the building, from an architectural and technical viewpoint, and the features of this kind of system. In this context, the reversibility of the retrofitting interventions is very important to preserve the structural and architectural features of the building (Maahsen-Milan and Fabbri 2013). The aim of this study is to focus on the architectural and executive aspects of the implementation of extensive green roofs so that they can be perfectly integrated with the existing building, and to quantify the possible energy benefits obtained. At present, the reference legislation in Italy is represented by UNI 11235:2015 (Guidelines for the design, execution, control and maintenance of green roofing) that defines the rules in relation to particular situations of use, climatic and building context.

2 METHODS

Speaking about retrofitting of existing buildings can appear easy since it is characterized by construction interventions based on dry assembly of several technological elements, but a careful evaluation of possible limitations to plan the strategies which permit to overcome them is needed. In this condition it is of considerable importance to define the field of action on the existing building heritage according to the construction period (in relation to both stylistic-architectural and technical-constructive canons) and to the presence of possible conservative limitations (the field of restoration for buildings with historical/landscape implies further verifications to check whether interventions are allowed). It is therefore necessary to confront ourselves with current, local and national laws and rules, i.e. binding legislation (laws) and voluntary rules (technical standards), in order to verify the legitimacy and the feasibility of retrofitting works; in this regard intervention guidelines were developed after the analysis of some implementations (Table 1).

3 THE CASE STUDY

This work describes the implementation made at the Frangerini Impresa S.r.l in Leghorn, illustrating the various phases of the installation and providing a cost estimation of that.

On June 2018, a green roof was implemented on building located in the industrial area of Leghorn. The building consists of two blocks: one destined to stock which develops completely on the ground floor and the other one characterized by two levels: the ground floor used as stock and the upper floor for offices. The area of the installation is the walkable terrace which is located on the first block adjacent to the office plan having dimensions 24.7m x 3m. The structure is made by prefabricated reinforced concrete and, in particular, the slab below the terrace consists of pre-fabricated Y concrete beams and concrete cast with electrowelded wire mesh; then there is an insulating panel in XPS, a waterproofing sheath and a floating floor.

Table 1. Logical frameworks of the problems inherent to retrofitting interventions on existing buildings.

Problems	Intervention strategies	Purposes
Higher costs for the recovery intervention	According to the level of complexity of the building that is going to be recovered, especially for those with a high historical-architectural value, the costs of the intervention may increase. A precise economic evaluation is necessary, even resorting to the Value Analysis, in order to determine the cost-benefit ratio of the intervention.	Economic sustainability of the intervention with value verification of the necessary categories of works.
Evaluation of the conservative importance of the historical elements that characterize the building.	Retrofitting interventions can involve both a single building and groups of buildings, but also any single intervention, if set in a historical context, conditions and is conditioned by the context. In-depth archival research is necessary to trace the evolutionary genesis of the settlement and geometric surveying campaigns in order to know the places and the building itself.	Verifying if the project interventions are integrated both at an architectural and technological level with the pre-existent building.
Verification of the presence of any regulatory protection limitations.	Checking the presence of any historical, artistic, archaeological, landscape limitations that could affect the area and the single building in the rules that regulate the construction activity in the area involved.	Consistency of the intervention with the regulations in force.
Verification of the technical feasibility of the intervention.	The retrofitting intervention must be verified during the design phase so that the achievement of the expected results in terms of energy, environmental and economic sustainability can be demonstrated. It is necessary to carry out analyses of the climatic and environmental context, of the evolutionary history of the building and of the context in which it rises, of the building type and of the specific bioclimatic properties, of the materials and construction techniques used, of the thermal performance and of the static system that connotes the building.	Verifying the correspondence of the expected performance of the intervention with the project requirements.
Poor knowledge of the results obtained.	The state of the art inherent to the retrofitting interventions both with direct interventions and with numerous case studies and simulations, illustrates the immediate and expected results, the durability and the problems that could arise between pre-existent buildings and new intervention. Given the relative recent field of application, they cannot be clearly determined, especially if we think about GSI. In the building dossier it will be appropriate to highlight the need to sample the efficiency of the intervention with pre-established deadlines according to its nature.	Durability of the intervention over time and sustainability of future maintenance interventions.

The intervention is characterized by the subdivision of the surface into two areas (5.6m x 6.56m): in the first one gray leaf species were planted (Lavandula angustifolia and Senecio, light green in Figure 1) and in the other one there are green leaf species (Rosmarinus officinalis prostratus and Thymus vulgaris, green in Figure 1). The existing floating floor was removed for a surface of 5.76m^2. Then this surface was cleaned, in order to obtain a surface suitable for laying the successive layers. Subsequently a perimeter border of 30 cm height was created for containing the whole system and in particular the substrate consisting of loose material; with the laying of this edge, the surface is reduced to 4.76m^2 due to the presence of flooring support systems.

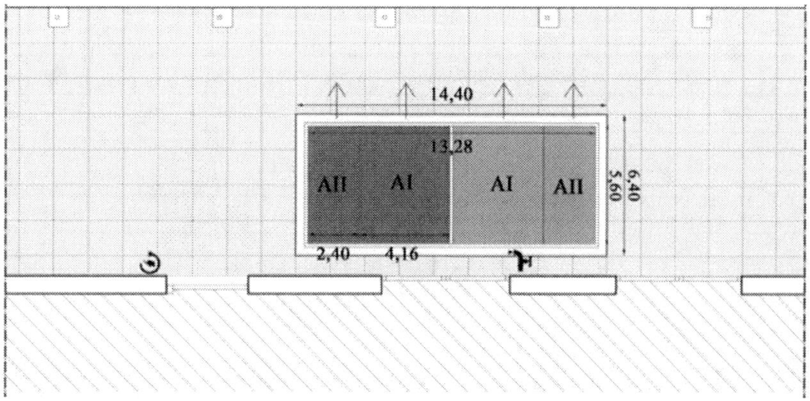

Figure 1. Installation diagram showing the position of the drains, the types of connections to water and electricity.

Figure 2. View of the main phases of the green roof implementation.

Then, a transparent polyethylene sheet was laid, taking care to create some openings in correspondence with the four types of green roof in order to convey the water in the collection tanks. Above the polyethylene, the insulating panels in expanded polystyrene foam were positioned at the edges of the surface and even the non-woven geotextile in polypropylene was laid there, whose function is to retain the substrate so that it is possible to avoid a leaching of the latter and a contamination of the waste water. Before laying the substrate, it was necessary to wet the non-woven geotextile for both having a damp surface in contact with the substrate and preventing dust formation. A separation panel was then placed, and the volcanic lapillus was positioned with an average thickness of about 6cm that works, as well as part of the substrate, as a loose mineral drainage too. The mix of pumice and compost was then placed above the lapillo and surmounted by a last pumice layer. The substrate was composed as follows by reference proportions of the Daku ROOF SOIL2, which provides a 60% of pumice and 40% of lapilli with addition of 20% of the total volume of compost. At this point it was possible to plant the plant species, previously in vase, subdividing them into gray-green leaf species and green leaf. Finally, the implementation was completed with the irrigation necessary for the system to start up.

4 FINDINGS

Table 2 shows the calculations concerning the loads acting on the coverage deriving from the installation of the green roof and those existing before the removal of the floating floor and the related supports.

Table 2. Comparison of the loads before and after the implementation of the green roof.

	ρ [kg/m³]	Spessore [mm]	Carico [kg/m²]
non-woven geotextile	200	2	0,4
insulating panels in XPS	30	40	1,2
Lapillo	900	30	27
Pumice	600	100	60
Compost	450	30	13,5
	[kg/pianta]	[piante/m²]	
Plants	1	10	10
Additional loads	-	-	**112,1**
	[kg]	**[quantità/m²]**	
Flooring	15	7,6	113,5
	[kg/supporto]	**[supporti/m²]**	
Floor supports	0,555	5	2,775
Removed loads	-	-	116,23
Resulting additional load	-	-	- 4,13

The metric estimate of the implemented system is shown in the Table 3. The cost per quantity and the cost per square meter, multiplied by the surface, provides the total cost of the complete installation to which the costs for labor, planning and annual maintenance are added up.

Table 3. Cost estimation of the green roof realized at Frangerini Impresa S.r.l.

	quantity	€/quantity	€/m²	m²	€
Plants in vase Φ 14 cm	10 plants/m²	2 €/plant	20	4,76	100
Lapillo	1 m²	7,90 €/35l	13,3	4,76	63,3
Pumice	1 m²	180 €/m³	18	4,76	85,7
Compost	1 m²	2,80 €/45l	2,5	4,76	11,9
non-woven geotextile	1 m²	49,50 €/100m²	0,5	4,76	2,4
insulating panels in XPS	1 m²	9,16 €/m²	9,2	1,5	13,8
polyethylene sheet	1 m²	5 €/20m²	0,25	5,5	1,4
Costs of stratigraphy	-	-	63,75	-	278,5
Specialized labor	3 workers (2hours)	30 €/h worker	-	-	180
Architectural project	-	600 €	-	-	600
Permission to build	-	900 €	-	-	900

Table 3 (contd). Cost estimation of the green roof realized at Frangerini Impresa S.r.l.

Maintenance	-	-	0,64	-	2,79
Total	-	-	64,39	-	1961,3
Tax relief	60% (65% with design costs)	-	- 38,2	-	1274,8
Total with tax relief	-	-	25,55	-	686,5

5 CONCLUSIONS

The simulation was conducted in Leghorn and involved four types of stratigraphy:

- TYPE AI, with rock wool insulation panel of 8 cm height;
- TYPE AII, with rock wool insulation panel of 8 cm height and XPS insulating panel of 4 cm height;
- TYPE BI, without any insulation panel;
- TYPOLOGY BII, with insulation panel in XPS of 4 cm height.

The comparison of these types shows that the one that guarantees the greater reduction of the annual primary energy requirement is the BII type (57% - 66%), followed by the AII with reductions ranging between 12% and 18%, then by BI (4% - 10%) and, finally, by the AI which increases the needs of 11% - 19%. From the point of view of the loads acting on the roof, an estimate was made of those currently present due to the pavement and those that were added with the implementation of the new stratigraphy: it was determined that, by removing the flooring and installing the green package, the loads on the floor are reduced by 0,04 kN/m^2 for this reason interventions on the structure to increase the load capacity are not necessary. In order to study the economic feasibility of the intervention, a cost analysis was conducted to identify the investment return period. Also, from the economic point of view the best type is BII with reductions of the costs for air conditioning between 53% and 56% and the best species, again, the Rosmarinus officinalis prostratus with a saving of 0,38 €/m^2 for the AII system and 3,97 €/m^2 for the BII system. Unfortunately, the AII typology does not allow to have an economic return because the costs to be incurred for maintenance over the years exceed the economic savings on energy.

References

Berardi, U., GhaffarianHoseini, A. H., and GhaffarianHoseini, A., State-of-the-Art Analysis of the Environmental Benefits of Green Roofs, *Applied Energy,* 115, 411–428, 2014.

Coutts, C., and Hahn, M., Green Infrastructure, Ecosystem Services, and Human Health., *International Journal of Environmental. Research and Public Health,* 12, 9768-9798, 2015.

European Commission, Green Infrastructure (GI), Enhancing Europe's Natural Capital, Bruxelles, 2013.

Francis, R. A., and Lorimer J., Urban Reconciliation Ecology: The Potential of Living Roofs and Walls, *Journal of Environmental Management*, 92, 1429-1437, 2011.

Galbrun, L., and Scerri, L., Sound Insulation of Lightweight Extensive Green Roofs, *Building and Environment*, 116, 130-139, 2017.

Maahsen-Milan, A., and Fabbri, K., Energy Restoration and Retrofitting. Rethinking Restoration Projects by Means of Reversibility/Sustainability Assessment, *Journal of Cultural Heritage*, 14, 41–44, 2013.

Tabares-Velasco, P. C., Zhao, M., Peterson, N., Srebric, J., and Berghage, R., Validation of Predictive Heat and Mass Transfer Green Roof Model with Extensive Green Roof Field Data, *Ecological Engineering*, 47, 165-173, 2012.

Interdependence between Structural Engineering and Construction Management
Edited by Ozevin, D., Ataei, H., Modares, M., Gurgun, A., Yazdani, S., and Singh, A.

ISBN: 978-0-9960437-6-2

ECOLOGY IN MULTI-STORY RESIDENTIAL BUILDINGS – INTERDEPENDENCY BETWEEN ECOLOGY AND COSTS OF BUILDING MATERIALS

DANIELA KOPPELHUBER and JOHANNES WALL

Institute of Construction Management and Economics, Graz University of Technology, Graz, Austria

More than 95% of multi-story residential buildings in Austria are currently predominantly constructed with conventional mineral construction materials. This fact combined with the increasing demands for a healthy residential living atmosphere demonstrates the great potential for using ecological materials. Life cycle assessments provide information on the ecological performance of buildings, but the corresponding economical aspects are not considered. Nevertheless, the economic aspects of a certain draft are important to clients and designers. Therefore, simplified assessment-tools are needed that take into account the ecological impact as well as the building costs. This paper presents the results of an investigation supplemented by a case study of a multi-story residential building, which was finished 2016 in Austria, illustrating the differences between the state-of-the-art material selection and ecologically optimized alternatives. The ecological impacts and the costs for the selected building-system were determined based on the case study. Subsequently, ecological optimization potentials were identified according to the environmental indicator OI3. Finally, the effects on component and construction costs were evaluated. The steps of this simplified process reveal the interdependency between ecological aspects and the costs of materials. This procedure represents a decision-making tool that can be used by clients as well as designers. The results of this research emphasize the large environmental impact improvements with little expenses when implementing sustainability in multi-story residential buildings as a crucial part of a green design.

Keywords: Construction materials, Green design, Environmental indicator OI3, Life cycle assessments, Sustainable development, Ecological performance.

1 INTRODUCTION

Environmental effects of building materials, components and the buildings themselves can be quantified and compared using the method of life cycle assessment (LCA). The results of an LCA provide information about the ecological performance of a building. Focusing on a holistic assessment, the interdependency between ecological aspects and costs of buildings are extremely important. However, these are the key factors in the planning process for decision-makers. Therefore, the issue directs the focus toward developing tools, which can be used to simplify and illustrate the relationship between the environmental impacts and corresponding financial aspects of building projects (Kovacic and Zoller 2015).

2 GREENING POTENTIALS OF HOUSING PROJECTS

Multi-story residential buildings have a great potential to reach the goals of sustainable development, as more than 95% of national housing projects are currently predominantly designed as conventional mineral-massive buildings in Austria (Teischinger *et al.* 2015), built with a low proportion of ecological building materials, especially insulation materials (Steinmann 2014) (Adamczyk and Dylewski 2017). This fact, together with increasing demands for healthy indoor climate conditions (Neubauer 2015), illustrates the great potential for the use of environmentally friendly building materials (Wall and Hofstadler 2016). Therefore, designers have to face the challenge of performing quick, reliable calculations of ecological and economic parameters to provide transparent data that help them select building materials for their clients.

3 CALCULATION OF THE CASE STUDY

The present research illustrates how the design process can involve ecological and economic aspects of different design options in an easy, understandable way for the clients.

3.1 Aim and Method

In a case study of a multi-story housing project finished in Austria in 2016, the higher and lower costs of ecological building materials and their effects on the building costs were calculated. Furthermore, the potential for greening in terms of optimizing the ecological performance was identified, and the level of interdependency between the costs and ecological aspects of several building materials was described. This method provides a simple way for designers to compare the ecological and financial qualities of building materials at an early point in the design process.

In the first step, the building costs were calculated with reference to the Austrian standard *ÖNORM B 1801-1* (2015). Second, the costs and ecological parameters for relevant multilayer components were determined, calculating the ecological indicator OI3. Subsequently, the ecological potential was identified, and ecological building materials with appropriate fire protection, sound insulation, thermal insulation and technical applicability were selected to carry out further calculations and for comparison purposes. Finally, the effects of an ecological optimization of the selected components and their construction costs were calculated and summarized.

3.2 Conditions of the Case Study

As a case study, a residential building with 36 apartments on six floors and a basement car park for 45 cars, finished in autumn 2016, was selected. The calculation of the construction costs and the OI3-value were based on the architectural drawings and the results of the calculation of the coefficient of thermal transmission (U-value). The project was designed and built with for multi-story-constructions using conventional mineral building materials as summarized in chapter 3.4.

3.3 Costs and Environmental Indicators for Conventional Building Materials

To calculate the costs, the unit prices from the German building cost database BKI were compared with the tender prices for the case study and supplemented by unit prices from similar construction projects. The ecological comparison of the components was based on the environmental indicator OI3, which can be used to assess the primary energy demand of the non-renewable energy, global warming potential and acidification potential of construction products from cradle-to-gate (Boogman and Mötzl 2010). Using the OI3-values of the individual materials, the ΔOI3-value for multilayer components was derived using the tool "*Baubook construction calculator*". The ΔOI3 illustrates the optimization potential of a construction element based on a score evaluation: The lower the score of a building material within the multilayer component, the better its ecological performance. The central theme, therefore, is to identify building materials, which have a high number of points and find appropriate alternatives. Finally, the calculated prices and ΔOI3-values of the conventional components were used to represent the basis for the comparison with ecological alternatives.

The eco-indicator OI3 is also the basis of calculations used for national subsidy programs as well as building certification systems such as *klimaaktiv* and *TQB 2010*. The selected method, which is used to assess the ecological improvement compared to the construction standard, is the principle of the ecological assessment in the *IBO Ökopass* certification system.

3.4 Selection of Ecological Building Materials

To select alternative ecological building materials, a number of parameters have been defined, in addition to technical serviceability and properties related to building physics. For example, the principles of construction and the U-value were not allowed to alter from those in the original design. Therefore, the thicknesses of the alternative thermal insulation material had to be adjusted. Finally, the ecological building materials listed below were selected for the comparative calculation with the conventionally constructed housing project:

- Exterior and structural walls: wood chip concrete instead of vertical coring bricks
- Apartment ceilings: prefabricated brick units instead of reinforced concrete
- Thermal insulation of walls: cork insulation slabs instead of EPS-F
- Thermal insulation of walls with fire protection: mineral foam instead of mineral wool
- Thermal insulation of flat roof: cork insulation slabs or mineral foam instead of XPS
- Thermal insulation of basement slab bottom: mineral foam instead of excelsior boards with a rock wool-core

3.5 Calculation of Costs and Environmental Indicators for Ecological Building Materials

After selecting the ecological building materials, the improvements in the ecological properties of the multilayer components were verified using the ΔOI3-value. Furthermore, the thermal insulation thickness was adjusted if necessary. The costs for the selected ecological material variants were calculated based on material price lists and added to the costs for delivery, wage assumptions, material offcut and, finally, the surcharge (Kropik 2016). While budgeting with conventional building materials is considered as a routine task for designers, the cost calculation for ecological building materials presents a challenge due to the lack of available databases for these materials.

4 RESULTS OF ANALYSIS

Applying the described method, a total of eight multilayer components with 20 ecological variants were examined for the use case. The results of the comparison as well as the ecological effects are summarized below.

4.1 Interdependency Between Ecology and Costs of Components

The outcome of the cost and environmental calculation is shown in Figure 1. The light bar on the left-hand side indicates the ecological improvements in the multilayer component, and the dark bar on the right illustrates the expenses of these components – compared with the conventional building elements.

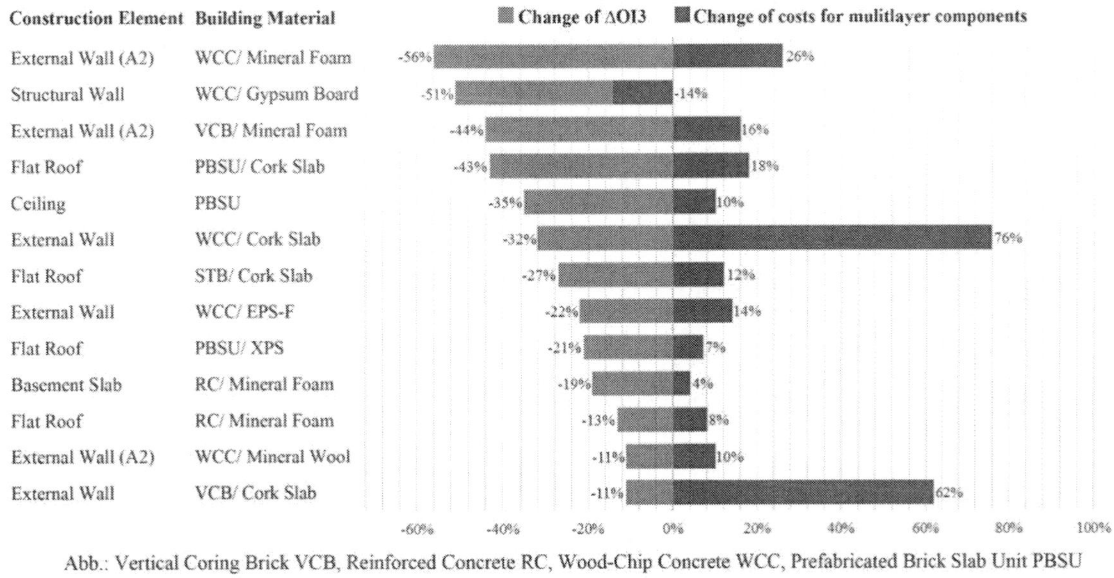

Figure 1. Changes in the costs and ecologically improvements in multilayer components.

It has been shown that structural alternatives cause fewer expenses, despite their exceptional ecological performance. The results of the comparison show that thermal insulation materials are more expensive despite their lower ecological potential to contribute towards the creation of a more ecological building.

4.2 Interdependency Between Ecology and Construction Costs

Perhaps more important than the results for the individual components were the illustration of the construction cost differences and the ecological improvement for the entire project. Therefore, the ΔOI3-value and the corresponding costs had to be multiplied by the surface of the components. To compare the construction costs, the costs of the shell and finish according to the national standard *ÖNORM B 1801-1* (2015) were considered. To illustrate the potential for greening, the ΔOI3-values of all components were summarized, based on their surfaces. For calculation purposes, design options with different ecologically optimized component structures had to be defined. Several calculated scenarios using different building materials and construction variants allowed the ecological and economic interdependency to be communicated transparently to the clients. Overall, the structural options were associated with fewer additional costs despite having excellent ecological performance values, as shown in Figure 2. On the contrary, insulation variants with lower ecological improvement values caused more expenses. The fact that the thermal conductivity of the alternative insulation material is higher than that of common materials is equally important to investors. This means that the ecological materials are thicker and results in a loss of usable area within the building up to 3% (Koppelhuber 2016).

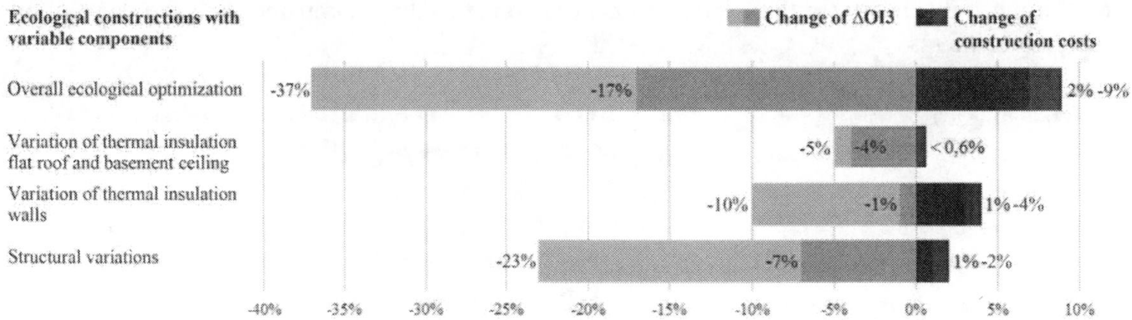

Figure 2. Changes in the construction costs and the ecologically improvement of the overall building.

4.3 Derived Benefits from Ecological Interventions

The outcome of this research illustrates the higher building costs and the loss of usable area affected by the use of ecological building materials. Although these are given facts, the designer has to motivate the client to choose environmentally friendly construction methods and materials for his or her project. The benefits of ecological-friendly materials to the environment and the indoor climate are well-known, but clients usually consider the financial benefits to be more relevant. Some financial rewards for the use of ecological building materials are listed in the following section, which illustrate selection criteria that supplement the environmental benefits:

- Improved evaluation results for building certifications: The results of an Austrian research project explain the influence of ecological building materials on the outcome of building certifications. The comparison of four frequently used national building certification systems illustrated that the selection of building materials could influence 29% to 59% of the overall assessment (Sölkner *et al.* 2014).
- Higher limits for construction costs for building certifications: In part, construction costs are prescribed for certifications. These cost limits may increase by using more ecological materials: In the case of the Swiss building certificate *Minergie*, this increase was up to 10% – or with *Minergie-Eco*, up to 15% – compared to the cost limits of a conventional residential building (Koppelhuber 2016).
- Higher limits for construction costs for promoting policies related to public housing: Maximum construction costs are generally defined for subsidized multi-story buildings. Depending on the funding systems, this upper limit increases by approximately 10% to 15% when ecological building materials are integrated into the construction of the building (Koppelhuber 2017).
- Higher subsidies for the owners of new apartments: Increasing the subsidy for first-time dwellers in the national subsidy scheme *Wohnbauscheck* by a maximum of 5% according to the ecological intervention would mean that cost increases caused by the use of environmentally friendly materials can be more easily covered by the new building owners (Koppelhuber 2017).

The conditions used in this project are typical for Central Europe, but the options presented clearly show that there are additional benefits for ecological building systems. These findings underline well-known arguments that support environmental protection and the promotion of a healthy indoor climate in residential buildings.

5 CONCLUSIONS AND OUTLOOK

The ecological and financial results of the calculations clearly illustrate the ecological potential and the associated low additional costs associated with the use of different combinations of alternative materials. It has been demonstrated that environmental improvements can already be achieved with little financial input. For example, if additional costs of 2% are accepted, an ecological improvement in the examined components of 23% is possible. A simplified and comprehensive cost calculation combined additionally with ecological indicators (such as the ΔOI3-value) can simplify the decision-making process for designers and strengthen the clients' understanding. This paper highlighted the challenges faced when budgeting ecological construction methods, which can be viewed as a major obstacle to making cost comparisons with ecological materials.

In conclusion, the results of this research can make it easier for designers to argue with the investors to bring about the acceptance of environmental improvements and illustrate the financial benefits of using ecological building materials. However, construction projects cannot only be evaluated on the basis of ecological properties; therefore, integrated, lifecycle-based strategies that include both economical and socio-cultural aspects in the decision-making process must be undertaken (Ofek *et al.* 2018).

References

Adamczyk, J., and Dylewski, R., The Impact of Thermal Insulation Investments on Sustainability in The Construction Sector, *Renewable and Sustainable Energy Reviews*, Volume 80, 421-429, 2017.

Boogman, P. and Mötzl, H, *IBO-Richtwerte für Baumaterialien – Wesentliche Methodische Annahmen. Version 2.2*, IBO, Vienna, 2010.

Koppelhuber, D., *Ökologie als Planungsaufgabe im Geschoßwohnbau – Aspekte der Nachhaltigkeit im Kostenvergleich der Baustoffe*, Graz University of Technology, Graz, 2016.

Koppelhuber, D., *Ökologie als Planungsaufgabe im Geschoßwohnbau – Vergleichende Betrachtung ökologischer Baustoffe unter Berücksichtigung von Bauherren- und Planungsaspekten*, Graz University of Technology, Graz, 2017.

Kovacic I, and Zoller V., Building Life Cycle Optimization Tools for Early Design Phases, 92(3), *Energy*, 409-419, 2015.

Ofek, S., Akron, S., Portnov, B.A Stimulating Green Construction by Influencing The Decision-Making of Main Players, *Sustainable Cities and Society*, 40, 165-173, 2018.

Kropik, A., *Baukalkulation und Kostenrechnung*, Perchtoldsdorf., 2016.

Neubauer, M.: *Immobilien Fokus – Die große Umfrage zum Thema nachhaltiges Wohnen*. Wien, Linz. Immobilien Fokus, 2015.

ÖNORM B 1801-1: Bauprojekt und Objektmanagement – Teil 1: Objekterrichtung. Wien. Austrian Standards, 2015.

Sölkner, P. J., Oberhuber, S., and Spaun et al., *Innovative Gebäudekonzepte im Ökologischen und Ökonomischen Vergleich über den Lebenszyklus*, BMVIT, Vienna, 2014.

Steinmann, H., Alternative Dämmstoffe – „Der Markt ist da", *a3 Das Baumagazin,* Oktober, 2014.

Teischinger, A., Stingl, R., and Berger, V., *Studie Holzbauanteil in Österreich und Wien*, proholz, 2015. Retrieved from www.proholz .at/news/news/detail/studie-holzbauanteil-in-oesterreich-und-wien/ on November 10, 2016.

Wall, J., and Hofstadler, C., *How Can Sustainability Issues be Considered in The Public Procurement Process?* Proceedings of the CIB World Building Congress 2016: Understanding Impacts and Functioning of Different Solutions. Nenonen, S. & Junnonen, J-M. (eds.). Tampere University of Technology, 4, 2016.

Interdependence between Structural Engineering and Construction Management
Edited by Ozevin, D., Ataei, H., Modares, M., Gurgun, A., Yazdani, S., and Singh, A.

ISBN: 978-0-9960437-6-2

REDUCTION OF CO_2 EMISSIONS IN CEMENT PRODUCTION BY USING PREHEATER

HAIYAN XIE[1], PRANSHOO SOLANKI[1], ALIREZA MOJADAM[1], and WENFANG LIU[2]

[1]*Dept of Technology, Illinois State University, Normal, USA*
[2]*School of Management, Guangxi University of Science and Technology, Liuzhou, China*

Cement has a pivotal role in the construction industry. However, cement is one of the key contributors to global CO_2 emission levels. This is due to the energy-intensive nature of cement production processes. This comparative-descriptive study focuses on the potential factors to reduce the CO_2 emission level in cement production and the decision-making process of adopting new environmental-friendly technology in production. Particularly, this study compares alternative technologies in cement manufacturing to reduce CO_2 emission. It collects both the industry data and the data from Texas, which is the biggest contributor to CO_2 emission in the US, to analyze how a shift in production technology could affect CO_2 emission and eventually improve the outcomes of environment protection and energy efficiency. This paper projects a possible improvement of implementing the method of preheater-precalciner in cement production in lieu of wet and long-dry process to upgrade kilns and reduce problematic CO_2 emission. This study suggests shifting from wet and dry kilns to preheater-precalciner systems to obtain the potential benefits of CO_2 emission reduction in the cement industry.

Keywords: Air pollution, Material production, Construction management, Sustainability, Investment analysis, Technology improvement and adoption.

1 INTRODUCTION

The Cement Industry subsumes the largest industrial non-combustion source of CO_2 emission by holding a record of 4.1% of the overall global emissions in 2014 (EPA 2015). Additionally, the cement manufacture process produces combustion-related CO_2 emissions, which also constitute an additional 4% of the total global emissions (Gibbs *et al.* 2000). Hence, the cement manufacture process by itself accounts for about 8% of the total global CO_2 emission. It is also in urgent need to control CO_2 emissions in the United States, which is the 3rd largest cement producer in the world (USGS 2015). The focus of this paper is on non-combustion related emissions in the cement industry. The purpose is to identify potential factors to reduce the CO_2 emission level in cement production. The factors can assist the decision-makers to smoothen the process to adopt new environmental-friendly technology in production. Specifically, this study collects both the industry data and the data from a particular area and compares alternative technologies.

The research goals include: (1) provide an up-to-date profile of the CO_2 emission levels of the cement industry; (2) detect and pinpoint the problematic cement production facilities to improve; (3) introduce potential improvement methods; and (4) provide in detail a transformation process for Portland cement manufacturing plants for successful adoption of new technologies.

The research seeks to answer the following questions: (1) What are the current CO_2 levels of the cement industry in the US?, (2) What are the potential CO_2 emission levels improvement rates in upgrading kilns to preheater? The shift in production technology could affect CO_2 emission and eventually improve the outcomes of environment protection and energy efficiency.

This comparative-descriptive study is conducted in a non-contrived setting. The purposes of study the production technologies affecting CO_2 emissions and eventually improve the outcomes of environment protection and energy efficiency. After compares alternative technologies in cement manufacturing to reduce CO_2 emissions in the literature review, this paper focuses on the influencing factors for new technology adoption. It collects both the industry data and the data from Texas, which is the biggest contributor to CO_2 emission in the US, to analyze the statuses and technical details of the controls on CO_2 emissions. This paper projects a possible improvement of implementing the method of preheater-precalciner in cement production in lieu of wet and dry process to upgrade kilns and reduce CO_2 emissions. In addition, this paper gives reasonable policy adjustments to help companies in the transformation processes.

2 LITERATURE REVIEW

2.1 Comparison of Various CO_2 Abatement Technologies

The CO_2 abatement technologies can be classified into five general categories: improving energy efficiency, alternative fuels, clinker substitution, innovative low-carbon cements, and carbon capture and storage. The clinker production is the most energy intensive step in the cement production (Hendriks *et al.* 1998). This step emits approximately 85% of overall CO_2 and consumes 80% of the energy used in the cement production. The average heat input required by wet, long dry, preheater, and precalciner were reported as 5.5, 4.1, 3.5, and 3.1, respectively (EPA 2015), which indicates that dry method is more energy efficient compared to wet method of cement production. Researchers demonstrated that preheater technology improved the energy efficiencies in cement production (Hanle *et al.* 2004 and Benhelal *et al.* 2013). The shift from wet to dry processes could reduce 50% energy usage and 20% CO_2 emission. For example, streamlining the kiln exhaust can save up to 20% of the energy input (Engin and Ari 2005). Galitsky and Price (2007) showed that installing an efficient kiln lead to 144 to 94,444 metric tons reduction in CO_2 emissions.

Coal is the most widely used fossil fuel in the cement production. It is possible to replace this fossil fuel with waste-derived fuel to save energy and reduce CO_2 emission (Hanle *et al.* 2004, Ali *et al.* 2011). The use of solid-waste-derived fuel might become the most ecologically and economically feasible method in lowering CO_2 emission in the cement industry (Mokrzycki and Uliasz-Bochenczyk 2003). However, the variation in properties of municipal solid waste is a significant drawback in their usage, which can lead to a low-quality final product (Kikuchi 2001).

Substitution of clinker with supplementary cementations materials (SCMs) is one of the most effective ways to reduce CO_2 emissions. The SCMs include fly ash, blast furnace slag and other pozzolans, which are able to form many types of blended cements (Gartner 2004, Hanle *et al.* 2004, Rehan and Nehdi 2005, Ali *et al.* 2011). For instance, Huntzinger and Eatmon (2009) evaluated CO_2 emission by substituting clinker with natural pozzolans and 100% cement kiln dust (CKD), which had insignificant impact on the cement's environmental footprint. Upon comparing 30 energy-efficient technologies, Worrell *et al.* (2001) concluded that the replacement of clinker with various additives was the most energy-efficient and cost-effective approach for the cement industry in the U.S. Kim and Worrell (2002) further concluded that such methods were viably duplicable on a global level. However, clinker substitution may result in low-quality final material (Schneider *et al.* 2011) and unacceptable or unmarketable products (Gartner 2004).

Low-carbon cements involve chemicals, which are completely different from the ordinary cement (IETD 2017). They offer significant reduction in CO_2 emissions. However, long-term sustainability and conformity of these cement products remain unproven. The innovative low-carbon cements need to gain customer acceptance.

Carbon capture and storage technologies are currently used for post-combustion CO_2 capture. Capturing and storing the CO_2 emission are effective solutions toward decreasing CO_2 emission levels, which were proven to reduce such emissions by 65-70% (Rehand and Nehdi 2005). Companies are interested in this business opportunity as well. In a recent study, an Australian company is planning to capture carbon emissions and convert it into solid carbonates (Guardian.com 2017). These carbonates have the potential to be used in building products, such as concrete and plasterboard, to create green construction materials.

2.2 Influencing Factors to New Technology Adoption

Many factors may influence the adoption of new technologies in control of pollutant emissions. Researchers found different influencing factors of decision-making to executives at strategic level, managers at tactical level, and technicians at technical level. Figure 1 shows the diverse influencing factors (Beggs 2000). Particularly, executives concerned on market orientation; managers focused on creativity, risks and usefulness; while technicians prefer easiness in learning and perceived benefits (Ishida *et al.* 2017, Lee 2009).

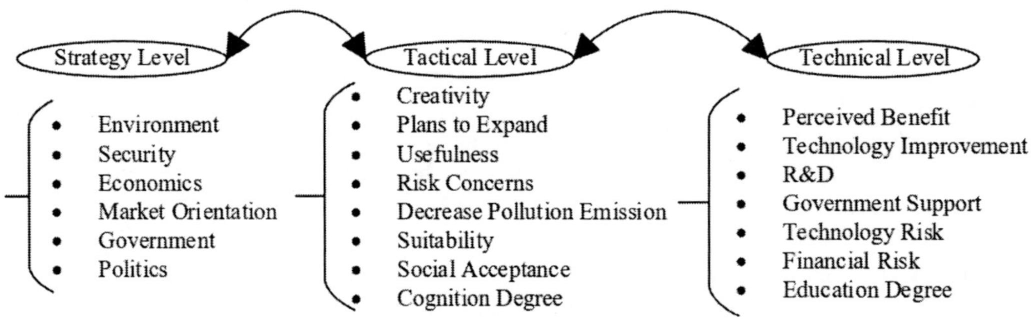

Figure 1. Diverse influencing factors.

3 DATA ANALYSIS AND DISCUSSION

Based on the database of the U.S. Environmental Protection Agency (EPA 2015), Texas accounted for 13.44% of the national total cement industry share in the non-combustion CO_2 emission and topped the list among all the states. There were 10 major and active cement production facilities in Texas (EPA 2015). Figure 2 shows their Greenhouse Gas (GHG) emission quantities over the past 6 years. Fortunately, the overall trend of the GHG emissions is decreasing. However, the GHG emission increased in the following plants: Ash Grove Cement Company, Holcim Texas Limited Partnership, Lehigh Cement Co LLLC/Waco Plant, Texas Lehigh Cement Company LP, and TXI Hunter Cement Plant. Although Ash Grove was pioneer in utilizing alternative fuels towards energy savings, it fell behind in terms of CO_2 emission levels pertaining to kiln types, because the company utilized 2 wet kilns and 1 dry kiln (USGS 2015). In terms of energy consumption, a conversion of wet or dry kiln to preheater-precalciner kiln can lead up to 40 percent increase in CO_2 emission (EPA 2015). Furthermore, such a conversion can cause a decrease in energy usage to 0.7 MMBtu /ton of cement production.

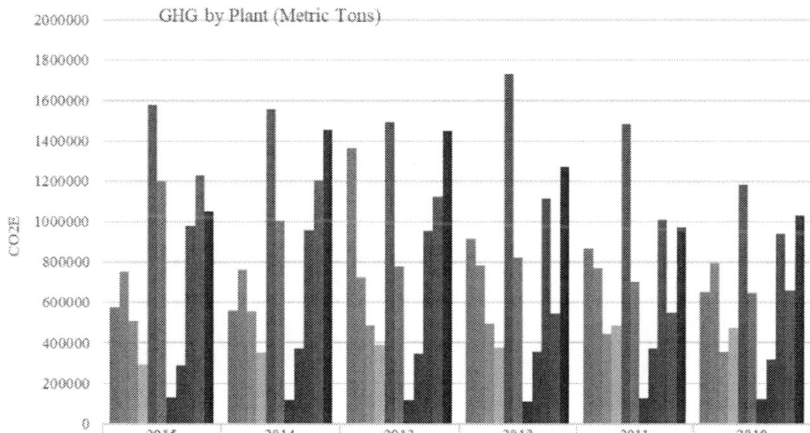

	2015	2014	2013	2012	2011	2010
Ash Grove Cement Company	576069	559764	1363796	915317	870671	651213
Alamo San Antonio Cement Plant	753290	762341	725541	786127	771784	797280
Capitol Cement Plant	506948	554230	485339	496935	445170	354667
Cemex Construction Materials South LLC	289694	350055	388469	376268	486803	473927
Cemex Construction Materials South, LLC	1580028	1557776	1494923	1729720	1482686	1182943
Holcim Texas Limited Partnership	1202046	1005020	781529	823891	702844	649166
Lehigh Cement Co LLC / Waco Plant	132123	116020	118315	110135	124505	122968
Lone Star Industries, Inc. /dba/ BUZZI UNICEM USA	288274	371051	344948	355097	371598	316778
Texas Lehigh Cement Company	978167	956245	953036	1113493	1010368	940974
TXI Hunter Cement Plant	1231790	1202912	1121981	546234	551893	662563
TXI Operations	1052468	1454146	1449259	1271535	971361	1031856

Figure 2. Greenhouse Gas (GHG) emission quantities (Data Source: EPA 2015).

4 CO_2 REDUCTION ESTIMATES

A conversion from wet or dry kiln to preheater-precalciner can lead to between 150 to 460 lb / (ton cement) reduction in CO_2 emission levels. Table 1 shows the differences that such conversion can make in the reduction of the emission levels of several Texan cement plants, including Ash Grove, Cemex, and Lehigh (EPA 2015). Figure 3 shows the results of these 3 companies when converting to 1 preheater and 1 precalciner in terms of reduction in CO_2 emission levels. Overall, this conversion can lead up to almost half million metric-ton of reduction in CO_2 emission levels, which accounts for almost 6% of the overall GHG emission quantity from the Texan cement industry.

The entire CO_2 reduction can be estimated in this method. Texas is responsible for 13.4% and almost 9 million metric tons of the overall industrial CO_2 emission in the U.S. If this conversion is used in the entire cement industry, there will be a reduction of approximately 13,840,000 metric tons of CO_2 emission. This amount leads to 20 percent of all industrial CO_2 emissions in the U.S. However, it is important to note that such a conversion calls for an investment of $96 million dollars /ton of cement production annually. In addition, operation and maintenance costs can be around $0.08 million dollars per ton cement (EPA 2015).

Table 1. Projections of Adoption of New Technology.

Facility	Kiln Type	Annual Production	GHG	GHG Reduction
Ash Grove	2 wet 1 dry	575,809	576,069	40,000-12,000
Cemex	1 preheater	1,585,163	1,580,208	108,000-330,000
Lehigh	1 wet	132,474	132,123	9,000-28,000
			Total	157,000-478,000

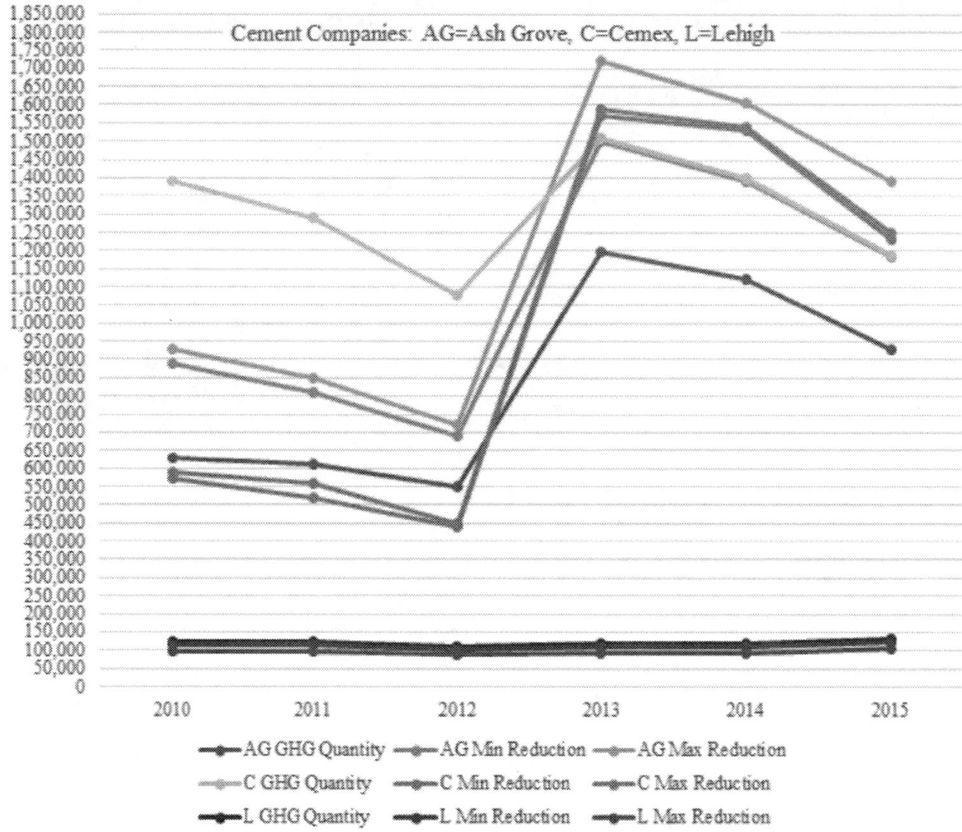

Figure 3. CO_2 emission projection (Data Source: EPA 2015).

5 CONCLUSIONS

This study focuses on the comparison between different cement production processes of different technologies. By capitalizing on the available data, this study evaluates the levels of CO_2 emissions among various cement production facilities. The results indicate that shifting from wet and dry kilns to preheater-precalciner systems has potential to reduce CO_2 emissions in the cement industry. In the final analysis, the data shows that the shift can reduce the CO2 emission levels up to almost 6% in the overall emission levels of the Texan cement industry. This saving is approximately 0.8% of the overall industrial CO_2 emission in the U.S. During this research, data confidentiality hindered the study greatly. For example, companies keep confidentiality on the data of the fuel consumption in a kiln and the production capacity of each kiln. This research is the start of a comprehensive look into the cement industry about technology improvement and adoption. The research would encourage the exploration of further methods towards more efficient cement-production with reduced CO_2 emission and energy consumption in the country.

Acknowledgments

This research is partially supported by the University Research Grant of Illinois State University (XIE2016) and the grant of Research on Multi Transportation network of Guangxi District Logistics on the Background of "One Belt and One Road", Guangxi Education Department (KY2016YB233).

References

Ali, M. B., Saidur, R., and Hossain, M. S., A Review on Emission Analysis in Cement Industries, *Renewable and Sustainable Energy Reviews*, 15(5), 2252-2261, 2011.

Benhelal, E., Zahedi, G., Shamsaei, E., and Bahadori, A., Global Strategies and Potentials to Curb CO_2 Emissions in Cement Industry, *Journal of Cleaner Production*, 51, 142-161, 2013.

Beggs, T. A., Influences and Barriers to the Adoption of Instructional Technology, 2000. Retrieved from http://files.eric.ed.gov/fulltext/ED446764.pdf on September 1, 2017.

Engin, T., and Ari, V., Energy Auditing and Recovery for Dry Type Cement Rotary Kiln Systems - A Case Study, *Energy Conversion and Management*, 46(4), 551-562, 2005.

EPA, United States Environmental Protection Agency 2015 Report, 2015.

Galitsky, C., and Price, L., *Opportunities for Improving Energy Efficiency, Reducing Pollution and Increasing Economic Output in Chinese Cement Kilns*, Proceedings of the 2007 ACEEE Summer Study on Energy Efficiency in Industry, 2007.

Guardian.com, Australian Firm Unveils Plan to Convert Carbon Emissions into 'Green' Concrete: Initiative to Convert CO_2 into Solid Carbonates Aims to Produce Building Materials on Commercial Scale by 2020, Guardian News and Media Limited. Retrieved from https://www.theguardian.com/environment/2017/aug/25/green-building-materials-carbon-capture-technology-ready-by-2020-says-manufacturer on September 1, 2017.

Gartner, E., Industrially Interesting Approaches to "Low-CO_2 Cements", *Cement and Concrete Research*, 34(9), 1489-1498, 2004.

Gibbs, M. J., Soyka, P., Conneely, D., and Kruger, M., CO_2 Emissions from Cement Production, Good Practice Guidance and Uncertainty Management in National Greenhouse Gas Inventories, 2000.

Hanle, L. J., Jayaraman, K. R., and Smith, J. S., *CO_2 Emissions Profile of the US Cement Industry,* Washington DC: Environmental Protection Agency, 2004.

Hendriks, C. A., Worrell, E., De Jager, D., Blok, K., and Riemer, P., *Emission Reduction of Greenhouse Gases from the Cement Industry,* Proceedings of the Fourth International Conference on Greenhouse Gas Control Technologies, 939-944, 1998.

Huntzinger, D. N. and Eatmon, T. D., A Life-Cycle Assessment of Portland Cement Manufacturing: Comparing the Traditional Process with Alternative Technologies, *Journal of Cleaner Production*, 17(7), 668-675, 2009.

IETD, Low-Carbon or Carbon-Negative Alternatives to Portland Cement, The Institute for Industrial Productivity, 2017. Retrieved from http://ietd.iipnetwork.org/content/low-carbon-or-carbon-negative-alternatives-portland-cement by September 1, 2017.

Ishida, S., Magnusson, M., and Nagahira, A., Factors Influencing Japanese Auto Suppliers' Predictions about the Future of New Technologies–An Exploratory Study of Electric Vehicles, *Futures,* 89, 38-59, 2017.

Kikuchi, R., Recycling of Municipal Solid Waste for Cement Production: Pilot-Scale Test for Transforming Incineration Ash of Solid Waste into Cement Clinker, *Conservation and Recycling*, 31(2), 137-147, 2001.

Kim, Y., and Worrell, E., CO_2 Emission Trends in The Cement Industry: An International Comparison, *Mitigation and Adaptation Strategies for Global Change*, 7(2), 115-133, 2002.

Lee, M. C., Factors Influencing the Adoption of Internet Banking: An Integration of TAM and TPB with Perceived Risk and Perceived Benefit, *Electronic Commerce Res. and Applns.*, 8(3), 130-141, 2009.

Mokrzycki, E., and Uliasz-Bocheńczyk, A., Alternative Fuels for The Cement Industry, *Applied Energy*, 74(1), 95-100, 2003.

Rehan, R., and Nehdi, M., CO_2 Emissions and Climate Change: Policy Implications for the Cement Industry. *Environmental Science & Policy*, 8(2), 105-114, 2005.

Schneider, M., Romer, M., Tschudin, M., and Bolio, H., Sustainable Cement Production-Present and Future, *Cement and Concrete Research*, 41(7), 642-650, 2011.

USGS, American Council for an Energy-Efficient Economy. PCA, Portland Cement Association Annual Reports USGS; US Geological Survey Annual and Quarterly Reports, 2015.

Worrell, E., Price, L., Martin, N., Hendriks, C., and Meida, L. O., CO_2 Emissions from the Global Cement Industry, *Annual Review of Energy and the Environment*, 26(1), 303-329, 2001.

Interdependence between Structural Engineering and Construction Management
Edited by Ozevin, D., Ataei, H., Modares, M., Gurgun, A., Yazdani, S., and Singh, A.

ISBN: 978-0-9960437-6-2

STUDY AND ANALYSIS OF INCENTIVES FOR STATE-OF-THE-ART NET-ZERO ENERGY HOUSES

DAVOUD SAHBAN RAFSANJANI, DEEPAK SHARMA, TUNG TRAN, AND PHOUNG NGUYEN

Dept of Civil and Environmental Engineering California State University Fullerton, Fullerton, USA

Over the past several years, the US Energy Information Administration (EIA) has reported that buildings consume about 40% of the total energy produced. As a consequence, buildings emit high volumes of carbon dioxide (CO_2). The ever-increasing population demand implies that buildings will continue their consumption and emit gasses in high proportions if appropriate steps are not taken. Federal and state agencies offer various financial incentives to motivate homeowners to adopt energy efficient alternatives. In this research, major cash inflows and outflows are identified for Net Zero Energy Houses (NZEH) and breakeven analysis is conducted by considering that the houses are constructed in two US states having significantly different climatic conditions. The cash inflows considered in this research are from Federal and State incentive programs and savings generated by adopting NZEH. Sensitivity analysis is also conducted to determine the impact of variation in in government incentives, market interest rate, electricity rates and rate of construction. Results show that when some of the incentives are utilized the breakeven for a NZEH construction could be less than 4 years. Sensitivity analysis shows that the breakeven is most sensitive to changes in government incentives and market interest rate.

Keywords: Energy conservation, Cash-flow analysis, Sensitivity analysis.

1 INTRODUCTION

Building energy efficiency has substantial impact on environment and emissions. Statistics show that buildings consume about 40% of the total energy production in the US and the EU (EIA 2018, Cao *et al.* 2016). So, any improvement in home energy efficiency will have significant impact on the overall environment. Improvements can be made by using energy efficient alternatives easily available in the market. Through this paper an attempt is made to present some of the latest state of the art systems used to achieve net-zero consumption and incentives that can help reduce energy intake in buildings. In addition, cash flows for houses in two US states are estimated to determine breakeven year for NZEH. The initial cost of NZEH being higher than traditional construction demoralizes homeowner to adopt green alternatives but a reasonable breakeven period can reverse this and motivate homeowners to adopt green alternatives. This is important because lack of proper adoption of energy efficient technologies will make it difficult to achieve any environmental benefits.

International Energy Agency (IEA) reported a historic increase in carbon emissions due to slowing of energy efficiency improvements and increase in energy demand (Chestney 2018). The EIA reported an increase of 1.7% carbon dioxide emissions (EIA 2017). Carbon emissions can be reduced in several ways. For example, some researchers focused on controlling emissions by focusing on embodied carbon (Pomponi and Moncaster 2018) while some others suggested adopting green contracting methods (Cui *et al.* 2011). The authors here suggest changing the way building energy is consumed. However, change is not easy, and a barrier exists when it comes to switching to energy efficient alternatives in buildings. A report by National Institute of Standards and Technology (NIST) stressed the importance of end users' acceptance of green technologies. It discusses financial, behavioral, policy level, and several other barriers faced when pursuing green alternatives. For example, the report mentions "*Investment costs are too high to encourage a transformation in consumer decisions*" (McNabb 2013). Similarly, social and cultural aspects were identified as barriers towards environment friendly options (Pomponi and Moncaster 2016). Clearly, when end-users 'see' such constraints ahead, the chances of large-scale adoption to green policies will be low. Agencies put tremendous amount of resources to implement green policies, but the existing barriers hinder their adoption and success.

The overall aim of this research is to lower some of the barriers towards green initiatives. To achieve this, Federal and State financial incentives supporting homeowners to adopt NZEH systems are documented. Government incentives such as tax credits and property tax are included in this paper. The paper also includes details of some of the state-of-the-art energy efficient systems that have been used and proven at 2017 Solar Decathlon competition. The details from incentives and the energy efficient systems are then collectively used to develop cash-flow for houses built in the states of California (CA) and New York (NY). The analysis is further extended by including sensitivity analysis which enabled identifying the most impactful parameters on breakeven. This research will help end users to understand that the NZEHs are not as expensive as they are perceived. As a result, agencies will be able to achieve higher end-user acceptance leading to achieving the desired green goals.

2 PREVIOUS STUDIES AND RESEARCH MOTIVATION

Energy efficient housing has several benefits. However, multiple barriers exist in making energy efficient housing a norm. Researchers have pointed out that one of the main barriers felt by home owners is cost. Li *et al.* (2018) surveyed 69 energy efficient housing stakeholders in Canada and found that half of the respondents were not ready to spend more than 5% additional money for NZEH. Similarly, Hast *et al.* (2015) found that cost was one of the main barriers that prevented end-users in China to adopt NZEH. However, Risholt *et al.* (2013), and Seto *et al.* 2016 found that the adoption barriers can be reduced by education and creating awareness.

Agencies are taking definite steps by promoting energy efficient houses and are gradually advancing towards making them a norm (Seto *et al.* 2016). For example, California Energy Commission conducts webinar to educate people about the various energy efficient options (Timothy 2017). Synchronously, researchers have advocated educating and bringing awareness about energy efficiency to overcome barriers for largescale adoption to green alternatives (Risholt *et al.* 2013). To support such a cause, the US DOE conducts Solar Decathlon competition every two years to promote and investigate affordable NZEH. In conjunction with all such awareness initiatives this research builds upon all the earlier works and attempts to bridge the gap between the technological advancements, incentives and financial analysis that can be used for promoting NZEH.

3 METHODOLOGY

In this research, the first task was to identify and study the designs of winning teams from Solar Decathlon 2017 competition. The designs used latest technologies and the teams made them work in a single-family home. Some of the innovative technologies are discussed in this paper. As a second task, the team identified information about incentives available from government agencies. Assuming that the energy efficient houses are constructed in CA and NY, applicable government incentives were identified for solar systems. Average utility bills for traditional houses were determined and potential energy savings were calculated. Cash flow statements were developed for all cash flow which became the basis of comparison between NZEH and traditional houses. The cash flow statement enabled conducting breakeven analysis. This was followed by sensitivity analysis that enabled determining influential parameters.

3.1 Studying NZEH Houses

Solar Decathlon 2017's competition documents of top three teams were studied. These teams were Swiss Team, Maryland and the UC Berkeley/ University of Denver team (UCB/UD). All these teams scored more than 800/1000. The teams were evaluated for excellence in architecture, market potential, engineering, communications, innovation, water usage, health & comfort, appliances, home life and energy. While all these teams presented great ideas, some of the main ones from UCB/UD are briefly described here.

The team used specialized glass having R value of 10 and paired them with frames of R value of 16. These items enabled having good lighting and insulation. The team chose to use water recycling system that treated grey water for reusing in toilet. Self-cleaning filters in the recycling system saved significant quantities of water. The team used an efficient, sustainable, and non-toxic wool that had a minimum R-value of 3.6 per inch. For lighting, the team used scrap optical fibers to let light enter from walls and thus increase the overall brightness in the house. The team chose to use moss matt panels to cover the entire north façade. This helped in generating a natural living surface contributing to the overall vegetation of the house. The team also used solar panels to transform sunlight to usable electrical energy. For a house that uses the above innovative methods can qualify for several government incentives, but for this research only solar panel incentives are included.

3.2 Identifying Federal and State Incentives

At the Federal level, a homeowner qualifies for a Federal subsidy in the form of tax credits. The tax credit for solar energy system covers 30% of the cost (including installation costs). Similarly, in the state of CA, property-tax incentive for solar photovoltaics and solar thermal covers 100% of proportionate system value. In NY, property tax credit for qualified solar energy system enables a credit of 25%. These incentives could help offset some of the expenses incurred from energy consumption in the states of CA and NY. For this research, single family houses with an area of 1048 sqft were assumed to be constructed in Westminster (CA) and New York City (NY).

Table 1. Federal and state incentives (source dsireusa.org).

Description of Incentive	**Westminster (CA)**	**New York City (NY)**
Federal Tax Credit	$5,357	$5,357
100% Proportionate Solar System for Property Taxes	$145	-
25% Proportionate Solar System for Property Taxes	-	$70
Total	**$5,502.28**	**$5,427.54**

3.3 Cost Comparison for Constructing and Consuming Energy

As per Zillow.com, for a single-family house in Westminster and NYC, the average cost of constructing a traditional house is \$406/sqft and \$630/sqft which translates to a total cost of \$425,488 (i.e. 1048×406) and \$659,999 (i.e. 1048×630) respectively. As expected, for NZEH the average cost/sqft is higher. With reference to UCB/UD team's construction the NZEH costs are estimated as \$434/sqft and \$646/sqft which totals to \$455,230 (i.e. 1048×434) and \$677,018 (i.e. 1048×646) respective. The differences in cost for pursuing NZEH in Westminster and NYC are \$29,742 (i.e. \$455,230-\$425,488) and \$17,020 (i.e. \$677,018-\$659,999) respectively. These differences are the financial barriers that can deter a decision maker to pursue NZEH but these must be supplemented with government's incentives to determine the actual financial burden.

3.4 Developing Cash Flow

Cash flow statements were developed to differentiate between the NZEH and traditional alternatives. Cash flow statements are developed by considering cash outflow and inflows. Annual electricity bills can have significant impact on the cash flow and so average annual electricity bills were determined. For Westminster, the average annual utility bills were \$2,763 and for NYC it was \$2,521 with an average increase of 20% every three years. A 3% interest rate was assumed for discount throughout the analysis. Putting together all the cash outflows and inflows following Table 2 was obtained.

Table 2. Cash outflow and inflow for breakeven analysis.

	Westminster, CA				**New York City, NY**			
Yr	**Utility Saving**	**Gov. Incentives**	**Total Savings**	**Total Difference**	**Utility Savings**	**Gov. Incentives**	**Total Savings**	**Total Difference**
0	2763	5502	8265	-29742	2521	5428	7948	-17020
1	2682	5342	8024	-21718	2447	5269	7717	-9303
2	2604	5186	7791	-13927	2376	5116	7492	-1811
3	3034	5035	8069	-5858	2768	4967	7735	5924
4	2946	4889	7834	1977	2688	4822	7510	13434
Note: All values are discounted to present time at 3%								

In Table 2, column "Utility Savings" represents discounted yearly electricity bill savings; "Gov. Incentives" represent all the solar incentives; and "Total Savings" is obtained by summing electricity bill saving and government incentives. The column "Total Difference" represents the cumulative difference between traditional and NZEH construction. The difference is seen reducing every in the table and the signs changes from negative to positive indicating breakeven. Sensitivity analysis was also conducted to determine the effectiveness of government benefits, interest rate, change in electricity rates and rate of construction in cost/sqft.

4 RESULTS

Results indicate that a CA homeowner can offset the cost of NZEH in three years and for the same construction in NY the offset occurs in four years as shown in Figure 1. Thus, after three years in CA and four years in NY the NZEH is equivalent to a traditional home in terms of cost.

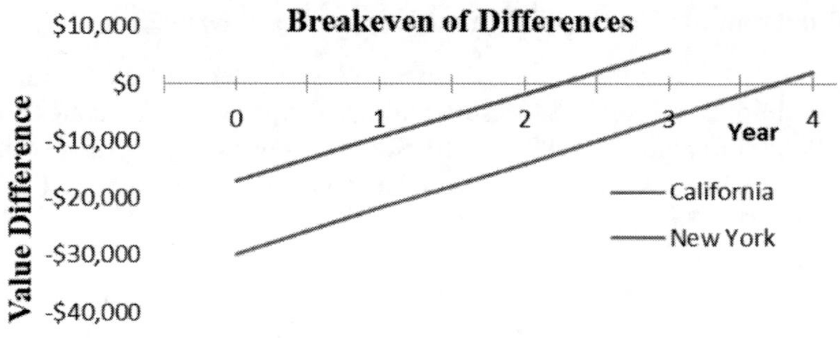

Figure 1. Breakeven analysis

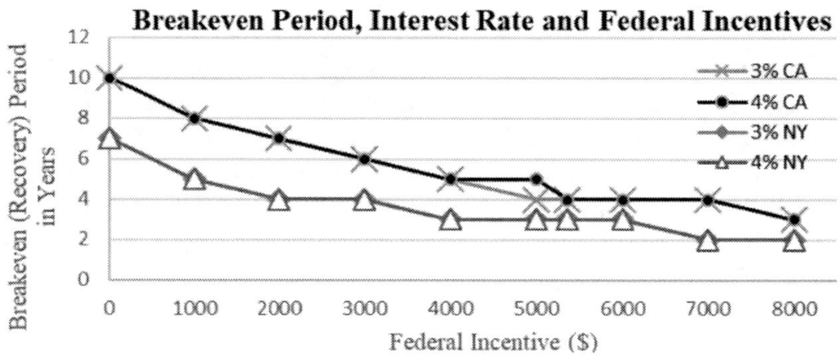

Figure 2. Sensitivity analysis results.

Sensitivity analysis showed that the breakeven period is sensitive to change in interest rate and the Federal incentives (Figure 2). It can be concluded that if Federal incentive is $5,000 the breakeven occurs within three years in NY even if the market interest rate increases from 3% to 4%. However, in CA the breakeven occurs in years when the market interest rate is 3% but it increases to five years if interest rate increases to 4%. Lastly, the breakeven period was found to be relatively insensitivity to changes in electricity rates and rate of construction in $/sqft.

5 CONCLUSION

This research was aimed at determining the impact of government incentives on adoption of NZEH. State-of-the-art systems used by top three teams at Solar Decathlon 2017 were studied and some of them were briefly described in this paper. Federal and State level solar system incentives were identified and used in this study. All the cash inflows and outflows were calculated and used to conduct breakeven analysis. The analysis showed that the difference between NZEH and traditional houses can be offset in three years in NY and four years in CA.

This study establishes relation between the additional cost incurred for NZEH housing and the time it takes to get even with traditional housing. The results indicate that while the NZEH construction appears expensive, the government incentives and utility bill savings enable to get even in less than four years in states like CA and NY that are the most expensive states in the US. Sensitivity analysis shows that breakeven periods are the most responsive to government incentives and market interest rate. This research can be further expanded to include other US states and other available incentives for sustainability.

References

Cao, X., Dai, X., and Liu, J., Building Energy-Consumption Status Worldwide and the State-of-the-Art Technologies for Zero-Energy Buildings During the Past Decade, *Journal of Energy and Buildings,* 128, 198–213, 2016.

Chestney, N., Global Carbon Emissions Hit Record High in 2017, Environment, Reuters, Published on 3/22/2018, 2018.

Cui, Q., Zhu, X., Whitten, L., and Deane, R., *Innovative Contracting Strategies for Combating Climate Change,* Report Number MD-11-SP009B4G, University of Maryland, 2011.

EIA, U.S. Energy-Related Carbon Dioxide Emissions, 2016, Report by U.S. Department of Energy, Washington DC, 2017.

EIA, How Much Energy is Consumed in US Residential and Commercial Building?, US Energy Information Administration. Retrieved from https://www.eia.gov/tools/faqs/faq.php?id=86&t=1 on Jan 08, 2018.

Hast, A., Alimohammadisagvand, B., Syri, S., Consumer Attitudes Towards Renewable Energy in China—The Case of Shanghai, *Sustainable Cities and Society,* 17, 69–79, 2015.

Li, H. X., Patel, D., Al-Hussein, M., Yu, H., and Gül, M., Stakeholder Studies and the Social Networks of NetZero Energy Homes (NZEHs), *Sustainable Cities and Society,* 38, 9–17, 2018.

McNabb, N., *Strategies to Achieve Net-Zero Energy Homes: A Framework for Future Guidelines,* Workshop Summary Report, NIST Special Publication 1140, 2013.

Pomponi, F., and Moncaster, A., Embodied Carbon Mitigation and Reduction in the Built Environment – What Does the Evidence Say? *Journal of Environmental. Management,* 181, 687–700, 2016.

Pomponi, F., and Moncaster, A., Scrutinising Embodied Carbon in Buildings: The Next Performance Gap Made Manifest, *Renewable and Sustainable Energy Reviews,* 81, 2431–2442, 2018.

Risholt, B., Time, B., and Hestnes, A. G., Sustainability Assessment of Nearly Zero Energy Renovation of Dwellings Based on Energy, Economy and Home Quality Indicators, *Journal of Energy and Buildings,* 60, 217–224, 2013.

Seto, B., Leahy, J., Herrschaft, B., Butterworth B., and Punjabi S., *Zero Net Energy Communities: Three Cities Leading the Way,* ACEEE Summer Study on Energy Efficiency in Buildings, 2016.

Timothy, N., *Upcoming Title 20 Lighting Standards Coming into Effect,* Appliances and Outreach and Education Office, California Energy Commission, 2017.

Interdependence between Structural Engineering and Construction Management
Edited by Ozevin, D., Ataei, H., Modares, M., Gurgun, A., Yazdani, S., and Singh, A.

ISBN: 978-0-9960437-6-2

REVIEWING GOVERNMENT INITIATIVES ON IMPLEMENTING SUSTAINABLE INFRASTRUCTURE CONSTRUCTION

DEBBY WILLAR and DAISY D. G. PANGEMANAN

Dept of Civil Engineering, Manado State Polytechnic, Manado, Indonesia

Sectoral development activities, lack of standards on eco-systematic approach, prioritizing short-term outcomes, and ignoring the holistic and comprehensive prospects of sustainability have led to environmental degradation. Therefore, the Indonesian government has realized the importance of implementing sustainable development concepts in every phase of infrastructure project life-cycle. This paper reviews the existing government regulations on sustainable construction implementation guidelines, which underlie the execution of infrastructure project throughout five phases of project life-cycle, namely (1) programming, (2) technical planning, (3) construction execution, (4) utilization, and (5) demolition. A relook at existing regulation in term of providing more technical concepts on implementing the sustainable principles throughout the infrastructure project life-cycle is necessary to help both the construction services users and providers in implementing sustainable construction principles in Indonesia. The outcome of this review can serve as a guide to develop a technical concept for the implementation of the sustainable construction in infrastructure projects in order to create sustainable development in Indonesia.

Keywords: Project life cycle, Sustainable development, Indonesia.

1 INTRODUCTION

The concept of sustainable construction is still new to the Indonesian construction industry. The government has realized that all this time, the infrastructure construction projects that have been undertaken merely rely on short-term achievements, less on systemic and ecosystemic approaches, ignoring the prospect of sustainability in a holistic manner, which has caused environmental damage. Sustainable construction concepts will have a positive impact on the global construction industry since the concepts aim to improve the environmental objectives (Abdul Jamil and Fathi 2016). Therefore, the implementation of infrastructure projects in the field of public works and housing must fulfill the provisions of environmental management and support sustainable development by taking into account the principles of benefit, safety, balance, and harmony of infrastructure and the environment.

In order to embrace the issue of 'sustainable construction that creates sustainable infrastructure,' the Indonesian government has enacted regulations regarding the implementation of sustainable construction in infrastructure projects owned by the Ministry of Public Works and Housing. The regulations contain guidelines to implement sustainable construction throughout the five phases of the execution of infrastructure project life-cycle, namely (1) programming, (2) technical planning, (3) construction execution, (4) utilization, and (5) demolition. The purpose of this guideline is to achieve sustainable development in Indonesia through the creation of sustainable infrastructure, which is constructed with a sustainable construction approach. The Brundtland Commission's 1987 report defines sustainable development as "development that meets the needs of the present without compromising the ability of future generations to meet their own needs" (WCED 1987). Whilst CIB and UNEP-IETC (2002) states that "sustainable development is not merely development that can be sustained, but rather the kind of development we need to pursue in order to achieve the state of sustainability". By these definitions, every activity related to the implementation of sustainable infrastructure should be more efficient, more effective, more productive, and more environmentally friendly. However, the concept of sustainable development that is supported by sustainable construction has not yet been well explored within Indonesia construction industry. The paper sets out to review the existing regulations from the viewpoints of related literature, to provide more technical concepts on implementing sustainable infrastructure construction.

2 SUSTAINABLE PRINCIPLES OF CONSTRUCTION PROJECTS

Sustainable development departs from a noble goal to achieve a better quality of life for society today and for future generations. This sustainable condition can be created if the development can meet four poles of sustainability at once, namely economic, social, environment, and technology (Agyekum-Mensah *et al.* 2012). By considering these four aspects, the construction industry should involve in an inclusively manner, not having a negative impact on the environment and with more efficient use of construction resources to provide the benefits to all communities.

Regarding sustainable construction implementation that support sustainable development, for those who have roles in the construction project design-based have agreed with the 1987 Brundtland report definition on sustainable development, while those who are working in project site-based emphasis most on material selection, recycling, life cycle analysis and cost analysis (Abrahams 2017). Moreover, construction projects that implement sustainable development principles should be proven throughout the project cycle from the planning, design, and construction of buildings and infrastructure, until their final deconstruction, concerning the extraction and beneficiation of raw materials and management of the resultant waste (CIB and UNEP-IETC 2002). Therefore, urgent issues on implementing sustainable principles in construction projects, particularly infrastructure projects that should be taken into account, since the project activities can change the condition and function of the environment, both in and social life, and will consume a lot of resources and have an impact on the surrounding environment.

Research studies on sustainable principles of construction projects have led to several definition of sustainable construction. Abdul Hamid and Kamar (2012) define sustainable construction as "the creation and responsible maintenance of a health built environment, based on ecological principles and by means of an efficient use of resources". In addition, according to Ismail *et al.* (2017), "a sustainable construction practice can be seen as ensuring all the construction activities are being carried out in a sustainable way, from the initial (planning) to the completion stages". The success of sustainable construction indeed depends on the attitude, awareness, and behavior of individuals and related groups of project stakeholders in making decisions based on a value system that encourages the formation of a balance between economic, social and environmental aspects.

Concerning the sustainable construction project life-cycle, according to Agyekum-Mensah *et al.* (2012) the project management principles are needed to sucessfully manage the phases and resources of sustainable construction while adhering to the sustanability pinciples. More practical ways to sustainable construction practices, according to Abdul Jamil and Fathi (2016), can be implemented with regard to the integration of social, economic, and environmental attributes, green design and procurement, technology and innovation to empower sustainability concepts, reorganized organizational structure and process, better education and training, measurement and reporting for improved environmental and social performance. Even, CIB and UNEP-IETC also identified the need for an initial condition or prerequisite that must be met in the implementation of the sustainable construction framework, especially for developing countries, like Indonesia. The prerequisites that must exist in the implementation of a sustainable construction framework consist of technology, institutions and value systems (CIB and UNEP-IETC 2002).

3 GOVERNMENT INITIATIVE ON SUSTAINABLE CONSTRUCTION

A basic regulation of Indonesian Ministry of Public Works and Housing No. 05/PRT/M/2015 relating to general guidelines for the implementation of sustainable construction in the infrastructure project execution was formulated in order to provide direction for the sustainable construction implementation that creates the sustainable infrastructure that eventually contributes to sustainable development. Concerning the practical issues on how the sustainable principles can be successfully implemented in infrastructure project life-cycle, the following sections discuss some technical concepts based on the literature review to help both the construction services users and providers in implementing sustainable construction principles in Indonesia.

3.1 Project Programming

Programming phase is a set of initial planning activities for determining aims, strategies, steps to be taken, schedule, as well as the needs of resources, especially funding to execute sustainable infrastructure. Sustainable infrastructure programming should be applied since the beginning by the technical unit of organization of the Ministry of Public Works and Housing to ensure the availability, continuity, and sustainability of resources fulfillment in achieving the distant targets.

The steps and techniques that should be applied in project programming phase are as follows: (1) determining the infrastructure object to be implemented; (2) determining the sustainable infrastructure implementation achievement; (3) identifying the interested parties and establishing cooperation among these parties since early stages; (4) equating the vision of sustainability for all the interested parties; (5) determining the method of project delivery system, which is appropriate with the sustainable infrastructure complexity and its ability; (6) determining the initial technical conception, choosing the technology and planning the funding for the sustainable infrastructure implementation in all stages by life-cycle cost approach; (7) analyzing the sustainable infrastructure feasibility; (8) selecting the professional service provider needed; (9) programming all stages of sustainable infrastructure implementation with a clear scope, funding and schedule in the form of work packages as well as risk management implementation; and (10) preparing the programming stage final report.

Operationally, the Indonesian government needs to start a holistic approach in the earliest phase of achieving sustainable infrastructure construction. Developing various indicators related to sustainability evaluation or assessment purposes (Abdul Hamid and Kamar 2012), and hiring life-cycle costing (LCC) professionals to evaluate the potential economic, social, and environmental benefits of proposed infrastrucure projects (Higham *et al.* 2015), might guide the government to determine the proposed infrastructure projects to meet the sustainability principles and criteria for Indonesian circumstances. Most importantly, the government as the projects owner and leaders need to establish the strategic and transformational leadership styles, which are leading the sustainability agenda (Opoku *et al.* 2015).

3.2 Project Technical Planning

The technical planning phase is defined as a sequence of activities which is a process of thought, creation, and engineering in realizing the expected sustainable infrastructure. Sustainable infrastructure technical planning must be implemented in accordance with the sustainable infrastructure technical requirements to ensure constructability in each phase of construction implementation and infrastructure performance in utilization and demolition phases.

Steps and techniques of management that must be applied in the technical planning stage are as follows: (1) identifying other interested parties in the technical planning phase; (2) selecting the needed professional technical planning service provider; (3) re-communicating the objectives, the scope and the target of sustainable infrastructure implementation to all parties involved in the technical planning; (4) assigning the criteria of sustainable infrastructure design referring to related technical requirement or using technical criteria of other agencies; (5) making technical planning process integrated with all parties involved; (6) analyzing and reviewing the technical planning results to ensure the quality of technical planning results; (7) preparing the final report of the technical planning phase. Given the important role of the technical planning phase in designing the sustainable infrastructure project execution, there is an urgent need for the Indonesian government to apply the project's life-cycle value through green design and the promotion of best construction procurement practice throughout the supply chain (Abdul Jamil and Fathi 2016, Abdul Hamid and Kamar 2012), and to benchmark and technology transfer of best practices from developed countries in implementing sustainable and green construction agenda (Abdul Hamid and Kamar 2012).

3.3 Construction Execution

In construction implementation phase, it is expected that this phase can contribute to the expected sustainable infrastructure accomplishment, sustainable infrastructure construction implementation that should be done using green construction approach and considering social and economic aspects in the location. The steps and techniques of management that should be done in the construction implementation phase are as follows: (1) identifying the interested parties and involve them in the construction implementation activities; (2) selecting professional construction implementation service provider needed; (3) obliging professional construction executor for conducting construction work at the site by applying green construction approach; (4) conducting activities of the sustainable infrastructure construction implementation; (5) ensuring that the results of every stage of construction implementation can be utilized efficiently and effectively by the users; (6) preparing the final report of construction implementation phase.

The roles of contractors organizations as the leading actor in this phase much depend on their knowledge and capabilities to implement sustainable construction principles. Upstill-Goddard *et al.* (2016) have urged that construction organizations must provide resources to their employees for supplementing knowledge, such as training programmes, to introduce sustainability policies and processes on their operations from environmental and social viewpoints. Education and training on the benefits of sustainable construction should reach the construction clients both public and private (Aigbavboa *et al.* 2017). In addition, Opoku and Ahmed (2014) recommend contractors to collaborate with their clients in adopting whole life costing (WLC), so that they do not only consider the initial cost of construction projects but rather considering the WLC of the project asset.

3.4 Utilization

Utilization phase is intended to utilize the sustainable infrastructure in accordance with its function and to maintain the sustainable infrastructure performance based on the possible actual conditions compared to the target and technical planning criteria. The sustainable infrastructure should be utilized optimally and should be maintained in accordance with its age of service for contributing to the goal accomplishment with asset management approach.

The steps and techniques of management to be applied in the utilization phase are as follows: (1) making the utilization plan of the sustainable infrastructure during the age of service; (2) conducting socialization, promotion and education for the users to support the sustainable infrastructure utilization; (3) managing the sequence of activities of utilization; (4) conducting the sustainable infrastructure feasibility test periodically and when approaching the end of the service period; (5) performing function enhancement for the sustainable infrastructure should be done on the request of the users based on the latest results of audit; (6) selecting a professional service provider for the utilization phase; and (7) documenting all processes and results of the sustainable infrastructure. Maintenance programme of infrastructure products is rarely highly considered by the infrastructure owner nor the users; this phase cannot be neglected since this is a part of sustainable principles. Schröpfer *et al.* (2017) have emphazised that improved knowledge transfer on how to build sustainably between all project participants could solve barriers towards delivering sustainable construction; this best practice should allow the Indonesian construction industry to implement sustainable principles in infrastructure projects from programming to project demolition.

3.5 Demolition

The sustainable infrastructure can be demolished at the end of its service period with a deconstruction approach in order to accomplish the sustainable infrastructure implementation goals in full. Deconstruction is the building demolition intended to get the building material or component for reuse and to get new material by recycling. One of the strategies in the infrastructure demolition is choosing the materials that still have values and throw away the unusable. Deconstruction can also be defined as a planned and careful infrastructure demolition intended to maximize the reuse of infrastructure material or components as well as minimizing the amount of material to be thrown away to the environment or nature (waste).

Steps to be taken in demolition stage of a sustainable infrastructure are as follows: (1) obtaining a license from the authorized official based on the feasibility test report of the sustainable infrastructure; (2) making the sustainable infrastructure demolition plans; (3) establishing a demolition service provider; (4) conducting a demolition process in accordance with the demolition plan; and (5) conducting orderly documentation in the demolition phase. It is clear that at the demolition phase, what has been programmed and planned for the infrastructure project execution will have an impact, for example, the use of construction materials. The way Malaysian construction industry addresses the issue of sustainable materials and green construction is by establishing an eco-labeling scheme for construction materials and promoting reduce, reuse and recycle on the various environmental, social, economic and even cultural factors (Abdul Hamid and Kamar 2012).

4 CONCLUSION

The purpose of this paper is to provide more technical concepts on implementing the sustainable principles throughout the infrastructure project life-cycle to help both the construction services users and providers in implementing sustainable construction principles in Indonesia. Based on the literature review several recommendations are proposed to comply with the Ministry of Public Works and Housing regulation (No. 05/PRT/M/2015). At the programming phase, the Indonesian government has to have strategies to lead the sustainability agenda including providing indicators that can be used to evaluate the proposed project that meets the sustainability criteria. The next phase is project planning, which should be based on green construction procrurement and processes. After that, the construction execution should be constructed by well-trained and competent contractors by considering not only the technical aspect but also environmental, social, and economic viewpoints. Finally, knowledge transfer about sustainable principles including the knowledge of sustainable construction materials among all project participants would support the execution of utilization and demolition phases. The results of the review can improve the understanding among the construction service providers to enable them to implement sustainable construction principles, and can prove their commitment on caring about environmental, economic, and social issues as well.

References

Abdul Hamid, Z., and Kamar, K. A. M., Aspects of Off-site Manufacturing Application Towards Sustainable Construction in Malaysia, *Construction Innovation*, 12(1), 4–10, 2012.

Abdul Jamil, A. H., and Fathi, M. Sy., The Integration of Lean Construction and Sustainable Construction: A Stakeholder Perspective in Analyzing Sustainable Lean Construction Strategies in Malaysia, in *Procedia Computer Science 100*, 634 – 643, 2016.

Abrahams, G., Constructing Definitions of Sustainable Development, *Smart and Sustainable Built Environment*, 6(1), 34-47, 2017.

Agyekum-Mensah, G., Knight, A., and Coffey, Ch., 4Es and 4 Poles Model of Sustainability: Redefining Sustainability in the Built Environment, *Structural Survey*, 30(5), 426-442, 2012.

Aigbavboa, C., Ohiomah, I., and Zwane, Th., Sustainable Construction Practices: "A Lazy View" of Construction Professionals in the South Africa Construction Industry, in *Energy Procedia,* 105, 3003–3010, Elsevier, 2017.

CIB and UNEP-IETC, *Agenda 21 for Sustainable Construction in Developing Countries*, CSIR Building and Construction Technology, Pretoria, 2002.

Higham, A., Fortune, Ch., and James, H., Life Cycle Costing: Evaluating Its Use in UK Practice, *Structural Survey*, 33(1), 73-87, 2015.

Ismail, F. Z., Halog, A., and Smith, C., How Sustainable is Disaster Resilience?: An Overview of Sustainable Construction Approach in Post-disaster Housing Reconstruction, *International Journal of Disaster Resilience in the Built Environment*, 8(5), 555-572, 2017.

Opoku, A., Ahmed, V., and Cruickshank, H., Leadership Style of Sustainability Professionals in the UK Construction Industry, *Built Environment Project and Asset Management*, 5(2), 184-201, 2015.

Opoku, A., and Ahmed, V., Embracing Sustainability Practices in UK Construction Organizations: Challenges Facing Intra-organizational Leadership, *Built Environment Project and Asset Management*, 4(1), 90-107, 2014.

Schröpfer, V. L. M., Tah, J., and Kurul, E., Mapping the Knowledge Flow in Sustainable Construction Project Teams Using Social Network Analysis, *Engineering, Construction and Architectural Management*, 24(2), 229-259, 2017.

Upstill-Goddard, J., Glass, J., Dainty, A., and Nicholson, I., Implementing Sustainability in Small and Medium-sized Construction Firms: The role of Absorptive Capacity, *Engineering, Construction and Architectural Management*, 23(4), 407-427, 2016.

WCED, *Our Common Future*, Oxford University Press, Oxford, 1987.

Interdependence between Structural Engineering and Construction Management
Edited by Ozevin, D., Ataei, H., Modares, M., Gurgun, A., Yazdani, S., and Singh, A.

ISBN: 978-0-9960437-6-2

DESIGN OF CERTIFIED GREEN BUILDINGS USING BUILDING INFORMATION MODELING

SENAY ATABAY, ASLI PELIN GURGUN, and ALPER GOKTUG DEMIRBAS

Dept of Civil Engineering, Yildiz Technical University, Istanbul, Turkey

Design of green buildings involves specific processes compared to conventional construction projects that require participation of several disciplines. Having particular activities, functions, and physical aspects, green buildings impose their own difficulties during design. Building Information Modelling systems (BIM) use different software for virtual simulation of characteristics of projects, which promotes early detection of potential problems even before production of the intended design details. Implementing BIM technology in design stage, not only helps to minimize design problems, but also to shorten the stage itself. With this effect of BIM combined with sustainable practices and energy efficient technologies, design stage alone can be considered as a sustainable application of construction activities. Therefore, principles and requirements of green building certification systems can easily be performed for projects with capabilities of BIM technologies and tools. In this study, redesign of Yildiz Technical University, Faculty of Civil Engineering Building based on LEED certification system using BIM tools is presented. The project is prepared as part of a capstone project by senior civil engineering students. Design stages are planned to promote sustainability and qualify the requirements of LEED v3 BD+C (Schools). The main objective is to introduce the students to two major trends in construction industry (green buildings and BIM applications) and to create awareness of sustainable construction, green building certification systems and emerging technologies.

Keywords: BIM, Sustainable construction, Green building certification systems, LEED, Civil engineering education.

1 INTRODUCTION

Sustainability is becoming an important factor due to depletion of natural resources and environmental pollution. With the introduction of sustainability into construction industry, design of environmentally-friendly "Green Buildings" is frequently practiced. Standards and certification systems have been developed in mostly developed countries are also used in other countries.

Construction sector constitutes an important role in the use of natural resources and energy consumption all over the world. It is reported that construction sector alone is responsible for 50% of the energy consumption and 42% of water consumption globally (Dikmen 2011). Management of construction and other activities in the whole life cycle buildings can be improved by the help of advances in information technology. Building Information Modeling (BIM) has become a significant technological tool that provides foresight about future problems and addresses solutions for multi-disciplinary working groups in projects (Akkaya and Basaraner 2012).

Although BIM is recognized as a tool to contribute positively to the management of projects, the current curriculum of civil engineering programs in Turkey merely provides theoretical and practical background of BIM. In this study, the effort to introduce BIM to senior students along with sustainable construction and LEED system by integrating all these concepts in a Capstone Project is presented. In the following parts, briefly sustainability, BIM and LEED are explained, followed by the implementation of them together on an existing building, namely Yildiz Technical University, Faculty of Civil Engineering Building.

2 SUSTAINABILITY, GREEN BUILDINGS AND LEED

Sustainability principles ideally should be applied throughout the life cycle of a building, including design and construction phases, while supporting decisions made in course (Kibert 2013). It should be noted that the life cycle starts from inception to the termination phases of a building. Principles of sustainable construction can be listed as follows:

With the increasing use of sustainable practices in construction industry, the efforts to establish standards started emerging, particularly in developed countries; followed or adapted by developing countries. Currently, there are several certification systems initiated, i.e., BREEAM (Building Research Establishment Environmental Assessment Method), LEED (Leadership in Energy and Environmental Design), DGNB (Deutshe Gesellschaft für Nachhaltiges Bauen e.V.), IISBE (International Initiative for Sustainable Built Environment), Greenstar (Environmental Rating System for Buildings), Casbee (Comprehensive Assessment System for Built Environment Efficiency).

Green buildings can be constructed with different features referring sustainable construction methods and practices. The design can provide low carbon emissions, efficient energy use, effective water management, automated and smart lighting systems, green material use, and other environmental-friendly technologies and methods (GYODER 2014).

3 BUILDING INFORMATION MODELING (BIM)

Building Information Modeling (BIM) can simply be described as a three-dimensional platform that aims for information sharing among architectural, structural, electrical, mechanical, construction design, and facility management teams. In this process, different software is used for the management of the projects. Project parties and members such as owners, contractors, engineers, and architects can be able to contribute, follow, and revise project information, and reach details whenever needed. BIM can be used for the entire construction process as well as for only a particular part in this process.

4 GREEN BUILDING MODELING BASED ON LEED CRITERIA USING BIM

The redesign of the Yildiz Technical University Faculty of Civil Engineering Building according to LEED certification system using BIM tools is presented. The design is performed with technologies using BIM provided by analysis steps and several implementation and practices supporting sustainable construction. Autodesk Revit is utilized as one of the BIM based programs. In the first step, architectural and structural drawings of the building were acquired in 2D, which were prepared in AutoCAD, and then these drawings were used in Revit to establish the 3D model. Figure 1 shows Autodesk Revit drawing and front entrance of the actual building.

Figure 1. Yildiz Technical University, Faculty of Civil Engineering Building.

4.1 Assessment of the Existing Building According to LEED

LEED certification system can be applicable on different type of buildings. In this study, the existing design of Yildiz Technical Faculty of Civil Engineering Building is evaluated according to LEED BD+C Schools v3 rating system. The objective is to identify the criteria and their corresponding points applicable to this building. Based on this evaluation, it is found that for the building in its existing form, 16 points could be achieved based on LEED BD+C Schools v3 rating system (Table 1).

Table 1. LEED points achieved with existing design (USGBC 2009).

LEED Criteria	Achieved point	Intent
Site selection	1	To avoid the development of environmentally sensitive lands and reduce the environmental impact from the location of a building on a site.
Joint use of facilities	1	To integrate the school with the community by sharing the building and its playing fields for non-school events and functions
Development density and community connectivity	4	To channel development to urban areas with existing infrastructure, protect greenfields, and preserve habitat and natural resources
Brownfield redevelopment	1	To rehabilitate damaged sites where development is complicated by environmental contamination and to reduce pressure on undeveloped land.
Alternative transportation- public transportation access	4	To reduce pollution and land development impacts from automobile use
Alternative transportation- bicycle storage and changing room	1	To reduce pollution and land development impacts from automobile use
Alternative transportation- parking capacity	2	To reduce pollution and land development impacts from automobile use
Site development- protect or restore habitat	1	To conserve existing natural areas and restore damaged areas to provide habitat and promote biodiversity
Process water use reduction	1	To increase water efficiency within buildings to reduce the burden on municipal water supply and wastewater systems

4.2 Redesigning the Building According to LEED System

Although with its existing condition, the building is insufficient to qualify for a level of certification through LEED system, it is possible to add some features that might help to achieve points adequate for certification, as well as acquiring additional sustainable practices in the building. New design options are searched to collect points based on LEED BD+C Schools v3. For this purpose, BIM libraries are reviewed for potential materials, which can be integrated to the new design. Once the convenient options are found, the building is redesigned Table 2 shows the additional 16 points that can be achieved after this stage.

Table 2. LEED points achieved after redesigning according to LEED BD+C Schools v3 (USGBC 2009).

LEED Criteria	Achieved point	Intent
Alternative transportation-low-emitting and fuel-efficient vehicles	2	To reduce pollution and land development impacts from automobile use
Site development-maximize open space	1	To promote biodiversity by providing a high ratio of open space to development footprint
Heat island effect- roof	1	To reduce heat islands1 to minimize impacts on microclimates and human and wildlife habitats
Heat island effect - nonroof	1	To reduce heat islands1 to minimize impacts on microclimates and human and wildlife habitats
Light pollution reduction	1	To minimize light trespass from the building and site, reduce sky-glow to increase night sky access, improve nighttime visibility through glare reduction and reduce development impact from lighting on nocturnal environment
Site master plan	1	To ensure that the environmental site issues included in the initial development of the site and project are continued throughout future development caused by changes in programs or demography.
Water use reduction	4	To increase water efficiency within buildings to reduce the burden on municipal water supply and wastewater systems
Enhanced refrigerant management	1	To reduce ozone depletion and support early compliance with the Montreal Protocol while minimizing direct contributions to climate change
Rapidly renewable materials	1	To reduce the use and depletion of finite raw materials and long-cycle renewable materials by replacing them with rapidly renewable materials
Certified wood	1	To encourage environmentally responsible forest management
Daylight and views - daylight	1	To provide building occupants with a connection between indoor spaces and the outdoors through the introduction of daylight and views into the regularly occupied areas of the building
Daylight and views - views	1	To provide building occupants a connection to the outdoors through the introduction of daylight and views into the regularly occupied areas of the building

4.3 Additional LEED Criteria Achieved Assumed for the Existing Building

In this section, the criteria that are assumed to have been qualified during the construction stage of the existing building are presented (Table 3) and sample drawings of the building can be seen in Figure 2. It is noted that 9 points can be achieved.

Table 3. LEED points achieved assumed during design stage.

LEED Criteria	Achieved point	Intent
Building Reuse-Maintain Existing Walls, Floors and Roof	2	To extend the life cycle of existing building stock, conserve resources, retain cultural resources, reduce waste and reduce environmental impacts of new buildings as they relate to materials manufacturing and transport
Construction Waste Management	2	To divert construction and demolition debris from disposal in landfills and incineration facilities. Redirect recyclable recovered resources back to the manufacturing process and reusable materials to appropriate sites
Building Reuse- Maintain Interior Nonstructural Elements	1	To extend the life cycle of existing building stock, conserve resources, retain cultural resources, reduce waste and reduce environmental impacts of new buildings as they relate to materials manufacturing and transport
Materials Reuse	2	To reuse building materials and products to reduce demand for virgin materials and reduce waste, thereby lessening impacts associated with the extraction and processing of virgin resources
Recycled Content	2	To increase demand for building products that incorporate recycled content materials, thereby reducing impacts resulting from extraction and processing of virgin materials

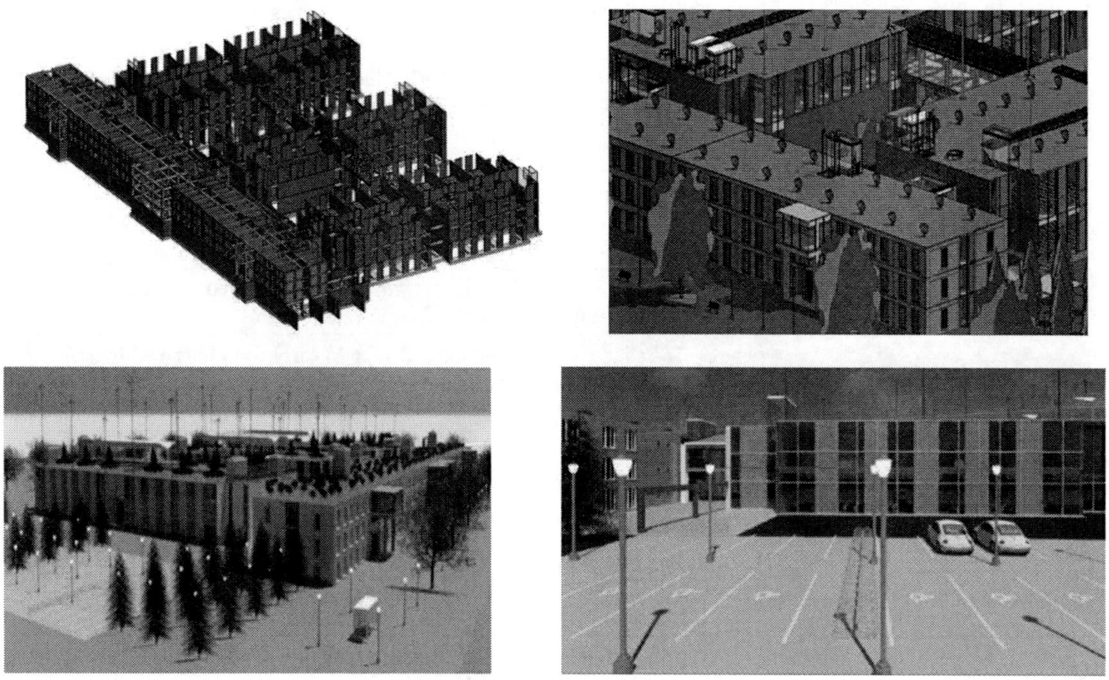

Figure 2. Sample drawings from the building.

5 CONCLUSIONS

In this study, Yildiz Technical University, Faculty of Civil Engineering Building was assessed and redesigned according to green building principles, using one of the BIM-based tools - Autodesk Revit.

Compliance with LEED certification system is analyzed through modeling by BIM applications. Analysis results showed that, existing design conditions exhibit 16 points. By redesigning the building, it is found that 16 more points can be achieved with additional sustainable construction practices. With the assumptions of points that are achieved during construction stage, the building score can reach to 41 points, which qualifies for LEED certification.

It is observed that BIM-based tools can be efficiently used during design stage of buildings, which aim to register for green building certification systems. Use of such tools can help to identify and prevent possible problems related to certification requirements that might occur in the future before construction stage.

BIM can be simply described as a system that provides a platform for all parties to participate in different phases of projects, which eventually influence the success of project management processes. Most of the undergraduate curriculum plans of civil engineering programs still lack related courses in Turkey. Therefore, students preparing a Capstone Project using means such as BIM could benefit from experiencing innovative approaches, which have become popular in construction industry before graduation. Inclusion of LEED green building certification system has also provided insights to students about sustainable construction and how they could utilize it as a means to achieve sustainability goals.

References

Akkaya, D., and Basaraner, M., *Yapı Bilgi Modellemesi (YBM) ve Coğrafi Bilgi Sistemleri (CBS)*, Akademik Bilisim'12- XIV. Akademik Bilişim Konferansı Bildirileri, Uşak Üniversitesi, Uşak, Türkiye. February 1-3, 2012.

Dikmen, C. B., Enerji Etkin Yapı Tasarım Ölçütlerinin Örneklenmesi, *Journal of Polytecnic*, 14(2), 121-134, 2011.

GYODER, Bir Değerleme Yaklaşımı: Yeşil Bina Sistemlerinin Gayrimenkul Değerine Etkisi, Gyoder.org.tr 2014. Retrieved from https://www.gyoder.org.tr/uploads/Yay%C4%B1nlar/sektorel_yayinlar/GYODER_PDF___20140605153004_2996yesil-degerleme.pdf in December, 2018.

Kibert, C. J., *Sustainable Construction: Green Building Design and Delivery*, John Wiley and Sons, New Jersey, 2013.

USGBC, U.S. Green Building Council, LEED 2009 for Schools New Construction and Major Renovations Rating System, http://www.usgbc.org/, 2009.

Interdependence between Structural Engineering and Construction Management
Edited by Ozevin, D., Ataei, H., Modares, M., Gurgun, A., Yazdani, S., and Singh, A.

ISBN: 978-0-9960437-6-2

CHALLENGES HINDERING THE ATTAINMENT OF SMART CITIES

DOUGLAS AGHIMIEN, CLINTON AIGBAVBOA, WELLINGTON THWALA, and IFIJE OHIOMAH

Faculty of Engineering and the Built Environment, University of Johannesburg, Johannesburg, South Africa

The concept of smart cities has become a buzzword in most developing countries. Most professionals within developing countries are quick to adopt this idea in their every day discuss on developing the state of their nation, without actually considering what this concept means to the average individual on the streets. Even lesser consideration is given to the readiness of these developing nations in terms of adopting the concept of smart cities. It is based on this notion that this study, through the review of existing studies, assessed the readiness of cities in Nigeria in adopting the concepts needed for achieving smart cities. The study gives an insight on the challenges impeding the attainment of city smartness in the country. It was observed that solving the problems caused by rapid urbanization within cities is the first step towards making these cities "smart ready". Therefore, if cities in Nigeria are to attain city smartness, first looking inward and solving the endemic problems within these countries is necessary before adopting concepts of smart cities that have been mastered in developed countries. Although the submissions of this paper form part of a much larger ongoing study, it contributes to the body of knowledge as it brings to light the challenges that must be solved if cities in Nigeria are to ever become smart.

Keywords: Digital technology, ICT, Sustainable cities, Sustainability.

1 INTRODUCTION

Govender (2018) noted that cities by nature are businesses on their own due to their ability to generate income for the government through taxes. Considering this important nature of cities to human existence and socio-economic development, coupled with the continuous growth in population and rapid urbanization, the call for smart cities has become popular in recent times. This call is with a view to handle the impact of urbanization on; the environment, the lifestyle of people within the environment, and governance (Silva *et al.* 2018). Although the definition of a smart city is believed to be a complicated one, several descriptions have evolved for a city that is smart (Sikora-Fernandez 2018). The simplest descriptions are that smart cities are connected cities (Silva *et al.* 2018), smart cities are safe cities in terms of humans and cyberspace (Gceza 2018), and smart cities are data-driven cities (Govender 2018). Drawing from these simple descriptions, it can therefore be said that a smart city is a city that is interconnected through the use of information and communication technology (ICT), provide adequate measure of safety for its occupants in terms of their physical and information space, and uses the enormous data being generated within the city to make informed decisions that will better the lives of its citizens. On a more comprehensive view, Harrison *et al.* (2010) described smart cities as cities that have the ability to connect the physical, social, business, information, and ICT infrastructure in a bid to increase the intelligence of the city.

A city is supposed to be advanced when compared to the rural areas. It is an environment that gives its dwellers a sense of belonging and some measure of fulfillment. Amer (2014) defined a city as a highly organized community. However, in some developing countries, particularly in Africa, most cities are far cry from what they should be. In fact, calling these places "cities" can be seen as a slur to areas with the true capabilities and characteristics of a city. Nigerian cities are no exception of this shortfall, with lack of necessary amenities characterizing major cities in the country. It is no gainsaying that rapid urbanization is one of the crucial issues in African cities in recent time. In fact, it is projected that by the year 2050, over half of Africa's population will be living in cities (Guneralp *et al.* 2017). Ogundare and Ogunbodede (2014) noted that in the case of Nigeria, there is a high rate of urbanization with a continuous rise in the population of the country. A major feature of the country's population is the increasing tendency for people to concentrate in cities, due to the city's ability to create settlement change and development for these individuals. This invariably puts severe pressure on available resources within these cities, with too many people depending on little infrastructure provisions. The resultant effect of this is poor and deteriorated infrastructures, poor living conditions, unhealthy environment, and the likes. According to Madikizela (2018) and Rees (2018), first solving these problems caused by rapid urbanization and making these cities "smart ready" is necessary, before trying to determine the route to take towards actually converting these existing cities into smart cities. It is based on this knowledge that this study through a review of literature assessed the challenges facing Nigerian cities in attaining city smartness with a view to proffering solutions that will increase the smart readiness of these cities.

2 OVERVIEWS OF SMART CITIES

Silva *et al.* (2018) described a smart city as one that applies the notion of the internet of things (IoT). In support, Kim *et al.* (2017) stated that IoT is one of the driving force behind smart city initiative all around the world as it provides the ability to monitor, manage and control devices, and to also create new information from a large sea of real-time data. In fact, it has been observed the smart cities adopt the use of ICT and IoT in city development, specifically in the aspect of government functionality, city operations, services deliveries, and intelligent analytics to optimize the services, production and usability (Kumar *et al.* 2018). It can, therefore, be seen that in describing what a smart city is, two aspects stand out, and they are the technological dimension and human dimension. The technological aspect deals with the networking of places using ICT deployment in every activity within a city in order to improve the standard of living within such city. However, it is clear that ICT alone will not give the desired improvement, hence the need for an enhancement in human capital and other forms of skill development among the citizens within the city (Kummitha and Crutzen 2017). Caragliu *et al.* (2011) also noted that the availability of ICT is not the only defining element of a smart city. The study went further to describe a smart city as one that invests in human and social capital and both traditional and modern communication infrastructure, and also allows for sustainable economic growth and high quality of life, through proper management of natural resources, and participatory governance.

3 CHALLENGES FACING CITY SMARTNESS IN NIGERIA

In 2016, the Governor of Lagos State signed a smart city deal with Dubai. The purpose of this deal was to transform Lagos, which is one of the major commercial cities in Nigeria, into a smart city – "the first in Africa". Unfortunately, aside mentioning of the signing of a memorandum of understanding (MoU) to this effect, no significant information was given for the commencement of this mouth-watering project (Ekwealor 2016). In a similar vein, the government at the federal level proposed a partnership in smart cities initiative with Huawei, a technology company. This was with the intent of reshaping and managing the delivery of public service (IT News Africa 2017). Siba and Sow (2017) further noted that in June 2017, the Nigerian government announced that the "Nigerian Smart City Initiative", which was aimed at increasing the use of ICT innovations in the delivery of improved physical infrastructure and service delivery. The results of these initiatives till today are yet to be evident within major cities in the country, with the continuous deterioration of infrastructures and poor city life still being dominant features in most cities. This can be attributed to the inherent challenges being faced within the country acting as barriers to the proper implementation of smart city concepts in cities within the country.

One of such challenges is poverty. BBC News (2012) reported that poverty in Nigeria has risen with almost 100 million people surviving on less than a dollar per day. Okhiria and Obadeyi (2015) observed that despite the increase in economic growth, the citizens have remained significantly poor. In 2018, a report from the Brookings Institution showed that Nigeria is now the nation with the highest number of people living in extreme poverty. The country overtook India as home to the world's greatest concentration of extreme poverty (Vanguard 2018). George and Ukpong (2013) stated that the major causes of poverty include lack of employment, high level of dependent population and overpopulation. With this overpopulation comes increased pressure on the meagre available infrastructure needed for survival. This leads to all other negative happenings within the society. In fact, Muhammad (2012) described poverty as the "octopus dragging along multiple criminal and nefarious activities among young and old, male and female folks".

The place of education in societal development cannot be overemphasised. Kazeem and Ige (2010) described education as the mechanism that helps the society generate the necessary knowledge and skills needed for the society's survival and sustenance. Moreover, Amzat (2010) noted that improving education is a good start towards alleviating poverty within a nation. In Nigeria, the education system today is a far cry from what it used to be. A continuous degradation in the system is evident in the last two to three decades. Public institutions are now characterised with poor and decayed infrastructures, leaving students to learn in unconducive environment and graduating with half-baked knowledge. The present democratic era in the country has suffered several industrial actions by academic bodies in the higher institutions. These bodies' demands among others include the need for standard and up-to-date infrastructures in the higher institutions (Edukugho 2013). In a country where the poverty level is high and education system is poor, having a digitalised city is closed to impossible as Malabi (2018) has noted that education plays a vital role in achieving smart cities in developing countries.

As observed by George and Ukpong (2013), lack of employment also contributes greatly to the high poverty level in Nigeria. In 2016, unemployment rose by 65% within the country (Noko 2016). This problem has been linked with the country's low economic growth rate, the desire to acquire white collar jobs only, high movement of people into urban areas, rapid population growth, and poor education system (Adesina 2013). Another rising issue is that of terrorism which is a global phenomenon and a major threat to national and international security and stability. It is also described as a "criminal act, against civilians, committed with the intent to cause death or serious bodily injury, or taking of hostages, with the purpose to provoke a state of terror in the general public or in a group of persons or particular persons, or compel a government or an international organization to do or to abstain from doing any act". Across the country's six geopolitical zones of Nigeria, ethnic militias and other violent non-state actors have emerged with different motivations and strategies to challenge state authority (Mustapha 2009). This situation is bound to scare away foreign investors that are ready to partner with the country in areas of infrastructural development and digital technologies needed for achieving city transformation. Issue of cybersecurity and digital safety is also a crucial issue that can be attributed to the adoption of technologies in a country like Nigeria where the cyber-crime rate is on the increase. Considering the importance of digital technologies and the internet in achieving of smart cities, the role of cyber insecurity in deterring the transformation of existing cities cannot be overlooked.

In addition, the epileptic power supply within the country is an issue that has been a reoccurring problem affecting digital technology adoptions (Oladapo 2007). This is coupled with the high cost, lack of access, and slow Internet connectivity in the country. In fact, Nigeria's Internet speed is so slow that Akamai (a global content delivery platform) ranked it as 114 out of 143 countries surveyed in May 2017 (Adepetun 2017). This is a major challenge for city transformation as a smart city has been described as a connected city (Friend 2018).

4 THE WAY FORWARD

The phrase "go digital or go home" is here to stay. Chambers (2017) noted that "the world will go digital: It will transform healthcare, enable people to live longer and be a huge disruption to society. Either you disrupt or you get left behind". Nigeria can no longer shy away from adopting current trends under the disguise of being a developing country. There is the need to adopt digital technologies in transforming of its cities and in providing better ways of living for its citizens. However, if this must be achieved, there is the need to first solve the inherent problems within the country that might serve as hindrances to the attainment of city transformation. No matter how laudable the proposed strategies for achieving smart cities within a country are, and the intent behind them, they are worthless if they are not contextualized. It is only by critically evaluating the problems within and solving them that the realization of smart ideas can be achieved (Madikizela 2018). This submission is further strengthened by the popular notion that "no two cities are alike" (Cowen 2017). Thus, understanding the problem peculiar to an area can help shape the solutions that will be attained, and only then can one think of bringing in smart notions to create a safe and happy city for its occupants.

Poverty alleviation programs are necessary for the adoption of new innovations. Through these programs, unemployment issues can be addressed. Similarly, more focus on the education is necessary. It has been observed that for smart cities to be achieved there must be cohesion, education, and utilization (Malabi 2018). It is only through proper education that the understanding in digital related practices and improvement on their utilization can be achieved. Through this improved education system, a reduction in crime rate can be attained as this will lead to acquiring of better jobs that will to a large extent reduce the rate of youths getting into criminal activities within cities and other areas of the country (Amzat 2010). There is the need for better partnering with countries and organizations that are excelling in combating issues surrounding cyber-crime. Increasing cybersecurity invulnerability will go a long way in increasing the chances of bringing the idea of the smart city into reality in the country. Increasing Government commitment towards delivering steady power supply is also crucial as this issue has been a reoccurring one within the country. Digital technologies needed to attain cities transformation require a steady power supply. In similar vein, creating strict sanctions on internet service providers is equally necessary as this is a major driver of the attainment of smart cities. These organizations with the support of the Government should be made to increase the quality of service delivery and by so doing create an enabling environment for cities transformation.

5 CONCLUSIONS

There is no shadow of doubt that cities in developing countries like Nigeria are in intolerable conditions. The need for a transformation of these cities into smart cities cannot be overemphasised. However, in order to conveniently attain this transformation, there is the need to evaluate the problems facing the country and proffer possible directions towards the attainment of a smart city in the country. Based on the review of existing literature, this study concludes that the challenges facing the transformation of Nigerian cities into smart cities are poverty, poor education, unemployment, terrorism which scares away foreign investors, cyber insecurity and poor digital information safety, poor power supply and poor internet connectivity. The study, therefore, recommends that the introduction of workable and realistic poverty alleviation programme to solve the issue of poverty and unemployment. Also, government and private investors need to place more focus on the education and intensify the fight again crimes both on and off the internet. Effort should also be made by the government to provide steady power supply and to create policies that will regulate the activities of internet service providers in ensuring better service delivery. It is believed that the submission of this study is a step towards achieving city readiness in the attainment of smart cities in Nigeria. However, the major limitation of the study is its dependence on existing literature. Hence, further studies hope to be conducted to harness empirical data within the study area to confirm the submissions of this study.

References

Adepetun, A., How Nigeria's Slow Internet Speed Affects Economic Growth, The Guardian Newspaper, 2017. Retrieved from https://guardian.ng/business-services/how-nigerias-slow-internet-speed-affects-economic-growth/ on March 2019.

Adesina, O. A., Unemployment and Security Challenges in Nigeria, *International Journal of Humanities and Social Science,* 3(7), 146-156, 2013.

Amer, S. B., *City definitions*, A Paper Presentation at DTU Management Engineering, Technical University of Denmark, May 25, 2014.

Amzat, I. H., The Effect of Poverty on Education in Nigeria: Obstacles and Solutions, *OIDA International Journal of Sustainable Development*, 1(4), 55-72, 2010.

BBC News, Nigerians Living in Poverty Rise to Nearly 61%, 2012. Retrieved from https://www.bbc.com/news/world-africa-17015873 in March 2019.

Caragliu, A., Del Bo, C., and Nijkamp, P., Smart Cities in Europe, *Journal of Urban Technology*, 18(2), 65-82, 2011.

Chambers, J., Go Digital or Go Home, and Other Thoughts from CISCO's, 2017. Retrieved from https://executive.mit.edu/blog/go-digital-or-go-home-and-other-thoughts-from-ciscos-john-chambers#.W6cAv2gzZaQ in March 2019.

Cowen, T., China's Megacities: No Two Cities are the Same, *Gulf News Thinkers,* 2017. Retrieved from https://gulfnews.com/opinion/thinkers/china-s-megacities-no-two-cities-are-the-same-1.2080002 in March 2019.

Edukugho, E., Education Sector Stinks! Infrastructure Bad, Now Worse, 2013. Retrieved from http://www.vanguardngr.com/2013/12/education-sector-stinks-infrastructure-bad-now-worse/ in March 2019.

Ekwealor, V., The Challenges to Lagos Becoming a Smart City Techpoint, 2016. Retrieved from https://techpoint.ng/author/suave/ on March 2019.

Friend, D., *DR and Cloud strategies, Data Protection and GDPR Compliance,* Smart Cities Africa Summit, Emperors Palace Hotel, Johannesburg, South Africa, July 3-4, 2018.

Gceza, N., *Cyber Security in Smart Cities: A People Perspective,* Smart Cities Africa Summit, Emperors Palace Hotel, Johannesburg, South Africa, July 3-4, 2018.

George, I. N., and Ukpong, D. E., Contemporary Social Problems in Nigeria and its Impact on National Development: Implication for Guidance and Counselling Services. *Journal of Educational and Social Research,* 3(2), 167-173, 2013.

Govender, V., *Laying Down the Groundwork for Smart Cities,* Smart Cities Africa Summit, Emperors Palace Hotel, Johannesburg, South Africa July 3-4, 2018.

Guneralp, B., Lwasa, S., Masundire, H., Parnell, S., and Seto, K. C., Urbanization in Africa: Challenges and Opportunities for Conservation, *Environmental Research Letters,* 13, 1-8, 2017.

Harrison, C., Eckman, B., Hamilton, R., Hartswick, P., Kalagnanam, J., Paraszczak, J., and Williams, P., Foundations for Smarter Cities, *IBM Journal of Research and Development*, 54, 1–16, 2010.

IT News Africa, Nigerian Government and Huawei Partner on Smart Cities initiative, IT News Africa – Africa's Technology News Leader, August 2017. Retrieved from http://www.itnewsafrica.com/2017/08/nigerian-government-and-huawei-partner-on-smart-cities-initiative/ in March 2019.

Kazeem, K., and Ige, O., Redressing the Growing Concern of the Education Sector in Nigeria. *Edo Journal of Counselling,* 3(1), 40-49, 2010.

Kim T., Ramos, C., and Mohammed, S., Smart City and IoT. *Future Generation Computer Systems,* 76, 159–162, 2017.

Kumar, H., Singh, M. K., Gupta, M. P., and Madaan, J., Moving Towards Smart Cities: Solutions That Lead to the Smart City Transformation Framework, *Technological Forecasting and Social Change,* 2018. Retrieved from https://doi.org/10.1016/j.techfore.2018.04.024 in March 2019.

Kummitha, R. K. R., and Crutzen, N., How do we Understand Smart Cities? An Evolutionary Perspective, *Cities,* 67, 43–52, 2017.

Madikizela, K., *How Smart Cities can Accelerate Transformation in Africa*, Smart Cities Africa Summit, Emperors Palace Hotel, Johannesburg, South Africa, July 3-4, 2018.

Malabi, K., *Bridging the Digital Divide,* Smart Cities Africa Summit, Emperors Palace Hotel, Johannesburg, South Africa July 3-4, 2018.

Muhammad, M., Poverty and National Security: A Review of Policies, 1960 – 2010, *Journal of Research in National Development,* 10(2), 243-251, 2012.

Mustapha, A. R., *Nigeria Since 1999: A Revolving Door Syndrome or the Consolidation of Democracy?* In A. R. Mustapha and L. Whitfield (eds.),. In T. p. Democracy. Oxford: James Currey, 2009.

Noko, E. J., Impact of Unemployment on Nigeria Economic Growth 1981-2015, *EduCacInfo,* 2016. Retrieved from https://educacinfo.com/impact-unemployment-nigeria/ on March 2019.

Ogundare, B. A., and Ogunbodede, E. F., Traffic Congestion and Parking Difficulties in Akure Metropolis, Nigeria, *IOSR Journal of Humanities and Social Science,* 19(8), 1-7, 2014.

Okhiria, A. O., and Obadeyi, J., Poverty, an African Epidemic: Empirical Evidence of Nigeria, *Developing Country Studies,* 5(6), 29-39, 2015.

Oladapo, A. A., An Investigation into the Use of ICT in the Nigerian Construction Industry, *Journal of Information Technology in Construction*, 12, 261-277, 2007.

Rees, G., *The Smart Water Utility of The Future*, Smart Cities Africa Summit, Emperors Palace Hotel, Johannesburg, South Africa, July 3-4, 2018.

Siba, E., and Sow, M., Smart City Initiatives in Africa, *Brookings - Africa in Focus*, 2017. Retrieved from https://www.brookings.edu/blog/africa-in-focus/2017/11/01/smart-city-initiatives-in-africa/ in March 2019.

Sikora-Fernandez, D., Smarter Cities in Post-Socialist Country: Example of Poland, *Cities,* 78, 52–59, 2018.

Silva, B. N., Khan, M., and Han, K., Towards Sustainable Smart Cities: A Review of Trends, Architectures, Components, and Open Challenges in Smart Cities, *Sustainable Cities and Society,* 38, 697–713, 2018.

Vanguard, Nigeria Overtakes India as World's Poverty Capital — Report, June 25, 2018. Retrieved from https://www.vanguardngr.com/2018/06/nigeria-overtakes-india-as-worlds-poverty-capital-report/ in March 2019.

Interdependence between Structural Engineering and Construction Management
Edited by Ozevin, D., Ataei, H., Modares, M., Gurgun, A., Yazdani, S., and Singh, A.

ISBN: 978-0-9960437-6-2

INNOVATION OF USING WASTE MATERIALS AS A REPLACEMENT OF CEMENT IN PCC

AHMED S. FARIED[1], MOSTAFA A. ABO-HASHEMA[1], ABDULRAHMAN E. ROUSHDY[2], and SAMEH A. GALAL[1]

[1]*Dept of Civil Engineering, Fayoum University, Fayoum, Egypt*
[2]*General Authority for Educational Buildings, Fayoum, Egypt*

Portland cement (PC) is used in the production of Portland cement concrete (PCC) to construct many infrastructure components such as rigid pavements. This research aims at studying the effect of using common waste materials in Egypt as a replacement of PC in the production of PCC. Using waste materials in concrete production could be an effective measure in maintaining the environment, improving the properties of PCC, and reducing cost of production. Nowadays, pavement technologist and researchers had put their interests on nanotechnology. This paper outlines the innovation of using Nanotechnology-based waste materials as a replacement of PC to enhance the physical and mechanical properties of PCC. The experimental work, using waste materials such as ceramic waste (CW) and ground-granulated blast-furnace slag (GGBFS), was conducted in five phases. In phase 1 and phase 2, different rations were used for CW and GGBFS by weight of cement, respectively. In phase 3, the best three ratios of CW were used with the best ratio of GGBFS. In phase 4, different ratios of CW and GGBFS were used with a standard ratio of Un-Hydrated lime (UHL). In phase-5, various mix of CW, GGBFS and UHL were suggested based on previous results. Modified PCC mixtures were produced, tested and compared to the conventional concrete mixture. Results showed that the cost of PCC production is reduced by 34% in average comparing to the control mix.

Keywords: Nanotechnology, Rigid pavement, Ground-granulated blast-furnace slag, Ceramic waste, GGBFS, Un-hydrated lime.

1 INTRODUCTION

Portland cement concrete (PCC) is a most versatile construction material because it was designed to resist loads and environment changes with adequate strength and durability. This necessitates the use of special combinations of performance and uniformity requirements that cannot always be achieved using conventional constituents and normal mixing. Another challenge is to reduce production cost of PCC without reducing the concrete performance, especially for big projects (Aly *et al.* 2012, Meyer 2009). Waste materials could be a solution for this challenge as a replacement of Portland cement (PC) and could be also an effective measure in maintaining the environment, improving the properties of concrete, and reducing cost of production (Khan 2011, Fwa and Wei 2005, Umapathy *et al.* 2014).

The Egyptian ceramics industry is producing around 3 million tons/year of waste (about 15%-25% of total waste material). Ceramic waste (CW) is of generally two types: waste earthenware and cracked during the manufacturing process. CW is considered as non-hazardous solid waste and possesses pozzolanic properties (Dayalan and Beulah 2014, Lopez *et al.* 2007). On the other hand, ground-granulated blast-furnace slag (GGBFS) or steel slag, a by-product of steel making, is produced during the separation of impurities from molten steel in steel-making furnaces. The slag occurs as a molten liquid melt and is a complex solution of silicates and oxides that solidifies upon cooling (Sultan *et al.* 2014, Birgisson *et al.* 2010).

Un-Hydrated lime (UHL) is a Calcium oxide (CaO) as in clinker and cement that has not combined with SiO_2, Al_2O_3, or Fe_2O_3 during the burning process, because of under burning, insufficient grinding of the raw mix, or the presence of traces of inhibitors. Lime has been used for a very long time in construction and buildings. Lime is an industrial product obtained by calcination of limestone in a limekiln (Oates 1998). Researchers have found that nanomaterials have significant effects in improving the strength and durability of PCC.

This paper outlines the innovation of using Nanotechnology-based waste materials (CW, GGBFS and UHL) as a replacement of PC in different dosages to enhance the physical and mechanical properties of PCC that could be used as rigid pavements. Modified PCC mixtures were produced, tested and compared to the conventional concrete in terms of Index Tests and Advanced Tests. Index tests include mix workability, compressive and flexural strength, density and base-carbonation. Advanced tests include X-ray diffraction (XRD) and Scanning Electron Microscope (SEM). The purpose is to evaluate the mechanical properties at 7 and 28 days to achieve the optimum percentage of the partial replacement of cement with waste materials.

2 MATERIALS AND METHODS

The materials involved in this research are cement, coarse aggregate, fine aggregate and waste materials such as CW, GGBFS, and UHL. The properties of these materials are tested and tabulated. The American code for concrete mix, ACI-211-11, was utilized provision for construction of PCC. Samples of wastes were collected manually and freshly at the beginning of the experimental work and stored as per standard specification. Wastes were broken into particles in nanometer (10^9 scale). The reduction in size was verified using SEM. Cement "AL-MUMTAZ" 52.5 grades (ASTM Type II) was used. Coarse and fine aggregates are used as follows: fractions of max size 10 mm of SUEZ crushed dolomite rock, and fractions of max size 2.70 of BANI-SWIF fine aggregate river sand, respectively.

2.1 Waste Materials

2.1.1 *Ceramic waste*

The CW was obtained from recycled ceramic tiles supplied by "AL-FARAANA" factory in Egypt. Cracked pieces were crushed. At the laboratory, these wastes were ground with an air jet mill to obtain powder. The resulting powders were sieved through a 75-lm (200 mesh) sieve.

The ceramic powder includes nearly 92% pozzolanic materials, where Al_2O_3 and SiO_2 were the main components with higher characteristic peak values during analysis. These two minerals could react with $Ca(OH)_2$ in the cement paste to produce crystalline C–A–H and low density C–S–H gel, which can fill nano pores in concrete, increasing the bond strength between the interface of aggregates, decreasing the permeability and improving the durability of the concrete.

2.1.2 *Ground-granulated blast-furnace slag*

The GGBFS was supplied from various steel industries in Egypt with a specific gravity of 2.40. As a replacement for cement, the GGBFS can be used from 5% to 70% depending upon compressive strength requirement. The same slag can be used effectively when it is converted to nano size. This would improve the performance of the slag as the surface area increases. The size reduction was achieved by loading raw slag of 17.15 microns into the planetary ball mill and ground for 2½ hours. The reduction in size was verified using Particle Size Analyzer and was found to be 370 nm after grinding (size has been reduced to around 45 times than original value).

2.1.3 *Un-hydrated lime*

In the lime industry, limestone is a general term for rocks that contain 80% or more of calcium or magnesium carbonates, including marble, chalk, and marl. UHL can be used as a replacement of cement from 10% to 20% depending upon compressive strength requirement and it could be used effectively when it is converted to nano size. An increase in clinker un-hydrated lime reduces the total silicates ($C_3S + C_2S$). Both C_3S and C_2S are involved in the hydration to produce the C-S-H gel, which is the main strength-forming phase in hardened Portland cement concrete. Un-hydrated lime hydrates to form more of Calcium Hydroxide $Ca(OH)_2$ crystals. More UHL leads to increase the proportion of C_2S and decreases the proportion of C_3S in the clinker, which is a more important contributor for 28-day strength. C_3S is much more reactive than C_2S.

2.2 Experimental Investigation

2.2.1 *Mix types and proportion*

The experimental work, using CW, GGBFS and UHL as a replacement of PC, was conducted in five phases as follows:

- **Phase 1**: CW was used with 20%, 25%, 30%, 35%, 40% and 45% by weight of cement;
- **Phase 2**: GGBFS was used with 5%, 10%, 15%, 20%, and 25% by weight of cement;
- **Phase 3**: The best three ratios of CW, based on the results, were used in conjunction with the best ration of GGBFS, based on the results;
- **Phase 4**: Three different ratios of CW and GGBFS (25%, 35% and 45%) were used in conjunction with the standard ration of UHL;
- **Phase 5**: Based on results of previous phases, different three ratios of CW (20%, 30% and 40%) were used in conjunction with the best ration of GGBFS and UHL.

In addition to the above five phases, a control mix was also designed with zero wastes.

2.2.2 *Mixture design*

The mixture was designed according to ACI-211, and ASTM C 595 for Blended Hydraulic Cements. The binder content (450 kg/m^3), fine aggregate content (892.8 kg/m^3), coarse aggregate content (717.8 kg/m^3) and water–cement ratio of 0.48 were chosen to be constant. The workability of the fresh concrete was measured with a standard slump cone using the slump test according to ASTM C 143. The coarse and fine aggregates were mixed first, followed by addition of cement, wastes and water. Nanoparticles are not easy to distribute uniformly due to their high surface energy. In order to solve this problem, before adding any nano waste to the mixture, nanoparticles were stirred separately for approximately 2-minutes at high speed using water. After making sure that the particles have been completely dissolved in water, they were added to the mixture.

2.2.3 *Compressive and flexural strength*

The test specimens were cast in steel cubic molds (100 x 100 x 100 mm) and steel beam molds (150 x 150 x 750 mm). Specimens were compacted on a vibrating table. After approximately 24 hours, the specimens were removed from the molds. The concrete specimens were cured in water at 21°C in cure tanks until the time of testing. Casting, compaction, and curing were accomplished according to ASTM C 192 and C 293. For each mix, cubic samples were tested to determine compressive strengths at 7 and 28 days of curing, and beams at 28 days of curing.

2.2.4 *Instrumental analyses*

The instrumental analyses tests were used to present the quality of concrete and changes in concrete dense after using wastes. The XRD and SEM give an indication about the mechanical properties of concrete functional in concrete chemical composition and pozzolanic reactions. X-ray diffraction (XRD) was carried out on a Philips PW3050/60 X-ray diffractometer using a scanning range from 2θ of 0° to 60° with a scanning speed of 1 sec/step and resolution of 0.05°/step. SEM can achieve resolution better than 1 nanometer.

3 RESULTS AND DISCUSSIONS

3.1 Index Tests

3.1.1 *Mixes workability*

The results of the flow table test showed that the workability of all mixes of phase 1 was decreased when compared to control mix. While the workability of all mixes of phase 2 was increased when compared to control mix. This is because of the water absorption nature. CW absorbs more water than GGBFS. For phase 3, when a part of GGBFS has been replaced by CW, the workability increases, as the water absorption property of CW is high compared to steel slag. Only lesser amount of water is left behind for the hydration of cement. For phases 4 and 5, UHL absorption is very low. Therefore, CW+UHL decreases workability when %CW increases; on the other hand, the workability is increased when using GGBFS+UHL. It is noteworthy that the slump value of control mix is 23 mm, where the specifications are in the range of 18-38 mm with average of 28 mm and standard deviation of 10 mm according to (ASTM C 143/ C143M -00). Consequently, all results are within the acceptable range with constant water per cement ratio.

3.1.2 *Compressive strength test*

The % change in compressive strength -- from an average of three cubic specimens -- for the best 40-70% replacement of PC in all phases is presented in Figure 1.

3.1.3 *Flexure strength test*

For each mix, cubic samples were tested to determine the flexure strengths at 28 days of curing. The flexure strength for each mixture was obtained from an average of three beams specimens. The % change in flexure strength for the best 40-70% replacement of PC in all phases is presented in Figure 2. As shown in the figures, the following conclusions can be drawn:

- For phase 1 mixes, compressive and flexure strengths decrease with increasing of CW, it is related to the CW works as a filler not as pozzolanic material as like GGBFS.
- For phase 2 mixes, strengths increase with 5% and 10% of GGBFS, then a reduction was noticed for 15% of GGBFS until 25% of GGBFS.
- Therefore, phase 3 mixes were selected to use 10%, 15% and 20% of GGBFS with three different percentages of CW as 20%, 30% and 40%. By increasing the percent of GGBFS with the same ratio of CW, strengths decrease.

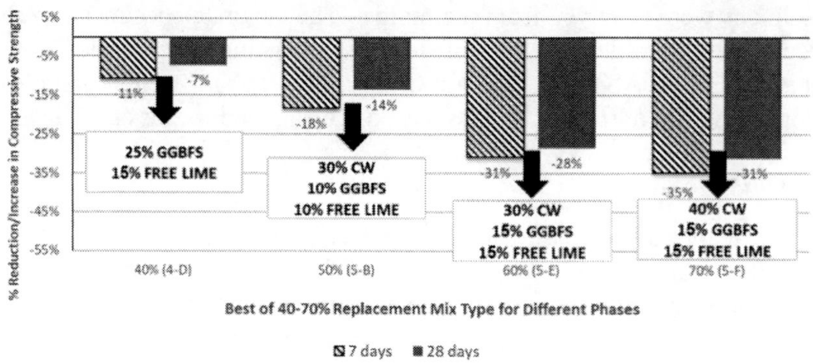

Figure 1. % change in compressive strength for the best 40-70% replacement of pc (all phases).

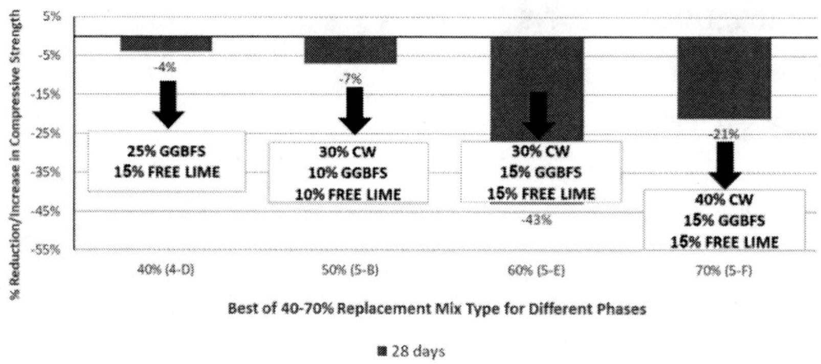

Figure 2. % change in flexure strength for the best 40-70% replacement of pc (all phases).

- Regarding phases 4 and 5, adding Un-hydrated lime, from 10 to 15 %, gives a good effect on both compressive and flexure strengths.
- For 70% cement replacement in (5-F) mix, only 31% reduction on compressive strength occurred by 20.9 N/mm2.
- For 40% cement replacement, mix has a minimum loss in compressive reading by only 14%. This returns to the percentage of CW as a pozzolanic relative to the percent of both GGBFS and Un-hydrated lime for continuing the interactions.
- The previous interpretation applies also on 50% and 60% replacement.
- Results of (5-E) and (5-F) mixes are very close despite 10% replacement difference.

3.2 Advanced Tests

To verify the mechanism predicted by concrete (compressive, flexural strength and base-carbonation) tests, X-ray diffraction and SEM Microscopy examinations were performed.

3.2.1 *XRD analysis*

Clearly, the (5-B) Mix showed the highest Portlandite {Ca (OH)2} peak intensity at 18.1° 2θ. This peak decreases with the addition GGBFS, CW and UHL suggesting the pozzolanic reaction. A low intensity Portlandite peak was observed in case of cement concrete containing higher percent of GGBFS and UHL. This proved the highest pozzolanic reactivity of GGBFS. The XRD results are in the agreement with that of mechanical properties.

3.2.2 *SEM microscopy*

The microstructure of concrete containing 10% of GGBFS is higher homogenous and compacted than that of other mixes. Also, the interfacial transition zone of this mix is stronger than that of net concrete and concrete containing ceramic waste. This indicates the higher pozzolanic activity of GGBFS compared to CW.

4 ECONOMIC ANALYSIS USING WASTE MATERIAL

Based on the previous results, there are four suggested mixes, which are 40%, 50%, 60%, and 70% replacement of PC with different dosages of CW, GGBFS and UHL. The cost of each material per ton was identified and consequently the cost of each mix comparable to the control mix was calculated. It can be concluded that production cost could be reduced up to 18.5% in case of 70% replacement of PC with waste materials. Another cost comparison was made with the traditional flexible pavement; where the results showed that the construction cost was decreased by nearly 34% when using PCC with waste materials for (5-F) mix.

5 CONCLUSIONS

In trying to find possibility of reducing cost of rigid pavement construction using Portland cement concrete (PCC), waste materials were used by means of partial replacement of cement in PCC. Ceramic Waste (CW), ground-granulated blast-furnace slag (GGBFS) and Un-hydrated lime were used as waste materials in nanoparticles through five different mixes phases. Results showed that up to 10% of GGBFS, as a replacement of cement, gives good results in increasing mechanical properties of PCC depending on the high pozzolanic reactivity of GGBFS. Finally, results showed that the cost of PCC production with waste materials is reduced by 18.5% comparable to the control mix; and is reduced by 34% comparable to flexible pavement.

References

Aly, M., Hashmi, M. S. J., Olabi, A. G., Messeiry, M., Abadir, E. F., and Hussain, A. I., Effect of Colloidal Nano-Silica on The Mechanical and Physical Behaviour of Waste-Glass Cement Mortar, *Materials and Design,* 33, 127-135, 2012.

Birgisson, B., Taylor, P., Armaghani, J., and Shah, S. P., American Road Map for Research for Nanotechnology-Based Concrete Materials, *Journal of the Transp. Research Board*, 130-137, March 2010.

Dayalan, J., and Beulah, M., Effect of Waste Materials in Partial Replacement of Cement Fine Aggregate and Course Aggregate in Concrete, *Intnl. Journal of Inventive Engineering and Sciences (IJIES),* 2014.

Fwa, T. F., and Wei, L., *Design of Rigid Pavements,* The Handbook of Highway Engineering, Fwa, T. W. (ed.), CRC Press, 2005.

Oates, J. A. H., *Lime and Limestone, Chemistry and Technology: Production and Uses,* Wiley-VCH verlag GmbH, Weiheim, 1998.

Khan, M. S., *Nanotechnology in Transportation: Evolution of a Revolutionary Technology,* TR News, 277, 3-8, November-December 2011.

Lopez, V, Llamas, B., Juan, A., Moran, J., and Guerra, I., Eco-Efficient Concretes: Impact of The Use of White Ceramic Powder on the Mechanical Properties of Concrete, *Biosystems Engineering,* 96(4), 559–564, 2007.

Meyer, C., The Greening of The Concrete Industry, *Cement and Concrete Composites,* 31, 601–605, 2009.

Sultan, A., Tarawneh, E. S, Gharaibeh, F., and Saraireh, M., Effect of Using Steel Slag Aggregate on Mechanical Properties of Concrete, *American Journal of Applied Sciences,* 11(5), 700-706, 2014.

Umapathy, U., Mala, C., and Siva, K., Assessment of Concrete Strength Using Partial Replacement of Coarse Aggregate for Waste Tiles and Cement for Rice Husk Ash in Concrete, *Intnl. Journal of Engineering Research and Applications*, 2014.

Interdependence between Structural Engineering and Construction Management
Edited by Ozevin, D., Ataei, H., Modares, M., Gurgun, A., Yazdani, S., and Singh, A.

ISBN: 978-0-9960437-6-2

SOCIAL AND ENVIRONMENTAL SUSTAINABILITY RATING SYSTEMS AND CERTIFICATION PROGRAMS

DIANA ANDRADE-RHOR, LAURA MONTALBÁN-DOMINGO, TATIANA GARCIA-SEGURA, and EUGENIO PELLICER

School of Civil Engineering, Universitat Politècnica de València, Valencia, Spain

The term sustainability has multidisciplinary use and meaning. Sustainability is typically described by many sources as development that meets the needs of the present generation without compromising the ability of future generations to meet their own needs. Sustainable construction projects are those that ensure the consideration of economic, environmental and social impacts that are created by their activities. It includes all the construction related tasks and the company's management. This research focuses mainly on the criteria of social and environmental sustainability. Social sustainability, whose criteria is based on the classification of company's stakeholders and the project's, and environmental sustainability, also called physical dimensions of sustainability, whose criteria are those related to resources and the environment. This paper aims to recognize the most known sustainability certification systems, how they classify their criteria and rate their metrics. After the research, we can conclude that the certification systems incorporate more metrics for criteria of environmental sustainability than for social sustainability.

Keywords: Certification systems, Construction project, Sustainability criteria.

1 INTRODUCTION

In the last couple of decades, the approach of responsible consumption, sustainable development and the inclusion of social and environmental aspects in construction projects have become increasingly important. The concept of sustainability refers, by definition, to the satisfaction of current needs without compromising the ability of future generations to satisfy their own needs, guaranteeing the balance between economic growth, care for the environment and social welfare (PNUD 2014).

Nowadays, sustainability must be considered a part of every decision we take, and the answer to each project should concern not only for the economic aspects but also for the social and environmental aspects of sustainability. It is said that construction is sustainable when it meets environmental challenges, responds to social and cultural demands and offers economic improvements (Bal *et al.* 2013).

The sustainability boom is having a positive impact on business certification and management systems. Organizations not only want to implement better management systems but also seek to certify them, in order to prove their stakeholders their commitment to the development of a sustainable project in a working environment adjusted to the objectives of the *triple bottom line* (economic, social and environment).

2 SOCIAL AND ENVIRONMENTAL SUSTAINABILITY METRICS AND CRITERIA

2.1 Social Sustainability Metrics

The social criteria result from the evaluation of the aspects that affect or may affect the stakeholders, each of them is benefited or harmed by the project in a different way, for that reason the social dimension includes specific criteria and metrics for each group of stakeholders. Within the construction projects, and the companies of the construction industry, we can differentiate four groups of stakeholders: the society, the local community, employees and users or customers (Bal *et al.* 2013; Benoît-Norris *et al.* 2013) (
Figure 1).

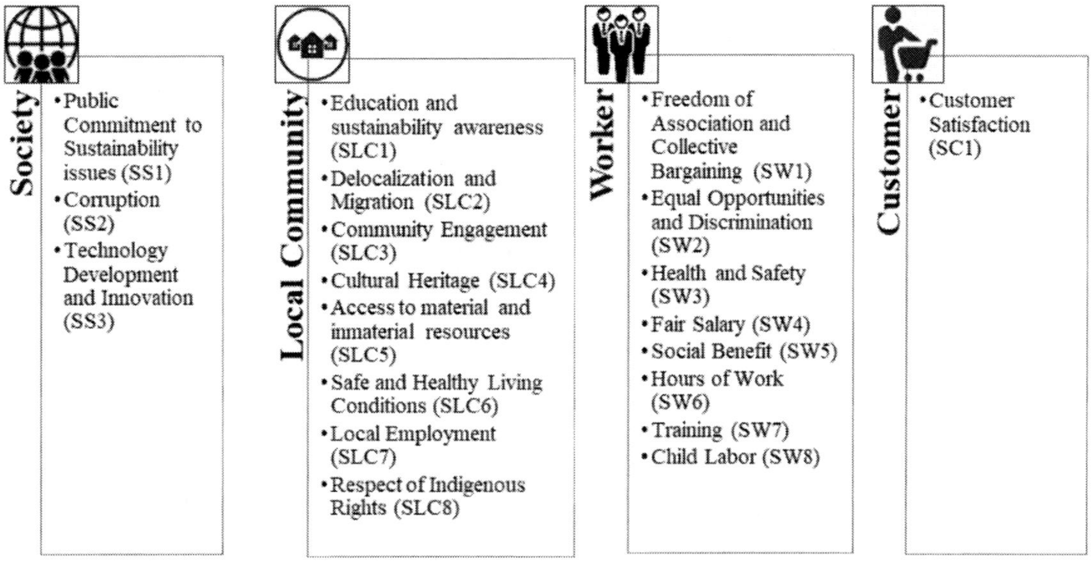

Figure 1. Social Sustainability Metrics by Stakeholders Categories.

2.2 Environmental Sustainability Metrics

Environmental indicators, also called physical dimensions of sustainability, are the traditional metrics of environmental sustainability (Cohen *et al.* 2014). The environmental sustainability metrics contemplate the impacts of the project or industry actions on the environment and human beings, any activity, that represents a change in the environment, must be identified, measured, mitigated and evaluated. The environmental metrics are classified into four categories (Figure 2), depending on the environment they affect; they can be related to human health, ecosystem, resources or water.

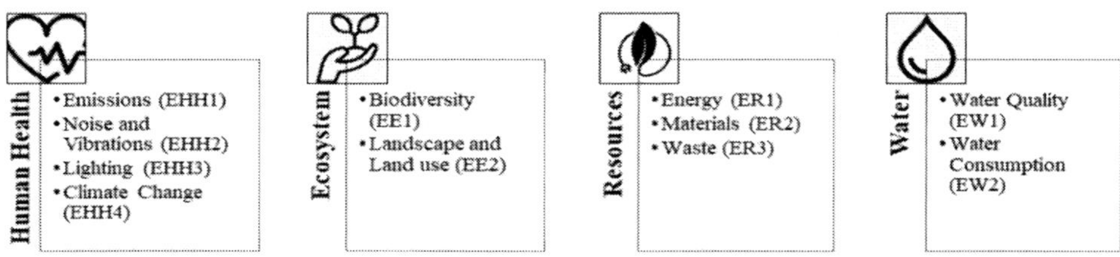

Figure 2. Environmental Sustainability Metrics by Categories.

3 RATING SYSTEMS AND CERTIFICATION PROGRAMS

The sustainability certification systems are a series of parameters or minimum standards in matters of social and environmental sustainability, carried out by specialized third parties, which are in charge of evaluating, classifying and certifying the organizations and/or their projects.

The construction industry has different certification systems (Table 1), ranging from evaluating the social management of employees within the organization, to standards to know if a building or infrastructure meets or does not meet the minimum requirements for energy performance. It means certification and rating systems that include the concept of sustainability in the management of the construction company and its projects.

Table 1. Rating Systems and Certification Programs.

#	SYSTEM - REFERENCE	SUBJECT - MATTER
01	Be^{2}st-in-highways (Recycled Materials Resource Center and University of Wisconsin-Madison 2010)	Social Requirements including regulation and Local Ordinances, Emissions, Energy, Waste, Water.
02	CEEQUAL (Beckwith *et al.* 2011)	Project Management, Land Use, Landscape, Biodiversity, Historic Environment, Water, Energy, Material, Waste.
03	ENVISION (ISI 2015)	Quality of Life, Leadership, Resource Allocation, Natural World, Climate and Risk.
04	GreenLITES (New York State Department of Transportation 2008)	Sustainable Sites, Water Quality, Material Resources, Atmosphere, Innovation.
05	Greenroads Manual (Muench *et al.* 2011)	Basic Project Requirements, Environment and Water, Access and Equity, Construction Activities, Materials and Resources, Pavement Technologies, Custom Credits.
06	ISO 26000 (Kritkausky and Schmidt 2011)	Organizational Governance, Human Rights, Labor Practices, Environment, Consumer Issues, Community Involvement and Development.
07	I-LAST (Illinois Department of Transportation 2012)	Planning, Design, Environmental Water Quality, Transportation, Lighting, Materials, Innovation.
08	INVEST (Bevan *et al.* 2012)	System Planning and Processes, Project Development, Operations and Maintenance.
09	LEED (U.S. Green Building Council 2002)	Sustainable Sites, Water, Energy and Atmosphere, Resources, Indoor Environmental Quality, Innovation, Regional Priority.
10	SA8000® (Social Accountability Internacional 2014)	Child Labor, Health and Safety, Freedom of Association and Right to Collective Bargaining, Discrimination, Working Hours, Salary, Management System.
11	STARS (North American Sustainable Transportation Council 2012)	Integrated Process, Access, Climate and Energy Ecological Function, Cost, Innovation.

4 SUSTAINABLE METRICS ANALYSIS

Once sustainability certification systems related to the construction industry and its projects are identified, we can determine which criteria is included by each one, and which are more representative according to each certification system. In
Table 2, those criteria that have been included in each system are marked down with an "X".

The social sustainability criteria that stand out above others, within the considerations of the certification systems analyzed, are Safe and Healthy Living Conditions (SCL6), Worker's Health and Safety (SW3) and Customer Satisfaction (SC1). ISO 26000 (06) and SA8000 (10) are, as expected, the certification systems that include the largest number of criteria in social sustainability terms.

Table 2. Consideration of criteria classified by certification system.

CRITERIA/SYSTEM	01	02	03	04	05	06	07	08	09	10	11
SOCIAL SUSTAINABILITY											
SS1	-	-	-	-	-	X	X	-	-	-	-
SS2	-	-	-	-	-	X	-	-	-	-	-
SS3	-	-	-	X	-	X	X	X	X	-	-
SLC1	-	-	X	-	X	-	X	X	-	-	-
SLC2	-	-	-	-	-	-	-	-	-	-	X
SLC3	-	X	X	-	X	X	X	-	-	-	-
SLC4	-	X	X	-	X	X	-	-	-	-	-
SLC5	-	-	-	X	X	X	X	-	X	-	-
SLC6	X	X	X	X	X	X	X	X	X	-	X
SLC7	-	X	-	-	-	X	X	-	-	-	-
SW1	-	-	-	-	-	X	-	-	-	X	-
SW2	-	-	-	-	-	X	-	-	-	X	-
SW3	-	X	-	-	X	X	-	X	X	X	-
SW4	-	-	-	-	-	X	-	-	-	X	-
SW5	-	-	-	-	-	X	-	-	-	X	-
SW6	-	-	-	-	-	X	-	-	-	X	-
SW7	-	-	-	-	X	X	-	-	-	-	-
SW8	-	-	-	-	-	-	-	-	-	X	-
SC1	-	-	X	-	-	-	X	-	X	-	X
ENVIRONMENTAL SUSTAINABILITY											
EHH1	-	X	X	-	X	X	X	X	X	-	X
EHH2	X	-	X	X	X	-	X	X	-	-	-
EHH3	-	-	X	X	X	-	X	-	X	-	-
EHH4	-	-	X	X	X	X	X	X	X	-	X
EE1	X	X	X	X	X	X	X	X	X	-	X
EE2	X	X	X	-	X	-	X	X	X	-	-
ER1	X	X	X	X	X	X	X	X	X	-	-
ER2	-	X	X	X	X	X	X	X	X	-	-
ER3	X	X	X	-	X	-	X	X	X	-	-
EW1	X	-	X	X	X	-	X	X	X	-	X
EW2	-	X	X	-	X	X	X	-	X	-	-

On the other hand, the environmental criteria related to material resources (ER2) and water quality (EW1) are the most important in the certification systems, and ILAST (07), ENVISION (03), INVEST (08) and LEED (09) are the systems that include the greatest number of environmental sustainability metrics. It is important to note that the two certification systems with the most metric criteria are ILAST, for infrastructure and transport, and LEED, for buildings. In general, it is possible to affirm that ISO26000 (06) is the most complete sustainability certification system, including most of the social and environmental criteria, and that could be applied in the construction industry despite of being focuses on social responsibility system.

We can specify there are some criteria that should be included in several certification systems, or that must receive more importance. For example, the social criteria related to corruption (SS2), employee training (SW7) and respect for indigenous rights (SLC8); and the environmental criteria that include lighting (EHH3), noise and vibration (EHH2) issues. Criteria directly or indirectly related to construction projects and their stakeholders.

5 CONCLUSIONS

The main functions of a sustainability rating system are to promote engineering awareness and integration into the construction process and at the project level, serve as a management tool for accountability and project evaluation, work for internal and external recognition for sustainable actions and to track sustainability program progress. Although the list of social sustainability criteria is greater than the environmental one, we can verify that the criteria of environmental sustainability are more incorporated in the certification systems. The certification systems ISO 26000 (06) and SA8000 are the most completed systems in terms of social sustainability, and ILAST (07) and LEED (09) includes a greater number of environmental sustainability metrics.

Acknowledgments

The authors acknowledge the financial support of the Valencian Regional Government (Project: GV/2018//085).

References

Bal, M., Bryde, D., Fearon, D., and Ochieng, E., Stakeholder Engagement: Achieving Sustainability in the Construction Sector. *Sustainability*, *5*(2), 695–710. https://doi.org/10.3390/su5020695, 2013.

Beckwith, P., Farrington, L., Ainsworth, A., Fox, J., Napier, L., Ball, S., … and Tomlinson, P., CEEQUAL The Assessment and Awards Scheme for Improving, Part 1 : Maintenance., 2011. Retrieved from http://www.ceequal.com/download/2355/ on May 3, 2018.

Benoît-Norris, C., Traverso, M., Valdivia, S., Vickery-Niederman, G., Franze, J., Azuero, L., … and Aulisio, D., *The Methodological Sheets for Sub-Categories in Social Life Cycle Assessment (S-LCA)*. Gothenburg: GreenDelta, 2013. Retrieved from https://www.lifecycleinitiative.org/wp-content/uploads/2013/11/S-LCA_methodological_sheets_11.11.13.pdf on May 3, 2018.

Bevan, T., Reid, L., Davis, A., Neuman, T., Penney, K., Seskin, S., … and Schulz, J., INVEST - *Sustainable Highways Self-Evaluation Tool*, October, 2012. Retrieved from https://www.sustainablehighways.org/INVEST_1.0_Compendium_Web.pdf on May 10, 2018.

Cohen, S., Bose, S., Guo, D., Miller, A., DeFrancia, K., Berger, O., … and Zhang, C., *The Growth of Sustainability Metrics*, May, 2014. Retrieved from http://spm.ei.columbia.edu/files/2015/06/SPM_Metrics_WhitePaper_1.pdf on May 20, 2018.

Illinois Department of Transportation, I-Last. *Illinois-Livable and Sustainable Transportation Rating System and Guide.*, *2.02*, 89, 2012. Retrieved from http://www.idot.illinois.gov/assets/uploads/files/transportation-system/reports/desenv/enviromental/i-last v 2 02.pdf on May 20, 2018.

ISI, *ENVISION Rating System for Sustainable Infrastructure*, Institute for Sustainable Infrastructure,. Washington DC, 2015. Retrieved from https://sustainableinfrastructure.org/wp-content/uploads/2017/02/Envision_Info_Packet.pdf on May 20, 2018.

Kritkausky, R., and Schmidt, C., *ISO 26000, Global Guidance Standard on Social Responsibility. Handbook for Implementers of ISO 26000*, Middlebury Vermont, 2011. Retrieved from http://www.ecologia.org/isosr/ISO26000Handbook.pdf on May 15, 2018.

Muench, S., Anderson, J., Hatfield, J., Koester, J., and Söderlund, M., *Greenroads Manual v1. 5*. Seattle, WA, University of Washington, 2011. Retrieved from https://www.greenroads.org/files/235.pdf, on May 15, 2018.

New York State Department of Transportation, *GreenLITES Project Design Certification Program*

Recognizing. New York, 2008. Retrieved from http://people.sunyit.edu/~barans/classes/ctc440/pdf/GreenLITES Certification Program Final.pdf, on May 15, 2018.

North American Sustainable Transportation Council, *Sustainable Transportation Analysis & Rating System - Pilot Plan Application Manual* (Vol. Version 1.). Portland, 2012. Retrieved from https://www.sccrtc.org/wp-content/uploads/2014/02/STARS-Pilot-Project-Application-Manual.pdf on May 10, 2018.

PNUD, *Estándares Sociales y Ambientales*. New York: Programa de las Naciones Unidas para el Desarrollo, 2014. Retrieved from http://www.undp.org/content/dam/undp/library/corporate/Social-and-Environmental-Policies-and-Procedures/UNDPs-Social-and-Environmental-Standards-SPANISH.pdf on May 10, 2018.

Recycled Materials Resource Center and University of Wisconsin-Madison, *Be 2 st-in-highways*. University of Wisconsin-Madison, 2010. Retrieved from http://rmrc.wisc.edu/wp-content/uploads/2012/09/BE2ST-Manual.doc on May 10, 2018.

Social Accountability Internacional, *Responsabilidad so cial 8000* - SA8000. SAI - Social Accountability International, 2014. Retrieved from http://www.iqnet-ltd.com/userfiles/SA8000/2008StdSpanish.pdf on June 5, 2018.

US Green Building Council, *LEED for New Construction and Major Renovations*, Green Building Rating System, 2002. Retrieved from https://www.usgbc.org/sites/default/files/LEED for New Construction v2.1 Rating System - March 2003.pdf on May 10, 2018.

Interdependence between Structural Engineering and Construction Management
Edited by Ozevin, D., Ataei, H., Modares, M., Gurgun, A., Yazdani, S., and Singh, A.

ISBN: 978-0-9960437-6-2

THE USE OF LIFE CYCLE ASSESSMENT TO ACHIEVE SUSTAINABLE CONSTRUCTION

IFIJE OHIOMAH, CLINTON AIGBAVBOA, THULISILE ZWANE, and DOUGLAS AGHIMEN

Sustainable Human Settlement and Construction Research Centre, University of Johannesburg, Johannesburg, South Africa

This study assesses the benefit of sustainable construction and the use of life cycle assessment (LCA) in the South African construction industry to achieve sustainable construction. The data used in this was derived from both primary and secondary resources. The primary data was collected through a questionnaire approach. The study revealed that LCA could be used for product comparison, product development, formulating of product eco-labels, evaluating construction processes and decision-making processes in the construction industry concerning materials. Furthermore, the study revealed that LCA has various strengths, challenges and weakness, this is due to the fact that there is a requirement of the high quality of knowledge and data when conducting a detailed LCA study. However, this study revealed that life cycle thinking is a cornerstone for developing policies and programs, which meet sustainability criteria and there is a great room for the development and use of LCA in the South African construction industry. Therefore, it is recommended that the South African Construction industry invest more in environmental tools such as LCA.

Keywords: Environmental impacts, Sustainability, Construction industry, Long-term company policy.

1 INTRODUCTION

The construction industry is a large employer of persons in economies, hence Hans *et al.* (2011) added that there are a few sectors in South Africa that have the kind of employment impact the construction industry has, particularly in relation to the employment of skilled and semi-skilled labor. Although the importance of the construction industry is clearly recognized worldwide, it is impossible to overlook the contribution it has on the environmental burden facing the country.

The construction industry plays an important role in the social and human development; at the same time, it is a contributor to environmental issues. Clements-croome (2000) stated that around half of all non-renewable resources that mankind consumes are used in construction, and therefore it is one of the least sustainable industries in the world. According to Baloi (2003), the traditional ways of the industry have been based on the fact that innovations and investments drive the economy and satisfy the consumer's needs. However, it is equally important to understand that construction activities make excessive use of natural resources, various sources of energy and water, and that the construction industries products and process have undesirable impacts on the environment. The evidence emanating from the study of Baloi (2003) has shown that raw material extraction, transportation, and manufacturing result in resource depletion and energy consumption. Further, construction processes' produced emissions contribute to global warming and wastes that pollute the air and water. Consequently, these problems have triggered the adoption of measures to mitigate them. It is becoming a key responsibility for construction organizations to deliver projects that address the environmental, economic, and social dimensions of sustainability.

In order to be able to manage the impact of construction activities to the environment, the use of Life Cycle Assessment (LCA) has been suggested to solve problems associated with the impacts of construction on the natural environment. LCA is described as a standardized method of tracking and reporting the environmental impacts of a product or process throughout its full life cycle. South Africa Bureau of Standards (SABS ISO 14040) 1998 also describes LCA as the calculation and evaluation of the environmentally relevant inputs and outputs and the potential environmental impacts of the life cycle of the product, material or service. Environmental inputs refer to the demand for natural and to emissions and solid waste.

The construction industry is a contributor to the environmental impacts that are experienced today. Sartori and Hestnes (2007) described it as an industry which consists of many phases, starting from mining, manufacturing, construction use, and demolition; within each phase, there is a large amount of energy consumption and considerable released emission as well. In addition, researchers (Asif *et al.* 2007, Zabalza *et al.* 2009) discussed that recent studies show that building is responsible for 30-40% of energy consumption and uses up to 40-50% of greenhouse emission. This impact on the environment has led to the conducting of this study to the reduction of environmental impacts through the use of Life Cycle Assessment.

Life Cycle Assessment is a technique used to evaluate the potential environmental and economic performances of building envelopes and products throughout their lifecycle (Stazi *et al.* 2012). According to Lindfors and Rosenquist, (1995) the European environment agency series states that Life Cycle Assessment is a process to evaluate the environmental burdens associated with a product system or activity and identifying and quantitatively describing the energy and material used and wastes released to the environment. The process includes the entire life cycle of the product or activity, encompassing extracting and processing of raw materials.

According to Klöpffer (2006), LCA has become a widely used methodology because of its integrated way of treating topics like framework, impact assessment and data quality. In addition, Klöpffer (2006) outlines that the LCA has four major stages which are: 1) Goal and scope definition where the purpose, objective, function and system boundaries are defined, 2) Inventory analysis which involves the collection of all data relating to the inputs, processes, emissions and others of the entire life cycle, 3) Impact assessment which involves environmental assessment and quantifying the input resources based on the inventory assessment, and 4) Interpretation wherein the result calculated from the impact is interpreted (Ortiz *et al.* 2009).

2 USE OF LIFECYCLE ASSESSMENT IN THE CONSTRUCTION INDUSTRY

Given the environmental impacts resulting from the construction industries processes, the use of the LCA tool becomes very important and without its use the sustainability and green building initiatives will be difficult to achieve. The LCA study has been used in different fields but its use in the construction industry is recent (Singh *et al.* 2011). Lifecycle assessment can be used in various ways and integrated environmental management information series nine (2004) stipulated the following uses:

By Manufacturers:

Product improvement – Lifecycle assessment can be used to reveal processes, components and systems to target environmental enhancement (Kataji 2003).

Product comparison – Lifecycle assessment can be used to identify and process systems that are major contributors to environmental impacts. For example, the Integrated Environmental Management Information Series Nine (2004) pointed out that BMW and Volvo use LCA to determine materials and methods to enable their vehicles to become more recyclable and reduce the generation of waste in the manufacturing and disposal phase.

Product development – New products are frequently developed from old designs and concepts. LCA makes use of that concept by taking "old information" and equating it with projections estimates for new product and service.

Development of product eco-label- Eco-labeling is a way of granting acknowledgement for products that attains a certain minimum standard in "environmental friendliness". In this case, LCA can be used to develop long-term policy regarding environmental impacts and risks posed by the materials and processes throughout the life cycle. LCA can also be used in evaluating resource conservation effects associated with source reduction and alternative waste management techniques. Lastly, LCA can be used to provide information to the public about the resource characteristics of products or materials. For example, Integrated Environmental Management Information Series Nine (2004) also pointed out that chemical companies like Unilever use LCA as a means of testing their products for eco-friendliness and legal compliance.

Product Information- Authorities might in some instance require product information for purpose of licensing or legal compliance, therefore the information produced from LCA can be used to supply information confirming the validity of data and product related decisions and choices.

By Public policymakers:

Formulating long-term company policy – LCA can contribute significantly to the development and adjustment of company policies in specific areas. For example, guidance on the choice of raw material use could directly affect the company's strategy for handling waste materials.

Application for construction systems and process evaluation- Evaluation of environmental impacts on construction includes more than just a simple combination of individual product and material assessment. However, LCA can also be used to assess this. In the study carried out by Keoleian *et al.* (2000)., evaluated life cycle energy use, greenhouse emission, and cost of the conventional residential home covered the pre-use (Material production and construction), use (Including maintenance) and demolition phases. It was discovered that the use phase accounted for approximately 91% of the overall life cycle energy usage over a 50-year life. The authors modelled a functionally equivalent energy-efficient house that combined 11 energy efficient strategies and discovered that these approaches led to a serious decrease in the total life cycle energy use.

Life cycle assessment is generally used as a research tool to analyze buildings or manufacturing processes that aid decision making (Simonen 2014). Moreover, the main application of LCA in the construction industry evaluating manufacturing process, developing product labels, comparing materials and methods and analyzing whole buildings.

When evaluating manufacturing, a manufacturer or research organization can commission an analysis of their manufacturing process, and this analysis will only focus on the "cradle-to-grave" impacts that a manufacturer can control. Again, Simonen (2014) stated that LCA application can be in developing product label, which includes a comprehensive report of LCA data of information of a product as well as rules.

As mentioned before, LCA can be used to compare materials and method. Through such comparisons, an environmentally preferable method or material can be identified and adopted

(Simenon 2014). Life cycle assessment can also be used in whole building to assess the impacts of manufacturing and construction in line with the operations and maintenance impacts. However, to do this, assumptions must be made about its life. Simonen (2014) explained that these assumptions would include the establishment life, the use of the building, maintenance, refurbishment and its end of life treatment. With the mentioned ways on how LCA can be used, it is important that in order to achieve the sustainable construction and healthy environment. LCA should be an integral part of any decision-making in the process of development in the construction industry.

3 METHODOLOGY

This study adopted a quantitative research approach in order to achieve the aim of the study, and a questionnaire was developed from an extant review of the literature. Practising construction professionals in the Gauteng Province construction industry in Johannesburg, South Africa were engaged in the collection of the primary data. The criteria for selecting a professional is that they had worked on a construction project in Johannesburg in the last year and are presently listed with a built environment professional council at any level. Fifty questionnaires were randomly distributed, among these selected construction professionals and 43 were retrieved with 40 ascertained to fit for analyses. The remaining three questionnaires were dropped due to poor completion and missing information. This represents an 80% response rate, and this was deemed adequate considering the limitation in the geographical scope of the study. The designed questionnaire adopted an agreement scale of 1 to 5 with 1 being strongly disagreed and 5 strongly agreeing. Data analysis was done using mean item score and standard deviation, and the mean were ranked in descending order. The ranking made it possible to cross compare the relative importance of the items as perceived by the respondents.

4 FINDINGS

The following results are based on the ranking (R) and the calculated standard deviation (SD) and the mean scores on the use of life cycle assessment. The following was observed to be among the dominant; making procurement decision and developing regulations (SD = 6.80; mean 4.10), making long-term policy regarding material use (SD = 7.10, mean 4.10), product improvement (SD = 7.30, mean = 4.10).

These findings are in agreement with a study conducted by Simonen (2014) stating that the main application of LCA in the construction industry includes evaluating manufacturing process, developing product labels, comparing materials and methods and analyzing whole buildings.

The data obtained from the questionnaires as answered by the respondents regarding the assessment of the use of life-cycle assessment to achieve sustainable construction in the South African construction industry was presented and analyzed in relation to research questions and literature review.

4.1 Results and Findings

Respondents were asked to rank the assessment of the use of Life Cycle Assessment to achieve sustainable construction. Table 1 below revealed that "Making procurement decision and developing regulations" was tied first with a mean score of 4.1 and SD of 6.80, "Making long term policy regarding material use" was tied first with the mean score of 4.1 and SD of 7.1, "Product improvement" was tied first with a mean score of 4.1 and SD of 7.3, and "Product development" was tied first with a mean score of 4.1 and SD of 7.3. The table further revealed that "Evaluating manufacturing processes" was ranked second with the mean score of 4.0 and SD 7.60, and "Natural resource conservation" was ranked third with a mean score of 3.9 and SD of 5.90. Lastly, "Aid decision-making process" was ranked fourth with a mean score of 3.8 and SD of 5.8.

Table 1. The use of LCA to achieve sustainable construction.

Lca uses	SD	MIS	Ranking
Making procurement decisions and developing regulations	6.80	4.10	1
Making long term policy regarding material use	7.10	4.10	2
Product analysis	7.10	4.10	2
Product improvement	7.30	4.10	3
Product development	7.30	4.10	3
Evaluate manufacturing process	7.60	4.00	4
Natural resource conservation	5.90	3.90	5
Product comparisons	6.20	3.90	6
Whole building analysis	6.20	3.90	6
Evaluating resource effect techniques	6.30	3.90	7
Aid decision-making process	6.70	3.80	8
Development of product eco-label	7.80	3.80	9

5 CONCLUSIONS AND RECOMMENDATIONS

This study assessed Life Cycle Assessment (LCA) as a tool to be used in the aim of achieving sustainable construction in the South African construction industry. Findings concerning the use of LCA in the construction industry were also in agreement with the reviewed literature in terms of it being fully adapted and used as a tool to mitigate environmental burdens imposed by the construction industry.

In line with the results from the study, there is a strong case for the construction industry to mitigate the impacts it imposes to environment; to achieve this, the construction industry must invest more into the development and the usage of tools such as life cycle assessment

From the findings, making procurement decision in the planning phase is important, as it shapes the purchase of sustainable materials for the construction phase of the project. This procurement decision cannot be done without the analysis of the building, which will include the pre-planning phase to post-completion phase; most clients and building professionals do not plan for the post-completion phase of the study. In addition, once the procurement decisions have been made, they will assist in the long-term development of policy that will guide the use of materials in the construction industry. Lastly the development of LCA will lead to the development of eco-label. Therefore, it is important that the implementation policies be carried out by construction professionals, and strict penalties be enforced if this is not followed. Incentives such as tax breaks and reduction of levies such as import tax should be offered to construction firms to ensure that practising LCA methodologies is not a burden.

References

Asif, M., Muneer, T. and Kelley, R., Life Cycle Assessment: A Case Study of a Home in Scotland, *Building Environment,* 42(3), 1391-1394, 2007.

Baloi, D., *Sustainable Construction: Challenges and Opportunities*, 9th Annual ARCOM Conference, University of Brighton, United Kingdom. 2003.

Clements-croome, D., (ed). Creating the Productive Workplace, Routledge, London, 2000.

Kataji, A. S., Life Cycle Assessment in Construction Practices, SETAC, 2003.

Keoleian, G. A., Blanchard, S., and Reppe, P., Life Cycle Energy, Costs, and Strategies for Improving a Single Family House, *Journal of Industrial Ecology,* 4(2), 135-156, 2000.

Klöpffer, W., The Role of SETAC in the Development of LCA, *The International Journal of Life Cycle Assessment,* 11(1), 116-122, 2006.

Lindfors, K. K., and Rosenquist, C. J., The Cost-Effectiveness of Mammographic Screening Strategies, *Jama,* 274(11), 881-884, 1995.

Ortiz, O., Castells, F., and Sonnemann, G., Sustainability in The Construction Industry: A Review of Recent Developments Based On LCA, *Construction and Building Materials,* 23(1), 28-39, 2009.

SABS ISO 14040, *Code of Practice: Environmental Management – Life Cycle Assessment – Principles and Framework,* SABS (South African Bureau of Standards), Pretoria, 1998.

Sartori, I. and Hestnes, A. G., Energy Use in The Life Cycle of Conventional and Low-Energy Buildings: A Review Article, *Energy and Buildings,* 39(3), 249-257, 2007.

Simonen, K., Life Cycle Assessment, Routledge, 2014.

Singh, A., Berghorn, G., Joshi, S., and Syal, M., Review of Life-Cycle Assessment Applications in Building Construction, *Journal of Architectural Engineering,* 2011.

Stazi, F., Mastrucci, A., and Munafò, P., Life Cycle Assessment Approach for the Optimization of Sustainable Building Envelopes: An Application on Solar Wall Systems, *Building and Environment,* 58278-288, 2012.

Zabalza, I., Scarpellini, S., and Aranda. A., Life Cycle Assessment in Buildings, 2009.

Interdependence between Structural Engineering and Construction Management
Edited by Ozevin, D., Ataei, H., Modares, M., Gurgun, A., Yazdani, S., and Singh, A.

ISBN: 978-0-9960437-6-2

LIFE CYCLE COST ANALYSIS OF ENERGY EFFICIENT SINGLE FAMILY RESIDENCE

AYUSHI HAJARE and EMAD ELWAKIL

School of Construction Management Technology, Purdue University, West Lafayette, USA

Residential and commercial buildings account for more than half of the electricity consumption in the United States. There are numerous practical solutions to make buildings more energy efficient and sustainable. Although it is well-established that green buildings are socially, environmentally, and economically beneficial, there is still a lack of green buildings in the residential sector. The installation and upfront costs for these houses are very high. This research aims to facilitate a broader understanding of the cost benefits of energy efficient and sustainable residences. The Life cycle cost analysis (LCCA) approach and energy simulation tools have been utilized and integrated for assessing the traditional single-family residence in the United States. A comparative study has been carried out including passive and net-zero energy through energy simulation software. This analysis will benefit academic researchers and industry practitioners to analyze and evaluate challenges and opportunities in energy efficient and sustainable residences.

Keywords: Residential, Upfront cost, Energy efficient, Energy simulation, Passive, Zero energy.

1 BACKGROUND

The life cycle energy used by buildings in their operation and maintenance phase is between 80-90% of the total life cycle energy (Ramesh *et al.* 2010). Thus, management and reduction of this energy during its operation phase should be of utmost importance and a major factor in play during the design phase of the project. There is growing awareness regarding the need for energy conservation and it is reflected in the way we now design buildings (Hoque 2007). However, Developers, owners, and tenants are reluctant to jump into the investments involved with Net zero energy buildings. New technologies often have high installation costs, which is one of the major reasons why there is a lack of energy-efficient buildings. Moreover, despite the numerous benefits of Life cycle cost analysis (LCCA), its adoption in the green building industry, especially for low rise residential buildings, is relatively slow. Most of the studies are limited to commercial building (Ryghaug and Sørensen 2009). There is need to expand LCCA into the residential sector, where its growth is hampered by inconsistent data and insufficient collaboration between stakeholders (Ramesh *et al.* 2010).

The purpose of this research is to explore the potential of current LCCA procedures as an evaluation tool for single-family residences; it will also seek to provide suggestions to improve its adoption for green building strategies.

The residential sector in the US accounts for a majority of energy consumption in the country. The supporting data has been published by the U.S. Energy Information Administration

(EIA) for the residential sector and the commercial sector (US Energy Information Administration 2015).

In recent studies conducted by U.S. Energy Information Administration (EIA), it was published that "Estimates from the most recent Residential Energy Consumption Survey (RECS), collected in 2010 and 2011 and released in 2011 and 2012, show that 48% of energy consumption in U.S. homes in 2009 was for heating and cooling (US Energy Information Administration 2017)". However, in 1993, the share of space heating and cooling in energy consumption in households was 58% (US Energy Information Administration 2017) suggesting that over the years, this share has declined. An explanation for this trend could be the increased use of newer heating systems or better insulation (Kansal and Kadambari 2010). On the other hand, there has been a steady increase in energy consumption through other electronic appliances (US Energy Information Administration 2017). Figure 1 gives us a better idea of the residential end use of electricity.

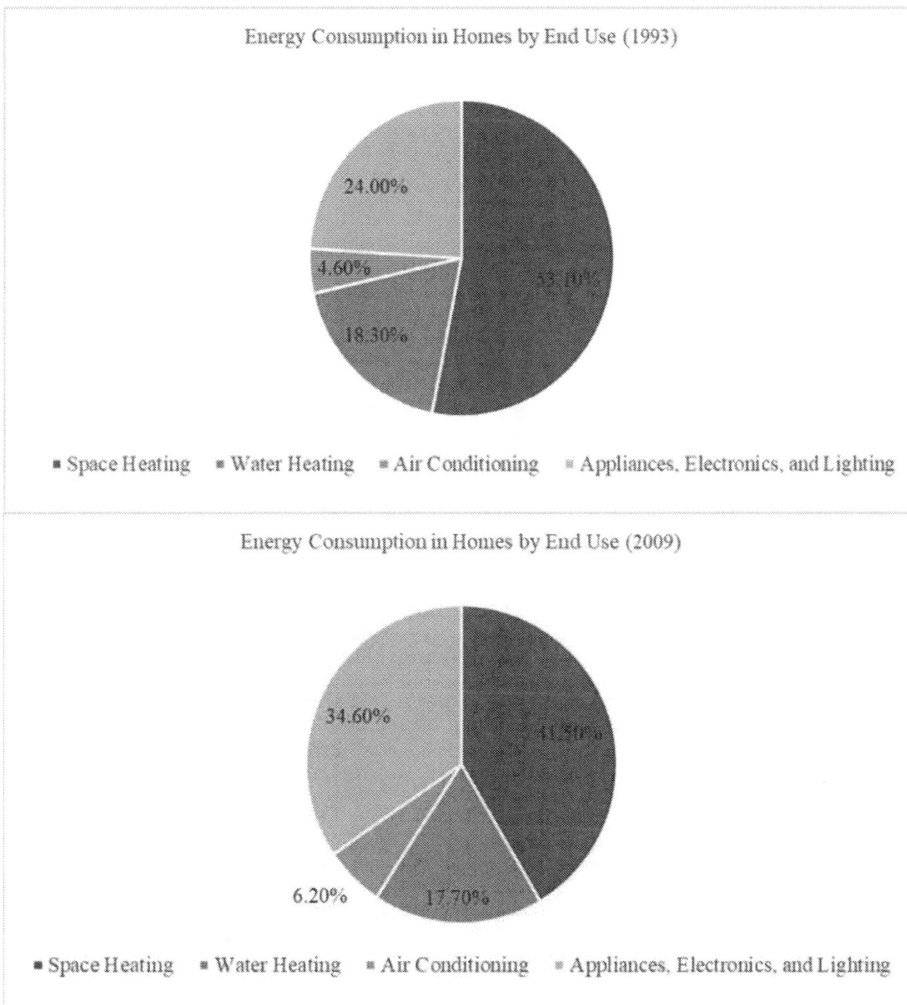

Figure 1. Energy consumption. (US Energy Information Administration, 2015).

3 CHALLENGES FACED BY PASSIVE AND ZERO ENERGY BUILDINGS

Today, only a few net-zero energy buildings (NZEB) exist around the world and they show that the design, construction, and operation have been challenging for many reasons (Butera 2013). We can conclude that energy efficiency is still deeply retrained in the construction industry. Some of the major reasons for this are:

- Lack of public policies to develop energy efficiency strategies (Ryghaug and Sørensen 2009)
- Limited or no government efforts in regulating green building and the construction industry (Ryghaug and Sørensen 2009)
- A rigid and conservative construction industry (Ryghaug and Sørensen 2009)
- Lack of knowledge of new technologies amongst the stalk holders (Kshirsagar *et al.* 2010)
- High installation costs (Marszal and Heiselberg 2011)
- Lack of consumer (home owner) understanding
- Lack of a quantitative approach to calculating the cost benefits of NZE houses (Perlova *et al.* 2015)
- Lack of renewable sources at some locations.

4 DESIGN OF NET ZERO ENERGY RESIDENCE

When we look at the construction of a building, it is often realized in a three-step process.

- Architecture design
- Mechanical/electrical system design
- Construction

In his work, Butera (2013) concluded that energy consultants must step in before any construction begins to provide optimum energy and cost-efficient alternatives. Thus, a new process was developed.

(i) Architecture design
(ii) Mechanical/electrical system design
(iii) Energy analysis
(iv) Construction

The process can be repeated until an appropriate design solution is proposed. This design process coupled with technological advances can be used to improve on strategies that make a building truly net-zero energy.

5 ASSESSMENT OF ZERO ENERGY BUILDINGS - LCCA

Buildings, in general, are increasingly evaluated to meet norms related to sustainability, cost-effectiveness, comfort and safety (Dwaikat and Ali 2014). However, zero energy buildings are a relatively new concept and haven't been fully developed. Thus, an optimum combination of measures has to be developed that are cost beneficial. Life cycle cost analysis (LCCA) is a great tool that helps the owner and stakeholders investigate the most cost-effective solution.

Life cycle cost analysis method was first developed by the Department of Defense (DOD) as an analytical method for evaluation of federal projects (Fernholz *et al.* 2013). "It is a method of determining the entire cost of a structure, product, or component over its whole life" (Snodgrass 2003). Because it includes operation and maintenance costs over a product's lifetime, it is considered a more appropriate evaluation measure when compared to only investments costs.

5.1 Net Present Value (NPV)

NPV is a method to calculate life-cycle cost (LCC) of a building while factoring in the time value of money. This is the most common tool used for life-cycle cost analysis (LCCA). However, it is not the most effective method for comparing buildings that have different lifespans. Additionally, this method does not include the user or owner's perspective (Marszal and Heiselberg 2011). The formula developed by Kaufman (1970), is the most widely used method for Life cycle cost calculation as shown in Eq. (1).

$$NPV = C + R - S + A + M \tag{1}$$

Where C is the investment cost, R is the Replacement costs, S is the Resale value A is the annually occurring costs and M is the non-annually occurring costs.

Life cycle cost analysis can serve as a great benchmarking tool for Green buildings, as it quantifies the benefits of one. Hoque (2007) writes, "It is useful for the comparison of different alternatives which satisfy the desired level of performance". Additionally, it is a great tool to access the performance of a green building.

6 RESEARCH METHODOLOGY

The researcher aims to develop green building strategies for a single-family residence through software and analyze the total cost of building through life-cycle cost analysis (LCCA).

The initial stage of the study would include a case study of a single family unit located in central Indiana. For this, energy data will be calculated from monthly energy bills. Energy modeling for a redesigned unit will be performed on an energy simulation software. The design would be of the following type:

- Passive Building design for a high-performance building
- Active Building equipment for a high-performance building
- A mixture of passive design and active building equipment for a net zero energy building.

The next steps in the research would be the cost and energy data calculations. Energy data from eQUEST would be used to evaluate operation and maintenance costs for the house. These costs along with other variables will be used to calculate the life cycle cost of the building using the net present value method.

7 DATA COLLECTION

The researcher will collect energy and cost data for the single-family residence and the energy modeling data. Based on previous studies and literature, following data variables will be included in this study.

(i) Occupancy Data- functionality, hours of use and specific features of building use
(ii) Physical Data- the number of floors, floor areas, HVAC systems, fenestration details, and water & sanitary data
(iii) Building performance data- Maintenance & cleaning cycles, electricity consumption, gas consumption
(iv) Cost Data- acquisition cost, taxes, inflation, management costs, replacement costs, salvage value

These variables have been determined from the larger pool of data depicted in the Figure 2.

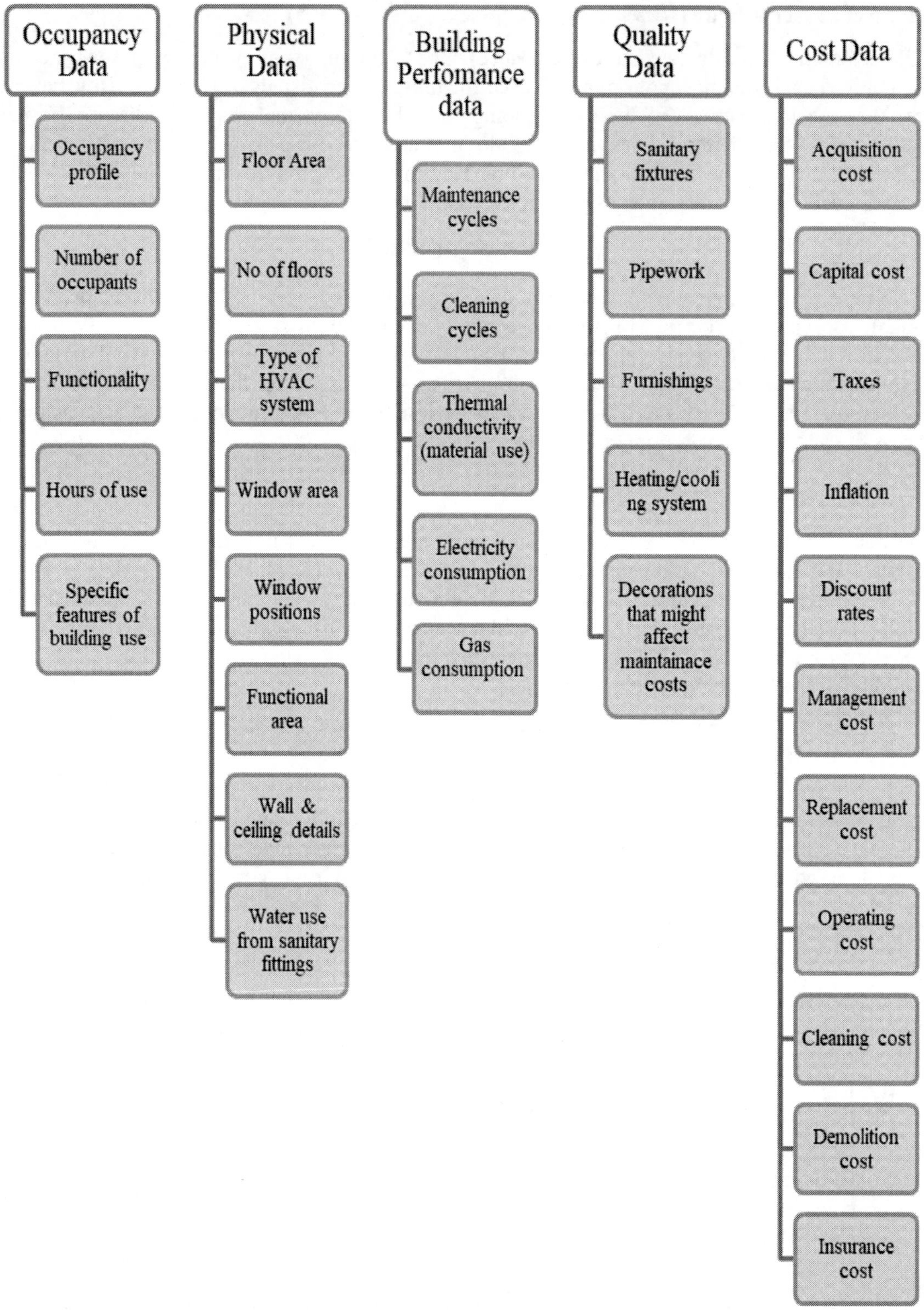

Figure 2. Data collection summary.

8 CONCLUSIONS

The study is ongoing. The data has been collected and analyzed data for a single residential home in Central Indiana.

While there has been tremendous growth in the design and implementation of sustainable practices, there is a dearth of efficient residential buildings in the United States. Energy efficient residential buildings form the core of implementable sustainability practices and a lack of research in the residential sector is prescient of bigger problems ahead. The cost benefits of energy efficient housing are far-reaching.

References

Butera, F. M., Zero-Energy Buildings: The Challenges, *Advances in Building Energy Research,* 7(1), 51–65, 2013.

Dwaikat, L. N., and Ali, K. N, *Green Buildings Actual Life Cycle Cost Control: A Framework for Investigation*, 13th Management in Construction Researchers' Association (MiCRA) Annual Conference and General Meeting, 06 November 2014, IIUM, (November). https://doi.org/10.1152/jn.00840.2006, 2014.

Fernholz, K., Bratkovich, S., Howe, J., Stai, S., and Frank, M., Life Cycle Cost Analysis of Non - Residential Buildings., Dovetail Partners, 2013.

Hoque, S., Net Zero Energy Homes: An Evaluation of Two Homes in the Northeastern United States, *Journal of Green Building*, 5(2), 79–90, https://doi.org/10.3992/jgb.5.2.79, 2007.

Kansal, R., and Kadambari, G., Green Buildings: An Assessment of Life Cycle Cost, *IUP Journal of Infrastructure*, 8(4), 50, 2010. Retrieved from http://search.ebscohost.com/login.aspx?direct=true&db=edb&AN=56642147&site=eds-live on March 2019.

Kaufman, R.J. (1970), "Life cycle costing: a decision-making tool for capital equipment acquisition", Cost and Management, March/April, pp. 21-28.

Kshirsagar, A. S., El-Gafy, M., and Abdelhamid, T. S., Suitability of Life Cycle Cost Analysis (LCCA) as Asset Management Tools for Institutional Buildings, *Journal of Facilities Management,* 8(3), 162–178, https://doi.org/10.1108/14725961011058811, 2010.

Marszal, A. J., and Heiselberg, P., Life Cycle Cost Analysis of A Multi-Story Residential Net Zero Energy Building in Denmark, *Energy*, 36(9), 5600–5609, https://doi.org/10.1016/j.energy.2011.07.010, 2011.

Perlova, E., Platonova, M., Gorshkov, A., and Rakova, X., Concept Project of Zero Energy Building, *Procedia Engineering*, 100(January), 1505–1514, https://doi.org/10.1016/j.proeng.2015.01.522, 2015.

Ramesh, T., Prakash, R., and Shukla, K. K., Life Cycle Energy Analysis of Buildings: An Overview, *Energy and Buildings*, 42(10), 1592–1600, https://doi.org/10.1016/j.enbuild.2010.05.007, 2010.

Ryghaug, M., and Sørensen, K. H., How Energy Efficiency Fails in the Building Industry, *Energy Policy*, 37(3), 984–991, https://doi.org/10.1016/j.enpol.2008.11.001, 2009.

Snodgrass, K., Life-Cycle Cost Analysis for Buildings is Easier than You Thought, United States Department of Agriculture, (August), 24, 2003. Retrieved from http://scholar.google.com/scholar?hl=en&btnG=Search&q=intitle:LifeCycle+Cost+Analysis+for+Buildings+Is+Easier+Than+You+Thought#1 on March 2019.

US Energy Information Administration, 2015. Retrieved from Energy Information Administration: https://www.eia.gov on March 2019.

US Energy Information Administration, 2017. Retrieved from Energy Information Administration: https://www.eia.gov on March 2019.

Interdependence between Structural Engineering and Construction Management
Edited by Ozevin, D., Ataei, H., Modares, M., Gurgun, A., Yazdani, S., and Singh, A.

ISBN: 978-0-9960437-6-2

BIOCLIMATIC ANALYSIS OF MECHE'S HOUSE: A CONSTRUCTION ALTERNATIVE

MAURO CEPEDA ORTIZ[1], SANTIAGO MORALES FLORES[2], and ENRIQUE VILLACIS[3]

[1]*School of Architecture and Design, University of the Americas, Quito, Ecuador*
[2]*Architecture with a Bioclimatic Approach, Quito, Ecuador*
[3]*School of Architecture, Arts and Design, Pontifical Catholic University, Quito, Ecuador*

Meche's House is an alternative post-disaster construction, and this is the study of its bioclimatic approach showing that social and post-disaster buildings also need this kind of research. Tropical climate conditions lead to buildings having a high energy consumption for cooling loads. In Ecuador, the energy consumption of the residential area is 28.78% of final demand. Also, there is very little relevant information on the analysis of bioclimatic design in buildings, as well as specific analysis of interior comfort. The carried-out analysis process considered methods of bioclimatic evaluation, which mainly focuses on building the user's comfort. For this reason, in the first place, the site climatic conditions and possible passive intervention strategies were determined. Followed by the evaluation of natural ventilation with which it was possible to evaluate the effectiveness of natural ventilation through simulations in Computational Fluid Dynamics program. Furthermore, thermal comfort analysis using an Energy Plus program is used for comparing the internal temperature ranges versus indoor natural ventilation. Finally, the data is discussed under an adaptive comfort and user perception of satisfaction. This research confirms the need to carry out bioclimatic evaluations of projects conceived under a good line of architectural design, since only in this way will it be possible to demonstrate that the proposed considerations and strategies have positive or negative outcomes.

Keywords: Indoor comfort, Tropical climate, Ecuador, Natural ventilation, Bamboo.

1 INTRODUCTION

Under tropical climate conditions, lead buildings having a high energy consumption, specifically for cooling loads (Al-Obaidi *et al.* 2014). They have several problems in terms of the materials that are applied. The energy consumption of non-residential and commercial buildings in countries that are in this region is approximately "a quarter of the total consumption of electric power" (Rattanongphisat and Rordprapat 2014). In Ecuador, the energy consumption of the residential area is 28.78% of final consumption and commercial, and public services are 21.78% (Iea.org 2018). This demand could be mitigated by a passive strategy well executed in buildings and, especially in a tropical climate. Especially with an adequate bioclimatic design, it is possible to easily save 30% of the energy produced by the energy demand by active cooling systems (Sadineni *et al.* 2011). Within the process of architectural design and construction process, the key elements to reach a suitable bioclimatic design are the materiality, orientation, and shape of the building. In terms of materiality in tropical climates, the most common and suitably used are: clay, wood, bamboo, and mainly other local materials (Al-Obaidi *et al.* 2014); unfortunately, today these materials are short in supply or non-existent. Cities that are developing in this geographic location have triggered the use of industrial materials such as steel, concrete, glass, light metals, and others which are more suitable for other contexts. With a correct passive design of the building in a said geographical area (Rattanongphisat and Rordprapat 2014), the performance could be substantially improved, and the demand for energy reduced to achieve a structure with a low carbon impact (Yeang and Yeang 2006) (Roslan *et al.* 2015). The buildings located within tropical climatic conditions generate a high energy consumption due to the cooling loads used (Al-Obaidi *et al.* 2014). In the same way, the materials used in the construction of low quality, allow a higher energy transfer. Therefore, the interior temperature is increased by thermal gains by radiation and convection, decanting in the use of mechanical cooling. The architectural design plays an important role when using appropriate processes framed in parameters of bioclimatic design that allow ensuring a proper application of passive strategies to regulate the conditions towards the interior of the architectural space (Sadineni *et al.* 2011) (Yeang and Yeang 2006). In Ecuador, there is poor relevant information on the analysis of bioclimatic design in buildings, as well as specific analysis of interior comfort with the use of specialized programs such as Energy Plus. This premise revalues this research because of the high degree of certainty that can be obtained from specialized models such as those developed under this process. Within the little that exists, the authors Calle *et al.* (2013) conclude that it is possible to successfully apply energy efficiency strategies in buildings in Ecuador whenever they are carried out in early stages of design, conclusion based on the study of a prototype building located in the province of Imbabura. A similar conclusion was presented when analyzing the variation of the materiality of the envelope of a minimum dwelling; where important comfort percentages were reached, which were between 79% and 95% of the total hours analyzed (Miño *et al.* 2013). Finally, from the study of a university building in the tropical zone of Ecuador, where the analysis of natural ventilation was included, it is concluded that it is possible to reduce energy consumption by 10% when combining the cross-ventilation strategy or solar chimney (Kastillo *et al.* 2015).

1.1 Geographic Location and Weather Conditions

The project is implemented in Ecuador in the province of Esmeraldas canton Muisne, specifically in the area called Pedro Carbo between the towns of Chamanga and Mompiche. Ecuador has four natural regions that have different climates due to their difference in altitude and proximity to the Pacific coast. In general, a large part of the Ecuadorian territory, except for the Sierra region, has warm climates typical of tropical zones. According to the climate classification Köppen-Geiger (Figure 1), it is stipulated that the area of intervention of the project is considered Cfa type, which means a temperate climate without dry seasons and with hot summers (Peel *et al.* 2007).

On the other hand, and with the aim to have a deep understanding of the site's climate, the meteorological data of the population of Muisne province of Esmeraldas was considered due to its proximity to the place. In this way, valuable information on climate data was obtained, as well as the sun's path and its different angles of inclination with the horizon during the year. The relative humidity of the air fulfills a preponderant factor when assimilating and evaluating the thermal sensation and levels of human comfort both inside and outside the buildings (Cheng *et al.* 2010).

According to the information obtained in the meteorological station closest to the site, the weather station CP M0153 INAMHI was used, the humidity of the air throughout the year remains at levels very close to 85%, in this way, the ranges of sensations. They vary from the oppressive type, which is present most of the time, to the unbearable type present during the months of higher temperature and solar radiation. Finally, it is essential to understand the behavior and velocity of the air in the sector, to include said variable in the thermal analysis and evaluation of the home. According to NASA SSE (2018), the annual air velocity at 50 meters above ground level is 2.98 m / s, with minimum and maximum speed peaks during March and August with 2.19 and 3.51 meters per second respectively.

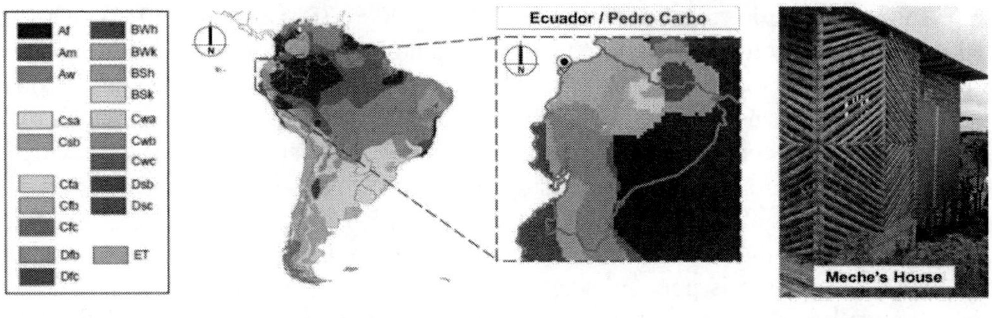

Figure 1. Climate classification Köppen-Geiger - Esmeraldas (Peel *et al.* 2007) and Meche's house façade.

2 METHODOLOGY

The analysis process that was carried out considered the methods of bioclimatic evaluation, which focuses mainly on the comfort of the user of the building. For this reason, in the first place, the site climatic conditions and the possible passive intervention strategies were determined. However, there was no data on the existence of files in format. EPW (EnergyPlus Weather file), which includes meteorological data in codes adapted for the analysis programs was used. In the same way, the tropical zones and particularly the latitude where the project is located have very little meteorological information loaded in this format. For this reason, it was necessary to generate a specific EPW file for the project considering climate data obtained from the meteorological station CP M0153 (INAMHI 2014) and mainly from NASA, to create results, close to the reality of the intervention site.

On the other hand, an internal thermal comfort analysis using the EnergyPlus program was used to compare the interior temperature ranges versus the presence of natural ventilation. The thermal study of the project was carried out based on related evaluation processes in such a way that it is possible to determine the incidence of the architectural and bioclimatic contributions in the thermal sensation of the users; this is how three scenarios modeled in the Energy Plus program were considered, which are identifiable. The first one evaluates the project considering a typical design and construction process in Ecuador with generic materials of frequent use and acceptance. The second scenario estimates the architectural project carried out by Ensusitio Arquitectura, including data and thermal capacities of local materials such as bamboo cane and soil-concrete. Finally, the third scenario starts from the base of the built project but includes natural ventilation considering the proposals and openings of the windows, as well as the bamboo framework spans. With the aim to establish similarity to reality, the cane framework was carried out in a more non-volumetric mathematical way, and lattice properties were included for each of the window openings. Natural ventilation is based on the direction and wind speed per hour determined by the weather station and loaded into the created EPW file. The evaluation of natural ventilation, with which it was possible to assess the effectiveness of natural ventilation through simulations in Computational Fluid Dynamics (CFD), was carried out based on the parameters of natural ventilation in CFD (Gan 2016). considering stable conditions. For representing and having a better understanding of the analysis process used in this research, an orthogonal chart was made where important and determining points in the continuity of each of the methods are highlighted. See Figure 2.

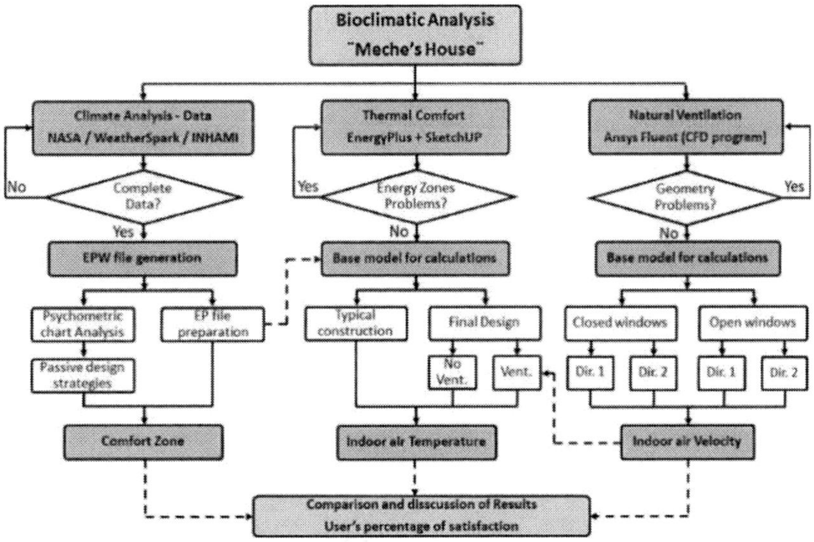

Figure 2. Orthogonal chart of the research methodology.

3 RESULTS

3.1 Natural ventilation analysis results with Ansys Fluent (CFD)

In the first analysis of the building (Figure 3), the direction of the wind is from North to South, and the building in terms of its openings is closed. There is evidence of natural ventilation even though the windows are closed, due to the materiality and configuration, which allows the passage of natural ventilation into the interior of the space with an average speed of 1.64 m/s.

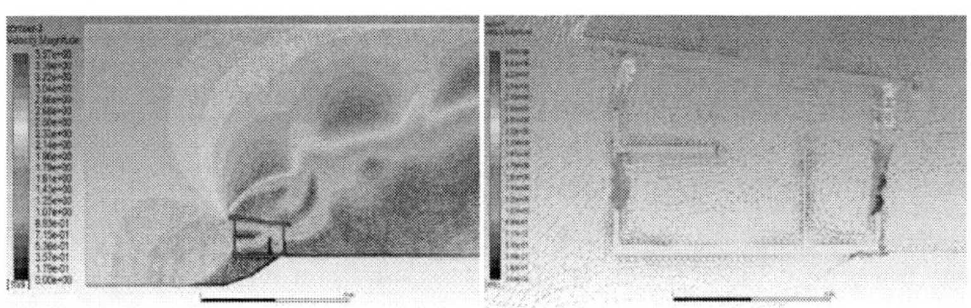

Figure 3. Natural ventilation on building CFD analysis 1.

3.2 Internal Thermal Analysis Results with Energy Plus

The average value of the interior air temperature of each space represents an excellent basis for comparing the thermal conditions of each scenario. With the aim to understand the annual behavior of them, the temperature is presented on an average day of each month. In the case of housing, the analysis shows that the construction of the project using typical materials from the area and without bioclimatic design strategies generates an interior temperature that is between 2 and 3.5 °C above the average room temperature that varies between 24 °C and 26.5 °C throughout the year. That is due to the heat gains through the materials and energy transfer from the outside to the inside. On the other hand, the same habitable space when it includes design strategies and architectural contributions decreases its temperature by at least 1.5 °C considering a traditional construction. Finally, when the passive design strategies are added plus the adaptability of the user using natural ventilation and window opening, a significant reduction of the temperature is obtained, and the internal thermal conditions are improved. Thus, the final project that includes natural ventilation manages to lower the temperature from the first scenario in a range between -1.9 °C to -3.4 °C throughout the year. This is evidenced in Figure 4.

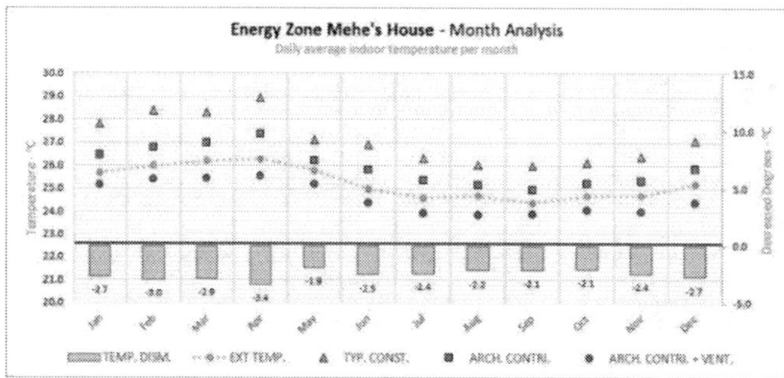

Figure 4. Indoor thermal comfort analysis (temperature) – house.

4 DISCUSSION OF RESULTS

With the aim to assess the results obtained from the natural ventilation and internal temperature analyzes, the PMVTool program developed by Marsh (2005) was used. It identifies the prediction of user satisfaction considering seven levels of complacency where 0 is comfortable, +3 is extremely hot and finally -3 is extremely cold, for this case study two scenarios were analyzed. The first evaluates average conditions during the day, which were obtained from the EnergyPlus and CFD modeling. Therefore, the inserted data is Ambient temperature 27 °C, radiant temperature 26 °C, relative humidity 80%, activity 1.61 (light), type of clothing 0.66 (Pants/short sleeve shirt) and finally a speed of wind of 2 m/s. As a result, 21% of the users could be dissatisfied with the conditions existing in the interior space, since the value obtained was +0.87, which falls outside the limit of what is considered comfortable with the beginning of heat sensations. The second analysis evaluates average conditions during the night, when the temperature of the environment is 23 °C, the radiant temperature 22 °C, the relative humidity 96%, the activity 0.77 (rest), the type of clothing 0.70 (Pants/shirt long sleeve) in case A and 1.59 (Thermal pajamas & blankets) in case B, finally a wind speed similar to 2 m/s. The variation of the cases (A/B) is generated to establish an adequate analysis with a possible adaptive comfort of the user. As a result, in case A 100% of the users could be dissatisfied since they obtained a score of -3 that represents a cold temperature sensation. On the other hand, case B with similar climatic conditions, but with the inclusion of user adaptability for the use of different clothes, a much lower percentage of dissatisfaction was obtained that was around 13.9%. It corroborates the hypothesis that the project has adaptive thermal comfort conditions.

5 CONCLUSIONS AND RECOMMENDATIONS

This work confirms the need to carry out bioclimatic evaluations to projects conceived under a good line of architectural design, since only in this way will it be possible to demonstrate that the considerations and strategies proposed have a positive result, as is the case of this project. It is evident that the morphology of the openings, as well as the materiality of the enclosure used, had a significant contribution in the reduction of thermal gains of the building and therefore the modification of internal temperature of the space; this guarantees a true adaptive comfort during the periods analyzed (between 20 °C and 28 °C) for the majority of users, as confirmed by temperature analysis and satisfaction percentage (higher than 70% of satisfied users) respectively. As could be seen, adaptive comfort (user adaptability) plays an essential role in the building in terms of thermal sensation. However, this process has a high degree of uncertainty about the average satisfaction obtained, since the conditions are assimilated differently by each user. Therefore, a more detailed analysis per hour could result in a proposal to improve the project, utilizing minimum changes in the morphology to ensure a greater interference of the volume in the internal temperature control and therefore a lower range of user dissatisfaction. Finally, as a project that has already been built, it has excellent potential to verify the results of the modeling described above and explained.

References

Al-Obaidi, K., Ismail, M., Malek, A., and Rahman, A., Passive Cooling Techniques Through Reflective and Radiative Roofs in Tropical Houses in Southeast Asia: A Literature Review, *Frontiers of Architectural Research, 3*(3), 283-297, 2014.

Calle, A. A., Toro, P. P., and Labus, J., Sustainable Buildings: Ecuador Case (In Spanish), *Technological Magazine ESPOL (In Spanish), 26*(2), 28-43. Retrieved from http://www.rte.espol.edu.ec/index.php/tecnologica/article/view/275/197 in 2013.

Cheng, V., Ng, E., and Givoni, B., Outdoor Thermal Comfort in Sub-Tropical Climate: A Longitudinal Study Based in Hong Kong, *Adapting to Change: New Thinking on Comfort,* Windsor: Network for Comfort and Energy Use in Buildings, 2010.

Gan, G., Natural Ventilation Driven by the Wind - CFD for Ventilation Design - Class, Nottingham, 2016.

Iea.org, International Energy Agency. Retrieved from IEA.org: http://www.iea.org/statistics/statisticssearch/report/?country=Ecuador&product=ElectricityandHeat&year=2013 on Feb 21, 2018.

INAMHI, National Institute of Meteorology and Hydrology - Meteorological Yearbook No. 51-2011 (In Spanish). Retrieved from http://www.serviciometeorologico.gob.ec/wp-content/uploads/anuarios/meteorologicos/Am%202011.pdf, 2014 on Jan 24, 2018.

Kastillo, J., Beltran, D., Ávila, C., Miño, I., and Aguinaga, Á., Energy Optimization for the Use of Natural Ventilation in Buildings in Warm Climates of Ecuador (In Spanish). Quito: ResearchGate. Retrieved from https://www.researchgate.net/publication/271703603 in 2015.

Marsh, A., *Predicted Mean Vote Calculator,* 2005.

Miño, I., Lobato, A., and Labus, J., Energy Thermal Simulations as a Tool for Social Housing Assessment of Ecuador: Case Yachay (in Spanish). Quito - Ecuador: Energy Efficiency and Renewable Energy Institute (INER - acronym in Spanish), 2013.

NASA SSE., (2018, January 15). *Surface Meteorology and Solar Energy (SSE).* Retrieved from https://eosweb.larc.nasa.gov/cgi-bin/sse/grid.cgi?email=skip%40larc.nasa.gov&step=2&lat=0.375&lon=-79.96&num=101091&p=grid_id&p=swvdwncook&p=swv_dwn&p=daylight&p=ret_tlt0&p=T10M&p=CDD18&p=HDD18&p=TSKIN&sitelev=&p=wspd10arpt&p=pct10m_wnd&p=vege_wnd&veg=6&h on Jan 15, 2018,

Peel, M., Finlayson, B., and Mcmahon, T., Updated World Map of The Köppen-Geiger Climate Classification, *HAL Id: hal-00298818, 4*(2), 439-473, 2007.

Rattanongphisat, W., and Rordprapat, W., Strategy for Energy Efficient Buildings in Tropical Climate, *Energy Procedia, 52*(1), 10-17, 2014.

Roslan, Q., Halipah Ibrahim, S., Affandi, R., Nasrun, M., Nawi, M., and Baharum, A., A Literature Review on The Improvement Strategies of Passive Design for The Roofing System of The Modern House in A Hot and Humid Climate Region, *Frontiers of Architectural Research, 5*(1), 126-133, 2015.

Sadineni, S., Madala, S., and Boehm, R., Passive Building Energy Savings: A Review of Building Envelope Components, *Renewable and Sustainable Energy Reviews, 15*(8), 3617-3631, 2011.

Yeang, K., and Yeang, L., *Ecodesign: A Manual for Ecological Design.* London: Wiley-Academy, 2006.

Interdependence between Structural Engineering and Construction Management
Edited by Ozevin, D., Ataei, H., Modares, M., Gurgun, A., Yazdani, S., and Singh, A.

ISBN: 978-0-9960437-6-2

ADOPTING NON-CONVENTIONAL MATERIALS IN AFFORDABLE HOUSING DELIVERY

IGNATIUS OMUH, PATIENCE TUNJI-OLAYENI, RAPHEAL OJELABI, ADEDEJI AFOLABI, and YIMIKA ERINLE

Dept of Building Technology, Covenant University, Ota, Nigeria

The second greatest human need is shelter and this indicates the necessity for everyone to be housed. Adequate housing provision reduces a host of other risk that the sustainable development goals are expected to reduce or even eliminate. However, due to policies, rapid population growth, rural-urban migration, housing provision for the homeless populace fall short. This paper highlights the prospects and problems of using non-conventional materials for low cost or affordable housing provision. The non-conventional materials investigated included earthworm cast as partial replacement of cement, stabilized earth bricks, bamboo, ash from waste incineration. A survey research was designed, and 200 questionnaires were distributed to both professionals and end users of housing. The findings showed that funding was a major obstacle to affordable housing provision, but this could be improved by employing non-conventional sustainable materials. The findings also showed that thought there was inadequate funding, lack of awareness of some of these materials were critical to the utilization of these materials. It was recommended that the potential users be sensitized to promote sustainable construction practices.

Keywords: Sustainable construction, Recycled waste, Low-cost housing, Shelter, Locally sourced materials, Perception.

1 INTRODUCTION

Aluko (2004) asserts that housing in the ranking of man's needs comes second only after food. Despite this importance of housing, 50.2 % of Nigerians still live in slums (United Nations 2015). Kabir and Bustani (2008) defined housing as dwellings, buildings or other shelters in which people live. Shelter is undeniably important for human survival, and with its great significance to man comes increasing demand for houses that are habitable. The basic shelter to be provided must be safe, healthy, and most importantly affordable for the occupants. According to the United Nations Habitat (UN Habitat), the population of families living in slums has increased by 18.62% between the year 2000 and 2020 (UN Habitat 2015). The number of families living in slums will continue to rise due to rural-urban migration, poverty, poor house planning and lack of infrastructure, and a host of other factors. It therefore is no surprise that housing is one of the sustainable development goals of the United Nations. To achieve this goal, stake holders in the construction industry and policy makers must work closely together to reduce the housing deficit being experienced worldwide. According to Olotuah and Bobadoye (2009), over 60% of the population of Nigeria are without shelter and over 75% dwell in slums.

In Nigeria, the housing deficit is upwards of 17 million houses (Rahimian *et al.* 2017). Aduwo *et al.* (2016) highlighted the efforts made by the Federal Government of Nigeria in trying to bridge the gap that is the deficit in housing provision. They showed that even though the Federal government proposed to build 653,271 housing units in forty-eight years (1962– 2010), they only succeeded in building only 95,594 units. This is a gross underachievement because even if the initial plan of 653,271 was achieved, there would still be a needed balance of 16,346,729 units that remain to be constructed because of the current housing deficit that sits at 17,000,000 (Ezeigwe 2015). This means that the government would have solved only 3.86% of the problem. Moreover, the Government's plan to provide "low-income" houses the low-income earners has not always been successful. Nigerian government in its quest to proving shelter for its populace admits that building materials take up about 60% of the total housing cost and the rising cost has been a major challenge.

Several researchers have investigated several aspects of affordable housing (Bajunid and Ghazali 2012, Puri *et al.* 2017, Gan *et al.* 2017, Ben-Shahar *et al.* 2018, Gohnert *et al.* 2018, Srivastava and Kumar 2018). Iwuagwu and Iwuagwu (2015) proposed the use of locally sourced materials in construction to reduce the cost of Building materials.

2 METHODOLOGY

This research was designed to investigate the utilization of indigenous materials for low-income housing delivery in Nigeria in the bid to bridge the house deficiency gap in the country. Hence, all the source of primary data found for this research is from structured questionnaires. The research was quantitative in nature and therefore a survey research design was adopted for the study. The survey research design was implemented the cross-sectional survey research design, which relied on data obtained from respondents. The sampling technique adopted was the random sampling technique. A reasonable amount of data was gotten from randomly selecting the subject matter considered.

The target population to be reached was that of the construction professionals and residents in the Federal Capital Territory, of Abuja, Nigeria. The population comprised of construction professionals such as architects, builders, quantity surveyors and structural/civil engineers, as well as residents who are government workers, entrepreneurs and private developers. These construction professionals were selected due to their experience in dealing with materials and residents were selected due to the importance of the client's preference.

The sample size selected from the population was two hundred (200): one hundred (100) construction professionals and one hundred (100) residents. The survey questions were designed as a 5-point Likert scale that ranged from strongly agree to strongly disagree. The data collected from the survey were analyzed and presented in descriptive format, with the aid of tables and charts. The different factors used in the ranking were obtained from other studies carried out by other researchers (Dayaratne 2010, Iwuagwu and Iwuagwu 2015, Gan *et al.* 2017, Nanyam *et al.* 2017). The ranking of the different factors considered was achieved using the relative importance index RII.

3 FINDINGS AND DISCUSSION

A total of 200 respondents answered the survey questions. Among the professionals in the construction industry, 12% of the respondents were females while 88% were male professionals in the construction industry. Among the residents, 40% were females, while 60% were male residents. This shows that the respondents were predominantly male. Both professionals and residents were asked about their age. From the result of respondents who were professionals in the construction industry, none were below the age of 21, 30% were between the ages of 20 – 40, 54% were between the ages of 40 – 60 and 16% were 60 and above. Among the residents, none were below the age of 21 or above the age of 60, 44% were between the ages of 20 – 40 and 56% were between the ages of 40 – 60.

From the result of residents, 22% were Government Workers, 4% were entrepreneurs, 34% were private developers and 40% responded as "others." From the result of professionals, those who were architects were 30%; those that were builders were 30%; those that were civil engineers were 30% and estate managers 10%. Also, from the result of residents, 10% had a higher national diploma, 48% had a BSc, 28% had an MSc and 4% had a PhD. The result showed that, 10% lived in self-owned houses, 60% lived in rented houses and 30% lived with family or engage in squatting.

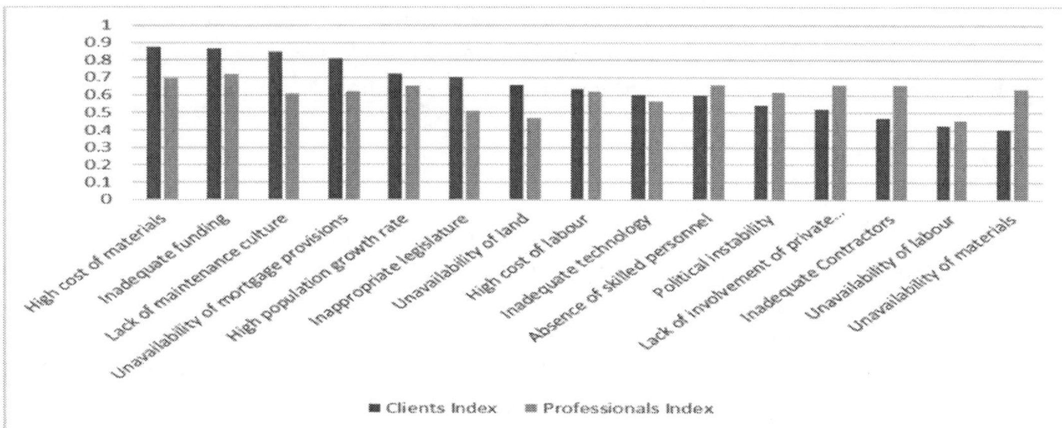

Figure 1. Evaluation of the housing deficit by Clients and Construction Professionals.

The chart in Figure 1 ranked the perceptions of both the clients and the construction professionals on the factors that lead to the housing deficit. According to the construction professionals, "Inadequate Funding" was the most important factor as it ranked 1st with an index score of 0.716 and the high cost of material ranked 2nd with an index score of 0.696. The clients on the other hand, perceived that high cost of materials was the most critical factor as it ranked 1st with an index score of 0.876 and inadequate funding ranked 2nd with an index score of 0.868. The least ranked factors highlighted include "Unavailability of Labour" and "Unavailability of Materials," which had index scores of 0.428 and 0.404 respectively. Both professional and clients both ranked inadequate funding and high cost of materials as their first two choices, showing that both categories of respondents shared the same perceptions towards housing deficit. While professionals ranked unavailability of land and labour as their least favourable factors, residents chose unavailability of labour and materials; these results are quite similar as well.

Table 1 shows the ranking of factors that influence the use of indigenous materials by Residents. The results obtained show that "Client Preference" was the most favoured benefit as it ranked 1st with an index score of 0.852 and the "Quality Superiority of Foreign Products" ranked 2nd with an index score of 0.808.

Both professionals and clients ranked client preference and quality superiority of goods as the most important factors that influence indigenous materials, but they ranked different factors as the least important.

Table 2 shows the construction professionals' ranking of the factors that influence the use of non-conventional building materials in construction. The table shows that "Client Preference" was the most favoured benefit as it ranked 1st with an index score of 0.736 and "Quality Superiority of Foreign Products" ranked 2nd with an index score of 0.68. The least ranked factors highlighted were the "Lack of Awareness" and "Unavailability of Materials," which had index scores of 0.64 and 0.608 respectively.

Table 1. Evaluation of factors that influence the use of indigenous materials by professionals.

Factors	Index	Rank
Client Preference	0.852	1st
Quality Superiority of Foreign Products	0.808	2nd
Unavailability of Technology to develop indigenous materials	0.76	3rd
Lack of Awareness	0.748	4th
Unavailability of Materials	0.712	5th
Inadequate Sustainability Culture	0.566	6th
Limitations from Building Codes and Regulations	0.444	7th

Table 2. Evaluation of factors that influence the use of indigenous materials by residents.

Factors	Index	Rank
Client Preference	0.736	1st
Quality Superiority of Foreign Products	0.68	2nd
Inadequate Sustainability Culture	0.668	3rd
Unavailability of Technology to develop indigenous materials	0.652	4th
Limitations from Building Codes and Regulations	0.648	5th
Lack of Awareness	0.64	6th
Unavailability of Materials	0.608	7th

Since the construction brief is expected to be based on the client's request and preference, it is only logical that both categories of respondents would agree on the Client's preference as being the most important factor influencing the use of non-conventional materials.

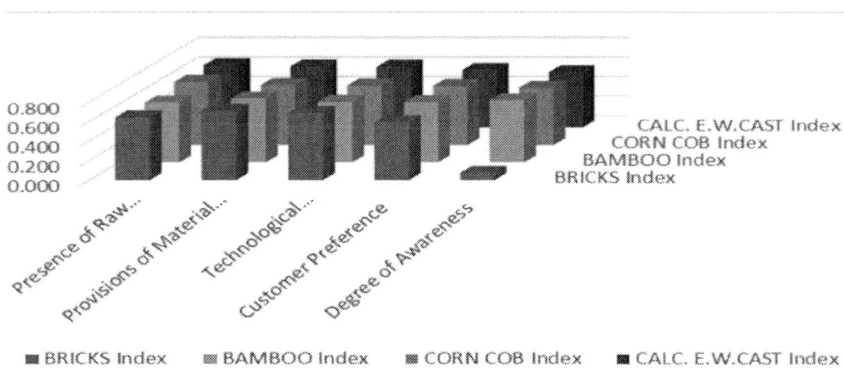

Figure 2. Evaluation of the utilization of different non-conventional materials.

Figure 2 displays the professional perceptions of the factors that influence the utilization of different non-conventional materials. The results obtained shows that “Degree of awareness” was the most favourable factor, in influencing the development of “Bricks”, as it ranked 1st with an index score of 0.72 and the least favourable factors highlighted was “Customer Preference” which had an index score of 0.596. The results also showed that “Provision for Material Development” was the most favourable factor, in influencing the development of “Bamboo”, as it ranked 1st with an index score of 0.656 and the least favourable factors highlighted was “Presence of Raw Materials” which had an index score of 0.596. One of the most common uses of bamboo in Nigeria has been for scaffolding. Few people have seen how its proper development and processing can enhance its aesthetic qualities. This could be why the respondents picked provision for material development as the major factor for the use of bamboo. Figure 2 also showed that the “Presence of Raw Materials” was the most favourable factor, in influencing the development of “Corn Cob”, as it ranked 1st with an index score of 0.64. The least favourable factor highlighted was the “Degree of Awareness” which had an index score of 0.568. The parts of the country where these raw materials are abundant are not readily accessible to those who would need them for construction. As for “Calcined Earthworm Cast,” the results showed that the “Presence of Raw Materials” was the most critical factor in influencing the development of “Calcined Earthworm Cast” as it ranked 1st with an index score of 0.648 and the least favourable factors highlighted was the Degree of Awareness which had an index score of 0.588.

4 CONCLUSIONS

From the findings, the high cost of materials and inadequate funding were the two most critical factors ranked by both professionals and residents in identifying the cause of Low-Income Housing Deficit in the Nigerian Construction Industry. “Inadequate funding” ranking high shows low-income housing projects are not funded properly.

The findings also showed that client preference and quality superiority of conventional materials were the two most critical factors ranked by both professionals and residents in analyzing factors that mitigate against the “Utilization of Non-Conventional Materials.” “Client preference” ranking high means that clients are identified as very important stakeholders in the material selection. Clients’ choice of materials needs to be influenced to ensure the use of non-conventional materials. “Quality Superiority of Conventional Materials” ranking high means that the quality of the conventional materials is perceived to be of higher standards than the quality of the non-conventional ones. Perceptions of quality superiority of foreign products greatly influence the choice of material by the client.

From the findings in examining the prospects and problems of developing non-conventional materials, “Provision of Material Development” ranked as the highest for “Bamboo” and second highest for “Ash Bricks.” This means professionals believe “Bamboo” and “Ash Bricks” have provision for material development. Also, in examining the prospects and problems of developing non-conventional materials, customer preference and degree of awareness ranked as the lowest factors for “Calcined Earthworm Cast” and “Corn Cob.” This implies that the professionals believe that the awareness on glulam and cob is very low and therefore, clients prefer conventional materials.

Based on the conclusions, it is therefore recommended that in order, to increase funding, the government should collaborate with the private sector towards implementing low-income housing schemes; this will aid in an incremental generation of financial resources for low-income housing projects. It is also recommended therefore that the government should endorse non-conventional materials, to create better price competition in the construction industry by implementing codes and regulations for the use of non-conventional materials. Seminars and workshops should be conducted to create awareness of the types of non-conventional materials and various benefits that use of non-conventional materials. Mass marketing of non-conventional materials can aid the perception of clients towards acquisition of non-conventional materials. The government should facilitate the development of industries that ensure the development of non-conventional materials to significantly increase the quality of non-conventional materials. Further research should be carried out on the psychological implications of the building materials in affordable housing and how this affects the stakeholders in the design and implementation of these non-conventional materials. Research into the effects of non-conventional materials on climate change and environment is also recommended.

References

Aduwo, B. E., Edewor, P., and Ibem, E. O., Urbanization and Housing for Low-Income Earners in Nigeria: A Review of Features, Challenges and Prospects, *Mediterranean Journal of Social Sciences*, *7*(3S1), 347-357, 2016.

Aluko, O., *Housing and Urban Development in Nigeria*, KINS, 2004.

Bajunid, A. F. I., and Ghazali, M., Affordable Mosaic Housing: Rethinking Low-Cost Housing, *Procedia-Social and Behavioral Sciences*, *49*, 245-256, 2012.

Ben-Shahar, D., Gabriel, S., and Golan, R., Housing Affordability and Inequality: A Consumption-Adjusted Approach, *Journal of Housing Economics*, 2018.

Dayaratne, R., Reinventing Traditional Technologies for Sustainability: Contemporary Earth Architecture of Sri Lanka, *Journal of Green Building*, 5(4), 23-33, 2010.

Ezeigwe, P. C., Evaluation of The Causes of Housing Problems in Nigeria: A Case Study of Awka The Capital City of Anambra State, *Journal of Economics and Sustainable Development*, *6*(20), 87-93, 2015.

Gan, X., Zuo, J., Wu, P., Wang, J., Chang, R., and Wen, T., How Affordable Housing Becomes More Sustainable? A Stakeholder Study, *Journal of Cleaner Production*, 162, 427-437, 2017.

Gohnert, M., Bulovic, I., and Bradley, R., A Low-Cost Housing Solution: Earth Block Catenary Vaults, *Structures*, July 2018.

Iwuagwu, U., B., and Iwuagwu, C. M. I. B., Local Building Materials: Affordable Strategy for Housing the Urban Poor in Nigeria, *Procedia Engineering*, *118*, 42-49, 2015.

Kabir, B., and Bustani, S. A., *A Review of Housing Delivery Efforts in Nigeria*, International Housing Conference, University of Glasgow, Scotland, September 2009.

Nanyam, V. N., Sawhney, A., and Gupta, P. A., Evaluating Offsite Technologies for Affordable Housing, *Procedia Engineering*, 196, 135-143, 2017.

Olotuah, A. O., and Bobadoye, S. A., Sustainable Housing Provision for The Urban Poor: A Review of Public Sector Intervention in Nigeria, *The Built and Human Environment Review*, *2*, 51-63, 2009.

Puri, V., Chakrabortty, P., Anand, S., and Majumdar, S., Bamboo Reinforced Prefabricated Wall Panels for Low Cost Housing, *Journal of Building Engineering*, *9*, 52-59, 2017.

Rahimian, F. P., Goulding, J., Akintoye, A., and Kolo, S., Review of Motivations, Success Factors, and Barriers to The Adoption of Offsite Manufacturing in Nigeria, *Procedia Engineering*, *196*, 512-519, 2017.

Srivastava, M., and Kumar, V., The Methods of Using Low Cost Housing Techniques in India, *Journal of Building Engineering*, *15*, 102-108, 2018.

UN Habitat, United Nations Habitat Almanac, U. H. S. 2015-2016, *Tracking Improvement in the Lives of Slum Dwellers,* Technical Report, 2015.

Interdependence between Structural Engineering and Construction Management
Edited by Ozevin, D., Ataei, H., Modares, M., Gurgun, A., Yazdani, S., and Singh, A.

ISBN: 978-0-9960437-6-2

UTILIZATION OF SUPERABSORBENT POLYMERS IN PLASTERS FOR MODERATION OF INDOOR RELATIVE HUMIDITY LEVEL

JAN FOŘT and MAGDALÉNA DOLEŽELOVÁ

Dept of Materials Engineering and Chemistry, Czech Technical University in Prague, Prague, Czech Republic

Nowadays, the energy efficiency and moderation of indoor environment of buildings became important topic in the field of building physics. Moisture occurrence in building construction poses a significant task for building designers and material engineers. Among other factors, a high level of relative humidity has negative effect on the work efficiency and health of building inhabitants. The presented paper is aimed on the analysis of chosen superabsorbent polymers (SAPs) and its influence on properties of newly developed interior plasters. The identification of optimal ratio between superabsorbent polymers and plaster mixture from the point of view of the workability of fresh mixtures is done. Also, the effect of water/dry mixture ratio is studied according to enhanced absorption capability of SAPs. Specific attention is paid to hygric properties, the determination of apparent moisture diffusivity, water vapor resistance factor and estimation of sorption isotherms. Obtained results directly point to significant effect of SAP admixture on material properties of developed plasters. Namely, addition of up to 3 wt.% of SAPs shifts the water absorption coefficient almost ten-times, while the total open porosity is increased only about 20%. Such materials can be applied for the damped areas or places where increased water transport parameters play an important role for the efficient moisture removal.

Keywords: Indoor climate, Moisture moderation, Hygric properties.

1 INTRUDUCTION

Moisture level significantly affects durability of constructions, their thermal performance and quality of indoor air. Although modern advanced ventilation systems provide sufficient air exchange rate, their wider application conflicts with sustainability development principles due to high energy demands (Koffi *et al.* 2011). Moreover, according to the European legislation related to the Nearly Zero Energy Buildings, air tightness of building envelopes leads to large moisture loads in building interiors. Especially in case of passive and low energy houses, the water vapor from numerous moisture sources and as a natural consequence of the changing weather conditions represent substantial moisture loads and can induce undesirable indoor conditions (Johansson *et al.* 2012). On the other hand, the low relative humidity level of the indoor air poses a problem, which have been struggling mostly during winter when indoor relative humidity often drops below 30% as the response to the relatively dry continental climate (Toftum *et al.* 1998). In addition, some specific buildings such as museums or galleries require specific indoor humidity levels to secure appropriate indoor conditions for historical artifacts.

The extensive research comprising utilization of mortars for ambient control by the combination of vermiculite, perlite and super absorbent polymers was made by Gonçalves *et al.* 2014. Mixtures with SAP achieved the most promising results in the field of humidity control in comparison with the other materials. The addition of perlite and vermiculite brings an increase of the moisture buffering compared to reference sample, but not so significantly. On the other hand, according to the high demands of SAP containing mixtures on the amount of batch water, the highest unrestrained shrinkage, weight changes and strong reduction of mechanical properties was obtained. Further research of application of SAP and porogene additives (aluminum powder-AP, sodium olefin-sulphonate-SOS) in the mortars was conducted in a study of Yang *et al.* (2011), which was focused on finding compromise between moisture buffer capacity and in particular mechanical properties. Here, the mixture with AP and SOS additives did not significantly shifted the moisture buffer properties of studied mortars. The lower SAP dosage resulted in decrease of the unrestrained shrinkage but also drop of capability to absorb water molecules. Senff *et al.* (2015) investigated the influence of difference dosages of SAP on rheological and mechanical properties. The combined effect of various dosages of SAP and w/c ratios on workability revealed importance of rheometry tests which play relevant role on the final properties. Necessity to increase of w/c ratio in line with SAP addition did not cause increase of the capillary index of mortars but negative impact on the flexural strength was obtained.

The superabsorbent polymers can provide a suitable environment for organisms such as fungi, yeasts, algae, lichens, protozoas or mites thank these properties (Haas *et al.* 2014). However, the presence of these organisms in indoor environment may pose the health problems for buildings occupants. Spores and metabolic products as enzymes and toxins may again get into the environment and sensitive persons may suffer from allergies and other diseases. Moreover, microbes under conditions when fungal peroxidases or cellobiose dehydrogenase are produced could mineralize polymers after solubilisation (Stahl *et al.* 2000).

Based on the literature survey, several attempts in the field of moderation and description of the relationship between indoor climate conditions, building inhabitants, and moisture performance of building materials can be found. The presented study is aimed at evaluation of influence of SAP admixtures on the basic physical and especially moisture transport parameters of newly designed plaster mixtures to outline possible ways for development of new type of plasters.

2 MATERIAL AND METHODS

2.1 Used Materials

Commercially produced cement-lime core plaster WEBER MV1 was used as a reference material, which was further enriched by particular SAP admixtures.

Table 1. Mixture composition.

Mixture	Dry plaster mixture (kg)	Water (kg)	SAP (kg)
PR	5	0.8	-
PS1	5	1.1	0.05
PS2	5	1.3	0.1
PS3	5	1.5	0.15

Free flowing white granules of SAP Favor Pac 300 with diameter about 500 – 2,000 μm from Evonik Industries (Germany) were used. Such material was applied for modification of reference plaster mixture in three grades. The detail description of particular mixtures enriched by 1, 2 and 3 wt.% of SAP is given in Table 1. Higher dosage of SAP was not tested because of negative effect of SAPs on the workability of fresh mixtures.

2.2 Determination Methods

Basic physical properties of studied plasters were characterized by meaning of measurement of the bulk density, the matrix density and the total open porosity. Performed measurement of the bulk density was done on five cubic samples of 50 mm side and determined from the measurement of sample sizes (using digital caliper) and its dry mass. The matrix density was accessed by helium pycnometry using apparatus Pycnomatic ATC (Thermo Scientific). The accuracy of the gas volume measurement using this device is ±0.01% from the measured value, whereas the accuracy of used analytical balances is ±0.0001 g. The measurement of bulk density uncertainty was 5.3% and 3% for matrix density.

The cup method in dry-cup arrangement was employed in the characterization of water vapor transport. It was based on measuring the diffusion water vapor flux through the specimen and partial water vapor pressure in the air under and above specific specimen surface. Water vapor transmission properties of a studied material were found by placing a specimen of the material on the top of a stainless-steel cup, whereas the specimen was in contact with the cup sealed by technical plasticine. For the measurement 5 samples with dimensions of 100 x 100 mm and 50 mm thickness were used. The cup contained a sorption material, namely silica gel. Measuring cups were placed in a controlled climate chamber and weighed periodically. The steady state values of mass gain were utilized for the determination of water vapor transport properties.

The ability of designed plasters to transport liquid water was described by the measurement of water absorption coefficient, which was defined as:

$$i = A \cdot t^{1/2} \tag{1}$$

where i (kg/m^2) is the cumulative mass of water, A (kg/m^{2}s$^{1/2}$) the water absorption coefficient and t the time (s). The water sorptivity test was used in the experiments.

For measurement of sorption and desorption isotherms, dynamic vapor sorption device DVS-Advantage was used, whereas the measurements were done at 21 °C. Before the measurement, the sample of studied material was dried at first and maintained in desiccator during cooling. Then, the sample was put into the climatic chamber of the DVS-Advantage instrument and hung on the automatic balances in the special steel tube. The instrument measures the uptake and loss of vapor gravimetrically, using highly precise balances having the resolution of 1.0 μg. Such a high resolution is obtained by hanging samples on the end of a beam where the position of the beam is measured by an optical sensor. The particular samples were exposed to the following partial water vapor pressure profile: 0; 20; 40; 60; 80 and 95% of relative humidity (RH). Each step in RH during the DVS measurement is incremented either when a stable mass is achieved with mass change less than 0.00004% /min or a maximum time interval of 400 min is reached. Because reaching of sample mass equilibrium at high RH was problematic, the maximum time interval of samples exposure to RH of 80% was 4000 min, and for RH of 98% it was extended up to 7000 min. The sample mass was 5–10 g.

3 RESULTS AND DISSCUSION

3.1 Basic Physical Properties

The basic physical properties of studied mixtures are shown in Table 2. Here, the bulk density, matrix density and total open porosity are given to highlight differences between individual mixtures and allow more coherent comparison of material parameters. The incorporation of 1 wt.% of SAPs did not result in fundamental decrease of the bulk density and the total open porosity was increased only about 3% to almost 41%. However, further addition of SAPs in case of PS2 and PS3 mixture significantly affected bulk density and total open porosity, while the matrix density remained almost constant. Here as one can see, a significant drop of the bulk density was obtained when the amount of SAP admixture exceeded 2 wt.%. To be specific, incorporation of 2 and 3 wt.% of SAPs led to the decrease of the bulk density to 1,225 kg/m^3 and 1,203 kg/m^3 respectively. The most distinct changes are visible for the total open porosity, which was shifted to 52.4 % and 53.1 % respectively.

Table 2. Basic material properties of designed mixtures.

Mixture	Bulk density (kg/m^3)	Matrix density (kg/m^3)	Total open porosity (%)
PR	1593	2547.9	37.5
PS1	1511.3	2549.6	40.7
PS2	1224.8	2574.2	52.4
PS3	1203.3	2563.2	53.1

3.2 Hygric Properties

The outstanding hygric storage properties of the pure SAPs were partially transferred to the newly developed mixtures and thus, the hygric properties of developed plasters were affected in great extent. The detail results of the water absorption coefficient and the moisture diffusivity coefficient are given in Table 3.

Table 3. Hygric properties.

Mixture	Moisture diffusivity (m^2/s)	Water absorption coefficient (kg/m^{2}s$^{1/2}$)
PR	9.93E-08	0.0379
PS1	3.76E-07	0.0779
PS2	6.80E-07	0.2425
PS3	1.02E-06	0.3737

Here, the hygric transport properties obtained for reference plaster were strongly changed when SAP admixture was added. The incorporation of 1 wt.% of SAP resulted in the doubled water absorption coefficient as well as for the moisture diffusivity coefficient. Obtained shift cannot be assigned only to increase of the total open porosity as was described latter and this notable change is linked to the very high water absorption of incorporated SAP admixture. The fundamental impact of the SAP addition was further confirmed by achieved results for other mixtures. The water absorption coefficient was significantly increased when SAP admixture exceeded 2 wt.%. Here, for PS2 mixture was eight times higher, while for PS 3 exceeded more than ten times initial value achieved by the reference plaster. Similar results were obtained also for the coefficient of moisture diffusivity, which was improved almost about three orders of magnitude.

3.3 Water Vapor Properties

The effect of increased moisture uptake was also studied by meaning of sorption isotherms and water vapor transport. Results of the cup method used for the determination of water vapor properties are given in Table 4 together with data of equilibrium moisture content (EMC) calculated from measurement of sorption isotherms. Here, similar behavior as was described in previous section was noted also for the water vapor transport and storage parameters. Looking at results of the water vapor diffusion coefficient, is possible to distinguish very strong effect of incorporated SAPs on material response when is exposed to the elevated humidity level. Namely, water vapor resistance factor was more than two times lower compared to the reference plaster. This fact can be assigned to the coupled effect of the increased total open porosity and moisture storage ability of SAP particles. EMC values were affected to a far greater extent and modified plaster was able to store more than five times higher amount (PS3) of water vapor compared to the reference plaster.

Table 4. Water vapor properties.

Mixture	Water vapor resistance factor (-)	EMC (%)
PR	16.8	0.9399
PS1	12.3	1.4961
PS2	9.8	3.9842
PS3	7.3	6.1631

4 CONCLUSIONS

The performed experimental analysis of the basic physical and hygric properties of cement-lime plasters modified by SAP admixture revealed an interesting behavior of such materials. Dosages of the SAP admixture is limited according the increased swelling capability of SAP materials. However even for small amounts of applied SAP is possible to distinguish very significant changes in material behavior, especially in case when is exposed to increased moisture level. While the mixture PS1 with SAP admixture only about 1 wt.% did not modified structure and texture of material to a greater extent, increase of SAP admixture over 2wt.% led to substantial increase of the total open porosity. Notwithstanding, massive increase of the water transport properties cannot be assigned only to this phenomenon. Major cause of these changes is related to outstanding absorption capacity of SAP admixture, which resulted in a multiple increase of all water transport parameters. Such material parameters can be easily utilized for design of new type of plaster applicable for the damped areas or places where are increased demands on moisture removal (from air or walls).

Acknowledgments

This research has been supported by the Czech Science Foundation, under project No. 18-03997S.

References

Gonçalves, H. B., Gonçalves, B., Silva, L., Raupp-Pereira, F., Senff, L., and Lanbrincha, J. A., The Influence of Porogene Additives on The Properties of Mortars Used to Control the Ambient Moisture, *Energy and Buildings*, 74, 61-68, May 2014.

Haas A., Habib, J., and Luxner, J., Comparison of Background Levels of Culturable Fungal Spore Concentrations in Indoor and Outdoor Air in Southeasetern Austria, *Atmospheric Environment*, 98, 640-647, December 2014.

Johansson, P., Ekstrand-Tobin, A., Svensson, T., and Bok, G., Laboratory Study to Determine the Critical Moisture Level for Mould Growth on Building Materials, *International Journal of Biodeterioration and Biodegradation*, 73, 23– 32, September 2012.

Koffi, J., Allard, F., and Akoua, J. J., Numerical Assessment of The Performance of Ventilation Strategies in a Single-Family Building, *International Journal of Ventilation*, 9, 337–349, March 2011.

Senff, L., Modolo, R. C. E., Ascensao, G., Hotza, D., Ferreira, V. M., and Lanbrincha, J. A., Development of Mortars Containing Superabsorbent Polymer, *Construction and Building Materials,* 95, 575-584, October 2015.

Stahl, J. D., Cameron, M. D., Haselbach, J., and Aust, S. D., Biodegradation of Superabsorbent Polymers in Soil, *Environmental Science and Pollution Research*, 7, 83-88, June 2000.

Toftum, J., Jorgensen, A. S., and Fanger, P. O., Upper Limits for Indoor Air Humidity to Avoid Uncomfortably Humid Skin. *Energy and Buildings*, 28, 1-13, August 1998.

Yang, H., Peng, Z., Zhou, Y., Zhao, F., Zhang, J., Cao, X., and Hu, Z., Preparation and Performances of a Novel Intelligent Humidity Control Composite Material, *Energy and Buildings*, 43, 386-392, March 2011.

Interdependence between Structural Engineering and Construction Management
Edited by Ozevin, D., Ataei, H., Modares, M., Gurgun, A., Yazdani, S., and Singh, A.

ISBN: 978-0-9960437-6-2

THEORETICAL VALUATION OF THE INFLUENCE OF CORPORATE GOVERNANCE AND SUSTAINABILITY ON LOW-COST HOUSING

GLADNESS MAKGOBI[1], NOKULUNGA MASHWAMA[1], WELLINGTON THWALA[1], and CLINTON AIGBAVBOA[2]

[1]*Department of Construction Management and Quantity Surveying, University of Johannesburg, Johannesburg, South Africa*
[2]*Faculty of Engineering and the Built Environment, University of Johannesburg, Johannesburg, South Africa*

The paper delivers a theoretical analysis of corporate governance and sustainability on low-cost housing. One of the constraints faced by low-cost housing projects is poor governance, which should be the development prospective of a company whose directive is to refine the economic potential of the countries in the process of transformation. Literature is the main method to investigate the research problem. Published, unpublished papers and journals were accurately reviewed to find solutions to the subject matter. The study revealed that it is a prerequisite in terms of the National Environmental Management Act for National and provincial department to formulate an Environmental Implementation Plan for any building construction activities that will have an impact on the environment. Green and sustainable are terms filled on company yearly reports, but the question is whether the Corporate Social Responsibility is valued or well thought of or highly regarded. The paper will assist the industry to realise that the achievement of the company does not only rely on the compliance of legislation and standards, but also by a broad series of economic, environment and social problems in a manner that is useful to global society. Organizational sustainability is an established requirement with many obligatory conditions enforcing ISO 14001 and environmental standards to be adhered to. Refining environmental performance through efficient use of energy and resources intensifies the success of companies.

Keywords: Compliance, Community, Society, Economy.

1 INTRODUCTION

Principles of corporate governance requires companies to fulfill their commitment to the society where they work and being accountable for their activity in the general public. Companies must not only focus on generating profit but ensure their conduct impact the community they are working in (Naidoo 2009). The prosperity of a company must reflect the transition from unsustainable to sustainable development, that is eliminating waste and pollution while using less resources and producing more products and services.

This simply means that the success of a company does not solely depend on short- and long-term capital but also a lengthy period ecosystem and social capital (Rothaus N.D.).

Lately companies are recognised as the primary source of social, economic and environmental issues. Companies are broadly observed to be successful by compromising the need of the society at large (Porter and Kramer 2011). An incredible challenge to regular business strategies with its attention on shareholder returns have developed movements like globalisation, climate change, societal unfairness and demographic transition (Wales 2013). CSR is a methodology that stimulate corporate governance, which bring about sustainability, hence the social responsibility of a company raises financial income (Blaga 2013).

Low cost housing is a lengthy investment cycle that must be advantageous to its beneficiaries as far as comfort, quality, and life expectancy is concerned (Jenkin *et al.* 2007). A financially effective construction industry, improves environmental sustainability by ensuring low cost techniques of development that support best distribution of resources and minimise losses. Economical viability within the construction business s requires that: social and ecological cost are approved and registered in the final cost of an item, (du Plessis & Landman 2002).

1 LITERATURE REVIEW

2.1 Corporate Social Responsibility and economic prosperity

Social sustainability does not rely on high ecological quality alone. All things should be considered including differences, which are brought together in areas of poor environmental quality, which may go as reactants for social support connection (Dempsey *et al.* 2011). Furthermore, bringing the community together to examine and strongly express their rights (Carrasco and Bilal 2016). Where members of the community join together, activate and campaign around such issues and make their own accumulative voices heard. They develop social capital in their own attempt to address unfairness and enhance their quality of life on the environment on which they live (Kwon and Addler 2014). At the point when residents unite against issues such as restricted access to services, for instance, regional equity might be served when access to services are corresponding equally (Demsy *et al.* 2011).

Green and sustainable economy are terminologies congested on yearly reports of companies, but the question is whether the Corporate Social Responsibility is valued or well thought of or highly regarded. If the answer is positive, about five hundred organizations phase illustrating directorship in carbon disclosure produced two times the median profit as an overall index (O'Mara *et al.* 2012). The achievement of the organizations does not only rely on compliance of legislation and standards, but also by a broad series of economic, environment and social problem in a manner that is useful to the whole community and society internationally. An extensive illustrative model stated three definitions (triple bottom line or 3p – model) which can be recognized as people, planet and profit (Elkington 1998). The three P's in the context of Corporate Social Responsibility refers to combination of social, environmental and economic factors into business model and strategic management (Jones *et al.* 2006).

2.2.1 *The Constitution*

The Constitution of Republic of South Africa Act No. 106 of 1996 in terms of section 24 is the primary obligatory Act pertaining to the environment right, and it states that everyone has a right to an environment that is not harmful to their health and well-being. The Act further declares that in order to accomplish this right the Constitution must also make provision to take reasonable statutory and other means in terms of section 24 (b) to: Promote conservation, secure ecologically sustainable development, use natural resources, and prevent pollution and ecological degradation while supporting substantial social and economic development (South African Government 1998).

2.2.2 *The National Environmental Management Act (No. 107 of 1998) (NEMA)*

NEMA is a regulation established in South Africa to enforce environmental rights. It contains the notion of sustainable development; it describes the incorporation of social, economic and environmental components into the organization to secure development that serves today and generation to come (South African Government 1998). It is a prerequisite in terms of the National Environmental Management Act (NEMA) for National and provincial department to formulate an Environmental Implementation Plan (EIP) for any building construction activities that will have an influence on the environment. A list of sustainable low-cost programs and projects undertaken by this plan is given in Table 1.

Table 1. Sustainable low-cost programs and projects undertaken by environmental implementation plan (The South African Human Rights Commision 2000 – 2002).

Department	Programs
Departments of Environmental Affairs and Tourism, Health, Housing, Minerals and Energy, and Water Affairs and Forestry in collaboration with Department of Housing	• A green financing feasibility study which examines the development of a financing system to enhance environmentally efficient housing development. • A Cabinet resolution to promote energy efficiency in the housing sector.
Department of Housing and Non-Government Organizations	• Green Professional Program of Sustainable Homes Initiative - Impacting the design of round 60 000 planned housing units nationwide to promote energy efficiency. • A National Sustainable Housing Training Course and other capacity building initiatives by the Sustainable Energy Environment and Development (SEED) program. • Sowing round 57 000 trees with new subsidized homes through the Trees for Homes program and the establishment of a National. • Urban Greening Fund which will proceed to assist residential greening projects.

2.2.3 *The environmental standard*

ISO 14001 standard is an accepted standard globally which outline obligations for Environmental Management System. The standard assists the company in enhancing environmental performance by minimising waste, effective utilisation of resources, acquiring credibility of associates, and competitive benefit (International Organization for

Standardization 2015). Environmental Management System is voluntarily accepted by the company under the ISO 14001 standard that incorporates effective reduction of environmental impacts (Directorate for Financial, Fiscal and Enterprise Affairs N.D.). The system gets audited at intervals to give the public a reasonable assurance that companies comply with the environmental standards.

The ISO14001 standard made a significant contribution which led to enhancement in water and air quality in Europe. The areas that were prominent leading contaminators of natural environment in Germany are the now good examples of clean environments. It is prerequisite for global businesses to fulfil the obligation of the standard (Ferguson 2014). The primary point in the Production of South African Standards and regulations is sustainable development, incorporated particularly in planning and housing policy (RSA, 1994). Currently, organizational sustainability is an established requirement with various mandatory prerequisites through survey enforcing environmental standards to be adhered to (Oke and Aigbavboa 2017).

3 METHODOLOGY

The study is literature review to investigate theoretical valuation of corporate governance and sustainability on low cost housing. General literature review published, unpublished, online journals and magazines by other researchers on similar study was critically assessed to investigate the research problem. The study adopted qualitative approach to gather and assess relevant information on corporate governance and sustainability on low cost housing. Qualitative research design is concerned with individuals'' own accounts of their attitudes, motivations and behavior. It offers richly descriptive reports of individuals'' perceptions, attitudes, beliefs, views and feelings, the meanings and interpretations given to events and things, as well as their behavior (Hakim 2000).

4 FINDINGS FROM LITERATURE

The literature revealed that triple bottom model known as the people, planet and profit comprises the integration of social, environmental and economic attributes into business aspects and good practice (Jones *et al.* 2006). Social sustainability should consider all elements that contribute to poor environmental quality, which may be a good response to promote social cohesion and not depend solely on ecological quality (Demsy *et al* 2011). Low cost housing is a long term investment which must benefit beneficiaries in terms of comfort, quality and life expectancy (Jenkin *et al* 2007). Businesses are currently viewed as the main causes of social, economic and environmental problems. Companies are neglecting the need of the broader society, but solely focus on profit (Porter and Kramer 2011). The main idea of sustainable development in the Production of South African Standards and regulations is extraordinarily included in planning and housing policy (RSA 1994).

Constitution of the Republic of South Africa is the regulatory framework regarding environmental rights. It states that everybody has the right to the environment that is not dangerous to their health and well-being. The success of a business must reveal the evolution from unsustainable to sustainable development by reducing waste and contamination whereas utilizing fewer assets and delivering more goods and services (Rothaus N.D.). It is a mandatory requirement to comply with environmental principles through assessments presently in organizational sustainability (Oke and Aigbavboa, 2017). Terms which are on businesses annual reports are green and sustainability, however the query is whether the Corporate Social Responsibility is treasured or well thought of or extremely considered

(O'Mara *et al.* 2012). The globally established standard which plans requirements for Environmental Management System is known as ISO 14001. It was designed to support businesses that improve environmental performance by reducing waste, effective use of resources, obtaining integrity of partners and competitive benefit (International Organization for Standardization, 2015). Regions that were primary polluters of natural environment in Germany are the now good examples of clean environments through the substantial influence of ISO14001 standard which resulted to improvement in water and air quality in Europe (Ferguson 2014).

According to Department of Housing 2001 – 2002, various sustainable low cost housing projects and programs were undertaken through Environmental Implementation Plan as an initiative, measure and compliance to impact lives of beneficiaries of low cost housing by:

- Influencing designs of about 60,000 houses to promote energy efficiency.
- Developing a financial system to enhance environmentally efficient and green low-cost housing projects.
- Planting trees to newly subsidised houses.
- Providing training courses on sustainable housing.

5 CONCLUSIONS AND RECOMMENDATIONS

Compliance to standards and legislative requirements forms the integral part of good corporate governance and sustainability on low cost housing and companies. Construction companies are obligated in terms of the National Environmental Management Act for national and provincial departments to formulate an Environmental Implementation Plan for any building construction activities that will have an influence on the environment. The constitution of the republic states that everyone has a right to an environment that is not harmful to their health and well-being. International companies are required to disclose any project that will have an environmental impact under the Carbon Disclosure Project.

The literature review revealed that the achievement of the organization does not only rely on compliance of legislation and standards, but also by a broad series of economic, environment and social problem in a manner that is useful to the whole community and society internationally. An extensive illustrative model stated three definitions (triple bottom line or 3p – model) which can be recognised as people, planet and profit. Currently, organizational sustainability is an established requirement with various mandatory prerequisites through the survey enforcing environmental standards to be adhered to. Therefore, it is highly recommended that construction companies should incorporate Environmental, Sustainability and Governance into their businesses, to ensure good business.

References

Blaga, S., Rethinking Business Sustainability, *Review of Economic Studies and Research,* 6(1), 5- 21, 2013.

Carrasco, M. A., and Bilal, U., A Sign of the Times: To have or to be? Social Capital or Social Cohesion?, *Social Science and Medicine*, 2016.

Constitution of the Republic of South Africa Act No. 106, 1996. Retrieved from www.gov.co.za on April 20, 2018.

Dempsey, N., Bramley, G., Power, S. and Brown, C., The Social Dimension of Sustainable Development: Defining Urban Social Sustainability, *Sustainable Development*, 2011.

du Plessis, C., and Landman, K., Sustainability Analysis of Human Settlements in South Africa, *Report Prepared for the Department of Housing by CSIR Building and Construction Technology*, 2002.

Elkington, J., Partners from Cannibals with Forks, *The Triple Bottom line of the 21st Century Business Environmental Quality Management*, 1998.

Ferguson, W., Environmental Compliance: Implications for South Africa Companies, 2014. Retrieved from http://m.engineeringnews.co.za/article/environmental-compliance-implications-for-south-africa-companies on September 15, 2018.
Hakim, C., Research Design: Successful Designs for Social Economics, 2000. Directorate for Financial, Fiscal and Enterprise Affairs (Undated). Retrieved from https://www.oecd.org/daf/inv/corporateresponsibility/18269204.pdf on March 20, 2018.
International Organization for Standardization, 2015. Retrieved from https://www.iso.org/files/live/sites/isoorg/files/archive/pdf/en/iso_14001__key_benefits.pdf on April 20, 2018.
Jenkins, P., Smith, H., and Ping Wang, Y., Planning and Housing in the Rapidly Urbanizing World, In *Housing, Planning and Design Series*, London, 2007.
Jones, P, Comfort, D, and Hillier, D., Corporate Social Responsibility and the UK Construction Industry, *Journal of Corporate Real Estate*, 8(3), July 2006.
Kwon, S. W., and Adler, P., Spacial Capital: Maturation of a Field of Research, *Academy of Management Review*, 39(4), 412-414, 2014.
Oke, A. E., and Aigbavboa, C. O., *Sustainable Value Management for Construction Projects*. Springer, Switzerland, 2017.
O'Mara, M., and Bates, S., *High-Performance Green Buildings Solutions, Global Segment Lead – Education & Smart Campus Solutions, Utilizing Smart Building to Reduce Environmental Impact, Ensure Occupant Satisfaction, Drive Better Business Outcomes, and Maximize Asset Value, Why Invest in High-Performance Green Buildings?*, 2012.
Naidoo, R., *Corporate Governance, an Essential Guide for African Companies*, Second Edition, 2009.
Porter, M., and Kramer, M., Creating Shared Value, *Harvard Business Review*, 89(1–2), 62–77, 2011.
RSA, *a New Housing Policy and Strategy for South Africa,* Department of Housing: White Paper, 1-57, 1994.
Rothaus, R., Return on Sustainability: Workforce Housing for People, Planet and Profit (ND), 2013.
South African Government, National Environmental Management Act No. 107, NEMA. Retrieved from www.gov.co.za, 1998. Accessed on April 20, 2018.
South African Human Rights Commission: The 4th Annual Economic and Social Rights Report (2000-2002) https://www.gov.za/sites/default/files/gcis_document/201409/4thesr0.pdf Accessed on April 20, 2018.
Wales, T., Organisational Sustainability: What Is It, and Why Does It Matter?, *Review of Enterprise and Management Studies,* 1(1), 2013.

Interdependence between Structural Engineering and Construction Management
Edited by Ozevin, D., Ataei, H., Modares, M., Gurgun, A., Yazdani, S., and Singh, A.

ISBN: 978-0-9960437-6-2

CASE STUDY COMPARISONS OF THE ECOLOGICAL FOOTPRINT ON SOCIAL HOUSING AFTER EARTHQUAKE

MICHAEL DAVIS[1], ENRIQUE VILLACIS[1], M. LORENA RODRIGUEZ[2], CYNTHIA AYARZA[2], DOMENICA DAVILA[1], CRISTINA FERNANDEZ[1], DANIELA JARRIN[1], JOSSELYN NACEVILLA[1], and JAVIER VIZCAINO[1]

[1]*School of Architecture, Design and Arts, Pontifical Catholic University, Quito, Ecuador*
[2]*Ensusitio Architecture, Quito, Ecuador*

This research seeks to distinguish which factors influence the ecological footprint and what types of construction have the least environmental impact in a post-disaster social housing building. The first case study is a government social housing design, built with bamboo and concrete masonry blocks, and another design by Ensusitio, a private practice approach to social housing built with bamboo and earth. These houses were granted to victims of the April 2016 earthquake in the Ecuadorian coastal region. The investigation process was carried out based on primary research, which was used to understand how Ensusitio carried out the construction process of Meche's house and also based on a secondary investigation of government social housing. With this information, a comparison is made between them to determine which of the two has the least ecological footprint.

Keywords: Local materials, Bamboo, Masonry block, Mass production, Meche's house.

1 INTRODUCTION

Wackernagel *et al.* (2002) state that the increase in consumption of energy and materials in the last 40 years has been greater than the growth of the human population, which translates into the ecological footprint of a population that according to this author is the biologically productive area required to generate the resources used and absorb the waste generated by said population. Within the industrial activities, the construction activity is the largest consumer of natural resources (Alavedra *et al.* 1997). On the other hand, the presence of natural disasters has been a constant contribution to the deterioration of the environment and cultural memory of a place. Disasters cause displacement, uprooting of the inhabitants, breakdowns, and destruction of habitable structures and public infrastructures (Gordillo 2004). This was the case after the 7.8 earthquake that occurred in Ecuador on April 16, 2016 (IG-EPN 2016), in which 35,264 buildings were destroyed or needed repairs (Ross 2017). To this end, the National Government through the Ministry of Urban Development and Housing/MIDUVI, international organizations and the private sector, MIDUVI (2014) including Ensusitio´s project, collaborated in the partial reconstruction of the affected areas. Within the MIDUVI programs, there are bamboo and concrete block houses. Ensusitio, on the other hand, is an architecture office that works with a deep concern for the environment and incorporates a thought of coherent social conscience

(Ensusitio 2017a). After the earthquake, Ensusitio chose to help by transferring knowledge. They held a hands-on workshop about good building practices while building Meches House and the house became a model for the community (Ensusitio 2017b).

The objective of this paper is to learn about the ecological footprint of two types of houses and compare them to know which has the least impact and see how the type of construction significantly influences these results.

2 METHODOLOGY

The research will be based on the ecological footprint in hectares per house of two types of post-earthquake houses: the traditional concrete block, self-standing, pile dwelling, and Meche´s House. Based on secondary research of scientific and academic articles we will obtain the ecological footprint of the materials that were necessary to build each of these houses in order to obtain the total ecological footprint of the construction and budgets in which the quantities of each material used for its construction are detailed. The ecological footprint of the transport was obtained on the basis that one hour of travel of a truck emits 0.18 tons of CO_2 (Bellart and Mesa 2009). Hours needed for travel in each house were quantified and calculation was made based on this figure. In case of Casa de Meche, there is more precise information regarding transportation of materials, therefore the information provided by Ensusitio office was taken into account.

3 ECOLOGICAL FOOTPRINT OF THE SELF STANDING CONCRETE BLOCK PILE-DWELLING (WITHOUT FOUNDATIONS)

According to Crespo (2015), the weight of a pressed block that is 40 cm long, 20 cm high, and 15 cm thick, is 10 kg. Therefore, 10 kg of mass concrete is required for its manufacture. Based on this data, we determined that the ecological footprint of the concrete required for a pressed block is 0.00064 ha/unit. The ecological footprint of the blocks used in this house is the multiplication of the ecological footprint of 1 block and the number of blocks required for construction of the pile dwelling, which is 1,150 units, data extracted from the tender issued by the MIDUVI (2014).

In general, conventional reinforced concrete has a range of unit weights ranging from 2240 to 2400 kg/m³ (De Guzmán 2001). According to this information, we proceeded to calculate the ecological footprint of this material when obtaining the volume of reinforced concrete from the tender issued by the MIDUVI (2014) in the pile dwelling, that is, 8.62 m³ and the weight of the concrete used in it, which is 19.82 tons and multiplying it by the ecological footprint of 1 m³ of reinforced concrete, which is 0.05 ha/ton, according to Bellart and Mesa (2009). The amount of mortar required in a square meter of the wall is 0.045 tons (Arroyo 2010). To this was added the weight of the plaster, which is fine plaster, so that both quantities were combined; this data was multiplied by the ecological footprint of 1 ton of mortar, obtaining the ecological footprint of the total material used in the dwelling, which is 12.47 tons.

According to the volumes of work extracted from the tender issued by MIDUVI (2014) for the construction of pile dwelling, 1288.32 kg of reinforcing steel are required. This figure is multiplied by the ecological footprint of 1 ton of this material (0.58 Ha/Ton); data abstracted from the thesis on *Environmental Impact and Life Cycle of Construction Materials* (Bellart and Mesa 2009) resulted in the ecological footprint emitted by the house.

The roof of the house is made of zinc; it has a thickness of 1.2 mm (MIPSA 2017) and it is required to cover 60 m². With data obtained from the MIDUVI tender (2014), the weight was calculated with the total of the material required for the roof, which is the area to be covered multiplied by the weight of the galvanized steel 18 gauge, which is 9.79 kg/m² (MIPSA 2017). Later, the calculation of the ecological footprint was made, which consists of multiplying the weight of the cover by the ecological footprint of 1 ton of this material. This data was abstracted from the thesis on *Environmental Impact and Life Cycle of the Materials of Construction* (Bellart and Mesa 2009).

Due to the lack of data about the transport of materials, an approximation of trips made in a truck with a capacity of 12 tons is assumed. Therefore, the estimated number of trips, taking into account the number of materials used to prepare the concrete block pile dwelling, is five. According to the Itec Bedec database (2008, 2009), a truck transporting 12 tons of material emits 189.18 kg of CO_2 into the atmosphere. In our calculation, we will consider that six hours of travel are necessary. Therefore, the final emission will be 1 135.08 kg of CO_2. Taking into account that the functional unit is tons of product, the emission per ton of CO_2 is 94.59 kg.

When estimating approximately 5 truckloads, the final emission is 5675.40 kg of CO_2. According to Bellart and Mesa (2009), to weigh the impact of the greenhouse effect, a certain number of hectares or square meters of the forest is needed to counteract the CO_2 that is emitted into the atmosphere annually. Due to photosynthesis, trees absorb CO_2 from the atmosphere, and store it as carbon, expelling oxygen. Growing trees absorb more CO_2 than old trees.

The Center for Ecological Research and Forestry Applications (CREAF 2009) specifies on its website that one hectare of forest absorbs some 5,000 kg of CO_2 annually, with an approximate density of 1000 trees per hectare. Therefore, to counteract the emission of CO_2 produced by the transport of materials from Quito to the province of Esmeraldas for the construction of the block house, 1.14 hectares of forest or an approximate density of 1140 trees are required. The ecological footprint has been calculated by a unit of material. The results are shown in Table 1.

Table 1. The ecological footprint of materials in the concrete block pile dwelling. MIDUVI (2014), Crespo (2015), De Guzmán (2001), Arroyo (2010), MIPSA (2017), Bellart and Mesa (2009).

Materials	Amount	Footprint	Total	Extraction (+)	Total
Mortar	12.47 Ton/m^2	0.084Ha/Ton	0.840 Ha	0.840 Ha	1.78 Ha
Concrete block	1.150	0.064Ha/Ton	0.074 Ha	0.074 Ha	0.15 Ha
Reinforced concrete	8.62m^3	0.050Ha/Ton	0.990 Ha	0.990 Ha	1.98 Ha
Reinforcing steel	1.28832kg	0.580Ha/Ton	0.750 Ha	0.750 Ha	1.50 Ha
Galvanized steel	60m^2	1.000Ha/Ton	0.580 Ha	0.580 Ha	1.17 Ha
				Total:	6.58 Ha
				Transport:	1.14 Ha
				TOTAL ECOLOGICAL FOOTPRINT	**7.72 Ha**

4 ECOLOGICAL FOOTPRINT OF MECHE´S HOUSE

Meche´s House was designed by Ensusitio following the 2016 earthquake. Ensusitio chose to contribute by transferring knowledge and creating a workshop of good building practices and skills, while at the same time building this house. It is mostly built with local materials; however, there are some that were transported from Quito, Atacames, etc., to August 10 in Esmeraldas (Ensusitio 2017c). These materials are industrial, so the carbon footprint represents 50% of the ecological footprint (Samaniego and Schneider 2009), the extraction of the material will constitute the other 50%, which is why in some materials the Ecological Footprint is doubled.

According to Ensusitio (2017a), the material that predominates in this house is bamboo; it needed 220 canes to construct. On the other hand, 660 canes can be harvested per hectare (Tandazo 2012), which shows that the productive area needed for 220 bamboo is 0.33 Ha. The bamboo does not emit CO_2 (Rea 2012), therefore, the ecological footprint of it would be only 0.33 Ha; the extraction process is not taken into account. Ensusitio states that they had to mobilize the cane 42.7 km so that it reaches the site in a van; if a truck emits 1 ton of CO_2 in 3000 km (Medina 2010), and it takes 1 Ha of forest to absorb 5 tons of CO_2 (Bellart and Mesa 2009), then to absorb what is produced by Ensusitio truck, 0.0028 hectares of forest is necessary.

Bahareque was used for another masonry with materials obtained in the area. According to Ensusitio (2017c), the materials they used were 7 parts sand, 3 parts clay, and 7 parts coconut fiber. They mention that the clay was obtained from the site, therefore, it does not have a representative ecological footprint. The extraction of the sand was not in large proportions; therefore, its ecological footprint was not taken into account. Nevertheless, it was necessary to transport it (Ensusitio 2017d). The sand transport was 5 minutes in a van; if a truck emits 1 ton of CO_2 in 33 hours at an average speed of 90 km/h (Medina 2010), in 5 minutes it emits 0.002 ton of CO_2. To absorb 5 tons of CO_2, one hectare of forest is needed (Bellart and Mesa 2009). Therefore, 0.0004 hectares of forest are needed to absorb what the truck produced. For coconut fiber, despite being a representative product of Esmeraldas, already crushed fibers were taken from Quito in a truck. A truck emits 1 ton of CO_2 in 3000 km (Medina 2010), and it takes 1 Ha of forest to absorb 5 tons of CO_2 (Bellart and Mesa 2009). To absorb what is produced by the Ensusitio truck in the 314 km from office to location of the house, 0.02 hectares of forest is necessary. It is important to mention that this could have been avoided since coconut is a material of the area. The ecological footprint is not taken into account since the coconuts used had been already considered waste. For bahareque masonry, it was also necessary to use wire mesh (Ensusitio, 2017a). According to material tables of Ensusitio, 8.32 kg of mesh was necessary. The ecological footprint is 0.58 Ha / Ton. Therefore, the footprint is 0.004 Ha. This is doubled by the production of the material (Samaniego and Schneider 2009), giving a total of 0.004 Ha. Transportation equivalent to 0.18 Ha is added, because the material was mobilized for 5 hours in a truck with 0.18 Ton/hour CO_2 emission and 1 Ha is needed to absorb 5 Ton of CO_2. Therefore, 0.19 Ha are necessary.

Table 2. The ecological footprint of materials in the concrete block pile dwelling. MIDUVI (2014), Crespo (2015), De Guzmán (2001), Arroyo (2010), MIPSA (2017), Bellart and Mesa (2009).

Materials	Amount	Footprint	Total	Extraction (+)	Transport	Total
bamboo	220 bamboo(b)	0,0015 Ha/b	0,840 Ha	-	0.0028 Ha	1.78 Ha
clay	-	-	-	-	-	0.15 Ha
sand	-	-	-	-	0.0004 Ha	1.98 Ha
coconut fiber	-	-	-	-	0.02 Ha	1.50 Ha
metal mesh	60m^2	0.58 Ha/Ton	0.580 Ha	0.580 Ha	0.18 Ha	1.17 Ha
wood	1.51 trees	0.009 Hs/tree	0.014Ha	-	-	0.014 Ha
galvanized roofing	39.6m^2	0.58 Ha/Ton	0.39 Ha	0.39 Ha	0.019 Ha	0.79 Ha
			TOTAL ECOLOGICAL FOOTPRINT			**1.35 Ha**

In the case of wood, 0.697 m^2 were necessary (Ensusitio 2017c). A tree is capable of producing 0.46 m^2 of wood (Nutto and Vázquez 2004), therefore, 1.51 trees are needed to obtain the amount of wood needed for the house. Nutto and Vázquez (2004) also mention that approximately 1100 trees can be harvested per hectare, therefore, productive area needed for a tree is 0.0009 Ha. Growing wood does not emit CO_2. Therefore, the ecological footprint of it would be 0.014 Ha. Wood was obtained from the same place, (Ensusitio 2017c), so there is no ecological footprint for its transport. For the roof, Ensusitio's budget reflects that 39.6 m^2 was needed; this is equivalent to 0.38 ton since weight of 1 m^2 is 9.79 kg / m^2 (MIPSA 2017). According to Bellart and Mesa (2009), the ecological footprint of this material is 0.58 Ha / Ton. Therefore, the footprint of the roof of Casa de Meche is 0.39 Ha, this is doubled by the production of the material (Samaniego and Schneider 2009). Added to this is the transportation that is

0.0194 Ha, since one hour of travel of a truck emits 0.18 Ton of CO_2 and 1 Ha is needed to absorb 5 Ton of CO_2 (Bellart and Mesa, 2009). Total ecological footprint is 1.35 Ha (Table 2).

5 RESULTS

There are several factors to consider in the calculation of the ecological footprint, such as the conditions of the land of implantation and consequently the foundations of the house as well as the mobilization of construction materials to the site. Due to the different terrain conditions in which each house is implanted, we have decided to separate the results.

Based on the research carried out, it is verified that the house with the highest environmental impact without considering the foundations is the MIDUVI Pile Dwelling; this is due to the materials used for its construction and the transportation of the same, resulting in an ecological footprint of 7.72 Ha. The reason for such a high level of impact of the materials is the industrial production of the same, which are manufactured in high quantities for the elaboration of a greater number of houses. The foundations used for this house are plinths that are built on four piles that separate the house from the soil and its ecological footprint is 0.59 Ha, which influences the total ecological footprint of the house, raising the result to 8.31 Ha.

Meche´s House is made mostly with local materials, which cause minimal impact and some of them avoid long-distance motorized transport, however, the site required the construction of a retaining wall and a more reinforced foundation for the house. This generates a great impact on the calculation of the ecological footprint since the concrete and the rods are the materials that have the most ecological footprint. If the foundation is not considered, Meche´s House would have the least ecological footprint, if the house were set on less complex terrain, the foundation would not have such a high ecological value. See Table 3.

Table 3. Ecological footprint comparison.

FOOTPRINT	Concrete block Pile dwelling	Meches house
With out foundation	7.72 Ha	1.35 Ha
Only foundation	0.59 Ha	2.18 Ha
TOTAL	**8.31 Ha**	**3.53 Ha**

6 CONCLUSIONS

Due to the different terrain conditions in which each of the houses is implanted, different types of foundations have had to be used, which significantly influences the calculation of the final ecological footprint. The materials that cause a greater environmental impact are those that require industrialization, as well as the transportation, these measures significantly increase the ecological footprint of a home. This is the reason why the Pile Dwelling presented by the MIDUVI has a greater ecological footprint than those in which local endemic materials or those with less environmental impact are used.

While Meche´s house, has a fairly high ecological footprint, due to the construction of a reinforced concrete retaining wall, so it should consider the possibility of implementing other strategies, replace the use of industrialized materials to save slopes. A retaining wall made with tires, which reduces the environmental impact due to the reuse of scrap tires, and is used in the lifting of slopes, this is an example of a strategy that generates a less impact than the use of reinforced concrete and satisfies the same need.

The use of local materials decreases the environmental impact because transportation is avoided. This is a practice that should always be considered when planning a construction project. Search for and implement different strategies, which reduce the ecological footprint, by replacing industrialized materials, by recycling or other options, such as the use of a retaining wall made with recycled tires, replacing the reinforced concrete retaining wall; this strategy reduces the environmental impact of the concrete and in turn reduces the monetary cost of the work.

References

Alavedra, P., Domínguez, J., Gonzalo, E., and Serra, J., *The Sustainable Construction: The State of the Issue,* Informes De La Construcción, 49(451), 41-47, 1997

Arroyo, C. J. A., Mortars Long Life: Design and Application. Retrieved from http://repositorio.usfq.edu.ec/handle/23000/752, 2010.

Bellart, M., and Mesa, S., Environmental Impact and Life Cycle of Construction Materials. Retrieved from https://upcommons.upc.edu/bitstream/handle/2099.1/7360/pfce%202009.110%20mem%C3%B2ria.pdf?sequence=1&isAllowed=y in 2009.

CREAF, Ecological Research and Forest Applications Center. Retrieved from http://www.creaf.uab.es in 2009.

Crespo, S. L., *Construction of Masonry Block, Through the Prefab of Macro Modular Elements*, 2015.

De Guzmán, D. S., *Concrete and Mortar Technology*, Pontifical Javeriana University, 2001.

Gordillo B, F., *Transitory Habitat and Emergency Housing.* Tabula Rasa, humanities magazine (2), 145-166, 2004. Retrieved from https://vdocuments.site/habitat-transitorio-y-vivienda-para-emergencias.html in 2017.

Ensusitio, Meche's House, (D. Jarrín, J. Vizcaino, Nacevilla, & Dávila, Interviewers), 2017a.

Ensusitio, La Casa de Meche: taller de buenas prácticas constructivas. Retrieved from http://ensusitioarq.com/casa-meche.html, on May 30, 2017b.

Ensusitio, Workbook of the Workshop: Good Practices in Construction, Meche's House, Quito, Ecuador. Retrieved from https://issuu.com/ensusitioarq/docs/manual_issuu in 2017c.

IGEPN, An earthquake of April 16, 2016. Special Seismic Report No. 19 - 2016. Geophysical Institute of the National Polytechnic School of Ecuador. Retrieved from http://www.igepn.edu.ec/servicios/noticias/1330-informe-sismico-especial-n-19-2016 in 2016.

MIDUVI, *Strategic Planning*, Ministry of Urban Development and Housing. Retrieved from: http://www.habitatyvivienda.gob.ec/el-ministerio/.2014 in 2014.

MIPSA, Smooth galvanized sheet. Retrieved from https://mipsa.com.mx/dotnetnuke/Productos/Lamina-galvanizada-lisa in 2017.

Medina, J., *The carbon dioxide diet (CO2).* Conciencia Tecnológica, 50-53, 2010.

Nutto, L., and Vázquez, M. C. T., *High-Quality Sierra Wood Production with Eucalyptus globulus*, CIS-Madera, 12, 6-18, 2004.

Rea, L. V., *Use of Bamboo as a Construction Material: Environmental Assessment against Traditional Construction Systems,* Technical architecture. Polytechnic University of Madrid, 2012.

Ross B., F. A. *Disaster Risk Management with an Emphasis on Prevention*; Case study: *Earthquake April 16, 2016, in Ecuador*, MSc Thesis, University of Barcelona, Barcelona, 2017.

Samaniego, J., and Schneider, H., *The Carbon Footprint in The Production, Distribution, and Consumption of Goods and Services,* Santiago, Chile: Economic Commission for Latin America and the Caribbean (ECLAC), 2009.

Tandazo J., Process of Industrialization of the Bamboo as an Alternative Material for The Construction and Design of One and Two-Story Type Houses, Using Bamboo in Its Structural Elements. Retrieved from https://repositorio.espe.edu.ec/bitstream/21000/5892/1/T-ESPE-034425.pdf.2012 on 2012.

Wackernagel, M., Schulz, N., Deumling, D., Callejas, A., Jenkins, M., Kapos, V., et al. *Tracking the Ecological Overshoot of the Human Economy.* In H. U. Edward O. Wilson (Ed.), Proceedings of the National Academy of Sciences USA, Cambridge: PNAS, 9266-9271, 2002.

Interdependence between Structural Engineering and Construction Management
Edited by Ozevin, D., Ataei, H., Modares, M., Gurgun, A., Yazdani, S., and Singh, A.

ISBN: 978-0-9960437-6-2

MECHE´S HOUSE OR "BEING IN THE WORLD": A POST-DISASTER COMMUNITY BUILDING EXPERIENCE

FERNANDO HINOJOSA[1], ENRIQUE VILLACIS[1], MARIA LORENA RODRIGUEZ[2], and CYNTHIA AYARZA[2]

[1]*School of Architecture, Design and Arts, Pontifical Catholic University, Quito, Ecuador*
[2]*Ensusitio Architecture, Quito, Ecuador*

On April 16, 2016, Ecuador had suffered a 7.8 earthquake. The experience designing and building of Meche´s house, a post-disaster alternative building process, is analyzed in this paper applying Martin Heidegger´s concepts of Dwelling, Being in the World and Being There. These allow better understanding to the idea of housing in the community where the object is built, to visualize its most symbolic space, also it enables to see how the users of the built dwelling make it their own, the construction, form and materials despite that in the beginning of the construction process they did not accept them. This article tries to explain two questions: The house of Meche reflects its habitat: Where is it built? And how Meche, who is the homeowner, and the participants make the house and experience their own, in order to replicate the appropriate construction and resilience.

Keywords: Architecture, Culture, Heidegger, Technic.

1 BUILDING IN COMMUNITY

In this article, the study approaches the experience towards what and how the object is produced. The study does not explain the construction but evaluate how the human being turns toward an authentic dwelling. The area is very rich and generates solid spatial experiences, such as the *casa montubia* (rural house on hillsides in the coastal area). These are pictured in the ceramic objects made by the ancestors and the form repeated in various spatial symbols and concepts such as the vertical relationship between heaven and earth. Duality in the dwelling space, with clearly defined active and passive keep clearly defined spaces in the local houses and seen in the built architectural object.

2 METHODOLOGY

To analyze the experience of the process of Meche's house we will use Heidegger´s concept "being-in-the-world" (Heidegger 1996). The proposed architectural object, the experience of construction with and by the community, the social environment of the house and the culture it reflects are very specific conditions. These will be analyzed objectively both for the user of the house and the community of the town 10th of August. Using Heidegger´s concepts and philosophy of living, allowing the study to visualize the experience from the "being", which is Heidegger's philosophy. The House and the culture it reflects are very specific conditions that enhance in an objective manner both the users of the House and the community of the town of 10th of August.

Heidegger's philosophy, his concepts, and his philosophical vision allow the study to analyze and interpret the experience of the construction of Meche's house in through the culture of the place, its technology, space configuration and materials used.

For Heidegger, the house is a "poetic dwelling", this concept permits approaching the art of building and landscape that is space, as a generator of human life, and as a work of art. The building makes "something", which is "truth", as well as a work of art, "it preserves the truth" (Heidegger 1994a).

When talking about "dwelling", Heidegger says, "it seems that we dwell, only by means of building. The aim of the act of building is to inhabit "(Heidegger 2015), this encourages direct observation of the built environment by the human being, and in turn results in better ways to understand and solve current architectural production. When Heidegger states that "Building is not the only way to inhabit, for a building is in itself already inhabiting" (Heidegger 2015) referring directly to the art of building, and Meche's House is that, construction by and for the user.

Heidegger's philosophy, his concept, and his vision indicate a way to analyze and interpret the experience of constructing Meche's House taking into consideration the local culture and the place and so incorporate the concept of "to be in the world".

3 HOW DOES MECHE'S HOUSE REFLECT THE ENVIRONMENT WHERE IT IS BUILT?

The house is located, replacing a space of sticks with plastic, where Meche had lived. As an earthquake victim, she had expected to get government assistance for a "normal" house for the area: concrete block, cement, iron rods, among other materials, all alien to the historical culture of the area. See Figure 1. Meche was reluctant to accept a house built with bamboo, perhaps the material was associated with the poor. There was disagreement due to the materials used due to shame of using traditional materials. She fought and got angry because she was ashamed to use traditional materials. However, after living in the house, the way of living was different, and a client specific environment developed. Meche created a space of her own a space for being her own being and her presence, which generates a space for being. "Because man becomes free just to the extent that he belongs to the region of fate and in this way becomes one who listens, but not a submissive and obedient listener." (Heidegger 1994b)

Figure 1. Left: Meche's House. Right: "typical" houses. Town of August 10. Access to Pedro Carbo.

The house articulates and generates unity composed of family life, the social environment as well as the natural and built environment. The way of life and customs strengthen the world of family and people who can meet and fulfill themselves there.

Several links can be observed between the house and the human being: it activates several connections that exist between the house and ancestral memory. The ancestral memory is incorporated in the present thus reserving, reviving, what was forgotten and revert a process of loss and oblivion. The house is the truth, and the logic and is neither consumerism nor a copy.
This house displays the raw materials increasing their natural value, see figure 2, it opens up a world and generates a history "the totality of the entity that changes "in time", it is understood that unlike nature, that it also moves "in time". The house follows the transformations and vicissitudes of individuals, groups and "culture". "But this would also occur in order to increase the dignity of being. For the sake of being-here that underlie human existence and not so man or civilization and culture would be valued through work." (Heidegger 2000)

Figure 2. Partially enclosed porch area is semipublic, and the back is private. All-natural materials.

This construction proposal is rooted in the historical cultural references of the place, such as the cocoa farmers, who joined into associations, have made cocoa a symbol of joint work effort and therefore an economic generator for society.

It must be stated that as cultural historical references are also a fundamental basis for fishermen who in their daily work bring back thousands of years of tradition and in their work have generated deep cultural practices that directly affect the cultural and social practices of the community.

Our question "What?" has four components. In the first place, the object makes the ancestral being present. Second, the object shapes the human destiny. Third, the object materializes a memory that is always waiting to manifest itself permanently. Finally, the object makes visible all the things of the earth: the stone, the clay, the sea, the air, the plants, the animals and even the light of day and the darkness of the night. In general, the architectural object and the technique "open a world and at the same time puts it back on the ground". In doing so, truth is put to work. "Technique is not a mere means, technique is a way out of the occult. If attention is paid to this, a totally different region will be opened for the essence of the technique. It is the region of disclosure, that is, of truth. (Heidegger 1994b)

The architectural - constructive proposal lays bridges to several traditional spaces, such as this space called Portal (porch) see Figures 2 and 3, where the inhabitants of the house meet to exchange experiences of life, it is a space that generates an opening towards the outside as a border between the public and the private, however in the architectural proposal Meche´s House, this space is covered in its limits, with a mesh of chopped cane, which generates a more intimate space and is used by Meche to generate her special space, and is in an indirect relationship with the House itself.

Figure 3. Porch: beauty parlor.

Though the porch has doors to make this space more private, Meche keeps them wide open during the day, so that her work activity, the hairdressing salon, is seen, and at night closes it to use it as a private room of the house.

The natural material; bamboo and clay following a course from nature toward the identity of a woman who has unknowingly been linked to what is natural since birth and brings us closer to the memory that lies in this "seeing" and goes into the intimacy of its presence. Heidegger defines this "something" as "truth", the building therefore "preserves the truth".

As shown in figure 4, the texture and color of the bamboo, open this experience toward memory makes this space visible. It contrasts with the surrounding buildings and this makes it worthy to be there, because it fulfills its role and appears as what it is: Meche´s family, space and community together are what the Greeks called *physis*, which is what establishes a human dwelling; Meche calls it her land, the *Pachamama* (mother earth in native kichwa language) that welcomes and protects us.

Figure 4. Rear Façade: opens to views.

The house opens a world, in it is found joy and security for Meche and her family, making them feel their land is their own. It is the temple, by the mere fact of rising there permanently; it is what gives things their appearance and people the vision of themselves. See figure 5. "Earth and sky, the divine and the mortal, forming a unity from themselves, belong to each other from the simplicity of the unitary Quaternity. Each of the Four reflects in its own way the essence of the others. With this, each one reflects himself in what is his and within the simplicity of the Four." (Heidegger 1994a).

Figure 5. Meche and her family: in their new home.

Through the analysis of the How, Meche´s house generates boundaries between the town's architectural works, forms with lack of both technical and climatic coherence, and the work itself that generates a place that approaches the self. The built place appropriates a hidden meaning that is revealed due to the built object, the place no longer is the same and the destiny of the town 10th of August, is already intimately linked to the created dwelling.

It is important to state that, the building site relates with the surrounding land. Erected in the sky, off the ground, this gesture allows to fully display the constructed object, which leads to the origin that, "meaning here from where and why a thing is what it is and as it is. What is something and how it is, is what is called its essence. "In the thinking of being, we never re-present only something real and we do not give as true what we have represented. To think "being" means to correspond to the interpellation of its essence " (Heidegger 1994a).

The origin of something is the source of its essence. It creates its own world. World and thing are, therefore, "Thinking the thing as a thing, we take care of the essence of the thing taking it to the region from which this essence. Making something is bringing the world closer. Approaching is the essence of closeness. To the extent that we take care of the thing as a thing, we inhabit the proximity" (Heidegger 1994a).

4 CONCLUSIONS

Meche´s house is the clear example of Heidegger´s ideas applied in the present. Meche speaks of changing attitudes toward the project soon after its completion. Now it is admired by the people of the town and is visited by local people. Tourists also visit it giving the house a prestige that becomes part of it. The truth manifested in spatial distribution, through the use of materials and their relationship to the environment. The initial reaction has been reversed. The local people rejected the house at first, now they accept it and so it is "in the world" completely.

Thus, the shape of the house, the handling of solid and open walls and the use of materials, give this building its own characteristics, some create more or less intimacy and others are linked to the land and landscape. Meche's life is generated in the earth and her house demonstrates this fact. The landscape is enriched by the natural materials of the house. The house contributes to the regeneration and destiny of this damaged environment. The utilitarian work, even books, has their "destinies"; while architectural works and institutions have their history.

Nature is also historical, although not precisely in the sense of "natural history", "History here does not mean the succession of certain events within time, however important they may be.

The house is a unit and unveils a world of combinations and it also expresses the un-concealment of being. Memory is the source of the strength of Meche´s house, it is a reminder to live fully, and it relates to Heidegger`s poetic dwelling, because the house is already an image and possesses the quality of dwelling.

Meche´s beauty parlor, part of the house, is her main source of income. It is named *Las Oyeritas* (clay pots or from the *oyerita*, a local bird that makes mud nests), according to Heidegger's thoughts, man lives between work and the word. The word opens the world, the work makes the world present. Human beings are alive in the word and are reflected in the essential Being that inhabits it and belongs to the dwelling.

Design work and construction requires technicians to have the absolute sensitivity in order to express the real needs of the user-owner, and the community. This almost causes the technician to disappear, so the human being appears in this process. This enables the construction process to be replicable and requires the efficient use of materials. Every nail counts; one must build houses of the XXI century out of local materials using ancestral techniques and meaning that challenges current architecture.

Meche´s House is of great importance for the family that inhabits it. It is maintained in a special way both in its form and in its essence, according to this philosophy. The space is built not only by the owner, Meche, but also by the team of architects and the community, that in their labor, transform and are transformed. Constructions, architectural objects, are those things built, that brings together a world in themselves and allow the full inhabitation of the human being.

References

Heidegger, M., *The Thing,* (E. Barjau, Trad.) Editorial Del Serbal, Spanish Translation, 1994a.

Heidegger, M., *The Question by Technique*. Barcelona, Editorial Del Serbal Spanish Translation. 1994b.

Heidegger, M., *Being and Time* (R. Editorial, Editor), Philosophia.cl: 1996. Retrieved from http://www.philosophia.cl. on December 2017.

Heidegger, M., *Letter on Humanism*. Alliance. 2000.

Heidegger, M., *Build Habitat Think*, In T. J. Escudero, & R. Editorial (Ed.), (J. Eduardo, Trad., Page 85). Madrid, Spain: Office of Art and Publishing. 2015.

Interdependence between Structural Engineering and Construction Management
Edited by Ozevin, D., Ataei, H., Modares, M., Gurgun, A., Yazdani, S., and Singh, A.

ISBN: 978-0-9960437-6-2

APPLICATION OF LEAN METHODS FOR PRODUCTIVITY IMPROVEMENT IN STEEL FABRICATION SHOPS: A CASE STUDY

MANUEL DUFRESNE, HO-YAN MOK, REZA SHAKERI, MOHAMED ELBARKOUKY, and VISHAL SHARMA

Northern Alberta Institute of Technology (NAIT), Edmonton, Canada

The objective of this paper is to identify tools and methodologies to increase productivity in commercial steel fabrication shops. The effectiveness of these tools and methodologies has been demonstrated through a case study in a Canadian steel fabrication shop. Three different productivity improvement concepts were identified as suitable for the steel fabrication environment. Lean, Six Sigma, and the Theory of Constraints (TOC). Lean was retained as being the most suitable to steel fabrication due to its low cost, focus on quality, and built in continuous improvement process. In the case study, Lean tools were identified and applied to address three top productivity issues. A baseline was established for existing productivity for a fitting station over a period of two weeks. Selected tools were implemented, and a session was conducted to educate the workers on the new processes and tools. Lastly, the productivity was re-calculated, the results were analyzed, and recommendations were made.

Keywords: Building construction, Six sigma, Theory of constraint, Optimization.

1 INTRODUCTION

There are approximately 2,500 steel fabricators in Canada. Due to expensive field labor and extreme weather on construction sites, shop fabrication is very common in Canada as the production can be carried out in a controlled environment (Ahmad *et al.* 2014). Steel fabrication shops are profit-oriented organizations, hence improving productivity is a major concern since it represents the effective and efficient conversion of resources into marketable products and determines business profitability (Sandbhor and Botre 2014). Since most of the local steel fabricators work with the same suppliers, subcontractors, and local conditions, the playing field is even. Hence, decreasing the amount of time spent on a piece of steel is the most effective way to help the fabricator increase productivity and win the next job. In addition, productivity is one of the most important factors to increase economic growth in a manufacturing environment, which is similar to a steel fabrication environment. Kwiatkowski (2016) concluded that decreasing the amount of time spent on a piece of steel is the most effective way to help the fabricator to increase productivity and win the next job. Because high productivity is one of the keys to financial success of steel fabricators (Kwiatkowski 2016), it is essential for the steel fabricator to understand the application of productivity improvement concepts and tools.

The aim of this research is to identify tools and methodologies to increase productivity in steel fabrication shops. This aim was achieved by implementing the research methodology in five steps. Step 1 involved the systematic identification of tools and techniques, which can be used to improve the productivity in the steel fabrication shop. Step 2 involved creation of fabrication shop process map (used as case study) and establish a current productivity baseline in selected work station. Step 3 involved analysis of selected work station to identify and rank the productivity issues. In this step, selected productivity improvement tools were also implemented. Step 4 involved re-measuring the productivity rate of the station after implementation of productivity improvement tools. Step 5 involved the analysis of productivity improvements, if any, based on collected data and make recommendations.

2 STEEL FABRICATION PROCESS

Construction steel fabrication industry, though it operates in a controlled environment like manufacturing industry, encounters the challenges similar to construction of a facility. For instance, every steel fabrication project is unique. Accordingly, steel fabrication projects manifest the hybrid nature between manufacturing and construction projects. All steel pieces are unique in steel fabrication projects, but they are fabricated in a shop under a controlled environment. A typical process map of a steel fabrication shop is in Figure 1.

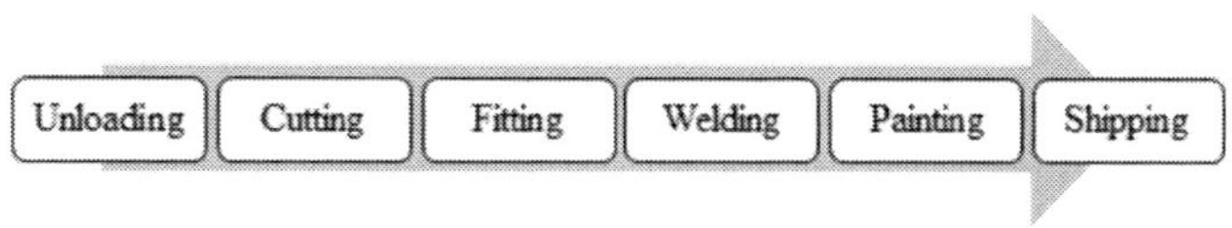

Figure 1. Current work flow process for a typical fabrication shop.

2.1 Productivity Measurement Units

According to Hofacker and Gehbauer (2010), hours per ton is the most common measure of steel fabrication productivity due to its simplicity. On the other hand, Song and AbouRizk (2008) argued that there is no consensus in how to measure productivity in steel fabrication environments since different shops will have different means of measuring productivity, which complicates the comparison process. In this research, "hours per ton" and "pieces per day" were retained as productivity measurement units.

2.2 Productivity Improvement Theories

2.2.1 *Theory of constraints (TOC)*

TOC is a methodology for identifying the most important limiting factor that stands in the way of achieving a goal and then systematically improving that constraint until it is no longer the limiting factor. In manufacturing, the constraint is often referred to as "a bottleneck." TOC takes a scientific approach to improvement. It hypothesizes that every complex system, including manufacturing processes, consists of multiple linked activities, one of which acts as a constraint upon the entire system such as the constraint activity is the "weakest link in the chain".

2.2.2 *Six sigma approach*

Six Sigma is a disciplined, data-driven approach and methodology for eliminating defects in any process by driving toward six standard deviations between the mean and the nearest specification limit. Six Sigma is a disciplined, data-driven approach and methodology for eliminating defects in any process by driving toward six standard deviations between the mean and the nearest specification limit. The fundamental objective of the Six Sigma methodology is the implementation of a measurement-based strategy that focuses on process improvement and variation reduction.

2.2.3 *Lean Thinking*

Lean Thinking is a set of principles and methodologies for improving cycle times and quality through elimination of waste, sometimes known by its Japanese name of Muda (waste). Lean Thinking allows businesses to distinguish between value-added and non-value-added activities. The objective is to improve cycle times, reduce waste, and increase value to the customer. Lean Thinking has been shown to gain dramatic benefits in organizations. Organizations are able to sustain production levels with half the personnel, improving quality and reducing cycle times from 50 percent to 90 percent (Womack and Jones 1996).

One of the similarities between these improvement theories is that their common first step is to identify the existing processes, but they have their own focus on how to improve those processes. Nave (2002) summarized that the selection of a process improvement methodology is dependent on the culture of the organization. Six Sigma will be a perfect program if the organization values analytical studies and the relationship of data. Lean Thinking is for organizations that values visual change. TOC suits organizations that value a system approach where total participation of all stakeholders is not required. Lean was selected as being the most appropriate to this organization because of its low initial implementation cost, speed of implementation, and overall simplicity.

3 CASE STUDY

The construction steel fabrication industry, though operating in a controlled environment like the manufacturing industry, still encounters challenges similar to the construction industry. For instance, every steel fabrication project is unique.

3.1 Process Mapping and Productivity Baseline

Interviews with the shop foreman revealed that the fitting station was the most time-consuming activity in the shop, where the fitter perform all of the coping, drilling, and fitting tasks. Process of current state was charted (Figure 2) and productivity was measured (Table 1) for two weeks, as baseline measure, by having the fitter fill out a production sheet on a daily basis.

Table 1. Average productivity of a fitting station (before implementation of process improvements).

	Weekly Average Productivity (Hrs/Ton)	Weekly Average Productivity (Pieces/day)
Week 1	4.7	9.25
Week 2	5.2	8.75

3.2 Identify and Rank the Productivity Issues

Based on collected data, observations and interviews, research team identified and ranked different productivity issues related to the fitting station. Top issues are listed below:

1. Wait for shared tools.
2. Time spent in searching for parts. Distance from parts storage location.
3. Shop culture: Resistant to change from the foremen to workers.
4. Wait for the crane because it is shared with other stations.

5. Wait for foreman for drawing clarifications.
6. Work Schedule: Dynamic changes in work schedule to meet emerging client needs.
7. Raw material supply delays: due to dynamic schedule
8. Machine or equipment downtime.
9. Lack of welding screens in the shop caused weld flashes, hence slowing the work.

3.3 Select and Implement Productivity Improvement Tools

Team identified and implemented three different lean tools in the steel fabrication shop to address the identified productivity issues namely, optimization of work process, 5S, and Kaizen. It was decided that some steps in the fitting station did not add value and would be better performed by general laborer. This was implemented in optimized work process map, shown in Figure 2.

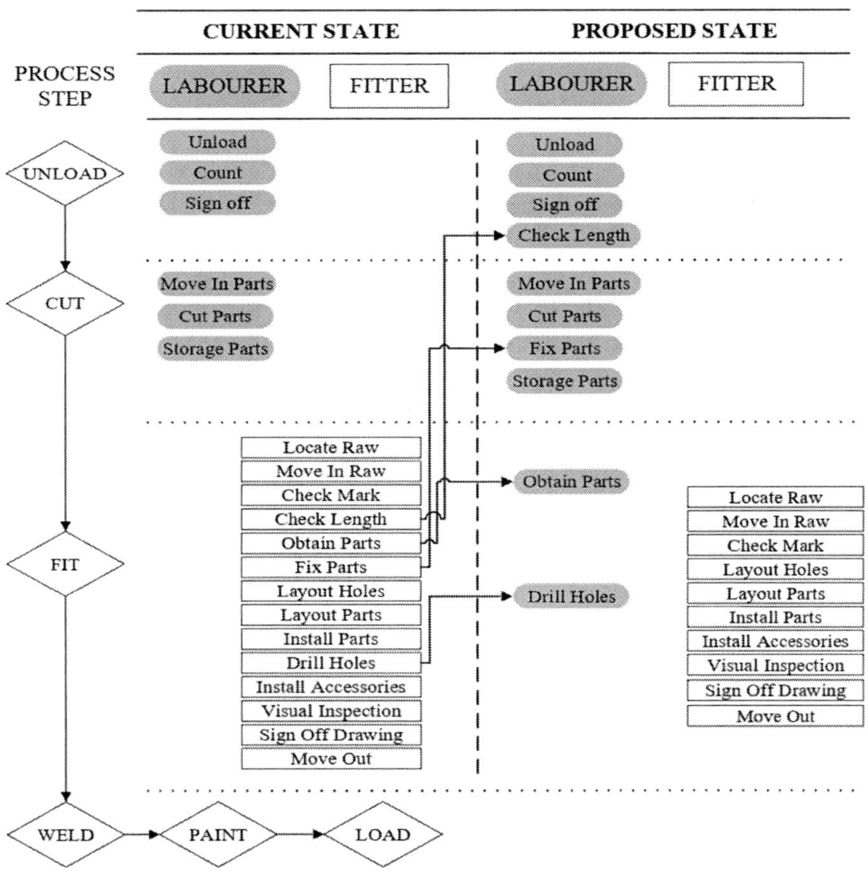

Figure 2. Current and optimized work flow process for Fitting Station.

Second productivity improvement tool used in this research was 5S. It was hypothesized that implementation of 5S will reduce lead times and cycle times of the fitting station, including the time to search the tools. Details of 5S implementation were eliminated due to space constraints.

Third productivity improvement tool used, to improve the culture to accept the change, was Kaizen. To make sure the office and shop employees understand the concept of Kaizen, the research team organized a training session with ten workers from different stations, one project manager, one foreman, and the vice president of operations. Topics of the training session included the origin of continuous improvement, the focus on small improvements the balance between quality and safety, and the continuous productivity improvement of their work.

3.4 Findings

After the implementation of Lean tools and methodologies, the productivity was re-calculated based on 'hours per ton' and 'pieces per day' (Table 2). When comparing the baseline with the new set of productivity data, two conclusions can be made. Firstly, the lean tools were successful in increasing the productivity by either increasing the pieces fitted per day or decreasing the hours spent on each ton of steel. Secondly, when the fitter is working on heavy pieces, the productivity is better harvested as 'hours per ton', but when the fitter is fabricating several light items, 'pieces per day' is more accurate measure.

Table 2. Average productivity of a fitting station (after implementation of process improvements).

	Weekly Average Productivity (Hrs/Ton)	Weekly Average Productivity (Pieces/day)
Week 3	4.1	7.25
Week 4	9.2	12.75

4 CONCLUSION

The Lean Thinking application is the focus of this research. It is recommended to combine its use with the other two improvement theories, Theory of Constraints and Six Sigma. Each of these improvement theories drives toward a common goal but from different perspectives. Combining the use of these theories can provide a larger set of tools that can be applied to each specific productivity improvement issues in a unique fabrication shop. This research validates the fact that lean tools can be used to increase productivity in a steel fabrication shop environment. Further study should be conducted to identify the best productivity measurement unit, which would take complexity of work in consideration. The implementation of lean tools in the stations must be made in a careful manner to not to create new bottlenecks.

References

Ahmad, K. I., Shrivastav, R. L., Pervez, S., and Khan, N. P., Analysing Quality and Productivity Improvement in Steel Rolling industry in Central India, *IOSR Journal of Mechanical and Civil Engineering*, 06-11, March 2014.

Hofacker, A., and Gehbauer, F., *The Tonnage-Flow*, Proceedings IGLC-18, 558-569, Technion, Haifa, Israel, July 2010.

Kwiatkowski, P., The Impact of Lean Manufacturing on Structural Steel Fabricating, thefabricator.com, January 2016. Retrieved from http://www.thefabricator.com/article/materialshandling/the-impact-of-lean-manufacturing-on-structural-steel-fabricating on Aug 01, 2018.

Nave, D., How to Compare Six Sigma, Lean and the Theory of Constraints, lean.org, March 2002. Retrieved from https://www.lean.org/Search/Documents/242.pdf on Aug 01, 2018.

Sandbhor, S., and Botre, R., Applying Total Interpretive Structural Modeling to Study Factors Affecting Construction Labour Productivity, *Australasian Journal of Construction Economics and Building*. 14(1), 20-31, 2014.

Song, L., and AbouRizk, S. M., Measuring and Modeling Labor Productivity Using Historical Data, *Journal of Construction Engineering and Management*, 134(10), 786-794, 2008.

Womack, J. P., and Jones, D. T., Beyond Toyota: How to Root Out Waste and Pursue Perfection, *Harvard Business Review*, 74(5), 140-158, 1996.

Interdependence between Structural Engineering and Construction Management
Edited by Ozevin, D., Ataei, H., Modares, M., Gurgun, A., Yazdani, S., and Singh, A.

ISBN: 978-0-9960437-6-2

3D PRINTING IN CIVIL ENGINEERING: A SYSTEMATIC LITERATURE REVIEW

IBRAHIM ENGIN OZTURK and GOZDE BASAK OZTURK

Dept of Civil Engineering, Aydin Adnan Menderes University, Aydin, Turkey

New technological developments determine the course of the construction industry. Three-dimensional (3D) printer technology has become one of the technologies that determine the production logic of the 21st century. The concept of Industry 4.0 gained importance in the construction industry with a 3D printer. In this study, a systematic literature review was performed. The publications related to the 3D printer have been examined within the scope of civil engineering implementations. The 3D printer researches associated with BIM, Industry 4.0, NASA, additive manufacturing, and point cloud technology subjects was reviewed through an intensive systematic literature review. As a result of the study, it has been understood that the studies in the field of civil engineering were insufficient compared to manufacturing industry. For this reason, the number of researches on 3D printer applications should increase in construction industry.

Keywords: BIM, Point cloud technology, Industry 4.0, NASA, Additive manufacturing.

1 INTRODUCTION

3D printer technology was a complex, expensive and predictable technology in 1984 when it was discovered. However, it is used in many industries such as medicine and manufacturing in today's newly revolutionized industrial environment (Hager *et al.* 2016). Successful applications medicine and other industries attracted the interest of different sectors and fields of science. The three-dimensional printing of concrete is currently being implemented in construction industry (Tay *et al.* 2017). Today, buildings can be printed with a three-dimensional printer with the developments in technology and engineering (Feng *et al.* 2015).

2 MATERIAL AND METHODS

3D printing technology researches related to BIM, Industry 4.0, NASA, additive manufacturing, point cloud technology have been reviewed in this study (Figure 1). The research trend that is connected to 3D printing in the field of civil engineering is assessed. In addition, NASA's 3D printing technology researches, which developed progressive technologies in the construction sector in order to build extraterrestrial habitats, have been reviewed. The Web of Science database was reviewed in all disciplines between 1983-2019 without any other restrictions. Literature Review; TI = (3D printing); TS = (3D printing AND BIM); TS = (3D printing AND Industry 4.0); TS = (3D printing AND NASA); TS = (3D printing AND additive manufacturing); TS = (3D printing AND point cloud technology). The screening was carried out in the English language papers. The publications were examined according to the disciplines and years.

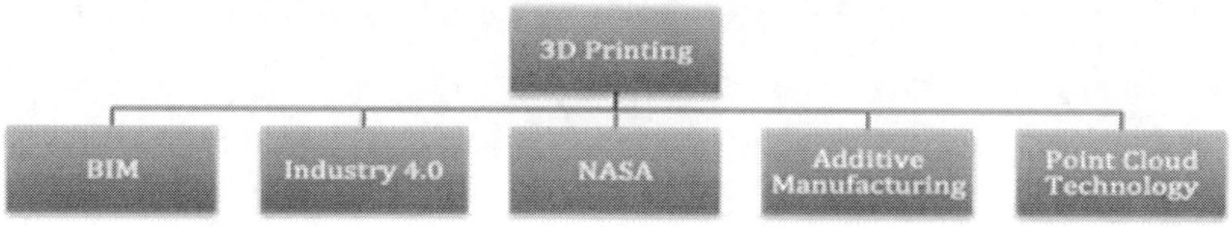

Figure 1. 3D printing related researches.

3 RESULTS AND DISCUSSION

5104 publications containing the word "3D printer" in title were found in the Web of Science scan. The first ten Web of Science categories with the largest number of publications are listed in Table 1. The number of publications containing "3D printing" phrase in the title in Web of Science through 1983 and 2019 is given in Figure 2. The number of publications in the field of "Civil Engineering and Construction Building Technology" is given in Table 2.

Table 1. The distribution of "3D printing" titled articles and proceedings in Web of Science categories.

Web of Science Categories	Number of Articles and Proceedings
Materials Science Multidisciplinary	1084
Electrical and Electronics Engineering	784
Physics Applied	514
Nano-science Nanotechnology	448
Manufacturing Engineering	371
Biomedical Engineering	368
Chemistry Multidisciplinary	353
Materials Science Biomaterials	302
Mechanical Engineering	299
Optics	290

When we consider all the disciplines on a global scale, the number of publications related to 3D printing increases compared to the years. The highest number of publications on 3D printing was in the field of "Materials Science Multidisciplinary." Globally published publications in "Civil Engineering and Construction Technology" were less than the other engineering areas. For example; the total number of publications in the field of "Civil Engineering and Construction Technology" is 93, and total number of publications in the field of "Electrical and Electronics Engineering" is 783, the ratio of latter to former is 11.8%.

Table 2. The distribution of "3D printing" titled articles and proceedings about civil engineering in Web of Science categories.

Web of Science Categories	Number of Articles and Proceedings
Construction Building Technology	48
Civil Engineering	45

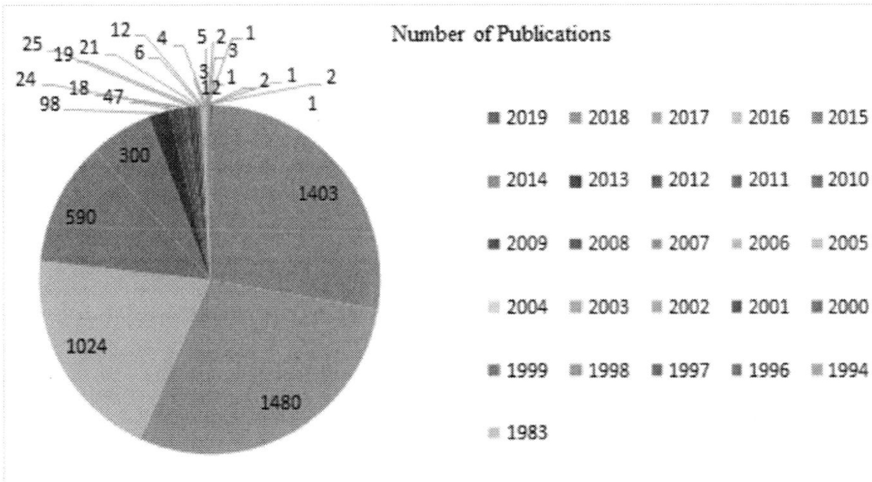

Figure 2. The number of publications containing "3D printing" phrase in the title in Web of Science categories through 1983 and 2019.

3.1 3D Printing and BIM

19 publications containing the word "3D printer" and "BIM" in topics together were found in the Web of Science scan. The first three "Web of Science" categories with the largest number of publications are listed in Table 3. The distribution of publications by year is given in Figure 3.

Table 3. The distribution of "3D printing" and "BIM" topics together in articles and proceedings in Web of Science categories.

Web of Science Categories	Number of Articles and Proceedings
Construction Building Technology	5
Education Educational Research	4
Civil Engineering	4

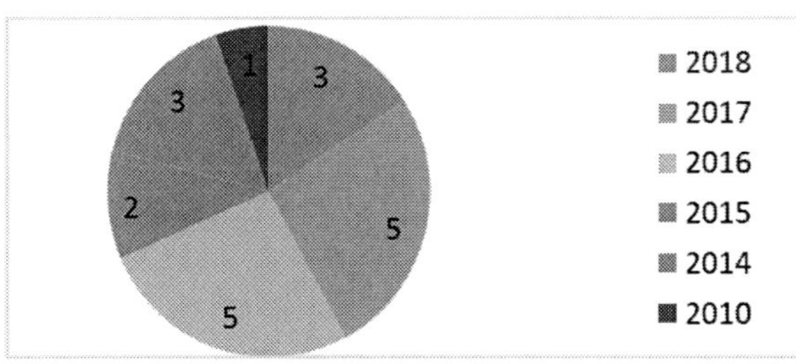

Figure 3. The number of publications containing "3D printing" and "BIM" topics in Web of Science categories through 2010 and 2018.

3.2 3D Printing and Industry 4.0

33 publications containing the word "3D printer" and "Industry 4.0" in topics together were found in the Web of Science scan. The first five "Web of Science" categories with the largest number of publications are listed in Table 4. The distribution of publications by year is given in Figure 4.

Table 4. The distributions of "3D printing" and "Industry 4.0" topics together in articles and proceedings in Web of Science categories.

Web of Science Categories	Number of Articles and Proceedings
Industrial Engineering	10
Manufacturing Engineering	9
Computer Science Artificial Intelligence	5

Figure 4. The number of publications containing "3D printing" and "Industry 4.0" topics together in Web of Science categories through 2014 and 2018.

3.3 3D Printing and NASA

15 publications containing the word "3D printer" and "NASA" in topics together were found in the Web of Science scan. The first four "Web of Science" categories with the largest number of publications are listed in Table 5. The distribution of publications by year is given in Figure 5.

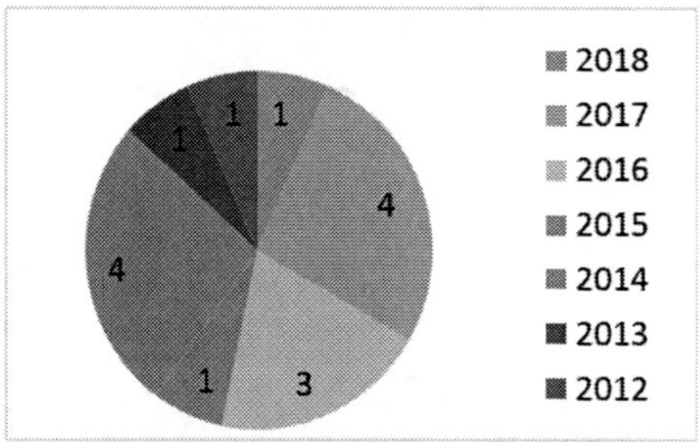

Figure 5. The number of publications containing "3D printing" and "NASA" topics together in Web of Science categories through 2012 and 2018.

Table 5. The distributions of "3D printing" and "BIM" topics together in articles and proceedings in Web of Science categories.

Web of Science Categories	Number of Articles and Proceedings
Electrical and Electronics Engineering	4
Optics	3
Physics Applied	3
Civil Engineering	2

3.4 3D Printing and Additive Manufacturing

2616 publications containing the word "3D printer" and "Additive Manufacturing" in topics together were found in the Web of Science scan. The first three "Web of Science" categories with the largest number of publications are listed in Table 6. The distribution of the publications by years is given in Figure 6. A total 63 publications containing the word "3D printer" and "Additive Manufacturing" in topics together were found in "Civil Engineering" and "Construction Building Technology" categories.

Table 6. The distributions of "3D printing" and "additive manufacturing" topics together in articles and proceedings in Web of Science categories.

Web of Science Categories	Number of Articles and Proceedings
Materials Science Multidisciplinary	696
Manufacturing Engineering	507
Electrical and Electronics Engineering	338

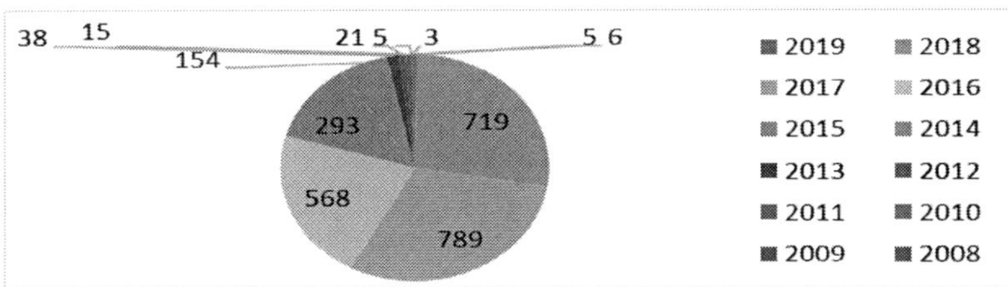

Figure 6. The number of publications containing "3D printing" and "additive manufacturing" topics together in Web of Science categories through 2008 and 2019.

3.5 3D Printing and Point Cloud Technology

24 publications containing the word "3D printer" and "Point Cloud Technology" in topics together were found in the Web of Science scan. The first three "Web of Science" categories with the largest number of publications are listed in Table 7. The distribution of publications by year is given in Figure 7.

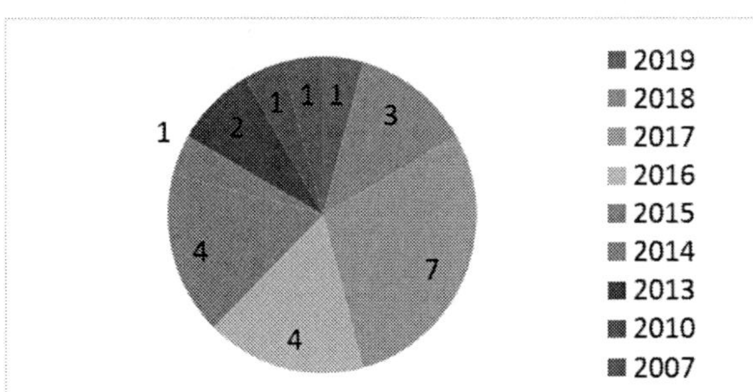

Figure 7. The number of publications "3D printing" and "point cloud manufacturing" topics together in Web of Science categories through 2007 and 2019.

Table 7. The distributions of "3D printing" and "point cloud technology" topics together in articles and proceedings in Web of Science categories.

Web of Science Categories	Number of Articles and Proceedings
Optics	4
Electrical and Electronics Engineering	3
Manufacturing Engineering	3

4 CONCLUSIONS

Scientific research on 3D printing is increasing over time. The print quality and diversity of the printing raw materials will increase by research in the field of Materials Science. The development of this technology alone is not enough to put pressure, for larger, more viable, more robust, more stable prints. Therefore, researchers must study raw materials. Most of the studies in the field of engineering are in the discipline of "Electrical and Electronics Engineering." BIM and 3D printing subjects together are studied more in the field of "Civil Engineering." The publications about the connection between 3D printing and BIM have increased over the years. Building Information Modeling is a prerequisite and necessary for the use of 3D print technology in civil engineering. The results of the systematic literature review on 3D printing, BIM, Industry 4.0, NASA, Additive manufacturing and point cloud technology shed light on the increase in research trends of the subject. NASA, as a pioneer in the 3D printing technology, will shape the future of research on subject. The adoption capacity on 3D printing of construction industry may affect the competitive edge of industry stakeholders. Therefore, an agile response in scientific research and industry adoption is needed in near future.

References

Feng, P, Meng, X, Chen, J-F, and Ye, L., Mechanical Properties of Structures 3D Printed with Cementitious Powders, *Construction Building Materials*, doi:10.1016/j.conbuildmat.2015.05.132, 2015.

Hager, I, Golonka, A, and Putanowicz, R., 3D Printing of Buildings and Building Components as the Future of Sustainable Construction, *Procedia Engineering,* doi:10.1016/j.proeng.2016.07.357, 2016.

Tay, Y. W. D., Panda, B., Paul, S. C., Noor Mohamed, N. A., Tan, M. J., and Leong, K. F., 3D Printing Trends in Building and Construction Industry: A Review, *Virtual Physical Prototyping,* doi:10.1080/17452759.2017.1326724, 2017.

Interdependence between Structural Engineering and Construction Management
Edited by Ozevin, D., Ataei, H., Modares, M., Gurgun, A., Yazdani, S., and Singh, A.

ISBN: 978-0-9960437-6-2

IDENTIFICATION OF CRITICAL INFRASTRUCTURE USING DATA MINING TECHNIQUES

SU-LING FAN[1], CHANG-SAAR CHAI[2], and KUMAR VIKRAM[1]

[1]*Dept of Civil Engineering, Tamkang University, New Taipei City, Taiwan*
[2]*School of the Built Environment, University of Reading Malaysia, Johor, Malaysia*

Critical Infrastructure (CI) is a term used to describe important national assets for producing or distributing a continuous flow of essential goods or services. They are marked by immense complexity, characterized predominantly by strong intra and interdependencies as well as hierarchies. These interconnections take many forms, including flows of information, shared security, physical flows of commodities, and others. Previous research has illustrated the relationship between the physical impacts of natural disasters and the social and economic factors on CI. Some research emphasized more the role of CI interdependencies and their importance and influence over the functioning of industries while others have looked the impacts due to disruption of CI after disasters. Nowadays comprehensive identification of all interdependency relationships of CI remains a challenge. As the complexity and interconnectedness of a country's CI evolve, threats and vulnerabilities increase. Thus, investigating how a set of CI interacts and identification of criticality of CI becomes an important topic. This research has made utilization of data mining techniques and proposes a method to identify the criticality of Critical Infrastructure so that to develop better disaster protection and prevention management.

Keywords: Critical infrastructure interdependency, General sequence pattern mining, Python.

1 INTRODUCTION

Critical Infrastructure (CI) are referring to the systems and assets, whether physical or virtual, so important to a particular country that the incapability or destruction of such systems and assets would have a debilitating impact on the security, national economy and public health and safety. These assets include, but are not limited to, facilities for transportation, communication systems, electric power systems, gas and oil storage facilities and pipelines, and government services (Chou and Tseng 2010).

Critical Infrastructure Interdependency (CII) defines the relationship between two elements of critical infrastructure as a bidirectional relationship in which the output of one is essential as the input of another (Rinaldi *et al.* 2001). Furthermore, these CIIs often create complex interdependency relationships that across system boundaries (Haimes 2005). As the complexity and interconnectedness of a country's critical infrastructure evolve, threats and vulnerabilities increase. Recently, it comes across many cases in which huge losses occur due to domino effects from some system failure. A typical example from a technical glitch of Taiwan Tatan Power Plant in 2017 which caused malfunctioning of six generators affecting the supply of 4 million

KWs of electricity. This incident brought a big impact to the local authority in transportation, manufacturing factories, hospitals, retails and many others (Chung-Han and Power 2017). Due to the great economy loses resulted from this incident, Taiwan National Security Office is prominently aware of the importance of CII in the context of homeland security. Therefore, a series of CII research and investigations are carried out to enhance the current performance (Chou and Tseng 2010).

This study is aimed to investigate the causal relationship of CII using data mining techniques. In order to achieve the aim, a case study using hospital system data were analyzed through General Sequencing Pattern and Python.

2 CRITICAL INFRASTRUCTURE INTERDEPENDENCY (CII)

There are different approaches demonstrating a CII relationship. For instance, Mendonça and Wallace (2006) employed a structure to represent the number of intersection between two critical infrastructure systems. In contrast, Mcdaniels *et al.* (2007) collected media reports and official ex-post-facto assessments to characterize the consequences of infrastructure failure interdependencies, in terms of impact and extend indices. They defined a CII relationship as a combination of its impact and extent indices. These indices can be further extended to impact score and extend score. What has heretofore been missing is an investigation of information contained in failure records. Such time-oriented failure for the Critical Infrastructure may be very difficult; using only the sequential information in the failure records may be a feasible alternative for analyzing CII relationships.

Other researchers have been concentrated efforts on the analysis of CII. Haimes and Jiang (2001) proposed the Leontief-based input-output model to investigate interconnectedness among various sectors of critical infrastructure. Although inability of one sector, called inoperability, can be precisely calculated, this model requires an input of a matrix structure denoting the relationship between each sector of critical infrastructure, and methods regarding how to generate the matrix have not been elaborated well in literature. Shih *et al.* (2009) used data warehousing and visualization techniques to explore CII for U.S. power generation. The study proposed a process to build a data warehouse, integrating data mining and geographic information system techniques so as to identify vulnerabilities in coal delivery to power plants. The results help to locate the need for redundancies in the coal supply chain.

3 BIG DATA

The Big Data concept is now receiving remarkable attention for tacking complex engineering problems. Among different fields, Big Data analytics is notably impacting the Civil Engineering domain. Currently, the operation and maintenance of Civil Engineering systems are undergoing a noticeable transformation as a result of huge amount of information provided by the emerging testing and monitoring systems. The key role of Big Data in this transformation is well understood. Despite the significance of the Big Data technologies to process large-scale data, current Civil Engineering information systems are still lacking in the successful implementation of them (Alavi and Gandomi 2017). The world is currently inundated with data, with fast advancing technology leading to its steady increase. The construction industry is not an exception to the pervasive digital revolution. The construction industry is responsible for undertaking some of the biggest and most expensive projects on Earth. Huge amounts of resources and work go into major construction projects and of course, this means that huge volumes of data are generated. The industry is dealing with significant data arising from diverse disciplines throughout the life cycle of a facility. Noticeably, this data in any form and shape has intrinsic value to the performance of the industry. With the advent of embedded devices and sensors, facilities have even started to generate massive data during the operations and maintenance stage, eventually leading to more rich sources of Big Construction Data. This vast accumulation of Construction data has pushed the construction industry to enter the Big Data era. (Bilal *et al.* 2016).

3.1 Data Mining

Data mining is the process of finding useful information from a large amount of data. The interesting patterns can be mined with the help of the several data mining techniques by (Suguna and Nandhini 2015). Data mining is defined as the nontrivial extraction of implicit, previously un-known and potentially useful information from data (Kreng and Yang 2013). Data Mining is concerned with the automatic or semiautomatic exploration and analysis, of large volumes of data, to discover meaningful patterns or rules. Data Mining has a broader scope than other traditional data analysis fields (such as statistics) since it tends to answer non-trivial questions (Wang and Leite 2013). For patterns discovery and extraction, Data Mining is primarily based on the technique(s) from statistics, machine learning, and pattern recognition (Linoff and Berry 2011). Several models are created and tested to assess the suitability of a particular technique(s) for solving the given business problem. Models with the highest accuracy and tolerance are chosen and applied to the actual data for generating predictive results (including predictions, rules, probability, and predictive confidence) (Bilal *et al.* 2016).

The development of Big Data analysis is eventually improving. However, General Sequencing Pattern (GSP) remained as one of the best technique concerning with finding statistically relevant patterns between data examples where the values are delivered in sequence.

3.2 General Sequences Pattern (GSP)

The sequential pattern is a sequence of item sets that frequently occurred in a specific order, the items in the same item sets are supposed to have the same transaction- time value or within a time gap. Usually, all the transactions are viewed as a sequence, usually called X-sequence (X can be renamed accordingly), where each transaction is represented as an item sets in that sequence, all the transactions are a list in a certain order with regard to the transaction-time. Sequences are important data, which occur frequently in many fields such as medical, business, financial, customer behavior, educations, security, and other applications.

Several algorithms have been devised for discovering such sequential patterns. The GSP algorithm was proposed by (Agrawal and Srikant 1994). The study aimed at mining sequential patterns over a large database of customer transactions. The GSP algorithm was subsequently applied in discovering various rules underlying a given set of sequences to predict a plausible sequence continuation.

Frequently occurring patterns ordered by time are found by sequential pattern mining. Sequential pattern mining has wide application since the data has a time component attached with them. For example, the medical domain can determine a correct diagnosis from the sequence of symptoms experienced; over customer data to help target repeat customers and with web-log data to better structure a company's website for easy accessibility of most popular links. There are several known methods for discovering general sequential patterns at present. Still, in a specific domain of web-log analysis more methods exist.

4 RESEARCH METHODOLOGY

Inoperability Input-Output Model (IIM) algorithm is adopted in this study. The algorithm has been designed to determine the criticality of the system and interdependencies within the system. It uses the sequences dataset generated to compute the matrix. With the power of enormous libraries contained in Python, it produces a quantitative output in the form of a matrix generated from the relationships obtained from the sequences. Through the CII matrix, the systems, which have the higher risks of failure can be identified and precaution measure could be proposed. This matrix also helps in determining how other systems behave if one of the systems fails, subsequently explains the impact of one system to another.

The data is collected from hospital Taipei Taiwan. Data format is in MySQL (.mdf) format of size 7.09 GB, contains 50,117,166 numbers with 30 days of data. Bad data contains 137,236, which falls in 536 system. The data information contains system details, IP address, date and time and system status (good or bad).

The collected data is fall in medium scale size, therefore Python is utilized in this study. It is a fast data mining approach, which is capable to create new outcome through the analyzed results. Python has emerged as a good option in data processing, and there is often a trade-off between scale and sophistication. Besides, it has excellent amounts of toolkits and features that gave advantage on data community.

5 RESULTS AND DISCUSSIONS

The collected data is first converted to ".csv" format then imported to the MySQL server in visual studio code. The excessive rows and columns will be deleted in this process to ensure that the data is clean and fit to analyze. Next, the data is filtered through "good" and "bad" status. The bad status data will be identified to further analyzed. In order to enhance the data analyzing the process, the distinct system names are renamed to "S01", "S02" accordingly as shown in Figure 2. Finally, the data will be sorted according to time frames (1 day, 5 days and 30 days time frame) and data with same time frames will be regrouped.

After the data screening process, min_count argument method is used to filter the low probability of occurrence dataset. This is to minimize results error due to the low occurrence rate. Then, the IIM matrix method is utilized by giving different time difference of failures and the added sequences were updated to the array containing all the sequences. The repeating systems within the same sequence will be eliminated. The formulated matrix is as in Eq. (1). The value of ith row and jth column of the matrix was calculated as:

$$A_{ij} = \frac{count(sys(i) \cap sys(j))}{count(sys(i))} \tag{1}$$

Where A_{ij} denotes the element corresponding to the ith row and jth column and count is used to calculate how many times the specific system has occurred in a sequence.

The result of day 3 is shown in Figure 3. The results are in the form of a CII matrix that represents the interdependencies between systems. From this matrix, the most critical system can be found out. The element (i, j) of the CII matrix indicates the dependent level of system i on j and impact of the system i. For example, in Figure 3 the element (2,4) indicates that the system "S04" will get 15% damaged if system "S02" is damaged. In a similar way, the entire matrix has been formulated. From the matrix, it can be observed that system "S01", "S02" and "S03" are highly interdependent on each other (100% dependent) which indicates that if any of the system S01, S02 or S03 damaged, the other two systems will be definitely affected. A similar pattern is found on systems 9 to 13 (i.e. "S09", "S10", "S11", "S12", "S13") which are 100 percent interdependent on each other.

	System_name	ID	Date_Time	Null	IP_address	status
0	/FACILITY/E-CH-1-SV.PV	-9999	29-11-2017 09:06	NaN	192.168.0.11	OPC_QUALITY_BAD
1	/FACILITY/E-CH-1-RA.PV	-9999	29-11-2017 09:07	NaN	192.168.0.11	OPC_QUALITY_BAD
2	/FACILITY/E-CH-1-RV.PV	-9999	29-11-2017 09:07	NaN	192.168.0.11	OPC_QUALITY_BAD
3	/FACILITY/E-CH-1-SA.PV	-9999	29-11-2017 09:07	NaN	192.168.0.11	OPC_QUALITY_BAD
4	/FACILITY/E-CH-1-SA.PV	-9999	29-11-2017 09:07	NaN	192.168.0.11	OPC_QUALITY_BAD
5	/FACILITY/E-CH-1-RA.PV	-9999	29-11-2017 09:07	NaN	192.168.0.11	OPC_QUALITY_BAD

Figure 1. Data screening process for bad status data.

	System_name	time
0	S02	32760
1	S01	32820
2	S02	32820
3	S03	32820
4	S05	32820
5	S04	32820
6	S05	32820
7	S04	32820
8	S03	32820

Figure 2. System renaming.

	SYS	S01	S02	S03	S04	S05	S06	S07	S08	S09	S10	S11	S12	S13	S14
0	S01	1.00	1.00	1.00	0.15	0.15	0.10	0.10	0.10	0.09	0.0	0.0	0.0	0.0	0.0
1	S02	1.00	1.00	1.00	0.15	0.15	0.10	0.10	0.10	0.09	0.0	0.0	0.0	0.0	0.0
2	S03	1.00	1.00	1.00	0.15	0.15	0.10	0.10	0.10	0.09	0.0	0.0	0.0	0.0	0.0
3	S04	0.54	0.54	0.54	1.00	1.00	0.11	0.11	0.11	0.12	0.0	0.0	0.0	0.0	0.0
4	S05	0.54	0.54	0.54	1.00	1.00	0.11	0.11	0.11	0.12	0.0	0.0	0.0	0.0	0.0
5	S06	0.57	0.57	0.57	0.16	0.16	1.00	1.00	0.99	0.00	0.0	0.0	0.0	0.0	0.0
6	S07	0.57	0.57	0.57	0.16	0.16	1.00	1.00	0.99	0.00	0.0	0.0	0.0	0.0	0.0
7	S08	0.56	0.56	0.56	0.16	0.16	1.00	1.00	1.00	0.11	0.0	0.0	0.0	0.0	0.0
8	S09	0.44	0.44	0.44	0.15	0.15	0.11	0.11	0.10	1.00	0.0	0.0	0.0	0.0	0.0
9	S10	0.00	0.00	0.00	0.00	0.00	0.00	0.00	0.00	0.00	1.0	1.0	1.0	1.0	1.0
10	S11	0.00	0.00	0.00	0.00	0.00	0.00	0.00	0.00	0.00	1.0	1.0	1.0	1.0	1.0
11	S12	0.00	0.00	0.00	0.00	0.00	0.00	0.00	0.00	0.00	1.0	1.0	1.0	1.0	1.0
12	S13	0.00	0.00	0.00	0.00	0.00	0.00	0.00	0.00	0.00	1.0	1.0	1.0	1.0	1.0
13	S14	0.00	0.00	0.00	0.00	0.00	0.00	0.00	0.00	0.00	1.0	1.0	1.0	1.0	1.0

Figure 3. Processed Data

6 CONCLUSION

The results that are obtained from the CII matrix are helpful in determining critical systems in the hospital. The matrix is able to predict the most likelihood of next system failure if one of the critical infrastructures is malfunctioning. This definitely helps to prevent domino effects of system failure in a particular facility. The function of this matric is not limited to compute the system in one particular facility, meanwhile, it can be extended to compute the national critical infrastructures to prevent Taiwan Tantan power plant incident to be reoccurred.

References

Alavi, A. H., and Gandomi, A. H. Big Data in Civil Engineering, *Automation in Construction*, *79*, 1–2, 2017.

Bilal, M., Oyedele, L. O., Qadir, J., Munir, K., Ajayi, S. O., Akinade, O. O., and Pasha, M., Big Data in the Construction Industry: A Review of Present Status, Opportunities, and Future Trends, *Advanced Engineering Informatics*, *30*(3), 500–521, 2016.

Chou, C.-C., and Tseng, S.-M., Collection and Analysis of Critical Infrastructure Interdependency Relationships, *Journal of Computing in Civil Engineering*, *24*(6), 539–547, 2010.

Chung-Han, B. L. Y., and Power, K., Outage across Taiwan Due to Malfunction at Power Plant, *Focus Taiwan News Channel*, 2017.

Haimes, Y. Y., and Jiang, P., Leontief-Based Model of Risk in Complex Interconnected Infrastructures, *Journal of Infrastructural Systems, 7*(1), 1–12, 2001.

Haimes, Y. Y., Infrastructure Interdependencies and Homeland Security, *Journal of Infrastructure Systems* 65–66, 2005.

Kreng, V. B., and Yang, S.-W., Data Mining of Hospital Characteristics in the Online Publication of Medical Quality Information, *Health*, 05(05), 931–937, 2013.

Linoff, G. S., and Berry, M. J., *Data Mining Techniques: For Marketing, Sales, and Customer Relationship Management,* John Wiley and Sons, 2011.

Mcdaniels, T., Chang, S., Peterson, K., Mikawoz, J., Reed, D., Empirical Framework for Characterizing Infrastructure Failure Interdependencies, *Journal of Infrastructure Systems*, 13(3), 175–184, 2007.

Mendonça, D., and Wallace, W. A., *Impacts of the 2001 World Trade Center Attack on New York City Critical Infrastructures*, 260–270, 2006.

Rinaldi, S., Peerenboom, J., and Kelly, T., Identifying, Understanding, and Analyzing Critical Infrastructure Independencies, *IEEE Control Systems Magazine,* 21, 11–25, 2001.

Shih, C. Y., Scown, C. D., Soibelman, L., Matthews, H. S., Jr, J. H. G., Dodrill, K., and Mcsurdy, S., Data Management for Geospatial Vulnerability Assessment of Interdependencies in U. S. Power Generation, *Journal of Infrastructure Systems*, 179–189, 2009.

Suguna, K., and Nandhini, K., Literature Review on Data Mining Techniques, *International Journal of Computer Technology and Applications*, 6, 583–585, 2015.

Wang, L., and Leite, F., *Knowledge Discovery of Spatial Conflict Resolution Philosophies in Bim-Enabled MEP Design Coordination Using Data Mining Techniques: A Proof-Of-Concept,* Computing in Civil Engineering, 2013.

Interdependence between Structural Engineering and Construction Management
Edited by Ozevin, D., Ataei, H., Modares, M., Gurgun, A., Yazdani, S., and Singh, A.

ISBN: 978-0-9960437-6-2

AMASS: ADVANCED MANUFACTURING FOR THE ASSEMBLY OF STRUCTURAL STEEL

ARTURO E. SCHULTZ[1], JIA-LIANG LE[1], MOHAMMAD E. SHEMSHADIAN[1], RAMZI LABBANE[1], DEBRA F. LAEFER[2,3], SALAM AL-SABAH[3], LINH TRUONG-HONG[3], MINH HUYNH[3], PATRICK MCGETRICK[4], TONY MARTIN[4], and PANTELIS MATIS[4]

[1]*Dept of Civil, Environmental and Geo-Engineering, University of Minnesota, Twin Cities, Minneapolis, USA*
[2]*Center for Urban Science and Progress and Dept of Civil and Urban Engineering, New York University, New York, USA*
[3]*School of Civil Engineering, University College Dublin, Dublin, Ireland*
[4]*School of Natural and Built Environment, Queen's University Belfast, Belfast, UK*

This paper describes a collaborative project between the US, Ireland, and Northern Ireland (UK) to investigate advanced manufacturing cutting techniques for the creation of a new class of intermeshed steel connections that rely on neither welding nor bolting. To date, advanced manufacturing equipment has only been used to accelerate traditional processes for cutting sheet metal or other conventional fabrication activities. Such approaches have not capitalized on the equipment's full potential. This project lays the groundwork to transform the steel building construction industry by investigating the underlying science and engineering precepts for intermeshed connections created from precise, volumetric cutting. The proposed system enhances the integration between design, fabrication, installation and maintenance through building information modeling platforms to implement advanced connections. Fully automated, precise, volumetric cutting of open steel sections introduces intellectual challenges regarding the load-transfer mechanisms and failure modes for intermeshed connections. The research activity addresses knowledge gaps concerning the load resistance and design of steel systems with intermeshed connections. Physical tests, finite element simulation and multi-scale modeling are being used to investigate the mechanics of intermeshed connections including stress and strain concentrations, fracture potential and failure modes, and to optimize connection geometry.

Keywords: Steel connection, Intermeshed, Plasma, Waterjet, Finite element modeling, Front-intermeshed connection, Side-intermeshed connection.

1 INTRODUCTION AND CONCEPTS

Despite field welding and bolting being time-consuming and/or expensive, the steel building market has not developed any new universally applicable structural steel connection systems since before World War II. To achieve improved construction efficiency and heightened material reuse, computer controlled, advanced manufacturing techniques in high-definition plasma, laser and water jet cutting (Figure 1) could be exploited (Ramakrishnan and Rogozinski 1997). This paper envisions the harnessing of those technologies to create an entirely new class of "intermeshed" steel connections.

Precise cutting of steel makes it possible to create the notches in beam flanges and web that can intermesh with other beam parts or external connectors, like puzzle pieces. This technique could radically change how structural steel is fabricated, assembled, deconstructed, and reused (Perreira *et al.* 1993). Without relying solely on bolting or welding to assemble a connection in the field, the simplicity and efficiency of the construction process may be significantly improved. To date, this class of fabrication equipment has only been used to accelerate traditional processes for cutting sheet metal or other conventional fabrication activities (e.g., cutting instead of drilling holes). Such approaches have not capitalized on the equipment's full fabrication potential.

Because assembly costs at a construction site are high for welded and bolted connections, the intermeshed connection offers the potential for a lower cost connection, even though manufacturing costs are likely to be higher. There is also the potential for life cycle savings as deconstruction costs would be significantly lower as an intermeshed connection can be designed specifically for disassembly and also for reuse. However, unlike traditional bolted or welded connections, where the industry has more than 100 years of experience, precise cutting of steel for an intermeshed connection is not yet part of the culture or its expectations.

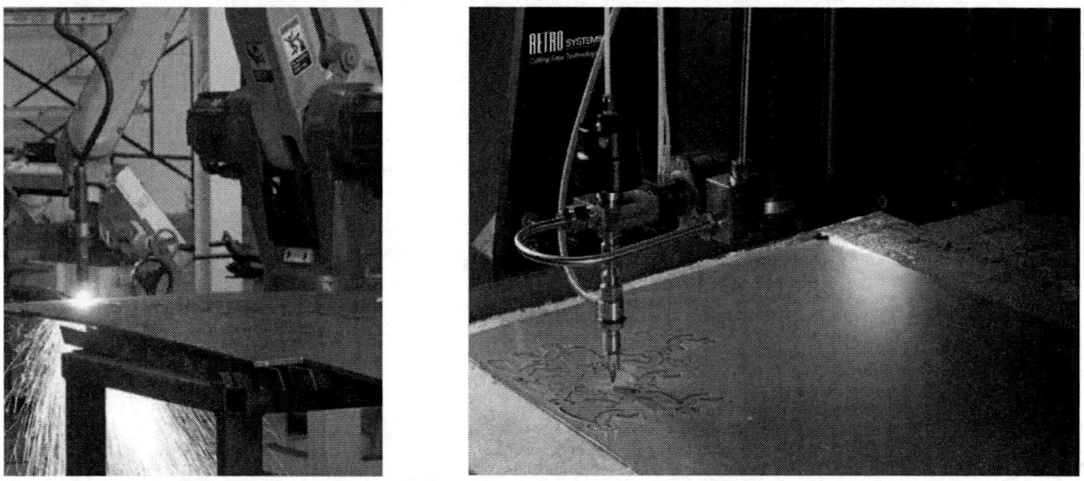

(a) Plasma cutting performed by an industrial robot (Wikipedia 2018a).

(b) Waterjet computer numerical control cutting machine. (Wikipedia 2018b).

Figure 1. Advanced cutting techniques.

While maintaining the original concept, different alternatives can be developed for the intermeshed connection. A variety of connection details were investigated and different connections were proposed. In this paper, two major ideas for intermeshed connection are presented in the following sections.

2 FRONT-INTERMESHED CONNECTION

The very first concept is a relatively simple intermeshed connection called the "Front-Intermeshed Connection" (Al-Sabah and Laefer 2017a). This connection is composed of three-dimensional interlocking through the top and bottom flanges and through the web. The flanges carry the tension and compression resulting from the bending moment at the connection, while the web carries the shear force. The connection transfers shear and compression from one beam section to the next through direct contact bearing of multiple, precisely shaped faces (Figure 2). The arrangement is ideally suited for connecting beams at or near ideal inflections points to create gravity load framing. For practical use of this new type of connection, understanding its mechanical behavior, especially in terms of the load carrying capacity under mixed mode loading scenarios, will be essential.

(a) Idealized connection in printed stainless steel.

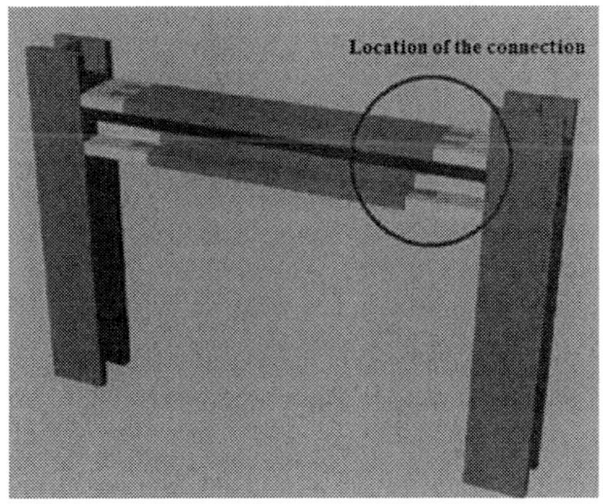

(b) Assembly of the connection in a frame.

Figure 2. Front-intermeshed connection.

Load is transferred between flanges through bearing and friction via the intermeshed teeth of the flanges. The connection has the advantages of simplicity and requires no additional parts to create the flange connection, although a locking mechanism can be added against uplift. However, field-assembled locking connections are unlikely to be able to fully transmit the flexural, axial, and shear capacities of the connected steel sections and may reduce the corresponding stiffness components of a continuous steel member, because of interruptions in the load path.

The stepped web connection allows for easy site assembly, as the middle beam part can be dropped from above, in a method similar to current practice. In Figure 2b, the ends shown in yellow are shop welded as stubs to the column. No other welding is required, and bolting is fully avoided in this configuration. The main drawbacks to this type of connection are the tightness of the tolerances and the lack of adjustability in the erection process.

2.1 Finite Element Analyses in Abaqus

To better understand how this type of precise interlocking performs structurally, initial lab testing was conducted for two-dimension meshed "fingers" (Matis *et al.* 2018). Based on the success of those initial tests, a three-dimensional numerical model was created in Abaqus (Abaqus 2013). The finite element analyses were performed under different load combinations to investigate the general behavior, failure modes, and peak capacity of the connection.

Under flexure, the connection shows a uniformly monotonic behavior with a relatively low flexural resistance. Failure occurs when the teeth on one side of the top flange slip out of the sockets on the other side (Figure 3a), so there is no capacity to carry tension force, which means no moment capacity can be developed. Moreover, in the case of combined tension and shear, the presence of the latter causes relative vertical movement between the two sides of the connection and, consequently, the flanges slip out of their intermeshed positions. When this phenomenon happens (Figure 3b), there is no component to resist tension, and the connection fails.

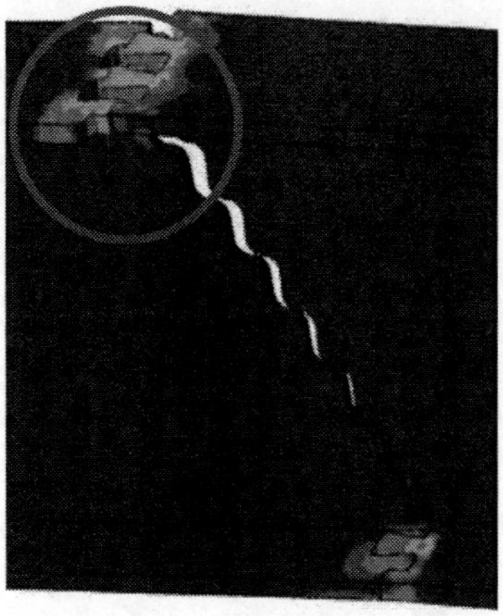

(a) In pure flexure top flange slips out.

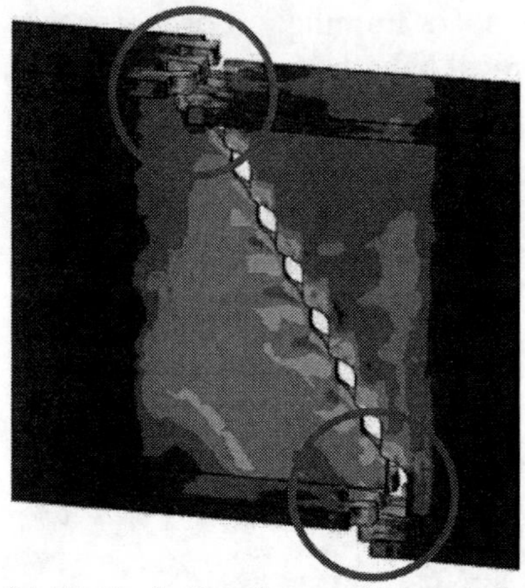

(b) Vertical slip of top and bottom flanges under combined tension and shear.

Figure 3. Performance of the front-intermeshed connection under different load conditions.

2.2 Connection Characteristics

Considering the connection configuration and the results of the finite element analyses, the front-intermeshed connection is generating insufficient strength and stiffness for effective use in practice. Therefore, it is helpful to characterize the front-intermeshed connection in terms of its performance characteristics in order to develop strategies for connection improvement.

Following current code standards and based on the results from the finite element studies, the front-intermeshed connection shows limited stiffness, strength, and ductility. So, it is classified as a non-ductile and partially restrained connection (AISC Specification for Structural Steel Buildings 2016, Eurocode 3 2005). This classification is due to the cuts in the flanges which reduce load carrying capacity in tension and, consequently, the moment significantly. Therefore, the front-intermeshed connection is not recommended for use where demands for 1) large moments, 2) large moments in combination with large shears, and 3) axial tension are possible.

3 SIDE-INTERMESHED CONNECTION

Given the limitations of the front-intermeshed connection in transferring loads, especially when in the presence of combined loads and strict tolerances, another alternative of the intermeshed connection is proposed (Al-Sabah and Laefer 2017b). The "Side-Intermeshed Connection" utilizes intermeshed external connectors to transfer flange tension and compression forces.

Two different versions of side-intermeshed connection were developed. The "original conception" (Figure 4) sought to meet the original goals of requiring no welding nor bolting at all, while maximizing erection speed and construction tolerance. Meanwhile, the second version (Figure 5) was modified for greater acceptance in the construction industry and includes minor bolting. This paper focuses on the modified version of the side-intermeshed connection.

In this connection, flange edges are cut to create a set of tooth-shape notches (i.e. "teeth"). To connect different sides of the beam, an angle is used on each edge with horizontal holes which match the teeth (see Figure 5). Having the beam flanges connected, the section will be able to transfer moments via the connector angles. However, for shear transfer, a pair of regular shear plates are bolted to the beam web.

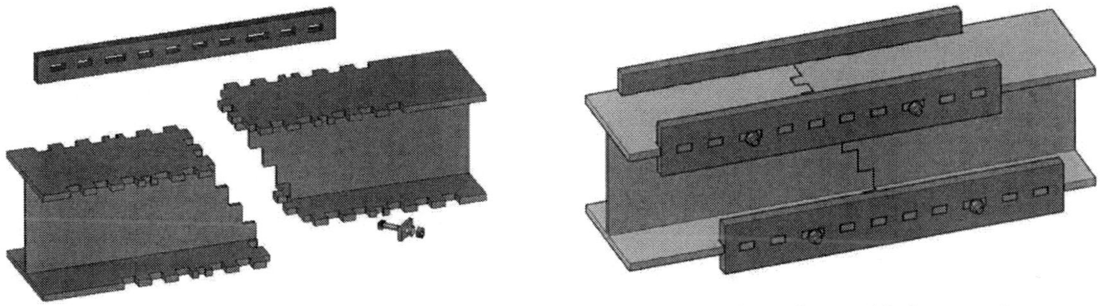

(a) 3D Exploded view. (b) 3D View of assembled connection.

Figure 4. Side-intermeshed connection (original conception).

A new design approach was adopted based on the concept of "Capacity Design" and the requirements of the current US steel design standard (AISC Steel Construction Manual 2017). In this approach, the angles are designed for applied loads (obtained from elastic analysis). Subsequently, the teeth are designed for maximum expected forces based on connector angle capacity. Using "Capacity Design" assures that failure occurs in the angles rather than the teeth. Such a mechanism is desirable since the replacement of the angles is easy and fast following a case of damaging overload. Given the number of variables needed to define connection resistance, an iterative approach for the proportioning of the elements is needed.

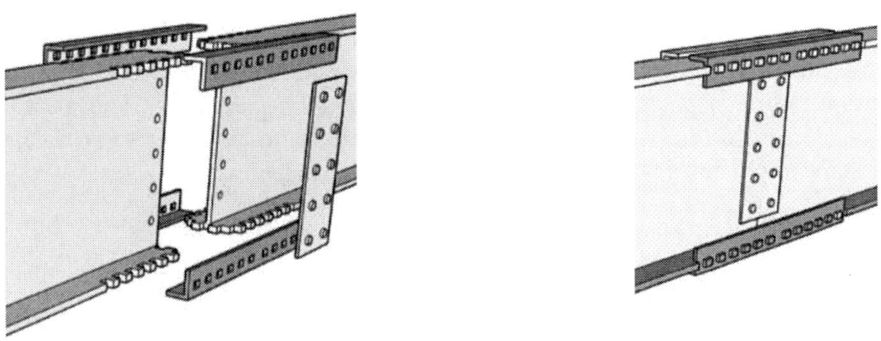

(a) 3D Exploded view. (b) 3D View of assembled connection.

Figure 5. Side-intermeshed connection (modified version).

The side-intermeshed connection allows larger tolerances and easier fabrication, which leads to a better potential for wider accepted in practice. However, one potential concern related to this connection is the stress concentration at the sharp corners of the angle holes. Depending on the length-to-width ratio of the angle holes, previous studies have shown that the stress concentration factor can be up to 5 (Sikora 1973) which can result in pre-mature rupture. To avoid any undesirable failure modes, circular holes were added to the sharp corners in the angle holes. A finite element model proved the effectiveness of such a change by reducing the stress concentration factor to 1.7. Figure 6 shows the configuration of the hole and the resulting stress. Further study of the side-intermeshed connection is ongoing, including cohesive modeling of the connection to enable efficient analysis, and experimental verification.

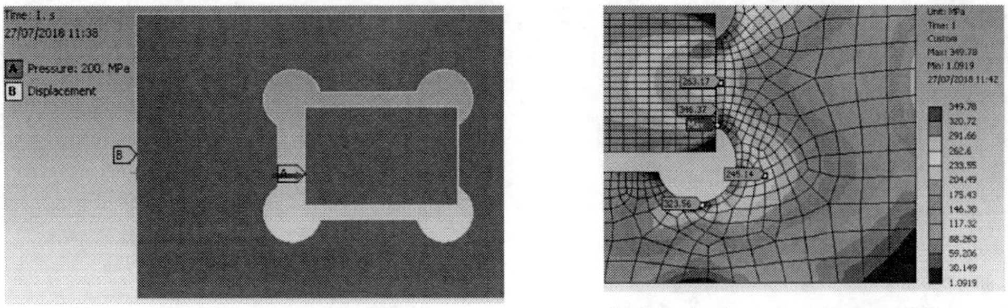

(a) Load and boundary conditions. (b) Equivalent stress around the corner.

Figure 6. Finite element analysis on the effect of the shape on the stress concentration around the hole.

4 CONCLUSIONS

This study analyzed a radically new connection for structural steel members, which uses multi-degrees of freedom, volumetric cutting to drastically reduce fabrication costs and vastly simplify and accelerate erection by avoidance of supplementary materials and methods. Based on this concept, two different alternatives of intermeshed connections were proposed and investigated.

Results from finite element modeling in Abaqus showed that the front-intermeshed connection exhibits excellent shear resistance even in the presence of moment or axial force. On the other hand, axial or flexural behavior is sensitive to the relative position of the flanges. The presence of shear can make the case more critical as it makes the flanges slip out, and the connection would lose the ability to hold together. Based on flexural characteristics, the front-intermeshed connection shows low stiffness and resistance rendering it a simple connection.

In keeping with the intermeshed connection concept, another alternative is proposed that eliminates the negative aspects of the front-intermeshed connection. The resulting side-intermeshed connection offers larger load capacity potential and larger erection tolerances. A special detail at the angle holes was adopted to avoid stress concentration.

References

Abaqus, *6.13 Manual,* SIMULIA, Providence, RI. 2013.

AISC Specification for Structural Steel Buildings, *American Institute for Steel Constructions,* 2016.

AISC Steel Construction Manual, *American Institute for Steel Constructions.* 2017.

Al-Sabah, A., and Laefer, D., *GB Patent Application No 1718744.4,* 2017a.

Al-Sabah, A., and Laefer, D., *GB Patent Application No 1718746.9,* 2017b.

Eurocode 3: Design of Steel Structures - Part 1-8: Design of Joints, *Eurocode,* 2005.

Matis, P., Martin, T., McGetrick, P. J., and Robinson, D., The Effect of Frictional Contact Properties on Intermeshed Steel Connections, *CERAI 2018 Dublin Ireland*, Paper 199, 2018.

Perreira, N. D., Fleischman, R. B., Viscomi, B. V, and Lu, L.-W., Automated Construction and ATLSS Connections; Development, Analysis, Experimentation, and Implementation of ATLSS Connections for Automated Construction, *ATLSS Reports,* Lehigh University, 1993.

Ramakrishnan, S., and Rogozinski, M. W., Properties of Electric Arc Plasma for Metal Cutting, *Journal of Physics D: Applied Physics*, 30(4), 636–644, 1997.

Sikora, J. A., Summary of Stress Concentrations in the Vicinity of Openings in Ship Structures, *Naval Ship Research and Development Center*, Report 3889, 1973.

Wikipedia, retrieved from https://en.wikipedia.org/wiki/Plasma_cutting, 2018a.

Wikipedia, retrieved from https://en.wikipedia.org/wiki/Water_jet_cutter, 2018b.

Interdependence between Structural Engineering and Construction Management
Edited by Ozevin, D., Ataei, H., Modares, M., Gurgun, A., Yazdani, S., and Singh, A.

ISBN: 978-0-9960437-6-2

THE APPLICATION OF WELD-BASED ADDITIVE MANUFACTURING STEEL TO STRUCTURAL ENGINEERING

VITTORIA LAGHI, MICHELE PALERMO, GIADA GASPARINI, STEFANO SILVESTRI, and TOMASO TROMBETTI

Dept of Civil, Chemical, Environmental and Materials Engineering, University of Bologna, Bologna, Italy

The present work aims at providing the first considerations upon the application of innovative manufacturing technology for civil engineering purposes. In particular, among the 3D printing processes currently available, Weld-Based Additive Manufacturing (WAM) results to be the most suitable technique for the realization of innovative structural forms in metal material. The great potential of taking the printing head "out of the box" allows for the construction of innovative shapes by adding layer upon layer of welded steel. In particular, the study is focused on the realization of the first 3D-printed steel footbridge by a Dutch company held in Amsterdam, called MX3D, and its Additive manufacturing process, which results in specific constraints and limitations to be taken into account for design purposes. First, the design issues are described, by considering the printing parameters to be adopted for the realization of large-dimensions structures, and then the implications in terms of specific geometrical and mechanical characteristics are studied. These first engineering evaluations are intended to pave the way towards the development of a ground-breaking technology for the fully-automated design and construction of novel 3D-printed building structures through innovative robotic manufacturing processes whose parameters are still not fully known.

Keywords: 3D printing, Mechanical characterization, Experimental tests, Stainless steel, Geometrical imperfections, Manufacturing techniques.

1 INTRODUCTION

Over the two last centuries, the progress in civil engineering always derived from advancements in technology and material science from the Industrial Revolution, which paved the way towards the realization of steel structures, to the studies of Hennebique at the beginning of XX century for the introduction of concrete in civil engineering. In the last decades, much attention has been brought to innovative manufacturing processes by means of Rapid Prototyping (RP), leading towards a new era for the construction field (Addis 2007).

The fundamental part of this process is the "digital turn" (Carpo 2013) in the design and planning process for civil engineering over the last 25 years, to emphasize the increasing influence of Computer-Aided Design (CAD) tools in the realization of innovative unexplored forms for architecture (i.e., complex, doubly-curved geometries, free-form design) (Schlaich and Schlaich 2000, Iwamoto 2013). Although many research efforts have been done since the 1950s on compressed and flux structures (Adriaenssens *et al.* 2014, Sasaki 2005), the construction technology at the time could not overcome the intrinsic issues related to the manufacturing process of such complex shapes. On the other hand, automation has been starting to prevail for the last decades in almost all production domains for aero-spatial, mechanical and biomedical fields, whereas for civil engineering it has still been a challenging task due to large dimensions of the products to be 3D-printed (Wong and Hernandez 2005).

Very recently, innovations in digital fabrication such as Additive Manufacturing (AM) processes present high potentials to be involved in the realization of structural elements without geometrical constraints, which would start a new trend towards automatic constructions.

2 WELD-BASED ADDITIVE MANUFACTURING PROCESS

The basic equipment of an Additive Manufacturing system consists of a motion part, a heat source and a feedstock (Williams *et al.* 2016). Several research activities aimed at providing appropriate terminology and classification among the widespread variation of technologies within this family (Sames *et al.* 2016), and a first distinction can be made based on the material adopted. In particular, for metal Additive Manufacturing processes, they can be divided into three main categories: (i) Powder-Based Fusion (PBF), (ii) Directed Energy Deposition (DED) and (iii) sheet delamination.

Although intense work has been done in the mechanical characterization of PBF metal outcomes (Buchanan *et al.* 2017, Song *et al.* 2015, Yap *et al.* 2015, Niendorf *et al.* 2013), this technology suffers from intrinsic dimensional constraints related to the volume of the box being part of the printing apparatus, therefore becoming highly challenging to realize real-dimension structural elements. Differently, among the DED processes, the so-called Wire-and-Arc Additive Manufacturing (WAAM), also referred to as simply Weld-based Additive Manufacturing (WAM), is able to provide larger outcomes due to the absence of the volume constriction and flexibility in the set up. In fact, the printing head is positioned on top of a robotic arm, literally taking 3D printing process "out of the box". On the other hand, larger dimensions of the outputs require also higher printing velocities, and therefore less accuracy in the geometry of the element realized.

3 THE FIRST APPLICATION OF WAAM IN STRUCTURAL ENGINEERING

The present work focuses on the application of Weld-based Additive Manufacturing process in the structural engineering field.

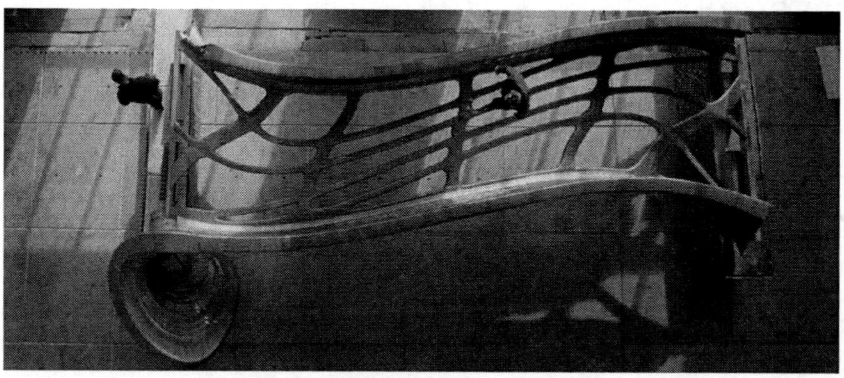

Figure 1. MX3D footbridge realized with the WAM process [courtesy of MX3D].

In particular, the first attempt to develop a new building trend has been done by MX3D (2019), a Dutch company based in Amsterdam which realized the first 3D-printed real-scale footbridge in Additive-Manufactured stainless steel (Figure 1).

The specifications of such manufacturing process in order to build a large-scale structure are crucial in the derivation of the structural response of the outcomes. First, the intrinsic lack of precision of the printing equipment has to be devoted to the higher velocities required to print large parts. Moreover, the welding process induces some non-negligible residual stresses due to heating, which might also alter the internal crystalized structure of stainless steel, to be objective of further research as well.

4 THE DESIGN ISSUES

In order to assess the design specification of a brand-new construction process, both geometrical and mechanical characterization should be performed, to properly quantify the peculiarities related to the WAAM process adopted in the realization of structural engineering parts.

4.1 Geometrical Imperfections

Figure 2 shows some of the crucial geometrical imperfections due to the manufacturing process: irregular cross-sectional area, surface roughness, and un-straightness.

Figure 2. The geometric issue of WAM tubular elements: irregular cross-section, un-straightness, surface roughness.

The cross-sectional area, of main importance to be considered in the derivation of the element structural response, is mainly affected by both surface roughness and un-straightness of the outcomes. In order to properly assess and quantify the external roughness and position of the longitudinal axis of the 3D-printed tubular elements studied, a high-resolution 3D scanning acquisition has been used to reproduce the surface by means of FE model, from which the irregularities have been studied and quantified.

The results from the geometrical characterization of Weld-based Additive Manufactured structural elements composing the MX3D's footbridge will be part of a more specific work. However, at first, it should be mentioned a discrepancy between the real and nominal cross-sectional areas, due to the irregular thickness of the welding portion, as well as the non-negligible roughness of the external surface. The same applies to the lack of straightness of some tubular elements, which alters their response in buckling under compressive load.

4.2 Mechanical Response

Regarding the structural response of Weld-based Additive Manufactured stainless-steel material, it is essential to perform some standard tests on scaled elements.

First, monotonic tensile and compressive tests are performed in order to assess the 0.2% proof stress, the ultimate stress, the Young modulus, and the ultimate deformation. Results of the first experimental campaign are briefly mentioned in recent work done by the authors (Laghi *et al.* 2018) and will be extensively described in a further study. Figure 3a shows the "dog-bone" shaped specimens used to derive the tensile strength, while Figure 3b shows the "stub columns" adopted for the compressive tests.

(a) (b)

Figure 3. (a) "dog bone" shaped specimens for tensile tests before and after the test; (b) "stub columns" specimens for compressive tests before and after the test.

The first experimental campaign shows some interesting results in terms of the mechanical properties of this innovatively processed material. As far as the 0.2% proof stress is concerned, the mean values resulting from both tensile and compressive tests are higher than the ones recommended for traditionally formed stainless steel (grade 316LSi), while for the ultimate stress the experimental results are on the lower side of the range requested by the EN 1993-1-4:2006+A1 (2015). The interesting parameter, which substantially differs from the traditional stainless-steel behavior is Young modulus, for which the first experimental tests show values around 100 GPa, corresponding to half the one commonly adopted for cold-formed stainless-steel material. This reduction might be due to several factors typical of the innovative manufacturing process adopted: some residual stresses coming from the heating treatment during the welding process, possible critical internal porosity or irregular crystalized microstructure, or else the geometric imperfections affecting the mechanical response of the element.

Further investigations on the overall structural response of WAM metal material will be the objective of some further specific work provided in a journal paper.

5 CONCLUSIONS

The advancements in digital fabrications have led over the last decade on the development of some innovative manufacturing processes for metal materials, which could be applied for civil engineering purposes. This new trend, paving the way towards a new era of construction technologies, has already been adopted in the realization of the first footbridge made by Weld-based Additive Manufactured stainless steel by a Dutch company, MX3D, held in Amsterdam.

The work focuses on the Weld-based Additive Manufacturing process adopted in the realization of the structure, and on its intrinsic design issues, concerning both the geometrical and the mechanical characterization of the 3D-printed steel structural elements.

From the studies conducted by means of 3D scanner acquisition models, both surface roughness and un-straightness are to be considered in the overall structural response of the 3D-printed structure. The results from tensile and compressive tests compared with EN 1993-1-4:2006+A1 (2015) for 316LSi stainless steel show a good match with respect to 0.2% proof stress and ultimate stress, while for the Young modulus a substantial discrepancy between the two manufacturing processes is visible. From these first considerations, further work is required to extensively characterize the WAM structural elements to be applied in the civil engineering field.

Acknowledgments

The authors acknowledge the support coming from MX3D Company by sharing the specifications of their work and sending the weld-based additive manufactured specimens to Structural and Geotechnical Engineering Laboratory (LISG) at the University of Bologna.

References

Addis, W, *Building: 3000 Years of Design Engineering and Construction*, Phaidon Press, London, 2007.

Adriaenssens, S., Block, P., Veenendaal, D., and Williams, C., (eds.), *Shell Structures for Architecture: Form Finding and Optimization*, Routledge, Abingdon, 2014.

Buchanan, C., Matilainen, V. P., Salminen, A., Gardner, L., Structural Performance of Additive Manufactured Metallic Material and Cross-Sections, *Journal of Constructional Steel Research,* 136, 35-48, 2017.

Carpo, M., (ed.*), The Digital Turn in Architecture 1992-2012*, John Wiley & Sons, Hoboken, 2013.

EN 1993-1-4:2006+A1, *Eurocode 3 Design of Steel Structures, Part 1–4: General Rules Supplementary Rules for Stainless Steel,* European Committee for Standardisation (CEN), 2015.

Iwamoto, L., *Digital Fabrications: Architectural and Material Techniques,* Princeton Architectural Press, 2013.

Laghi, V., Palermo, M., Pragliola, M., Girelli, V. A., Van Der Velden, G., and Trombetti, T., *Towards 3D-Printed Steel Grid-Shells: The Main Idea and First Studies*, Proceedings of IASS Annual Sympoia, 2018(9), 1-9, 2018.

MX3D, Retrieved from http://mx3d.com/projects/bridge/ on March 2019.

Niendorf, T., Leuders, S., Riemer, A., Richard, H. A., Tröster, T., and Schwarze, D., Highly Anisotropic Steel Processed by Selective Laser Melting, *Metallurgical and Materials Transactions B,* 44(4), 794-796, 2013.

Sames, W. J., List, F. A., Pannala, S., Dehoff, R. R., and Babu, S. S., The Metallurgy and Processing Science of Metal Additive Manufacturing, *International Materials Reviews,* 16(5), 315-360, 2016.

Sasaki, M., *Flux Structures*, Toto, Tokyo, 2005.

Schlaich, J., and Schlaich, M., *Lightweight Structures, Widespan Roof Structures*, Thomas Telford Publishing, 177-188, 2000.

Song, B., Zhao, X., Li, S., Han, C., Wei Q., Wen, S., et al., Differences in Microstructure and Properties Between Selective Laser Melting and Traditional Manufacturing for Fabrication of Metal Parts: A Review, *Frontiers in Mechanical Engineering*, 10 (2): 111–125, http://dx.doi.org/10.1007/s11465-015-0341-2, 2015.

Williams, S. W., Martina, F., Addison, A. C., Ding, J., Pardal, G., and Colegrove, P., Wire+Arc Additive Manufacturing, *Materials Science and Technology*, 32(7), 641-647, 2016.

Wong, K. V. and Hernandez, A., A Review of Additive Manufacturing, *ISRN Mechanical Engineering*, 1-10, 2012.

Yap, C. Y., Chua, C. K., Dong, Z. L., Liu, Z. H., Zhang, D. Q., Loh, L. E., and Sing, S. L., Review of Selective Laser Melting: Materials and Applications, *Applied Physics Reviews,* 2(4), 041101, 2015.

Interdependence between Structural Engineering and Construction Management
Edited by Ozevin, D., Ataei, H., Modares, M., Gurgun, A., Yazdani, S., and Singh, A.

ISBN: 978-0-9960437-6-2

CONTENTS UNDER PRESSURE: DISTINGUISHING IMPREGNATED FIRE-RETARDANT-TREATED WOOD FROM SURFACE-APPLIED TREATMENTS

MIKE ECKHOFF

Hoover Treated Wood Products, Inc., Thomson, USA

Increasing populations combined with decreasing developable plots are leading to higher real estate prices. Communities are attempting to address these higher prices by increasing the number of residential units per acre. One way to increase the supply is to use fire-retardant-treated wood (FRTW) to build larger and taller structures. However, as FRTW is increasingly being used, substitute products that do not meet the definition of FRTW in the building codes are also being increasingly used. How does one distinguish the two? By relying on knowing how FRTW is manufactured, where it can be used, and on building code requirements, responsible parties will better understand why FRTW is different from these products and how to verify that only code-compliant, chemically impregnated FRTW is used in larger, taller structures.

Keywords: Pressure-impregnated, Intumescent, Paints, Cementitious Coatings, Steiner Tunnel Test, Extended UL 723, 2018 International Building Code.

1 INTRODUCTION

As populations increase, the supply of housing in proximity to workplaces and desired amenities declines, in turn leading to rising real estate prices. Urban areas have responded to this price pressure by spreading, but with decreasing amounts of outlying undeveloped areas, these urban areas are now responding by consolidating; in other words, urban areas are becoming increasingly densified by building up instead of building out. This trend is expected to continue. By the year 2040, two-thirds of the United States' population will be concentrated in just 10 large-scale, regional "megapolitan" clusters (Nelson and Lang 2013). The question becomes how do we build sustainably in densifying, urban areas while maintaining standards for life safety?

One answer is to use more fire-retardant-treated wood (FRTW). FRTW products are defined in the 2018 International Building Code (IBC 2018) as "wood products that, when impregnated with chemicals by a pressure process or other means during manufacture, exhibit reduced surface-burning characteristics and resist the propagation of fire" (International Code Council [ICC] IBC 2018). FRTW can contribute to a sustainable housing solution while helping to mitigate environmental impacts. FRTW can be manufactured to meet standards required by independent, third-party, green building certification systems (e.g., FSC, SFI, etc.) It can also be used as a substitute for untreated wood in any rated assembly without altering its fire resistance, as stated in references such as UL's Fire Resistance Directory (Bueche 2013). Through this substitution, FRTW could be used in applications where untreated wood is not allowed, like the exterior walls in a Type III building or the interior, nonbearing partitions of a Type I or II building i.e., FRTW could be used in certain applications in larger, taller buildings where the use of untreated wood is prohibited. Finally, because extracting, manufacturing, and constructing with wood releases substantially less greenhouse gas and consumes much less energy than other building materials with (e.g., steel, concrete), using sustainably harvested FRTW provides a more environmentally friendly solution to the housing crisis (Malmsheimer *et al.* 2008).

One unintended consequence stemming from this increased use of FRTW currently is the installation of products that do not meet the FRTW definition in the IBC (2018) but are touted as fully code-compliant substitutes for FRTW. Often these products are less expensive than FRTW and appear to be properly credentialed; if building inspectors or other officials are not familiar with the FRTW provisions in the building codes, they could easily overlook the discrepancies. The way to identify legitimate FRTW starts with knowing how the product is manufactured.

2 HOW FRTW IS PRODUCED

Most FRTW will be chemically impregnated using a pressure process, although other, non-pressurized methods are possible; pressure impregnation, however, is the only effective means for fire-retardant treating wood (Bueche 2010). This pressure impregnation, first patented in 1893, occurs through a simple six-step process, as shown in Figure 1.

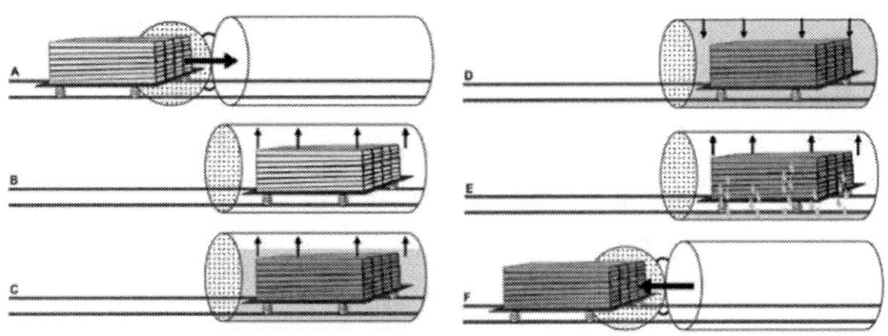

Figure 1. The pressure-impregnation process used by FRTW. Adapted from USDA Forest Service, Forest Products Laboratory (2010), 15-19.

First, untreated wood is loaded into a horizontal cylinder (A); then, the cylinder door is sealed and a vacuum is applied to remove air from the cylinder and the wood (B); next, the treatment is then pumped into the cylinder (C); the pressure in the cylinder is then increased, forcing the treating solution deep into the wood (D); after pressurizing, the treating solution is then pumped out, and a second vacuum removes any excess treatment (E); finally, the treated wood is moved to a kiln to be dried using carefully controlled moisture conditions (F). Once kiln-dried after the treatment to moisture content levels prescribed by the IBC, the FRTW is ready to be labeled and is then subsequently ready for use.

3 HOW FRTW WORKS

Once through the chemical impregnation process, fire-retardant treatment of wood improves fire performance by greatly reducing the amount of flammable gases released, thus reducing the rate at which flames spread over the surface. Treatments reduce the amount of heat available or released by the volatiles during the initial stages of fire and result in the wood self-extinguishing once the primary source of external fuel is exhausted.

To ensure that the treatment is effective, FRTW is required to pass the UL 723 (2018) "Test for Surface Burning Characteristics of Building Materials." This test is designed to gauge the fire-spread over the surface of a material and the smoke developed due to the burning. The test apparatus, originally developed by Steiner (1936), consists of a 25-foot-long rectangular tunnel with two gas burners at one end that direct a 4.5-foot flame under controlled conditions of draft and temperature onto the surface of a 24-foot-long by 20-inch-wide specimen. Flame spreads along the surface of the material as the test progresses. The distance of flame travel and rate at which the flame front advances during a 10-minute exposure determine the calculated flame spread index (FSI).

However, UL 723 (2018) realized that this 10-minute period only demonstrated delayed ignition and gave little indication of non-combustibility. At the request of several insurance rating bureaus, UL wrote a performance standard for pressure impregnated FRTW that requires the test period to be extended an additional 20 minutes to a total of 30 minutes. Malcomson and Bono (1967) wrote that during the 20-minute extension, "the test samples are observed to determine that the flame spread does not exceed the equivalent of 25 (5 ft. beyond the standard igniting flame) and that there is no evidence of significant progressive combustion."

Over the years, changes to the standard affected the calculation procedures used to determine FSI values from raw data. In order to allow listed FRTW from older tests to remain compliant, an additional criterion was added to the codes to address what was already required by the UL Listings. This additional code criterion requires that "the flame front shall not progress more than 10-½ feet beyond the centerline of the burners at any time during the test."

4 FRTW APPLICATIONS

Because of its working properties stemming from the chemical impregnation, FRTW has myriad applications. Table 1 lists these applications next to their code citation in the 2018 IBC. These applications identify applications where an architect may decide to use FRTW or where a building or fire inspector may encounter FRTW in the field. These applications can be significant to address the housing crisis. For instance, UL Design Number V314 is a two-hour rated exterior wall assembly permitted by the code. It consists of two sheets of 5/8" Type X Gypsum Board on an interior-facing side, with 2x4 FRTW studs and either mineral wood or fiberglass insulation behind them, with FRTW plywood on the exterior-face side, and finished with any exterior cladding as permitted by the authority having jurisdiction.

5 DISTINGUISHING FRTW FROM SURFACE-APPLIED TREATMENTS

With so many applications, how specifically can a person distinguish FRTW from surface-applied treatments. Three main approaches exist: how these wood products are manufactured, how they are listed, and how they are labeled.

5.1 Means of Manufacture

For the 2018 IBC (2018), clarifying language was added to the FRTW provisions in Chapter 23 (Section 2303.2.2.): "*The use of paints, coating, stains or other surface treatments is not an approved method of protection as required in this section*" (emphasis added). In other words, materials such as intumescent paints, cementitious coated wood panel products (e.g., oriented strandboard), stains, and other surface-applied or topical treatments do not meet the definition of FRTW in the 2018 IBC.

To reiterate, this added code language, while new, was more clarifying in nature; these surface-applied treatments never met the definition of FRTW since the initial IBC's promulgation in 2000. The primary reason that these surface treatments do not meet the definition of FRTW is that FRTW is "impregnated with chemicals" (ICC 2018, 41). These other treatments are applied topically to the surface of a wood product typically on only one side, or in the case of the cementitious coated products, possibly two (front and back)—but never all six sides. FRTW, because of its pressurized chemical impregnation, provides passive protection for all sides of a wood product, including lumber and plywood, instead of relying on a vulnerable surface barrier.

Table 1. Selected FRTW applications contained in the 2018 IBC.

Code Application	**Citation**
Architectural trim, exterior wall coverings	1405.1.1#3
Attics, elimination of sprinklers in residential occupancy	903.3.1.2.3
Awnings & canopies	3105.2
Balconies and similar projections	603.1#1.4
Bay and oriel windows	705.2.4
Children playground structures in malls	424.2#1
Combustible projections	705.2.3
Exterior bearing & nonbearing walls: Type III const.	602.3
Exterior bearing & nonbearing walls: Type IV const.	602.4.1
Exterior nonbearing walls in Types I & II construction	603.1#1.2
Enclosed combustible spaces in sprinklered buildings of all types of construction: Sprinklers not required	**NFPA 13:** 1999 ed. 8-13.1.1#9; 2002 ed. 8.14.1.2.11; 2007, 2010, 2013, 2016 ed. 8.15.1.2.11
Fire barrier: See partitions Types I & II construction	603.1#1.1
Fuel dispensing station (marine and motor vehicle)	406.7.2
Interior finish with flame spread index < 25 (Class A)	803.1.2
Kiosks in covered and open mall buildings	402.6.2
Liquid storage rooms (shelving, racks, and wainscotting)	415.11.5.2#3
Mechanical equipment screens	1510.6.2#2
Parapet not required when using FRTW sheathing:	
Exterior walls	705.11#5.1
Fire and party walls in Types III, IV, and V	706.6#4.3
Townhouses: Exterior and common wall use within 4ft of such walls	**International Residential Code:** R302.2.4
Partitions (2 hr or less) in Types I & II construction	603.1#1.1
Partitions (fixed) establishing corridors in buildings with one tenant serving no more than 30 people	603.1#11
Pedestrian walkways	3104.3#2
Platforms in Types I, II, and IV construction	410.3
Plenums in all types of construction	**International Mechanical Code:** 602.2.1
Roof construction in Types I & II construction	603.1#1.3
Roof construction in Types I, II, III, & VA construction when > 20 ft. above the floor	Table 601, Footnote b
Rooftop structures (penthouses)	1510.2.4
Shakes and shingles: Wood	1505.6
Wood veneer	1404.5.1
Walls and ceiling furred & dropped more than 1-3/4"	803.15.2.1

The ramifications for an entity approving a product with a surface-applied treatment as a substitute for FRTW could be severe. Imagine a cementitious coated wood panel product with both the front and back sides protected by the coating. These panels are typically installed, per the manufacturer's recommendations, with a minimum one-eighth inch gap. Should a fire find its way into the gap, then that fire has unfettered access to the panel product's exposed, combustible, engineered wood core. Similarly, if an intumescent painted wood panel product is damaged during installation or if it is not maintained or reapplied per the manufacturer's recommendations, then its fire resistance is jeopardized, to say nothing about its likely exposed edges and unprotected, opposite-facing side.

5.2 Listing

Defined in the IBC (2018), a listed product is a product "listed by an organization acceptable to the building official and concerned with evaluation of products or services that maintains a periodic inspection" of those products to ensure that they meet "identified standards or [have] been tested and found suitable for a specified purpose." To earn a UL listing, FRTW must meet the requirements of the UL 723 (2018) test as discussed previously.

Often, proponents for materials that do not meet the building code's definition of FRTW will state that their products do conform as they pass the UL 723 test. Remember, however, that UL 723 is only a 10-minute test. For a product to meet the standard for the purposes of a UL listing, it must pass the extended UL 723 (2018) test i.e., it must also endure the 20-minute extension for a total of 30 minutes.

Once an FRTW product is listed, the listing indicates that the FRTW has been tested for conformance to an applicable standard and is subject to a third-party inspection quality assurance (QA) program. The QA verifies that the minimum level of quality required by the appropriate standard is maintained. The listing agency performing the third-party inspection must be approved by the code official and the basis for this approval may include, but is not limited to, the capacity and capability of the agency to perform the specific testing and inspection.

5.3 Labeling

Closely linked to the FRTW listing is the FRTW label. FRTW, for interior and exterior applications, is required by the IBC to be labeled with all the information shown in Figure 2. Label provisions in the IBC for both interior and exterior FRTW applications are essentially identical except for two key requirements. First, the exterior label must have the text "No increase in the listed classification when subjected to the standard rain test," signifying that the FRTW product has successfully passed the ASTM D2898 (2017) "Standard Practice for Accelerated Weathering of Fire-Retardant-Treated Wood for Fire Testing" test, which is a 12-week test that simulates FRTW exposed to 80 inches of rain per year for a decade. Second, the label must feature the word "Exterior" to signify the FRTW is to be used in a weather-exposed as opposed to a weather-protected application.

Two additional features of the label link it to the listing. First is the mark of an approved agency, in this case the UL Classified Mark. Through its Classification and Follow-Up Services program, UL verifies a material's quality and provenance of production through countercheck and on-site surveillance. As part of the FRTW label, this Classified Mark provides assurances that FRTW products are "listed," and have been tested and produced under the UL Classification and Follow-Up Services program. Second, the phrase "30 Minute Test" reiterates that this material has successfully undergone the code-required extended UL 723 (2018) test.

6 CONCLUSIONS

With the ever-increasing pressures to increase the supply of housing nationwide, the use of FRTW is rapidly increasing as are products that do not meet the definition of FRTW in the building code but are being touted as such. The way to distinguish code-compliant FRTW from these other surface-applied treatments is for architects and building and fire officials to know and understand how these products are manufactured, how they are listed and labeled, and be familiar with the FRTW provisions in the building codes. Only by taking these steps can all involved stakeholders ensure that the pressure to build is used to create larger, taller structures that are code-compliant and safer for the public to occupy.

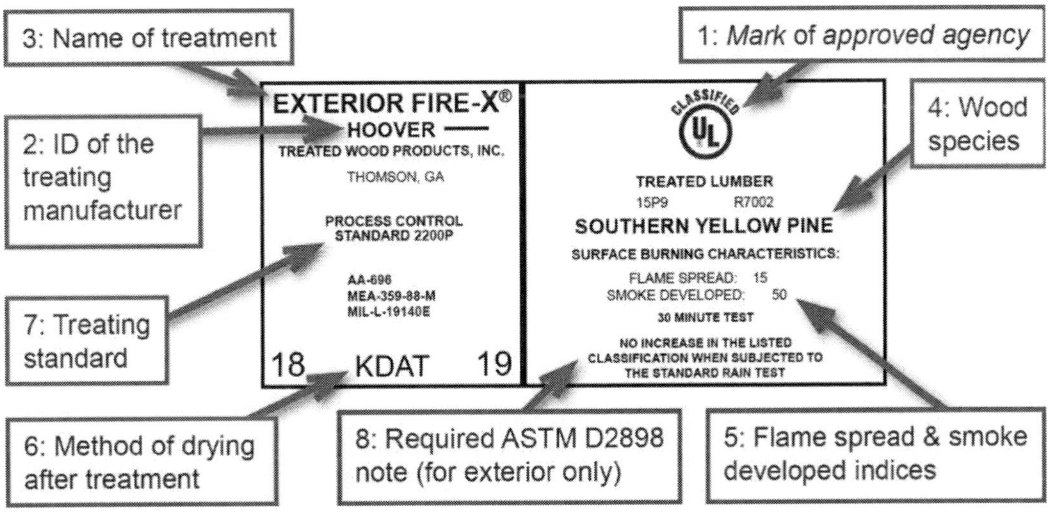

Figure 2. An example of a code-complaint FRTW label for an exterior application.

References

ASTM D2898, *Standard Practice for Accelerated Weathering of Fire-Retardant-Treated Wood for Fire Testing*, ASTM International, West Conshocken, PA, 2017.

Bueche, D., Wood Use in Type I and II (noncombustible) Construction, *Challenges, Opportunities and Solutions in Structural Engineering and Construction*, Ghafoori, N. (ed.), 581-585, CRC Press, Boca Raton, FL, 2010.

Bueche, D., NFPA Code Provisions and Fire-Retardant-Treated Wood, *New Developments in Structural Engineering and Construction*, Yazdani, S., and Singh, A. (eds.), 245-250, Research Publishing Services, Singapore, 2013.

IBC, *International Building Code,* International Code Council (ICC) Publications, Country Club Hills, IL, 2018.

Malcomson, R.W. and Bono, J.A., Underwriters' Laboratories, Inc. Issues new labels for FRTW, *Wood Preserving News*, September 1967.

Malmsheimer, R.W., Hefferman, P., Brink, S., Crandall, D., Deneke, F., Galik, C., Gee, E., Helms, J.A., McClure, N., Mortimer, M., Ruddell, S., Smith, M., and Stewart, J., Forest Management Solutions for Mitigating Climate Change in the United States, *Journal of Forestry*, Oxford Univ. Press, 106(3), 115-173, April 2008.

Nelson, A., and Lang, R., *Megapolitan America: A New Vision for Understanding America's Metropolitan Geography*, Routledge, New York, NY, 2013.

Steiner, A.J. *Investigation of Effectiveness of Fireproofed Red Oak and Maple Lumber*, MS Thesis, Armour Institute of Technology, Chicago, IL, 1936.

UL 723, *Standard for Test for Surface Burning Characteristics of Building Materials*, 11th Edition, UL Publications, Northbrook, IL. 2018.

USDA Forest Service, Forest Products Laboratory, *Wood Handbook: Wood as an Engineering Material, General Technical Report FPL-GTR-190*, U.S. Department of Agriculture, Forest Service, Forest Products Laboratory, Madison, WI, 2010.

Interdependence between Structural Engineering and Construction Management
Edited by Ozevin, D., Ataei, H., Modares, M., Gurgun, A., Yazdani, S., and Singh, A.

ISBN: 978-0-9960437-6-2

MULTI-CRITERIA DECISION MAKING IN EARLY STAGE DESIGN: CAPTURING THE DYNAMICS USING UTILITY THEORY

JOAS SERUGGA and MICHAIL KAGIOGLOU

School of Art, Design and Architecture, University of Huddersfield, Huddersfield, U.K

This paper introduces a conceptual framework that captures the key processes in early stage design (ESD) processes mapping out Decision Making (DM) to support value delivery using utility theory. A lot of construction waste in downstream processes can be traced back to the design stage according to research. Yet there's currently little research into interdependences key to value generation in this most dynamic of construction processes. This research identifies the role of the end-user as a key element in the process of value generation in ESD. Understanding the user requirements is just as important a step as it is for the design process to engage in requirement forecasting. This paper explores new ground in understanding of DM processes in ESD in defining and understanding value. The research first explores the theoretical concepts in multi-criteria DM, UT and value generation in ESD. Design science research methodology is used to review literature to inform the framework that's validated with case studies in two design firms.

Keywords: Front end design, Complexity and uncertainty, Value design.

1 INTRODUCTION

There's an increasing need for new paradigms to deliver improved value in construction projects. Yet at the same time, construction continues to be dominated by a lot of the waste both in the physical and process spaces (Bølviken and Koskela 2016). Halttula *et al.* (2017) attribute a lot of this waste to the limited integration in construction processes. Specific focus has been drawn to the role of early stage design (ESD) processes in value management (VM) as a result (Kpamma *et al.* 2017). VM in ESD is information intensive and dominated by a lot of decision making (DM) in the process of devising design solutions. Much of the waste at this stage is process based though ultimately, this translates into material waste if decisions have to be changed. DM shapes the whole process of value generation from (1) defining the project purpose and objectives, (2) establishing the project attributes (3) trading off attributes by weighing their consequences, (4) defining utilities of attributes and finally, (5) value definition and requirement forecasting. Moreover, DM in this case will be influenced by the project's context including sociocultural, economic, political, technological or environmental constraints. The compounding constraining effect added to the nature of the decision maker does influence the nature of decisions that underpin value delivery in addition to the often conflicting nature of design requirements (DR). On this basis, understanding and underscoring the process of DM is needed. Design typically utilizes attributes (design goals transformed into requirements) and alternatives (models and solutions) to design problems. Good decisions can contribute to value generation through reducing waste. Won *et al.* (2016) traced a lot of waste in construction processes back to DM in ESD through errors or need for reworks among other wastes. Multi Attribute Decision Making (MADM) theory of which Utility Theory (UT) is one, has been used successfully in supporting DM in arriving at decisions among alternatives in many industries processes including construction in value generation (Turskis and Juodagalvienė, 2016). However, none of the applications have been specific to ESD to date. Moreover, UT extends the MADM conceptualization by mapping the expected utility value (EUV) of choice decisions of the DM among the many alternatives (Mintz 2016). This is especially important in ESD when decisions made impact on processes later on in the project cycle. The use of UT therefore presents a step change in the understanding of stakeholder DM during design as design decisions drawn from user goals can be assessed on their EUV. Value can be mapped from UT conceptualization. This paper therefore contributes towards knowledge of ESD DM by presenting a framework that maps value using UT.

2 VALUE MANAGEMENT (VM) AND DM IN EARLY STAGE DESIGN (ESD)

According to Trevisan and Brissaud (2016) process dynamics and interaction of the various constructs that underpin value generation can be understood from system down to subsystem level for the purpose of value definition. The importance of such an approach is the ability to identify any gaps in processes that either are merely non value adding (in which any mitigations or management can be devised); or sheer waste (in which case they have to be addressed). This sets a useful basis for process level analysis in ESD to any improvement propositions. The conceptual basis by extension is important in capturing any dynamics in the flow of tasks and activities that constitute value definition in relation to DM.

In reality however, Kpamma *et al.* (2017) remark on the limited participation of the end user in DM of ESD processes. Additionally, Luo *et al.* (2017) say that understanding the project context is a crucial part of this DM. Zhu and Mostafavi (2017) add that such understanding has to build with it any emergent properties relating to *'absorptive, adaptive, and restorative capacities'* of the project. This requires a structured decision support mechanism. It's important to point out that indeed there has been a proliferation of decision support tools aimed at supporting complex decision making (Nielsen *et al.* 2016). The conceptual UT considered is that by Keeney and Raiffa (1976).

During requirements capture many attributes will present potential for many alternatives in the process of design DM. Mapping the nature of DM on one hand while defining the user benefits on the other is therefore an important part of VM. UT affords both these benefits (Keeney and Raiffa, 1976). Transitivity in UT is a key element in ensuring consistency in the process. UT transitivity during DM therefore provides further support in the way decision makers go about capturing user requirements and the value propositions. The adoption of UT for this research therefore strongly supports efforts for wider need for support tools to bridge the gap in decision support in current dispensations. This paper forms part of the wider research into DM in ESD processes to understand how this contributes to value delivery. The discussion that follows establishes a framework integrating the key concepts of DM, UT and value generation in ESD from a utilitarian perspective; with firstly, a look at the methodological approach.

3 METHODOLOGY

A mixed methods approach of case studies and literature review is adopted Guetterman and Fetters (2018). Design science research was adopted where knowledge was built to inform the design and implementation of the framework (Van Aken *et al.* 2016). A detailed literature review was first carried out across a section of relevant journals and academic articles to develop the conceptual basis of the framework in terms of design practice. The literature from some seminal works like Keeney and Raiffa (1976) supported the conceptual basis of UT. Two general spaces in design process were identified i.e. – the integrated front end design (FED) and implementation/post implementation space. In the former, literature helped identify the attribute definition/requirement capture and tradeoffs spaces. In these stage, information is gathered, outline designs developed, requirements analyzed and alternatives can be compared. In the latter stage, there are opportunities for requirement and utility forecasting, evaluation of attribute utilities that ultimately define the value space. Two architectural firms in Porto Alegre in Brazil dealing in a cross section of housing products, from small residential to public and commercial design were choice selected for the action research. The choice in Brazil presented the element of a dynamic context a key part of the wider research. Through a series of interviews with senior designers, the processes mentioned above were cross examined with ongoing design work to validate the processes integrated in the framework.

4 DECISION MAKING IN UTILITY THEORY

Design processes capture user goals to transform these into design requirements. These effectively are high level goals that are essentially difficult to quantify (Keeney and Raiffa, 1976). The goal of utility theory (UT) is in part to interpret these high level goals into measurable attributes; in manner represented by a Utility function (UF) to support any tradeoffs process. A goal of *'nice views'* or *'quality workspace'* for an office space is qualitative. UT allows each attribute to be considered for its utility by defining a UF; essentially a representation of a DM's preferences when presented with a series of consequence trade-offs. EUV is the aggregation of all expected utilities of a given criterion. It's suggested that defining a UF for any given criterion takes the following three steps for an attribute X; i.e., 1) determining the upper and lower scales of the criterion (X, X_L). A minimum of two is needed for a function to be derived, 2) determining the threshold (X_T) - the neutral point between the two which is given a value zero; and the most preferred (X) point that's set to 1. i.e. $U(x_r)_j = 0 \; and \; U(x_M)_j = 1$ and finally 3) anchoring the points to define a cardinal utility and connecting the points to define a UF either with a straight line as $U_j(x_j) = A_j y_j + B_j$ or exponential function in Eq. (1).

$$U_j(x_j) = A_j e^{B_j y_j} + C_j \tag{1}$$

Where; $U_j(x_j)$ = utility of the criterion j while $A_j, B_j \; and \; C_j$ are constants.

Figure 1 an illustration of such two attributes x, y. Keeney and Raiffa (1976) have demonstrated that the UF (Figure 1) marks interface of attribute points with equal utilities. The UF is equally important in informing the nature of DM; whether the decision maker's risk averse, prone or neutral. Their work is quite important in underscoring DM in ESD when design solutions have to be drawn from a raft of alternatives and from sometimes conflicting attributes. Firstly, they describe a decision maker's risk premium as $X_i - \hat{X}_i$ to support the decision makers risk position. Secondly, they demonstrate that the optimum quantifiable and qualitative variables for the objectives for example can better be captured by understanding the intricate nature of the DM process via the UF. In ESD for example, it's as much important to understand the underlying expectations of the stakeholders as it is to map the DM processes that defines the value. In Figure 1, the utilities of A_3 (Y_3, Z_1),A_2 (Y_2, Z_2), A_1 (Y_1, Z_3) are equal.

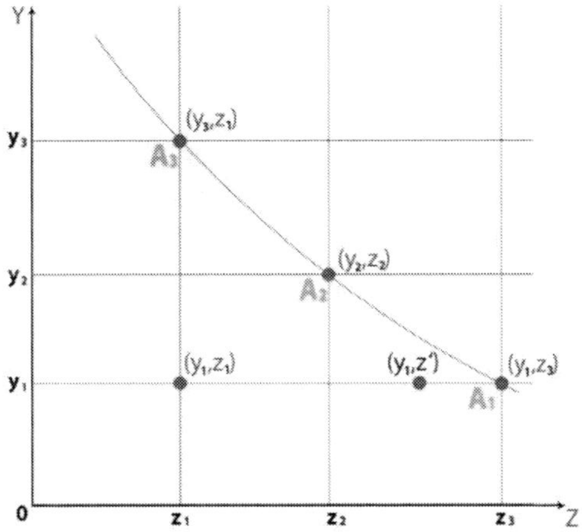

Figure 1. Mapping the decision maker's DM with UF.

In an effort to bring about convergence of DM, a stepwise process is proposed by Keeney and Raiffa (1976) as; drawing up any preliminaries of an assessment that draws focus between process analysis and DM. Fully defining the relevant qualitative parameters relating to the project such as – time frames, cost structures and any incentives; quality or environmental requirements that define the intended objectives and specifying quantitative limits based on lower level attributes relating to these broad high level goals/objectives. Finally, the analyst chooses a UF of the parameters and carries out consistency checks to ensure any results align with not only the objectives but also DM. Again Keeney and Raiffa (1976) suggest a five step process for analysis of a UF as; 1) pre analysis involving examining the nature of the functions, 2) structuring a problem, 3) Assessing the judgmental probability distributions, 4) Carrying out an assessment for all consequences relating to the DM process and preferences and 5) Harnessing opportunities in maximizing the EUV. The basis of UT is thus shown in the attributes S, Q, R, T, W (see Figure 2). The attribute space is bounded by specifying the limits of the attribute X in terms of X_0 and X^*, such that the expected utility X is reflected by $X_0 \leq X \leq X^*$. As buildings are elements of space and time requiring trade-offs during DM, the role of the end-user is important right from the process of purpose of definition from Figure 2. For example, attributes S, Q, R, T, W can be traded off taking into account the costs and benefits of each. This delicate process if integrated with all stakeholders, including the end-user is key in supporting value generation through the attribute utility transformation seen in Figure 2. Defining user requirements – Q - great views – R-optimal workspace, S - Ease of maintenance – T – low energy use, Q – flexible space; and that Utility (U)assumes an additive function defined in Eq. (2).

$$U_{Q,R,S,T,W} = U_Q(Q) + U_R(r) + U_S(S) + U_T(T) + U_W(W) \tag{2}$$

This can be summarized in Eq. (3).

$$U_{Q,R,S,T,W} = \sum_{i=1}^{n} U_i(X_i) \tag{3}$$

5 DISCUSSION AND CONCLUSIONS

The major part of ESD DM is in fact the process of resolving the often conflicting requirements. Collyer and Warren (2009) discuss the sources of these conflicting goals and purposes as stemming from defining design parameters, interpreting design needs and defining concept processes.

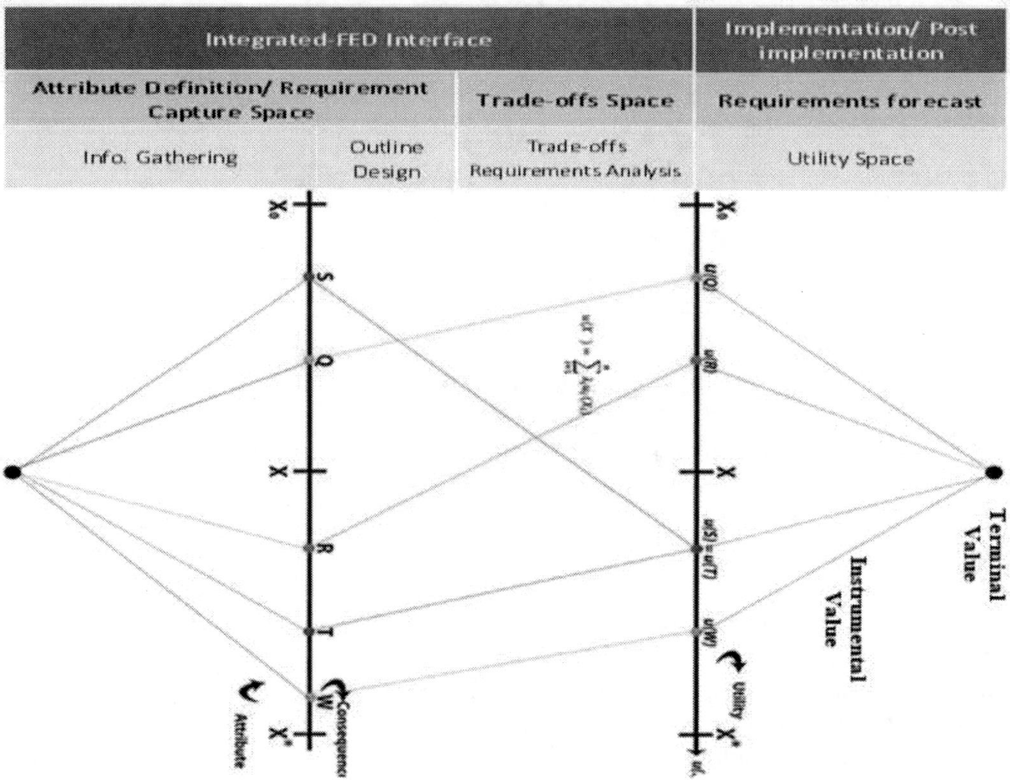

Figure 2. Framework to map value evolution and capture in ESD.

In the framework in Figure 2, this mismatch between interpretations is seen with the utilities of transformed attributes $Q\ and\ R$ on one hand and S on the other; so that in the former, their utilities $(U_Q\ and\ U_R)$ were lower than what was captured in the attribute space and interpreted in the trade-off space. Similarly, the attribute S is seen to have a higher utility (U_S), equal to $T, (U_T)$, than could supposedly be assumed in the previous spaces. According to Pich *et al.* (2002), DM has to adopt some level of ambiguity in this case built around a core purpose, so that any uncertainty arising out of these misunderstandings can be accommodated. Zhu and Mostafavi (2017) reinforce this position arguing that the right approach should be *'absorptive, adaptive, and restorative'* so as to have any change capacity. This stage seeks to reinforce the capacity of the DM process to accommodate any changing needs in the design process.

From the foregoing, UT provides a strong basis for analysis of the decision maker's propensity to risk which can influence their ability to make decisions in ESD. It's important to underscore the decision maker's preference structure from Figure 1 if various decisions carry the same utility especially if that decision is key to value definition. Perhaps some previous criticisms of UT do not adequately reflect the fact that in some settings, value trade-offs are a fundamental part of DM like it's illustrated crucially in this ESD framework. Some of these criticisms relate to transitivity, one of the central tenets of UT. Critics argue that transitivity may not necessarily be present in some real world DM scenarios such as expecting a professional boxer A to beat boxer B simply because boxer C beat boxer B. This perhaps is important in supporting further investigations into the limits of UT as a MADM tool by looking dynamic contexts in a more detail study as part of the wider research.

References

Bølviken, T., and Koskela, L., Why Hasn't Waste Reduction Conquered Construction?, 2016.

Collyer, S., and Warren, C., Project Management Approaches for Dynamic Environments, *International Journal of Project Management,* 27(4), 355-364, 2009.

Guetterman, T. C., and Fetters, M. D., Two Methodological Approaches to the Integration of Mixed Methods and Case Study Designs: A Systematic Review, *American Behavioral Scientist,* 62(7), 900-918, 2018.

Halttula, H., Haapasalo, H., Aapaoja, A., and Manninen, S., Early Involvement and Integration in Construction Projects: The Benefits of DfX in Elimination of Wastes, *International Journal of Management, Knowledge, and Learning,* 6(2), 215-237, 2017.

Keeney, R. L., and Raiffa, H., *Decisions with Multiple Objectives: Preferences and Value Trade-Offs*, Cambridge University Press, 1976.

Kpamma, Z. E., Adjei-Kumi, T., Ayarkwa, J., and Adinyira, E., Participatory Design, Wicked Problems, Choosing by Advantages, *Journal of Engineering, Construction, Architectural Management,* 24(2), 289-307, 2017.

Luo, L., He, Q., Jaselskis, E. J., and Xie, J., Construction project complexity: Research Trends and Implications, *Journal of Construction Engineering Management,* 143(7), 2017.

Mintz, A., *Integrating Cognitive and Rational Theories of Foreign Policy Decision Making: The Polyheuristic Theory of Decision*, Springer, 2009.

Nielsen, A. N., Jensen, R. L., Larsen, T. S., and Nissen, S. B., Early Stage Decision Support for Sustainable Building Renovation–A Review, *Building Environment,* 103, 165-181, 2016.

Pich, M. T., Loch, C. H., and Meyer, A. D., On Uncertainty, Ambiguity, and Complexity in Project Management, *Management Science,* 48(8), 1008-1023, 2002.

Trevisan, L., and Brissaud, D., Engineering Models to Support Product–Service System Integrated Design. *CIRP Journal of Manufacturing Science and Technology,* 15, 3-18 2016.

Turskis, Z., and Juodagalvienė, B., A Novel Hybrid Multi-Criteria Decision-Making Model to Assess A Stairs Shape for Dwelling Houses, *Journal of Civil Engineering Management,* 22(8), 1078-1087, 2016.

Van Aken, J., Chandrasekaran, A., and Halman, J., Conducting and Publishing Design Science Research: Inaugural Essay of the Design Science Department of the Journal of Operations Management, *Journal of Operations Management,* 47, 1-8, 2016.

Won, J., Cheng, J. C., and Lee, G., Quantification of Construction Waste Prevented by BIM-Based Design Validation: Case Studies in South Korea, *Waste Management,* 49, 170-180, 2016.

Zhu, J., and Mostafavi, A., Discovering Complexity and Emergent Properties in Project Systems: A New Approach to Understanding Project Performance, *International Journal of Project Management,* 35(1), 1-12, 2017.

Interdependence between Structural Engineering and Construction Management
Edited by Ozevin, D., Ataei, H., Modares, M., Gurgun, A., Yazdani, S., and Singh, A.

ISBN: 978-0-9960437-6-2

AUTOMATING BACKHOE OPERATION USING FUZZY LOGIC

DEEPAK K. SHARMA

Dept of Civil and Environmental Engineering, California State University, Fullerton, Fullerton, USA

During the past decade, automation has significantly increased in all forms of engineering. With the fast-paced research and development in automation, it is expected that automation will be commonplace and many operations will be performed using robots. Automation of heavy construction equipment is in its early stages. Attempts are persistently made to push the frontiers of science and technology to automate heavy equipment but, limited progress has been made. This research presents a fuzzy logic modeling approach for moving a back-hoe bucket from a fill point to a dump point. The fuzzy logic model is designed to accept all feasible points formed by boom and stick movements in 3D space. Inputs to the model are coordinates of fill and dump spots, as well as stick and boom dimensions. The model provides output signals to automate the bucket movement by swinging the bucket, lowering (or raising) boom, and stretching (or retracting) stick. The paper also includes three site scenarios to demonstrate the performance of the model. The three scenarios are varied from a very basic to a very complex maneuver. All three scenarios are tested for error, and the errors are within acceptable limits.

Keywords: Heavy equipment, Artificial intelligence, Excavator, Construction.

1 INTRODUCTION

Construction projects are characterized by dynamically changing site conditions. As a result, construction sites present challenges when it comes to any innovation and that has impacted the overall productivity of the industry. This is applicable to horizontal and vertical construction projects, but it is more prominent in case of horizontal construction projects like highways, canals, runways, and other similar projects. Many tasks on horizontal construction projects are executed by manually operated heavy equipment and as a result, project outcome is dependent on operators' skill. A highly competent operator will be able to lessen a cycle time as compared to an average operator (Holt and Edwards 2015). The availability of skilled operators is thus critical for overall construction industry. However, US is facing labor shortage (FRD 2018) and the issue doesn't seem to be abating soon, mandating that we seek ways to get around it.

Automation is known to improve productivity (Chen *et al.* 2018) and can also help in mitigating the impacts of labor shortage. This paper presents a fuzzy logic model for automating backhoe excavator operation. The approach will enable agencies, and equipment manufacturers to develop autonomous equipment. Using trigonometry, spherical geometry concepts and fuzzy theory, a model is here presented that aims to automate backhoe bucket movement. Model inputs consist of 3D coordinates of fill and dump points as well as stick and boom dimensions. As an output, the model provides a set of signals that will operate the excavator control joysticks. These commands will move excavator bucket from filling point to dumping point. The movement will include bucket swinging, lowering (or raising) boom, and stretching (or retracting) stick.

2 PREVIOUS STUDIES AND RESEARCH MOTIVATION

Construction productivity is lagging in terms of benefiting from technological advancements. As a result, the industry has lagged when it comes to productivity improvements in comparison to other industries (Fulford and Standing 2014). Researchers have demonstrated that construction automation can improve its productivity (Chen *et al.* 2018). While other industries have transformed into fully automated industries, construction and earthmoving operations could not be automated fully till date. Several researchers have identified challenges faced while automating earthmoving equipment (Dadhich *et al.* 2016) and these challenges require that the research community attempts to meet the automation needs.

Researchers are persistently trying to push the frontiers of science and technology to make construction automation a norm (Naskoudakis and Petroutsatou 2016). A remote-controlled robotic arm has been developed that can control an equipment from a distance (Sasaki and Kawashima 2008). Similarly, a tele-operation system that enables an operator to mount the system on arm and control the excavator system remotely is also now available (Kim *et al.* 2009). Recently, the tele-operated system was equipped with additional systems consisting of head mounted display (Le *et al.* 2017). Agencies are in the process of automating construction processes. CALTRANS is developing Automated Machine Guidance (AMG) technology to improve safety, efficiency and accuracy during construction (CALTRANS 2018). On the same lines, FHWA's e-construction group reports about various technologies that use mobile phones, GPS and other advance technologies during construction. Despite these efforts, a recent review implies that equipment automation is only limited on controlling heavy equipment by remotes (Naskoudakis and Petroutsatou 2016) mandating continuous human intervention. As a result, there is a need to fully automate equipment operation. This research work builds on the previous research and presents a model for achieving comprehensive heavy equipment automation.

3 RESEARCH METHODOLOGY

A model is developed in this research that will help contemporary researchers to pursue heavy equipment automation. The model focuses on a back-hoe excavator's (excavator) displacements to move a bucket automatically. The overall process has been broken down in to small steps. As a first step, the research boundaries were created. This helped in defining the tasks and extent of this research. The second step included using feasible coordinates for filling and dumping spots followed by determining resulting angular position for excavator's boom and stick. The third step requires calculating 3-dimensional displacement for boom and stick to be able to move the bucket from fill to dump. Finally, the model outputs were analyzed to determine accuracy of the model. Scenario analysis is also included to study model performance in varying situations.

Automating excavator operation imposes variety of challenges due to its parts' movement. Boom and stick operate in coordination (swings and angularly moves) to move bucket between desired points. The bucket controls allow scooping and releasing earth. Also, crawlers move the whole excavator on ground. Considering all the activities makes the problem extremely large and complex. So, in this research the scope is restricted to the coordinated boom and stick movements along with the swinging action, thus covering many points scattered in a 3-dimensional space. The focused and limited research enables us to concentrate this paper only on one of the most unique movement seen only in excavators. The approach proposed here can be easily extended to consider bucket operations, other attachments and ground movement.

The next step focuses on locational, trigonometric and geometrical aspects of the excavator. We start with the assumption that the 3D Cartesian coordinates for fill and dump points are available. This assumption is made because accurate GPS devices are easily available and these are frequently in earthwork industry. GPS devices can be attached to the end of stick and accurate coordinates of fill and dump points can be obtained. Based on the coordinates, the angles between boom and stick are calculated. Figure 1 shows the layout of boom and stick.

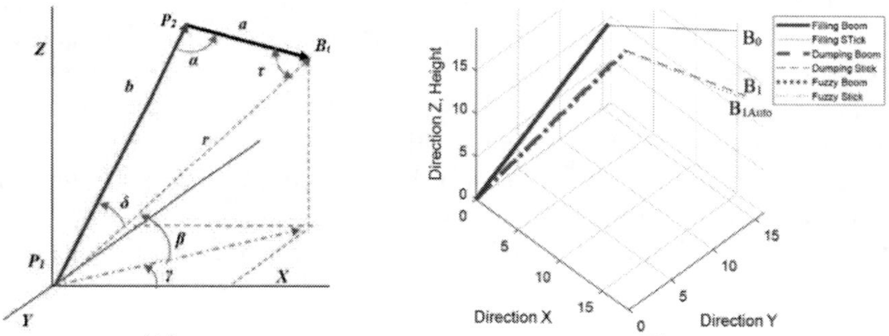

(A) (B)

Figure 1. (A) Geometry and angular relations between boom and stick, (b) Base case scenario results showing the resulting fuzzy boom and stick.

P_1 represents center of the equipment with coordinates [0,0,0]. The equipment swings at point P_1 about the Z-axis and boom is hinged at P_1. Segments P_1P_2 *and* P_2B_0 represents boom and stick with dimensions *b* and *a* respectively. Point B_0 is tip of the stick where bucket is attached. Boom makes an angle θ with the XY plane. It is obtained by adding β (angle made by the resultant of boom and stick (P_1B_0) with the XY plane) and δ (angle made by the resultant P_1B_0 and boom). The projection of P_1B_0 on XY plane makes an angle γ with the positive X axis. α represents angle between boom and stick. The stick makes angle ϕ with the negative Z-direction and τ with the resultant P_1B_0. MATLAB was used to establish the relation between coordinates and angular relations of boom and stick. Calculations started with determining angular position and length of P_1B_0. Matlab's cart2sph function was used to calculate γ and β as well as distance P_1B_0 using coordinates of B_0. Cosine rule was then used to calculate the angles α, δ, τ in triangle $P_1P_2B_0$ (Eq. 1-3).

$$\alpha = \mathrm{Cos}^{-1}(b^2 + a^2 - r^2)/(2ab) \tag{1}$$

$$\tau = \mathrm{Cos}^{-1}(a^2 + r^2 - b^2)/(2ar) \tag{2}$$

$$\delta = \mathrm{Cos}^{-1}(b^2 + r^2 - a^2)/(2br) \tag{3}$$

This allowed us the calculate θ (not shown in figure) as per Eq. (4):

$$\theta = \beta + \delta \tag{4}$$

Using geometrical relations, angle ϕ was obtained using Eq. (5):

$$\phi = \theta + \alpha - 90^0 \tag{5}$$

The coordinates of point P_2 were obtained by using MATLAB's sph2cart feature. This completed one set of calculation that enabled determining all the necessary geometrical details for the filling spot. Using the same steps again, angles and coordinates were recalculated for dumping spot. The dumping spot is represented as B_1 in this research. This enabled determining the stick and boom movement to move bucket from fill to dump location. All these data became input to the fuzzy model.

Fuzzy logic is an Artificial Intelligence (AI) modeling approach that allows using linguistic variables (Zadeh 1996), a property that can be conveniently applied to the current research. It is also useful when a process is complex, non-linear and operator dependent (Chen and Kairys 1993). Its ease of converting linguistic variables to desirable outputs has been exploited here to develop a fuzzy model for excavator control.

During excavation, an operator observes bucket fill and dump locations and moves the bucket between the points. The same process can be repeated through fuzzy modeling. First the visual observations are replaced by precise (crisp) coordinates. Using the geometrical relations, the locations of boom and stick are fed to the fuzzy inputs which are then fuzzified. The operator maneuver skills in fuzzy modeling are represented by several if-then rules. For example, several rules can be created that could look like *"if the dump location is too far, stretch the stick and forwarding the boom to reach the distant point"* to cover all possible scenarios. All this information collectively provided the desired outcome to be able to operate joystick controls.

The fuzzy logic model was developed in MATLAB using the approach discussed in The Basic Tipping Problem (Matlab 2017). It has three inputs and three outputs. Mamdani Fuzzy Inference System was used. The inputs included the displacements needed by boom and stick in X, Y and Z directions to move a bucket from fill to dump location. The outputs were designed to be able to provide push-pull commands for left and right joystick controls. The excavator was assumed to have standard SAE controls which means that as the left-hand joystick control is moved right the equipment swings right; left movement swings it left; forward movement pushes boom down and a pull lifts the boom up. Similarly, when the right joystick is moved forward the stick moves away and when pulled back moves the stick comes closer. All these details were used to create outputs. All inputs and outputs were assigned triangular membership function, a method detailed in Tipping Problem. Finally, validation of this approach is necessary to determine if the automated system provides acceptable outcome. Validation was done by comparing outputs from the fuzzy model with the expected outputs obtained through the trigonometric equations and geometrical relations (Eqs. (1) to (5)).

4 BASE CASE, SCENARIO ANALYSIS AND RESULTS

To test the above model a simple scenario was considered. This scenario represents an easy maneuver for an operator. In this scenario B_0 was assigned [15, 16, 19] and B_1 was assigned [18, 14, 15] coordinates. For any excavator operator this will be an easy maneuver because the overall bucket movement is 3 ft along the X direction, -2 ft along the Y direction and 4 feet in the Z direction, making a linear movement of just 5.3852 ft. Since both the points are above ground (Z coordinates are positive for both points) both points are clearly visible to operator, making it a simple maneuver. The fuzzy model developed in this research gave outputs that enabled moving the bucket from B_0 to B_{1Auto} (B_{1Auto} is the dump point obtained from the fuzzy model). The coordinates of the B_{1Auto} point came out as [18.137, 13.850, 14.792] which is very close to the desired B_1 coordinates [18, 14, 15]. The linear distance between points B_1 and B_{1Auto} represents the error and it comes out to be 0.2909 ft. Figure 1 B shows the details of this scenario in 3D view. In the figure the filling position is represented by solid lines. The thick solid line is boom and the thin solid line is stick. At some angle the dashed lines represent stick and boom reaching the dump position. The dashed lines can be observed with closely following dotted lines, representing the stick and boom reaching the dump location through the fuzzy model. Figure 1(B) shows the fill and dump spots and the corresponding boom and stick positions in a 3D view.

In the base case the overall movement of boom and stick was limited to a few feet in X, Y and Z directions. This represented a very simple maneuver. To test the model performance for some difficult scenarios the model was tested for two more scenarios representing increasingly challenging maneuvers. The results of FL model are given in Table 1.

Table 1. Results obtained using the fuzzy logic model.

Sr	Location Coordinates		Maneuver Complexity	Automatic Dump Coordinates	Error (ft)
	Pick Up	Dump			
1	[15, 16, 19]	[18, 14, 15]	Low (Base)	[18.137, 13.850, 14.792]	0.2909
2	[15, 22, 3]	[21, 5, 19]	Medium	[20.851, 4.4534, 19.436]	0.7148
3	[10, 22, -10]	[21, -18, 19]	Severe	[21.0597, -17.6105, 19.2689]	0.4771

The results indicate that the fuzzy model was able to move the bucket within a very close range of the targeted dump location. The errors in the three cases were 0.2909 ft (3.491 in) for the base case, 0.718 ft (8.616 in.) for the medium severity case and 0.4717 ft (5.6604 in.) for the extreme case. Given that the dump location will be above a dump-truck and dimension of a standard dump-truck in the US is 59×37 in^2, the maximum error of 8.616 in. can be considered acceptable. This is also considered acceptable because a human operator empties a bucket approximately above the 59×37 in^2 truck-bed and while doing so relies on visual judgement.

All these results indicate that if an excavator is provided with fill and dump coordinates and the designed fuzzy model is used, a bucket can be autonomously moved between desired locations. Repeating this several times will enable autonomous earth moving. While the current system does not include bucket and crawler operation, the learning from this paper can be extended to include bucket and crawler in the model. Having a fully automated excavator will have significant impact on productivity and safety.

5 CONCLUSIONS

Automating heavy equipment has a tremendous opportunity but a very limited advancement could be seen in terms of fully automating heavy equipment. Through this paper, a fuzzy logic model is proposed for automating excavator operation. The insights gained through this work can be extended to other excavation operations and equipment. This work will enable researchers, agencies and equipment manufacturing companies to pursue fully automated equipment.

The results of the research showed that fuzzy logic modeling enabled automating excavator operation. This was demonstrated by the fact that the model was tested for three scenarios and gave good results. Visually, the results are very good and the automated excavator will be able to move a bucket autonomously. The results are comparable to the work by other researchers (Shi *et al.* 1996). The authors used fuzzy logic to operate a dozer replica and the trajectory of equipment was analyzed. Similarly, in this research the visual analysis (Figure 1B) indicates a very good outcome. Quantitatively the error ranged from 3.491 in. to 8.616 in. which is acceptable considering the size of equipment employed on excavation sites. The errors are likely due to the suboptimal membership function design. Additional research, which will be largely based on trial and error and thus could be intensive, is expected to improve the results further.

Based on the findings of this research it is expected that researchers extend this excavator problem by including bucket operation and movement on ground. Future researchers will be able to build upon this work by considering other heavy equipment for automation. An immediate application is to automate front shovels and cranes. From an application perspective, it is suggested that the automation model be extended further and used for on mass earthwork sites such as canals, dams, highways and others. At an initial phase these excavators can work as a part of other equipment fleet where the fleet is dumping dirt near an automated excavator and the excavator keeps repeating the loading operation. Such a deployment will enable studying the model's real-life performance and improve it. The deployment is expected to improve on-site productivity and thus enable meeting the project's overall goals efficiently.

5.1 Future Works

The modeling approach discussed in the paper provides a basic groundwork that can be pursued further and one of them is productivity. Automating backhoes will have a noticeable impact on productivity because the operations will be very precise. An operator maneuvers a backhoe by moving the left joystick forward and backward to move the boom down and up respectively. Similarly, moving right joystick forward and backward moves the stick away and brings it closer respectively. Manual control of the joysticks will move bucket along a non-linear path and thus it takes more time than a movement on a linear path. As a result, the production will increase, because, as per Peurifoy *et al.* (2011) the basic production is inversely proportional to cycle time and thus any reduction in the cycle time will increase the overall productivity of the backhoe.

References

CALTRANS, Implementation of Automated Machine Guidance, CALTRANS.com, (n.d.). Retrieved from http://www.dot.ca.gov/hq/construc/amg/Implementation.pdf on May 15, 2018.

Chen, G., and Kairys, G., *When and When Not to Use Fuzzy Logic in Industrial Control,* Conference Record of the 1993 IEEE Ind. App. Conf. Twenty-Eighth IAS Annual Meeting, 2035-2040, 1993.

Chen, Q., Soto, B. C., and Adey, B. T., Construction Automation: Research Areas, Industry Concerns and Suggestions for Advancement, *Journal of Automation in Construction,* 94, 22–38, 2018.

Dadhich, S., Bodin, U., and Anderson, U., Key Challenges in Automation of Earth-moving Machines, *Journal of Automation in Construction,* 68, 212–222, 2016.

FRD, *The Beige Book - Summary of Commentary on Current Economic Conditions,* Published by Federal Reserve District, 2018.

Fulford, R., and Standing, C., Construction Industry Productivity and the Potential for Collaborative Practice, *International Journal of Project Management,* 32, 315–326, 2014.

Holt, G., and Edwards, D., Analysis of Interrelationships Among Excavator Productivity Modifying Factors, *International Journal of Productivity and Performance Management,* 64(6), 853–869, 2015.

Kim, D., Kim J., Lee, K., Park, C., Song, J., and Kang, D., Excavator Teleoperation System Using a Human Arm, *Journal of Automation in Construction,* 18, 173–182, 2009.

Le, Q. H., Lee, J. W., and Yang, S. Y., Remote Control of Excavator Using Head Tracking and Flexible Monitoring Method, *Journal of Automation in Construction,* 81, 99-111, 2017.

Matlab, The Basic Tipping Problem, www.mathworks.com (n.d.). Retrieved from https://www.mathworks.com/help/fuzzy/building-systems-with-fuzzy-logic-toolbox-software.html#brzqs45 on October 15, 2017.

Naskoudakis, I., and Petroutsatou, K., A Thematic Review of Main Researches on Construction Equipment Over the Recent Years, *Procedia Engineering,* 164, 206–213, 2016.

Peurifoy, R. L., Schexnayder, C. J., Shapira, A., and Schmitt, R. L., *Construction Planning, Equipment and Methods,* 8th Ed., McGraw Hill, ISBN, 978-0-07-340112-6, 2011.

Sasaki, T., and Kawashima, K., Remote Control of Backhoe at Construction Site with a Pneumatic Robot System, *Journal of Automation in Construction*, 17, 907–914, 2008.

Shi, X., Lever, P. J., and Wang, F. Y., Fuzzy Behavior Integration and Action Fusion for Robotic Excavation, *IEEE Transactions on Industrial Electronics*, 43(3), 1996.

Zadeh, L. A., Fuzzy Logic = Computing with Words, *IEEE Transactions on Fuzzy Systems*, 4(2), 103-111, 1996.

Interdependence between Structural Engineering and Construction Management
Edited by Ozevin, D., Ataei, H., Modares, M., Gurgun, A., Yazdani, S., and Singh, A.

ISBN: 978-0-9960437-6-2

ESTIMATING THE RUNNING COSTS OF COMMERCIAL BUILDINGS: ARTIFICIAL NEURAL NETWORK MODELING

DEVINDI GEEKIYANAGE and THANUJA RAMACHANDRA

Dept of Building Economics, University of Moratuwa, Moratuwa, Sri Lanka

Running costs of a building is a substantial share of its total life-cycle cost (LCC) and it ranges between 70-80% in commercial buildings. Despite its significant contribution to LCC, investors and construction industry practitioners tend to mostly rely on construction cost exclusively. Though the early stage estimation of running costs is limited due to the unavailability of historical cost data, several efforts have been taken to estimate the running costs of buildings using different cost estimation techniques. However, the prediction accuracy of those models is still challenged due to less quality and amount of data employed. This study, therefore, developed an artificial neural network (ANN) model for running costs estimation of commercial buildings with the use of building design variables. The study was quantitatively approached and running costs data together with 13 building design variables were collected from 35 commercial buildings. The ANN model developed resulted in a 96.6% perfect correlation between the running cost and building design variables. The testing and validation of the model developed indicate that there is greater prediction accuracy. These findings will enable industry practitioners to make informed cost decisions on implications of running costs in commercial buildings at its early stages, eliminating excessive costs to be incurred during the operational phase.

Keywords: Cost modeling, Operations cost, Maintenance cost, Building design variables, Decision-making, LCC.

1 INTRODUCTION

Usually, costs incurred during the operational phase of a building responsible for a substantial share of its Life Cycle Cost (LCC). Some buildings have inherently higher running cost than others, such as commercial buildings. For example, the running costs of commercial buildings account for over 69% of the total LCC (Wang *et al.* 2014). Similarly, Wong *et al.* (2010) revealed that the running cost of an office building varies between 72 to 81% of its total LCC. Despite its contribution to the LCC structure, often running cost is given less focus in investment decision making and investors tend to mostly rely on initial cost alone.

A recent study on the review of existing models for LCC estimation revealed that there is no simple model for estimating the running cost of buildings to date (Krstić and Marenjak 2017). The application of available methods and models for the running cost estimation of buildings are also limited to the later stage of building life cycle as these models require an extensive set of operational cost data (Krstić and Marenjak 2017). For example, Al-Hajj and Horner (1998) have presented a running costs model for institutional buildings, with eleven cost elements and to an accuracy of 1.13%. Similarly, Kirkham *et al.* (2002) and El-Haram *et al.* (2002) have developed

WLCC models for hospital buildings where cost components such as facilities management costs, energy costs, maintenance costs, residual costs, and discount rate were determinants of WLCC. Early-stage supportive running cost estimation models are therefore essential as it provides implications of costs to be incurred during the operating phase of buildings at early design stages of building constructions.

Estimation of cost of a product, system, or service based on its determinants is a well-known and approved method for cost estimation over the years. For example, Kirkham *et al.* (1999) have developed an energy cost model for sports centres based on building design variables such as the number of users and floor area. However, Krstić and Marenjak (2017) stressed that these models are not based on adequate historical cost records and not based on the available cost structure, rather than standard cost structure. Authors further indicate that the models developed so far ignore some important factors such as the age, location, level of occupancy, and standards of operation.

Deciding through which type of building to include in a forecasting model is not the only problem. The choice of modelling technique is also important (Boussabaine *et al.* 1999). Among the statistical approaches, regression techniques deserve attention due to relative ease of implementing and requirement of less computational power than other statistical approaches (i.e. genetic algorithms, neural networks, support vectors machine) (Fumo and Biswas 2015). However, the application of purely parametric cost estimation methods is limited due to the lack of reliable historical cost data and building design variables, which have a direct influence on its LCC. In contrary, Boussabaine *et al.* (1999) opined that statistical models have been used for some time but in present, artificial intelligence is proposed as a more reliable and accurate modelling technique. Providing professionals with accurate forecasting techniques will enable them to make informed and reliable estimates of likely running cost in commercial buildings, as well as other forms of buildings. Therefore, this study introduces an early-stage supportive running cost estimation model for commercial buildings with use of the artificial neural network (ANN) modelling.

2 RESEARCH METHODS

The research was primarily approached quantitatively to develop early-stage supportive running cost estimation models for commercial buildings with the use of ANN modelling. The documents including architectural drawings, bills of quantities, historical cost records, and monthly utility bills were reviewed to collect the required data. The case buildings selected for the study was limited to 35 out of the population of 117 commercial buildings, which were recorded in Sri Lanka due to the time constraints and limited access to cost data. Generally, a sample size of more than 30 at 5% confidence level is sufficient for many types of research. Though it is said that a big sample of data is required to run an ANN, an ANN tool including a particular training-validation-test procedure for small datasets has been developed some years ago and recently refined in order to obtain not only realistic regression laws, but also reliable ones (One can refer to Pasini and Potestà (1995) and Pasini *et al.* (2001) for the fundamentals of this tool) (Pasini 2015). Accordingly, the commercial buildings selected for the study consists of 49% of office buildings and 37% of banks while remaining include educational institutes, retails, and multi-purpose (i.e. hotel + apartment) buildings. Further, a majority of the selected buildings (63%) consists of three to 12 while remaining 26% and 11% are 13 to 25 and above 25 storied buildings respectively.

Based upon statistical pre-analysis, 13 variables (i.e. building design variables), which are quantitative in nature and convertible (nominal data) were selected for predicting the running cost of commercial buildings. The influence of variables on the running cost and ease of availability of data were the primary factors in the selection of the variables. Further, the running cost data were collected in accordance with the standards of BCIS, BS ISO 15686-5:2008 standard, and NRM3, for three consecutive financial years: 2014, 2015, and 2016.

Initially, the collected dataset was subjected to the 'Multiple imputation" technique to impute the missing values within the data set. Next, the target variable was normalized using the gross internal floor area and obtain the normalized target variable called running cost/sq. ft. The ANN model was developed with the aid of Neural Designer machine learning software and the Feedforward neural network with backpropagation training was administered as it is commonly used with linear activation function. Finally, the prediction accuracy of the developed model was evaluated with use of the mean absolute percentage error (MAPE) and Theil's U value.

3 DATA ANALYSIS AND FINDINGS

In order to proceed with the neural network analysis, there are three basic assumptions to be satisfied. Firstly, both dependent and independent variable s should be the continuous form of data. In this study, the dependent variable, which is running costs/sq. ft and independent variables including working days/week, working hours/day, building age, GIFA, net floor area, circulation area, height, number of floors, window area, Window-to-Floor-Ratio, and number of occupants are scale data. In addition, two dummy variables namely, the grouping of buildings (1=Detached, 2=Attached), and type of structure (1=Concrete, 2=Steel, 3=Pre-fabricated) were added to the analysis to represent the nominal data collected. Therefore, satisfied the first assumption. Next, the Shapiro-Wilk normality test was conducted to explore the normal distribution of residual values. As observed from Table 1, the significance of the standardized residual (ZRESI) is greater than 0.5 indicates that the ZRESI is normally distributed.

Table 1. Test of normality: Shapiro-Wilk.

	Statistic	**df**	**Sig.**
Standardized Residual	0.954	30	0.211

Next, the relationship between the dependent variable and the independent variables are needed to be linear, both for each independent variable and globally. Accordingly, a scatterplot analysis was conducted between each independent variable and the dependent variable and the charts derived are presented in Figure 1. As shown in the scatterplot matrix, five continuous independent variables namely GIFA, NFA, CA, building height, and the number of floors out of 11 have strong linear relationships with the dependent variable: running cost. Although other six independent variables don't represent strong linear relationships with the dependent variable as the points are more scattered and it is observed that the points are trying to gather along the diagonal. Therefore, it is concluded that all the independent variables have linear relationships with the dependent variable, thus satisfied the third assumption. As shown in the scatterplot matrix, five continuous independent variables namely GIFA, NFA, CA, building height, and the number of floors out of 11 have strong linear relationships with the dependent variable: running cost. Although other six independent variables don't represent strong linear relationships with the dependent variable as the points are more scattered and it is observed that the points are trying to gather along the diagonal. Therefore, it is concluded that all the independent variables have linear relationships with the dependent variable, thus satisfied the third assumption.

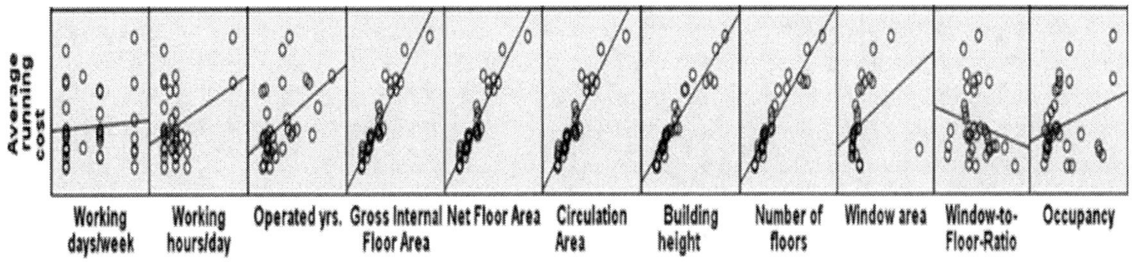

Figure 1. The relationship between the running cost and building design variables.

3.1 The Artificial Neural Network Model for Running Cost Estimation of Buildings

Initially, the pretreated data (with multiple imputation and normalization) was input to the "Neural Designer" software and the data set consists of 35 instances was itself divided into three types of instances such as 60% of training instances, 20% of selection instances, and 20% of testing instances. Next, the data set was further pretreated by setting all instances, which includes constant variables, repeated variables and univariate outliers and multivariate outliers as unused instances. Fortunately, the results derived indicate that there are no any constant or repeated variables in the data set and only 2 instances were set as unused due to outliers. Then, the neural network was developed and it represents the predictive model. Accordingly, the size of the scaling layer is 13, the number of inputs. The scaling method for this layer is the "MinimumMaximum". Further, the neural network was designed with three layers. Table 2 depicts the size of each layer and its corresponding activation function. The architecture of this neural network can be written as 13:10:7:1.

Table 2. The summary of the neural network.

	Input number	Neurons number	Activation function
1	13	10	Linear
2	10	7	Linear
3	7	1	Linear

The statistics of the parameters shown in Table 3 depict information about the complexity of the model. In general, it is desirable that all minimum, maximum, mean and standard deviation values are not very big as shown for the developed model.

Table 3. Parameters statistics of the natural network model.

	Minimum	Maximum	Mean	Deviation
Parameters	-0.990295	0.990173	-0.0601779	0.57255

The loss index plays an important role in the use of a neural network. It defines the task the neural network is required to do and provides a measure of the quality of the representation that it is required to learn. The normalized squared error (NSE) is used here as the error method. If the NSE has a value of unity then the neural network is predicting the data 'in the mean', while a value of zero means a perfect prediction of the data. In this network, the NSE is 3.28.

The procedure used to carry out the learning process is called training (or learning) strategy. The quasi-Newton method was applied as the training strategy of the neural network in this study in order to obtain the best possible loss. It is based on Newton's method but does not require calculation of second derivatives. Instead, the quasi-Newton method computes an approximation of the inverse Hessian at each iteration of the algorithm, by only using gradient information. Accordingly, the initial value of the training loss is 3.18746, and the final value after 230 iterations is 0.00238913 whereas the initial value of the selection loss is 5.59346, and the final value after 230 iterations is 0.208116.

A standard method to test the loss of a model is to perform a linear regression analysis between the scaled neural network outputs and the corresponding targets for an independent testing subset. This analysis leads to three parameters for each output variable. The first two parameters, a and b, corresponding to the y-intercept and the slope of the best linear regression relating scaled outputs and targets. The third parameter, R^2, is the correlation coefficient between the scaled outputs and the targets. If we had a perfect fit (outputs exactly equal to targets), the slope would be 1, and the y-intercept would be 0. If the correlation coefficient is equal to 1, then there is a perfect correlation between the outputs from the neural network and the targets in the testing subset. Accordingly, Table 4 lists the linear regression parameters for the scaled output running cost/sq. ft.

Table 4. The linear regression parameters for the scaled output running cost/sq. ft.

Regression parameters	Value
Intercept	-0.0355
Slope	1.14
Correlation	0.966

The mathematical expression represented by the neural network inputs working days/week, working hours/day, attached/detached, age, gross internal floor area, net floor area, circulation area, height, no. of floors, type of structure, window area, window to floor ratio and occupancy to produce the output Running cost/sq. ft. For function regression problems, the information is propagated in a feed-forward fashion through the scaling layer, the perceptron layers and the unscaling layer.

3.2 Model Testing

The purpose of model testing is to evaluate the performance of the developed ANN model in estimating a functional form that relates the design variables of commercial buildings to the running cost. Table 5 presents the prediction accuracy of the developed ANN model with use of the MAPE and Theil's U statistic, which commonly used to measure the performance.

Table 5. Results of test statistics for model accuracy.

Test	ANN
MAPE	-4.9%
Theil's U value	0.049

As shown in Table 5, the average MAPE of the ANN model is -4.9%, indicates that the ANN model has been achieved a high accuracy. The Theil's U value for the ANN model is 0.049 (where the U value indicates greater accuracy as $U \rightarrow 0$). Further, the neural network model recorded a correlation of 0.966, this accuracy is better than that recorded by the so far developed parametric regression models for LCC estimation.

4 CONCLUSIONS

The paper has highlighted the importance of different modelling techniques for predicting cost to be incurred during the operation phase of buildings particularly, commercial. The level of MAPE for the ANN model can be considered acceptable in most real applications, depending on the phase of application of the model. It is clear from this limited experiment that ANN was able to extract a functional form (i.e., a function) that represents the problem under investigation. The study has also shown that ANN models may prove as a good alternative to parametric cost modelling. Within the limits of this study, ANN models have been shown to be able to model data that strongly exhibit noise and achieve reasonable accuracy.

Acknowledgments

This work was supported by the Senate Research Committee of University of Moratuwa under the Grant No. SRC/LT/2017/21.

References

Al-Hajj, A., and Horner, M. W., Modelling the Running Costs of Buildings, *Construction Management and Economics*, Taylor and Francis, 16(4), 459-470, 1998.

Boussabaine, A. H., Kirkham, R. J., and Grew, R. G., Estimating the Cost of Energy Usage in Sport Centres: A Comparative Modelling Approach, *15th Annual ARCOM Conference,* Hughes, W. (ed.), 481-488, Liverpool, UK, 1999.

El-Haram, M. A., Marenjak, S., and Horner, M. W., Development of A Generic Framework for Collecting Whole Life Cost Data for The Building Industry, *Quality in Maintenance Engineering*, Emerald, 8(2), 144-151, 2002.

Fumo, N., and Biswas, R. M. A., Regression Analysis for Prediction of Residential Energy Consumption, *Renewable and Sustainable Energy Reviews*, Elsevier, 47(C), 332-343, 2015.

Kirkham, R. J., Boussabaine, A. H., and Awwad, B. H., Probability Distributions of Facilities Management Costs for Whole Life Cycle Costing in Acute Care NHS Hospital Buildings, *Construction Management and Economics*, Taylor and Francis, 20(3), 251-261, 2002.

Kirkham, R. J., Boussabaine, A. H., Grew, R. G., and Sinclair, S. P., Forecasting the Running Costs of Sport and Leisure Centres, *08th International Conference on Durability of Building Materials and Components,* Lacasse, M. A., and Vanier, D. J. (eds.), 1728-1738. Institute for Research in Construction, 1999.

Krstić, H., and Marenjak, S., Maintenance and Operation Costs Model for University Buildings, *Technical Gazette*, 24(1), 193-200, 2017.

Pasini, A., Artificial Neural Networks for Small Dataset Analysis, *Journal of Thoracic Disease*, AME Publication Company, 7(5), 953, 2015.

Pasini. A., Pelino, V., and Potestà, S., A Neural Network Model for Visibility Nowcasting from Surface Observations: Results and Sensitivity to Physical Input Variables, *Journal of Geophysical Research,* American Geophysical Union, 106(D14), 14951-14959, July, 2001.

Pasini, A., and Potestà, S., Short-Range Visibility Forecast by Means of Neural-Network Modelling: A Case-Study, *Il Nuovo Cimento C*, Springer, 18(5), 505-516 September, 1995.

Wang, N., Wei, K., and Sun, H., Whole Life Project Management Approach to Sustainability, *Management in Engineering*, American Society of Civil Engineers, 30(2), 246-255, March, 2014.

Wong, I. L, Perera, S., and Eames, P. C., Goal Directed Life Cycle Costing as A Method to Evaluate the Economic Feasibility of Office Buildings with Conventional and TI-Façades, *Construction Management and Economics*, Taylor and Francis, 28(7), 715-735, 2010.

Interdependence between Structural Engineering and Construction Management
Edited by Ozevin, D., Ataei, H., Modares, M., Gurgun, A., Yazdani, S., and Singh, A.

ISBN: 978-0-9960437-6-2

ANALYSIS OF IMPEDIMENTS TO A SUCCESSFUL CONSTRUCTABILITY PROCESS IN HIGHWAY CONSTRUCTION

R. EDWARD MINCHIN[1], REIHANEH SAMSAMI[1], DAN TRAN[2], DAN D'ANGELO[3], SIDNEY SCOTT[4], YIFENG TIAN[1], HAEKYUNG IM[1], and JEFFREY RUSSELL[5]

[1]*M.E. Rinker Sr. School of Construction Management, University of Florida, Gainesville, USA*
[2]*Dept of Civil, Environmental and Architectural Engineering, University of Kansas, Lawrence, USA*
[3]*Applied Research Associates, Inc, Champagne, USA*
[4]*HKA, Philadelphia, USA*
[5]*Dept of Civil and Environmental Engineering, University of Wisconsin, Madison, USA*

The US Federal Highway Administration (FHWA) Every Day Counts Program (EDC) has resulted in state Departments of Transportation (DOTs) putting evermore emphasis on speeding up the delivery of highway and bridge construction projects for use by the driving public. This has resulted in an increase in the use of integrated project delivery methods and adding alternative technical concepts (ATCs) to traditional design-bid-build (DBB) contracts. ATCs have exhibited great potential for delivering substantial benefits like cost savings, increased constructability, and quicker project delivery. Previous research has found that knowledge of project constructability was lacking in state DOT planning, programming, and environmental staffs. At the same time, the permitting process for several government agencies has become increasingly restrictive. The intent of this paper is to report on the research team's progress in an ongoing effort to furnish the US government with a uniform set of guidelines for the application of the constructability process during all phases of project development and delivery. The research uses surveys, focus groups and interviews to determine which states have implemented formal programs to ensure that the constructor is furnished with a set of contract documents that affords said constructor with the best possible opportunity to successfully construct the project with the highest quality standards, within the contract duration and without exceeding the construction budget.

Keywords: Project management, Alternative technical concepts, Project planning and design, Highway and road design, NCHRP10-99.

1 INTRODUCTION

The concept of constructability was introduced into the construction industry in the 1980s. The research directed by the Construction Industry Institute (CII) was the driving force behind applying and implementing constructability. CII defines constructability as "the optimum use of construction knowledge and experience in planning, design, procurement, and field operations to achieve [the] overall project objective" (CII 1986). In the highway sector, several state Departments of Transportation (DOTs) applied constructability to their projects in the early

1990s. During this period, Constructability Review (CR) systems were developed for Florida DOT (Ellis *et al.* 1992) and Wisconsin DOT (Russell *et al.* 1994). The Wisconsin constructability work process enables construction input in design, creation of constructability teams, continuity and communication between design and construction, and continuous improvement of highway projects.

The intent of CRs is to identify unclear specifications, refine project design, increase project construction efficiency, and potentially reduce disputes, cost overruns, and delays. Research has documented several key benefits of conducting CRs, including reduced cost, shorter schedules, improved quality, better risk-control, fewer claims, and fewer change orders. NCHRP Reports 390 and 391 are considered key references for implementing CRs for DOTs because of their systematic approach to integrating constructability into the project development process (Anderson and Fisher 1997).

DOTs across the nation are constantly dealing with ever-increasing technical complexities, increasing regulatory restrictions, and pressures to deliver quality products on time and within budget. This can create a schedule-driven environment during the project development process that can lead to errors, omissions, and constructability problems. To overcome such issues, DOTs are increasingly using alternative contracting methods, including design-build (D-B) and construction-manager-as-general-contractor (CM/GC), along with applying alternative technical concepts (ATCs) to procurement in several project delivery systems.

The Federal Highway Administration (FHWA) Every Day Counts Program (EDC) brought national visibility to implementing ATCs for incorporation into transportation projects. ATCs have well-documented potential to accrue sizable benefits in cost savings, increased constructability, and schedule reduction. Among the many documented benefits of alternative contracting is early construction input during project development and design. In fact, some practitioners believe that CM/GC inherently builds constructability into project delivery, eliminating the need for a separate constructability process.

DOTs often seek contributions from industry representatives to identify constructability issues and improve their environmental, design, construction, and maintenance processes. The objective of the NCHRP 10-99 research is to develop a decision-making Framework scalable to all contract delivery methods that complies with all environmental statutory requirements, in order to assist DOTs in evaluating savings from using CRs and industry input during planning, design, and permitting.

2 LITERATURE REVIEW

Although constructability is an important concept in construction projects, it was rarely applied to construction projects until relatively recently. In 1989, the Texas DOT sponsored a project named "Investigation of Highway Project Constructability Improvement" implemented by Hugo *et al.* (1989). In this project, constructability concerns related to highway specifications were studied. The results of the analysis show that information and communication, project scoping, and processes and methods are the major constructability problems. The resistance by owner, contractor, and designers hinders the integration between design and construction, as well as other factors like contractual limitations and time shortage.

Fischer and Tatum (1997) studied the design-relevant characteristics of constructability knowledge. They demonstrated that lack of formal constructability knowledge is the primary reason for lack of constructability inputs. Availability and level of detail are two concerns of designers. Access to sufficient knowledge at the proper time was known as one of the factors that play an enormous role in understanding constructability issues. The other element was the level of detail of CRs, which provide designers with feasible alternatives that would lead them to constructible designs. Fischer and Tatum showed that an appropriate level of detail at the right time will enable companies to improve their performance.

Anderson *et al.* (1997) studied the CR process for transportation facilities through the NCHRP 390 research project. As stated in this study, State Transportation Agencies (STAs) are interested in initiating the CRs in early stages of projects to maximize their benefits and according to them, lack of feedback to the design team, lack of input from construction team, low clarity of the specifications and plans, and lack of available reviewers are the main issues in conducting CR processes. In addition, the authors stated that paradigm shifts in both project-level and agency-level are necessary to implement the constructability process. While strategies like a team approach and pilot projects benefit the STAs at the project level, organizing for constructability and a systematic lessons-learned process help them improve the CR processes at the agency-level.

Delay in delivering highway projects is a great source of conflict. Ford *et al.* (2004) studied this issue and recommended formal CRs as a potential solution to reduce construction project delays. The effects of these reviews were studied at the design and construction phases to determine the optimal amount of constructability to achieve the greatest effect on projects. For this purpose, a model was designed to relate CRs to project duration as a management tool. Their study showed that there is a positive correlation between the time put into CRs and the design phase duration, and a negative correlation between the time invested in CRs and construction duration. Based on these results, the authors illustrated that an intermediate-sized CR is the most efficient solution to reduce delays in comparison to small or large reviews.

The design phase has been targeted through different studies on constructability because it is known as the most obvious phase within which to implement CRs. Lam *et al.* (2006) gathered the major factors affecting constructability of the designed projects. Those authors demonstrate that three major factors of through-site and ground investigation before the actual design process starts, coordinating the design documents and sequences, and standardized and safe designs are the most critical factors. By considering these factors, it is possible to improve the efficiency of projects and save time and cost while delivering a high-quality project.

Tauriainen *et al.* (2014) studied constructability through assessing Building Information Modeling (BIM). After discussing the system used in Finland as a constructability appraisal model, those authors developed an experimental constructability assessment method (ECAM) as a constructability information source, using BIM. By applying ECAM, constructability scores of the building and building elements were calculated. These scores could forecast the potential constructability issues of the projects in the future and support decision making on them.

One of the latest research studies reviewing the past, present and anticipated future practices of constructability was conducted by Kifokeris and Xenidis (2017). Their study presented research regarding constructability implementation, tools, and current trends, to evaluate its applications. The success performance parameters for any project considered within this article are cost, time, quality, and client satisfaction.

3 RESEARCH METHODOLOGY

The research is broken into two phases. These phases are named "Benchmark the Current State of Practice" and "Decision-Making Framework". The main objective of the first phase is to define the state of the practice regarding the use of CRs. For this purpose, designing a rigorous Research Plan is required to ensure the collection of high-quality data that can yield meaningful trends and the detection of any gaps and barriers in the current state of CRs. Special attention will be given to impacts on NEPA and the environmental permitting process.

In Phase I, "Benchmark the Current State of Practice," five tasks are defined as: literature review, define the current state of practice, develop conceptual Framework, prepare interim report, and face-to-face meeting. Task 1 is literature review, through which a comprehensive content analysis will be conducted to cover both historic and contemporary writings, covering the current environmental requirements for public transportation agencies, use of BIM in the CR process, and the codes that impact the CRs. All tasks related to defining the current state of the practice are covered under Task 2. Every effort is being made to include a clear description of the best practices discovered in Phase I, and then to utilize what is learned in Phase II of the project. Gathering data on the current state of practice is done utilizing a survey, focus groups, case studies, and a workshop. Task 3 is focused on developing a conceptual Framework to provide transportation agencies with sound constructability practices that will help in evaluating cost and time savings, project quality, and other metrics across the entire project development process. The decision-making Framework that is finally recommended by the team is apt to resemble the Conceptual Framework shown in Figure 1. Task 4 is preparing the interim report, which will be started toward the end of completion of Tasks 1 and 2 and will include the following: the findings of the state of practice review; a preliminary design of the proposed decision-making Framework; and a Work Plan for Phase II. The last task of the first phase is a face-to-face-meeting between project team, which will include the Work Plan for Phase II, the preliminary proposed decision-making Framework, and plans for the Workshop.

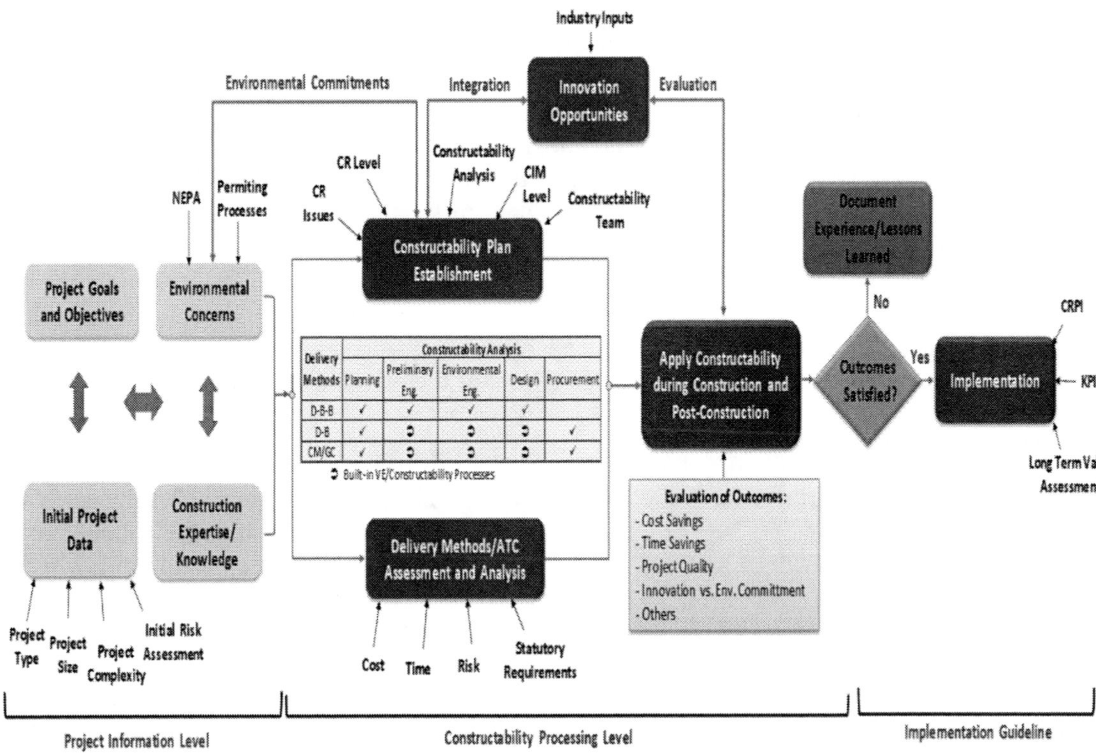

Figure 1. Decision-making framework.

In Phase II, "Decision-Making Framework," the following tasks will be performed: develop Framework, workshop, develop training materials, develop an electronic presentation, final report, and prepare final deliverables. Task 1 is assigned to Developing a Decision-Making Framework. This new Framework will result from changes made to the Conceptual Framework (Figure 1), based on information obtained from the data analysis and the comments of the NCHRP 10-99 Panel in the face-to-face Meeting. Under Task 2, a Workshop will be developed and convened with representatives from DOTs, panel members, and the research team to demonstrate the proposed decision-making Framework and gather feedback. The next step is Developing Training Materials for the use of DOTs in training their personnel in the implementation of CRs and industry input during project development. This will be done as Task 3. In conjunction with Task 3, Task 4 is to Develop an electronic presentation to be used as part of the training effort. Once Tasks 3 and 4 are completed, a final report will be prepared as an outcome for Task 5.

4 RESEARCH PROGRESS

Tasks 1 and 2 were worked on for the time period spanning from the last week of May 2018 to the end of August 2018. The literature review (Task 1) began during proposal preparation and has been completed. Through this task, over 60 writings, consisting of papers, National Cooperative Highway Research Program (NCHRP) projects, and guidelines were reviewed. Several areas such as the definition of constructability, CR implementation and issues, integrating constructability into the project development process, design-relevant constructability knowledge, risk management for addressing constructability issues, incorporating maintainability in the CR process, the role of CRs in conflict resolution, the relationship between total quality management and CR, lessons-learned systems for constructability, and the benefits of BIM / virtual reality in constructability are covered by Task 1.

A data gathering process is underway under Task 2, contacting the Central Design and Central Construction Offices of each of the 52 State DOTs have been contacted separately and key personnel from each office interviewed. This task is 99% completed. The team has successfully interviewed 103 offices (of the 104) with the purpose of identifying the DOTs which have formal constructability processes. Besides, this task was focused on gathering data on whether the DOTs include contractor/industry input during their process. Interviewees have provided the names/contacts of consultants and contractor who are experienced in conducting CRs, and the team has successfully interviewed 25 out of the 49 additional contacts.

5 RESULTS AND SUMMARY

The next step after gathering data is the data analysis. Table 1 has the results of the questions asked of 103 Construction and Design employees from 52 DOTs. This initial telephone survey consists of three main questions and two follow-up questions. The questions and the results are in Table 1

Each state is ranked using its Design and Construction Offices' answers to the basic questions of the survey. Using this ranking criterion, a number ranging from 1 to 5 is assigned to each state. States ranking higher than 4.0 are the selected states. Through the telephone interview process, the team has recognized ten DOTs as most knowledgeable and experienced in conducting CRs. These state DOTs are Arizona, California, Colorado, Maryland, New Mexico, New York, Oregon, Pennsylvania, and Virginia.

The research team plans to interview the 1 remaining DOT office and 20 referred additional contacts in the coming month. The data gathering process will last through the end of 2018.

With the completion of data gathering, data analysis and conceptual Framework development will start at the beginning of 2019.

Table 1. Answers to questions in initial telephone interview.

No.	Question	YES	Participants	Percentage
1	Do you have a formal constructability process?	73	98	74%
1a	Does your formal constructability process change when you use alternative delivery?	36	73	49%
1b	Does your agency's effort to improve constructability involve / include industry input?	60	73	82%
2	Does your agency have established processes that allow contractor input into the design?	49	98	50%
3	Has contractor input to the design ever been hampered by a fear on the part of the parties involved of having to re-open the NEPA process, environmental permitting process, or any other pre-construction process?	23	98	23%

References

Anderson, S. D., and Fisher, D. J., *Constructibility Review Process for Transportation Facilities,* No. 390. Transportation Research Board, 1997.

CII, *Constructability: A Primer,* Construction Industry Institute, University of Texas at Austin, 1986.

Ellis, R. D., Kumar, A., and Ahmed, Q. A., Engineering and Industrial Experiment Station: Life Cycle Reconstructibility, University of Florida, 1992.

Fischer, M., and Tatum, C. B., Characteristics of Design Relevant Constructability Knowledge, *Journal of Construction Engineering and Management*, 123(3), 253-260, 1997.

Ford, D. N., Anderson, S. D., Damron, A. J., de Las Casas, R., Gokmen, N., and Kuennen, S. T., Managing Constructability Reviews to Reduce Highway Project Durations, *Journal of Construction Engineering and Management*, 130(1), 33-42, 2004.

Hugo, F., O'Connor, J. T., and Stamm, E. M., Specification Improvements for Enhanced Constructibility, *Research Project,* 3-6-88/0, 1989.

Kifokeris, D., and Xenidis, Y., Constructability: Outline of Past, Present, and Future Research, *Journal of Construction Engineering and Management,* 143(8), 04017035, 2017.

Lam, P. T., Wong, F. W., and Chan, A. P., Contributions of Designers to Improving Buildability and Constructability, *Design Studies*, 27(4), 457-479, 2006.

Russell, J. S., Swiggum, K. E., Shapiro, J. M., and Alaydrus, A. F., Constructability Related to TQM, Value Engineering, and Cost/Benefits. *Journal of Performance of Constructed Facilities*, 8(1), 31-45, 1994.

Tauriainen, M., Puttonen, J., Saari, A., Laakso, P., and Forsblom, K., The Assessment of Constructability: BIM Cases, *eWork and eBusiness in Architecture, Engineering and Construction*, 55-61, 2014.

Interdependence between Structural Engineering and Construction Management
Edited by Ozevin, D., Ataei, H., Modares, M., Gurgun, A., Yazdani, S., and Singh, A.

ISBN: 978-0-9960437-6-2

IMPROVING MARKETING FUNCTIONS THROUGH SOCIAL MEDIA PRESENCE AMONG CONSTRUCTION SMEs

ADEDEJI AFOLABI[1], RAPHEAL OJELABI[1], PATIENCE TUNJI OLAYENI[1], IGNATIUS OMUH[1], and OPEYEMI OYEYIPO[2]

[1]*Building Technology, Covenant University, Ota, Nigeria*
[2]*Quantity Surveying, Covenant University, Ota, Nigeria*

With over 3.5 billion people connected to the internet, small and medium-scale (SME) construction businesses have the opportunity to enhance their customer base. The study assessed ways by which marketing functions of SME construction businesses can be improved through social media presence of the firms. The study used a quota-purposive sampling of construction businesses in Nigeria's mega city, Lagos State. The small and medium scale construction businesses comprised of consulting and contracting firms. Using a questionnaire instrument, the social media arm of the marketing unit of SME construction businesses was measured. Presentation of the data was carried out using pie charts and principal component analysis. The study identified two main critical success factors (CSFs) – Firm/Social Media characteristics and Security/Investment factor enabling the Social Media infrastructure for SME construction businesses. A focus on these entities would ensure the successful integration of social media marketing strategies in the traditional marketing functions of construction SMEs. The survival of SME construction business is hinged on the availability of customers which social media exposure helps attract and interact with potential customers. SME Construction businesses of the 21st century must align with the 'global village' trend of incorporating social media applications in their traditional marketing functions.

Keywords: Construction industry, Critical success factors, Customers, Facebook, Performance, Nigeria.

1 INTRODUCTION

The performance of small and medium scale enterprises (SMEs) is appreciated in their ability to create employment and wealth for the nation (Sandberg *et al.* 2002). Osotimehin *et al.* (2012) noted that SMEs engage the use of mainly local raw materials while employing local labor from the urban and rural centers thereby balancing the wealth even to the rural parts. Despite the increased budgetary allocations and loan acquisition schemes available locally, financial and technical aid internationally, Osotimehin *et al.* (2012) reported that SMEs have not been able to accelerate employment and poverty alleviation in order to foster the Nigerian economic development. Apart from the financial concerns, which a large number of SMEs find it difficult to access due to stringent measures and conditions pegged by financial houses (Chukwuemeka 2009). Osotimehin *et al.* (2012) asserted that SMEs are faced with other constraints in management, technology, competition, the price of available local raw materials, infrastructural facilities and multiplicity of policies resulting in double taxation and so on. The distinct nature of SMEs makes them easy prey to unpalatable business environment that larger firms may easily manoeuver. Coad (2007) opined that the growth of SMEs can be somewhat unpredictable. However, this has not deterred the high frequency of new firms been formed daily but with low survival ratings over a few years (Asma *et al.* 2015). To buttress this, Bartelsman *et al.* (2005) revealed that 20-40 percent of new firms fail within two years post-entry, while only about 40-50 percent survive beyond the seventh year. This can be attributed to the numerous hurdles each new firm has to face over time.

One of the ways SMEs can survive in the harsh terrain of doing business is by increasing its share of marketing functions. This can be improved by using several innovative tools and techniques order than the traditional methods of marketing. The construction industry has been mostly traditional in its operation, although, it is fast changing and embracing new innovative information and communication technologies (ICTs) (Afolabi and Oyeyipo 2017, Ojelabi *et al.* 2018a). An innovative marketing function available to construction SMEs is the use of social media platforms. In today's world, every organization must as a necessity exist in an online platform. This is because, online platforms such as the use of websites, social media platforms and other internet-based systems that link customers to products have changed the way businesses should be run. The global impact of web-based systems cannot be over-emphasized for the far-reaching benefits they engender on businesses (Ibem and Laryea 2015, Afolabi *et al.* 2017a). Companies need customers to grow and survive in an economic environment, while a large percentage of these customers' time is increasingly being spent on social media platforms. Therefore, getting consumers to purchase the company's products is hinged on meeting customers in the market place of online media. With more than 3.5 billion people users of one form of social media app, the likelihood of attracting a user to a product is fairly high (Iblasi *et al.* 2016). Lee *et al.* (2014) noted that other firms aside construction SMEs are judiciously increasing their marketing budgets to include advertising on social networking sites such as in developed countries where spending on social media marketing has increased from 8.4% to about 22% within the last five years. A social media presence is able to showcase construction SMEs' strength, past and ongoing projects, which can go a long way in assuring clients of their prowess. Therefore, the study assessing dimensions to which marketing functions of construction SMEs can be improved through social media presence. In order to achieve this, the study showed the critical success factors (CSF) enabling the social media infrastructure for SME construction businesses.

2 RESEARCH METHOD

This study is based on related works that have measured the use of social media platforms in the construction section. The study by Ojelabi *et al.* (2018b) measured the management of the social relationship that exists between clients and construction companies using social media platforms. While this study focused on the marketing functions – the enablers and limiting factors, which are referred to as the critical success factors (CSFs). The study area is centered in the Southwestern part of Nigeria in Lagos State. The state has a high number of registered and unregistered construction SMEs within the study location involved in contracting and consulting services. A survey research design was used in carrying out the study, which entailed the design of a questionnaire instrument. Due to the lack of a comprehensive list, the sample size was selected using a quota-purposive sampling technique. A total of seventy-nine (79) construction SMEs participated in the social media study. This represented a 63% response rate and is deemed adequate for the study. The study used a 5-point Likert scale in measuring the critical success factors (CSF) enabling the social media infrastructure for SME construction businesses. The data collected was presented using pie charts and factor analysis.

3 RESULTS AND DISCUSSION

This section contains the result and discussion on the critical success factors (CSF) enabling the social media infrastructure for SME construction businesses. Firstly, the SME construction information was described to understand the characteristics of the participants. Section two (2) contains an analysis of the critical success factors (CSF) enabling the social media infrastructure for SME construction businesses.

3.1 Characteristics of the Construction SMEs

The study showed the company details of construction SMEs that participated in the survey. Figures 1 and 2 showed the pie-chart breakdown of the organization type and the years of existence of the companies in the Nigerian business sector. The pie-chart in Figure 1 showed the organization type, which indicated that 46 (58.2%) of the construction SMEs were into contracting, 29 (36.7%) of them were mainly into consulting works on construction projects, and 4 (5.1%) were into other types of jobs within the construction sector. In Figure 2, the years of the company's existence in the construction sector showed the 14 (17.7%) of the construction SMEs have been in operation for 1 to 5 years, 34 (43%) have been in the construction sector for 6 to 10 years, 10 (12.7) of the construction SMEs have been in operation for 11 to 15 years, 8 (10.1%) of them had 16 to 20 years' experience in the construction industry and 13 (16.5%) of the construction SMEs had above 20 years of expertise in the construction sector.

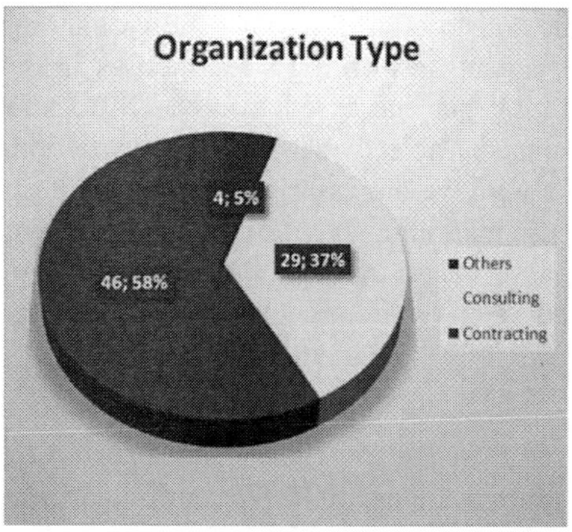

Figure 1. Organization type.

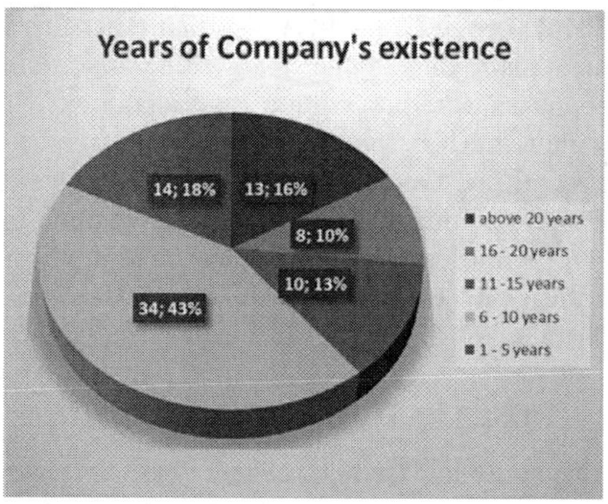

Figure 2. Years of Company's existence.

3.2 Critical Success Factors (CSF) Enabling the Social Media Infrastructure

In order to classify the critical success factors (CSFs) enabling the social media infrastructure by SME construction businesses a factor analysis test was carried out. The thirteen (13) CSFs tested in this study were adapted for the study by Preece *et al.* (2014) and Ojelabi *et al.* (2018a). These CSFs focused on the use of the information and communication technologies (ICTs), the process involved and the human involvement in the process and acquisition. The factor analysis is a statistical technique to explore relationships among variables of the CSFs. By this, it would reduce or summarize the data of the variables into groups. Principle factor extraction analysis (PCA) with Varimax rotation was performed using SPSS 21.

The eigenvalues associated with each linear component after rotation was generated with the CPA indicating 10 components. In this light, a scree plot as shown in Figure 3 was resorted to and this showed a clear break and point of inflection at the second factor. Therefore, only two (2) components are significant for exploring the critical success factors (CSFs) enabling the social media infrastructure by SME construction businesses.

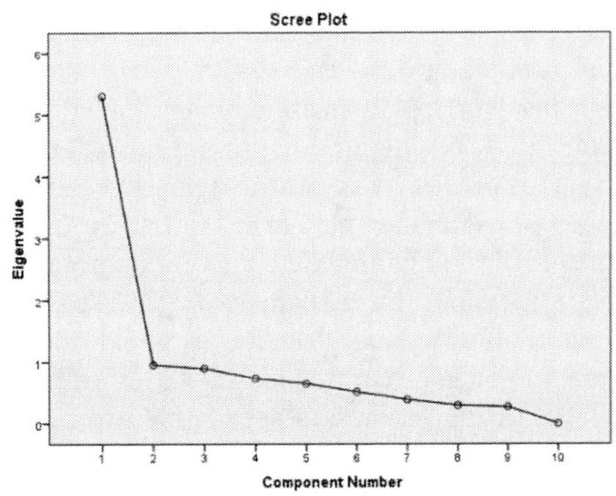

Figure 3. Scree plot of the Critical success factors (CSFs) enabling the social media infrastructure.

There are two (2) factors shown from the scree plot which accounted for 63% (factor 1 = 53%; factor 2 = 10%) of the total variance from the variables tested. Table 1 showed the extracted rotation component matrix of the factor loading for Critical success factors (CSFs) enabling the social media infrastructure.

Table 1. Factor loading for Critical success factors (CSFs) enabling the social media infrastructure (Rotated Component Matrix).

	Critical Success Factors	**Component 1**	**Component 2**
Firm/ Social media Characteristics	Management's Commitment	.819	.348
	Capacity of firm	.819	.384
	Optimism about social media marketing strategy	.795	.209
	Social media capacity as a business tool	.653	.256
	Ease of navigating social media platforms	.644	.199
	Understanding of social media metrics	.609	.516
Security/ Investment Factor	Secured company/clients' information	.176	.816
	Investment in social media infrastructure	.255	.762
	Control over social media use	.421	.694
	Competitors' influence	.271	.582

Extraction Method: PCA.
Rotation Method: Varimax with Kaiser Normalization.

From Table 1, there are two main factors that reflect in the loading and have factor loadings above 0.50. The result revealed two main categories of factors, which are "Firm/Social Media Characteristics" and "Security/Investment Factor". In the first factor, which is "Firm/Social Media Characteristics", there is a need for critical success factor of management commitment and optimism about social media marketing strategy. For SME construction businesses, social media platforms should be marketed as a business tool and therefore the required investment and infrastructure should be put in place. Another CSF is "Capacity of Firm" which can be viewed from the size of the firm. Large construction companies adequately use social media platforms and are visible to different clients. This means that SME construction businesses can tap from the available clients on social media platforms and also showcase their strength and business profile through pictures of their head offices, construction equipment, qualified personnel, past and ongoing construction projects. It is essential that social media app developers and actors should try and ensure that their platforms are easy to navigate by construction SMEs. Furthermore, metrics of how the profiles of each construction SMEs can be provided on potential clients and those that have visited the profiles including the construction products they are most interested in, can be provided such as in the study by Afolabi *et al.* (2017b). The second CSF classified as "Security/Investment Factor" highlighted that for a social media marketing strategy to be successful there is a need to generate confidence in the information stored and communicated through the platform. Construction businesses need some level of control over their project and client information so as to avoid undue influence from other competitors. "Competitors' influence" would also enable social media marketing functions of construction SMEs. The world has changed, and no business should want to be left behind due to the prospects and benefits social media marketing strategy has to offer every sector of the economy.

4 CONCLUSIONS AND RECOMMENDATION

The study assessed ways by which marketing functions of SME construction businesses can be improved through social media presence of the firms. Essentially, there are some critical success factors (CSFs) that enable the social media infrastructure for SME construction businesses. The study identified two main CSFs: "Firm/Social Media characteristics" and "Security/Investment factor". A breakdown of the factors showed that Firm/Social Media characteristics comprised of "Management's Commitment", "Capacity of Firm", "Optimism About Social Media Marketing Strategy, Social Media Capacity as a Business Tool", "Ease of Navigating Social Media Platforms" and "Understanding of Social Media Metrics", while "Security/Investment Factor" comprised of "Secured Company/Clients' Information", "Investment in Social Media Infrastructure", "Control Over Social Media Use" and "Competitors' Influence". These entities would ensure the successful integration of social media marketing strategies in the traditional marketing functions of construction SMEs. The survival of SME construction business is hinged on the availability of customers which social media exposure helps attract and interact with potential customers. SME Construction businesses of the 21st century must align with the "global village" trend of incorporating social media applications in their traditional marketing functions. It is recommended that on the part of construction SMEs, there is a need for increased commitment, optimism and investment towards social media infrastructure as a business tool. On the part of social media developers/actors, there is also a need to increase the security of information and provision of social media business metrics so as to measure the performance of the tool by construction SMEs.

Acknowledgments

The researchers appreciate Covenant University for funding the open access publication of this manuscript.

References

Afolabi, A., Owolabi, D., Ojelabi, R., Oyeyipo, O., and Aina, D., Construction Professionals' Perception of a Web-Based Recruiting System for Skilled Labour, *Journal of Theoretical and Applied Information Technology*, 96(10), 2885 – 2899, 2017a.

Afolabi, I. T., Ezenwoke, A. A., and Ayo, C. K., Competitive Analysis of Social Media Data in The Banking Industry, *International Journal of Internet Marketing and Advertising,* 11(3), 183 – 201, 2017b.

Afolabi, A., and Oyeyipo, O., The Perception of Future Decision-Makers on The Building Profession, *Malaysian Construction Research Journal*, 21(1), 55 – 73, 2017.

Asma, B. B., Diabate, A., and Othman, A., Establishing the Factors Affecting the Growth of Small and Medium-Sized Enterprises in Algeria, *American International Journal of Social Science*, 4(2), 101-115, 2015.

Bartelsman, E., Scarpetta, S., and Schivardi, F., Comparative Analysis of Firm Demographics and Survival: Evidence from Micro-Level Sources in OECD Countries, *Industrial and Corporate Change*, 14, 365-391, 2005.

Chukwuemeka, E. O., Poverty and the Millennium Development Goals in Nigeria: The Nexus, *Educational Research and Review*, 4(9), 405-410, 2009.

Coad, A., *Firm Growth: A Survey*, Papers on Economics and Evolution, 2007.

Ibem, E. O., and Laryea, S., E-Procurement Use in The South African Construction Industry, *Journal of Information Technology in Construction*, 20, 364-384, 2015.

Iblasi, W. N., Bader, D. M. K., and Al-Qreini, S. A., The Impact of Social Media as a Marketing Tool on Purchasing Decisions (Case Study on SAMSUNG for Electrical Home Appliances), *International Journal of Managerial Studies and Research,* 4(1), 14–28, 2016.

Lee, D., Hosanagar, K., and Nair, H. S., *The Effect of Social Media Marketing Content on Consumer Engagement: Evidence from Facebook*, WITS Conference, 1–50, 2014.

Ojelabi, R. A., Oyeyipo, O., Afolabi, A.O., and Amusan, L. M., Presence of Social Client Relationship Management within the Nigerian Construction Industry, *Buildings*, 1-15, 2018a.

Ojelabi, R.A., Afolabi, A.O., Oyeyipo, O., Tunji-Olayeni, P. F., and Adewale, B. A., Data Exploration of Social Client Relationship Management (CRM2.0) Adoption in The Nigerian Construction Business, *Data in Brief*, 18, 1471-1476, 2018b.

Osotimehin, K. O., Jegede, C. A., Akinlabi, B. H., and Olajide, T. O., An Evaluation of the Challenges and Prospects of Micro and Small Scale Enterprises Development in Nigeria, *American International Journal of Contemporary Research*, 2 (4), 174–185, 2012.

Preece, C., Chong, H.Y., Golizadeh, H., Rogers, J., A Review of Customer Relationship (CRM) Implications: Benefits and Challenges in Construction Organizations, *International Journal of Civil Engineering,* 13, 362–371, 2014.

Sandberg, K., Vinberg, S., and Pan, Y., *An Exploratory Study of Women in Microenterprise: Owner Perceptions of Economic Policy in A Rural Municipality: Gender-Related Differences*, CD-Proceedings of 12th Nordic Conference on Small Business Research, 1-14, 2002.

Interdependence between Structural Engineering and Construction Management
Edited by Ozevin, D., Ataei, H., Modares, M., Gurgun, A., Yazdani, S., and Singh, A.

ISBN: 978-0-9960437-6-2

NEURAL NETWORK AND PROBABILTY BASED COST EXPECTATION LIMIT MODEL FOR RESIDENTIAL BUILDING COST

LEKAN AMUSAN, JOSHUA OPEYEMI, ADEYEMI EBUOLUWA, and IGNATIUS OMUH

Dept of Building Technology, Covenant University, Ota, Nigeria

The aim of this study is to develop a Cost Expectancy Limit Model that could assist clients in having proactive information about construction cost expectation of a particular building type with a view of assisting the client in proactive determination of expected construction cost of a building under predetermined conditions. Two population frames were used in this context. First, a population frame of 1500 samples of actual construction cost of residential building in Lagos state Nigeria out of which 1000 samples of As-built cost (Actual cost) of residential buildings were used, in artificial neural network data training and model development using MATLAB Neuro tools. The second population sample was 250 samples of construction professionals, out of which 200 samples was picked for purpose of questionnaire administration to capture data on factors that could influence building cost expectancy. Mean Item Score, Simple Percentage, and Relative Agreement Index of SPSS package was used to analyze and process the data. Cost expectancy limit was developed with parameters trained with Artificial Neural Networks, while factors that influence the accurateness of the expectancy model were articulated, such as economic factors, political factors, activity of maestros, macro and micro economic variables, and corruption factors, among others. The study recommends the use of the model and strategy for effectiveness in accurate prediction of construction cost among other things.

Keywords: Parameters, Prediction, Neuro tools, Agreement, Information.

1 COST IN BUILDING

Contributions Cost is one of the major challenges faced during each phase of construction. Many construction projects are started and later cannot be completed because of many reasons, including financial challenges. This is because clients do not have proper information concerning the project and do not know how to plan financially for the project, leading to project abandonment and other challenges like time and cost overrun, which causes a waste of resources, and it becomes a nuisance to the environment, especially when it keeps reoccurring (Efi and Andrea 2011). Cost forecasting if done poorly also contributes highly to this; it will lead to estimates being lower than the actual value and improper funding of the project. This requires cost expectancy limit to be developed by reliable means, as there are many approaches to developing this. The causes of project abandonment are numerous, and most of them are directly or indirectly related to cost. The professionals in the industry of construction have great concerns about the costs, i.e., the actual cost which is the final cost, the estimated cost or the initial cost at the beginning of the project and the tender figure. Professionals are concerned in particular with the wide gap that exists between these costs (Boussabaine 2010, Offei–Nyako *et al.* 2016).

2 RESEARCH METHODOLOGY

Relative Agreement Index (R.A.I.) is a metric that is usually used for carrying out of rating scale evaluation and derives an index that can serve as a ratio of measurement of opinion of the respondents. An agreement scale of 1 to 5 was used: Strongly, Agree, Agree, Strongly Disagree, Disagree, and Neutral. In this section, population sample, sample size, and methods of analysis were presented. Also, the method of analysis to be used for the analysis of the various parameters that could help in the development of cost expectancy limit which could lessen or eliminated cost related challenges on site that sometimes lead to project abandonment, was presented using the relative agreement index (R.A.I.) The R.A.I. method was used to derive the results manual where SA=5, A=4, N=3, D=2, SD=1 (Jafarzadeh *et al.* 2013). The population constituents for the purpose of this research work was categorized along the line of clients which include the type of buildings which can be residential in nature, religious, academic, office, health facilities and special buildings. The population frame for this study includes private buildings that were used for residential purpose. The projects are those completed within the last 5 years. The initial cost and as-built cost extracted from the project documents of these projects were used as modeling parameters for the neural network-based model. Two samples were used in this study. One of these were samples for identifying parameters that influences project abandonment on account of cost related problems and information of importance of cost limit determination in project execution. The second sample was the sample of cost data of completed project used in model development with the aid of neural network data training. In the context of this study, a population frame of 250 was used, of which 200 samples were used for analysis through questionnaires distributed for response collation on parameters that influence project abandonment and importance of development of cost expectancy limit in building. However, for the purpose of training the data with Neural Networks, large samples are needed for modeling. To this end, sample frame of 1500 was used, while a sample of 1000 of actual cost of residential building was used for data training.

Table 1 indicates the parameters that influence project abandonment. "Funds mismanagement" was ranked 1st with a R.A.I. of 0.874 after which "poor construction planning" was ranked 2nd with a R.A.I. of 0.854 and "land or legal disputes" was ranked third with R.A.I. of 0.850. "Improper planning and design and inception of project" is ranked 4th with R.A.I. of 0.844 with "improper project estimates" ranking 5th with R.A.I. of 0.836. Also, "economic recession" ranked 6th with R.A.I. of 0.830, "death of client" ranked 7th with a R.A.I. of 0.814. Closely following is "lack of adequate fund allocation and payment delay" which ranked 8th with R.A.I. of 0.808. "Inflation and bankruptcy of contractor" also ranked 8th with R.A.I. of 0.808 as well. Ranking 10th was "lack of proper assessment" with a R.A.I. of 0.788; following this is "delay in payment" ranking 11th with R.A.I. of 0.780 and ranking 12th is the "project manager incompetence" with R.A.I. of 0.752. Table 2 indicates the importance of developing cost expectancy limit as "prevention of project abandonment" was ranked 1st with a R.A.I. of 0.862 after which "to enable the client determine if the project can continue" was ranked 2nd with a R.A.I. of 0.838 and "keep expenditure within cost limit" was ranked third with R.A.I. of 0.832 (Elfaki *et al.* 2014).

Table 1. Parameters that influence project abandonment.

S/N	Project Abandonment Parameters	Mean	R.A.I	Ranking
1	Fund mismanagement	4.37	0.874	1
2	Poor construction planning	4.27	0.854	2
3	Land or legal disputes	4.25	0.850	3
4	Improper planning and design and inception of project	4.22	0.844	4
5	Improper project estimates	4.18	0.836	5
6	Economic recession	4.15	0.830	6
7	Death of client	4.07	0.814	7
8	Lack of adequate fund allocation and payment delay	4.04	0.808	8
9	Inflation and bankruptcy of contractor	4.04	0.808	8
10	Lack of proper assessment	3.94	0.788	10
11	Delay in payment	3.90	0.780	11
12	Project manager incompetence	3.76	0.752	12
13	Lack of project risk assessment	3.75	0.750	13
14	Change of priority	3.70	0.740	14
15	Inability to adhere to specifications and building codes	3.67	0.734	15
16	Poor quality control by regulatory bodies	3.45	0.690	16
17	Poorly developed clients brief	3.43	0.686	17
18	Negligence of quantity surveyor resulting in wrong estimation	3.37	0.674	18
19	Lack of available skilled personnel	3.33	0.666	19
20	Weakened market conditions	3.28	0.656	20

R.A.I--- Relative Agreement Index

Table 2. Importance of development of cost expectancy limit in building.

S/N	Importance of development of cost expectancy limit	Mean	R.A.I	Ranking
1	Prevention of project abandonment	4.31	0.862	1
2	To enable the client determine if the project should proceed	4.19	0.838	2
3	Keeps expenditure within cost limit	4.16	0.832	3
4	To ensure adequate funding for project delivery	4.15	0.830	4
5	To enable the contractor exercise effective control over resources	4.06	0.812	5
6	Gives rise to effective cost plan	4.04	0.808	6
7	Awareness of financial expectation	4.01	0.802	7
8	Improved building quality and performance	4.00	0.800	8
9	Used as cost prediction	4.00	0.800	8
10	Clients satisfaction with better value for money	3.96	0.792	10
11	Determine overall feasibility of the project	3.91	0.782	11
12	Helps to manage cash flow	3.90	0.780	12
13	Prevention of cost overruns and time overruns	3.90	0.780	12
14	Help target cost to be distributed in a balanced way over the different parts of the building	3.84	0.768	14
15	Enables designers arrive at practical and balanced designs	3.78	0.756	15
16	Allows client to secure a business plan	3.72	0.744	16
17	Promotes effective communication among project team	3.67	0.734	17
18	Help contractor determine if they will bid	3.57	0.714	18
19	To enable the contractor maximize his profitability	3.51	0.702	19
20	Helps to determine source of funds	3.15	0.630	20

R.A.I--- Relative Agreement Index

2.1 Neural Network Training of Sampled Cost Data for Model Training

As a result of the training, the graph in Figure 1 was obtained and the values at the bottom left end. The blue dot graph denotes the true values and the yellow dot on the graph denotes the predicted values. The red line joining them denotes the errors between them. On the bottom left end, the following result was given from the training of the network architecture which is the table above; RMSE= 3.3117e+06; -Square= 0.00; MSE= 1.0967e+13; MAE= 2.7536e+06; The prediction time= ~60obs/sec; The training time=12.494seconds. (Jafarzadeh *et al.* 2013). After the training of the network, a single value was predicted from the trained network by using a coding formula: yfit= trainedmodel.predict. Fcn (X) where X represents the values that were used in the training. A single value was gotten from this prediction, which is 3,050200 naira. Some of the above contains some basic assumptions in application of neural network application.

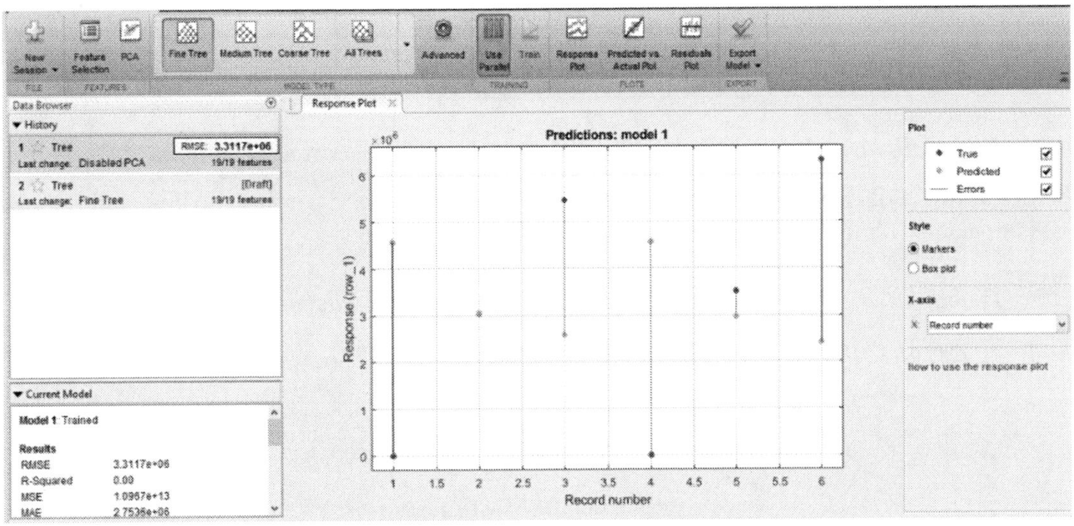

Figure 1. Neural network cost data training pane 1. Source: Mat-lab R2017b (2018)

2.2 Developing Cost Expectancy Limit to Prevent Cost Related Challenges on Site

2.2.1 *Probability estimating*

The PERT (Program Evaluation and Review Technique) was used for the probability estimation technique. The value predicted by the neural network was used to get the range of values that were used in the PERT technique. The nearest cost from the original costs was selected and the optimistic cost, pessimistic cost, and most-likely cost were selected from the nearest cost to the predicted cost.

From Table 1, 2- and 3-Bedroom Bungalows were used as type of building that could be easily built and that can be planned towards. From Table 3, the most likely construction cost for 1-Bedroom bungalow is N4,567,820[$12,723]; 2-Bedroom bungalow is N9,588,000 [$26,707] while 3-Bedroom bungalow project is N13,235,612[$36,868]. For the 1-Bedroom bungalow, the predicted cost was N3,050,200 naira within a close range to the B.O.Q. value. The cost was used as the optimistic cost and the resultant as-built cost as the pessimistic cost. The same trend emerged in the cases of 3-Bedroom Bungalow and 2-Bedroom Bungalow. The expected cost was calculated using the formula in Eq. (1) and the value derived was close to the most-likely cost (Efi and Andrea 2011, Amusan 2012, Buratti *et al.* 2014, Amusan *et al.* 2018b). The Hedonic cost equation used for probability estimation of most-likely cost generated is shown in Eq. (1):

$$ECR= [OP[0.1667]+0.667MC+0.1667PES] \quad (1)$$

where OP=Optimistic Cost; MC= Most Likely Cost and PES is Pessimistic Cost.

Table 3. Table of probability estimating calculation of expected cost limit.

Building Types	Optimistic cost [BOQ Value]	Most likely cost	Pessimistic cost	Expected cost [(0+4M+P)0.1667]	Expected cost Region
1-Bedroom Bungalow	3,050,55 [$8,497]	4,500,000 [$12,535]	6,356,370 [$17,705]	**4,567,820** [$12,723]	4,500,000 [$12,535]
2-Bedroom Bungalow	9,674,000 [$2,6947]	9,450,000 [$38,441]	10,654,000 [$29,676]	**9,588,000** [$26,707]	9,450,000 [$26,323]
3-Bedroom Bungalow	13,543,670 [$37,726]	13,800,000 [$38,441]	10,670,000 [$29,721]	**13,235,612** [$36,868]	13,800,000 [$38,441]

2.2.2 *Probability cost hedonic model for different probability cost expectations*

The break down of different cost expectation of most-likely cost, pessimistic cost and optimistic cost is presented alongside the Hedonic model in the next sections.

2.2.3 *Probability regression cost hedonic model for expected cost and most-likely cost*

Using Eq. 2(a-d), expected cost and most-likely cost were both calculated:

$$y = mx + c \quad (2a)$$

$$\mathbf{4500000} = 3050550 \text{ x } 0.1576 + 4019233. \quad (2b)$$

$$y = 4500000 - \textbf{Most Likely Cost} \quad (2c)$$

$$m = 3050550 - \text{Optimistic Cost ---}x = 0.1576 - \text{Inflation Rate } c = 4019233 \quad (2d)$$

2.2.4 *Probability cost hedonic model for optimistic cost*

Using Eq. 3(a, b), optimistic cost was calculated:

$$y = mx + c \quad (3a)$$

$$3050550 = 6356370.02 \text{ x } 0.1576 + 2048786.08 \quad (3b)$$

2.2.5 *Probability cost hedonic model for pessimistic cost*

Using Eq. 4(a-c), pessimistic cost was calculated:

$$y = mx + c \quad (4a)$$

$$6356370.02 = 3050550 \text{ x } 0.1576 + 5875603.34 \quad (4b)$$

$$y = 6356370.02 - \textbf{Pessimistic Cost} \quad (4c)$$

These equations represent various hedonic models generated for the purpose of determining the cost dichotomies such as optimistic cost, most likely cost, and expected cost for all categories of different residential building projects.

3 CONCLUSIONS

Factors that influence building cost and building cost expectancy have been reviewed in order to know the factors that affect the cost of building so measures can be taken in order to avoid them or reduce them. The most influential factor was chosen by the respondents and the order of importance was stated in detail in the tables presented in the text. Finally, Probability estimating model that could be used in determining cost dichotomy of projects for purpose of planning was developed which cut across cost determination that covers Most likely, Pessimistic and Most likely cost (Kim *et al.* 2004, Efi and Andrea 2011, Afolabi *et al.* 2018, Amusan *et al.* 2018a, Amusan *et al.* 2018b). The study has generated an hedonic form of a regression model that could capture the Pessinmistic cost optimistic cost and most likely cost of a building type. The hedonic model is described using Eq. (1) as given earlier in Section 2.2.1.

Acknowledgements

The support of Covenant University and Covenant University Center for Research and Innovations (CUCRID) is acknowledged for sponsoring this research and funding the publication of the research article.

References

Afolabi, A. O., Oyeyipo, O. O., Ojelabi, R. A., and Amusan, L. M., Professionals' Perception of a Web-Based Recruiting System for Skilled Labour, *Journal of Theoretical and Applied Information Technology*, 96 (10), 2885-2899, 2018.

Amusan, L. M., *Neural Network-based Cost Predictive Model for Building Works,* PhD Thesis, Covenant University, Ota, Nigeria, 2012.

Amusan L. M, Charles, A. K., Adeyemi, E., Joshua, O., and Raphael, O. A., Data on Expert System-Econometric Entropy Informatics Model for Adjudicating Residential Building Project Costs, *Data Brief*, 20, October 2018, Pages 1721-1729, 2018a.

Amusan, L., Kehinde, A.-Y., Ayodeji, O., Tunji-Olayeni P., Adedeji, A., and Owolabi, J., Work Order Building Informatics Platform for Planning Residential Building Projects and Maintenance Work, *International Journal of Civil Engineering and Technology*, 9(9), 135-146, 2018b.

Boussabaine, A. H., The Use of Artificial Neural Networks in Construction Management: A Review, *Construction Management and Economnics,* 14(5), 427-436, 2010.

Buratti, C., Lascaro, E., Palladino, D., and Vergoni, M., Building Behavior Simulation by Means of Artificial Neural Network in Simmer Conditions, *Sustainability*, 2014.

Efi , P., and Andrea, A., Sized Based Software Cost Modelling with Artificial Neural Network and Genetic Algorithms, *Artificial Neural Networks - Application*, Limassol, Cyprus, doi: 10.5772/15851, 2011.

Elfaki , A. O., Alatawi, S., and Abushandi, E., Using Intelligent Techniques in Constuction Project Cost Estimation: 10-Year Survey. (S. Mdanat, Ed.) *Advances in Civil Engineering, 2014*, 11. Retrieved from https://www.hindawi.com/journals/ace/2014/107926/ on Dec. 2, 2014.

Kim, G.-H., An, S.-H., and Kang, K.-I., Comparison of Construction Cost Estimating Models Based on Regression Analysis, Neural Networks, and Case-Based Reasoning, *ScienceDirect,* 39(10), 1235-1242, 2004.

Jafarzadeh, R., Ingham, J. M., Wikinson, S., Gonzalez, V., and Aghakouchak, A. A., Application of Artificial Neural Network Methodology for Predicting Seismic Retrofit Construction Costs, *Journal of Construction Engineering ans Management*, 140(2), 2013.

Mat-lab R2017b, MatLab Neuro Tool. 3rd Edition. United States of America, 2018.

Offei-Nyako K., Tham, L. C. O., Bediako, M., Adobor, C. D., and Asamoah, R. O., Deviations between Contract Sums and Final Accounts: The Case of Capital Projects in Ghana, *Journal Of Construction Engineering*, Vol. 2016, 2016.

Interdependence between Structural Engineering and Construction Management
Edited by Ozevin, D., Ataei, H., Modares, M., Gurgun, A., Yazdani, S., and Singh, A.

ISBN: 978-0-9960437-6-2

THE IMPORTANCE OF THE PROPERTY MANAGEMENT COMPANIES AS CONDOMINIUM JURIDICAL PERSONS

KANYAPA ARAMRAKS[1] and TOMOYUKI GONDO[2]

[1]*Dept of Architecture and Building Engineering, Tokyo Metropolitan University, Tokyo, Japan*

[2]*Dept of Architecture, The University of Tokyo, Tokyo, Japan*

This study examined Bangkok's condominium management system with a particular focus on the role of property management companies as members of juridical persons. The number of condominiums in Bangkok has been proliferating rapidly, particularly in areas near the mass rapid transit system. The Thailand Condominium Act requires that a condominium juridical person be registered upon transfer of ownership of the first condominium unit. This study's analysis of data collected from interviews with 16 condominium juridical persons found that there are three types: 1) property management companies; 2) juridical person employees; and 3) a combination of 1) and 2). All but one condominium—which used the mixed management type—had hired property management companies to oversee management and maintenance. Reasons cited for this overwhelming preference included insufficient management and maintenance knowledge among juridical employees, and limited beneficial connections with outsourcers and suppliers among juridical person employees. The interview results demonstrate that the responsibilities of property management companies extend beyond their job description and include caring for residents, seeking solutions for condominium problems, and creative thinking for innovative proposals for alternative management practices.

Keywords: Management and maintenance, Problem, Interview, Duty, Bangkok.

1 INTRODUCTION

1.1 Background

Before 1979, there were no condominiums in Thailand. Residences were limited to detached houses, row houses, apartments, and leaseholds (a type of building similar to a condominium), and no laws or regulations existed to protect the rights of residents living in the same building. In 1979, the government enacted the Condominium Act, which led to the development of a condominium construction industry in Bangkok, with over 900 buildings during the first 15 years (Real Estate Business Promotion Bureau 2018). The construction of a mass rapid transit system improved accessibility and increased travel convenience, thus further stimulating condominium construction. Figure 1 shows the proliferation of condominiums constructed in Bangkok from 1979 to the present. In January 2018, there were 748,123 condominium units in the city.

The Condominium Act requires a condominium juridical person to manage and maintain in common property areas and a real estate company must register a condominium juridical person upon transfers in the first unit's ownership (Suvisit and Roachanakanan 2011). A condominium juridical person refers to a group of people who registers under this Act as a non-human legal entity, in other words any organization that is not a single natural person but is authorized by law with duties and rights and is recognized as a legal person and as having a distinct identity. A group consists of one manager and committees.

Real estate companies have had a dominant role in condominium unit sales; however, they often have little involvement in management or maintenance services due to the smaller profits available to them in these domains and the lack of staff with the required skills and experience. In response, numerous property management companies have been founded, and there are currently 632 such companies in Bangkok (Table 1) (Department of Business Development 2018).

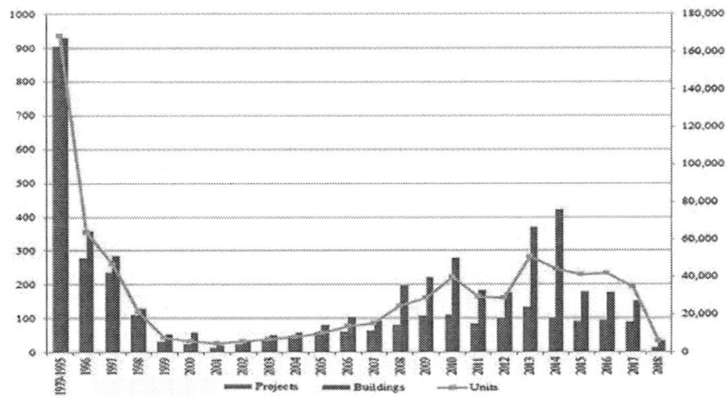

Figure 1. Condominiums in Bangkok.

Table 1. The number of property management companies in Thailand.

Region	Numbers		
	2016	2017	2018
Bangkok	601	620	632
Central region	146	157	162
Eastern region	452	479	480
Northeastern region	28	30	32
North region	33	35	39
South region	336	426	459
Western region	24	29	28
Total	1,620	1,776	1,832

Previous researches show that in the past, the owner of the project is the person who manages the condominium. The first property management company was hired to manage the condominium in 1981 (Sangsuwan 1999), the management system for mid-income condominiums in Bangkok can be classified into four types and all types have hired the property management companies to support in condominium management (Meesiri 1999), the duty of the property management company has changed to manage the relationship between residents (Rattanaprichavej 2016).

However, it is often difficult to find a person with the required maintenance and management skills to serve as both the manager and a committee board member of the condominium juridical person. Therefore, the condominium juridical person is often required to hire a property management company to maintain the condominium building and manage its operations. This paper aims to explain the condominium management system in Bangkok and better understand the significance of property management companies as condominium juridical persons.

1.2 Objectives

This study aims to investigate the management of condominium juridical persons and examine the management process of property management companies. This paper focuses on the mid-range condominium segment, which comprises condominiums priced between THB 70,000–100,000 per square meter (1 USD equals about 32 THB), and the city condominium segment, which consists of condominiums priced lower than THB 70,000 per square meter (The Thai Real Estate Association 2017)

1.3 Methodology

This research involved the distribution and analysis of interviews focusing on the importance of property management companies as condominium juridical persons. First, we reviewed the previous research about condominiums in Bangkok, including journal articles, books, dissertations, and studies performed by the Real Estate Information Center and the Department of Lands. Second, information was collected from interviewees living and working in condominiums, namely six committee members, three juridical person managers, seven area managers, 14 building managers, four assistance-building managers, and two management staffers. Interview topics were related to the management systems, maintenance plans and problems that occurred in the management of condominiums.

2 THE CONDOMINIUM JURIDICAL PERSON

As mentioned above, Section 31 of the Thailand Condominium Act requires that a real estate company must register a condominium juridical person at the Department of Lands upon transferring the first condominium unit to the customer, and Section 33 stipulates that the duly registered condominium juridical person "shall have the objectives to manage and maintain the common property with the power to do any reasonable acts for the benefits of the said objectives" (Tantikulanan 2008). A condominium juridical person is a team consisting of one manager who may be a natural person or a juridical person and a committee board comprised of co-owners not less than three persons but not exceeding nine persons. The paragraphs below elucidate some important elements of the condominium juridical person system.

Information from interviews with 16 condominium juridical persons found that there are three categories of condominium juridical persons in Bangkok: 1) property management companies 2) condominium juridical person employees and 3) a combination of a property management company and condominium juridical person employee. All but one condominium—which used the third type—used the first type to oversee management and maintenance of the condominium. The first type is by far the most common form because it is difficult to find personnel with the required expertise and experience in management and maintenance to perform all the functions of condominium juridical person employees. Co-owners such as committee board members and residents often lack sufficient knowledge in these areas. When these co-owners hire a property management company, they are procuring a management team to support the condominium juridical person by providing staffing at the condominium building. Core members of such a management team include a building manager, administrator, accountant, and technician. Survey data identified the number of staff in a management team as ranging from 4 - 15 people per site, depending on the budget of the condominium juridical person and the scale of the project. Table 2 displays the number of members and duties of each member in a condominium juridical person, and Table 3 presents information on management team numbers gleaned from the interviews.

Table 2. Condominium juridical person members and duties.

Member	Source	Number	Duty
Committee board	Co-owner	3–9	-Monitoring and control the work of management team -Arranging the meeting at least once every six months
Juridical person manager	-Co-owner -PMC -Natural person	1	-Sign an important and transaction documents -Providing security operations or taking actions in maintaining peace and order within the condominium
Management team:			
Building manager	PMC	1	-Manage the condominium in good condition -Monitoring the staff team working
Administrator	PMC	1–2	-Make the documents and keep the input-output documents -Coordinate about equipment and supplies purchasing
Accountant	PMC	1–2	-Make a document about revenue accounts and expense accounts -Keeping all accounting documents
Technician	PMC	2–5	-Controlling, checking and repairing the equipment and mechanical systems -Make a memo of maintenance systems

Table 3. Information of condominiums and property management gleaned from the interviews.

Condos Name	Developer	Project Area (m^2)	Property management company	Number of management team	Type of property management company
A	Pruksa	3,466	OPM	4	Independent
B	Pruksa	3,688	One Property	4	Independent
C	Pruksa	12,640	SMC	10	Independent
D	Noble	2,328	VPC Asia Pacific	7	Independent
E	NHA	n/a	CEMCO	4	Independent
F	NHA	n/a	QPM	10	Independent
G	NHA	25,651	Royal Fame	10	Independent
H	Asia Stars	3,992	Grand Estate	5	Independent
I	Rojana	3,920	JLL	12	Independent
J	SC Asset	6,140	U Life	7	Independent
K	Supalai	16,360	PMC	10	Independent
L	Land&House	10,348	Vertical	7	Independent
M	LPN	4,932	LPP	5	Subsidiary
N	LPN	12,800	LPP	7	Subsidiary
O	AP	3,276	Smart	7	Subsidiary
P	Ananda	26,017	The Works	15	Subsidiary

3 THE PROPERTY MANAGEMENT COMPANY

The property management company (PMC) is responsible for taking care of housing, the condominium, and the office building. It also manages the customer service and public utilities, including organizing personnel to provide assistance to the residents, manage budgeting, and systematically allocate income and expenses. Services performed by property management companies encompass the areas of facility management, environment management, community management, and financial and service operations (Rattanaprichavej 2016).

Table 3 identified two main types of property management companies, including companies that are subsidiaries of the real estate company that initially constructed the condominium project.

The advantages of using this type of property management include expedience because the design and construction are known to the company, which facilitates operations and communication between the condominium juridical person, the management team, and the real estate company. The latter also tends to have its own technician teams to support condominium maintenance. However, residents perceive this system as unfair because the subsidiary organization belongs to the real estate company. The second type is the independent company, which includes Thai companies as well as those based overseas. The main advantage of using this type is that it reduces the chance of corrupt relations between the property management and the real estate companies. However, some smaller companies will not have a technician team to support maintenance work, and they might be unable to spare extra time for solving additional problems in the condominium unrelated to maintenance.

4 RESPONSIBILITIES OF PROPERTY MANAGEMENT COMPANIES

The working system of property management companies is divided into supporting teams, namely the accounting and technician divisions, which stand by at the head office, and the onsite management team, which consists of a building manager, administrators, accountants, and technicians. These teams are coordinated by a juridical person manager who liaises between the head office and the condominium site. The onsite management team works with the committee board as a representative of the co-owners and has a duty to attend committee board meetings, which the committee board generally arranges to be convened monthly as a means to follow up on the work of the management team.

The main duty of the management team is to manage and maintain the common property of the condominium; however property management also assists residents in addressing issues related to their private property. The management team is responsible for purchasing maintenance materials and equipment, as well as creating the condominium's budget and procuring outsourced services and suppliers such as security guards, housekeepers, gardeners, and pest control services for the approval of the committee board. Therefore, the management team needs to have knowledge of financial management as well understand the physical structure of condominium and building systems. The committee board has a duty to give comments and approve the budget and procurements. For example, Figure 2 outlines the procurement process to hire supplier companies and purchase equipment in condominiums.

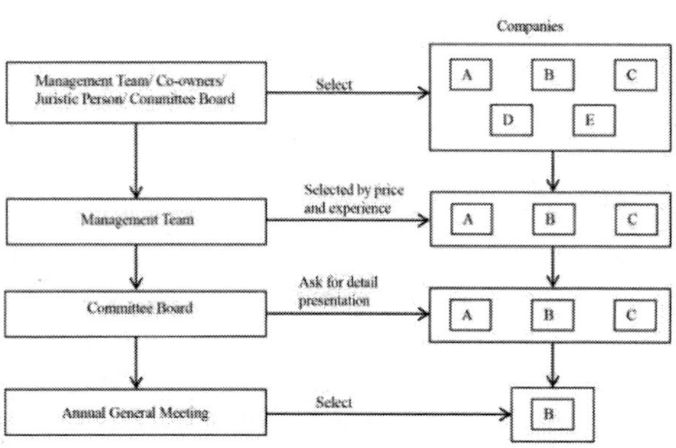

Figure 2. Procurement process in condominiums.

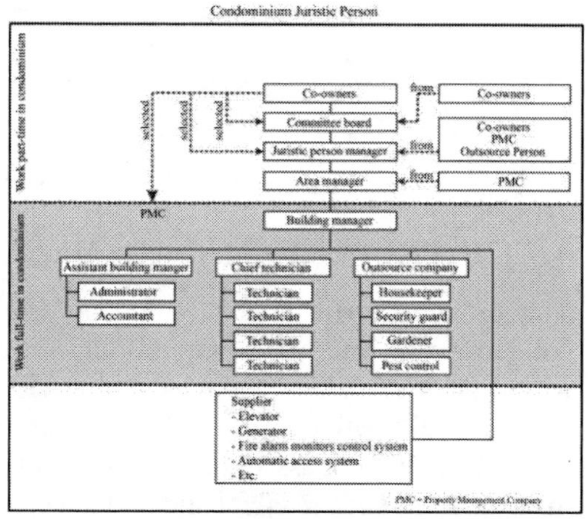

Figure 3. Structure of a condominium juridical person.

Figure 3 shows the structure of a condominium juridical person; one group comprising people who are not onsite on a daily basis, namely the committee board, juridical person manager, and area manager. The second group encompasses daily (full-time) onsite staff engaged in condominium operations of management and maintenance, specifically the management team of the property management company, the outsourced companies, and the suppliers. The outsourced companies and suppliers do not belong to the property management company; however, the management team has a duty to control all operations and services, which requires it to coordinate with the head offices of external companies, include summarizes the various work reports for co-owners at the annual general meeting.

5 CONCLUSION

Condominium owners lack management and maintenance knowledge and skills, do not have specialists in these fields inside the condominium, and lack connections with outsourcers and suppliers. These limitations in their ability to manage and maintain condominiums can result in long-term problems as the building ages. Therefore, it is necessary to hire a property management company with the necessary knowledge and experience to support the daily operations of a condominium to enhance the effectiveness and longevity of their use. Condominium owners thus avoid having to spend time managing and solving problems that they are ill-equipped to address, including creating and overseeing maintenance and repair budgets and managing the building's equipment, services and external relationships. Furthermore, the management team of the property management company is part of the condominium juridical person that works at the condominium daily and takes care of its residents. Such advantages have resulted in property management companies becoming essential elements of the condominium juridical system, and property management companies have continued to proliferate in Bangkok. However, there is no license of professional practice in property management, including no standards and each company uses its own standard to manage condominium. The government cannot monitor and control the management of condominiums. Therefore, the government or related departments must have a supporting role and set the condominium management standards for working and monitoring.

References

Department of Business Development, The Statistics of Business Registration, 2018. Retrieved from http://datawarehouse.dbd.go.th/bdw/menu/est/1.html on July 24, 2018.

Meesiri, P., *Mid-income Condominium Management in Bangkok CBD*, Thesis, Chulalongkorn University, Bangkok, 1999.

Rattanaprichavej, N., *Principles and Concepts in Real Estate Business*, Thammasat, Bangkok, 2016.

Real Estate Business Promotion Bureau, The Statistics of Real Estate Business, January 31, 2018. Retrieved from www.m-society.go.th/article_attach/22335/21559.pdf on August 6, 2018.

Sangsuwan, S., *Condominium Management*, Wiwat Publishing, Bangkok, 1999.

Suvisit, V. and Roachanakanan, K., *Land Allocation and Condominium Laws for Real Estate in Thailand*, Amarin Printing & Publishing, Bangkok, 2011.

Tantikulanan, V., *The Description of Condominium Act*, Pimaksorn Publishing, Bangkok, 2008.

The Thai Real Estate Association, Real Estate 2017, January 11, 2017. Retrieved from http://www.realist.co.th/blog/อสังหา-2560/ on July 30, 2018.

Interdependence between Structural Engineering and Construction Management
Edited by Ozevin, D., Ataei, H., Modares, M., Gurgun, A., Yazdani, S., and Singh, A.

ISBN: 978-0-9960437-6-2

BEST PRACTICES FOR THE DESIGN PROCESS FOR THE HORIZONTAL DESIGN-BUILDER

GIOVANNI C. MIGLIACCIO[1], R. EDWARD MINCHIN[2], UMBERTO C. GATTI[1], LOURDES PTSCHELINZEW[2], KENNETH E. ATKINS[3], and GREGORY A. HOSTETLER[4]

[1]*College of Built Environment, University of Washington, Seattle, USA*
[2]*M.E. Rinker Sr. School of Construction Management, University of Florida, Gainesville, USA*
[3]*Strategic Solutions Partners, LLP, Winter Haven, USA*
[4]*Infrastructure Engineers, Inc., Edmond, USA*

In early 1990s, the American driving public insisted that planned highway and bridge projects be completed quicker than was possible using the Design-Bid-Build (DBB) construction project delivery system, which had dominated the industry since 1930s. This led state Departments of Transportation (DOTs) to explore fast-track methods of construction. In late 1980s, some DOTs had experimented with Design-Build (D-B) delivery system. Forty-two state DOTs and numerous county and municipal transportation agencies now use the system and it remains the most popular integrated construction project delivery system in US transportation construction despite Federal Highway Administration efforts through the Every Day Counts initiatives to popularize newer methods such as Construction-Manager-as-General-Contractor and Alternative Technical Concepts. Its popularity is due to the speed with which projects move from conception through to completion and the ability through early contractor participation to implement innovative ideas that improve quality and further enhance the speed of the project. With all this speed, however, the design process has struggled to stay ahead of construction, especially when the Design-Builder is faced with new responsibilities such as Right-of-Way acquisition, the National Environmental Policy Act (NEPA) process, the environmental permitting process, utility re-location, etc. This paper recounts the research performed by a team led by the University of Florida to produce a guidebook to help both design-builder and owner with the design management process for transportation infrastructure construction.

Keywords: Design management, Delivery system, Fast-track, DOT.

1 INTRODUCTION

Beginning with the implementation of the Intermodal Surface Transportation Efficiency Act in 1991, Design-Build (D-B) has grown in popularity as a faster and in many applications, better alternative to the traditional Design-Bid-Build (DBB) construction project delivery system popularized in the first half of the 20th century. Along with the anticipated advantages of the new system, came inevitable disadvantages. The shorter project durations and advantages from contractor input into the design were welcomed. However, owners do not like the loss of control over the design process, and managing the design so as to keep pace in the evermore popular fast-track systems has been problematic. It is the latter problem that prompted the National Cooperative Research Program to sponsor research project NCHRP 15-46, "Guide for Design

Management of Design-Build and Construction-Manager-as-General Contractor Projects." This paper focuses on the research on the D-B side of the guidebook.

An agency self-assessment outlined in the guidebook identifies elements within the agency that could hinder the successful D-B implementation. This assessment is particularly useful to agencies that only recently have adopted or are considering adopting an innovative project delivery method, since it allows them to address certain issues before experiencing their effects on projects. It also benefits agencies already using D-B to help them identify issues they might previously have overlooked.

One of the most difficult questions an agency can ask is whether it is prepared to adopt and implement an innovative project delivery approach, since it forces the agency to face the reality of its organizational culture. Because the decades-long use of the DBB method has so fundamentally shaped employee perceptions and organizational structures and practices, implementing a new delivery approach constitutes a major paradigm shift for the state agencies adopting it (Miller *et al.* 2000). Early studies have found that "as agencies attempt D-B for the first time, they are constrained by the low-bid culture in their organizations" (Molenaar and Gransberg 2001). The USDOT has acknowledged these difficulties, reporting that "states not accustomed to this method of procurement can find it difficult to oversee these types of projects" (USDOT-FHWA 2004). Moreover, although the D-B method's combined procurement of services is expected to reduce transactional costs for delivering a project (Pietroforte and Miller 2002), this approach usually prompts state personnel to spend considerable time experimenting and developing new organizational routines to support the change (USDOT-FHWA 2004). This extra time is often justified by a wider concern that safeguards embedded in traditional approaches will be lost in the change process (USDOT-FHWA 2004). However, these concerns often appear with respect to the agency's approach to design management (DM) under D-B, since losing control of design is one of the major agency concerns when D-B is implemented for the first time.

Challenges to changing the project delivery approach are common when an agency adopts any innovative delivery method (e.g., D-B or CM/GC) and often depend on an agency's formal and informal cultures. When an agency is procedurally rooted in traditional means and methods, it is likely to face opposition to innovative delivery approaches. To mitigate this, the agency's formal culture should be open to innovation, risk-taking, and improvement.

2 LITERATURE REVIEW

The two most commonly employed methods of delivery of infrastructure construction are now DBB and D-B. Gransberg and Senadheera (1999) conducted a national survey of 15 DOTs; the results showed D-B to be the alternative method commissioned by all surveyed DOTs. Through further analysis, Gransberg was able to identify three different systems that were used to advertise and award D-B. Low-bid, Adjusted Score and Best Value were then compared to identify their strengths and weaknesses.

Scott (2006) pointed out the problem associated with low bid in D-B; instead of getting the benefits of cost control, most of the time it will result in a decrease in the quality of the final product. Scott focused on the Best Value approach, which places the emphasis not only on the price but also on other factors. Investigation into the legislative regulation and nature of this contracting method was analyzed to help develop the Best Value procurement method.

Shr (2004) studied the growing popularity of incentive/disincentive bidding. This method was found to shorten the contract time by making it difficult for the contractor to not accelerate the project. Shr developed a model to establish reasonable incentive or disincentive rates based on construction cost and time. But incentive/disincentive contracting may cause the quality of the final product to decline if the incentives are aimed only at time reduction.

Molenaar *et al.* (1999) studied the emergence of D-B into the public segment of the industry. Their work analyzed each party's responsibility with regard to the delivery system and also explains the procurement process and the structure of this particular method. Marwa *et al.* (2006) went more in depth than Molenaar by studying 76 D-B projects and identifing correlations between the procurement processes and the projects' overall performance. Chan (2002) sought specific project conditions that could help increase success rates of projects when using D-B. He noted that measures of success are defined by three factors—time, cost, and quality—but he believes that a more comprehensive metric should be established.

Gransberg and Sendheera (1999) conducted a survey aimed at all the DOTs in the United States. With the fifteen DOTs that employed D-B, they reviewed the three main methods of D-B—Low-bid D-B, Adjusted-Score D-B, and Best-Value D-B. Gransberg concluded that each of the methods can be utilized with different types of highway construction, depending on the nature of the project. Gransberg *et al.* (2008) also addressed the issue of quality assurance concerning D-B as it relates to transportation projects. One of the disadvantages of D-B is the lack of control over the detailed components of construction, which requires the agency to form a more comprehensive method to ensure the quality of the work. The report demonstrated the different ways in which DOTs have successfully controlled quality by focusing on all aspects of the construction phases; but this does not seem to be the case for all agencies. In the same year, Gransberg and Windel (2008) pinpointed the issue of communicating the quality requirements of public agencies on D-B projects. The study found that some owners tend to rely on the qualification evaluation process rather than being proactive on the issue.

Touran *et al.* (2009) presented on an evaluation process to help agencies identify the suitable delivery system to use for specific projects. They identified 24 key concerns that will determine the ideal delivery system. The paper includes a beneficial example that demonstrates how the evaluation should be conducted. However, the best option is not always available to the owner. Ghavamifar and Touran (2008) investigated all regulations that had been enacted by state legislation regarding public transportation projects. Sate statutes that address different innovative delivery processes were also provided.

3 RESEARCH METHODOLOGY

Initial research included a survey (Task 1) to determine how many state agencies have used D-B and when they began to implement it. About 80 percent of state agencies have used D-B, but many agencies have still not adopted this delivery method. Additionally, among the agencies that have used D-B, about half have used it sporadically and for fewer than ten projects. In light of these findings, the purpose of this research was twofold: while focusing on providing DM guidelines under D-B, it also summarized the results of previous research on D-B implementation so that all agencies can benefit from the experience of agencies around the country.

3.1 Level 1 Interviews (65 Agencies)

In Task 1, 52 DOTs (including Puerto Rico and DC) and 13 non-DOT agencies were interviewed by telephone, and all answered basic questions about their knowledge of, and experience with, design-build (D-B) and construction-manager-as-general-contractor (CMGC). These were designated as Level 1 interviews. Data from these interviews were compiled into a comprehensive spreadsheet denoting each state's knowledge and application of D-B; their frequency of use; and when the system was implemented.

It was found that D-B systems are recognized by all DOT agencies and 85% of non-DOT agencies. D-B is used by 81% of DOTs and 69% of non-DOTs, and for states that did not use D-B, 8% claimed to not have enabling legislation. The earliest use of D-B by DOTs dates to 1983 (Kentucky). For non-DOT agencies, D-B projects were implemented first in 2002. For both types of agencies, the majority influx was in the 2000s. Table 1 shows the level of implementation for agencies as of the time of the interviews. The first number in the body of the table is the number of agencies that have completed the number of D-B projects on that row (one, 2-10, etc.); the two percentages are the percent represented by that first number of total agencies interviewed, and of the total number of agencies interviewed that have executed D-B projects.

Table 1. Number of D-B projects executed by agency.

	DOT Agency Total (% of 52, of 42 using D-B)	**Non-DOT Agency Total (% of 13, of 9 using D-B)**
One Project:	3 (6%, 7%)	2 (15%, 22%)
2-10 Projects:	22 (42%, 52%)	7 (54%, 78%)
11-20 Projects:	8 (15%, 19%)	
21-50 Projects:	2 (4%, 5%)	
51-100 Projects:	2 (4%, 5%)	
Over 100 Projects:	4 (8%, 10%)	

3.2 Level 2 Interviews (9 Agencies)

The Level 2 interviews also were conducted by telephone and participants answered questions of a more intense and, at times, project-specific nature. Those who had a particularly high level of experience and knowledge were asked to fill out a supplemental form that asked for budget information and other particulars. The goal of this phase was to identify agencies to conduct a scan that would allow researchers to identify emerging features of DM under D-B.

After reviewing the responses to Level 1 interviews, the following criteria were identified for selecting states/agencies to interview in Level 2:

- An agency has used D-B within the last five years.
- An agency has used D-B on more than five projects.
- An agency has submitted the additional information on candidate case studies OR a member of the research team has identified potential case studies delivered by the state/agency (e.g., Maryland, Washington, North Carolina, Oregon).

Once the Interview Instruments were completed, the interview process was initiated by contacting all the chosen DOTs (10 total). Initial communications were conducted through e-mail. If a response was not received within a few days, a phone call was used to follow up. Since the project schedule was tight, researchers decided to consider this phase satisfactorily completed once a response rate of at least 50% (five agencies) was achieved. However, each agency was contacted at least three times independently after the research team received a commitment from at least five DOTs. The research team was able to complete the scan of the nine agencies listed below. Level 2 information was collected during the onsite visit for the D-B portion of the UDOT data collection. Level 2 information was collected on nine DOTs for a response rate of 90%. The agencies interviewed as part of Level 2 are Maryland, Massachusetts, Minnesota, Mississippi, Missouri, New Mexico, North Carolina, Oregon, Utah, and Washington.

3.3 Case Studies

Four Case Studies were conducted with three agencies. Case Studies were conducted through research team site visits and face-to-face interviews. Different research team members interviewed a completely different set of officials in each state. Some Case Studies were performed on an agency's program as a whole and some were performed on individual projects.

4 RESULTS

The team learned many lessons from the project. Among those are:

- Understand the importance of the project planning and procurement phase. The agency must understand that D-B is a more sophisticated project delivery method than DBB. The more work the agency performs up front, the less likely will be the occurrence of issues and disputes after contract award. In particular, an agency is expected to carefully define the project scope and develop contractual documents while involving all relevant project stakeholders in this process, including local government, public agencies, and utility companies.
- Perform a risk analysis to identify the project risks and allocate each risk to the entity (i.e., agency or design-builder) that can better manage it. The integration of design and construction services (and sometimes right-of-way acquisition and utility relocation) makes the risk allocation process under D-B more fluid than for DBB. This information will be used to develop a draft risk allocation that will undergo an industry review phase or be included in the RFP.
- Educate the agency and all project stakeholders about the D-B process that has been adopted. In general, D-B is a project delivery method. In practice, there are different ways to implement D-B within an agency. Some agency departments—such as regional offices—and project stakeholders—such as local government, public agencies, and utility companies—may not be familiar with D-B in general and with the way your agency is implementing it in particular. Thus, the agency must actively involve all interested agency departments and project stakeholders as soon as possible to educate them about D-B procedures and peculiarities. Contrary to DBB, the project scope must be defined clearly before the design phase in D-B projects.
- Clearly communicate to the proposers the selected payment method and how it can affect design activities. Since many proposers consist of constructors that contract out the design to design firms, they may not fully understand how the selected payment method can affect design activities. If the agency determines to pay each line of work when it is 100% complete and the proposer does not break down the design activities into multiple lines of work to obtain payments consistent with the accomplished design activities, the proposer and/or designers will have to finance some of the design activities. Since this may create adversarial relationships, the selected payment method should be spelled out clearly in the RFP documents, and its potential impact on design should be understood by the agency and communicated to the proposers during the procurement phase.
- Develop a process for pre-award value engineering (VE). The agency can highly benefit in terms of quality improvement, cost savings, and/or schedule reduction from pre-award VE concepts. When these innovations are outside the scope outlined in the RFP documents, they are submitted as alternative technical concepts. Furthermore, pre-award VE allows the agency to retain all cost savings while the cost savings from post-award VE generally are shared with the design-builder. On the other hand, the process for soliciting and handling pre-award VE concepts should be transparent and not detract from

the objectivity of the competition. Thus, the agency should develop a process that supports the proposers in proposing innovations.

5 SUMMARY AND CONCLUSIONS

At the program level, it is crucial how the agency allocates DM responsibilities among its units, and how the project delivery process is managed by units that deal with phases that deal with post-award design. At the project level, the approach to deal with pre-award design activities substantially affects post-award DM. During post-award DM, an agency's approach to DM is mostly shaped by how it establishes a collaborative partnering environment, how it handles communications and coordination, how it handles VE proposals, how it handles interdependencies between design and other activity, and especially how it handles formal DM processes.

References

Chan, A. P., Scott, D., and Lam, E., Framework of Success Criteria for Design/Build Projects, *Journal of Management in Engineering*, 18(3), 120-128, 2002.

Ghavamifar, K., and Touran, A., Alternative Project Delivery Systems: Application and Legal Limits in Transportation Projects, *Journal of Professional Issues in Engineering Education and Practice*, 134(1), 106-111, 2008.

Gransberg, D. D., Datin, J., and Molenaar, K., *NCHRP Synthesis 376: Quality Assurance in Design-Build Projects*, Transportation Research Board of the National Academics, Washington, D.C., 2008.

Gransberg, D. D., and Senadheera, S. P., Design-Build Contract Award Methods for Transportation Projects, *Journal of Transportation Engineering*, 125(6), 565-567, 1999.

Gransberg, D. D., and Windel, E., Communicating Design Quality Requirements for Public Sector Design/Build Projects, *Journal of Management in Engineering*, 24(2), 105-110, 2008.

Marwa, A., Messner, J. I., Horman, M. J., Comparing Procurement Method for Design-Build Project, *Journal of Construction Engineering and Management*, 132(3), 230-238, 2006.

Miller, J. B., Garvin, M. J., Ibbs, C. W., and Mahoney, S. E., Toward a New Paradigm: Simultaneous Use of Multiple Project Delivery Methods, *Journal of Management in Engineering*, 16(3), 58-67, 2000.

Molenaar, K., and Gransberg, D., Design-Builder Selection for Small Highway Projects, *Journal of Management in Engineering*, 17(4), 214-223. 2001.

Molenaar, K. R., Songer, A. D, Barash, M., Public-Sector Design/Build Evolution and Performance, *Journal of Management in Engineering*, 15(2), 54-62, 1999.

Pietroforte, R., and Miller, J. B., Procurement Methods for US Infrastructure: Historical Perspectives and Recent Trends, *Building Research and Information*, 30(6), 425-434, 2002.

Scott, S., III, *NCHRP Report No. 561: Best-Value Procurement Methods for Highway Construction Projects*, Transportation Research Board of the National Academies, Washington, D.C., 2006.

Shr, J. F., and Chen, W. T., Setting Maximum Incentive for Incentive/Disincentive Contracts for Highway Projects, *Journal of Construction Engineering and Management*, 130(1), 84-93, 2004.

Touran, A., Molenaar, K. R., Gransberg, D. D., Ghavamifar, K., Decision Support System for Selection of Project Delivery Method in Transit, *Transportation Research Record*, 2111(1), 148-157, 2009.

USDOT-FHWA, 2004, Report to Congress on Public-Private Partnerships. Retrieved from http://www.fhwa.dot.gov/reports/pppdec2004/pppdec2004.pdf on September 01, 2018.

Interdependence between Structural Engineering and Construction Management
Edited by Ozevin, D., Ataei, H., Modares, M., Gurgun, A., Yazdani, S., and Singh, A.

ISBN: 978-0-9960437-6-2

REHABILITATION OF MARINE TERMINAL PIERS

MUANGSANGOP SENIWONGSE

Bermello Ajamil and Partners, New York, USA

This paper presents practical methods and techniques for rehabilitation and strengthening marine terminal piers in the USA. These piers are waterfront structures for berthing of large cargo vessels and cruise vessels. Typically, the piers consist of timber or concrete deck supported on multiple pile bents. The piers are subjected to large impact forces from the mooring of the vessels acting at the fenders and bollards, in addition to the other normal vertical dead and live loads. Depending on the size of the vessels, the design service loads on the bollards are in the range of 100 tons to 200 tons. Other possible severe loads are from the winds and waves during the hurricanes. The piers are continuously exposed to outdoor, marine and coastal environments, which are corrosive in nature. For these reasons, all piers of 15 years or more are normally up for repair/rehabilitation. Typical deteriorations found are: cracked and spalled concrete, rust in exposed reinforcing bars, broken timber piles, timber piles with reduced cross sections, cracked concrete piles, corroded steel piles, and displaced out-of-plumb piles. The paper presents the methods for repairing the above-mentioned defects. For concrete crack and spall repair: sealing with epoxy, concrete patching and shotcrete are used. For pile repair: concrete jacketing of piles is used for splicing or strengthening of piles.

Keywords: Concrete, Deck, Jacket, Pile, Steel, Timber.

1 INTRODUCTION

This paper presents the rehabilitation techniques that are commonly used in the United States to extend the service life of marine terminal piers. These methods can equally be used for similar structures around the world. These piers are basically bridge structures located on the waterfront along the shoreline or extended into the sea or river channel for berthing of the cargo or cruise vessels. The marine piers, however, are subjected to a much larger lateral force than the design lateral force on highway bridges. The impact forces on the fenders and bollards due to mooring of the cargo or large cruise vessel are in the range of 100 tons to 200 tons. In addition to subjecting to large impact forces, the main structural components, and the piles and decks are constantly exposed to contact with water and a corrosive environment. The materials for the deck and piles are timber, steel and concrete. The following sections give rehabilitation techniques for repairing the existing deck and piles.

2 MARINA PIER STRUCTURES

Marina piers for mooring of cargo and cruise ships generally consist of apron slab deck supported on pile columns. The pile columns are foundation piles that extend to support the deck. Timber piles are seen in many piers built 30 to 40 years ago. Concrete piles and steel piles are used in a more recent design.

The deck is of timber or concrete. Timber deck is generally used with timber piles as seen in Figure 1(a). Most cruise and cargo pier decks are concrete decks supported on steel beams and steel pipe piles as shown in Figure 1(b).

(a) (b)

Figure 1. Piers with a) timber deck on timber piles, and b) concrete deck on steel pipe piles.

2.1 Piles in Grid Pattern

In the so-called flat slab deck, the deck is supported on pile columns arranged in a regular square or rectangular grid pattern. This type of pile configuration is found mostly in timber pile foundation of the pier. The spaces between piles are about five to six feet in both x- and y-directions as determined by the strength of timber piles.

2.2 Piles in Row Pattern

Most pier superstructures are multiple span concrete deck supported on transverse pile bents which are spaced at 30 feet to 50 feet span range. The piles are arranged in one- or two- row configuration with or without pile cap. The deck may be a one-way slab or a beam and slab deck.

3 REHABILITATION AND REPAIR

All exterior structures are subject to environmental weathering and corrosion. Maintenance and repair are performed to extend the service life of these structures. Common repair procedure on the deck and piles of marine piers are given in the following sections.

3.1 Repair of Timber Deck

Timber deck can be readily repaired by replacing the damaged or deteriorated planks. Replacing an old deck with the new deck is not uncommon.

3.2 Repair of Concrete Deck

Deteriorated concrete decks have cracks and spall areas with exposed reinforcement. Figure 2 shows a repair method of concrete cracks. The cracks are V-grooved to a depth of approximately half an inch and blown clean. The groove is then filled with neat epoxy.

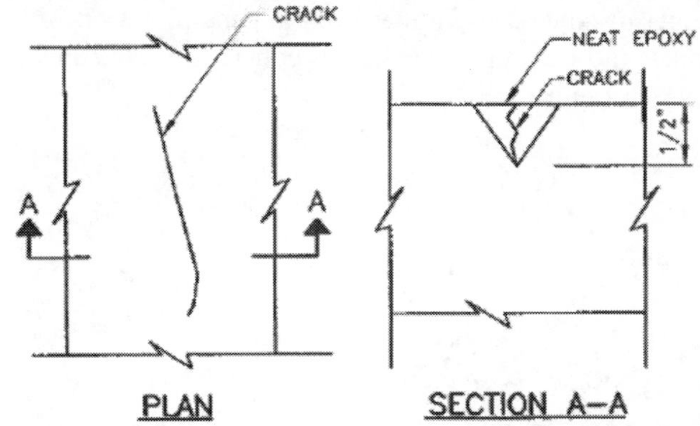

Figure 2. Repair of concrete cracks (Seniwongse 2007).

Figure 3 shows suggested patching repair methods for disintegrated or spalled concrete at the top surface of the deck. As shown, epoxy mortar is used to patch a shallow depth repair, and non-shrink mortar for deeper depth. In cases where steel bars have lost more than one quarter of their original diameter, supplemental reinforcement is provided before patching.

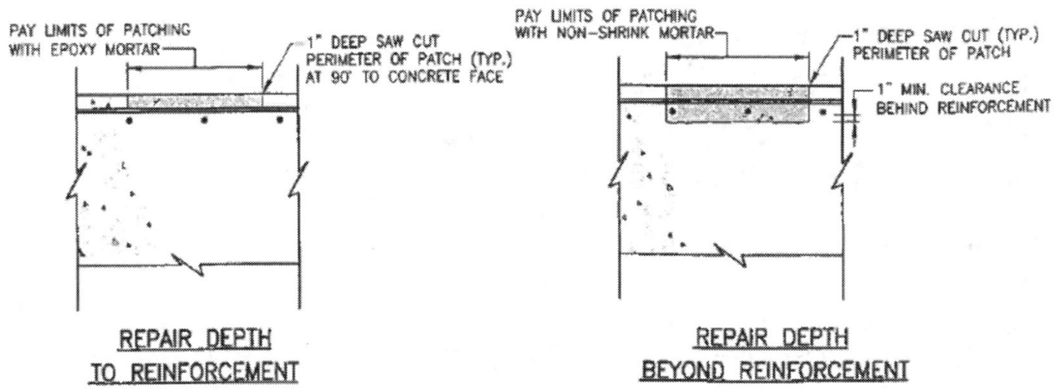

Figure 3. Repair of concrete cavity (Seniwongse 2007).

For repairing of spalled concrete area at the bottom surface (soffit) of the deck, shotcrete method of repair is recommended as shown in Figure 4.

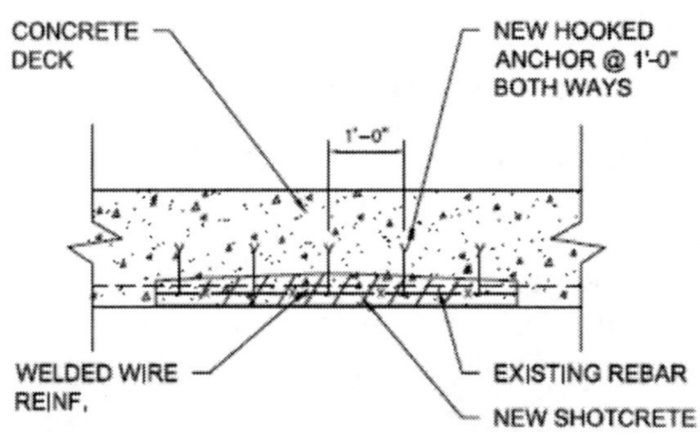

Figure 4. Shotcrete repair of deck soffit.

Areas of soffit of concrete deck with spalling concrete, honeycombed concrete and exposed reinforcement are repaired by removing loosened concrete to sound concrete, adding reinforcing steel as required and applying new shotcrete. Shotcrete application and surface preparation are given in specification for shotcrete (ACI 506.2-13 2013).

3.3 Repair of Timber Piles

Two methods of repairing of broken or severely deteriorated timber piles are shown in Figures 5 and 6. In Figure 5 the derelict timber pile is encased in concrete. Figure 6 shows the splicing of new timber pile to existing timber pile with concrete jacketing.

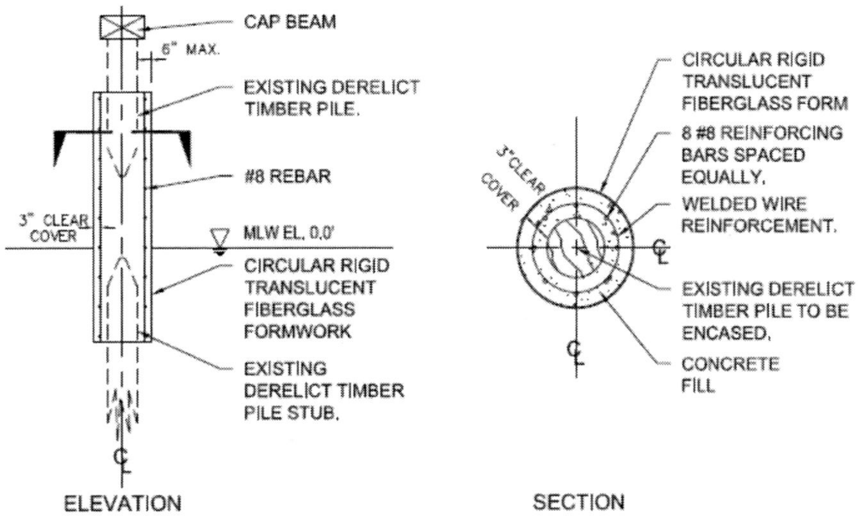

Figure 5. Concrete jacketing of existing derelict timber pile.

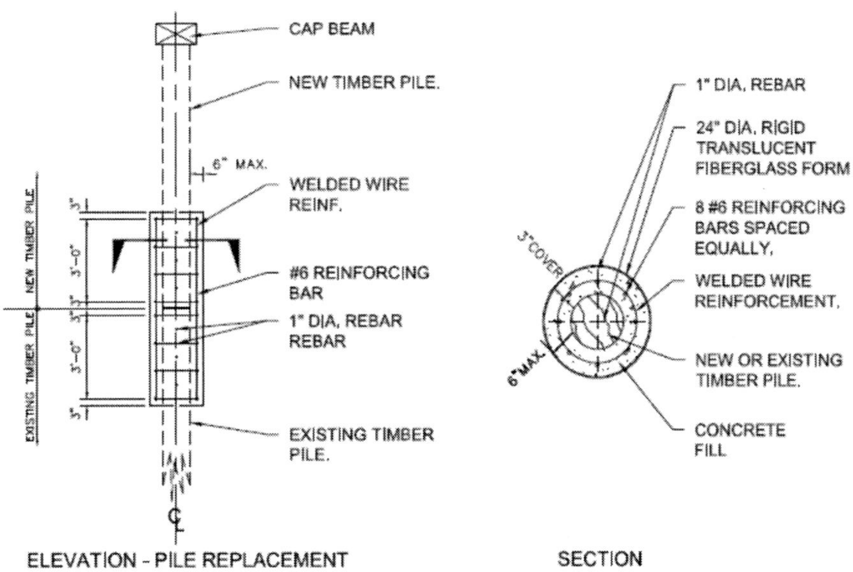

Figure 6. Partial timber pile replacement.

3.4 Repair of Steel Piles

Rehabilitation of steel pipe piles by jacketing is shown in Figures 7 and 8. The diameter of the steel pipe caissons and the length of the jackets shown are indicative only. As shown the jackets start from top of the pile (caisson) and extend 18 ft. to below the aggressive corrosion zone, normally a few feet below the lower low water level. The diameter of the jacket must be larger than the outside diameter of the existing piles as dictated by the design thickness of the concrete jacket, and at least must provide adequate gap for flowing of concrete grout. In Figure 7, the 3-inch concrete grout jacket is designed to protect the existing steel caisson from further corrosion. The steel jacket acts as permanent form and protection of the concrete grout.

The 5/16" wall thickness of the steel jacket will be adequate for corrosion allowance for the remaining service life of the pier.

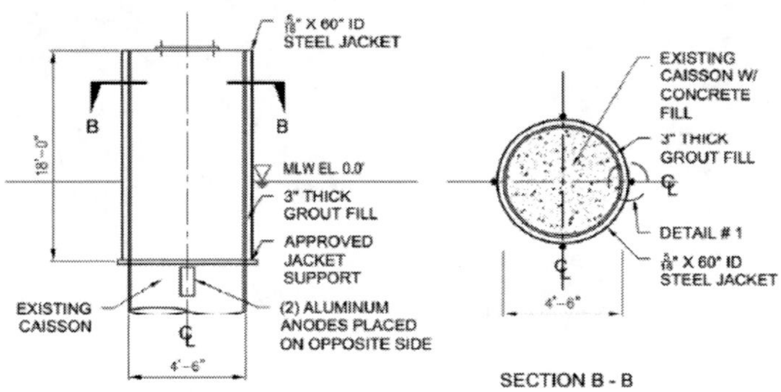

Figure 7. Steel jacketing of existing steel pipe pile.

Figure 8 shows the situation where the existing caisson is more corroded than that of Figure 7 and has to be strengthened by encasing in the reinforced concrete jacket. The fiberglass form is large enough to provide 8 inches gap for the circular reinforced concrete column section. The vertical rebar and the hoop rebar are designed for the required column strength.

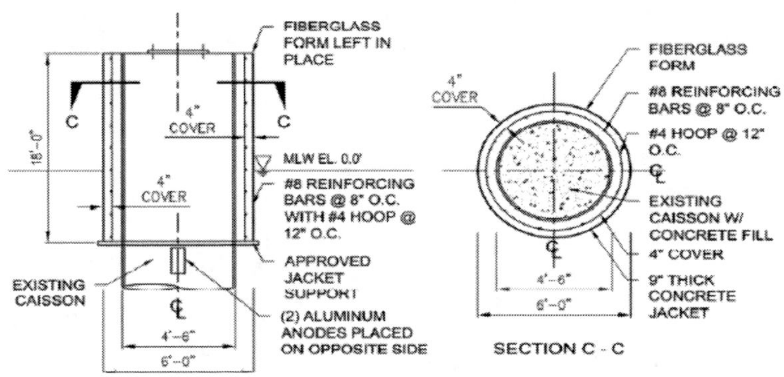

Figure 8. Fiberglass jacketing of existing steel pipe pile.

Note that in Figures 7 and 8, sacrificial aluminum anodes are attached on opposite faces of the steel caissons as parts of the cathodic protection system to protect the steel components from corrosion.

3.5 Repair of Concrete Piles

Rehabilitation of existing concrete piles can be done by jacketing similar to those shown for steel piles.

4 CONCLUSIONS

This paper has described methods of repair and rehabilitation of marine terminal piers that have been used successfully in the US. It can be concluded that:

- Timber deck is replaced by new timber partially or for the entire span.
- Concrete deck is repaired for cracks and spalled areas by sealing with epoxy and filling of cavity by patching using concrete or non-shrink mortar for top deck surface. For the soffit, shotcrete is more practical. Reinforcing bars are added as required before patching.
- For repair and strengthening of timber, concrete and steel piles, the common technique is by wrapping the repaired or strengthening portion by reinforced concrete, the concrete jacketing technique. In this technique, the steel tubes or fiberglass tubes of 6 inches to 10 inches larger radius than that of the existing piles are used as permanent formwork, reinforcing bars installed and concrete is pumped in to fill the gap.

The methods given in this paper would be applicable for rehabilitation of similar projects in the US as well as in any other countries.

References

ACI 506.2-13, *Specification for Shotcrete*, American Concrete Institute, USA, 2013.

Seniwongse, M., *Evaluation and Rehabilitation of Concrete Bridges in USA,* 4th International Structural Engineering and Construction Conference, Melbourne, Australia, 2007.

Interdependence between Structural Engineering and Construction Management
Edited by Ozevin, D., Ataei, H., Modares, M., Gurgun, A., Yazdani, S., and Singh, A.

ISBN: 978-0-9960437-6-2

KNOWLEDGE ENABLED TECHNOLOGY SELECTION FOR SEWER CONDITION ASSESSMENT

JEHAN ZEB

Dept of Public Works, City of Topeka, Topeka, USA

To improve the quality of life of communities, local government provides municipal services, including the sewer collection, effectively and efficiently employing asset management best practices. The condition assessment of assets (including the sewer collection system) is one of the core elements of implementing effective asset management across the utility organizations. Availability of sophisticated and complex condition assessment technologies, lack of common understanding about them and related high cost associated with sewer condition assessment requires the technologies and related knowledge to be explicitly defined in a neutral format—the ontology, to support the development of applications for technology selection. An ontology of condition assessment technologies for sewer network, CATS_Onto was developed following a seven-step approach at two levels of abstraction: meta-model and detailed ontology. This paper presents the development and verification of the meta-model. The proposed ontology will be used to develop a tool for the selection of the most appropriate technology for sewer condition assessment. The knowledge representation was verified while validation is underway.

Keywords: Ontology, Criteria, Inspection, Application, Asset management.

1 INTRODUCTION

The local government provides basic municipal services to communities through infrastructure systems (assets), which plays a vital role in the local, state and national economic development. Infrastructure assets are critical for the safety, welfare, and public health of communities (ASCE 2017). To provide uninterrupted services, these assets need to be maintained and managed to employ a comprehensive asset management planning. One of the core elements of effective asset management planning is to assess the condition of assets on a regular basis using appropriate technology to support maintenance and rehabilitation decision making. Although asset management of all types of assets is equally important; however, this paper focuses on wastewater (sanitary and storm) assets because most of these assets are at least 60 years old and many communities have sewers that are older than 100 years (US EPA 2015).

A number of technologies are currently available to assess the condition of the sewer network. Each technology has specific strengths and weaknesses and due to the complexity of these technologies, the utility management experts find it difficult to select a suitable, feasible and cost-effective technology. According to Agarwal (2010), these experts choose technologies that are not suitable to collect the data required for accurate condition assessment. This emphasizes the need to develop a tool that the utility experts can use to select an appropriate technology for sewer condition assessment using an ontological approach. The underlying knowledge model was created in a neutral format (the Ontology of Condition Assessment Technologies for Sewer network, CATS_Onto). An "ontology is an explicit specification of a conceptualization." (Gruber 1995). Following a layered architecture proposed by Gomez-Perez (1996), the CATS_Onto was developed at two levels: (i) an abstract Meta-Model of Condition Assessment Technologies for Sewer Network, CATS_MM, and detailed CATS_Onto. This paper discusses the meta-model and is divided into 7 sections. The background information and related literature are discussed in section 1 and 2 respectively. Section 3 discusses the methodology and section 4 explains the development of the meta-model. A potential area of application is discussed in section 5 and evaluation is depicted in section 6. Finally, conclusions are discussed in section 7.

2 LITERATURE REVIEW

2.1 Infrastructure Management and Asset Management Ontologies

In the area of infrastructure management, Osman (2007) developed the Infrastructure Product Ontology to represent infrastructure product knowledge in the water, wastewater, electrical, telecommunication, and gas sector. Zeb and Froese (2014, 2015b) created a Tangible Capital Asset Ontology to represent Tangible Capital Assets in the transportation, water, wastewater, and solid waste management to develop an Asset Information Integrator System (Zeb *et al.* 2015). These ontologies represent engineered or man-made assets in the built-environment. In the natural environment, an Eco Asset Ontology was built to represent natural assets (e.g., ditch, storm channel, ponds, etc.) (Zeb 2017). El-Gohary (2008) developed the Infrastructure and Construction Process Ontology to represent construction processes. To support the seamless exchange of information between the systems of the utility organizations, a Transaction Domain Ontology was developed (Zeb and Froese 2012, 2016, 2017). These ontologies represent product and process knowledge lacking the knowledge required to support the development of applications for the selection of an appropriate technology for sewer condition assessment. In asset management domain, Frolov *et al.* (2009) developed an Asset Management Ontology to integrate product (asset) and process (best practices) knowledge. In the condition assessment domain, a model was developed to monitor the condition of industrial assets (Campos 2007).

2.2 Condition Assessment Ontologies and Technology Selection Considerations

To assess sewer networks, a set of condition assessment technologies and selection considerations are available. The utility experts don't have adequate knowledge about the capabilities and limitations of these technologies (Agarwal 2010). Tuccillo *et al.* (2010) classified technologies into four categories: screening, internal pipe surface, wall integrity, and pipe bedding and void conditions and identified selection criteria including; inventory of pipes and operating conditions, data needs, cost and implementation issues. Similarly, Lee (2017) categorized technologies into eight sub-classes: pitting depth measurement, visual inspection, electromagnetic inspection, acoustic inspection, ultrasonic testing, laser profiling, flow meters, and innovative technologies. Presently, the technology selection process is manual, which is a very time-consuming task and requires a lot of technology-related know-how. The lack of such a knowledge results in poor decision-making regarding technology selection. This requires the need to develop a tool to select an appropriate technology for sewer condition assessment.

The existing literature lacks; (i) a formal classification system of condition assessment technologies for sewer network; (ii) explicit description of the technologies and selection criteria; and (iii) a framework for condition assessment technology selection. The proposed CATS_Onto will address these issues.

3 METHODOLOGY

The following seven-step approach was used to develop the proposed ontology. A similar methodology was adopted for the development of Infrastructure Product Ontology (Osman 2007), Infrastructure and Construction Process Ontology (El-Gohary, 2008), Tangible Capital Asset Ontology (Zeb and Froese 2015a and 2015b) and Eco Asset Ontology (Zeb 2017).

- Step 1: Define scope —The scope represents the purpose, use, and users of the ontology
- Step 2: Develop a meta-model—The meta-model represents the knowledge at the abstract level to ease knowledge categorization and improve understanding of the knowledge.
- Step 3: Define taxonomy—The abstract concepts represented in the meta-model were extended to develop detailed taxonomies of concepts.
- Step 4: Code ontology—The knowledge was formally coded in the Ontology Web Language (OWL) using the Protégé Ontology Editor (Protégé 2018).
- Step 5: Capture ontology—All concepts were explicitly defined in the plain English language (soft axioms) and OWL Description Logic Syntax (hard axioms).
- Step 6: Evaluate ontology—As part of the evaluation, the knowledge was verified using the built-in Protégé Reasoners and validation is underway through industry experts.
- Step 7: Document ontology—The knowledge was documented for future use.

4 ONTOLOGY DEVELOPMENT – META-MODEL

According to Gomez-Perez (1996), ontologies are developed using a layered architecture with the top layer representing generic concepts, which are further specialized at the lower levels. The CATS_Onto was developed using this approach with the top layer representing the core concepts related to technology selection at the abstract level and is called as CATS_MM. A detailed CATS_Onto was developed through specialization of core concepts. This paper discusses the development and verification of the knowledge representation at the meta-model level. Figure 1 shows the key concepts (entities, things) and relationships between them to formalize the knowledge required for the technology selection.

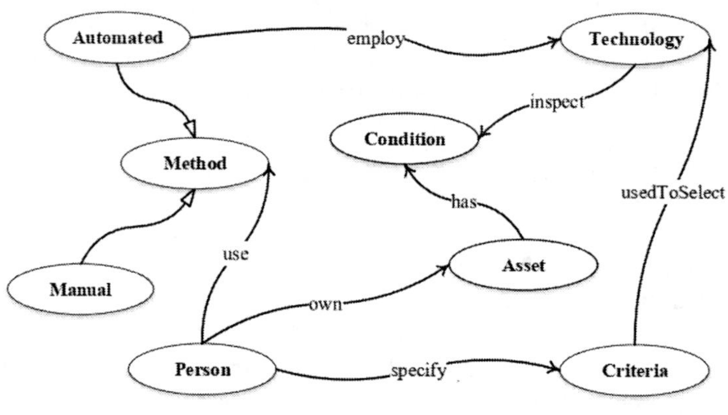

Figure 1. Meta-model of condition assessment technologies for sewer network.

4.1 Key Concepts

The "concepts are key entities in the production, services and manufacturing systems", (Fox and Gruninger 1998). A brief definition of each concept is as follows:

Asset is defined "as an item, thing or entity that has potential or actual value to an organization", (ISO55000 2014). The asset refers to the sanitary and stormwater pipe network, including collectors, interceptors, and force main.

Condition is a measure of the physical, capacity and operational health of the sanitary and stormwater pipe network. The utility organizations usually employ automated technologies to inspect the condition of the pipe network that depends on the age, environment, maintenance, history and how well it is used by the community.

Person includes individuals and organizations who own, operate and manage sanitary and stormwater systems. The organization refers to public and private utility agencies and corporations responsible for providing uninterrupted municipal services.

Method is a generic concept covering all such means, mediums, and techniques used to assess sewer condition. These methods are either manual or automated. The manual methods include physical observation of the pipe network through visual inspection. The automated methods include physical observation of the pipe network using a set of fully automated machines and devices without the aid of human beings.

Criteria are rules, principles, aspects, elements, and factors used to identify the most appropriate technology for sewer inspection. According to Agarwal (2010), technology selection criteria include; affordability, detectability, feasibility, suitability, and usability.

Technology refers to the capabilities given to a machine, device or a piece of equipment created as a result of the practical application of knowledge (Merriam-Webster 2018) for sewer inspection. Tuccillo *et al.* (2010) categorized technologies into the following four types: Bedding and Void Inspection Technologies, Internal Surface Inspection Technologies, Screening Technologies, and Wall Integrity Inspection Technology.

The knowledge in the meta-model was conceived based on the notion that the person (individuals and organizations) own, operate, and manage infrastructure systems including sewer systems to provide services effectively and efficiently. These persons inspect and assess condition of assets as part of effective asset management using manual and automated method. In an automated method, a technology is employed to inspect the sewer network. To select a technology, an evaluation criteria is needed. The utility organizations specify the criteria to select the most appropriate technology for sewer condition assessment.

5 ONTOLOGY APPLICATION

The meta-model was used to develop the detailed CATS_Onto. The software developers will use the CATS_Onto to develop applications for the selection of appropriate technology. Such applications are lacking in the asset management domain, which need to be developed for effective decision-making regarding technology selection. The CATS_Onto provides a common understanding of the terms in the area of asset management and ensures applications are developed consistently across the industry. The proposed applications will be implemented using a multi-step exclusion protocol in which technologies will be excluded based on technical feasibility and technical suitability criteria. After screening, the selected technologies will be compared against a set of performance and cost parameters to determine performance and cost. Work is in progress to develop the application and is beyond the scope of this paper.

6 EVALUATION

The ontology was verified and validated using a set of criteria. In verification, the knowledge is verified against a set of criteria to assess that the knowledge model is built right. In ontology validation, the knowledge is validated through industry experts to assess that the right knowledge is built. The following criteria were used to evaluate the ontology including; consistency, conciseness (Gomez-Perez 1996), clarity, completeness (Yu *et al.* 2007), and correctness (Guarino 1998). Consistency measures uniformity in the knowledge representation. Conciseness measures redundancy in the knowledge representation. Clarity measures understandability of knowledge representation. Completeness measures the extent of coverage of knowledge representation. Correctness measure accuracy of knowledge representation. The CAT_Onto was verified for consistency and conciseness using Protégé Reasoners: ELK 0.4.3, FaCT++ 1.6.5, HermiT 1.3.8, Mastro DL-Lite Reasoner, Ontop 1.18.1, Pallet and jcel (Protégé 2018) as shown in Figure 2. These reasoners were run to check consistency and conciseness of the knowledge representation. The reasoning analysis results are shown in Figure 2 (image on right), where a superclass "Nothing" under the inferred class hierarchy (an automatically generated class hierarchy) represents classes with errors. There was no class found under the superclass "Nothing", indicating that the knowledge representation is consistent and concise.

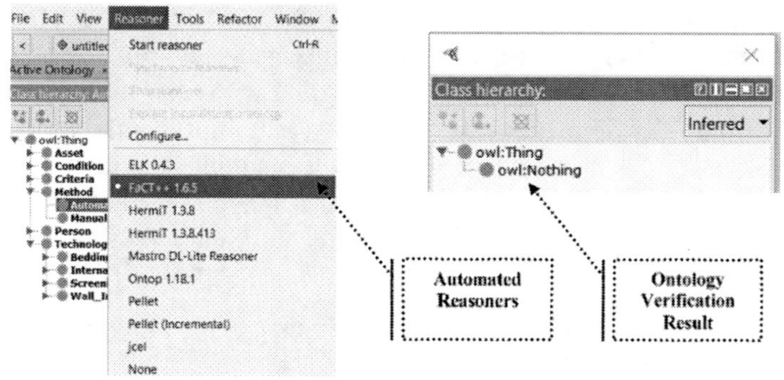

Figure 2. CATS_Onto verification.

7 CONCLUSIONS

The CATS_Onto was developed at two levels of abstraction: meta-model and detailed ontology using a seven-step approach. This paper discusses the development and verification of the meta-model. The meta-model represents the core concepts at the abstract level to ease knowledge categorization and navigation. As part of the evaluation, the knowledge in the meta-model was verified using a set of criteria. The ontology verification was accomplished using automated reasoners in the Protégé Ontology Editor. The verification results indicate that the knowledge representation was consistent and concise. The ontology validation is currently underway and beyond the scope of this paper. The proposed CATS_Onto will be implemented in a prototype stand-alone or web-based application to demonstrate the validity of the approach. From a theoretical perspective, the CAT_Onto has a significant contribution to the body of knowledge in the area of asset management. From a practical perspective, the knowledge representation; (i) provides a common understanding of the terms for software developers and industry experts; (ii) eases consistent implementation in applications; and (iii) eases knowledge base extensions in case more sewer technologies and criteria are added in future. The research limitations include the lack of a full fledge implementation in an application to test and validate the proposed approach. In the current state, the most common technologies are represented in the CATS_Onto. In future, the knowledge represented in the CATS_Onto will be implemented in a stand-alone or web-based application. The proposed ontology will be extended once more technologies are identified.

References

Agarwal, M., *Developing a Framework for Selecting Condition assessment Technologies for Water and Wastewater Pipe*, M.Sc. Thesis, Virginia Polytechnic Institute, Virginia, USA, 2010.

ASCE., Infrastructure Report Card, *A Comprehensive Assessment of America's Infrastructure*, American Society of Civil Engineers, USA, 2017.

Campos, J., *An Ontology for Asset Management*, 1st IFAC Conference on Cost Effective Automation in Networked Product Development and Manufacturing, IFAC Proceedings, 40(19), 36-41, 2007.

El-Gohary, N., *Semantic Process Modeling and Integration for Collaborative Construction and Infrastructure Development*, Ph.D. Thesis, Dept. of Civil Eng., Univ. of Toronto, Canada, 2008.

Fox, M. S. and Gruninger, M., *Enterprise Modeling*, AI Magazine, AAAI Press, 109-121, 1998.

Frolov, V., Megel. D., Wasana, D., Yong, S., and Lin, M., *Building an Ontology and Process Architecture for Engineering Asset Management*, Proceeding of 4th World Congress on Engineering Asset Management, Athens, Greece, 2009.

Gómez-Pérez, A., Towards a Framework to Verify Knowledge Sharing Technology, *Expert Systems with Applications*, 11(4), 519-529, 1996

Gruber, T. R., Towards Principles for the Design of Ontologies Used for Knowledge Sharing. *International Journal of Human-Computer Studies*, Academic Press Ltd, 43(5-6), 907–928, 1995.

Guarino, N., *Some Ontological Principles for Designing Upper-Level Lexical Resources*, Proceedings of the 1st International Conference on Lexical Resources and Evaluation, 1998.

ISO55000., *Asset Management—Overview, Principles, and Terminology*. International Standard, 1st Edi., Ref. #., ISO 55000:2014(E), International Organization for Standardization, USA, 2014

Lee, A., *Condition Assessment Technologies for Water Transmission and Sewage Conveyance Systems*, A Collaborative Project of the University of British Columbia Sustainability Scholar Program 2017 and Metro Vancouver, Vancouver, Canada, 2017

Merriam-Webster., *Definition of Technology from Merriam-Webster*, retrieved from, https://www.merriam-webster.com/dictionary/ technology on September 18, 2018.

Osman, H. M., *A Knowledge-Enabled System for Routing Urban Utility Infrastructure*, Ph.D. Thesis, Department of Civil Engineering, University of Toronto, 2007

Protégé., *Protégé is a National Resource for Biomedical Ontologies and Knowledge*, Developed by Stanford Center for Biomedical Informatics Research, Stanford University, USA, 2018.

Tuccillo, M. E., Jolley, J., Matel, K., and Boyd, G., *Report on Condition Assessment Technology of Wastewater Collection Systems*, National Risk Management Research Laboratory–Water Supply and Water Resources Division, U.S. Environmental Protection Agency, Ohio, USA, 2010.

US EPA., *Condition Assessment of Underground Pipes*, Excerpt from Condition Assessment of Wastewater Collection Systems, EPA/600/R-09/049, Water Infrastructure Outreach, US Environmental Protection Agency, Cincinnati, Ohio, USA, 2015.

Yu, J., Thom, J. A., and Tam, A., *Ontology Evaluation Using Wikipedia Categories for Browsing*, Proc. of 6th ACM Conf. on Information and Knowledge Mgt., Lisbon, Portugal, 223–232, 2007.

Zeb, J., An Ontology of Eco Assets Towards Effective Eco Asset Management, *Built Environment Project and Asset Management Journal*, Emerald Publishing Ltd., UK, 7(4), 388-399, 2017.

Zeb, J. and Froese, T. Transaction Ontology in the Domain of Infrastructure Management, *Canadian Journal of Civil Engineering*, Published by NRC Research Press, 39(9), 1-12, 2012.

Zeb, J. and Froese, T., Tangible Capital Asset Ontology in Infrastructure Management, *Infrastructure Asset Management Journal*, Institution of Civil Engineers, ICE Publishing, UK, (3), 81-92, 2014.

Zeb, J. and Froese, T., *Transaction Formalization in the Infrastructure Management Using an Ontological Approach*. Chapter 8, Ontology in the AEC Industry: A Decade of Research and Development in the Architecture, Engineering, and Construction, ASCE, USA, 169-192, 2015a.

Zeb, J. and Froese, T., *An Ontology-Supported Transaction Formalism Protocol in Infrastructure Management*, 5th Intl./11th Construction Specialty Conference, CSCE, Vancouver, Canada, 2015b

Zeb, J. and Froese, T., An Ontology-Supported Infrastructure Transaction Management Portal in Infrastructure Management, *Journal of Infor. Technology in Construction*, 21, 100-118, 2016.

Zeb, J. and Froese, T., Transaction Formalism Protocol Tool in Infrastructure Management, *Construction Innovation Journal*, Emerald Group Publishing Ltd., UK, 17(2), 180-203, 2017.

Zeb, J. Froese, T., and Vanier, D., An Ontology-Supported Asset Information Integrator System in Infrastructure Management, *Built Environment Project and Asset Mgt.*, 5(4), 380– 397, 2015.

Interdependence between Structural Engineering and Construction Management
Edited by Ozevin, D., Ataei, H., Modares, M., Gurgun, A., Yazdani, S., and Singh, A.

ISBN: 978-0-9960437-6-2

ANALYSIS OF THE MAINTENANCE WORK ORDER DATA IN EDUCATIONAL INSTITUTIONS

DENIZ BESIKTEPE[1], MEHMET E. OZBEK[2], and REBECCA A. ATADERO[1]

[1]*Dept of Civil and Environmental Engineering, Colorado State University, Fort Collins, USA*
[2]*Dept of Construction Management, Colorado State University, Fort Collins, USA*

The growing number of aging buildings is of critical importance to the economic and social well-being of a country. For institutional organizations, particularly those with large building portfolios, an effective facility management approach is required to ensure these buildings function properly for their missions. As a part of facilities management, building maintenance activities occupy a significant role in reaching the goal of delivering an acceptable level of performance while minimizing costs and failures. Historical work order data may potentially include a substantial value for assessing the condition of building systems by helping to identify frequent and common building maintenance activities. As an exploratory study on building maintenance activities in educational institutions, this paper conducts an analysis of historical work order data collected from six educational institutions in the State of Colorado and Connecticut in the United States between 2008 and 2018. Within this context, a total of 877,668 work orders have been analyzed to identify possible trends of the maintenance activities with building age and building type. The results of this study provide a preliminary understanding of the frequent maintenance activities in the data sets with their relations of building age and building type. The need for the comprehensive statistical analysis is apparent to better understand any possible trends in the maintenance activities and their correlations between building age and building type.

Keywords: Facilities management, Building maintenance, Condition assessment, Educational facilities.

1 INTRODUCTION

Educational facilities are a fundamental part of educational institutions, providing the necessary environment and services for the institutions to meet their objectives. As the average age of educational institutions increases, facilities management practices and their effectiveness become important to sustain their educational missions. The Sightlines Report (2017) states that the average age of buildings in higher educational facilities as of 2007 was 34.2 years; and with no new building additions, the average age would have been 44.7 years. The database of the Sightlines Report (2017) tracks 52,200 buildings over 400 campuses in 44 U.S. states and four Canadian provinces for a total of 1.5 billion gross square feet (GSF) of space. The spaces for educational facilities grow and their ages increase faster than the resources allocated to maintain the spaces. Subsequently, limited budgets and resources highlight the need for effective maintenance management.

The main purpose of this study is to identify possible trends in maintenance activities such as the relation between the maintenance activity and building age as well as the maintenance activity and building type. The findings of this paper provide a basis for a resource efficient condition assessment framework in educational institutions. The operation and maintenance of the facilities include many different daily activities. Work order management systems provide a significant amount of data including activity types, crew/trade types, building ages, building types, dates, and etc. This paper conducts an analysis of work orders collected from six educational institutions in the State of Colorado and Connecticut in the United States between 2008 and 2018. The preliminary analysis of 877,668 work orders suggests that electrical, heating, ventilation and air conditioning (HVAC), and plumbing activities are the most frequent and common building maintenance activities. Therefore, this study mainly focuses on electrical, HVAC, and plumbing work orders.

2 BACKGROUND

Cost constraints, budget cuts, and the lack of proper management are the main reasons of lower performance of facilities, which result in the need for effective facilities management approaches (Eweda *et al.* 2013). Kaiser (2016) states that almost 75 percent of the higher educational facilities are in the range of 30 to 40 years old and they are facing the challenge of deferred maintenance backlog and inadequate funding for their maintenance needs. The same study mentioned that the deferral of major maintenance activities causes significant issues for system and infrastructure reliability, health and safety, and functional use of space (Kaiser 2016).

The aging buildings and insufficient funding of maintenance needs highlight the importance of the condition assessment process, which is a fundamental step of maintenance decisions. The limited resources have a negative impact on the condition assessment practices of the higher educational institutions. Therefore, most of the higher educational institutions are not able to perform the visual inspection part of the condition assessment process.

The historical work orders include comprehensive information related to the maintenance activities. Over the life of the facilities, the analysis of work orders may provide a preliminary feedback about the buildings' current conditions. There are some studies in the literature using the work order analysis for fault detection and maintenance budget calculations. Bossmann and Lennerts (2013) studied the maintenance data sets of 140 federal buildings in Germany to identify the key influencing factors of accurate and reliable budget calculations for the building maintenance. Their study emphasizes the influence of the age of facilities, the type of use, the size of the building and the number of stories on the budget calculations. Additionally, Yang *et al.* (2018) proposed a failure mode and effect analysis method for HVAC issues in the building maintenance. Their method uses a data mining approach for the analysis of work orders, though, it is limited to a specific type of maintenance activity. A more comprehensive approach is needed to analyze the historical work order data for obtaining meaningful information that will help better understand the condition of the building systems.

3 DATA COLLECTION AND METHODOLOGY

The work order data sets used in this study are collected from six educational institutions in the state of Colorado and Connecticut in the United States between 2008 and 2018. Among six educational institutions, three are state universities, two are private universities and one is a school district. The GSF of these six institutions is 34,964,771 square footage (SF) with a total of 1,671 buildings. The characteristics of the institutions with the average age of their buildings are shown in Table 1.

Table 1. The characteristics of the educational institutions of the study.

Institutions	State	GSF of Buildings	Total # of Buildings	Average Age of Buildings	Time Interval of Work Orders	Work Order Management Software
State #1	Colorado	4,076,953	60	63	2009-2018	TMA
State #2	Colorado	12,361,537	748	43	2008-2017	FAMIS
State #3	Colorado	8,186,982	409	67	2013-2017	FAMIS
Private #1	Colorado	3,567,470	74	51	2012-2018	TMA
School District #1	Colorado	3,979,365	70	43	2011-2017	SchoolDude
Private #2	Connecticut	2,792,464	310	105	2013-2017	SchoolDude

The work order data sets were exported to Microsoft Excel from the work order management software of the institutions. Table 1 shows the names of their work order management software. The primary analysis of the work orders focused on identifying the distribution of the maintenance activities for each institution.

The total of 877,668 work orders of six institutions contain all types of maintenance activities. However, the keywords and classification of the work orders were not constant for all institutions the names of crew or trade types were similar, namely architectural, carpentry, electrical, HVAC, mechanical, plumbing, structural, etc. Therefore, the work orders were sorted according to the crew or trade keywords. The content of the keywords was confirmed with the facilities management departments of each institution.

The first step of the analysis is focused on identifying any possible relations between maintenance activity types and building ages. On the second step of the analysis relations between maintenance activity types and building types are revealed. The researchers acknowledge that the analysis of this study provides a preliminary understanding of relations between the maintenance activity types and building age as well as the maintenance activity type and building type.

4 RESULTS

The preliminary data analysis of this study revealed that electrical, HVAC, and plumbing maintenance activities comprise the majority of all work orders as presented in Table 2. The "Others" category of Table 2 includes the variety of maintenance activities namely access control, carpentry, construction, custodial, key and lock services, fire safety, grounds, painting, pest management, operational services, structural, security, telecommunications, utility, welding, and etc. The contribution of these maintenance activities is less than 6% to the "Others" category individually. Subsequently, this analysis confirmed the similar feedback given by the facilities management executives who pointed out the higher frequencies in these three categories. Therefore, further analysis of the study focuses on these three maintenance activities.

The first step of the analysis explored the relationship between the maintenance activity types and building age. The corrective maintenance work orders including electrical, HVAC, and plumbing activities were extracted from the data and used in this analysis. The average age of the buildings was calculated per maintenance activity type for each institution considering only the year of construction, and renovations were not included in the calculations. It is important to note that, some of the buildings' ages were not provided by the facilities management departments. Since the focus of the analysis is exploring the relationship between the maintenance activity types and building age, the data of these buildings could not be included in the analysis, and their data was disregarded. The percentage of the disregarded data varies between 20% and 3% of the total work orders per each institiution.

Table 2. The distribution of the electrical, HVAC, and plumbing work orders of the institutions.

Institutions	Electrical	HVAC	Plumbing	Others
State #1	17.54%	16.76%	11.94%	53.76%
State #2	9.62%	22.58%	10.12%	57.68%
State #3	14.56%	12.96%	9.31%	63.17%
Private #1	7.31%	23.57%	5.15%	63.97%
School District #1	14.55%	7.86%	16.50%	61,09%
Private #2	13.87%	13.39%	14.12%	58.62%

The distribution of work orders for the maintenance activities was relatively estimated for each individual building using Eq. (1).

$$Dwb_i = \frac{1}{Nw} \sum_{i=1}^{n} Nwb_i \tag{1}$$

where:

Dwb = The distribution of work orders for each maintenance activity at building *i*;

Nw = The total number of work orders for each maintenance activity; and

Nwb = The number of work orders for each maintenance activity at building *i*.

These distributions are grouped into three clusters in each maintenance activity for all institutions: higher, lower, and zero work orders. The higher cluster represents five percent and greater of total work orders (>5%). The lower cluster represents the buildings that have the work orders less than one percent of total work orders (< 1%). Lastly, the zero work order cluster represents the buildings that have zero work orders assigned (= 0). The higher and lower clusters point out the two extreme ranges in the data set. Table 3 shows the average age of the buildings of each institution in these clusters per maintenance activity type.

Table 3. The average age of buildings with higher #, lower #, and zero # of work order clusters (% of work orders > 5%, < 1%, = 0).

Maintenance Activity Type	State #1			State #2			State #3			Private #1			School District #1			Private #2		
	H	**L**	**Z**	**H**	**L**	**Z**	**H**	**L**	**Z**	**H**	**L**	**Z**	**H**	**L**	**Z**	**H**	**L**	**Z**
Electrical	36	73	47	53	43	32	49	62	48	51	55	46	38	44	-	53	108	64
HVAC	37	77	44	50	47	34	28	63	52	51	51	53	38	47	-	51	107	104
Plumbing	40	76	45	51	45	32	79	54	44	27	51	55	45	48	-	41	109	141

H = Higher # of Work Orders, **L** = Lower # of Work Orders, **Z** = Zero # of Work Orders

The comparison of the maintenance types and building ages revealed that the average age of the buildings in each cluster is more than the average age of 34.2 years provided on the Sightlines

Report (2017). Table 4 shows the average age distribution of the buildings in this study. The majority of the buildings in all clusters of the analysis are in the range of 40 and 60 years. The comparison of the same institutions in the three clusters revealed that the average age of the buildings in the the lower cluster is larger than the average age of the buildings in the higher cluster. In other words, the analysis does not show that older buildings receive a higher number of work orders as might be expected.

Table 4. The average age distribution of buildings in higher #, lower #, and zero # of work order clusters (% of work orders > 5%, < 1%, = 0).

Average Age	**H**	**L**	**Z**
<40	33%	0%	20%
40 - 60	61%	56%	60%
>60	6%	44%	20%

Classroom, office, and research buildings have the largest square footage among the facilities of this study; thus, only those building types are used for the subsequent analysis. However, based on the data it was only possible to classify the types of buildings of two educational institutions within classroom, office, and research. Therefore, the two educational institutions used for the second step of the analysis are State #2 and Private #1. Table 5 shows the distribution of buildings in classroom, office, and research types of these two institutions.

Table 5. The distribution of classroom, office, and research buildings in State #2 and Private #1.

Institutions	**C**	**O**	**R**	**OT**
State #2	7%	24%	14%	55%
Private #1	22%	19%	12%	47%

C = Classroom, **O**= Office, **R** = Research, **OT**= Others

The distribution of the building types is grouped into the same clusters of building age distributions namely higher, lower, and zero work orders. Table 6 shows the percentage distribution of building types per maintenance activity and per institution within each cluster. The comparison of the maintenance activity types and building types revealed that classroom, office and research types receive a higher number of work orders. This finding needs further analysis to better understand the correlation between the maintenance activity type and building type.

Table 6. The percentage distribution of building types with higher #, lower # and zero # of work orders. (% of work orders > 5%, < 1%, = 0).

Maintenance Activity Type	**H**								**L**								**Z**							
	State #2 (%)				Private #1 (%)				State #2 (%)				Private #1 (%)				State #2 (%)				Private #1 (%)			
	C	O	R	OT	C	O	R	OT	C	O	R	OT	C	O	R	OT	C	O	R	OT	C	O	R	OT
Electrical	0	33	33	34	100	0	0	0	7	21	12	60	7	9	1	83	0	14	5	81	0	1	0	99
HVAC	33	33	0	34	100	0	0	0	12	30	15	43	4	15	3	78	1	9	6	84	0	0	0	100
Plumbing	25	0	25	50	0	0	100	0	8	26	11	55	8	9	3	80	1	7	10	82	0	0	0	100

Additionally, the data analysis performed for this study helped to identify some common issues in the work order management practices faced by the higher educational institutions. One issue is that the general trend of identifying the maintenance activity depends on the crew or trade type in the work order management systems. None of the data sets used in this study include a category of the maintenance activity type. For that reason, the maintenance issues need to be identified with some additional information. The data sets include task descriptions that may provide useful information to better identify the maintenance issues. However, there is no consistency on the keywords of those descriptions and the variety of the keywords make the descriptions complicated to be analyzed.

5 CONCLUSIONS

The facilities management executives of the educational institutions in this study advised the higher frequency of electrical, HVAC, and plumbing maintenance activities throughout the data collecting process. Subsequently, the findings of the study confirm the opinion of these executives.

The analysis of the study does not reveal a specific relation between the maintenance activity and building age as well as the maintenance activity and building type. The results of this study provide a preliminary understanding of the frequent maintenance activities in the data sets with their relations of building age and building type. Consequently, the need for the comprehensive statistical analysis is apparent to better understand any possible trends in the maintenance activities and their correlations between building age and building type.

Although the data sets include the maintenance types (corrective, preventive, predictive, etc.) and crew/trade types, they do not represent the maintenance activity types properly. The maintenance activities need a subcategory that includes the basic definitions of the frequent failures. For instance, the electrical maintenance activity category requires subcategories such as light bulb changing, electrical panel issues, outlet/switch issues, etc. The subcategory of the maintenance activities may increase the effective use of work order data in the condition assessment process.

References

Bossmann, J., and Lennerts, K., Maintaining Technical Building Facilities: A Growing Challenge, *International Conference on Construction and Real Estate Management: Construction and Operation in the Context of Sustainability*, 298-305, Karlsruhe, Germany, 2013.

Eweda, A., Zayed, T., and Alkass, S., Space-Based Condition Assessment Model for Buildings: Case Study of Educational Buildings, *Journal of Performance of Constructed Facilities,* 29(1), 1-12, June, 2013.

Kaiser, H., Capital Renewal and Deferred Maintenance, Association of Physical Plant Administrators (APPA), Leadership in Educational Facilities, 2016. Retrieved from http://bokcms.appa.org/pdfs/131-05281612.pdf on Jan 2019.

Sightlines Report, The State of Facilities in Higher Education: 2017 Benchmarks, Best Practices, & Trends Annual Sightlines Report, Sightlines a Gordian Company, 2017. Retrieved from https://www.sightlines.com/insight/2017-state-of-facilities-in-higher-education/ on Jan 2019.

Yang, C., Shen, W., Chen, Q., and Gunay, B., A Practical Solution for HVAC Prognostics: Failure Mode and Effects Analysis in Building Maintenance, *Journal of Building Engineering,* 15(2018), 26-32, January, 2018.

Interdependence between Structural Engineering and Construction Management
Edited by Ozevin, D., Ataei, H., Modares, M., Gurgun, A., Yazdani, S., and Singh, A.

ISBN: 978-0-9960437-6-2

TECHNOLOGICAL AND SOCIAL RETROFIT STRATEGIES FOR THE PUBLIC RESIDENTIAL BUILDINGS OF THE SECOND POST-WAR

MICHELE DI SIVO[1], DANIELA LADIANA[2], GIOVANNI SANTI[1], and LEDIANA RRJOLLI[1]

[1]*Dept of Energy, Systems, Territory, and Constructions Engineering (D.E.S.T.e.C.), University of Pisa, Pisa, Italy*
[2]*Dept of Architecture, University G. D'Annunzio of Chieti-Pescara, Pescara, Italy*

This experimental study on the renewal of post-World War II suburban areas addresses the theme of the technological retrofitting of buildings, focusing on typological features, construction techniques and, more in general, of all aspects that fail to meet the needs of contemporary living. An initial examination of Italian and international case studies helped to identify possible guidelines for the optimization of urban renewal activities. The guidelines were tested by applying them to a case study: a 1950s-era neighborhood in the city of Pisa. The study undertaken highlights the need to consistently integrate all of the activities that affect the building structures - consolidations and static adjustments, improvement of energy efficiency, functional and spatial adaptation - and that work towards strengthening social integration by taking into account the specific needs of the individual. The work is implemented on individual buildings and urban areas, with the aim of improving the local identity characteristics. The suggested approach is that of a circular economy, which helps to reduce the impact on the environment, mainly through the extensive use of recycled materials. The proposed experiment thus aims to test the resilience of buildings following their seismic, energy and functional updating, including sociological and environmental psychology assessments, with particular attention to vulnerable users, in order to find solutions capable of promoting social inclusion.

Keywords: Suburbs, Urban and social renewal, Social housing, Integrated retrofitting, Seismic adaptation.

1 URBAN RENEWAL AND SOCIAL HOUSING

Statistics show that around 75% of the European population currently lives in urban areas, a number that is expected to increase to 80% by 2020; the number could exceed 90% in as many as seven Member States. Furthermore, since the mid-1950, the total area of urban areas in the EU has increased by 78%, compared to a population growth of only 33%. According to the UN-Habitat 2016 World City Report (UN-Habitat 2016), 70% of urban land is occupied by housing. This modality of expansion, defined by experts as the "occupation of decoupled land", is attributable to profound changes in lifestyles and to the obsolescence of housing, which is no longer capable of fitting in with new paradigms of contemporary housing (Di Sivo and Cellucci 2016), rather than to population growth. It is, therefore, apparent that urgent and critical action be taken to slow down this phenomenon. There must be a shift from the tendency towards the indefinite and relentless expansion of cities (Latouche 2012) to that of urban renewal, not only, of the historical town centers but also of those early suburbs that have been subsumed by the rapid expansion of recent decades (Landolfo *et al.* 2013).

Within this framework lies an important area of work that has been identified as a priority for intervention: the social housing built in the period after the Second World War. Built following the destruction of the war and the subsequent steady urbanization of the population, it is unthinkable that this housing can be demolished on a large scale (Giacchetta 2012) and, although it must be adapted to the changed housing standards and the new energy/environmental constraints, it may prove to be an important resource for residential policies.

The modern implications of all this is that the significant functional and technological obsolescence of these structures is also accompanied by marked and evident physical deterioration. In particular, social changes with regard to the composition of households and the age of the occupants, together with changes in energy availability and technological progress, tend to modify the uses and habits of residents (Jacobs 1961) and accelerate the process of obsolescence.

It therefore seems appropriate that research in the field of technological and environmental design aim towards determining possible approaches, methods, and tools to facilitate the optimization of urban renewal activities for these settings. This involves implementing strategies, policies, and interventions aimed at increasing urban *resiliency*, by adapting city systems so that, under variable conditions, these are capable of systemic adaptation and flexibility so as to render it possible to simulate behaviors and alternatives of response (Habraken 1998).

An experimental research initiative is underway at the University of Pisa, where a working group is carrying out studies and projects in a neighborhood of Pisa with the aim of coming up with a valid approach for the types of building found mainly in post-war Italian housing that characterizes the earliest suburbs of Italian cities.

1.1 A Possible Approach

The research began with a study of the current housing stock, making a thorough examination of the buildings – including the typological features, the construction techniques – as well as on the building stock in Italy and in Europe, upon which qualitative/quantitative assessments were made of the technological and energy deficits encountered. See Table 1.

An assessment of these experiences confirmed the necessity of taking an integrated approach that takes into account the object as a whole and makes use of all the elements available to enhance its potential (Bellomo and Pone 2011). The traditional approach of making timely interventions aimed at responding to individual problems is in fact not completely effective in terms of reformulating building features and architecture.

The aim of retrofitting is to renew the building system in question by operating both on the environmental and technological sub-system utilizing an architectural project capable of implementing, enhancing or improving (depending on current needs) the energy, structural, seismic, and functional features of the building. Retrofitting consists in the introduction of new features not included in original project phase or, in any case, not provided (Ascione and Bellomo 2012); it may involve the entire building or only certain parts of it; it may be superficial or volumetric depending on the elements involved; it may involve integrating or replacing the shell, structure or systems, or removing or adding entire volumes or parts of the building. Utilizing retrofitting strategies brings various types of issues into play that are found in operational practices of multiple technical fields such as *feedback* and adaptation through the application of innovative solutions, new technologies, new integrative elements or in modification of physical and functional characteristics and features of existing elements. Being a relatively recent approach, retrofitting is a realm that is still open to new areas and possibilities for development. Implemented mainly to achieve the improvement/updating of energy and functional features, more recently it has kept in stride with the progress of increasingly less-standardized market offerings (Di Sivo and Angelucci 2012), moving towards *custom-fitting*, in attempt to respond to the specific needs of users. This implies greater attention towards the final user of the "improved" product which adds not only to the technological connotations of retrofitting, but also to social ones aimed at increasing the quality of the housing.

Table 1. Project strategies extrapolated from the assessment of Italian and European case studies.

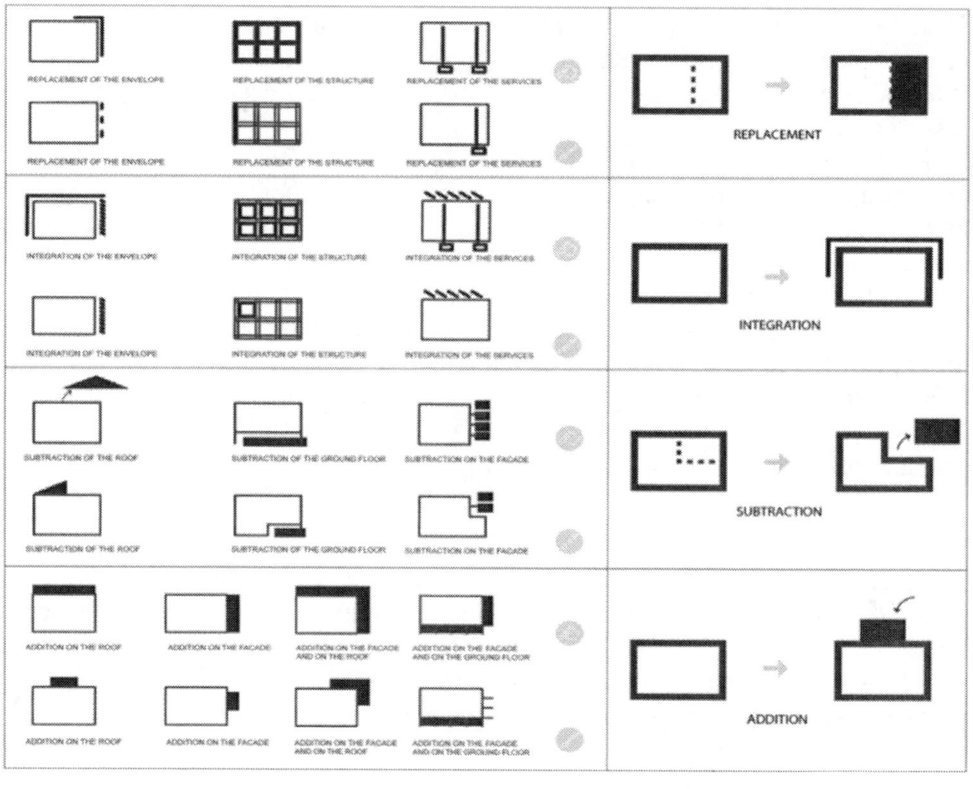

The retrofitting approach, aimed at the upgrading of the features and characteristics of buildings through the use of innovative technologies, is of particular interest for its social and environmental repercussions, and operates within the logic of a circular economy. In the case of work on post-World War II public buildings, retrofitting should be grasped as a powerful opportunity not only to remedy the physical and social deterioration found in the intermediate suburbs, but also to operate with critical capacity, combining the energy-related, seismic, and functional renewal of the buildings, in the dialectic between conservation and innovation of architecture within the logic of an economy that works in harmony with the environment.

2 A CASE STUDY

In order to carry out a preliminary experiment on a category of building commonly found throughout the municipal territory, we mapped out the residential areas of the city of Pisa and performed a detailed analysis of the distinctive character for each of these areas.

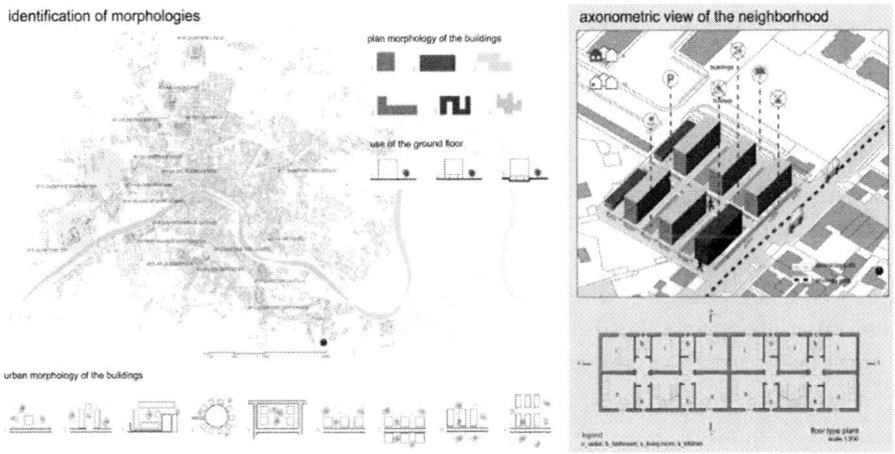

Figure 1. Preliminary study on the types of buildings and households.

The neighborhoods were classified by year of construction, size, types of buildings and property developer. The morphological layout, number of floors above ground, the construction technique, and the type of users were identified.

These considerations led to the selection for the case study being the neighborhood of Ciro Ravenna, a block consisting of six almost identical buildings three-story condominium built with load-bearing walls. See Figure 1. The examinations allowed the areas for renewal to be defined: social, technological, energy, and architectural. It was on this basis that the work schedule was planned. See Figure 2.

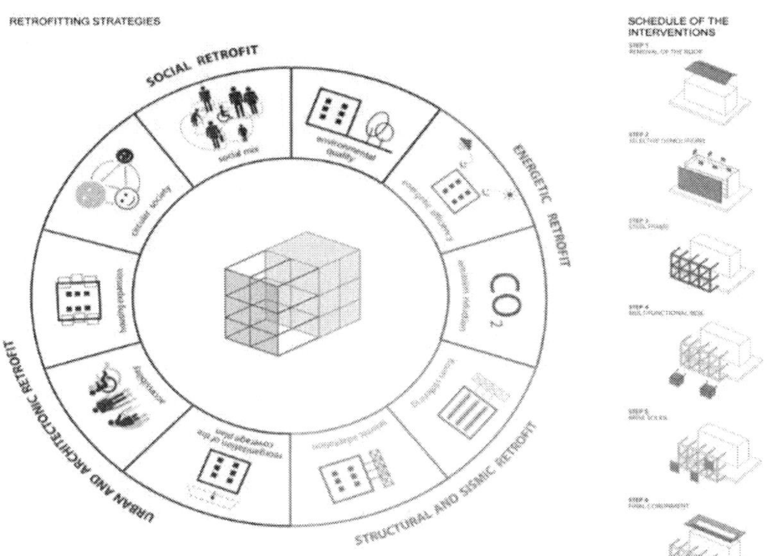

Figure 2. Retrofitting strategies and work diagrams.

As for the outdoor spaces, the entire parcel was redesigned to encourage soft mobility through the inclusion of a cycle path and rest areas. Three recreational areas have been set out in the green spaces that meet the needs of three different age groups: the elderly, young people, and small children. In order to create green spaces that are also educational (Stein 2010), it was decided to make one of the existing green spaces a place for gardening. A driveway was created with parking areas surrounding the block, creating a green, pedestrian center.

Since the inhabitants are at present mainly elderly people and some immigrant families (Regione Toscana 2017), the social renewal will include the integration of other types of users, such as students and young couples. There are, therefore, "4 different sizes" of residence that have been designed: XS, S, M and L. These may be combined in different configurations to meet the needs of the newer and older resident user types. Furthermore, study of the relationship between the organization of the residential layout and the furnishings made it possible to render the space more open and flexible. Particular attention was paid to vulnerable users), in order to ensure their autonomy with the provision of home automation systems. See Figure 3.

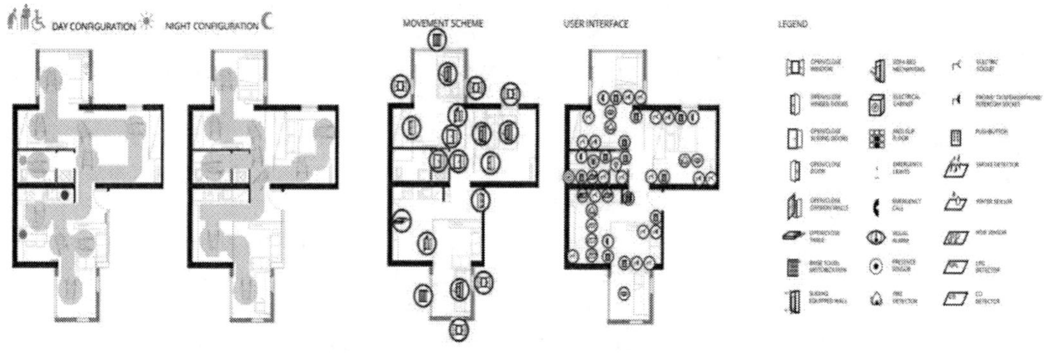

Figure 3. Day and night home automation configuration. Movement and installation interface.

The structural safety of the building has been ensured by installing a steel exoskeleton with floor and wall braces. This structural solution was enhanced from a functional and architectural point of view: in order to better integrate with the existing building (Muselli *et al.* 2017), a box was created, a sort of pre-fabricated wooden building *prosthesis*. T his multi-functional volumetric addition can take on different functions, depending on the different times of the day and the type of user.

The external structure, in addition to ensuring the safety of the building, serves to support the integrated solar panel and photovoltaic panel crowning system for the production of energy for use in the residences and outdoor spaces. The roof has been refurbished as a meeting place to encourage socializing and social integration for the whole building; to this end, it will include an outdoor kitchen, a relaxation area, and a central "agora" as a multi-purpose area. See Figure 4. In order to increase energy efficiency, the external facades have been integrated with insulating panels of different materials depending on the exposure (north with rock wool and south with wood fiber). These additions made it possible for the building to achieve Class A rating. Rainwater is planned to be collected in a collection and purification tank located in the basement for use by the residences or to service the outdoor spaces.

3 CONCLUSIONS

The technological research process in the retrofitting project described that included the addition of new technologies and new functionalities to a group of public housing buildings in order to prolong their life cycle will require further in-depth research to explore the definition of the realization costs, which are not only economic in nature, but also environmental and social, in order to be able to evaluate the feasibility and advantages that this approach offers. This is why the application of circular economy policies aimed at limiting waste and environmental costs, as well as the reintegration of waste materials into the production cycle play a fundamental role.

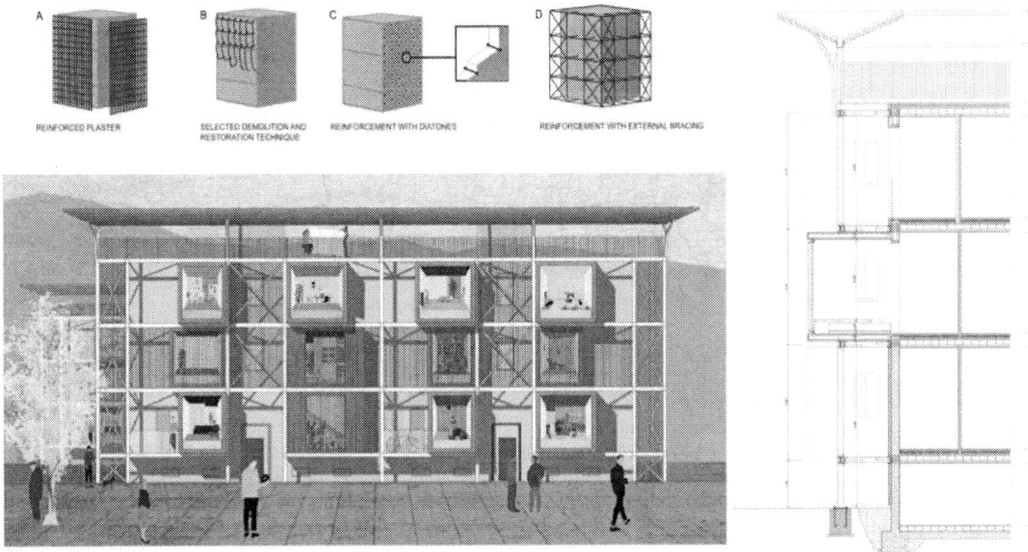

Figure 4. Structural reinforcement techniques and layout of exterior spaces.

References

Ascione, P., and Bellomo, M., *Retrofit Per La Residenza*. Clean Edizioni, Napoli, 2012.

Bellomo, M., and Pone, S., Il Retrofit Tecnologico Degli Edifici Esistenti: Qualità Dell'abitare, Sostenibilità Ambientale, Rilancio Economico, *Techne*, Firenze University Press, 1, 82-87, 2011.

Cellucci, C., and Di Sivo, M., *Habitat Contemporaneo, Flessibilità Tecnologica E Spaziale*, Franco Angeli Editore, Milano, 2016.

Di Sivo, M., and Angelucci, F., Il Mass Costumization Process Per L'housing Sociale, *Techne*, Firenze Univercity Press, 4, 132-137, 2012.

Giacchetta, A., Retrofitting Solare Di Edilizia Sociale: Un Progetto A Savona, *Techne*, Firenze Univercity Press, 4, 366-373, 2012.

Habraken, N. J., *The Structure of the Ordinary. Form and Control in the Built Environment*, The MIT Press, London, 1998.

Jacobs, J., *The Death and Life of Great American Cities*, Random House, New York, 1961.

Landolfo, R., Losasso, M., and Pinto, M. R., *Innovazione E Sostenibilità Negli Interventi Di Riqualificazione Edilizia. Best Practice Per Il Retrofit E La Manutenzione,* Alinea, Firenze, 2013.

Latouche, S., *Per Un'abbondanza Frugale*, Bollati Boringhieri, Torino, 2012.

Muselli, E., Tartaglia, A., and Di Pasquale, J., Typological and Technological Innovation for the Application of Hybrid Systems to Housing Construction: Between Technological Culture and Application Testing, *Techne*, Firenze University Press, 13, 287-294, 2017.

Regione Toscana, Abitare in Toscana – Anno 2017, Sesto Report Sulla Condizione Abitativa. Retrieved from http://www.regione.toscana.it/-/abitare-in-toscana-anno-2017-sesto-rapporto-sulla-condizione-abitativa on January 4, 2018.

Stein, C., *Greenning Modernism. Preservation, Sustainability, and the Modern Movement*, W.W. Norton & Company, New York, NY, 2010.

UN-Habitat, Urbanization and Development: Emerging Futures, *World City Report 2016*, UN-Habitat, 2016.

Interdependence between Structural Engineering and Construction Management
Edited by Ozevin, D., Ataei, H., Modares, M., Gurgun, A., Yazdani, S., and Singh, A.

ISBN: 978-0-9960437-6-2

PRIORITIZATION OF PIPELINE REPLACEMENT BASED ON FAILURE RATE

AMARJIT SINGH and XI SONG

Dept of Civil Engineering, University of Hawaii at Manoa, Honolulu, USA

The purpose of this paper is to prioritize water pipelines for replacement through the analysis of pipeline failure rates and reliability. By using actual pipe condition record data from Honolulu Board of Water Supply (BWS), the constant failure rate is recognized through Bathtub profile and used to calculate reliability. This paper provides the utility manager a quantitative method to evaluate and arrange pipeline renewal. It assists in determining maintenance schedules depending on the age and type of the pipe by plotting the corresponding failure rate against time. Reliability helps determine whether a certain group of pipes are within the acceptable practice threshold to signal replacement. In this paper, the analysis of four major types of pipe systems from BWS are conducted. With the data from fiscal year (FY) 2008, the failure rates are calculated, and the reliability of all pipelines are seen to be above 80%.

Keywords: Reliability, Bathtub curves, Water supply, Main breaks, Length in ground.

1 INTRODUCTION

According to the AWWA (2001), water is the most capital intensive of all utility services. As the pipelines need to be buried underground, the costs of management and maintenance are always high. It is very important to catalogue the pipe type, age and length in the area under scrutiny, and undertake prediction analysis for pipe replacement. As such, a proactive approach of pipe asset management is crucial for determining the optimal time to replace a pipe.

In this article, the actual pipe condition in fiscal year (FY) 2008 provided by the Honolulu Board of Water Supply (BWS) is used. BWS serves approximately one million customers with over 150 million gallons of water daily (BWS Report 2010), maintaining more than 2,000 miles of pipelines, which break at a rate of approximately one per day. By analyzing the failure rate of pipelines, the paper can provide the utility manager to prioritize the pipe replacement.

2 METHODOLOGY

The study of pipeline replacement prioritization has been conducted and published by various researchers and scholars. For example, by matching the threshold and predicted break rates, Loganathan *et al.* (2002) derived the method to predict optimal renewal time for pipe systems. Rogers and Grigg (2009) combined the probability models and the ranking systems to determine the possibility of pipe failure. Wood and Lence (2009) used statistical deterministic equations to estimate the pipe failure rate.

In this study, pipes with similar characteristics (i.e. type and age) are grouped and used to form reliability distributions. Furthermore, the constant failure rate period is identified by using the bathtub curve for reliability calculations.

2.1 Failure Patterns for Complex Products

Complex products often follow a familiar pattern of failure (Juran and Gryna 1993). When the failure rate is plotted against a continuous time scale, the resulting graph (Figure 1) known as the "bathtub curve", exhibits three distinct periods. The first period, known as the early life phase, is characterized by high failure rate right after installation, due to faulty installations or defective pipe. The second period, or useful life phase, is characterized by a constant low failure rate. The failure rate increase in the third period, wear-out life phase, is caused by pipe deterioration and ageing.

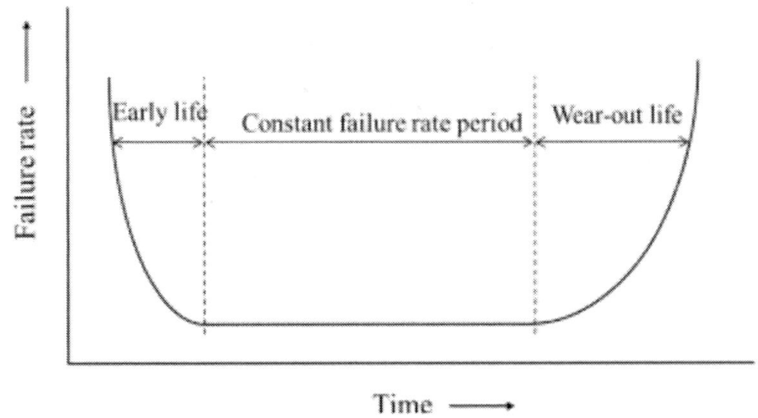

Figure 1. The theoretical bathtub curve.

"The bathtub curve is a fundamental tool in quality engineering for analyzing the failure process. When the failure rate is constant, the distribution of time between failures is distributed exponentially." (Juran and Gryna 1993) This is the basis of the exponential formula for reliability, requiring a comprehensive work log to verify.

2.2 Reliability Formula

The distribution of time between failures indicates the chance of failure-free operation for the specified time period. By plotting the bathtub curve, the early and wear-out periods can be easily filtered; only the constant failure rate λ in the useful life period is considered; the probability of reliability can be estimated by the exponential formula in Eq (1) (Juran and Gryna 1993).

$$R = e^{-t\lambda} \qquad (1)$$

Where, R is probability of failure-free operation for a time period equal to or greater than t
λ is failure rate in the constant period
t is a specified period of failure-free operation

3 BWS PIPE SYSTEM OBSERVATION

To gain an understanding of the water supply pipe system at BWS, data was collected from inventory records, annual reports, statistical summaries, and main break data from fiscal year (FY) 2008. Based on total length of pipe in the ground (LIG), the four major types of pipe are, in order of prevalence, cast iron (CI), ductile iron (DI), polyvinyl chloride (PVC), and concrete cylinder (CC). These four types account for 90% of the pipe system; the other 10% are asbestos concrete (AC), steel (STL), copper (CU), as illustrated in Figure 2.

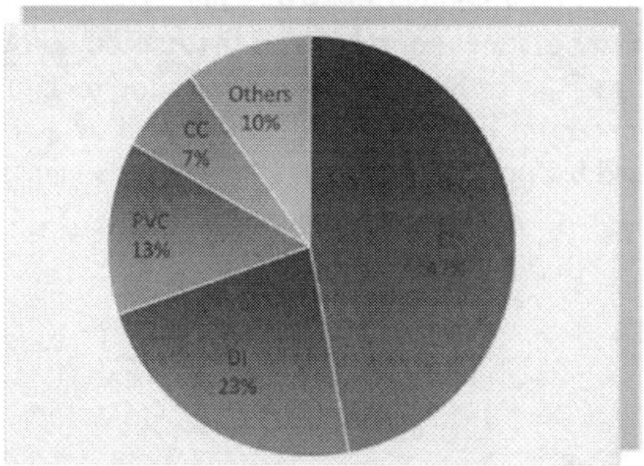

Figure 2. Percent of total length of all types of pipes in FY2008.

The pipes ranged in age from new to over 100 years, with the average age of all pipes in FY 2008 being approximately 32.2 years. The oldest pipe in the system was installed in 1898; pipes between 40 to 50 years were the most prevalent, comprising approximately 16.1 percent of the total population. For each pipe type, the average age varies. By FY 2008, the average age of CI pipes was 46.7 years; 16.6 years for DI pipes; 9.5 years for PVC pipes; and 32.3 years for CC pipes, shown in Figure 3.

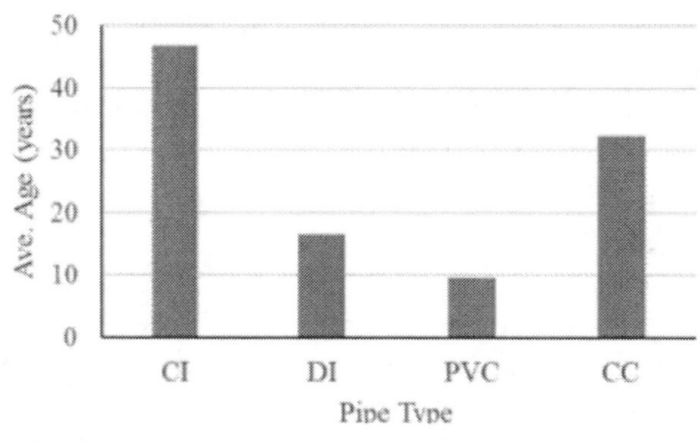

Figure 3. Average age of all types of pipes in FY2008.

3.1 Failure Rate and Future Breaks

The following equation calculates failure rate:

$$Failure\ rate = (Number\ of\ breaks)/LIG/mile \quad (2)$$

The failure rate was calculated by using number of breaks in a specific age group divided by the length of pipe in the ground (LIG). The unit of failure rate is breaks/year/mile.

Tables 1-4 summarize the number of failures and the LIG for each four-major pipe types in FY 2008. The number of failures and the LIG for each pipe type was grouped into ten-year intervals, with the exception of PVC pipes, which were grouped into subsequent five-year intervals, because they are relatively new and installed only recently in the system. For each pipe type, assuming the failure rate of each age group remains constant, the expected annual failure rates after 20 years are forecast based on the current statistical data from FY2008. It is noticeable that, after 20 years, CI and DI pipes will face a larger volume of failures; the number of annual failures will increase by 70% and 863%, respectively.

Table 1. Failure rate for CI pipes by age in FY 2008 (Modified from Singh and Adachi 2013).

Age	Number of failures	LIG (miles)	Failure rate (breaks/yr/mile)	LIG (miles) after 20 years	Expected annual failures after 20 years
0-10	0	0.0	-	-	-
10-20	0	0.0	-	-	-
20-30	1	0.2	6.311	0.0	0.0
30-40	30	163.2	0.184	0.0	0.0
40-50	49	244.0	0.201	0.2	0.0
50-60	43	222.4	0.193	163.2	31.5
60-70	23	90.9	0.253	244.0	61.7
70-80	26	97.6	0.266	222.4	59.2
80-90	13	31.6	0.411	90.9	37.4
90-100	1	1.5	0.647	97.6	63.1
100-110	12	4.3	2.809	31.6	88.8
Sum	198	-	-	-	341.9

Table 2. Failure rate for DI pipes by age in FY 2008 (Modified from Singh and Adachi 2013).

Age	Number of failures	LIG (miles)	Failure rate (breaks/yr/mile)	LIG (miles) after 20 years	Expected annual failures after 20 years
0-10	1	100.8	0.010	-	-
10-20	13	200.6	0.065	-	-
20-30	6	123.5	0.049	100.769	4.9
30-40	9	33.8	0.266	200.578	53.4
40-50	1	1.0	1.031	123.473	127.3
50-60	2	0.6	3.621	33.783	122.3
Sum	32	-	-	-	308.0

Table 3. Failure rate for PVC pipes by age in FY 2008 (Modified from Singh and Adachi 2013).

Age	Number of failures	LIG (miles)	Failure rate (breaks/yr/mile)	LIG (miles) after 20 years	Expected annual failures after 20 years
0-5	4	87.9	0.046	-	-
5-10	3	78.8	0.038	-	-
10-15	0	65.9	0.000		0.0
15-20	0	24.3	0.000	78.8	0.0
20-25	1	13.1	0.076	65.9	5.0
Sum	8	-	-	-	5.0

Table 4. Failure rate for CC pipes by age in FY 2008 (Modified from Singh and Adachi 2013).

Age	Number of failures	LIG (miles)	Failure rate (breaks/yr/mile)	LIG (miles) after 20 years	Expected annual failures after 20 years
0-10	0	3.9	0.000	-	-
10-20	1	25.6	0.039	-	-
20-30	1	19.6	0.051	3.9	0.198
30-40	2	40.6	0.049	25.6	1.261
40-50	2	52.2	0.038	19.6	0.750
Sum	6	-	-	-	2.209

It is predictable that the replacement of certain pipes is very necessary. Due to the relative recent installations of PVC and CC pipes, the failure rate of these two pipe types will become constant in 20 years.

3.2 Bathtub Curves

Taking a glance at the Tables 1-4, CI pipes have the longest age range; while PVC pipes have the shortest. Due to the relatively new pipeline system and material of the pipe, the number of failures were only eight and six for PVC and CC pipes, respectively. Detailed observations and comparisons from the Bathtub curves are discussed below.

Bathtub curves for four types of pipes are shown in Figure 4. It is observed that the curves of CI and DI were the only two that fell into the theoretical Bathtub curve, where the constant failure rate periods were recognized in the graphs. Some portions of the Bathtub curve for PVC pipes were recognized as the Bathtub profile. Since the PVC pipes were relatively new and the total numbers of failure were low in FY 2008, the results are inconclusive and will not be considered for replacement at this moment. Finally, the Bathtub curve for CC pipes was inverted to the opposite of the bathtub profile. Since CC pipes have the lowest failure rate and shortest LIG, CC pipes will also have low priority for replacement compared to CI and DI pipes.

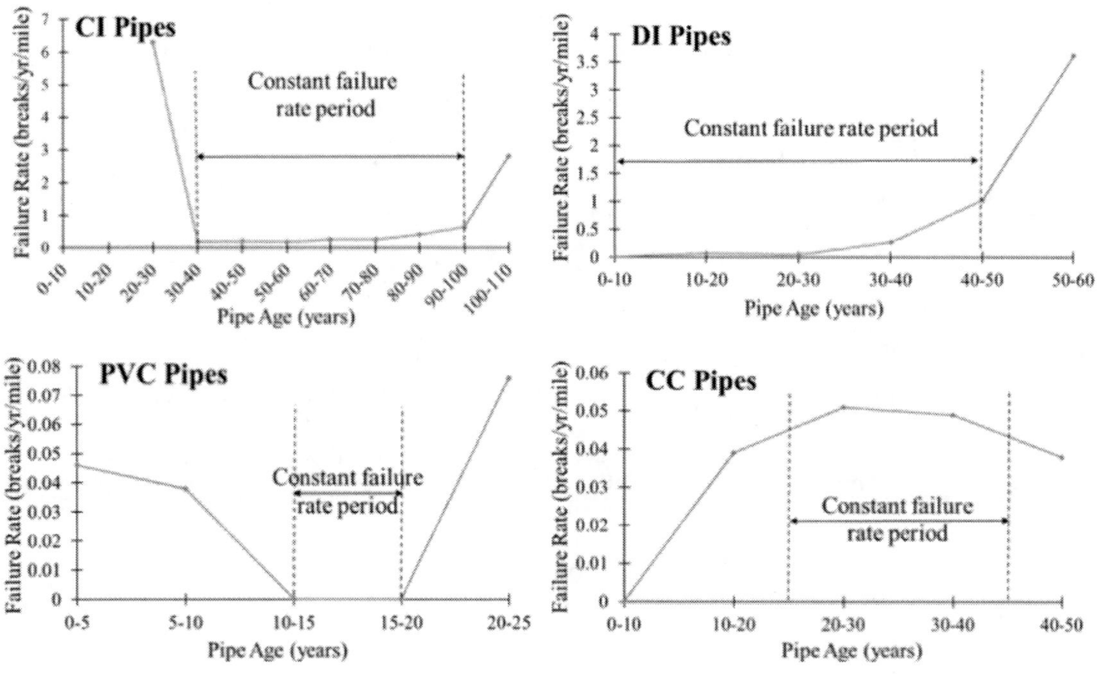

Figure 4. Failure rate vs. Time plot for four types of pipes in FY 2008.

4 RELIABILILTY FOR BWS PIPE SYSTEM

Once the constant failure rate periods are identified for CI and DI pipes, the average failure rates are computed by using total numbers of failure divided by the total LIG in the age groups within the constant failure rate, which is 30-100 year for CI pipes and 0-40 year for DI pipes. Furthermore, this average failure rate is the λ in Eq. (1).

The reliabilities for CI and DI pipes were calculated using Eq. (1). Set t equal to 1 for the FY 2008 reliability rate calculations. Results are shown in Table 5. It can be seen from Table 5, in

FY 2008, CI pipes have a 19.5% rate of failure in one year, while DI pipes have as low as 6% of the chance.

Table 5. Reliability in FY 2008.

Pipe type	Constant failure rate period	Average failure rate	Reliability ($R = e^{-t/\lambda}$)	Probability of failure
CI	30-100	0.217	0.805	0.195
DI	0-40	0.063	0.939	0.061
PVC	10-20	0	1	0
CC	25-45	0.046	0.955	0.045

After the calculation, the BWS utility manager will use the calculated reliability to make the pipe replacement decision. Furthermore, an experienced manager will also cross-reference the previous record to evaluate the necessity of pipe replacement for certain types of pipes in particular field areas. For budgeting purposes within that fiscal year, based on the reliability, the manager is now able and ready to recommend the priority of pipe replacements.

5 CONCLUSIONS AND LIMITATIONS

The overall goal of this paper was to provide an assessment of the pipe system at BWS and a guidance for the utility manager to prioritize the replacement of pipelines. The methodology is straightforward and can be done by anyone even without an engineering background. Based on the data collected in FY 2008, the major observations are as follow:

- The failure rate v. time plot of CI pipes fits the Bathtub profile quite typically, establishing a visible constant failure rate period, allowing reliability to be calculated.
- In FY 2008, it was discovered that CI pipes had the highest average failure rate, followed by DI and CC pipes; PVC pipes showed zero average failure rate.
- High breakage of CI and DI pipes is predicted for FY 2028, and these two types of pipes are recommended to have the highest priority for replacement; PVC and CC pipes have low probability of failure and have low priority for replacement.

The reliability analysis provides a guideline and sound methodology for prioritization of pipeline replacement. Due to the lack of data from BWS, this paper provided an approximate method in 2008 to predict the future performance of pipelines. Therefore, this study needs to be continually undertaken every year for multiple years to achieve adequate statistical data and set a reference database for more accurate future breakage predictions.

References

AWWA, *Dawn of the Replacement Era: Reinvesting in Drinking Water Infrastructure*, American Water Works Association Research Foundation, Denver, CO, 2001.

BWS Report, Department and Agency Reports, *Board of Water Supply*, Honolulu, HI, 2010.

Logannathan, G. V., Park, S., and Sherali, H. D., Threshold Break Rate for Pipeline Replacement in Water Distribution Systems, *Journal of Water Resources Planning and Management*, 128(4), 271-279, 2002.

Juran, J. M., and Gryna, S. M., *Quality Planning and Analysis*, McGraw-Hill Inc, New York, NY, 1993.

Rogers, P. D., and Grigg, N. S., Failure Assessment Modeling to Prioritize Water Pope Renewal: Two Case Studies, *Journal of Infrastructure Systems*, 15(3), 162-171, 2009.

Singh, A., and Adachi, S., Bathtub Curves and Pipe Prioritization Based on Failure Rate, *Built Environment Project and Asset Management*, Emerald, 3(1), 105-122, 2013.

Wood, A., and Lence, B. J., Using Water Main Break Data to Improve Asset Management for Small and Medium Utilities: District of Maple Ridge, B.C., *J. of Infrastructure Systems*, 15(2), 111-119, 2009.

Interdependence between Structural Engineering and Construction Management
Edited by Ozevin, D., Ataei, H., Modares, M., Gurgun, A., Yazdani, S., and Singh, A.

ISBN: 978-0-9960437-6-2

ASSET CONDITION ASSESSMENT MODEL FRAMEWORK FOR HEALTHCARE FACILITIES

DALIA SALEM and EMAD ELWAKIL

School of Construction Management, Purdue University, West Lafayette, USA

The poor healthcare facilities performance costs the USA tens billions of dollars each year. There are several factors affect the critical assets of healthcare facilities such as the physical condition of the facility, infection prevention, life safety, and hospital revenue loss. The objective of this research is to develop an asset condition assessment model for healthcare facilities. The model identifies several factors, especially the condition, capacity, and code compliance of a facility's infrastructure. The asset condition assessment model has been developed based on the mechanical, electrical and plumping systems and evaluates the asset condition based on its age, life expectancy, recent condition assessment and energy efficiency among other factors. The data was collected from different hospitals in Indiana. This model has been developed using the Regression analysis technique. The model has been validated through analytical and mathematical methods. Results showed average validity at 88.32%. The developed model will help the healthcare facilities managers to prioritize the budget allocation and to reduce the impact of consequences of asset failure, which will improve the healthcare infrastructure performance.

Keywords: Asset management, Critical assets, Hospitals, Regression analysis.

1 INTRODUCTION

Healthcare environment quality is linked to patient safety, clinical output, patient stress, patient outcome and staff stress. Healthcare facility directors need to ensure that complex systems are working properly as a part of maintaining the physical environment within healthcare settings (Lucas 2012). Effective facility directors take actions to ensure that healthcare facilities are clean and well-maintained.

2 BACKGROUND

Facility management (FM) is "the integral planning, realization and management of buildings and accommodation, services and resources which contribute towards the effective, efficient and flexible attainment of organizational goals in a changing environment" (Regterschot 1990). Alexander *et al.* (2002) suggest that FM is "the management of non-core company assets to support and increase the efficiency of the main business of the organization."

The exponential growth of healthcare FM has greatly improved healthcare effectiveness and quality. Healthcare facilities management is one of the most vital components of successful healthcare services. These services can be broken down into two categories: hard and soft facilities management. Hard facilities management (FM) includes property maintenance and management; soft facilities management (FM) is the management of support services (Reuvid and

Hinks 2001). Hard FM includes the built environment: the estate, electricity, indoor air, property, water supply and structural fabric. On the other hand, cleaning, catering, security, laundry and waste management comprise soft FM (Liyanage and Egbu 2008).

Previous studies developed a facility failure analysis model that links healthcare delivery processes to facility failures at a hospital. The model assists in ranking and identifying the failures that highly affect service delivery to prioritize maintenance activities (Mohammadpour and Leila 2015, Ali and Hegazy 2013). A clear, precise knowledge of asset performance and condition is necessary for all management decisions in terms of assessing renewal needs, maintenance, and rehabilitation. Premature failure due to a lack of asset knowledge necessitates replacing the asset, an action that has major consequences related to business potential loss and risk.

To address the above issues, a tool for assessing asset criticality is necessary in capital needs decision-making. According to this body of knowledge, critical asset condition assessment in terms of criticality was addressed by very limited research. This is considered a major research gap, as maintenance strategy is a main part of the healthcare facilities management.

3 RESEARCH OBJECTIVES AND POINT OF DEPARTURE

The main objective is to develop a model framework to evaluate and prioritize the asset critical condition for the healthcare facilities. This main objective can be broken down into the following sub-objectives:

- Identify the asset criticality for healthcare facilities.
- Identify and determine the main factors and sub factors influencing asset criticality for healthcare facilities.
- Develop a model framework of asset condition assessment.
- Validate the developed model.

4 FACTORS AFFECTING THE ASSET CRITICALITY

Based on the literature review and the experts' interviews, asset condition assessment (ACA) factors are identified and selected. Four main categories incorporated in the research include fifteen other sub factors (Salem and Elwakil 2018a) as shown in Table 1.

Table 1. Asset criticality factors in healthcare facilities (Salem and Elwakil, 2018a).

Physical Condition	Environmental Aspects	General Safety	Revenue Loss
Age/Life Expectancy	Lead	Life Safety	Labor Hours
Condition Assessment	Asbestos	Physical Safety	No. of Work Orders
Parts Availability	Mold		Repair Cost
Code Compliance	Water Contamination		
Energy Efficiency	Indoor Air Quality		

5 RESEARCH METHODOLOGY

The research methodology used in this research consists of several steps started with literature review, followed by collecting data, ranking the asset condition assessment (ACA) factors, developing the ACA regression analysis, results, conclusion and recommendations.

The methodology used to evaluate the facility was based on its spaces. The patient risk space, the principal element evaluated, was measured according to numbers provided by patient risk groups and Infection Control Risk Assessment (ICRA) standards. The physical aspects, quality of the environment of care, general safety procedures, and hospital revenue loss impact were automated and integrated using several modeling techniques to develop an ACA critical rating model for the healthcare facility equipment. Regression analysis is a robust statistical methodology that can be used to determine the relationship between two or more quantitative or qualitative independent variables to predict the dependent variable (Al-Zwainy *et al.* 2013). The regression analysis technique was conducted to measure the relationship between these factors and the asset risk score. The rationale for using this methodology rather than the traditional AHP methodology was its use of the model factor weights in the representation of their behaviors.

6 DATA COLLECTION

Data collected in this research was qualitative and quantitative as shown in Figure 1.

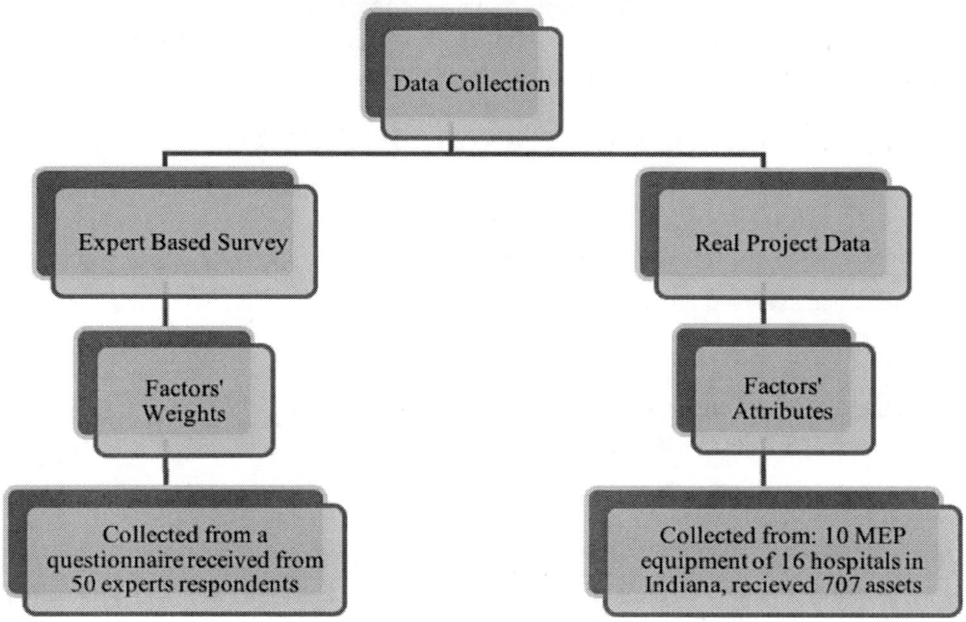

Figure 1. Main data collection process.

The study population was limited to Indiana healthcare facilities. The primary sample was based on the convenience and availability of the healthcare directors and professionals, including professionals from related disciplines such as infection prevention, life safety, and administration. The participants selected were 83 healthcare professionals in Indiana. The population samples varied by position and affiliated department. Valid case study data were collected from 707 MEP equipment types operated in 16 Indiana hospitals. The MEP equipment types utilized in this study were chillers, boilers, air handling units, cooling towers, condensate storage tanks (return and deaerator), generators, electrical switch gear, fire pumps, medical air/gas/vacuums, and water softeners. The MEP equipment is considered one of the main assets in hospitals.

7 DEVELOPMENT OF ASSET CONDITION ASSESSMENT MODEL

Salem and Elwakil (2018b) used the analytic hierarchy process (AHP) to rank critical asset condition assessment factors. A pair-wise comparison matrix of the main factors and their subfactors was used to obtain the factors' relative weights, or their importance among the other factors. The logical consistency CR of the overall priority weights was used to verify the reliability and consistency of the expert survey results. The results showed that the General Safety Factors with the highest priority and effect on the Asset Condition Assessment (0.31), followed by Environmental Considerations (0.27), Physical Factors (0.24), and Revenue Loss Factors (0.19). The Life Safety Sub factor had the highest weight of the General Safety Factors (0.65), followed by physical safety (0.11). The Code Compliance score was the highest of the Physical Factors (0.37), and the Repair Cost Sub factor was the highest of the Revenue Lost Factors (0.46). At 0.03, Age/Life Expectancy had the smallest contribution, as shown in Figure 2. Finally, the overall weights were decomposed to calculate the critical average weight of each factor in the framework (Saaty 1980, 1994).

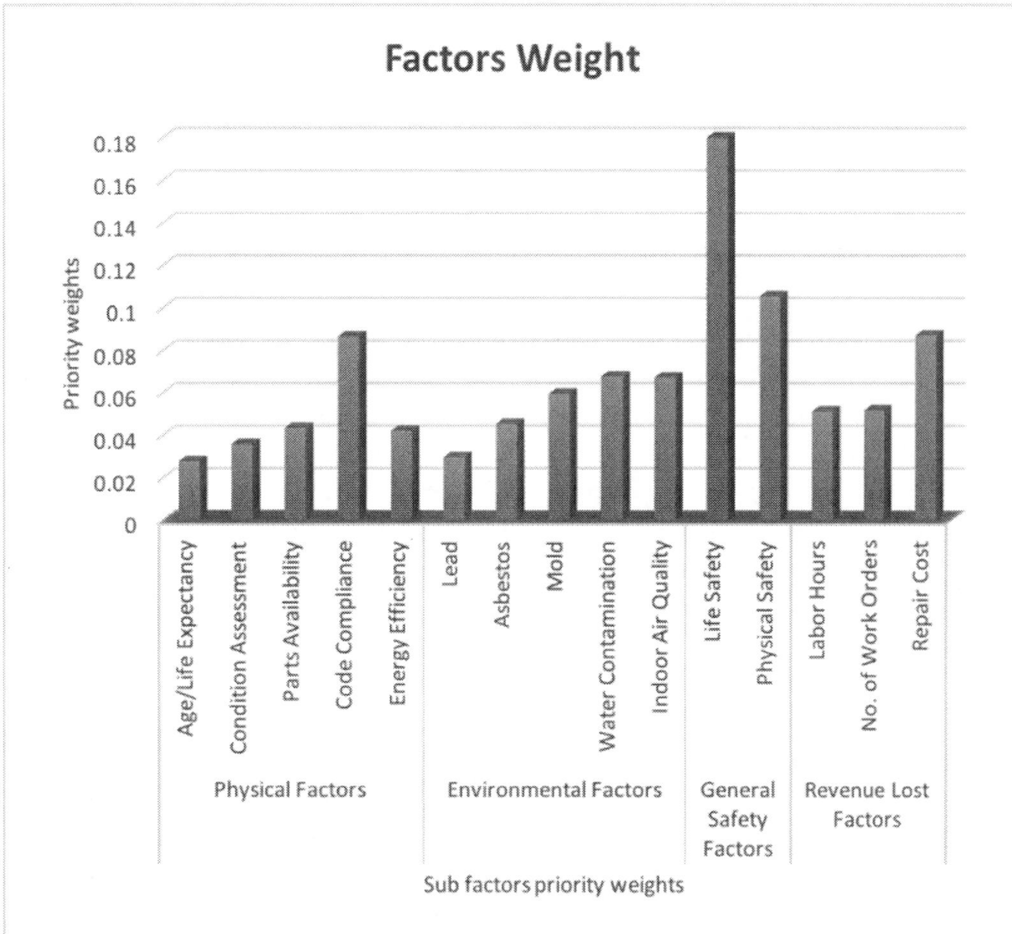

Figure 2. Critical asset factors relative weights.

The next step was to combine the collected factors' weights and effect attribute factors into one model representing the experts' knowledge that could be used in the ACA Critical Score model. Due to the multiple scales used to determine the case study attribute values, the attribute effects were normalized to a scale from 0 to 1.

To develop the AHP/Regression methodology for the ACA rating model, the factors' weights and normalized effect attributes were multiplied. The regression analysis technique was conducted to measure the relationship between these factors and the asset risk score. The rationale for using this methodology rather than the traditional AHP methodology was its use of the model factor weights in the representation of their behaviors.

The R^2, adjusted R^2, mean square error (S or MSE), and Mallow's Cp values were used to select the best ACA model based on the highest R^2, highest adjusted R^2, minimum MSE, and Cp closest to the number of independent variables can be summarized in Table 2.

Table 2. ACA model summary.

S	R-sq	R-sq(adj)	R-sq(pred)
0.0573766	90.56%	90.44%	90.24%

The Analysis of Variance (ANOVA) shows that the P-value (statistical significance) was 0.000 for the developed regression model. The null hypothesis was rejected at a significance level of 0.05. The F-test results demonstrated that the developed model was statistically sound. The P-value and the F-test results showed that the model reflected satisfactory results. Table 3 features the model's statistical analyses, including the constants, coefficients, F-values, and P-values.

Model verification was completed to test the validation of the model. The verification used the 141 case studies to compare the results of the developed model with the facility directors' rating scale. The results of the validation process show that AVP was 88.32%, which indicates that the developed ACA critical rating model was robust and its results satisfactory.

Table 3. Coefficients and statistical analyses.

Term	Coef	SE Coef	T-Value	P-Value	VIF
Constant	0.0152	0.0133	1.15	0.252	
Redundancy	0.29027	0.00816	35.57	0.000	1.60
Energy Efficiency	-0.562	0.270	-2.08	0.038	3.41
Environmental Aspects	0.0421	0.0223	1.89	0.059	1.09
Life Safety	1.4327	0.0451	31.75	0.000	2.92
Physical Risk	2.027	0.137	14.84	0.000	2.73
Labor Hours	3.108	0.219	14.20	0.000	2.34
Work Orders	2.290	0.243	9.44	0.000	1.72

8 CONCLUSION

The present research aims to develop an asset condition assessment model of the healthcare facility asset in terms of its criticality, which will assist in the decision-making process related to capital needs. This model will improve the asset management of healthcare buildings, one of the largest national infrastructure sectors. The patient risk space was the main factor considered by the developed ACA model. The ACA critical rating model framework consists of four main factors: Physical, Environmental, General Safety, and Revenue Loss. Regression analysis technique was used to develop the asset condition assessment model based on the aforementioned factors. The model validation process showed that the robustness of the developed ACA critical rating model at 88.32%. The developed ACA model has the potential to reduce the amount of time and money spent on decision making processes, allowing leaders to allocate funding in ways that better serve hospital communities and patients.

References

Alexander, J. M., Nelson, C. A., Van Berkel, V., Lau, E. K., Studts, J. M., Brett, T. J., and Fremont, D. H., Structural Basis of Chemokine Sequestration by a Herpesvirus Decoy Receptor, *Cell,* 111(3), 343-356, 2002.

Ali, A., and Hegazy, T., Multicriteria Assessment and Prioritization of Hospital Renewal Needs, *Journal of Performance of Constructed Facilities,* 28(3), 528–538, 2013.

Al-Zwainy, F. M. S., Abdulmajeed, M. H., and Aljumaily, H. S. M., Using Multivariable Linear Regression Technique for Modeling Productivity Construction in Iraq, *Open Journal of Civil Engineering*, 3(3), 127-135, 2013.

Hinks, J., Reuvid, J., and Hinks, J., Measuring the Value of Facilities and Facilities Management, in Reuvid, J. and Hinks, J. (Eds), *Managing Business Support Services: Strategies for Outsourcing and Facilities Management, Building on Experience*, Kogan Page/Institute of Directors, London, 3, 177-87, 2001.

Liyanage, C., and Egbu, C., A Performance Management Framework for Healthcare Facilities Management, *Journal of Facilities Management*, 6(1), 23–36, 2008.

Lucas, J. D., *An Integrated BIM Framework to Support Facility Management in Healthcare Environments,* Ph.D. Thesis, Virginia Polytechnic Institute and State University, Blacksburg, VA, 2012.

Mohammadpour, A., and Leila P., Global Fault-Tolerant Control Technique for Multiphase Permanent-Magnet Machines, *IEEE Transactions on Industry Applications,* 51.1, 178-186, 2015.

Regterschot, J., *Facility Management in Changing Organizations*, In Proceedings of the International Symposium on Property Maintenance Management and Modernization, CIB International Council for Building Research Studies and Documentation Working Commission, Singapore, 70(1), 146-155. 1990.

Saaty, T. L., *The Analytical Hierarchy Process, Planning, Priority, Resource Allocation*, McGraw-Hill, New York, 1980.

Saaty, T. L., Highlights and critical Points in The Theory and Application of the Analytic Hierarchy Process, *European Journal of Operational Research*, 74(3), 426-447, 1994.

Salem, D., and Elwakil, E., Toward a Sustainable World Institutional Buildings, *Architecture, Engineering and Construction*, 5(2), 61-71, 2018a.

Salem, D., Elwakil, E., *Develop an Assessment Model for Healthcare Facilities: A Framework to Prioritize the Asset Criticality for the Capital Renewals,* International Conference on Construction and Real Estate Management, Charleston, South Carolina, US, August 9-10, 2018b.

Interdependence between Structural Engineering and Construction Management
Edited by Ozevin, D., Ataei, H., Modares, M., Gurgun, A., Yazdani, S., and Singh, A.

ISBN: 978-0-9960437-6-2

RECTIFICATION THEORY AND REINFORCEMENT PRACTICE BASED ON THE PRINCIPLE OF COUNTERFORCE CENTER

YANPENG ZHU, KUIBIN YANG, JUN TAO, and XIULI WANG

School of Civil Engineering, Lanzhou University of Technology, Lanzhou, China

Like buildings with other types of foundations, buildings with pile foundations can tilt because of irrational designs, imperfections of foundation treatment, poor construction qualities, even the change of water environment during usage, etc. By studying and analyzing the causes for the inclination, this paper has proposed a theory of rectification and reinforcement for buildings with pile foundations based on the principle of counterforce center, which makes building be rectified by changing the counterforce center of pile foundation so that it can be in the same vertical line with the gravity center of upper structure. Then the calculation method was given. Combining with an engineering project, it has been proved that the principle of counterforce center is a good guidance for the rectification and reinforcement for buildings with pile foundations.

Keywords: Pile foundation, Building rectification, Foundation reinforcement, Pile supplement.

1 INTRODUCTION

Although the traditional conception believes that differential settlement and tilting rarely happens on structures with pile foundations, structures with pile foundations frequently tilt with the influence of many factors including design, construction, or use, etc. The inclination of the pile-foundation-structures brings a hidden trouble to the safe usage of structures and a threat to the safety of users' lives and properties.

Some scholars have made a conclusion from many engineering practices that one of the important reasons for the structures' inclination is that the center of gravity of upper structure and the centroid of foundation are not in the same vertical line (Gao 1991, Wen and Wei 2004, Wang *et al.* 2006). But these conclusions were not made from force perspective to illustrate inclination from its essential reason. Moreover, according to Technical Code for Improvement of Soil and Foundation of Existing Buildings (JGJ123 2012), the work for rectification is usually considered specifically in two methods: forced settlement and jacking. The principles of the two methods are stress-releasing principle (Liu 1990) or breaking soil structure and stress-transferring principle (Zhu *et al.* 2005). However, relating construction methods cannot satisfy the requirement of diverse types of structures, and especially they focus little on the method that can rectify pile-foundation-structures.

Through analysis about the tilting causes of pile-foundation-structures, this journal puts forward a new rectification and reinforcement theory based on the principle of counterforce center. The theory was applied in a project, and the building in this project was rectified successfully.

2 THE RECTIFICATION THEORY BASED ON THE PRINCIPLE OF COUNTERFORCE CENTER OF PILE FOUNDATION

2.1 The Principle of Counterforce Center of Pile Foundation

The counterforce of pile foundation addressed in this paper is not actual bearing force. It is a hypothesis during force analysis. How much load that a single pile can actually bear is related to the ultimate bearing capacity, so the counterforce of pile foundation was calculated by assuming that every single pile bears the value of ultimate bearing capacity. Therefore, the ultimate bearing capacity was applied to calculate the counterforce of pile foundation.

The counterforce of pile foundation is the maximum stress set provided by all the single pile, whose direction is straight up, and the value is the sum of bearing capacity of all piles. The counterforce center is located at the point of bearing capacities' resultant force of all piles.

In projects, the counterforce center of pile foundation and the gravity center of upper structure sometimes are not in same vertical line, as shown in Figure 1. With the eccentric effect, the bearing stress of pile foundation would be nonuniform, and the more settlement would happen on the side bearing more. As a result, the whole structure would tilt. Moreover, the tilting of the whole structure causes more eccentric increment of upper loads, which leads to new moment increment increasing the tilt of whole structure in turn. Higher the structure is, more obvious the effect is (Lou *et al.* 2000).

To make pile foundation bear more loads from upper structure, the deformation of each pile should be made as much concerted with each other as possible, and the center of gravity of upper structure and the counterforce center of pile foundation should be kept as in one vertical line as possible. It is the basic principle to rectify tilting. According to the skewing degree between the counterforce center of pile foundation and the gravity center of upper structure, based on considering eccentric moment, the bearing loads on piles are rearranged. Then additional piles are added according to the rearranged bearing situation. Therefore, the gravity center of upper structure and counterforce center of pile foundation act on one vertical line and the structure is rectified.

2.2 The Calculation for Counterforce Center of Pile Foundation

The counterforce center is located at the point of action of bearing capacities' resultant force of all piles. Now how to locate the counterforce center of pile foundation is introduced by the example in Figure 2. The counterforce of a single pile is assumed to be the bearing capacity of the pile, which is calculated according to Eq. (1) from *Technical Code for Building Pile Foundations*.

$$Q_{uki} = u\sum q_{sik} l_i + q_{pk} A_p \tag{1}$$

where Q_{uki} is the standard value of single pile's bearing capacity, u is the perimeter of pile body, q_{sik} is the standard value of single pile's limited side friction in the ith layer, l_i is the length of single pile body buried in the ith layer, q_{pk} is the standard value of single pile's limited tip resistance, and A_p is the area of pile tip.

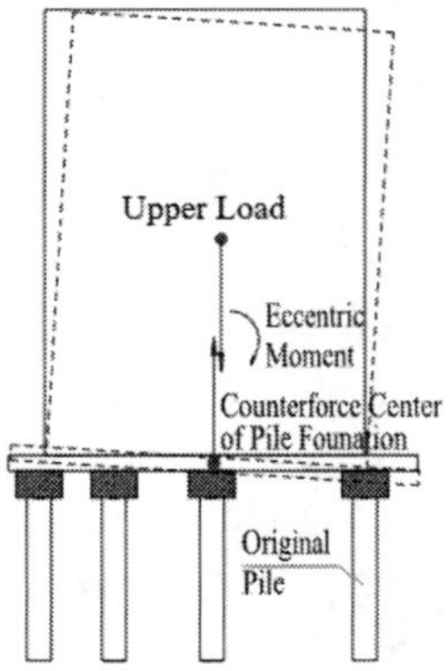

Figure 1. Diagram of inclined building.

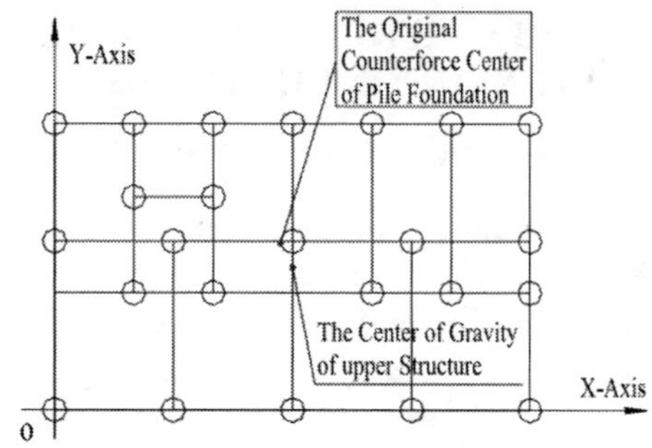

Figure 2. Layout plan of pile foundation.

After the calculation about the standard value of bearing capacity of a single pile, by establishing a rectangular plane coordinate system as a reference, the location of counterforce center of pile foundation can be calculated according to the principle of moment balance from Eq. (2):

$$X=\frac{\sum_{i=1}^{n}Q_{uki}x_i}{\sum_{i=1}^{n}Q_{uki}},\quad Y=\frac{\sum_{i=1}^{n}Q_{uki}y_i}{\sum_{i=1}^{n}Q_{uki}} \tag{2}$$

where X and Y are the coordinate figures of the counterforce center in the rectangular plane coordinate system, x_i and y_i are the coordinate figures of the ith pile.

2.3 The Calculation for the Bearing Capacity of Pile Foundation Considering Additional Eccentric Moment

When the counterforce center of pile foundation and the center of gravity of upper structure are not in the same vertical line, eccentric effect would produce the additional moment, which results in that the bearing load is rearranged. Considering the nonuniformity of the bearing load on pile foundation brought by eccentric effect, the rearranged bearing capacity of piles is recalculated from Eq. (3).

$$N_i=\frac{G}{n}\pm\frac{G(X-X')y_i'}{\sum y_j'^2}\pm\frac{G(Y-Y')x_i'}{\sum x_j'^2} \tag{3}$$

where N_i is the eigenvalue of bearing capacity of a single pile considering additional moment, G is the total weight of the whole structure, n is the number of piles, X' and Y' are the coordinate figures of the gravity center of the upper structure in the rectangular plane coordinate system, x_i', x_j', y_i', y_j' are the distances of the ith and jth piles to the y-axis and x-axis, respectively.

2.4 The Calculation for Adding Piles

Comparing the standard value of bearing capacity of a single pile calculated from Eq. (1) and the bearing capacity of a single pile considering the additional eccentric effect calculated from Eq. (3), when the former is smaller than the double value of the latter, supplement for piles should be designed, which means adding one or several piles around the original pile to satisfy the requirement for bearing capacity. In the end, the center of gravity of upper structure and the counterforce center of pile foundation are in the same vertical line by this way. The number of additional piles is calculated from Eq. (4),

$$n = \frac{2N_i - Q_{uk}}{N} \tag{4}$$

where n is the number of addition piles around the original pile, Q_{uk} is the standard value of ultimate bearing capacity of original pile, N is the standard value of ultimate bearing capacity of new single pile added.

3 THE PRACTICE OF RECTIFICATION AND REINFORCEMENT BASED ON THE PRINCIPLE OF COUNTERFORCE CENTER

In recent years, the authors have made some rectification and reinforcement practices with the guidance of principles and technologies mentioned above, which proved to have a great effect. Hopefully the following example can provide any thought and reference for the future similar rectification and reinforcement engineering for pile-foundation structure.

3.1 Overview of the Project

With a total weight of 16,000 tons, the pile-foundation-building is 49.0m long, 13.0m wide, and 35.5m high. There are 1 floor underground and 12 floors on the ground of this frame-shear wall structure. The foundation type was bored concrete piles. The building started to tilt after 5 years from being built, shown in Figure 3 and Figure 4.

Figure 3. The tilting building.

Figure 4. The crack of the tilting building

The displacement measured from top to bottom of the building varies in the range of 225mm to 357mm. The whole building deformed and inclined to south, and the maximum gradient reached 12.75‰ beyond the allowable value for high-rise building required in *Code for Design of Building Foundation*, which threatens the safety of whole structure.

3.2 The Calculation for Rectification and Reinforcement Design Based on the Principle of Counterforce Center of Pile Foundation

The pile foundation was divided into two parts according to the tilting direction and settlement difference. The north part was induced settlement zone while the south was adjustment zone. The design integrated both rectification and reinforcement for the whole building by applying the new grout-steel pipe composite piles. By means of installing the bearing platforms on the top of the new grout-steel pipe composite piles, connected with original piles and bottom plate, the induced settlement was made by drilling the original bored concrete piles to fulfill rectification and reinforcement, shown in Figure 5.

The information of different layers is determined by geological survey materials. The vertical ultimate bearing capacity standard value of every single pile was calculated by Eq. (1). The calculated minimum bearing capacity standard value was 2054kN while the maximum bearing capacity standard value was 3081kN, and the total bearing capacity of pile foundation was 131466kN, which was smaller than the total weight of the building, 160000kN. So the original piles did not satisfy the requirement of bearing capacity. The counterforce center of pile foundation was at (-290.6, 980.9), calculated using Eq. (2). Obvious excursion happened comparing to the center of gravity (0, 0) of upper structure.

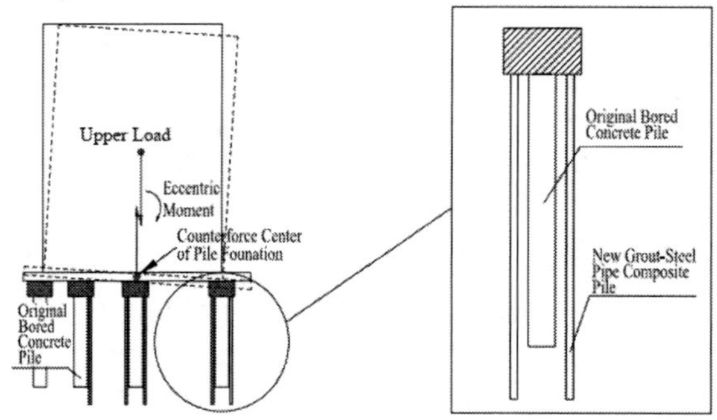

Figure 5. Figure of rectification.

The bearing capacity of every pile was calculated according to Eq. (3) considering the additional eccentric moment. The minimum counterforce of single pile was 2126.1kN appearing at northwest part, which was smaller than the actual bearing load of this pile, 2558.8kN. It did not satisfy the requirement of bearing capacity, and more piles need to be added to reinforce the foundation.

The selected pile type was grout-steel pipe composite pile, whose pile tip enters the intermediary weathered sandstone to the depth of 2m. The bearing capacity eigenvalue of single pile was 800kN. The amount of piles needing to be added was determined with Eq. (4). There would be 58 new grout-steel pipe composite piles and 36 bearing platforms being needed in total by the whole building through calculation. The arrangement for bearing platforms is shown in Figure 6. After adding new piles, the new counterforce center of foundation was calculated to be (-31.9, 37.5), with Eq. (2), which means the additional eccentric moment decreased a lot. So the rectification requirements were satisfied and the rectification is accomplished.

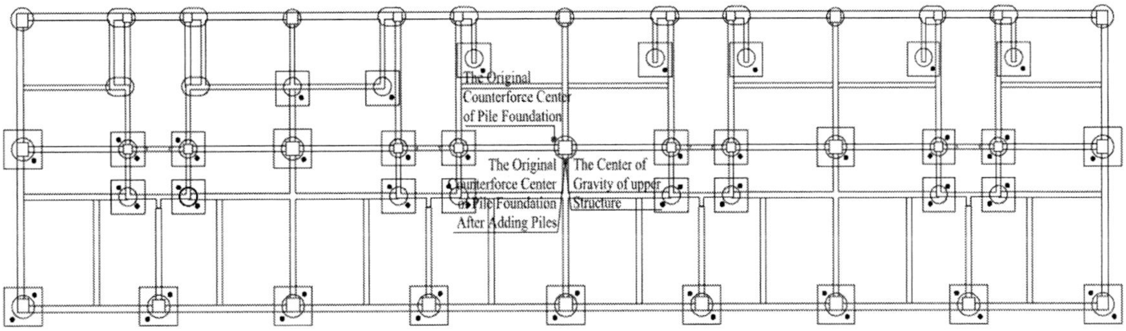

Figure 6. Plan of pile foundation after supplementary piles.

4 CONCLUSIONS

(1) Differential settlement of pile foundation would happen because of eccentric effect caused by that the gravity center of upper structure and the counterforce center of pile foundation are not in a vertical line, which is an important reason for the inclination of the whole structure.

(2) By decreasing the nonuniformity of additional pressure on foundation based on the thought of rectification and reinforcement applying the principle of counterforce center of pile foundation, the loads on pile foundation can be redistributed, which can be a good guidance for the rectification and reinforcement for buildings with pile foundation.

(3) In the practice of rectifying and reinforcing the buildings with pile foundations, firstly the eccentric distance between the counterforce center and the gravity center of upper structure is calculated. Then more piles are added according to the load redistributed considering the eccentric moment. In this way, the counterforce center of pile foundation and the gravity center of upper structure can be moved in the same vertical line so that the rectification of the structure is fulfilled.

References

JGJ123, *Technical Code for Improvement of Soil and Foundation of Existing Buildings*, China Academy of Building Research, China Architecture and Building Press, Beijing, 2012.

Gao, Y. C., Differential Settlements of Residential Buildings on Soft Foundation, *Chinese Journal of Geotechnical Engineering*, 13(4), 59-65, 1991.

Liu, Z. D., Tilt-Correcting of Foundation Stress Relieve Method, *Soil Engineering and Foundation*, 4(1), 1-6, 1990.

Lou, X. M., Liu, J. H., and Hu, Z. X., Systematic Analysis on the Inclination of Box Foundations for High-Rise Buildings, *Journal of Tongji University*, 28(6), 646-650, 2000.

Wang, H., Chen, J. P., and Que, J. S., Research on Design and Application of Root Piles to Strengthening Foundations, *Rock and Soil Mechanics*, 27(Supp.), 1290-1294, 2006.

Wen, X. G., and Wei, G., Rectification and Reinforcement of an Inclined Building on Soft Soil Foundation, *China Civil Engineering Journal*, 37(8), 61-65. 2004.

Zhu, Y. P., Wang, X. L., and Zhou, Y., Theoretical Analysis and Practice of Rectifying Reinforcement Using Expansive Method for Incline Building on Collapsible Loess, *Chinese Journal of Rock Mechanics and Engineering*, 24(15), 2786-2794, 2005.

Interdependence between Structural Engineering and Construction Management
Edited by Ozevin, D., Ataei, H., Modares, M., Gurgun, A., Yazdani, S., and Singh, A.

ISBN: 978-0-9960437-6-2

A STRUT AND TIE MODEL FOR FRCM-STRENGTHENED REINFORCED CONCRETE DEEP BEAMS

TADESSE WAKJIRA and USAMA EBEAD

Dept of Civil and Architectural Engineering, College of Engineering, Doha, Qatar

Exiting literature revealed that fabric reinforced cementitious matrix (FRCM) is a promising material for the strengthening of shear deficient reinforced concrete (RC) beams. However, most of the available experimental studies are devoted to the use of FRCM system for the strengthening of slender beams and limited literature is available on the strengthening of deep beams using FRCM system. Moreover, there is no available literature on the analytical modelling of FRCM-strengthened deep beams. In this paper, a simple strut and tie model (STM) has been used to predict the ultimate load carrying capacity of RC deep beams strengthened in shear using FRCM system. The model accounts for the internal transverse reinforcement ratio and axial rigidity of the FRCM system. The proposed model is validated against an experimental results of RC deep beams strengthened with different types of FRCM system available in the literature.

Keywords: Fabric reinforced cementitious matrix, Strengthening, Shear capacity.

1 INTRODUCTION

One of the major problems faced in construction industries worldwide is an early deterioration of reinforced concrete (RC) structures encountering their life cycle (Younis *et al.* 2018).

Fiber reinforced polymer (FRP) system has been widely used for the strengthening of deteriorated structures owing to their favorable properties over the traditional strengthening system (Aly *et al.* 2006, Ebead 2015, Ebead and Saeed 2013, 2014). However, FRP possess some drawbacks, such as incompatibility with the concrete substrate and failure at high temperature due to reliance on epoxy adhesives. To overcome these problems, fabric reinforced cementitious matrix (FRCM) has been introduced by replacing the epoxy adhesives with cementitious mortar and fibers with the textile fabrics. Exiting literature revealed that FRCM can be used to significantly enhance the flexural and shear capacity of RC beams (Ebead *et al.* 2017, Ebead and Wakjira 2018a, 2018b, Elghazy *et al.* 2017, 2018, Pino *et al.* 2017, Tetta *et al.* 2016, Wakjira and Ebead 2018c, 2018a, Younis *et al.* 2017) possessing a good bond with the concrete (Younis and Ebead 2018). However, most of the available experimental studies are devoted to the use of FRCM system for the strengthening of slender beams and the literature available on the FRCM-strengthened deep beams are scarce (Azam *et al.* 2018, Tetta *et al.* 2018, Wakjira and Ebead 2018b, 2019). Moreover, there is no available literature on the analytical modelling of FRCM-strengthened deep beams. ACI-ASCE Committee 445 (Joint ACI-ASCE Committee 445 1998) defined deep beams as beams in which the ratio of the critical shear span to effective depth *(a/d)* is less than 2.50. In this type of beams, unlike slender beams in which the combined shear and bending action transfers the load to the support, the load is transferred to the support through the compressive stresses by arch action (He *et al.* 2012). A strut-and-tie model (STM) has shown to be an effective design method for this type of beam.

In this paper, a strut and tie model is used to predict the ultimate load carrying capacity of FRCM-strengthened RC deep beams. The model is then validated against an experimental data of FRCM-strengthened deep beams available in the literature.

2 STM FOR FRCM-STRENGTHENED RC DEEP BEAMS

The failure in deep beams is approximated to follow the shortest possible path (Tuchscherer *et al.* 2016). Figures 1a and 1b show the proposed STM for FRCM-strengthened beams in which the load is assumed to be transferred directly to the support in a form of a compressive strut. The diagonal compressive strut may fail because of concrete splitting caused by the diagonal tension stress or the crushing of concrete caused by formation of diagonal compressive stress. The latter is resisted by concrete compressive strength of the diagonal strut, while the diagonal tension stress is mainly resisted by transverse reinforcement; i.e., internal transverse reinforcement (ITR) and external transverse reinforcement (FRCM) as shown in Figure 1b.

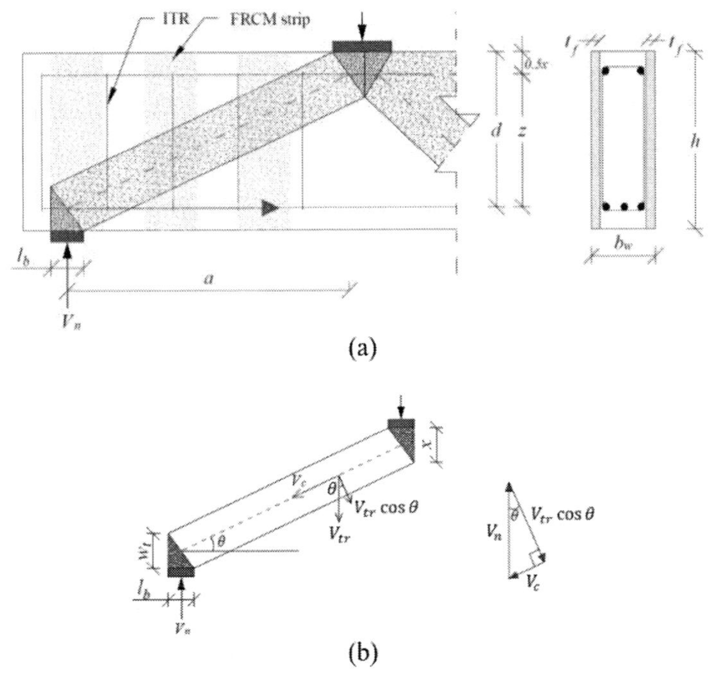

Figure 1. Detail of STM for FRCM-strengthened beams within the critical shear span: (a) layout of STM and (b) load transfer and equilibrium of forces.

From the equilibrium of forces on the strut, Figure 1b, the nominal shear capacity (V_n) of FRCM-strengthened beam is given in Eq. (1).

$$V_n = V_c \sin\theta + V_{tr} \cos^2\theta \quad (1)$$

where, V_c and V_{tr} are the shear force in the concrete compressive strut and the shear strength provided by the transverse reinforcement, which are ITR and FRCM reinforcement. The inclination (θ) of the compressive strut from the horizontal can be determined as follows.

$$\theta = \tan^{-1}(z/a) \tag{2}$$

Where

a is the critical shear span;

z is the lever arm determined from the cracked rectangular section of singly reinforced RC beam as given in Eq. (3).

$$z = d - \frac{kd}{3} \tag{3}$$

Where

d is the effective depth of the section;

kd is the position of the neutral axis from the extreme compression side of the beam.

The factor *k* is determined from the moment-curvature relationship of singly reinforced RC beams as given in Eq. (4) (Park and Kuchma 2007).

$$k = \sqrt{(n\rho)^2 + 2n\rho} - n\rho \tag{4}$$

Where ρ is the flexural reinforcement ratio (A_s/b_wd) and *n* is the steel-concrete modular ratio, which is the modulus of elasticity of steel (E_s=200,000 MPa) to concrete (E_c). The modulus of elasticity of concrete is given by Eq. (5) according to (ACI Committee 318 2014).

$$E_c = 4700\sqrt{f_c'} \tag{5}$$

The force in the concrete compressive strut (V_c) is given by Eq. (6).

$$V_c = f_{ce}A_{st} \tag{6}$$

Where,

f_{ce} is the effective concrete strength;

A_{st} is the smaller of cross-sectional area of the strut at the two ends.

According to (ACI Committee 318 2014) the effective concrete strength is the smaller of the effective compressive strength of the concrete in strut ($f_{ce,st}$) and nodal zone ($f_{ce,n}$) as given in Eq. (7) and Eq. (8), respectively.

$$f_{ce,st} = 0.85\beta_s f_c' \tag{7}$$

Where, β_s is the factor that depends on the shape of the strut, which is unity for prismatic strut considered in this study.

$$f_{ce,n} = 0.85\beta_n f_c' \tag{8}$$

Where, β_n is given by Eq. (9) below as per (ACI Committee 318 2014).

$$\beta_n = \begin{cases} 1.0, & \text{nodal zone bounded by strut only} \\ 0.80, & \text{nodal zone anchoring one tie} \\ 0.60, & \text{nodal zone anchoring more than one tie} \end{cases} \tag{9}$$

Thus, β_n is 0.80 for the beams considered in this study. Area of the strut is given by Eq. (10).

$$A_{st}=b_w a_{st} \tag{10}$$

Where, a_{st} is the minimum of the width of the strut at the two ends given by Eq. (11).

$$a_{st}=\min\left(l_b \sin\theta + w_t \cos\theta,\ l_p \sin\theta + x \cos\theta\right) \tag{11}$$

Where,

l_b is the with of the support plate;

l_p is the width of the loading plate;

w_t is the depth of the tie, which is taken as twice the distance between the centroid of flexural reinforcement and extreme tension side of the beam;

x is the depth of the top horizontal strut, which is taken as twice the distance between the centroid of the compression zone in the moment-curvature and extreme compression side of the beam (2kd/3). The shear strength provided by the transverse reinforcements (V_{tr}) is the sum of the shear provided by the ITR (V_{itr}) and FRCM reinforcement (V_f) as given in Eq. (12).

$$V_{tr}=f_{itr}A_{itr}+f_f A_f \tag{12}$$

Where,

f_{itr} and A_{itr} are the tensile stress and area of ITR;

f_f and A_f are the effective tensile stress and area of FRCM reinforcement.

The effective tensile stress of the FRCM system is determined as the product of its elastic modulus (E_f) and effective strain (ε_f) as given in Eq. (13).

$$f_f=E_f\varepsilon_f \tag{13}$$

The effective strain in FRCM system is limited to the minimum of 0.004 and 75% of its ultimate strain as per (ACI Committee 549 2013). Therefore, the shear capacity of FRCM-strengthened beam is given by Eq. (14).

$$V_n=f_{ce}A_{st}\sin\theta+f_{itr}A_{itr}\cos^2\theta+E_f\varepsilon_f A_f\cos^2\theta \tag{14}$$

Finally, the ultimate load carrying capacity of the beam can be determined based on the loading condition.

3 VERIFICATION OF THE MODEL

The proposed model is validated against an experimental database of RC deep beams strengthened with FRCM system available in the literature (Azam *et al.* 2018, Tetta *et al.* 2018, Wakjira and Ebead 2018b). Azam *et al.* (2018) tested rectangular RC beams of dimensions 1700 × 250 × 400 mm (length × width × height) with critical shear span to effective depth (*a/d*) ratio of 1.62. The beams were strengthened with externally bonded cement-based composites and internally reinforced with or without internal ITR. The beams tested by (Wakjira and Ebead 2018b) were of 2100 mm long rectangular beams with cross-sectional dimensions 150 × 330 mm (width × height) tested under three-point bending with *a/d* ratio of 1.96. The average 28-day compressive strength of concrete used to cast the beams was 30 MPa. The beams were strengthened with both continuous and discontinuous configurations of three different types of FRCM composites (carbon, glass, PBO) applied using a near surface embedded technique. The beams tested by (Tetta *et al.* 2018) had rectangular cross section of dimensions 1677 × 102 × 203 mm) (length × width × height) tested under three-point loading with *a/d* ratio of 1.6. The beams were strengthened with an externally bonded carbon FRCM. However, the slender beams with *a/d* ratio greater than 2.50 tested by (Tetta *et al.* 2018) are not included in this study.

The ratio of the theoretically predicted to experimental load carrying capacity (P_{th}/P_{ex}) is depicted in Figure 2. This ratio ranged between 0.81 and 1.11 with an average of 0.94 and standard deviation of 0.09 as shown in Figure 2. The correlation of determination (R^2) between the experimental and theoretically predicted ultimate load carrying capacity was 0.96. This observation shows that the proposed STM can predict the load carrying capacity of the FRCM-strengthened deep beams with an acceptable accuracy.

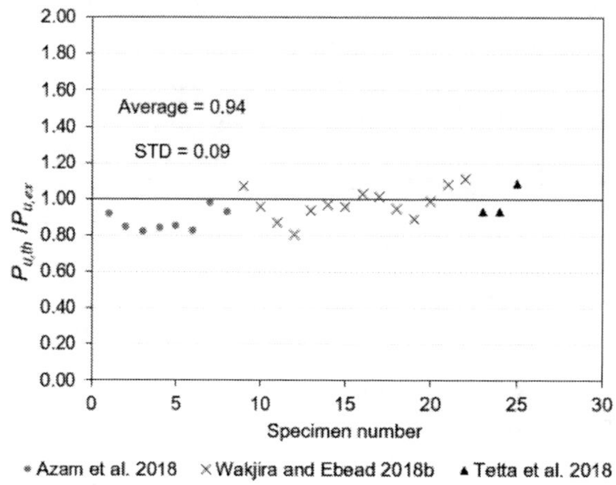

Figure 2. Verification of the model against experimental data.

4 CONCLUSIONS

A simple strut and tie model was used to predict the ultimate load carrying capacity of FRCM-strengthened RC deep beams. The model accounts for internal transverse reinforcement ratio and axial rigidity of the FRCM system. An experimental database of FRCM-strengthened deep beams available in the literature was used to validate the model and a reasonable agreement between the experimental and theoretically predicted ultimate load of the beams was obtained. The correlation of determination (R^2) between the experimental (P_{ex}) and theoretically predicted (P_{th}) ultimate load carrying capacity was 96%, while the average of P_{th}/P_{ex} ratio was 0.94 with a standard deviation of 0.09.

Acknowledgments

This paper was made possible by NPRP grant # NPRP 9-110-2-052 from the Qatar National Research Fund (a member of Qatar Foundation) and internal grant # QUST-CENG-SPR-14/15-15. The findings achieved herein are solely the responsibility of the authors.

References

ACI Committee 318. *Building Code Requirements for Structural Concrete (ACI 318-14)*. American Concrete Institute, Farmington Hills, MI, USA, 2014.

ACI Committee 549. 2013. *Guide to Design and Construction of Externally Bonded Fabric-Reinforced Cementitious Matrix (FRCM) Systems for Repair and Strengthening Concrete and Masonry Structures (ACI 549.4R-13), American Concrete Institute, Farmington Hills, MI, USA, 2013.*

Aly, R., Brahim, B., and Usama, E., Tensile Lap Splicing of Fiber-Reinforced Polymer Reinforcing Bars in Concrete, *ACI Structural Journal,* 103(6), 857–864, 2006.

Azam, R., Khaled, S., Jeffrey, W., and Martin, N., Shear Strengthening of RC Deep Beams with Cement-Based Composites, *Engineering Structures,* 172(June), 929–937, 2018.

Ebead, U., Inexpensive Strengthening Technique for Partially Loaded Reinforced Concrete Beams: Experimental Study, *Journal of Materials in Civil Engineering*, (10), 2015.

Ebead, U., and Saeed, H., Hybrid Shear Strengthening System for Reinforced Concrete Beams: An Experimental Study, *Engineering Structures,* 49(April 2013), 421–433, 2013.

Ebead, U., and Saeed, H., Flexural and Interfacial Behavior of Externally Bonded/ Mechanically Fastened Fiber-Reinforced Polymer Strengthened Reinforced Concrete Beams, *ACI Structural Journal,* 111(4), 2014.

Ebead, U., Ebead, U., Shrestha, K. C., Afzal, M. S., El Refai, A., and Nanni, A., Effectiveness of Fabric-Reinforced Cementitious Matrix in Strengthening Reinforced Concrete Beams, *Journal of Composites for Construction,* 21(2), 04016084, 2017.

Ebead, U., and Wakjira ,T., Behaviour of RC Beams Strengthened in Shear Using near Surface Embedded FRCM, *IOP Conference Series: Materials Science and Engineering*, 431:072001, 2018a.

Ebead, U., and Wakjira ,T., FRCM/Stirrups Interaction in RC Beams Strengthened in Shear Using NSE-FRCM, *IOP Conference Series: Materials Science and Engineering*, 2018b.

Ebead, U., and Younis, A., Pull-off Characterization of FRCM/Concrete Interface, *Composites Part B* 165(November 2018), 545–553, 2019.

Elghazy, M., Refai, A., Ebead, U., and Nanni, A., Effect of Corrosion Damage on the Flexural Performance of RC Beams Strengthened with FRCM Composites, *Composite Structures,* 180, 994–1006, 2017.

Elghazy, M., El Refai, A., Ebead, U., and Nanni, A., Post-Repair Flexural Performance of Corrosion-Damaged Beams Rehabilitated with Fabric-Reinforced Cementitious Matrix (FRCM), *Construction and Building Materials*, 166(January), 732–744, 2018.

He, Z. Q., Zhao, L., and Zhongguo, J. M., Investigation of Load-Transfer Mechanisms in Deep Beams and Corbels, *ACI Structural Journal,* 109(4), 467–476, 2012.

Joint ACI-ASCE Committee 445, Recent Approaches to Shear Design of Structural Concrete, *Journal of Structural Engineering*,124(12), 1374–1417, 1998.

Park, J. W., and Kuchma, D., Strut-and-Tie Model Analysis for Strength Prediction of Deep Beams, *ACI Structural Journal,* 104(6), 657–661, 663–666, 2007.

Pino, V., Hadad, H., Basalo, F., Nanni, A., Ebead, U., and Refai, A., Performance of FRCM-Strengthened RC Beams Subject to Fatigue, *Journal of Bridge Engineering,* 22(10), 4017079, 2017.

Tetta, Z., Koutas, L., and Bournas, D., Shear Strengthening of Full-Scale RC T-Beams Using Textile-Reinforced Mortar and Textile-Based Anchors, *Composites Part B: Engineering,* 95, 225–239, 2016.

Tetta, Z., Koutas, L., and Bournas, D., Shear Strengthening of Concrete Members with TRM Jackets: Effect of Shear Span-to-Depth Ratio, Material and Amount of External Reinforcement, *Composites Part B: Engineering,* 137(February 2017), 184–201, 2018.

Tuchscherer, R. G., Birrcher, D. B., and Bayrak, O., Reducing Discrepancy between Deep Beam and Sectional Shear-Strength Predictions, *ACI Structural Journal,* 113(1), 3–16, 2016.

Wakjira, T., and Ebead, U., A New Approach for Predicting the Shear Capacity of FRCM Strengthened RC Beams in Shear, *IOP Conference Series: Materials Science and Engineering,* 2018a.

Wakjira, T., and Ebead, U., FRCM/Internal Transverse Shear Reinforcement Interaction in Shear Strengthened RC Beams, *Composite Structures,* 201(June), 326–339, 2018b.

Wakjira, T., and Ebead, U., Hybrid NSE/EB Technique for Shear Strengthening of Reinforced Concrete Beams Using FRCM: Experimental Study, *Construction and Building Materials*, 164, 164–177, 2018c.

Wakjira, T. G., and Ebead, U., Internal Transverse Reinforcement Configuration Effect of EB/NSE-FRCM Shear Strengthening of RC Deep Beams, *Composites Part B: Engineering,* 166, 758–772, 2019.

Younis, A., and Ebead, U., Bond Characteristics of Different FRCM Systems, *Construction and Building Materials,* 175(June), 610–620, 2018.

Younis, A., Ebead, U., and Judd, S., Life Cycle Cost Analysis of Structural Concrete Using Seawater, Recycled Concrete Aggregate, and GFRP Reinforcement, *Construction and Building Materials,* 175, 152–160, 2018.

Younis, A., Ebead, U., and Shrestha, K. C., Different FRCM Systems for Shear-Strengthening of Reinforced Concrete Beams, *Construction and Building Materials*, 153, 514–526, 2017.

Interdependence between Structural Engineering and Construction Management
Edited by Ozevin, D., Ataei, H., Modares, M., Gurgun, A., Yazdani, S., and Singh, A.

ISBN: 978-0-9960437-6-2

ASSESSMENT OF THE TARGET RELIABILITY DEPENDING ON STRUCTURAL PROPERTIES

LILITA OZOLA and DANA ZIRNOVA

Dept of Structural Engineering, Latvia University of Life Sciences and Technologies, Jelgava, Latvia

Although the establishment of the maximum acceptable failure probability for structures is not a mandatory requirement in design, it is important to develop a unified system for safety differentiation of bearing structures in public buildings so that people can feel equally safe in all the countries of the European Union. Differences from country to country in safety requirements declared by national codes may be permissible regarding the economic losses only. This study is aimed to clarify the understanding of how to examine a structure regarding the weakest link (element, fastener, system) in the limit state. In this study the authors propose a method for differentiation of structures regarding the variation of bearing capacity, which variates depending on a number of characteristics involved, and the failure mode anticipated when overloading takes place. Variation of bearing capacity is expressed in terms of reliability index variating correspondingly. Differentiation of the safety regarding failure modes anticipated has been developed using the toughness indices as a decisive criterion for the comparable assessment of structural compositions regarding the ability to sustain in the limit state for some period of time. Also, a criterion of possible redistribution of internal forces in the limit state is taken into account. The developed example of the assessment of the target reliability of structures is presented.

Keywords: Forensic engineering, Structures, Bearing capacity, Failure mode, Reliability index.

1 INTRODUCTION

An acceptable framework for the assessment of structural redundancy, moreover a set of criteria for sufficiently safe structures is a widely discussed issue in professional circles nowadays. The additional mandatory requirements for the redundancy of structures vary considerably from country to country. It is important to develop a unified system for safety differentiation of bearing structures in public buildings so that people can feel equally safe in every country of the European Union. Differences in safety requirements declared by national codes may be permissible regarding economic losses only, not the loss of human life.

Reliability, redundancy, toughness, robustness, and plastic failure mode of a structure- these are the properties in mutually interdependent complexity correlating closely with the safety level to be ensured. Highly sophisticated theoretical methods have been developed for the reliability analysis of structures (Zhao and Ono 2001, Ditlevsen and Madsen 2005, Nowak and Collins 2012, Schneider and Vrouwenvelder 2017) and the results sensitive to every influencing data group involved may be obtained quickly. The problem pertaining to reliability analysis has been constant due to the lack of input data samples for influencing factors. However, the uncertainty of the results of reliability analysis is mainly due to a complex interaction of factors both numerically measurable and logically quantified. A more comprehensive review on reliability-based performance criteria for structural systems has been completed by Ghosn *et al.* (2016).

In this study a proposal for the structural safety assessment incorporating essentially different indices of properties has been examined. The method proposed is based on the findings and examination of the behavior of the weakest link in structural system (element, fastener, whole system) when the limit state occurs.

2 BACKGROUND FOR THE SAFETY ASSESSMENT FRAMEWORK

Comprehensive structural design has been built upon three mutually interdependent categories:

- Code-based design
- Reliability and/or risk analysis
- Design for robustness, redundancy and plastic failure mode of a system

Most of the structural design codes are established in partial factor format. It is provided that load and resistance sets are located in a distance safe enough by means of partial factor values. Normally larger distance leads to a safer but more expensive structure. Also, some overlap zone is unavoidable in any case because of indeterminate factors peculiar to structural system in its complexity. Moreover, code conditions are developed to provide desirable safety margins for individual members and joints of a structure. The assessment of a safe behavior of an entire system remains in the competence of the professionals. Usually the approach to safety assessment is quite formal – basically done following some generalized safety provisions stated by national authorities. Normally these safety assessment procedures involve an uncertainty that should be taken into account.

Uncertainty is the main problem in structural design and construction. Variation of bearing capacity is generated by a large variety of input data (Keskküla and Ozola 2003). It has been proven by a large number of sample tests that material resistance variables may be assumed as normally distributed and mutually independent on load in terms of statistics. In such a situation the simplest linear limit state function has been applied for the definition of the reliability index as it is stated in Annex C of Eurocode EN 1990:2002/A1:2005/AC (2010). The reliability index in its often-used definition includes statistical properties of both data sets- those of resistance and load. However, it is doubtful that two variable reliability index reflects adequately a safety zone of the structure, since a large number of variables involved come from the load data set and thereby hide away some portion of resistance variation, which may occur more often. Let us consider the probabilities of negative performance function values in the case of combination of statistically distinctive load and resistance data set, illustrated by normally distributed probability diagrams in Figure 1. Case A is characterized by coefficient of variation (COV) for action (E) variables of 0.22 and COV=0.06 for resistance (R) variables. In Case B corresponding values of COV are 0.18 and 0.14. It is worthwhile to draw attention to big gap between the probabilities of the negative performance function (R-E) value in Case A, when P(R-E=-0.3)= 7.66E-18, and in Case B, when for the same (R-E) value the probability is P(R-E=-0.3)= 2.27E-06 due to different overlap zones.

In order to put focus on the variation of resistance effect being studied, it is assumed that the action effect is a certain fixed value (e_o), which may be adopted as the fractile of the resistance (R), and probability of failure (P_f) is expressed by Eq. (1). But one variable reliability index (β) is defined by Eq. (2), where μ_R is the average value of bearing capacity data set and σ_R is the standard deviation.

$$P_f = P(R < e_o) = \Phi_R(e_o) \quad (1)$$

$$\beta = (\mu_R - e_o)/\sigma_R \quad (2)$$

Redundancy, robustness and the anticipated failure mode of a structure, when overloading takes place, are essentially important properties in safety analysis not measurable by unambiguously defined characteristics. During recent years, the topic of the complex interaction of the above-mentioned important phenomenon in limit state has been discussed by Fang and Fan (2011) and many others.

As a result of the current study and previous research (Ozola 2013) a logically built-up bridge has been created between the reviewed properties of a structure in the final stage of decision making. Both the measurable and logic estimates proposed are presented in Table 1.

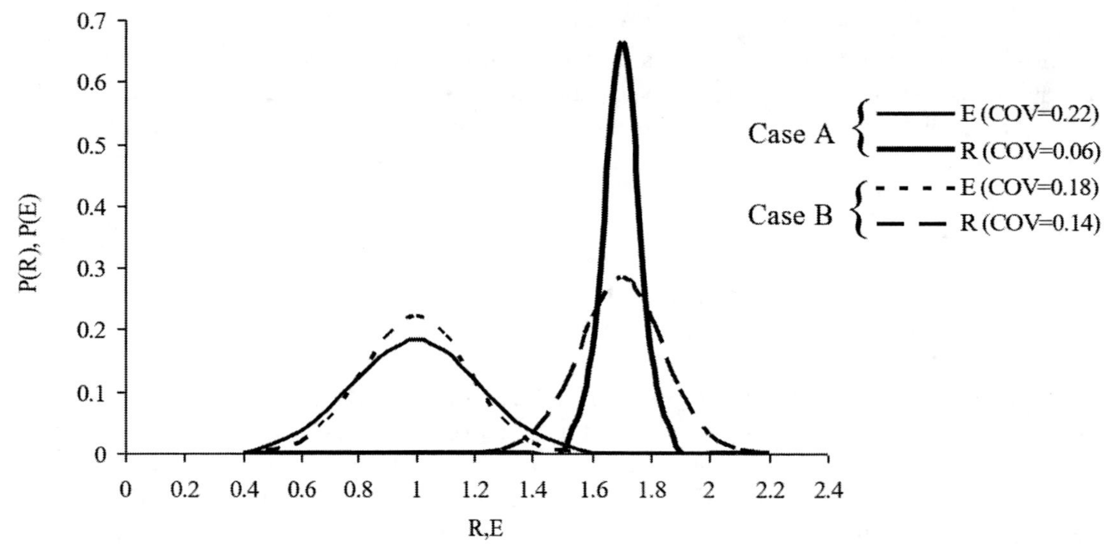

Figure 1. Randomness of action effect E and resistance R variables.

Table 1. Criteria proposed for the assessment of target reliability.

Description	Criteria corresponding to buildings' failure severity			Assessment mode
	high consequence for loss of human life (RC3)	**medium consequence (RC2)**	**low consequence for loss of human life (RC1)**	
Minimum value of reliability index recommended by EN 1990	4.3	3.8	3.3	Calculation
One variable reliability index	3.7	3.2	2.8	According test
Characteristic value of bearing capacity by design conditions	X	X	X	Code based design
Characteristic value of bearing capacity by tests	X	X	X	Experimental test
Coefficient of variation (COV)	COV≤0.15	COV≤0.20	COV≤0.25	Experimental test
Redundancy/ failure mode/ robustness	Robust/ redundant/ plastic failure	Robust/ plastic failure	Brittle failure	Professional judgement

3 ASSESSMENT OF TARGET RELIABILITY ACCORDING TO TEST DATA

The floor structure named by manufacturers as "Easi-Joist" has been tested and analyzed with the purpose to assess the target reliability. The experimental and design model inspected are shown in Figure 2. The span of simply supported model is 1.64 m, the span to depth ratio – 9.6. Section sizes of solid timber (C24) elements are 70 x 45 mm. Structural steel profiles of class S275 (U-profile of thickness 1.2 mm, depth 30 mm, width 12 mm). Ten "Easi-Joist" models were tested up to failure using the universal testing machine "Instron" under static load of a rate of 50 N/s. Buckling of compressed steel elements were observed over thin section around upper chord as predominant failure mode during the static tests of ten models (see Figure 3).

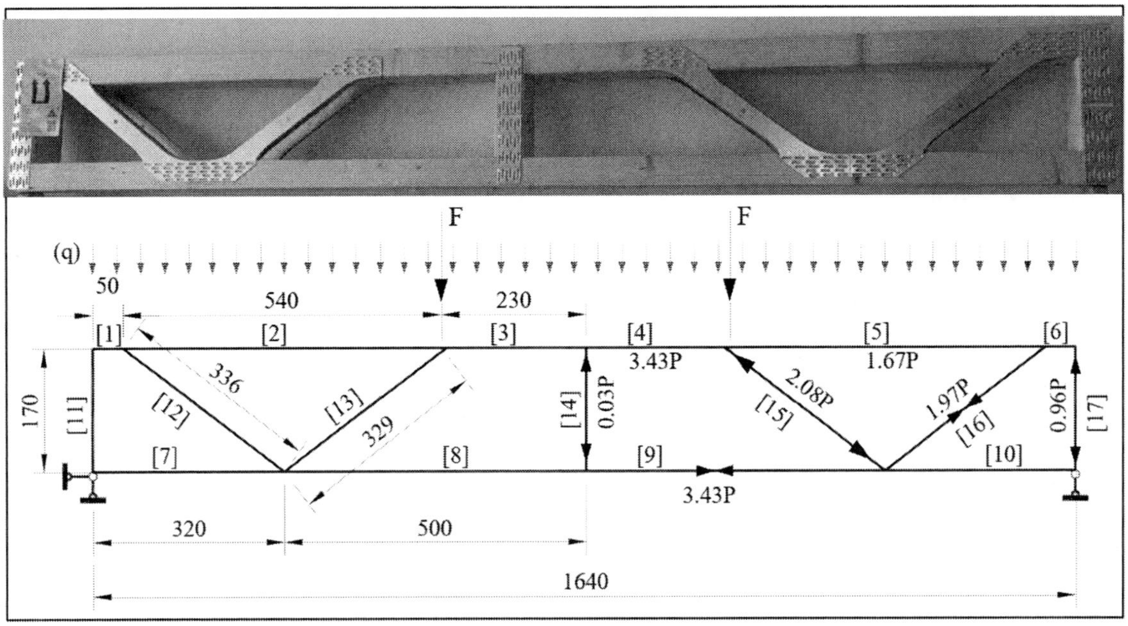

Figure 2. "Easi-Joist" model, geometric shape and axial force multipliers.

Figure 3. Characteristic failure mode of "Easi-Joist" model.

The safety assessment procedure of "Easi Joist" was completed with the results listed as follows:

- Characteristic value of shortterm point load is F_k= 6 kN according to Eurocode-based design methodology determined as a minimal value from nine limit state conditions of EN 1995-1-1 (2004) and EN 1993-1-1 (2005) referable to the timber-steel composition inspected.
- Processing of experimentally determined bearing capacities data set using the EN 14358 (2006) method “Acceptance criteria for a sample” results to value very close $F_{k,r}$= 6.04 kN. Capacities data set exhibited good fitness with lognormal distribution, see Figure 4, and the proper value of the coefficient of variation estimated is COV= 0.078.
- Experimental tests are useful to reveal the weakest link in the system. For system inspected it is a section of buckled steel element near upper chord where flanges of U-profiles are cut off, and the whole compression force remains to be transferred by a thin (1.2 mm) web section only. Normally this weakest section may be overlooked by designer in routine design procedure.
- Regarding the anticipated failure mode, the “Easi Joist” structure may be characterized as one capability, which depends on the behavior of the elements in the longitudinal buckling. That means an inherent trend of the structures toward a plastic failure mode in the ultimate limit state.
- The part of the area under force – displacement diagrams reproduced by the INSTRON software is indicative for the limit state toughness of a structure. It was estimated as a positive feature in terms of safety of the structure as an onset of the limit state is expected to be accompanied by progressive growth of plastic deformations.
- Assuming design capacity F_d= 5.4 kN (F_d= F_k/γ_f) equal to fixed maximal value in load side of performance diagram (see Figure 1), a value of one variable reliability index is β= $(\mu_R$-$e_o)/\sigma_R$= (7.14-5.45)/0.56 =3.0.
- Redundancy of the “Easi Joist” structure was assessed by gradually excluding elements assumed to be broken. Large deformations are expected but not a brittle failure.

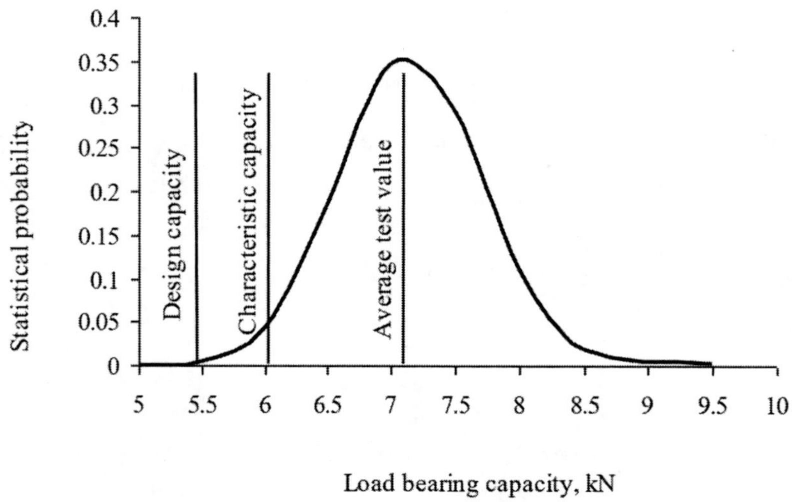

Figure 4. Lognormal distribution curve superimposed on static test data of bearing capacity and its codified values.

4 CONCLUSIONS

- Both measurable and logically examined variables shall be considered in the structural safety assessment.
- It is necessary to establish a codified system for the requirements of experimental tests for newly implemented structures with the purpose to obtain: (1) statistical characteristics of bearing capacity, (2) predominant failure mode including some indices for toughness, and (3) professional assessment of robustness and redundancy of structure desirable with a trend to the plastic failure mode of higher toughness indices.
- It is recognized that one variable reliability index computed according to the experimental test data sets is more adequate for the characterization of structural safety.
- The initial framework for the safety assessment of newly designed structures has been created, but further development is needed for the implementation in practice.

Acknowledgments

The authors sincerely appreciate the material support from the manager Mr M.Karlsons of Ltd "JMR-Frame" with the test models.

References

Ditlevsen, O., and Madsen, H.O., *Structural Reliability Methods*, Technical University of Denmark, 2005.

EN 14358, *Timber Structures - Calculation of Characteristic 5-Percentile Values and Acceptance Criteria for A Sample,* European Committee for Standardization, 2006.

EN 1990:2002/A1:2005/AC, *Eurocode - Basis of Structural Design,* European Committee for Standardization, 2010.

EN 1993-1-1, *Eurocode 3: Design of Steel Structures - Part 1-1: General Rules and Rules for Buildings,* European Committee for Standardization, 2005.

EN 1995-1-1, *Eurocode 5: Design of timber structures - Part 1-1: General - Common Rules and Rules for Buildings,* European Committee for Standardization, 2004.

Fang, Z.X., and Fan, H.T., Redundancy of Structural Systems in the Context of Structural Safety, *Procedia Engineering*, 14, 2172-2178, 2011.

Ghosn, M., Frangopol, D.M., McAllister, T.P., Shah, M., Diniz, S.M.C., Ellingwood, B.R., Manuel, L., Biondini, F., Catbas, N., Strauss, A., and Zhao, X.L., Reliability-Based Performance Indicators for Structural Members, *Journal of Structural Engineering*, 142(9), F4016002, F4016002-1: F4016002-13, September, 2016.

Keskküla, T., and Ozola, L., Codified Design and Reliability Aspects of Structural Timber Elements, *System-Based Vision for Strategic and Creative Design*, Bontempi, F. (ed.), 25-30, Balkema, Netherlands, 2003.

Nowak, A. S., and Collins, K. R., *Reliability of Structures*, CRC Press, New York, 2012.

Ozola, L., Some Considerations for Safety and Robustness of Structures, *New Developments in Structural Engineering and Construction*, Yazdani, S. and Singh, A. (eds.), 1505-1510, Research Publishing, Singapore, 2013.

Schneider, J., and Vrouwenvelder T., *Introduction to Safety and Reliability of Structures*, International Association for Bridge and Structural Engineering - IABSE, Zurich, 2017.

Zhao, Y-G., and Ono, T., Moment Methods for Structural Reliability, *Structural Safety*, 23, 47-75, 2001.

Interdependence between Structural Engineering and Construction Management
Edited by Ozevin, D., Ataei, H., Modares, M., Gurgun, A., Yazdani, S., and Singh, A.

ISBN: 978-0-9960437-6-2

INCREASING THE HEIGHT OF A GUYED STEEL STACK WHICH IS SUPPORTED ON AN INDUTRIAL BUILDING

ROMAN OKELO and BRADFORD RUSSELL

BR Architects and Engineers, Richardson, USA

The paper studied the rehabilitation and increase in height of an existing 30 ft tall, industrial, guyed steel stack, which was built around 1977. The stack is supported on an industrial building. Stacks are an essential part of the industrial facilities. They allow the discharge of chemical waste gases from industries to the atmosphere, under stringent pollution control guidelines. The height of the stack is increased by 40 ft to comply with the new guidelines. Analytical and numerical simulations were performed using MecaStack software (local model) and SAP2000 V.20 software (global model). Analysis and design are based on the ASME STS-1-2016 standard code. Discontinuity around the manhole at the bottom of the stack is investigated with SAP2000 V.20 finite element software. Results in terms of principle stresses, and Von-Mises stress, were used to prepare recommendations, repair and construction documents for the project.

Keywords: Dynamic wind loading, Slender structures, Stability, Analysis, Stress.

1 INTRODUCTION

Guyed steel stacks provide a more economical solution for taller stacks compared to self-supporting ones, depending on available space. The main component of these structures is usually a slender, cylindrical stack, circular in shape, which is pinned at its base. Sets of inclined guy cables support laterally the stack at several levels along its height. These guy cables are pretensioned and, generally, spaced at equal angles around the stack. The structural behavior of guyed stacks is complex due to the following reasons: geometric and material nonlinearities in conjunction with the sagging tendency of the guy cables; the interaction between the cables and the stack; the slenderness of the stack, and the boundary conditions. This paper describes the analytical and numerical research, which was carried out to increase the height of the existing steel stack, so as to comply with new pollution guidelines. The steel stack is designed in accordance with the American Society of Mechanical Engineers (ASME STS-1-2016 2016) standard. The design loads are based on the Minimum Design Loads for Buildings and Other Structures (ASCE/SEI 7-16 2016). Furthermore, the International Committee of Industrial Chimneys (CICIND 2010) model code, as well as ASCE Chimney and Stack Inspection Guidelines (ASCE 2003) provided supplemental data for the design.

The behavior of guyed structures has been investigated by several authors. Most of their work has been based on a "local model" where, the stack is independent of any adjacent or supporting structure. However, most design codes require that, structural interactions due to any supporting framework, geometric nonlinearities and potential interactions between guy wires and stack, should be taken into consideration in a detailed analysis of guyed stacks by correct modelling of inertia properties of the guy wires, the stack properties, and the boundary conditions. In essence, depending on the given geometric configuration, this amounts to determining the forces in the stack by using a global 2D or 3D model; especially, for unstiffened chimneys with a low aspect ratio. This may be explained by the fact that, the interactions between these members have a significant influence on overall performance of the whole system. A nonlinear analysis should be performed, if it is determined that the magnitude of cable displacements is such that the equilibrium equation for the structure should be based on the geometry of the displaced structure (P-Delta effect). Cables should maintain a minimum tensile force under all loading conditions to minimize visible sag and potential for induced vibrations.

Prinz and Roels (1998) performed the design of strengthening of a brick-built chimney structure by adding a guyed steel stack onto the structure. The added steel section was guyed with three steel ropes. During a period of four years, in which the system was in use, routine maintenance inspection, every two months revealed no defects. Bhowmik and Custodis (2001) performed a challenging chimney retrofit project for a steel producing company in Ontario, Canada. The stack was 160 ft tall with ten ft ten in inside diameter, and with approximately the bottom 30 ft inside a building. The authors used a Beam model to help retrofit the chimney.

This paper deals with analysis, design and rehabilitation of a guyed steel stack. The most important parameters considered in the evaluation and practical design of the stack will be presented. The performance of the stack is examined by using MecaStack software (2016) and through the finite element computer program SAP2000 V. 20 (2018). The effects of some important structural and material parameters on the guyed steel stack are evaluated and discussed.

2 AS-BUILT STRUCTURE

The project is located in Dallas, Texas, U.S. The as-built steel stack has a diameter of 4.25 ft, and a current height of 30 ft. It is supported on a platform, 20 ft x 15 ft, at an elevation of about 44 ft from the pavement, in conjunction with a set of three guy wires, with a diameter of 3/8 inches at about 120 degrees around the stack. The thickness of the shell is 0.167 in along the whole height; and the steel plate on the platform is 5/8 in thick. It is a single wall construction with fixed ladders. A steel-braced frame supports the platform and transfers the loads over four columns to the 30-in diameter, nine ft deep concrete shafts. The shafts are reinforced with eight #7 vertical rebars, and #3 stirrups at 18 in on center. The platform configuration consists of typical wide-flange beams: W16x26 sections in the north-south direction, at four ft spacing (approx.); and W12x16 in the east-west direction, at three ft spacing (approx.). The columns are square, hollow structural sections, HSS8x8x1/4 at the lower levels and HSS5x5x1/4 section at upper levels. X-diagonal bracings are ¾ in dia. steel rods. Additionally, some W16x26 beam sections are used as horizontal bracings below the platform level. There is a two ft by 7.5 ft flue opening at the bottom of the stack. The stack was built around 1977. The as-built steel is assumed to be A-36.

3 REHABILITATION

Based on the evaluation of the inspection reports, and the need to increase the height of the existing stack, a retrofit of the supporting steel frame was deemed necessary. Thus, the exiting stack shell is reinforced with a total of eight WT4x5 sections as vertical stiffeners. Additionally, two WT7x11 sections are welded adjacent to the flue opening. In the east-west direction, the platform is reinforced with three additional beams, W16x26, reducing the spacing of the beams to about two ft; and, the existing W8x10 beams are reinforced with WT4x5 section. The perimeter beams, W12x16 in the north-south direction are reinforced with WT6x7. These built-up beams are then boxed-up with two 1/8 in thick plates, one on each side. Furthermore, in the north-south direction, W16x26 and W12x14 beams have been added. The bracing configuration, just beneath the platform has been retrofitted with pipe sections PIPE 4, in combination with ¾ in diameter diagonal rods. The structural system is supported additionally by four new columns, HSS5x5x3/8 at the top and HSS8x8x3/8 below. The load from each column is transferred to a 40-ft deep, 30-in diameter concrete drilled shafts, with an embedment of seven ft into rock. The shafts are reinforced with eight #9 vertical rebars, and #3 spirals at 6 in pitch. The new steel is one of the following: A36, A53, A475 (7 Wire), A500, A572. Concrete has a minimum compressive strength of 3 ksi.

4 INCREASING THE STACK HEIGHT

The stack height has been increased by 40 ft, using ¼ in thick steel cylinder sections. The diameter is kept constant as in the as-built section, at a value of 4.25 ft. Angle sections, L3x3x1/4, at five ft spacing are used as horizontal rings. The whole rehabilitated stack, including the new section, is tied onto the platform by three sets of four, 1.5 in. diameter guy wires at different heights of 20 ft; 40 ft and 60 ft above the platform. The guy wire is attached onto the stack and onto the platform by using eye lug and a dead man. This is done in conjunction with turnbuckle, which allows the adjustment to the pretension force. Additional platform has been added to the stack.

5 LOCAL ANALYSIS WITH MECASTACK SOFTWARE

A local analysis has been carried out using MecaStack software (2016), in accordance with ASME STS-1-2016 and the Minimum Design Loads for Buildings and other Structures, ASCE/SEI 7-16. Fig. 1a) shows the model used for the analysis. All geometric configuration described above is included in the model. The stack has been analyzed under constant wind load of 41 psf. The platform live load is 50 psf. The pretension guy wire force at the platform is normally a value between 6% and 15% of the wire breaking strength. This analysis used a value of 6%. Furthermore, the analysis takes the following parameters into consideration: a) nonlinear cable effects; b) wind loads in different directions; c) thermal expansion of the stack; d) vortex shedding of guyed stack, and e) fatigue.

The results of the local analysis, in terms of guy wire tension, F; axial compression stress, σ_a; bending stress, σ_b; circumferential stress, σ_c, in the stack shell are provided in Table 1. The table also includes the allowable design values per ASME STS-1-2016. It can be seen that, for the proposed configuration, the calculated values are less than the allowable values. Thus, the design is safe.

6 GLOBAL ANALYSIS WITH SAP2000 V. 20 (2018) – FEA MODELLING

A 3D finite element model has been developed using SAP2000 V.20 (2018) commercial software, for the steel stack, as shown in Fig. 1b. This model is used to capture any other features of the structure, which could not be effectively simulated with the local model, including the supporting steel-braced frames. The ladders and platforms are not simulated in this global model. However, the dead and live loads due to the platform are included in the model. The guy wires are modelled as tension-only elements. The shell element SHELL is used in modelling the stack wall and platform plates. This is a three- or four node formulation that combines membrane and plate bending behavior. There is a rigid connection between the stack shell and the platform plates.

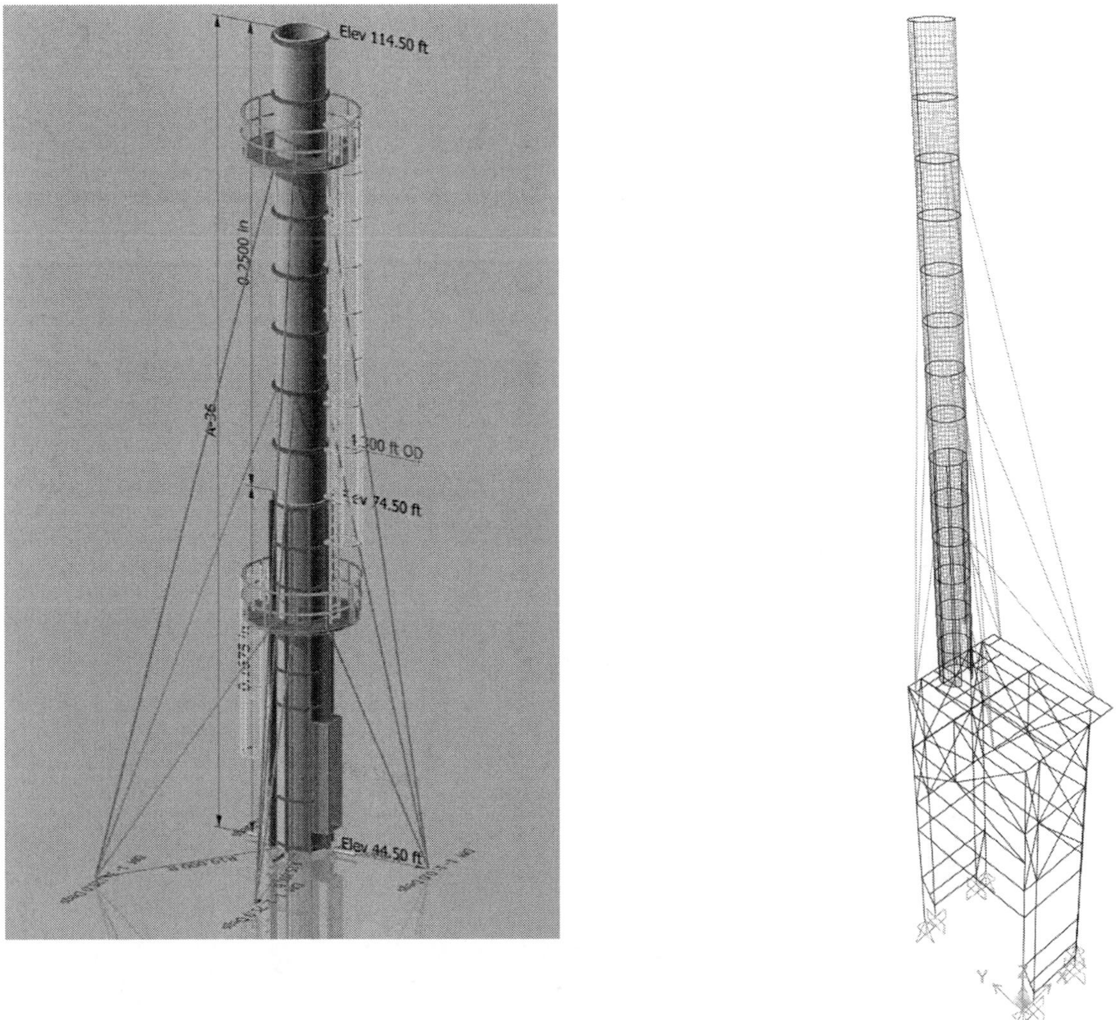

a) b)

Figure 1. Analysis models: a) Local model with MecaStack; b) Global model with SAP2000 V.20 (2018).

Table 1. Comparison of analysis results with allowable values (ASD).

Item	MecaStack		SAP2000	
	Calc.	Allow.	Calc.	Yield
Guy Wire Force, [kips]	37.74	54.72	17.4	54.72
Shell Stress at 44.5 ft				
Axial Compr., σ_a [ksi]	4.22	9.88		
Bending, σ_b [ksi]	0	9.88		
Circumferential, σ_c [ksi]	0.10	19.9		
Von-Mises, σ_{vM}			26.0	36.0
Shell Stress at 74.2 ft				
Axial Compr., σ_a [ksi]	6.6	9.88		
Bending, σ_b [ksi]	2.49	9.88		
Circumferential, σ_c [ksi]	0.1	19.9		
Von-Mises, σ_{vM}			5.9	36.0

Wind loading is a constant value of 41 psf., as determined under section 5. Additional dead and live loads due to the upper platforms, as well as the loads on the lower platform have been included. Thus, wind, live and dead loads have been considered in this model. As depicted on Figure 1b), a fixed boundary condition is used at the pavement level. The analysis is based on allowable stress as per code AISC 325-05 (2006). SAP2000 (2018) takes the following parameters into consideration: a) nonlinear cable effects, b) P-Delta effects with small and large deformations; and c) wind loads in different directions.

The results show plausible distribution of the principle stresses, and the need for the retrofitting of the bracing system configurations. Generally, the maximum principal stresses and Von-mises stress are used to evaluate the capacity of shell structures. The results presented in Figure 2, show realistic distribution of the Von-Mises stress, σ_{VM} [ksi] on the stack shell and on the platform plate.

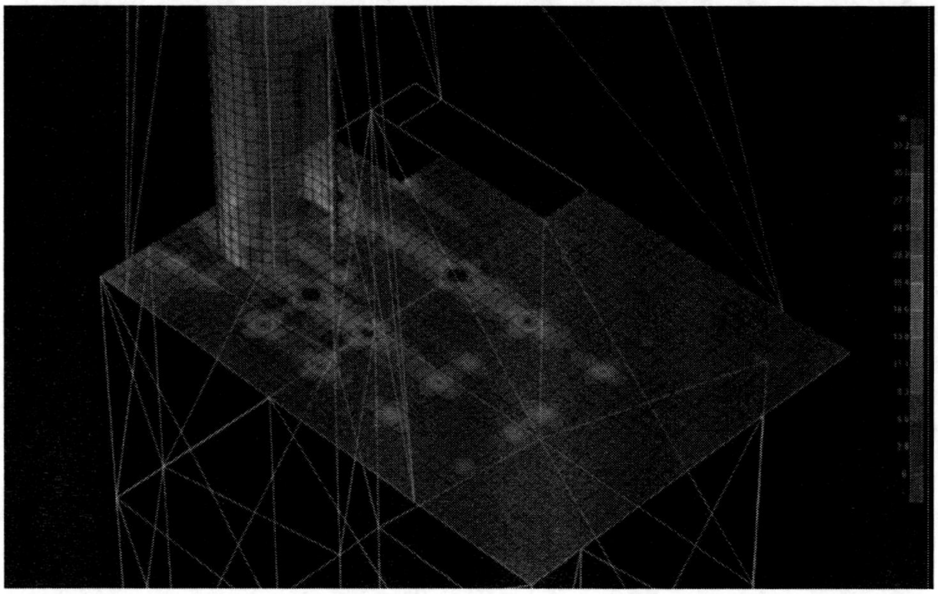

Figure 2. Global model analysis results – computed Von-Mises stress [ksi].

Due to the flue opening at the bottom of the stack, the stiffness of the cross-section is reduced significantly, and high concentration of stresses could be expected, when no reinforcement is provided. However, the reinforcement provided section 3 above seems to be adequate. Except for singularity cases, the plotted values are less than the yield strength, which is 36 ksi. Furthermore, the results of the combined stress in the frame members show that their maximum stress ratios are less than unity. The additional outputs of the analysis are not presented in this paper due to space limit; but, they will be presented in the conference. Therefore, the selected geometric configuration meets the design safety requirements.

As shown in Table 1, the local analysis provides stresses that are lower than the ones based on the global model. However, the results from both models are used to determine the necessary stack shell thickness to resist flexural loading; the need, size and spacing of ring and vertical stiffeners; the size and location of any additional bracings, and the size and number of guy wire cables. Furthermore, in adherence to ASME STS-1-2016, section 4.11.3.1 [2], this study took into consideration the "structural interaction between the guyed steel stack and its supporting structure."

7 CONCLUSIONS

Simplified, local models may not be adequate for slender stacks with large opening. Geometric and material nonlinearities and cable-stack-framework interactions are important and should be simulated properly by using appropriate type and number of elements in the cable model, suitable formulation, and also correct modelling of inertia properties of the guy cables, the stack and any supporting framework. Thus, it can be concluded that, the analysis, design and performance of a guyed steel stack may be accurately predicted by the use of an efficient and effective software such as MecaStack (2016) and SAP2000 V.20 (2018).

References

AISC 325-05, *Steel Construction Manual.* 13th Edition, American Institute of Steel Construction, Inc., Chicago, IL, 2006.

American Society of Civil Engineers (ASCE). *Chimney and Stack Inspection Guidelines: Design and Construction of Steel Chimney Liners.* American Society of Civil Engineers. Reston, VA, 2003.

ASCE/SEI 7-16, *Minimum Design Loads for Building and Other Structures*, American Society of Civil Engineers. Reston, VA, 2016.

ASME STS-1-2016, *Steel Stacks*, American Society of Mechanical Engineers, New York, NY, 2016.

Bhowmik, A. and Custodis, H., A Challenging Chimney Retrofit Project for a Steel Producing Company in Ontario, Canada, in *CICIND Report*, 17(2), September 2001.

CICIND, *Model Code for Steel Chimneys,* The International Committee of Industrial Chimneys Zurich, Switzerland, 2010.

Prinz, M., and Roels, V. K., Heightening of a Brick-Built Chimney by a Guyed Steel-Top, Munich, Germany, in *CICIND Report*, 14(1), June 1998.

MecaStack Software, A Program for the Analysis and Design Stacks, July 2016.

SAP2000 Version 20, Integrated Software for Structural Analysis and Design, Computer and Structures Inc, Berkeley, CA, USA, 2018.

Interdependence between Structural Engineering and Construction Management
Edited by Ozevin, D., Ataei, H., Modares, M., Gurgun, A., Yazdani, S., and Singh, A.

ISBN: 978-0-9960437-6-2

LOAD PATH REFORMULATION AROUND MULTIPLE RESTRICTIONS FOR AN INDUSTRIAL RETROFIT

TIM HOGUE

Hargrove Engineers + Constructors, Mobile, USA

Process, operational, and electrical restrictions inherent in the design of a regulatory compliance upgrade for an industrial plant, and overburdened existing structure, required an unusual structural retrofit. There were multiple restrictions on placement of structural members and multiple needs for additional strength in the existing members. To manage the constraints, the load path was reformulated to bypass problem areas and focus upgrade work on just a few locations. The key in selecting upgrade locations was determining where there was reserve capacity in the existing foundation. Two such locations were identified and super-bents were formed there using existing moment frames. Horizontal trusses were built to bridge between the super-bents. A super-bent included a new grade beam, with end bulbs surrounding existing column pedestals, column section augmentation, new heavy anchor bolts and new diagonals to convert two parallel beams into a deep truss. Embedded structural steel was utilized in the grade beam end bulbs to distribute anchorage forces and develop reliable load paths around obstructed columns. Unusual uses of structural steel, work in obscured areas, unconventional structural systems and construction schedule interuptions led to an increase in construction management and inspection scope and complexity. CM lessons learned from the project involved better interaction of disciplines during design and greater input from engineering just prior to and during construction.

Keywords: Concrete, Grade beams, Embedded structural steel, Strengthening, Super-bent, Construction management lessons.

1 INTRODUCTION

Historically, the typical industrial pipe rack has comprised of a series of parallel moment frames or bents, each approximately 20 to 30 feet wide and one spaced from the next at about the same distance, resisting lateral load in the short (or transverse) direction, with bracing every few bent-to-bent bays resisting lateral loads in the long (or longitudinal) direction (Arya *et al.*, 1979). Typical pipe rack height is in the range of 15 to 30 feet. For racks built prior to 2001, when OSHA began requiring a minimum of 4 anchor bolts for columns (Safety Standards 2001), column anchorages often consisted of only two anchor bolts near the column center, meaning no significant moment resistance could be developed in the base connection. Further, anchorage capacity of older anchor bolt assemblages has been found by the author to usually have relatively little capacity computed according to the latest generation of strength formulas (ACI 318-02 2002).

One such pipe rack, the original construction drawings for which bear the date of 1981 (Figure 1), was to be retrofitted with new structure, equipment and piping installed on top of it as part of a regulatory compliance upgrade for the plant process unit. The added objects would increase wind and seismic load. That the retrofit would be on top of the existing framing would magnify the overturning effect of those loads. Further, it was likely that current code wind and seismic load provisions (ASCE/SEI 7-10 2010), especially when applied in conjunction with more modern insight on the effect of wind on industrial open structures (Wind Loads 2011), would lead to larger unit seismic loads and wind pressures and/or (for wind) effective tributary areas for all parts of the structure, whether original or retrofit.

Figure 1. Pipe rack prior to retrofit.

Beside load effects, sources of challenges for retrofit design and construction were electrical, structural, piping and process constraints and construction management demands. To overcome these, the load path for the transverse direction of the pipe rack had to be reformulated relative to the typical concept for such structures.

2 LOAD PATH REFORMULATION CONCEPT

2.1 Analysis of Existing Structure

As originally designed, the moment frames had to be self-sufficient relative to transverse forces. Each moment frame must support the gravity and transverse lateral forces tributary to it. There is no means for redistribution of these loads to adjacent bents. Thus, if the moment frame capacity were inadequate, each and every moment frame would, keeping the original load path concept, have to be strengthened.

And it was indeed found that the bents, once the new loads were applied, were inadequate. Analysis showed that the two-bolt column anchorage could not meet demands. Frame members were shown to be overstressed. Story drift was too much.

2.2 Load Path Constraints

Under most columns, the foundation comprised three "composite" piles: driven timber pile topped by a cast concrete column stub. The lateral capacity of these piles was indeterminate and considered unreliable at any rate.

Foundations under columns to which a vertical, longitudinally oriented brace was attached, though, were more substantial: four 12-inch diameter steel piles. These foundations were designed primarily to react against load in the longitudinal direction. Therefore, they had considerable capacity for transverse lateral loads beyond the tributary load for a single moment frame.

Strengthening of column anchorages would have required enlargement of the concrete pedestals on which the columns rest, additional base plate and more anchor bolts. This rather involved proposition, which would have to be employed at every column of every bent, was deemed impractical from a schedule and construction cost standpoint.

Further, some of the columns rested not on an isolated pedestal but a corbel integral with a wall of an underground tank. Plant operations did not want to apply further loads or add structural elements to this wall due to concerns over its structural condition. Moreover, there were any number of pipes, machines and instruments that could not be moved which prevented access to concrete in the area to which additional anchorage elements would have to be affixed.

It was proposed to add bracing in the lower "story" of each moment frame. This would take care of the overstress of the frame members but not the anchorage inadequacy. However, bracing was disallowed by plant operations because it would prevent passage of maintenance vehicles down the "hallway" of the pipe rack.

Consideration was given to adding plates to flanges of beams and columns in order to increase strength and stiffness. The amount of plate required, though, was so great as to be impractical from the standpoint of material cost, weld placement and constructability. Constructability problems stemmed not only from handling heavy pieces but also from having to work around multiple and frequent blockages of access to structural members by electrical conduit, cable tray, instruments and piping that are typically clustered around and supported off of structural columns rather than separate, dedicated vertical support members.

The attachment of WT sections to the faces of wide flange beams and columns was considered as well. They required less free surface area of column and beam flanges but nevertheless, there were still too many items attached to members to allow this to be feasible.

2.3 Super-Bents

In order to satisfy the above noted constraints, the only feasible approach perceived, short of building a whole new, separate structural system, was to turn two existing moment frames into super-bents that would collect essentially the entire lateral load applied to the area where the new regulatory compliance infrastructure was installed. The bents selected were ones whose columns sat on steel pile-supported foundations so that there would be sufficient foundation lateral capacity. A rendering of the structural model of the retrofit super-bent structural system is shown in Figure 2. The super-bents are on column grid lines 3 and 9.

2.3.1 *Bridge trusses*

To affect the super-bent system, a number of structural retrofits were required. One of these was means to collect transverse lateral loads otherwise tributary to typical moment frames and deliver them to the super-bents. To do this, two horizontal trusses were added to the structure. The lower of the two (Figure 2) was composed of new steel threaded in amongst the existing framing. To avoid conflict with and disturbance of the existing framing, the chords of the lower truss were set outside the footprint of the original pipe rack. The upper bridge truss was composed entirely of new framing.

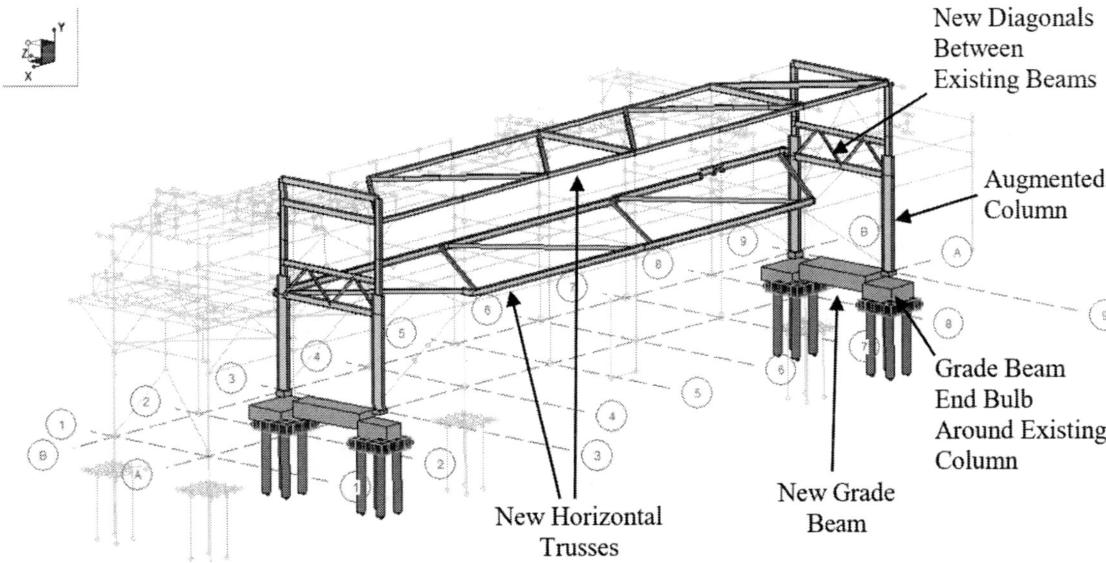

Figure 2. Super-bent structural model (with essentially same viewpoint as in Figure 1). Note, members not part of the super-bent system appear as faint wireframe elements.

2.3.2 *Frame strengthening*

The superstructure framing of the super-bents had to be strengthened in order to support the lateral load. This strengthening comprised three main aspects:

(1) Diagonal elements were added to the space between two existing parallel beams in order to form, a stiff, deep truss. The diagonals were placed without interfering with existing piping.
(2) Most of the height of the original columns was augmented with WT spines welded to the outsides of the columns. The outside faces of the columns at the super-bent locations were essentially free of obstruction. With the spines, the columns as strengthened were double the depth of the original columns.
(3) Anchorage capacity was added to the columns such that the base connections would be moment-resisting.

2.3.3 *Foundation strengthening*

It was determined that the steel pile groups had sufficient strength to resist the applied lateral loads. However, the existing foundation system did not have enough moment capacity to develop a moment-resisting reaction at the bases of the columns. To address this, a concrete grade beam was added between the two columns of each super-bent. The beam was tied into the column anchorage assemblage in order to transfer moment but surrounded the existing pedestals with such girth that only minimal strain could be developed in the pedestal before the grade beam-column interaction arrested joint rotation. Thus, moment was effectively prevented from transferring into the pedestal-pile cap-pile assemblage.

3 COLUMN-TO-GRADE BEAM DETAILING

One final constraint affecting the retrofit structural system concerned the presence of a vertical duct bank extending up alongside some of the super-bent pedestals. It was required by plant operations that the duct bank not be touched in any way. Therefore, the grade beam could not follow a direct path between the two columns of a bent. In essence, stresses had to "turn" several corners and "go around" the duct banks to transmit column base moments into the body of the grade beam. This is illustrated in Figure 3.

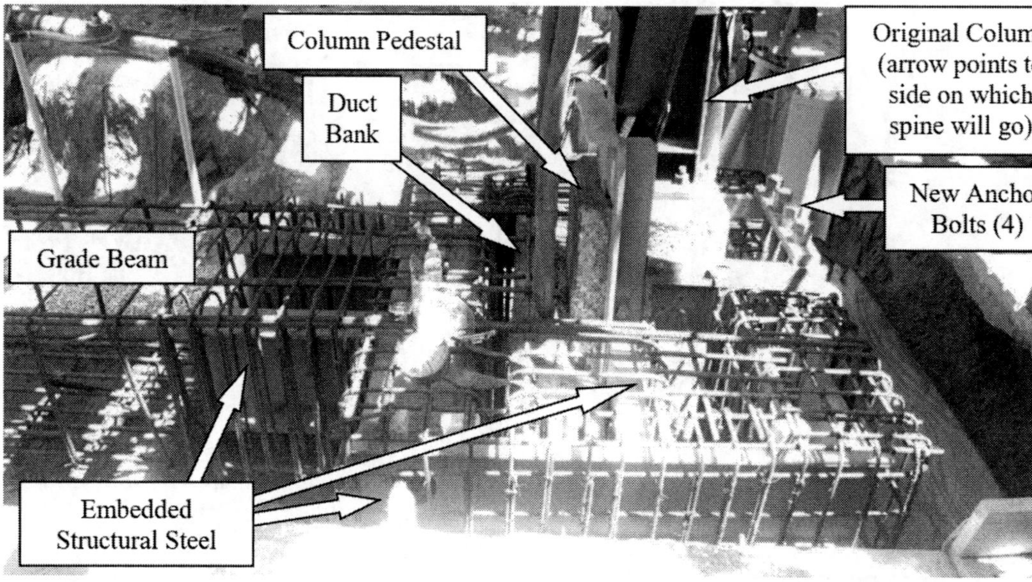

Figure 3. Super-bent grade beam on column grid line 3, looking along grid line B towards line 2. Note, column spine not yet installed.

To direct stresses through the concrete grade beam bulb surrounding the existing column and stabilize stress transfer around the duct bank, structural steel was embedded within the bulb and enclosed in ties. This included a pair of channels at the back or outside half of the bulb, against which the new anchor bolts could bear and with which anchor bolt forces could be reliably distributed into the rest of the bulb. Usual limit states for cast-in-place anchor bolts, except for steel tension failure, were considered not applicable in this case.

To design the embedded structural steel, an analysis was done with only the steel in the model, no bulb concrete. The steel had to carry all the load. Then, for checking lateral drift, the analysis was repeated with the bulb concrete included in the model. The embedded structural steel was continued into the main body of the grade beam far enough for grade beam longitudinal rebar to be developed.

4 CONSTRUCTION MANAGEMENT LESSONS LEARNED

The structural detailing concrete as well as embedded steel and the superstructure around the column and column pedestal, was unconventional and, more pointedly, awkward. It was important that construction management activities involve not just routine inspection personnel and schedule. The engineer of record needed to perhaps not continuously inspect the work but did need to frequently and regularly inspect the construction work and communicate with field personnel. It is important that both engineering and construction management personnel insist on this high level of field observation through to the end of the project. There is a tendency to bypass detailed observation as the project progresses and a sense of familiarity is developed. However, not one grade beam bulb was completely identical to any of the others so familiarity can be misleading, and the sense of familiarity should always be held suspect.

The principals of frequent inspection by the engineer of record and engineering-construction management communication holds as well for superstructure installation where details are unusual, especially as they are made so by construction scheduling. A case in point was one of the column spines. Some electrical conduit interfered with the spine. The conduit was scheduled for relocation regardless of structural work but could not be worked on until the turnaround for the process involved. Therefore, the spine had to be installed in three pieces; two prior to turnaround, one after. This odd detail was well known to engineering who had been intimately involved in developing it and making sure it would be structurally effective. But by the time the last piece of the spine could be installed, it had been forgotten and was left out—engineering had not been involved enough in construction inspection.

5 CONCLUSIONS

(1) Key to managing major structural strengthening is establishing a suitable load path.

(2) The first task is to determine availability of reserve foundation strength. Every effort must be made to exploit excess foundation capacity before resorting to requiring new foundations or major foundation strengthening.

(3) As in the case at hand, a likely source for excess foundation capacity needed for lateral load in one direction is a foundation relatively lightly loaded in that direction but designed for large lateral loads in the other direction.

(4) Once sources of sufficient foundation strength are identified, if they can be identified, the superstructure must be configured and strengthened to transmit loads to the new foundation locations. Load path reformulation will likely be necessary. The load path reformulation scheme ultimately selected will be sifted out from among several possibilities by the various spatial, process, structural, electrical, instrumentation and mechanical constraints presenting.

(5) Close and frequent field observation by the engineer of record and contact between the engineer of record and the construction management team and contractor must be insisted on and carried out when load path reformulation is unconventional, whether due to structural configuration, schedule or both.

References

ACI 318-02, *Building Code Requirements for Structural Concrete*, American Concrete Institute, Farmington Hills, MI, 2002.

Arya, Suresh C., Feng, Edward G., and Pincus, George, Optimum Design of Steel Pipe Racks, *Engineering Journal*, American Institute of Steel Construction, 16(3), 84-97, 3rd quarter, 1979.

ASCE/SEI 7-10, *Minimum Design Loads for Buildings and Other Structures*, American Society of Civil Engineers, Reston, VA, 2010.

Safety Standards, *Safety Standards for Steel Erection,* Department of Labor, OSHA, Final Rule for 29 CFR Part 1926, Subpart R, Fed Register # 66:5317-5325, January 18, 2001. Retrieved from https://www.osha.gov/laws-regs/federalregister/2001-01-18-0 on August 31, 2018.

Wind Loads, *Wind Loads for Petrochemical and Other Industrial Facilities,* Task Committee on Wind-Induced Forces of the Petrochemical Committee of the Energy Division of the American Society of Civil Engineers, Reston, VA, 2011.

Interdependence between Structural Engineering and Construction Management
Edited by Ozevin, D., Ataei, H., Modares, M. Gurgun, A., Yazdani, S., and Singh, A.

ISBN: 978-0-9960437-6-2

EXPERIMENTAL EVALUATION OF SOIL REINFORCEMENTS IN UNPAVED ROAD

NAHLA SALIM

Dept Civil Engineering, University of Technology, Baghdad, Iraq

In this study, a series of 24 laboratory tests were conducted on a footing resting on crushed stone with 17.68 kN/m^3 dry unit weight overlying sandy soils of two relative densities corresponding to (60% and 80%). The subbase layer is of crushed stone with a thickness of 5, 7.5 and 10 cm. Ten tests were conducted under static load with and without geogrid. All the other 14 model tests were carried out under harmonic load which was applied in a sequence determined prior (40% of static load). Tests were conducted at (2) Hz frequency according to the loading value. The process of the loading was continued until the number of cycles reached 10^4. The results indicated that, for static load and with the inclusion of the geogrid, as the thickness of the subbase layer increases, the percentage of increase in bearing capacity was reduced. In general, using geogrid reinforcement with subbase thickness of 7.5 and 5 cm causes an increase in bearing capacity approximately 1.5 to 2 times greater than for unreinforced respectively. This means that by using geogrid reinforcement, the thickness of subbase can be reduced which causes a reduction in construction cost.

Keywords: Geogrid, Static, Dynamic loading, Crushed stone, Sandy soil, Bearing capacity.

1 INTRODUCTION

In general, unpaved roads consist of an unbound crushed stone as a subbase overlying an existing, typically weak subgrade. The types of fill used for the subbase layer vary from locally available materials or waste material to good quality crushed stone. Generally, the sub grade has a poor bearing capacity or may range from medium strength to very low strength. The road may be not paved for the following reasons:

- In developing countries, the cost may be the main factor.
- For temporary access road, so the road has a very brief design life.

The unpaved roads usually are poor quality, potholed, unable to with stand loads of heavy vehicle or equipment. One way to improve soil properties for pavement construction is to reinforce the material with geosynthetic. Geosynthetic used in unpaved roads are essentially geotextile and geogrid. This work focuses on the use of geogrid as reinforcement, which improved interface resistance due to interlocking as compared to geotextile. This interlocking between the crushed stone and geogrid provided the confinement which minimizes lateral movement of crushed stone particles and increases the modulus of the base course, which causes a wider vertical stress distribution over the subgrade and consequently a reduction of the vertical and lateral deformation.

In recent years, the use of geosynthetic materials in road construction has increased drastically. The first use of fabrics for reinforcing roads was investigated by the South Carolina Highway Department in 1926 as reported by Koerner (2005). Giroud and Noiray (1981) used an improvement factor of $(\pi + 2)$ for geotextile reinforced unpaved road and $(3\pi/2 + 2)$ for geogrid reinforced unpaved road. Krishnaswamy and Sudhakar (2005) carried out a test to study the effect of using geotextile in increasing the California Bearing Ratio. Gali and Nair (2014) performed laboratory and field studies on unreinforced and reinforced unpaved low volume roads section constructed over weak subgrade using different types of geosynthetic reinforcement. Like geotextile biaxial geogrid and geocell layer at the interface of subgrade and base course in order to increase the load carrying capacity and reduction in rut depth. The main conclusions were drawn, that using Geosynthetic are effective in increasing the load carrying capacity and reducing rut depth. Researchers like Elleboudy *et al.* (2017), Calvarano *et al.* (2017), and Mousavi *et al.* (2017) use numerical analysis to study the performance of geogrid reinforced of the unpaved road under static and repeated load through ABAQUS and 3D-Plaxis Software. From the literature, it was evident that the application of geosynthetic reinforcement in the pavement has not been extensively studied. In this paper, the experimental study was carried out to evaluate the effect of using geogrid reinforcement in improving the load carrying capacity and deformation of the unpaved road under the static and dynamic load. In the model tests, a load of 40% of the static load was applied repeatedly for 10^4 cycles and the settlement of the plate was measured.

2 EXPERIMENTAL SETUP

The model test apparatus consisted of a test tank, loading system, and instrumentation. The test tank has inner dimensions of 1000mm length, 750 mm width and 700mm depth. Each part of the container is made of steel plates of 5 mm thick. A load was applied through a steel plate of 200 mm width, 400mm length and 5 mm thickness. Different materials were used in performing the experimental model which is subgrade soil, crushed stone, and geogrid.

2.1 Subgrade Soil

Sand used in the present study was brought up from Bedra City at Wasit Province, southern East of Baghdad in Iraq. The subgrade soil used was classified as poorly graded sand (SP) according to the Unified Soil Classification System. The maximum unit weight was 19.4 kN/m^3 and the minimum unit weight was 15.8 kN/m^3. The angle of internal friction was 31.5^0 for 60% relative density and 38.7^0 for 80% relative density.

2.2 Crushed Stone

The subbase layer rests over a sandy soil with maximum dry unit weight equal to 17.68 kN/m^3. The crushed stone of various sizes was collected and sampled according to grading I of granular sub-base design as given by R6E (Iraqi Standard Specification 1983). The subbase is classified as GP according to the Unified Soil Classification System (USCS). Figure 1 depicts the grain size distribution of the crushed stone used in the present study.

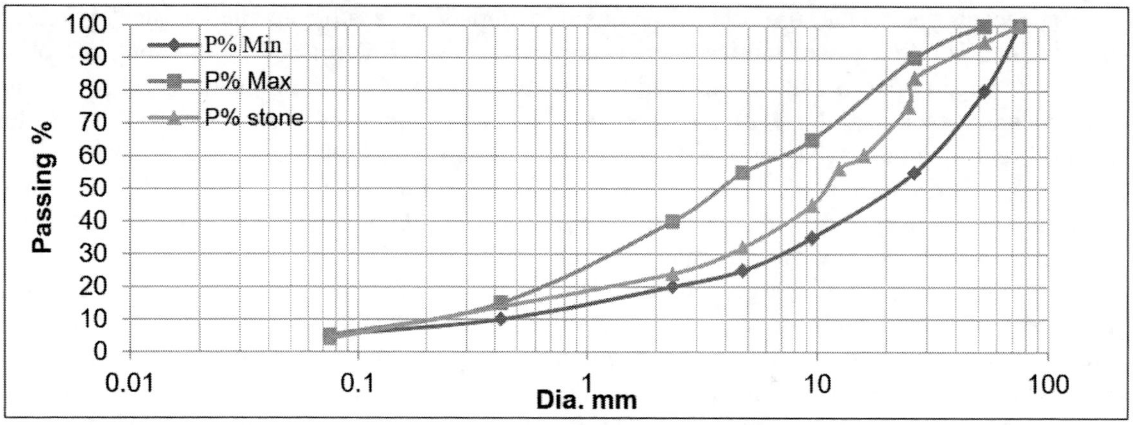

Figure 1. Grain size distribution of crushed stone.

2.3 Reinforcing Materials

The geosynthetic reinforcement usually placed above weak subgrade to improve the performance of this layer. These improvements can be listed into four functions: Separation, filtration, drainage, and reinforcement. The most affected functions are separation and reinforcement. The Separation function of the reinforcing element prevents the base course from sinking in the subgrade soil. Thus, the base course thickness remains constant without deterioration through the road life. This means that it will be able to distribute vehicle loads inefficient way without causing distress in the subgrade. In this study, the used geogrid was manufactured by Al-Latifia Factory, the plastic mesh having engineering properties; mass per unit area is 700 g/m^2, tensile strength at 2% strain is 5.1 kN/m and tensile strength at 5% strain is 9.1 kN/m. One layer of geogrid is used, and it is placed at the base course/subgrade soil interface.

3 TESTING PROGRAM

The experimental program included a series of model tests on two-layered soil system, these tests were carried out under static and harmonic load. The static load was applied gradually through the hydraulic jack that actuating at the controlled displacement of 0.03 mm/sec. Dynamic loads applied according to the testing program, the harmonic loading was applied in a sequence determined prior (40% of static load). Tests were conducted at (2) Hz frequency according to the loading value. The process of the loading was continued until the number of cycles reached 10^4.

4 RESULTS AND DISCUSSIONS

Table 1 summarizes the results of unpaved road bearing capacity under static load. For all models tests, the failure is defined as the load causing a settlement corresponding to 10% of the footing width (Terzaghi 1943). It is clearly shown that the soil with a relative density of 60%, the geogrid reinforcement causes an increase in the bearing capacity from 1.5-2 times greater than unreinforced for 7.5 cm and 5 cm of subbase layer respectively. This is due to the improvement of load distribution and lateral restraint of base material which is related to the geogrid/crushed stone material interaction which prevents lateral spreading of the base material and imparts tensile stiffness to the base. For subbase layer thickness of 10 cm, a small increment in bearing capacity, this is may be due to higher deformations are required for the mobilization of reinforcing tensile strength due to higher subbase layer thickness. It can be seen that for soil with

80% relative density, there is no significant increase in the bearing capacity due to small deformation. Table 2 summarizes the results of unpaved road surface settlement under harmonic load took at 1000 cycle. Figures A-1 to A-8 in Appendix A has presented the test results of the unpaved road under static and dynamic test.

Table 1. Bearing capacity of unpaved road under static load with and without geogrid reinforcement.

Relative Density of subgrade %	Subbase thickness (cm)								
	h = 5 cm			h= 7.5 cm			h= 10 cm		
	Bearing Capacity (kPa)								
	Without Geogrid	With Geogrid	% of increase in bearing capacity	Without Geogrid	With Geogrid	% of increase in bearing capacity	Without Geogrid	With Geogrid	% of Increase in bearing capacity
60	152	300	97	225	375	67	329	375	14
80	373	382	2.4	446	545	22	653	710	9

Table 2. Surface settlement under cyclic load with and without geogrid reinforcement at the end of 1000 cyclic.

Relative Density of subgrade %	Subbase thickness (cm)								
	h = 5 cm			h= 7.5 cm			h= 10 cm		
	Settlement in mm at 1000 cycle								
	Without Geogrid	With Geogrid	Set. Redn%	Without Geogrid	With Geogrid	Set. Redn%	Without Geogrid	With Geogrid	Set. Redn%
60	6.5	5.8	10	5.4	4.5	16	4.8	4.3	10
80	4.0	3.9	2.5	3.6	3.2	11	3.3	2.9	12

5 CONCLUSIONS

The main conclusions from the inclusion of geogrid at the interface subbase/subgrade are:

- Unpaved road under static load provides an increase in the bearing capacity and reduced the settlement. The maximum increment varies from 1.5-2 times greater than unreinforced one. While in the case of unpaved road subjected to a dynamic load, the maximum reduction in the settlement is only 16% using geogrid reinforcement compared to the unreinforced case.
- The thickness of the subbase layer has a significant effect on the performance of geogrid. Specific deformation is required for geogrid to have fully mobilization of reinforcing tensile strength.
- The geogrid prevents the intermix of base and subgrade by a geogrid performing the function of separation. Finally, geogrid reduces construction cost by reducing the thickness of the subbase layer.

Appendix A. Experimental Test Results

The experimental test results for the unpaved road under static and harmonic load with and without geogrid are shown in Figures A.1 to A.8.

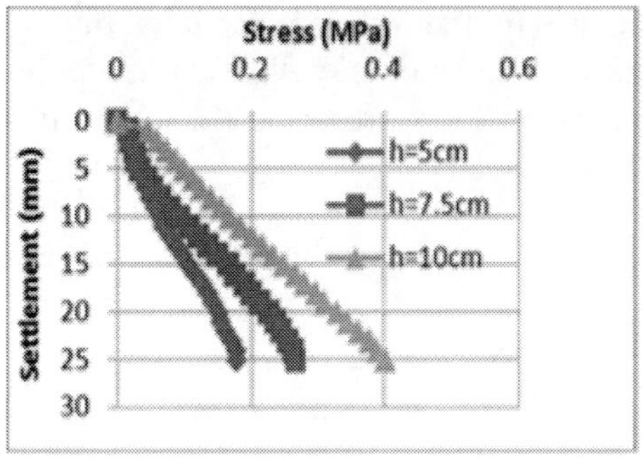

Figure A.1. The effect of the subbase layer thickness on the bearing capacity (RD. = 60%) without geogrid under static load.

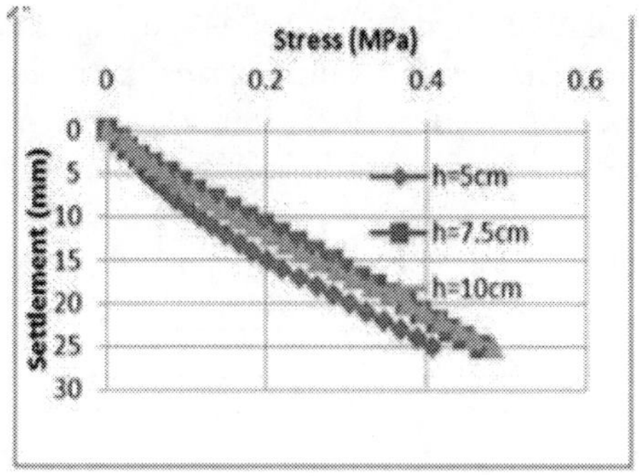

Figure A.2 The effect of geogrid reinforcement on the bearing capacity with different subbase layer thicknesses (RD = 60%) under static load.

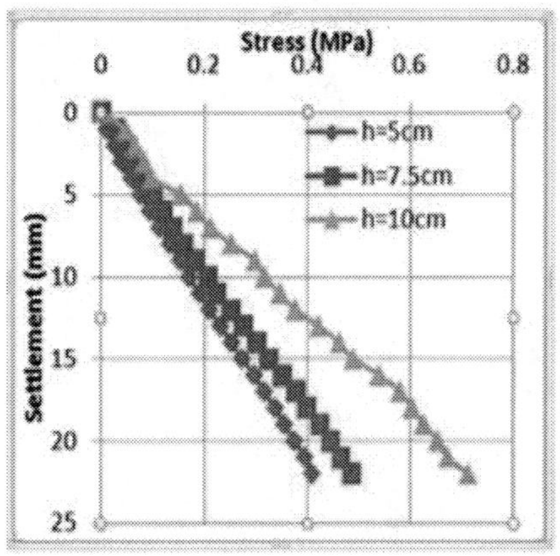

Figure A.3. The effect of the subbase layer on the bearing capacity with different subbase layer thicknesses (RD = 80%) under static load.

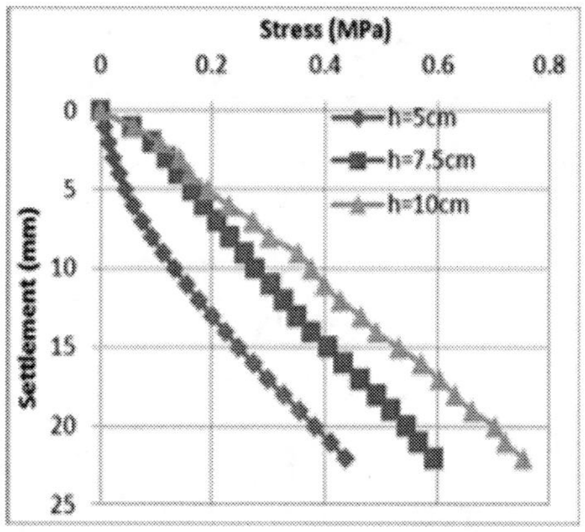

Figure A.4. The effect of the subbase layer on the bearing capacity with different subbase layer thicknesses using geogrid reinforcement under static load.

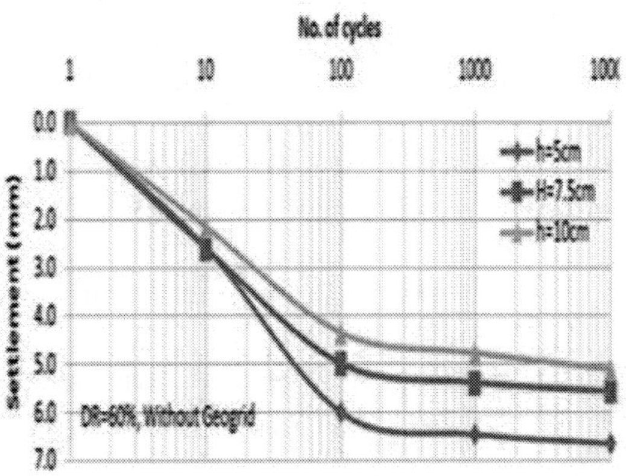

Figure A.5. The effect of dynamic load on the bearing capacity with different subbase layer thickness (RD = 60).

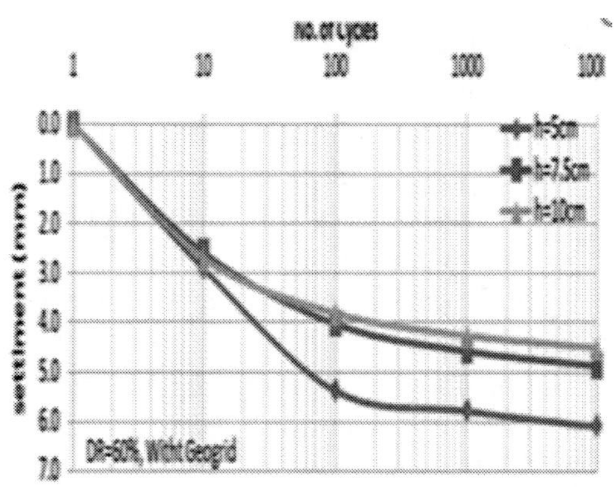

Figure A.6. The effect of dynamic load on the bearing capacity with different subbase layer thickness (RD = 60).

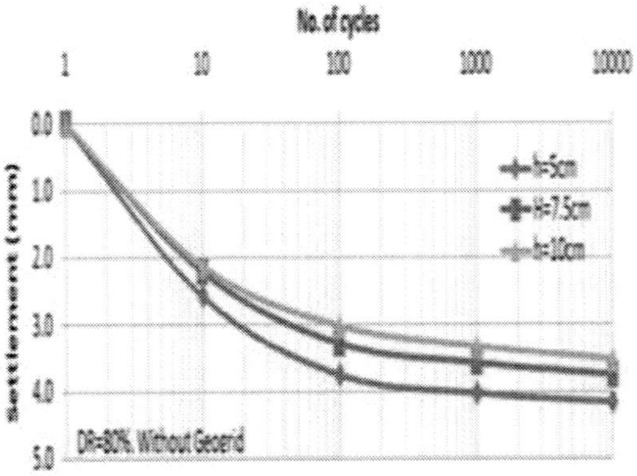

Figure A.7. The effect of dynamic load on the bearing capacity with different subbase layer thickness (RD = 60).

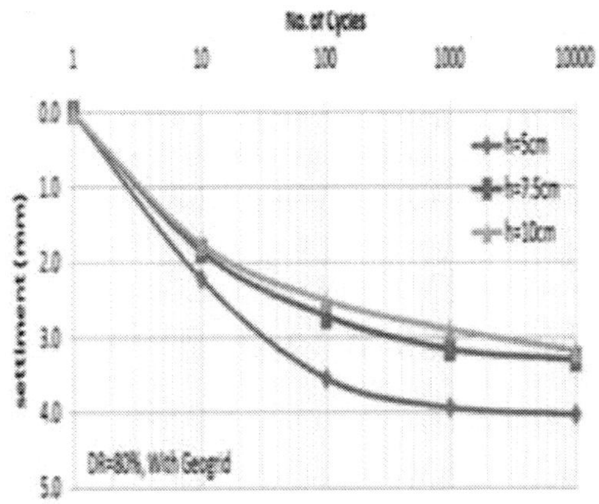

Figure A.8. The effect of dynamic load on the bearing capacity with different subbase layer thickness (RD = 60).

References

Calvarano, L.S., Palamara, R., Leonardi, G., and Moraci, N., 3D FEM Analysis on Geogrid Reinforced Flexible Pavement Roads, *Earth and Environmental Science,* 95, 2017.

Elleboudy, A. M., Saleh, N. M., and Salama, A. G., Assessment of Geogrid in Gravel Roads under Cyclic Loading, *Alexandria Engineering Journal (Hosted by ELSEVIER),* 56(3), 319-326, 2017.

Gali, M. L., and Nair, A. M., *Geosynthetic in Unpaved Roads,* 2014.

Giroud, J. P., and Noiray, L., Geotextile-Reinforced Unpaved Roads. *Journal of Geotechnical Engineering Division,* American Society of Civil Engineers, 107(GT9), 1233-1254, 1981.

Krishnaswamy, N. R. and Sudhakar, S., Analytical and Experimental Studies on Geosynthetic Reinforced Road Subgrade, *Journal of Indian Road Congress,* 66(1), 151-200, 2005.

Koerner, R. M., *Designing with Geosynthetic,* 5th Edition, Prentice-Hall Inc. Englewood Cliffs, NJ, 2005.

Mousavi, S. H., Gabr, M. A., and Borden, R. H., Optimum Location of the Geogrid Reinforcement in Unpaved Road, *Canadian Geotechnical Journal,* 2017.

Iraqi Standard Specification, Standard Specification For Roads and Bridges/Republic of Iraq, Ministry Of Housing and Construction State Organization of Roads and Bridges R6E, 1983.

Terzaghi, *K., Theoretical Soil Mechanics,* Wiley, New York. *(1943).*

Interdependence between Structural Engineering and Construction Management
Edited by Ozevin, D., Ataei, H., Modares, M., Gurgun, A., Yazdani, S., and Singh, A.

ISBN: 978-0-9960437-6-2

INVESTIGATION ON THE EFFECT OF COUNTERMEASURES FOR SUBSIDENCE AT THE APPROACHING AREAS OF ABUTMENTS

YUKIHIDE KAJITA[1], KAZUKI ONODA[1], TAIJI MATSUDA[1], KUNIHIKO UNO[2], and TAKESHI KITAHARA[3]

[1]*Dept of Civil and Structural Engineering, Kyushu University, Fukuoka, Japan*
[2]*Institute of Technology, Penta-Ocean Construction Co. Ltd., Tochigi, Japan*
[3]*Dept of Science and Engineering, Kanto Gakuin University, Yokohama, Japan*

The purpose of this study is to confirm the effect of soil improvement methods on preventing ground subsidence at the back of abutments. Earthquake seismic analysis is performed for three models. One is a model with no ground improvement. Next is a model with deep mixing method. The third is the model with lightweight banking method. As a result, from the perspective of the passage possibility of the emergency vehicles, both the deep mixing method and lightweight banking method are effective in preventing ground subsidence at the approaching area of abutments. However, in the case of the deep mixing method, it is found that the maximum bending moment of the pile under the liquefaction layer increases because a lump of improved rigid soil that falls down toward the piles.

Keywords: Ground subsidence, Liquefaction, Structure-soil interaction, Deep mixing method, Lightweight banking method, Effective stress analysis.

1 INTRODUCTION

The subsidence of the soil at the approaching area of the abutment and the horizontal movement of the abutment were observed as shown in Figure 1 in 2011 off the Pacific Coast of Tohoku earthquake and the 2016 Kumamoto earthquake. This is because the lateral flow was generated in the soft ground due to the severe ground excitation (Japan Society of Civil Engineers 2017). If the amount of the subsidence was large, the emergency cars would not be able to pass through the bridge just after a severe earthquake. Some types of ground improvement methods have been introduced to improve the ground stiffness hardness or reduce the surcharge load. However, previous studies have not mentioned how the abutment and the back embankment of the abutment behave when an earthquake occurs. The purpose of this study is to confirm the effect of soil improvement methods for prevention of ground subsidence at the back of the abutment. The deep mixing method which improves the liquefaction layer and the lightweight banking method which reduces surcharge load and soil pressure is adopted in this study. In this research, the effect of the deep mixing method and the lightweight banking method for subsidence of soil at the approaching area of the abutment is investigated by performing seismic response analysis.

Figure 1. Ground subsidence at the back of abutment.

2 THE OUTLINE OF TWO-DIMENSIONAL EFFECTIVE STRESS ANALYSIS

2.1 The Modelling of Structures and Ground

In this analysis, FLIP (Finite element analysis program of liquification process/response of soil-structure systems during an earthquake), which is a dynamic effective stress analysis program is used (Morita *et al.* 1997, Iai *et al.* 1992). Figure 2 shows the numerical model in this analysis. The girder (orange line) is connected to the pier rigidly. The clearance between the girder and the abutment is set to be 10 cm. The ground consists of 5 layer types. The physical properties of each layer are shown in Table 1.

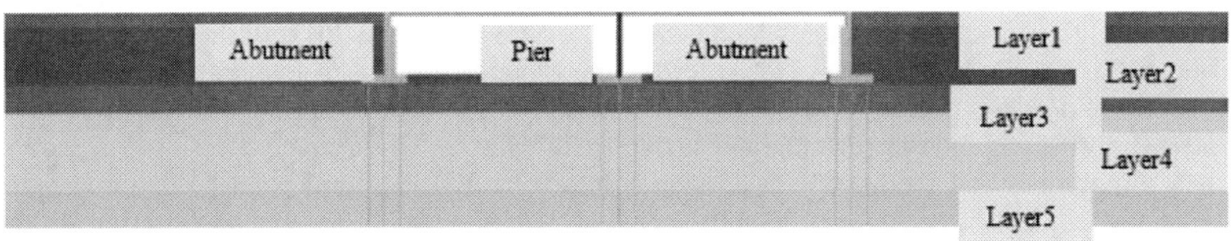

Figure 2. Numerical model in this analysis.

In Table 1, Layer 1 and Layer 2 are the ground at the back of the abutment. Layer 3 is the ground just under the foundation of the abutments and the pier. In this analysis, it is assumed that only Layer 3 is a liquefaction layer. So, the excess pore water pressure ratio at Layer 3 is calculated. The excess pore water pressure ratio has defined the ratio of the excess pore water pressure to the initial effective mean stress. This analysis is conducted under the undrained condition. The moment-curvature relationship of the piles is defined as the bi-linear model as shown in Figure 3. The foundation piles are steel tube piles whose diameter is 800mm, and thickness is 12mm. The foundation piles are arranged three in the bridge-axial direction, four in the orthogonal direction. The foundation piles of the abutment and the ones of the pier is the same material and size. The bottom of the numerical model in Figure 2 is the surface of the engineering base. In this model, the node of the end member for the steel pile is located at the bottom of the numerical model. That means the movement of the end member for steel pile is the same as the movement of the engineering base surface.

Table 1. The physical properties of each layer.

	Thickness of a layer	Average effective constraing pressure	Density	N-value	Internal friction angle
	m	kPa	ton/m^3		degree
layer1	2.1	13.90	1.8	5	38
layer2	7.9	80.12	1.8	10	37
layer3	5	147.15	2	2	30
layer4	10	215.82	2	10	33
layer5	5	353.16	2	20	36

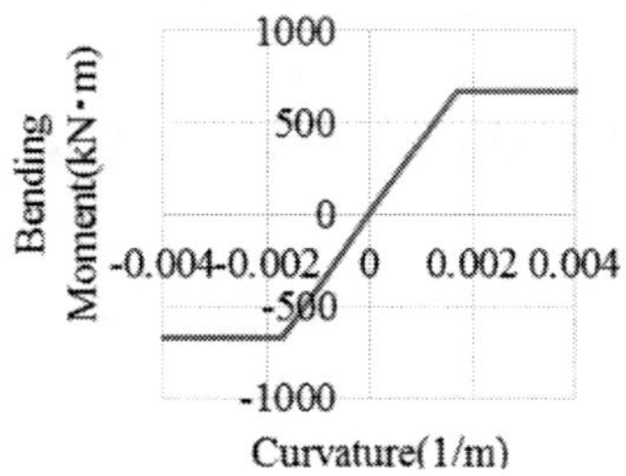

Figure 3. Moment-curvature relationship of piles.

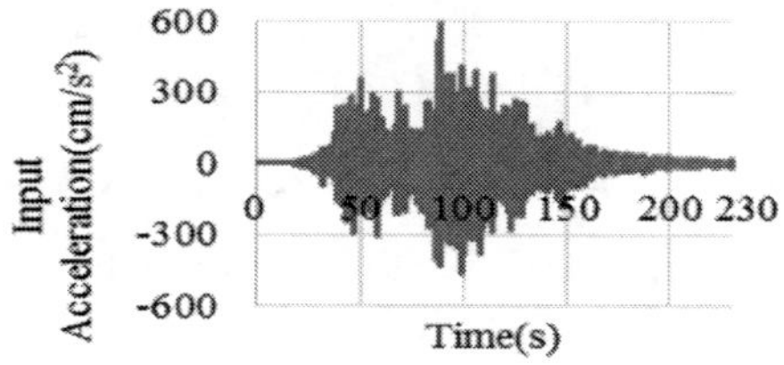

Figure 4. Acceleration at the engineering base.

2.2 Input Earthquake Motion at the Engineering Base

Firstly, the ground surface motion, which is the observation record in 2011 off the Pacific coast of Tohoku Earthquake is prepared. This record is the ground surface acceleration of the east-west direction and this is recorded in Soma city, Fukushima Prefecture. The Observation point is FKS001 by Strong-motion seismograph networks in Japan. The characteristic of this ground surface motion is that the maximum acceleration is over 300cm/s^2 and both the low and high-frequency components are considerably included. In this analysis, the acceleration at the engineering base is needed. So, the acceleration at the engineering base is calculated by using one-dimensional wave propagation analysis. Figure 4 shows the input wave at the engineering base.

2.3 The Case of Analysis

Two types of ground improvement methods are investigated in this study. One is the deep mixing method and the other is the lightweight banking method. So, in this paper, 3 cases of analytical models are prepared. Case1 is the model with no ground improvement. Case2 is the model with the deep mixing method. Case3 is the model with the lightweight banking method. The Deep mixing method is a ground improving method for solidifying ground by mixing the improved material and the soft soil. By conducting a lightweight banking method, reduction of surcharge load and soil pressure to the back of abutment prospects. Therefore, the reduction of subsidence at the approaching area of the abutment is also prospected by reduction of lateral movement of the abutment.

2.4 The Model of Deep Mixing Method

In this research, the deep mixing method is conducted to liquefaction layer for reducing the subsidence of soil at the approaching area of the abutment. Improving area is determined by the past technical report (Public Works Research Institute 2010). The improving area at the back of the left side abutment is shown in Figure 5. In Figure 5, the green area surrounded by red line is the improving area by deep mixing method. In this study, the improving area is modeled as a solid material like cement mortar. This is because the improving area is solidified. So, the improving area of the deep mixing method is modeled by the liner plane element. The physical property is decided from the result of the past research (Yamamoto *et al.* 2006).

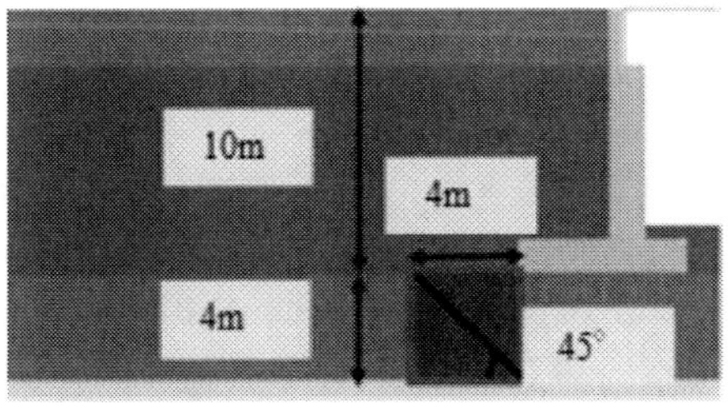

Figure 5. The improving area of deep mixing method.

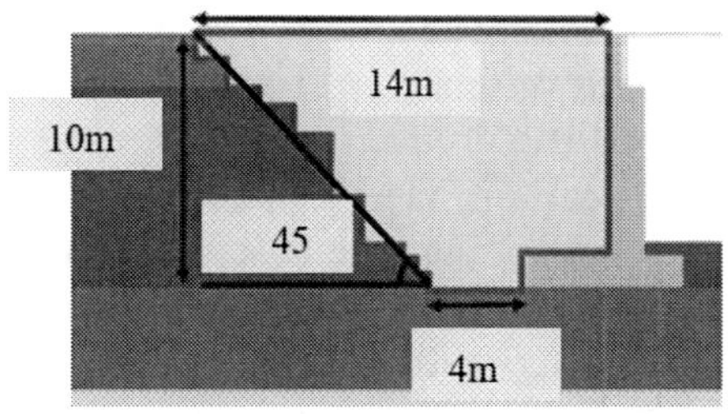

Figure 6. The improving area of lightweight banking method.

2.5 The Model of Lightweight Banking Method

The physical properties of the lightweight banking method are decided from the test data of the private company (Taiheiyo Cement Corporation 2018). The density of the lightweight banking is about 60% of the original ground. The place replaced by lightweight banking is shown in Figure 6. In the case that the lightweight banking method is conducted, the density and the stiffness are changed from the original ground. In this analysis, we assumed that the lightweight soil ground exists from the beginning. In real site situation, when the lightweight banking method is adopted, the original soil ground has already existed, and then we replace the original soil with the lightweight soil. Comparing these two cases, that means the case that the lightweight soil is existed from the beginning and the case that the lightweight soil is replaced the original soil with, the surcharge load at the liquefaction layer might be different. This difference is ignored in this research.

3 ANALYTICAL RESULTS AND DISCUSSIONS

3.1 The Amount of Vertical Gap

The contour figure of the excess pore water pressure ratio of Case 1(no ground improvement) is shown in Figure 7. When the excess pore water pressure ratio is close to 1, the color turns to be from blue to red. Figure 8 shows the time history of the excess pore water pressure ratio in the middle of Layer 3. As shown in Figure 8, the excess pore water pressure ratio is rising sharply from 40 seconds in all cases which is the time that the acceleration of the input wave shown in Figure 4 starts to be large. The excess pore water pressure ratio is closer to 1 at the end of analysis in all case. In other words, the liquefaction occurs in Layer 3. The yellow points in Figure 7 are the node of the top of the abutment and the node of ground which is located at the 4.5m back of the abutment. In this study, the vertical gap is defined as the vertical relative displacement between two points. Figure 9 presents the time history of the vertical gap at the right-side abutment. If the node of back fill soil ground is located below against the node of the abutment, the vertical gap is a positive value. From Figure 8 and Figure 9, when the excess pore pressure ratio becomes 0.8, the amount of vertical gap increases sharply. And it is confirmed that the horizontal displacement of the right abutment also becomes the maximum value. This is because the stiffness of the liquefaction layer is starting to be reduced. Figure 10 shows the vertical gap at the end of the analysis. As shown in Figure 10, the node of backfill soil is located above against the top of the abutment in Case 3 (The lightweight banking method). This is because the surcharge load of liquefaction layer under the replaced area is reduced. The soil just under the abutment foundation might be easy to move to the area just under the replaced area. So, the abutment might sink in contrast with the backfill soil of the abutment. It is reported that the traffic speed of the emergency vehicle is limited when the vertical gap is 10cm and more (Tokida and Oda 2009). Therefore, in this research, 10 cm is defined as the limiting vertical gap. The vertical gap exceeds the limiting vertical gap in Case1, which is the model with no ground improvement. In the other two cases, the vertical gap is smaller than the limiting vertical gap. The absolute value of the vertical gap of the model with deep mixing method (Case 2) is smaller than the absolute value of gap of the model with lightweight banking method (Case 3). In this analysis, the liquefaction is the main factor of the occurrence of the vertical gap. The deep mixing method improves the liquefaction layer directly. Therefore, the deep mixing method has a great effect on preventing ground subsidence at the approaching area of abutments.

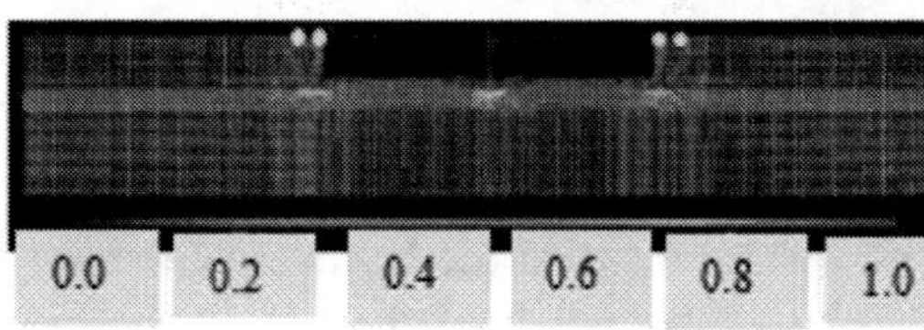

Figure 7. The excess pore water pressure ratio in Case 1 (no ground improvement).

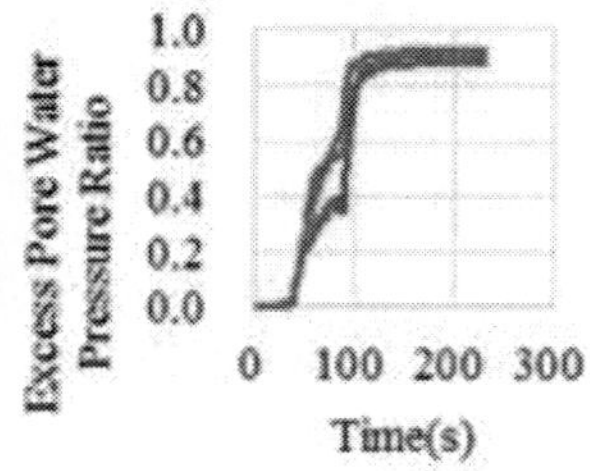

Figure 8. The time history of excess pore water pressure ratio.

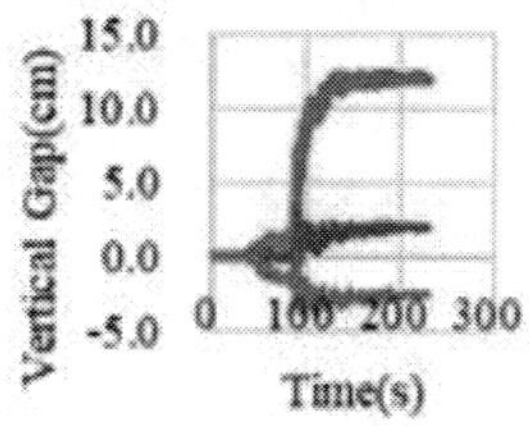

Figure 9. The time history of vertical gap.

3.2 The Response of Piers

Figure 11 and Figure 12 present the maximum bending moment and the maximum horizontal movement of the foundation pile which is located at the center of the right-side abutment. The value of the bending moment and horizontal movement is the absolute value. The yield bending moment is set to be 670 kNm in this model. The bottom surface of the abutment foundation is located at 20m upward from the bottom of this numerical model. And the thickness of the liquefaction layer (Layer 3) is 4m. So, 16m from the bottom of the numerical model indicates the boundary between Layer 3 (Liquefaction layer) and Layer 4 (Non-liquefaction layer). As shown in Figure 11, the bending moment reaches the yield value from 12.5m to 15.5m and from 18.5m to 19.5m from the bottom. However, the bending moment doesn't reach the yield value from 15.5m to 18.5m from the bottom. This decrease in bending moment is caused by liquefaction. From Figure 12, the horizontal displacement is almost the same from the bottom to 16m. The displacement increases sharply from 16m due to the liquefaction layer (Layer 3). The displacement at the top of the pile in Case 2 (the deep mixing method) is smaller than the other two cases. Therefore, it is thought that the deep layer mixing which improves liquefaction layer directly has an effect on the response of the piles. As shown in Figures 11 and 12, the bending moment in Case 2 is larger than the other two cases and the displacement is a little larger under the liquefaction layer. It is confirmed that the improving area moves to diagonally forward downward due to the surcharge load. It is thought that this displacement causes the increasing of the bending moment of the pile.

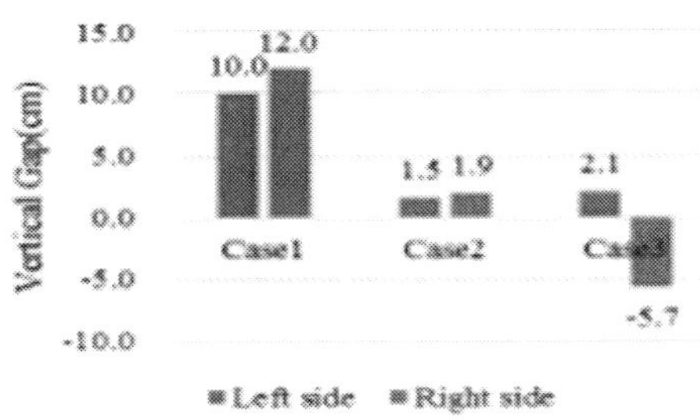

Figure 10. Final vertical gap.

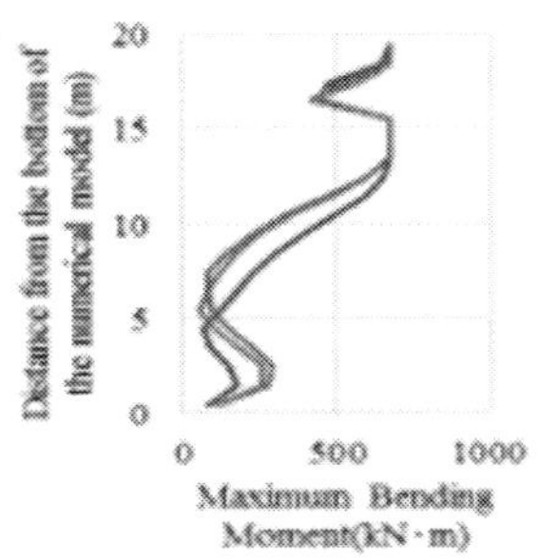

Figure 11. Maximum bending moment.

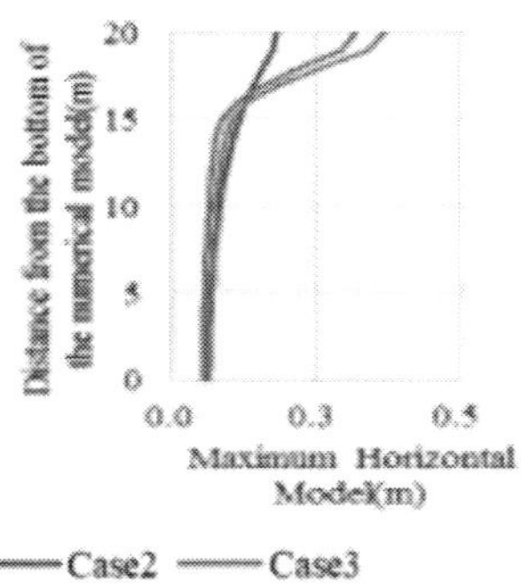

Figure 12. Maximum horizontal displacement.

4 CONCLUSIONS

In this research, the seismic response analysis which is considered the interaction between ground and bridge is performed for investigating the behavior of abutment and background and confirming the effect of ground improvement.

(i) The vertical gap in no ground improvement model is larger than 10 cm. However, the vertical gap in other cases is smaller than 10 cm. From the perspective of the passage possibility of the emergency vehicles, these two improvement method has the effect for prevention of ground subsidence at the back of the abutment.

(ii) In the deep mixing method model, the bending moment of the pile and the maximum horizontal displacement increases under the liquefaction layer. This is because the improvement area (the rigid solid area) moves to diagonally forward downward and then the improvement area pushes the soil and pile under the liquefaction layer.

(iii) In this research, input earthquake acceleration data is one case. Therefore, the increase of input earthquake data and three-dimensional analysis are future topics of discussion.

Acknowledgments

This work was supported by JSPS KAKENHI Grant Number JP25KO6182.

References

Iai, S., Matsunaga, Y., and Kameoka, T., Analysis of Undrained Cyclic Behavior of Sand under Anisotropic Consolidation, *Soils and Foundations*, 32(2), 16-20, June 1992.

Japan Society of Civil Engineers, *Report on Damage Surveys and Investigations Following the 2016 Kumamoto Earthquake* (in Japanese), Japan, 2017.

Morita, T., Iai, S., Liu, H., Ichii, K., and Sato, Y., Simplified Method to Determine Parameter of FLIP (in Japanese), *Technical Notes of The Port and Harbor Research Institute*, 869, 1-36, 1997.

Public Works Research Institute, Investigation on The Guideline of Countermeasures for Lateral Flow of Abutments (in Japanese), 2, *Technical notes of PWRI 4174*, Japan, 2010.

Taiheiyo Cement Corporation, Artificial Light Weight Filling Material (in Japanese), 2018. Retrieved from http://www.taiheiyo-cement.co.jp/service_product/aggregate/pdf/keiryo-morido2018.pdf.

Tokida, K., and Oda, K., Future Views on Seismic Performance-Based Design Reinforcement and Management of Road Embankment (in Japanese), *Journal of Japan Society of Civil Engineers*, 65, 857-873, November 2009.

Yamamoto, Y., Takahashi, N., and Kurokawa, Y., Analytical Study on Liquefaction Countermeasure Effect of the Pile Type Deep Mixing Method (in Japanese), *Technical Report of The Research Institute of Mitsui-Sumitomo Construction*, 45-53, Japan, 2006.

Interdependence between Structural Engineering and Construction Management
Edited by Ozevin, D., Ataei, H., Modares, M., Gurgun, A., Yazdani, S., and Singh, A.

ISBN: 978-0-9960437-6-2

SLOPE STABILIZATION USING LOW DENSITY POLYETHYLENE

HARSHIT VARMA, DARSHAN SHAH, DEEPAK SHARMA, and SAAHIL WADHWANI

Dept of Civil and Environmental Engineering, California State University Fullerton, Fullerton, USA

Landslides still concern transportation agencies. In 2017, the Big Sur coastline roadway in Monterey, California was buried under a landslide that displaced over 6 million cubic yards of mud. Though slope stabilization techniques have been successfully used in the past, the Big Sur landslide has highlighted the necessity for innovation. Through this paper, a new slope stabilization technique is proposed using Low Density Polyethylene (LDPE) pins, used tires and vetiver grass. The paper includes a comparison of a retaining wall (a traditional slope stability alternative) with the proposed new technique. This includes stability analyses considering traditional and the proposed method. The analysis includes checking for overturning, sliding and bearing capacity failure for retaining wall and slope stability analysis for the proposed slope stabilization. The paper also includes cost comparison between the two alternatives. The results indicate that the proposed technique is stable and very comparable to traditional methods. Because the method uses waste material the proposed method costs significantly less than traditional approaches.

Keywords: Retaining walls, Vetiver grass, Used tire, Landslide.

1 INTRODUCTION

Rockfall, slope failure, shallow debris movement, or a combination of all three disturbances can be deemed a landslide. Natural and man-made factors can trigger landslides, including but not limited to, gravity, earthquakes, weather changes, erosion, and mining. Landslides can inflict significant economic losses, damage to infrastructure, loss of life and property, altering natural landscapes and impacts to various ecosystems.

In this research, special emphasis has been made on promoting green alternatives. Millions of tires need to be managed to be recycled as reusable products, but a significant portion of those unrecycled used tires end up in landfills. Similarly, plastic waste needs treatment to be transformed in to reusable form to ensure that they do not end up untreated and eventually affect the environment. Agencies are concerned about dealing with such waste. In this research, the waste materials are being proposed for use to help agencies deal with waste and replace the traditional material to reduce cost.

In this research, an innovative approach is proposed to stabilize slopes in landslide-prone areas. This paper includes stability analyses of the proposed technique to ensure stability and appropriateness of the proposed method. The proposed method uses Low-Density Polyethylene (LDPE) pins and vetiver grass to stabilize slopes and tire wall for retaining slopes. Big Sur landslide is used as a case study which occurred between Carmel Highlands and Santa Lucia Mountain moving about a million tons of rock and dirt on a coastal highway and into the sea.

The results indicate that the proposed method is stable and economically cheaper than the traditional method.

1.1 Components of the Proposed Technique

The proposed method has three components as follows:

(i) LDPE Pins: LDPE is made out of waste plastic. The material is inert and does not react easily with the natural environment. The pins can be nailed using special equipment such as drilling rigs (by Davey-Kent) and hammer drills (by Krupp HB28A Hydraulic). This equipment have been used in Texas for Red and Beaumont clays in Texas (Khan, 2013) Driving LDPE pins displace soil particles to densify treated soil mass (Loehr *et al.* 2013).

(ii) Tire Walls: A wall made by arranging used tires against the sloped surface can withstand earth pressure from behind the wall (lwhome.org). Constructing tire retaining wall will require filling tires with high-density clay and then compacting each tire with a sledgehammer or pneumatic tamper.

(iii) Vetiver Grass: These are a special species of grass and characterized by deep roots. These grasses can be helpful in several ways including sloe stabilization (Truong *et al.* 2008, Gordon and Sherar 2018) that can be used to benefit the engineering community. These grasses will be able to hold the top layers of soil and prevent soil erosion and slope stability (Ni *et al.* 2018).

2 LITERATURE REVIEW

Literature review shows that several slope stabilization methods are proposed. Harabinova (2017) conducted a slope stability assessment and found the most unfavorable slip surface. The researcher proposed reinforcing method using geogrids and ground anchors. Pons *et al.* (2018) assessed the stability of earth retaining wall and suggested gabion walls for heights 1m to 6m. For retaining earth with heights greater than 4.5m, the authors recommended cantilever retaining walls. Zamiran *et al.* (2018) analyzed the seismic response of cantilever retaining wall. They used the finite difference method and Newmark sliding block method for their research. They found that the displacement of a retaining wall is dependent on the cohesive property of backfill material and even a small variation in cohesion will influence displacement of retaining walls. Tang *et al.* (2018) investigated stabilizing slopes using porous concrete and implanting grass and found that the method was very effective under tension and stresses, which prevented slope failure. Broda *et al.* (2016) researched the effectiveness of wool geotextile for protecting steep slopes. The geotextiles provided protection against water erosion and landslides. However, degradation of geotextile caused the reduction of mechanical strength and led to the deterioration of geotextile. This research builds upon the existing body of knowledge and proposes a method that replaces traditional retaining walls with the used tired wall, LDPE and vetiver grass.

3 STABILITY ANALYSIS

In this research, the analysis is based on the details provided by Arora (2005) and Das (2016). The overall analysis can be divided into three parts; a) overturning, b) sliding, and c) bearing capacity failure analysis for tire wall retaining wall. The analysis is also conducted for slope stability considering natural slope and reinforced slope. The overturning analysis ensures that the structure overcomes the overturning moment created by the soil pressure behind it. The sliding analysis enables ensuring that a retaining structure does not slide due to backfill pressure. The bearing capacity failure analysis ensures that the vertical load exerted from the wall and the backfill is not excessive. The slope stability analysis tells us if the slope is stable against failure.

For conducting the analysis, soil is tested for index properties, and slope profile is obtained. This includes determining the internal angle of friction (ϕ), cohesion (c) and the angle of backfill slope (i). Further, horizontal and vertical stresses in the backfill soil mass are required. This needs determining horizontal stress (σ_h) and vertical stress (σ_v) on backfill soil particles. As per Das (2016), vertical stress can be calculated as $\sigma_v = \gamma Z Cos(i)$ where γ is the unit weight of the soil, Z is the height and i is the angle of backfill slope. Horizontal stress can be calculated as $\sigma_h = K_a \sigma_v$, where K_a is the coefficient of lateral earth pressure and can be calculated as $K_a = \left(\frac{1-\sin\phi}{1+\sin\phi}\right)$ for backfill slope greater than 15^0 (Das 2016). For overturning analysis, moments must be calculated that could topple a retaining structure. All the overturning moments ($M_{O1} + M_{O2} + ... + M_{On}$) acting on a retaining structure are added to obtain Overturning Moment M_O (i.e., $M_O = M_{O1} + M_{O2} + ... + M_{On}$) (Das 2016). Similarly, resisting moments are calculated and added to obtain M_R (i.e., $M_R = M_{R1} + M_{R2} + ... + M_{Rn}$). Taking a ratio as M_R/M_O provides a Factor of Safety for Overturning (FOS_O) (Das 2016). As per Das (2016), FOS_O of 2 to 3 is considered safe. For bearing capacity analysis, the distance ($\overline{X}$) can be calculated as $\overline{X} = \frac{\sum M - P_h * \bar{y}_n}{\sum V}$, where M is summation of total moments, P_h is horizontal pressure acting on the retaining structure ($P_h = P_a Cos(i)$), $\bar{y}_n$ is distance of load acting from the base and $\sum V$ is summation of all vertical forces. Further, eccentricity (e) can be obtained as $e = \frac{B}{2} - \frac{M_R - M_o}{\sum P_V}$, where B = top width of the retaining wall and P_v is the vertical pressure (Das 2016). The eccentricity obtained is compared with B/6 to determine if the bearing capacity is less or greater than desired. If it comes less, the allowable maximum bearing capacity (q_{max}) can be calculated as $q_{max} = \frac{\sum P_V}{B}\left(1 + \frac{6e}{B}\right)$ (Das 2016). The FOS for bearing capacity (FOS_B) can be obtained as FOS_B = (Safe Bearing Capacity)/q_{max} and a value of FOS_B between 2 to 3 is considered safe (Das 2016).

The sliding analysis requires calculating all the resisting forces (F_R). The total F_R forces collectively resist all the driving forces (F_D) generated by the backfill in the horizontal direction. This can be done by using the formula $F_R = \sum F_r$ generated from different weight component taken into consideration. The factor of safety for sliding can be calculated as $FOS_S = \frac{\sum F_R}{\sum F_D}$. As per Das (2016), the FOS_S should be greater than 1.5.

Lastly, the stability analysis of natural slopes includes finding FOS using infinite slope method. FOS against sliding can is given by $FS = \frac{s}{\tau.} = \frac{c + ((\gamma H * COS2i) \tan\phi)}{\gamma H * \cos i * \sin i}$ (Arora 2005). For determining FOS for reinforced stabilized slopes the approach proposed by Bhuiyan (2014) has been used $\frac{Fs = cL + h L \gamma \cos 2i * \tan\phi + (\frac{L}{s} + 1) * P}{hL * \sin i * \cos i * \gamma sat}$ where L is the length of slip surface, s is the LDPE spacing, P is mobilized load and h is depth of slip surface $= \frac{D}{\cos i}$

For this research, data was assumed to match the Big Sur area that failed in 2017. For the tire wall stability analysis ϕ was assumed to be 30°. Other assumptions were made considering that a traditional RCC retaining wall is constructed in the area to prevent landslides. The assumptions include backfill slope angle (i) = 16, soil-wall friction angle (δ) = 25, top width of the retaining wall (W) = 0.984 ft, height of retaining wall (H) = 16.404 ft, width of base slab (2/3H) = 10.941 ft, depth of cantilever base (0.1H) = 1.640 ft, unit weight of soil (γ) = 375.93 psf, unit weight of concrete = 501.25 psf, width of stem at base = 1.6404 ft, extension of base slab = 1.6404 ft, depth (D) of base wall below ground = 2.64 ft, extension of base slab at toe = 7.660 ft, height of backfill beyond structure = 2.076 ft, allowable soil pressure = 10445 psf. The assumed details are displayed in Figure 1.

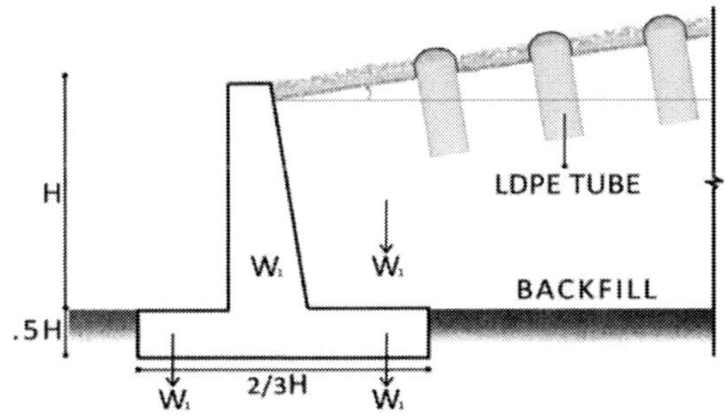

			Moments	
Item	Forces	Leverage	Overturning	Resisting
W1	36	0.85		30.60
W2	12	0.63		7.60
W3	40.02	1.67		66.73
W4	210.15	2.17		455.50
W5	13.36	2.56		34.16
Pv	23.53	3.34		78.48
Ph*	82.07	1.88	154.17	
		Total	154.17	673.08

* represents horizontal force, all others are vertical forces

Figure 1. Traditional retaining wall and forces calculation.

As an alternative, a new soil stabilization method was proposed for the same location. The assumptions made for the proposed rectangle shaped tire-wall slope stabilization include height of tire wall = 16.404 ft, base of tire wall = 16.404 ft, backfill slope angle (j) = 16°, top width of tire wall = 2.833 ft, depth of tire base = 1.6404 ft, unit weight of soil for filling tire = 375.93 psf, angle of tire wall friction (δ') = 27, allowable soil pressure (qna) = 10445 psf.

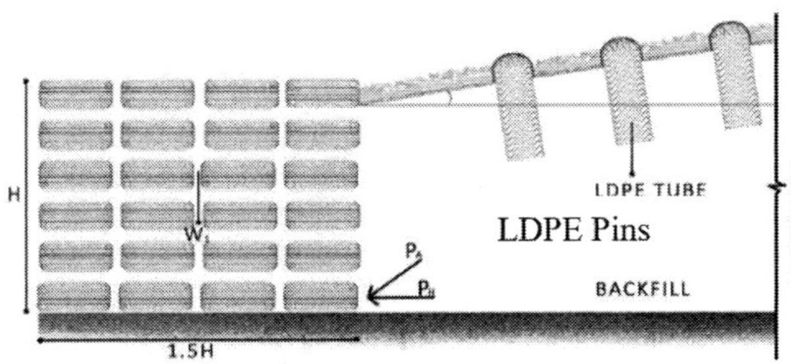

FOR RECTANGULAR SECTION

			Moments	
Item	Forces	Leverage	Overturning	Resisting
W1	450.00	2.50		1125.00
Pv	23.53	5.00		117.66
Ph*	82.07	1.83	150.45	
Total			150.45	1242.66

* represents horizontal force, all others are vertical forces

Figure 2. Proposed rectangular tire wall & force calculation.

For the slope stability analysis slope height is assumed to be 40 ft giving us L=126.5 ft, s is considered to be 3 ft, and P is 725 lb/ft.

4 RESULTS

Both the proposals were compared by conducting the overturning, sliding and bearing capacity calculations. These are detailed in Table 1. Also, the slope stability analysis of natural slope gave us a FOS of 0.76 against sliding, but with the RPP reinforcement, the FOS improved to 3.99. Table 1 shows that the proposed tire wall with a rectangular cross-section (shown in Figure 2) passes the required checks. When implemented, the wall will appear as shown in Figure 3.

The tire wall can be constructed by placing used tires in any masonry bond patterns and filled with dense earth. This approach of constructing tire retaining wall pursued at a few places in the US (Source: lwhome.org) and in order to ensure that the tires act monolithically the tire walls can be treated with methods such as shotcrete, stucco, vegetative covering, geo-fabric, concrete blocks, and concrete blocks (Hossain and Jayawickrama 2000).

From calculations in Table 1, the tire retaining wall will support the backfill material. Also, it is believed that the backfill soil will be loosely packed and will require densification. This can be achieved by nailing LDPE pins into the loose soil. This approach has been used successfully by (Loehr *et al.* 2013). In addition, vetiver grass (a deep rooted grass and a xerophyte that requires relatively low water to sustain (Mekonnen, 2000)) is proposed to be sown to impart extra strength by holding the soil particles and to be able to prevent soil erosion in case of precipitation (Truong *et al.* 2008). The grass will also act as a natural barrier to soil erosion due to the wind.

Table 1. Comparing the RCC retaining wall with the proposed tire retaining wall.

Steps	RCC Retainting Wall	Rectangular Section, Tyred Retaining Wall
Determining Ka	0.379	0.38
Overturning Check Calculations		
Active Pressure (kN), Pa = 0.5KaγH^2 =	85.373	85.37
Active Pressure (kN), Ph = Pacos(i)=	82.066	82.07
Active Pressure (kN), Pv = Pasin(i)=	23.532	23.53
Overturning Moments =	154.170	150.45
Resiting Moments =	673.076	1242.66
Factor of Safety (FOS_O) =	4.366	8.26
Sliding Check Calculations		
Summation of Vertical Forces =	335.064	473.53
Summation of Horizontal Forces =	82.066	82.07
Factor of Safety (FOS_S) =	1.904	2.94
Bearing Capacity Check Calculations		
Xbar = ∑M/∑V =	1.549	2.31
e = (b/2) - Xbar =	0.119	0.19
b/6 =	0.556	0.83
e < b/6	Yes => OK	Yes => OK
pmax =	121.947	116.70
pmin =	78.991	72.72
Factor of Safety (FOS_B) =	4.100	4.28

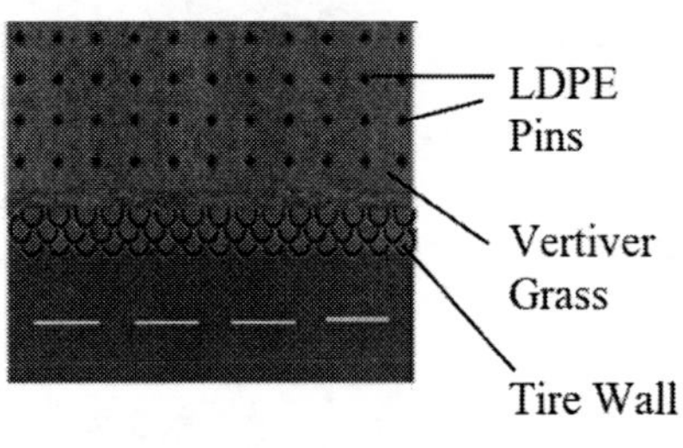

Figure 3. Top view and axonometric view of the proposed slope stabilization.

The cost of the tire wall is found to be lower than traditional retaining wall. For a one mile long and 16.404 ft high RCC retaining wall, the cost of construction will be $2,956,800 (i.e. 5280 ft x 16.404 ft x $35/sq ft). Conversely, a used tire costing $1 (sold at scraptire.net) will cost $74,895 for the same area needing 74,895 tires (i.e. (5280 ft /1.41 ft dia)* 20 layers). Assuming that we use 70,000 LDPE pins at a spacing of 3 feet in XY direction and considering a slope height of 40 ft, an area of approximately 211,200 sq. ft can be stabilized along the 5280 ft stretch of tire retaining wall. This will cost us $340,900 at the cost of $4.87 per sqft. Lastly planting vetiver grass will cost $7000 (at $3 per pound (available at Molokai Seed Company) and used at a rate of 10 pounds per 1000 sq ft at an area of 211,200 sqft). This totals to $422,795 (i.e. 74,895 +

340,900 + 7,000) which is just about 80% cost saving when compared with the cost of traditional retaining wall.

5 CONCLUSIONS

A landslide can trigger tremendous loss to life and property. Government agencies have successfully stabilized slopes in the past. However, some landslides like the case of Big Sur have highlighted the need for innovation. The new method should be reliable and cost-effective and should be able to deal with all the issues. In this research, a new method is proposed which uses LDPE, used tires and vetiver grass. The proposed design was primarily focused on preventing landslides using waste materials. Theoretically, the research results suggest that the proposed retaining wall will be able to safely retain slopes safely. It can be observed from the provided calculations that the stability and factor of safety are relatively good when compared with traditional retaining walls. The proposed model in the research is cost effective and can be used as an alternative to the traditional methods.

While the method is theoretically acceptable, further research is required to determine the effect of pore pressure on overall stability. Also, the use of tires could be unpleasant for some and may require additional methods to cover it. Weathering effects on tire such as erosion, excessive heating, rubber smell, and leaching are yet to be studied. All these will require conducting a pilot project and testing the outcomes in real life. From an application perspective, the proposed method has a lot of future potential

References

Arora, K. R., *Soil Mechanics and Foundation Engineering,* Sixth Edition Reprint, Standard Publishers Distributors, Delhi, ISBN: 81-8014-028-8, 2005.

Bhuiyan, H. R., *Group Resistance of Recycled Plastic Pin in Sustainable Slope Stabilization*, Thesis, University of Texas (Arlington), TX, USA. 2014

Broda, J., Grzybowska-Pietras, J., Gawlowski, A., Rom, M., Przyblo, S., and Laszczak, R., Biodegradation of Sheep Wool Geotextiles, *J. of International Biodeterioration & Biodegradation,* 115, 31-38, 2016.

Das, B., *Principles of Foundation Engineering,* Eighth Edition, ISBN-13: 978-1-305-08155-0, 2016.

Gordon, K., and Sherar, J., Best Management Practices Field Guide, Low Volume Roads Engineering Program, US Forset Service. Retrieved from https://www.fs.fed.us/t-d/programs/forest_mgmt/projects/lowvolroads/ on April, 11, 2018.

Harabinova, S., Assessment of Slope Stability on the Road, *J. of Procedia Engrg.,* 190, 390-397, 2017.

Hossain, S., and Jayawickrama, P. W., *Use of Whole Tires in Earth Retaining Structures,* Research Report Number 0-1876, Center for Multidisciplinary Research in Transportation, Texas Tech Uni., 2000.

Loehr, J. E., Bowders, J. J., Owen, J. W., Sommers, L., and Liew, W., Stabilization of Slopes Using Recycled Plastic Pins, *Journal of the Transportation Research Board,* Washington D.C. 2013.

lwhome.org, Tire Retaining Wall. Retrieved from https://www.lwhome.org/tire-retaining-walls-1 on November 4th, 2018.

Mekonnen, A., Hand Book on Vetiver Grass Technology, 2000, GTZ IFSP S/GONDER, Ethiopia. Retrieved from https://www.vetiver.org/ETH-handbook.htm on February 25th, 2019.

Ni, J. J, Leung, A. K, Ng, C. *W. W,* and Shao, W., Modelling Hydro-Mechanical Reinforcements of Plants to Slope Stability, *Computer and Geotechnics,* 95, 99-109, 2018.

Pons, J. J., Pia, V. V, Yepes, V., and Marti, J. V., Life Cycle Assessment of Earth-Retaining Walls: An Environmental Comparison, *Journal of Cleaner Production,* 192, 411-420, 2018.

Tang, W., Mohseni, E., and Wang, Z., Development of Vegetation Concrete Technology for Slope Protection and Greening, *Construction and Building Materials,* 179, 605-613, 2018.

Truong, P., Van, T. T., and Pinners, E., *Vetiver Systems Application,* Technical Reference Manual, The Vetiver Network International, 2008.

Zamiran, S., and Osouli, A., Seismic Motion Response and Fragility Analyses of Cantilever Retaining Walls and Cohesive Backfill, *Journal of Soils and Foundation,* 58(2), 412-426, 2018.

Interdependence between Structural Engineering and Construction Management
Edited by Ozevin, D., Ataei, H., Modares, M., Gurgun, A., Yazdani, S., and Singh, A.

ISBN: 978-0-9960437-6-2

CLAYEY SOIL TREATED BY CEMENT AND QUICKLIME BY MICROSTRUCTURE: A MINERALOGICAL ANALYSIS

HUSAM HIKMAT BAQIR, AQEEL AL-ADILI, KAWTHER AL- SOUDANY, and ALI SHAREEF

Dept of Civil Engineering, University of Technology, Baghdad, Iraq

Soft clayey soil was treated by a combination of cement (PC) and Quicklime (LQ) in order to modify and stability. This study shows an improvement of clay soil brought from Garma Ali site in the Al Basra governorate, Iraq. The PC was added in percentages of 0, 2, 4, 6, 8, and 10%, and LQ was added to 2 and 4%, of dry weight. Also, this research used Microstructure Analysis by Scanning Electron Microscope (SEM) testing and Mineralogical Analysis by X – Ray Diffraction (XRD) testing on the examination soil treated with mix between cement (PC) and Quicklime (LQ) for the purpose of knowing the reasons for the increase in the shear strength and decrease maximum dry density. Through the micrographs that result from the scanning electron microscoping and the curves of X-ray that demonstrate presentence, the formation of the hydration reaction product (CSH gel) shows tiny bristle (rod) crystals. The CSH gel and the tiny bristle (rod) crystals worked on coating and contact the particles together. The micrographs for soils treated with (2% LQ + 10% PC) show an increase in the amount hydrated gel (CSH) compared to the soil treated with 6% PC and 2% LQ and 8% PC and 4% LQ, but the number of bristle–like crystals decreases compared to soil treated by 8% PC and 4% LQ.

Keywords: Scanning electron microscoping (SEM), X – Ray Diffraction (XRD), UUU test, CSH (calcium silicate hydrate), Dolomite (D).

1 INTRODUCTION

Random data collected from several site investigation reports demonstrated the value of undrained shear strength at less than 30 kPa in Basrah governorate and less than 40 kPa in Maysan and Dhi Qar governorates; compression indices as high as 0.3 were also reported (Rahil 2007). Lime is used primarily to dewater the soil in order to improve the workability and its bearing capacity. This Lime modification is widely used for building embankment and subgrade of clayey soil since the effect is rapid and modifies the geotechnical characteristics of the soil like the plastic limit, the shear strength and the soil compaction characteristics. Later the pozzolanic reaction between the soil minerals and lime in the presence of water leads to the formation of secondary cementitious products (C-S-H, C-A-S-H) increasing the soil cohesion and its resistance (Boardman *et al.* 2001, Muller 2005, Maubec 2010, Nichloson 2014). Cement stabilization is quick, does not need mellowing time and provides a non-leaching platform (Sariosseiri and Muhunthan 2009). Furthermore, in practice, a mixed treatment with lime and cement is used since the mixture allows one to facilitate the workability without disturbing the effect of the cement and the gain of resistance in the long term. The addition of lime and cement exerts impacts on the material microstructure, and consequently, influences pore distribution over a very short term. Yet to this day, none of these efforts have provided a complete description of the full set of physicochemical mechanisms and their impacts on both the microstructure and macroscopic behavior. With this backdrop, the objective of the present study is to describe the effects of adding lime and cement on the characteristics of the treated softy soil by employing a multi-scale approach. To explain performance trends, the effects of adding lime and cement on the soil's physical chemistry and microstructure will be characterized.

2 EXPERIMENTAL OF INVESTIGATION

2.1 Materials Used

2.1.1 *Soils*

In this study, soil samples used were obtained from the Garma Ali site about 538 km south of Baghdad in Al Basra city. The engineering properties of clayey soils are presented in Table 1, which shows the grain size distribution of soils used.

Table 1. Physical and chemical properties of natural soils used.

Index Property	**Test Standard**	**Index Valve Garma Ali Soil**
Liquid Limit (L.L) (%)	ASTM D 4318	58
Plastic Limit (P.L) (%)	ASTM D 4318	27
Plasticity Index (P.I) (%)	ASTM D 4318	31
Specific Gravity (G.s)	ASTM D 854	2.75
Gravel (larger than 4.75mm) (G) %	ASTM D 422	0
Sand (0.075 to 4.75 mm) (S) %	ASTM D 422	2
Silt (0.005 to 0.075 mm) (M) %	ASTM D 422	35
Clay (less than 0.005mm) (C) %	ASTM D 422	63
Classification (USCS)	ASTM D 2487	CH
Calcium Oxide (CaO) (%)	Chemical Analysis	21.12
SO3 Content (%)	Chemical Analysis	0.38
Total Dissolved Salt (TDS %)	Chemical Analysis	1.73
Total Solved Salt (TSS %)	Chemical Analysis	6.89
PH Value (%)	ASTM D 4972	8.69
MDD (KN/m3)	ASTM D 698	16.5
OMC (%)	ASTM D 698	20.5

2.1.2 *Types and properties of additives*

The cement used to be sulfate resistance Portland cement (type V) manufactured by (Al Jessir) factory made in Iraq. The type of lime used was quicklime (unhydrated) manufactured from limestone by the Alnoora factory in Kerbella governorate.

2.2 Laboratory Tests

A series of laboratory tests consisting of, compaction, UUU test, scanning electron microscope (SEM) and mineralogical analysis by x – ray diffraction (XRD) were conducted on clayey soils. The combinations of Portland cement and quicklime (PC and LQ) were used for stabilization of the two soils. The percentages of PC were 0, 2, 4, 6, 8 and 10%, while the percentages of LQ were 0, 2 and 4%.

3 RESULT AND DISCUSSION

3.1 Microstructure Analysis by SEM and Mineralogical Analysis by XRD

The microstructure refers to fabric that is the arrangement of the particles, particle group and pore spaces in the soil as well as cementation. The changes of microstructural development in the two tested soils due to PC and LQ addition play a significant role in the geotechnical properties and the mechanical behavior in these stabilized soils. It was noted in this study that the cohesion strength (C) in soils treated with proportions 8% PC and 4% LQ is higher than that of soils treated with 10% PC and 2% LQ. Therefore, Microstructure Analysis by Scanning Electron Microscope (SEM) and Mineralogical Analysis by X – Ray Diffraction (XRD) tests were performed for the purpose of knowing the cause of this convergence, Also, soils treated by 6% PC and 2% LQ in order to be compared with higher ratios from soils treated were examined. The scanned image was obtained in the magnification range from 50 to 20,000 times.

3.1.1 *Treated stabilized soil with 6% PC and 2% LQ*

Figure 1 illustrates a micrograph of treated soil. The micrographs of the treated soil with 6% PC and 2% LQ show crumbs of floccules with a porous nature and cementitious compounds (calcium silicate hydrate) coating the clay particles. Additionally, the reaction of the composite (PC and LQ) with clay led to the formation of aggerates of various sizes. It can be seen in both treated soils that there is a micro fabric silt – fine sand like structure. Figure 2 shows XRD pattern for untreated soil and Figure 3 illustrates XRD pattern for treating soil and with 6% of PC and 2% of LQ, respectively. For treating soil with 6% PC and 2% LQ, calcite ($CaCO_3$) shows a small increase in quartz (SiO_2) content, while feldspar and dolomite ($Ca(MgCO_3)_2$) show a slight reduction in its content. CSH has intense peaks centered on 26° and 28° for treated soil. SEM micrograph was the best technique to trace and confirm the CSH existence inside soil.

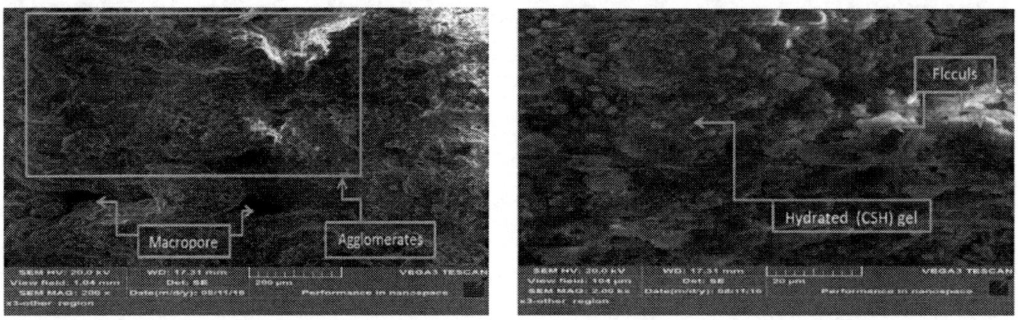

(A) (B)

Figure 1. SEM micrograph of treated soil with 6% PC and 2% LQ (A) at a magnification 200 times. (B) At a magnification of 2,000 times.

3.1.2 *Treated stabilized soil with 8% PC and 4% LQ*

Figures 4a and 4b illustrate micrographs of treated soil stabilized with 8% PC and 4% LQ. The micrograph shows aggregated arrangement due to flocculation and the formation of the hydration reaction (CSH) product (CSH gel) coating and cementing the soil particles together. In addition to treated soil, the micrograph illustrates the formation of more new bristle-like crystals. The result from XRD reveals that crystals are mainly composed of Si and Ca. The SEM images of specimens illustrate that the stabilized clay particles are coated with massive, fibrous, calcium silicate hydrated (CSH) gel. The magnified figure for treated soil shows highly fabric compared to treated soil with 6% PC and 2% LQ and 10% PC and 2% LQ. These bristles fill the voids inside treated soil. The XRD of treated soil with 8% PC and 4% LQ is shown in Figure 5. Figure 4 shows a huge increase in quartz (SiO_2), in addition to the increase in calcite ($CaCO_3$), where this increase is much higher than treated soil with 6% PC and 2% LQ. So, the CSH present intense peaks centered on 27° for treated soil. The increase in silt– sized particle content leads to increase in the cohesion strength value of the stabilized soil is observed. Quartz (SiO_2) and Calcite contains hard minerals so the increasing in their contents made the cohesion strength to gain up. A reduction can be noticed in the content of montmorillonite, Feldspar, illite and kaolinite.

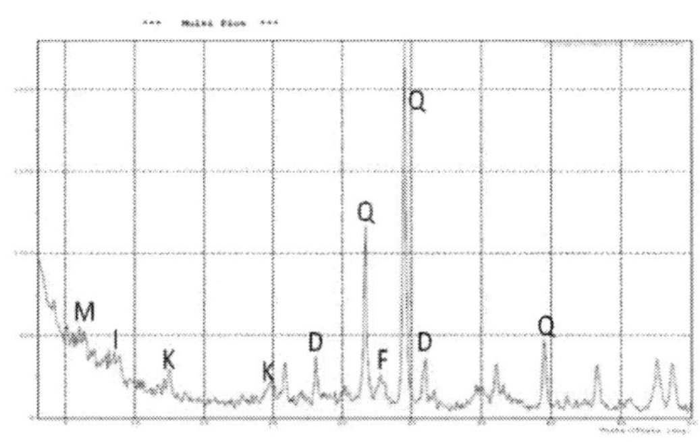

Figure 2. XRD of Natural Soil.

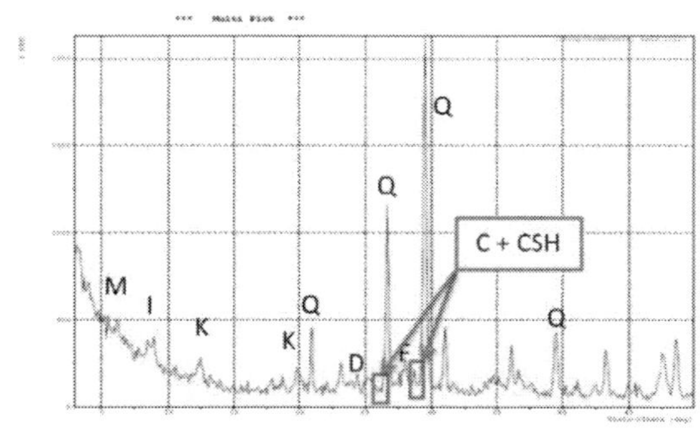

Figure 3. XRD analysis of treated soil with 6% PC and 2% LQ.

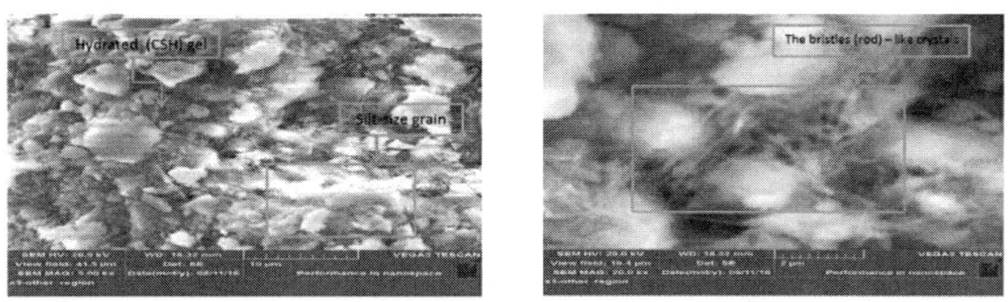

(A) **(B)**

Figure 4. SEM micrograph of treated soil with 8% PC and 4% LQ (A) at a magnification 5,000 times. (B) at a magnification of 20,000 times.

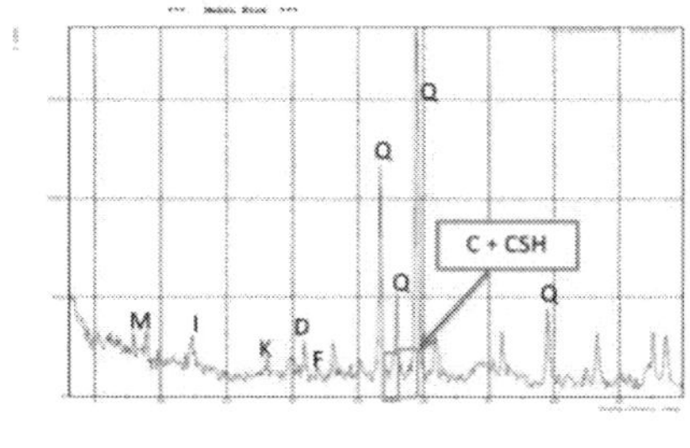

Figure 5. XRD analysis of treated soil with 8% PC and 4% LQ.

3.1.3 *Treated stabilized soil with 10% PC and 2% LQ*

Figure 6 illustrates treating sample of soil with 10% PC and 2% LQ, and Figure 7 shows an increase in the amount of hydrated gel (CSH) compared to soil treated with 6% PC and 2% LQ and 8% PC and 4% LQ, but the number of bristle-like crystals is decreased compared to soil treated by 8% PC and 4% LQ in the last two figures (as a result of pozzolanic reaction); it developed more aggregates. This implies that the sample has a high amount of hydration cement and pozzolanic products, and bonding among the clay particles, where it is noticed in the SEM image of treated soil, which represents a gathering of hydration cement and pozzolanic reaction products. XRD pattern of the treated soil with 10% PC and 2% LQ, shows a huge increasing in quartz content compared to the natural soil, while Figure 7 illustrates XRD for treated soil with 10% PC and 2% LQ increase in calcite, and an increase in quartz compared to treated soil where the increase is in calcite and then quartz. Calcium silicate hydrated (CSH) appears in both images of SEM and XRD of treated soil with 10% PC and 2% LQ as follows.

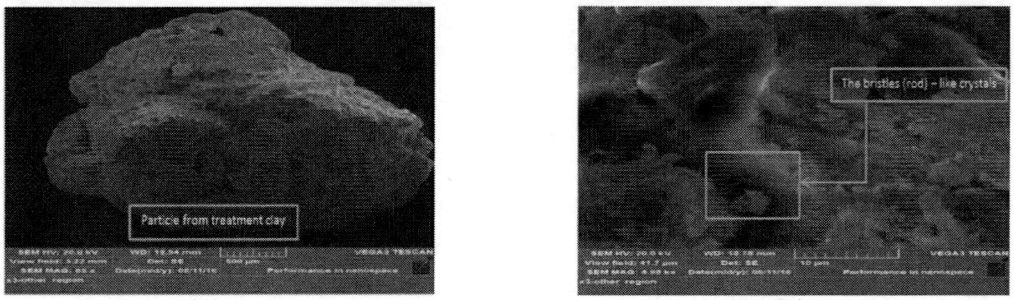

(A) (B)

Figure 6. SEM micrograph of treated soil with 10% PC and 2% LQ (A) at a magnification 50 times. (B) at a magnification of 5,000 times.

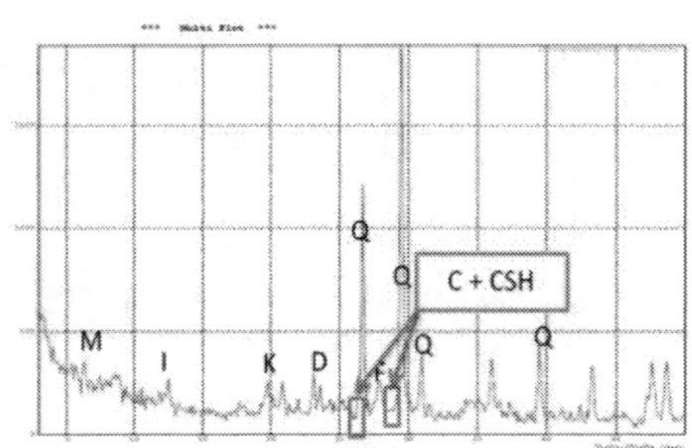

Figure 7. XRD analysis of treated soil with 10% PC and 2% LQ.

4 CONCLUSIONS

This study has demonstrated the effect of PC and LQ on compaction, UUU test and interpretation of some results by microstructure and mineralogical analysis of soft clay

A- The SEM-micrographs of treated soil with different PC and LQ content show crumbs of floccules with a porous nature and cementitious compounds coating clay particles. It can be seen that this soil has a microfabric silt-fine sand like structure.

B- Mineralogical analysis by X – ray diffraction (XRD) shows an increase in quartz and calcite. The increase in silt-sized particle content leads to an increase in the pore voids size. Hence, an increase in the cohesion strength (C) value of the stabilized soils is observed. Quartz and calcite contain hard minerals, so the increase in their contents made the cohesion of treated soil increase as well. A reduction can noticed in the content of montmorillonite, illite, and kaolinite clays compared to natural soil.

References

Boardman, D. I., Glendinning, S., and Rogers, C. D. F., Development of Stabilization and Solidification in Lime–Clay Mixes, *Geotechnique,* 51(6), 533–543, 2001.

Maubec, N., *Multi-scale Approach of Soils Treated with Lime — Interaction Study with Clays,* PhD Thesis, Université de Nantes, 2010.

Muller, C. J., *Pozzolanic Activity of Natural Clay Minerals with Respect to Environmental Geotechnics.* PhD Thesis, University of Karlsruhe, 2005.

Nichloson, P. G., *Soil Improvement and Ground Modification methods,* Elsevier Scientific Publish Company, Amsterdam, 2014.

Rahil, F. H., *Improvement of Soft Clay Underneath a Railway Track Model Using Stone Columns Technique,* PhD Thesis, University of Technology, Iraq, 2007.

Sariosseiri, F., and Muhunthan, B., Effect of Cement Treatment on Geotechnical Properties of Some Washington State Soils, *Journal of Engineering Geology,* 104,119-125, 2009.

Interdependence between Structural Engineering and Construction Management
Edited by Ozevin, D., Ataei, H., Modares, M., Gurgun, A., Yazdani, S., and Singh, A.

ISBN: 978-0-9960437-6-2

STUDY OF A MULTIPLE-TUNED MASS DAMPER FOR EXCESSIVE FLOOR VIBRATIONS MITIGATION

MEHDI SETAREH and STEPHANIE RENARD

School of Architecture and Design, Virginia Tech, Blacksburg, USA

Excessive vibration of architectural structures due to human movements has become an important serviceability issue. Tuned Mass Dampers or Vibration Absorbers have been used to mitigate the problem. This paper presents studies on an easy to fabricate, assemble and tune Multiple Tuned Mass Damper (MTMD) that can alleviate vibrations due to the excitations of several modes of a floor system. Using a test floor at the Virginia Tech Vibration Testing Laboratory, a number of experiments were conducted with the MTMD active and inactive. Different excitations generated by an electrodynamic shaker, and people walking on the floor were used. The vibration tests were repeated at various times to check the consistency of the MTMD performance. The measured reduction in floor vibrations were compared to those predicted based on the analytical equations. The results have shown that the device is effective in reducing floor vibrations due to human movements in spite of its small mass compared to the mass of the test floor.

Keywords: Multiple Tuned Vibration Absorber (MTVA), Vibration serviceability, Vibration control, Civil engineering structures, Walk tests, Human vibrations.

1 INTRODUCTION

Vibration problems due to human movements such as walking in civil structures have become more prevalent in recent years due to several factors such as: reduction in the structural mass resulting from the use of higher strength construction materials, reduction in damping due to less office furniture and fewer partitions, reduction in stiffness due to the use of optimized computer-aided design approaches (which generally result in less stiff supporting members), and reduction in the natural frequency of floor system due to the requirements for fewer support members such as columns or walls.

Tuned mass dampers (TMDs) or tuned vibration absorbers (TVAs) have been used to control excessive vibrations of various structural and mechanical systems. A TMD is a vibrating system made of a mass, spring, and damper which is usually installed within a structural system. It is tuned to the natural frequency of the structure such that it counteracts the movements of the structure. Frahm (1911) proposed the first application of TMDs to an engineering problem. Ormondroyd and Den Hartog (1928), Brock (1946), and Den Hartog (1947) contributed significantly to the theory of vibration reduction of undamped systems using TMDs.

Setareh *et al.* (2006, 2007) used pendulum and semi-active TMDs for floor vibration control. In both cases a TMD consisting of a frame hinged at one end and free at the other end supporting steel plates was used. The semi-active damper consisted of a magnetorheological damping device. Varela and Battista (2011) conducted laboratory tests on TMDs made of steel plates directly hung from the supporting beams of a test floor and showed that they were effective in reducing floor vibrations.

The common feature of the devices discussed above is their relatively large size, which require them to be installed within the floor structure. Also, the assembly and tuning of the devices are not generally easy tasks and require involvement of highly technical personnel. In addition, there are usually several modes of vibration that need to be controlled in multi-bay floor systems. These result in the application of TMDs as a retrofit option to be difficult and cost-prohibitive. Therefore, this paper introduces an easy to fabricate, install, and tune multiple tuned mass damper (MTMD) to control several modes of vibration of floor systems. It presents the results of the studies using the MTMD on a laboratory test floor susceptible to excessive vibrations due to human movements.

2 THEORETICAL BACKGROUND

The theoretical aspect of the performance of TMDs or MTMDs to control vibration is well-established. However, for clarity of discussion, the theoretical development of the application of a TMD on a single-degree of freedom (SDOF) system is briefly discussed here.

Figure 1 shows a TMD represented by a mass (m_T), stiffness (k_T), and damping (c_T) connected to a SDOF structure represented by a mass (M_s), stiffness (K_s), and damping (C_s). X_s and x_T represent the movement of the structure and TMD, respectively. F is the frequency-dependent harmonic force acting on the structure.

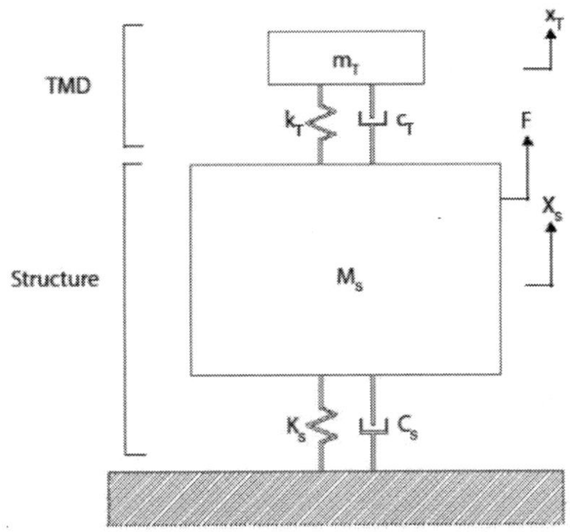

Figure 1. SDOF Model with TMD.

The acceleration response of the structure, $\ddot{X}_s$, is normalized and derived in Eq. (1) in terms of non-dimensional parameters of frequency ratio, f (natural frequency of the TMD to the natural frequency of the structure); mass ratio, μ (mass of TMD to the mass of the structure); force frequency ratio, g (excitation frequency to the natural frequency of the structure); ξ_s (damping ratio of the structure); and ξ_T (damping ratio of the TMD).

The relationship between the physical properties of the dynamic system and the above non-dimensional parameters are given below in Eq. (1):

ω_s= Natural frequency of structure (M_s)

ω_T= Natural frequency of TMD (m_T)
$f = \omega_T/\omega_s$ = Frequency ratio
$\mu = m_T/M_s$= Mass ratio
$g = \omega/\omega_s$= Force frequency ratio
$K_s = M_s\omega_s^2$= Stiffness of the structure
$k_T = m_T\omega_T^2$= Stiffness of the TMD
$C_s = 2M_s\omega_s\xi_s$ = Damping of the structure
$c_T = 2m_T\omega_T\xi_T$= Damping of the TMD
$\xi_s = C_s / 2M_s\omega_s$= Damping ratio of the structure
$\xi_T = c_T / 2m_T\omega_T$= Damping ratio of the TMD

$$R = \left|\frac{\ddot{X}_s}{\frac{F}{M_s}}\right| = g^2\sqrt{\frac{(f^2-g^2)+(2\xi_T f g)^2}{[(1-g^2)(f^2-g^2)-g^2 f(\mu f+4\xi_s\xi_T)]^2+4g^2\,[\xi_T f(1-g^2-g^2\mu)+\xi_s(f^2-g^2)]^2}} \qquad (1)$$

In the above equation, R, is the normalized acceleration response of the structure. In general, for a given μ and ξ_s the TMD non-dimensional parameters, f , and ξ_T are computed to minimize the maximum response of the structure (R_{max}).

3 DESCRIPTION OF THE MTMD

The MTMD is entirely made of steel parts which are screwed to each other. This facilitates the fabrication and assembly of the unit. It consists of five parts:

- Base: a steel plate to provide a flat surface for the stand components to be attached to.
- Stand: steel angles to provide support for the cantilevered plates.
- Wings: double-cantilevered uniform thick steel plates providing the required stiffness for the device.
- Steel Weights: steel plates providing the required mass for the MTMD.
- Dampers: these are standard dampers to provide the required damping for the device.

4 TUNING TESTS

Tests were conducted at the Virginia Tech Vibration Testing Laboratory. The facility houses a two-story full-scale test structure that was used for the studies conducted here. Figure 2 shows the MTMD on the test floor used for this study, which consists of a 30ft by 30ft concrete floor supported by steel beams and columns (at the corners).

Figure 2. MTMD on the test structure at the Vibration Testing Laboratory.

Table 1. Measured natural frequencies and damping ratio of the test floor.

Mode Number	f_s (Hz)	ξ_s (%)
1	5.225	0.51
2	9.225	0.36
3	10.65	0.42

First, the natural frequencies (f_s) and damping ratios (ξ_s) of the test floor with the MTMD inactive (locked) were measured as shown in Table 1. The MTMD wings were tuned to the measured floor natural frequencies, and subsequently placed on the test floor. The location of the MTMD placement was selected such that it can simultaneously provide vibration reduction to the first three modes of the floor. An electrodynamic shaker (APS 113) and a number of accelerometer (PCB 393C) connected to an OROS signal analyzer were used for this purpose. The total steel weights used on each wing to tune the device to the first three modes were as follows: W_1 = 54 lb W_2 =40.5 lb, and W_3 =27 lb.

The optimum MTMD frequency ratio and damping ratio based on the mass ratios and measured floor damping ratios were computed analytically. The mass ratios for the first three modes are: 0.0018, 0.00325, and 0.00167, respectively. Using the equation (1), the predicted reduction in the floor's first three modes vibration due to the application of the MTMD were 74%, 85%, and 78%, respectively. Figure 3 shows the frequency response functions (FRF) of the floor comparing the response of the floor with the MTMD inactive (locked wings) and active (unlocked wings). A burst chirp excitation was used to obtain the FRFs. The measured average reduction in the response of the first three vibration modes of the floor were 67% (Range: 58 – 73%), 50% (Range: 39– 70%), 48% (Range: 39-54%), respectively.

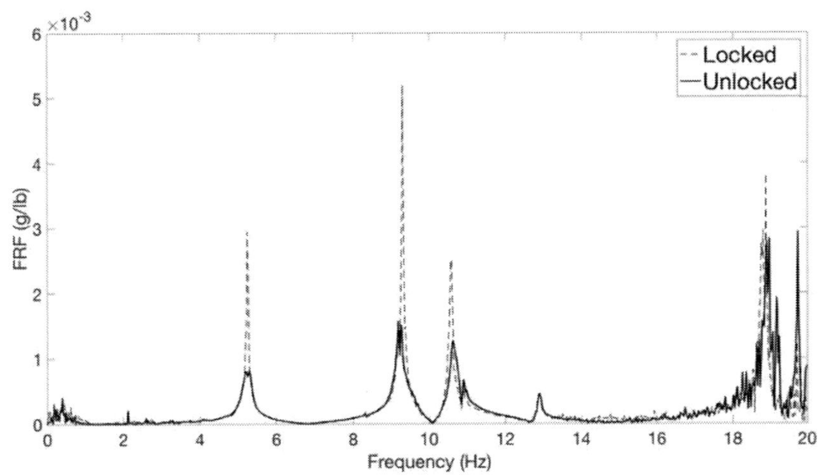

Figure 3. Comparison of the floor frequency response function with the MTMD active and inactive.

5 WALK TESTS

To evaluate the performance of the MTMD to reduce floor vibrations due to walking, 30 human subjects were asked to walk individually on the floor at the corresponding sub-harmonics of the first and second mode natural frequencies of the floor. Each person walked from one end of the floor along the center line to the opposite end and waited until the measurement was complete. The selected step frequency for walking at the first mode sub-harmonic was 158 spm (steps per minute) and for the second mode was 139 spm. Figure 4 shows a sample of the measured floor response in the time and frequency domains when the subject walked at the first mode sub-harmonic (158 spm). The average reduction in the vibration for the first mode was 59% and for the second mode was 42%. It has to be noted that typically walking excitations cannot cause resonance at the frequencies above 10 Hz since the dynamic load factor at such high frequencies is normally insignificant, which was the case for the third mode of vibration of the floor.

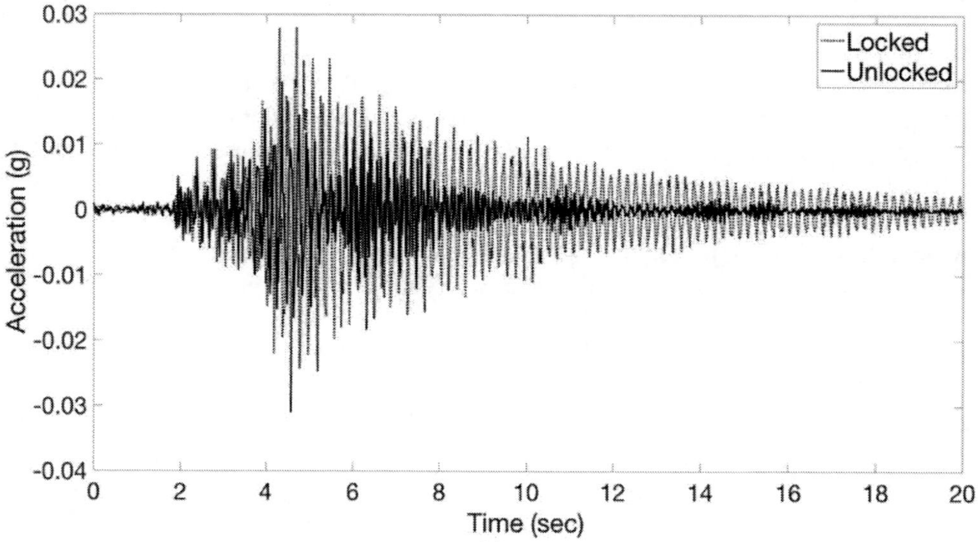

(a)

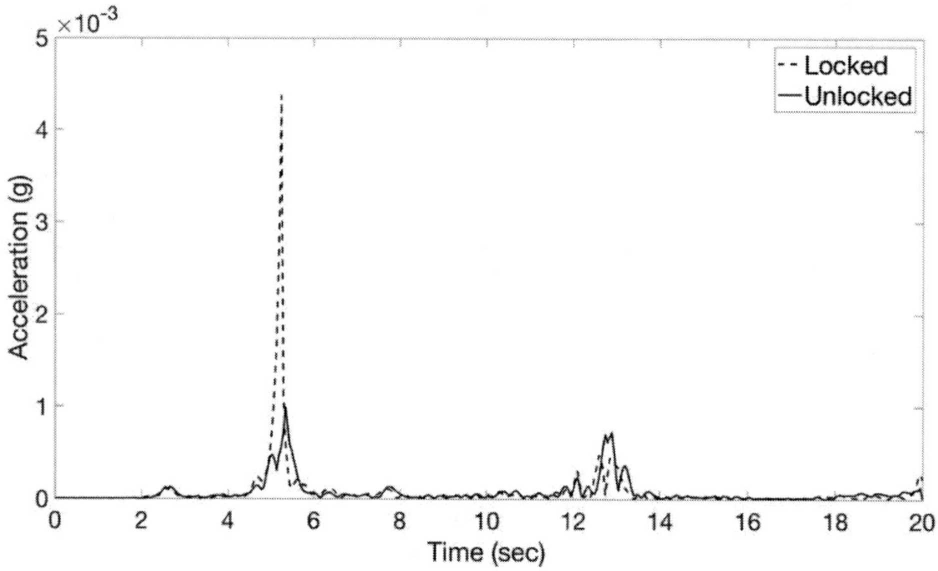

(b)

Figure 4. Comparison of the floor response when the subject walked at the first mode Subharmonic (158 spm): (a) time-history, (b) frequency spectrum.

6 CONSISTENCY TESTS

To check the consistency in the performance of the device due to the variations in the environmental conditions and the possibility of presence of non-linearities in the floor and device, the FRF curves were measured on different days while the ambient temperature was also recorded. The results showed that the performance remained consistent, which was demonstrated by the small coefficients of variation for the vibration reduction for the FRFs to be 7%, 15%, and 9% for the first three modes, respectively. The total number of measurements for each case was 40.

7 CONCLUSION

This paper provided the details of a multiple-tuned-mass-damper (MTMD) which is easy to fabricate, assemble, and tune to control vibrations of different modes of a floor system. The results of the tests conducted showed that even though the device had very small mass compared to the floor mass, it could provide good reduction in the floor vibration due to human walks at the subharmonics of the natural frequencies of a building floor system. It also showed that the device performs consistently and the system non-linearities are insignificant.

Acknowledgements

The research presented here was supported by the National Science Foundation under grant number CMMI-1335004. This support is gratefully acknowledged. Any opinions, findings, and conclusions expressed in this paper are those of the writers and do not necessarily reflect the views of the National Science Foundation.

References

Brock, J. E., A Note on the Damped Vibration Absorber, *Journal of Applied Mechanics*, 13, A-284, 1946.

Den Hartog, J. P., *Mechanical Vibrations*, 3rd ed., McGraw-Hill Book Co., Inc., New York, New York, 1947.

Frahm, H., *Device for Damping of Bodies*, U.S. Patent No. 989,958, 1911.

Ormondroyd, J., and Den Hartog, J. P., The Theory of the Dynamic Vibration Absorber, *Transactions of the American Society of Mechanical Engineers*, 50, 9-22, 1928.

Setareh, M., Ritchey, J. K., Murray, T. M., and Baxter, A., Application of a Pendulum Tuned Mass Damper for Floor Vibration Control, *Journal of Performance of Constructed Facilities*, 20 (1), 64-73, 2006.

Setareh, M., Ritchey, J. K., Murray, T. M., Koo, J. H., and Ahmadian, M., A Semi-Active Tuned Mass Damper for Floor Vibration Control, *Journal of Structural Engineering*, 133 (2), 242-250, 2007.

Varela, W. D., and Battista, R. C., Control of Vibrations Induced by People Walking on Large Span Composite Floor Decks, *Engineering Structures*, 33, 2485-2494, 2011.

Interdependence between Structural Engineering and Construction Management
Edited by Ozevin, D., Ataei, H., Modares, M., Gurgun, A., Yazdani, S., and Singh, A.

ISBN: 978-0-9960437-6-2

RETROFITTING METHODS AGAINST FATIGUE CRACKING IN STEEL GIRDER WEB PENETRATION

CHIHIRO SAKAMOTO[1], MASAHIRO SAKANO[1], HIDEYUKI KONISHI[2], and TAKASHI FUJII[3]

[1]*Dept of Civil and Environmental Engineering, Kansai University, Osaka, Japan*
[2]*Japan Bridge Association, Osaka, Japan*
[3]*Ministry of Land, Infrastructure and Transport, Himeji, Japan*

Fatigue cracking in steel girder web penetration details is so dangerous that it can break steel girders. Since a number of highway bridges have such web penetration details in Japan, it is of urgent importance to grasp these fatigue strength properties and develop effective retrofitting methods. In a previous report, we investigated the stress distributions around web penetration details, and fatigue cracking behavior, using steel girder specimens with web penetration details. In this study, we investigate effects of retrofitting methods against fatigue cracking in web penetration details through fatigue tests using large girder specimens with web penetration details in which cross beam lower flanges are connected to lower surface of a slot by welding. Principal results obtained through this study are as follows: (1) Weld toe grinding can extend fatigue life more than 5 times, (2) Two-face attachment can extend fatigue life more than 10 times, and (3) Two-face attachment with weld toe grinding can extend fatigue life more than 25 times.

Keywords: Fatigue test, Weld toe grinding method, Two face attachment method.

1 INTRODUCTION

Fatigue cracking in steel girder web penetration details is so dangerous that it can break steel girders (Fisher 1984). A one-meter-long crack was detected in Yamazoe Bridge in 2006 (Nara National Highway Office 2006). Since a number of highway bridges have such web penetration details in Japan, it is of urgent importance to grasp these fatigue strength properties. However, few fatigue tests have been reported on steel girder web penetration details (Sakano *et al.* 1995 and Hanshin Expressway Co. Ltd. 2012).

In previous studies, we investigated stress distributions around web penetration details and fatigue cracking behavior through fatigue tests using girder specimens with web penetration details (Yoshida *et al.* 2014 and Sakamoto *et al.* 2018a). Then we investigated the stress reduction effect by three face attachment retrofit through static loading tests using a large girder specimen with web penetration details where cross beam lower flanges are connected to lower surface of a slot by welding (Sakamoto *et al.* 2018b).

In this study, we investigate effects of two face attachment retrofitting methods against fatigue cracking in web penetration details through fatigue tests using a large girder specimen with web penetration details in which cross beam lower flanges are connected to lower surface of a slot by welding.

2 EXPERIMENTAL PROCEDURE

2.1 Specimen

Figure 1 shows configurations and dimensions of the specimen. Dimensions of the specimen were increased from the previous one (Yoshida *et al.* 2014 and Sakamoto *et al.* 2018a), so that two face attachments can be fixed in the space between the cross beam flange and the main beam flange. The specimen had 3 web penetration details, and 6 test parts. Test parts A, B, and C were as weld, and A', B', and C' were weld toe ground. Two face attachments were applied to test parts A and A'.

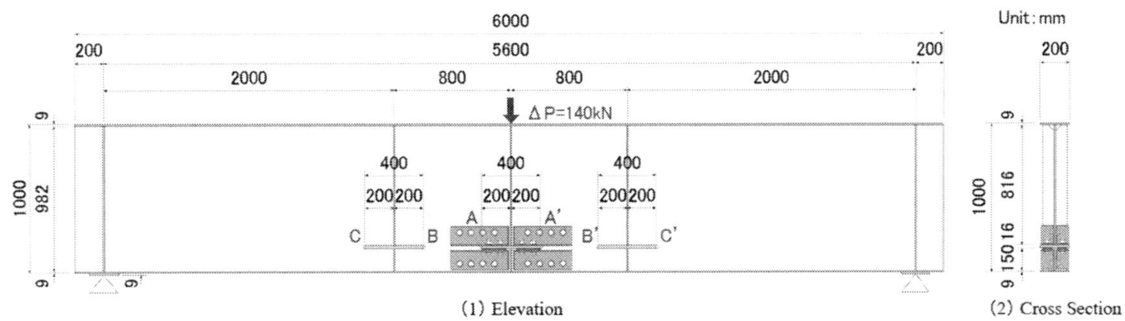

Figure 1. Configurations and dimensions of the specimen.

2.2 Weld Toe Grinding Method

Weld toes were ground to depth of 0.1 – 1.2mm and radius of 5.1 – 6.7mm, using a bar grinder in test parts A', B', and C'. After weld toe grinding, weld cracks were detected in the weld in all test parts as shown Figure 2.

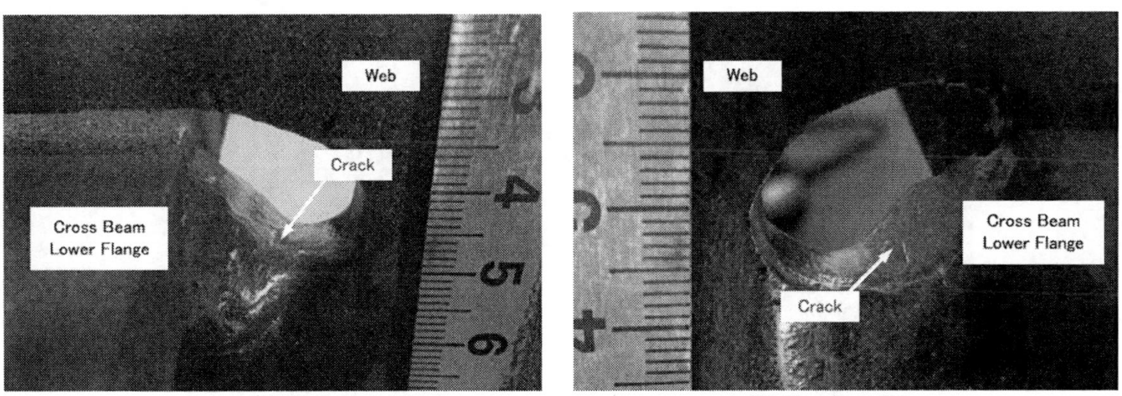

(1) Test Part A' (2) Test Part B'

Figure 2. Ground weld toe.

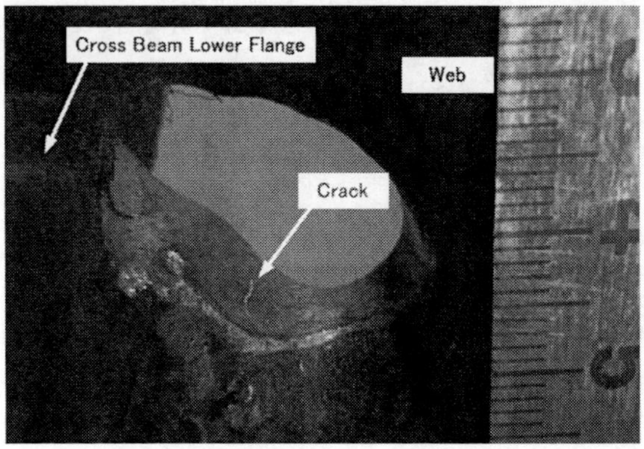

(3) Test Part C'

Figure 2. Ground weld toe (contd).

2.3 Two Face Attachment Method

From the previous study (Sakamoto *et al.* 2018b), the upper and lower side attachments were most effective on stress reduction, and there was very little difference in stress reduction effects between two and three face attachments. Therefore, we adopted the upper and lower side two face attachments.

2.4 Fatigue Test Procedures

Figure 3 shows the loading conditions. We conducted fatigue tests in the 3-point bending conditions similarly as static loading test (Sakamoto *et al.* 2018b). The load range was set to 140kN (Pmin=260kN, Pmax=400kN) so that the maximum tensile stress of about 170MPa could be generated in the bottom flange (JIS SM490Y steel), considering the dead load stress. Magnetic Particle Test (MT) was applied to detect fatigue cracks at weld toe along cross beam lower flange edges. When fatigue cracks were propagated, a stop hole (SH) was drilled and a high tension bolt (HTB) was tightened. Then fatigue tests were continued.

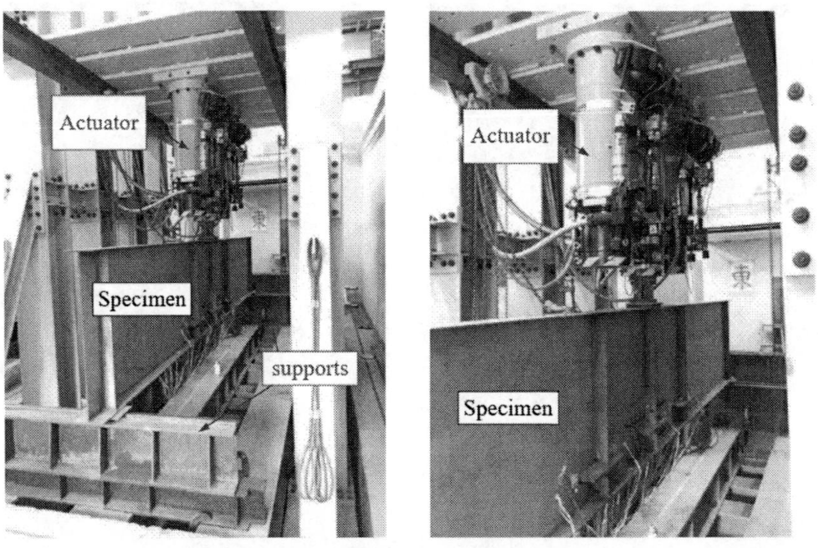

Figure 3. Loading conditions.

3 EXPERIMENTAL RESULT

Figure 4 shows principal stresses distributions. Considering directions of principal stresses, there may be low possibility of fatigue cracking in test parts B and B'.

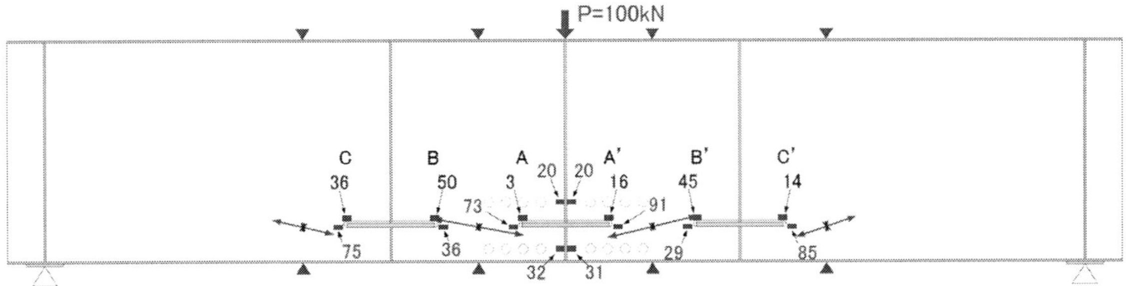

Figure 4. Principal stresses distributions.

Figure 5 shows locations of fatigue cracks. In the test part C (as weld), fatigue crack was detected after 0.2Mcycles loading. In the test part C' (weld toe grinding), fatigue crack was detected after 1.0Mcycles loading. After 1.4Mcycles loading, these crack length were reached to 30mm, SH was drilled and HTB was tightened. In the test part A (as weld with attachments) a fatigue crack was detected after 2.0Mcycles loading. In the test part A' (weld toe grinding with attachments), no fatigue cracks were detected after 3.0Mcycles loading with upper and lower side two face attachments and further after 3.0Mcycles loading with lower side attachments.

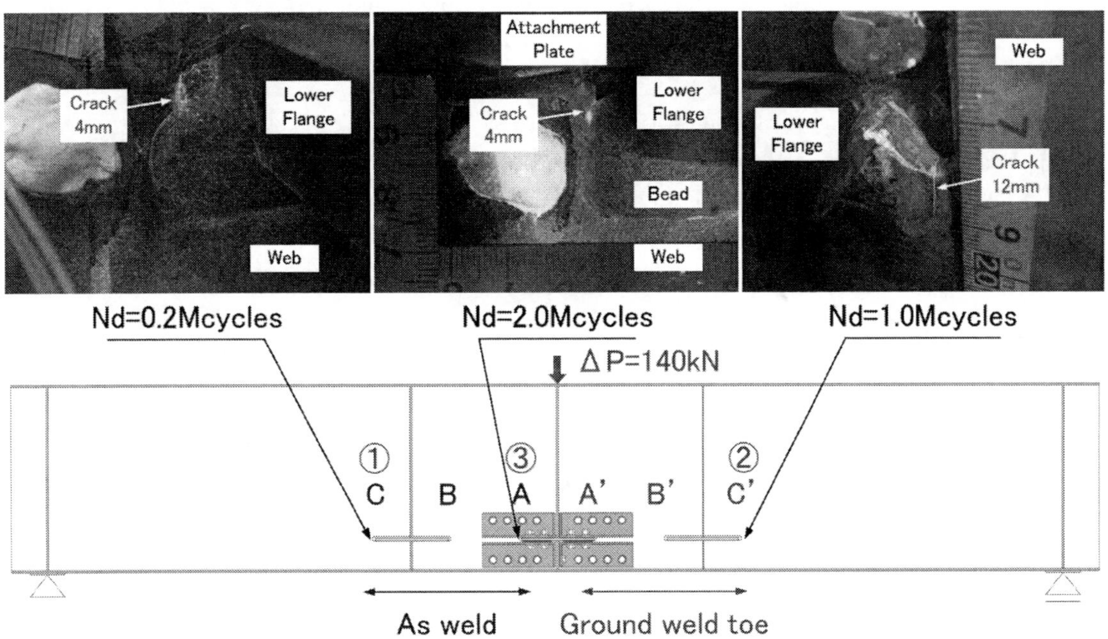

Figure 5. Locations of fatigue cracks.

Figure 6 shows stress range vs. fatigue life relationship. Comparing test parts C (as weld) and C' (weld toe grinding), weld toe grinding can extend fatigue life more than 5 times. Comparing test parts A (as weld with attachments) and C, two face attachment can extend fatigue life more than 10 times. Comparing test parts A' (weld toe grinding with attachments) and C, two face attachment with weld toe grinding can extend fatigue life more than 25 times.

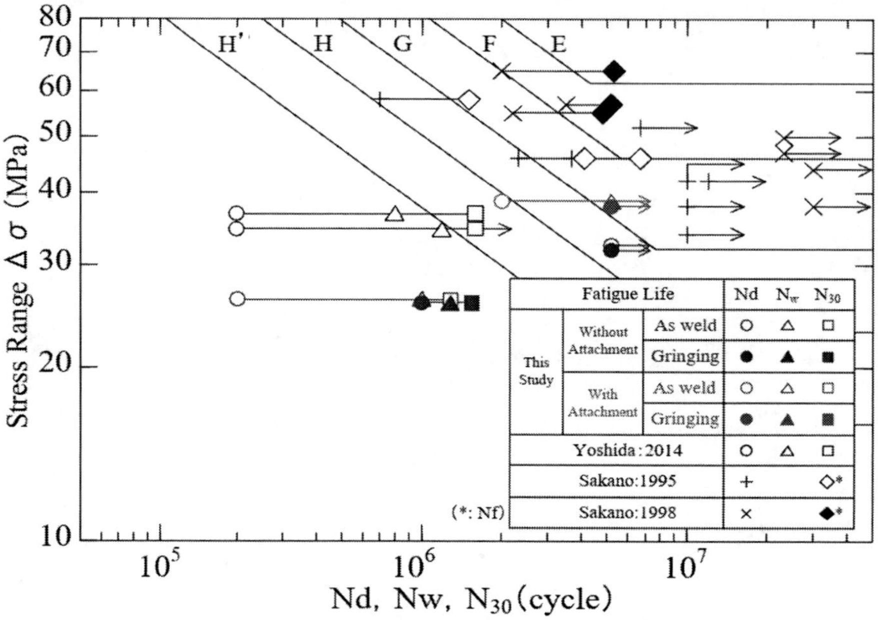

Figure 6. Stress range vs. fatigue life relationship.

4 CONCLUSIONS

The principal results obtained through this study are as follows;

(1) Weld toe grinding can extend fatigue life more than 5 times.
(2) Two face attachment can extend fatigue life more than 10 times.
(3) Two face attachment with weld toe grinding can extend fatigue life more than 25 times.

Future work is required on more effective and smart retrofitting measures.

References

Fisher, J. W., Fatigue and Fracture in Steel Bridges, Chapter 6, Wiley-Interscience, 1984.

Hanshin Expressway Co. Ltd., Fatigue Counter Measures for Steel Bridges in Hanshin Expressway, Chapter 7, (in Japanese), Hanshin Expressway Management Technology Center, 2012.

Nara National Highway Office, Retrieved from http://www.kkr.mlit.go.jp/nara/, (in Japanese), 2006.

Sakamoto, C., Sakano, M., Konishi, H., and Fujii, T., *Retrofit against Fatigue Cracking at Web Penetration Details with a Slot in Steel Girder*, 9th International Symposium on Steel Bridges, No.3.4, Prague, 2018a.

Sakamoto, C., Sakano, M., Konishi, H., and Fujii, T., *Three Face Attachment Retrofit against Fatigue Cracking at Steel Girder Web Penetration,* The Fourth Australasia and South-East Asia Structural Engineering and Construction Conference, No.11.6, Brisbane, Australia, 2018b.

Sakano, M., Mikami, I., Murayama, H., and Moriwaki, K., Fatigue Strength of Railway Plate Girder with Web Penetration (in Japanese), *Journal of Structural Engineering*, 41A, 965-973, 1995.

Yoshida, N., Sakano, M., Konishi, H., and Fujii, T., *Fatigue Test of Steel Girder Web Penetration Details with a Slit*, Sustainable Solutions in Structural Engineering and Construction, 95-100, ISEC Press, 2014.

Interdependence between Structural Engineering and Construction Management
Edited by Ozevin, D., Ataei, H., Modares, M., Gurgun, A., Yazdani, S., and Singh, A.

ISBN: 978-0-9960437-6-2

STRENGTHENING OF RC BEAMS USING NEAR SURFACE EMBEDDED-FRCM

USAMA EBEAD and HOSSAMELDIN EL-SHERIF

Dept of Civil and Architectural Engineering, College of Engineering, Qatar University, Doha, Qatar

Fibre reinforced cementitious matrix (FRCM) systems are mostly externally bonded (EB) for the strengthening of reinforced concrete (RC) and masonry structures. In this paper, the relatively new concept of near-surface embedded (NSE) FRCM, has been introduced for the flexural strengthening of beams. The process of the application of NSE-FRCM strengthening technique involves the removal of the concrete layer at beam soffit, being the most deteriorated in actual practices. Experimental evidence of the flexural strengthening efficacy of this technique is provided here. Eight RC beams were prepared and tested under four-point loading with the consideration of two test parameters: (a) FRCM material (polyparaphenylene benzobisoxazole (PBO)/carbon/glass); and (b) the reinforcement ratio (0.5% representing flexure-deficient beams and 1.28% representing typical under-reinforced beams). The strengthening led to gains in ultimate loads that ranged between 31.4% and 84.3%.

Keywords: Reinforced concrete, Flexure, Repair, Roughening.

1 INTRODUCTION

Research on the strengthening of RC beams deals mostly with new prepared specimens without signs of concrete deterioration (Elghazy *et al.* 2017, 2018a, Pino *et al.* 2017, Tetta *et al.* 2017, Younis *et al.* 2017). This does not simulate real practices where the beams would have suffered deterioration prior to repair or strengthening. This deterioration can be due to steel reinforcement corrosion that leads to degradation in the steel/concrete bond and cross-sectional loss of steel bars (Zhang *et al.* 2018). More recently, fabric reinforced cementitious matrix (FRCM) has emerged as a viable solution for RC strengthening. FRCM includes dry fibers impregnated in a cementitious matrix, and is surface-applied for the aim of strengthening RC and masonry structures. Researchers focused on the effect of corrosion on the mass loss of steel reinforcement through long-term exposure to severe artificial conditions (Elghazy *et al.* 2018b). The beam soffit is subjected to cracking due to concrete deterioration, and these cracks affect the integrity of the concrete and the concrete/FRCM bond if not addressed. Therefore, any problems in the concrete surface have to be addressed prior to any strengthening work according to the ACI 549 guideline (ACI Committee 549 2013). To better study real applications, some studies tested beams strengthened after being pre-damaged (Ebead 2015) or exposed to accelerated corrosion environments, thus, examined the effect of strengthening corroded beams (Elghazy *et al.* 2017, 2018a). Elghazy *et al.* (2018a) reported on the success of FRCM in the flexural strengthening of corrosion damaged RC beams.

A new concept of substrate roughening, proposed by Wakjira and Ebead, included the removal of the concrete cover and embedding the fabric layers within the created grooves (Wakjira and Ebead 2018a, 2018b). They referred to the installed FRCM system in this technique as near-surface embedded (NSE) FRCM. This technique allows the surface to be prepared without the safety hazards associated with sand blasting. FRCM in the NSE roughening method replaces the damaged concrete preserving the shape and dimensions of the original beam conforming with any architectural restrictions. Moreover, the fact that FRCM is embedded in the concrete cover rather than applied to the surface was reported to also improve the FRCM/concrete interaction for shear strengthened beams (Wakjira and Ebead 2018a). This can be further supported by other research reporting that debonding was observed between the FRCM layer and concrete in beams strengthened with externally bonded FRCM (EB-FRCM) (Younis *et al.* 2017). This NSE roughening technique, however, has not yet been investigated for flexural strengthening.

2 EXPERIMENTAL PROGRAM

The test matrix of the experimental program is given in Table 1. Eight un-strengthened and strengthened RC beams were tested in this study. The beams included flexural-deficient beams (with ρ_s = 0.5%) and typical under-reinforced beams (with ρ_s = 1.28%) strengthened using FRCM with two plies of PBO, carbon, or glass fabrics. The behavior and effectiveness of FRCM with one ply was assessed elsewhere by the authors for characterization tests and EB-FRCM flexural strengthening applications (Ebead *et al.* 2017), and NSE- and EB-FRCM shear strengthening applications (Wakjira and Ebead 2018a, 2018b, Younis *et al.* 2017). The letters 'P', 'C' and 'G' in the beam identification shown in Table 1 refer to PBO, carbon, and glass fabrics, respectively. 'N' refers to NSE strengthening. 'L' and 'H' refer to low and high reinforcement ratios of ρ_s = 0.5% and 1.28%, respectively.

Table 1. Test matrix.

Fabric type	ID	Fabric type	Flexural Reinforcement
1	R-0.5%	-	Low, ρ_s = 0.5%
2	R-1.28%	-	High, ρ_s = 1.28%
3	P-0.5%	PBO	Low, ρ_s = 0.5%
4	C-0.5%	Carbon	Low, ρ_s = 0.5%
5	G-0.5%	Glass	Low, ρ_s = 0.5%
6	P-1.28%	PBO	High, ρ_s = 1.28%
7	C-1.28%	Carbon	High, ρ_s = 1.28%
8	G-1.28%	Glass	High, ρ_s = 1.28%

2.1 Specimens

The geometry and reinforcement details of the test specimens are shown in Figure 1. All beams were 2.5 m long with a rectangular cross-section of 150×260 mm. The tensile steel bars were placed at a depth of 210 mm measured from the top surface of the beam. All beams were reinforced with closed steel stirrups of 8 mm diameter spaced at 100 mm to prevent premature shear failure.

2.2 Test Setup

All beams were four-point loaded at a loading rate of 1 mm/min, which is a common testing speed in static tests (Akbari Hadad *et al.* 2018, Ebead and Marzouk 2002, Wakjira and Ebead 2018a). The beams had an effective span of 2200 mm with shear spans measuring 825 mm as shown in Figure 1. All beams were instrumented with 60 mm strain gauges bonded to the top surface of the concrete in the compression zone and with two 5 mm strain gauges bonded to the tensile steel bars at midspan. Deflections were measured using two linear variable differential transducers (LVDTs) located at midspan.

2.3 Materials

Ready-mixed concrete was used to cast all beams. The concrete proportions per cubic meter were 780 kg of sand, 950 kg of gravel, and 350 kg of ordinary Portland cement with water/cement ratio of 0.45. ASTM C39/C39M (2009) compression tests were conducted on 150×300 mm concrete cylinders revealing an average 28-day compressive strength of 39.5 MPa with a standard deviation of 1.6 MPa. In addition, ASTM C1609 (2012) flexural tests were conducted on six prisms 100×100×500 mm revealing modulus of rupture of concrete of 5.65 MPa with a standard deviation of 0.6 MPa. The yield stress of grade 500B steel bars (ISE/104 Committee 2005) used in this study ranged between 594 and 627 MPa as tested by the authors.

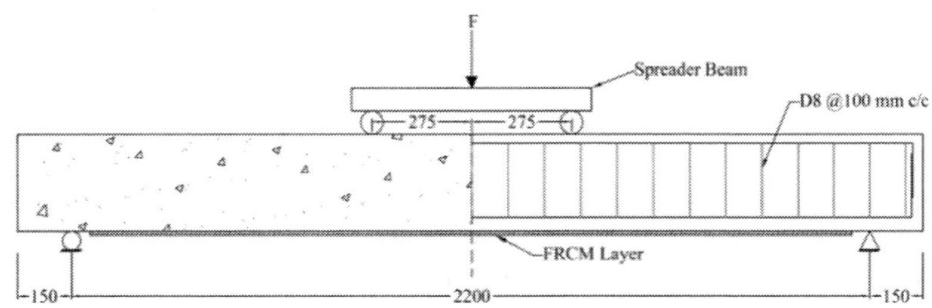

Figure 1. Dimensions and reinforcement details of the longitudinal section of the beams.

Three different types of commercially available FRCM fabrics namely, Carbon, PBO, and Glass were used in this study. The mechanical and geometric properties of each fabric in the warp and weft directions can be found elsewhere (Younis *et al.* 2017). The fabrics were adhered to the concrete surface using their associated mortars as per the manufacturers' recommendations. Mortars had 28-day compressive strengths of 30±2.4, 20±1.4, and 40±2.3 MPa for Carbon-, PBO- and Glass-FRCM, respectively, as tested by the authors according to ASTM C109/C109M (2005) provisions. In addition, FRCM characterization of coupon tests of size 410×50×10 mm was conducted in accordance with AC 434 (ICC (International Code Council) 2016).

2.4 Strengthening Procedure

Prior to FRCM installation for the NSE-FRCM strengthened beams, Group 1 specimens, a 90 mm wide and 15 mm deep groove over the full loaded length of the beam was created using a concrete slitting machine so that the beam did not have the groove at the support location. The depth of the groove was determined as 10 mm for both PBO-FRCM and Carbon-FRCM and 15 mm for Glass FRCM to accommodate the two layers of FRCM plies. The cut concrete was manually chipped off leaving a rough surface as shown in Figures 2a and 2b. The hand lay-up method recommended by ACI 549 (ACI Committee 549 2013) was adopted during the installation of FRCM. The roughened surface was first saturated with water for one hour before applying the first layer of the cementitious matrix with a thickness of 3 to 4 mm. The first fabric ply was then installed in place and was gently impregnated into the matrix as shown in Figure 2c. A second layer of the matrix was immediately applied followed by the second fabric ply and a finishing matrix layer as shown in Figure 2d.

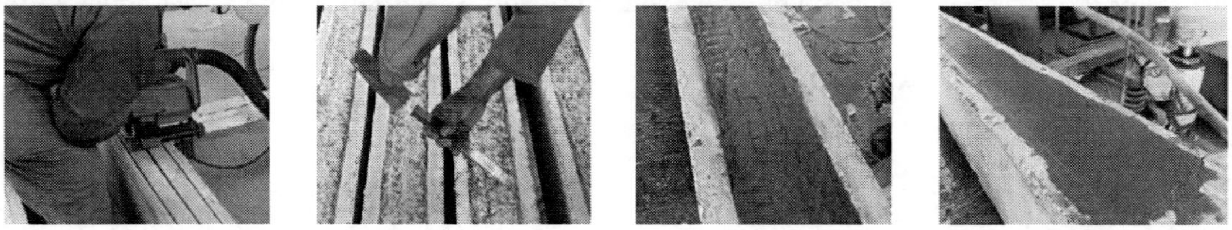

(a) (b) (c) (d)

Figure 2. FRCM Strengthening: (a) grooving the beam, (b) chipping the concrete and clearing the groove, (c) applying first layer of mortar and impregnating the fabric, (d) finishing after FRCM application.

3 RESULTS AND DISCUSSION

Table 2 lists the ultimate loads, P_u, recorded for all beams. Significant gains in the load-carrying capacities were observed in the strengthened specimens. The gain in the load carrying capacity ranged between 31.4% for Specimen G-0.5% and C-1.28% for Specimen C-1.28%. The strengthening effect was more noticeable for specimens with lower reinforced ratio; i.e., the flexure-deficient beams than that for the typical under reinforced counterparts. Specimens with ρ_s= 0.5% showed an increase in P_u that ranged between 31.4% and 57.1% and between 30.1% and 72.8%, respectively, compared to that of the corresponding reference specimen R-0.5%.

Table 2. Summary of the test results.

Specimen	ρ_s (%)	Initial stiffness, S_I (kN/mm)	Post cracking stiffness, S_c (kN/mm)	Ultimate capacity, P_u (kN)	P_u gain (%)	Ductility index,ΔI
R-0.5%	0.5	15.1	3.53	39.9	-	15.31
	1.28	26.8	7.85	99.2	-	7.31
P-0.5%	0.5	26.8	7.85	62.7	57.1	5.73
C-0.5%	0.5	26.5	5.03	59.2	48.4	3.21
G-0.5%	0.5	16.2	4.11	52.4	31.4	5.56
P-1.28%	1.28	22.9	9.42	123	83.5	3.6
C-1.28%	1.28	23.7	9.74	124	84.3	2.45
G-1.28%	1.28	25.2	8.79	110	70.4	2.87

It was found that both carbon and PBO fabrics provided similar enhancement in the load carrying capacities of the strengthened beams. Specimens strengthened with G-FRCM showed considerably lower strength gains. Specimens with C-, PBO-, and G-FRCM systems showed average gains in the ultimate loads of 1.66, 1.70, 1.51 times the average load capacity of the control specimens. It was observed that G-FRCM strengthened beams presented the highest FRCM fabric utilization. All G-FRCM strengthened beams failed due to fabric rupture.

The load-deflection relationships for the specimens are shown in Figures 3a and 3b. The initial stiffness (S_I) and post cracking stiffness (S_c) are shown in Table 2 for all the tested specimens.

The average S_I for the control beams was 21 kN/mm. This value was increased due to strengthening to 23.6 kN/mm. The average value of S_I was very close to that of the control specimens. The average S_I values for specimens with ρ_s=0.5% (P-0.5%, C-0.5%, and G-0.5%) was 23.2 kN/mm compared to S_I of the control specimen 15.1 kN.mm showing an initial stiffness increase. For higher reinforced (ρ_s=1.28%) NSE specimens (P-1.28%, C-1.28%, and G-1.28%), the average initial stiffness was found to be 23.9 kN/mm that is lower than S_I for the reference specimen R-1.28%. Inspecting the values of the initial stiffness for all NSE specimens shows a slight scatter of the values without a clear trend; however, there are, in average, similar to the average of the reference specimens which is 21 kN/mm. Post crack stiffness, S_C, shows a different behavior to the initial stiffness. The S_C average values were 7.5 kN/mm. The S_C values showed an increase between 12% (Specimen G-1.28%) and 123% (Specimen P-0.5%) compared to their corresponding benchmark Specimens R-0.5% and R-1.28%. The average S_C value for beams with ρ_s=0.5% (P-0.5%, C-0.5%, and G-0.5%) was 5.67 kN/mm. As for the beams with ρ_s=1.27%, all specimens showed similar trend and average at very close values at 9.32 kN/mm. This supports that flexure deficient beams are the most affected in terms of post cracking stiffness by the strengthening scheme.

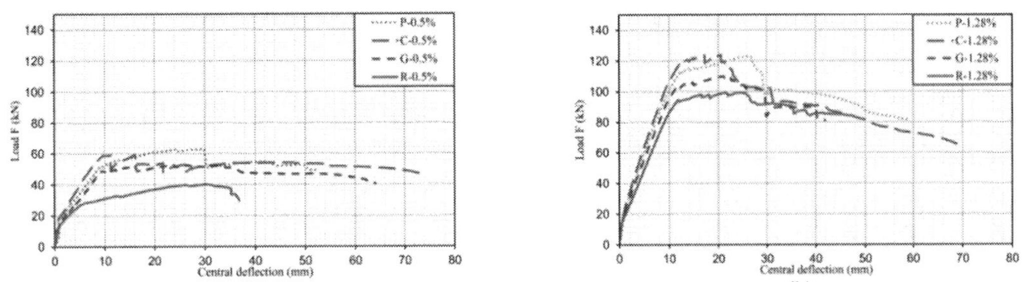

(a) (b)

Figure 3. Load-Deflection relationships for beams with a) low reinforcement and b) high reinforcement.

The ductility index (ΔI) is defined as the ratio between deflections at the ultimate and yield loads, respectively (Aly *et al.* 2006, Ebead and Saeed 2010, 2013, 2014) and is listed in Table 2. Generally, strengthening lead to a reduction in ΔI when comparing the strengthened beams to their reference counterparts. The strengthened specimens showed an average ΔI that is 35% of that shown by the reference specimens.

4 CONCLUSIONS

Based on the results of this investigation, the following conclusions have been made:

- The average gain in the load carrying capacity was 62.5 % for the strengthened beams.
- It was observed that G-FRCM strengthened beams presented the highest FRCM fabric utilization. All G-FRCM strengthened beams failed due to fabric rupture.
- increase in the initial stiffness was shown for the strengthened specimen with an average of 23.6 kN/mm compared to 21 kN/mm for the reference specimen.

Acknowledgments

This paper was made possible by the NPRP grant # NPRP 7-1720-2-641 from the Qatar National Research Fund (a member of Qatar Foundation) and by Qatar University grant # QUST-CENG-SPR-13/14. The findings achieved herein solely the responsibility of the authors.

References

ACI Committee 549, *Guide to Design and Construction of Externally Bonded Fabric-Reinforced Cementitious Matrix (FRCM) Systems for Repair and Strengthening Concrete and Masonry Structures (ACI 549.4R-13),* American Concrete Institute, Farmington Hills, MI, USA, 2013.

Akbari Hadad, H. A., Nanni, A., Ebead, U. A., and El Refai, A., Static and Fatigue Performance of FRCM-Strengthened Concrete Beams, *Journal of Composites for Construction*, 22(5), 2018.

Aly, R., Benmokrane, B., and Ebead, U., Tensile Lap Splicing of Bundled CFRP Reinforcing Bars in Concrete, *Journal of Composites for Construction*, 10(4), 287–294, 2006.

ASTM C39/C39M, *Standard Test Method for Compressive Strength of Cylindrical Concrete Specimens,* ASTM International, West Conshohocken, PA, 2009.

ASTM C109/C109M, *Standard Test Method for Compressive Strength of Hydraulic Cement Mortars. Annual Book of ASTM Standards*, 9, 2005.

ASTM C1609, *Standard Test Method for Flexural Performance of Fiber-Reinforced Concrete (Using Beam with Third-Point Loading)*, ASTM International, West Conshohocken, PA., 2012.

Ebead, U., Inexpensive Strengthening Technique for Partially Loaded Reinforced Concrete Beams: Experimental Study, *Journal of Materials in Civil Engineering*, 11(10), 1–11, 2015.

Ebead, U., and Marzouk, H., Strengthening of Two-Way Slabs Subjected to Moment and Cyclic Loading, *ACI Structural Journal*, 99(4), 435–444, 2002.

Ebead, U., and Saeed, H., Modeling of reinforced Concrete Slabs Strengthened with Fiber-Reinforced Polymer or Steel Plates, *ACI Structural Journal*, 107(2), 218–227, 2010.

Ebead, U., and Saeed, H., Hybrid Shear Strengthening System for Reinforced Concrete Beams: An Experimental Study, *Engineering Structures*, 49(4), 421–433, 2013.

Ebead, U., and Saeed, H., Flexural and Interfacial Behavior of Externally Bonded/Mechanically Fastened Fiber-Reinforced Polymer Strengthened Reinforced Concrete Beams, *ACI Structural Journal*, 111(4), 741–752, 2014.

Ebead, U., Shrestha, K. C., Afzal, M., El Refai, A., and Nanni, A., Effectiveness of Fabric-Reinforced Cementitious Matrix in Strengthening Reinforced Concrete Beams, *Journal of Composites for Construction*, 21(2), 04016084, 2017.

Elghazy, M., Refai, A., Ebead, U., and Nanni, A., Effect of corrosion damage on the flexural performance of RC beams strengthened with FRCM composites, *Composite Structures*, o180, 994–1006, 2017.

Elghazy, M., El Refai, A., Ebead, U., and Nanni, A., Post-Repair Flexural Performance of Corrosion-Damaged Beams Rehabilitated with Fabric-Reinforced Cementitious Matrix (FRCM), *Construction and Building Materials*, 166, 2018a.

Elghazy, M., El Refai, A., Ebead, U., and Nanni, A., Corrosion-Damaged RC Beams Repaired with Fabric-Reinforced Cementitious Matrix, *Journal of Composites for Construction*, 22(5), 04018039, 2018b.

ICC (International Code Council), *Acceptance Criteria for Masonry and Concrete Strengthening Using Fabric-reinforced Cementitious Matrix (FRCM) and Steel Reinforced Grout (SRG) Composite Systems - AC434,* ICC Evaluation Service, 2016.

ISE/104 Committee, *BS 4449:2005: Steel for The Reinforcement of Concrete. Weldable Reinforcing Steel. Bar, Coil and Decoiled Product,* BSI, 2005.

Pino, V., Hadad, H., Basalo, F., Nanni, A., Ebead, U., and Refai, A., Performance of FRCM-Strengthened RC Beams Subject to Fatigue, *Journal of Bridge Engineering*, 22(10), 2017.

Tetta, Z., Koutas, L., and Bournas, D., Shear Strengthening of Concrete Members with TRM Jackets: Effect of Shear Span-to-Depth Ratio, Material and Amount of External Reinforcement, *Composites Part B: Engineering*, Elsevier, 137, 184–201, 2017.

Wakjira, T. G., and Ebead, U., Hybrid NSE/EB Technique for Shear Strengthening of Reinforced Concrete Beams Using FRCM: Experimental Study, *Construction and Building Materials*, 164, 164–177, 2018a.

Wakjira, T. G., and Ebead, U., FRCM/Internal Transverse Shear Reinforcement Interaction in Shear Strengthened RC Beams, *Composite Structures*, Elsevier, 201, 326–339, 2018b.

Younis, A., Ebead, U., and Shrestha, K. C., Different FRCM Systems for Shear-Strengthening of Reinforced Concrete Beams, *Construction and Building Materials*, 153, 514–526, 2017.

Zhang, W., Zhang, H., Gu, X., and Liu, W., Structural Behavior of Corroded Reinforced Concrete Beams Under Sustained Loading, *Construction and Building Materials*, Elsevier Ltd, 174, 675–683, 2018.

Interdependence between Structural Engineering and Construction Management
Edited by Ozevin, D., Ataei, H., Modares, M., Gurgun, A., Yazdani, S., and Singh, A.

ISBN: 978-0-9960437-6-2

BEHAVIOR OF RC BEAM RETROFITTED USING ULTRA HIGH-PERFORMANCE CONCRETE UNDER IMPACT LOADS

MOHAMMED ALI AL-OSTA

Dept of Civil and Environmental Engineering, King Fahd University of Petroleum and Minerals, Dhahran, Saudi Arabia

Several new types of materials have recently been used as retrofitting materials for structural elements such as ultra-high performance concrete with steel fiber reinforcement (UHPFRC). These materials are used as jacking to enhance the strength and ductility reinforced concrete (RC) beams. Considerable attention has been focused on the response of retrofitted RC beam under static loads but the behavior of such beam under impact loading is somewhat lacking. Therefore, in this study, a 3-D finite element model (FEM) of retrofitted RC beams under impact loading using non-linear finite element software (ABAQUS) was investigated. Since experimental work on this topic is scarce, the FEM is validated using the results of retrofitted RC beam under static loads. The impact load was applied in ABAQUS as equivalent to an initial velocity of 2500 mm/s. A parametric study was carried out to study the flexural response of RC beams retrofitted with different thicknesses and strengthening configurations of UHPFRC under impact loading.

Keywords: UHPFRC, Finite element model, ABAQUS.

1 INTRODUCTION

Several techniques have been reported in the literature for the strengthening of reinforced concrete (RC) elements. Some of the conventional methods of strengthening and retrofitting of damaged RC beams include removal or replacement or RC jacketing, the addition of fiber reinforced polymer composites and external steel elements. Recently, several new techniques for the repairing of RC beams have been reported including using Carbon Fiber Reinforced Polymer (CFRP). Some of these techniques are labor-intensive and require a different level of artful detailing, high cost, disruption of building occupancy in terms of considerable loss of floor space and limited range of applicability. A major inadequacy of CFRP sheets is its early brittle failure due to the loss of bond. Mechanical anchorage was provided to fully enhance the bond of CFRP layers, by avoiding the early failure. The use of anchoring the FRP sheets in concrete may damage the anchorage zone and it will result in stress concentration zone and early brittle failure. Therefore, an ultra-high performance fiber reinforced concrete (UHPFRC) repairing system is another technique to strengthen or restore the damaged RC beams. Recently, Martinola *et al.* (2007) studied the use of a precast layer with a thickness of 40 mm and made of high-performance fiber reinforced concrete (HPFRC) to enhance RC beams. In order to overcome the above-mentioned issues and shortcomings, the non-seismically designed joint will be repaired and strengthened by using the UHPFRC jacket. Mahmud *et al.* (2013) studied the size effects of

UHPFRC on the flexural response of the beams. These researchers developed a numerical modeling using ABAQUS software. Both the experimental and numerical results showed that the size-effect in the flexural strength of the beams is almost negligible. Chalioris *et al.* (2014) investigated the use of thin reinforced self-compacting concrete (SCC) for the strengthening of conventional RC beams. Their results showed an increase in the strength with improved ductility and favorable failure behavior. It was concluded that the high strength self-compacting concrete (SCC) is a quick option for rehabilitation or strengthening the existing RC beams. Ruano *et al.* (2015) carried out an experimental work and a numerical modeling of the behavior of RC beams strengthened in shear with high-performance self-compacting concrete. Both studies from the experimental work and numerical simulation demonstrated that the fiber content not only prevented debonding from the concrete substrate but also enhanced the failure load of strengthened RC beams. Lampropoulos *et al.* (2016) investigated the efficiency of using UHPFC for strengthening of conventional RC beams. Different types of configuration of UHPFC layers were used, in the tensile, in the compressive sides and with a three-side jacket. The results showed a significant moment increase when three sides jacket was used. Al-Osta *et al.* (2017) have recently studied numerically and experimentally the performance of RC beams retrofitted with layers of UHPFC that had a thickness of 30 mm only. Different types of configuration of UHPFC layers were used, bottom side, 2 sides and 3 sides. The results indicated that beams retrofitted on three-side showed the highest failure load. Nasrin *et al.* (2017) studied numerically the effect of impact load on the behavior of bridge girder made of ultra-high performance concrete (UHPC) and reinforced concrete beams strengthened with CFRP. The results demonstrated that UHPC beams showed less deflection as compared to RC beams made of normal concrete.

The need for retrofitting the existing structures has been enormous last years. The most popular technique is using CFRP laminates for upgrading the deteriorated structures. CFRP possesses desired properties such as high-strength, corrosion protection, easy to apply and minimal size change. Besides all these advantages, CFRP system has some shortcomings, which are mainly related to the bond and the incompatibility problems Al-Osta (2018). Therefore, in this study, the UHPFRC panels attached with different thicknesses to the sides of RC beams were numerically investigated under impact loads.

2 EXPERIMENTAL TEST PROGRAM

A simply supported RC beams were tested under static four-point loading as can be seen in Figure 1. The vertical deflection at the mid-span and the corresponding load were measured and plotted. The results obtained from the experimental work and the finite element models were compared to validate the developed model.

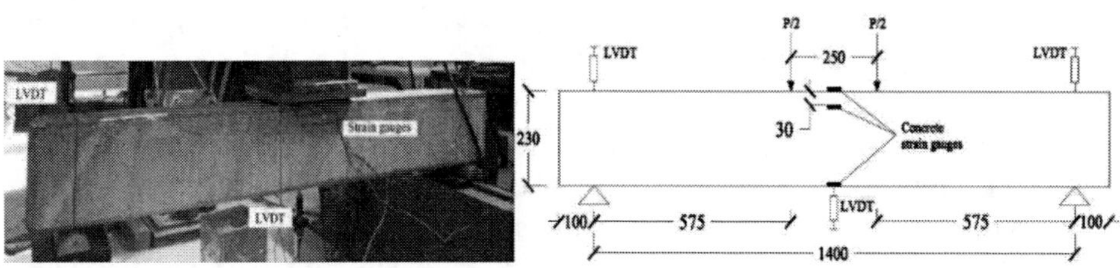

Figure 1. Beam test set-up for static load and instrumentation of beam flexural strength test.

3 FINITE ELEMENT MODELLING

The performance of normal concrete (NC) and UHPFRC is non-linear. In this study, a 3D finite element model (FEM) of beams strengthened with different configuration was developed using ABAQUS software to predict the response of the RC beams specimens under impact loads. The linear performance of materials was defined using the elastic modulus, E=34 GPa; and poisson's ratio, v =0.15 for NC and E= 46 GPa; v=0.18 for UHPFRC. The concrete damage plasticity model (CDPM) developed by Lubliner *et al.* (1989) and Lee and Fenves (1998) was utilized to capture the nonlinear behavior of NC and UHPFRC. The behavior of the materials in CDPM was defined using the parameters presented in Table 1 and nonlinear experimental data given by the author in Ref. Al-Osta *et al.* (2017). The damage parameters d_c and d_t for compressive and tension loads, respectively, in Eqs. (1) and (2) were used below Birtel and Mark (2006). The behavior of steel rebars and stirrups was modeled as an elastic-perfectly plastic with information obtained experimentally; f_y = 590MPa, E=200GPa.

Table 1. Parameters of CDPM for NC and UHPFRC Al-Osta *et al.* (2017).

Ψ (°)	*Eccentricity*	σ_{b0}/σ_{c0}	*K*	Viscosity parameter
NC/ UHPFRC				
36	0.1	1.16	0.667	0

where:

$\frac{\sigma_{b0}}{\sigma_{c0}}$ is the ratio of initial biaxial compressive yield stress to initial uniaxial compressive yield stress;

K is the ratio of the second stress invariant on the tensile meridian (TM) to that on the compressive meridian (CM).

Ψ is dilation angle.

$$d_c = 1 - \frac{\sigma_c E_c^{-1}}{\epsilon_c^{pl}\left(\frac{1}{b_c}-1\right)+\sigma_c E_c^{-1}} \quad (1)$$

$$d_t = 1 - \frac{\sigma_t E_c^{-1}}{\epsilon_t^{pl}\left(\frac{1}{b_t}-1\right)+\sigma_t E_c^{-1}} \quad (2)$$

where:

d_c and d_t = Damage parameters for concrete in compression and tension, respectively;
σ_c and σ_t = Compressive and tensile stresses, respectively;
E_c = Modulus of elasticity of concrete;
ϵ_c^{pl} and ϵ_t^{pl} = Plastic compressive and tensile strains, respectively;
b_c and b_t = constant values with range $0 < b_c$ and $b_t < 1$.

3.1 Type of Element

A 3D- eight noded linear brick element was utilized to model NC and UHPFRC. The steel rebars and stirrups were modeled as 2-noded linear 3D truss element. The contact between UHPFRC and NC was considered as a perfect bond. Figure 2 shows the geometry of models.

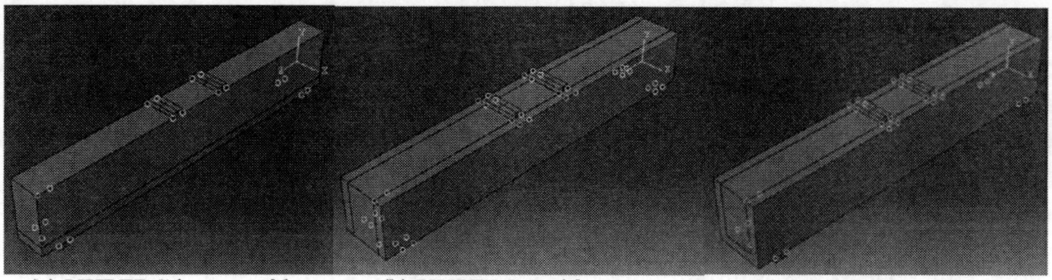

(a) UHPFRC bottom side (b) UHPFRC 2-sides (c) UHPFRC 3-sides

Figure 2. Geometry of models.

3.2 Validation of Model

The proposed FE model was validated with the results obtained from the experimental work for the beams tested under static loading. These beams were strengthened by different layers of UHPFRC (bottom, two-side and three-side). Figure 3 shows that the FEM can be used to predict the failure load and the performance of RC beams strengthened with layers of UHPFRC. More information about the validations of the FE model is available elsewhere Al-Osta *et al.* (2017).

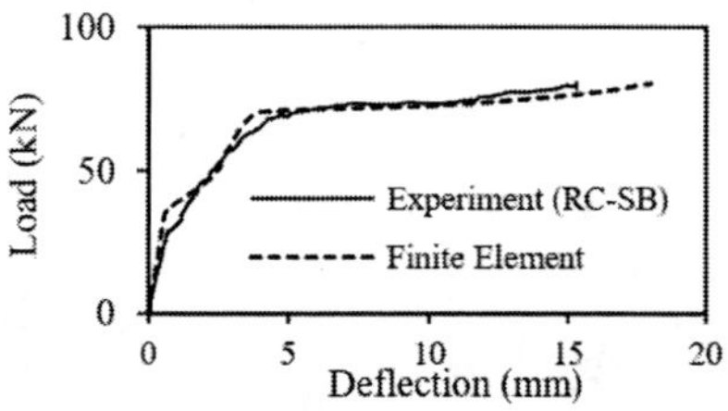

(a) UHPFRC bottom side- Static load

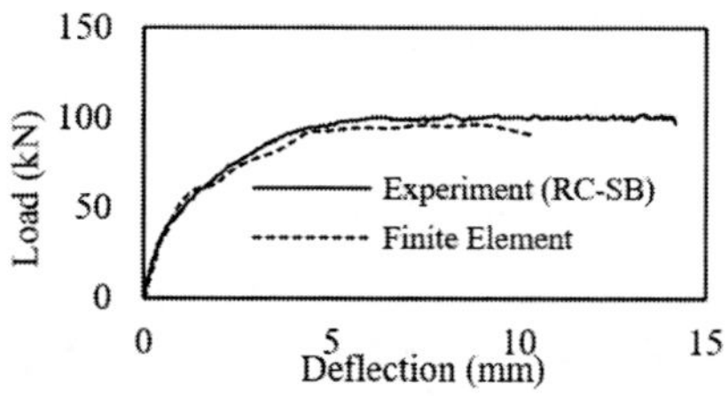

(b) UHPFRC 2-side- Static load

Figure 3. Experimental and FE load deflection behavior (a) RC-BS static load (b) RC-2S static load Al-Osta et al. (2017).

4 RESULTS AND DISCUSSION

In this study, the specimens retrofitted with different thicknesses of UHPFRC layers are subjected to an impact load characterized by an initial velocity of 2500 mm/s. that was applied at the mid-span. The performance of strengthened RC beams with different configuration and thicknesses of UHPFRC layers under impact load is presented. The beams were named FE-control, FE-BS-t1, FE-2S-t1, and FE-3S-t1. For example, FE-BS-20, FE stands finite element, BS is for strengthening from the bottom side and 20 mm stands for the thickness of the layer of UHPFRC. 2S and 3S indicate the strengthening from two and three sides, respectively. Figure 4 indicates that FE-control without strengthening was subjected to more deflection than RC beam specimens strengthened with layers of UHPFRC in the bottom, two-side or in three sides under the same impact load. It can be seen from the figure that increasing the thickness of UHPFRC layers results in decreasing the mid-span deflection. This could be attributed to the dense structure and high modulus of elasticity of UHPFRC layers used as compared to NC thereby enhancing the stiffness and resisting the strengthened beams to the applied load. For RC beam, the mid-span deflection of FE-BS-20mm, FE-2S-20mm, and FE-3S-20mm decreased by 15%, 29% and 37%, respectively, compared to the FE-control. Furthermore, the beams in Figure 4 are experiencing non-linear deflection behavior of beams with the time due to impact loading. Therefore, it is evident that the impact energy is initiating/triggered the non-linear material models. Figure 5 shows the relationship between the energy and time history of FE-control, FE-BS-20mm, FE-2S-20mm, and FE-3S-20mm. It can be confirmed from Figure 5(a) that the kinetic energy is constant as expected for all beams. It is maximum when the steel block bits the concrete surface of the beams and becomes zero when it is away from the beams. Figure 5(b) shows that the total energy for all beams is almost close to the kinetic energy, which indicates the accuracy of the model according to the conservation law of energy.

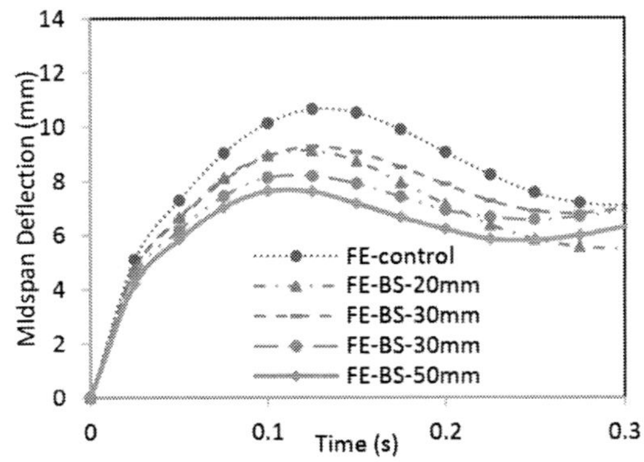

(a) UHPFRC bottom side- Impact load.

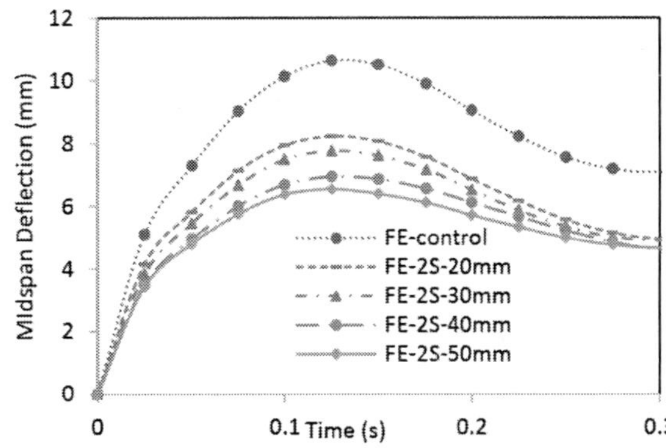

(b) UHPFRC 2-side- Impact load.

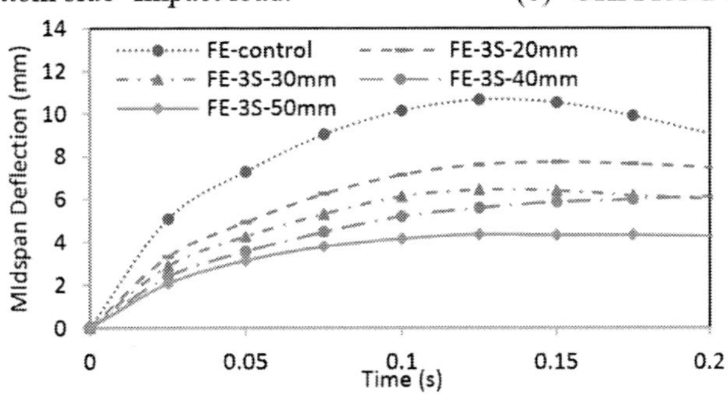

(c) UHPFRC 3-side- Impact load.

Figure 4. Deflection time history at mid-span.

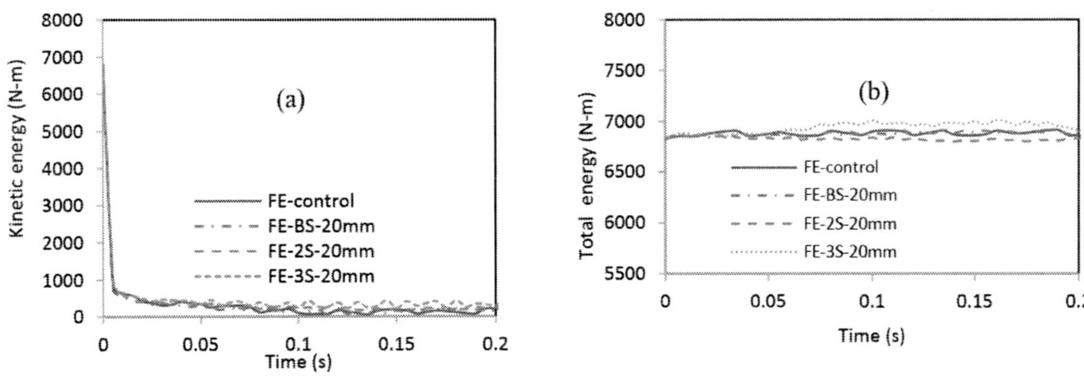

Figure 5. Energy -time relationship (a) Kinetic energy (b) Total energy in the model.

5 CONCLUSIONS

Based on the numerical simulation of the flexural performance of RC beams retrofitted with different thicknesses of UHPFRC layers under impact load, the following conclusions could be drawn:

- Since no experimental works could be found in the literature to study the effect of impact load on the performance of RC beams strengthened by layers of UHPFRC, the ABAQUS software package was utilized to preliminary investigate the effect of impact load on the behavior of retrofitted RC beams.
- It was noted that it is possible to reduce the mid-span deflections of RC beam by 15% through strengthening the beams with a bottom layer of UHPFRC with a thickness of 20 mm only. While the amount of mid-span deflection could be reduced to 37% by using layers of UHPFRC with a thickness of 20 mm that surround the beam from three sides.
- The total energy stored for all beams is almost close to the kinetic energy during the impact loading history, which indicates the accuracy of the model.

Acknowledgements

This study was made possible with the support of Department of Civil and Environmental Engineering and Research institute at King Fahd University of Petroleum and Minerals.

References

Al-Osta, M. A., Isa, M., Baluch, M. H., and Rahman, M. K., Flexural Behavior of Reinforced Concrete Beams Strengthened With Ultra-High Performance Fiber Reinforced Concrete, *Construction and Building Materials*, Elsevier, 134(1), 279-296, March, 2017.

Al-Osta, M. A., Exploitation of Ultrahigh-Performance Fibre-Reinforced Concrete for the Strengthening of Concrete Structural Members, *Advances in Civil Engineering*, 2018, 1-12, January, 2018.

Birtel, V., and Mark, P., *Parameterised Finite Element Modelling of RC Beam Shear Failure*, Ababqus User's Conference, 95-108, 2006.

Chalioris C. E., Thermou, G. E., and Pantazopoulou S. J., Behaviour of Rehabilitated RC Beams with Self-Compacting Concrete Jacketing - Analytical Model and Test Results, *Construction and Building Materials*, Elsevier, 55(31), 257-273, March, 2014.

Lampropoulos, A. P., Paschalis, S. A., Tsioulou, O. T., and Dritsos, S. E., Strengthening of Reinforced Concrete Beams Using Ultra High Performance Fibre Reinforced Concrete (UHPFRC), *Engineering Structures*, Elsevier, 106(1), 370-384, January, 2016.

Lee, J., and Fenves, G. L., Plastic-Damage Model for Cyclic Loading of Concrete Structures, *Journal of Engineering Mechanics*, 124(8), 892-900, August, 1998.

Lubliner, J., Oliver, J., Oller, S., and Oñate, E., A Plastic-Damage Model for Concrete, *International Journal of Solids and Structures*, Elsevier, 25(3), 299-326, January, 1989.

Mahmud, G. H., Yang, Z., and Hassan. A .M . T., Experimental and Numerical Studies of Size Effects of Ultra High Performance Steel Fibre Reinforced Concrete (UHPFRC) Beams, *Construction and Building Materials*, Elsevier, 48, 1027-1034, November, 2013.

Martinola, G., Meda, A., and Plizzari, G. A., *An Application of High Performance Fiber Reinforced Cementitious Composites for R/C Beams Strengthening*, FraMCos-6 Catania (Italy), Proceedings, 2007.

Nasrin, S., Ibrahim, A., Al-Osta, M., Khan, U., *Behavior of Retrofitted UHPC Beams Using Carbon Fiber Composites under Impact Loads,* in Structures Congress 2017.

Ruano, G., Isla, F., Sfer, D., and Luccioni, B., Numerical Modeling of Reinforced Concrete Beams Repaired and Strengthened with SFRC, *Engineering Structures*, Elsevier, 86(1), 168-181, March, 2015.

Interdependence between Structural Engineering and Construction Management
Edited by Ozevin, D., Ataei, H., Modares, M., Gurgun, A., Yazdani, S., and Singh, A.

ISBN: 978-0-9960437-6-2

IMPROVEMENT OF REINFORCEMENT METHOD OF R/C CYLINDRICAL SHELL WITH OPENING

TAKASHI HARA

Dept of Civil Engineering and Architecture, National Institute of Technology, Tokuyama College, Shunan, Japan

Reinforced concrete (R/C) chimney with a hole is prone to decrease the strength. Therefore, the particular reinforcements have been placed around the hole. Both meridional and hoop reinforcements have been placed additionally equivalent to the loss amount of reinforcements at the hole and the diagonal reinforcements have been placed to prevent the corner failures as well. Diagonal reinforcements in the shell were curved shape. It is the laborious work to place them. In this analysis, the improved reinforcement method was proposed instead of placing the conventional diagonal reinforcements. In the proposed methods, only the hoop and the meridional reinforcements were arranged around the corners instead of diagonal reinforcements. FEM analyses were done to evaluate the proposed reinforcing method. Numerical model was the cylindrical panel with a rectangular opening. From the numerical analyses, it was concluded that the R/C panel with proposed reinforcement method showed the same behavior as that with conventional reinforcement method.

Keywords: R/C Shell, Chimney, Inlet hole, Crack, FEM.

1 INTRODUCTION

A reinforced concrete (R/C) cylindrical chimney has a large opening at the lower portion to take in a flue-gas. The region around the hole is prone to decrease the strength. Therefore, particular reinforcements have been supplied around the hole. Figure 1 shows the Tupras stack collapsed under Kocaeli earthquake in 1999 (Huang *et al.* 2004). The dotted line denotes the collapsed part, and the chimney had a rectangular hole at the lower part. It was concluded that the cause of collapse was introduced by the wide opening and a lack of reinforcement around the corners.

Figure 1. Collapse of Tupras chimney (Photo by P.L. Gould 2004).

To improve the panel strength with hole, many researchers have been studied (e.g. Doh *et al.* 2006). In the same manner, the design code of the chimney with hole was recommended (ACI Committee 307 2008). The thickness of the concrete around the hole is increased and the meridional and hoop reinforcements were placed equivalent to the loss amount of reinforcements at the hole. In addition, the diagonal reinforcements were placed to prevent corner failures. However, when placing diagonal reinforcements in the shell, the reinforcing bars have a curved shape (Figure 2). Sami stated the difficulties to place them and the requirement of improving these works. (Sami *et al.* 2016).

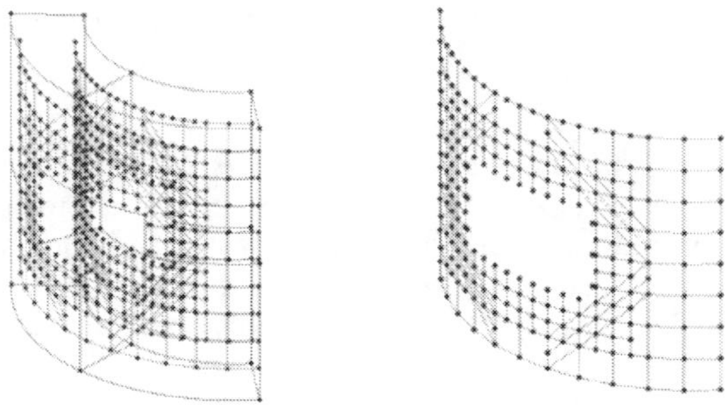

Figure 2. Current reinforcement around the Hole.

In this paper, the improved reinforcement method was proposed instead of placing the conventional diagonal reinforcements. In the proposed methods, only the hoop and the meridional reinforcements were arranged around the corners. The reinforcements equivalent to the diagonal reinforcements were placed at the same region. FEM analyses were done to evaluate the proposed reinforcing method. In numerical analyses, four types of model were examined. One model was R/C panel without hole and three models were R/C panels with hole. Specimens were cylindrical panel supported on hoop edges and were free along the meridional edges. The stiffness, the ultimate strength and the crack propagations were compared among these models.

2 NUMERICAL MODEL

Figure 3 shows the configuration of the numerical model. The model is a cylindrical panel. The radius and the opening angle are 693.75mm and 83.08°, respectively. The height and the thickness are 900mm and 40mm, respectively. The shell has the rectangular (264mm x 240mm) opening in the middle of the shell panel. The reinforcements are placed in both hoop and meridional directions. The reinforcement ratio is 0.35% in each inner and outer surface (ACI-ASCE Committee 334 2017). The concrete cover is 5mm. The reinforcing methods adopted in this paper are shown in Figure 4. Figure 4(a) shows the panel without hole. Figure 4(b) shows the panel with hole and without reinforcing around the hole edges. In Figure 4(c), the panel with hole has conventional reinforcements (ACI Committee 307 2008). Finally, in Figure 4(d), diagonal reinforcements are replaced to meridional and hoop reinforcements with the same reinforcement ratio as shown in Figure 4(c). The material properties adopted in these analyses are shown in Table 1.

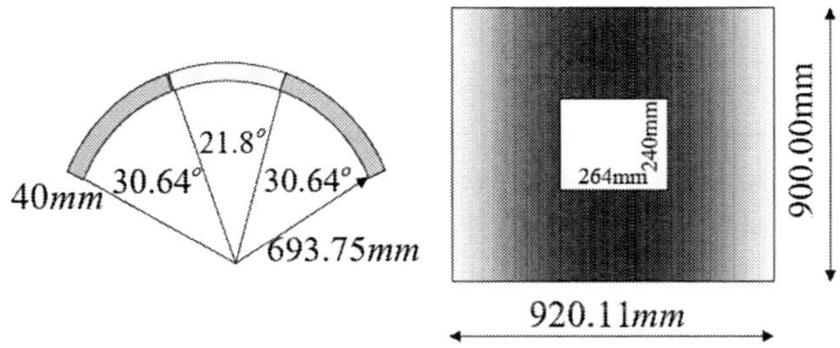

Figure 3. Configuration of R/C panel.

Table 1. Material properties.

CONCRETE		STEEL	
Elastic Modulus E_c	$22.3GPa$	Elastic Modulus E_s	$208GPa$
Poisson's ratio ν	0.20	Hardening Parameter H'	0.01
Compressive Strength f_c'	$34.4MPa$	Yield Strength f_y	$235MPa$
Tensile strength f_t	$3.44MPa$	Tensile Strength f_b	$402MPa$

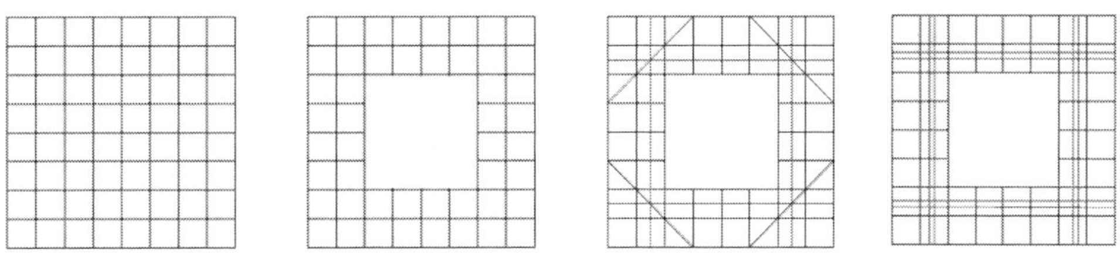

(a)Without hole. (b)With hole. (c) With hole+m+h+d. (d)With hole+2m+2h.

Figure 4. Reinforcing patterns.

3 NUMERICAL ANALYSIS

3.1 Numerical Method

In numerical analysis, Finite Element Method was applied. The concrete and the reinforcements were represented by twenty nodes hexahedral element and the thin surface model, respectively. The inelastic behavior of concrete possesses the recoverable strain components and irrecoverable strain components. Under tri-axial stress state, the yield function depends not only on the mean normal stress but also on the second deviatoric stress invariant (Hinton 1988). The yield condition of tri-axial compressive concrete is expressed by Drucker Prager criterion (Hinton 1988 and Hara 2017). Parameters adopted in this criterion were defined by the Kupfer's experiment (Kupfer and Hilsdorf 1969). It is assumed that the initial yield begins when the equivalent stress exceeds $0.3f'_c$ (f'_c: compressive strength of concrete). Also, the crushing condition of concrete is described as a strain control phenomenon and the crushing condition is defined as like as the yield function (Hinton 1988 and Hara 2017).

The response of concrete in tension is modeled as a linear-elastic brittle material; maximum tensile stress criteria are employed. After cracking, to evaluate the stiffness of concrete, the stress reduction normal to the crack plane is assumed as an exponential degradation curve.

The reinforcing bars are considered as steel sheets. The bilinear idealization is adopted in order to model the elasto-plastic stress strain relationship and both the tensile and the compressive states are governed by the same relationship.

3.2 Numerical Model

Figure 5 shows the numerical model. R/C panels are supported on hoop edges and are free on meridional edges. Supporting points are equally placed five portions on the hoop edges. In the case of the chimney, the structure is subjected to wind load. If the shell thickness is small comparing with the shell diameter, the out of plane bending is quite important. However, in this numerical model, the axial tensile and the compressive loads are applied to the meridional direction of R/C panel. The model is divided into 32 elements in hoop direction and 28 elements in meridional direction, respectively. Also, the R/C panel is divided into four elements through the thickness.

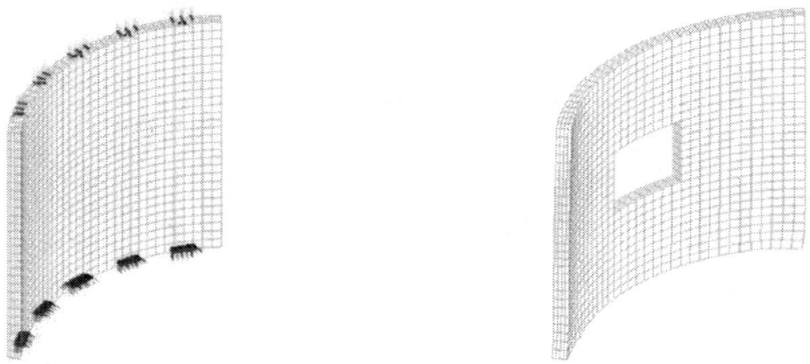

(a)R/C panel without hole. (b) R/C panel with hole.

Figure 5. Numerical model.

4 NUMERICAL RESULTS

4.1 R/C Cylindrical Panel in Compression

There is the possibility of the buckling of R/C cylindrical panel under compression. Figure 6 shows the numerical results of the eigenvalue analysis.

In the case of the R/C panel without hole, the local deformation arises at the end of the bottom hoop. However, in the case of the panel with hole and without reinforcement, the edge around the hole shows the buckling (see Figure 6(b)). It is important to reinforce the opening edge. When the reinforcing steels are placed around the hole (see Figure 4(c) (d)) the buckling mode is the same as that of the R/C panel without hole (see Figure 6(c)).

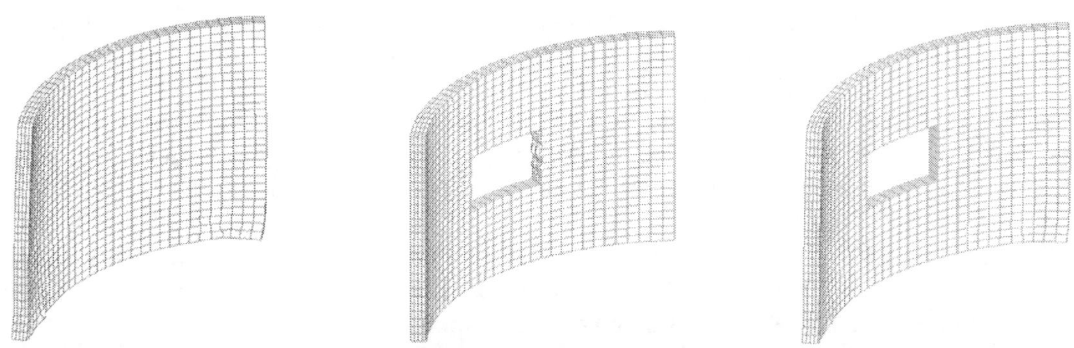

(a)Panel without hole. (b)Panel with hole (no reinf.). (c) Panel with hole (with reinf.).

Figure 6. Buckling mode.

Figure 7 shows the cracks and the crushing points depending on the reinforcing method under compressive load 140kN. Marks (*) show the crushing point and the rectangles show the cracking of the concrete. These are represented at each Gaussian integration point. At the loading points and supporting points, the crushing is detected and the portion between loading points shows the cracking. Also, upper and lower edges of the hole show the cracking. There are few differences between the results for each reinforcing method.

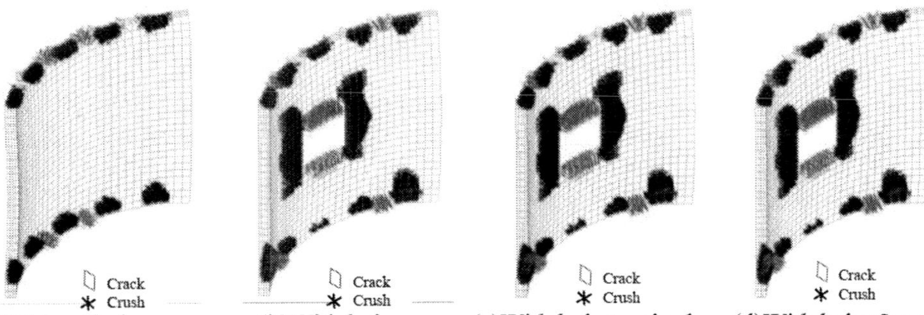

(a)Without hole. (b)With hole. (c)With hole+m+h+d. (d)With hole+2m+2h.

Figure 7. Cracks and crush in compression.

4.2 R/C Cylindrical Panel in Tension

Figure 8 shows the cracks when the tensile force is applied. In all cases, the element around the hoop edges shows the cracks. In the case of R/C panel with hole, tensile cracks appear at the hole corners. It is important to prevent the tensile crack propagation around the corners using the diagonal reinforcement.

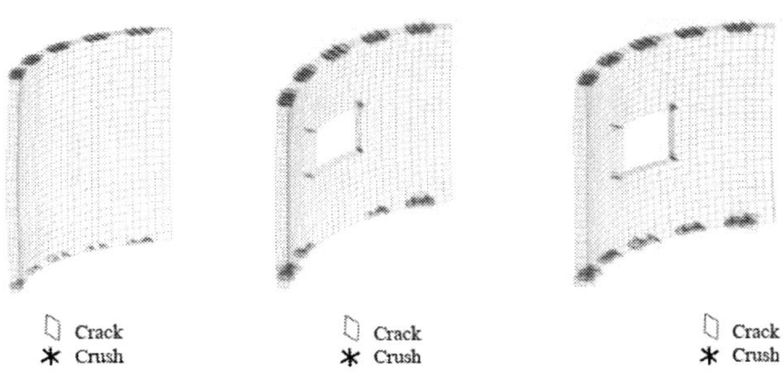

(a)Panel without hole. (b)Panel with hole (no reinf.). (c) Panel with hole (with reinf.).

Figure 8. Cracks in tension.

4.3 Load Deflection Curve

Figure 9 shows the load deflection relation of R/C panel in compression (a) and in tension (b). The abscissa and the ordinate show the displacement at the top center of the opening and the total applied load, respectively. Each mark shows the numerical results depending on the reinforcing method (see Figure 4).

In the case of the panel without hole, the stable load bearing behavior is shown up to failure. The stiffness of the panel with hole under the compressive force decreases gradually and the panel fails. The ultimate strength of R/C panel is almost the same between R/C panel with and without hole. In case of the R/C panel with hole, the models adopted in this analysis show the same ultimate strength. However, replacing of diagonal reinforcements with meridional and hoop reinforcements shows the larger strength.

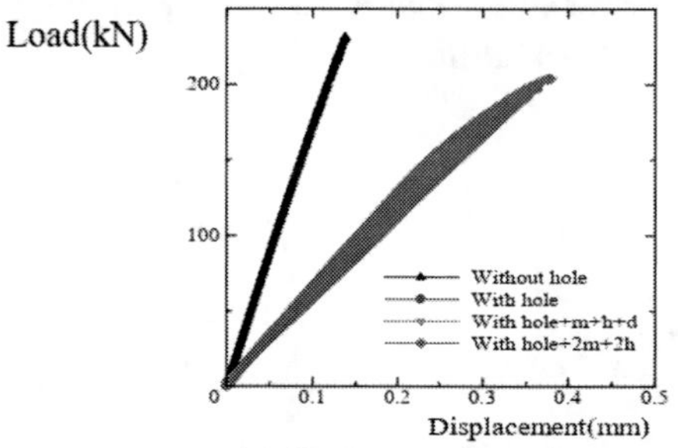

(a) Under compression.

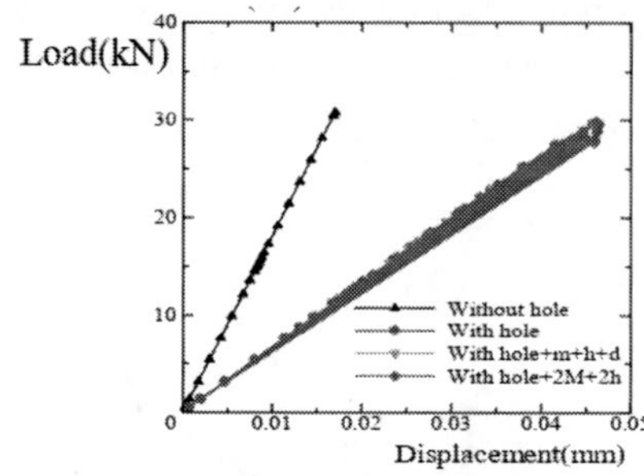

(b) Under tension.

Figure 9. Load deflection relation.

5 CONCLUSIONS

The deformation and the strength of R/C panel with hole was investigated by replacing the diagonal reinforcement with both meridional and hoop reinforcements. From the numerical analyses following conclusions are obtained.

(i) R/C cylindrical panel with hole degraded the stiffness under compressive load
(ii) The stiffened R/C panel with hole and R/C panel without hole show the same strength.
(iii) The ultimate strength of R/C panel with proposed reinforcements and a hole showed the same or the larger ultimate strength than that with conventionally reinforced R/C panel and hole.

The proposed reinforcement method was effective under the loading and the supporting conditions shown in this analysis. It is required to investigate the behavior of R/C panel with hole under other conditions combining with the experimental analyses.

Acknowledgements

This research work was done under the supports of Electric Technology Research Foundation of Chugoku.

References

ACI Committee 307, Code Requirements for Reinforced Concrete Chimneys and Commentary, ACI, 2008.

ACI-ASCE Committee 334, Concrete Shell Structures Practice and Commentary, 334.2, ACI, 2017.

Doh, J. H., and Fragomeni, S., Ultimate Load Formula for Reinforced Concrete Wall Panels with Openings, *Advances in Structural Engineering,* 9, 103-115, 2006.

Hara, T., Behavior of Reinforced Concrete Cooling Tower Shell under Lateral Load on Nonuniform Foundation, *Heat Transfer Engineering,* 38(11-12), 1127-1134, 2017.

Hinton, E., *Numerical Methods and Software for Dynamic Analysis of Plates and Shells*, Pineridge Press, Swansea, U.K, 1988.

Huang, W., Gould, P. L., Martinez, R., and Johnson, G., Non-linear Analysis of a Collapsed Reinforced Concrete Chimney, *Earthquake Engineering and Structural Dynamics,* 33, 485-498, 2004.

Kupfer, H., and Hilsdorf, K. H., Behavior of Concrete under Biaxial Stress, *ACI Journal*, 66, 656-666 1969.

Sami, A. K., Akhyniyazov, D., and Altay, S., An Investigation of Opening Reinforcement Configurations for Reinforced Concrete Chimneys, *ICCT 2016 International Conference on Industrial Chimneys and Cooling Towers, Rotterdam*, 213-218, 2016

Interdependence between Structural Engineering and Construction Management
Edited by Ozevin, D., Ataei, H., Modares, M., Gurgun, A., Yazdani, S., and Singh, A.

ISBN: 978-0-9960437-6-2

A PREVENTIVE STRENGTHENING STRATEGY FOR AGED STEEL STRUCTURES

NOZOMU TANIGUCHI[1], WEIWEI LIN[2], and MASAAKI HIROE[3]

[1]*Maebashi Institute of Technology, Gunma, Japan*
[2]*Waseda University, Tokyo, Japan*
[3]*Kobayashi Institute of Physical Research, Tokyo, Japan*

Steel columns are widely used in buildings, bridge piers, and railway platform roofs etc. With aging, those steel columns are vulnerable to corrosion and fatigue, and can deteriorate for a variety of reasons, especially in great earthquakes. In order to avoid the unrecoverable damage of existing steel columns, the high cost of the structure owners and the great impact on the public use, effective preventive maintenance methods on the aged steel piers are necessary. On this background, a strengthening method for aged steel columns by using new construction materials such as glass fiber reinforced polymer (GFRP) plates, rapid hardening concrete, rubber-latex mortar, and reinforcing bars, is introduced in this study. Depending on possible applied load directions and corrosion conditions, two specimens were used in the loading tests. Static loading tests were performed on steel columns with and without strengthening. Applied load and deflection relationship and strain distribution on original and strengthened columns were measured and compared. Moreover, three-dimensional FE models were built, and the numerical results were compared with the test results. Both experimental and numerical results indicate that the present strengthening method can significantly enhance both rigidity and ultimate load carrying capacity of aged steel columns.

Keywords: Steel columns, Rubber-latex mortar, Loading test, FE model.

1 INTRODUCTION

Steel column structures are used in buildings, bridges, railway platforms and other civil engineering structures around the world, due to their suppleness in both design and construction. In recent years, however, severe damages of aged steel columns were observed in earthquakes. For instance, a large amount of steel bridge piers damaged in the 1995 Hyogoken-Nanbu earthquake mainly because of either local buckling or rupture of welded corners (Chu 2011). During the 2011 Tōhoku earthquake, severe damage was caused to electrification poles on viaducts of Tōhoku Shinkansen. Based on the existing experience, Japanese design specifications have been revised and updated a lot, but they are mainly used for designing new columns. For aged steel piers constructed with old design methods, they are generally not satisfied to new design codes, thus they should be strengthened appropriately. On this background, seismic retrofit on existing steel column structures has become a major concern.

With the purpose of improving the seismic performance of existing steel columns, a preventive seismic retrofit method for existing steel column structures is proposed in this paper. Both static loading tests and numerical simulations were conducted to investigate the real effectiveness of the present strengthening method and its improvement on the seismic performance of existing column structures.

2 STRENGTHENING STRATEGY

In this study, a strengthening method using GFRP plates, rubber-latex mortar, rapid hardening concrete and reinforcing bars is proposed to strengthen the existing column structures. Rubber-latex mortar shows various abilities especially in adhesion bonding, waterproofing, shock absorption and abrasion resistance (Lin and Yoda 2013, Lin *et al.* 2014a), and significant contribution of rubber-latex mortar in enhancing bonding strength on steel-concrete interface was confirmed in previously performed bonding tests. In this method, the rubber-latex mortar is mainly used to enhance the integrity between aged structural steel and newly added surrounding concrete. Rapid hardening concrete is used for resisting the external loads together with the existing steel column, thereby increasing the rigidity and load carrying capacity. FRP plates are used as the formwork for concrete casting, and its contribution in the load carrying capacity is not considered in the design. Similar to other RC structures, reinforcements are used for controlling the crack width after concrete cracking.

This method was originally proposed by authors of this paper for strengthening superstructures of aged short-span railway bridges, and both laboratory loading tests and field loading tests were performed by authors' group (Lin *et al.* 2014b, 2014c, 2017, and Taniguchi 2016). Considering it outstanding effects in improving the structural performance of bridge superstructures, this study was performed to investigate its possibility in strengthening existing column structures.

3 EXPERIMENTAL PROGRAM

3.1 Test Specimens

Two steel columns with same cross-section (an H-shape section of 194×150×6×9) were designed. Each of the specimens was 1.42m in height and the loading point was decided at the height of 1.2m. Stiffeners were welded at the loading point to prevent buckling and crippling of the web before flexural failure. A 700mm square steel plate with thickness of 30 mm was used as the base plate and connected with the steel column by welding. To strengthen the original steel column, the rubber-latex mortar with the thickness of 5mm was used on the surface of the aged structural steel. Reinforcement with a nominal diameter of 10mm for longitudinal reinforcement and 6mm for stirrups were used. Finally, rapid hardening concrete was casted. The test specimens were fixed through connecting the base plate with the ground by using 4 bolts with a circular diameter of 60mm. The geometry dimension of the test specimen is shown in Figure 1.

3.2 Static Loading Test Set-up

The load is applied by a jack in horizontal direction to simulate the earthquake load. For columns like electrification poles or columns in railway platform roofs, the axial force can be ignored in the design due to their negligible deadweight. In terms of steel pier in bridge structures, the axial force is generally designed as no larger than 20% of its yield load. Therefore, the axial force in the column is not considered in order to simplify the loading test. However, the influence of the axial force on mechanical behavior of steel columns will also be investigated in this study, by using the numerical results.

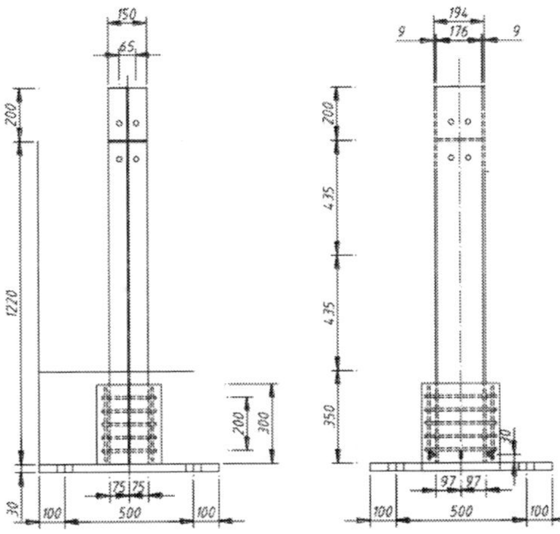

(a) Front view (b) Side View

Figure 1. Size dimensions of test specimen (unit: mm).

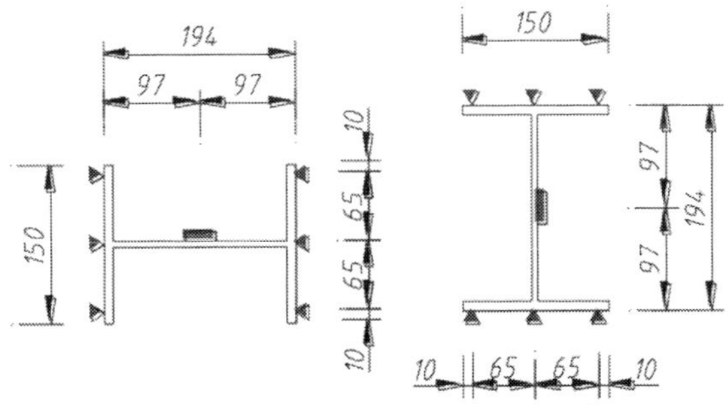

(a) Strong axis. (b) Weak axis.

Figure 2. Strain gauge locations (unit: mm).

As shown in Figure 2, there are strong axis and weak axis in a typical column with an H-shape cross-section. In order to investigate the effects of the proposed method in two different directions, the horizontal load was applied in both strong and weak axis directions, respectively. For this reason, two specimens were used in this study. Before strengthening, static loading tests were performed on the two original steel columns, including SC-S-1 (original steel column under load in strong axis direction) and SC-W-1 (original steel column under load in weak axis direction). The applied load was controlled in a low level to avoid any possible damage on columns. After strengthening, SC-S-2 (strengthened steel column under load in strong axis direction) and SC-W-2 (strengthened steel column under load in weak axis direction) were tested again to compare with the original steel columns without strengthening.

In order to confirm the sectional strain distribution, 7 strain gauges were used to measure the normal strain in the web and steel flanges near the bot-tom section, as shown in Figure 2. LVDTs were also used to measure the horizontal displacement at the loading section and two sections.

4 NUMERICAL MODELING

The simulation of test specimen was carried out using the finite-element method and DIANA software. Solid elements were used to simulate the concrete, and shell elements were employed to model the steel column. Numerical models of steel column before and after strengthening were shown in Figure 3. To simulate the bond connection between the steel column and surrounding concrete, and the connection between the ground and the base plate, interface elements used in the numerical analysis, as shown in Figure 4. Rebar elements were used for modeling the reinforcement, as shown in Figure 4.

Material tests were performed, and the nominal compressive strengths of concrete achieved after 7 days and 11 days (test day) were 36.2 N/mm^2 and 39.3 N/mm^2, respectively. Experimental tensile stress-strain curve proposed by Nakasu *et al.* (1996) was used to simulate the tension softening behavior of concrete. The compression behavior of concrete was referred to JSCE (2002). The stress-strain curve considering the strain hardening was used to simulate the structural and reinforcing bars (JSCE 2002). The yield strength of the H-shape column was 309 N/mm^2 according to the material test report. The yield strength of reinforcement was yield strength: 295 N/mm^2. The simplified interface model developed from Dörr's bond-slip model is used to model the steel-concrete interface (Dörr 1980).

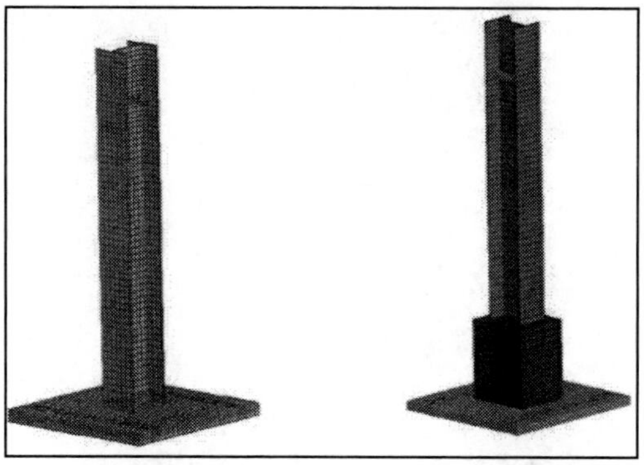

(a) Before strengthening. (b) Strengthened.

Figure 3. Numerical models for columns.

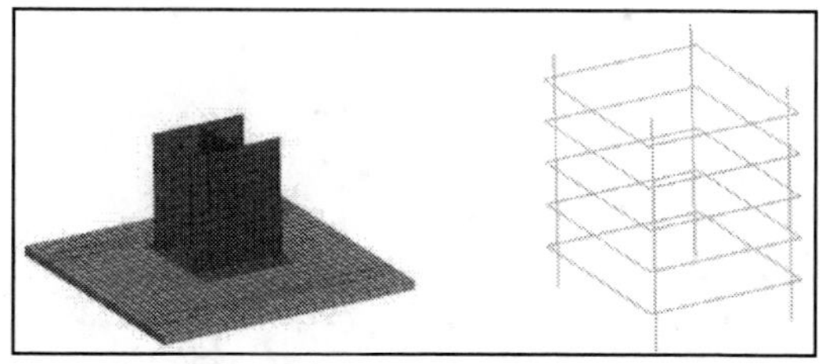

Figure 4. Simulation of interface and reinforcement.

5 RESULTS AND DISCUSSION

5.1 Applied Load in the Strong Axis Direction

When the steel column is subjected to a load in the strong axis direction (SC-S), the load-displacement curve obtained from the numerical analyses is compared to the experimental results as shown in Figure 5. For retrofitting of aged steel columns, increase of rigidities in service state and increase of load carry capacity in the ultimate state are major concerns for engineers. On the basis of the obtained results, both experimental and numerical results demonstrate increase of the column stiffness after strengthening. Taking the horizontal deflection of 6.3mm (the maximum displacement in loading test for SC-S-1) as an example, the test load changed from 24.9kN in SC-S-1 to 31.4kN in SC-S-2, in-creased by 26%. For the strengthened column, the numerical model was stronger than the test model, which might be due to the ignored residual stress at the welded section in the numerical analyses. The comparison between the numerical results also indicates that the ultimate load carrying capacity of the column has increased obviously after strengthening. Taking the horizontal displacement of 61mm as an example, the corresponding load increased from 79.8 kN in SC-S-1 to 98kN in SC-S-2, thus 22.8% increase was confirmed.

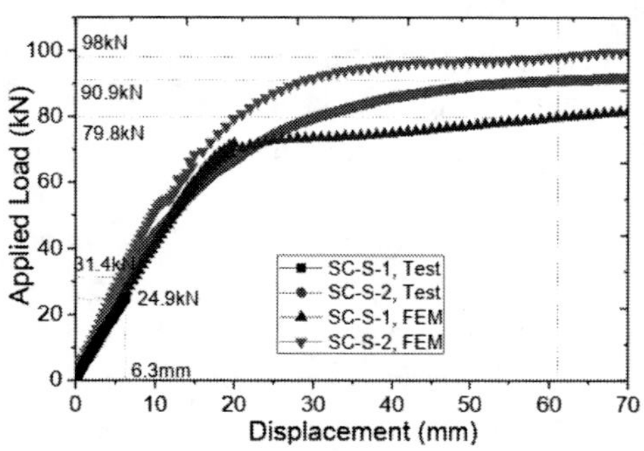

Figure 5. Load-deflection curve for SC-S.

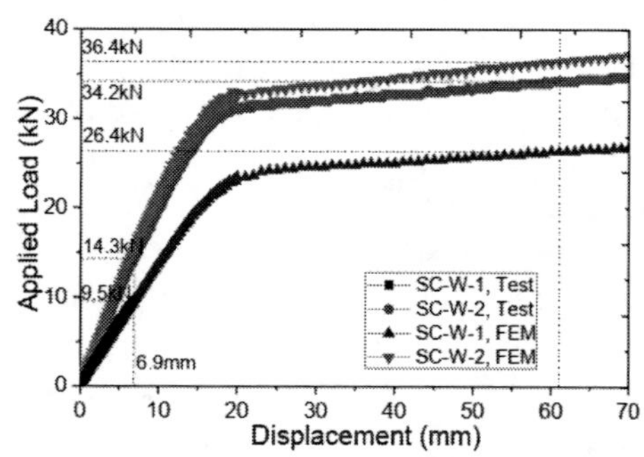

Figure 6. Load-deflection curve for SC-W.

5.2 Applied Load in the Weak Axis Direction

For the steel column under load in the weak axis direction (SC-W), the applied load and deflection relationship was given in Figure 6. According to the comparison, significant increase of the column stiff-ness and ultimate load carrying capacity in the weak axis direction were also confirmed. Taking the horizontal deflection of 6.9mm (the maximum displacement in loading test for SC-W-1) as an example, the test load changed from 9.5kN in SC-W-1 to 14.3kN in SC-W-2, increased by 51%. The ultimate load was increased from 26.4kN in SC-W-1 to 36.4kN in SC-W-2, thus 38% increase can be confirmed. The significant increase of the rigidity and load carrying capacity demonstrate the improvement of the seismic performance during earthquakes. Also, the comparison indicates that the numerical results agree well with the test results, indicating the validity of the numerical models used in this study.

5.3 Effects of Axial Force

In this study, the axial force for columns was ignored in order to simplify the loading tests and numerical analyses. This assumption will not affect the behavior for steel columns with negligible axial force, e.g. electrification poles or columns in railway platforms. For other columns subjects to unneglectable axial forces, e.g. steel piers, the effect of due axial force on the structural behavior of column structure is still known and needs to be investigated.

Phase analyses (two steps) were performed in the numerical analyses. In phase-1, the axial force considering the loads due to a bridge superstructure was applied at the top of the column. The axial force was taken as 20% of its yield load, according to the existing studies in Japan. Thereafter, the horizontal load was applied in phase-2 until the ultimate state of the column models.

For a column subjected to a horizontal load applied in strong axis, the load-displacement curves of both original and strengthened columns with and without considering axial force were shown in Figure 7 (a) and Figure 7 (b), respectively.

It is found that if the axial force was considered, the rigidities of columns (both original and strengthened columns) were increased slightly. However, no matter the axial force is taken into account or not, the effects of the proposed strengthening method can be confirmed. In addition, the ultimate load was affected little by the applied axial force. Therefore, it can be concluded that the present strengthening method is still effective even if a certain level (≤20%) of axial load is considered. Thus, this method can also be used for strengthening steel columns with axial force, like steel piers.

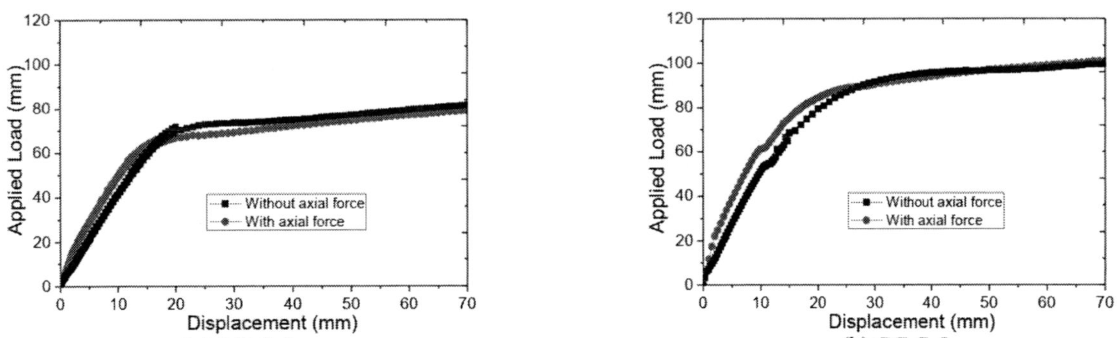

(a) SC-S-1. (b) SC-S-2.

Figure 7. Load-displacement curve between numerical models with and without axial force.

6 CONCLUSIONS

A strengthening method for existing steel columns was introduced in this paper. In order to confirm the effect of this method, static loading tests were performed on two original steel columns and columns strengthened with rubber-latex mortar, rapid hardening concrete and reinforcing bars. Detailed load-deformation response, and strain distribution on the steel columns were reported in this paper. Easily obtained construction materials and simple operated construction methods make this method competitive in comparison with other repairing methods. Static loading tests were performed to confirm the effects of the present strengthening method. On the basis of the test results, the present strengthening method was proved to be effective for increasing both rigidity and load carrying capacity of the aged steel columns, resulting in improvement of seismic performance of aged steel columns.

A 3-D nonlinear finite element model was established for simulating the present test specimens. The ultimate load carrying capacity, load-displacement curves and strain development process predicted by FE models agree well with those obtained from the experiments, which demonstrates the accuracy and efficiency of proposed FE models. In addition, the influence of axial force in steel columns was investigated in the numerical analyses. For a column subjected to an axial force in service condition (where the axial force is relatively small compared with its yield strength), the proposed strengthening method is still effective.

References

Chu, K., *Study on Seismic Retrofitting Method for Existing Thin Steel Members*, PhD Thesis, 2011.

Dörr, K., *Ein Beitrag zur Berechnung von Stahlbetonscheiben unter besonderer Berücksichtigung des Verbundverhaltens*. PhD Thesis, University of Darmstadt, Darmstadt, 1980.

JSCE, *Standard Specification for Concrete Structures*-2002, Japan Society of Civil Engineers, 2002.

Lin, W., and Yoda, T., Experimental and Numerical Study on Mechanical Behavior of Composite Girders under Hogging Moment, *International Journal of Advanced Steel Construction,* 9(4), 309-333. 2013.

Lin, W., Yoda, T., Taniguchi, N., Kasano, H., and He, J., Mechanical Performance of Steel-Concrete Composite Beams Subjected to a Hogging Moment, *Journal of Structural Engineering,* ASCE. 140(1), 04013031, 2014a.

Lin, W., Yoda, T., and Taniguchi, N., Rehabilitation and Restoration of Old Steel Railway Bridges: Laboratory Experiment and Field Test, *Journal of Bridge Engineering*, ASCE. 19(5), 04014004, 2014b.

Lin, W., Yoda, T., Taniguchi, N., Satake, S., and Kasano, K., Preventive Maintenance on Welded Connection Joints in Aged Steel Railway Bridges, *Journal of Constructional Steel Research*, 92, 46-54, 2014c.

Lin, W., Yoda, T., Taniguchi, N., Hansaka, M., Satake, S., and Sugino, Y., Renovation of Existing Steel Railway Bridges: Field Test and Numerical Simulation, *Advances in Structural Engineering*, 1-15. 2017.

Nakasu, M., and Iwatate, J., Fatigue Experiment on Bond Between Concrete and Reinforcement, *Transaction of JSCE,* 426, 852-853. 1996.

Taniguchi, N., Fujiwara, Y., Lin, W., Yoda, T., Matsuo, H., and Kubo, T., On the Evaluation of Actual Stiffness of a Railway Composite Bridge with the Help of FE Analysis, *Journal of Japan Society of Civil Engineers, Ser. A2 (Applied Mechanics), JSCE,* 71(2), 785-793. 2016.

Interdependence between Structural Engineering and Construction Management
Edited by Ozevin, D., Ataei, H., Modares, M., Gurgun, A., Yazdani, S., and Singh, A.

ISBN: 978-0-9960437-6-2

STRENGTHENING METHOD OF UNREINFORCED MASONRY WALLS BY BRACE ELEMENTS

ARIUNAA GANBAATAR and HARUYUKI YAMAMOTO

Development Technology, Hiroshima University/IDEC, Hiroshima, Japan

The ductility and strength properties of lightweight steel brace element for structural strengthening of unreinforced masonry (URM) wall is reported in this study. The performance of URM wall without brace and with brace is evaluated by an elastoplastic nonlinear analysis including the yield criterion, flow rule and hardening rule. Masonry constituent is implemented into two-dimensional finite element model developed in Fortran code. The brittle failure of URM wall is analyzed using by an interface element to model bricks and joints. Two types of steel brace elements which can resist only tension and both tension and compression have been considered to increase the ductility of URM wall. The comparison between experimental and numerical results adopted in simplified-micro masonry modeling is illustrated with the relationship between force and displacement. The ductility and strength of masonry wall with the tension and the tension-compression brace element were different results of URM wall by in-plane pushover analysis. As for the simplification for a calculation of masonry wall, equivalent masonry model is discussed and besides diagonal brace elements, the additional vertical brace element is assumed for the high masonry wall by the numerical analysis.

Keywords: Masonry modeling, Joint element, Finite element, Elastoplastic analysis.

1 INTRODUCTION

Most of the low-rise buildings in Ulaanbaatar city, Mongolia covers the clay brick masonry buildings. After the Govi-Altai earthquake occurred in a remote area of Mongolia in 1957 (Mw 8.1), seismic building code for masonry structure of Soviet Union was being adopted in Mongolia until 1990. The unreinforced masonry buildings constructed before the Govi-Altai are still existing without strengthening throughout Mongolia. There is a need to enhance performance of their buildings under seismic load. Scientists noted that various influences of masonry structures depend on two different material properties and quality of workmanship (Lourenço and Rots 1997). As such, masonry modeling is complex. Also, the weak connection between brick and mortar is assumed to be the ability to not endure the tension force. To increase the ultimate strength and ductility of masonry wall, diverse strengthening methods are employed all over the world.

The purpose of this study is to develop a low cost and effective strengthening method. A suitable strengthening method is considered by the steel brace element. In our study, two types of brace elements are analyzed to increase the ductility capacity of masonry wall in the numerical analysis. The first type of the brace element only resists the tension force and the second type is the tension-compression brace element. Also, if the bending moment occurs in the wall, additional steel braces are installed in the vertical direction of the wall. Finally, the force-displacement relation of the numerical and experimental analysis was compared with braces and without braces for the brick wall.

2 FINITE ELEMENT ANALYSIS

In this study, the elastoplastic nonlinear analysis of a two-dimensional load-bearing masonry wall is highlighted in the finite element program. Simplified micro modeling is utilized in the finite element analysis to consider the failure mechanism (see Figure 1). The connection between brick and mortar is assumed together in the model because of presuming zero thickness of mortar (Lourenço 1996).

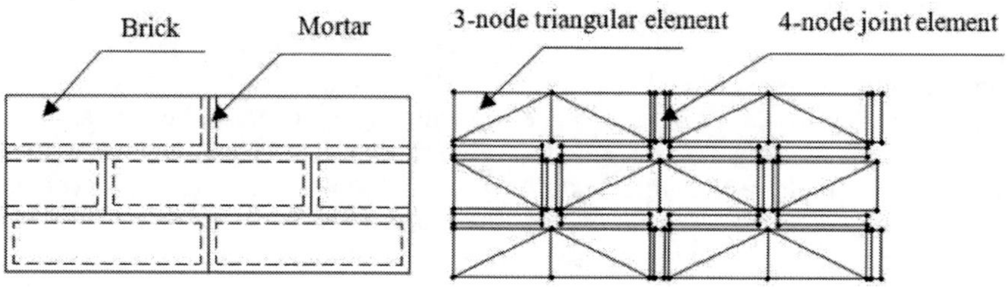

Figure 1. Finite element modeling of masonry structure.

Brick is discretized into 3-node triangular elements and joint elements in the vertical and horizontal direction are developed by 4-node interface elements in the finite element analysis to calculate a masonry wall. The interaction between brick and mortar displays the stress and displacement relation. It shows in Eq. (1) used in an elastic case.

$$\sigma = D^e \times \varepsilon \tag{1}$$

For a 2D formulation, D^e = diag[K_s, K_c], ε=(Δu, Δv)^T. k_s and k_c denote shear and compression stiffness of joint element, Δu and Δv are relative displacements. When defining the relation between stress and displacement, yield function for slip and potential function related to non-associated flow rule are adopted in the elastoplastic case. This means that the adhesion of brick and mortar is assumed in the present study. The non-associated flow rule presents in shear and compressive stress. Eq. (2) shows stress-displacement relation based on elastoplastic theory.

$$D^{ep} = \left[D^e\right] - \frac{\left[D^e\right]\cdot\left\{\frac{\partial Q}{\partial \sigma}\right\}\cdot\left\lfloor\frac{\partial F}{\partial \sigma}\right\rfloor\cdot\left[D^e\right]}{\left\lfloor\frac{\partial F}{\partial \sigma}\right\rfloor\cdot\left[D^e\right]\cdot\left\{\frac{\partial Q}{\partial \sigma}\right\}} \tag{2}$$

F is yield function for slip, *Q* is a potential function. Stiffness matrices of 3-node triangular brick elements and 4-node interface joint elements state the superposition principle in Fortran program developed on finite element analysis whereas global stiffness matrices for a system consider integrating the stiffness matrices of each element. It shows in Eq. (3).

$$[K]^{global} = \Sigma[K]^{brick} + \Sigma[K]^{mortar} \tag{3}$$

$[K]^{global}$ – global stiffness matrices, $[K]^{brick}$ – stiffness matrices of triangular elements and $[K]^{mortar}$ – stiffness matrices of interface joint element.

3 STRENGTHENING METHOD

The purpose of this study is to choose a low cost and effective strengthening method. As for the unreinforced brick masonry wall, the shear capacity is low due to lateral loads. In order to increase the ductility of the brick wall, we chose the steel brace element to strengthen the unreinforced brick masonry wall. The tension brace and the tension–compression brace were computed for the unreinforced brick wall in the numerical analysis. Brace is installed by diagonal direction on the wall. As for the numerical analysis we considered a masonry sample developed by Raimakers and Vermelfoort. This wall size is 953 mm x 950 mm x 100 mm, a brick size is 204 mm 98 mm x 50 mm and mortar thickness is 10 mm in Figure 2. An analytical model of the masonry wall sample is exhibited with brace in Figure 3.

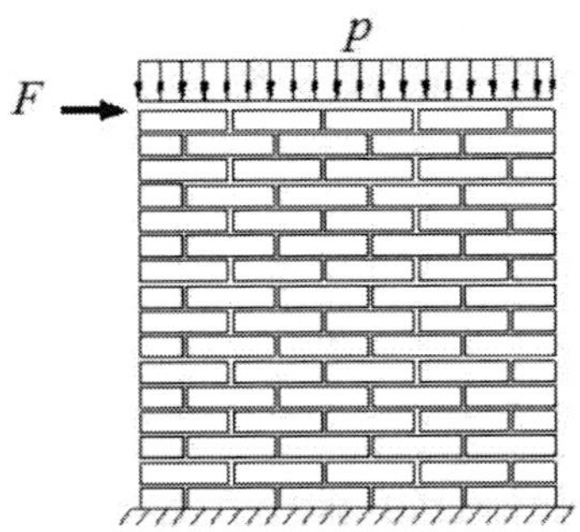

Figure 2. A masonry wall sample.

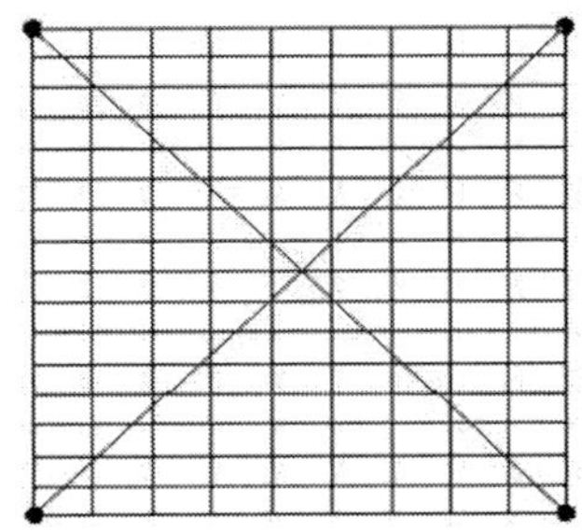

Figure 3. An analytical model of the masonry wall with braces.

The unreinforced masonry wall with 10cm^2 cross-section area of brace element was computed and compared with the brick wall without brace elements. The result of the numerical analysis with diagonal brace elements shows the load-displacement diagram by 300 kN/m^2 initial stress in Figure 4. Figure 4 emphasizes a relationship between the load-displacement of the wall compared with both braces and without braces. Masonry is brittle material. Therefore, the brick wall without braces are suddenly broken as shown in Figure 4.

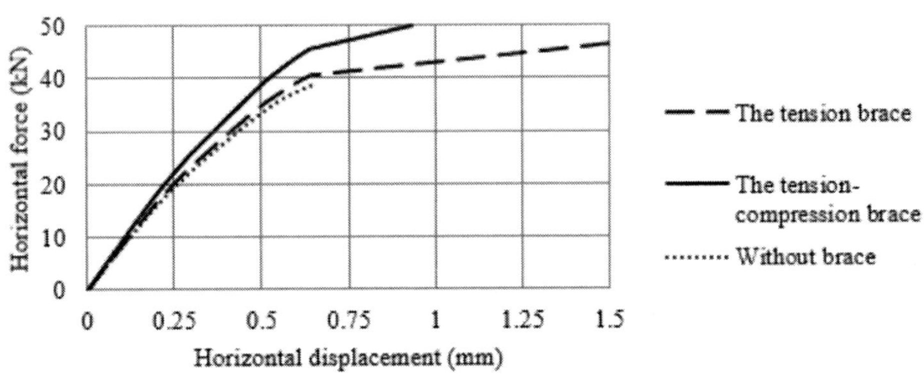

Figure 4. The load-displacement diagram of the brick wall with braces and without braces of the numerical analysis.

The brick wall with the tension braces can be more capable of bearing. The principal goal of brace elements is to enhance the ductility of the masonry wall. But the tension-compression brace element is not enough to increase the ductility capacity of the wall than the tension brace elements. If size of masonry wall increases in the vertical direction, the bending moment besides shear forces will happen in the wall under the lateral force.

Also, the data preparing is the time consuming for the simplified-micro modeling in the numerical analysis and a memory area of the personal computer needs a more large capacity. To avoid the circumstance, we considered the equivalent model of the brick masonry wall. The equivalent model is shown in Figure 5. In this model, three different types of wall (case A, case B and case C) were used to compare the effectiveness of an additional vertical bar element. Case A is the brick wall without brace elements. Case B is the brick wall with only diagonal brace elements. Case C is the brick wall with both diagonal and vertical brace elements.

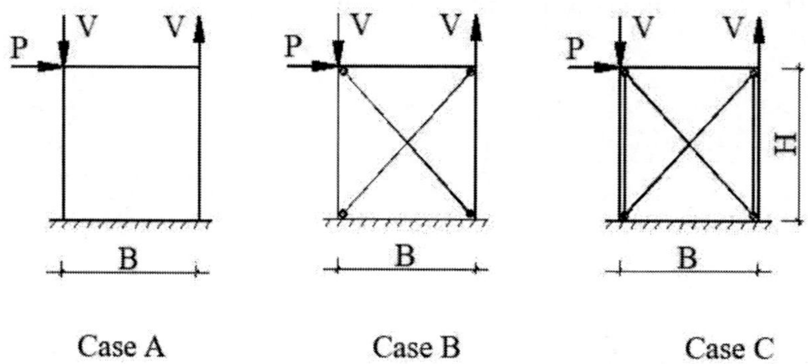

Figure 5. The equivalent model of the high masonry wall without braces and with braces.

If the solid brick wall (953 mm x 950 mm x 100 mm) is assumed one story, we increased it until five stories. Figure 6 shows the load-displacement relation of the top point of the five stories masonry wall (n=5) compared with cases A, B and C. As shown in Figure 6, the ultimate horizontal force of case C is four times more than case A and two times more than case B. Therefore, the vertical bar elements are effective to perform the lateral loads.

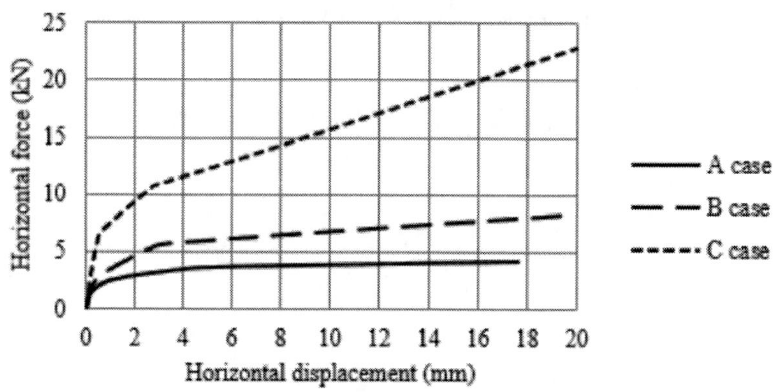

Figure 6. The load-displacement diagram of cases A, B and C (the story number n=5).

4 EXPERIMENT

4.1 Material Test

Experiments were conducted to define the elastic and inelastic properties of masonry material employed in this study. Cement sand ratio of mortar is 1:10 in this study. The test specimens were constructed by a brick size 210 mm x 100 mm x 60 mm. Elastic properties of the material (Young's module and Poisson's ratio of brick, and shear and compression stiffness of the joint element) were acquired in accordance with brick and mortar tests. Inelastic properties (frictional coefficient and cohesion of the joint element) were obtained from the brick–mortar interface test. Material test results are presented in Table 1.

Table 1. Elastic and inelastic properties of brick and joint element.

Element	Elastic				Inelastic	
	Young's module (MPa)	Poisson's ratio	Shear stiffness (kN/m^3)	Compressive stiffness (kN/m^3)	Frictional coefficient	Cohesion (kPa)
Brick	$1.567 \cdot 10^4$	0.13	-	-	-	-
Joint	-	-	$1.50 \cdot 10^6$	$3.45 \cdot 10^6$	0.54	40.5

4.2 Wall Test

To simulate empirical condition, the solid clay brick walls were tested under compression and an incremental lateral loads. The brick wall size is 650 mm x 480 mm x 100 mm and mortar thickness is 10 mm. Two types of wall (with brace and without brace) are tested to characterize the load-displacement relation. The steel brace used for this experiment is 9 mm diameter bar on both side of the wall face in Figure 7. An actual diameter of a bar element to perform due to lateral force is 7 mm. A load-displacement relationship in Figure 8 from finite element analysis have been compared with experimental values. The load-displacement diagram from the finite element analysis is an approximate with the experimental values of the URM wall. A result of the numerical analysis is slightly stiffer than the experimental result. This is related to model like homogeneous material for the finite elements as compared with the non-homogeneous real wall. The ultimate force presumed by the numerical analysis is 3.7 kN which is more than the force of 3 kN from the experiments.

Figure 7. The brick wall specimen without braces.

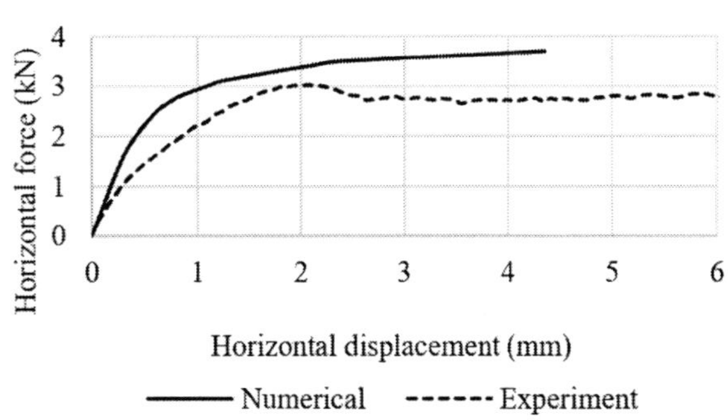

Figure 8. Load-displacement diagram of the brick wall.

The failure mechanism of the brick wall without brace element occupied the sliding of the bottom part of the wall (see Figure 9). This sliding shear failure usually reveals in case of the weak mortar, stiff brick and low vertical load (Tomaževič 1999). As for the wall with brace elements, there are not any damages to both joint element and brace element.

Figure 9. The brick wall specimen with braces.

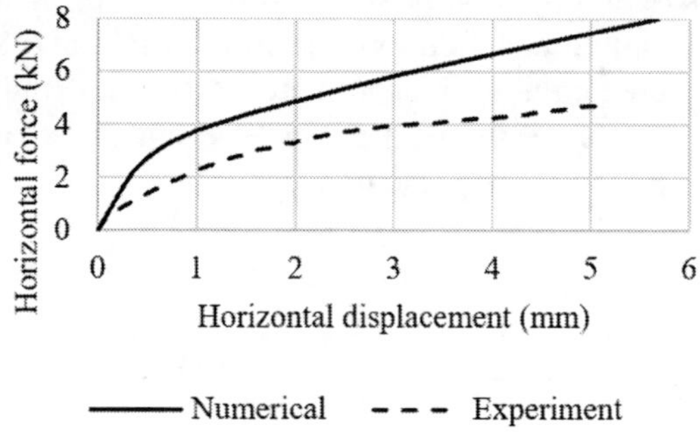

Figure 10. Load-displacement diagram of the brick wall with braces.

Figure 10 shows that load-displacement diagram of URM wall strengthened by braces. The numerical analysis is little higher than the experimental result. The ultimate force of the finite element analysis is 8 kN which is higher than 4.3 kN from the experimental data. It is also related to assume as homogenous modeling for the finite element analysis. Strengthening technique used by steel brace elements is significant to increase in-plane strength and ductility of the unreinforced brick wall.

5 CONCLUSION

This research emphasized the numerical analysis and experimental study of the unreinforced brick masonry wall without braces and with braces under the static vertical and horizontal forces. An elastoplastic non-linear finite element analysis was dealt with to presume the calculation of URM wall and URM wall with brace elements. The numerical results were received employing finite element software Fortran. Selecting the simplified micro modeling of masonry structure is appropriate to assess the failure mechanism and also, the simple finite elements is a reasonable model for assessing in the calculation. URM wall with the tension brace element is effective to increase the ductility capacity of the brick wall. But the tension-compression brace element increases the ultimate strength of the brick wall. The vertical bar elements reduce the bending moment influence than only using diagonal braces on the brick wall.

References

Lourenço, P. B, *Computational Strategies for Masonry Structures*, PhD Thesis, Delft University of Technology, Delft, https://doi.org/ISBN 90-407-1221-2, Faculty of Civil Engineering, 1996.

Lourenço, P. B., and Rots, J. G, Multisurface Interface Model for Analysis of Masonry Structures, *Journal of Engineering Mechanics-Asce*, 123(7), 660-668, 1997.

Tomaževič, M., *Earthquake-Resistant Design of Masonry Buildings (Vol. 1),* Imperial College Press, https://doi.org/10.1142/9781848160835, 1999.

Interdependence between Structural Engineering and Construction Management
Edited by Ozevin, D., Ataei, H., Modares, M., Gurgun, A., Yazdani, S., and Singh, A.

ISBN: 978-0-9960437-6-2

MODELING CONSTRUCTION ORGANIZATION PERFORMANCE IN POST-CONFLICT ENVIRONMENT

QAIS AMARKHIL and EMAD ELWAKIL

School of Construction Management, Purdue Polytechnic Institute, West Lafayette, USA

In the field of the construction industry, research work has widely focused on identifying Key performance indicators and critical success factors without assessing the impact of conflict environment factors. This study focusses on the impact of post-conflict environment factors on local construction organization performance. This paper presents a framework for improving construction organization performance in a post-conflict environment. The proposed framework consists of four stages: identify post-conflict environment impacting factors, determine critical success factors (CSFs), determine key performance indicators (KPIs), and adopt the best strategy to improve performance. Analytical hierarchy process (AHP) and multiple linear regression (MLR) modeling has been used to analyze quantitative and qualitative variables obtained from the literature and expert opinion through comprehensive literature search, meetings, and survey to determine critical success factors and to identify performance improvement strategy. The study finding suggests that twenty factors from the questioner have a critical impact on the identified five performance measures. The presented CSFs helps the organization management team to consider the impact of these factors on their firm and to formulate a competitive strategy in a post-conflict environment.

Keywords: Performance improvement framework, Critical success factors (CSFs), Key performance indicators (KPIs), Analytical hierarchy process (AHP), Multiple linear regression analysis (MLR).

1 INTRODUCTION

Environmental uncertainty and constraints have the potential to impact any organization performance (Grewal and Tansuhaj 2001). The external environment provides the organizations with the inputs which impact the internal process of an organization where the external environment impacting factors are not in direct control of organization management team (Farmer and Richman 1964, Schein 1965). The external environment is a source of constraints, opportunities, and uncertainties, which affect the organization performance with respect to its business form (Bourgeois 1980). The Cambridge Business Dictionary (2017) define the external environment as" the conditions and events outside a company that affects the way its operations." Lee and Choi (2003), and Elenkov (2002) describe organization performance the degree to which the organization attained its business objectives. Therefore, organization success and failure are profoundly impacted by the external environment.

Construction organization performance assessment and measurement have received significant attention in recent years to meet the construction industry challenges and competency. Many researchers have developed performance prediction models and methodologies to help construction organization to achieve profit and success in the market. For instance, Horta *et al.*

(2009) studied Portuguese companies to develop a framework to assess construction organization performance, Elwakil *et al.* (2009) have determined 18 CSFs for the organization performance assessment in developed and developing countries such as the USA, Canada, and Egypt. Chan and Chan (2004) has studied construction companies in Hong-Kong to develop a set of key performance indicators for the construction industry success. And, Abraham (2003) has studied top 400 U.S. companies to identify the critical success factors methodology to enhance construction organization success. However, questions can be probed that are these all performance frameworks and critical success factors applicable to a different environment? Likewise, are these identified factors have the same impact in another environment on the organizational performance, for instance, what would be the impact of these developed success strategies on organizations performance in the post-conflict environment? The post-conflict countries environment is significantly different from the developed countries or developing countries environment.

Post-conflict countries are that have suffered from the civil war or other internal conflicts which must embark for reconstruction and economic recovery, and on social and political reforms to provide the foundation for peace and democracy (Del Castillo 2001). There is a high level of uncertainty, various constraints, and some unstable opportunities in the post-conflict environment (Haughton 1998). In addition to the political and security instability, there are ubiquitous features of the post-conflict environment economies which differentiates it from the non-conflict environment. For example, some of the post-conflict environment features are as: poor infrastructure, high inflation, weak financial system, small abnormal industry sector, undermined institutions, and as well as there are: financial support from donors and the diaspora pool for the country development (Haughton 1998). Thus, all these features resulting in numerous critical factors that impact organizational success and failure. It also makes the organizational management team perceive performance indicator differently in such an environment than the non-conflict environment.

On the other hand, there is limited research has been done to model construction organization performance in a post-conflict environment. Consequently, there is an essential need to develop a framework to be able to determine (CSFs) and key performance indicators (KPIs) for construction organization and to identify performance improvement strategy in a post-conflict environment such as Afghanistan.

2 METHODOLOGY

Organization profit and success are based on the impact of many factors. Identifying and determining these critical impacting factors help organizations to concentrate on the areas of performance that needs improvement (Elwakil *et al.* 2009). From a comprehensive literature review, total twenty-nine post-conflict environment impacting factors were identified, and five performance KPIs were determined by balanced scorecard model (BSC) the survey questioner research method is applied to collect the data for this study. The survey questioners were sent to 500 construction companies which were selected randomly from the list of 20013 registered companies with The Afghanistan Investment Support-Agency (AISA) in all over the country from 2001 to August 2016. Most of these registered companies are headquartered in Afghanistan big cities, Kabul, Kandahar, Jalal-Abad, Herat, and Balkh. A total of 51 filled survey questionnaires were received from those distributed.

In this study, organization performance was analyzed using post-conflict environment impacting factors as the independent variables and using performance indicators as dependent variables to determine critical success factors and develop performance improvement strategy.

Critical success factors are important to know to improve organizational effectiveness and efficiency. Organizations must understand critical success factors and its impact on the different divisions of an organization to achieve long-term success, (Kaplan and Norton 1995). CSFs apply to any organization operating in an industry (Rockart 1979). Also, Rockart, (1979) defines the critical success factors (CSFs) as "the critical success factors are areas of performance that should receive constant and careful attention from management."

Similarly, understanding what parameter or key performance indicator (KPIs) must be monitored and gauged is crucial, since, the KPIs are general indicators of performance concentrating on output or outcome (Collin 2002). The KPI working group (2000) describes KPIs as the enabler of measurement of the construction project and organizational performance. Therefore, it is essential to know what metrics or KPIs to be selected to analyze and evaluate the impact of CSFs on it. In this study five essential KPIs were shortlisted in the survey questioner, the participants made the KPIs pairwise comparison and rated them by Likert five-degree scale, and also the participants ranked the impact of twenty-nine shortlisted post-conflict impacting factors on these indicators by the seven-degree Likert scale.

To analyze the obtained data, Analytic hierarchy process (AHP) decision-making technic is applied to weight the selected KPIs. Saaty (2008) describes the Analytic hierarchy process (AHP) a non-complicated tool for human decision making. Also, AHP is a fixable multi-criteria decision-making process which can be easily integrated with other modeling technics such as multiple linear regression, fuzzy logic, artificial neural network and others (Elwakil 2017). Consistency index (CI) and consistency ratio (CR) were calculated to verify the validity of KPIs in a pairwise comparison. The multiple linear regression (MLR) modeling technics is used to determine the impact of critical success factors on the KPIs and to develop an organizational performance prediction model. The MLR can determine that how well the impacting factors work together to predict construction organization performance (the best set of CSF) and in the same time the MLR technic makes it possible to determine which factors contribute more to predict construction organization performance (CSFs). The t-test was run to evaluate the result significance. And also, the validly of the developed model was checked through 10% of the test data.

3 ANALYSIS OF THE RESULTS

3.1 Analytical Hierarchy Process (AHP)

The AHP technic is applied to this study to determine the weights of the KPIs measures. The Table 1 presents the weights for the pairwise compared KPIs from the survey questioner. The determined KPIs measures are:

- Average cost predictability per project (R_C)
- Average time predictability per project (R_T)
- The volume of annual work growth (R_W)
- Contractor satisfaction (R_S)
- Annual growth in the number of biding projects (R_B)

The consistency analysis for this study shows that the consistency index is CI = 0.000132< 0.1 which mean there are logical consistency and reliability between the compared measures. The weighted measure is used as a coefficient of each developed model predicting factors to determine the final set of critical success factors for the construction organization performance.

3.2 Multiple Leaner Regression Model (MLR)

Regression analysis is performed to determine CSFs and to develop a performance prediction model. The predicting factors in each model are selected according to the best subset of performance prediction model presented in the Table 1.

Table 1. Performance prediction model for each individual measure.

Weighted value of KPI measure	KPI	0.282	0.074	0.335	0.234	0.074
Discerption of impacting factors	Factor	Time predictability	Bid growth	Cost predictability	Work growth	Contractor satisfaction
Anticompetitive behavior	X1		X	X	X	
Monitory uncertainty	X2	X	X	X		X
Not being paid	X3			X		X
Lack of regulatory policy	X4		X			X
Project & warranty failure (instability)	X5	X		X	X	X
Corruption	X6			X	X	X
Theft and crime	X7	X		X	X	X
Overall security -conflict	X8		X	X		
Lack of access to finance (banking)	X9		X	X	X	
Tax-admins, the tax rate	X10		X			X
Lack of legal & judicial system	X11	X			X	
Lack of skilled & educated workforce	X12	X			X	
Lack of access to land	X13		X			
Poor-infrastructure	X14	X			X	X
Government financial aid dependency	X15		X	X	X	X
Lack of internet and technology	X16		X	X		X
Market structure & competition	X17	X	X			X
International financial support	X18		X		X	
Local expenditure of international agencies	X19	X		X	X	X
Government & nongovernment training	X20	X				X
Government invest support	X21					
Lack of construction materials availability	X22	X	X	X		X
Lack of admins technical capabilities	X23	X	X	X		X
Bureaucratic process	X24	X		X		X
International co-investment	X25	X				
Diaspora investment and tech support	X26		X			
Uncompetitive quality and price	X27	X	X			X
The rate on return on investment	X28			X		
Lack of government risk reduction policy	X29			X	X	

The t-test result for the developed model at the 0.1 significance level shows that all impacting factors are significantly predicting construction organization performance. Table 2 shows the rank of top twenty critical success factors that impact construction organization performance in a post-conflict environment.

Table 2. Rank of construction organization performance in a post-conflict environment.

Rank	Factor	Critical Success Factors (CSFs) description
1	X1	Anti-competitive behavior
2	X12	Lack of skilled & educated workforce
3	X6	Corruption
4	X20	Governmental & nongovernmental organizations support& training
5	X27	Uncompetitive quality and price
6	X11	Lack of legal & judicial system
7	X28	Rate of return on investment
8	X17	Market structure & competition
9	X2	Monitory uncertainty
10	X7	Theft and crime
11	X8	Overall security -conflict
12	X29	Lack of government risk reduction policy
13	X15	Government-financial-aid dependency
14	X9	Lack of Access to finance-banking
15	X3	Not being paid
16	X18	International financial support
17	X19	Local expenditure of international agencies
18	X5	Project failure& warranty failure because of instability
19	X24	Bureaucratic process
20	X14	Poor-Infrastructure

4 CONCLUSION

The purpose of this study was to develop a framework to assess the impact of the post-conflict environment on construction organization performance. Previous research has widely focused on developing critical success factors and success strategies for organization performance improvement without considering the impact of the environment where these organizations perform. This study evaluated the impact of external environment on organization performance and examined five performance measures which are shown in the Table 1 to determine the impact of post-conflict environment factors on these identified KPIs. The study result indicates that top twenty factors from the questioner which are shown in the Table 2 have a critical impact on the organization performance and considering these impacting factors can help to formulate a successful strategy and improve organizational performance. The finding of this study suggests that there is a significant association between construction organization performance and post-conflict environment factors. However, the study did not consider the organizational differences such as the type of construction, years of experiences, and the size of the firm. Thus, the developed framework will benefit the academic researchers and industry practitioners to analyze and evaluate challenges and opportunities caused by different external environmental factors in the post-conflict construction industry.

References

Abraham, G. L., *Critical Success Factors for The Construction Industry*, Construction Research Congress: Wind of Change: Integration and Innovation, 1-9, 2003.

Bourgeois III, L. J., Strategy and Environment: A Conceptual Integration, *Academy of Management Review, 5(1), 25-39,* 1980.

Collin, J., Measuring The Success of Building Projects–Improved Project Delivery Initiatives, *Report for the Queensland Department of Public Works,* Australia, 2002.

Chan, A. P. C., and Chan, A. P. L., Key Performance Indicators for Measuring Construction Success, *Benchmarking: An International Journal*, 11(2), 203-221, 2004.

Cambridge Business Dictionary, retrieved from https://dictionary.cambridge.org/us/dictionary/english/external-environment on November 24, 2017,

Del Castillo, G., Post-Conflict Reconstruction and The Challenge to International Organizations: The Case of El Salvador, *World Development*, 29(12), 1967-1985, 2001.

Elenkov, D. S., Effects of Leadership on Organizational Performance in Russian Companies, *Journal of Business Research*, *55*(6), 467-480, 2002.

Elwakil, E., Ammar, M., Zayed, T., Mahmoud, M., Eweda, A., and Mashhour, I., In *Construction Research Congress 2009: Building a Sustainable Future*, 350-359, 2009.

Elwakil, E., Integrating the Analytical Hierarchy Process and Regression for Assessing Construction Organizations' Performance, *International Journal of Construction Management*, 17(1), 76-88, 2017.

Farmer, R. N., and Richman, B. M., A Model for Research in Comparative Management, *California Management Review*, *7*(2), 55-68, 1964.

Grewal, R., and Tansuhaj, P., Building Organizational Capabilities for Managing Economic Crisis: The Role of Market Orientation and Strategic Flexibility, *Journal of Marketing,* 65(2), 67-80, 2001.

Haughton, J., The Reconstruction of War-Torn Economies, *Harvard Institute for International Development (June)*, 1998.

Horta, I. M., Camanho, A. S., and Da Costa, J. M., Performance Assessment of Construction Companies Integrating Key Performance Indicators and Data Envelopment Analysis, *Journal of Construction Engineering and Management*, 136(5), 581-594, 2009.

Kaplan, R. S., and Norton, D. P., Putting the Balanced Scorecard, *Performance Measurement, Management, and Appraisal Sourcebook*, 66, 1995.

Lee, H., and Choi, B., Knowledge Management Enablers, Processes, and Organizational Performance, *Journal of Management Information Systems*, 20(1), 179-228, 2003.

Rockart, J. F., Chief Executives Define Their Own Data Needs, *Harvard Business Review*, 57(2), 81-93, 1979.

Schein, E. H., Organizational Psychology, 1965.

Saaty T. L., Decision Making with The Analytic Hierarchy Process. *International Journal of Services Sciences,* 1(1), 83-98, 2008.

Interdependence between Structural Engineering and Construction Management
Edited by Ozevin, D., Ataei, H., Modares, M., Gurgun, A., Yazdani, S., and Singh, A.

ISBN: 978-0-9960437-6-2

EXPLORING APPLICATIONS OF BLOCKCHAIN TECHNOLOGY IN THE CONSTRUCTION INDUSTRY

ALIREZA SHOJAEI

Building Construction Science, Mississippi State University, Mississippi State, USA

Blockchain provides a secure decentralized information management system that can solve many common problems facing the construction industry. The loose structure of the construction industry, the way that public and private projects are tendered, and the supply chain system it uses for material and service delivery provide unique challenges and problems. New information technology management systems such as BIM and RFID are used to address some of these issues, though not completely. Blockchain technology can be used to further improve the information management systems in construction, provide more automation and mitigate many possible legal conflicts by default. Implementation of blockchain technology in the construction industry can also result in the use of smart contracts with fewer administrative struggles, improve the flow of the project, material, and service delivery, and increase the security and currentness of BIM or project documents. This study aims to explore the applications of blockchain technology in improving the construction industry's information management systems. It is concluded that not only the blockchain technology has potential in addressing some of the common problems in the construction industry but also it is adaptable to the construction industry structure and the way it is practiced. Thus, blockchain technology is a viable option for adaptation in the construction industry.

Keywords: Smart contract, Information technology, Supply chain management, BIM, Circular economy, Facility management.

1 INTRODUCTION

The construction industry is often critiqued because of its inefficiency and low productivity. The disaggregated structure of the construction industry, its sequential nature where works need to be done in a sequential and chain resembling system, and the number of stakeholders with different interests involved in each project are named as root causes of its problems. Coordinating all the necessary tasks, contract administration, handling claims, and supply chain management through manual paperwork is proved to be troublesome and inefficient. Blockchain is an information technology that inherently adapts to these structural problems. Blockchain can provide a direct solution to each of these problems or provide a comprehensive platform where project design and management happens in a blockchain backed Building Information Model (BIM) integrated with a self-enforcing smart contract supplied by vendors chosen and managed by a blockchain network in a larger scale of circular economy. This technology is capable of bridging the lack of trust between the stakeholders (Mathews *et al.* 2017), automate a lot of currently manual processes, provide a secure and reliable infrastructure for collaboration and information exchange while increase transparency and provide a reliable chronological record keeping.

Blockchain is an implementation of distributed ledger technology, where peers are the data storage nodes. In such a system, the information is shared with everyone completely and verifiably with complete traceability and chronological order. The technology itself is free and implementing a blockchain system is far cheaper than any other shared database system such as common cloud solutions. There have been three generations of blockchain so far, that is, Blockchain 1.0 for digital currency, Blockchain 2.0 for digital finance, and Blockchain 3.0 for digital society. Blockchain 1.0 is solely regarding the decentralize transaction of money and payments. Blockchain 2.0 works in a more general way and covers transactions related to any kind of asset. For instance, smart contracts and smart properties are examples of the second generation's implications. Blockchain 3.0 goes beyond the concept of asset transaction recording and covers areas such as government, health, science, and culture. Blockchain implementations in supply chain management and banking systems show it is highly secure and reliable (Morabito 2017). Public, consortium, and private blockchains provides different information management systems frameworks that can address different needs based on the level of openness and access it is needed (Morabito 2017). A public blockchain is accessible by anyone, and the data can be read or be written by any user, whereas, in a private system only the selected few can access the system. A hybrid or consortium blockchain allows exchange between separate blockchain networks. The absence of administrative power in blockchain and the power of crowed over the data validity is the source of its power.

Blockchain is identified by Wang *et al.* (2017) as one of the tools that can help to overcome the acknowledged lack of trust and inadequate information sharing technologies in Architecture, Engineering, and Construction (AEC) industry (Lau and Rowlinson 2010). Blockchain, due to its working system, would eliminate the need for trust between the parties or the need for confrontational contractual relationships due to the lack of trust. Five areas of blockchain technology's applications in the built environment sector is investigated in this study. Namely, smart contracts, supply chain management and circular economy, BIM, facility management, and sustainability. Finally, the challenges of using blockchain technology in the built environment sector are discussed, and directions for future research are suggested.

2 SMART CONTRACT

A smart contract can be defined as a computer program consist of if/then statements dividing the work into smaller measurable work packages and automating the process of compliance and payment (Lamb 2018). The conditions where each work package or milestone is considered complete is defined, and the completion of each one would trigger the predefined compensation automatically. This approach of breaking the project into smart contracts would provide a new type of work breakdown structure, which would help all the stakeholders to better understand their obligations, requirements, and liabilities as well as how other work packages is going to affect theirs or vice versa. As a result, a better holistic view of the project before execution and also during execution would be provided, where the real-time progress of the project can be tracked and provide a well-recorded track of developments during project execution. Controlling the contract governance with a computer program would decrease the number of uncertainties involved in project execution as the outcomes and the triggers to each outcome is completely predefined and expected. Signing off payments would become automatic dependent on the proof of work and compliance and reduce the need for administrative support in that aspect of the work. The if/then system would ultimately reduce the disputes from contract stipulations as the conditions and compensations of the work is clearly stated by if/then statements and is governed automatically by a computer program. Blockchain can be the platform for implementation of smart contracts as a digital protocol. The information recorded in blockchain would clearly show who is responsible for what work at what time. Any tempering and change in the project data would make that block of information invalid, and it is fully traceable to the party at fault. In other words, the data recorded in the blockchain network is secured and tempered proof.

The underlying assumption of construction contracts is that most of the rights and obligations of stakeholders are against each other. This environment is in direct contrast to the new collaborative project delivery methods, such as BIM requires. The shared data environment provided by blockchain would allow a seamlessly automated compliance evaluation and provide an instrumental data record for dispute resolution. Consequently, using smart contracts and blockchain would eventually change the landscape of use of escrows, liens, and other sorts of bonds in the construction industry. Smart contracts provide extreme transparency and Potentially reduce the number of conflict points and also provide an unalterable track of project progress which can be used for conflict resolution in case it becomes necessary.

Cryptocurrencies such as Bitcoin are suggested to be used as the form of payment or collaterals in smart contracts (Cardeira 2015), but it is not necessary and not practical with current fluctuations in cryptocurrencies values. Payments and collaterals could still be hard currencies handled through banks by integrating existing payment accounts and the blockchain network. The end result would be a reduction in the number of people involved in contract administration and a better record keeping of the procedures followed. Reduction of intermediary parties and paper processes such as payment applications would increase the efficiency of the industry, not because of the increase in productivity but because it would smooth the administrative process such as compliance and financial procedures and automate many of the processes involved, which would increase the speed of project governance. Ultimately, such a system would smooth the project delivery process and provide a transparent and traceable payment system. Auditing the project cost, progress, labor practices, and any other project's record would be much faster and more reliable compared to the conventional project delivery systems. Consequently, the industry would become more efficient and inherently more ethical. The increased transparency would also result in better accountability and better project governance.

Implications of blockchain in construction and contract administrations go beyond the scope of smart contracts. Issues such as temporal equipment lease (Wang *et al.* 2017) and temporal insurance policies (Kakavand *et al.* 2017) are already being discussed where the user only pays for the time construction workers are present at the job site, and the validity of this and other conditions under question can be securely verified through blockchain network. Having a complex construction project executed though smart contract and blockchain might not be feasible in the near future (Gabert 2018). However, the viability of the concept is evident and implementing such a system in simple projects or subcontracts would be an appropriate first step to evaluate its performance in the real world.

3 SUPPLY CHAIN MANAGEMENT AND CIRCULAR ECONOMY

The decentralized and fragmented structure of the construction industry's supply chain very well pair with the decentralized ledger system of blockchain. Blockchain can provide the infrastructure needed to securely and reliably advance material traceability (Hultgren and Pajala 2018, Petersson and Baur 2018) and promote the circular economy (Rudolphi 2018). Smart contracts can play a role in supply chain management and material tracing too. Both supplier and buyer can be assured that they are going to get compensated for their money/product by using a smart contract. As a result, the purchase could happen more directly between the source and the end user without the need for local suppliers. The payments can be sequential and proportionate to the status of material/product delivery, and the final payment could be tied to the final inspection/commission of the material/product.

There is also a high potential for integrating internet of things or RFID tags with blockchain to provide real-time material monitoring system resulting in better site management practice and increase in construction efficiency. Using a blockchain network for supply management would help vendors to easily show their certification of identity from authorities. The buyers can easily validate those certifications of identity and also see the track record of the vendor to check their reputation and capacity from previous works. As a result, a buyer and a vendor without knowing and trusting each other can engage in a transaction.

On a grander scale, a true circular economy is achievable through blockchain. When a raw material is extracted, its information can be stored on a blockchain network, storing its source and characteristics. Then producers can use that network to order and obtain their raw materials. Each order is also stored, and it is traceable that each product consists of what raw materials from where. This chain of information continues to the end user who is buying the product and installing it in their project, which would increase the material transparency in the construction industry. Furthermore, during the lifetime of a product when maintenance is necessary or if there is a question regarding its source or its materials' source it is clearly traceable. There will be a chain of information from raw material source to factories, vendors, sub-contractors, contractors, and the final project. That would allow a fully transparent material usage in the industry, where planning for reusing of materials are possible through pre-planning and current knowledge of materials status and their background.

4 BIM

The introduction of BIM to the AEC industry provided a shared data environment where all the information is stored in a shared project file. The location and maintenance of this shared data environment presented a new challenge. Blockchain technology is a promising solution to such a problem (Turk and Klinc 2017). An always current distributed ledger system with high security, which can be used as an infrastructure for maintaining up to date BIM models during the project Lifecycle between the involving stakeholders. BIM provides a single project environment where all the information regarding the project is created and stored there. Blockchain provides a single platform to maintain and update that project environment and connect it with the reality of site work. This would provide a time-stamped, tampered proof data. The peer-to-peer structure of blockchain aligns well with the collaborative way that a BIM model is developed by different stakeholders, and ultimately, it would improve collaboration. The tampered proof track record of changes to the model by each user can be used to find a party at fault in design or miscommunication. Any change to the model is recorded and communicated to other people linked in the blockchain network. Consequently, the BIM related claims during the project lifecycle can be solved much more natural.

Connecting a project's BIM model to a smart contract would need an information system to link project elements to reality and reflect the project progress on the model. Blockchain can be used separately or in adjunction of governing the smart contract to link the BIM model to the smart contract and update the BIM model according to the project progress. The smart contract program can be applied to the BIM model elements and linked to the project's reality through blockchain. As a result, the actual construction should match the model to get compensated. If the contractor finds a clash or an error, he can send a change order or RFI, and it would be securely time stamped and recorded. The granular work progress can be tracked on the model with or without reality capture technologies. The inspection process can be done by manual inspection or automatically through reality capture technologies such as 3D laser scanning or Lidar. The end result would be a streamlined contract administration and better project governance.

5 FACILITY MANAGEMENT

Integration of blockchain and BIM or Building Maintenance System (BMS) would provide a reliable integrated system which can provide the complete history of the project and also trace every detail of the building to its source (Mathews *et al.* 2017). Furthermore, this integration can stretch to the future and use smart contracts when maintenance is needed to automatically place a work order and upon the verification of completeness, release the payment to the contractor. The concept of Decentralized Autonomous Organization (DAO) is introduced as an organization, which is governed via multiple smart contracts. A DAO can be attributed to a building through its lifecycle where everything from design and construction to operation, maintenance and demolition is done by smart contracts cohesively and autonomously. Blockchain is one of the few technologies can bear the burden of supporting such complex interactions through time. Longitudinal health record (Angraal *et al.* 2017) is an example of a similar concept from a different industry.

6 SUSTAINABILITY

The material transparency discussed in supply chain section would have an impact in sustainability in areas such as whole life cycle cost, carbon emission estimates, and raw material verification. For instance, the designers or users can make a sustainable choice by using material traceability through blockchain up to the source of any product's raw materials. Typically, the supply chain would provide the specific information required by the clients. A blockchain platform would enable not only the direct suppliers to provide the required information but also the indirect suppliers such as the raw material providers to a prefabrication factory can also put their information in the database for more accuracy and verifications. This process would provide consistent and structured asset information. This database can be used not only for decision making during the design, procurement, and construction but also would be beneficial for the post-occupancy management of the facility. A blockchain network can also help energy management on a grand scale to achieve a smart grid (Mengelkamp *et al.* 2018). First, both energy consumption and production should be tracked using a blockchain. Then, this could provide a basis for a better supply and demand control and ultimately a true dynamic pricing for energy.

7 CONCLUSION

Successful implementation of blockchain in other industries such as accounting, financial technology, and commodity market shows the viability of this technology. Blockchain can hypothetically provide a platform for supporting the link between the physical world and the digital one. It can also cover the whole Lifecycle of a project from material sourcing, contract administration to the operation, maintenance, and eventually demolition and material reusability. Blockchain would help to smooth the project development processes and reduce the need for intermediary parties. One necessary step for this technology to become prevalent is becoming prevalent as much as possible so the software providers would enable their software to use push and pull data by integrating with blockchain. A common standard and template for each application area is crucial. Thus a loosely structured industry such as construction can use this system. Consequently, everybody in the supply chain would be able to input their data and use the other stakeholders' input in a meaningful manner. One important issue raised in this study is while cryptocurrency can play a role in each discussed section, it is not necessary to be part of the system and blockchain can be used regardless of the cryptocurrency state.

The disaggregated structure of the construction industry makes it a suitable match for using blockchain. However, this loose structure is a disadvantage in implementing innovation and new technologies. This problem is more noticeable when a grand scale change is requiered. As a result, the administrative gap within the industry is a hindrance to blockchain technology implementation in this industry. Even though computer aided design and BIM is digitizing the construction industry, there is an apparent gap between the framework envisioned in this paper and the current state of digitization of the construction industry. A looped two-step approach is necessary to, first study the industry to find the requirements and applicable areas. Then, the second step is to test the viability of the solutions and adjust the framework based on the empirical findings. This study is a contribution to the first step, which is going to be followed by testing each explored area with test cases in future works.

References

Angraal, S., Krumholz, H. M., and Schulz, W. L., Blockchain Technology: Applications in Health Care, *Circulation: Cardiovascular Quality and Outcomes*, 10(9), e003800, 2017.

Cardeira, H., Smart Contracts and Possible Applications to the Construction Industry, in *The New Perspectives in Construction Law Conference,* Bucharest, Romania, 2015.

Gabert, H., *Blockchain and Smart Contracts in The Swedish Construction Industry - An Interview Study on Smart Contracts and Services,* MSc Thesis, Department of Real Estate and Construction Managament, Royal Institute of Technology in Stockholm, Stockholm, Sweden, 2018.

Hultgren, M., and Pajala, F., *Blockchain Technology in Construction Industry- Transparency and Traceability in Supply Chain,* MSc Thesis, Department of Real Estate and Construction Managament, Royal Institute of Technology in Stockholm, Stockholm, Sweden, 2018.

Kakavand, H., Kost De Sevres, N., and Chilton, B., The Blockchain Revolution: An Analysis of Regulation and Technology Related to Distributed Ledger Technologies, *Available at SSRN: https://ssrn.com/abstract=2849251*, 2017.

Lamb, K., *Blockchain and Smart Contracts : What The AEC Sector Needs to Know, Centre for Digital Built Britain*, CDBB_REP_003, 2018.

Lau, E., and Rowlinson, S., Trust Relations in The Construction Industry, *International Journal of Managing Projects in Business*, 3(4), 693–704, 2010.

Mathews, M., Robles, D., and Bowe, B., BIM+ Blockchain: A Solution to the Trust Problem in Collaboration? In *CITA BIM Gathering 2017*, 11, 2017.

Mengelkamp, E., Notheisen, B., Beer, C., Dauer, D., and Weinhardt, C., A Blockchain-Based Smart Grid: Towards Sustainable Local Energy Markets, *Computer Science - Research and Development*, 33(1–2), 207–214, 2018.

Morabito, V., Business Innovation Through Blockchain, *Cham, Springer International Publishing*, 2017.

Petersson, E., and Baur, K., *Impacts of Blockchain Technology on Supply Chain Collaboration, MSc Thesis, Business Administration, Jönköping University, Jönköping, Sweden, 2018.*

Rudolphi, J. T., Blockchain for a Circular Economy: Explorative Research Towards The Possibilities for Blockchain Technology to Enhance The Implementation of Material Passports, *MSc Thesis, Department of Construction Management and Engineering, Eindhoven University of Technology, Eindhoven, Netherlands, 2018.*

Turk, Ž., and Klinc, R., Potentials of Blockchain Technology for Construction Management, *Procedia Engineering*, 196(June), 638–645, 2017.

Wang, J., Wu, P., Wang, X., and Shou, W., The Outlook of Blockchain Technology for Construction Engineering Management, *Frontiers of Engineering Management*, 4(1), 67, 2017.

Interdependence between Structural Engineering and Construction Management
Edited by Ozevin, D., Ataei, H., Modares, M., Gurgun, A., Yazdani, S., and Singh, A.

ISBN: 978-0-9960437-6-2

REVISITING SYSTEMS AND APPLICATIONS OF ARTIFICIAL NEURAL NETWORKS IN CONSTRUCTION ENGINEERING AND MANAGEMENT

ALIREZA SHOJAEI[1] and AMIRSAMAN MAHDAVIAN[2]

[1]*Building Construction Science, Mississippi State University, Mississippi State, USA*
[2]*Civil and Environmental Engineering, University of Central Florida, Orlando, USA*

Artificial neural networks have been widely used for modeling and simulation of different problems in the construction industry, including, but not limited to, regression, clustering, and classification. They provide solutions for complex problems where other modeling methods often fail. For instance, they can capture nonlinear and complex relationships between the variables while many traditional modeling methods fail. However, they have their own limitations. They often can only be trained for a specific problem with a predetermined number of inputs and outputs. As a result, any change that requires an update in the architecture of the network cannot be automatically done and require human intervention. The recent developments in the field of artificial neural networks resulted in new concepts such as neural architecture search, reinforcement learning, and neuroevolution. These new areas can provide new methods for solving past and existing problems facing the construction industry in a more efficient, elegant, and versatile manner. One of the main contributions of the recent developments is networks that can optimize their own architecture and networks that are able to evolve and change their architecture. This paper aims to briefly review the application areas of the artificial neural networks in construction engineering and management and discuss how the recent developments in this field can be applied and provide better solutions.

Keywords: Reinforcement learning, Neuroevolution, Artificial intelligence, Neural architecture search.

1 INTRODUCTION

Artificial Neural Networks (ANNs) are an abstract form of biological neural systems. They have been a pioneer modeling approach in Artificial Intelligence (AI) and machine learning field. ANNs are implemented for a variety of tasks such as regression, clustering, and classification. They are especially powerful in complex pattern recognition and nonlinear relationship modeling, and image processing. Figure 1 illustrates a sample architecture of a feed-forward neural network. The first layer shows the neurons of the input layer; the second layer is called the hidden layer, which here includes five neurons and the last layer is the output layer, which here includes only one neuron. The developer of an ANN model needs to optimize two set of details, first, the architecture of the model and second, optimization of the models' parameters. Tuning the parameters of the model such as learning rate, epoch, initial weights, and biases can be done through different methods such as backpropagation, particle swarm optimization, and many other methods discussed in the following sections.

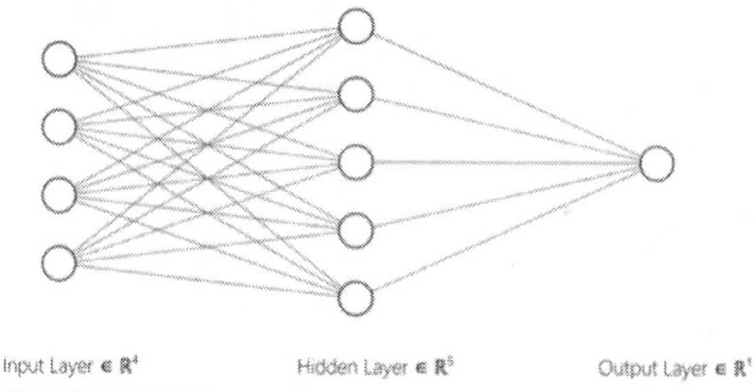

Figure 1. An illustration of an ANN's architecture.

The main problem in developing ANN models is decision making regarding the architecture of the model. Until recently, the decisions regarding the architecture of the ANN models were more of guesswork, where the developer based on experience or testing different values would come up with a not necessarily optimal architecture for the ANN under development. The following sections would provide a review on applications of ANNS in construction engineering and management (CEM), followed by the discussion and conclusion section where the recent developments in ANNS field and their implications in CEM is discussed.

2 COST

Cost prediction is one of the significant areas of study where ANN is applied in CEM field. McKim (1993) successfully used ANN to predict tender price, Williams (1994) used an ANN for construction cost prediction. Al-Tabtabai *et al.* (1998) developed an ANN model for forecasting of percentage increase in the cost of a highway project from a reference estimate. Boussabaine and Kaka (1998) applied an ANN to project cash flow forecasting. Cheng *et al.* (2009) suggested web-based conceptual cost estimates for the construction projects using an evolutionary fuzzy neural inference model that calculates the whole life cycle costs of construction projects by considering substantial items' costs. Feng and Li (2013) proposed an optimization method for cost estimation. Their model integrates the Genetic Algorithm (GA) and Back Propagation (BP) techniques. This study concluded that GA-BP model could get lower predict error which takes longer run time compared to the conventional BP models. El-Sawah and Moselhi (2014) used Back Propagation Neural Network, Generalized Regression Neural Network, Probabilistic Neural Network, and regression analysis models for forecasting an approximate cost estimate of low-rise structural steel buildings and their relevant cost. Yadav *et al.* (2016) developed a cost estimating technique based on the principles of ANN to forecast structural cost of residential buildings. Lee *et al.* (2016) developed a model using the ant colony optimization algorithm to optimize the selection of ANN parameters that can be used to address the cost overruns in projects and improve construction waste management.

3 SCHEDULING AND PRODUCTIVITY

Scheduling and productivity is another area where ANN seems to be widely adopted by researchers. Chao and Skibniewski (1994) developed an ANN regarding the Labor productivity in the construction industry. Boussabaine (1995) developed a neural network system for forecasting construction planning and productivity. Lu *et al.* (2000) developed an ANN regarding the labor production rate for pipe installation activity. It was predicted that the duration of concrete operations via an artificial neural network by focusing on supply chain parameters. Ezeldin *et al.* (2006) estimated the productivity within a developing market for formwork assembly, steel fixing, and concrete pouring activities. Golpayegani *et al.* (2007) presented a framework for planning work breakdown structure (WBS) of construction projects based on an ANN. Bai *et al.* (2009) used four successively arranged neural networks to develop the WBS of the projects using ANNs. Cheng and Kou (2003) developed an object-oriented evolutionary fuzzy neural inference system in order to predict a subcontractor's performance, and duration estimation of slurry walls. Yahia *et al.* (2011) used the number of change orders to develop a project time prediction model.

4 DECISION MAKING

Decision making due to its complex nature is another area where ANN has shown significant contributions. Kamarthi *et al.* (1992) used two layers BP neural network in order to select the formwork systems. Murtaza and Fisher (1994) developed an ANN model that facilitates decision-making on whether to use a conventional or modularization method to build an industrial process plant. Chua *et al.* (1997) proposed an ANN model for the identification of leading measurement factors that impact budget performance to forecast various management strategies in a project. Lou *et al.* (2001) used the past conditions records to predict the short-term future conditions using an ANN. Lam *et al.* (2001) used an ANN as a decision-making tool to evaluate the prequalification of contractors. Chew and Tan (2003) proposed a maintainability grading system which helps in improving decision-making of wet areas design that ultimately leads to minimum lifecycle maintenance cost using an ANN. Al-Sobiei *et al.* (2005) developed a decision-making tool to employ the services of a contractor that can allow the use of a rational and effective policy by using ANN and GA. Khalafallah (2008) developed a model for predicting housing market performance in order to advocate home developers and real estate investors using an ANN method. Vahdani *et al.* (2011) used a neuro-fuzzy model to estimate the qualitative and quantitative performance of the projects accurately. Sharmik *et al.* (2016) applied a BP-NN in order to estimate the soil characteristics. El-Sawalhi and Hajar (2016) used an ANN to investigate the election of the best-qualified contractor in the Gaza strip. Apanaviciene and Juodis (2005) applied an NN method to propose a model which can help as a decision-making tool for competitive bidding and for evaluating the management risks of the projects. Jamil *et al.* (2009) showed the feasibility of fitting ANN in the development of a simulator and an intelligent system for the forecasting of workability and compressive strength of high performance concrete.

5 LEGAL DOMAIN

Legal domain due to its qualitative nature is not as developed as other areas investigated in this study. Cheung *et al.* (2000) reported using ANN to classify projects in accordance with project "dispute resolution satisfaction (DRS)" which also recognized the sensitive variables that determine projects with adverse DRS and favorable DRS. Fatima *et al.* (2014) develop an ANN model based on identified significant qualitative parameters for minimizing construction disputes and reduce the cost of construction.

6 RISK ASSESSMENT AND ANALYSIS

Risk assessment and analysis due to its quantitative nature is a good candidate for employing ANN. McKim (1993) used an ANN regarding risk quantification. ANN was for cost-based risk prediction and identification of project cost drivers. Al-Tabtabai and Alex (2000) used ANN system to recognize the cost deviations that happen, regarding the political risk involved in a construction project. Odeyinka *et al.* (2002) developed a cost flow risk assessment model employing a BP neural network. The model was tested on 20 new projects with a satisfactory prediction of variation between forecast and actual cost flow at 30%, 70%, and 100% stages. Maria-Sanchez (2004) applied a neural network approach to evaluate the impact of environmental risk in the construction projects. Xiang and Luo (2012) proposed a principal project parties' behavioral risk evaluation model. The BP was employed to avoid subjectivity factors in the risk assessment process. Polat (2012) proposed a contingency estimation model based on ANN to enable managers to evaluate the level of risk in their projects more objectively and systematically. Liu and Guo (2014) used ANN to propose a risk assessment method using rough sets to decrease uncertainties. Moayed *et al.* (2011) applied an ANN and Logistic Regression to model the occupational safety and health of construction workers. Goh and Chua (2013) used a neural network to study the relationship between elements of accident severity and safety management on proactive management. Mohammadfam *et al.* (2015) modeled the factors affecting the health of the workforce and predicting the harshness of occupational injuries using chain analytical approach including ANN modeling and rough set theory.

7 DISCUSSION AND CONCLUSION

The conventional ANN architecture engineering often involves guesswork and trial and error. Beside the user experience, a series of trial and error are needed to get to the point of satisfactory architecture, and even that is not a guaranteed optimal solution. Most of the ANNs reviewed in this paper used gradient descent for optimizing their network. Some studies have used other soft computing methods such as ant colony, particle swarm optimization, and GA. However, their implementations lack the novelty in architecture search, and the optimization only applies to certain parameters of the network. Recent developments (Such *et al.* 2017, Conti *et al.* 2017) opened the door to ANNs, which can evolve and automatically change their architecture and topology to find the optimal architecture by combining evolutionary algorithms and ANNs. The main concept is that an AI that is designed for stochastic gradient descent can change its structure to a better performing structure through reinforcement learning and/or neuroevolution. Zoph and Le (2016) and Zoph *et al.* (2017) used a recurrent ANN to conduct a neural architecture search to generate the optimal architecture of an ANN for the desired task. In other words, they used an AI to build another AI with an optimized structure. The reviewed literature in previous sections shows the versatility and importance of ANN in CEM field. However, considering the possibility of finding the optimal network architecture through computation and not trial and error and the possibility of evolving AI is very promising. Consequently, revisiting the previous work and employing the new ANN systems on the same problems, theoretically, would significantly increase the performance of the previous generations of ANNs in CEM.

References

Al-Sobiei, O. S., Arditi, D., and Polat, G., Predicting the Risk of Contractor Default in Saudi Arabia Utilizing Artificial Neural Network and Genetic Algorithm Techniques, *Construction Management and Economics*, 23(4), 423–430, 2005.

Al-Tabtabai, H., Alex, P., and Tantash, M., Preliminary Cost Estimating of Highway Construction using Neural Network, *Cost Engineering*, 41(3), 19–24, 1998.

Al-Tabtabai, H. M., and Alex, A. P., Modeling the Cost of Political Risk in International Construction Projects, *Project Management Journal,* 31(3), 4-13, 2000.

Apanaviciene, R., and Juodis, A., *Modelling of Construction Project Management Effectiveness by Applying Neural Networks,* Symposium on Sustainability and Value Through Construction Procurement, Salford, United Kingdom, 2005.

Bai Y, Zhao Y, Chen Y, and Chen L., *Designing Domain Work Breakdown Structure (DWBS) Using Neural Networks*, 6th International Symposium on Neural Networks, Wuhan, China. Springer Berlin Heidelberg,1146–1153, 2009.

Boussabaine, A. H., *A Neural Network System for Productivity Forecasting,* Proceedings of the 12th ISARC, Warsaw, Poland, 1995.

Boussabaine, A. H., and Kaka, A. P., A Neural Networks Approach for Cost Flow Forecasting, *Construction Management and Economics,* 16(4), 471–479, 1998.

Chao, L. C., and Skibniewski, M. J., Estimating Construction Productivity: Neural-Network-Based Approach, *Journal of Computing in Civil Engineering*, 8(2), 234–251, 1994.

Cheng, M. Y., Tsai, H. C., and Hsieh, W. C., Web-Based Conceptual Cost Estimates for Construction Projects Using Evolutionary Fuzzy Neural Inference Model, *Automation in Construction*, 18(2), 164–172, 2009.

Cheng, M., and Kou, C., Object-Oriented Evolutionary Fuzzy Neural Inference System for Construction Management, *Journal of Construction Engineering and Management*, 129(4), 461-469, 2003.

Cheung, S., Tam, C., and Harris, F., Project Dispute Resolution Satisfaction Classification through Neural Network, *Journal of Management in Engineering,* 16(1), 70-79, 2000.

Chew, M. Y. L., and Tan, S. S., Neural Network Approach for Grading of Maintainability of Wet Areas in High-Rise Buildings. *International Journal of Architectural Science*, 4(4), 150–167, 2003.

Chua, D. K. H., Kog, Y. C., Loh, P. K., and Jaselskis, E. J., Model for Construction Budget Performance-Neural Network Approach, *Journal of Construction Engineering and Management*, 123(3), 214–222, 1997.

Conti, E., Madhavan, V., Such, F. P., Lehman, J., Stanley, K. O., and Clune, J., Improving Exploration in Evolution Strategies for Deep Reinforcement Learning Via a Population of Novelty-Seeking Agents, arXiv *preprint arXiv:1712.06560,* 2017.

El-Sawah, H. and Moselhi, O., Comparative Study in the Use of Neural Networks for Order of Magnitude Cost Estimating in Construction, *Journal of Information Technology in Construction*, 19, 462–473, 2014.

El-Sawalhi, N., and Hajar, Y.A., Development of Awarding System for Construction Contractors in Gaza Strip Using Artificial Neural Network (ANN), *KICEM Journal of Construction Engineering and Project Management*, 6(3), 1-7, 2016.

Ezeldin, A., and Sharara, L., Neural Networks for Estimating the Productivity of Concreting Activities, *Journal of Construction Engineering and Management*, 132(6), 650-656, 2006.

Fatima, A., Sekhar, T. S., and Hussain, S. A. M., Analysis of Construction Dispute Resolution Process Using Artificial Neural Networks, *International Journal of Innovative Research and Developments,* 3(7), 81–86, 2014.

Feng, G. L., and Li, L., Application of Genetic Algorithm and Neural Network in Construction Cost Estimate, *Advanced Materials Research*, 756–759, 3194–3198, 2013.

Golpayegani, S. A. H., and Emamizadeh, B., Designing Work Breakdown Structures Using Modular Neural Networks, Decision Support Systems, 44(1), 202–222, 2007.

Goh, Y., and Chua, D., Neural Network Analysis of Construction Safety Management Systems: A Case Study in Singapore, *Construction Management and Economics*, 31(5), 460-470, 2013.

Jamil, M., Zain, M. F., and Basri, H. B., Neural Network Simulator Model for Optimization in High Performance Concrete Mix Design, *European Journal of Scientific Research*, 34(1), 61–68, 2009.

Kamarthi, S. V., Sanvido, V. E., and Kumara, S. R. T., Neuroform-Neural Network System for Vertical Formwork Selection, *Journal of Computing in Civil Engineering*, 6(2), 178–199, 1992.

Khalafallah, A., Neural Network Based Model for Predicting Housing Market Performance, *Tsinghua Science and Technology,* 13(1), 325–328, 2008.

Lam, K. C., Hu, T., Thomas Ng, S., Skitmore, M., and Cheung, S., A Fuzzy Neural Network Approach for Contractor Prequalification, *Construction Management and Economics,* 19(2), 175–188, 2001.

Lee, D., Kim, S., and Kim, S., Development of Hybrid Model for Estimating Construction Waste for Multifamily Residential Buildings using Artificial Neural Networks and Ant Colony Optimization, *Sustainability*, 8(9), 870–883, 2016.

Liu, J., and Guo, F., Construction Quality Risk Management of Projects on the Basis of Rough Set and Neural Network, *Computer Modeling & New Technologies,* 18(11), 791-794, 2014.

Lou, Z., Gunaratne, M., Lu, J. J., and Dietrich, B., Application of Neural Network Model to Forecast Short-Term Pavement Crack Condition: Florida Case Study, *Journal of Infrastructure Systems*, 7(4), 166–171, 2001.

Lu, M., AbouRizk, S., and Hermann, U., Estimating Labor Productivity Using Probability Inference Neural Network, *Journal of Computing in Civil Engineering*, 14(4), 241-248, 2000.

Maria-Sanchez, P., Estimation of Environmental Risks in Construction Projects in Puebla (Mexico): A Neural Network Approach, *WIT Transactions on Ecology and the Environment*, 77, 2004.

McKim, R. A., Neural Networks and Identification and Estimation of Risk, *AACE International Transactions*, 5.1–5.10, 1993.

Moayed, F.A., and Shell, R.L., Application of Artificial Neural Network Models in Occupational Safety and Health Utilizing Ordinal Variables, *Ann. Occup. Hyg*., 55(2), 132– 142, 2011.

Mohammadfam, I., Soltanzadeh, A., Moghimbeigi, A. and Alizadeh, S.B., Use of Artificial Neural Networks (ANNs) for the Analysis and Modeling of Factors that Affect Occupational Injuries in Large Construction Industries, *Electron Physician*, 7(7), 1515-1522, 2015.

Murtaza, M. B., and Fisher, D. J., A Neural Network Model for Decision Making with Application in Construction Management, *Journal of International Information Management*, 3(2), 27–38, 1994.

Odeyinka, H., Lowe, J., and Kaka, A., *A Construction Cost Flow Risk Assessment Model,* ARCOM 18th Annual Conference, Northumbria, United Kingdom, 2002.

Polat, G., ANN Approach to Determine Cost Contingency in International Construction Project, *Journal of Applied Management and Investments,* 1(2), 195–201, 2012.

Sharmik, S., Lekha, G., and Kanshik, S., Cost and Time Effective Prediction of Soil Characteristics Using Artificial Neural Networks Model, *International Journal of Innovative Research in Science, Engineering and Technology*, 5(3), 3829– 3834, 2016.

Such, F. P., Madhavan, V., Conti, E., Lehman, J., Stanley, K. O., and Clune, J., Deep Neuroevolution: Genetic Algorithms are a Competitive Alternative for Training Deep Neural Networks for Reinforcement Learning, arXiv preprint arXiv:1712.06567, 2017.

Vahdani, B., Mousavib, S., Mousakhanic, M., Hashemi, Time Prediction using a Neuro-Fuzzy Model for Projects in the Construction Industry, *Journal of Optimization in Industrial Engineering*, 19, 97-103, 2011.

Williams, T. P., Predicting Changes in Construction Cost Indexes Using Neural Networks, *Journal of Construction Engineering and Management*, 120(2), 306–320, 1994.

Xiang, P. C., and Luo, K., The Evaluation for the Behavioral Risk of Participants in Construction Projects Based on Back Propagation Neural Network, *Advances in Information Sciences and Service Sciences*, 4(14), 97, 2012.

Yadav, R., Vyas, M., Vyas, V., and Agrawal, S., Development of Cost Estimating Method for Bricklayer Cost, *International Journal of Engineering Research and Technology,* 5(1), 430–432, 2016.

Yahia, H., Hosny, H., and Abdel Razik, M., Time Contingency Assessment in Construction Projects in Egypt using Artificial Neural Networks Model, *International Journal of Computer Science*, 8(4), 523-531, 2011.

Zoph B, Vasudevan V, Shlens J, and Le Q. V., Learning Transferable Architectures for Scalable Image Recognition, *arXiv preprint arXiv:1707.07012. 2017.*

Zoph, B., and Le, Q. V., Neural Architecture Search with Reinforcement Learning, *arXiv preprint arXiv:1611.01578,* 2016.

Interdependence between Structural Engineering and Construction Management
Edited by Ozevin, D., Ataei, H., Modares, M., Gurgun, A., Yazdani, S., and Singh, A.

ISBN: 978-0-9960437-6-2

MULTISYSTEMIC MODELING TO IMPROVE FORECAST ACCURACY IN CONSTRUCTION MANAGEMENT

CHRISTIAN HOFSTADLER

Institute of Construction Management and Economics, Graz University of Technology, Graz, Austria

Process modeling and simulations provide indispensable support during project preparation, construction process planning, and execution. Realistic correlation modeling in construction management and/or economics is crucial to inform meaningful, beneficial process simulations. Inductive approach to reflecting reality in a (computation) model enables prospective consideration of the object or process under study and related analyses. This method aids decision making in construction management related projects and contributes to developing new computation models. Linking the management of chances and risks to construction process modeling is thus an essential tool for systematically making and implementing decisions in construction management and economics. Forecast-based conclusions regarding future developments or events require computation models to be developed based on hands-on experience and practical feasibility whilst relying on sound theoretical assumptions, required abstraction levels, simplifications, and considering their mathematical implementation. The resulting model is linked to model objects for which it was developed and to which it is applied. Model objects may include entire buildings or structures, contract sections, or individual structural components, such as those for which construction processes and logistics as well as construction times and costs are determined. Monte Carlo simulations are applied to systematically account for uncertainties. Derived models should be closely related to the aspects and societal challenges of interdisciplinarity, simulation, and digitization. This paper outlines the requirements that a multisystemic model should fulfill to increase forecast accuracy.

Keywords: Project preparation time, Construction costs, Construction time, Productivity, Management of chances and risks.

1 INTRODUCTION

Forecast-based conclusions regarding future developments or events require computation models to be developed based on hands-on experience and practical feasibility whilst relying on sound theoretical assumptions, required levels of abstraction, and simplifications, and considering their mathematical implementation.

Reflecting reality in a (simplified) computation model is an inductive thought process that enables indirect consideration and analysis of the object under study (Töllner 2010). As a rule, real systems can never be completely identified or represented, which is why it is usually inevitable to only consider certain sections of such a real system. Models are generally characterized by three main criteria or features:

- Representation feature: models are representations of real or theoretical systems or originals (social, technical, socio-technical systems, or cyber-technical systems).
- Reduction feature/simplification: models do not capture all attributes of the represented system but only those relevant to fulfilling the intended purpose.
- Pragmatic feature: links between models and originals are purpose-driven; the model replaces the original for limited periods of time. Use of the model is restricted to certain theoretical or real operations (Töllner 2010).

The use of (computation) models can be pictured as a relational triangle involving the model subject (i.e., its developer or user), the model itself, and the model object, or real system. In this setting, the model subject intends to achieve specific targets related to the model object, such as forecasting construction costs, and develops or uses a model for this purpose. This model then represents the model object in an idealized or abstract form. We refer to Jockisch and Rosendahl (2010) for a classification of various models in business administration, economics, and the engineering sciences as well as in an interdisciplinary context. This paper refers to, and discusses, classifications proposed by different authors.

Calculating construction costs and times requires an abstraction of reality or of prevailing cause-effect relationships. This level of abstraction is achieved by introducing a computation model, which represents real processes whilst inevitably incorporating simplifications. "Desktop tests"[1] are carried out to revise, expand, and improve the computation model in order to rectify linkage and model errors. Validation is applied to prove the applicability of the computation model to reality, resulting in calibration and adjustments if and when required. To enhance validity, it is useful to integrate into the model mathematically described interactions that were derived from practice (including from expert surveys), such as non-linear curves representing losses of productivity or increases in labor consumption rates.

2 MODELING IN PROJECT MANAGEMENT, CONSTRUCTION MANAGEMENT AND ECONOMICS

Any model will interact with the model subject and the model object. The model subject is the developer of the model, but it can also be its user. Provided the model can also be used by third parties, individuals who were not involved in the modeling stage can also apply it in specific ways. The model is linked to model objects for which it was developed and to which it is applied. Model objects may include entire buildings or structures, contract sections, or individual structural components, such as those for which construction processes and logistics as well as construction times and costs are determined (see Figure 1).

Any modeling will necessitate a related understanding in order to fully comprehend systems and to reflect the required processes or existing dependencies. A key aspect of modeling is to delimit the model to other systems and processes. Such limits must be defined and documented in a clear and plausible manner so as to enable users who were not involved in the modeling exercise to apply the model according to its intended purpose.

The type and focus of models to reflect interdependencies in construction management and economics will vary depending on the individual project phases. Modeling requirements essentially depend on whether we are dealing with forecasting or planning processes and on the

---

[1]A desktop test is the check of a computation algorithm or of a routine for its correctness. For this purpose, a deterministic computation model must be developed in which results are known, and can be subsequently reviewed, for a defined input.

project phase(s) considered. Generally speaking, models exist for *ex ante, inter actio* and *ex post* analyses. Figure 1 shows the multisystemic model as a hexagonal pyramid that forms the basis for analyses, forecasts, comparisons, decisions etc.

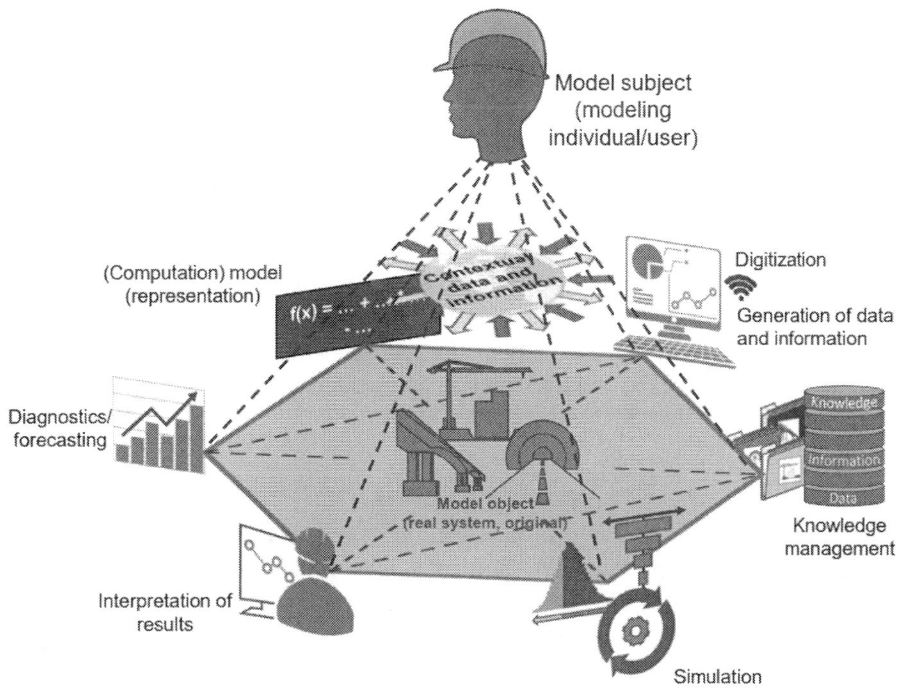

Figure 1. Multisystemic modeling.

"One cannot not model!" The author of this paper has modeled this statement on the famous Paul Watzlawick[2] quote "One cannot not communicate!" Why is this statement, "One cannot not model", correct? Any human being will use models, either consciously or unconsciously, for their actions, decisions, forecasts, plans, etc. Likewise, organizations require and apply models for corporate forecasting, planning, and decision-making processes.

Furthermore, models should be clear, plausible, and capable of being communicated, particularly if model developer and user are not one and the same person. A modeling exercise is at the very beginning of any analysis or development process.

But how can we systematically proceed from models to accurate, usable outputs? A hexagonal pyramid is a good means to represent this progress. This multisystemic pyramid should create the basis for achieving individual project break-even and thus ensuring long-term corporate profitability. The individual triangles forming the pyramid are described below.

2.1 First Triangle

The model subject, i.e., the human being, initially faces the challenge of having to develop a model suitable for a specific task. Such a model can be a check list, an equation, or a simulation

[2] Paul Watzlawick was an Austrian/American communications scientist, psychotherapist, sociologist, philosopher, and author.

– and it will inevitably strongly interact with the model object because the model itself is developed as a reflection of the model object (Kummer 2015). At this stage, the following important questions should be asked: What should the model capabilities include? To which project phases should it be applied? What data and information base is available, and what degree of detail will be possible or useful in the specific phase? Will a deterministic model suffice, or should a semiprobabilistic or probabilistic model be developed?

2.2 Second Triangle

Valid data and information is required to arrive at solid, usable outcomes as a result of applying the model. Where does this data and information come from, and how should it be created for a new project? To some extent, data is generated in automated processes, such as weather data that provides information on temperature trends or humidity curves. Humans present on the construction site must first identify information on the effects of such variables on the site's production system and then document it in an accurate and plausible manner. This process is essentially determined by the specific mode of understanding and subjective judgment of situations by the individuals performing related actions.

2.3 Third Triangle

Digitization should enable the capture of real-time data and information. Such information can be utilized for the current project and should also be systematically entered in the knowledge base. Knowledge is a crucial, and in some areas most important, factor within the planning and production system. Contextual application of models should be improved based on existing knowledge and performance indicators taken from the knowledge base.

2.4 Fourth Triangle

The next triangle is formed by the model subject, model object, and simulation. This setting should enable the following:

- Systematic consideration of uncertainties
- Analyses of a variety of scenarios
- Large number of computing operations (iterative steps)
- Optimization
- Forecasts
- Consideration of correlations etc.

2.5 Fifth Triangle

The fifth triangle comprises the model subject, the model object, and the interpretation of results. In digitization and simulation, the type and representation of generated results are important. Distribution functions are an appropriate means to visualize and communicate the chance/risk ratio, for example.

2.6 Sixth Triangle

The last triangle involves the human being, the model object, and diagnostics/forecasting. Diagnostics are used to replicate the course of past events. Forecasts attempt to simulate the future as realistically as possible.

This multisystemic consideration of the hexagonal pyramid is crucial for holistic modeling. This pyramid is highly relevant not only to research but also to practical applications. Contextuality is essential for ensuring high-quality results when applying the model to construction projects.

In project management, construction management and economics, models are developed and used for various purposes. They are created, for example, for organizational, selection, decision-making, assessment, management, control, safety and security, forecasting, and comparison processes.

The model subject defines the targets to be achieved with the model under development. In this respect, it is crucial to break down targets into "essential", "priority", "nice to have", and non-targets.

3 UNCERTAINTIES IN MODELS

Any abstraction or simplification performed as part of developing such a model will be associated with uncertainties, which can be grouped into aleatoric and epistemic uncertainties in the case of engineering problems (see Figure 2).

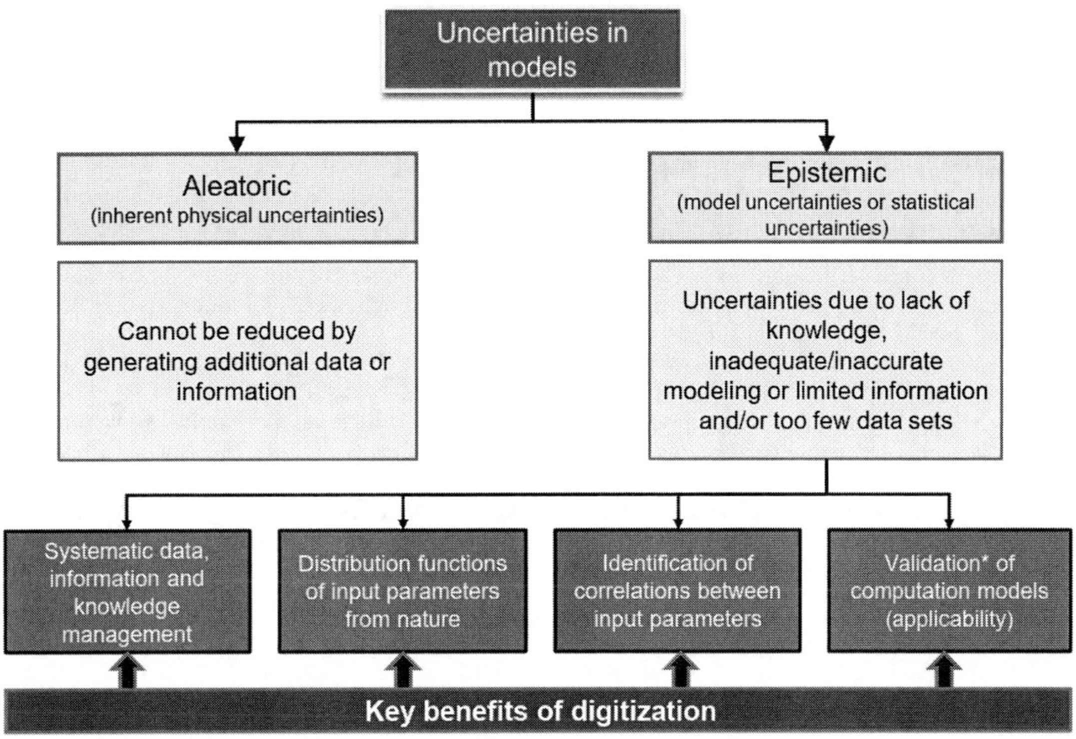

* Method validation designates the verification of the fitness of an analytical method for an intended purpose.
Validation means to verify if the behavior of a model is essentially identical to that of the original. Cf. Töllner et al. (2010): Modelle und Modellierung – Terminologie, Funktionen und Nutzung; p. 19

Figure 2. Uncertainties in models and simulations.

Aleatoric uncertainties result from the fact that the world is (co)determined by the element of chance. Capturing additional information will not reduce existing aleatoric uncertainties. On the other hand, epistemic uncertainties result from lack of knowledge, inadequate or inaccurate modeling, or limited information and/or too few data sets.

This is where the huge benefit of digitization comes into play: systematic data, information and knowledge management makes it possible to represent systemic or natural spreads of input parameters by data fitting using distribution functions, and to integrate these spreads into the calculations. Furthermore, interdependencies of input parameters can be captured in the form of correlations and also be incorporated in the computation model (Hofstadler and Kummer 2017). In-depth analyses also use copulas to reflect dependencies between input parameters.

4 CONCLUSIONS

Realistic modeling essentially requires a thorough understanding and intimate knowledge of the subject matter or area under study. System and process limits should be defined relatively generously at the beginning of the modeling exercise. Modeling is then restricted to the main aspects of the analysis on the basis of precisely defined targets.

Calculations and forecasting require abstraction of reality or of prevailing cause-effect relationships by establishing a computation model, which represents real processes but also inevitably includes simplifications. "Desktop tests" are carried out to revise, expand, and improve the computation model in order to rectify linkage and model errors. Validation is applied to prove the applicability of the computation model to reality, resulting in calibration and adjustments if and when required.

To enhance validity, it is useful to integrate into the model mathematically described interactions, such as non-linearities, correlations, copulas, or uncertainties expressed as ranges and distribution functions, that were derived from practice (including from expert surveys, experiments, and site surveys).

References

Hofstadler, C., and Kummer, M., *Chancen- und Risikomanagement* in der Bauwirtschaft – Für Auftraggeber und Auftragnehmer in Projektmanagement, Baubetrieb und Bauwirtschaft, Springer-Verlag, Berlin, Heidelberg, 2017.

Jockisch, M., and Rosendahl, J., *Klassifikation von Modellen,* In: Das ist gar kein Modell!, Bandow, G., Holzmüller, H. (eds.), Gabler, GWV Fachverlag GmbH, Wiesbaden, 2010.

Kummer, M., *Aggregierte Berücksichtigung von Produktivitätsverlusten bei der Ermittlung von Baukosten und Bauzeiten – deterministische und Probabilistische Betrachtungen*, PhD Thesis, Graz University of Technology, Graz, 2015.

Töllner, A., *Modelle und Modellierung – Terminologie, Funktionen und Nutzung,* In: Das ist gar kein Modell!, Bandow, G., Holzmüller, H. (eds.), Gabler, GWV Fachverlag GmbH, Wiesbaden, 2010.

Interdependence between Structural Engineering and Construction Management
Edited by Ozevin, D., Ataei, H., Modares, M., Gurgun, A., Yazdani, S., and Singh, A.

ISBN: 978-0-9960437-6-2

QUANTITY UNCERTAINTIES IN REINFORCEMENT WORKS – COMPARISON OF PUBLIC VERSUS PRIVATE CLIENTS

MARKUS KUMMER

Institute of Construction Management and Economics, Graz University of Technology, Graz, Austria

Uncertainties are systematically considered and dealt with by applying probabilistic calculation methods, such as Monte Carlo simulations. When selecting appropriate distribution functions for input parameters, users are constantly faced with the issue of having to choose the "right" distribution function for the relevant parameter. Quantities of individual works play a crucial role for costing and pricing, but also for construction process and logistics planning purposes. Quantities stated by the client in its structural specifications are fraught with uncertainties owing to, for instance, incomplete plans at the time of specification, inaccurate calculations, or mere estimates. This is why actual quantities can either be greater or smaller than the specified quantities. This paper demonstrates how distribution functions can be derived from expert surveys delivering responses from actual construction practice. Specific reference is made to reinforcement works whilst distinguishing between public and private clients. The outcomes of the survey presented and discussed in this paper include descriptive data analyses as well as violin plots and fitted distribution functions.

Keywords: Distribution functions, Expert survey, Data fitting, Management of chances and risks, Monte Carlo simulation.

1 INTRODUCTION

Any probabilistic calculation requires the selection of distribution functions for uncertain input parameters. When trying to choose the "right" distribution, users are constantly confronted with the issue which of the many available theoretical (mathematically described) distributions would best reflect the characteristics of the relevant parameter. After selecting the suitable type or kind of distribution, the corresponding shape parameters must be defined to be able to perform a Monte Carlo simulation (Goulet and Smith 2011, Raaber 2003, Sander 2012, Mayer 2013).

Three methods and their combinations are generally available for selecting distribution functions: theoretical considerations based on known or assumed characteristics, gathering data from construction practice, and expert surveys.

In this context, theoretical considerations refer to the basic characteristics of the relevant distribution function, such as discrete vs. continuous, open vs. closed boundaries, mode, and skewness. The second basis for selecting an appropriate distribution function is to gather data directly from construction practice. In this case, an additional distinction is made between historical data (*ex post* – after the construction phase, for instance information taken from final costing or daily work reports) and data collected during the construction phase (*inter actio*[1] – such as from direct site observations). When continuously gathering data during the construction phase, appropriate boundaries or limits must be defined in terms of space, time, and content whilst also ensuring data quality. Furthermore, collected data must be assessed to find out if data can be applied directly or if there is any need for fitting. Only if data has been put in, or linked to, the appropriate context information can be generated from it for the purpose of making this data usable for additional analyses and calculations. Another key criterion that determines data quality is the number of data points, or number of observations. The smaller the sample and the greater the standard deviation of gathered data, the greater the inaccuracy of the derived estimation of the parameters of the underlying population (Hofstadler and Kummer 2017).

AbouRizk and Halpin (1992) state that it is not possible to categorize distributions on the basis of fewer than 20 observations. It should be noted, though, that 20 responses taken from expert surveys provide a higher quality of information than 20 data points taken from observations, which, for instance, can be extracted from various projects using related logs and records. This principle holds true particularly if special emphasis is placed on the selection of appropriate experts, the drafting and preparation of a standardized questionnaire, and the discussion of answers with respondents (similar to the Delphi method).

For the purpose of this research, expert surveys (*ex ante* surveys) were used to determine distribution functions for quantity deviations. Our preference for expert responses over individual measurements taken from construction practice is based on the fact that any expert statement already incorporates experiential evidence from several past projects. Moreover, respondents are capable of linking objective data to (known) complex contextual conditions in construction practice as well as to heuristics, which adds value to their responses. Any identification of quantity deviations between specification and invoicing will also involve the issue of defining appropriate boundaries with respect to contract addenda and/or cancelled works.

2 EXPERT SURVEY

Experts are defined as individuals who possess specific knowledge and intellectual skills and competencies in a clearly delineated field and who serve as a source of specific knowledge for the purpose of a survey. Expert knowledge usually comprises exceedingly large amounts of information, including simplifications, lesser-known facts, rules of thumb, and smart practices (i.e. heuristics) that enable efficient problem solving (Gläser and Laudel 2010). In a preselection process, a total of about 130 experts from Austria and Germany with experience in the fields of costing, process planning, construction, final costing, and invoicing were contacted in writing and asked to participate in the survey. In total, 27 experts were recruited for the survey. They provided 23 responses with respect to quantity deviations in reinforcement works.

The majority of respondents (i.e., approx. 63%) worked for large companies with more than 250 employees, about 26% worked for medium-sized businesses (50 to 249 employees), and about 11% came from small businesses (10 to 49 employees). In the survey presented in this paper, experts had an average professional experience of 17.7 years; this experience ranged from 5 to 41 years. Questions were designed and developed together with social researchers (i.e., sociologists) in several revision steps, applying the principles of simplicity, clarity, impartiality, and specificity. On average, each respondent was interviewed for about 45 minutes either on the phone or in a face-to-face session to overcome ambiguities, collect missing information, or obtain

[1] Latin for: during (work) execution

background information and justifications of responses (Hofstadler and Kummer 2017).

3 DATA FITTING

The issue of which distribution functions to apply to individual calculation parameters essentially determines the quality of the output of a Monte Carlo simulation. There is no uniform answer to this question. In the literature, both considerations regarding the "nature" and potential characteristics of individual parameters, the incorporation of historical data, as well as the use of expert opinions are discussed.

The transfer of such data records (e.g. from *ex ante*, *ex post* or *inter actio* surveys or observations) into distribution functions that can be described on a theoretical level is referred to as "data fitting". In this process, existing data is first analyzed graphically in histograms and then assessed using mathematical/statistical methods in order to identify the theoretical distribution function that best reflects existing data (i.e. with the smallest deviation). Since distributions need to be adjusted to project-specific circumstances or conditions, the shape or curve of distributions of uncertain calculation parameters is of greater interest than absolute values.

An expert survey is recommended as the method to select the underlying data because the *inter actio* and *ex post* gathering of data is likely to present definition and allocation problems in the course of data processing. Data processing is followed by actual data fitting, which involves a comparison and check for consistency of adjusted data with theoretical distributions that can be described mathematically. This theoretical distribution can then either be applied directly to the uncertain input parameter, or the input is optimized for the fitted distribution. This optimization step provides the advantage that the boundaries of the distribution can be modified on the basis of theoretical considerations and characteristics, which significantly simplifies the input (of minimum and expected values, for example) for the user. Furthermore, optimization makes it possible to define a conversion scheme to scale and adjust the derived distribution function to individual/project-specific conditions (Hofstadler and Kummer 2017). Fitting aims to determine the parameters of a distribution function (displacement and shape parameters) in a numerical process so that the distribution reflects gathered raw data as accurately as possible.

4 QUANTITY DEVIATIONS IN REINFORCEMENT WORKS

An expert survey conducted by Kummer and Hofstadler at Graz University of Technology in 2015/16 collected responses with respect to the magnitude of deviations from quantities stated in the specifications of reinforcement works expected for public and private clients (Kummer 2015, Hofstadler and Kummer 2017). No absolute quantities but percentage deviations from the specified quantity were requested, with the latter being defined as 100%. Experts determined the lower limit of quantity deviations (quantity reduction) by indicating a percentage value smaller than or equal to 100% and the upper limit (quantity increase) by indicating a percentage value greater than or equal to 100%.

In the expert survey, 23 responses were received with respect to quantity deviations in reinforcement work (see Table 1). The figures stated for public and private clients differed only to a minor extent, which permits the conclusion of a similar overall degree of accuracy of specifications for reinforcement works. Differences were primarily found for extreme values (i.e. minimum and maximum values). A mean quantity reduction to up to about 91 and 92% was observed for reinforcement works (see Table 1, line 2, cells B and D, and line 14, cells B and D).

The greatest specified quantity reduction declines to 70% in the case of public and to 60% in the case of private clients (line 10). Mean expected quantity increases amount to 7 to 8% (see Table 1, line 2, cells C and E, and line 14, cells C and E). A slightly larger quantity increase should be assumed for public-sector clients relative to the mean values and robust M-estimators (according to Huber). This tendency is also confirmed when considering the maximum values stated for quantity increases. Experts stated a quantity increase of up to 35% for private clients, whereas this increase amounted to up to 40% for public clients (line 11). The calculated range of M-estimators is roughly identical for public and private clients.

Table 1. Descriptive representation of data pertaining to quantity deviations in reinforcement work.

No.	Parameter	Reinforcement – steel bar – public clients		Reinforcement – steel bar – private clients	
		Negative deviation	Positve deviation	Negative deviation	Positve deviation
0	A	B	C	D	E
1	N	23	23	23	23
2	Mean	91.57	108.30	91.30	108.70
3	Standard error of mean	1.57	1.81	1.77	1.90
4	Standard deviation	7.53	8.66	8.48	9.10
5	Coefficient of variation	8.22 %	7.99 %	9.28 %	8.37 %
6	Median	95.00	110.00	95.00	110.00
7	Mean absolute deviation	5.00	5.00	5.00	6.00
8	Robust coefficient of variation	5.26 %	4.55 %	5.26 %	5.45 %
9	Mode	95.00	110.00	90.00	100.00
10	Minimum	70.00	100.00	60.00	100.00
11	Maximum	100.00	140.00	100.00	135.00
12	Spread	30.00	40.00	40.00	35.00
13	Skewness	-1.25	2.24	-2.45	1.26
14	M-estimator (H12)	92.51	107.26	92.65	107.54
15	Standard deviation (H12)	6.69	6.73	5.78	7.99
16	Robust coefficient of variation	7.24 %	6.27 %	6.24 %	7.43 %
17	Normal distribution (Shapiro-Wilk)	No	No	No	No

Figure 1 shows the ranges of expert responses but also the violin plots relating to them. Violin plots are an extension to the known box plot representation. They combine the information of a density function (or of a smoothed histogram) with that of a box plot in a single diagram. Their advantage is that the graphical representation includes not only box plot information but also distribution characteristics. In the diagram, the density curve is plotted symmetrically above and below the (vertical) box plot. The two halves of the violin plot are thus mirrored along the axis on which the relevant parameter is shown.

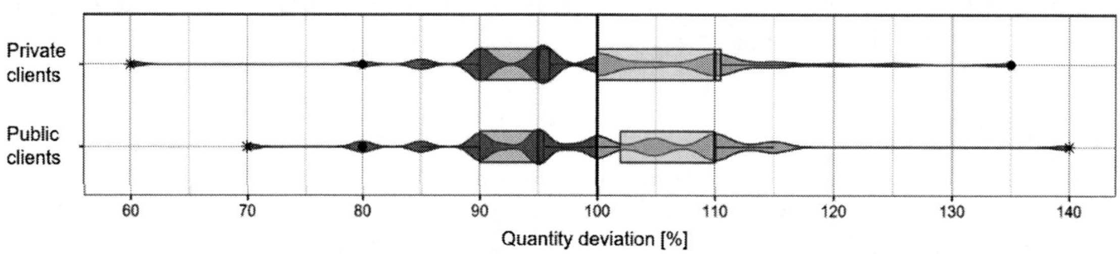

Figure 1. Violin plots of quantity deviations in reinforcement work – public vs. private clients.

Another benefit of violin plots is that the shape of the distribution or the presence of any accumulations or clusters within certain value ranges is easier to visualize. This type of information can get lost in conventional box plots. The greater the frequency of tapered zones in the violin plot, the larger the likelihood of a greater frequency of expert responses within certain value ranges. For instance, responses got clustered at quantity reductions of 90% and 95% in relation to reinforcement works both for public and private clients. Between these two percentages, the violin plot exhibits only a small width, which permits the conclusion that fewer responses lie within this range. If no responses exist within a defined value range, the associated area of the violin plot reduces to zero and appears as a line (see, for example, quantity increase in reinforcement works – public clients – near the 130% mark).

Figure 1 uses circles/dots for outliers and stars for extreme values. The number of responses cannot be derived directly from the violin plot, which is why descriptive data analyses are listed in Table 1. The @Risk software suite was used to perform data fitting on the basis of expert responses for public and private clients (see Figure 2) to enable realistic recommendations for potential distribution functions that would appropriately represent quantities in reinforcement works. A LogLogistic distribution for public clients and a Logistic distribution for private clients reflect the expert statements very accurately, as confirmed by a Kolmogorov-Smirnov test.

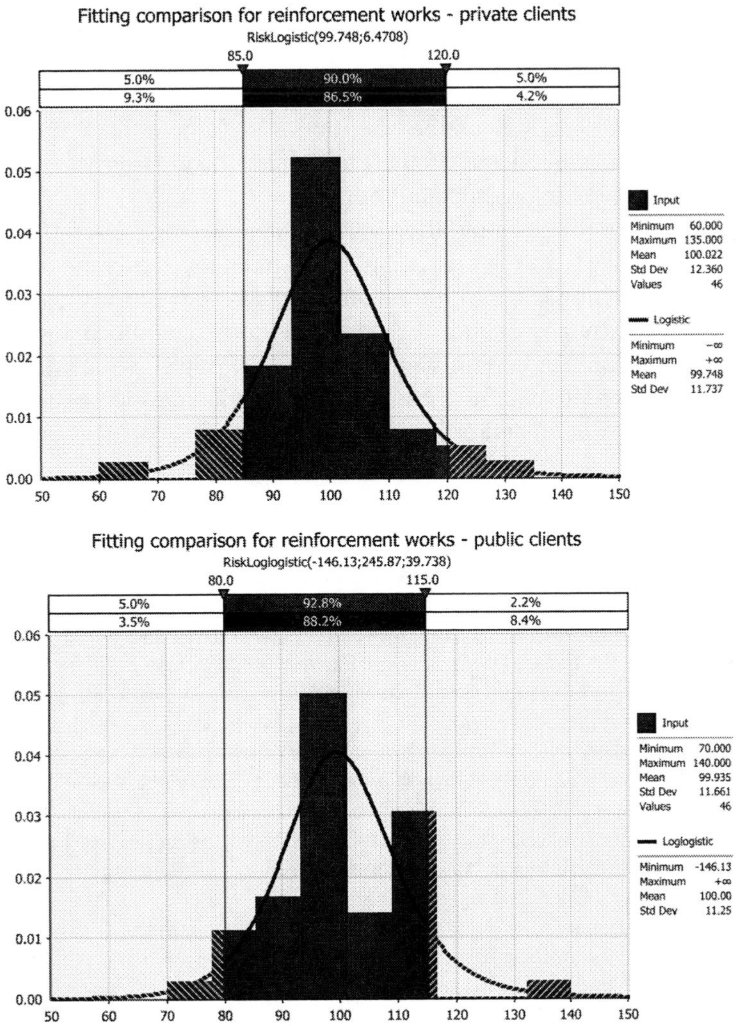

Figure 2. Data fitting – quantity deviations in reinforcement works.

5 DISCUSSION AND CONCLUSIONS

From the bidder's/contractor's point of view, the above-mentioned quantity uncertainties constitute a risk (for instance with respect to the contribution margin), but also a chance (such as in unit-priced contracts in the case of quantity increases at a high unit price) and should thus be reflected by distribution functions of relevant quantities in the probabilistic costing exercise. Considering quantity uncertainties is also of interest to clients in order to investigate, during the phase of selecting the best bidder, how prices quoted by bidders would move when specified quantities change in the case of unit-priced contracts. Overall, the results of the expert survey show that, quantity uncertainties in reinforcement works are considered to be slightly greater for private clients than those assumed for public clients. A LogLogistic distribution for public clients and a Logistic distribution for private clients reflect the best approximation. However, the open ends of these theoretical types of distribution should be viewed critically. If probabilistic calculations are performed using such distributions (with no upper and/or lower limit), this may lead to exceedingly implausible individual results within the simulation. It is thus useful to combine the findings of the survey and the results of mathematical/statistical data fitting with theoretical considerations, particularly with respect to possible limits to individual parameters. This broader analysis includes both data fitting and theoretical considerations and can be used to establish a polygonal representation of the distribution's shape and to apply limit values, thus making it possible to retain the shape of the distribution taken from the expert survey and to adjust the value range to specific project conditions.

Quantity uncertainties will not only affect costing and pricing. Even more importantly, they will also have an influence on construction process planning and logistics as well as on the scheduling of resources. Knowledge of potential/expected ranges thus helps project stakeholders to include outliers and extremes in their project considerations and planning exercises and to develop a variety of solutions for certain scenarios from the outset. Generally speaking, knowing the possible ranges of specified quantities contributes to achieving greater transparency in process planning and project organization and/or management.

References

AbouRizk, S., and Halpin, D., Statistical Properties of Construction Duration Data, *Journal of Construction Engineering and Management,* 118(3), 525-544, 1992.

Gläser, J., and Laudel, G., *Experteninterviews und Qualitative Inhaltsanalyse als Instrumente Rekonstruierender Untersuchungen,* Springer Fachmedien GmbH, Wiesbaden, 2010.

Goulet, J., and Smith, F., *Extended Uniform Distribution Accounting for Uncertainty of Uncertainty,* Proceedings of the First International Conference on Vulnerability and Risk Analysis and Management (ICVRAM 2011), Hyattsville, Maryland, ASCE, April 11-13, 2011.

Hofstadler, C., and Kummer, M., *Chancen- und Risikomanagement in der Bauwirtschaft – Für Auftraggeber und Auftragnehmer in Projektmanagement, Baubetrieb und Bauwirtschaft,* Springer-Verlag, Berlin, Heidelberg, 2017.

Kummer, M., Aggregierte Berücksichtigung von Produktivitätsverlusten bei der Ermittlung von Baukosten und Bauzeiten – deterministische und probabilistische Betrachtungen, Doctoral Thesis, Graz University of Technology, Graz, 2015.

Mayer, F., *Kostensicherheit zum Zeitpunkt der Realisierungsentscheidung – Entwicklung eines Kosten-Prognose-Modells für Bauwerkskosten im Hochbau,* Ph.D. Thesis, Lehrstuhl für Bauprozessmanagement und Immobilienentwicklung der Technischen Universität, München, 2013.

Raaber, N., *Beitrag zu Ausschreibung und Vergabe von Bauleistungen. Schriftenreihe des Instituts für Baubetrieb und Bauwirtschaft der Technischen Universität Graz,* Verlag der TU Graz, Graz, 2003.

Sander, P., *Probabilistische Risiko-Analyse für Bauprojekte – Entwicklung eines branchenorientierten softwaregestützten Risiko-Analyse-Systems,* Ph.D. Thesis, Leopold-Franzens-Universität Innsbruck, 2012.

Interdependence between Structural Engineering and Construction Management
Edited by Ozevin, D., Ataei, H., Modares, M., Gurgun, A., Yazdani, S., and Singh, A.

ISBN: 978-0-9960437-6-2

COMPARISON OF OWNER AND CONTRACTOR PERSPECTIVES IN ESTABLISHING COST CONTROL STRUCTURE

RUVEYDA KOMURLU[1] and AKIN ER[2]

[1] *Dept of Architecture, Kocaeli University, Kocaeli, Turkey*
[2] *TUPRAS General Management, Kocaeli, Turkey*

Cost control is a part of cost management which is carried out by the project owner and the contractor throughout a project. However, the structures of the control function developed by each party generally differ since the purpose of the function and the level of the structure are not exactly the same. Contractors have several advantages while building a cost control system such as more detailed information about the project, more background, more dedicated personnel, specifically developed software etc. Therefore, contractors have a broader vision on the issue. Owners need to utilize cost control systems not only for common reasons but also due to some specific necessities. Unlike contractors, owners have to capitalize projects just before they put their investments into operation. This paper intends to focus the necessities prioritized by the owners and contractors on the issue. A comparison will be made to distinguish the differences between the cost perspectives of project owners and contractors. An assessment will be done of the matters that owners pay attention while establishing a cost control structure in light of the experiences practiced in oil and gas projects. Thus, the important points which should be better to consider by the owners' project control teams will be advised.

Keywords: Cost management, Budget compliance, Cost breakdown structure, Capitalization, Project control team.

1 INTRODUCTION

Cost management is simply defined as "the processes involved in planning, estimating, budgeting and controlling costs so that the project can be completed within the approved budget" (PMI PMBOK 2008). Following, the definition is enhanced to include financing, funding and managing processes (PMI PMBOK 2013). Herein, the cost control part of the cost management will be reviewed regarding the necessities prioritized by the owners and contractors.

Cost control refers to project procedures which are applied to monitor actual expenditures against the progress and compute deviations from the approved budget to allow effective measures to be taken to achieve minimum costs (AACE 10S-90 2017). This perception is valid for all parties involved in a project since the value of money has been increased in the recent decades. Contractors' profit margins have been reduced considerably due to increased competition in the construction industry, because of which they have greater cost-consciousness to be able to compensate possible losses (Ashworth *et al.* 2015). Thus, owners and contractors are highly motivated to follow a cost control process. In the absence of a control system, projects proceed without an adequate oversight and a clear understanding of status, so an analytic action plan cannot be implemented to bring the project back on track in case of a deviation (Venkatamaran and Pinto 2008).

On the other hand, outputs of cost control process could provide valuable data for further projects. This data could be used as a library during cost estimating if recorded and classified in a proper way. A cost control system can also be established to cooperate with the accounting system, so those records could be organized to allow the owner complete capitalization of the project with ease.

Project stakeholders should set up a cost control system which satisfies their expectations and necessities. That system should be built up on a project breakdown structure which specifically called as cost breakdown structure (CBS). CBS has a similar frame as work breakdown structure (WBS) however indicative factor is the cost monitoring philosophy. It allocates the project several levels beginning from project overall to final cost elements required to be followed. With this hierarchical breakdown, occurred cost could be addressed to cost elements which enables project parties to roll up project costs to each cost level of CBS systematically (Er and Komurlu 2017). Such kind of methods are also used to calculate the physical progress of projects. (Komurlu and Er 2018a).

This paper intends to compare cost perspectives of owners and contractors. Since, there are limited number of research studies analyzing owner procedures on the subject, a CBS is studied to introduce owners' point of view to readers.

2 COST PERSPECTIVES OF OWNERS AND CONTRACTORS

The main objective of cost control is to measure variances from the approved budget. The quality of the budget depends on the accuracy of cost estimates, while the accuracy of estimates is directly related to the level of information at hand.

On the owner's side, it is necessary to realize some project development steps to reach a maturity level, where the project owner can reckon a more reliable cost estimation and decide whether to invest or not. However, even at this stage, many details may not be clear yet. Therefore, cost estimation of owners generally could be done with high level cost elements. It also includes a certain level of contingency. Typically, project is divided into three main phases i.e. engineering, procurement and construction. Project budget mainly stands on those phases.

However, previous experiences of authors show that on the contractor's side, project budgets are generally built up on two main categories; direct costs and indirect costs. Since the contract price is fixed, any cost increase results with a loss of profit. That reality motivates the contractors to establish an apparent and easily traceable cost structure as given in Table 1. It is much more detailed and CBS level of cost elements are lower compared to the owners' cost structure.

Regarding the project life cycles owners' ability to influence project expenditures rapidly decreases after the final approval of the investment decision. However, on the contractors' side, it is much more dynamic. Lower level of cost elements allows contractors to focus on and analyze each work item in detail. For example, in a proper cost control structure a work item like concrete pouring can be deeply analyzed. The occurred cost of one cubic meter of concrete can be compared with budget unit price and total variance in the total cost of concrete pouring can be calculated. Moreover, root cause of unit cost variance could be determined (such as material, labor etc.) or cost and quantity variance amounts in the total variance could be calculated. Thus, a well-designed and controllable cost control system allows contractors to take effective actions to achieve minimum costs. That system necessarily contains three main components;

i) a proper cost control structure based on an accepted cost control method,
ii) sufficient information to reckon budget costs as per requirements of the structure,

iii) sufficient organization to monitor and control the actual data.

Table 1. Direct and Indirect Costs of Project Budgets on the Contractor's Side.

Direct Costs	**Indirect Costs**
Personnel Costs • Direct Personnel	Personnel Costs • Indirect Personnel • Head Office Personnel
Equipment and Materials • Direct Equipment • Direct Materials	Equipment and Materials • Indirect Equipment • Construction Machinery

Subcontractor Costs	Administration Costs
	• Facilities
	  o Site Facilities
	  o Camp Facilities
	• Site Running Costs

	Financial Costs
	Insurance Costs
	Taxes and Duties

Owners experience difficulties to gather detailed cost information at the beginning of the project because the detailed design is rarely completed at that stage. Therefore, to set up a higher level of CBS is an obligation rather than a preference for owners. In parallel, they set up more compact but effective cost control organizations.

Contractors have much more information at the beginning of the project since they have a complete organization dealing with engineering, procurement and construction specifically. They have in-house data due to their experience and a steady cost control system which was improved during previous projects. Therefore, good implementations of tailor-made or packaged cost control software at contractors' organizations are quite common. Generally, Earned Value methodology is the basis for these systems, because this method provides an objective measurement, earlier identification of problems, and fast feedback on any corrective action taken (Gerson 2013).

3 A CBS PROPOSAL REGARDING OWNERS' REQUIREMENTS

Main objective of owners for cost control is to prevent deficits in the budget, and maintain consistency of the feasibility regarding cost part. Second, they would like to have a reliable cost register for their future investments. Third, they would like to separate the investment cost in a compatible way to help the capitalization process.

Uniform accounting system requests owners to record their investment expenditures in "construction work in progress" account. They have to classify and transfer those expenditures below capital asset accounts just before putting them into operation.

- Land
- Buildings
- Infrastructure
- Equipment and Furniture
- Intangible Assets
- Other Fixed Assets

This process is called capitalization which is the final step of recording of a cost as an asset rather than an expense. Since capitalization regulations stipulate specific rules while registering costs at the inventory, collaboration between owners' project and financial affairs teams becomes an inevitable necessity. A proper cost breakdown structure, which divides the project costs into agreed components, enables financial affairs teams to apply requirements of capitalization with ease.

In Table 2., a cost breakdown structure of a plant project is presented as an example considering an oil and gas project. This structure is recommended by taking into account various experiences of the authors in different investment projects built in the TUPRAS refineries.

Table 2. Cost Breakdown Structure of a Plant Project.

Level 0	**Level 1**	**Level 2**	**Level 3**	**Level 4**
PROJECT	**UNITS**	**PHASES**	**SUB-PHASES**	**UNIQUE ELEMENTS**
Project Overall	Unit 1	Engineering	Basic Engineering	
			Detailed Engineering	
			Procurement Engineering	
		Procurement	Equipment	Equipment – 1
				Equipment – 2
				...
			Bulk Material	
		Construction		
		Others	Consultancy	
			Freight & Insurance	
			Financial Costs	
			Services	
			Utilities	
			Other Fixed Assets	Inventory – 1
				Inventory – 2
				...
			Intangible Assets	Asset – 1
				Asset – 2
				...
			Others	
	Unit 2			
	...			

The logic behind the structure is as follows: Firstly, the project is divided into units. Those units will be the cost centers in the accounting system during the operation phase. For the following level, main cost categories such as engineering, procurement, construction etc. could be defined. By that way, total cost of main phases could be calculated and recorded.

Equipment is the most important cost element of the structure since it will be the base of capitalization. Maintenance or replacement costs should be addressed to equipment during operation. Additionally, each equipment may have different depreciation periods which directly affect the company's annual financial statement. Therefore, each itemized equipment should be considered in the lower level of equipment and their cost shall be assigned to those cost elements. Each equipment also has engineering, construction, freight, customs etc. costs. However, it is difficult to determine them equipment by equipment. Therefore, those costs could be recorded at their specific categories. Financial groups can distribute those costs to each equipment proportionally by calculating the weight of the equipment in the total equipment cost, while implementing the capitalization process.

Buildings, tanks, infrastructure and interconnecting lines are also other important structures required to be followed preferably as a separate unit. If the owner assigns some personnel specifically for the management of the project, expenditures of those personnel could also be accepted as a part of the project cost.

4 CONCLUSIONS

In the light of the practices experienced in both owner and contractor organizations, it is concluded that cost control perspectives of owners and contractors generally differ since their purpose and cost control levels are not exactly the same. Listed below are the similar and different cost perspectives of owners and contractors:

i) Owners generally set their budget on project phases, whereas contractors prefer to focus on direct and indirect costs.
ii) Contractors have much more detailed project information and experience than owners, so that they can build up more detailed cost control structures.
iii) Project organizations of contractors consist of more personnel dedicated to cost control, so that they can monitor actual costs in details. As a result, contractors can set up more complex cost control systems and use Earned Value methodology more efficiently.
iv) Both parties utilize cost control systems to measure variance from approved budgets and allow effective action to be taken to achieve minimum costs.
v) Both parties record cost data for future projects.
vi) Unlike contractors, owners, using a cost control system, would like to separate the investment cost in a compatible way to help the capitalization process.

Considering the limited number of research studies analyzing owner procedures on the subject, a CBS is recommended to convey authors' approach to readers who work as a cost controller at owner organizations. The example is selected from an oil and gas project, however the approach could be utilized in several types of industrial projects. Main idea is to divide project into levels corresponding to units, phases, sub-phases, unique elements respectively, and form a structure not only useful for cost control but also for capitalization process of financial affairs departments. In addition, such kind of CBS could be utilized by estimation teams during open book cost estimate (OBCE) period of EPC convertible contracts (Komurlu and Er 2018b) which allows to calculate project cost within a systematic and traceable structure.

References

AACE International, *Cost Engineering Terminology*, Recommended Practice 10S-90, 2017.
Ashworth, A., and Perera, S., *Cost Studies of Buildings*, Routhledge, UK, 2015.

Er, A., and Komurlu, R., Importance of Planning in Refinery Projects and Recommendations for Successful Planning, *Journal of Architecture and Life*, 2(2), 253-264, ISSN: 2564-6109, doi: 10.26835/my.349372, 2017.

Gerson, M., Using Earned Value Analysis to Manage Projects, *Journal of Applied Business and Economics*, 15(1), 11-14, 2013.

Komurlu, R., and Er, A., *A Simplified Method for Physical Progress Measurement,* The Second European and Mediterranean Structural Engineering and Construction Conference, *Responsible Design and Delivery of the Constructed Project*, Beirut, Lebanon, doi: 10.14455/ ISEC.res.2018.2, 2018a.

Komurlu, R., and Er, A., *Evaluation of an OBCE Conversion Contract in a Large-scale Oil and Gas Project,* 9th International Conference on Engineering, Project, and Production Management (EPPM 2018) Conference, Nazarko, L. (ed.), Cape Town, South Africa, Nelson Mandela University and Central University of Technology, 2018b.

Project Management Institute, *A Guide to the Project Management Body of Knowledge (PMBOK Guide),* Fourth Edition, 2008.

Project Management Institute, *A Guide to the Project Management Body of Knowledge (PMBOK Guide),* Fifth Edition, 2013.

Venkatamaran, R. R., and Pinto, J. K., *Cost and Value Management in Projects*, Jon Wiley & Sons, New Jersey, 2008.

Interdependence between Structural Engineering and Construction Management
Edited by Ozevin, D., Ataei, H., Modares, M., Gurgun, A., Yazdani, S., and Singh, A.

ISBN: 978-0-9960437-6-2

TOWARDS DIGITIZING THE CONSTRUCTION INDUSTRY: STATE OF THE ART OF CONSTRUCTION 4.0

MD ASLAM HOSSAIN and ABID NADEEM

Dept of Civil and Environmental Engineering, Nazarbayev University, Astana, Kazakhstan

Having a significant economic impact in country's GDP and being a major workforce, construction industry is yet characterized by low production rate, low technological advancement, minimum automation and robotic usage, and so on. With the visionary idea of Industry 4.0 that focuses on digitization of the value chain of a product and improving productivity through a variety of technologies and automated manufacturing environment, this research aims to develop a framework to adopt Construction 4.0 within a construction company. Current state of the art of the technologies in construction associated with the notion of Industry 4.0 (e.g., Building Information Modelling, virtual reality, augmented reality, Drone, etc.) is explored through extensive literature studies. The proposed framework incorporates current technological advancement related to construction industry, legislative requirements, barriers, enterprise transformation requirements and so on. Construction 4.0 would make a great impact in construction industry through improved value chain of construction projects, productivity improvement, and safe and sustainable construction. Indeed, the proposed framework would contribute to the advancement of new knowledge in the worlds' construction companies and provide potentials of new research focuses.

Keywords: Productivity improvement, BIM, Value chain, Automation, Industry 4.0, Internet of things.

1 INTRODUCTION

The construction industry accounts for 9% world's GDP (gross domestic product) share (Crosthwaite 2000) which indicates its significant impact on the country's economic development. As reported in (McKinsey 2018), global spending in construction was $11 trillion in 2017 and is projected to increase by $14 trillion in 2025. However, the industry is characterized by low productivity with low technological advances, minimum mechanization/computerizations, and robotic usage etc. The digitization, automation, and integration provide opportunities for productivity improvement as well as design and construction quality (Global Industry 4.0 Survey 2016). Nevertheless, embracing new technologies and automation in the construction industry is very slow. For instance, Building Information Modelling (BIM) is becoming the core of information management in the Architecture, Engineering and Construction industry that allows all stakeholders to exchange and manage information throughout the lifecycle of a built facility. Yet, BIM adoption rate in the construction industry has been much slower than anticipated (Walasek and Barszcz 2017).

Because the construction industry is one of the major contributors of the country's economy, the industry must adopt the emerging technologies for effective management of construction projects and improve the productivity and value chain. In recent time, Industry 4.0 has received much attention in manufacturing industries towards increasing digitization, automation and ICT use with an aim to create digital value chain of the lifecycle of the product starting from concept to development, manufacturing, use and maintenance, and recycling (Lasi *et al.* 2014). This aids producing quality product with a decrease in cost and time-to-market as well as overall improvement in enterprise performance (Brettel *et al.* 2017). Despite the benefits that Industry 4.0 offers, construction industries are yet to incorporate these innovative technologies. It is important to understand the major challenges specific to the construction projects that restrain integrating new technologies and be addressed accordingly.

With the visionary idea of Industry 4.0, this study explores the current state of the art of the technologies associated with construction (Construction 4.0) and proposes a framework to incorporate new technologies within construction projects. The framework would help to understand different goal sets of digitization and automation within a construction company.

2 INDUSTRY 4.0 AND CONSTRUCTION INDUSTRY

Industry 4.0 (the fourth industrial revolution) encapsulates future industry development trends to achieve more intelligent manufacturing processes, including reliance on Cyber-Physical Systems (CPS), construction of Cyber-Physical Production Systems (CPPS), Internet of things (IoT) and implementation and operation of smart factories (Zhou *et al.* 2015). The applications of the new concepts related to Industry 4.0 can be applied to the construction arena. The manufacturing of the construction machinery industry can take help from virtual reality to produce construction machinery in a short time, with good quality and at lower cost (Wu *et al.* 2013). The construction industry is quickly adopting Industry 4.0 compliant technologies in the form of BIM, virtual reality (VR), augmented reality (AR) and mixed reality (MR). The maturity level of such technologies is at different levels. BIM, cloud computing, mobile computing, and modularization technologies are relatively mature due to their wide acceptance in the industry whereas AR, VR and MRs are in their formative years.

3 STATE OF THE ART OF CONSTRUCTION 4.0

Industry 4.0 has four main design principles as identified by Gregor (2002) and further explained by Herman *et al.* (2016). Following the same notion, Figure 1 depicts the design principles of Construction 4.0. Though the characteristics of the project in the construction industries are much different than the products in manufacturing industries, the concept of Industry 4.0 can be well matched in the perspective of construction industries. As explored by Oesterreich and Teuteberg (2016), major components of Industry 4.0 such as IoT, IoS (Internet of Services), CPS, Big Data, Cloud Computing, Robotics, etc. are now getting attention to enable a digitized construction environment though the practical applications of Construction 4.0 are still at its infancy. Following subsections describe design principles to understand the current state of the art of Construction 4.0.

The idea of digitization deals with seamless communication technologies that enable interconnected objects and people share information to attain the common goals collaboratively (Hermann *et al.* 2016). BIM has been found to be the prominent digital planning method in construction (Oesterreich and Teuteberg 2016). Along with IoT and IoP (Internet of People), BIM can be used to set the common communication standards between project participants. Though there is a lack of standardized data schema in BIM and lack of protocol that define the responsibility for information usage (Chen *et al.* 2018), buildingSMART is working towards interoperable standards of BIM.

Sensor data and digitalized plant models are linked together to create a virtual copy of the physical world that helps IoT participants to make informed decisions (Hermann *et al.* 2016). In the context of construction, VR and AR are becoming increasingly popular (Meža *et al.* 2014, Wu *et al.* 2013) that can minimize the gap between the digital (e.g. 3D model, simulation model etc.) and real world (position or condition of materials, equipment).

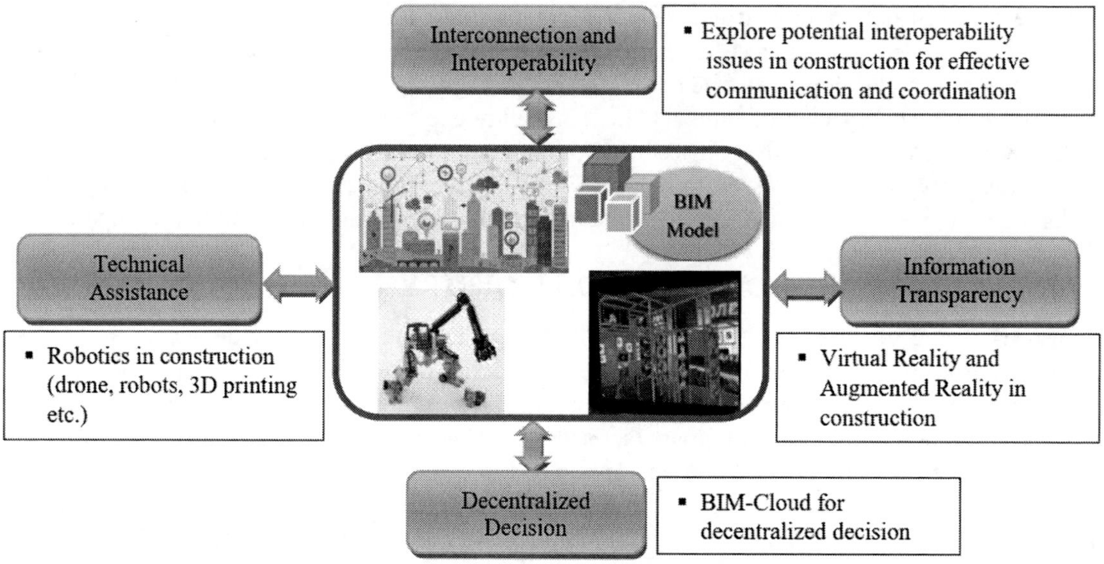

Figure 1. Construction 4.0 design principles.

Decentralized decision making involves the interconnection of objects and people utilizing local and global information to make a better decision and improve productivity (Hermann *et al.* 2016). Construction projects often require collaboration between parties who are geographically placed in diverse locations. Moreover, different participants use different BIM platforms for their models which create interoperability issues (Hossain and Yeoh 2018). There is a growing interest in using Cloud technologies as a means of team collaboration throughout a building's lifecycle in a BIM environment. And, researchers have been investigating to develop specifications for a cloud-based BIM platform (Alreshidi *et al.* 2016).

In an Industry 4.0 environment, the humans' role is changing towards a strategic decision maker instead of the operator of machines (Hermann *et al.* 2016). Unsafe, unpleasant and exhaustive works are assisted by robots for which humans are needed to be trained properly for effective human-machine collaboration (Awais and Henrich 2013). Being a labor-intensive industry, the construction industry has huge potential to improve its productivity by technical assistance (e.g., robot usage), especially for hazardous and unsafe works for a human. In digital construction platform, there are limited uses of robots such as 3D printing, fabricating walls, rebar placement, welding, drones etc. (Chen *et al.* 2018, García de Soto *et al.* 2018, Keating *et al.* 2017, Zhang *et al.* 2018) and all of them are single-application robots. Use of multi-functional and multiple robots should be investigated considering the dynamic site conditions in construction projects.

4 IMPLEMENTING THE CONCEPT OF CONSTRUCTION 4.0

As can be seen from the previous discussions, the design principles of Industry 4.0 can be well aligned with the construction industry's principles and practices. However, the construction industry has not utilized Industry 4.0 on a widespread scale. Digitization and automation of the construction industry to improve the value chain, productivity and sustainable construction are far isolated, focusing on a single point solution. Following the blueprint for digital success (Global Industry 4.0 Survey 2016), Figure 2 presents a framework depicting a step-by-step procedure to implement the concept of Construction 4.0 among construction companies.

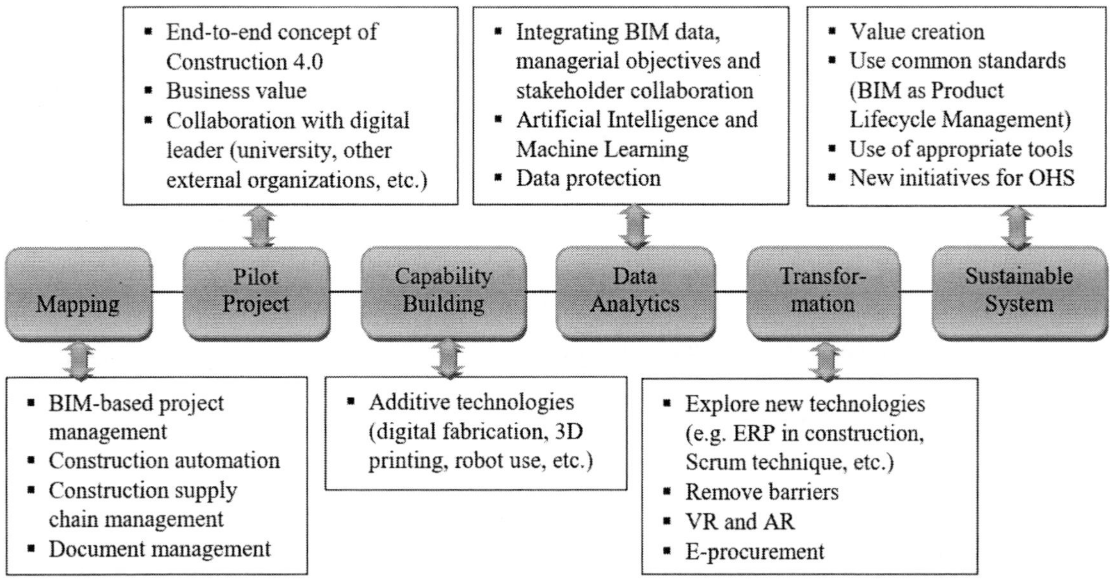

Figure 2. Framework to implement the concept of Construction 4.0.

The implementation steps start with mapping out the Construction 4.0 strategies that consist of evaluating the current digital maturity within the company and setting targets for the next five years. Particularly, the company should have well set BIM-based project management system since BIM is a digital planning method in the construction industry (Li and Yang 2017, Oesterreich and Teuteberg 2016). Other measures are automation in the construction process (Chen *et al.* 2018), Industry 4.0 enabled Construction Supply Chains (Dallasega *et al.* 2018), and a standardized document management system which is key to digitization (Wright 2013).

Next, select pilot projects to establish proof of concept. The company should highlight the end-to-end concept of Construction 4.0 following the notion of Industry 4.0, demonstrate business value and work in collaboration with outside digital leaders and universities to accelerate the digital innovation (Global Industry 4.0 Survey 2016).

While working with the pilot projects, it is important to comprehend lessons learned, develop strategies to improves processes by incorporating new/additive technologies such as digital fabrication (García de Soto *et al. 2018)*, robot usage (Keating *et al.* 2017) etc., and train people to put digitization into place.

Data analytics is the key to decision-making and intelligent systems design by integrating BIM data, managerial objectives, and stakeholder collaborations. In such, IoT enabled Artificial Intelligence (AI), Machine Learning (ML) and other intelligent techniques can be used

(Woodhead *et al.* 2018) while carefully considering the data protection and cyber security issues (Chen *et al.* 2018).

To gain the competitive advantage the company should transform into a digital enterprise with clear leadership and commitment from the top management and financial stakeholders. Continue to explore new technologies such as ERP in construction, Scrum technique, E-Procurement etc. (Costa and Grilo 2015, PRNewswire 2018, Streule *et al.* 2016). The metrics for the transformation are for the positive influences in the construction industry in achieving improved productivity, efficiency, quality and construction processes integration. Furthermore, remove barriers that may hinder digitization of the construction company. As for any business transformation, this will occur under the barriers of political, economic, socio-cultural, technological, environmental, ethical and legal influences (World Economic Forum 2016). For example, the adoption of BIM in the construction industry since the beginning of this millennium has witnessed such influences in terms of the organizational culture and work practices and lack of clarity about the costs and benefits of such adoptions.

Finally, the company should plan for a sustainable system in developing complete product and solutions for customers. Use appropriate tools and standards (e.g., BIM for lifecycle management) (Woodhead *et al.* 2018), build suitable partnerships, and develop a new safety culture for Construction 4.0 (Badri *et al.* 2018). This, in turn, will need more regulatory requirements and compliances to be devised. Several organizations including the government championed BIM in their respective countries; Construction 4.0 needs similar or better support.

5 CONCLUSIONS AND DISCUSSION

There is a huge potential for construction companies to improve productivity through digital value creation, using the notion of Industry 4.0. Nevertheless, the relevant technologies such as CPS, IoT, cloud computing have not been comprehensively considered in construction projects though the demonstration of such technologies to illustrate the best practices exists. This study illustrated the design principles for Construction 4.0 and presented a systematic framework to adopt the concept of Construction 4.0 within a construction company. The construction companies can use the framework to develop capabilities and transform enterprises towards digitalization. Nevertheless, the biggest challenge in attaining the goals of digitization and automation in construction are lack of digital culture and training, especially among SMEs who are the major contributors in the construction industries. Considering the high impact of the construction industry towards the country's economic development, the government should provide adequate support to SMEs who have limited capabilities to invest in new technologies. Further investigations are to be made while the companies use the framework and identify the good practices, lesson learned, and challenges encountered to attain an ecosystem approach.

References

Alreshidi, E., Mourshed, M., and Rezgui, Y., Cloud-Based BIM Governance Platform Requirements and Specifications: Software Engineering Approach Using BPMN and UML, *Journal of Computing in Civil Engineering*, 30(4), 04015063, 2016.

Awais, M., and Henrich, D., Human-Robot Interaction in an Unknown Human Intention Scenario, *Proc., 11th International Conference on Frontiers of Information Technology*, 89-94, 2013.

Badri, A., Boudreau-Trudel, B., and Souissi, A. S., (2018). Occupational Health and Safety in the Industry 4.0 Era: A Cause for Major Concern?, *Safety Science*, 109, 403-411, 2018.

Brettel, M., Friederichsen, N., Kcroseller, M., and Rosenberg, M., (2017). How Virtualization, Decentralization and Network Building Change the Manufacturing Landscape: An Industry 4.0 Perspective, *FormaMente*, 12, 47-62, 2017.

Chen, Q., García de Soto, B., and Adey, B. T., Construction Automation: Research Areas, Industry Concerns and Suggestions for Advancement, *Automation in Construction*, 94, 22-38, 2018.
Costa, A. A., and Grilo, A., BIM-Based E-Procurement: An Innovative Approach to Construction E-Procurement, *The Scientific World Journal*, 2015, 905390-905390, 2015.
Crosthwaite, D. The Global Construction Market: A Cross-Sectional Analysis, *Construction Management and Economics,* 18(5), 619-627, 2000.
Dallasega, P., Rauch, E., and Linder, C., Industry 4.0 As an Enabler of Proximity for Construction Supply Chains: A Systematic Literature Review, *Comp. in Ind.,* 99, 205-225, 2018.
García de Soto, B., Agustí-Juan, I., Hunhevicz, J., Joss, S., Graser, K., Habert, G., and Adey, B. T., Productivity of Digital Fabrication in Construction: Cost and Time Analysis of a Robotically Built Wall, *Automation in Construction*, 92, 297-311, 2018.
Global Industry 4.0 Survey, Engineering and Construction Key Findings, 2016. Retrieved from https://www.pwc.nl/nl/assets/documents/industry-4-0-engineering-and-construction-key-findings.pdf in October, 2018.
Gregor, S., A Theory of Theories in Information Systems, *Proc., Inf. Sys. Found. Workshop,* 1-19, 2002.
Hermann, M., Pentek, T., and Otto, B., *Design Principles for Industrie 4.0 Scenarios*, Proc., 49th Hawaii International Conference on System Sciences (HICSS), 3928-3937, 2016.
Hossain, M. A., and Yeoh, J. K. W., BIM for Existing Buildings: Potential Opportunities and Barriers, *IOP Conference Series: Materials Science and Engineering*, 371(1), 012051, 2018.
Keating, S. J., Leland, J. C., Cai, L., and Oxman, N., Toward Site-Specific and Self-Sufficient Robotic Fabrication on Architectural Scales, *Science Robotics*, 2(5), 2017.
Lasi, H., Fettke, P., Kemper, H.-G., Feld, T., and Hoffmann, M., Industry 4.0, *Business and Information Systems Engineering*, 6(4), 239-242, 2014.
Li, J., and Yang, H. A., *Research on Development of Construction Industrialization Based on BIM Technology under the Background of Industry 4.0,* MATEC Web of Conferences, Vol 100, p 02046 (2017), 02046, 2017.
McKinsey, Construction Industry Spending Worldwide from 2014 to 2025 (in Trillion U.S. Dollars). Retrieved from www.statista.com/statistics/788128/construction-spending-worldwide/ in October, 2018.
Meža, S., Turk, Ž., and Dolenc, M., Component Based Engineering of a Mobile BIM-Based Augmented Reality System, *Automation in Construction*, 42, 1-12, 2014.
Oesterreich, T. D., and Teuteberg, F., Understanding the Implications of Digitisation and Automation in The Context of Industry 4.0: A Triangulation Approach and Elements of a Research Agenda for The Construction Industry, *Computers in Industry*, 83, 121-139, 2016.
PRNewswire, *Alliance Precast Industries Selects Epicor ERP to Support Digital Transformation,* PR Newswire Association LLC, 2018.
Streule, T., Miserini, N., Bartlomé, O., Klippel, M., and de Soto, B. G., Implementation of Scrum in the Construction Industry, *Procedia Engineering*, 164, 269-276, 2016.
Walasek, D., and Barszcz, A., Analysis of the Adoption Rate of Building Information Modeling [BIM] and its Return on Investment [ROI], *Procedia Engineering*, 172, 1227-1234, 2017.
Woodhead, R., Stephenson, P., and Morrey, D., Digital Construction: From Point Solutions to Iot Ecosystem, *Automation in Construction*, 93, 35-46, 2018.
World Economic Forum, Shaping the Future of Construction: A Breakthrough in Mindset and Technology, 2016. Retrieved from http://www3.weforum.org/docs/WEF_Shaping_the_Future_of_Construction_full_report__.pdf in October 2018.
Wright, S., Why would a Construction Business have a Document Management System?, *Credit Control,* 34(10/11), 70, 2013.
Wu, Y.-F., Zhang, Y., Shen, J.-W., and Peng, T., The Virtual Reality Applied in Construction Machinery Industry, Virtual, Augmented and Mixed Reality, *Systems and Applications*, 340, 2013.
Zhang, X., Li, M., Lim, J. H., Weng, Y., Tay, Y. W. D., Pham, H., and Pham, Q.-C., Large-scale 3D Printing by A Team of Mobile Robots, *Automation in Construction*, 95, 98-106, 2018.
Zhou, K., Taigang, L., and Lifeng, Z., Industry 4.0: *Towards Future Industrial Opportunities and Challenges*, Proc., 12th International Conference on Fuzzy Systems and Knowledge Discovery (FSKD), 2147-2152, 2015.

Interdependence between Structural Engineering and Construction Management
Edited by Ozevin, D., Ataei, H., Modares, M., Gurgun, A., Yazdani, S., and Singh, A.

ISBN: 978-0-9960437-6-2

QUANTIFYING THE IMPACT OF CONGESTION ON CONSTRUCTION LABOR PRODUCTIVITY USING AGENT-BASED MODELING

MOHAMMAD RAHAL and HIAM KHOURY

Civil and Environmental Engineering, American University of Beirut, Beirut, Lebanon

Several findings from the construction field stipulate that productivity falloffs are primarily management-related; however, this notion does not consider the direct impact of these same management decisions on the workers themselves. For instance, the planning of the workspace layout delves in a spatial configuration which if not properly managed can potentially result in congestion that, in turn, directly affects labor productivity. Previous research efforts developed models to analyze the effect of congestion on labor productivity but failed to capture all the complexities of this mechanism and its dynamics. Therefore, this paper puts forward the groundwork of an agent-based simulation model (ABM) and presents work targeted at quantifying the impact of congestion on the productivity of construction crews. More specifically, the ABM model takes into account two construction trades working in the same area and tackles five scenarios each depicting different congestion and interaction levels. At the heart of this simulation is a quantitative model that defines essential congestion metrics and outputs space interference values. Experiments were conducted and results highlighted that the higher the space interference values the less productive the crews become. Additionally, these values will constitute an integral part in future work when studying the impact of congestion on the crews' learning curve, whereby the latter being a major gauge for levels of productivity.

Keywords: Management, Site layout, Spatial configuration, Space interference, Crews' interaction, ABM.

1 INTRODUCTION AND BACKGROUND

Site congestion has mainly been associated with stacking of trades, lack of space on confined sites, ineffective workforce management, mismanagement of material delivery and storage, out-of-sequence work, and poor waste management on site (Thomas and Ellis Jr 2017). For instance, stacking of construction trades happens whenever multiple tradespeople work simultaneously in a single work area. It is a common secondary action taken by contractors using overmanning to speed up the schedule, although it sometimes takes place under normal schedule (Reichard and Norwood 2001). However, crews need enough room to comfortably maneuver and work with their necessary tools, material, and equipment. Therefore, stacking of trades creates congestion and crew interferences negatively impact productivity (Hanna *et al.* 1999).

On the other hand, prior research efforts studied the issue of workspace availability as a new decision factor for establishing the construction schedule (Thabet 1992). A scheduling model was developed to quantify workspace parameters for any activity (i.e. space demand and space availability/supply). The developed model also provided a procedure to compare the space demand and supply in order to sequence construction activities. Furthermore, Mallasi (2006) and Dawood and Mallasi (2006) introduced the Critical Space-Time Analysis (CSA) concept to quantify the congestion degree between the overlapping workspaces, and visualize workspace congestions in architectural projects. Watkins *et al.* (2009) used agent-based modeling to study construction labor productivity as an emergent property of individual and crew interactions. The goal of the latter work was to efficiently utilize construction space and develop plans and schedules that account for congestion arising from crew interactions in a given space. Another research work discussed spatial demand and supply from the perspective of construction operators and a modeling methodology based on spatiotemporal utilization was proposed (Chua *et al.* 2010). Pradhananga and Teizer (2014) developed a cell-based continuous simulation model to quantify congestion for operations in which equipment move continuously.

Therefore, previous research efforts developed models to either quantify congestion or analyze its effect on labor productivity. However, no single work attempted at quantifying the impact of congestion on labor productivity while capturing all the complexities of this mechanism and its dynamics. This paper, on the other hand, uses the agent-based modeling (ABM) technique (Bonabeau 2002) to simulate spatiotemporal interactions of activities and study how resulting congestion can impact construction productivity. Additionally, the proposed ABM model studies the interactions in a granular fashion for each activity block rather than for the whole activity.

2 METHODOLOGY: IMPLEMENTING THE ABM MODEL

2.1 Underlying Quantitative Congestion Model

The quantitative model of congestion adopted throughout this study is inspired from the work by Yeoh and Chua (2010) that quantifies workspace congestion based on the concept of space and time utilization or space-time-volume. In this case, the congestion factor, represented in Eq. (1) by an index called "dynamic space interference" (DSI), measures the extent to which a work operator can accommodate the interferences due to other activity workspaces.

$$DSI_A = \rho_A + \sum_i \rho_i \times \frac{s_{iA} t_{iA}}{s_A t_A} \quad (1)$$

More specifically, Eq. (1) consists of two parts whereby the first part (ρ_A) is the individual spatio-temporal utilization factor of the activity, while the second part is an incremental component that includes the effect of interfering space entities. More details on the adopted quantitative congestion model can be found in Chua *et al.* (2010). In this study, it was assumed that the absence of congestion corresponds to a DSI value of 0.85 and its presence corresponds to DSI values greater than 0.85 and, as such, require greater attention from the construction manager (Yeoh and Chua 2012).

2.2 Agents, Environment, and Interaction Rules

In this study, two different trades were considered to perform and execute tiling and false ceiling activities in a research office area of (8m x 40m). As such, two active agents, namely the ceiling and tiling crews, were modeled to interact in this environment. The tiling crew consists of one skilled worker and one helper performing 20 m^2/day, while the false-ceiling crew consists of two skilled workers completing 25 m^2/day. The durations of both tiling and false-ceiling were set to 13 and 10 days respectively. This construction environment was chosen as it can witness workspace conflicts and interferences.

Furthermore, in order to better mimic the various management decisions that can possibly take place in reality, the model was designed to allow the user to select out of five different scenarios that may occur on the construction field. In this case, each scenario delineates the path along which each trade moves to complete the tasks and, as such, presents a different congestion level whenever the two activities get close to each other from any direction. These scenarios were established by consulting with construction professionals while taking into consideration work continuity. Figure 1 illustrates the layout adopted for one of the designed scenarios, whereby both crews follow the same path/direction of work. More specifically, the tiling crew goes first and the false-ceiling crew, having higher productivity, is assumed to overlap with the other crew and create congestion around blocks H and I and towards the end.

Scenario 1: Tiling and False-Ceiling paths.

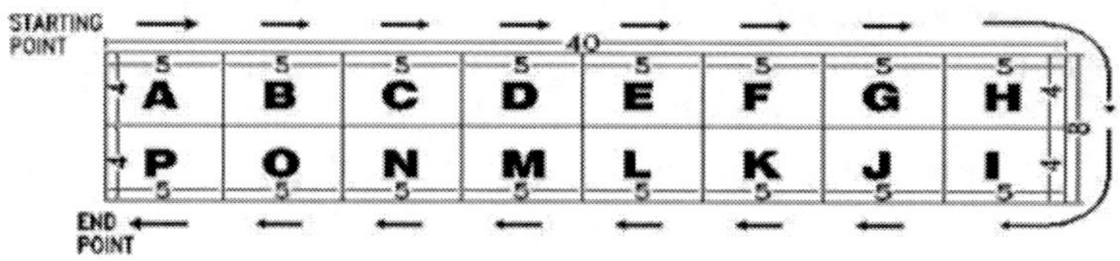

Figure 1. Case study layout.

It is worth mentioning that, within the proposed model, the user can opt to select the operating space and the total boundary comfort space for both activities. The operating space is the actual space needed to perform the activity while the comfort space is the total area depicting the entire activity (i.e. operating space) in addition to path spaces, on-site inventory (i.e. material storage area), and a safe/comfortable distance between the current activity's crew and all other neighboring crews. In fact, beyond that safe distance, comfort space interferences are likely to happen and trigger congestion at different locations which would impact work efficiency and, in turn, would negatively impact labor productivity. As such, in this study, reasonable operating space for a certain working area could be taken as a 0.7m offset from the working area boundaries while the comfort space can be assumed to extend an offset of 2m from the working area boundaries. Nonetheless, the model is flexible enough to allow the user to compare results based on various offset values.

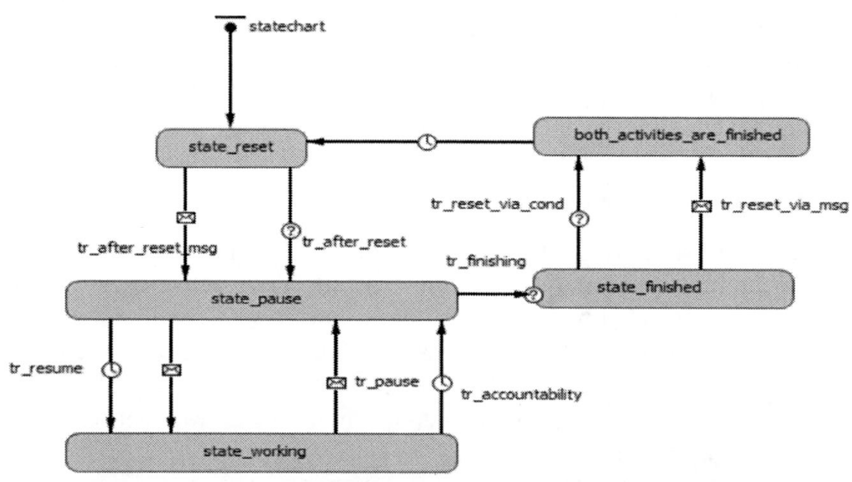

Figure 2. State-chart of the false-ceiling agent.

Figure 2 illustrates the ABM state-chart of the ceiling agent. A similar state-chart is also constructed for the tiling agent to allow interaction between the two activities depending on the selected scenario.

The state-chart provides a (*state_pause-state_working*) loop that is created to complete all iterations of a certain path before the activity finishes (i.e., *state_finished*). The ABM simulation generates specific congestion DSI values for each unit block in the layout (Figure 1) given a particular activity and at a specific time and location. When both activities are completed (i.e. *both_activities_are_finished*), DSI values generated, and congestion levels analyzed, the user can then run another scenario or path.

It is good to note again the contribution lying in the proposed ABM model that represents each activity block by a generalized DSI value to portray overall congestion, unlike previous studies that only represented the whole activity (Chua *et al.* 2010).

3 RESULTS AND ANALYSIS

Table 1 illustrates the DSI values output from the ABM model for each activity (i.e., Tiling, T or Ceiling, C) and varying according to the degree of congestion established under each of the five designed scenarios (i.e., S1-S5). Furthermore, every experimental scenario includes sixteen iterations for each activity, whereby each iteration represents a different block depending on the path followed in each scenario.

Table 1. DSI values output from the ABM simulation.

	DSI Values									
	S(1)		S(2)		S(3)		S(4)		S(5)	
Iteration	T	C	T	C	T	C	T	C	T	C
1	0.799	0.799	1.331	1.331	0.799	0.799	0.799	0.799	0.799	0.799
2	0.746	0.746	1.203	1.241	0.746	0.746	0.746	0.746	0.799	0.799
3	0.746	0.746	1.162	1.173	0.746	0.746	0.746	0.746	0.746	0.746
4	0.746	0.746	1.121	1.137	0.898	0.746	0.898	0.746	0.746	0.746
5	0.746	0.746	1.080	1.101	0.746	1.023	0.799	0.799	0.746	0.746
6	0.746	0.746	1.040	1.066	0.746	0.746	0.746	0.746	0.746	0.746
7	0.746	0.746	0.999	1.030	0.746	0.746	0.746	0.746	0.829	0.746
8	0.799	1.331	0.799	1.118	0.799	0.799	1.175	0.746	0.746	0.746
9	0.816	1.012	1.018	0.799	0.799	0.799	0.746	0.802	0.746	1.078
10	0.835	0.995	0.876	0.746	0.746	0.746	0.746	0.912	0.974	0.746
11	0.990	1.078	0.835	0.898	0.829	0.746	0.746	0.746	0.746	0.746
12	1.078	1.161	0.794	0.852	0.760	0.746	0.799	0.799	0.746	0.746
13	1.182	1.244	0.753	0.817	0.746	0.746	0.746	0.746	0.746	0.746
14	1.285	1.327	0.746	0.781	0.746	1.078	0.746	0.746	0.746	0.746
15	1.389	1.410	0.746	0.746	0.746	0.746	0.746	0.746	0.799	0.799
16	1.597	1.597	0.799	0.799	0.799	0.799	0.799	0.799	0.799	0.799

In order to analyze the obtained DSI values and relate them to labor productivity, several interviews were conducted with construction project managers and practitioners, resulting in the fifth-degree polynomial curve (R^2 value= 0.998) depicted in Figure 3 that mainly reflects a decline in productivity for a construction crew as congestion increases (i.e., DSI increases).

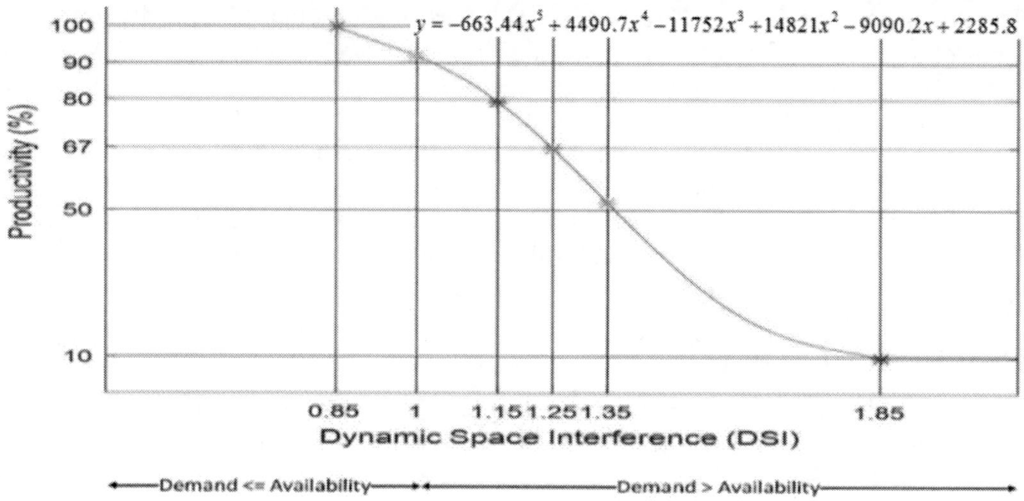

Figure 3. Productivity-DSI curve.

Accordingly, obtained DSI values (Table 1) ranging from 0.85 to 1.15 reflect acceptable levels of congestion that typically take place in the construction industry and a productivity level ranging from 80-100%. On the other hand, values ranging from 1.15 to 1.35 are considered critical and alarming values that require serious attention from site engineers. The upper bound of 1.35 is considered the maximum tolerable congestion value beyond which productivity is reduced by 50% compared to its initial value. Furthermore, DSI values above 1.35 represent severe congestion levels and drop in productivity that may lead to inefficiencies in workflow/labor-flow, bottlenecks, interruptions, etc.

4 CONCLUSIONS AND FUTURE WORK

This paper aimed at quantifying the impact of workspace congestion on construction labor productivity using agent-based simulation. More specifically, the ABM model takes into account two consecutive trades working in the same area and tackles five scenarios depicting different congestion and interaction levels based on spatial interferences among activities. The outcome of this paper supports the fact that unnecessary congestion and workspace conflicts can be minimized or alleviated by taking better planning/management actions, which, in turn, positively affect labor productivity.

Future work will further develop the ABM model to:

1. Show the impact of congestion on the crew learning curve through a mathematical function.
2. Calculate the individual duration timeouts for all iterations using different learning curve models and consequently obtain the overall cumulative timeout for each scenario.
3. Conduct different simulation experiments using factors that affect positively and negatively the learning rate.

Acknowledgements

The presented work has been supported by AUB's University Research Board (URB). The authors gratefully acknowledge URB support. Any opinions, findings, conclusions, and recommendations expressed by the authors in this paper do not necessarily reflect the views of URB.

References

Bonabeau, E., Agent-Based Modeling: Methods And Techniques for Simulating Human Systems, *Proceedings of the National Academy of Sciences, 99*(suppl 3), 7280-7287, 2002.

Chua, D. K., Yeoh, K., and Song, Y., Quantification of Spatial Temporal Congestion in Four-Dimensional Computer-Aided Design, *Journal of Construction Engineering and Management, 136*(6), 641-649, 2010.

Dawood, N., and Mallasi, Z., Construction Workspace Planning: Assignment and Analysis Utilizing 4D Visualization Technologies, *Computer- Aided Civil and Infrastructure Engineering,* 21(7), 498-513, 2006.

Hanna, A. S., Russell, J. S., Nordheim, E. V., and Bruggink, M. J., Impact of Change Orders on Labor Efficiency for Electrical Construction, *Journal of Construction Engineering and Management,* 125(4), 224-232, 1999.

Mallasi, Z., Dynamic Quantification and Analysis of The Construction Workspace Congestion Utilising 4D Visualisation, *Automation in Construction,* 15(5), 640-655, 2006.

Pradhananga, N., and Teizer, J., Congestion Analysis for Construction Site Layout Planning Using Real-Time Data and Cell-Based Simulation Model, *Computing in Civil and Building Engineering,* 681-688, 2014.

Reichard, D. D., and Norwood, C. L., Analyzing the Cumulative Impact of Changes, *AACE International Transactions*, CD61, 2001.

Thabet, W. Y., *A Space-Constrained Resource-Constrained Scheduling System for Multi-Story Buildings.* Virginia Tech, 1992.

Thomas, H. R., and Ellis Jr, R. D., Construction Site Management and Labor Productivity Improvement, American Society of Civil Engineers, Reston, 2017.

Watkins, M., Mukherjee, A., Onder, N., and Mattila, K., Using Agent-Based Modeling to Study Construction Labor Productivity as an Emergent Property of Individual and Crew Interactions, *Journal of Construction Engineering and Management,* 135(7), 657-667, 2009.

Yeoh, K., and Chua, D. K., *Mitigating Workspace Congestion: A Genetic Algorithm Approach,* EPPM 2012 Conference, 2012.

Interdependence between Structural Engineering and Construction Management
Edited by Ozevin, D., Ataei, H., Modares, M., Gurgun, A., Yazdani, S., and Singh, A.

ISBN: 978-0-9960437-6-2

THE BREXIT AND ITS ECONOMIC IMPACTS ON THE CONSTRUCTION INDUSTRY

HOSSEIN ATAEI and FARNAZ TAHERKHANI

Dept of Civil and Materials Engineering, University of Illinois at Chicago, Chicago, USA

In June 2016, the United Kingdom voted to depart from the European Union (EU) with 51.9% of the votes versus 48.1%. A short-term increase in the overall construction project costs is anticipated due to shortage of less-expensive skilled workers and more costly construction equipment imports from the EU. This may result in slower growth for commercial and residential industry sectors. The long-term issue of Brexit will affect private investment in British infrastructure. As demand increases globally on construction of new civil infrastructure or rehabilitation of existing systems, the global investors and Engineering, Procurement and Construction (EPC) firms will be more mindful and conservative in investing in UK projects. The UK government will seek more private financing for the country's capital projects as easier access to the EU funds - as existed prior to Brexit - may no longer be an alternative. The UK will be a riskier destination for private investors or Public-Private Partnerships (PPP's) because of the developing instability and complexities in the new post-Brexit econo-political environment. As a result, a slower turnout in major infrastructure projects is anticipated. This paper investigates the economics of the post-Brexit UK and the opportunities and threats to the construction industry. It addresses the role of the government and the availability of public funding and how construction regulations and standards will be impacted as a result.

Keywords: United Kingdom, Economics, Skilled labor, Training, Equipment and materials costs, Housing and infrastructure projects.

1 INTRODUCTION

The process of the United Kingdom (UK) leaving the European Union (EU) formally began on March 29, 2017. It was agreed that the UK has roughly two years from that date to negotiate and implement a new relationship with the EU. On Dec. 8, British and European Nation negotiators announced an agreement regarding the country's separation settlement: Britain will pay an exit bill of £35 billion to £39 billion; there will be no official "hard border" between Northern Ireland (which is currently part of the UK) and the Republic of Ireland (an official EU member), and EU citizens in Britain – and vice-versa – will retain a number of rights (Kenton 2019).

The "Brexit bill," or financial settlement the UK owes Brussels following its withdrawal, has been a major point of discussion for both sides of the spectrum. During the campaign, a large financial settlement would therefore irritate many Brexit supporters; most of who believe that the UK should not have to pay anything. "Leave" voters base their support for Brexit on several factors, including the European debt crisis, immigration, terrorism and Brussel's bureaucracy. Britain has long been wary of the European Union's projects, which Leavers feel threatens the UK's sovereignty: the country never opted into the European Union's monetary union, meaning that it uses the pound instead of the euro. It also refused to share open borders with a number of other European nations. Opponents of Brexit also cite a number of reasons for their position. One is the risk involved in pulling out of the EU's decision-making process, given that it is by far the largest destination for British exports. Another is the economic and societal benefits of the EU's free movement of goods, services, capital and people across borders. A common thread in both arguments is that leaving the EU would destabilize the British economy in the short term and make the country poorer in the long term.

Construction in the United Kingdom accounts for a significant proportion of output and employment. The construction industry is vital to building homes in the United Kingdom where housebuilding has been below demand for decades. The construction industry is also vital to delivering the infrastructure that the UK needs (Dromey *et al.* 2017).

The construction industry in the UK already faces significant challenges such as low productivity; slower rate of incorporating innovative methods and also dysfunctional skills system. The Brexit threatens to increase these significantly by triggering a crisis for the industry. This could have serious consequences for the UK economy in general.

Businesses in general, with few exceptions, mainly wanted to stay in EU for facilitated financial transactions and easier access to global services, goods and skilled labor. Brexit invoked fears of losing jobs, properties and conventional travel benefits for many. However, the "Leave" supporters campaigned for it mainly for the following reasons: (1) there was a perception in the UK that the EU imposed too many rules on business and charged billions of pounds a year in membership fees for much less in return; (2) the UK wanted to make all its own laws again, rather than being created through shared decision-making with other EU nations; and (3) immigration was a major issue for Brexit supporters who wanted UK to take back full control of its borders and reduce the number of people coming to UK to live and/or work.

2 THE UNITED KINGDOM ECONOMY AND CONSTRUCTION INDUSTRY

The construction industry contributes 90 billion pounds to the UK economy which accounts for 6.7% of the total UK economy (Construction 2025 2013). There are 2.9 million jobs filled in the construction industry, which approximately is 10% of the entire UK job market. The construction industry in the UK is a key contributing factor to other economic sectors mainly the housing market; commercial property; transportation; water; energy and digital communication infrastructure. It is expected that 1.7 – 2.5 million new homes will be built by 2025. The UK Government is planning to invest £110 billion in the energy sector in the next decade and another £100 billion in water infrastructure systems over the next 20 years (Construction 2025 2013).

The UK Government's Strategic Priorities for the construction sectors are through further focusing on: (1) Smart Construction and Digital Design; (2) Sustainable Construction; and (3) Improved Trade Performance (Construction 2025 2013). To realize the above Strategic Priorities, the construction industry relies heavily on educated skilled workforce for improving the construction industry image and increases the workforce capability. This is also vital in improving more-complex stakeholder demands through delivering stronger, sustainable and more resilient supply chain that is made possible via effective research and development (Construction 2025 2013).

After a significant slowdown during the Global Financial Crisis of 2017 and 2008, the construction industry in the UK grew steadily between 2012 and 2017. In Third Quarter of 2018, the construction industry output was 14% higher than that of the First Quarter of 2008 while the whole UK economy output grew by 15% over the same period (Rhodes 2018). Housing projects accounted for 25% of the 2017 total UK construction output, of which 4% was publicly-funded projects and 21% private, while infrastructure accounted for 12% of total construction projects (Rhodes 2018).

According to the UK Office of National Statistics (2019), the construction sector's consolidated growth was only 0.7% (compared to 7.1% in 2017). There was a 1.1% increase in all new work (versus 7.4% for 2017), which was mainly due to a 2.9% increase in private new housing and 2.6% increase in infrastructure output.

3 SHORT-TERM EFFECTS OF BREXIT ON THE UK CONSTRUCTION INDUSTRY

The residential and commercial construction industry in the United Kingdom felt the effects of Brexit when the decision to rescind from the European Union was first made public. Between the date of the referendum (June 2016) and August 2017, the sterling pound fell approximately 17.6% of its value compared to the Euro (Dromey *et al.* 2017). It is important to note that the construction industry in the United Kingdom is heavily reliant on migrant workers, and these migrant workers make up for the disparity between the native UK workers who are retiring within the next decade and those native UK workers entering the construction field. The non-native UK workers had mainly saw opportunities of higher income in the UK and strong currency as they save towards the retirement. The pound depreciating in value essentially reduced the savings of the non-native UK workers. The immediate depreciation in value of the pound rendered the United Kingdom far less attractive to non-native UK workers, relative to other EU nations. According to Dromey *et al.* (2017), even before the Brexit announcement, non-native UK workers had already begun to leave the United Kingdom mainly because of these two reasons: (1) the pound depreciated in value, and (2) non-native UK workers had no guarantees of being permitted to stay and continue working in the post-Brexit United Kingdom.

Employers in the construction sector are concerned with retaining these individuals rather than recruiting new workers. The infrastructure projects in the United Kingdom are dependent on private funding where the private investors may not decide to withdraw from funding such infrastructure despite the economic and political uncertainty as a result of Brexit. This is mainly due to the fact that the infrastructure projects are usually low risk investments and they have minimal correlation with the foreign exchange and equity markets. It is expected, however, that Direct Construction Costs (labor, equipment and materials) may see an increase in prices.

Few of the current large infrastructure projects in the UK that may be directly impacted by Brexit are: (1) Thames Tideway which is a sixteen mile long tunnel under the Thames River to capture, store and transport sewage and rainwater discharges that currently flow into the Thames River and (2) High Speed 2 which is a high-speed railway system responsible for connecting London, Birmingham, the East Midlands, Leeds, and Manchester. Such infrastructure projects are mainly made possible in part by private investors, the European Investment Bank (EIB), and the European Investment Fund (EIF). The United Kingdom needs investment in infrastructure due to ever-increasing demands for such with large-scale private funds as the UK is lacking sufficient national funds for such capital projects.

As a member of the European Union, the UK has had access to the EIB and the EIF which collectively invested billions of euros in the United Kingdom's economy. If the roles of the EIB and EIF are diminished, the government, moving forward, in the medium to long term, has to look for alternative options of inexpensive financing for its infrastructure projects mainly through the private investors who still have confidence in the UK's economy. The United Kingdom is known for its transparency and fair procurement practices, which are attractive features for private investors in governmental infrastructure projects. These transparency features and protocols are expected to remain un-changed in the post-Brexit United Kingdom.

4 LONG-TERM EFFECTS OF BREXIT ON THE UK CONSTRUCTION INDUSTRY

The long-term forecast of Brexit cannot be accurately determined. The most prevalent concern, however, will be the potential restriction of free movement of people and goods. Construction industry will be one of the main market sectors that may be heavily impacted by lower investor confidence stemming from the uncertainty over the post-Brexit UK economy (Timetric 2016).

The UK construction market is reliant on foreign labor from within the EU due to the insufficient numbers of new and existing skilled domestic workers. The UK construction industry is also reliant on imports for most construction materials and equipment from other EU countries, especially Germany, Sweden and Italy. Therefore, depending on how the new trade agreements are negotiated between the UK and the EU, it could have significant impacts on construction costs particularly if new taxes and tariffs are imposed on construction materials and equipment imports. In general, a sharp deceleration in economic growth is expected for the UK economy.

The UK is one of the EU's biggest beneficiaries of funds. Once the United Kingdom officially leaves the EU, the levels of the EU direct investments in the UK will be significantly plummeted squeezing industry margins even further. This will highly affect the commercial and residential construction industries as many contracting companies will have a hard time with labor and material costs. This may, in turn, drive multi-national companies out of the UK into other countries where it is easier to invest in new construction projects.

5 THE FUTURE OF BREXIT

According to Dromey *et al.* (2017), the United Kingdom (UK) currently has demographic challenges within the construction industry that will ultimately lead to shortages of skilled workers. It is expected that the Brexit exacerbates this shortage particularly within London and south-east Britain. These regions are heavily reliant on migrant European-born (EU-born) workers who have moved to the United Kingdom for work. The native-UK construction workforce is rapidly ageing with 43.4% over the age of 45, and nearly 387,000 construction workers who will approach retirement in the next decade, and approximately 1 million workers approaching retirement in the next two decades (Dromey *et al.* 2017). EU-born workers, and workers born outside of Europe (Non-EU-born), make up the majority of construction workers aged 25-44. Currently, the UK's skilled worker shortage is masked by the foreign workers, as they fill the void not filled by the UK-born workers. With the United Kingdom closing its borders and restricting movements as a result of Brexit, the future of the UK's construction industry needs to be looked at from various perspectives of the current construction workforce: (1) from the perspective of workers who already had some sort of construction industry; (2) from the perspective of young people entering the workforce; (3) construction workers who qualify for further education or other type of additional training, and (4) workers who have already completed some sort of apprenticeship who qualify for regular employment.

According to Dromey *et al.* (2017), only 57% of construction employers arranged or funded training for their staff in 2016 which is the second lowest of all industries, and far lower than the economywide average of 66%. This may be because it is more expensive to provide training in construction industry compared to the other sectors of the economy. Moreover, there may be low or non-existent incentives for construction workers in the industry to receive additional non-mandatory training to boost their careers with newer technologies and innovations in construction practices. The construction workers and industry leaders may not see a reason to change what is not broken. This is also amplified by recent reform that has limited public funding for further education.

The UK vocational education sector does not give the construction workers the best of required skills and experience needed by employers so that they can progress into a sustainable career in the industry. Apprenticeships are also not efficient since many apprentices are unable to find permanent employment after completing a Level 2 apprenticeship. To combat this, the United Kingdom government has implemented two levies: (i) Construction Industry Training Board (CITB) Levy to advance and improve further education and skillsets for the construction workforce and (ii) the Apprenticeship Levy which is a payroll tax applicable to companies with a payroll higher than 3-million euro to boost employer investment in raining with a target goal of three-millions apprenticeships by the year 2020 (Dromey *et al.* 2017). Faced with post-Brexit uncertainties in construction industry, many small or medium-size UK construction firms would rather not to contribute and pay theses levies and would look for ways to reclaim their levy payments rather than using the funds pooled from the levies to provide training for their employees. This is also due to the fact that larger employers get proportionately more of the CITB grants than they contributed in the CITB levy while small and medium size employers are simply not incentivized to recuperate their funds for training of their employers" (Dromey *et al.* 2017).

These tax laws need to be reformed so that companies become incentivized to train their employees, provide apprenticeships or hire apprentices, and so that companies align their practices in such a way that helps the United Kingdom achieve its goals as outlined in Construction 2025 (2013), using advanced construction technology and modern methods of construction. To achieve technological advances and modernization in the construction industry, education is key. The construction industry is often not the first career choice for those entering the workforce, because it may be seen as dirty, and laborious. Therefore, education on the latest technological advancements is essential, not only for productivity, but also for retention of construction personnel, and to increase the numbers of UK-born workers entering the construction sector. By using the levy funds for effective education and efficient training, construction can be promoted as a career of choice. Currently, there is no guarantee from the government as to whether the EU nationals will be allowed to remain in the United Kingdom once the country formally exits the union as the UK Government also plans to limit the annual immigration.

As a member of the European Union, the United Kingdom allowed for free movement of goods within the EU, and this ultimately benefited the construction industry. By removing itself from the European Union, the United Kingdom may now face limits on the quantities of materials it can import or export, which could lead to economic uncertainty and inflation in the price of construction materials, equipment, and construction labor. Moreover, as a member of the European Union, the United Kingdom has had access to the EU funds that were invested in the country's major infrastructure projects. In the post-Brexit era, the UK Government may have to look for alternative inexpensive construction financing options. According to Catterall (2017), while the United Kingdom government may end up saving money on the EU membership fees, however, these savings will not be able to offset the total construction costs for current and future UK infrastructure projects.

6 SUMMARY AND CONCLUSION

The construction industry in the United Kingdom is already dealing with a shortage of technical talent. With approximately half of the UK-born workforce to retire within the next decade, and a shortage of UK-born workers entering the construction industry, the construction industry is heavily reliant upon migrant workers. Brexit only complicates matters as it restricts the free movement of skilled construction workers from both EU nations and other countries into the UK. There are also uncertainties present pertaining the work visas and labor rights for non-native workforce to remain in the post-Brexit UK. Moreover, establishment of hard borders between the UK and the EU will restrict free movement of construction material and equipment imports into the UK and may lead to higher construction costs. This will also be combined with potential tax hikes and new tariffs on the UK imports and exports.

The lack of UK-born native construction workers is a result of stereotypes and disinterest of the construction industry by younger generations, as well as poor recruitment and retention practices. This is also exacerbated partly by the historical hesitance of the construction industry to become modernized and implement the latest technological advancements.

The UK Government is looking into boosting the interests of the British construction workers who are ready to enter the workforce through tax and education reforms. However, the government would need to be cognizant of the post-Brexit ramifications with which the construction industry may deal with by closing its borders and limiting access to international private

References

Catterall, O., *To Leave or not to Leave - How will Brexit Affect the Construction Industry?*, LABC, The United Kingdom, June 2017.

Construction 2025, *Industrial Strategy: government and industry in partnership*, HM Government, United Kingdom, The National Archives, London, UK, July 2013.

Dromey, J., Morris, M., and Murphy, L., *Building Britain's Future? The Construction Workforce After BREXIT*, The Institute for Public Research, London, UK, November 2017.

Kenton, W., (Reviewer), *Brexit*, Investopedia, https://www.investopedia.com/terms/b/brexit.asp, Updated May 2019.

Rhodes, C., *Construction Industry: Statistics and Policy*, Briefing Paper, House of Commons Library, The United Kingdom, December 2018.

Office of National Statistics (ONS), *Construction Output in Great Britain: February 2019,* (Latest Release April 2019), Newport, The United Kingdom.

Timetric, *Brexit and the Impact on Construction in the UK*, Construction Intelligence Center, London, England, July 2016.

Interdependence between Structural Engineering and Construction Management
Edited by Ozevin, D., Ataei, H., Modares, M., Gurgun, A., Yazdani, S., and Singh, A.

ISBN: 978-0-9960437-6-2

PRACTICE EXPERIENCE IN ASSESSING THE PROJECT MANAGERS' LEADERSHIP SKILLS

SILVIA MAZZETTO

School of Architecture and Interior Design, Lebanese American University, Byblos, Lebanon

In construction, the achievement of a project is mainly due to the work of the project manager and the contribution they make, as team leader, of knowledge, experience, and coordination skills. This paper presents a practical approach in assessing the students' leadership skills at the College of Engineering at Qatar University, by running a multidisciplinary collaborative project. The students acted as the project managers, involved in leading several working groups. The main aim of the exercise was to propose a practical approach to assessing the extent to which the knowledge and skills of a leader are important for ensuring that a project is completed successfully. The project manager's ability to lead a team through the different work stages of a project is seen as a fundamental contributor to its success. This practical approach to the appraisal of leadership brings the theoretical teaching of Project Management closer to its practical applications, by encouraging students to learn the leadership techniques and tools commonly used in the professional setting. The paper concludes by suggesting that there is a need to focus more attentively on assessing leadership skills when selecting a project manager, in either an academic or a professional context.

Keywords: Multidisciplinary collaboration, Entrepreneurship, Practical approach, Teamwork, Responsibilities.

1 INTRODUCTION

Many research exercises and studies have described the importance of project managers for the success of all projects. To attaining that success, the project managers usually have to demonstrate their knowledge of management tools, techniques, and their leadership skills when leading the project team. These skills allow the project managers to guide the team members in the right direction, resolving collaboration problems, coordinating the different tasks, assessing and reducing the risks, and building reciprocal respect, and trust.

The paper describes a practical approach for assessing the leadership skills of the project manager as he/she applies at Qatar University for two consecutive academic years (Fall 2015 and Fall 2016) in the form of a multidisciplinary collaborative project (Mazzetto 2016a).

The primary research objective was to simulate a real-work scenario opportunity for the students to address the real problems related to professional projects, such as the need for leadership skills in multidisciplinary teams, and the need to cope with the pressure of the imposed deadlines. The new practical experiences enabled the students to develop a correct professional attitude, highlight their leadership skills, and improve their positioning in the world of work.

2 LITERATURE REVIEW

Numerous past studies have presented detailed research evaluating the impact of the project manager's skills and performance on the successful completion of a project.

In the literature (Muller and Turner 2010) to achieve the projects' success, the project managers have to demonstrate a solid knowledge base, and good leadership skills to guide their team during the progress of the projects, towards attaining the expected outcomes. In fact, the project manager's leadership skills significantly affect the success of every project, for this reason, such skills should be assessed at the selection stage to choose the appropriate leader for the project implementation.

In the literature, many studies have measured the skills, knowledge, and professional characteristics of project managers in the construction sector, particularly concerning the evaluation of the managerial abilities that are associated with leadership skills. Moreover, the ability to take the lead in managing and coordinating all the roles involved in the project.

Finally, there is a growing debate in the academic world about the contemporary needs of students' practical training that is required more and more in the professional project management sector.

Many authors (Edum-Fotwe and McCaffer 2000, O'Donnell *et al.* 2008) have explored learning methods that could be applied in the educational sphere to develop project management capabilities and skills in students, in compliance with the characteristics that the construction industry most requires in a project manager, so that these can be developed at the academic training stage.

Recent studies in the educational field (Major 2016, Muller *et al.* 2012, Shamir 2011) have highlighted the issue of bridging the gap between the professional and academic worlds, especially in the project management field. This issue has opened many critical questions, particularly about contemporary project management techniques, practices, standards, and approaches that are currently applied to address the increasing request for professionalism in the academic sector. The paper's outcomes provide an innovative assessment approach that could reduce the disconnection between theory and practice, particularly concerning the project managers' leadership ability (Mazzetto 2016b, Mazzetto 2017).

3 THE LEADERSHIP ASSESSMENT METHODOLOGY

This paper describes an assessing approach that was applied at Qatar University, College of Engineering, involving both the Department of Architecture and Urban Planning (AUP) and the Department of Industrial and Systems Engineering (ISE). The applied multidisciplinary collaborative project brought together students from both the Departments in the Project Management courses, assembling them into working groups. That assessing methodology is shown schematically in Figure 1.

Developing the project required data to be collected; each student was required to make personal contributions and interactions within the groups, assessing their skills and their innate attitude to leadership, and thereby selecting the group leaders. Then the project managers during the multidisciplinary collaboration simulated a real-life working experience that was very important for teaching students what to expect in their working lives. In the end, the achieved results of the project managers were assessed and described in the final report to demonstrate the contribution of their leadership.

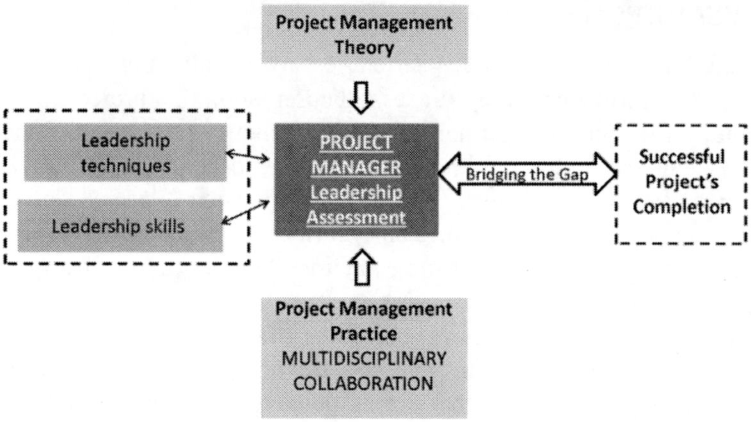

Figure 1. Multidisciplinary assessing methodology for leadership skills and techniques. Framework for collaborative projects.

3.1 Data Collection and Final Reports: Tools for Leadership Assessment

The group reports showed the collected data for assessing leadership skills including: (1) general information about the students and the projects carried out within the groups (2) a checklist of skills for assessing the leadership abilities of the project managers (3) an evaluation of the projects placed on the leadership abilities of the project managers. However, the results presented here only show the data for evaluating the leadership, to show how a possible study method was applied to multidisciplinary university collaboration.

3.2 Project Managers' Leadership Skills Assessment

The students selected the most critical leadership skills used by the project managers for carrying out their project, and they scored them (Figure 2), using a standard set of ratings (Table 1 and Table 2). The results (Figure 2) show the characteristics that were voted as the most important for the success of the leaders during the two academic years.

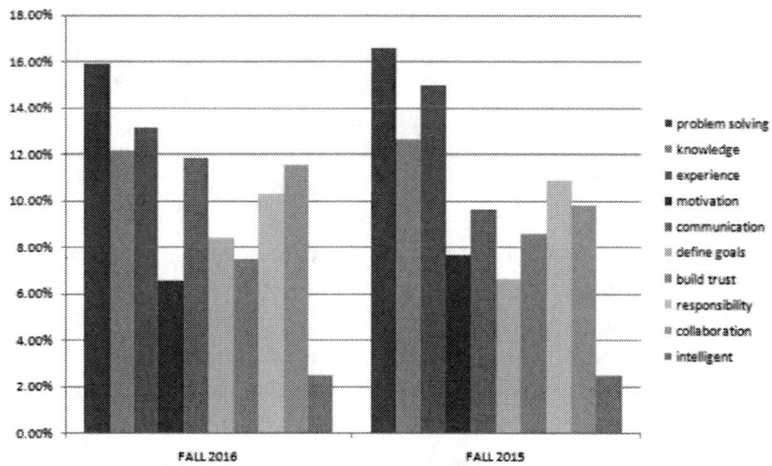

Figure 2. Leadership skills assessment for the multidisciplinary collaboration projects, average values (Fall 2015 and Fall 2016).

3.3 Assessment of the Project Managers' Leadership Skills and Techniques

After completing the process of assessing the project managers' leadership skills (Table 1 in Fall 2015 and Table 2 in Fall 2016), the data were compared, to show the correspondence between the skills and the project's success. There was a clear correspondence showing that a low percentage score for the leadership skills of project managers corresponds to a similarly low percentage score

Table 1. Leadership's skills and techniques assessment, total scores in Fall 2015.

Leadership Assessment Fall 2015	**Understanding of the leadership techniques and skills leaders use to work collaboratively in their team and communities.**		
Criterion	Adequate (Moderate understanding) Low (61-70) /100	Accomplished (Clear Understanding) Average (71-80) /100	Mastery (Excellent Understanding) Best (81-90) /100
There is understanding of techniques Leaders use to work collaboratively with the team	10%	15%	7%
There is understanding of the leadership skills which leaders use to provide for the community	11%	15%	8%
There is understanding of the leadership principles and rules	10%	16%	8%
TOTAL VALUES	31%	46%	23%

Table 2. Leadership's skills and techniques assessment, total scores in Fall 2016.

Leadership Assessment Fall 2016	**Understanding of the leadership techniques and skills leaders use to work collaboratively in their team and communities.**		
Criterion	Adequate (Moderate understanding) Low (61-70) /100	Accomplished (Clear Understanding) Average (71-80) /100	Mastery (Excellent Understanding) Best (81-90) /100
There is understanding of techniques Leaders use to work collaboratively with the team	5%	20%	6%
There is understanding of the leadership skills which leaders use to provide for the community	5%	25%	7%
There is understanding of the leadership principles and rules	3%	22%	7%
TOTAL VALUES	13%	67%	20%

for the project outcomes. The same values were repeated during the two multidisciplinary collaboration experiences both for average and high percentage scores, showing how the outcomes of a project are very often directly affected by the manager's leadership skills.

4 RESULTS

The presented multidisciplinary experience for assessing the leadership skills was conducted two years, and generated 28 complete projects, enabling the leadership skills of the project managers to be assessed by evaluating them against the results they obtained managing the final projects. The skills, interactions, and responsibilities that most significantly influenced the performance of the project managers were identified from the values they obtained in the leadership assessment.

The results showed that the outcomes of a project are very often directly affected by the manager's leadership skills, both regarding project quality and the completeness of the results obtained. There was a direct correspondence between the project outcomes and the leadership ratings of the project managers: low, medium, and high assessment scores for the projects matched with similarly low, medium, and high scores for the leadership skills of the project managers. The overall results of the students' leadership assessment have increased by about 21%, and the evaluation of the multidisciplinary projects has increased by about 1% of average values, in Fall 2016. The values of the best performances have approximately been in the same range across the two semesters (around 20% of the total values). Some common leadership styles were also identified that have not yet been analyzed in detail, but that did make it possible to identify common outcomes.

5 CONCLUSIONS

From the evaluations made, it can be concluded that the fundamental skills for good leadership by a project manager are an excellent problem-solving attitude, substantial practical experience, an adequate store of knowledge, and good communication skills.

The roles of the different participants in a project are often challenging to manage, but the ability and responsibility of the leader, as the guide, is precisely to know how to create a fair balance between the parties involved, and an adequate level of communication and collaboration. Some important conclusions can be drawn from the outcomes obtained, of which one is the importance of assessing the leadership skills of a project manager, both in the academic and the professional fields. A correct evaluation would make it possible to select the most appropriate project manager for a multidisciplinary project, also taking account of their skills and the level of authority they attained. The skills of a project manager, including their leadership style, can directly affect the outcome of how a project is managed, and its success.

Some limitations were imposed on this research, for example only a small sample of students was involved, due to the fact the multidisciplinary experience, done in coordination between different departments, was repeated for only two semesters. For that reason, the outcomes cannot be extrapolated to make an exhaustive assessment of the leadership requirements in the field of project management in higher education. That will require subsequent confirmation by similar exercises.

References

Edum-Fotwe, F. and McCaffer, R., Developing Project Management Competency: Perspectives from The Construction Industry, *International Journal of Project Management,* 18, 111-124, 2000.

Major, D., Models of Work-Based Learning, Examples, and Reflections, *Journal of Work Applied Management*, 8(1), 17-28, 2016.

Mazzetto, S., Practice Experience and Multidisciplinary Collaboration in Project Management: A Case Study, in *Integrated Solutions for Infrastructure Development* Lau, H. H.; Tang, F. E., Ng, C. K.; and Singh, A. (eds.), Third Australasia and South-East Asia Structural Engineering and Construction Conference ASEA-SEC-3, Kuching, Sarawak, Malaysia, Oct 31-Nov 4, 2016a.

Mazzetto, S., Theory, and Practice in Project Management. A Multidisciplinary Approach, in *Interaction between Theory and Practice in Civil Engineering and Construction,* Komurlu, R., Gurgun, A. P., Singh, A.; and Yazdani, S. (eds.). First European and Mediterranean Structural Engineering and Construction Conference, EURO-MED-SEC-1, Istanbul, Turkey. doi: 10.14455/ISEC.res.2016.100. May 24-29, 2016b.

Mazzetto, S., Leadership and Collaboration in Project Management Education: A Case Study, in *Resilient Structures and Sustainable Construction* Pellicer, E., Adam, J. M., Yepes, V., Singh, A., and Yazdani, S. (eds.), Ninth International Structural Engineering and Construction Conference Valencia, Spain July 24-29, 2017.

Muller, R., and Turner, R., Leadership Competency Profiles of Successful Project Managers, *International Journal of Project Management*, (28)5, 437-448, 2010.

Muller, R., Geraldi, J., and Turner J. R., Relationship between Leadership and Success in Different Types of Project Complexities, *IEEE Transactions on Engineering Management*, 59(1), 77-90, 2012.

O'Donnell, H., Karallis, T., and Sandelands, E., Reflecting on the Skills Agenda: A Construction Industry Perspective, *Education and Training,* 50, 59-63, 2008.

Shamir, B., Leadership Takes Time: Some Implications of (Not) Taking Time Seriously in Leadership Research, *The Leadership Quarterly*, 22(2), 307-315, 2011.

Interdependence between Structural Engineering and Construction Management
Edited by Ozevin, D., Ataei, H., Modares, M., Gurgun, A., Yazdani, S., and Singh, A.

ISBN: 978-0-9960437-6-2

THE PROJECT BOUNDARY AND OCCUPATIONAL STRESS

KEN FARNES

School of Property, Construction and Project Management, RMIT University, Melbourne, Australia

Despite advances in project management practices and profession, projects continue to have an unacceptably high failure rate, which may be influenced by project manager being highly stressed and their decision-making capabilities becoming narrowed. The project manager's role necessitates that they occupy the boundary between sponsoring organisation, clients, and other actors involved in the project environment, where they are often subjected to time pressures, project uncertainties, conflicting demands, complex and often dynamic social structure involved in the project environment. In this social constructionist study thirty six senior executives participants were interviewed. It was observed that when discussing the person-environment fit and competencies of a project manager, occupational stress and stress management were overlooked by executive project sponsors as a possible contributing factor for poor project manager performance and poor project outcomes. Despite the growing body of research evidence that has identified occupational stress across many different professions as having an adverse impact on an individual's performance and health, there has been little recognition within project management literature of the impacts of stress or coping strategies for managing stressors within the project environment. This strongly suggests that additional research on stress focused on the project environment is warranted, and that both project sponsors and project managers need to be made aware of the antecedents to stress and consequences of stress both in the personal and business sense.

Keywords: Workplace stress, Project manager, Job performance, Decision making.

1 INTRODUCTION

A project manager commences a project with an underlying belief that with human endeavour, a combination of a supportive sponsor, skilled resources, coupled with suitable prescriptions and process, the planned results can be achieved (Gaddis 1997, Hamel and Prahalad 2013). Nevertheless, due to unexpected events, poor planning, inexperience, and a variety of other reasons, many projects fail to achieve the expected outcomes. In response to the dilemma of high failure rates, industry and academia have been focused on the technocratic and rationalistic aspects of project management, namely the development of prescriptive books and education processes that articulate the hard skills of managing a project (Atkinson *et al.* 2006). However, the prescriptions and process are ineffective at facilitating stakeholder "sense-making" within a project environment, failing to address many of the conflicting issues facing the project manager within the project environment.

Project environments are 'human systems' operating in socially complex environments where a decision or action from one individual or group may affect related individuals and groups. While hard skills may be essential in managing a project, it is the soft skills such as management of the interface boundary between the project, the organisation, suppliers, and the clients that provides the most important contribution that a project manager brings to a project

(Langer *et al.* 2008). In this role the project manager faces countless conflicting and ambiguous pressures as internal and external, and formal and informal interfaces make demands on the project manager which results in the project environment becoming unpredictable and uncontrollable, generating conflict, tension, and workplace stress for the project manager.

2 RESEARCH METHOD

This research is premised on Appreciative Inquiry (Whitney and Trosten-Bloom 2003) and explores the project manager from the perspective of executive management. Appreciative inquiry is an alternative approach to traditional problem solving centred or inquisitional research style of uncovering "what went wrong and who is at fault", it instead seeks "what can be done to make things better". The study is qualitative and based on postmodern social constructionist / interpretivist philosophy (Gergen 1978, Gergen 2009, Lewis *et al.* 2008). The author interviewed thirty-six senior executives who sponsor strategic projects within their respective organisations, the organisations range from telcos, utility providers, postal services, ICT providers, and other large organisations. The executives were selected through a purposive sampling approach (Miles *et al.* 2013) to ensure that the selected participants came from a broad range of industries and thereby enabling the opportunity to 'generalise' the study outcomes to other situations (Lincoln and Guba 1985, Saldaña 2015).

3 LITERATURE REVIEW

Stress in the workplace can impact an organisation's productivity through absenteeism and poor employee productivity, it can negatively impact an individual's mood states and health. Occupational stress can be viewed as the individuals' reactions to the characteristics of the workplace which appear threatening (Kahn *et al.* 1964). It points to a poor "person-environment" fit between an individuals' capabilities, competencies, and the workplace environment, or that the individual is not fully equipped to handle a particular work situation with the "misfit" between the individual and the environment being the basis for an individual's adverse behavioral and psychological reactions (French 1963, Bussing and Glaser 1999).

Since the early 1970s there has been considerable research into occupational stress in professions that are thought to be subjected to dynamic working environments, such as nursing, teaching, and law enforcement. Reseach by organisational behavioural scientists has provided a growing body of evidence that suggests stress is linked to an individual's physical and mental health (Cobb 1976, Morris and Snyder 1979) to coronary heart disease (House 1974, Matteson and Ivancevich 1979); to absenteeism (Margolis *et al.* 1974, Gupta and Beehr 1979); to staff turnover (Porter *et al.* 1974, Gupta and Beehr 1979); and to job dissatisfaction (Lyons 1971, Miles 1975, Beehr 1976). Although project management is a dynamic working environment there has been little research into the implications of stress within the project environment.

Dr. Hans Selye postulated in the 1930s that stress is a non-specific response of human body to any demand on it (Selye 1959, Selye 1976, Selye, 1993). This is supported by more recent research that postulates stress occurs whenever a demand exceeds the regulatory capacity of an individual (Dickerson and Kemeny 2004). The project environment is often unpredictable, resulting in perplexing problems for the project manager and making them susceptible to conflict, tension, and stress which results in many decisions being made under stressful conditions. The decision making of an individual who works under stressful situations is more rigid, simplistic, and superficial (Friend 1982, Cherrington 1994) and research has shown that decisions made under stressful conditions tend to be less well-thought out and much more irrational (Keinan *et al.* 1987, Lazarus 2000, Starcke *et al.* 2008, Galvan and Rahdar 2013) thus indicating that stress has a negative impact on an individual's performance.

A project manager's role spans the boundary between the sponsoring organisation, the clients, and other actors involved in the project environment. This spanning role exacerbates the project manager's tendency to experience incompatible demands and conflicting expectations from people involved in the project environment resulting in decision making under stressful conditions. Research has demonstrated that an individual under stress will not fully consider the situation and all of the possible options, tending to make decisions in a rushed and unsystematic manner (Janis and Mann 1977, Mather and Lighthall 2012, Galvan and Rahdar 2013) leading to a number of undesirable consequences, including a restriction or narrowing of attention, increased distraction, increases in reaction time, and deficits in the individual's working memory (Driskell *et al.* 1999) and a tendency to develop poor interpersonal relationships (Leung *et al.* 2005). This is supported by researchers who have investigated the impacts of stress on the performance of a variety of groups including: managers (Joiner 2001), police officers (Tang and Hammontree 1992, Collins and Gibbs 2003), students (Wolk and Bloom 1978), nurses (Dailey *et al.* 1986), teachers (Byosiere 1987, De Heus and Diekstra 1999), and human service workers (Wieclaw *et al.* 2006).

4 DISCUSSION

Project management literature is de-personalised and primarily focused on processes and prescriptions and as a consequence information into the impact of stress on the performance of project managers is sadly lacking. Project managers are often confronted with complex organisational and project environments. The complexity of the project environments and its associated occupational stressors, surprisingly was directly addressed by a single participant when they acknowledged that "we've got to be very careful that we don't burn them out". Several other participants obliquely referred to the drivers and impacts of occupational stress by employing expression like "it can be quite bullish at times" and "the job is demanding" when discussing the project environment within their respective organisations. Previous studies of stress have indicated that there is an optimal amount of occupational stress in terms of its effects on an individual's performance (Anderson 1976).

Stress that is higher or lower than the optimal amount results in steadily decreasing performance of the project manager and consequently an increasing risk of a poor project outcome. Performance under stress follows an inverted-U-shaped function known as the Yerkes-Dodson law (Anderson 1976). Figure 1 shows the inverted U-shape function of the Yerkes-Dodson law's curvilinear relationship between the amount of stress felt by workers and their level of performance. Nearly two thirds of the participants were anxious about various aspects of the project manager's person-job fit attributes and their ability, as described in different ways by several participants, 'to get stuff done'. Many of the participants were keen on increasing the job demands on the project manager by introducing more challenges, constraints, and a need to drive creative solutions to resolve problems within the project, there was no recognition that these additional demands may result in additional stressors. The Yerkes-Dobson curve suggests that increasing the demand will tend to increase the project manager's performance as they experience eustress up to a certain level of stress which corresponds to a range of stress for optimal performance.

Increasing the level of stress beyond this point, the project manager's performance will begin to deteriorate. Anecdotal evidence from project managers suggest that this happens quite often when a project sponsor has unreasonable expectations and imposes unrealistic budget, time, and other constraints on the project. The level of motivation to solve a problem under high levels of stress becomes so high that the project manager's perception narrows to only obvious cues, and he/she ignores relevant information and may focus on reducing anxiety rather than performing project management tasks. When experiencing excessive levels of stress, the project manager experiences distress and may become agitated, or emotionally upset, or experience health problems, thus causing a significant reduction in their level of performance and also have negative impacts on their health. At this point the project manager will switch from class 1 problem solving behaviours and focus on class 2 behaviours which emphasise emotional and defensive coping mechanisms leading to a reduction of task (project management) performance.

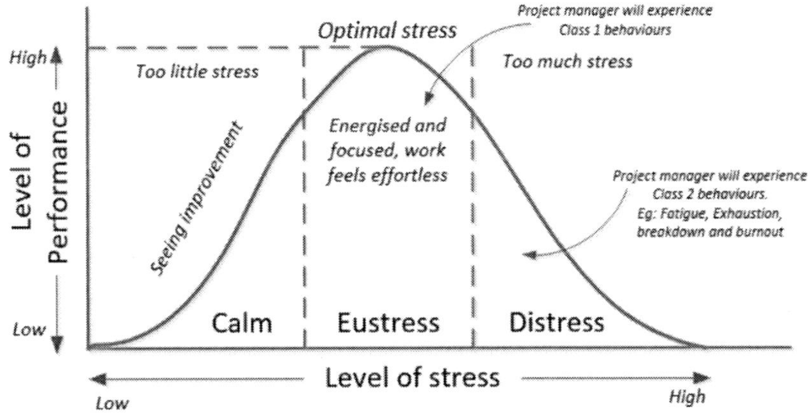

Figure 1. Shows the typical relationship between an individual's performance and stress. Yerkes-Dobson's curve, adapted from Anderson (1976).

Increasing the job demands on a project manager, as suggested by the Yerkes-Dobson curve, tends to increase the project manager's performance as they experience eustress up to a certain level of stress which corresponds to a range of stress for optimal performance. If the level of stress increases beyond this point, the project manager's performance will begin to deteriorate. As previously noted, the level of motivation to solve a problem under high levels of stress becomes so high that the project manager's perception narrows to only obvious cues, and he/she ignores relevant information and may focus on reducing anxiety rather than performing project management tasks.

5 CONCLUDING REMARKS

All participants are project sponsors and it was not clear that they appreciated the relationship between the project and the project manager's performance and occupational stress. It is disturbing that only one participant considered that the welfare of the project manager to be a factor in their description of a competent project manager. It is a little reassuring that several other participants obliquely addressed stress, but it is difficult to explain why virtually every participant overlooked workplace stress. It is possible that they overlooked stress due the organisational cultural factors, perhaps they consider stress to be a normal element within the workplace, or perhaps the topic is taboo, or perhaps the lines of communication between the participants and the project managers are not open and the signs of stress are missed

The project environment is essentially a complex social environment and considering the complexity of project environments, it is surprising that little attention has been given to work-induced stress within the project management industry although there are many studies in other disciplines on stress that have produced a long list of possible antecedents of burnout, the theoretical progress within the project environment has been limited. This lack of acknowledgement strongly suggests that additional research on stress focused on the project environment is warranted and that both project sponsors and project managers need to be made aware of the antecedents to stress and the consequences of stress both in the personal and business sense. In recognising that stress is unavoidable within the project environment, a logical extension of this study will be to investigate the project manager's response to various stress inducers, their resilence, and their ability to regulate their emotions.

References

Anderson, C. R., Coping Behaviors as Intervening Mechanisms in The Inverted-U Stress-Performance Relationship, *Journal of Applied Psychology,* 61, 30, 1976.

Atkinson, R., Crawford, L., and Ward, S., Fundamental Uncertainties in Projects and The Scope of Project Management, *International Journal of Project Management,* 24, 687-698, 2006.

Beehr, T. A., Perceived Situational Moderators of The Relationship Between Subjective Role Ambiguity And Role Strain, *Journal of Applied Psychology,* 61, 35, 1976.

Bussing, A., and Glaser, J., Work Stressors in Nursing in The Course of Redesign: Implications for Burnout and Interactional Stress, *European Journal of Work and Organizational Psychology,* 8, 401-426, 1999.

Byosiere, P. H. R., *Effects of Societal, Organizational, and Individual Factors on Job Performance, Job Satisfaction, and Job Strain: Multiple Structural Equation Modeling in A Three Wave Longitudinal Panel Study of New Teachers,* 1987.

Cherrington, D. J., *Organizational Behavior: The Management of Individual and Organizational Performance*, Allyn and Bacon Boston, 1994.

Cobb, S., Social Support as A Moderator of Life Stress, *Psychosomatic Medicine,* 38, 300-314, 1976.

Collins, P., and Gibbs, A., Stress in Police Officers: A Study of The Origins, Prevalence And Severity of Stress - Related Symptoms within A County Police Force, *Occupational Medicine,* 53, 256-264, 2003.

Dailey, R. C., Ickinger, W.and Coote, E. Personality and Role Variables as Predictors of Tension Discharge Rate in Three Samples, *Human Relations,* 39, 991-1003, 1986.

De Heus, P., and Diekstra, R. F., Do Teachers Burn Out More Easily? A Comparison of Teachers With Other Social Professions on Work Stress and Burnout Symptoms, 1999.

Dickerson, S. S., and Kemeny, M. E., Acute Stressors and Cortisol Responses: A Theoretical Integration and Synthesis of Laboratory Research, *Psychological Bulletin,* 130, 355, 2004.

Driskell, J. E., Salas, E., and Johnston, J., Does Stress Lead to a Loss of Team Perspective? *Group Dynamics: Theory, Research, and Practice,* 3, 291, 1999.

French, J. R., The Social Environment and Mental Health, *Journal of Social Issues*, 1963.

Friend, K. E., Stress And Performance: Effects of Subjective Work Load and Time Urgency. *Personnel Psychology,* 35, 623-633, 1982.

Gaddis, P. O. , Strategy Under Attack, *Long Range Planning,* 30, 38-45, 1997.

Galvan, A.and Rahdar, A., The Neurobiological Effects of Stress on Adolescent Decision Making, *Neuroscience,* 249, 223-231, 2013.

Gergen, K. J., Toward Generative Theory, *Journal of Personality and Social Psychology,* 36, 1344, 1978.

Gergen, K. J., *Realities And Relationships: Soundings In Social Construction*, Harvard University Press, 2009.

Gupta, N., and Beehr, T. A., Job Stress and Employee Behaviors, *Organizational Behavior And Human Performance,* 23, 373-387, 1979.

Hamel, G., and Prahalad, C. K., *Competing for The Future*, Harvard Business Press, 2013.

House, J. S., Occupational Stress and Coronary Heart Disease: A Review and Theoretical Integration, *Journal of Health and Social Behavior*, 12-27, 1974.

Janis, I. L,.and Mann, L., *Decision Making: A Psychological Analysis of Conflict, Choice, and Commitment*, Free Press, 1977.

Joiner, T. A., The Influence of National Culture and Organizational Culture Alignment on Job Stress and Performance: Evidence from Greece, *Journal of Managerial Psychology,* 16, 229-242, 2001.

Kahn, R. L., Wolfe, D. M., Quinn, R. P., Snoek, J. D., and Rosenthal, R. A., Organizational Stress: Studies in Role Conflict and Ambiguity, 1964.

Keinan, G., Friedland, N., and Ben-Porath, Y., Decision Making under Stress: Scanning of Alternatives under Physical Threat, *Acta Psychologica,* 64, 219-228, 1987.

Langer, N., Slaughter, S. A., and Mukhopadhyay, T., *Project Managers' Skills and Project Success in It Outsourcing*, Icis 2008 Proceedings, 147, 2008.

Lazarus, R. S., Toward Better Research on Stress And Coping, 2000.

Leung, M. Y., Ng, S. T., Skitmore, M., and Cheung, S. O., Critical Stressors Influencing Construction Estimators in Hong Kong, *Construction Management And Economics,* 23, 33-44, 2005.

Lewis, S., Passmore, J., and Cantore, S., *Appreciative Inquiry for Change Management: Using Ai To Facilitate Organizational Development*, Kogan Page Publishers, 2008.

Lincoln, Y. S., and Guba, E. G., *Naturalistic Inquiry*, Sage, 1985.

Lyons, T. F., Role Clarity, Need For Clarity, Satisfaction, Tension, and Withdrawal. *Organizational Behavior and Human Performance,* 6, 99-110, 1971.

Margolis, B. L., Kroes, W. H., and Quinn, R. P., Job Stress: An Unlisted Occupational Hazard, *Journal of Occupational and Environmental Medicine,* 16, 659-661, 1974.
Mather, M., and Lighthall, N. R., Risk and Reward Are Processed Differently in Decisions Made Under Stress, *Current Directions In Psychological Science,* 21, 36-41, 2012.
Matteson, M. T., and Ivancevich, J. M., Organizational Stressors and Heart Disease: A Research Model, *Academy of Management Review,* 4, 347-357, 1979.
Miles, M. B., Huberman, A. M., and Saldana, J., *Qualitative Data Analysis: A Methods Sourcebook*, Sage Publications, Incorporated, 2013.
Miles, R. H., An Empirical Test of Causal Inference Between Role Perceptions of Conflict and Ambiguity and Various Personal Outcomes, *Journal of Applied Psychology,* 60, 334, 1975.
Morris, J. H., and Snyder, R. A., A Second Look at Need for Achievement and Need for Autonomy as Moderators of Role Perception–Outcome Relationships, *Journal of Applied Psychology,* 64, 173, 1979.
Porter, L. W., Steers, R. M., Mowday, R. T., and Boulian, P. V., Organizational Commitment, Job Satisfaction, and Turnover Among Psychiatric Technicians, *Journal of Applied Psychology,* 59, 603, 1974.
Saldaña, J., *The Coding Manual for Qualitative Researchers*, Sage, 2015.
Selye, H., Perspectives in Stress Research, *Perspectives in Biology and Medicine,* 2, 403-416, 1959.
Selye, H., Forty Years of Stress Research: Principal Remaining Problems and Misconceptions, *Canadian Medical Association Journal,* 115, 53, 1976.
Selye, H., History of The Stress Concept, 1993.
Starcke, K., Wolf, O. T., Markowitsch, H. J., and Brand, M., Anticipatory Stress Influences Decision Making Under Explicit Risk Conditions, *Behavioral Neuroscience,* 122, 1352, 2008.
Tang, T. L.-P., and Hammontree, M. L., The Effects of Hardiness, Police Stress, and Life Stress on Police Officers' Illness and Absenteeism, *Public Personnel Management,* 21, 493-510, 1992.
Whitney, D. K., and Trosten-Bloom, A., *The Power of Appreciative Inquiry: A Practical Guide to Positive Change*, Berrett-Koehler Publishers, 2003.
Wieclaw, J., Agerbo, E., Mortensen, P. B., and Bonde, J. P., Risk of Affective and Stress Related Disorders Among Employees in Human Service Professions, *Occupational and Environmental Medicine,* 63, 314-319, 2006.
Wolk, S., and Bloom, D., The Interactive Effects of Locus of Control and Situational Stress Upon Performance Accuracy and Time, *Journal of Personality,* 46, 279-298, 1978.

Interdependence between Structural Engineering and Construction Management
Edited by Ozevin, D., Ataei, H., Modares, M., Gurgun, A., Yazdani, S., and Singh, A.

ISBN: 978-0-9960437-6-2

THE POTENTIAL OF KNOWLEDGE MANAGEMENT ON CONSTRUCTION SITES

CORNELIA NINAUS[1] and DAVID KNAPP[2]

[1]*Institute of Construction Management and Economics, Graz University of Technology, Graz, Austria*
[2]*Graz University of Technology, Graz, Austria*

Knowledge management is the process of creating, sharing, using, and managing knowledge, one of the most valuable organizational resources. This approach is well-known in Austria's industrial sector, but applied only in major construction companies. It is mainly used to share knowledge between different departments, but it is not commonly found on construction sites. During the construction phase a variety of separate firms build a temporary multidisciplinary organization, to produce investment goods. To show the potential of knowledge management on building sites in Austria, 78 interviews were conducted. Construction sites of different types (new construction and refurbishment of buildings) were taken into account in order to guarantee a representative outcome. The highest cost-benefit ratio for knowledge management can be seen in knowledge intensive processes. The execution phase is characterized by craftwork which often includes many routine steps. But the survey shows that almost a quarter of the daily business is about knowledge intensive processes while the amount doesn´t correlate with the working experience. Furthermore, on construction sites with many trades the lack of information and knowledge transfer is the cause of nearly a quarter of the problems faced. The findings indicate the need for knowledge management on construction sites and the potential grows with the number of trades. The teambuilding process can be seen as the most important step for an efficient knowledge management during the execution phase.

Keywords: Knowledge transfer, Tacit knowledge, Explicit knowledge, Communication, Construction phase, Connectivity, Teambuilding, Austria.

1 INTRODUCTION

Knowledge Management is a strategy for getting the right knowledge to the right person at the right time and helping them to share it and put information into action to improve organizational performance (O'Dell and Grayson 1998). Additional knowledge is the theoretical and practical skill a person owns and uses to solve problems. It is based on data and information, but in contrast it is linked to a person (Probst *et al.* 2006).

The origins of this go back to the year 400 BC when the philosophers Plato and Aristotle discussed the acquisition, the forms and transfer of knowledge (Ritter *et al.* 2007). Michael Polanyi (1966) was the first to divide knowledge into implicit and tacit, one of the most important basics for further research in this field. Explicit knowledge can be measured, captured, examined and codified into documents. In contrast tacit knowledge is very hard to transfer because it relies on a person's experiences (Polanyi 1966, Pathirage *et al.* 2008). The scientific peak of knowledge management was in the late 20th to the first decade of the 21st century and resulted from the publication of the 'Knowledge Creating Company' (Nonaka and Takeuchi 1995) in which the knowledge spiral was presented, which is the fundament for many of the following knowledge models. The knowledge management approach is well-known in Austria's industrial sector, but applied only in major construction companies even though it can improve every phase of a construction project from the acquisition to the handover and operation (Wais 2006). It is mainly used to share knowledge between different departments, but it is not commonly found on construction sites (Rathswohl 2014). An obstacle for the implementation of knowledge management in the construction industry is the lack of pragmatic and simple measures and tools (Cüppers 2006).

During the project execution phase, the various trades working on site form a temporary organization (Grover and Froese 2016). The problem in this specific phase is that these different working parties do not have any contractual relationships and are not familiar with each other and as a result they do not share any knowledge. The thoughts about knowledge management often lead to traditional information and communication technologies (e.g. data bases), but beneath the technical surface the human element (willingness for knowledge sharing) and organizational aspect (conditions for knowledge management provided by the company) are also pillars essential for success (Bullinger *et al.* 1997). The affects for knowledge sharing in project based organizations are trust, organizational culture (how people behave), organizational structure (coordination and supervision within the organization), IT infrastructure, top management support, communication and furthermore willingness to share knowledge (Ju *et al.* 2017).

A survey in Germany showed that knowledge transfer in the construction industry is inhibited through temporal, spatial, contextual and social barriers. It indicated that Mobile devices like smartphones or tablets could increase the connectivity and thus also knowledge transfer (Waizenegger *et al.* 2016). According to a survey conducted in the construction industry in Turkey the common driver for knowledge management is reducing rework effort (Kivrak *et al.* 2008), while a study in the United Kingdom points to the sharing of tacit knowledge as one of the most important drivers (Carrillo *et al.* 2004). Subsequent to reporting on all of these findings the study determines the knowledge management behavior on construction sites in Austria during the execution phase. The preconditions for multidisciplinary knowledge exchange as well as the potential of knowledge management are identified.

2 METHOD

In order to identify the potential of knowledge management at the trade layer on Austria´s construction sites a situation analysis through semi-structured interviews were conducted to ensure the maintenance of consistency over the concepts covered in each interview (Corbin and Strauss 2015). Six constructions sites of different types in the Province of Styria where taken into account for the investigation. These were: (1) a new residential construction project with more than 150 units, (2) the demolition and new construction of a three storied apartment building, (3) the refurbishment of a brewery, (4) the renovation of a residential project with five units, (5) the new construction of a detached house and (6) the renovation of a detached house. (1) This project was located in Graz, the provincial capital of Styria, while the others where located in Murau, a town in Upper Styria. 78 semi-structured interviews were conducted with experts which were selected following a study of the site documents. The chosen interviewees had to work on the site in a trade which had at least one intersection with another company during the execution phase. The 78 participants aged between 23 and 72 represented 29 different professional groups of the 6 investigated construction sites to guarantee a holistic output. With the age mix chosen it was aimed to identify correlations with the ages or work experience and the answers. The professional groups considered are distributed as follows: (1) 16 carcass workers, (2) 42 interior construction workers, (3) 19 workers for building services and (4) one worker for the demolition process. The aim was to include at least one person of every trade to record all interfaces and interactions. The semi-structured interviews were carried out by two researchers face to face or by telephone and were audio recorded. The audio recordings were subsequently transcribed, cleaned, coded and analyzed through a qualitative content analysis. The interview guideline was divided into four research parts: (1) knowledge intensive processes, (2) problems during the execution phase, (3) multidisciplinary knowledge exchange and (4) digital tools. To achieve a profitable knowledge management with a justified cost-benefit ratio, the implementation should concentrate on (1) knowledge intensive processes. In contrast to routine processes they create, transfer and use knowledge within the work steps. Knowledge intensive processes are characterized by high complexity, few repetitions, a high degree of novelty and little available data. With the section (2) problems during the execution phase, it was aimed to identify if any problems are caused by the lack of information and knowledge transfer. In the third category (3) multidisciplinary knowledge exchange it was clarified if the different trades could learn something from each. Regarding to Bullinger *et al.* (1997) a holistic knowledge management consists of IT-Infrastructure, supportive organization and keen humans. The last part (4) should evaluate the current usage of digital tools and the potential of further IT.

3 RESULTS

3.1 Knowledge Intensive Processes

In this part the interviewees should estimate their daily work in the categories routine processes or knowledge intensive processes. Figure 1 shows the distribution over all investigated sites categorized by the professional experience of the interviewee.

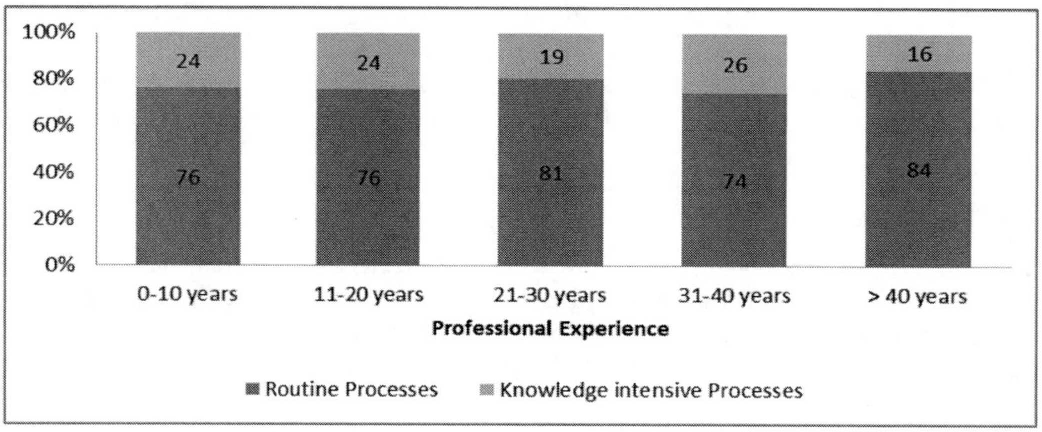

Figure 1. Distribution of knowledge intensive processes in dependence on age.

According to the interviewees on average 78% of the daily business on a building site are routine while 22% are knowledge intensive processes. The knowledge intensive processes are result from: (1) 41% first use of materials, technologies or building methods, (2) 32% detailed solutions, (3) 11% construction process, (4) 10% project type, (5) 4% technical standards and law and (6) 2% project organization.

3.2 Problems During the Execution Phase

This section identified the main triggers for problems during the execution phase. 72% interviewees of the construction sites (2), (3), (4), (5) and (6) (see chapter 2 Method) told us they had not noticed any disturbances, while no one involved in project (1) indicated this in their answers. On average in all the investigated construction sites 38% of the problems where bound to construction disturbances (e.g. weather) while 30% were caused by planning problems. Both are very common problem issues for the construction industry. The remaining problems could be divided into 17% caused by the client and 15% due to the lack of communication and knowledge transfer.

3.3 Multidisciplinary Knowledge Exchange

In order to evaluate the potential of multidisciplinary knowledge exchange in the construction industry participants were asked if the different trades could learn something from each other and what would force them to exchange their knowledge with other companies. 90% of the interviewees believed the trades definitely could learn from each other and would benefit from this exchange. The other 10% are divided in 5% who tend to agree and 5% who see no benefit in knowledge sharing.

The following preconditions for knowledge exchange were reported: (1) 64% personal contact, (2) 20% interest, (3) 9% trust and (4) 7% same language.

3.4 Digital Tools

People were asked about their willingness to use digital tools to support the knowledge and information exchange on the construction site. Figure 2 shows the distribution of the answers.

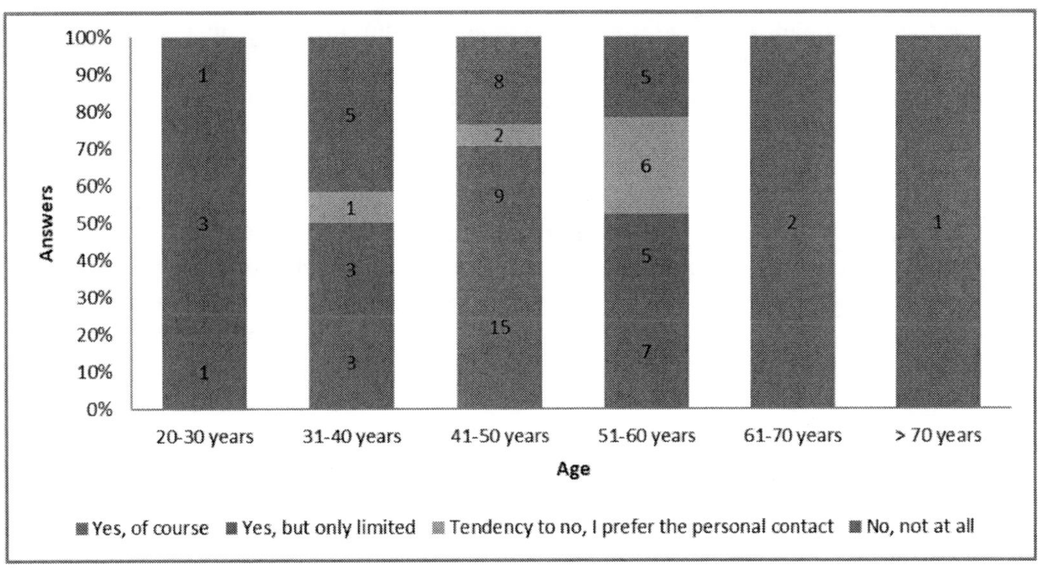

Figure 2. Willingness to use digital tools.

On average across all age groups the answers were distributed as follows: (1) 36% Yes, of course, (2) 26% Yes, but only limited, (3) 12% Tendency to no, I prefer the personal contact, (4) 26% No, not at all.

4 DISCUSSION

Even if the building industry is characterized by routine processes, as a result of the study it was established that almost a quarter of the processes are knowledge intensive, which indicates potential for knowledge management. It can be seen that there is no significant correlation between the age and the amount of knowledge intensive processes. The reason is because the building industry is faced with new materials, technologies, building methods and detailed solutions on a daily basis. An apprentice will naturally face knowledge intensive processes virtually each working day, because every method or material is something new. The challenge – for the new and experienced worker - is to react swiftly and efficiently on this unforeseen situation. Knowledge management can support the knowledge finding processes, which makes the execution phase more efficient. According to the survey the major portion of all problems on the investigated sites were related to the planning process and construction disturbances (e.g. weather), which are very common issues for the construction industry. A surprise result of the detailed analysis, however, was that the answer "communication and knowledge transfer" was only given on construction site (1) where it caused almost a quarter of the problems. An explanation could be the bigger project size and the location in a larger city. In fact, the interviewees mentioned that the communication on construction sites in non-urban areas is always easier because the people know each other from other projects. This conforms with the outcome of the section on multidisciplinary knowledge exchange. Future research is necessary to identify the differences of communication and knowledge sharing in urban and non-urban sites. It can be seen as very positive that the majority of the interviewees think the different trades could both learn something from each other and would benefit from doing so. The preconditions "personal contact" and "trust" referred to indicate that the laborers today do their daily tasks without having any contact to the people working next to them and furthermore they neither want to communicate nor to share information and knowledge. Knowledge management thus has the potential to reduce problems during the execution phase by linking the people at work. But it also indicates that not every building site is faced with problems caused by the lack of communication and knowledge transfer. The biggest effort can be seen on complex construction sites involving trades, which have never worked together previously, a situation that arises frequently in the bigger cities. Kivrak et al. (2008) indicated reducing rework as a common driver for knowledge management, while 20% of the interviewees mentioned "interest" as their precondition. This category means that the worker will only be willing to share knowledge when seeing own advantages from doing so (e.g. saving time). An efficient knowledge management is based on a supportive organizational structure, open-minded corporate culture and an appropriate IT-infrastructure (Bullinger *et al.* 1997). Use of a digital tool is an essential for connecting the different companies at work on the site. The majority (Fig. 2) are willing to use IT to support the knowledge exchange. A detailed look at the answer "Yes, of course" shows that the majority of those who gave this answer are over 41 years old. In this the prejudice that older people are not willing to use digital tools cannot be validated. This conclusion is also borne out by the finding that the younger generation is more skeptical and frequently does not believe that digital tools can be the miracle cure for every problem. The IT-Infrastructure could bring a solution for the language problem mentioned in chapter 3.3 by the inclusion of a translation function.

5 CONCLUSIONS

The qualitative interviews resulted potential for knowledge management on the construction site. But it also can be seen that not every building project needs this approach. The number of problems, especially the ones related to communication and knowledge transfer, increase with the size and complexity of a construction project. This indicates that large projects stand to gain more from knowledge management than small ones (e.g. detached houses). Contrary to expectations the working experience does not correlate with the number of knowledge intensive processes. The lack of well-structured multidisciplinary organization and open minded workers can be seen as the most important barrier for knowledge management on the construction site as well as its highest potential. Linking the people working on the trade layer will help reduce problems as a result of heightened communication and knowledge sharing. The opinions of interviewees on the subject of digital tools confirms this finding. They would use a digital tool to support the knowledge sharing, but the organizational boundary conditions must be set in advance. The different trades create a temporary organization during the execution phase but they are not building a team. Further research is necessary to identify processes in order to break down the barriers described above and to push through and establish a satisfactory multidisciplinary knowledge management.

References

Bullinger, H., Wörner, K., and Prieto, K., *Wissensmanagement Heute: Daten, Fakten, Trends*, Frauenhofer-Institut für Arbeitswirtschaft und Organisation, Stuttgart, 1997.

Corbin J., and Strauss, A., *Basics of Qualitative Research*, SAGE Publication, California, US, 2015.

Cüppers, A., *Wissensmanagement in Einem Baukonzern. Anwendungsbeispiele bei Bauprojekten*, VDI-Verlag, Düsseldorf, Germany, 2006.

Grover, R., and Froese, T., *Knowledge Management in Construction using SocioBIM Platform: A Case Study of AYO Smart Home Project,* International Conference on Sustainable Design, Engineering and Construction, 145, 1283-1290, 2016.

Ju, J., Ren, X., and Hao, S., *Modeling Enablers for Successful Knowledge Sharing in Project-Based Organizations,* International Conference on Construction and Real Estate Management, Wang, Y., Pang, Y., Shen, G., Zhu, Y. (eds), Guangzhou, China, 158-167, 2017.

Kivrak, S., Arslan, G., Dikmen, I., and Birgonul, T., Capturing Knowledge in Construction Projects: Knowledge Platform for Contractors, *Journal of Management in Engineering*, 24(2), 87-95, 2008.

Nonaka, I., and Takeuchi, H., *The Knowledge Creating Company*, Oxford University Press, New York, 1995.

O'Dell, C., and Grayson, C. J., If only We Knew What We Know: The Transfer of Internal Knowledge and Best Practice, *The Free Press*, New York, 1998.

Pathirage, C., Amaratunga, D., and Haigh, R., The Role of Tacit Knowledge in the Construction Industry: Towards A Definition, *International Conference on Building Education and Research (BEAR)*, School of Built Environment, University of Salford, UK, 204-217, 2008.

Polanyi, M., *The Tacit Dimension*, MA: Peter Smith, Gloucester, 1966.

Probst, G., Raub, S., and Romhardt, K., *Wissen Managen*, Betriebswirtschaftlicher Verlag Dr. Th. Gabler, Wiesbaden, Germany, 2006.

Ritter, J., Gründer, K., and Gabriel, G., *Historisches Wörterbuch der Philosophie Gesamtwerk Bd 1-13*, Schwabe Verlag, Basel, Germany, 2007.

Rathswohl, S., *Entwicklung eines Modells zur Implementierung eines Wissensmanagement-Systems in kleinen und mittleren Bauunternehmen*, Kassel University Press, Kassel, Germany 2014.

Wais, A., *Wissensmanagement zur Unterstützung von Baustellen in Bauvertragsfragen*, Books on Demand GmbH, Norderstedt, 2006.

Waizenegger, L., Thalmann, S., Sarigianni, C., Eckhardt, A., and Kolb, D., *From Isolation to Collaboration How the Increasing Diffusion of Mobile Devices Has Changed Practices of Knowledge,* Twenty-Fourth European Conference on Information Systems (ECIS), AIS eLibrary, Istanbul, Turkey, 2016.

Interdependence between Structural Engineering and Construction Management
Edited by Ozevin, D., Ataei, H., Modares, M., Gurgun, A., Yazdani, S., and Singh, A.

ISBN: 978-0-9960437-6-2

STRATEGIC SYNERGY OF SERVICE OFFERINGS WITHIN A PROFESSIONAL SERVICES FIRM

JOHN A. KUPRENAS[1] and LARA H. JENNINGS[2]

[1]*Harris and Associates, Irvine, USA*
[2]*Program and Construction Management, Harris and Associates, San Diego, USA*

The architecture, engineering, and construction (AEC) industry has traditionally been structured based upon a firm concentrating on select markets and specific service offerings. More recently, however, the industry has moved away from the piece by piece process of developing a project into a more holistic view. This paper suggests the concept that holistic strategies can create positive business synergies. Two types of synergies are identified – synergies related to practice and knowledge and synergies related to service and delivery. Examples of these synergies are provided with cases provided for engineering, architecture, and management firms. The increase in strength of the firm as a result of the strategy is more than the net additive growth or change; hence, it is deemed synergistic. Conclusions identify the synergies outcomes and discuss the need for future research in this area, particularly data-based research to test the theory.

Keywords: Project delivery, Engineering, Architecture, Design, Construction management.

1 INTRODUCTION

The growth of a typical business can have many paths. Much research has been conducted around the concept. The body of research in area does not have a clear consensus on factors that influence firm growth. Gibart's law suggests that firm growth is independent of firm size (Evans 1987). Mansfield (1961) suggests that smaller firms have higher and more variable growth rates than larger ones. Evans, however, found that firm growth decreases with age and with firm size (1987). Other research suggests that rapid and widespread growth among small firms is not to be expected at all. Smaller firms have higher and more variable growth rates than larger ones (O'Farrell and Hitchens 1988). Creating alliances to grow seems to be an established growth practice, however, alliances were found to be created mostly by desire of the firm's management (BarNir and Smith 2002). Competitive advantage is recognized as a key to growth. Dynamic core skills can create competitive advantages and technological learning creates the ability to develop, maintain, and exploit dynamic core competencies (Hitt *et al.* 2000).

Given these varied ideas, a logical question is how best to structure an AEC firm to succeed in today's evolving industry. This paper will introduce the concept of strategic synergies. Strategic synergies is term coined to capture the nonlinear consequences of a strategy. As an example of a non-linear strategy, rather than adding 10 staff to a firm of 10 to create the ability of 20, strategic synergies theorize that the now 20-person firm has the power of more than 20. This power is generated through the synergy of the combination of the parts. This power could be manifested in many ways. It may mean more revenue (higher professional services billing rates), less competition (greater volume and revenue), or better quality (less rework or write offs). Not every combination, however, will result in such synergy. These following sections will detail examples of two types of synergies – synergies related to practice and knowledge and synergies related to services and delivery.

2 PRACTICE AND KNOWLEDGE BASED SYNERGIES

Practice and knowledge-based synergies are those created by the growth is expertise around a domain of knowledge. Figure 1 shows a typical civil engineering firm on the left side of the figure. The firm may do basic design work related to streets, water transmission, or storm drains. As the firm grows, they would take on more staff with higher levels of expertise. These staff would help the firm win larger projects. As the firm wins larger projects, there influence grows and they can develop more expertise around more sophisticated design work, for example pump stations and water treatment. They may develop sub-specialties, for example reverse osmosis as part of water treatment. Throughout this growth, the firm would use subconsultant designers as needed, depending on the work requirements.

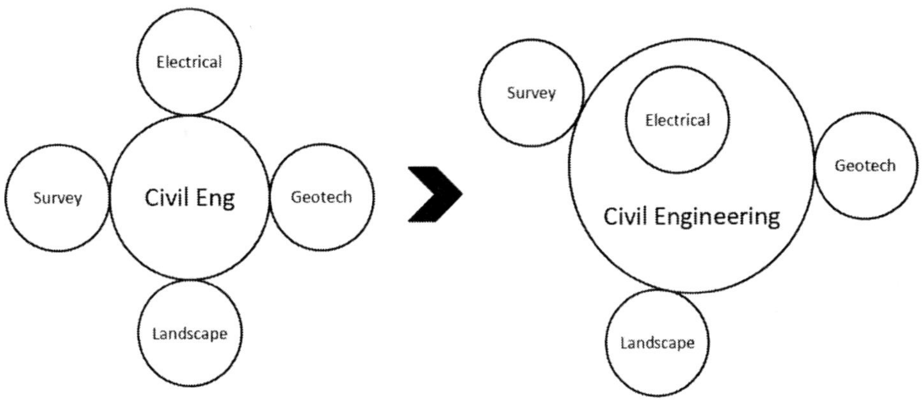

Figure 1. Civil engineering strategic synergy – initial action.

Looking at the right side of Figure 1, a strategic synergy is shown. As the civil engineering design firm grows, one strategy would be to acquire electrical design expertise. This could be done through a group hire or an acquisition. In executing this strategy, the firm will be able to more completely execute a segment of their core work, for example pump station design. By bringing the electrical design in-house, they will be able to present a more competitive fee (no mark up on the sub) and given their office locations and home office structure, they may not need to increase corporate overheard. The synergy of the two design groups working together in the same offices would also improve quality of the design through better coordination. Insurance costs should not be significantly more. Finally, the electrical group could pursue work on their own apart from civil collaborations. This will open up the firm to new clients. Downsides to this action would be the loss of cash for the acquisition and the cultural considerations of bringing on essentially a new team.

This example can advance even more. Figure 2 shows the potential evolution of this strategic synergy. Over time, the firm finds that electrical work is more profitable, perhaps less competitive, and provides better billing rate multipliers. The firm begins to hire more and more electrical designers. Slowly the identity of the firm changes. As less civil design work is won, the firm concentrates more on electrical work. The firm becomes an engineering firm specializing in in electrical design with a group that could do in-house civil design. As an example, the firm might specialize in power distribution. In this case, the strategic synergy demonstrates a firm better evolved to capture work in their environment. Over time, a nimble firm would experience multiple such evolved actions.

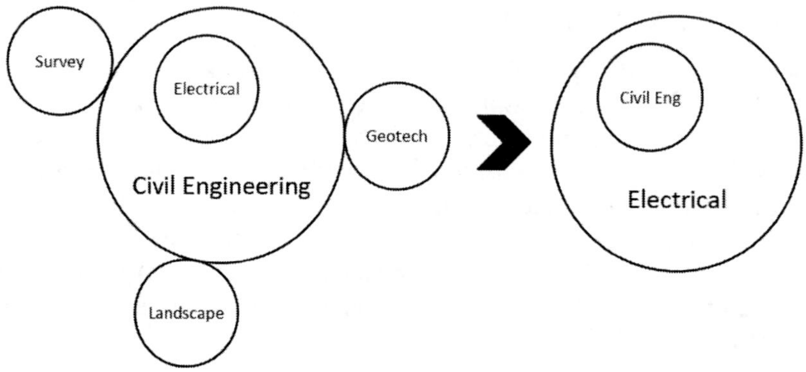

Figure 2. Civil engineering strategic synergy – evolved action.

Similar to the civil engineering design firm, an architecture firm typically uses subconsultants. As shown in Figure 3, in order to design a building project, an architect may team with MEP engineers, structural engineers, civil engineers. As the firm grows, some of this expertise might be brought into the firm. Synergy is created by closer collaboration and immediate access. Creativity is enhanced by shared experiences created by working together and through a better understanding of the architect's vision or style. Efficiency could also be enhanced. For example, unique structural details required to execute one design could potentially be improved and used in future designs.

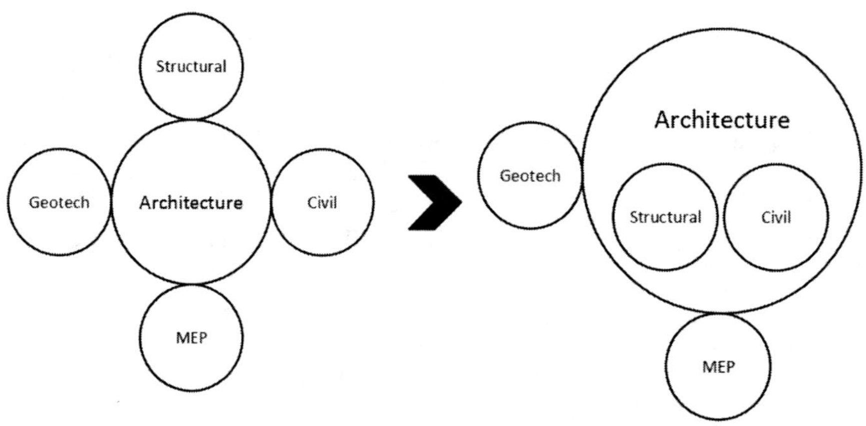

Figure 3. Architecture strategic synergy – initial action.

Eventually, this architecture firm could become a firm with multiple specialties and become a design firm as shown in Figure 4. Each sub specialty could work individually or collaboratively on more complex projects. As the design firm grows, as shown in figure 4, additional specialties can be added. A synergy is created to allow the firm to undertake not only different kinds of projects, but also projects with different delivery methods (perhaps design-build or a public private partnerships).

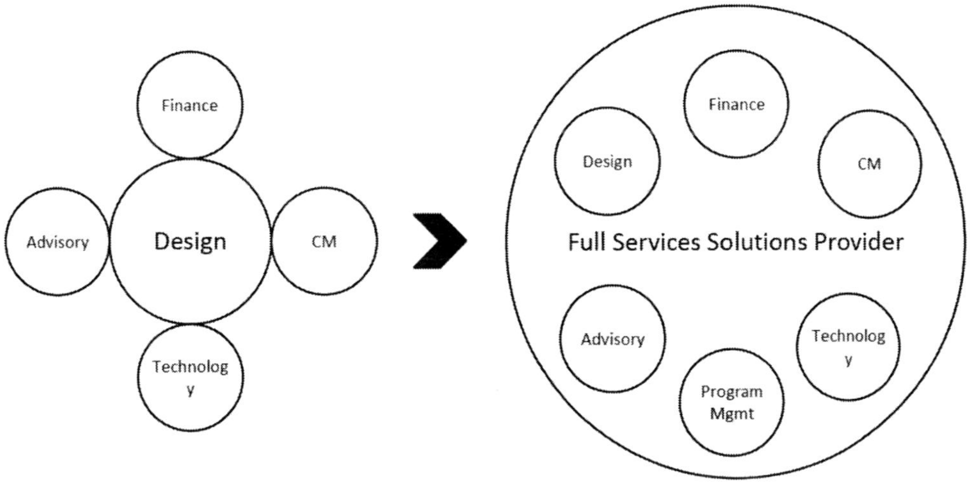

Figure 4. Architecture strategic synergy – evolved action.

In the case of the right side of Figure 4, the increasing expertise of the firm gives synergy related to design efforts, but also potentially synergy related to the types of services across a variety of delivery methods. These are called service and delivery-based synergies.

3 SERVICE AND DELIVERY BASED SYNERGIES

Strategic synergies are not exclusive to design. In fact, many service and delivery based synergies would be related to AEC management organizations. Consider a construction management firm as shown in Figure 5.

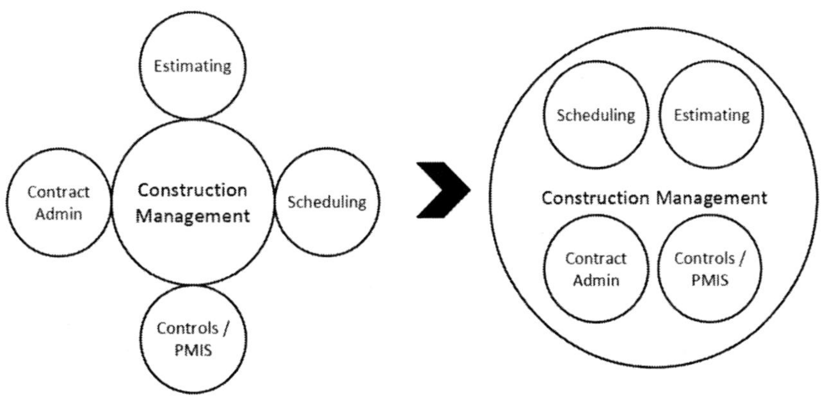

Figure 5. Construction management strategic synergy – initial action.

The firm may start with a few partners who have good connections in the industry and strong core construction expertise. As they advise their clients, depending on the engagement, their scope of work may require technical work related to construction. Examples might be complex scheduling for a multiple phase industrial project or utilizing strong controls / project management information systems for a program with multiple contractors. In each case, this service would be done through an external subconsultant. Looking at the right side of the diagram, the strategic synergy would bring these services into the firm as the firm grows.

Figures 6 shows how a program a management firm could grow even more. Many construction management firms decide to stay lean, but by developing in house expertise, the firm could now compete and win program management contracts. The synergistic advantage of program management is not related to fee, but rather related to volume. A program manager could have dozens, if not hundreds of staff, billing on a project. This strategy allows the construction management firm to move beyond the billability of its principals, into a role that provides orders of magnitude larger volumes.

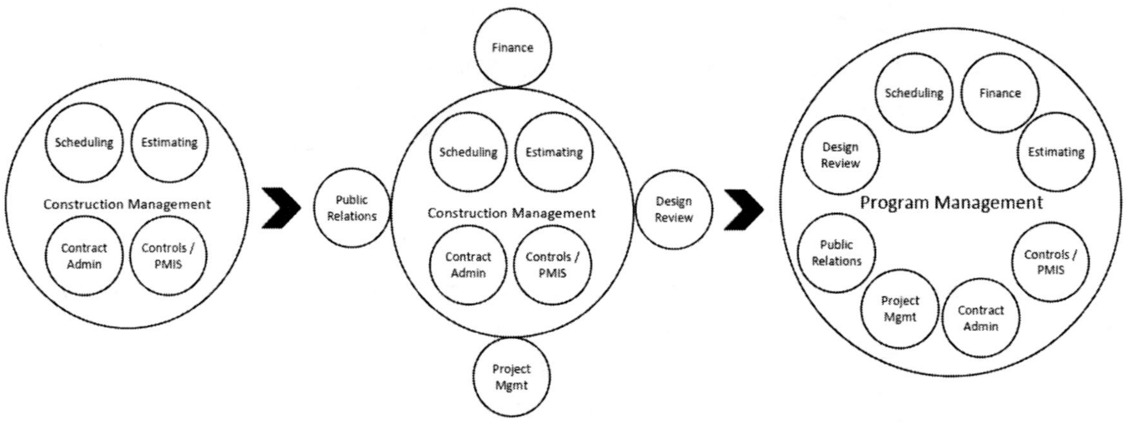

Figure 6. Construction management strategic synergy – evolved action.

Consider the growth in size of a firm on the far-right hand side of Figure 6. Beyond the economics of billable hours, a unique strategic synergy would be created. This firm would start to look less like an AEC firm and more like a solutions provider. The firm could become a large or mega size firm and acquisitions of expertise would become a routine course of business. These firms have very limited competition on huge value programs, very limited competition on their portfolio of expertise, and the ability to generate tremendous fees through there large staffing models.

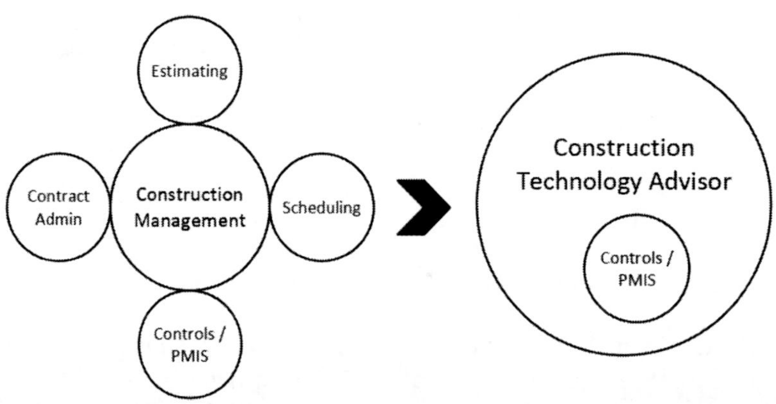

Figure 7. Construction management strategic synergy – evolved specialty action.

Not every service-based strategy needs to entail growth of the firm. Another service based strategic approach might be to specialize the construction management firm from Figure 5 into a single area. One current trend in Construction Management is leveraging technology. Figure 7 shows how a firm might accomplish this by concentrating in the PMIS work. Rather than growing expertise around technical skills associated with construction, like scheduling or estimating, a firm might concentrate on the technology of project management information systems (PMIS) and project controls. This strategy creates a potential synergy in that the firm opens themselves to potentially more types of clients (owners, contractors, suppliers, etc.) and positions them ahead of the technology curve that is pressing forward on construction management.

4 CONCLUSIONS

Based on the above analyses, strategies can be focused in two ways – on a practice / knowledge basis and on a service / delivery basis. This article has summarized examples of each type of strategic synergy. In most cases, immediate benefits can be realized at what is termed evolved actions. In some cases, further potential is identified in secondary evolved actions. These can be manifested through enlarging the size of the effort or concentrating in a specific area. Benefits from the concept of strategic synergy are numerous but focus around increased profitability (from higher fees or larger projects) and / or greater margin (from more specialization / fewer competitors. Benefits can also be manifested through better positioning of the firm for future work and opportunities for employees to advance their careers.

Future research in this area of AEC business strategies is needed. Given the upcoming changes in the industry around technology and technology-based tools, as well as the new project delivery models, sound AEC strategy is critical to a firm's success. Additional research to attempt to quantity the value of these types of synergies needed. Case study-based research with actual financial and performance-based results over long periods of time would be of value for validation of the concepts. Additional conversation and exploration around the theory of strategic synergies would also be of value.

References

BarNir, A., and Smith, K. A., Interfirm Alliances in the Small Business: The Role of Social Networks, *Journal of Small Business Management,* 40(3), 219-232, 2002.

Evans, D. S., Tests of Alternative Theories of Firm Growth, *Journal of Political Economy*, 95(4), 657-674, 1987.

Hitt, M. A., Ireland, R. D., and Lee, H., Technological Learning, Knowledge Management, Firm Growth and Performance: An Introductory Essay, *Journal of Engineering and Technology Management,* 17(3-4), 231-246, 2000.

Mansfield, E., *Entry, Gibrat's Law, Innovation, and the Growth of Firms*, No. 126. Cowles Foundation for Research in Economics, Yale University, 1961.

O'Farrell, P. N., and Hitchens, D. M., Alternative Theories of Small-Firm Growth: A Critical Review, *Environment and Planning A,* 20(10), 1365-1383, 1988.

Interdependence between Structural Engineering and Construction Management
Edited by Ozevin, D., Ataei, H., Modares, M., Gurgun, A., Yazdani, S., and Singh, A.

ISBN: 978-0-9960437-6-2

CHALLENGES AND RISKS RELATING TO MEP ENGINEERING IN LARGE BUILDING CONSTRUCTION PROJECTS

MICHAEL MONSBERGER

Institute of Construction Management and Economics, Graz University of Technology, Graz, Austria

MEP (mechanical, electrical, plumbing) systems are integral parts of modern buildings. Adequate conception, design, and installation of MEP systems are an essential prerequisite for a successful project. Experience has shown that the practical implementation of this aspect is often a challenge for the involved project stakeholders due to the increasing complexity of MEP systems. This raises the question of whether the field of MEP engineering is particularly sensitive to potential conflicts and disruptions in complex construction projects compared to other trades. In order to investigate these questions, a survey based on a standardized questionnaire was carried out among Austrian experts with experience in large building construction projects. Overall, 515 experts participated in this survey, with 365 completely answering the questionnaire. The results show that the participants consider proper project integration of MEP engineering as a critical success factor in complex construction projects. They also show that inadequate consideration of this aspect poses a substantial risk for project disruptions.

Keywords: Time and cost overruns, Project disruptions, MEP planning and coordination, Online survey, Stakeholder-specific analysis.

1 INTRODUCTION

Large-scale construction projects facing substantial time and cost overruns are often subject of public interest and debates. In Germany, for example, projects such as those analyzed in Kostka and Fielder (2016) led to the establishment of a high-level commission in 2013 with the aim of preparing reform proposals for the development and implementation process of large-scale construction projects. In its report (BMVI 2015), the commission calls for a new culture in such projects. The topic has also attracted interest in the scientific community. Flyvbjerg *et al.* (2003), for example, investigated the frequency and magnitude of cost overruns of transportation infrastructure projects. They conclude that cost escalation is the rule rather than the exception, that it is a global phenomenon and that the situation has not improved over the last 70 years. Iyer and Jha (2005) investigated factors affecting cost performance of Indian construction projects. They argue that coordination among project participants is the most significant factor with respect to positive influence on cost performance. According to the analysis of Sambivasan and Soon (2007), who investigated causes for time overruns in Malaysian construction projects, improper planning, poor site management, and inadequate contractor experience are the top three causes for delays.

The relevance of MEP engineering has not been studied in this context so far. For this reason, a survey was conducted in Austria in order to assess the importance of MEP in complex construction projects such as airports, hospitals, laboratory buildings etc. and to investigate causes of project disruptions associated with MEP. The results are summarized in Monsberger and Fruhwirth (2018). The aim of this study is to help managers and decision makers in large-scale building projects, who are often not MEP specialists, to recognize and assess risks associated with MEP.

2 METHODOLOGY

A standardized questionnaire containing 57 questions was compiled and sent to 1527 Austrian experts, who can be assigned to one of the following stakeholder groups: clients, architects, consultants (MEP), consultants (others), contractors (MEP), contractors (construction), facility managers and others. Stakeholder group specific evaluations were made for all questions. The questionnaire was implemented using the online tool LimeSurvey, and the software IBM SPSS Statistics version 25 was used for evaluation. The answers were analyzed by means of descriptive statistics and cross tabulation. In addition, the Kruskal-Wallis test (H test) and the SPSS built-in Dunn-Bonferroni post-hoc test were used in case of ordinal questions to evaluate differences between the stakeholder groups. Overall, 515 experts responded and 365 completed the questionnaire in full. The number of respondents is listed for each result in total and per stakeholder group. The figures and tables presented in the following are adapted from Monsberger and Fruhwirth (2018).

3 SELECTED RESULTS

Figure 1 shows an assessment of works with respect to project disruptions. About 77% of the respondents identify "MEP installations" as a domain that frequently causes disruptions in complex construction projects, followed by "interior work" (excluding MEP installations) with a score of 53%. The number of respondents considering "structural work", "earthwork," and "facade assembly" as a frequent cause of disruption is substantially smaller.

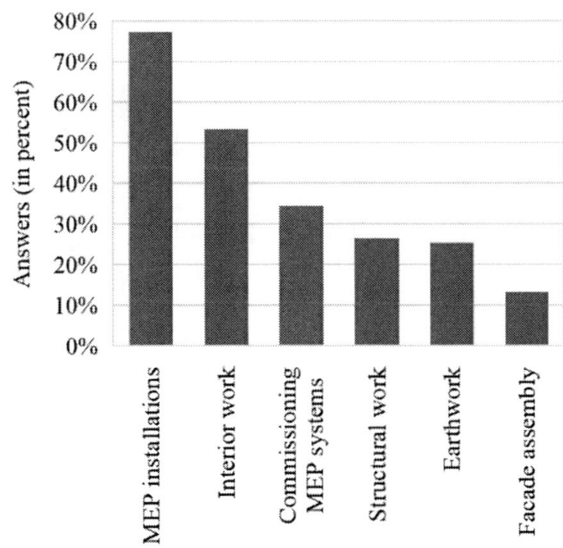

Figure 1. Works that frequently cause disruptions in complex construction projects (n=378).

Table 1 shows the stakeholder-specific result. All stakeholder groups except "MEP consultants" and "MEP contractors" consider "MEP installations" as a frequent source of project disruptions. It seems comprehensible that these groups do not see their own profession as a predominant disturbing factor in construction projects. However, even in these two groups "MEP installations" are on the second position. In case of "MEP contractors," the difference between "interior work" (73.7%) and "MEP installations" (71.1%) is as small as 2.6 percentage points.

Table 1. Works that frequently cause disruptions in complex construction projects – stakeholder-specific analysis.

	Clients	Architects	Consultants (MEP)	Consultants (others)	Contractors (MEP)	Contractors (construction)	Facility managers	Others
	n=82	n=32	n=84	n=40	n=38	n=45	n=34	n=23
MEP installations	79.3%	87.5%	56.0%	85.0%	71.1%	91.1%	85.3%	91.3%
Interior work	58.5%	28.1%	72.6%	50.0%	73.7%	31.1%	35.3%	43.5%
Commissioning of MEP systems	39.0%	43.8%	20.2%	47.5%	23.7%	33.3%	50.0%	30.4%
Structural work	20.7%	15.6%	36.9%	27.5%	31.6%	22.2%	17.6%	34.8%
Earthwork	30.5%	28.1%	19.0%	25.0%	7.9%	44.4%	20.6%	26.1%
Facade assembly	13.4%	18.8%	7.1%	17.5%	15.8%	15.6%	8.8%	17.4%

Moreover, participants were asked to assess MEP in terms of the client's risk for additional costs (Figure 2). The respondents consider the risk for additional costs caused by MEP comparatively high, with the majority (42.5%) of respondents rating it with 7 or 8 on a 10-point scale.

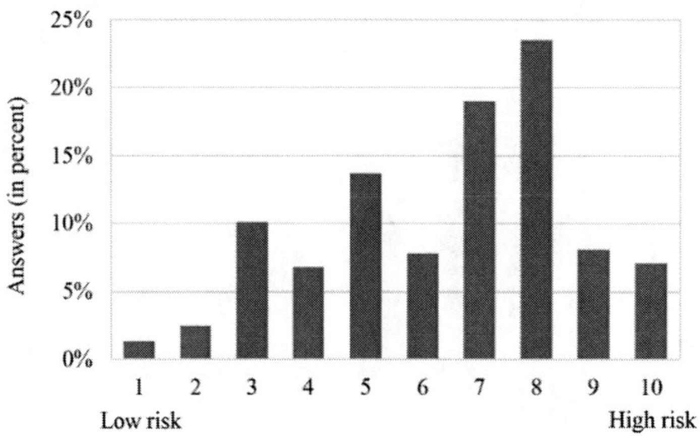

Figure 2. Assessment of the client's risk for additional costs caused by MEP (n=395).

The stakeholder-specific analysis in Table 2 reveals that the assessment of the stakeholder group 'MEP consultants' differs from the other groups. A Kruskal-Wallis test (H=47.502; p=0.000) shows that this difference is statistically significant. Pairwise comparison of the stakeholder groups via a Dunn-Bonferroni post-hoc test confirms that the answer pattern of the group "MEP consultants" differs from all other stakeholder groups except "facility managers" with statistical significance. For the latter the calculated p-value of the post-hoc test is $p=0.384$ and hence above the pre-defined threshold of 0.05.

Table 2. Assessment of the client's risk for additional costs caused by MEP - stakeholder-specific analysis.

	Clients	**Architects**	**Consultants (MEP)**	**Consultants (others)**	**Contractors (MEP)**	**Contractors (construction)**	**Facility managers**	**Others**
Scale (clustered)	n=88	n=33	n=86	n=41	n=42	n=46	n=35	n=24
1 - 3	11.4%	12.1%	31.4%	2.4%	9.5%	2.2%	14.3%	12.5%
4 - 7	50.0%	33.3%	53.5%	56.1%	35.7%	43.5%	51.4%	41.7%
8 - 10	38.6%	54.5%	15.1%	41.5%	54.8%	54.3%	34.3%	45.8%

A further goal of the survey was to identify causes of project disruptions and additional costs associated with MEP. Figure 3 shows an assessment of nine different causes. Almost 70% of the respondents consider "incomplete planning when call for tender is released" as the main cause of additional costs caused by MEP, followed by "late decision making" (64.5%) and "lack of coordination between trades" (60.3%). According to this result, the factors "poor contracts" (4.7%), "regulations" (11.2%), and "lack of cooperation between contractor and subcontractors" (18.3%) are of minor significance.

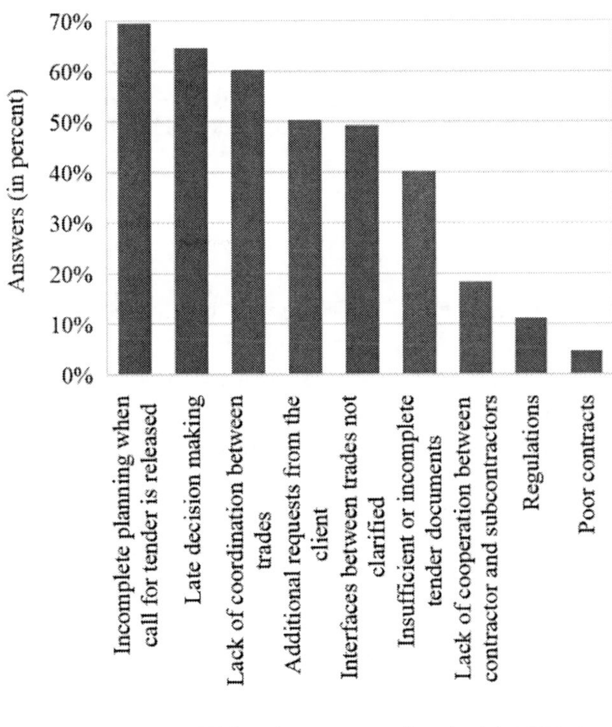

Figure 3. Causes of additional costs associated with MEP (n=383).

Table 3. Causes of additional costs associated with MEP – stakeholder-specific analysis.

	Clients	Architects	Consultants (MEP)	Consultants (others)	Contractors (MEP)	Contractors (construction)	Facility managers	Others
	n=84	n=31	n=86	n=40	n=39	n=45	n=34	n=24
Incomplete planning when call for tender is released	73.8%	64.5%	48.8%	77.5%	84.6%	88.9%	67.6%	62.5%
Late decision making	45.2%	64.5%	81.4%	65.0%	76.9%	64.4%	55.9%	62.5%
Lack of coordination between trades	73.8%	61.3%	47.7%	62.5%	53.8%	44.4%	70.6%	79.2%
Additional requests from the client	33.3%	51.6%	76.7%	50.0%	53.8%	46.7%	35.3%	37.5%
Interfaces between trades not clarified	54.8%	45.2%	37.2%	60.0%	46.2%	55.6%	52.9%	50.0%
Insufficient or incomplete tender documents	45.2%	29.0%	23.3%	30.0%	59.0%	57.8%	35.3%	58.3%
Lack of cooperation between contractor and subcontractor	19.0%	12.9%	27.9%	17.5%	5.1%	8.9%	29.4%	12.5%
Regulations	15.5%	6.5%	10.5%	12.5%	7.7%	8.9%	14.7%	8.3%
Poor contracts	2.4%	6.5%	10.5%	0.0%	5.1%	2.2%	5.9%	0.0%

The stakeholder evaluation gives an inhomogeneous picture. MEP contractors consider "late decision making" as the main reason for additional costs. "Facility managers:" and "others" see "lack of coordination between trades" as the main cause. All other stakeholder groups regard "incomplete planning when call for tender is released" as the main reason for additional costs. It should be noticed that "clients" and "architects" consider "lack of coordination between trades" and "late decision making" equally important with "incomplete planning when call for tender is released", respectively.

4 DISCUSSION

Figure 1 shows that the vast majority of respondents consider MEP as a domain that frequently causes disruptions in complex construction projects. This indicates that the field of MEP is more sensitive to project disruptions than other areas. This reasoning is supported by other results in Monsberger and Fruhwirth (2018), which show, for example, that MEP is also an area with a high potential for conflicts. Furthermore, the risk of additional costs caused by MEP is considered to be relatively high (Figure 2). The stakeholder-specific analysis in Table 2 shows that the risk assessment by the group 'MEP consultants' substantially differs from the other groups. This picture is also partially reflected in the result in Table 1, in which 'MEP consultants' chose 'MEP installations' noticeable less often than the other groups. This pattern can also be observed in other results in Monsberger and Fruhwirth (2018). This finding is particularly relevant for clients. They should take into account that consultants might downplay project risks related to MEP. Clients should therefore be aware of the importance of MEP in complex construction projects themselves and ensure a thorough project integration and coordination of MEP.

The result shown in Figure 3 and further results in Monsberger and Fruhwirth (2018) underline the importance of planning, decision-making, and coordination. These factors are success criteria for construction projects in general. However, they are particularly important for the MEP sector. The scope and complexity of MEP systems has increased rapidly over the past decades and will continue to increase (e.g., due to the digitalization of system components and developments such as the Internet of Things). The design and implementation of MEP systems has thus become a highly multidisciplinary and interactive task with numerous interfaces between different project domains requiring increased efforts in terms of planning and coordination. Underestimation of this complexity by decision makers and project managers is a substantial source of conflicts and disturbances resulting in time and cost overruns.

5 SUMMARY AND CONCLUSIONS

This paper discusses the significance and impact of MEP engineering in complex construction projects and identifies causes of additional costs caused by MEP. The results derive from a stakeholder survey in Austria with more than 365 participants from eight different stakeholder groups. The findings are useful for project managers and decision makers, as they can use them to make their own assessment of the relevance of MEP in a project and the risks involved. The stakeholder-specific analysis gives insight into how different project participants assess risks associated with MEP and corresponding factors causing cost overruns. Such analyses can help project managers to identify and understand differing viewpoints of stakeholder groups and to take measures to create consistent views between the client, consultants, and contractors, which is vital for project success. According to the results, project managers and decision makers should recognize that MEP engineering has become a success critical factor in complex building projects and ensure that adequate expertise and resources are available in both, management and technical project domains across all stakeholder groups. In addition, they should give MEP planning and coordination a high priority and make sure that decisions regarding MEP are adequately taken. The presented results particularly apply to the situation in Austria. They may also apply to countries where the construction sector is comparable to that in Austria (e.g., Germany). However, the increasing complexity and relevance of MEP in construction projects must be taken into account whenever modern buildings are built today.

References

BMVI, Reformkommission Bau von Großprojekten: Komplexität beherrschen - kostengerecht, termintreu und effizient, Bundesministerium für Verkehr und digitale Infrastruktur, 2015. Retrieved from https://www.bmvi.de/SharedDocs/DE/Publikationen/G/reformkommission-bau-grossprojekte-endbericht.pdf?__blob=publicationFile on September 21, 2018.

Flyvbjerg, B., Skamaris Holm M. K., and Buhl, S. L., How Common and How Large are Cost Overruns in Transport Infrastructure Projects?, *Transport Reviews,* Taylor and Francis Group, 23(1), 71-88, 2003.

Iyer, K. C., and Jha, K. N., Factors Affecting Cost Performance, *International Journal of Project Management,* Elsevier, 23(4), 283-295, 2005.

Kostka, G., and Fiedler, J., *Large Infrastructure Projects in Germany: Between Ambition and Realities*, Palgrave Macmillan, 2016.

Monsberger, M., and Fruhwirth, M., *Die Gebäudetechnik im österreichischen Bauprozess: Eine Studie über Herausforderungen in komplexen Hochbauprojekten aus Sicht unterschiedlicher Stakeholder am Bau,* Verlag der Technischen Universität Graz, Graz, 2018. Retreived from https://dx.doi.org/10.3217/978-3-85125-604-8 on September 21, 2018.

Sambivasan, M., and Soon, Y. W., Causes and Effects of Delays in Malaysian Construction Industry, *International Journal of Project Management*, Elsevier, 25(5), 517-526, 2007.

Interdependence between Structural Engineering and Construction Management
Edited by Ozevin, D., Ataei, H., Modares, M., Gurgun, A., Yazdani, S., and Singh, A.

ISBN: 978-0-9960437-6-2

PROJECT MANAGERS' RESPONSIBILITIES IN MULTIDISCIPLINARY COLLABORATION PROJECTS: A CASE STUDY IN EDUCATION

SILVIA MAZZETTO

School of Architecture and Interior Design, Lebanese American University, Byblos, Lebanon

The success of a project is not only built upon the performance of the team involved but also on the contribution of the project manager and the other participants such as the client, stakeholders, consultant, suppliers, and authorizing bodies. In particular, the project managers' contribution in coordinating and leading the participants is fundamental to the project success. The research hypothesis addressed in this paper is to investigate how a multidisciplinary collaboration experience between two departments, using a Work-Based Learning (WBL) method, can improve students' training while handling responsibilities in project management. This paper presents a practical approach to the teaching of Project Management, based on multidisciplinary collaboration protocol developed by the College of Engineering at Qatar University. Collaboration between two Project Management courses was repeated for two years, involving students of both disciplines in a joint assignment (a collaborative project). During the experience, the project managers' responsibilities were analyzed and evaluated by the students. Outcomes of the study show that because there is a close correspondence between project manager's skills, interpersonal action, and readiness to take responsibility, the success of a project is strongly influenced by the capabilities of its leader. The paper meditates on the contribution of the multidisciplinary experience in developing students' capabilities and responsibilities when working on a collaborative project, improving entrepreneurship's performance and building new ways of collaboration in higher education.

Keywords: Leadership, Practice experience, Entrepreneurship, Practical approach, Teamwork.

1 INTRODUCTION

This paper describes a practical approach for assessing the responsibilities of the project managers during a multidisciplinary collaboration experience that was applied in the Department of Architecture and Urban Planning (AUP) and the Department of Industrial and Systems Engineering (ISE) at the College of Engineering at Qatar University. The project brought together students from the Construction and Project Management course (ARCT 530, AUP) and the Project Engineering course (IENG 481, ISE), mixing them into working groups (Mazzetto 2016a).

The fundamental aspect of the multidisciplinary approach was the organization and management of a shared project, and the primary intention was to simulate a real project and enable the students to understand the dynamics of group work and experiment with leadership techniques by applying the theoretical knowledge they learned on the course.

A first multidisciplinary collaboration project started in the Fall 2015 semester and was repeated in the following Fall 2016 semester, with a total of 183 students, divided into 28 multidisciplinary groups.

The primary research objective was to simulate a real-life scenario that helped students to address the roles and problems usually arising in professional projects, such as the need for collaboration and leadership responsibilities in multidisciplinary teams.

2 LITERATURE REVIEW

In the literature, it is well known that it is crucial to understand the importance of the project manager in the achievement of construction projects (Cleland 1995, Duncan 1993, Chileshe and Ali 2009), due to the fact that the manager's duties and responsibilities have an immediate impact on the completion of such projects. Usually, the project manager is one of the critical people that lead and guide projects in the right direction and has three clearly defined managerial responsibilities (Lientz and Rea 2001, Kerzner 1998) such as resource manager, control manager, and coordinator. Additionally, the project manager is responsible for providing the achievement of group and organizational objectives and targets.

In compliance with the European Commission contributions to reinforce Work-Based Learning (WBL) elements in Vocational Education and Training (VET) (European Union EU's Charter of Fundamental Rights 2000), there are many WBL models have been developed in Europe to enhance the students practical experience and integrating into a school-based program. The Work-Based Learning approach, used in the present research, provides students with "real-work" experiences in which they can apply their knowledge and expertise to enhance their employability.

The WBL approach of the proposed multidisciplinary collaboration was integrated in the Qatar University-based program and the aim was to create "real project" work environments, establish contacts and cooperation with real clients, develop students' entrepreneurship competencies and skills while working in multidisciplinary teams (Mazzetto 2016b, Mazzetto 2017) and focusing on work-process orientation, innovation, and creation process.

3 MULTIDISCIPLINARY COLLABORATION METHODOLOGY

In particular, the present paper focuses on the collaborative WBL model adopted by the Department of Architecture and Urban Planning (AUP) and the Department of Industrial and Systems Engineering (ISE). The collaborative WBL model used and developed during the multidisciplinary experience enabled the individual learners to work in collaboration both with other learners and with the tutors (instructors), coordinating, developing and adapting their projects with the needs of private organizations and governmental institutions.

With this model, through a formal protocol, the College of Engineering effectively involves the students from different disciplines to focus on multidisciplinary activity through a joint assignment, guided by the instructors. The groups are requested to work as part of a collaborative team. The results are the collaborative design and delivery which carry academic credits.

The collaborative model's primary aim is to build up a discussion between the students' learning experiences in the multidisciplinary collaboration project and to develop the competence of meditative practice. The collected data and observations enabled the implementation of the tailor-made WBK model.

The advantage of the adopted approach of WBL is that it permits students to get an empirical assessment of the project management responsibilities in the workplace, during the coordinated multidisciplinary classes and thereby meeting the standard requirements.

In the end, the contributions of individual students concerning responsibility, either as members of groups or as project managers, were described in a final activity report in which the progress of the work was shared, to facilitate the development of multidisciplinary interaction and permit assessment by tutors.

3.1 Data Collection and Final Reports: Tools for Leadership Assessment

The collected data were divided into sections as part of the final report requirements: (1) general data about the students and the projects carried out within the groups (2) an interaction diagram for evaluating the interpersonal activities of the project managers (3) a responsibility matrix for evaluating the responsibilities of the project managers.

The multidisciplinary projects enabled a significant amount of other data to be collected relating to project management techniques and tools, and the collaboration between the participants in the groups. The results presented here are only related to evaluating the participant responsibilities while working into multidisciplinary university collaboration.

3.2 Collaborative Interaction Diagram

The data presented by the students in the final reports showed the degree of interaction between the participants in the multidisciplinary collaborative exercise, and the value of the project managers- responsibilities, by creating a collaborative interaction diagram.

The students evaluated the interaction between the participants by establishing which values and roles had been most involved. The data in Figure 1 shows the averages of the results obtained across the two years of the multidisciplinary collaboration exercise divided into high, medium, and low levels of participant interaction.

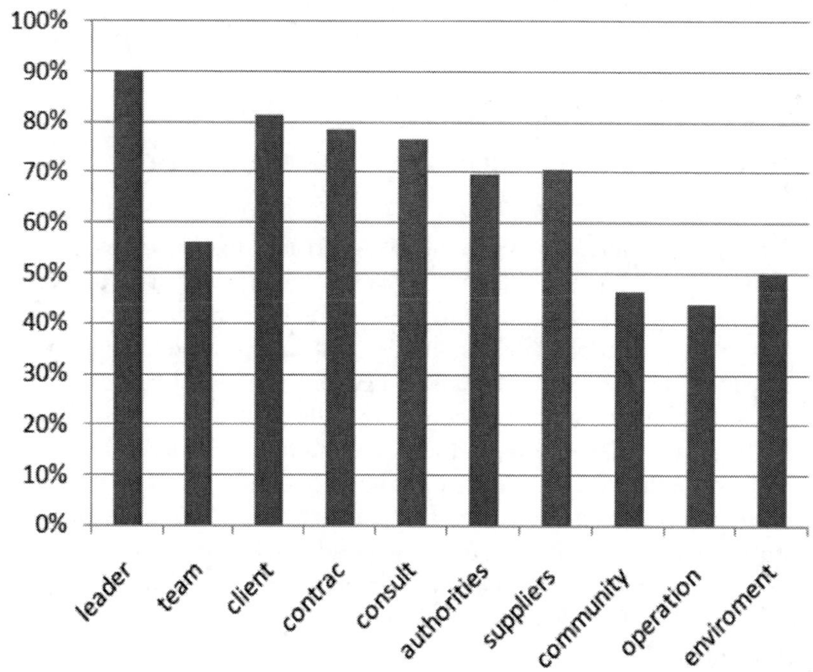

Figure 1. Multidisciplinary collaboration' interaction diagram, average scores (Fall 2015 and Fall 2016). Source: Author.

The highest levels of evaluation for interaction were assigned to the project manager being the most important key person in the multidisciplinary collaborative exercise. However, given the low evaluation results obtained, we deduce that the level of interaction within the multidisciplinary team and the operating team could have been better if greater involvement and collaboration between the members had been encouraged during the various project development stages.

3.3 Project Managers' Responsibility Matrix

The data presented in the third section of the final reports showed the assessed level of responsibility of the project manager, using a responsibility matrix. These responsibilities were divided into the four phases of the project lifecycle: (1) Planning (2) Design (3) Execution (4) Completion.

Table 1 gives the average values obtained for the assessments, across both semesters (Fall 2015 and Fall 2016) of the collaborative projects, showing the values of each member contribution for achieving the project's completion. It is noteworthy from the reached values that the highest level of responsibility was assigned to the project manager, particularly at the initial "project development" stage, followed by the "time scheduling" whereas the project manager's responsibility was scored as low concerning the "purchase orders" responsibility. In fact, such responsibility was usually carried out by independent group of students in the multidisciplinary teams.

Thanks to these evaluations made by the students, the managers' responsibility matrix shows that during the multidisciplinary collaboration, the most responsible phase for the project managers was during the initial phase of the project. Although some of the activities in Table 1 are not the direct responsibility of the project manager, they are usually indirectly influenced by the leaders' skills and responsible approach.

Table 1. Project Managers' responsibility matrix, average values (R= responsibility role, S=Supportive role, I=informative role) (Fall 2015 and Fall 2016). Source: Author.

	PLANNING (%)			DESIGN (%)				EXECUTION (%)						COMPLETION (%)		
	Develop Project	Establish Budget	Time schedule	Project Responsibility	Develop Design	Perform Project	Coordinate Integration	Review Submittal	Develop Checklist	Purchase Orders	Perform Quality	Progress Reports	Hold Meetings	Supervise Corrections	Address Concerns	Final Report
R	84	57	64	48	55	11	14	16	4	0	43	5	21	9	2	13
S	0	11	0	23	39	23	54	23	43	64	9	20	16	55	45	27
I	16	32	36	29	5	66	43	61	54	36	48	75	63	36	54	61

4 RESULTS

Following completion of the multidisciplinary projects, the project manager responsibilities were assessed and compared with the project outcomes showing a high level of correspondence. It can be seen that a low percentage score for the responsibility of project managers corresponds to a similarly low percentage score for the project outcomes, in the same way in which a medium/high percentage score for the responsibility corresponds to a similarly medium/high percentage score for the projects.

To enable assessment of the ability of the project managers, most of the projects required them to take a high level of responsibility during the initial phase of project development, team selection, subdivision, and assignment coordination, followed by overseeing the project planning, risk assessment, budget management, and the quality control of the project, particularly during the completion phase. From these experiences of multidisciplinary collaboration, it emerges that the outcomes of a project are very often directly affected by the manager's responsibilities, both concerning project quality and the completeness of the results obtained.

5 CONCLUSIONS

The research presented in this paper considers the competencies, interactions, responsibilities, and leadership experience that were evaluated as the most important regarding their contribution to the success of many multidisciplinary collaborative projects carried out in the academic experience.

The roles of the different participants in a project are often difficult to control, but the ability and responsibility of the leader, is precisely to know how to create a fair balance between the parties involved, and an adequate level of communication and collaboration.

So far as the possible implications for academic research are concerned, the characteristics highlighted here as the most important for correct responsibility by project managers could be included in specific degree programs, to construct the basics of a reliable method for teaching the project managers of the future.

References

Cleland, D. I., Leadership and the Project-Management Body of Knowledge, *International Journal of Project Management,* 13(2), 83-88, 1995.

Chileshe, N., and Ali, M. A. M., *The Influence of the Project Manager on the Success of the Construction Project,* Sixth International Conference on Construction Project Management, Korea, Oct. 11-14, 2009.

Duncan, W. R., The Process of Project Management, *Project Management Journal,* 24(3), 5-10, 1993.

European Commission, EU's Charter of Fundamental Rights, 2000. Retrieved from https://www.ec.europa.eu/info/aid-development-cooperation-fundamental-rights/your-rights-eu/eu-charter-fundamental-rights_en on October 10, 2017.

Kerzner, H., *Project Management: A Systems Approach to Planning, Scheduling, and Controlling,* Wiley, New York, 1998.

Lientz, P. B., and Rea, P. K., *Project Management for the 21st Century*, Taylor & Francis, Abingdon, United Kingdom, 2001

Mazzetto, S., Practice Experience and Multidisciplinary Collaboration in Project Management: A Case Study, in *Integrated Solutions for Infrastructure Development,* Lau, H. H.; Tang, F. E., Ng, C. K.; and Singh, A. (eds.), Third Australasia and South-East Asia Structural Engineering and Construction Conference ASEA-SEC-3, Kuching, Sarawak, Malaysia, Oct 31-Nov 4, 2016a.

Mazzetto, S., Theory and Practice in Project Management: A Multidisciplinary Approach, in *Interaction between Theory and Practice in Civil Engineering and Construction,* Komurlu, R., Gurgun, A. P., Singh, A.; and Yazdani, S. (eds.), First European and Mediterranean Structural Engineering and Construction Conference, EURO-MED-SEC-1, Istanbul, Turkey. doi: 10.14455/ISEC.res.2016.100. May 24-29, 2016b.

Mazzetto, S., Leadership and Collaboration in Project Management Education: A Case Study, in *Resilient Structures and Sustainable Construction,* Pellicer, E., Adam, J. M., Yepes, V., Singh, A., and Yazdani, S. (eds.), Ninth International Structural Engineering and Construction Conference Valencia, Spain July 24-29, 2017.

Interdependence between Structural Engineering and Construction Management
Edited by Ozevin, D., Ataei, H., Modares, M., Gurgun, A., Yazdani, S., and Singh, A.

ISBN: 978-0-9960437-6-2

GOVERNMENT ROLES IN PUBLIC-PRIVATE PARTNERSHIPS PROCUREMENT OF INFRASTRUCTURES: THEORETICAL APPROACH

RAPHEAL ABIODUN OJELABI, OLABOSIPO ISHOLA FAGBENLE, LEKAN MURITALA AMUSAN, and ADEDEJI OLUSHOLA AFOLABI

Dept of Building Technology, Covenant University, Ota, Nigeria

Social and economic infrastructures provision has been the sole responsibility of the government in the time past. However, due to the geometric demand in human infrastructures needs, the government supply capacity has been constrained. The inability of the government to close the infrastructural gaps is due to the inequality in financial capacity and the financial worth of social and economic infrastructures. Despite the paucity of the fund required for infrastructures provisions in government, the need for social and economic infrastructures cannot wait due to its relevance in nation-building. The innovative approach to meeting the social and economic infrastructures is engineered through the adoption of a Public-Private Partnership (PPP) procurement option. The procurement option is a sure route through which infrastructural provision can be sustained. This paper reviews literature on Public-Private Partnership by buttressing on the expected roles of the government through the Governance theory concept.

Keywords: Innovation, Procurement option, PPP, Developing country.

1 INTRODUCTION

There has been an assertion that governments across the globe are faced with limitations in meeting the demand for infrastructural supply and management (Babatunde *et al.* 2015). United Nations (2009) however revealed that one of the key limitations to the supply of infrastructures in developing economies can be attributed to a funding issue. This warranted the Government from the developing countries of the world to solicit for private sector collaboration in infrastructural provisions through the platform of Public-Private Partnership (Ibem 2010).

The Public-Private Partnership (PPP) infrastructure procurement option has been a trending issue of debate across the globe since its adoption (Ojelabi *et al.* 2017). It has generated mixed reactions from stakeholders as to its adoption as a tool for infrastructure procurement. The controversy surrounding the PPP procurement option was traced to different perceptions of stakeholders on PPP across the globe. Monibot (2010) opined that PPP adoption in infrastructure procurement has enriched some contractors due to the excessive profit made from the execution of public works. Sambrani (2014) also viewed the Public-Private Partnership as a medium through which governments are shifting infrastructure provision responsibility to the private sector. However, Robinson *et al.* (2010) opined that it is a medium through which sluggishness in the provision of infrastructure facilities by the public sector can be managed. Arewa and

Farrell (2011) however asserted that PPP has come to stay because infrastructure provision is gradually overwhelming the government.

The multi-dimensional application of PPP has diversified such that it cannot be capped into a single definition. The World Bank Institute (2012) defined PPP as a long-term contractual relationship between private party and government agency towards public asset or service provision. Zhang and Cheng (2013) also referred to Public-Private Partnership as a contractual agreement in which the private sector is responsible for provision and management of public works and services in a long-term for public sector acquisition. For this study, Public-Private Partnership is defined as a synergy between the public sector (government) and the private sector towards infrastructure provision in which the private sector is saddled with the management responsibility while the public sector is to provide the enabling environment.

2 APPRAISING THE THEORY SUPPORTING PUBLIC-PRIVATE PARTNERSHIP IN INFRASTRUCTURE DELIVERY

The emergence and acceptance of the Public-Private Partnership (PPP) procurement route in infrastructures delivery are not without a root, it stems out from some the existing theories which include new public management theory and governance theory among others.

One of the theories that give credence to the Public-Private Partnership (PPP) procurement is the new public management theory. Gruening (2001) asserts that the theory is based on the reinvention of government. Osborne and Plastrik (2000) viewed New Public Management as a practical management approach to controlling operational challenges in government. Aucoin (1990) reinforced that reinventing government is only feasible under flexible management and customer's oriented style rooted in private sector operational techniques.

Likewise, the governance theory supports the advent of Public-Private Partnership in the procurement of social and economic infrastructures. Montenegro and Bulgacov (2014) explained that governance and government do not mean the same as the former signifies the change in the latter. Pierre (2000) referred to governance as a system that is capable of sustaining coordination and togetherness among diverse actors with different purpose and objectives. Stoker (1998) make some propositions on governance one of which it refers to governance as the set of institutions drawn from actors beyond the government toward solving social and economic issues.

3 PUBLIC-PRIVATE PARTNERSHIP PRACTICES ACROSS THE GLOBE

Public-Private Partnership procurement option is very crucial in infrastructural delivery, as such that developing and developed countries across the globe are engaging in practices that can encourage the attraction of the private investors and as well grantee the procurement acceptance by the people. It is not debatable to state that the developed and developing nations differs in infrastructure strength and financial capacity. These two factors to a larger extent dictate the varying practices between the two regions in Public-Private Partnership from research findings. Winch *et al.* (2012) revealed that developed countries like United Kingdom, France and Japan immediate demand for infrastructure development is not as high as for countries in the developing world as they have reached maturity level in infrastructure development and public welfare services. The researchers revealed that the available infrastructures in the developed world are aging and likewise the population which is likely to impact on tax revenue in the future and increases social services expenditure in medical and health infrastructures. Despite the visible need for infrastructures development and maintenance, the acceptance of PPP for its procurement among the people is not allowed if they are to pay for the services. As such, the government mostly engaged PPP for social projects that it has the capacity to make a unitary charge to the private contractor. Irrespective of the source of financing the PPP projects, value for money appraisal is of uttermost important in the use of PPP in public infrastructures in the developed world.

On the other hand, the need for infrastructures development in the developing world is very high and the financial means of closing the infrastructural gap is not readily available. The countries in this region are mostly dependent on private finance in ensuring equity in their social and economic infrastructure supply. Winch *et al.* (2012) explained the practice in the use of PPP in the procurement of infrastructures through a case study of some selected countries like Indonesia and India from the developing world. The researchers revealed that the governments of Indonesia and India attract private investors by providing financial assistance to limit the private party risk despite the certainty of the users paying for the facility. This is necessary because the cash flow from the users are not sufficient to augment the initial investment due to the political constraint in the charges to the users. The government support grant is called viability gap fund schemes in India and modified BOT schemes in Indonesia. The for value for money assessment is not practiced in this region due to a heavy commitment of the private investors and primary concern for the use of the procurement option is mainly for infrastructural development.

4 GOVERNMENT ROLE IN PUBLIC-PRIVATE PARTNERSHIP PROCUREMENT

PPP sustainability in any country is highly dependent on the parties involved which are mainly the government and the private sector. It is evidenced by PPP's description that it takes two to make PPP functional. However, Namblard (2000), emphasized that the parties (governments and the private sector) are not altogether equal as much premium is placed on one than the other.

Wibowo and Alfen (2015) however, emphasized that the government is inevitable in the implementation of the PPP procurement option. Therefore, the researchers identified government role critical to the development of PPP under some key categories to include; legal, and regulatory provisions, policy framework, public sector capacity, and projects preparation and planning.

The PPP capacity to deliver is dependent on the credible legal and regulatory framework. The legal and regulatory tool is an essential anchor required to keep PPP from drifting from its expectations. Tafesse (2014) further buttressed that the existence of legal and institutional environment is very crucial prior to the adoption of PPP tool. The need for an established legal and regulatory tool in PPP affairs is to ensure that operations of PPP are not totally been dictated to by the government.

Policy development has been the norms for all going-concerns in every area of life. It has been found to be the driving force behind the success and sustainability of business and organization around the world. An organization that lacks policy development is without control. Georghiou *et al.* (2014) identified the role of policy as a tool used in developing capacity and also aids management of an organization to achieve the expected outcome. It is evidenced by the foregoing that policy development is highly critical to the continuous operation of organizations or businesses. Therefore, the sustainability of Public-Private Partnership (PPP) as a form of business is dependent on sound policy development. Srivastav and Shakaya (2012) opined that trust is one of the fundamental requirements for partnership between the Governments and the private investors to thrive under the platform of PPP and such can only be achieved under a sound policy development.

Public-Private Partnership (PPP) procurement system has been lauded as a sustainable and realistic medium of infrastructural procurement by stakeholders around the globe. The laudable compliment of this procurement option will not have been without the contributions of the parties involved. The private and the public sectors have been affirmed to be the bedrock of this procurement system, however, Namblard (2000), asserted that the parties are not on the same pedestal in the discharge of their duties under the procurement system. Irrespective of their varying positions, lack of competency of either of this party can compromise the delivery system. European PPP Expertise Centre (2015) consented that deficiency in skill required for PPP operation can affect the PPP process even under the right framework. It was further stated that the weakness of public sector capacity has been one of the challenges contributing to the failure of the PPP system. Public sector capacity is very crucial to the success of PPP as it dictates the knowledge base and ability of the participants. The foregoing fact has warrants the need to address the human capacity of the public party being one of the key players under the procurements option.

Preparation and planning are the inseparable twins that are crucial to the success of any proposed project. Rui *et al.* (2015) affirmed that preparation and planning are highly essential in managing time delay, cost overrun and quality defects that plague most construction projects. Preparation and planning are drivers of projects successful delivery and are not limited to a procurement option. It encompasses all procurement routes including PPP that is used for the delivery of social and economic projects. Inwa *et al.* (2014) also consented to the need to develop a workable plan for procurements of construction projects due to its uniqueness in terms of complexity compared to other forms of general procurements. Among other forms of construction projects procurements delivery route, PPP has been found to be more demanding in term of management. The need for management of PPP procurement option can be hinged on the complexity of the system as a result of the parties involved and the construction process. World Economic Forum (2016) further linked the complexity of PPP procurements option to the large team and multiple stakeholders which include ministries, banks, investors, construction firms, regulators and users among others that are crucial to the operation of this procurement option. However, management of these multiple stakeholders for effective delivery of PPP projects cannot be achieved without proper preparation and planning of the procurement option. Therefore, proper planning of PPP projects are dependent on some key determinants as revealed by World Economic Forum (2016) and it includes thorough project preparation process, bankable feasibility study, balanced risk allocation, and regulation and ensuring a conducive enabling environment is guaranteed, among others.

5 CONCLUSIONS

This paper reviewed the government role in the implementation of PPP procurement in the delivery of social and economic infrastructures. The findings from the study revealed that the practices in the implementation of PPP in the developed and developing countries differ. Governments from the developed world only embark on PPP projects that have the capacity to finance due to their maturity in infrastructural strength and the resistance of the people to pay for tariffs. On the other hand, governments in the developing countries pushing for PPP implementation are very many; however, private investors are very careful to invest in the regions due to the need for the certainty of social and economic infrastructure, the risk in doing business in the regions and the uncertainty of governments. The study, therefore, revealed that the attraction of the private investors to engage in PPP is dependent on the roles of the governments, which include managing the inherent risk, encouraging private investor, enhancing public sector capacity in PPP, establishing sound legal and regulatory framework for PPPs, developing sound policies to promoting PPP and ensuring adequate preparation and planning.

Acknowledgments

The researchers appreciate the publication support fund given by Covenant University towards the publication of this article.

References

Arewa, A., and Farrell P., *The Private Finance Initiative: Why Value for Money Remain Elusive*, CIB.TG72/ARCOM Doctoral research workshop, University of Central Lancashire, United Kingdom, 2011.

Aucoin, P., Administrative Reform in Public Management: Paradigms, Principles, Paradoxes and Pendulums. Governance, *International Journal of Policy and Administration*, 3, 115-137, 1990.

Babatunde, S. O., Perera, S., Zhou, L., and Udeaja, C., Barriers to Public Private Partnership projects in Developing Countries: A Case of Nigeria, *Engineering, Construction and Architectural Management*, 22(6), 669-691, 2015.

European PPP Expertise Centre, PPP Motivations and Challenges for the Public Sector, 2015. Retrieved from www.eib.org/epec/resources/publications/epec_PPP_motivations_and_challenges_en in July, 2018.

Georghiou, L., Elder J., Uyarra, E and Yeow J., Policy Instruments for Public Procurement of Innovation: Choice, Design and Assessment, *Technological Forecasting and Social Change*, 86, 1-12, 2014.

Gruening, G., Origin and Theoretical Basis of New Public Management, *International Public Management Journal*, 4(1), 1-25, 2001.

Monibot, G., The UK'S Odious Debts: Many PFI Deals Are Undemocratic, It's Time We Stop Honoring Them, 2010. Retrieved from http://www.monbiot.com/2010/11/22/the-uks-odious-debts/ in July, 2018.

Montenegro, L. M., and Bulgacov, S., Reflections on Actor-Network Theory, Governance Networks, and Strategic Outcomes, *BAR-Brazilian Administration Review*, 11(1), 107-124, 2014.

Namblard, C., *In Favour of Pragmatic Approach Towards Public-Private Partnership*, In Perrot, J.Y and Chatelus G. Financing of Major Infrastructure and Public Service Projects: Public Private Partnership, Presses Points, 2000.

Ibem, E., An Assessment of the Role of Government Agencies in Public-Private Partnerships in Housing Delivery in Nigeria, *Journal of Construction in Developing Countries*, 15(2), 23-48, 2010.

Inwa, I. I., Gthae, W., and Stephen, D., Application of Project Planning Techniques in Construction Procurement: The Case of Nigerian Indigenous Contractors, *International Journal of Economic Research and Investment*, 5(1), 31-47, 2014.

Ojelabi, R. A., Oyeyipo, O., and Afolabi, A. O., Built Environment Professionals' Perception of the Effectiveness of Building Control Measures in Lagos State, *Journal of Construction In Developing Countries*, 22(1), 41-54, 2017.

Osborne, D, and Plastrik, P., *The Reinventors Field Book: Tools for Transforming Your Government*, San Francisco, CA, 2000.

Pierre, J., Introduction: *Understanding Governance in Debating Governance. Authority Steering and Democracy*, Oxford University Press, 1-12, 2000.

Robinson, H., Carrillo, P., Anumba, C. J., and Patel, M., *Governance & Knowledge Management for Public-Private Partnerships*, Wiley-Blackwell, Oxford, 2010.

Rui, L. M., Ismail, S., and Hussaini, M., Professional Development of Project Management for Contractor in The Construction Project. A Review, *Procedia-Social and Behavioral Sciences*, 174, 2940-2945, 2015.

Sambrani, V. N., PPP for Asia and African Perspective Towards Infrastructure Development: A Case Study of Greenfield Bangalore International Airport, India, *Social and Behavioral Sciences*, 157, 285-295, 2014.

Srivastav, H., and Shakya, P. L., *Policy Framework for Public-Private Partnership (PPP) Projects in Nepal*, 3rd Ministerial Meeting on PPP for Infrastructure Development, Tehran, Iran, May 11-14, 2012.

Stoker, G., Governance as Theory: Five Prepositions. *International Social Science Journal*, 50(55), 17-28, 1998.

Tafesse, T., Public-Private Partnership in Development: Lesson in Devising Legal and Institutional Framework from South Korea, *Public Policy and Administration Research*, 3(4)50-57, 2014.

United Nations, What is Good Governance? 2009. Retrieved from www. gdrc.org/u-gov/escap.governance.html in June 2018.

Wibowo, A and Alfen, H. W. Government-Led Critical Success Factors in PPP Infrastructure Development, *Built Environment Project and Asset Management,* 5(1) 121-134, 2015.

Winch, G.M., Onishi, M., and Schmidt, S., *Taking Stock of PPP and PFI Around the World,* Certified Accountant Education Trust, London, 2012.

World Bank Institute, Public-Private Partnerships, *Reference Guide Version 1.0. International for Reconstruction and Development Association*, Washington D.C, USA, 2012.

World Economic Forum. *The Global Competitiveness Report, 2016-2017*, 2016. Retrieved from www.weforum.org/gcr in August 2018.

Zhang, X., and Cheng S. A Systematic Framework for Infrastructure Development Through Public-Private Partnerships, *International Association of Traffic and Safety Sciences (IATSS) Research,* 36, 88-97, 2013.

Interdependence between Structural Engineering and Construction Management
Edited by Ozevin, D., Ataei, H., Modares, M., Gurgun, A., Yazdani, S., and Singh, A.

ISBN: 978-0-9960437-6-2

QUALITY MANAGEMENT IN CONSTRUCTION PROJECT: EMPIRICAL STUDY OF COVENANT UNIVERSITY SPORTS COMPLEX

SOLOMON OYEBISI[1], ANTHONY EDE[1], FESTUS OLUTOGE[2], BEN NGENE[1], OLATOKUNBO OFUYATAN[1], and JOHN OLUWAFEMI[1]

[1]*Dept of Civil Engineering, Covenant University, Ota, Nigeria*
[2]*Dept of Civil Engineering, University of Ibadan, Ibadan, Nigeria*

The acceptable level of quality in construction projects has long been a problem. Significant resources are wasted on construction projects because of inefficient or non-existent quality management procedures. Therefore, this study evaluates the quality management and the quality grading practices in the construction project with emphasis on the Covenant University Sports Complex. The quantitative research strategy and questionnaire survey were used as the main data collection instrument for soliciting information from the project participants in order to determine the best key for quality management practice. Moreover, an ultrasonic pulse velocity tester was engaged to assess and evaluate the conditions of concrete structural members. It was found out that management commitment to quality is the most key for effective quality management practices in the construction project. In addition, the ultrasonic pulse velocity (UPV) test carried out on the randomly selected structural members varied in the range 4.2km/sec to 4.6km/sec. From this, it is inferred that the quality grading and homogeneity of concrete in the tested reinforced concrete columns, beams and slabs fall in the range of "good to excellent concrete". The findings would assist project participants in implementing an efficient quality grading and management while executing construction projects.

Keywords: Ultrasonic pulse velocity, Reinforced concrete, Questionnaire, Client, Consultant, Contractor.

1 INTRODUCTION

A construction project according to Ashokkumar (2014) goes through different phases such as conceptual planning, feasibility study, design, procurement, construction, acceptance, operation, and maintenance in its life span; and one of the important factors that determines the success of construction projects is quality. The effective management of construction project in all phases of its life cycle is linked to quality. Tan and Abdul-Rahman (2005) opined from the perspective of a construction firm that following the specified standard and maintaining the quality of construction works for the accomplishment of client's satisfaction resulting in the firm's long-term competitiveness and enduring business growth is referred to as "Quality Management in Construction Projects."

This predicates that faulty construction takes place if the project's quality is not in conformity to the specified standards and specifications. The wastages due to corrective action against shoddy jobs, lately discovered during construction and maintenance are found to be 6-15% and

5% of construction cost respectively (Akinci and Boukamp n.d. 2004, Fagbenle *et al.* 2009). In addition to this, Ogunde *et al.* (2017) stated that quality control, quality assurance, quality improvement, and quality standards are concepts encompassed and embedded in the term "quality management" in the construction firm.

2 LITERATURE REVIEW

2.1 Quality Management Systems in Construction Industry

Schlickman (2003) generally defines quality as "the degree to which a set of inherent characteristics fulfill requirements." This connotes that quality seems to be achieved in the construction firm whenever the client's needs and project participants are accomplished. Therefore, implementing the practice of quality management system in a construction company as a strategic management instrument requires the understanding of quality concepts in order to derive benefits from it. The principles of Quality Management Systems regarding quality include the establishment of policies and objectives by construction company to manage resources, the delegation of responsibilities, roles and authority to personnel, and the development of a company's structure among the personnel. The International Standard Organization has specifically incorporated the following eight quality management principles namely customer focus, leadership, involvement of people; process approach, system approach to management, continual improvement, factual approach to decision making, and mutually beneficial supplier relationships into its requirements (Schlickman 2003).

2.2 Application of Quality Management in Construction Project

The implementing concepts of quality management in construction project require the identification of quality standards (quality planning), the evaluation of overall project performance (quality assurance), and the monitoring of specific project results (quality control) (Chandrasena 2010). Quality assurance (QA) and quality control (QC) are importantly and mostly used in construction project. Tender documents, specifications, and working drawings are based on the quality control procedure in construction projects.

3 METHODOLOGY

3.1 Principle of Approach Choice

The study was posited in the positivist paradigm, which recognized and produced quantitative data that fit the survey approach, objectively gathered data as a small sample, and made an independent observation.

3.2 Research Approach

The study used the deductive research design as a result of substantial data and adopted the use of quantitative methods for analysis. The theory was tested by empirical observation.

3.3 Research Strategy

The study used quantitative strategy and questionnaires as the main data collection techniques.

3.4 Data Collection Method

Desk survey and field survey were adopted as approaches for collecting data in this study.

3.5 Data Collection Instruments

3.5.1 *Questionnaire*

The researchers used a questionnaire adopted from (Schlickman 2003, Rumane 2011) and researchers' own experience in engineering field and construction management to achieve the study objectives. Generally, the questionnaire was developed by the researchers and appraised by specialists in academia and construction project management to gain a comprehensively statistical information from project engineers (client representatives), contractors, and consultants. The questionnaire was composed of two sections: Section one was focused on the respondents' profile such as education, position, and experience, while section two sought to determine whether the contractor is committed to quality implementation in their deliveries. The grading point was evaluated according to five-point Likert scale Level of Quality from excellent (5) to poor (1).

3.5.2 *Administration of questionnaire*

Pilot tests were subsequently conducted on the questionnaire for the contractor, the consultant, and the project engineers (client representatives) for the identification and the removal of potential obscurity to enhance the reliability and validity of the questionnaire. The questionnaires were directly administered on one-on-one basis.

3.6 Data Analysis

The data was analysed using MINITAB 17 software computer program. Reliability of internal consistencies was studied using Multivariate – Item Analysis while One-Way Analysis of Variance (ANOVA) was used to determine whether the means of two or more groups differ. Descriptive Analysis was used to compare relationships between variables.

3.7 Reliability of Internal Consistency

The data comprised 16 responses from experienced personnel within the construction company, with a 100% valid response rate. Cronbach's alpha value was used to determine the internal consistency of the construction of the section. If the value is closer to 1, the internal consistency is higher. A value greater than 0.80 was considered to be good for significant consistency; value between 0.70 and .80 was acceptable for significant consistency while value lesser than 0.70 was considered poor (Sekaran and Bougie 2009). As shown in Table 1, the internal consistency of the measures adopted in this study was therefore considered to be acceptable and sufficiently reliable for the measurement of quality management in construction project.

Table 1. The reliability of internal consistency.

Section	Variable/Measure	No. of Question	Cronbach's alpha
2	Quality implementation level in the firms	20	0.8715

4 DATA ANALYSIS AND DISCUSSIONS

4.1 Respondents' Demographic Profile

The test proportion (TP) according to Goodness-of-Fit Test (One variable) of the respondent details in each class from Table 2 was 0.25, and P-value was 1.00 ($> \alpha = 0.05$) with no contribution to chi-square. This showed that respondent details had a 100% valid response rate and both observed values and expected values was equally fit in proportion.

Table 2. Respondents' demographic profile.

Respondent Information	Class	Frequency (n=16)	Percent (%)	Cum. Percent (%)
Education Level	OND	1	6.3	6.3
	HND	4	25.0	31.3
	Bachelor's Degree	9	56.3	87.6
	Master's Degree	2	12.5	100
	PhD	0	0	100
Education Background	Architecture	2	12.5	12.5
	Civil Engineering	9	56.3	68.7
	Project Management	3	18.8	87.5
	Quantity Surveying	1	6.3	93.8
	Other	1	6.3	100
Management Position	Top level	6	37.5	37.5
	Middle level	7	43.8	81.3
	Low level	3	18.8	100
Experience	1-5 years	2	12.5	12.5
	6-10 years	3	18.8	31.3
	11-15 years	6	37.5	68.8
	16 years & above	5	31.3	100

Source: Field Survey 2018

4.2 Quality Implementation Level

The quality assurance implementation level of the construction firm was adopted from (Schlickman 2003) and evaluated in this part of the study. The respondents were given 20 questions of quality indices. The rank result of One-Way ANOVA is presented in Table 3.

The result of the level of quality implementation as presented in Table 3 indicated that the means of the quality indices are not significantly different. Also, the analysis of variance table statistics gave Adj SS, Adj MS, F-Value and P-Value as 10.31, 0.5426, 1.06 and 0.392 respectively. Since p-value (**0.392**) is larger than the α-level selected (**0.05**), it shows that the means are not significantly different. According to Table 3, the respondents ranked "Management Commitment to Quality" as the most important indices in achieving the quality management. This is in conformity with (Project Management Institute 2013) that keeping an accepted and consistent level of quality in the project and top management commitment to examine the needed experiment is very essential. "Client Satisfaction" was ranked 2nd. This confirms (Project Management Institute 2013)'s finding that the condition of an effective quality management system implementation is to ascertain that work is done according to specifications throughout the project life cycle and ensures that clients' satisfactions are met. Respondents' ranks of 4th, 5th, 6th 7th and 8th also satisfied (Project Management Institute 2013). "Plan, Schedule and Assign Independent Personnel to Investigate Internal Quality Audits" was ranked as the least indices for implementing quality assurance by the respondents.

Table 3. The company quality implementation Level (N =16).

S/N	Quality Indices	Mean	StDev	95% CI	Rank
Q1	Evaluate, select and appropriately exercise control over the capabilities for quality delivery by subcontractors	4.500	0.632	(4.148, 4.852)	3
Q2	Control of quality records	4.375	0.500	(4.023, 4.727)	5
Q3	Verification control, storage, and maintenance of customer-supplied product	4.188	0.655	(3.836, 4.539)	8
Q4	Adherence to certified quality specifications such as the Nigeria FMW&H, Revised 2007; ISO 9000	4.500	0.673	(4.148, 4.852)	3
Q5	Plan design activities by identifying, reviewing, verifying, and validating design input	4.313	0.704	(3.961, 4.664)	6
Q6	Inspection control, measuring and testing of equipment prior to use and after use	4.375	0.500	(4.023, 4.727)	5
Q7	Review, approve and control purchasing documents prior to release	4.250	0.856	(3.898, 4.602)	7
Q8	Management commitment to quality	4.625	0.500	(4.273, 4.977)	1
Q9	Training for team members on quality	4.438	0.727	(4.086, 4.789)	4
Q10	Review variation order prior to acceptance and transmit corrected requirements to required functions	4.000	1.155	(3.648, 4.352)	11
Q11	Verification of subcontracted work by the client representative is allowed at the premises where specified in contract document	4.500	0.673	(4.148, 4.852)	3
Q12	Conduct receiving inspection and testing of incoming materials and components in accordance with quality plan	4.438	0.727	(4.086, 4.789)	4
Q13	Corrective and preventive action	4.313	0.704	(3.961, 4.664)	6
Q14	Client satisfaction	4.563	0.629	(4.211, 4.914)	2
Q15	Handling, storage, preservation and delivery	4.125	0.719	(3.773, 4.477)	9
Q16	Plan, schedule and assign independent personnel to investigate internal quality audits	4.000	1.033	(3.648, 4.352)	11
Q17	Conformity or nonconformity of work on the basis of inspection and performance	4.375	0.500	(4.023, 4.727)	5
Q18	Definition, establishment and documentation of quality planning on a project or contract, and also preparation of a quality plan	4.500	0.673	(4.148, 4.852)	3
Q19	Identify, implement and control the application of statistical techniques in quality control	4.375	0.500	(4.023, 4.727)	5
Q20	Verification of servicing according to the required specifications	4.063	0.929	(3.711, 4.414)	10

Source: Field Survey 2018

Table 4. The result of the UPVT.

Wing	RC Member	Pulse Velocity by Cross Probing (PVCP) (Km/sec)	Average PVCP (Km/sec)	Concrete Quality Grading IS: 13311-1 (1992)
1	Column	4.5, 4.8, 3.9, 4.4, 4.4, 4.5	4.4	Good
	Beam	4.8, 4.6, 4.9, 3.9, 4.6, 4.6	4.6	Excellent
	Slab	4.0, 4.7, 3.8, 4.5, 4.5, 4.0	4.3	Good
2	Column	4.6, 4.5, 4.6, 4.7, 4.6, 4.5	4.6	Excellent
	Beam	3.9, 4.4, 4.6, 4.5, 4.2, 4.0	4.3	Good
	Slab	3.9, 4.5, 4.2, 4.4, 4.3, 3.8	4.2	Good

Source: Empirical Study 2018

4.3 Non-Destructive Testing on the Structural Members

An Ultrasonic Pulse Velocity Test (UPVT) was carried out at 28 days curing on RC columns, beams and slabs using an Ultra Pulse Analyzer (58-E4800 UPV tester) at randomly selected members in both wings. The experimental test on quality grading and homogeneity of concrete at randomly selected structural members varied from 4.2 km/sec to 4.6 km/sec. The results (Table 4) showed that condition of concrete structures fall in the range of "Good to Excellent Concrete".

5 CONCLUSIONS, LIMITATIONS, AND RECOMMENDATIONS

Quality practices and client's needs and satisfaction were discovered as the key quality management techniques. Moreover, UPV test proved the concrete as good and excellent in term of quality grading. However, the study was limited to a certain geopolitical, cultural, and societal zone. Work commitment and pressure did not permit some respondents to have enough time to respond to the questionnaires, but the researchers always contacted and worked with them on site. Thus, the study recommended the quality management techniques in the execution of projects by construction engineers to build a safer society. Also, control and regulatory bodies in the construction environment should institute an award scheme for the best firm which strictly adheres to quality practices as a way of motivating them to use quality management techniques.

Acknowledgments

The authors appreciate the management of Covenant University for her financial support and the physical planning development for her technical support in course of the study.

References

Akinci, B., and Boukamp, F., *Towards Automated Defect Detection: Object-Oriented Modeling of Construction Specifications,* Carnegie Mellon University, Pittsburgh, USA, 1-12, 2004.

Ashokkumar, D., Study of Quality Management in Construction Industry, *International Journal of Innovative Research in Science Engineering and Technology,* 3(1), 36-43, 2014.

Chandrasena, G. L., *Importance of Quality Assurance in the Building Construction Industry of Sri Lanka,* National Building Research Organization, Sri Lanka, 2010.

Fagbenle, O. I., Olabosipo, I., and Owolabi, J. D., Developing a Realistic Budget for Construction Projects: Lessons from Nigeria, *International Review of Business and Social Sciences,* 1(3), 17-23, 2009.

IS: 13311-1, Method of Non-Destructive Testing of Concrete, Part 1: Ultrasonic Pulse Velocity (Ced 2: Cement and Concrete), *Bureau of Indian Standards*, New Delhi, 1992.

Ogunde, A. O., Olaolu, O., Afolabi, A., Owolabi, J., and Ojelabi, R., Challenges Confronting Construction Project Management System for Sustainable Construction in Developing Countries: Professionals Perspectives (A Case Study of Nigeria), *Journal of Building Performance*, 8(1), 1-11, 2017.

Project Management Institute, *A Guide to the Project Management Body of Knowledge* (5th Ed.), Project Management Institute Inc., Newtown Square, Pennsylvania, USA, 2013.

Rumane, A. R., *Quality Management in Construction Projects*, Boca Raton, FL: CRC Press, 2011.

Schlickman, J., ISO 9001:2000, *Quality Management System Design*, Artech House, London, Boston, 2003.

Sekaran, U., and Bougie, R., *Research Methods for Business: A Skill-Building Approach* (5th ed.), Wiley, Chi Chester-UK, 2009.

Tan, C. K., and Abdul-Rahman, H., *Preliminary Research into Overcoming Implementation Problems in Construction Projects,* 15-18, Micra Conference, Faculty of the Built Environment, University Malaya, 2005.

Interdependence between Structural Engineering and Construction Management
Edited by Ozevin, D., Ataei, H., Modares, M., Gurgun, A., Yazdani, S., and Singh, A.

ISBN: 978-0-9960437-6-2

CONCEPTUAL STAKEHOLDER THEORY IN PROJECT MANAGEMENT

PAULIN TCHUMTCHA WEMBE

School of Construction Economics and Management, University of the Witwatersrand, Johannesburg, South Africa

Stakeholder management is an important part of management notion particularly in project management today. In recent years, the concept of stakeholder grows comprehensively among academics and professionals. Despite the widespread stakeholder management, the theory is not straightforward to understand and apply. One of the reasons is the number of theories in literature. Literature reviews reveal the following as stakeholder management theories: normative, instrumental, descriptive, convergent stakeholder and social science stakeholder theories. The research employs existing literature to develop a contemporary understanding of stakeholder management theory of stakeholder management. This paper contributes to the stakeholder management theory by presenting the new concept of stakeholder management theory using two cultures (Western and Chinese cultures in Africa). These management approaches can be classified as Western and Chinese models. The Western model is contractual by nature, while the Chinese model is characterized by informality (guanxi). The new stakeholder management theory called "Accustom theory". The new stakeholder management theory developed – "Accustom theory" – had a lot to do with the stakeholder management in project management. It deals with hard skills, soft skills, and the power structure skills. This paper contributes to stakeholder management theory by presenting accustom theory as a new perspective of stakeholder management theory. The new theory aims to improve the practice of project management by presenting how formality, participation and communication levels in projects improve stakeholders' satisfaction.

Keywords: Leadership, Culture, Influence, Decision making, Construction management.

1 INTRODUCTION

Stakeholder management is a prominent element of management practice. In recent years, the concept of stakeholder has achieved widespread attention among academics and managers. In Project Management, stakeholder management has evolved among scholars and professionals over the decades. Friedman and Miles (2006), and Freeman *et al.* (2018) argue that stakeholders appear to be more knowledgeable about project management than in the past. With the increase of stakeholders understanding of their role on the project, an appropriate approach is required to management them. Project Management Body of Knowledge (PMI 2013) has added project stakeholder management as a new project management knowledge area. This is to help the Project Manager in managing projects to manage stakeholders' satisfaction. However, there is still confusion and miss understanding around stakeholder theory. There is a question around stakeholder management theory (What is stakeholder theory?). This paper attempts to make a contribution to stakeholder management theory by expanding the current theory and presenting the conceptual theory developed using two cultures (Western and Chinese cultures in Africa). These management approaches can be broadly classified as Western and Chinese models (Wembe 2016). The Western model is contractual by nature, while the Chinese model is characterized by informality (guanxi).

According to Stawicki *et al.* (2007), a Chinese approach to stakeholder management is based on three pillars embedded in the philosophy of Confucianism, Taoism, and Buddhism. These three approaches also reflect the plan-do-check-act circle: Confucianism – how to make plans and how to manage; Taoism – how to act in accordance with nature; Buddhism – how to review everything. In contrast, Western management philosophy and approach is based on thoughts and beliefs deeply rooted in Western civilization which is closely connected with rules and law. Therefore, the legal contract is the key to doing business (Stawicki *et al.* 2007). Chinese Project Managers (PMs) pay greater attention to maintaining personal relationships within the project stakeholders, and to group harmony and lasting relationships when resolving conflicts than Western project managers. For Chinese PMs, tasks can only be accomplished if a moral and close relationship within the group is achieved (see Table 1). Therefore, Chinese performance is measured morally, whereas Western performance is measured on a task-based platform. Both approaches to project management have their advantages (Chen and Partington 2003, Stawicki *et al.* 2007).

Table 1. Comparison of Chinese and Western management approach.

Chinese approach	Western approach
Collectivism	Individualism
Large power distance	Small power distance
Strong uncertainty avoidance	Weak uncertainty avoidance
Long-term orientation	Short-term orientation, Inner-directed
Outer-directed	Inner-directed
Relationship	Contractual
Conservatism, the tension between hierarchy and harmony	Autonomy, the tension between mastery and egalitarian commitment/ harmony

1.1 Stakeholder Management

The purpose of this paper is to attempt to deal with conflicting opinions on stakeholder management theory. Stakeholder theory is timely, yet young; it is controversial, whilst at the same time important (Laplume *et al.* 2008). It is timely because of the emergence of formal organizations as the dominant institutions of our time, increasing reports of ethical misconducts, particularly issues around the harmful impacts of corporate negligence with regards to the natural environment. It is young because empirical validity is yet to be established on several of its key propositions (e.g., Jones 1995). Stakeholder theory is also controversial because it questions the conventional assumption that the pursuit of profits is the preeminent management concern. Jensen (2002) identifies profit-making as the "single-valued objective" of a corporation. It is also important because stakeholder theory seeks to address the often-overlooked sociological question of how organizations affect society (Hinings and Greenwood 2003, Stern and Barley 1995).

The origins of stakeholder management theory come from the field of business and strategic management theory. Other scholars since Freeman (1984) have further developed multiple theories of stakeholder management, although most of these theories are somewhat confusing.

2 THEORY OF STAKEHOLDER MANAGEMENT

There exists extensive literature on the topic of stakeholder management research, but the theories are not well developed. This paper presents a conceptual stakeholder management theory developed by Wembe (2016), using two cultures (Western and Chinese cultures in Africa), which derive from refutable propositions for empirical testing.

Extensive discussion of stakeholder theory has, to date, mainly applied to the normative and instrumental perspectives (Donaldson and Preston 1995, Jones 1995, Jones and Wicks 1999, Jawahar and McLaughlin 2001, Mellahi and Wood 2003, Butterfield *et al.* 2004). Consequently, the descriptive aspects of stakeholder theory have been largely neglected. Egels-Zanden and Sandberg (2010) claim that the tendency is at least partly related to Donaldson and Preston's (1995) framing of normative stakeholder theory as the core interpretation of the theory (Berman *et al.* 1999). Freeman (1999) and Kaler (2003) have criticized the separation of these three aspects into distinct parts. However, Donaldson and Preston (1995) argue that there is a good reason to separate the theory into three distinct parts. This taxonomy of normative, instrumental and descriptive stakeholder theory shadows the theory with a CSR understanding.

Some researchers have combined Donaldson and Preston's (1995) normative and instrumental theory to develop other theories, including Jones and Wicks (1999) and Gibson (2000). Jones and Wicks (1999) developed what they called "convergent stakeholder theory". Convergent stakeholder theory is descriptive and instrumental, but it is more fundamentally normative. Stakeholders are identified by their interests and all stakeholder interests are intrinsically valuable. The convergent stakeholder theory was dismissed as unsound by Freeman (1999); we do not need more theories that converge but more narratives that are divergent – that shows us different but useful ways to understand organizations in stakeholder terms. Next, social science stakeholder theory developed by Gibson (2000) focuses on the concepts of justice, equity, and social rights having a major impact on the way that stakeholders exert moral reason over project development or change initiatives. Bourne and Walker (2006) use three different stakeholder theories to develop their tools. They use social science stakeholder theory by Gibson (2000), instrumental stakeholder theory developed by Donaldson and Preston (1995), and convergent stakeholder theory by Jones and Wicks (1999). However, the use of these three theories is a repetition of the same theories, because social science stakeholder theory and convergent stakeholder theory are a combination of normative and instrumental theory by Donaldson and Preston (1995).

It is generally accepted that stakeholders are not product, liability, or service. Stakeholders are people and groups of people whom PMs must manage during the life cycle of the project (Freeman 2008, Bourne and Walker 2006).

2.1 What is a Stakeholder Management Theory?

Stakeholder management theory is a combination of both rational and affective theory. First of all, stakeholder management involves the optimization of processes which is one set of skills and falls into rational theory. Second, the complex processes involving human emotion and behaviors, which is another set of skill, falls into affective theory. Third, the line in between rational and affective theory is not static and should not be static – it should be flexible to facilitate the understanding of the theory and to allow a non-ideological, neutral approach, which is also the willingness of a PM to operate in different environments (Wembe 2016). If a PM takes any set of skill principles to the extreme, there will be problems. So, PMs have to guide stakeholders from deception by adding a willingness to the set of skills required to manage stakeholders.

Two sets of skills are needed for stakeholder management, namely, hard skills and soft skills. The latter includes informal and the ability to read the power structures of the organization and the willingness to operate in this environment. When the two sets of management skills are exercised differently, stakeholder satisfaction levels will change. The contractual arrangement model is transactional, and according to Miles and Ballard (1997), in contractual arrangements, the object of the exchange is clearly understood. The skills required to manage contractual arrangements are hard skills and some soft skills. Freedman (2006), argues that the idea of stakeholders, or stakeholder management, or a stakeholder approach to strategic management, is to formulate and implement processes which satisfy all and only those groups who have a stake in the project, the central task in this process is to manage and integrate the relationships and interests of shareholders in a way that ensures the long-term success of a project.

The informal approach is the ability to read the power structures of the organization, and the willingness to operate in this environment. The recommendation of Freedman (2006) seems to be an idea from the informal management approach. The informal management approach to strategic management model is to manage and integrate the relationships and interests of shareholders in a way that ensures the long-term success of the project. In an informal management approach relational arrangement, goals and objectives are much more undefined and the results uncertain, such that "rules" change with the life of the contract (Miles and Ballard 1997).

3 STRUCTURE OF PROPOSITIONS

There are three variables of stakeholder management in project management determined in the literature review. These can be considered as playing an important role in forming stakeholder satisfaction: level of formality, level of participation and level of communication. What is the bottom line of stakeholder satisfaction? Previously, stakeholder satisfactions were measured based on the three constraints (Time, Cost and Scope). It further moved to expectation, process and information (Hartmann and Hietbrink 2013, Walker *et al.* 2008). These researchers identified the three variables that will make stakeholders even are more satisfying.

1. The level of formality relates to the type of contract and relationship PMs have with the stakeholders; stakeholders can have the first choice of the relationship.
2. The level participation addresses the involvement of PMs in stakeholder activities or the co-operation between PMs and stakeholders, and again stakeholders can have expectations about the involvement of the PMs.
3. The level of communication relates to the frequency and quantities of communication the PM has with the stakeholders and the number of feedback stakeholders provide to the PM.

The research analyzes the formality level, participation level and communication level in projects in Africa. The main idea is to think about the responsibility of the PMs with regard to stakeholders; how the model will enhance stakeholder satisfaction; how more effective communication, either formal or informal, will improve and hence stakeholder satisfaction; and what the impact of the formal contractual relationships and informal contractual will have on the projects; how the stakeholder satisfaction will improve the project successfully?

4 NEW CONCEPTUAL OF STAKEHOLDER MANAGEMENT THEORY

This new perspective of stakeholder management theory is seen as the 'best' part of both Western and Chinese way of stakeholder management. Although no other scholars appear to have defined the stakeholder management theory, the latter is a key to stakeholder management. The development of the theory is underpinned by assumptions. The research assumptions consider the relationships between PMs and project stakeholders, qualitatively and quantitatively.

This paper proposes a contribution to the stakeholder management theory base on the formality, participation and communication level of project managers. The contribution to the stakeholder management theory derives from the substantive propositions developed in section 3. It presents a significant contribution to the stakeholder management conceptual theory that can be classed as an "accustom theory". Accustom theory is a new stakeholder management conceptual theory. It deals with the hard skills (which is the craft of stakeholder management) which are the processes, the soft skills (described as the art of project leadership) which are the attitude and behaviors of PMs, and the power structure skills (the ability to read the power structure of the stakeholder ecology and the willingness to operate in that environment) which is the understanding of the complexity and the creativity of the PMs when managing stakeholders.

Hard skills are a process in which project managers use to potentially deliver a successful project while soft skills are the emotional, intuition, communication, interpersonal skill and leadership project managers use to manage the process which deliver the project successfully. The important skill is the line between hard skills and soft skills which is defined as "guanxi", the power structure skills. The power structure skills are the caution, values, critical, understanding why some stakeholders do not fit within the group and take action to make them feel acceptable for the purpose of the project, and the benefits stakeholders convey and gain within the project stakeholders' group.

5 CONCLUSIONS

This paper has presented an account of the inconsistencies in stakeholder management theory and demonstrated how corporate social responsibility theory shadows stakeholder management theory. This paper forecasted on the three variables to elaborate the new perspective theory of stakeholders' management which is the new conceptual theory of stakeholders' management coming from the best part of both Chinese and Western stakeholders' management. A new stakeholder management approach based on best practices from Western and Chinese approaches aims to increase stakeholders' satisfaction and improve project success.

References

Berman, S. L., Wicks, A. C., Kotha, S., and Jones, T. M., Does Stakeholder Orientation Matter? The Relationship between Stakeholder Management Models and Firm Financial Performance, *The Academy of Management Journal*, 42(5), 507-525, 1999.

Bourne, L., and Walker, D. H. T., Visualizing Stakeholder Influence- Two Western Examples, *Project Management Journal*, 37(1), 5–22, 2006.

Butterfield, K. D., Reed, R., and Lemak, D. J., An Inductive Model of Collaboration from the Stakeholder's Perspective, *Business and Society*, 43(2),162-195, 2004.

Chen, P., and Partington, D., An Interpretive Comparison of Chinese and Western Conceptions of Relationships in Construction Project Management Work, *International Journal of Project Management*, 22, 397–406, 2003.

Donaldson, T., and Preston, L. E., The Stakeholder Theory of Corporation: Concepts, Evidence and Implication, *Academy of Management Review*, 20(1), 65-91, 1995.

Egels-Zanden, N., and Sandberg, J., Distinctions in Descriptive and Instrumental Stakeholder Theory: A Challenge for Empirical Research, *Business Ethics: A European Review*, 19(1), 2010.

Freeman, R. E., *Strategic Management: A Stakeholder Approach*, Pitman, Boston, MA, 1984.

Freeman, R. E., Response Divergent Stakeholder Theory, *Academy of Management Review*, 24(2), 233-236, 1999.

Freeman, R. E., Managing for Stakeholders, In Donaldson and Wehane (eds.), *Ethical Issue in Business: A Philosophical Approach*, 8th ed., Englewood Cliffs, NJ: Prentice Hall. 39-53, 2008.
Freeman, R. E., Harrison, J. S., and Zyglidopoulos, S., *Stakeholder Theory Concepts and Strategies*, Cambridge University Press, ISBN: 9781108539500, 2018.
Friedman, A. L., and Miles, S., *Stakeholders: Theory and Practice*, Oxford University Press, 2006.
Gibson, K., The Moral Basis of Stakeholder Theory, *Journal of Business Ethics,* 26, 245-257, 2000.
Hartmann, A., and Hietbrink, M., An Exploratory Study on the Relationship Between Stakeholders' Expectations, Experiences and Satisfaction in Road Maintenance, *Construction Management and Economics*, 31(4), 345-358, 2013.
Hinings, C. R., and Greenwood, R., Disconnects and Consequences in Organization Theory?, *Administrative Science Quarterly*, 47, 411-422., 2003.
Jawahar, I. M., and McLaughlin, G. L., Toward a Descriptive Stakeholder Theory: An Organizational Life Cycle Approach, *Academy of Management Review,* 26(3), 397-414, 2001.
Jensen, M. C., Value Maximization, Stakeholder Theory and the Corporate Objective Function, *Business Ethics Quarterly*, 12(2), 235-256, 2002.
Jones, T. M., Instrumental Stakeholder Theory: A Synthesis of Ethics and Economics, *Academy of Management Review*, 20(2), 404–437, 1995.
Jones, T. M., and Wicks, A. C., Convergent Stakeholder Theory, *Academy of Management Review*, 24(2), 206-221, 1999.
Kaler, J., Differentiating Stakeholder Theories, *Journal of Business Ethics*, 46(1), 71-83, 2003.
Laplume, O. A., Sonpar, K., and Litz, A., Stakeholder Theory: Reviewing a Theory That Moves Us, *Journal of Management*, 34, 1152, 2008.
Mellahi, K., Wood, G., The Role and Potential of Stakeholders in "Hollow Participation": Conventional Stakeholder Theory and Institutionalist Alternatives, *Business and Society Review*, 108(2), 183-202, 2003.
Miles, R., and Ballard, G., *Contracting for Lean Performance*, The Annual Conference of the International Group for Lean Construction, Griffith University, Gold Coast, Australia, 1997.
PMI, *A Guide to the Project Management Body of Knowledge*, 5th ed., Published by: Project Management Institute, Pennsylvania, USA, 2013.
Stawicki, J., Ding, R., and Zieglmeier, O., *What Western Project Management Can Learn from Chinese Culture, Philosophy and Management Approach*, Proceedings of the 21 IPMA World Congress, Poland/Cracow, 2007.
Stern, R. N., and Barley, S. R., Organizations and Social Systems: Organization Theory's Neglected Mandate, *Administrative Science Quarterly*, 41(1), 146-162, 1995.
Walker, D. H. T., Bourne, L., and Rowlinson, S., Stakeholders and the Supply Chain, in *Procurement Systems – A Cross Industry Project Management Perspective,* Walker, D. H. T. and Rowlinson, S. (eds.), *Taylor and Francis*, Abingdon, 70-100, 2008.
Wembe, P. T., *A New Stakeholder Management Framework for EPCM Projects: Analysis of Approaches Used by Australian and Chinese Project Managers in Africa*, PhD Thesis, School of Built Environment, Curtin University, 2016.

Interdependence between Structural Engineering and Construction Management
Edited by Ozevin, D., Ataei, H., Modares, M., Gurgun, A., Yazdani, S., and Singh, A.

ISBN: 978-0-9960437-6-2

IMPLEMENTING CONTRACTOR KNOW-HOW IN EARLY PHASES OF COMPLEX INFRASTRUCTURE PROJECTS

LENA PAAR, CORNELIA NINAUS, and DETLEF HECK

Institute of Construction Management and Economics, Graz University of Technology, Graz, Austria

The importance of bringing contractor know-how into early construction project phases has long been recognized by the construction industry. Although experts and practitioners have considered new forms of procurement and management for construction projects, no significant solutions have been identified until now. It should be mentioned in this context that the Austrian government initiated the implementation of a new procurement approach in 2015 by reducing the opportunity for public clients to award the contract to the bid with the lowest price in conformity with the European guidelines. Austria's infrastructure industry is thus seeking new approaches that are able to implement the know-how of all parties involved in a project. The aim of this paper is to identify possibilities for early contractor involvement and its potential for optimizing the project management in the Austrian infrastructure industry.

Keywords: Project delivery model, Early contractor involvement, Project alliancing, Competitive dialog.

1 INTRODUCTION

The construction industry is a project-based industry, which utilizes a variety of separate firms in a temporary multidisciplinary organization, to produce investment goods such as buildings, roads and bridges (Kamara *et al.* 2002). Every construction project must fulfil its own unique requirements in dependence on budget, time and technical conditions and has to meet client's specifications. In order to meet these requirements, the design and planning phase of infrastructure projects need to insure an orderly course of construction. In Austria the project phases of design and construction are traditionally separated: based on a more or less finished design the client awards the construction contract to the contractor based on lowest price principle. Specific knowledge gained during a construction project is frequently lost after completion of the construction phase. Figure 1 shows the typical planning and design phases of traditional infrastructure projects. The client hires designers who are responsible for the whole planning process (from basic evaluation to operational design). Contractors are only involved in the last phase of operational design, which makes it very difficult to implement practical knowledge and innovative ideas from them. It is an acknowledged fact that teamwork and a collaborative relationship among project parties are essential for a successful construction project (Black *et al.* 2000, Mosey 2009, Bellini *et al.* 2016). When thinking of alternative delivery models for infrastructure projects in Austria, aspects of cooperation among project participants should also be taken into consideration.

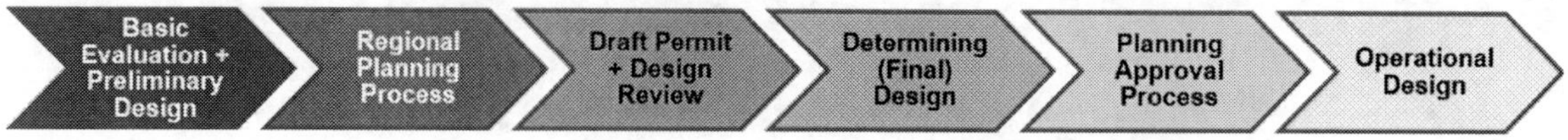

Figure 1. Planning and design phases of traditional infrastructure projects.

The aim of this paper is to identify possibilities for implementing contractors' know-how in the initial project phases of infrastructure projects and its potential for optimizing the project management in order to imply approaches of collaborative delivery models. This paper therefore describes the existing limitations and hindrances in the current situation for public clients of complex infrastructure projects and proceeds to the question of how this process can be optimized.

2 METHOD

In order to assess the possibilities for implementing contractor know-how in order to optimize the project management of construction projects and to enable innovation within construction projects a qualitative survey was carried out. The aim of this survey was to establish the existing hindrances and limitations for the implementation of contractor know-how in public infrastructure projects in Austria. For that purpose, six qualitative interviews were conducted, in a first phase with seven experts in the field of infrastructure construction. The selection criteria for these experts were: (1) number of years experience in the field of construction, (2) working experience in the field of infrastructure construction in Austria, (3) working experience in the field of infrastructure construction outside Austria (e.g. UK; Finland, Sweden, US, AUS) and (4) an open-minded attitude to alternative delivery models (e.g. participation in committees on this topic, publication of papers regarding some form of alternative delivery models). Three experts had a client background, which means that they work for a public client and are involved in infrastructure projects on the client's side as project managers. Three of the experts were employees of contractors and one expert had a neutral background. The interviews were conducted by the first author between June 2016 and April 2017. The length of the interviews was between 1.5 and 2.0 hours. The interviews were semi-structured and can be described as mostly narrative. A transcript of all the interviews from a tape recording (conforming to defined transcription rules and with no consideration given to latent content) provided the foundation for the qualitative content analysis. In a first step the text of the transcript was subdivided into logical units (from 1 up to 3 sentences) and summarized into paraphrases. In a second phase these paraphrases were converted and clustered into 7 main categories with different subcategories on the basis of which mutual statements were summarized. The next step was to find out possibilities for improving the current situation of infrastructure projects. The aim was to analyze alternative forms of project delivery models (e.g. Early Contractor Involvement, Project Alliance, Integrated Project Delivery, Competitive Dialogue, Project Partnering), which are already in use in other countries of the European Union and define their field of application. The necessary data came from a literature review that also considered case studies (e.g. Ross 2003, Mosey 2009, Schlabach 2013, HS2 2014, Gee *et al.* 2018). Based on specific literature for the infrastructure market in Austria and Germany (Schlabach 2013), in a third step an investigation was made into the applicability of these alternative delivery models in the context of the Austrian legal system. Based on the findings, recommended courses of action for adapting a new delivery model for infrastructure projects in Austria were developed.

3 RESULTS

The analysis of the conducted qualitative survey shows the essential role of a cooperative relationship between the parties in order to capture know-how from the contractor's perspective and implement this in the construction project phases. The main findings can be sum up as followed: (1) Project parties are dissatisfied with the current atmosphere of infrastructure construction projects in Austria and are confirming a lack of cooperative behavior on site. The construction market is therefore seeking new (alternative) delivery models for e procurement and contractual phases of infrastructure projects. (2) Infrastructure projects in Austria face a wide range of different legal restrictions, concerning procurement law and the law of environmental impact assessment. There is also general reservation against the binding application of best value criteria for the awarding of contracts. Against this background it is essential to establish "intelligent" and suitable best value criteria in the procurement regulations in order to find the best tender. (3) The key success factor is in the domain of the persons involved in the project (key staff members) who should have full decision-making powers in order to react immediately to disruptions, delays and claims. In order to avoid opportunistic behavior in the relations between the parties it is necessary to implement more collaborative contracting tools, such as project alliancing, integrated project delivery, project partnering and early contractor involvement. Regarding the analysis of the literature review, the joint characteristics of the investigated alternative delivery models (Project Alliancing, Early Contractor Involvement, Integrated Project Delivery, Project Partnering) can be systematized in the following findings: (1) Contractors are involved in the early project phases of construction and their specific knowledge can thus be implemented in the design phase. The client defines the exact scope of work in collaboration with the main contractors. (2) The contractual relationships between client and contractor are on an "open book" basis and "no blame, no dispute"-principles. It is essential to formulate project principles that also include the common goals of all the project participants. (3) During procurement and execution, the emphasis lies on co-operational behavior between client and contractor, which is achieved by means of mutual workshops and interviews. (4) Moving on to the organizational characteristics, it is significant that a leadership team and a separate management team are responsible for controlling and implementation. Both teams consist of at least one member from every project party (client, designer and contractor) who should work together as a team with the same shared interests. This also leads on logically to the principle that each project party will either win or lose ("gainshare/painshare"), if the project costs are higher than estimated.

4 RECOMMENDED COURSES OF ACTION

To establish an alternative delivery model for complex infrastructure projects in Austria, two fundamental aspects must be described beforehand: the current situation of delivery models in Austria (see chapter 4.1) and a definition of "complex" infrastructure projects (see chapter 4.2).

4.1 Traditional Delivery Model of Infrastructure Projects in Austria

As mentioned above nearly all the design work for infrastructure projects in Austria comes from the client side, which means that based on a more or less finished design, the public contractor awards the contract by an open procurement procedure. In the past, the tender was awarded in most cases to the lowest priced bid. The regulation was introduced in 2015 that public clients must now also consider best value aspects in procurement procedures.[1] These circumstances

[1]This is attributable to European procurement regulations: Directive 2014/24/EU and 2014/25/EU.

made it necessary to adapt procurement principles in order to enhance the selection of the "best" instead of the "cheapest" tender. However, the current procurement model is still based on approach of design-bid-build. Figure 2 shows the traditional delivery model for infrastructure projects in Austria.

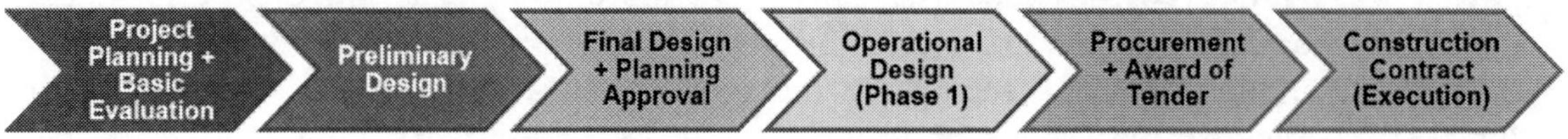

Figure 2. Traditional delivery model of infrastructure projects.

Starting with project planning and basic evaluation, developing the preliminary and afterwards final design, the procurement process begins on the basics of operational design. Contractors are involved only in the last two project phases (procurement and contractual phase), which make it very difficult to implement practical knowledge and innovative ideas from them.

4.2 Project Classification for Infrastructure Projects

Although construction can be defined as a complex and dynamic task (Bertelsen 2004), it is not necessary to consider every construction project as a complex infrastructure project in this paper. A classification was necessary for making a distinction between non-complex and complex infrastructure projects. Table 1 shows the project classification matrix, which refers to 9 different criteria (K1 to K9) for defining complexity: type (K1) and scale of project (K2), schedule (K3), duration (K4) and specifications of work (K5), structure of tender market (K6), risk of approval (K7), budget preferences (K8) as well as the strategic role of the project (K9). The different criteria have been deduced from the literature (e.g. Bertelsen 2004, Schlabach 2013).

Table 1. Project classification matrix.

	Assessment [points]							
K1: Type of Project	small redevelopment	0	new construction (open terrain)	1	rebuilding (with restraints)	3	rebuilding (during full operation)	6
K2: Scale of Project	< 5,548M EUR	0	> 5,548M EUR < 50M EUR	1	> 50M EUR < 150M EUR	3	> 150M EUR	6
K3: Schedule preferences	no priority	0	moderate schedule	1	tight schedule + penalty	3	fast track-project	6
K4: Duration of Construction	< 1 year	0	> 1 to 5 years	1	> 5 years	3	> 10 years	6
K5: Specifications of works	standardised specification of works	0	new type of contruction method (partly)	1	new type of contruction method (mostly)	3	contractors know-how necessary	6
K6: structure of tender market	many (>10) potential tenderers	0	some (>3) potntial tenderers	1	< 3 potential tenderers	3	unknown	6
K7: environment (approval)	low risk	0	moderate risk	1	high risk	3	very high risk	6
K8: budget preferences	no budget defined	0	increase of budget possible	1	increase of budget difficult	3	increase of budget impossible	6
K9: strategic role of project	low	0	moderate	1	high	3	very high	6

Clients of infrastructure projects should use this project classification matrix to assess the complexity in an initial phase. The sum of all points divided by 6 results in a score of between 0.0 and 9.0. Roughly speaking all projects with a score below 6.0 can be characterized as "non-complex" and above as "complex" (see Figure 2).

Figure 2. Score of project classification matrix.

This paper as also the recommended alternative delivery model refers to projects that reach a score of at least 6.0 or higher and can therefore be qualified as a complex infrastructure project.

4.3 Recommended Delivery Model for Complex Infrastructure Projects

Figure 3 shows the developed alternative delivery model for infrastructure projects. Contractors in this recommended delivery model are involved much earlier in the project: after the client specifies the preliminary design, a competitive dialogue (CD) must be accomplished. The client finally awards the contract to the "best" tender and concludes the construction contract using project alliancing approaches.

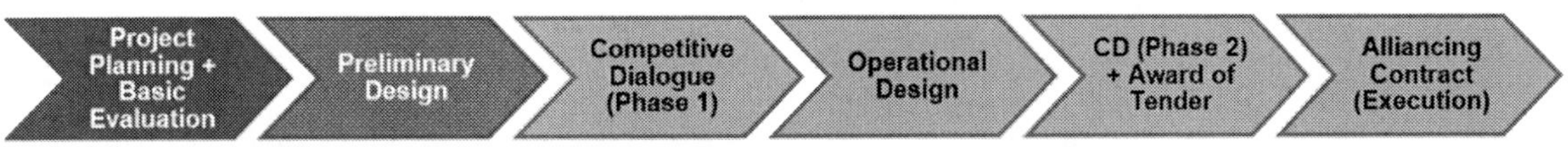

Figure 3. Recommended alternative delivery model for infrastructure projects.

For a successful implementation of this alternative delivery model it is necessary to consider the following recommendations: (1) The competitive dialogue should take into consideration the legal restrictions under public procurement law. In Austria this means the client must select at least 2 (better 3) bidders for a tender using an objective prequalification system. In phase 2 of the competitive dialog the bidders develop the exact scope of work in order to find the best solution for the project and to define the target costs. It is not possible to select the bidder using a non-price procedure under the procurement regulations. The definition of the target price must thus be taken into consideration when awarding the contract. (2) During the project execution phase the client and the contractor should work together based on an alliancing contract, installing a leadership and management team that works on the basis of "no blame, no dispute" and "best for project" principles.

5 CONCLUSION

The results of the qualitative interviews and the literature review show that there is a need for more cooperative delivery models in the field of infrastructure projects. To address this collaborative behaviour it is necessary to implement contractors earlier in project phases than in the current procurement and contract model. The recommended alternative delivery model presented in this paper has the advantage of implementing contractor know-how early in project phases. Selecting the bidder for a tender based on a competitive dialogue, as part of the recommendations, is also in compliance with public procurement law. In Austria, however, it has not been possible to also include the designer in the project team until now, due to legal restrictions. It is highly recommended that the designer should also be included in the project team in order to achieve a more cooperative working atmosphere. Whether and how this recommended delivery model will find a field of application in practice will need to be investigated in the course of further research.

References

Bellini, A., Aarseth, W., and Hosseini, A., Effective Knowledge Transfer in Successful Partnering Projects, *Energy Procedia*, Elsevier, 96, 218-228, 2016.

Bertelsen, S., *Construction as a Complex System*, Proc., Int. Group for Lean Construction 11th Annual Conf. (IGLC-11), IGLC, Blacksburg, May 15, 2004. Retrieved from http://leanconstruction.dk/ August, 1, 2018.

Black, C., Akintola, A. and Fitzgerald, E., An Analysis of Success Factors and Benefits of Partnering in Construction, *International Journal of Project Management*, Elsevier, 18, 423-434, 2000.

Gee, R.A., Parker, C. J., and Cuttler R. J., Nortside Storage Tunnel, Sydney: Investigation, Design and Construction. Retrieved from http://www.ats.org.au on August 01, 2018.

HS2, Early Contractor Involvement (ECI) Guidance, High Speed Two Limited, October 17, 2014. Retrieved from https://www.gov.uk/government/publications/early-contractor-involvement on August 01, 2018.

Kamara, J. M., Augenbroe, G., Anumba, C. J., and Carrillo, P. M., Knowledge Management in The Architecture, Engineering and Construction Industry, *Construction Innovation*, Emerald, 2, 53-67, 2002.

Mosey, D., *Early Contractor Involvement in Building Procurement*, Wiley-Blackwell, Chichester, West Sussex, United Kingdom, 2009.

Ross, J., Introduction to Project Alliancing (on Engineering and Construction Projects), April 2003. Retrieved from https://iccpm.com on August 01, 2018.

Schlabach, C., *Untersuchungen zum Transfer der Australischen Projektabwicklungsform Project Alliancing auf den Deutschen Hochbaumarkt*, Kassel University Press, Kassel, Germany, 2013.

Interdependence between Structural Engineering and Construction Management
Edited by Ozevin, D., Ataei, H., Modares, M., Gurgun, A., Yazdani, S., and Singh, A.

ISBN: 978-0-9960437-6-2

REACTIVE DEVALUATION AS A BARRIER AGAINST AMICABLE CONSTRUCTION DISPUTE SETTLEMENT

KEYAO LI, ON YI CHOW, and SAI ON CHEUNG

Dept of Architecture and Civil Engineering, City University of Hong Kong, Kowloon, Hong Kong

Disputes are common in construction projects. Protracted unresolved dispute destroys contracting parties' desire to fulfil the project objectives. In this regard, promoting amicable and efficient construction project dispute resolution (CPDR) has been one of the key recommendations of almost every construction industry review. However, disputants' irrational behaviors would derail trajectory to settlement. Notable example of irrational decision is failing to seize settlement option that provides utility to both disputing parties. This can be the result of a deep-rooted psychological barrier called reactive devaluation (RD). RD describes the behavior of habitually devaluing the proposal raised by the counterpart. Disputants would reject a proposal irrespective of the quality and reasonableness of the proposal, just because it is raised by the counterpart. The occurrence of RD in CPDR would block proposal exchanges that are inevitable if a settlement is to be reached. This study aims to examine the application of RD in CPDR. Possible manifestations of RD in CPDR scenario were summarized and a conceptual framework of RD was proposed. Five types of RD were identified: reluctance to change, doubts about counterpart's ability, overconfidence, biased information processing, and mistrust of the counterpart. Furthermore, suggestions to alleviate RD were also discussed.

Keywords: Conceptual framework, Irrational decisions, Manifestations, Loss aversion, Attitude polarization, Naïve realism.

1 INTRODUCTION

Conflicts are inevitable in modern construction business, in particular where projects are increasingly complex with project team members coming from diversified background (Cheung and Yiu 2006). This diversity is conflict laden with the ultimate outcome of construction dispute. The happenings of construction dispute will hinder the project schedule and cause delay. Therefore, almost all construction industry reviews have called for efficient and amicable construction project dispute resolution (CPDR).

Facing the enormous cost of solving the dispute in court, many researchers and practitioners welcome the contractual use of cost-friendly alternative dispute resolution (ADR) methods. Nevertheless, barriers arising from one's cognition and self-perception hinder amicable settlement because these may lead to irrational decisions of not choosing the most economical and speedy solutions. In other words, irrational decisions are attributed to some deep-rooted psychological barriers against dispute settlement (Ross and Stillinger 1991, Mnookin and Ross 1995). The topic of psychological barriers against dispute settlement has been discussed widely in business decisions (Korobkin and Guthrie 1994, Ross and Ward 1995, Korobkin 2005, Spangler 2012). The study on psychological barriers against CPDR was inspired by Korobkin and Guthrie (1994), in which three types of phenomena affecting disputants' decision making in litigation were identified: framing, equity seeking and reactive devaluation. Framing explains that disputants' choice to accept or reject a settlement offer is dependent on their frames on the offer in relation to a certain reference point. These frames can be positive or negative. Equity seeking describes that disputants' desire to seek perceived equity may allow personal feelings to override economically rational considerations. Reactive devaluation (RD) is a cognitive bias happening when a suggestion is always devaluated simply because it is proposed by a counterpart. In extreme cases, one would reject a proposal regardless of its quality and reasonableness (Ross and Stillinger 1991). Among the three psychological barriers, reactive devaluation was studied in-depth for its applicability in CPDR.

The objective of this research study is to identify the conceptual framework of RD in CPDR. Relevant literatures about theoretical explanations and underpinnings of RD have been reviewed. Applications on how to overcome RD were also discussed. The study is reported in four parts: i) Theoretical background of the study; ii) Conceptual framework of RD in CPDR; iii) Suggestions to overcome RD; and iv) Concluding remarks of the study.

2 THEORETICAL BACKGROUND OF THE STUDY

The theoretical basis of this study is supported by a literature review on RD. In this part, the theoretical bases of RD are discussed in terms of its initiated studies and explanations.

2.1 Pioneer Studies of Reactive Devaluation

Professor Lee Ross and his colleagues (Ross and Stillinger 1991) first initiated RD. They proposed that RD has three major observations (Mnookin and Ross 1995): i) The proposal raised by the counterpart is rated less positively than the one raised by himself or by a neutral third party; ii) proposed plan is rated less positively than plan withheld; and iii) proposal raised unilaterally is rated less positively than the one that was brainstormed by the parties together (Ross and Stillinger 1991, Ross and Ward 1995). The first observation was supported by the study about the arm reduction proposal between United States and Soviet Union (Ross and Ward 1995). It was found that the proposal was seen unfavorable by US subjects when Soviets were the originator, more favorable if a neutral third party is the putative originator and plan with US putative attributor was considered most favorable. Ross and Stillinger (1991) and Ross and Ward (1995) proposed another phenomenon that the presence of availability of choices causes the occurrence of RD. Even the choices are from non-hostile sources (Korobkin and Guthrie 1994). Ross and Stillinger (1991) and Ross and Ward (1995) reported the tendency that people devaluate the 'proposed' plan announced to the one yet to be announced. RD happens that people consider the withheld choice is better, even there is neither negative relationship nor prior assurance existed among the choices. The third observations described that unilateral offer caused the occurrence of RD effect. Mono-directional concession will render the recipient to believe that it is always possible to bargain (Ross and Ward 1995). As a result, although the offeror attempted to build goodwill in dispute resolution, the proposal is lamentably devaluated (Ward *et al.* 2011).

2.2 Explanations of Reactive Devaluation

2.2.1 Loss aversion

Loss aversion is reported as one of the contributors of RD (Kahneman and Tversky 1984, Korobkin and Guthrie 1994, Johnson *et al.* 2007). Just as Kahneman and Tversky (1984) described, the adverse feeling caused by a loss is deeper than the attractive feeling caused by a gain of same amount. In other words, the attractiveness of possible gain cannot commensurate the aversion of possible loss. The possibility in identifying an event as "gain" or "loss" usually depends on how the event was framed. People are usually reluctant to give up something that they consider as their endowment or personal assets (Thaler 1980, Kahneman and Tversky 1984, Ericson and Fuster 2011, Morewedge and Giblin 2015). People tend to avoid taking losses thus it is much more reluctant to give up something they own than to purchase them. If accepting the resolution offer is considered as losing one's interest, it is believed that loss aversion will take priority and will encourage the party to devaluate and reject the offer (Johnson *et al.* 2007). As the feeling of loss outweighs the perceived gains, the offeree will resist to proceed with the negotiation (Kahneman and Tversky 1984). In CPDR, both contracting parties would try to 'hide' their weaknesses and acceptable trade-offs. This inclination would make resolution more difficult than anticipated. Affected by loss aversion in CPDR, a party may be very conscious about not exposing to potential losses.

2.2.2 Attitude polarization

Attitude polarization is highly believed to be a cause of RD (Ross and Ward 1995). It is a phenomenon that belief polarization escalates after reviewing evidences on the issue found by the parties (Lord *et al.* 1979). People will selectively pay attention to the information that supports their views and dismiss those contradicting (Lord *et al.* 1979, Klayman 1995). Tesser (1978) claimed that the longer the time one exposes to supportive information, the more confirmation he/she has towards the information. As one gains more positivity towards his/her opinion, he/she tends to believe that the counterpart would also consider the same (Ross *et al.* 1977). As a result, when one is exposed to information that supports his dispute resolution proposal, he will find his proposal more convincing and therefore should be accepted by his counterpart (Ross *et al.* 1977, Ross and Ward 1996, Ericson and Fuster 2011). Being self-enhanced, one will lose interest in negotiating (Johnson-Cartee and Copeland 1997). Hence, biased information processing strengthens parties' already held opinions and makes attitudes from two sides more polarized. Not providing the whole picture for the negotiation, attitude polarization destroys the trust among parties and thus adds difficulty to amicable settlement (Lord *et al.* 1979, Morewedge and Giblin 2015).

2.2.3 Naïve realism

Naïve Realism (NR) is also highly recognized as a source of RD (Ross and Ward 1996). There are three tenets of NR according to Ross and Ward (1996). The first one is about objectivity. People with NR think that they are able to see the world as how it is without subjective interpretations. They believe their perception is objective without any bias. These disputants believe their proposals are the most practical in solving the dispute (Pallier *et al.* 2002). As a result, they would challenge and devaluate the effectiveness of the proposal put forward by the counterpart (Ross and Stillinger 1991, Ross and Ward 1995, Bland *et al.* 2012). Another tenet deals with homogeneity. People tend to believe the counterpart would draw same conclusion as theirs with the same piece of information if the counterpart were reasonable and open-minded

(Ross and Ward 1996). They expect the other side to act in the same way as they do because they believe they are correct (Ross *et al.* 1977). Final tenet is that people consider their counterparts as lazy, unintelligent and biased. The counterparts are expected to put in more effort and work harder to reach the same judgment. Therefore, facing a different conclusion drawn by others, disputants influenced by RD would attribute the differences to the incapability and biases of others. Therefore, they tend to devaluate counterpart's proposals without thorough consideration.

3 CONCEPTUAL FRAMEWORK OF REACTIVE DEVALUATION IN CPDR

Literature review was conducted to elaborate the conceptual underpinnings of RD in CPDR. RD behaviors were collected from the reported studies on the three causes of RD. Twenty-five RD manifestations were operationalized in CPDR. These manifestations were summarized in Table 1. The manifestations were organised into five types according to their nature. As a result, a conceptual framework of RD in CPDR is proposed.

Table 1. Conceptual framework of RD in CPDR.

Manifestations of RD in CPDR	References	Types of RD
1. I want to maintain my decisions therefore I am uninterested in negotiating (Attitude Polarization; Loss Aversion).	Johnson-Cartee and Copeland (1997), Kahneman and Tversky (1984)	Type 1: **Reluctance to change**
2. Before negotiation, I believe my proposal will be accepted (Naïve Realism; Attitude Polarization).	Ross *et al.* (1977), Ross and Ward (1996), Ericson and Fuster (2011)	
3. After forming a reasonable opinion, I tend to devaluate other possibilities (Attitude Polarization; Naïve Realism).	Klayman (1995), Ross and Ward (1995,1996)	
4. When rethink about my decision, I unconsciously recall reasons why my decision can hold (Attitude Polarization).	Lord *et al.* (1979), Klayman (1995)	
5. I think the resolution proposal raised by the counterpart is based on incomplete information (Naïve Realism; Attitude Polarization).	Ross *et al.* (1977), Ross and Ward (1995, 1996)	Type 2: **Doubts about counterpart's ability**
6. I think the proposal from the counterpart cannot resolve the project dispute (Naïve Realism).	Ross and Stillinger (1991), Ross and Ward (1995), Bland *et al.* (2012)	
7. I think the counterpart is biased (Naïve Realism).	Ross and Ward (1996)	
8. I doubt the capability of the counterpart (Naïve Realism, Attitude Polarization).	Ross *et al.* (1977), Ross and Stillinger (1991), Bland *et al.* (2012)	
9. I doubt the effectiveness of the suggestions raised by the counterpart (Naïve Realism).	Ross and Stillinger (1991), Ross and Ward (1995), Bland *et al.* (2012)	
10. I think the counterpart provides premature decisions (Naïve Realism; Attitude Polarization).	Ross *et al.* (1977), Ross and Ward (1995), Bland *et al.* (2012)	
11. I think my proposal is more practical (Naïve Realism, Attitude Polarization).	Ross and Ward (1996), Pallier *et al.* (2002), Ericson and Fuster (2011)	Type 3: **Overconfidence**
12. I think my proposal is the best solution to the project dispute (Naïve Realism, Attitude Polarization).	Ross and Ward (1996), Pallier *et al.* (2002), Ericson and Fuster (2011)	
13. I believe the counterpart will agree with us if they put more effort in decision making (Naïve Realism; Attitude Polarization).	Ross and Ward (1996), Ross *et al.* (1977)	
14. I always think there is still room for bargaining with the proposal from the counterpart (Naïve Realism).	Ross and Ward (1995, 1996), Ross *et al.* (1977)	
15. I think only my proposal can tackle the project dispute (Naïve Realism, Attitude Polarization).	Ross and Ward (1995, 1996), Lord *et al.* (1979), Pallier *et al.* (2002)	

Table 1 (contd). Conceptual framework of RD in CPDR.

16. I endorse information that supports my decisions (Attitude Polarization).	Lord *et al.* (1979), Klayman (1995)	Type 4: **Biased information processing**
17. I search for information that confirms my assessment (Attitude Polarization).	Lord *et al.* (1979), Klayman (1995)	
18. After forming a decision, I tend to pay more attention to the information that supports my decision (Attitude Polarization).	Lord *et al.* (1979), Morewedge and Giblin (2015)	
19. My information searching process terminates after I consider I have found enough supporting information for my decision (Attitude Polarization; Naïve Realism).	Pallier *et al.* (2002), Ericson and Fuster (2011)	
20. I tend to think the choices provided by the counterpart are disadvantageous to my side (Loss Aversion; Naïve Realism).	Korobkin and Guthrie (1994), Johnson *et al.* (2007), Bland *et al.* (2012)	Type 5: **Mistrust of the counterpart**
21. I tend to believe the choices not provided by the counterpart are more advantageous to my side (Loss Aversion).	Kahneman and Tversky (1984), Ross and Stillinger (1991), Ross and Ward (1995)	
22. I tend to think the counterpart wants to gain interests from me with his offer (Loss Aversion).	Kahneman and Tversky (1984), Johnson *et al.* (2007), Morewedge and Giblin (2015)	
23. I think the compromise of the counterpart is just their negotiation strategy (Loss Aversion).	Korobkin and Guthrie (1994), Ross and Ward (1995)	
24. I tend to think the counterpart is behaving opportunistically (Loss Aversion; Naïve Realism).	Ross and Ward (1995, 1996)	
25. I tend to devaluate the proposal if I know that it was based on information out of my knowledge (Loss Aversion).	Korobkin and Guthrie (1994), Ross and Ward (1995)	

4 SUGGESTIONS TO OVERCOME REACTIVE DEVALUATION

Suggestions to overcome RD were categorized in two types—raising one's self-awareness of irrationality and enhancing communication. Raising awareness of irrational behavior is the first step for performance improvement. It is expected that after disputants realized their biased reluctance to change, overconfidence and biased information processing, more efforts will be paid to improve their performance and decision-making approaches in CPDR. In this regard, raising disputants' self-awareness is believed to be useful in mitigating RD for type 1, type 3 and type 4. Enhancing communication is the recommended approach to alleviate the RD attitudes in type 2 and type 5. Seamless communication and partnering relationship among the disputing parties would alleviate hostile RD attitudes and facilitate the settlement procedures (Bayliss *et al.* 2004).

5 CONCLUDING REMARKS

Disputes are common in construction projects. Amicable and efficient CPDR is welcomed by all contracting parties. RD is a well-known psychological behavior that describes the tendency to devalue the proposal raised by counterpart. It may happen among construction disputants and threaten rational dispute resolution. This study examined the conceptual framework of RD in CPDR. A comprehensive literature review was conducted. Underpinning theoretical background was explored. Twenty-five possible RD manifestations in CPDR were summarized from literatures. Five types of RD behaviors in CPDR were identified. These are: i) reluctance to change; ii) doubts about counterpart's ability; iii) overconfidence; iv) biased information processing, and v) mistrust of the counterpart. Furthermore, raising one's self-awareness of irrationality and enhancing communication are suggested as possible means to curb RD.

Acknowledgments

The study was fully supported by a CityU Strategic Research Grant (No. 7004870) and a HKSAR RGC General Research Fund project (No. 11209118).

References

Bayliss, R., Cheung, S. O., Suen, H. C., and Wong, S. P., Effective Partnering Tools in Construction: A Case Study on MTRC TKE Contract 604 in Hong Kong. *International Journal of Project Management,* 22(3), 253-263, 2004.

Bland, B., Powell, B., and Ross, L., Barriers to Dispute Resolution: Reflections on Peacemaking and Relationships between Adversaries, *Understanding Social Action, Promoting Human Rights*, Oxford University Press, 269-291, 2012

Cheung, S. O., and Yiu, T. W., Are Construction Disputes Inevitable? *IEEE Transactions on Engineering Management*, 53(3), 456-470, 2006.

Ericson, K. M. M., and Fuster, A., Expectations as Endowments: Evidence on Reference-Dependent Preferences from Exchange and Valuation Experiments, *The Quarterly Journal of Economics,* 126 (4), 1879-1907, 2011.

Johnson, E. J., Keinan, A., and Haubl, G., Aspects of Endowment: A Query Theory of Value Construction, *Journal of experimental psychology: Learning, memory, and cognition*, 33 (3), 461-474, 2007.

Johnson-Cartee, K., and Copeland, G., *Inside Political Campaigns: Theory and Practice,* Westport, CT: Praeger, 1997.

Kahneman, D., and Tversky, A., Choices, Values, and Frames, *American Psychologist,* 39 (4), 341–350, 1984.

Klayman, J., Varieties of Confirmation Bias, *Psychology of Learning and Motivation*, 32, 385-418, 1995.

Korobkin, R., and Guthrie, C., Psychological Barriers to Litigation Settlement: An Experimental Approach, *Michigan Law Review*, 93 (1),107-192, 1994.

Korobkin, R., Psychological Impediments to Mediation Success: Theory and Practice. *Ohio State Journal on Dispute Resolution*, 21, 281-307, 2005.

Lord, C. G., Ross, L., and Lepper, M. R., Biased Assimilation and Attitude Polarization: The Effects of. Prior Theories on Subsequently Considered Evidence, *Journal of Personality and Social Psychology*, 37(11), 2098- 2109, 1979.

Mnookin, R., and Ross, L., Strategic, Psychological, and Institutional Barriers: An Introduction, In *Barriers to the Negotiated Resolution Conflict,* New York: Norton, 3-27, 1995.

Morewedge, C. K., and Giblin, C. E., Explanations of The Endowment Effect: an Integrative Review, *Trends in Cognitive Sciences,* 19(6), 339- 348, 2015.

Pallier, G., Wilkinson, R., Danthir, V., Kleitman, S., Knezevic, G., Stankov, L., and Roberts, R. D., The Role of Individual Differences in the Accuracy of Confidence Judgments, *The Journal of General Psychology,* 129(3), 257-299, 2002.

Ross, L., and Stillinger, C., Barriers to Conflict Resolution, *Negotiation Journal,* 7(4), 389-404, 1991.

Ross, L., and Ward, A., Psychological Barriers to Dispute Resolution, In *Advances in Experimental Social Psychology*, 255-304. San Diego, CA: Academic Press, 1995.

Ross, L., and Ward, A., Naive Realism in Everyday Life: Implications For Social Conflict and Misunderstanding, In *Values and Knowledge*, Hillsdale, NJ: Erlbaum, 103–135, 1996.

Ross, L., Greene, D., and House, P., The "False Consensus Effect": An Egocentric Bias in Social Perception and Attribution Processes, *Journal of Experimental Social Psychology*, 13(3), 279-301, 1977.

Spangler, B. M., Heads I Win, Tails You Lose: The Psychological Barriers to Economically Efficient Civil Settlement and a Case for Third-Party Mediation, *Wisconsin Law Review*, 1435-1466, 2012.

Tesser, A., Self-Generated Attitude Change, *Advances in Experimental Social Psychology*, 11, 289-338, Jan, 1978.

Thaler, R., Toward a Positive Theory of Consumer Choice, *Journal of Economic Behavior and Organization,* 1(1), 39-60, 1980.

Ward, A., Atkins, D. C., Lepper, M. R., and Ross, L., Affirming the Self to Promote Agreement with Another: Lowering a Psychological Barrier to Conflict Resolution, *Personality and Social Psychology Bulletin,* 37(9), 1216-1228, 2011.

Interdependence between Structural Engineering and Construction Management
Edited by Ozevin, D., Ataei, H., Modares, M., Gurgun, A., Yazdani, S., and Singh, A.

ISBN: 978-0-9960437-6-2

A STUDY OF ENDOWMENT EFFECT IN CONSTRUCTION PROJECT DISPUTE RESOLUTION

KEYAO LI, YUI HANG NG, and SAI ON CHEUNG

Dept of Architecture and Civil Engineering, City University of Hong Kong, Kowloon, Hong Kong

It has been reported that the average dispute value (the additional entitlement included in the contract to resolve disputes) was US $84 million and average length of dispute (the period between the time a dispute is formalized and it is settled) was 14.6 months in Asia in the year of 2016 (Arcadis Global Construction Dispute Report 2017). These figures are not impressive in terms of the non-productive use of valuable resources. Endowment effect (EE), which was firstly coined by 2017 Nobel laureate in economics science professor Richard Thaler, is a behavioral phenomenon that people attribute more value to the things they own. EE would derail making rational decisions. This study explores the application of EE on disputants' behavior in construction project dispute resolution (CPDR). In this regard, four sources of EE in CPDR were identified. These are i) ownership, ii) loss aversion, iii) status quo bias and iv) bargaining habit.

Keywords: Psychological barrier, Ownership, Loss aversion, Status quo bias, Bargaining habit, Overcome.

1 INTRODUCTION

In June 2017, the seventh annual Arcadis Global Construction Dispute Report (2017) was published and it highlighted the financial values and time spent on resolving dispute in recent years (Table 1). From Table 1, it can be seen that many additional resources and time have been recurrently spent in resolving construction dispute. Especially, Asia recorded higher average dispute value (the additional entitlement included in the contract to resolve disputes) than the global average. In 2016, it was even almost twice of the global average. Therefore, considerable resources would be saved if construction disputes could be resolved amicably and speedily.

Table 1. Financial values and time input for dispute resolution.

	Average Dispute Value (US$ Millions)					Average Length of Disputes (Months)				
Year	2012	2013	2014	2015	2016	2012	2013	2014	2015	2016
Global Average	31.7	32.1	51.0	46.0	42.8	12.8	11.8	13.2	15.5	14.0
Asia	39.7	41.9	85.6	67.0	84.0	14.3	14.0	12.0	19.5	14.6

Barriers against dispute resolution were categorized into three forms by Ross and Ward (1995): institutional barriers, strategic barriers and psychological barriers. Institutional barriers relate to the institutional and organizational factors that discourage settlement and are often imposed by a company's senior management. Restricted channels of communication and information, varied interests among the groups and political considerations are other forms of institutional barriers (Ross and Ward 1995). If effective communication is present, the impact of institutional barriers against settlement may be minimized (Spangler 2012). Strategic barriers are the negotiation strategies used by parties with the belief that these would help to bring benefits for them from the negotiation. Miscalculations or misinterpretations may lead to opportunistic negotiation tactics (Raiffa 1982, Mnookin 2003). Refusing to share information is one of the most common strategies applied on the negotiation table. However, refusal to share information would easily lead to negotiation deadlocks (Raiffa 1982). To deal with strategic barriers, enhancing mutual trust is the solution. Trust would minimize unnecessary suspicion among the disputing parties (Wong *et al.* 2005). The third barrier to successful dispute settlement is psychological in nature. Psychological barriers will govern individual's decision-making activities, including interpreting information, evaluating risks and setting priorities (Ross and Ward 1995). Psychological barriers arise from human instinctive tendency to avoid or delay making decisions when the level of uncertainty is high. Among the three types of barrier in dispute resolution, it is believed that institutional and strategic barriers could be minimized by designing appropriate resolution procedures, assistance from neutral third party and deploying experienced negotiators (Spangler 2012). However, it is more difficult to deal with psychological barriers. Endowment effect (EE) has been reported to be one of psychological barrier that hampers the successful settlement of dispute (Thaler 1980, Tversky and Kahneman 1991, De Dreu and Van Knippenberg 2005, Morewedge and Giblin 2015). This study aims to explore the application of the concept of EE in construction project dispute resolution (CPDR).

2 CONCEPT OF ENDOWMENT EFFECT

2.1 Definition of Endowment Effect

It was generally believed that human decision making can be well explained by Coase theorem (Coase 1988). Coase theorem described the economic allocation of resources in negotiation or similar trade-offs settings (Kahneman *et al.* 1991). Coase theorem predicts the final allocation of resources in negotiations shall be based on an ultimate assumption whereby individuals will value an entitlement the same when consider it as an acquisition or a loss. It was later found that Coase theorem was not always correct: people usually ask for much more as a compensation for giving up an entitlement than they would be willing to pay to acquire the same—a phenomenon named endowment effect (EE) by Thaler (1980). The 2017 Nobel laureate Richard Thaler firstly introduced the concept of EE in 1980, which he referred to as the underweighting of foregone gain of adding goods to individual's endowment (Thaler 1980). Morewedge *et al.* (2009) also explained that the amount people demanded to relinquish something they own was larger than the amount they were ready to spend to get the same. There are two situations of EE: (1) psychological ownership effect: people who own goods will value it greater than people who do not (Beggan 1992, De Dreu and Van Knippenberg 2005, Shu and Peck 2011); (2) In economic exchange situations, the maximum amount of money people are willing to pay (WTP) to acquire goods is much lower than the minimum amount of money people are willing to accept (WTA) to forgo it (Carmon and Ariely 2000, Morewedge and Giblin 2015).

It is found that the endowment effect is not confined to exchange of goods or services, it is also applicable in the domain of decision making. De Dreu and Knippenberg (2005) claimed that ownership attachment of EE also applied to beliefs, ideas and positions people use in decision making, people had the tendency to defend the arguments and positions they owned. Liedtka

(2015) also proposed that in the process of problem solving, EE would lead to attachment to first solutions and reduction in options considered. This study aims to investigate the behavior and influence of EE in construction project dispute resolution (CPDR).

2.2 Sources of Endowment Effect and Respective Influence in CPDR

After the concept of EE was first coined by professor Thaler, extensive follow-on studies have been conducted to explore the theories and sources of EE (Tversky and Kahneman 1991, De Dreu and Van Knippenberg 2005, Shu and Peck 2011, Morewedge and Giblin 2015). Four widely proposed sources of EE and their influence in CPDR were discussed in this section and they were summarized with respective key references in Figure 1.

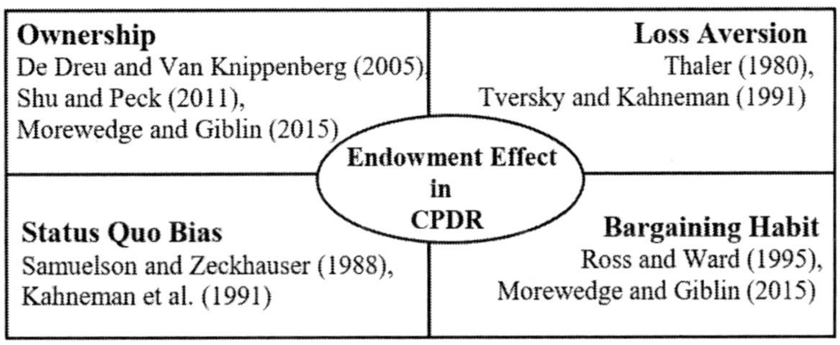

Figure 1. Four sources of EE and respective key references.

2.2.1 Ownership

Psychological ownership explains the tendency of EE (De Dreu and Van Knippenberg 2005, Shu and Peck 2011, Morewedge and Giblin 2015). Beggan (1992) and Morewedge *et al.* (2009) claimed that ownership would increase the perceived value of goods people own. Not confined to goods, but values of beliefs and ideas would also be raised as well (De Dreu and Van Knippenberg 2005). Ownership creates an association between owner and the good and it becomes part of the self-concept of the owner (Ariely *et al.* 2005, Weiss and Johar 2013). Therefore, the positive nature of human's self-enhancement renders a positive evaluation of the goods people owns (Carmon *et al.* 2003, Ariely *et al.* 2005). It has been pointed out that once a psychological attachment has been formed, the potential loss of the good is seen as a threat to the individual (Beggan 1992, Dommer and Swaminathan 2013). One will unconsciously ask for more compensation if he has to give up his owning goods, leading to a WTA-WTP gap. Besides, Carmon and Ariely (2000) stated that ownership would activate and increase the owners' accessibility to the positive features of the goods they own, which results in a higher value of their goods compared to other people. In CPDR, under the influence of ownership effect, disputants would attribute irrational value to their assessments and opinions about the dispute matter. Their further decision making would be affected to defend their previous assessments.

2.2.2 Loss aversion

Thaler (1980) described EE as a form of manifestation of loss aversion. Loss aversion is a concept derived from prospect theory and it is the tendency to prefer avoiding loss over acquiring gains of the same amount (Tversky and Kahneman 1991). Loss aversion implies that decision makers will attach more weights to prospective loss than to equivalent prospective gains, where losses are the outcome below the reference states while gains are the outcome above the reference state (Kahneman and Tversky 1979). Receiving an entitlement will move an individual from a state of not-owning to owing the entitlement, whereas forgoing an entitlement will move and individual from owing to not owing the entitlement (Novemsky and Kahneman 2005, Morewedge and Giblin 2015). People are generally more sensitive to loss than gain, loss looms larger than its corresponding gain (Tversky and Kahneman 1991). Therefore, he would ask for more compensation if he had to sell his ownings, compared to the price he would pay to acquire the identical amount of goods. In CPDR, taking their own assessment about the dispute as a reference state, disputants would consider comprising as receiving losses. Disputants' loss aversion tendency makes them become more sensible to their compromises than their prospective gains from settlement. Hence, disputants would be reluctant to step out and accept the offer from the counterpart.

2.2.3 *Status quo bias*

Korobkin (2003) claimed that EE is developed from status quo bias. Samuelson and Zeckhauser (1988) first introduced the term status quo bias and they conducted a series of decision-making experiments and subjects were found to be inclined to stick with their original status quo. Status quo bias is a psychological bias that individuals have a greater tendency to remain status quo when making decisions under risks (Kahneman *et al.* 1991). Status quo is the baseline of the decision makers when confronted with great uncertainties over the circumstances of the decision. Any changes from this baseline will be seen as a loss. Under the influence of status quo bias, individuals are reluctant to make changes. They will only do so if the expected outcome offers a significant higher gain. Thus, their WTA would be higher than the corresponding WTP. This shows the presence of EE in exchange situations. In CPDR, bias towards status quo would make disputants keep their original assessments about the dispute. They are unwilling to search alternative information or consider evidence from the opposite side to construct new positions.

2.2.4 *Bargaining habit*

It was suggested that strategic bargaining habit was one of the factors that contribute to the formation of a WTP-WTA gap (Kahneman *et al.* 1991). When people have to make trade-offs in negotiations, individuals were found to misrepresent their valuations by over-stating or under-stating their true value in order to receive maximum profit. Knez *et al.* (1985) discovered that sellers often over-claimed their WTA in order to maximize profits. On the other hand, buyer will tend to under-claim their WTP when he attempts to acquire something. Morewedge and Giblin (2015) related this bargaining habit as a strategic misrepresentation. In negotiations, people will strategically misrepresent their valuation of the goods, either over-value it or under-value it, so as to achieve a more favorable outcome. This phenomenon is not only observed in goods exchange situations, it is also presented in business negotiations (Ross and Ward 1995). Parties' unfaithful offers and responses are important reasons leading to negotiation failures (Ross and Ward 1995). In CPDR, disputants may use opportunistic bargaining strategies to maximize their potential interests, which will devastate the relationship among the parties and makes the resolution process become more difficult.

3 SUGGESTIONS TO OVERCOME ENDOWMENT EFFECT IN CPDR

Involvement of a third party neutral to assist CPDR can be a feasible way to narrow the impact of EE (Ross and Ward 1995, Elliott 1999). Although the presence of a third party cannot completely eliminate the psychological barriers as parties are still affected by their own thinking

(Korobkin 2005). A third-party neutral can guide the negotiations in the "right" direction to a mutual consented settlement (Ross *et al.* 1975, Lord *et al.* 1984). A third-party neutral helps disputants to unveil their stereotypes, discover their common goals and re-build mutual trust (Clark and Marshall 1981, Ross and Ward 1995). Awareness is the first step for EE minimization. Besides, when the common goals and mutual trust are developed, communication would be enhanced among the disputing parties. This will in turn reduce opportunistic and aggressive bargaining strategies. In these regards, the impact of EE would be alleviated with the facilitation of third-party neutral advisors in CPDR. It is therefore worthwhile and economical to incorporate a third party neutral not only just after a dispute has arisen, instead, having the expert assistance of a third-party neutral like dispute resolution advisor throughout the whole project duration would be instrumental in reducing the chance of developing a EE-based dispute resolution attitude. The improvement in communication and early detection of contracting parties' irrational behaviors together would reduce the chance of developing EE. Timely advice and intervention by a standing third-party neutral would steer the disputing parties back to rational courses.

4 CONCLUSIONS

Dispute is inevitable due to the complex nature of construction project (Cheung and Yiu 2006). Extensive resources would be saved if construction dispute could be resolved efficiently. Identifying the barriers to amicable CPDR is the first step. This study focuses on one of the psychological barriers—endowment effect (EE). EE was first proposed by the 2017 Nobel laureate professor Richard Thaler and it is an irrational behavior underpinned by the tendency of people attributing higher value to the things they own. The existence of EE would bring difficulties to amicable communication among the disputing parties and hamper successful CPDR settlement. This study examined the influence of EE in CPDR. The nature and theoretical background of the EE were explored. Four sources of EE and their respective influence in CPDR were identified: ownership; loss aversion; status quo bias and bargaining habit. The involvement of third-party neutral was further highlighted as an instrumental way to minimize the influence of EE.

Acknowledgements

The work described in the paper was supported by a CityU Strategic Research Grant (no. 7004870) and a HKSAR RGC General Research Fund Project (no. 11209118).

References

Arcadis, *Global Construction Dispute Report 2017: Avoiding the Same Pitfalls*, 2017.

Ariely, D., Huber, J., and Wertenbroch, K., When Do Losses Loom Larger Than Gains?, *Journal of Marketing Research,* 42(2), 134-138, 2005.

Beggan, J. K., On the Social Nature of Nonsocial Perception: The Mere Ownership Effect, *Journal of Personality and Social Psychology*, 62(2), 229, 1992.

Carmon, Z., and Ariely, D., Focusing on the Forgone: How Value Can Appear So Different to Buyers and Sellers, *Journal of consumer research*, 27(3), 360-370, 2000.

Carmon, Z., Wertenbroch, K., and Zeelenberg, M., Option Attachment: When Deliberating Makes Choosing Feel Like Losing, *Journal of Consumer research,* 30(1), 15-29, 2003.

Cheung, S. O., and Yiu, T. W., Are Construction Disputes Inevitable?, *IEEE Transactions on Engineering Management,* 53(3), 456-470, 2006.

Clark, H. H., and Marshall, C. R., Definite Reference and Mutual Knowledge, *Linguistic Structure and Discourse Settings*, Cambridge University Press, 10-63, 1981.

Coase, R. H., *The Firm the Market and the Law*, The University of Chicago Press, Chicago and London, 1988.

De Dreu, C. K., and Van Knippenberg, D., The Possessive Self as a Barrier to Conflict Resolution: Effects of Mere Ownership, Process Accountability, and Self-Concept Clarity on Competitive Cognitions and Behavior, *Journal of Personality and Social Psychology*, 89(3), 345-357, 2005.

Dommer, S. L., and Swaninathanm, V., Explaining the Endowment Effect Through Ownership: the Role of Identity, Gender, and Self-Threat, *Journal of Consumer Research,* 39(5), 1034-1050, 2013.

Elliott, M. L. P., The Role of Facilitators, Mediators, and other Consensus Building Practitioners, *The Consensus Building Handbook: A Comprehensive Guide to Reaching Agreement,* Sage Publications, Inc: California, 199-239, 1999.

Kahneman, D., and Tversky, A., Prospect Theory: An Analysis of Decision under Risk, *Econometrica*, The Economic Society, 47(2), 263-292, 1979.

Kahneman, D., Knestsch, J. L., and Thaler, R. H., Experimental Tests of the Endowment Effect and the Coase Theorem, *Journal of Political Economy*, The University of Chicago, 98(6), 1325-1348, 1991.

Knez, P., Smith, V. L., and Williams, A. W., Individual rationality, market rationality, and value estimation, *The American Economic Review*, 75(2), 397-402, 1985.

Korobkin, R., Psychological Impediments to Mediation Success: Theory and Practice, *Ohio State Journal on Dispute Resolution*, 21, 281-307, 2005.

Korobkin, R., The Endowment Effect and Legal Analysis, *Northwest University Law Review*, 97(3), 1227, 2003.

Liedtka, J., Perspective: Linking Design Thinking with Innovation Outcomes Through Cognitive Bias Reduction, *Journal of Product Innovation Management*, 32(6), 925-938, 2015.

Lord, C. G., Lepper, M. R., and Preston, E., Considering the Opposite: A Corrective Strategy for Social Judgement, *Journal of Personality and Social Psychology*, 47(6), 1231-1243, 1984.

Mnookin, R., Strategic Barriers to Dispute Resolution: A Comparison of Bilateral and Multilateral Negotiations, *Harvard Negotiation Law Review*, 8(1), 1-27, 2003.

Morewedge, C. K., and Giblin, C. E., Explanations of the Endowment Effect and Integrative Review, *Trends in cognitive sciences*, Cell Press, 19(6), 339-348, 2015.

Morewedge, C. K., Shu, L. L., Gilbert, D. T., and Wilson, T. D., Bad Riddance or Good Rubbish? Ownership and Not Loss Aversion Causes The Endowment Effect, *Journal of Experimental Social Psychology*, 45(4), 947-951, 2009.

Novemsky, N., and Kahneman, D., The Boundaries of Loss Aversion, *Journal of Marketing Research*, 42(2), 119-128, 2005.

Raiffa, H., *The Art and Science of Negotiation*, Harvard University Press, Cambridge, 1982.

Ross, L., and Ward, A., Psychological Barriers to Dispute Resolution, *Advances in Experimental Social Psychology,* Academic Press, 27, 255-304, 1995.

Ross, L., Lepper, M. R., and Hubbard, M., Perseverance in Self-Perception and Social Perception: Biased Attributional Processes in The Debriefing Paradigm, *Journal of Personality and Aocial Psychology*, 32(5), 880-892, 1975.

Samuelson, W., and Zeckhauser, R., Status Quo Bias in Decision Making, *Journal of Risk and Uncertainty,* 1(1), 7-59, 1988.

Shu, S. B., and Peck, J., Psychological Ownership and Affective Reaction: Emotional Attachment Process Variables and The Endowment Effect, *Journal of Consumer Psychology,* 21(4), 439-452, 2011.

Spangler, B. M., Heads I Win, Tails You Lose: The Psychological Barriers to Economically Efficient Civil Settlement and a Case for Third-Party Mediation, *Wisconsin Law Review*, 1435-1466, 2012.

Thaler, R., Toward a Positive Theory of Consumer Choice, *Journal of Economic Behavior and Organization*, 1(1), 39-60, 1980.

Tversky, A., and Kahneman, D., Loss Aversion in Riskless Choice: A Reference Dependent Model, *The quarterly journal of economics,* Oxford University Press, 106(4), 1039-1061, 1991.

Weiss, L., and Johar, G. V., Egocentric Categorization and Product Judgment: Seeing Your Traits in What You Own (and Their Opposite in What You Don't), *Journal of Consumer Research*, 40(1), 185-201, 2013.

Wong, P. S., Cheung, S. O., and Ho, P. K., Contractor as Trust Initiator in Construction Partnering Prisoner's Dilemma Perspective, *Journal of Construction Engineering and Management,* 131(10), 1045-1053, 2005.

Interdependence between Structural Engineering and Construction Management
Edited by Ozevin, D., Ataei, H., Modares, M., Gurgun, A., Yazdani, S., and Singh, A.

ISBN: 978-0-9960437-6-2

AN INVESTIGATION OF THE CONTRACTUAL ROADWAY SAFETY TERMS IN TRANSPORTATION PPP PROJECTS

LUMING SHANG and AHMED M. ABDEL AZIZ

Dept of Construction Management, University of Washington, Seattle, USA

Roadway safety is an important aspect of all transportation facilities. With operation and maintenance becoming contracted roles under public-private partnerships (PPP), it is expected that safety rates will improve beyond those experienced in traditional delivery. The international experience shows that PPPs experienced mechanisms designed to provide safer roads. In the United States, PPP is gaining momentum and several states initiated PPP projects. While studies show the potential of PPPs for improving cost efficiency and expediting the delivery time, few researches investigated the correlation between PPP and roadway safety. To fill the gap, this study investigated seventeen PPP projects through content analysis of their agreement and procurement documents. The findings show that PPP projects were lacking further consideration toward safety beyond what is normally available in traditional delivery. Safety was an objective in most (76%) of the projects, however, without direct mechanisms to implement such objective. Around half of the projects have mentioned safety as part of proposal evaluation but only two projects have assigned points/weights in the evaluation. None of the projects provided for a direct link between the compensation to the achievement of improvements in the safety rates. Generally, PPPs are meant to provide public agencies with means to achieve particular objectives. It is recommended to use the particulars of PPPs in performance measures and payment mechanisms to enforce roadway safety in highways infrastructure.

Keywords: Transportation systems, Evaluation, Accident management, Safety incentives, Documents, Design.

1 ROADWAY SAFETY AND PUBLIC-PRIVATE PARTNERSHIP

Roadway safety is an important aspect of safe and reliable transportation systems. Poor roadway design and poor pavement conditions are among the factors that would render a highway unsafe with an increasing number of accidents, injuries, and fatalities. During the life cycle of a road, from its preliminary design to operation, there are decisions that affect roadway safety. In this respect, the operation stage offers great opportunities to improve roadway safety (Vassallo *et al.* 2009). Well-maintained roads are expected to have lower accident rate than roads with pavement cracks, rutting, potholes and other pavement distresses. Therefore, it is fair to assume that roads with sufficient funding for operation and maintenance would have fewer accidents.

Although state and federal agencies have continuously provided capital for roadway design and construction, funding for operation and maintenance is always in shortage. Innovative procurement approaches are increasingly used to fill the financial gap by involving the private sector in managing and financing infrastructure through Public-Private Partnership (PPP). In performance-based PPP contracts, road maintenance and financing could be provided as part of the contractual delivery system (Abdel Aziz 2007). While there have been several studies showing the potential of PPPs for improving cost efficiency and expediting the delivery time, research on the relationship between roadway safety and PPP are scarce with regard to the contractual safety terms and safety performance. Geddes (2014) conducted research regarding the effects of PPP on traffic safety based on the Mexico federal toll roads and found that private management of federal roads in Mexico had no improvement on safety performance. In another research, Vassallo *et al.* (2009) concluded that PPPs with safety payments or incentives is effective in encouraging the contractor to implement safety approaches and finally could improve the roadway safety performance.

To fill the research gap between PPP and roadway safety performance in the US, this paper aims to investigate the contractual terms related to roadway safety of transportation PPP projects in the US and to develop recommendations to articulate contractual terms for safety improvement.

2 RESEARCH METHODOLOGY

States that have (1) PPP enabling legislation, (2) availability of sufficient number transportation PPP projects, and (3) a wealth of experience and history of using PPPs are identified as targets for the research purpose. Table 1 shows such target states and projects used in this research.

Table 1. The information of the selected PPP projects.

Project	State	PPP Contract Model	Year of Financial Closure	Status
The Port of Miami Tunnel	Florida	DBFOM	2009	In Operation
I-595 Improvement	Florida	DBFOM	2009	In Operation
I-4 Ultimate	Florida	DBFOM	2014	Under Construction
SR125	California	DBFOM	2003	In Operation
The Presidio Parkway	California	DBFOM	2012	In Operation
IH 635 Managed Lanes	Texas	DBFOM	2008	In Operation
North Tarrant Express	Texas	DBFOM	2009	In Operation
SH 183 Managed Lanes	Texas	DBFOM	2014	Under Construction
SH 130 Seg. 5 & 6	Texas	DBFOM	2007	In Operation
Elizabeth River Tunnels	Virginia	DBFOM	2012	Under Construction
I-495	Virginia	DBFOM	2008	In Operation
I-95	Virginia	DBFOM	2012	In Operation
Pocahontas Parkway	Virginia	Lease-Develop-Operate	2007	In Operation
Indiana Toll Road	Indiana	75-year long-term lease	2006	In Operation
PR-22 and PR-5 Toll Road	Puerto Rico	40-year long-term lease	2011	In Operation
Northwest Parkway	Colorado	Long-term lease	2007	In Operation
Chicago Skyway	Illinois	99-year long-term lease	2005	In Operation

A content analysis of the procurement and contract documents is conducted to investigate how traffic safety was managed and accounted for in the target projects. The documents of each project has been divided into four categories, including bidding document, contract document – Part I Design & Operation, contract document – Part II Safety Incentives and Payments, and contract document – Part III Safety Accident Management. A set of questions has been designed for each category to fully explore how roadway safety is managed in these projects.

3 CONTRACTUAL ANALYSIS

The objective of the content analysis is to establish the understanding of the roadway safety management in current US PPP projects.

3.1 Safety Objectives in Procurement Documents

This part aims to answer two questions – 1) whether roadway safety is considered as a significant goal in the project, and 2) how well the safety considerations have been reflected in bidding evaluation criteria. The results of the analysis are in Figure 1.

Regarding the first question, only those projects that explicitly include safety as one of the project objectives are marked as "Yes". For example, this study marked the I-595 Improvement project as "Yes" since its RFP claimed "*the primary objective throughout the Term of The Project are: ... Maintain a high level of quality and safety provisions in the engineering, construction, maintenance and operations services provided by the Concessionaire.*" A total of 13 projects (76%) included safety in their project objectives, as shown in Figure 1(a).

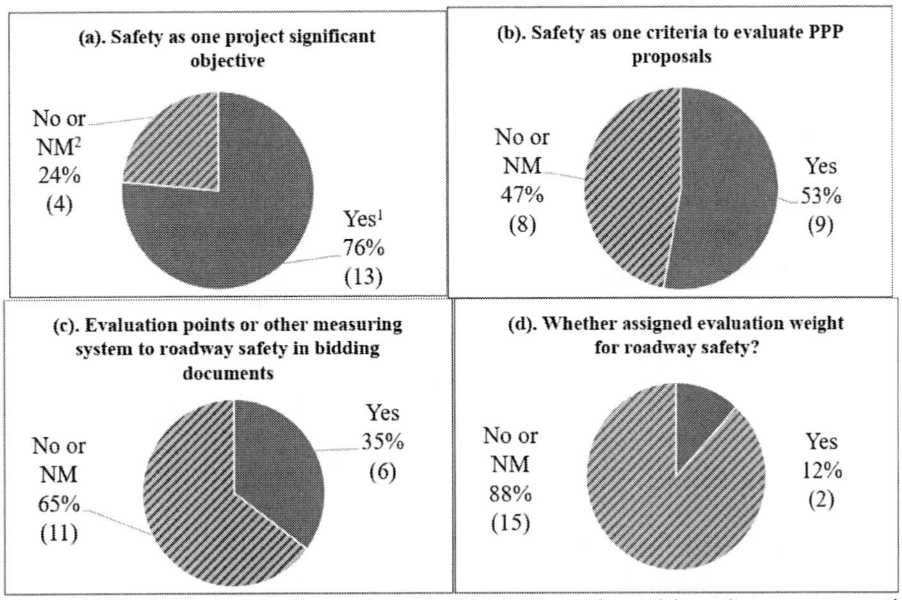

Notes: [1]Projects marked as "Yes" only if containing certain contractual provisions that can answer the proposed questions; [2]NO or NM (Not Mentioned) for projects not containing any provisions related to the proposed questions.

Figure 1. Safety considerations in the bidding documents.

In terms of the selection criteria, about half of the selected projects (9 out of 17) included safety as one of the project objectives and also contained safety in their Evaluation Criteria, Figure 1(b). In these nine projects, a common method is to include safety management as an element to evaluate the technical capability of the bidding team. In these projects, safety is categorized as part of technical capability, however, a fewer number of projects specifically assigned evaluation points or rating weights to safety; this is presented in Figure 1(c) and Figure 1(d). Usually, the evaluation weights are only assigned for team's Technical Capability as a whole. Only Pocahontas Parkway and SH 183 assigned evaluating points for safety.

Although most PPP transportation projects considered safety as one of important project goals, there is a much smaller portion that clearly reflected the safety objective on the selection criteria. This is clearly noted where the percentages go down from 76% to 53% to 35% and finally to 12% in Figure 2. For instance, I-595 Improvements, the Presidio Parkway, the Elizabeth River Tunnel, I-495 Express Lanes, Indiana Toll Road, and PR-22 and PR-5 have similar contracts to include safety as part of evaluation criteria for the Technical Capacity, but none of them assigned any weighting points specifically on safety evaluation.

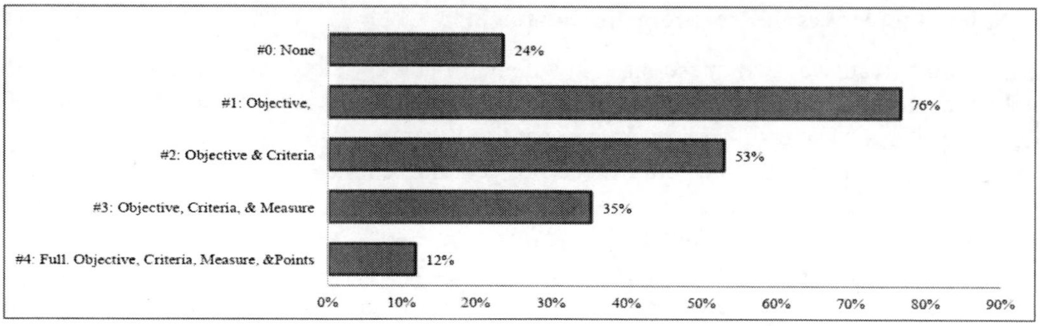

Figure 2. Relative safety considerations.

3.2 Contractual Document Part I: Design and Operation

The PPP Technical Requirements commonly contain safety-related contractual terms, such as the road slope, lighting, and materials. Also, the concession agreements require their concessionaires to provide a comprehensive Safety Plan. However, since these safety-related terms are inherited from the traditional project delivery methods, they are considered as traditional terms in this study. This subsection aims to investigate if PPP agreements include any special safety terms, beyond what are included in traditional contracts in terms of structural design and operational requirements.

In terms of special design or extra roadway elements for safety improvement, as shown in Figure 3(a), 8 of the 17 (47%) target projects contain at least one contractual design term that differs from the traditional contractual template. These special design-related provisions are heterogeneous and vary from project to project. For example, I-595 Improvements required a custom design should meet or exceed all of the safety requirements provided by the standard signature gantry, which is a set of particular design requirements in the Agreement; LBJ 635 adopted a geotechnical instrumentation program to monitor the safety and adequacy of the design and construction approach; I-95 Express Lanes required a reliability assessment shall be made for safety purpose, and furthermore, the project also required an operational analysis to demonstrate the concessionaire's revised design did not have a significant adverse impact on the safety and operation of the existing facility; Indiana Toll Road developed a particular traffic plan, named Safety Initiative, for roadway safety improvement; and PR-5 / PR-22 implemented a program called "the accelerated safety upgrades" to improve the performance and safety conditions.

In terms of operation, as presented in Figure 3(b), 12 of the 17 (71%) projects include special provisions to improve roadway safety in two ways. The first requires the concessionaire to conduct periodic safety inspection to monitor the safety condition. The second strategy includes a requirement for the concessionaire to regularly update the safety plans.

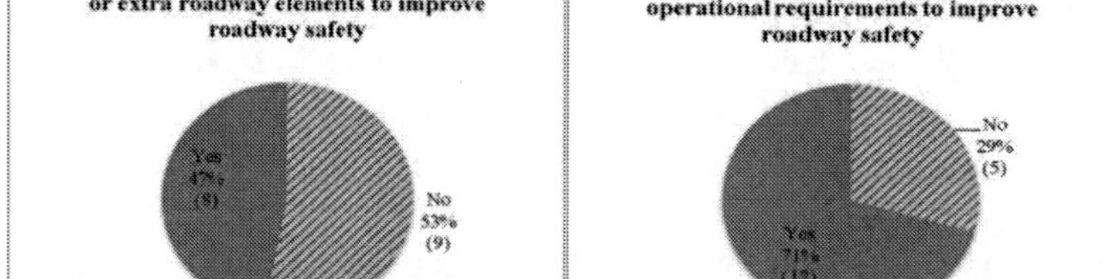

Figure 3. Special design and operation safety terms in target projects.

3.3 Contractual Document Part II: Safety Incentives and/or Payments

This part aims to identify the usage, if any, of monetary incentives for managing roadway safety. According to the payment mechanisms, none of the 17 projects contained any payment that links a monetary compensation to the traffic safety performance.

All the selected projects contained traditional provisions regarding roadway safety precautions (Figure 4). Two safety-related contractual chapters, Safety Compliance and Remedies for Failure to Meet Safety Standards or perform Safety Compliance, are widely applied to all PPP projects. In addition, these two chapters also included some terms potentially involved in a safety-related deduction or compensation. For example, I-4 Ultimate required if the concessionaire failed to be compliant with safety standards or any other regulations, the public agency has the right to take actions directly to ensure the safety and quality of the project and the concessionaire is responsible for all the expenses and potential losses due to such actions. However, the safety mentioned in most of the provisions does not clearly refer to the roadway or traffic safety. Although some contracts provided for deductions of the general O&M performance of which safety-related requirements is indirectly included (*e.g.* sealing of pot holes and cracks), none of the 17 projects required the evaluation of any safety performance measure (*e.g.*, number of crashes, injuries, or fatalities). None have provided any safety payment or incentive for a better safety performance.

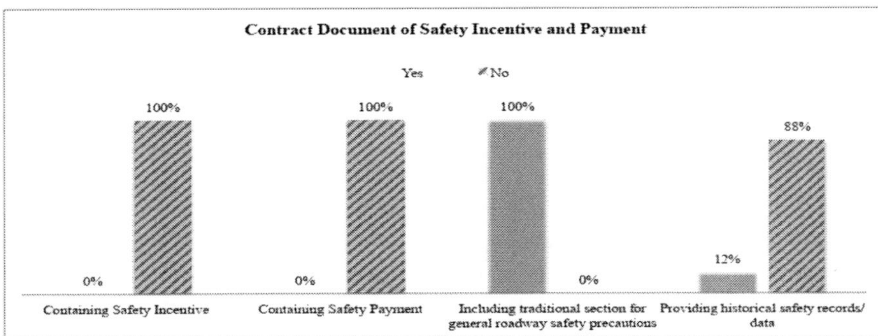

Figure 4. The adoption of safety-related payments and incentives.

Additionally, few projects required the use of historical safety records to assess roadway safety performance. Only two projects (I-4 Ultimate and Port of Miami Tunnel) required in the Technical Requirements utilizing traffic data for project analysis. However, Port of Miami Tunnel only used AADT to establish a revised traffic operational analysis for congestion management rather than for improving traffic safety. I-4 Ultimate is the only one of the 17 projects providing historical traffic safety records in the contract. It adopted a safety ratio to assess roadway safety performance. However, this ratio was only used for identifying the hazard location of the project instead of evaluating the operational performance regarding safety.

3.4 Contractual Document Part III: Roadway Accident Management

In terms of traffic accident/incident reporting and management, this research found that collecting and keeping safety data (*e.g.* number of accidents was not a major contractual requirement) which can be seen in Figure 5. The majority of the 17 projects (94%) required their concessionaires to immediately report any safety-related incidents to the public law enforcement agency. There is one project (6%), LBJ 635, allowed an independent agency to do this task. Only four (25%)

projects required the concessionaire to collect safety data and only the Elizabeth River Tunnel) required the concessionaire to manage information identified through incident reports.

While collecting and keeping safety data was not a major requirement for the concessionaire, all of 17 projects required the concessionaire to develop Incident Management/Response Plans.

All of the selected projects required development of an incident management plan to immediately response and manage the roadway incidents. Additionally, most (81%) of the 17 projects required the concessionaires to be responsible for the maintenance and repair, and 11 projects required the concessionaires to pay for the repair works. Some agencies authorized pursuing claims against any responsible third party for reimbursement of expenses incurred.

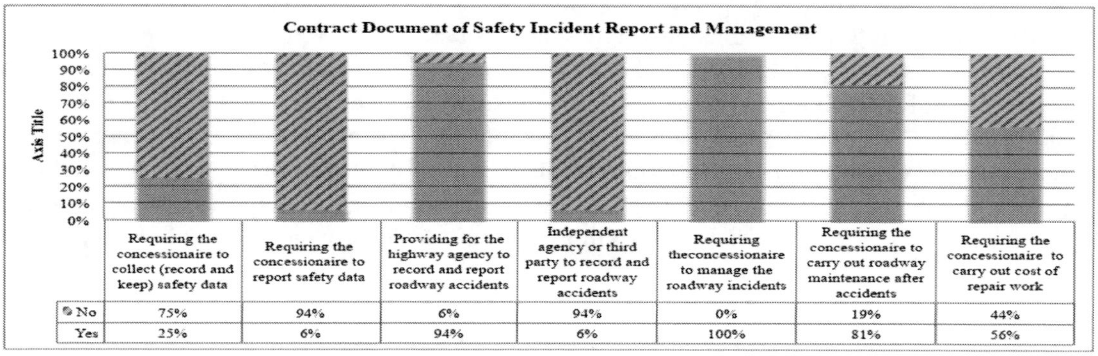

Figure 5. Safety incident report and management.

4 SUMMARY AND CONCLUSIONS

Following to the analysis of the 17 PPP projects, it can be fairly concluded that the US take a lenient stance toward using PPPs to enforce better roadway safety performance. By default, all the selected projects account for the traditional safety requirements as set by the states/federal governments. The analysis shows that safety was an important objective, however, there was no particular mechanism to implement that objective. Half of the projects (53%) include safety in proposal evaluation, but only two projects assigned points/weights to the safety requirements.

All of the agreements required the concessionaire to provide regular safety plans. However, for the operation phase, 12 out of the 17 projects (71%) called for carrying inspections and updating of the roadway safety plans. Only four projects required collect and report traffic and crash data. The PPP projects had no safety incentives, payment, or other mechanism. No provisions were made to assess any safety performance indicators except for one project.

It is suggested that public highway agencies to investigate proactive systems that enforce and incentivize higher roadway safety standards. Along with the traditional contractual safety terms, public agencies would be advised to investigate the broad range of options for proactive systems, e.g. link the concessionaire compensation to the achievement of particular safety objectives.

References

Abdel Aziz, A. M., A Survey of The Payment Mechanisms for Transportation DBFO projects in British Columbia, *Construction Management and Economics*, 25(5), 529-543, 2007.

Geddes, R., The Effects of Public-Private Partnerships on Traffic Safety: Evidence from Mexico, 2014. Retrieved from http://www.utrc2.org/sites/default/files/pubs/Final-PPP-Traffic-Safety.pdf on Jan 2019.

Vassallo, J. M., Rangel. T., de Vilar, P.P., and Areans, B., Do PPP Contracts Improve Road Safety, *Research Paper, European Investment Bank University Research Sponsorship Program*, European Investment Bank and Universidad Politecnica de Madrid, Madrid, 2009.

Interdependence between Structural Engineering and Construction Management
Edited by Ozevin, D., Ataei, H., Modares, M. Gurgun, A., Yazdani, S., and Singh, A.

ISBN: 978-0-9960437-6-2

AN INVESTIGATION OF FATAL ACCIDENTS IN DEMOLITION WORK OF BUILDINGS

HIROKI TAKAHASHI

Country Construction Safety Research Group, National Institute of Occupational Safety and Health, Kiyose, Japan

Buildings constructed in Japan during a period of high economic growth, 1954–1973, have deteriorated, and many need to be demolished. In the course of that demolition work, accidents have occurred, but the causes and circumstances of many of them are largely unknown. This study investigates accidents that occurred during building demolition work in Japan between 2010 and 2014, with the goal of preventing future accidents. The results show that fatal accidents while demolishing buildings accounted for 6.5% of all fatal accidents on construction sites. The greatest number of fatal accidents can be attributed to falls and collapsing walls. The causes of falls were found to be two-fold. First, in the process of demolishing a building, frequently, a hole is cut in the floor and scrap materials are thrown into it to dispose of them; however, workers have been known to fall into this hole and be injured fatally. It was also found that many workers do not use full harnesses. The cause of collapsing walls was identified as follows. When a wall is demolished, its lower part is often cut through to weaken it, at which point the wall can collapse, crushing workers. Analysis of the study results shows that greater attention must be paid to creating a system that promotes fall prevention and safe wall-cutting on demolition sites.

Keywords: Fall, Collapse, Japan, Occupational injuries.

1 INTRODUCTION

The buildings constructed in Japan during a period of high economic growth, 1954–1973, have deteriorated, and many need to be demolished. In addition, accidents have occurred during restoration work after the 2011 Great East Japan Earthquake. But, the causes and circumstances of many of them are largely unknown.

There is a difference in competence among demolition firms in Japan. Highly competent demolition firms can safely demolish high buildings. But, many accidents occur during work by incompetent demolition firms. Qualification programs exist to improve skills and safety (JCTC 2018 and JDCA 2018). These programs cover the duties of workers, the relevant laws, demolition methods, and the practical business of ensuring a safe work environment. Workers who complete the qualification programs become demolition site supervisors. In addition, the qualification examination exists to improve construction management ability (JDCA 2018). The examination tests knowledge of demolition methods and equipment, ability to create construction plans and quotations for demolition works, and ability to manage demolition works and teach workers. Workers who pass the qualification examination become technical managers of demolition works. But these programs and examinations do not address the present situation regarding accidents during demolition works.

There are papers on the destruction of concrete (Suzuki *et al.* 2018 and Yuasa 2018), a study on supporting machines on the floor (Aoki *et al.* 2018), and research on asbestos (Horie *et al.* 2017). However, there are no scholarly articles on the present situation of accidents during demolition work. There are safety prevention guidelines (PBA 2013), recommendations (JCOSHA 2012a, JCOSHA 2012b and JCOSHA 2016), and technical books (SGDM 2017) on demolition work, but they do not address the current situation of accidents during demolition work.

This study investigated fatal accidents that occurred during building demolition work in Japan over a five-year period, with the goal of preventing future accidents.

2 METHOD OF INVESTIGATION

The fatal accident data used were public data from the government of Japan (Ministry of HLW 2018) and included the types and causal agents of the accidents. From the fatal accident data, the study surveyed a sample of building demolition accidents that occurred during a five-year period, between 2010 and 2014.

3 RESULTS OF INVESTIGATION

3.1 Occurrences by Year

Figure 1 shows the fatal accidents occurring during building demolition work by year. The figure shows that the number of fatal accidents has decreased gradually since 2012. For comparison, Figure 2 shows fatal accidents at all construction sites by year. As that figure shows, the number of fatal accidents occurring annually at all construction sites remained consistent until increasing in 2014. The fatal accidents during building demolition work were about 6.5 % of all fatal accidents on construction sites during the period investigated.

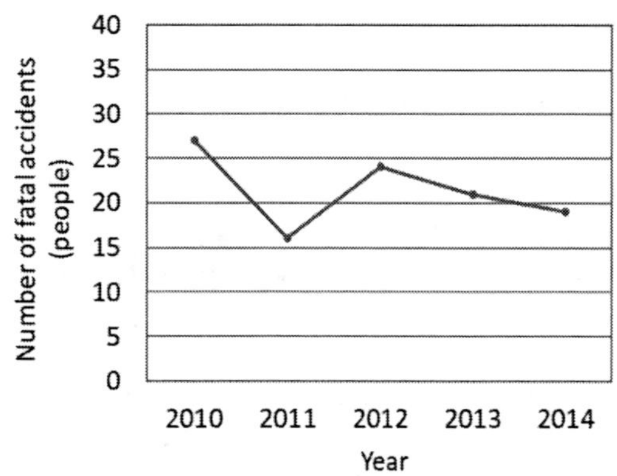

Figure 1. Fatal accidents during building demolition work by year.

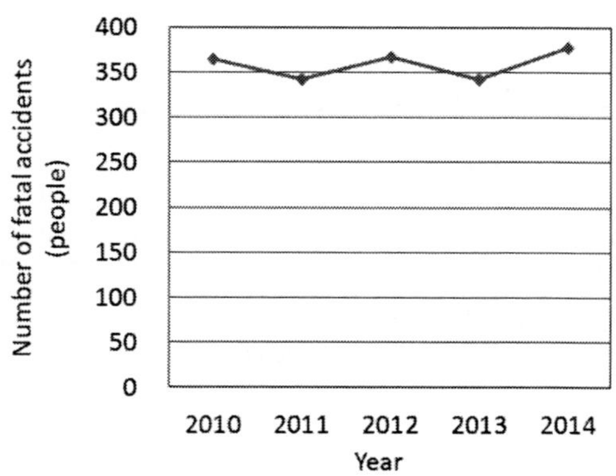

Figure 2. Fatal accidents at all construction sites by year.

3.2 Occurrences by Type

Figure 3 shows fatal accidents during building demolition work by type. The highest number, 56, were attributed to "Fall," which accounted for 52% of fatal accidents during demolition work, followed by "Collapse," which killed 20 workers and accounted for 19% of fatalities.

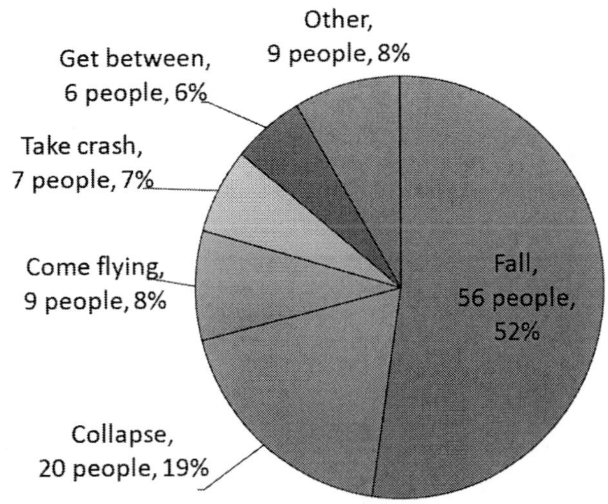

Figure 3. Fatal accidents during building demolition work by type.

3.3 Fatal Fall

Figure 4 breaks down fatal falls by the cause of the fall. Slate roof, wooden house, and steel-framed building each caused a significant number of fatal falls.

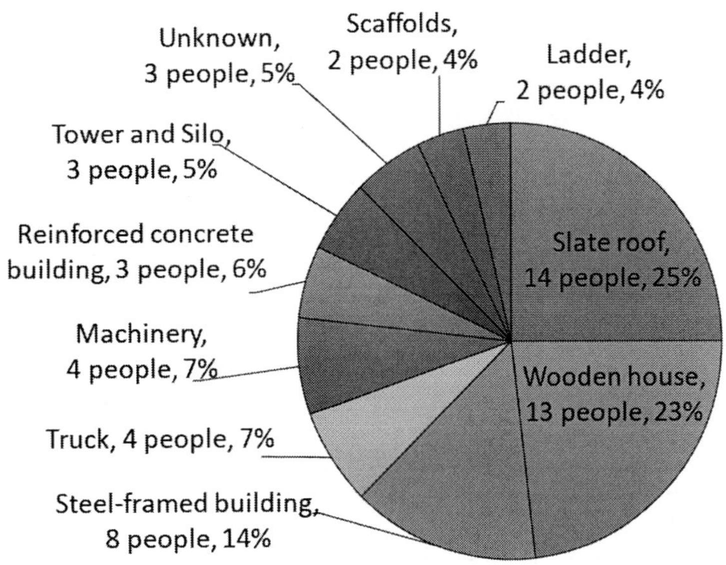

Figure 4. Fatal falls during building demolition work by causal agents.

Figure 5 breaks down the three biggest causes of fatal falls, identified in Figure 4, by the major causal agents for each. Figure 5(a) shows that "Roof tile gave way" is the major cause of falls from slate roof. Figure 5(b) shows that "Falling from roof" is the major type of fatal fall associated with wooden houses. Figure 5(c) shows that "Beam" are the most frequent site of fatal falls from steel-framed building. Figure 5(d) shows that "Opening" are the most frequent site of fatal falls in reinforced concrete building. The workers fall through these openings, into which they throw pieces of broken concrete to dispose of them. Workers must also set up guardrails around excavations to prevent falls.

The rates of these fatal falls are increased by the fact that workers are not using full harnesses during demolition work (JCOSHA 2018). Given this, these accidents cannot be prevented just by knowing the causal agents of the falls. Fall prevention must include making sure that every worker wears a full harness every time that it is warranted. However, the set of main rope is difficult during demolition work because the strength of an unstable structure is unclear. There is no fall prevention method during demolition work. Fall prevention methods must be examined during demolition.

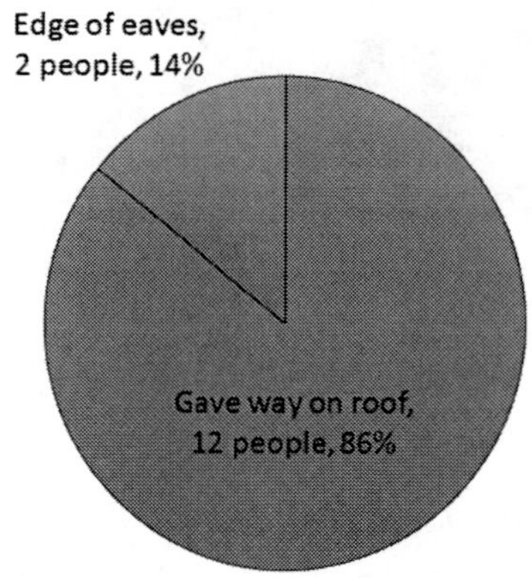

(a) Slate roof.

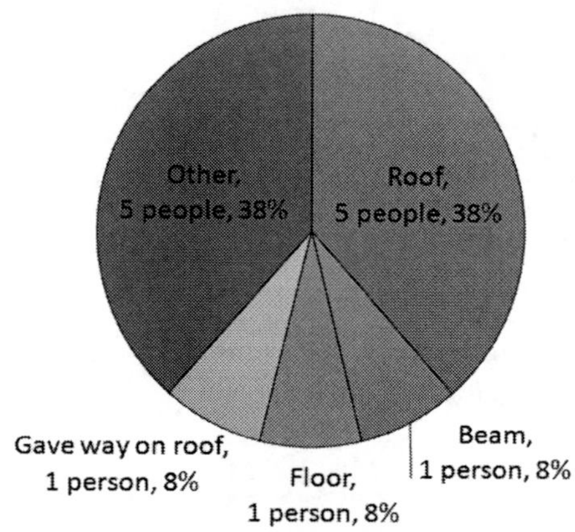

(b) Wooden house.

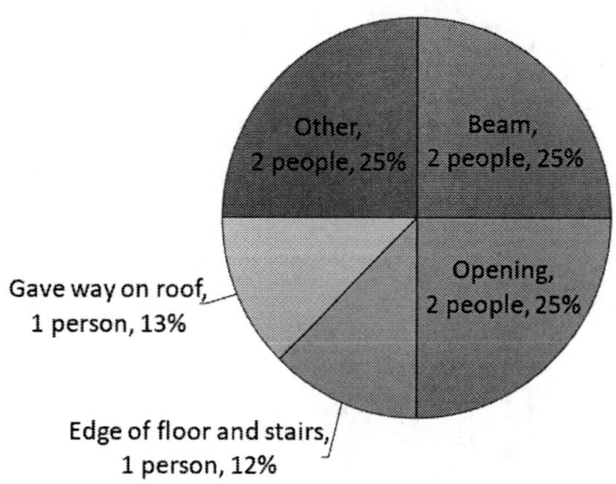

(c) Steel-framed building.

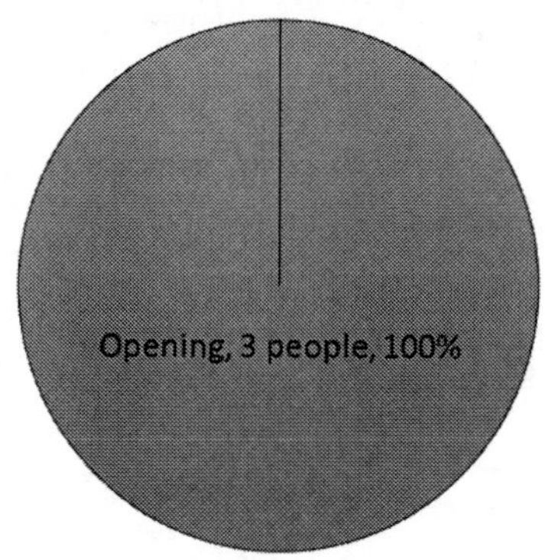

(d) Reinforced concrete building.

Figure 5. Fatal falls during building demolition work by part of the structure involved.

3.4 Collapse

Figure 6 breaks down fatalities from collapsing walls, the second most frequent cause of demolition site fatalities, by causal agent.

In Japan, when workers demolish concrete and wooden walls, they knock them down (JCOSHA 2012a, JCOSHA 2012b). Before doing this, they cut through the lower part of each wall to weaken it. If the cut is too deep, the wall may collapse during cutting and crush the worker. If a concrete wall is reinforced with steel, the worker cannot tell how much of the steel has been cut through, which may increase the chances that the wall collapse and crush the worker. Guidelines (PBA 2013), recommendations (JCOSHA 2012a, JCOSHA 2012b and JCOSHA 2016), and technical books (SGDM 2017) do not discuss the extent to which a wall should be cut. Because there is no uniform standard regarding how deeply to cut into walls, a worker knocking down a wall does not know how to minimize the danger of the wall collapsing. Another cause of crushing fatalities is workers entering areas of unsteady walls during the demolition process. Research is needed into how deeply to cut into walls. Such research would enable the proposal of uniform technical standards regarding wall demolition that would improve safety.

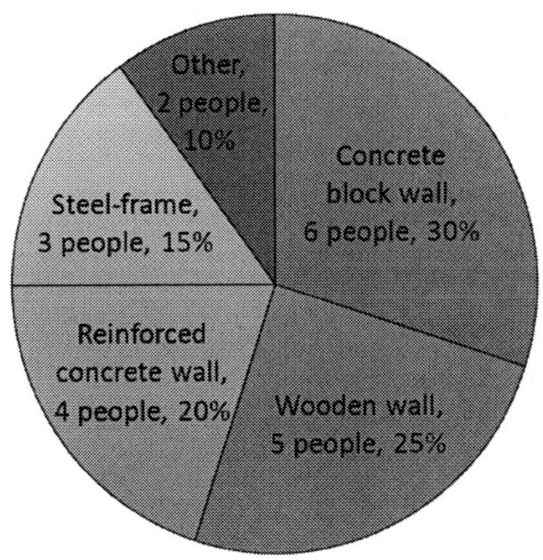

Figure 6. Fatalities due to collapsing walls by wall type.

4 CONCLUSIONS

This study investigated fatal accidents that occurred during building demolition work in Japan between 2010 and 2014. The results showed that fatal accidents while demolishing buildings accounted for 6.5% of all fatal accidents on construction sites during that period. The greatest number of fatal accidents can be attributed to workers falling and walls collapsing.

In the process of demolishing a building, frequently, a hole is cut in the floor into which workers throw scrap materials to dispose of them. A significant number of fatal injuries is caused by workers falling into this hole. Another significant cause of fatal falls is that workers fail to wear full harnesses. Furthermore, when a wall is demolished, its lower part is often cut into to weaken it, but if the cut is too deep, the wall may collapse, crushing the worker.

Given these major causes of fatalities on demolition sites, greater consideration must be given to promoting harness use to prevent fatal falls, and safe wall-cutting to prevent collapses that crush workers. Finally, we must evaluate the fall-prevention methods, and safety-management methods used during wall demolition.

References

Aoki, H., Aoki, K., Choyo, K., and Yusawa, N., *Study on Safety of Structure on Demolition Work*, Summaries of Technical Papers on Demolition Work, Japan Demolition Contractions Association, 7, 6-10, Tokyo, Japan, 2018.

JCOSHA, *Recommendation on Demolition Work of Concrete Building*, Japan Construction Occupational Safety and Health Association, Japan, 2012a.

JCOSHA, *Recommendation on Demolition Work of Steel Building*, Japan Construction Occupational Safety and Health Association, Japan, 2012b.
JCOSHA, *Safety on Demolition Work of Wooden Building*, Japan Construction Occupational Safety and Health Association, Japan, 2016.
JCOSHA, *Annual on Safety and Health of Construction Industry*, Japan Construction Occupational Safety and Health Association, Japan, 2018.
JCTC, Japan Contractions Training Center, Web site of Japan Contractions Training Center. Retrieved from http://www.zenkaikouren.or.jp/ on 2018.
JDCA, Japan Demolition Contractions Association, Web site of Japan Demolition Contractions Association. Retrieved from http://www.jctc.jp/ on 2018.
Horie, Y., Ichihara H., Nagai K., and Furuya S., *Study of the Dust Control of Asbestos Included Spraying Rock Wool*, Summaries of Technical Papers of Annual Meeting, Architectural Institute of Japan, 8, 1239-1244, Hiroshima, Japan, 2017.
Ministry of HLW, Ministry of Health, Labour and Welfare, Web site of the Ministry of Health, Labour and Welfare. Retrieved from http://anzeninfo.mhlw.go.jp/index.html on 2018.
PBA, *Standard Specification for Demolition Work of Building*, Public Buildings Association, Houbundo, 2013.
SGDM, *New Demolition Method and Estimating*, Study Group on Demolition Method, Economic Research Association, 2017.
Suzuki, S., Ishimaru, T., Miwa, A., Nakamura, Y., and Ueki, K., *Studies on the Crushing Technique of Concrete Structures Utilizing Pressure Rise due to Freezing of Water Filled in a Circular Tube*, Summaries of Technical Papers of Annual Meeting, Architectural Institute of Japan, 1164, 327-328, Tohoku, Japan, 2018.
Yuasa, N., Study on demolition system of High-Strength Concrete Structure, *Summaries of Technical Papers on demolition work*, Japan Demolition Contractions Association, 7, 1-5, Tokyo, Japan, 2018.

Interdependence between Structural Engineering and Construction Management
Edited by Ozevin, D., Ataei, H., Modares, M., Gurgun, A., Yazdani, S., and Singh, A.

ISBN: 978-0-9960437-6-2

DISPUTED RISKS IN CONSTRUCTION

RITA AWWAD and JOE KARAM

Dept of Civil Engineering, Lebanese American University, Byblos, Lebanon

The construction industry is perceived as one of the most crucial fields in risk management studies due to its inherent risks and complexity. Despite the large amount of research presented in this area, experts still prefer relying on their own knowledge and experience to manage project risks. This approach leads to various disputes among construction stakeholders over the ownership of these risks and their mitigation. Such disputes normally affect the project's goals in terms of budget and schedule. Many researchers have addressed risk factors in the construction field; however, none has tackled the disputed risks among construction parties. This research aims at identifying the different disputed risk factors among project stakeholders along with their respective mitigation strategy. It is based on an empirical questionnaire sent to construction practitioners with various expertise. An analysis of the results is conducted using descriptive statistics and one-on-one interviews. This research provides experts with a list of the most disputed risks in construction, along with their mitigation approach. In the end, construction parties will be able to meet the project's goals and steer clear of any delay or cost overrun resulting from the occurrence of disputed risks.

Keywords: Risk management, Risk mitigation, Risk analysis, Construction industry, Owner, Contractor.

1 INTRODUCTION

Risk is defined as an uncertain event that can produce severe consequences on an organization. It can happen in various environments and with diverse characteristics and impacts (Serpell *et al.* 2015). In construction projects, stakeholders usually perceive the risk at an advanced level because of the inherent challenges and tough working conditions faced during construction (Hanna *et al.* 2013). This results in controversies in risk allocation, with each party attempting to transfer it to others, leading to disputes between them.

However, despite having a lot of studies about risk management, experts still prefer using their own knowledge and experience for approaching risks in construction (Jung and Han 2017). This subjective approach in handling risks often leads to conflicts between the different parties in a construction project.

Despite that, the construction management literature agrees on the identification, severity and likelihood of risks in construction; every party still has its own perspective about risk responsibility and sharing. Thus, the aim of this research is to identify disputed risks between construction stakeholders through collecting their different perspectives on risk responsibility and sharing as well as mitigation strategies. This will provide practitioners in the field with a clear idea on the commonly disputed risks in construction and their most adequate mitigation strategies.

2 LITERATURE REVIEW

Research studies tackling risk management can be mainly divided into two categories: risk assessment and risk mitigation. This section will elaborate on these two groups of studies.

Many researchers have addressed risk assessment. Some have tackled this topic using empirical approaches where they surveyed construction experts' opinions through questionnaires or interviews. For example, Konior (2015) calculated a correlation factor that links the likelihood of occurrence and the impact of a risk for different project risks using direct data derived from a survey. Nan *et al.* (2016) analyzed project risks as random variables in order to estimate the probability distributions of the attributes of perceived risk and analyze their criticality. In this article, risk criticality is the significance of the risk, determined by both its probability of occurrence and level of impact. Furthermore, researchers have also used modeling in order to address this topic. For instance, Jannadi and Almishari (2003) developed a model called Risk Assessment Model (RAM) that determines, based on specific input data, the risk score of an activity, and generates a justification factor that gives an estimate of the efficiency of the proposed solution.

On the other hand, risk mitigation has also gained interest among researchers. In these studies, researchers have widely used empirical approaches to assess the appropriate mitigation strategies based on experts' opinions. For example, Abdul-Rahman *et al.* (2012) ranked 57 different risk factors in the Gulf Region, as to their own severity and likelihood of occurrence, and then associated risk responses for every risk factor, as a solution to mitigate it. Albalate *et al.* (2015) studied the different risk mitigation strategies and sharing in Public-Private Partnerships for different countries across the world. He related the country's own economic situation to the company's approach for risk strategies.

Despite the several studies about risk assessment and mitigation, experts still rely on their own experience and knowledge to identify, assess and mitigate risks. In fact, stakeholders do agree on the identification of risks and their magnitude and likelihood in construction but are still in constant dispute over the ownership of a specific risk and who is responsible for mitigating it. This study addresses this issue through surveying construction stakeholders and identifying the main risks that are disputed among them. Also, this study presents mitigation strategies for the same risks, in order to provide practitioners with a complete perspective on the issue of risk management.

3 RESEARCH METHODOLOGY

3.1 Data Collection

To achieve the objective of this paper, an empirical questionnaire was drafted and sent to contractors, developers and architects in the construction field. A literature review was done to gather all the information published on the most important risk factors in the body of knowledge. Hariharan *et al.* (2012), Abd El-Karim *et al.* (2017), Chan *et al.* (2011), Shen *et al.* (2001), Abdul-Rahman *et al.* (2012) and many more scholars have assessed the different construction risk factors. In this study, the risk factors presented by the aforementioned publications were grouped and organized in a way to limit the number of factors in the study in a concise manner, but still cover all the aspects of risk management. Table 1 shows all the risk factors that were used.

Table 1. List of investigated risk factors.

ID	Risk
R1	Change or Ambiguity in Scope of Work:
R2	Missing items or insufficient design at tender stage/Intense Competition/tender methodology (lowest bid)
R3	High complexity of Project (Many uncertainties / Different stakeholders / High expectation of owner)
R4	Ambiguous, incomplete and one-sided contract
R5	Unforeseeable ground conditions
R6	Lack of experience of contracting parties / poor quality of Work
R7	Accidents and safety issues on site
R8	Unsatisfactory performance and frequent change of Subcontractors
R9	Delayed site investigation, handing over and mobilization
R10	Poor planning, estimating and scheduling of critical activities
R11	Delays in solving contractual disputes
R12	Incompetence of Project management / organizational issues:
R13	Inflexibility of Project consultant (delays in testing and revising alternative design etc..)
R14	Poor/rushed Project design and Poor Constructability
R15	Project Finance Problems and Delayed Payments
R16	Political, Governmental, and Regulatory Issues
R17	Resource Price fluctuation, inflation, and changes in Interest Rate
R18	Unfavorable social and weather conditions
R19	Poor quality of material and equipment
R20	Changes in material quality and specification
R21	Delays in Resource Availability / Low productivity of Labor

In addition, a list of mitigation strategies was extracted from the literature and was included in the questionnaire. Chan *et al.* (2012) and Abdul Rahman *et al.* (2012) presented well elaborated risk mitigation strategies which were used in this study as well.

3.2 Design of Empirical Questionnaire

The questionnaire is composed of two parts to specifically address the aim of this research. The first part targets the respondents' professional profile and experience in the field. It is crucial to differentiate between the different backgrounds of the respondents in order to evaluate their response in relation to their expertise. A total of 95 respondents have completed the survey. Table 2 presents the distribution of the respondents based on their field of expertise and type of projects. The results of this research will also be based on a comparison between the respondents based on their type of project experience. The second part is considered the main and most important part of the survey. Respondents were given the list of all the risk factors presented previously where they were asked to choose whether the risk factor is the owner's responsibility, the contractor's or shared. Then, they had to pick one or more mitigation strategies that they deemed adequate to each investigated risk factor.

Table 2. Respondent profile.

Field of Expertise	Percentage	Study Group	Type of Projects	Percentage	Group
Owner/Development Firm	7.37%	Owner Group	Residential	29.32%	G1
Consultancy Firm	48.42%		Commercial and Institutional	26.70%	
Design Firm	12.63%		Highway Construction	9.95%	
Real Estate Agency	1.05%				
Contracting Firm	26.32%	Contractor Group	Industrial	17.80%	G2
Subcontracting Firm	4.21%		Heavy Construction	16.23%	

4 RESULTS

After gathering all the responses, three analyses were completed in this research. An overall analysis was done for all responses and two others based on the grouping shown in Table 2 for the type of project experience. For all three analyses, each group of answers (Contractor and Owner) are considered separately. The collected responses are summarized, and percentages are calculated for each group. The response with the highest percentage within the same group was considered for that specific risk. If both groups share a different response for the same risk, then it would be considered a disputed risk. Otherwise, that risk would be considered as an agreed upon risk. As for the mitigation part, both groups were considered as one. The mitigation strategy with the highest number of responses is chosen for each risk. Table 3 presents a summary of the descriptive statistics, showing the type of risk (whether disputed or agreed) for all respondents. It also shows the results for each group of projects, along with their corresponding mitigation strategy.

The results obtained in Table 3 were analyzed and discussed during one-on-one interviews with experts in the construction field. The results of this research are interpreted as follows:

(i) Overall Responses - Both owner and contractor groups agree on the ownership of 17 out of 21 risks. This observation proves that they have a clear agreement on the responsibility and ownership of most risks in construction when asked in an objective setting. However, this is not the case in the field, where they are subjectively involved in construction risks.

(ii) G1 Responses – Results show that the two groups are in dispute over 4 out of 21 risks. The same interpretation applies as that of the overall responses.

(iii) G2 Responses – The two study groups within these types of projects show an overall dispute over 7 out of 21risks. The types of projects within this group are known to be more complex than those within G1 because of the more intricate construction activities and traits involved aside the higher number of stakeholders. When subjected to a complex working environment, project stakeholders tend to be in larger disagreement over risk ownership and responsibility.

Table 3. Risk ownership and mitigation.

Risk ID	G1	G2	Overall	Mitigation
R1	Agreed	Agreed	Disputed	Clearly defined scope of works in client's project brief
R2	Disputed	Agreed	Disputed	Tender interviews and tender briefings to ensure tenderers gain a clear understanding of scope of works involved and necessary obligations to be taken in the project
R3	Agreed	Agreed	Agreed	Right Selection of Project Team
R4	Agreed	Disputed	Agreed	Development of standard contract clauses in connection with project's schemes or methodology
R5	Disputed	Disputed	Agreed	Prompt evaluation and agreement on any variations as they are introduced
R6	Agreed	Agreed	Agreed	Right Selection of Project Team
R7	Agreed	Agreed	Agreed	Adhere to the international occupational health and safety (OH and S) management system specification
R8	Agreed	Agreed	Agreed	Work with familiar subcontractor, supplier or client with thorough background checks
R9	Agreed	Agreed	Agreed	Prompt evaluation and agreement on any variations as they are introduced
R10	Agreed	Agreed	Agreed	Sufficient time given to interested contractors to submit their bids for consideration
R11	Agreed	Agreed	Agreed	Establishment of adjudication committee and meetings to resolve potential disputed issues
R12	Agreed	Disputed	Agreed	Right selection of project team
R13	Agreed	Disputed	Agreed	Right selection of project team
R14	Disputed	Disputed	Disputed	Employing a third party to review the project design in compliance with prevailing building regulations and buildability at tender stage
R15	Agreed	Agreed	Agreed	Insert provision in the contract to allow contractor to suspend work in the event of non-payment
R16	Disputed	Disputed	Agreed	Engage local partner or local manager
R17	Agreed	Agreed	Agreed	Application of price fluctuation clause in the contract
R18	Agreed	Agreed	Agreed	Implementation of relational contracting within the project team
R19	Agreed	Agreed	Agreed	Counter propose materials and equipment that are obtainable in local areas
R20	Agreed	Disputed	Disputed	Adapt and adopt: the standards in accordance to the specifications
R21	Agreed	Agreed	Agreed	Prepare standbys for the equipments according to priority of age, frequency of usage, and the likelihood of failure

5 CONCLUSIONS

The dynamic and complex nature of construction projects reflects high uncertainties and risks which causes construction experts to be in constant dispute over the ownership of these risks.

This paper aims at identifying the risks that are being disputed in the field. A survey was sent out to experts after a comprehensive literature review of the most common risk factors and mitigation strategies. Descriptive statistics and one-on-one interviews were conducted to analyze results of the survey. The study showed that experts tend to share an understanding on risk ownership, but have larger disagreements when subjected to a more complex environment. Furthermore, this paper lists proper mitigation strategies for each risk, as agreed by both parties. That way, experts will get an early chance to agree on the ownership of disputed risks and their mitigation strategies, prior to the start of the project and avoid possible delays that the project may face due to these disputes.

References

Abd El-Karim, M. S. B. A., El Nawawy, O. A. M., and Abdel-Alim, A. M., Identification and Assessment of Risk Factors Affecting Construction Projects, *HBRC Journal,* 13(2), 202-216, doi://doi.org/10.1016/j.hbrcj.2015.05.001, 2017.

Abdul-Rahman, H., Loo, S. C., and Wang, C., Risk Identification and Mitigation for Architectural, Engineering, and Construction Firms Operating in The Gulf Region, *Canadian Journal of Civil Engineering,* 39(1), 55-71, 2012.

Albalate, D., Bel, G., Bel-Piñana, P., and Geddes, R. R., Risk Mitigation and Sharing in Motorway PPPs: A Comparative Policy Analysis of Alternative Approaches, *Journal of Comparative Policy Analysis: Research and Practice,* 17(5), 481-501, doi:10.1080/13876988.2015.1010788, 2015.

Chan, D. W. M., Chan, A. P. C., Chan, J. H. L., Lam, P. T. I., and Yeung, J. F. Y., Risk Ranking and Analysis in Target Cost Contracts: Empirical Evidence from The Construction Industry, *International Journal of Project Management,* 29(6), 751-763, doi:10.1016/j.ijproman.2010.08.003, 2011.

Chan, J. H. L., Chan, D. W. M., Chan, A. P. C., and Lam, P. T. I., Risk Mitigation Strategies for Guaranteed Maximum Price and Target Cost Contracts in Construction: A Factor Analysis Approach, *Journal of Facilities Management,* 10(1), 6-25, doi:10.1108/14725961211200379, 2012.

Hanna, A. S., Greg, T., and Swanson, J. R., Construction Risk Identification and Allocation: Cooperative Approach, *Journal of Construction Engineering and Management,* 139(9), 1098-1107, doi:10.1061/(ASCE)CO.1943-7862.0000703, 2013.

Hariharan, S., Sawant, P. H., and Vandana, B., Construction Project Risk Assessment: Development of Model Based on Investigation of Opinion of Construction Project Experts from India, *Journal of Construction Engineering and Management,* 138(3), 409-421, doi:10.1061/(ASCE)CO.1943-7862.0000435, 2012.

Jannadi, O. A., and Almishari, S., Risk Assessment in Construction, *Journal of Construction Engineering and Management,* 129(5), 492-500, doi:5(492), 2003.

Jung, W., and Han, S. H., Which Risk Management is Most Crucial for Controlling Project Cost? *Journal of Management in Engineering,* 33(5), 04017029, doi:10.1061/(ASCE)ME.1943-5479.0000547, 2017.

Konior, J., Enterprise's Risk Assessment of Complex Construction Projects, Archives of Civil Engineering, 61(3), 63-74, 2015.

Nan, L., Dongping, F., and Yu, S., Cognitive Psychological Approach for Risk Assessment in Construction Projects, *Journal of Management in Engineering,* 32(2), 04015037, doi:10.1061/(ASCE)ME.1943-5479.0000397, 2016.

Serpell, A., Ferrada, X., Rubio, L., and Arauzo, S., Evaluating Risk Management Practices in Construction Organizations, *Procedia, Social and Behavioral Sciences,* 194, 201-210, 2015.

Shen, L. Y., Wu, G. W., and Ng, C. S., Risk Assessment for Construction Joint Ventures in China, *Journal of Construction Engineering and Management,* 127(1), 76-81, doi:1(76), 2001.

Interdependence between Structural Engineering and Construction Management
Edited by Ozevin, D., Ataei, H., Modares, M., Gurgun, A., Yazdani, S., and Singh, A.

ISBN: 978-0-9960437-6-2

MODELING CONTRACT PRACTICES RESULTING IN RISK MISALLOCATION

CHIPOZYA TEMBO-SILUNGWE

Dept of Construction Economics and Management, Copperbelt University, Kitwe, Zambia

Risk misallocation has been a topic of discussion in the extent literature for some time now. The literature points out that risk management is the key to ensuring that optimal risk allocation is achieved. While the focus on risk allocation has been about the party best suited to carry a given risk, the issue of how misallocation occurs is rarely dealt with. This paper modeled how risk misallocation can result from contract practice. This was done through a literature review and 15 purposive heterogeneous semi-structured interviews conducted with clients, project managers, architects, engineers and quantity surveyors in the Zambian building sector. The derived model suggests that misallocation could result from inappropriate selection of risk mitigation mechanisms provided for in the contract, poor selection of contracts and use of inappropriate procurement routes. These findings advance knowledge that necessitates the allocation of risk appropriately as the areas of contract practice and risk management are rarely researched with regards to risk misallocation.

Keywords: Buildings, Casual network, Documentation, Resources, Zambia.

1 INTRODUCTION

Contracts are the main mode of communication between the two contracting parties (client and contractor), providing documentary evidence of what has been agreed. Contracts set out the risks and liabilities to be carried by each contracting party and they generally affect project delivery. It has been argued that fair and carefully drafted contracts with appropriate risk allocation normally result in preferred construction project outcomes of delivering projects within time, cost, with minimal disputes (Mooney and Mooney 2014). However, when the contract is not viewed as fair, this results in projects shortfalls of poor quality, cost overruns, schedule overruns, onerous claims, disputes and risk misallocation (Alsalman and Sillars 2013). Risk misallocation in this context refers to allocations of risks wrongly or inappropriately (Hanna *et al.* 2013). Misallocation could arise from some combination of imperfect methods of allocation, imperfect response methods, allocation to the inappropriate party between client and contractor, allocation of imperfect resources or simply to a lack of allocation.

The construction industry in Zambia and especially in the public sector has evidence of risk misallocations. These are evidenced by rampant claims (Sibanyama *et al.* 2013), poor quality works as well as time and cost overruns (Kaliba *et al.* 2013, Auditor General's Office 2015). However, it is unclear what kinds of risk misallocations are involved. This paper documents an investigation of the various forms of misallocation that can result from contract practice and further displays the practices in a causal network.

2 CONTRACT PRACTICE AND RISK ALLOCATION IN THE CONSTRUCTION INDUSTRY

Risk allocation in the construction industry is mainly done through the various contracts used. Various contract practices aspects achieve the desired risk allocation. Contract practice is exercised at different levels (RICS 2015): firstly, the demonstration of knowledge and understanding of the various forms of contract used in the area of business involved; secondly, application of the knowledge in the use of the various standard forms of contract at project level including the implications and obligations applying to the contracting parties and lastly, provision of evidence of reasoned advice, preparation and presentation of reports on the selection of the appropriate form of contract and warranties for a chosen procurement route. This last practice includes advising on the most appropriate contractual procedure at the various stages of construction or other contract. The construction contract is basically used to allocate risks that are, to assign responsibility for a given occurrence to mainly the contractor and/or client.

Risk allocation can be done using various means namely representations & warrantees, indemnification, limitation of liability, express contractual remedies, payment terms, product warranties, and force majeure clauses (Practical Law Commercial 2013). The aforementioned are allocated in the various contract clauses. To understand the link between contracts and risk allocation, one needs to appreciate that various forms and types of contract allocate risk differently; that different risk mechanisms, warranties and responsibilities have different bearings on risk allocation and lastly that the risk apportionment is different in different types of procurement systems.

Contract forms are legal agreements that are designed to be used by employers and contractors in many situations without modification, however, in practice, they are commonly modified (Mooney and Mooney 2014). Modification may take various forms namely omission of clauses, change in working of clauses and addition of more clauses. Contract modification may make the risk allocation undesirable for one of the contract parties. Since the client in normally in a position to modify, the modification is normally to the detriment of the contractor. By nature, some contract forms are designed with a specific work type in mind, for example, small or large buildings, roads, dams or bridges. The contract procurement may be a standard form or bespoke contract (Smith *et al.* 2014). Correspondingly, risk apportionment may depend on the complexity of the works and the size of the project. Furthermore, the form of contract may be suitable for works with quantities or without. In the Zambian construction industry, the forms of contract used include the various Zambia Public Procurement Agency (ZPPA) forms of 2013 (open international, open national and small works contract), Joint Contracts Tribunal contract forms of 1972, New Engineering Contract 4 of 2017 and the FIDIC forms of contract of 2005.

Forms of contract that have definite quantities presented in bills of quantities usually have a fixed or lump sum payment mode where most of the risk is allocated to the contractor, while contracts without definite quantities are normally priced on a cost-plus basis where the least risks are allocated to the contractor. Nevertheless, Hackett *et al.* (2007) note that hybrid payment modes can exist on one contract so that some risks can be paid for using a different payment mode than the principal contract payment mode. In ideal situations, the risk allocation is supposed to be negotiated before the signing of a contract. Nonetheless, contracts of adhesion have been known to be formulated using a take it or leave it a model (Kanamugire 2013) and since contractors are always looking for work in the competitive construction industry, they are more likely to engage in contracts that have a risk allocation unfavorable to them. Additionally, projects are on occasion procured using unsuitable procurement routes.

Procurement routes range from conventional design bid build to unconventional types such as design and build, management-oriented routes (management contracting and construction management) and on to framework agreements. The various procurement routes have different risk allocation inherent in them. Design-bid build types have more risk allocated to the contractor with the design expected to be relatively complete while the management types normally have more risk allocated to the client (Smith *et al.* 2014).

3 METHODOLOGY

The research used a pragmatic approach using a sequence of semi-structured interviews, a questionnaire survey, contract documents, and a Delphi as solving of a practical problem (risk misallocation) was envisioned. However, for this paper, only interviews are reported. The interviews were with 15 heterogeneously and purposively selected respondents. The 15 comprised of quantity surveyors (2), architects (2), civil engineers (2), project managers (2), contractors (3), procurement officers (2) and clients (2). The respondents were drawn from both the public and private sector each having at least more than 10 years of experience in the building sector on various types of buildings ranging from residential, industrial and commercial. Prior phone calls were used to ascertain the required parameters and obtain consent for the interviews. Public sector respondents were identified through various government ministries and private sector respondents from professional bodies. The interviews lasted between 30 to 70 minutes. Respondents had between 10 and 30 years of experience in the building sector, on various types of building such as commercial, industrial and residential. The interviews were recorded using a digital recorder or using written notes when permission to record was not given. Recordings were then transcribed, and content analysis was done to make meaning of the data. Interpretive judgment on the transcriptions was used to formulate the nature of the casual relationships between contract practice and misallocation of risk. Miles *et al.* (2014) describe a causal network as an abstract, inferential picture organizing field study data in a coherent way showing how one thing led to another in a linear yet interwoven pattern. The constructing of a causal network involves a list of the antecedent (start variables), mediating variables and outcomes (ibid). The causal network was then mainly based on the researchers' interpretation of the interview data of contract practices identified leading to various types of misallocations such as resource misallocation, inappropriate allocation or use of inappropriate risk response, mechanism were then mapped.

4 FINDINGS

The misallocations resulting from contract practice are presented under three themes as shown below and the derived causal network in Figure 1 sums up the findings.

4.1 Contract Selection

Contract forms are selected by the client, as indicated by respondent PRC2 in section 4.3. The findings show that a common contract used in the building sector is ZPPA suites: Open National, Open International, and Small Works contract (13/15 respondents). These contracts are fixed or lump sum types except Open International, which can be used as an ad measurement contract. The Open International contract is based on the International Federation for Consulting Engineers (FIDIC) harmonized version of 2005. Another common contract used in the private sector is the Joint Liaison tribunal (JLC) contract popularly known as the ZIA (Zambia Institute of Architects) contract (6/15 respondents). Respondents identified several problems associated with the types and forms used. Firstly, it was highlighted that the selection of contract form in the public sector is based on estimates rather than actual project characteristics, sometimes resulting in the use of an inappropriate form. Secondly, the lack of contracts based on unconventional procurement types and payment modes other than fixed price (which is used for a contract that exceeds 12 months) sometimes leads to unfavorable bids and inappropriate risk allocation as the contracts based on traditional procurement are modified for use in unconventional procurement.

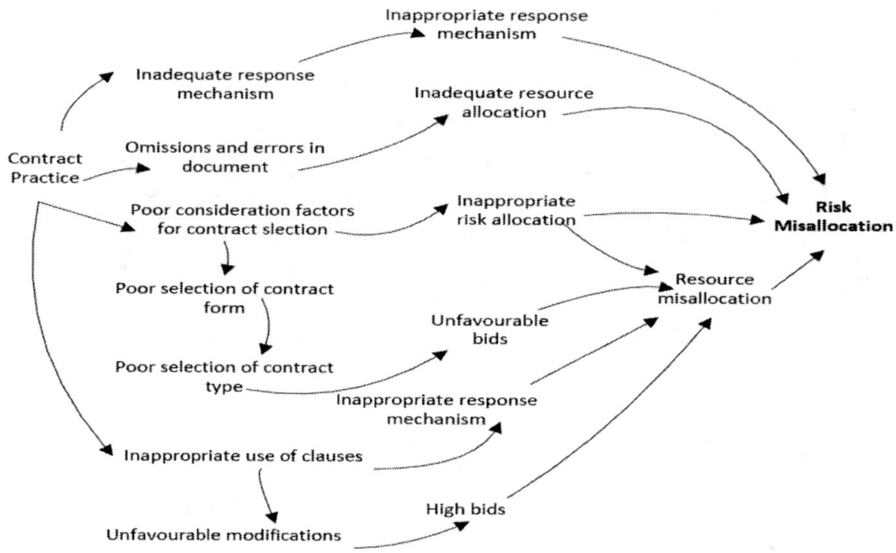

Figure 1. Modeling of risk misallocation resulting from contract practice.

4.2 Mechanisms

"Mechanisms" refers to treatment used to mitigate given risks, for instances the use of testing and sub-contracting. These and more mechanisms can be provided for in the contract. The findings were that on occasion inadequate mechanisms are put in place. An example given was that the 20% subcontracting limit was often inadequate as more than 20% of the works would need to be done by specialist subcontractors. This was indicated by one of the respondents "*there are times when the specialized work is more in value compared to the rest of the works e.g. combining plumbing, electrical works, and mechanicals such as lifts, escalators, air-conditioning"- PUCL.* Inappropriate use of clauses could also result in an inappropriate allocation. Fifty-six percent (8/15) of the respondents had used integrated procurement modes, yet the traditional contract form was used, resulting in inappropriate risk allocation when contract modifications were not made.

4.3 Contract Preparation

Standard contract forms in the building sector are seldom used without modification (11/15 respondents). Due to the modifications, unfavorable allocations are experienced (See comment by PRC2 below). Additionally, inadequate response mechanisms are put in place due to the inappropriate use of clauses, which also results in inappropriate risk allocation. Ultimately, unfavorable modifications result in high bids, which then result in resource misallocation.

"...because they know if those clauses are there, their risks will be high and they would be found wanting and because they are the ones who have the sole privilege of coming up with the form of contract that they want to have. Us as contractors have no input in that if anything those contracts that we use are just shoved down our throats...."- PRC2

Omissions and errors are also common in contract documents as indicated by respondent PUQS below. These result in resource misallocation and lack of cover for a given risk provided by the contract.

"Sometimes they do not indicate the retention period or omit the retention clause entirely, other times the insurance minimum cover is not indicated. Liquidated and ascertained damages are not indicated at times. There are many deficiencies". - PUQS

5 DISCUSSION

Misallocations in the Zambian building sector arising from contract practice are mainly from contract documentation, followed by contract selection and lastly mechanisms provided for. The most common mode of misallocation is resource misallocation. In line with the understanding of contract practice as laid out in RICS 9+/(2015), it appears that professionals in the building sector and particularly in the public sector need to improve their practice in certain areas while practice in other areas could be considered to be moderately adequate

The professionals demonstrate familiarity with and understanding of the various forms of contract used in the construction business. There is an appreciation of the various forms through their selection is commonly flawed because it is dependent on estimates (amounts). Additionally, different procurement modes that are supposed to be accompanied by their respective contract forms instead rely on the traditional contract form. This implies a need for the contract portfolio to include contracts based on different procurement systems.

At project level in the building sector, application of knowledge about the use of the various standard forms of contract, including their implications and obligations for the contract parties, is upheld to an extent. However, modifications to clauses change the obligations and implications for the parties. The client mainly initiates these changes and it is unclear to what extent the contractor is allowed to negotiate them. This negates the predictability of standard forms of contract.

The area that seems to need the most work is the provision of evidence of reasoned preparation and selection of the appropriate form of contract and a chosen procurement route, seeing that a variety of misallocations stem from contract selection. The findings also point to problems resulting from procurement routes. This is an area of practice that needs to be worked on to improve contract practice and reduce misallocations in the building sector. Use of appropriate procurement routes should be promoted in line with the selection of the appropriate form of contract.

6 CONCLUSION

Contract practice involves various aspects that all contribute to the successful fruition of a project. In the Zambian building sector contract selection, contract preparation and mechanisms stipulated in the contract document have been found to contribute to a variety of risk misallocations which the contract is supposed to guard against. These misallocations are illustrated in a causal network derived from the interview data. Eliminating such misallocations would be a basis for improving contract practice in the Zambian building sector. This study focused particularly on the building sector and future research might reveal how far its findings also apply in other construction sectors such as roads.

References

Alsalman, A. A., and Sillars D., Modelling the Effects of Sub-Optimal Risk Allocation in the Construction Industry, *Engineering Project Organization Conference, Devils Thumb Ranch,* Colorado, USA, July 9-11, 2013.

Auditor General's Office, *Report of The Auditor General on The Accounts of The Republic for The Financial Year Ended 31st December 2015, for The Period 2015, Zambia,* Lusaka, 2015. Retrieved from www.ago.gov.zm on September 12, 2016.

Hackett, M., Robinson, I., and Statham, G., *Procurement, Tendering and Contract Administration,* The Aqua Group ed. Oxford: Blackwell Publishing, 2007.

Hanna, A., Thomas, G., and Swanson, J., Construction Risk Identification and Allocation: A Cooperative Approach, *Journal of Construction Engineering and Management,* 139(9), 1098-1107, 2013.

Kaliba, C., Muya, M., and B Sichombo, Causal Factors of Cost Escalation, Schedule Overruns and Quality Shortfalls in Construction Projects in Zambia, *International Journal of Construction Management*, 13, 53-68, 2013.

Kanamugire, C. J., The Future of Standard Form of Contracts in South Africa with Particular Reference to Recent Developments in Law, *Mediterranean Journal of Social Sciences,* 4(13), 335-352, November 2013.

Miles, M. B., Huberman, A. M., and Saldana, J., *Qualitative Data Analysis.* London: Sage, 2014.

Mooney, C., and Mooney, E., Risk Allocation in Construction Contracts: Irish Public Works, *Proceedings of the Institution of Civil Engineers- Management, Procurement and Law,* 167(2), 68-74, 2014.

RICS, *APC: Technical Competencies,* London: RICS, 2015.

Practical Law Commercial, Using Contractual Risk Allocation Provisions to Minimize Rise and Maximize Reward, Thompson Reuters, 2013. Retrieved from https://content.next.westlaw.com on April 12, 2018.

Smith, N., Merna, T., and Jobling, P., *Management of Risk in Construction Projects.* 3rd ed., Wiley Blackwell Publishing, West Sussex, 2014.

Interdependence between Structural Engineering and Construction Management
Edited by Ozevin, D., Ataei, H., Modares, M., Gurgun, A., Yazdani, S., and Singh, A.

ISBN: 978-0-9960437-6-2

PROJECT PERFORMANCE INDEX FOR CAPITAL INTENSIVE CONSTRUCTION PROJECTS

SHOEB SYED and EMAD ELWAKIL

Building Construction Management, Purdue University, West Lafayette, USA

Holistic construction performance measure has been an elusive and highly sought-after goal. The stochastic nature of construction processes and the number of stakeholders involved in these processes have added to the complexity of the problem. Large-scale infrastructure projects produce big data in form of actual site parameters, safety data, risk analysis data, cost data, financial data, quality assurance data and procurement data. This research aims at developing a project performance index. The factors influencing project performance have been grouped together as Critical Success Factors (CSFs) then through survey and expert opinions, relative weights have been collected for these factors. The factors and weights have been modeled through Regression Technique and validated through analytical and mathematical methods. The developed index will benefit the academic researchers and industry practitioners to assess the performance of the project and to increase the efficiency of project performance.

Keywords: Critical success factors, Construction project, Project performance, Schedule, Cost, Quality, Safety, KPI, Balanced scorecard.

1 INTRODUCTION

Construction projects are prone to time delays and cost overruns. (Mirza and Ehsan 2017). There are several measures that have been devised to gauge project performance based on numerous studies of critical project success factors, but these are mostly quantitative factors that have been calculated based on formulae developed by researchers. Project complexity factors have been characterized in the past in five broad categories: time/schedule, scope, cost/budget, quality, resources and risk (Mirza and Ehsan 2017)

The construction industry has come a long way in incorporating more advance technology on and off the site. There is an influx of newer information systems and more robust reporting systems but the major hurdle still happens to be reluctance of project managers to share data. Most companies have measuring systems to monitor their performance and while companies are entitled to their privacy, a generic rating system built for a specific industry can go a long way in improving efficiency – this is the aim of this research. Even with the intra-company measures, staff and executives have difficulty triaging the available factors to come up with an effective action plan.

2 BACKGROUND

Nguyen *et al.* (2013) performed an exploratory analysis by listing project success measures from literature published in the past decade or so. Though the list isn't comprehensive it provides a great start at how researchers have looked at measuring project success in the past. As a summary of the findings of Nguyen *et al.* (2013) the following areas were identified that success measures may be categorized into:

- Time Performance
- Cost Performance
- Quality Performance
- Customer Satisfaction
- Safety Performance
- Organizational Objectives

Several quantitative measures can be devised to measure the performance of a project in the above-mentioned areas. Alternatively, Young and Poon (2013) researched the effect on project performance of 'Top Management Support' – a qualitative factor. Maqbool *et al.* (2017) focused their efforts on studying the importance of leadership traits such as emotional intelligence, competency, and of different leadership styles and their impact on project performance. There is merit to both qualitative and quantitative factors. One of the major pursuits of this paper is to select factors from a mix of qualitative and quantitative factors and to identify factors that may have been overlooked because they couldn't be described quantitatively and, subsequently, measure their contribution towards project success.

2.1 Critical Success Factors

Critical Success Factors (CSFs) are defined by Toor and Ogunlana (2009) as, "elements which significantly contribute to, and are vital for, the success of a project". In theory, CSFs have been employed by several industries – Medicine, Manufacturing, IT and Construction. Pinto and Slevin (1987) are among the first researchers to classify factors as 'critical success factors' in an attempt to quantify the performance of construction projects.

Other researchers have also compiled lists of critical success factors and categorized them into roughly the same areas as mentioned previously under project success areas (Belassi and Tukel 1996, Fortune and White 2006). There has been a marked shift in research focus recently; where construction performance was measured in terms of time, cost, quality and safety (de Wit 1988, Pinto and Slevin 1987, Sanvido *et al.* 1992) there is a move towards quantifying the effect of human aspects such as emotional intelligence (Maqbool *et al.* 2017), upper-management support (Young and Poon 2013), and clear definition of goals (Toor and Ogunlana 2009).

2.2 Existing Project Performance Metrics

The Project Excellence (PE) model used by the German Project Management Association is one of the earliest nation-wide performance measurement models (Westerveld 2003). The model was derived from acceptable industry practices such as those developed by the EFQM (European Foundation for Quality Management) and the PDCA (Plan-Do-Check-Act) cycle of continuous development.

The Balanced Scorecard (BSC) is another popular performance measurement metric devised by the Harvard Business School (HBS) Professor Mr. Robert Kaplan and Renaissance Solutions President, David Norton. A versatile metric used across various industries, it identifies "leading and lagging" measures for various factors affecting a process to ascertain whether the process is achieving its desired goal or not. Kagioglou *et al.* (2001) proposed the "Performance Measurement Process (conceptual) Framework (PMPF)" inspired by the Balanced Scorecard method as a metric tailored for the construction industry. The European KPI system and Canadian metric program are both similar performance measurement models built for construction companies operating within these regions (Bassioni *et al.* 2004, Costa *et al.* 2007, Nasir *et al.* 2012, Rankin *et al.* 2008).

KPIs are a part of the Constructing Excellence Productivity program and were designed in 2003 as an initiative to benchmark construction performance in the UK. The process involved extensive reviews of construction projects by a panel of experts that rated the project on a comprehensive scale of factors. The KPI was implemented as a national program with companies volunteering to participate and making use of the results for planning their own projects.

With the aim to holistically measure construction project performance, the first step is to decide on all the spheres of influence that may contribute towards that performance. The methodology employed by the researcher is somewhat similar to Elwakil's (2017) investigation of organizational performance index. Namely, shortlisting success factors from existing literature, spreading them across functional units in a project and performing mathematical modelling to determine their ideal weightages in calculating the performance index. The modelling approach will be tailored according to the groups of factors out of the available tools such as analytical hierarchy process, fuzzy modelling, and/or multiple linear regression. The mathematical models will be validated against data collected from actual projects and their score on an industry-accepted metric like the ones described in the next section.

3 METHODOLOGY

The steps followed in this research are summarized in a step-wise fashion as below:

- The process begins with identifying and shortlisting factors that contribute to the performance assessment of construction projects. This step has been accomplished by undertaking a review of relevant literature in the body of knowledge. Both qualitative and quantitative factors have been documented for further analysis.
- The factors are documented against their sources and clustered under broad areas in Table 1 for better analysis. This list has been subjected to further analysis to yield a list of Critical Success Factors (CSFs). The proposed Project Performance Index (PPI) takes these Critical Success Factors as primary inputs to the model.
- The process of shortlisting has been accomplished by designing a questionnaire aimed at deducing the impact of these factors as estimated by industry professionals. The survey also notes the respondent's experience, designation and organization to add perspective to the response. More about the survey design has been discussed in the section on Survey Design.
- Simultaneously, a database has been started of performance factors available in the industry to assess performance of existing projects. This will serve as a validation tool for the Project Performance Index developed as part of this research and a way to translate industry generated data as inputs to this research model.
- Responses from the survey shall be subjected to further mathematical analysis (Multiple Linear Regression/Analytical Hierarchy Process) to gain insight on weights of Critical Success Factors, correlation between factors and interaction among factors.

3.1 CSF Matrix

Work on CSFs available in the literature is summarized in Table 1.

Table 1. Critical Success Factors shortlisted from literature review.

S no.	Factor	Source
1	Top Management Support	(Young and Poon 2013), (Martin 1976), (Cleland and King 1983), (Toor and Ogunlana 2009), (Pinto and Slevin 1987), (Morris and Hough 1987), (Fortune and White 2006)
2	Regular Client Involvement	(Young and Poon 2013), (Cleland and King 1983), (Toor and Ogunlana 2009), (Sayles and Chandler 1971), (Pinto and Slevin 1987), (Fortune and White 2006)
3	Clear Statement of Requirements	(Young and Poon 2013), (Toor and Ogunlana 2009), (Cleland and King 1983)
4	Proper Planning and Project Controls	(Young and Poon 2013), (Toor and Ogunlana 2009), (Martin 1976), (Cleland and King 1983), (Fortune and White 2006), (Locke 1984), (Cleland and King 1983), (Sayles and Chandler 1971), (Baker *et al.* 1983), (Morris and Hough 1987)
5	Smaller Project Milestones	(Young and Poon 2013), (Martin 1976)
6	Realistic Expectations	(Young and Poon 2013), (Toor and Ogunlana 2009)
7	Clear Vision and Objectives	(Young and Poon 2013), (Toor and Ogunlana 2009), (Martin 1976), (Fortune and White 2006) , (Locke 1984), (Baker *et al.* 1983), (Morris and Hough 1987)
8	Team Composition and Competency	(Young and Poon 2013), (Toor and Ogunlana 2009), (Martin 1976), (Fortune and White 2006), (Cleland and King 1983), (Baker *et al.* 1983), (Pinto and Slevin 1987)
9	Change Management	(Young and Poon 2013), (Toor and Ogunlana 2009), (Fortune and White 2006)
10	Project Manager Competence	(Toor and Ogunlana 2009), (Fortune and White 2006), (Locke 1984), (Sayles and Chandler 1971), (Baker *et al.* 1983)
11	Contractors and Sub Contractors	(Toor and Ogunlana, 2009), (Fortune and White 2006)
12	IT Support	(Toor and Ogunlana 2009), (Fortune and White 2006), (Cleland and King 1983)
13	Monitoring and Feedback	(Toor and Ogunlana 2009), (Fortune and White 2006), (Cleland and King 1983), (Sayles and Chandler 1971), (Morris and Hough 1987)
14	Stakeholder Support and Involvement	(Toor and Ogunlana 2009)
15	Continued Financial Support	(Young and Poon 2013), (Fortune and White 2006), (Martin 1976), (Cleland and King 1983), (Baker *et al.* 1983), (Pinto and Slevin 1987), (Morris and Hough 1987)
16	Communication between all parties	(Toor and Ogunlana 2009), (Martin 1976), (Locke 1984), (Pinto and Slevin 1987)
17	Risk Management	(Fortune and White 2006)
18	Staff Training	(Fortune and White 2006)

3.2 Data Collection

The outputs from the Model are validated by the Average Validity Percentage (AVP) and Average Invalidity Percentage (AIP) as a first layer of validation. The second layer is to validate against the database of industry-accepted performance assessment database.

4 CONCLUSIONS

A Project Performance Assessment Model (PPAM) for construction projects shall be developed using a Hierarchical Fuzzy Expert System (method still under consideration).

The outputs from the Model will be validated by the Average Validity Percentage (AVP) and Average Invalidity Percentage (AIP) as a first layer of validation. The second layer will be to validate against the database of industry-accepted performance assessment database. The researcher is currently in the process of data collection and hopes to deduce conclusions in the coming months.

References

Baker. B. N., Murphy, D. C., and Fisher, D., Factors Affecting Project Success, *Project Management Handbook,* Van Nostrand Reinhold Co., New York, 1983.

Bassioni, H. A., Price, A. D. F., and Hassan, T. M., Performance Measurement in Construction. *Journal of Management in Engineering, 20*(2), 42–50, https://doi.org/10.1061/(ASCE)0742-597X(2004)20:2(42), 2004.

Belassi, W., and Tukel, O. I., A New Framework for Determining Critical Success/Failure Factors in Projects, *International Journal of Project Management, 14*(3), 141–151, https://doi.org/10.1016/0263-7863(95)00064-X, 1996.

Cleland, D. I., and King, W. R., *Systems Analysis and Project Management,* McGraw Hill, New York 1983.

Costa, D. B., Formoso, C. T., Kagioglou, M., Alarcón, L. F., Caldas, C. H., and Asce, M., Benchmarking Initiatives in the Construction Industry : Lessons Learned and Improvement Opportunities, *Journal of Mnagement in Engineering, ASCE, 22*(4), 158–167, https://doi.org/10.1061/(ASCE)0742-597X(2006)22:4(158), 2007.

de Wit, A., Measurement of Project Success, *International Journal of Project Management, 6*(3), 164–170, https://doi.org/10.1016/0263-7863(88)90043-9, 1988.

Elwakil, E., Integrating Analytical Hierarchy Process and Regression for Assessing Construction Organizations' Performance, *International Journal of Construction Management, 17*(1), 76–88, https://doi.org/10.1080/15623599.2016.1187247, 2017.

Fortune, J., and White, D., Framing of Project Critical Success Factors by a Systems Model, *International Journal of Project Management, 24*(1), 53–65, https://doi.org/10.1016/j.ijproman.2005.07.004, 2006.

Kagioglou, M., Cooper, R., and Aouad, G., Performance Management in Construction: A Conceptual Framework, *Construction Management and Economics, 19*(1), 85–95, https://doi.org/10.1080/01446190010003425, 2001.

Locke, D., *Project Management,* St Martins Press, New York, 1984.

Maqbool, R., Sudong, Y., Manzoor, N., and Rashid, Y., Transformational Leadership on Project Success : An Empirical Perspective, *Project Management Journal, 48*(3), 58–75, Retrieved from http://ezproxy.liberty.edu/login?url=http://search.ebscohost.com/login.aspx?direct=true&db=bah&AN=123094185&lang=es&site=ehost-live, 2017.

Martin, C. C., *Project Management,* Amaco, New York, 1976.

Mirza, E., and Ehsan, N., Quantification of Project Execution Complexity and its Effect on Performance of Infrastructure Development Projects, *EMJ - Engineering Management Journal, 29*(2), 108–123, https://doi.org/10.1080/10429247.2017.1309632, 2017.

Morris, P., Hough, G., 1987. *The Anatomy of Major Projects: A Study of the Reality of Project Management,* vol. 1. Chichester, John Wiley & Sons, Ltd., UK.

Nasir, H., Haas, C. T., Rankin, J. H., Fayek, A. R., Forgues, D., and Ruwanpura, J., Development and Implementation of a Benchmarking and Metrics Program for Construction Performance and Productivity Improvement, *Canadian Journal of Civil Engineering,* https://doi.org/10.1139/l2012-030, 2012.

Nguyen, T. A., Chovichien, V., and Takano, S., Quantitative Weighting for Evaluation Indexes of Construction Project Success by Application of Structural Equation Modeling, *International Journal of Construction Engineering and Management, 2*(3), 70–84, https://doi.org/10.5923/j.ijcem.20130203.05, 2013.

Pinto, J. K., and Slevin, D. P., Critical Factors in Successful Project Implementation, *IEEE Transactions on Engineering Management, EM-34*(1), 22–27, https://doi.org/10.1109/TEM.1987.6498856, 1987.

Rankin, J., Fayek, A. R., Meade, G., Haas, C., and Manseau, A., Initial Metrics and Pilot Program Results for Measuring The Performance of The Canadian Construction Industry, *Canadian Journal of Civil Engineering, 35*(9), 894–907, https://doi.org/10.1139/L08-018, 2008.

Sanvido, V., Grobbler, F., Parfitt, K., Guvenis, M., and Coyle, M., Critical Success Factors For

Construction Projects, *Journal of Construction Engineering and Management, 118*(1), 94–111, https://doi.org/10.1061/(ASCE)0733-9364(1992)118:1(94), 1992.

Sayles, L. R., and Chandler, M. K., *Managing Large Systems*., Harper & Row, New York, 1971.

Toor, S., and Ogunlana, S. O., Construction Professionals' Perception of Critical Success Factors for Large Scale Construction Projects, *Construction Innovation*, *9*(2), 149–167, https://doi.org/10.1108/14714170910950803, 2009.

Westerveld, E., The Project Excellence Model®, *Linking Success Criteria and Ciritcal Success Faktor*, 21(6), 411–418, 2003. Retrieved from https://www.wiso-net.de/document/BLIS__20030864110833351919318920518911 on March 2019.

Young, R., and Poon, S., Top Management Support-Almost Always Necessary and Sometimes Sufficient for Success: Findings From a Fuzzy Set Analysis, *International Journal of Project Management*, 31(7), 943–957, https://doi.org/10.1016/j.ijproman.2012.11.013, 2013.

Interdependence between Structural Engineering and Construction Management
Edited by Ozevin, D., Ataei, H., Modares, M., Gurgun, A., Yazdani, S., and Singh, A.

ISBN: 978-0-9960437-6-2

COST OPTIMIZATION OF MULTIPLE RESOURCES ALLOCATION FOR MULTI-PROJECTS

SARAH ABOUL FOTOUH and A. SAMER EZELDIN

Dept of Construction Engineering, American University in Cairo, Cairo, Egypt

As the construction industry grows year by year, optimization of resources is becoming essential to reduce their required number, their costs and as a consequence the total cost of the project. Project managers have to face problems regarding management of cost, time, and available resources for single projects. What is more challenging is to optimize the available resources for multiple projects, which would result in appreciable savings. Most of the companies in the construction industry; commonly; optimize the resources for single project. However, with the presence of several mega projects in many developing countries running at the same time, there is a need for a model to enhance the efficiency of available resources among multiple simultaneous projects. This paper discusses a numerical model of cost optimization and allocation of up to nine resources for up to three projects for a given company, taking into consideration the transportation of resources from one project to another and the cost of unused resources. The model was developed using a genetic algorithm, and it is used on the identified critical resources. It calculates the cost of each resource, minimizes the cost of extra resources, cost of unused resources, and generates the schedule of each project within a selected overall program.

Keywords: Construction management, Genetic algorithm, Modeling, Multiple projects.

1 INTRODUCTION

Construction industry is becoming more sophisticated every day, and there is usually deficiency in the utilization of resources. Multi-national companies face this problem more often as they might have several mega projects running at the same period of time. Therefore, such companies have to utilize available resources efficiently to cover all its projects, and minimize the cost of any extra resources needed from outside the company.

1.1 Managing Multiple Projects

In multi-national companies, projects need to be viewed as an integrated portfolio rather than separate projects. In managing multiple projects, project managers are required to maintain control over different types of construction projects, balance conflicting requirements of resources and coordinate the project portfolio to ensure the optimum outcome for the organization is achieved (Dooley *et al.* 2005).

The main challenges that have to be addressed when managing multiple are:

- Projects are overlapping with other projects and day-to-day operations, sharing common deliverables, resources, information or technology across those overlapping projects.
- Prioritization of resources on daily basis over all running projects.

- Meeting the deadlines of the projects, which contribute to the overall development objectives of the parent organization (Turner and Speiser 1992).

1.2 Resources Management

Resource management plays an important role in project management, in which the project manager tries to avoid unnecessary resources overload. There are three objective functions well established in the literature to cope with resource management (Rieck and Zimmermann 2015):

- The minimum moment has to be minimized (Harris 1973);
- The *total overload cost problem*, where costs are generated when the pool of a given resource is exceeded (Easa 1989);
- The *total adjustment cost problem*, where one is concerned with the minimization of the cumulative costs arising from increasing or decreasing the utilizations of resources (Bianco *et al.* 2016).

1.3 Previous Models

Many researches have been performed on resource levelling, with different approaches. Harris (1973) developed one of the earliest simple heuristic processes called the "Minimum Moment Algorithm" for the resource leveling problem. Later, Hiyassat (2000) modified Harris's process by taking into consideration the activities' free float and the resources needed in the selection criteria.

In addition to the above exact formulations for the resource leveling problem, several authors proposed meta-heuristic procedures to generate near-optimal schedules. Many studies proposed a model based on genetic algorithm (GA) for resource scheduling that performs resource leveling along with resource allocation simultaneously (Aboul Fotouh and Ezeldin 2017, Jun and El-Rayes 2011). Another study proposed a GA based optimization system to minimize the weighted total deviations of resources' requirements (Leu *et al.* 2000). In 2011, a model was built for resource levelling by splitting the activities (Hariga and El-Sayegh 2011).

Recently in 2017, a multiple resources levelling and allocation model was made using genetic algorithm but not taking into consideration the cost of transportation of resources not the cost of unused resources (Aboul Fotouh 2017).

1.4 Objective

This paper generates a multi-objective genetic algorithm-based model for multiple resources levelling and allocation for multiple projects taking into consideration the following:

1- Relationships between the activities.
2- Cost of each resource within company.
3- Cost of each out-sourced resource.
4- Cost of transportation of resources.
5- Cost of unused resources.

This model is tested on three different over-lapping projects and nine different resources. The model can be adjusted to fit the needs of the project manager by increasing the number of projects and resources as desired.

2 METHODOLOGY

Evolutionary algorithms (EAs) are stochastic search methods that simulate the natural biological evolution and/or the social behavior of different species. Such algorithms have been developed to arrive at near-optimum solutions to large-scale optimization problems, when precision is not the highest priority and the optimal solution would be exhaustive or difficult to find (Elbeltagi *et al.* 2005).

The first evolutionary-based technique introduced in the literature was the genetic algorithms (GAs) (Elbeltagi *et al.* 2005). GAs were developed based on the Darwinian principle of the 'survival of the fittest' and the natural process of evolution through reproduction (Elbeltagi *et al.* 2005). It is a method for solving both constrained and unconstrained optimization problems by generating feasible solutions for optimization problems by natural evolution, such as mutation, selection, and crossover (Elbeltagi *et al.* 2005).

2.1 Model's Input

The user has to input some data;

- The start date of each project.
- The predecessors of each activity.
- Duration of each activity.
- The number of resources needed by each activity.
- The number available of each resource (Pool of available resources).
- The cost of each extra resources needed.
- The transportation cost of each resource.
- Cost of unused resources

2.2 Model's Development

2.2.1 The Model's objective function (fitness function) is shown in Eq. (1):

$$Min \sum_{i}^{1} ECR + TC + CUR \qquad (1)$$

Where;

- "ECR" is the Extra Cost of Resources,
- "TC" is the Transportation Cost of resources,
- "CUR" is the Cost of Unused Resources
- "i" is the number of projects.

2.2.2 Model's variable

The variable cells in this model is the number of days needed to shift each activity to reach the minimum cost of extra resources needed taking into consideration the transportation cost. These variables will be added to the start date, so that the activity will be shifted by the number written in the variable cells.

2.2.3 Model's constraints

This model has two types of constraints:

Hard Constraints: These are the constraints that the model cannot break, in this model it is the deadline of each project. This means the model cannot extend the duration of any project and only able to use the available float for each activity (Eq. (2)).

$$Max.\ of\ FDP \leq DLP \tag{2}$$

Where,

- FDA is the Finish Date of Project
- DLP is the Deadline of Project

Soft Constraints: These are the constraints that the model will try to abide by it, but if it cannot, then they can ignore it but add a certain penalty. In this model the soft constraints are the maximum number of resources needed per period of time should be less than the pool of available resources. If the model cannot reach a near optimum solution that all the resources needed are within the pool limit, then an additional cost will be added depending on which resource is exceeding the pool limit and its associated cost. The model will automatically select the resource with the least additional cost instead of the resource with the higher cost to minimize the cost as much as possible (Eq. (3)).

$$max\ NRT \leq Pool\ of\ available\ resources \tag{3}$$

Where,

- NRT is the number of resources needed per unit time.

2.3 Model's Outputs

The model generates the following outputs;

I- Updated schedule of each project.
II- Extra cost of extra resources such as in Figure 1.
III- Cost of resources' transportation from one project to another such as in Figure 1.
IV- Cost of un-used resources such as in Figure 1.
V- Histograms of resources needed per unit time before and after optimization such as in Figure 2.

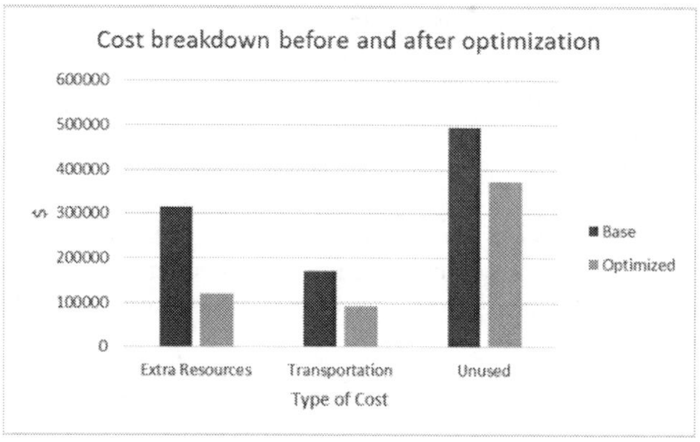

Figure 1. Cost breakdown before and after optimization.

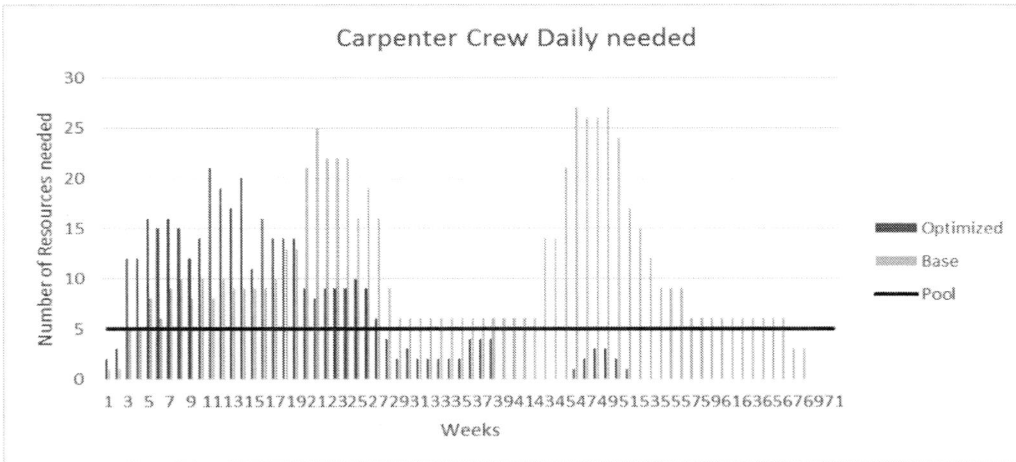

Figure 2. Sample of resources needed per unit time before and after optimization.

3 RESULTS AND DISCUSSION

The model is able to reduce the cost of extra resources, transportation of resources and the cost unused resources for all three projects and nine resources. In these specific data the cost of extra resources needed was reduced the most in comparison with cost of un-used resources and transportation cost as shown in Figure 1.

Figure 2 shows a sample of one resource's (Carpenter) bar chart before and after optimization that is in the model. Figure 2 shows that the extra resources needed at the end of the projects were shifted to the beginning of the projects.

Figure 3 shows that the total cost of resources decreased by almost 40% for all three projects together using nine resources.

The results obtained shows how the model succeeded to decrease the cost of projects and eventually increase the company's mark-up. This model is best to use in the scheduling process and resource allocation, but it can also be used after the start-up of the projects to be able to utilize efficiently the use of available resources and maintaining the deadlines of the projects.

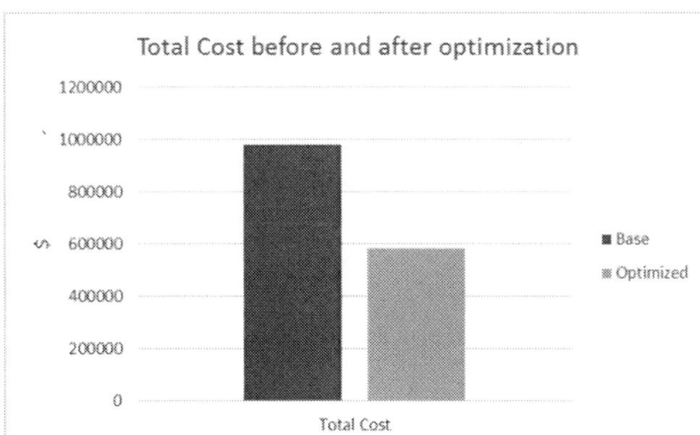

Figure 3. Total Cost of Resources before and after optimization.

4 CONCLUSION

This paper discusses the allocation of resources for multiple projects and multiple resources, taking into consideration the cost of extra resources needed, the cost of unused resources, and the transportation of resources from one project to another. A genetic algorithm-based model was built, and it was able to reduce all three types of costs, which at the end can save the company a large percentage of the total cost of resources. The results generated by the model indicate a good validation of the model.

References

Aboul Fotouh, S., *Optimization of Multiple Resources for Multi-Projects (Text)*. The American University in Cairo, Cairo- Egypt, 2017.

Aboul Fotouh, S., Ezeldin, A.S., *Genetic Algorithm Optimization of Multiple Resources for Multi-Projects, in: Resiliet Structures and Sustainable Construction,* Presented at the ISEC-09, 2017, ISEC Press, Valencia, Spain, https://doi.org/10.14455/ISEC.res.2017.44, 2017.

Bianco, L., Caramia, M., and Giordani, S., Resource Levelling in Project Scheduling with Generalized Precedence Relationships and Variable Execution Intensities, *Spectrum* 38, 405–425, 2016.

Dooley, L., Lupton, G., and O'Sullivan, D., Multiple Project Management: A Modern Competitive Necessity, *Journal of Manufacturing Technology Management* 16, 466–482, https://doi.org/10.1108/17410380510600464, 2005.

Easa, S. M., Resource Leveling in Construction by Optimization, *Journal of Construction Engineering and Management,* 115, 302–316, 1989.

Elbeltagi, E., Hegazy, T., and Grierson, D., Comparison Among Five Evolutionary-Based Optimization Algorithms, *Advanced Engineering Informatics,* 19, 43–53, https://doi.org/10.1016/j.aei.2005.01.004, 2005.

Hariga, M., and El-Sayegh, S. M., Cost Optimization Model for the Multiresource Leveling Problem with Allowed Activity Splitting, *Journal of Construction Engineering and Management,* 137, 2011.

Harris, R. B., *Precedence and Arrow Networking Techniques for Construction*. University of Michigan, 1973.

Hiyassat, M. A. S., Modification of Minimum Moment Approach in Resource Leveling. *J. Constr. Eng. Manag.* 126, 278–284, https://doi.org/10.1061/(ASCE)0733-9364(2000)126:4(278), 2000.

Jun, D. H., and El-Rayes, K., Multiobjective Optimization of Resource Leveling and Allocation during Construction Scheduling, *Journal of Construction Engineering and Management,* 137, 1080–1088, https://doi.org/10.1061/(ASCE)CO.1943-7862.0000368, 2011.

Leu, S.-S., Yang, C.-H., Huang, J.-C., Resource Leveling in Construction by Genetic Algorithm-Based Optimization and Its Decision Support System Application, *Automation in Construction*, 10, 27–41, 2000.

Rieck, J., and Zimmermann, J., *Exact Methods for Resource Leveling Problems*, in: Schwindt, C., Zimmermann, J. (Eds.), Handbook on Project Management and Scheduling, Springer International Publishing, Cham, 1, 361–387, https://doi.org/10.1007/978-3-319-05443-8_17, 2015.

Turner, J., and Speiser, A., Programme Management and Its Information Systems Requirements, *International Journal of Project Management,* 10, 196–206, https://doi.org/10.1016/0263-7863(92)90078-N, 1992.

Interdependence between Structural Engineering and Construction Management
Edited by Ozevin, D., Ataei, H., Modares, M., Gurgun, A., Yazdani, S., and Singh, A.

ISBN: 978-0-9960437-6-2

CHANCES AND RISKS OF COMBINED PRICING

MARKUS KUMMER and CHRISTIAN HOFSTADLER

Institute of Construction Management and Economics, Graz University of Technology, Graz, Austria

Appropriate costing creates a plausible, reliable basis for determining the bid price. One of the key requirements that applicable laws, standards, and guidelines stipulate for (structural) specifications is that works should be specified clearly, completely and objectively so as to ensure bid comparability. Furthermore, cost estimates should be possible for the specified works (excluding non-quantifiable risks). The specifying party usually requires combined pricing because works related to various structural components, construction trades, subsoil types, rebar diameters etc. that necessitate a variety of types and combinations of production factors are merged into a single item. Combined pricing occurs whenever a unit price is to be derived from different cost estimates with respect to the utilization of production factors. This means that bidders intending to develop an activity-based costing model would have to know the type of work or service, the underlying conditions for performing the work, and the associated timeframe (i.e. project start, binding milestones, project end). This paper investigates if combined pricing items meet these requirements whilst putting a particular focus on the relevance of combined pricing to construction management and economics, including construction expert opinions on the chances and risks of this approach. For this purpose, this paper primarily deals with reinforcing works.

Keywords: Expert survey, Management of chances and risks, Reinforcing works, Reinforcement ratio, Building specification.

1 INTRODUCTION

Construction management is characterized by a combination of production factors. Pricing is essentially the result of how and when which production factors are combined (Hofstadler 2014). Only production factors of an identical type and pricing may be merged into a single item as part of a structural work specification. This principle provides the basis for appropriate costing. In practice, however, deviations occur relatively frequently and inevitably necessitate a combined pricing approach.

The client can create the prerequisites for appropriate pricing by disclosing its bases of costing, which enables appropriate costing/pricing by the contractor. Furthermore, such a pricing exercise can also be supported and made plausible by the contractor disclosing its own pricing assumptions. As a matter of course, the client's position will also depend on the degree of design detail reached at the time of preparing the specification. If detailing has progressed only to a limited extent, no reinforcement plans or drawings will usually be available for specifying reinforcing works, which forces the client to start from its own assumptions. Nonetheless, at this project stage, the client will be in a better position than the bidder(s) to come up with fairly accurate assumptions, for instance on the basis of a preliminary structural analysis.

The client has much more detailed project knowledge and will have analyzed the project more thoroughly at this stage. It is worthy of note that the client is in possession of a larger amount of information than the contractor with respect to project timing and planning. No bidder would be in a position to overcome this deficit within the specified submission period; nor are bidders under the obligation to do so.

Clients can protect themselves against speculative bids by preparing an appropriate quantity survey on the basis of a sufficiently accurate and detailed planning status. If, however, the client discloses its bases for pricing only partially or not at all, the bidder (and future contractor) is bound to make own assumptions to be able to conduct a pricing exercise at all. If the bidder also fails to disclose its assumptions as part of the submission process, this will very likely give rise to disputes over any changes in quantities or the allocation of combined pricing components.

The contractor cannot know or derive the quantity structure that the client used for its specification or request for proposals. Conversely, the client is not in possession of any information with respect to the assumptions made by the contractor for the purpose of costing and pricing.

Yet another distinction needs to be made regarding the bidder's disclosure, namely if its disclosure is based on its own assumptions or assumptions made by a subcontractor, which may result in different mean wage costs, for example. If the bidder discloses its assumptions, the client will know the bases of pricing.

If these are not disclosed, however, they will subsequently be submitted by the bidder in the event of a dispute, and will then have to be checked for plausibility and, if and when required, critically compared with relevant information provided in the literature or, as far as reasonably possible, recalculated from the bid price.

Generally speaking, however, the contractor is forced to come up with free pricing assumptions if the client fails to supply related detailed information in its specification. It should be noted, though, that such free assumptions must not be unrealistic whilst also taking account of any applicable legislation, such as on minimum wages to prevent dumping (Hofstadler and Kummer 2015).

The following sections of this paper outline the relevance of combined pricing for construction management and economics in the context of structural building specifications.

1.1 Relevance for Construction Economics

For the bidder (and future contractor), the bill of quantities forming part of the structural building specification is the most important basis upon which to proceed with costing and pricing. In this document, the client must describe all circumstances and conditions for works to be performed so as to enable the bidder to prepare its proposal in accordance with the objective explanatory content of the bid documents. This means that it is the client's responsibility to describe the works to be performed in sufficient detail.

If the client merges several types of work into single items and if the bases for costing (quantity structure, work proportions etc.) of such merged items are not disclosed, the bidder will be forced to come up with free cost assumptions to be in a position to proceed with the pricing exercise at all. Bidders that do not, or only partially, realize the complexity of the work to be performed will usually be more inclined to take risks than other bidders with prior experience in performing complicated or complex works.

Any change in the proportions of the works included in the merged item compared to the bidder's original pricing assumptions will lead to a changed pricing structure associated with potentially higher or lower costs compared to the bid price.

If the bidder's (realistic) costing assumptions are disclosed in the bid submission process, they can be used for updating the construction contract accordingly if and when required. If, on the other hand, no such disclosures were made, and if additional costs are claimed, contractors are always confronted with the issue of having to prove on which assumptions they have based their pricing or if they have conducted a costing/pricing exercise at all (or rather resorted to mere price estimates).

1.2 Relevance for Construction Management

In addition to construction economics, combined pricing is also relevant to some aspects of construction management. Free pricing assumptions made by the bidder or contractor are also used as a basis for process planning, for instance for the scheduling and utilization of production factors. Related tasks include the preparation of high-level and detailed timelines as well as the allocation of construction equipment required on-site, such as defining the number of cranes as one of the key equipment categories for building construction, and of the number of required workers. In the event of any changes to quantity proportions within a combined pricing item, such as varying proportions of different soil or contamination classes in the case of earthworks or merging various rebar diameters and/or bend shapes when pricing reinforcing works, these will inevitably also lead to modifications of combined labor consumption rates or outputs. If the same timeframe is specified for the related works despite these changes, such a scenario may result in acceleration measures including, for example, the use of additional workers or pieces of equipment. Such measures may result in losses of productivity due to suboptimal team sizes, underrunning the minimum required workspace, or increased daily working hours.

2 COMBINED PRICING FOR REINFORCING WORKS

For reinforcing works, bidders are requested to apply combined pricing whenever no individual rebar diameters are specified and/or no details regarding reinforcement ratios and/or bend shape complexity and/or allocation to specific structural components (such as columns, walls, floors) are provided. Any source-based costing would at least require knowledge of the type of work or service, the conditions or circumstances under which work is to be performed, the prevailing site conditions, and the timeframe for work completion.

In Austria, for example, the standard specification or bill of quantities relating to building construction works includes one item for rebar and another for mesh reinforcement. Unit prices must be stated per kilogram or tonne of reinforcement and broken down to individual structural components. Furthermore, the standard specification may include connecting elements. Thus, the standard bill of quantities makes it very difficult for bidders to arrive at accurate reinforcement ratios, mean rebar diameters, and bend shapes, particularly in the case of engineering structures. Accurate information can only be taken from comparisons with similar, previously completed projects and/or from details provided in the literature. Bidders (or the future contractor) thus tend to be exposed to a high risk of stating "exceedingly optimistic" labor consumption rates, i.e., rates that are too low.

If the client does not provide additional details, such as on the distribution of rebar diameters or bend shapes, the bidder is again bound to make free pricing assumptions for the purpose of performing its pricing exercise. Such assumptions also include diameter and bend shape distributions. As a result, the following cost and price components of reinforcing works are interrelated: cost of materials (including costs of various specified dimensions), costs of rebar cutting and bending, and costs of installing the reinforcement.

An expert survey was conducted in 2015/16 at Graz University of Technology to determine the various types of information required for reinforcing works (specified as a combined pricing item) in order to perform a source-based costing exercise.

Respondents were given the following variables in the survey:

- Mean rebar diameter
- Diameter distribution
- Bend shape distribution
- Number of different reinforcement items
- Rebar lengths
- Reinforcement ratio
- Details regarding threaded connection bars
- Details regarding rebend connections
- Number and position of construction joints
- Number and position of expansion joints

A sound basis for the comparability of bids exists only if sufficient details are provided with respect to the reinforcement. The larger the related information gaps in the specification, the more significant the need for free pricing assumptions, with an associated adverse effect on the comparability of bids.

3 EXPERT SURVEY

Experts are defined as individuals who possess specific knowledge and intellectual skills and competencies in a clearly delineated field and who serve as a source of specific knowledge for the purpose of a survey. Expert knowledge usually comprises exceedingly large amounts of information, including simplifications, lesser-known facts, rules of thumb, and smart practices (i.e., heuristics) that enable efficient problem solving (Gläser and Laudel 2010, Springer Gabler 2017). In a preselection process, a total of about 130 experts from Austria and Germany with experience in the fields of costing, process planning, construction, final costing, and invoicing were contacted in writing and asked to participate in the survey. In total, 29 experts were recruited for the survey. They provided 28 responses with respect to details that must be stated for appropriate pricing of reinforcing works.

The majority of respondents (i.e., approx. 66%) worked for large companies with more than 250 employees, about 24% worked for medium-sized businesses (50 to 249 employees), and about 10% came from small businesses (10 to 49 employees). In the survey presented in this paper, experts had an average professional experience of 17.4 years; this experience ranged from 5 to 41 years. Questions were designed and developed together with social researchers (i.e., sociologists) in several revision steps, applying the principles of simplicity, clarity, impartiality, and specificity. On average, each respondent was interviewed for about 45 minutes either on the phone or in a face-to-face session to overcome ambiguities, collect missing information, or obtain background information and justifications of responses (Hofstadler and Kummer 2017).

4 ANALYSIS

Figure 1 shows the analysis of the relevance of individual details pertaining to reinforcing works. This diagram reveals that the reinforcement ratio is by far the most important factor to enable source-based costing. This variable describes the amount of reinforcement to be installed per cubic meter of concrete, thus indirectly providing information on how tight the spacing of the reinforcement should be. For instance, installing rebar in a wall at a reinforcement ratio of 215 kg/m³ is much easier than at a reinforcement ratio of 510 kg/m³ (see Figure 2).

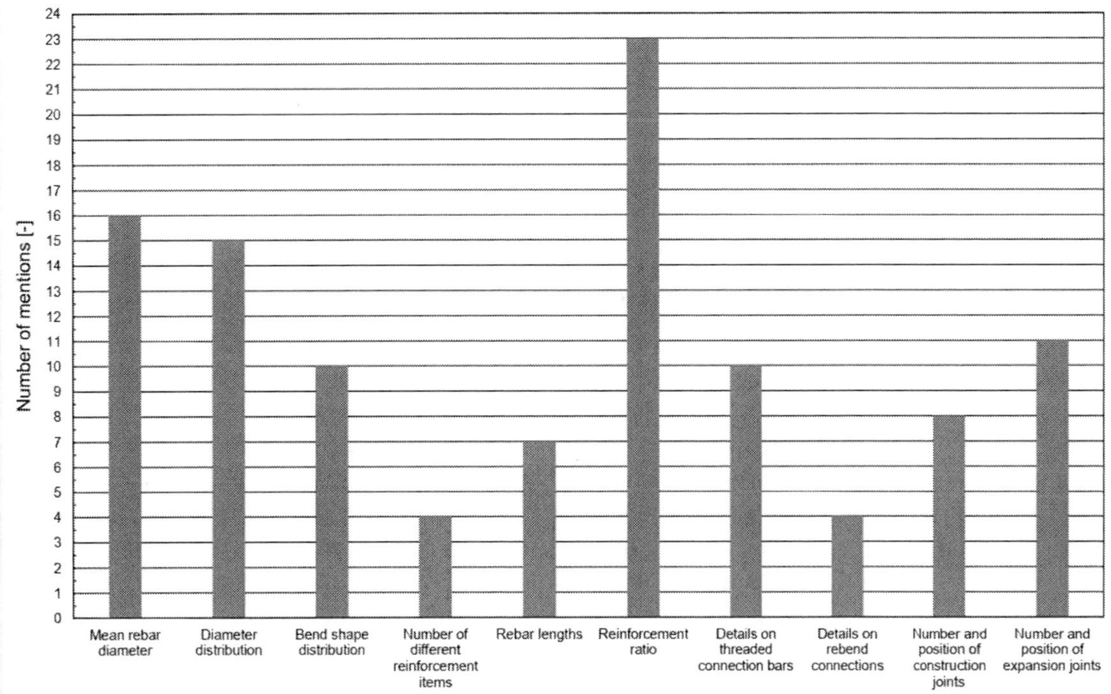

Figure 1. Number of mentions of details that must be stated for appropriate pricing.

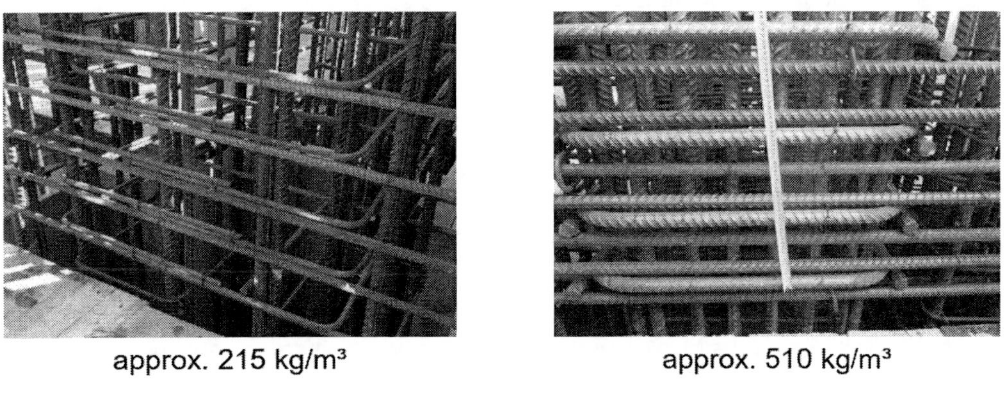

Figure 2. Different reinforcement ratios for core walls of a high-rise building.

Other significant factors for performing a comparable, source-based costing exercise include the mean rebar diameter (i.e., the weighted average of different rebar diameters and quantity proportions) and diameter distribution. Together with the distribution of bend shapes, these factors relate to the quantity structure of the combined pricing item and are thus crucial for costing and pricing (Wanninger 2014).

Any change in the quantity structure on which costing is based will have a direct effect on the assumed labor consumption rates and costs and will subsequently lead to cost over- or underruns. Experts also considered the number and position of construction and expansion joints as well as details regarding threaded connection bars to be significant, whereas rebar lengths, details on rebend connections, and the number of different reinforcement items were considered to be less significant.

5 CONCLUSIONS

This paper described the issues that may arise from structural building specifications if the client does not disclose accurate details regarding the reinforcement. In such as case, it is difficult to prove if additional cost claims are justified on their merits alone; yet it is equally difficult to determine the correctness of the claimed amount. The fewer the details provided by the client in its structural specifications, the greater the freedom for bidders at the submission stage. However, the question arises if bidders are actually comfortable with such a large room to maneuver in terms of pricing. Ways and means to reduce or prevent uncertainties in costing and pricing as well as in the related handling and invoicing processes are highly relevant to construction management and economics, far beyond reinforcing works. These issues can be resolved by preparing a clearer, more comprehensive specification combined with provisions that stipulate how to proceed in the event of any changes to quantities, the reinforcement ratio, or distributions at the detailed design stage.

The client's specification should include designated items for component groups that belong together, such as foundations, columns, walls, or floors, and specify separate items for individual rebar diameters. These are necessary to describe the work clearly, completely and objectively, to ensure price comparability, and, ultimately, to prevent non-quantifiable risks from being shifted to the bidder(s). Moreover, further items would have to be created for additional costs of different reinforcement ratios, quantity over- or underruns, or bend shape distributions. This approach would enhance the quality of the specification, thus promoting the plausibility and comparability of bids, whilst also making a major contribution to reducing conflicts in the course of managing and completing the project.

References

Gläser, J. and Laudel, G., *Experteninterviews Und Qualitative Inhaltsanalyse Als Instrumente Rekonstruierender Untersuchungen,* Springer Fachmedien GmbH, Wiesbaden, 2010.

Hofstadler, C., *Produktivität im Baubetrieb – Bauablaufstörungen und Produktivitätsverluste,* Springer-Verlag, Berlin, Heidelberg, 2014.

Hofstadler, C., and Kummer, M., Risiken und Chancen der Mischpreiskalkulation, *Proceedings of 13. Grazer Baubetriebs- und Bauwirtschaftssympsoium – Preisermittlung und Vergabe in der Bauwirtschaft – Baubetriebliche, bauwirtschaftliche und rechtliche Aspekte,* Heck, D., Mauerhofer, G., Hofstadler, C., (eds.), Graz University of Technology, 49-92, 2015.

Hofstadler, C., and Kummer, M., *Chancen- und Risikomanagement in der Bauwirtschaft – Für Auftraggeber und Auftragnehmer in Projektmanagement, Baubetrieb und Bauwirtschaft,* Springer-Verlag, Berlin, Heidelberg, 2017.

Springer Gabler Verlag (editor), Gabler Wirtschaftslexikon, Keyword: Expertenwissen. Retrieved from http://wirtschaftslexikon.gabler.de/Archiv/55229/expertenwissen-v10.html on November 18, 2017.

Wanninger, R., Fairness vs. Praxis – Risikoallokation am Beispiel Bewehrung, Festschrift Univ.-Prof. Dr.-Ing. Josef Zimmermann zum 60.Geburtstag und zur 145-Jahrfeier des Lehrstuhls Bauprozessmanagement und Immobilienentwicklung, Technical University of Munich, Munich, B19-1-B19-8, 2014.

Interdependence between Structural Engineering and Construction Management
Edited by Ozevin, D., Ataei, H., Modares, M., Gurgun, A., Yazdani, S., and Singh, A.

ISBN: 978-0-9960437-6-2

INFLUENCE OF PROJECT PREPARATION QUALITY ON THE PROJECT OUTCOME: A QUALITATIVE APPROACH

CHRISTIAN HOFSTADLER and MARKUS KUMMER

Institute of Construction Management and Economics, Graz University of Technology, Graz, Austria

To achieve the targets of all project key stakeholders, construction projects, in particular, require sufficiently long preparation and execution periods that go beyond ensuring project break-even. Assessing the break-even, or commercial viability, of a project directly depends on its complexity and underlying conditions. Time and funds are necessary to achieve a sufficiently high quality in construction project management, planning, and specification. At the project planning stage, the budget associated with specific use requirements should be calculated whilst also determining the quantities and quality standards achievable within a pre-defined budget. To successfully plan and execute construction projects, it is crucial to identify available resources in terms of funds and time in advance, thus making it possible, at a very early stage, to strike a realistic balance between the (financial) project targets and the available budget. If these targets exceed the budget, works are often specified incompletely, or specified quality standards are lowered with a view to arrive at lower costs in early project phases in order to ensure that the required permit is issued. This paper adopts a qualitative approach to highlight the influence of project preparation (i.e. project lead time and project preparation budget) on achieving the client's/owner's project targets.

Keywords: Project preparation time, Construction costs, Construction time, Productivity, Management of chances and risks.

1 INTRODUCTION

Sound preparation and sufficient time allotted to construction are important environmental factors of project target achievement. Besides construction time, however, project success will essentially depend on the way the project is prepared, and how its completion is managed and coordinated (i.e. project lead time, project working hours, project organization, project coordination, and project interfaces).

It is not only the contractor commissioned with performing construction works that will be influenced by the construction time specified by the client when selecting and combining its production factors. Likewise, the client itself will be dependent on its own specifications. Project break-even is essentially determined by the client's allocation of an appropriate time period in the construction contract. Any short (or extremely short) period for project completion specified by the client will keep the pace of the work at a high level (Hofstadler 2017). Both the contractor and the client are influenced by this speed: the former will be forced to work faster, while the latter will have to significantly accelerate their planning, control, management, coordination, inquiry, and decision-making processes. Accordingly, the client's agents must also be able and willing to keep up with this high pace.

It is well-known from road traffic, for example, that exceedingly high speeds may also result in hefty penalties if you are caught by a speed camera (e.g. Road Traffic Act and Driver's License Law for Austria). On the other hand, not only speed limits must be complied with; driving speeds also need to be adjusted to prevailing circumstances, such as traffic flow, road conditions, weather, and visibility, as well as to road characteristics such as slopes, gradients, bend radii, and lane widths, and one's own driving skills. Speeds may be too high and lead to accidents even in the case of compliance with the permissible maximum speed. In road traffic, exceedingly high speeds can be associated with major risks.

Likewise, all stakeholders involved in a construction project are more or less penalized if the speed or pace of project preparation and completion is too high. The magnitude of such "penalties" will essentially depend on the provisions of the construction contract and the associated rights and, in particular, obligations of the parties thereto.

Generally speaking, any analysis of past and current projects boils down to the following statement: "If you want to achieve a high quality standard of the building, you need to come up with detailed plans and specifications whilst keeping work speed at a reasonable level and considering, or adhering to, required setting, drying, and evaporation times, timing of formwork stripping, pull-out and adhesive strength values etc.". This paper refers to the major influence of project preparation and construction time in the context of overall project success, or viability.

2 TARGET SYSTEM FOR CONSTRUCTION PROJECTS

The well-known "project management triangle" establishing a relationship between cost, quality, and time is no longer sufficient to achieve project and operational targets. From a holistic point of view, reaching project break-even essentially requires the triangle to be transformed into a hexagon to integrate planning and forecasting. This enlarged system integrates the additional targets of process quality, insensitivity to disruption, and quantity; it covers not only the construction but also the operational phase.

Such an integrated approach is mainly driven by the decision-making/design variables of quality, time, cost, quantity, sensitivity to disruption, and process quality. These factors need to be analyzed more thoroughly to arrive at a sufficiently comprehensive assessment of the construction project. Project quality is the determining success driver because it will define the market price achievable for the project. Moreover, the quality of the project constitutes the lever with which to control follow-up costs because higher quality usually implies longer maintenance intervals, for instance owing to the robustness and easy maintenance of surfaces (such as those of metal or copper roofs or created by emulsion or mineral paints), thus optimizing building operation and use. Any time-related analysis will be guided by the key principle that ever-shorter product life cycles lead to corresponding cost increases. In this context, potential costs are obviously limited by the credit lines provided by banks. As holistic considerations are becoming increasingly popular in the construction industry, such views require enlargement of the classic "magic triangle" of cost, time, and quality beyond the project target to also cover newly introduced operational targets. The latter focus on life cycle costs, sustainable quality standards, and time-bound long-term targets such as warranties and service agreements and contracts. It thus becomes obvious that current project targets can be considered "spot on" (or narrow) in terms of quality and deadlines because costs will usually have been defined by the client. Furthermore, a life cycle approach is necessary to prevent this rigid focus on targets and to develop a strategic sourcing strategy with long-term targets, as shown in Figure 1 (target system for the life cycle of construction projects). Moreover, process quality and sensitivity to disruptions in the construction and operational phases of the building should be assessed (Hofstadler and Kummer 2017).

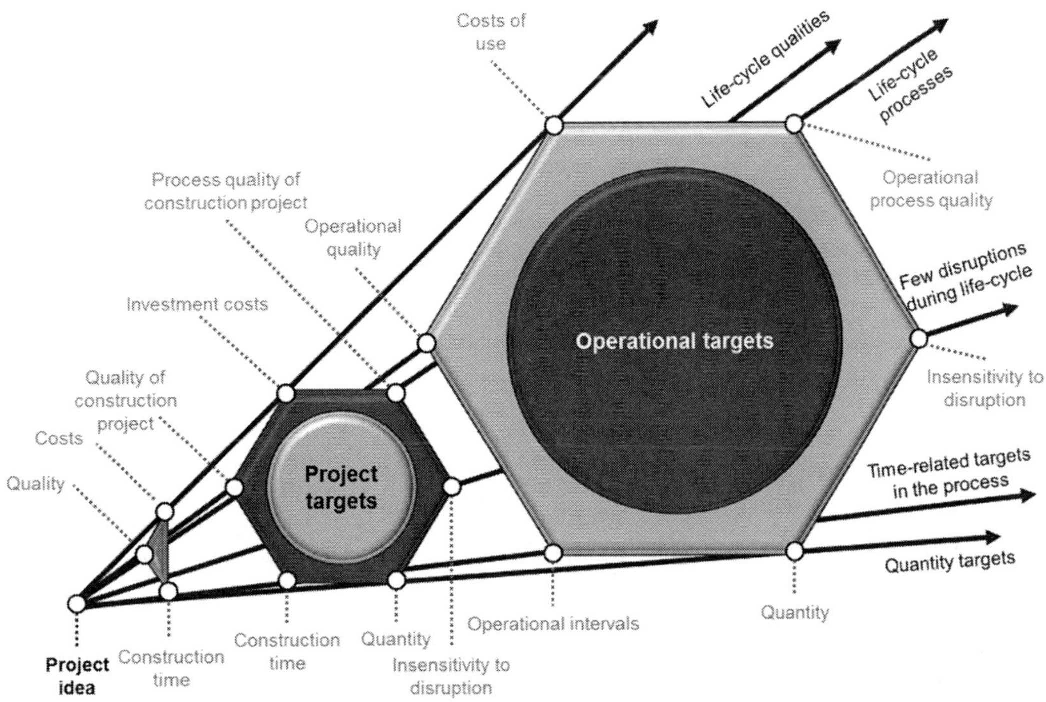

Figure 1. Target system for the life cycle of construction projects (Hofstadler 2014).

The defined (selected) chance/risk ratio will have a major influence on target achievement. If, for instance, a very short period and tight budget are chosen for project preparation, there will be a correspondingly high probability for project preparation costs to be relatively low. However, such an approach will inevitably be associated with a relatively high risk of underachievement (or no achievement at all) of defined project targets.

3 PROJECT PREPARATION TIME AND PROJECT TARGETS

Sufficient time and funds are required for achieving a high-quality standard in the management, planning, and specification of construction projects. Moreover, significant additional resources are needed especially for the systematic, structured preparation of highly complex projects to be completed within a short timeframe in a sophisticated work environment whilst involving a large number of stakeholders with an influence on specific structural parameters and having to comply with comprehensive requirements in terms of their environmental impact. Add to this the impact of political lobbying activities, and the need for specific yet comprehensive preparatory works will increase even further.

Time is money! This famous Benjamin Franklin quote is included in his essay entitled “Advice to a Young Tradesman” and suggests that we will save money (i.e., costs) if we expend less time (i.e., effort). However, this approach can also be fallacious, particularly in construction projects where it may lead to disastrous time and cost overruns whilst having an adverse effect on building quality and use.

Construction projects require sufficient preparation and completion times aligned with their degree of complexity to achieve the targets defined by all project stakeholders. Any assessment of viability thus directly depends on the complexity of the project and its underlying conditions. For construction projects, achieving a high-quality standard in project management, planning, and specification requires time and funds to be aligned with the necessary effort.

Based on a requirements analysis, project planning should derive the budget necessary for certain use requirements. Another plausible approach would be to start from an available budget (cost cap or ceiling) to define possible quantities and qualities for the building or structure. Any successful project planning and implementation effort will inevitably necessitate prior definition of required resources in terms of time and funds, which makes it possible to strike a realistic balance between the project targets and the available budget at a very early stage. More often than not, works are specified in the absence of sufficiently detailed planning information, or quality targets are initially reduced to adhere to the specification so as to arrive at lower costs in early project phases in order to ensure that the required permit is issued (e.g. Lechner 2018, Wall 2018). These cost caps result, for instance, from public-sector constraints and potential co-funding by government agencies.

Cost estimates are initially kept at a low level particularly in the case of public buildings with the associated political influence because decision makers fear public resistance or outrage and thus do not want to put the project at risk before it has even started – a prime example of politicians' fear of losing voter support!

Furthermore, the structural framework is sometimes specified to a lesser extent at the contract award stage, as opposed to the degree of complexity shown in the detailed design or construction phases. Only after the contract has been awarded will the selected bidder find out that the structural framework and associated works are much more complex than originally disclosed by the client in the bidding phase. If the contract was awarded to the seemingly "cheapest" bid (i.e., the lowest contract amount), the quality and quantity targets will subsequently be increased by the required additional works identified post-award, as well as by subtly different interpretations of the works owed under the construction contract.

This approach to project handling leads to much higher costs – as if the actual requirements for the building or structure and the circumstances and conditions under which work should be performed had been incorporated in the specification prior to the award of the (original) contract. Client's requests or needs that become known only after the contract award will almost always result in cost increases, which can be avoided only if such requirements are stated as clearly and unambiguously as possible before awarding the original contract. Such cost increases are mainly due to the fact that the competitive setting has changed and "forgotten" works need to be procured at a higher price level. Prior to entering into the construction contract, the bidder is bound to compete with other bidders and is thus under severe competitive pressure. After the award, the contractor is the only counterpart of the client when it comes to requested changes or alterations, and is thus in a stronger negotiating position. Pricing of changed or additional works does no longer (completely) take place in the competitive environment of the original contract with corresponding prices of originally agreed works or services. Any simplified or incomplete work specification must be firmly rejected because it will almost always give rise to a disaster from a construction management and/or economics point of view because the originally agreed cost, timeframe, and quality of the building or structure can no longer be adhered to.

Besides sufficient funds, it is crucial to allocate sufficient time in line with project complexity to ensure networked project preparation and effective work execution. In an integrated planning exercise, this step will enable the feedback loops needed for a sufficient degree of planning detail. Planning quality has a direct influence on specification quality and thus on the achievable planning and specification maturity as a basis for bid costing and pricing exercises that reduce pricing risks for bidders. The resulting high degree of bid comparability will also enable the client to commission construction works exactly as specified and on budget. In early project phases, estimation methods play a prominent role in order to arrive at initial values and ratios that are then replaced with more solid calculations as the project is progressing. Such calculations lead to predominantly deterministic results. In these project phases, there are no quantitative options (yet) for interpretation of risks taken or of chances that may potentially arise.

Figure 2 shows the qualitative correlation between estimated project lead time and achievable project, structural, operational and use qualities. The horizontal axis represents time, whereas the vertical axis shows the degree of target (under)achievement. A "normal" project lead time should be aimed at, depending on the type and nature of the project. The normal project lead time area describes the range of lead times sufficient for preparing or processing the bases for planning and decision-making to a sufficiently high-quality standard. The 100% mark shown on the vertical axis indicates that defined targets can be achieved at minimal costs, disruptions, disputes etc.

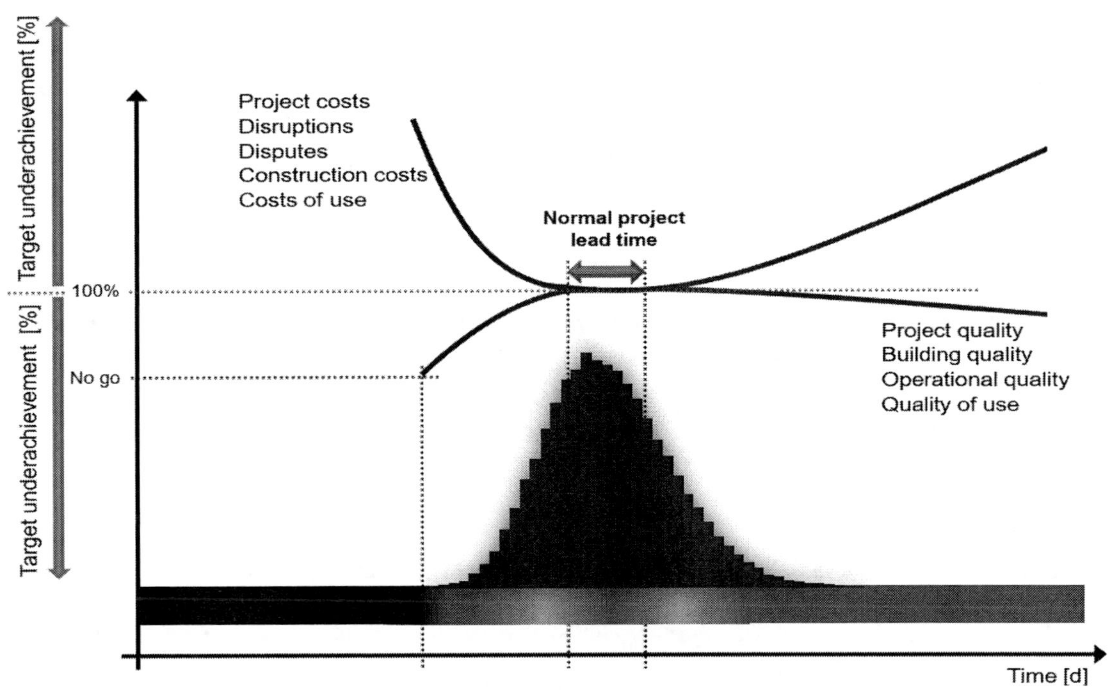

Figure 2. Correlation between project preparation time and project targets (Hofstadler 2014).

Any project lead time that is too short or too long will lead to cost increases. The number and severity of disruptions or disputes will rise particularly in the case of an exceedingly short (and continuously shortened) lead time, resulting in higher construction and project costs as well as higher costs of use at a later stage. Although the client will often have prepared a very stringent construction contract, its provisions may not be sufficient to resolve all issues arising in the ensuing project phases due to deficient project organization (structural and workflow organization), planning, and specification. The key factor to be considered in this context is that any planning, organization and coordination activity will still be performed by humans requiring adequately allotted time periods in order to achieve normal productivity. All too often, important steps to coordinate and clarify responsibilities at project interfaces are skipped because of the existing "project pressure" and exceedingly tight time budget.

Follow-up costs potentially incurred by an exceedingly short project lead time are hard to quantify *ex ante*, which is also associated with the major issue of having to "discount" such costs to the corresponding point in time according to the net present value method. This approach also makes it impossible to compare total costs in the case of normal project lead time to those associated with project lead times that are too short.

Even in the (rather rare) event of an exceedingly long project preparation time, costs will increase as a result of unnecessary planning and organizational activities (such as considering compliance with new regulations and guidelines, changes in the client's ownership structure, or modified financing agreements) that do not lead to a further improvement of results. Quite on the contrary, exceedingly long project lead times can reduce quality standards, for instance by overstretched planning. At the execution stage, the production system will be in the center of considerations. Achievable productivity is essentially determined by the efficiency and effectiveness of the production system.

4 CONCLUSIONS

It is not only "chaos projects" that highlight the importance and necessity of integrating aspects related to project lead times into organizational, management, planning, and construction processes more thoroughly and systematically. A forward-looking, smart approach to managing chances and risks in order to define the related targets and methods will pave the road to success.

Key aspects of the correlation between project preparation time and the degree of project target achievement are outlined in a transparent and plausible manner.

This paper provides a particularly realistic basis for demonstrating the relevance of project preparation time in project and construction management and economics, thus laying a sound foundation for drawing the right conclusions in the interest of ensuring the commercial success of the project. Consideration of the aspects outlined in this paper makes the steps of preparing and making decisions more systematic and target-driven whilst also enhancing decision certainty and transparency.

References

Hofstadler, C., *Methoden zur Ermittlung von Lebenszykluskosten*. In: Nachhaltig Bauen mit Beton: Werkstoff und Konstruktion, Beton Graz'14 – 2. Grazer Betonkolloquium; 25. / 26. September 2014. Nguyen, V., Maydl, P., Freytag, B., and Santner, G. (eds.). 139-152. Graz, Verlag der Technischen Universität Graz, 2014.

Hofstadler, C., *Calculation of "Normal" Construction Time for Building Construction Projects Whilst Considering Uncertainties*, Resilient Structures and Sustainable Construction, Proceedings of the Ninth International Structural Engineering and Construction Conference, Valencia, Spain, July 24-July 29, 2017.

Hofstadler, C., and Kummer, M., *Chancen- und Risikomanagement in der Bauwirtschaft – Für Auftraggeber und Auftragnehmer in Projektmanagement, Baubetrieb und Bauwirtschaft*, Springer-Verlag, Berlin, Heidelberg, 2017.

Lechner, H., *Systematische Kostenschätzungen und –berechnungen – Wo sind die Lücken im Kostentrichter?*. In: Kostenschätzung, Kostenberechnung, Kostenanschlag – Baubetriebliche, bauwirtschaftliche und rechtliche Aspekte, Tagungsband zum 16. Grazer Baubetriebs- und Bauwirtschaftssymposium, Verlag der TU Graz, Graz, 2018.

Wall, J., *Lebenszyklusorientierte Modellierung von Planungs-, Ausschreibungs- und Vergabeprozessen*, PhD Thesis, Graz University of Technology, Graz, 2018.

Interdependence between Structural Engineering and Construction Management
Edited by Ozevin, D., Ataei, H., Modares, M., Gurgun, A., Yazdani, S., and Singh, A.

ISBN: 978-0-9960437-6-2

LOOKING FOR PROJECT SUCCESS IN INFRASTRUCTURE PROJECTS: HOW CAN WE ACHIEVE IT?

KONRAD SPANG[1], HEINZ EHRBAR[2], and AHMED ELBAZ[1]

[1]*Dept of Project Management, Kassel University, Kassel, Germany*
[2]*Dept of Construction, Environment, and Geomatics, ETH Zurich, Zurich, Switzerland*

Even though the main aim of project management (PM) is to assure successful projects, the PM literature still lacks a clear definition of project success (PS), since it highly depends on the stakeholders' expectations and the point in time in which the project was evaluated. In this context and due to their specialties e.g. long duration or a large number of stakeholders, the infrastructure projects are subjected to time delays and cost overruns and therefore, criticized in the society and described as failed or unsuccessful projects. This paper hypothesizes the need to adapt our PM approach to follow more systemic methods and presents the main principles of systematic project management (SPM) as a success factor (SF) for infrastructure projects. In addition, a case study about the Gotthard Base Tunnel is presented to verify the literature findings.

Keywords: Systematic project management, Gotthard base tunnel, Success factors, Success criteria.

1 INTRODUCTION

Despite the major technical development in the last decades, the well-known written research results about PM, the growth and acceptance of PM in the last forty years, or the rapid growth in membership of PM bodies, projects still fail to satisfy their stakeholders (Spang 2016), especially those still associated with massive cost overruns and time delays (Flyvbjerg 2005).

In order to phrase a holistic definition of PS for infrastructure projects that include multiple stakeholders with multiple perceptions of PS, we need to consider the following points:

- The difference between PS and successful projects: According to Pinto and Slevin (1987) a successful project is defined as one that resulted in organizational change. PS has several pragmatic definitions, including meeting schedule, budget, and performance.
- The difference between PS and PM success: According to de Wit (1988), PM success can be measured in terms of time, cost and quality. PS can be measured by comparing the project outcomes and the planned project targets.
- The specialties of infrastructure projects: Besides the specialties of construction projects, infrastructure projects, especially transportation projects, acquire others (e.g., strong interaction with the environment, big budgets that to be provided from taxes, technical challenges, special regulation for the procurement process, long-lasting projects, limited ability to predict the construction condition (e.g., subsoil, weather, problematic for public acceptance (Spang 2016))).

PM has always been presented as a big support for PS (Spang 2016, de Wit 1988), moreover, many authors such as Spang (2016), and Wirkus (2016) claimed that by adapting of the PM approach, better outputs in terms of cost, time and quality can be achieved in infrastructure projects. Consequently, this means higher PS.

2 DEFINITION OF PROJECT SUCCESS IN INFRASTRUCTURE PROJECTS

In order to define the PS, the researchers have been facing many challenges such as coping with the different expectation of the stakeholders as well as choosing the right point in time to evaluate the project. Therefore, a great focus has been given to identifying PS criteria or dimensions as a means to assess PS, however, these criteria can vary from project to project, from stakeholder to stakeholder or from a point in time to another.

2.1 Multidimensional Concept

In order to overcome these problems, many authors have been using multidimensional PS assessment models and adopting its core concept to fit with different industry sectors. Shenhar *et al.* (1997) presented a four-dimensions project assessment model. The first dimension "project efficiency" defines success according to three criteria on-time, within budget, and within specification. Second dimension "impact on customer" addresses client-oriented criteria e. g. meeting performance measures and functional requirements. Third dimension "business success" deals with the direct impact on the organization in terms of market shares and total improvement of the organization. Fourth dimension "preparing for the future" addresses the preparation of the organization for the future e.g. new opportunities for further markets or innovation capabilities.

Bannerman (2008) presented a multi-level assessment model for IT-Projects and this model consists of five levels to evaluate PS. The first two levels are processed success and PM success and in these levels, the project will be evaluated against its performance such as time, cost and quality. The third level reflects the perceptions of users and emphasis their acceptance and satisfaction. The last two levels are future-oriented levels to address the impact of the project on its own organization (business success) and on the market (strategic success).

2.2 Multidimensional Framework for Infrastructure Projects

In order to integrate the specialties of infrastructure projects and the big number of stakeholders, Elbaz and Spang (2018) presented a multidimensional PS assessment model consisting of six dimensions distributed over the project life cycle and adaptable to acquire different stakeholders' success criteria. These dimensions are:

- <u>Function Success</u>: This dimension outlines the fact that a project must deliver its supposed functions. Unfortunately not all infrastructure projects achieve this goal e. g. Berlin Airport BER or Big Dig Boston. Even with a project delay or overruns, once the project delivers its supposed function this dimension can be considered as achieved.
- <u>Management Success</u>: This dimension testes if the project is completed in time, within budget and according to specifications. A very few numbers of projects achieve this dimension by 100% and the literature has shown that an overrun of 30% in infrastructure projects is considered "normal" (Kostka and Fiedler 2016).
- <u>Investment Success</u>: Addresses the financial effects of projects and can be measured by Pay-Back-Period.

- Organization Success: This dimension addresses the improvement of the internal process of the organization (e.g., organizational structure, culture, and the ability of the organization to manage parallel projects).
- Business Success: Addresses the project's effect on the business of its stakeholders especially the owner. This dimension covers financial criteria such as market shares as well as non-financial criteria (e.g., sense of pride and achievement).
- Strategic Success: Infrastructure projects are parts of a strategic plan; this dimension assesses the strategic effects of the project on the long run and introduces many criteria to measure it (e.g., economic growth, socioeconomic trends and ecological trends).

3 ACHIEVING PROJECT SUCCESS IN INFRASTRUCTURE PROJECTS

Despite the significant number of SFs in the literature, stakeholders still cannot achieve a harmonic cooperation atmosphere in their projects and are not satisfied with the output of their projects. In order to facilitate the management of these complex projects, this paper presents SPM as a key SF and therefore claims that projects can be seen and considered as systems.

3.1 Project as a System

A system is a set of things working together as parts of a mechanism or an interconnecting network; a complex whole. This definition fits our understanding of projects very well but lacks the temporary nature, that's why projects can be seen as temporary systems (Heinrich 2014) yet meant to achieve predefined goals.

3.2 Systematic Project Management

Systematic means "done or acting according to a fixed plan or system; methodical" or as "according to an agreed set of methods or organized plan". Accordingly, PM has a systematic aspect by its definition.

This paper defines SPM as a collaboration of various yet linked and standardized management practices and techniques which have to be implemented as a whole set to ease and facilitate the implementation of SFs and to elevate the chances of PS.

3.3 Systematic Project Management as a Success Factor

PM has been approved as a key SF for project-oriented industries such as construction (Chan *et al.* 2004). Yet, the implementation of an efficient PM faces many difficulties, (e.g., the maturity of PM culture in companies, interlacing the project to its program or portfolio and further company's strategic vision, linking different PM knowledge areas to each other (e.g., risk management and stakeholder management or even connecting different project phases)).

The traditional PM approach assumes that the sponsor/customer can define project objectives, time and cost required to achieve these objectives as well as the way to get the work done (Wirkus 2016). But due to the complexity of infrastructure projects, this assumption can be very hardly and rarely achieved. In order to be more successful in today's projects, the PM needs to adopt more systematic approach in which: a) Project objectives and targets are agreed among project stakeholders and comply with the strategic goals and objectives of their organizations. b) PS definition covers all project phases from initiating to closing. c) The purport of PS is unified between the different management levels. d) The PM processes are integrated in a nonlinear approach, that facilitates the flow of feedbacks between PM processes, project phases, and project stakeholders.

4 SYSTEMATIC PROJECT MANAGEMENT AS A SUCCESS FACTOR IN INFRASTRUCTURE PROJECTS

The SPM is a great support for the PS as discussed in section 2.2.

Function success: Infrastructure projects are predicted to have several changes, that affect their functionality or lead to a functional failure e.g Berlin Airport. Due to the complexity of these projects, changes can not be avoided by 100%. The project must have change management, that systematically proves and integrates these changes in the project plan, re-estimates the scope, cost, time, quality, resources, risks and sometimes even the procurement and contract management.

Management success: The SPM helps to achieve management success in many ways by:

- Linking the different PM elements and implement them as a whole set to manage the resources more properly during the project life cycle.
- Aligning project objectives with the company's strategic goals and linking the PS to the company success to insure continuous support of the top management.
- Complying the stakeholders to the PS by integrating their motivations and objectives into the project goals.

Investment success: since the project sponsors might have a different understanding of PS, the SPM aims to integrate the investment criteria into the project plan and provides the sponsors with regular reports about their investment to ensure more compliance and transparency.

Organization success: SPM helps organizations to better manage parallel projects by setting the organizational objectives in the first place and highlighting the outcomes of the project to the organization's assets. It also enables the organization to maintain a better relationship to customers, partners, and suppliers by addressing their goals and benefits in its proper understanding of the PS.

Business success: SPM links the project to its business case and hypothesizes the effect of the project on maximizing the value for the project's owner, sponsors, and stakeholders.

Strategic success: SPM provides a long term tracing to assess and evaluate the strategic effects of the projects and compare them to the strategic plan.

The next section presents the Gotthard Base Tunnel as a successful case study to verify the usability of the SPM.

5 CASE STUDY GOTTHARD BASE TUNNEL, SWITZERLAND (1992 – 2016)

The Gotthard Base Tunnel in Switzerland is the longest railway tunnel in the world with a total length of 57.1 km, total underground system of 151.8 km, approved by Swiss voters in a plebiscite in 1992 and in operation since 2016. It can be considered as a successful major project.

The New Rail Link through the Alps (NRLA) at the Gotthard posed an unusual challenge to all involved considering the sheer size, the political significance and the exceptional length of the base tunnel. Moreover, the project owners and sponsors had to be able to rely on the project being constructed as it had been decided by the Swiss Parliament and the Swiss people in the agreed target (cost, time and quality)

Whereas cost and time could be defined quite accurately by target values with a range of variation (-10%/+40% for costs) already in an early stage, the definition of the quality requirements is more challenging. In the case of the Gotthard Base Tunnel the quality requirements were defined project specifically (based on the quality management standard ISO 8402 (1994)).

The project's owner developed a Sponsor-Contractor delivery model as shown in Figure 1, where the sponsor (Swiss Federal Government) charged AlpTransit Gotthard Ltd (ATG) with a design-build mandate as a constructor (sponsor constructor model). The ministry (DETEC) and the federal railway authority (Federal Office of Transport, FOT) represented the federal government in the project organization. The system of project requirements was divided into "fixed requirements without freedom of action" and "open requirements with freedom of action" as shown in Table 1. At the Gotthard Base Tunnel, stakeholders had a different view on the project requirements (i.e., the sponsor had his focus mainly on fixed requirements and the requirements related to final object).

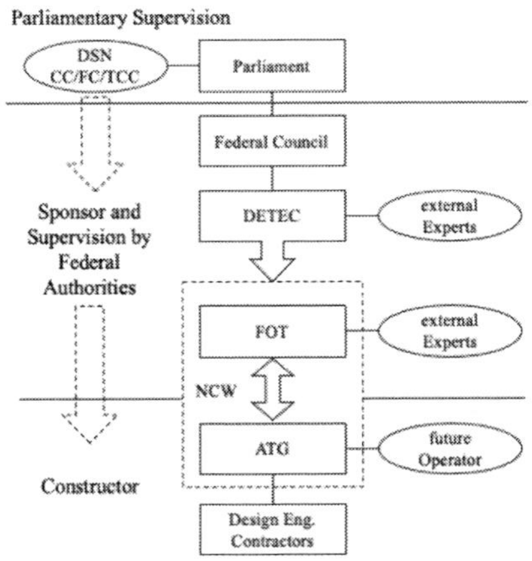

Figure 1. The sponsor constructor delivery model (Ehrbar *et al.* 2016).

Table 1. Project Requirements at the Gotthard Base Tunnel (Ehrbar *et al.* 2016).

Fixed Project Requirements without Freedom of Action	Open Project Requirements with Room of Action	
	Object-related Project Requirements	Process-related Project Requirements
Agreed Project Requirements: (Agreement Swiss Confederation / ATG) Completion • Of scope of work and agreed standers • Within budget • Within time schedule	Quality / Function Fulfillment required levels of • Reliability • Availability • Maintainability • Safety	Occupational Health and Safety • Highest OHS-Standers • Reasonable health protection
	Costs Minimum of • Investment costs • Operation costs • Maintenance costs	Project Organization • High reliability: minimize errors and contradictions • Organization: clear structures, explicit interfaces, unambiguous tasks. • Processes: leadership, effcient performance and controlling • Information (internally / externally): at the right time, at the appropriate level
Presupposed Project Requirements Compliance with • Laws, regulations, codes • Project approval with obligations • Duty of care	Schedule / Milestones Minimum • Planning period • Period of approval • Construction period Environment • Optimum environmental safety and resources management	

The Project Management System at the Gotthard Base Tunnel was created with a best practice approach and was an important key SF. The key SFs for this project were systematic and very professional project management (refer to Table 1) from the earliest beginning, based on a strategic plan (shifting the heavy traffic from the road to rail), taking into account all the requirements of the various stakeholders (public opinion, affected people, environmental organizations) as well as the market. The successful realization of the project was only possible thanks to the project-specific organization (refer to Figure 1), which was supported by a project-specific legislation.

6. SUMMARY

This Paper highlights the specialties of the infrastructure projects links the effects of these specialties with the project performance in terms of time, cost and quality as main project success criteria. These three criteria have been widely criticized to be the only ones used with infrastructure projects, thereupon this paper presents the six-dimensional Project success from Elbaz and Spang (2018) and illustrates the importance of systematic project management to each dimension.

The more complicated our projects are going to be, the more we need to adapt our way to manage things around them, and therefore this paper criticizes the traditional approach of project management and presents the SPM as an approach to achieve better results from infrastructure projects by a) Link project's objectives to organization's objectives to ensure more support from the top project management; b) Discuss and unify the purport of project success between all projects phases and project stakeholders; c) Reconstruct our project management process to ensure more agility and feedbacks among project stakeholders. In the end, it verifies the usability of SPM by presenting Gotthard-Base-Tunnel as a successful case study.

References

Bannerman, P., Defining Project Success a Multilevel Framework, PMI® Research Conference: Defining the Future of Project Management. Warsaw, Poland, 2008.

Chan, Albert P. C., Scott, D., and Chan, A P. L., Factors Affecting the Success of a Construction Project, Journal of Construction Engineering and Management, 130(1), 2004.

de Wit, A., Measurement of Project Success, International Journal of Project Management, 6(3), 164–170, 1988.Ehrbar, H., Gruber, L. R., and Sala, A., Tunnelling the Gotthard - Gotthard Base Tunnel, Swiss Tunnelling Society, Bauverlag Gütersloh, 2016.

Elbaz, A. E. M. and Spang, K., Mapping the Success Dimensions of The Infrastructure Projects in Germany. International Project Management Association Research Conference 2017, 2018.

Flyvbjerg, B., Policy, and Planning for Large Infrastructure Projects: Problems, Causes, Cures, World Bank Policy Research Working Paper 3781, 2005.

Heinrich, H., Systemisches Projektmanagement: Grundlagen, Umsetzung, Erfolgskriterien, Carl Hanser Verlag GmbH Co KG, 2014.

ISO 8402, Quality Management and Quality Assurance - Vocabulary, International Organization for Standardization, Geneva, Switzerland, 1994.

Kostka, G., and Fiedler, J., Large Infrastructure Projects in Germany: Between Ambition and Realities, Cham: Palgrave Macmillan, 2016.

Pinto, J. K., and Slevin, D. P.., Critical Factors in Successful Project Implementation, IEEE Transactions on Engineering Management EM, 34(1), 22–27, 1987.

Shenhar, A. J., Levy, O., and Divr, D., Mapping the Dimensions of Project Success, Project Management Journal, 28(2), 5–13, 1997.

Spang, K., Projektmanagement von Verkehrsinfrastrukturprojekten, Berlin, Heidelberg: Springer Berlin Heidelberg, 2016.

Wirkus, M., Adaptive Management Approach to an Infrastructure Project, Procedia - Social and Behavioral Sciences, 226(Juli), 414–422, 2016.

Interdependence between Structural Engineering and Construction Management
Edited by Ozevin, D., Ataei, H., Modares, M., Gurgun, A., Yazdani, S., and Singh, A.

ISBN: 978-0-9960437-6-2

CAUSES OF DEVIATIONS IN PROJECT EXECUTION AND THEIR EFFECTS

FLORIAN MÜLLER[1], ALEXANDER TUPI[2], and DETLEF HECK[1]

[1]*Institute of Construction Management and Economics, University of Technology, Graz, Austria*

[2]*Granit GmbH, Graz, Austria*

Over the past few years the German and Austrian construction industries have been confronted with increasing complexity in projects, a situation, which has resulted in a significantly higher number of construction claims. The lack of an effective documentation system and a standardized claim management system has led to a growing number of disputes and legal cases. To develop a standardized claim management system, it is essential to understand and analyze deviations in the as-planned vs. the as-built occurring during project execution. A standardized system may assist in the quantification of impacts of different deviations and change procedures in future projects. This paper focuses on deviations and their effects occurring during project execution, as well as possible disputes arising from them. Further research is necessary in order to develop an effective standardized claim management system.

Keywords: Claim management, Claim process, Dispute resolution, Conflict management, Standardized processes.

1 INTRODUCTION

Rapidly changing social demands have a major impact on the technical and legal requirements for construction projects. The complexity of construction projects has been continually rising over the past few years. A precise definition of the owner and user objectives has become a major challenge. Projects suffer more and more from insufficient design (Mueller and Stempkowski 2015). Shortage of time and cost efficiency put additional pressure on them (Burr 2016). In Austria, the design and execution phase are traditionally separated. Recently, a shift towards planning during project execution has been observed. This is making the preliminary work preparation and cost calculations inaccurate. The Austrian construction industry has experienced a considerable increase in claims as a result. Contractors often lack sufficient time and human resources to prepare a proper claim. Additionally, a substantial documentation and claim examination is considered a secondary activity, which aggravates the situation (Kangari 1995). The increase of claims and inadequate claim documentation can cause a loss of confidence between the owner and contractors (Zaghloul *et al.* 2002). Specific knowledge, obtained through previous successful claim processes, is rarely used in following claims. This leads to an uncontrolled development of numerous documentations and claim process systems. A legal opinion of an Austrian high court judge (Kodek *et al.* 2017) regarding requirements on documentation for construction claims shows that the construction industry in Austria is facing a change of heart regarding claim management. Based upon this legal opinion, a stringent causal evidence for each single event linked to its effects is required, making documentation for large scale and complex projects exceptionally difficult. Global claims with generic cost calculation models, based on few specific documented events, should be avoided. The idea is to eliminate the possibility of *cum hoc ergo propter hoc* (correlation does not imply causation). The objectives of this research are to identify the most common deviations causing a claim during execution and their effects on the workflow, as well as problems occurring during the claim process. Understanding them may be beneficial for an optimization of the organization processes and for a possible standardization of the documentation and claim process system in Austria.

2 OBJECTIVE AND METHODOLOGY

A research project on the subject of claim management, conflict management and settlement of disputes has been carried out in order to investigate in detail relevant deviations during execution, which are the cause of a claim and their effects on the workflow. Based on the findings (Bakhary *et al.* 2015, Jaffar *et al.* 2011, Mohamed *et al.* 2014, Odeh and Battaineh 2002, Werkl and Heck 2013, Zaneldin 2006) a questionnaire was prepared in consultation with claim experts from the Austrian construction industry. The questionnaire is divided into five sections.

(i) The first section of the questionnaire contains questions regarding the background of participants.
(ii) The second section of the questionnaire contains questions related to common causes of deviation during work execution.
(iii) The third section of the questionnaire contains questions related to the effects on the construction process caused by disruptions.
(iv) The fourth section of the questionnaire contains questions related to problems occurring during different stages of the claim process.
(v) The fifth section of the questionnaire contains questions related to conflicts arising from claims.

Experts that took part in the questionnaire were selected by (1) background, (2) work experience in construction industry and (3) work experience with claim management. The survey consisted of closed-ended questions with a Likert scale from 0 (likelihood = never respectively no effect) to 10 (likelihood = always respectively severe effect). They could add extra options or contribute opinions.

3 RESULTS

3.1 Expert's Profile

Data of 55 claim experts from the Austrian construction industry was obtained. Due to diverging interests of stakeholders of a construction project, it is necessary to obtain data from the main parties dealing with a claim. This is why the experts are divided into three groups: owner, contractor and consultant. Table 1 shows the profile of the experts in terms of profession and experience.

Table 1. Expert's profile.

Characteristics	Owner	Contractor	Consultant
Number of experts	27.16%	34.57%	38.27%
Years of experience in the construction industry	21.65	20.07	18.71
Years of experience in claim management	17.06	15.89	12.76

The share of different groups is well balanced. Multiple answers were possible regarding the profession group. Most of the experts have 20 years of experience in the construction industry and 15 years in claims management. Information gathered is to be considered reliable.

3.2 Causes of Deviation

In the questionnaire, experts scaled in total 23 different deviations during project execution causing a claim according to their frequency. Table 2 shows the five most frequent deviations during project execution causing a claim. According to owners and consultants, "inadequate definition and or specification of scope of work" is the most common deviation causing a claim. Consultants rank "lack of decision-making by the owner" as an equally frequent deviation. Contractors consider a "delay of deployment of plans by the owner" to be the most frequent deviation.

Table 2. Deviations causing a claim.

Deviations causing a claim	**Total**	**Owner**		**Contractor**		**Consultant**	
	Ranking	Mean Score	Ranking	Mean Score	Ranking	Mean Score	Ranking
Incomplete and/or inadequate design	1	6.12	2	7.33	3	6.88	3
Inadequate definition and/or specification of scope of work	2	6.53	1	7.11	5	7.12	1
Delay of deployment of plans by owner	3	5.24	4	7.52	1	6.35	5
Lack of decision-making by owner	4	4.94	7	7.26	4	7.12	1
Changes of owner requirements (quantity)	5	5.47	3	6.22	6	6.47	4

3.3 Effects on the Workflow

Deviations occurring during project execution may have a negative impact on the workflow. To develop a standardized claim management system, it is essential to understand similarities of effects on the workflow caused by deviations during project execution. In collaboration with claim experts, 16 recurring effects were detected. Table 3 shows the five most frequent effects on the workflow. According to the results of all participating groups a "productivity loss due to cumulative effects" is the most common effect.

Table 3. Effects on execution.

Effects on execution	**Total**	**Owner**		**Contractor**		**Consultant**	
Table 3 (Continued)	Ranking	Mean Score	Ranking	Mean Score	Ranking	Mean Score	Ranking
Cumulating effects	1	5.81	1	7.04	1	7.21	1
Change of conditions	2	5.35	3	6.81	2	6.44	2
Change of quantities	3	5.59	2	6.78	3	6.38	3
Disruption of workflow	4	3.94	8	6.30	4	5.87	5
Acceleration of work	5	5.00	4	5.37	8	5.25	6

3.4 Problems Related to Claim Process

Austrian Standards define a two-stage model for the claim process. The first stage is the notification of an event causing a possible increase of cost and an extension of time. The second stage is the submission of claim documents. Several steps are required to successfully settle a claim. Several claim process models for the achievement of this exist in the German-speaking area (e.g. Mueller and Stempkowski 2015, Duve 2008). A five-stage model was defined for this questionnaire, starting with the identification of a claim-causing event, followed by notification, documentation, quantification and negotiation. Table 4 shows the three most common issues related to each stage of the claim process.

Table 4. Identification of a claim-causing event.

		Total	Owner		Contractor		Consultant	
		Ranking	Mean Score	Ranking	Mean Score	Ranking	Mean Score	Ranking
Identification	Insufficient documentation to identify a claim event	1	6.27	1	7.00	2	7.71	2
	Poor communication between site and head office	2	6.07	3	7.12	1	7.71	2
	Insuf. contract knowledge to identify a claim event	3	5.75	4	6.38	5	7.76	1
Notificatio	Ambiguity of legal basis	1	6.00	1	4.73	1	6.71	3
	Poor communication on site	2	5.50	4	4.46	3	7.29	1
	Insufficient documentation for notification	3	6.00	1	4.19	4	6.76	2
Documentatio	Incomplete documentation	1	6.88	1	7.22	1	7.82	1
	Lack of time for sufficient documentation	2	6.60	2	6.85	2	7.35	2
	Insufficient documentation due to verbal instructions	3	5.56	6	6.67	3	7.24	3
Quantificati	Unrealistic idea about losses	1	7.25	1	5.19	2	7.88	1
	Lack of time to prepare claim	2	6.56	2	5.63	1	7.29	2
	Unclear legal and contractual basis to prepare a claim	3	5.20	3	5.15	3	7.18	3
Negotiation	Disagreement regarding costs increased by claim event	1	7.88	1	7.93	1	8.76	1
	Insufficient documentation of claim event	2	6.94	2	6.56	5	6.47	2
	Insufficient owner knowledge about constr. processes	3	4.06	6	6.93	2	5.71	5

3.5 Reasons for Conflicts During Claim Process

In the last section of the questionnaire, experts ranked conflicts arising during the claim process. They were grouped into three categories: contractual, behavioral and operational conflicts. Table 5 shows the five most common reasons for conflicts arising during the claim process. Three out of five most common conflicts come from the group of contractual conflicts.

Table 5. Reasons for conflicts during claim process.

Conflicts during claim process	Total	Owner		Contractor		Consultant	
	Ranking	Mean Score	Ranking	Mean Score	Ranking	Mean Score	Ranking
Incomplete design	1	8.22	1	5.75	3	7.35	2
Diverging interests of project partners	2	7.67	3	5.75	3	7.88	1
Inadequate design	3	7.78	2	5.81	2	6.71	3
Insufficient description of scope of work	4	6.52	4	6.00	1	6.65	4
Change of quantities	5	6.37	5	5.06	5	5.59	5

4 DISCUSSION

Most deviations during project execution and conflicts during claim process arise due to inadequate design. This indicates that the owner is unable to define his project goals, which leads to an insufficient definition of the scope of work for the contractor. Design amendments during execution are necessary and result in multiple workflow disruptions. Contractors suffer from productivity losses due to numerous disruptions and are unable to document the increasing number of claim events effectively. Consequently, claim quantification is based on insufficient documentation. Inaccurate quantification, pared with the (frequent) over-optimistic idea about losses on the part of the contractor, finally results in a conflict with the owner. The results suggest that several factors can effectively decrease the occurrence of claim events during execution. A high planning standard avoids later changes in the design. Sufficient time for work preparation allows the contractor to plan their workflow effectively and to cope with changes. A systematization of processes prevents errors and makes it easier to introduce workflow amendments due to changes. A standardized documentation and claim process makes a claim more accurate and comprehensible.

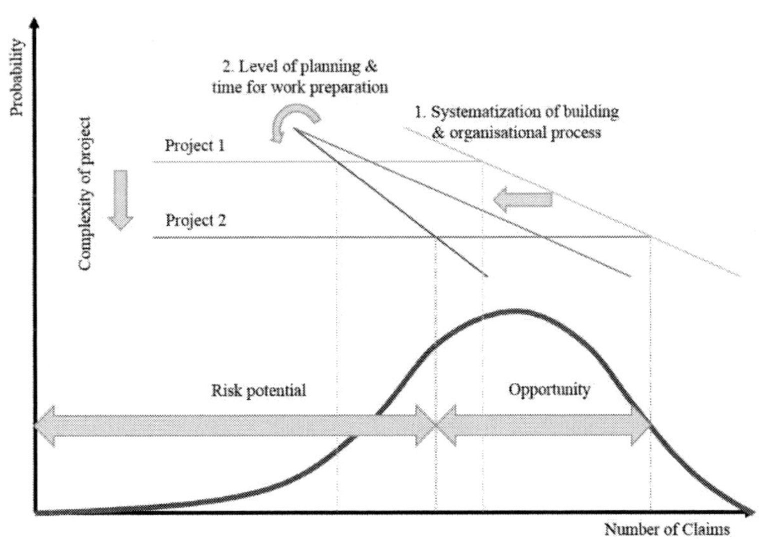

Figure 1. Assessment of risk due to good project organization.

Figure 1 shows the decrease of risk regarding the occurrence of claims. A project is more or less likely to face claims according to its complexity. The horizontal line defines the complexity of a project. The inclined line defines the level of planning and the time for work preparation. The steeper the line, the more accurate the project design and the more time a contractor has for work preparation. The line can be shifted to the left if processes are standardized. The intersection of the two lines defines the probability of the occurrence of a claim.

5 CONCLUSION

The results of the questionnaire indicate that there is a strong need for a project delivery model that allows the design and execution phases to merge. Traditional contracts and claim management cannot accommodate such a project delivery model. The recommendations given in this paper may reduce the risk of claims during project execution. Further research is necessary to propose a standardized claim management system customized for the Austrian construction market.

References

Bakhary, N. A., Adnan, H., and Ibrahim, A., A Study of Construction Claim Management Problems in Malaysia, *Procedia Economics and Finance*, Elsevier, 23, 63-70, 2015.

Burr, A., *Delay and Disruption in Construction Contracts*, 5th Edition, Informa Law from Routledge, Abingdon, 2016.

Duve, H., Nachweis von Bauablaufstörungen, *Baubetriebs- & Baurechtsseminar*, TU Graz, 2008.

Jaffar, N., Tharim, A. H. A., and Shuib, M. N., Factors of Conflict in Construction Industry: A Literature Review, *Procedia Engineering*, Elsevier, 20, 193-202, 2011.

Kangari, R., Construction Documentation in Arbitration, *Journal of Construction Engineering and Management,* 121(2), 201-208, 1995.

Kodek, G., Plettenbacher W., Draskovits A., Kolm R., Mehrkosten Beim Bauvertrag, Linde, Vienna, 2017.

Mohamed, H. H., Ibrahim, A. H., and Soliman, A. A., Reducing Construction Disputes Through Effective Claims Management, *American Journal of Civil Engineering and Architecture*, Science and Education Publishing, 2(6), 186-196, 2014.

Mueller, K., and Stempkowski, R., *Handbuch Claim-Management*, 2nd Edition, Linde, Vienna, 2015.

Odeh, A. M., and Battaineh, H. T., Causes of Construction Delay: Traditional Contracts, *International Journal of Project Management*, Pergamon, 20, 67-73, 2002.

Werkl, M., and Heck, D., Risiko- und Nutzenverhalten in der Bauwirtschaft, TU Graz, Graz, 2013.

Zaghloul, R., and Hartman, F. Construction Contracts: The Cost of Mistrust, *International Journal of Project Management*, Elsevier, 21, 419-424, 2003.

Zaneldin, E. K., Construction Claims in United Arab Emirates: Types, Causes, and Frequency, *International Journal of Project Management*, Elsevier, 24, 453-459, 2006.

Interdependence between Structural Engineering and Construction Management
Edited by Ozevin, D., Ataei, H., Modares, M., Gurgun, A., Yazdani, S., and Singh, A.

ISBN: 978-0-9960437-6-2

THEORETICAL ASSESSMENT OF IMPACTS OF PROPERTY RIGHTS ON EXISTING RECONSTRUCTION AND DEVELOPMENT PROGRAM HOUSES

GLADNESS MAKGOBI[1], NOKULUNGA MASHWAMA[1], and CLINTON AIGBAVBOA[2]

[1]*Department of Construction Management and Quantity Surveying, University of Johannesburg, Johannesburg, South Africa*

[2]*Sustainable Human Settlement and Construction Research Centre, University of Johannesburg, Johannesburg, South Africa*

The paper assesses the impacts of property right on low-cost housing and the solutions to address the challenges of property rights on low-cost housing. A literature review is the primary research method on the subject matter to find answers to impacts of property rights on affordable housing. The basis of the study is a general literature review (locally and internationally) published by other researchers on the similar study. The study indicated that enhanced political voice and political accountability has an impact on improving the prospects for secure property rights. Moreover, influence and supports the notion that the rule of law, in providing legal security for property rights, is an enabling condition for economic growth. Furthermore, if governments of developing countries provide real property ownership with clear titles and rights enforceable by law, then poor people will be able to use their assets to obtain credit which can be employed in productive activities, and countries could lever themselves and their poor inhabitants out of poverty. The research will make contribution to type of land reform, which lead to sustainable development through land titling, a reduction in the probability of being evicted, or otherwise losing land rights and provide land users with greater assurance that they will enjoy fruits of their labor and investment, thereby encouraging them to make long-term investments for sustainable use of their properties.

Keywords: Insecure tenure, Housing, Regulatory framework, Governance.

1 INTRODUCTION

The major cause of failure in land administration systems in many countries is poor management and a lack of good governance. Furthermore, land governance has been identified as an integral part of poverty reduction and the provision of tenure security (Zakout *et al.* 2009). Moreover, good governance in land administration aims to protect the property rights of individuals, enterprise as well as state, by introducing principles such as transparency, accountability, rule of law equity and participation into land related public sector.

Rights are described as socially or legally recognized entitlements to access, use, and control areas of land and related natural resources (Zakout *et al.* 2009). Formal improvement refers to a house that has a legitimate formal title, conforms with local planning regulations and building codes, structurally sound, and can be used as a security for a lengthly term home loan. Tenure security is defined as the preservation in conceptual contrast to eviction, the probability of saleability of property, exchange of ownership through the estate, pledging as collateral for mortgage and access to a loan under a certain state of affairs (De Soto 2003). Household income, housing policies, legal ownership of property with correct title deeds and standards enacted by regulation are elements which have a direct impact on property rights of houses occupied by the poor (Mattingly 2013).

2 LITERATURE REVIEW

2.1 Notion of Reconstruction and Development Program Houses

The National Housing policy has an appropriate framework which provides free houses to poor household members who earns between R0 to R3500, 00 monthly (Department of Housing 2009). The Housing White Paper indicates that the free low-cost housing is secured by individuals who are not financially able to qualify to purchase a house in the open market. It was also determined that the housing subsidy assistance is once off capital sponsorship, which is registered in the name of the beneficiary, yet paid out to the developers responsible for developing the subsidized houses (RDP) (The National Department of Housing 1994).

2.2 Governance on Low-Cost Properties (Land and Buildings)

Poor governance, whether in the formal land administration or customary tenure arrangements, implies that the land rights are not protected (Olujimi and Iyanda 2013). The absence of reliable approaches to record ownership of land leaves all spheres of government with no answers as to who are the legal owners of the land, in which side and the purpose for which land has been utilized (Ghebru and Okumo 2016). The essence of honesty and accountability favorable to functional trading of land will be nonexistence due to lack of uncomplicated regulatory implementation, transparent legislation, affordable regulatory procedures, secure type of tenure, responsive potential departments (Burns 2007).

Governments must initiate processes which are much more positive and supportive to the poor community including the provision of land, upgrading shelter and services, and help with building sustainable livelihoods (Turok and McGanahan 2013). An enhanced political voice and political accountability have an impact on improving the prospects for secure property rights (Domingo 2013). This perspective draws on the assumption that the rule of law is necessary to secure property rights. As such, it focuses on understanding the enabling governance conditions needed for the rule of law to be strengthened as a precondition for stabilizing property rights (Domingo 2013).

2.3 Authenticating Security of Tenure System

The formalization of land ownership, Land Titling program were recommended by the World Bank, IMF, and USAID (UN-Habitat 2003). A land right is secured when the owner is free of all encumbrances on the title of such land (Ndukwe 2001). In many developing countries, poor people are essentially excluded from formalizing their land and property rights and, consequently, a black market exists outside of the formal economy (De Soto 2003). If the governments of developing countries provide real property ownership with clear titles and rights enforceable by law, then poor people will be able to use their assets to obtain credit which can be employed in productive activities, and countries could lever themselves and their poor inhabitants out of poverty (Fabiyi 1990). The reasoning implies that more secure and well-defined property may lead individuals to substitute away from unproductive conflict over property towards productive activities (Trebilcock and Veel 2014).

Various researchers opposed the proposition that formal ownership to land occupied by the poor can bring about enhanced housing system. Evidence has been drawn up in Peru Land Titling program where immense land titling project presumed that land tenure formalization does not have a positive effect on household income or ability to obtain financial service including credit, but it may have a less effective impact on existing low-cost housing development (Kawaga 2001). It has been further discovered in Peru's Land Titling Program that acquiring formal property rights for land does not promote access to housing finance, but instead, the primary principle to obtain a mortgage is through family's labor remuneration (Cockburn 2001). Evidence from an impact evaluation study conducted on nationwide titling programme in Peru shows that titling led to a limited reduction of credit scoring. But the same analysis found that formal property ownership had no effect on approval rates for private sector loans (Rodrik 2000).

2.4 Government Intervention on Land Reform

Land reform as regulation requires appropriate capacity and the organizations to work feasibly and land sales have been politically deceived under the hands of disreputable workers, hence the poor have land outside the formal structure, yet don't have title deed (Cousins *et al.* 2005). Government financed units are regularly informally sold/exchanged, with no relating record being kept of the change in ownership. The underlying recipients would then not be able to further access to housing subsidy as they are registered on the National Housing Subsidy Database as a recipient (Umhlaba Consulting Group 2013).

Property holders are not allowed to apply for building plan or land use endorsement since they don't hold the lawful status of ownership (SPLUMA 2013). Adherents of this type of land reform justify their argument based on reduction in the probability of being evicted, or otherwise losing land rights, provides land users with greater assurance that they will enjoy the fruits of their labor and investment, thereby encouraging them to make long-term investments for sustainable use of properties (Kasimbazi 2017).

3 METHODOLOGY

The study was conducted with reference to the existing theoretical literature review. Published, unpublished, online journals and magazines by other researchers on similar research were critically examined to investigate the research problem. The qualitative approach provides perceptions to social components of the procedures amongst agencies (Denzin and Lincoln 1994). The qualitative technique is regarded as a suitable method to investigate impacts of property rights on existing low-cost housing. The approach will assist the manner on which role players (beneficiaries, land reform, government, and policy makers) better understand what is happening in their respective authorities with reference to property rights on existing low-cost housing.

4 FINDINGS FROM LITERATURE

The literature review showed that property rights are lawfully known entitlements to right to use and manage areas of land and related natural resources. Rights in the land are identified as an essential measure to poverty alleviation and establishment of tenure security (Zakout *et al.* 2009). Tenure security is the preservation in theoretical distinction to eviction, likelihood of selling the property, change of ownership by the estate, used as surety for a home loan and access to credit under a definite state of affairs (De Soto 2003). Preserving property rights of individuals, enterprise as well as state, by presenting standards such as transparency, accountability, rule of law, equity and contribution into land related public sector is the primary objective of good regulation in land administration (Zakout *et al.* 2009).

Poor regulation in land administration or normal tenure provisions entails that the land rights are not secured. An economic regulator, which delays effectiveness and prevents private investment, originates from urban overcrowding (Olujimi and Iyanda 2013). Properties are sold illegally by the poor in many countries and they are not incorporated in formalizing land use and property rights (De Soto 2003). There is no record kept for a change of ownership of low-cost houses, which are frequently sold illegally (Umhlaba Consulting Group 2013). In order for institutes to function efficiently, land reform as a by-law needs proper expertise to deal with sales of land that has been politically deceived under the hands of dishonest employees, hence the poor have and use land informally, and are not in possession of title deeds (Cousins *et al.* 2005).

All levels of government have no solution as to who are the legitimate owners of land and which side and for what purpose the land has been used due to lack of consistent methods to record ownership (Ghebru and Okumo 2016). The poor reside on condition, which is harmful to their well-being (Turok and McGanahan 2013). Building plan applications of low-income residents cannot be processed because they are not legal owners of properties (SPLUMA 2013). It is the role of the government to develop measures to assist the poor to establish a maintainable standard of living and access to services and infrastructure (Turok and McGanahan 2013). The aim of Urban Land Management directive is to preserve, develop investments and prevent land use that negatively affects the value of the property (Kitchin and Ovens 2013). Moreover, the need of the poor must be directly addressed by the Land Use Management system through developments of appropriate systems (Görgens and Denoon-Stevens 2012). Conforming to this type of by law reduces the probability of eviction, losing the land right and reassure land and building users value for money on their investment (Kasimbazi 2017).

A case study drawn up from Peru Land Titling program recognized that strengthening land tenure does not yield a positive household income or guarantee a home loan or credit approval from financial institutions, but it can have a less effective influence on the construction of existing low-cost housing (Kawaga 2001). Furthermore, it has been revealed in Peru's Land Titling Program that formalizing property right does not stimulate access to housing finance, instead the key standard to acquire a home loan is through the household income (Cockburn 2001). An assessment conducted on Peru titling program indicated that titling led to an enormous decrease in credit scoring. But the similar investigation indicated that formal property ownership has no implications on loan approval rates by financial institutions (Rodrik 2000).

5 CONCLUSIONS AND RECOMMENDATIONS

The study investigated the impacts of property rights on existing low-cost housing. Governance on property right is the contributing factor to control the way and purpose, which land and buildings occupied by the poor can be used. There should be some point of accountability and intervention to reinforce formal property rights to existing low-cost houses in order to enhance the quality of life to beneficiaries and socio-economic development of the country. There is a need to find suitable standards for a land management system that react to methods of the urban land market that clearly address the need of the poor. The objective of the current legislative (Spatial Land Use Management) framework in South Africa is to protect and promote investment but the framework does otherwise for the poor. The Act prevents unauthorized land use that has a negative impact on property values.

A literature review revealed that in various developing countries, poor people are essentially excluded from formalizing their land and property rights and, consequently, a black market exists outside of the formal economy. Moreover, if the government of developing countries provides real property ownership with clear titles and rights enforceable by law, then poor people will be able to use their assets to obtain credit, which can be employed in productive activities so that countries could get themselves and their poor inhabitants out of poverty. Other researchers found that acquiring formal property rights for land does not promote access to housing finance, but instead, the household income is the primary principle to obtain a mortgage. The study recommends a balanced socio-economic development for beneficiaries of low-cost housing holistically. It means the type of land reform that regulates low-cost housing should be interlinked. It must not only provide secure tenure without sustainable employment. Secure tenure can only give partial security, whereas a combination of security of tenure and employment make it possible to alleviate poverty and promote investment. Beneficiaries will be able to apply for building plans in the event they want to extend their properties, to formally sell properties in the open market, use the property as collateral to secure a mortgage, have properties registered in their names at deeds office and be able to formalize all property related transactions.

References

Burns, T., *Land Administration Reform: Indicators of Success and Future Challenges*, 2007.

Cockburn, J. A., *Official Registration (Formalization) of Property in Peru (1996-2000),* ESF N AERUS International Workshop on Coping with Informality and Illegality in Human Settlements in Developing Cities, 2001.

Cousins, B., Cousins, T., Hornby, D., Kingwill, R., Royston, L., and Smit, W., Will Formalizing Property Rights Reduce Poverty in South Africa's 'Second Economy'?, Questioning The Mythologies of Hernando De Soto, in Programme for Land and Agrarian Studies – PLAAS, Debating Land Reform, Natural Resources and Poverty, 2005.

Denzin, N. K., and Lincoln, Y. S., *Handbook of Qualitative Research.* Sage Publications, Thousand Oaks, California, 1994.

De Soto, H., The Mystery of Capital: *Why Capitalism Triumphs in The West and Fails Everywhere Else*, Basic Books, New York, 2003.

Department of Housing, National Housing Code, 2009. Retrieved from http://www.dhs.gov.za/Content/The%20Housing%20Code%202009/index.htm in March 2019.

Domingo, P, *Property Rights and Development: Property Rights and Social, Political and Economic Empowerment*, 2013.

Fabiyi, Y. L, Land Policy for Nigeria: Issues and Perspectives, An Inaugural Lecture, Obafemi Awolowo University, Ile-Ife, June, 1990.

Ghebru, H., and Okumo, A., *Land Administration Service Delivery and Its Challenges in Nigeria,* 2016.

Görgens, T., and Denoon-Stevens, S., Towards Facilitative and Pro-poor Land-Use Management in South African Urban Areas: Learning from International Experience, *Urban Forum*, 24, 85-103, 2012.

Kagawa, A,. Policy Effects and Tenure Security Perceptions of Peruvian Urban Land Tenure Regularization Policy in the 1990s, *ESF N AERUS International Workshop on Coping with Informality and Illegality in Human Settlements in Developing Cities*, 2001.

Kasimbazi, E., *Land Tenure and Rights for Improved Land Management and Sustainable Development*, 2017.

Kitchin, F., and Ovens, W., Land Governance in South Africa: Implementing the Land Governance Assessment Framework, Urban Landmark, 2013.

Mattingly, M., *Briefing Paper Property Rights and Development: Property Rights and Urban Household Welfare*, 2013.

Ndukwe, O. U., *Case and Commentaries on Land Use Act 1978*, Calabar: University of Calabar Press, 2001. North, D.C. *Institutions, Institutional Change, and Economic Performance.* New York: Cambridge University Press, 1990.

Olujimi, J. A., and Iyanda, A., *Physical Planning Implications of Access to Residential Land and Legal Security of Tenure in Lagos Metropolis, Nigeria, FIG Working Week 2013* Environment for Sustainability Abuja, Nigeria, 6 – 10 May 2013. TS04B - Planning and Development – 6438, 2013.

Rodrik, D., *Trade Policy Reform as Institutional Reform in Handbook on Developing Countries and the Next Round of WTO Negotiations*, Bernard Hoekman, (ed.), 2000.

SPLUMA, Spatial Land Use Management Act 16 of 2013. Retrieved from https://www.customcontested.co.za/laws-and-policies-/the-spatial-planning-and-land-use-management-act-spluma/ on August 20, 2017.

The National Department of Housing, *The Housing White Paper: A New Housing Policy and Strategy for South Africa,* 1994. Retrieved from http://www.info.gov.za/whitepapers/1994/housing.htm on March 2019.

Trebilcock, M., and Veel, P. E., *Property Rights and Development: The Contingent Case for Formalisation*, 2014.

Turok, I., and McGranahan, G., Urbanization and Economic Growth: The Arguments and Evidence for Africa and Asia, *Environment and Urbanization*, 25(2), 465-482. doi:10.1177/0956247813490908, 2013.

Umhlaba Consulting Group, Housing Project Life Cycle, *Land Settlement Development Research*, 2013.

UN-Habitat, *Handbook on Best Practices Security of Tenure and Access to Land: Implementation of the Habitat Agenda*, 2003.

Zakout, W., Wehrmann, B., and Törhöne M. P., *Good Governance in Land Administration Principles and Good Practices*, 2009. Retrieved from http://www.fao.org/docrep/011/i0830e/i0830e00.ht in March 2019.

Interdependence between Structural Engineering and Construction Management
Edited by Ozevin, D., Ataei, H., Modares, M., Gurgun, A., Yazdani, S., and Singh, A.

ISBN: 978-0-9960437-6-2

BUILDING INFORMATICS NEURAL NETWORK AND REGRESSION HEURISTICS PROTOCOL FOR MAKING DECISIONS IN BUILDING CONSTRUCTION PROJECTS

LEKAN M. AMUSAN, IGNATIUS O. OMUH, and TIMOTHY O. MOSAKU

Dept of Building Technology, Covenant University, Ota, Nigeria

Building Informatics is a body of knowledge that uses the ICT computer system, digital systems, building information modeling, and state-of-the-art software in solving technical and management issues in building and construction fields. One of the modern methods used in data forecasting and modeling is Artificial Neural Networks, considering its advantage over traditional regression method. A data sample was taken of 1500 residential building projects' completion costs. Regression analysis was carried out and model validated with functionality and Jackknife re-sampling technique. 150 Questionnaires were used to capture data on factors influencing application of heuristics protocol for decisions in residential building construction projects and data samples were analyzed using severity index, ranking, and simple percentages. Analysis of data brought up some factors that influence effective application of heuristic protocol in solving decision problems in construction decision process. The linearity analysis was carried out on the model and results indicated high level of tolerance and -0.0876 lowest variation prediction quotients to 0.9878 highest variation quotients. Also, 0.069 regression model fitness coefficient (R-square) was generated with 0.9878 highest variation quotients with standard error of 0.045. The results data attests to the stability of the model generated and the model is flexible in accommodating new data and variables, thus, allows for continuous updating.

Keywords: Coefficient, Collinearity, Fitness, ICT, Updating, Expert system.

1 INTRODUCTION

The construction sector has been noted for activities that are cost intensive; since ancient times construction of building and facilities has been carried out with various cost components. Often, cost components of a building or construction work need to be packaged adequately for proper interpretation. Sometimes, systems are often used to interpret and decode cost packages in construction works in form of models. These models refer to as "building models." Skitmore (2003) and Brandon (1987) defined "building cost model" as the "symbolic representation of a building system expressing the content of the system in terms of cost influencing parameters." According to Skitmore (2003), the first model that was used in solving building related problems is called the cube method. The cube method was used about around 14 BC alongside methods which were used in solving cost related problems, such as the floor area method and the story enclosure method. Among all the model types, the story enclosure method was discovered to be more accurate in structure, form, and application (Rosen 1974). This was documented as one of the early works in the system of Hedonic equations and regression models.

1.1 Heuristic Cost Models Applications in Solving Construction Related Challenges

In the context of this study, different areas of application of cost models are presented. Their application could be summarized into two categories: the process application and application by product. There are two different schools of thought in cost modeling: product-based and process based. Skitmore (2003) defined product-based model as a system that usually models finished product, while process-based model on the other hand creates a model through synchronization of building process to synthesize a model. Similarly, there are areas of defect in the application of cost models in construction operations. For instance, Rosen (1974) opined that the area method of estimating item costs is deficient in being influenced by factors other than floor area. Also, it has overdependence on system antecedents to model real situation as contained in the Monte Carlo simulation methods. Finally, expert–based systems have been found to have strength in the areas of deficiency of regression models. It was discovered that their output has less error when compared with output of regression analysis; it generate less error between input and expected output, also, and tends to have variation error within the range of 2% to 4% while the parametric model (regression model) often has a variation error greater than 7% (Kim *et al.* 2004, Cattel, Bowen and Kaka, 2008).

2 REVIEW OF RELATED WORKS ON NON-TRADITIONAL MODELS [ANNs]

Artificial Neural Networks have been used in modeling cost variables in almost all areas of human endeavor. In this section, some of the articles reviewed cover the following areas: highway cost modeling, final construction cost modeling, risk analysis, and risk estimating among others. In Picard and De-Palma (2010), risk was identified on project sites, with the aid of Artificial Neural Networks, in the study, percentage change in the estimated cost from final cost prediction was carried out as an index of risk analysis and measurement. Also, in Kim *et al.* (2004), the estimating model in use at the early stage of construction works was examined and analyzed; between regression analysis and neural network, neural network was found in the study to have better prediction ability. Also, in Spooner and Setyawati (1974), a combination of regression and neural network models to generate a regression-based model was developed; likewise, a stock market return network model was developed by Thawornnwong and Enke (2004) using neural network. Moreover, Picard and De-Palma (2010) carried out research on a econometric model with the aim of developing hedonic price model. Regression model was used to generate hedonic regression model. The hedonic model was used in estimating demand and value of a specific good by classifying it into its constituent characteristics (Amusan 2012). Hedonic models are used in several applications such as real estate application, real estate appraisals, computation of consumer price index (CPI) and relative price index (RPI), among others. According to Amusan (2012), and Elfaki *et al.* (2014), in real estate economics, hedonic model is applicable in solving the problems of price determination and price adjudication (Amusan 2012). Hedonic models have ability to treat variables based on their separate identity or elasticity in case of a log model. To this end, the econometric model developed in this study toes the line of submissions of Picard and De-Palma (2010). Also, the hedonic related model adopted cost entropy and econometric approach to generate a model that incorporates heterogeneous variable of residential project for price and cost judgment.

3 RESEARCH METHODOLOGY

3.1 Source of Data Information

Two groups of samples were used in this study: 150 samples for qualitative aspect of data training parameters, that is, the criterion for effective prediction ability and Factors that influences the Application of Heuristic Protocol and Expert System in Construction Decision and samples for data training for Heuristic proposed model. One thousand (1000) samples were picked at random from 1500 samples of residential building projects completed within the past four (4) years at selected locations: Ogun State, Lagos State, and Federal Capital Territory (FCT) in Nigeria. These areas are regarded as economic nerve centers and regions of high construction activity. The initial and final costs of the sampled projects were extracted and adjusted with price index to 2017. The 1000 samples were adjusted with price and prevailing inflation index to be able to capture economic variable that influences building cost and was further used to train the Multi Layer Perceptron Neural network with Back Propagation system and Levenberg Marqua, this was for network configuration. From Table 1, 500 (50 percent) of the samples was used in model testing, while another set of 750 samples of the earlier adjusted/modified data was used in output prediction.

Table 1. Active cross validation performance for building projects.

Parameters	Active Cross Validation Performance	Cross Validation Performance
Mean Square Error	0.0583	0.000465
Normal Mean Square Error	0.0878	4546521.80
Regression Value ‘r’	0.9790	0.0465

The building project costs were selected for validation. The cross-validation performance of the cost samples treated was presented in Table 1 and 3. Regression value of 0.9790 was obtained while the Mean Square Error is 0.0583 and the Normal Mean Square Error is 0.0878. The error of 6% between the original values relative to predicted value is adjudged low enough for acceptance. The skewness of the data is normal and positive at 0.9790 (Afolabi *et al.* 2018).

Table 2. Regression coefficients of the developed model.

Parameters	Unstandardized Coefficients	Standard Coefficient		Collinearity Statistics
B(Constant)	Standard .Error	Beta	t	Significance
	4.1398	4.1587	9.953	.000

Notes: Dependent Variable: Neural Networks

3.2 Re-sampling

Re-sampling test was conducted on the model in order to ascertain the stability and the influence of outliers on the models' stability. The results are presented in Tables 2 and 3; two models are presented here: model of as-built sum and neural network model. Neural model has standard error of 0.197 while as-built sum model has 0.312. Generally, the two models showed stability with high level of tolerance.

Table 3. Resampling test.

Dimension	Eigen value Index	Condition (Constant)	Variance Proportion As Built Sum	Neural Network Sum	
1	3.923	1.000	.01	.00	.000
2	.088	8.759	.58	.018	.045
3	.087	16.995	.42	.030	.099

Factors that influence the application of heuristic protocol and expert system in construction decision are presented in Table 4. 150 questionnaires designed in Likert scale were administered on some respondents to harvest user perception on factors that influence the application of heuristic protocol and expert system in construction decision. "Preparation of cost budget limits for heuristic protocol investment" was ranked 1st, "inadequate knowledge about the profit on heuristic protocol application investment" was ranked 2nd, "high cost of employing ICT professionals" was ranked 3rd, and "lack of staff with appropriate skill and knowledge in ICT" was ranked 4th. Also, "cost of training professionals in cost expert application like neural network" was ranked 5th, while "inadequate education in ICT in educational institutions" and "lack of commitment by firms management towards ICT application in solving basic construction problems" were ranked 6th and 7th respectively, among others (Afolabi *et al.* 2018).

Table 4. Data on factors influencing application of heuristic protocol and expert system in construction decision.

Determinants/ Parameters	R.A.I	Rank
Budget limits for heuristic protocol investment	0.836	1st
Inadequate knowledge about the profit on heuristic protocol application investment	0.788	2nd
High cost of employing ICT professionals	0.776	3rd
Lack of staff with appropriate skill and knowledge in ICT	0.764	4th
Cost of training professionals in cost expert application like neural network	0.756	5th
Inadequate education in ICT in educational institutions	0.756	6th
Lack of commitment by firms management towards ICT application in solving basic construction problems	0.748	7th

4 CONCLUSIONS

After configuration of a neural network system that could be used for processing of residential building cost, the cost data was used to train neural network system and generate a suitable algorithm as shown in Figure 1. The neural output generated represents a predicted cost range for the office projects with regards to prevailing economic situation like inflation and building price index. This was factored into the as-built cost of the project and predicted upward for the period of six months. Thus, the specified range of prediction expressed for the model in the context of this study is six months, subject to constant economic variables. However, if economic variables change before the six-month prediction window period, the cost should be adjusted with the current economic variables. Cross validation analysis indicates -0.07403 lowest variation prediction quotients to 0.66639 highest variation quotients. Also, the regression coefficient (R-square) value for determining the model fitness is 0.035 with standard error of 0.397. Model fitness and standard error are often used as index to measure the nature of a regression model.

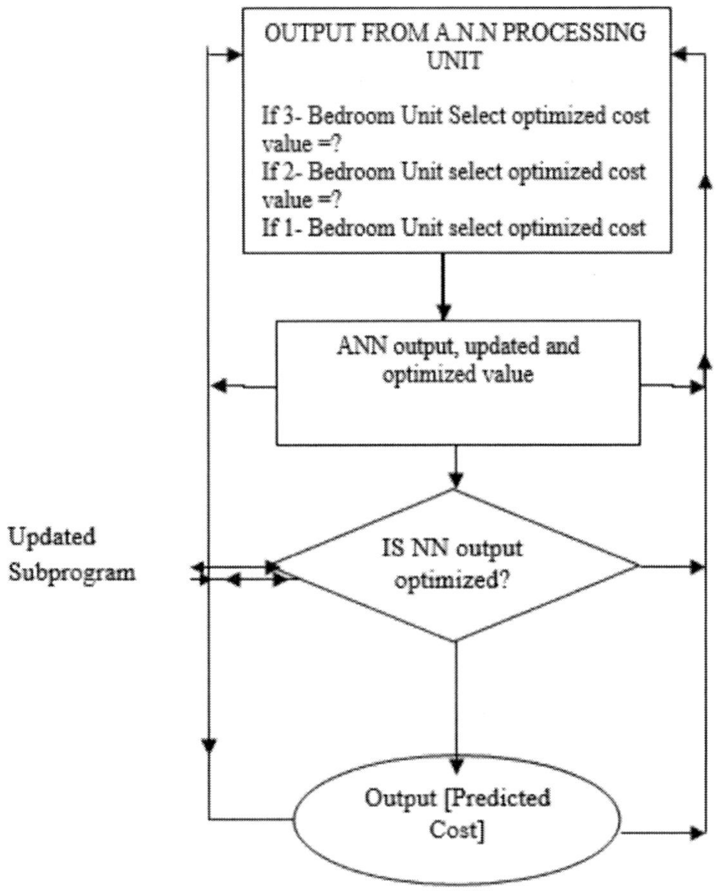

Figure 1. Algorithm for cost processing.

References

Afolabi, A., Fagbenle, O., Ojelabi, R., Amusan L., Tunji-Olayeni, P., and Daniyan, V., Development of a Web-Based Human Resource Sourcing System for Labour Only Contracts, in *Construction Research Congress 2018*, Wang, C., Harper, C., Lee, Y., Harris, R., and Berryman, C. (eds.), ASCE Press, 736-746, 2018.

Amusan, L. M., *Neural Network–Based Cost Predictive Model for Building Works*, PhD Thesis Covenant University, Department of Building Technology, Covenant University, Nigeria, 215-220, 2012.

Brandon, A., *Cost Modeling for Estimating,* The Royal Institution of Chartered Surveyors, London, 1987.

Cattel, W. D., Bowen, P. A., and Kaka, A. P., A Simplified Unbalanced Bidding Model, *Construction Management and Economics*, 26(10-12), 1291-1302, 2008.

Elfaki , A. O., Alatawi, S., and Abushandi, E. Using Intelligent Techniques in Constuction Project Cost Estimation: 10-Year Survey. (S. Mdanat, Ed.) *Advances in Civil Engineering, 2014*. Retrieved from https://www.hindawi.com/journals/ace/2014/107926/ on December 2, 2014.

Kim, G.-H., An, S.-H., and Kang, K.-I., Comparison of Construction Cost Estimating Models Based on Regression Analysis, Neural Networks, and Case-Based Reasoning, *Building and Environment*, ScienceDirect, 39(10), 1235-1242, 2004.

Picard, A., and De Palma, A., Neuromodex-Neural Network System for Modular Construction Decision Making, J*ournal of Computing in Civil Engineering*, Proc. Paper No. 5708, 8(2), 221-233, 2010.

Rosen, H. S., Hedonic Prices and Implicit markets: Product Differentiation in Price Competition, *Journal of Political Economy*, 82, 1974.

Spooner, J, Setyawati A, Probilistics Estimating. Journal of the Construction Division. 100(1).65-77, 1974.

Skitmore, R. M, and Ng, S. T., Forecast Models for Actual Construction Time and Cost, *Building and Environment,* 38(8), 1075, 2003.

Thawornwong, S.,wang and Enke, D., Forecasting Stock Returns with Artificial Neural Networks, *Intelligent Engineering Systems through Artificial Neural Networks,* 12, 695-705, 2004.

Interdependence between Structural Engineering and Construction Management
Edited by Ozevin, D., Ataei, H., Modares, M., Gurgun, A., Yazdani, S., and Singh, A.

ISBN: 978-0-9960437-6-2

IMPROVING COMPETITIVENESS IN INTERNATIONAL CONSTRUCTION

SUAT GUNHAN

Dept of Construction Science, The University of Texas at San Antonio, San Antonio, USA

Historically, U.S. contractors have a successful track record in international markets. They built that track record by bringing advanced technologies and applying efficient project management methodologies to projects. According to ENR data from the early 1980s to early 2008, U.S. contractors led the international construction arena. However, starting from 2009, Chinese construction firms lead the international construction with increasing revenue volume. The study analyzes the U.S. construction firms' competitiveness in international markets between 1980 and 2017. The results of the analysis indicate that the U.S. contractors need to be competitive in creative financing. Creating joint ventures from companies that are active in Middle East, and Asia can improve the U.S. activities in those markets. To improve capabilities in delivering full-cycle services, U.S. construction companies can acquire facilities management companies. Involvement in emerging delivery methods and contracts such as build-operate-transfer, design-build-finance-maintain-operate, and public-private-partnerships can improve the competitiveness of U.S. contractors in international markets.

Keywords: Build-operate-transfer, Public-private-partnership, Mergers and acquisitions, Design-build-finance-operate-maintain, Chinese contractors, U.S. contractors.

1 INTRODUCTION

The competitiveness of construction companies in international markets is affected by many factors. Historically, U.S. contractors have had a successful track record in international markets. They built that track record by bringing advanced technologies and applying efficient project management methodologies to projects (Strassman and Wells 1988). However, since 2009, Chinese contractors lead the international construction arena in terms of revenue volume. This paper analyzes the Engineering News Record (ENR) data on international construction over the period of 1982-2017 to explore the factors that affected the competitiveness of U.S. international contractors. The study has identified the strengths and weaknesses that affected the performance of U.S. contractors in international markets.

According to Mahalingam *et al.* (2005), partnerships, contracting practices, knowledge of regional standards, and knowledge of the existing market conditions affect the performance of international contractors. The global monetary policies affect the perception of owners. Starting from the early 2000s, the market trends indicate that construction entities are expected to include financing in their bids. Recently, contractors are expected to assume responsibilities for the full cycle services to become competitive. This requires contractors to provide design, procure, build and manage facilities (Han *et al.* 2010). The paper provides an insight into the U.S. contractors' journey in international markets and explores how U.S. contractors should position themselves to be surrounded by the new market dynamics. The findings of the review provide strategic options to restore U.S. leadership in the international construction arena.

The existing literature in international construction related to positioning of U.S. construction companies mainly concentrates on the analysis of periods where U.S. contractors had superior competitive advantages. There is a gap in literature analyzing the periods where U.S. contractors are experiencing difficulties in maintaining the leadership status. The findings of this study are expected to identify major factors affecting the competitiveness of U.S. international contractors in the last two decades.

2 THE REVIEW OF THE FACTORS AFFECTING COMPETITIVENESS

The technological skills and project management expertise of U.S. contractors made them the most competitive international contractors between 1980 and 1993. The ENR Magazine publishes the Top International 225/250 contractors annually. The annual list is highly recognized by international construction researchers. The findings of Lu (2014) indicate that the ENR data can be used with a high degree of confidence. The ENR ranked the international contractors based on their new contract amounts in international markets until 1993. The average percentage of new international contract amounts won by U.S. contractors is approximately 35% (annually) of all international contract amounts from 1982 to 1993. This is a successful period for U.S. contractors as they led the international construction arena with substantially high percentage new contract ratios. Despite the fact that U.S. contractors had disadvantages on the tax issues and export financing, their technological advantage such as expertise on petro-chemical projects and project management provided major advantages to secure projects in oil and gas producing nations. The only decline of the market share by U.S. contractors took place from 1986 to 1989, and this can be attributed to the decline in oil prices. Despite the decline, U.S. contractors continued to have the largest share (in terms of revenue volume) in international construction markets during that time frame (1986-1989). The market share of U.S. contractors declined after 2003; since 2009, Chinese contractors lead the international construction arena (Zilke and Taylor 2014). The data starting from 1994 provides rankings of international construction revenues annually (Figure 1).

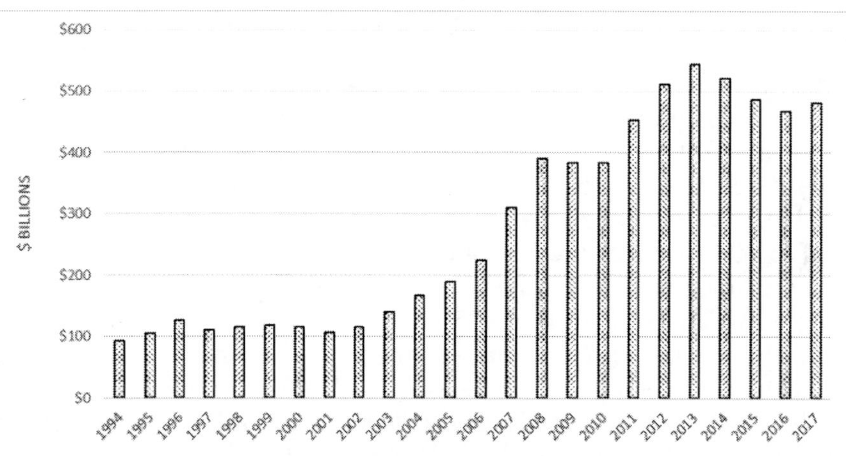

Figure 1. The total international construction revenues between 1994 and 2017.

The Japanese contractors had the largest revenues from 1994 to 1996 followed by U.S. contractors. However, U.S. contractors took the lead back in 1997 and continued to have the major share until 2008. The peak years with top percentage ratios were 1998 and 1999. The U.S. leadership continued until 2008 with a decreasing percentage ratio (Figure 2). The weakening dominance of U.S. contractors in international construction, starting from 1999, can be attributed to the trend that the contractors in developing countries started to deliver technology driven projects. During the early 2000s, more and more Chinese companies became competitive in international construction. According to Reina and Tulacz (2002), cultural issues also play a role in international construction; eastern culture can be more appealing to developing countries. The declining trend is not aligned with the growing international revenues within the same time period (Figure 1). Despite the decline, U.S. contractors were still the leading contractors until 2008 in international markets. At the same time period, the domestic construction markets in the U.S. were very active; this could be one of the reasons that contractors concentrated in domestic projects (Reina *et al.* 2000).

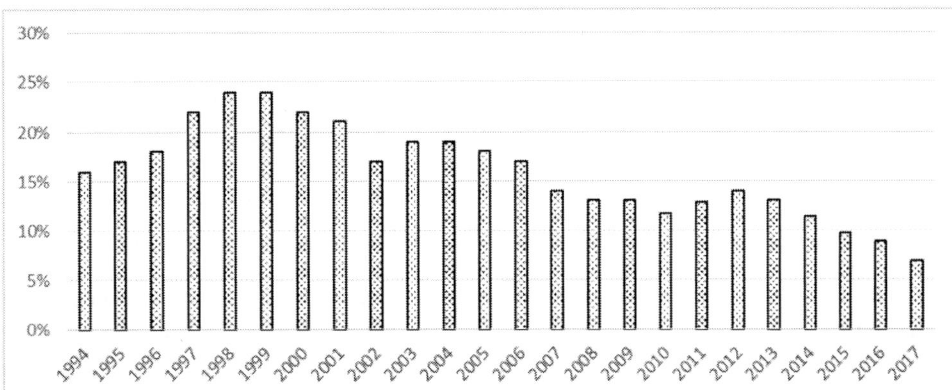

Figure 2. The percentage ratio of the revenues by U.S. construction companies to the total revenues in international markets between 1994 and 2016.

Even though the leadership of U.S. contractors continued, some factors did not favor U.S. contractors, such as relatively less support from U.S. government compared to the other nations' governments. More and more sophisticated firms from developing countries joined international construction. The advancements in communication technologies started to make an impact on the pace and effectiveness of projects in the beginning of this period (Reina and Tulacz, 1995). Being able to bring financing also became more critical in this period and the build-operate-transfer (BOT) method was highly effective in getting new contracts. Mergers and acquisitions (M&A) also became an important trend, especially as a method of entry into established markets such as the U.S. (Reina *et al.* 2000). BOT became a very natural process for some European contractors, and it became a competitive advantage against U.S. contractors. In addition to BOT and M&A, facilities management (FM) became a trend used by some large international contractors. FM, especially in the airport segment, became an alternative venue stream for European contractors (Reina and Tulacz 2001, 2002). In 2001, Spain's *Dragados Obras y Proyectos* SA reported that 40-45% of their international work account was BOT. French Bouygues also reported that in addition to BOT, design-build (DB) constituted large portion of their international work (Reina and Tulacz 2001). Political factors are always critical (Ashley and Bonner 1987). Some western companies (excluding the U.S.) had some advantages in Northern Africa and in some parts of the Middle East (Reina *et al.* 2000).

Fluctuating currency values can make an impact on competitiveness of international contractors (Gunhan and Arditi 2005). Japanese and European contractors increased their revenues because of the increasing value of yen and euro starting from early 2000s through late 2000s (Reina and Tulacz 2003). It is common to have companies from newly industrialized countries (NIC) that built power plant projects in developed countries with BOT capabilities. For example, the large Turkish firm GAMA built a power plant in Ireland with BOT method. The same company built a power plant in Jordan by joint venturing with Black and Veatch, an American firm (Reina and Tulacz 2004). Besides its technical expertise, BOT provided a competitive advantage to get a project in Europe, and the American firm benefited from the joint venture (JV) since the Turkish firm has cultural advantages that affect the project performance while operating in the Middle East. Chinese firms continued to increase their share not only by becoming competitive in technology intensive projects but also, they have continued to use their competitive labor rate labor advantages especially in Africa. Companies from NICs such as China, Korea, and Turkey, which were historically competitive in labor-intensive projects, became competitive in technology intensive projects. They developed expertise by establishing JVs with companies from the developed world. They brought cultural advantages and price competition with lower bids (Reina and Tulacz 2005). Another competitive advantage of Indian and Chinese contractors is that they developed a major technical expertise in their domestic markets, and they used it in international projects. They were transformed from labor exporters to engineer-procure-construct (EPC) and BOT exporters (Reina and Tulacz 2007). The market growth was significant, especially with petroleum production facilities and power plants, major infrastructure projects, and signature buildings in 2000s. Starting from 2009, Chinese companies had the largest share (excluding the years of 2012 and 2013 due to Spanish companies' leadership) until 2017. In this period, government supported contractors' competitive advantage continued. Chinese companies expanded their operations in African market, and they had great success with Chinese financial aid package (Reina and Tulacz 2015, 2017). They continue to increase their share in the Middle East. Similar to Chinese contractors, Korean contractors have increased their market share in the Middle East's oil- and gas-plant market. Projects in Central and South Africa are also emerging due to developments and natural resources (Reina and Tulacz 2010). According to some construction executives, government-supported methods may create political and financial unfairness during the bidding process. Starting from the mid-2010s, technology is not a pure competitive advantage anymore. According to Reina and Tulacz (2015), clients are in demand for PPP and BOTs. Risk transfer is the major trend by using emerging delivery methods such as Design-Build (DB), Design-Build-Finance (DBF), and Design-Build-Finance-Operate-Maintain (DBFOM). Companies are in an era where they need to consider revenue streams by considering the overall life cycle either with delivery methods (BOTs and PPP, DB, offering full life-cycle services) or FM. Public-private-partnership (PPP) is seen as a long-term revenue generating stream in a competitive construction industry (Reina and Tulacz 2011).

Companies invest through acquisitions to become more competitive and create new niche segments (Reina and Tulacz 2014). In order to minimize risks through diversification hence improve competitive strengths, firms acquire technically capable firms (Reina and Tulacz 2017). Cultural similarities clearly make a difference. One of the success factors to working in Commonwealth of Independent States and Russia is the ability to bring technical expertise but use local resources at the same time; understand how to operate, and find manpower (Reina and Tulacz 2014, 2015). The Middle East's oil and gas construction market was crowded in 2010s with too many contractors. Many contractors took contracts with unreasonable prices. Companies with robust health and safety programs and quality standards often cannot afford to bid against these new competitors (Reina and Tulacz 2015). This can be one of the major reasons why U.S. contractors have a declining ratio in the Middle East. Low cost mentality may cause an increase in the number of failed contracts in the near future.

3 CONCLUSIONS AND RECOMMENDATIONS

The study analyzed the factors affecting U.S. companies' competitiveness in international construction. U.S. companies have always been very competitive in international construction. U.S. leadership continued until 2008 with a declining ratio, and starting from 2009, mainly Chinese construction firms lead the international construction with increasing percent ratios to the total international construction volume. U.S. contractors dominated international construction in 1980 – 2000 period with their technical skills in petro-chemical, petroleum processing, and power plant projects accompanied with their project management expertise despite the high salaries of their employees. Contractors from developing nations were also involved in international construction at the same time period but they were building labor-intensive, low-technology projects. Starting from the mid-90s though, companies from developing nations started to build technically complex projects as well. Despite the continuing leadership of U.S. companies, their percent ratio to the overall international revenues continued to decline until 2008. Starting from the mid-90s, the companies from developing nations such as China, Korea and Turkey were able to build technically complex projects while still being competitive with labor costs. Owners in the international markets became more demanding on the financing side. Companies, which are backed by their governments that subsidize respective companies with export credits obtained competitive advantage against U.S. contractors. They developed expertise in BOT, PPP projects as well as DBFOM. During the period from 2009 to date, Chinese contractors continue to lead international construction markets. Government support especially for Chinese contractors continues. The major markets such as the Middle East, Asia, and Africa are with full of contractors from China, Korea, Turkey, and recently India (in the Middle East market). The major competitive advantage of those companies is that they all have eastern culture, which becomes a major competitive advantage to win contracts in developing countries. One reason, that the Middle East, Asia, and Africa became tough markets for U.S. contractors is that it is hard to compete with robust health and safety standards. Owners continue to be demanding, and companies that can offer full life-cycle services obtain competitive advantage. In order to stay competitive, most non- U.S. companies were active in acquisitions.

U.S. contractors need financial backing from their government to be competitive in an arena where creative financing is crucial for success. Creating joint ventures with companies that are active in Middle East, and Asia can improve the U.S. involvement in those markets. FM is a niche market for U.S. contractors, but more U.S. construction companies can acquire FM companies to improve their competitiveness in international construction. Involvement in emerging delivery methods and contracts such as BOT, DBFMO, and PPPs is crucial to succeed in international markets.

For future studies, the rise of construction firms from the NICs in international construction should be analyzed. There are many studies on Chinese contractors analyzing their competitiveness, however the companies from the NICs other than China such as India, South Korea, and Turkey should also be analyzed to explore their strengths contributing to their competitiveness in the last two decades. The future research should also focus on the additional success strategies other than the recommended ones for U.S. construction firms in this paper. The future research should contribute to the exploration of unique and innovative ways of improving U.S. competitiveness in international construction.

References

Ashley, D. B., and Bonner, J. J., Political Risks in International Construction, *Journal of Construction Engineering and Management*, 113(3), 447-467, 1987.

Gunhan, S. and Arditi, D., Factors Affecting International Construction, *Journal of Construction Engineering and Management*, 131(3), 273-282, 2005.

Han, S. H., Kim, D. Y., Hyoun, S. J., and Choi, S., Strategies for Contractors to Sustain Growth in the Global Construction Market, *Habitat International*, 34(1), 1-10, 2010.

Lu, W., Reliability of Engineering News-Record International Construction Data, *Construction Management and Economics*, 32(10), 968-982, 2014.

Mahalingam, A., Levitt, R. E., and Scott, W. R., Cultural Clashes in International Infrastructure Development Projects: Which Cultures Matter?, in *Project Procurement for Infrastructure Construction*, International Council for Building, Rotterdam, Netherlands, 254-257, 2005.

Reina P., and Tulacz, G. J., The Top 225 International Contractors, *ENR*, 235(9), 36, 1995.

Reina P., and Tulacz, G. J., The Top 225 International Contractors, *ENR*, 247(8), 66, 2001.

Reina P., and Tulacz, G. J., The Top 225 International Contractors, *ENR*, 249(9), 26, 2002.

Reina P., and Tulacz, G. J., The Top 225 International Contractors, *ENR*, 251(8), 28, 2003.

Reina P., and Tulacz, G. J., The Top 225 International Contractors, *ENR*, 253(8), 34, 2004.

Reina P., and Tulacz, G. J., The Top 225 International Contractors, *ENR*, 255(8), 40, 2005.

Reina P., and Tulacz, G. J., The Top 225 International Contractors, *ENR*, 259(7), 30, 2007.

Reina P., and Tulacz, G. J., The Top 225 International Contractors, *ENR*, 265(6), 44, 2010.

Reina P., and Tulacz, G. J., The Top 225 International Contractors, *ENR*, 267(6), 45, 2011.

Reina P., and Tulacz, G. J., The Top 250 International Contractors, *ENR*, 273(5), 1, 2014.

Reina P., and Tulacz, G. J., The Top 250 International Contractors, *ENR*, 275(5), 33, 2015.

Reina P., and Tulacz, G. J., The Top 250 International Contractors, *ENR*, 279(6), 33, 2017.

Reina P., Tulacz, G. J., and Rosenbaum, D., B., The Top 225 International Contractors, *ENR*, 245(7), 60, 2000.

Strassman, P., and Wells, J., *The Global Construction Industry*, Croom Helm, London, UK, 1998.

Zilke, J. P., and Taylor, J. E., Shifting Sands and Shifting Grounds: Analysis and Implications of Shifting Dynamics in the Global Construction Industry, *Journal of Management in Engineering*, 31(5), 04014076-1 - 04014076-7, 2014.

Interdependence between Structural Engineering and Construction Management
Edited by Ozevin, D., Ataei, H., Modares, M., Gurgun, A., Yazdani, S., and Singh, A.

ISBN: 978-0-9960437-6-2

THE INFLUENCE OF BAMBOO REINFORCEMENT AND ADDITION OF FIBERS ON HOLLOW BEAMS

FAISAL ANANDA, HENDRA SAPUTRA, and ARMADA SUKRI

Dept of Civil Engineering, Politeknik Negeri Bengkalis, Riau, Indonesia

The development of the artistic value of a building with reinforced concrete material continues to grow, one of which burrows the pipeline so as unexposed so that the condition of a building looks beautiful. This condition makes a weakening of the strength of concrete. As for the weakness of the concrete caused by a small crack due to deflection or the other anticipated by the addition of fiber one of the rebar tie wire (bendrat). Based on this condition, the beam testing was done with size by 150mm x 200mm x 1200mm with using one-point loading. Concrete mix design using SNI 03-2834-2000. The load obtained without fibers using the steel reinforcement by 12.9% greater than the use of bamboo reinforcement. This is because the tensile strength of bamboo is smaller than the tensile strength of iron. Likewise, on the beam that added fiber as much as by 2% on the bamboo reinforcement, there is a difference of loading is smaller than the steel reinforcement by 25%. The interesting result is found in the 0.5 inch diameter beam which has an increase in load with the initial crack conditions close to the same as the beam without a hole.

Keywords: Initial crack, Steel fiber, Flexural strength.

1 INTRODUCTION

Concrete is one of the main materials of infrastructure development. Concrete is formed by hardening of cement mixture, water, fine aggregate, coarse aggregate (crushed or gravel), air, and sometimes other additives (Nawy 2010). The concrete has a compressive strength value comparatively higher than its tensile strength. A 28-day strength concrete strength ranges from approximately 10-65 MPa. The value of tensile strength of concrete is only about 9% to 15% of its compressive strength (Mulyono 2005). The function of reinforcement in the concrete is to withstand the tensile forces carrying the loads acting on the concrete. Over time, the use of reinforcement continues to be developed to find a replacement, considering this material is a material that cannot be renewed by nature. One of the developments was to replace using natural materials such as bamboo. Bamboo is one of the oldest building materials used by mankind. The bamboo culm, or stem, has been made into an extended diversity of products ranging from domestic household products to industrial applications. In Asia, bamboo is quite common for bridges, scaffolding and housing, but it is usually a temporary exterior structural material (Latief *et al.* 1990). In Asia, bamboo is quite used for bridges scaffolding and housing, but it is usually temporary exterior structural material. In many overly populated regions of the tropics, certain bamboo supplies are the one suitable material, i.e., sufficiently cheap and plentiful. In response to global warming issues and sustainable society, manufacturing using natural material has become actively in developing countries. Bamboo is low cost, fast growing and broad distribution of growth is expected to contribute significantly in earthquake-resistant construction and seismic retrofit technology. On the other hand, plants and fibers are annually reproducible clean resources. Bamboo is a unique group of gigantic grasses the culm of which originates in underground rhizomes. It grows naturally in many parts around the world country but some species are artificially planted. Bamboo forests are found across tropic and sub-tropic zones between latitudes of about 40°south, i.e., areas with mean annual temperatures of from 20°C to 30°C. Compared with other fibers such as steel fiber and carbon fiber, bamboo fiber has significant advantages: high tensile strength (its tensile strength could reach 180Mpa (Huang *et al*). Concrete weak to tensile forces makes the initial crack of concrete in the tensile area. In addition to reinforcement, we can also add fiber. Whether in the form of stiff fiber (steel fibre) or not stiff. The addition of fiber can increase tensile strength (Ananda 2016a, Ananda 2016b, Ananda *et al.* 2017b), and minimize crack width (Ananda *et al.* 2017a).

2 MATERIALS AND METHODS

The study was conducted at the Laboratory of Building Materials and Concrete, Civil Engineering Department, State Polytechnic of Bengkalis. The test object was a beam with a dimension of 150mm x 200mm x 12mm with one-point loading (Figure 1). According to ACI Committee 224(2001), maximum crack width allowable for the protective condition is 0.41 mm (Table 1). This experiment was in full-scale test specimen, 1:1. It tested until it reached the first crack condition and used the simple support on both sides.

Table 1. Maximum crack width allowable.

Exposure Condition	Width Crack	
	mm	in
Dry air or protective membrane	0,016	0,41
Humidity, Moist air, Soil	0,012	0,30
Deicing Chemicals	0,007	0,18
Seawater and seawater spray, wetting and drying	0,006	0,15
Water-retaining structures	0,004	0,10

ACI Committee 224

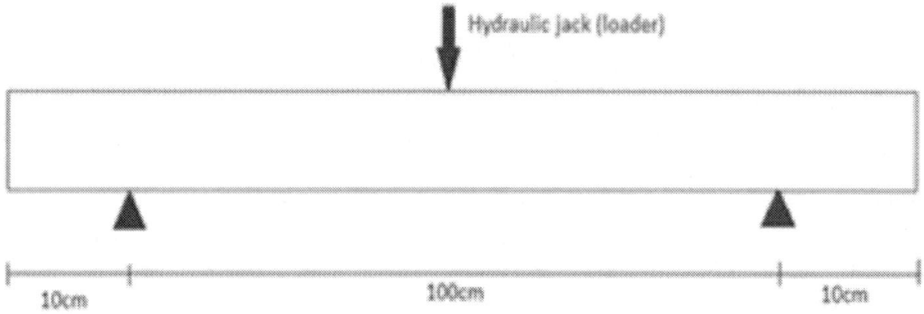

Figure 1. Loading scheme and testing.

The concrete mix design based on SNI 03-2834-2000, fc' = 15 MPa (T03-2834-2000, 2000). Steel fiber used in this research is the conventional steel fiber (bendrat) with length of ± 6cm.

Deflection testing and crack width use the fiber mixed with concrete by the initial slump 13 cm ± 2 cm. The experiment samples of compressive strength and tensile created using a cylinder.

After the specimen reaches the age of 28 days, the test was carried out to examine the static deflection, width crack and strain that occurs. Reinforcement uses fy = 240 MPa and bamboo reinforcement as a comparison. Specimens were subjected to a load line, started from zero to the first crack. Load (P) can be read on proving ring at intervals of 50 kg per one strip reading.

3 RESULT AND DISCUSSION

From the results of laboratory tests, it was found that the influence of the main reinforcement using bamboo reinforcement has a larger crack width size exceeding the maximum limit of crack allowed (Tables 2-5).

The interesting thing is found in this research is the increase of load on the 0.5-hole beam to the beam is not hollow with the initial crack conditions are more or less the same. The increase occurred on the beam of steel and bamboo reinforcement ± 3%, while on the beam of steel and bamboo reinforcement with additional 2% fiber by 25% - 35%.

In addition to the hollow block beam fibers to the hollow beam, there is no effect of fiber, where the initial crack that occurs close to the same magnitude despite the difference in the bending load that occurs on the beam (see Figures 2-5).

Table 2. Bending and initial first cracking loads of steel reinforcement beam (Tbi) without conventional fibers (bendrat).

Fiber (%)	Beam Holes (inch)	Initial First Cracks (mm)	Bending Loads (kg)
0	0	0,2	1550
	0,5	0,2	1600
	0,75	0,3	1300
	1	0,4	1250

Table 3. Bending and initial first cracking loads of steel reinforcement beam with additional conventional fibers (bendrat) 2%.

Fiber (%)	Beam Holes (inch)	Initial First Cracks (mm)	Bending Loads (kg)
2	0	0,1	1000
	0,5	0,2	1550
	0,75	0,2	1200
	1	0,1	1250

Table 4. Bending and initial first cracking loads of bamboo reinforcement beam (Tba) without conventional fibers (bendrat).

Fiber (%)	Beam Holes (inch)	Initial First Cracks (mm)	Bending Loads (kg)
0	0	1	1350
	0,5	0,8	1400
	0,75	1	1400
	1	0,6	1250

Table 5. Bending and initial first cracking loads of bamboo reinforcement beam with additional conventional fibers 2%.

Fiber (%)	Beam Holes (inch)	Initial First Cracks (mm)	Bending Loads (kg)
2	0	1	750
	0,5	0,9	1000
	0,75	1	800
	1	0,9	950

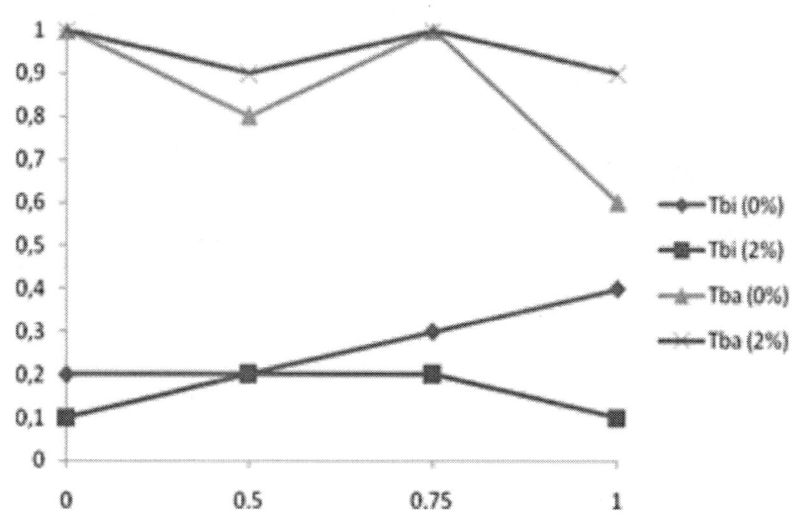

Figure 2. The width of cracking beam without holes and hollowed using bamboo and reinforcement.

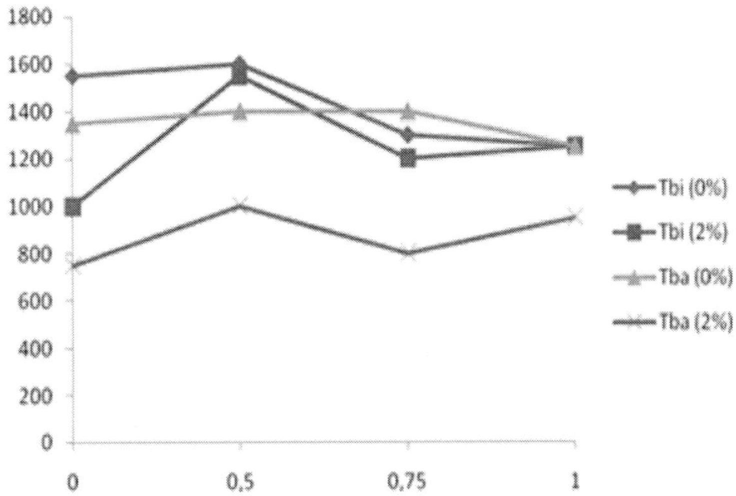

Figure 3. Bending loads of beam without holes and hollowed using bamboo and reinforcement.

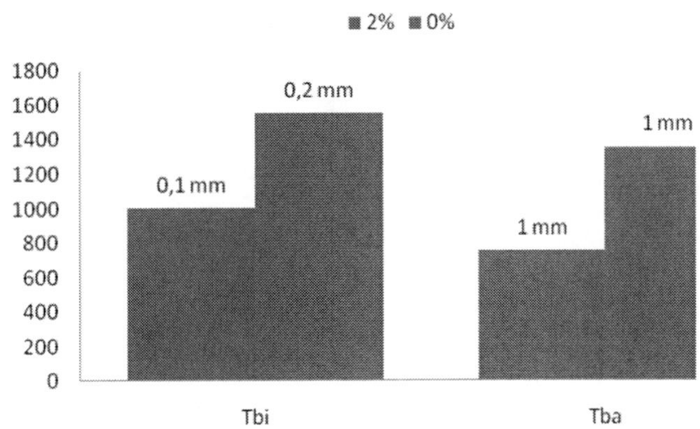

Figure 4. Comparison of load and initial cracks width in concrete reinforced beams of steel and Bamboo.

4 CONCLUSIONS

From the results of analysis and discussion, it can be concluded that the use of bendrat fibers on the bending test beam can affect the initial crack value although the bending load does not continue to increase with the beam enlargement. The 0.5-hole beam has better flexural strength than the 0.75 inches and 1-inch holes since the resilient load is greater than that of a hollow block, both of which are 2% or without fibers.

References

ACI Commitee 224,. *Control of Cracking in Concrete Structures (224R01)*, Farmington Hills, Mich: American Concrete Institute, 2001.

Ananda, F., *Pengaruh Metode Perawatan Dengan Penambahan Serat Baja Terhadap Kuat Tekan dan Tarik Beton*, Seminar Nasional Ketekniksipilan Bidang Vokasional IV, Bali, 155-158, 2016a.

Ananda, F., *Perbandingan Nilai Kuat Tekan Terhadap Proses Perawatan (curing) Dengan Penggunaan Serat Konvensional (bendrat) dan Serat Dramix*, Seminar Nasional Industri dan Teknologi, Bengkalis-Riau, 363-367, 2016b.

Ananda, F., Soehardjono, A., Zacoeb, A., and Gunawan, S., *The Reinforced Concrete Beam Deflection and Cracking Behavior with Additional Fiber Steel*, Resilient Structures, and Sustainable Construction, ISEC 9, Valencia, 1-5, 2017a.

Ananda, F., Gunawan, S., and Saputra, *H., Pengaruh Penambahan Serat Dramix dan Perawatan Terhadap Kuat Tekan, Kuat Tarik dan Biaya Beton*, Seminar Nasional Teknik Sipil 2, Bali, 49-54, 2017b.

Latif, Tarmeze and. Fauzidah, *Anatomical features and mechanical properties of three Malaysian bamboos*. Journal Tropical Forest Science 2(3): 227-234, 1990.

Mulyono, T. *Teknologi Beton.* Yogyakarta: C.V ANDI OFFSET, 2005.

Nawy, E. G., *Beton Bertulang Suatu Pendekatan Dasar*. Terjemahan Ir. Bambang Suryoatmono., M.Sc. Bandung: PT Refika Aditama, 2010.

T03-2834-2000, *Tata Cara Pembuatan Rencana Beton Normal*, BSN, Puslitbang Teknologi Pemukiman, Balitbang Kimpraswil Departemen Kimpraswil, Bandung, 2000.

Z. Huang, Y.J. Wu and C. Zhang: *Bamboo Strengthened with Carbon Fiber Reinforced Polymer for the Substitutes of Steel Substructures*, Key Engineering Materials, Vol. 517. p.233, 2012

Interdependence between Structural Engineering and Construction Management
Edited by Ozevin, D., Ataei, H., Modares, M., Gurgun, A., Yazdani, S., and Singh, A.

ISBN: 978-0-9960437-6-2

SCRAP TIRE RUBBER-BASED AGGREGATE IN LIGHTWEIGHT CONCRETE

MARTINA ZÁLESKÁ[1], DAVID ČÍTEK[2], MILENA PAVLÍKOVÁ[1], VOJTĚCH BAZGIER[1], and ZBYŠEK PAVLÍK[1]

[1]*Dept of Materials Engineering and Chemistry, Faculty of Civil Engineering, Czech Technical University in Prague, Prague, Czech Republic*

[2]*Dept of Experimental Methods, Klokner Institute, Czech Technical University in Prague, Prague, Czech Republic*

Reusing rubber particles from used tires is good for the environment and, moreover, the world's population is becoming progressively conscious that a very high demand for natural resources is unsustainable. As the concrete industry consumes high amounts of natural resources, both for clinkering and gravel aggregate production, one must focus on its sustainability, considering environmental and economic issues. Therefore, reuse of waste tire rubber in concrete or in other composite materials is a logical solution for sustainable production of construction materials. Hence, the paper is aimed at the development and testing of lightweight concrete composed of a high volume of crushed waste tire rubber used as partial replacement of natural silica aggregate. In order to access the effect of incorporation waste tire rubber-based aggregate in concrete composition, reference concrete mix based on silica aggregate only is studied as well. The crushed waste tire rubber is characterized by its powder density, specific density, and particle size distribution. Specific attention is paid to thermal transport and storage properties of waste rubber that are examined in dependence on compaction time. For the developed lightweight concrete, mechanical, hygric, and thermal properties are tested. The tested lightweight concrete is found to be alternative and environmentally friendly construction material possessing improved thermal insulation function, interesting hygric parameters and sufficient mechanical resistance.

Keywords: Waste rubber, Mechanical resistance, Thermal insulation, Hygric properties, Construction material, Sustainability.

1 INTRODUCTION

Construction materials in which concrete is modified by the replacement or addition of different constituents have been constantly sought due to both performance and environmental considerations (Bompa *et al.* 2017). According to the U.S. Tire Manufacturers Association (2017), 246.4 million of scrap tires were generated in the United States in 2015, which is equivalent to circa 4.0 million tons. Accordingly, In the UK, it is estimated that 37 million car and truck tires are being discarded annually and this number is set to increase with the growth in road traffic by 63% by 2021, as reported by Cairns *et al.* (2004). The scrap tires are disposed of using various methods like landfilling, to produce carbon black, use as fuel and for pyrolysis. However, the easiest and cheapest method of disposal discarded tires is their burning, which causes a serious fire hazard and moreover emits dense black smoke with toxic gases. One of the most environmentally friendly solutions for the treatment of scrap tires is their use in cement concrete to replace the natural aggregate (Thomas and Gupta 2016). Most of the previous studies investigated the utilization of waste rubber in concrete as aggregate with the results indicated a significant reduction in all mechanical parameters, but the improvement of the thermal and sound insulation properties and reduction of the unit weight (Aliabdo *et al.* 2015, Iqbal 2017, Bompa *et al.* 2017).

The present study aims to evaluate the mechanical, thermal and hygric properties of lightweight concrete containing a different proportion of rubber aggregate as partial replacement of coarse natural aggregate. Developed lightweight concrete with rubber can find utilization in non-structural applications with thermal insulation function.

2 EXPERIMENTAL

Rubber particles of fraction 4/8 mm (Figure 1) were obtained from the crushing of waste tires. First, assessment of basic physical and thermal properties of both natural and tire rubber-based aggregate were done. Then the lightweight concrete with different amount of rubber aggregate was prepared. Portland slag cement CEM II/A-S 52.5 N (Spenner Zementwerk Berlin GmbH & Co. KG, Germany) was used as a binder and silica sand of particle size fraction 0/4 mm and 4/8 mm from gravel-pit Dobříň, Czech Republic, was used as natural aggregate. Silica sand of fraction 4/8 mm has been substituted by scrap tire rubber-based aggregate of fraction 4/8 mm in a proportion of 10, 20 and 30 mass %. Reference concrete samples without rubber aggregate were prepared as well. The concretes composition is shown in Table 1. The water/cement ratio was kept 0.5 for all prepared composites. The casted prismatic samples 40 × 40 × 160 mm and cubic samples with a side dimension of 100 mm were left for 1 day at laboratory conditions and then they were unmolded and cured for 28 days in water.

Table 1. Mix composition of lightweight concrete with waste rubber aggregate.

Mixture	Mass (g)					w/c
	Cement	Natural aggregate 0/4	Natural aggregate 4/8	Rubber aggregate 4/8	Water	
Ref	450	675	675	-	225	0.5
10%	450	675	540	135	225	0.5
20%	450	675	405	270	225	0.5
30%	450	675	270	405	225	0.5

Figure 1. Rubber aggregate of fraction 4/8 mm.

2.1 Characterization of Used Natural and Rubber-Based Aggregate

In order to assess used aggregate in terms of physical properties, its specific density, powder density, and particle size distribution were determined. Automatic helium pycnometer Pycnomatic ATC (Thermo-Scientific) was used for specific density measurement. The grain-size curve was obtained using the standard sieve method with sieves of mesh dimensions 0.063; 0.125; 0.25; 0.5; 1.0; 2.0; 4.0; 8.0; 16.0; 31.5 and 63.0 mm. For the purpose of evaluation of possible changes in thermal properties of both natural and rubber aggregates of fraction 4/8 mm during sample preparation, their thermal conductivity λ (W/m·K), volumetric heat capacity C_v (J/m^3·K) and thermal diffusivity a (m^2/s) as well as their powder density were measured in dependence on compaction time. For these tests, a high-frequency vibrating table VSB-15 (Brio) was used. For thermal parameters determination, the commercially produced device ISOMET 2114 (Applied Precision, Ltd.) was used as a representative of transient impulse methods (Záleská *et al.* 2017). In the thermal parameter tests, needle probe was applied.

2.2 Basic Physical Properties

Lightweight concrete specimens were first dried in a vacuum drier at 60 °C. Matrix density measurement was performed using an automatic helium pycnometry as described above. The bulk density was accessed on a gravimetric principle following the standard EN 12390-7 (2009). This method is based on the measurement of sample size and its dry mass. Calculation of the total open porosity was then done from the bulk and matrix density values. The relative expanded uncertainty of the applied measuring method was approx. 5 %.

2.3 Mechanical Resistance

After 28 days of water curing were on the prismatic specimens of dimension 40 × 40 × 160 mm applied the strength tests. The flexural strength was measured according to the standard EN 12390-5 (2009). Halves of broken specimens were used for compressive strength testing following the standard EN 12390-3 (2009). The loading area was 40 × 40 mm. The Young´s modulus was accessed on the dynamic principle using the pulse ultrasonic method on the prismatic samples dried at 60 °C. Device DIO 562 (Starman Electronic) was used for the measurement.

2.4 Thermal Properties

In order to understand the influence of the rubber aggregate on the improvement of thermal insulation properties of the developed lightweight concretes, their thermal conductivity λ (W/m·K) was accessed using commercially produced device ISOMET 2114 (see above) equipped with a surface probe. The measurement was conducted in dependence on moisture content at the dry state and at the full water saturated state on the cubic samples with a side dimension of 100 mm.

2.5 Water Transport Parameters

To access performance of the developed lightweight concrete with rubber aggregate in terms of capillary suction, experimental test for determination of water absorption coefficient A (kg/m^2·s$^{1/2}$) was carried out. Dried prismatic samples were cut into half and water and vapor-proof insulated by epoxy resin on all lateral sides. Then they were exposed by their 40 × 40 mm face to the distilled water. The weight change of the samples in time was monitored and registered. On the basis of experimental data, the capillary water absorption coefficient was calculated as originally proposed by Hall (1989).

3 RESULTS AND DISCUSSION

The grain-size curves of used both natural and rubber-based aggregates are given in Figure 2. Rubber aggregate was found to be slightly finer compared to the natural aggregate of fraction 4/8, especially for grain size ranging from 1 to 3 mm.

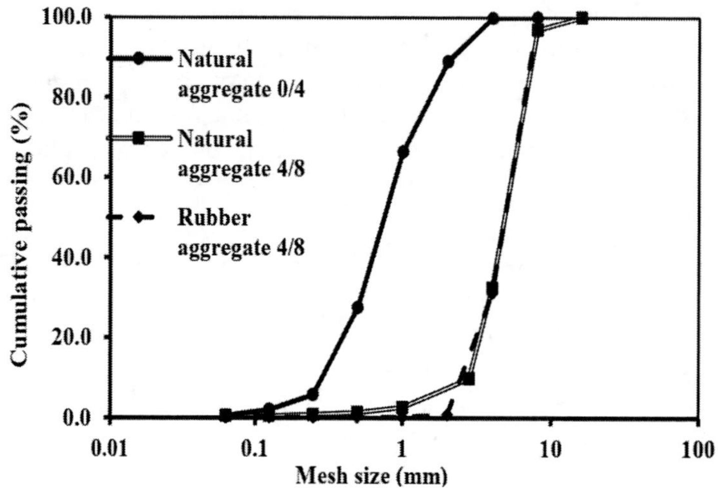

Figure 2. Grain-size analysis of used natural and rubber aggregates.

Thermophysical properties of both natural and rubber-based aggregates of fraction 4/8 are given in Table 2. The specific density was 1174 kg/m^3 for the rubber aggregate and 2662 kg/m^3 for the natural aggregate of fraction 4/8 respectively. We can see that the thermal performance of rubber particles is better in comparison with silica sand. As expected, with compaction time the thermal conductivity, as well as the volumetric heat capacity, increased for both rubber and natural aggregates due to the reduced amount of air gaps between aggregates particles.

Table 2. Aggregate properties in dependence on the compaction time.

Compaction Time (s)	Rubber Aggregate 4/8				Natural Aggregate 4/8			
	Powder Density (kg/m^3)	λ (W/m·K)	$C_v \times 10^6$ (J/m^3·K)	$a \times 10^{-6}$ (m^2/s)	Powder Density (kg/m^3)	λ (W/m·K)	$C_v \times 10^6$ (J/m^3·K)	$a \times 10^{-6}$ (m^2/s)
0	0.445	0.0951	0.2580	0.3684	1.408	0.315	1.4040	0.2246
10	0.460	0.0954	0.2542	0.3753	1.539	0.348	1.4195	0.2451
20	0.465	0.0957	0.2558	0.3741	1.581	0.369	1.4432	0.2559
180	0.465	0.0963	0.2623	0.3671	1.603	0.388	1.4260	0.2720

In Table 3, there are presented results of bulk density, matrix density, and open porosity tests performed for lightweight concretes with incorporated rubber aggregate. The bulk density values of prepared concrete containing rubber are between 1500-2000 kg/m^3, they can be therefore classified as lightweight concretes in classes LC 1.6, LC 1.8 and LC 2.0 respectively, according to the standard EN 206-1 (2014). Rubber has much lower specific density than natural aggregate, so its replacement with rubber consequently reduces the density of the developed rubber concrete (Shu and Huang 2014). The porosity of developed lightweight concrete increased with increased rubber content, this can be explained by the higher content of entrained air in concrete mixes and pores present in rubber particles (Girskas and Nagrockienė 2017).

Table 3. The basic physical characteristic of lightweight rubber concrete.

Material	Bulk density (kg/m³)	Matrix density (kg/m³)	Open porosity (%)
Ref	2134	2453	13.0
10%	1914	2249	14.9
20%	1760	2101	16.3
30%	1543	1902	18.9

Results of mechanical strength tests are summarized in Table 4. We can see that mechanical resistance significantly decreased with increasing amount of rubber-based aggregate in the mixture. These results are in accordance with the findings of other authors (Shu and Huang 2014, Thomas and Gupta 2016, Girskas and Nagrockienė 2017). The bond between the non-wetting rubber surface and cement paste is probably the main parameter that controls the reduction in mechanical concrete properties (Aliabdo *et al.* 2015).

Table 4. Mechanical properties of lightweight rubber concrete.

Material	Compressive strength (MPa)	Flexural strength (MPa)	Young´s modulus (GPa)
Ref	61.5	6.8	39.0
10%	26.0	5.1	17.0
20%	13.0	2.7	5.2
30%	4.5	1.7	3.1

The thermal conductivity values in the dry and full water-saturated state are presented in Table 5. The thermal conductivity of concrete is sensitive to many factors, such as concrete density, which is primarily related to the type and density of the contained aggregates and moisture condition of specimens (Iqbal 2017). With the increased rubber substitution rate, the thermal conductivity value of rubberized concrete significantly decreased, this observation was also reported in many recent studies (Aliabdo *et al.* 2015, Iqbal 2017). With increased moisture content the thermal conductivity value increases for all samples. Rubber particles in present of moisture still reduce the thermal conductivity value compared to the reference sample but in opposite direction is the higher porosity associated with higher moisture presence in concrete porous space.

Table 5. The thermal conductivity of lightweight rubber concrete in the dry and fully water saturated state.

Material	λ (W/m·K)	
	Dry state	Saturated state
Ref	1.776	2.378
10%	1.275	1.899
20%	0.857	1.205
30%	0.653	1.123

Results of the water absorption coefficient test are given in Table 6. The water transport behavior of lightweight rubber concrete is mainly influenced by the non-wetting character of the used rubber aggregate and by the porosity of the sample, which acts against each other.

Table 6. The water absorption coefficient of lightweight rubber concrete.

Material	A (kg/m^2·s$^{1/2}$)
Ref	0.0197
10%	0.0168
20%	0.0187
30%	0.0171

4 CONCLUSIONS

This study aimed at the evaluation of scrap tire rubber-based aggregate with different replacement levels in lightweight concrete in terms of mechanical, thermal, hygric and basic physical properties of the developed materials. The obtained results have shown that application of the rubber aggregate in the building industry can lead to an environmentally friendly material with improved thermal insulation function that can increase the energy efficiency of buildings whereas its mechanical parameters are suitable for non-structural applications.

Acknowledgments

The authors gratefully acknowledge the financial support received from the Czech Science Foundation, under project No 17-04215S - Thermal insulation composites containing waste plastic-based fillers.

References

Aliabdo, A. A., Elmoaty, A. E. M. A., and AbdElbaset, M. M., Utilization of Waste Rubber in The Non-Structural Application, *Construction and Building Materials*, 91, 195-207, 2015.

Bompa, D. V., Elghazouli, A. Y., Xu, B., Stafford, P. J., and Ruiz-Teran, A. M., Experimental Assessment and Constitutive Modelling of Rubberised Concrete Materials, *Construction and Building Materials,* 137, 246-260, 2017.

Cairns, R. A., Kew, H. Y., and Kenny, M. J., The Use of Recycled Rubber Tyres in Concrete Construction, *In Proceedings of the International Conference of the Concrete and Masonry Research Group 2004: Sustainable Waste Management and Recycling: Used/Post Consumer Tyres*, Thomas Telford Ltd., 135–142, 2004.

EN 12390-3, *Testing Hardened Concrete – Part 3: Compressive Strength of Test Specimens, CEN,* 2009

EN 12390-5, *Testing Hardened Concrete – Part 5: Flexural Strength of Test Specimens,* CEN, 2009.

EN 12390-7, *Testing Hardened Concrete – Part 7: Density of Hardened Concrete,* CEN, 2009.

EN 206-1, *Concrete – Part 1: Specification, Performance, Production and Conformity,* CEN, 2014.

Girskas, G., and Nagrockienė, D., Crushed Rubber Waste Impact of Concrete Basic Properties, *Construction and Building Materials*, 140, 36-42, 2017.

Hall, C., Water Sorptivity of Mortars and Concretes: A Review, *Magazine of Concrete Research,* 41(147), 51-61, 1989.

Iqbal, M., Thermal Conductivity of Hybrid Recycled Aggregate – Rubberized Concrete, *Construction and Building Materials*, 133, 516-524, 2017.

Shu, X., and Huang, B., Recycling of Waste Tire Rubber in Asphalt and Portland Cement Concrete: An Overview, *Construction, and Building Materials*, 67(Part B), 217-224, 2014.

Thomas, B. S., and Gupta, R. Ch., A Comprehensive Review of the Applications of Waste Tire Rubber in Cement Concrete, *Renewable and Sustainable Energy Reviews*, 54, 1323-1333, 2016.

Záleská, M., Pokorný, J., Pavlíková, M., and Pavlík, Z., *Thermal Properties of Light-Weight Concrete with Waste Polypropylene Aggregate*, AIP Conference Proceedings 1866 (1), AIP Publishing, 2017.

U.S. Tire Manufacturers Association, 2015 U.S. Scrap Tire Management Summary, May 2017. Retrieved from USTires.org on July 2018.

Interdependence between Structural Engineering and Construction Management
Edited by Ozevin, D., Ataei, H., Modares, M., Gurgun, A., Yazdani, S., and Singh, A.

ISBN: 978-0-9960437-6-2

SOME PROPERTIES OF REACTIVE POWDER CONCRETE MADE WITH RECYCLED AGGREGATE

SHATHA HASAN

Dept of Civil Engineering, University of Technology, Baghdad, Iraq

An experimental study was carried out to investigate, the strength properties, of recycled aggregates for use in reactive powder concrete (RPC), to better understand the properties of reactive powder concrete with recycled aggregates used as a substitution material for normal sand in reactive powder concrete. The effects of variable parameters on these properties were carefully studied which are the percent of recycled fine aggregate (0%, 25%, 50%, and 75%) as a partial replacement by weight of normal sand, recycled fine aggregate (50%) as a partial replacement by weight of glass sand, and two different curing (20, 90 °C). The inclusion of using reactive silica sand powder leads to a considerable increment in both compressive and tensile strength for recycled reactive powder concrete (RRPC) and the heat treatment improves the properties of the RPC and RRPC considerably. Finally, the results indicated that it is possible to produce reactive powder concrete from recycled aggregate depending on the strength, the results show to suggest only a gradual lowering in compressive and splitting tensile strength.

Keywords: Strength, Silica sand, Silica fume.

1 INTRODUCTION

Reactive powder concrete (RPC) is a relatively new family of Portland cement – based material with high strength, high elastic modulus, very low porosity, ductility with ultimate elongation, and high impact strength (Richard and Cheyrezy 1995, Hassan 2006, and Abdelalim *et al.* 2008). RPC is obtained by an optimization of its grain size distribution by scaling down the maximum size of aggregate to 600 microns and by applying some very simple heat treatment. The properties of RPC are achieved by combining Portland cement, silica fume, fine silica sand, high range water reducing admixture, water, and steel fiber (Richard and Cheyrezy 1995, Perry and Zakariasen 2004).

Concrete is one of the most widely used structural materials in the world, and continues to evolve. The main purpose behind its popularity is its characteristics such as high strength and durability. For this reason, it is important to use crushed concrete from demolition waste as concrete aggregate instead of natural aggregate to preserve natural resources of materials in order to reduce the amount of demolition waste (Weimann *et al.* 2003, Puri *et al.* 2013). During the last few years, many studies and projects have been carried on the utilization of Recycled Aggregate Concrete (RAC) in ordinary concrete. These researches have explained that RA from building demolition can be utilized instead of normal aggregates. Prior investigations on the replacement of natural sand by recycled sand from building demolition in several types of concrete are studied. In this study, it was, therefore, examined the possibility of replacing normal sand with recycled concrete sand in RPC to produce recycled reactive powder concrete, RRPC.

Demiss *et al.* (2018) carried out an experimental study on the mechanical and microstructural properties of RRPC containing waste glass powder and fly ash at standard curing. The experimental results indicated that replacing the silica fume fully by finely dispersed local waste glass powder and fly ash is a promising approach for local structural construction applications. In this study, some properties of the RPC incorporating local powders as a pozzolanic material and replacing natural aggregate by recycled concrete aggregate were investigated.

2 EXPERIMENTAL PROGRAM

2.1 Materials

2.1.1 *Cement*

Ordinary Portland cement, manufactured in Iraq, was used throughout this study. Its chemical and physical properties conform to Iraqi specification No.5.

2.1.2 *Fine aggregate*

Two types of normal fine aggregate are used in this study. The first one is Al-Ekhaider sand (NS) and the second type is silica sand (SS) known as glass sand. Glass sand used throughout this study was crushed silica rock brought from Al- Ramadi Glass factory. The two types of fine aggregate with maximum size 600 μm were tested and the physical and chemical properties were within the requirements of the Iraqi specification No. 45.

2.1.3 *Recycled aggregate*

The fine recycled concrete aggregate was gained from buildings destruction. Large parts of demolition were brought to the Laboratory and broken up by a laborer into small parts and the parts passing through a 4.75mm sieve were utilized as a fine recycled concrete aggregate (RCS). The particle size distribution of fine recycled aggregate was thus similar to the particle size distribution of normal fine aggregate.

2.1.4 *High range water reducing admixture*

High range water reducing admixture (SP) used in this study is modified polycarboxylic ether, which has been primarily developed for applications where the highest strength, durability, and performance is required. It is free from chlorides and complies with ASTM C494.

2.1.5 *Mineral admixtures*

Condensed Silica fume (CSF) is a highly reactive pozzolanic material and is an extremely fine powder, with particles about 100 times smaller than an average cement grain. The chemical composition and physical requirements show that the silica fume conforms to the chemical and physical requirements of ASTM C1240 specifications.

2.1.6 *Steel fibers*

Hooked steel fibers (S_f) used in this work and each steel fiber has a diameter of 0.4mm and length of approximately 20mm with aspect ratio = 50, density= 7800 kg/m^3 and ultimate tensile strength of 1600 MPa.

2.1.7 *Reactive silica sand powder*

Reactive silica sand powder (RSSP), which is used in this work, brought from Al- Ramadi Glass factory and conformed to the physical and chemical requirements of ASTM C618 class N Pozzolan.

2.2 Mix Proportions

Many mix proportions according to pervious researches (Hassan 2006, Gao *et al.* 2005) were tried in this study to have maximum compressive strength and flow of (75+5%) of RPC and recycled RPC. Nine mixes were prepared in the laboratory. Ordinary RPC (mix1) was used as a reference mix, then eight additional mixes were prepared to investigate the effects of replacing the normal fine aggregate with fine recycled aggregate on the properties of RPC. Table 1 shows the mix proportions for all mixes. Fine silica sand known as glass sand is used in mixes eight and nine to compare them with mixes made with natural fine sand in other mixes.

Table 1. Normal and RRPC mixes used in the present research.

		Series 1		Series 2		Series 3		Series 4	
Mix No.	**Mix 1**	**Mix 2**	**Mix 3**	**Mix 4**	**Mix 5**	**Mix 6**	**Mix7**	**Mix 8**	**Mix 9**
C	970	970	970	970	970	970	970	970	970
NS	1060	795	795	530	530	265	265		
SS								1060	530
CSF	175	175	135	175	135	175	135	175	135
RSSP			40		40		40		40
RCS		265	265	530	530	795	795		530
W/C	0.187	0.19	0.191	0.192	0.194	0.199	0.204	0.192	0.194
Flow Table %	78	77	77	76	74	74	73	74	73
SP	All mixes the *SP* 8.5 % by weight of cement								
S_f	All mixes the steel fibers 2% by volume								

2.3 Mixing, Casting and Curing Procedure

All normal and recycled RPC mixes were performed in a rotary mixer of 0.1m^3. The compaction was affected by means of a vibrating table for a period of twenty seconds/ layer for (5 cm cubes) and (10cm × 20 cm cylinder). After twenty-four hours from casting, the specimens were molded. Then concrete specimens were moist cured in two different conditions:

A. 20 ° C (Room temperature).
B. Steam curing at 90 ° C after preliminary curing at room temperature for twenty-four hours.

3 TEST RESULTS AND DISCUSSION

3.1 Compressive Strength

The compressive strength tests of normal and recycled RPC were carried out in accordance with B.S: 1881: part 116. Five cm cube specimens were used to define compressive strength using a standard testing machine with a capacity of 2000 kN. The loading was used at a rate of fifteen Mpa per minute. Cubes were removed from curing solution at age of (14, 28, and 90) days. The average of three specimens was recorded for each testing age. Generally, test results demonstrate a considerable decrease in compressive strength of all recycled reactive powder mixes made by replacing normal fine sand by fine recycled concrete aggregate compared to the mixes of ordinary RPC made with normal sand as shown in Table 2.

Figures 1 and 2 exhibits the growth of compressive strength for all RPC when tested at different ages. The test results of normal and recycled RPC showed that replacing normal fine aggregate with recycled demolished concrete aggregate at the levels of 25%, 50%, and

75% had only a slight effect on the compressive strength. In general, the decrease of compressive strength may have been due to the goodness of recycled fine aggregate, as the fine recycled aggregate was gained from the demolition of unknown quality building concrete. Similar results have been reported by other researchers (Bo-Tsun *et al.* 2010, Vaishali and Rao 2012).

Results given in this Table 2 show that RPC with glass sand (Mix 8) can also reach the high compressive strength. On the other hand, the compressive strength of the mix 9 in series 4, which replaced the silica sand with recycled fine aggregate, were greater than mix 5 in series 2, which replaced the natural fine aggregate with recycled fine aggregate, which may be due to the better properties of silica sand and powerful bonding between old mortar.

Steam curing at 90°C gave a better performance of RPC and RRPC with silica fume and RSSP in terms of higher strength than the curing at room temperature (20°C) as shown in Figures 1 and 2.

Table 2. The compressive strength of RPC and RRPC at two different curing.

Series		**1**	**2**		**3**		**4**		**5**	
Curing	**Days**	**Mix 1**	**Mix 2**	**Mix 3**	**Mix 4**	**Mix 5**	**Mix 6**	**Mix7**	**Mix 8**	**Mix 9**
20° C	**14**	78.5	73.8	72.6	68.1	66.8	58.3	57.1	80.1	76.3
	28	120.4	114.2	110.9	111.5	110.3	95.7	90.3	123.3	126
	90	128.2	123.9	121.7	118.7	119.2	102.6	98.4	130.9	130.2
90° C	**14**	85.2	82.1	81.5	74.6	72.5	66.3	62.2	82.3	79.6
	28	129.9	123.8	123.1	125.6	123.1	104.5	99.8	132.8	131.9
	90	140.5	136.8	133.5	126.6	126.1	119.4	115.7	139.8	138.2

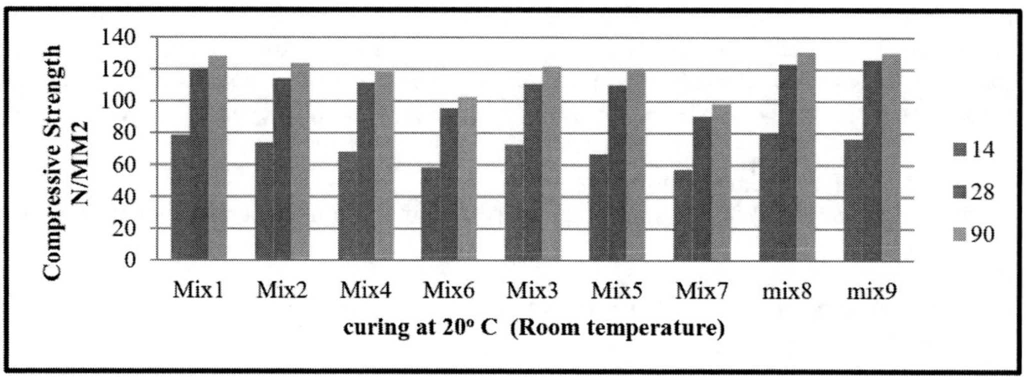

Figure 1. The compressive strength of RPC and RRPC at 20 °C.

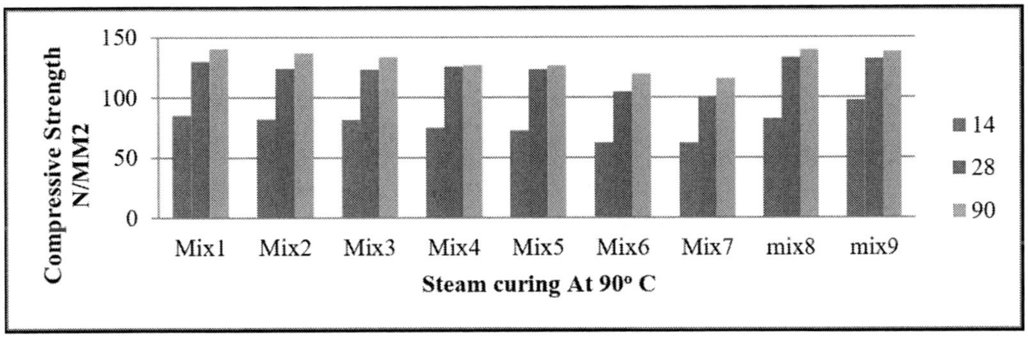

Figure 2. The compressive strength of RPC and RRPC at 90°C.

3.2 Splitting Tensile Strength

This test was done according to ASTM C496-86 specification. Cylinders of ten cm diameter and twenty cm height were utilized and tests were completed using a standard testing machine with a capacity of 2000 kN. The load was gradually raised up to a point of failure. The loading was applied at a rate of 1.5 MPa per minute. The test results of the splitting tensile strength are plotted in Figure 3 which presents the splitting tensile strength behavior of RPC and RRPC mixes at 28 days with different curing condition. The heat treatment has a considerable influence on the splitting tensile strength, for instance, the splitting tensile strength of mixes 5 and 9 rises from (9.2 to 11.7) MPa and from (11.9 to 14.7) MPa respectively. The results also indicate that the splitting tensile strength increase due to the heat treatment is less than the increase in compressive strength of the same mixtures. Since the heat treatment affects the w/cm ratio, it can be concluded that the compressive strength is significantly affected more than the splitting tensile strength. The test results of normal and recycled RPC showed that replacing fine aggregate with recycled demolished concrete aggregate had only a slight effect on the splitting tensile strength.

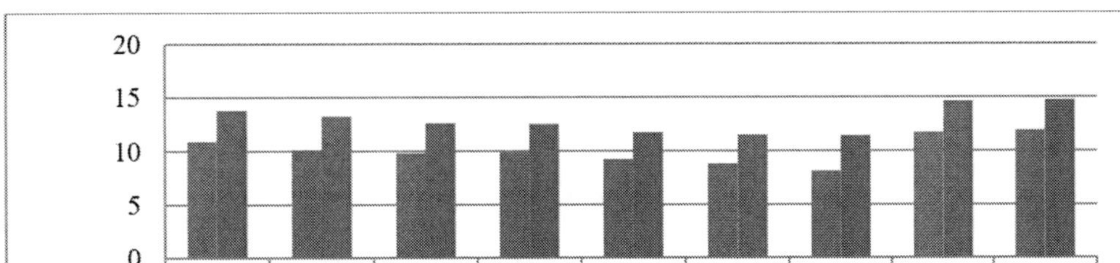

	mix1	mix2	mix3	mix4	mix5	mix6	mix7	mix8	mix9
2o o C	10.9	10.1	9.8	9.9	9.2	8.8	8.1	11.7	11.9
9o o C	13.8	13.2	12.6	12.5	11.7	11.5	11.4	14.6	14.7

Figure 3. Splitting tensile strength of RPC and RRPC at 28 days.

4 CONCLUSIONS

Depending on the results of this study on RPC and RRPC the following conclusions can be drawn:

1. It is possible to produce recycled reactive powder concrete from recycled fine concrete aggregate as a partial replacement of normal fine sand or glass sand.
   The heat treatment improves the properties of the RPC and RRPC considerably.
2. It is possible to produce RPC and RRPC from glass sand as a fine aggregate.
3. The effect of the use of recycled fine aggregate on the strength of RRPC depends on

the percentage of recycled fine aggregate utilized

4. Concrete mixes designed with a group of recycled fine aggregate as a partial replacement for glass sand offered higher compressive and splitting tensile strengths than concrete mixes designed with a group of recycled concrete aggregate used as partial replacement normal sand.
5. The inclusion of reactive silica sand powder as a partial replacement of mineral admixture leads to a considerable increase in strength.

References

Abdelalim, A., Ramadan, M., Bahaa, T., and Halawa, W., Performance of Reactive Powder Concrete Produced Using Local Materials, *HBRC Journal*, 4 (3), December 2008.

Bo-Tsun, C., Ta-Peng, C., Tzong-Ruey, Y., Ya-Chu, C., and Tien-Chin, H., Statistical Study on Properties of High-Performance Recycled Aggregate Concrete, *Korea Concrete Institute*, ISBN978-89-5708-182(2), 2010.

Demiss, B., Oyawa, W., and Shitote, S., Mechanical and Microstructural Properties of Recycled Reactive Powder Concrete Containing Waste Glass Powder and Fly Ash at Standard Curing, *Cogent Engineering*, 5, 1464877, 2018.

Gao, R., Stroeven, P., and Hendriks, C., Mechanical Properties of Reactive Powder Concrete Beams, *ACI Special Publication*, 228(79), 1237-1252, 2005.

Hassan, S., *Static and Impact Properties of Reactive Powder Concrete*, Ph.D. Thesis, University of Technology, Baghdad, 2006.

Perry, H., and Zakariasen, D., First Use of Ultra-High-Performance Concrete for an Innovative Train Station Canopy, *Concrete Technology Today*, 25(2), August 2004.

Puri, N., Kumar, B., and Tyagi, H., Utilization of Recycled Wastes as Ingredients in Concrete Mix, *International Journal of Innovative Technology and Exploring Engineering (IJITEE)*, 2(2), January 2013.

Richard, P. and Cheyrezy, M., Composition of Reactive Powder Concretes, *Cement and Concrete Research*, 25(7), 1501-1511, 1995.

Vaishali, G., and Rao, H., Strength and Permeability Characteristics of Fiber Reinforced High-Performance Concrete with Recycled Aggregates, *Asian Journal of Civil Engineering (Building And Housing)*, 13(1), 55-77, 2012.

Weimann, K., Giese, L., Mellmann, G., and Simon, R., Building Materials from Waste, *Materials Transactions*, 44(7), 1255-1258, 2003.

Interdependence between Structural Engineering and Construction Management
Edited by Ozevin, D., Ataei, H., Modares, M., Gurgun, A., Yazdani, S., and Singh, A.

ISBN: 978-0-9960437-6-2

STRENGTH PROPERTIES OF DURABLE CONCRETE MADE WITH VARIOUS ALTERNATIVE CEMENTITIOUS MATERIALS

ISAMU YOSHITAKE[1], SHUN INOUE[1], KEISUKE MIYAMOTO[2], and KOICHIRO YAMATO[3]

[1]*Dept of Civil and Environmental Engineering, Yamaguchi University, Ube, Yamaguchi, Japan*
[2]*Hagimori Industries, Ube, Yamaguchi, Japan*
[3]*Cement and Construction Materials Company, Ube Industries, Ube, Yamaguchi, Japan*

Many researchers investigated various concrete made with the cementitious materials, hence the effects of various admixture are well known. Most researches focused on an alternative cementitious material, tested concrete properties by varying the cement-replacement ratio. The optimum cement-replacement ratio may be different by each admixture. The present study examined the strength properties of concrete incorporating various admixtures. The cementitious materials tested in this study were fly-ash, blast furnace slag powder, silica-fume, limestone powder, and an artificial admixture which has high resistance to chloride attack. First, fundamental properties of the concrete incorporating the admixture were summarized referring to previous durability tests. Second, the strength properties of concrete made with some cement-replacement ratios by the cementitious materials were examined. In addition, alternative cement efficiency factors (***k***) of each cementitious material were estimated.

Keywords: Admixture, Durability, Alternative cement efficiency factor, Chloride resistance admixture.

1 INTRODUCTION

Alternative cementitious materials such as fly-ash and blast-furnace slag are often used in concrete to improve strength and durability in addition to the reduction of CO_2 emission from cement production. Many researchers investigated various concrete made with the cementitious materials, hence the effects of various admixture on strength and durability are well known. Most researches focused on an alternative cementitious material, tested concrete properties by varying the cement-replacement ratio. The optimum cement-replacement ratio may be different by each admixture. Hence it is not easy to compare the effect of admixture under a certain test condition. To compare and quantify the effect of the admixture, the present study examined strength properties and durability of concrete incorporating various admixtures. The cementitious materials tested in this study were fly-ash, blast furnace slag powder, silica-fume, limestone powder. In addition, the study tested on an artificial admixture developed in recent years. The admixture has unique properties which are high-resistance for chloride attack and high-early-strength (Chloroguard 2018).

Yamato *et al.* (2017, 2018) conducted various durability tests using artificial admixture. They examined chloride penetration resistance, carbonation, sulfate resistance, and freeze-thaw resistance of the concrete incorporating the admixture. Their test results confirmed that the durability properties are comparable with or higher than that of conventional concrete without the admixture. For example, effective diffusion coefficients of steam-cured and water-cured concrete incorporating the admixture are presented in Figure 1 (a) (b), respectively. In addition, Figure 2 shows the relative dynamic moduli of elasticity and mass-change of the water-cured concrete subjected to freeze-thaw cycles. In these graphs, N or OPC was used for concrete using ordinary Portland cement, and BB was used for concrete made with blast-furnace slag cement. As well, it is well known that general concrete made with admixtures such as fly-ash, blast-furnace slag powder and silica-fume indicates excellent durability (Berry and Malhotra 1980, Özbay *et al.* 2016, Khan and Siddique 2011). The study focuses on strength properties of concrete containing such mineral admixtures which improve concrete durability and indicates fundamental properties for the designing of optimum mixture proportion of concrete made with admixture.

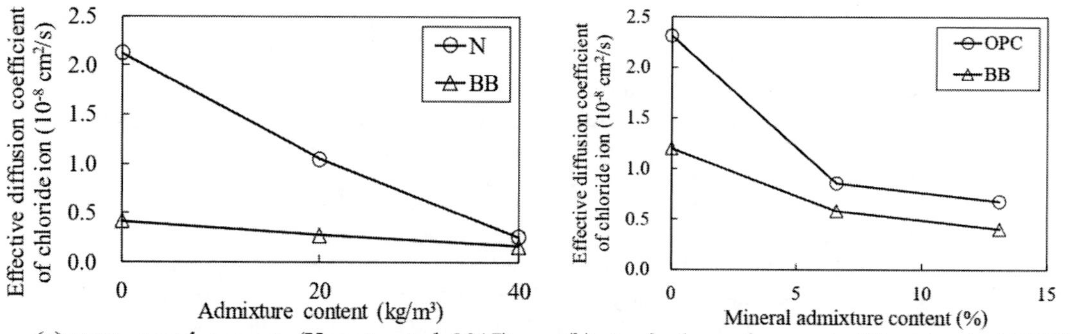

(a) steam-cured concrete (Yamato *et al.* 2017). (b) standard-cured concrete (Yamato *et al.* 2018).

Figure 1. Effective diffusion coefficients of concrete incorporating the developed admixture.

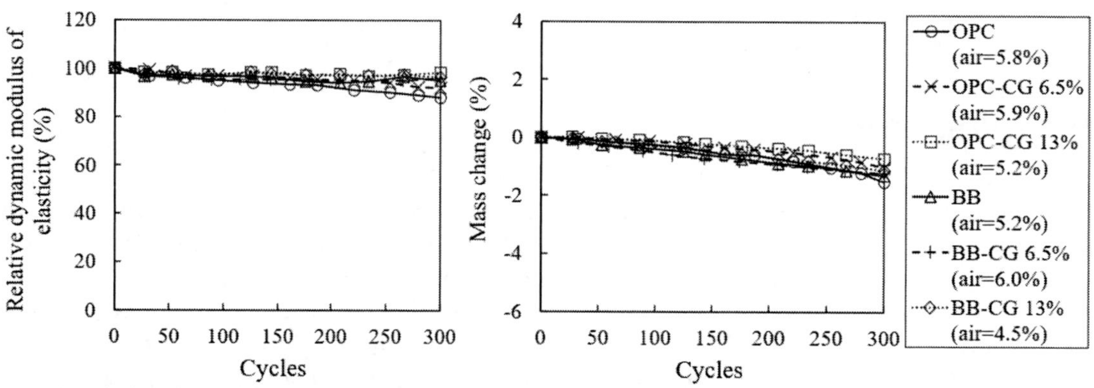

(a) Relative dynamic moduli of elasticity. (b) Mass change.

Figure 2. Freeze-thaw test of concrete incorporating the admixture (Yamato *et al.* 2018).

2 MATERIALS AND CONCRETE MIXTURE

2.1 Chloride Resistance Admixture

The chloride-resistance admixture (C) is a pozzolanic mineral powder incorporating SiO_2 and Al_2O_3, and it has a Brunauer–Emmett–Teller (BET) specific surface area of 13000 m^2/kg or higher. It was confirmed that the admixture contributes to improve concrete durability and strength development at an early age in the previous investigations (Yamato *et al.* 2017, Yamato *et al.* 2018, Yamato 2018). The fundamental properties of the admixture are given in Table 1. It is remarkable that the admixture (C) has higher Al_2O_3 than silica-fume (S). Further details of the admixture (C) cannot be released here because of a commercial contract with the manufacturer.

Table 1. Chloride resistance admixture.

	Density (kg/m^3)	Fineness (m^2/kg)	SiO_2 (%)	Al_2O_3 (%)	Na_2O (%)	K_2O (%)	Cl^- (%)
C	2360	13300	71.2	23.3	0.27	0.27	0.029
S	*2240*	*16900*	*94.0*	*0.26*	*0.21*	*0.28*	*0.014*

2.2 Materials

The materials are summarized in Table 2. Ordinary Portland cement defined in a Japanese Industrial Standard (JIS R 5210) was used. Fine and coarse aggregates are general aggregate in western Japan. Chemical admixtures given in Table 2 were used for suitable fresh properties.

Table 2. Materials used in the study.

Material	Type	Property
Water	Tap-water (W)	Density 1.00 g/cm^3
Cement	Ordinary Portland cement (OPC)	Density 3.16 g/cm^3, Blaine fineness 3280 cm^2/g
Admixture	Limestone powder (L)	Density 2.71 g/cm^3, Blaine fineness 3640 cm^2/g
	Fly-ash (F)	Class II (JIS A 6201), Density 2.22 g/cm^3, Blaine fineness 3530 cm^2/g
	Blast furnace slag powder (B)	Density 2.90 g/cm^3, Blaine fineness 4840 cm^2/g
	Silica-fume (S)	Density 2.21 g/cm^3
	Chloride resistance admixture (C)	Density 2.36 g/cm^3, BET fineness 13.3 m^2/g
Fine aggregate	Sea sand (S1)	Density 2.60 g/cm^3, FM 2.91, Size 5-0 mm
	Crashed stone sand (S2)	Density 2.60 g/cm^3, FM 2.90, Size 5-0 mm
Coarse aggregate	Crashed stone (G1)	Density 2.71 g/cm^3, Size 20-15 mm
	Crashed stone (G2)	Density 2.70 g/cm^3, Size 15-5 mm
Chemical admixture	Air-entraining and water reducing agent (WRA)	Lignin sulfonate, oxy-carboxylate, and poly-carboxylic acid-based compound
	Air-entraining and high-range water-reducing agent (HRWRA)	Poly-carboxylic acids-based compound
	Antifoaming agent (AFA)	Nonionic surfactant

2.3 Mixture Proportions of Concrete

Mixture proportions of tested concrete are summarized in Table 3. The water-cementitious material ratios (*w/cm*) were designed as 0.45, 0.55 and 0.65 for control mixtures and concrete containing the chloride resistance admixture (C). All admixtures tested herein were used as an alternative cementitious material for Portland cement. The cement-replacement ratios were 0.1, 0.2 and 0.4 by mass of cement. In the mixture I.D. of admixture concrete, the first number shows the water-cementitious material ratios (*w/cm*), the following letter presents the admixture and the last number shows the cement-replacement ratios. The mixed volume ratios of aggregate were S1: S2 = 0.4 : 0.6, and G1 :G2 = 0.5 : 0.5, respectively.

Table 3. Mixture proportions and fresh properties.

Mixture I.D.	Unit weight (kg/m^3) W	OPC	Admixture	S1	S2	G1	G2	Chemical admixture (cm×%)		Slump -cm-	Air -%-
Control-45	162	360	0	581	249	488	486	1.00 [a)]	0.0015 [c)]	14.0	3.4
Control-55	162	295	0	618	265	488	486	0.40 [a)]	0.0020 [c)]	14.0	4.5
Control-65	162	250	0	644	276	488	486	0.80 [a)]	0.0018 [c)]	15.0	4.5
45-C10	162	324	36	574	246	488	486	1.00 [b)]	0.0026 [c)]	11.0	4.7
45-C20	162	288	72	567	243	488	486	1.05 [b)]	0.0009 [c)]	9.0	5.0
45-C40	162	216	144	552	237	488	486	1.70 [b)]	0.0090 [d)]	14.0	4.7
55-C10	162	266	30	312	262	488	486	1.10 [a)]	0.0014 [c)]	12.0	4.2
55-C20	162	236	59	607	260	488	486	1.40 [a)]	0.0023 [c)]	12.5	5.8
55-C40	162	177	118	595	255	488	486	1.75 [b)]	0.0002 [c)]	11.0	5.7
65-C10	162	225	25	639	274	488	486	1.00 [a)]	0.0	13.5	5.4
65-C20	162	200	50	634	272	488	486	1.60 [a)]	0.0	11.0	4.7
65-C40	162	150	100	624	268	488	486	1.30 [b)]	0.0125 [d)]	9.0	5.5
55-F10	162	266	30	610	261	488	486	0.70 [a)]	0.0112 [c)]	14.0	4.5
55-F20	162	236	59	604	259	488	486	0.40 [a)]	0.0090 [c)]	19.0	5.9
55-F40	162	177	118	589	253	488	486	0.40 [a)]	0.0210 [c)]	16.5	4.2
55-B10	162	266	30	616	264	488	486	0.55 [a)]	0.0019 [c)]	14.0	4.9
55-B20	162	236	59	615	264	488	486	0.50 [a)]	0.0021 [c)]	16.0	4.7
55-B40	162	177	118	612	262	488	486	0.45 [a)]	0.0012 [c)]	12.0	4.6
55-L10	162	266	30	615	263	488	486	0.50 [a)]	0.0027 [c)]	12.5	4.3
55-L20	162	236	59	612	262	488	486	0.40 [a)]	0.0018 [c)]	17.5	5.2
55-L40	162	177	118	607	260	488	486	0.40 [a)]	0.0018 [c)]	11.0	4.0
55-S10	162	266	30	610	261	488	486	1.40 [a)]	0.0	13.5	5.0
55-S20	162	236	59	603	259	488	486	1.50 [a)]	0.0125 [d)]	9.5	4.7
55-S40	162	177	118	589	252	488	486	1.80 [b)]	0.0275 [d)]	9.0	5.0

[a)]WRA, [b)]HRWRA, [c)]AEA, [d)]AFA

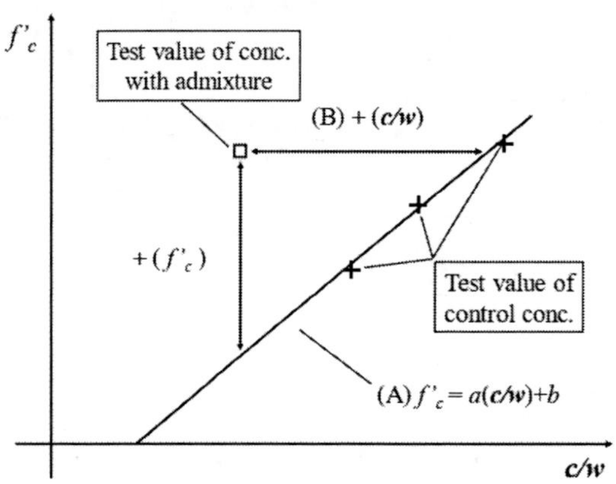

Figure 3. Alternative cement efficiency factor (k) based on $f'_c - c/w$.

2.4 Alternative Cement Efficiency Factor (*k*) Based on Compressive Strength

To determine the effectiveness of each alternative cementitious material for Portland cement, the study evaluated the cement efficiency factor based on the compressive strength test. The factor can be calculated by applying the compressive strength of admixture concrete into a compressive strength (f'_c) – cement-water ratio (c/w) relation of the reference concrete (control mixtures). Figure 3 illustrates the estimation method for the cement efficiency factor (k).

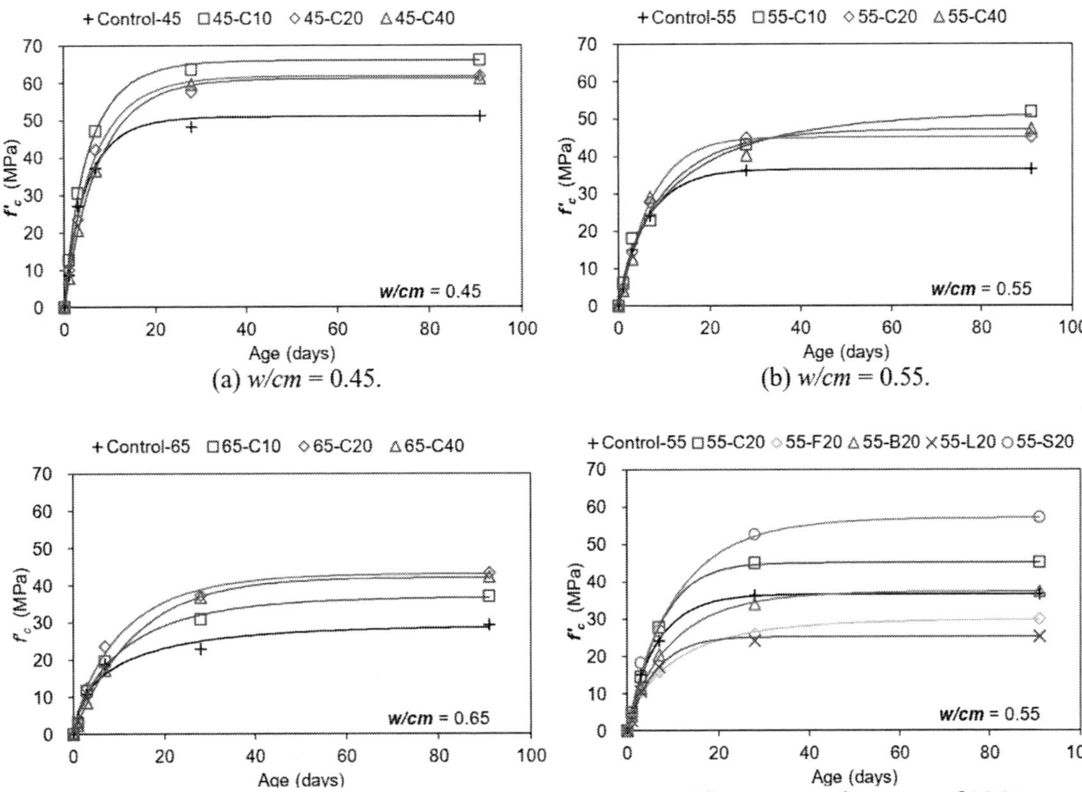

(a) w/cm = 0.45. (b) w/cm = 0.55.

(c) w/cm = 0.65. (d) cement-replacement of 20 %.

Figure 4. Compressive strength.

3 RESULTS AND DISCUSSION

3.1 Compressive Strength

Compressive strength test was conducted in accordance with the Japanese standard (JIS A 1108). Three cylindrical specimens, 100 x 200 mm, were used at each test age. Figure 4 (a) (b) (c) presents compressive strengths of the chloride resistance admixture (C) concrete of w/cm = 0.45, 0.55, 0.65, respectively. These test results confirm that the chloride resistance admixture (C) concrete achieved higher strength than the control concrete at all test ages. It is noteworthy that the concrete has higher early-age strength through the admixture is a kind of pozzolanic powder material. In addition, Figure 4 (d) shows compressive strengths of various admixture concrete made with the cement-replacement ratio of 20 %. The result shows that the chloride resistance admixture (C) concrete indicated always higher strength than fly-ash (F), blast-furnace slag powder (B), limestone powder (L) concretes. The concrete indicated high early-age strength as well as silica fume (S) concrete. The strength property assures that chloride resistance admixture (C) is a useful material to improve concrete quality in addition to excellent durability.

3.2 Alternative Cement Efficiency Factor (*k*)

Figure 5 (a) (b) summarizes the alternative cement efficiency factors (*k*) of the chloride resistance admixture and other admixtures, respectively. Regardless of the water-cementitious ratios and the cement-replacement ratio, the factor (*k*) of the chloride resistance admixture (C) achieved 1.0 or higher as well as silica fume (S). The result implies that strength contribution of these admixtures (C, S) was higher than Portland cement.

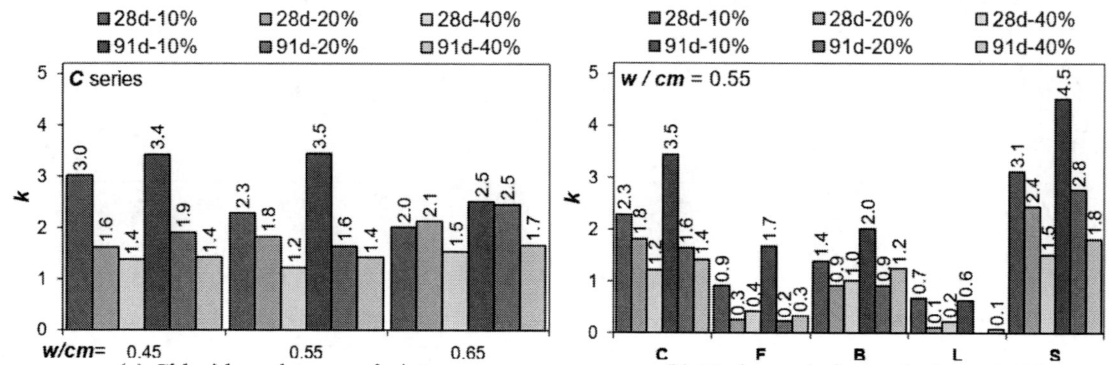

(a) Chloride resistance admixture. (b) Various admixture (*w/cm* = 0.55).

Figure 5. Alternative cement efficiency factor (***k***).

4 SUMMARY

This paper reported the effect of various admixtures on strength properties in addition to durability of concrete based on the previous investigations. It can be concluded that the chloride resistance admixture (C) is an effective material to improve strength and durability of concrete.

Acknowledgments

The study was sponsored financially by the Japan Society for the Promotion of Science (JSPS KAKENHI Grant Number 18K04303). The authors would like to acknowledge Mr. Mizushima J. and Ms. Yamamoto K. for their support throughout this work.

References

Berry, E. E., and Malhotra, V. M., Fly Ash for Use in Concrete - A Critical Review, *Journal of the American Concrete Institute*, 77(2), 59-73, 1980.

Chloroguard. retrieved from http://www.chloroguard.jp/ on Aug. 2018.

Khan, M., and Siddique, R., Utilization of Silica Fume in Concrete: Review of Durability Properties, *Resources, Conservation and Recycling*, 57, 30-35, 2011.

Özbay, E., Erdemir, M., and Durmuş, H. I., Utilization and Efficiency of Ground Granulated Blast Furnace Slag on Concrete Properties – A review, *Construction, and Building Materials*, 105, 423-434, 2016.

Yamato, K., Ishida, T., Yamaji, N., Tsugo, S., and Yoshitake, I., Durability of Steam-cured Concrete Incorporating a High-resistance Admixture for Chloride Attack (in Japanese), *Journal of the Society of Materials Science, Japan*, 66(5), 328-333, 2017.

Yamato, K., Yamaji, N., and Yoshitake, I., Properties of Concrete Incorporating a High-Resistance Admixture for Chloride Attack (in Japanese), *Cement Science and Concrete Technology*, 71, 667-673, 2018.

Yamato, K., *Experimental Study on Properties of Concrete Mixed with a Fine Admixture Incorporating High-Volume of SiO_2 and Al_2O_3 (in Japanese)*, PhD dissertation, Yamaguchi Univ., 2018.

Interdependence between Structural Engineering and Construction Management
Edited by Ozevin, D., Ataei, H., Modares, M., Gurgun, A., Yazdani, S., and Singh, A.

ISBN: 978-0-9960437-6-2

POTENTIAL UTILIZATION OF GRANITE WASTE WITH METAKAOLIN IN SUSTAINABLE CONCRETE

RAVI K. SHARMA, TRILOK GUPTA, and DIGVIJAY S. CHOUHAN

Dept of Civil Engineering, College of Technology and Engineering, MPUAT, Udaipur, India

Cement production is a crucial factor responsible for climate change and recognized as major environmental challenge for humankind. It causes around 5% of global CO_2 emissions resulting in lowering of air quality and human health. Production of cement is highly energy intensive process requiring consumption of depleting limited natural resources. Granite stone is famous for strength, elegance and durability. Enormous quantity of granite slurry waste is generated during processing of granite stone blocks, which adversely affect the fertility of land, environment and society. Metakaolin is an ultrafine pozzolonic and economical alternative to silica fume. It has positive influence on concrete properties because of filling effect and additional pozzolanic reaction. In this experimental investigation replacement level of both granite slurry waste and metakaolin are taken up to 10% as partial replacement of cement. Mechanical and durability properties of modified concrete have been evaluated. It has been shown that optimum replacement level of both materials enhances the compressive and flexural strength. Also, improvement in water permeability and resistance to acid attack has been observed. This study demonstrates that utilization of granite slurry waste and metakaolin as partial replacement of cement will reduce the consumption of cement, which will conserve the natural resource, reduces CO_2 emission, energy demand and waste accumulation. Hence, the production of new composite concrete will be sustainable and advantageous to the environment and society.

Keywords: Cement replacement, Granite slurry waste, Mechanical, Durability, Sustainability.

1 INTRODUCTION

Cement production is highly energy intensive process, which involves intensive fuel consumption for the clinker making and other production process. In cement production green house gases emitted is the root cause of the global warming. Furthermore, cement production requires extraction of natural resources, which results in depletion of these limited natural resources. Hence, there is a vital need of eco-efficient sustainable material, which can overcome the adverse issues related with cement production by reducing its consumption.

The granite industry produces about 15% to 20% of total waste during the cutting process of granite blocks. Granite slurry waste produced in powder form after drying is a fine material, which can be easily airborne and has detrimental effect on fertility of land, open source water, environment and society (Mashaly *et al.* 2018). Further, accumulation of this waste reduces the porosity of land, prevents ground water recharge and also decreases the quality of underground water. Metakaolin is an ultrafine pozzolonic and economical alternative to silica fume. It has positive influence on concrete properties because of filling effect and additional pozzolanic reaction.

Elmoaty (2013) assessed the impact of replacement of granite dust as cement alternative in concrete. The outcome of the result shows that granite dust can be used up to 5% as cement without compromising the compressive strength property of the concrete. Sharma *et al.* (2016) conducted an experimental investigation to determine strength properties of concrete by replacing cement with granite slurry waste. Test results shows increase in compressive and flexural strength of concrete as compared to reference concrete depending upon replacement level and w/c ratio. Dinakar *et al.* (2013) reported that incorporating 10% metakaolin as substitution of cement enhances the mechanical and durability properties of concrete.

It is evident from the available literature review that granite slurry waste and metakaolin modifies the properties of concrete depending upon various parameters. In the present experimental study granite slurry waste and metakaolin are used at w/c ratio 0.40. In this study the replacement level of granite slurry waste is taken up to 10% and metakaolin also up to 10% as partial replacement of cement. Various concrete properties such as workability, compressive strength, flexural strength, water permeability and loss of compressive strength in acid attack have been evaluated.

2 MATERIALS AND PROPORTION

2.1 Materials

In this research work Ordinary Portland cement confirming IS 8112 (2013) was used. The locally available Kharka river sand passing 4.75 mm sieve as per IS 383 (2016) was used and coarse aggregates were procured from the local supplier confirming the gradation as per IS 383 (2016). The granite slurry waste was collected from the dump yard of Udaipur, Rajasthan, India and Metakaolin procured from the local supplier. Physical properties and chemical composition of various materials are shown in Table 1 and Table 2 respectively.

Table 1. Physical properties of materials used in this research work.

Material	Physical Property
Cement	Consistency-32%, Initial setting time-130minutes, Final setting time-213 minutes, specific gravity-3.12
Fine aggregate	Specific gravity-2.71
Coarse aggregate	Specific gravity-2.80
Granite slurry waste	Form-fine powder, colour-red, water absorption-7.6%, consistency-37%, specific gravity-2.17
Metakaolin	Form-fine powder, colour-off white, consistency-45%, specific gravity- 2.42

Table 2. Chemical composition of materials (%).

Material	SiO_2	Fe_2O_3	Al_2O_3	TiO_2	CaO	MgO	LOI	Na_2O	K_2O
Cement	18.70	3.55	8.31	-	61.60	0.80	3.34	-	-
Granite slurry waste	71.22	0.56	12.48	0.16	1.40	0.81	1.20	6.16	4.56
Metakaolin	57.08	1.42	32.72	0.16	0.56	0.20	5.60	1.45	0.37

2.2 Mix Proportions

The concrete mix design of M25 grade was made as per IS 10262 (2009) and IS 456 (2000). Concrete specimens for w/c ratio 0.40 with granite slurry waste and metakaolin as a partial replacement of cement were cast and cured for 28 days standard curing. Mix proportions of various ingredients used in the study are presented in Table 3.

Table 3. Mix proportion of concrete containing granite slurry waste and metakaolin.

Sample	Cement kg/m^3	% of cement replacement by GSW	GSW as partial replacement of cement Kg/m^3	% of cement replacement by MK	MK as partial replacement of cement Kg/m^3	FA Kg/m^3	CA Kg/m^3
CC	372.0	0	0.0	0	0.0	586	1279
G5M0	353.4	5	18.6	0	0.0	586	1279
G10M0	334.4	10	37.2	0	0.0	586	1279
G0M5	353.4	0	0.0	5	18.6	586	1279
G5M5	334.4	5	18.6	5	18.6	586	1279
G10M5	316.2	10	37.2	5	18.6	586	1279
G0M10	334.4	0	0.0	10	37.2	586	1279
G5M10	316.2	5	18.6	10	18.6	586	1279
G10M10	297.6	10	37.2	10	37.2	586	1279

GSW: Granite slurry waste, MK: Metakaolin, FA: Fine aggregates, CA: Coarse aggregates

3 RESULTS AND DISCUSSION

3.1 Workability

Slump test is used to determine the workablity of concrete and has been performed as per Indian standard IS 1199 (1959). Variation of slump of control concrete and other concrete mixes has been presented in Figure 1. It can be seen from the figure 1 that slump value decreases with an increase in replacement level of cement by granite slurry waste and metakaolin. Slump value of control concrete mix is highest (60mm) whereas minimum slump value (30mm) is observed at 10% replacement of both granite slurry waste and metakaolin. Both granite slurry waste and metakaolin consist of very fine particle, which increases the surface hydration of the modified concrete produced, leading to the greater water absorption which intern decreases the workability of mix.

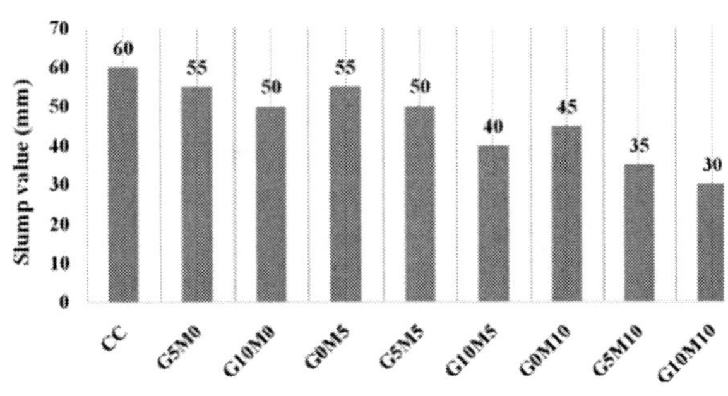

Figure 1. Slump value for granite slurry waste concrete with and without metakaolin.

3.2 Compressive Strength

Compressive strength test has been conducted according to IS 516 (1959) on cube specimen of size 150mm. It can be observed from the Figure 2 that the 28 days compressive strength of concrete increases with increase in the replacement level up to 10% of granite slurry waste without use of metakaolin G10M0 (35.82 N/mm^2). However, inclusion of metakaolin leads to increases the compressive strength of the granite slurry concrete. This behavior is due to the chemical reaction between $Ca(OH)_2$ and metakaolin, which results in the formation of additional CSH gel. It is observed that the maximum compressive strength was found for the mix containing maximum utilization of metakaolin without granite slurry waste (G0M10). It can be also seen that the compressive strength for the mix G10M10 was observed as 34.62 N/mm^2, which was more than the control concrete mix 29.82 N/mm^2. Hence, to accomplish maximum replacement of cement and also at the same time maximum utilization of granite slurry waste, optimum replacement combination is G10M10.

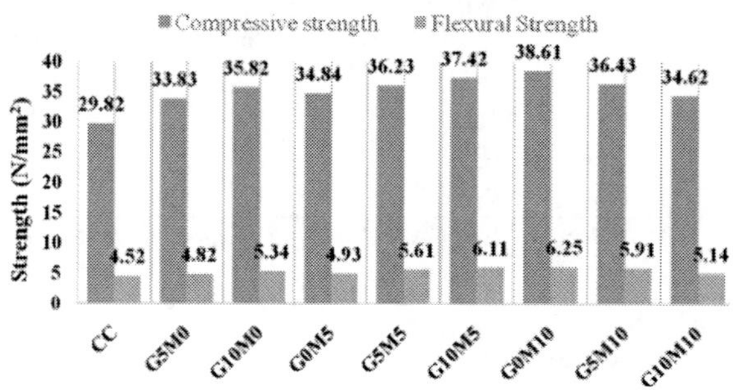

Figure 2. Compressive and Flexural strength for granite slurry waste concrete with and without metakaolin.

3.3 Flexural Strength

The flexural strength of concrete beam samples (100mmx100mmx500mm) was evaluated by two point loading test as per IS 516 (1959). Figure 2 illustrates the results of flexural strength test. It can be observed from the test results that variation of flexural strength of concrete samples was similar to that of compressive strength. In optimum replacement combination G10M10, the value of flexural strength is higher (5.14 N/mm^2) than that of control concrete mix (4.52 N/mm^2).

3.4 Water Permeability Test

The water permeability of cube specimens of size 150mm was determined as per guidelines of DIN 1048 (1981). It can be observed from the Figure 3 that the depth of water penetration increases with increase in replacement of granite slurry waste. However, addition of metakaolin in the granite slurry concrete reduces the water penetration depth. Dinakar *et. al.* (2013) also reported similar behavior of metakaolin. In optimum replacement level combination G10M10, the value of water penetration depth is lower (23 mm) as compared to that of control concrete mix (24 mm).

3.5 Loss of Compressive Strength in Acid Attack

Acid resistance of concrete samples was tested after immersing the samples in 5% H_2SO_4 acid solution for 28 days. It can be observed from the Figure 4 that loss in compressive strength increases with an increase in the percentage of granite slurry waste. Increase in loss of compressive strength was due to increase in the permeability of water through the grains of granite slurry waste. However, inclusion of the metakaolin in the granite slurry concrete further decreases the loss of compressive strength. In optimum replacement level combination G10M10, loss of compressive strength shows lesser value (23.56%) as compared to the value (24.46%) for control concrete mix.

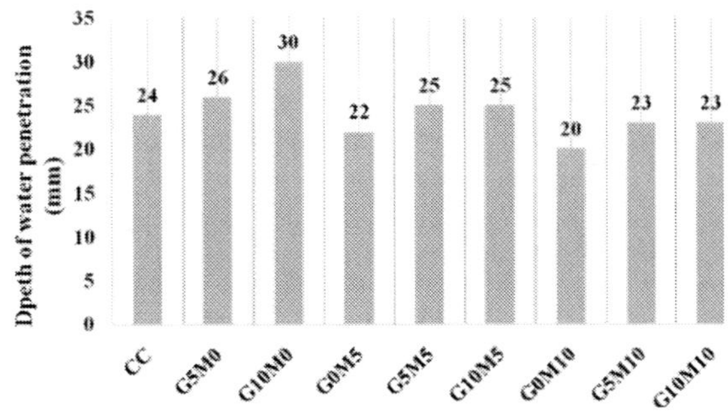

Figure 3. Depth of water penetration for granite slurry waste concrete with and without metakaolin.

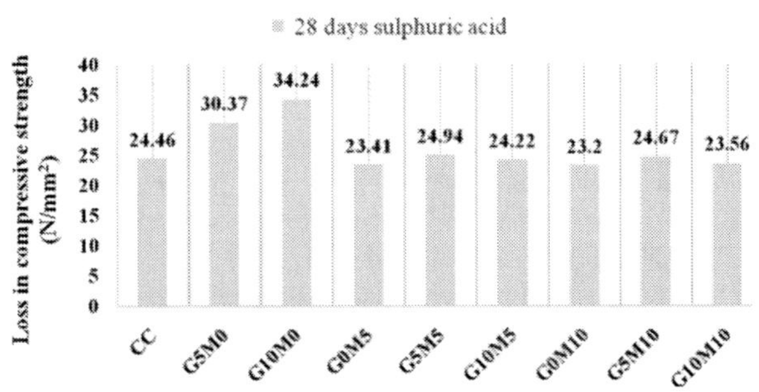

Figure 4. Loss of compressive strength in acid attack for granite slurry waste concrete with and without metakaolin.

4 CONCLUSIONS

Suitability of granite slurry waste and metakaolin for the production of concrete has been evaluated in this research work. Cement is partially replaced by granite slurry waste and metakaolin at varied replacement level ranging (0%, 5%, and 10%). The behavior of granite slurry waste concrete was evaluated first, followed by the addition of metakaolin at varied replacement level in the concrete. The following conclusions can be drawn based on the experimental results:

- The slump value of concrete decreases with the increasing replacement level of granite slurry waste. Further, decrease in the slump value was also observed with increasing percentage of metakaolin.
- The compressive and flexural strength of concrete increases with increasing replacement level of granite slurry waste. Further, addition of metakaolin improves the compressive and flexural strength of the concrete.
- Depth of water penetration increases with the increasing replacement level of granite slurry waste. Conversely, inclusion of metakaolin in granite slurry concrete decreases the depth of water penetration.

- The acid attack test demonstrated that loss in compressive strength increase with the increasing replacement level of granite slurry waste. However, inclusion of metakaolin shows better resistance to acid attack.

Based on the various test results, it was found that the optimum replacement combination level is G10M10. This combination is eco-friendly and sustainable because of the maximum replacement of cement and also maximum utilization of granite slurry waste in the concrete.

The development of modified concrete will lower down the consumption of cement significantly which intern will conserve the natural resources and reduces the emission of green house gases. Furthermore, it will also solve the disposal and accumulation problem associated with the granite slurry waste. Hence, the production of new composite concrete will be sustainable and advantageous to the environment and society.

Acknowledgements

This experimental study was sponsored by Department of Science and Technology, New Delhi under sanctioned no.SB/S3/CEE/042/2013. The authors, therefore, acknowledge with thanks to DST, New Delhi for this support.

References

DIN 1048, *Testing Concrete: Testing of Hardened Concrete Specimens Prepared in Mould, Part 5,* Deutsches Institut fur Normung, Germany, 1981.

Dinakar, P., Sahoo, P. K., and Sriram, G., Effect of Metakaolin Content on the Properties of High Strength Concrete, *International Journal of Concrete Structures and Materials*, 7(3), 215-223, 2013.

Elmoaty, A. E. M. A., Mechanical Properties and Corrosion Resistance of Concrete Modified with Granite Dust, Construction *and Building Materials*, 47, 743-752, 2013.

IS: 1199, *Indian Standard Methods of Sampling and Analysis of Concrete,* Bureau of Indian Standards, New Delhi, India, 1959.

IS: 383, *Indian Standard Coarse and Fine Aggregate for Concrete – Specification,* Bureau of Indian Standards, New Delhi, India, 2016.

IS: 8112, *Indian Standard Ordinary Portland cement 43 Grade-Specification,* Bureau of Indian Standards, New Delhi, India, 2013.

IS: 10262 *Recommended Guidelines for Concrete Mix Design,* Bureau of Indian Standards, New Delhi, India, 2009.

IS: 456, *Indian Standard Plain and Reinforced Concrete-Code of Practice,* Bureau of Indian Standards, New Delhi, India, 2000.

IS: 516, *Indian Standard Methods of Test for Strength of Concrete,* Bureau of Indian Standards, New Delhi, India, 1959.

Mashaly, A. O., Shalaby, B. N., and Rashwan, M. A., Performance of Mortar and Concrete Incorporating Granite Sludge as Cement Replacement, *Construction and Building Materials*, 169, 800-818, 2018.

Sharma, S., Gupta, T., and Sharma, R. K., Assessment of Mechanical Properties of Concrete Containing Granite Slurry Waste, *International Journal of Engineering*, 29(5), 599-605, 2016.

Interdependence between Structural Engineering and Construction Management
Edited by Ozevin, D., Ataei, H., Modares, M., Gurgun, A., Yazdani, S., and Singh, A.

ISBN: 978-0-9960437-6-2

USING ARABIC GUM IN PRODUCTION OF REFRACTORY BONDING MORTAR

SHATHA HASAN[1], OSAMA EEDAN[1], and KATREEN AZWZ[2]

[1]*Dept of Civil Engineering, University of Technology, Baghdad, Iraq*
[2]*Implementation of Transport Projects, Ministry of Transport, Baghdad, Iraq*

Refractory bonding mortar is a specially-engineered blend consists of non-metallic materials that give the mortar unique heat-resistance and thermal characteristics for use in high-temperature applications. The goal of this study was to determine the performance of refractory mortar that produced from local raw materials in this study; Iraqi clays were utilized as raw materials while Arabic gum was used as adhesive materials. Three proportion mixes were applied, (96:4), (93:7), and (90:10) as (grog: kaolin binding material) and the burning temperatures for specimens were 1500°C. In the present research, some physical test such as the percentage of apparent porosity, bulk density, and specific heat are studied. Also, some mechanical properties such as compressive strength, binding strength, and dimensional stability are investigated. Finally, thermal conductivity and thermal shrinkage as thermal properties are also surveyed. All refractory bonding mortar produced in this study which has been used local materials have a good quality and it has all the characteristics of an economical product.

Keywords: Kaolin, Strength, Grog, Thermal properties, Dimensional stability, Iraq.

1 INTRODUCTION

Refractories are made of inorganic, nonmetallic, porous and heterogeneous materials composedof thermally stabilized mineral aggregates, a binder phase, and additives. The principal raw materials used in the production of refractory are normally the oxides of silicon, aluminum, magnesium, and calcium (Agbajelola 2011, Kadhum and Jaffer 2013).

Refractories are ceramic materials designed to withstand very high temperatures (in excess of 540 °C) (Chukwudi 2008).

According to previous researches (Simon *et al.* 2003, Kadhum and Jaffer 2013) refractory materials must be chemically and physically stable at high temperatures. Depending on the effect of the environment, they must be resistant to thermal shock, be chemically inert, and/or have specific ranges of thermal conductivity and of the coefficient of thermal expansion.

2 THE OBJECTIVES OF RESEARCH

This study consists of the following objectives and significance:

A. Producing refractory bonding mortars from local materials such as Iraqi clay and Arabic gum that have properties comfortable to standard specification for refractory mortar which is utilized in fusion kilns building and lining.

B. This study is carried out to investigate some mechanical, physical and thermal properties of local refractory bonding mortar to encourage using Arabic gum as an adhesive material in refractory bonding mortar.

3 EXPERIMENTAL PROGRAMS

3.1 Materials

White kaolin was used as a granular product in this study. This clay was produced by crushing and grinding calcined refractory material (Al_2O_3-SiO_2 composition).

Kaolin clays were also utilized as a binder due to its good properties during molding without appearing of cracks. The white kaolin, which is utilized in this work, conformed to the chemical analysis requirements of (Iraqi Chemical Standard 2004), as shown in Table 1.

Table 1. Chemical analysis of white-kaolin.

Composition	Al_2O_3%	SiO_2 %	MgO%	SO_3%	CaO%	Fe_2O_3%	L.O.I%
White-Kaolin	42.90	42.96	Nil	Nil	0.76	0.32	13.06
Limit of (Iraqi Chemical Standard 2004)	32 (min.)	50 (Max)				1.4 (Max)	

Arabic gum was used as adhesive materials in this study, which were brought from local markets. Arabic gum is an organic material from the Iraqi acacia tree, which has adhesive properties when mixed with water and in high temperature.

Arabic gum is a natural exudate of acacia trees that is widely used in the food industry and has a unique combination of excellent emulsifying properties, low solution viscosity, and high molecular weight.

3.2 Preparation and Testing of Specimens

The procedures of preparation and testing of specimens processes include preparation of grog as refractory aggregates from kaolin clays and preparation of adhesive materials from Arabic gum solution or powder.

The following process results in the production of refractory bonding mortar:

A. Raw materials are crushed by the grinder and the powder was sieved to obtain the required sizes of the mixture. The purpose of the grinding process is to improve the physical and mechanical properties of materials.

B. The white kaolin clay formed in a rectangle mold at 20°C (room temperature) for one day to loss water, and later drying in dryer oven at (110°C) also for one day. The purpose of burnt clay was to reduce shrinkages cracks and deformations throw green stage.

C. The rectangular samples were burned for (2 hrs) at (1400°C) in a special kiln.

D. The burning samples were crushed by disc and finally, sieved by utilizing sieve No. (200) of seventy-five microns according to ASTM – C64 (1972) to make the grog.

E. The Arabic gum was crushed, sieved and dissolved in hot water to produce a solution with 1.5 gm/cm^3 density by using hot water.

F. All materials (Arabic gum with water as an adhesive material, kaolin grog, and kaolin as a binding material) were mixed by the electrical mixer. The details of the mix proportion are shown in Table 2.

G. After mixing, the mixtures were casted in the 2 cm cubic mold, pressing the mixtures to gain high density and lower porosities, placing at 20°C for one day, and then drying the specimens at 110°C for one day.

H. Finally, the specimens were burned at 1500°C in a special kiln for two hours.

Table 2. Mix proportion of all Iraqi refractory bonding mortar samples used in the study.

Mix No	Arabic Gum	Description (grog: binding material)	Mix proportion (grog: binding material)
Mix 1			96:4
Mix 2	(5%) % of water	Kaolin: Kaolin (K: K)	93:7
Mix 3			90:10

4 TEST RESULTS AND DISCUSSION

4.1 Mechanical properties

4.1.1 *Compressive strength (Crushing strength)*

The average value of compressive strength of the hardened samples was specified by using 2cm cubes according to ASTM C64 (1972).

The compressive strength of the Iraqi refractory bonding mortar was determined before and after burning at 1500°C and the results shown in Table 3.

Table 3. The cold and hot compressive strength of Iraqi refractory bonding mortar specimens.

Mix No	Mix proportion (grog: binding)	Mix Description (grog: binding)	Cold compressive strength (MPa) at 20°C before burning	Hot compressive strength (MPa) after burning
Mix1	(96:4)	Kaolin: Kaolin	4.47	85.29
Mix2	(93:7)	(K: K)	4.13	90.38
Mix3	(90:10)		3.16	116.61

Table 3 shows that the cold crushing strength of the mix (3) gives a minimum value of (3.16Mpa); which is within an acceptable range suggested for refractory. However, ASTM C64 (1972) recommended 1.3 Mpa as a minimum value of compressive strength for refractory materials. This shows that kaolin grog and local adhesive material can be used in refractory bonding mortar. It is also noticed that the cold compressive strength of specimens before burning was decreased whenever the kaolin grog content decreased, while the hot compressive strength of specimens after burning up 1500°C was increase when the kaolin grog content decreased.

4.1.2 *Bonding strength test*

This test was measured by using a flexural / tensile testing machine of ELE international company. The preparation of the specimens, included of cutting the fire brick into (40*40*160) mm^3 prism dimension by electrical hacksaw for concrete sawing, then bonding the two parts by refractory mortar products in this study according to ASTM C 198-02 (2002). The specimens were tested before burning and after burning and the results showed in Table 4.

From the Table 4 that mentioned below, it was noticed that the mix (2) gave higher cold bonding strength and followed by mix (1) and mix (3) respectively. While when comparing the values of cold bonding strength with hot bonding strength the mix (1) gave higher hot bonding strength and followed by mix (2) and mix (3).

Table 4. The cold and hot bonding strength of Iraqi refractory bonding mortar specimens.

Mix No	**Mix proportion (grog: binding)**	**Mix Description (grog: binding)**	**Cold bonding strength (MPa) at 20°C before burning**	**Hot bonding strength (MPa) after burning**
Mix1	(96:4)	Kaolin: Kaolin	2.38	12.31
Mix2	(93:7)	(K: K)	3.39	12.13
Mix3	(90:10)		1.19	9.39

4.2 Physical Properties

The physical properties of refractory mortars included bulk density, specific gravity, the percentage of the apparent porosity and water absorption. The results of physical properties of Iraqi refractory mortars are shown in Figure 1. The bulk density of the mixes that burned at 1500°C showed an increment whenever the kaolin (binding material) content increased while specific density, percent of apparent porosity, and percent of water absorption decreased whenever the kaolin (binding material) content increased.

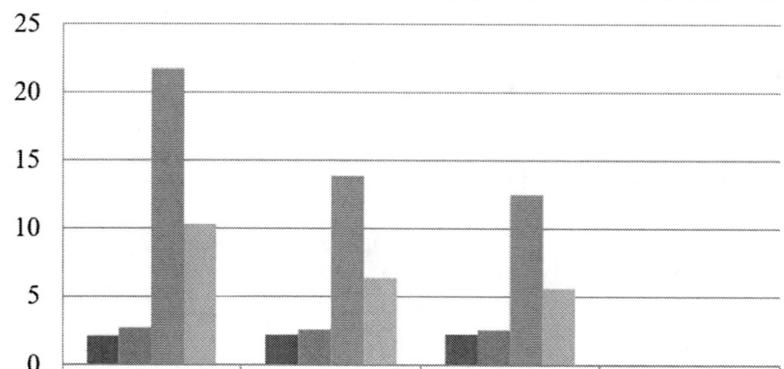

	mix 1	mix 2	mix 3	
■ Bulk density(gm./cm3)	2.1	2.19	2.23	
■ Specific density(gm./cm3)	2.69	2.56	2.55	
■ % apparent porosity	21.73	13.87	12.5	
■ % Water absorption	10.33	6.37	5.6	

Figure 1. Physical properties of Iraqi refractory mortars burning at 1500°C.

4.3 Dimensional Stability Tests

These tests included linear thermal shrinkage and reheating expansion which were measured according to ASTM C 179–04 (2004) by using the electronic digital caliper tool before and after burning the specimens to 1500°C. The linear thermal, shrinkage of the refractory mortar was evaluated for three mixes after burning the specimens to 1500°C.

The results were illustrated in Table 5 to range from (5% to 10 %). It was also noticed from Table 5 that re-burning the specimens at 1500°C was led to expand only mix (2) and mix (3), that's due to the phase of Mullite is seen as a binding phase in most of the refractory brick and it has a high resistance to melting and minimum thermal expansion as well as low thermal conductivity as stated by (Al-Amer and Al-Kadhemy 2015).

Table 5. Linear thermal shrinkage and reheating expansion of specimens after burning to 1500°C.

Mix No	Mix proportion (grog: binding)	Mix Description (grog: binding)	linear Thermal shrinkage%	Reheating expansion %
Mix1	(96:4)	Kaolin: Kaolin	5	0
Mix2	(93:7)	(K: K)	9.5	5.3
Mix3	(90:10)		10	5.9

4.4 Thermal Properties

The thermal conductivity and specific heat of the mixes that burned at 1500°C showed an increment whenever the kaolin (binding material) content increased while thermal diffusivity decreased whenever the kaolin (binding material) content increased as shown in Figure 2. Moreover, these results have a lower than the value was stated (2.1w/m.K) by (Rasin *et.al.* 2009).

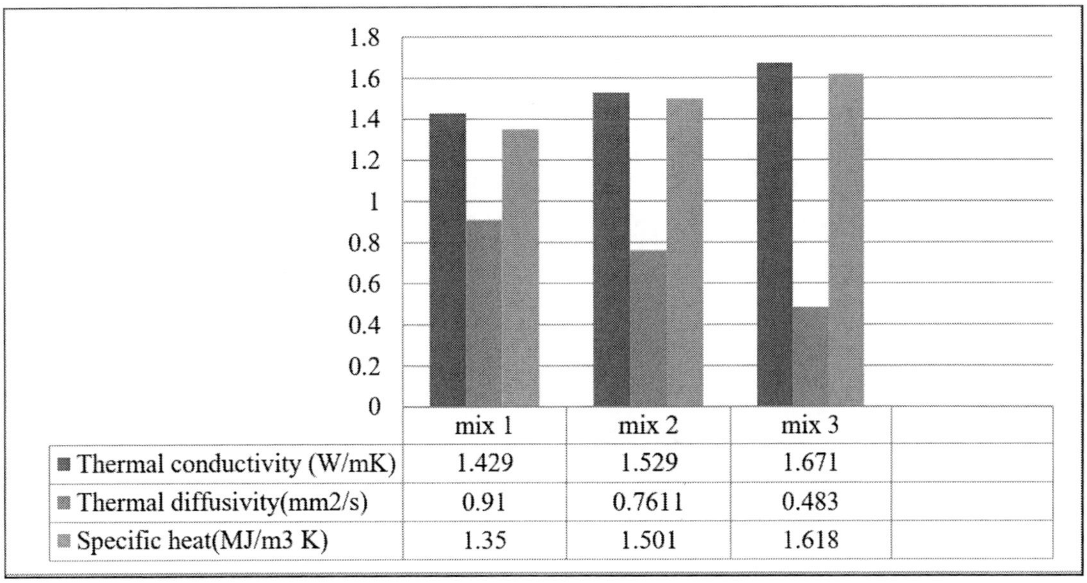

Figure 2. Thermal properties of Iraqi refractory mortars burning at 1500°C.

5 CONCLUSIONS

From the experimental results presented in this research, the following some conclusions are drawn:

A. Results indicated that hot compressive strength increased when the kaolin grog decreased while the cold compressive strength decreased when the kaolin grog decreased for all mixes.
B. Increasing binding material content caused a considerable increase in the bulk density, thermal conductivity, and specific heat while specific density, percent of apparent porosity, thermal diffusivity, and percent of water absorption decreased with the increment of kaolin (binding material) content.
C. Finally, refractory bonding mortar can be produced by using local materials such as Iraqi clay and Arabic gum as adhesive material with good properties.

References

Agbajelola, D., Production of Refractory Binding Material for Cement Kiln, *Journal of Innovative Research in Engineering and Science*, 2(1), 44-49, 2011.

Al-Amer, E., Al-Kadhemy, M., Improving the Physical Properties of Iraqi Bauxite Refractory Bricks, *Journal of Al-Nahrain University*, 18(3), 67-73, 2015.

ASTM C 64, Standard Test Methods for Refractories for Incinerators and Boilers, *American Society for Testing and Material Internationa*l, 1-6, 1972.

ASTM C 198-02, Standard Test Methods for Cold Bonding Strength of Refractory Mortar, *American Society for Testing and Material Internationa*l, 1-4, 2002.

ASTM C 179-04, Standard Test Methods for Drying and Firing Linear Change of Refractory Plastic and Ramming Mix Specimens, *American Society for Testing and Material International*, 1-2, 2004

Chukwudi, B., Characterization and Evaluation of the Refractory Properties of NSU Clay Deposit in Imo State Nigeria, *The Pacific Journal of Science and Technology*, 9(2), 487-494, 2008

Kadhum, M., and Jaffer, H., Study of the Thermal Durability of Refractory Mortar Prepared from Local Clay Mixed with Different Percentage of Silica, *Iraqi Journal of Science*, 54(4), 1096-1101, 2013.

Iraqi Chemical Standard, Iraqi Chemical Specification of Clay, Ministry of Industry and Mineral, *State Company for Mining Industry*, Iraq, 2004.

Rasin, F., Harby, K., Moauyad, S., and Salam, S., Preparation of Refractory from Local Materials, *Ibn Al-Haitham Journal for Pure and Applied Science*, 22(2), 11, 2009.

Simon, F., Adamczyk, B., and Kley, G., Refractory Materials from Waste, *Journal of The Materials Transaction*, 44(7), 1251-1254, 2003.

Interdependence between Structural Engineering and Construction Management
Edited by Ozevin, D., Ataei, H., Modares, M., Gurgun, A., Yazdani, S., and Singh, A.

ISBN: 978-0-9960437-6-2

RHEOLOGY OF SELF-COMPACTING CONCRETE USED FOR SUPERTALL BUILDINGS

MIEN TRAN and YEN CU

Dept of Materials Construction, Ho Chi Minh City University of Technology, Ho Chi Minh City, Vietnam.

Nowadays, self-compacting concrete (SCC) is widely being used for supertall building construction in Vietnam. With high workability, SCC allows us to speed up pumping, placing, and surface finishing, which results in a reduction in building cost. Popular methods such as slump flow, T_{500}, and V funnel test are used to indirectly determine rheological properties of SCC. However, these methods do not present yield stress and plastic viscosity, which are the most parameters affecting the pumping property. In this paper, the SCC rheology is determined directly through yield stress and plastic viscosity with the flow curve test and stress growth test by using the rheometer equipment. Thereby, limitation of the yield stress and the plastic viscosity is recommended for vertically pumping to supertall building project without segregation and bleeding.

Keywords: Yield stress, Plastic viscosity, Workability, Pumping, Rheology.

1 INTRODUCTION

In 2018, Vietnam topped the "The Landmark 81" building with a height of 461.2m, which is the tallest building in Vietnam and the seventh highest in Southeast Asia. This shows that the demand for concrete pumped for high buildings in Vietnam is huge. There have been many types of research on concrete mixes for tall buildings as well as the relation between rheological parameters and the pumping process. Few studies have addressed the problem of pumping for buildings over 300m high. In technical terms, self-compacting concrete (SCC) is suitable for the requirements for concrete pumped for high-rise buildings because minimum size of coarse aggregate is under 12.5mm helping to reduce the pressure when pumping SCC mixture up (Riding *et al.* 2016). In addition, due to the continuous grading of the SCC mixture, the molecular bonding force is large, therefore, under high pumping pressure, the SCC mixture is highly stable compared to the normal concrete mix with the same flow rate (Wallevik 2003). The ACI 237R (2007), the EN 12350 (2010) and the European Guidelines for SCC (2005) have specifications of the test procedure, classification, and evaluation of SCC mixtures. Besides, the rheological property of the SCC mixture is a useful tool for assessing, designing and managing SCCs through static yield stress, dynamic yield stress, and plastic viscosity. Rheological property can be optimized to ensure the workability of the SCC (Koehler and Fowler 2007). The rheological parameters of the SCC mixture, including yield stress and viscosity, can be determined indirectly through field experiments such as slump flow, T_{500} (s) and T_v (s) as shown on Table 1.

Table 1. Methods of indirect evaluation of rheological parameters of the SCC mixture.

Method	Measurement value	Rheological parameters
Slump flow	Measure the slump diameter	Yield stress
T500	Measure the time for the flow to reach the 500mm diameter	Viscosity
V funnel	Time of V-shaped funnel flow	Viscosity

The concrete ingredients affect the rheological parameters of the concrete mix. Specifically, the higher the water content in the mix, the lower the stress yield and viscosity of the concrete mixture. The superplasticizer increases leading to the decrease of yield stress but the viscosity of the concrete mix changed insignificantly (Wallevik 2003).

Rheometer used to measure the rheological property of SCC is useful, thus adjusting the water, air, silica fume content and different dispersants for each SCC. Becomes easier based on the yield stress value and plastic viscosity, the rheological graph divides the areas for use by SCC (Wallevik and Wallevik 2011). Furthermore, the author also gives the lowest slump flow value which concrete mixture can be considered as SCC on Figure 1, for each slump flow range there will be a viscosity range and yield stress range simultaneously. If the viscosity is less than 40 Pa.s, the SCC will have significant yield stress depending on its viscosity. On the other hand, if the SCC is too viscous with plastic viscosity greater than 70 Pa.s, the yield stress is close to zero. The authors also recommended that concrete mixtures with yield stress greater than 40 Pa and viscosity less than 50 Pa.s are completely pumpable, with the corresponding rheometer.

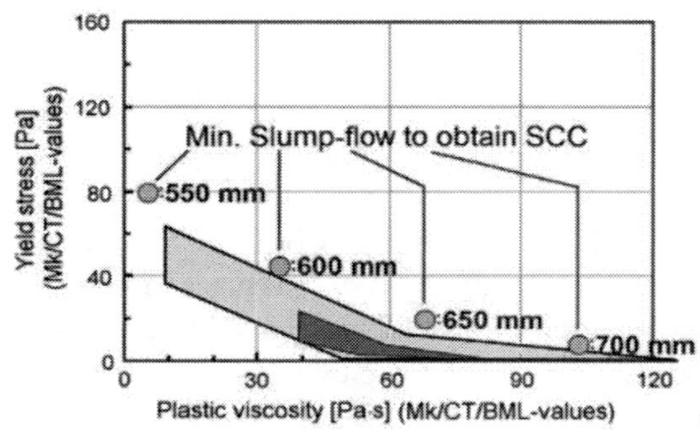

Figure 1. Proposed rheography for SCC (Wallevik 2002).

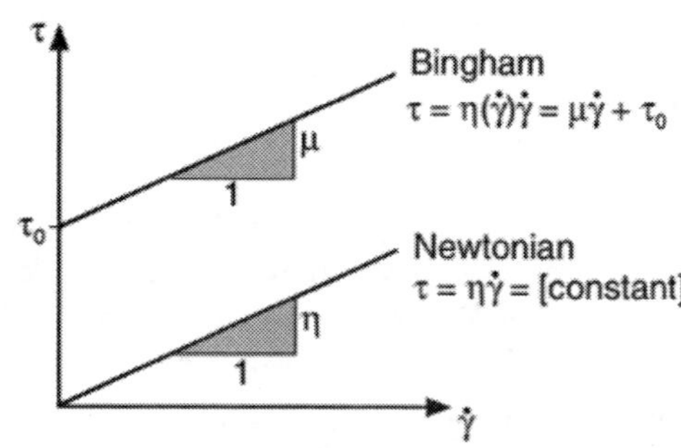

Figure 2. Flow curve for Newtonian and Bingham fluids.

There are many rheological models, but the Bingham model in Figure 2 is the best fit (Riding *et al.* 2016 and Schmidt 2014). Concrete mixtures always have initial tremendous yield stress due to the flocculation of the cement particles. It is necessary to provide the concrete mixture initial shear stress greater than the yield stress to break the flocculation structure so that produce the initial flow in the concrete mix (Ferraris 1999). Therefore, the Newton model is not suitable for describing the rheological behavior of the concrete mix. Bingham suggested the model in Eq. (1) as follows:

$$\tau = \tau_0 + \mu \times \gamma \quad (1)$$

where :

τ = *shear stress (Pa)*; τ_0 = *yield stress (Pa)*; μ = *plastic viscosity (Pa.s)*; γ = *shear rate (1/s)*

In this study, the SCC mixture design with a grade of 60MPa, water to binder ratio (W/B) from 0.34 to 0.37, fly ash varying from 15 to 25% and silica fume from 4 to 8%, is investigated in terms of slump flow, yield stress and viscosity.

2 EXPERIMENTAL PROGRAM

Concrete mixture design with a grade of 60MPa which has total cementitious content of 540kg, including cement, fly ash and silica fume. The coarse aggregate with a maximum size of 12.5 mm is used. Sand is mixed between crushed sand and river sand where river sand occupies 65% by volume. The concrete mixing proportions are shown in Table 2.

The workability of the SCC mixture was measured by slump flow, T_{500} and V funnel flow T_v tests at the mixing time and at every 30 minutes after mixing until the SCC lost the ability to maintain the flow. At these times, the yield stress and viscosity were determined by the flow curve test with four-bladed propeller ICAR (Koehler and Fowler 2005). The results of SCC rheology properties is in terms of slump flow, V flow, yield stress, and viscosity as shown in Table 3, and the Rheometer ICAR is shown in Figure 3.

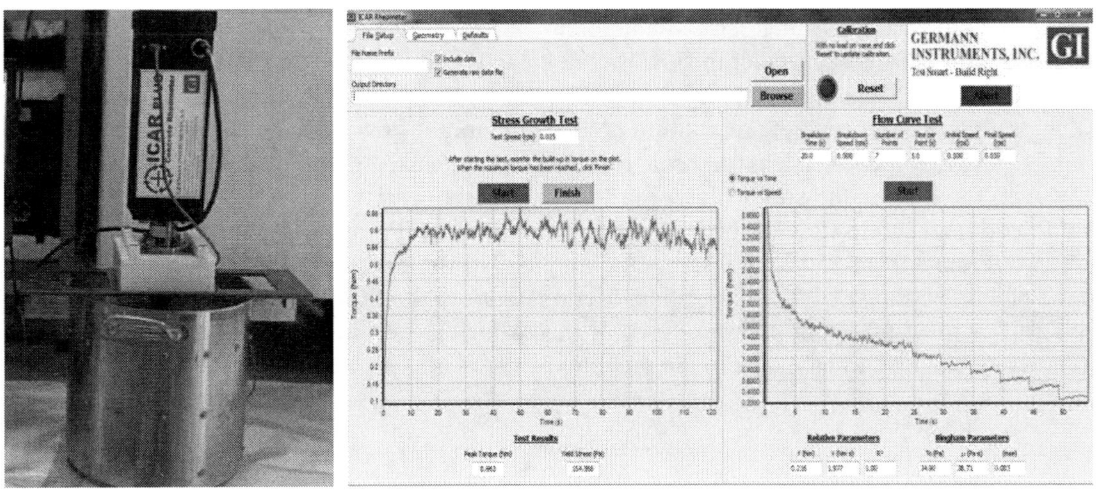

Figure 3. Rheometer ICAR.

Table 2. SCC mix proportions.

Sample	Cement (kg/m^3)	Water (kg/m^3)	Fly ash (kg/m^3)	Silicafume (kg/m^3)	Admixture (l/m^3)	W/B
SCC1	389	195	108	43.2	8.370	0.36
SCC2	400	188	108	32.4	8.370	0.35
SCC3	410	188	108	21.6	7.830	0.35
SCC4	373	186	135	32.4	8.370	0.34
SCC5	362	196	135	43.2	7.830	0.36
SCC6	383	190	135	21.6	7.830	0.35
SCC7	416	198	81	43.2	7.830	0.37
SCC8	427	198	81	32.4	7.830	0.37
SCC9	437	194	81	21.6	7.830	0.36

Table 3. Results of the rheology of SCC.

Sample	Slump flow (mm)	T_{500} (s)	T_v (s)	τ_0 (Pa)	η (Pa.s)
SCC1	760	2.4	4.5	15.21	31.12
SCC2	780	2.2	4.67	17.13	31.39
SCC3	730	1.86	4.7	31.37	32.59
SCC4	740	2.4	5.1	26.9	32.7
SCC5	780	1.88	5.47	18.89	29.1
SCC6	770	2.19	4.72	16.88	33.2
SCC7	750	1.57	3.72	32.07	30.46
SCC8	720	2.51	4.82	27.78	32.59
SCC9	720	2.56	4.78	30.2	35.3

3 TEST RESULTS AND DISCUSSIONS

The relation between the yield stress and slump flow on Figure 4 is inversely proportional, i.e., the yield stress decreases as slump flow increases and vice versa. This relation is appropriate as the SCC mixture has a higher slump flow value, the higher the flowability so that the shear stress value provided to the concrete mixture to produce the initial flow is smaller (yield stress is small). This is perfectly beneficial as the concrete mixture is easily pumped up under low pressure. However, the greater the slump flow, the less segregation resistance of the concrete mixture, so that the optimal slump flow of the SCC mixture pumped up for actual structures is recommended within the range from 650 to 780 mm. As the yield stress increases, the SCC mix hardly moves in the pipeline because of the tremendous friction between the SCC mixture and the pipeline wall. This causes great pumping pressure to compensate for the loss in the pipe. Inadequate pumping up pressure can lead to blockage in the pipeline.

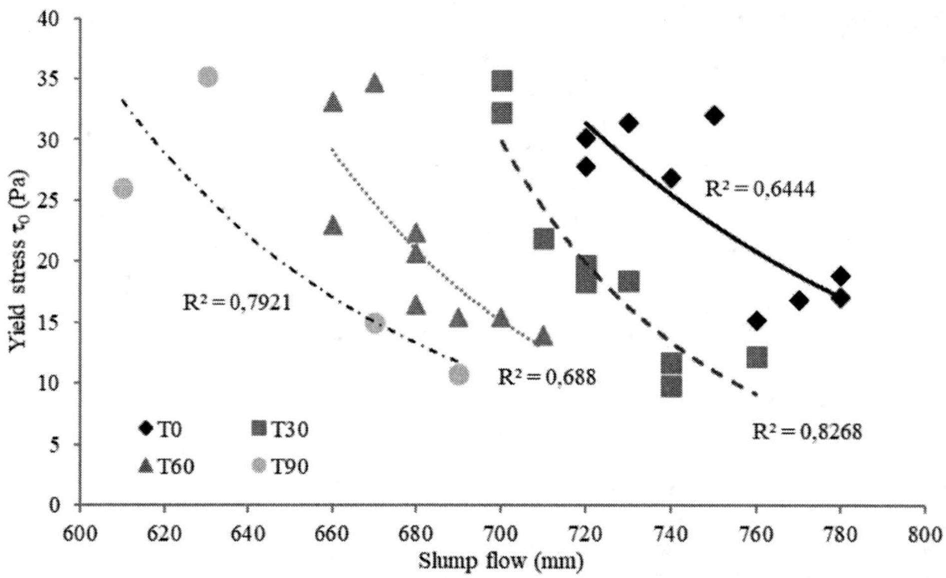

Figure 4. The relation between yield stress τ_0 (Pa) and slump flow (SF) (mm).

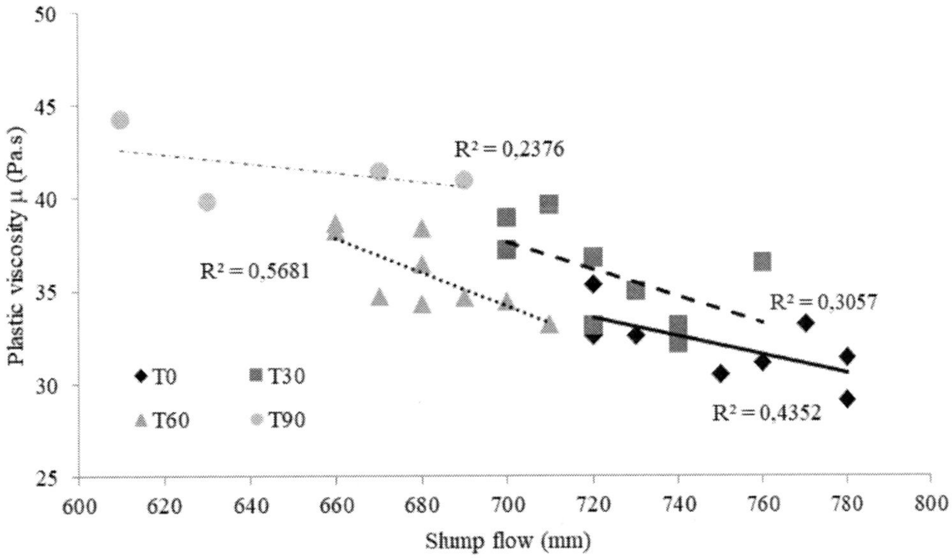

Figure 5. The relation between viscosity η (Pa.s) and slump flow (mm).

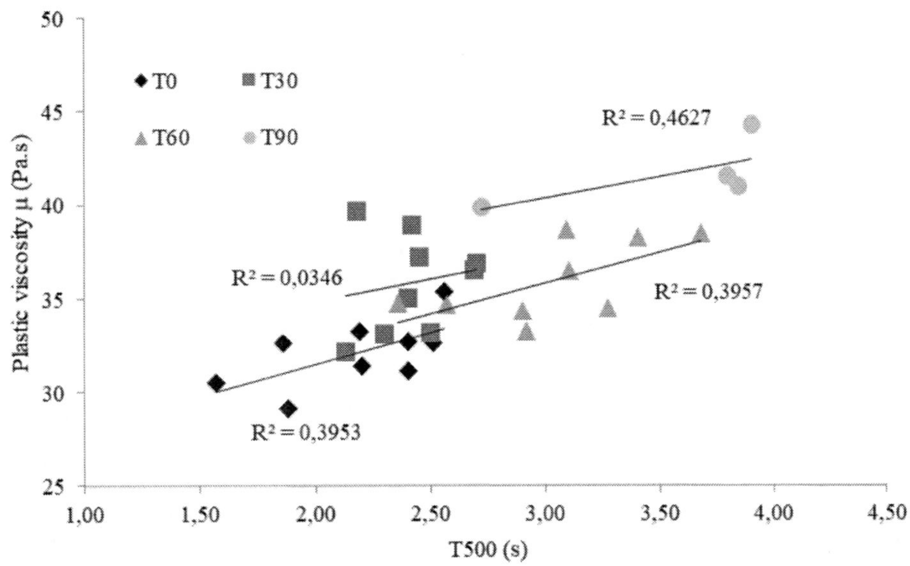

Figure 6. The relation between viscosity η (Pa.s) and T_{500} (s).

The viscosity characterizes the resistance of material flow so that when the SCC has a high viscosity value corresponding to the small SF value on Figure 5. The viscosity depends on the internal friction and molecular force between the particles. As viscosity is greater, the concrete mixture becomes more stable (hard to be segregation, bleeding during the pumping up process). However, too high viscosity will cause the concrete mixture to move hardly in the pipe due to the large friction between the concrete mixture and the pipe, which means that a huge pumping pressure is required.

As stated in Figure 6, the high T_{500} value reducing flowability of concrete mixtures is one of the causes of blockage in the pipeline. This does not mean that low viscosity will benefit the pumping process. With low viscosity, the concrete mixture is susceptible to segregation and bleeding. Therefore, a suitable range of viscosity values at which the SCC mixture can be pumped should be selected.

Workability is also demonstrated by the ability to maintain the SCC, most of the grades are maintained up to 90 minutes, corresponding to the time from mixing to pumping up. The relations between the experimental parameters at the survey times have the same trend, which illustrates the stability, homogeneity of SCC.

4 CONCLUSIONS

The rheological property of the SCC mixture is determined directly through yield stress and plastic viscosity. From the results, conclusions are drawn as follows:

- The relationship between the yield stress and the slump flow is non-linear. The lower the yield stress as the slump flow increases and vice versa.
- The relationship between the viscosity and the slump flow is linear as well. The SCC has a high viscosity value corresponding to the low slump flow value.
- The SCC mixture with W/B from 0.34 to 0.37, fly ash content from 15 to 25% and silica fume content from 4 to 8% has the rheological parameters such as yield stress τ_0 ranging from 20 Pa to 40 Pa and viscosity η ranging from 25 Pa.s to 40 Pa.s respectively. These values are suitable to the SCC to avoid segregation and ensure the stability together with the homogeneity when directly pumped up to the supertall buildings in the vertical direction.

References

ACI 237R-07, *Self-Consolidating Concrete*, American Concrete Institute, 1-30, 2007.

EN 12350-8, Testing Fresh Concrete. Slump-Test, European Norms, 1-9, 2010.

Ferraris, F., Measurement of the Rheological Properties of High-Performance Concrete: State of the Art Report, *Journal of Research of the National Institute of Standards and Technology*, 104 (5), 461-478, 1999.

Koehler, E., and Fowler, D., *A Portable Rheometer for Self-Consolidating Concrete*, SCC 2005 Conference, 2005.

Koehler, E., and Fowler, D., ICAR Mixture Proportioning Procedure for Self-Consolidating Concrete, 2007.

Riding, A., Vosahlik, J., Feys, D., Malone, T., and Lindquist, W., Best Practices for Concrete Pumping, *Report No. K-TRAN*, KSU-14-2, December 2016.

SCC, *The European Guidelines for Self Compacting Concrete*, SCC028, European Project Group, 1-63, May 2005.

Schmidt, W., *Design Concepts for the Robustness Improvement of Self- Compacting Concrete : Effects of Admixtures and Mixture Components on the Rheology and Early Hydration at Varying Temperatures,* PhD Thesis, Technische Universiteit Eindhoven, Eindhoven, January 2014.

Wallevik, O., *Practical Description of Rheology of SCC*, SF Day at the Our World of Concrete, Singapore, 2002.

Wallevik, O., *Rheology – A Scientific Approach to Develop Self-Compacting Concrete,* 3rd International Symposium on SCC, Rilem, Reykjavik, Iceland, August 23–31, 2003.

Wallevik, O., and Wallevik, E., Rheology as a Tool in Concrete Science : The Use of Photographs and Workability Boxes, *Cement and Concrete Research,* 41, 1279–1288, January 2011.

AUTHOR INDEX

AUTHOR INDEX

AUTHOR INDEX

AUTHOR INDEX

AUTHOR INDEX

ISEC PRESS
PO Box 1334
Fargo, North Dakota 58107
USA

CURRAN ASSOCIATES INC.
proceedings.com

ISBN 978-1-5108-8867-8

9 781510 888678

International Structural Engineering and Construction Conference (ISEC-10)

Interdependence of Structural Engineering and Construction Management

Chicago, Illinois, USA
20 - 25 May 2019

Volume 2 of 2

International Structural Engineering and Construction Conference (ISEC-10)

Interdependence of Structural Engineering and Construction Management

Chicago, Illinois, USA
20 - 25 May 2019

Volume 2 of 2

ISBN: 978-1-5108-8867-8

Printed from e-media with permission by:

Curran Associates, Inc.
57 Morehouse Lane
Red Hook, NY 12571

Some format issues inherent in the e-media version may also appear in this print version.

Printed by Curran Associates, Inc. (2019)

For permission requests, please contact The International Structural Engineering and Construction Society (ISEC) at the address below.

ISEC PRESS
PO Box 1334
Fargo, North Dakota 58107
USA

isec.sec@gmail.com

Additional copies of this publication are available from:

Curran Associates, Inc.
57 Morehouse Lane
Red Hook, NY 12571 USA
Phone: 845-758-0400
Fax: 845-758-2633
Email: curran@proceedings.com
Web: www.proceedings.com

TABLE OF CONTENTS

VOLUME 1

CHAPTER 3: INNOVATIVE HOUSING TECHNOLOGIES

CHAPTER 4: MANUFACTURING SYSTEMS FOR CONSTRUCTION

CHAPTER 5: DECISION MAKING AND NEURAL NETWORK MODELS IN CONSTRUCTION

CHAPTER 6: REHABILITATION AND RETROFITTING METHODS – PART 1

CHAPTER 7: REHABILITATION AND RETROFITTING METHODS – PART II

CHAPTER 8: RETROFIT OF COMPOSITE AND METAL STRUCTURES

CHAPTER 9: MODELING IN CONSTRUCTION MANAGEMENT

CHAPTER 10: PROJECT MANAGEMENT

CHAPTER 11: LAW AND DISPUTE RESOLUTION

CHAPTER 12: COST AND DURATION OPTIMIZATION IN PROJECT MANAGEMENT

CHAPTER 13: ALTERNATIVE MATERIALS FOR CONCRETE

VOLUME 2

CHAPTER 14: CONCRETE DESIGN

CHAPTER 15: GEOPOLYMER CONCRETE

CHAPTER 16: DISASTER PREVENTION STRATEGIES

CHAPTER 17: EDUCATIONAL COMPONENT AND PROFESSIONAL ETHICS IN CIVIL ENGINEERING

CHAPTER 18: BRIDGES

CHAPTER 19: REINFORCED CONCRETE STRUCTURES

CHAPTER 20: STEEL STRUCTURES PART I

CHAPTER 21: STEEL STRUCTURES PART II

CHAPTER 22: DYNAMIC RESPONSE OF STRUCTURES

CHAPTER 23: COMPOSITE SYSTEMS

CHAPTER 24: STRUCTURAL MEASUREMENTS AND COMPUTATIONAL METHODS

CHAPTER 25: FOUNDATION AND PAVEMENT DESIGN

Interdependence between Structural Engineering and Construction Management
Edited by Ozevin, D., Ataei, H., Modares, M., Gurgun, A., Yazdani, S., and Singh, A.

ISBN: 978-0-9960437-6-2

STRENGTH, SHRINKAGE, AND PERMEABILITY PERFORMANCE OF SEAWATER CONCRETE

ADEL YOUNIS[1], USAMA EBEAD[1], PRANNOY SURANENI[2], and ANTONIO NANNI[2]

[1]*Dept of Civil and Architectural Engineering, Qatar University, Doha, Qatar*
[2]*Dept of Civil, Architecture and Environmental Engineering, University of Miami, Coral Gables, USA*

Given the increasing global concern of freshwater scarcity, the use of seawater in concrete mixtures appears to be a way forward towards achieving sustainable concrete, especially in the case of non-reinforced concrete applications or with the use of non-corrosive reinforcement. This paper reports on the results of an experimental study to compare the freshwater- and seawater-mixed concretes in terms of their strength, shrinkage and permeability performance. The experimental program included the following: (i) compressive strength test (at 3, 7, 28, and 56-day ages); (ii) concrete shrinkage test (at Days 4, 7, 14, 21, 28, and 56 following mixing); and (iii) permeability tests (rapid chloride permeability and water absorption at Days 28 and 56 following mixing). As for the study results, seawater concrete showed a slightly higher early-age (i.e., till Day 7) strength performance than that of freshwater-mixed counterpart, followed by a strength performance that is 7–10% inferior to the freshwater concrete after 28 days or later. Also, the shrinkage of seawater concrete was slightly higher than that of freshwater concrete, with a difference of 5% reported after 56 days following mixing. Finally, the permeability performance of hardened concrete in seawater and freshwater mixtures was similar.

Keywords: Sustainability, Seawater-mixed concrete, Compressive strength, Shrinkage test, Rapid chloride permeability, Water absorption.

1 INTRODUCTION

Concrete is the most commonly-used construction material worldwide, typically produced by mixing freshwater, cement, aggregates, and admixtures. The environmental impact of concrete production from its raw ingredients is deemed significant (Miller *et al.* 2016). Consequently, there has been a recent growing interest in other sources of raw materials to reduce energy consumption and save natural resources; i.e., to achieve a more ''green'' concrete (Rahal 2007; Saeed *et al.* 2012; Younis *et al.* 2018b). In this context, seawater has increasingly emerged among researchers as an alternative mixing water for concrete, bearing in mind the increasing global concern of freshwater scarcity (Mekonnen and Hoekstra 2016) as well as the development of non-corrosive reinforcement such as fiber-reinforced polymer (FRP) to counter possible corrosion (Aly *et al.* 2006, Benmokrane *et al.* 2006). Therefore, combining seawater and FRP reinforcement in structural concrete appears to be plausible from both technical (Younis *et al.* 2017) and economic (Younis *et al.* 2018a) perspectives.

Significant research efforts were devoted in the past few decades to comprehend the effect of seawater mixing on the resulted concrete, evidently from the recent literature surveys of Nishida *et al.* (2013) and Xiao *et al.* (2017) in this regard. As for the strength performance, most studies suggest that the use of seawater results in a slight negative effect on the long-term concrete strength (i.e., 28 days or later) (Kaushik and Islam 1995; Wegian 2010). In contrast, some studies showed a strength increase not only at early ages but also in the long term (Griffin and Henry 1964, Shi *et al.* 2015); this imposes the need for further research in this regard. Moreover, the majority of previous studies were focused on the basic physical and mechanical characteristics of concrete such as workability, setting time, compressive and tensile strength, etc. Studies on the shrinkage or permeability performance of seawater-mixed concrete are scarce (Modupeola and Olutoge 2014, Park *et al.* 2011) and therefore, these topics remain relatively poorly understood. In view of that, the current paper establishes a comparison between freshwater- and seawater-mixed concretes in terms of their strength, shrinkage and permeability characteristics. At first, materials and test methods adopted in this study will be explained. After that, test results will be presented and discussed, from which the conclusions of this comparison will be drawn.

2 EXPERIMENTAL PROGRAM

2.1 Materials

Two concrete mixtures were considered: (i) the conventional freshwater-mixed concrete which is regarded as a reference; and (ii) seawater-mixed concrete. The difference between the two mixtures was only in the mixing water. Seawater was delivered from Al-Khor coastal area in Qatar to the concrete plant, and then fabric-filtered and used for concrete mixing. Chemical characteristics were obtained for both types of mixing water as shown in Table 1. The pH and alkalinity measurements were within the acceptable limits for both water types. The sulfates, chlorides, and dissolved solids in seawater were (as expected) significantly higher than those of the freshwater or even the allowable limits.

Table 1. Chemical characterization of the two mixing waters.

Measure (unit)	Method	Max. limit	Freshwater results	Seawater results
Chloride (mg/L)	BS 1377-3	1000	14.1	18600
Sulfate (mg/L)	BS 1377-3	2000	20.9	2359
Alkalinity (mg/L)	BS 6068-2.51	500	69.5	149
Total dissolved solids (mg/L)	BS 1377-2	2000	62.0	30300
pH at 25 °C (-)	BS 6068-2.5	6.5–9.0	8.1	8.2

Ready-mix concrete with a 28-day 60-MPa design compressive strength was considered. The mixture proportions per cubic meter of concrete as per BS EN 206 (2013) were 1144 kg of gravel, 756 kg of sand, 158 kg of ordinary Portland cement, 292 kg of blast furnace slag, and 164 kg of water; for which the water-to-cementitious material (w/cm) ratio was 0.34. Blast furnace slag was considered at 65% replacement level as a supplementary cementitious material, being generally observed to enhance the performance of seawater concrete (Nishida *et al.* 2013, Xiao *et al.* 2017). A commercial superplasticizer was added at a dosage of 3.8 kg/m^3 to both concrete mixtures.

2.2 Test Methods

The compressive strength test was performed on concrete cylinders (of 150 mm in diameter and 300 mm in height) in accordance with ASTM C39/C39M (2016). Samples were tested to investigate the effect of two variables: (a) mixing water (seawater/freshwater); (b) testing age (3, 7, 28, and 56 days). Three specimens were considered for each test point.

Concrete shrinkage was measured for both concrete mixtures as per ASTM C157/C157M (2014). For each mixture, three concrete prisms ($100 \times 100 \times 500\ mm$) were cast and then, after one day, immersed in water for 30 minutes. The initial length measurements were taken, and then specimens were kept in a room with air-drying conditions (i.e., temperature of $25 \pm 1°C$ and humidity less than or equal 50%). The concrete shrinkage measurements were obtained at days 4, 7, 14, 21, 28, and 56 following mixing. For each measurement, the length difference (ΔL) was determined between a reference bar and the concrete prism. The concrete shrinkage at time t, S_t, can be calculated as the change in ΔL_t with respect to that initially measured (ΔL_0) divided by the gauge length ($G = 25.4\ mm$) as shown in Eq. (1):

$$S_t\ (\%) = \frac{\Delta L_t - \Delta L_0}{G} x100 \quad (1)$$

Rapid chloride permeability (RCP) test was performed according to ASTM C1202-71a (2017) as shown in Figure 1. In this test, the concrete specimen (100-mm diameter and 50-mm deep cylinder) was kept under a potential difference of 60 V for six hours between its two ends; the first end was immersed in an NaCl solution and the second in an NaOH solution. During the immersion period, the total amount of electrical current passing through the specimen was measured in Coulombs, which indicates the resistance of the specimen to the chloride penetration and thus the concrete quality.

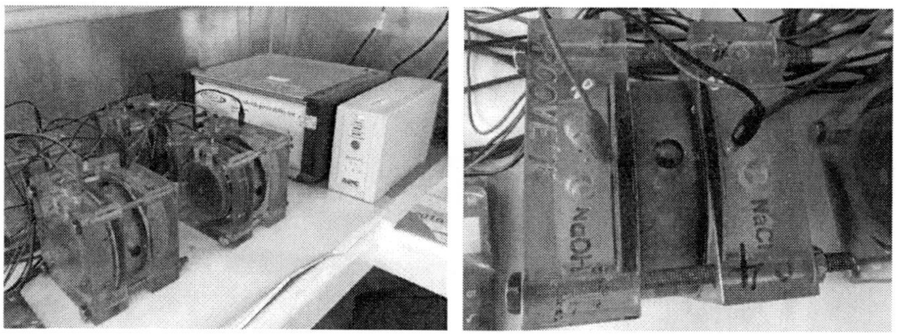

Figure 1. Rapid chloride permeability test.

Water absorption (WA) test was performed according to BS 1881-122 (2011). Three cylindrical cores (75 mm in diameter and 47 mm in depth) were obtained from the top, middle, and bottom of a 150 mm concrete cube. After that, the samples were dried by a 110 °C oven for three days and then cooled for one day using a dry airtight vessel. The initial weight (W_1) of each sample was first measured. After that the sample was immersed in water for 30 minutes, then surface-dried and weighed again (W_2). The water absorption can be calculated in Eq. (2):

$$WA\ (\%) = \frac{W_2 - W_1}{W_1} \times 100 \quad (2)$$

3 RESULTS AND DISCUSSION

3.1 Strength

Figures 2 depicts the compressive strength test results. As shown in the figure, the use of seawater led to a slight increase (around 5%) in the concrete compressive strength at the early stage (up to Day 7). However, after 28 days or longer following mixing, the compressive strength of seawater concrete was 7-10% lower than that of freshwater concrete. These results are consistent with previous research (Kaushik and Islam 1995, Nishida *et al.* 2013, Wegian 2010, Xiao *et al.* 2017). The higher early strength of seawater concrete can be attributed to the lower porosity as a result of hydration acceleration, while the lower long-term strength could be attributed to the leaching of hydration products as suggested by Kaushik and Islam (1995).

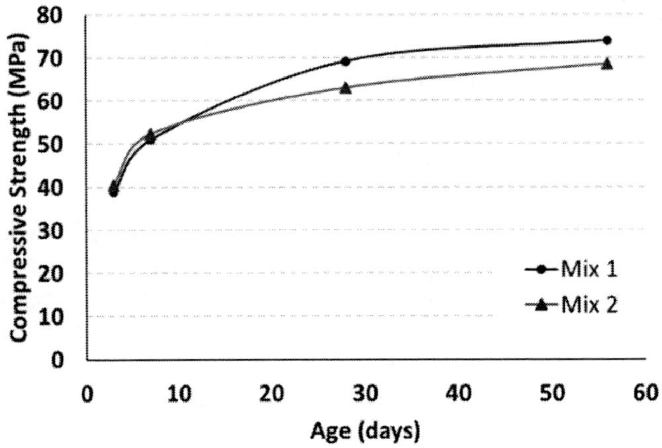

Figure 2. Compressive strength test results (standard deviations are 0.40 MPa for 3-day, 0.95 MPa for 7-day, 1.17 MPa for 28-day, and 1.47 MPa for 56-day measures on average).

3.2 Shrinkage

Figure 3 shows the concrete shrinkage (%) as a function of time for both concrete mixtures. Each test point is the average of three measurements.

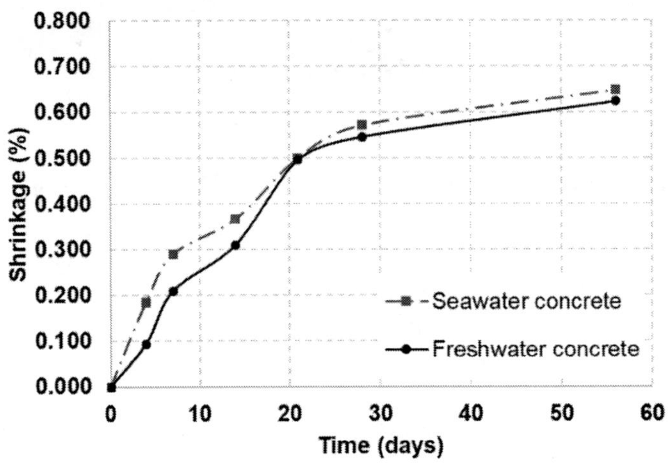

Figure 3. Shrinkage test results (standard deviations are 0.038% for 3-day, 0.076% for 7-day, 0.065% for 14-day, 0.076% for 21-day, 0.096% for 28-day, and 0.092% for 56-day measures on average).

As shown in the figure, the concrete shrinkage exhibited a rapid increase until Day 21 following mixing, after which both curves showed a gentle increase. At Day 56, seawater-mixed concrete showed a slightly higher drying shrinkage (within 5%) than that of the conventional mix. The observed influence of seawater on the drying shrinkage can be related to the presence of chloride in the pore solution, or to the formation of a finer pore structure in seawater concrete (Park *et al.* 2011, Shi *et al.* 2015).

3.3 Permeability

Table 2 presents the results of rapid chloride permeability (RCP) and water absorption (WA) tests. Each concrete mixture was tested at two different periods after mixing, namely, 28 and 56 days. Each test result is the average of two samples. It can be realized that the measurements of RCP and WA at Day 56 are lower than those of Day 28. This can be attributed to the enhancement of concrete due to the ongoing hydration. Besides, the RCA and WA measurements obtained for both mixtures are comparable, which indicates that using seawater had almost no effect on the permeability performance of hardened concrete.

Table 2. Permeability performance test results.

Specimen	**RCP**		**WA (%)**
	Charge passed (coulombs)	Permeability class (as per the standard)	
Freshwater concrete (Day 28)	407	Very low	1.79
Freshwater concrete (Day 56)	369	Very low	1.58
Seawater concrete (Day 28)	439	Very low	1.69
Seawater concrete (Day 56)	349	Very low	1.56

4 CONCLUSIONS

This paper aimed at establishing a comparison between seawater and freshwater concretes in terms of strength, shrinkage, and permeability characteristics. Based on the study results, the following conclusions have been drawn:

- Using seawater resulted in an early slight increase in the compressive strength of hardened concrete (till Day 7), followed by a decrease of around 7–10% as compared to that of freshwater concrete after 28 days or later.
- The shrinkage of hardened concrete in case of seawater mixing was slightly higher than that of the conventional freshwater concrete (about 5% difference as of Day 56 following mixing was observed).
- Using seawater in concrete mixtures showed almost no effect on the permeability and/or resistance to chloride ingression of hardened concrete. Results of RCP and WA tests were almost the same for the two mixtures under comparison.

Acknowledgments

This paper was made possible by the NPRP grant # NPRP 9-110-2-052 from the Qatar National Research Fund (a member of Qatar Foundation). The findings achieved herein solely the responsibility of the authors. Special thanks are due to Readymix Qatar (as a part of LafargeHolcim) for providing the expertise that greatly assists the authors in their research.

References

Aly, R., Benmokrane, B., and Ebead, U. A., Tensile Lap Splicing of Bundled CFRP Reinforcing Bars in Concrete, *Journal of Composites for Construction*, 10(4), 287–294, 2006.

ASTM C1202-17a, *Standard Test Method for Electrical Indication of Concrete's Ability to Resist Chloride Ion Penetration*, ASTM International, West Conshohocken, PA, USA, 2017.

ASTM C157/C157M-08, *Standard Test Method for Length Change of Hardened Hydraulic-Cement*

Mortar and Concrete, ASTM International, West Conshohocken, PA, USA, 2014.
ASTM C39/C39M-16b, *Standard Test Method for Compressive Strength of Cylindrical Concrete Specimens*. ASTM International, West Conshohocken, PA, USA, 2016.
Benmokrane, B., Aly, R., and Ebead, U. A., Tensile Lap Splicing of FRP Reinforcing Bars in Concrete, *ACI Structural Journal*, ACI, 103(6), 857–864, 2006.
BS 1377-2, *Methods of Test for Soils for Civil Engineering Purposes. Classification Tests*. BSI, 1990.
BS 1377-3, *Methods of Test for Soils for Civil Engineering Purposes. Chemical and Electro-Chemical Tests*, BSI, 1990.
BS 1881-122, *Testing Concrete. Method for Determination of Water Absorption*, BSI, 2011.
BS 6068-2.50, *Water Quality. Physical, Chemical and Biochemical Methods. Determination of pH*, BSI, 1995.
BS 6068-2.51, *Water Quality. Determination of Alkalinity. Determination of Total and Composite Alkalinity*, BSI, 1996.
BS EN 206, *Concrete Specification, Performance, Production and Conformity*, BSI, 2013.
Griffin, D. F., and Henry, R. L., *The Effect of Salt in Concrete on Compressive Strength, Water Vapor Transmission, and Corrosion of Reinforcing Steel*, Port Hueneme, California, 1964.
Kaushik, S. K., and Islam, S., Suitability of Sea Water for Mixing Structural Concrete Exposed to a Marine Environment, *Cement and Concrete Composites*, 17(3), 177–185, 1995.
Mekonnen, M. M., and Hoekstra, A. Y., Four Billion People Facing Severe Water Scarcity, *Science Advances*, 2(2), e1500323, 2016.
Miller, S. A., Horvath, A., and Monteiro, P. J. M., Readily Implementable Techniques Can Cut Annual CO_2 Emissions from The Production of Concrete by Over 20%, *Environmental Research Letters*, IOP Publishing, 11(7), 74029, 2016.
Modupeola, A. G., and Olutoge, F. A., The effect of seawater on shrinkage properties of concrete, *International Journal of Research Engineering and Technology*, 2(10), 1–12, 2014.
Nishida, T., Otsuki, N., Ohara, H., Garba-Say, Z. M., and Nagata, T., Some Considerations for Applicability of Seawater as Mixing Water in Concrete, *Journal of Materials in Civil Engineering*, American Society of Civil Engineers, 27(7), B4014004, 2013.
Park, S. S., Kwon, S. J., and Song, H. W., Analysis Technique for Restrained Shrinkage of Concrete Containing Chlorides, *Materials and Structures*, 44(2), 475–486, 2011.
Rahal, K., Mechanical Properties of Concrete with Recycled Coarse Aggregate, *Building and Environment*, 42(1), 407–415, 2007.
Saeed, H. A., Tagnit-Hamou, A., Ebead, U. A., and Neale, K. W., Stoichiometric Study of Activated Glass Powder Hydration, *Advances in Cement Research*, 24(2), 91–101, 2012.
Shi, Z., Shui, Z., Li, Q., and Geng, H., Combined Effect of Metakaolin and Sea Water on Performance and Microstructures of Concrete, *Construction and Building Materials*, 74, 57–64., 2015.
Wegian, F. M., Effect of Seawater for Mixing and Curing on Structural Concrete, *The IES Journal Part A: Civil and Structural Engineering*, Taylor and Francis, 3(4), 235–243, 2010.
Xiao, J., Qiang, C., Nanni, A., and Zhang, K., Use of Sea-Sand and Seawater in Concrete Construction: Current Status and Future Opportunities, *Construction and Building Materials*, Elsevier Ltd, 155, 1101–1111, 2017.
Younis, A., Ebead, U. A., and Nanni, A., *A Perspective on Seawater/FRP Reinforcement in Concrete Structures*, Proceedings of the Ninth International Structural Engineering and Construction Conference, Resilient Structures and Sustainable Construction, ISEC Press, Valencia, Spain, St-38, 2017.
Younis, A., Ebead, U., and Judd, S., Life Cycle Cost Analysis of Structural Concrete Using Seawater, Recycled Concrete Aggregate, and GFRP Reinforcement, *Construction and Building Materials*, 175, 152–160, 2018a.
Younis, A., Ebead, U., Suraneni, P., and Nanni, A., Fresh and Hardened Properties of Seawater-Mixed Concrete, *Construction and Building Materials*, 190(C), 276–286, 2018b.

Interdependence between Structural Engineering and Construction Management
Edited by Ozevin, D., Ataei, H., Modares, M., Gurgun, A., Yazdani, S., and Singh, A.

ISBN: 978-0-9960437-6-2

PERFORMANCE EVALUATION OF VARIOUS CONCRETE MIXES SUBJECTED TO SULFURIC ACID

MUHAMMAD KALIMUR RAHMAN[1], HASSAN ALI ALKHALIFAH[2], ALI HUSAIN AL-GADHIB[3], SALAH OTHMAN ALDULAIJAN[3], and MOHAMMED SALIHU BARRY[1]

[1]*Research Institute, King Fahd University of Petroleum and Minerals, Dhahran, Saudi Arabia*
[2] *Consulting Services Dept, Saudi Aramco, Dhahran, Saudi Arabia*
[3]*Civil Engineering Dept, King Fahd University of Petroleum and Minerals, Dhahran, Saudi Arabia*

Concrete is an essential structural material that has long been used for the construction of buildings, bridges, tanks, pavements and variety of the other types of structures. Due to its physical properties and economical use, concrete is used extensively in the country's infrastructure. In the Oil and Gas sector, concrete infrastructures are more challenging structures, which are exposed to a highly aggressive environment. A special type of reinforced concrete structure exposed to different forms of sulfur attack is the sulfur storage structure, typically referred to as "Sulfur Pit". Sulfur Pit is an essential part of oil and gas processing facilities, where the sulfur after extraction from the hydrocarbons in Sulfur Recovery Units is stored and maintained in the liquid phase at temperatures ranging from 130 °C to 160 °C. The gas sweeting process results in the formation of acid gas consisting of H_2S, water vapor in addition to residual sulfuric acid. Reinforced concrete exposed to this environment is subject to deterioration and corrosion of the reinforcing steel. This paper presents the experimental investigation of four different concrete mixes exposed to 5% sulfuric acid at ambient temperature. These mixes include normal OPC, sulfate resistant (Type V) cement, and two mixes with supplementary cementitious materials blast furnace slag (GGBFS) and Class-F fly ash. The investigation is focused on mechanical properties and mass loss of the concrete samples.

Keywords: Sour crudes, Sulfur pit, Corrosion, Service life, GGBFS, Fly ash.

1 INTRODUCTION

Concrete is an essential structural material that has long been used for the construction of buildings, bridges, foundations, tanks, pavements, tunnels, dams and variety of the other types of structures. Concrete is subject to various forms of deterioration mechanisms, which affects its ability for the intended use, and limit its service life. In general, concrete has a low resistance to chemical attacks. The most common forms of chemical attack on concrete are the corrosion of steel reinforcement arising from chloride attack and due to carbonation, sulfate attack, alkali-aggregate reactions, and acid attacks.

Sulfate attacks on the concrete manifest in terms of extensive cracks-based deterioration mechanisms in concrete structures. Sulfates present in soils, underground water, sewage, and seawater ingress through the concrete, react with its components and form expansive compounds.

The reaction leads to an increase in the volume of the products resulting in the disintegration of the concrete matrix. In industrial applications and sewage transport and processing concrete structures, another mechanism of concrete deterioration involving sulfate ions is the sulfuric acid attack. The sulfuric acid has a two-prong attack, with the sulfate ions forming expansive products and the acid reducing the concrete alkalinity, thereby, destroying the concrete passivity which protects steel from corrosion.

A special type of reinforced concrete structure exposed to a different form of sulfate attack is the sulfur storage structure, referred to as "Sulfur Pit", which is an essential part of oil and gas processing facilities, where the sulfur extracted from the hydrocarbons in Sulfur Recovery Units is stored and maintained in the liquid phase at temperatures ranging from 130°C to 160°C.

The reinforced concrete sulfur pits are exposed to a very corrosive environment and subject to frequent deterioration in a short span of time. Corrosion of reinforcing steel and sulfate attack are prominent forms of deterioration leading to delamination spalling of concrete cover in the walls and the roof of the sulfur pit. Heavy deterioration is more prominent in the soffit of the roof slab and the upper part of the walls (vapor zone). The extracted molten sulfur contains some moisture in addition to the ingress of moisture from external sources as well as from the steam coil used to heat the sulfur to maintain the liquid phase. The sulfuric acid and fumes attack the reinforcing steel causing corrosion (Rahman *et al.* 2016).

Corrosion of concrete due to sulfuric acid can generally be characterized by the following reactions, (Salek *et al.* 2016, Vincke *et al.* 2002):

1. Gypsum Formation $Ca(OH)_2 + H_2SO_4 \rightarrow CaSO_4.2H_2O$
2. Decalcification $C_3S_2H_3 + H_2SO_4 \rightarrow CaSO_4.2H_2O + C_2S_2H_2$
3. Ettringite Formation $3CaO.Al_2O_3.12H_2O + 3(CaSO_4.2H_2O) + 14H_2O \rightarrow 3CaO.Al_2O_3.3CaSO_4.32H_2O$

For improving the chemical resistance of concrete to sulfuric acid, many researchers have studied the effect of cement type, cement content, water-to-cementitious materials ratio (w/cm) (Ehrich *et al.* 1999, Fattuhi and Hughes 1988), supplementary cementitious materials (SCMs) (Torii and Kawamura 1994, Roy *et. al* 2001), and polymeric materials (Monteny *et. al* 2003, Vincke *et. al* 2002) on improving the resistance of mortar or concrete to sulfuric acid attack.

In a series of chemical tests with different sulfuric acid concentrations of 1–3%, Fattuhi and Hughes (1988) showed that sulfate resistant portland cement (SRPC) did not offer marked improvement compared to that of ordinary portland cement (OPC) in reducing the mass loss of mortar or concrete specimens.

There is a lack of a consensus concerning the effectiveness of substituting cements with supplementary cementitious materials (SCMs) and nanoparticles in resisting attack caused by acidic media. For instance, Durning and Hicks (1991) and Mehta (1985) reported that the incorporation of silica fume increased the resistance of concrete to 1% sulfuric acid attack due to reduced calcium hydroxide content and lower permeability. Conversely, Monteny *et al.* (2003) reported a negative effect of silica fume incorporation in concrete specimens exposed to 0.5% sulfuric acid. Monteny *et al.* (2003) reported that the highest resistance to a 0.5% sulfuric acid solution was achieved by a binary binder mixture comprising more than 60% ground granulated blast furnace slag. Chang *et al.* (2005) reported that binary binder concrete mixtures prepared with 60% slag and ternary binder mixtures with 56% slag and 7% silica fume had inferior performance compared to that of a 100% OPC mixture when immersed in a 1% sulfuric acid solution with a Ph of 1.27.

Bassuoni and Nehdi (2007) in their study explained the beneficial impacts of SCMs in resisting acid attacks and attributed the better performance of SCMs to the reduction of calcium hydroxides and increase in C-S-H. They explained further that the secondary C-S-H has lower

C/S ratio, less reactive to acid attack and forms a protective layer by reducing the diffusion of acids. Girardi and Di Maggio (2011) concluded that in general, all the beneficial effects of SCMs in resisting aggressive acidic attack are supposed to be diminished with an increase in the severity of the chemical attack.

2 EXPERIMENTAL INVESTIGATION

In this experimental program, four concrete mixes were investigated subjected to sulfuric acid attack. These include concrete mixes with (1) ordinary portland cement (M1), (2) Sulfate resistant portland cement (M2), (3) Glass Granulated Blast Furnace Slag (GGBFS) (M3), and (4) Fly Ash and Silica Fume (M4). Table 1 illustrate the mix design proportions. To evaluate the performance of various concrete mixes to sulfuric acid, samples representing 4 different mixes were placed in 5% by volume sulfuric acid for 12 weeks. Samples were cast and cured for 28 days and tested for initial mechanical properties including compressive and tensile strength, mass, absorption, density, air voids. Exposed as well as cured samples retrieved at 2, 4, 8 and 12 weeks were tested for compressive and tensile strength and mass was measured.

The concentration of the sulfuric acid was monitored through pH measurements as well as through titration with a base (NaOH). The titration process is based on the chemical reaction between an acid (H2So4) and a base (NaOH) in the presence of a colour indicator (phenolphthalein). The base is added to the acid until colour change to pink indicating a neutralized solution. Acid was added regularly to keep the sulfuric acid concertation constant at 5%. The solution-to-specimen volume ratio was kept constant at 2.5.

Table 1. Mix proportions (kg/m^3).

Mix No.	Cement	GGBSF	FA	SF	W/C ratio	Sand	Aggregate	Superplasticiser (L)
M1	370				0.4	776	1121	3.7
M2	370 Type V				0.4	776	1121	3.7
M3	165	340		45	0.3	696	1006	5.5
M4	335		160	37	0.3	629	1119	6

3 RESULTS AND ANALYSIS

3.1 Density, Absorption and Voids

Density, water absorption and volume of permeable voids were measured according to ASTM C-642 (2013). Oven-dry mass was determined after drying specimens at a temperature of 100°C to 110°C for 24 hrs. Saturated mass was determined after immersing the specimen in water at approximately 21°C for 48 h. Saturated mass after boiling was then determined after placing the specimen in water and boil for 5 h. Apparent mass of concrete specimens was measured for samples immersed in water. Table 2 shows the computed values for the four mixes.

3.2 Mass Reduction

Mass was measured for samples before and after exposure to sulfuric acid. Mass loss as a percentage of original mass at different exposure time for specimens from all mixtures exposed to the 5% sulfuric acid solutions are calculated and presented in Table 3. A sample representing the different mixes after exposure for 90 days are shown in Figure 1. After 90 days of exposure, the mass loss was 21% for OPC and SRPC mixes, 22% for the GGBFS and SF mix and 28% for Fly

Ash and SF mix. During the first two weeks of exposure, a notable mass loss of about 6% was observed for all the specimens, except for the GGBFS mix, which showed an increase in mass by 3%. The rate of mass loss mass loss was slower over the period between 2-4 weeks due to neutralization of solution and the increase of the pH in the solution up to 2.0 (Acid is added after titration every 4 weeks). The rate of mass loss continued from the fourth week up to the end of the exposure.

Table 2. Calculated absorption, density, and voids.

Mix	Absorption after immersion, %	Absorption after Immersion and Boiling	Bulk Density, Dry kg/m^3	Bulk Density After Immersion kg/m^3	Bulk Density after immersion and boiling	Apparent density kg/m^3	Volume of permeable pore space (voids), %
M1	4.25%	4.50%	2.292	2.390	2.396	2.556	6.51%
M2	3.81%	4.55%	2.338	2.427	2.444	2.616	7.23%
M3	2.12%	2.37%	2.369	2.419	2.425	2.510	3.61%
M4	3.09%	3.27%	2.341	2.413	2.418	2.535	4.81%

Table 3. Mass reduction after 2, 4, 8 and 12 weeks of exposure.

Mix No.	% mass reduction 2W	% mass reduction 4W	% mass reduction 8W	% mass reduction 12W
M1	6.47%	6.17%	12%	21%
M2	5.80%	5.88%	12%	21%
M3	-3.14%	3.28%	11%	22%
M4	6.35%	6.46%	15%	28%

Figure 1. Samples after exposure to sulfuric acid (From left M1, M2, M3 and M4).

3.3 Compressive Strength

Compressive strength tests were conducted in accordance with ASTM C39 (2018). Average of 3 cylinders 3" x 6" size were considered. Samples were capped with sulfur to ensure a smooth surface of the acid deteriorated surfaces. Samples tested for 7 days, 28 days before starting the exposure to sulfuric acid. Exposed samples were retrieved and tested for periods of 2, 4, 8 and 12 weeks. Each retrieved sample includes both water cured and acid exposed in order to make a comparison. Compressive strength results of the water cured and exposed samples are shown in Figure 2.

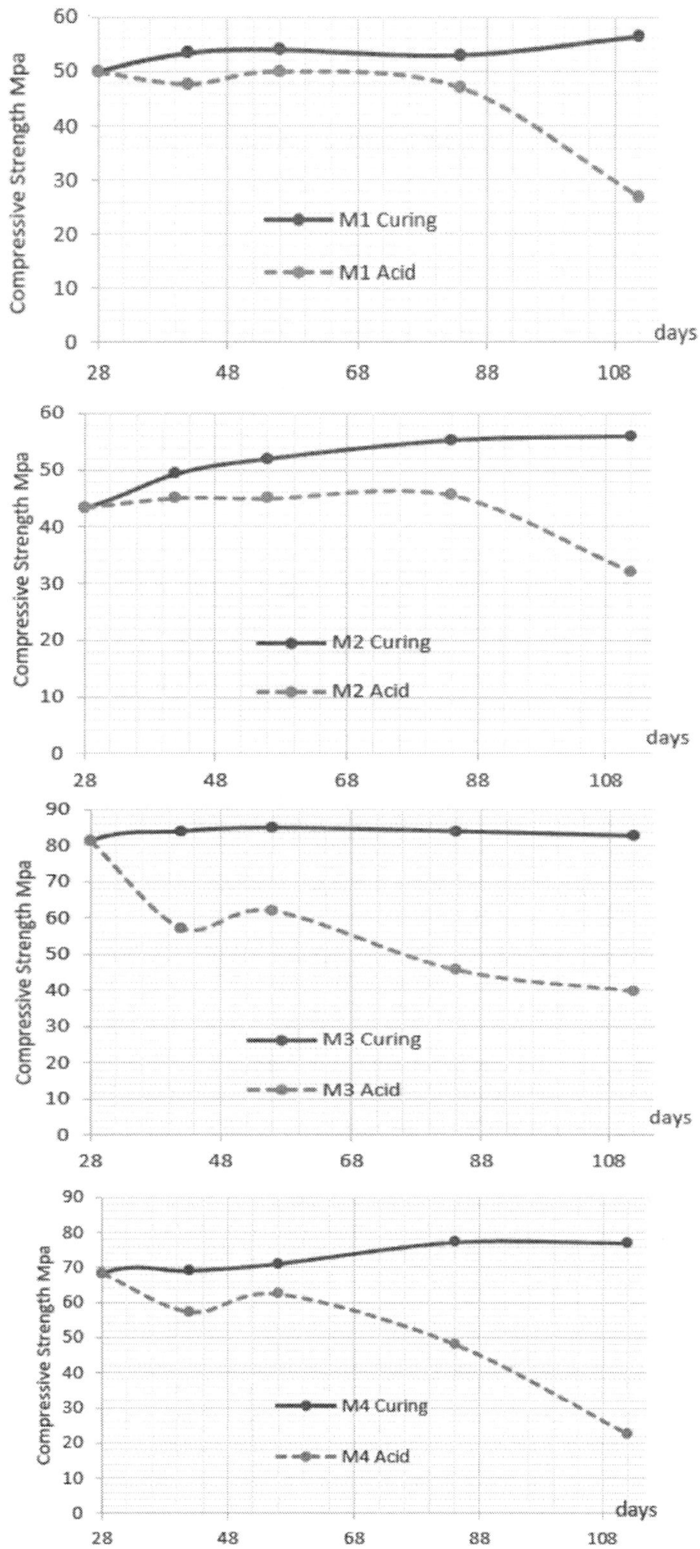

Figure 2. Compressive strength of water cured and acid exposed samples for M1, M2, M3, and M4.

4 CONCLUSION

Sulfuric acid reacts with calcium-bearing phases (portlandite and CSH in absence of portlandite) in hydrated cement paste to form expansive gypsum, resulting in gradual disintegration of paste matrix, consequential loosening of aggregate, and precipitation of outer paste.

After 90 days of exposure to sulfuric acid, the compressive strength OPC mix decreased by 51% as compared to 39% for SRPC. The SCM mixes on the other hand show similar or higher loss of strength compared to OPC. The strength of GGBFS+SF mix decreased by 52%. For FA+SF mix the reduction was 67%. Prominent reduction in compressive strength started after 8 and 12 weeks of exposure. The mass loss for OPC and SRPC mixes and the GGBFS+SF mix were both at about 21-22%, whereas, the FA+SF mix lost about 28%, reflecting in lower strength.

The mass stability and compressive strength of concrete mixes subjected to sulfuric acid attack depend on the chemical composition of the cementitious materials as well as the pH of the acid under consideration. In this research addition of SCMs didn't help improve the performance of the concrete as has been reported by many researchers. The SRPC mix did offer marked improvement compared to that of OPC mix in reducing the mass loss of concrete specimens.

References

ASTM C39 / C39M-18, Standard Test Method for Compressive Strength of Cylindrical Concrete Specimens, ASTM International, West Conshohocken, PA, 2018.

ASTM C-642, Standard Test Method for Density, Absorption, and Voids in Hardened Concrete, ASTM International, West Conshohocken, PA, 2013.

Bassuoni M. T., and Nehdi, M. L., Resistance of Self-Consolidating Concrete to Sulfuric Acid Attack with Consecutive Ph Reduction, *Cement and Concrete Research*, 2007.

Chang, Z., Song, X., Munn, R., and Marosszeky, M., Using Limestone Aggregates and Different Cements for Enhancing Resistance of Concrete to Sulphuric Acid Attack, *Cement and Concrete Research*, 35(8), 1486–1494, 2005.

Durning, T., and Hicks, M., Using Microsilica to Increase Concrete's Resistance to Aggressive Chemicals, *Concrete International,* 13(3), 42–48, 1991.

Ehrich, S., Helard, L., Letourneux, R., Willocq, J., and Bock, E., Biogenic and Chemical Sulfuric Acid Corrosion of Mortars, *Journal of Materials in Civil Engineering*, 11(4), 340–344, 1999.

Fattuhi, N., and Hughes, B., Ordinary Portland Cement Mixes with Selected Admixtures Subjected to Sulfuric Acid Attack, *Materials Journal*, 85(6), 512–518, 1988.

Girardi, F., and Di Maggio, R., Resistance of Concrete Mixtures to Cyclic Sulfuric Acid Exposure and Mixed Sulfates: Effect of The Type of Aggregate, *Cement and Concrete Composites*, 2011.

Mehta, P. K., Studies on Chemical Resistance of Low Water/Cement Ratio Cements, *Cement and Concrete Research*, 15(6), 969–978, 1985.

Monteny, J., De Belie, N., and Taerwe, L., Resistance of Different Types of Concrete Mixtures to Sulfuric Acid, *Materials and Structures,* 36(258), 242–249, 2003.

Rahman, M. K., Khalifah, H. A., Ammar, M. K., and Abu-Aisheh, E., *Condition Assessment and Finite Element Modelling of a Sulfur Pit Structure in The Sulfur Recovery Unit of a Gas Plant,* 4th SCMT Conference, 2016.

Roy, D., Arjunan, P., and Silsbee, M., Effect of Silica Fume, Metakaolin, And Low-Calcium Fly Ash on Chemical Resistance of Concrete, *Cement and Concrete Research*, 31(12), 1809–1813, 2001.

Salek, S., Samali, B., Murphy, T., Wuhrer, R., and Adam, G., *Comparative Study Between Microstructure of a Novel Durable Concrete and Normal Concrete Subjected to Harsh Environments,* 9th International Conference on Fracture Mechanics of Concrete and Concrete Structures, 2016.

Torii, K., and Kawamura, M., Effects of Fly Ash and Silica Fume on The Resistance of Mortar to Sulfuric Acid and Sulfate Attack, *Cement and Concrete Research*, 24(2), 361–370, 1994.

Vincke, E., Wanseele, E., Monteny, J., Beeldens, A., De Belie, N., Taerwe, L., Van Gemert, D., and Verstraete, D., Influence of Polymer Addition on Biogenetic Sulfuric Acid Attack of Concrete, *International Biodeterioration and Biodegradation,* 49(4), 283–292, 2002.

Interdependence between Structural Engineering and Construction Management
Edited by Ozevin, D., Ataei, H., Modares, M., Gurgun, A., Yazdani, S., and Singh, A.

ISBN: 978-0-9960437-6-2

CHALLENGES OF USING NON-DEGRADABLE WASTE MATERIAL POLYSTYRENE PACKING IN REINFORCED CONCRETE DESIGN AND CONSTRUCTION

BEN NGENE, GIDEON BAMIGBOYE, and OSATO ASEMOTA

Dept of Civil Engineering, Covenant University, Ota, Nigeria

Weight and Environmental concerns are two elements that have made the search for alternative materials of construction critical in today's construction industry. The importance of finding a solution to the problem has given rise to the use of non-degradable materials. This study examines the challenges of making such material as polystyrene used in product packaging a part of the structural element in construction. Such adventure it is envisaged reduced the volume and number of landfill sites in Nigeria where air pollution arising from dump sites affects the health of citizens. To achieve this aim, Styrofoam was used as a partial replacement for coarse aggregate (granite) by volume. The replacement was carried out in varying percentages of 10, 20 and 30% Styrofoam concretes respectively with a concrete mix ratio of 1:1:2. The result obtained showed that concrete produced using Styrofoam as the alternative aggregate possess lesser values of compressive strength when compared with the control of conventional concrete without Styrofoam. The optimal percentage of a replacement for Styrofoam concrete was 10% with a compressive strength of 21.33 N/mm^2 while the control concrete had a compressive strength of 33.26 N/mm^2. The loss of strength of Styrofoam concrete is compensated by the reduction of the requirement for landfill sites and the attendant pollution generated while the concrete can be used for non-structural elements in construction.

Keywords: Environment, Non-degradable waste, Air-pollution, Compressive strength, Styrofoam concrete, Conventional concrete.

1 INTRODUCTION

Zhang (2011) defined concrete as a man-made "stone" which is produced by the combination of materials, granular coarse and fine aggregates, cement and water in a standardized ratio, after which the mix is consequently coagulated and hardened. As concrete is a composite material and is made of various components, its properties and characteristics are functions of the aggregated properties of each constituent or component as well as the proportions in which each component is combined (Bamigboye *et al.* 2017). Mamlouk and Zaniewski (2006) described aggregates as a granular material such as sand, gravel, crushed stone, blast-furnace slag that generally occupies roughly 60-75% of the total volume of concrete. The issue of waste management has local, national and international dimensions and arises due to the inefficiencies of municipal solid waste management strategies adopted as stated by (Otitoju 2014). The word "Styrofoam" is a commercial name ascribed to a material known technically as expanded polystyrene or polystyrene foam. Polystyrene is produced from styrene, which can be found in naturally occurring substances or can be synthesized from petroleum by-products such as benzene according to Friend (2005). Polystyrene is a vinyl polymer and can be defined according to its chemical structure, as a long hydrocarbon chain, consisting of a phenyl group attached to every other carbon atom. Expanded polystyrene (EPS) however, can be described as polystyrene in the form of raw beads being steam-heated, causing expansion (Ismail *et al.* 2003). Production and use of Styrofoam (expanded polystyrene) have considerably harmful effects on the environment, according to the Environmental Protection Agency and Styrofoam is the fifth largest source of hazardous waste in the United States (Kuruppu and Chandratilake 2012). This is due to the difficulty experienced in the recycling of Styrofoam as a result of its lightweight and the high economic costs of transporting and degreasing the petroleum-based material. Also, on the weight side of the argument, the construction of high-rise buildings and other structures, the cost of construction of a foundation can usually be attributed to the large amounts of the dead loads incident on it, due to the self-weight of the structural members. It is therefore important to investigate the production of lightweight concrete which may possess the same strengths as structural normal weight concrete and may be produced using alternative materials as aggregates. Conventional aggregates are naturally occurring materials and are obtained from the environment. The process of aggregate extraction is harmful to the environment and may contribute to damaging geological phenomena such as landslides and erosion. According to the Cement and Concrete Institute Australia (2008), there is a critical shortage of natural aggregates for the production of new concrete. On both weight and environmental fronts, the case for alternative lighter and sustainable material cannot be over-emphasized.

The objectives of the study, therefore, is to determine the suitability of Styrofoam as aggregates for the production of structural concrete and also the percentage of coarse aggregate required to be replaced with Styrofoam for optimal concrete strength after 28 days of curing. The scope of the work is limited to comparing the compressive strength of conventional concrete and that produced using Styrofoam as coarse aggregate. The use of alternative aggregates in the production of concrete contributes to sustainable construction, observed by Kuruppu and Chandratilake (2012). The partial or full replacement of natural aggregates mitigates the need for quarrying and consequently, the environmental impact of aggregate extraction is eliminated. Therefore, the use of waste Styrofoam as alternative aggregates for the production of lightweight concrete is a practical means of diverting materials which would have otherwise been disposed of, into useful products.

2 MATERIALS AND METHODS

Ordinary Portland cement with grade 42.5N, which is in line with the requirement of ASTM C150 (2007) was used. The fine aggregates used were natural sharp river sand obtained from river Ogun. The sand is free of silt or any deleterious materials and did not contain particles exceeding 5mm in size as shown in Figure 1a. Graduated granite with a maximum size of 20 mm was then used as a coarse aggregate, the supply was free of dust and other impurities as shown in Figure 1b. The Styrofoam used was procured from packaging for electronics such as air conditioners, laptops, and the likes. It was then carefully shredded into coarse aggregate sizes, to be used as a partial replacement for granite at various percentages as shown in Figure 1c. The water utilized for the concrete mixing have the following properties: It is free from harmful amounts of acids, soils, alkalis or other inorganic or organic impurities; It does not contain iron, vegetable matter or any other type of substances which are likely to have inauspicious effects on concrete or reinforcements; It is potable. The plasticizer used in the experimental procedure is

Conplast SP430. It complies with Type 'F' as a high range water reducing admixture. Sieve analysis was carried out on both the fine and coarse aggregates to determine aggregates grading in line with ASTM C136 (2007). The water used for the experimental procedure was potable and was observed to be free from organic or inorganic impurities, which might likely have harmful effects on the resulting concrete. The mix ratio adopted for the study is 1:1:2. Batching by volume was adopted for this study with 0.45 water-cement ratio. The specimens produced were classified into groups and sub-groups based on the material used for replacement, the percentage of replacement and the curing duration as follows Group 1(GRP1), which serve as control specimens in which conventional aggregates are used for concrete production, Group 2(GPR2), consists of specimens produced with 10% coarse aggregates replacement with Styrofoam by volume, Group3 (GPR3) consists of the specimens produced with 20% coarse aggregate replacement with Styrofoam by volume and Group 4(GPR4) consists of the specimens produced with 30% coarse aggregate replacement with Styrofoam by volume. The fresh concrete tests carried out on the concrete produced were a slump and compaction factor. The water-cement ratio of each mix was constant at 0.45 and a super-plasticizer was utilized in adequate proportions for each group mix. The fresh concrete was filled in 150x150x150mm cubic molds Figure 1d, while the specimens were de-molded after 24hrs and cured in water for 28 days in line with ASTM C31 (2015). The compressive strength of specimen was tested on the bearing surface of the Universal Testing Machine (UTM) in line with ASTM C39 (2003), which has a loading capacity of 100 tones without eccentricity and a uniform rate of loading of 550 kg/cm^2 per minute, which is applied until failure of the cube.

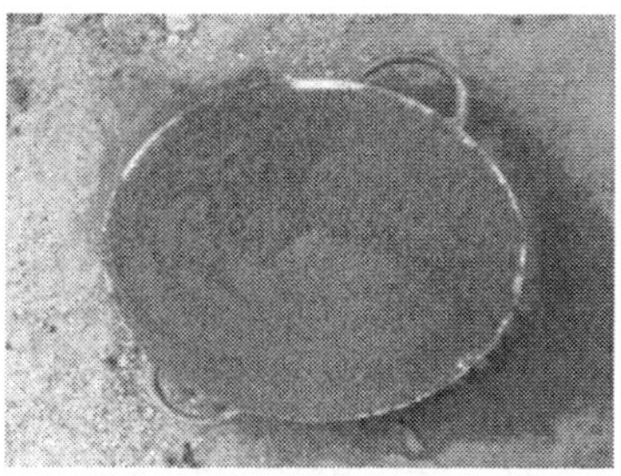

(a) Fine aggregate used.

(b) Coarse aggregate used.

(c) Styrofoam used as a replacement for granite.

(d) The casting of the specimen.

Figure 1. Fine (a) and Coarse (b) aggregates, Styrofoam (c) and Cast Concrete Specimen (d) Samples

3 RESULTS AND DISCUSSION

3.1 Sieve Analysis

Figures 2 and 3 show the sieve analysis results of both the fine and coarse aggregates to ascertain the particle size distribution of aggregates used. From the results obtained the aggregates are well graded.

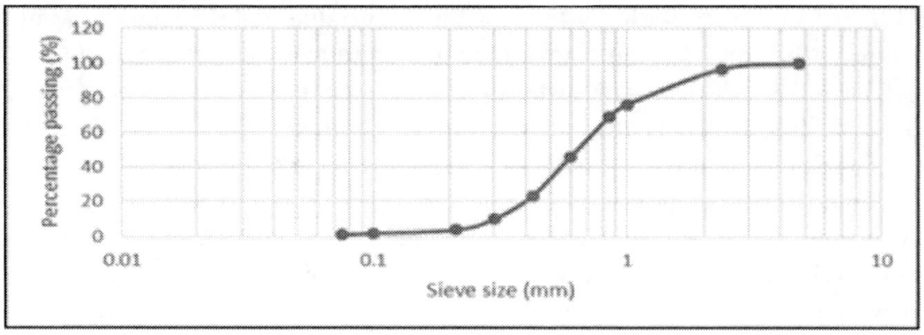

Figure 2. Particle size distribution curve for Fine aggregates.

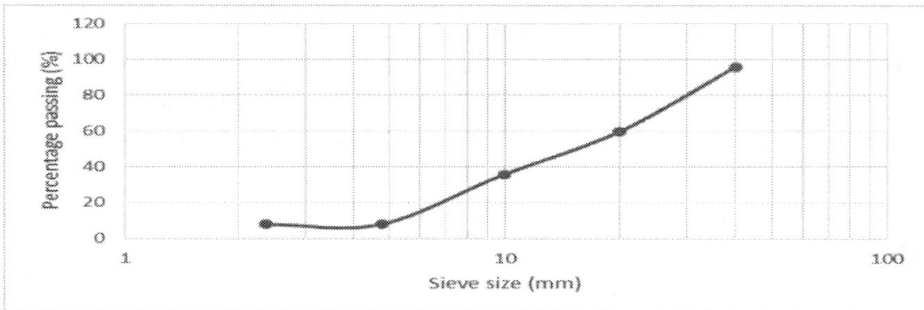

Figure 3. Particle size distribution curve for coarse aggregates.

3.2 Slump Test Result

Slump test was carried out to determine the workability of each specimen group mix as shown in Figure 4. From test results, it was observed that the highest slump value is obtained by the group 4 specimen, which consists of concrete produced with 30% coarse aggregate replacement with Styrofoam by volume. This can be attributed to the hydrophobic nature of the Styrofoam, which would reduce the absorptive properties of the specimens and allow for more water to be available within the mix thereby increasing the slump. The lowest value of slump is attained by the group 1 specimens, which are the control group 1. This is because of the roughness of the granite surface, which increases the aggregate interlocking ability of the mix.

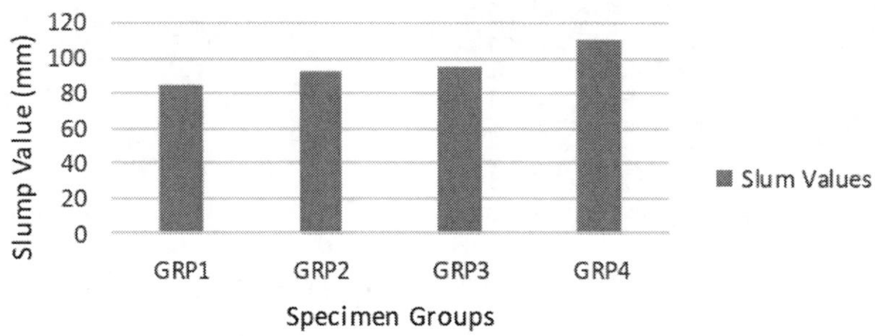

Figure 4. Slump value for each specimen.

3.3 Compressive Strength

Compressive strengths of concrete produced using Styrofoam as partial replacement of conventional aggregates were determined. The specimens produced were classified into four groups and sub-groups based on the material used for replacement, the percentage of replacement and the curing duration. Example of such grouping consists of the control specimens in which the conventional aggregates are used for concrete production. This group is further classified into the following sub-groups: GRP1-7, GRP1-14, and GRP1-28 where the numbers after the hyphen are the duration for which the specimens within each group were cured. The other groups include five for specimens produced with 10% coarse aggregate replacement with Styrofoam by volume; six for specimens produced with 20% coarse aggregate replacement with Styrofoam by volume and seven for specimens produced with 30% coarse aggregate replacement with Styrofoam by volume. It is also important to note that the following conditions were varied in the production of the specimens: curing period, the volume of granite and volume of Styrofoam. On the other hand, the following conditions were kept constant during the experimental procedure: volume of cement, the volume of sand, the volume of water, water-cement ratio, grade of cement and mix ratio.

From Figure 5, it can be observed that the highest values of compressive strengths are obtained by the group 1 specimens. The graphs show a consistent decrease in the compressive strengths of the specimen with an increase in the percentage replacement of the granite. The largest values of compressive strength Styrofoam values were obtained after 28 days of curing. It can be observed from Figure 5, that the values of compressive strength of the Styrofoam specimen are substantially smaller than those of the control specimen with the lowest value occurring at 30% replacement of coarse aggregate. This reduction can be attributed to segregation, which occurs during mixing of the Styrofoam specimens due to its hydrophobic nature. This would lead to a reduction in the bond experienced within the concrete matrix between the constituents and consequently lead to a reduction in strength.

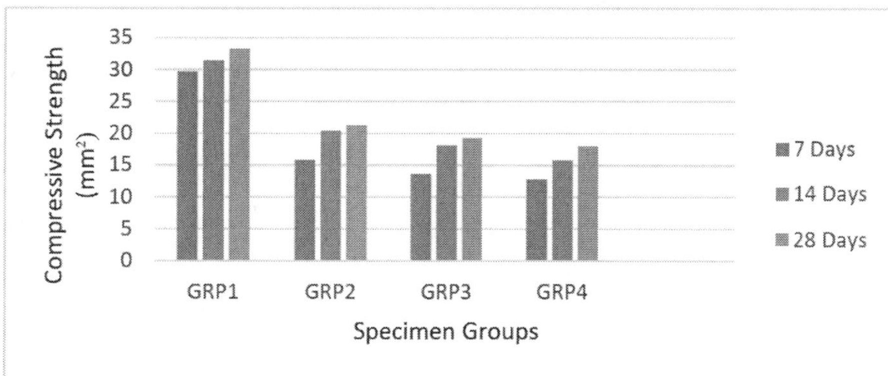

Figure 5. Bar chart showing strength development of Styrofoam specimens for control, 10, 20 and 30% coarse aggregate replacement for varying curing period.

4 CONCLUSION

The results of the experimentation provided significant information for the application of lightweight concrete in construction. The comparison between the compressive strengths of concrete produced using conventional aggregates and those produced using Styrofoam as a partial coarse aggregate replacement showed that the control specimens of conventional concrete were found to have the highest values of compressive strengths for the curing period of 7, 14 and 28 days. The optimal replacement percentage for Styrofoam as a replacement for natural granite in concrete was obtained to be 10% as this gave the highest compressive strength value of 21.33

Mpa after 28 days of curing. The values obtained at 20 and 30% were found to be considerably low and therefore not recommended to be used for structural purposes. To further improve the possibility of field application, the bond between the resulting concrete and conventional steel reinforcement need to be investigated also.

Acknowledgments

The authors wish to thank the chancellor and the management of Covenant University for the platform made available for this research.

References

ASTM C31/C31M, *Practice for Making and Curing Concrete Test Specimens in The Field,* American Society for Testing and Materials, West Conshohocken: ASTM International, 2015.

ASTM C39, Standard Test Methods for Compressive Strength of Concrete Specimens, American Society for Testing and Materials, West Conshohocken: ASTM International, 2003.

ASTM C136/C136M-14, *Standard Test Method for Sieve Analysis of Fine and Coarse Aggregates,* American Society for Testing and Materials, West Conshohocken: ASTM International, 2007.

ASTM C150, *Standard Specification for Portland Cement,* American Society for Testing and Materials. West Conshohocken: ASTM International, 2007.

Bamigboye, G. O., Adedeji, A. A, Olukanni, D. O., and Jolayemi, J. K., Reliability of Gravel in Place of Granite in Concrete Production, *Journal of Engineering and Applied of Sciences*. 12(20), 5121-5128, January 2017.

Cement and Concrete Institute Australia, Use of Recycled aggregates in construction, Australia, 2008.

Friend, D., *Conserving Natural Resources in Illinois,* Land and Water, 2005.

Ismail, I., Saim, A. A., and Saleh, L. A., *Properties of Hardened Concrete Bricks Containing Expanded polystyrene beads*, The 5th Asia-Pacific Structural Engineering and Construction Conference *(ASPEC 2003)*, 171-179, 2003.

Kuruppu, G., and Chandratilake, R., *Use of Recycled Glass as a Coarse Aggregates in Concrete*, Global Challenges in Construction Industry, 221-227, Colombo: World Construction Conference, 2012.

Mamlouk, M., and Zaniewski, *Materials for Civil and Construction Engineers,* Pearson Education Inc., New Jersey, 2006.

Otitoju, T., Individual Attitude toward Recycling of Municipal Solid Waste in Lagos, Nigeria, *American Journal of Engineering Research,* 78-88, March 2014.

Zhang, D. H., *Building Materials in Civil Engineering*. Woodhead Publishing, Cambridge, 2011.

Interdependence between Structural Engineering and Construction Management
Edited by Ozevin, D., Ataei, H., Modares, M., Gurgun, A., Yazdani, S., and Singh, A.

ISBN: 978-0-9960437-6-2

MAGNETIC FIELD TREATED WATER INFLUENCE ON STRUCTURAL CONCRETE ELEMENTS: ENVIRONMENTAL ISSUES OF CONCERN

BEN NGENE, GIDEON BAMIGBOYE, CHIDIEBERE AGOMO, KENECHUKWU ENEH, and OLAYINKA OLAJIDE

Dept of Civil Engineering, Covenant University, Ota, Nigeria

The impact of magnetized water on the strength properties of concrete calls for a holistic appraisal of its effect on the environment of production and use. In this study, the influence of the Magnetic Field Treated Water (MFTW) on the strength of concrete produced under various environments is considered. The work considered three environments of use of magnetized water in concrete with respect to partial replacement of fine aggregate with waste glass; quality of curing medium on mechanical properties of concrete and effect on the mechanical properties of self-compacting concrete. The three scenarios presented involved preparing and testing concrete cubes made with magnetized water, cement, fine (sand and glass) and coarse (granite) aggregates at various levels of combination and environment of curing. The concrete cubes were cured for 7, 14, 28, and 56 days and the strength and other parameters, such as slump and workability, were compared with conventional concrete in normal environment. It was observed that concrete under self-compacting concrete made with MFTW had a 9.1% increase in compressive strength than normal; for concrete made with clay brick powder and MFTW, both the workability and compressive strength was higher than normal. The magnetized water in concrete with waste glass as partial replacement had higher slump value and 24% increase in compressive strength than conventional concrete. MFTW therefore improves strength development of concrete.

Keywords: Waste glass, Curing medium, Compressive strength, Magnetized water, Conventional concrete.

1 INTRODUCTION

Cost considerations, environmental concerns are driving the search for new ways and means of attaining sustainable concrete production and use. The use of magnetic field treated water (MFTW) is one means of utilizing local and accessible material for the development of cost-effective concrete members in the construction of civil works (Matawal and Danladi 2011). In his work, Jamal (2017) described the various types of concrete such as normal, high strength, high performance, air entrained, lightweight, self-compacting and lime Crete concrete depending on the method of production and use. Notwithstanding the type of concrete, water is required in the preparation and curing to attain the necessary or expected strength of the concrete mix. In the case of the production of self-compacting concrete (SCC) for example, the challenge of availability and cost of skilled labor in the construction industry lead to its discovery. This type of concrete is a low flow stress, high deformability, good segregation (prevents separation of particles in the mixture), low viscosity concrete mix, capable of consolidating under own weight. It is widely used in developed countries but in developing countries such as Nigeria, its use especially in constructing pavement is observed to be rare (Busari *et al.* 2017).

Portland cement, fine aggregates such as sharp sand, crushed waste glass and coarse aggregates such as gravel and crushed rocks are usually mixed with (clean) water in predetermined proportions in the preparation of concrete for building elements (Bamigboye *et al.* 2017). The medium that activates the processes necessary for the required bond formation between the cement and constituent aggregates mixture is water, which implies that water is basic in the making of concrete. Research has recently shown that magnetized water can be used in the fields of health, agriculture, construction etc. (Olugbenga 2014). The quality of water employed in concrete production can affect the qualities of such concrete according to Olugbenga (2014). Bellis (2006) reported that several factors influence the performance of concrete, such as water, the binder percentage, the moisture content, the type of aggregate used to make concrete, extra binders and fibres. Production of concrete all through the years especially in developing countries has been traditionally with the utilization of ordinary water. Current effort is to look at alternative source of water such as magnetized water, which is gotten when faucet water is passed through a magnetic field with the purpose of altering its structure or magnetized water (MW) as stated by (Hamza *et al.* 2017). A more specific name however is magnetic field treated water (MFTW) as this helps to clarify that the water in no way has properties associated with magnetism. Magnetic field's magnetic force improves water activity by breaking down the molecules into single or finer form. Diamagnetism of the hydrogen bond results in higher stability under a magnetic field, which is the major factor behind some adjustment in the water properties (Hosoda *et al.* 2004, Alwediyani *et al.* 2015). Concrete can solidify when cast in place and develop high compressive strength, its value depends on its grade, level of workmanship in the casting of concrete process and nature of vibrations which may constitute errors that increase porosity and decrease strength. For a large project requiring great precision, such accumulated errors can create a big problem. This underscores the need for a more flowable concrete mix according to Walraven (2013).

The aim of this study is to assess and compare the development of strength of concrete made with both magnetic field treated water and normal water under varied environmental conditions. The specific objectives being to obtain various mixes using magnetized water and then ordinary water for the same mix design of concrete; to compare the results of the tests for characteristics of the fresh state and compare the strength of the hardened properties of concrete made with normal water and magnetized water with both cured in either of magnetized and normal water and to finally compare the results of the test of the mechanical properties of hardened concrete such as compressive and tensile strength at the ages of 7, 14, 28, and 56 days.

2 MATERIALS AND METHODS

The materials used for this study are ordinary Portland cement (OPC) in line with ASTM C150 (2007). The fine aggregates used were natural sharp river sand obtained from river Ogun, Ogun state, Nigeria, in line with ASTM C33 (2003). The sand is free of silt or any deleterious materials and did not contain particles exceeding 5mm in size. Coarse aggregate with maximum size of 20mm was used in line with ASTMC 136 (2014), the supply was free of dust and other impurities. Figure 1 and 2 shows sample of both coarse and fine aggregates respectively. A brown, liquid, chlorine-free superplasticizer that instantly disperses in water was used to increase the strength, durability and maneuverability of concrete without increasing the demand for water.

Figure 1. Coarse aggregate.

Figure 2. Fine aggregate.

Portable water was used in this study for the production of the concrete control mix. The source of the portal water is municipal water supply from the University laboratory while magnetic water was prepared by passing water from source through a magnetic field. In the process, magnetic water is obtained by connecting a hose from an open tap from which the water flows into the magnetic field produced by the solenoid. The water, which comes out through the other end of the solenoid is then stored in a plastic drum. The magnetic field is generated by passing current of about 32 amps through the field which is made up of a coil (turns) wound round a pipe of 600 mm length and about 100mm diameter. The current is generated from an input of 220 volts which is been step down by a transformer of 150volts. During this time the water passing through the pipe is magnetized. Figure 3 shows the setup of the magnetic device.

Figure 3. Magnetic field simulator.

Tests carried out on the cement include: Initial setting time, final setting time, and consistency test. The grading of aggregates (fine and coarse) into various sizes was determined by sieve analysis. Other tests carried out on the aggregates include; water absorption and specific gravity. The following water hardness tests were carried out on water; calcium, magnesium and calcium carbonate using a spectrophotometer. Other tests include pH of water to show the level of alkalinity or acidity of a sample. For this work, all the itemized tests were carried out to determine the fresh properties of concrete, that is, the slump flow test and the V-funnel test. The

Normal concrete was designed using grade M30 for 42.5N cement grade, that is, 1:1.5:3. Ordinary water and magnetic field treated water was used in mix 1 and mix 2 respectively. The same water/cement ratio of 0.46 at a cement content of 417.15 Kg/m^3 was utilized for all mixes. Compressive strength and tensile strength of concrete after curing for 7 days, 14 days, 28 days, and 56 days were determined with a compression-testing machine. Concrete mix design was established for the various types and mixes using normal and magnetized water.

3 RESULTS AND DISCUSSION

3.1 pH Test

In this test, it was observed that pH value of the MFTW increased from that of ordinary water by 24%, that is, the pH of ordinary water collected for the test was 5.55 while the pH for the water after passing through the magnetic field increased to 6.88. The test was carried to confirm that the ordinary water, which passed through the magnetic field, was actually magnetized.

3.2 Fine and Coarse Aggregates Properties Results

The fineness modulus for fine aggregates shows the value of 2.94, which implies that most of the aggregates are found between 150 μm and 300 μm sieves. The aggregate used is a zone-2 fine aggregate. The value is also between 2.9 and 3.2 and therefore coarse sand as shown in Figure 4. The fineness modulus found for combined or all-in-aggregate samples vary between 3.5 and 6.5 for it to be coarse aggregate. The fineness modulus of the coarse aggregate is determined using the sieve analysis result from the result the fineness modulus value of 6.10 implies that most of the aggregates are found between 10 mm and 20 mm. The maximum size of the coarse aggregates is 20 mm as the fineness modulus value fits between the ranges of 6.1 and 6.9. The values for specific gravity are: sand 2600 kg/m^3, fine crushed glass 2300 kg/m^3, granite coarse 2630 kg/m^3; bulk density is 1680 kg/m^3 and for fineness modulus the result is 6.10.

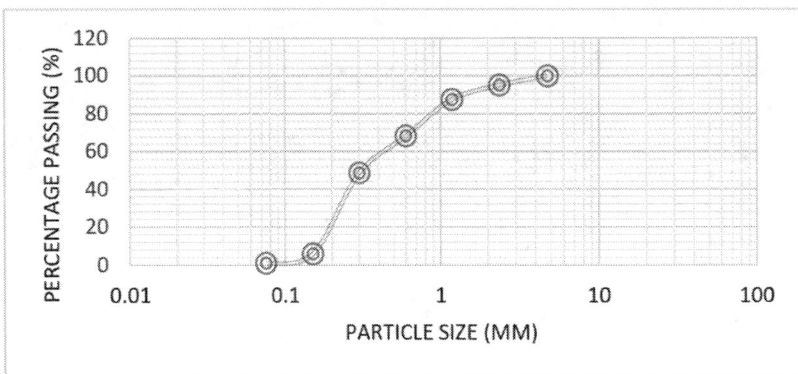

Figure 4. Graph of sieve analysis of fine aggregates.

3.3 Slump Test

The result of the slump test is shown in Table 1 for various types of concrete and environments. From the test result, the general workability of the self-compacting concrete types is higher than those of the normal (consolidated) concrete types. The slump value of MFTW is 29.5mm, which is higher than 18.5mm of NC. Based on the classification of slump given in the standard, both slumps are class S1.

Table 1. Various slump test results (mm).

MIX TYPE	Normal Concrete	Crushed Waste Glass at 15% Replacement	Self-Compacting Concrete	Crushed Brick Dust at 20% Replacement
NC	45	140	18.5	70
MFTW	260	220	29.5	140

NC = Normal Concrete, MFTW = Magnetic Field Treated Water

3.4 Compressive and Split Tensile Strength Tests

The 7, 14, 28, and 56 days compressive and tensile strength of the different concrete mixes were obtained, however only the strength test results for 28 days are shown in Table 2. Generally, the Mix for concrete manufactured and cured in MFTW gave a higher compressive and tensile strength for each curing day than those, which was mixed and cured with ordinary water. The 28 days and 56 days compressive strengths of MFTW are higher than those of Normal concrete by about 10% and 8.32% respectively. For the tensile strength, the increase is 7% of MFTW above that of normal concrete. This implies that at the same cement content and water/cement ratio, magnetized water gives rise to a higher compressive and tensile strength of concrete. The trend of higher compressive and tensile strength values for all curing periods of 7, 14, 28 and 56 days between MFTW and Normal water in concrete for all concrete types, that is, normal concrete, concrete with crushed waste glass at percentages, self-compacting concrete and concrete with brick dust at percentage replacement of fine were observed as indicated in the table.

Table 2. Various compressive and tensile test results (MPa).

MIX TYPE	Normal Concrete		Crushed Waste Glass at 15% Replacement		Self-Compacting Concrete		Crushed Brick Dust at 20% Replacement		Sea Water Cured Environment		HCL Cured Environment	
	C	T	C	T	C	T	C	T	C	T	C	T
NC	32.55	3.27	33.02	3.67	32.55	3.27	25.45	2.94	-	-	-	-
MFTW	36.23	3.50	34.99	3.67	36.23	3.50	30.27	3.09	32.77	2.94	36.51	1.87

NC = Normal Concrete, MFTW = Magnetic Field Treated Water, Compressive Test, Tensile Test

After curing in salt-water, concrete cubes have a darker surface than the reference and magnetized water concrete cubes. When cured in salt water a deposit of salt formed with a whitish appearance all over the cube.

4 CONCLUSIONS

From the above it evident that magnetized water increases the slump of normal concrete by at least 15mm for SCC. Magnetized water reduces the yield stress of SCC and therefore increases the flowability and the rate of flow of SCC. Magnetized water gives rise to a higher compressive strength of normal concrete by about 10% at 28 days. SCC made and cured in MFTW gives a higher strength of about 9.1% than that made and cured in ordinary water. SCC made with ordinary water but cured in MFTW gives little increase in strength when compared to that cured in ordinary water. SCC made and cured in MFTW starts showing significant higher strength (2%) when compared with that made using MFTW but cured in normal water from the 28th day.

SCC made using MFTW gains strength more slowly than its normal concrete counterpart until the 28th day. Study may be carried out by subjecting the concrete made with MFTW to some form of environment conditions such as elevated temperature and presence of pollutants to determine the long-term carrying capacity of elements built with it. The results indicate that the environment of production and use of concrete affects its strength considerably.

Acknowledgments

The authors thank the Chancellor and management of Covenant University for the platform made available for this research.

References

Alwediyani, H., Almasoudi, A., Abdulrahman, A., Kenkarr, N., Alsaidi, S., Khalofa, H., and Bjafar F., The Change in Physical Properties of Magnetic Water, *International Journal for Water Management in India,* 7(9), 45-55, 2015.

American Society for Testing and Materials ASTM C150, Standard Specification for Portland Cement. West Conshohocken: ASTM International, 2007.

American Society for Testing and Materials, ASTM C33, Standard Specification for Concrete Aggregates. West Conshohocken: ASTM International, 2003.

American Society for Testing and Materials, ASTM C136/C136M-14, Standard Test Method for Sieve Analysis of Fine and Coarse Aggregates. West Conshohocken: ASTM International, 2014.

Bamigboye, G. O., Olukanni, D. O., Adedeji, A. A., and Jolayemi J. K, Effect of Granite/Gravel (washed) Combination on Self-Compacting Concrete, *International Journal of Applied Engineering Research*, 12(15), 4799-4805, January 2017.

Bellis, M., The History of Concrete and Cement, The Inventors. Retrieved from http://theinventors.org/, 2006.

Busari, A. A., Akinmusuru, J. O., Dahunsi, B., Ofuyatan, O. M., and Ngene, B. U., Pavement Construction Using Self-Compacting Concrete: Mechanical Properties. *International Journal of Advanced and Applied Sciences,* 4(8), 50-55. 2017.

Hamza, B. A., Anis, E., Nahla, B. S., and Khaled, E., Effect of Magnetic Treatment on Surface Tension and Water Evaporation. *International Journal of Advance Industrial,* 120, 2017.

Hosoda, H., Mori, H., Sogoshi, N., Nagasawa, A., and Nakabayashi, S., Refractive Indices of Water and Aqueous Electrolyte Solutions under High Magnetic Fields, *The Journal of Physical Chemistry A,* 108(9), 1461-1464, 2004.

Jamal, H., Common Types of Concrete and Their Characteristics. Retrieved from https://www.aboutcivil.org/types-of-concrete.html on 2017.

Matawal, D. and Danladi, S., Linking Academia to Industry: A Case Study of the Building and Construction Industry, *Nigerian Building and Road Research Institute, NBRRI.* Retrieved from http://notap.gov.ng on September 20, 2011.

Olugbenga, A., Effects of Different Sources of Water on Concrete Strength: A Case Study of Ile-Ife. *Civil and Environmental Research,* 6(3), 39-43, 2014.

Walraven, J., Self-Compacting Concrete: Challenge for Designer and Researcher. Retrieved from https://www.masterbuilder.co.in/data/edata/Articles/September2013/72.pdf on 2013.

Interdependence between Structural Engineering and Construction Management
Edited by Ozevin, D., Ataei, H., Modares, M., Gurgun, A., Yazdani, S., and Singh, A.

ISBN: 978-0-9960437-6-2

EXPERIMENTAL INVESTIGATION ON SELF CURING RECYCLED AGGREGATE CONCRETE MEMBERS

V. BHIKSHMA and J. RAVI KUMAR

Dept of Civil Engineering, Osmania University, Hyderabad, India

Self-curing concrete is an important aspect in strength development, which is neglected due to various reasons such as shortage of water, inaccessibility of structures during curing. Self-curing agents play important role in heat of hydration and strength development. The present study involves the use of self-curing agent polyethylene glycol (PEG 6000). In this investigation, M20 to M50 grade concrete has been considered to evaluate strength characteristics and flexural behavior using natural and recycled aggregate concrete. The various percentages of PEG 6000 have been used. The test result indicates self-curing concrete has indicated strength enhancement compared to conventional curing concrete. The strength for normal coarse aggregate concrete with a self-curing agent (PEG 6000) at 2% with 10% GGBS is decreased when compared with recycled coarse aggregate concrete with 1% (PEG 6000) with addition 10% GGBS. Mechanical properties are 7% to 20% higher for self-curing concrete when compared to natural concrete. Similarly, the flexural strength of normal coarse aggregate concrete was found to be varying from 4% to 16% more than with recycled coarse aggregate concrete at 28 days. Ultimate deflection of RC beams is 10% greater than the NC beams and ultimate moments of RC beams were 1% to 3% less than the corresponding NC beams. The moment-curvature relationship and load-deflection characteristics of the NC beams and RC beams show a similar trend.

Keywords: PEG 6000, GGBS, Recycled coarse aggregate, Moment-curvature, Deflection.

1 INTRODUCTION

Curing is the process of controlling the rate and extent of moisture transport from concrete during cement hydration. It may be either or after it has been placed in position thereby providing time for the hydration of the cement to occur. Since the hydration of cement does take time in days and even weeks thus curing must be undertaken for a reasonable period of time in order to concrete achieve potential strength and durability. Curing may also encompass the control of temperature since this affects the rate at which cement hydrates. The curing period may depend on the properties concrete, the purpose for which it is used and ambient conditions, i.e. the temperature and relative humidity of the surrounding atmosphere. Curing is designed primarily to keep the concrete moist, by preventing the loss of internal moisture during the period in which it gains strength. Curing of concrete shall be maintained satisfactorily during its early stages in order to develop the desired properties.

2 NEED FOR THE PRESENT STUDY

A self-curing concrete is provided to absorb water from the atmosphere to achieve better hydration of cement in concrete. It solves the problem that the degree of cement hydration is lowered due to no curing or improper curing. It is now widely accepted that there is a significant potential for reclaiming and recycling demolished debris for use in value-added applications to maximize economic and environmental benefits.

3 OBJECTIVES

- To determine the mechanical properties of natural coarse aggregate and recycled coarse aggregate with self-curing concrete
- To study the flexural behavior of self-curing concrete of RC beams and comparing with NC beams.

4 LITERATURE REVIEW

El Dieb (2007) has investigated to find out the water retention capacity and degree of hydration and moisture transport by using a self-curing agent and compared to conventional curing concrete. El Dieb *et al.* (2012) carried out a research on synthesized water-soluble polymers viz., polyethylene glycol (PEG) and polyacrylamide (PAM) as self-curing agents and its effect on the degree of hydration, water absorption, permeable pores and microstructural characteristics of Portland cement mixtures without and with 8% silica fume replacement. Kumar *et al.* (2012) carried out of a research on the mechanical characteristics of concrete such as compressive strength, splitting tensile strength and modulus of rupture by varying the percentage of PEG from 0% to 2% by weight of cement for both M20 and M40 grades of concrete. Kholia *et al.* (2013) carried out a research on the working and efficiency of curing methods, which are generally adopted in the construction industry and compared with the conventional water curing method. Laboratory tests show that concrete in a dry environment can lose as much as 50 percent of its potential strength compared to moist cured. Evangeline (2014) carried out a research on specific water-soluble chemicals such as Polyvinyl alcohol added during the mixing which can reduce water evaporation from and within the set concrete, making it "self-curing".

5 EXPERIMENTAL WORK

The materials used in the investigation are cement, fine and coarse aggregate, recycled coarse aggregate, Polyethylene glycol (PEG 6000), Ground granulated blast furnace slag (GGBS). Polyethylene glycol is a condensation polymer of ethylene oxide and water with the general formula H(OCH2CH2)n OH. For obtaining mechanical properties, cubes of 150 mm x 150 mm x 150 mm, cylinders 150 mm diameter x 300 mm length and prisms 100 mm x 100 mm x 500 mm were cast. For flexural behavior reinforced concrete beams 150 mm x 230mm x 1500 mm size has been adopted. The properties of recycled aggregate and PEG 6000 are presented in Table 1 and Table 2 respectively. Properties of GGBS are presented in Table 3 and Table 4. The reinforcement details of beams used for flexural analysis shown in Table 5. The compressive and tensile strains are recorded and these strains are used to plot strain diagram, which is used to find the depth of neutral axis, which is further used for calculation of curvature.

Table 1. Physical properties of recycled coarse aggregate.

S.No.	Specification	Value
1	Fineness modulus	1.37
2	Bulk density (kg/ m^3)	1325
3	Specific gravity	2.3

Table 2. Properties of PEG 6000.

S.No.	Specification	Value
1	Molecular weight	5500-6500
2	Appearance	white flake
3	Colour, Boha	10 max
4	Moisture	0.5% max
5	Hydroxyl Value	16-23 (mg KOH/g)
6	PH	5 – 7
7	Specific Gravity	1.08 - 1.09
8	Dioxane	1ppm max

Table 3. Physical properties of ground granulated blast furnace slag (GGBS).

S.No	Description	Value
1	Color	Off white
2	Specific gravity	3.15
3	Fineness (m^2/kg)	420
4	Glass Content (%)	75

Table 4. Chemical Composition of GGBS.

S.No	Description	Mass (%)
1	Loss of Ignition	2.6
2	Calcium, CaO	42.5
3	Silica, SiO_2	34.4
4	Alumina, Al_2O_3	12.9
5	Iron, Fe_2O_3	1.12
6	Magnesium, MgO	4.87
7	Sodium, Na_2O	0.45
8	Sulphur, SO_3	3.10
9	Potassium, K_2O	0.66

Table 5. Reinforcement details of the test specimens.

S.No	Beam Designation	Grade of Concrete	Reinforcement Provided		Spacing of stirrups
			Tension zone	Compression zone	
1	NC-M20	M20	2No's -12ø	2No's-8ø	8mm @130mm c/c
2	RC-M20	M20	2No's -12ø	2No's-8ø	8mm @130mm c/c
3	NC-M30	M30	2No's-16ø	2No's-8ø	8mm @130mm c/c
4	RC-M30	M30	2No's-16ø	2No's-8ø	8mm @130mm c/c
5	NC-M40	M40	4No's-12ø	4No's-8ø	8mm @130mm c/c

6 RESULTS AND DISCUSSIONS

6.1 Workability

Series of tests were carried out on normal coarse aggregate (NC) and recycled coarse aggregate (RC) specimens of M20, M30, M40, and M50 concrete grade mixes. Slump values of recycled aggregate concrete are low compared to that of natural aggregate concrete due to high water absorption of recycled aggregate concrete. As per test result, the highest slump obtained was 140mm for NC mix and the lowest slump was 30mm for RC mix and the average slump for each batch of the mix was 84mm. Thus, target slump of 50mm to 120 mm has been achieved. It was found that M20, M30, M40 concrete grade with natural aggregate show medium workability and whereas M50 grade of concrete indicates low workability because of increase in water cement ratio. On the other hand, low workability has obtained for M20 and M30 grade concrete of recycled aggregate and for M40 and M50 even very low. This is because of the high water absorption capacity of recycled aggregate.

6.2 Mechanical properties

The test results of compressive strength for NC and RC are presented in Table 6. The compressive strength of the recycled aggregate concrete is 2.5% lesser than the normal aggregate concrete for all mixes. Further, it is found that for NC-M20, NC-M30, NC-M40 and NC-M50 grade concretes has an average splitting tensile strength of 2.25MPa, 3.5MPa, 3.78MPa, and 4.9MPa respectively. Average Flexural strengths for NC-M20, NC-M30, NC-M40 and NC-M50 grade concrete obtained as 4.24MPa, 4.84MPa, 5.58MPa, and 6.07MPa respectively. Similarly, for RC-M20, RC-M30, RC-M40 and RC-M50 grade concretes has an average flexural strength of 3.93MPa, 4.15MPa, 5.1MPa, and 5.8MPa respectively.

Table 6. Compressive strength (N/mm^2).

S.No.	Grade of concrete	Natural aggregate (NC)	Recycled aggregate (RC)
1	M20	34.33	32.79
2	M30	43.45	42.27
3	M40	52.27	51.33
4	M50	62.59	62.17

6.3 Flexural Behavior of Beams

The ultimate failure load and deflections were obtained and presented in Table 7 and Table 8. It was observed that the ultimate failure load of recycled aggregate beams were lesser than that of the natural aggregate beams. Also, the ultimate deflection of the recycled aggregate beams is more than the natural aggregate beams. Strain characteristics was found to be same for self-curing concrete of normal aggregate and recycled aggregate for all the mixes of grades M20, M30, M40 and M50 respectively and the moment resistance of beams of normal aggregate concrete is found to be 3% to 8% more than the recycled aggregate concrete for all the mixes.

Table 7. Ultimate carrying capacity (kN).

S.No.	Grade of concrete	Natural aggregate (NC)	Recycled aggregate (RC)
1	M20	167.64	162.56
2	M30	243.84	238.70
3	M40	269.24	248.92
4	M50	330.20	264.16

Table 8. Deflections (mm).

S.No.	Grade of concrete	Natural aggregate (NC)	Recycled aggregate (RC)
1	M20	6.70	7.20
2	M30	7.40	8.30
3	M40	7.45	8.40
4	M50	7.93	7.95

6.4 Moment Curvature Relation of Self-Curing Concrete of NC (Mix-A) and RC (Mix-B)

In the experimental investigation, the moment corresponding to every incremental load was computed using the solid mechanics principle, for all 8-mix cases. Deformations at the level of compression fibers and tension fibers were recorded through dial gauges corresponding to various load increments. These observations were recorded as mean values of the corresponding deformation measured for two beams. The data of strain measurements were then used to plot moment curvature graphs. With reference to these tables, the moment of resistance was found to be 3% more for NC-M20, 2% more for NC-M30, 8% more for NC-M40 and 25% more for NC-M50 respectively.

7 CONCLUSIONS

- The compressive strength was approximately 1% to 4% more for self-curing concrete of normal aggregate concrete compared to recycled aggregate concrete for mixes M20, M30, M40 and M50 respectively.
- The splitting tensile strength was approximately 7% to 20% more for self-curing concrete of normal aggregate concrete compared to recycled aggregate concrete for the mixes considered above.
- The flexural strength approximately observed to be 4% to 16% is more for self-curing concrete compared to recycled coarse aggregate for the mixes indicated above.
- The Ultimate deflections of RC-beams resulted 7% to 12% greater for self-curing concrete when compared normal coarse aggregate for the same mixes.
- The moment of resistance of beam for normal aggregate concrete is found to be 3% to 8% more than the recycled aggregate concrete for all the mixes.

References

El Dieb, A. S., Self-Curing Concrete: Water Retention, Hydration and Moisture Transport, *Construction and Building Materials*, 21, 1282-1287, June 2007.

El Dieb, A. S., El Maaddawy, T. A., and Mahmoud, A, A. M., Water-Soluble Polymers as Self-Curing Agent in Cement Mixes, *Advances in Cement Research,* 24(5), 291-299, October 2012.

Evangeline, S., Self-Curing Concrete and its Inherent Properties, *International Journal of Engineering and Application*, 8(4), 66-71, August 2014.

Kholia, N. R., Vyas, B. A., and Tank, T. G., Effect on Concrete by Different Curing Method and Efficiency of Curing Compounds – A Review, *International Journal of Advanced Technology*, 2(4), 57-60, June 2013.

Kumar, J. M. V., Srikanth, M., and Rao, K. J., Strength Characteristics of Self-Curing Concrete, *International Journal of Research in Engineering and Technology*, 1(1), 51-57, September 2012.

Interdependence between Structural Engineering and Construction Management
Edited by Ozevin, D., Ataei, H., Modares, M., Gurgun, A., Yazdani, S., and Singh, A.

ISBN: 978-0-9960437-6-2

BEHAVIOR OF M 50 GRADE SELF-COMPACTING CONCRETE DEVELOPED USING PORTLAND SLAG CEMENT AND METAKAOLIN

SRAVYA NALLA[1], JANARDHANA MAGANTI[1], and DINAKAR PASLA[2]

[1]*Jawaharlal Nehru Technological University, Hyderabad, India*
[2]*School of Infrastructure, Indian Institute of Technology, Bhubaneswar, India*

Self-compacting concrete (SCC) is a revolutionary development in concrete construction. The addition of mineral admixtures like metakaolin, which is a highly reactive pozzolana to the SCC mixes, gives it superior strength and durability. The present work is an effort to study the behavior of M50 grade SCC by partial replacement of Portland Slag Cement (PSC) with metakaolin. Its strength and durability aspects are comparable with a controlled concrete (without replacement of cement). In the present work, a new mix design methodology based on the efficiency of metakaolin is adopted. The optimum percentage replacement of cement with metakaolin is obtained based on compressive strength test results. The influence of metakaolin on the workability, compressive strength, splitting tensile strength and flexural strength of SCC and its behavior when subjected to elevated temperature was investigated through evaluation against controlled concrete and non-destructive testing. From the test results, it was observed that incorporation of metakaolin at an optimum dosage satisfied all the fresh properties of SCC and improved both the strength and durability performance of SCC compared to controlled concrete.

Keywords: Pozzolana, Optimum percentage replacement, Effective cementitious materials, Efficiency concept, Non-destructive testing.

1 INTRODUCTION

Self-Compacting Concrete (SCC) is a highly flowable, yet stable concrete that can spread readily into place and fill the formwork without compaction and without undergoing any significant separation of material constituents (Khayat and Schutter 2014). SCC is usually produced using supplementary cementitious materials (SCM) such as silica fume, fly ash or slag. Metakaolin, a highly reactive pozzolona was successfully used as a mineral admixture for the development of SCC. Unlike other mineral admixtures, metakaolin is not industrial waste. It is manufactured by calcination of kaolinitic clay under controlled conditions of 650 to 800°C. Not only metakaolin is economical over silica fume but also due to higher silica and alumina content, the pozzolanic action of metakaolin is more significant as compared to silica fume and fly ash (Badogiannis *et al.* 2015). The particle size of metakaolin is not as fine as silica fume, hence the requirement of high range water reducing admixture (HRWRA) is less in case of metakaolin as compared to silica fume for the same slump flow (Balogh 1995, Ding and Li 2002).

In the present study, the effect of Indian metakaolin on the fresh concrete properties, strength, and durability of SCC has been investigated by adopting a specific mix design methodology proposed by Dinakar and Manu (2014). From the study, it was observed that as the percentage of metakaolin in SCC mixture increases, the 28-day compressive strength of SCC, the demand for high range water reducing admixture (HRWRA) and the rheological parameters (plastic viscosity and yield stress) increases (Hassan *et al.* 2010).

2 EXPERIMENTAL PROGRAM

2.1 Materials

Portland slag cement (PSC) of 53 grade conforming to IS 455 (1989) was used. Commercially available metakaolin was collected from a company in Gujarat, India, having a specific gravity of 2.5. Table 1 indicates the properties of the metakaolin used in this study. The Coarse aggregates of size 10 mm and down (granite rock) and well-graded river sand of maximum size 4.75 mm were used as fine aggregate. The specific gravity of coarse and fine aggregates was found to be 2.7 and 2.62 respectively. Polycarboxylic ether (PCE) based superplasticizer (SP) conforming to IS 9103 (1999) is used as a chemical admixture. The SP contains 0.01-0.02% of viscosity modifying agent (VMA) as reported by the manufacturer.

Table 1. Properties of Metakaolin (supplied by manufacturer).

Properties	Metakaolin
Physical Form	Off white powder
Specific Gravity	2.5
Bulk Density	300 ± 30 gm/ltr
Average Particle Size	1.5 μ
Residue 325 #	0.5 % max
Pozzolan Reactivity - mg $Ca(OH)_2$	>1000
Loss of ignition (LOI)	0.68
Fineness {Blaine} (m^2 /kg)	15,000

2.2 Mix Proportions

The mix design was carried out according to the mix design methodology proposed by Dinakar and Manu (2014) based upon the efficiency concept. It was also taken care that all mixes satisfied the regulations given in EFNARC (2002) guidelines. Trial mixes were prepared to arrive at optimum percentage replacement of PSC with metakaolin based on compressive strength test results. The optimum 'water to effective cementitious materials content' (ratio) was arrived at like 0.27 and the SP dosage was kept constant to satisfy fresh properties of SCC. The final mix proportions of SCC containing different percentages of metakaolin are shown in Table 2.

3 RESULTS AND DISCUSSIONS

Based on the tests carried out, properties of fresh and hardened concrete are reported in the following section.

3.1 Fresh Properties

The fresh properties of SCC were determined using slump flow, U-box, and L-box tests as per EFNARC (2002) guidelines and the results are shown in Table 3.

From the above results, it is evident that with the increase in metakaolin content, the slump flow decreased and the T_{50cm} time increased. This may be due to the higher surface area of metakaolin which increases the water demand and thereby SCC loses fluidity. All the fresh properties of SCC satisfied the regulations given in EFNARC guidelines.

Table 2. Details of the mix proportions.

$w/(c + k_{28}\times m)$	P (%)	k_{28}	C (kg/m^3)	m (kg/m^3)	W (kg/m^3)	F.A (kg/m^3)	C.A (kg/m^3)	SP (kg/m^3)	SP (%)
0.27	0	0	550	0	148.5	783.89	873.478	5.5	1
0.27	8	4.8346	506	44	194.057	725.42	808.323	5.5	1
0.27	10	4.4059	495	55	221.2	713.302	794.825	5.5	1
0.27	15	3.5375	467.5	82.5	205.022	710.663	791.851	5.5	1
0.27	20	2.9596	440	110	206.701	707.868	788.76	5.5	1
0.27	25	2.6729	412.5	137.5	210.606	702.266	782.54	5.5	1

$w/(c + k_{28}\times m)$ - water to effective cementitious materials content ratio, *c* - Cement content, k_{28}- Efficiency of metakaolin, *m* – Metakaolin, *p* - Percentage replacement of cement with metakaolin, *F.A* - Fine aggregate, *C.A* - Coarse aggregate, *SP* – Superplasticizer. SCC without metakaolin is referred to as control concrete.

Table 3. Fresh Properties of SCC.

Replacement of cement with metakaolin (%)	Flow time T_{50cm} (sec)	Slump flow (mm)	L-Box	U-Box (mm)
0	2.06	780	0.98	1
8	2.21	765	0.94	5
10	2.36	745	0.89	9
15	3.08	720	0.873	15
20	3.9	690	0.832	21
25	5.01	646	0.813	27

3.2 Strength Studies

3.2.1 *Compressive strength*

The unconfined compressive strength of 100 mm cubes was obtained, by subjecting to a compressive loading at the rate of 2.5 kN/s, at the age of 7 and 28 days on 3,000 kN machine and the test was performed as per IS 516 (1959). The strength variation with varying percentage of metakaolin is illustrated in Figure.1.

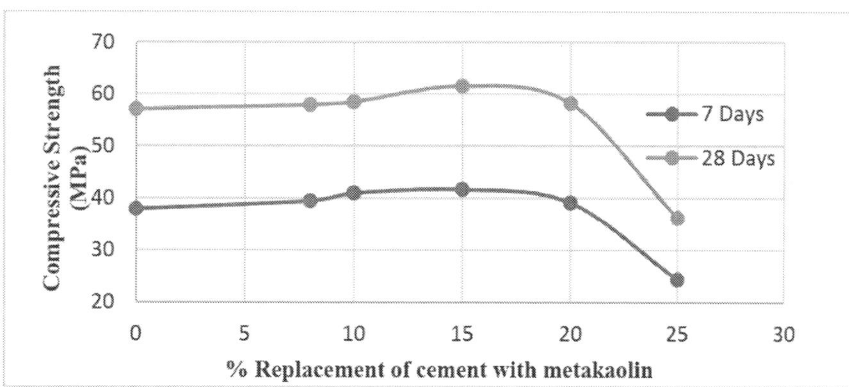

Figure 1. Variation of compressive strength with respect to metakaolin percentage.

From the above figure, it can be observed that compressive strength of concrete is maximum for 15% replacement of cement with metakaolin and hence, it is chosen as optimum dosage. Further, when the dosage of metakaolin was increased, it was observed that the compressive strength was decreased, and this might be due to clinker-dilution effect.

3.2.2 *Flexural strength and splitting tensile strength*

The split tensile strength test was conducted on 150×300 mm cylindrical specimens at the age of 28 days, and the test was performed as per IS 5816 (1999) and the flexural strength test was conducted on a beam of size 100×100×500 mm subjected to pure bending and the test was performed as per IS 516 (1959). Table 4 indicates the flexural and splitting tensile strength test results for control concrete and SCC having an optimum replacement of PSC with metakaolin.

Table 4. Flexural and Splitting Tensile Strength Test Results.

Replacement of cement with metakaolin (%)	Flexural strength at 28 Days (MPa)	Splitting tensile strength at 28 Days (MPa)
0	8.042	4.576
15	8.209	4.844

From the test results, it was observed that the flexural and split tensile strengths for SCC having an optimum replacement of PSC with metakaolin was slightly higher than the values obtained for control concrete.

3.3 Durability Studies

3.3.1 *Elevated temperature test*

For elevated temperature testing, the cube specimens are subjected to 100°C temperature for 24 hours at the age of 28 days and tested for compressive strength. The average compressive strength of specimens subjected to 100° C is compared with average compressive strength of specimens before heating, for both control concrete and SCC having optimum replacement of PSC with metakaolin and the results are shown in Table 5.

Table 5. Elevated Temperature Test.

Replacement of Portland slag cement with metakaolin (%)	Compressive Strength (MPa) at 28 Days	
	At Room Temperature	After subjected to 100^{0}C for 24 hour and cooled to room temperature
0	57.05	66.81
15	61.53	72.29

From the test results, it was observed that the compressive strength was increased for both the concretes after subjected to 100^{0}C for 24 hours and cooled to room temperature. The increase in compressive strength is slightly higher in case of SCC having an optimum replacement of PSC with metakaolin when compared to control concrete. This might be due to further densification of pore structure, due to the addition of metakaolin to concrete (Mehta and Monteiro 2006). Increase in temperature beyond a certain value can result in explosive spalling due to the build-up of pore pressure by steam, so the rate of increase in compressive strength decreases for concrete with metakaolin due to densified pore structure. Thermal cracks develop due to the evaporation of physically absorbed water which starts at 80°C and such concretes may show inferior performance as compared to concretes without metakaolin at elevated temperatures (Phan 1996).

3.4 Non-Destructive Tests

3.4.1 *Rebound hammer and ultrasonic pulse velocity tests*

The non-destructive tests on 100 mm cube specimens were performed as per IS 13311 (1992). A summary of the results of non-destructive tests of control concrete and SCC with optimum dosage is presented in Table 6 respectively.

Table 6. Non-Destructive Test Results of Control Concrete and SCC with optimum dosage.

Sample at 28 Days (Average)	Rebound number	Ultrasonic pulse velocity (m/s)	Compressive Strength from combined UPV and Rebound hammer test (MPa)
0	38	4326.67	53.83
15	49.4	4370	57.5

From the results, it was observed that both the rebound number and ultrasonic pulse velocity values are increased for SCC having an optimum replacement of PSC with metakaolin as compared to control concrete and this increase is due to the pozzolanic properties of metakaolin i.e., densified pore structure. The compressive strength of concrete is predicted from combined ultrasonic pulse velocity and rebound number to have better accuracy in estimating the compressive strength of concrete (Malhotra 1976). The compressive strength obtained from NDT was found to be lesser than the actual compressive strength for both the control concrete and SCC having an optimum replacement of PSC with metakaolin, but the percentage variation is within the range (± 20 %) as specified in IS 13311 (1992).

4 CONCLUSIONS

Based on the experimental investigation of M 50 grade SCC developed by partially replacing Portland slag cement with metakaolin (having an optimum replacement of PSC with metakaolin), following conclusions are drawn:

- With the increase in the percentage of metakaolin, the slump value decreased and T_{50cm} time was found to increase.
- The 'water to effective cementitious materials content' ratio was fixed as 0.27 as it satisfied all the fresh concrete properties. By keeping this constant, optimum percentage replacement of cement (PSC) with metakaolin was found to be 15% for SCC.
- The increase in compressive strength of self-compacting concrete mix with optimum percentage replacement of cement with metakaolin is 9.918% for 7-days and 7.846% for 28-days age of the sample.
- There is an increase in flexural strength and splitting tensile strength by 2.07% and 5.86% respectively by replacing cement (PSC) with optimum content of metakaolin (15%).
- The compressive strength was found to increase after 24 hours of exposure to an elevated temperature of 100°C, and this increase in compressive strength is relatively higher for SCC with optimum percentage replacement of metakaolin content when compared to controlled concrete.

- The results obtained from the correlation of in-situ compressive strength with rebound number and ultrasonic pulse velocity showed lower values of concrete compressive strength, compared to results obtained from actual compressive strength test for the same mixes, for both controlled concrete and SCC with optimum percentage replacement of metakaolin content.

References

Badogiannis, E., Sfikas, I. P., Voukia, D. V., and Trezos, K., Durability of Metakaolin Self-Compacting Concrete, *Construction and Building Materials*, 82, 133-141, 2015.

Balogh, A., *High-Reactivity Metakaolin,* The Aberdeen Group, Publication #C950604, 1995.

Ding, J. T., and Li, Z., Effects of Metakaolin and Silica Fume on Properties of Concrete, *Materials Journal,* 99(4), 393-398, 2002.

Dinakar, P., and Manu, S. N., Concrete Mix Design for High Strength Self-Compacting Concrete using Metakaolin, *Materials and Design*, 60, 661–668, March 2014.

EFNARC, *Specification, and Guidelines for Self-Compacting Concrete,* Surrey, UK, 2002.

Hassan, A. A. A., Lachemi, M., and Hossain, K. M. A., Effect of Metakaolin on The Rheology of Self-Consolidating Concrete, In: Design, *Production, and Placement of Self-Consolidating Concrete.* Rilem Bookseries, 1(3), 103–112, July 2010.

IS 455, *Portland Slag Cement Specification,* Bureau of Indian Standards, New Delhi, India, 1989.

IS 516, *Method of Tests for Strength of Concrete,* Bureau of Indian Standards, New Delhi, India, 1959.

IS 5816, *Splitting Tensile Strength of Concrete-Method of Test,* Bureau of Indian Standards, New Delhi, India, 1999.

IS 9103, Concrete Admixtures Specification, Bureau of Indian Standards, New Delhi, India, 1999.

IS 13311, *Non-Destructive Testing of Concrete-Methods of The Test,* Bureau of Indian Standards, New Delhi, India, 1992.

Khayat, K. H., and Geert, De, Schutter., *Mechanical Properties of Self-Compacting Concrete State-of-the-Art Report of the RILEM Technical Committee 228-MPS on Mechanical Properties of Self-Compacting Concrete,* RILEM, 14, Springer, 2014.

Malhotra, V. M., Testing of Hardened Concrete: Non-destructive Methods, Monograph No. 9, *American Concrete Institute*, Detroit, MI, 52, 1976.

Mehta, K., and Monteiro, P. P. J. M., *Concrete-Microstructure, Properties and Materials,* Third Edition, *McGraw-Hill,* United States of America, 2006.

Phan, L.T., Fire Performance of High-Strength Concrete: A Report of The State-of-The-Art. *National Institute of Standards and Technology*, Gaithersburg, MD, 105, 1996.

Interdependence between Structural Engineering and Construction Management
Edited by Ozevin, D., Ataei, H., Modares, M., Gurgun, A., Yazdani, S., and Singh, A.

ISBN: 978-0-9960437-6-2

NOVEL CONCRETE MIXTURE USING SILICA RICH AGGREGATES: WORKABILITY, STRENGTH AND MICROSTRUCTURAL PROPERTIES

PAUL AWOYERA, JOSEPH AKINMUSURU, ANTHONY EDE, and JOSHUA JOLAYEMI

Dept of Civil Engineering, Covenant University, Ota, Nigeria

The persistent reliance on traditional construction materials is of no gain to the future generation. The rate at which the natural aggregate sources are explored is alarming, and as a result, the threat of depletion of the natural materials has inspired interest in sustainable construction materials, focusing on construction and demolition wastes and local materials. In this study, an experimental insight on modified concrete, based on workability, strength and microstructural properties, is provided, in an attempt to ascertain the suitability of silica-rich aggregates (ceramic industry wastes and laterite) as a replacement for conventional fine and coarse aggregates. Various mix proportions were considered, and material batching was done by weight for concrete casting. The workability test, using slump, indicates that the flowability of the modified concrete mixes is achievable at a water-binder ratio of 0.6. The strength properties of the concrete increased with the increasing ceramic substitution for granite while increasing laterite content beyond 10% negates the strength gain by the concrete. A concrete mix containing 90% ceramic fine and 10% laterite, as fine aggregate, and 100% of cement and ceramic coarse, as binder and coarse aggregate, respectively, gave higher compressive strength (22.5 MPa), and split-tensile strength (3.6 MPa), and these results were found as comparable to the conventional concrete.

Keywords: Compressive strength, Hydration, Interfacial transition zone, Morphology, Recycled aggregate, Split-tensile strength.

1 INTRODUCTION

The high growth rate in world population and increasing urbanization, are major, among many factors that are contributing to the scarcity of construction materials, and this has caused the cost of natural aggregates to rise in some parts of the world (Karthik *et al.* 2017). Thus, more volume of aggregates is needed to be able to meet the current world demand. The attention of researchers are drawn towards the use of solid wastes resulting from industrial and construction activities for construction, so as to prevent an event of total depletion of the natural aggregate sources, and moreover, to reduce the environmental nuisance caused by the disposal of solid wastes (Murthi *et al.* 2018). Ceramic wastes sourced from production (Juan *et al.* 2010), and those sourced from construction and demolition (García-González *et al.* 2015, Awoyera *et al.* 2018), have been considered as a partial replacement for conventional aggregates and cement. Those studies showed that ceramic aggregate addition to concrete could improve both its physical and mechanical properties. There are also investigations that showcased the pozzolanic reactivity of ceramic products (Elkhadiri and Puertas 2008).

On the other hand, laterite as a local soil material has been studied (Awoyera *et al.* 2017) to explore its potential as a replacement for river sand; most studies reported that the strength properties of concrete containing laterite were near that of conventional concrete. Laterite is often regarded as a marginal material, and only a 10 – 30% of it can be added to concrete (Awoyera *et al.* 2016). However, despite the extensive studies on the use of ceramics and laterite in concrete, there is still scanty information regarding combined addition of these two materials in concrete. Therefore, this study investigates the use of ceramic tiles sourced from construction and demolition sites, and laterite, an abundant soil material in sub-Saharan African countries, as a replacement for natural aggregates in concrete. The strength properties and some microscale properties of the concrete are determined.

2 EXPERIMENT

2.1 Materials and Methods

The ceramic tiles, comprising of mostly floor and wall tiles, used in this study was sourced from construction and demolition sites in Ota, Ogun State, Nigeria. The ceramics were crushed using a hammer mill, and subsequently pulverized and graded to reflect natural aggregates, using British standard (BS) sieves, into coarse (12.7 mm) aggregate and fine (0 – 4 mm) aggregate categories. Laterite was sampled at a depth of 10 meters in a borrow pit located within the precinct of Covenant University, Nigeria. The soil has been exposed to several borrowing activities by land fillers and construction workers. The coarse ceramic aggregate was substituted for granite, while its fines, and laterite were substituted for river sand. Ordinary Portland cement conforming to standards (BS 12 1989) was used as a binder. The aggregates were air-dried for a few weeks before use. A mix proportion of 1:1.5:3 (binder, fine aggregate: coarse aggregate), and a constant water/binder ratio of 0.6 was used. The mix proportions for concrete samples are presented in Table 1, and Table 2 shows the physical properties of the aggregates. The properties of both laterite and ceramics are in close range of the recommended standards for natural aggregates.

The morphology of pulverized ceramic and cement was obtained using a Scanning electron microscopy (SEM - Hitachi S4100 equipped with energy dispersion spectroscopy, EDS – Rontec) at 20 kV. The SEM micrographs of ceramic particles and cement are shown in Figures 1a and 1b respectively. These images were captured at high magnification, and the analysis was performed with X-ray microanalysis (EDS). The dominant oxides identified in the ceramics were silica (SiO2) while calcium oxide (CaO) was seen in the cement. Concrete samples produced include cubes of 150 mm dimensions and cylinders of 100 mm x 200 mm. The batching was done by weight while casting the concrete samples. The samples were removed from the mold after 24hrs of casting and cured in water for maximum 28 days. During casting, workability of the concrete mixes, using slump (as per the requirements of BS EN 12350-2 2009), was randomly assessed before placing the concrete in the molds. The slump results for all fresh concrete mixes were in the range 50-60 mm, which represent a true slump condition. Concrete that falls into this category are usually considered for the purpose of casting structural elements such as beam, slabs, columns and foundations.

Strength properties were determined (as per the requirement of BS EN 12390-6 2006), after 3, 7, 14, and 28 days curing, using triplicate samples from each mix. The microstructure of selected samples (obtained from the crushed cubes), comprising of the reference mix, and a sample with optimal strength from the modified mixes, were examined using SEM techniques in backscattered mode. This approach helps to visualize the interfacial transition zone between aggregates and bulk cement pastes in concrete.

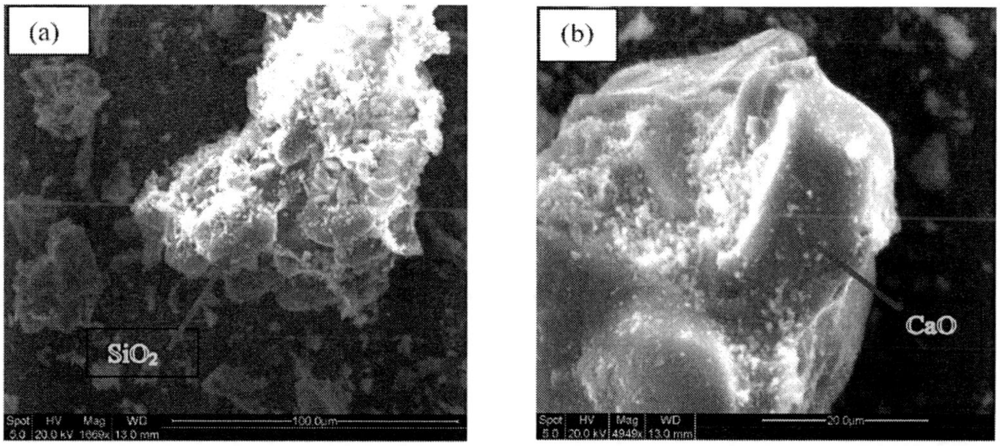

Figure 1. SEM micrograph of (a) Ceramic particles (b) cement.

Table 1. Mix proportion design (%) for concrete samples.

Mix ID	Binder	Fine Aggregates			Coarse Aggregates		w/c
	Cement	Sand	Ceramic fine	Laterite	Granite	Ceramic coarse	
Ref	100	100	-	-	100	-	0.6
M10	100	90	-	10	100	-	0.6
M20	100	80	-	20	100	-	0.6
N0	100	-	100	-	100	-	0.6
N10	100	-	90	10	100	-	0.6
N20	100	-	80	20	100	-	0.6
L0	100	-	100	-		100	0.6
L10	100	-	90	10		100	0.6
L20	100	-	80	20		100	0.6

Table 2. Physical properties of aggregates used.

Properties	**River sand**	**Laterite**	**Ceramic fine**	**Ceramic coarse**	**Granite**
Specific gravity	2.61	2.13	2.26	2.31	2.87
Water absorption	2.24	4.70	2.52	0.55	0.23
Fineness modulus	2.24	1.80	2.20	6.88	6.95
Aggregate crushing value	-	-	-	20.86	34.00
Aggregate impact value	-	-	-	27.00	24.00

3 RESULTS AND DISCUSSIONS

Figures 2 and 3 show the compressive strength and the split-tensile strength respectively, gained through 28 days curing of the samples. The axial compressive strength of all the concrete samples tested increased as the curing age increased, however, the reference sample demonstrated higher strength than the modified mixes. The samples M10, N10, and L10 (containing 10% laterite), gave maximum compressive strength which represented an approximately 11% strength gain over the mixes without laterite. The increased compressive strength could be influenced by the pozzolanic influence of fragments of ceramic aggregate or due to chemical effect of laterite in concrete containing minimal laterite (10%), which apparently contributed to the hydration of the concrete and consequently enhanced the compressive strength.

There was no significant increase in strength as laterite was increased beyond 10%, this could be as a result of retarding effect of laterite on hydration because laterite contains large clay minerals, which may alter the rate of hydration. This effect might be taken care of by curing concrete containing laterite for a longer duration in order to allow it to complete its hydration.

When 28-day strength was considered, sample L10 from the modified mix gave the highest strength among these set, however, this sample had a 9% less strength than the reference mix. The increased strength in the reference concrete suggests that the hydration within the reference concrete matrix reached its peak within 28-days curing. This suggests that an improved interfacial transition zone (ITZ) exist between ceramic aggregate, laterite, and hardened cement paste. Similar to the compressive strength tests, the split-tensile strength of all the samples increased with increasing curing age. Also, mixes M10, N10, and L10 (containing 10% laterite) gave the higher values of split-tensile strength, which represented an approximate 8% strength gain when compared with the samples without laterite. The split-tensile strength also decreased as the laterite content was increased beyond 10%.

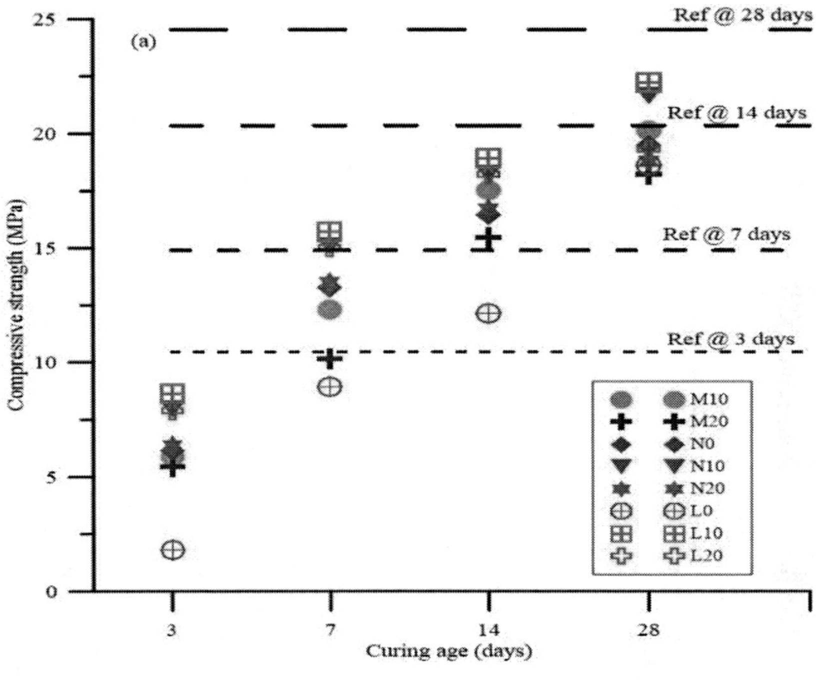

Figure 2. Compressive strength development with curing age.

This increase in the splitting strength of the mixes N10 and L10 may be a result of the considerably higher water content used (Liu *et al.* 2015) because there will be enough water for hydration despite absorptions by ceramic aggregate and laterite aggregates prior to the hydration process.

The SEM micrographs of the mix L10, having higher strength from the modified samples, and the reference concrete is shown Figures 4a and 4b respectively. A relatively well-compacted ITZ can be seen in the reference concrete when compared with the mix L10. Therefore, it can be inferred that the rigidity of the ITZ in concrete depends largely on the hydration mechanism, which in turn influences the strength properties.

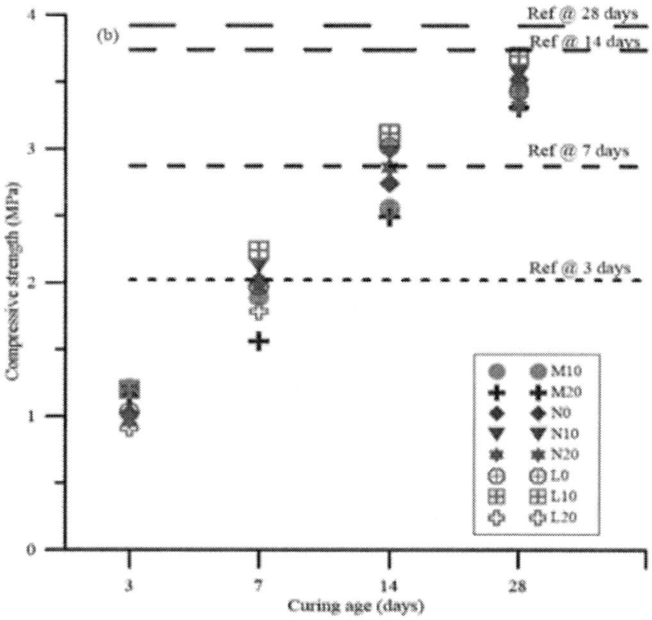

Figure 3. Split-tensile strength development with curing age.

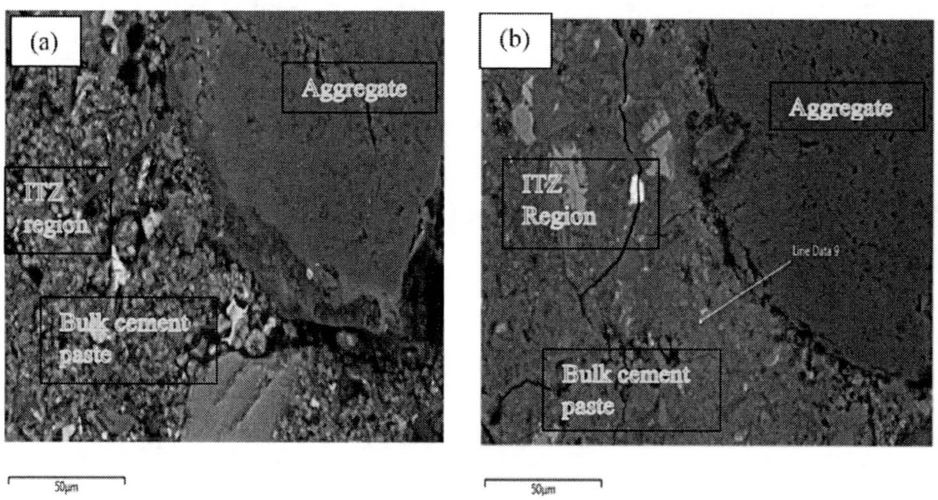

Figure 4. SEM micrographs of (a) optimal modified mix (b) reference concrete.

3 CONCLUSION

The concrete mixes considered in this study demonstrated an increase in both compressive and splitting tensile strengths with increasing curing age. The reference concrete has higher strength properties than the mixes containing alternative aggregates, and this strength increase was attributed to rapid hydration in the concrete. Furthermore, the reference concrete produced higher strength as a result of an improved ITZ.

For the modified mix categories, a ceramic fine and laterite substitution in the proportion of 90% and 10% respectively, for river sand, and 100% coarse ceramics for granite, enhanced both the compressive and splitting tensile strengths.

Acknowledgments

This research was supported by Covenant University, Ota, Nigeria.

References

Awoyera, P. O., Akinmusuru, J. O., Dawson, A. R., Ndambuki, J. M., and Thom, N. H., Microstructural Characteristics, Porosity and Strength Development in Ceramic-Laterized Concrete, *Cement and Concrete Composites,* 86, 224-237, February, 2018.

Awoyera, P. O., Akinmusuru, J. O. and Ndambuki, J. M., Green Concrete Production with Ceramic Wastes and Laterite, *Construction and Building Materials,* 117, 29–36, August, 2016.

Awoyera, P. O., Dawson, A. R. Thom, N. H. and Akinmusuru, J. O., Suitability of Mortars Produced Using Laterite and Ceramic Wastes: Mechanical and Microscale Analysis, *Construction and Building Materials*, 148, 195-203, September, 2017.

BS EN, 12390-6, *Testing Hardened Concrete – Part 6: Tensile Splitting Strength of Test Specimens*, British Standard, London, UK, 2006.

BS EN 12350-2, *Testing Fresh Concrete - Slump-Test,* British Standard, London, UK, 2009.

BS 12, Specification for Ordinary and Rapid-Hardening Portland Cement. *British Standard, London, UK*, 1989.

Elkhadiri, I., and Puertas, F., The Effect of Curing Temperature on Sulphate-Resistant Cement Hydration and Strength, *Construction and Building Materials*, 22(7), 1331–1341, July, 2008.

García-González, J., Rodríguez-Robles, D. Juan-Valdés, A., Pozo, J. M. M., and Guerra-Romero, M. I., Ceramic Ware Waste as Coarse Aggregate for Structural Concrete Production, *Environmental Technology*, Taylor & Francis, 36(23), 3050–3059, September, 2015.

Juan, A., Medina, C., Guerra, M., Morán, J., Aguado, P., Sánchez de Rojas, Frías, M., and Rodríguez, O., Re-Use of Ceramic Wastes in Construction, In *Ceramic Materials*, Ed. Wunderlich, W., InTech, London, 2010.

Karthik, S., Rao, P. R. M., and Awoyera, P. O., Strength Properties of Bamboo and Steel Reinforced Concrete Containing Manufactured Sand and Mineral Admixtures, *Journal of King Saud University-Engineering Sciences,* 29(4), 400 – 406, October, 2017.

Liu, F., Liu, J., Ma, B., Huang, J., and Li, H., Basic Properties of Concrete Incorporating Recycled Ceramic Aggregate and Ultra-Fine Sand, *Journal Wuhan University of Technology, Materials Science Edition,* 30(2), 352–360, April, 2015.

Murthi, P., Awoyera, P., Selvaraj, P., Dharsana, D., and Gobinath, R., Using Silica Mineral Waste as Aggregate in a Green High Strength Concrete: Workability, Strength, Failure Mode, and Morphology Assessment, *Australian Journal of Civil Engineering,* (in press), 1–7, 2018.

Interdependence between Structural Engineering and Construction Management
Edited by Ozevin, D., Ataei, H., Modares, M., Gurgun, A., Yazdani, S., and Singh, A.

ISBN: 978-0-9960437-6-2

EFFECTS OF DIFFERENT SOURCES OF WATER ON THE PROPERTIES OF CONCRETE

IGNATIUS OMUH, RAPHEAL OJELABI, ADEDEJI AFOLABI, PATIENCE TUNJI-OLAYENI, CHUKWUMA OBI, and TIMOTHY MOSAKU

Dept of Building Technology, Covenant University, Ota, Nigeria

Water is vital to human existence and life can only be sustained by it. Concrete is a widely used construction material and water is an important part of its composition. Potable water is what is recommended for concrete works, but unfortunately, some places do not have access to this. Places that do not have access to potable water might have access to other water sources that might be used for concrete works. This study was undertaken to investigate the effects of water from different sources on concrete mechanical properties. This study evaluates the characteristics of concrete produced with river water, well water, and potable tap water. Compressive strength and Density, were used to evaluate the characteristics of concrete specimens of mix ratios 1:2:4 and $1{:}1^1/_2{:}3$ produced with water from the different sources. The results showed that concrete specimens produced with tap water had the highest mean compressive strength at 28 days. While well water had the lowest compressive strength, it was concluded that well water was not suitable for concrete works even though it is already being used on some sites that can't access tap water.

Keywords: Compressive strength, Concrete density, Hard water, Mix ratio, Potable water, Water-cement ration, Turbidity, Acidity.

1 INTRODUCTION

Concrete is a major component of infrastructural facilities in the 21st century because of the versatility in its use. According to the European commission, concrete is the most popularly used construction material (Montero and Laserna 2017). Concrete is a mixture of cement, fine aggregates, coarse aggregates, and water combined in the correct proportions to give a strong, dense, homogeneous material which can be easily molded into desired shapes and is ideally suited for the manufacture of structural components (Ekmekyapar and Orung 2001). Water is essential of concrete practice for both mixing the concrete and curing the hardened concrete (Omuh *et al.* 2018). According to Kulkarni *et al.* (2014), the concrete industry is consuming annually 1 billion tons of mixing water in the world. The water performs two main functions. First, it allows the cement to hydrate, which is the chemical reaction between cement and water which results in the setting of the concrete. Secondly, water is also included in a concrete mix to provide workability to the mix. This enables the constituent's materials to be mixed together easily, to be placed in the position and compacted properly. The presence of water in any concrete mix is a major factor that controls the strength development and other properties of concrete (Aitcin 2016). However, not every potential concrete site is situated near a water source. For instance, concrete production in some rural areas may compel the use of water from alternative sources such as wells and even rivers. These waters from alternative sources that are readily accessible might or might not be fit for concreting works therefore the purpose of this research. ACI 318M (2015) specifies that water suitable for concrete should be potable, colorless, odorless water fit for drinking. However, ASTM C94 (2016) permits the use of non-potable water for ready mix concrete. Although according to Neville (2000) there is no standard for mixing water in concrete, most other researchers would disagree because it is usually specified that water fit for drinking is most appropriate for concrete.

The most important property of concrete is strength which determines the quality of the concrete according to Rakesh and Dubey (2014). The compressive strength of concrete is the most common performance measure used by engineers and builders for testing of concrete cubes and cylinders. The compressive strength of concrete can be calculated from the failure load divided by the cross-sectional area resisting the load.

2 MATERIALS AND METHODS

2.1 Materials

2.1.1 *Ordinary Portland cement*

Portland cement was used in the production of the concrete; cement acts as a binder for the rest of the materials, specifically Dangote Cement, which was obtained from a local dealer of cement at Ota, Ogun State. The grade to be employed is 42.5R.

2.1.2 *Aggregates*

The fine aggregate used for the study was bought from a local source at Ota, Ogun State. The fine aggregate is sharp river sand with sizes not greater than 5mm diameter. The aggregates conform to the ASTM C136 (2006).

The coarse aggregate used for this experiment is granite from crushed quarried rocks. The size of the granite is not greater than 19 mm and it was bought from local granite seller at Ota, Ogun State. This also conforms to the ASTM C136 (2006) (Method for sieve analysis of fine and coarse aggregates).

2.1.3 *Water*

The water samples were obtained from three sources. The sources of the water are listed:

1. River water which was obtained from the Atuara River, Ota Ogun State.
2. Open well water which was obtained from a residence in ilogbo, Ota Ogun State.
3. Tap water which was obtained from Building Technology's Concrete Laboratory, Covenant University, Ota, Ogun State.

The water used in this study was analyzed in the Chemistry Laboratory of Covenant University, Ota, Ogun State.

The properties that were analyzed were the physical and chemical properties of the water. The physical properties determined included color, odor, pH at 20° C, turbidity and conductivity. The chemical properties determined includes acidity, alkalinity, total hardness, chloride, sulphite, sulphate, nitrite, ammonia, and silica.

2.1.4 *Preparation of concrete cube specimens*

A total of 54 cubes were cast for concrete of mix ratio 1:2:4 and $1{:}1^1/_2{:}3$. The concrete cubes were of size 150 mm x 150 mmx 150 mm. the design for the experiment had three samples for each mix and three hydration periods as shown in Table 1. Steel molds were used for casting the cubes. The materials will be mixed to produce a total of 54 concrete cubes with the specifications:

a. The grade of concrete to be achieved are of characteristic strength 25 MPa, and 30 MPa respectively with their respective mix ratio's 1:2:4 and $1:1^{1}/_{2}:3$
b. The mixing method was the same for all the concrete samples with different water sources. Manual mixing was employed because of the small volume of the concrete to be mixed.
c. The curing method employed was ponding for 7 days, 21 days and 28 days.

2.1.5 *Mix Ratio*

For each water sample used, there were two mix ratios. The cube samples were marked to differentiate the test conditions (A1, A2, A3), (B1, B2, B3), (C1, C2, C3), (D1, D2, D3,), (E1, E2, E3), (F1, F2, F3). Table 1 represents the sample identification and Table 2 shows details about concrete cube samples.

Table 1. Concrete specimen identification.

S/N	Source of water	Mix Ratio	Symbol
1	River water	1:2:4	**A**
2	River water	$1:1^{1}/2:3$	**B**
3	Well water	1:2:4	**C**
4	Well water	$1:1^{1}/2:3$	**D**
5	Tap water	1:2:4	**E**
6	Tap water	$1:1^{1}/2:3$	**F**

Table 2. Batching summary for concrete cubes samples.

Sample	Cement (Kg)	Sand (Kg)	Granite (Kg)	Water (Kg)	Water/ Cement Ratio
A	13	26	52	8.5	0.60
B	17	25.5	51	11.1	0.65
C	13	26	52	8.5	0.60
D	17	25.5	51	11.1	0.65
E	13	26	52	8.5	0.60
F	17	25.5	51	11.1	0.65

The concrete cubes were removed from the molds after 24 hours and placed into the curing tanks. There were three hydration periods within which samples were tested for the compressive test to examine the strength development of the concrete samples.

3 FINDINGS AND DISCUSSION

From Table 3, River water tested showed a slightly clear appearance, pale yellow in color. The odor was objectionable, which may indicate the presence of organic matters, and a pH of 6.46 which is acidic. For the Well Water tested, it showed a clear sample, colorless and odorless, and a pH of 5.91 which is acidic. For Tap Water, the water was clear in appearance, odorless, and a pH of 5.23 indicating acidity. The WHO standard for water states: clear, odorless, and has a pH ranging from 6.50 - 9.20. The pH value of River Water did not satisfy the WHO and Nigerian Industrial Standards (NIS) standard for drinking water but fails to satisfy the conductivity standard. The pH of Tap Water and Well Water used does not satisfy the WHO and NIS standard at 5.23.

Table 3. Physical characteristics of the water samples.

S/No	Parameter	River water	Well water	Tap water	WHO Standard for Drinking water		NIS Standard for Drinking Water
					Minimum acceptable	Maximum acceptable	Maximum acceptable
1	Appearance	Slightly Clear	Clear	Clear	Clear	Clear	Clear
2	Color	Pale Yellow	-	-	-	-	-
3	Odor	Objectionable	Odorless	Odorless	Odorless	Odorless	Odorless
4	pH at 28°C	6.46	5.91	5.23	6.5	9.2	6.5-8.5
5	Turbidity (mg/l)	<50	<50	<50	5	50	5
6	Conductivity (µScm-1)	192.6	132.6	30	0.9×10^{-4}	1.20×10^{-1}	1000

Table 4. Chemical characteristics of water sample.

S/No	Parameter	River water	Well water	Tap water	WHO Standard for Drinking water		NIS Standard for Drinking Water
					Minimum acceptable	Maximum acceptable	Maximum permitted
1	Acidity-P (mg/l $CaCaO_3$)	40	20	20	Nil	Nil	Nil
2	Alkalinity-M (mg/l $CaCaO_3$)	70	40	40	30	500	Nil
3	Total Hardness (mg/l $CaCaO_3$)	60	30	20	30	200	150
4	Chloride, Cl^- (mg/l)	3.55	1.78	1.78	200	600	250
5	Sulphate, SO_2^{-4} (mg/l)	<50	<50	<50	200	400	200
6	Silica, SiO_2 (mg/l)	ND	ND	ND	-	-	-

Note: ND= Not Detected

From Table 4, River Water tested showed chlorine content of 3.55 mg/l, total hardness of 60 mg/l $CaCO_3$ and sulphate was less than 50 mg/l. For Well Water tested, it showed a total hardness of 30 mg/l $CaCO_3$, chloride content of 1.78 and sulphate was less than 50 mg/l. For Tap Water, the water showed total hardness of 20 mg/l $CaCO_3$, chloride content of 1.78 mg/l and sulphate was less than 50 mg/l. The WHO standard for water states a total hardness of 30 – 200 mg/l $CaCO_3$, chloride content of 200 – 600 mg/l and a sulphate content of 200 – 400 mg/l. The total hardness and the chlorine content of River Water fails to satisfy the WHO and NIS standard for drinking water. The Tap Water, Well Water and River Water used fail to satisfy the WHO and NIS standard in terms of acidity with an acidity of 20 mg/l $CaCO_3$ as compared to the standard which is Nil. All three water samples fall below W.H.O's minimum standard for chloride. The total hardness of Tap Water falls below W.H.O's minimum standard for total hardness. Figure 1 shows the compressive strength of 1:2:4 concrete mixed with water from different sources. It shows that the concrete that was mixed with river water possessed the lowest compressive strength at 28 days of 17.8 MPa.

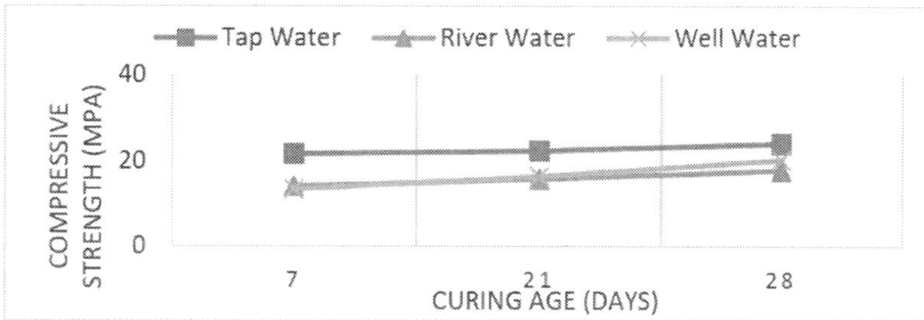

Figure 1. Compressive strength development for concrete with different mixing water types (1:2:4).

This could be because the water possessed impurities that may have been detrimental to the concrete strength development. The concrete mixed with well water yielded concrete of compressive strength of 20.24 MPa while the concrete mixed with tap water (which is supposed to be potable water) yielded the highest compressive strength of 24.01 MPa. It was also noted that the concrete mixed with well water possessed a lower 7-day strength than the concrete mixed with river water. The cube sample yielded a compressive strength at 28 days that was higher than the compressive strength of the samples mixed with river water.

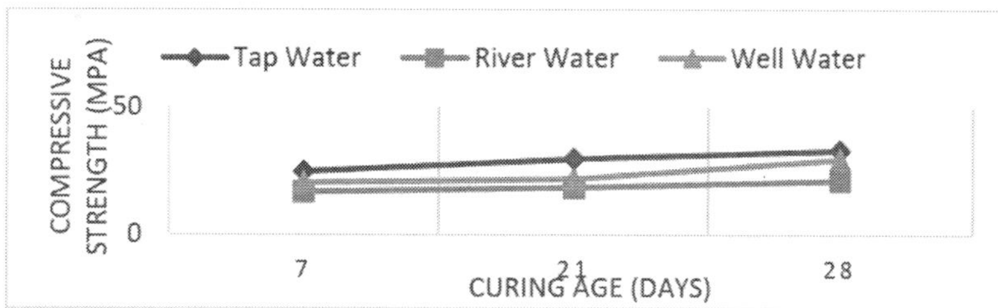

Figure 2. Compressive strength development for concrete with different mixing water types (1:1$^1/_2$:3).

Figure 2 shows the compressive strength of 1:1$^1/_2$:3 concrete mixed with water from different sources. The trend of the results is like that for concrete of mix ratio 1:2:4. The expected strength of mix 1:$^1/_2$:3 concrete is 30 MPa but both the concrete mixed with river water and well water respectively did not attain the 28-day strength of 30 MPa.

The Tap Water produced the highest compressive strength. The Well Water produced lower compressive strength while the river water used produced the lowest compressive strength, with slow strength development as observed from the Figures 1 and 2. This can be attributed to the organic nature of the water which can be deduced from the physical properties of been pungent in odor and pale color and the acidity of the water can be attributed for low strength recorded. It was determined by De Melo *et al.* (2016), that River Eater possessed some form of humic acid that is harmful to concrete (Kucche *et al.* 2015). This may have been responsible for the low compressive strength obtained.

4 CONCLUSIONS AND RECOMMENDATIONS

The Tap and Well Water samples tested did not meet the WHO and NIS standards for acidity but were within the standards for all the other parameters. The River Water was within the limits for acidity/pH value but possessed color while both the WHO and NIS standards specify that it must be clear. The findings also showed that Tap Water, which is the most potable to drink, is the most suitable for concrete works which agrees with what was indicated from literature (Shetty

2000, Neville and Brooks 2010). While the use of potable water is generally safe, water not fit for drinking may also be satisfactorily used in making concrete. As a rule, any water with pH of 6.0 to 8.0 which does not taste saline or brackish is suitable for use (Kucche *et al.* 2015). Color and odor do not indicate that deleterious substances are present in water. Natural waters that are slightly acidic are harmless, but water containing humic or other organic acids may adversely affect the strength of concrete. The Well Water used was tested to be acidic therefore not complying with W.H.O and NIS standard for drinking water in terms of the pH value. Concrete specimens produced with Well water had densities. Tap Water (or potable water) is the most appropriate for concrete works but in the circumstance where it is not available, Well Water can still be used. It should be noted that the strength might be reduced but not significantly. It is also recommended that River Water should be avoided by all means in concrete works.

Acknowledgments

The researchers wish to acknowledge the Covenant University Centre for Research, Innovation and Discovery (CUCRID) for their support in this study.

References

ACI 318M, Building Code Requirements for Structural Concrete (ACI 318M-14): an ACI Standard: Commentary on Building Code Requirements for Structural Concrete (ACI 318M-14), ACI Committee. *Farmington Hills, MI: American Concrete Institute*, 2015.

Aitcin, P. C., Water and its Role on Concrete Performance, *Science and Technology of Concrete Admixtures*, 75-86, 2016.

ASTM, C136, Standard Test Method for Sieve Analysis of Fine and Coarse Aggregates, American Society for Testing and Materials, West Conshohocken, USA, 2006.

ASTM, C94, Standard Specification for Ready Mixed Concrete, American Society for Testing and Materials, West Conshohocken, USA, 2016.

De Melo, B. A. G., Motta, F. L., and Santana, M. H. A., Humic Acids: Structural Properties and Multiple Functionalities for Novel Technological Developments, *Materials Science and Engineering: C*, 62, 967-974, 2016.

Ekmekyapar, T., and Orung, I., Buildings Materials, *Ataturk Univ. Agricultural Faculty Publication*, 145, 70-135, 2001.

Kucche, M. K., Jamkar, S. S., and Sadgir, P. A., Quality of Water for Making Concrete: A Review of Literature, 2015.

Kulkarni, V., Patil, S. G., Shivasharanappa, Study on Compressive Strength of Concrete by using Treated Domestic Waste Water as Mixing and Curing of Concrete, *International Journal of Research in Engineering and Technology*, 03(12), 152-156, 2014. Retrieved from https://ijret.org/volumes/2014v03/i12/IJRET20140312021.pdf on June 22, 2018.

Montero, J., and Laserna, S., Influence of Effective Mixing Water in Recycled Concrete, *Construction and Building Materials*, *132*, 343-352, 2017.

Neville, A., Water Cinderella Ingredient of Concrete, *Concrete International*, *22*(9), 66-71, 2000.

Neville, A. M., and Brooks, J. J., *Concrete Technology*. Second edition, Prenticehall, 269, 276, 2010.

Omuh, I. O., Mosaku, T. O., Joshua, O., Ojelabi, R. A., Amusan, L. M., Afolabi, A. O., and Arowolo, A. O., Data on Mixing and Curing Methods Effects on the Compressive Strength of Concrete, *Data in Brief*, 18, 877-881, 2018.

Rakesh, A. M., and Dubey, S. K., Effects of Different Types of Water on Compressive Strength of Concrete, *International Journal of Emerging Technologies*, 5(2), 40-50, 2014.

Shetty, M. S., *Concrete Technology*, S. Chand and Company Ltd., New Delhi, India, 2000.

Interdependence between Structural Engineering and Construction Management
Edited by Ozevin, D., Ataei, H., Modares, M., Gurgun, A., Yazdani, S., and Singh, A.

ISBN: 978-0-9960437-6-2

RESPONSE SURFACES FOR PROPERTIES OF CONCRETE WITH CLAY BRICK POWDER AS SUBSTITUTE FOR CEMENT

OLUWAROTIMI OLOFINNADE[1], ANTHONY EDE[1], KAYODE JOLAYEMI[1], KEHINDE OYEYEMI[2], and JULIUS NDAMBUKI[3]

[1]*Dept of Civil Engineering, Covenant University, Ota, Nigeria*
[2]*Dept of Applied Geophysics, Covenant University, Ota, Nigeria*
[3]*Dept of Civil Engineering, Tshwane University of Technology, Pretoria, South Africa*

This study examined the application of pulverize clay brick wastes as alternative constituent for Portland cement replacement in green concrete production using the response surface methodology (RSM). The adopted response surface approach is central composite design (CCD). The statistical models were developed between the concrete constituents (clay brick powder and water cement ratio) and their response variables (slump, compressive and split tensile strength). Relationships were established and mathematical models in terms of actual factors from predicted responses were developed. The influence of the considered factors on the properties of response were visually observed from the contour and response surface plots. The statistical models of experimental values clearly depict that the obtained experimental values are in close agreement with the predicted values, which validates the response surface models with desirability value of 1. The results show that pulverized clay brick waste have significant influence on the properties of concrete than the water cement (w/c) ratio, however, a declining trend was seen for all analyzed concrete properties. In addition, this study showed that the pulverized waste brick can be used as alternative substitute for Portland cement up to 20% in the production of sustainable concrete.

Keywords: Mathematical modeling, Waste clay brick, Cement, Sustainable concrete.

1 INTRODUCTION

Sustainability is a key issue for the construction industry because the industry relies more on nonrenewable resources for construction materials (Calkins 2009). Moreover, these materials constitute major components for the construction of engineering infrastructures; hence, limiting the impact of these materials on the environment is now very crucial. Concrete is one of the commonly used construction materials and finding ways to reduce its production impact on the environment is important (Akinwumi *et al.* 2016, Olofinnade *et al.* 2018b). Moya *et al.* (2010) reported that the process of producing concrete demands a considerable amount of raw materials and energy thus resulting in the emission of significant quantity of greenhouse gases (GHG) into the atmosphere. Previous studies emphasized on the impact of CO_2 emission on the environment (Shakir *et al.* 2014, Olofinnade *et al.* 2016). Meanwhile, studies also depicted that the impact of cement on the environment can be limited through the use of supplementary cementitious materials such as metakaolin, fly ash and slag used as alternative material to partially replace cement (Bektas *et al.* 2008, Schneider *et al.* 2011). Moreso, additional benefits include energy and cost saving (Olofinnade *et al.* 2016, 2017, 2018a). With the continuous advancement in knowledge, new materials are being researched upon as alternative materials for cement replacement. One of such materials is clay brick wastes generated mostly from brick and ceramic industry or as construction and demolition (C&D) wastes (Bektas *et al.* 2008, Olofinnade *et al.* 2018a). The clay bricks are produced from natural clay by calcination at very high temperature not less than 900°C (Olofinnade *et al.* 2018a). The clay brick contains relevant chemical compositions used to classified pozzolanic materials and exhibit amorphousness due to the high temperature exposure (Olofinnade *et al.* 2018a). Consequently, when the clay brick waste is finely ground into powder, it can be deployed as pozzolana to partially replace cement in concrete or mortar production instead of discriminate disposing as reported by (Olofinnade *et al.* 2018a). Previous studies by researchers have focused on using clay brick wastes as partial replacement for cement (Bektas *et al.* 2008, Olofinnade *et al.* 2016, 2018a) and natural aggregates (Aliabdo *et al.* 2014) in concrete composites. This study aims at developing mathematical models using the response surface methodology (RSM) to model the response of concrete properties containing pulverize clay brick wastes as alternative constituent for Portland cement replacement in green concrete.

2 MATERIALS

In this study, the pulverized clay brick wastes were sourced as generated wastes from brick factory in Lagos Metropolis, and used to partially replace Portland cement in concrete at dosage percentages of 0%, 10%, 20%, 30%, and 40%. The Portland cement (Grade 42.5) with specific gravity of 3.15 was used in all the concrete mixes in accordance to NIS 444 (2003). The natural gravel with a maximum size of 19 mm and specific gravity of 2.70, while the sand with a particles size ranging from 0.075 to 4.75 mm and specific gravity of 2.62 were used in this study. The chemical compositions of the clay brick powder and cement are presented in Table 1. Example of the concrete mix proportioning is presented in Table 2. Tests were carried out in accordance with BS EN 12350-2 (2009) and BS EN 12390 (2002) respectively.

Table 1. Chemical composition of cement and clay brick powder.

Composition%	SiO_2	Al_2O_3	Fe_2O_3	CaO	MgO	K_2O	Na_2O	TiO_2	P_2O_5
Cement	24.08	19.40	6.28	74.25	3.96	0.85	1.74	0.62	1.21
Clay brick	60.64	14.23	4.93	0.27	1.72	1.44	1.94	0.98	0.90

2.1 Experimental Design

The RSM comprises of statistical and mathematical procedures of analyzing and modelling problems involving a response influenced by several factors. In this study, RSM was used to the analyzed and developed relationship for fresh and hardened properties of concrete at 28 days as the response of interest. The response surface design approach used is the face-centred central composite design (CCD) with $\alpha = 1$ and full quadratic model for each response. The dosage percentage of clay brick (%) is represented as A and water-cement ratio (w/c) is coded as B; both considered as factors. The Eq. (1) depicts the full quadratic model used in this study.

$$Y=\beta_0+\beta_1A+\beta_2B+\beta_{12}AB+\beta_{11}A^2+\beta_{22}B^2 \quad (1)$$

where Y is predicted response, β_0 is the intercept, β_1 and β_2 are linear influence coefficients, β_{11} and β_{22} are quadratic influence coefficients, and β_{12} is interaction coefficient.

Table 2. Concrete mix proportion.

Mixture	Control Mix	CB10%	CB20%	CB30%	CB40%
w/b ratio	0.5	0.5	0.5	0.5	0.5
Water, kg/m^3	197	197	197	197	197
Cement, kg/m^3	385	347	308	269.5	231
Sand, kg/m^3	790	790	790	790	790
Gravel, kg/m^3	1073	1073	1073	1073	1073
Clay brick, kg/m^3	0	38	77	115.5	154
Target strength, MPa	25	25	25	25	25

3 RESULTS AND DISCUSSION

3.1 Response Surface for Fresh Properties: Slump

The obtained values from the experiment show that the slump shows a reduction trend as the percentage of the clay brick content increases. The obtained values were used to evaluate the influence of w/c ratio and clay brick (%) in fresh concrete mixes using RSM. Table 3 depicts the results of the analysis of variance (ANOVA) carried out for slump response surface quadratic model. The results show that the clay brick content % effect ($P < 0.0001$) and the quadratic clay brick content% effect ($P < 0.0001$) were statistically significant. The mathematical model for the slump is presented in Eq. (2). Furthermore, Table 4 shows the actual and predicted values of slump (mm) of fresh concrete.

$$\text{Slump (mm)}(Y) = 7185.81990 + 0.26998\,A - 28794.78610\,B - 0.056409\,AB - 0.024390A^2 + 29085.26267\,B^2 \quad (2)$$

Table 3. ANOVA for slump (mm) response surface quadratic model.

Source	Sum of Squares	Degree of freedom	Mean Square	F Value	p-value Prob > F
Model	1653.08	5	330.62	115.75	< 0.0001
Clay Brick, A (%)	1374.13	1	1374.13	481.08	< 0.0001
w/c ratio, B	3.817E-004	1	3.817E-004	1.336E-004	0.9911
AB	1.328E-004	1	1.328E-004	4.649E-005	0.9948
A^2	254.62	1	254.62	89.14	< 0.0001
B^2	1.56	1	1.56	0.55	0.4839
Residual	19.99	7	2.86		
Lack of Fit	19.99	5	4.00		
Pure Error	0.000	2	0.000		
Cor Total	1673.08	12			

3.2 Response Surface for Compressive and Split Tensile Strength

The obtained values from the experiment show that the compressive strength and the split tensile strength of the hardened concrete measured reduce as the percentage of the clay brick content increases. However, the results show a significant compressive strength increase at lower dosage level of cement replacement with clay brick powder (%). The obtained values were used to evaluate the influence of w/c ratio and clay brick (%) in fresh concrete mixes using RSM.

Table 4. Actual and predicted values of slump, compressive and split tensile strength (MPa) at 28 days.

Run	Slump (mm)		Compressive Strength (MPa)		Split tensile strength (MPa)	
	Actual Value	Predicted Value	Actual Value	Predicted Value	Actual Value	Predicted Value
1	55.00	54.84	23.00	23.44	3.80	3.78
2	55.00	59.00	24.89	24.02	3.66	3.63
3	60.00	59.74	20.00	20.02	3.34	3.35
4	30.00	29.68	20.00	20.02	3.34	3.27
5	55.00	54.10	20.00	19.83	2.55	2.51
6	55.00	54.10	23.56	23.58	3.34	3.35
7	30.00	30.42	23.00	23.44	3.34	3.35
8	30.00	30.39	23.67	23.63	3.80	3.80
9	55.00	54.82	23.00	22.96	3.80	3.83
10	55.00	54.10	22.00	22.04	3.34	3.35
11	45.00	45.05	23.00	23.02	3.34	3.35
12	60.00	59.02	23.50	23.79	2.80	2.87
13	60.00	59.75	23.60	23.44	2.55	2.56

Table 4 also shows the actual and predicted values of compressive and split tensile strength of concrete. Table 5 depict the results of the ANOVA carried out for strength response surface quadratic model respectively. The results show that for compressive strength, the quadratic clay brick content % effect ($P < 0.0036$), were statistically significant. Similarly, for split tensile strength, the quadratic clay brick content % effect ($P < 0.0005$), were statistically significant. The mathematical model in terms of actual factors for the compressive strength and split tensile strength are presented in Eqs. (3) and (4) respectively.

Compressive Strength,

$$Y\ (\text{MPa}) = -4755.32581 + 164.46331\ \mathbf{A} + 19253.54710\ \mathbf{B} - 666.74000\ \mathbf{AB} + 0.044192\ \mathbf{A}^2 - 19391.46172\ \mathbf{B}^2 - 0.097500\ \mathbf{A}^2\mathbf{B} + 676.00000\ \mathbf{AB}^2 \quad (3)$$

Split Tensile Strength,

$$Y\ (\text{MPa}) = -390.89076 + 0.23793\ \mathbf{A} + 1592.56093\ \mathbf{B} - 0.51178\ \mathbf{AB} - 4.4137E\text{-}004\ \mathbf{A}^2 - 1606.34270\ \mathbf{B}^2 \quad (4)$$

Furthermore, the contour and response for strength properties are depicted in Figures 1 and 2.

Table 5. ANOVA for split tensile strength (STS) and compressive strength (CS) MPa quadratic model.

Source	Sum of Squares		Degree of freedom		Mean Square		F Value		p-value Prob > F	
	STS	CS	STS	CS	STS	CS	STS	CS	STS	CS
Model	2.25	29.09	5	7	0.45	4.16	227.06	16.16	< 0.0001	0.0037
Clay brick, A (%)	2.14	9.78	1	1	2.14	9.78	1079.39	38.04	< 0.0001	0.0016
w/c ratio, B	8.686E-003	0.22	1	1	8.686E-003	0.22	4.38	0.87	0.0746	0.3931

Table 5 (contd). ANOVA for STS and CS MPa quadratic model.

Source	Sum of Squares		Degree of freedom		Mean Square		F Value		p-value Prob > F	
	STS	CS	STS	CS	STS	CS	STS	CS	STS	CS
AB	7.338E-003	0.078	1	1	7.338E-003	0.078	3.70	0.30	0.0957	0.6046
A^2	0.071	6.80	1	1	0.071	6.80	35.89	26.44	0.0005	0.0036
B^2	4.902E-003	0.062	1	1	4.902E-003	0.062	2.47	0.24	0.1598	0.6439
Residual	0.014	1.29	7	5	1.981E-003	0.26				
Lack of Fit	0.014	1.05	4	3	3.468E-003	0.35		2.90		0.2665
Pure Error	0.000	0.24	3	2	0.000	0.12				
Cor Total	2.26	30.37	12	12						

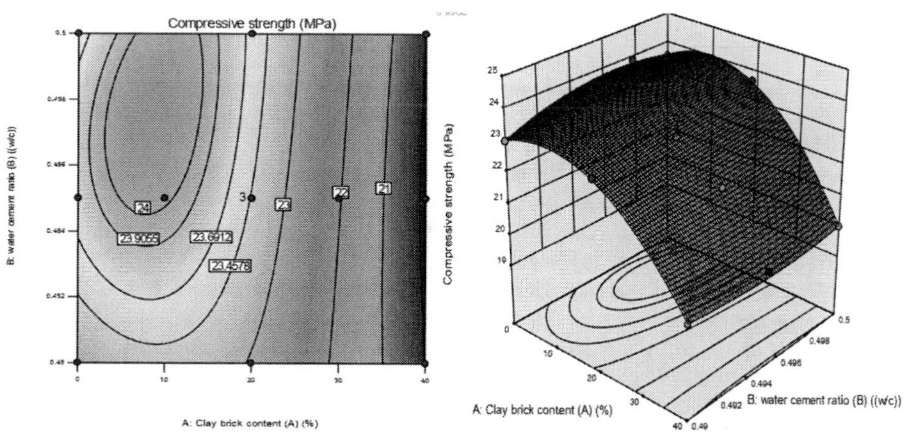

Figure 1. Contour plot and response surface of compressive strength.

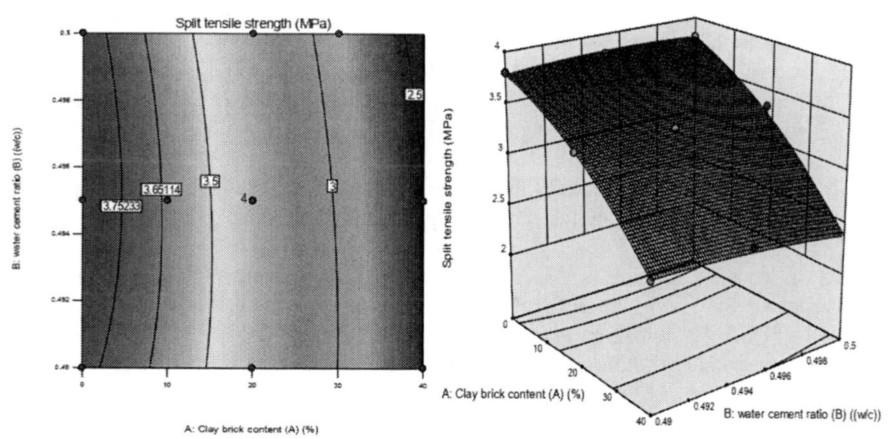

Figure 2. Contour plot and response surface of split tensile strength.

4 CONCLUSION

In this study, pulverized clay brick waste was used as partial substitute for cement in concrete of moderate strength. The experimental study clearly indicated a significant decrease in the slump with increasing clay brick powder content. Moreover, the compressive and split tensile strength of the hardened concrete specimens were found to decrease as the replacement quantity of clay brick increases compared to control. Using the response surface method, the measured values were analyzed by the face-centered composite surface design (CCD) employing the full quadratic model to determine the effect of clay brick and water-cement variables on the response, it was found that the models were significant. The results showed that pulverized clay brick waste had significant effect on the properties of concrete compared to water-cement (w/c) ratio, however, a declining trend was seen for all analyzed concrete properties. However, linear influence of clay brick content (%) was noticed to be more significant in the models.

Acknowledgments

The authors would like to appreciate the Centre for Research, Innovation and Discovery, Covenant University, Ota, Nigeria for supporting this research.

References

Akinwumi, I. I., Awoyera, P. O., Olofinnade, O. M., Busari, A. A., and Okotie, M, Rice Husk as A Concrete Constituent: Workability, Water Absorption and Strength of The Concrete, *Asian Journal of Civil Engineering*, 17, 887–898, 2016.

Aliabdo, A. A., Abd-Elmoaty, A. M., and Hassan, H. H., Utilization of Crushed Clay Brick in Concrete Industry, *Alexandria Engineering Journal,* 53(1), 151–168, 2014.

Bektas, F., Kejin, W., and Ceylan, H., Use of Ground Clay Bricks as A Pozzolanic Materials in Concrete, *Journal of ASTM International,* 5, ISSN 1546-962X, 2008.

BS EN 12350-2. *Testing of Fresh concrete,* European Committee for Standardization, 2009.

BS EN 12390. *Testing Hardened concrete,* European Committee for Standardization, 2002.

Calkins, M., *Materials for Sustainable Sites: A Complete Guide to The Evaluation, Selection, and Use of Sustainable Construction Materials.* Hoboken: Wiley, 2009.

Moya, J. A., Pardo, N., and Mercier, A., *Energy Efficiency and CO_2 Emissions: Prospective Scenarios for The Cement Industry*, JRC Scientific and Technical Report, 2010.

NIS 444: Part 1, *Composition, Specifications and Conformity Criteria for Common Cements*, Nigeria Industrial Standards Center, Standard Organisation of Nigeria (SON), Abuja, 2003.

Olofinnade, O. M, Ndambuki, J. M, Ede, A. N., and Booth, C, Application of Waste Glass Powder as a Partial Cement Substitute Towards More Sustainable Concrete Production, *International Journal of Engineering Research in Africa*, 31, 77–93, 2017.

Olofinnade, O. M., Ede, A. N., and Booth, C. A., *Sustainability of Waste Glass Powder and Clay Brick Powder as Cement Substitute in Green Concrete*, Handbook of Environmental Materials Management, Springer, 2018a.

Olofinnade, O. M., Ede, A. N., Ndambuki, J. M, and Bamigboye, G. O., Structural Properties of Concrete Containing Ground Waste Clay Brick Powder as Partial Substitute for Cement, *Materials Science Forum,* 866, 63–67, 2016.

Olofinnade, O. M., Ede, A. N., Ndambuki, J. M., Ngene, B. U., Akinwumi, I. I., and Ofuyatan, O., Strength and Microstructure of Eco-Concrete Produced Using Waste Glass as Partial And Complete Replacement for Sand, *Cogent Engineering*, 5, 1483860, 1-19, 2018b.

Schneider, M., Romer, M., Tschudin, M., and Bolio, H, Sustainable Cement Production –Present and Future, *Cement and Concrete Research,* 43, 642–650, 2011.

Shakir, A. A., Naganathan, S., and Mustapha, K. N, Effect of Quarry Dust and Billet Scale Additions on The Properties of Fly Ash Bricks, *Iranian Journal of Science and Technology. Transactions of Civil Engineering*, 38(C1), 51–60, 2014.

Interdependence between Structural Engineering and Construction Management
Edited by Ozevin, D., Ataei, H., Modares, M., Gurgun, A., Yazdani, S., and Singh, A.

ISBN: 978-0-9960437-6-2

MICROSTRUCTURE INVESTIGATION OF SEAWATER VS. FRESHWATER CEMENT PASTES

ADEL YOUNIS[1], USAMA EBEAD[1], PRANNOY SURANENI[2], and ANTONIO NANNI[2]

[1]*Dept of Civil and Architectural Engineering, Qatar University, Doha, Qatar*

[2]*Dept of Civil, Architectural and Environmental Engineering, University of Miami, Coral Gables, USA*

Recently, seawater has emerged as viable mixing water for concrete, especially in the case of non-reinforced concrete applications or with the use of non-corrosive reinforcement. Previous studies concerning seawater-mixed concrete mostly revealed an initial slight increase in its strength performance (i.e., till Day 14 following mixing), followed by a strength reduction of 7–15% (i.e., after 28 days or longer) as compared to the conventional freshwater-mixed concrete. With an attempt to explain such observations, this paper aims at comparing the microstructure of freshwater- and seawater-mixed cement pastes. Scanning electron microscopy was utilized to observe the microstructure of freshwater and seawater pastes at Days 3 and 28 following mixing. At Day 3, seawater paste was observed to have more densified microstructure as compared to that of the freshwater counterpart, resulting in relatively higher strength performance. At Day 28, the microstructure was almost similar for the two cement pastes. However, seawater paste was observed to have salt impurities as a result of seawater ions, which possibly cause a slightly lower strength performance as compared to the freshwater paste.

Keywords: Sustainability, Seawater-mixed concrete, Scanning electron microscopy, Secondary electrons, Back-scattered electrons, Energy dispersive X-ray.

1 INTRODUCTION

More often than not, there has been a growing interest in using alternative resources of raw materials in concrete to achieve sustainability goals (Saeed *et al.* 2012, Suraneni *et al.* 2018, Xiao *et al.* 2017), given the significant environmental impacts of producing concrete using the current practices (Miller *et al.* 2016). With the increase of global concern of freshwater scarcity (Mekonnen and Hoekstra 2016), seawater is being increasingly proposed as an alternative mixing water for concrete, especially after the development of non-corrosive reinforcement such as fiber-reinforced polymer (FRP) to counter possible corrosion (Aly *et al.* 2006, Benmokrane *et al.* 2006). In principle, using seawater concrete in association with FRP reinforcement seems to be viable from both technical and economic standpoints (Younis *et al.* 2017, 2018a).

Significant research efforts were devoted in the past few decades to grasp, as possible, the effects of seawater mixing on the strength performance of the resulted concrete (Kaushik and Islam 1995, Nishida *et al.* 2013, Wegian 2010, Xiao *et al.* 2017, Younis *et al.* 2018b). In summary, an initial slight increase was observed in the strength performance of seawater concrete as compared to the freshwater-mixed counterpart (until 7–14 days following mixing), followed by a slight reduction of 7–15% after 28 days. The current paper aims at providing some explanations for such observations through the lens of microstructure investigation. Scanning electron microscopy was used to study the microstructure of both seawater and freshwater pastes at Days 3 and 28 following mixing. At first, materials and test methods adopted in this study will be explained. After that, test results will be presented and discussed, from which the conclusions of this comparison will be drawn.

2 METHODOLOGY

2.1 Materials

A comparison was established between freshwater and seawater cement pastes of the same water-to-cement ratio (0.34). The difference between the two mixtures was only in the mixing water. Chemical characteristics were obtained for both types of mixing water and stipulated in Table 1. The pH and alkalinity measurements were within the acceptable limits for both water types. However, the sulfates, chlorides, and dissolved solids in seawater were (as expected) significantly higher than those of the freshwater or even the allowable limits. Table 2 presents the chemical composition of the ordinary Portland cement used in this study, obtained by BS EN 196-2 (2013). The fineness obtained using Blaine air permeability test (by BS EN 196-6 2010) was measured as 3350 cm^2/g.

Table 1. Chemical characterization of the two mixing water types.

Measure (unit)	Method	Max. limit	Freshwater results	Seawater results
Chloride (mg/L)	BS 1377-3 (1990)	1000	14.1	18600
Sulfate (mg/L)	BS 1377-3 (1990)	2000	20.9	2359
Alkalinity (mg/L)	BS 6068-2.51 (1996)	500	69.5	149
Total dissolved solids (mg/L)	BS 1377-2 (1990)	2000	62.0	30300
pH at 25 °C (-)	BS 6068-2.5 (1995)	6.5–9.0	8.1	8.2

Table 2. The chemical composition of the cement used.

Chemical	Composition (% wt.)
SiO_2	19.90
Al_2O_3	4.30
Fe_2O_3	3.21
CaO	61.84
MgO	4.49
$Na_2O + K_2O$	0.51
S	2.70
Cl	0.051
Loss on Ignition	2.34
$C_2S + C_3S$	69.98
C_3A	5.96
C_4AF	9.76

2.2 Observation Methods

Scanning electron microscopy was performed, in accordance with ASTM C1723 (2016), for seawater and freshwater pastes at Days 3 and 28 following mixing. Three analysis techniques were utilized (Winter 2012):

- Secondary electron (SE) imaging: in which the secondary electrons (generated as a result of the collision with the passing electron beam) were detected. This provides information about the morphology of the specimen's surface, and therefore, the specimens used in this technique were surface-fractured.

- Backscattered electron (BSE) imaging: in which the backscattered electrons were detected. BSE imaging provides a general indication of the atomic distribution using the discrepancy in the surface brightness. In accordance with ASTM C1723 (2016), the specimens analyzed by BSE imaging had a polished surface.
- Energy-dispersive X-ray microanalysis (EDX): in which the X-rays (produced as a result of the collision with the projected electron beam) were detected. Analyzing the resulting X-ray spectra is regarded as a useful tool to identify the chemical composition.

3 RESULTS AND DISCUSSION

3.1 Early-Age Microstructure

Figure 1 shows the difference in BSE images between freshwater and seawater pastes at Day 3 following mixing. From a qualitative viewpoint, seawater paste generally showed a more densified microstructure as compared to that of the freshwater counterpart. In the case of freshwater paste, it was observed that pores (black-colored) and anhydrous cement (white-colored) occupy a larger area as compared to that of the seawater-mixed counterpart. Likewise, the higher early-age strength of seawater paste can be evidenced by comparing the SE images of both 3-day-aged pastes (Figure 2). By observing the morphological behavior, it can be realized that the C-S-H needles (Franus *et al.* 2015) were less densified and more distributed in the freshwater paste (Figure 2-a); whereas a more solid microstructure can be observed in case of seawater paste (Figure 2-b). Such observations may explain the relatively higher early-strength performance of seawater concrete as reported in previous studies (Kaushik and Islam 1995, Nishida *et al.* 2013, Wegian 2010, Xiao *et al.* 2017). The higher early-age strength of seawater concrete can be attributed to the lower porosity due to the acceleration of cement hydration (Kaushik and Islam 1995). It is also suggested that ettringite and gypsum crystals fill in the voids in the seawater paste and thus further densify the microstructure (Katano *et al.* 2012).

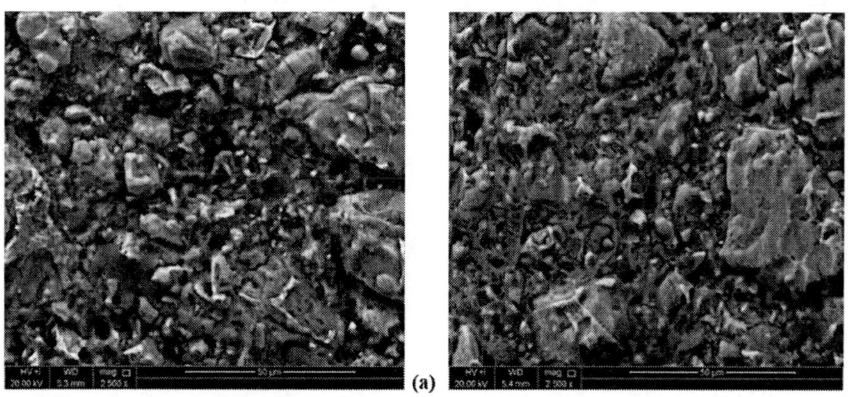
(a) (b)

Figure 1. BSE images for (a) freshwater paste and (b) seawater paste at Day 3.

(a) (b)

Figure 2. SE images for (a) freshwater paste and (b) seawater paste at Day 3.

3.2 Later-Age Microstructure

Figure 3 shows the BSE images for both pastes at Day 28 following mixing. As intuitively expected, the microstructure of the freshwater and seawater pastes at later ages was more densified than that of the early-age pastes (in Figure 1). The microstructure solidness of seawater and freshwater pastes, however, appeared to be the same when observed through the lens of Figure 3. With a more in-depth insight (i.e., SE imaging accompanied by EDX microanalysis), one can observe the formation of salt impurities in seawater paste (see Figure 4), that are most likely to be gypsum as shown from the EDX and also suggested by Weerdt and Justnes (2015). It is suggested that part of the calcium in the pore solution reacted with the sulfate ions (that are abundant in seawater) to form such phases. In this mechanism, magnesium sulfates ($MgSO_4$) in seawater react with calcium hydroxide ($Ca(OH)_2$) in the pore solution to form soluble magnesium hydroxide ($Mg(OH)_2$) and gypsum ($CaSO_4.2H_2O$), yielding an expansive crystallization pressure that decreases concrete strength (Wegian 2010). This can be regarded as a possible explanation for the slight reduction in the strength performance of seawater mature concrete as compared to that of freshwater-mixed counterpart (Nishida *et al.* 2013, Xiao *et al.* 2017).

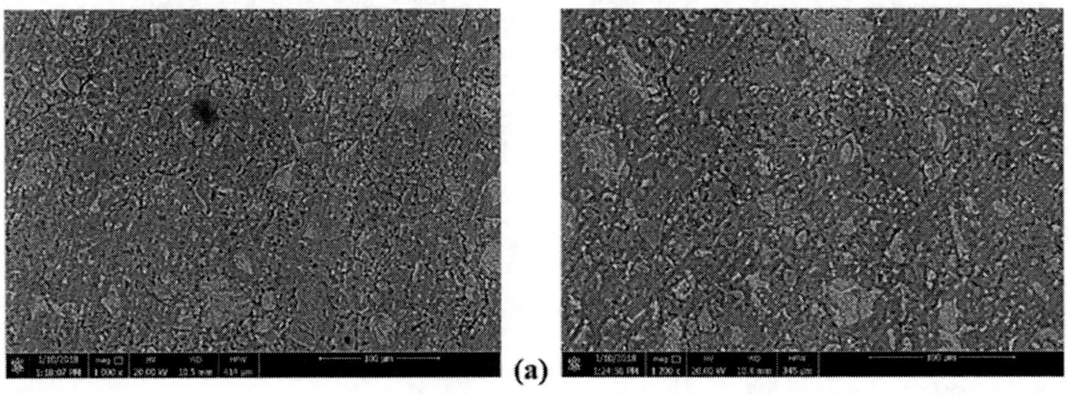

(a) (b)

Figure 3. BSE images for (a) freshwater paste and (b) seawater paste at Day 28.

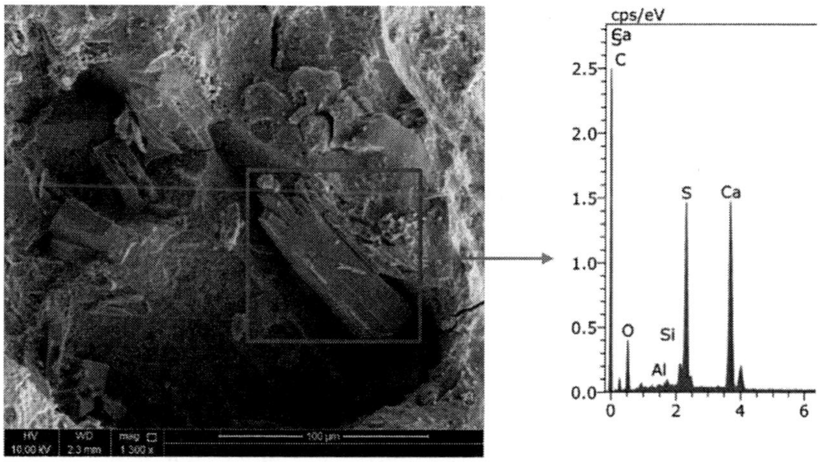

Figure 4. Crystallization of salt impurities in seawater paste.

4 CONCLUSION

This paper aimed at investigating the difference in the microstructure between seawater and freshwater cement pastes. To achieve this, scanning electron microscopy was utilized as an analytical technique to study the microstructure of both pastes using secondary electron imaging, back-scattered electron imaging, and energy dispersive X-ray microanalysis. Each paste was observed at Day 3 and Day 28 following mixing. Based on the study results, the following conclusions have been drawn:

- The more densified microstructure was observed in seawater paste at early ages as compared to that of the freshwater counterpart. This explains the relatively higher early-strength performance of seawater concrete (i.e., within 5% higher as of 14 days following mixing) as reported in the literature.
- At later ages, the microstructure was almost similar for the two cement pastes. However, salt impurities were observed in seawater paste as a result of seawater ions. Such phases can be regarded as a possible explanation to the slightly inferior strength performance of seawater concrete at later ages (i.e., within 7–15% lower after 28 days or later) as reported in the literature.

Finally, it should be emphasized that, given its high contents of chloride, the use of seawater in concrete mixtures is deemed most practical in concrete structures with non-reinforced applications or with the use of non-corrosive reinforcement (e.g., FRP).

Acknowledgments

This paper was made possible by the NPRP grant # NPRP 9-110-2-052 from the Qatar National Research Fund (a member of Qatar Foundation). The findings achieved herein solely the responsibility of the authors. Special thanks are due to Readymix Qatar (as a part of LafargeHolcim) for providing the expertise that greatly assists the authors in their research.

References

Aly, R., Benmokrane, B., and Ebead, U. A., Tensile Lap Splicing of Bundled CFRP Reinforcing Bars In Concrete, *Journal of Composites for Construction*, 10(4), 287–294, 2006.

ASTM C1723-16, *Standard Guide for Examination of Hardened Concrete Using Scanning Electron*

Microscopy, ASTM International, 2016.

Benmokrane, B., Aly, R., and Ebead, U. A., Tensile Lap Splicing of FRP Reinforcing Bars in Concrete, *ACI Structural Journal*, ACI, 103(6), 857–864, 2006.

BS 1377-2, *Methods of Test for Soils for Civil Engineering Purposes, Classification Tests*, BSI, 1990.

BS 1377-3, *Methods of Test for Soils for Civil Engineering Purposes, Chemical and Electro-Chemical Tests*, BSI, 1990.

BS 6068-2.50, *Water Quality, Physical, Chemical and Biochemical Methods, Determination of pH.* BSI, 1995.

BS 6068-2.51 *Water Quality. Determination of Alkalinity, Determination of Total and Composite Alkalinity*, BSI, 1996.

BS EN 196-2, *Method of Testing Cement, Chemical Analysis of Cement*, BSI, 2013.

BS EN 196-6, *Methods of Testing Cement, Determination of Fineness, British Standards Institution, London*, 2010.

Franus, W., Panek, R., and Wdowin, M., *SEM Investigation of Microstructures in Hydration Products of Portland Cement*, Springer Proceedings in Physics, 164, 105–112, 2015.

Katano, K., Takeda, N., Ishizeki, Y., and Iriya, K., *Properties, and Application of Concrete Made with Sea Water and Un-Washed Sea Sand*, Third International Conference on Sustainable Construction Materials and Technologies, 2012.

Kaushik, S. K., and Islam, S., Suitability of Sea Water for Mixing Structural Concrete Exposed to A Marine Environment, *Cement and Concrete Composites*, 17(3), 177–185, 1995.

Mekonnen, M. M., and Hoekstra, A. Y., Four Billion People Facing Severe Water Scarcity, *Science Advances*, 2(2), e1500323, 2016.

Miller, S. A., Horvath, A., and Monteiro, P. J. M., Readily Implementable Techniques Can Cut Annual CO_2 Emissions from The Production of Concrete by Over 20%, *Environmental Research Letters*, IOP Publishing, 11(7), 74029, 2016.

Nishida, T., Otsuki, N., Ohara, H., Garba-Say, Z. M., and Nagata, T., Some Considerations for Applicability of Seawater as Mixing Water in Concrete, *Journal of Materials in Civil Engineering*, 27(7), B4014004, 2013.

Saeed, H. A., Tagnit-Hamou, A., Ebead, U. A., and Neale, K. W., Stoichiometric Study of Activated Glass Powder Hydration, *Advances in Cement Research*, 24(2), 91–101, 2012.

Suraneni, P., Azad, V. J., Isgor, O. B., and Weiss, J., Role of Supplementary Cementitious Material Type in the Mitigation of Calcium Oxychloride Formation in Cementitious Pastes, *Journal of Materials in Civil Engineering*, 30(10), 04018248, 2018.

Weerdt, K. De, and Justnes, H., The Effect of Sea Water on The Phase Assemblage of Hydrated Cement Paste, *Cement and Concrete Composites*, 55, 215–222, 2015.

Wegian, F. M., Effect of Seawater for Mixing and Curing on Structural Concrete, *The IES Journal Part A: Civil and Structural Engineering*, Taylor and Francis, 3(4), 235–243, 2010.

Winter, N. B., *Scanning Electron Microscopy for Cement and Concrete*, WHD Microanalysis, 2012.

Xiao, J., Qiang, C., Nanni, A., and Zhang, K., Use of Sea-Sand and Seawater in Concrete Construction: Current Status and Future Opportunities, *Construction and Building Materials*, 155, 1101–1111, 2017.

Younis, A., Ebead, U. A., and Nanni, A., *A Perspective on Seawater/FRP Reinforcement in Concrete Structures,* Proceedings of the Ninth International Structural Engineering and Construction Conference, Resilient Structures and Sustainable Construction, ISEC Press, Valencia, Spain, St-38, 2017.

Younis, A., Ebead, U., and Judd, S., Life Cycle Cost Analysis of Structural Concrete Using Seawater, Recycled Concrete Aggregate, and GFRP Reinforcement, *Construction, and Building Materials*, Elsevier Ltd, 175, 152–160, 2018a.

Younis, A., Ebead, U., Suraneni, P., and Nanni, A., Fresh and Hardened Properties of Seawater-Mixed Concrete, *Construction and Building Materials*, 190(C), 276–286, 2018b.

Interdependence between Structural Engineering and Construction Management
Edited by Ozevin, D., Ataei, H., Modares, M., Gurgun, A., Yazdani, S., and Singh, A.

ISBN: 978-0-9960437-6-2

CONSTITUTIVE MODELS FOR PREDICTING MECHANICAL PROPERTIES OF NATURAL POZZOLAN-BASED ALKALI ACTIVATED CONCRETE

MOHAMMED IBRAHIM[1], MEGAT AZMI MEGAT JOHARI[2], MUHAMMAD KALIMUR RAHMAN[1], MOHAMMED MASLEHUDDIN[1], and HATIM DAFALLA MOHAMED[1]

[1]*Center for Engineering Research, Research Institute, KFUPM, Dhahran, KSA*
[2]*School of Civil Engineering, Universiti Sains Malaysia, Pulau Pinang, Malaysia*

A large variety of mix design variables and environmental conditions in which alkali activated concretes (AAC) are cured, influences the nature and intensity of the binder formed. Equations developed for OPC-based concrete in various codes and published work may not accurately predict engineering properties of these AAC. AAC in this study was synthesized utilizing natural pozzolan (NP) in the presence of alkaline activators. Nano-SiO_2 was added for enhancing the strength development at room temperature curing as NP is a low calcium precursor material. Development of compressive strength, flexural strength and modulus of elasticity were investigated. Using the data generated, two constitutive models, relating the compressive strength to flexural strength and modulus of elasticity were developed and compared with the equations specified in international codes for OPC-based concrete and previous studies in the area. The results show that the ACI 318, underestimates the flexural strength and overestimates the modulus of elasticity of alkali activated concrete. However, constitutive models developed in this study are in good agreement with the equations proposed in the previous works for AAC. Only the limited data available to date on the engineering properties of AAC cannot be used to establish robust constitutive relationships and more research is required in this area.

Keywords: Curing, Strength, Modulus of elasticity.

1 INTRODUCTION

Alkali activated binders (AABs) are evolving as the possible alternative to OPC due to its multiple benefits (Rovnanik 2010). Thus far, industrial byproduct, such as fly ash, has been extensively used as source material in synthesizing AABs and cured at temperatures between 40 to 80 °C (Hardjito *et al.* 2005). However, the engineering properties and microstructure of AAB depends on several factors, including; composition of alkaline activators and their ratio, fineness and constituents of source materials as well as curing conditions (Duxson *et al.* 2007). The strength gain of these binders is rapid when cured at elevated temperature, while, it is slow at room temperature curing (Pangdaeng *et al.* 2014). In order to widen the practical application of AAB in this study natural pozzolan (NP) was partially replaced with nanosilica (NS) to enable it to be cured at room temperature. Considering the fact that the nature of the binder formed during alkali activation is quite different than that of OPC, the behavior of AAC under different loading conditions could possibly be different than OPC-based concrete. Hence, constitutive models specified for OPC concrete by the international standards and codes, such as ACI 318-10 (ACI 318 2010), as well as equations proposed in the previous studies may not accurately predict engineering properties of NP-based AAC. According to the results of a study (Diaz-Loya *et al.* 2011), ACI 318-10 (ACI 318 2010) underestimates the flexural strength of AAC, while, it overestimates the modulus of elasticity for similar grade concrete. For instance, based on Diaz-Loya *et al.* proposal, the flexural strength estimated was 11% more than that computed by ACI 318-10 (ACI 318 2010). On the contrary, the modulus of elasticity of AAC reported by several researchers (Olivia and Nikraz 2012) was less than that estimated by ACI 318-10 equation for similar strength concrete. Therefore, the data obtained on the engineering properties in this study were used to develop constitutive models describing the relationship between compressive strength with flexural strength and modulus of elasticity using a regression analysis.

2 MATERIALS AND METHODS

2.1 Materials

Natural pozzolan (NP), in a powdered form of volcanic rock, was the main precursor material used in the study. The chemical composition of NP is given in Table 1. The physical properties of the NS used in the study are shown in Table 2. 14 M NaOH solution and Na_2SiO_3 of 3.3 silica modulus were used as alkaline activators. The composition of sodium silicate includes: H_2O: 62.5%, SiO_2: 28.75% and Na_2O: 8.75%. Crushed limestone, having specific gravity of 2.56, was used as coarse aggregate while dune sand having specific gravity of 2.62 was utilized.

Table 1. Chemical composition of NP.

Oxides	SiO_2	Al_2O_3	Fe_2O_3	CaO	MgO	K_2O	Na_2O	P_2O_5	TiO_2	LOI
Weight, %	40.48	12.90	17.62	11.83	8.33	1.67	3.60	1.37	0.60	1.6

Table 2. Physical properties of NS.

Property	Solid content, %	Avg. Particle size, nm	Bulk density, g/cm^3	Specific surface area, m^2/g	Na_2O content, %	Viscosity cps	pH
Value	50	35	1.4	80	0.2	15	9.5

2.2 Methods

Table 3 shows the quantities of constituent materials of mixes prepared by incorporating NS. All AAC mixtures were prepared with a constant Na_2SiO_3/NaOH weight ratio of 2.50. Additionally, OPC concrete was prepared using mix proportions generally adopted in the construction industry.

Table 3. Quantities of AAC mix constituents.

Mix #	Legend	NP, kg/m^3	NS, kg/m^3	Na_2SiO_3, kg/m^3	NaOH, kg/m^3	FA, kg/m^3	CA, kg/m^3
M0	0%-NS	400	0	150	60	650	1206
M1	1%-NS	396	8	150	60	646	1200
M2	2.5%-NS	390	20	150	60	640	1188
M3	5%-NS	380	40	150	60	630	1170
M4	7.5%-NS	370	60	150	60	620	1152

The concrete specimens were de-molded after 1 day of casting, placed in plastic bags to avoid evaporation of moisture, and kept in the laboratory that was maintained at 23±2 °C until the predetermined curing period. OPC concrete was cured under wet burlap. The compressive strength of concrete was measured after 3, 7, 14, 28, 56, 90 and 180 days of curing on 50 mm cube specimens. Prismatic specimens measuring 50 x 50 x 200 mm were prepared to determine the flexural strength of concrete using third point loading in accordance with ASTM C78 (ASTM C78 2010). The modulus of elasticity of concrete was measured on 75 mm diameter and 150 mm high cylindrical concrete specimens in accordance with ASTM C469 (ASTM C469 2010). In order to ensure the accuracy of results, three specimens for each period were prepared and tested.

3 RESULTS AND DISCUSSION

3.1 Compressive Strength

The compressive strength development in the AAC prepared with NS varying from 0 to 7.5% by weight and OPC concrete is depicted in Figure 1. The compressive strength was low at the onset of curing in the concrete specimens prepared with NS compared to those prepared without it, particularly after three days of curing. However, as the period of room curing continued, the strength gain was remarkably high in the concrete mixes containing NS than in the control specimens.

According to these results, there was about 117% and 110% increase in the strength in concrete containing 5% and 7.5% NS, respectively, compared to the control samples (0% NS), at 90 days. The remarkable improvement in the performance of the AAC containing NS, in terms of compressive strength, particularly with 5% NS, could be attributed to the transformation of source materials to the polymeric gel in the presence of highly reactive NS as well as due to possible particle packing effect of nanoparticles in the binder structure (Yip *et al.* 2005).

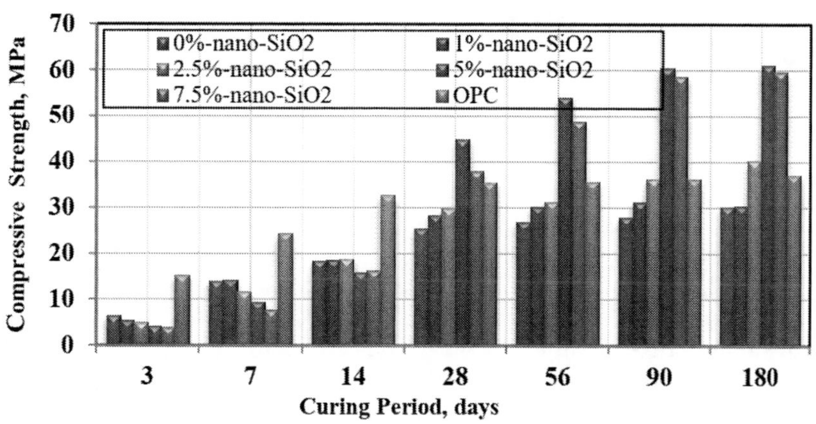

Figure 1. Compressive strength development of the concrete.

3.2 Flexural Strength

The flexural strength of the AAC mixtures varied from 3.88 to 6.20 MPa and 4.92 to 7.59 MPa, after 28 and 90 days of curing, respectively. The maximum value was observed in the AAB with 7.5% NS. After 90 days of curing, there was about 4.47%, 11.99%, 53.25% and 54.27% increase in the flexural strength in the AAC mixtures incorporating 1%, 2.5%, 5% and 7.5% NS, respectively, over the control mixture prepared without NS. These findings were consistent with that of compressive strength results, which exhibited a similar trend, as reported by other researchers, where partial replacement of binders with nanoparticles resulted in an improvement in the mechanical properties (Phoo-Ngernkham *et al.* 2014). The compressive and flexural strength after 28 and 90 days of curing were utilized to develop a correlation between them by plotting square root of compressive strength vs. flexural strength (Figure 2). The correlation between compressive and flexural strength can be expressed by Eq. (1) obtained by regression analysis. In order to compare the relationship developed in this study with other statistical models, the flexural strength was also computed using relationships proposed by Phoo-ngernkham *et al.* and Nath and Sarker, Eqns. (2) and (3), respectively, also plotted in Figure 2. The flexural strength was calculated using Eq. (4) as specified in ACI 318-10 and plotted in Figure 2, where ft is the flexural strength and f'c is the compressive strength, both in MPa.

$$ft=1.246\sqrt{f'c}-2.060 \qquad (1)$$

$$ft=1.430\sqrt{f'c}-4.223 \qquad (2)$$

$$ft=0.93\sqrt{f'} \qquad (3)$$

$$ft=0.6\sqrt{f'c} \qquad (4)$$

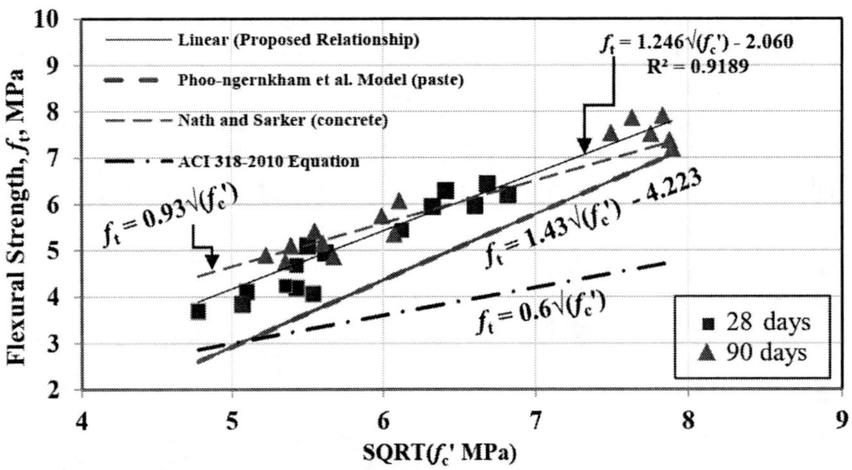

Figure 2. Correlation between flexural strength and compressive strength.

The flexural strength of AAC after 28 and 90 days of room curing computed, as a function of compressive strength, was greater than that estimated using the ACI 318-10 equation, particularly for those AAC mixes containing higher NS content. The flexural strength predicted using the model developed by Phoo-Ngernkham *et al.* was also less than that obtained in this study, which could be attributed to the absence of coarse and fine aggregates in the mixtures. However, the flexural strength calculated using the statistical model proposed by Nath and Sarker gives values of similar order. Based on these results, it can be concluded that ACI 318-10 equation did not accurately predict the flexural strength of AAC.

3.3 Modulus of Elasticity

The modulus of elasticity increased with increasing curing period as well as the quantity of NS in the AAC. It ranged between 13.80 to 24.00 GPa and 14.53 to 28.70 GPa, after 28 and 90 days of curing, respectively. AAC prepared with 5% NS resulted in the highest modulus of elasticity at both curing periods, whereas, it was the lowest in the concrete mixture prepared without NS. In this study, an attempt was made to develop a statistical model for computation of modulus of elasticity and to compare it with the models available in the literature using the obtained data. Figure 3 provides the correlation between compressive strength with that of modulus of elasticity using regression analysis. The equation for predicting modulus of elasticity for OPC as per ACI 318-10 and the equations proposed by Phoo-ngernkham *et al.*, Noushini *et al.*, and Nath and Sarker for AAC are given in Eqns. (5-8), respectively. The moduli of elasticity predicted by these equations utilizing compressive strength results obtained experimentally in this study are also plotted in Figure 3.

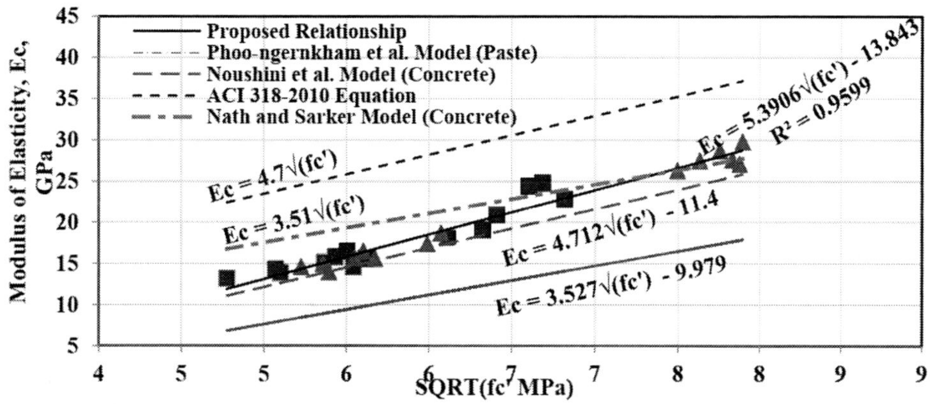

Figure 3. Correlation between modulus of elasticity and compressive strength.

Based on the data obtained in this study a relationship between elastic modulus with the square root of compressive strength is proposed in Eq. (9) for AAC with NP as precursor material, which was partially replaced with NS.

$$Ec = 4.7\sqrt{(fc')} \quad (5)$$

$$Ec = 3.527\sqrt{(fc')}-9.979 \quad (6)$$

$$Ec = 4.712\sqrt{(fc')}-11.4 \quad (7)$$

$$Ec = 3.51\sqrt{(fc')} \quad (8)$$

$$Ec = 5.390\sqrt{(fc')} - 13.843 \quad (9)$$

Where; Ec is the modulus of elasticity in GPa and f'c is the compressive strength in MPa. It is evident from the figure that the modulus of elasticity calculated using ACI 318-10 is significantly more than that predicted by other models as well as proposed model by the authors. With regard to the model presented by Phoo-ngernkham *et al.*, it under estimates the modulus of elasticity of AAC compared to the value obtained in this study. However, statistical models proposed by Noushini *et al.* and Nath and Sarker match very well with the experimental results of this investigation. Based on the results presented here, the equation given in ACI 318-10 overestimates the modulus of elasticity and is not suitable for predicting these values for AAC.

4 CONCLUSIONS

The compressive strength, flexural strength and modulus of elasticity of NP and NS-based AAC increased with the period of curing. At the onset of curing, the compressive strength of AAC was less than that of OPC. However, it significantly improved as the room curing continued. The partial replacement of NP with 5% and 7.5% NS significantly improved the mechanical properties due to the enhanced transformation of source material to C-S-H or C/N-A-S-H gel. Constitutive models developed in this research utilizing modest data generated on the engineering properties were in good agreement with some of the other models proposed by previous researchers. However, ACI 318-10 underestimated flexural strength compared to experimental value; it was contrary in the case of modulus of elasticity. It is important to highlight that considering the large number of mix design variables and limited availability of engineering properties of AAC, the proposed statistical model in this study as well as suggested by other researchers, mentioned earlier, cannot be used as a general correlation between the compressive strength and other engineering properties of AAC. Consequently, there is a need to develop models correlating the engineering properties of AAC by undertaking future research studies.

Acknowledgments

The authors gratefully acknowledge the Center for Engineering Research (CER) of the Research Institute, King Fahd University of Petroleum and Minerals and the School of Civil Engineering, Universiti Sains Malaysia for providing the necessary support for this work.

References

ACI 318-10, Building Code Requirement for Structure Concrete Practice and Commentary, *American Concrete Institute*, 2010.

ASTM C496-10, Standard Test Method for Splitting Tensile Strength of Cylindrical Concrete Specimens, *ASTM International*, 2010.

ASTM C78-10, Standard Test Method for Flexural Strength of Concrete (Using Simple Beam with Third-Point Loading), *ASTM International*, 2010.

Diaz-Loya E. I., Allouch F. N., and Vaidya S., Mechanical Properties of Fly-Ash Based Geopolymer Concrete, *ACI Materilas Journal*, 3, 108, 2011.

Duxson P., Fernández J. A., Provis J. L., Lukey G. C., Palomo A., and van Deventer J. S. J., Geopolymer Technology: The Current State of the Art, *Journal of Material Science*, 42(9), 2917-2933, 2007.

Hardjito, D., Wallah S. E., Sumajouw D. M. J., and Rangan, B. V., Fly Ash-Based Geopolymer Concrete, *Australian Journal of Structural Engineering*, 6(1), 77-85, 2005.

Nath P., and Sarker P.K., Flexural Strength and Elastic Modulus of Ambient-Cured Blended Low-Calcium Fy Ash Geopolymer Concrete, *Construction and Building Materials*, 130, 22–31, 2017.

Noushini A., Aslani F., Castel A., Gilbert R. L., and Uy B., Compressive Stress-Strain Model for Low-Calcium Flyash-based Geopolymer and Heat-Cured Portland Cement Concrete, *Construction and Building Materials*, 73, 136–146, 2016.

Olivia M., and Nikraz H., Properties of Fy Ash Geopolymer Concrete Designed by Taguchi Method, *Materials and Design.*, 36, 191-198, 2012.

Pangdaeng S., Phoo-ngernkham, T., Sata V., and Chindaprasirt P., Influence of Curing Conditions on Properties of High Calcium Fly Ash Geopolymer Containing Portland Cement as Additive, *Materials and Design*, 53, 269-274, 2014.

Phoo-Ngernkham T., Chindraprasirt P., Sata V., Hanjitsuwan S., and Hatanaka S., The Effect of Adding Nano-SiO_2 And Nano-Al_2O_3 on Properties of High Calcium Fly Ash Geopolymer Cured at Ambient Temperature, *Materials and Design*, 55, 58–65, 2014.

Rovnanik, P., Effect of Curing Temperature on The Development of Hard Structure of Metakaolin-Based Geopolymer, *Construction and Building Materials*, Elsevier, 24(7), 1176-1183, 2010.

Yip C. K., Lukey G. C., and Van Deventer J. S. J., Coexistence of Geopolymeric Gel and Calcium Silicate Hydrate at The Early Stage of Alkali Activation, *Cement and Concr. Research*, 35, 1688-1697, 2005.

Interdependence between Structural Engineering and Construction Management
Edited by Ozevin, D., Ataei, H., Modares, M., Gurgun, A., Yazdani, S., and Singh, A.

ISBN: 978-0-9960437-6-2

MECHANICAL AND DURABILITY CHARACTERISTICS OF GGBS DOLOMITE GEOPOLYMER CONCRETE

AIKOT PALLIKKARA SHASHIKALA, PRAVEEN NAGARAJAN, SARANYA PARATHI

Dept of Civil Engineering, National Institute of Technology Calicut, Calicut, India

Production of Portland cement causes global warming due to the emission of green-house gases to the environment. The need for reducing the amount of cement is necessary from sustainability point of view. Alkali activated and geopolymeric binders are used as alternative to cement. Industrial by-products such as fly ash, ground granulated blast furnace slag (GGBS), silica fume, rice husk ash etc. are commonly used for the production of geopolymer concrete. This paper focuses on the development of geopolymer concrete from slag (100% GGBS). Effect of different cementitious materials such as lime, fly ash, metakaolin, rice husk ash, silica fume and dolomite on strength properties of slag (GGBS) based geopolymer concrete are also discussed. It is observed that the addition of dolomite (by-products from rock crushing plants) into slag based geopolymer concrete reduces the setting time, enhances durability and improves rapidly the early age strength of geopolymer concrete. Development of geopolymer concrete with industrial by-products is a solution to the disposal of the industrial wastes. The quick setting concrete thus produced can reduce the cost of construction making it sustainable also.

Keywords: Marine resistance, Setting time, Sorptivity, Strength, Sulphate resistance, Water absorption.

1 INTRODUCTION

Production of cement is harmful for environment as it is the main source for the development of green-house gases to the atmosphere (Li *et al.* 2011). Many researchers have focused their studies on the use of cementitious material as substitute for cement to protect the environment from global-warming. Later, Geopolymer concrete (GPC) was developed by Davidovits (1984), in which industrial by-products were used for the complete replacement of cement (Davidovits 1991). GPC is a sustainable concrete, which has high strength, durability and less carbon footprint (Juenger *et al.* 2011). Industrial by-products such as fly ash, Ground Granulated Blast Furnace Slag (GGBS) and metakaolin are the main source material for the development of geopolymer concrete (Duxson *et al.* 2007).

GGBS is a by-product from steel industries and its chemical composition is similar to cement. It can react with water at slow rate of hydration and can be activated with an alkali solution for producing high strength (Nath *et al.* 2014). Dolomite is a by-product from rock crushing plant. Use of dolomite powder in self-compacting concrete had been reported earlier (Barbhuiya *et al.* 2011). No work has been reported on the effects of dolomite on geopolymer concrete. In this study, effect of dolomite on strength and durability properties of GGBS geopolymer concrete was investigated.

2 MATERIALS USED

2.1 Ground Granulated Blast Furnace Slag (GGBS)

GGBS and dolomite are the by-products from steel industries and rock crushing plants respectively. Physical and chemical properties of GGBS and dolomite are shown in Table 1.

Table 1. Physical and chemical properties of GGBS, dolomite and OPC.

	Cement	GGBS	Dolomite
Source	Ultra Tech, 53 grade OPC	Mangalore Steel Industries (Pvt. Limited)	Rock crushing plant, Jaipur
Colour	Grey	Off white	Pure white
Specific gravity	3.14	2.9	2.85
Blaine's fineness (cm^2/gm)	2435	4032	3500
CaO	61.53	38.9	35.58
SiO_2	20.36	33.5	20.78
Al_2O_3	4.31	10.68	8.54
Fe_2O_3	5.98	2.35	2.45
MgO	1.36	9.45	20.58
LOI	6.46	5.12	12.07

2.2 Aggregate

The property of aggregates used in the study, satisfying IS 2386-3 (1963), is shown in Table 2.

Table 2. Properties of aggregates.

Property	Fine aggregate	Coarse aggregate
Size	Passing through 1.18 mm sieve	Maximum size 12 mm
Water absorption	1.4%	0.93%
Fineness modulus	2.75	6.8

2.3 Alkaline Solution

Combination of sodium hydroxide and sodium silicate was used as an alkaline medium for the test. Sodium hydroxide pellets having purity 97% were dissolved in water one day before casting, since it is an exothermic reaction. Sodium silicate in solution form composed of SiO_2 (27.2%), Na_2O (8.9%) and H_2O (63.9%). Solutions of sodium silicate and sodium hydroxide were mixed together at least one hour before casting (Davidovits 1991). Sulphonated naphthalene based super plasticizer also added to improve the workability.

3 MIX PROPORTIONING OF GEOPOLYMER CONCRETE

There are no proper mix design guidelines available for GGBS geopolymer concrete under ambient curing condition. Mix proportion for M60 concrete was carried out based on trial and error basis for GPC and is shown in Table 3. GPC was filled in each mould in three layers and compacted on a vibrating table. All the specimens were demolded within 24 hours and cured under ambient condition (25-28°C).

Table 3. Mix proportioning of GPC.

Sl. No	Material	Quantity (kg/m^3)
1	Binder content	400
2	Alkaline solution	175
3	Fine aggregate	618
4	Coarse aggregate	1196
5	Super plasticizer	6

4 EFFECT OF DOLOMITE ON STRENGTH PROPERTIES OF GGBS GPC

4.1 Preparation of Specimens

Specimens were prepared as per the mix proportions given Table 3. GGBS was replaced with dolomite in different proportions (Table 4).

Table 4. Proportions of GGBS (G), Dolomite (D) and OPC.

MIX	GGBS	Dolomite	Cement
G100:D0	100	0	0
G90:D10	90	10	0
G80:D20	80	20	0
G70:D30	70	30	0
G60:D40	60	40	0
G50:D50	50	50	0
OPC100	0	0	100

4.2 Setting Time

Setting time of binder was found as per IS 4031 Part IV (1988) using Vicat apparatus. Initial and final setting time of geopolymer paste with different proportions of GGBS and dolomite are shown in Table 5.

Table 5. Setting time of OPC and GPC concrete.

Mix	Initial setting time (min)	Final setting time (min)
G100:D0	47	62
G90:D10	44	58
G80:D20	42	53
G70:D30	36	47
G60:D40	34	46
G50:D50	32	41
OPC100	37	265

Initial setting time for 100 % GGBS was obtained as 47 minutes, which is 27 % more than OPC. Replacement of GGBS with dolomite reduces the initial setting time. When mix ratio of GGBS and dolomite become 50:50, the setting time was reduced by 16 % less than OPC.

4.3 Workability of GPC

Table 6 gives the slump values based on the workability tests conducted as per IS 1199:1959 (1959). GGBS geopolymer concrete has 14% more slump than OPC concrete. 38% more slump was observed in G50:D50 than that of G100:D0. As the percentage of dolomite increases, workability also increases due to the spherical shape of dolomite particles.

4.4 Compressive Strength

150 mm ×150mm × 150 mm cube concrete specimens were tested at 1, 3, 7, 14, 28 and 56 days of curing as per the procedure specified in IS 516:1959 (1959), which is shown in Table 6.

Table 6. Compressive strength of GGBS-dolomite GPC.

Mix	Slump (mm)	Compressive strength (N/mm^2)				
		1 day	3 day	7 day	28 day	56 day
G100:D0	120	42.5	57	62.8	64.5	67.8
G90:D10	132	47.5	65	67.5	68.4	69
G80:D20	145	50.4	65	69.5	70	72
G70:D30	155	53.0	63	71	72.5	73.6
G60:D40	160	45.2	52	64	65.7	68.2
G50:D50	166	30.5	48.5	58	60.4	64
OPC100	105	15.7	23.5	35.6	71.5	73

Addition of dolomite up to 30% increases the compressive strength and decreases thereafter. The optimum percentage of GGBS and dolomite is obtained as 70:30 in terms of strength. Ambient curing provided 95% of the compressive strength of geopolymer concrete within 7 days whereas the strength gaining process of OPC was completed after 28 days of curing in water.

5 TESTS FOR DURABILITY

5.1 Water Absorption

Water absorption test was done as per ASTM C 642-82 (1994) by using cubes of size 100 × 100 × 100mm. Test results are shown in Table 7.

Table 7. Water absorption test results.

MIX	Initial weight (kg)	Weight after 30 minutes (kg)	Weight after 120 hours (kg)	Initial water absorption (%)	Final water absorption (%)
G100:D0	2.62	2.645	2.675	0.954	2.09
G90:D10	2.63	2.653	2.68	0.874	1.91
G80:D20	2.62	2.641	2.65	0.802	1.14
G70:D30	2.61	2.629	2.64	0.728	1.15
G60:D40	2.64	2.66	2.67	0.776	1.13
G50:D50	2.67	2.69	2.75	0.749	2.23
OPC100	2.491	2.525	2.628	1.36	5.49

From Table 7, OPC concrete has 2.63 times more water absorption than G100:D0. Replacement of GGBS with dolomite (30%) decreases water absorption 1.82 times than that of G100:D0. G70:D30 mix has less initial water absorption compared to others, which is 1.87 times and 1.31 times less than OPC100 and G100:D0 respectively.

5.2 Sorptivity

Sorptivity refers to the rate of water absorption through capillary pores. It is measured as per ASTM C 642-82 (1994) using cylinder specimens having diameter 150mm and height 50mm (Figure 2). Sorptivity test results are given in Table 8. OPC concrete has 2.68 times more sorptivity than GGBS geopolymer concrete. Addition of dolomite makes the concrete denser than that of GGBS geopolymer concrete.

Table 8. Sorptivity of geopolymer concrete.

MIX	Sorptivity (cm/min$^{1/2}$) x 10^{-3}
G100:D0	2.45
G90:D10	2.34
G80:D20	2.13
G70:D30	2.14
G60:D40	2.12
G50:D50	2.18
OPC100	6.56

Figure 2. Test setup for sorptivity.

5.3 Sulphate Attack

Oven dried cube specimen having 100 x 100 x 100 mm size were used were immersed in a 3% sodium sulphate solution for 180 days. Loss in weight and compressive strength were measured after 180 days and are shown in Table 9.

Table 9. Acid and sulphate resistance of concrete.

MIX	Acid attack (after 180 days)		Sulphate attack (after 180 days)	
	% weight loss	% loss in compressive strength	% weight loss	% loss in compressive strength
G100:D0	1.21	10.5	0.26	6.45
G90:D10	1.51	10.6	0.20	6.28
G80:D20	1.32	9.51	0.18	6.15
G70:D30	1.47	9.45	0.12	6.18
G60:D40	1.46	9.40	0.11	6.17
G50:D50	1.58	11.58	0.11	6.78
OPC100	35.64	48.56	0.68	20.45

OPC concrete has 2.62 times more weight loss than G100:D0. GGBS-dolomite geopolymer concrete showed reduced weight loss of 2.36 times less than that of G100:D0.

5.4 Marine Attack

Artificial marine water with composition was prepared in the laboratory as per the procedure given in ASTM D1141 (2013). Loss of weight and percentage reduction in compressive strength of 100 x 100 x 100 mm cube specimens was found out after 180 days is given in Table 10.

Table 10. Marine water resistance of GGBS-dolomite GPC.

MIX	% weight loss	% loss in compressive strength
G100:D0	0.95	10.24
G90:D10	0.92	9.84
G80:D20	0.94	9.80
G70:D30	0.92	9.81
G60:D40	1.45	9.64
G50:D50	1.89	10.85
OPC100	8.21	25.78

OPC 100 has 8.6 times more weight loss than that of G100:D0 during marine water attack. 2.52 times compressive strength loss was also observed for OPC concrete than that of GPC.

6 CONCLUSIONS

- Addition of dolomite decreases the setting time of GGBS geopolymer concrete, thus reducing the total construction time.
- Maximum compressive strength was obtained at GGBS and dolomite proportion 70:30.
- GGBS-Dolomite geopolymer concrete attained 95% of its compressive strength within 7 days of ambient curing, while that of conventional concrete completed its strength gaining process after 28 days of water curing.
- GGBS-Dolomite GPC has improved water absorption, sorptivity and high resistance towards sulphate and marine attack. Hence, it can be used for construction in marine environments.

Acknowledgements

The authors thankfully acknowledge the financial support provided by Kerala State Council for Science, Technology and Environment [TDAP/01/2017/KSCSTE], Kerala, India.

References

ASTM C 642-82, Standard Test Method for Specific Gravity, Absorption and Voids in Hardened Concrete, *Annual Book of ASTM Standards*, Vol. 4.02, 1994.

ASTM-D1141-98: Standard Practice for the Preparation of Substitute Ocean Water, *Annual Book of ASTM Standards*, Vol. 11.02, 2013.

Barbhuiya, S., Effects of Fly Ash and Dolomite Powder on the Properties of Self-Compacting Concrete, *Construction and Building Materials*, 25(8), 3301-3305, 2011.

Davidovits, J., Geopolymers: Inorganic Polymeric New Materials, *Journal of Thermal Analysis and Calorimetry*, 37(8), 1633-1656, 1991.

Davidovits, Synthetic Mineral Polymer Compound of the Silicoaluminates Family and Preparation Process, *US Patent No. 4*, 172, 199, 1984.

Duxson, P., Provis, J. L., Lukey, G. C., and Van Deventer, J. S., The Role Of Inorganic Polymer Technology in The Development of 'Green Concrete', *Cement and Concrete Research*, 37(12), 1590-1597, 2007.

Juenger, M. C. G., Winnefeld, F., Provis, J. L., and Ideker, J. H., Advances in Alternative Cementitious Binders, *Cement and Concrete Research*, 41(12), 1232-1243, 2011.

IS 516-1959: Methods of Test for Strength of Concrete, *Bureau of Indian Standards*, New Delhi, P.11, 1959.

IS 1199-1959: Methods of Sampling and Analysis of Concrete, *Bureau of Indian standards*, New Delhi, P.11, 1959.

IS 2386 (Part III)-1963: Methods of Test for Aggregate for Concrete Part 3-Specific Gravity, Density, Voids, Absorption and Bulking, *Bureau of Indian Standards*, New Delhi, P.17, 1963.

IS 4031-1988: Methods of Physical Tests for Hydraulic Cement, *Bureau of Indian Standards*, New Delhi, P.1, 1988.

Li, C., Gong, X. Z., Cui, S. P., Wang, Z. H., Zheng, Y., and Chi, B. C., CO_2 Emissions Due to Cement Manufacture, *Materials Science Forum,* Trans Tech Publications, 685, 181-187, 2011.

Nath, P., and Sarker, P. K., Effect of GGBFS on Setting, Workability and Early Strength Properties of Fly Ash Geopolymer Concrete Cured in Ambient Condition, *Construction and Building Materials*, 66,163-171, 2014.

Interdependence between Structural Engineering and Construction Management
Edited by Ozevin, D., Ataei, H., Modares, M., Gurgun, A., Yazdani, S., and Singh, A.

ISBN: 978-0-9960437-6-2

EFFECT OF INCORPORATING NANO ZEOLITE ON THE STRENGTH OF NATURAL POZZOLAN-BASED GEOPOLYMER CONCRETE

MUHAMMAD KALIMUR RAHMAN[1], MOHAMMED IBRAHIM[1], MOHAMMED SALIHU BARRY[1], and HASSAN AL KHALIFA[2]

[1]*Center for Engineering Research, Research Institute, KFUPM, Dhahran, Saudi Arabia*
[2]*Consulting Services Department, Saudi Aramco, Dhahran, Saudi Arabia*

Natural pozzolan is a green and sustainable material, which has a strong potential for utilization in the geopolymer concrete. At room temperature, the development of strength in natural pozzolan based geopolymer concrete is fair, however, it is desirable, to have higher strength and to enhance the early age strength development. The effect of nano zeolite (NZ) as an additive to the mix by incorporating it at 3% and 5% by weight of natural pozzolan on the strength development and microstructural properties of natural pozzolan based geopolymer concrete mixes are presented. Scanning electron microscopy (SEM) was utilized to determine the morphology of the paste. The results indicated that 5% nanozeolite replacement resulted in superior strength and denser microstructure compared to other mixes. The addition of nanozeolite improved dissolution of source material in the alkaline activators and enhanced transformation of it to form polymeric compounds.

Keywords: Natural pozzolan (NP), Geopolymer concrete, Nano zeolite, Compressive strength, Morphology.

1 INTRODUCTION

The utilization of nano materials in the building material research is growing due to their ability to enhance the engineering properties of binders (Jayapalan *et al.* 2013). The hardened properties of these materials are improved due to the development of compact microstructure as a result of high reactivity to form compounds (Balaguru and Chong 2006). Also, there is a particle packing effect (Sanchez and Sobolev 2010). Currently, alkali activated binders are being extensively studied due to their multiple benefits including potential alternative to ordinary Portland cement (Rovnaník 2010). These benefits are: lower environmental impact, high early strength, and better durability performance (Yusuf *et al.* 2014, Brough and Atkinson 2002). However, strength development of these binders depends on the chemical composition of precursor materials as well as curing conditions. Thus far, industrial byproduct, such as fly ash, has been extensively used as source material in synthesizing AABs and cured at temperatures between 40 to 80 °C (Hardjito *et al.* 2005). The strength development is rapid when these binders are cured at elevated temperature because of the fact that heat accelerates the reactivity. On the contrary to improve the reaction kinetics, finer materials can be added to the primary precursor materials to improve the reactivity. Most significantly, the fineness of the precursor material has a profound effect on the rate of reactivity (Wang *et al.* 2005). The finer materials dissolve easily in the alkaline solution leading to a higher conversion of the source materials to polymeric compounds which enhances the strength of these binders (Phoo-ngernkham *et al.* 2014). It is estimated that the dissolution of source materials having particle size of more than 20 μm is rather slow, whereas, materials with particle size of less than 2 μm react rapidly (Wang *et al.* 2005). In this regard nano materials with enormous specific surface area are believed to be potentially useful in improving the strength development of these binders. Earlier studies have used nanosilica and nano alumina to improve the properties of these binders. Phoo-ngernkham *et al.* (2014) studied the influence of incorporating NS and nanoalumina (NA) on the strength and microstructure of high calcium fly ash-based AAB cured under ambient conditions at a given alkaline activator concentration and ratio. It was reported that 2% nano material was sufficient to obtain enhanced strength due to the formation of calcium silicate hydrate (C-S-H) or calcium alumina silicate hydrate (C-A-S-H) gel along with sodium alumina silicate hydrate (N-A-S-H) gel (Phoo-ngernkham *et al.* 2014). In another study (Adak *et al.* 2014), fly ash was replaced with up to 10% NS in developing alkali activated mortar even though the source material was rich in silica and alumina. It was reported that the strength increased by 15% due to 6% replacement of fly ash with NS (Adak *et al.* 2014). In this study, natural pozzolan based binder was developed and to improve its properties nano zeolite (NZ) addition was proposed. The developed binder was evaluated by conducting mechanical tests and studying its microstructure.

2 MATERIALS AND METHODS

2.1 Materials

Natural pozzolan (NP), the main precursor material used in the study, was in a powdered form of volcanic rock. The chemical composition of NP is given in Table 1. The chemical composition of NZ used in the study is shown in Table 2. 14M NaOH solution and Na_2SiO_3 of 3.3 silica modulus were used as alkaline activators. The composition of sodium silicate includes: H_2O: 62.5%, SiO_2: 28.75% and Na_2O: 8.75%. Crushed limestone, having specific gravity of 2.56, was used as coarse aggregate while dune sand having specific gravity of 2.62 was utilized as fine aggregate in the concrete mixtures.

Table 1. Chemical composition of NP.

Oxides	SiO_2	Al_2O_3	Fe_2O_3	CaO	MgO	K_2O	Na_2O	P_2O_5	TiO_2	LOI
Weight, %	40.48	12.90	17.62	11.83	8.33	1.67	3.60	1.37	0.60	1.6

Table 2. Properties of nano zeolite.

Property	Solid content, %	Avg. Particle size, nm	Bulk density, g/cm^3	Specific surface area, m^2/g	Na_2O content, %	Viscosity cps	pH
Value	50	35	1.4	80	0.2	15	9.5

2.2 Methods

Table 3 shows the quantities of constituent materials of mixes prepared by incorporating NZ. All the AAC mixtures were prepared with a constant Na_2SiO_3/NaOH weight ratio of 2.50. The concrete specimens were de-molded after 1 day of casting, divided into two groups and placed in plastic bags to avoid evaporation of moisture. For curing, one group of concrete specimens were kept in the laboratory which was maintained at 23$\pm$2 °C, while, the other group was cured at elevated temperature in the oven that was maintained at 60$\pm$2 °C. The compressive strength of concrete was measured after 1/2, 1, 3, 7, 14, and 28 days of elevated temperature curing, while, it was measured after 3, 7, 14, 28 and 90 days in case of room temperature curing (23±2 °C) on 50 mm cube specimens according to ASTM C109-10 (2010). Alkali activated paste was prepared and cured at elevated temperature for 28 days after which morphology was determined using a JEOL model 5800 LV instrument.

Table 3. Quantities of AAC mix constituents.

Mix #	Legend	NP, kg/m^3	NZ, kg/m^3	Na_2SiO_3, kg/m^3	NaOH, kg/m^3	FA, kg/m^3	CA, kg/m^3
M0	0%-NZ	400	0	150	60	650	1206
M1	3%-NS	388	12	150	60	646	1200
M2	5%-NS	380	20	150	60	640	1188

3 RESULTS AND DISCUSSION

3.1 Compressive Strength

The compressive strength development in the AAC cured in the oven prepared with 0%, 3%, and 5% NZ is depicted in Figure 1. Figure 2 gives the strength development in the AAC cured at room temperature. The compressive strength in both the curing regimes increased with period of curing. However, the strength development was faster in case of specimens cured in the oven compared to the ones at room temperature. 7 days of heat curing was sufficient to gain maximum strength in the AAC prepared with various NZ contents, while, the strength development was continued in the AAC cured at room temperature even after 28 days. The maximum compressive strength measured in case of AAC cured in the oven was 37.52 MPa, 41.54 MPa and 43.62 MPa synthesized with 0%, 3% and 5% NZ, respectively, while, it was 27.93 MPa, 32.56 MPa and 33.97 MPa in the room temperature cured batch of concrete specimens. Based on these results, the strength gain in the concrete prepared with 3% and 5% NZ addition cured at elevated temperature over control samples was 10% and 16%, respectively; however, this value was 16% and 21% in case of specimens cured at room temperature. It is to be noted that maximum compressive strength was gained in case of concrete specimens cured at room conditions was at 90 days. The improvement in the performance of AAC containing NZ, in terms of compressive strength, particularly with 5%, could be attributed to the enhanced transformation of source materials to the polymeric gel in the presence of highly reactive NZ as well as due to the possible particle packing effect of nanoparticles in the binder structure (Yip *et al.* 2005).

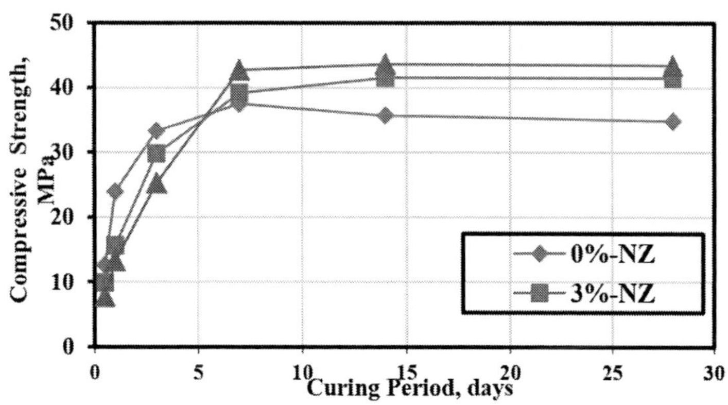

Figure 1. Compressive strength development in the concrete cured at elevated temperature.

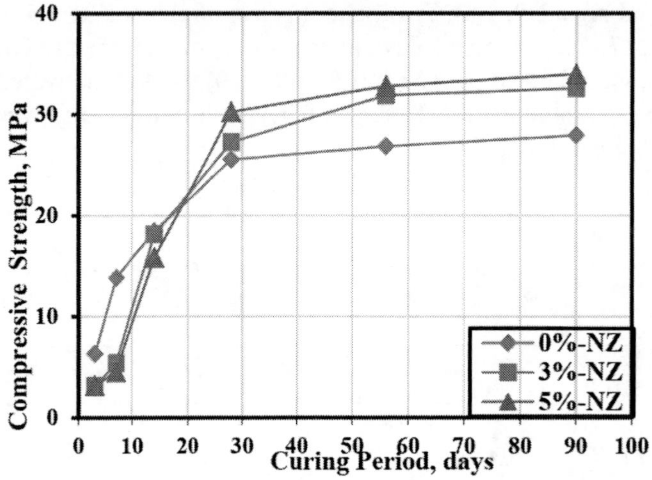

Figure 2. Compressive strength development in the concrete cured at room temperature.

3.2 Morphology

Figure 3 presents the SEM images of geopolymer paste incorporating NZ cured at elevated temperature for 7 days. Morphology of alkali activated paste was not carried out for the specimens cured at room temperature. Microstructure of the specimen prepared without NZ was very porous in nature with widespread cracks. When NZ was added to the mixture, microstructure started to densify with reduction in cracks and voids. In the mixtures prepared by incorporating 3% and 5% NZ, micrographs showed denser microstructure with homogenous gel-like matrix. The microstructure of AAP modified with 3% and 5% NZ was more or less similar. The more homogenous gel in case of AAP modified with 3% and 5% NZ with undefined boundaries most probably consisted of polymeric compounds. Based on the several micrographs observed while taking the images, there was partial filling of voids in the specimens prepared with 3% and 5% NZ which resulted in higher compressive strength. On the contrary, in the specimens prepared without NZ, the solid area was less, which did not beneficially helped in increasing the strength of developed concrete. These outcomes are consistent with the compressive strength results wherein 3% and 5% replacement of natural pozzolan with NZ lead to superior strength. However, the initial sluggishness in the development of strength in the concrete mixes containing higher NZ was due to the increase in the silica content for a given amount of alkaline solution which may have contributed in delayed polymerization, particularly in the specimens cured at room temperature (Panias *et al.* 2007, Oh *et al.* 2010). However, there was considerable strength gain in these mixes when cured at elevated temperature. Nonetheless, as the curing continued due to the availability of additional soluble silica accelerated the polymerization process, which resulted in enhanced transformation of source materials to the polymeric gel (Nath and Sarker 2017). These results are indicative of the fact that in the process of polymerization due to the availability of soluble silica, particularly in the mixes incorporating micro silica, resulted in the formation of CSH or CASH gels along with the NASH products (Juhyuk *et al.* 2014).

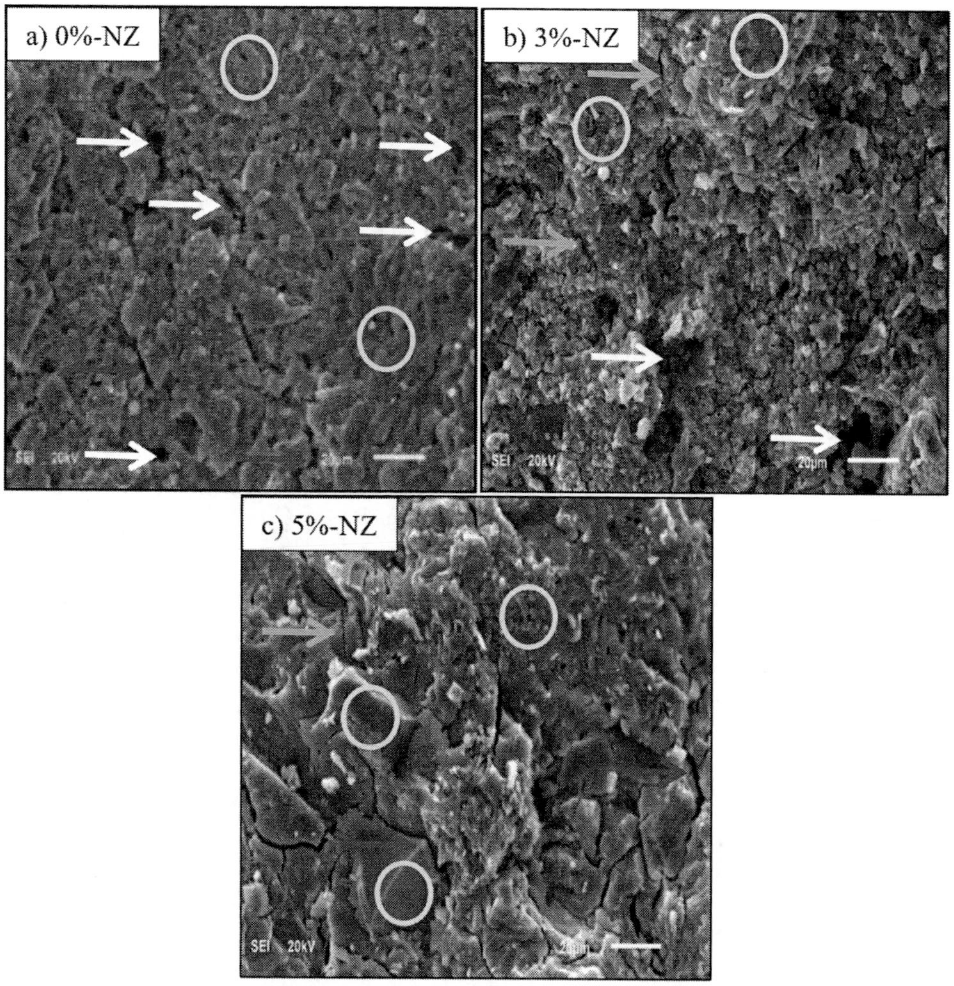

Figure 3. SEM images of alkali activated paste modified with NZ. (white arrows: voids, red arrows: wide cracks, yellow arrows: fine cracks, yellow circles: uniform gel).

4 CONCLUSIONS

The maximum compressive strength of thermally cured AAC prepared with and without NZ was higher than room temperature cured specimens. The strength development was faster when concrete was cured at elevated temperature, while it was slower in room conditions. However, the AAC prepared with 3% and 5% exhibited enhancement in strength compared to control mixture when cured at both the regimes. According to these outcomes, the concrete developed in this study has potential to be used for structural purposes, as the strength was more than 40 MPa and 30 MPa when cured in the oven and room temperature, respectively. The microstructure of AAC modified with 3% and 5% NZ showed denser matrix and compact probably due to enhanced transformation of source material in to polymeric compounds. On the other hand, the microstructure of AAC prepared without NZ was relatively discrete and non-uniform.

Acknowledgments

The authors gratefully acknowledge the Center for Engineering Research (CER) of the Research Institute, King Fahd University of Petroleum and Minerals, for providing the necessary support for this work.

References

Adak D., Sarkar S., and Mandal S., Effect of Nano Silica on Strength and Durability of Fly Ash Based Geopolymer Mortar, *Construction and Building Materials*, 70, 453–459, 2014.

ASTM C109-10, Standard Test Method for Compressive Strength of Hydraulic Cement Mortars (Using 2-in. or (50-mm) Cube Specimens), *ASTM International*, 2010.

Balaguru P., and Chong K., *Nanotechnology and Concrete: Research Opportunities*, Proceedings of ACI Session on Nanotechnology of Concrete: Recent Developments and Future Perspectives, Denver, USA, November 7, 2006.

Brough A.R., and Atkinson A., Sodium Silicate-Based Alkali-Activated Slag Mortars: Part I. Strength, Hydration and Microstructure, *Cement and Concrete Composites*, 32, 865–879, 2002.

Hardjito, D., Wallah, S. E., and Sumajouw, D. M. J., and Rangan, B. V., Fly Ash-Based Geopolymer Concrete, *Australian Journal of Structural Engineering*, 6(1), 77-85, 2005.

Jayapalan, A. R., Lee B. Y., and Kurtis K. E. Can Nanotechnology Be 'Green'? Comparing Efficacy of Nano and Microparticles in Cementitious Materials, *Cement and Concrete Composites*, 36, 16-24, 2013.

Juhyuk, M. A., Sungchul, B. B., Celik, K. B., Yoon, S. B., Ki-Hyun, K. B., Kang, S. K. C., Paulo, J. M., and Monteiro, B., Characterization of Natural Pozzolan-Based Geopolymeric Binders, *Cement and Concrete Composites,* 53, 97-104, 2014.

Nath P., Sarker P.K., Flexural strength and elastic modulus of ambient-cured blended low-calcium fly ash geopolymer concrete, *Construction and Building Materials*, 130 (2017) 22–31.

Oh, J. E., Monteiro, P. J., Jun, S. S., Choi, S, and Clark, S. M., The Evolution of Strength and Crystalline Phases for Alkali-Activated Ground Blast Furnace Slag and Fly Ash Based Geopolymers, *Cement and Concrete Research,* 40(2), 189–196, 2010.

Panias, D., Giannopoulou, I. P., and Perraki, T., Effect of Synthesis Parameters on The Mechanical Properties of Fly Ash-Based Geopolymers, *Colloids and Surfaces, A,* 301(1), 246–254, 2007.

Phoo-ngernkham, T., Chindraprasirt, P., Sata, V., Hanjitsuwan, S., and Hatanaka, S., The Effect of Adding Nano-SiO_2 and Nano-Al_2O_3 on Properties of High Calcium Fly Ash Geopolymer Cured at Ambient Temperature, *Materials and Design,* 55; 58–65, 2014.

Rovnaník, P., Effect of Curing Temperature on The Development of Hard Structure of Metakaolin-Based Geopolymer, *Construction and Building Materials,* 24(7), 1176–1183, 2010.

Sanchez F., and Sobolev, K., Nanotechnology in Concrete – A Review, *Construction and Building Materials,* 24, 2060–2071, 2010.

Wang, P. Z., Trettin, R., and Rudert, V., Effect of Fineness and Particle Size Distribution of Granulated Blast Furnace-Slag on The Hydraulic Reactivity in Cement Systems, *Advanced Cement Research,* 17, 160-166, 2005.

Yip, C. K., Lukey, G. C., Van Deventer, J. S. J., Coexistence of Geopolymeric Gel and Calcium Silicate Hydrate at The Early Stage of Alkali Activation, *Cement and Concrete Research*, 35, 1688-1697, 2005.

Yusuf, M. O., Megat Johari, M. A., Ahmad, Z. A., and Maslehuddin, M., Strength and Microstructure of Alkali-Activated Binary Blended Binder Containing Palm Oil Fuel Ash and Ground Blast-Furnace Slag, *Construction and Building Materials,* 52, 504–510, 2014.

Interdependence between Structural Engineering and Construction Management
Edited by Ozevin, D., Ataei, H., Modares, M., Gurgun, A., Yazdani, S., and Singh, A.

ISBN: 978-0-9960437-6-2

FLEXURAL BEHAVIOR OF HIGH STRENGTH GEO-POLYMER CONCRETE BEAMS

VANKUDOTHU BHIKSHMA, KANDIRAJU PROMODKUMAR, and PUTTA PANDURANGHIAH

Dept of Civil Engineering, Osmania University, Hyderabad, India

The demand for concrete is increasing day by day. As the consumption of cement is increased, environmental issues arise due to the release of CO_2 during the manufacturing of cement. The objective of this research work is to produce a pollution free concrete with a combination of fly ash and GGBS (Ground granulated blast furnace slag) and without the use of cement. In this paper an attempt was made to study the mechanical properties of high strength geo-polymer concrete of grade M60 using GGBS, fly ash and micro silica. The testing program was planned for the mechanical properties of geo-polymer concrete and flexural behavior of corresponding beams. The experimental results indicated that the geo-polymer concrete M60 grade has a compressive strength of 70.45 MPa at the age of 28 days cured at ambient condition. Further, flexural strength and split tensile strengths for M60 grade high strength geo-polymer concrete at 28 days were observed to be 5.45 MPa and 3.63 MPa respectively. The modulus of elasticity was higher than the theoretical value proposed by IS 456-2000. It was also observed that the load carrying capacity of M60 grade high strength geo-polymer concrete found to be more than corresponding grade conventional concrete. The load-deflection, moment-curvature relationships were studied. The experimental results were encouraging to continue for further research in the area high strength geo-polymer concrete.

Keywords: GGBS, Fly ash, Micro silica, Compressive strength, Splitting tensile strength, Flexural strength, Modulus of elasticity, Mechanical properties.

1 INTRODUCTION

Concrete is the second most used material after water. Ordinary Portland Cement (OPC) becomes an important material in the production of concrete which acts as a binder for the ingredients of concrete. Utilization of cement causes pollution to the environment and reduction of natural resources. The current contribution of greenhouse gas emission from Portland cement production is about 1.5 billion tons annually.

The purpose of this study to produce environmental friendly concrete, replacing cement with the industrial by products such as fly ash, GGBS (Ground granulated blast furnace slag) etc. and to develop cost-efficient and eco-friendly concrete. To produce high strength concrete using geo-polymers and without ordinary Portland cement.

The objective is to determine the mechanical properties of geo-polymer concrete of M60 grade. To investigate the flexural behavior of geo-polymer concrete beams and comparing with conventional reinforced concrete beams.

2 SCOPE OF THE WORK

Three Cubes (150mm x 150mm x 150mm) each for 7 days and 28 days were tested for compressive strength. Three cylinders (150mm diameter x 300mm length) for splitting tensile strength and three cylinders of the same size were used for determination young's modulus of elasticity at 28 days. Six prisms (100mm x 100mm x 500mm) were tested for flexural strength at 28 days. Two numbers geo-polymer concrete reinforced concrete beams (150mm x 230mm x 1500mm) were cast and another two beams of the same size of conventional concrete were cast for flexure behavior.

3 LITERATURE REVIEW

Naidu *et al.* (2012) with 28.57% replacement of fly ash with slag, achieved compressive strength of 57MPa for 28 days. For the same mix strength obtained was 43.56 MPa after exposure to 500°C for 2 hours. Vignesh and Vivek (2015) made an attempt to study strength properties of geo-polymer concrete using low calcium fly ash replacing with slag in five different percentages. Ramujee and Potharaju (2014) aimed at making eco-friendly concrete and increasing the strength of the concrete. Sagawa *et al.* (2013) investigated the chloride diffusion coefficient of the concrete mixed with ground granulated blast-furnace slag (GGBS). Islam *et al.* (2014) concluded that experimental study on the development of compressive strength of ground granulated blast furnace slag-palm oil fuel ash-fly ash based geo-polymer mortar.

4 EXPERIMENTAL PROGRAM

After several trial mixes, mix proportions for M60 grade geo-polymer concrete are determined. Materials fly ash, GGBS), alkaline liquids, sodium hydroxide (NaOH), sodium silicate (Na_2SiO_3), fine aggregates, coarse aggregates, micro silica, super plasticizers are used. Physical and chemical properties of fly ash and GGBS are presented in Table 1 to Table 4.

Table 1. Physical properties of fly ash.

S.No	Description	Value	As per IS 3812-2007
1	Specific gravity	2.2	---
2	Fineness (m^2/kg)	369	Minimum - 320
3	Lime Reactivity	4.9	Minimum - 4.5
4	Soundness(Autoclave Test)	0.034	Maximum - 0.8

Table 2. Chemical composition of fly ash.

S.No	Description	Mass (%)	Requirement as per IS 3812-2007
1	Loss of Ignition	1.90	12.0 (Max)
2	Silica, SiO_2	52.16	($SiO_2+Al_2O_3+Fe_2O_3$)
3	Alumina,Al_2O_3	31.97	70
4	Iron, Fe_2O_3	9.32	(Max)
5	Magnesium MgO	0.71	5.0 (Max)
6	Sodium Na_2O	0.89	1.5 (Max)
7	Sulphur SO_3	0.5	2.75 (Max)

Table 3. Physical properties of GGBS.

S.No	Description	Value
1	Colour	Off white
2	Specific gravity	3.15
3	Fineness (m^2/kg)	420
4	Glass Content (%)	75

Table 4. Chemical composition of GGBS.

S.No	Description	Mass (%)
1	Loss of Ignition	2.6
2	Calcium, CaO	42.5
3	Silica, SiO_2	34.4
4	Alumina,Al_2O_3	12.9
5	Iron, Fe_2O_3	1.12
6	Magnesium MgO	4.87
7	Sodium Na_2O	0.45
8	Sulphur SO_3	3.10
9	Potassium K_2O	0.66

4.1 Alkaline Liquid

The most common alkaline liquid in geo-polymerization is a combination of a sodium silicate solution and sodium hydroxide. Sodium based solutions were chosen because it is a low cost than potassium based solutions. The solution is prepared 24 hours prior to testing.

4.2 Sodium Hydroxide (NaOH)

In the present investigation for the preparation of alkaline solution sodium hydroxide of 12M molarity was considered. The Sodium Hydroxide solids of a laboratory grade in pellets form with 99 percent purity. The sodium hydroxide (NaOH) solution was prepared by dissolving the pellets in water.

4.3 Sodium Silicate (Na_2SiO_3)

Sodium silicate solution obtained from local chemical suppliers was used. The chemical composition of the Sodium silicate solution was Na2O =14.7%, SiO2 = 29.4% and water 55.9% by mass. The PH value of the liquid was 11.7. The mix proportions for mix design for M60 grade geo-polymer concrete are shown in Table 5.

4.4 Specification of Reinforced Concrete Beam

An under reinforced concrete beam was designed to determine the flexure strength of the geo-polymer concrete beam. Also two conventional under reinforced RC beams were cast with same grade i.e., M60 and compared the flexural strength. Main reinforcement was formed by 4 no's of 16mm diameter in tension zone (bottom) and 2 no's of 12mm diameter in compression zone (top). Further, the stirrups of 8mm diameter at 125mm center to center are proposed.

Table 5. Mix proportions of geo-polymer concrete for M60 grade.

S.No	Description	Qty (kg/Cum)
1	Fly ash	274
2	GGBS	206
3	Fine aggregate	767
4	Coarse aggregate	948
5	Na2 SiO3	171
6	NaOH	18
7	Water	51

5 TEST SPECIMENS AND RESULTS

Mechanical test results are tabulated in Table 6 to Table 9.

Table 6. Test results of the Compressive strength of geo-polymer concrete (GPC).

S.No.	Compressive strength at 7 days for GPC (N/mm^2)	Compressive strength at 28 days for GPC (N/mm^2)	Compressive strength at 28days for conventional concrete (N/mm^2)
1	41.86	69.51	73.00
2	44.00	71.74	72.45
3	42.40	70.11	69.67
Average Strength	42.75	70.45	70.70

Table 7. Splitting tensile strength of M60 grade geo-polymer concrete.

S No	Sample	Splitting tensile strength (N/mm^2)
1	M-60-A	3.60
2	M-60-B	3.50
3	M-60-C	3.78
Average value		

Table 8. Flexural strength of geo-polymer concrete.

S.No.	Sample	Flexure strength (N/mm^2)
1	M60-A	5.18
2	M60-B	5.58
3	M60-C	4.78
4	M60-D	5.18
5	M60-E	5.98
6	M60-F	5.98
Average value		5.45

Table 9. Test results of modulus of elasticity.

S.No	Sample	Modulus of Elasticity (GPa)
1	1	41.00
2	2	42.00
3	3	47.00
Average value		43.00

Crack patterns and failure mode were observed for tested beams. The load carrying capacity of the geo-polymer concrete beam is little higher than convention concrete. The load and corresponding deflections were recorded for both geo-polymer concrete beam and convention concrete beams. The results are presented in Figure 1. Also, the moment-curvature relationship is shown in Figure 2.

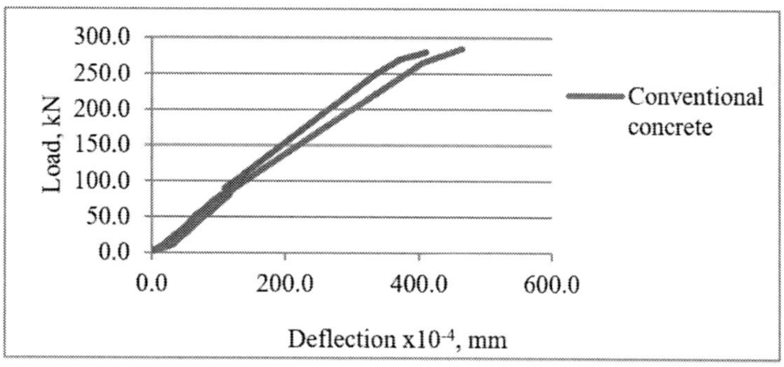

Figure 1. Comparisons of Load vs Deflection for M60 grade between Geo-polymer concrete beam and a conventional concrete beam.

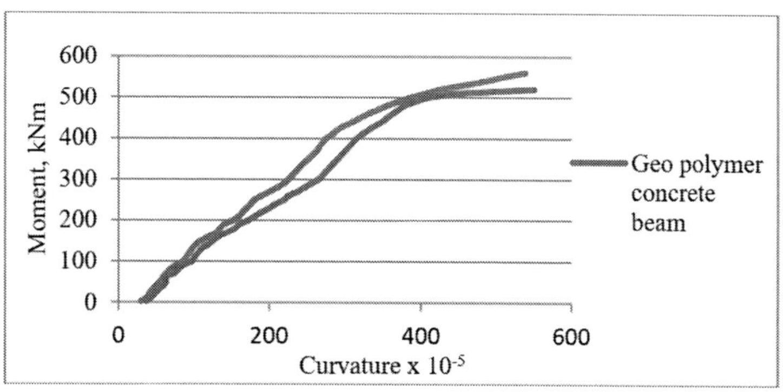

Figure 2. Comparisons of Moment vs Curvature for M60 grade between Geo-polymer concrete beam and a conventional concrete beam.

6 CONCLUSIONS

An attempt was made for the development of high strength geo-polymer concrete beams. Based on the investigation reported in this paper the following conclusions were drawn:

- The average density of fly ash based geo-polymer concrete is observed to be similar to that of conventional concrete.

- The compressive strength of fly ash-GGBS-Micro silica-based high strength geo-polymer concrete of grade M60 under the ambient temperature at 28 days is almost equal to the compressive strength of conventional concrete.
- The modulus of elasticity and load carrying capacity of high strength geo-polymer concrete of grade M60 are observed to be slightly higher side when compared with conventional concrete of the same grade.
- Hence, geo-polymer concrete mix with 47.5 % fly ash and GGBS, 20% of Micro Silica and 12M give required compressive strength in terms of M60 grade of concrete.
- Geo-polymer concrete can be used to replace high strength normal concrete since it does not only satisfy all the properties but also eco-friendly.

References

Islam, A., Alengaram, U. J., Jumaat, M. Z., and Bashar, I. I., The Development of Compressive Strength of Ground Granulated Blast Furnace Slag-Palm Oil Fuel Ash-Fly Ash Based Geo-Polymer Mortar, *Materials and Design*, 56(10), 833–84, Dec 2014.

Ramujee, K., and Potharaju, M., Development of Low Calcium Fly Ash Based Geo-Polymer Concrete, *International Journal of Engineering Research and Applications*, 6(1), 91-95, February 2014.

Naidu, G., Prasad, A. S. S. N., Adiseshu, S., Satayanarayana, P.V.V., A Study on Strength Properties of Geo-Polymer Concrete with The Addition of GGBS, *International Journal of Engineering Research and Development*, e-ISSN: 2278-067 X, p-ISSN: 2278-800X, 2(4), 19-28, July 2012.

Vignesh, P., and Vivek, K., An Experimental Investigation on Strength Parameters of Fly Ash Based Geo-Polymer Concrete with GGBS, *International Research Journal of Engineering and Technology*, 2(2), 135-142, May 2015.

Sagawa, Y., Yamamoto, D., and Henzan, Y., *Properties of Concrete with GGBS and Its Applications for Bridge Superstructures,* 3rd International Conference On Sustainable Construction Materials and Technologies, Japan, August 2013.

Interdependence between Structural Engineering and Construction Management
Edited by Ozevin, D., Ataei, H., Modares, M., Gurgun, A., Yazdani, S., and Singh, A.

ISBN: 978-0-9960437-6-2

BLENDED METAKAOLIN AND WASTE CLAY BRICK POWDER AS SOURCE MATERIAL IN SUSTAINABLE GEOPOLYMER CONCRETE

MAHMOOD F. AHMED, WASAN I. KHALIL, and QAIS J. FRAYYEH

Dept of Civil Engineering, University of Technology, Baghdad, Iraq

Recently, sustainability and ecological related problems have attracted more attention around the world. The construction sector incorporates directly and indirectly in global warming, natural resources depletion, and environmental pollution. This study aims, firstly; to identify the optimum mix of metakaolin (MK) geopolymer concrete required to achieve high compressive strength with respect to the concentration of the alkaline solution and curing system. Secondly, to reduce the impact of brick waste on the environment, by producing geopolymer concrete based on blended MK and waste clay brick powder (WBP). The compressive strength, splitting tensile strength and flexural strength of MK-based geopolymer concrete specimens were studied. Different contents of waste clay brick powder (WBP) (0%, 10%, 15%, and 20%) as a replacement by weight of (MK) were investigated. The results appear that it is possible to produce MK-based geopolymer concrete with a compressive strength of 44.03 MPa, while it was 34.76 MPa at 28 days for specimens with 15% WBP replacement of main source binder. Finally, it could be concluded that green moderate strength geopolymer concrete can be produced and used in different civil engineering applications.

Keywords: Alkaline solution, CO_2 emission, Environmental pollution, Heat curing, Mix proportions, Molarity.

1 INTRODUCTION

The sustainability has become an important issue in recent years as the growth of waste production associated with the awareness of the environmental problems and the exhaustion of natural resources has increased at a rapid rate for producing materials such as concrete (Brito and Saikia 2013). In fact, the greatest challenge facing the sustainability of concrete as a building material comes from the production of cement, which is the key ingredient in concrete. Worldwide, high consumption of natural raw materials (1.6 tons) is required to produce one ton of Portland cement (Toniolo and Boccaccini 2017). Moreover, the cement industry has an extraordinary contribution to global CO_2 emission (Haberta and Plamondonb 2016, Ken *et al.* 2015, Karthika *et al.* 2017). Hence, considerable efforts have been made in order to get new innovations in the concrete industry by using more eco-friendly construction materials such as "geopolymers". Geopolymer concrete can be produced by a polymeric reaction of alkaline liquid; usually is Na_2SiO_3 and NaOH, with aluminosilicate-based powder materials such as Metakaolin (Verma and Dev 2018, Mehta and Siddique 2016). On the other hand, the generation of clay brick waste is increasing rapidly due to a spike in construction activities around the world. According to the United States Environmental Protection Agency, about 44 million tons of brick waste was produced during construction and demolition actions just between 2012 and 2014 (Wong *et al.* 2018). Most of the demolished bricks are disposed of in landfills, which causes economic losses and serious environmental problems. The main aims of this research are: [1] determine the optimum trial mix of geopolymer concrete using Iraqi metakaolin (MK) as a source material with respect to higher compressive strength and less curing energy consumption, and [2] investigate the effect of substitute the (MK) with different contents of waste clay brick powder (WBP) on mechanical properties of geopolymer concrete.

2 EXPERIMENTAL WORK

2.1 Materials

The source material Metakaolin was obtained from the calcination of Iraqi Kaolin clay at 700°C for 2 hours, while the waste clay brick powder was produced by crushing waste bricks to small pieces, then grinding it by cyclone machine. The physical properties and chemical composition of the (MK) and (WBP) according to X-Ray Fluorescence results are given in Table 1. The alkali activator was a mix solution of NaOH (pellets form with 98% purity dissolved in distilled water) and commercial grade Na_2SiO_3 which contents 54% H_2O, 32.5% SiO_2, and 13.5% Na_2O. For all mixes, the coarse and fine aggregate were within the requirements of the Iraqi Specification No.45/1984. The coarse aggregate was natural and crushed with size 5–14 mm, dry rod density of 1595 kg/m^3, specific gravity of 2.61 and absorption of 1.3%. The fine aggregate was natural sand within zone 1 with a fineness modulus of 3.3, dry rod density of 1787 kg/m^3, specific gravity of 2.58 and absorption of 1.6%. A superplasticizer (Conplast SP2000) and extra tap water were used to get the applicable workability and improving the mixing process.

Table 1. Chemical composition and physical properties of metakaolin and waste bricks powder

Oxide%	SiO_2	Al_2O_3	Fe_2O_3	CaO	MgO	SO_3	Na_2O	K_2O	LOI
MK	54.20	39.0	0.92	1.37	0.15	0.45	0.22	0.27	0.71
WBP	56.82	11.36	2.36	20.20	3.02	0.83	0.86	0.86	1.19

Physical Properties	MK	WBP
Specific gravity	2.64	2.84
Specific surface area (m^2/kg)	14300	462
7 days pozzolanic activity index, (%)	113.3	89.2

2.2 Preparation of Geopolymer Concrete

The alkali activator solution was prepared one day before casting to be coolede at room temperature. The coarse aggregate, sand, MK and WBP were mixed together in electrical pan mixer for four minutes. Next, the alkali activator solution, extra water, and superplasticizer were added to the dry mixture and the mixing process continues 6 to7 minutes. The fresh geopolymer concrete was cast in molds and compacted by tamping each layer 30 times then followed by 15-20 second on a vibration table. After demolded, the specimens curried according to the selected curing method. The compressive, splitting tensile and flexural strengths were determined according to BS 1881: Part 116 (1983) using 100 mm cubes, ASTM C496/C496M (2017), and ASTM C78/C78M (2018), respectively.

2.3 Mixture Proportioning

Geopolymer concrete attains most of its strength at early age, and the rate of strength development beyond 7th day is not significant Joseph and Mathew (2012). Thus, in this study, the trail mixes were prepared to select the optimum mix with higher compressive strength at 7 days. Table 2 shows the details of mix proportions and compressive strength results for trial mixes which have the same content of extra water (10% by weight of MK) and oven cured at 50°C for 5 hours. The results indicate that mix (M4) has the highest compressive strength at 7 days. To investigate the effect of alkaline liquid concentration, ratio of Na_2SiO_3 to NaOH and curing regime, other trials were carried out on mix (M4) to select the optimum MK-based geopolymer concrete which considered as reference mix. Four mixes that have 0, 10, 15, and 20% waste brick powder as a partial replacement by weight of metakaolin were prepared.

Table 2. Details of MK-based geopolymer concrete trial mixes

Mix	MK, (kg/m^3)	Coarse Agg., (kg/m^3)	Fine Agg., (kg/m^3)	Alkaline solution, by wt. of MK	SP by wt. of MK (%)	Molarity	NaOH to Na_2SiO_3	Compressive Strength at 7days (MPa)
M1	400	1100	720	0.45	3	12	1:3.5	6.4
M2	400	1100	720	0.45	3	12	1:2.5	5.83
M3	400	1100	720	0.52	3	12	1:2.5	18.0
M4	415	1240	475	0.65	2	12	1:2.5	26.56
M5	414	1136	660	0.52	3	12	1:2.5	23.78
M6	400	1185	630	0.55	3	12	1:2.5	20.16
M7	400	1300	500	0.5	3	12	1:2.5	13.03
M8	450	1217	608	0.55	2	12	1:3.5	9.8
M9	450	1267	558	0.55	2	12	1:3.5	12.7
M10	400	1185	630	0.65	2	14	1:2.5	12.36
M11	400	1185	630	0.65	2	14	1:2	17.5
M12	500	1150	575	0.65	2	12	1:2.5	22.5
M13	500	1481	787.5	0.65	2	12	1:2.5	24.5

3 RESULTS AND DISCUSSION

3.1 Effect of Curing Method

For geopolymer concrete, the heat curing is very important for geopolymerization of aluminosilicate gel, which enhances polycondensation process to obtain high early strength gain Ken *et al.* (2015). Table 3 illustrates the compressive strength at 7 days for mix (M4) after subjected to different curing methods. It is found that the optimum curing method was the oven treatment for 4-5 hours at 60°C then under sunlight in hot weather. The compressive strength of the specimens subjected to the optimum curing method was 32.6 MPa at 7 days, with an increase up to 49.38% compared with those subjected to sunlight curing alone. This curing method provides a continuous geopolymerization process to produce a more homogenous and dense structure of MK- geopolymer concrete. Al-Shathr *et al.* (2018) used curing in oven at 45 °C for 24 hrs. with molds followed by 48 hrs. without molds. They obtain MK-geopolymer concrete with a compressive strength of 22.7 MPa at 7 days. However the optimum combined curing method used in this research had an advantage in produce more sustainable geopolymer concrete with higher strength and less curing energy consumption, which can be applied in wide areas of civil engineering.

Table 3. Effect of curing method on compressive strength of mix M4

No.	Curing Regime	Compressive strength at 7 Days (MPa)
1	Inside Laboratory (12- 25 °C)	14.5
2	Outdoor subjected to sunlight at summer (35-49°C)	16.5
3	In oven at 50°C for 4-5 hours, then under sunlight at summer (35-49°C)	26.5
4	In oven at 60°C for (4-5) hours, then under sunlight at summer (35-49°C)	32.6
5	In oven at 70°C for (4-5) hours, then under sunlight at summer (35-49°C)	31.1
6	In oven at 80°C for (4-5) hours, then under sunlight at summer (35-49°C)	30.2
7	In oven at 100°C for (3) hours then, under sunlight at summer (35-49°C)	31.5

3.2 Effect of Alkaline Activator Concentration

The results in Figure 1 show that for the same (NaOH: Na_2SiO_3) ratio, the compressive strength of mix M4 increases with the increase of molarity from 12M up to14M. This behavior is attributed to the influence of the concentration of NaOH solution in geopolymerization as it is responsible for dissolving, hydrolyzing and leaching the (Si and Al) to form the precursor gel (Joseph and Mathew 2012, Livi and Repette 2017). Also, the (NaOH: Na_2SiO_3) ratio at (1:2) enhances the workability better than (1:2.5), therefore, the compressive strength was higher (Al-Shathr *et al.* 2018). The highest compressive strength obtained in this study was 41.4 MPa at 7 days for (1:2) ratio and 14 molarity, so it's selected as a reference mixture.

3.3 Mechanical Properties for Geopolymer Concrete with WBP

3.3.1 *Compressive strength*

The results of compressive strength for all mixes at 7 and 28 days are shown in Figure 2. Generally, the compressive strength at 28 days is higher than that at 7 days for all mixes. Geopolymer concrete containing WBP shows a reduction in compressive strength. The percentages reduction were 24.5, 21.1, and 29.8% for specimens with 10, 15, and 20% WBP respectively as compared with reference specimens (not containing WBP) at 28 days. This reduction may be attributed to the difference in fineness and strength activity index between MK and WBP.

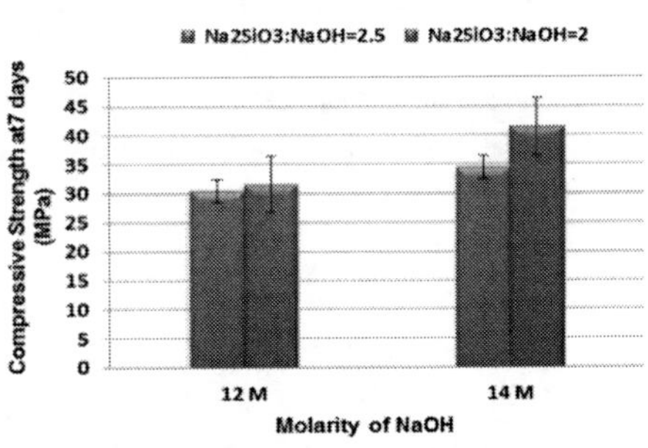

Figure 1. Effect of alkali solution concentration on the compressive strength of mix M4.

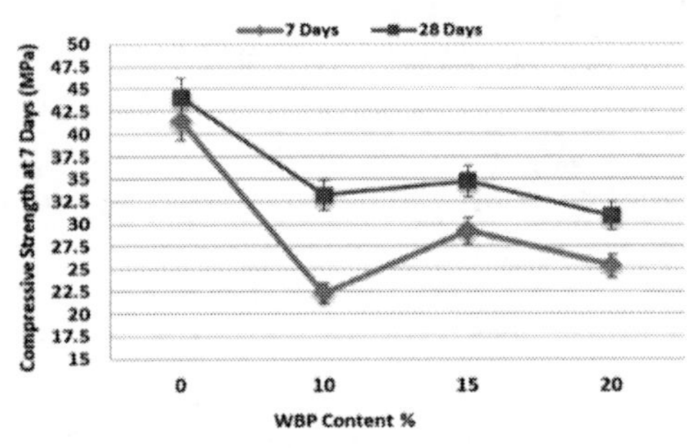

Figure 2. Compressive strength of geopolymer concrete with different waste bricks powder.

In addition, the incorporation of WBP altered the ideal balance of silicon/aluminum ratio and other elements, especially CaO during the geopolymerization (Wongsa *et al.* 2017). Also, the mismatching in optimal curing method (temperature and period) for MK based and WBP-based geopolymer causes reduction in compressive strength. According to Tuyan *et al.* (2018) the maximum compressive strength of waste clay brick powder-based was gained upon curing at 90 °C for 5 days. The reduction in compressive strength results is approved with the results presented by Rovnaník *et al.* (2016). This is related to the microstructure of alkali-activated brick powder, which is less compact and more porous. The optimum dosage of WBP was 15% by weight of MK, that gained 29.26 and 34.76 MPa strength at 7 and 28 days respectively.

3.3.2 *Splitting tensile strength*

Figure 3 presents the results of splitting tensile strength at 7 and 28 days. As in compressive strength, the splitting tensile strength increases with age, and the incorporation of WBP as partial replacement of MK decreases the results. The reduction in splitting strength reaches its maximum value of 37.9% at 28 days for mixture with 10% WBP. This declining in splitting results related to the same causes discussed in compressive strength previously. The minimum reduction in splitting strength was 6.66% for 20%WBP replacement of MK.

3.3.3 *Flexural tensile strength*

As shown in Figure 4, the blended binder geopolymer concrete mixtures exhibited better flexural strength than the mixture with 0%WBP at 28 days age. This enhancement is related to the feature of the microstructure of WBP-geopolymers, which consist of small particles of different sizes and shapes, and a few fiber-forms like crystals (Allahverdi and Kani 2013).

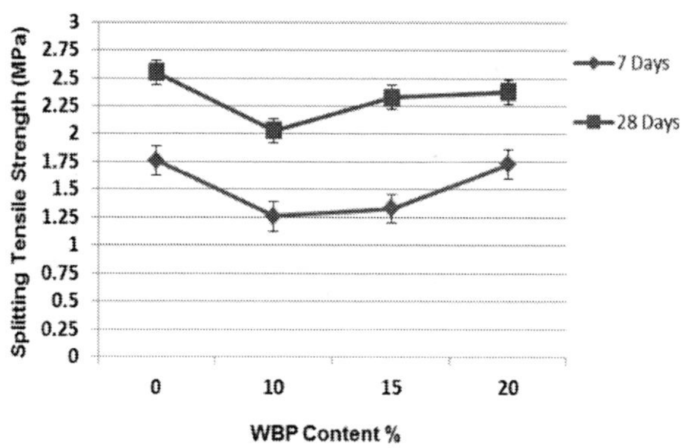

Figure 3. Splitting tensile strength of geopolymer concrete with different WBP contents.

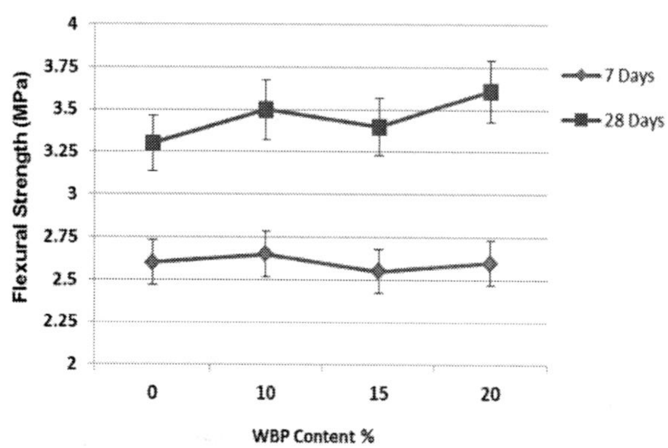

Figure 4. Flexural strength of geopolymer concrete with different WBP contents.

4 CONCLUSIONS

Based on the results obtained in this research, there is a direct relationship between the alkaline solution concentration, the curing method and characteristics of geopolymer concrete. Thus, it is possible to produce MK-based geopolymer concrete with higher than 40 MPa compressive strength at 7 days depending on accurate mix proportions, increased the alkaline activator concentration up to 14 molarity, and adopting efficient curing method. The combined curing method (in oven at 60°C for 4 - 5 hours, then exposure to the sunlight in hot weather) provides the best performance of MK-geopolymer with respect to the strength properties, sustainability and economically, because it reduces the energy consumption and the time of production especially in regions that have hot weather like Iraq. In addition, the incorporation of WBP in MK-based geopolymer concrete reduces both compressive and tensile strength, while the flexural strength showed a slightly increase. Related to compressive strength, the 15% WBP is the optimum dosage, while the 20% WBP gives the highest flexural strength.

The outcomes of this research confirm that the geopolymer concrete synthesized from blended MK and WBP had satisfactory mechanical properties for the concrete structures like multi stories buildings. It has the ability to be used in building industry such as precast elements, interlocking paving units, masonry units, sidewalk, and rigid pavement. Further studies on microstructure, long-term durability characteristics and time dependent deformations (creep and shrinkage) are suggested to be investigated.

References

Allahverdi, A., and Kani, E. N., Use of Construction and Demolition Waste (CDW) for Alkali-Activated or Geopolymer Cements- 18, *Woodhead Publishing Series in Civil and Structural Engineering*, 2013.

Al-Shathr, B., Shamsa, M., and Al-Attar, T., *Relationship between Amorphous Silica in Source Materials and Compressive Strength of Geopolymer Concrete*, MATEC Web of Conferences, 162, 02019, 2018.

Brito, J., and Saikia, N., *Recycled Aggregate in Concrete,* Springer- Verlag, London, UK, 2013.

Haberta, G., and Plamondonb, C. O., Recent Update on the Environmental Impact of Geopolymers, *RILEM Technical Letters*, 1, 17-23, April, 2016.

Joseph, B., and Mathew, G., Influence of Aggregate Content on the Behavior of Fly Ash Based Geopolymer Concrete, *Scientia Iranica, Transactions A: Civil Engineering*, Elsevier, 19(5), 1188-1194, July, 2012.

Karthika, A., Sudalaimanib, K., and Vijayakumarc, C. T., Durability Study on Coal Fly Ash-Blast Furnace Slag Geopolymer Concretes with Bio-Additives, *Ceramics International,* Elsevier, 43, 11935–11943, June, 2017.

Ken, P. W., Ramli, M., and Ban, C. C., An Overview on The Influence of Various Factors on The Properties of Geopolymer Concrete Derived from Industrial By-Products, *Construction and Building Materials*, Elsevier, 77, 370–395, January, 2015.

Livi, C. N., and Repette, W. L., Effect of NaOH Concentration and Curing Regime on Geopolymer, *IBRACON Structures and Materials Journal*, 6(10), 1174-1181, December, 2017.

Mehta, A., and Siddique, R., An Overview of Geopolymers Derived from Industrial by-Products, *Construction and Building Materials,* Elsevier, (127), 183-198, October, 2016.

Rovnaník, P., Rezník, B., and Rovnaníková, P., Blended Alkali-Activated Fly Ash / Brick Powder Materials, *Procedia Engineering*, Elsevier, 151, 108-113, 2016.

Toniolo, N., and Boccaccini, A. R., Fly ash-based geopolymers containing added silicate waste: A review, *Ceramics International*, Elsevier, 43, 14545–14551, July, 2017.

Tuyan, M., Çakir, Ö. A., and Ramyar, K., Effect of Alkali Activator Concentration and Curing Condition on Strength and Microstructure of Waste Clay Brick Powder-Based Geopolymer, *Composites Part B*, Elsevier, 135, 242-252, 2018.

Verma, M, and Dev, N., Geopolymer Concrete: A Way of Sustainable Construction, *International Journal of Recent Research,* 1(5), 201-205, March, 2018.

Wong, C. L., Mo, K. H., Yap, S. P., Alengaram, U. J., and Ling, T. C., Potential Use of Brick Waste as Alternate Concrete-Making Materials: A Review, *Journal of Cleaner Production*, Elsevier, 195, 226-239, May, 2018.

Wongsa, A., Boonserm, K., Waisurasingha, C., Sata, V., and Chindaprasirt, P., Use of Municipal Solid Waste Incinerator (MSWI) Bottom Ash in High Calcium Fly Ash Geopolymer Matrix, *Journal of Cleaner Production*, Elsevier, 148, 49-59, 1 April, 2017.

Interdependence between Structural Engineering and Construction Management
Edited by Ozevin, D., Ataei, H., Modares, M., Gurgun, A., Yazdani, S., and Singh, A.

ISBN: 978-0-9960437-6-2

BUILDING A SUSTAINABLE WORLD: ECONOMY INDEX OF GEOPOLYMER CONCRETE

SOLOMON OYEBISI[1], ANTHONY EDE[1], FESTUS OLUTOGE[2], OLATOKUNBO OFUYATAN[1], and TOLULOPE ALAYANDE[1]

[1]*Dept of Civil Engineering, Covenant University, Ota, Nigeria*
[2]*Dept of Civil Engineering, University of Ibadan, Ibadan, Nigeria*

Geopolymer concrete offers a considerable solution not only to the environmental problem but also to the structural deterioration confronting the world. But, limited or no study is found on its cost implications. Consequently, this study evaluates the production cost and the economy index of geopolymer concrete (GPC) and compares it with the Portland cement concrete (PCC). Corncob ash (CCA) and ground granulated blast furnace slag (GGBFS) were used as source materials in the production of geopolymer concrete. Alkaline liquids were prepared to obtain 12 molar concentrations. The concentration was used to activate the source materials. Grade 30 concrete (M30) was adopted as a mix design proportion. GGBFS was replaced by CCA in varying percentages as 0%, 20%, 40%, 60%, 80%, and 100%. The research findings reveal that GPC is 27.71% lesser than the PCC in terms of production cost while the economy index of GPC is higher than the PCC for the same grade of concrete. The results infer that GPC is cheaper and more viable than the PCC. Thus, geopolymer concrete proves to be an innovative product and appears to be a feasible solution not only to the environmental and structural deteriorating problems but also to the problem of high cost of Portland cement in the construction industry.

Keywords: Corncob ash, Ground granulated blast furnace slag, Sodium hydroxide, Sodium silicate, Compressive strength, Regression model.

1 INTRODUCTION

The high price of Portland cement and failure to utilize the truly indigenous and eco-friendly materials have not solved the consistent yearning for sustainable housing delivery. Thus, a more eco-friendly and sustainable construction process can be implemented with the introduction of geopolymer technologies that allow the production of concrete without Portland limestone cement. Geopolymer concrete is an inorganic chemical reaction between the source materials that are rich in silica and alumina (such as fly ash, metakaolin, and ground granulated blast furnace slag) and the alkaline activators (Lloyd and Rangan 2010). The most commonly used alkaline activators are sodium hydroxide (NaOH) and sodium silicate (Na_2SiO_3) (Davidovits 2013, Lloyd and Rangan 2010). The interest in the application of eco-friendly concrete is further necessitated by the global concern for the environment, in terms of sustainable development, reduction in energy consumption and greenhouse gas emissions (Raheem *et al.* 2010, Oyebisi *et al.* 2017, Ofuyatan *et al.* 2018a, Ofuyatan *et al.* 2018b, Oyebisi *et al.* 2018a, Oyebisi *et al.* 2018b, Oyebisi *et al.* 2018c). Portland cement (PC) production contributes 7% of the total greenhouse gas emissions to the earth's atmosphere. In comparison, there are about 70-80% less carbon dioxide emissions and 43-59% less energy required in the production of geopolymer cement, slag by-product (Davidovits 2013). Hence, the application of geopolymer concrete can significantly reduce the emissions of carbon dioxide into the atmosphere.

Many studies have been conducted on geopolymer concrete two decades ago with limited or no study on its cost implications. Even though, the available ones with cost analyses remain as "grey box". Malathy (2009) estimated that GPC is 10-25% cheaper than the PCC but mix design ratio, curing type, and concentration of alkaline activators was not stated. Thaarrini and Dhirva (2016) evaluated the comparative study on the production cost of Portland cement concrete and geopolymer concrete. It was discovered that cost of production of geopolymer concrete for grade 30 mix proportion is 1.7% marginally higher than the Portland cement concrete of the same grade. Also, the production cost of geopolymer concrete for grade 50 mix proportion is 11% higher than Portland cement concrete of the same grade. Based on these results, type of alkaline activators, the rate of sodium hydroxide concentration, and curing type were not specified. Saurabh and Yogesh (2017) compared the geopolymer concrete with Portland cement concrete based on the strength and cost. The study concluded that production cost per cubic meter (m^3) of geopolymer concrete is 34.75% less than the Portland cement concrete of the same grade. But, the rate of sodium hydroxide concentration was not clearly expressed.

The prime yardstick which determines the structural behavior of any concrete is a compressive strength (Rajarajeswari and Dhinakaran 2016) and this can be measured with the cost of concrete to evaluate feasibility. This is referred to as an economy index. Therefore, this study presents the economy index of geopolymer concrete and compares its result with the Portland cement concrete using the characteristic compressive strength as a strength yardstick and the current market prices relating to the material costs to determine the total cost of the concrete. 12 molar concentration of sodium hydroxide and grade 30 mix design ratio were selected and the prediction equations suitable for the relationship between compressive strength and economy index of geopolymer concrete was presented. The optimum scope of the chemistry laboratory procedures, mixture proportions and designs were chosen based on the relevant standards and codes. All the experimental works in this study were carried out at the Civil Engineering Department, Covenant University, Ota, Nigeria.

2 MATERIALS AND METHODS

2.1 Materials

The following materials were used as fresh concrete mixture constituents namely, Dangote 3X Portland-limestone cement, corncob ash, granulated blast furnace slag (GBFS), fine aggregate (FA), 12.5mm and 19mm coarse aggregates (CA), sodium hydroxide (NaOH) pellet with 99% purity, sodium silicate (Na_2SiO_3) solution, naphthalene-based superplasticizer (Conplast- SP 430) and water.

2.2 Experimental Methods

2.2.1 *Design of concrete mix proportion for grade 30Mpa*

Both the geopolymer concrete and the Portland limestone cement concrete mix proportions were designed in accordance to American Concrete Institute (ACI 214R 2011 and ACI 211.1 2002) to arrive at initial mix proportions taking into considerations, the specific gravities, water absorption capacity, and the moisture contents of materials used. The mix replacement levels are presented in Table 1 while the quantity of mixture proportion for the concrete ingredients is presented in Table 2.

Table 1. The mix proportions for the concrete.

S/N	Ingredient Proportions	Mix No.	Remarks
1	PCC	PCC	Control sample
2	100% GGBFS + 0% CCA	GPC 1	
3	80% GGBFS + 20% CCA	GPC 2	
4	60% GGBFS + 40% CCA	GPC 3	
5	40% GGBFS + 60% CCA	GPC 4	
6	20% GGBFS + 80% CCA	GPC 5	
7	0% GGBFS + 100% CCA	GPC 6	

Note: PCC (Portland cement Concrete); GGBFS (Ground Granulated Blast Furnace Slag); CCA (Corncob Ash); GPC (Geopolymer Concrete)

Table 2. Quantity of concrete ingredients (kg/m^3).

Mixture ID	PCC	GGBFS	CCA	CA 1	CA 2	FA	SS	SH	SP	AL/B	W/S
PCC	390	0	0	516	516	675	NA	NA	3.9	0.54	-
GPC 1	0	390	0	516	516	675	150	60	-	0.54	0.26
GPC 2	0	312	78	516	516	675	150	60	-	0.54	0.26
GPC 3	0	234	156	516	516	675	150	60	-	0.54	0.26
GPC 4	0	156	234	516	516	675	150	60	-	0.54	0.26
GPC 5	0	78	312	516	516	675	150	60	-	0.54	0.26
GPC 6	0	0	390	516	516	675	150	60	-	0.54	0.26

Note: CA 1 (12.5mm Coarse Aggregate size); CA 2 (19mm Coarse Aggregate size); FA (Fine Aggregate); SS (Sodium Silicate solution); SH (Sodium Hydroxide solution); SP (Superplasticizer); AL/B (Alkali Liquid/Binder- for GPC, and Water/Binder- for PCC); W/S (Water-to-Geopolymer Solids ratio).

2.2.3 *Preparation of alkaline liquids, mixing of concrete and testing*

The alkaline liquids were prepared 24 hours prior to the casting of concrete based on the standard chemistry laboratory procedures established by (Rajamane and Jeyalakshmi 2014). The mixing was done for about 5 minutes until a homogenous mixture was obtained. Superplasticizer was administered at 1.0% by mass of binder (PCC). As in the case of Portland cement concrete, superplasticizer is not required for the production of geopolymer concrete because it has no significant effect on its workability and strength properties (Chindaprasirt 2014, Daniel *et al.* 2010). All samples were cured at room temperature in ambient condition (23 ± 5 °C; 60% ± 5% RH). For each mixture, three samples were prepared for each testing age. Compressive strength was carried out in accordance with British Standards (BS EN 12390-3 2000) and the test on the concrete specimens was done at 28 days curing using a digital compressive testing machine of 2000KN maximum capacity. The economy index of the concrete is calculated using Eq. (1).

$$Economy\ Index = \frac{Compressive\ Strength}{Total\ cost\ of\ production} \quad (1)$$

3 RESULTS AND DISCUSSIONS

3.1 Production Cost per cubic meter (m^3) of PCC and GPC

The production costs per cubic meter for both PCC and GPC are presented in Table 3 and Table 4 respectively. From Table 3, it is shown that the total cost of production for an M30 grade of Portland cement concrete is \$ 109.59 with cement appropriating a significant percentage of 53.38 in the production cost. Similarly, Table 4 indicates the total cost of production for an M30 grade of geopolymer concrete as \$ 79.22 with sodium hydroxide solution appropriating a significant percentage of 37.26 in the production cost. By comparison, it is inferred that the cost of production per cubic meter for an M30 grade of geopolymer concrete is 27.71% lesser than the Portland cement concrete of the same grade.

Table 3. Production cost for M30 per m^3 of PCC.

S/N	Constituent	Quantity (kg/m^3)	Unit	Rate ($)	Amount ($)	Constituent Cost (%)
1	Cement	390	Kg	0.15	58.50	53.38
2	FA (SSD)	675	Kg	0.01	6.75	6.16
3	CA (SSD)	1031	Kg	0.02	20.62	18.82
4	Water	204.15	Kg	0.01	2.04	1.86
5	Conplast SP-430	3.90	Kg	5.56	21.68	19.78
	Total				**109.59**	**100**

Table 4. Production cost for M30 per m^3 of GPC.

S/N	Constituent	Quantity (kg/m^3)	Unit	Rate ($)	Amount ($)	Constituent Cost (%)
1	GGBFS/CCA	390	Kg	0.01	3.90	4.92
2	FA (SSD)	675	Kg	0.01	6.75	8.52
3	CA (SSD)	1,031	Kg	0.02	20.62	26.03
4	NaOH solution	21.24	Kg	1.39	29.52	37.26
5	Na_2SiO_3 solution	66.15	Kg	0.26	17.20	21.71
6	Water	122.61	Kg	0.01	1.23	1.55
	Total				**79.22**	**100**

3.2 Economy Index (EI)

Table 5 shows the economy index of both PCC and GPC, and from the results, control concrete (PCC) has an economy index of 0.320. All geopolymer concretes produced indicate higher economy index than the PCC except GPC 5 and GPC 6. This is primarily due to the cost difference between the cement and the source materials (GGBFS and CCA) and the higher strength of GGBFS and CCA-based GPC. The cost of source materials is very less (around 15-20% of cement) (Rajarajeswari and Dhinakaran 2016). Both GGBFS and CCA are waste materials from steel industries and agricultural produce respectively.

Table 5. Economy index of concretes produced.

Mix ID	Total Cost/m^3 ($)	Compressive Strength (MPa)	EI	% Increase/Decrease
PCC	109.59	35.12	0.320	-
GPC 1	79.22	43.17	0.545	+41.28
GPC 2	79.22	40.13	0.507	+36.88
GPC 3	79.22	35.54	0.449	+28.73
GPC 4	79.22	28.77	0.363	+11.85
GPC 5	79.22	22.89	0.289	-10.73
GPC 6	79.22	20.07	0.253	-26.48

3.3 Regression Model for Compressive Strength and Economy Index of GPC

General model power 1 in Matlab (2017a) was employed and the result of the regression equation is presented in Figure 1. With respect to compressive strength, the coefficients of determination (R^2) is 99.99% for the economy index at 28 days curing. This concludes that the model is 100% significantly fit to predict the relationship and also, compressive strength largely depends on the economy index at 95% confidence bounds.

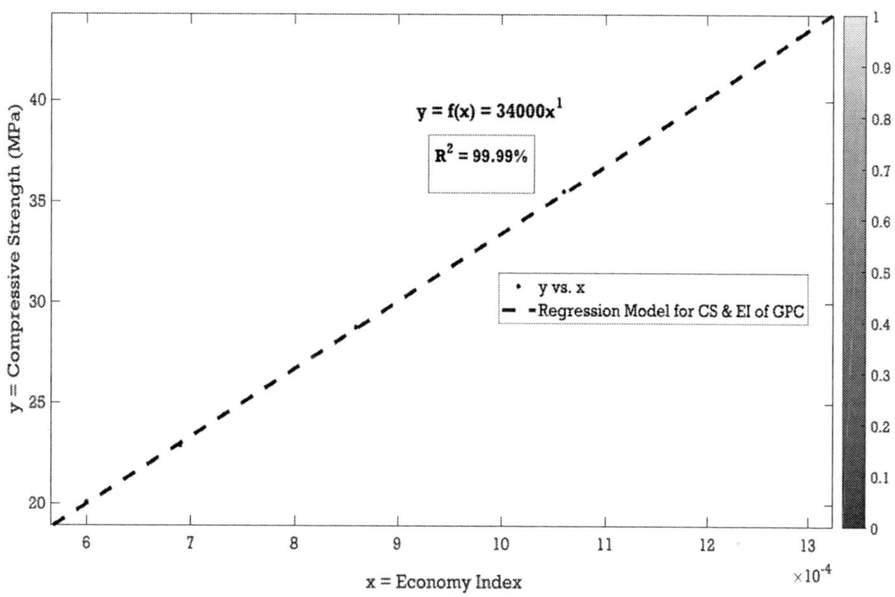

Figure 1. Regression model for compressive strength (MPa) against an economy index.

4 CONCLUSIONS AND RECOMMENDATIONS

Based on the above findings, it is concluded that a higher economy index and a lesser cost of production exist in the production of Grade M30 for GPC activated with sodium hydroxide (NaOH) and sodium silicate (Na_2SiO_3) solutions at ratio 2.5:1 of Na_2SiO_3-to-NaOH for 12 molar concentration of NaOH when compared with PCC of the same grade of concrete. And accordingly, the study contributed to the evaluation of sustainable concrete production and cost implication in comparison with Portland cement concrete. Moreover, the study developed a regression model equation and predicted the relationship between the compressive strength and the economy index. It is therefore recommended that standard specifications and procedures should always be followed in the course of evaluating the production cost for the geopolymer concrete. Economy index of a higher grade of concrete may also be determined to get the higher economy index.

Acknowledgements

The authors value the Covenant University management for the technical and financial supports in course of the study.

References

ACI 211.1, Standard Practice for Selecting Proportions for Normal, Heavyweight, and Mass Concrete, *American Concrete Institute Committee,* Farmington Hills, USA, 2002.

ACI 214R, Guide to Evaluation of Strength Test Results of Concrete, *American Concrete Institute Committee,* Farmington Hills, USA, 2011.

BS EN 12390-3, Testing Hardened Concrete: Compressive Strength of Test Specimens, *British Standard Institution*, 2 Park Street, London. 2000.

Chindaprasirt, P. W. C., Effect of Sodium Hydroxide Concentration on Chloride Penetration and Steel Corrosion of Fly Ash-Based Geopolymer Concrete Under Marine Site, *Construction and Building Materials*, 63, 303-310, 2014.

Daniel, G., Kong, L., and Sanjayan. J., Effect of Elevated Temperatures on Geopolymer Paste, Mortar And Concrete, *Cement and Concrete Research*, 40, 334-339, 2010.

Davidovits, J., *Geopolymer Cement: A Review,* 2013. Retrieved from http://www.geopolymer.org/library/technical-papers/21-geopolymer-cement-review on August 20, 2018.

Lloyd, N., and Rangan, B. V., *Geopolymer Concrete with Fly Ash,* Second International Conference on Sustainable Construction Materials and Technologies, 2010.

Malathy, R., Fresh and Hardened Properties of Geopolymer Concrete and Mortar, 2009. Retrieved from www.schleibinger.com/cmsimple/?download=r2009_malathy_geopolymer.pdf on March 5, 2018.

Ofuyatan, O., Adeola, A., Sulymon, N., Ede, A., Oyebisi, S., Alayande, T., and Ogundipe, J., Pseudo-Dynamic Earthquake Response Model of Wood-Frame with Plastered Typha (Minima) Bale Masonry-Infill, *International Journal of Civil Engineering and Technology,* 9(2), 27-35, February, 2018a.

Ofuyatan, O., Ede, A., Olofinnade, R., Oyebisi, S., Alayande, T., and Ogundipe, J., Assessment of Strength Properties of Cassava Peel Ash-Concrete, *International Journal of Civil Engineering and Technology,* 9(1), 965-974, January 2018b.

Oyebisi, S. O., Olutoge, F. A., Ofuyatan, M. O., and Abioye, A. A., Effect of Corncob Ash Blended Cement on the Properties of Lateritic Interlocking, *Progress in Industrial Ecology-An International Journal*, 11(4), 373-387, 2017.

Oyebisi, S., Ede, A., Ofuyatan, O., Oluwafemi, J., and Akinwumi, I., Comparative Study of Corncob Ash – Based Lateritic Interlocking and Sandcrete Hollow Blocks, *International Journal of Geomate*, 15(51), 209-216, November 2018a.

Oyebisi, S., Ede, A., Olutoge, F., Ofuyatan, O., and Oluwafemi, J., Modeling of Hydrogen Potential and Compressive Strength of Geopolymer Concrete, *International Journal of Civil Engineering and Technology,* 9(7), 671-679, July 2018b.

Oyebisi, S., Ede, A., Olutoge, F., Ofuyatan, O., and Oluwafemi, J., Influence of Alkali Concentrations on the Mechanical Properties of Geopolymer Concrete, *International Journal of Civil Engineering and Technology,* 9(8), 734-743, August 2018c.

Raheem, A. A., Oyebisi, S. O., Akintayo, S. O., and Oyeniran, M. O., Effects of Admixture on the Properties of Corncob Ash Cement Concrete, *Leonardo Electronic Journal of Practices and Technologies,* 16, 13-20, January-June 2010.

Rajamane, N. P., and Jeyalakshmi, R., Quantities of Sodium Hydroxide Solids and Water to Prepare Sodium Hydroxide Solution of given Molarity for Geopolymer Concrete Mixes, *Indian Concrete Institute Technical Paper*, SRM University, India, 2014.

Rajarajeswari, A., and Dhinakaran, G., Compressive Strength of GGBFS based GPC under Thermal Curing, *Construction and Building Materials*, 126, 552-559, 2016.

Saurabh D., and Yogesh, S., Comparison of Geopolymer Concrete based on Strength and Cost with Concrete, *Imperial Journal of Interdisciplinary Research (IJIR),* 3(9), 1026-1029, 2017.

Thaarrini, J., and Dhivya, S., Comparative Study on the Production Cost of Geopolymer and Conventional Concretes, *International Journal of Civil Engineering Research,* 7(2), 117-124, 2016.

Interdependence between Structural Engineering and Construction Management
Edited by Ozevin, D., Ataei, H., Modares, M., Gurgun, A., Yazdani, S., and Singh, A.

ISBN: 978-0-9960437-6-2

REUSE OF GLASS AND CLAY BRICK WASTES AS SUBSTITUTE FOR NATURAL AGGREGATE IN GREEN CONCRETE

OLUWAROTIMI OLOFINNADE[1], ANTHONY EDE[1], JULIUS NDAMBUKI[2], and AYANFE OPEYEMI[1]

[1]*Dept of Civil Engineering, Covenant University, Ota, Nigeria*
[2]*Dept of Civil Engineering, Tshwane University of Technology, Pretoria, South Africa*

Utilization of recycled waste materials in making concrete is identified as a sustainable means of managing wastes, thereby reducing the energy consumption, preserving the environment and conserving of natural resources from depletion. Researchers referred to this type of concrete as "green" concrete. This study examines the possible reusing of crushed waste glass as partial and complete substitute for natural aggregates in production of moderate strength green concrete with the addition of ground clay brick as admixture. The clay bricks were obtained as generated wastes from the ceramic and brick producing factory, while the glass wastes were sourced from dump sites and waste collection points within Ota, Nigeria. The waste glass varied from 25% – 100% in steps of 25%, and the ground clay brick was added in 10%, 15% and 20% by mass of Portland cement into the concrete mixes. Tests, which include workability and characteristics strength were carried out on the concrete specimens. Microstructural examination was performed on selected concrete specimens. Results indicate reduction in workability with increased waste glass and clay brick powder content. Moreover, the characteristic compressive strength of the concrete specimens increased with curing age, however, concrete mixes containing 10% clay brick powder and 25% waste glass aggregate showed significant improvement in strength at curing age of 28 days than the control concrete.

Keywords: Compressive strength, Sustainability, Recycling, Eco-concrete, Microstructure, Waste management.

1 INTRODUCTION

The concept of sustainability in the construction industry can be attributed to sustainable development. The report of World commission in Environment and Development on sustainable development mentioned that the idea of sustainable development implies the present generation meeting their needs without compromising the needs of the future generation (WCED 1987). Furthermore, sustainability in the construction industry is now being considered a major concern due to the identifiable environmental impact associated with the amount of natural resource depleted for production of building materials such as cement and concrete for buildings and engineering structures (Bilodeau and Malhotra 2000). The industry is reported to be contributing more CO_2 emission into the atmosphere compare to other industries (Kline and Barcelo 2012). Calkins (2009) mentioned that the construction industry depends more on nonrenewable materials as resources for construction, which in turn aid the destruction of the environment. Of recent, there are continuous research efforts on using the construction industry as a means of managing generated solid wastes. It was reported that the industry provides an effective means of removing large amounts of generated solid wastes from the environment (Calkins 2009, Ling *et al.* 2013). Moreover, over the years there is an increased emphasis on sustainability in the construction industry. That is, limiting the huge impacts of materials on the environment as well as waste materials (Calkins 2009). Some solid waste materials such as waste glass can be reuse and recycled for use in sustainable production of concrete (Siddique 2008, Ali and Al-Tersawy 2012, Tan and Du 2013, Rashad 2014, Olofinnade *et al.* 2017, 2018a). Crushed waste glass was reported to provide an alternative as a substitute material for traditional aggregate in concrete (Rashad 2014, Olofinnade *et al.* 2016a, 2016b, 2018b). However, many studies have pointed out the restraint that is limiting the utilization of waste glass aggregate in concrete. This restraint is referred to as the alkali-silica reaction (ASR) gel formed when the glass material is exposed to attack from the high alkali from the cement paste causing cracking of concrete. Meanwhile, some positive results have pointed out the suppressing effect of ASR through the use of pozzolanic admixtures, low alkali cement in concrete, and lithium compounds (Tan and Du 2013, Carsana *et al.* 2014, Afshinnia and Rangaraju, 2015). The aim of this study is to examine the possibility of using waste glass as a substitute for sand with the addition of clay brick powder as pozzolanic admixture in the production of medium strength green concrete.

2 METHODOLOGY

2.1 Materials: Ordinary Portland Cement

The Ordinary Portland cement used for this study was sourced locally complying with the standard specification of ASTM Type I for cement used for general purpose. Moreover, the chemical constituents of the cement were determined using the X-ray Fluorescence (XRF), and the results are presented in Table 1, including some physical properties of cement.

2.2 Clay Brick Powder (GCB)

The clay bricks were obtained as generated wastes from a ceramic and brick producing factory located within the Lagos metropolis, Nigeria. The bricks were manufactured by calcination of raw clay under high temperature not less than 900 °C. However, before crushing the clay brick wastes into powder, the waste was properly separated to remove unwanted impurities. Thereafter, the wastes bricks were crushed and milled to powder size. The powder specimen was then sieved through the BS mesh 75 μm sieve opening in order to ensure uniform particle size of the same fineness. The chemical compositions of the clay brick powder as shown in Table 1 were obtained using the X-ray Fluorescence (XRF).

2.3 Crushed Waste Glass (CWG)

The glass wastes used in this study were sourced from dump sites and waste collection points within Ota environment, Ogun state. The sourced glass waste debris comprises mainly disposed flat glasses (windows) and container glasses of mixed coloured. After collection of the waste glasses, they were all thorough soaked in water for 48 hours for the purpose of removing unwanted dirt's, contaminants and residues. The waste glasses were then properly washed and openly air-dried. However, due to the nature of glass, safety precaution was taken to protect the hands and eyes by wearing rubber gloves and eye google protection. The waste glasses were crushed to the required particle sizes similar to natural sand. The chemical compositions of the waste glass as shown in Table 1 were determined through X-ray Fluorescence (XRF).

2.4 Natural Fine and Coarse Aggregate

The fine aggregates (river sand) used in this study are sourced locally and commercially. The maximum size of the river sand is 4.75 mm, while that of coarse aggregate (gravel) is 19 mm.

Table 1. Chemical composition of cement, clay brick powder and crushed waste glass materials.

Chemical composition	Mass (%)			Physical properties			
	Cement	Clay brick powder	Waste glass		cement	Clay brick powder	Waste glass
SiO_2	19.38	60.64	64.31	Specific	3.15	2.52	2.50
Al_2O_3	.00	14.23	19.98	gravity	0.52		
Fe_2O_3	4.14	4.93	6.25	Soundness (%)			
CaO	3.19	0.27	10.61	Setting time	68		
MgO	2.442	1.72	0.63	Initial (min)	473		
SO_3	.00	1.20	0.25	Final (min)		Brownish	Light
K_2O	2.44	1.44	0.74	Colour			grey
Na_2O	1.59	1.94	12.52				
TiO_2	0.21	0.98	0.61				
Cr_2O_3	0.28	0.90					
Na_2Oeq =	0.04						
$Na_2O + 0.658K_2O$	0.02		0.02				
Loss on ignition	0.18		1.47				

Table 2. Proportion of concrete mixtures for this study.

Concrete mixtures		Binder (kg/m^3)		Aggregate (kg/m^3)				
		Cement	Clay brick powder	Crushed waste glass	Sand	Gravel	Water (kg/m^3)	Water/binder ratio
Control	100%	275	0	0	550	1100	138	0.5
Glass sand-CWG	25%CWG	275	0	138	413	1100	138	0.5
	50%CWG	275	0	275	275	1100	138	0.5
	75%CWG	275	0	413	138	1100	138	0.5
	100%CWG	275	0	550	0	1100	138	0.5
CWG/ 10%GCB	25%CWG	247.5	27.5	138	413	1100	138	0.5
	50%CWG	247.5	27.5	275	275	1100	138	0.5
	75%CWG	247.5	27.5	413	138	1100	138	0.5
	100%CWG	247.5	27.5	550	0	1100	138	0.5
CWG/ 15%GCB	25%CWG	233.75	41.25	138	413	1100	138	0.5
	50%CWG	233.75	41.25	275	275	1100	138	0.5
	75%CWG	233.75	41.25	413	138	1100	138	0.5
	100%CWG	233.75	41.25	550	0	1100	138	0.5
CWG/ 20%GCB	25%CWG	220	55	138	413	1100	138	0.5
	50%CWG	220	55	275	275	1100	138	0.5
	75%CWG	220	55	413	138	1100	138	0.5
	100%CWG	220	55	550	0	1100	138	0.5

2.5 Methods: Mixing Procedures, Casting and Testing of Concrete Specimens

Batching summary for proportioning of the various concrete constituents is presented in Table 2. All mixtures were batched by weight at ratio of 1:2:4 (cement: sand: gravel) to achieve a 28-day strength of 20 MPa. The mixing procedures involve firstly; the sand, waste glass and gravels were admixed together for two (2) minutes. Secondly; cement and clay brick powder used thoroughly mixed together for two (2) minutes, and then added to the aggregate mixture, and all were properly mixed in the dry state for over 5 minutes. Thirdly; the required quantity of water was then added while the mixing process continues for about 15 -18 minutes until a uniform paste was achieved. The slump of the freshly prepared concrete mixes was tested. Concrete cubes specimens were produced using a 150 × 150 × 150 mm dimensions. After curing by total immersion in water, compressive strength of the hardened concrete specimens at the ages of 3, 7 and 28 days were tested (BS EN 12390 2009). Microstructural examination was carried out on selected concrete from the tested concrete specimens.

3 RESULTS AND DISCUSSION

The particle size distribution analysis indicates close similarity for the natural sand and CWG materials. In addition, the workability of the freshly prepare green concrete mix shows a declining trend as the percentage of the glass and GCB increases relative to the control.

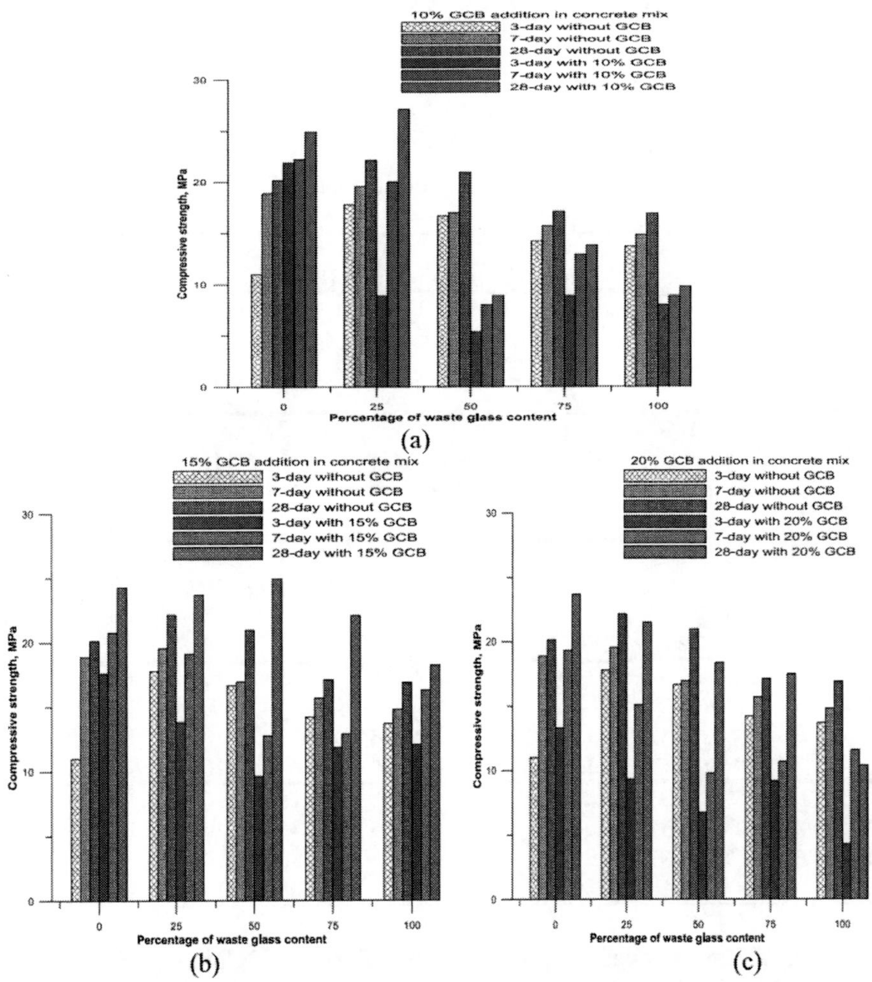

Figure 1. Compressive strength of concrete specimens containing CWG at (a) 10% GCB; (b) 15% GCB; (c) 20% GCB.

Results on the chemical composition of clay brick powder as presented in Table 1 shows that the GCB material can function as pozzolanic substitute for the cement in accordance with the recommendation of ASTM C618 (2003) for pozzolanic material. According to the ASTM C618 specification, that a potential pozzolana should have the additions of SiO_2 + AL_2O_3 + Fe_2O_3 to be minimum of 70% by mass of the chemical compounds. Consequently, for the GCB, the recorded minimum percentage of is 79% by mass. The results on the compressive strength test for the cube specimens are presented in Figures 1 (a – c).

The evolution of compressive strength development for the green concrete containing CWG show a decrease in strength as the CWG contents increases. However, the results clearly indicate that addition of GCB has significant influence on the strength development of the CWG modified green concrete. This can be attributed to the pozzolanic nature of the GCB reacting with the calcium hydroxide $Ca(OH)_2$ from the cement to form a compound of hydrated C-S-H. Moreso, resulting in modification of the green concrete properties resulting in strength enhancement as observed for the 10% GCB addition (He *et al.* 1994, Olofinnade *et al.* 2018a). In addition, it was observed that the compressive strength increases proportionally with the curing ages. Significant improvement in compressive strength was noticed at 10% GCB and 25% waste glass as a partial substitute for cement and natural sand respectively. Figures 2(a – c) depicts the scanning electron microscope (SEM) images taken on the fractured surface of concrete specimens showing pore structure for the control (Figure 2a), containing CWG aggregate (Figure 2b), and a further improved structure with a low number of pores and cracks (Figure 2c).

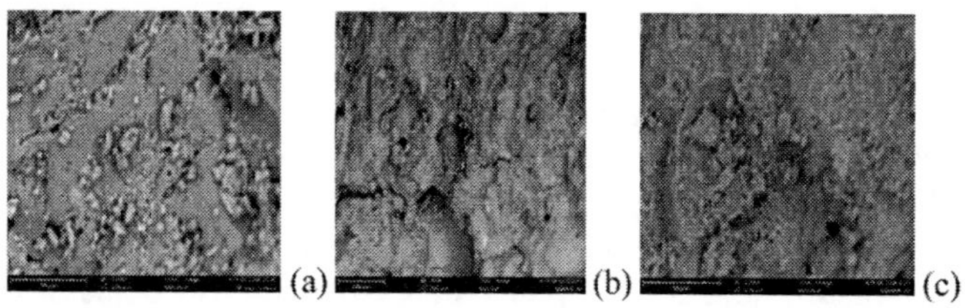

Figure 2. (a) SEM image on control concrete; (b) SEM image on concrete specimen containing CWG particles; (c) SEM image with the addition of GCB.

4 CONCLUSIONS

This study examines the possible production of green concrete using the crushed waste glass and clay brick powder as a substitution material for natural sand and cement respectively. The study shows that the workability of freshly prepared concrete decreases as percentage of CWG and GCB content increases. The addition of GCB into concrete mixes containing CWG improves the compressive strength of the hardened concrete compared to the control. The improvement in strength due to addition of clay brick powder was also corroborated by the microstructure of concrete specimens. Based on findings, crushed waste glass of about 25% replacement for sand with the addition of clay brick powder of about 10% is suggested as a substitute for cement in eco-concrete production.

Acknowledgments

The authors would like to appreciate the Centre for Research, Innovation and Discovery, Covenant University, Ota, Nigeria for supporting this research.

REFERENCES

Afshinnia, K., and Rangaraju, P. R., Influence of Fineness of Ground Recycled Glass on Mitigation of Alkali–Silica Reaction in Mortars, *Construction and Building Materials,* 81, 257–267, 2015.

Ali, E. E., and Al-Tersawy, S. H., Recycled Glass as a Partial Replacement for Fine Aggregate in Self-Compacting Concrete, *Construction and Building Materials,* 35, 785–791, 2012.

ASTM C618. *Standard Specification for Coal Fly Ash and Raw or Calcined Natural Pozzolan for Use in Concrete,* American Society for Testing and Materials, West Conshohocken, PA, ASTM International, 2003.

Bilodeau, A., and Malhotra, V. M., High Volume Fly Ash System: Concrete Solution for Sustainable Development, *ACI Materials Journal,* 1, 41-47, 2000.

BS EN 12390: 2009, *Testing Hardened Concrete*, British Standards Institute, BSI, London, 2009.

Calkins, M., *Materials for Sustainable Sites: A Complete Guide to the Evaluation, Selection, and Use of Sustainable Construction Materials.* Wiley, Hoboken, NJ, 2009.

Carsana, M., Frassoni, M., and Bertolini, L., Comparison of Ground Waste Glass with Other Supplementary Cementitious Materials, *Cement and Concrete Composites,* 45, 39–45, 2014.

He, C., Makovicky, E., and Osaeck, B., Thermal Stability and Pozzolanic Activity of Calcined Kaolin, *Applied Clay Science*, 9, 165–187, 1994.

Kline, J., and Barcelo, L., Cement and CO_2, A Victim of Success!, *Cement Industry Technical Conference, IEEE-IAS/PCA,* 53, 1–14, 2012.

Ling, T. C., Poon, C. S., and Wong, H. W., Management and Recycling of Waste Glass in Concrete Products: Current Situations in Hong Kong, *Resource, Conservation and Recycling,* 70, 25–31, 2013.

Olofinnade, O. M., Ede, A. N., and Booth, C. A., *Sustainability of Waste Glass Powder and Clay Brick Powder as Cement Substitute in Green Concrete*, Handbook of Environmental Materials Management, Springer, 2018a.

Olofinnade, O. M., Ede, A. N., Ndambuki, J. M., and Bamigboye, G. O., Structural Properties of Concrete Containing Ground Waste Clay Brick Powder as Partial Substitute for Cement, *Material Science Forum,* 866, 63-67, 2016a.

Olofinnade, O. M., Ede, A. N., Ndambuki, J. M., Ngene, B. U., Akinwumi, I. I., and Ofuyatan, O., Strength and Microstructure of Eco-Concrete Produced using Waste Glass as Partial and Complete Replacement for Sand, *Cogent Engineering*, 5, 1483860, 1-19, 2018b.

Olofinnade, O. M., Ndambuki, J. M, Ede, A. N., and Olukanni, D. O., Effect of Substitution of Crushed Waste Glass as Partial Replacement for Natural Fine and Coarse Aggregate in Concrete, *Material Science Forum,* 866, 58-62, 2016b.

Olofinnade, O. M., Ndambuki, J. M., Ede, A. N and Booth, C., Application of Waste Glass Powder as a Partial Cement Substitute Towards More Sustainable Concrete Production, *International Journal of Engineering Research in Africa,* 31, 77 – 93, 2017.

Rashad, A. M., Recycled Waste Glass as Fine Aggregate Replacement in Cementitious Materials Based on Portland Cement, *Construction and Building Materials,* 72, 340–357, 2014.

Siddique, R., *Waste glass*, Chap. 4 in *Waste Materials and by-Products in Concrete.* Springer, 2008.

Tan, K.H., and Du, H., Use of Waste Glass as Sand in Mortar, Part I: Fresh, Mechanical and Durability Properties, *Cement and Concrete Composites,* 35, 109–117, 2013.

WCED, *Our Common Future*: Report of the World Commission in Environment and Development. Oxford University Press, Oxford, 1987.

Interdependence between Structural Engineering and Construction Management
Edited by Ozevin, D., Ataei, H., Modares, M., Gurgun, A., Yazdani, S., and Singh, A.

ISBN: 978-0-9960437-6-2

EFFECTS OF BAMBOO FIBERS AND LIMESTONE POWDER ON FRESH PROPERTIES OF SELF-COMPACTING CONCRETE

ANTHONY NKEM EDE, OBATARHIE OSHOGBUNU, OLUWAROTIMI MICHAEL OLOFINNADE, KAYODE JOSHUA JOLAYEMI, SOLOMON OLAKUNLE OYEBISI, OLUWASEUN GRACE MARK, and PAUL OLUWASEUN AWOYERA

Dept of Civil Engineering, Covenant University, Ota, Nigeria

Self-compacting concrete (SCC) flows through densely steel reinforced elements and consolidates under self-weight without need for vibration or compaction. This helps in complex and densely reinforced structures. The integration of fibers and fillers in concrete improves its general properties. The addition of fibers in particular can regulate the flow and workability of the concrete; hence, the high workable nature of SCC can be an ideal mix for the incorporation of fibers. This research investigates the effect of bamboo fibers and limestone powder on the fresh properties of self-compacting concrete. Bamboo fibers of an aspect ratio of 50 and varied volumes of 0.25%, 0.5%, 0.75% and 1% were adopted for this research. The workability of the mix was assessed by slump flow test and V- funnel test. For fiber volumes of 0.25%, 0.5%, 0.75%, it was observed that the coarse aggregate was evenly distributed across the spread, indicating good viscosity and stability of the mix. The presence of 10% percent limestone powder improved the workability of the concrete mix. This can be attributed to filler properties of limestone powder, which, affecting the cement particle system, changed the ordinary distance between them and modified the water quantity available for the hydration process. These results proved that the bamboo fiber and limestone powder can be sustainably adopted to regulate the flow-ability of SCC without compromising desired properties.

Keywords: Natural fiber, Workability, Flowability, Congested reinforcement, Slump test, V-funnel test.

1 INTRODUCTION

Concrete is a revolutionized material over the years of human existence and both open to accommodate new inclusions and give rise to new types. Vast use of concrete made up of lime and volcanic ash was verified during the era of Roman Empire (Li 2011). Then came the era of concrete using the blend of Portland cement, fine and coarse aggregates, admixtures and water. Currently, concrete is the primarily material adopted by the construction industry for it the ability to be molded into different shapes and for affordability. In today's sustainability-driven world, guided by the ever-increasing concern for the deteriorating environment, efforts are made to create more ecological concrete with many constituent materials substitution (Olofinnade *et al.* 2016 and 2017, Ede *et al.* 2017 and 2018). The use of concrete worldwide is twice as much as wood, steel, aluminum, and plastics put together, and are only exceeded in the present world by the usage of naturally occurring water (Koehler and Fowler 2007). Self-compacting concrete

(SCC) is an emerging technique of concrete technology, to tackle the problem of concreting through congested reinforcement. The uniqueness of SCC is due to properties, like filling ability, flow ability, pump ability, durability, and the ability to make production of concrete more mechanized. Modern concrete made with Portland cement generates about 5% of the total Co_2 emitted today to the environment, while normal SCC requires an increased powder content which leads to a notable upsurge in material cost and other associated negative effects such as increase in thermal stress and shrinkage. This necessitated the need to provide suitable filler materials to assuage such undesirable effects. The fillers materials or supplementary cementitious materials (SCM) can be pozzolanic or hydraulic. Because concrete is a material open to inclusions, the SCMs may consist of locally available materials such as rice husk ash, fly ash, pulverized fuel ash (PFA), silica fume, granulated ground blast furnace slag (GGBS), and lime stone powder (Zhu and John 2005).

The addition of limestone powder in self-compacting concrete significantly enhances the fresh properties due to the increase of the workability derived from the ability of limestone powder to retain water in fresh mix and improve the stability and deformability of fresh SCC mix. The enriched fresh properties can be attributed to the improve particle packing system, better water retention of fresh mix, and arguably possible chemical reaction involving cement and Calcium carbonate ($CaCO_3$) to form carbon-aluminate ($C\text{-}Al_2O_3$) (Surabhi *et al.* 2009). The introduction of fibers as reinforcements in concrete is not new, it was first developed by the Egyptians, where straws and the hair of animals were adopted (Bindu 2016). Resent research and design of fiber reinforced concrete began to increase in the 1970s and various types of fibers have been developed such as steel fiber, glass, Polypropylene, Nylon and Polyester, etc. (ACI Committee 2002). In most of the developing countries, agricultural plants are utilized for development and production of useful composites. Bamboo is one of these natural materials and it is found in abundance in Africa, Asia and Southern part of America, yet not adequately exploited for concrete design technology.

Self-compacting concrete was developed in Japan in the late 1980's by Professor Okamura and his team in a quest for a more durable and sustainable concrete in highly congested reinforcement environment without the need for vibration or compaction (Okamura and Ouchi 1999). Use of admixtures to improve the workability of SCC made with natural fiber and supplementary cementitious material such as limestone, fly ash, silica fume is documented in literature (Li, 2011). Although, the introduction of natural fibers in concrete can be found in literature, the use of bamboo fiber in concrete has not been adequately explored. These problems brought about this research. Self-compacting concrete, containing environmentally friendly bamboo fibers and limestone as a supplementary material is being considered in this research. The aim of this research is to determine the effect of limestone powder and bamboo fibers on the fresh properties of the fiber reinforced self-compacting concrete. This will be done by performing fresh property tests of concrete specimen. The type of concrete to be obtained will be referenced as Bamboo Fiber-Reinforced Self Compacting Concrete (BFR-SCC).

2 METHODOLOGY

2.1 Material Selection

The material selection and methodology implemented affect the results obtained. The materials adopted, their various properties, and the methodologies embraced for analyzing the fresh properties of BFR-SCC are discussed in this section. All tests were carried out at Covenant University. An Ordinary Portland cement brand available in Nigeria (Dangote cement), of grade 42.5N was adopted for this research. The filler material used is locally available, finely grinded and sieved limestone powder. CONPLAST SP430 super-plasticizer, which disperses fine particles in concrete mix and enables water content of the concrete to perform more effectively was used. Sharp sand of maximum particles size of 5 mm and crushed granite of 19 mm were used as aggregates. A water to cement ratio of 0.5 was adopted. Figure 1 shows the limestone powder and the aggregates adopted for the research.

(a) (b) (c)

Figure 1. (a) Limestone powder; (b) Coarse Aggregate; (c) Fine Aggregate

The bamboo fibers used had a maximum length of 50 mm with varying diameters. Bamboo sticks were cut into pieces with the back scrapped off and soaked in water for at least a day. The soaked bamboo was beaten into strips then combed into fibers. These fibers were placed in 2% solution of sodium hydroxide for a day as treatment. Water absorption test is also carried out. Bamboo tree and fiber are shown in Figure 2.

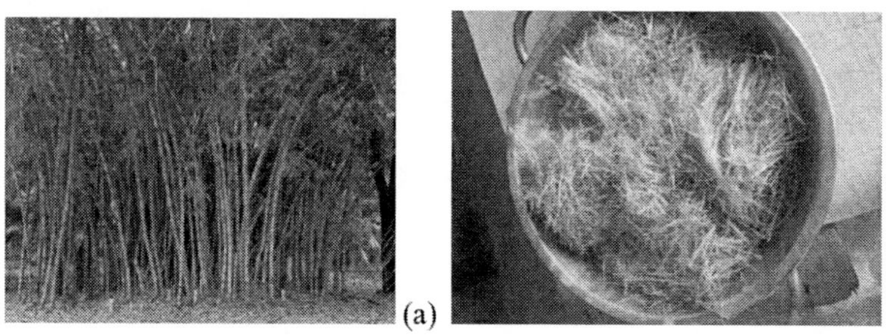
(a) (b)

Figure 2. (a) Bamboo tree; (b) Bamboo fiber.

2.2 Mix Design and Test Methods

For this study, various tests were carried out to determine the fresh properties of the BFR-SCC samples such as the slump test, L-box test, and the V- funnel test. For 0, 0.25, 0.5 and 0.75% of fiber to cement weight with 10% of limestone powders and fiber length (50mm) kept constant, fresh properties and microstructural test of hardened concrete were determined. Different mix design of self-compacting concrete was tried by volume in accordance with EFNARC (2002). Based on test results, adjustments are made, and then the mix is re-batched with further testing until the required properties are achieved. The curing process was done by submerging the specimen in clear fresh water in a curing tank. Prior to this, the samples were stored in the laboratory at room temperature for 24 hours, and then they were marked and removed from the molds.

2.3 Slump Test and V-Funnel Test

The slump test was carried out to assess the flowability of SCC. It gives some indication of the segregation of the sample but doesn't give any indication on the ability of the concrete to pass through dense reinforcements. The V-funnel test was used to measure the flowability of concrete with a maximum aggregate size of 19 mm. The results of this test can be influenced by the properties of the concrete such as volume of aggregate in the mix. Figure 3 shows slump test and the V-funnel test procedures.

(a) (b)

Figure 3. (a) Slump flow test; (b) V funnel test.

2.4 Preparation of Cubes

The molds were cleaned and oiled and then filled with BFR-SCC concrete without compacting and the top surface is leveled and smoothened with trowel. The test sample are stored in moist air under room temperature for 24 hours and after then marked and removed from the mold. The removed cubes are cured in a curing tank prior to the compressive test.

2.5 Microstructural Properties

A scanning electron microscope (SEM) was used to analyze the microstructural properties of selected BFR-SCC mixtures in order to get a more detailed understanding of the fresh properties.

3 RESULTS

3.1 Fresh Property Test Results

3.1.1 *Flow and T_{50}*

EFNARC LIMITS: slump flow – (550- 800) mm; T50- < 12 seconds.

From Figure 4, the SCC mixtures had a slump flow diameter of 490-690 mm, with SCC2 (10% limestone) having the highest slump flow. All SCC mixtures, where within the acceptable limits for SCC design except BFRSCC4 (1% fiber content), which had poor flowing ability due to the high presence of bamboo fibers. SCC1, SCC2 and BFRSCC1 had good deformability and are considered as class 1 slump flow while BFRSCC3 (0.75% fiber content) falls under class 2. It can also be observed from the bar chart that an increase in fiber content causes a decrease in the filling ability of SCC. There is a 25% reduction in slump flow between BFRSCC (0.75% fiber) and plain SCC containing limestone. After the slump flow was carried, it could also be observed that the coarse aggregate got to the edge of the slump flow and did not remain in the center except for the case of BFRSCC4. The presence of uniform distribution of larger particles across the spread indicates good viscosity and stability of the mix.

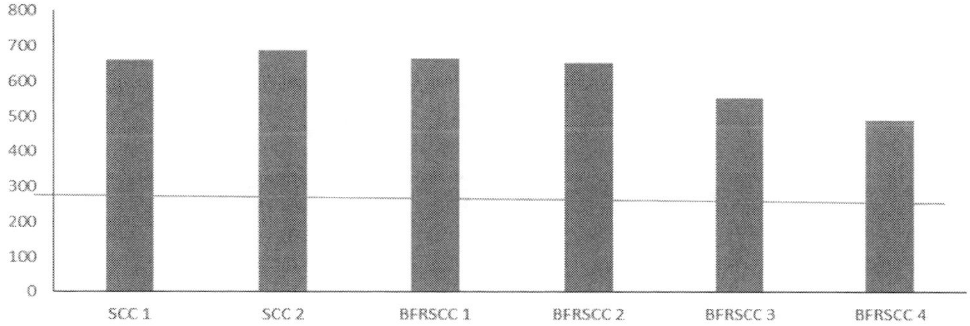

Figure 4. Slump flow for the various samples.

3.1.2 *V- funnel test result*

EFNARC LIMITS: V- funnel – (6 - 12) sec.

Table 1. V-funnel test results.

MIX	V-funnel (sec)
SCC1	8.08
SCC2	7.06
BFRSCC1	9.52
BFRSCC2	11.48
BFRSCC3	18
BFRSCC4	0

From Table 1, the flow times for the various SCC samples were within 8-18 seconds. All samples were within specified limits except BFRSCC3 and BFRSCC4, which is an indication that both mixes have poor passing ability. It could also be seen that SCC 2 containing 10% percent limestone required the least time, indicating that the addition of limestone powder improved the workability of the concrete mix. This could be attributed to filler properties of limestone powder that affected the cement particle system, changed the ordinary distance between them, and modified the water quantity available for the hydration process. The increase in fiber gave rise to a reduction in flowability.

3.1.3 *Microstructural properties*

Figure 5(a) and (b) represent control 1 (plain self-compacting concrete) and 2 (10% limestone powder) after 28 days of curing. From the scan as magnified by the SEM, Figure 5(a) shows a dense interfacial structure. The SEM scan shows the presence of pores on the interface, this is due to the lack of fine particles to fill the pore spacing. Compared to the control 1, sample control 2 shows a tight pore structure. From the SEM, it can be shown that the pores and particle size distribution have a significant effect on the physical strength of the concrete.

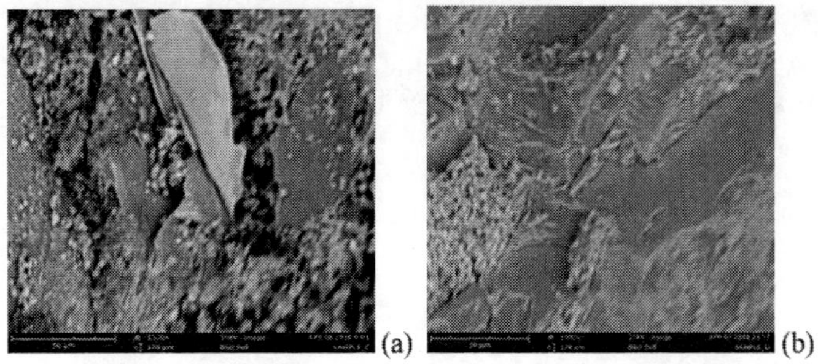

Figure 5. (a) SEM features of control 1 mix; (b) SEM features of control 2 mix.

4 CONCLUSIONS

This research determined the influence of bamboo fibers and limestone powder on the fresh and microstructural properties of self-compacting concrete. It can be seen that an increase in fiber content enhanced the cohesiveness and internal resistance of the fresh mix, thereby reducing the workability of the mix, while the optimal bamboo fiber content that can be used is 0.75%, as it had a slump flow of 552 mm, which is within the required limit.

Acknowledgments

The authors would like to deeply appreciate the Centre for Research, Innovation and Discovery, Covenant University, Ota, Nigeria for supporting this research.

References

ACI Committee 544.1R-96, ACI Technical Report on Fiber Reinforced Concrete (FRC), 2002.

Bindu, D. N., Bamboo Fiber Reinforced Concrete - A Review, *International Journal of Science Technology and Management*, 5(06), pp 189-193, 2016.

Ede, A.N., Olofinnade, O.M., Bamigboye, G.O., Shittu, K.K., and Ugwu, E.I., Prediction of fresh and hardened properties of normal concrete via choice of aggregate sizes, concrete mix-ratios and cement, *International Journal of Civil Engineering and Technology* 8(10), 288-301, 2017.

Ede, A. N., Olofinnade, O. M., Ugwu, E. I., and Salau, A. O, Potentials of Momordica Angustisepala fiber in Enhancing Strengths of Normal Portland Cement Concrete, *Cogent Engineering*, 5, 1431353, 1-17, 2018.

EFNARC, Specification and Guidelines for Self-Compacting Concrete, European Federation, 2002.

Koehler, E. P., and Fowler, D. W., Inspection Manual for Self-Consolidating Concrete in Precast Members, *TxDOT Project 0-5134*, 2007.

Li, Z., *Advanced Concrete Technology*, John Wiley & Sons, Inc., New Jersey, 2011.

Okamura, H., and Ouchi, M., Self-compacting concrete: Development, Present Use and Future, *First International RILEM Symposium on Self-Compacting Concrete, Rilem Publications*, 3–14, 1999

Olofinnade, O. M., Ndambuki, J. M., Ede, A. N., and Olukanni, D. O., Effect of Substitution of Crushed Waste Glass as Partial Replacement for Natural Fine and Coarse Aggregate in Concrete, *Materials Science Forum*, 866, 58-62, 2016.

Olofinnade, O. M., Ede, A. N., and Ndambuki, J. M., Sustainable Green Environment through Utilization of Waste Soda-Lime Glass for Production of Concrete, *Journal of Materials and Environmental Sciences* 8(4), 1139-1152, 2017.

Surabhi, C. S., Mini, S., and Syam, P., *Influence of Limestone Powder on Properties of Self-Compacting Concrete*, 10th National Conference on Technological Trends *(NCTT09)*, 161–164, 2009.

Zhu, W., and John, C. G., Use of Different Limestone and Chalk Powders in Self-Compacting Concrete, *Cement and Concrete Research*, *35*, 1457–1462, 2005.

Interdependence between Structural Engineering and Construction Management
Edited by Ozevin, D., Ataei, H., Modares, M., Gurgun, A., Yazdani, S., and Singh, A.

ISBN: 978-0-9960437-6-2

MECHANICAL AND MICRO-STRUCTURE CHARACTERIZATION OF STEEL FIBER-REINFORCED GEOPOLYMER CONCRETE

SAID ELKHOLY and HILAL EL-HASSAN

Dept of Civil and Environmental Engineering, United Arab University, Al Ain, UAE

This paper aims to investigate the mechanical performance of steel fiber-reinforced geopolymer concrete made with fly ash and ground granulated blast furnace slag as blended aluminosilicate source material. To activate the binding phase, combinations of sodium silicate (SS) and sodium hydroxide (SH) solutions with three different molarities (8M, 10M, and 14M) were used. Steel fibers were added to the geopolymer concrete mix in varying proportions up to 3%, by volume. Constant binder, activator solution, and aggregate contents were adopted for all 13 mixes. Samples were cast and cured at ambient conditions for measuring the rheological and mechanical properties, including slump, modulus of elasticity, compressive strength, tensile splitting strength, and flexural strength. Experimental test results show that geopolymers made with higher molarity of SH were less workable but had improved mechanical performance. The effect of adding steel fibers on the mechanical performance was more apparent at an early age and in weaker geopolymer concretes. Additionally, scanning electron microscopy, differential scanning calorimetry, and Fourier transform infrared spectroscopy highlighted the co-existence of calcium aluminosilicate hydrate and sodium aluminosilicate hydrate gels.

Keywords: Rheological properties, Mechanical properties, Analytical models.

1 INTRODUCTION

To reduce the greenhouse gas emissions and consumption of natural resources, scientists and environmentalists recommend the replacement of cement by supplementary cementitious material (SCMs). Fly ash and ground granulated blast furnace slag (GGBS) are respective industrial by-products of the combustion of coal and production of steel. Fly ash and GGBS are highlighted as primary replacements to ordinary Portland cement (OPC) in concrete due to their pozzolanic properties and global abundance. Moreover, fly ash served as the main binder in an inorganic geopolymer concrete. Typically, it was activated through mixing with an alkaline solution of sodium silicate (Na_2SiO_3) and potassium or sodium hydroxide (KOH or NaOH). Research findings showed good compressive strength, low creep, acid resistance and low shrinkage (García-Lodeiro *et al.* 2007). It could also be used in severe environmental conditions with marine or heat exposure (Chanh *et al.* 2008, Kong and Sanjayan 2008, Zhang *et al.* 2010). Despite their superior performance, geopolymers have shown little resistance to cracking due to their brittle nature. For application in the construction industry, the tensile and flexural properties should be improved. Several studies, such as Al-Majidi *et al.* (2017), Aydın and Baradan (2013),

Bernal *et al.* (2010) and Gao *et al.* (2017), investigated the mechanical properties of steel fiber-reinforced geopolymer composites.

This research investigates the mechanical performance of steel fiber-reinforced slag/fly ash blended geopolymer concrete. Fibers were added up to dosages of 3%, by volume. Three different molar concentrations of sodium hydroxide were used to assess the combined effect. Being locally abundant, desert dune sand served as a sustainable fine aggregate. The performance of geopolymer concrete mixes was evaluated in terms of compressive, tensile, and flexural strengths.

2 EXPERIMENTAL PROGRAM

2.1 Materials

Class F (ASTM C618 2015) fly ash and ground granulated blast furnace slag (GGBS) were used as the geopolymer binding material. Crushed stone was used as coarse aggregate with a nominal maximum size of 10 mm, specific gravity of 1660 kg/m^3, and water absorption of 1.5%. The aggregates were prepared to surface-saturated dry (SSD) condition. Desert dune sand served as fine aggregate. Its specific gravity and unit weights were 2.57 and 1670 kg/m^3, respectively. An alkali-activator solution was prepared as a mixture of sodium silicate (SS) and sodium hydroxide (SH). The mass chemical composition of the grade N SS solution was 26.3% SiO_2, 10.3% Na_2O, and 63.4% H_2O. The SH solution was formulated to a molarity of 14 (14M) by dissolving 97-98% pure NaOH flakes in tap water. Hooked-end steel fibers (specific gravity 7.9) with an aspect ratio of 65 and length of 35 mm were used. To ensure sufficient workability for mixes with high steel fiber content, a polycarboxylic ether polymer-based superplasticizer (SP) was used.

2.2 Geopolymer Concrete Mix Design

Thirteen slag/fly ash blended geopolymer concrete mixes with varying Sodium Hydroxide (SH) molar concentrations and fiber volume fractions were prepared. A 3:1 ratio of GGBS: fly ash was selected to eliminate the need for heat curing and to provide the concrete with sufficient workability for proper placing and finishing. The incorporation of steel fibers ranged from 0 to 3%. Three different molar concentrations of SH were used to assess the combined effect as shown in Table 1.

Table 1. Mix proportion of geopolymer concrete (kg/m^3).

Binder		**Aggregate**		**Activator**		**SP**	**Steel fiber volume (%)**
Fly Ash	GGBS	Dune Sand	Coarse	SS	SH (Molarity)		
125	375	550	1100	143	57(8M, 10M& 14M)	10	0.0~3%

2.3 Sample Preparation

The alkali-activator solution was prepared two days prior to casting to allow for dissipation of heat associated with the exothermic chemical reactions of SH flakes with water and SH solution with SS solution. The dry components were mixed in a pan mixer for 3 minutes. The prepared solution was gradually incorporated into the dry components and mixed for 3 minutes to ensure homogeneity and uniformity. Superplasticizer was added a few seconds after the activator solution. For each mix, freshly-mixed geopolymer concrete was prepared as fifteen ϕ100 x 200 mm cylinders and three 100 x 100 x 500 mm prisms for subsequent mechanical testing. Specimens were cast into two layers, compact-vibrated for 10 seconds on a vibrating table, left to rest for 24 hours at ambient conditions, and then demolded. Subsequently produced geopolymers were kept at ambient conditions until testing age.

3 RESULTS AND DISCUSSION

3.1 Compressive Strength

The compressive strength development of geopolymer concrete mixes with respective 8M and 14 M SH is illustrated in Figure 1. On average, 81 and 97% of the 28-day compressive strength were obtained at 1 and 7 days of age, respectively for 8M geopolymers. Similarly, for 14M geopolymers, 78 and 89% of the respective 28-day compressive was reached within 1 and 7 days. The addition of steel fibers led to an increase in the load carrying capacity of geopolymers. An increase between 7 to 22% was observed after 1 day when 0.5 to 3% of steel fibers were added. At later ages of 7 and 28 days, up to 25% improvement in compressive strength was recorded with 3% steel fibers.

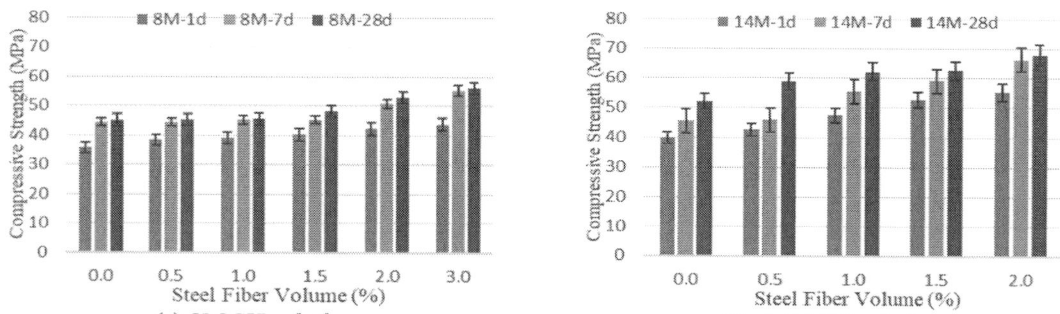

(a) 8M SH solution (b) 14M SH solution

Figure 1. Strength development of geopolymer made with different molarity of SH solution.

3.2 Indirect Tensile Strength

Figure 2 presents the splitting tensile strength as a function of steel fiber volume fraction incorporated for 8M, 10M, and 14M geopolymers at 28-day age. It can be seen that 14M geopolymer concrete was consistently superior to counterparts made with lower molarity of SH. This is well-aligned with compressive strength behavior. Geopolymers with 8M-SH increased by 7, 41, 55, 65, and 147% when 0.5, 1, 1.5, 2, and 3% of steel fibers were added, respectively. Clearly, the contribution of steel fibres to compressive strength was more significant at higher fiber dosages. On the other hand, 14M counterparts increased by 7, 22, 28, and 60% for similar steel fibers volume fractions.

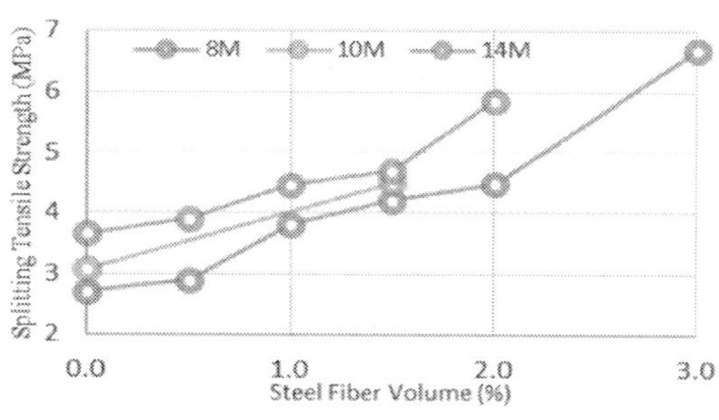

Figure 2. Splitting tensile strength fiber reinforced geopolymer concrete.

3.3 Flexural Strength

Figure 3 presents the flexural strength of plain and steel fiber-reinforced geopolymer concrete at 28-day age. An increase in flexural strength can be noted upon the incorporation of steel fibers. For 8M-geopolymers, a volumetric addition of fiber up to 1.5% could only enhance the flexural capacity by 5%. However, the effect of fiber addition was more pronounced at higher dosages of 2 and 3%. The associated increase in strength was 20 and 32%, respectively. Apparently, crack propagation through geopolymer concrete was retarded or arrested due to the incorporation of 2 to 3% steel fibers. An increase of 9% was also reported for 10M-geopolymers upon the addition of 1.5% steel fiber volume fraction. In contrast, a near-linear relation between the volume fraction of steel fibers and flexural strength increase was noted for 14M-geopolymers. On average, the flexural strength increased by 8% for every 0.5% volume fraction of steel fiber added. A comparison between geopolymers with constant steel fiber content and different molarity of SH (8M-1.5SF, 10M-1.5SF, and 14M-1.5SF) showed that higher molarity resulted in further increase in flexural strength.

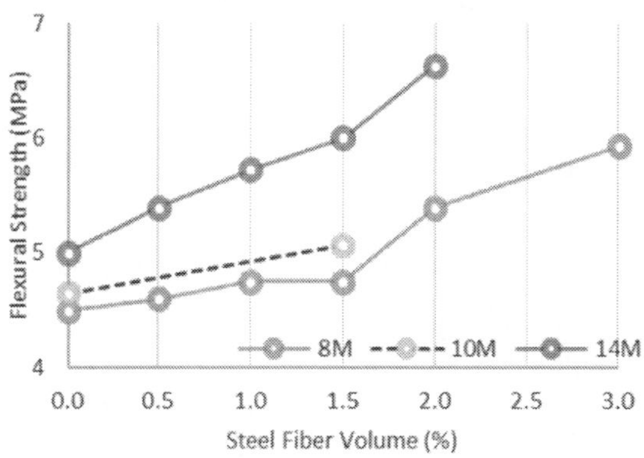

Figure 3. Flexural strength of 8M, 10M, and 14M geopolymer concrete.

3.4 Failure Modes

A comparison of the failure mode of plain and steel fiber-reinforced concrete cylindrical specimens after undergoing compression tests are shown in Figure 4. Plain geopolymer concrete made with slag and fly ash did not undergo an explosive failure at peak load as in fly ash-based geopolymers. Rather, it retained its original shape after the peak load. The crack propagation in the geopolymer binder was restricted under compression due to the transverse confinement effect of steel fibers and adequate bond with the geopolymeric matrix. Effectively, the addition of steel fibers led to higher compressive strength.

(a) plain (b) 1.0% SF (c) 2.0% SF (d) 3.0% SF

Figure 4. Compression failure modes of 14M-geopolymer concrete

Figure 5 shows the splitting tensile failure modes of plain and fiber reinforced geopolymers. It is clear from the figure that the extent of damage decreases as more fiber reinforcement is incorporated into the geopolymeric mixture. The failure transformed from a brittle to a more ductile behavior due to the inclusion of steel fibers. While the plain geopolymeric cylinder split into two parts, the cylindrical sample incorporating 3% steel fibers, by volume, remained completely intact.

(a) plain (b) 1.0% SF (c) 2.0% SF (d) 3.0% SF

Figure 5. Splitting tensile failure modes of 14M-geopolymer concrete.

Figure 6 shows the flexural failure modes of plain and fiber-reinforced geopolymer concrete prisms. All cracks formed in the middle third and propagated nearly vertically upward. Plain geopolymer concrete of Figure 6(a) experienced fracture and separation of the prism into two parts. On the contrary, fiber-reinforced counterparts remained intact; higher load capacity was observed with higher volumetric fiber fractions.

(a) Plain (b) 1.0% SF

(c) 2.0% SF (d) 3.0% SF

Figure 6. Flexure failure modes of 14M-geopolymer concrete.

3.5 SEM and FTIR Analyses

The microstructure of plain geopolymer concrete was analyzed by 100 differential scanning calorimetry, 101 scanning electron microscopy), and Fourier transforms infrared spectroscopy. The micrographs of geopolymer concrete samples after 7 and 28 days did not significantly change. The FTIR spectra of concrete samples after 28 days showed larger quantities of reaction products and few water peaks.

3 CONCLUSIONS

The results of this study show that the mechanical performance of geopolymer concrete composed of Class F fly ash and slag depends on the molarity of SH. The mechanical properties of steel fiber-reinforced geopolymer composites were investigated. An increase in compressive, tensile and flexural strength was obtained using steel fibers in dosages up to 3%, by volume. The addition of steel fibers led to an increase in compression, tensile, and flexural strength. The effects of steel fibers in dosages on compression, splitting tensile and flexural failure modes were also reported.

Acknowledgments

The authors are grateful to the United Arab Emirates University for financial support through grants 31N305 and 31N282. The assistance of students and laboratory staff at the United Arab Emirates University is gratefully acknowledged.

References

Al-Majidi, M. H., Lampropoulos, A., and Cundy, A. B., Steel Fibre Reinforced Geopolymer Concrete (SFRGC) with Improved Microstructure and Enhanced Fibre-Matrix Interfacial Properties, *Construction and Building Materials*, 139, 286-307, 2017.

ASTM C618, *Standard Specification for Coal Fly Ash and Raw or Calcined Natural Pozzolan for Use in Concrete,* ASTM International, West Conshohocken, PA, 2015.

Aydın, S., and Baradan, B., The Effect of Fiber Properties on High Performance Alkali-Activated Slag/Silica Fume Mortars, *Composites Part B: Engineering*, 45(1), 63-69, 2013.

Bernal, S., De Gutierrez, R., Delvasto, S., and Rodriguez, E., Performance of an Alkali-Activated Slag Concrete Reinforced with Steel Fibers, *Construction and Building Materials*, 24(2), 208-214, 2010.

Chanh N V, Trung B D, and N, V. T., *Recent Research on Geopolymer Concrete*, Proc., the 3rd ACF International Conference-ACF/VCA, 2008.

Gao, X., Yu, Q. L., Yu, R., and Brouwers, H. J. H., Evaluation of Hybrid Steel Fiber Reinforcement in High Performance Geopolymer Composites, *Materials and Structures*, 50(2), 165, 2017.

García-Lodeiro, I., Palomo, A., and Fernández-Jiménez, A., Alkali–Aggregate Reaction in Activated Fly Ash Systems, *Cement and Concrete Research*, 37(2), 175-183, 2007.

Kong, D. L. Y., and Sanjayan, J. G., Damage Behavior of Geopolymer Composites Exposed to Elevated Temperatures, *Cement and Concrete Composites*, 30(10), 986-991, 2008.

Zhang, Z., Yao, X., and Zhu, H., Potential Application of Geopolymers as Protection Coatings for Marine Concrete: II. Microstructure And Anticorrosion Mechanism, *Applied Clay Science*, 49(1–2), 7-12, 2010.

Interdependence between Structural Engineering and Construction Management
Edited by Ozevin, D., Ataei, H., Modares, M., Gurgun, A., Yazdani, S., and Singh, A.

ISBN: 978-0-9960437-6-2

RESEARCH ON THE MEASURES AGAINST LARGE EARTHQUAKES AND TAKING REFUGE AT ONE'S OWN HOME

AKIE HISAGI[1] and TAKASHIGE ISHIKAWA[2]

[1]*Dept of Architecture and Interior Design, Bunka Gakuen University, Tokyo, Japan*
[2]*Dept of Housing and Architecture, Japan Women's University, Tokyo, Japan*

The possibility of a huge earthquake striking Tokyo in the future is high. In the event of such an earthquake, many people will face problems such as the inability to return to their homes, which will cause complete disorder and confusion in the city as it will be impossible for everyone to access emergency shelters. If buildings remain safe, it will be desirable for residents to take refuge at home. In this study, we investigated the disaster awareness of 461 residents in central Tokyo using an internet survey. Our results reveal that approximately 60% of residents expect to take refuge at their own homes. However, almost none of the respondents have emergency supplies such as water, food, and other daily necessities. This lack of preparedness could result in many people having to move to emergency shelters when disaster strikes. By analyzing these results, we aim to develop a method for avoiding confusion in the aftermath of a disaster and educate building residents on their needs and the preparatory actions they must take in such situations.

Keywords: Disaster preparedness, Awareness of disaster mitigation, Evacuation activity, Shelter, Emergency supplies, Residents.

1 INTRODUCTION

If a huge earthquake strikes Tokyo, enormous damage will occur, owing to its high earthquake risk and large population. Damage to humans and buildings, which are direct consequences of the earthquake, are expected to be extensive, but a major obstacle can also hamper evacuation activities in the aftermath of the earthquake. In the center of Tokyo where the population density is high, the number of evacuation shelters is not enough; it will be impossible to accommodate all the residents. Therefore, in Tokyo, it is desirable for residents who are ensuring the safety of their home buildings to take refuge at home rather than at evacuation centers in case of an earthquake. This idea is also recommended by the government. However, while many of the citizens wish to spend a stress-free life at home following an earthquake, they also expect to secure their own space in evacuation centers and to receive supplementary supplies. In the Tokyo metropolitan area, where stockpiling and evacuation spaces cannot be secured, it is essential to maintain a reserve of food and beverages that will last more than a week with one's own stockpiles. For citizens to understand how to cope with the changes to lifestyle while taking refuge at home in the aftermath of a disaster, their awareness of evacuations and the actual state of their preparedness has been investigated in our previous studies (Shirai *et al.* 2011, Hirata *et al.* 2011, Takahashi *et al.* 2013, Hisagi *et al.* 2013).

As a result, it was revealed that many residents in the metropolitan area wish to remain at home after a disaster; however, their preparedness in terms of stockpiling. is insufficient. Even though they have knowledge about the necessary amounts of stockpiles to be maintained, they have not taken any action in terms of being prepared. One of the influencing factors of these results may be people's optimism regarding damage estimation, such as their disregard for damage to life, expectation of prompt recovery after a disaster, and excessive expectation of administrative support.

Therefore, in this research, we investigated people's current perceptions of the damage estimate with the aim of broadening their knowledge so that can they can properly prepare for evacuation or taking refuge at home in central Tokyo in the aftermath of an earthquake.

2 OUTLINE OF INVESTIGATIONS

Our survey respondents were 248 residents of high-rise condominiums (10 or more stories) and 213 residents of detached houses in the 23 special wards constituting central Tokyo. They were people in a position to manage stockpiles at home.

The survey was conducted between November 1 and November 7, 2016. The investigation assessed the respondents' awareness of disaster mitigation, earthquake preparedness, consciousness of refuge-preparation needs, etc. The investigation outline is given in Table 1.

Table 1. Outline of the investigation.

Investigation date	November 1–7, 2016
Investigation method	Internet questionnaire
Respondents	461 individuals: 248 residents of high-rise condominiums (10th to 74th floor) and 213 residents of detached houses. Respondents were between 25 and 85 years of age
Investigation topic	Awareness of issues and needs for disaster mitigation, preparedness for an earthquake, consciousness of refuge-building and supplies, knowledge about formal shelters, ability or willingness to take refuge in one's own home

3 SURVEY RESULTS AND CONSIDERATIONS

In the event of a major earthquake, we investigated whether ordinary citizens preferred to take refuge at their own homes or be evacuated to a shelter. The results are shown in Figure. 1.

Most citizens would prefer to take refuge at home than at evacuation shelters. It is noteworthy that few people, less than 5% of the respondents, wish to live in evacuation centers for more than a week.

Next, we investigated the extent to which the residents considered the damage estimate. Figure 2 shows the responses of residents to being asked how many restoration days are required for the resumption of their usual lifestyles and the restoration of important facilities that have been damaged by the earthquake.

Most citizens think that support and communication networks should be restored within three days, and they assume that they will be able to resume their normal lifestyles within a week. The number of days for restoration according to the assumption of experts and officials in Tokyo, the number of days it would take for complete restoration is one week for electricity, one day for water supply and sewerage, one month for gas, and two weeks for communication networks.

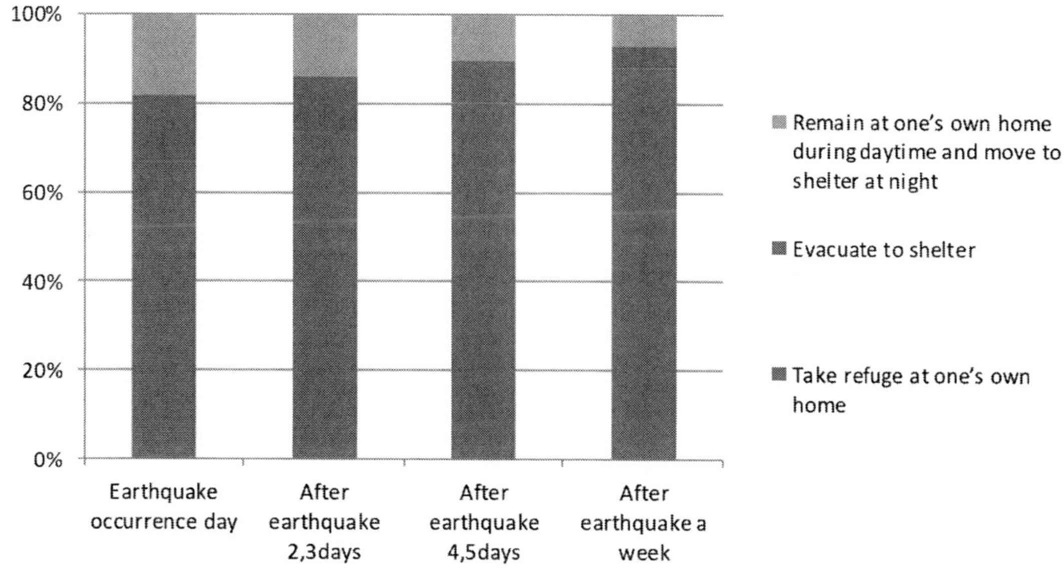

Figure 1. Places to which to evacuate after a huge earthquake.

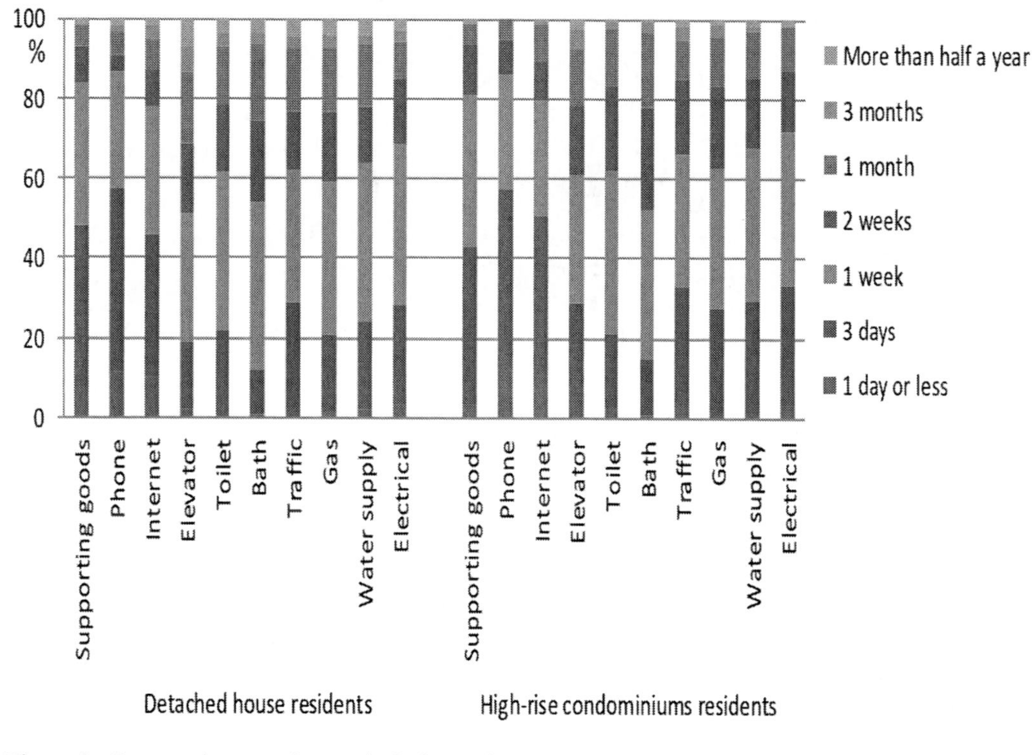

Figure 2. Expected restoration periods for various essential services that could suffer damage in the aftermath of a huge earthquake.

Therefore, when comparing these pieces of information, the citizens imagine the restoration period to be far shorter than it actually is in reality, indicating that their damage estimate is very optimistic. There was also a tendency for residents of houses to assume shorter recovery days than detached residents. It was found that about 80% of all residents were not realistically aware of the difficulties of life after an earthquake.

This tendency to disregard the assumption of damage is considered to be one of the factors why people do not follow appropriate stockpiling behavior. Respondents who wished to live in evacuation centers after the earthquake estimated that restoration would be briefer than estimated by those who wanted to stay at home. Figure 3 shows the responses to inquiring about the reasons for respondents' estimation of the number of days it would take for restoration of essential services.

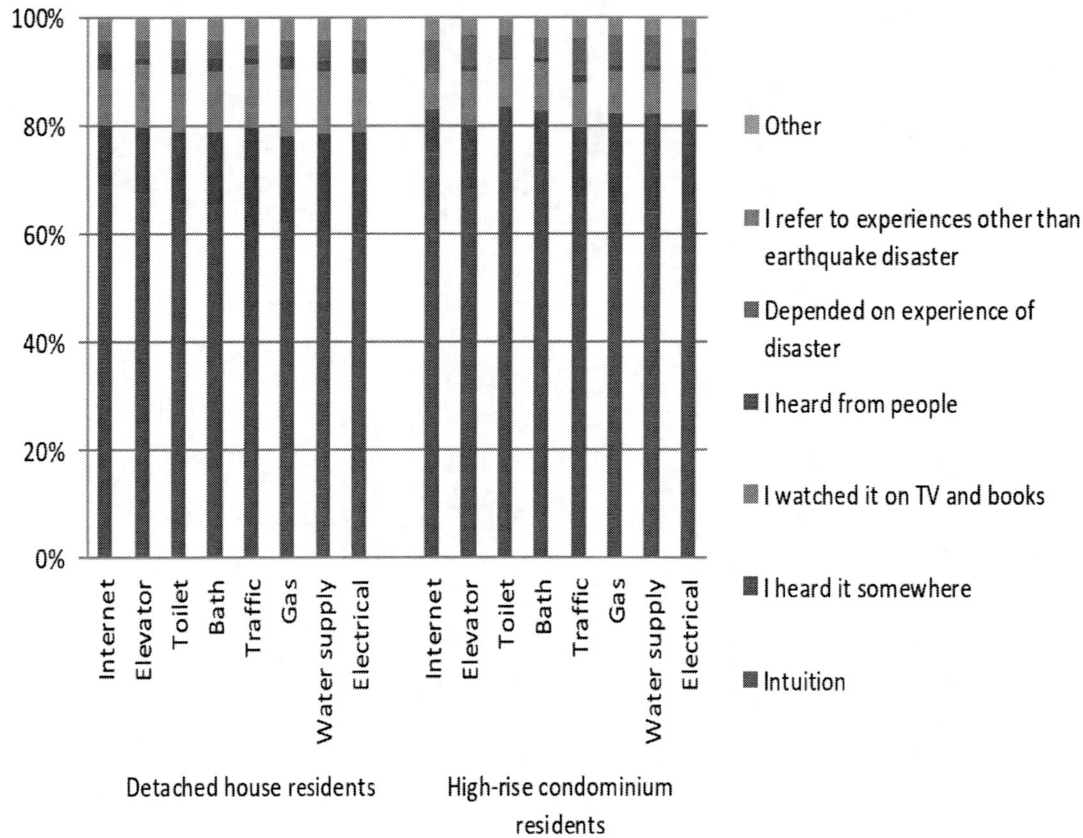

Figure 3. Reasons for respondents' estimation of the number of days for restoration of essential services.

"Intuition" constitutes approximately 60% or more of all responses, and the percentages of high-rise condominium residents are higher than those of detached house residents. In addition, the majority of respondents who expected that the number of days for restoration would be short made this judgment based on their intuition.

Furthermore, we investigated scenarios that would make respondents feel uneasy if a large-scale earthquake were to occur in the capital. The results are shown in Figure 4.

Anxiety about the complete destruction of one's home and concern about the breaking of furniture are more common among detached house residents than among high-rise condominium residents. In addition, similar shortages of stockpiles are found for both respondents who feel very worried and those who are not at all worried. The results of the survey revealed that most citizens were prepared for only about three days with their stockpiles of drinks and food. In the case of the Tokyo metropolitan area, although stockpiles are required to last for seven days, this shows a clear shortage. However, many respondents did not feel worried.

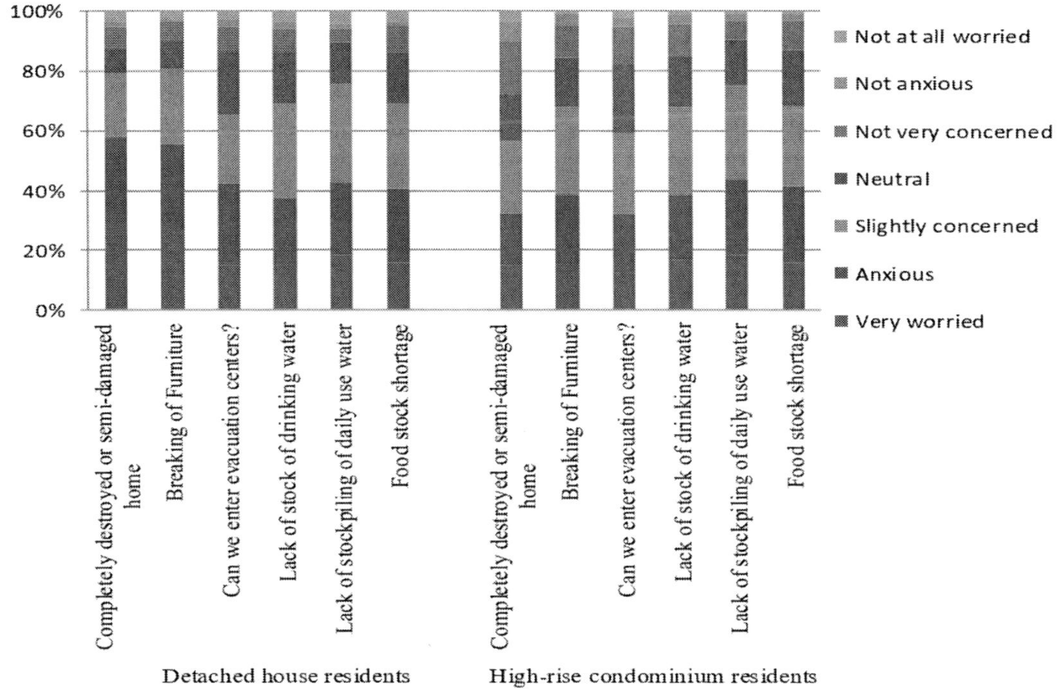

Figure 4. Major worries when an earthquake occurs.

4 CONCLUSIONS

Residents in central Tokyo are aware of the possibility of an earthquake that may directly hit the Tokyo area, but they tend to assume little damage, and they judge such possible earthquakes intuitively. In particular, residents of high-rise condominiums tend to estimate less damage than residents of detached houses. Most of the damage estimates are judged by intuition. These are considered to be some of the current factors that result in residents not maintaining sufficient stockpiles in order to help themselves in the case of a future earthquake.

In the future, it is necessary to consider investigating disaster information that makes people aware of accurate damage estimations.

Acknowledgments

We graciously thank the people who took part in this study. This work was supported by JSPS KAKENHI Grant Numbers 26560035.

References

Hirata K., Hisagi A., Ishikawa T., and Saito T., Development of Safe and Secure Environment in High-Rise Residential Buildings against Large Earthquakes- Part 7 Consciousness Survey on Means of Escape for Residents Lived in High-rise Condominiums, *Architectural Research Meeting, the A.I.J. (Academic Journal)*, 957-958, 2011.

Hisagi A., Hirata K., Ishikawa T., and Saito T., Development of Safe and Secure Environment in High-Rise Residential Buildings against Large Earthquakes- Part 12 The Current Status on Preparedness of Refuge Life in High-rise Condominiums Derived from a Resident's Survey, *Architectural Research*

Meeting, the A.I.J. (Academic Journal), 1143-1144, 2013.

Shirai M., Ishikawa T., and Saito T., Development of Safe and Secure Environment in High-Rise Residential Buildings against Large Earthquakes- Part 6 A Resident's Attitude Survey about Earthquake Prevention in High-rise Condominiums, *Architectural Research Meeting, the A.I.J. (Academic Journal)*, 955-956, 2011.

Takahashi M., Hisagi A. Ishikawa T., and Saito T., Development of Safe and Secure Environment in High-Rise Residential Buildings against Large Earthquakes- Part 11 Possibility of Preparedness and Evacuation Life in High-Rise Condominiums Based on the Terms of the Building, *Architectural Research Meeting, the A.I.J. (Academic Journal)*, 1141-1142, 2013.

Interdependence between Structural Engineering and Construction Management
Edited by Ozevin, D., Ataei, H., Modares, M., Gurgun, A., Yazdani, S., and Singh, A.

ISBN: 978-0-9960437-6-2

MODEL DISASTER PREVENTION LESSONS IN PRIMARY SCHOOL BASED ON DISASTER PREVENTION EDUCATION POLICIES

NORIKO IMURA[1] and TAKASHIGE ISHIKAWA[2]

[1]*Dept of Architecture, Musashino University, Tokyo, Japan*
[2]*Dept of Housing and Architecture, Japan Women's University, Tokyo, Japan*

This study deals with disaster mitigation education in primary schools and as a final goal aims to improve citizens' capabilities to mitigate disaster situations in society. This report summarizes the current national policy on safety education in Japan and follow it up with a summarization of the Tokyo Metropolitan Government policy, which is based on the national policy. In addition, it compares Japan's education policy with that of New Zealand. Analysis revealed the following three points. First, the content of safety education in Japanese primary schools consists of six fields (traffic safety, daily life safety, disaster safety, etc.). Just before, and after, the Great Tohoku Earthquake in 2011, the main educational content changed from daily life safety to disaster safety. Second, by focusing on lessons, it was found that the content of disaster safety was taught not as part of various subjects but largely during "homeroom activities". Third, it became clear that the subject of earthquakes now accounts for half of the disaster prevention lesson contents in Japanese primary schools and has been included in disaster prevention lesson contents in all school grades.

Keywords: Disaster mitigation education, Curriculum, Educational program, Safety education, Japan, Tokyo Metropolitan Government.

1 INTRODUCTION

This study deals with educational programs for disaster mitigation in elementary school education, with the aim of improving the disaster mitigation management ability of each citizen and thus ultimately contributing to disaster mitigation. Japanese education policy prescribes six years of compulsory primary school for 7–12-year-old children and three years of middle school for 13–15-year-olds. This report summarizes the Japanese government's established disaster mitigation education policy for primary schools and the safety education program established by the Tokyo Metropolitan Government for primary schools, in line with the nation's policy, as a case study for prefectures. Then, it analyzes what kinds of educational contents are exemplified as model lessons in the national guidelines for disaster education, in the Tokyo Metropolitan Government safety education program guidelines, and the model classes in the research conferences organized over the last 12 years.

2 NEW DISASTER MITIGATION EDUCATIONAL PROGRAM IN SCHOOLS AFTER THE 1995 GREAT HANSHIN-AWAJI EARTHQUAKE

In recent years, as a result of the 1995 Great Hanshin-Awaji Earthquake, schools have begun implementing disaster education programs. The Ministry of Education, Culture, Sports, Science, and Technology (MEXT) announced guidelines for disaster mitigation programs in school education in 1998 (MEXT 1998). These guidelines were revised in 2013 (MEXT 2013) after the 2011 Great East Japan Earthquake. Table 1 shows these changes.

Table 1. Revisions in the disaster mitigation program after the Great Hanshin-Awaji Earthquake.

Date	Events
January 15, 1995	The Great Hanshin-Awaji Earthquake
March 1998	"Reference documents of disaster mitigation educational program"
November 2001	"Reference documents of safety programs", first edition
March 2010	"Reference documents of safety programs", second edition
March 11, 2011	The Great East Japan Earthquake
March 2013	"Reference documents of safety programs", revised and extended version

According to the 2013 version, the overall aim of disaster mitigation education is to separately facilitate three main learning areas: "knowledge, thinking, judgment", "crisis prediction, subjective action", and "social contribution, support foundations". Table 2 shows these aims organized by school year. Among them, the colored parts are attached to particularly important items. Targets have been set according to the child's developmental stage.

Table 2. Disaster education goals of each grade (as established by MEXT).

	Knowledge, Thinking, Judgment	**Crisis prediction, Subjective action**	**Social contribution, Support foundations**
1st & 2nd grade	The student will gain an interest in disasters and be able to think about safe behavior in case of disaster.	The student will be equipped to understand the dangers posed by disasters, follow instructions by adults, and take appropriate action.	In the event of a disaster, the student will be able to avoid danger and contact adults.
3rd & 4th grade	The student will be equipped to understand the fundamental aspects of disasters and think ingeniously about disaster prevention.	The student will gain an interest in understanding the dangers posed by disasters and will be able to think of ways to avoid danger.	In the event of a disaster, the student will be able to cooperate with family, friends, and surrounding people so as to avoid danger.
5th & 6th grade	The student will be enabled to understand the characteristics of disasters, as well as organizing disaster mitigation.	The student will be able to predict the risks associated with disasters, and in case of disasters can take the necessary action to avoid danger.	In the event of a disaster, the student will be able to take the safety of surrounding people into consideration and behave in a way that will prove useful to others.

3 THE TOKYO METROPOLITAN GOVERNMENT'S SAFETY EDUCATION PROGRAM

The Tokyo Metropolitan Government has established a "Safety Education Program" (Tokyo Metropolitan Board of Education 2016) as a foundation for safety education. This program's purpose is not limited to disasters. It includes acquiring the ability to predict and avoid other dangers such as crime and accidents and acquiring the skills and qualities necessary for protecting others.

MEXT defines safety education in schools as consisting of three aspects: "daily life safety", "traffic safety", and "disaster safety". Disaster mitigation is classified as "disaster safety". The

Tokyo Metropolitan Safety Education Program has identified "basic matters that must be taught" with regard to these safety aspects, and the necessary contents that children have to learn are classified based on each type of safety; a one-year teaching plan has been created for these activities. The Tokyo Metropolitan Government has independently created supplementary readers.

4 ANALYSIS OF MODEL LESSONS AT THE TOKYO METROPOLITAN GOVERNMENT SCHOOL SAFETY EDUCATION RESEARCH CONFERENCE

The Tokyo Metropolitan Government School Safety Education Research Conference is an annual seminar where the Tokyo Metropolitan area teachers, as well as teachers from all over the country, gather for study. Every year, the host school is selected from among the public elementary schools in Tokyo. Based on the guidance plan established by the national and Tokyo Metropolitan governments, the teachers at the host school are required to create and demonstrate their model lessons.

To understand the trends in the model lessons, we surveyed the contents of the open model class held at 12 elementary schools over a period of 12 years from 2005 (the 29th convention) to 2016 (the 40th convention) (National-Tokyo Metropolitan School Safety Education Study Research Conference 2016). Table 3 summarizes the class contents by subject and grade. Safety education in these contents includes the above-mentioned daily life safety, traffic safety, and disaster safety. In addition to the contents of information safety, general safety and morals are introduced. Table 3 shows all these contents of safety fields.

Table 3. Classification of the model classes' contents by each "safety".

Subject	Grade	the 29th 2005	the 30th 2006	the 31st 2007	the 32nd 2008	the 33rd 2009	the 34th 2010	the 35th 2011	the 36th 2012	the 37th 2013	the 38th 2014	the 39th 2015	the 40th 2016
Living Environment Studies	1				▲								●
	2				■								▲▲
Social Studies	3										▲		
	4										●		
Science	4	■			■								
	5		■										
	6						▲▲		▲				
Physical Education	4		■■										
	5	■				■							
	6	■		■	■	■							
Moral Education	1	◆◆		▲	◆	◆							
	2	◆◆					▲						
	3	◆●											
	4	◆◆											
	5	●◆	▲		▲								
	6	◆◆	▲										
Art and Handicraft	2	■		■	■	■	■						
	4				■								
Japanese	2										▲		▲
Home Economics	6				■								
Period for Integrated Study	3				■								▲
	4			■	▲		■				▲	■	
	5			■▲	▲■						■		▲▲
	6			▼	■	■						▲	▲★
Homeroom Activities	1	●	■■	●	●	●	■■	■■	■■	■	■▲	▲■	■■●●▲▲▲▲▲
	2		●▲		▲	●		●●	▲▲	▲	▲	●■▲	●▲
	3	■	■■	●	■	■■	■■	▲■	▲	●	●	●	■▲▲
	4	■	▲		■	▲	■	▲	▲▲	▲		▲	■●▲
	5		■			▼	▼▼▼	●▼	●	▲▲	▲	▲	■●▲▲▲
	6	▲	▼					■▲	▼	▲▲	▲▲	■	■

Legend ●Traffic safety ■Daily life safety ▲Disaster safety
▼Information safety ★Safety in general ◆Morals

The survey results demonstrate that, as shown in Table 3, until the 35th convention (held before the Great East Japan Earthquake), most of the class contents were based on moral thinking and daily life safety such as personal dangers. These classes were centered on the subjects of the department. However, after the 36th convention—held after the Great East Japan Earthquake—and at those held after it, classes on moral topics disappeared and daily life safety classes decreased, replaced by disaster safety classes. The contents of these disaster safety classes include lesson materials related to preparation for future disaster situations, such as emergency earthquake early warning systems, and cardiopulmonary resuscitation methods.

5 ISSUES FROM THE VIEWPOINT OF THE MEXT PROGRAM, THE TOKYO METROPOLITAN GOVERNMENT PROGRAM, AND THE DISASTER SAFETY MODEL CLASSES

Focusing on the contents of the disaster safety classes, we investigated the types of disasters that these classes have discussed or dealt with and then classified each disaster type by subject and grade. The classes targeted for analysis were picked from examples in the nationwide guidelines prescribed in Chapter 2 of this paper, from examples of practical lessons provided in the Tokyo Metropolitan Government's "Safety Education Program" in Chapter 3 of this paper, and from the 12-year model classes provided in the "Tokyo Metropolitan Government School Safety Education Research Conference Reference Collection" described in Chapter 4 of this paper. The results are shown in Table 4. In addition, the black cell originally had no subject setting, and the orange cell was a subject with setting, but there was no lesson subject in Table 4. Figure 1 shows the graph of class contents in the model schools.

Table 4. Model class contents classified by disaster type.

Grade	Living Environment Studies	Social Studies	Science	Physical Education, Health and Physical Education	Home Economics	Moral Education	Period for Integrated Study	Homeroom Activities	School Event	Other
1	◇■★					■		●■●■ ★■■● ■■	●★■ ★▽★	■■■ ★
2	■■					■★		●■★■ ★■■■ ★■■★ ■		
3		▽					■■	■▽★◇ ●■		
4		★					■■	■▽■■ ▽◇★○ ■▽■		
5			▽	★		■■	▽▲◇■ ★■▽■	■★■▽ ★○▽★ ■●▲■		
6			■■■			◇■	◇★■	■★■◇ ◇◎■■ ○		

Legend ●Fire ■Earthquake ▲Volcano ▽Meteorological disaster ○Nuclear disaster ◇Shelter ★Preparing for disaster ◎General disaster safety

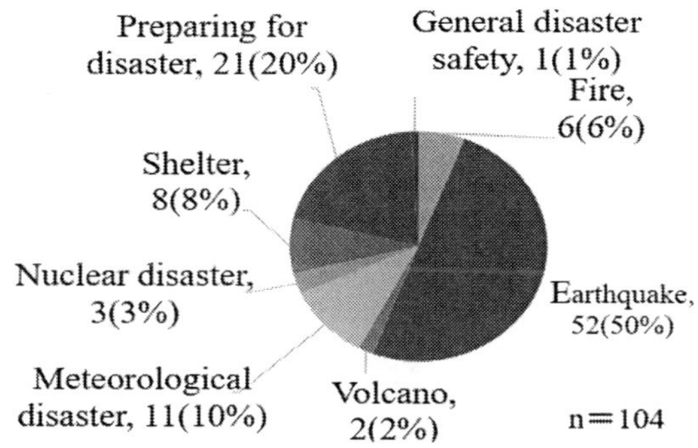

Figure 1. Class contents of the model schools organized by safety education program.

From this study, we found that half of all the classes dealt with contents concerning earthquakes, which were taught in all grades. Regarding safety in case of fire, there were classes in almost all grades from the first grade of elementary school upward. There were few classes that dealt with volcanoes and nuclear disasters because it seems difficult for students to understand these disasters. These contents can thus be set only for the upper grades in primary school. By class subjects, the study found that disaster safety classes were conducted mainly as "homeroom activities" in all the grades. Furthermore, it turned out that these classes were hardly conducted during home economics and physical education. Even though physical education classes are held in all grades, there was only one example of a model class using the subject of physical education.

6 A COMPARISON OF DISASTER MITIGATION EDUCATION IN JAPANESE AND NEW ZEALAND SCHOOLS

In New Zealand, the Ministry of Civil Defence & Emergency Management (CDEM) is responsible for disaster prevention. "Get Ready" (CDEM 2018a) on CDEM's web page consists of published contents for the disaster prevention education of the public, and Table 5 shows its items.

Table 5. The contents of "Get Ready".

• At Home – Get Ready Get Thru
• At School (Primary School & Intermediate School)
• Early Childhood Education Centres
• At Work
• People with Disabilities
• Get involved
• Civil Defence Emergency Management Alerts & Warnings

Among them, the disaster prevention education contents for children, equivalent to Japanese elementary school disaster mitigation education, is "At School" (CDEM 2018b). The educational contents/materials are published online, and the collection, called "What's the Plan Stan?" has the following features:

(i) It deals with six kinds of disasters (floods, storms, earthquakes, tsunamis, eruptions, and landslides).

(ii) Explanations for each disaster are rich in content.
(iii) From this webpage one can access links to other useful services, such as the Met service and the EQC, if required; thus, it can be a rich database for teaching material.
(iv) There are teaching pointers for teachers and explanations for school operators, in addition to commentary for children and their families.

Compared to Japanese teaching materials, these specialize in natural disasters, are content-rich, and distributed through the web. This study's interview survey (conducted at Wellington's C elementary school in November 2017) showed that teachers in New Zealand have greater discretionary powers as compared to Japanese teachers in the field in terms of selecting teaching contents and lesson management with regard to disaster prevention education.

7 CONCLUSION

As a result of the analysis, this paper arrives at the following three points:

(i) The main contents of safety education shifted from daily life safety to disaster safety just before and after the Great East Japan Earthquake.
(ii) By focusing on lessons, it was found that disaster safety was not included in the classes of various subjects; rather, it was found to be taught mainly during time allocated to homeroom activities.
(iii) The contents with regard to earthquakes were taught using only half of the total class time. Furthermore, disaster safety classes were held for all grades.

In the future, it is necessary to foster an educational environment where there is no bias regarding grade and the implementation of contents. In addition, it is important to make comparisons with New Zealand regarding teaching materials and disaster mitigation education classes in school.

Acknowledgments

The authors sincerely thank Maho ANDO, who was a member of the authors' laboratory during the study period.

References

CDEM, Get Ready, Ministry of Civil Defence & Emergency Management, New Zealand, 2018a. Retrieved from https://www.civildefence.govt.nz/ on March 2018a.
CDEM, At School, Ministry of Civil Defence & Emergency Management, New Zealand, 2018b. Retrieved from https://www.civildefence.govt.nz/get-ready/at-school/ on March 2018b.
MEXT, School Safety Reference Materials Safety Education at Schools That Will Make "Living Power", MEXT, Japan, 1998.
MEXT, Reference Documents of the Safety Programs added and Revised Version, MEXT, 2013. Retrieved from http://www.mext.go.jp/a_menu/kenko/anzen/1289310.htm on May 2016.
National-Tokyo Metropolitan School Safety Education Study Research Conference, National-Tokyo Metropolitan School Safety Education Study Research Conference Material CD, National-Tokyo Metropolitan School Safety Education Study Research Conference, Japan, 2016.
Tokyo Metropolitan Board of Education, The Safety Educational Program<No.8 Edition>, Tokyo Metropolitan Government, 2016. Retrieved from http://www.kyoiku.metro.tokyo.jp/pickup/.tokyo.jp/pickup/p_gakko/anzenkyoikuprogram/28anzenkyoikupro.pdf on May 2016.

Interdependence between Structural Engineering and Construction Management
Edited by Ozevin, D., Ataei, H., Modares, M., Gurgun, A., Yazdani, S., and Singh, A.

ISBN: 978-0-9960437-6-2

CONCEPTUAL BIG DATA CASE STUDY TO IDENTIFYING RISKS OF NEW NUCLEAR TECHNOLOGIES

JAEHEUM YEON[1], MARK CZARNY[1], JOHN WALEWSKI[2], and JULIAN KANG[3]

[1]*Dept of Engineering & Technology, Texas A&M University-Commerce, Commerce, USA*
[2]*Dept of Civil Engineering, Texas A&M University, College Station, USA*
[3]*Dept of Construction Science, Texas A&M University, College Station, USA*

New technologies associated with nuclear power plants are being introduced regularly. However, many of the risks and uncertainties associated with these new nuclear technologies have yet to be identified. In this study, the risks related to newly-developed nuclear technologies were determined through an extensive review of the extant literature. A qualitative visual content analysis was selected as the research method employed to identify words repeatedly occurring in 147 journal articles. Through this conceptual "big data" approach, frequently mentioned words were identified using a co-occurrence map. The analysis results were then grouped into four categories: fuel resources, operational system designs, nuclear reactor cooling systems, and steam generators. Words used repeatedly to reference these four key categories were determined to also represent potential causes of risk factors. Many texts could be analyzed in a short period of time through the use of visual content analysis software. Frequently emphasized correlating words were then identified. This big data approach can also be applied to additional nuclear power practices to identify other uncertainties. Last, the limitations of a visual content analysis employed as a risk identification approach were revealed through this study.

Keywords: Risk identification, Visual text mining, Content analysis.

1 INTRODUCTION

Nuclear power is a major worldwide energy resource that accounted for 2,571 billion kWh, or about 11% of the global electricity production, in 2015. This immense amount of electricity was generated by approximately 450 nuclear reactors that are currently being operated in 30 countries (WNA 2018a). Governments looking for sustainable energy resources are aggressively investing in nuclear power in their respective countries. In 2018, approximately 50 nuclear power reactors were under construction in 13 countries (WNA 2018a). Moreover, nuclear power plant construction is just one representative international construction projects because seven countries – France, Japan, USA, Russia, Canada, South Korea, and Argentina – are capable of exporting nuclear reactors to other countries (NPP Exporters 2018). Although nuclear power is attractive as an energy resource for the next generation, the risks and uncertainties associated with it are inherent in nuclear power plant construction projects. These risks usually originate from non-recurring uncertainties that cannot be processed statistically such as social trends, technological developments, and economic changes (Hertz and Thomas 1984). The owners and contractors of international construction project must assess the political, geographic, economic, environmental, regulatory, and cultural risk factors before they engage in implementing alternatives if they hope to achieve business success (Walewski *et al.* 2006). Poor construction project performance and unsuccessful delivery are more common in international construction projects than in domestic efforts (Walewski *et al.* 2006). Thus, risk identification is crucial, especially in terms of nuclear power plant construction, if issues are to be avoided and proactive responses planned. Also, to comprehensively identify the risks of construction projects, both internal and external dangers must be considered at the same time (see Figure 1) (Khodeir and Mohamed 2015).

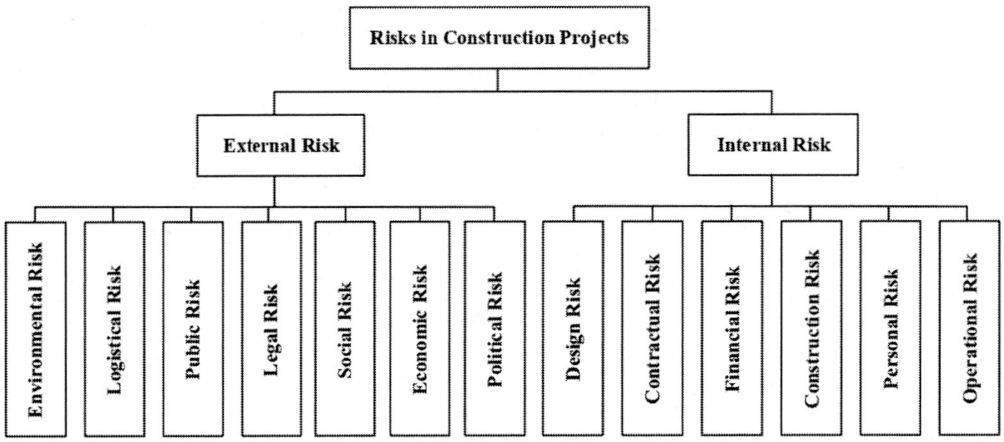

Figure 1 . General risk framework for construction projects (Khodeir and Mohamed 2015).

2 PROBLEM STATEMENTS

Various traditional and advanced technologies are employed in nuclear power plant construction projects. Even among the same types of construction, huge differences can exist in terms of schedule, cost, quality, and resources demanded, depending on the applied technology. In addition, countries that export nuclear reactors to other countries continue to develop new technologies in order to remain competitive in the market. However, new technology means additional risk because related uncertainties have yet to be investigated from a long-term perspective. To prevent and resiliently respond to risks, these issues should be classified systematically and their causes identified. Uncertainties associated with nuclear power can cause irreversible disasters. Thus, this study explored the causes of potential risks associated with new nuclear power plant technologies.

3 RESEARCH METHOD

The risks were identified by quantitative and qualitative risk analysis methods; the general process of risk identification is shown in Figure 2. Quantitative research methods were applied when numerical data existed. These methods included multivariate statistical models, event trees, system dynamics models, sensitivity analyses, project simulations, stochastic simulation models, and additive models. In contrast, the methods of qualitative risk assessment were applied to situations involving risks that could not be quantified, were beyond the scope of the analytical models, or involved uncertainties that could not be processed statistically. These uncertainties were caused by unique situations that were not repeatable or by technology development, economics, or social changes. The qualitative risk analysis methods included project definition rating indexes, pareto diagrams, and failure modes and effects analyses (National Research Council 2005).

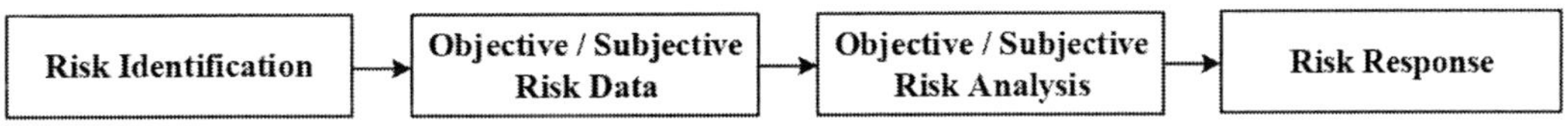

Figure 2. Risk identification process.

In this study, the research method selected to identify risk factors and their causes, as derived from the texts, was qualitative conventional content analysis (Hsieh and Shannon 2005). Identification of risk factors from repeated mention in an extensive literature review is not easy if one attempts to manually read each study. Therefore, in this study an extensive literature review was conducted via a visual text analysis method. Many journal articles and technical papers associated with nuclear power plant construction can be obtained online. However, newly developed nuclear power technologies cannot be researched in old publications. To search for trends in cutting-edge nuclear technology, the entirety of the Journal of Nuclear Energy and Technology (JNET) online was gathered for submission to the visual content analysis software because JNET provides articles associated with nuclear energy, training, science, and technology (NET 2018). The total number of articles collected was 169. Among these 169, 22 were excluded after reviewing the abstracts; their topics were outside the scope of this research. Thus, 147 articles from JNET were analyzed by KH Coder, the visual content analysis software was illustrated in Figure 3.

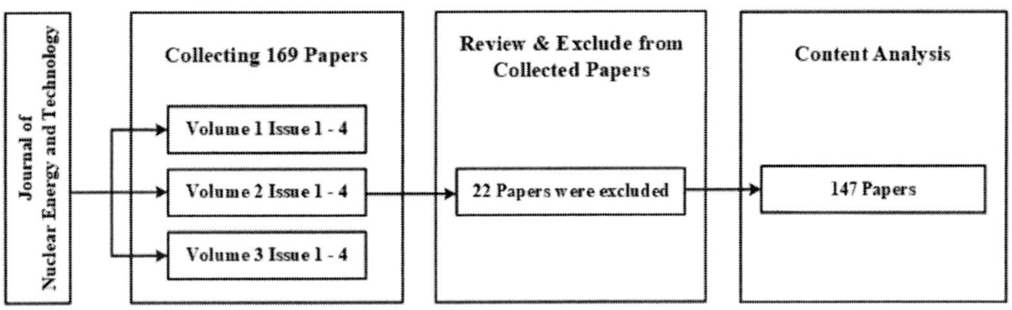

Figure 3. Data collection process.

This study was based on a conceptual big data approach. Through big data, new correlations were found among words mentioned repeatedly in an extensive literature review (Boyd and Crawford 2011). The advantage of visual content analysis software is that many occurrences of repeated words in a sizeable collection of literature can be counted in a short period of time (Weber 1990). KH Coder, the software selected for this study, is advantageous for counting word frequency and generating co-occurrence maps (Higuchi 2018). Co-occurrence maps are a powerful visualization tool for examining emphasized and frequently-connected words at a glance (Ryan and Bernard 2003). The repeated / emphasized words in this research were determined to be important nuclear energy risk factors that require attention. Words related to the emphasized words were identified as elements of risk.

4 RESULTS AND DISCUSSION

The results of the visual content analysis are shown in Table 1. The repeated words of greatest importance among the 30 most frequent words associated with new nuclear energy technologies, as obtained from 147 journal papers, were: reactor, temperature, neutron, coolant, heat, steam, design, uranium, and plutonium. To identify risks based on these repeated words, words related to the most frequently mentioned words needed to be identified. Thus, co-occurrence maps provided by a function of the visual content analysis software were generated.

Table 1. Consistency analysis results.

Count	Word	Frequency	Count	Word	Frequency
1	fuel	2,041	16	rate	662
2	reactor	1,857	17	parameter	618
3	system	1,300	18	process	593
4	power	1,150	19	core	557
5	value	1,061	20	heat	554
6	temperature	977	21	steam	543
7	calculation	870	22	material	520
8	result	825	23	method	520
9	neutron	814	24	energy	496
10	model	751	25	level	490
11	datum	750	26	water	486
12	flow	743	27	design	478
13	coolant	712	28	uranium	478
14	time	709	29	plutonium	450
15	operation	701	30	unit	438

The generated co-occurrence map is shown in Figure 4. Through this map, risk factors associated with new nuclear power plant technologies were categorized. The first risk factor was related to fuel resources such as uranium and plutonium that are used to generate energy in nuclear reactors. The risk factor related to uranium is human health dangers when uranium is mined and processed because doing so emits radiation (National Research Council 2011). Also, plutonium is a radioactive substance and dangerous when inhaled (Voelz 2000). The second risk factor was related to the design of a reactor's operational system. Uncertainties associated with the design of newly developed nuclear technologies could include design defects stemming from a lack of specification, differences in site conditions, or communication problems between designers and installers (Perry and Hayes 1985). The third risk was connected to the temperature of a nuclear reactor. Most of the new technologies introduced in the literature dealt with coolants used to control the temperature in a reactor and keep it under 350°C (WNA 2018b); this reduces the likelihood that an explosion will occur. The fourth most frequently mentioned word was related to steam generators. Many new technologies associated with stream generators have been introduced. Electricity is generated when a steam turbine is made to spin by steam produced from heated water (Trigilio 2006). This type of generator is not directly related to a risk. However, many studies have been conducted on steam pressure generators, meaning that technical change factors such as new laws and designs related to new steam generator technology could produce risk during a long-term project lifecycle.

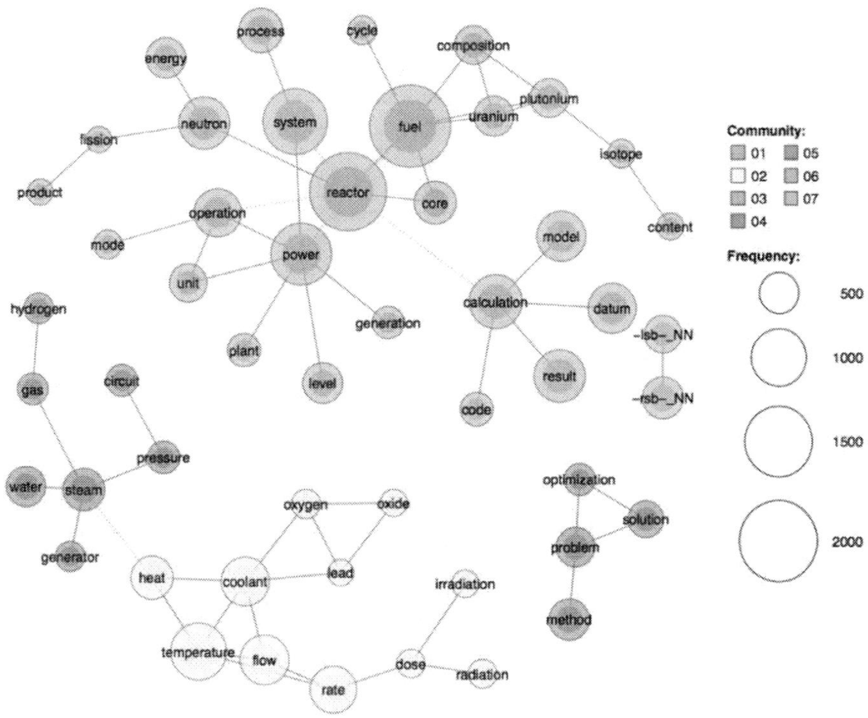

Figure 4. Co-occurrence map.

5 CONCLUSION

In this study, the risks related to newly developed nuclear technologies were identified using visual context analytics (i.e. the KH Coder). The texts of 147 papers published in the Journal of Nuclear Energy and Technology were analyzed through a big data approach. Four potential risk categories associated with nuclear technologies were identified: fuel resources, operational system designs, nuclear reactor cooling systems, and steam generators. Through a co-occurrence map, words correlating with the four categories were determined. These words could be causes of risks or newly determined key words. However, a deep-level analysis to determine the exact causes of these uncertainties is difficult. Thus, these texts should be manually analyzed by readers after specific uncertainties are identified by the visual context analysis software. This limitation should be addressed as a topic of future work.

References

Boyd, D., and Crawford, K., *Six Provocations for Big Data.* A Decade in Internet Time: Symposium on the Dynamics of the Internet and Society, 2011.

Hertz, D. B., and Thomas, H., *Practical Risk Analysis*, Wiley, New York, 1984.

Higuchi, K., *KH Coder*. Retrieved from http://khcoder.net/en/ on August 22, 2018.

Hsieh, H-F, and Shannon, S., Three Approaches to Qualitative Content Analysis, *Qualitative Health Research*, 15(9) 1277-1288, 2005.

Khodeir, L. M., and Mohamed, A. H., Identifying the Latest Risk Probabilities Affecting Construction Projects in Egypt according to Political and Economic Variables, *HBRC Journal*, 129-135, 2015.

National Research Council, *The Owner's Role in Project Risk Management*, National Research Council, The National Academies Press, Washington DC, 2005.

National Research Council, *Uranium Mining in Virginia: Scientific, Technical, Environmental, Human Health and Safety, and Regulatory Aspects of Uranium Mining and Processing in Virginia,* National Academies Press, Washington DC, 2011.

NET, Nuclear Energy and Technology, Retrieved from National Research Nuclear University: https://www.journals.elsevier.com/nuclear-energy-and-technology on August 22, 2018.

NPP Exporters, The Nuclear Power Plant Exporters' Principles of Conduct, Retrieved from http://nuclearprinciples.org/participants/ on August 22, 2018.

Perry, J. G., and Hayes, R. W., *Risk an its Management in Construction Projects*, Proceeding of the Institution of Civil Engineers, Institution of Civil Engineers, 78(3), 499-521, 1985.

Ryan, G. W., and Bernard, H. R., Techniques to Identify Themes, *Field Methods*, 15(1), 85-109, 2003.

Trigilio, S., *Applying the Principles of Risk Management to Nuclear Power Plant Safety,* Naples: International Foundation for Protection Officers, 2006.

Voelz, G., Plutonium and Health: How Great is the Risk?, *Los Alamos Science*, 26, 78-79, 2000.

Walewski, J., Gibson, G. E., and Vines, E. F., Risk Identification and Assessment for International Construction Projects, In D. I. Cleland, & R. Gareis, *Global Project Management Handbook*, 1-17, McGraw-Hill, New York, 2006.

Weber, R. P., *Basic Content Analysis,* Sage, 1990.

WNA, Plans For New Reactors Worldwide,World Nuclear Association. Retrieved from http://www.world-nuclear.org/information-library/current-and-future-generation/plans-for-new-reactors-worldwide.aspx on August 22, 2018a.

WNA, Nuclear Power Reactors, World Nuclear Association. Retrieved from http://www.world-nuclear.org/information-library/nuclear-fuel-cycle/nuclear-power-reactors/nuclear-power-reactors.aspx on August 22, 2018b.

Interdependence between Structural Engineering and Construction Management
Edited by Ozevin, D., Ataei, H., Modares, M., Gurgun, A., Yazdani, S., and Singh, A.

ISBN: 978-0-9960437-6-2

STUDY ON THE SEISMIC VULNERABILITY OF A VIADUCT SUBJECTED TO NEAR-FAULT GROUND MOTIONS

YONG LI, MENGFEI XIE, and LIJUN MENG

School of Civil Engineering, Shijiazhuang Tiedao University, Shijiazhuang, China

Piers, abutments and bearings of viaducts may suffer severe damage during earthquakes, so it's not insufficient to evaluate the seismic vulnerability of a bridge system only by plastic hinge curvature, which is adopted in seismic design guidelines. In this paper, the seismic vulnerability evaluation of a viaduct is conducted by incremental dynamic analysis under 30 near-fault ground motions, which are selected from PEER database. Then several damage measures are recommended to make an overall estimation for the seismic vulnerability of the viaduct, including plastic hinge curvature, shear failure and sliding displacement failure of bearings and pounding force between abutments and the girder. The analysis results show that the transversal seismic excitations may lead to more severe damage than the longitudinal ground motions. No matter in which direction the ground motions are inputted, the bearings' seismic vulnerability resulted by shear force or sliding displacement is higher than the plastic hinge of piers, which indicates that the seismic vulnerability of the bridge system is determined by the bearings to an extent. As a result, bearings should be designed according to both static and seismic analyses to guarantee the safety during earthquakes.

Keywords: Near-fault earthquake, Incremental dynamical analysis, First-order reliability principle, PGA, Damage measure.

1 INTRODUCTION

Near-fault ground motion records which always refer to the fault distance no more than 20 km, have a distinct pulse type ground motion caused by rupture directivity effect and fling-step effect (Mavroeidis and Papageorgiou 2003, Ambraseys and Douglas 2003). With a higher PGV/PGA than far field ground motions, near-fault earthquake records express a relative wide sensitive area of their acceleration response spectrums. A near-fault earthquake record with velocity pulses has longer period of response spectrum platform and its response spectra value is larger in the long period range than that without the pulse. Additionally, the incremental spectral value in the long period can enlarges the deformation of long-period bridges, which may lead to pounding response even dropping of superstructure (Loh *et al.* 2002). However, there is not a seismic design code for bridge supplying an explicit provision to take the near-fault ground motion effect in to account. Therefore, the incremental dynamic analysis (IDA) method is adopted to conduct a seismic vulnerability analysis of viaducts under near-fault earthquakes.

IDA method can reflect the structural seismic performance under different intensity for a same ground motion by the elasto-plastic time-history analysis (Vamvatsikos and Cornell 2002,

Vamvatsikos and Cornell 2004). It can overcome disadvantages resulted in pushover method which execute a static process analysis instead of a dynamic one. Generally speaking, IDA method with enough ground motions is able to comprehensively analyze the seismic performance of bridges as long as the finite element model is reasonably and correctly built.

In general, IDA method is a reasonable way to take study on the seismic vulnerability on viaducts under near-fault ground motions. Based on the PEER Ground Motion Database, a set number of typical near-fault earthquake records can be selected to conduct IDA and analyze failure probability of viaducts. The correspondence between intensity measures of ground motions and damage measures of viaducts can be built, which can contribute to seismic vulnerability analysis of viaducts under near-fault earthquakes (Andreas *et al.* 2005).

2 STRUCTURAL FINITE ELEMENT MODEL AND NEAR-FAULT GROUND MOTIONS SELECTION

A continuous viaduct with spans 4×40 m is selected to conduct the IDA and evaluate the seismic vulnerability. The superstructure is single box with three cells, and the substructures are double-column piers and gravity abutments, as shown in Figure 1. The finite element model includes several nonlinear characteristics, such as plastic hinge of piers, pounding behavior between the girder and abutments, and hysteretic behavior of sliding bearings.

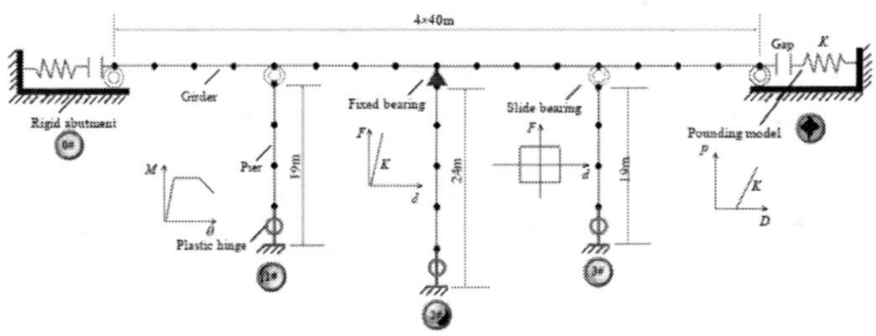

Figure 1. Finite element model of the viaduct.

30 near-fault ground motion records with impulse are chosen in PEER Database. Suppose the bridge site type is II, characteristic period is 0.45s, and seismic design intensity is Ⅶ degree. The distances between records site and earthquake rupture are all within 25km. The acceleration spectrum of selected records and target spectrum are shown in Figure 2.

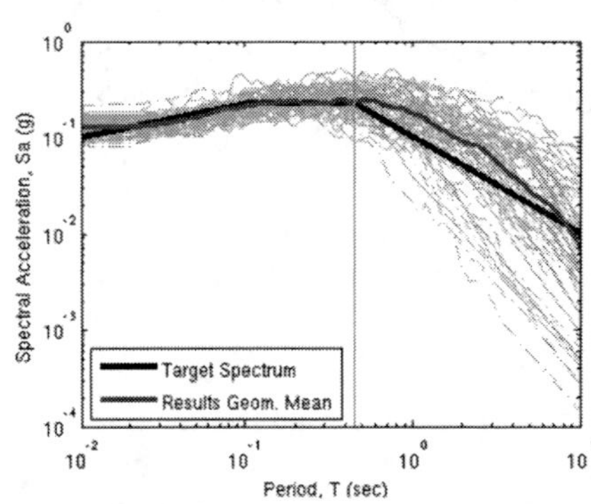

Figure 2. Comparison between target spectrum and the average spectrum of selected near-fault earthquake records.

3 SEISMIC VULNERABILITY ANALYSIS FOR TRANSVERSAL GROUND MOTIONS

3.1 IDA results

Viaduct piers are always flexible with the capacity design for shear performance under seismic excitations, so plastic hinges of piers are often adopted for failure evaluation in seismic design codes. However, under transversal ground motion input, the main seismic damage of a viaduct includes plastic hinge forming on piers and shear failure of all bearings. For plastic hinges at the bottom of piers, the recommended damage state division determined by curvature limit is shown in Table 1.

Table 1. Recommended damage states for plastic hinges.

Damage state	Description	Curvature limit
No damage (ND)	First yield	0.005
Slight damage (SD)	Crack	0.007
Moderate damage (MD)	Core concrete failure	0.015
Extreme damage (ED)	Collapse limit	0.025
Complete damage (CD)	Collapse	0.05

Bearings arrangement of the viaduct is shown in Figure 3. For shear damage of bearings, the damage status is determined by shear force limit. The first damage state should be the design limit in the transversal direction, and the second damage state represents complete shear failure of bearings, which refers to the transversal restriction between piers and the girder suffers destruction, corresponding to the failure of anchoring bolts of bearings.

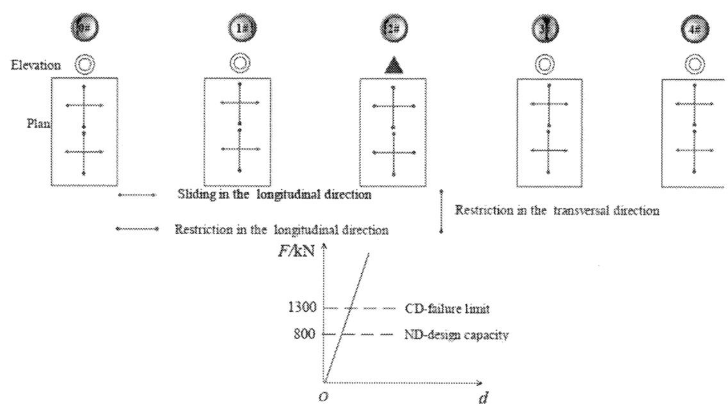

Figure 3. Finite element model of the bridge.

IDA results of the viaduct according to different DMs under transversal near-fault earthquake input are shown in Figure 4. Because the viaduct is a symmetrical structure, results of 1# and 3# piers are almost the same and the seismic damage of them is lower than 2# pier. ED state of 1# and 3# piers is reached at 1.0 g while 2# pier at 0.7 g. Transversal shear fragility of all bearings is reached more easily than plastic hinges of piers. Additionally, the sliding bearings at abutments reach the failure state at 0.3 g, which is lower than the bearings of 2# pier at 0.6 g.

Figure 4. IDA results for transversal near-fault ground motions.

3.2 Seismic Vulnerability Analysis

The seismic vulnerable curves of piers and bearings corresponding to each DM are shown in Figure 5. Probability distribution rules can be deduced from ND to CD according to PGA levels. Meanwhile it demonstrates the transversal shear vulnerability of bearings is higher than plastic hinges' failure of piers. For a certain level of transversal ground motions, the bearings at abutments are much easier to reach failure state, which may lead to the transversal restriction failure on the girder even unpredictable severe damage, for example girder falling in the traversal direction.

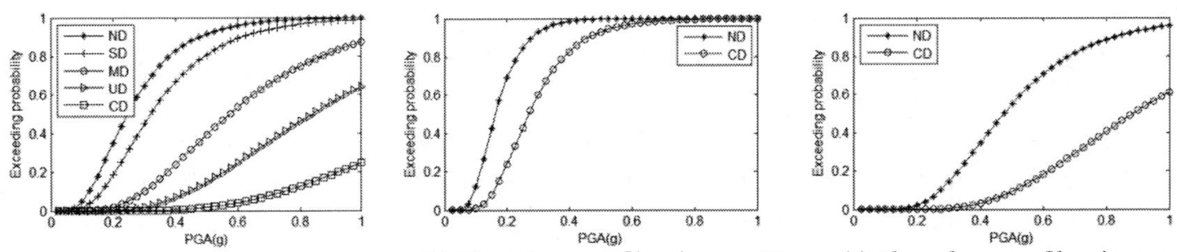

(a) Plastic hinge damage of 2# pier

(b) Shear damage of bearings on 0# abutment

(c) Shear damage of bearings on 1# pier

Figure 5. Failure probability of piers and bearings under transversal seismic excitations.

Under transversal earthquake excitations, damage of piers and bearings both may lead to seismic failure of a bridge system. So, it is necessary to evaluate the damage probability of the bridge system by the joint fragility probability method. The first-order reliability principle also called wide bounds method is adopted to evaluate the failure probability of the bridge system.

4 SEISMIC VULNERABILITY ANALYSIS FOR LONGITUDINAL GROUND MOTIONS

IDA results of the viaduct for longitudinal near-fault ground motions are shown in Figure 6. Considering the plastic hinge curvature, the seismic vulnerability of 1# and 3# piers is slightly higher than 2# fixed pier's. Compared with Figure 4, the seismic vulnerability of piers is higher than that under transversal excitations. When PGA reaches 1.0 g, the piers are damaged slightly under longitudinal excitations while may be damaged extremely in the transversal direction. Seismic vulnerability of the fixed bearing is higher than the plastic hinge of piers under longitudinal seismic excitations. The fixed bearing can reach the complete damage state at 0.3 g, indicating that its seismic vulnerability is higher than that under transversal excitations. The sliding bearings are hard to be destroyed because of the displacement restriction possibly by the pounding effect between abutments and the girder. Additionally, the pounding response is unable to generate the concrete damage of the girder or abutments.

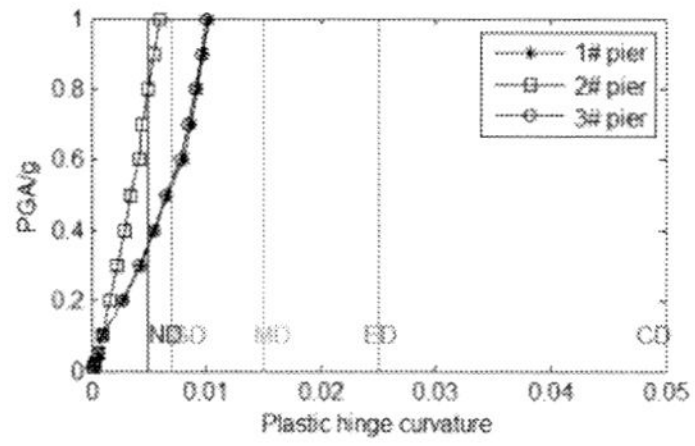

(a) Plastic hinges of piers.

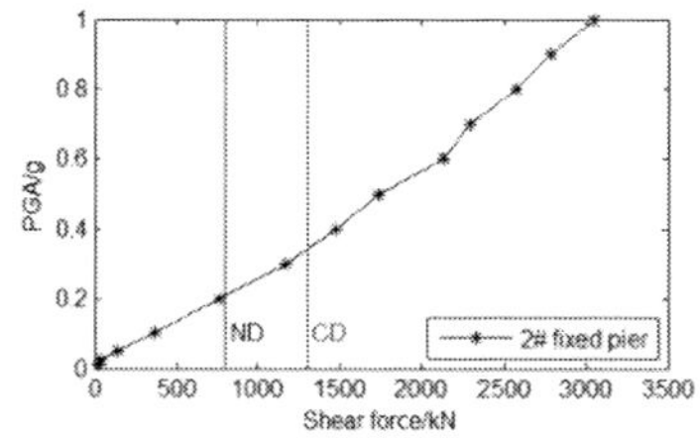

(b) Shear force of the fixed bearing.

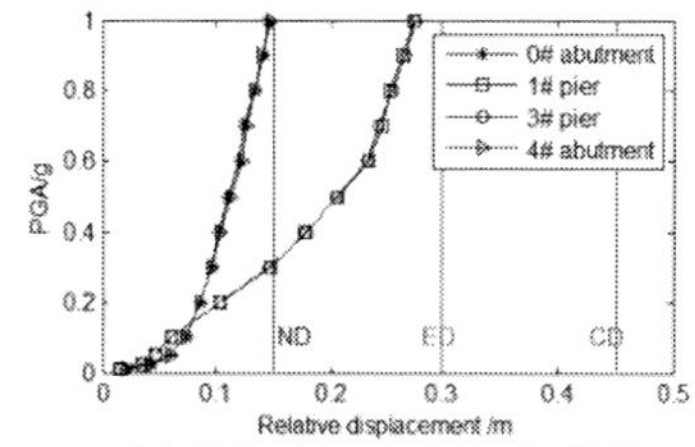

(c) Displacement of sliding bearings.

Figure 6. IDA results according to different DMs under the longitudinal seismic excitations.

The seismic vulnerability curves of each member under longitudinal near-fault earthquake excitations are shown in Figure 7. It can be deduced that the shear failure vulnerability of the fixed bearing is higher than plastic hinges of piers. The failure probability of sliding bearings on piers is higher than that on abutments, and the failure probability of unfixed piers is higher than the fixed pier.

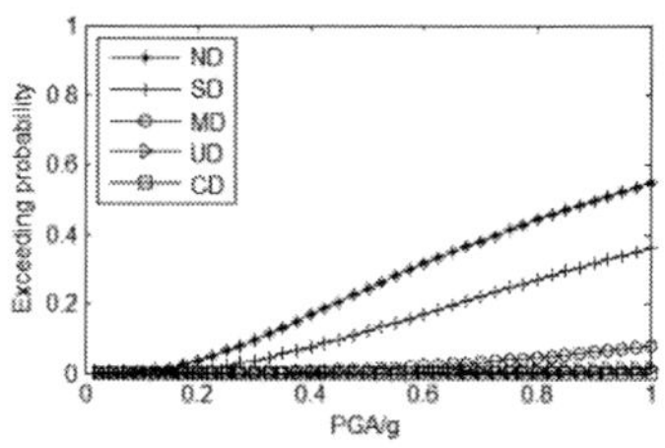

(a) Plastic hinge damage of 2# pier.

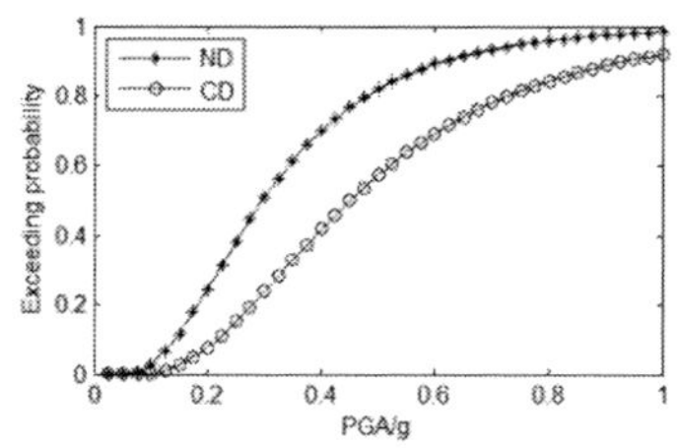

(b) Shear failure of the 2# fixed bearings.

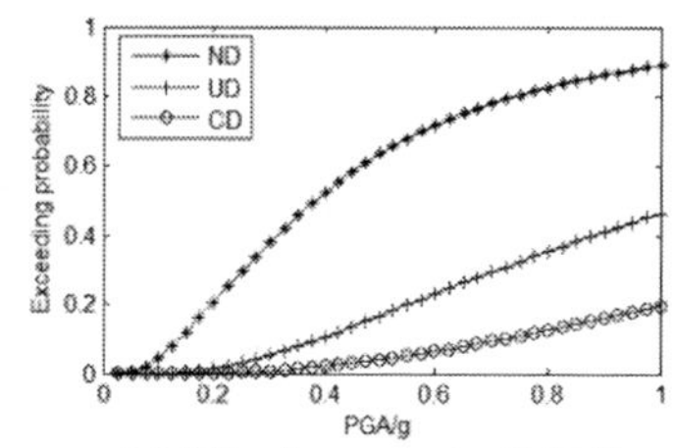

(c) Displacement of sliding bearings at 1# and 3# piers.

Figure 7. Failure probability of piers and bearings under longitudinal seismic excitations.

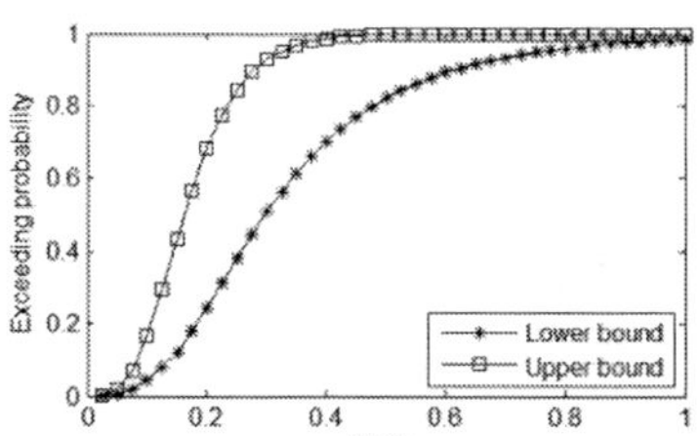

(a) No damage.

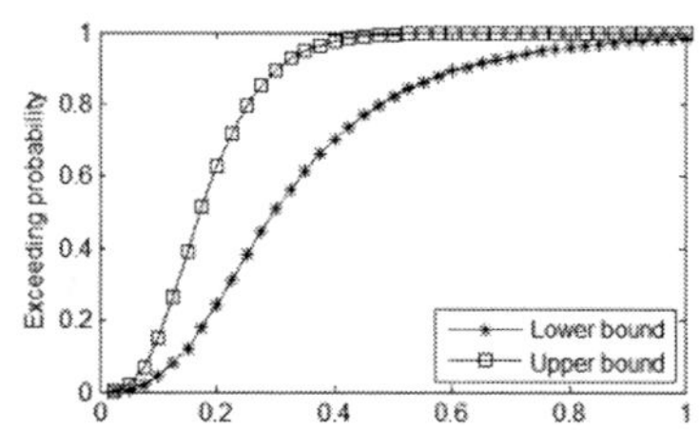

(b) Slight damage.

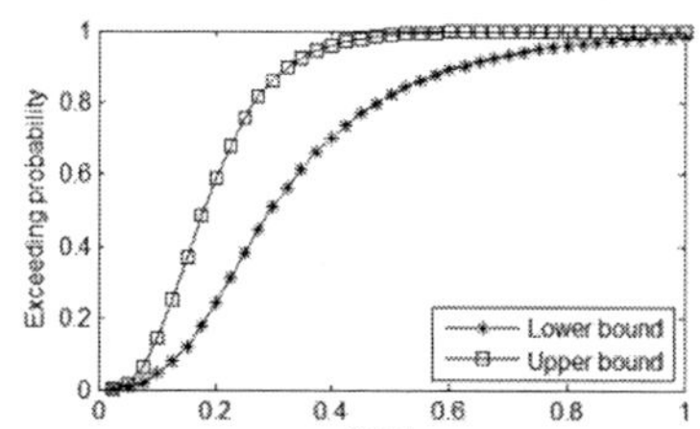

(c) Moderate damage.

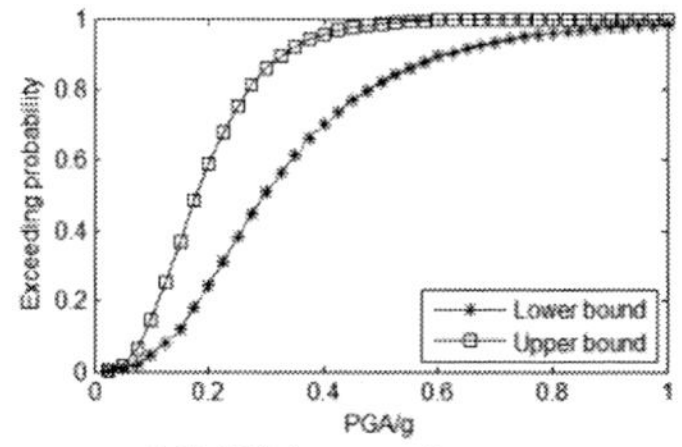

(d) Ultimate damage.

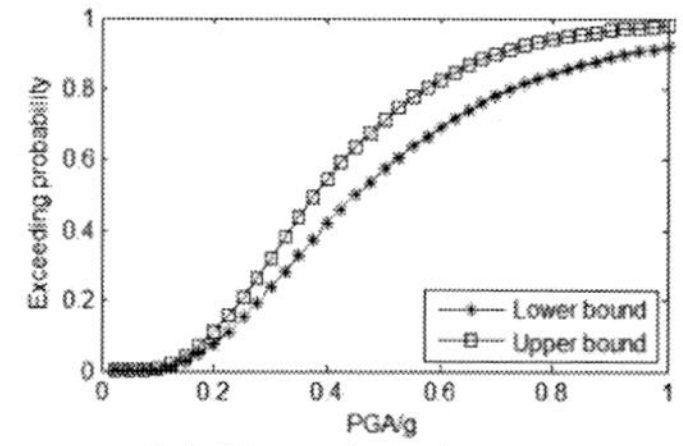

(e) Complete damage.

Figure 8. Seismic vulnerability probability of the bridge system.

The failure probability of the bridge system is shown in Figure 8. Comparing with Figure 5 and Figure 7, it can be reflected that the seismic vulnerability probability of the bridge system under longitudinal seismic excitations is higher than that under transversal seismic excitations. Overall the seismic vulnerability of the viaduct in the longitudinal direction greatly depends on the shear failure of the fixed bearing.

5 CONCLUSIONS

(1) The seismic vulnerability under transversal earthquake inputs is higher than that under longitudinal seismic excitations. No matter the excitation directions, the probability of shear failure or the displacement failure of bearings is higher than that of plastic hinge of piers. The seismic vulnerability of the bridge system depends on the failure probability of bearings.

(2) Under longitudinal earthquake inputs, the plastic hinges' curvature of all piers and displacement response of all sliding bearings are small because of the pounding effect between abutments and the girder, but the shear force response of the fixed bearing

(3) Under transversal earthquake inputs, the seismic vulnerability of bearings' shear failure is much larger than that of piers' plastic hinge failure. As a result, the seismic vulnerability of the bearings on the abutments is the largest, secondly the fixed bearing, and the bearings on the other piers in the end.

(4) The bearings on the abutments are suitable for bi-direction sliding because of the rigid stiffness meanwhile the side restrainers are rigid enough to avoid pounding failure, which may lead to phenomenon of girder falling in transversal direction. All bearings of the bridge system should have preferable shear performance, which should be higher than that of piers.

Acknowledgements

This study is supported by Natural Science Foundation of Hebei Education Department in China (Grant No: BJ2018048), the Natural Science Foundation of China (Grant No: 51508350, 51508351) and Natural Science Foundation of Hebei Province in China (Grant No: E2016210087)

References

Ambraseys, N., and Douglas, J., Near-field Horizontal and Vertical Earthquake Ground Motions, *Soil Dynamics and Earthquake Engineering,* 23(1), 1-18, 2003.

Andreas, F., Stempniewski, L., and Mrichwalski, S., A Combined Geophysical/Engineering Approach for The Seismic Safety of Long-Span Bridges, *Lecture Notes in Earth Sciences,* 105(3), 455-468, 2005.

Loh, C. H., Liao, W. I., and Chai, J. F., Effect of Near-Fault Earthquake on Bridges: Lessons Learned from Chi-Chi Earthquake, *Earthquake Engineering and Engineering Vibration,* 1, 86-93, 2002.

Mavroeidis, G. P., and Papageorgiou, A. S., A Mathematical Representation Of Near-Fault Ground Motions, *Bulletin of the Seismological Society of America,* 93(3), 1099-1131, 2003.

Vamvatsikos, D., and Cornell, C. A., Incremental Dynamic Analysis, *Earthquake Engineering and Structural Dynamics*, 31, 491-514, 2002.

Vamvatsikos D, and Cornell C. A., Applied Incremental Dynamic Analysis, *Earthquake Spectra,* 20(2), 523-553, 2004.

Interdependence between Structural Engineering and Construction Management
Edited by Ozevin, D., Ataei, H., Modares, M., Gurgun, A., Yazdani, S., and Singh, A.

ISBN: 978-0-9960437-6-2

BURIED TRANSPORTATION STRUCTURES: FUZZY-MARKOV COMPUTATION CASE STUDY

KOOROSH GHAREHBAGHI and KEN FARNES

School of Property, Construction and Project Management, RMIT University, Melbourne, Australia

This paper will focus on how to improve buried structures' resilience and sustainability so that they hold up over time, withstand extreme weather events, and minimize the use of natural and financial resources. Generally, buried Transportation structures are those which have some portion of their elements and configuration dormant, such as culverts, bridges, and their foundation systems. These structures are a significant part of transportation networks and their related systems. These structures can be disrupted by extreme weather events such as seismic activities, extreme wind gusts, tornado, and tropical cyclones. Repairing the damage after such events can challenge resources and negatively impact the area's economy by disrupting traffic. Moreover, a common problem with buried structures is the lack of understanding about each individual structural element. Accordingly, as a part of this paper's focus, careful attention is given to such important structural integrity. Nevertheless, a number of Australian case studies were undertaken to further improve the buried structures' resilience and sustainability. To support this paper's aim a Fuzzy-Markov computation using a modified *FuzzyStatProb* for buried assets were also explored. The Fuzzy-Markov computation accurately identified the appropriate treatment regimes for selected bridges; which ultimately reduced the natural and financial resources usage.

Keywords: Resilience and sustainability, Fuzzy-Markov process.

1 INTRODUCTION

Universally, buried Transportation structures are those which some of their elements and configurations are concealed, more likely underground. These assets include bridges, where some of their structural elements are dormant (beneath the natural ground level), including columns, pillars, and so on. Such buried Transportation structures conventionally have their footing systems, along with their connections and joints, sealed under the *natural ground level*. Since all of these structures are an important part of the transportation network, their dormant and concealed components need to be regularly monitored and examined.

Although these components typically have low failure rates, when they fail the consequences can be quite severe. Their low rate of failure, coupled with the high cost of inspection and condition assessment contributes to the situation where there is a lack of data necessary to model the deterioration rates of these assets and subsequently to make rational decisions regarding their repair or renewal (Gharehbaghi and Georgy 2015). Nevertheless, inspections determine the maintenance requirements and subsequent maintenance regimes (Hurt and Schrock 2016). Gharehbaghi and Chenery (2017) noted that, while structural resilience includes sturdiness and regaining abilities (under extreme loads), sustainability on the other hand, involves the holistic integration of economic, social, and in particular environmental factors. They also noted, although structures are subject to conventional corrosion and rustis, they also can be disrupted by extreme weather events such as seismic activities. In addition, they also highlighted that efficient maintenance regimes for structures are paramount to prolong these asset's life span.

Since repairing the damage after extreme weather events can challenge resources and negatively impact and disrupt the general traffic, an understanding of buried assets and their individual structural elements is vital. Although, such investigations have been considered and examined widely, the aim of this paper is to provide a number of recent Australian case studies to further improve the buried structures resilience and sustainability. Numerous computation methods such as Artificial Neural Network (ANN), Genetic Algorithms (GA), and Bayesian Networks (BN) have been proposed for the measurement of the deterioration of buried assets, the most prominent approach has been the Markovian processes which will be used be explore the case studies. Lorenzo (2014) rightly highlighted that the Fuzzy-Markov process is one of the most effective computational methods for such structures. Such a proclamation is due to the Fuzzy-Markov process's ease of use and more importantly, accurately validating some of the more common considerations for the buried structures.

2 COMMON CONSIDERATIONS FOR THE BURIED STRUCTURES

Gharehbaghi and Sagoo (2016) argued that the primary consideration for buried Transportation structures consists of understanding how to improve resilience and sustainability of such assembly. Nevertheless, Hurt and Schrock (2016) highlighted that to successfully achieve this, a checklist needs to be developed and maintained that reemphasizes the importance of minimization of natural resource demands and environmental impacts. Such a checklist would highlight environmental mitigation and durability, along with identifying climate change risks and subsequent areas. On the other hand, Shi *et al.* (2012) noted that an essential factor to accomplish more sustainable buried structures also comprises comprehension of when buried structures are viable and appropriate. However, to achieve such an outcome capabilities of buried structures and their installed elements needs to be precisely identified and carefully validated. Fuzzy-Markov process provides an appropriate apparatus to accurately validate such considerations for buried structures.

Furthermore, Gharehbaghi and Rahmani (2017) also underlined the importance of materials selection and usage as a part of structural Resilience and Sustainability. The structural Resilience and Sustainability is paramount for stable buried Transportation structures (Stewart *et al.* 2011). Such material application would further strengthen the buried structures against extreme weather events such as seismic activities, extreme wind gusts, tornado, tropical cyclones, and so on. Accordingly, the buried Transportation structures resilience and sustainability are paramount factors to ensure these assets retain their structural integrity and therefore their life-span is prolonged. For Australia, tropical cyclones are common phenomenon and spread throughout the northern, north-east and north-west coast regions. Accordingly, this paper will examine a number of case studies in Australia to further improve the buried structures' resilience and sustainability.

3 AUSTRALIAN CASE STUDIES

Four bridges (including their structural elements) were selected as the basis of the case studies to further investigate the buried structures resilience and sustainability. These four sites are within the state of New South Wales (NSW). These sites all faced extreme wind gusts which were part of the category 4 Cyclone Debbie which occurred in 2017 and was the strongest tropical cyclone since 2015 (Tropical Cyclone Marcia). Cyclone Debbie had the highest wind speed of 195 km/h and wind gusts of 250 km/h. It affected Queensland, New South Wales, and New Zealand. The location of the four bridges and the actual Cyclone Debbie are presented in Figure 1.

As can be seen from an overview of the bridges in Figure 1, bridge NE-12-BBK2 was closest to the NSW and the Queensland border. In addition, this bridge was also closest to Cyclone Debbie's radius and subsequently would have faced more force and thus additional damage.

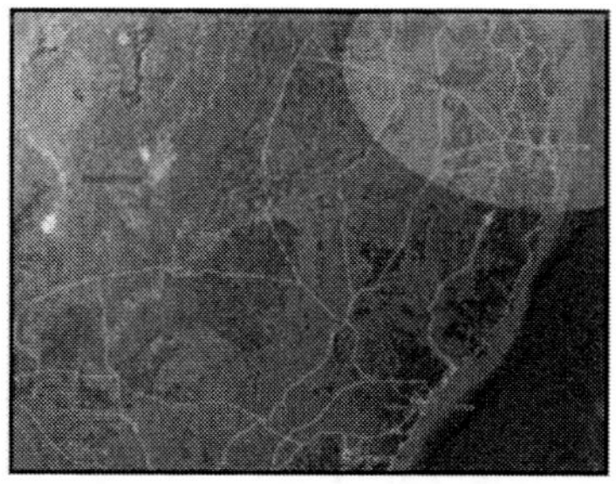

Figure 1. Represents overview of Cyclone Debbie and the four case studies.

Although, various damages were quite visible, however, appropriate Structural Health Monitoring (SHM) method was utilized as the basis of determining the level of impairment for each bridge. As noted, only one of the bridges required significant repair due to considerable structural damage. NE-12-BBK2 a Banded Beam Bridge sustained severe structural displacement at its base. The damage was centered on the connecting point of the third column base and its foundation system. The damage to the other three bridges was minimized and thus enabled them to be in-use. Nonetheless, to further validate the damage to these buried Transportation structures a detailed Fuzzy-Markov computation was carried out. The overview of this analysis is provided below in Table 1.

Table 1. Overview of the four Australian (NSW) case studies.

Bridge Id #	Bridge Type	Buried Element	Damage Detected	Structural Integrity and Performance
N-24-BB31	Banded Beams	Column and base connection area	Minor - surface (Micro) crack	Operational - require minimum maintenance
NW-43-CS5	Cable Stayed	Foundation	Minor - trivial crack	Operational - require minimum maintenance
N-09-BBC54	Banded Beams	Foundation	Minor - surface crack	Operational - require minor maintenance
NE-12-BBK2	Banded Beams	Column and base connection area	Sever - structural displacement	Non-operational - require replacement due to rapid deteriorate

4 FUZZY-MARKOV COMPUTATION (OVERVIEW) FOR THE CASE STUDIES

Traditionally, the Fuzzy-Markov technique uses the condition of the asset in a distinct manner in a few discrete states (Anastassiou 2016). Nonetheless, this technique utilizes Computational Theorem and Discrete Mathematics to predict the transition of the asset from one state to the next (Gharehbaghi 2016). Figure 2 represents the schematics of this technique. As shown in Figure 2, stage 1 prescribes the algorithm development to determine the deterioration rates, and stage 2 imposes specific state of high, intermediate, and low. While these states are based on predicted asset conditions, the general algorithm developed further validates and filters these states. As Gharehbaghi (2006) argued, the Fuzzy-Markov computation, specifically complements the complexities of the Buried Transportation Structures. This is supported by Chmielowski (2016) who demonstrated that the computation needs the ability to be flexible and adaptable when computing to prolong the asset's life.

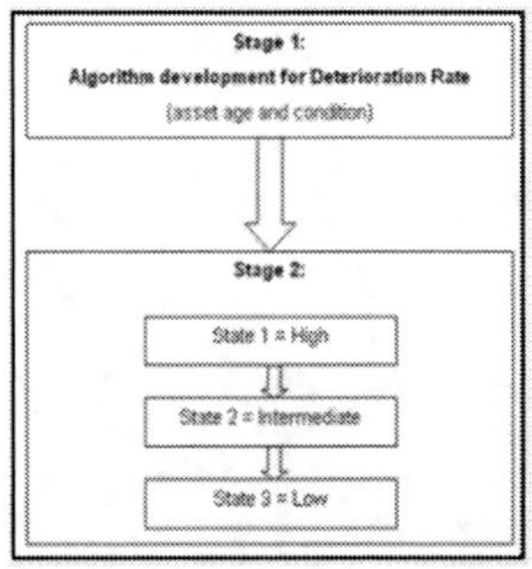

Figure 2. Fuzzy-Markov for buried Asset - Schematics (Gharehbaghi 2006).

Such an approach is fundamental to effectively validate the irregular variables such as design life, rate of use, maintenance and rehabilitation regimes and so on, of the Buried Transportation Structures. Appropriately, the Fuzzy-Markov computation (modified *FuzzyStatProb*) was used for the analysis of the four case studies. This process was selected (buried structures) since there was a probability of inaccurate inspection regimes and data acquisitions. The primary data was sourced from "updated records" along with the "validated and verified archives." These specific records were maintained by the relevant transportation authorities, such as Transport for NSW, and were also summarized as a part of the "preliminary condition assessment". In addition, the encapsulated data was then grouped into specific categories (discussed below) for advance measurement and evaluation processing. The finalized data was next collaborated into the *algorithm sequencing* for pre-modeling purposes. The deterioration process was then modeled based on *fuzzy rule*, using the Markov method. Subsequently, this process was applied each time via two definite stages:

- Stage one: As a part of the input, this involved the design of fuzzy rule-based algorithm for the deterioration rate of specific *time steps* from asset's age and condition. As already noted, the deterioration rate was thus finalized using SHM method.
- Stage two: As a part of the procedure stage, the condition rating of the asset was defined based on present conditions and deterioration rate. In other words, the deterioration process of an asset gradually moved from a high (good) state to lower condition states.

The process was then formulated to reflect what happens in reality where the condition of an asset can only be in transition between two or three states (as per Figure 2). Moreover, the Fuzzy-Markov computation for the Case Studies was carefully developed along with their guidelines. These are shown in Figure 3 and Table 2. While Figure 3 portrays the interaction between "each stage and state", Table 2 on the other hand represents the commentary for all the case studies. Whereas S1 (commencement of the segmentation) and S2 are the primary Fuzzy Markov fields, $S2^{-1}$ to $S2^{-3}$ are the secondary and ultimately the *taxonomy grounding*. Moreover, the taxonomy fields represent the 'sorting process' through the connecting lines as a part of the actual chain mechanism. Correspondingly, all the nodes were part of this chain system. The model ultimately uses a simple language to describe the existing problems, identify objects, and verify the main elements of such objects. Moreover, such objects consist of the specific elements including Stage 1, and so on. The Fuzzy-Markov computation was a complex model which also determined the interaction between elements. Nonetheless, the Fuzzy-Markov computation model also possessed multiple solutions, and thus allowed for the best solution to be selected. Based on the best outcome, a new strategy was then issued to the actual Buried Transportation Structures. Nevertheless, upon successful Fuzzy-Markov computation, a holistic inspection, maintenance and rehabilitation regimes were carefully developed. These regimes included:

- Cumulative waiting time's distributions were calculated using ANNs to observe effects.
- As the basis of Structural resilience, specific matrices for age-dependent transition probability were compiled using conditional survival probabilities in the various structural states.
- As the basis of Structural sustainability, environmental and economic schemes were then determined for each structure (for schemes of inspection, intervention and failure phases).
- Scheduled inspection and maintenance intervals were then calculated to reduce the overall cost to a minimum.
- Warning signs were also formulated for immediate intervention when the structural integrity of the asset was compromised.

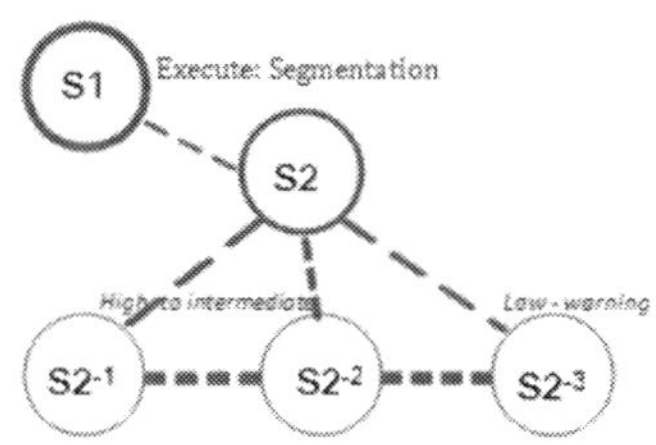

Figure 3. Shows the simplified Fuzzy-Markov computation model.

Table 2. Represents the guidelines of the Fuzzy-Markov computation model.

Bridge Id #	Stage 1	Stage 2 ^{S1}	Stage 2 ^{S2}	Stage 2 ^{S3}	Solution and suggestion
N-24-BB31	1.00	**0.99**	0.01	0.00	Structure performance high
NW-43-CS5	1.00	**0.93**	0.07	0.00	Structure performance intermediate to high
N-09-BBC54	1.00	**0.81**	0.19	0.00	Structure performance intermediate to high
NE-12-BBK2	1.00	0.00	0.00	**1.00**	Structure performance low - inefficient: require immediate repair

Finally, the Fuzzy-Markov computation process validated that, while assets N-24-BB31 and NW-43-CS5 were deemed totally safe and operational however, minimum maintenance was required. On the other hand, although N-09-BBC54 was operational, it needed minor maintenance. Conversely, it was established that the asset NE-12-BBK2 was in total despair and thus required immediate replacement. As a result, these four Australian case studies clearly highlighted the significance of the Fuzzy-Markov process to further improve the buried structures' resilience and sustainability through better identification and thus proper treatment recommendation for such assets. Such resilience and sustainability mechanisms were the outcome of transition probability, using conditional survival possibilities along with early warning signs for prematurely interventions. Accordingly, Fuzzy-Markov process highlighted the structural resilience and sustainability of the four case studies and in-turn specific maintenance regimes were also developed to further prolong these asset's lives. Ultimately, while to ensure

Structural resilience matrices for age-dependent probability was accumulated, to warrant sustainability, environmental and economic schemes were then determined for each bridge. Such an outcome would eventually lead to minimizing the use of natural and financial resources.

5 CONCLUSIONS

The aim of this paper was to review strategies for the improvements of the buried transportation structures' resilience and sustainability. To achieve this, four different types of bridges (including their structural elements) were selected all within the state of NSW, Australia. All of these bridges were subject to Cyclone Debbie, which occurred in 2017. These four bridges were selected since as they each showed different signs of structural damage. Nevertheless, to support this paper's aim, a Fuzzy-Markov process for buried assets was also explored. The examination of the Fuzzy-Markov process for four case studies validated the advantage of such computational process to further augment the buried structures' resilience and sustainability. As already noted, such improvements were through superior identification using Fuzzy-Markov processes and thus the appropriate treatment was recommended for the assets; which in-turn would ultimately point to curtailing the use of natural and financial resources. Consequently, the utilization of the Fuzzy-Markov computation, precisely distinguished the most suitable rehabilitation regimes for the said bridges. Ultimately, the instituted Fuzzy-Markov computation led to reduction of the resources required for the purpose of the four bridge's rehabilitation regimes.

Acknowledgements

The authors wish to dedicate their gratitude to Linda Watson (Senior Transportation Engineer) who assisted in retaining some of the necessary primary data needed for this research.

References

Anastassiou, G., *Intelligent Systems II: Complete Approximation by Neural Network Operators,* Springer, 51-59, 2016.

Chmielowski, W., *Fuzzy Control in Environmental Engineering, Springer Books*, USA, 38-44. 2016.

Gharehbaghi, K., *Infrastructure Asset Diagnostics: Operations and Maintenance*, Proceedings of the Asset Management and Engineering International Conference, Bristol, England, 8–11 March, 2006.

Gharehbaghi, K., *Artificial Neural Network for Transportation Infrastructure Systems*, MATEC Web of Conferences, 1-5, 2016.

Gharehbaghi, K., and Chenery R., *Fiber Reinforced Concrete (FRC) for High Rise Construction: Case Studies*, IOP Conference Series: Materials Science and Engineering, Institute of Physics, United Kingdom, 272, 223-229, 2017.

Gharehbaghi, K., and Georgy, M., Utilization of Infrastructure Gateway System (IGS) as a Transportation Infrastructure Optimization Tool, *International Journal of Traffic and Transportation Engineering, SAP,* USA, 4(1), 8-15, 2015

Gharehbaghi, K., and Rahmani, F., Viscoelastic Materials: Innovation and Development Countenance, *Solid State Phenomena - Key Engineering Materials, Trans Tech Publications,* Switzerland, 266, 165-171. 2017.

Gharehbaghi, K., and Sagoo, A., Reducing Deterioration and Corrosion of Reinforcements in Transportation Structures, *J. of Civil Engineering and Environmental Technology*, 2(1), 19-22. 2016.

Hurt, M., and Schrock, S., *Highway Bridge Maintenance Planning and Scheduling,* Springer, 264-284, 2016.

Lorenzo, F., *Methodologies and Techniques for Advanced Maintenance*, Springer, 29-34, 2014.

Shi, X., Xie, N., Fortune, K., and Gong, J., Durability of Steel Reinforced Concrete in Chloride Environments: An Overview, *Construction and Building Materials*, 30, 125–132, 2012.

Stewart M. H., Wang, X., and Nguyen, M. N., Climate Change Impact and Risks of Concrete Infrastructure Deterioration, *Engineering Structures*, 33, 1326–1337, 2011.

Interdependence between Structural Engineering and Construction Management
Edited by Ozevin, D., Ataei, H., Modares, M., Gurgun, A., Yazdani, S., and Singh, A.

ISBN: 978-0-9960437-6-2

INVESTIGATION OF DAMAGE ON DERBENDIKHAN DAM DURING EARTHQUAKE EXCITATION

CAGRI MOLLAMAHMUTOGLU[1] and IDRIS BEDIRHANOGLU[2]

[1]*Civil Engineering Faculty, Yildiz Technical University, Istanbul, Turkey*
[2]*Dept of Civil Engineering, New York University Abu Dhabi, Abu Dhabi, UAE*

In this study, the performance of a damaged dam was evaluated through a three-dimensional finite element model. The dam is located in Derbendikhan city of Northern Iraq and damaged during a 7.3 magnitude earthquake which was happened 30 kilometers south of Halabja city. Derbendikhan dam which was built between the years 1956-1961 is a clay-core rock fill dam. The damage of the dam was investigated at the site right after the earthquake and some cracks were observed in the main body of the dam. The main goal of this work is to present the results of the survey which was conducted at the site and investigating the damage development mechanism through a realistic three-dimensional finite element model of the dam. As complying with the observations at the site, the finite element analysis has shown that the primary failure mechanism is due to the separation of the core and rock fill sections at the downstream side of the dam.

Keywords: Damage development mechanism, Three-dimensional finite analysis, Google Earth, Mohr-Coulomb Model, Plastic analysis, Hydrodynamics.

1 INTRODUCTION

A 7.3 Mw earthquake occurred on 12.11.2017 with local time at 18:18:17 around the part of Bitlis-Zagros mountain range near the Iraq-Iran border, which is 30 Km south of Halabja province. A scientific team was formed from the members of Dicle University College of Engineering in order to investigate the effects of the earthquake on-site. The team, which included researchers Assist. Prof. Dr. M. Sefik Imamoglu from the Dept. of Geology of Mining Engineering, Assoc. Prof. Dr. Idris Bedirhanoglu from the Dept. of Structures of Civil Engineering, and CE Nihat Noyan from the Chamber of Civil Engineers/Diyarbakır branch, arrived at the earthquake affected region on 14.11.2017, and started investigations just after three days. Regions with significant damages were identified via consulting with the local authorities and examining earthquake records in Suleymaniyah Earthquake Center. Detailed reports on the earthquake, damage to the structures and geological aspects are covered in a field report (Bedirhanoglu and Imamoglu 2017), and also in two conference proceedings (Imamoglu *et al.* 2018, Bedirhanoglu *et al.* 2018). Eventually, the city of Derbendikhan and especially the dam near the city were found to sustain major damages. Thus, investigation efforts were first directed to the dam and later to the heavily damaged structures in Derbendikhan city. On-site investigation efforts on the dam revealed a large lateral crack at the top of the dam structure running through the entire width of the dam. The crack is on the downstream side and becomes wider as one moves towards to the center of the span of the dam. The crack indicates a suspected case of separation of the clay core and the surrounding rock-fill sections. In this study, a 3D finite element model with infinite boundary conditions and hydrodynamic effects is used to analyze the case. Mohr-Coulomb material model is used with characteristic values encountered in rock-filled clay core structures. Moreover, actual schematics of the structure are unavailable, so Google-Earth measurements and characteristic slope values with some information from a photo of the dam obtained from the site investigation, will be used to model the geometry of the dam. Nevertheless, it will be shown that observed damage is likely a result of the partial separation of core and surrounding filling.

2 DERBENDIKHAN DAM AND SURROUNDING AREA

Derbendikhan (Derbendixan) dam is located inside a narrow valley near Derbendikhan district along north-northeast direction (Figure 1). Derbendikhan district in turn is located at the near south of Suleymaniyah province. The 535 m long dam is constructed on a thick limestone layer which is split off deeply by the streams feeding the reservoir and which sits on an alternate formation of mudstone and silt stone. It was constructed between the years 1956-1961 as a rock fill dam with a clay impermeable core. It has a height of 128 m and has a reservoir capacity of 3 billion m^3 (Figure 1). Lack of information required to obtain data from open sources and as a result geometrical data were measured from Google Earth as seen in Figure 1 left side. The exact dimensions of the clay core are unknown and assumptions made on the slope and bottom width of the core sections.

Figure 1. Close-up view of Derbendikhan Dam view from right bank and geometrical dimensions.

3 EFFECT OF EARTHQUAKE ON DERBENDIKHAN DAM

The earthquake focused in this study occurred on 12.11.2007 at UTC 18:18:17 (21:18:17 local time) at 30 Km south of the city of Halabja, Suleymaniyah/Iraq on the Bitlis-Zagros rupture zone near Iraq-Iran border. The faults of the rupture zone caused an earthquake with a magnitude of 7.3 Mw. Due to the first announcement by USGS, the epicenter is located at 34.957°N 45.792°E with a depth of 25 km. Later the location is revised as farther to east 34.905°N 45.956°E and the depth is revised as 19 km. Thus the territorial address of the epicenter was shifted from Suleymaniyah/Iraq to Kermanshah/Iran. The earthquake was felt in a vast region and caused major damage to the dam structure. A prominent crack running through the span of the dam is visible on the crest of the dam as can be seen in Figure 2. This crack is on the downstream side and reaches up to 30 cm in the middle. Moreover, end cracks appeared on both sides of the dam.

These findings suggest rocking of the entire dam body along the stream direction and resulting separation of the core and filling sections. Also, a depression occurred on the downstream side of the crest.

Figure 2. Lateral downstream crack (green) and end cracks (red) occurred on Derbendikhan dam structure during the earthquake.

4 FINITE ELEMENT MODEL

We analyzed the dam structure with static (dead weight, hydrostatic pressure) and dynamic (ground motion, hydrodynamic) loads. The model consisted of a 3D finite element model of the structure with Mohr-Coulomb plasticity. The exact dimensions of the dam structure, the clay core section, exact material parameters, water level during the earthquake and ground acceleration records were unable to obtain. Thus, a series of assumptions were made. Assumptions about the geometry were made partly due to the Google Earth measurements as indicated before (Figure 1) and assumptions on the material parameters were made due to characteristic values for the Mohr-Coulomb model. They are selected as in (Table 1) similar to given values in the literature (Ghafari *et al.* 2014).

Table 1. Material properties.

	E (Mpa)	v	c (kPa)	φ	φ	ρ (kg/m^3)
Clay core	40	0.32	80	26	0	1600
Rock fill	80	0.28	20	37	8	2000
Ground	3000	0.28	5	42	7	2250

Assumed dimensions are approximately taken from the Google Earth measurements (Figure 3). Thus base width of the model is 570 m and crest width is assumed to be 20 m. The dam is assumed to be holding a water level of 120 m during the earth quake. The clay core's base width is assumed to be 70 m. The slopes of the filling and the core are around 25° (slope ratio ≡ 1/2) and 79° (slope ratio ≡ 5/1) respectively.

4.1 Hydrodynamic Effects and Modeling Considerations

Hydrodynamic effects are taken into consideration via an approximate method. The analytical solution of pressure distribution over the downstream slope of a 2D rigid dam with a height of H under constant acceleration of a_0 can be given per Chwang (1978) in Eq. (1),

$$p = C_p \rho_w H a_0. \tag{1}$$

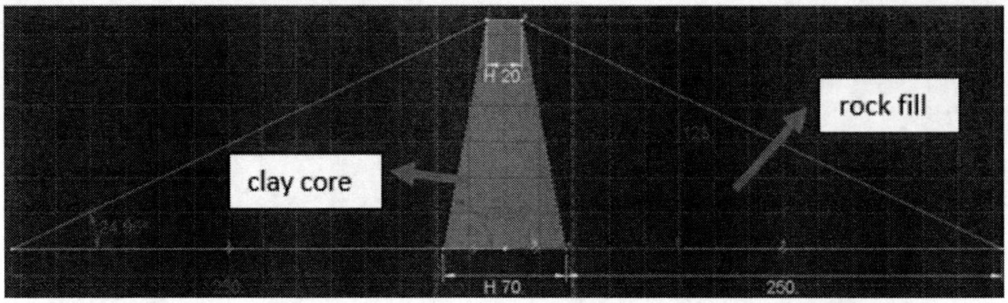

Figure 3. Assumed cross section for the Derbendikhan Dam used in the analysis.

Here C_p is the depth varying pressure coefficient (Figure 4) and ρ_w is the water mass density (1000 kg/m^3). In the assumption of the theory a certain portion of the water body moves with the dam thus an added associated mass is considered. In the analysis this contribution is considered over a thin section on the slope of the downstream side. The 120 m depth is divided into 10 equal sub-sections with a strip of 2.5 m wide. Each sub-section is then assigned with the average value of the increased density.

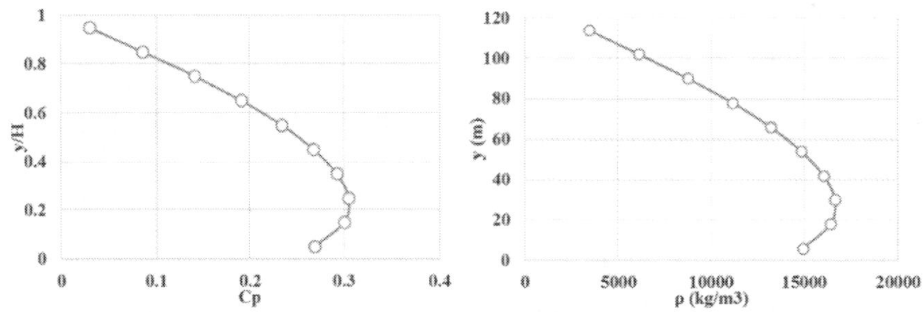

Figure 4. Variation of C_p Increased mass density due to hydrodynamic effects on downstream slope.

The body of the dam is modeled with 60039 C3D8R type 3D elements (Figure 5). Moreover, the ground and immediate surroundings are modeled with infinite elements in order to mitigate the effects of rigid boundary conditions (ABAQUS™ Analysis User Guide 2013). For this purpose 832 CIN3D8 type elements were used (green section in Figure 5). Out of plane dimensions are 535 m at the crest and 328 m at the base. Perfect bonding is considered between the dam body and the ground as well as between the rock-fill section and the clay core. Loads consist of the deadweight of the entire assembly with respect to the given densities in Table 1, hydrostatic water pressure on the downstream slope up to 120 m, hydrodynamic effects due to increased mass during the earthquake and finally the ground accelerations. As it was not possible to obtain the actual records ground accelerations are taken directly factorized (0.5) versions of the 1999 Kocaeli Earthquake whose components are shown in Figure 6. The load is applied in three steps with explicit dynamic analysis. In the first step, dead weight is applied due to self-weight of the dam structure only. In the second step hydrostatic pressure is superimposed over the first step. Finally in the third step ground acceleration is given. In the first and second steps, although explicit analysis is used, a quasi-static approach is taken and loads are linearly ramped to their final values with a total loading time of 400 secs.

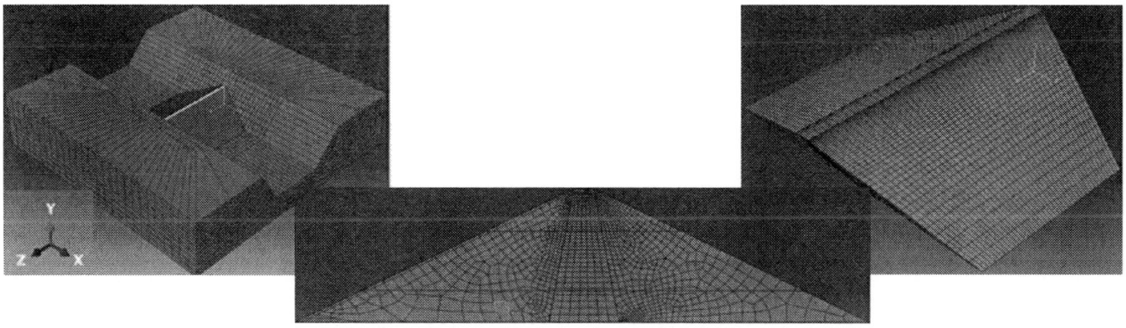

Figure 5. Complete dam assembly with the ground (infinite elements) and meshed dam cross-section and 3D body.

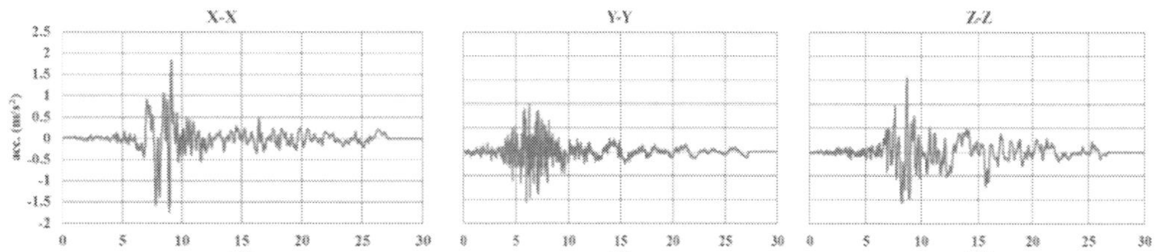

Figure 6. Components of ground acceleration used in the analysis.

5 RESULTS AND DISCUSSION

Analysis results are reasonable with the field observations (Figure 2). Figure 7 shows the vertical displacement field before (left) and after (right) the earthquake. It is seen that due to self-weight,

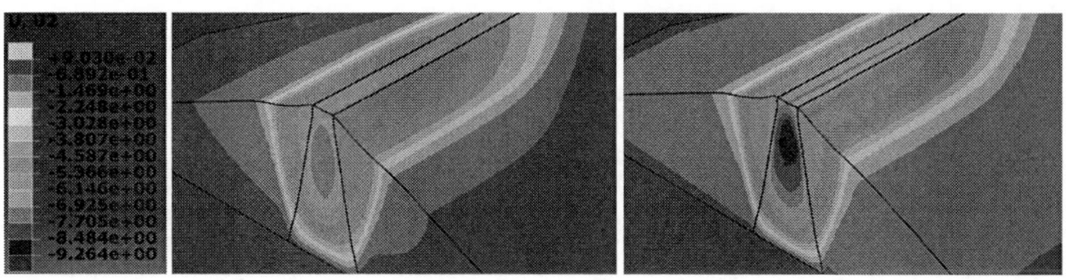

Figure 7. Vertical displacement (m) field before (left) and after the (right) earthquake.

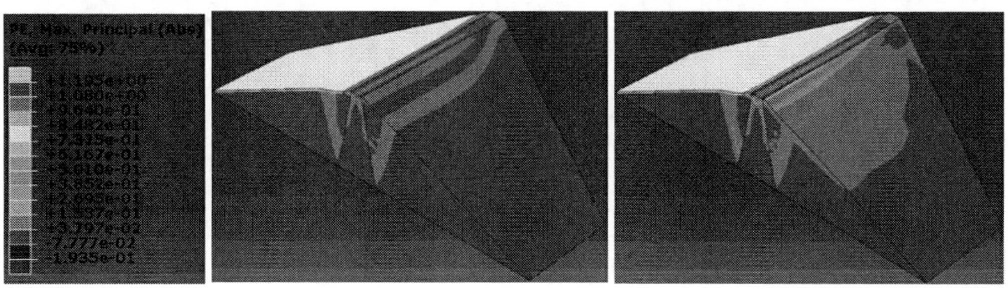

Figure 8. Absolute values of max principal plastic strains before (left) and after (right) earthquake.

consolidation type settling occurs and as observed in the field an additional depression around 1 m occurs around the crest region after the earthquake.

In plastic analysis, a measure of the concentration of max plastic strains can be assumed as probable fault (crack) zones. Figure 8 shows the absolute value of max plastic strains which also indicates the end slip bands which manifested as end cracks and the lateral downstream crack running along the length. Mohr-Coulomb material model seems to be reasonable when elaborate data on material parameters are not available. As it is seen that this simple nonlinear material model with assumed parameters and partly assumed dimensions are able to capture the observed damage mechanism. Thus we can conclude a possible separation of the core and rock fill sections which is prominent at the downstream side. The water pressure seems to limit the separation on the upstream side. Overall the damage mechanism fits with the on-site outer observations. Thus it can be concluded that the dam structure is non-negligibly damaged and care must be taken.

Acknowledgments

We deeply appreciate the support of Diyarbakir Branch of Chamber of Civil Engineers and its member Civil Engineer Nihat Noyan who took part in our investigation group; and Dicle University Physics Department's graduate student Mr. Qadri Ahmed Yusif who accompanied us with his vehicle and also acted as our translator and guide. We also thank Dr. Ramazan Hamza Muhammed from Duhok University, Department of Geology, and members of Duhok Seismology and Meteorology Institute; Head Dara Hasan Faraj, seismology Bakir M Saeed Ali, Aras Muhammed Tofig, Omid Muhammed Mahsun, Borhan İbrahim Salih for their companionship during the on-site investigations. We need to also mention the help of Union of Engineers Duhok Chamber members Abdul Emir and Diler Abo. We also want to thank Dicle University Engineering department and Dean of Engineering Faculty Prof. Dr. Abdullah Toprak, who made this site visit possible. Finally, we appreciate the financial support provided by MASAKA General Trading – Import and Export LTD Company, KRG, Iraq.

References

ABAQUS™ 6-13 Analysis User Manual, 2013. Retrieved from http://dsk.ippt.pan.pl/docs/abaqus/v6.13/books/usb/default.htm on April 2019.

Bedirhanoglu, I., and Imamoglu, M. S., 12 Kasım 2017 Halepçe Depremi Ön Değerlendirme Raporu, *Dicle Üniversitesi Mühendislik Fakültesi Maden Mühendisliği Bölümü Genel Jeoloji Anabilim Dalı Başkanlığı*, Nov 2017.

Bedirhanoglu, I., Mollamahmutoglu, C., and Imamoglu, M. S., *Damage of the Darbandikhan Dam During the Last 7.3 Halabja Earthquake*, 5th International Symposium on Dam Safety and Exhibition, 27-31 October 2018 (Accepted for publication).

Chwang, A. T., Hydrodynamic Pressures on Sloping Dams During Earthquakes. Part 2. Exact Theory, *J. Fluid Mech*, 87(2), 343-348, 1978.

Ghafari, A., Nikraz, H. R., and Sanaeriad, A., Finite Element Analysis of Deformation and Arching Inside the Core of Embankment Dams During Construction, *Australian Journal of Civil Engineering*, 14(1), 13-22, 2014.

Imamoglu, M. S., Bedirhanoglu, I., Mollamahmutoglu, C., *Evaluation of 12 November 2017 Mw 7.3 – 30 km south of Halabja Iraq-Iran Border Earthquake and its Effects on Darbandikhan Dam from A Geological Perspective,* 5th International Symposium on Dam Safety and Exhibition, 27-31 October 2018 (Accepted for publication).

Interdependence between Structural Engineering and Construction Management
Edited by Ozevin, D., Ataei, H., Modares, M., Gurgun, A., Yazdani, S., and Singh, A.

ISBN: 978-0-9960437-6-2

OPTIMISING COLLABORATIVE LEARNING AND GROUP WORK AMONGST TERTIARY STUDENTS

MEHRDAD ARASHPOUR[1], JULIA LAMBORN[1], and PARISA FARZANEHFAR[2]

[1]*Dept of Civil Engineering, Monash University, Melbourne, Australia*
[2]*Melbourne Medical School, University of Melbourne, Australia*

Team work and group assessments have become very common in the higher education context with the aim of enhancing communication skills and student engagement. However, group member accountability introduces a new challenge and constraint to groups in achieving their best performance. To investigate this issue, the current research analyzes the relationship between performance of individual members and group composure to test their relationship and dependency. Towards this research aim, individual and group marks in several units in were analyzed statistically. Empirical results show that there is tendency amongst same-gender students to team up together and this has a significant impact on group results. The current research contributes to improvement of theory and practice in higher education and identifies opportunities to enhance effectiveness of group work.

Keywords: Collaborative learning, Free-riding, Gender equality, Group assignments, Individual assessment, Performance and output, Project-based learning, Self-selected groups, Student experience, Higher education pedagogy.

1 INTRODUCTION

Previous research has shown the interdependence between group learning and graduate attributes related to employability and resilience (Arashpour *et al.* 2015). These justify the investment of higher education providers in developing modern learning infrastructure that facilitates 'learning in the round'. Interactive lecture spaces are often equipped with cutting-edge technology to enable peer presentation and group collaboration (Asok *et al.* 2017, Stache *et al.* 2017). Sometimes design of learning spaces can be self-configured by students to better suit collaborative activities (see Figure 1).

Although modern and interactive spaces provide the required infrastructure for collaborative work, the role of instructors in facilitating 'learning in the round' is still crucial (Arashpour and Aranda-Mena 2017, Khuzwayo 2018). Educators should encourage student groups to work efficiently on collaborative assignments with the aim of developing decision-making skills, communication and critical thinking (Adwan 2016). Towards this aim, important factors in designing group activities include but are not limited to optimal group size, member selection methods, proper scheduling of teamwork, and setting challenging tasks (Arashpour *et al.* 2018b)

Figure 1. Collaborative learning spaces for facilitating group works.

The aforementioned factors can significantly influence dynamics of teamwork; however, they do not necessarily optimize member interdependence, involvement, and fair division of labor (Caple and Bogle 2013). One variable that has been largely overlooked by the higher education literature is the influence of gender-inclusiveness. This variable can provide essential intellectual and social resources to groups and encourage both female and male members to develop a sense of responsibility towards the team success (Magogwe *et al.* 2015, Arashpour *et al.* 2018a). The focus of current research will be on impacts of gender inclusiveness on collaborative learning.

2 BACKGROUND

The main objectives of group work in classrooms are to develop skills including communication, cooperation, collaboration, and compromise (Arashpour *et al.* 2017a). Considering the importance of group work and its objectives, tertiary educators need to make every effort to understand dynamics of group activity in order to enhance student experience. The following sections provide background information from the literature on two important variables of member assignment and gender differences that have major influence on group dynamics.

2.1 Group Composition

There are several methods by which students can be assigned to working groups including self-selection, random assignment, hybrid methods, and purposeful assignment. Self-selected and instructor-selected groups are both very common in the tertiary education context (Ficano 2018). In self-selected groups, students can freely select the members they want to work with. Previous research has shown the superior performance of self-selected groups in terms of communication between members, confidence in the collective competence, and enthusiasm to work together (Khuzwayo 2018). Self-selected groups, however, can face problems with regard to time management, completing work of other members, and failure to follow commitments (Woolard 2018).

Instructor-selected groups are usually formed by random assignment of students to different groups. Furthermore, group composition can be determined by adopting a hybrid approach by which students first have the opportunity to form groups themselves and then instructors assign remaining students to existing or new groups (Woolard 2018). Another approach to form instructor-selected groups is to purposefully assign students based on some assignment-related criteria. Instructor-selected groups minimize the problems of free riding (Ali *et al.* 2018), and boredom caused by working with same circles of friends (Nasirian *et al.* 2018). Instructor-selected groups, however, are not usually favored by students, and members are less likely to work together in future group activities (Thrall *et al.* 2016). Reviewing the tertiary education literature suggests that member assignment methods have significant impacts on group dynamics and should be chosen based on course and cohort characteristics (Stache *et al.* 2017).

Self-efficacy beliefs can explain many gender differences in learning and academic performance. Self-efficacy determines whether or not female and male students believe in their learning capability (Stache *et al.* 2017). Theorists contend that past research related to gender differences in academic learning has frequently centered on individual performance and has often neglected group activities and their dynamics (Zitzelsberger *et al.* 2015). The objective and contribution of the current study is to bridge the existing gap by focusing on group dynamics in higher education under influence of important variables such as gender differences and member selection.

3 RESEARCH METHOD

Different higher education courses were selected in which there were both group assessments and individual assessments including projects and final exams. Collected student performance data were analyzed to test the three developed research hypotheses. Altogether, 60 students comprising 25 males and 35 females participated in this study. The students were in year two to four of their undergraduate studies in two public Australian universities. Table 1 shows the median and interquartile ranges (IQR) of assignment marks for the study population sample.

Table 1. Median and interquartile ranges of assignment marks for the study population.

N= 60 **Female 41.6% Male 58.4%**	**Marks Median** (IQR 25%-75%)
Individual Project	**85** (75.6-92.5)
Group Assignment	**80** (76-88)
Final Exam	**80** (68.5-86)
Total Score	**81.5** (72.2-87)

4 ANALYSIS AND RESULTS

Previous research has shown the importance of motivational beliefs on academic performance of tertiary students (Caple and Bogle 2013). Among the others, self-efficacy, recognition of task value, and learning goal orientation are the most influential motivational beliefs. On this basis, pedagogical interventions to enhance motivational beliefs can be beneficial. However, such interventions have not been customized to reflect unique characteristics of learning in female and male students (Ali *et al.* 2018). Some research in higher education literature claims that motivational beliefs and therefore performances of female and male students in group activities are not significantly different. Towards the aim of this research, the validity of the above claim is tested by developing the following research hypothesis:

Null Hypothesis: Performances of female and male students in group activities are not significantly different.

As discussed, individual and group marks of 60 students were analyzed across a variety of assignments including projects, quizzes and exams. The first null hypothesis proposes that performance of female and male students in group activities are not significantly different. In the first step to test the hypothesis, descriptive statistics including median and interquartile ranges (25%- 75%) were used to summarize data (see Table 2). As can be seen, for the studied population sample, the median (50th percentile) of achieved marks has been higher for female students in both individual and group assignments.

Table 2. Effect of gender difference on assignment outcomes.

	Female Median (IQR)	**Male Median (IQR)**	**Effect Size**	**P-value (t-test)**
Individual Project	**90** (77.5-92.5)	**82.5** (75-92.5)	0.7	0.1
Group Assignment	**88** (80-92)	**80** (72-84)	0.8	0.05*
Final Exam	**82** (70-90)	**77** (64-84)	0.5	0.1
Total mark	**85** (73.5-89.5)	**78** (70-85)	0.75	0.02*

Although descriptive statistics are helpful in measuring central tendency and variability of data, inferential statistics are required for an in-depth interpretation of trends (Arashpour *et al.* 2018c). On this basis, t-test was deployed to measure the significance of difference between achieved marks by female and male students. Furthermore, Cohen's effect size was calculated to translate the difference between the two groups into standard deviation units.

As Table 2 shows, the p-value derived from t-statistics indicates the significant difference between performance of female and male students in group activities. More importantly, the effect size of 0.8 suggests that performance of the two groups differ by 0.8 standard deviations. Both descriptive and inferential statistics result in rejection of the null hypothesis and confirm that performance of female and male students in group activities are significantly different.

5 CONCLUSIONS

The optimization of collaborative learning and student experience in group work has long been the focus of research in higher education literature. However, only few studies have adopted a holistic approach that considers important performance variables such as gender inclusiveness and selection interventions in collaborative learning (Arashpour *et al.* 2017b). In order to bridge this gap, the current research analyzed individual and group performances of a large population sample of tertiary students to shed light on dynamics of group work in higher education. Results of analysis were used to test three research hypotheses around performance of different genders in group work, relationships between individual performance of associated members and group outcomes, and finally the role of instructor interventions to form gender-inclusive teams.

Using both descriptive and inferential statistics, it was found that female students generally outperform their male counterparts in group assignments. This superiority creates a tendency to form all-female groups with the aim of achieving higher marks in group exercises. Therefore, instructors' interventions will be required to maximize gender balance in all groups. Considering the diversity of motivational beliefs and self-regulation in male and female cohorts, such interventions will optimize learning goal orientation, task value, and self-efficacy that are key ingredients for group output and performance.

References

Adwan, J., Dynamic Online Peer Evaluations to Improve Group Assignments in Nursing E-Learning Environment, *Nurse Education Today,* 41, 67-72, 2016.

Ali, M. F., Tahir, L. M., Said, M. N. H. M., Junaidi, J., Atan, N. A., and Hahsan, A., *Promoting Staged Self-Directed Learning (SSDL) Among Malaysian Tertiary Learners Through Online Discussion in Completing Group Assignment,* 2017 IEEE International Conference on Teaching, Assessment and Learning for Engineering, TALE 2017, Institute of Electrical and Electronics Engineers Inc, 2018.

Arashpour, M., Sagoo, A., Wingrove, D., Maqsood, T., and Wakefield, R., *Single Capstone or Multiple Cornerstones? Distributed Model of Capstone Subjects in Construction Education,* Proceedings of 8th International Structural Engineering and Construction Conference, ISEC 2015, ISEC Press, 2015.

Arashpour, M. and Aranda-Mena, G., *Curriculum Renewal in Architecture, Engineering, and Construction Education: Visualizing Building Information Modeling Via Augmented Reality,* 9th International Structural Engineering and Construction Conference: Resilient Structures and Sustainable Construction, ISEC 2017, ISEC Press, 2017.

Arashpour, M., Too, E., and Le, T., *Improving Productivity, Workflow Management, and Resource Utilization in Precast Construction,* 9th International Structural Engineering and Construction Conference: Resilient Structures and Sustainable Construction, ISEC 2017, ISEC Press, 2017a.

Arashpour, M., Bai, Y., Aranda-Mena, G., Bab-Hadiashar, A., Hosseini, R., and Kalutara, P., Optimizing Decisions in Advanced Manufacturing of Prefabricated Products: Theorizing Supply Chain Configurations in Off-Site Construction, *Automation in Construction*, 84, 146-153, 2017b.

Arashpour, M., Bai, Y., Kamat, V., Hosseini, R., and Martek, I., *Project Production Flows in Off-Site Prefabrication: BIM-Enabled Railway Infrastructure*, 35th International Symposium on Automation and Robotics in Construction and International AEC/FM Hackathon: The Future of Building Things, ISARC 2018, International Association for Automation and Robotics in Construction I.A.A.R.C), 2018a.

Arashpour, M., Kamat, V., Bai, Y., Wakefield, R., and Abbasi, B., Optimization Modeling of Multi-Skilled Resources in Prefabrication: Theorizing Cost Analysis of Process Integration in Off-Site Construction, *Automation in Construction*, 95, 1-9, 2018c.

Arashpour, M., Miletic, M., Williams, N., and Fang, Y., *Design for Manufacture and Assembly in Off-Site Construction: Advanced Production of Modular Façade Systems,* 35th International Symposium on Automation and Robotics in Construction and International AEC/FM Hackathon: The Future of Building Things, ISARC 2018, International Association for Automation and Robotics in Construction I.A.A.R.C), 2018b.

Asok, D., Abirami, A. M., Angeline Cv, N., and Lavanya, R., *Active Learning Environment for Achieving Higher-Order Thinking Skills in Engineering Education,* 4th IEEE International Conference on MOOCs, Innovation and Technology in Education, MITE 2016, Institute of Electrical and Electronics Engineers Inc., 2017.

Caple, H., and Bogle, M., Making Group Assessment Transparent: What Wikis Can Contribute to Collaborative Projects, *Assessment and Evaluation in Higher Education*, 38(2), 198-210, 2013.

Ficano, C. K. C., Identifying Differential Benefits from A Flipped-Group Pedagogy in Introductory Microeconomics, *International Review of Economics Education*, 2018.

Khuzwayo, M. E., Assessment of Group Work in Initial Teacher Education and Training, *South African Journal of Education,* 38(2), 2018.

Magogwe, J. M., Ntereke, B., and Phetlhe, K. R., Facebook and Classroom Group Work: A Trial Study Involving University of Botswana Advanced Oral Presentation Students, *British Journal of Educational Technology,* 46(6): 1312-1323, 2015.

Nasirian, A., Arashpour, M., and Abbasi, B., *Multiskilled Human Resource Problem in Off-Site Construction,* 35th International Symposium on Automation and Robotics in Construction and International AEC/FM Hackathon: The Future of Building Things, ISARC 2018, International Association for Automation and Robotics in Construction I.A.A.R.C), 2018.

Stache, J. M., Barry, B. E., Klosky, J. L., and Freisinger, G. M., *Assessing Individual Temperament and Group Performance in a Project-Based Learning Experience,* 124th ASEE Annual Conference and Exposition, American Society for Engineering Education, 2017.

Thrall, G. C., Coverdale, J. H., Benjamin, S., Wiggins, A., Lane, C. J., and Pato, M. T., A Randomized Controlled Trial of Team-Based Learning Versus Lectures with Break-Out Groups on Knowledge Retention, *Academic Psychiatry,* 40(5), 755-760, 2016.

Woolard, N. A., Rethinking Management Group Projects with The Adaptive Leadership Model: The Lesson is The Process, *Journal of Education for Business*, 2018.

Zitzelsberger, H., Campbell, K. A. , Service, D., and Sanchez, O., Using Wikis to Stimulate Collaborative Learning in Two Online Health Sciences Courses, *Journal of Nursing Education,* 54(6), 352-355, 2015.

Interdependence between Structural Engineering and Construction Management
Edited by Ozevin, D., Ataei, H., Modares, M., Gurgun, A., Yazdani, S., and Singh, A.

ISBN: 978-0-9960437-6-2

CURRENT PRACTICES OF EXPERIENTIAL LEARNING IN CONSTRUCTION AND TECHNOLOGY PROGRAMS

AHMED ABDELATY[1] and K. JOSEPH SHRESTHA[2]

[1]*Dept of Construction, Southern Illinois University Edwardsville, Edwardsville, USA*
[2]*Dept of Engineering, Engineering Technology, and Surveying, East Tennessee State University, Johnson City, USA*

Construction education is dynamic and practice oriented. As such, effective construction programs require significant collaboration with the construction industry. This collaboration, in the form of internship or cooperative programs, increase the student readiness for the job market by providing valuable field experience. Construction programs in the United States (US) established several internship requirements that range from being optional to multiple required internships. This study focuses on scanning the current internship requirements set by construction and engineering technology programs in the US by gathering information including; 1) Number and length of required internships, 2) Internship prerequisites, 3) Internship deliverables, 4) assessment method. The outcome of this study is expected to help construction programs improve their internship or cooperative requirements by considering the prevailing practices developed by other schools. Additionally, the study provides recommendations to enhance the effectiveness of internship for positive experiential learning.

Keywords: Internship, Field-experience, Cooperative education, Teaching methodology, United States.

1 INTRODUCTION

The construction industry is characterized by being a dynamic and transformative to new and effective practices. As such, preparing future construction professionals requires the inclusion of a dynamic and flexible curriculum. One of the best practices to create such a curriculum is to include experiential learning as one of the learning instruments. Experiential learning happens when the learner gains knowledge through the transformation of experience (Kolb 2014). In a construction management curriculum, this transformation can be effectively achieved through internships and cooperative education (co-ops) experiences. Many construction management programs require students to complete at least one internship or co-op before graduation. However, the requirement varies from one program to another.

Maintaining a continuous connection between construction management programs and the industry is vital to enhance the quality of future graduates. Some programs have realized the importance of this continuous connection and includes an extensive co-op or internship requirement in their curriculum while other programs are still lagging. One of the advantages of including a strict internship or co-op requirement is exposing students to actual construction experience outside the classroom. This allows students to apply knowledge gained in the classroom to real-life construction projects. It further enhances students' classroom learning experience as they will be better equipped at identifying and understanding the potential applications of classroom knowledge in their future work environment.

From the university's perspective, programs benefit from keeping an active link with the industry to validate and improve their curriculum. However, requiring several credit hours internship or co-op requirement may pose a challenges on some students to graduate on time especially since several programs require such internships or co-op to be completed one or two semester before graduation. Additionally, some programs may struggle with distributing their required credit hours for graduation between experiential learning-based courses and other required traditional courses.

The objective of this study is to survey the internship and co-op requirements of construction engineering, technology, and management programs from Southeast region and Great Lakes region of Associated Schools of Construction (ASC) which includes the state of Alabama, Florida, Georgia, Mississippi, North Carolina, South Carolina, Virginia, and Tennessee, Illinois, Ohio, Indiana, Michigan, Kentucky, and Wisconsin. Based on the data collected from public websites, the authors analyze the advantages and disadvantages for different internship requirements. The authors also compare the similarities and differences in the internship requirements between the programs in Southeast region and Great Lakes region. Additionally, recommendations for programs that plan to include an internship or co-op requirement are presented.

2 LITERATURE REVIEW

The experiential learning model is not something new. The theory has been proposed and implemented more than 40 years ago. Kolb (2014) reformulated the experiential learning theory. There are 4 structural dimensions underlying the process of experiential learning. Kolb's model involves four areas of experiential learning 1) reflective observation, 2) active experimentation, 3) abstract conceptualization and 4) concrete experience. These areas would be encountered through an experiential learning activity such as an internship or a co-op experience. An internship has innumerous definitions; however, it can be defined as a supervised work experience with an intention of encountering an experiential learning strategy (O'Neill 2010). According to O'Neill (2010), an internship should be a deliberative form of learning that has a cycle of feedback for improvement to support learning goals and objectives.

Tener *et al.* (2001) analyzed the effectiveness of experiential learning from internships in construction engineering and management. The study examined reports submitted by 170 construction engineering and management students after finishing their internships. The authors analyzed the language in the reports and match it to the four learning modes suggested by Kolb's experiential learning model. Active experimentation and reflective observation are the highest learning modes encountered during the internship. The study also reported a significant development in student's self-efficacy because of the internship experience.

In disciplines close to construction management, McCarthy and McCarthy (2006) integrated experiential learning into business curricula. Experiential learning as a learning pedagogy gives an opportunity to the students to learn from personal encounter and actual work experience. This learning process occurs through cooperative education placements, internships, job shadowing, and classroom-based hands-on activities. McCarthy and McCarthy (2006) investigated the effectiveness of incorporating job shadowing in a junior level business communication course. Sixty-eight students participated in an 8-hour job shadowing assignment. Students were asked to write a reflective paper on what they experienced. An exit survey indicated that 100% of the students found that job shadowing was helpful or very helpful activity (McCarthy and McCarthy 2006). Similarly, Gault *et al.* (2000) investigated the relationship between the student success and exposure to experiential learning in the form of internship. The study shows that experiential learning plays an important role in preparing students for entry job-level market. Through a survey, students who interned before were able to secure jobs faster than students who never interned. Students who previously had an internship reported securing higher monetary compensation from their first job compared to students with no internship experience.

Doel (2008) describes an approach that aims at maximizing the benefits gained through industry internships. The approach adopted the School of Engineering at the University of Queensland enhances the students' reflection on their internship experience. Students are required to report logs and critical learning activities. Afterwards, students are required to analyze their significant learning activities by analyzing the situation, self-impact, interpretation, and decision. Doel (2008) also notes an important observation related to experiential learning: students actually learn from reflecting on their experience and taking responsibility.

3 OBJECTIVES AND METHODOLOGY

The main objective of this study is to investigate the adoption and implementation of internship and co-op education in the US construction engineering, technology, and management programs. This objective is achieved by: 1) collecting data on the internships and co-op requirements of the construction programs in Southeast and Great Lakes regions, 2) analyzing the variation of such requirements using descriptive statistics, 3) compare the requirements among the programs from the two regions, 4) providing recommendations based on the prevailing internship or co-op requirements.

4 DATA

The authors used ASC's regional directory to identify the colleges and universities in Southeast and Great Lakes regions that offer associate or bachelor's degrees in construction engineering, technology, or management. The Southeast region consist of eight states (Figure 1): Alabama, Florida, Georgia, Mississippi, North Carolina, South Carolina, and Tennessee; the Great Lakes Region consists of six states: Illinois, Indiana, Kentucky, Michigan, Ohio, and Wisconsin. The authors studied programs from 15 colleges and universities from Southeast region and 23 from Great Lakes region.

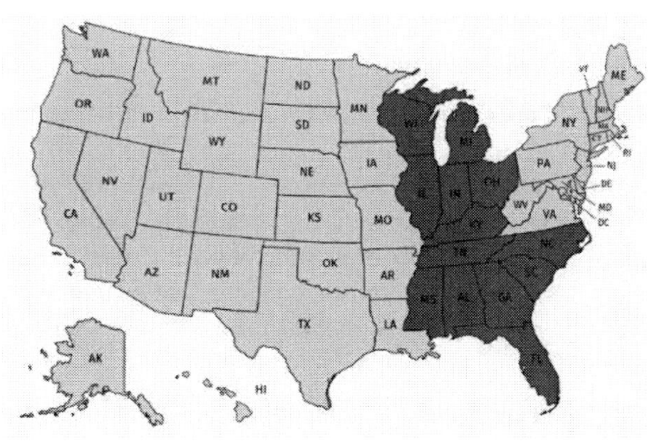

Figure 1. States covered in the survey.

Most of the data were obtained from their course catalog and program overview pages from their websites. The data collected include a) number and length of required internships, b) internship prerequisites, c) internship deliverables, and d) assessment methods.

5 INTERNSHIP REQUIREMENTS

The results of the survey are summarized in Table 1. More than half the programs surveyed have internship and/or co-op requirements for graduation. More percentage of programs in Great Lakes region require internship and/or co-op. While median number of internship and/or co-op hour requirement in Great Lakes regions is higher, the average is higher in the Southeast region. Overall, the median number of hours required by the programs in both regions combined is 400 hours.

Table 1. Descriptive statistics of internship and co-op requirements.

Descriptions	Southeast	Great Lakes	Combined
Total Number of Programs Surveyed	27	23	50
Average Internship and/or Co-op Hours	530	401	438
Median Internship and/or Co-op Hours	400	490	400
Percentage of Programs Requiring Internships or Co-Op	48%	83%	64%

Figure 2a shows that most of the programs do not require any internship and/or co-op while 16 programs surveyed requires at least one. Some programs required up to 5 internships. Figure 2.b shows that most of the programs prefer internships over co-op.

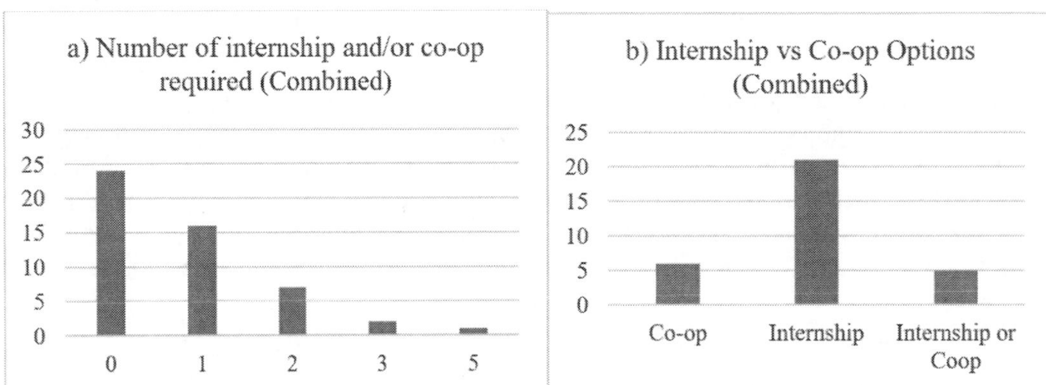

Figure 2. Number of internship and/or co-op requirements in the combined region.

Figure 3 shows the distribution of the number of hours required for graduation in the construction programs in Southeast and Great Lakes regions and their combination. The number of hours required varies widely in the programs surveyed. In all cases, 400 hours is the most common number of hours for internship and/or co-op. Those that require any such hours, the requirement is as low as 80 hours to as high as 800 hours. The second most common number of such hours is 300.

The study further noticed variation in requirements to take the internship/co-op positions. Some programs required junior or senior standings, others have various other course as prerequisites before enrolling for the class, and others require departmental approval. Further the total number of credit hours for the internship also varied from 1 to 12 credits when it is required.

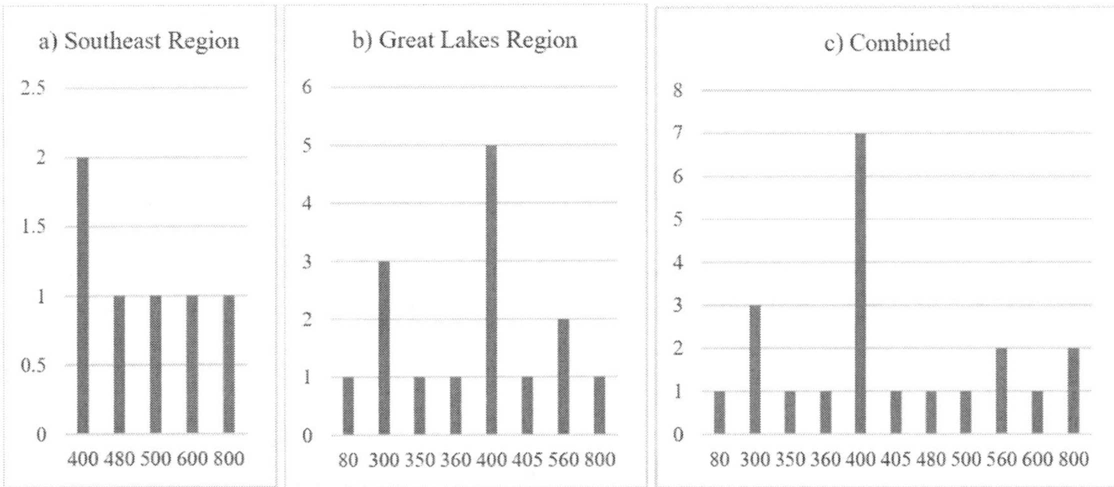

Figure 3. Distribution of number of internship and/or co-op required for graduation.

Many programs require a formal presentation and/or reports from the student and documentations from the company once internships and/or co-op are completed. This is a good practice to have.

6 DISCUSSION AND CONCLUSION

Based on the data collected, the authors believe that a 400 hours internship is a reasonable and important requirement for graduation. Requiring more than one internship or co-op can create graduation hurdles especially when the construction market is not growing. Students may face difficulties finding internship opportunities to graduate on time. In addition, academics have noted that many states in the United States are pushing universities to lower the number of credit hours required for finishing a degree. This means that some schools may face challenges to redesign their curriculum.

Few programs require finishing 12 credit hours of internship or co-op experience. This requirement is considered one of the toughest requirements observed among all programs surveyed. On the other hand, not requiring any internship to graduate can significantly lower the level of readiness of some students with no prior construction experience. Based on the data collected, the prevalent requirement is one internship for 400 hours.

Most of the programs require interns to be in a junior or senior standing. Few other programs require specific coursework on safety or specific certification such as OSHA 10. The authors believe that having a safety coursework or certification is a good practice to ensure the interns safety in construction jobsites. Some programs have a publicly available internship handbook which describes the intern, industry mentor, and academic supervisor responsibilities and roles. The authors believe that this a good practice to identify the responsibilities and expectations before the start of the internship.

One major concern observed in regards with the effectiveness of internships is how the number of hours are counted. There is no discussion in any of the internship handbooks on this topic. The number of hours required for the internship completion should be tied to the productivity and type of tasks assigned to the intern. Only one university have segregated the maximum number of hours for counted from various types of works such as management, trades, labor, and volunteering experience. As such, it's important to communicate the level of expectation with the interns and their industry mentors regarding the level of work and tasks assigned. It is also important to note that Hajshirmohammadi (2017) indicated that experiential learning activities should be assessed by the level of understanding compared to control group. While this may be challenging to incorporate in construction and technology internships, the idea of deeper level of internship assessment should be explored.

As for future research, it is recommended to conduct a nationwide survey of the internship requirement and practices. In addition, a questionnaire survey that would target academic supervisors, students, and industry mentors would greatly reveal important insights regarding the best practices of experiential learning.

References

Doel, S., *Measuring Student Reflection During Engineering Internships*, WACE Asia Pacific Conference, 2008.

Gault, J., Redington, J., and Schlager, T., Undergraduate Business Internships and Career Success: Are They Related?, *Journal of Marketing Education,* 22(1), 45-53, 2000.

Hajshirmohammadi, A., Incorporating Experiential Learning in Engineering Courses, *IEEE Communications Magazine*, *55*(11), 166-169, 2017.

Kolb, D. A., Experiential Learning: Experience As The Source of Learning and Development, *FT Press,* 2014.

McCarthy, P. R., and McCarthy, H. M., When Case Studies Are Not Enough: Integrating Experiential Learning into Business Curricula, *Journal of Education for Business,* 81(4), 201-204, 2006.

O'Neill, N., Internships as a High-Impact Practice: Some Reflections on Quality, *Peer Review,* 12(4), 4-8, 2010.

Tener, R. K., Winstead, M. T., and Smaglik, E. J., *Experiential Learning from Internships in Construction Engineering,* Proceedings of the 2001 ASEE Annual Conference & Exposition, Albuquerque, NM, June 24 - 27, 2001.

Interdependence between Structural Engineering and Construction Management
Edited by Ozevin, D., Ataei, H., Modares, M., Gurgun, A., Yazdani, S., and Singh, A.

ISBN: 978-0-9960437-6-2

ASSESSMENT OF LEVEL OF STUDENTS' SATISFACTION WITH QUALITY OF MANAGEMENT SERVICES IN PRIVATE STUDENT ACCOMMODATION

GETRUDE MAMPURU and CLINTON AIGBAVBOA

Dept of Construction Management and Quantity Surveying, University of Johannesburg, Johannesburg, South Africa

This paper presents findings on students' satisfaction with one of the biggest private student accommodation in Johannesburg, South Africa. Management services play an important role in the level of student satisfaction. Without good management services and a poor interaction between student and management, the quality of services provided will have an adverse effect on the level of student satisfaction. In assessing overall student satisfaction, the quality of services offered to students is the significant element to be considered. Hence the purpose of this study was to determine the level of student satisfaction with the quality of management services provided to them in private student accommodations. Data that was received was analyzed by means of frequencies and tables, which employed the use of descriptive statistical procedures and Mean Item score. Analysis was based on the 58 useable questionnaires obtained from students out of 60 questionnaires issued. Findings from the study revealed that electricity supply, Wi-Fi connection, and general rules are among the most important aspect of student satisfaction/dissatisfaction with quality of management services. Thus, the study recommends that the management engages with students regarding their problems, and timely deal with arising complains. Furthermore, management should provide high quality Wi-Fi connection that will be suitable for all student accommodated, and also provide proper supply of electricity as well as increase the supply of hot water.

Keywords: University, Education, Housing, Performance, Higher institutions.

1 INTRODUCTION

Student accommodation is housing built for students or to accommodate solely students while pursuing their educational endeavors. Private student accommodation refers to housing that is not directly provided by a specific higher learning institution but that is provided by private individuals, foundations, and corporations (Bekurs 2007). The main purpose for a student accommodation is to create an environment where students can learn and live (Fields 2011). Student accommodation can be referred to by such terms as: student housing, dormitory, units, student apartments, hostel, student halls, and college housing. Charlton (2004) described housing as much more than a physical structure, according to him management services, environmental services, culture are part of housing. Housing is one of the most important needs and ranks on top, hence Omole (2001) says is the most important thing in human existence.

According to Amole (2005), student residence satisfaction simply refers to the fulfilment of students' needs, whether it is through the provision of management services or physical features of the residence. Amole (2005) continued to explain that satisfaction is when their rooms have sense of privacy. Another meaning of student satisfaction with residence is an environment that promotes positive social process, and adequate amenities which students respond well to. A proper guidance for residence management and university housing department is lacking because previous studies do not focus on student satisfaction residence especially in South Africa.

The nature in which management interacts with students determines the level of student satisfaction with a student accommodation. According to Amole (2009), one of the factors in student satisfaction with housing is management services. He included features such as fees, attitude of residence management, and accommodation rules. Management services such as dealing with student's problems, staff availability, staff competence also contribute to students' satisfaction with private student accommodation (Abdullah 2009). The quality of management services is the most important factor presented by private student accommodation owners to students' that defines overall satisfaction (Torbica and Stroh 2001). Yet, this is the part that most private student accommodation owners show the poorest performance, hence

The quality of management services determines the level of students' satisfaction. Although student accommodation is a significant element toward student academic performance, it has not received sufficient attention, thus most private students' accommodations continue to deliver poor quality of management services. The quality of management services plays an important role in housing satisfaction, even though housing preferences and satisfaction are subjective and influenced by a number of factors such as financial situation, cultural and social background, and different phases of life (Thomsen and Eikemo 2010).

In terms of good management services in a student accommodation, proper accommodation will encourage a study environment that is quiet, privacy and provide security, and assist the university housing department satisfy students' expectations for private student accommodation (Hassanain 2008). Student housing managers and university administrators are unable to provide students with quality of management services, mainly because previous studies do not address wide range of satisfaction with student housing. Danso and Hammond (2017) emphasized that management of services in private student accommodation play a significant role in addressing the housing challenges in Higher Educational Institutions. According to Amole (2005) a comfortable good and well managed student housing contributes to good and academic performance of students.

Ajayi *et al.* (2015) further explained that the improvement in academic excellence of students is dependent on students' satisfaction with the quality of management services provided in private student accommodations, hence it is important that more research is directed to finding out student satisfaction with residence. Due to poor management services, there had been reported cases of students getting sick in the residence because of poor sanitary conditions Ajayi *et al.* (2015). Aigbavboa *et al.* (2014) emphasized that health and performance in every way are important factors in a well-managed student accommodation. In order for students to learn effectively, a decent student accommodation needs to be provided whether by private or public student housing (Mugambwa *et al.* 2016).

2 METHODOLOGY

This study was a quantitative research, a well-structured questionnaire was used as means of collecting data for this research study, questionnaire was developed from the extensive literature reviewed. The researcher approached all relevant participants, which were students living in a privately-owned residence in Johannesburg, South Africa. The variables in the table were identified through the review of literature and were given to students to rate. A Likert scale of 1 to 5 was used were respondents were asked to mark the questions.

The data obtained from the completed questionnaires was presented and analyzed by means of frequencies, cross-tabulations and tables. The analysis employed the use of descriptive statistical procedures and Mean Item Score to interpret and draw conclusions from the data collected. This study was designed to assess the hypothesis that management services in student accommodation will affect student satisfaction with the residence. The rankings of the mean item score made it possible to compare the literature. Since one cannot really measure satisfaction, using descriptive analysis to rank the level of satisfaction is the best way.

The Mean Item Score (MIS) was used to convert the responses from the Likert Scales into easily analyzable data.

The mean item score (MIS) was calculated for each item as in Eq. (1);

$$\text{MIS} = \frac{1n_1+2n_2+3n_3+4n_4+5n_5}{\sum N} \quad (1)$$

Where;

n1 =number of respondents for Not satisfied

n_2=number of respondents for Slightly satisfied

n3 =number of respondents for Neutral

n4 =number of respondents for Satisfied

n5 =number of respondents for Very satisfied

N =Total number of respondents

Following the mathematical calculation of the MIS, the criterions were ranked in descending order (from highest to lowest)

3 FINDINGS AND DISCUSSION

3.1 Background Demographic Information

Based on background information, respondents from rural areas ranked the highest three as: “24hour security guard system” (MIS=4.14), “Fire safety system” (MIS=4.00), and “Electrical lighting in the apartment” (MIS=3.79). The lowest three ranked were: “Quality of Wi-Fi connections” (MIS=2.21), “Working air-conditioning in the apartment” (MIS=1.79), and “ATM machines in the apartment” (MIS=1.21).

Based on respondents from the township, the highest three included: "Electrical lighting in the apartment” (MIS=4.11), “24hour security guard system” (MIS=4.00), and “Electricity supply” (MIS=3.78). The lowest three consisted of: “ATM machines in the apartment” (MIS=1.28), “Quality of Wi-Fi connections” (MIT=2.17), and “Cold and hot water supply” (MIT=2.03).

Results obtained from informal settlement respondents revealed that the highest three rankings were: “24hour security guard system” (MIS=5.00), “Electrical lighting in the apartment” (MIS=5.00), and “Conveying systems” (lifts and or stairs) (MIS=5.00). The three lowest ranked were: “Least Measures taken with non-rent payment” (MIS=1.00), “Public Washroom” (MIS=1.00), and “Maintenance management practices” (MIS=1.00).

Respondents’ from the urban area ranked the highest three as: “Electricity supply” (MIS=4.57), “Electrical lighting in the apartment” (MIS=4.14), “24hour security guard system” (MIS=3.86). The lowest three were: “Rules regarding visitors” (MIS=2.00), “Working air-conditioning in the apartment” (MIS=1.57), and “ATM machines in the apartment” (MIS=1.00).

Referring to Table 1, the total shows that "Electrical lighting in the apartment" had the highest rank score of (MIS=4.05), the second highest ranked was the "24hour security guard system" with (MIS=4.03), and "Electricity supply" (MIS=3.86,) ranked the least three included: Rules regarding visitors (MIS=1.98), Working air-conditioning (MIS=1.86), and lastly at number 22 ATM machines in the apartment (MIS=1.22).

The basis of this study was area of origin, to measure if students from different backgrounds had the same level of expectations.

Table 1. Management services of a private student accommodation.

Management services	TOTAL	
	MIS	Rank
Electrical lighting in the apartment	4.05	1
Electricity supply	3.86	3
24hour security guard system	4.03	2
Fire safety	3.76	4
Conveying systems (lifts and or stairs)	3.74	5
Mini markets in the apartment	3.31	6
Quality of repaired works	3.26	8
First aid kit	3.19	10
Maintenance of appliances	3.28	7
Public Washroom	2.81	11
Garbage disposal	3.22	9
Maintenance management practices	2.76	14
Measures taken with non-rent payment	2.78	13
Cleaning services	2.81	11
Laundry rooms	2.79	12
Management response to complaints	2.64	15
Recreation room and support services	2.59	16
Quality of Wi-Fi connections	2.22	18
Cold and hot water supply	2.31	17
Rules regarding visitors	1.98	19
Working air-conditioning in the apartment	1.86	20
ATM machines in the apartment	1.22	21

3.2 Discussion

Ranked at the top five were: "Electrical lighting in the apartment", "24hour security guard system", "Electricity supply", "Fire safety", and "Conveying systems (lifts and or stairs)", ranked at the bottom 5 were: "ATM machines in the apartment", "Working air-conditioning in the apartment", "Rules regarding visitors", "Quality of Wi-Fi connections", and "Cold and hot water supply."

These results agreed with the study done by Ajayi (2015) who ranked "Electricity supply" as the most important aspect of student satisfaction with quality of management services, as this study also ranked "Electricity supply" at the top ten of management services. Amole (2009) also agreed with the study when his findings were that rules were the most important factors in determining student satisfaction, on this study however students were not satisfied with the rules set by the management team, rules are in fact factors determining the level of student satisfaction.

Muslim *et al.* (2013) study agrees with this study as it ranked "Internet connection" as one of the important factors to determine student satisfaction. They were also similar to the study done by Khozaei *et al.* (2010) who ranked "Good internet connection" as one of the most important factors that influence student accommodation. However, in this study students were not satisfied with the quality of Wi-Fi connection, which means management failed to satisfy students.

In contrary to these findings, Danso and Hammond (2017) ranked "Garbage disposal" as the highest on the study done on Knus campus in Ghana, however these findings ranked "Garbage disposal" in the middle with no effect to the student's satisfaction, because it was not in the highest five neither was it on the bottom five of the results.

Whether it is Wi-Fi connection or network connection it is a requirement for public and private student accommodations to have good quality of internet connection for students, as it is well known by scholars that student do and submit documents via internet. Supply of electricity is very significant for any student accommodation, because student study at night and during the day, in order to do that they need sufficient electricity and water supply. Management team needs to provide 24hours security systems in order for the student to study with peace knowing that laps, phones and other belongings are safe inside the accommodation.

4 CONCLUSIONS

The finding of the study revealed that "24hr security guard system" provided by the management was the most satisfying service, followed by "Electricity supply", "Fire safety systems", "Conveying systems", and "Electrical lighting in the apartment", which were the major features which influenced student satisfaction and ranked at the top five of the study. The study's findings suggest that students were not satisfied with "Cold and hot water supply", "Quality of Wi-Fi connections", "Rules regarding visitors", "Working air-conditioning in the apartment" and "ATM machines in the apartment." Therefore, the objective has been successfully acknowledged from both literature and the structured questionnaire.

5 RECOMMENDATIONS

The management team needs to work hand in hand with the house committee to solve student problems and deal with students complains, for example:

- Implement suggestion boxes, which will be checked every now and then.
- Pay for quality Wi-Fi connection that is suitable for the number of students who are accommodated in this residence.
- Review rules regarding visitors especially the one that says no visitors two days before and after month end.
- Add more boilers for hot water supply and maintain the supply of water; cold water needs to be cold and hot water needs to be hot.

References

Abdullah, R. B., A Descriptive Study on Students' Satisfaction Towards the Services Provided by Universitiutura Malaysia. 1-76, 2009.

Aigbavboa, C. O, Thwala, W. D., and Eke, C. C., *An Evaluation of Students' Satisfaction with University Hall of Residence*, Yang, D and Qian, Y (eds.), Proceedings of the 18th International Symposium on Advancement of Construction Management and Real Estate, University of Johannesburg, 703-711, 2014.

Ajayi, M, Nwosu, A., and Ajani, Y., Students' Satisfaction with Hostel Facilities in Federal University of Technology Akure Nigeria, *European Scientific Journal,* 11(34), 402415, 2015.

Amole, D., Coping Strategies for Living in Student Residential Facilities in Nigeria, *Environment and Behaviour,* 200-219, 2005.

Amole, D., Residential Satisfaction in Students' Housing, *Journal of Environmental Psychology,* 76-85, 2009.

Bekurs, G., Outsourcing Student Housing in American Community Colleges: Problems and Prospects, *Community College Journal of Research and Practice,* 31, 621-636, 2007.

Charlton, S., An Overview of the Housing Policy and Debates, Particularly in Relation to Women (or Vulnerable Groupings), *Centre for the Study of Violence and Reconciliation,* 2004.

Danso, A. K., and Hammond, S. F., Level of Satisfaction with Private Hostels Around Knust Campus, 719-726, 2017.

Fields, T., *A Hedonic Model for Off-Campus Student Housing: The Value of Location, Location,* Muncie, Bale State University, 2011.

Hassanain, M. A., On the Performance Evaluation of Sustainable Student Housing Facilities, *Journal of Facilities Management,* 6(3), 212-225, 2008.

Khozaei, F., Hassan, A. S., and Khozaei, Z., Undergraduate Students' Satisfaction with Hostel and Sense of Attachment to Place: Case Study of University Sains Malaysia, *American Journal of Engineering and Applied Sciences,* 3(3): 516-520, 2010.

Mugambwa, J., Mugerwa, G. W., Mutumba, W. W., Muganzi, C., Namubiru, B., Waswa, Y., and Kayongo, I. N., Privately Provided Accommodation Service Quality and Customer Satisfaction, 2016.

Muslim, M. H., Karim, H. A., Abdullah, I. C., and Ahmad, P., Students' Perception of Residential Satisfaction in the Level of Off-Campus Environment, *Procedia - Social and Behavioral Sciences,* 105, 684696, 2013.

Omole, F. K., Basic Issues in Housing Development, Femo Bless Publications, 2001.

Thomsen, J., and Eikemo T. A., Aspects of Student Housing Satisfaction: A Quantitative Study, 2010.

Torbica, M. Z., and Stroh, R. C., Customer Satisfaction in Home Building, *Journal of Construction Engineering and Management,* January/February, 2001.

Interdependence between Structural Engineering and Construction Management
Edited by Ozevin, D., Ataei, H., Modares, M., Gurgun, A., Yazdani, S., and Singh, A.

ISBN: 978-0-9960437-6-2

ASSESSMENT OF BLENDED LEARNING EDUCATION MODEL: UPTAKE AND REASONS FOR ITS IMPLEMENTATION IN CONSTRUCTION EDUCATION

RAPHEAL ABIODUN OJELABI, ADEDEJI OLUSHOLA AFOLABI, IGNATIUS OWOICHO OMUH, and PATIENCE FIKIYEMO TUNJI-OLAYENI

Dept of Building Technology, Covenant University, Ota, Nigeria

The emergence and integration of information technology in modern age education training have changed the face of education across the globe. The need for the blend of information and communication technology tools in construction education training is of the high essence as it enhances students learning capacity through visuals and other means. Therefore, the study investigates the level of the blend of ICT with the traditional method of learning among lecturers within the built environment departments in Covenant University in Nigeria and reasons for its use were also examined. A survey method was adopted for the study; 125 structured questionnaires were administered to the randomly selected lecturers and 100 were retrieved. The data collected was analyzed with Mean item score, percentage, and Mann-Whitney U respectively. The result revealed that the adoption of the ICT tools, which include interactive whiteboards PowerPoint, a learning management system, and video clips among the lecturers is very high. The reasons attributed to its use is due to the visual platform it provides for the students while weblogs, web conferencing, and mobile phone were the lesser used tools. In conclusion, the study recommends that Covenant University management should increase the funding of ICT tools and also enhance the lecturers training in the use of blended learning style.

Keywords: ICT, Covenant University, Built environment, Nigeria.

1 INTRODUCTION

The 21st century education system has taken a new dimension with the advent of information and communication technology (Afolabi *et al.* 2017). Buran and Evseeva (2015) asserted that the world has become digital and likewise the education process as a result of the introduction of ICT tools into modern education training. Matukhin and Zhitkova (2015) also affirmed the integration of technology in all areas of life including education in the modern age. It is evidenced from the foregoing that modern age awareness and adoption of information and communication technology is second-to-none. The integration of ICT tools into traditional educational space as birth a new model of education known as 'blended learning model'.

Blended learning according to Culatta (2011) is described as the merger of e-learning together with traditional face-to-face teaching aided by an instructor. Likewise, Mason and Rennie (2006) asserted that blended learning is the integration of online learning into the traditional model of education. The acceptance of the blended learning model in the modern age can be traced to the need for the integration of information and communication technology in all fields of human endeavors, in which the education field is not an exception. Tapscott (2008) however buttressed that the 21st century students attitude towards information and communication technology tools have made the need for blended learning education model inevitable. Coiro (2003) further elucidate that blended learning model adoption is not an option in the modern age because literacy in education has been redefined beyond students' ability to read, write, learn, and comprehend; it also now includes students' ability to interact with technology. Undoubtedly, the students expected learning outcomes in the 21st century cannot be achieved under traditional teaching environment alone, therefore, there is a need for integration of active virtual learning environment, which necessitates the introduction of the blended learning model in education training.

2 REASONS FOR BLENDED LEARNING MODEL IN CONSTRUCTION EDUCATION

The blended learning model has been identified as the trending education model in the 21st century education model as it has caused a paradigm shift in education training across the globe. The adoption of the modern education model (blended learning) in modern education has been supported by most education stakeholders across the globe due to its deliverables that are evidenced in various field of human education training. The benefits accrued to the use of the modern education model has formed the basis for its wide adoption among education stakeholders. Despite the education model benefits, reasons for its use differ among educators. Different educators adopted the blended learning education model for different reasons in a bid to create the most effective learning environment for students. Its integration into construction education is been driven by reasons which are very germane to students' education training in the construction field.

Eydelman (2013) revealed that one of the reasons for the adoption of the blended learning education model in education training is because of its capacity to induce motivating and friendly learning environment for students. The relevance of blended learning cut across all field of human education training including construction education. The 21st century students are been driven by technology presence. Therefore, the introduction of the blended learning education model will no doubt stimulate students to learn due to their perceived friendliness of the learning environment. Another convincing reason as cited by Pape (2010) for the use of the modern education model across the globe is linked to its ability to cause reduction of time in class teaching in all field human education training. The capacity of the blended learning education model to cause a reduction in time in class during teaching session lies in its effectiveness as an education training model. Its effectiveness is linked to its combined strength in online teaching together with the traditional method face-to-face approach. In construction education training, the use of a blended learning model can reduce class time due to the enhancement of face-to-face teaching method with media presence. Pearcy (2009) also identified the reason for blended learning education model adoption in education training to include its capacity to facilitate interesting teaching and learning during the class session. The capacity of the education model to facilitate interesting class can be linked to the affinity of the modern age students for a technologically driven learning environment. Vernadakis *et al.* (2011) also revealed that the reason for the blended learning model adoption in education training is aligned to its capacity to create a student-centered learning environment. The blended education model ability to create a learner-centered focus was affirmed by Morgan (2002) as he attested that the model has the capacity to shift the education system from a teacher-centered approach to more of student-centered focus. Under the learner-centered approach, the ability of the students or learners to develop independent learning skills cannot be compromised (Aborisade 2013).

3 APPRAISAL OF BLENDED LEARNING EDUCATION MODEL

Ojelabi *et al.* (2017) asserts that sustainability of education capacity is dependent on the teaching and learning effectiveness. The demand for effectiveness in teaching and learning in the modern age have necessitated for a shift in the existing education model (traditional education style) to a contemporary education model with outcomes, which include learning and innovation, critical thinking and collaboration, and media and technology skills. Hubactova and Semradova (2016) however, opined that blended learning education model is the contemporary education model with capacities to meet the demand of the 21st century. The blended learning education model according to Kosar (2016) is the combination of instructional modalities (e.g., onsite, self-paced learning, and web-based), delivery media (e.g., internet, PowerPoint presentations, textbooks) and instructional methods (e.g., face-to-face or technology-based sessions) and web-based technologies (e.g., wikis, chat rooms, blogs, textbooks, online courses). Garrison and Kanuka (2004) claimed that the hybrid of face-to-face and online technology education model (blended learning model) have the capacity to reinforce both interactive and communicative learning environment critical to expected learning outcomes capable of sustaining education deliverables in the modern age.

4 RESEARCH METHODOLOGY

The aim of the study was to assess the uptake of the blended learning model in a higher institution in Nigeria. In a bid to achieve the aim, the study assesses the level of adoption of blended learning tools in construction education and reasons for its implementation in construction education training. Relevant data was sourced from both primary and secondary sources. Primary data was collected with the aid of structured questionnaires administered to academics and students within the in the built environment departments in Covenant University. The sample was selected using purposive sampling technique one of the non-probabilistic techniques. A total of 125 questionnaires were distributed among the academics in the built environment and 100 were retrieved. The questionnaire was divided into three sections: section one examined the characteristics of the respondents, the second section addresses the level of use of blended learning tools and they are placed on the likely scale of 1-5 which are; 5 = use very often, 4 = use often, 3 = sometimes use, 2 = rarely use, 1 = not used, and the third section addresses the reasons for implementation of the blended learning tools in construction education using the criteria; 5 = strongly agree, 4 = agree, 3 = slightly agree, 2 = disagree, 1 = strongly disagree.

5 ANALYSIS AND RESULT

Table 1 shows the breakdown of the respondent's characteristics. The respondent's academic status in Table 1 shows that 74% of the respondents were students while 26% were lecturers. Also, the gender distribution of the respondent's shows that 54% represents the male while 46% represents the female. The breakdown down of the respondent's department shows that Building Technology and Architecture have the highest percentage score with 42% and 22% respectively.

5.1 Adoption Level of Blended Learning Tools in Construction Education Training

Table 2 shows the level of adoption of different blended learning tools in construction education training. It was evidenced that the interactive whiteboards and PowerPoint rank very high with mean scores of 4.51 and 4.48 respectively. The next three blended learning tools are often used in construction education training are tablets PCs, learning management system (module) and laptop with the mean score of 4.25, 4.24 and 4.14 respectively. Other blended learning tools, which include video clips, e-mail, audio response system, video conferencing, group collaboration software, web forms and audio file all have a mean score above 3.0, which simply signifies that their use is not that often but sometimes. Web conferencing and mobile phone both have mean score slightly above 2.0, which signifies that they are rarely used in construction education training.

Table 1. Respondent characteristics.

Characteristics	Frequency	Percentage
Respondent Academic Status		
Lecturer	26	26
Student	74	74
Total	**100**	**100**
Gender		
Male	54	54
Female	46	46
Total	**100**	**100**
Respondent Department		
Architecture	22	22
Building Technology	42	42
Estate Management	20	20
Civil Engineering	16	16
Total	**100**	**100**

Table 2. Level of adoption of blended learning tools in construction education.

Blended Learning Tools	Mean	Remarks
Interactive whiteboards	4.51	Use very often
PowerPoint	4.48	Use often
tablets PCs	4.25	Use often
Learning management system	4.24	Use often
Laptop	4.14	Use often
Video clips	3.56	Use often
e-mail	3.51	Use often
Video Conferencing	3.50	Use sometimes
Audio Response System	3.39	Use sometimes/use often
Group collaboration software	3.38	Use sometimes/use often
Web forms	3.28	Use sometimes/use often
Audio Files	3.22	Use sometimes/use often
Virtual world	2.98	Use rarely/ sometimes use
Web conferencing	2.18	Use rarely/ sometimes use
Mobile phone	2.16	Use rarely/sometimes use
weblogs	1.52	Rarely use/ never use

5.2 Reasons for Blended Learning Tools Adoption in Construction Education

Findings from the results presented in Table 3 revealed the educationist assessments of the level of agreement to the reasons identified for blended learning tools adoption in construction education are all above mean score of 3.50, which signifies agree on the Likert scale.

Table 3. Reasons for blended learning tools adoption in construction education.

Reasons	Mean	Mean
Enhance visual aid.	4.48	Agree
Increase the accessibility of learning materials.	4.43	Agree
Creates a student-centered learning environment.	4.40	Agree
Creates interactive learning classroom	4.39	Agree
Enhance students' knowledge	4.36	Agree
Support face-face learning approach	4.35	Agree
Creates a motivating learning environment	4.29	Agree
Support collaboration	4.25	Agree
Enhances students learning skills	4.22	Agree
Creates an interesting lesson	3.85	Agree

5.3 Research Hypothesis

The study hypothesis tested for the significant difference between lecturers and students on the reasons for adopting blended learning tools in construction education is presented in Table 4 using the Mann Whitney U Test. The results revealed that there is no significant difference between the lecturers and the students on the reason for blended learning model adoption at a significant value of 0.205.

Table 4. Mann Whitney U test statistics.

	Reasons for Blended Learning Adoption
Mann-Whitney U	802.500
Wilcoxon W	1153.500
Z	-1.267
Asymp. Sig. (2-tailed)	.205

6 CONCLUSIONS AND RECOMMENDATIONS

This study examined the level of adoption of blended learning tools and reasons for its implementation in Covenant University in Nigeria. Findings revealed that the institution affinity for blended learning tools in construction education is fairly good with evidence in the adoption of tools like interactive whiteboards, presentation software (PowerPoint), tablets Pcs, learning management system, laptop and video clip among others in education training. The reasons for its implementation from the study include its capacity to enhance visual aid, increase accessibility of learning materials and student-centered learning environment created among others. The reasons identified in the study for technological tools integration in education training are supported by Eydelman (2013), Pape (2010), Pearcy (2009), and Morgan (2002). However, despite the institution encouraging levels of adoption in some of the ICT tools in education training, the study revealed that its adoption in tools like a virtual world, web conferencing, mobile phones and weblog for teaching are poor. Therefore, the study suggested the need for the institution to enhance funding education ICT tools and also encourage the lecturer readiness to the use of a blended learning model through relevant training.

Acknowledgments

The researchers appreciate the publication support fund given by Covenant University towards the publication of this article.

References

Aborisade, P. A., Blended Learning in English for Academic Purposes Courses: A Nigerian Case Study, 2013. In B. Tomlinson and C. Whittaker (Eds.), *Blended Learning in English Language Teaching: Course Design and Implementation*. London British Council, UK.

Afolabi, A., Ojelabi, R., Oyeyipo, O., Tunji-Olayeni, P., Omuh, I., and Amusan, L., Integrating Software Development Courses in The Construction Curriculum, *Turkish Online Journal of Education Technology*, Special Issue, 2017.

Buran, A., and Evseeva, A., Prospects of Blended Learning Implementation at Technical University, *Procedia-Social and Behavioral Sciences,* 177-18, 2015.

Coiro, J., Reading Comprehension on the Internet: Expanding Our Understanding of Reading Comprehension to Encompass New Literacies, *The Reading Teacher,* 56, 448-464, 2003.

Culatta, R., Distance Learning. Innovative Learning, 2011. Retrieved from http://innovativelearning.com/online-learning/distance_learning.html in July 2018.

Eydelman, N., A Blended English as A Foreign Language Academic Writing Course, 2013. In B. T. C. Whittaker (Ed.), *Blended Learning in English Language Teaching: Course Design and Implementation*. UK: London British Council.

Garrison, D. R., and Kanuka, H., Blended Learning: Uncovering Its Transformative Potential in Higher Education, *The Internet and Higher Education,* 7, 95–105, 2004.

Hubactova, S., and Semradova, I., Evaluation of Blended Learning, *Procedia-Social Behavioral Sciences,* 217, 551-557, 2016.

Kosar, G., A Study of EFL Instructors, Perceptions of Blended Learning, *Procedia-Social and Behavioral Sciences,* 232, 736-744, 2016.

Mason, R., and Rennie F., *E-Learning: The Key Concepts,* London: Routledge, 2006.

Matukhin, D., and Zhitkova, E., Implementing Blended Learning Technology in Higher Professional Education, *Procedia-Social Behavioral Sciences,* 2016, 183-188, 2015.

Morgan, K. R., *Blended Learning: A Strategic Action Plan for A New Campus,* University of Central Florida, 2002.

Ojelabi, R., Afolabi, A., Tunji-Olayeni, P., Amusan, L., and Omuh, I., Learning Adequacy of Nigerian Tertiary Educational System for Sustainable Built Environmental Course, *Turkish Online Journal of Education,* Special Issue 2017.

Pape, L., Blended Teaching and Learning. American Association of School Administrators, 2010. Retrieved from http://facultycenter.ischool.syr.edu/files/2012/02/blended-learning-and-teaching.pdf in June, 2018.

Pearcy, A. G., Finding the Perfect Blend: A Comparative Study of Online, Face-To-Face, and Blended Instruction, PhD Thesis, University of North Texas, 2009. Retrieved from http://search.proquest.com/docview/304963133?accountid =4488, on June, 2018.

Tapscott, D., *Grown Up Digital: How The Net Generation is Changing Your World,* New York: McGraw-Hill, 2008.

Vernadakis, N., Giannousi, M., Derri, V., Michalopoulos, M., and Kioumourtzoglou, English. The Impact of Blended and Traditional Instruction in Students' Performance, *Procedia Technology,* 1, 439-443, 2011.

Interdependence between Structural Engineering and Construction Management
Edited by Ozevin, D., Ataei, H., Modares, M., Gurgun, A., Yazdani, S., and Singh, A.

ISBN: 978-0-9960437-6-2

CRITICAL FACTORS INFLUENCING BUILDING GRADUATES' EMPLOYABILITY IN A DEVELOPING ECONOMY

ADEDEJI AFOLABI[1], RAPHEAL OJELABI[1], PATIENCE TUNJI-OLAYENI[1], IGNATIUS OMUH[1], and OPEYEMI OYEYIPO[2]

[1]*Building Technology, Covenant University, Ota, Nigeria*
[2]*Quantity Surveying, Covenant University, Ota, Nigeria*

Schools of higher learning are endowed with the prerogative of churning out employable graduates to the workplace. This can only be achieved by measuring the need-thermometer in skills and capacities required in the industry in comparison with the education provided by the higher institutions. The study was aimed at evaluating the critical factors that influence building graduates' employability in a developing economy. Using an employer-academia perspective, a cross-section survey system through a questionnaire instrument provided information on graduate employability in the built environment industry. Top management officials in construction industry and academia in the built environment in schools of higher learning in Lagos and Ogun State provided the data for the study. Statistical tools such as stacked bars and Linear Regression was used in presenting the data. The study revealed major skills (communication, teamwork, professional expertise/problem-solving, self-management, planning and organizing, ICT, life-long learning, and initiative/enterprise) required by employers in the construction industry. The study showed that teaching strategies and curriculum content can influence building graduates' employability. In conclusion, the study developed a framework for increasing building graduates' employability to align with the skills needed by employers in the built environment in a developing economy. It is recommended that schools of higher learning need to continually measure the needs of the industry and incorporate findings into a robust construction curriculum. Work-learning settings should be encouraged for construction students.

Keywords: Construction industry, Education, Higher institution, Industry, Skill.

1 INTRODUCTION

Unemployment remains a clog in the wheels of progress of most developing economies. This continues to be an issue despite the upsurge number of institutions of higher learning; increment in new intakes and those that eventually graduate (Akinyemi *et al.* 2012). This study opined that it is not a question of unemployment but whether the graduates produced in developing economies are employable. Therefore, it is a case of structural unemployment, whereby, as the economy is developing, the industry's needs and technologies are also changing. Whereby, once the educational institutions producing the graduates cannot adapt, the result is massive structural unemployment. With over 200 tertiary institutions churning up unemployable graduates in Nigeria for instance (Anyadike *et al.* 2012), Uwaifo (2009) argued that the educational system must move away from the era of chalk and talk; rather it should focus on the industry needs to build employable graduates. According to Anyadike *et al.* (2012), unemployment in Nigeria has worsened due to the outdated school curricula and lack of employable skills. Researchers such as Kakwagh and Ikwuba (2010) and Olokundun *et al.* (2014) noted that unemployment has been the most socio–economic challenge gripping developing nations such as Nigeria. A sad aspect of unemployment was felt in March 2014, where 522,650 youths gathered to fill 4,556 vacant positions advertised by the Nigerian Immigration Service, resulting in the death of several young graduates. The youth (15 – 35 years) unemployment and underemployment rate for the fourth quarter (Q4) of 2017 were put at 52.65 percent, which is 22.64 million people (National Bureau of Statistics 2017). Rather than waiting on the government in developing countries, the onus is on schools of higher learning to re-strategize on how to ensure that their graduates are employable and create employment for others. Whereas, in developed countries, the higher institutions have deeply taken initiatives to create graduates that are ready for the workplace as a result of listening to the needs of industries (Jackson 2013). Adeyemo *et al.* (2010) opined that this made possible in developed countries, whereby their university researchers are able to observe the need trends of industries and examine their importance in improving the state of the nation's economy and technological advancements. The case for building graduates' employability in a developing economy is built around their relevance in oiling the necessary housing and urban infrastructure needed in developing nations. Nigeria, for instance, struggles with housing deficit, for it has an over 180 million population and over-pressured infrastructure. The building graduates also find their skills relevant in other sectors from oil and gas, financial, manufacturing, and so on. Through planning, design, construction, maintenance and repair, and operation, building graduates are able to transform various resources into habitable and usable facilities. The construction sector is looking for raw talents that are able to assume the required positions within organizations with little or no need of retraining (Afolabi *et al.* 2017). These employable recruits will be able to fill the skill gap that exists in the construction sector. SME construction businesses are looking for individuals that would enhance the productivity and profit of the firms rather than becoming liabilities and increasing the employment turnover, ones they do not fit. For employers of labor, Jackson (2013) noted that a skill deficient graduate will lead to a loss and poor output for the firm. The construction industry is so critical in using graduates that have the required skills because of the nature of the products in terms of housing and other civil infrastructure (Afolabi and Oyeyipo 2017). Therefore, it is paramount that the graduates produced in this sector are of utmost quality as mandated by the industry. For schools of higher learning in developing countries, building graduates that have the required employability skills should be seen as a means of building competitive advantage and attracting more students. In some cases, employers of labor approach universities to supply their graduates based on the competence and reputation of the universities that have been tested in their graduates (Falola *et al.* 2018). The role of universities should not end in producing academically sound graduates but additionally ensuring that they are desirable to employers. Therefore, the study intends to evaluate the critical factors influencing building graduates' employability in a developing economy.

2 RESEARCH METHOD

This pilot study was carried out from a two-way perspective of the employers in the construction industry and academics in the built environment. Using the two (2) groups of respondents afforded, the study looked at a wide view of the industry needs and the curriculum review of the employability needs of building graduates. The two (2) group comprised of the population of the study. To narrow the respondents that participated in the study, a quota-purposive sampling technique was used to select the employers-academics respondent. A total of 50 respondents in each group were selected. The survey research design used in this study required the use of a well-structured questionnaire data instrument, which was used in receiving responses on the subject of the study. Questionnaires were delivered by hand and through electronic means and retrieved via the same medium. A total of 30 completed questionnaires from each group were retrieved from the field survey. The study was carried out in Lagos and Ogun State. These two states have a vibrant construction industry with large volumes of construction activities ongoing. The states also have world-class universities that have a robustly built environment program. The data collected were analyzed using SPSS v.21 and presented using stacked bars and categorical regression.

3 RESULTS AND DISCUSSION

This section contains the result and discussion on the factors influencing building graduates' employability in a developing economy. The first section analyzed using stacked bars the employability skills perceived to have been covered in the curriculum by the academics and employability skills required in the construction firms by the employers. The second section showed the categorical regression of the factors influencing building graduates' employability in a developing economy.

3.1 Perceived Employability Skills

This section covered the employability skills perceived by the tutors in university and employers in the construction field. Academics provided the level of employability skills covered in the University curriculum in the built environment while employers of labor assessed the employability skills to conform to their requirement based on positions in their construction firms. The employability skills identified in this study are adapted from the Department of Education, Science and Training in Australia (DEST 2006). Figure 1 showed the perceived level of employability skills by academics and employers. In Figure 1, academics placed emphasis on the communication skills, professional expertise/problem-solving skills and teamwork skills. Although this is at par with the view of employers of labor in the construction sector, there is some form of gap in creating a well-rounded graduate fit for the needs of the industry. Other employability skills perceived important by the employers above the academics include planning and organizing skills, life-long learning skills, and initiative/enterprise skills. These are skills that are can be learnt in the classroom setting but perfected in a work integrated learning environment.

3.2 Factors Influencing Building Graduates' Employability

This section examines the factors influencing building graduates' employability in the construction sector. The factors considered are basically factors from teaching strategies and curriculum content. The factors analyzed using linear regression include "Meeting Deadlines for Assignment Submission", "Leadership Courses/Public Speaking Courses", "Writing and Presenting Assignments And Report", "Simulations", "Work Placement Programmes", "Town and Gown Sessions", "Entrepreneurship Courses", "Use of Case Studies", "Presenting & Participation in Class Discussion", "Mentoring and Coaching Activities", "Business Development Training", "Designing Change Processes", "Research Project", "Brainstorming Sessions", "Attendance in Class", "Role Plays/Demonstration", "Group Discussion", "Student Membership of Professional Bodies", "Start-Up Boot Camp Series", "Developing a Career Plan", "Planning and Organizing an Event", "Developing and Designing Models", "Data Collection for

Research", "Searching for Internship", "Sourcing for Library & Online Materials", "Involvement in Student Bodies", "Society or Sporting Activities", "Use of Log-Books", "Working on Group Assignments", "Practical Experience in Courses", "Decision Making Activities", "Utilizing Specialized ICT Packages" and "Development of Portfolios".

The null hypothesis; H_0 was constructed as teaching strategies and curriculum content cannot influence building graduates' employability. While the alternate hypothesis; H_1 stated that teaching strategies and curriculum content influence building graduates' employability.

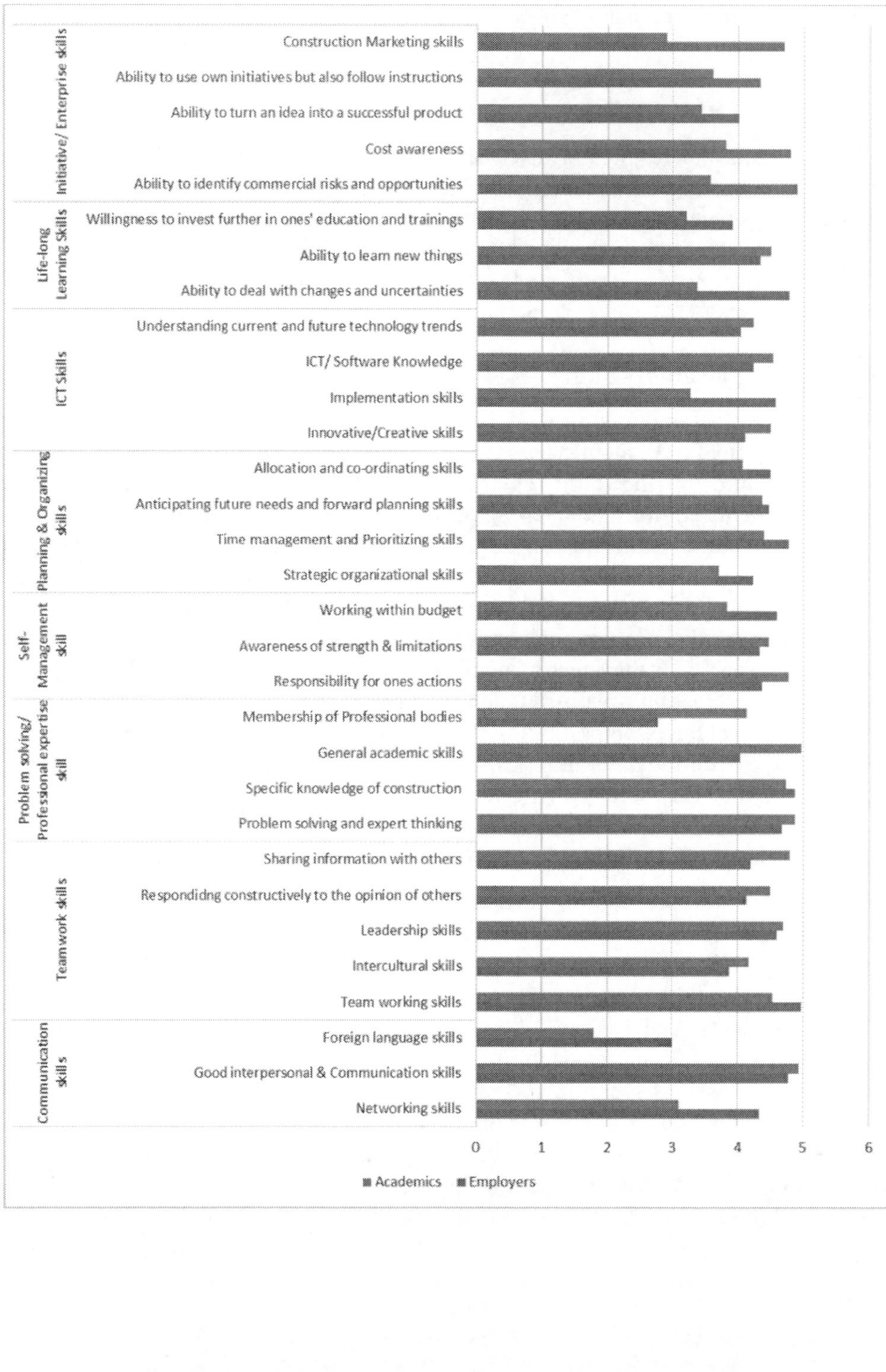

Figure 1. Employability skills perceived by academics vs employers.

With the significance level below 0.05 ($F (32, 27) = 3.972$; $p = 0.000$) as shown in Table 1, the alternate hypothesis is accepted and the null hypothesis is rejected, which means that teaching strategies and curriculum content significantly influence building graduates' employability.

Table 1. Analysis of variance[a].

Model		Sum of Squares	df	Mean Square	F	Sig.
	Regression	12.372	32	.387	3.972	.000[b]
1	Residual	2.628	27	.097		
	Total	15.000	59			

a. Dependent Variable: Category of respondents
b. Predictors: (Constant), Teaching strategies & Curriculum content

Table 2 showed the model summary of the regression analysis. From Table 2, it showed that teaching strategies and curriculum content has an influence (R^2) of 82.5% on building graduates' employability. This finding is corroborated by the study in Olokundun *et al.* (2018).

Table 2. Model summary.

Model	R	R Square	Adjusted R Square	Std. Error of the Estimate
1	.908[a]	.825	.617	.312

a. Predictors: (Constant), Teaching strategies & Curriculum content

The result of the findings helped to develop a framework for increasing building graduates' employability in Figure 2. The framework highlighted the importance of having students go through a work-learning environment at different levels from their first year coupled with a rich curriculum content and effective teaching strategy, employability skills can be imparted on the Building students.

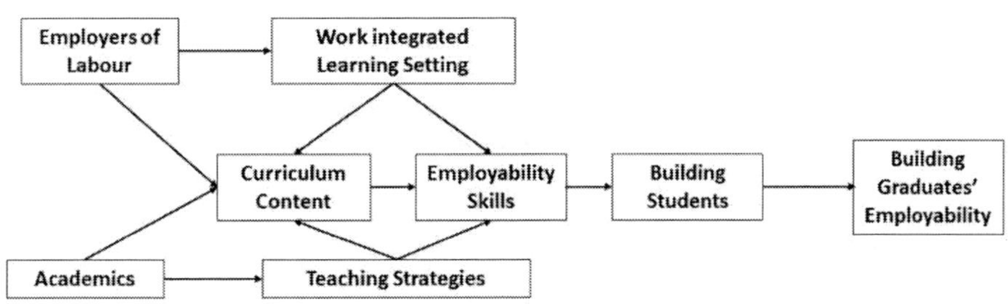

Figure 2. Framework for increasing building graduates' employability.

4 CONCLUSIONS AND RECOMMENDATIONS

The study was aimed at evaluating the critical factors that influence the building graduates' employability in a developing economy. The study revealed the major skills (communication; teamwork, professional expertise/problem-solving; self-management; planning & organizing; ICT; life-long learning; initiative/enterprise) required by employers in the construction industry.

There are specific skills that employers perceived as important in the construction industry. However, academics are of the opinion that it is not adequately captured in the construction curriculum. Even though academics and employers of labor are on par with their importance placed on communication skills, professional expertise/problem-solving, teamwork skills and ICT skills, there are still deficiencies in the construction curriculum to impact the knowledge and experience of planning and organizing skills, life-long learning skills and initiative/ enterprise skills, which are needed in the construction industry. The study showed that teaching strategies and curriculum content can influence building graduates' employability. In conclusion, the study developed a framework for increasing building graduates' employability to align with the skills needed by employers in the built environment in a developing economy. It is recommended that schools of higher learning need to continually measure the needs of the industry and incorporate the findings in a robust construction curriculum. The employers' role should not be neglected in building a well-rounded graduate for the construction industry. Work-learning settings should be encouraged for construction students as this would influence the skills that will engender a competitive and quality Building graduate in the construction industry.

Acknowledgments

The researchers appreciate Covenant University for funding the open access publication of this manuscript.

References

Adeyemo, S. A., Ogunleye, O. A., Oke, C. O., and Adenle, S. O., A Survey of Factors Determining the Employability of Science and Technology Graduates of Polytechnics and Universities in The Nigerian Labour Market, *Journal of Science and Technology Education Research*, 1(5), 99-106, 2010.

Afolabi, A. O., and Oyeyipo, O., The Perception of Future Decision-Makers on the Building Profession, *Malaysian Construction Research Journal*, 21(1), 55-73, 2017.

Afolabi, A. O., Ojelabi, R. A., Oyeyipo, O., Tunji-Olayeni, P. F., and Omuh, I. O., Integrating Software Development Courses in the Construction Curriculum, *The Turkish Online Journal of Educational Technology*, Special Issue, 215-225, 2017.

Akinyemi, S., Ofem, I. B., and Ikuenomore, S. O., Graduate Turnout and Graduate Employment in Nigeria, *International Journal of Humanities and Social Science*, 2(14), 257-265, 2012.

Anyadike, N., Emeh, I. E. J., and Ukah, F. O., Entrepreneurship Development and Employment Generation in Nigeria: Problems and Prospects, *Universal Journal of Education and General Studies*, 1(4), 88-102, 2012.

Department of Education, Science and Training (DEST), *Employability Skills from Framework to Practice, an Introductory Guide for Trainers and Assessors*, Canberra, Australia, 2006.

Falola, H. O., Oludayo, O. A., Olokundun, M. A., Salau, O. P., Ibidunni, A. S., Igbinoba, E., Employee Resourcing Strategies and Universities' Corporate Image: A Survey Dataset, *Data in Brief*, 18, 1378-1382, 2018.

Jackson, D., The Contribution of Work-Integrated Learning to Undergraduate Employability Skills Outcomes, *Asia-Pacific Journal of Cooperative Education*, 14, 2, 99-115, 2013.

Kakwagh, V. V., and Ikwuba, A., Youth Unemployment in Nigeria: Causes and Related Issues, *Canadian Social Science*, 6(4), 231-237, 2010.

National Bureau of Statistics, NBS, Labour Force Statistics Vol. 1: Unemployment and Underemployment Report (Q1 – Q3 2017), FOS, Nigeria, 2017.

Olokundun, A. M., Falola, B. H., and Ibidunni, A. S., Agro Business as a Remedy for Youth Unemployment Towards the Achievement of Sustainable Development in Nigeria: Comparative Perspectives from the Kwara State Agro Business Economy, *Journal of Economics and Sustainable Development*, 5(3), 46-57, 2014.

Olokundun, M., Ibidunni, S., Ogbari, M., Peter, F., Borishade, T., Falola, H., Salau, O., Kehinde, O., Survey Data on Teaching Strategies and Product Innovation: A Focus on Selected University Students in Nigeria, *Data in Brief*, 18, 248-254, 2018.

Uwaifo, V. O., Industrializing the Nigerian Society Through Creative Skill Acquisition Vocational and Technical Education Programme, *International NGO Journal*, 4(4), 142-145, 2009.

Interdependence between Structural Engineering and Construction Management
Edited by Ozevin, D., Ataei, H., Modares, M., Gurgun, A., Yazdani, S., and Singh, A.

ISBN: 978-0-9960437-6-2

ROLE OF INDUSTRY EMPLOYERS IN FOSTERING EMPLOYABILITY SKILLS

JOHN ALIU[1], CLINTON AIGBAVBOA[2], and WELLINGTON THWALA[2]

[1]*Dept of Civil Engineering, University of Johannesburg, Johannesburg, South Africa*
[2]*Faculty of Engineering and the Built Environment, University of Johannesburg, Johannesburg, South Africa*

Graduate skills and employability discussions have continued to be a contentious issue for both international and local employers. The concerns of industry employers stem from their inability to acquire graduates with the requisite skills and competencies to effectively solve arising industry problems. This concern has led employers to cast blames on higher education institutions (HEIs), who they feel are responsible for the development of these skills which employers continuously seek from fresh graduates. Subsequently, present-day employers have sought innovative ways to train and re-train new employees to help ease their transition into the world of work. This paper identifies the various approaches and ways by which industry employers develop the required skills of graduates. An extant review of literature was conducted from databases such as Academic Search Complete, ASCE conference proceedings, EBSCOhost, Emerald, ERIC, Science Direct, Scopus, Springer, Taylor and Francis online, amongst others. One of the primary findings stemming from this study revealed that in-house training, job rotation, coaching and mentoring, and cross-functional teams are some of the key approaches by which industry employers can develop skills and competencies among fresh graduates. This study concludes that HEIs are not solely responsible for the development of employability skills among graduates, but reckons that both the employers of labour and graduates themselves have a crucial role to play in the construction education discourse.

Keywords: Mentoring, Construction education, Construction industry, Graduate skills, Higher education.

1 INTRODUCTION

The importance of HEIs in equipping students with the right skills has positioned it as a catalyst for economic growth (Munap *et al.* 2015). Its roles in the areas of research and development have further earmarked it as a critical influence in the construction education discussion. However, in recent years, the world of work has also gained significant recognition in the development of skills and knowledge among graduates or employees. This is because employers have continuously expressed their dissatisfaction at the quality of graduates coming through the HEIs. This frustration stems from the fact that present-day employers are seeking to recruit graduates based on their qualities, abilities and discipline-specific knowledge. This situation has led employers to regularly design several opportunities to train and re-train graduates to hit the ground running in the world of employment. Several scholars also suggest that the industry provides ample opportunity for sound learning which is referred to as work-place learning

(Beckett and Hager 2002, Boud and Middleton 2003, Collin *et al.* 2011). Hence, this paper reviews the significance of these opportunities provided by industry employers.

In the employment sector, learning can be embedded in the context of social practice. This is as a result of the opportunities provided for employees to attain the required knowledge, which connects theoretical and practical activities realistically and efficiently (Lave and Wenger 1991, Billett 1996). Various ways through which graduates can benefit from work-place activities include informal learning (experience-based learning and incidental learning); self-directed learning and formal organizational learning (Foley 1999, Marsick and Volpe 1999, Senge 1990, Marsick and Watkins 1999, Hager and Halliday 2006). According to Mallon *et al.* (2005) and Doornbos *et al.* (2008), the world of work is more concerned with informal learning rather than the formal education and academic qualification. Hence, this study will cast more focus on the informal ways by which graduates' skills can be developed through an extensive literature survey. The acquisition and development of new skills and knowledge ensure the productivity and satisfaction of graduates cum employees.

2 LITERATURE REVIEW

2.1 Informal Learning Process

This form of learning process dates back as far as 1926 in the early writings of Lindeman (1926), Dewey (1938) and Knowles (1970). These scholars suggested that matured learners are apprised of their learning experiences through self-direction. Further studies by Marsick and Volpe (1999) and Dale and Bell (1999) further expressed the correlation between employees and the work environment and acknowledged that a considerable part of the learning process occurs through interaction with others. This learning process could be regarded as a by-product of everyday activities, which are random, implicit, idiosyncratic, opportunistic, unstructured, unintended and unsystematic and driven by serendipity (Sawchuk 2003, Eraut 2004, Hager and Halliday 2006). This form of learning is not limited to a body of knowledge and can be attained via a range of strategies including social interaction, coaching, self-directed learning, random conversation, teamwork and mentoring. Simply put, this form of learning does not lead to formal qualifications. When juxtaposed with the formal learning process, the informal learning process suggests greater flexibility and recognizes the significance of socialization and learning from other people. Hence, this learning process can improve certain values, attitudes, skills and competencies among employees. These include: improved interpersonal skills, problem-solving skills, creative thinking abilities, communication skills, leadership effectiveness, and teamwork skills. Knowledge and skills gained via this learning process can be described as tacit knowledge (Foley 2004).

Furthermore, this form of learning is also referred to as incidental learning (Marsick and Watkins 1990) and can be used interchangeably with informal learning. Just like informal learning, incidental learning is tacit and unintended and can occur by learning from poor judgements, mistakes, trial and error and even a hidden agenda of an organization's culture or set-up (Marsick and Watkins 1999, Tusting 2003).

2.2 Forms of Informal Learning

2.2.1 *In-house training*

In this form of learning, employers of labor provide intentional and planned activities, courses, symposiums and seminars that provide a learning element. In this case, trainers and facilitators are usually from the organisation, industry, professional bodies or external bodies. According to

Hameed and Waheed (2011) and Kumar *et al.* (2017), training of recruits can be classified as behavioral and cognitive approaches. The behavioral methods are also called on-job training methods. These form of training permits the individuals to practice behavior in real or simulated fashion. These types of training stimulate learning through behavior, which can lead to skill and knowledge development. On the other hand, cognitive methods, which are also referred to as off-job training methods provide verbal, written and conceptual information on how to carry out activities (Hameed and Waheed 2011, Kumar *et al.* 2017). Benefits of in-house training include: improved performance and productivity of the employee, development of technical skills of employees, understanding of what the industry entails, amongst others.

2.2.2 *Job rotation*

This is another strategy through which industry can improve the performance and productivity of employees (Schultz and Schultz 2010). Job rotation refers to the process of switching an individual from one role to another, which increases their value to an organization. It can also occur when individuals are introduced to a schedule of tasks or assignments, which are aimed at providing a holistic breadth of exposure. Two categories of job rotation were proposed, which includes: within-function rotation (rotation between jobs of similar responsibility and orientation) and cross-functional rotation (rotation between jobs in different aspects of the organisation over a period). As a learning approach, recruits are provided ample opportunities to learn the necessary skills needed to thrive in the industry (Adjei 2012). Job rotation also provides the following: adequate work experience, career development, career management and motivation, career satisfaction, understanding of strategy issues, exposure to various functions and professionals, quality learning experience, increased network of contacts, self-confidence, increased transfer of company, increased responsiveness to change, self-efficacy, greater job variety, personal development, skill diversity amongst others (Earney and Martins 2009, Khan *et al.* 2014).

2.2.3 *Coaching and mentorship*

This form of learning is primarily concerned with the improvement of graduate skills, attitudes and performance within their present job role. Coaching can also be regarded as a task-oriented form of personally tailored training. Carter (2001) further described coaching as an interactive process, which provides employees with the opportunity to develop rapidly, which in turn increases their productivity. Yeo (2008) argue that mentorship provides the opportunities for young employees to master their craft to become experts. This can be achieved by making plans for a trainee to work alongside an experienced head to observe and learn the trends of achieving tasks. Several scholars have identified several roles mentors can play in the development of skills among graduates. These include: monitoring the progress of employees, facilitating their participation in various industrial activities and functions, and providing guidance through the performance of tasks. These mentor-mentee relationships can improve: interpersonal skills, self-image, cultural skills, psychological skills, job satisfaction, amongst others (Ragins and Kram 2007, Abedin *et al.* 2012). In summary, while coaching help trainees to carry out tasks effectively, mentorship deals with the preparation of trainees for future activities and purposes.

2.2.4 *Cross-functional teams*

Over time, the ability of recruits to work together with other employees in other departments or functions has become key competency employers seek. This has prompted most employers to encourage cross-functional teams as an effective initiative to further improve recruits. Kettley and Hirsh (2000) state that cross-functional teams are composed of individuals who have a specific domain base but work collaboratively on issues or processes requiring diverse resources. Several benefits of working in cross-functional teams includes timely and effective delivery of tasks, promotion of teamwork, improved communication, new knowledge about job roles, self-development, direct transfer of skills and knowledge from experts and collective problem-solving (Lovelace *et al.* 2001).

3 RESEARCH METHODOLOGY

This research study was conducted with reference to extant works of literature published in peer-reviewed conference papers and journals articles in order to review the various informal strategies by which the construction industry can develop new employees. In understanding these various activities, extensive literature searches were carried out several weeks in July and August 2018, covering the major academic databases including ASC, EBSCO, Emerald, Science Direct, ISI Web of Knowledge, and GoogleScholar. In developing an initial shortlist of these approaches, a preliminary survey of nine sources was done - Academic Search Complete, ASCE conference proceedings, EBSCOhost, Emerald, ERIC, Science Direct, Scopus, Springer, Taylor and Francis online which identified a number of key approaches. The initial search generated a total of 167 articles which was later filtered down to 72. These 72 articles were then examined individually which helped in the realization of this research study. In order to be included in this review, the scholarly article had to comply with certain criteria. One of the conditions is that the article should have been published in a peer-reviewed journal or conference proceedings (which implied that newspaper reports, government articles and unpublished works) were not considered. The other condition was that the article discussed explicitly various approaches by which industry employers can further improve the skills of employees. Furthermore, the literatures utilized for this study was extracted based on the search criteria that included the following keyword combinations: 'employability', 'mentoring', 'formal AND learning', 'informal AND learning', 'construction industry' and 'graduate AND skills'.

4 LESSONS LEARNT AND RECOMMENDATIONS

The construction industry of today is constantly seeking the services of skilled graduates to meet up with its ever-changing and challenging needs. The benefits of owning a pool of skilled employees with a myriad of skills cannot be over-emphasized. In fact, employers who demand skills from recruits have a responsibility to support and develop such graduates when they are employed. This study suggests various strategies and initiatives by which employers can achieve this. Employers can also improve graduates by implementing a system, which ensures progressive graduation of work activities. These learning curves can be further enhanced by providing a regular feedback mechanism, which strengthens their awareness and abilities. It is also worthy to note that effective learning can be strengthened if recruits are proactive, willing, reflective and creative. This study recommends that employers need to improve on the following to achieve a holistic development of their recruits. These include: developing quality time and resources to train individuals, monitoring of individual interactions, designing task-teams to foster teamwork, ensuring adequate supervision, production of learning maps for trainees amongst others. In summary, it is germane for industry employers to understand their role in further developing their new recruits and employees to ensure a seamless transition into the world of work.

5 CONCLUSIONS

In conclusion, industry employers can further improve the quality of graduates cum employees through a variety of strategies and perspectives. These can be achieved through formal learning activities, which are intentionally planned, informal activities, which are unintentional and self-directed directed learning which can be accomplished by establishing a rapport with their fellow professionals. Through a literature review, the aim of this study, which was to examine the various informal approaches by which industry employers can develop skills and competencies among fresh graduates were achieved. The studies revealed that in-house training, job rotation, coaching and mentoring, and cross-functional teams are some of the key approaches by which industry employers can achieve this. The study found out that, in-house trainings involve symposiums, workshops and seminars, which can lead to improved employee performance and productivity. Job rotation was found to provide employees with adequate work experience and exposure to various functions within the industry. The benefits of coaching and mentoring as well as cross-functional teams were also examined from literature as these activities lead to improved communication and provide adequate knowledge about job roles. Furthermore, informal learning has the potential to occur more frequently than the formal learning process and can also be regarded as incidental learning. The approaches discussed in this study are extremely beneficial to graduates and reckons that the employers of labour have a crucial role to play in the construction education discourse.

Acknowledgments

This research was supported by grants received from the Global Excellence and Stature (GES) Postgraduate Scholarships and supported by the Faculty of Engineering and the Built Environment, University of Johannesburg, South Africa.

References

Abedin, Z., Biskup, E., Silet, K., Garbutt, J. M., Kroenke, K., Feldman, M. D., McGee, Jr R., Fleming, M., and Pincus, H. A., Deriving Competencies for Mentors of Clinical and Translational Scholars, *Clinical and Translational Science,* 5(3), 273-280, June, 2012.

Adjei, D., *The Impact of Job Rotation on Employees' Performance a Case Study: Utrak Financial Services Limited,* MSc Thesis, School of Business, KNUST, 2012.

Beckett, D., and Hager, P., *Life, Work and Learning-Practice in Postmodernity*, Routledge, UK, 2002.

Billett, S., Constructing Vocational Knowledge: History, Communities and Ontogeny, *Journal of Vocational Education and Training,* 48(2):141-54, Jan, 1996.

Boud, D., and Middleton, H., Learning from Others at Work: Communities of Practice and Informal Learning, *Journal of Workplace Learning,* 15(5), 194-202, Sep, 2003.

Carter, A., Executive Coaching: Inspiring Performance at Work, *Institute for Employment Studies*, Brighton UK, June 2001.

Collin, K., Sintonen, T., Paloniemi, S., and Auvinen, T., Work, Power and Learning in A Risk Filled Occupation, *Management Learning*, 42(3), 301-18, July, 2011.

Dale, M., and Bell, J., Informal Learning in the Workplace, *Department for Education and Employment,* London, Aug, 1999.

Dewey, J., *Experience and Education*, New York: Collier Books, 1938.

Doornbos, A. J., Simons, R. J., and Denessen, E., Relations Between Characteristics of Workplace Practices and Types of Informal Work-Related Learning: A Survey Study Among Dutch Police, *Human Resource Development Quarterly,* 19(2), 129-51, June, 2008.

Earney, S., and Martins, A., Job Rotation at Cardiff University Library Service: A Pilot Study, *Journal of Librarianship and Information Science*, 41(4), 213-226, 2009.

Eraut, M., Informal Learning in The Workplace, *Studies in Continuing Education*, 26(2), 247–273, 2004.

Foley, G. (ed.), *Dimensions of Adult Learning: Adult Education and Training in the Global Era*, Allen and Unwin, Maryborough, Victoria, 2004.

Foley, G., *Learning in Social Action: A Contribution to Understanding Informal Education*, Zed Books, London, 1999.

Hager, P., and Halliday, J., *Recovering Informal Learning: Wisdom, Judgement and Community*, Lifelong Learning Book Series, Vol. 7, Dordrecht: Springer, 2006.

Hameed, A., and Waheed, A., Employee Development and its Affect on Employee Performance A Conceptual Framework, *International Journal of Business and Social Science*, 2(13), 2011.

Kettley, P., and Hirsh, W., *Learning from Cross-Functional Teamwork*, Institute for Employment Studies, Brighton, 2000.

Khan, F., Rasli, A. M., Yusoff, R. M., Ahmed, T., ur Rehman, A., and Khan, M. M., Job Rotation, Job Performance, Organizational Commitment: An Empirical Study on Bank Employees, *Journal of Management Info*, 3(1), 33-46, 2014.

Knowles, M., *The Modern Practice of Adult Education: Andragogy Vs Pedagogy*, Associated Press, New York, 1970.

Kumar, A., Singh, S. K. and Kumar, G., Effectiveness of In-House Training on Technical Employees in Biotech Industry, *Journal of Technical Education and Training*, 9(1), 2017.

Lave, J., and Wenger, E., *Situated Learning: Legitimate Peripheral Participation,* Cambridge University Press, Cambridge, 1991.

Lindeman, E. C., *The Meaning of Adult Education*, New Republic, New York, 1926.

Lovelace, K., Shapiro, D. L. and Weingart, L. R., Maximizing Cross-Functional New Product Teams' Innovativeness and Constraint Adherence: A Conflict Communications Perspective, *Academy of Management Journal*, *44*(4), 779-793, 2001.

Mallon, M., Bryson, J., Pajo, K., and Ward, R., *Learning at Work, Organizational Opportunities and Individual Engagement: A Case Study of a New Zealand Wine Company*, 8th conference on International Human Resource Management, Cairns, *Australia*, June, 2005.

Marsick, V. J., and Volpe, M., The Nature and Need for Informal Learning, *Advances in Developing Human Resources,* 1, 1–9, 1999.

Marsick, V. J. and Watkins, K., *Informal and Incidental Learning in The Workplace*, Routledge, London, 1990.

Marsick, V. J., and Watkins, K., "Envisioning New Organizations for Learning," in D. Boud. and J. Garrick (eds.), *Understanding Learning at Work,* Routledge, London, 199–216, 1999.

Munap, R., Badrillah, M. I. M., and Mokhtar, A. R. M., *Graduates Employability Skills: Hard and Soft Skills towards Employee Productivity from the Perspective of Malaysian Employers*, Proceedings of the International Symposium on Research of Arts, Design and Humanities *(ISRADH 2014),* Springer, 151, 2015.

Ragins, B. R., and Kram, K. E., eds. *The Handbook of Mentoring at Work: Theory, Research, and Practice*, Sage, 2007.

Sawchuk, P. H., Informal Learning as a Speech-Exchange System: Implications for Knowledge Production, Power and Social Transformation, *Discourse and Society*, 14(3), 291–307, 2003.

Schultz, D., and Schultz, S. E., Psychology and Work Today, Person, Boston, 136, 144, 2010.

Senge, P. M., *The Fifth Discipline: The Art and Practice of the Learning Organization*, Random House Sydney, Australia, 1990.

Tusting, K., *A Review of Theories of Informal Learning*, Lancaster Literacy Research Centre, Lancaster University, Lancaster, 2003.

Yeo, R. K., How Does Learning (Not) Take Place in Problem-Based Learning Activities in Workplace Contexts?, *Human Resource Development International*, 11(3), 317-330, 2008.

Interdependence between Structural Engineering and Construction Management
Edited by Ozevin, D., Ataei, H., Modares, M., Gurgun, A., Yazdani, S., and Singh, A.

ISBN: 978-0-9960437-6-2

STUDENTS' HOUSING SATISFACTION IN PRIVATE HIGHER INSTITUTIONS

DOUGLAS AGHIMIEN[1], CLINTON AIGBAVBOA[1], JOSHUA ADAMA[2], and WELLINGTON THWALA[1]

[1]*Faculty of Engineering and the Built Environment, University of Johannesburg, Johannesburg, South Africa*
[2]*Dept of Estate Management, Kaduna State University, Kaduna, Nigeria*

Student housing is essential to the attainment of quality education, most especially in higher institutions were students come from far distance to study. While research abounds on happenings with regards to student housing in public higher institutions, there is a paucity of information on this situation within the private higher institutions. This study, therefore, presents the result of an investigation of the satisfaction of students with housing facilities in private higher institutions in Nigeria. The study adopted a survey design and quantitative data were gathered through the use of a questionnaire administered to 216 students selected through stratified random sampling. The students' residence satisfaction was assessed from the physical, and social/management stands. Data gathered were analysed using percentage, mean score and Mann-Whitney U–Test. The study revealed that on a general level, while private institution students are more satisfied with the social and management characteristics of the residence than the physical characteristics. The study observed that although overall satisfaction is derived, there is still adequate room for improvement in order to deliver better housing for students. The outcome of this study provides a possible direction for the management of private higher institutions in improving their students' housing delivery in order to achieve higher students' satisfaction.

Keywords: On- and off-campus, Private university, Student residence.

1 INTRODUCTION

Price *et al.* (2003) observed that student housing is an important feature students tend to consider when enrolling in higher institutions. Unfortunately, the inadequacy and dissatisfaction among students with regards to housing in higher institutions, especially the public institutions in developing countries like Nigeria, have been observed in recent times (Ajayi *et al.* 2015, Akinpelu 2015, Amole 2009, Sawyerr and Yusof 2013). Egunyenga (2009) mentioned that the original idea of student's housing which led to the construction of hostels within the first-generation higher institutions in the country was that of providing comfortable accommodation that will enable students to carry out their primary function of learning. However, the increased number of candidates seeking admission into higher institutions has made the situation of hostels a far cry from what they used to be.

This study recognizes the fact that past research has dwelt on issues relating to students housing in Nigeria (Ajayi *et al.* 2015, Amole 2007 and 2009, Najib *et al.* 2011a, Najib *et al.* 2011b, Oke *et al.* 2017, Oyetunji and Abidoye 2016, Sawyerr and Yusof 2013). However, more focus has been placed on government-owned institutions. Understanding issues surrounding the satisfaction of students from the private institutions is also important, since the growth of private institutions in Nigeria is becoming overwhelming (Adama *et al.* 2018). As of 2012, Nigeria was said to have a total of one hundred and twenty-two universities, and 43% of them are privately owned (Okojie 2012). This number today has undoubtedly increased with more students being admitted into higher institutions in the country. Thus, the satisfaction of students with the housing system in these private institutions requires research attention.

Weidemann and Anderson (1985) introduced the residence satisfaction index, which contains an overall satisfaction question, and three loyalty behavior questions. This satisfaction index has over time been modified to suit different situations (Amole 2009, Najib *et al.* 2011a, Najib *et al.* 2011b, Sawyerr and Yusof 2013). This approach of measuring satisfaction alongside loyalty has been favoured in recent research because it is believed that satisfaction and loyalty levels are linked, with extremely dissatisfied customers engaged in slightly more disloyal behaviour than extremely satisfied ones. Najib *et al.* (2011a) adopted this approach in assessing students' residential satisfaction in Malaysia. Sawyerr and Yusof (2013) while assessing the satisfaction of students with hostel facilities in a federal polytechnic in Nigeria, adopted this approach, but dropped the aspect of loyalty as it was stated that students' payment of hostel fee in Nigeria is made on session basis. Hence, students have no say on whether or not they will be staying on or living the hostel within the session.

A similar approach was adopted in this study in assessing students' residence satisfaction in a private higher institution in the country. However, while previous studies assessed students' housing satisfaction through the physical and social characteristics of the students' residence, this study included some management characteristics. This was done since private institutions are owned by individuals and they are majorly profit-oriented; it is, therefore, possible that certain management factors such as hostel policies are bound to play a vital role in the determination of students' satisfaction. This is in line with Azeez *et al.* (2016) submission that factors influencing residential satisfaction can be broadly categorised as physical, social and management factors.

2 METHODOLOGY

This study assessed students' satisfaction with on and off-campus residence, in a private University in Osun State, Nigeria. A quantitative survey approach was adopted due to the need to gather first-hand information from students within the institution, and a stratified random sampling method was employed. Students were first divided into strata based on their academic level, before being randomly selected. Students of 200, 300, and 400 levels took part in the study, while the 100 level students were exempted based on the fact that they are made to stay in hostels provided by the institution. As result, these set of students do not have a choice as regards their place of residence. Preliminary investigation reveals that these three levels of students cover a population of about 2000. Using the Yamane 1967 formula of $n = N/1+N(e^o)$, with a significance level (e^o) of 0.05, a sample size of 333 was derived for the study.

The instrument for data collection was a structured questionnaire, with an equal number distributed to students in the three levels. The questionnaire used was designed in two parts and variables measured were gathered from extensive review of existing literature. Part 1 dwelt on the background information of the students, while Part 2 addressed the satisfaction of students with their respective residence. The students were provided with some satisfaction requirement and were asked to rate them on a 5-point Likert scale based on their level of satisfaction. In the scale, 1 represents highly dissatisfied, 2 being dissatisfied, 3 being slightly satisfied, 4 being satisfied, and 5 being highly satisfied. In order for the respondents to have a clear range for their answer, a scale of between 0 - 20% was set for highly dissatisfied, 21 - 40% for dissatisfied, 41 - 60% for slightly satisfied, 61 – 80% for satisfied, and above 80% for a highly satisfied. From the 333 questionnaires distributed, 242 were retrieved out of which 216 were ascertained fit for analyses. The remaining 26 were dropped due to incorrect filing and missing vital details. Analyses of data were done using percentage, frequency, mean item score, and Mann-Whitney U-Test which is a non-parametric test used in testing the statistically significant difference in the view of two groups. The reliability of the questionnaire was analysed using Cronbach's alpha test. The alpha value of 0.778 was derived for the satisfaction variables and this shows that the instrument is reliable since the derived value is closer to 1 (Moser and Kalton 1999).

3 FINDINGS AND DISCUSSIONS

3.1 Background Information

Background information revealed that 98 male students and 118 female students took part in the survey, 94 of which stays in hostels provided by the institution within the campus. The remaining 122 students reside in accommodations off campus. Also, the majority of those staying on campus are the female students (63), while more male students (67) prefer to stay off campus. Majority of the students (112) fall within the age range of 15 to 20 years, while 95 of them fall within the range of 21 to 25 years. The age range with the lowest number of students is 26 to 30 years. Also, most of the younger students (15 to 20 years) stay on campus while the older ones (21 to 30 years) stay off campus. This implies that some measure of restriction exists among the younger students in terms of staying alone outside the school vicinity. More students from the 200 level (82) participated in the study. This is followed by the 300 level students (76), with the least level being 400 levels (58). Only a few students tend to stay on campus when they are in their final year, as only 19 students out of the 58 stated that they stay on campus. More students in 200 level can be found staying on campus. This can be linked to the fact that it is mandatory for them to stay on campus in their first year in the institution, hence, the tendency of wanting to stay there in their second year might be high. However, as they move higher, they tend to outgrow the school environment and the urge to stay off campus increases.

3.2 Students' Housing Satisfaction

Results in Table 1 show that students in the private institution are satisfied with the physical requirements of their hostels, as an overall satisfaction level of 62% was derived. More satisfaction is derived in the aspect of water supply, electricity supply, security, bedroom, bathroom, and waste disposal facilities. Slight satisfaction is derived in the aspect of the toilet, reading room and internet facilities, with considerable dissatisfaction with common room facilities. This implies that the school management and off-campus residence providers need to improve in these areas in other to increase their students' satisfaction level.

It is worthy to note that the Mann-Whitney U-test conducted shows that there is no statistically significant difference in the view of students staying on campus, and those staying off-campus with regards to their level of satisfaction with 9 out of the 10 identified physical requirements. A significant p-value of above 0.05 was derived for these 9 requirements. However, a significant p-value of 0.003 was derived for toilet facilities. This implies that there is a significant difference in the level of satisfaction derived by both sets of students as regards this facility in their respective residences. Reason for this disparity can arguably be linked with the fact that most students staying off campus tend to stay in accommodations where they have their toilet facilities to themselves or at worse have a limited number of people to share with. The same situation does not apply to those on campus as they have to share the toilet facilities with other students within their hostel.

Table 1. Students' satisfaction with the physical characteristics of their residence.

Physical Characteristics	MIS	Rank	%	Remark	Z.	Sig.
Water Supply	3.71	1	74	*Satisfied*	-2.807	0.005
Electricity Supply	3.67	2	73	*Satisfied*	-0.970	0.332
Security facility	3.63	3	73	*Satisfied*	-1.179	0.238
Bedroom facility	3.54	4	71	*Satisfied*	-1.529	0.126
Bathroom facility	3.17	5	63	*Satisfied*	-1.809	0.070
Waste disposal facilities	3.10	6	62	*Satisfied*	-1.230	0.219
Toilet facilities	2.86	7	57	*S. Satisfied*	-3.015	0.003**
Reading room facility	2.76	8	55	*S. Satisfied*	-1.933	0.053
Internet facilities	2.54	9	51	*S. Satisfied*	-0.500	0.617
Common room	1.84	10	37	*Dissatisfied*	-1.962	0.053
Overall Satisfaction	**3.08**		**62**	***Satisfied***		

Note: *MIS = Mean Item Score, %= Percentage, S. Satisfied =Slightly Satisfied*

For the students' satisfaction with the social and management characteristics of their residents, some social and management features were identified from literature and respondents were asked to rate them based on their level of satisfaction. Results in Table 2 show that students are satisfied with the social and management aspects of their residence as a 68% satisfaction level was derived. Higher satisfaction can be seen in the aspect of the behavior of the landlord or officer-in-charge of the hostel, and the availability of transportation. Mann-Whitney U-Test conducted shows that there is no statistically significant difference in the satisfaction of students staying on and off campus. A significant p-value of above 0.05 was derived for all the assessed requirements.

Table 2. Students' satisfaction with Social and Management characteristics of their residence.

Social and Management Characteristics	MIS	Rank	%	Remark	Z.	Sig.
The behavior of the landlord or officer-in-charge of the hostel	3.59	1	72	*Satisfied*	-0.739	0.460
Availability of transport from the hostel/house to lecture halls	3.55	2	71	*Satisfied*	-0.587	0.557
Affordability of the housing fee	3.46	3	69	*Satisfied*	-1.336	0.182
Interpersonal relationship among tenants/students at the hostel	3.34	4	67	*Satisfied*	-1.193	0.233
Pleasantness and quietness of the environment to allow for easy study	3.28	5	66	*Satisfied*	-1.280	0.201
Population of students using the facility	3.27	6	65	*Satisfied*	-0.365	0.715
Overall Satisfaction	**3.41**		**68**	***Satisfied***		

Note: *MIS = Mean Item Score, %= Percentage*

Results in Table 3 show the overall residential satisfaction level of students of private higher institutions. From the table, it is evident that a higher satisfaction level of 68% is derived in the aspect of social and management characteristics as against that of the physical requirement of 62%. The general students' housing satisfaction is 64%, which shows that the students are generally satisfied with their residence. However, this satisfaction level can still be improved upon to get a higher residence satisfaction among the students.

Table 3. Overall student's housing satisfaction.

Students' Residential Satisfaction	MIS	%	Remark
Physical	3.08	62	*Satisfied*
Social and Management	3.41	68	*Satisfied*
Overall Satisfaction	**3.25**	**64**	***Satisfied***

3.3 Discussion of Findings

Findings of this study reveal that students are satisfied with their residence, with more satisfaction being derived in the aspect of social and management criteria than the physical aspect. This satisfaction level shows that the situation of facilities in a private institution in the country is somewhat better than what is obtainable in government institutions as observed in the studies of Ajayi *et al.* (2015), Amole (2009) and Sawyerr and Yusof (2013). However, more focus is needed in the area of physical characteristics, as this area tends to have lower satisfaction level when compared to that of the social and management characteristics. This finding further corroborates Danso and Hammond (2017) submission that students in KNUST, Ghana are generally satisfied with their campus residence. It is also in tandem with Ifaturoti (2017) submission that students are generally satisfied with residence in Akoka, Lagos State, Nigeria.

The general students' residential satisfaction level shows that students' in private institution in Nigeria are satisfied with their housing facilities. This is in contrast with what is obtainable among students in public institutions in the country. The findings also show that private institutions are gradually taking their student residence standard to a satisfactory level as obtained in other countries around the world. This is because a similar level of satisfaction has been recorded in countries like Ghana (Danso and Hammond 2017), South Africa (Oke *et al.* 2017), Malaysia (Khozaei *et al.* 2014, Najib *et al.* 2011a), and Norway (Thomsen 2010). To maintain overall improvement, the government, therefore need to follow suit by improving their delivery of housing facilities for students in the public institutions. This will also assist them to retain and attract more students to government-owned institutions.

4 CONCLUSIONS

Based on the findings, this study concludes that both the physical features and the social and management aspect of students housing in private higher institutions can still be improved upon to attain better housing satisfaction for students within these institutions. On a more unified view, a satisfactory level was derived for the students' housing satisfaction in private higher institutions. This study has contributed to the body of knowledge as it brings to light housing satisfaction level of students in private higher institution in the country; an aspect that has been devoid of adequate research attention in the discussion of students housing satisfaction. It is believed that the findings of this study will help private institutions management to improve on their students' residence delivery in order to achieve higher students' satisfaction, better student's performance and increase in profit. As for the government, the findings of this study will help them understand the need to improve students housing facilities being provided in the public institutions in a bid to compete favourably with their private counter who seems to be delivering better housing to their students. The major limitation of this study lies in the use of a single private higher institution in the country, thus further studies can be carried out in other private institutions around the country in other to compare findings.

References

Adama, J. U., Aghimien, D. O., and Fabunmi, C. O., Students' Housing in Private Universities in Nigeria: Influencing Factors and Effect on Academic Performance, *International Journal of Built Environment and Sustainability*, 5(1), 12-20, 2018.

Ajayi, M., Nwosu, A., and Ajani, Y., Students' Satisfaction with Hostel Facilities in Federal University of Technology, Akure, Nigeria, *European Scientific Journal,* 11(34), 402-415, 2015.

Akinpelu, O. P., Students' Assessment of Hostel Facilities in The Polytechnic Ibadan, Ibadan, Nigeria: Realities and Challenges, *Research on Humanities and Social Sciences*, 5(17), 74-81, 2015.

Amole, D., A Study of the Quality of Student Residential Facilities in Nigeria, *Planning for Higher Education*, 35(4), 40-50, 2007.

Amole, D., Residential Satisfaction in Students' Housing, *Journal of Environment Psychology*, 29(1), 76-85, 2009.

Azeez, T., Taiwo, D., Mogaji-Allison, B., and Bello, A., Comparative Assessment of Students' Satisfaction with Hostel Accommodation in Selected Private Universities in Ogun State, Nigeria, *European Scientific Journal*, 12(32), 410-425, 2016.

Danso, A. K., and Hammond, S. F., Level of Satisfaction with Private Hostels around Knust Campus, *International Journal of Science and Technology*, 6(3), 719-727, 2017.

Egunyenga, E. J., *Room – Mate Conflicts in Nigerian Universities: Causes and Management Strategies*, Department of Educational Administration and Policy Studies, Delta University, Abraka, Nigeria, 2009.

Ifaturoti, A. S., Residential Satisfaction in Private Tertiary Student Housing Facilities in Akoka, Lagos State, Nigeria, *Advancement in Science and Technology Research,* 3(2), 8-14, 2017.

Khozaei, F., Hassan, A. S., Al Kodmany, K., and Aarab, Y., Examination of Student Housing Preferences, Their Similarities and Differences, *Facilities,* 32(11/12), 709-722, 2014.

Moser, C. A. and Kalton, G., *Survey Methods in Social Investigation,* 2nd Edition. Gower Publishing Company Ltd, Aldershot, 1999.

Najib, N. U. M., Yusof, N., and Abidin, N. Z., Student Residential Satisfaction in Research Universities, *Journal of Facilities Management*, 9(3), 200-212, 2011a.

Najib, N. U. M., Yusof, N., and Osman, Z., Measuring Satisfaction with Student Housing Facilities, *American Journal of Engineering and Applied Sciences*, 4(1), 52-60, 2011b.

Oke, A. E, Aigbavboa, C. O., and Raphiri, M. M., Students' Satisfaction with Hostel Accommodations in Higher Education Institutions, *Journal of Engineering, Design and Technology*, 15(5), 652-666, 2017.

Okojie, J., *The Punch Newspaper*, Wednesday, February 29th, 9, 2012,

Oyetunji, A. K., and Abidoye, R. B., Assessment of the Factors Influencing Students' Choice of Residence in Nigerian Tertiary Institutions, *Sains Humanika,* 8(2), 39–47, 2016.

Price, I., Matzdorf, F., Smith, L., and Agahi, H., The Impact of Facilities on Student Choice of University, *Journal of Facilities,* 21(10), 212-222, 2003.

Sawyerr, P. T., and Yusof, N., Student Satisfaction with Hostel Facilities in Nigerian Polytechnics, *Journal of Facilities Management*, 11(4), 306-322, 2013.

Thomsen, J., *Student Housing Student Homes? Aspects of Student Housing Satisfaction,* PhD Thesis, Norwegian University of Science and Technology, Trondheim, 2008.

Weidemann, S., and Anderson, J. R., A Conceptual Framework for Residential Satisfaction, *in Altman, I. and Werner, C.M. (eds), Home Environments: Human Behaviour and Environment*, Plenum Press, New York, NY, 8, 153-81, 1985.

Interdependence between Structural Engineering and Construction Management
Edited by Ozevin, D., Ataei, H., Modares, M., Gurgun, A., Yazdani, S., and Singh, A.

ISBN: 978-0-9960437-6-2

GLOBALIZATION AND CONSTRUCTION ENGINEERING MANAGEMENT EDUCATION

HOSSEIN ATAEI

Dept of Civil and Materials Engineering, University of Illinois at Chicago, Chicago, USA

Globalization is the study of dramatic global changes and their consequences in different fields of economy, technology and communications during the past two to three decades. Similar to other industry sectors or professional practices, globalization has enabled the construction industry to take advantage of cross-border markets and benefit from the opportunities that the globalized economy has provided. Advances in transportation, data sharing and communication have given construction professionals and the labor force unprecedented access to markets that were previously insulated. As the political borders have turned to be increasingly irrelevant and the national economies have become more interdependent, the socio-cultural differences are hence accentuated and further exposed due to historically-existing dissimilarities in societal cultures and business norms. In this paper, the main goal is to understand the globalization and its implications for the world economy specifically focusing on the construction industry. This paper studies the impacts of globalization in construction engineering and management profession and further investigates on the opportunities and threats that a globalized economy and more-synchronized business practices will bring for industry and how Construction Engineering educators shall successfully address those unprecedented issues in teaching Construction Engineering Management courses to the next generation of the industry leaders.

Keywords: Global mindset, Cross-cultural project management, Stakeholder diversity, Training, Academic, International markets.

1 INTRODUCTION

In the past decades, the global construction industry has grown rapidly as the demand for residential and commercial construction as well as civil infrastructure systems has grown considerably. The demand for construction output is projected to hit record highs by 2021 with an output of US$10.1 Trillion (Timetric 2017). Much of this is made possible through globalization in the construction industry. As the desire for bigger cities increases in developing countries, but limitations of resources such as knowledge, labor force, and materials become a challenge, well-established companies in developed countries seek to capitalize on these opportunities. Yet, the risk has grown parallel to the output as companies face new challenges in the international construction industry.

This research focuses on the changes in construction industry over time due to the global influences as well as the impacts and consequences that it brings. It looks into the role of the Construction Engineering and Management academic curricula to better educate and prepare students for new future challenges that they will face in foreign markets.

The global market in the construction industry outside of the United States has opened up a wider network of opportunities for companies from different countries who have gained advantages in revenues and the new market exposure. However, projects carried out in the United States have experienced much smaller effects of globalization than that of projects carried out in foreign markets.

With its geographical location and strict foreign work and investment policies, domestic companies - many of which are multi-national - carry much of the construction work in the US. However, as more opportunities open in foreign markets, companies from United States have been motivated to participate and offer their services abroad. The lack of awareness and experience in dealing with multicultural work force puts many entry-level or mid-career US construction engineering professionals at higher risk and uncertainty when dealing with large and complex projects carried out in other countries. In an ever-changing industry, future engineering graduates who specialize in construction management face more challenges ahead compared to their peers of the past.

This can also be attributed to the fact that much of the academic curricula in construction engineering and management in the universities and institutions across the US focuses mainly on technical aspects and issues that are seen at a domestic scale. While these are very important fundamentals to understand so that the students are prepared when entering the workforce, such curriculum designs disregard many of the challenges that engineering students will face in foreign markets through their career journeys.

2 GLOBALIZATION AND THE ACADEMIC EDUCATION

Globalization requires a proper education for humanity to understand its ideologies. As it involves social evolution on a worldwide level, collective cooperation is a necessity, and therefore the weight of education becomes most apparent (Schneider and Barsoux 2003). Local or national market boundaries no longer restrain the construction projects; supply and demand flows through a larger, singular market where any of the parts and pieces - ranging from bricks and mortar to human capital and design concepts - can come from corners of the world and used to build a project in other places.

The extensive construction projects that followed the advent of modern construction materials formed the background of what emerged as modern construction industry. Most building construction remained in the hands of small and medium-sized local contractors. However, civil engineering projects required much larger scale operations and, hence, formed a launching pad for international construction (Ngowi *et al.* 2005). In construction industry, the essence of the market focus and position is targeting the right business at the right moment in the right market. The industry's leading players have bolstered their competitiveness by securing strategic positions in high-growth emerging markets such as Brazil, Russia, India, China and South Africa (commonly referred to as BRICS countries), taking advantage of domestic infrastructure opportunities while securing lucrative contracts overseas.

In global construction market, financial success in project delivery is sought by securing global competitiveness, diversifying revenue sources and project portfolios. The volume of international contracts has contributed to firms' sustained growth by mitigating the impact of the domestic market's cyclical nature. Reduction in the influence of market's cyclical nature ensured a level of financial stability for multi-national construction firms through increased global activities (Low and Lim 2000, You and Zi 2007). For construction industry, globalization has been a trend of augmented exposure to the opportunities and competition in international markets (Toor and Ofori 2008). The larger contractors in the United States, for example, must compete alike against the multi-billion-dollar Engineering-Procurement-Construction firms worldwide. This competition, however, directly and adversely affects mid-sized and smaller, local firms which do not have the means, methods or logistics refuse to adjust towards the ever-growing opportunities in international market (NRC 2009, Scholte 2005, Tomlinson 1999, Gunhan and Arditi 2005).

International construction can be riskier than domestic construction because of difficulties in communication management, understanding new cultures and obtaining market wisdom, avoiding local politics, as well as supervising a diverse group of professionals. According to Abdul-Aziz (1994), in a study of nine large US and Japanese construction companies, it was concluded that long-term profitability and balance of growth are the two prime concerns for going global. Moreover, technological advantages, sophisticated management systems for scope, cost, schedule, quality, risk and procurement for construction projects are among the main factors for success of a construction company in international markets. In order to enter new markets, construction companies must assign a much higher importance on client relationship management within their organizations vis-à-vis the culture and business norms of the country in which they are seeking to bid on a new project.

2.1 The Global Mindset

The Global Construction Perspectives and Oxford Economics Report (2009) predicts that construction in emerging markets, including Asia, Latin America, the Middle East and North Africa account for approximately 55% of global construction output. In such increased demand for new construction in these markets, the US companies will face a fierce competition with new players from other economic giants such as China and India. This demand will prepare the next generation of the US construction engineering and management professionals with a "Global Mindset".

In today's fast-pacing global construction industry, the barriers to global trade have been reduced significantly which enables capital, materials, labor, and technology to flow freely across borders. However, globalization has considerably increased the complexity of the engineering and construction business environment: Increased project demand and new funding mechanisms drive the stakeholders to diversify and seek mergers in newly-developed international markets to gain further and uninterrupted access to new opportunities.

To compete in today's uncertain, dynamic and interactive business environment, design and construction companies, domestic and foreign alike, need to focus on key trends that shape their external operating environment. New generation of construction project managers must be equipped with technical competencies and soft skills to anticipate and precisely observe and act upon emerging trends to find market opportunities, test the inherent risks and seize on new opportunities. In a world of rapidly changing of resources, competition and customer demands, the successful future leaders are the ones who look beyond traditional performance standards and the short-term thinking to explore, shape and implement inclusive multi-cultural financial strategies to address the complex sets of project stakeholders' needs.

2.2 Globalization and Construction Education Challenges

According to Han *et al.* (2010), these are the critical changes that have challenged the global construction engineering and management industry: (1) competition paradigm has been shifting from conventional price competition to a more complex competitive framework where intangible factors are gaining more traction; (2) the steady rise of private investment schemes and the growing size of projects requires a more comprehensive scope management in planning, executing and monitoring the construction projects; (3) financing capability plays a key role in acquiring more opportunities in the international construction market as more international construction projects require newer financing arrangements; (4) price fluctuations and unpredictable increased costs of materials, equipment and labor in emerging markets that result in more complicated risk factors stemming from economic instability and cross-cultural business deal making.

International construction can be riskier than domestic construction. Increased project delays, scope or budget overruns may be resulted by managing projects in an unfamiliar business environment and/or operating and complying with different sets of regulations, business norms, and cognitive-cultural beliefs of the diverse project stakeholders.

The main challenge in front of the construction engineering and management programs in the globalized economy is to train the next generation of industry leaders and project management professionals to be creative thinkers and problem solvers. These professionals shall possess all or as many as of the following attributes: (a) technically competent in their field of expertise with a broader knowledge of the global economy, opportunities and threats in international markets; (b) lifetime learners of cross-cultural communication skills; (c) interdisciplinary problem solvers in managing broader and more diverse project stakeholders; (e) comfortable in managing projects with high levels of risk and uncertainty; (f) creative global mindset to identify and research on market capabilities for future business development purposes.

3 TRAINING THE NEXT GENERATION OF CONSTRUCTION ENGINEERING AND MANAGEMENT PROFESSIONALS AS GLOBAL COLLABORATORS

In order to adapt to emerging global trends and surviving the highly dynamic future markets, the construction engineering and management firms need to hire, train and retain the next generation of project management professionals who will be advancing the industry and sustaining the long-term profits of the firm through incorporating the following strategies: (1) developing a comprehensive understanding of financing structuring options for managing construction projects in global scale; (2) practicing niche marketing by identifying and focusing on prosperous international target markets; (3) collaborating with various multi-background stakeholders of complex international projects; (4) adapting to changing social environment and platforms; (5) enhancing the comprehensiveness of the Engineering-Procurement-Construction (EPC) services of the firm in undertaking global construction projects; (6) investing in new technologies, skill-sets and expertise; (7) incorporating human resources practices to train next generation of the industry leaders with globalization mindset for the future of the firm.

As new competitors enter the scene of domestic and international competition in construction, industry, the local and global engineering, design, and construction companies must response with new business solutions to target greater involvement of clients and reassessing their risk. Emphasis on greater efficiency and innovation, in particular, will ensure that a larger number of stakeholders can rise to the challenge of the new competitive fields both at home and abroad. Today's successful industry firms are providing professional services through incorporating human resources practices that understand the importance of hiring the best professionals with risk management capabilities, market knowledge and a deep understanding of cross-cultural client management especially when performing under pressure and tight deadlines in international markets.

3.1 Best Practices in Achieving Cross-Cultural Leadership Potential Realization

As the global engineering, procurement and construction (EPC) firms would be implementing newer and more holistic strategies to be more prepared to face the business challenges of globalization, acquiring and retaining the next generation of industry professionals with global mindset will play a vital key role.

Here are a few best practices for design and construction firms in achieving cross-cultural leadership potential realization that focuses on high-skilled global multi-background workforce and engineering teams.

3.1.1 *View competitors as potential collaborators*

In a globalized competition to win and deliver on more complex construction projects in international markets, both larger and smaller design and construction firms can work together towards successful project planning, executing and delivery.

Larger companies usually bring large-scale capabilities, financial resources, higher-end technologies or knowledgeable staff to the project, while smaller companies may offer a particular specialized staff or a niche set of expertise to an international project in which a larger firm may be interested.

3.1.2 *Understand the cost and risk barriers to entry*

A comprehensive risk management plan and sophisticated financial structuring strategies are among the main considerations for construction firms in doing business in international markets. The cost of entering international markets is usually much higher for construction firms as it requires detailed strategic planning, highly-skilled professional workforce with global mindset and new technologies to benefit from economies of scale in executing overseas projects. Therefore, risk management becomes a major factor in strategic and future planning.

Allocating the intricacies of risk usually present a new ground for contractors to cover when doing business outside of traditional borders.

Public Private Partnership (PPP) Projects specifically can be very challenging and often bring greater risk in terms of longer terms, larger scale of liability and increased vulnerability to the changes in external dynamics as the project progresses.

3.1.3 *Communicate and innovate to create new adaptation strategies*

In today's dynamic global business market, almost all international engineering and construction jobs are noticeably complex and collaborative. The project engineering teams are more diverse, and seldom can an engineering project be run in an isolated and closed manner.

Newly introduced technologies and increasingly sophisticated demands of various project stakeholders are pushing construction and design professionals to work in multi-background diverse project teams that consist of people from different cultures, ethnicities and religions. Therefore, managing communications among multitude of project stakeholders is a key factor towards successful project delivery. This important goal may be achieved, as more companies will be hiring, training and retaining construction professionals with a global mindset in order to build reliable networks to work in innovative ways.

4 CONCLUSIONS

The construction engineering and management curricula need to be accordingly adjusted to incorporate the effects of globalization in construction engineering industry. This will certainly include increased focus on interdisciplinary subjects such as (but not limited to): ethical and cultural awareness, cross-cultural business deal-making, international market strategies, integrated project management in foreign markets, multi-cultural stakeholder management, knowledge of negotiation techniques, developing leadership and inter-personal skills and learning one or more foreign (i.e., non-English) languages.

The construction engineering education in the era of globalized economies has to be conducted with a global mindset. The next generation of construction project managers need to be trained through a globalized education in their field of expertise to successfully master essential attributes and the problem-solving skills to meet the demands of the evolving global markets facing challenges of sustainability and green energy alternatives, increased global population, moral and environmental awareness and the global climate change issues. US universities may achieve such goals not only by revisiting and modifying their current construction engineering and management curricula but also through establishing academic collaborations with international counterparts in the form of faculty exchange and/or student exchange programs. Moreover, inclusion of international internship or co-op opportunities within the construction engineering curricula, across the nation, can serve the students who are seeking to acquire cross-cultural awareness experience in starting their careers to be better prepared for future challenges.

References

Abdul-Aziz, A. R., Global Strategies: A Comparison Between Japanese and American Construction Firms, *Construction Management and Economics*, 12(6), 473-484, 1994.

Gunhan, S., and Arditi, D., Factors Affecting International Construction, *Journal of Construction Engineering and Management*, 131(3), 273-282, 2005.

Han, S. H., Kim, D. Y., Jang, H. S., and Choi, S., Strategies for Contractors to Sustain Growth in the Global Construction Market, *Habitat International*, 34, 1-10, 2010.

Low, S. P., and Lim, N. H., The Strategic Responses of Construction Firms to the Asian Financial Crisis in 1997–1998, *International Journal of Construction Marketing*, 1(2), 1-12, 2000.

NRC, *Advancing the Competitiveness and Efficiency of The US Construction Industry*, National Research Council, National Academies Press, 2009.

Ngowi, A. B., Mbachu J., Pienaar E., Talukhaba A., The Globalisation of The Construction Industry- A Review, *Building and Environment*, 40(1), 135-141, 2005.

Schneider, S. C., and Barsoux, J. L., *Managing Across Cultures*, Pearson Education, 2003.

Scholte, J. A., *Globalization: A Critical Introduction*, Palgrave Macmillan, 2005.

Timetric, *Key Highlights from Global Construction Outlook 2021*, Construction Intelligence Center (CIC), 2017

Tomlinson, J., *Globalization and Culture*, University of Chicago Press, 1999.

Toor, S.-ur-R., and Ofori, G., Developing Construction Professionals of the 21st Century: Renewed Vision for Leadership, *Journal of Professional Issues in Engineering Education and Practice*, 134(3), 279-286, 2008.

You, T., and Zi, H., The Economic Crisis and Efficiency Change: Evidence from the Korean Construction Industry, *Applied Economics*, 39(14), 1833-1842, 2007.

Interdependence between Structural Engineering and Construction Management
Edited by Ozevin, D., Ataei, H., Modares, M., Gurgun, A., Yazdani, S., and Singh, A.

ISBN: 978-0-9960437-6-2

ADOPTION OF CLOUD BASED LEARNING IN INSTITUTES OF HIGHER LEARNING

OLAWALE KUKOYI and CLINTON AIGBAVBOA

Faculty of Engineering and the Built Environment, University of Johannesburg, Johannesburg, South Africa

Every developing nation is moving away from the typical classroom teaching to the evolution of the 21st century. Most developed countries and nations have progressed beyond traditional learning methods in classrooms to virtual environment means of learning, which includes cloud-based learning. However, the same cannot be said of the academic system in institutes of higher learning in Nigeria. Based on this knowledge, this study would aim to look at the challenges faced in the adoption of cloud-based learning and teaching service delivery in institutes of higher learning in Nigeria. It will also consider the benefits and drivers for cloud-based learning in the country. This study shall be conducted using systematic reviews from relevant literature published on cloud-based learning and indexed on top rated databases like Springer, Emerald insight and Scopus. Findings from this study would show the benefits and drivers for adopting cloud-based learning and its impact on those promoting academic productivity in the country, which includes the learners in institutes of higher learning in Nigeria. Further findings from this study will highlight ways to deal with challenges observed from the adoption of cloud-based learning in institutes of higher learning in the country and how governmental support can aid in using cloud-based learning to improve the quality of education delivered in institutes of higher learning in Nigeria as benefited by developed nations across the world.

Keywords: Benefits, Challenges, Teaching service delivery, Nigeria.

1 INTRODUCTION

In the advent of every nation moving away from typical classroom teaching to that of the 21st century, the use of diverse information technologies has become essential. Balasubramanian *et al.* (2009) observed that information technology strengthens a nation's higher education institutes, in the areas of teaching, research and service to the community. One of such information technology innovations is cloud computing, whose immense benefits have been noted in recent times in quality education delivery. According to Thomas (2011), the use of cloud computing will, to a large extent, enhance engagement by educators to better understand and improve teaching practice. This will increase quality of student learning outcome via advancing the scholarship of teaching and learning. Cloud computing is within the domain of recent evolution across the information technology world and academia. Gital and Zambuk (2011) defined cloud computing as a model for enabling on demand network access to a shared pool of resources or computing resources.

In 2010, a learning-based system called a cloud-based program that interfaces with all databases containing information from user courses and content was created for use Pina (2012).

This can be used as an educational means for students irrespective of their locations. Information and resources are available without searching for them manually via catalogued books, which may not be available physically. This saves both time and solves problems alluded to this feature (Kats 2013).

Recent trends in teaching and learning strategies shows that the presentation of information has changed, and this is evident in the way students interact with information using technological innovation. This innovation, such as cloud-based learning, demands educators to use new technologies in the delivery of teaching services (Chang and Guetl 2010). This has led to a better learning process and ease in teaching. Despite this innovations and obvious benefits inherent in the use of cloud base learning, developing countries around the world (Nigeria inclusive) is still lagging in its adoption. Eze *et al.* (2014) carried out an investigation on the adoption of cloud-based learning in Nigeria and observed that cloud-based learning has not been fully exploited due to poor regulatory policies and lack of governmental support. This is rather disheartening considering the poor education system within the country. Based on this knowledge, this study through the review of relevant literature assessed the possible ways of adopting cloud-based learning within institutes of higher learning in Nigeria. The study assessed the challenges of cloud-based learning, the factors that can drive its adoption and its inherent benefits to the education system in the country.

2 TEACHING SERVICE DELIVERY IN INSTITUTES OF HIGHER LEARNING IN NIGERIA

Teaching service delivery in institutions of higher learning in Nigeria requires a need to meet up with counterpart nations in the world. In 2010, Arong and Ogbadu realized that instructional materials needed in teaching and student learning process in the country need be enhanced in Nigeria due to its lack in the country (Arong and Ogbadu 2010). This led to a review by Oye *et al.* (2011) who advocated for the use and acceptance of information and communication technology for teachers and learners of higher education in Nigeria thereby adding substance to education in the country. Asiyai (2015) realized that for Nigeria to meet and make up for the deficiency in teaching service delivery, there was a need for more training on the latest method of teaching service enhancements used by developed nations. This would require teaching enhancement developments like cloud-based learning and e-learning methods in improving teaching effectiveness in the Nigerian teaching service curriculum and the country which will enable the country meet up in part with teaching service delivery in the 21st century. For Nigeria as a country to improve in service delivery in education, stakeholders both in the governmental control end, teachers, and learners would have to support each other and get trained in the use of digital and online resources to meet present trend from nations globally. This will increase quality teaching and learning service delivery in the nation (Asiyai 2013).

In Nigeria today, advancements in the use of educational technology tools and gadgets are required to bridge the gap of providing online resources for a central teacher/learner instructional environment. Students of the present generation in the country need a means to meet up with the social networked resources for training and educational delivery in the nation (Laleye 2015).

3 OVERVIEW OF CLOUD BASED LEARNING

Developed nations have moved away from the traditional need to be in classrooms only to access good means of education to online platforms as a means of additional education via using growing technologies to redirect learners to get educated from any location without being in a classroom and yet achieve set educational goals (Kotoua *et al.* 2015). The problem faced from this is that most educators who are skilled in the art of imparting this online means of education are mostly in developed nations, as reiterated by (Adeyinka 1975, Kotoua *et al.* 2015). To enhance the study of this form of online education, online-learning tools to aid teaching and learning processes were created as a means of informal learning. Cloud technologies are being used for easy access of information and materials by both learners and institutions worldwide and its services include exchange of data and information using online social means to achieve the purpose of knowledge and information gathering by learners, institutions and workers (Garcia-Penalvo *et al.* 2014).

The use of social means of learning has overtaken the traditional means of information gathering, hence the reason why a social means of approach to bolster the educational learning system as advocated by Barrantes and Yagüe (2015), which considers the effectiveness and sustainability of a social approach to learning through cloud-based learning. Conclusions were reached after collating data for three years by Wannapiroon and Nookhong, that innovations based on a social means of learning are more effective and sustainable to the learning process. In 2015 social media and e-education, strategies used in a model developed in Thailand spoke on the basic skill for working and learning in the 21st century educational system as a means of survival, this model was used for solving problems on cloud-based technology literacy for effective educational management in the future. This result showed that the use of cloud-based learning helped learners in learning both fast and effectively (Nookhong and Wannapiroon 2015).

A systematic approach to the innovation of the traditional mode of teaching in classrooms and the use of virtual learning environment combined in a form called blended learning. This was done incorporating the principle behind being present in class and getting instructions note and data guidelines from an online source and social means of interaction in higher educating institutes as described by Graham, Woodfield and Harrison in 2012. The study adopted ways to implement blended learning of this type into higher institutes of learning using six institutes in the United States to achieve this as a case study (Porter *et al.* 2014).

To meet present global market excellence in developing economic countries, the government of these countries are involving their resources into polices to provide adequate internet and information technology to their countries. To safeguard this, there must be the creation of adequate policies and infrastructure that would make it possible for cloud-based learning in their country via provision of adequate funding for education with this form of learning (Ezenwoke *et al.* 2013). As a continual part of integration of cloud-based learning in developing countries especially in Nigeria, it is both efficient and cost saving for students of higher education irrespective of their certification to submit dissertations and research thesis. To access this for knowledge virtual access would be created for further study or access via databases in the cloud through their digitalization (Ifijeh 2014).

Cloud based learning cannot fully be accepted if we do not understand its advantages from the learner's perspective as said by Ashtari and Eydgahi (2017), who captured the learner's perspective in their research to the effectiveness of cloud-based learning. They termed it an attractive way in sharing, editing and storage of huge amount of information and wealth of knowledge for use yet, also described it as having ease of access with internet usage via using their own equipment's which are mobile devices or institution equipment's. Most times, even without leaving their location or taking on cost for excessive printing or time management to get resource and data (Ashtari and Eydgahi 2017). This method of data storage is used in Saudi Arabian universities already due to their amount of low-cost information technology resources in teaching and researching learning. They moved all this information together into a cloud-based database for access to everyone for use and continual upgrading. A downsize of this was noted to be from security of data and lack of legal policy by the government for implementation (Alharthi *et al.* 2017).

The necessity of having a cloud-based system of learning increases every day due to reduction of learning cost by less purchase of hard copy materials and the use of stored materials in the cloud for continual education into the future which has aided learners in their academic growth using e-learning (Bouyer and Arasteh 2014).

4 CHALLENGES OF CLOUD BASED LEARNING ADOPTION

A study done by Akin *et al.* using south western public universities in Nigeria showed that cloud computing in Nigeria has a good impact on low environment, cost effectiveness, IT complexities being reduced, and reduced investment in physical assets. They spoke on the challenges facing adoption being security of data, non-regulatory compliance and privacy on cloud-based learning being a concern. The study showed a minute percentage of not yet adopting cloud-based learning compared to a larger amount adopting it due to its advantages and effectiveness in learning (Akin *et al.* 2014). A similar study was done in Malaysia where the major drawback in Africa is its vulnerability to hacking and data security plus the addition of who owns control (Wakunuma and Masika 2017). Further study by Olanrewaju *et al.* in their systematic review described the challenges facing cloud-based learning as poor network infrastructure, high cost of ICT, lack of cloud awareness, and unreliable internet, which is a major hindrance in Africa (Olanrewaju *et al.* 2017).

5 DRIVERS AND BENEFITS OF CLOUD-BASED LEARNING

Bora and Ahmed (2013) identified crucial benefits of cloud-based learning. These benefits include low cost to learners, wherein learners do not have to be looking for anything heavy, because the cloud can be accessed through their mobile devices, laptops, or PCs at any location via the cloud storage with no expense to them, instant software updates, where learners get updates immediately as soon as there is an upgrade, and improved document format capability where cloud based applications open directly in the cloud and does not require the use of other applications to format documents to different compatibility mode. Learners benefit from cloud-based learning by the abilities to make use of online feedback and courses, and sending of assignments and projects without physical presence, time or distance to location. Teacher's benefits include the abilities to make preparatory note available earlier, manage student and content reports via feedback immediately and send assignments online. It also aids in the keeping of each student record to be able to monitor them systematically on improvements or otherwise, and security of data, where data stored in the cloud rarely crashes and has security benefits of permanence due to an unexpected system or server crash. In Malaysia, Hew and Abdul Kadir (2016) discovered that the use of cloud-based learning via knowledge sharing; easy access to teaching resources over the same community and ability to retrieve information has scaled up the acceptance of cloud-based learning in a virtual environment. Cloud based learning has economic benefits in terms of cost effectiveness and reduction in hard copies or printing for learners. It aids scalability in terms of storage increase or reduction depending on space required. It also aids data security on stored material for future use and limited to crashing and disappearance of stored data. It allows for e-learning also called distance or online learning via easy access from any location and real time via devices. It also promotes shared knowledge of materials and resources at a click away which ensures equality in what learners have access to irrespective of which side of the country or Ivy-league school (Chung *et al.* 2013)

6 LESSONS LEARNT

Challenges related to the adoption of cloud-based learning in Nigeria, as observed by the literal reviews discussed above, can be dealt with by creating data encrypted clouds to reduce if not protect from hacking which would resolve issues on data security and vulnerability to hacking. Provision of cheaper and available internet network coverage across the country, especially in institutes of higher learning, plus continual upgrades and updates of the cloud servers would solve challenge on poor network infrastructure, high cost of ICT, and unreliable internet. The government can control ownership by using a body or corporate organization to ensure central control access by institutions, thereby making it universal to all. Government policies need be created to monitor and protect cloud-based learning in institutions of higher learning in Nigeria, which would aid and put better protection on its management in the country.

7 CONCLUSIONS AND RECOMMENDATIONS

Cloud based learning is being adopted in the Nigerian educational system of higher learning, but it has not yet been optimized to its fullest extent due to the challenges of poor network and internet coverage, which is key in the utilization of cloud-based learning and its relative benefit. The Nigerian government so far is doing all it can to evolve and yet benefit from the technological advancement via knowledge sharing and ease of access by supporting institutions of higher learning to start use of this innovation. Due to the ongoing progress of cloud-based learning in Nigeria, further studies can be done to take statistical data on improvements in cloud-based learning in higher institutions in Nigeria including governmental support at secondary level education process as time goes. Simulations or models can as well be created to look at the compares between before cloud-based learning and after cloud-based learning.

References

Adeyinka, A. A., Current Problems of Educational Development in Nigeria, *The Journal of Negro Education,* 44, (2), 177-183, 1975.

Akin, O., Matthew, F., and Comfort., D., The Impact and Challenges of Cloud Computing Adoption on Public Universities in Southwestern Nigeria, *International Journal of Advanced Computer Science and Applications,* 5, 8-17, 2014.

Alharthi, A., Madini, O., Robert, A., Walters, J., and Wills, G. B., An Exploratory Study for Investigating the Critical Success Factors for Cloud Migration in The Saudi Arabian Higher Education Context, *Telematics and Informatics,* 34(2), 664-678, 2017.

Arong, F. E., and Ogbadu, M. A., Major Causes of Declining Quality of Education in Nigeria from Administrative Perspective: A Case Study of Dekina Local Government Area, *Canadian Social Science,* 6(3), 183-198, 2010.

Asiyai, R. I., Challenges of Quality in Higher Education in Nigeria in the 21st Century, *International Journal of Educational Planning and Administration,* 3(2), 159-172, 2013.

Asiyai, R. I., Improving Quality Higher Education in Nigeria: The Roles of Stakeholders, *International Journal of Higher Education,* 4(1), 61-70, 2015.

Ashtari, S., and Eydgahi, A., Student Perceptions of Cloud Applications Effectiveness in Higher Education, *Journal of Computational Science*, 23, 173-180, 2017.

Balasubramanian, K., Clarke-Okah, W., Daniel, J., Ferreira, F., Kanwar, A., Kwan, A., Lesperance, J., Mallet, J., Umar, A., and West, P., *ICTs for Higher Education*, 2009.

Barrantes, C., and Yague, J. L., Adults' Education and Agricultural Innovation: A Social Learning Approach, *Procedia-Social and Behavioral Sciences,* 191, 163-168, 2015.

Bora, U. J., and Ahmed, M., E-learning Using Cloud Computing, *International Journal of Science and Modern Engineering*, 1(2), 9-12, 2013.

Bouyer, A., and Arasteh, B., The Necessity of Using Cloud Computing in Educational System, *Procedia-Social and Behavioral Sciences,* 143, 581-585, 2014.

Chang, V., and Guetl, C., Generation Y Learning in the 21st Century: Integration of Virtual Worlds and Cloud Computing Services, In *Global Learn*, 1888-1897, Association for the Advancement of Computing in Education (AACE), 2010.
Chung, C. H., Pasquini, L. A., and Koh, C. E., Web-Based Learning Management System Considerations for Higher Education, *Learning and Performance Quarterly,* 1(4), 24-37, 2013.
Eze, S. C., Awa, H. O., Okoye, J. C., Emecheta, B. C., and Anazodo, R. O., Determinant Factors of Information Communication Technology (ICT) Adoption by Government-Owned Universities in Nigeria: A Qualitative Approach, *Journal of Enterprise Information Management,* 26(4), 427-443, 2013.
Ezenwoke, A., Omoregbe, N., Ayo, C. K., and Sanjay, M., NIGEDU CLOUD: Model of a National E-Education Cloud for Developing Countries, *IERI Procedia,* 4, 74-80, 2013.
Garcia-Penalvo, F. J., Johnson, M., Alves, G. R., Minović, M., and Conde-González, M. A., Informal Learning Recognition through A Cloud Ecosystem, *Future Generation Computer Systems,* 32, 282-294, 2014.
Gital, A. Y., and Zambuk, F. U., Cloud Computing: Solution to ICT in Higher Education in Nigeria, *Advances in Applied Science Research*, 2(6), 364-369, 2011.
Hew, T. S., and Abdul Kadir, S. L. S., Predicting the Acceptance of Cloud-Based Virtual Learning Environment: The Roles of Self Determination and Channel Expansion Theory, *Telematics and Informatics,* 33(4), 990-1013, 2016.
Ifijeh, G., Adoption of Digital Preservation Methods for Theses in Nigerian Academic Libraries: Applications and Implications, *The Journal of Academic Librarianship,* 40(3-4), 399-404, 2014.
Kats, Y., *Learning Management Systems and Instructional Design: Best Practices in Online*, 2013.
Kotoua, S., İlkan, M., and Kılıç, H., *The Growing of Online Education in Sub Saharan Africa: Case Study Ghana*, 2015.
Laleye, A. M., Educational Technology for Effective Service Delivery in Educational Training and Research in Nigeria, *Procedia-Social and Behavioral Sciences,* 176, 398-404, 2015.
Nookhong, J., and Wannapiroon, P., Development of Collaborative Learning Using Case-Based Learning Via Cloud Technology and Social Media for Enhancing Problem-Solving Skills and ICT Literacy within Undergraduate Students, *Procedia-Social and Behavioral Sciences*, 174, 2096-2101, 2015.
Oye, N. D., Lahad, N. A., and Ab Rabin, Z., A Model of ICT Acceptance and Use for Teachers in Higher Education Institutions, *International Journal of Computer Science and Communication Networks,* 1(1), 22-40, 2011.
Pina, A. A., An Overview of Learning Management Systems, *Virtual Learning Environments: Concepts, Methodologies, Tools and Applications*, 33-51, 2012.
Porter, W. W., Graham, C. R., Spring, K. A., and Welch, K. R., Blended Learning in Higher Education: Institutional Adoption and Implementation, *Computers and Education*, 75, 185-195, 2014.
Thomas, P. Y., Cloud computing: A Potential Paradigm for Practicing the Scholarship of Teaching and Learning, *The Electronic Library,* 29(2), 214-224, 2011.
Wakunuma, K., and Masika, R., Cloud Computing, Capabilities and Intercultural Ethics: Implications for Africa, *Telecommunications Policy,* 41(7-8), 695-707, 2017.

Interdependence between Structural Engineering and Construction Management
Edited by Ozevin, D., Ataei, H., Modares, M., Gurgun, A., Yazdani, S., and Singh, A.

ISBN: 978-0-9960437-6-2

SHEAR STRENGTH OF DETERIORATED STEEL GIRDERS IN MULTI-GIRDER BRIDGES

AMANDA BAO, CALEB GUILLAUME, and ALANA MORAES

Civil Engineering Technology, Rochester Institute of Technology, Rochester, USA

Bridge deterioration is mainly caused by repeated traffic loads and adverse environmental exposure. According to the 2017 American Society of Civil Engineers' infrastructure report card, more than 9% of the bridges in the United States were labeled structurally deficient. For steel bridges, the most dominant deterioration form is corrosion, which is characterized by the metal area loss resulting in structural capacity reduction. Corrosion is very common in steel multi-girder bridges because of moisture exposure, leakage through bridge joints as well as the frequent use of deicing detergents during the winter season in cold regions. Over the years, the rust can be serious enough to disconnect the web from the flanges of the girder, which poses significant concerns for load capacity especially at girder ends. This research studies shear strength loss in deteriorated steel multi-girder bridges by 3-D finite element models built in ABAQUS. Our analysis is focused on web area loss and web thinning due to corrosion, and their consequences for shear and web buckling capacity reduction. Area loss will be modeled by removing materials from the web, and web thinning will be simulated by reducing the web thickness. The numerical models resemble real steel corrosion forms by changing the shape, size and location of the area loss. A load rating method will be proposed based on the analyses.

Keywords: Corrosion, Structural capacity, Abaqus, Finite element.

1 INTRODUCTION

According to the ASCE 2017 Infrastructure Report Card, more than 9% of bridges in the United States were labeled structurally deficient, indicating some degree of deterioration of structural elements (ASCE 2017). In cold regions, frequent use of deicing chemicals in the winter season and bridge joint leakage are the main causes of corrosion in steel girder bridges. Figure 1 shows the typical girder corrosion forms. Excessive rust accumulation over the years can be serious enough to disconnect the girder web from the flanges, which poses significant concerns for structural capacity especially at girder ends (Al Badran 2013, Bao *et al.* 2018, Kayser and Novak 1989, Kulicki *et al.* 2017, Sharifi 2012, Van de Lindt and Ahlborn 2005).

Figure 1. Typical steel girder corrosion at web and flange (Bao *et al.* 2018).

In this study, the residual shear strength of corroded steel girder bridges is evaluated by 3-D finite element modeling. Our analysis and testing are focused on the effects of web area loss and web thinning on shear and web buckling capacity reduction.

2 3D FINITE ELEMENT MODELING

It is costly to perform full scale physical experiments on bridges. Finite element modeling has been used for decades to do research on large scale structures. In this research, ABAQUS is used as a tool to analyze the mechanical behaviors of deteriorated girders (Abaqus 2016). 3D finite element models are set up to study web buckling and shear capacity of corroded girders. The load cases and the controlling forces are selected according to the load combinations specified in AASHTO LRFD Bridge Design Specifications (AASHTO 2016, NYSDOT 2017). Translation and rotation at the girder ends are modeled properly to simulate the support conditions. Corrosive area loss is simulated by removing part of web area, and web thinning is modeled by reducing the web thickness uniformly (Bao *et al.* 2018). Web thinning due to rust mainly reduces the girder's shear strength and web buckling strength. The structural capacity of the deteriorated girders is analyzed.

The prototype bridge is a simple span I-Plate steel girder bridge with a cast-in-place concrete deck. The span length of the bridge is 130 feet. The overall width of the bridge is 98 feet, and the bridge carries six lanes of traffic. The design live load on the bridge is the typical AASHTO HL-93 Live Load. The cast-in-place deck is 9 inches thick. The bridge framing has 10 girder lines with 10 feet girder spacing. Figure 2 and Figure 3 shows the bridge elevation and the bridge typical section, respectively.

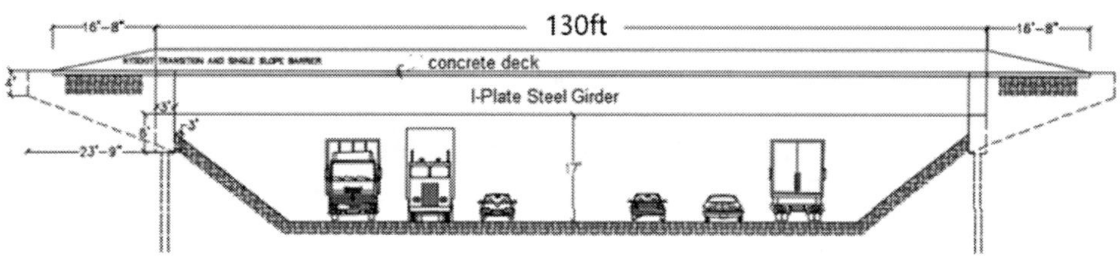

Figure 2. Bridge elevation.

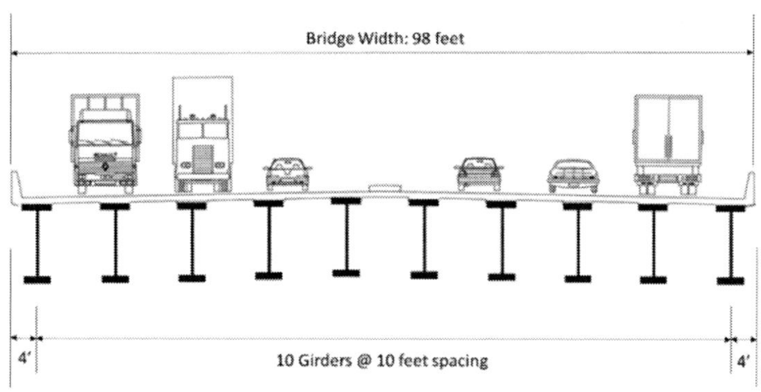

Figure 3. Bridge typical section.

The I-plate steel girder has a top flange of 12 inches wide and 2 inches thick, a web of 48 inches deep and ½ inch thick and a bottom flange of 12 inches wide and 2 inches thick. The overall depth of the steel girder is 52 inches. The length of the girder end model is 100 inches. The boundary conditions are pinned at one end of the top and bottom flanges to simulate the real girder end embedded into the abutment diaphragm. The pinned support is 6 inches long on both the top and the bottom flanges to represent the real bearing length of a simple span girder. The other end is free in vertical movement but restricts horizontal movement along the girder line. The girder end finite element model in ABAQUS is shown in Figure 4. The steel girder material uses Grade 50 steel with a yield strength of 50 ksi, and an ultimate strength of 65 ksi. The steel has an elastic modulus of 29,000 ksi and a Poisson's ratio of 0.30 (American Institute of Steel Construction 2017).

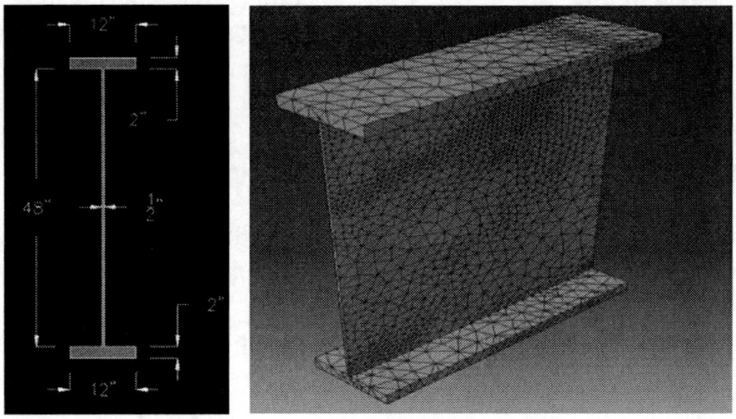

Figure 4. Girder end FEM model in ABAQUS.

3 MODELING RESULTS

A typical interior girder is used for calculating the shear force at the girder end. The unfactored non-composite and composite dead load (DC) on the girder is 2.0 kip/ft, and the unfactored wearing surface (DW) on the girder is 0.40 kip/ft. The live load (LL) on the girder is AASHTO HL-93 live load, the dynamic impact factor of the design truck HS20-44 is 1.33, and the live load distribution factor for shear is 0.952, which is calculated according to AASHTO LRFD Bridge Design Specifications as shown in Eq. (1) (AASHTO 2016).

Live load distribution factor for shear =

$$\text{Max}\begin{cases} \text{one lane loaded: } 0.36 + \frac{S}{25} = 0.36 + \frac{10}{25} = 0.760 \\ \geq 2 \text{ lanes loaded: } 0.2 + \frac{S}{12} - \left(\frac{S}{35}\right)^2 = 0.2 + \frac{10}{12} - \left(\frac{10}{35}\right)^2 = 0.952 \end{cases} \quad (1)$$

where S = girder spacing.

The maximum design shear force at the girder end is obtained by Strength I load combination specified in Eq. (2) according to AASHTO LRFD Bridge Design Specifications (AASHTO 2016).

$$\text{Strength I Load Combination: } 1.25DC + 1.5DW + 1.75LL \quad (2)$$

Figure 5 shows the load diagram for the single girder analysis. The calculated design shear force according to the single girder analysis is 420 kips at the girder end.

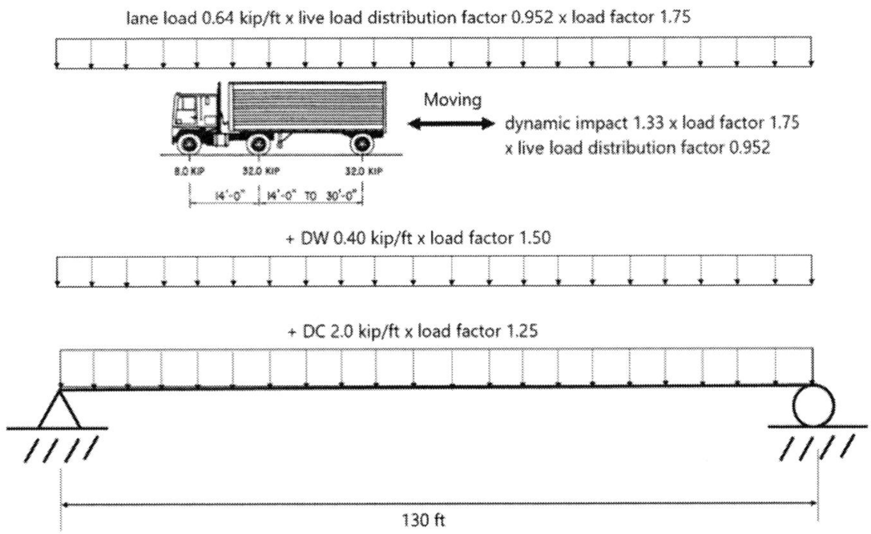

Figure 5. Single girder analysis.

4 RESULTS

The original web size is 48 inches deep by ½ inch thick. The web thinning due to rust is modeled by reducing the web thickness in the section geometry input. The web bucking capacity of the intact web is 678 kips according to the ABAQUS analysis. The web bucking capacity vs. web thinning is plotted in Figure 6. The web thicknesses in the plot are 0.50 inch (original thickness), 0.475 inch (5% thickness reduction), 0.45 inch (10% thickness reduction), 0.425 inch (15% thickness reduction), 0.40 inch (20% thickness reduction) and 0.375 inch (25% thickness reduction), respectively.

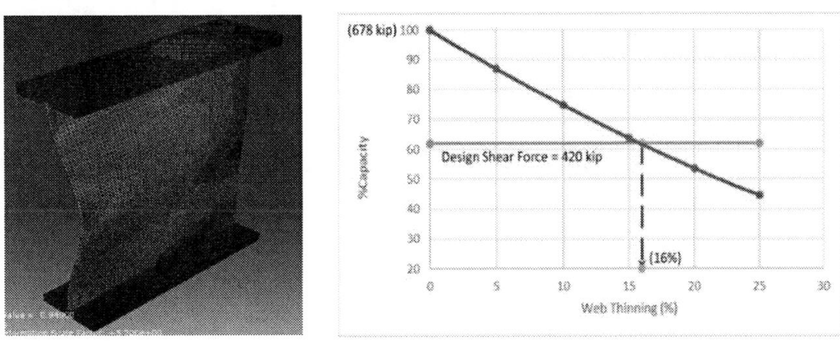

Figure 6. Web buckling capacity vs. web thinning.

The area loss due to corrosion is modeled by removing portions of the web. In this paper, the shape of the area loss is modeled by a rectangular hole with the longer side parallel to girder line. The results of web buckling capacity vs. sizes of area loss are shown in Figure 7 and Figure 8. The effect of varying hole location on the web buckling strength is illustrated in Figure 9. The area loss size used in Figure 9 is 6"x18" = 108 in^2.

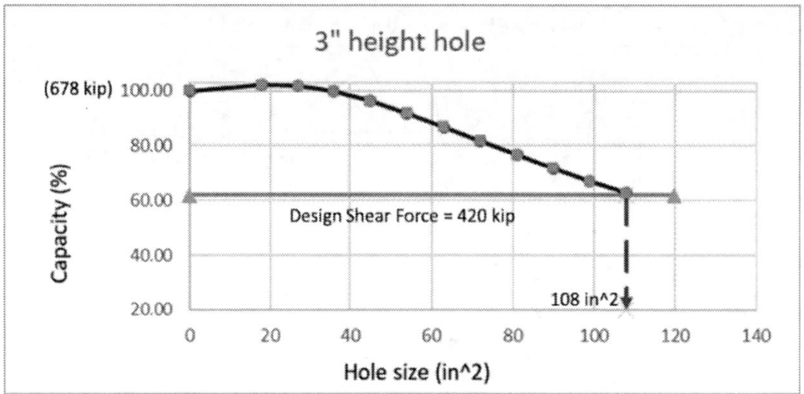

Figure 7. Web buckling capacity vs. area loss – hole height 3".

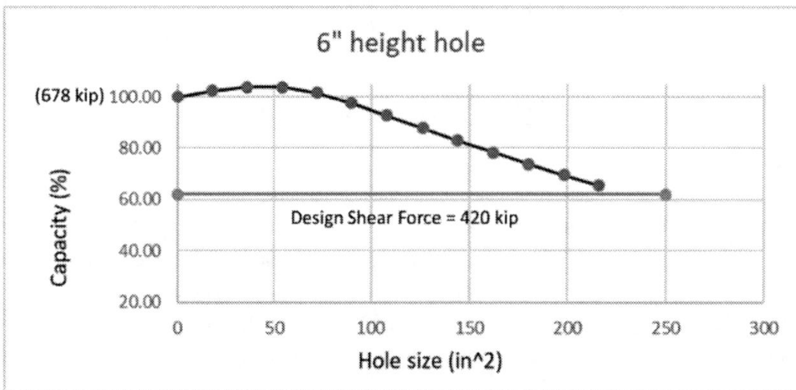

Figure 8. Web buckling capacity vs. area loss – hole height 6".

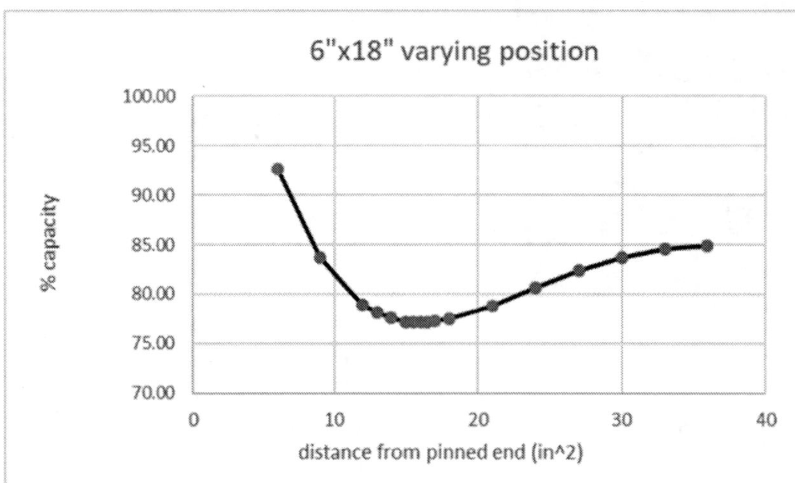

Figure 9. Web buckling capacity vs. hole location – hole size 6"x18".

5 CONCLUSIONS

Common corrosion types including web thinning and web area loss and their effects on the structural capacity of deteriorated girders are investigated in this paper. According to the results of 3D finite element analyses using ABAQUS, the conclusions of this research are:

1) Web thinning due to corrosion has a significant effect on girder shear and web buckling strength. Development of extensive rust can dramatically reduce the girder's structural capacity. 10% reduction in web effective thickness may result in 25% or more buckling strength and 20% reduction of web thickness may lead to 50% structural capacity loss due to decreasing effective web cross sectional area as well as increasing width to thickness ratio of the web.
2) Figure 7 and Figure 8 show consistently that the larger size of area loss, the lower the structural capacity. The strength may drop below the girder design load when the size of hole keeps growing, which represents structural deficiency.
3) Comparing Figure 7 with Figure 8, we can find that when extending area loss in the direction parallel to the girder line, the structural capacity decreases at a higher rate than extending area loss in the direction parallel to the web depth.
4) Figure 9 shows that the location of the area loss significantly affects the structural capacity, when the hole starts about 1/3 of the web depth from the beam end support, the structural capacity reaches its lowest point.

Acknowledgement

This study is supported by Western New York Association for Bridge Construction and Design Research Grant (2017-2018) and Rochester Institute of Technology College of Applied Science and Technology Scholarship Incentive Grant (2017).

References

AASHTO, *AASHTO LRFD Bridge Design Specifications,* 7th Edition, American Association of State Highway and Transportation Officials, 2016.

ABAQUS, CAE User's Guide, 2016.

Al Badran, M.S., Structural Reliability Analysis of Corroded Steel Girder Bridge, MSc Thesis, University of Nebraska-Lincoln, 2013.

American Institute of Steel Construction, *Steel Construction Manual,* 15th Edition, 2017.

ASCE, *2017 Infrastructure Report Card,* American Society of Civil Engineers, 2017.

Bao, A., Gulasey, M., Guillaume, C., Levitova, N., Moraes, A., and Satter, C., *Structural Capacity Analysis of Corroded Steel Girder Bridges,* Proceedings of the 3rd International Conference on Civil, Structural and Transportation Engineering, Paper ID 118, Niagara Falls, Canada, June 10-12, 2018.

Kayser, J., and Nowak, A., Reliability of Corroded Steel Girder Bridges, Elsevier Science Publishers: *Structural Safety*, 6, 1989.

Kulicki, J., Prucz, Z., Sorgenfrei, D., Mertz, D., and Young, W., *Guidelines for Evaluating Corrosion Effect in Existing Steel Bridges,* National Cooperative Highway Research Program, Report 333, 1990.

NYSDOT, *Bridge Manual,* 2017.

Sharifi, Y., Residual Web Bearing Capacity of Corroded Steel Beams, Advanced Steel Construction, 8(3), 242-255, 2012.

Van de Lindt, J., and Ahlborn, T., *Development of Steel Beam End Deterioration Guidelines,* Michigan Technological University, 2005

Interdependence between Structural Engineering and Construction Management
Edited by Ozevin, D., Ataei, H., Modares, M., Gurgun, A., Yazdani, S., and Singh, A.

ISBN: 978-0-9960437-6-2

VIBRATION RESPONSE OF A FOOTBRIDGE DUE TO PEDESTRIANS' MOVEMENTS

MEHDI SETAREH, RUTHVIK KADAM, and MOHAMMAD BUKHARI

School of Architecture and Design, Virginia Tech, Blacksburg, USA

Failure of footbridges in England and France due to excessive vibrations from marching soldiers in the nineteenth century resulted in many fatalities and injuries. The unexpected lateral vibrations of the Millennium Bridge over the Thames River in London was a watershed event as large number of research studies were conducted on this subject following this incidence. However, there are still a number of issues related to vibration serviceability of footbridges that require further studies. Among these are: the amplification of structural response when a group of pedestrians cross a footbridge as compared to a single individual (group effects). This is important as footbridge vibrations are usually computed when subjected to one pedestrian's crossing and are magnified to estimate those generated by a group of people. Therefore, it is important to conduct vibration testing of footbridges to better define the group effects. This paper presents an experimental study of a footbridge susceptible to vibrations due to human movements. With the help of a group of volunteers, the group effects were computed as a function of the group size. It was found that the results were not completely consistent with those in the literature. Conclusions are made based on the results of the data analysis.

Keywords: Footbridge vibration, Group effects, Vibration serviceability, Enhancement factor, Human-structure interactions.

1 INTRODUCTION

Footbridges can be susceptible to large vibrations due to human movements. The excessive vibrations of the Millennium footbridge over the Thames River in London (Dallard *et al.* 2001) initiated large number of research studies focused on evaluating the vibration serviceability issues of footbridges.

To evaluate the dynamic response of footbridges, the structural designer needs to have a good estimate of the applied loads. It is customary to compute the response of a footbridge subjected to a single pedestrian, and increase the results when a group of people moves over the structure. Therefore, few researchers have studied the vibration amplification when a group of pedestrians cross a footbridge as compared to that due to an individual. This amplification is represented by the "enhancement factor".

Matsumoto *et al.* (1978) proposed an enhancement factor of $\sqrt{N}$ to estimate the vibration amplitude of a footbridge when crossed by a group of N pedestrians from that attributed to one person, assuming a Poisson arrival probability distribution. Wheeler (1982), Bachmann and Ammann (1987), and Barker (2007) conducted analytical studies to define the enhancement factors. Unfortunately, all conducted investigations on this subject consisted of analytical studies with very limited experimental verifications.

Setareh (2016, 2017) conducted studies on the group effects in footbridges, which are to the knowledge of the authors the only experimental investigation conducted on this topic. Setareh (2016) conducted a number of dynamic tests with a group of human subjects and concluded that when the pedestrians walked slowly or at normal speed in unison, the enhancement factor for group effects increased approximately by the square root of the group size as proposed in the literature for the random walks. However, when the group speed approached the first mode natural frequency of the footbridge, the enhancement factor became closer to the group size.

Setareh (2017) conducted an experimental study of another footbridge and concluded that when people moving in unison on a footbridge at the structure's fundamental frequency, the enhancement factor is significantly larger than the recommended values in the literature. He also observed that the enhancement factors for when the people move in unison at the footbridge fundamental frequency is close to the number of people for smaller group size. This reduces with an increase in group size, which can be attributed to an increase in the footbridge damping due to human-structure dynamic interactions and the fact that it is more difficult for a larger group to move in unison.

To add to the available literature on this issue, this paper presents an experimental study of the group effects on a footbridge using a number of human subjects. It uses different evaluation parameters and computes the enhancement factors. It provides the results of the data analysis and concludes on the outcome.

2 DESCRIPTION OF THE STRUCTURE

The footbridge used in the study connects a multi-story building to an adjacent parking structure located in Blacksburg, Virginia (see Figure 1). It is made of a steel structure consisting of two trusses at 8 ft apart with a standard Pratt configuration. The total length of the footbridge is 116'-10" and has a clear depth of 7'-6". All the truss members are made of Hollow Structure Shapes (HSS), which are welded together. The top chord is made of HSS6x6x3/8, the bottom chord consists of HSS8x6x3/8, the verticals are HSS6x6x3/8 and the diagonals are made of HSS4x3x1/4. The truss panel points are 9'-0" apart. The footbridge floor deck consists of a 20 gage, 2-inch thick form metal deck with 4 inches of concrete topping (total thickness=6"). The concrete has a compressive strength of 3,500 psi.

Figure 1. Footbridge used in the study.

3 DESCRIPTION OF THE TESTS

To measure the resonance frequency of the structure, an electrodynamic shaker was placed on the footbridge and the frequency response function (FRF) was found. This resulted in a resonance (natural) frequency for the vertical modes of 4.08 Hz and 11.43 Hz.

Subsequently, up to 32 volunteers participated in a series of walk tests on the footbridge. Four speeds were selected: (1) 122 spm (steps per minute) to represent the half-harmonic of the measured first vertical mode natural frequency, (2) 110 and 115 spm to investigate the possibility of reduction in the structure's natural frequency due to the presence of pedestrians; and (3) random movements. PCB 393C accelerometers were placed at the center of the footbridge on both sides to record the vibrations along the three orthogonal directions. However, since the largest vibrations occurred in the vertical direction, this study will focus on these measurements only.

Using a metronome to synchronize their movements, different groups of volunteers (1, 4, 8, 14, 20, 26 and 32 people) crossed the footbridge from one end to another and stopped while the measurements were completed. Figure 2 shows a group of pedestrians crossing the footbridge during one of the tests. An OROS signal analyzer was used to record and analyze the data.

Figure 2. People walking on the footbridge during the test.

4 GROUP EFFECT STUDIES

Different vibration evaluation parameters were computed for the measurements made during the conducted tests. These include the peak acceleration (a_p), maximum transient vibration value (MTVV) and vibration dose value (VDV). For detailed definition of these parameters, please refer to Setareh (2017). For computing the values of the MTVV and VDV, the frequency weighting function, W_k, from the International Standard ISO2631-1 (1997) was used.

The enhancement factors for a_p, MTVV and VDV for different groups of pedestrians were computed by dividing the values for each group by its counterpart for an individual pedestrian. To provide more consistent values and minimize the effects of noise and biases during the tests, each data point was computed by averaging the values from two accelerometers on the two sides of the footbridge. In addition, each test was conducted twice, once the volunteers crossed the footbridge from left to right and again from right to left. The results for these tests were also averaged for each step frequency. Therefore, the results used for the analysis were based on the average of four data sets.

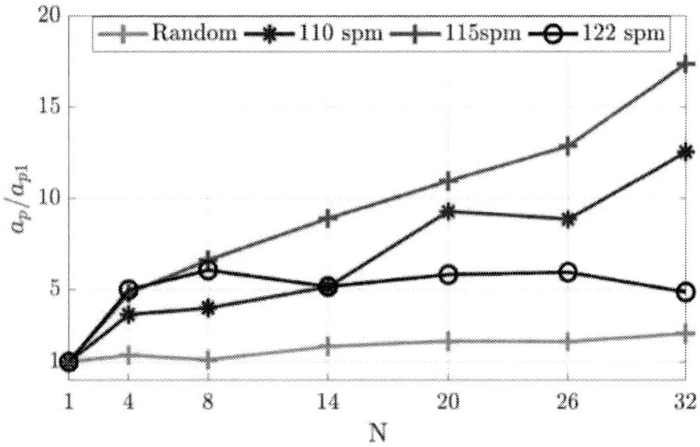

(a)

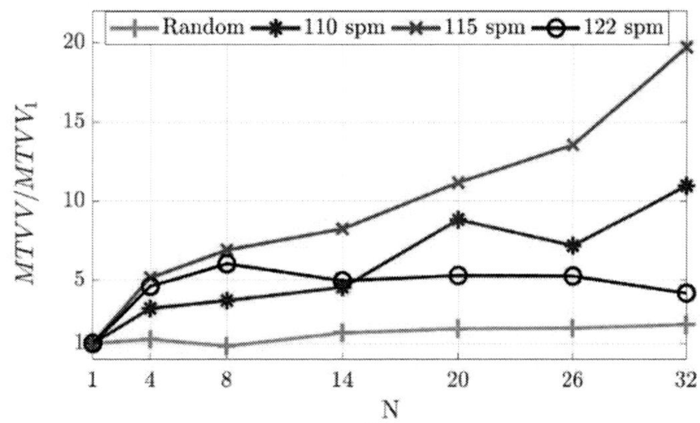

(b)

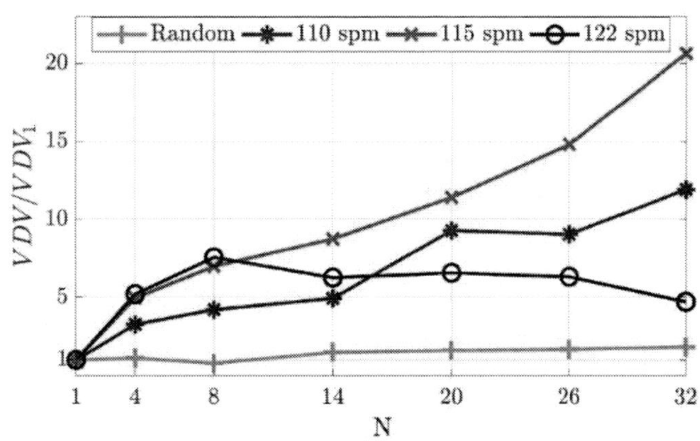

(c)

Figure 3. Variation of different group enhancement factors with group size (N):
(a) a_p/a_{p1} vs. N; (b) MTVV/ $MTVV_1$ vs. N; (c) VDV/VDV_1 vs. N.

Figure 3 shows the enhancement factors for different evaluation parameters versus the number of pedestrians for different step frequencies. As expected, the enhancement factors for the random walks are small. However, the unexpected observation is the reduction of these values when more than eight pedestrians walked at the sub-harmonic of the natural frequency of the footbridge (122 spm) even though for the 110 and 115 spm the enhancement factors increased with an increase in the number of people.

To better comprehend the results, the peak acceleration and VDV versus the number of people for each walk speed is shown in Figure 4. As can be observed, the peak acceleration and VDV for the case when the pedestrian moved at the sub-harmonic of the natural frequency were much larger than when they walked at other speeds. This may have resulted in some of the pedestrians not being able to keep their pace with the beat from the metronome. However, the results from the prior studies conducted by Setareh (2017) show that at the measured vibration levels this could not cause the level of reduction in the footbridge response observed here.

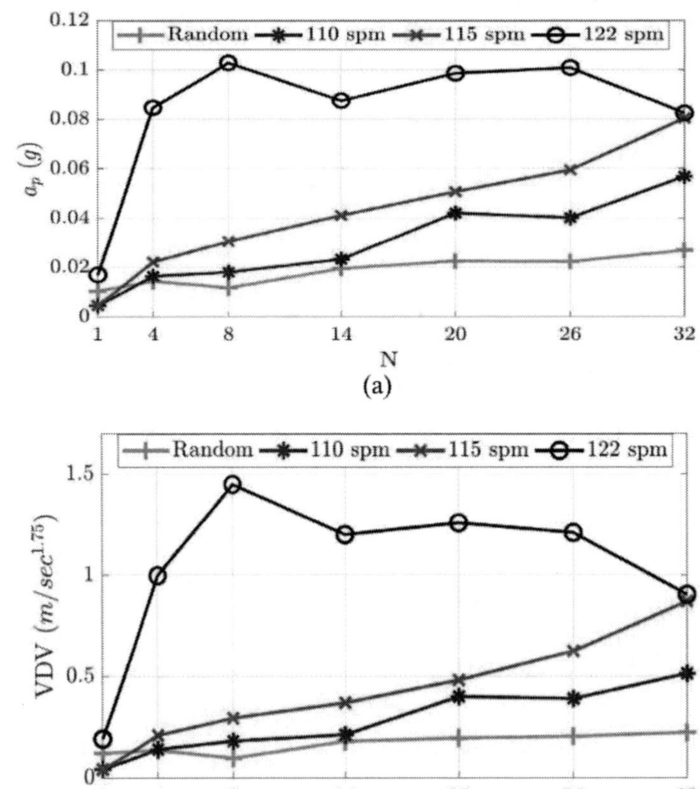

(a)

(b)

Figure 4. Peak acceleration and VDV values for different group size and various walking speeds: (a) a_p vs. N; (b) VDV vs. N.

The other major factor in reducing the enhancement factor, when the people walked at 122 spm, can be attributed to the effects of human-structure interactions. Based on the measured dynamic properties of the footbridge, if the natural frequencies of pedestrians were close to the vertical mode of vibration, they could act as tuned mass dampers (TMD), absorbing the vibration energy, and result in a reduction in footbridge response. As the mass ratio between the pedestrians and the footbridge increases (with more pedestrians), the required equivalent TMD damping ratio increases. Since human body has inherently large damping levels, as the number of people increases the human bodies become more tuned to the structure and result in larger vibration reduction. This can provide an explanation of the observed phenomenon. More studies are needed to provide a quantitative evaluation of the effects of the observed human-structure interactions.

5 CONCLUSIONS

This paper provided the results of studies conducted on a footbridge to evaluate the group effects. The presented results show a reduction in the enhancement factors when a number of pedestrians (beyond a limit) walked at the sub-harmonic of the footbridge natural frequency. This was attributed mainly to the effects of human-structure interaction and to a lesser degree to the inability of the pedestrians to keep pace with each other when the footbridge had large levels of vibration.

Acknowledgments

The research presented here was supported by the National Science Foundation under grant number CMMI-1335004. This support is gratefully acknowledged. Any opinions, findings, and conclusions expressed in this paper are those of the writer and do not necessarily reflect the views of the National Science Foundation.

References

Bachmann, H., and Ammann, W., Vibration in Structures Induced by Man and Machines, *International Association for Bridge and Structural Engineering*, Switzerland, 1987.

Barker, C., Footbridge Pedestrian Vibration Limits Background to Response Calculations, *International Journal of Space Structures,* 22(1), 35-43, 2007.

Dallard, P., Fitzpatrick, A. J., Flint, A., Le Bourva, S., Low, A., Ridsdill-Smith, R. M., and Willford, M., The London Millennium Footbridge, *Structural Engineering*, 79(22), 17-33, 2001.

ISO2631-1, *Mechanical Vibration and Shock – Evaluation of Human Exposure to Whole-Body Vibration – Part 1: General Requirements – ISO 2631-1.* Second ed., International Organization for Standardization, Geneva, Switzerland, 1997.

Matsumoto, Y., Nishioka, T., Shiojiri, H., and Matsuzaki, K., Dynamic Design of Footbridges, *International Association for Bridge and Structural Engineering,* 1-15, 1978.

Setareh, M., Vibration Serviceability Issues of Slender Footbridges, *Journal of Bridge Engineering, ASCE,* 21(11), 04016084-1-12, 2016.

Setareh, M., Study of the Group Effects on the Vibration Serviceability of Slender Footbridges, *Footbridge 2017 Berlin*, Berlin, 2017.

Wheeler, J., Prediction and Control of Pedestrian-Induced Vibration in Footbridges. *Journal of the Structural Division,* 108(9), 2045-2065, 1982.

Interdependence between Structural Engineering and Construction Management
Edited by Ozevin, D., Ataei, H., Modares, M., Gurgun, A., Yazdani, S., and Singh, A.

ISBN: 978-0-9960437-6-2

INFLUENCE OF RAILING STIFFNESS ON WHEEL LOAD DISTRIBUTION IN TWO-SPAN CONCRETE SLAB BRIDGES

SARAH JABER[1], MOUNIR MABSOUT[1], and KASSIM TARHINI[2]

[1]*Dept of Civil and Environmental Engineering, Amer. Univ. of Beirut, Beirut, Lebanon*
[2]*Dept of Civil Engineering U. S. Coast Guard Academy, New London, USA*

The American Association of State Highway and Transportation Officials (AASHTO) Standard Specifications or LRFD do not account for the presence of railings in the analysis and design of concrete slab bridges. This paper presents a parametric investigation of the influence of railing stiffness on the wheel load distribution in simply-supported, two-equal-span, and one- and two-lane reinforced concrete slab bridges using the finite-element analysis (FEA). A total of 160 bridge cases were modeled and bridge parameters such as span lengths and slab widths were varied within practical ranges. Various railing stiffness were investigated by assuming railings built integrally with the bridge deck and placed on both edges of the bridge. The FEA wheel load distribution and longitudinal bending moments were compared with reference bridge slabs without railings as well as to the AASHTO design procedures. Accordingly, the presence of railings reduced the FEA negative moments by a range of 54% to 72% and the FEA positive moments by a range of 40% to 61% depending on the railing stiffness. This reduction in slab moments due to the presence of railings could be considered an increase in the bridges load carrying capacity. The results of this investigation will assist bridge engineers in better designing and/or evaluating concrete slab bridges in the presence of railings. This could also be considered an alternative for strengthening existing concrete slab bridges.

Keywords: Concrete slab bridges, Railing stiffness, Finite-element analysis, AASHTO procedures, Load-carrying capacity.

1 INTRODUCTION

A significant number of highway bridges are short-span reinforced concrete slabs that are owned and maintained by local and state governments. The design of highway bridges in the United States conforms to the American Association of State Highway and Transportation Officials (AASHTO) Standard (2002) or Load and Resistance Factor Design (LRFD) Bridge Design Specifications (2012). The current AASHTO procedures do not consider the effect of railings that are built integrally with bridge deck in the analysis and design or evaluating the load-carrying capacity of concrete slab bridges.

A published parametric study by Mabsout *et al.* (2004) investigated single-span, simply-supported reinforced concrete slab bridges using finite-element analysis (FEA). Results indicated that AASHTO Standard Specifications moments overestimate the FEA moments for short spans and one-lane bridges but agreed with FEA moments for short-span bridges with two or more lanes. It was also found that AASHTO Standard Specifications (2002) underestimated the FEA moments for longer spans. However, for AASHTO LRFD (2012) procedure, it overestimated the FEA moments for all bridge cases. In addition, several published studies were conducted by investigating the influence of sidewalks and railings on wheel load distribution in steel and prestressed girder bridges which was shown to increase the stiffness of the superstructure and increase the load-carrying capacity of these bridges (Akinci *et al.* 2008, Chung *et al.* 2006, Conner and Huo 2006, Eamon and Nowak 2002, Mabsout *et al.* 1997). Recently, a parametric investigation by Fawaz *et al.* (2017) studying the influence of one standard railing size on straight concrete slab bridges was performed. The results indicated that by placing two railings on straight bridges, AASHTO Standard Specifications (2002) procedures overestimated the FEA moments by about 100% for one-lane bridges, and by about 20% for bridges with two-lanes. AASHTO LRFD (2012) overestimated the FEA moments in all bridge cases by about 150% for one-lane, and about 70% for two-lanes when placing two railings on slab bridges. This paper builds on the recent study to investigate the influence of railings stiffness on the increase in load carrying capacity in two-span continuous reinforced concrete slab bridges.

2 AASHTO BENDING MOMENTS

For simply-supported concrete slab bridges, AASHTO Standard Specifications (2002) suggest three approaches in determining the live-load bending moment but only one procedure is used in this study that was compared with the finite-element analysis results.

$$M = 13{,}500S \; for \; S \leq 15m \tag{1}$$

or

$$M = 1{,}000(19.5S - 90) \; forS > 15m \tag{2}$$

where S is the span length (m) and M is the longitudinal moment per unit width (N-m/m).

AASHTO LRFD Section 4.6.2.3 (2012) provides an equivalent strip width procedure to design reinforced concrete slab bridges that is comparable to procedures specified in the Standard Specifications. However, the AASHTO LRFD Section 3.6.1.2 requires the use of HL93 live loading (addition of HS20 Truck plus lane loading). This approach is to divide the total bending moment from a line of wheel and lane load by an equivalent width defined by AASHTO LRFD (2012) to obtain a statically design moment per unit width.

3 DESCRIPTION OF BRIDGE CASES

Typical simply-supported two-equal-spans, one- and two-lane reinforced concrete slab bridge cases were analyzed in this investigation. Four span lengths were considered in this parametric study: 7.2, 10.8, 13.8, and 16.2 m (24, 36, 46, and 54 ft) with corresponding slab thicknesses of 450, 525, 600, and 675 mm (18, 21, 24, and 27 inches), respectively, to control deflection. The overall slab widths were assumed to be 4.2 m (14 ft) for one lane and 7.2 m (24 ft) for two lanes.

The base case for the standard railings size was adopted from previous research by assuming 200 mm (8 in) wide and 760 mm (30 in) high railing above slab, as reported by Fawaz *et al.* (2017). Five stiffness factors are considered including X0, X1, X2, X3, X4, and X0.5, assuming X0 as the reference case with no railings and X1 to be the base case with the standard railings. The railing moment of inertia (I) was computed at the bottom of the railing section as per Eq. (3).

$$I_{(bottom)} = I_{(center)} + Ad^2 = \frac{bh^3}{12} + bh(\frac{h}{2})^2 = \frac{bh^3}{3} \qquad \therefore I_{(bottom)} = 4I_{(center)} \tag{3}$$

Therefore, the railings stiffnesses considered have the following moment of inertia with X0 (I = 0), X1 (I = 4Ic), X2 (I = 8Ic), X3 (I = 12Ic), X4 (I = 16Ic), and X0.5 (I = 2Ic). The various railings sizes corresponding to (X0.5 – X4) are shown in Figure 1. Also, Figure 2 (right) shows typical plan views and cross-section of 10.8 m (36 ft) per each span, two-lane bridge case with railings (base case, X1), with HS20 trucks loading placed longitudinally and transversely to cause maximum longitudinal positive moments, respectively, as explained in the following section.

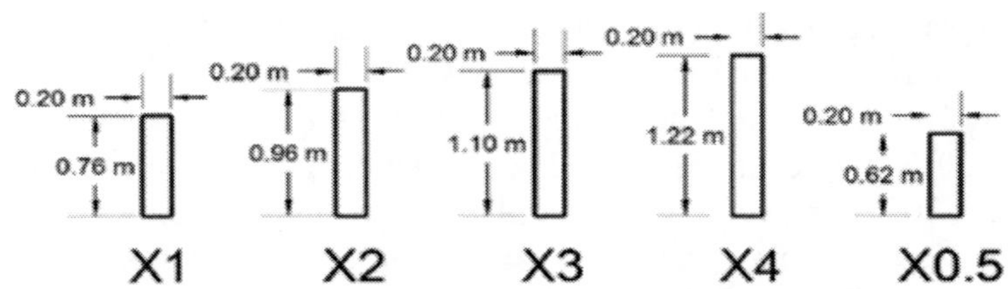

Figure 1. Various railing sizes corresponding to stiffnesses (X1, X2, X3, X4, X0.5).

4 BRIDGE LOADING

The bridge cases considered in this study were subjected to AASHTO HS20 standard design trucks. The AASHTO HS20 truck, with two line of wheel spaced at 6 ft (1.8 m) consisting of loads of 4, 16, 16 Kips spaced at 14 ft (or 18, 72, 72 KN spaced at 4.2 m), were placed longitudinally on the bridge deck to produce maximum positive or negative bending moments in the slab. Transversally, as per Fawaz *et al.* (2017), Edge loading condition was adopted in this study, which consists of positioning the center of the left wheel of the left most truck at 0.3 m (1 ft) from the left edge of the slab, and the other trucks placed side-by-side with a distance 1.2 m (4 ft) between the adjacent trucks. Figure 2 (left) illustrates the Edge loading conditions for the 10.8 m (36 ft) span, two-lane bridge cases for maximum positive moments.

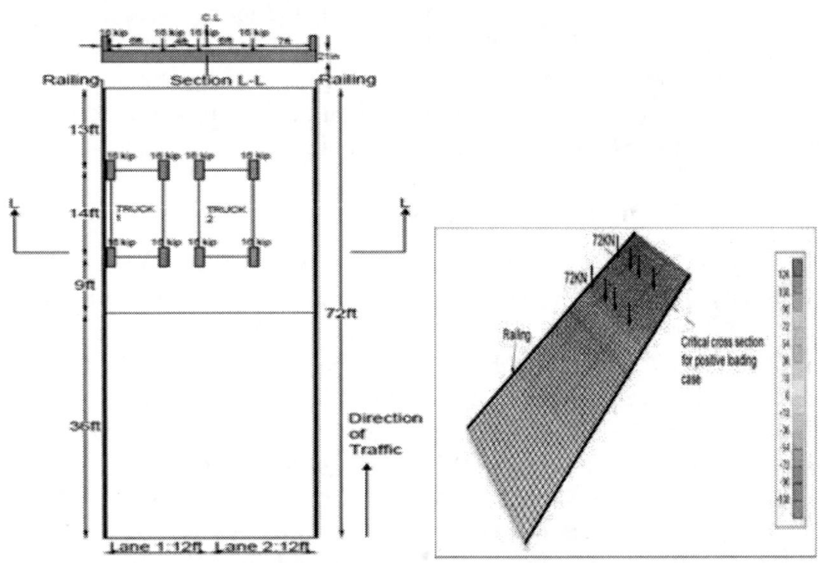

Figure 2. (Left) Typical plan view and cross-section for a 10.8 m (36 ft) per span, two-lane bridge, subjected to positive Edge loading condition with base case railings (X1); and (Right) FEA corresponding longitudinal positive bending moments (KN-m/m).

5 FINITE ELEMENT MODELING

A total of 160 concrete slab bridge cases were investigated in this study using the FEA. The computer program SAP2000 (version 19) was used to discretize the bridge into four-node shell elements with six degrees of freedom at each node. The simple eccentric beam element was adopted in this investigation to model the railings, as per Fawaz *et al.* (2017). Figure 2 (right) illustrates typical finite element models with the corresponding positive moment contours, for a 10.8 m (36 ft) span, two-lane bridges, subject to HS20 Edge loading condition, and in the presence of two standard base case railings.

6 FINITE-ELEMENT ANALYSIS RESULTS

The FEA results are reported in terms of the maximum longitudinal bending moments at critical section in the concrete slab bridges. The FEA results for bridges with railings of different stiffness factors were compared with reference bridge cases without railings, and with AASHTO Standard Specifications (2002) as per Eqs. (1) or (2), and with LRFD procedures (2012).

6.1 FEA Results vs. AASHTO

Figure 3 (left) shows sample plots of the FEA longitudinal positive moment plots at the critical sections for all the two-lane bridge cases in combination with the four span lengths (S) with base case railings (X1). Figure 3 (right) shows the positive moment plots for all the two-lane bridges with 10.8 m (36 ft) span length, with different railing stiffness configurations, along with the AASHTO Standard Specifications (2002) and LRFD procedures (2012) moments. The maximum FEA moments for the concrete slabs were defined as the first peak value occurring after the maximum value at the leftmost edge; the latter "edge" moment is assumed to be resisted by an edge beam. Similar figures were obtained for the negative moments.

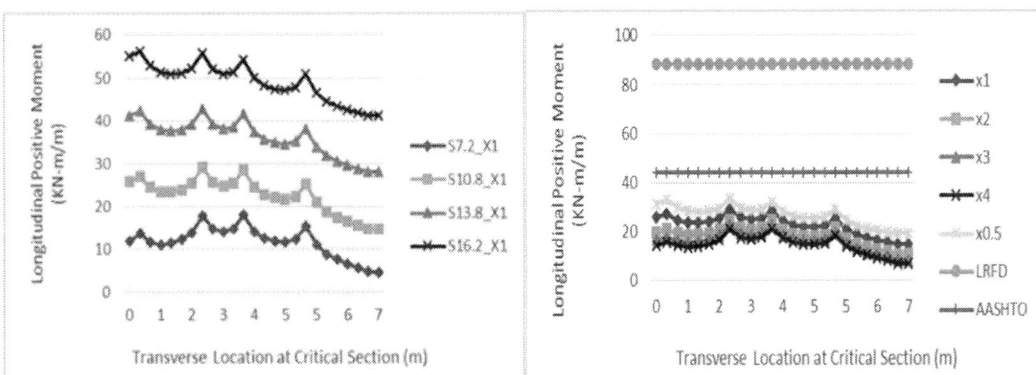

Figure 3. FEA positive moments for two-lane bridges with base case railings (X1) for various spans (left); and for a 10.8 m (36 ft) span with various railings sizes compared with AASHTO moments (right).

The maximum FEA positive and negative moments were summarized and compared with the AASHTO Standard Specifications (2002) and LRFD (2012) moments for all the bridge cases. It was observed that, for bridge cases with no railings (X0), AASHTO Standard Specifications (2002) generally tends to give similar results to the FEA negative moments, with the exception of one-lane with spans less than 12 m (40 ft) where the AASHTO overestimates the FEA negative moments by about 30%. This is more pronounced with more lanes and longer spans, where AASHTO underestimation of the FEA moments reaches up to 26% for two lanes with spans greater than 12 m (40 ft). When base case railings (X1) are present in a concrete slab, the FEA slab negative moments decrease significantly and AASHTO overestimates or gives similar moments in almost all cases, reaching about 68% for the one-lane bridges with spans less than 12 m (40 ft), and gives similar moments for two-lane bridges with spans longer than 12 m (40 ft). Also, as the stiffness factor of railings increases, the FEA negative moments decrease and a more significant AASHTO overestimation is observed. This overestimation reaches about 81% for one-lane bridges and 72% for two-lane bridges with (X4) stiffer railings. It can be observed that, for bridge cases with no railings (X0), AASHTO Standard Specifications generally tends to overestimate the results of the FEA slab positive moments, with the AASHTO overestimation reaching about 35%. As the number of lanes and span length increase, this overestimation decreases. When base case railings (X1) are present in a concrete slab, the FEA slab positive moments decrease significantly and AASHTO overestimates moments in all cases, reaching about 55% for the one-lane bridges with spans less than 12 m (40 ft), and this overestimation decrease to about 18% for two-lane bridges with spans longer than 12 m (40 ft). Also, as the stiffness factor of railings increases, the FEA positive moments decrease and a more significant AASHTO overestimation is observed. This overestimation reaches about 70% for one-lane bridges and 53% for two-lane bridges with (X4) stiffer railings.

AASHTO LRFD (2012) overestimates the FEA slab negative moments and positive moments in almost all bridge cases with or without railings. AASHTO LRFD overestimates the FEA slab negative moments by about 30% and positive moments by about 40% for one-lane bridges. For two-lane bridges, this overestimation increases to about 48% for the negative moment and about 55% for the positive moment. This overestimation decreases with the increase in span length. When base case railings (X1) are present, the AASHTO LRFD overestimation of the FEA slab moments becomes more significant reaching an average high of 66% for the negative moment and 61% for the positive moment in one-lane bridges, and 67% for the negative moment and 70% for the positive moment in two-lane bridges. This overestimation is further increased as the railings stiffness factor increases where it reaches about 81% for the negative moment and 74% for the positive moment for one-lane bridges and about 82% for the negative moment and 77% for the positive moment for two-lane bridges with (X4) stiffer railings.

Table 1. Comparison of FEA positive moment with railings to reference case without railings.

Number of Lanes	Span Length (m)	Ratio of FEA Maximum Longitudinal Moment with Railings to Reference Case without Railings						Reference Moment X0
		Stiffness Factor						
		X0		X1		X2		
1	7.2	19.1	1.00	15.6	0.82	14.0	0.73	19.1
	10.8	32.9	1.00	19.8	0.60	15.9	0.48	32.9
	13.8	46.6	1.00	30.1	0.65	23.7	0.51	46.6
	16.2	57.8	1.00	40.5	0.70	32.4	0.56	57.8
2	7.2	23.7	1.00	18.0	0.76	16.9	0.71	23.7
	10.8	40.5	1.00	29.2	0.72	24.7	0.61	40.5
	13.8	56.4	1.00	42.7	0.76	35.8	0.63	56.4
	16.2	69.4	1.00	55.7	0.80	47.5	0.68	69.4
		X3		X4		X0.5		
1	7.2	12.9	0.68	12.3	0.64	17.1	0.90	19.1
	10.8	14.1	0.43	13.0	0.40	23.9	0.73	32.9
	13.8	20.2	0.43	18.0	0.39	36.1	0.77	46.6
	16.2	27.5	0.48	24.4	0.42	47.3	0.82	57.8
2	7.2	16.3	0.69	16.1	0.68	19.4	0.82	23.7
	10.8	22.2	0.55	20.7	0.51	33.2	0.82	40.5
	13.8	31.6	0.56	28.7	0.51	48.1	0.85	56.4
	16.2	42.0	0.61	38.1	0.55	61.6	0.89	69.4

6.2 FEA Results: Railings vs. No Railings

The maximum slab positive bending moments are summarized in Table 1 for all bridge cases in terms of ratios of FEA results for cases with various railings stiffness factors as compared to the reference cases X0 without railings; a similar table was obtained for negative moments. It can be observed that the presence of railings reduces the maximum moments, and this becomes significant as the railings stiffness factor increases, and more for one lane vs. two lanes. For one-lane bridges, the maximum longitudinal moment was reduced by about 40 to 55% when adding railing with stiffness factor (X1) and reduced by about 60 to 70% with (X4) railing stiffness factor. As for two-lane-bridges, the slab moment was reduced by about 30 to 40% with (X1) railing stiffness factor, and a higher reduction was observed for (X4) stiffer railing of about 50 to 65%.

7 SUMMARY AND CONCLUSIONS

AASHTO Standard Specifications (2002) and AASHTO LRFD (2012) procedures do not account for the presence of railings as integral parts of a bridge slab. Based on this study, it is evident that these integral elements increase the capacity of the concrete slab bridges. This reduction in the slab moment decreases with the increase in the number of lanes and increases with the change in the railing stiffness. These railings can be used as an alternative strengthening technique to upgrade or rehabilitate existing bridges and allow permit vehicles on the bridge.

Acknowledgments

This research was supported by a grant from the University Research Board (URB) at the American University of Beirut to whom the authors are indebted and thankful.

References

AASHTO, *Standard Specifications for Highway Bridges*, 17th ed., American Association of State Highway and Transportation Officials (AASHTO), Washington, D. C., 2002.

AASHTO, *LRFD Bridge Design Specifications*, 5th ed., American Association of State Highway and Transportation Officials (AASHTO), Washington, D. C., 2012.

Akinci, N. O., Liu, J., and Bowman, M. D., Parapet Strength and Contribution to Live Load Response for Super Load Passages, *Journal of Bridge Engineering*, 13(1), 55-63, 2008.

Chung, W., Liu, J., and Sotelino, E. D., Influence of Secondary Elements and Deck Cracking on the Lateral Load Distribution of Steel Girder Bridges, *Journal of Bridge Engineering*, 11(2), 178-187, 2006.

Conner, S., and Huo, X. S., Influence of Parapets and Aspect Ratio on Live-load Distribution. *Journal of Bridge Engineering*, 11(2), 188-196, 2006.

Eamon, C., and Nowak, A., Effects of Edge-Stiffening Elements and Diaphragms on Bridge Resistance and Load Distribution, *Journal of Bridge Engineering*, 7(5), 258-266, 2002.

Fawaz, G., Waked, M., Mabsout, M., and Tarhini, K., Influence of Railings on Load Carrying Capacity of Concrete Slab Bridges, *Bridge Structures*, IOS Press, Vol. 12, No. 3-4, 85-96, 2017.

Mabsout, M., Tarhini, K., Frederick, G., and Kobrosly, M., Influence of Sidewalks and Railings on Wheel Load Distribution in Steel Girder Highway Bridges, *Journal of Bridge Engineering*, 2(3), 88-96, 1997.

Mabsout, M., Tarhini, K., Jabakhanji, R., and Awwad, E., Wheel Load Distribution in Simply Supported Concrete Slab Bridges, *Journal of Bridge Engineering*, 9(2), 147-155 (2004).

SAP2000 (version 19), *Computers and Structures Inc.*, Berkeley, California.

Interdependence between Structural Engineering and Construction Management
Edited by Ozevin, D., Ataei, H., Modares, M., Gurgun, A., Yazdani, S., and Singh, A.

ISBN: 978-0-9960437-6-2

CASE STUDY OF 3D SEISMIC PUSHOVER ANALYSIS OF INTEGRAL ABUTMENT BRIDGE

XINYA LIU and CHUNG C. FU

Dept of Civil and Environmental Engineering, University of Maryland, College Park, USA

Integral abutment bridges (IABs) have a continuous deck monolithically encased into abutment stem, and typically using single row of piles to carry vertical loads and accommodate longitudinal thermal deformation. Except for smooth pavement and low maintenance cost, IABs have also outperformed conventional seat-type abutment bridges in seismic performance due to increased redundancy, higher damping, and smaller displacements. However, lack of information on their seismic design and performance may have discouraged their use in high seismic zones. In this study, current research and implementation of IABs are comprehensively reviewed. IABs with steel-concrete girders provided by NYDOT are chosen for intensive seismic case study. Three-dimensional finite element models of IABs for nonlinear seismic analysis are elaborated to capture the behavior of components of superstructure, abutment stem, piles, backfill, etc. Pushover analyses are carried out to obtain the capacity curves. Through parametric studies, the effects of bearing are outlined. Conclusions and some recommendations are made for seismic evaluation and design practice of IABs.

Keywords: Three-span bridge, Jointless, Finite element models, Plastic hinges, Capacity curve, Parametric study.

1 INTRODUCTION OF INTEGRAL ABUTMENT BRIDGES

Integral abutment bridges (IABs) are designed without any expansion joints between spans and abutments. Resistance to longitudinal thermal movements and braking loads is provided by the stiffness of the soil abutting the end supports and, in some cases, by the stiffness of the intermediate supports.

IABs are usually considered as a prime alternative to conventional jointed bridges. IABs have recently become very popular in North America and Europe as they provide many economical and functional advantages (Spyrakos and Loannidis 2003, Briseghella and Zordan 2015). Modern IABs are known to have performed well in recent earthquakes due to the increased redundancy, larger damping resulting from cyclic soil–pile-structure interaction, smaller displacements and elimination of unseating potential (Itani and Sedarat 2000). The monolithic construction of IABs also provides better transfer of seismic loads to the backfill and pile foundations.

In 2005, the integral abutment-backfill behavior on sand soil was study by pushover analysis on a 2-D model. A study of earthquake resistance of IABs was conducted by Purdue University in 2009, in which a time-history analysis was done on 2-D models. A study by Far *et al.* (2015) combined seismic and actual thermal loads at the time of an earthquake is considered in the analysis of 2-D IAB model.

However, a research of seismic capacity of the IABs based on 3-D finite model and pushover analysis has yet been provided. Accordingly, this research study is aimed at investigating the seismic capacity using capacity curves resulted from the pushover analysis.

2 PROPERTIES OF THE THREE-SPAN INTEGRAL BRIDGE

Three IABs designed by NYDOT were analyzed as case studies. In this presentation, one of the three cases, the I-87 South Bound Bridge, is focused on. The I-87 South Bound Bridge over Megsville is a straight three-span semi-integral abutment bridge with a total length of 330ft and a central span of 130ft as shown in Figures 1(a).

The bridge deck is composed of a 9.5 in.-thick, 520-in-width reinforced concrete slab supported by five I-shaped steel girders spaced at 110in. from center to center as shown in Figure 1(b). The abutments are 3-ft-thick and 43.33-ft-long each, supported by eight HP12x84 piles in a single row spaced 69 in center to center. The layout of piles at the beginning abutment is shown in Figure 1(c); the semi-integral abutment detail is shown in Figure 1(d).

The two piers are supported on single columns with a height of 50ft and 42ft, respectively. Each pile cap is 330in by 276in and 72in thick and supported by 36 HP12x84 steel piles with the length of 50ft. All piles are in their weak axes in the longitudinal direction.

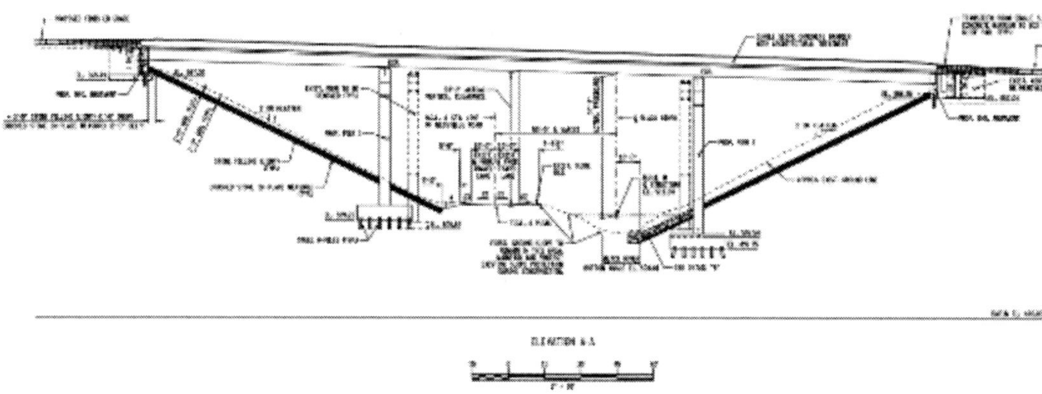

(a) Elevation view.

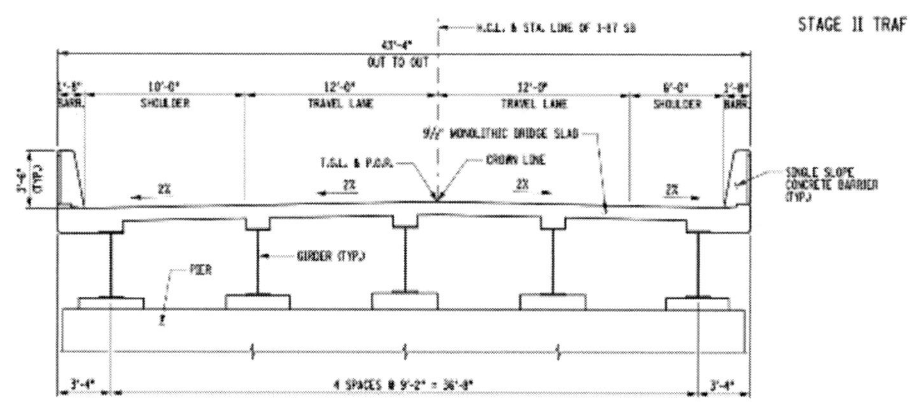

(b) Deck cross section.

Figure 1. Construction details of I-87 South Bound Bridge.

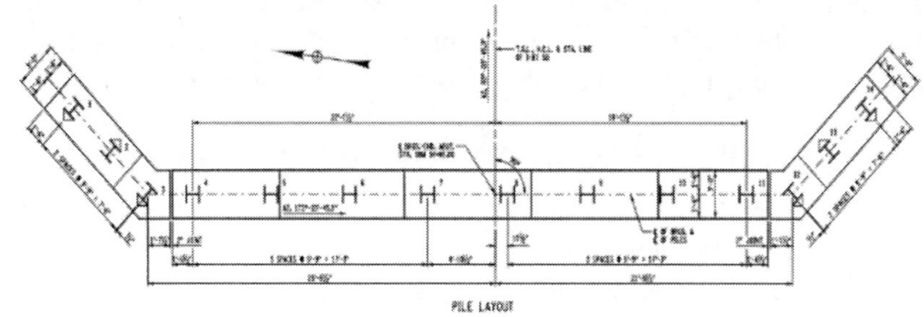

(c) Piles layout at abutment.

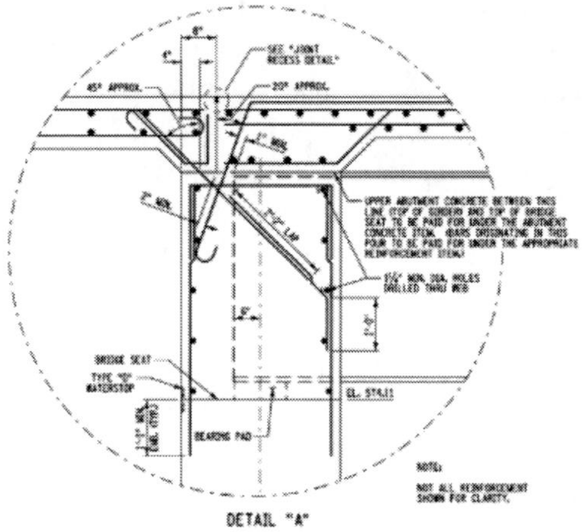

(d) Semi-integral abutment.

Figure 1. Construction details of I-87 South Bound Bridge (contd).

3 STRUCTURAL MODELING

The three-dimensional nonlinear finite element model was established by CSiBridge and incorporates the entire bridge structure, including the bridge superstructure, substructure and foundation as well as the soil behind the abutments and around the piles.

3.1 Modeling of Structure

The bridge superstructure was modelled using 3-D shell elements. Full composite action between the slab and the girders was assumed. Abutments were modeled by thick shell elements. The bearing pads at the semi-integral abutment were simulated by links that fixed y and z translational DOFs. The piles, piers and cap beams were modelled by beam elements. The 2ft embedded length of piles was considered in the model, allowing full moment transfer between piles and abutments.

3.2 Modeling of Soil-Structural Interaction

The soil-structural interaction can be defined by a nonlinear force (P)-displacement(Y) curve, where P is the lateral resistance of soil and Y is the lateral displacement. In this study, the actual nonlinear P-Y curves of soil are simplified with an elastic-plastic force-displacement curve relating the ultimate resistance of the soil as shown in Figure 2.

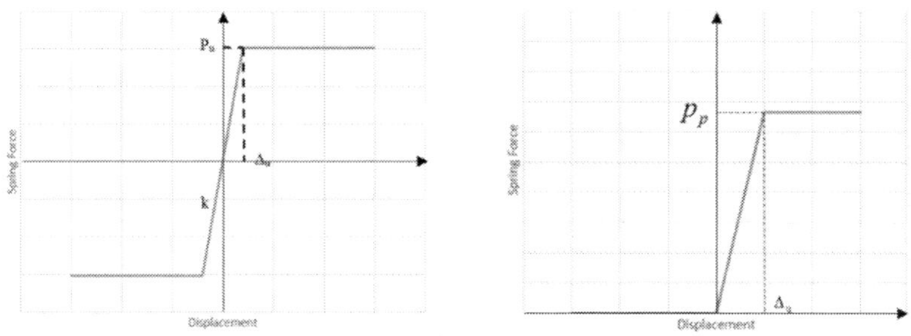

Figure 2. Simplified P-y curve of pile-soil interaction and abutment-soil interaction.

The soil around the piles was assumed sand according to the soil information used in the design given in the general notes. According to the Broms method (Broms 1964), the equivalent maximum force Pu (lb) for each spring on pile at the depth of z is shown in Eq. (1):

$$P_u = 3k_p \gamma D \cdot z \cdot s \tag{1}$$

For the abutment, according to the Rankine's earth pressure, the backfill horizontal-passive earth pressure the depth of z is determined as in Eq. (2):

$$p_p = \gamma z k_p \tag{2}$$

The soil spring stiffness k (lb/ft) at the depth of z can be obtained as in Eq. (3):

$$k = n_h \cdot z \cdot s \tag{3}$$

The maximum displacement is as follows as in Eq. (4):

$$\Delta_u = \frac{P_u}{k} \tag{4}$$

Where kp is the coefficient of passive earth pressure, γ is the unit weight of soil, D is the width or diameter of pile, s is the space of soil springs, and nh is the constant of the subgrade reaction. The force- displacement relationship keeps linear before the displacement reaches the maximum displacement Δ_u. After reaching Δ_u, the displacement keeps constant while force increases.

The pile-soil springs and abutment-soil springs were modelled by non-linear joint link elements and area springs respectively in CSiBridge. The soil reaction is linearly increased with depth; for the soil spring stiffness k is proportional to Pu, the soil spring stiffness varies with depth.

4 SEISMIC ANALYSIS

4.1 Eigenvalue Analysis

An eigenvalue analysis was conducted to determine the natural period Tn of the bridges. For the I-87 South Bound Bridge, the first mode is in the transverse direction and the second one in the longitudinal direction. The frequencies and periods of the first five modes are listed in Table 1.

Table 1. Modal periods and frequencies of three cases.

Modes No.	T_s(sec)	f_s(Hz)	Mode shape description
1	0.6141	1.6283	1st transverse
2	0.6047	1.6538	1st longitudinal
3	0.4499	2.2225	1st vertical bending
4	0.4241	2.3581	1st transverse antisymmetric torsion
5	0.4203	2.3793	1st vertical symmetrical torsion

4.2 Pushover Analysis

In the model, plastic hinges were defined as default PMM, PM2 and PM3 in the program for hinges formed under pushover action in different directions. The yield rotation factors are in accordance with ASCE 41-13 Table 9-6.

The relatively high stiffness of integral abutments would attract most of the longitudinal and transverse seismic forces and the expansion bearings at the top of piers limit the forces passed to the pier columns in the longitudinal direction. Thus, the piles and piers in IABs are allowed to act as "weak links" during seismic events and limit the seismic forces. These piles will subject to large flexural moments that cause sections to yield and eventually form plastic hinges at some positions (Monzon *et al.* 2014). The static pushover analyses were performed in the longitudinal and the transverse direction; then the plastic hinges were assigned at locations with the largest positive or negative moment.

Plastic hinges were formed on piles in the longitudinal direction as shown in Figure 3. In the transverse direction, the stiffness of piers is relatively high, thus it attracted most of the pushover loads and contributes to the form of plastic hinges at the bottom of the piers first as shown in Figure 4. Also, because of different heights of piers, plastic hinges occurred in sequence when the pushover loads were symmetrically applied.

Figure 5 shows the total base shear plotted against the deck displacement from the longitudinal and transvers pushover analyses respectively. In the figure, the displacements where the hinges start to yield are marked. It can be observed that the system remains generally elastic when the piles start yielding.

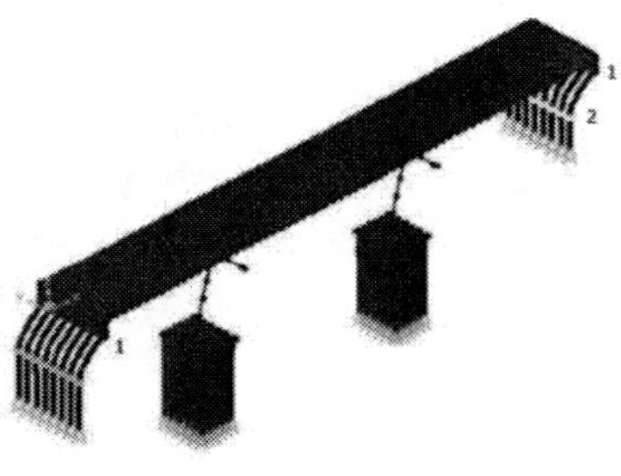

Figure 3. Plastic hinge occurrence in the x-dir.

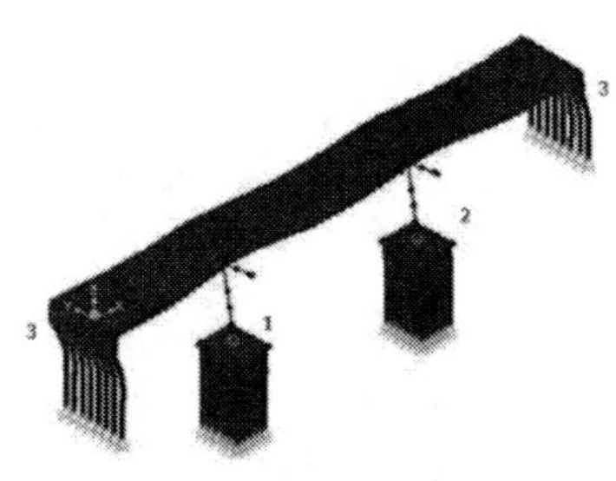

Figure 4. Plastic hinge occurrence in the y-dir.

5 PARAMETRIC STUDY

A parametric study was performed on the variation of bearing at the abutment stem. The three-span bridge was originally designed as a semi-integral abutment bridge (SIAB). In this study, by changing the bearing restraint, bridges with integral abutment and semi-integral abutment were analyzed and compared.

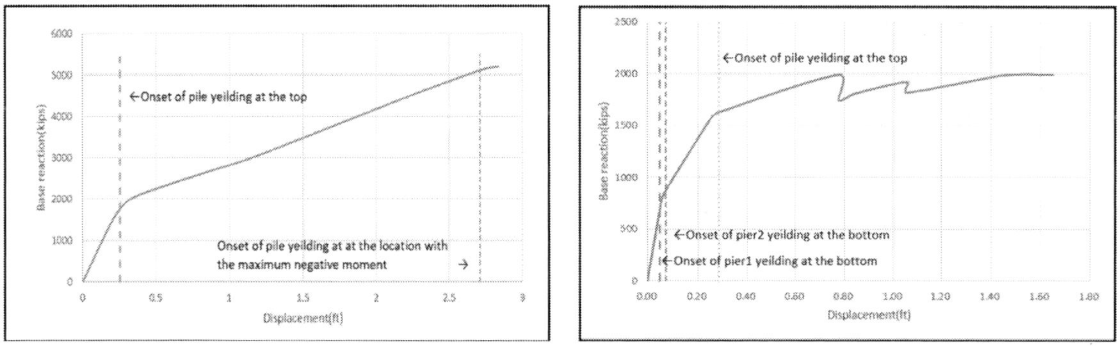

Figure 5. X and y-direction pushover curve with the mark of hinges yielding.

The modal shapes of IAB and SIAB are similar in the same mode. The period and frequency of each mode are slightly different, for the bridges with fully integral abutment have a higher stiffness, the periods of each mode become lower.

Comparing the pushover curves of IAB and SIAB in the x direction, the curve of IAB shows a higher force demand when reaches the same displacement. For the bearing pad not restrained in the longitudinal direction in SIAB, it is reasonable for the structure to have larger flexibility in this direction. In the y direction, the effect is not so obvious.

6 CONCLUSIONS

By performing the pushover analysis on the structures, the elastic design of the structure can be checked and the potential failure mechanism of structure under severe earthquake determined.

The location and sequence of plastic hinge occurrences were obtained from the pushover analysis. The sections at the top of piles went into plastic stage first and then the sections at the location with maximum negative moment followed. Locations of maximum negative moment on piles are different in the longitudinal and transverse directions.

In the comparison of integral abutment bridges and semi-integral abutment bridges, as expected, IABs have higher stiffness and smaller displacement under the same magnitude of earthquake compared to the SIABs.

References

Briseghella B, and Zordan T., An Innovative Steel-Concrete Joint for Integral Abutment Bridges. *Journal Traffic and Transportaion Engineering,* 2(4), 209–222, 2015.

Broms, M., Lateral Resistance of Piles in Cohesionless Soils. *Journal of the Soil Mechanics and Foundations Division © ASCE*, May, 1964.

Far, N. E., Maleki, S., and Barghian, M., *Design of Integral Abutment Bridges for Combined Thermal and Seismic Loads*, World Congress on Advances in Structural Engrg. and Mechanics (ASEM15), 2015.

Itani, A. M., and Sedarat, H., *Seismic Analysis and Design of the AISI LRFD Design Examples of Steel Highway Bridges.* Center for Civil Engineering Earthquake Research, University of Nevada, 2000.

Monzon, E. V., Itani, A. M., and Pekcan, G., Seismic Behavior and Design Of Steel Girder Bridges With Integral Abutments, *Bridge Structures,* 10(4), 117-128, 2014.

Spyrakos C., and Loannidis G., Seismic Behaviour of Post-Tensioned Integral Bridge Including Soil-Structure Interaction, *Soil Dynamics and Earthquake Engineering,* 23(1), 53–63, 2003.

Interdependence between Structural Engineering and Construction Management
Edited by Ozevin, D., Ataei, H., Modares, M., Gurgun, A., Yazdani, S., and Singh, A.

ISBN: 978-0-9960437-6-2

INFLUENCE OF CURVED WEB CLOSED STEEL SECTIONS IN BRIDGE DESIGN

GILLES VAN STAEN, HANS DE BACKER, and PHILIPPE VAN BOGAERT

Dept of Civil Engineering, Ghent University, Ghent, Belgium

A bridge is nowadays more than a structure that connects people over an obstacle. When a bridge has to be built in an urban area, either it has to be a landmark or it should blend away in the environment. The use of curved steel panels is one option to obtain these requirements. However, due to a lack of knowledge, engineers end up with a conservative design to implement these structural elements. For that reason, a Finite Element Model is made of a railway bridge, where the outer webs of the main girders have a varying web curvature. Six different models are made and compared. The most important parameters that are compared are the deformations and the stresses in the webs. The study finds that curved webs have an equal or even better behavior than flat webs, even with smaller web thickness. This makes that designers can use curved webs in their design, without needing extra steel to make their design safe.

Keywords: Railway bridge, Buckling, Stress, Plates, FEM, Eurocode, Deformation.

1 INTRODUCTION

Through the centuries, the design of bridges has evolved a lot from using more efficient materials, to changing the purpose of it. A bridge is nowadays more than a structure that connects people over an obstacle. When a bridge has to be built in an urban area, either it has to be a landmark or it should blend away in the environment. The use of curved steel panels is one option to obtain these requirements. Examples of the implementation of these elements are a railway bridge in Zemst (Van Bogaert 2010) and a bicycle bridge in Bruges (Flour 2012), both in Belgium (Figure 1).

Figure 1. Railway bridge in Zemst and bicycle bridge in Bruges.

Due to a lack of knowledge about curved webs suitable for bridge design, the examples in Figure 1 ended up with a rather conservative design. However, research is done on the buckling behavior of such panels as an isolated element (Batdorf *et al.* 1947, Featherstone 2003). These investigations are done for aerospace engineering purposes and thus made use of very thin steel plates of only a few millimeters, which is rather small for bridge design. Van Bogaert (2014) and Mozhdeh *et al.* (2011) expended the research towards bridge design with using curved webs either in a box section or by comparing the finite element approach with the theoretical formulas.

Curved panels do not only affect bridges on a local level, but also have an influence on the general bridge behavior. In order to find this influence, a typical railway bridge in Belgium is modified in such a way that the webs of the boxes of the main girder have a varying curvature radius. These curvature radii are based on previous research (Van Bogaert 2014) and exist in six different steps between a flat plate and an almost perfect cylindrical plate.

The influence of the modified webs is verified by several parameters. The first one is the deflection of the main girder under the serviceability limit state combination of only live loads. Next to that, the stress along the main girder in ultimate limit state in the webs is compared for each radius with each other. The last parameter that is compared is the web thickness of the most curved model in order to have an equal amount of steel use as for the model with flat webs.

2 FINITE ELEMENT MODELLING OF ARCH BRIDGE

2.1 Model Description

A finite element model of the Schaarbeek railway bridge, which is visible in Figure 2, is used as the basics in order to find the influence of curved web panels on the structural behavior of the arches and the main girders. The bridge is part of a high-speed railway that is connecting Brussels and Leuven, and equals a span of 140 m while crossing other railway tracks. The arch has a height of 22.36 m and is connected to the deck with 20 inclined hangers at each arch. The deck has a width of 10.5 m and is an orthotropic steel deck, carrying two railway tracks. The cross-sections of the arch and the main girder are shaped as rectangular boxes and the girder will have varying web curvatures as main parameter for this research, whereof only the outer web will vary. At the bottom of Figure 3 are the cross section with a flat web and the most curved web visible. Furthermore, the bridge is supported by neoprene bearings.

Figure 2. Schaarbeek railway bridge with span of 140 m.

In order to find the influence of the varying webs curvatures, the finite element model is kept as simple as possible. Therefore, a beam-plate model is created in Scia Engineer (NEMETSCHEK 2018). The software makes it easy to combine the loads, as described in the Eurocode, and to modify the properties of the bridge. A disadvantage is that at the intersecting points between plate elements and beam elements peak stresses can occur, due to different degrees of freedom the elements have. However, these are encountered by the use of averaging strips, which covers the area the beam and the plate does have in common.

All elements of interest for this paper are modelled as plate elements. This includes the box for the arches and the box for the main girders. Next to that, the hangers are connected to the box's flanges with the use of plate elements, as visible in Figure 3. At last, all internal stiffeners are plate elements as well. These are placed in the main girders at the locations where a cross beam is located and in the arches at the locations where a hanger is connecting with the box. All other elements, which are the hangers, the crossbeams and the wind bracings, are modeled as beam elements. A mesh size of 0.2 m is used for all elements.

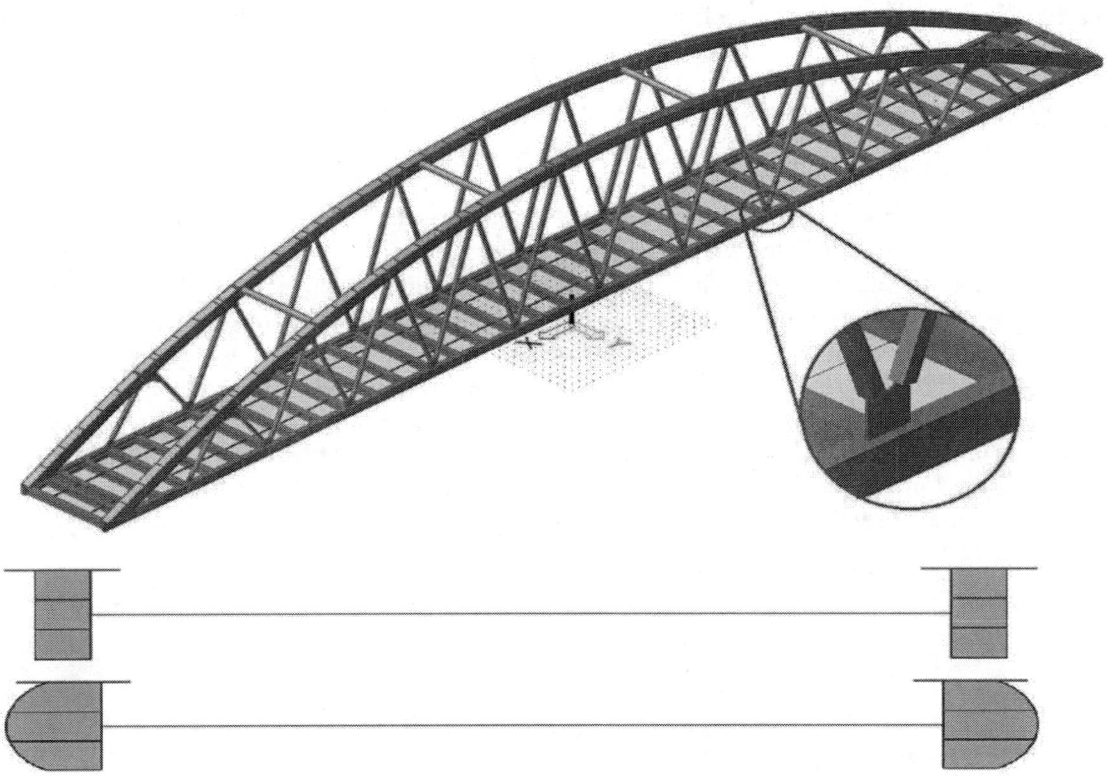

Figure 3. Finite element model of bridge and cross-section of the main girders.

The basic model with straight webs is verified with a comparison of previous research on the same bridge. De Backer *et al.* (2014) used the Schaarbeek bridge in order to find the critical normal force that leads to out-of-plane buckling of the arch. However, the bridge was modelled in detail; it is possible to compare the results. The detailed model gave a normal force of 13,500 kN for the dead loads. This simplified model in the Scia software returns a normal force of 11,900 kN for the dead loads. Despite of this small difference, the behavior of both bridges is similar.

2.2 Loads

2.2.1 *Self-weight and permanent load*

The self-weight of the bridge is generated by the software itself, while the ballast on which the railways supports is accounted with a mass density of 17 kN/m³ with a height of 0.6m.

2.2.2 *Train load model 71 (EN1991-2)*

Load model 71 from Eurocode EN 1991-2 (2003) is applied on the bridge. The combination of point loads and line loads is placed over the full span and half span of the bridge, with the point loads in the middle of the loaded field. The bridge has a double railway track, which is loaded together and separately. The loads are multiplied with the factor α, which is taken as 1.33 since this value is recommended for international railway traffic.

2.2.3 *Wind load*

Eurocode EN 1991-1-4 (2003) describes the loads due to wind on a structure. Chapter 8 describes the loading on a bridge structure. For the wind load transverse on the structure, a simplified method is chosen as described in the code. For the wind load in the z-direction, this value has to be multiplied with a factor of 0.90.

3 INFLUENCE OF CURVATURE

The basic model is modified in such a way that the moment of inertia increases, especially around the z-axis of the cross-section. It also implies the use of an extra amount of steel, which increase the dead weight of the bridge. To compare the behavior of the different web configurations, the first three fields are ignored. This is because due to the model simplification, the first fields have to be stiffened to create a stiff base for the transverse stiffness of the arch. With this in mind, the deformation of the girders due to live loads and the stresses along the girder are studied.

3.1 Girder Deformation

The deformation of a bridge tells a lot about the structural behavior of the structure. Table 1 gives the deformation values for the main girder in z- and y- direction. As one can expect, the reduced deformation is relatively higher in the transvers direction, due to the larger stiffness of the curved beam. However, these differences are all quite small because the global deformation is mostly determined by the arch stiffness.

Table 1. Deformation of main girder.

Curvature ratio r/h	Uz (mm)	Uy (mm)
1500	80	13.6
3	79.9	13.5
1.667	79.9	13.4
1	79.7	13.3
0.7	79.6	13.1
0.55	79.2	12.7

This additional stiffness also influences the out of plane buckling behavior of the arches. The critical buckling factor increases with increasing web curvature. For the model with a curvature ratio of r/h equal to 1.667, the buckling factor increases from 1.27 to 1.28 (for all loads in ULS).

This indicates that the critical buckling load increases with a few hundred kilo Newton's with an increasing web curvature. The next step in the investigation is to find the influence of curving the webs of the arch, where this parameter has a direct influence on the buckling stiffness.

3.2 Stress Variation in Web

The design of a web plate is mostly determined by its local buckling stiffness. The stress pattern in the web and the stress magnitude both have their influences. For that reason, it is interesting to verify if the curvature changes the stress pattern or magnitude. Figure 4 gives the stress patterns of the web fields in the middle of the bridge. The rectangle in blue is the steel plate connecting the upper flange with the hanger. The figure shows the flat web above and has an increasing curvature radius while descending in the figure.

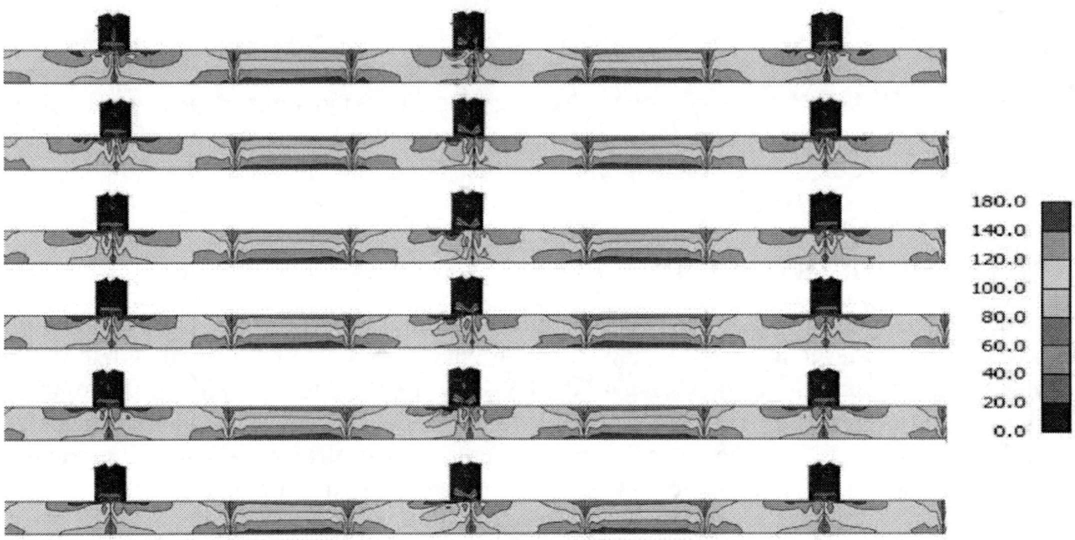

Figure 4. Stress along the girder for flat webs (above) to almost cylindrical web (down).

One can see that the stress pattern in the central web between two hangers does not change much for all models. In addition, the web fields next to the hanger supports do not change in value but has a small reduction in stress magnitude with increasing web curvature. This same behavior makes that the design procedure of a flat plate or a curved plate does not change. Next to that, Batdorf *et al.* (1947) found that the buckling coefficient used for the formula of the critical elastic buckling stress is increasing with increasing curvature, which makes that a curved panel can bear more stress. However, due to the thickness of 25 mm and a web height of 1.14 m, the plates can be considered as stocky. This means that they will yield before buckling occurs, and thus the yield stress is of more importance.

All stresses vary between 60 MPa and 180 MPa in Figure 4, which is obtained by the ultimate limit state for a railway bridge, using both permanent and variable loads. This is less than the yield strength of the S235 steel, resulting in a safe design.

3.2 Varying Web Thickness

With giving the web a large curvature and the same steel thickness as a flat web, more steel is used due to the increased length of the plate. With decreasing the web thickness, the same amount of steel can be obtained. For the model with the most curved web, a thickness of 20 mm instead of 25 mm results in the same use of steel as the model with the straight webs.

With a smaller web thickness, the moment of inertia about both axis decrease. This is immediately seen when the deformations of the girders are checked. The deformation in z-direction is equal with the one for the flattest web (80 mm). On the other hand, the deformations in the y-directions are 13.2 mm, which is still better than most other models. Due to the increased transverse stiffness, the out of plane buckling factor is larger than for the model with the flat webs. Next to that lies the stress pattern and values in line with the ones displayed in Figure 4.

These results indicate that with a same amount of steel; equally or even better structural requirements can be reached with curved panels. This makes that designers can use curved webs in their design, without forcing the engineer into a conservative and thus expensive design.

4 CONCLUSIONS

A finite element model is used to find the influence of curved webs on the behavior of the bridge. A median span railway bridge in Schaarbeek is modelled and modified in such a way that the outer webs of the main girders have varying curvatures. It is found that by giving the webs a curvature and thus with increasing the moment of inertia around both axis, the bridge has an equal and even a slightly better behavior. The curved webs even increase the critical buckling factor for out-of-plane buckling of the arch. A next step in this research is modifying the webs of the arch boxes with curved webs.

In the second phase is the web thickness decreased in order to have a same amount of steel as used for the bridge with flat webs. The girder has still a better performance when the webs are curved. This makes it so the designer can use curved webs in their design, without forcing the engineer into a conservative and thus expensive design.

References

Batdorf, S. B., Stein, M., and Schildcrout, M., Critical Shear Stress of Curved Rectangular Panels. *Langley Memorial Aeronautical Laboratory Langley Field*, Va. Washington, 1947.

De Backer, H., Outtier, A., and Van Bogaert, Ph., Buckling Design of Steel Tied-Arch Bridges, *Journal of Constructional Steel Research* 103, 159-167, 2014.

Eurocode EN 1991-1-4, *Actions on structures - Part 1-4: General actions - Wind actions*, European Committee for Standardization CEN; Brussels, 2003.

Eurocode EN 1991-2, *Actions on Structures – Part 2: Traffic Loads on Bridges*, European Committee for Standardization CEN; Brussels, 2003.

Featherston, C. A., Imperfection Sensitivity of Curved Panels under Combined Compression and Shear, *International Journal of Non-Linear Mechanics* 38, 225-238, 2003.

Flour, P., Bicycle Bridge Brugge, 2012. Retrieved from pietflour.wordpress.com on December 2018.

Mozhdeh, A., Edlund, B. L. O., and Alinia, M. M., Buckling and Postbuckling Behaviour of Unstiffened Slender Curved Plates under Uniform Shear, *Thin-Walled Structures*, 49, 1017-1031, 2011.

NEMETSCHEK – Scia Engineer. Retrieved from http://help.scia.net/nl/support/dowload/scia-engineer-18 on 2018.

Van Bogaert, Ph., *Design and Construction of a Double-Curved Railway Overpass and 3-Track Tubular Arch*, Proc. IABSE Conference Large Structures and Infrastructures for Environmentally Constrained and Urbanised Areas, IABSE, Venice, 564-565, 2010.

Van Bogaert, Ph., *Buckling Strength of Curved Bridge Girder Web Panels, Including Residual Stress*, 37th IABSE Symposium, IABSE, Madrid, 58-59, 2014.

Interdependence between Structural Engineering and Construction Management
Edited by Ozevin, D., Ataei, H., Modares, M., Gurgun, A., Yazdani, S., and Singh, A.

ISBN: 978-0-9960437-6-2

ACCURATE MODELING OF CURVED STEEL I-GIRDER BRIDGE

YIFAN ZHU, CHAORAN XU, and CHUNG C. FU

Dept of Civil and Environmental Engineering, University of Maryland, College Park, USA

A curved and/or skewed steel I-girder bridge, in addition to the basic vertical shear and bending effects, will be subjected to torsional and warping effects. Thus, simplified hand calculation and line girder methods, might not be enough when bridges are to be analyzed. Refined methods, termed by AASHTO, have to be adopted. This paper has investigated the closeness and difference between curved bridge finite element models using 2-D gird and 3-D shell elements of I-girders, both are part of AASHTO refined method. Moreover, the results are calibrated by comparing analysis result with various two-dimensional and three-dimensional computations with varied curvature effects. It is concluded that when introducing torsional effects to finite element models, the modified torsional constant J with consideration of warping effect should be taken into the 2-D grid model as a refined model. When using 3-D shell elements as the refined model, stiffeners and connection plates play an important role of global model stiffness and should not be ignored, especially for sharp curved steel I-girder bridges.

Keywords: Torsional effect, Warping effect, Finite element analysis, Model optimization, Approximate method, Refined method, Stiffener, Connection plate.

1 INTRODUCTION

Steel bridges that designed and built in early times are primarily straight and simple-span, which can be analyzed and calculated by hand. Indeterminate structures, such as continuous span bridges, can now be easily handled by computer software. Those bridges, including simple span and continuous span bridges, are still mainly straight bridges that subject to major-axis shear and bending moment effects of the main girders. However, in addition to the vertical shear and bending effects, a curved and/or skewed girder bridge is subjected to torsional effects, which cause both normal stresses and shear stresses in steel girders. For I-shape girders, the torsions they carried are primarily by means of warping other than St. Venant torsion due to open cross-sectional geometry. Furthermore, the St. Venant torsional shear flow around the perimeter of the cross section can only develop relatively small force couples. The total normal stress in an I-shaped girder is a combination of any axial stress, major axis bending stress, lateral bending stress, and warping normal stress (Figure 1a). The total shear stress is the sum of vertical shear stress, horizontal shear stress, St. Venant torsional shear stress, and warping shear stress (Figure 1b). For non-skewed straight bridges, only the major axis bending stress and vertical shear stress are dominant; the other factors can be ignored in the design phase but have to be included in other load combinations for code checking.

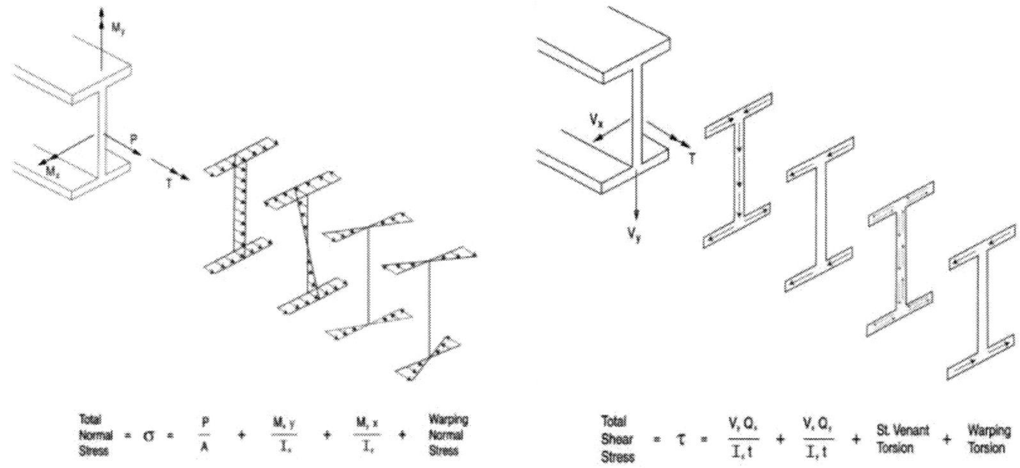

(a) Normal stress (b) Shear stress

Figure 1. Illustration of the general I-girder normal and shear stresses which can occur in a curved or skewed I-shaped girder.

Generally, I-girders carry torsion through combination of pure torsion and restrained warping and the total torsional resistance can be expressed in Eq. (1). For the calculation of section properties, including C_w, refer to AISC Design Guide 9: Torsional Analysis of Structural Steel Members (AISC 2003).

$$T = GJ\theta' - EC_w\theta'' \quad (1)$$

Where:
G = shear modulus of elasticity of steel; J = torsional constant of cross-section and can be approximated using Eq. (2) for rolled and built-up I shapes; E = modulus of elasticity of steel; C_w = warping constant of cross section and can be approximate as $\frac{I_y h^2}{4}$ for rolled and built-up I shapes; I_y = lateral moment of inertia about Y-axis, and h = distance between centerlines of top and bottom flanges.

2 2D MODELING AND COMPUTATION

In two-dimensional grid bridge modeling, the distribution of forces through the system is highly dependent on member stiffness parameters such as EI_x, EI_y, GJ, and EC_w. The warping stiffness parameter EC_w, is not used in a generic structural analysis method base on the beam theory with six degrees of freedom per node. For special analyses, cross sectional warping deflection, the seventh degree of freedom can be included to consider the warping of thin-wall cross sections. Thus, the additional warping stiffness is required. For curved structure, EC_w is often the dominant contributor to the individual girder torsional stiffness. Without consideration of EC_w, the local twisting responses of the girders cannot be modeled accurately. On the other hand, a full 3D FEM analysis, in which the thin-wall sections are modeled by plane shell elements, bypasses the need for the modeling of warping stiffness within the single beam element used to model the girder in 2D grid analysis approaches (Fu and Wang 2014).

A rigorous solution of grid analysis to take care of the warping problem of a thin-wall beam requires an additional degree of freedom. Several researchers (e.g., Hsu *et al.* 1990, Fu and Hsu

1995) have included the warping deflection as the seventh degrees of freedom, in addition to the regular six DOFs, at each node for the curved beam analysis to consider the warping effect. For the case of partial warping restrained, an effective torsional constant, K_{eff}, was proposed by Fu and Hsu (1994) and later improved by Elhelbawey and Fu (1998) to consider warping effects in a regular six DOFs analysis. A simple, easy-to-apply effective torsional constant for the rotational stiffness of a restrained open section was developed to count for both the pure torsion and the warping torsion into account. This effective (equivalent) torsional constant, K_{te},, can be easily calculated and used for any generic finite-element structural analysis program.

The original torsional constant for most common structural shapes, *J*, can be approximated by Eq. (2)

$$J = \sum bt^3/3 \tag{2}$$

where *b* and *t* = width and thickness of the thin-wall elements, respectively. The effective (equivalent) torsional constant, K_{te}, developed by Fu and Hsu (1994), can be expressed as in Eq. (3)

$$K_{te} = J\cosh\frac{\lambda}{2} / \left\{\cosh\frac{\lambda}{2} - 1.0\right\} C \tag{3}$$

where $\lambda^2 = GJ/EC_w$, (λ = l/a, where a is used in AISC documents), *C* = correction factor that equals $\{1.0/[1.0 + 2.95\,(b/l)^2]\}$, *l* = unbraced length, and *b* = flange width.

Once the effective torsional constant is determined, the stiffness matrix for a grid structure can be derived by using the traditional straight beam method with three DOFs (torsional rotation, bending rotation and deflection) per node. The stiffness matrix of an element in grid model with warping partially restrained is as following in Eq. (4):

$$[K_e] = \begin{bmatrix} \frac{GK_{te}}{l} & 0 & 0 & -\frac{GK_{te}}{l} & 0 & 0 \\ 0 & \frac{4EI_y}{l} & \frac{-6EI_y}{l^2} & 0 & \frac{2EI_y}{l} & \frac{6EI_y}{l^2} \\ 0 & \frac{-6EI_y}{l^2} & \frac{12EI_y}{l^3} & 0 & \frac{-6EI_y}{l^2} & \frac{-12EI_y}{l^3} \\ -\frac{GK_{te}}{l} & 0 & 0 & \frac{GK_{te}}{l} & 0 & 0 \\ 0 & \frac{2EI_y}{l} & \frac{-6EI_y}{l^2} & 0 & \frac{4EI_y}{l} & \frac{6EI_y}{l^2} \\ 0 & \frac{6EI_y}{l^2} & \frac{-12EI_y}{l^3} & 0 & \frac{6EI_y}{l^2} & \frac{12EI_y}{l^3} \end{bmatrix} \tag{4}$$

A similar study was done years later by the NCHRP Project 12-79 Report 725 (2012) with two equivalent equations with warping fixity at each end of a given unbraced length L_b (Eq. 5a) and warping fixity at one end and warping free boundary conditions (Eq. 5b) where J_{eq} is equivalent to K_{te} in Eq. (3).

$$J_{eq(fx-fx)} = J[1 - \frac{\sinh(pL_b)}{pL_b} + \frac{[\cosh(pL_b)-1]^2}{pL_b\sinh(pL_b)}]^{-1} \tag{5a}$$

$$J_{eq(s-fx)} = J[1 - \frac{\sinh(pL_b)}{pL_b\cosh(pL_b)}]^{-1} \tag{5b}$$

3 3D FINITE ELEMENT MODELING

The category 3D FEM analysis method is meant to encompass any analysis/design method that includes a computerized structural analysis model where the superstructure is modeled fully in three dimensions, including: modeling of girder flanges using line/beam elements or plate/shell/solid type elements; modeling of girder webs using plate/shell/solid type elements; modeling of cross frames or diaphragms using line/beam, link/truss, or plate/shell/solid type elements (as appropriate); and modeling of the deck using plate/shell/solid elements. This method is time-consuming and complicated, thus is arguably deemed most appropriate for use for complicated bridges (e.g., bridges with severe curvature, skew, or both, unusual framing plans, unusual support/substructure conditions, or other complicating features). 3D analysis methods are useful for performing refined local stress analysis of complex structural details (AASHTO/NSBA 2011).

When and how to use a refined 3D FEM analysis for engineering design is a controversial issue, and in the US such an approach has not been fully incorporated into the AASHTO specifications to date (AASHTO 2013). The typical AASHTO methodology for design is generally based on assessment of nominal (average) stresses calculated by simplified methods, such as P/A or Mc/I, and not localized peak stresses obtained by shell- or solid-based finite element models. Refined analysis can provide substantially more detailed and accurate information about the stress state of the structure. This could allow for more cost effective and reliable design but often comes with increased engineering effort and increased potential for error. However, if properly modeled, in the forensic or load test cases, such refined analysis is commonly adopted due to its refinement and accuracy. Researchers have done several modeling and analysis over the years. Hays Jr. *et al.* (1986) and Mabsout *et al.* (1997) modeled the deck slab using quadrilateral shell elements in plane with five degrees of freedom per node and the steel girders using 3D beam elements with six degrees of freedom per node (In-plane shell -beam model). Tarhini and Frederick (1992) et. al. used eight node linear solid brick elements with three displacement degrees of freedom in each node to model the concrete deck as a 3D Brick-shell model. Fu and Lu (2003) idealized bridge deck with isoperimetric quadrilateral shell elements and the reinforcement was modeled as a smeared two-dimensional membrane layer with isoperimetric plane stress element, and called 3D shell-shell model. In addition, 3D shell-beam model (Tabsh and Tabatabai 2001, Issa *et al.* 2000) and 3D brick-beam model (Ebeido and Kennedy 1996, Barr *et al.* 2001, Chen 1999, Sebastian and McConnel 2000) were used for three-dimensional finite element analysis of steel bridges (Fu and Wang 2014).

4 CASE STUDY

Bridge No. 27W18 is a two-span curved steel bridge with skewed angle and currently under design process. The general plan of the bridge is shown in Figure 2. Engineers and researchers had done structural analysis under dead load in the non-composite stage, i.e., the concrete deck is assumed with no stiffness but has weight on steel girders. In this study, seven analyses including four in two-dimensional and three in three-dimensional are computed and analyzed. For two dimensional calculations, engineers used two software, *DESCUS-I* and *STAAD*, based on various calculation methods, to compute reaction forces at bearings, moments and deflections for each girder in the bridge. *CsiBridge*, *ANSYS*, and *STAAD* are three software that used in 3D finite element analysis in this case and choose the 3D shell-shell model as the original model. It should be noted that in 3D finite element modeling for this case, modeling stiffeners and connection plates has significant influence on the results due to torsional effect by the curvature and skews.

The results of reaction forces and moments are listed and compared in Table 1, the deflection shapes of each girder are similar, and the maximum values are pretty close:

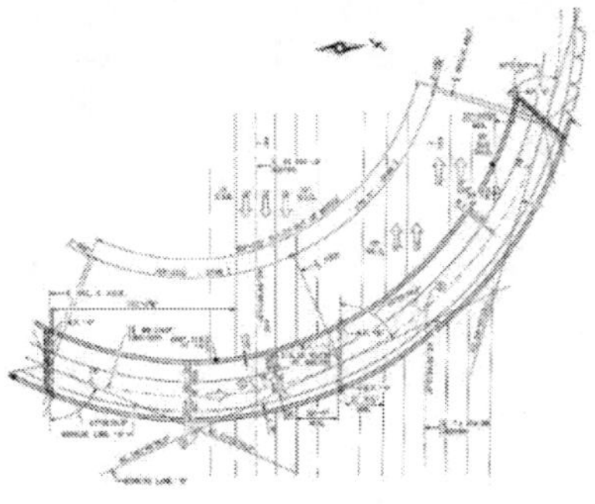

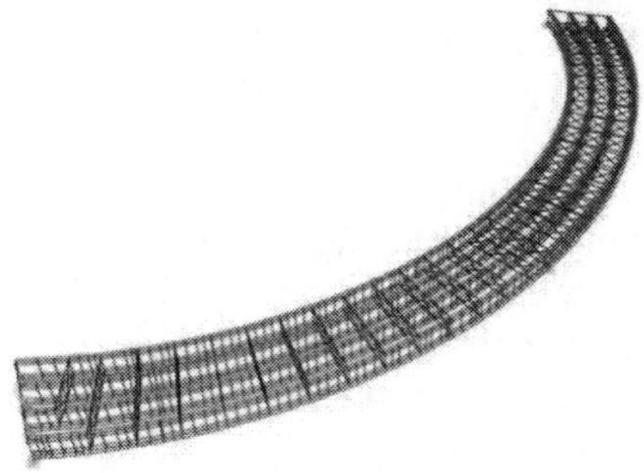

Figure 2. General plan and 3D FEM model of the bridge.

Table 1. Results of model analysis for bridge 27W18.

Girder No.	Analysis software and method	Moment (kip-ft)		Unfactored non-composite reaction (kips)		
		Maximum	Minimum	South Abutment	Pier	North Abutment
1	STAAD - NCHRP 725	5550.159	-9039.145	192.6	198.7	243.5
	STAAD - 2D grid	5678.432	-9113.4	200.5	184.2	243.2
	DESCUS-I	6086.24	-9639.28	202.9	207.3	247.1
	DESCUS-I - adjusted	5810.24	-9210.16	193.4	193	235.2
	STAAD - 3D	7719.243	-9549.975	187.5	198.1	223.9
	CsiBridge	6322.346	-9631.13	174.35	149.06	215.05
	ANSYS	5290.61	-9681.7	162.04	159.24	209.82
2	STAAD - NCHRP 725	4018.314	-9108.697	115.9	515.6	99.4
	STAAD - 2D grid	4048.466	-9248.765	101.1	525	100.3
	DESCUS-I	4252.88	-10077.28	113.5	534.6	113.9
	DESCUS-I - adjusted	4058.72	-9625.28	108.2	521.3	108.6
	STAAD - 3D	4697.415	-9928.611	118.7	492.8	128
	CsiBridge	4049.07	-9835.96	106.89	545.03	108.12
	ANSYS	4094.77	-9460.32	119.62	529.97	111.82
3	STAAD - NCHRP 725	2377.601	-7242.275	47.4	461.4	38.7
	STAAD - 2D grid	2311.716	-7320.26	51.1	472.1	33.4
	DESCUS-I	2324.08	-7635.52	53.7	479.8	35.4
	DESCUS-I - adjusted	2216.24	-7289.84	51.3	456.1	34
	STAAD - 3D	1919.312	-8300.873	49.8	471.4	34.2
	CsiBridge	1904.45	-7433.59	56.35	438.11	34.15
	ANSYS	1889.64	-7983.89	52.46	434.52	31.22
4	STAAD - NCHRP 725	1125.841	-5709.293	47.2	370.7	43.2
	STAAD - 2D grid	1113.356	-5651.61	49.2	367.5	46.4
	DESCUS-I	1059.44	-5828.4	51	402.6	46.9
	DESCUS-I - adjusted	1008.32	-5559.52	48.6	384.5	44.7
	STAAD - 3D	642.579	-6590.325	45.3	410.7	35.3
	CsiBridge	748.93	-5460.8	50.96	406.06	49.67
	ANSYS	898.74	-6723.443	46.8	407.22	47.41

5 CONCLUSIONS

The curved skew I-girder bridges have different behavior under loading compared with straight bridges. This difference would be caused most due to torsional effects on the open section of I girder. Thus, for accurate modeling, the effective torsional constant should be taken into consideration for 2D analysis, and the connection plates and stiffeners should be modeled and added to the steel girders when doing 3D finite element analysis for curved and/or skewed bridge.

References

AASHTO, *AASHTO LRFD Bridge Design Specifications*, American Association of State Highway and Transportation Officials, Washington, DC, 6th Edition with 2013 Interim.

AASHTO/NSBA, *Guidelines for Steel Girder Bridge Analysis*, Steel Bridge Collaboration Task Group 13, Document G13.1, 1st Edition, 155 pp. 2011.

AISC, *Design Guide 9: Torsional Analysis of Structural Steel Members*, Chicago, IL, USA, 2003.

ANSYS Mechanical APDL. ANSYS Academic Teaching Introductory. ANSYS, Inc.

Barr, P. J., Eberhard, M. O., and Stanton, J. F., Live-Load Distribution Factors in Prestressed Concrete Girder Bridges, *Journal of Bridge Engineering*, 6(5), 298-306, 2001.

Chen, Y., Distribution of Vehicular Loads on Bridge Girders by The FEA Using ADINA: Modeling, Simulation, and Comparison, *Computers and Structures*, 72(1-3), 127-139, 1999.

CSiBridge, Structural Bridge Design Software, Computer and Structures, Inc.

DESCUS-I, Design and Analysis of Curved I-Girder Bridge Systems, *Users' Manual*, PSI.

Ebeido, T., and Kennedy, J. B., Girder Moments in Simply Supported Skew Composite Bridges, *Canadian Journal of Civil Engineering*, 23(4), 904-916, 1996.

Elhelbawey, M. I., and Fu, C. C., Effective Torsional Constant for Restrained Open Section, *Journal of Structural Engineering*, 124(11), November, 1998.

Fu, C. C., and Hsu, Y.T., Bridge Diaphragm Elements with Partial Warping Restraint, *Journal of Structural Engineering*, 120(11), November, 1994.

Fu, C. C., and Hsu, Y. T., The Development of An Improved Curvilinear Thin-Walled Vlasov Element, *Computer and Structures*, 54(1), 147–159, 1995.

Fu, K. C., and Lu, F., Nonlinear Finite-Element Analysis for Highway Bridge Superstructures, *Journal of Bridge Engineering*, 8(3), 173-179, 2003.

Fu, C. C., and Wang, S. Q., *Computational Analysis and Design of Bridge Structures, ISBN*-13 978-1466579842, book published by CRC Press, December 2014.

Hays Jr., C., Sessions, L. M., and Berry, A. J., Further Studies on Lateral Load Distribution Using a Finite Element Method, *Transportation Research Record*, 6-14, 1986.

Hsu, Y.T., Fu, C.C. and Schelling, D.R., An Improved Horizontally Curved Beam Element, *International Journal of Computer and Structures*, 34(2), 313-316, 1990.

Issa, M. A., Yousif, A. A., and Issa, M. A., Effect of Construction Loads and Vibrations on New Concrete Bridge Decks, *Journal of Bridge Engineering*, 5(3), 249-258, 2000.

Mabsout, M. E., Tarhini, K. M., Frederick, G. R., and Tayar, C., Finite-Element Analysis of Steel Girder Highway Bridges, *Journal of Bridge Engineering*, 2(3), 83-87, 1997.

Sebastian, W. M., and McConnel, R. E., Nonlinear FE Analysis of Steel-Concrete Composite Structures, *Journal of Structural Engineering*, 126(6), 662-674, 2000.

STAAD, Structural Analysis and Design Program, Bentley Systems, Incorporated.

Tabsh, S. W., and Tabatabai, M., Live Load Distribution in Girder Bridges Subject to Oversized Trucks, *Journal of Bridge Engineering*, 6(1), 9-16, 2001.

Tarhini, K. M., and Frederick, G. R., Wheel Load Distribution in I-Girder Highway Bridges, *Journal of Structural Engineering*, 118(5), 1285-1294, 1992.

White, D. W., Coletti, D., Chavel, B. W., Sanchez, A., Ozgur, C., Chong, J. M. J., Leon, R. T., Medlock, R. D., Cisneros, R. A., Galambos, T. V., Yadlosky, J. M., Gatti, W. J., and Kowatch, G. T., Guidelines for Analysis Methods and Construction Engineering of Curved and Skewed Steel Girder Bridges. *NCHRP Report, 725*.Transportation Research Board, 2012.

Interdependence between Structural Engineering and Construction Management
Edited by Ozevin, D., Ataei, H., Modares, M., Gurgun, A., Yazdani, S., and Singh, A.

ISBN: 978-0-9960437-6-2

EFFECT OF TRANSIENT THERMAL GRADIENT ON STRESSES IN COMPOSITE BRIDGES

CAESAR ABI SHDID[1] and OMAR EL-MASRI[2]

[1]*Dept of Civil Engineering, Lebanese American University, New York, USA*
[2]*Popli Design Group, Penfield, USA*

Composite steel-concrete bridges experience higher thermally induced stresses than their concrete and steel cousins. These thermal stresses which can result from support restraints, debris accumulating in expansion joints, or from non-uniform thermal gradients, can lead to significant damage in the concrete deck. Conventional heat transfer theory in solids, in three-dimensional finite element formulation, is used to perform a sequentially coupled thermal-stress analysis in a selected single-span case study bridge. Actual environmental boundary conditions for a selected geographical region are used to develop the thermal profile. The vertical thermal gradient is shown to be largely non-linear as opposed to existing models, such as AASHTO. The thermally induced tensile stresses in the concrete are shown to be significant compared to service load stresses and constitute 60% of the tensile strength of the concrete deck.

Keywords: Temperature profile, Finite element analysis, Cold regions, Construction cracks, Steel, Concrete.

1 INTRODUCTION AND JUSTIFICATION

Bridges undergo continuous heat exchange with their surroundings thus developing a transient thermal profile within their cross section. The thermal gradient is largely affected by the thermal diffusivity of the constituent materials. It is this difference between the low and high thermal diffusivity of concrete and steel, respectively, that results in a significant thermal gradient within the transverse cross-section of composite bridges. This thermal gradient, coupled with different coefficients of thermal expansion for concrete and steel and the presence of shear connectors that prevent slip between the steel girders and concrete deck, make the thermal stresses in composite bridges particularly high. Design codes, such as the AASHTO LRFD Bridge Design Specifications prescribe hypothetical vertical temperature profiles in various types of bridges located in different geographical regions (AASHTO 1989).

The provided bridge expansion mechanisms can only accommodate overall deformations due to annual variations in temperatures and uniform strain components. The vertical thermal gradient resulting from diurnal variations produces three components of thermal stresses: (1) a uniform component (the average strain) which produces axial movement without stress if the movement is unrestrained; (2) a linear component (curvature-inducing strain), which produces vertical deflections and curvature without stress if the vertical movement is completely unrestrained; and (3) a nonlinear component (a stress-inducing strain) that leads to non-uniform self-equilibrating stresses. These self-equilibrating stresses result from the continuity of the cross section and the assumption that plane sections remain plane. These stresses will lead to considerable damage in the structure in the form of deck cracks that lead to the corrosion of the steel reinforcement (Ramey *et al.* 1997). Limited studies have been dedicated to investigating the thermal gradients and the resulting thermal stresses in composite bridges (Fu *et al.* 1990).

Previous studies—on which the proposed AASHTO gradient is based—have failed to take into account the presence and effect of a longitudinal thermal gradient that will produce self-equilibrating stresses in the transverse direction, regardless of whether the vertical thermal gradient is uniform or not, or whether the section is restrained or not. This is mainly due to the continuity of the section, the Poisson's effect, and displacement compatibility at the interface of consecutive sections. While some previous studies considered a two-dimensional model (Berwanger and Symko 1975, Emanuel and Hulsey 1978), most existing research is based on a one-dimensional model (Emerson 1973, Priestley and Thurston 1979), environmental boundary conditions synthesized using predictive sinusoidal models rather than actual measured data—and in some cases—steady state conditions, and simplistic finite element models. The various theoretical and experimental studies that had been conducted on composite bridges were synthesized by Kennedy and Soliman (1987), who proposed a simple one-dimensional vertical temperature distribution within the section that is uniform through the depth of the steel beam and linear through the concrete deck. This distribution led to simple formulation of thermal stresses in simple and continuous spans.

2 BRIDGE MODEL

The Colquitz River Bridge near Victoria in British Columbia, Canada was used as a case study because of its perfectly symmetrical cross section that makes it typical of composite bridges. The bridge length is 83 meters divided into five spans of varying lengths, and its width is 11.9 m. The bridge cross-section is shown in Figure 1 and the material properties are summarized in Table 1. The thermal coefficient of expansion of concrete varies between 7.2 E-6/°C and 11.7 E-6/°C. A constant value was assumed near the higher end—closer to the thermal coefficient of expansion of steel. Lower values would produce more dramatic thermal stresses. A single span of 18.3 meters was modeled in this study having pinned and fixed supports—boundary conditions that would produce the highest thermal stresses of all possible support conditions as was shown by Dilger *et al.* (1983). The City of Fargo, ND (Latitude: 46°, 52', 38"; Longitude: 96°, 47', 22"; Elevation: 275 m), with its extremely cold winters and hot summers, was selected for the purpose of highlighting the thermal stresses produced by diurnal cycles through large thermal gradients.

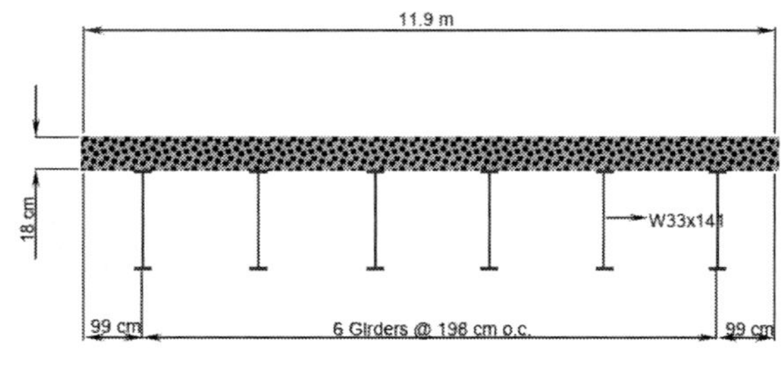

Figure 1. Cross section of Colquits river bridge.

When performing thermal stress computations for bridges, two cases of temperature profiles have to be considered in order to estimate the full range of critical stresses for design. The first is when a positive thermal gradient results from having the top of the concrete deck at a temperature higher than the bottom of the deck, which is at the same temperature as the steel beam. The second case is when the top of the concrete deck is at a temperature lower than the bottom of the deck and steel beam, and a negative thermal gradient ensues. The first case can occur on a sunny day in the summer when the top of the deck is heated by the sun radiation faster than the shaded steel beams that remain at a uniform temperature—equal to that of the ambient air temperature. The first case can also take place on a sunny winter day, but with much lower temperature difference, *ΔT*, between the top of the deck and the steel beams due to the lower intensity of solar radiation. The second case can occur on a winter or a summer day when the top of the deck is suddenly drenched with cold rain or snow.

Table 1. Bridge model material properties.

Bridge Component	Unit Weight (Kg/m^3)	Modulus of Elasticity (MPa)	Poisson's Ratio	Coefficient of Thermal Expansion (cm/cm/°C)	Grade	Compressive Strength (MPa)
Concrete Deck	2,400	26,435	0.2	10.8 E-6	NA	28
Steel Girders	7,850	200,000	0.3	11.7 E-6	ASTM A992	NA

The most critical case is when the thermal and service load stresses are of an additive nature. In a fixed-pinned single span bridge as considered, a positive thermal gradient similar to the one depicted in Figure 2(a) will produce the associated shown stresses that would be additive to dead and live load stresses. A negative thermal gradient, as the one depicted in Figure 2(b), will produce thermal stresses that are of reverse sign to those of service loads, and are consequently not detrimental. Since larger differential temperatures would obviously induce higher stresses (although not proportionally), the case of a hot sunny day in summer would therefore constitute the most critical case of stresses. The time domain chosen for this transient analysis consists therefore of a 24-hour time span occurring over June 4. The selection was based on the highest radiation intensity day of the year for Fargo, ND, since it has been shown that the solar radiation has the highest influence on thermal gradients in bridges (Tong 2002). A transient time step of one hour, divided into four (4) equal increments, was used for the transient thermal simulation.

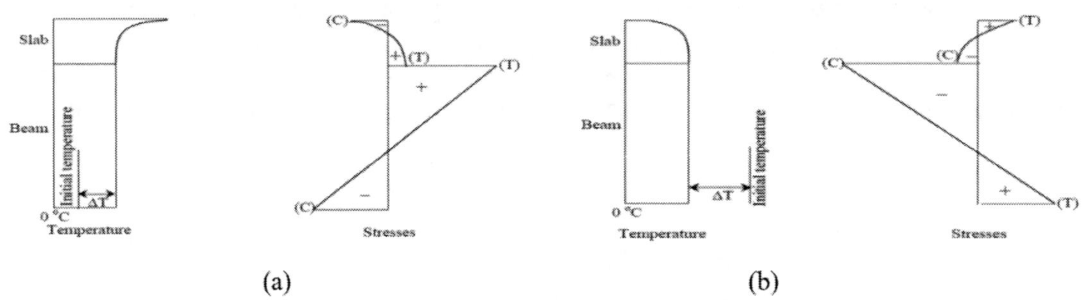

Figure 2. Thermal gradient and corresponding thermal stresses: (a) positive gradient; (b) negative gradient.

The finite element model of the bridge section was constructed using homogenous isotropic solid elements. Both concrete deck and steel girders were modeled using 3D 8-node linear hexahedral heat transfer mesh elements with temperature as a single degree of freedom at each node, and second-order (quadratic) interpolation in three dimensions. The model (shown in Figure 3) comprised the full span length, width, and depth of the bridge with all six supporting steel girders. The refined mesh resulted in nine integration points along the depth of the concrete deck, thus allowing accurate representation of non-linear behavior in bending and retrieval of thermal profile, stresses, and strains at each integration point. Shear connectors, in the form of "tie" connectors, were continuously provided between the bottom of the concrete deck and the top flange of the steel girders to ensure full composite action zero slip. Translation and rotation displacement restraints were provided at the end of the model as appropriate. The stress-strain behavior of the concrete is modeled to include compressive crushing, tensile softening, and cracking. The stress-strain behavior for the concrete deck in tension is shown in Figure 4.

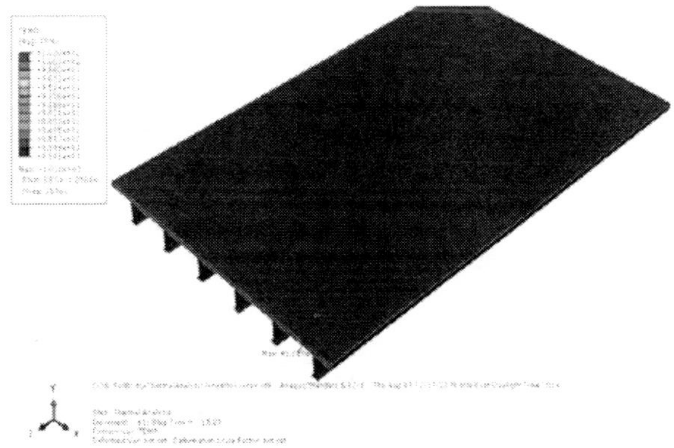

Figure 3. Finite element model of the bridge.

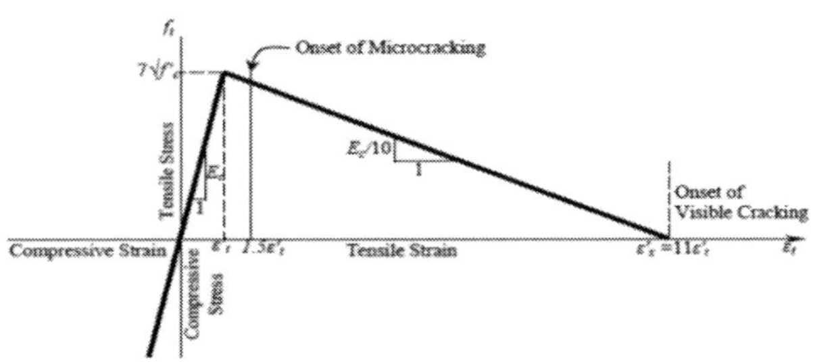

Figure 4. Concrete stress-strain model in tension.

3 RESULTS

Sequentially-coupled non-linear transient thermal-stress analysis is conducted in which the stress/deformation field in the bridge depends on the temperature field. This is performed by first conducting an uncoupled heat transfer analysis, and then a stress/deformation analysis of the model. The simulation time domain spanned over a period of 24-hours that represents a full day (June 4). Results of the 3D finite element analysis show that the vertical temperature distribution in the web of the steel girder is almost uniform, which agrees with previous studies (Kennedy and Soliman 1987). Figure 5 provides the vertical temperature distribution at midspan of the interior girder at the time of maximum thermal gradient, along with the thermal profile as calculated using the AASHTO and the Kennedy and Soliman models. Due to solar radiations reflected from the ground, the temperature at the bottom surface of the concrete deck tends to be slightly cooler right over the two steel girders than the regions between the two girders and under the overhang.

The absolute maximum temperature differential between the concrete deck and the steel girder is critical for calculating the bridge thermal stresses resulting from the various components of thermal strains (uniform, linear, and non-linear). On June 4, the temperature differential reached a maximum value of 10.7 °C (19.3 °F) at the location of the interior girder when the temperature in the concrete deck reached a maximum value of 38.6 °C (102 °F) at time 15:15. This value is less than half that suggested by Kennedy and Soliman, which is 22.3 °C.

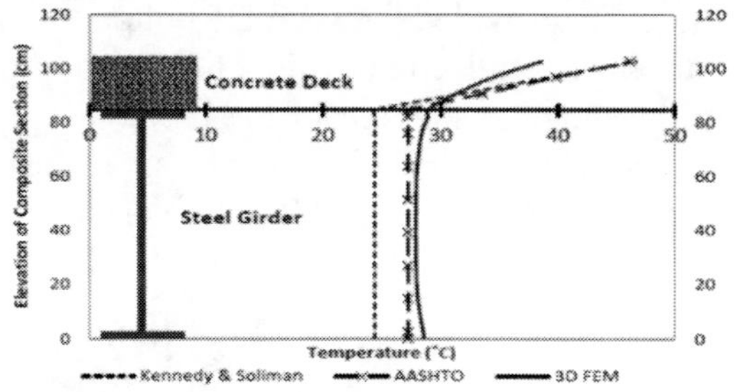

Figure 5. Vertical temperature distribution at interior girder (15:15; June 4).

AASHTO recommends that the vertical temperature distribution be uniform in the steel girders and linear in the superimposed concrete deck. The temperature differential in the concrete deck is based on a map of solar radiation zones in the United States, in which the city of Fargo, ND (located in zone 2) has a maximum positive vertical temperature differential of 18.9 °C between the top and bottom surfaces of the concrete deck. Figure 5 shows that the AASHTO maximum temperature differential is 8.2 °C higher than the one obtained here. This difference reaches 11.6 °C for the Kennedy and Soliman model. This conservative approximation by AASHTO leads to considerable error in assessing the thermal stresses. Another issue is the nonlinear thermal profile which produces a nonlinear thermal strain component whose effects on the design is unknown.

The vertical thermal profile proposed by AASHTO for the case study bridge is highly conservative. The reason is that AASHTO is based on a 1D conduction analysis and ignores the reflected solar radiation which decreases the positive thermal gradient by heating the bottom surface of the deck. Also, the AASHTO model does not consider the depth of the concrete deck or the overhang-to-depth ratio, which produces a shading effect on the exterior beam. AASHTO also assumes a linear vertical temperature distribution in the concrete deck at all times.

The highest positive value for the maximum principal stress is used as the maximum tensile stress in the concrete deck, and the highest negative value of the minimum principal stress is used as the maximum compressive stress in the concrete deck. As for ductile materials such as steel, the Von Mises effective stress is used for the determining the stress in the steel girders; these stresses are only positive since they predict the yielding failure of the material.

Figure 6 shows the stress/deformation results of the steel girders. The maximum Von Mises effective stress is shown to be at around 98 MPa (14.5 ksi), which is well within the yield strength of any structural grade steel. This maximum stress occurs at the top of the exterior girder near the supports of the bridge. Figure 6 shows the thermal stresses to produce lateral torsional buckling in the steel girders that increase from the centerline of the bridge towards the edges.

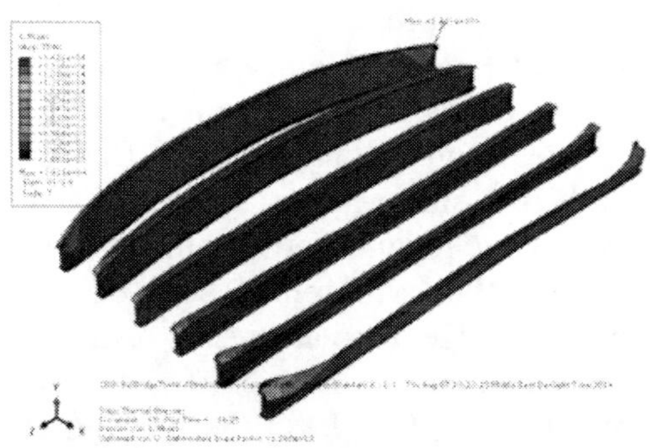

Figure 6. Maximum Von Mises stresses and deformation of steel girders.

The thermal stresses are compared to those obtained by previous studies in Table 2. The tensile stresses in the concrete deck are reasonably comparable with those obtained by Emanuel and Taylor (1985), and much higher than those obtained by Chen (2008) and Fu *et al.* (1990). This high discrepancy in the value of tensile stresses can be attributed to the fact that previous studies used the principle stresses to compute the thermal stresses in the steel girders, and the fact that these models have not considered the 3D effect of the thermal gradient on the bridge stresses.

Table 2. Thermal stresses comparison with previous models.

Maximum Stresses (MPa)	3D FEM	Chen (2008)	Fu *et al.* (1990)	Emanuel and Taylor (1985)
Compressive in Concrete Deck	−3.8	--	-1.87	−1.98
Tensile in Concrete Deck	2.16	0.61	1.14	2.83
Compressive in Steel Girder	--	--	-18.62	−16.25
Tensile in Steel Girder	97.98	--	6.76	0

4 CONCLUSIONS

The AASHTO and Kennedy and Soliman models are overly conservatively for bridges in cold regions, leading to overestimation in predicting the thermal stresses. A nonlinear thermal profile develops in the concrete deck as compared to the linear profile proposed by AASHTO and other models. This nonlinearity creates a nonlinear strain component that significantly affects the thermal stresses in the bridge. The stresses in the deck constitute 60% of the available tensile strength of concrete and are thus substantially large to warrant consideration in the design of steel reinforcement. Existing 1D and 2D models fail to account for the lateral torsional buckling stresses that the thermal gradient produces in the steel girders. They also fail to model the effect of the transverse thermal gradient on the transverse curvature and thermal stresses of the bridge.

References

AASHTO, *AASHTO Guide Specifications, Thermal Effects in Concrete Bridge Superstructures,* American Association of State Highway and Transportation Officials, Washington, DC, 1989.

Berwanger, C., and Symko, Y., Thermal Stresses in Steel-Concrete Composite Bridges, *Canadian Journal of Civil Engineering*, 2(1), 66-84, 1975.

Chen, Q., *Effects of Thermal Loads on Texas Steel Bridges,* PhD Thesis, Retrieved from ProQuest Dissertations and thesis database. (UMI No. 3320677), 2008.

Dilger, W. H., Ghali, A., Chan, M., Cheung, M. S., and Maes, M. A., Temperature Stresses in Composite Box Girder Bridges, *Journal of Structural Engineering*, 109(6), 1460-1478, 1983.

Emanuel, J. H., and Hulsey, J. L., Temperature Distributions in Composite Bridges, *Journal of the Structural Division*, 104(1), 65-78, 1978.

Emanuel, J. H., and Taylor, C. M., Length-Thermal Stress Relations for Composite Bridges, *Journal of Structural Engineering*, 111(4), 788-804, 1985.

Emerson, M., *The Calculation of the Distribution of Temperature in Bridges* (No. TRRL LR561 R&D Rept.), 1973.

Fu, H. C., Ng, S. F., and Cheung, M. S., Thermal Behavior of Composite Bridges, *Journal of Structural Engineering,* 116(12), 3302-3323, 1990.

Kennedy, J. B., and Soliman, M. H., Temperature Distribution in Composite Bridges, *Journal of Structural Engineering*, ASCE, 113(3), 475-482, 1987.

Priestley, M. J. N., and Thurston, S. J., Discussion of the Paper Titled Thermal Calculations for Bridge Design by Hunt et al., *Journal of Structures*, Elsevier, 504(102), 1277-1279, 1979.

Ramey, G. E., Wolff, A. R., and Wright, R. L., Structural Design Actions to Mitigate Bridge Deck Cracking, *Practice Periodical on Structural Design and Construction*, 2(3), 118-124, 1997.

Interdependence between Structural Engineering and Construction Management
Edited by Ozevin, D., Ataei, H., Modares, M., Gurgun, A., Yazdani, S., and Singh, A.

ISBN: 978-0-9960437-6-2

VISIBLE TEST ON BUGHOLE GENERATION OF FLUIDITY CONCRETES FOR TUNNEL LINING

ISAMU YOSHITAKE[1], MASAHIRO HIEDA[1], KENTA OKAMOTO[2], and TOMOYUKI MAEDA[2]

[1]*Dept of Civil and Environmental Engineering, Yamaguchi University, Ube, Yamaguchi, Japan*

[2]*Penta-Ocean Construction, Tokyo, Japan*

Bughole (surface air-void) is a concern for surface quality though it rarely decreases strength properties of the concrete structure. In particular, sidewall of NATM tunnel lining which generally has negative angle is a significant concern in the bughole problem. To reduce the surface imperfection of concrete, the authors examined the bughole properties and discussed on construction methods in previous investigations. The present study focuses on the relation of bughole generation and concrete viscosity. A visible test of bughole generation was conducted using high and medium fluidity concretes in addition to the conventional concrete. Bughole distribution and properties were analyzed by the image analysis which was developed in our previous study. This paper reports the effect of concrete fluidity on decrease of bugholes on concrete surface. A remarkable observation in the experimental study is that number and area-ratio of bugholes increased with concrete viscosity while large voids can be decreased by using high/medium fluidity concrete.

Keywords: Surface air-void, Surface quality of concrete structures, Water-soluble cellulose thickener, WSCT.

1 INTRODUCTION

To improve surface quality of concrete structures, bugholes should be reduced as much as possible. Bughole problem is not only aesthetic of concrete, it may influence on maintenance of concrete structures. Smooth surface without bugholes can contribute to easy clean for concrete surface and to check-up of durability such as gas-permeability test.

Numerous construction engineers have tried to decrease the imperfection of the concrete surface. However, we have little effective reduction method of the surface air-voids even now. The author's research group has conducted various experiments to observe bughole generations and to reduce the surface imperfection (Maeda *et al.* 2014, Harada *et al.* 2015, Maeda *et al.* 2016). The previous experimental study focused on bughole problems of conventional concrete in road tunnel constructed by New Austrian Tunnel Method (NATM). The sidewall of NATM tunnel lining which generally has a negative angle is a significant concern in bughole problem.

In Japanese highways, medium fluidity concrete has been often used in NATM tunnels for improvement of concrete-construction. Voids at concrete-casting, a cause of bugholes, can be reduced in case of fluidity concrete. It should be noted that thickener is often used in such fluidity concretes. In such high viscosity concrete, it may be hard to remove surface air-voids under the vibration of the concrete.

The focus of this study is to examine the effect of thickener on concrete bugholes. A laboratory test was conducted to observe bughole generation of the fluidity concrete. The experimental study used a transparent acrylic form to observe concrete voids before/after vibration. The observed bugholes were quantified in an image analysis developed in a previous study (Yoshitake *et al.* 2018). This paper reports bughole distributions and properties based on the image analysis.

2 METHODOLOGY

2.1 Materials and Concrete Mixtures

The present study used ordinary Portland cement defined in Japanese Industrial Standard (JIS) R5210 for the fluidity concretes. In addition, the study used sea-sand for fine aggregate and crushed sandstone for coarse aggregate. The chemical admixtures used in the study were a poly-carboxylic acid high-range water reducing agent (HRWRA) and a water-soluble cellulose thickener. The materials are given in Table 1.

Table 1. Materials used in the study.

Materials	Symbol	Properties
Cement	C	ordinary Portland cement, density: 3.15 g/cm^3
Fine aggregate	S	sea sand, density: 2.60 g/cm^3
Coarse aggregate	G_1	crushed sandstone, 20-15 mm density: 2.70g/cm^3
	G_2	crushed sandstone, 15-5 mm density: 2.70g/cm^3
Admixture	HRWRA	a poly-carboxylic acid high-range water reducing agent
	WSCT	water-soluble cellulose thickener

Table 2 gives concrete mixtures. The thickeners (WSCT) in the range of 0 to 0.45 % of water were added to the concrete. Fresh properties such as slump-flow and air content are also summarized in Table 2.

Table 2. Mixture proportions and fresh properties.

I.D.	*w/cm*	W kg/m^3	C kg/m^3	S kg/m^3	G_1 kg/m^3	G_2 kg/m^3	HRWRA g/m^3	WSCT g/m^3	Air %	Slump Flow cm
T00	0.512	174	340	869	458	458	1.7	0	5.3	45×44
T05	0.512	174	340	869	458	458	2.7	87	3.3	49×43
T10	0.512	174	340	869	458	458	3.4	174	3.4	48×40
T15	0.512	174	340	869	458	458	3.4	261	4.4	47×47
T30	0.512	174	340	869	458	458	5.1	522	4.9	45×43
T45	0.512	174	340	869	458	458	6.8	783	4.9	44×43

2.2 Visible Test

To observe bughole generation under vibration, the study used a transparent acrylic form shown in Figure 1. Dimensions of the inner space of the form were 300-mm thickness × 300-mm width × 750-mm height. The 300-mm thickness was based on the design thickness of the lining concrete of an NATM tunnel. The vibrator used in the present study had a diameter of 28-mm and a frequency of 220–280 Hz. Fresh concrete was cast into the form with an angle of 30 degree and consolidated for 60 seconds using the vibrator. The negative angle form was designed to simulate sidewall of tunnel linings which often appeared bugholes. Front and sides surface conditions of concrete were photographed by using three commercial digital still cameras which have approximately 18 million (5184 × 3456) usable pixels.

Figure 1. Transparent acrylic form and photography condition.

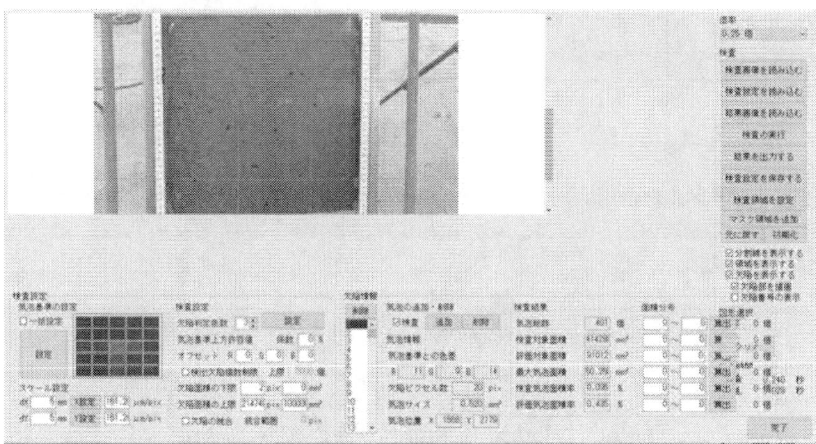

Figure 2. Interface of the image analyzing software.

2.3 Image Analysis

Image analyses based on thresholding are conducted to detect and evaluate the bugholes randomly distributed in concrete surface (Liu and Yang 2017, Liu *et al.* 2017). The present study also conducted an image analysis; however, it analyses colored photographs for detection of bugholes. The image analysis can detect invisible bugholes and calculate automatically bughole size. The image analysis is demonstrated in Figure 2. Further details of the image analysis are available in the previous paper (Yoshitake *et al.* 2018).

3 RESULTS AND DISCUSSION

3.1 Air-Voids and Bugholes

Air voids before vibrating are frequently observed during general concrete construction. It is well known that the air voids include entrapped air at concrete casting. Vibrators are generally used to remove such air voids in the construction of tunnel lining concrete. To improve concrete quality, medium fluidity concrete (MFC) has been often used in NATM tunnel construction in recent years. Harada *et al.* (2017) performed the image analysis of air-voids of various concretes including a self-consolidating concrete (SCC) and MFC. They also used the transparent acrylic form shown in Figure 1. The purpose of this study is to examine the air-voids of various MFC in addition to the previous investigation (Harada *et al.* 2017).

Figure 3 (a) (b) present surface photographs of conventional concrete and the MFC (T15) tested in this study, respectively. These photos show concrete conditions before vibrating for consolidation. It can be confirmed that MFC has significantly lower air-voids than the conventional concrete tested in the previous study (Harada *et al.* 2017).

(a) conventional concrete (slump:16.5 cm, air:5.5%). (b) MFC (T15) (S.F.: 47 × 47cm, air: 4.4%).

Figure 3. The surface condition of concrete without vibration.

To quantify the air-voids of the tested MFC, the study uses an empirical equation given in Eq. (1) by referring to the previous study (Harada *et al.* 2017).

$$r_a = \frac{S_f \times \sqrt{(S_{sl} + S_{sr})/2}}{V} \times 100 \qquad (1)$$

where r_a is the pseudo-air-void volume ratio (%), S represents area of air-voids (mm^2) and V is concrete volume (mm^3). The subscripts f, sl, sr represent front, left-side and right-side of concrete specimen, respectively. Each area can be calculated automatically by summing the number of pixels in the photograph.

Bugholes generated under vibration were detected and evaluated by using the image analysis mentioned above. Bugholes of hardened concrete may be slightly different from the bugholes of fresh concrete, however, the differences were negligible. The present study focuses on the bugholes detected in the image analysis of tested fresh concrete (MFC) and quantified the bughole characteristics. To compare the bugholes of each tested concrete (MFC), the study employs the bughole area ratio defined in Eq. (2) as well as previous studies (Harada *et al.* 2015, Maeda *et al.* 2016, Harada *et al.* 2017, Yoshitake *et al.* 2018).

$$r_b = \frac{S_f}{S} \times 100 \qquad (2)$$

where r_b is the bughole area ratio of front concrete surface (%). S_f and S are area of bugholes and concrete area for evaluation, respectively.

3.2 Image Analysis of Air-Voids

Table 3 summarizes the pseudo air-void volume ratio (r_a) obtained from the image analysis. For comparison, the test data reported in the previous investigation (Harada *et al.* 2017) was added to the table. The note is that the "Conventional" in this table means general purpose ready-mixed concretes which include various water-cement ratios (0.42 – 0.70) and designed slump (8, 15, 21 cm). Further information such as mixture proportions are available in the previous study (Harada *et al.* 2017).

Table 3. Pseudo air-void volume ratio (r_a) and bughole area ratio (r_b).

I.D.	WSCT -g/m^3-	r_a %	r_b %
T00	0	0.227	0.954
T05	87	1.061	1.082
T10	174	2.342	1.753
T15	261	0.031	1.570
T30	522	0.229	1.521
T45	783	0.335	2.686
MFC[a]	N/A	2.410	1.060
SCC[a]	N/A	0.011	1.330
Conventional[a]	N/A	13.746 – 44.287	0.346 – 1.425

a: previous test data (Harada *et al.* 2017)

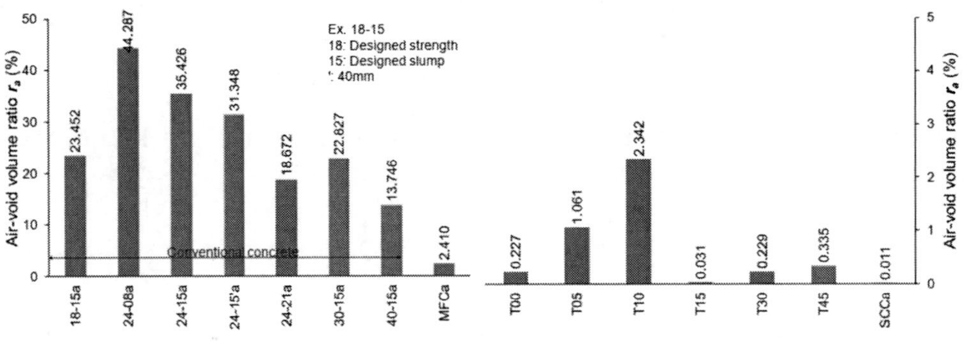

Figure 4. Variation of the pseudo air-void volume ratio (r_a).

Figure 4 presents the pseudo air-void volume ratio (r_a) varied by addition of the thickener (WSCT). Obvious tendency due to the thickener was not observed in this study while the volume ratios were significantly lower than the ratio of previous MFC. Different from the self-consolidating concrete (SCC), even the medium fluidity concrete (MFC) needs vibration to remove air-voids. The initial condition of such air-void distribution may be due to the casting concrete, so the air-void volume ratios (r_a) were not so different by the thickener.

3.3 Image Analysis of Bugholes

Figure 5 (a) shows the variation of area-ratio (r_b) of bughole having 5 mm or larger diameter. As well, Figure 5 (b) presents the area-ratio (r_b) of bughole having 5 mm or smaller diameter.

Noteworthy is that the area-ratio of large bugholes is proportional to the volume of thickener (Figure 5 (a)). Meanwhile, the difference of the area-ratio of small bugholes was small by the thickener as shown in Figure 5 (b).

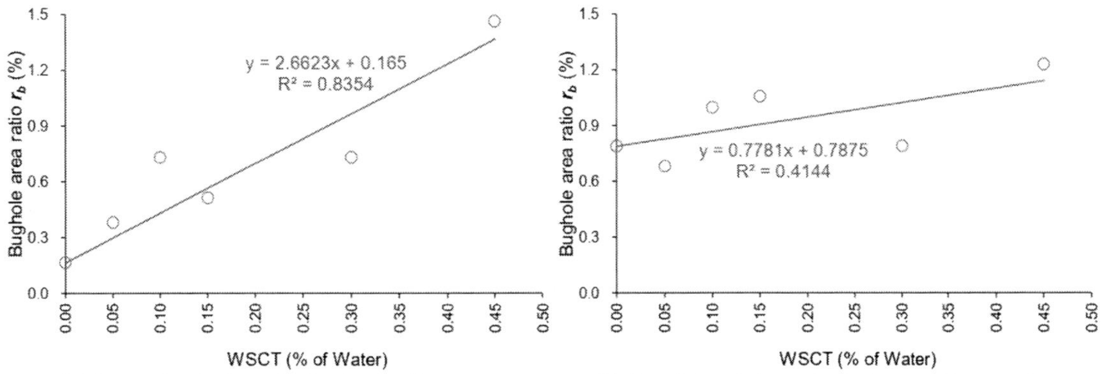

(a) Large bughole ratio (> 5 mm diameter). (b) Small bughole ratio (< 5 mm diameter).

Figure 5. Bughole area ratio (r_b) – thickener.

4 CONCLUSIONS

The present study focused on a relation of bughole generation and viscosity of medium fluidity concrete (MFC) including thickener. The visible test was conducted to observe bughole generation during vibration. The conclusions of the study are as follows.

- The air-void volume ratios of MFC using the present thickener were lower than the ratio of previous MFC and conventional concrete.
- The area-ratio of large bugholes (> 5 mm dia.) was proportional to the volume of thickener, while the area-ratio of small bugholes (< 5 mm dia.) was hardly varied by the thickener.

References

Harada S., Maeda T., Hirano M., and Yoshitake I., *Reducing Bug-Holes on Tunnel Lining Concrete by Using Covering Sheets,* Proceedings of ISEC-8, 689-694, Sydney, 2015.

Harada S., Hieda T., Maeda T., Hirano M., Miyamoto K., and Yoshitake I., Experimental Investigation on Characteristics of Bughole Generation Based on a Visualization Test for Various Concretes (in Japanese), *Journal of the Society of Materials Science, Japan*, 66(8), 582-587, Aug. 2017.

Liu B., and Yang T., Image Analysis for Detection of Bugholes on Concrete Surface, *Construction and Building Materials*, Elsevier, 137, 432-440, April 2017.

Liu B., Yang T., and Xie Y., Factors Influencing Bugholes on Concrete Surface Analyzed by Image Processing Technology, *Construction and Building Materials*, Elsevier, 153, 897-907, Oct. 2017.

Maeda T., Honma H., Hirano M., and Yoshitake I., *Permeability of Tunnel Lining with Air/Water Bubbles on Concrete Surface,* Proceedings of ASEA-SEC-2, 321-325, Bangkok, 2014.

Maeda T., Harada S., Moriuchi M., and Yoshitake I., *Reevaluation of the Effect of Covering Sheets for Reducing Bugholes on Tunnel Lining Concrete,* Proceedings of EURO-MED-SEC-1, 79-84, Istanbul 2016.

Yoshitake I., Maeda T., and Hieda M., Image Analysis for the Detection and Quantification of Concrete Bugholes in a Tunnel Lining, *Case Studies in Construction Materials*, Elsevier, 8, 116-130, June 2018.

Interdependence between Structural Engineering and Construction Management
Edited by Ozevin, D., Ataei, H., Modares, M., Gurgun, A., Yazdani, S., and Singh, A.

ISBN: 978-0-9960437-6-2

EXPERIMENTAL ASSESSMENT OF TRC CYLINDRICAL TUBE-SHAPED UNITS

DIDEM DONMEZ[1], MUSTAFA GENCOGLU[2], and AHMET ABDULLAH DONMEZ[2]

[1]*Dept of Earthquake Engineering, Istanbul Technical University, Istanbul, Turkey*
[2]*Dept of Civil Engineering, Istanbul Technical University, Istanbul, Turkey*

The purpose of this study is to investigate the behavior of tube-shaped TRC ductility experimentally via monotonic loading. Ar-Glass and PVA as textile materials are used in the production of composite tube parts. These parts were tested in four and six layers and in different thicknesses. As a result of the mortar thickness, textile layers play the major role in ductility and strength. As the tube thickness increases, the strength increases, but the ductility decreases. However, ductility increases as the tube thickness decreases and the energy absorption increase by making the parts more flexible. This paper analyzes how much ductile the 10 cm long tube can be under pressure and tensile forces. The use of different textile materials, different thicknesses, and different numbers of textile layers has been examined. According to the test results, it is concluded that cylindrical tube shaped units produced by using PVA fabric material is more ductile than units produced by using Ar-Glass fabric material. Also, with the wall or slab elements or pipeline obtained by using these units, the structure system can be strengthened against earthquake loads.

Keywords: TRC composite, Ductile, Flexural behavior, Tube/pipe-shaped system, Fabric, Strength.

1 INTRODUCTION

The development of high strength and ductile composite materials in seismic areas has become a major research subject in recent years. These types of materials have experienced high strength and displacement capacity under compression and tension loadings. One of the most widely used composite materials is textile-reinforced concrete (TRC), which consists of a fine-grained concrete matrix and high-strength textile fabric reinforcement made of alkali-resistant glass fibers or polyvinyl alcohol (PVA) fibers. TRC can be used as a structural element in addition to being used generally as structural strengthening material such as fiber reinforced polymer (FRP). The combination of fabrics and fine-grained concrete, which in fact resembles more a mortar than a concrete, allows architects and engineers to design thin and lightweight structures, characterized by a high load-bearing capacity. The main advantages of this material are related to its durability and strength performance since no cover against corrosion is required, while it is also possible to align fibres with the load direction (Colombo *et al.* 2013, Mobasher 2012). Structural elements made of TRC material show much more ductile behavior because of the high form flexibility of the textile reinforcement (Pakravan *et al.* 2016). In order to increase ductility and flexibility, cylindrical tube form was chosen, and experimental studies have been associated with it. Understanding the link between the basic damage effects in the material structure (matrix cracks, debonding, etc.) and the corresponding shape of the load-deflection curve is essential for the targeted design of the composite material structure (Chudoba *et al.* 2016). This paper investigates the behavior of ductility under the compression and tensile forces of the structural components of the cement-based textile composite tube-shaped units with different thicknesses and layers. The behavior of large-scale shell structures can be predicted with the investigated TRC members (Sharei *et al.* 2017).

2 EXPERIMENTAL STUDIES

2.1 Specimen Production

One-meter-long cylindrical tube specimens are produced using a PPR (Pull Pour Roll) method, a new method similar to the winding method (Figure 1). The production is carried out by this method, which has less production effort than the "hand laying method". The samples are obtained by pulling and wrapping the fabric material according to the desired strength without applying any heat treatment for the molding process (Daskiran 2018).

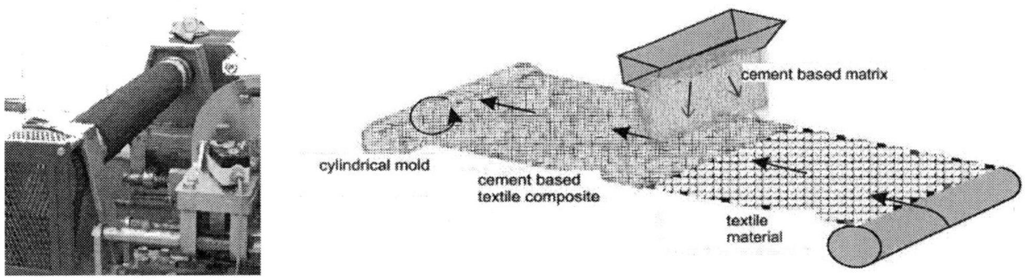

Figure 1. Production setup - PPR method.

Ar-Glass and PVA, as textile materials, are chosen in the production of composite tube parts. These parts were produced in 4 and 6 layers, with various thicknesses. As a result of that, the textile layers and the thickness of mortar plays the major role on ductility and strength. These parts were cut into 10 cm pieces. Thus, the cylindrical tube-shaped test specimens are obtained (Figure 2).

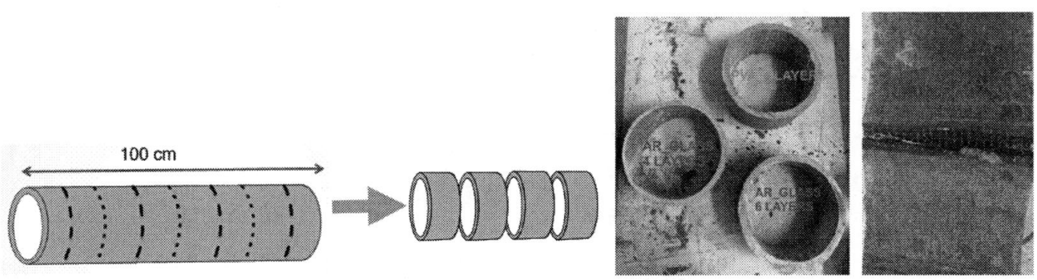

Figure 2. Cement based textile composite tube-shaped units.

The energy absorption capacity is the main parameter for deciding the shape and used materials. Therefore, the cylindrical tube shape is chosen for the experimental investigation. PVA as a textile material is more ductile than Ar-Glass, but the strength of PVA is less than Ar-Glass. TRC specimens are produced by using both PVA and Ar-Glass and polymer-containing cement matrix. The properties of the units are shown in Table 1.

Table 1. Properties of the TRC cross-section.

Description	Length (mm)	Thickness (mm) (average)	Cross Section Diameter (mm)	Number of Textile Layers
Ar_Glass 4 Layer	100	8-10	150	4
Ar_Glass 6 Layer	100	18	150	6
PVA 4 Layer	100	15	150	4

2.2 Compression Test Setup

The load vs. displacement relationship of cylindrical tube-shaped specimens under monotonic loading is investigated by using a servo-hydraulic jack. MTS brand loading jack with 600 kN capacity, located in the ITU construction materials laboratory, is used for this purpose. The load is applied with a displacement-controlled setup. Loading rate for all the tests is 2 mm/min. A sample of compression loading test setup is shown in Figure 3.

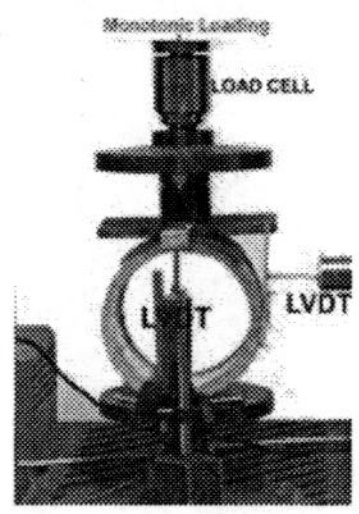

Figure 3. Compression test setup - MTS machine.

Three samples for each type of specimens are tested, which are Ar_Glass 4 Layer, Ar_Glass 6 Layer, and PVA 4 Layer. Two LVDTs for vertical and horizontal directions are used for measurements. It is intended to measure the displacement capacities in orthogonal directions.

2.3 Tensile Test Setup

Special head brackets for the tensile tests were prepared and the tensile test was applied to three specimens of each unit (Ar_Glass 4 Layer, Ar_Glass 6 Layer, and PVA 4 Layer).

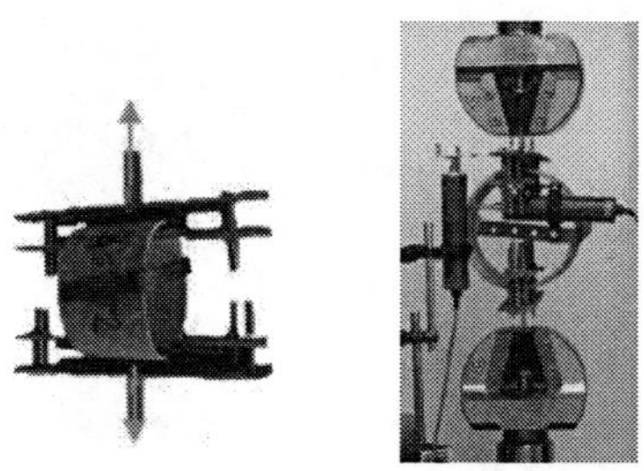

Figure 4. Tensile test setup - MTS machine.

Load-displacement relationships were investigated by applying monotonic tensile loading to specimens. It is expected that the load-displacement curves and maximum load capacities are similar to the load-displacement curves and load capacities obtained by the compression test. The test setup and LVDT placements for tensile loadings are shown in Figure 4.

3 COMPARISON OF TEST RESULTS

Three specimens of each unit (Ar_Glass 4 Layer, Ar_Glass 6 Layer, and PVA 4 Layer) were subjected to compression and tensile tests. Then the load-deflection curves and load bearing capacities are determined with the average of the results. In Figure 5, it can be deduced that TRC tube cracking pattern has a staggered shape in load vs displacement curves. The averages of three samples are shown in red in Figure 5. The testing materials with their experimental setups are shown on the graphs as well.

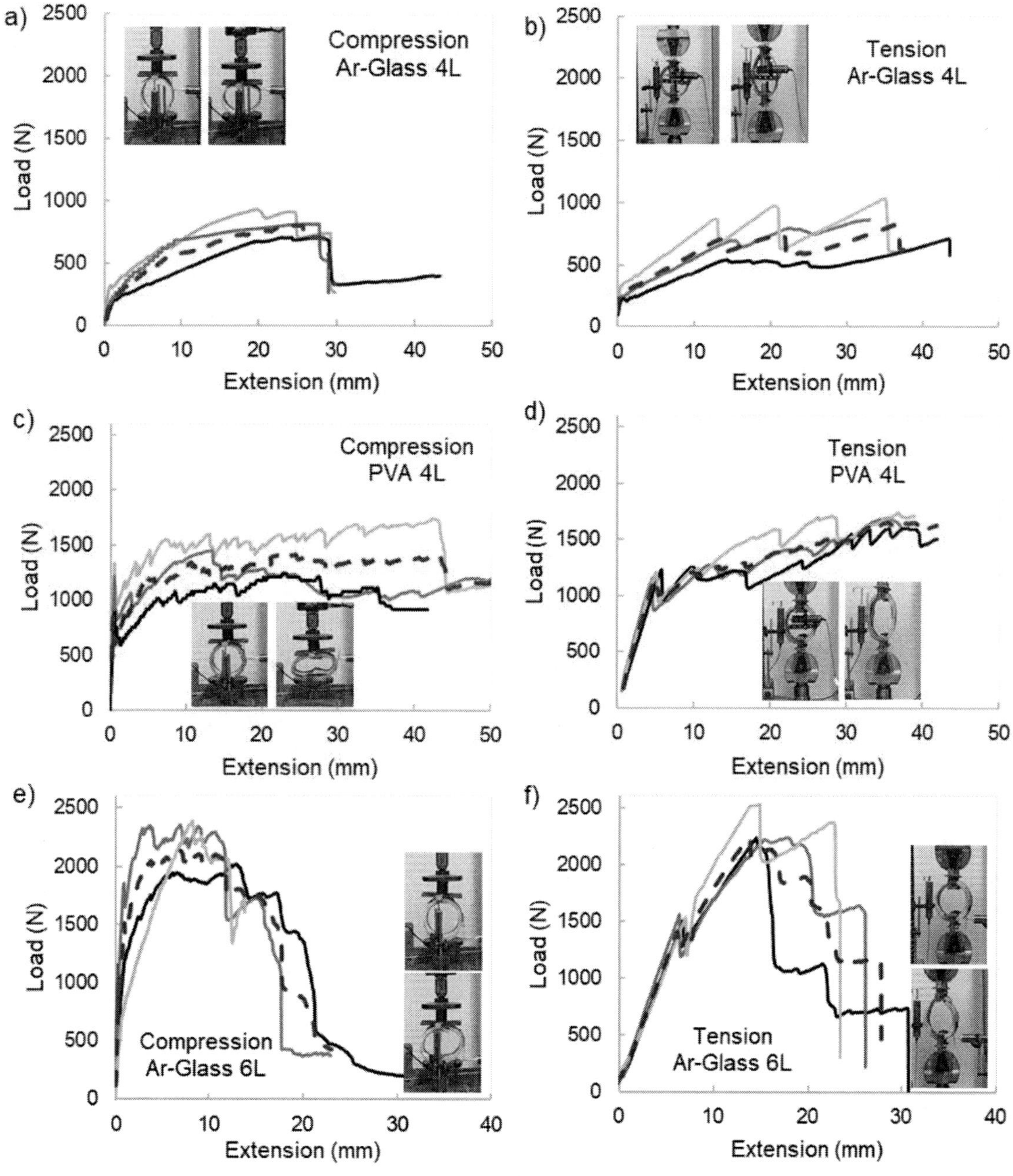

Figure 5. Compression and tensile load (N) –deflection (mm) curves.

The load-bearing capacity of 4 layered cement based PVA textile composites is measured as 1400N while the load bearing capacity of 4-layer Ar-glass textile composites is around 800 ~ 1000N. The strength of the cement-based 6-layer Ar-glass textile composite is about 2150N. When the number of cement-based Ar-glass textile composites increases from four to six, the load capacity doubles and the displacement decreases. The cement-based Ar-glass 6L cylindrical textile composite units have less ductility and consume less energy than Ar-glass 4L. The initial crack of PVA 4L cylindrical textile composite is exhibited at around 1000N load. Delamination of layers in the PVA reinforced TRC specimens is the main observed failure type. On the other hand, Ar-Glass 4L and 6L cylindrical textile composite specimens mainly exhibit transverse cracking. As the number of layers (or thickness) increases, the brittleness increases, and the displacement capacity drops sharply. The ultimate load and displacement forms are similar for compression and tension tests as expected. Table 2 shows the maximum loads and displacements under tension and compression experiments.

Table 2. Experimental results of the cylindrical tube shaped TRC units.

Description	The Average Ultimate Compression Load (N)	The Average Ultimate Compression Deflection (mm)	The Average Ultimate Tensile Load (N)	The Average Ultimate Tensile Deflection (mm)
Ar_Glass 4 Layer	800 ~ 1000	29	800 ~ 1000	35~40
Ar_Glass 6 Layer	2100~ 2300	20	2200~ 2500	28
PVA 4 Layer	1500	50	1800	45

4 CONCLUSIONS

In this paper, the experimental results of the textile reinforced concrete cylindrical tube-shaped specimens are presented. The load-bearing behavior of TRC is influenced by material, amount and orientation of the textile reinforcement and the fine concrete matrix. It has been observed that the strength of the Ar-Glass is high and that the 4-layer cement-based Ar-Glass textile composite behaves in sufficient ductility. The mesh spacing of Ar-Glass is higher than the PVA textiles. This feature provides great convenience during production. Moreover, the cement-based mortar of Ar-Glass TRC provides better adherence with the Ar-Glass textile than PVA material. Compression and tension tests result in textile-mortar delamination of TRC produced by using PVA textile material. In this content, TRC produced by using PVA textile material can absorb high strain energy. However, the ductility of TRC produced by using Ar-Glass textile material depends on layer number. If the number of layers increases, ductility, and deflection of TRC produced by using Ar-Glass textile material decreases. Moreover, the failure modes of the specimens change from delamination type to flexural failure with an increasing number of textile layers (Du *et al.* 2018).

In addition, the cylindrical tube-shaped form is effective in strengthening structural elements. In addition, this form provides system flexibility. By using tube-shaped TRC members, high energy absorbent slab or wall elements can be produced. Also, ductility and resistance of load-bearing structures against seismic loads can be increased.

References

Chudoba, R., Sharei, E., and Scholzen, A., A Strain-Hardening Microplane Damage Model for Thin-Walled Textile-Reinforced Concrete Shells, Calibration Procedure, and Experimental Validation, *Composite Structures*, 152, 913-928, June, 2016.

Colombo, I. G., Magri, A., Zani, G., Colombo, M., and Di Prisco, M., Erratum to: Textile Reinforced Concrete: Experimental Investigation on Design Parameters. *Materials and Structures*, 46(11), 1953-1971, 2013.

Daskiran, G., *Development of High Performance Cementitious Laminated Composites in Terms of High Strength and High Durability*, Ph.D. Thesis, Istanbul Technical University, 2018.

Du, Y., Zhang, X., Liu, L., Zhou, F., Zhu, D., and Pan, W., Flexural Behavior of Carbon Textile-Reinforced Concrete with Prestressing and Steel Fibres, *Polymers*, 10(1), 98, 2018.

Mobasher, B., *Mechanics of Fiber and Textile Reinforced Cement Composites*, Chapter 8, 2012.

Pakravan, Hamid R., Masoud Jamshidi, and Hojatolah Rezaei, Effect of Textile Surface Treatment on The Flexural Properties of Textile-Reinforced Cementitious Composites, *Journal of Industrial Textiles*, 46(1), 116-129, 2016.

Sharei, E., Scholzen, A., Hegger, J., and Chudoba, R., Structural Behavior of a Lightweight, Textile-Reinforced Concrete Barrel Vault Shell, *Composite Structures*, 171, 505-514, March, 2017.

Interdependence between Structural Engineering and Construction Management
Edited by Ozevin, D., Ataei, H., Modares, M., Gurgun, A., Yazdani, S., and Singh, A.

ISBN: 978-0-9960437-6-2

FLEXURAL DUCTILITY OF STRUCTURAL CONCRETE MEMBERS SUBJECTED TO LIMITED CYCLES OF REPEATED LOADING

NAZAR OUKAILI and MOHAMMED KHATTAB

Dept of Civil Engineering, University of Baghdad, Baghdad, Iraq

For structural concrete members that may expose to serious earthquake, overload or accident impact, the design of ductility must be given the same importance as the flexural strength. The aim of this investigation is to study the change in ductility of structural concrete flexural members during their exposure to limited cycles of repeated loading. Twenty full-scale beam specimens have been fabricated in to two identical groups; each group consisted of ten specimens. The first group was tested under monotonic static loading to failure and regarded as control beams, while the specimens of the second group were subjected to ten cycles of repeated loading with constant load interval, which ranged between 40% and 60% of ultimate load. Specimens in each group were categorized as follows: two traditional reinforced concrete specimens with different intensity of tension reinforcement; three partially prestressed specimens with bonded strands; three partially prestressed specimens with unbonded strands; and two fully prestressed concrete specimens. The main variable, which was considered for all specimens was the partial prestressing ratio (*PPR*). It was observed that, the ductility of reinforced concrete beams was insignificantly increased during subjecting to limited repeated loading. For fully prestressed and partially prestressed concrete beams with high level of *PPR*, the ductility was significantly enhanced, while, it was decreased for specimens with small level of *PPR*.

Keywords: Curvature, Deflection, Rotation, Flexural members, Bonded strands, Unbonded.

1 INTRODUCTION

Ductility assures gradual failure rather than brittle failure to provide a warning to the occupiers before total collapse. "The ductile structure is that one, which is capable to experience large inelastic deformations at near maximum load carrying capacity without brittle failure" (Thompson and Park 1980). For structural members, ductility is expressed usually in terms of deformation ductility ratio, which is described in terms of rotation, curvature and deflection (Naaman *et al.* 1986, Thompson and Park 1980 and Giannini *et al.* 1986). The ductility ratio is defined as the ratio of the ultimate deformation to the yield deformation (ACI 423-5R-99 1999).

The displacement ductility factor which is the value commonly used in nonlinear dynamic analyses is expressed as in Eq. (1):

$$\mu_\Delta = \Delta_u / \Delta_y \quad (1)$$

Where Δ_u = maximum deflection of the structure and Δ_y = deflection of the structure at first yielding.

Some analyses of structures have used the rotational ductility factor of members which is defined as in Eq. (2):

$$\mu_\theta=\theta_u/\theta_y \quad (2)$$

Where θ_u = maximum plastic hinge rotation of the end of member and θ_y = rotation at the end of member at first yielding.

The essential information wanted by the designer concerns the required member section behavior at the plastic hinge expressed by the curvature ductility factor as in Eq. (3):

$$\mu_\varnothing=\varnothing_u/\varnothing_y \quad (3)$$

Where $\varnothing_u$= maximum curvature at the section and $\varnothing_y$= curvature at the section at first yielding.

When calculating ductility factors for partially prestressed members, the description of the first yielding deformation (displacement, rotation or curvature) often causes difficulty when the load or moment-deformation curve is not elasto-plastic. In addition, it is not distinctive because the section contains both types of steel (prestressed and nonprestressed).

Cohn and Bartlett's (1982) anticipated that the yielding curvature accompanying with yielding of ordinary reinforcement. Thompson and Park (1980) proposed that, the yielding point is the intersection of the tangent of elastic portion of load-deflection curve and the horizontal line at the ultimate load. Park and Falconer (1983) took the yielding deformation at the intersection of the secant from zero to 75% of the ultimate moment capacity and the post elastic slope. Naaman *et al.* (1986) defined the yielding point as the intersection of secant from zero to the proportional limit of prestressed steel and the post-elastic slope.

All the above-mentioned methodologies are based on the ductility of the member at specific section rather than the yielding of specific components in the cross section.

2 THE EXPERIMENTAL PROGRAM

The experimental program was undertaken to study the ductility of partially prestressed concrete beams under monotonic static (S) and limited repeated loading (R).

This research included testing twenty simply supported structural concrete beams that have the same geometrical layout, same transverse reinforcement and variable longitudinal reinforcement. All beams were with (200 mm width × 300 mm height) rectangular cross section simply supported on a clear span of (3000 mm) length, where the span to total section depth ratio was 10. All beams were loaded in four-point loading scheme using two symmetrical monotonic concentrated static loads applied at one-third of the span length. These beams were divided into two identical categories. The first category was tested under monotonic static loading up to failure and regarded as controlled beams. The second category was exposed to repeated loading according to the following stages:

- Stage one: the minimum and maximum cyclic loads P_{min} and P_{max} were taken, respectively, as 40% and 60% of the ultimate load of the accompanying beams of first category. The load level was selected to simulate self-weight and self-weight plus superimposed dead load and service overload, respectively. Ten cycles of loading and unloading were implemented.
- Stage two: after ten cycles of repeated loading, the load was released to zero.
- Stage three: during this stage, beams were subjected to monotonic static loading until failure.

Each category consisted of ten beams divided into four groups as follows:

First Group – consisted of two conventional reinforced concrete beams (FR), one under-reinforced (UN) and the other with maximum reinforcement ratio (MAX). The Partially Prestressed Ratio ***PPR*** (Naaman *et al.* 1986) for these beams is zero.

Second Group – included three partially prestressed concrete beams (PP) with bonded strands (B). Each beam is reinforced with two Ø12.7 mm strands and the initial prestressing stress was equal to 0.7 of the characteristic strength (f_{pu}) of low-relaxation strand. Different amount of nonprestressing reinforcement was provided for each beam so as to get different value of ***PPR***. According to ACI-318-14 (2014), the first beam is categorized as tension-controlled (TC), while the second is transition-controlled (TRC) and the third is compression-controlled (CC).

Third Group – this group is identical to Second Group but with un-bonded strands (U). It comprised also three partially prestressed concrete beams with wide range of ***PPR*** value.

Fourth Group – consisted of two fully prestressed concrete beams (FP) with bonded strands, one is tension-controlled, and the other is compression-controlled. The first is reinforced with two strands and the other with three strands. ***PPR*** for these beams is one.

Figure 1 shows the typical reinforcement section and the transverse steel distribution along the member axis.

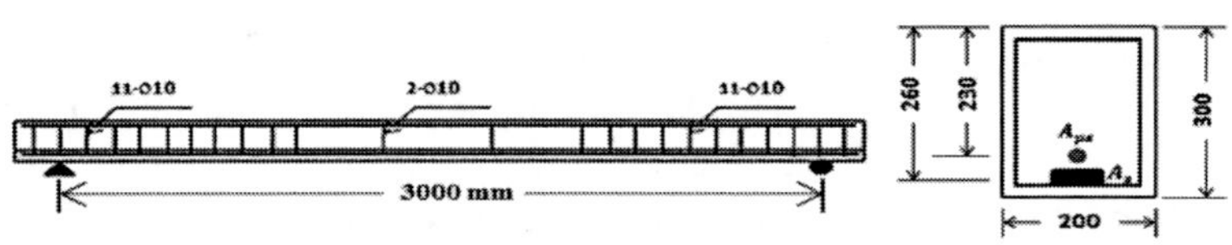

Figure 1. Transverse reinforcement distribution along member's axis and typical beam section.

The longitudinal reinforcement details of the tested beams are illustrated in Table 1. The beams were designed in such a way that the expected failure should occur due to flexure rather than shear, therefore steel stirrups of (Ø10 mm @ 100 mm) c/c were used in shear spans to guarantee that the beams will not fail under shear. The steel stirrups were tied to two longitudinal bars of (10 mm) diameter at the top (except FP-B-CC-S and FP-B-CC-R beams, they were provided with one strand and with initial prestressing stress equal to *0.5* f_{pu}) and to different number of mild steel bars at the bottom depending on whether the beam is tension controlled, transition or compression controlled. Yield stress of the longitudinal and the transverse bars was (570 MPa).

Table 1. Geometrical properties of steel used for reinforcing experimental beam specimens.

Beam's labelling	A_s (mm^2)	A_{ps} (mm^2)	A'_s or A'_{ps} (mm^2)	ω	ω_{ps}	$\acute{\omega}$	ϖ	*PPR*
FR-UN-S,R	4Ø12	-	2Ø10	0.043	-	0.043	0.000	0.000
FR-MAX-S,R	4Ø12+5Ø10	-	2Ø10	0.245	-	0.045	0.200	0.000
PP-B-TC-S,R	2Ø10	2 strands	2Ø10	0.043	0.185	0.043	0.185	0.771
PP-B-TRC-S,R	2Ø10+2Ø12	2 strands	2Ø10	0.124	0.178	0.043	0.259	0.529
PP-B-CC-S,R	6Ø12	2 strands	2Ø10	0.259	0.169	0.044	0.384	0.358
PP-U-TC-S,R	2Ø10	2 strands	2Ø10	0.043	0.152	0.043	0.152	0.743
PP-U-TRC-S,R	2Ø10+4Ø12	2 strands	2Ø10	0.172	0.143	0.044	0.271	0.409
PP-U-CC-S,R	2Ø10+5Ø12	2 strands	2Ø10	0.224	0.138	0.044	0.318	0.339
FP-B-TC-S,R	-	2 strands	2Ø10	0.000	0.187	0.049	0.138	1.000
FP-B-CC-S,R	-	3 strands	1 strand	0.000	0.240	0.000	0.240	1.000

Note: $\omega=\rho_s f_y/f'_c$; $\omega_{ps}=\rho_{ps} f_{ps}/f'_c$; $\acute{\omega}=\rho' f_y/f'_c$; $\varpi=\omega+\omega_{ps}-\acute{\omega}$ where $\rho_s=A_s/bd_s$; $\rho_{ps}=A_{ps}/bd_{ps}$; $\rho'=A'_s/bd_s$

The characteristic strength of prestressed low-relaxation strand was (1862 MPa) and the initial prestressing stress for all strands in all prestressed concrete beams was $0.7f_{pu}$ except for FP-B-CC-S and FP-B-CC-R, which was $0.6\,f_{pu}$. The compressive strength of the concrete cylinder at age 28 days used for all the beams was (40MPa).

The main objective of the present study is to investigate the effect of limited cycles of repeated loading on the ductility of reinforced, partially prestressed and fully prestressed concrete beams. Two parameters were studied with respect to ductility ratio; they were the Partial Prestressing Ratio (***PPR***) and the global reinforcing index (ϖ).

3 THE EXPERIMENTAL RESULTS

The load-deflection diagrams for each two identical beams, (one from category 1 and the other from category 2), were compared (Figure 2).

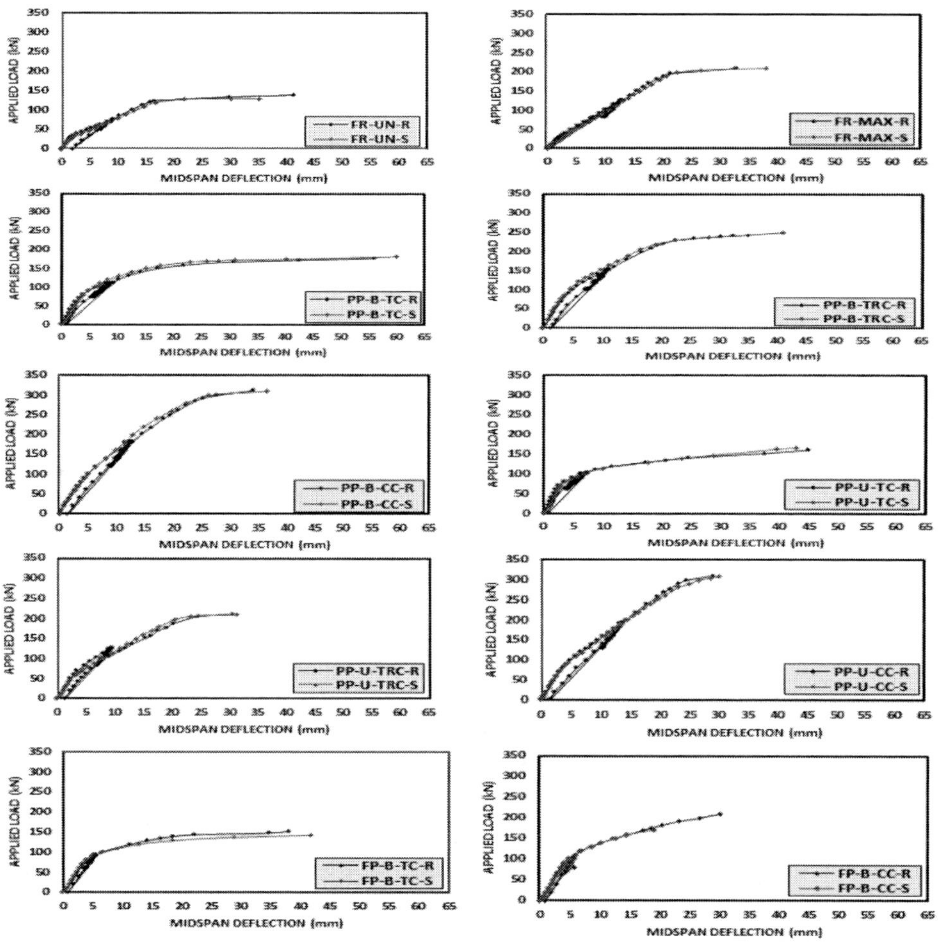

Figure 3. Load-deflection curves for the tested beams.

Reloading the beam with monotonically increasing static load, after releasing the tenth cycle of the repeated load, showed that the behavioral response of members under the applied load was approximately the same as for the identical counterpart beam in category 1 with little differences except for beams (FP-B-CC-S and FP-B-CC-R). It can be seen that for beams (PP-B-TC-S, R,

PP-B-TRC-S, R, PP-U-TC-S, R and PP-U-TRC-S, R), three distinctive points could be observed. These points are characterized cracking, yielding and ultimate loads. This behavior is expected since they are designed as tension and transition-controlled beams, respectively. However, for (PP-B-CC-S,R and PP-U-CC-S,R) beams, where the behavioral response has different character regarding the load and the deflection at ultimate, only two points which characterized cracking and ultimate loads can be observed. Prior to cracking all beams behaved in an elastic manner.

Thompson and Park approach (1980), which is the most commonly used approach, has been adopted to determine the yielding deformation in the current study. Yield deflection, ultimate deflection and displacement ductility factor are illustrated in Table 2.

Table 2. Yield deflection, ultimate deflection and displacement ductility factor.

Beam's labelling	*PPR*	ϖ	Yield deflection, Δ_y (mm)	Ultimate deflection, Δ_u (mm)	Displacement ductility factor, μ_Δ	$\mu_{\Delta\ Repeated}/\mu_{\Delta\ Static}$
FR-UN-S FR-UN-R	0.000	0.000	8.0 7.5	30.0 41.5	3.75 5.53	1.47
FR-MAX-S FR-MAX-R	0.000	0.200	12.0 13.5	38.0 32.5	3.16 2.41	0.76
PP-B-TC-S PP-B-TC-R	0.771	0.185	7.5 7.0	60.0 56.0	8.0 8.0	1.00
PP-B-TRC-S PP-B-TRC-R	0.529	0.259	9.0 10.0	41.0 35.0	4.56 3.50	0.77
PP-B-CC-S PP-B-CC-R	0.358	0.384	15.0 16.0	36.0 34.0	2.40 2.13	0.89
PP-U-TC-S PP-U-TC-R	0.743	0.152	6.0 5.1	43.0 45.0	7.17 8.82	1.23
PP-U-TRC-S PP-U-TRC-R	0.409	0.271	11.0 9.0	31.0 30.5	2.82 3.40	1.21
PP-U-CC-S PP-U-CC-R	0.339	0.381	12.0 12.5	30.0 28.5	2.50 2.28	0.91
FP-B-TC-S FP-B-TC-R	1.00	0.138	6.0 7.0	42.0 38.0	7.00 5.43	0.78
FP-B-CC-S FP-B-CC-R	1.00	0.240	8.0 8.5	18.5 30.0	2.31 3.53	1.53

It was found that, the ductility ratio is inversely proportional to the reinforcing index (ϖ) for beams of the same type. In other words, the ductility factor increased when the reinforcing index decreased for FR, PP (bonded and unbonded) beams and for FP beams. That means increasing reinforcement (nonprestressed or prestressed) in the beam resulted in a considerable reduction in the ductility ratio. The case is different concerning the relation between ***PPR*** and the ductility ratio. The results show that the ductility ratio is directly proportional to ***PPR***. By studying the relation between the ductility ratio and the bonding properties, another observation can be distinguished. As shown from the Table 2, the ductility ratio is very close for both beams (PP-B-TC-S) and (PP-U-TC-S). The two beams are reinforced with the same prestressed and non-prestressed steel. The difference between the two beams is that the first one has bonded strands while the other has unbonded strands. Therefore, the ductility ratio is not affected by the bonding properties. The comparison between (PP-B-TC-S) and (FP-B-CC-S), which have approximately the same ultimate load strength is very interesting. Both beams have bonded strands but the first one is partially prestressed, whereas the second is fully prestressed. The partially prestressed beam (PP-B-TC-S) is more ductile than the fully prestressed beam (FP-B-CC-S). It means that presence of nonprestressed steel enhanced the ductility of prestressed concrete members. Ductility ratio was the same for beams PP-B-TC-S and PP-B-TC-R but it was increased due to repeated loading for beam PP-U-TC-R comparing with PP-U-TC-S by about 23%. It seems that the effect of repeated loading on beams with unbonded strands is more significant rather than beams with bonded strands. Repeated loading enhanced the ductility ratio of (FP-B-CC-R) beam comparing with the counterpart beam (FP-B-CC-S) by about (53%). The situation is different concerning ductility of beam (FP-B-TC-R). The ductility is decreased by about (22%) when the beam is subjected to repeated loading.

4 CONCLUSIONS

The following conclusions can be drawn:

- For all types of beams (FR, PP and FP), displacement ductility ratio (μ_Δ) is inversely proportional to the reinforcing index (ϖ) regardless of the type of testing (static or repeated). On the other hand, for partially prestressed beams, the ductility ratio is directly proportionated to Partially Prestressed Ratio (***PPR***).
- The effect of bonding between strands and concrete has little effect on ductility ratio for PP-beams type when they subjected to static loading.
- For FR beam with little amount of reinforcing index, the ductility ratio has been enhanced when this beam was subjected to repeated loading. The same thing can be said for tension-controlled with unbonded strands PP beam. The effect of repeated loading on tension-controlled with bonded strands PP beam is approximately zero.
- The repeated loading led to decreased ductility ratio for both types of PP compression-controlled beams (bonded and unbonded).
- Unexpected phenomenon has been noticed for compression-controlled FP beam. Although the prestressed steel area is increased comparing with tension-controlled FP-beam but the ductility ratio is increased when it was subjected to repeated loading. It seems that, decreasing the initial prestressing stress from $0.7f_{pu}$ to $0.6f_{pu}$ was the reason for that unexpected phenomenon.

References

ACI-318-14, *Building Code Requirements for Structural Concrete and Commentary*, American Concrete Institute, Farmington Hills, 2014.

ACI 423.5R-99, State-of-the-Art Report on Partially Prestressed Concrete, Reported by Joint ACI-ASCE Committee 423, 1999.

Cohn, M. Z., and Bartlett, M., Nonlinear Flexural Response of Partially Prestressed Concrete Sections, *Journal of the Structural Division,* 108(ST12), 2747-2765, Dec, 1982.

Giannini, R., Menegotto, M., and Nuti C., Influence of Prestressing on Seismic Response of Structures: A Numerical Study, Partial Prestressing, from Theory to Practice, *NATO ASI Series,* 2, Martinus Nijhoff Publishers, Dordrecht, The Netherlands, 255-274, 1986.

Naaman, A. E., Harajli, M. H., and Wight, J. K., Analysis of Ductility in Partially Prestressed Concrete Flexural Members, *PCI Journal*, 31(3), 64-87, May-June, 1986.

Park, R., and Falconer, T. J., Ductility of Prestressed Concrete Piles Subjected to Simulated Seismic Loading, *PCI Journal*, 28(5), 111-144, Sept.-Oct., 1983.

Thompson, K. J., and Park, R., Ductility of Prestressed and Partially Prestressed Concrete Beam Sections, *PCI Journal,* 25(2), 46-70, Mar 1980.

Interdependence between Structural Engineering and Construction Management
Edited by Ozevin, D., Ataei, H., Modares, M., Gurgun, A., Yazdani, S., and Singh, A.

ISBN: 978-0-9960437-6-2

MACROMECHANICAL ANALYSIS OF SPHERICALLY VOIDED BIAXIAL CONCRETE SLABS

GIRUM URGESSA[1], WONDWOSEN ALI[2], and JAVAD ESFANDIARI[1]

[1]Sid and Reva Dewberry Dept of Civil, Environmental and Infrastructure Engineering, George Mason University, Fairfax, USA
[2]BEL Engineering LLC, Temple Hills, USA

The use of spherically voided biaxial concrete slab (SVBS) system, which uses hollow plastic balls as infill material, has increased widely because of its reduced weight-to-strength and weight-to-stiffness ratios when compared to solid concrete slabs. However, SVBS is a heterogeneous composite structure in which building a representative continuum model poses a significant challenge. To mitigate this challenge, the feasibility of determining the macromechanical structural behavior of spherically voided biaxial concrete slabs is studied using plate theories, aided by mechanical properties that were determined from a homogenization process of the representative volume element (RVE). This paper presents numerical analysis results of SVBS using both Mindlin-Reissner (thick) and Kirchhoff-Love (thin) plate theories. The results from both theories predicted the slab behavior reasonably well and they were within 10% of each other with the exception of the prediction of the twisting moment. Possible explanation of this deviation is provided in the paper.

Keywords: Plate theory, RVE, Mindlin-Reissner, Kirchhoff-Love.

1 INTRODUCTION

The use of spherically voided biaxial concrete slab (SVBS) system, which uses hollow plastic balls as infill material, has increased in recent years because of its reduced weight-to-strength and weight-to-stiffness ratios when compared to solid concrete slabs. The inherently high flexural rigidity of SVBS systems mostly comes from the two solid facing plates separated by a hollow core (Voros 2011, Mota 2013). Figure 1 shows a typical biaxially reinforced concrete flat slab system that uses grids of hollow plastic balls as void formers.

A great deal of analytical research on modeling of solid, ribbed, and hollow core reinforced concrete slabs has been conducted over the past decades. However, research on the analytical modeling of spherically voided biaxial concrete slabs is relatively limited. Generally, there are three fundamentally different approaches to obtain continuum mechanics based constitutive equations for predicting the structural behavior (responses) of an SVBS system. These include phenomenological, numerical, and micromechanics approaches.

The phenomenological approach is based on conducting experiments on test specimens whose dimensions are large enough for determining stiffness properties. The stiffness properties obtained from the experiments are then used to determine parameter estimates for an assumed constitutive model in predicting the macroscopic/structural behavior of the slab. This approach certainly offers the best advantage for isotropic materials, because the response characteristics of the materials can be well-described based on few tests (Ali and Urgessa 2012, Ali and Urgessa 2014a). However, the anisotropic and non-homogeneous nature of the SVBS system greatly magnifies the challenge involved in the experimental determination of the stiffness properties. To describe a full 3-dimensional constitutive model of the SVBS, its mechanical properties in all orthogonal planes have to be obtained, which is difficult to obtain using testing results only.

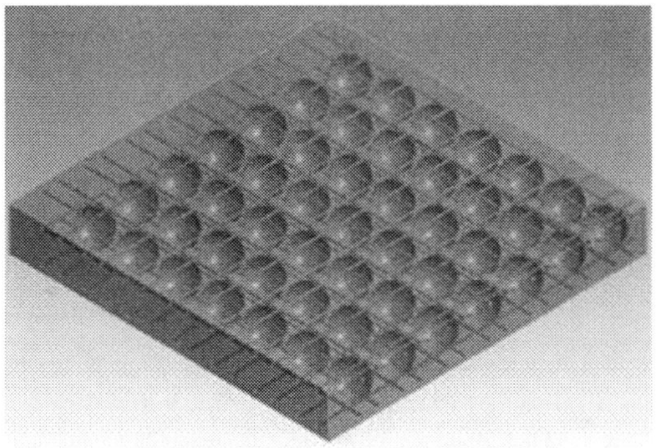

Figure 1. SVBS system.

The numerical approach is based on a detailed finite element analysis, which is a powerful technique for carrying out structural analysis. However, conducting 3D finite element analysis on a large (full-scale) SVBS has its own challenges because the geometry is complex and it needs contact algorithms between different constituents. Limited finite element simulations that investigated the structural capacities and failure mechanisms of SVBS systems under different load configurations, without conducting expensive experimental tests, are reported in Schnellenbach-Held and Pfeffer (2002), Abramski *et al.* (2010), and Ali and Urgessa (2014b). The results from these studies confirmed that SVBS systems are cost-effective when compared to conventional solid concrete slabs.

The computational effort of analyzing an SVBS system using a detailed three dimensional finite-element can be greatly reduced by the third analytical approach: modeling it as an equivalent thick-orthotropic plate through the process of micromechanical homogenization. Figure 2 illustrates this computational approach. First, a representative volume element (RVE) containing all the geometric and constitutive information of the SVBS is extracted. Then the axial, flexural and torsional properties of the RVE are determined based on the application of suitable boundary conditions that generate characteristic deformation and rotation modes. Once the homogenization process is complete, the mechanical properties of the RVE are used in conducting macromechancial analysis of the entire SVBS system.

2 MICROMECHANICAL HOMOGENIZATION

The objective of micromechanical homogenization is to establish the macroscopic behavior of a heterogeneous system by capturing the effective (overall) characteristics of its heterogeneity. The heterogeneous material is then replaced by a homogeneous material whose global characteristics are a good representation of the heterogeneous system (Urgessa and Casey 2013). In the homogenization process used here, the RVE is subjected to eight linearly independent deformations (including displacements and rotations). For each of the eight deformation/rotation cases, the equivalent reaction forces and moments at the constrained boundaries of the RVE are computed from the FE-model analysis using ANSYS (Figure 3).

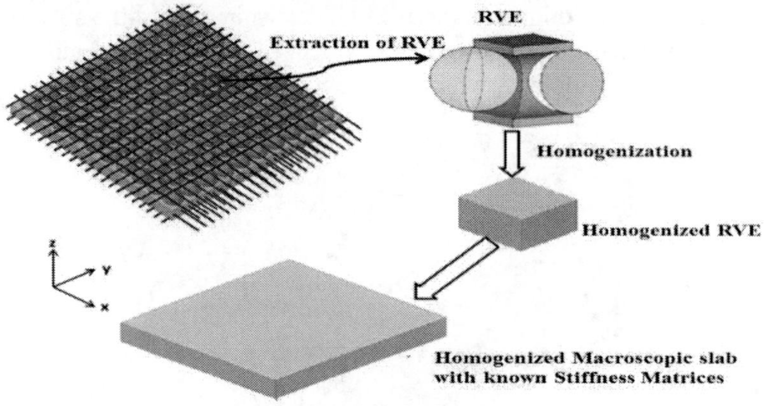

Figure 2. Framework for homogenization.

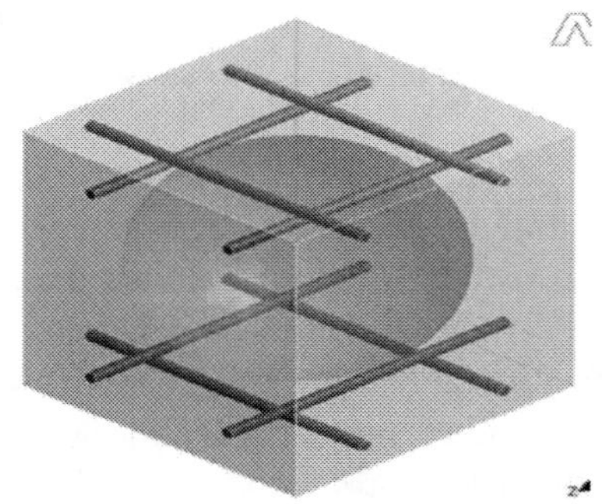

Figure 3. RVE showing concrete, rebar, and hollow composite core.

Then, based on the FE analysis results, the equivalent material properties such as the stiffness coefficients of the RVE are determined. These coefficients are similar to what are commonly referred to as ABDE stiffness matrices in mechanics of composite materials (Challagulla *et al.* 2010).

3 MACHROMECHANCIAL ANALYSIS

A macromechanical analysis of an example SVBS system was conducted using ABDE stiffness matrices determined from micromechanical modeling of RVE. A 340 mm thick SVBS slab having a length of 12 m and a width of 10 m was subjected to a vertical uniform distributed load of 20 kPa. The RVE selected for micromechncial characterization was 300 mm by 300 mm in the planar directions. Once the ABDE stiffness matrices of the RVE were characterized, a macromechancial analysis of the slab was conducted using classical plate theories.

Based on the dimensions of SVBS, the typical thickness to span ratio can be as high as 4%. Therefore, the displacement based first order shear deformation theory (FSDT), also known as Mindlin-Reissner or thick plate theory, was incorporated in the analysis. In FSDT, the effect of the transverse shear force on the deformation of the SVBS is included and the transverse shear strain distribution is assumed to be constant through the plate thickness (Cecchi *et al.* 2007). A shear correction factor was applied to account for the strain energy due to shear deformation. The other widely cited plate theory, Kirchhoff-Love theory (KLT) or thin plate theory, which was expected to under-predict deflections and stresses was also included in the study for comparison purposes (Reddy 2006).

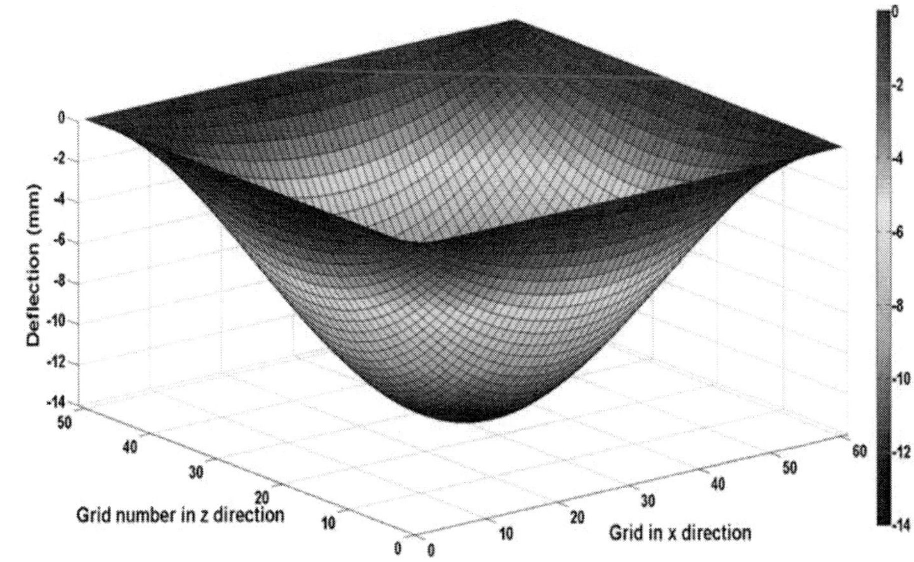

Figure 4. Mid-plane deflection using FSDT: Maximum deflection = 14 mm.

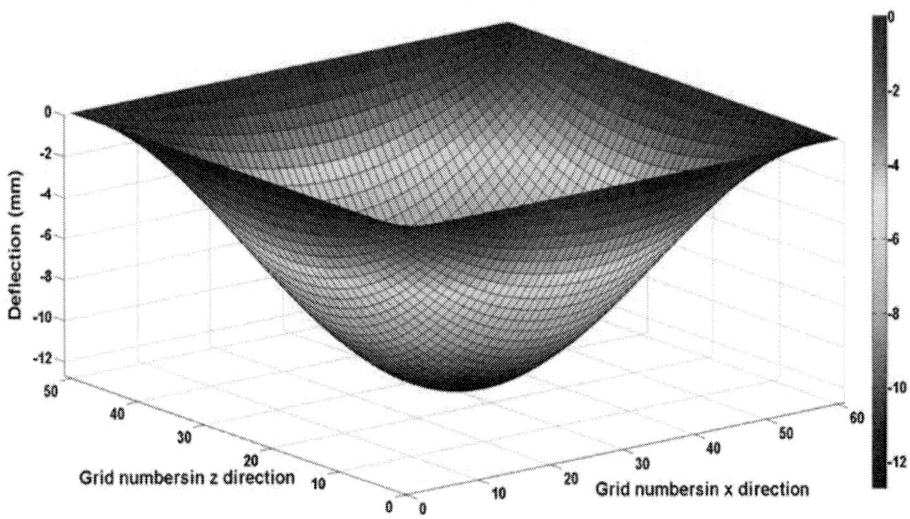

Figure 5. Mid-plane deflection using KLT: Maximum deflection = 12.5 mm.

A MATLAB program was written that determines the mid-plane deflection, bending moments in both planar directions, twisting moment, maximum bending stresses in both planar directions, and in-plane shear stress of the homogenized slab using both FSDT and KLT. Figure 4 and Figure 5 show sample outputs for the mid-plane deflection of the SVBS system using FSDT and KLT, respectively.

Results from KLT were found to be smaller than FSDT by about 10% because of the fact that KLT does not take into account the effect of transverse shear and transverse normal stresses.

This observation was true for all response parameters (deflection, bending moment and bending stress) with the exception of twisting moment in which the observed difference was about 20%. The latter can be explained by the fact that in KLT elements were assumed to remain perpendicular to the mid-plane, yet equilibrium requires that planar stress components would be present causing these elements to twist substantially.

4 CONCLUSIONS

This paper presented macromechanical analysis of a SVBS system through the process of micromechanical homogenization of RVE. The heterogeneity of the SVBS was captured by mechanical properties extracted from a homogenous RVE. Once the homogenization process was complete, two of the classic plate theories were used to determine structural responses. The results from both theories predicted the slab behavior reasonably well and results were within 10% of each other with the exception of the prediction of the twisting moment.

Acknowledgments

This research was supported by the Presidential Scholarship at George Mason University. The authors are grateful for the support.

References

Abramski, A., Albert, A., Nitsch, A., and Schnell J., Experimental and Numerical Investigations of the Load-bearing Behavior of Reinforced Concrete Slabs using Spherical Void Formers, *Beton- und Stahlbetonbau*, 105(6), 1-13, 2010,

Ali, W., and Urgessa, G., Numerical Prediction Model for Temperature Distributions in Concrete at Early Ages, *American Journal of Engineering and Applied Sciences*, 5(4), 282-290, 2012.

Ali, W., and Urgessa, G., Computational Model for Internal Relative Humidity Distributions, *Journal of Computational Engineering*, 7, 2014a.

Ali, W., and Urgessa, G., *Structural Capacities of Spherically Voided Biaxial Slabs (SVBS),* Proceedings of the 2014 Structures Congress, American Society of Civil Engineers, 2048-2059, 2014b.

Cecchi, A., and Sab, K., A Homogenized Reissner–Mindlin Model for Orthotropic Periodic Plates: Application to Brickwork Panels, *International Journal of Solids and Structures*, 44, 6055-6079, 2007.

Challagulla, K. S., Georgiades, A. V., and Kalamkarov, A. L., Asymptotic Homogenization Modeling of Smart Composite Orthotropic Grid Reinforced Shells: Part I Theory, *European Journal of Mechanics A/Solids*, 29, 530-540, 2010.

Mota, M., *Voided "Two-Way" Flat Slabs*, American Society of Civil Engineers Structures Congress, 1640-1649, 2013.

Reddy, J. N., *Theory and Analysis of Elastic Plates,* CRC Press, 2006.

Schnellenbach-Held, M., and Pfeffer, K., Punching Behavior of Biaxial Hollow Slabs, *Cement and Concrete Composites*, 24, 551-556, 2002.

Urgessa, G., and Casey, M. J., Temporary Support for Rebar Column Cages during Construction, *Practice Periodical on Structural Design and Construction*, 18(3), 145-148, 2013.

Voros, S., *Analysis of Bi-Axial Hollow Core Slabs,* Proceedings of the JUNIORSTAV 2011 Conference, Brno University of Technology, Czech Republic, 2011.

Interdependence between Structural Engineering and Construction Management
Edited by Ozevin, D., Ataei, H., Modares, M., Gurgun, A., Yazdani, S., and Singh, A.

ISBN: 978-0-9960437-6-2

BEHAVIOR OF CONTINUOUS CONCRETE BEAMS REINFORCED WITH BASALT BARS

SALVIO ARAGAO ALMEIDA JUNIOR and AZADEH PARVIN

Dept of Civil and Environmental Engineering, The University of Toledo, Toledo, USA

The use of fiber-reinforced polymer (FRP) for reinforcement and retrofit of structures has become common in recent years. Although considerable research exists on carbon, glass, and aramid fibers, new materials continue to emerge, requiring new knowledge to optimize their use and safe design. This study analyzed the behavior of continuous beams reinforced with FRP bars. The parameters studied were concrete type (normal-strength concrete (NSC) and ultra-high-performance fiber-reinforced concrete (UHPFRC)), fiber material (carbon, glass, and basalt), and environmental exposure (acid, alkaline, and saline environments). Finite element (FE) beam models were developed and validated with published experimental data. The validated models were used to study the aforementioned parameters. Although the basalt fibers provoked higher displacements when compared to carbon and glass fibers, they also provided better bond with the concrete and higher tensile strength, allowing the beams to reach higher load capacity. Exposure to aggressive environments diminished the adhesion between GFRP and BFRP bars and the concrete, but the reduction was not sufficient to initiate the debonding and the failure was governed by FRP bars rupture. However, it was concluded that the load-displacement response was not affected much by the environmental exposure. The UHPFRC provided superior bond strength between concrete and the reinforcement bars, which assured rupture of the fibers in these beams as well. The use of UHPFRC also resulted in an increase in the beams' load carrying capacity.

Keywords: UHPFRC, BFRP, FRP-concrete bond, Acid environment, FRP bars, Strengthening.

1 INTRODUCTION

Externally bonded FRP reinforcement is often inefficient due to debonding of the sheets (Khalifa 2016). A better alternative is the use of FRP reinforcement bars. Such bars are used mostly in simply-supported beams and there is limited research regarding their application in continuous beams due to complex phenomena of moment redistribution (Kara and Ashour 2013). Consequently, the behavior of continuous beams reinforced with FRP bars is still not well-understood (Sakr *et al.* 2015). Another point in need of further research is the structural applications of new materials, such as basalt fiber-reinforced polymer (BFRP) and ultra-high-performance fiber-reinforced concrete (UHPFRC) for continuous beams. To this date, only limited investigations have been performed on simply-supported beams reinforced with BFRP bars (Elgabbas *et al.* 2015, High *et al.* 2015) and on their bond with the concrete (Shen *et al.* 2015), but no research was found examining their application in continuous beams. Similarly, simply-supported UHPFRC beams were studied by Ferrier *et al.* (2015) and Yoo *et al.* (2015) and the UHPFRC's bond with CFRP and GFRP bars was analyzed by Islam *et al.* (2015) and Firas *et al.* (2011), but no study covered continuous beams reinforced with BFRP.

This study consists of an original numerical investigation of continuous beams, cast with normal-strength concrete (NSC) and UHPFRC reinforced with three types of FRP bars. Finite element (FE) beam models were developed and validated with data published by Ashour and Habeeb (2008). The validated model was used to examine the behavior of NSC and UHPFRC continuous beams reinforced with CFRP, GFRP, and BFRP bars. Furthermore, the response of the beam when GFRP and BFRP bars were exposed to acid, alkaline, and saline conditions for 90 days was also investigated.

2 PROPOSED FINITE ELEMENT MODEL

Figure 1 shows the geometry of developed FE models, which represent half of the actual beam and contain rollers at the axis of symmetry (left end). The models were validated through experimental data of two of the continuous beam specimens, CC5 and SC6, tested by Ashour and Habeeb (2008). The beams had 35 mm cover to the center of the bars and 8 mm stirrups with 140 mm spacing, which prevented the shear failure. Reinforcement details and material properties of tested specimens are given in Tables 1 and 2, respectively.

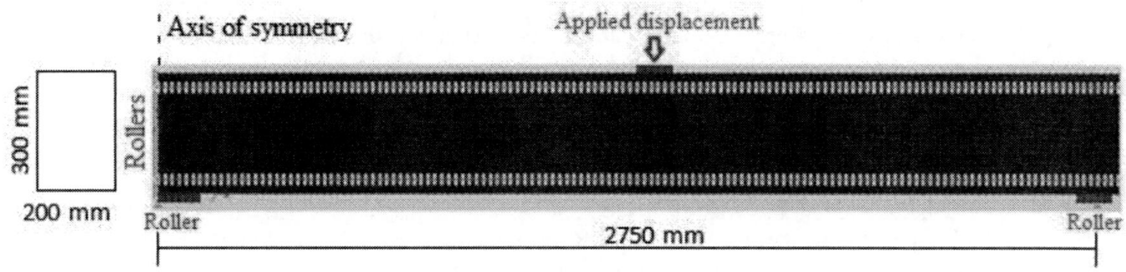

Figure 1. Specimens CC5 and SC6: FE beam model.

Table 1. Reinforcement details of tested beams.

Specimen	f'c (MPa)	Type	n_{top}	ϕ_{top} (mm)	n_{bottom}	ϕ_{bottom} (mm)
CC5	28.0	CFRP	2	12	2	12
SC6	26.3	Steel	4	12	4	12

Table 2. Material properties of tested beams.

Material	E (GPa)	f_y (MPa)	f_u (MPa)	ε_u	ε_{fu}
CFRP (12 mm)	200	-	1061	-	0.0053
Steel (8 mm)	206.8	525.5	611.6	0.15	-
Steel (12 mm)	200	510.8	594.4	0.15	-

The pre-peak segment of concrete stress-stress curve was modeled as proposed by Popovics (1973) and its bond with the FRP bars as a tri-linear stress-slip curve, shown in Figure 2. The bond parameters were extracted from stress-slip curves given in Gravina and Smith (2008), Elgabbas *et al.* (2015), Yan and Lin (2016), Firas *et al.* (2011), and Yoo *et al.* (2015). The steel was modeled as elastic-plastic, with a nonlinear strain hardening effect and rupture at the ultimate strain (ε_u). The FRP fibers were modeled as linear-elastic, with rupture at the ultimate strain (ε_{fu}).

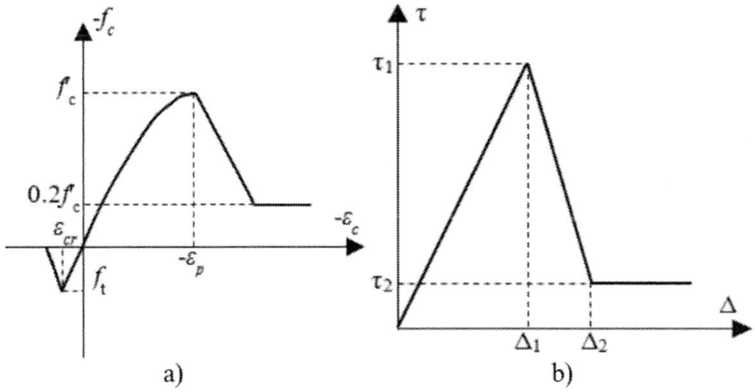

Figure 2. FE Material models: a) concrete stress-strain and b) bond stress-slip curves.

The beam models were created using a FE analysis software program called VecTor2 (VT2), specific to reinforced concrete structures. 2-D rectangular elements were employed to model the reinforced concrete (in blue) and the steel plates (in red), which are placed at load and supports locations. The stirrups were uniformly distributed in the beam using a smeared reinforcement approach and are not visible (Figure 1). The rebars were represented by 1-D truss elements (in red), however, they are mostly covered by the link elements (in yellow), generated to represent the bond between reinforcement bars and concrete. After validation of the FE beam model, the same geometry was used to study three parameters: fiber type, environmental exposure, and concrete strength.

3 DISCUSSION OF RESULTS

3.1 Model validation

The load versus displacement curves of the CC5 and SC6 specimens tested by Ashour and Habeeb (2008) were compared with the ones obtained from the FE model (Figure 3). Since a good agreement was observed between the FE analysis and the experimental results, the model was considered validated and was used to study the aforementioned parameters.

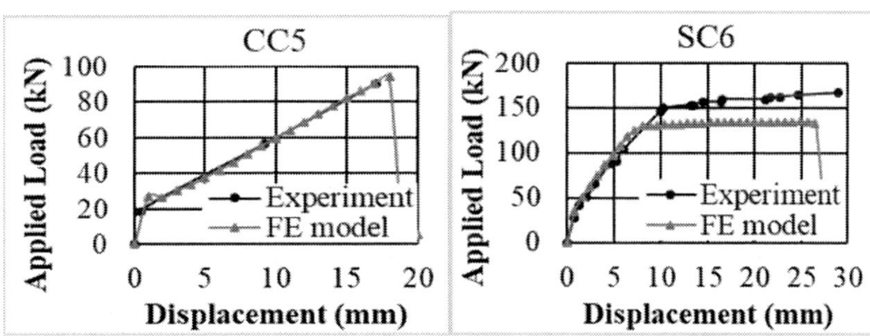

Figure 3. Numerical and experimental load-displacement curves for continuous beams.

3.2 Fiber types

Table 3 shows the fiber composite material properties used in the present parametric study. The bond stress-slip parameters of the reinforcement bars were obtained from Gravina and Smith (2008), Elgabbas *et al.* (2015), and Yan and Lin (2016). Figure 4 shows the load versus displacement curves for various bar fiber types. The BFRP bars provided the highest load capacity for the beam, at the cost of higher deflection. This is due to basalt's higher tensile strength and lower modulus of elasticity as compared to CFRP or GFRP materials.

Table 3. Fiber composite material properties.

Fiber	E (GPa)	f_u (MPa)	τ_1 (MPa)	Δ_1 (mm)	τ_2 (MPa)	Δ_2 (mm)
Carbon	200	1061	11.6	1.23	7.79	1.26
Glass	60	1000	8.0	1.29	1.51	9.45
Basalt	59.5	1567	28.9	0.69	27.4	1.27

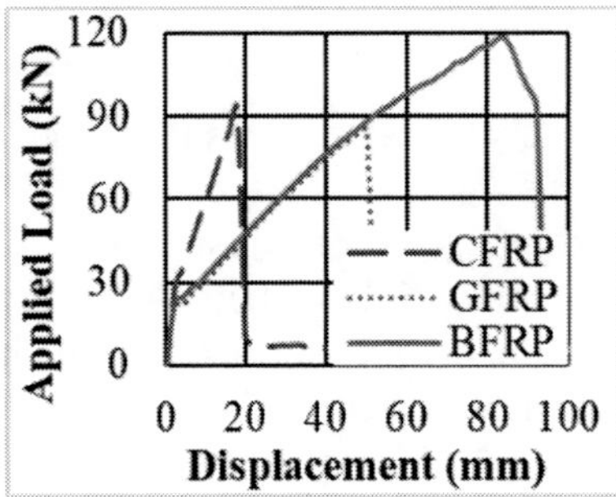

Figure 4. Load-displacement curve of specimen CC5 (NSC) reinforced with CFRP, GFRP, and BFRP bars.

3.3 Ultra-High-Performance Fiber-Reinforced Concrete

The use of UHPFRC (f'_c = 170 MPa) in place of NSC (f'_c = 28 MPa) not only enhanced the capacity of the beam, but also the bond between the concrete and reinforcing bars (Figure 5). The mechanical properties of the UHPFRC were obtained from Firas *et al.* (2011) and used with the Popovics stress-strain curve. According to stress-slip data from Firas *et al.* (2011) and Yoo *et al.* (2015), the CFRP and GFRP bars bond with UHPFRC was superior to NSC. This finding was also confirmed in the present FEA study. The stronger bond can prevent bar slippage in other scenarios where this is an issue. Since no data was found regarding the bond strength between BFRP bars and UHPFRC, it was assumed to follow the trend observed in the other fiber types.

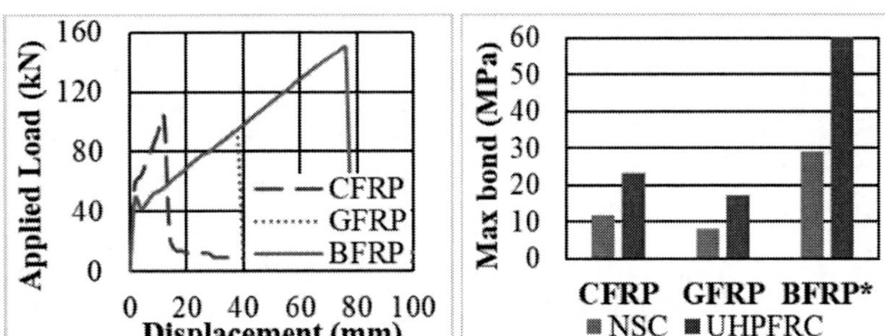

a) b)

Figure 5. a) Load-displacement response of CC5 cast with UHPFRC and b) bond strength comparison.

3.4 Environmental Exposure

The exposure of the reinforcement bars to acid, alkaline, and saline conditions reduced their bond with the concrete (Figure 6). The exposure time was taken as 90 days for all bars. The data for GFRP and BFRP bond stress-slip curves were obtained from Altalmas *et al.* (2015) but was not found for the CFRP bars. The original data referred to 60 MPa concrete and was adjusted to 28 MPa by multiplying the values by $\sqrt{28}/\sqrt{60}$. The potential environmental effects on concrete properties were not examined in this study.

The effect of the environmental exposure did not cause a significant change in the load capacity of the beam since the reduction in the bond strength was not sufficient to initiate the debonding. Subsequently, despite the long exposure time, the rupture of the FRP bars was observed. This indicates that the FRP bars are appropriate to be used in aggressive conditions in which steel would likely be subjected to corrosion; especially BFRP bars since they have the strongest bond with the concrete.

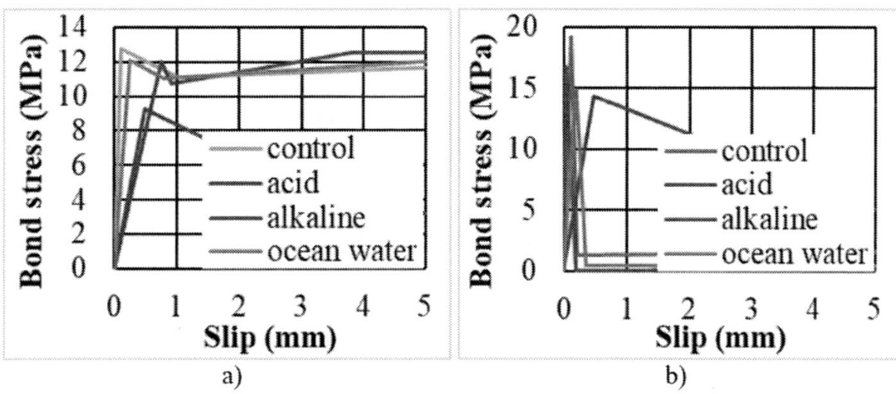

Figure 6. Bond strength of a) GFRP and b) BFRP bars subjected to aggressive environment for 90 days.

4 CONCLUSIONS

This study investigated the effect of using carbon, glass, and basalt fiber-reinforced polymer bars as the reinforcement for continuous beams. A finite element model was created, validated, and used to analyze new scenarios. The effects of bars exposure to acid, alkaline, and saline environments were examined, as well as the influence of using ultra-high-performance fiber-reinforced concrete on the load capacity of the beam and the bond between concrete and reinforcing bars.

The major conclusions obtained in this study are as follows:

- The use of BFRP bars led to higher load capacity and deflection than CFRP and GFRP bars. This was due to the basalt's higher tensile strength and lower modulus of elasticity.
- The debonding was prevented due to the adequate embedment length of the bars.
- The use of UHPFRC in the place of NSC increased bond and the load capacity of the beam due to higher concrete compressive and tensile strengths. The bond enhancement was not observable in the load versus displacement response because debonding never occurred.
- Exposure of the bars to acid, alkaline, and saline environments lowered their bond with the concrete, but not enough to initiate debonding.

References

Altalmas, A., El Refai, A., and Abed, F., Bond Degradation of Basalt Fiber-Reinforced Polymer (BFRP) Bars Exposed to Accelerated Aging Conditions, *Construction and Building Materials*, Elsevier, 81, 162-171, February, 2015.

Ashour, A. F., and Habeeb, M. N., Continuous Concrete Beams Reinforced with CFRP Bars, *Proceedings of the Institution of Civil Engineers–Structures and Buildings*, University of BardFord, 161(6), 349-357, 2008.

Elgabbas, F., Ahmed, E. A., and Benmokrane, B., Physical and Mechanical Characteristics of New Basalt-FRP Bars for Reinforcing Concrete Structures, *Construction and Building Materials*, Elsevier, 95, 623-635, July, 2015.

Ferrier, E., Michel, L., Zuber, B., and Chanvillard, G., Mechanical Behaviour of Ultra-High-Performance Short-Fibre-Reinforced Concrete Beams with Internal Fibre Reinforced Polymer Bars, *Composites Part B: Engineering*, Elsevier, 68, 246-258, January, 2015.

Firas, S. A., Gilles, F., and Robert, L. R., Bond between Carbon Fibre-Reinforced Polymer (CFRP) Bars and Ultra High Performance Fibre Reinforced Concrete (UHPFRC): Experimental Study, *Construction and Building Materials*, Elsevier, 25(2), 479-485, February, 2011.

Gravina, R. J., and Smith, S. T., Flexural Behaviour of Indeterminate Concrete Beams Reinforced with FRP Bars, *Engineering Structures*, Elsevier, 30(9), 2370-2380, September, 2008.

High, C., Seliem, H. M., El-Safty, A., and Rizkalla, S. H., Use of Basalt Fibers for Concrete Structures, *Construction and Building Materials*, Elsevier, 96, 37-46, October, 2015.

Islam, S., Afefy, H. M., Sennah, K., and Azimi, H., Bond Characteristics of Straight- and Headed-End, Ribbed-Surface, GFRP Bars Embedded in High-Strength Concrete, *Construction and Building Materials*, Elsevier, 83, 283-298, May, 2015.

Kara, I. F., and Ashour, A. F., Moment Redistribution in Continuous FRP Reinforced Concrete Beams, *Construction and Building Materials*, Elsevier, 49, 939-948, December, 2013.

Khalifa, A. M., Flexural Performance of RC Beams Strengthened with Near Surface Mounted CFRP Strips, *Alexandria Engineering Journal*, Elsevier, 55(2), 1497-1505, June, 2016.

Popovics, S., A Numerical Approach to the Complete Stress-Strain Curve of Concrete, *Cement and Concrete Research*, Elsevier, 3(5), 583-599, September, 1973.

Sakr, M. A., Khalifa, T. M., and Mansour, W. N., External Strengthening of RC Continuous Beams Using FRP Plates: Finite Element Model, *International Journal of Structural Analysis and Design- IJSAD*, Seek Digital Library, 2(1), 65-71, April, 2015.

Shen, D., Ojha, B., Shi, X., Zhang, H., and Shen, J., Bond Stress-Slip Relationship Between Basalt Fiber-Reinforced Polymer Bars and Concrete Using a Pull-Out Test, *Journal of Reinforced Plastics and Composites*, Sage Journals, 35(9), 747-763, January, 2016.

Yan, F., and Lin, Z., Bond Behavior of GFRP Bar-Concrete Interface: Damage Evolution Assessment and FE Simulation Implementations, *Composite Structures*, Elsevier, 155, 63-76, November, 2016.

Yoo, D. Y., Kwon, K. Y., Park, J. J., and Yoon, Y. S., Local Bond-Slip Response of GFRP Rebar in Ultra-High-Performance Fiber-Reinforced Concrete, *Composite Structures*, Elsevier, 120, 53-64, February, 2015.

Interdependence between Structural Engineering and Construction Management
Edited by Ozevin, D., Ataei, H., Modares, M., Gurgun, A., Yazdani, S., and Singh, A.

ISBN: 978-0-9960437-6-2

OPTIMIZATION PARAMETER SETS FOR SUSTAINABLE CONCRETE IN TUNNELS

LARISSA STRÖMBERG[1], MALIN LÖFSJÖGÅRD[2], ANDERS ANSELL[2], and STAFFAN HINTZE[1]

[1]*NCC Nordic Construction Company, Infrastructure Business Area, Solna, Sweden*
[2]*KTH Royal Institute of Technology, Division of Concrete Structure, Stockholm, Sweden*

Public and private clients are beginning to set complex optimization requirements, taking into consideration environmental and cost-efficiency parameters over the built construction's lifetime. The early design process is currently irreversible, and this makes it difficult to change a concrete structure in the later detailed design stage, when more accurate information is available regarding environmental impact and life-cycle costs. There is a dilemma in complying with existing standards to achieve technical requirements while optimizing a concrete structure in order to reduce the climate impact. The long-term goal of the project is to develop a new theoretical concept for dynamic optimization strategies which can be applicable to the early design, the client-requirement preparation, the detailed design, the production and the follow-up stages. This paper presents the results of the up-start phase of the project. The work has focused on the identification of current practice regarding clients' requirements for technical, environmental and cost-efficiency parameters. An analysis of these requirements with sprayed concrete (shotcrete) in a number of ongoing projects has led to the identification of optimization parameter sets. The project has also shown how the physical values of those parameters can be collected from existing statistics, experience recovery databases and previous project requirements, or calculated according to standardized methods and tools. The concept developed will be used in a demonstrative modeling in the next project step.

Keywords: Multiple client requirements, Markov chain analysis, Life cycle assessment, Life cycle cost, Concrete structures.

1 INTRODUCTION

The European Commission has developed guidelines on climate change (EU commission 2011) and challenges the Member States to redesign their industries and community-building process to be more climate-neutral by 2050. Sweden has taken initiatives aiming to achieve a fossil-free transport sector by 2030 and a climate-neutral nation by 2045 (The Swedish Parliament 2017). The European directive (EU directive 2014/24/EU) on public procurement in the Member States proposes a supplement to the traditional tender evaluation model in which competitive bidding proposals are no longer evaluated on the basis of price. Sweden adopted the directive as the new Public Procurement Act, PPA (The Swedish Parliament 2016), which proposes that alternative tenders are evaluated on the basis of the total life-cycle costs, LCC. PPA allows a contracting authority to refer to an eco-labeling as proof that the procurement object meets certain requirements, e.g. an Environmental Product Declaration, EPD.

Flexibility in dimensioning concrete structures based not only on traditional technical parameters but also on "new" parameters such as LCC and Life Cycle Assessment, LCA is becoming increasingly important. The main obstacle to the development of sustainable structures and the creation of business models with these is the fact that environmental engineering is not integrated with current technical design practice.

Several Swedish public clients such as the Swedish Transport Administration, STA (2015), Stockholm County Council, SCC (Stockholm County Council 2016) and Swedish municipalities (Common Environmental Requirements for Construction Work 2018) have begun to require that infrastructural projects be evaluated with respect to climate impact based on the total life-cycle perspective as well as price. When the STA tightens its demands by 2020, with a 30% reduction in climate impact in each infrastructural project, it will be very difficult to meet them without altering the current technical design. The early design process is irreversible, and this makes it difficult to change a concrete structure in the later detailed design stage to take into consideration not only a contract price but also environmental parameters.

There is a growing interest in developing digital optimization tools based on inter-disciplinary aspects that can be used to support the evaluation of design options for the creation of more sustainable concrete structures. Almeida *et al.* (2016) have developed an optimization method for bridges based on technical performance and LCC with numerical modeling based on Markov chains theory and it is used to calculate a compound performance index for each design option based on a five-level scale. In another North American project, an attempt has been made to create an optimization strategy to identify the most sustainable blend of Portland cement, PCC for road pavements. The aim was to find the option with the best performance based on three aspects: LCC, technical performance and LCA, and a matrix of 10 different strategies including different mixes of Portland cement, road designs and maintenance actions has been developed (Alauddin Ahammed 2017). In a Norwegian project, the LCA modeling tool has been used in the Scandinavian ETSI model (Bridge Life Cycle Optimization) to optimize multiple sustainability aspects for bridges. The tool enables the most sustainable bridge structure to be designed based on LCA, LCC and technical functions (Brattebø 2012). Sen and Roesler (2017) have developed a digital modeling concept for the entire life cycle of concrete road pavements.

The present research project aims to further develop existing models for optimizing individual concrete structures into a more flexible and interactive modeling concept for all concrete structures. The focus of the project is on creating ways of making changes in technical design based on LCA and LCC aspects, regardless of the stage of the project. An earlier KTH study of the parameters reflecting the technical performance of concrete structures and a numerical modeling matrix based on Markov chains theory (Ansell *et al.* 2002) will be further developed in the present project. A previously developed digital LCA calculation interface (Strömberg 2017) will also be used.

2 DESCRIPTION OF THE FRAMEWORK

The purpose of this research project is to develop a dynamic optimization concept to complement traditional structural design and construction practice, public procurement routines, product development processes and client follow-up control in order to create more sustainable concrete structures. The aim is to build a mathematical model that indicates which optimization parameters determine that one concrete structure is more climate-neutral than another for the same technical performance and price. The model can then be used to identify the most effective and direct actions, from the early planning phase to the final demolition of the structure.

The developed concept is based on international Standards and building practice, and it complies with current research in this area. The international Standard, ISO 15686-5 (2008), has been used to define in-depth aspects of the sustainability of a construction project. Environmental, economic and technical performance are the sustainability aspects described in the Standard, which gives a basic outline of how the different sustainability aspects can be integrated into the entire design process taking into consideration the whole life-cycle of the building or engineering work, see Figure 1. The Standard describes the general framework for the application of LCA and LCC, along with technical parameters to different types of construction and to different process stages: early planning, preparation of bidding requirements, design, construction, maintenance and end of life.

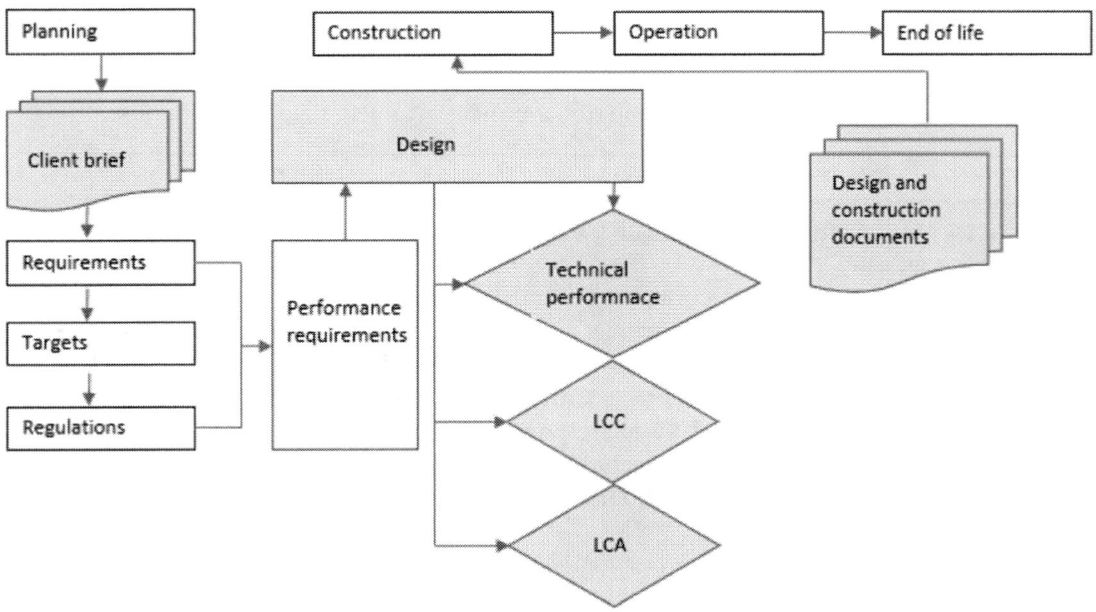

Figure 1. Performance requirements according to ISO 15686-5 (2008) (with some modifications).

The next step will include an overall demonstration of the optimization matrix and the modeling concept. To be able to use the Markov chains theory, a matrix with optimization parameters for each aspect and the physical values of all possible random parameters is required. These values can be generated by measurement or taken from literature sources or feedback from previous infrastructural projects. The probability vector with different life-cycle scenarios for the design of a concrete structure can then be modeled, making it possible to optimize the most sustainable option based on multiple aspects such as technical or functional performance, LCA and LCC. Optimization according to Markov chains theory makes it possible to include the times when different parameters which are a part of the optimization matrix may affect the optimization results.

The developed concept will be demonstrated on shotcrete used as rock reinforcement in traffic tunnels with optimization parameters based on typical requirement specifications in Swedish infrastructural projects. In the future, it will be possible to generalize the concept and apply it internationally to different types of concrete structures. The climate impact of the spray concrete reinforcements in STA's tunnel projects accounts for nearly 33% of the overall climate impact of the whole project calculated on a single tunnel section (Gröndal 2018). Shotcrete is interesting because the Norwegian Public Roads Administration has banned the use of plastic fiber reinforcement in sprayed concrete for environmental reasons (Myren *et al.* 2018). Plastic fiber reinforcement is nevertheless still considered in some countries to be an environment-friendly alternative to steel reinforcement due to its lower weight.

3 IDENTIFICATION OF PARAMETERS AND CONSTRAINTS

The large number of project documents describing the requirements of Swedish clients with regard to technical performance, LCA and LCC have been analyzed, and project requirements from STA, SCC and several Swedish municipalities have been evaluated. The project has developed a matrix with optimization parameters covering most of current clients' requirements regarding climate impact reporting, the calculation of life-cycle costs and functional requirements for the shotcrete reinforcements. The need for information and the sources for this information, e.g. some physical values for the optimization parameters in order to perform a mathematical modeling with the concept in the next step of the project, was also mapped.

3.1 Parameters for Technical Performance

The current practice of using technical parameters makes it possible to compare alternative designs independently of the client or the performer. The industry-wide acceptance of the same standardized requirements in accordance with the European standards and Swedish building regulations have led to a uniform working method. Many parameters are in use to define the technical performance of shotcrete reinforcement according to Standard EN 1992-1-1, STA's requirements (STA 2016) and project-specific requirements, but three parameters of crucial importance for this type of concrete structure: overlay, stock thickness and material consumption (Dalmalm 2018) have been chosen. According to Dalmalm (2018), optimizing these parameters leads to the greatest economic and environmental savings.

3.2 Parameters for LCA

The analysis has shown that there are different variants in the requirements for calculation, reporting, optimization and follow-up of the LCA in the client's project specifications. A full LCA contains several environmental categories, but the primary requirement in the evaluated inquiries from the STA, SCC and Swedish municipalities is the climate impact. This may lead to sub-optimization when a design is optimized to be climate neutral but may cause other adverse environmental impacts, such as waste generation, eutrophication, acidification, utilization of non-renewable resources etc. In the long term, more extensive LCA client requirements must be developed.

ISO 15686-5 (2008) sets out the system perspective for the definition of sustainability for a building or a civil engineering work, but in order to elaborate some explicit parameters for LCA in this study we use the EN 15804 (2012). Standard which specifies an industry-approved method for the development of an EPD for a construction product. An EPD should always be reviewed and approved by an official EPD-program operator so that it can be used to compare the environmental performances of similar products in, for example, public procurement. An EPD contains seven mandatory environmental impact categories, including climate impact, acidification, eutrophication, etc. The information in an EPD can be used as a reliable measure of environmental performance and can be accepted for use in reporting, marketing, procurement and product development.

This present project plans to use only climate impact as an optimization parameter for LCA, since this is judged to be sufficient to meet the current reporting requirements in the Swedish construction industry. In order to be able to perform an LCA for a shotcrete structure, the LCA-software Gabi (2018) will be used with an associated database of emission factors.

3.3 Parameters for LCC

The analysis has shown that claiming an LCC is not as widespread among Swedish public clients although both the European directive 2014/24/EU and the PPA (The Swedish Parliament 2016) recommend applying life-cycle costs for the evaluation of tender proposals instead of the traditional procurement based on price.

The project proposes that the ISO 15686-5 (2008) standard be used to calculate the LCCs. Both the STA (2017) and the recommendation of the Swedish authorities regarding the implementation of the PPA (The Swedish Parliament 2016) refer to this Standard for the conduct of an LCC. The Standard defines life-cycle costs for a built structure as the sum of the investment, operation, maintenance and final settlement costs. An LCC can be calculated on the basis of present values and as annual costs. With the present values, investment costs can be weighed against the future costs and benefits. The annual costs are mainly useful when different maintenance costs are weighed against each other or against the benefits gained.

The STA has developed and implemented an Excel-based calculation tool for the LCC for road structures, the ambition being to conduct LCC for the evaluation of all infrastructural projects within a few years. For the LCC calculations in the next stage of this project, the STA's LCC tool (STA 2017) and the STA' Bridge and Tunnels Management System, BaTMan (STA 2018) will be used.

4 CONCLUSIONS

Firstly, theoretical grounds for a modeling concept have been developed based on a review of international standards, previous research, current design practice, and European Unit's and Sweden's goals in a sustainable construction sector. Secondly, optimization parameters and input data requirements for such modeling have been identified based on mapping of a large number of client specifications in Swedish infrastructural projects.

The next step will include an overall demonstration of the optimization matrix and the modeling concept. The developed concept will be demonstrated on shotcrete used as rock reinforcement in traffic tunnels with optimization parameters based on typical requirement specifications in Swedish infrastructural projects.

It is a major socio-economic challenge to convert the current construction process into a more climate-neutral and cost-effective process taking into consideration the whole life-cycle perspective and achieving the high-level climate policy goals at the European Unit level. The concept developed in the present project will hopefully create more flexibility in the inter-disciplinary optimization of concrete structures and allows changes to be made during the detailed design and construction. The overall result of this project is to increase knowledge accumulation about sustainability aspects at least in 20% of the Swedish construction sector and to adopt the developed modeling concept at least for 20% of all public infrastructural projects with budget over EUR 500,000 by 2021.

Acknowledgments

The project is financed by The Development Fund of the Swedish Construction Industry (SBUF), Sweden's Innovation Agency (Vinnova) through Strategic Innovation Program InfraSweden2030 and project participants: Swedish Transport Administration (STA), Stockholm County Council (SCC), Svensk Byggtjänst, entrepreneurs (NCC and Svevia) and KTH Royal Institute of Technology.

References

Alauddin Ahammed, M., Concrete Pavement Life Cycle Environmental Assessment and Economic Analysis: A Manitoba Case Study, *Pavement Life-Cycle Assessment conference*, Al-Qadi, I. L., Ozer, H. and Harvey, J. (eds.), Illinois, USA, 2017.

Almeida, J. O., Delgado, R. M., and Teixeira, P .F., A Bridge Life-Cycle Cost Optimization Methodology, *International Association for Life-Cycle Civil Engineering*, 2016.

Ansell, A., Racutanu, G., and Sundquist H., *A Markov Approach in Estimating the Service Life of Bridge Elements in Sweden,* 9th International Conference on Durability of Building Materials and Components, Brisbane, Australia 2002.

Brattebø, H., *ETSI Stage 3 Task group 4: Life Cycle Assessment of Bridges*, Norwegian University of Science and Technology (NTNU), 2012.

Common Environmental Requirements for Construction Work (Gemensamma miljökrav för entreprenader), Published by Stockholms stad, Malmö stad, Göteborgs stad and Trafikverket, (in Swedish), 2018.

Dalmalm, T., Certifiering Med Hjälp Av Simulator, *Aktuellt om Produktivitet,* Trafikverket, (in Swedish), June 2018.

EN 15804, Sustainability of Construction Works, Environmental Product Declarations, Core Rules for The Product Category of Construction Products, European Committee for Standardization, 2012.

EN 1992-1-1, *Eurocode 2: Design of concrete structures,* European Committee for Standardization, 2005.

EU commission, *A Roadmap for moving to a competitive low carbon economy in 2050*, 2011.

EU directive 2014/24/EU, on public procurement, February 2014.

Gabi, Software. Retrieved from https://www.thinkstep.com/ on September 13, 2018.

Gröndal, T., Om betong vore guld: Klimat- och energieffektiviseringsarbete i planering och projektering av infrastruktur, *Trafikverkets Resultatkonferens om Klimatkrav och Klimatkalkyl*. Retrieved from https://www.trafikverket.se/contentassets/205e449539334db1b42cd5246f8f5986/resultatkonferens_180614.pdf, (in Swedish) on September 13, 2018.

ISO 15686-5, Buildings and Constructed Assets – Service-Life Planning – Part 5: Life-Cycle Costing, 2008.

Myren, S. A., Hagelia, P., and Bjøntegaard, Ø., The Ban of Polymer Fibre In FRS In Norwegian Road Tunnels, *Eight International Symposium on Sprayed Concrete*, 252-255, Trondheim, Norway, June 11-14, 2018.

Sen, S., and Roesler, J., Contextual Life Cycle Assessment Framework for Pavement Preservation, *International Journal Pavement Engineering*, 2016.

STA, TDOK 2016:0231, Trafikverket, Krav Tunnelbyggande, Version 1.0, (in Swedish), 2016.

STA, Trafikverkets riktlinje 2015:0480: Klimatkrav I Planläggning, Byggskede, Underhåll Och På Tekniskt Godkänt Järnvägsmateriel, (in Swedish), 2015.

STA, Stöd För Arbete Med Livscykelbedömningar Och LCC Vid Planering Och Projektering Av Vägar Och Järnvägar, Version 1.1, (in Swedish), 2017.

STA, Öppen Databas BaTMan. Retrieved from https://www.trafikverket.se/tjanster/system-och-verktyg/forvaltning-och-underhall/BaTMan/ on September 13, (in Swedish), 2018.

Stockholm County Council (Stockholms Läns Landsting, SLL), *Miljöprogram 2017–2021*, (in Swedish), 2016.

Strömberg, L., Conceptual Framework for Calculation of Climate Performance with Pre-verified LCA-Tools, *Journal of Civil Engineering and Architecture*, 11, 29-37, DOI:10.17265/1934-7359, 2017.

The Swedish Parliament (Sveriges Riksdag), Lag (2016:1145) Om Offentlig Upphandling, (in Swedish), 2016.

The Swedish Parliament (Sveriges Riksdag), Ett klimatpolitiskt ramverk för Sverige, (in Swedish), 2017.

Interdependence between Structural Engineering and Construction Management
Edited by Ozevin, D., Ataei, H., Modares, M., Gurgun, A., Yazdani, S., and Singh, A.

ISBN: 978-0-9960437-6-2

ICONIC STRUCTURES: CASE STUDY OF A HISTORIC MUSEUM WITH NOTABLE SPANS DESIGNED IN CONCRETE

ANTHONY EDE[1], ADEOTI WILLIAMS[1], OLATUNJI ONI[2], OLUWAROTIMI OLOFINNADE[1], SOLOMON OYEBISI[1], TOLULOPE ALAYANDE[1], OLUWASEUN MARK[1], OLATOKUNBO OFUYATAN[1], and JOHN OLUWAFEMI[1]

[1]*Dept of Civil Engineering, Covenant University, Ota, Nigeria*
[2]*Dept of Architecture, Covenant University, Ota, Nigeria*

The emergence of iconic structures around the world increased tourism and economic status of the of host cities. Iconic structures stand out and attract the attention of the general public due to their daring structural limits and aesthetic forms. Hence, the design idea of an iconic historic museum is conceptualized in this research. This paper models an art center, analyses and designs the critical elements according to BS 8110, trying to determine the feasibility of a large-cantilevering structure with numerous curves being achieved. The activities carried out in this research involved a systematic modeling of the structure with Building Information Modeling (BIM) tools, analysis and design in reinforced concrete with particular emphasis on the cantilevered wing of about 37.5m. Numerous load combinations were applied, and various member section properties were experimented. The results obtained from the analysis of the designed reinforced concrete model identified the structural efficiency of certain critical members in the west wing of the building and how they were made to adhere to structural limits in the code of practice by ensuring deflections of critical cantilevered members did not exceed the limit. A unique arrangement of structural systems has been combined to solve the problem of deflection of the seemingly impossible cantilever of 37.5m.

Keywords: Design and analysis, Deflection, Structural limit, Long cantilever, Reinforced concrete, BIM.

1 INTRODUCTION

Various regions and countries have astonishing structures, but a handful have become instantly recognizable and easily symbolize a destination or an epoch in time and a location as they develop the skyline of such areas. Such structures can be classified as iconic structures.

An iconic structure is an architectural masterpiece that stands out and is captivating (Ede 2014a, Ede and Udoh 2015a, 2015b). Numerous structures around the world belong to this category of eye-catching structures. Examples are the Sydney Opera House, Tower of Pisa, Eiffel Tower, the Burj Al Arab, CCTV Headquarters, the Millennium Dome, Pyramids of Egypt, British Airways London Eye and Malaysia's Petronas Towers. These structures are easily recognizable and have come to symbolize their host cities. Their presence has been immensely beneficial to their host cities and nations as they boast the value of tourism and therefore, inspires the design of more outstanding eye-catching structures.

This research considers the structural analysis and design of a proposed museum building according to BS 8110 (1997), conceptualized by architect Tunji. The museum is an ingenious artifice testing the limits of spatial requirements and extents of long-span with a daring curved extensive cantilever. Museums are structures in which historical, scientific, artistic, or cultural objects of interest are stored and exhibited. Other museums buildings across Nigeria, like the National Museum in Lagos and the Civic Centre in Victoria Island, Lagos amongst others, are unique structures and this proposed model is to bring a new touch to the city by exploring new architectural and structural limits.

The economic benefits of the world's most iconic structural wonders have been immensely beneficial to the countries in which they are present. Hence, a ripple effect in the generation of more outstanding structures. If this structure is implemented, it could contribute immensely the value of internal and external tourism in Nigeria as there will be an influx of tourists from within and from other countries of the world.

2 METHODOLOGY

Museums buildings serve a large segment of society and therefore, to achieve end-user satisfaction, there is a need to adopt innovative design philosophies as they relate to accessibility and usability requirements (Ibem *et al.* 2017). They require spaces with aesthetic priorities to properly relay the essence of historical objects. Hence the need for open rooms with large areas and spans. The traditional role of museums is to gather and store objects and materials of cultural, religious and historical importance, preserve them and present them to the public for the purpose of education and entertainment (EGMUS 2017).

This museum building consisting of three storey, has a 22m long cantilever, which supports a 488 square meter atrium with a critical wing of 37 m cantilever, children interpretation class, curator's lounge and interpretation space, which carry large significant distributed loads. The approach consisted of the development of the structural computer model, the analysis of the critical wing of the building and the design of selected elements.

Building Information modeling (BIM) tools were adopted for various part of the works. The importance of BIM tools in present day buildings cannot be over-emphasized (Ede 2014b, Ede *et al.* 2017). BIM tools are used to model and manage a whole lot of information; the building teams need to make proper decisions especially in the design process, thereby minimizing errors and enhancing structural qualities. In this museum design, BIM tools were extensively used by the creator of the design and the supervising team of engineers and architects of Covenant University, Ota, Nigeria in order to ensure that the structural model is feasible.

The architectural model was realized in Autodesk REVIT 2017 while Autodesk Robot was adopted for the analysis and design. Studies on the deflection and moment envelopes of the concrete elements of the west-wing of the structure containing the cantilever were meticulously carried out. The west side of the building gradually progresses in height for three levels and then stretched out from this level upwards. The cantilever extended for a distance of 22m about its curved perimeter to the supported end. Based on structural efficiency considerations, tension cables members of 275 MPa were adopted to reduce the deflection of the cantilever. Two Core columns were used to support the cables at each half of the cantilever.

The museum has a spatial requirement of about 450 square meters, which is required on all four floors. The topmost floor consists a curved glazed glass roof, which can be effectively accounted for in numerous iconic structures by application of steel structural systems forming portal frames, space frames, trusses or space trusses. The elevation and skeleton views are shown in Figure 1.

Figure 1a. Architectural model displaying the extensive cantilever (Arc. Tunji 2017).

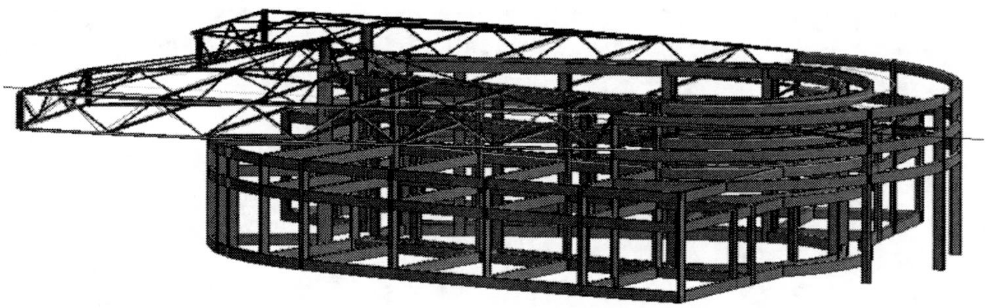

Figure 1b. Skelton of left wing cantilever (Arc. Tunji 2017).

In today's world, being guide by the ever-increasing concern for sustainability, vast combinations of materials are available for construction (Olofinnade *et al.* 2017, Ede *et al.* 2018a). For this project, normal reinforced concrete, strengthened with cables were adopted. The estimated dead and imposed loads values are in accordance with the BS 6399 (1996). Wind load was generated in Autodesk robot in the x and y directions, assuming wind velocity of 25m/s, terrain level one meter above sea level and load generation deviation factor of 0.5%.

The analysis considered the shear stress in beams and columns, bending moment in beams and columns, deflection of cantilever and tension in cables. The analysis was based on the decomposed beam loads since all the slab loads are transferred to the beams. The slabs were defined as the rigid diaphragm while the tension cables were considered as steel full sections.

3 RESULTS AND DISCUSSIONS

The works carried out consisted of the architectural modeling of the structure via BIM tools, structural analysis of the model with simultaneous consideration of various load interaction curves of critical elements, with strong consideration of the deflection limits as prescribed in BS 8110 (1997). A sketch of the structure with its least resisting moment compared to the overturning moment is shown in Figure 2.

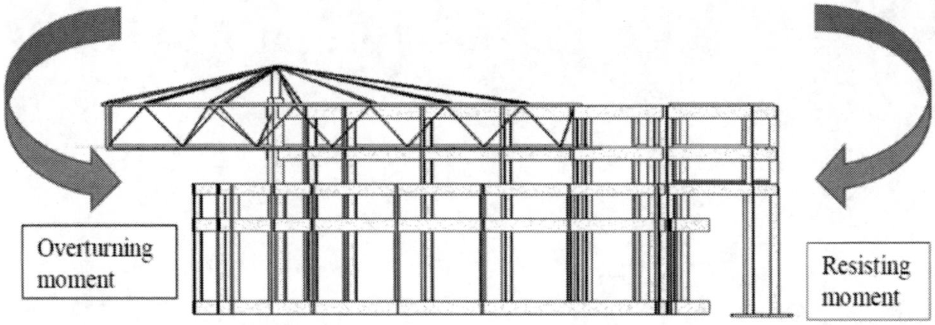

Figure 2. Elevation of building with graphic aids of entire moment directions.

The member sections were selected based on functional and aesthetic feasibility, bearing in mind the best options that will not encroach into functional space. Certain members most susceptible to the effects of the loads were identified and carefully analyzed. A section of the longest cantilever B of daring length is shown in Figure 3.

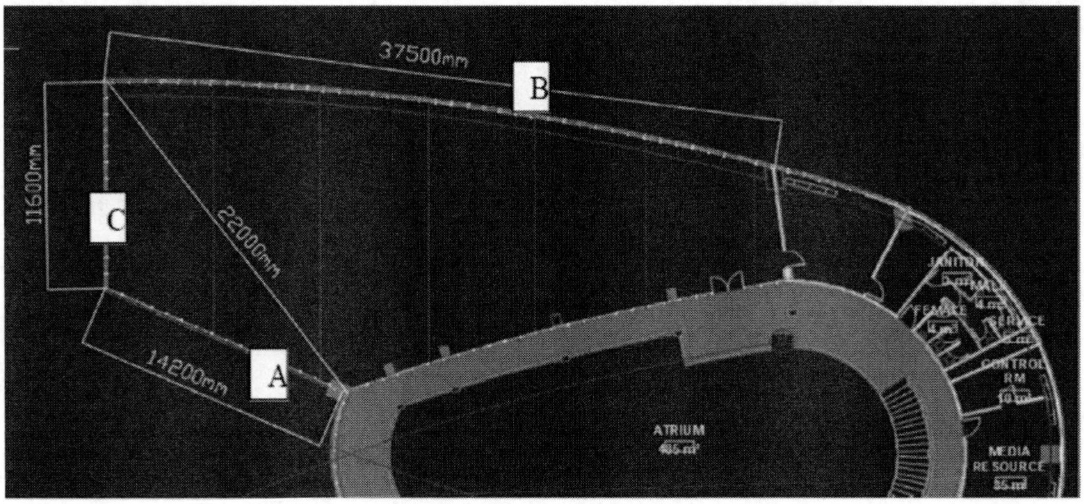

Figure 3. Plan view of cantilever (AutoCAD drawing).

Deflection is an essential criterion to be considered for structural members (especially in a cantilever) due to the risks that excessive deflection may likely lead to damage of finishes, glass panels and unnecessary slope on ground surface. In this research, the deflections of the cantilevers were vigorously analyzed. Numerous concrete sections were experimented on the various cantilevering members during the analysis. It was observed that an increase in member section was directly related to a decrease in deflection but with continuous increase, the self-weight of the steadily increasing members leads to an increase in deflection. Variation of beam A with section sizes is shown in Figure 4.

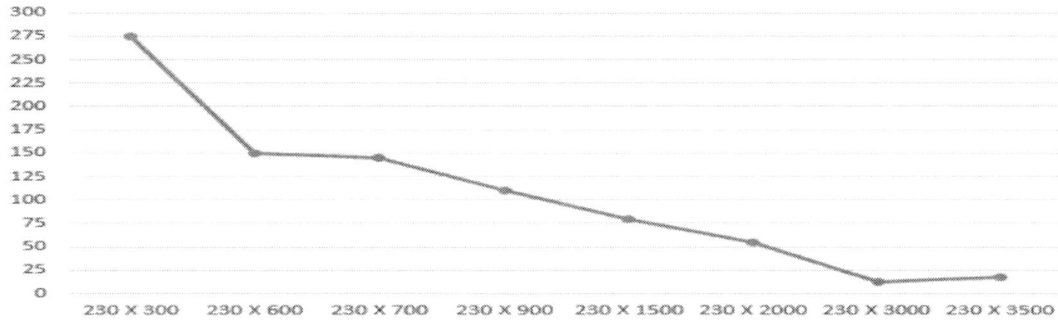

Figure 4. Graph showing variation of beam A with section size.

Two principal types of C35 concrete sections were eventually adopted: a 300X600 mm rectangular section and a T-section with flange breadth of 600mm, flange height of 175 mm, overall height of 900 mm and web breadth of 230 mm. The deflection of beam A for section types 1 and 2 are shown in Figures 5a and 5b.

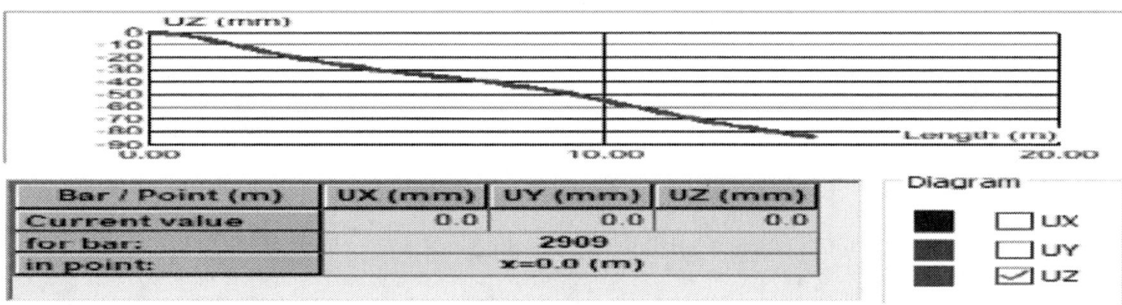

Figure 5a. Deflection graphs of members A for section type 1.

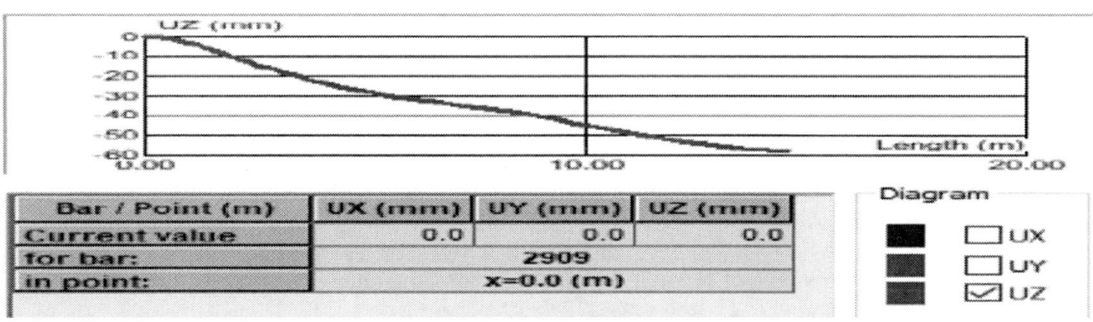

Figure 5b. Deflection graphs of members A for section type 2.

It can be seen that by applying section 2, deflection reduced considerably from 90 mm to 60 mm in beam A.

4 CONCLUSIONS AND RECOMMENDATIONS

The structural model of this eye-catching museum building was successfully generated and subjected to various load cases. With the arrangement of structural members, it can be said that the negative outcomes of the actual long cantilever were greatly catered for. The research was tasking and helped to realize systems unique to the structure, which helped to reduce the typical outcome of such complex structure and define the limits of the architect's concept. With further analysis and study of such iconic structures, more efficient solutions may be developed, which will lead to the realization of more amazing structures, which may boost tourism and transform the skylines of Nigeria and other nations of the world. As can be seen, a unique arrangement of structural systems has been combined to solve the problem of the deflection of the seemingly impossible cantilevers, thereby reducing safety risks (Ede *et al.* 2018b). Also, variations of deflection extents with member sizes were studied as member sections were increased. Hence, this work serves as a strong basis for further research into other extensive cantilevering structures.

Based on the challenges encountered in translating the architectural drawing to functional structural model, it is recommended to engage reputable contractors who experienced in complex construction projects. This is particularly important because excessive deflection after installation of finishes, cladding and perimeter glass can lead to cracks and damages to supplementary installations and pose safety risk for the entire structure.

Acknowledgements

The authors wish to acknowledge the management of Covenant University for supporting this research.

References

BS 6399-1, *Loading for Buildings-Part 1: Code of Practice for Dead and Imposed Loads*, British Standard Institute, London. 1996.

BS 8110-1, *Structural use of Concrete-Part 1: Code of Practice for Design and Construction,* British Standard Institute, London. 1997.

Ede, A. N., Building Information Modeling: Case Study of a Duplex Building Project in Nigeria, *International Journal of IT, Engineering and Applied Sciences Research*, 3(4), 25-28, 2014a.

Ede, A. N., Challenges Affecting the Development and Optimal Use of Tall Buildings in Nigeria, *The International Journal of Engineering and Science*, 3(4), 12-20, 2014b.

Ede, A. N., and Udoh, E. G., Modelling, Analysis and Design of a Bottle-Shaped Building, *IISTE Civil and Environmental Research,* 7(1), 114-121, 2015a.

Ede, A. N., and Udoh, E. G., Modelling, Analysis and Design of a Water Drop-Shaped Building, *The International Journal of Scientific and Engineering Research*, 6(4), 14-19, 2015b.

Ede, A. N., Olofinnade, O. M., and Sodipo, J. O., *Use Of Building Information Modelling Tools For Structural Health Monitoring*, Proceedings of the IEEE International Conference on Computing, Networking and Informatics, ICCNI, 1-4, 2017.

Ede, A.N., Olofinnade, O. M., and Awoyera, P. O., Structural Form Works and Safety Challenges: Role of Bamboo Scaffold on Collapse of Reinforced Concrete Buildings in Nigeria, *International Journal of Civil Engineering and Technology,* 9(9), 1675-1681, 2018a.

Ede, A. N., Olofinnade, O. M., Ugwu, E. I., and Salau, A. O., Potentials Of Momordica Angustisepala Fiber In Enhancing Strengths of Normal Portland Cement Concrete, *Cogent Engineering*, 5(1), 1-17, 2018b.

EGMUS - *The European Group on Museum Statistics,* 2017.

Ibem, E. O., Oni, O. O., Umoren, E., and Ejiga, J., An Appraisal of Universal Design Compliance of Museum Buildings in Southwest Nigeria, *International Journal of Applied Engineering Research,* 12(23), 13731-13741, 2017.

Olofinnade, O. M., Ede, A. N., and Ndambuki, J. M., Sustainable Green Environment through Utilization of Waste Soda-Lime Glass for Production of Concrete, *Journal of Material and Environmental Sciences,* 8(4), 1139-1152, 2017.

Interdependence between Structural Engineering and Construction Management
Edited by Ozevin, D., Ataei, H., Modares, M., Gurgun, A., Yazdani, S., and Singh, A.

ISBN: 978-0-9960437-6-2

SHEAR BEHAVIOR OF RC BEAMS STRENGTHENED WITH DIFFERENT TYPES OF FRCM: EFFECT OF STIRRUPS' CONFIGURATION

TADESSE WAKJIRA and USAMA EBEAD

Dept of Civil and Architectural Engineering, Qatar University, Doha, Qatar

Fabric-reinforced cementitious matrix, (FRCM) system has shown to be promising for the strengthening of reinforced concrete (RC) beams. However, the available experimental investigation on the shear strengthening efficacy of FRCM system is limited, particularly for deep beams. Moreover, to the authors' knowledge, no literature is available on the effect of the stirrups' configuration relative to the FRCM strips on the shear capacity of FRCM-strengthened beams. Studying this effect will aid in a better understanding of the FRCM/stirrups interaction. Thus, in this paper the experimental study on the shear behavior of RC deep beams strengthened in shear using FRCM system is presented. The test matrix involved two unstrengthened and six FRCM-strengthened deep beams tested under three-point bending. The primary test variable was the effect of stirrups' configuration relative to the FRCM strips. The other test variable includes the effect of different types of FRCM fabric (made of carbon, glass, and polyparaphenylene benzobisoxazole, PBO). Experimental results demonstrated an effective application of the FRCM in improving the load capacities of RC deep beams, up to 40.3% increase in the load capacity was achieved.

Keywords: Reinforced concrete, Deep beam, Strengthening, Fabric-reinforced cementitious matrix.

1 INTRODUCTION

Fiber reinforced polymer (FRP) system has been widely implemented as an effective strengthening technique for reinforced concrete (RC) structures due to its light weight, ease of installation, an excellent resistance to corrosion, and high tensile strength (Aly *et al.* 2006, Ebead 2015, Ebead and Saeed 2013, 2014). However, some drawback associated with FRPs such as their incompatibility with the concrete substrate, low fire resistance, and high cost of resins limits their use. In order to overcome this problem, fabric reinforced cementitious matrix (FRCM) has been introduced by replacing the epoxy adhesives with cementitious mortar. FRCM system has successfully been used for the strengthening of RC beams both in shear (Ebead and Wakjira 2018a, Elghazy *et al.* 2018, Marcinczak and Trapko 2019, Ombres 2015, Wakjira and Ebead 2018c, 2019, Younis *et al.* 2017) and flexure (Ebead *et al.* 2017, Ebead *et al.* 2019, Ebead and El-sherif 2019, Elghazy *et al.* 2017, Pino *et al.* 2017) owning a better bond with the substrate concrete (Ebead and Younis 2019, Younis and Ebead 2018) relative to FRP. However, the available experimental investigation on the shear strengthening efficacy of FRCM system is limited. The shear strengthening efficacy of such system depends on various factors including but not limited to the application methods, wrapping schemes, and geometric configurations.

Moreover, previous experimental and analytical results has shown that the stirrups reinforcement ratio significantly affect the efficacy of the FRCM system (Awani *et al.* 2015, Ebead and Wakjira 2018b, Wakjira and Ebead 2018a, 2018b). Nonetheless, unlike FRP strengthening system, the study on the interaction of the FRCM/stirrups is scarce. Moreover, to the authors' knowledge, no literature is available on the effect of the stirrups' configuration relative to the FRCM strips on the shear capacity of the strengthened beams. Studying this effect will aid in a better understanding of the FRCM/stirrups interaction.

Thus, the present study investigates the structural performance of RC beams strengthened with an externally bonded FRCM. The strengthening was performed with three fabric plies embedded in the cementitious matrix. The effects of the stirrups' configuration (aligned and unaligned with the FRCM strips) and types of fabrics (carbon, glass, and polyparaphenylene benzobisoxazole (PBO)) were investigated.

2 EXPERIMENTAL PROGRAM

A total of eight medium-scaled RC beams, two unstrengthened/control beams and six strengthened beams, were used. Figure 1 shows the geometric and reinforcement details of the test beams. The specimens were internally reinforced with two 16 mm diameter tensile bars and two 8 mm diameter compressive bars with yield strengths of 595 MPa and 535 MPa, respectively. The shear reinforcement was done using 8 mm diameter bars spaced at 100 mm outside the shear span, while 6 mm diameter bars spaced at 215 mm were used within the shear span as shown in Figure 1. The yield strengths of the stirrups within and outside the shear span were 298 MPa and 234 MPa, respectively.

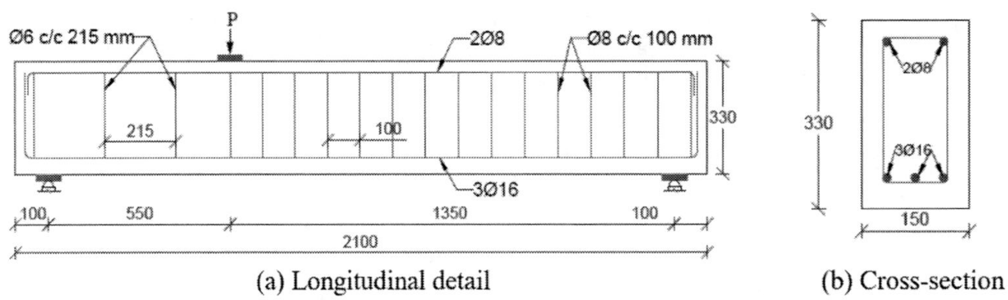

Figure 1. Geometric and reinforcement detail of the reference beam (all dimensions are in mm).

The beams were cast using normal weight ready-mix concrete with an average compressive strength of 30 MPa. Three different types of FRCM fabrics were used; viz., glass, carbon, and PBO fabrics as shown in Figures 2a through 2c along with their geometric and mechanical properties.

The strengthening is performed using 0.12 m wide FRCM strips spaced at 0.095 m as shown in Figures 3a and 3b. The test variables were the stirrups' configuration relative to the FRCM strips; namely, stirrups aligned versus unaligned with the FRCM strips as shown in Figures 3a and 3b, respectively, and the type of fabrics (glass, PBO, and carbon) as listed in Table 1. The test specimens are classified in to two groups based on the configuration of the stirrups relative to the FRCM strips. Group-1 beams are specimens internally reinforced with stirrups aligned with the FRCM strip, Figure 3a, while in Group-2 beams the stirrups are unaligned with the FRCM strips, Figure 3b.

Grid spacing: 18.2 × 14.2 mm
Ultimate elongation: 0.0325
Elastic modulus (E_f): 80 GPa
Tensile strength: 2600 MPa
Area per unit width in the warp direction: 0.0477 mm

(a) Glass fabrics

Grid spacing: 10 × 10 mm
Ultimate elongation: 0.0180
Elastic modulus, E_f: 240 GPa
Tensile strength: 4800 MPa
Area per unit width in the warp direction: 0.0477 mm

(b) Carbon fabrics

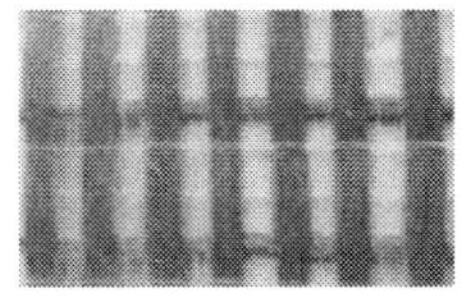

Grid spacing: 10 × 17 mm
Ultimate elongation: 0.0215
Elastic modulus, E_f: 270 GPa
Tensile strength: 5800 MPa
Area per unit width in the warp direction: 0.0455 mm

(c) PBO fabrics

Figure 2. Fabric types used in this study and fiber properties.

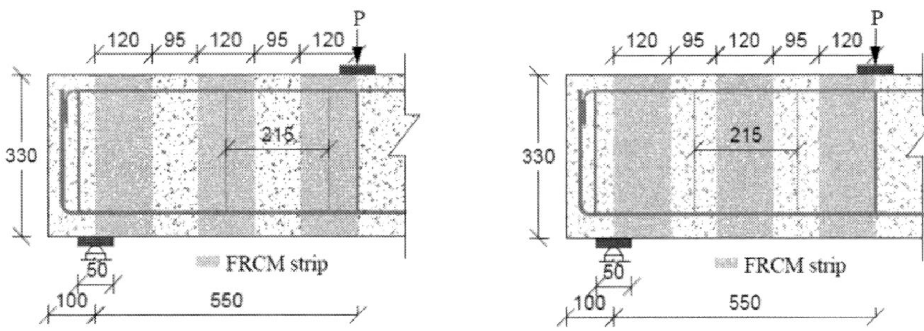

(a) Group-1 (b) Group-2

Figure 3. Strengthening configuration details (all dimensions are in mm).

3 TEST RESULTS AND DISCUSSION

The test results are summarized in Table 1 in terms of the ultimate load, P_u, and its corresponding deflection, δ_u, and the gain in P_u relative to the reference beams.

Table 1. Experimental test results.

Group	Beam ID[a]	FRCM fabric	P_u (kN)	Gain in P_u (%)	K_{ft} (kN)	δ_u (mm)
1[b]	B1	–	130	–	–	5.12
	BC1	Carbon	182	40.3	13705	7.54
	BG1	Glass	158	21.5	6091	6.36
	BP1	PBO	163	25.7	11007	7.52
2[c]	B2	–	138	–	–	6.02
	BC2	Carbon	188	36.6	13705	8.28
	BG2	Glass	161	16.6	6091	6.73
	BP2	PBO	175	26.8	11007	7.45

[a]B: beam, C: carbon, G: glass, P: PBO; [b]stirrups aligned with the FRCM strips;
[c]stirrups unaligned with the FRCM strips.

3.1 Ultimate Load and Failure Modes

The ultimate load (P_u) and the gain in P_u for FRCM-strengthened beams relative to the corresponding reference beam are listed in Column 4 and Column 5 of Table 1. The reference beam of Group-1 specimens, B1, failed at the ultimate load of 130 kN. The strengthened beams of the same group; namely, Specimens BC1, BG1, and BP1 failed at the respective ultimate loads of 182 kN, 158 kN, and 163 kN representing percentage gain in P_u of 40.3%, 21.5%, and 25.7% relative to B1 as listed in Table 1.

The strengthened specimens of Group-2; namely, BC2, BG2, and BP2 failed at the ultimate loads of 188 kN, 161 kN, and 175 kN, respectively. These load values represent 36.6%, 16.6%, and 26.8% increase in the ultimate load relative to B2 (138 kN) for Specimens BC2, BG2, and BP2, respectively as listed in Table 1.

For the FRCM type, carbon FRCM strengthened specimens showed higher gain in the load capacity relative to the glass and PBO counterparts regardless of the stirrups' configuration. This can be explained in terms of the FRCM axial rigidity (E_f), which is the product of the axial stiffness (E_f) and effective area (A_f) of the FRCM reinforcement as given in Table 1. For instance, Specimen BC1 from Group-1 with higher E_f value failed at the ultimate load of 182 kN representing 40.3% gain in the load capacity. The glass and PBO counterparts of the same specimen, BG1 and BP1, showed lower gain in P_u of 21.5% and 25.7%.

The effect of the stirrups' configuration on the percentage gain in the load capacity is illustrated in Figures 4a through 4c for glass, PBO, and carbon FRCM strengthened beams, respectively. As can be seen in these figures, generally, aligned FRCM/stirrups resulted in a slightly higher gain in the load capacity. However, this difference in the load capacity is not significant, less than 5%.

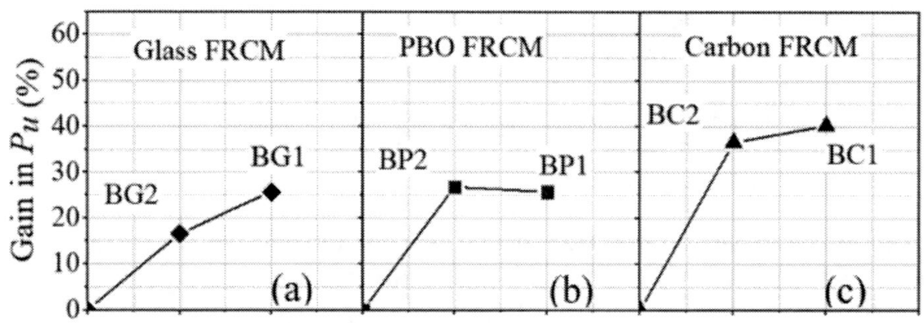

Figure 4. The effect of stirrups' configuration on the gain in the ultimate load.

All tested beams failed in shear with the formation of a diagonal shear crack. The strengthened beams failed in shear with a premature delamination of the FRCM laminate off the concrete substrate as shown in Figures 5a and 5b for Specimens BC1 and BP1, respectively. This is a typical mode of failure of specimens strengthened with an externally bonded FRCM system.

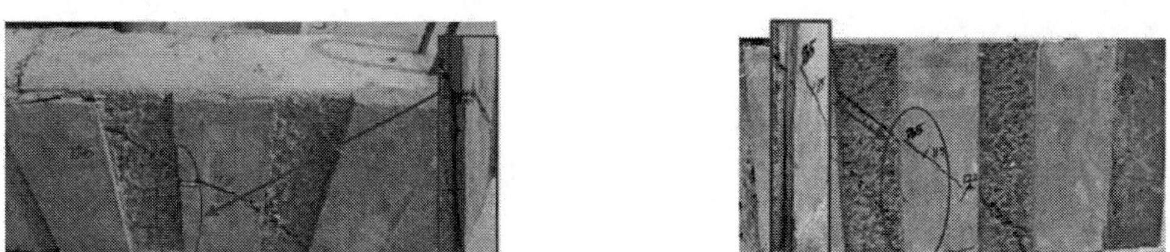

(a) BC1 (b) BP1

Figure 5. Typical failure modes of the strengthened beams.

3.2 Deformational Characteristics

Figures 6a and 6b show the load–deflection responses of the tested beams, while the ultimate deflection (δ_u), i.e., the deflection at the ultimate load is given in Table 1. In general, all the tested beams showed a bilinear load–deflection response with a partial linear curve before and after failure. The strengthened beams experienced higher deflection at the ultimate load relative to the unstrengthened beams as shown in Figures 6a and 6b. The maximum deflection of up to 8.28 mm, was observed for the strengthened specimens as can be seen in Figure 6 and Table 1. This observation indicates the efficacy of the FRCM system in enhancing the load–deflection response of the strengthened beams.

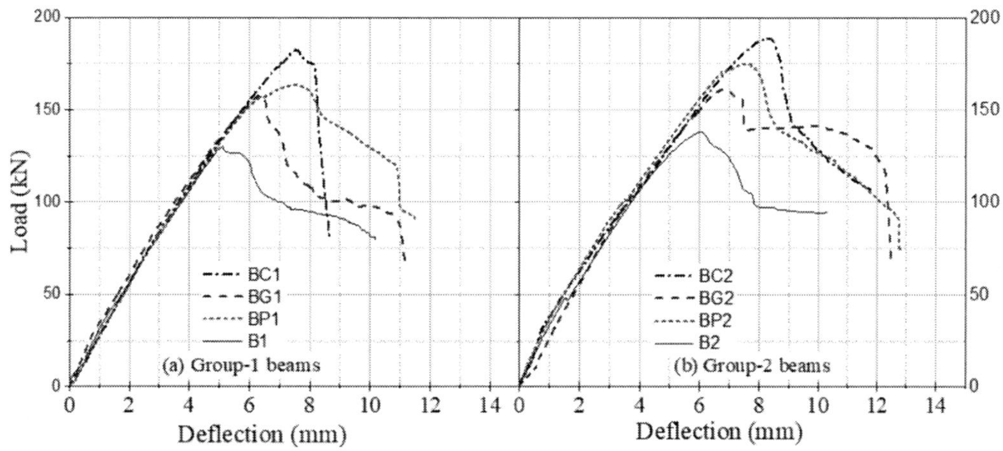

Figure 6. Load–deflection diagram.

4 CONCLUSIONS

In this paper, an experimental study on the shear behavior of fabric-reinforced cementitious matrix, FRCM-strengthened reinforced concrete (RC) beam was presented. Different types of commercially available FRCM fabrics were used along with their associated manufacturer recommended mortars including glass, carbon, and polyparaphenylene benzobisoxazole fabrics. The effect of stirrups' configuration relative to the FRCM strips and the type of the FRCM fabrics on the load capacity, failure modes and deformational characteristics of the strengthened beams were studied. All beams failed in shear with a formation of main diagonal shear crack running from the loading point to the support. The following conclusions can be drawn from this study:

Regardless of the stirrups' configuration, the FRCM system has considerably increased the load capacities of the strengthened beams. The gain in the ultimate load ranged from 16.6% to 40.3% relative to the reference beams and this enhancement depends on the tested variables.

The stirrups' configuration relative to the FRCM strips does not show a significant effect on the percentage gain in the load capacities of the strengthened beams. A maximum of 4.9% gain in the ultimate load was observed by aligning the stirrups with the FRCM strips compared to the unaligned FRCM/stirrups configuration.

The specimens strengthened with carbon FRCM showed a higher gain in the load capacity than the PBO and glass FRCM counterparts due to its the higher axial rigidity compared to the others.

Acknowledgments

This paper was made possible by NPRP grant # NPRP 9-110-2-052 from the Qatar National Research Fund (a member of Qatar Foundation). The findings achieved herein are solely the responsibility of the authors.

References

Aly, R., Benmokrane, B., and Ebead, U., Tensile Lap Splicing of Fiber-Reinforced Polymer Reinforcing Bars in Concrete, *ACI Structural Journal,* 103(6), 857–864, 2006.

Awani, O., El Maaddawy, T., and Refai, A., Numerical Simulation and Experimental Testing of Concrete Beams Strengthened in Shear with Fabric-Reinforced Cementitious Matrix, *Journal of Composites for Construction,* 20(6), 1–11, 2015.

Ebead, U., Inexpensive Strengthening Technique for Partially Loaded Reinforced Concrete Beams : Experimental Study, *Journal of Materials in Civil Engineering,* (10), 2015.

Ebead, U., and El-sherif, H., Near Surface Embedded-FRCM for Flexural Strengthening of Reinforced Concrete Beams, *Construction and Building Materials,* 204, 166–176, 2019.

Ebead, U., and Saeed, H., Hybrid Shear Strengthening System for Reinforced Concrete Beams: An Experimental Study, *Engineering Structures,* 49, 421–433, 2013.

Ebead, U., and Saeed, H., Flexural and Interfacial Behavior of Externally Bonded/ Mechanically Fastened Fiber-Reinforced Polymer Strengthened Reinforced Concrete Beams, *ACI Structural Journal,* 111(4), 2014.

Ebead, U., Shrestha, K. C., Afzal, M. S., El Refai, A., and Nanni, A., Effectiveness of Fabric-Reinforced Cementitious Matrix in Strengthening Reinforced Concrete Beams, *Journal of Composites for Construction,* 21(2), 04016084, 2017.

Ebead, U., Shrestha, K. C., and Saeed, H., Soffit and U-Wrap Fabric-Reinforced Cementitious Matrix Strengthening for Reinforced Concrete Beams, *ACI Structural Journal,* 116(2), 267–278, 2019.

Ebead, U., and Wakjira, T., Behaviour of RC Beams Strengthened in Shear Using near Surface Embedded FRCM, *IOP Conference Series: Materials Science and Engineering* 431, 072001, 2018a.

Ebead, U., and Wakjira, T., FRCM/Stirrups Interaction in RC Beams Strengthened in Shear Using NSE-FRCM, *IOP Conference Series: Materials Science and Engineering*, 2018b.

Ebead, U., and Younis, A., Pull-off Characterization of FRCM/Concrete Interface, *Composites Part B* 165(November 2018), 545–553, 2019.

Elghazy, M., Refai, A., Ebead, U., and Nanni, A., Effect of Corrosion Damage on the Flexural Performance of RC Beams Strengthened with FRCM Composites, *Composite Structures,* 180, 994–1006, 2017.

Elghazy, M., El Refai, A., Ebead, U., and Nanni, A., Post-Repair Flexural Performance of Corrosion-Damaged Beams Rehabilitated with Fabric-Reinforced Cementitious Matrix (FRCM), *Construction and Building Materials,* 166(January), 732–744, 2018.

Marcinczak, D., and Trapko, T., Shear Strengthening of Reinforced Concrete Beams with PBO-FRCM Composites with Anchorage, *Composites Part B: Engineering,* 158, 149–161, 2019.

Ombres, L., Structural Performances of Reinforced Concrete Beams Strengthened in Shear with a Cement Based Fiber Composite Material, *Composite Structures,* 122, 316–329, 2015.

Pino, V., Hadad, H., Basalo, F., Nanni, A., Ebead, U., and Refai, A., Performance of FRCM-Strengthened RC Beams Subject to Fatigue, *Journal of Bridge Engineering,* 22(10), 4017079, 2017.

Wakjira, T., and Ebead, U., A New Approach for Predicting the Shear Capacity of FRCM Strengthened RC Beams in Shear, *IOP Conference Series: Materials Science and Engineering*, 2018a.

Wakjira, T., and Ebead, U., FRCM/Internal Transverse Shear Reinforcement Interaction in Shear Strengthened RC Beams, *Composite Structures* 201(June), 326–339, 2018b.

Wakjira, T., and Ebead, U., Hybrid NSE/EB Technique for Shear Strengthening of Reinforced Concrete Beams Using FRCM: Experimental Study, *Construction and Building Materials,* 164, 164–177, 2018c.

Wakjira, T., and Ebead, U., Internal Transverse Reinforcement Configuration Effect of EB/NSE-FRCM Shear Strengthening of RC Deep Beams, *Composites Part B: Engineering* 166, 758–772, 2019.

Younis, A., and Ebead, U., Bond Characteristics of Different FRCM Systems." *Construction and Building Materials* 175(June), 610–620, 2018.

Younis, A., Ebead, U., and Shrestha, K. C., Different FRCM Systems for Shear-Strengthening of Reinforced Concrete Beams, *Construction and Building Materials,* 153, 514–526, 2017.

Interdependence between Structural Engineering and Construction Management
Edited by Ozevin, D., Ataei, H., Modares, M., Gurgun, A., Yazdani, S., and Singh, A.

ISBN: 978-0-9960437-6-2

DETERIORATION OF CONCRETE SULFUR STORAGE TANK UNDER ACIDIC VAPOR ATTACK

HASSAN A. KHALIFAH[1], MUHAMMAD K. RAHMAN[2], OMAR M. HAMDAN[4], MOHAMMED ABU HAID[4], MOHAMMED IBRAHIM[2], and ALI H. GADHIB[3]

[1]*Dept of Consulting Services, Saudi Aramco, Dhahran, Saudi Arabia,*
[2]*Research Institute, King Fahd University of Petroleum and Minerals, Dhahran, Saudi Arabia,*
[3]*Dept of Civil Engineering, King Fahd University of Petroleum and Minerals, Dhahran, Saudi Arabia,*
[4]*UGP Inspection Unit, Saudi Aramco, Dhahran, Saudi Arabia*

Sulfur storage tanks are integral parts of refineries and gas plants. During the gas sweetening process, liquid sulfur is separated in the sulfur recovery units and is stored in underground reinforced concrete tanks in liquid phase by steam-generated temperatures of 315 °F. Roof of the sulfur pit is about 300 – 400 mm thick reinforced concrete structure with two layers of reinforcement. Inside the sulfur pit, the ceiling and walls are subjected to high temperatures and highly corrosive acidic vapors above liquid sulfur line which leads to severe deterioration of concrete and corrosion of reinforcement in the roof slab. This paper presents the NDT based investigations of the roof slab of the tank using the Ground Penetrating Radar (GPR) in a typical sulfur pit with 30 years in service. GPR survey is compared with ultrasonic pulse echo and half-cell potential measurements on the slab. Compressive strength and petrographic examination of selected concrete cores and SEM/EDS and XRD tests on powder samples, from inside the pit are presented.

Keywords: Corrosion, Delamination, Service life, Condition assessment, Ground penetrating radar (GPR), Ultrasonic, SEM/EDS, Petrography.

1 INTRODUCTION

Natural gas obtained from Saudi Arabian sour gas fields contains significant amount of sulfur in the form of hydrogen sulphide. As a part of the gas treatment process, it is sent to the sulfur recovery units (SRU) to recover Sulfur. Typical SRU consist of a thermal reaction furnace, waste heat boilers, and a series of catalytic reactors (converters) and condensers. The sulfur generated from the process is stored in below-grade reinforced concrete sulfur storage tank (sulfur pits). (GPSA 2004).

Sulfur pits reinforced concrete structures are subjected to aggressive environment involving exposure to sulfur, sulfuric acid, vapor and moisture and elevated operating temperature (150°C). The roof slab of the sulfur pit is the primary element, which suffers from major deterioration due to reinforcing steel corrosion. Corrosion of reinforcement and spalling of concrete on the underside of the slab have been observed, resulting in severe reduction in load carrying capacity (Kline 2004, Kline 2006, Dowling 1992, Schwabenlander *et al.* 2004).

The sulfur pit under assessment has been in service for the past 30 years. It is a reinforced concrete structure constructed below grade level, consisting of retaining wall on the periphery covered with a reinforced concrete roof slab. The overall length of the sulfur pit is about 206.5 feet (63 m) and the width of the sulfur pit is about 48 feet (14.7 m). The roof slab is a 15 inches flat plate slab supported on column (with capital) at the center and retaining walls at the edges. The clear height of the sulfur pit is 7.5 feet. Walls and columns are supported on a raft foundation about 18 inches thick. The main objective of the study was to assess the potential of Ground Penetrating Radar (GPR) and Ultrasonic Impact Echo (UIE) techniques for NDT of roof slab of the sulfur pit. In addition to these NDT laboratory testing conducted includes compressive strength, petrographic examination and microstructural characterization.

2 RESULTS AND FINDINGS

2.1 Visual Inspection and Delamination Survey

Visual inspection of the roof slab was carried out to record distresses such as delamination and spalling. All defects observed during the visual inspection were measured and transferred to the drawings, as illustrated in Figure 1. The visual survey also included photographic evidence of the distresses found in the structural components of the sulfur pit. The delamination survey on the top of the roof slab was conducted by hammer sounding.

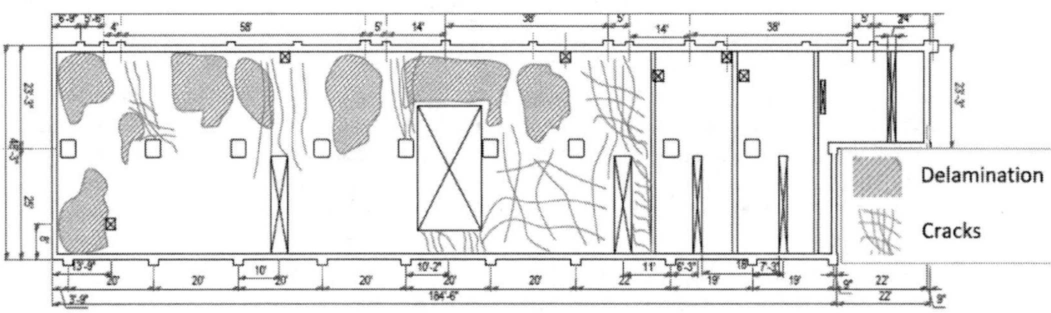

Figure 1. Cracking and delamination on the top side of the roof.

2.2 Ultrasonic Pulse Echo or Impact Echo

Ultrasonic pulse echo technique was used to detect flaws in the hardened concrete such as, delamination, cracks, honeycombing and voids. The ultrasonic pulse echo measurements were carried out on roof slab at a predetermined grid spacing of 2 feet in both horizontal and vertical directions. The technique is based on the propagation and reflection of shear waves using 50 kHz transducers. The average signal recorded by receiving transducers is presented as a 2D picture. The results of impact echo tests were plotted in a 2D format as shown in Figure 2. Irregularities in the concrete were identified based on the shear wave intensity as detailed in Table 1.

Table 1. Concrete quality based on shear wave speed.

Speed (m/s)	Concrete Quality
X<2000	Bad
2000>X>3000	Moderate
3000>X>4000	Very Good

The contour shown in Figure 2 indicates very weak reflection of the shear waves around the openings. On the contrary, as we move from the openings the shear wave reflection was very strong. This shows that the concrete around the opening is weak or delaminated. The ultrasonic pulse echo results are in good agreement with the hammer survey carried out to determine delamination of concrete.

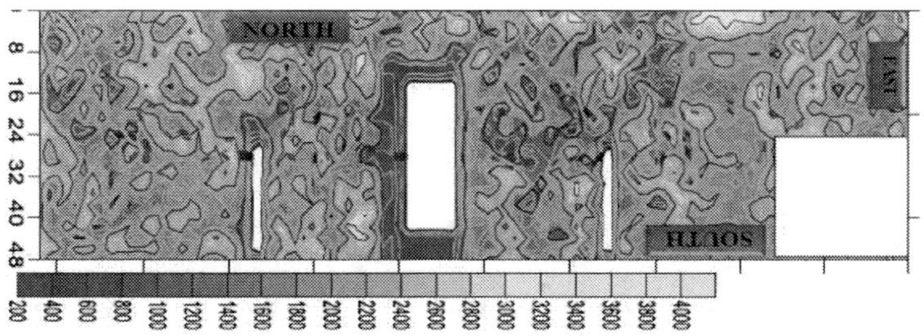

Figure 2. Ultrasonic pulse echo contours.

2.3 Corrosion Potential Measurements

Corrosion potential for detecting reinforcement corrosion and state of passivity of the reinforcing steel in concrete structures was based on ASTM C876 (2009). Potentials were measured at a grid spacing of 2 x 2 feet, in longitudinal and transverse directions on the top of the roof slab. A reference electrode was placed on the concrete surface, which is connected via a high impedance Voltmeter (> 109 Ohms) to the reinforcement cage. Interpretation of the half-cell potential values, vis-a-vis the corrosion status, is shown in Table 2. Corrosion potential mapping for the roof slab is shown in Figure 3. It can be seen that the corrosion potential at several locations slab particularly around openings are higher than threshold value indicating that corrosion of steel is in progress in a large portion of the slab with some areas indicating high corrosion.

Table 2. Corrosion potential and their interpretation according to ASTM C876 (2009).

Potential V (Vs $Cu/CuSO_4$)	Probability of Corrosion
<-0.35	> 95 %
-0.20 to -0.035	approx. 50 %
>-0.020	< 5 %

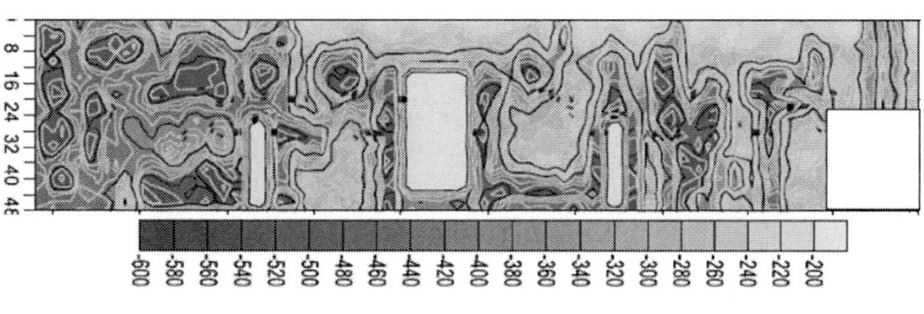

Figure 3. Corrosion potential mapping.

2.4 Ground Penetration Radar Survey

The GPR employed in this project is a high frequency 1600 MHz GPR (model GSSI), which can penetrate into the concrete to a depth of 0.5 meters. The overall thickness of the roof slab is 400 mm. GPR was also used to locate reinforcement in the concrete slab before extracting cores. All accessible rectangular areas were marked to conduct GPR survey. In each area GPR line scans were run two feet apart and analyzed to identify any anomalies in the substrate. These scans were taken using identical settings for the GPR equipment, data processing procedures, and scanning location to eliminate the impact of any parameter other than the concrete quality. Dielectric constant was assumed as 6.0, which corresponds to that of a dry concrete. Data was analyzed using the software RADAN by GSSI. The subsurface anomalies in the form of corrosion of steel reinforcement are shown in the Figure 4. It can be seen from this figure if the reinforcement is clean and not corroded there is a clear reflection of waves from the top reinforcement.

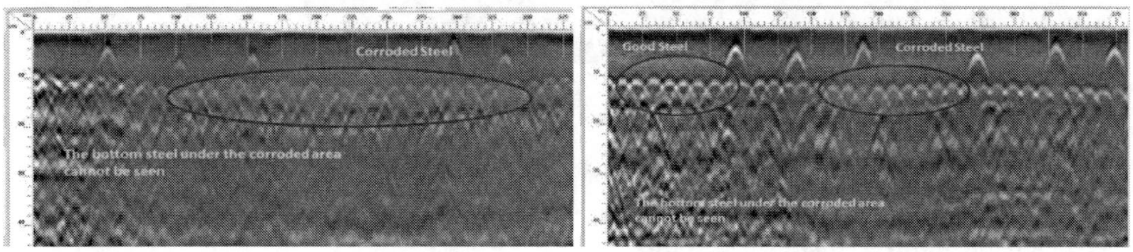

Figure 4. Reflection from top slab corroded reinforcing steel with GPR.

The estimated corrosion obtained from GPR analysis is shown in Figure 5. Severe to moderate corrosion of reinforcing steel is noted on the North side of the slab. Apart from this area, corrosion of steel is identified in a number of patches around the openings as well as away from them. The GPR results agree with the corrosion potential and impact echo tests.

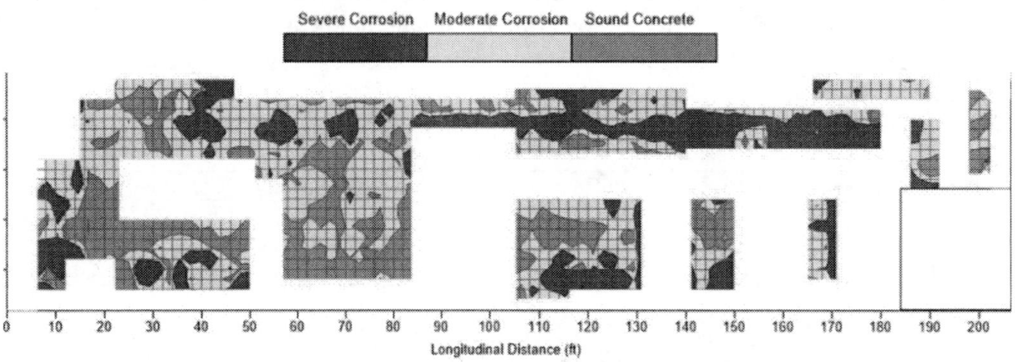

Figure 5. Estimated Corrosion based on GPR survey.

2.5 Compressive Strength of Concrete

Compressive strength of cored samples was determined according to ASTM C39 (2012) and ASTM C42 (2012). The compressive strengths of concrete cores obtained from the top slab at two locations were 25.7 MPa and 16.9 MPa. Sulfate resistant cement was used with a minimum compressive strength of 24 MPa required after 28 days of curing.

2.6 Petrographic Analysis of Concrete

Petrographic examination according to ASTM C856 (2011) "Standard Practice for Petrographic Examination of Hardened Concrete" was carried out on concrete specimens cored from the sulfur pit. Figure 6 shows a typical photomicrograph obtained from the concrete cores. The petrography of concrete core from the slab shows that the acidic sulfur vapors have affected the soffit of slab. The outer surface of the concrete core from the slab was worn and exhibited evidence of sulfur deposits. The concrete mix has nominal 19mm, crushed limestone/dolomite coarse aggregate and quartzitic sand fine aggregate, bound by a probable Portland-type cement matrix. The concrete is apparently well mixed, exhibiting good compaction. The excess voidage is about 0.5%. The petrography of concrete core from the slab shows that the acidic sulfur vapors have affected the soffit of slab. Top 10 mm seems to be heavily affected. Petrography shows weak, friable and cracked concrete in top 7 mm. There is a moderate level of sulfur penetration in the slab, up to a depth of 30 mm with voids and microcracks filled with sulfur. The maximum depth to which the sulfur has infiltrated into the slab is up to 60 mm. Alteration of the cement matrix and increased densification can be observed.

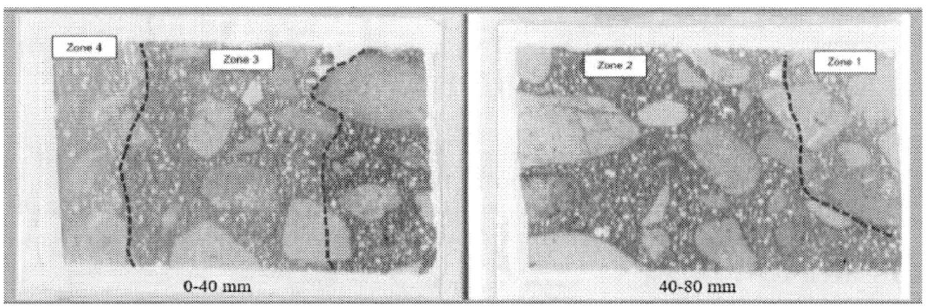

Figure 6. View of thin section showing zones of alteration within the concrete.

2.7 Microstructural Characterization

Scanning electron microscopy (SEM) and Energy dispersive x-ray spectroscopy (EDS) were used to determine the elemental composition of various phases in the specimens retrieved from underside of the roof slab at different depths. At each depth, elemental composition of different spots in the matrix was taken (Fig. 7). In the specimen extracted from 0 to 5 mm, 29.4% and 31.7% sulfur were noted at two different spots. Other elements such as Ca, Si, and Al were observed. Sulfur content was less at depth exceeding 5 mm from surface. High sulfur at the surface propagates the sulfate attack for the acids, resulting in formation of expansive material which causes cracking and delamination at the underside of roof slab.

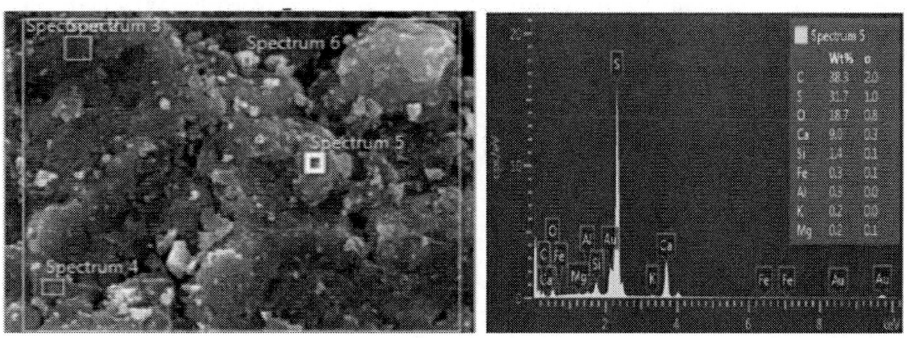

Figure 7. SEM and EDS of specimen extracted underside of roof from 0 to 5 mm depth.

2.8 X-Ray Diffraction (XRD)

Powder samples extracted from different structural components were analyzed. XRD for a powder sample collected from 0 to 5 mm from the surface (Figure 8) indicate presence of sulfur, anhydrite (calcium sulfate) and calcite peaks. Calcite, calcium silicate hydrate (CSH), anorthite, quartz and portlandite were also noted in the XRD patterns.

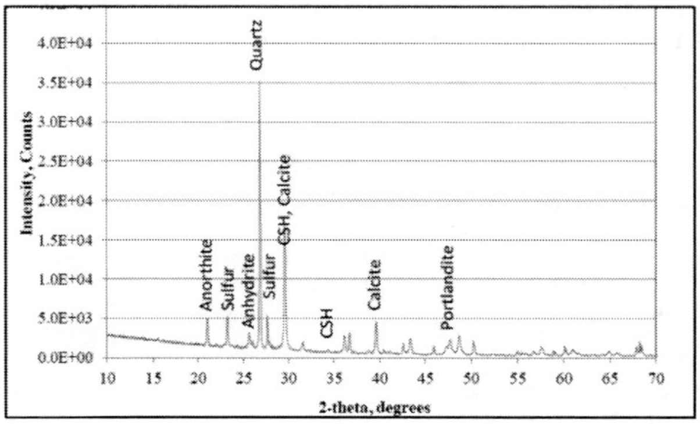

Figure 8. X-Ray diffraction of samples collected from underside of roof 0 to 5 mm from the surface.

3 CONCLUSIONS

GPR survey, impact echo and corrosion potential mapping suggested concrete delamination and to reinforcement corrosion on the top of the roof slab. GPR provided an excellent indication of corrosion severity at the top of the roof slab. However, the condition of the bottom reinforcement with the pit being operational cannot be diagnosed due to severe corrosion in top steel. Microstructural characterization and petrographic studies reveal the extent of deterioration.

References

ASTM C39, Standard Test Method for Compressive Strength of Cylindrical Concrete Specimens, ASTM International, West Conshohocken, PA, 2012.

ASTM C42, Standard Test Method for Obtaining and Testing Drilled Cores and Sawed Beams of Concrete, ASTM International, West Conshohocken, PA, 2012.

ASTM C856, Standard Practice for Petrographic Examination of Hardened Concrete, ASTM International, West Conshohocken, PA, 2011.

ASTM C876. Standard Test Method for Corrosion Potentials of Uncoated Reinforcing Steel in Concrete, ASTM International, West Conshohocken, PA, 2009.

Dowling, N. I., Corrosion of Materials Used in Storage and Handling of Solid Elemental Sulphur, Alberta Sulphur Research, Ltd. Sulphur and Energy, 103-115, 1992.

GPSA, Engineering Data Book, FPS Version Volume I & II, 12th ed., Gas Processing Suppliers Association, 2004.

Kline, T., Sulfur Pit Assessment and Repair Strategies, *Brimstone Sulfur Symposium*, Vail, Colorado, September 2004.

Kline, T., Repair of Subsurface Molten Sulfur Containment Structures, *Brimstone Sulfur Symposium*, Banff, Alberta, CA, 2006.

Schwabenlander, R., and Kline, T., Sulfur-Recovery Operations Pose Formidable Challenge to Concrete Infrastructure, *World Refining,* 12(4), 30-31, May 2002.

Interdependence between Structural Engineering and Construction Management
Edited by Ozevin, D., Ataei, H., Modares, M., Gurgun, A., Yazdani, S., and Singh, A.

ISBN: 978-0-9960437-6-2

THIN-WALLED STEEL TUBULAR COLUMNS WITH UNIFORM AND GRADED THICKNESS UNDER CYCLIC LOADING

QUSAY AL-KASEASEBH and IRAJ H.P MAMAGHANI

Dept of Civil Engineering, University of North Dakota, Grand Forks, USA

Thin-walled steel tubular circular columns are widely used as cantilever bridge piers due to their geometric efficiency, aesthetic appearance, and high earthquake resistance. However, local buckling, global buckling, or interaction between both is usually the main reason of significant strength and ductility loss in these columns, which eventually leads to their collapse. This paper investigates the behavior of uniform circular (C) and graded-thickness circular (GC) thin-walled steel tubular columns under constant axial and cyclic lateral loading. The GC column with size and volume of material equivalent to the C column is introduced and analyzed under constant axial and cyclic lateral loading. The analysis carried out using a finite-element model (FEM), which considers both material and geometric nonlinearities. The accuracy of the employed FEM is validated based on the experimental results available in the literature. The results revealed that, significant improvements in strength, ductility, and post-buckling behavior of thin-walled steel columns obtained using the GC column.

Keywords: Circular section, Buckling, Ductility, Strength.

1 INTRODUCTION

Civil engineering structures are exposed to increased earthquake risks in severe seismic regions. Their integrity is always challenged due to extreme uncertainties of severe earthquakes (Miller 1998, Nakashima *et al.* 1998). Thin-walled steel tubular circular columns are widely used in modern buildings, offshore platforms, elevated storage tanks, and transmission towers (Ucak and Tsopelas 2006). In addition, they can be used in wind turbines and as cantilever bridge piers in seismic regions due to their geometric efficiency, aesthetic appearance, and high earthquake resistance (Ucak and Tsopelas 2014). Thin-walled steel tubular columns may be a superior to their counterparts of reinforced concrete due to their stiffness to cross-sectional area ratio, light-weight, and ductility, especially when limited construction space is preferable. However, local buckling, global buckling, or interaction between both is usually the main reason for a significant loss in strength and ductility in these members, which eventually leads to their failure under cyclic lateral loading (Mamaghani *et al.* 1996). Thin-walled steel tubular columns are susceptible to damage when subjected to strong earthquakes (i.e. the Kobe earthquake (1995), the Sichuan earthquake (2008), and the East Japan earthquake (2011)) and severe local buckling was reported in many studies (Bruneau 1998). As a result, several analytical and experimental studies were carried out to identify the factors, which might improve the strength and ductility of these columns under axial and cyclic lateral loading. Numerically, factors that affect the strength and ductility of circular tubular columns have been investigated (Gao *et al.* 1998b). Finite element

(FE) studies concluded that ductility of tubular circular columns is sensitive to the radius-to-thickness ratio parameter (R_t). Moreover, decreasing R_t and the column slenderness ratio parameter (λ) resulted in improvements on both strength and ductility of the columns (Gao *et al.* 1998a).

In this paper, a tested thin-walled steel tubular circular column (C) is numerically studied under constant axial and quasi-static cyclic lateral loading. The FE results are compared to the experimental results from literature to confirm the validity of the FE analysis. Moreover, a graded-thickness thin-walled steel tubular circular column (GC) is used with equivalent size and volume of material to C column in attempt to improve the strength, ductility, and post-buckling behavior. The main reason for the improved behavior of GC columns is their ability to eliminate severe local buckling near the base of the column where the buckling most likely occurs.

2 FINITE-ELEMENT ANALYSIS

There is no doubt that full-scale testing results in a better insight into understanding the structures' behaviors, however, physical experimentation is expensive and time consuming. For this purpose, FE analysis on the cyclic behavior of thin-walled steel tubular circular columns is carried out using ABAQUS Ver 6.14 (Hibbit *et al.* 2014). The FEM takes into account both material and geometric nonlinearities. The accuracy of FEM is validated based on the experimental results available in the literature (Nishikawa *et al.* 1998). The key parameters considered in the practical design of thin-walled steel tubular columns are radius-to-thickness ratio parameter (R_t), and column slenderness ratio parameter (λ) (Mamaghani and Packer 2002). R_t affects the local buckling behavior of thin-walled steel tubular circular columns, while λ controls the global buckling (Mamaghani 2008, Mamaghani and Packer 2002). R_t and λ parameters are determined as in Eq. (1) and Eq. (2):

$$R_t = \frac{D}{2t}\frac{\sigma_y}{E}\sqrt{3(1-\nu^2)} \tag{1}$$

$$\lambda = \frac{2h}{r}\frac{1}{\pi}\sqrt{\frac{\sigma_y}{E}} \tag{2}$$

Local buckling effect usually occurs near the base of C columns when they are subjected to constant axial load and cyclic lateral loading (Nishikawa *et al.* 1998). For this purpose, as shown in Figure 1, the upper part of the column is modeled using two-node beam element (B31), whereas four-node shell element (S4R), which considers accurately the local buckling, is used for the lower part of the column. All the elements are available in the ABAQUS library (Hibbit *et al.* 2014). The interface between the S4R and B31 elements is modeled using multi-point constraint (MPC). For computational efficiency, the bottom half of the lower part (*D*) is divided into 26 shell elements, while the remaining height (*D*) is only divided into 14 shell elements. The upper part of the column is divided into 14 beam elements. The above stated mesh divisions are determined by trial-and-error. It is found that such mesh density gives accurate results. The initial geometrical imperfection and residual stresses are not considered in the analysis, as their effects were not measured in the experiment (Hibbit *et al.* 2014). For thin-walled steel tubular columns, cyclic lateral load is more dominant than the axial load, which implies that the effect of the initial imperfections is negligible (Goto *et al.* 1998).

2.1 Cyclic Loading Program

The displacement-controlled unidirectional cyclic loading is illustrated in Figure 1d and adopted as a lateral loading program. The quasi-static cyclic loading is applied to the top of the column with the presence of the constant axial load (*P*) throughout the loading history. The cyclic displacement amplitude is increased as a multiple of the yield displacement (δ_y), which is calculated by Eq. (3):

$$\delta_y = \frac{H_y h^3}{3EI} \tag{3}$$

Where $H_y = (\sigma_y - P/A)\, Z/h$ = lateral yield load and *A*, *h*, *EI*, and *Z* = cross-sectional area, the height, the bending stiffness, and the section modulus of the column, respectively (Goto *et al.* 2010). The yield displacements and lateral yield loads for the columns are listed in Table 1. All the analyzed columns are made of carbon steel SS400 (equivalent to ASTM A36 (2014)).

3 PROPOSED GC COLUMN

Conventional C columns experience premature buckling behavior in either local or global buckling form, near the base of the column, under constant axial and cyclic lateral loading. Under this buckling behavior, C columns are unable to fully utilize their strength and ductility capacities. To overcome these limitations, GC column is proposed as alternative for the counterpart C column. The column height and diameter are kept same for both C and GC columns. The GC column is divided into three segments of constant cross sections. The first and second segments have the same height that is equal to the diameter of the circular section from the base. The third segment has a height of (*h-2D*). As shown in Figure 2, a thicker cross section (t_1=1.25*t*) is used along the first segment, and the original thickness (t_2= *t*) is kept for the second segment. Finally, the remaining material volume is distributed on the third segment with (t_3=0.86*t*). The above configurations of GC sections are chosen based on which achieved better behavior. Table 1 shows material and geometrical properties of C and GC columns. As can be seen, same material and geometrical properties (except the plate thickness) are used for both C and GC columns.

Table 1. Geometric and material properties of the C and GC columns.

Properties	**C Column** (Goto *et al.* 1998)	**GC Column**
Steel material	SS400	SS400
h (mm)	3403	3403
D (mm)	900	900
$t_1/t_2/t_3$ *(mm)*	9	11.25/ 9/ 7.75
λ	0.26	0.26
R_t	0.115	0.115
H_y *(KN)*	414.9	414.9
δ_y*(mm)*	10.6	10.6
$P/\sigma_y A_s$	0.124	0.124
σ_y *(MPa)*	298.6	298.6
σ_u *(MPa)*	495	495

4 COMPARISON OF NUMERICAL AND EXPERIMENTAL RESULTS

In this section, the numerical cyclic behavior results of the C columns are compared with the experiment results that were obtained by the Public Works Research Institute (PWRI) of Japan (Nishikawa *et al.* 1998).

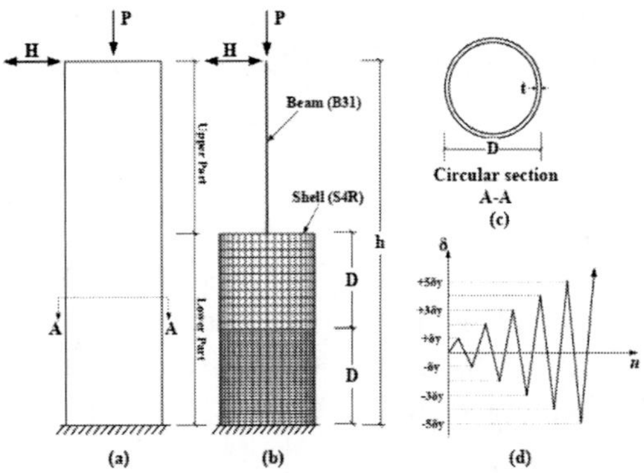

Figure 1. C Analytical model: (a) Bridge pier, (b) FE model, (c) Cross section, and (d) Loading program.

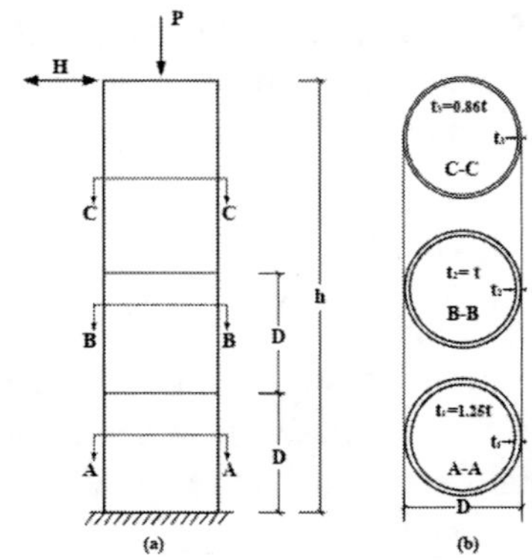

Figure 2. GC Analytical model: (a) Bridge pier, and (b) GC sections.

4.1 Hysteresis Curves of C and GC Columns

The analysis normalized lateral load vs. lateral displacement hysteresis curves, are shown in Figure 3. H_y and δ_y denote the lateral yield load and yield displacement, respectively. First, the FE analysis hysteresis loops are compared with experimental results available in the literature (Nishikawa *et al.* 1998). From Figure 3a, the comparison of hysteresis curves of C column, shows a reasonable agreement with the experimental results. This indicates that FE analysis, using kinematic hardening material behavior, gives a reasonable accuracy to describe the material behavior with regard of local buckling of thin-walled steel tubular columns. As shown in Figure 4, the deformed shape of C column (Figure 4b) at the end of the FE analysis is compared to the deformed shapes at the end of the experiment (Figure 4a) (Nishikawa *et al.* 1998). Based on this comparison, the deformed shape of C column is captured relatively well in the FE analysis. Using the same validated FEM, a comparison study has been performed between the behavior of the GC column and their counterpart C columns under the same axial and cyclic lateral loading. Figure 3b compares the hysteretic behavior of C and GC columns obtained from the FE analysis.

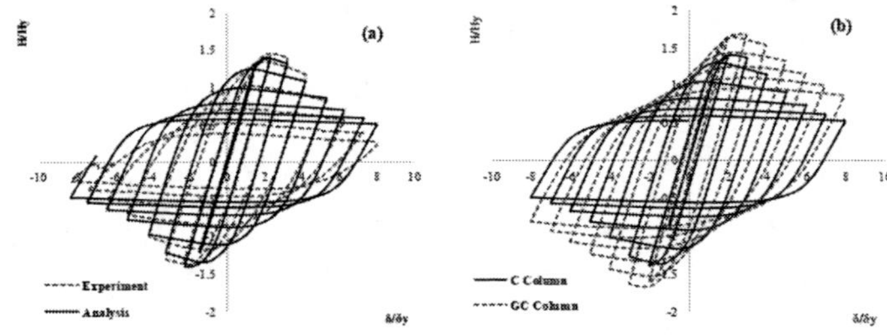

Figure 3. Horizontal load vs. horizontal displacement hysteretic curve.

In C column, the buckling initiates when the displacement is between $2\delta_y$ and $3\delta_y$. A strength drops of 17.6% of the ultimate column strength (observed at $\delta = +2.16\delta y$) occurs at $\delta = +4\delta y$. As the displacement increases, the column strength decreases at a rapid pattern to only 38% of its ultimate strength at $\delta = +8\delta y$. On the other hand, GC column shows a maximum load capacity at δ = +3δy. Only 8.4% strength drop of the ultimate strength took place at δ = +4δy, which gives an indication that local buckling started between 3δy and 4δy. As the displacement amplitude is increased, more strength deterioration is observed. The residual strength of the GC column is 52% of its ultimate strength at δ = +8δy. It is worth mentioning that strength significantly drops at δ = -4δy, and δ = -8δy for both columns. Generally, strength capacity improvement of 20% is obtained with the proposed GC column in comparison with the C column. Figure 4 shows the final deformed shapes of GC column compared to C column at δ = +8δy. For C column, buckling occurs near base as excepted, while buckling shifted upward from the base in case of GC column.

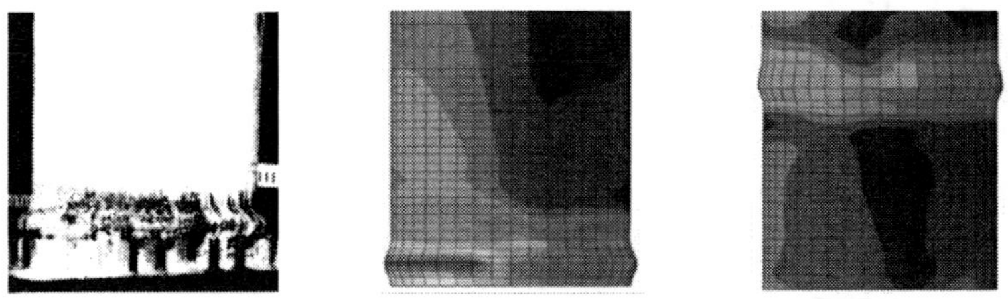

(a) Experiment (Nishikawa *et al.* 1998) (b) C Column (c) GC Column

Figure 4. Deformation of columns.

5 ENERGY ABSORPTION CAPACITY

Ductility is the ability of a structure to resist large deformations. The dissipated energy is a measure of the cyclic behavior of the structures. Therefore, the energy absorption capacity of the C and GC columns has been studied. A normalized energy absorption (*E*) is defined in Eq. (4) (Mamaghani *et al.* 1996):

$$E = \frac{2}{H_y \delta_y} \sum_{i=1}^{n} E_i \tag{4}$$

In Eq. (4), E_i = energy absorption in the *i-th* half-cycle, n = number of half-cycles (one half-cycle is defined from any zero-lateral load to the subsequent zero-lateral load). Using the Eq. (4), Figure 5a compares the normalized cumulative energy absorption vs. *n*, obtained from the experiment and analysis of C column. The normalized energy absorption curves vs. *n* obtained from the analysis fit very close to the experimental results. For GC column, as shown in Figure 3b the strength of GC column decreases in a controlled rate from cycle to cycle compared to the C columns, which is expected to dissipate larger energy than the C column under cyclic lateral loading. Instead and by determining the area under lateral load vs. lateral displacement curves, Figure 5b shows that the dissipated energy of the GC column is larger than that of C column, which is expected to experience higher ductility in the case of proposed GC column.

6 CONCLUSIONS

In this paper, FE analysis has been carried out to evaluate the cyclic behavior of the C column. In addition, graded-thickness column (GC) equivalent to the C column is introduced in an attempt to improve the strength, ductility, and post-buckling behavior. First, the accuracy of the employed FEM was validated based on the experimental results available in the literature for the C column. The relatively good agreement between the FE analysis and experiment confirms the ability of the FEM to capture the structural behavior with regard of the local buckling of thin-walled steel tubular circular columns. GC column shows a superior strength and ductility performance in comparison with the C column. An improvement of 20% in the ultimate strength was achieved using the GC column. Furthermore, improvements in the residual strength were observed in the case of GC column at the end of the analysis. T he dissipated energy of the GC columns was higher which exhibited higher ductility. Buckling behavior of the C column was captured relatively well by the employed FEM. GC column delays the local buckling occurrence under cyclic lateral loading. Moreover, the buckled shape of the C column occurred near to the base as expected, while buckling shifted upward from the base in the GC column.

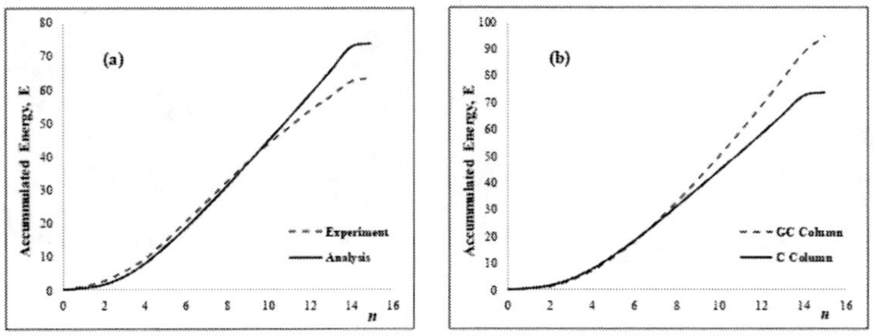

Figure 5. Energy absorption capacity.

References

ASTM A36 / A36M - 14 Standard Specification for Carbon Structural Steel, *ASTM International, West Conshohocken, PA*, 12–14, 2014.

Bruneau, M., Performance of Steel Bridges During the 1995 Hyogoken–Nanbu (Kobe, Japan) Earthquake—a North American Perspective, *Engineering Structures*, *20*(12), 1063–1078, 1998.

Gao, S., Usami, T., and Ge, H., Ductility Evaluation of Steel Bridge Piers with Pipe Sections, *Journal of Engineering Mechanics*, *124*(3), 260, 1998a.

Gao, S., Usami, T., and Ge, H., Ductility of Steel Short Cylinders in Compression and Bending. *Journal of Engineering Mechanics*, *124*(2), 176–183, (1998b).

Goto, Y., Kumar, G., and Kawanishi, N., Nonlinear Finite-Element Analysis for Hysteretic Behavior of Thin-Walled Circular Steel Columns with In-Filled Concrete, *Journal of Structural Engineering*, *136*(11), 1413–1422, 2010.

Goto, Wang, Q., and Obata, M., FEM Analysis for Hysteretic Behavior of Thin-Walled Columns, *Journal of Structural Engineering*, *124*(11), 1290–1301, 1998.

Hibbit, D., Karlsson, B., and Sorensen, P., *Abaqus 2014 Documentation,* Dassault, 2014.

Mamaghani, I. H. P., Seismic Design and Ductility Evaluation of Thin-Walled Steel Bridge Piers of Box Sections, *Transportation Research Record: Journal of the Transportation Research Board*, *2050*(1), 137–142, 2008.

Mamaghani, I. H. P., and Packer, J. A., *Inelastic Behaviour of Partially Concrete-Filled Steel Hollow Sections,* 4th Structural Specialty Conference, 1–10, 2002.

Mamaghani, I., Usami, T., and Mizuno, E., Cyclic Elastoplastic Large Displacement Behaviour of Steel Compression Members, *Journal of Structural Engineering*, *42*, 135–145, 1996.

Miller, D. K., Lessons Learned from the Northridge Earthquake, *Engineering Structures*, *20*(4–6), 249–260, 1998.

Nakashima, M., Inoue, K., and Tada, M., Classification of Damage to Steel Buildings Observed in the 1995 Hyogoken-Nanbu Earthquake, *Engineering Structures*, *20*(6), 271–281, 1998.

Nishikawa, K., Yamamoto, S., Natori, T., Terao, K., Yasunami, H., and Terada, M., Retrofitting for Seismic Upgrading of Steel Bridge Columns, *Engineering Structures*, *20*(4–6), 540–551, 1998.

Ucak, A., and Tsopelas, P., Cellular and Corrugated Cross-Sectioned Thin-Walled Steel Bridge-Piers/Columns, *Structural Engineering and Mechanics*, *24*(3), 355–374, 2006.

Ucak, A., and Tsopelas, P., Load Path Effects in Circular Steel Columns under Bidirectional Lateral Cyclic Loading, *Journal of Structural Engineering*, *141*(2009), 1–11, 2014.

Interdependence between Structural Engineering and Construction Management
Edited by Ozevin, D., Ataei, H., Modares, M., Gurgun, A., Yazdani, S., and Singh, A.

ISBN: 978-0-9960437-6-2

SEISMIC DESIGN EVALUATION OF LIQUID-FILLED STIFFENED STEEL CYLINDRICAL TANKS

WIRIYACHAI ROOPKUMDEE and IRAJ MAMAGHANI

Dept of Civil Engineering, University of North Dakota, Grand Forks, USA

This paper aims to estimate dynamic buckling loads of cylindrical liquid storage tanks. Finite element analysis is performed using ANSYS computer program. Twelve different geometries of the cylindrical tanks are analyzed with height to diameter (*H/D*) ratios of 0.5, 1.0, 1.5, and 2.0 and the diameter to thickness (*D/t*) ratios of 1000, 1500, and 2000 to cover tall and short cylindrical tanks. The dynamic buckling capacities of cylindrical tanks filled with water up to 90% of their height are investigated in this study. The transient dynamic analysis is performed to find the dynamic buckling loads. Applied dynamic loads in this study are horizontal earthquake excitations in terms of acceleration (g) due to gravity. The transient dynamic analysis results indicate that the dynamic buckling loads decrease when the H/D ratios increase, and the dynamic buckling loads decrease when the D/t ratios increase. Design curves for the cylindrical tanks of various geometries subjected to earthquake accelerations are generated.

Keywords: Finite element, Buckling, Earthquake, Transient analysis, Stability, Shell.

1 INTRODUCTION

Liquid storage tanks are subjected to the horizontal and vertical ground accelerations during the earthquakes. These earthquake ground accelerations may cause damages to the liquid storage tanks. Spillage of toxic liquid from the liquid storage tanks could cause a serious threat to human health and the environment. Additionally, failure of tanks, which contain an inflammable substance such as petroleum has frequently led to an uncontrolled fire which occurred during Niigata and Alaska earthquakes in 1964 (Priestley *et al.* 1986).

The damages of petroleum storage tanks were reported due to earthquakes of 1933 Long Beach, 1952 Kern County, 1964 Alaska, 1971 San Fernando, 1979 Imperial Valley, 1983 Coalinga, 1989 Loma Prieta, 1992 Landers, 1994 Northridge, and 1995 Kobe (Cooper and Wachholz 1999). Damages of cylindrical tanks due to earthquake loading can occur in several forms. The American Lifelines Alliance (ALA) reported the failure modes that have occurred to steel storage tanks (Edinger *et al.* 2001). These failure modes are shell buckling mode, roof, and miscellaneous steel damage, anchorage failure, tank support system failure, foundation failure, hydrodynamic pressure failure, connecting pipe failure, and manhole failure. This study is interested in the shell buckling mode subjected to the horizontal earthquake accelerations. Housner (1963) reported that the hydrodynamic behavior between water and the storage tanks, which are subjected to horizontal accelerations can be distinguished into two kinds. First, impulsive mass, a mass of water is rigidly attached to the tank at the proper height. Second, convective mass, the horizontal accelerations from the tank, excites a mass of water into oscillations. Malhotra *et al.* (2000) presented that the design of fixed-based cylindrical tank is a simple and efficient way to design for the convective component, compulsive component, and shell deformation due to interaction effects from the impulsive component. Sezen *et al.* (2008) used ANSYS computer program to study liquefied gas-structure interaction and a simplified model of three tanks in Turkey that experienced an earthquake on August 17, 1999, and they reported that shear and bending moments are overestimated if the fluid is modeled as a single rigid mass.

2 FINITE ELEMENT MODELS

Twelve different geometries of the tanks are analyzed with height to diameter (H/D) ratios of 0.5, 1.0, 1.5, and 2.0 and the diameter to thickness (D/t) ratios of 1000, 1500, and 2000 to investigate the buckling behavior of various sizes of the cylindrical tanks. These twelve cylindrical tanks are modeled as aboveground storage tanks (AST) that are open at the top. The cylindrical tanks are considered fixed at the bottom and free on the top. Summary of twelve different geometries of the cylindrical tanks adopted in this study is listed in Table 1.

Table 1. Summary of twelve different geometries of the cylindrical tanks.

Model	H (m)	D (m)	t (mm)	H/D	D/t
1	4.573	9.144	9.144	0.5	1,000
2	4.573	9.144	6.096	0.5	1,500
3	4.573	9.144	4.572	0.5	2,000
4	9.144	9.144	9.144	1.0	1,000
5	9.144	9.144	6.096	1.0	1,500
6	9.144	9.144	4.572	1.0	2,000
7	13.716	9.144	9.144	1.5	1,000
8	13.716	9.144	6.096	1.5	1,500
9	13.716	9.144	4.572	1.5	2,000
10	18.288	9.144	9.144	2.0	1,000
11	18.288	9.144	6.096	2.0	1,500
12	18.288	9.144	4.572	2.0	2,000

The material for all cylindrical storage tanks is steel with a modulus of elasticity, E = 200,000 MPa, Poisson's ratio, ν = 0.3, and the mass density, ρ = 7,850 kg/m^3. Bilinear isotropic hardening of the steel is included with the yield stress of 344.74 MPa and the tangent modulus of 13,790 MPa. The liquid filled inside the cylindrical tanks is water with the bulk modulus of 2,068.4 MPa, and the mass density of 1,000 kg/m^3.

The finite element analysis (FEA) computer program, ANSYS, is used to carry all computations. Due to the symmetric geometries, all FEA models are modeled with only half of cylinder to reduce the computation time. From the element types in ANSYS, SHELL181 element is used to be the element for the steel cylindrical tanks. SHELL181 is a four-node element with six degrees of freedom at each node (translation in x, y, and z directions, and rotation about x, y, and z-axes). FLUID80 element is used to be the element for water filled inside the cylindrical tanks. FLUID80 is an eight-node element with three degrees of freedom at each node (translation in x, y, and z directions).

3 TRANSIENT DYNAMIC ANALYSIS

The transient dynamic analysis is the technique for the response of a structure subjected to a time-dependent loading. In addition, inertia and damping effects are considered for the transient dynamic analysis. The equation of motion, Eq. (1), is solved by the transient structure simulation in ANSYS program (ANSYS Inc 2011).

$$[M]\{\ddot{u}\} + [C]\{\dot{u}\} + [K]\{u\} = \{F_{(t)}\} \tag{1}$$

where: $[M]$ = mass matrix, $[C]$ = damping matrix, $[K]$ = stiffness matrix, $\{\ddot{u}\}$ = nodal acceleration vector, $\{\dot{u}\}$ = nodal velocity vector, $\{u\}$ = nodal displacement, $\{F_{(t)}\}$ = load vector, t = time

3.1 Earthquake Accelerations

Earthquake loads of El Centro of May 18, 1940, are used in this study. Numerical values of these earthquakes are in the unit of g, the accelerations due to gravity are presented in Figure 1. Since the PGA is 0.319g at time = 2.02 s, the accelerogram from 0 s to 8 s is used in this study to reduce CPU time-consuming.

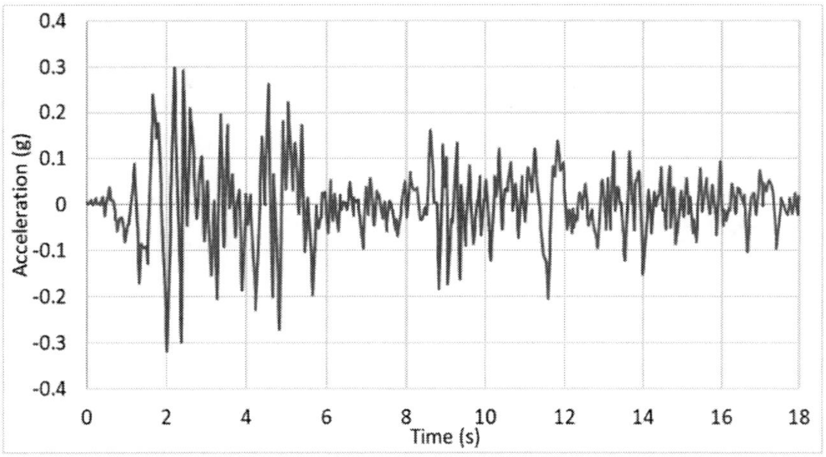

Figure 1. Accelerogram of North-South component of El Centro earthquake, 1940.

4 PSEUDO-EQUILIBRIUM PATHS

Budiansky and Roth's criterion are used to generate pseudo-equilibrium paths in this study. Buckling instability occurs when a small increase in the pulse intensity causes a strong increase rate of the deflection (Budiansky and Roth 1962). The maximum displacements are plotted in as a function of PGA level. These curves have been identified as pseudo-equilibrium paths. The pseudo-equilibrium path is used to indicate the dynamic buckling along the path of transient displacements. Figure 2, 3, 4, and 5 illustrate the deformed shape of model 5, 7, 8, and 10, respectively. Figure 6 and 7 represent pseudo-equilibrium paths for model 5 and 7, respectively.

For von-Mises stresses at the dynamic buckling loads, it is found that von-Mises stresses for all models are less than the yield stress of the steel, which is 344.74 MPa. The highest von-Mises stress occurs with Model 12 at 255.80 MPa.

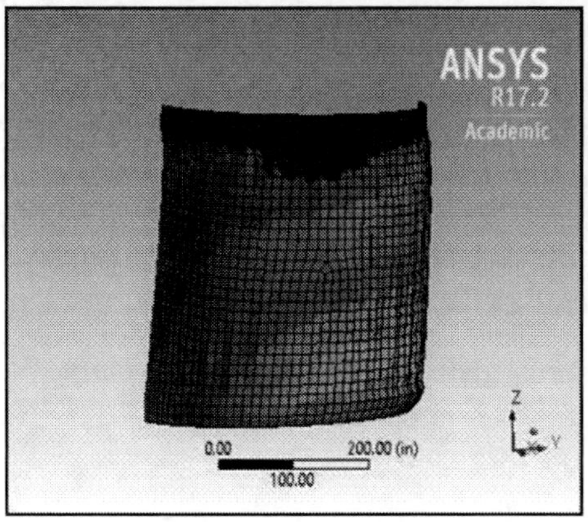

Figure 2. Deformation of model 5.

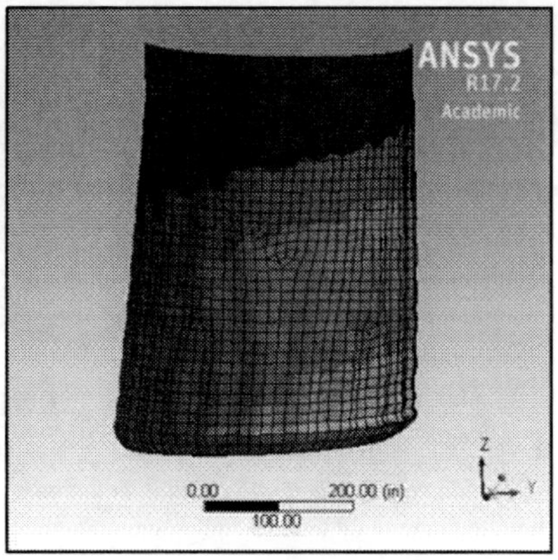

Figure 3. Deformation of model 7.

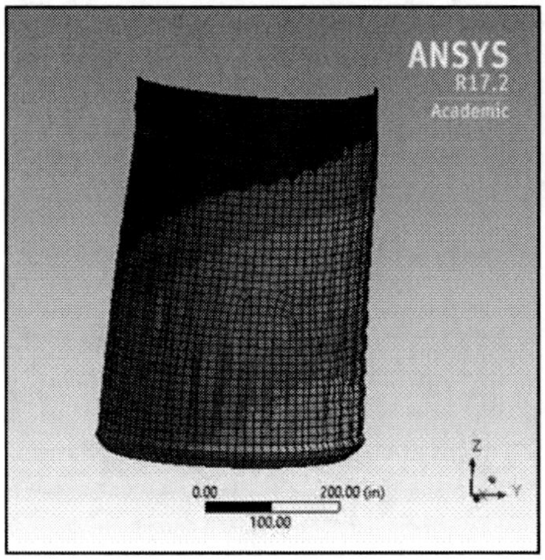

Figure 4. Deformation of model 8.

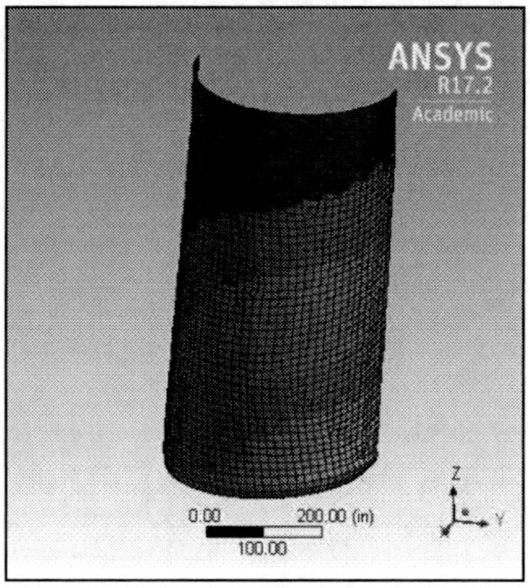

Figure 5. Deformation of model 10.

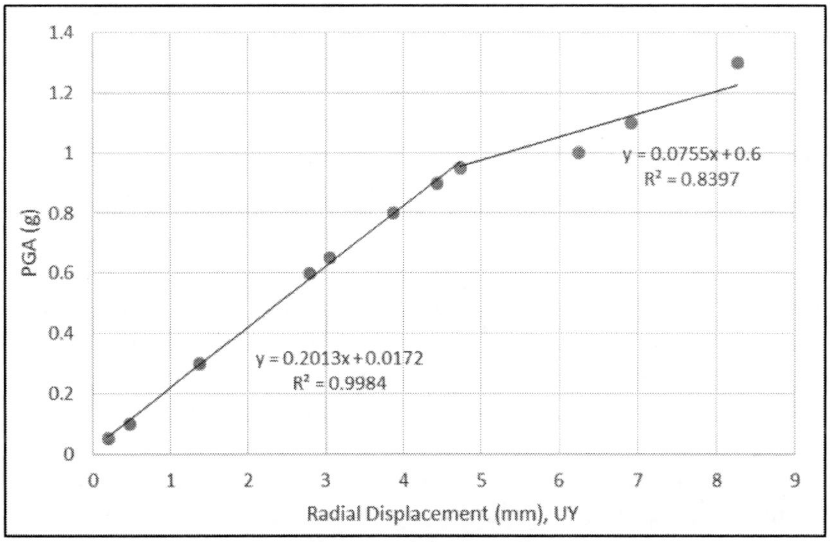

Figure 6. Pseudo-equilibrium paths for model 5.

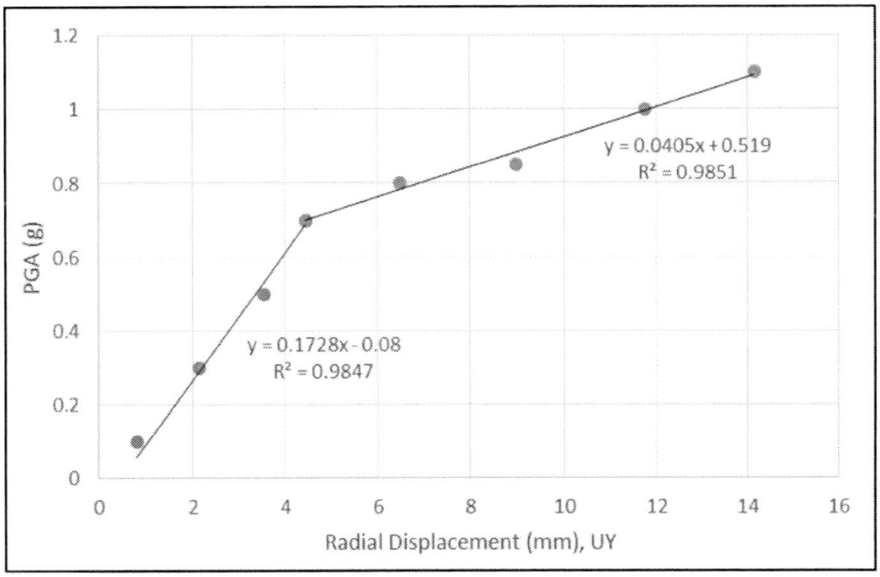

Figure 7. Pseudo-equilibrium paths for model 7.

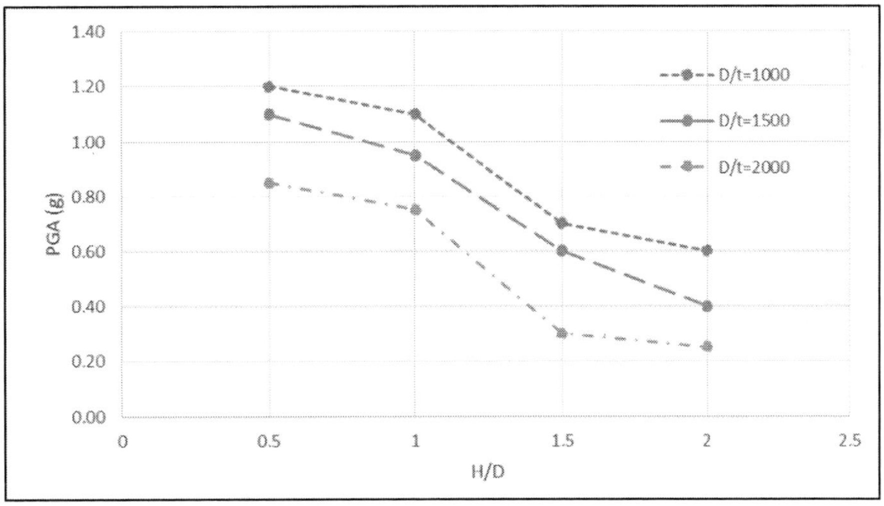

Figure 8. Dynamic buckling capacities.

After different analyses of the structure for several load, levels have been performed, and the value for which there is a significant jump in the response for a small increase in the load indicates that the structure passes from a stable state to a critical state as shown in Figure 6 and 7. In terms of shell buckling for the liquid-filled cylindrical tanks, this study found that the interaction effects between H/D and D/t ratios are significant when the cylindrical tanks are subjected to horizontal accelerations. The dynamic buckling capacities of the cylindrical tanks decrease when H/D ratios increase, and the dynamic buckling capacities decrease when D/t ratios increase. The dynamic buckling capacities of liquid-filled cylindrical tanks can be summarized as illustrated in Figure 8.

5 CONCLUSIONS

This paper dealt with dynamic buckling of steel tanks with fixed support subjected to the horizontal component of the real earthquake. The interaction effects of D/t and H/D ratios on the dynamic buckling were investigated. The buckling loads of the cylindrical tanks filled with water up to 90% of their height subjected to the El Centro earthquake range between 0.25g and 1.20g. The results from the transient analysis show that, for the cylindrical tanks filled with water up to 90% of their height, the dynamic buckling loads decrease when the H/D ratios increase, and the dynamic buckling loads decrease when the D/t ratios increase.

References

ANSYS, Inc., *ANSYS Mechanical APDL Element Reference,* 2011.

Budiansky, B., and Roth, S., Axisymmetric Dynamic Buckling of Clamped Shallow Spherical Shells, *NASA Collected Papers on the Stability f Shell Structures*, TN-1510, 597–606, 1962.

Cooper, T. W., and Wachholz, T. P., Optimizing Post-Earthquake Lifeline System Reliability, in *Proceedings of the 5th US Conference on Lifeline Earthquake Engineering,* 16, 878-86, 1999.

Edinger, J. M., Avila, E. A., Ballantyne, D., Cheng, L., Kiureghian, A. Maison, B. F., O'Rourke, T. D., and Power, M., Seismic Fragility Formulations for Water Systems: Part 1-Guideline, American Lifelines Alliance, 2001. Retrieved from https://www.americanlifelinesalliance.com/pdf/Part_1_Guideline.pdf on January 2019.

Housner, G. W., The Dynamic Behavior of Water Tanks, *Bulletin of The Seismological Society of America*, 53(2), 381-387, 1963.

Malhotra, P. K., Wenk, T., and Wieland, M., Simple Procedure for Seismic Analysis of Liquid-Storage Tanks, *Structural Engineering International*, 10(3), 197-201, 2000.

Priestley, M. J. N., Davidson, B. J., Honey, G. D., Hopkins, D. C., Martin, R. J., Ramsey, G., and Wood, J. H., Seismic Design of Storage Tanks, *Recommendations of a Study Group of the New Zealand National Society for Earthquake Engineering*, 3, 1986.

Sezen, H., Livaoglu, R., and Dogangun, A., Dynamic Analysis and Seismic Performance Evaluation of Above-Ground Liquid-Containing Tanks, *Engineering Structures*, 30(3), 794-803, 2008.

Interdependence between Structural Engineering and Construction Management
Edited by Ozevin, D., Ataei, H., Modares, M., Gurgun, A., Yazdani, S., and Singh, A.

ISBN: 978-0-9960437-6-2

STRESS INTENSITY FACTOR ANALYSIS OF AN EDGE CRACK PERPENDICULAR TO THE INTERFACE IN BI-MATERIALS

YONGKANG HOU, SHUO ZHANG, SHUJIN DUAN, and RUIMEI AN

Research Institute of Structural Engineering, Shijiazhuang Tiedao University, Shijiazhuang, China

The purpose of this paper is to investigate the stress intensity factor in the problem of a crack perpendicular to the interface in bi-materials. Based on the theory of an edge crack perpendicular to the interface between two dissimilar isotropic half-planes, the stress intensity factor was yield out from the stress singularity eigen-equation. The stress intensity factor for a composite beam segment with an edge crack under bending and tension was computed by the ordinary finite element and the singular finite element, and it was found that the singular element method is more accurate and applicable than the ordinary element method. The influence of the distance from the crack tip to the interface and the material's mechanical properties to the stress intensity factor were analyzed. The results show that the stress intensity factor increases first, and then decreases with decreasing of the distance from the crack tip to the interface, and the stress intensity factor increases with increasing of the elastic modulus of the cracked material.

Keywords: Fracture mechanics, Ordinary element method, Singular element method.

1 INTRODUCTION

Composite materials have been increasingly used in practical engineering applications, the fracture and failure of composite materials have attracted much attention of investigators (Xu 2006), but the most researches are devoted to the crack that along the interface or the tip at the interface (Williams 1959, Duan *et al.* 1989, Xu 2006). The problem of crack perpendicular to the interface was researched by Zak (1963) firstly, he used the eigenfunction expansion method to analyze the stress singularity ahead of a crack tip, which is perpendicular to and terminating at the interface. Cook and Erdogan (1972) used the Mellin transform method to derive the governing equation of a finite crack perpendicular to the interface and obtained the stress intensity factors. Qin and Noda (2002) used supersingular integral equation to analyze the problem of 3D crack perpendicular to the interface and obtained the numerical solution of stress intensity factor. Bouhala *et al.* (2013) studied the crack with the tip at the interface by the finite element method and computed the stress intensity factor by the physical method.

All the above studies are almost about the calculation method of stress intensity factor of crack perpendicular to the interface, the influence factor to the stress intensity factor needs to be studied further. In this paper, the stress intensity factor is yield out by the by the ordinary finite element and the singular finite element, and the influence of the distance from the crack tip to the interface and the material's mechanical properties to the stress intensity factor are analyzed.

2 THE STRESS INTENSITY FACTOR OF CRACK PERPENDICULAR TO THE INTERFACE IN BIMATERIALS

The model of crack perpendicular to the interface in bi-materials is established, as shown in Figure 1 (here, E is elastic modulus, v is Poisson's ratio).

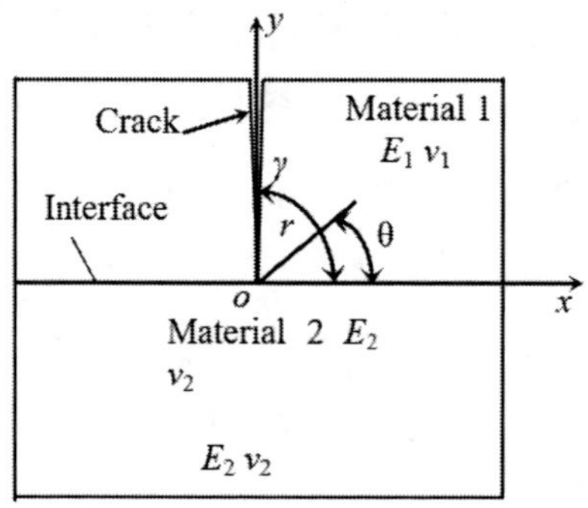

Figure 1. Schematic of crack perpendicular to the interface in bi-materials.

Following Zak and Williams (1963), we assume a stress function $\Phi(r, \theta)$, which satisfies the biharmonic equation, Eq. (1):

$$\nabla^2\nabla^2\Phi(r,\theta)=0 \tag{1}$$

The function of materials 1 and materials 2 take the form as shown in Eq. (2) and Eq. (3):

$$\Phi_1(r,\theta)=r^{\lambda+1}\left[A_1\cos(\lambda+1)\theta\text{-}B_1\sin(\lambda+1)\theta+C_1\cos(\lambda-1)\theta\text{-}D_1\sin(\lambda-1)\theta\right] \tag{2}$$

$$\Phi_2(r,\theta)=r^{\lambda+1}\left[A_2\cos(\lambda+1)\theta-B_2\sin(\lambda+1)\theta+C_2\cos(\lambda+1)\theta-D_2\sin(\lambda-1)\theta\right] \tag{3}$$

The stress and displacement components can be obtained from Eq. (2) and Eq. (3). By satisfying the continuity conditions at the interface and the stress-free boundary conditions at the crack surfaces, the obtained homogeneous equations about λ should be zero to determining λ:

$$\begin{aligned}&\left[1-\beta+2\alpha\left(1+\alpha-\alpha\lambda^2+\beta\left(\lambda^2-1\right)\right)\right]\sin\pi\lambda-\alpha(\beta-\alpha)\lambda_2\cos(\pi\lambda-2\gamma)\\&-\left(\alpha+\alpha^2-\alpha\beta\right)\sin(\lambda\pi-2\lambda\gamma)-\beta\lambda\sin2\gamma-(1+\alpha)(\alpha-\beta)\sin(2\lambda\gamma+\lambda\pi)\\&-\left(\alpha\beta\lambda^2-\alpha2\lambda^2\right)\sin(\pi\lambda+2\gamma)=0\end{aligned} \tag{4}$$

When the crack perpendicular to the interface, $\gamma=\pi/2$, the Eq. (4) can be written as:

$$\sin\lambda\pi\left[\lambda^2\left(-4\alpha^2+4\alpha\beta\right)+2\alpha^2-2\alpha\beta+2\alpha-\beta+1+\left(2\alpha^2-2\alpha\beta+2\alpha-2\beta\right)\cos\lambda\pi\right]=0 \tag{5}$$

Where $\alpha=(\mu_1/\mu_2\text{-}1)/(1+\kappa_1)$, $\beta=(\mu_1/\mu_2)(1+\kappa_2)/(1+\kappa_1)$, μ is shear modulus, $\kappa=3\text{-}4v$ for plane strain and $\kappa=(3\text{-}v)/(1+v)$ for plane stress, v is Poisson's ratio.

Refer to the definition of stress intensity factor in material stress singularity, using the stress of crack perpendicular to the interface in bi-materials, the K_I can be written as:

$$K_{\mathrm{I}} = \lim_{r \to 0} \sqrt{2\pi} r^{1-\lambda} \sigma_\theta (r,\theta) \quad (\theta = -\frac{\pi}{2}) \tag{6}$$

Substituting the expression of stress into Eq. (6), the expression of stress intensity factor is as follow in Eq. (7):

$$K_{\mathrm{I}} = \lim_{r \to 0} \sqrt{2\pi} r^{\lambda-1} \lambda(\lambda+1) F_\lambda \quad (\theta = -\frac{\pi}{2}) \tag{7}$$

3 THE SIMULATION MODEL

A partial beam segment of 1 m long with a central edge-crack is cut from a beam structure under tension and negative bending for analysis, and the half plane is selected for modeling, as shown in Figure 2.

Where the width of material 1 and material 2 are 0.3 m and 0.7 m respectively, and the length is 0.5 m. E_1, E_2 are elastic modulus, n_e is the elastic modulus ratio ($n_e = E_1/E_2$ and $n_e \leqslant 1$), the material 1 is weaker; v_1, v_2 are Poisson's ratio, η is the ratio of Poisson's ratio ($\eta = v_1/v_2$ and $\eta \leqslant 1$). The crack lies in material 1 and expands to the interface. N is the distance from the crack tip to the interface, and H is the distance from the crack tip to the top surface ($H + N = 0.3$ m).

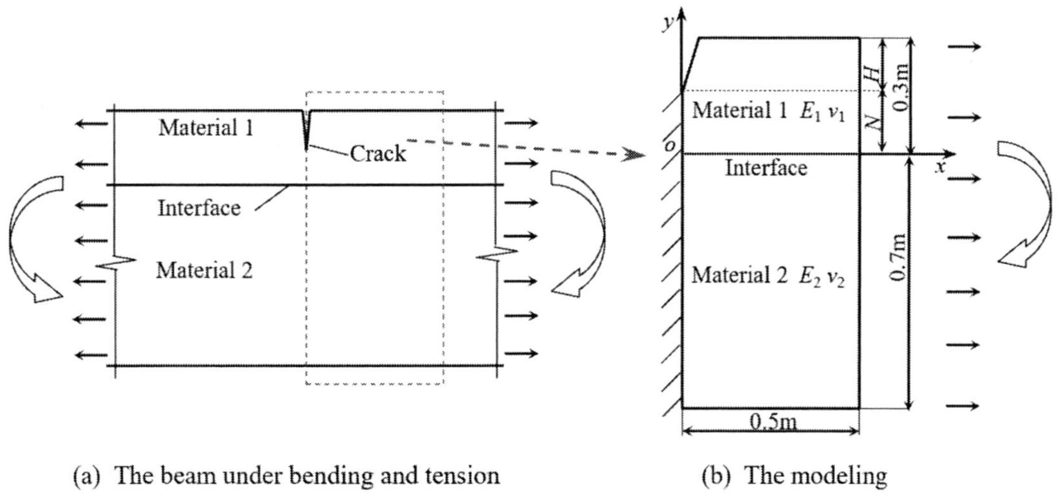

Figure 2. The problem under consideration.

The above problem of a crack perpendicular to the interface in bi-materials under bending or tension is examined respectively, and the stress intensity factors are obtained by using the ordinary finite element and the singular finite element.

3.1 Stress Intensity Factor Analysis under Tension

In the present study, the model under tension is analyzed, H is taken as 0.10 m, 0.20 m, 0.25 m, 0.27 m and 0.3 m respectively, and the $K_{\mathrm{I}}/\sigma_{\mathrm{I}}$ is calculated, one can obtain the tendency of $K_{\mathrm{I}}/\sigma_{\mathrm{I}}$ to change with H, as shown in Figure 3.

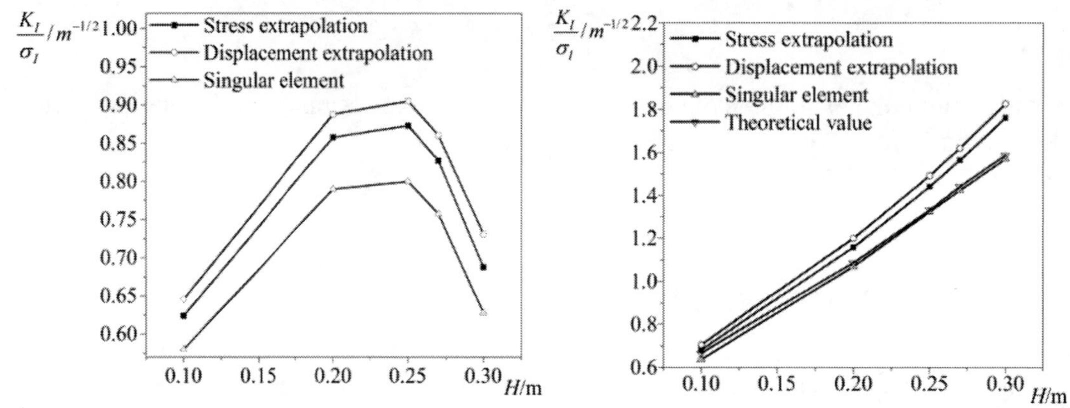

(a) $n_e = 1.0$ $\eta = 1.0$ (b) $n_e = 0.2$ $\eta = 1.0$

Figure 3. The stress intensity factor versus the *M* under tension.

From Figure 3, it is shown that the error of the calculation result between the finite element method and the theory method is smaller, in which the singular element method is more accurate than the ordinary element method; when n_e =1.0, or n_e =0.2, but the crack tip is far away from interface, the model can be seen as homogenous material, the stress intensity factor increases with increasing of the *H*; when n_e = 0.2, and the crack lies in weaker material, if the crack tip is closer to the interface, due to the influence of the stiffer material, the stress intensity factor decrease with increasing of the *H*.

Continue to analyze the model, and the elastic modulus ratio is taken as 0.1, 0.2, 0.5, and 1.0 respectively, calculated the value of K_I/σ_I, to obtain the tendency of K_I/σ_I to change with n_e, as shown in Figure 4.

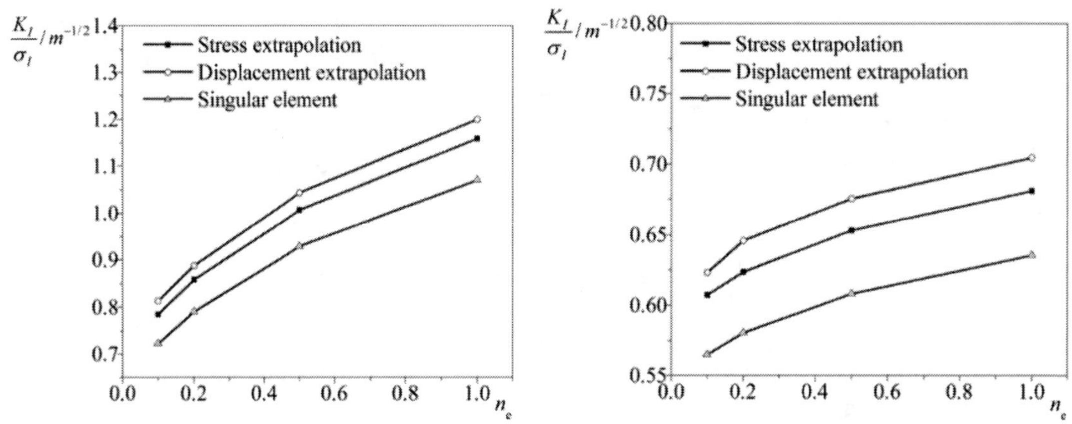

(a) $N = 0.1$ $\eta = 1.0$ (b) $N = 0.2$ $\eta = 1.0$

Figure 4. The stress intensity factor versus the n_e under tension.

From Figure 4, it is shown that the stress intensity factor increases with increasing of the elastic modulus of the cracked material, and the stress intensity factor decrease with the increasing of the difference of the elastic modulus between constituent materials.

3.2 Stress Intensity Factor Analysis under Bending

The stress intensity factor of crack perpendicular to the interface in bi-materials under bending, is analyzed, the parameters are the same as the tension.

Figure 5 shows the tendency of K_I/σ_I to change with H. From Figure 5, it is shown that the error of the calculation result between the finite element method and the theory method are smaller, in which the singular element method is more accurate than the ordinary element method. The tendency of K_I/σ_I to change with H is the same as the tension.

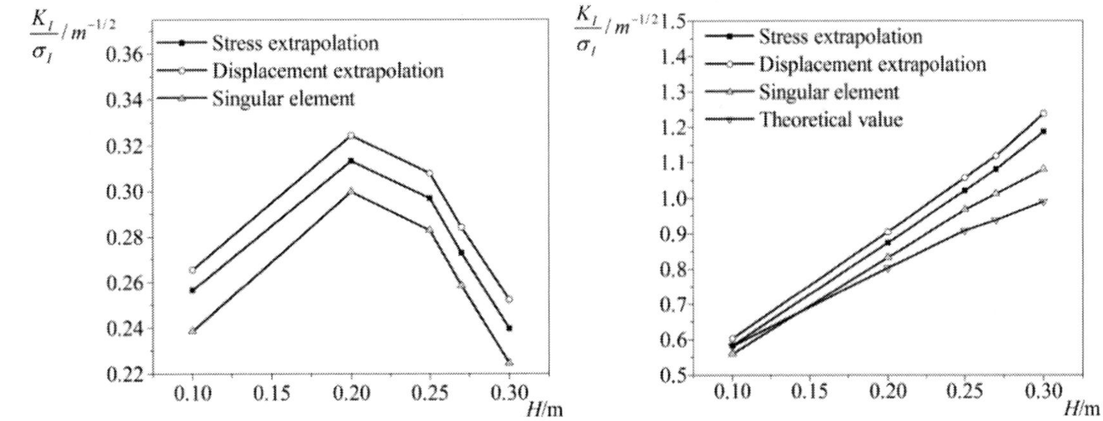

(a) $n_e = 1.0$ $\eta = 1.0$ (b) $n_e = 0.5$ $\eta = 1.0$

Figure 5. The stress intensity factor versus the M under bending.

Figure 6 plots the tendency of K_I/σ_I to change with n_e. From Figure 6, one observes that the stress intensity factor increases with increasing of the elastic modulus of the cracked material, and the law of change is the same as the tension.

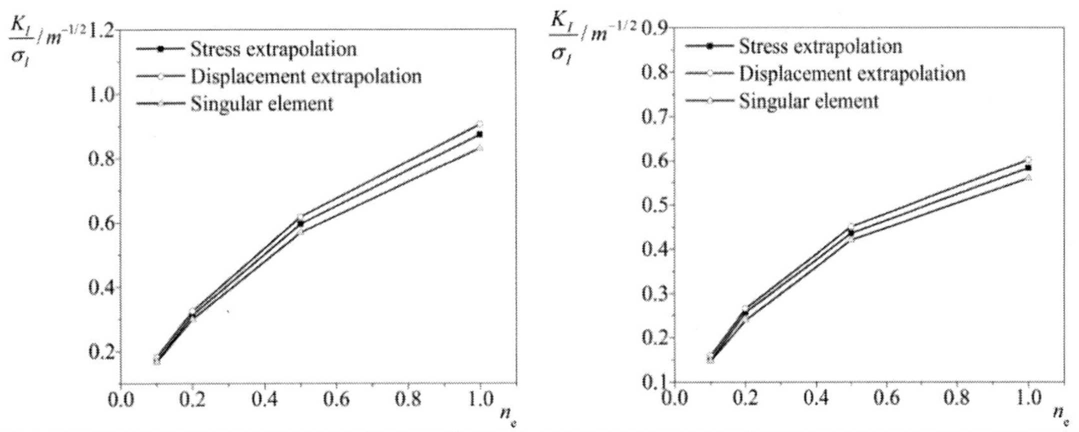

(a) $N = 0.1$ $\eta = 1.0$ (b) $N = 0.2$ $\eta = 1.0$

Figure 6. The stress intensity factor versus n_e under bending.

4 CONCLUSIONS

In the paper, the stress intensity factor of crack perpendicular to the interface in bi-materials was yielded out theoretically, and the stress intensity factor was analyzed by the ordinary finite element and the singular finite element, and from the numerical results some conclusions can be drawn as follows:

(i) Compared with ordinary element method, the result of the singular element method is more accurate and close to theoretical value;

(ii) When the edge crack lies in a weaker material, if the crack tip is far away from interface, the model can be considered as homogenous material, the stress intensity factor increases with decreasing of distance from the crack tip to the interface; if the crack tip is closer to the interface, the stress intensity factor decreases with decreasing of distance from the crack tip to the interface due to the effect by the stiffer material.

Acknowledgments

The authors greatly acknowledge the work support from the Hebei Provincial Natural Science Foundation, P. R. China (No. A2015210029) and the Hebei Provincial Graduate Innovation Fund Project, P. R. China (No CXZZBS2017132).

References

Bouhala L., Shao Q., Koutsawa Y., Younes, A., Nunez, P., Makradi, A., and Belouettar, S., An XFEM Crack Tip Enrichment for a Crack Terminating at a Bi-Material Interface, *Engineering Fracture Mechanics*, 102(2), 5164, 2013.

Cook, T. S., and Erdogan, F., Stresses in bonded materials with a crack perpendicular to the interface. *International Journal of Engineering Science*, 10(8), 677697, 1972.

Duan, S., Yazaki, H., and Nakagawa, K., A Crack at the Interface of an Elastic Half Plane and a Rigid Body, *Engineering Fracture Mechanics*, 32, 573-580, 1989.

Qin, T. Y., and Noda, N. A., Analysis of a Three-Dimensional Crack Terminating at an Interface Using a Hyper-Singular Integral Equation Method, *ASME Journal of Applied Mechanics*, 69(5), 626631, 2002.

Williams, M. L., The Stresses Around a Fault of Cracks in Dissimilar Media, *Bulletin of The Seismological Society of America*, 49, 199-204, 1959.

Xu, J., *The Mechanics of Interface,* Beijing: Science Press, 17, 2006.

Zak, A. R., and Williams, M. L., Crack Point Stress Singularities at a Bi-Material Interface, *Journal of Applied Mechanics*, 30(1), 142-143, 1963.

Interdependence between Structural Engineering and Construction Management
Edited by Ozevin, D., Ataei, H., Modares, M., Gurgun, A., Yazdani, S., and Singh, A.

ISBN: 978-0-9960437-6-2

DESIGN OF BOLTED CONNECTIONS SUBJECT TO TORSION AND SHEAR IN THE ULTIMATE LIMIT STRESS STATE

MOHAMED S. ABU-YOSEF[1], EZZELDIN Y. SAYED-AHMED[2], and EMAM A. SOLIMAN[1]

[1]*Dept of Structural Engineering, Ain Shams University, Cairo, Egypt*
[2]*Dept of Construction Engineering, The American University in Cairo, Cairo, Egypt*

Steel connections transferring axial and shear forces in addition to bending moment and/or torsional moment are widely used in steel structures. Thus, design of such eccentric connections has become the focal point of any researches. Nonetheless, behavior of eccentric connections subjected to shear forces and torsion in the ultimate limit state is still ambiguous. Most design codes of practice still conservatively use the common elastic analysis for design of the said connections even in the ultimate limit states. Yet, there are some exceptions such as the design method proposed by CAN/CSA-S16-14 which gives tabulated design aid for the ultimate limit state design of these connections based on an empirical equation that is derived for ¾ inch diameter A325 bearing type bolts and A36 steel plates. It was argued that results can also be used with a margin of error for other grade bolts of different sizes and steel of other grades. As such, in this paper, the performance of bolted connection subject to shear and torsion is experimentally investigated. The behavior, failure modes and factors affecting both are scrutinized. Twelve connections subject to shear and torsion with different bolts configurations and diameters are experimentally tested to failure. The accuracy of the currently available design equations proposed is compared to the outcomes of these tests.

Keywords: Eccentric connection, Shear center, Instantaneous center, Shear failure, Steel structures.

1 INTRODUCTION

The majority of steel connections are eccentrically loaded. Thus, design of these eccentric connections has become the focal point of many researches: for such connections, the moment-induced stresses must be taken into considerations besides the stresses induced due to normal and/or shear force. By large, design of bolted concentric connection subject to bending moment has been extensively investigated. In contrast, behavior and design of eccentric connections subjected to shear forces and torsion in the ultimate limit state are still ambiguous: current researches still focus on using the elastic analysis in its design. Most codes of practice (e.g. ECP 205 (2008) and BS EN 1993-1-1 (2005)) still use this elastic analysis for design of the said connections even in the ultimate limit states. Yet, there are some exceptions such as the design method proposed by CAN/CSA-S16-14 (2014) and AISC (2017) which give tabulated design aid for the ultimate limit state design of these connections based on an empirical equation that was derived for ¾ inch diameter Grade 4.8 bearing type bolts and A36 steel plates; it was argued that results can also be used with a margin of error for bolts of different sizes/grades and other type steel. As such, in this research, the performance of bolted connection subject to shear and torsion is experimentally investigated in order to either verify the currently (and barely) available limit states design methods or recommend new ones. An ongoing research experimentally scrutinizes the behavior, failure modes and factors affecting the capacity of these connections. Herein, the results of this experimental program are used to investigate the accuracy of the currently available limit state method proposed for by CAN/CSA S-16-14 (2014).

2 ELASTIC ANALYSIS

Figure 1 shows a schematic of a bolted connection subject to shear and torsion with bolts' shear areas and loads shown separately from the column and bracket plate. The eccentric load P can be replaced with the same load value acting at the bolts' centroid plus a couple M = PL, where L is the load eccentricity (Figure 1). As such, each bolt is assumed to resist an equal share of the load, which is given by $P_V = P/n$, where n is the number of bolts. Each bolt's force resulting from the couple M can also be assumed based on the distance between this bolt and the bolts' centroid. Based on this assumption, the forces acting on each bolt due to the couple M can be found from Eq. (1) (Sayed-Ahmed and Elserwi 2017),

$$P_{mx} = \frac{M \cdot X_i}{\sum_1^n \left(X_i^2 + Y_i^2\right)} \quad and \quad P_{my} = \frac{M \cdot Y_i}{\sum_1^n \left(X_i^2 + Y_i^2\right)}$$
$$d_i = \sqrt{X_i^2 + Y_i^x} \tag{1}$$

Where d is the distance from the centroid of the bolt to the bolts centroid. The total force acting on any bolt due to the shear force and the couple M is thus given by Eq. (2);

$$P_i = \sqrt{\left(P_{mxi} + P_{vx}\right)^2 \left(P_{myi} + P_{vy}\right)^2} \tag{2}$$

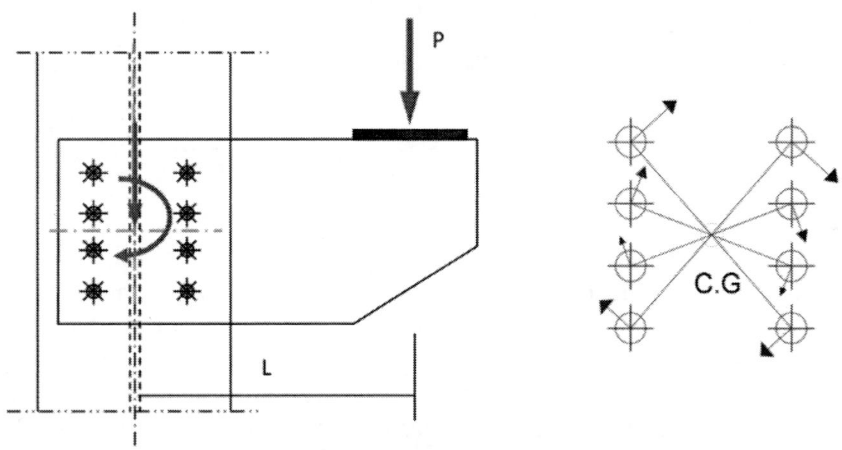

Figure 1. Parameters adopted in the elastic analysis of connections subject to shear and torsion.

3 INELASTIC ANALYSIS

A method of analysis for eccentric connections subject to shear and torsion is described by Kulak *et al.* (1987) and Kulak and Gilmore (2011) and adopted by CAN/CSA-S16 (2014). At the ultimate load, it is assumed that bolt furthest from the instantaneous center (IC) just reaches its failure load (Figure 2); Brandt (1982) presented in details the theoretical approach behind the IC method. Each bolt resistance is assumed to act on a line perpendicular to the radius joining the bolt to the instantaneous center; and, displacement Δ of each bolt is assumed to vary linearly with the length of that radius. The resistance of each bolt is calculated according to the load-deformation relationship of the bolt as Eq. (3):

$$R = R_u(1 - e^{-\mu\Delta})^{\lambda} \tag{3}$$

Where R is the bolt load at any given deformation, R_u is the ultimate bolt load, Δ is the shear, bending and bearing deformation of the considered bolt, μ and λ are regression coefficients and e is the base of natural logarithms. At the ultimate load, $\Delta = \Delta_{max}$ for the bolt furthest away from the instantaneous center.

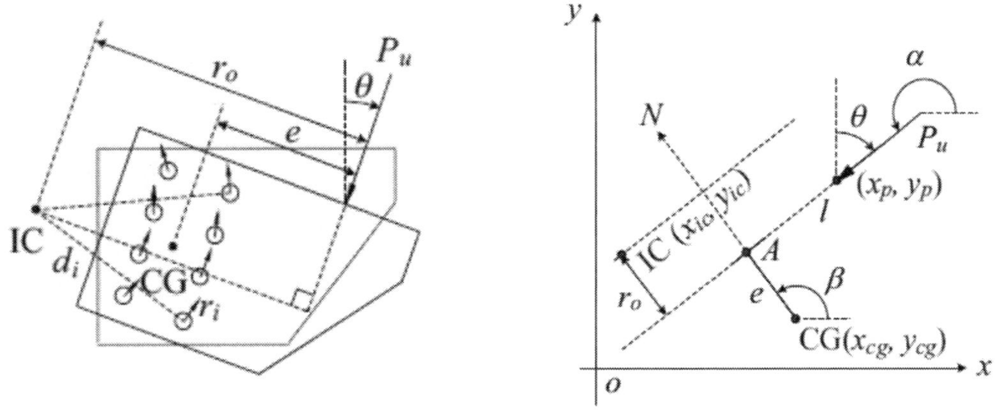

Figure 2. IC method (left) and perpendicular distance r_o from IC to a line *l*.

Tables have been developed using this method with normalized tabulated values for a coefficient C, which may be used for bolts of any diameter and/or grade. In determining C, the following values were used: R_u=329 kN, μ=10, λ=0.55, Δ_{max}=8.64 mm; these values were obtained experimentally for ¾ inch diameter A325 bolts and are reported by Crawford and Kulak (1971). Thus, the ultimate load for each bolt group and eccentricity can be computed from the tabulated values and then divided by the maximum value of R when $\Delta=\Delta_{max}$ to obtain the corresponding value of C.

AISC (2017) design manual provides two practical methods for the design of these bolted connections. The first method is essentially an elastic method and is regarded as a conservative method; the second method is based on the IC concept providing more realistic results. The C coefficient is listed for six inclination angles of the load (θ = 0°, 15 °, 30 °, 45 °, 60 °, and 75 °). Design Engineers tend to interpolate linearly the C coefficient for a nonspecific θ value. However, doing so is not entirely justified. Additionally, the direct implementation of the IC method is difficult because it involves a tedious trial and error process.

4 THE EXPERIMENTAL PROGRAM

Twelve specimens were designed to investigate behavior and failure mode of bolted connections subject to shear and torsion. Two and three rows of bolts (Grade 4.8) with 10 mm, 12 mm, and 16 mm bolt diameters were adopted in the tested connections. Test set-up and instrumentation are shown in Figure 3 while Table 1 shows details of the tested specimen with L and S symbols in the specimen names indicating long (80 mm) and short (60 mm) bolts, respectively.

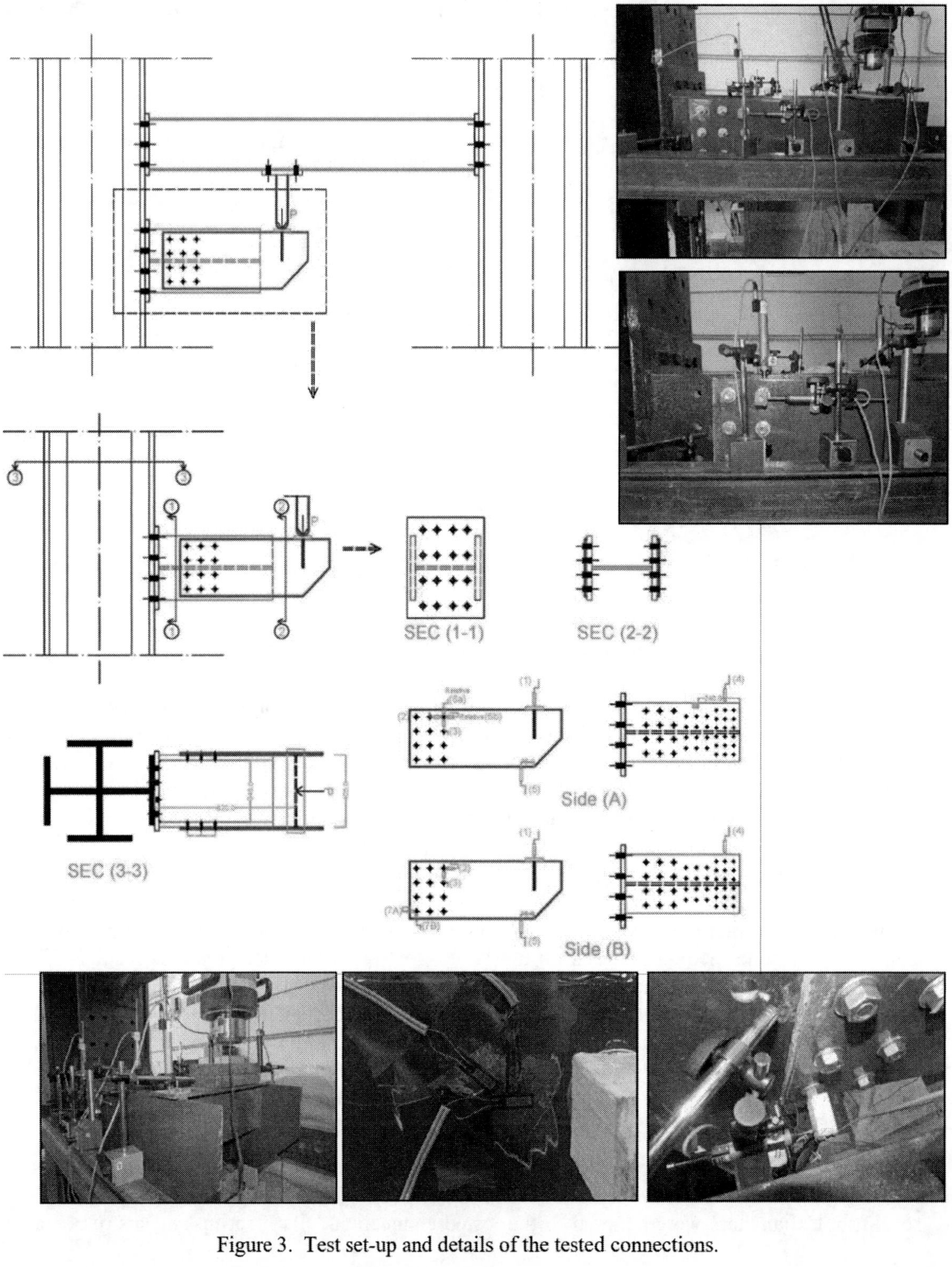

Figure 3. Test set-up and details of the tested connections.

The moving plates (brackets) were bolted directly to a bracket (fixed part), which was rigidly connected to the testing frame column (as a cantilever) using 27 mm diameter G8.8 bolts. Instrumentations were set, as shown in Figure 3 via 14 LVDTs and six strain gauges. The load was applied on a top plate connected the two brackets by an automatically controlled hydraulic jack, which was adjusted to maintain constant loading rate. All instrumentations and load cell were connected to an automatic data acquisition system, which was connected to a computer to record all data acquired by the system software.

Table 1. Details and results of all the tested connections.

Specimen ID	Bolt Dia.(mm)	No of rows	Exp. failure load P_{Exp} (kN)	Failure load Eq. 3 $P_{Eq.3}$ (kN)	$P_{Exp}/P_{Eq.3}$	Elastic Load Eq. 2 $P_{Eq.2}$ (kN)	$P_{Exp}/P_{Eq.2}$
M(10)-2R-e1-S	10	2	60	70	0.86	43	1.39
M(10)-2R-e1-L	10	2	62		0.89		1.43
M(10)-2R-e2-S	10	2	52	60	0.87	35	1.47
M(10)-2R-e2-L	10	2	53		0.88		1.50
M(12)-2R-S	12	2	60	70	0.86	43	1.40
M(12)-2R-L	12	2	60		0.86		1.40
M(12)-3R-S	12	3	90	98	0.92	61	1.47
M(12)-3R-L	12	3	83		0.85		1.36
M(16)-2R-S	16	2	95	101	0.94	65	1.46
M(16)-2R-L	16	2	100		0.99		1.53
M(16)-3R-S	16	3	132	145	0.91	95	1.39
M(16)-3R-L	16	3	130		0.90		1.37
Average ± St. Dev.					0.9±0.04		1.4±0.05

5 TEST RESULTS

Table 1 provides a summary of the failure loads of all tested connections and compares these loads to both the elastic design approach and the inelastic design one adopted CAN/CSA-S16-14 (2014). Samples of failed connections are shown in Figure 4.

Table 1 reveals that the experimental failure load of the tested connections is about 90%±4% of that predicted by the inelastic design Eq. (3) while it is about 140%±5% of that predicted vial the elastic design approach Eq. (2). As such the elastic design tends to be very conservative and significantly underestimate the failure load of connections subjected to shear and torsion. On the other hand, the inelastic design predicted closer value to those recorded experimentally; however, it still needs some adjustment as it overestimates the failure load by about 10%.

6 CONCLUSIONS

An experimental program was conducted on twelve steel bolted connections, which are subjected to shear and torsion. The results of the experimental investigation were compared to the currently adopted elastic and inelastic design techniques.

The comparative investigation revealed that the elastic approach significantly underestimates the connection capacity and tends to be very conservative and uneconomic. On the other hand, the inelastic approach almost correctly predicted the connection capacity with about 10% error, which indicates that it is still in need of some adjustment since it currently overestimates the connection capacity by the said marginal error.

Figure 4. Sample of failed connections.

References

AISC, *Steel Construction Manual*, 15th ed., American Institute of Steel Construction, Chicago, Illinois, USA, 2017.

Brandt, G., Rapid Determination of Ultimate Strength of Eccentrically Loaded Bolt Groups, *Engineering Journal,* 19(2), 94-100, 1982.

BS EN 1993-1-1, *Eurocode 3: Design of Steel Structures*, British Standards Institution (BSI), London, UK, 2005.

CAN/CSA-S16-14, 1-M89. *Design of Steel Structures.* Association Canadienne Normalisation, Canadian Standards Association, Toronto, Canada, 2014.

Crawford, S. F., and Kulak, G. L., Eccentrically Loaded Bolted Connections, *ASCE Journal of Structural Engineering*, 97(ST3),738–65, 1971.

ECP 205, *Egyptian Code of Practice for Steel Construction Load and Resistance Factor Design (LRFD),* Ministry of Housing, Utilities and Urban Development, Egypt. 2008.

Kulak, G. L., Fisher, J. W., and Struik, J. A., *Guide to Design Criteria for Bolted and Riveted Joints*, 2nd Edition, John Wiley, New York, USA. 1987

Kulak, G. L., and Gilmore, M. I., *Limit States Design in Structural Steel*, Canadian Institute of Steel Construction CISC, 9th edition, 2011.

Sayed-Ahmed, E. Y., and Elserwi, A. A., *Limit States Design of Steel Structures*, Lambert Academic Publishing (LAP), Germany, 2017.

Interdependence between Structural Engineering and Construction Management
Edited by Ozevin, D., Ataei, H., Modares, M., Gurgun, A., Yazdani, S., and Singh, A.

ISBN: 978-0-9960437-6-2

STRESS STATE OF STEEL TUBE OF SQUARE CFT COLUMNS WHICH REACHED SHORT-TERM ALLOWABLE STRENGTH

MAO LIU[1], RYO HANDA[2], MASAE KIDO[2], and KEIGO TSUDA[2]

[1]*Dept of Civil Engineering and Architecture, National Institute of Technology, Tokuyama College, Shunan, Japan*
[2]*Dept of Architecture, The University of Kitakyushu, Kitakyushu, Japan*

The allowable strength was compared with the yield strength of CFT short columns subjected to constant axial force and horizontal force in previous studies. The yield strength of CFT columns is determined by 2/3 of compressive force of concrete in most cases. And the yield strength of CFT columns is smaller than the allowable strength. However, as an important index representing the damage state of CFT column, the stress state of steel tube when the CFT column reaches the short-term allowable strength is not clarified. The objective of this study is to make clear whether the steel tube yielded when the short-term allowable strength of square CFT column is reached by an analytical method. The analytical parameters are the effective length to depth ratio, axial force ratio and width to thickness ratio. The range of parameters in the cases which the steel tubes yield before the CFT columns reaches the allowable strength are shown and the decrease of stiffness in these cases are discussed.

Keywords: Steel concrete composite structure, Beam-column, Damage conditions, Length-to-width ratio, Width-to-thickness ratio, Initial stiffness.

1 INTRODUCTION

As a current type of structure, the concrete filled steel tube columns (CFT columns for short) are widely used especially for the high-rise buildings. In Japan, the Recommendations for Design and Construction of Concrete Filled Steel Tubular Structures (CFT Recommendations for short) published by Architectural Institute of Japan is one of the standards that show principles and design methods for CFT members (AIJ 2008). According to CFT Recommendations, the yield strength of CFT column should be determined by the smallest value among the compressive yield strength of steel tube, the tensile yield strength of steel tube and 2/3 of the compressive strength of concrete. However, it is cumbersome in calculation. For the allowable design method that bases on the limit state of safety and serviceability when the materials are elastic, the yield strength of CFT column is calculated by adding the strengths of concrete column and steel tube.

In the previous studies (Liu *et al.* 2011, Miyazaki *et al.* 2014), we have shown the relation between the short-term allowable strength (Short-term means the effects of occasional loads such as earthquake and wind are considered.) and the yield strength of CFT column, as well as the relationship with rotational angles. According to the relation of horizontal force and rotational angle, the stiffness of CFT column doesn't decrease sharply as soon as the yield strength is reached. Because whether the yield strength of CFT column is early reached or not depends on axial force ratio, the design method using the yield strength as a short-term allowable strength is considered too safe in some cases. There are no studies to clarify the stress state of steel tube when the yield strength of CFT column reaches the short-term allowable strength. Therefore, as an important index representing damage state of CFT column, it is necessary to clarify whether the steel tube yielded when the strength of CFT column reaches the short-term allowable strength.

The purpose of this paper is to analyze the relation between horizontal force and horizontal deflection when a square CFT member is subjected to constant axial force and horizontal force, and to clarify whether the steel tube yielded before the strength of CFT member reaches the short-term allowable strength. Length-to-width ratio l_k/D (l_k is the buckling length and D is the width of cross-section.), axial force ratio n and width-to-thickness ratio D/t (t is the thickness of steel tube.) are chosen as parameters for analysis, and the ranges of these parameters when the steel tube yielded before the strength of CFT column reaches the short-term allowable strength are investigated. In addition, the tangential stiffness corresponding to the yield strength of steel tube and the short-term allowable strength is calculated respectively to evaluate the decrease of stiffness.

2 ANALYSIS

2.1 Analytical Model and Loading Conditions

As shown in Figure 1(a), a constant axial force N and a horizontal force H act on a cantilever CFT member whose length is L. The cross section of CFT member is a square. The width and the thickness of steel tube are D and t shown in Figure 1(b).

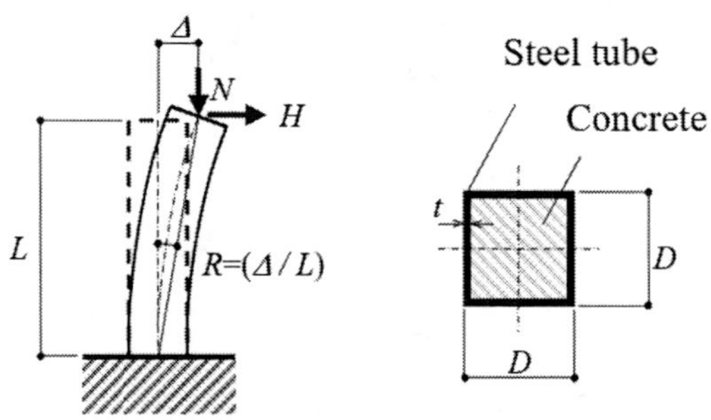

(a) Loading conditions (b) Cross section

Figure 1. Analytical model.

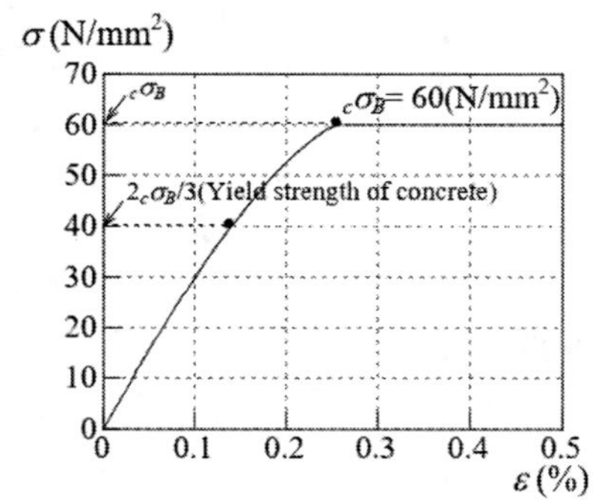

Figure 2. Stress-strain relation for concrete.

2.2 Analytical Method

The relation between horizontal force and rotational angle is obtained by the Column Deflection Curve method (CDC method for short) which is shown in CFT Recommendations. Firstly, the relation between bending moment M and curvature ϕ is calculated under the assumptions that the cross section does not change after deformation and the deflection is quite small. Then, based on the M-ϕ relationship, the relation between horizontal force H and rotational angle R is obtained. The second-order effects are considered in the calculation of CDC method.

2.3 Stress-Strain Relations

The stress-strain relation of concrete is calculated by Eq. (1) which is presented in State-of-the-Art Report on High-Strength Concrete edited by AIJ (2009). The stress-strain model of Fafitis and Shah is used. However, the strength is set to be constant to equal ${}_c\sigma_B$ if the strength σ is over the compressive strength of concrete ${}_c\sigma_B$.

$$\left.\begin{array}{ll}\sigma=\left\{1-\left(1-\dfrac{\varepsilon}{\varepsilon_m}\right)^a\right\}{}_c\sigma_B & (\varepsilon<\varepsilon_m)\\ \sigma={}_c\sigma_B & (\varepsilon\geq\varepsilon_m)\end{array}\right\} \tag{1}$$

In which $a=\dfrac{E_c\cdot\varepsilon_m}{{}_c\sigma_B}$ and $\varepsilon_m=0.93\cdot{}_c\sigma_B{}^{1/4}\cdot10^{-3}$.

The elastic modulus of concrete E_c is calculated by Eq. (2).

$$E_c=\left(3.32\times\sqrt{{}_c\sigma_B}+6.9\right)\times10^3 \tag{2}$$

In which, ${}_c\sigma_B$ is the compressive strength of concrete.

The tensile strength of concrete is ignored in this study. The stress-strain relation of concrete is shown as Figure 2. The stress-strain relation of steel tube is assumed to be elastic-perfectly plastic model. The elastic modulus of steel tube is equal to 2.05×10^5N/mm^2.

2.4 Short-Term Allowable Strength

In the CFT recommendations, the CFT members with different lengths are divided into three kinds for structural design, short column (i.e., $l_k/D\leq4$), intermediate column (i.e., $4<l_k/D\leq12$) and slender column (i.e., $l_k/D>12$). The design equations are different by the kinds of CFT members. The short-term allowable strength of CFT members which is subjected to axial force and bending moment is calculated by adding the strengths of concrete column and steel tube. For short column, second-order effects are ignored in the calculation of allowable strength.

2.5 Horizontal Force H_1 Corresponding to Short-Term Allowable Strength

In this paper, we define the horizontal force when the first-order moment ($H\times L$) of CFT column is equal to short-term allowable strength M_a as H_1, and the rotational angle is R_1 corresponding to H_1. Figure 3 shows the relation between horizontal force H and rotational angle R. The value of H_1 is marked by • in this figure.

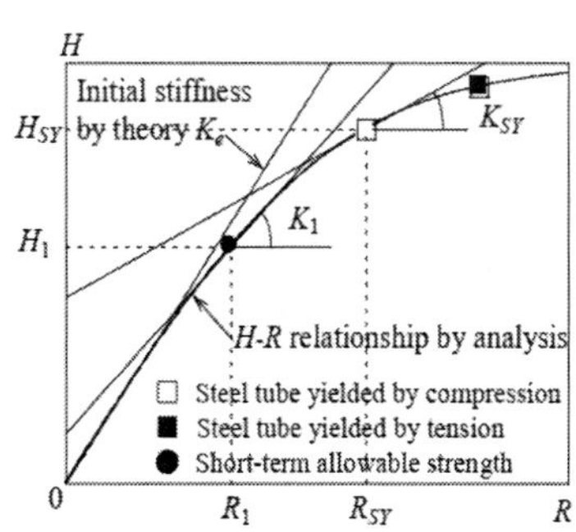

Figure 3. Relation of H-R.

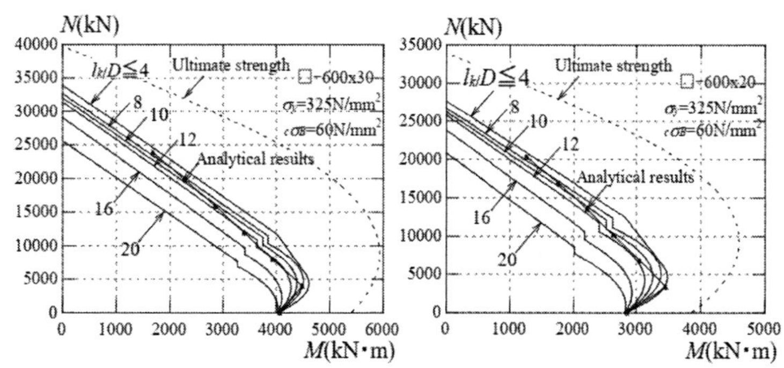

Figure 4. Relation of M-N with different l_k/D.

2.6 Horizontal Force H_{SY} Corresponding to Yield Strength of Steel Tube

The bending moment when the steel tube yielded by compression or tension (i.e., the stress of the edge of cross-section equals the yield stress of material σ_y) is obtained from the M-ϕ relationship. Then, making the bending moment at the bottom of CFT column equal to this value, the horizontal force in this situation is obtained from the H-R relationship. In Figure 3, the points when the steel tube yielded by compression and tension are marked respectively by □ and ■. The smaller one of □ and ■ is defined as the horizontal force H_{SY}. The rotational angle corresponding to H_{SY} is named as R_{SY}. According to Figure 3, because the value of H corresponding to □ is smaller than that corresponding to ■, the horizontal force H_{SY} is determined by the compressive yielding.

Figure 4 shows the short-term allowable strength mentioned in Section 2.4 by M-N relationship. The bending moment for short column (i.e., $l_k/D \leq 4$) when steel tube yielded obtained by analysis is shown by ▲. According to this figure, it is understood that the values by analysis are almost smaller than the values of short-term allowable strength.

3 RESULTS AND DISCUSSION

3.1 Analytical Parameters

The parameters used in this paper are shown as follows.

(i) Length-to-width ratio l_k/D: 4, 8, 10, 12, 16, 20

(ii) Axial force ratio $n=N/N_0$ ($N_0={}_sA\sigma_y+{}_cA_c\sigma_B$): 0, 0.1, 0.2, 0.3, 0.4, 0.5, 0.6

(iii) Width-to-thickness ratio D/t: 20, 30

The width of cross-section D is set to equal 600mm. The compressive strength of concrete ${}_c\sigma_B$ and the yield strength of steel σ_y is 60N/mm^2 and 325N/mm^2 respectively.

3.2 Comparison of H_1 and H_{SY}

Figure 5 shows the relation between the axial force ratio n and the value of H_{SY}/H_1 obtained by analysis, with the length-to-width ratio l_k/D as a parameter. When the value of H_{SY}/H_1 is smaller than unity, the steel tube yielded before the short-term allowable strength is reached. Table 1 shows the data used in Figure 5. The range of $H_{SY}/H_1<1$ is enclosed with bold line. Except the cases of n=0 and n=0.1 of D/t=30, the yield strength of steel tube is determined by compressive yield strength.

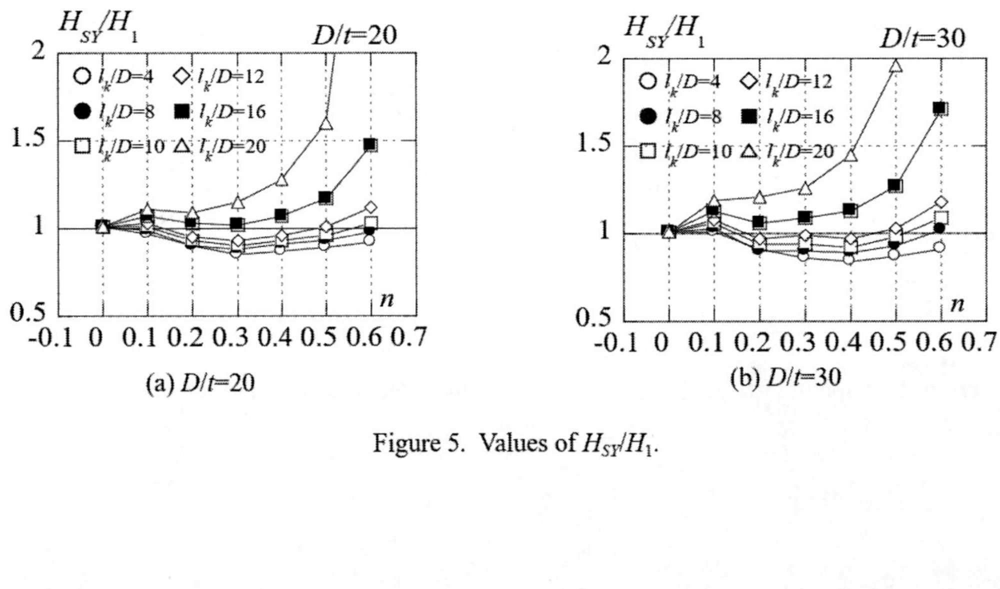

Figure 5. Values of H_{SY}/H_1.

According to Figure 5, the value of H_{SY}/H_1 increases as the length-to-width l_k/D increases. This is probably because the short-term allowable strength decreases as the value of l_k/D increases. In cases of l_k/D =16 and 20, the value of H_{SY}/H_1 is larger than unity whatever the axial force ratio n is. It is understood that steel tube doesn't yield in case of slender column (l_k/D >12) when the short-term allowable strength of CFT member is reached. In addition, when the length-to-width l_k/D equals 12 or less, the smallest value of H_{SY}/H_1 appears in case of n=0.3 or 0.4.

Table 1. (a) Values of H_{SY}/H_1 when D/t =20.

n	l_k/D					
	4	8	10	12	16	20
0	1.01	1.01	1.01	1.01	1.01	1.01
0.1	0.97	0.99	1.01	1.03	1.07	1.11
0.2	0.90	0.90	0.93	0.95	1.03	1.09
0.3	**0.85**	0.88	0.90	0.93	1.02	1.15
0.4	0.87	0.91	0.93	0.96	1.07	1.28
0.5	0.89	0.93	0.96	1.01	1.17	1.60
0.6	0.92	0.98	1.03	**1.12**	1.47	**3.57**

Table 1. (b) Values of H_{SY}/H_1 when D/t =30.

n	l_k/D					
	4	8	10	12	16	20
0	1.01	1.01	1.01	1.01	1.01	1.01
0.1	1.01	1.03	1.06	1.08	1.13	1.19
0.2	0.91	0.90	0.94	0.97	1.06	1.21
0.3	0.86	0.90	0.94	0.99	1.09	1.26
0.4	**0.84**	0.89	0.92	0.97	1.13	1.45
0.5	0.87	0.93	0.98	1.03	1.27	1.96
0.6	0.91	1.02	1.09	**1.18**	1.71	**9.82**

Table 2. (a) Values of R_{SY}/R_1 (l_k/D=10, D/t=20).

n	R_{SY}(%)	R_1(%)	R_{SY}/R_1	H_{SY}/H_1
0	0.478	0.473	1.01	1.01
0.1	0.509	0.503	1.01	1.01
0.2	0.428	0.466	0.92	0.93
0.3	0.359	0.402	**0.89**	**0.90**
0.4	0.300	0.322	0.93	0.93
0.5	0.241	0.248	0.97	0.97
0.6	0.181	0.172	**1.05**	**1.05**

Table 2. (b) Values of R_{SY}/R_1 (l_k/D=10, D/t=30).

n	R_{SY}(%)	R_1(%)	R_{SY}/R_1	H_{SY}/H_1
0	0.455	0.450	1.01	1.01
0.1	0.517	0.486	1.06	1.06
0.2	0.428	0.462	0.93	0.94
0.3	0.352	0.378	**0.93**	**0.94**
0.4	0.293	0.319	0.92	0.92
0.5	0.234	0.240	0.98	0.98
0.6	0.171	0.158	**1.09**	**1.09**

3.3 Comparison on Rotational Angle

The rotational angles R_{SY} and R_1 corresponding to H_{SY} and H_1 are shown in Tables 1 and 2. According to Table 2, the difference between the values of R_{SY}/R_1 and the values of H_{SY}/H_1 shown in Section 3.2 is almost 0, i.e., the values of H_{SY}/H_a and R_{SY}/R_1 are almost the same.

3.4 Decrease of Stiffness

In this section, the decrease of stiffness when the steel tube yielded before the short-term allowable strength is reached is discussed. K_{SY} and K_1 represent the tangential stiffness corresponding to H_{SY} and H_1, and the values of K_{SY} and K_1 are obtained by the relation of H-R like Figure 3. The initial stiffness by theory K_e is calculated by Eq. (3).

$$K_e = \frac{H}{R} = \frac{ZN}{\tan Z - Z} \tag{3}$$

in which $Z = L\sqrt{N/EI}$ and $EI = E_c I_c + E_s I_s$.

The comparison of the initial stiffness K_e to the stiffness K_{SY} and K_1 is presented in Figure 6. The values of K_{SY}/K_e and K_1/K_e are shown with dotted lines and solid lines respectively. According to

Figure 6, the value of K_{SY}/K_e is 0.77 to 0.97, which is almost the same regardless of the value of l_k/D. In regard to the effect of width-to-thickness ratio, the value of K_{SY}/K_e with the same axial force ratio is smaller when the width-to-thickness ratio is larger. On the decrease of stiffness when the allowable strength is reached, when the value of l_k/D is smaller, the value of K_1/K_e decreases at a larger rate in most cases. Comparing the values of K_{SY}/K_e and K_1/K_e, the maximum difference is 0.214 which appears in the case when l_k/D=4 and n=0.4 in Figure 6(a), i.e., after the steel tube yielded, the stiffness of CFT column decreases by about 21% until the short- term allowable strength is reached.

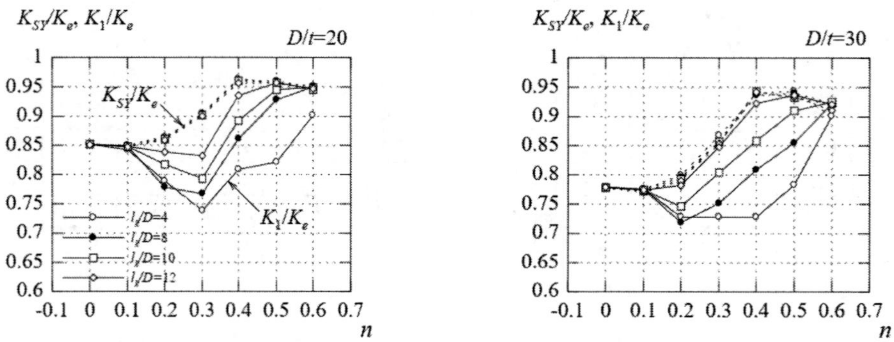

(a) D/t=20 (b) D/t=30

Figure 6. Decrease of stiffness.

4 CONCLUSIONS

In this paper, the relation between horizontal force and deflection when a square CFT member is subjected to constant axial force and horizontal force is shown by analytical method. The horizontal force H_{SY} and rotational angle R_{SY} when the steel tube yielded are compared with the lateral load H_1 and rotational angle R_1 when the short-term allowable strength is reached by taking length-to-width ratio l_k/D, axial force ratio n and width-to-thickness ratio D/t as parameters. The main findings are shown as follows.

(i) The ranges of these parameters when the steel tube yielded before the CFT column reaches the short-term allowable strength are investigated. According to Figure 5, the value of H_{SY}/H_1 increases as the length-to-width l_k/D increases. The steel tube doesn't yield in the case of slender column (l_k/D>12) when the short-term allowable strength of CFT member is reached.

(ii) According to Table 2, the values of H_{SY}/H_1 and R_{SY}/R_1 are almost the same.

(iii) Comparing the values of K_{SY}/K_e and K_1/K_e, the stiffness of CFT column decreases by about 21% until the short-term allowable strength is reached.

References

Architectural Institute of Japan (AIJ), *Recommendations for Design and Construction of Concrete Filled Steel Tubular Structures*, AIJ, Japan, 2008.

Architectural Institute of Japan (AIJ), *State-of-the-Art Report on High-Strength Concrete*, AIJ, 2009.

Liu, Q. S., Kido, M. and Tsuda, K., *Study on Relationship between Superposed Strength and Yield Strength of the Square CFT Columns,* ISEC-6, Zurich, Switzerland, 21-26 June 2011.

Miyazaki, T., Nai, A. and Kido, M., Comparison Between the Rotation Angle at the Yield Strength and the Allowable Strength of Square CFT Columns, *J. of Constructional Steel*, JSSC, pp.569-576, Nov., 2014.

Interdependence between Structural Engineering and Construction Management
Edited by Ozevin, D., Ataei, H., Modares, M., Gurgun, A., Yazdani, S., and Singh, A.

ISBN: 978-0-9960437-6-2

MECHANICAL PROPERTIES OF STRUCTURAL STEEL AND HIGH-TENSION BOLT AT ELEVATED TEMPERATURE

KYUNG-JAE SHIN, HEE-DU LEE, SO-YEONG KIM, DA-SOM CHU, and JONG-HUN WOO

Dept of Architectural Engineering, Kyungpook National University, Daegu, Korea

This paper presents the test results of the mechanical properties of three types of structural steel at high temperature, which are generally used for the Pre-Engineered Building (PEB) system. The PEB system is generally used for non-residential buildings, such as factories and warehouses. The structural steel members are installed without fire resistance protection, which means they are very weak in the case of fire. The end-plate connection could be critical in the case of fire because most of the moment is resisted by the tensile force of the bolts. Therefore, the mechanical properties of bolts at elevated temperatures are tested. Coupon test specimens for SS400(SS275), (SM490(SS355), and F10T bolts were tested according to ASTM E8M. The high-temperature coupon tests were performed at 20°C, 400°C, 500°C, 600°C, 700°C, and 800°C. The test results were compared with the design reduction factors obtained from the American and European standards. The yield strength and tensile strength satisfied the minimum strength of the specified standards at 20°C. However, the reduction factor for yield strength obtained at a high temperature was lower than that of the standard value suggested by the code. In particular, the reduction factors for the high-strength bolt (F10T) were lower than those of the structural steel members (SS400(SS275), SM490(SM355)).

Keywords: High temperature, Thermal properties, Reduction factor, Tensile test, Extensometer.

1 INTRODUCTION

The Pre-Engineered Building (PEB) system is an efficient structural system for long-span structures, such as factories or warehouses. If this system is adopted for non-residential buildings, the structural members are not fire-protected, which results in early collapse in the case of fire. Figure. 1 shows a PEB system and end-plate connection detail. The depth of the main steel girder is higher than that of the H-shaped beam and its slenderness ratio is very high. The member is critical for lateral torsional buckling in the case of fire because the flexural strength is governed by reduced yield strength and the modulus of elasticity. Especially in cases of fire, PEB, made of slender elements, loses strength at temperature more easily and collapses more quickly than general steel structure systems. The strength of the end-plate connection is governed by the reduced tensile strength of the bolts and end plate. This paper shows the material test results at high temperature for three main materials used in the PEB system. Three types of material, SS400(SS275) (F_y = 235MPa, F_u = 400MPa) for the web, SM490(SM355) (F_y = 315MPa, F_u = 490MPa) for the flange, and a F10T bolt (F_u = 1000MPa), were tested at 20°C,

400°C, 500°C, 600°C, 700°C and 800°C according to ASTM E8M (*ASTM, 2010a*). The test results were compared with the design reduction factors obtained from the American and European standards (*AISC 2005, CEN 2005, Choi et al. 2014, and Chen et al. 2006*).

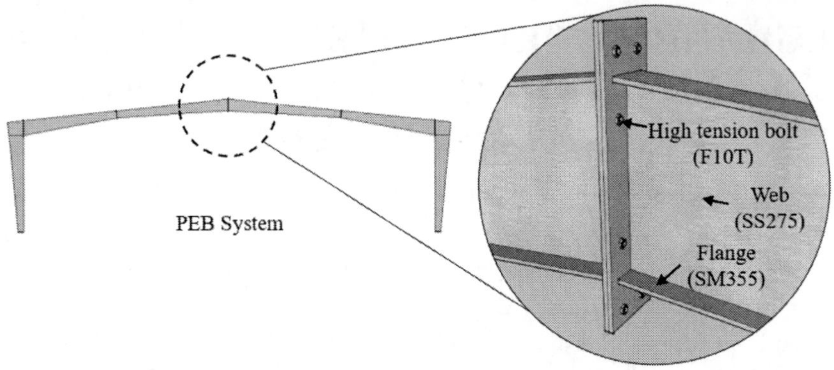

Figure 1. PEB system components.

2 TEST COUPON INFORMATION

High-strength steel specimens for tensile strength tests were made according to American standards (*ASTM 2010a*). Table 1 shows the detail of test coupons. The tensile tests at elevated temperatures were carried out by preparing the specimens as shown in Figure 2 to be subjected to tensile testing at high temperatures between 400°C and 800 °C at 100 °C increments.

Table 1. Detail of test coupons.

Steel	Target Temp. (°C)	Standard	Section size (mm)
SS275 SM355 F10T	20 (Room Temp.)	Specimen 3 according to ASTM E8M	Φ6
	400		
	500		
	600		
	700		
	800		

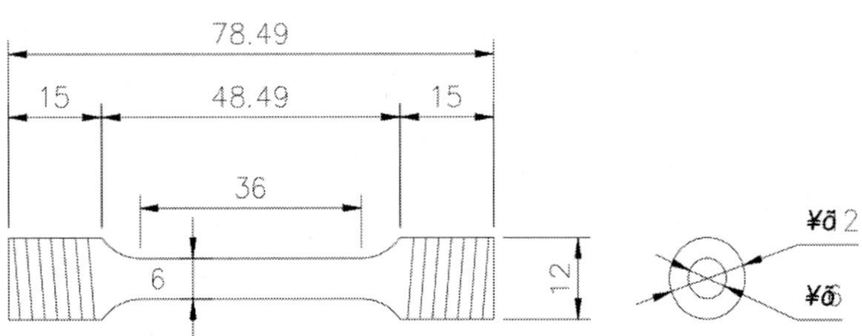

Figure 2. Dimension of tensile test specimen.

The displacement measurement equipment for the high-temperature tensile test was made as shown in Figure 3. It is composed of two extended bar sets, an inside bar connected to the upper gauge point, and an outside bar connected to the lower gauge point. The location of the gauge point that holds the specimen using a knife edge is inside the furnace, and the other side of the gauge point is exposed to the outside of the furnace. The inside bar connected to the upper gauge point slides into the outer tube, which is connected to the lower gauge point. Two extensometers were attached using the knife edge of one at each extended bar. The average value of the two extensometers is used as the strain for the coupon test. Many small holes outside the tube were made to compensate for the strain change of the inside bar induced by the temperature change. A thermocouple (K-type) was installed at the surface of the steel in the heating furnace for the tensile test at a precise target temperature. The heating velocity was 5°C/min due to the heating furnace capacity. In order to equalize the heat distribution, the experiment was performed after more than 20 minutes from reaching the target temperature (*ASTM 2010b*). Two extensometers were installed external to the heating furnace, as shown in Figure 3(a), (b). The average value of the measured strains was calculated.

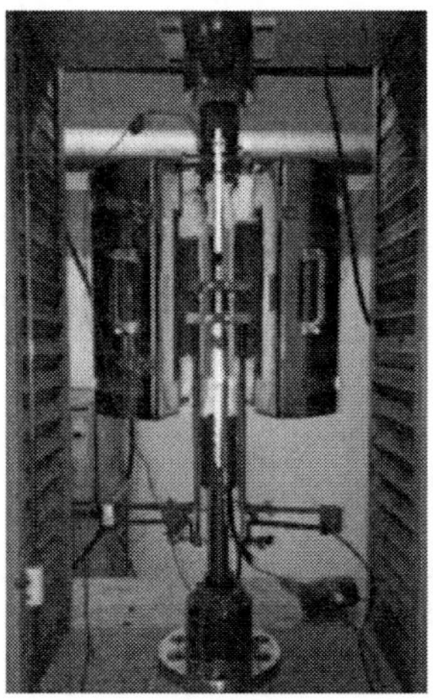

(a) Test in foreground

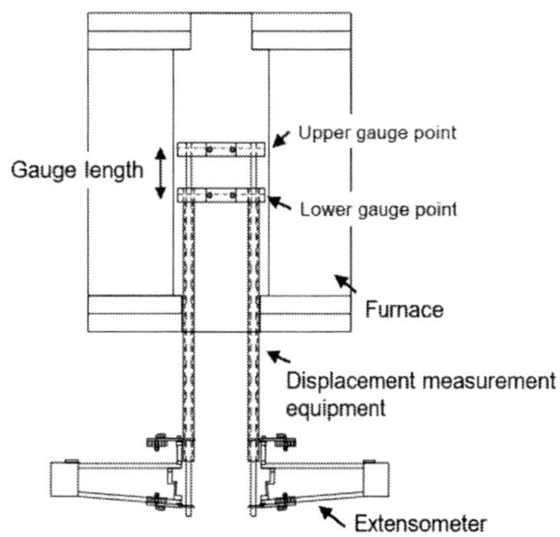

(b) Displacement measurement equipment

Figure 3. High-temperature tensile test.

3 TEST RESULTS

3.1 Material Properties of Steel

Table 2 indicates the material properties of steel at high temperature. The modulus of elasticity was not measured because the measurement device is not sensitive enough to measure small displacements. As shown by the general high-temperature behavior, as the temperature increased, the yield strength and tensile strength decreased, whereas elongation increased.

Table 2. Material properties of steel at high temperature.

Steel	Temp (°C)	Yield strength (MPa)	Tensile strength (MPa)	Elongation (%)
SS400 (SS275)	20	287.5	432.6	36.6
		300.6	450.7	36.6
		295.4	432.7	28.8
	400	221.6	391.1	32.0

Table 2 *(contd).* Material properties of steel at high temperature.

SS400 (SS275)	400	205.9	373.1	39.2
		220.1	344.5	31.9
	500	164.1	236.0	36.2
		177.6	256.7	34.5
		165.1	184.2	33.3
	600	133.7	159.2	36.0
		116.3	150.3	33.1
		146.0	171.1	28.7
	700	83.5	103.9	35.0
		58.7	101.2	36.3
		57.8	106.4	41.1
	800	38.5	52.0	45.3
		32.2	64.9	57.4
		23.3	64.9	56.1
SM490 (SM355)	20	310.1	486.8	26.0
		317.8	476.6	28.4
		321.7	474.1	38.3
	400	222.3	419.6	25.3
		228.7	435.1	28.5
		252.7	391.1	29.0
	500	179.6	248.8	34.8
		209.7	264.3	35.6
		156.1	217.6	40.3
	600	141.0	186.7	39.1
		138.2	155.7	36.0
		107.1	155.7	41.5
	700	85.6	85.7	39.3
		75.4	98.6	34.8
		55.5	83.1	39.8
	800	70.0	83.1	30.2
		34.8	83.0	35.1
		36.3	64.9	42.3
F10T	20	981.5	1,037.7	16.2
		1,002.1	1,048.1	14.9
		996.2	1,040.2	10.3
	400	792.9	888.2	13.3
		748.5	865.0	15.5
		837.3	880.7	12.7
	500	532.1	640.5	13.8
		415.1	609.6	15.3
		508.2	625.1	16.5
	600	179.2	271.5	38.6
		142.9	287.0	42.0
		159.3	302.4	35.3
	700	73.4	106.2	37.0
		35.1	90.0	38.4
		67.7	98.6	37.8
	800	61.4	64.8	49.7
		59.0	71.1	47.2
		72.6	78.8	49.5

Figures 4–6 show the stress–strain curves at different temperatures. The strain-hardening section between yield strength and tensile strength decreased as the temperature increased. Additionally, as the temperature increased, the difference between yield strength and tensile strength decreased. Yield strength was obtained by 0.2% offset of the stress–strain curve.

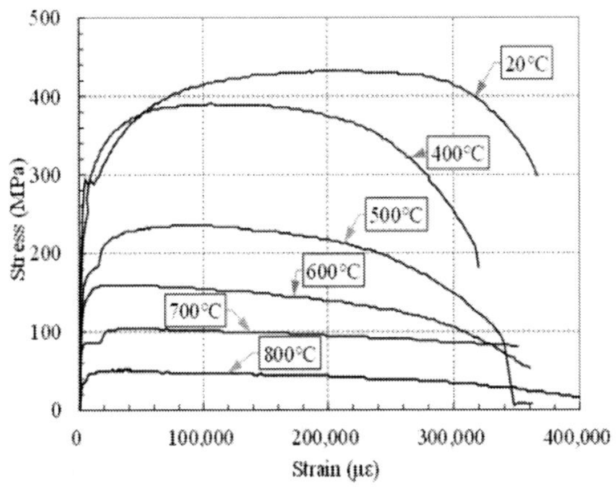

Figure 4. Stress–strain relationship (SS400(SS275)).

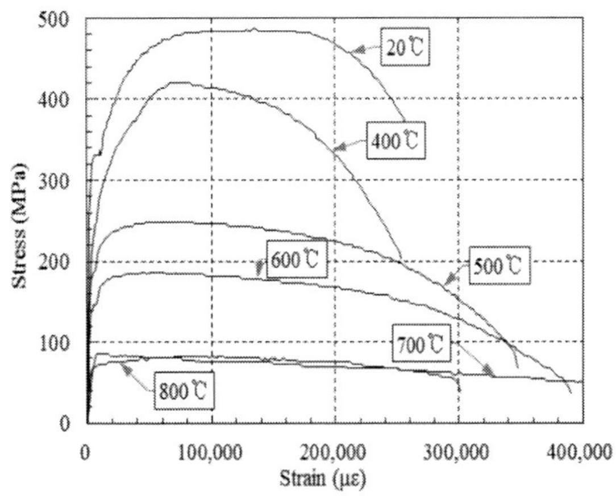

Figure 5. Stress–strain relationship (SM490(SM355)).

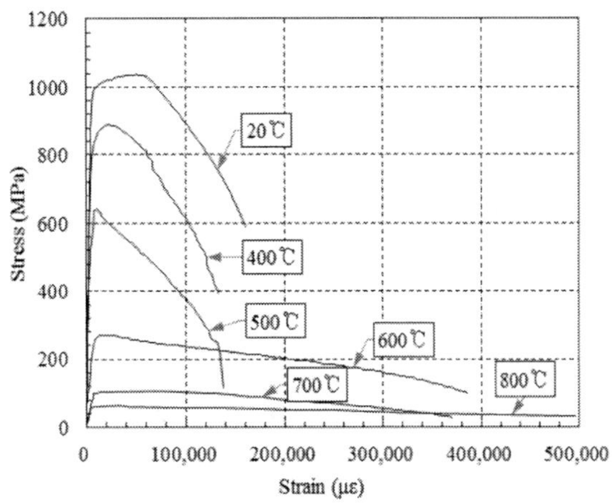

Figure 6. Stress–strain relationship (F10T).

3.2 Comparison of Reduction Factor with Existing Standards

Figure 7 shows the reduction factors of the yield strength. The reduction factors of the yield strength in AISC and Eurocode 3 were plotted using linear interpolation methods. Findings showed that the reduction factors of all the coupons were lower than the specified AISC and Eurocode 3 standards. In particular, the reduction factors of high-tension bolts (F10T) decreased more sharply than those of SS400 and SM490 at 500°C. When the heating temperature reached 400°C, the yield strengths of all coupons were similar, each with 73–83% of yield strength from normal temperature. However, when the heating temperature reached 500°C, the yield strengths of all coupons decreased sharply.

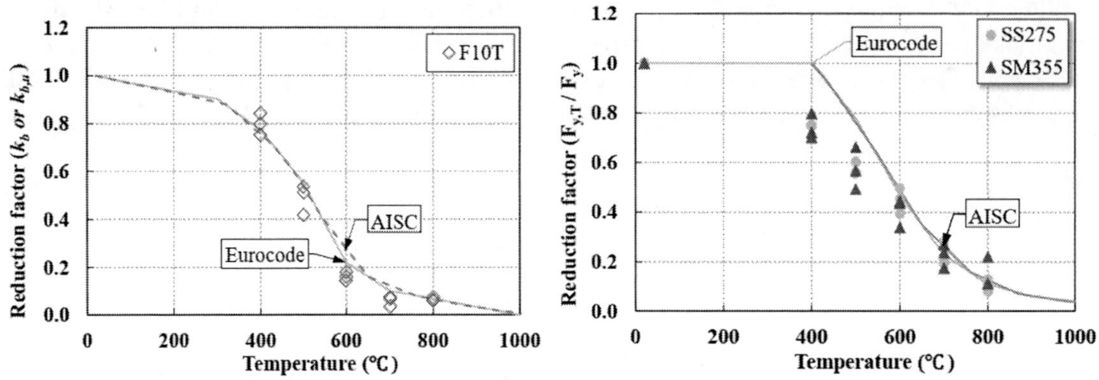

Figure 7. Reduction factor of yield strength compared with existing standards.

4 CONCLUSION

The tensile test of steels (SS400, SM490, and F10T) used for the PEB system was performed at an elevated temperature. The strength of steel and reduction factor obtained from the tests were compared with the existing standard values at high temperature. The reduction factor of yield strength obtained from the tests was lower than those of AISC and Eurocode 3. Fire protection may be required at the bolted area even though the PEB system may not require fire protection according to the design code. The measured reduction factors data of steel used in the PEB system at an elevated temperature are not enough to make a concrete conclusion. More test data on high-strength bolts are especially needed.

Acknowledgments

This research was supported by the Fire Fighting Safety & 119 Rescue Technology Research and Development Program funded by the Ministry of Public Safety and Security ("MPSS-fire safety-2015-72")

References

ASTM, Standard Test Methods for Tension Testing of Metallic Materials – ASTM E8/E8M, ASTM International, West Conshohocken, USA, 2010a.

ASTM, Standard Test Methods for Elevated Temperature Tension Tests of Metallic Materials – ASTM E21, ASTM International, West Conshohocken, USA, 2010b.

AISC, Specification for Structural Steel Buildings-ANSI/AISC 360-05, American Institution of Steel Construction (AISC), Chicago, USA, 2005.

CEN, Eurocode 3: Design of Steel Structures - Part 1-2: General Rules - Structural Fire Design, European Committee for Standardization, Brussels, Belgium, 2005.

Choi, I. R., Chung, K. S., and Kim, D. H., Thermal and Mechanical Properties of High-Strength Structural Steel HSA800 at Elevated Temperatures, *Materials & Design*, 63(2014), 544-551, 2014.

Chen, J., Young, B., and Uy, B., Behavior of High Strength Structural Steel at Elevated Temperatures, *Journal of Structural Engineering*, 132(12), 1948-1954, 2006.

KSA, KS D 3503: Rolled Steels for General Structure, Korean Standards Association (KSA), Seoul, Korea, 2014a.

KSA, KS D 3515: Rolled Steels for Welded Structures, Korean Standards Association (KSA), Seoul, Korea, 2014b.

Interdependence between Structural Engineering and Construction Management
Edited by Ozevin, D., Ataei, H., Modares, M., Gurgun, A., Yazdani, S., and Singh, A.

ISBN: 978-0-9960437-6-2

RESEARCH ON WIND PERFORMANCE AND FIELD EXPERIMENT OF LARGE SCALE COMPLEX SURFACE SPACE TRUSS WITH STEEL PIPE

XIULI WANG[1,2], YONGHONG RAN[1,2], HUVUE ZHANG[1,2], and YANPENG ZHU[1,2]

[1]*School of Civil Engineering, Lanzhou University of Technology, Lanzhou, China*
[2]*Western Center of Disaster Mitigation in Civil Engineering, Lanzhou University of Technology, Lanzhou, China*

The wind-induced performance response of large-span truss curved roofs is extremely complex and changes obviously under wind load. This paper has taken the large-span steel roof of *Liu Zhaike* highway toll station as an example to analyze by both numerical simulations and field tests of responses of the steel canopy under wind load. In this case, simulated analysis results using finite element software were calibrated and verified by the field test results. In addition, a new method of large structure field testing was proposed which obtained the internal force and displacement of the canopy structure. Moreover, this paper analyzed and determined the actual stress state of the rectangular pipe truss. Finally, the safety condition of the roof was evaluated based on the monitoring data and the simulation of the finite element model. And the analysis methods provides references for similar engineering field tests, as well as guidance for the operation and maintenance for this project.

Keywords: Rectangular steel tube truss, Numerical wind tunnel, Coefficient of wind pressure, Shape factor, Field test.

1 INTRODUCTION

Large-span structures are attributed to wind-induced sensitive ones, and the wind load tends to be the main control load of the structural design (Xiang 1997). For example, *Zhengzhou* stadium suffered from disaster caused by a windstorm in 2003, with more than 4000 m^2 cantilevered roof whipped by strong winds, and the direct loss of about ¥240,000 (Pan 2008). Although the structural wind resistance analysis of the numerical wind tunnel method has been carried out for such structures (Ma *et al.* 2007), and the computer finite element technology is employed to analyze precisely various stress conditions of structures, there may be a great deviation between the calculated results and the actual force of the structure due to the errors in load assumption (Chen *et al.* 2010, Shen and Wu 2010). Therefore, it is of great theoretical significance and practical value to carry out field experiments under special wind load conditions of large-scale projects. However, the field load tests of large space structures are few at home and abroad (Zhou *et al.* 2012). Taking the large roof structure of *Liu Zhaike* expressway toll station at as an example, the numerical simulation of wind tunnel and field load test are carried out to study the structural response of rectangular steel truss with space-curved surface under wind load. Meanwhile, a loading method for field test of large structure is presented, and the results of numerical simulation and field test are compared and analyzed.

2 PROJECT OVERVIEW

The new toll station of *Liu Zhaike* is located in *Wuhe* Township of *Jingyuan* County, *Gansu*, China, which is also the juncture of *Gansu* and *Ningxia*. The average altitude is 1693 m, with extensive terrain and strong wind. The average maximum wind speed is 1.8 m/s, and the instantaneous maximum wind speed reaches 23.2 m/s over the years. The region is also full of windy days, and the average annual number of days with wind speed exceeding 5 m/s reaches 160 days. Therefore, wind loads are the main factor to be considered for this structure. The total height of the new toll station roof in *Liu Zhaike* is 23.8 m, with the width of 21 m, and the total length is 112.4 m, and the projected area of the building 2360.40 m^2. The toll booth is made up of two parts: steel structure and membrane structure. And the steel structure is made up of space plane rectangular steel tube truss structure, the whole building surface formed by vertical and horizontal crossing (Figure 1).

Figure 1. Construction site.

3 SIMULATION OF NUMERICAL WIND TUNNEL

3.1 Computational Model

3.1.1 *Wind direction*

The geometric entity model is shown in Figure 2. Since the toll station is a biaxial symmetric structure, the wind load under 7 wind directions (0°, 15°, 30°, 45°, 60°, 75° and 90°) are chosen as the simulated condition (Figure 3). Among them, the 0° wind direction is the wind load parallel to the north south direction, with the smallest windward area; The 90° wind direction is the wind load parallel to the east-west direction, with the biggest windward area.

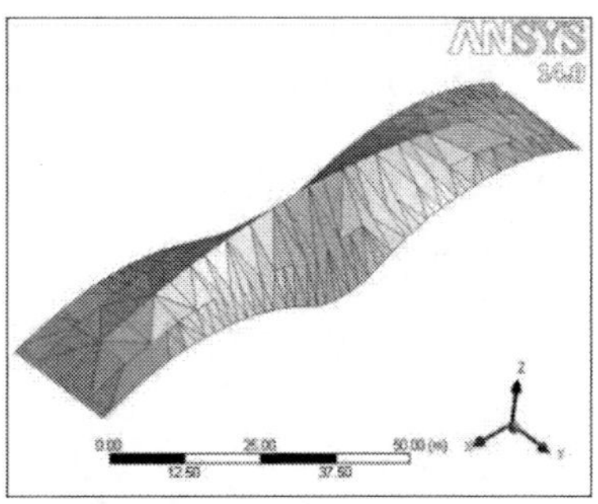

Figure 2. Geometry model of the toll station.

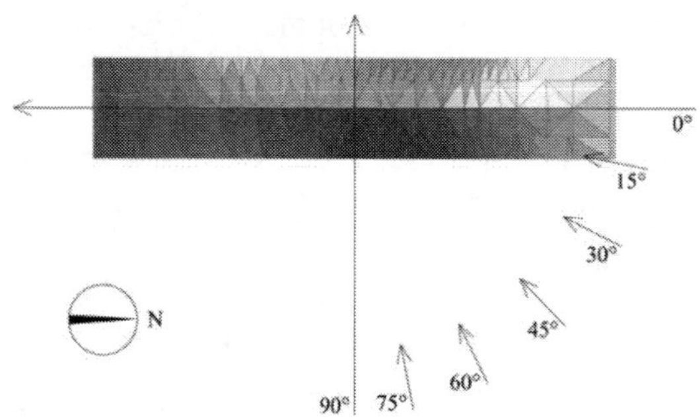

Figure 3. Wind direction diagram.

3.1.3 *Computing watershed model*

Dimension of computing watershed in the structure refers to the design parameters of large span structures in reference (Li 2013). *L*1, advanced position of computing watershed, is 190 meters according to dimension of building surface ($l_x \times l_y \times l_z = 21\text{m}\times 112.4\text{m}\times 23.8\text{m}$), and *L*2, as the backward position of computing watershed, is 285 m, with its width B 450 m. The dimension of computing watershed is $B \times L \times H = 450\,\text{m} \times 500\,\text{m} \times 238\,\text{m}$.

3.1.4 *Boundary condition*

Entrance boundary condition: Surface roughness of the structure is class B, and basic wind pressure $\omega_0 = 0.45\,\text{kN} \cdot \text{m}^{-2}$, reappearing period 50 years. The average wind speed at the standard reference height 10 m can be calculated as: $u_0 = 28.77\,\text{m} \cdot \text{s}^{-1}$. The characteristics of free-stream turbulence are defined by the method of directly giving the turbulent kinetic energy k and turbulent dissipation rate ε.

Exit boundary condition: Boundary conditions of building surface and ground: Adopting no free sliding wall.

3.2 Calculation Results and Analysis

According to (Wang and Wang 2015), it can be seen that the SST k-ω shear stress transport model has better convergence than the BSL-RSM Reynolds stress model. Therefore, SST k-ω shear stress transport model has been chosen. Within the B type geomorphic wind field, the model was simulated under the 7 wind directions. The surface of the structure is divided into four regions, A, B, C and D. Each region is divided into 15 sub-regions, totaling 60 sub-regions, as shown in Figure 4. The average wind pressure coefficient of measuring points corresponding to the 60 partitions on the roof surface is also monitored.

D1	D2	D3	D4	D5	D6	D7	D8	D9	D10	D11	D12	D13	D14	D15
C1	C2	C3	C4	C5	C6	C7	C8	C9	C10	C11	C12	C13	C14	C15
B1	B2	B3	B4	B5	B6	B7	B8	B9	B10	B11	B12	B13	B14	B15
A1	A2	A3	A4	A5	A6	A7	A8	A9	A10	A11	A12	A13	A14	A15

Figure 4. Number of measuring points on the structure surface.

The model is a semi open membrane structure, with different wind pressure on the upper and lower surface of the structure, which is subject to the wind load. Therefore, after obtaining wind pressure on the upper and lower surface, the resultant force of the structure and the corresponding wind pressure coefficient after superposition are obtained by the block superposition treatment of the surface of the structure.

Take the 90° wind direction as an example, the average wind pressure coefficient and velocity streamline of the structure surface are analyzed (Figure 5 and 6). Because the central section of the windward of the structure is concave and it is a larger area, the impact and the stagnation point will occur when passing through the air. Thus, the maximum point of positive pressure is generated, which is concentrated in the middle of windward side, with the maximum positive pressure coefficient 1.10. When the airflow is impeded in the central section of the windward side, it begins climbing to the top and sides of the roof. When climbing to the ridge, the airflow separated apparently forming the tiny discrete high-speed vortex and falling off the back ridge, thus, generating great wind suction near the separation point. Therefore, from the lower to the upper edge of the roof, the average wind pressure coefficient decreases gradually. The positive pressure is gradually transformed to the negative pressure when the airflow nears the ridge. The maximum negative pressure area is concentrated in the central of two wings of the ridge, with the maximum negative pressure coefficient -2.304.

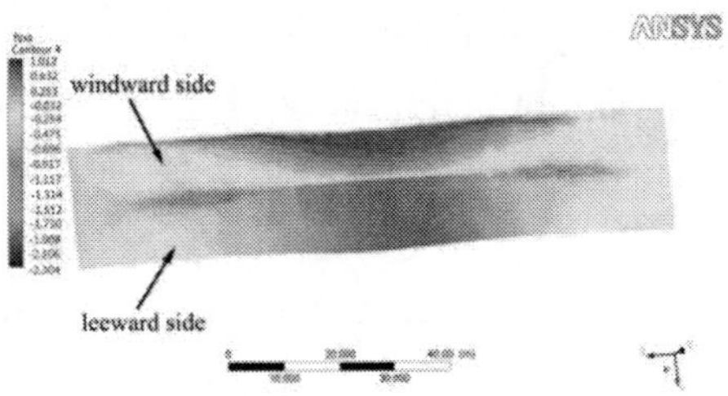

Figure 5. Average wind pressure coefficient (90°).

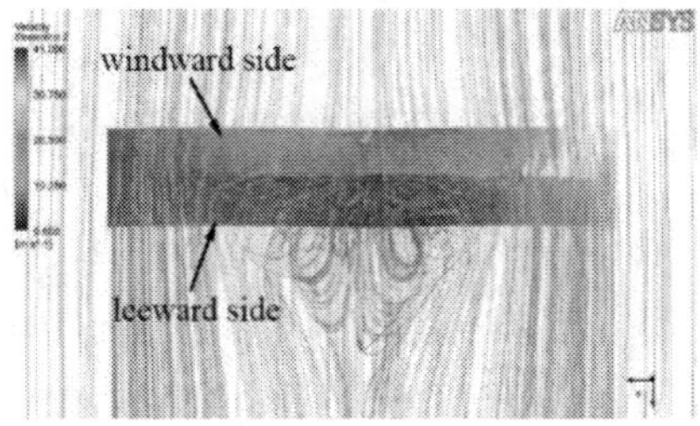

Figure 6. Velocity streamline.

Due to the adhesion and separation of airflow, greater suction is produced. The negative wind pressure in the middle part of leeward of the structure is smaller, and it increases with the extension to both sides. The negative pressure begins to decrease when it is 1/3 meter away from the edge of the wing, until reduced to the minimum at the edge of the wings. The negative pressure at the entire leeside is obviously smaller than that at the ridge.

The results show that the maximum wind pressure area on the windward side is concentrated on the areas of A8, A9 and A10 that are located in the middle part of the windward. And the average wind pressure coefficient of the three areas is between 0.75 and 1.25. The negative pressure area on the windward side is mainly concentrated on B1 and B2 that are located in the ends of the windward's two wings. The negative pressure area on the windward side is mainly concentrated on the areas of B1 and B2 that are located the ends of the windward's two wings. The maximum negative wind pressure on the leeside is mainly concentrated on the areas of C11 and C12 that are located in the ridge of the middle part of the two wings. And the average wind pressure coefficient of the two areas is between -1 and -2. The D area in the leeside is the least affected by the airflow and its average wind pressure coefficients are very small under the 7 wind directions.

To sum up, when the wind angles are between 60°-90°, the absolute value of the average wind pressure coefficient of the structure surface is larger, with wider distribution. The maximum positive and negative mean wind pressure coefficients have little difference, which shows that the structure is greatly influenced by the wind load in the range of the wind angle.

3.3 Shape Factor of Wind Load

The average wind pressure coefficient is converted into the wind load shape coefficients (Table 1), which is used as the load parameter of static analysis. The average wind pressure coefficient C_{pi} obtained from numerical wind tunnel simulation is reference to the wind pressure at 10m height. The shape coefficient of wind load μ_{si} at different heights is corrected by the change of wind pressure height.

4 FIELD LOADING TEST

4.1 Loading Scheme

Adopting grading loading system, the loads from the first to third grade are 40%, 80%, and 100% respectively of the set load. Preloading is needed before test, and the load is 10% of set load.

After unloading, formal load starts (Wang *et al.* 2012). Each load should be stabilized for 30 minutes and then proceed to next level loading.

Table 1. Structural shape factor of wind load (90°).

Area	Factor	Area	Factor	Area	Factor	Area	Factor
A1	-0.270	B1	-0.213	C1	-0.157	D1	-0.199
A2	-0.235	B2	-0.194	C2	-0.180	D2	-0.201
A3	-0.213	B3	-0.180	C3	-0.314	D3	-0.219
A4	-0.128	B4	-0.156	C4	-0.583	D4	-0.449
A5	0.162	B5	-0.103	C5	-0.572	D5	-0.447
A6	0.270	B6	-0.054	C6	-0.574	D6	-0.448
A7	0.676	B7	-0.026	C7	-0.579	D7	-0.427
A8	1.080	B8	0.178	C8	-0.578	D8	-0.407
A9	1.082	B9	0.173	C9	-0.572	D9	-0.415
A10	0.276	B10	-0.009	C10	-0.570	D10	-0.446
A11	0.161	B11	-0.101	C11	-0.563	D11	-0.449
A12	-0.126	B12	-0.158	C12	-0.489	D12	-0.451
A13	-0.210	B13	-0.178	C13	-0.310	D13	-0.220
A14	-0.234	B14	-0.196	C14	-0.178	D14	-0.199
A15	-0.270	B15	-0.211	C15	-0.135	D15	-0.162

When loading, the tension of multiple cables simulates the multi-point loads on the structure under wind load. The test system consists of DH3816 static strain acquisition instrument, steel surface strain mete and strain gauge, adopting high precision total station to record the displacement in three directions of each measuring point. According to the simulation results of the numerical wind tunnel, the key parts of the structure are determined and arranging the measuring point. Considering the influence of wind speed at the test site, the real-time wind speed and direction are monitored by the wind speed indicator and wind direction indicator. The wind speed and wind direction data are used to convert the required tension, and the test load is adjusted in real time.

4.2 Test Results

The first loading grade: Start loading from the third group and stop loading when 40% of the set load is reached. The first group starts loading, when the loading reaches at 40% of the set load, stops loading. The second group starts loading, when the loading reaches at 40% of the set load, stops loading. Thus, the first-grade load ends. At this point, the stress value of each measuring point is -60~80MPa, and the overall structure is in an elastic state.

The second loading grade: Start loading from the second group and stop loading when 80% of the set load is reached. The third group starts loading, when the loading reaches at 80% of the set load, stops loading. The first group starts loading, when the loading reaches at 80% of the set load, stops loading. Thus, the second-grade load ends. The stress of the measuring point is -60~60MPa, and the number of the measuring points increases when the stress is -60MPa.

The third group loading: Start loading from the first group and stop loading when the set load is reached. The second group starts loading, and stops loading, when the reaching the set load. The third group starts loading and stops loading when reaching the set load. Thus, the third-grade load ends and all the process of loading is completed. At this point, the stress value of each measuring point is -120~20Mpa, and the stress of some measuring points increases obviously, with the maximum value close to -120MPa. The structure is still in an elastic state.

After all loading is finished, keep the loading steady for 30 minutes and then start unloading. After unloading, keep the structure steady for 60 minutes and then start measuring data. At this point, the stress of all the measuring points tends to zero, which shows that the structure in the whole experiment is in the elastic stress state fully.

5 CONCLUSIONS

In this paper, the toll station of *Liu Zhaike* expressway is taken as an example, the numerical simulation and field load test of the complex curved rectangular tube truss structure are carried out, and the main conclusions are drawn as follows:

When the wind direction is between 60° and 90°, the absolute value of the average wind pressure coefficient of the structure surface is greater, with a wider range of distribution.

During the test, the maximum stress is far less than the design strength of the component. After unloading, the stress and displacement of the structure can be restored to the initial state, indicating that the structure is working in an elastic state.

Under the asymmetric load working condition, the stress of some web member in the structure appears inverse phenomenon, which means the tension and compression state of the member is converted reciprocally. If the load increases to a certain value, it will lead to the destabilizing of the belly bar and affect the safety of the structure.

Acknowledgments

The authors would like to thank the financial support from the National Natural Science Foundation of China (51778273).

References

Chen, B., Yang, Q., and Wu, Y., Multi-Objective Equivalent Static Wind Loads for Large-Span Structures (in Chinese), *China Civil Engineering Journal*, 43(3), 62-67, 2010.

Li, X., Numerical Simulation Study of Wind Load on Structures Based on CFD (in Chinese), *Guangzhou: School of Civil Engineering and Transportation, South China University of Technology*, 2013.

Ma, J., Zhou, D., and Li, H., Numerical Wind Tunnel Technique for The Wind Resistance Analysis of Long Span Spatial Structures (in Chinese), *Engineering Mechanics*, 7(24), 77-93, 2007.

Pan, F., Refinement Theory of Random Wind-Induced Dynamic Response for Long-Span Roof Structures, (in Chinese), *Hangzhou: Institution of Architecture and Civil Engineering,* Zhejiang University, 2008.

Shen, S., and Wu, Y., *Present Situation and Prospect of Wind Resistant Research on Long-Span Space Structures (in Chinese),* in Proceedings of the 30th Anniversary of The General Assembly on Structural Wind Engineering Research in China, 94-103, China Civil Engineering Society, 2010.

Wang, X., Li, Y., and Yang, W., Health Monitoring and Simulation Analysis Over Construction Loading Process for Long-Span Steel Structure (in Chinese), *Spatial Structures*, 18(4), 49-54, 2012.

Wang, X., and Wang, X., Numerical Simulation of Wind Pressure Distribution on the Cantilever Steel Truss Roof Structure with Multi-Zigzag Plane Shape (in Chinese), *Spatial Structures*, 21(1), 26-33, 2015.

Xiang, H., State of the Art and Prospect in Studies on Structural Wind Engineering (in Chinese), *Journal of Vibration Engineering*, 10(3), 258-263, 1997.

Zhou, X., Gan, M., and Li, G., Grouping Response Method for Equivalent Static Wind Loads Based on a Modified LRC Method, *Earthquake Engineering and Engineering Vibration*, 11(1), 107-119, 2012.

Interdependence between Structural Engineering and Construction Management
Edited by Ozevin, D., Ataei, H., Modares, M., Gurgun, A., Yazdani, S., and Singh, A.

ISBN: 978-0-9960437-6-2

LOAD BEARING STRUCTURAL ASSEMBLY

BRADFORD RUSSELL

BR Architects & Engineers, Richardson, USA

This article is based on an invention focused on absorbing and redirecting forces two dimensionally or three dimensionally in any mechanism that relies on force resolving members, so more benefits of the mechanism can be achieved. This Assembly is focused on new ideas for framing and connections to resolve forces so the rigidity between members can be minimized and the Assembly can sustain performance under longer durations and more unique loadings, i.e., seismic loading on structure. The Assembly uses several techniques and fundamentals for efficient force dissipation with, or without deflection while creating a controlled rotation when loaded. The allowable arrangement of the Assembly within itself (web members within outer and inner rings and arcs), and as an assortment of assemblies, can be limitless. This arrangement, configuration of web members, numbering, width and flexibility, can be arranged for a predictable and controlled deflection or rotation in any direction, due to loading from any direction and type, thus creating a load bearing and load dissipating Assembly with greater efficiency, and advance the life of interacting parts under loadings, i.e., seismic loading of structures. This Assembly, in particular the web members, can be an arrangement and properties of arcs to offer a desired response, as well as create a rotation when loaded.

Keywords: Absorbing, Redirecting, Extending, Force manipulation, Controlling.

1 INTRODUCTION

The Assembly is focused on new ideas for framing and connections to resolve forces so the rigidity between members can be minimized and the Assembly can sustain performance under a longer duration and more unique loadings. The Assembly is used to change linear (horizontal or vertical) acceleration and loading into a rotation about a point thus controlling the eccentricity on the critical load bearing members of the structure, i.e., columns (United States Patent and Trademark Office 2013, 2014).

2 DESCRIPTION OF ASSEMBLY

As shown in Figure 1b, the Assembly includes an outer loop (arc) member; an inner loop (arc) member spaced apart from and sized smaller than the outer loop (arc) member; a web assembly coupled to and extending between the outer loop (arc) member and the inner loop (arc) member, the web assembly comprising a plurality of arcuately formed web members.

The arcuately formed web members and end connections allow the members to go through increased curvature during loading while creating a controlled rotation in excited events, similar to the loads generated from seismic forces on structures. The properties of each outer and inner ring (arc), as well as web members and their end connections, i.e., fixed, pinned, can be selected for a controlled performance – that of controlled directional movement or rotation, or lack thereof.

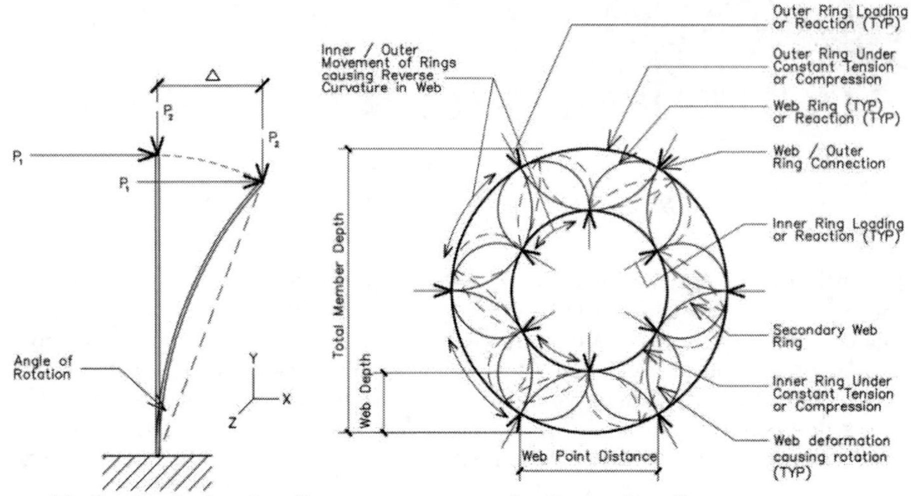

a. Similar Equivalent Loading. b. System Loading.

Figure 1. Load bearing structural assembly, i.e., 'ForceGenie' component section / elevation schematic.

The Assembly is literally a flat or 3-dimensional spring, which can be designed for a confined space and provide an array of outcomes or interact with an array of forces. The circumferential rings have minimal deformations because of the bracing effect of the web members, and the extensions of the web members through bending causing rotation of the Assembly rings and compression / tension in the circumferential rings (arcs) (Russell 2015).

3 BENEFITS OF ASSEMBLY

The Assembly's, i.e., ForceGenie's, design is a response to structural failures during extreme events on the structural connections and created by dissipating of the energy from the events. When the inner or outer ring is loaded, it generates a rotation (spin) in relation to the other by bending of the web members (Ref. Figure 1b). This rotation about a center will have the tendency to control the eccentricity of the vertical loading and dissipate energy in movement and heat. As well, in this bending of the Assembly's web members, a web member can be placed into single curvature, or double curvature through the addition of mid rings (Figure 2) and increase the performance of the web member section along the length.

The loading of the Assembly will cause a rotation (spin) of the inner and outer rings simply because the web members are much stronger in compression than in bending, and the web members are pre-formed in a bent shape (arc). In ideal conditions, the rotation (spin) of the inner and outer rings, in a cyclical event, will put the web members into full tension and bending, and rings into tension and braced compression. In addition, the amount of rotation (spin) can be predicted by simple structural analysis and angle of rotation set by the design of the members, i.e. Moment of Inertia, the quantity of web member arcs, the spacing of the arcs around the parameter of the rings, etc. (Russell 2015).

4 FUNCTION OF THE ASSEMBLY

As depicted in Figure 1, one can see the analogy with the bending of column and a possible configuration of the members. With this, one can recognize the efficiency of the circle (arc) can be used to generate efficiency in strength, in particular in bending, but also in compression, tension, and shear of the web members. The allowable arrangement of the Assembly (Figure 1, 2 and 3) within itself (web members within outer and inner rings (arcs), and as an assortment of Assemblies, can be limitless (Refer to Figures 2 and 3 above for possible configurations). In Figure 1b is shown an Assembly with six ring (arc) groups of web members. The web members can be of any arc arrangement separating the inner and outer rings. (Additionally, the outer and inner rings can be of an assortment of arc configurations, i.e., elliptical, tear-shaped, etc. for a specific required response.)

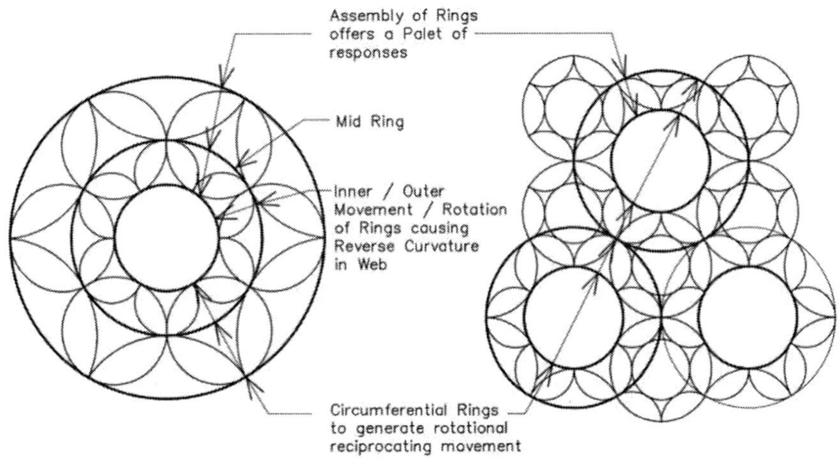

Figure 2. Circumferential rings (arcs) elevation.

Figure 3. Assortment of rings (arcs) elevation.

This arrangement, configuration of web members, web numbering, web width, web flexibility, etc. can be arranged for a predictable rotation (controlled deflection) in any direction, due to loading from any direction (Figure 1b) and type, all while keeping the members within the elastic and inelastic range and thus creating a load bearing, a load dissipating Assembly with greater efficiency in a cyclical event. This will tend to enhance the life of interacting parts under repetitive or reciprocating loadings, i.e. seismic loading on structures. The Assembly rotates due to the ease of displacing a web member (bending) rather than compressing the member length. Refer to Figure 4 for a possible connection integrating the Assembly.

The rotation of a ring in relation to the other (inner or outer) is achieved by directional displacement (rotational) from the differing moment of inertia of member directions under loadings, i.e., bending about the axis (Ref. Figure 1a for axis orientation). That is, in this case the moment of inertia about the 'Z' axis is much less than about the 'X' axis and 'Y' axis, and the kl/r is sufficient to cause bending of the member rather than failure in member compressive stress. The dimensions of the cross section in the members are set so bending is assured in one direction (i.e. about the 'Z' axis) and thus rotation of the rings occurs.

5 ENGINEERING PRINCIPLES OF THE ASSEMBLY

The Assembly makeup has the effect of isolating the excited loading, buffering the impact of seismic (or other cyclic event), and may prevent collapse. Figure 1 earlier shows directional forces on the Assembly; however, the forces on the inner and / or outer rings can be in any direction from predictable loadings and cause the rotation between the rings (arcs) unique to the Assembly.

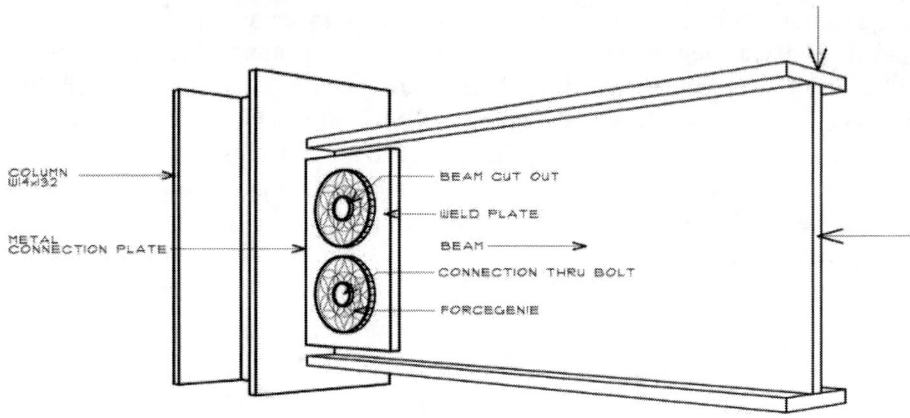

Figure 4. Assembly flange connection perspective.

$$F \quad \leftrightarrow \tau = r\,F\perp \quad (\text{N, lbf} \leftrightarrow \text{rad/sec}^2) \tag{1}$$

$$F_{net} = M\,a \leftrightarrow \tau_{net} = I\,\alpha\,(\text{N, lbf} \leftrightarrow \text{rad/sec}^2) \tag{2}$$

Eq. (1) and Eq. (2) show engineering formulas for Translation (X, Y, Z) and Rotation. One can see the F_{net} generated from the acceleration (i.e. seismic forces on structure) is converted into a rotational acceleration. This acceleration occurs to all mass, but because of the loading points of the Assembly occur around the parameter (inner or outer rings), the web members move into reverse curvature from the current position and are put into tension, and the adjoining web member into bending. This web member bending and ring rotation can be uniquely designed so that when it is loaded and the web arch ends are fixed or pinned, the inner / outer rings (arcs) move through a predictable angular deflection (rotation).

Table 1. Modified Mercalli intensity scale.

Instrumental Intensity	Acceleration (g)	Velocity (cm/s)	Perceived shaking	Potential damage
I	< 0.0017	< 0.1	Not felt	None
II–III	0.0017 – 0.014	0.1 – 1.1	Weak	None
IV	0.014 – 0.039	1.1 – 3.4	Light	None
V	0.039 – 0.092	3.4 – 8.1	Moderate	Very light
VI	0.092 – 0.18	8.1 – 16	Strong	Light
VII	0.18 – 0.34	16 – 31	Very strong	Moderate
VIII	0.34 – 0.65	31 – 60	Severe	Moderate to heavy
IX	0.65 – 1.24	60 – 116	Violent	Heavy
X+	> 1.24	> 116	Extreme	Very heavy

The Modified Mercalli Intensity Scale gives us a range of the seismic events. The Assembly can be designed accordingly by the respective ranges of acceleration and velocity along with the loading of the specific structures to determine the respective forces causing displacement or rotation. The number, properties, and arrangement of the web members are selected so to generate a designed and predictable rotation before the web members go into tension and bending (Refer to Figure 1b for deflected shape).

The Assembly reaction will have the effect of dissipating the forces, distributing the moment and shear through the members with respect to inflection points, and will tend to limit the maximum values accordingly. Or the web members can move, with angular deflection (rotation), through single curvature with the use of hinge connections on the ends of the arc to the inner and outer rings.

With respect to the assemblies shown in the figures, the loading of one ring (either inner, outer, or both) can be from any direction and have the same twisting affect because of the nature of the arc web members being set in curvature. The Assembly can be as big as the Main Force Resisting System of the structure (i.e. many feet) and / or as small as a connection component (i.e. less than an inch). The strength of the Assembly is achieved when it is loaded it causes rotation and loss of energy into the system through force dissipation, ring and mass displacement.

Potential energy can be stored in either masses or springs. In a spring the potential energy, U, is given in Eq. (3):

$$U = \int f(x) * dx = \int k * x * dx = ½ k * x^2 \tag{3}$$

(k = spring constant, x = displacement (rotation)

For a mass moved in a constant gravitation field as in the Assembly is given as Eq. (4):

$$U = \int f(x) * dx = \int m * g * dx \tag{4}$$

(m = mass of structure, g = acceleration of gravity, i.e. 9.8 m/s^2, 32.2 ft/s^2)

In the ideal sense, we want the friction forces between the Assembly and the structure to never be realized so energy is not transmitted to the structure. The energy dissipation is created by movement of the web members in, and through the elastic, and into, and through the inelastic range. There are significant energy losses with this movement when converted into heat energy. The mechanical energy is then transported to the rings, which increases the kinetic energy of the Assembly. Some of this kinetic energy is lost to heat energy from the friction between the Assembly and the structure, and some is lost to the Assembly and the web members (arcs) in the act of rotating.

6 RATIONALIZATION OF THE ASSEMBLY

The Assembly only accounts for the above grade forces as they interact with the various levels of the structure (in early development of foundation interaction with the structure). Additionally, the Assembly should be used throughout the structure so there is no induced rotation in the structural diaphragms and the energy of the event is released in the Assembly. The two-dimensional nature of the Assembly shown in Figures works nicely with the horizontal and vertical accelerations of the seismic event, and the orthogonal framing of structure will tend to protect multi-directional of the possible seismic events.

Numerous other arc web arrangements for the Assembly can be set for specific spatial allowances, specific loading requirements, and specific rotational displacements needed. The effect of the Assembly on the building is that of changing Natural Frequency of the structure. A structure, which is positioned for optimal performance in an earthquake is that of a box, which moves with a stiff earth surface. If the period of ground motion matches the natural resonance of a building, it will undergo the largest oscillations possible and suffer the greatest damage. Inversely speaking, the increased ductility of the Assembly increases the building's natural frequency of the structure, which makes it more difficult for the building's frequency to be reached in a seismic event. The Assembly increases the ductility without increasing the displacement. The web assembly configuration can be adapted to individual reaction needs, both directional, amount of rotation and length of deflection (rotation), needed from the Assembly.

7 CONCLUSIONS

The primary structural parameters, which are used in the Assembly are that of member curvature and member confinement. If a member can be allowed or forced to move into curvature, each length of the member can be allocated a portion of the moment and shear in reverse, and the PΔ effect of the structure is minimized and controlled. This makes each web member act more efficiently along the member length and will tend to control forces from the Assembly to that being connected and dissipate energy accordingly.

The uses, which could benefit from the actions created in the Assembly are structural in nature but can go beyond the building environment. Additionally, this may include aeronautics, medical, athletics, and other incorporating mechanically loaded components, which could benefit by causing rotational displacement from linear forces as a form of energy transposition. The author is certain you can look around, where you currently stand and see numerous other connections and structures, which could benefit by the strengths generated from the use of the Assembly.

Acknowledgements

The views expressed herein are those of the author and do not necessarily reflect those of the advisors. The author is grateful to BRAE and for their advice and input during the preparation of this paper.

- Dr. Ali Abolmaali, Professional Engineer, Chair Department of Civil Engineering and Director of Center for Structural Engineering Research and Simulation, University of Texas at Arlington, Arlington, Texas (email: abolmaali@uta.edu).
- Dr. Bijan Mohraz, Professional Engineer, PhD advisor, Southern Methodist University, Dallas, Texas (email: bmohraz@engr.smu.edu).
- Dr. Roman Okelo, Professional Engineer (Structural), BR Architects & Engineers, Richardson, Texas 75080 (email: romano@brarchitects.com).

References

Russell, B., Framing Systems for Cyclical Loading Events, Unpublished report, Southern Methodist University, December 2015.

United States Patent and Trademark Office, Load Bearing Structural Assembly, Patent Appl. Serial No. 13/969,529, filed August 17, 2013.

United States Patent and Trademark Office, Load Bearing Structural Assembly, Patent Appl. Serial No. 14/450,425, filed August 4, 2014.

Interdependence between Structural Engineering and Construction Management
Edited by Ozevin, D., Ataei, H., Modares, M., Gurgun, A., Yazdani, S., and Singh, A.

ISBN: 978-0-9960437-6-2

FATIGUE LIFE ANALYSIS OF RAIL-WELDS USING LINEAR ELASTIC FRACTURE MECHANICS

MEQUANENT M. ALAMNIE[1] and YALELET ENDALEMAW[2]

[1]*Department of Engineering Science, University of Agder, Grimstad, Norway*
[2]*College of Engineering and Technology, Wolkite University, Wolkite, Ethiopia*

The initiation and growth of fatigue cracking is mainly due to high stress concentration, heterogeneity and poor quality of weld. The detection and rectification of such weld defects are major concerns of rail network managers to reduce potential risk of rail breaks and derailments. To estimate the fatigue life of welded joints and to analyze the progress of fatigue cracks, a fracture mechanics-based analysis and fatigue models were developed using Finite Element Analysis. The initial flaw is obtained from a sample weld using ultrasonic flaw detecting machine test. Linear Elastic Fracture Mechanics (LEFM) approach based on the Paris law was applied to determine critical crack size and the number of cycles to failure using FRANC3D software. The inspection interval of rail welds before fracture (failure) was suggested based on reliability and life cycle analysis that correspond with minimum overall cost and frequency interval. It is recommended that fracture-based models in combination with reliability analyses can be a sustainable infrastructure decision-making algorithm.

Keywords: Ultrasonic test, Paris law, FRANC3D, Inspection interval.

1 INTRODUCTION

Railway infrastructures are high investment assets. They are designed to work in a very demanding safety conditions and must display a very low occurrence of failures. Rails are the most significant and basic components of railway systems. However, different factors, which influence the rail degradation process gradually reduce the performance, reliability and safety of railway infrastructure. Railway rails are manufactured in sections of 25 - 120 m length, which are joined in track by either bolting or welding. The joints and track irregularities are the weak sections associated with high risk of failure (Ringsberg *et al.* 2005). The service performance of rail-welds is affected by their ability to support the service load without fatigue damage. Fatigue and fracture behavior are important considerations in determining the condition of metal structures subjected to cyclic loads. Specifically, the expected life of a rail-weld subjected to random, variable-amplitude traffic cycles are highly dependent upon damage accumulation caused by various fatigue mechanisms. Structural failure is seldom attributed to load considerations; the occurrence of stresses exceeding those predicted by the designer is rare (Peter and Thomas 2011). This paper is aimed at determining the response of rail-welds to fatigue load using linear fracture mechanics approach. A Cost model is also employed to address the trade-offs among various factors related to rail defect and inspection frequency (Sauragh 2008).

2 RAIL WELDING

According to the American Society of Welding, welding is a localized coalescence of metal where coalescence is produced by heating to suitable temperature, with or without the use of filler metal. Alumino-thermic (thermite) welding is a process that produces coalescence of metals by heating them with a powdery mixture of metallic aluminum (Al~22%) and iron scale (Fe_3O_4~78%) as a thermite (Sergejevs and Mikhaylovs 2008, Napoleon 1997, Ringsberg *et al.* 2005). Weld sample used in this paper is a thermite weld. Visual examination of the weld for visible defects by the welding personnel such as geometry is first test. *Ultrasonic, Hardness, Slow Bends Test, Fracture Surface Examination, and Micro-and Macro Structure Examination* are other tests for weld quality (Liu *et al.* 2014). Ultrasonic Testing (UT) uses high frequency sound waves (typically 0.5 and 15 MHz) to conduct examinations and make measurements. It has wide use in engineering applications (*flaw detection/evaluation, dimensional measurements, material characterization, etc.*). A typical pulse-echo UT inspection system consists of several functional units, such as the pulser/receiver, transducer, and a display device. Angle beam transducers and wedges are typically used to introduce a refracted shear wave into the test material. An angled sound path allows the sound beam to come in from the side to improve detectability of flaws in weld areas.

3 FRACTURE MECHANICS

Fracture mechanics is broadly classified into two types: Linear Elastic Fracture Mechanics (LEFM) and Elastic Plastic Fracture Mechanics (EPFM). LEFM assumes small deformation and minimal yielding at the crack tip, while EPFM can account large deformation and plastic effects. In LEFM, stress intensity factor (SIF) is a measure of the stress field at the crack tip. In general, the stress intensity factor is proportional to the remote stress σ, crack size a, and the geometry of the mechanical component with crack and serve as a measure of the severity of the crack tip for different crack configurations. Paris's law (Eq. (1)) is used for fatigue crack growth analysis (Paris and Erdogan 1963).

$$\frac{da}{dN} = C(\Delta K(a))^m \tag{1}$$

Where C and m are material parameters in Paris region (3 and $1*10^{-11}$ respectively) (David 2001), $\Delta K = K_{max} - K_{min} = \Delta\sigma\sqrt{\pi a}$, K_{max} (maximum) and K_{min} (minimum) stress intensity factors.

The number of cycles to failure, N_f, can be obtained by integrating (Eq. (1)) with respect to crack size a, from initial crack size, a_i, to final crack size, a_f, as indicated in Eq. (2) and Eq. (3).

$$N_f = \int_{a_i}^{a_f} \frac{da}{c[\Delta\sigma\sqrt{\pi a}]^m} \tag{2}$$

$$a_f = \frac{1}{\pi}\left[\frac{K_{IC}}{c[\Delta\sigma\sqrt{\pi a}]^m}\right]^2 \tag{3}$$

Where, K_{IC} is the fracture toughness.

3.1 Fracture Behavior of Rail-Welds

For thermite welds after crystalline rearrangement Poisson's ratio and modules of Elasticity are in the range of -17% to +13% and ∓5%, respectively (IACS 2016). Since the material at the running surface of the weld, and some millimeters beneath, is strongly deformed and compressional residual stresses are generated in that region, and since the global bending stresses are highest in the foot, fatigue cracks in the welded rails tend to grow from underneath rather than from the railhead (Zerbist *et al.* 1978).

3.2 Flaw Detection of Weld Sample

Flaws are detected on the rail weld at the rail base using a "crack tip diffraction" technique. An angled transducer of 60° and using the shear velocity of sound in steel as 3130mm/s are used so that, it could be used as a tip for discontinuity/flaw (Det Norske Veritas 2004). The flaw size can be determined as half the velocity of sound in the rail, 3130mm/s multiplied by the product of the time taken by the sound to traverse the flaw, which is 100 µsec and cosine of angle of transducer used in the determination (Figure 2a). Thus, the flaw size is shown in Eq. (4) as:

$$a = \frac{dtV}{2}\cos(\theta_R) = \frac{1}{2}(100\mu\text{sec}) * (3130\text{mm/s}) * \text{Cos}\,(60°) \cong 0.07825\text{mm} \quad (4)$$

3.3 Optimal Inspection Interval

Preventive replacement and periodical inspections are two major approaches used for improving system availability and reducing maintenance cost. The Association of American Railroads (AAR) developed a Rail Defect Detection Cost Model that estimates the total cost of repairing detected defects and broken rails (Orringer 1990). Based on a model developed by AAR, the long-term cost of Rail-welds is given by Eq. (5) (Liu *et al.* 2014).

$$K\frac{L}{V}C_{hr} + \frac{S(K)L}{h\left(\frac{T}{K}-q\right)}(DDC + C_{DDT}) + S(K)L(S_{DC} + C_{SDT}) \quad (5)$$

K is annual inspection interval; L is track length (mi); V is average inspection vehicle speed (mph); C_{hr} is inspection cost per hour per vehicle($/hr/veh); S(K) is annual number of broken rail per mile; h is parameter related to inspection frequency (≈ 0.0108); T is annual traffic density(MGT); DDC is cost of repairing detected weld defect($); C_{DDT} is train delay cost due to fixing a detected defect($); S_{DC} is cost of repairing broken rail($); C_{SDT} is train delay cost due to fixing a broken rail($).

4 WHEEL-RAIL CONTACT STRESS ANALYSIS

The forces arising between wheel and rail contact generate contact stresses in a local volume of the two bodies. Hertzian-Contact-Model describes the local stresses with good accuracy for the most common wheel-rail contact problems (Kalker 1991). The distribution of the contact pressure in this elliptical area represents a semi-ellipsoid, which can be expressed as Eq. (6):

$$P(x,y) = \sqrt{1 - \frac{x^2}{a^2} - \frac{y^2}{b^2}} \times P_o \quad (6)$$

Where, a and b are semi axes, and x and y are the required coordinates to specify the point of contacts on the rail surface based on the lateral rail surface parameter. The location of the contact point depends on the relative position of the wheelset with respect to the rail and the two bodies' profile. If $x = 0$ and $y = 0$ that is if the point of contact is on the centerline of the railhead, the stress is maximum, which is equal to Eq. (7):

$$P(x,y) = P_o = \frac{3N}{2\pi ab} \quad (7)$$

The dynamic effect of static wheel load, F_V^{dyn} is modeled as a statistical distribution by multiplying the static load, N with a magnification factor (K_{dyn}) (Eq. (8) to (10)).

$$F_y^{dyn} = K_{dyn} \times N \quad (8)$$

$$K_{dyn} = 1 + 3n\varphi \quad (9)$$

$$\varphi = \begin{cases} 1 & \text{for } V \leq \frac{60Km}{hr} \\ 1 + \frac{0.5(V-60)}{190} & \text{for } 6o \leq V \leq \frac{300Km}{h} \text{ for Passenger train} \\ 1 + \frac{0.5(V-60)}{80} & \text{for } 60 \leq V \leq \frac{140Km}{h} \text{ for freight train} \end{cases} \quad (10)$$

Where n= 0.15 to 0.25 for different types of track, 0.2 for normal wheel-rail contact and φ is coefficient due to speed (Peter and Thomas 2011), V = train speed.

4.1 Finite Element Modelling

The track components, namely, the rail, the sleepers, the wheel, and the weld are modeled assigning their respective material properties and in their respective positions. Meshing the assembly of those parts as seen in Figure 1 (a and b), the job is submitted for analysis and is then exported to FRANC3D for further analysis (ABAQUS 6.12 1998), (FRANC3D 2011).

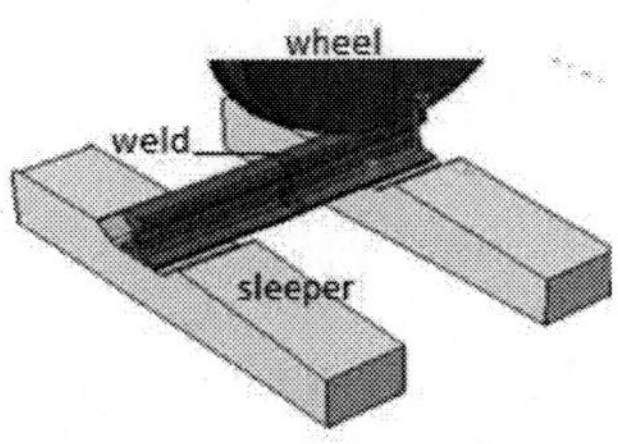

(a)

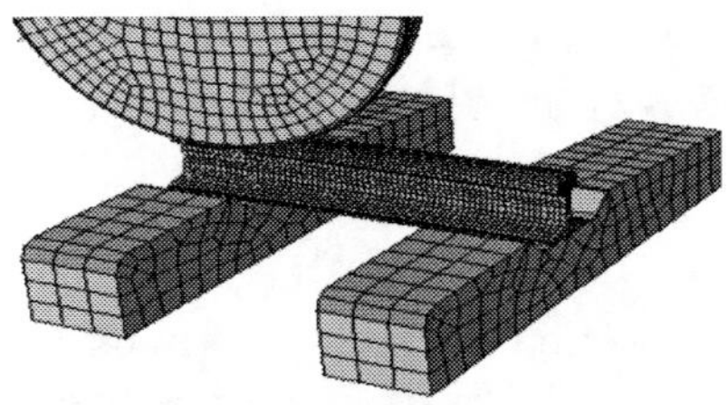

(b)

Figure 1. ABAQUS FE Modeling of welded track.

5 RESULT AND DISCUSSIONS

5.1 Contact Stress Analysis

In rail welds, as the molten metal cools with crystalline re-arrangement, there is a wide variation in modules of elasticity and Poisson's ratio (IACS 2016). Two cases considered (maximum and minimum) Case 1; E_{max} and μ_{min} and Case 2; E_{min} and μ_{max} the response of these Weldments to contact stress is depicted for different train speed. The contact stress is verified from both empirical and FEA.

Table 1. Contact stress comparison of the Base and Welded rail with different crystalline rearrangement.

Speed (Km/hr.)	For base rail			For Weld (Case 2)			For Weld (Case 1)		
	a (mm)	b (mm)	P_o (N/mm^2)	a (mm)	b (mm)	P_o (N/mm^2)	a (mm)	b (mm)	P_o (N/mm^2)
60	8.983	6.747	1546.58	8.92	6.7	1566.73	9.0923	6.829	1509.49
70	9.052	6.799	1558.57	8.99	8.99	1185.86	9.1628	6.8819	1521.19
80	9.121	6.85	1570.38	9.06	6.8	1590.84	9.2322	6.9341	1532.72
90	9.188	6.901	1582.01	9.12	6.85	1602.63	9.2322	6.9341	1567.03
100	9.255	6.951	1593.48	9.19	6.9	1614.24	9.368	7.0361	1555.26
110	9.321	7.000	1604.78	9.26	6.95	1625.69	9.3206	7.0005	1604.78
120	9.385	7.049	1615.92	9.32	7.00	1636.98	9.4999	7.1352	1577.17

5.2 Fatigue Life Computation

FRANC3D is a program that inserts and extends a crack and/or voids in a pre-existing finite element meshes to see a fracture response of material. In this paper, the ABAQUS/CAE model is exported and the initial crack (0.07825mm) found from the UT is inserted and the model is re-meshed with the inserted crack as shown in Figure 2.

(a) (b) (c)

Figure 2. (a) Echograph of UT (b) and (c) Re-meshed Model with elliptical initial crack geometry.

The analysis of a fracture-based fatigue model in FRANC3D estimated the crack growth rate versus number of cycles to failure, and the maximum crack growth versus maximum stress intensity facture to fracture (See Figure 3 (a) and 3(b)). With different initial crack sizes, the response of weldment to remaining number of cycles to failure is determined.

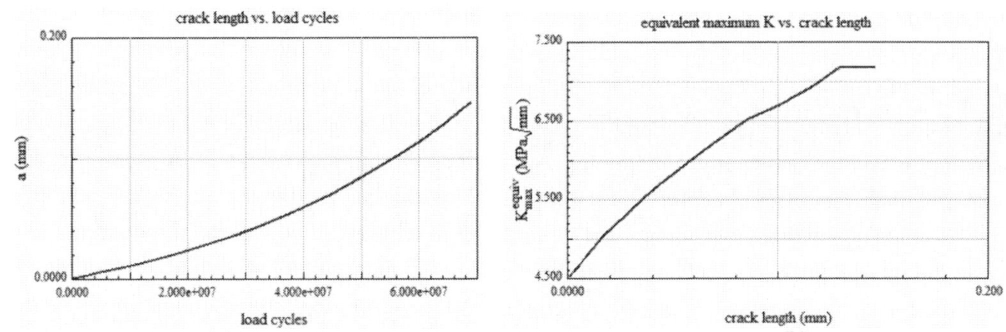

(a) (b)

Figure 3. *(a)* crack growth Vs. number of cycles *(b)* Crack length Vs. K_{max}.

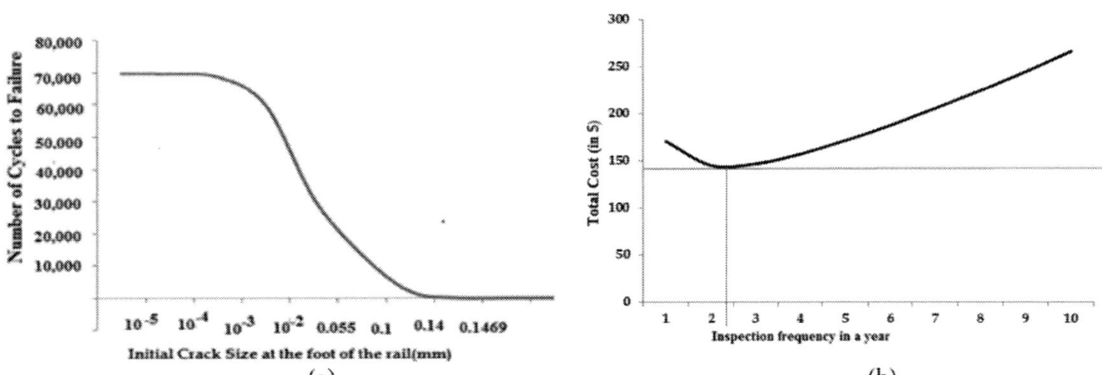

(a) (b)

Figure 4. (a) Load frequency and crack growth (b) Total cost ($) and annual inspection frequency.

5.3 Inspection Interval

Using a model developed by AAR (the cost items are modified based on case study data), the calculated total long-term cost is shown for different inspection frequencies.

As can be seen in Figure 4, the inspection interval which balances the cost of inspection from frequent inspection and failure cost from an infrequent inspection can be balanced at an inspection interval of at least two times a year or nearly every five-month interval.

6 CONCLUSIONS

In this paper, a fracture mechanics based approach was applied to analyze the performance of rail weld joint from Ethiopian railways project. After detecting a flaw in the thermite weld, a finite element method based on Paris law is develop to analyze the fatigue response of the weld subjected to 12.5 ton wheel load at different speed until the critical crack reached. Based on cost model, the inspection interval is estimated that corresponds to minimize the overall defect prevention cost. The remaining number of cycles to failure and fatigue crack growth rate is determined to decrease with a very small increment of an initial crack size.

In addition, as the material property of welds is slightly different from the base rail for a weld with a lower elastic modulus and higher Poisson's ratio (μ), the contact stress at the weldment exceeds the base rail. However, as the crystalline of the weld is rearranged in a manner with larger E and smaller μ the contact stress slightly lessen the contact stress between the base rail and the wheel. The result of this paper can be a starting point to develop a methodology for further detail repair and replacement of the rail, i.e., Maintenance strategy Models.

References

Abaqus 6.12, *Dassault System Center of Simulation Excellence,* Theory Manual, 1998.
David R., *Fatigue*, Department of Material Science and Engineering, MIT, Cambridge MA, 2001.
Det Norske Veritas, *Introduction to Non Destructive Testing Techniques*, Ultrasonic Testing, 2004.
Franc3D, *Fracture Analysis Consultant,* Inc., Reference Manual, 2011.
IACS, *Requirements Concerning Materials and Welding*, International Association of Classification Societies, 2016.
Kalker J., Wheel-Rail Rolling Contact Theory, *Wear*, 144, 243-261, 1991.
Liu, X., Dick, C. T., and Saat, M. R., *Optimizing Ultrasonic Rail Defect Inspection to Improve Transportation Safety and Efficiency,* The Second T&DI Congress in Orlando, FL, June 8-11, 2014.
Napoleon, *Railway Applications-Track Alumino-Thermic Welding of Rails*, OH, 1997.
Orringer, O., *Control of Rail Integrity by Self-Adaptive Scheduling of Rail Tests*, Federal Railroad Administration, Washington, D.C., 1990.
Paris, P., and Erdogan, F., A Critical Analysis of Crack Propagation Laws, *Journal of Basic Engineering*, 85(4), 528-533, 1963.
Peter J., and Thomas, T., *Fatigue Behavior of Steel Highway Bridge Detail*, Virginia, 2011.
Ringsberg, J. W., Josefson, B. L., and Skyttebol, A., *Fatigue Crack Growth in Welded Rails*, Göteborg, Sweden, 2005.
Sergejevs, D., and Mikhaylovs, S., Analysis of Factors Affecting Fractures of Rails Welded by Alumino-Thermic Welding, *Transport Problems*, 3(4), 33-37, 2008.
Sauragh, K., *Reliability Analysis and Cost Modeling of Degrading Systems*, Ph.D. Thesis, Division of Operation and Maintenance Engineering, Luliya University, 2008.
Zerbist. U., Lauden, R., Eden L., and Smith, R., *Introduction to the Damage Tolerance Behavior of Railway Rails*, Philadelphia, PA., 1978.

Interdependence between Structural Engineering and Construction Management
Edited by Ozevin, D., Ataei, H., Modares, M., Gurgun, A., Yazdani, S., and Singh, A.

ISBN: 978-0-9960437-6-2

IMPACT OF INTRODUCING SEMI-RIGID MOMENT FRAMES ON SEISMIC RESPONSE OF BRACED FRAMES

MAHMOUD FAYTAROUNI[1], ONUR SEKER[2], BULENT AKBAS[3], and JAY SHEN[1]

[1]*Dept of Civil, Construction and Environmental Engineering, Iowa State University, Ames, USA*
[2]*Dept of Civil Engineering, MEF University, Istanbul, Turkey*
[3]*Dept of Civil Engineering, Gebze Technical University, Kocaeli, Turkey*

Maximum seismic inelastic drift demand in a steel building with braced frames as primary seismic-force-resisting (SFR) system tends to concentrate in few stories without considering inherent participation of designed gravity-force-resisting (GFR) system in actual structural stiffness and strength. The influence of GFR system on stiffness and strength can be taken into account by considering the composite action in beam-to-column shear connections that exist in modern steel building construction to form actual semi-rigid moment-resisting frames. Therefore, modeling semi-rigid moment frames as an equivalent to the GFR system in braced frame buildings could be utilized as a representative to the strength provided by gravity frames. This paper presents a seismic evaluation of a six-story chevron braced frame, with and without semi-rigid moment frame. Four different cases are investigated under a set of ground motions and results are discussed in terms of story drift distribution along the height. The results pointed out that the current findings lay a foundation to conduct further investigation on the seismic performance of braced frames as designed SFR system together with GFR system.

Keywords: Braced frame, Semi-rigid moment frame, Pushover analysis, Non-linear time history analysis, Seismic drift demand.

1 INTRODUCTION

Concentrically braced frames (CBFs) are considered among the most efficient lateral force resisting systems that absorb energy through buckling and yielding of braces. Their high lateral stiffness and strength placed them among the most commonly used structural systems for resisting earthquakes. Despite the beneficial features that this system exhibit, it presents some major shortcomings in the seismic performance. When braces buckle at any given story level, the post-buckling stiffness and strength of that story drops significantly causing a rapid increase in story drift demand, which leads to an unanticipated ductility demand concentration on braces. Eventually, this drift concentration in isolated stories give rise to brace fracture while upper stories moves similar to a rigid body without being affected from the inelastic deformation demand. In addition, recent research (Uriz 2008) has also found that the concentration of post-buckling deformations in a single story induces significant flexural demand on columns that may lead to a column fracture. More recently, a study by Shen *et al.* (2017) have also found that braces in CBFs are very likely to fracture prior to the design story drift due to large ductility demand on braces that have been mutually influenced by the inelastic deformation in beams and columns. In view of these shortcomings, considerable attempts have been devoted to obtain a more desirable and uniform inelastic deformation distribution either by providing a more redundant and ductile backup system like moment resisting frames (Goel 1992) or by controlling buckling in braces (Seker *et al.* 2017) or both (Kiggins and Uang 2006).

Further, the inherent participation of GFR system can contribute significantly to the lateral stiffness and strength of the steel braced building, and capable of limiting the drift concentration subsequent to brace buckling. In fact, the impact of GFR system on the stiffness and strength can be taken into account by considering the composite action in beam-to-column shear connections that exist in modern steel building construction to form actual semi-rigid moment-resisting frames. Majority of existing studies consider the effect of GFR system by modeling the gravity columns only. Nevertheless, very few studies can be found in the literature investigating the impact of gravity beams with shear tab connections together with gravity columns on the seismic performance of CBFs, predominately when they act compositely with the concrete floor system. Although no study has been found on special CBFs, studies sponsored by AISC on non-ductile CBF buildings (i.e., Shen *et al.* 2017 and Wen *et al.* 2017) have found that beam-to-column shear connections are capable of resisting substantial moments, about 30% to 50% of the beam plastic moment. Thus, the focus herein is on the seismic behavior of braced frames coupled with semi-rigid moment frames, employed as an equivalent to the strength provided by the GFR system in steel braced building. Accordingly, six-story braced frame with chevron configuration was evaluated to study the impact of semi-rigid moment frames, referred to as MFs hereafter.

2 BUILDING DESCRIPTION

The structure considered in this study is one of the six-story braced frames designed by Goel and his research associates in late 1980s (Goel 1992), namely F4MR. This structure presents an opportunity to evaluate the impact of MFs combined with CBFs for several reasons: (a) the response of this structure together with a group of other structures, have influenced the development of CBFs design requirements based on which early seismic design provisions were recommended; (b) braces in CBF were designed with width-to-thickness ratios (b/t) that are comparable to the current seismic provisions (ANSI/AISC 341-16 2016) to ensure adequate ductility; and (c) it reflects the design philosophy of the early days and represents a good example of the existing structures. The six-story floor plan and the elevation view of the lateral force resisting frames including the steel section sizes are shown in Figure 1.

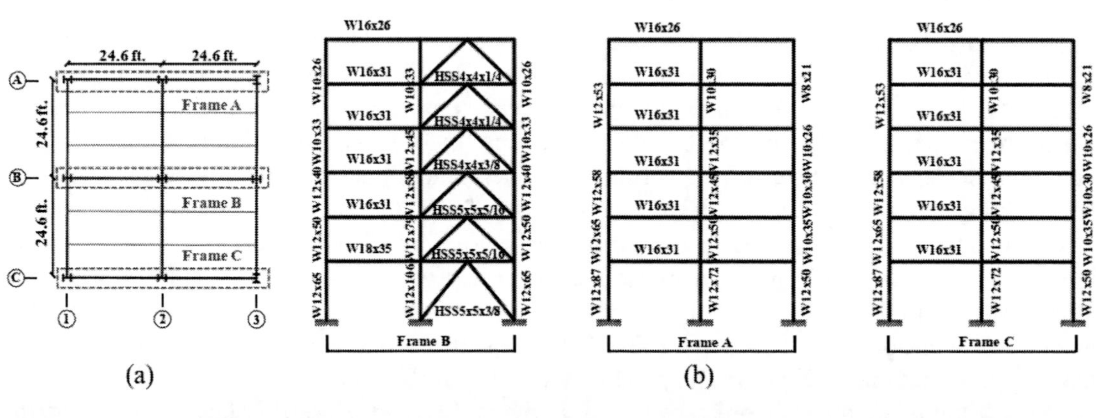

Figure 1. Six-story dual structural system: (a) floor plan view; and (b) elevation of braced bay (i.e., Frame B) and unbraced bays (i.e. Frames A and C) with cross-sectional dimensions.

As illustrated in Figure 1(a), three SFR systems were located on grids A, B, and C. The braced frame with chevron configuration was on one bay of Frame B and the moment frame on the other unbraced bay, whereas in both bays of Frames A and C, SFR systems were moment frames, as seen in Figure 1(b). The first story height was 14.8 feet, while all remaining stories were 11.2 feet. Detailed information on the structure can be found in Hassan (1991). Note that the girder-to-column connections of the chevron-braced bay presented in Figure 1 were moment connections. Thus, this braced bay (in Frame B) is also acting as a moment frame. In a typical design of braced frames, however, these connections are often deemed simple connections. In addition, the base supports of the SFR in all bays were considered perfectly fixed which is also not common in design.

3 CASES OF STUDY AND NONLINEAR PUSHOVER ANALYSIS

Several cases were derived from the above-described six-story structure to establish the impact of MFs on the CBFs inelastic behavior. Maintaining the original structural configuration and cross-sectional sizes of Frames A, B, and C in Figure 1, four cases were considered, as shown in Figure 2. These cases represent MFs integrated with CBF system with different base shear proportions. The case definitions and their corresponding design base shear fractions are as follows: (a) Case 1, CBF only; (b) Case 2, CBF with one bay of MF, with strength equivalent to 5% of the CBF base shear; (c) Case 3, CBF with two bays of MF, with strength equivalent to 10% of that of CBF; and (d) Case 4, CBF with five bays of MF, with strength equivalent to 25% of the CBF.

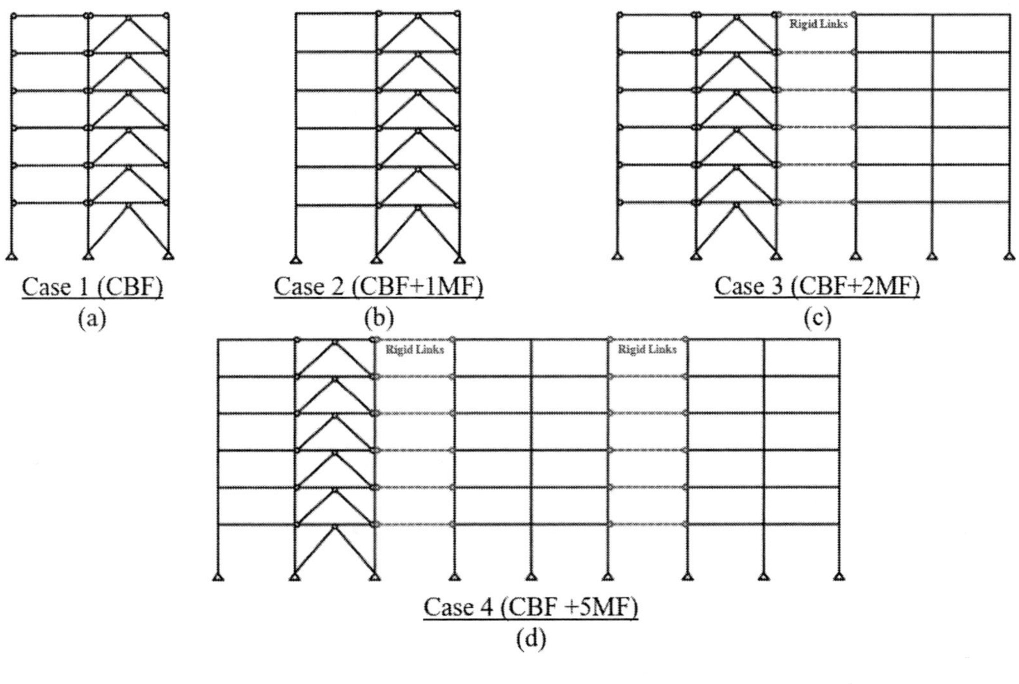

Figure 2. Frame cases considered in this study: (a) Case 1 regards CBF only; (b) Case 2, CBF with one MF; (c) Case 3, CBF with two MFs; and (d) Case 4, CBF with 5 MFs.

These base shear proportions were determined by means of pushover analyses that were exclusively performed on the CBF system, the moment frame next to the braced bay, and one of the two-moment frames located on grids A and C in Figure 1(a), respectively. Note that the six-story frame was modified by incorporating simple connections in the chevron-braced bay, and the base support of the SFR systems in all bays were assumed to be pinned to be consistent with the current design practice. The results of the pushover analyses are illustrated in Figure 3. From the markers shown at the initial stiffness in Figure 3, one can notice that moment frame in the braced bay has an elastic base shear capacity of 6.0 kips, which is about 5% of that of CBF. This capacity was increased to 10 kips, about 10% of the CBF base shear, when the two-moment frames located on grids A or C were incorporated. Also, a few observations can be made by investigating Figure 3: (a) in the CBF, the contribution of framing action to the strength after braces buckled appears to be significant compared to the scenario when the connections are assumed to be perfectly pinned (without framing action), which is generally the case in the traditional design practice; and (b) the residual post-buckling capacity of CBF w/o framing action is almost equivalent to the post-yielding strength of one moment frame. Note that, the seismic weight was 1356 kips, and the numerical models used for the pushover analyses were built in the Open System for Earthquake Engineering Simulation Platform, *OpenSees* (McKenna *et al.* 2000) following an experimental validation discussed in the next section.

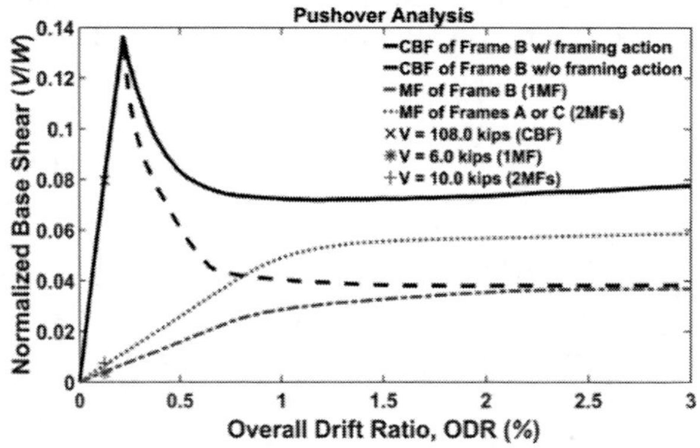

Figure 3. Pushover response of CBF and MFs of Frames B and A (or C).

4 NUMERICAL MODELS CALIBRATION

The numerical models of Cases 1, 2, 3, and 4, illustrated in Figure 2, were built as 2D models in *OpenSees*. The reliability of the models was validated by the pseudo-dynamic test results of a full-scale six-story steel building tested at the Building Research Institute in Japan (Foutch *et al.* 1986). In fact, the structural scheme of the building designed by Goel (2002) in Figure 1 and the one considered in this study, were originated from the latter test. Under a cooperative research program between the U.S. and Japan in 1982, the building shown in Figures 4(a) and (b) was tested to investigate the seismic behavior of steel structures with different LFRSs, and to provide design recommendations accordingly. The experimental program consisted of four phases. The focus herein is on Phase I program that included a concentrically braced frame with chevron (inverted-V) configuration tested under the 1978 Miyagi-Ken-Oki (N-S) earthquake. As Figure 4(c) compares the simulated displacement histories of each story level with that reported from the experimental test up to fracture of braces that occurred at 8.2 seconds. As can be interpreted from the figure, the tested and simulated displacement histories are in good agreement. Therefore, one can notice that analyzing the developed analytical models (i.e., frames of Cases 1,2,3, and 4) under several ground motions can be considered adequately reliable. Note that a fracture criterion was not introduced in the analytical models, and hence, response beyond 8.2 seconds was not illustrated as discrepancies become significant.

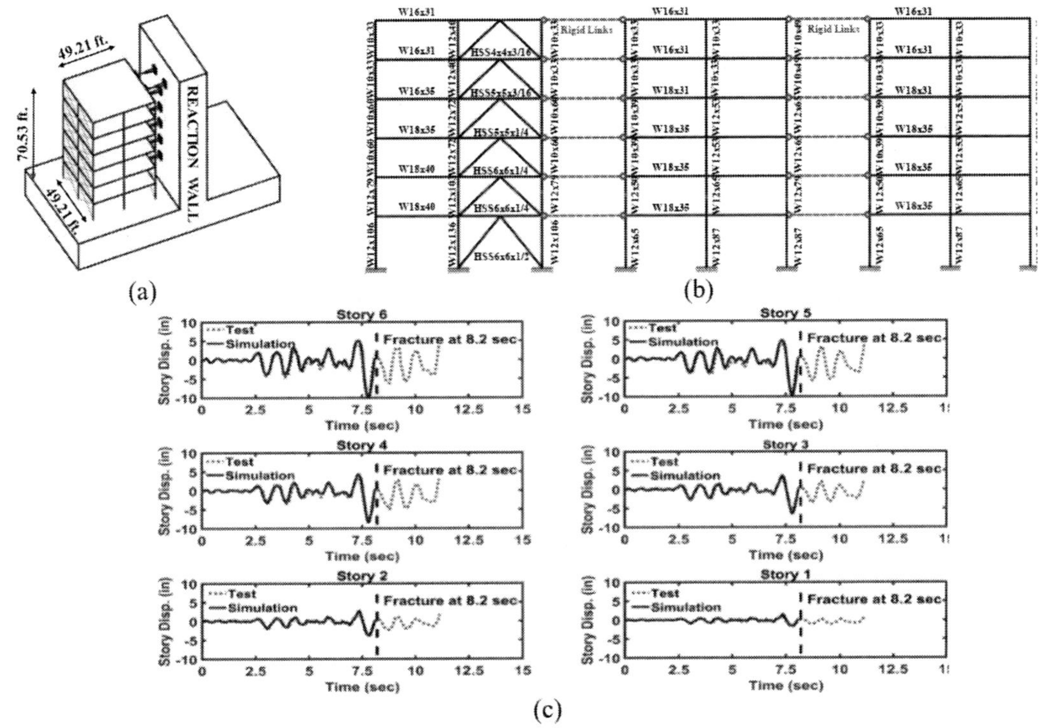

Figure 4. Comparison between the results of the analytical simulations of the tested six-story structure of Phase I experimental program with the test response: (a) tested six-story structure; (b) elevation of the lateral force resisting systems; and (c) simulated story displacement versus that of the test.

5 ANALYSES, RESULTS, AND CONCLUSIONS

Nonlinear dynamic responses of Cases 1, 2, 3, and 4 subjected to four different ground motion records are studied in terms of story drift ratio (SDR). This was performed to investigate the impact of incorporating MFs designed with three different strength levels on the seismic performance of CBFs. The first fundamental period (T_1) of the considered cases are 0.9, 0.82, 0.81, and 0.77 seconds. The unscaled response spectra of the four records are shown in Figure 5(a) and the drift response of the considered cases to these records are plotted in Figures 5(b) through (e). Accordingly, one can observe the following:

(a) The deformation patterns of Case 1 (CBF only) and Case 2 (CBF+1MF) were similar under all applied ground motion excitations. The seismic drift demands were concentrated in the first story level and no apparent trend was observed that MFs designed for 5% of the CBF base shear yield to a uniform drift distribution along the structural height.

(b) As the number of rigid bays and accordingly the participation of MFs to the lateral strength and stiffness increase (i.e., Cases 3 and 4), the mitigation in the inter-story drift demands becomes significant on the order of 50 (Figures 5b, c and d) to 75% (Figure 5e) compared with the other two cases. Although the strength of MFs in Case 3 (CBF+2MF) was based on slight participation of about 10% of the CBF base shear, the peak SDR demand obtained from Case 3 was equal or less than half of that of Case 2 under all ground motions. Thus, incorporating the MFs seemed to have a significant impact on the CBF behavior since a substantial reduction in the demand and a more uniform drift demand distribution throughout the height of the CBF were achieved even when such weak and low-redundant moment frames were introduced.

(c) The results pointed out that the current findings lay a foundation to conduct further investigation on the seismic performance of braced frames together with GFR system.

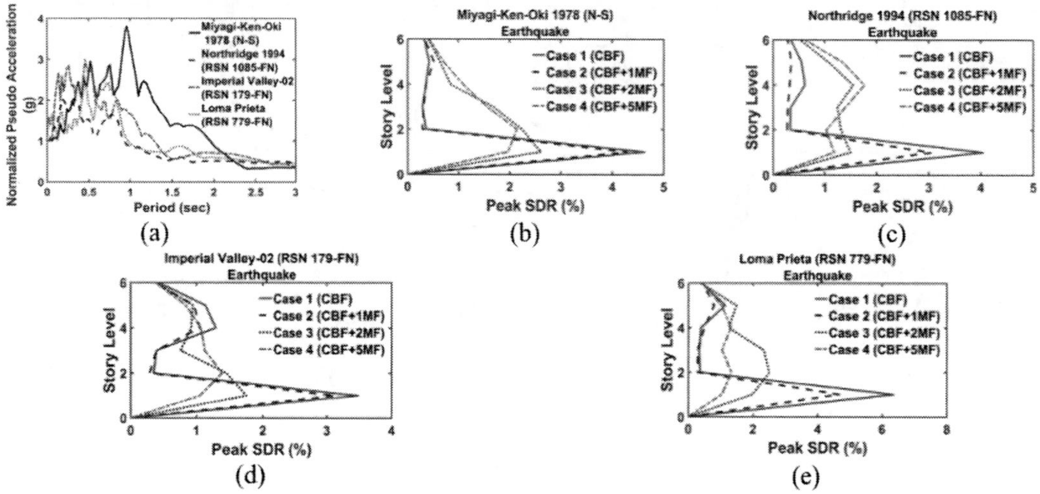

Figure 5. Applied ground excitations and response of the considered cases: (a) response spectra; (b) peak SDR response to Miyagi-Ken-Oki earthquake; (c) peak SDR response to Northridge earthquake; (d) peak SDR response to Imperial Valley earthquake; and (e) peak SDR response to Loma Prieta earthquake.

References

ANSI/AISC 341-16, *Seismic Provisions for Structural Steel Buildings,* Chicago (IL): American Institute of Steel Construction, 2016.

Foutch, D. A., Goel, S. C., and Roeder, C. W., *Preliminary Report on Seismic Testing of a Full-Scale Six-Story Steel Building*, University of Illinois Engineering Experiment Station, College of Engineering. University of Illinois at Urbana-Champaign, 1986.

Goel, S. C., Earthquake Resistant Design of Ductile Braced Steel Structures, *Stability and Ductility of Steel Structures under Cyclic Loading*, 297-308, 1992.

Hassan, O. F., *Modeling of Bracing Members and Seismic Behavior of Concentrically Braced Steel Structures*, PhD Thesis, Horace H. Rackham School of Graduate Studies, University of Michigan, 1991.

Kiggins, S., and Uang, C.-M., Reducing Residual Drift of Buckling-Restrained Braced Frames as a Dual System, *Engineering Structures,* 28 (11) 1525–1532, 2006.

McKenna, F., Fenves, G. L., and Scott, M. H., Open System for Earthquake Engineering Simulation, *University of California, Berkeley, CA.*, 2000.

Seker, O., Akbas, B., Seker, P.T., Faytarouni, M., Shen, J. and Mahamid, M., Three-Segment Steel Brace for Seismic Design of Concentrically Braced Frames, *Journal of Constructional Steel Research, 137*, 211-227, 2017.

Shen, J., Seker, O., Akbas, B., Seker, P., Momenzadeh, S. and Faytarouni, M., Seismic Performance of Concentrically Braced Frames with and without Brace Buckling, *Engineering Structures*, 141, 461-481, 2017.

Shen, J., Rou, W., Akbas, B., Seker, O. and Uckan, E., 2015. Near-collapse behavior of steel buildings with non-ductile concentrically braced frames. *Journal of Constructional Steel Research, 113*, pp.101-114.

Uriz, P., *Toward Earthquake-Resistant Design of Concentrically Braced Steel-Frame Structures,* Pacific Earthquake Engineering Research Center, 2008.

Wen, R., Akbas, B. and Shen, J., 2013. Practical moment–rotation relations of steel shear tab connections. *Journal of Constructional Steel Research, 88*, pp.296-308.

Interdependence between Structural Engineering and Construction Management
Edited by Ozevin, D., Ataei, H., Modares, M., Gurgun, A., Yazdani, S., and Singh, A.

ISBN: 978-0-9960437-6-2

ANALYSIS OF BUCKLING AND BENDING FAILURE OF ALUMINIUM COLUMN SECTIONS USED IN EMERGENCY RESTORATION TOWERS

SARAH RAMJI[1], CIRO PASINI[1], YONGJIAN CHANG[2], and OYA MERCAN[2]

[1]*Tower Solutions Inc., Markham, Canada*
[2]*Dept of Civil Engineering, University of Toronto, Toronto, Canada*

This paper presents an experimental analysis of buckling and bending failure modes of a 2.58 m long, 460×460 mm^2 6061-T6 Aluminum alloy column section used in emergency restoration towers. The main objective is to determine the bending and buckling load capacities of the column section through experiments, as these values are critical in an emergency tower (guyed mast) design. Within the context of this overarching goal, a secondary objective is to ascertain whether the presence of certain manufacturing non-conformance affects the loading capacity of the section significantly. Finally, finite-element analysis (FEA) simulations are conducted in order to compare the experimental data with numerical results. The results show that the ultimate bending and buckling load capacities of the column section are 383 kN and 3,868 kN respectively. Furthermore, the results indicate that the presence of manufacturing non-conformances such as air bubbles and delamination do not have a detrimental effect on the load capacity of the column. Of the two non-conformances studied, the specimen with bubbles had a 1% difference from the good specimen, and the delaminated specimen had a 10% deviation. Comparison of experimental data with FEA simulation results shows that the numerical solution tends to overestimate the stiffness of the column, and that the FEA approach may require further calibration.

Keywords: Modular construction, Buckling load, Bending behavior, Finite element analysis.

1 INTRODUCTION

High voltage transmission lines are vulnerable to breakage and service disruption when extreme weather conditions impose heavy loads on permanent towers, causing them to collapse. Emergency restoration systems (ERS) are used to temporarily restore the conductors so that electrical power can be resumed quickly, while the permanent towers are replaced. Although an extreme weather event can pass by quickly, the temporary tower must be able to withstand even extreme conditions, particularly when they are used for maintenance applications. These emergency restoration structures are essentially guyed mast structures. The tower column, or mast, consists of a 2.58 m long, 460×460 mm^2 6061-T6 square aluminum section, having a corner rail that permits the travel of a safety tether and a hoisting apparatus, allowing line workers to remain attached at all times. These masts can be joined together to form a column of any desired height, held in place by guy wires. These masts are essential to the strength of the emergency tower, and determination of the number and position of guy wires required for the tower design, and hence accurate determination of the properties of these emergency tower sections is beneficial. The main objective of this paper is to analyze the bending and buckling failure of an aluminum section used in emergency restoration towers through experiments. A secondary objective of the present study is to determine whether the presence of manufacturing non-conformances such as bubbles and delamination has a detrimental effect on the loading performance of the column section. Finally, a comparison between FEA results and experimental data is made for both bending and buckling to gain a better understanding of the validity of the numerical simulations.

2 LITERATURE REVIEW

Some previous studies have presented testing results for aluminum alloy hollow section columns. Zhu and Young (2006) investigated the behavior of circular hollow aluminum column sections. They studied the effects of welding on aluminum alloy columns and observed that the experimental ultimate load of welded columns was 54-76 % of the ultimate load corresponding to the non-welded columns. They also developed a non-linear finite element model and found that the numerical models developed matched the experimental results. Zhu and Young (2008) also conducted compression and bending tests for circular, square and rectangular section aluminum beam columns. Fan *et al.* (2013) tested a series of aluminum alloy H-type sections fabricated using 6082-T6 heat-treated aluminum alloy subject to axial compression between two pinned ends. The overall buckling and axial load bearing capacities were obtained. Dimopoulos and Gantes (2012) investigated the experimental and numerical behavior of cantilevered shells with opening and stiffening, which reflected the geometric characteristics of wind turbine tunnels.

3 EXPERIMENTAL SETUP, RESULTS, AND DISCUSSION

3.1 Test Procedure and Setup

The bending and buckling tests on the tower section (mast, or column) were conducted using the Baldwin Universal Testing Machine at the Department of Civil Engineering, University of Toronto. This is a large-scale test machine capable of applying both tension and compression loads up to 5400 kN, on specimens up to 6 m high. For both these tests, the loads were applied using a pin connection at the top – a 610 mm diameter plate that can rotate to a maximum of 4 degrees. The load cases, maximum load applied before failure and the nature of the specimens tested are listed in Table 1.

For the buckling test shown in Figure 1, the load is applied on the top along the vertical axis of the tower modular section specimen via a pivoting head connection. A 600×600×50 mm^3 steel plate is fixed to the bottom of the modular tower section by 8 ASTM A324 bolts in order to fix the end of the column on the ground. The bending test was conducted by setting the specimen on two roller supports 100 mm distance away from the ends, as shown in Figure 2. The bending load is applied at the middle of the modular section via a pivoting head connection and a steel plate.

Table 1. Load case, specimen type, and maximum load for all tests.

No.	Load Case	Specimen type	Max load (kN)
1	Bending1	Rejected specimen (with bubbles)	383.88
2	Bending2	Good specimen	386.08
3	Buckling1	Rejected specimen (with bubbles)	3868.4
4	Buckling2	Good specimen	3801.5
5	Buckling3	Good specimen	3822.6
6	Buckling4	Good specimen	3735.4
7	Buckling5	Delaminated specimen	3351.6

Local strain and stress were determined using strain gauges. These gauges were positioned at areas predetermined to be critical from an initial FEA simulation. Based on the results of these tests, the number of strain gauges was adjusted for the subsequent tests. The gauges were also subsequently concentrated on areas where failure was determined to occur from the initial experimental tests, particularly on the heat affected zones (HAZ) of the aluminum welds. Figure 3 and 4 show the location of strain gauges for the bending and buckling tests, respectively.

The load for the buckling tests was increased at a rate of approximately 1.75 kN/s before the yield point, and 0.65 kN/s after. For the bending test, the load was initially increased at a rate of approximately 0.4-0.6 kN/s, and the load rate was dropped to approximately 0.2 kN/s once the welds cracked. The load-displacement relationship, load-local strain relationship, and the failure modes of all the specimens were captured during the experiments. In addition, the maximum load carrying capacity, the ultimate compressive displacement, and the initial stiffness of the tower modular section specimens were determined.

Figure 1. Buckling test experimental setup.

Figure 2. Bending test experimental setup.

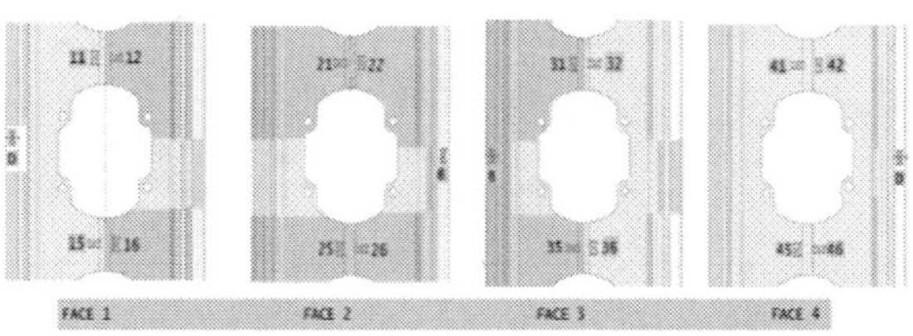

Figure 3. Strain gauge locations for bending tests.

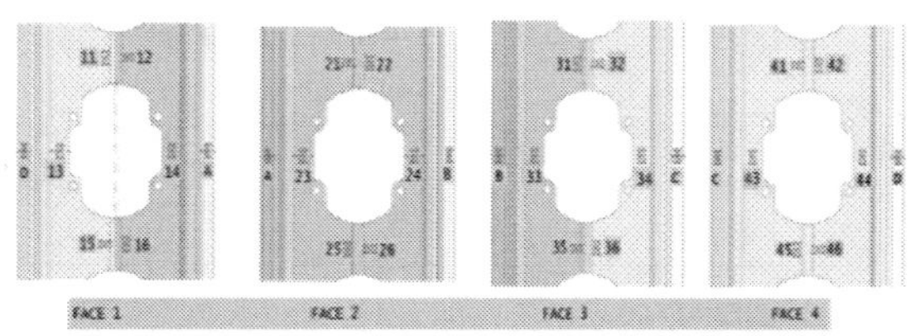

Figure 4. Strain gauge locations for buckling tests.

3.2 Bending Tests

Two bending tests were conducted, and the results were compared with the FEA simulations. The maximum loads reached before failure for first (rejected) and second (good specimen) experimental tests before failure were 383 kN and 386 kN, respectively, as shown in Figure 5. In the corresponding FEA simulation, the maximum load reached before failure was almost two times greater than the experimental values, and the initial stiffness was marginally higher in the

FEA. Comparisons between the experimental results of the two specimens showed that the maximum displacements for the rejected specimen were 23.5 mm and for the good specimen was 28.5 mm. Comparing failure modes, the reject specimen had six 'drop' points, each of which indicates cracking in the aluminum welds. As the load increased over the load bearing capacity of the welds, they cracked, causing the applied load to decrease. This is evidenced by the drop in the load-displacement curves. For the test of the 'good' specimen, only two of the welds fail, implying that the aluminum section had carried a greater load beyond the ultimate tensile strength of the material. Figure 6 shows the two specimens at the completion of the tests, and Table 2 provides the stress values as determined by the strain gauges. It is observed that the values are mostly below the yield stress of 275 MPa.

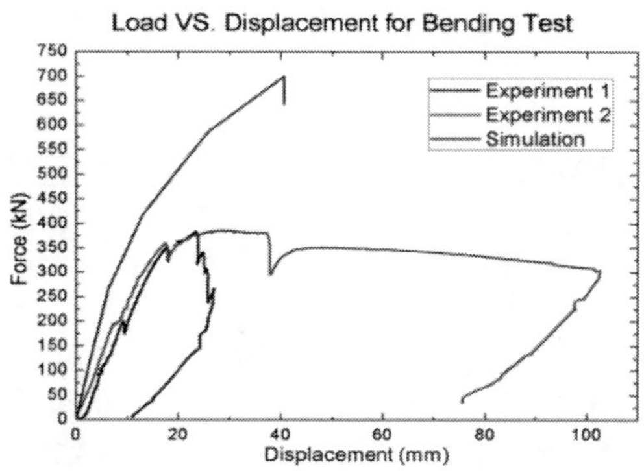

Figure 5. Load-displacement curves for the bending tests.

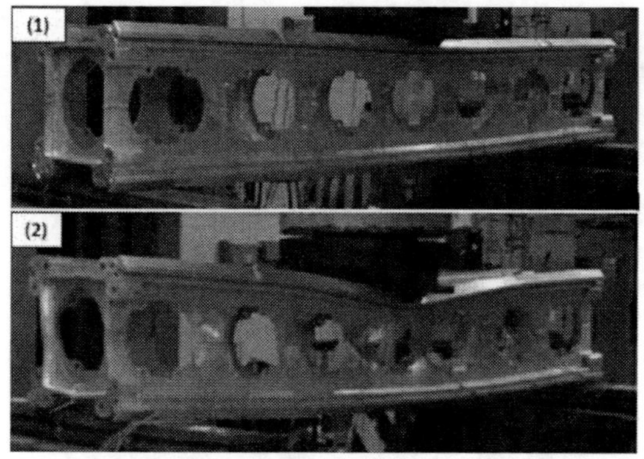

Figure 6. The specimens at the completion of bending test 1 and 2.

Table 2. Results of the bending tests.

	Corner D Stress (MPa)	Face 1 Stress (MPa)				Face 2 Stress (MPa)				Corner B Stress (MPa)
Location		11	12	15	16	21	22	25	26	
Simulation	203.4	13.4	92.8	197.7	4.1	-145.3	-18.9	-23.3	-121.4	204.0
Bending 1		-42.2	167.6	-32.7	200.1	-284.5	43.0	60.2	168.7	130.4
Bending 2		289.9	-111.0	281.9	-141.2	-88.4	-558.0	152.9	157.9	139.5
	Corner B Stress (MPa)	Face 3 Stress (MPa)				Face 4 Stress (MPa)				Corner D Stress (MPa)
Location		31	32	35	36	41	42	45	46	
Simulation	204.0	5.3	-30.6	-4.9	3.3	-127.5	-62.8	-20.8	-158.0	203.4
Bending 1	130.4	-8.6	29.8	-7.6	41.3	149.9	-77.1	-275.8	42.5	
Bending 2	139.5	165.4	-41.6	193.2	-44.8	272.1	204.6	-72.1	-558.0	

3.3 Buckling Tests

As mentioned earlier, five specimens were used for the buckling tests, and the loads applied before failures are listed in Table 1. The load-displacement curves for these five specimens are shown in Figure 7. The displacements ranged from 30 mm to 38 mm, and the initial stiffness for each test is almost the same. The FEA simulation shown in Figure 8 indicated a higher initial stiffness, but the same maximum load, as shown in Figure 7. As mentioned before, the first two buckling tests used the maximum number of strain gauges, which were strategically positioned near the welds between each hole. Figure 4 shows the initial location of the strain gauges, and Table 3 shows the corresponding stress values. Stress values obtained for locations 11, 12, 21, and 22 are omitted from the table since the measured stress is not critical at those locations. It can be seen that most values measured by the strain gauges are in the elastic range, and only a few are beyond the yield strength. It is further observed that the strain gauges showing readings in the plastic range are predominantly in the corners of the specimen.

The first two tests showed that the critical portions of the section are around the third hole of the tower section, where global buckling was observed. Strain gauges were thus increased at this location. Table 3 shows the results of these tests, which are similar to the first two buckling tests. However, it can be observed that the strain gauges concentrated around the third hole of the tower section have reached their saturation point, which corresponds to a stress of 558 MPa. Finally, considering the two manufacturing non-conformances studied, the specimen with bubbles did not have any observed strength reduction, while the delaminated specimen failed at a load approximately 10% lower than the other specimens.

Table 3. Results of the buckling tests at the locations and corners shown for each of the 6 simulations

	D (MPa)	Face 1 Stress (MPa)				A (MPa)	Face 2 Stress (MPa)				B (MPa)
Location		13	14	15	16		23	24	25	26	
Simulation	-244.4	-315.9	-298.0	-22.9	-2.9	-253.0	-296.7	-293.9	2.5	-14.6	-176.4
Buckling 1	/	/	/	87.9	281.6	-351.1	/	/	180.5	-100.1	/
Buckling 2	-558.0	/	/	176.7	130.8	/	/	/	95.8	-284.6	268.6
Buckling 3	-240.0	-558.0	-558.0	22.0	12.6	-198.3	-558.0	-558.0	36.9	31.6	-179.9
Buckling 4	278.4	-558.0	-558.0	95.1	-558.0	-558.0	-558.0	-558.0	103.3	39.6	-345.7
Buckling 5	-82.2	-558.0	-558.0	/	/	281.1	-327.6	-558.0	/	/	217.0
	B (MPa)	Face 3 Stress (MPa)				C (MPa)	Face 4 Stress (MPa)				D (MPa)
Location		33	34	35	36		43	44	45	46	
Simulation	-176.4	-298.1	-303.3	-13.2	12.2	-180.5	-317.6	-314.2	12.7	-15.7	-244.4
Buckling 1	/	/	/	-277.6	77.1	-266.2	/	/	48.5	-74.5	/
Buckling 2	268.6	/	/	-72.1	48.3	/	/	/	-53.8	98.9	-558.0
Buckling 3	-179.9	-558.0	-558.0	-28.3	57.1	-182.1	-558.0	-558.0	33.5	-49.6	-240.0
Buckling 4	-345.7	-558.0	-558.0	143.0	202.0	287.9	-558.0	-558.0	-63.9	9.4	278.4
Buckling 5	217.0	-558.0	-558.0	-295.3	67.5	-558.0	-558.0	-558.0	/	/	-82.2

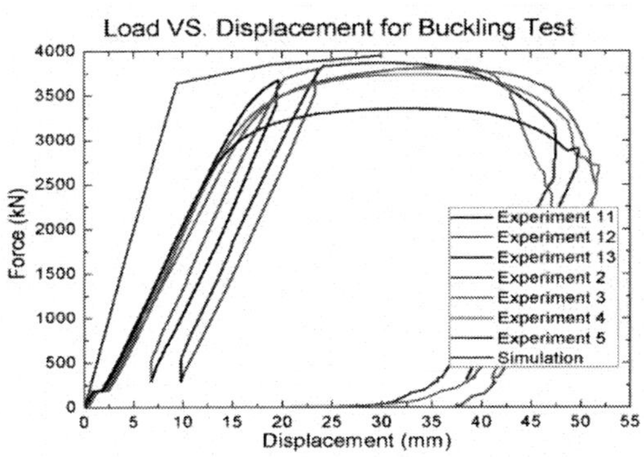

Figure 7. Load-displacement curves for the buckling tests.

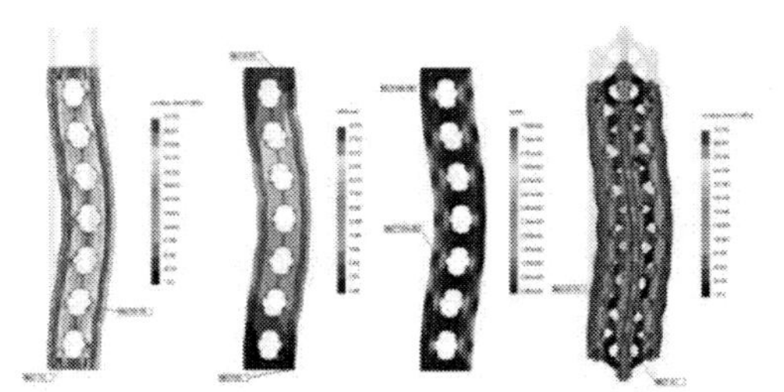

Figure 8. FEA simulation results for the buckling test.

3.4 Stress-Strain Curve

Using load-displacement curves discussed above and the local strain measurements from the strain gauges obtained during the experiment, the strain values can be translated into stresses given the stress-strain relationship of the aluminum alloy. The elastic modulus is 69,000 MPa; the yield strength is 275 MPa, and the ultimate tensile strength is 310 MPa. A simplified stress-strain relationship was also obtained from an FEA analysis and is shown in Figure 9. The local stresses are obtained from this curve.

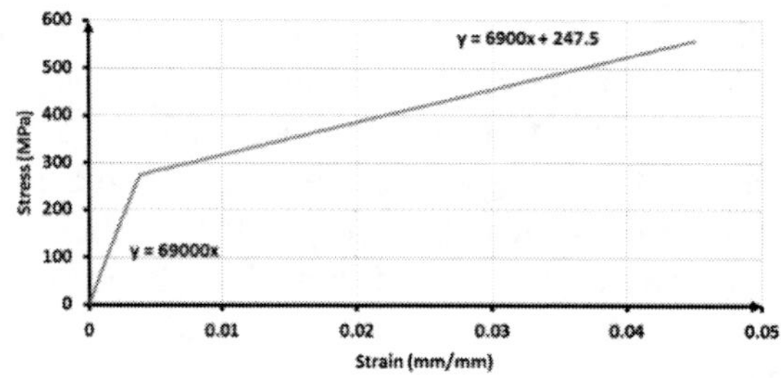

Figure 9. Simplified stress-strain curve for the aluminum alloy.

4 CONCLUSIONS

This paper investigates the bending and buckling load capacity of an aluminum column section used in Emergency Restoration Towers. The maximum bending and buckling load capacities of the column section are 383 kN and 3,868 kN, respectively. Moreover, the presence of certain manufacturing non-conformances does not reduce the maximum load capacity of the section significantly. The bubble non-conformances reduce the bending load capacity by less than 1% without reducing the buckling capacity. The delaminated specimen shows an approximately 10% reduction in buckling capacity. Finally, a comparison of experimental results with FEA simulations shows that the numerical stiffness tends to be underestimated by the FEA approach, and that the FEA approach requires further calibration.

Acknowledgments

The authors acknowledge financial support from the Government of Canada's NSERC ENGAGE program and thank Harry Sildva and Lino Borroto for their early contributions.

References

Dimopoulos, C. A., and Gantes, C. J., Experimental Investigation of Buckling of Wind Turbine Tower Cylindrical Shells with Opening and Stiffening Under Bending, *Thin-Walled Structures*, 54, 140-155, March 2012.

Fan, F., Wang, Y., Lin, L., and Qian, H., Experimental Study on Axial Compression Performance of High-Strength Aluminum-Alloy Columns with H-Section, *Applied Mechanics and Materials*, 274, 459-462, 2013, January 2013.

Zhu, J. H., and Young, B., Experimental Investigation of Aluminum Alloy Circular Hollow Section Columns, *Engineering Structures*, 28, 207-215, January 2006.

Zhu, J. H., and Young, B., Numerical Investigation and Design of Aluminum Alloy Circular Hollow Section Columns, *Thin-Walled Structures*, 46, 1437-1449, December 2008.

Interdependence between Structural Engineering and Construction Management
Edited by Ozevin, D., Ataei, H., Modares, M., Gurgun, A., Yazdani, S., and Singh, A.

ISBN: 978-0-9960437-6-2

A PULL-DOWN DYNAMIC ANALYSIS OF TWO-SPAN STEEL FRAMES SUBJECTED TO PROGRESSIVE COLLAPSE

MOHAMMADJAVAD ESFANDIARI and GIRUM URGESSA

Dept of Civil, Environmental, and Infrastructure Engineering, George Mason University, Fairfax, USA

Progressive collapse, also known as disproportionate collapse, describes a chain reaction of structural element failures in which a primary structural element failure results in the failure of adjoining structural elements. It eventually causes widespread structural damages and a disproportionate collapse. While high level finite-element models incorporating non-linear dynamic analysis will produce more realistic results in progressive collapse scenarios, they are computationally time consuming. Therefore, the development of a non-linear time history pull-down model that is validated with experimental results would be beneficial for producing acceptable and efficient design solutions, particularly for practicing structural engineers. In this paper, a non-linear time history pull-down model of a two-span steel frame is analyzed in ETABS. The ETABS model results are compared with experimental results of two steel frames with two-spans conducted by the National Institute of Standards and Technology (NIST). The NIST experiments include beam-column assemblies from the second-floor framing system of a ten-story building and each span is 20 feet long. The numerical results from ETABS pull-down analysis showed good agreement with the results from the NIST experimental study.

Keywords: Pull-down analysis, Finite element model, Time-history, Disproportionate collapse.

1 INTRODUCTION

Progressive collapse is defined as a chain reaction of structural system failure initiated from a local failure of a structural member due to terrorist attacks, gas explosions, major earthquakes, or fires. Failure in a local structural element leads to an increase in internal forces and an overload in nearby structural elements, ultimately causing a progressive collapse of the entire or a disproportionately large part of the structure. The collapses of the Alfred P. Murrah Federal Building in 1995 in Oklahoma City, the Twin Towers of the World Trade Center in 2001 in New York City, and recently the Plasco Building in 2017 in Tehran, demonstrate the disruptive nature of progressive collapse. The consequences include economic losses and casualties of human lives. Even though it is impossible to predict the timing of natural or man-made causes that can lead to progressive collapse, structures can be designed to better withstand progressive collapse.

In recent years, particularly after the events of September 11, 2001, many studies consisting of experimental programs and numerical studies have been conducted on progressive collapse of structures. Furthermore, the General Services Administration (GSA 2003) and the Department of Defense (Department of Defense 2005) have published guidelines aiming at reducing the potential for progressive collapse when designing new or renovated buildings. Byfield *et al.* (2014), Wang *et al.* (2014) and Qian and Li (2015) have provided extensive reviews on this subject.

NIST (Sadek *et al.* 2011) generated two finite element models of structural frames in DYNA: a detailed model with a large number of elements, primarily solid and shell elements, and a reduced model with a limited number of elements, primarily beam and spring elements. DYNA accounts for both geometrical and material nonlinearities. The center stub column of the frame was pushed down under displacement control until failure occurs in both models. Moreover, an effective stress versus effective plastic strain curve was specified. For this purpose, the engineering stress-strain curve was converted to a true stress vs. plastic strain curve and several coupon tests were conducted for each steel type. Iterative finite element analyses of the tensile test were carried out, and the failure strain and the extrapolation of the true stress-strain curve were adjusted until quantitative agreement of the measured and calculated engineering stress-strain behavior in the softening region beyond the maximum stress was reached. An arrangement of spring elements was used to represent the behavior of diagonal braces in all models. The results of the numerical study were compared with the result from the experimental study and found to be in an excellent agreement (Sadek *et al.* 2011).

Kaewkulchai and Williamson (2004) created a beam-column element with multi-linearization and lumped plasticity. They studied the significance of dynamic load redistribution following the failure of one or more elements. They also created a simplified finite element model to assess the impact effect due to falling failed members on the remaining structure. Their results indicated that dynamic redistribution of loads is a significant feature of the progressive collapse problem, which should be accounted for in order to avoid estimates of capacity that are not conservative (Kaewkulchai and Williamson 2004, 2006).

While general purpose finite element software programs, such as ABAQUS, DYNA, and ANSYS are utilized for understanding progressive collapse mechanism in a research setting, structural design and capacity checks per building codes are not performed in these programs due to the expertise needed to post-process and interpret results (Lallotra and Singhal 2017). In fact, Hartmann *et al.* (2008) showed that the computations associated with the simulation of collapses of real-world structures based on conventional methods are very costly. Consequently, progressive collapse is rarely considered in conventional design limiting engineers to rely only on structural integrity considerations, such as providing standard rebar details for minimizing progressive collapse in reinforced concrete building. On the other hand, design-oriented software programs, such as ETABS and SAP2000, can handle large multi-story design projects in less time when compared to general purpose programs.

This research utilizes a design-oriented program for facilitating progressive collapse analysis for a wider structural engineering community in the design world. The model can offer engineers the capability to efficiently evaluate a wide range of column removal scenarios in new building designs. It can also aid to strengthen existing structures against progressive collapse through retrofitting techniques (Esfandiari *et al.* 2018a, Urgessa and Esfandiari 2018).

2 NIST EXPERIMENTAL TEST DESCRIPTION

One of the most important large scale studies on progressive collapse was conducted by the National Institute of Standards and Technology (NIST) (Sadek *et al.* 2011). For this experimental study, two experimental models were built: one frame designed to meet the requirements of Seismic Design Category (SDC) C with welded unreinforced flange-bolted web (WUF-B) connections, and another frame satisfying the requirements of SDC D with reduced beam section

(RBS) connections. These specimens were beam-column assemblies from the second-floor framing system derived from a ten-story building and each span measured 20 ft in length. For both the WUF-B and RBS test specimens, the beam-column assembly initially remained in the elastic range at small displacements of the center column at about 2 in (50 mm). In that early stage of the response, the behavior was shown to be dominated by flexure. With increasing the vertical displacement, yielding occurred at the beam-to-column connections, and axial tension developed in the beams, indicating catenary behavior. The axial tension in the beams increased until the connections could no longer sustain the combined bending and axial stresses, and the beam-column assemblies failed. NIST utilized these experimental results and developed computational models for determining reserve capacity and robustness of steel structures. The specification of one of the NIST test setups is shown in Figure 1.

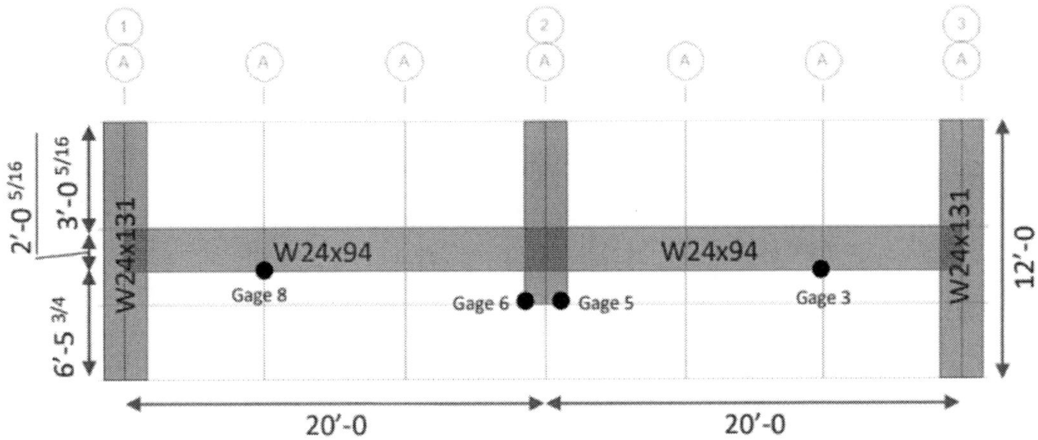

Figure 1. Details of RBS specimen from NIST Technical Note 1669 [34].

3 NUMERICAL MODELING AND RESULTS

For this study, a nonlinear dynamic analysis method is used to compare our results with the experimental test values of NIST. First, a finite element model is built in ETABS. The default hinge properties, which corresponds to the hinge definitions in FEMA 356 (2000), were used. Then, a vertical load is applied at the place of the hydraulic ram in NIST model. While in the NIST experimental model, the specimen was unloaded and reloaded twice. However, only one-half load cycle is applied in progressive collapse. As a result, in this numerical study the maximum load is assigned to be the maximum load at the first unloading in the experimental model excluding the effects of unloading and reloading. It should be noted that having comparable results for the linear response as well as the first stage of the yielding is considered enough for design practices. A time history function is utilized to increase the vertical load until the structure collapses (i.e., vertical "pushover analysis"). Figure 2 illustrates the applied vertical load versus vertical displacements at 1/3 span of beams and center column for the RBS specimen.

Figure 3 shows the comparison of the NIST experimental test with our ETABS numerical study as far as vertical displacement profiles of beams corresponding to vertical loads for the RBS specimen are concerned. As it can be seen, there is a good agreement between numerical and experimental results. The difference can be justified by the fact that connections are not explicitly modeled in ETABS and they are considered as fixed connections. The results are close enough that can be used for future studies on progressive collapse.

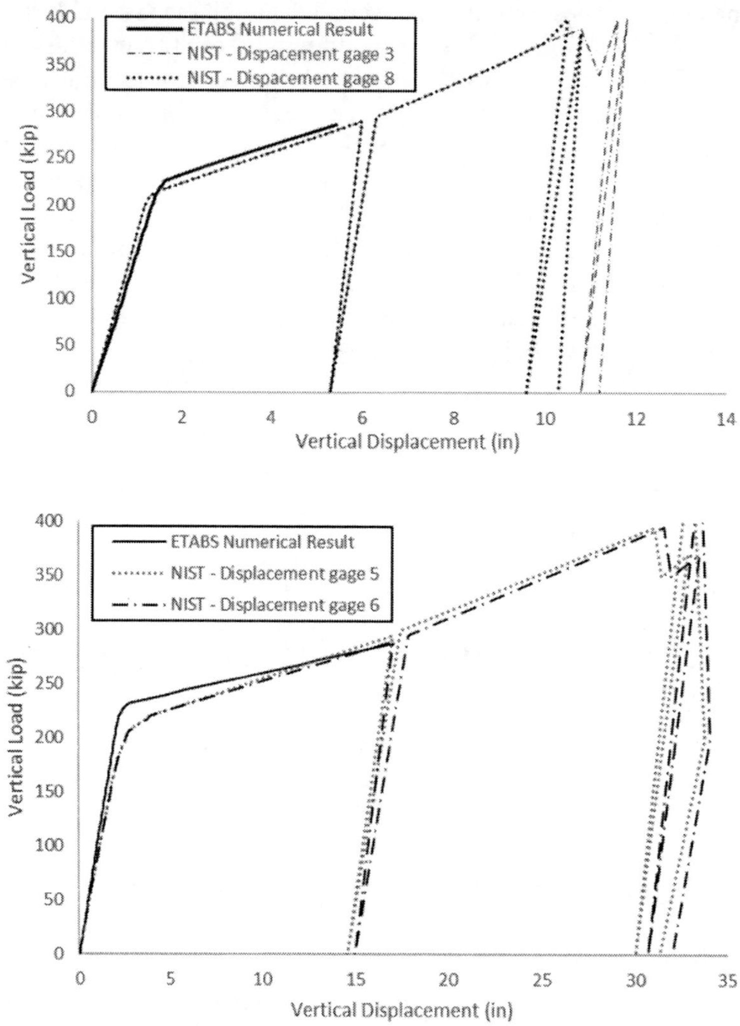

Figure 2. Applied vertical load versus vertical displacements at (a) 1/3 span of beams and (b) center column for the RBS specimen.

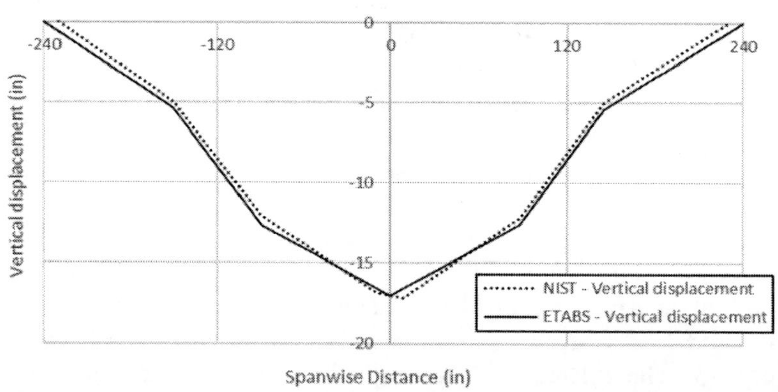

Figure 3. Vertical displacement profiles of beams corresponding to vertical loads for the RBS specimen at 286 kip.

Figures 2 and Figure 3 demonstrate that results from numerical studies using ETABS pull-down analysis are in good agreements with the NIST experimental studies.

4 CONCLUSIONS AND FUTURE RESEARCH

This paper presented a pull-down dynamic analysis of two-span steel frames subjected to progressive collapse. The accuracy of the proposed numerical study was verified using experimental results available in the literature. While the model proposed in this article facilitates progressive collapse analysis, it can be used for addressing practical issues such as combining it with optimization algorithms for developing an integrated computational framework that is capable of incorporating robust and fast progressive collapse analysis for designing new structures or retrofitting existing structures. The progress of our optimization algorithm is reported in (Esfandiari *et al.* 2018b, Esfandiari *et al.* 2018c, Esfandiary *et al.* 2016) and our research is underway to integrate it with progressive collapse analysis of structural frames.

References

Byfield, M. P., Mudalige, W., Morison, C., and Stoddart, E., *A Review of Progressive Collapse Research and Regulations*, Proceedings of the ICE - Structures and Buildings, 167, 447–456, 2014.

Department of Defense, Unified Facilities Criteria: Design of Buildings to Resist Progressive Collapse (UFC 4-023-03), *Department of Defense,* Washington, DC, USA, 2005.

Esfandiari, M. J., Bondarabadi, H. R., Sheikholarefin, S., and Manshadi, S. H. D., Numerical Investigation of Parameters Influencing Debonding of FRP Sheets in Shear-Strengthened Beams, *European Journal of Environmental and Civil Engineering*, 22(2), 246–266, 2018a.

Esfandiari, M. J., Urgessa, G. S., Sheikholarefin, S., and Manshadi, S. H. D., Optimization of Reinforced Concrete Frames Subjected to Historical Time-History Loadings Using DMPSO Algorithm, *Structural and Multidisciplinary Optimization*, 1–16, 2018b.

Esfandiari, M. J., Urgessa, G. S., Sheikholarefin, S., and Manshadi, S. H. D., Optimum Design of 3D Reinforced Concrete Frames Using DMPSO Algorithm, *Advances in Engineering Software*, 115, 149–160, 2018c.

Esfandiary, M. J., Sheikholarefin, S., Bondarabadi, R., and A, H., A., Combination of Particle Swarm Optimization and Multi-Criterion Decision-Making for Optimum Design of Reinforced Concrete Frames, *International Journal of Optimization in Civil Engineering*, 6(2), 245–268, 2016.

FEMA-356., Prestandard and Commentary for the Seismic Rehabilitation of Buildings. Retrieved from FEMA.gov on August 14, 2018

GSA, Progressive Collapse Analysis and Design Guidelines for New Federal Office Buildings and Major Modernization Projects, Washington, DC, 2003.

Hartmann, D., Breidt, M., Nguyen, van V., Stangenberg, F., Höhler, S., Schweizerhof, K., Liebscher, M., Structural Collapse Simulation Under Consideration of Uncertainty – Fundamental Concept and Results, *Computers and Structures*, 86(21), 2064–2078, 2008.

Kaewkulchai, G., and Williamson, E. B., Beam Element Formulation and Solution Procedure for Dynamic Progressive Collapse Analysis, *Computers and Structures*, 82(7), 639–651, 2004.

Kaewkulchai G., and Williamson E. B., Modeling the Impact of Failed Members for Progressive Collapse Analysis of Frame Structures, *Journal of Performance of Constructed Facilities*, 20(4), 375–383, 2006.

Lallotra, B., and Singhal, D., State of the Art Report - A Comparative Study of Structural Analysis and Design Software - STAAD Pro, SAP-2000 and ETABS Software, *International Journal of Engineering and Technology*, 9(2), 1030–1043, 2017.

Qian, K., and Li, B., Research Advances in Design of Structures to Resist Progressive Collapse, *Journal of Performance of Constructed Facilities*, 29(5), B4014007, 2015.

Sadek, F., Main, J. A., Lew, H. S., Robert, S. D., Chiarito, V., and El-Tawil, S., An Experimental and Computational Study of Steel Moment Connections under a Column Removal Scenario, *Technical Note (NIST TN)* - 1669 Rev2, 2011.

Urgessa, G. S., and Esfandiari, M., Review of Polymer Coatings Used for Blast Strengthening of Reinforced Concrete and Masonry Structures, *In International Congress on Polymers in Concrete (ICPIC 2018),* Springer, Cham, 713–719, 2018.

Wang, H., Zhang, A., Li, Y., and Yan, W., A Review on Progressive Collapse of Building Structures, *The Open Civil Engineering Journal*, 2014.

Interdependence between Structural Engineering and Construction Management
Edited by Ozevin, D., Ataei, H., Modares, M., Gurgun, A., Yazdani, S., and Singh, A.

ISBN: 978-0-9960437-6-2

COLLAPSE AND BUCKLING OF STAINLESS STEEL PRESTRESSED STAYED COLUMNS

JOSEF MACHACEK and RADEK PICHAL

Faculty of Civil Engineering, Czech Technical University in Prague, Prague, Czech Republic

Buckling and collapse of stainless steel elements loaded in compression are studied experimentally and numerically. Based on four tests of single crossarm stayed columns the numerical analysis using ANSYS software package is validated and presented in a detail. First, for a completeness, the study refers to columns with one central crossarm, but the main emphasis is devoted to columns with two crossarms located in the thirds of the element length. The analysis employed geometrically and materially nonlinear analysis (GMNIA) to respect a change of inner energy during buckling of an "ideal" (perfect) column, initial deflections of an "imperfect" column (covering various initial deflection modes and amplitude values) and nonlinear stress-strain relationship belonging to stainless steel material. The results cover both critical buckling and maximal collapse loads of the columns in compression. Finally, the important comparisons of the load capacities concerning stayed columns in compression with one/two crossarms and ratios of critical/maximal loadings, elastic/inelastic material and fixed/sliding support of the stays at the crossarms are provided. Conclusions comprise evaluation of these results and principal recommendations for the design of stayed columns.

Keywords: ANSYS, Buckling modes, Crossarm, Imperfection, Inelastic, Prestressing.

1 INTRODUCTION

The prestressed stayed columns are used to achieve a sufficient collapse strength together with an attractive appearance in cases of very slender compression elements, see Figure 1. The most common arrangement consists of a central steel tube, single, double or triple tube crossarms and prestressed cable or rod stays. The crossarms may have two, three or four arms (arms in the angle of 180°, 120°, or 90°). Material of all these elements may be from common or stainless steel.

The principal research on the stayed columns with one central crossarm was presented by Smith *et al.* (1975) and Hafez *et al.* (1979). Using an analytical approach concerning "ideal" (perfectly straight) columns they discovered and described three zones of buckling behavior under an arbitrary prestress level of the stays including "optimal prestressing" T_{opt} leading to maximal critical load $N_{cr,max}$. Subsequent research of stayed columns with one central crossarm concerned influence of various buckling modes and, in particular, also "imperfect" columns (i.e. with initial deflections), e.g. by Chan *et al.* (2002), Saito and Wadee (2008, 2009), Pichal and Machacek (2018). The essential results, demonstrating the reduction of critical loads N_{cr} with respect to maximal collapse loads N_{max} of imperfect columns (with reasonable initial deflections), were published by Wadee *et al.* (2013) and are roughly illustrated in Figure 1.

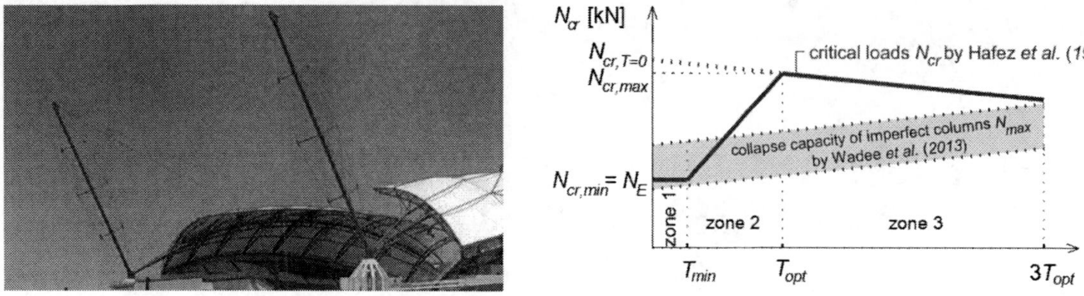

Figure 1. Stadium in Faro (Portugal) with triple crossarms, each with three in-space arms (photo J. Machacek), critical and collapse loads for stayed columns with just one central crossarm.

The theoretical research was accompanied by laboratory tests by Araujo *et al.* (2008), Servitova and Machacek (2011), Osofero *et al.* (2012), and Serra *et al.* (2015). Stayed columns with two and three crossarms were investigated in recent years, following some first attempts in the 80s of the 20th century. The stability of these "ideal" columns was investigated by Martins *et al.* (2016), Yu and Wadee (2017) and Lapira *et al.* (2017). Lastly, Machacek and Pichal (2018) investigated numerically both "ideal" and "imperfect" stayed columns with two crossarms.

This paper will focus on comparisons of critical and collapse loads of prestressed stayed columns with either one or two crossarms and effectiveness of the second crossarm in a case of otherwise the same column (with the same geometrical and material properties). The numerical values were received using ANSYS software, followed by proper and successful validation of the modelling via four stayed column tests.

2 NUMERICAL ANALYSIS

The analyzed stayed columns are shown in Figure 2 and correspond to the tested ones with just one central crossarm by Servitova and Machacek (2011). The entry characteristics (length, area and second moment of area) are as follows:

- Central stainless steel tube Ø 50x2 [mm]: $L = 5000$ mm, $A_c = 302$ mm^2, $I_c = 87009$ mm^4,
- Crossarm stainless steel tubes Ø 25x1.5 [mm]: $a = 250$ mm, $A_a = 111$ mm^2, $I_a = 7676$ mm^4,
- Stays as cables or rods Ø 4 mm: $L_s = 2513$ mm, $A_s = 12.6$ mm^2.

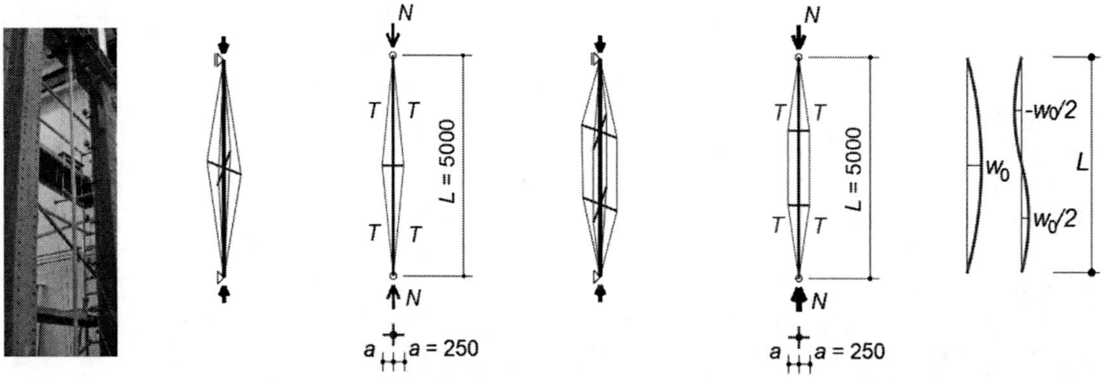

Figure 2. Photo of tested stayed columns, geometry of analyzed columns with just one central crossarm or two crossarms located in the thirds of the column length and considered initial deflection modes.

Tensile tests of full tube cross section stainless steel (grade 1.4301) specimens for both central column and crossarm tubes resulted into the average nonlinear stress-strain relationship as shown in Figure 3, with the initial elastic modulus of 184 GPa.

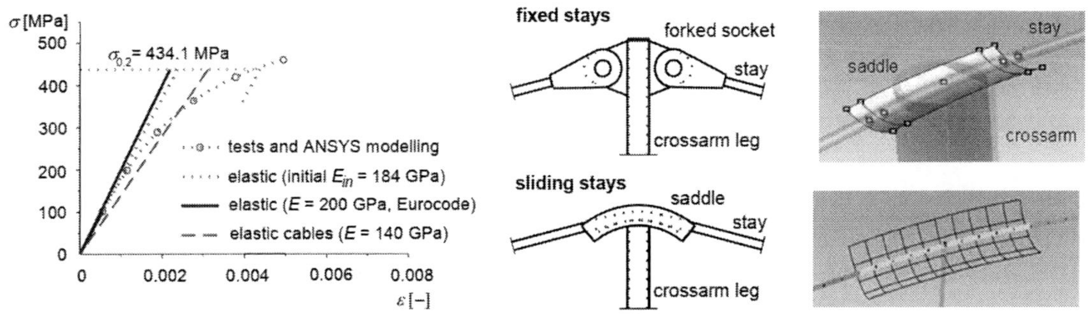

Figure 3. Stress-strain relationships used in the following analyses. Arrangements of stay-crossarm connections (commonly fixed, sliding over saddles at tests) and FEM modeling of the saddles in ANSYS.

Three-dimensional (3D) FEM in ANSYS used elements BEAM188 for the central column and all crossarms (2x6 = 12 DOF, embodying large deflections and material nonlinearity) and LINK180 for cable/rod stays (2x3 = 6 DOF, large deflections, material nonlinearity and adopting tension forces only). In the case of sliding stays the cylindrical and toroidal parts of the saddles was formed from SHELL281 elements according to Figure 3, with friction coefficients between stay and saddle ranging from v = 0.01 to 0.1. The contact between the saddle and the stay provided elements TARGE170 and CONTA175. The stay's prestress was imposed by a relevant thermal change and following external loading via axial displacement of the column, using standard Newton-Raphson iteration. The FE meshing was tested for various divisions with resulting satisfactory element lengths $L/250$, $a/25$ and area of shell elements 23.0 mm^2.

Finite element (FE) model was validated using results of the four tests by Servitova and Machacek (2011). The tests with otherwise identical arrangement differed in initial deflections of the central column and level of the stays prestress. The stays, each with the prestress of T [kN], were continuous over the saddles and friction coefficient 0.1 was considered. Geometrically and materially nonlinear analysis with imperfections in 3D (GMNIA) was employed. The validation for all four tests was successful, but here only the tests No. 1 and No. 2 are shown in Figure 4.

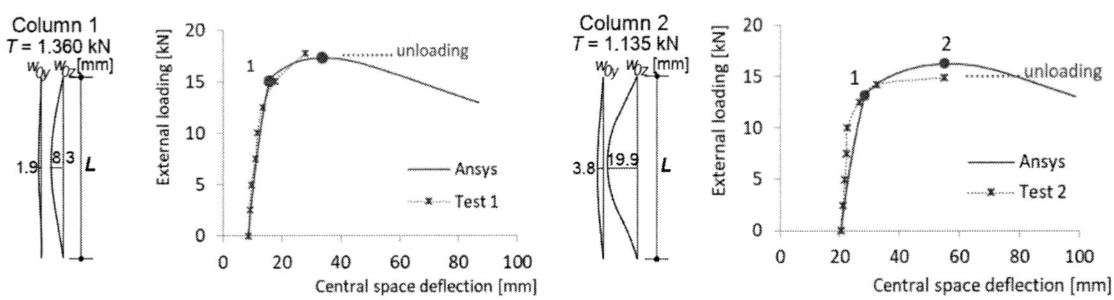

Figure 4. Initial deflections along the columns lengths and comparison of tests and 3D GMNIA results.

3 COMPARATIVE STUDIES

Columns with symmetrical and antisymmetrical initial deflections were investigated (Figure 2).

Saito and Wadee (2008) confirmed that in prestressed stayed columns linear buckling analysis (LBA) doesn't give the relevant critical loads in the full range of prestressings due to a sudden change of the column inner energy under buckling. Therefore, initial deflections have to be introduced and GMNIA or in an elastic region GNIA need to be used. In the current studies the infinitesimal initial deflection amplitude $w_0 = L/500000 = 0.01$ mm was used for the "ideal" columns to determine critical loads and value of $w_0 = L/200 = 25$ mm for the "imperfect" columns to determine maximal (collapse) loads. The latter value corresponds to Eurocode EN 1993-1-1 for cold-formed tubes and elastic analysis and covers all kinds of initial imperfections.

3.1 Columns with One Central Crossarm

The spatial direction of initial deflections (in between the crossarm arms) and up to 20 different prestressings were imposed to the columns. The results of 3D GMNIA are shown in Figure 5. Note that in a low prestress the maximal load is higher than the critical one due to activating of stays on convex sides of the buckling column (while at concave sides the stays slacken). The maximal critical load $N_{cr,max} = 31.6$ kN is given by antisymmetric buckling (with antisymmetric initial deflections), while maximal collapse load for the imperfect column $N_{max} = 19.6$ kN can be reached for a great prestress only. The buckling modes at collapse for different prestress levels between point 1 and 2 are changing from antisymmetric to interactive and back to antisymmetric ones.

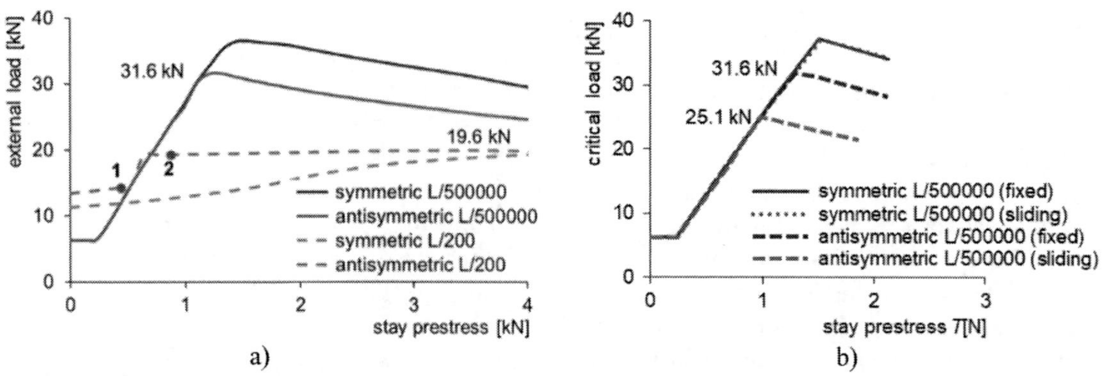

Figure 5. a) GMNIA results for critical load (with $w_0 = L/500000$) and maximal collapse load (with $w_0 = L/200$). b) Influence of sliding stays over crossarm saddles.

Using 3D GNIA (i.e. elastic analysis with $E = 200$ GPa recommended by Eurocode) gives results on unsafe side with $N_{cr,max} = 36.2$ kN and $N_{max} = 22.7$ kN (not shown in this paper). Influence of the support arrangement of stays at crossarms (see Figure 3) was studied with several friction coefficients between stays and saddles. Results for critical loads under very low friction $v = 0.01$ are shown in Figure 5. The sliding support may give more economical solution (less stay pieces and forked sockets), but the antisymmetric buckling results in substantial decrease of both critical and maximal loads.

3.2 Columns with Two Crossarms

The similar 3D GMNIA studies concerning "ideal" and "imperfect" columns with two crossarms were performed (see Figure 6). For the given entry data the antisymmetric initial deflections are deciding for both critical (40086 N) and maximal collapse (25040 N) loads. Again, in a low prestress the maximal load is higher than the critical one due to activating of the convex stays.

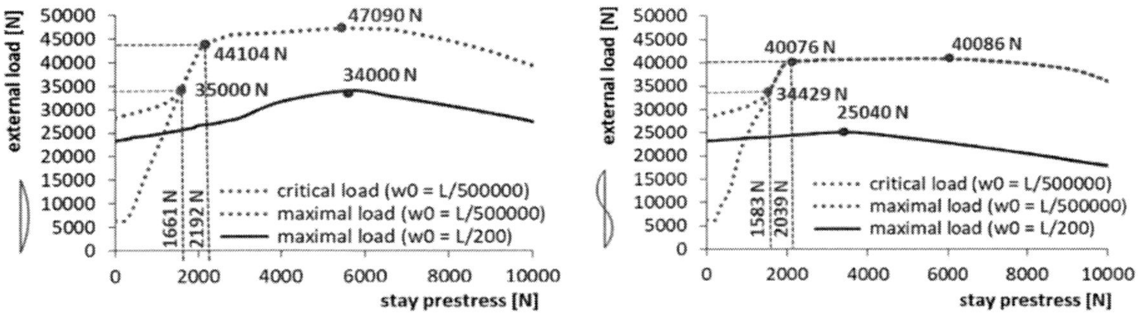

Figure 6. GMNIA results for critical and maximal collapse loads of columns with two crossarms: left for symmetric initial deflections, right for antisymmetrical initial deflections.

Rather interesting is behavior of the "ideal" column in a low prestress. While the critical load is given by buckling resulting in slackening of the concave side stays, the column may carry higher maximal load by activating of the convex side stays, see Figure 6 (this is valid also for unprestressed column, with the prestress $T = 0$). For higher prestress both the critical and maximal loads of the "ideal" column merge together.

The significance of the material nonlinearity concerning stainless steel follows from the results of elastic geometrically nonlinear analysis with imperfections (GNIA) shown in Figure 7. In the analysis the elastic Young's modulus E =200 GPa was employed (see also Figure 3).

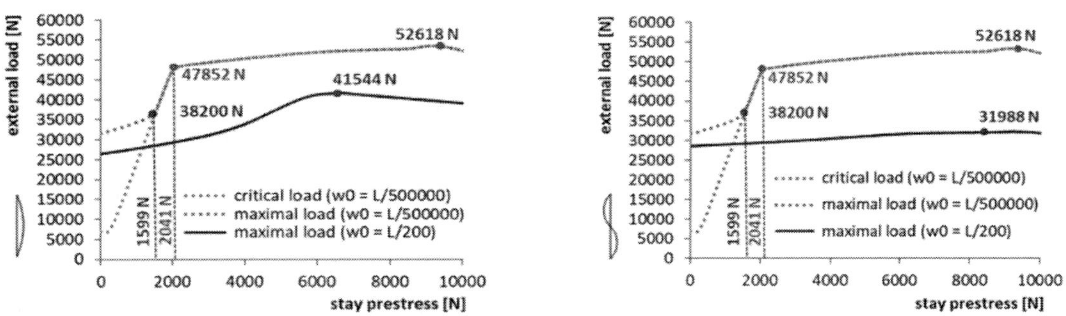

Figure 7. GNIA results for critical and maximal collapse loads of columns with two crossarms: left for symmetric initial deflections, right for antisymmetrical initial deflections.

Comparison of both critical and maximal loads indicates the substantial relevance of proper modeling of the material nonlinear behavior.

4 CONCLUSIONS

The investigation of the prestressed stayed columns with one central crossarm and two crossarms located in the thirds of the column length are presented. The study covers both "ideal" (perfectly straight) and "imperfect" (real) columns to find critical or maximal collapse loads. While the uniform main geometrical entry data are used, the initial imperfections and level of prestress are variable. The results therefore enable comparison of the relevant factors influencing the column behavior and recommend taking them into account in a practical design.

- Adding the second crossarm into otherwise the same prestressed stainless steel stayed column increases significantly both critical and maximal collapse loads. In the investigated column the increase of critical load was 27 % (40086 N/31600 N) and that of maximal collapse load was 28 % (25040 N/19600 N).

- The critical loads of "ideal" columns can't be used in a practical design as the influence of initial imperfections is enormous. In the studied columns the decrease due to imperfections was to 62 % for both columns with one or two crossarms.
- Material nonlinearity of stainless steel needs to be carefully taken into account. In the investigated column the decrease of maximal load due to the nonlinearity is to 78 %.
- Sliding of stays over saddles decreased the maximal load to 79 %.

Acknowledgments

The work was supported by the Czech Grant Agency under grant GACR No. 17-24769S.

References

Araujo, R. R., Andrade, S. A. L., Vellasco, P. C. G. S., Silva, J. G. S., and Lima, L. R. O., Experimental and Numerical Assessment of Stayed Steel Columns, *Journal of Constructional Steel Research*, 64(9), 1020-1029, 2008.

Chan, S. L., Shu, G., and Lü, Z., Stability Analysis and Parametric Study of Pre-Stressed Stayed Columns, *Engineering Structures*, 24(1), 115-124, 2002.

Hafez, H. H., Temple M. C., and Ellis, J. S., Pretensioning of Single-Crossarm Stayed Columns, *Journal of the Structural Division*, 14362(ST2), 359-375, 1979.

Lapira, L., Wadee, M. A., and Gardner, L, Stability of Multiple-Crossarm Prestressed Stayed Columns with Additional Stay Systems, *Structures*, 12, 227-241, 2017.

Machacek, J., and Pichal, R., *Stability and Strength of Stainless Steel Stayed Columns with One and Two Crossarms,* Proceedings 8th International Conference on Thin-Walled Structures 2018 (ICTWS 2018), Camotin, D. (ed.), 1-12, Lisbon, 2018.

Martins, J. P., Shahbazian, A., Silva, L. S., Rebel, C., and Simoes, R., Structural Behavior of Prestressed Stayed Columns with Single and Double Cross-Arms Using Normal and High Strength Steel, *Archives of Civil and Mechanical Engineering*, 16, 618-633, 2016.

Osofero, A. I., Wadee M. A., and Gardner, L., Experimental Study of Critical and Post-Buckling Behaviour of Prestressed Stayed Steel Columns, *Journal of Constructional Steel Research*, 79, 226-241, 2012.

Pichal, R., and Machacek, J., Single-Crossarm Stainless Steel Stayed Columns, *Advances in Technology Innovation*, 3(1), 9-16, 2018.

Saito, D., and Wadee, M. A., Post-Buckling Behaviour of Prestressed Steel Stayed Columns, *Engineering Structures*, 30, 1224-1239, 2008.

Saito, D., and Wadee, M. A., Numerical Studies of Interactive Buckling in Prestressed Steel Stayed Columns, *Engineering Structures*, 31(2), 432-443, 2009.

Serra, M., Shahbazian, A., Silva, L., S., Marques, L., Rebelo, C., and Vellasco, P. C. G. S., A Full Scale Experimental Study of Prestressed Stayed Columns, *Eng. Structures*, 100, 490-510, 2015.

Servitova, K., and Machacek, J., *Prestressed Stainless Steel Stayed Columns*, Proceedings EUROSTEEL 6th European Conf. on Steel and Composite Structures, Dunai, L. *et al.* (eds.), 1767-1722, Budapest, 2011.

Smith, R. J., McCaffrey, G. T., and Ellis, J. S., Buckling of a Single-Crossarm Stayed Column, *Journal of the Structural Division*, 11071(ST1), 249-268, 1975.

Wadee, M. A., Gardner, L., and Osofero, A. I., Design of Prestressed Stayed Columns. *Journal of Constructional Steel Research* 80, 287-298, 2013.

Yu, J., and Wadee, M. A., Mode Interaction in Triple-Bay Prestressed Sayed Columns, *International Journal of Non-Linear Mechanics*, 88, 47-66, 2017.

Interdependence between Structural Engineering and Construction Management
Edited by Ozevin, D., Ataei, H., Modares, M., Gurgun, A., Yazdani, S., and Singh, A.

ISBN: 978-0-9960437-6-2

THE BASICS OF FIXED CONNECTION IN ANALYSIS AND DESIGN OF STEEL FRAMES

MUANGSANGOP SENIWONGSE

Bermello Ajamil and Partners, New York, USA

This paper is concerned with the out-of-date design criteria in the assumption of end boundary conditions in computer analysis and design of steel building frames in some sectors of heavy industry. The two major concerns are addressed: one is the use of pinned end for the steel column base plates whose details are fixed end, and the second is the use of pinned connection design of steel beam to steel column in lieu of using moment connection. A typical frame is shown for demonstration purposes. The typical connection details used are shown. The author presents the pinned and fixed connection definition with reference to AISC Specification for Structural Steel Buildings. Examples of actual pinned end and actual fixed end are shown with reference to actual structures. A simple structure of a single column traffic sign post with a cantilever beam at the top of the column was selected to demonstrate a fixed end column base and a beam to column moment connection. It is the aim of this paper to clarify this simple but significant concept of fixed connections of column base plates and moment connection of beam to column, which in turn will lead to significant savings in the construction of steel buildings.

Keywords: Baseplate, Bolt, Computer, Moment, Pinned, Rotation, Welding.

1 INTRODUCTION

This paper refers to the structural steel analysis and design practice in some sectors of heavy industry in USA using computer software. The concerns are on at least two major assumptions in computer modeling that should be reconsidered and updated to reflect the actual details and structural behavior. In a certain specialized heavy industry, the design and build is usually exclusively awarded to a few big well-known companies which have been in the business for more than 30 - 40 years. This criterion of prequalification has its own advantages and disadvantages. Using the argument that the structures built with existing design criteria for the past 40 years are still standing up and hence any suggested revision or improvement are ignored. The author points out herein three modeling criteria that would improve the future design, which would result in a more successfully outcome to all concerned parties, leading to easier- to-build and cheaper structures. The cost reduction can then be passed on to the consumers. Most importantly, it concerns the basic understanding of fixed and pinned connections of professional engineers, which needs to be addressed and corrected.

2 OUT-OF-DATE DESIGN CRITERIA AND GROUP THINK

Without naming the industries, the typical interior steel frame of the industrial buildings housing heavy machinery is shown in Figure 1. The buildings usually have one or two mezzanine levels.

The length of the buildings is in the order of 300 ft. The portal frames are spaced at 30 ft to 40 ft across the entire length of the building. The ground floor is usually of concrete mat foundation on grade or on piles depending on the soil foundation. The mezzanine floors are either concrete slab on steel beams or steel grating on steel beams.

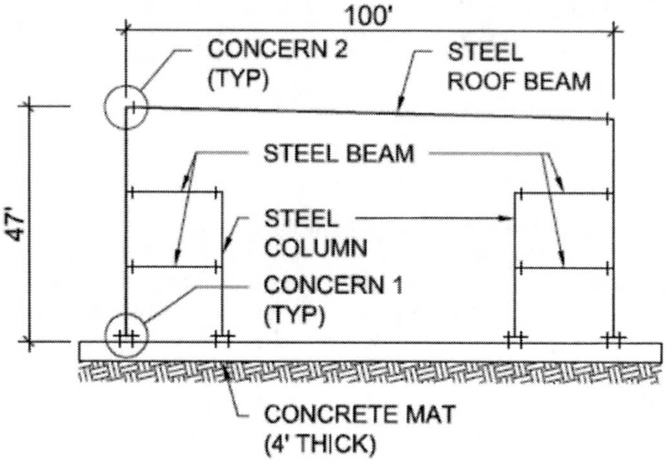

Figure 1. Typical steel frame of concern.

The concerns in this paper are the Out-of-Date Design Criteria and the Group Think Mentality. The Establishment imposed the following design criteria.

1) All steel column bases are assumed pinned ends, while the design detail used are actually of fixed ends.
2) No field welding is allowed in any steel structural members. All steel connection design for field connection must be bolted connection only, with no exception, while much smaller steel pipes and equipment supporting steel angles inside the buildings are allowed to be field welded.
3) The steel beams supporting concrete floor slab must be designed as non-composite sections, while the composite design would have reduced the size of the steel beams.

The reason given by the Group Think was these criteria have been used repeatedly for more than 40 years and there have been no problems. The first two of the above three criteria will be discussed in this paper for education sake and for saving of steel from wrong design assumptions.

Figure 2 shows a typical column base plate similar to those used for the steel frame in Figure 1. It is shown for demonstration purpose only. Typical column sizes are W36 and the base plates 2" to 3" thick, 44" x 48" in plan with 10 to 16 high strength bolts of 2" to 2 ½" dia. The bolts are cast-in-place and embedded some 3 ft deep in the concrete mat foundation. The W36 column is fully welded all round to the steel base plate. It is obvious that there is moment resistance at the welds at the W36 base, and there is moment resistance at the bolt group at the mat concrete foundation. The steel base plate itself has moment of inertia for moment resistance. Yet, all column bases of this type were input in the structural analysis programs as pinned ends, that is the computer was told the column bases have no moment resistance and are all free to rotate, which is not the case.

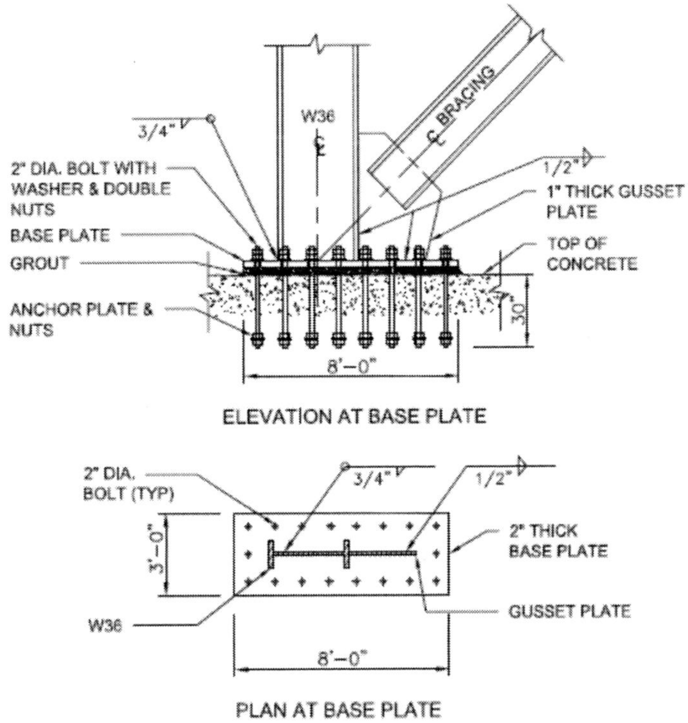

Figure 2. Typical column base plate and anchor bolt details.

Figure 3 shows typical steel connection between the roof beam and the column. The beam is W36 and the column is W36. Shear tab pinned connection was used as mandated by the design criteria. The primary purpose of the out-of-date criteria given by group think was to avoid field welding.

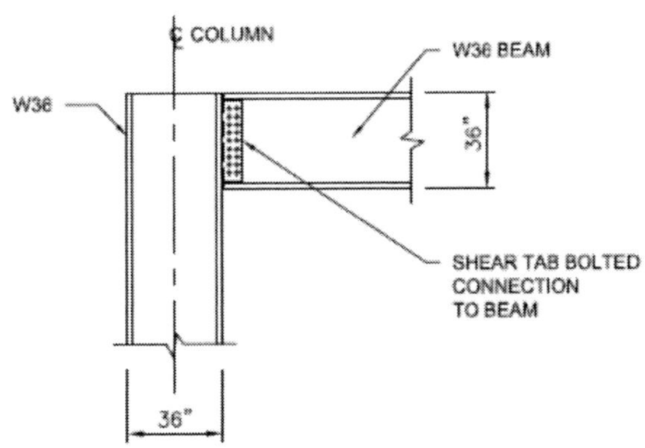

Figure 3. Shear tab pinned connection of steel beam to column.

It must be pointed out that the shear tab connection in Figure 3 can be easily improved to moment connections as shown in Figure 4(a) with field bolting and 4(b) with full penetration welding at the top and bottom flanges. These details or similar are commonly used around the world. But, the details were prohibited by the establishment. The sub-consultants and the contractors did not question the design criteria as long as they got paid for what they were told to do. Hence, the cycles continue.

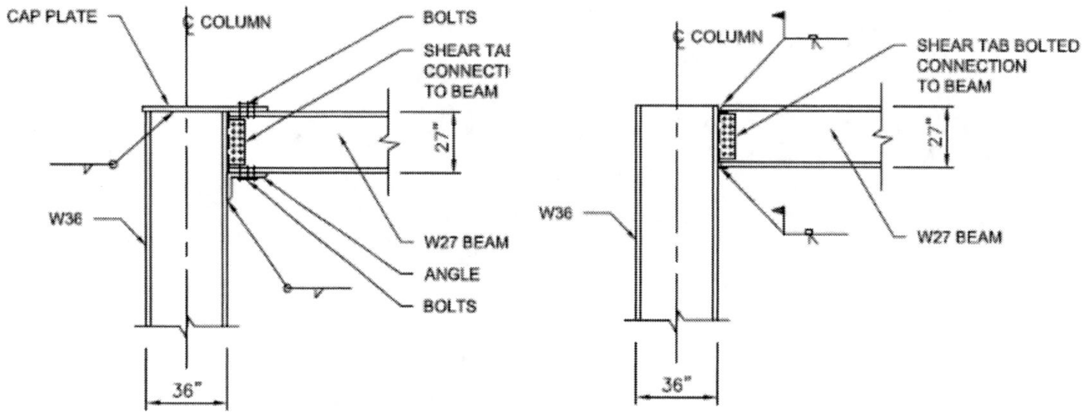

Figure 4. Moment connections: (a) by bolting at the flanges, (b) by field welding at the flanges.

3 PINNED SUPPORT VERSUS FIXED SUPPORT

Figure 5 shows the legend and definition of fixed and pinned supports taken from AISC (1989)[1]. Pinned support is free to rotate, having no resistance against rotating forces or moments. Fixed end is fixed against rotation and translation and is the most stable type of support. The structural details for pinned support and fixed support are different and are shown in the next sections.

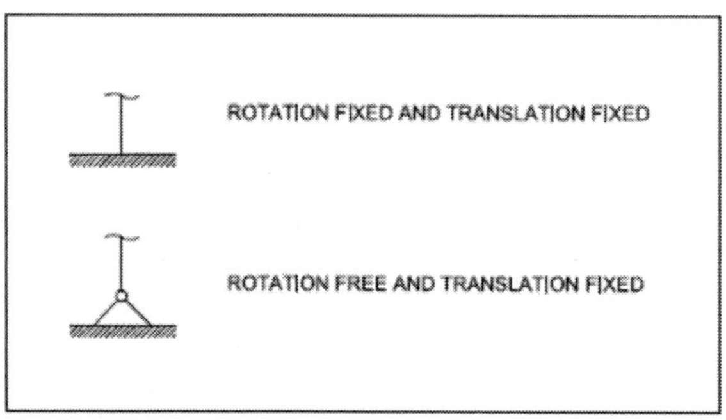

Figure 5. Legend and definition of fixed against rotation and free to rotate (AISC 1989).

3.1 Pinned-end Detail

The examples of pinned-base details are shown in Figures 6 and 7. The support is provided with an elastomeric bearing (Figure 6a) or a pin or roller (Figure 6b) to behave as a pined-end, free to rotate.

[1] Part 5 Chapter C, Table C-C2.1, page 5-135.

Figure 7a shows the actual pin used at the bottom of the right column, which the design engineer wanted it to behave as pinned-end. It was there for a reason, not by chance. The picture was taken in an International airport in Europe. Figure 7b shows a pinned connection detail of the steel columns that support the roof of Turkey International airport (Celtikci 2001).

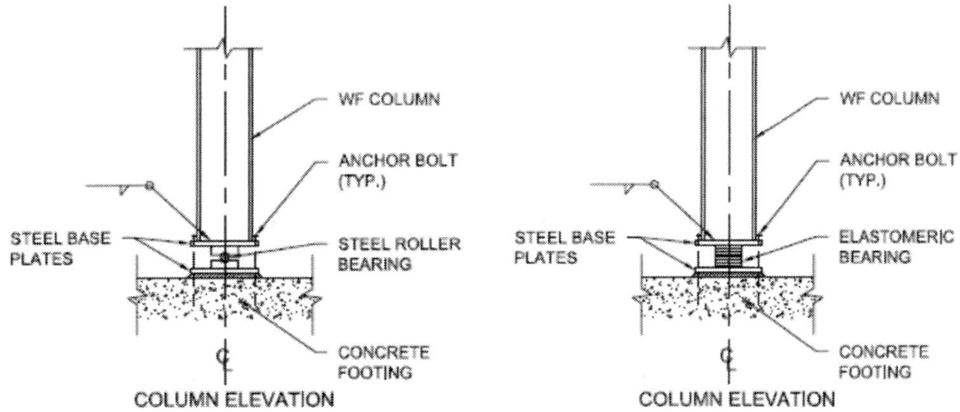

Figure 6. Column base details for pinned support – free to rotate.

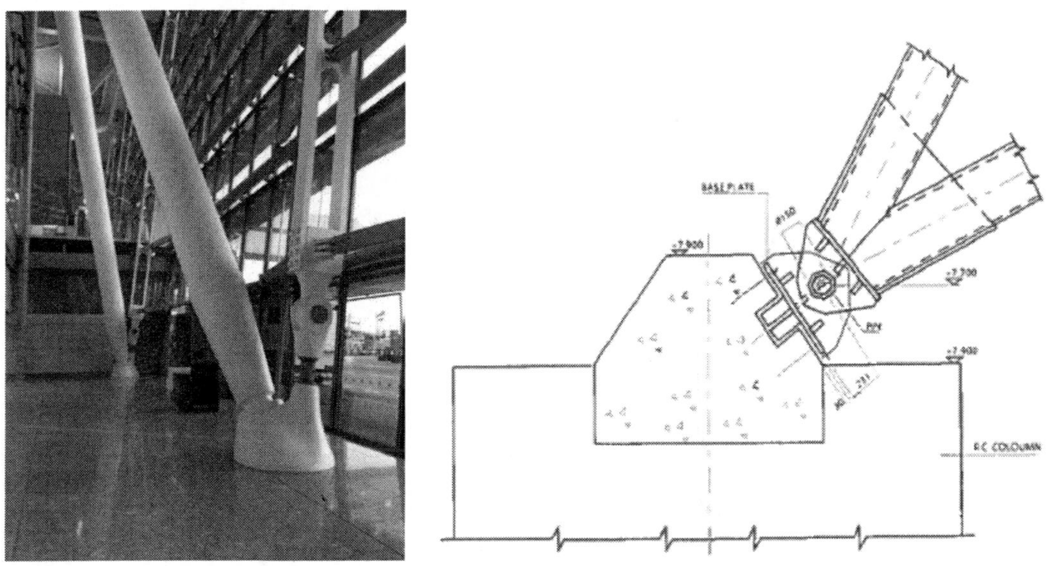

Figure 7. Pin connection (a) at the base of the right column, (b) at the base of steel columns at Turkey International Airport (Celtikci 2001).

3.2 Fixed- end Detail

Fixed support by direct definition is fixed. Fix means no movement. Fixed support is a stable support and can resist forces and moments that act on it or transferred to it. The typical fixed-end support for steel column is shown in Figure 2. The base plates and the anchor bolts do not have to be oversized to behave as a fixed base as shown in Figures 8 and 9. Figure 8a shows a traffic sign post with overhanging cantilever beam on a small steel base plate with 4 anchor bolts (Figure 8b). This support is a fixed or moment connection, the post is stable against the lateral forces and moments that act on it. Figure 8a also shows the cantilever steel beam is connected to the top of the post with four bolts. It is obviously a moment connection, otherwise the post and the cantilever beam would have rotated and collapsed under lateral wind loads. Note that the base plate and bolts in Figure 2 are much larger than the base plate in Figure 8b. It is very obvious Figure 2 detail is a fixed support detail, not a pinned support.

(a) (b)

Figure 8. Base of traffic post moment connection (fixed connection).

3.3 Beam to Column Moment Connection

Refer to Figure 3 of the roof beam connection to top of column. It is a pinned connection for the beam. If the fixed or moment connection (Figure 4) is used, the beam size would be reduced in this case from W36 to W27. The mid-span deflection will also be reduced. The portal frame would be a rigid frame and can resist horizontal transverse forces due to the joint rigidity without the need for the horizontal diagonal bracings. Hence, much construction time and cost would have been saved. Figure 8a above demonstrates beam to column moment connection is simple to achieve.

4 CONCLUSIONS

This paper has addressed two major concerns of the incorrect assumption in the existing design criteria of steel building frames of some major heavy industry. One concern is the use of pinned end column computer input instead of using the fixed end coding. The other is the use of pinned connections to columns for all the beams in the entire buildings. These two incorrect or in appropriate assumptions have led to grossly over designed steel frames. Due to its grossly over-design, nothing has ever happened to the structures. But this does not mean a good engineering practice, besides engineers have the duty to provide "Duty of Care" for the work performed, and to attempt to give the Clients the most optimum structure.

References

AISC, *Manual of Steel Construction*, Allowable Stress Design, Ninth Edition, 1989.

Celtikci, N., Istanbul Sabiha Gokcen Airport, Turkey International Terminal, *Modern Steel Construction,* 55, August 2001.

Interdependence between Structural Engineering and Construction Management
Edited by Ozevin, D., Ataei, H., Modares, M., Gurgun, A., Yazdani, S., and Singh, A.

ISBN: 978-0-9960437-6-2

PERIODIC BOLT PLACEMENT IN STEEL STRUCTURES FOR GUIDED DAMAGE DETECTION

LU ZHANG, DIDEM OZEVIN, and GORKEM OKUDAN

Civil and Materials Engineering, University of Illinois at Chicago, Chicago, USA

Connections form the weakest link in structural systems. Bolted connections are especially susceptible to stress corrosion cracking due to stress-riser points at the bolt edges and the intrusion of water between plates leading to galvanic corrosion. Bolts are conventionally placed based on minimum and maximum spacing requirements guided by American Institute of Steel Construction. Unfortunately, the detectability of crack and corrosion is currently not a design variable. In this study, bolts are placed in periodic pattern such that they exhibit unique frequency response. The periodic placement allows modeling only unit cell in the shape of hexagon with periodic boundary conditions to obtain the frequency response. When damage occurs, the periodic pattern is broken, and the frequency response behavior changes. The influence of crack to vibration modes is numerically modeled, and the damage detection ability with unique bolt placement is demonstrated. It is shown that the spatial distribution of bolt can assist the damage detection ability, which should be considered as a criterion in the design of bolted connections.

Keywords: Frequency response, Crack detection, Corrosion detection, Spatial distribution, Periodic pattern, Failure.

1 INTRODUCTION

Welded or bolted connections are usually considered as the weakest elements in structural systems that impact the performance and the resistance to normal service as well as extreme events. For seismic regions, bolted connections are preferred over welded connections as welds are brittle and do not have good energy dissipation mechanism to reduce the dynamic load. The design of bolted connections is based on strength limit states depending on loading (axial, shear or combined). Once number of bolts is determined, they are placed based on minimum and maximum spacing requirements of American Institute of Steel Construction (AISC). The distance between centers of standard, oversized, or slotted holes shall not be less than 2 2/3 times the nominal diameter of bolt while it is preferred to use 3 times the nominal diameter of bolt. The longitudinal spacing between bolts shall not exceed 24 times the thickness of the thinner part or 12 in for painted members or unpainted members not subjected to corrosion, or 14 times of the thickness of the thinner part or 7 in for unpainted members of weathering steel subjected to atmospheric corrosion. The purpose of maximum spacing is to prevent water intrusion between layers and provide good contact. The purpose of minimum spacing is to facilitate construction.

In this paper, the concept of "inspectability" will be considered as another design variable such that the presence of corrosion, hidden cracking and stress increase can be detected at the earliest state. Typical inspection methods to assess the presence of damage in bolted connections are visual inspection, ultrasonics, vibration analysis, and fiber optic sensing (Ryan *et al.* 2012, Fu 2005, Graybeal *et al.* 2002, Hong-Nan *et al.* 2004 and Fritzen 2005). Visual inspection fails as there may be hidden defects. With current design practice, the performance of ultrasonics and vibration methods is limited due to unstructured bolt distribution as shown in Figure 1(a).

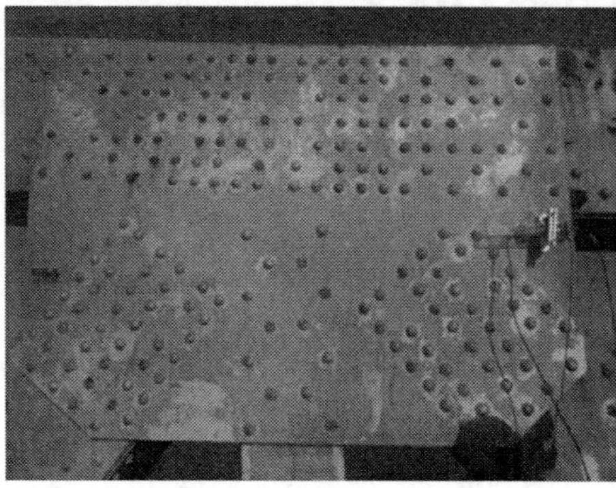
(a)

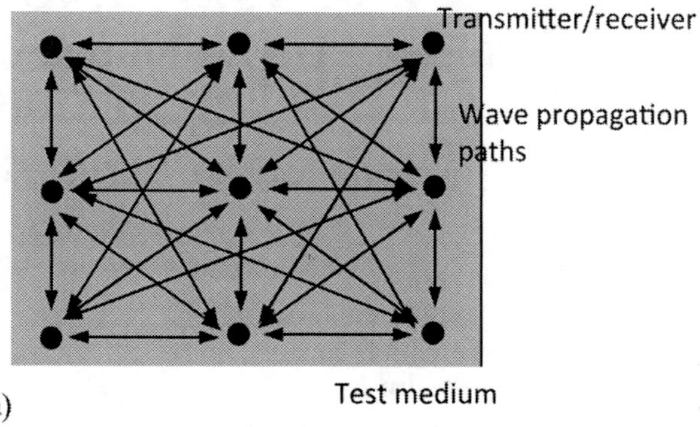

(b)

Figure 1. (a) Typical bolt placement in steel bridges and (b) an example of guided wave UT on plate-like structure.

2 GUIDED WAVE ULTRASONICS IN PERIODIC PERFORATED PLATED

Ultrasonic Testing (UT) is an active nondestructive evaluation method based on exciting structural systems by a transmitter and analyzing the received waveform. This is a high frequency dynamic method. In bulk ultrasonics, frequencies above 1 MHz are selected; however, the method is limited spatial resolution. To inspect larger volumes, lower frequencies are selected such that guided waves are generated in plate-like structures.

2.1 Guided Wave Ultrasonic Testing

Guided wave ultrasonic testing is applied to structures with bonded media such as plates. Multiple wave modes are generated due to reflections and refractions at boundaries leading to multi-modal and dispersive waves as opposed to single mode and non-dispersive bulky waves (Mitra and Gopalakrishnan 2016). Figure 1(b) shows an array of transmitter/receiver placed on plate-like structure and wave propagation paths. The transducers are typically made of piezoelectric material with reciprocal property functioning as transmitter and receiver. Therefore, the same transducers in the array are used as either transmitter or receiver. Using the waveforms recorded through multiple wave paths, damage location, type and size are measured by applying signal processing methods to the received signals. Figure 2 shows two examples of waveforms recorded from pristine condition (no damage) and damage condition. Various information such as arrival times of multiple wave modes, phase shift, amplitude and energy changes is utilized to assess the damage of structure. The method is successful in plain plates without bolts. However, when bolts are placed between the paths of transmitter and receivers, ultrasonic waves scatter, which complicates the measurement process. Therefore, bolts should be placed strategically such that their effect on ultrasonic waves is known at the design stage.

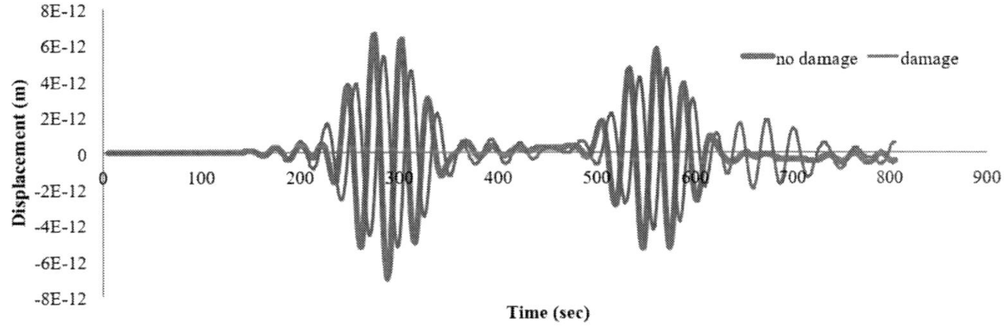

Figure 2. Typical guided wave ultrasonic signal.

2.2 Periodic Perforated Geometries

In recent studies on periodic structures, it is recognized that structures with periodic design have unique effective mechanical properties due to geometric pattern rather than constituents. Metamaterials and phononic crystal (PC) structures are of growing interest due to their ability to manipulate and control the propagation of elastic waves in ways that are not possible in conventional materials. Elastic metamaterials are artificial composite, man-made periodic systems creating new responses through physical constraints in the constituent materials (Craster and Guenneau 2013). The superior property of the acoustic metamaterial is attributed to its engineered geometry and artificially induced inhomogeneity (Pai 2010). The periodicity or the lattice constant, and composite behavior of resonating unit cells control the scattering of elastic waves, as a consequence, the wave properties (e.g., energy, direction). The metamaterial medium exhibits band gaps, or frequency ranges with negative modulus for acoustic waves that cannot propagate into the medium (Lee *et al.* 2009). The band gap of the metamaterial can be controlled by the contrast of elastic constants, the filling fraction and the lattice (i.e., periodicity) of the constructed element, and the thickness of the base plate (Hou *et al.* 2004, Charles *et al.* 2006). In this paper, the bolts are placed periodically to exhibit unique frequency response, which is influenced by the presence of damage such that single measurement becomes sufficient to change from the pristine state of the connection.

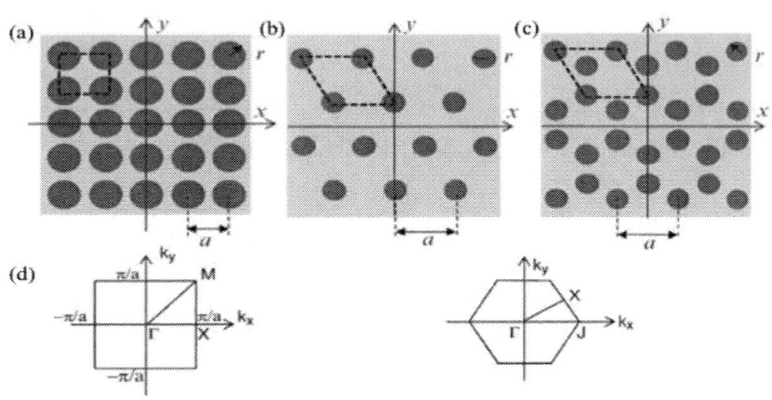

Figure 3. Three binary composite metamaterial designs, (a) square lattice, (b) hexagonal lattice, (c) honeycomb lattice and (d) first Irreducible Brillouin zone (IBZ) (Pennec and Djafari-Rouhani 2016).

3 NUMERICAL ASSESSMENT OF DAMAGE IN HONEYCOMB PLACED BOLTS

Binary composite materials are defined as steel-air and hexagonal placement of unit cells is considered. Figure 3 shows examples of binary composite materials in square, hexagonal and honeycomb lattice. Considering the effective placement of bolts with minimum spacing, hexagonal placement and the influence of crack on vibration modes are studied.

3.1 Numerical Models

When the gusset plate is subjected to loading, crack may initiate perpendicular to the loading direction. The studied gusset plate with and without crack is shown in in Figure 4(a). The bolt layout with the gusset has a repeated pattern; therefore, the edges of base plates are defined using the Bloch-Floquet (periodic) boundary condition. One unit cell is sufficient for creating the dispersion curve, which represents the solution of entire structure such that changes in wave modes due to the presence of damage can be detected with minimum effort. The unit cell models with and without crack are built. Then, the transmission analysis is conducted to simulate guided wave ultrasonics. The models are built using COMSOL Multiphysics Software. The typical gusset with the bolted connection is selected. The schematic of periodic unit cell is shown in Figure 4(b). The bolt units are arranged periodically in a hexagonal lattice. The diameter of the bolt is 5/8 inch, the spacing between each bolt is 2 2/3 times of its diameters, and the thickness of gusset is 3/8 inch.

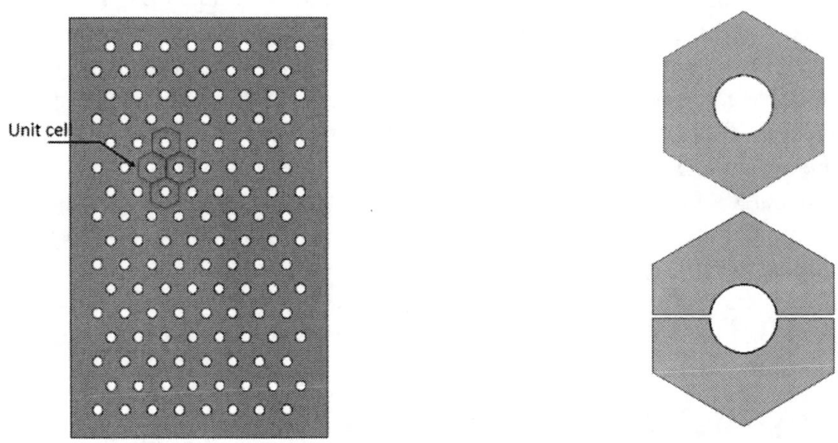

(a) (b)

Figure 4. (a) The gusset plate model and (b) the unit cells with and without crack.

3.2 Results

The influence of presence of crack on the wave propagation within the periodic structure is studied. The fundamental frequency is the main concern since it has largest contribution to the structural response. The dispersion relations between frequency and wave number is calculated for the ten lowest bands from 0 to 80 kHz and plotted in Figure 5. The changes in the dispersion curve occur, mostly in the high frequency range. In addition, the band gap (no vibration occurs) of the crack case moves to the frequency range of 34.5-37.7 kHz. The width of band gap increases from 1.6 kHz to 3.2 kHz. The mode shape within the band gap of pristine gusset plate is shown in the figure. If that particular wave mode is excited with the guided wave ultrasonics, it is expected that the ultrasonic amplitude will decay with the presence of crack due to the formation of band gap.

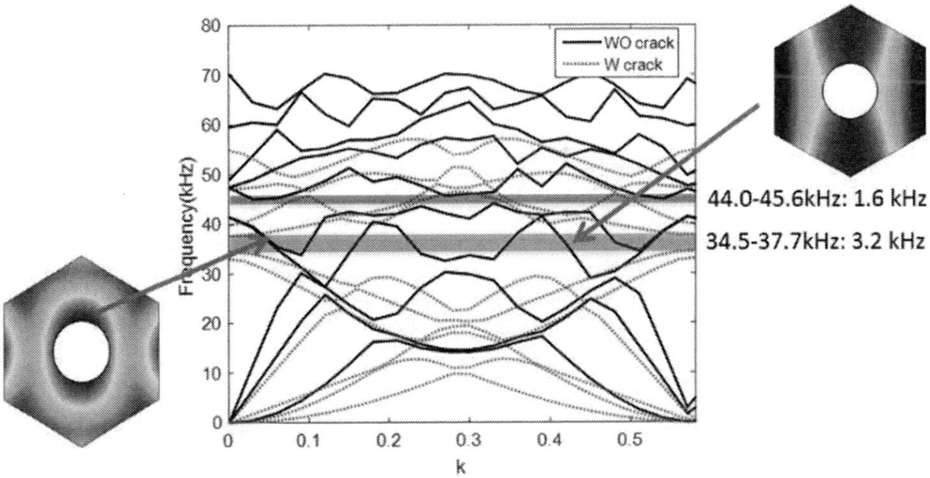

Figure 5. The dispersion curve with and without crack indicating two mode shapes of pristine case within the band gap of cracked case.

To demonstrate the influence of crack to guided wave ultrasonics, the transmission analysis is conducted using the entire gusset plate (Figure 4(a)) within the bandgap of the cracked plate. The excitation signal is selected as 6-cycle sine wave with 36 kHz excitation, and applied to the edge

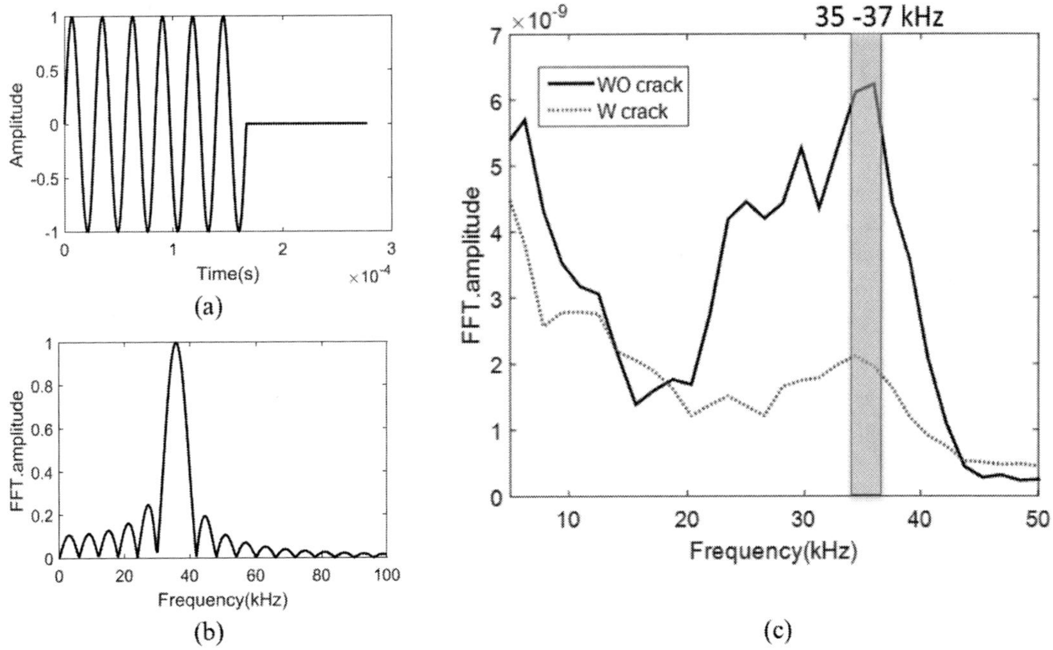

Figure 6. The transmission model result (a) excitation signal in time domain (b) excitation signal in frequency domain, (c) the spectra responses without (WO) and with (W) crack.

of plate. The displacement response at the other edge is extracted from the model. The displacement histories without and with crack are shown in Figure 6. Significant reduction in amplitude is observed, which verifies the unit cell mode.

4 CONCLUSIONS

In this paper, a novel nondestructive evaluation method for detecting the crack of bolt connections is proposed by taking advantage of periodic pattern of bolt layout. The typical gusset plate with repeated hexagonal placement of bolts is studied. The band gap is observed due to the repeated bolt pattern and crack. The presence of crack significantly changes the periodicity of bolt layout, leading the change of vibration and ultrasonic wave propagation, which can be used for identifying the crack. The concept of 'inspectability' within design is verified numerical results, which will be validated with experimental results as future work.

Acknowledgments

This research is based upon work supported by the National Science Foundation under Award No. CMMI 1552375 entitled "CAREER: Engineered Spatially Periodic Structure Design Integrated with Damage Detection Philosophy". The support from the sponsoring organization is gratefully acknowledged. Any opinions, findings, and conclusions or recommendations expressed in this paper are those of the authors and do not necessarily reflect the views of the organizations acknowledged above.

References

Charles, C., Bonello, B., and Ganot, F., Propagation of Guided Elastic Waves in 2D Phononic Crystals, *Ultrasonics*, 44, 209–13, 2006.

Craster, R. V., and Guenneau, S., *Acoustic Metamaterials: Negative Refraction, Imaging, Lensing and Cloaking*, Springer, 2013.

Fritzen, C. P., Vibration-Based Structural Health Monitoring – Concepts and Applications, *Key Engineering Materials,* 293-294, 3-20, 2005.

Fu, G., *Inspection and Monitoring Techniques for Bridges and Civil Structures*, Woodhead Publishing Series in Civil and Structural Engineering, 2005.

Graybeal, B. A., Phares, B. M., Rolander, D. D., Moore, M., and Washer, G., Visual Inspection of Highway Bridges, *Journal of Nondestructive Evaluation,* 21(3), 67-83, 2002.

Hong-Nan, L., Dong-Sheng, L., and Gang-Bing, S., Recent Applications of Fiber Optic Sensors to Health Monitoring in Civil Engineering, *Engineering Structures* 26, 1647–1657, 2004.

Hou, Z., Wu, F., and Liu, Y., Phononic Crystals Containing Piezoelectric Material, *Solid State Communications*, 130, 745–749, 2004.

Lee, S. H., Park, C. M., Seo, Y. M., Wang, Z. G., and Kim, C. K., Acoustic Metamaterial with Negative Density, *Physics Letters A*, 373, 4464–4469, 2009.

Mitra, M., and Gopalakrishnan, S., Guided Wave Based Structural Health Monitoring: A Review, *Smart Materials and Structures*, 25, 053001, 2016.

Pai, P. F., Metamaterial-based Broadband Elastic Wave Absorber, *Journal of Intelligent Material Systems and Structures*, 21, 517-528, 2010.

Pennec, Y., and Djafari-Rouhani, B., *Fundamental Properties of Phononic Crystal.* Khelif A., Adibi A. (eds) Phononic Crystals. Springer, New York, NY, 2016.

Ryan, T. W., Mann, J. E., Chill, Z. M., and Ott, B. T., Bridge Inspector's Reference Manual, U.S. Department of Transportation Federal Highway Administration, Publication No. FHWA NHI 12-049, 2012.

Interdependence between Structural Engineering and Construction Management
Edited by Ozevin, D., Ataei, H., Modares, M., Gurgun, A., Yazdani, S., and Singh, A.

ISBN: 978-0-9960437-6-2

DEVELOPING A METHODOLOGY FOR USING THE CONTINUOUS STRENGTH METHOD FOR BUILT-UP I-SECTION

RAQUIBUL HOSSAIN, SHADMAN SAKIB, and MALEHA ALAM

Dept of Civil Engineering, Bangladesh University of Engineering and Technology, Dhaka, Bangladesh

Built-up steel sections are widely used in the construction industry due to their wide range of compression resistance. The main objective of the study is to introduce an efficient and economic design methodology for built-up I-sections by using the Continuous Strength Method (CSM) equations. These I-sections are made from two C-sections with bolts spaced at a specific interval. This methodology includes the calculation of individual capacity of C-sections using the CSM base curve and conversion that to bolted I-sections by using I-section properties. The predicted capacities show that the modified application of CSM equations can predict compressive resistance with high accuracy for built-up I-sections. Though the effect of bolt spacing is not considered in this study, the proposed methodology paves the path for deriving CSM equations for built-up sections.

Keywords: Bolted, Base curve, Cross-section, CSM, Steel.

1 INTRODUCTION

Built-up sections are comprised of two or more standard sections. They are required when the desired capacities and configuration cannot be achieved by a standard or hot rolled section. Besides, it is sometimes difficult to produce required sizes in standard sections. In that case, built-up sections offer alternative options against standard sections.

Built-up sections can be bolted or welded. For achieving economy and ease in construction, bolted sections are preferable compared with welded sections. It can provide acceptable strength, durability and safety. Besides, it is more effective in consideration of time, cost and installation.

This paper focuses on the study of compressive resistance of the bolted steel sections using the Continuous Strength Method (CSM) equations. This study may provide greater significance in increasing the design efficiency of bolted steel sections than standard CSM (Ahmed *et al.* 2016).

2 THE CONTINUOUS STRENGTH METHOD (CSM)

The Continuous Strength Method (CSM) is a strain-based design approach. Gardner (2008) first introduced the term 'Continuous Strength Method' and made the technique more generalized for easy applications. It has already been validated for stocky and slender standard sections (Gardner and Afshan 2013, Ahmed *et al.* 2016).

The CSM has two major parts: a base curve and a material model. A design base curve establishes a continuous relationship between the normalized cross-section deformation capacity

$\epsilon_{csm} / \epsilon_y$ and the cross-section slenderness ($\bar{\lambda}_p$) (Gardner and Afshan 2013). It is suggested to determine the cross-section slenderness ($\bar{\lambda}_p$) using the elastic buckling capacity $\sigma_{cr,cs}$ and 0.2% proof stress $\sigma_{0.2}$ (Gardner and Afshan 2013). To tackle the observed significant post-buckling of a slender cross-section, a new parameter called Equivalent Elastic Deformation Capacity $\epsilon_{e,ev}$ was also introduced recently by modifying the CSM equations by Ahmed *et al.* (2016). Equivalent elastic deformation capacity $\epsilon_{e,ev}$ is a function of ϵ_{csm}.

Material Model is used to determine the cross-section resistance in combining with the strain measure (Ahmed *et al.* 2016). Using the material model, the limiting buckling stress f_{csm} for the cross-section can be calculated. The final cross section resistance in compression $N_{c,rd}$ is estimated as recommended in *EN 1993-1-4, Eurocode 3*.

3 CUFSM SOFTWARE

CUFSM (Li and Schafer 2010) is finite strip elastic buckling analysis application. The use of finite strip analysis has the unique capability to provide complete and relevant stability. Design and hand methods that are traditionally used for "plate" structures often ignore compatibility at plate junctures and usually provide no means to calculate a variety of important buckling modes such as distortional buckling. CUFSM allows all elastic buckling modes of a structure to be quantified and examined. This software was used by Ahmed *et al.* (2016) to obtain $\sigma_{cr,cs}$ in order to calculate f_{csm} for slender sections.

4 EXPERIMENTAL STUDIES

A total number of 21-collected test results (Stone and LaBoube 2005, Lue *et al.* 2006, Liu *et al.* 2009) for both stocky and slender I-sections (bolted) were used to conduct the study.

5 PROPOSED METHODOLOGY

The compression resistance $N_{standard\text{-}csm}$ of the built-up I-section was first calculated using the CSM formulas proposed by Ahmed *et al.* (2016). It was observed that there was large variation between test results and estimated values as shown in Figure 1 ($N_{standard\text{-}csm}/N_{test}$).

To improve the prediction capability, a new methodology is developed. Firstly, the critical stress $\sigma_{cr,cs}$ of each C-section is determined using CUFSM software (Li and Schafer 2010) instead of calculating the value for the I-section. Based on these $\sigma_{cr,cs}$, cross-section slenderness ($\bar{\lambda}_p$), strain ratio $\epsilon_{csm}/\epsilon_y$, equivalent elastic deformation $\epsilon_{e,ev}$ and buckling stress f_{csm} were calculated based on CSM formulation (Ahmed *et al.* 2016).

Cross-section compression resistance for individual channel section $N_{c,rd\ (C-section)}$ was determined considering only C-section by multiplying f_{csm} and the area of C-section $A_{(c-section)}$. Then cross-section compression resistance for built-up I-section $N_{c,rd\ (I-section)}$ was calculated using Eq. (1).

$$N_{c,rd(I-section)} = 2 \times N_{c,rd(c-section)} \tag{1}$$

Buckling resistance of columns was calculated according to the proposed equations of *EN 1993-1-4, Eurocode 3*. Elastic critical buckling capacity N_{cr} was determined using the moment of inertia of the whole I-section using Eq. (2). $\bar{\lambda}_{csm}$ was then calculated using the area of C-section $A_{(c-section)}$ and elastic critical buckling capacity N_{cr} using Eq. (3). The buckling reduction factor (χ) was calculated for both X-axis and Z-axis. Buckling resistance of the column $N_{b,rd\ (I-section)}$ was determined for both axis using Eq. (4). Minimum of $N_{c,rd\ (I-section)}$, $N_{b,rd-x\ (I-section)}$ and $N_{b,rd-z\ (I-section)}$ was taken as the cross-section resistance of the built-up section $N_{modified-csm}$.

$$N_{cr} = \pi^2 \times E \times \frac{I_{(I\text{-section})}}{L_e^{\,2}} \tag{2}$$

$$\bar{\lambda}_{csm} = \sqrt{\frac{A_{(c\text{-section})} \times f_{csm}}{N_{cr}}} \tag{3}$$

$$N_{b,rd(I\text{-section})} = \chi \times N_{c,rd(I\text{-section})} \tag{4}$$

6 PARAMETER SELECTION

In the modified CSM, the value of material partial safety factor γ_{M0} was taken as 1.0 that is less than 1.10 according to *EN 1993-1-4, Eurocode 3* and the imperfection factor α was taken as 3.0 that was proposed by Lecce and Rasmussen (2016). The values of a and b for calculating equivalent elastic deformation $\epsilon_{e,ev}$ were taken as 3.05 and 3.00 for I-section under axial compression proposed by Ahmed *et al.* (2016).

7 DATA ANALYSIS WITH THE CSM

The collected test data mentioned in section 4 were used for evaluating the proposed methodology ($N_{modified-csm}/N_{test}$) as summarized in Table 1 and shown in Figure 1. Comparison between estimated compression resistance of built-up columns by respected method and test values in Figure 1 shows that variations are within 20% mostly that are more accurate and less scattered than the values calculated from standard CSM (Ahmed *et al.* 2016). In this study, effect of bolt spacing was ignored which may cause non-conservative result for proposed methodology. Further research will be conducted to include the effect of bolt spacing to make the methodology more accurate and reliable.

Table 1. Performance evaluation of proposed methodology.

Section ID	Specimen Description	$N_{modified-csm}/N_{test}$	$N_{standard-csm}/N_{test}$	Reference
1	2C75*6.9, bolted 4@52.5cm	1.03	0.59	*Liu et al.* (2009)
2	2C75*6.9, bolted 2@105.0cm	1.18	0.67	
3	2C100*9.4, bolted 4@52.5cm	1.03	0.71	
4	2C100*9.4, bolted 2@105.0cm	1.18	0.82	
5	2C180 × 21.4/B-B 4 bolted @ 50	1.01	0.88	*Lue et al.* (2006)
6	2C180 × 21.4/B-B 3 bolted @ 67	1.03	0.90	
7	1.372-152 @305	0.94	0.61	
8	1.372-152 @610	0.95	0.61	
9	1.372-152 @762	0.95	0.62	

Table 1 (contd). Performance evaluation of proposed methodology.

10	1.372-152 @914	1.10	0.72	
11	1.372-152 @1016	0.96	0.62	
12	1.372-152 @1067	0.95	0.62	
13	1.155-92 @305	1.06	0.73	*Stone and LaBoube* (2005)
14	1.155-92 @610	1.27	0.87	
15	1.155-92 @914	1.25	0.86	
16	0.88-92 @305	1.11	0.85	
17	0.88-92 @610	0.99	0.76	
18	0.88-92 @914	1.01	0.78	
19	0.84-152 @305	1.01	0.63	
20	0.84-152 @610	0.88	0.55	
21	0.84-152 @914	1.06	0.66	

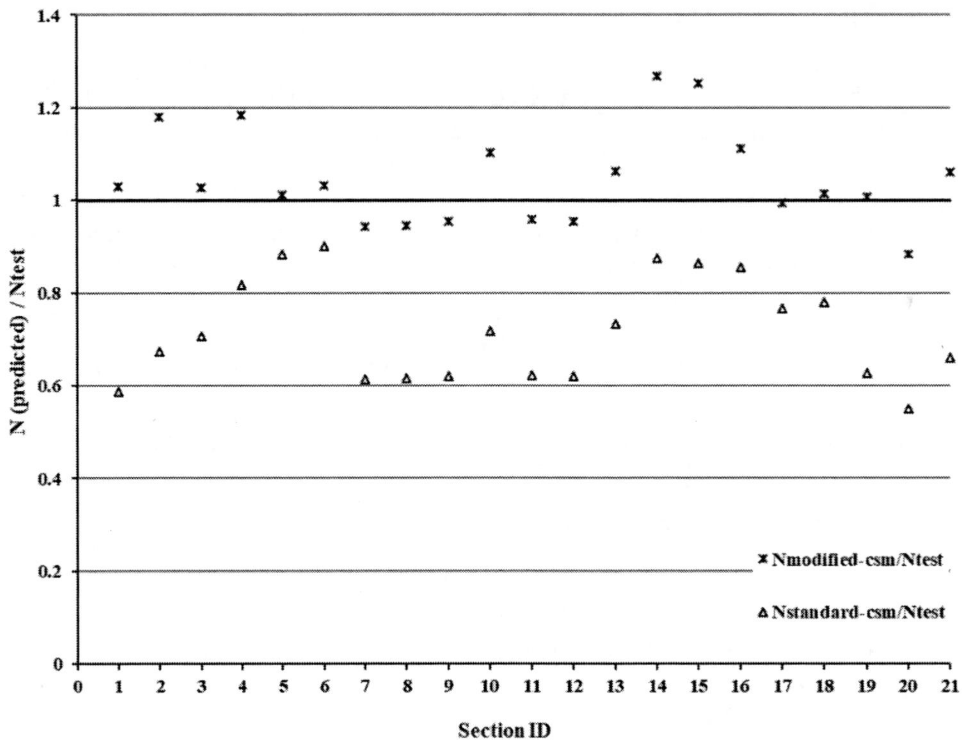

Figure 1. Performance evaluation of proposed methodology.

8 RELIABILITY ANALYSIS

A statistical analysis is presented in Table 2 to show the reliability of the standard CSM to calculate the compression resistance of built-up I-sections. The proposed methodology has the co-efficient of variance (COV) of 0.10 that indicates the better accuracy than standard CSM (Ahmed *et al.* 2016) having the COV of 0.16.

Table 2. Statistics of the predictions.

	$N_{modified\text{-}csm}/N_{test}$	$N_{standard\text{-}csm}/N_{test}$
Mean	1.05	0.72
Std. Deviation	0.10	0.11
COV	0.10	0.16

9 CONCLUSIONS

In this paper, a modified CSM methodology was proposed for built-up (bolted) I-sections. Comparison between test data and predicted values shows that the method offers improved cross section compression resistance than standard CSM. Though the effect of bolt spacing is not considered in this study, the proposed methodology paves the path for deriving CSM equations for built-up sections. Currently a study is being conducted to introduce bolt spacing consideration to make the proposed methodology more effective.

Acknowledgments

The authors would like to express their gratitude to Dr. Mahmud Ashraf, Associate Professor, School of Engineering from Deakin University, Melbourne, Australia and Dr. Shameem Ahmed, Assistant Professor, Department of Civil Engineering from Bangladesh University of Engineering and Technology (BUET), Dhaka, Bangladesh for their valuable suggestions in this research. The study was conducted as an undergraduate thesis in Bangladesh University of Engineering and Technology.

References

Ahmed, S., Ashraf, M., and Anwar-Us-Saadat, M., The Continuous Strength Method for Slender Stainless Steel Cross-sections, *Thin-Walled Structures*, 107, 362-376, October, 2016.

EN 1993-1-4: Eurocode 3: Design of Steel Structures - Part 1-4: General Rules - Supplementary Rules for Stainless Steels, *European Committee for Standardisation*, October, 2006.

Gardner, L., The Continuous Strength Method, *Proceedings of the Institution of Civil Engineers - Structures and Buildings*, ICE Publishing, 161(3), 127-133, June, 2008.

Gardner, L., and Afshan, S., The Continuous Strength Method for Structural Stainless Steel Design, *Thin Walled Structures*, 68, 42-49, July, 2013.

Lecce, M., and Rasmussen, K., Distortional Buckling of Cold-Formed Stainless Steel Sections: Finite-Element Modeling and Design, *Journal of Structural Engineering*, 132(4), April, 2006.

Li, Z., and Schafer, B., Buckling Analysis of Cold-formed Steel Members with General Boundary Conditions Using CUFSM: Conventional and Constrained Finite Strip Methods, Missouri University of Science and Technology, 2010.

Liu, J., Lue, D. M., and Lin, C. H., Investigation on Slenderness Ratios of Built-up Compression Members, *Journal of Constructional Steel Research*, Elsevier, 65(1), 237-248, January, 2009.

Lue, D. M., Yen, T., and Liu, J., Experimental Investigation on Built-up Columns, *Journal of Constructional Steel Research*, 62(12), 1325-1332, December, 2006.

Stone, T., and LaBoube, R., Behavior of Cold-formed Steel Built-up I-sections, *Thin-Walled Structures*, 43(12), 1805-1817, December, 2005.

Interdependence between Structural Engineering and Construction Management
Edited by Ozevin, D., Ataei, H., Modares, M., Gurgun, A., Yazdani, S., and Singh, A.

ISBN: 978-0-9960437-6-2

EXPERIMENTAL AND NUMERICAL STUDIES OF STRUCTURAL PERFORMANCE OF NOTCH DAMAGED STEEL BEAMS

SAID ELKHOLY and BILAL EL-ARISS

Dept of Civil and Environmental Engineering, United Arab University, Al Ain, UAE

Performance of notch damaged unrepaired steel I-shaped beam specimens was examined experimentally and numerically simulated using finite element modeling. Nine U-shaped notch damaged specimens and one control beam specimen were considered. The nine unrepaired laterally unsupported I-shaped full-scale steel specimens were tested in two-point loadings and the parameters considered in the study were the size and location of the notch. The notch size was 30 mm, 40 mm, and 60 mm in depth with a constant opening width of 20 mm. The location of the notch was at one-quarter, one third, and middle of the specimen clear span. All pre-damaged specimens had notches on one side of the tension flange with respect to the beam web except for two specimens where two notches were cut at the middle of the specimen clear span on both sides of the tension flange. The test specimens were modeled analytically using three-dimensional models of the bare-steel I-shaped specimens for comparison of the analytical predictions with experimental results. The test results showed the strength of the notched beams was hardly affected by the notch size or location. However, the stiffness of the beams was slightly decreased as the notch got closer to the mid-span with increased ductility. The corresponding analytical results were in good agreement with the test data.

Keywords: Laboratory testing, FE simulation, I-shaped members, Stiffness, Load bearing capacity.

1 INTRODUCTION

Kim and Brunell (2011) and Kim and Harries (2012) explored the CFRP restoration approaches for notch damaged steel beams. The impact of the notch shapes was considered numerically to find the relations between CFRP restoration approaches and initial damage in the steel beams tested in flexure. Siwowski and Siwowska (2018) investigated the structural behavior of CFRP plate-strengthened steel beams. Four different shapes of strengthening steel plates with a notch under fatigue load. They showed the possibility of extending the fatigue life of steel plate approximately four times with non-prestressed plates and, in some cases, fully stopping the crack growth by means of prestressed plates. Elchalakani (2016) offered investigational results of corroded steel circular hollow sections (CHS) repaired with CFRP sheets and tested under large displacements. The author showed that corrosion level can affect the percentage in combined flexural and bearing strength of the CHS. Kadhim (2012) studied the behavior of CFRP plate-repaired continuous steel beam. The analytical results specified that for CFRP repairing plate lengths in hogging and sagging regions of the beam of around 60% and 40% of beam span length, respectively, the ultimate strength increase rate diminished. Yu *et al.* (2011) established a bond slip model to predict the debonding of FRP laminates used to strengthen steel beams considering the effects of adhesive thickness on the bond strength, laminate thickness, and bond length and strength. It was determined that the bond strength between the strengthening FRP laminates and steel beams can be increase as a result of decreasing the laminate thickness and modulus, minimizing the adhesive thickness, or increasing the bond length. Chen *et al.* (2018) experimentally observed the fatigue performance of the damaged steel rectangular hollow section (RHS) beams. The test data indicated that the used CFRP sheets with high modulus increased the RHS beam fatigue life. Tavakkolizadeh and Saadatmanesh (2003a and 2003b) and Photiou *et al.* (2004) studied the behavior of damaged steel-concrete composite steel girders repaired with CFRP sheets under static loading. Test results revealed that epoxy bonded CFRP sheets could restore the stiffness and ultimate load-carrying capacity and of repaired girders. Comparison between experimental and numerical modeling showed that the traditional methods of analysis of composite beams were conservative. Deng *et al.* (2018) applied a mixed-mode cohesive approach to imitate CFRP- strengthened notch damaged steel beams. The study revealed that although the beam strength was increased, the ductility reduced as the notch extent increase and debonding mode of failure premature occurred. It is clear that previous works show that researchers have mostly proposed retrofitting techniques to damaged steel members and investigated the rehabilitated member performance. The objective of this study is to examine the flexural behavior of unrepaired laterally unsupported notch damaged steel beams.

2 EXPERIMENTAL PROGRAM

A program for testing a total of ten W-shaped steel beams was prepared. One undamaged (control) and nine artificially U-shaped notch damaged beams were instrumented and tested in flexural. The notch was cut in the beam bottom flange and all beams were left unrepaired and laterally unsupported while tested. The notch size was 30 mm, 40 mm, and 60 mm in depth with a constant width of 20 mm and was cut on one side of the beam bottom (tension) flange with respect to the beam web. All of the beams had a clear span of 1800 mm between the support and a total length of 2000 mm, Figure 1. The location of the notch was at one-quarter, one-third, and middle of the specimen clear span, as shown in Figure 2 for the notch at the middle of the beam.

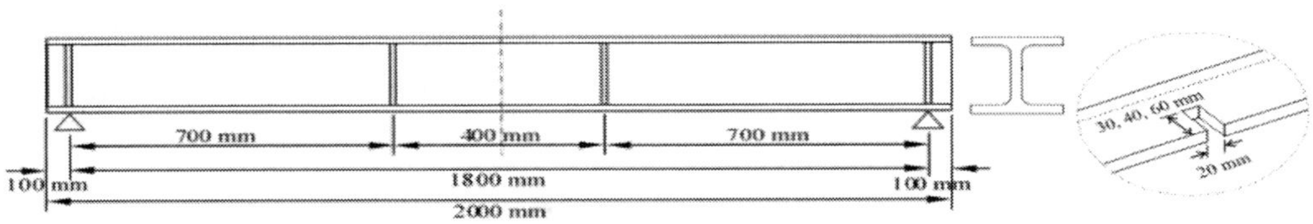

Figure 1. Details of a typical notch damaged I-beam specimen.

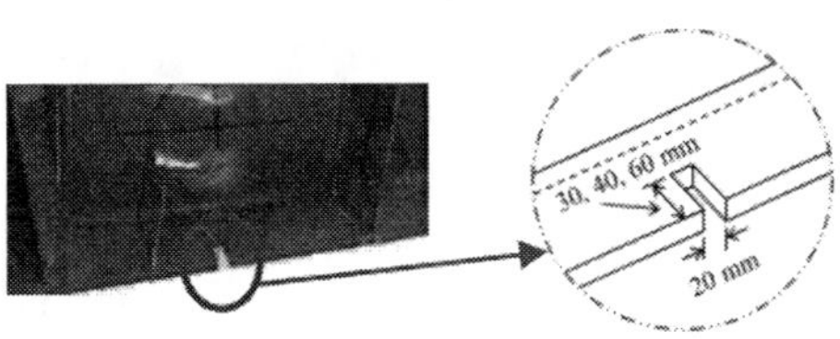

Figure 2. Location of the notch in the beam bottom flange.

The w-shape steel beams were of W 203 x 203 x 52 sections with a web height and thickness of 206 mm and 7.9 mm, respectively, and a flange width and thickness of 204 mm and 12.6 mm, respectively, and 46.1 kg/m mass. The beam specimens were web-stiffened to avoid web crippling at the support locations and at the two loading point locations which are 400 mm apart and 700 mm away from supports, as presented in Figures 1 and 3.

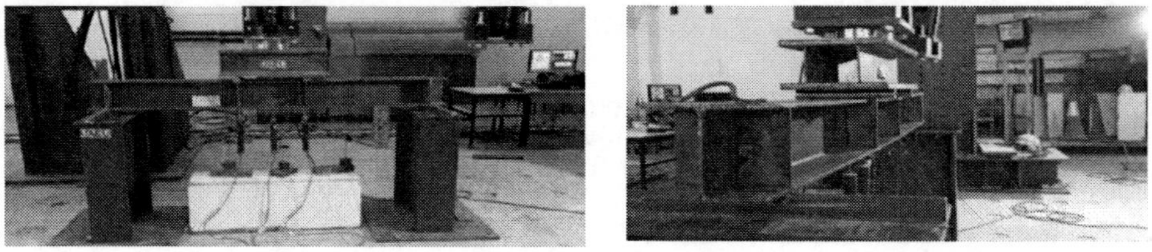

Figure 3. The beams with vertical stiffeners over supports and under loading points.

2.1 Material Properties and Test Matrix

The mechanical properties of the studied steel W-shaped specimens were obtained from standard tensile coupon test. The average ultimate and yielding strength values obtained were 441 MPa and 294 MPa, respectively. The damaged beams were divided into three groups and are shown in Table 1.

Table 1. Test matrix.

Group (Location of Notch)*	Length of Notch (mm)	Specimen Name
C	0	C-NG-0
	30	QS-30
Q	40	QS-40
	60	QS-60
	30	TS-30
T	40	TS-40
	60	TS-60
	30	MS-30
M	40	MS-40
	60	MS-60

C=control beam (undamaged); Q=notch at beam quarter span; T=notch at beam one-third span; M=notch at beam mid-span. NG-0=no notches (0 depth); S=single notch on one side of the bottom flange.
Specimen Name - example: *QS-30=single notch of 30 mm depth on one side of the bottom flange at beam quarter span.*

2.2 Instrumentation of Test Setup

The steel W-shaped beams were tested in four-point bending set-up with two point loads applied. The beams had of the clear span between the supports of 1800 mm. The two applied point loads were 200 mm on each side of the centerline of the beam and 700 mm from the supports, Fig. 1. The control beam and notch damaged beams were left unrepaired; however, to avoid web crippling of the beams during testing, the beams were web-stiffened with vertical stiffeners under the load application points and above the supports. All steel beam specimens were loaded in a displacement-control mode at a rate of 2 mm/min and were kept laterally unsupported during testing, Figures 2 and 3. The beam was equipped and instrumented with strain gauges, for strain readings, and Linear Variable Differential Transformers (LVDT), for deflection measurements. The strain gauges were bonded to the outer surface of top and bottom flanges and the web of the beams at locations. The LVDTs were located below the bottom flange at the load application points and at the centerline of the beam, as shown in Figure 4.

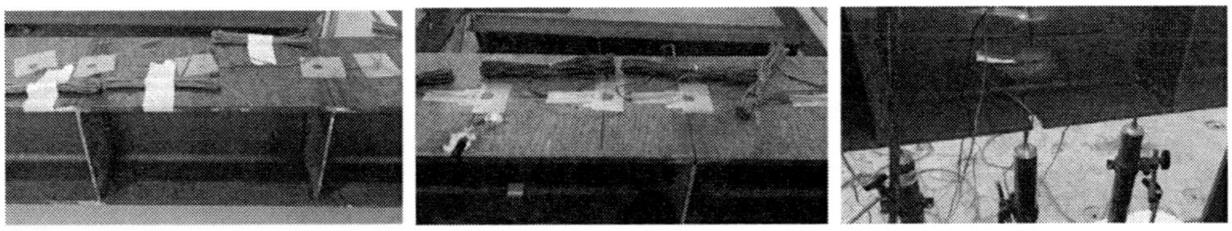

Figure 4. Installation of strain gauges and placement of LVDT's.

2.3 Finite Element Simulation

Finite element simulation of the tested beams was carried out. Three-dimensional models were generated to simulate the tested W-shaped steel beams numerically up to starting of lateral torsional buckling as the beams were not laterally supported. The models were produced using finite element (FE) analysis program, ATENA. Three types of three-dimensional solid elements are provided by ATENA-3D: tetrahedral elements (CCIsoTetra), brick elements (CCIsoBrick), and wedge elements (CCIsoWedge). After several trials of tuning the FE mesh, the Tetrahedral 3-D (T3D) solid element was found to be the most appropriate element that best explained the behavior of the specimens and; therefore, was used to generate the FE models. A displacement-controlled incremental loading method was employed in the FE analysis. The standard Newton-Raphson iterative solution method implemented in ATENA was adopted in the analysis. A schematic sketch of the beams modeling is shown in Figure 5.

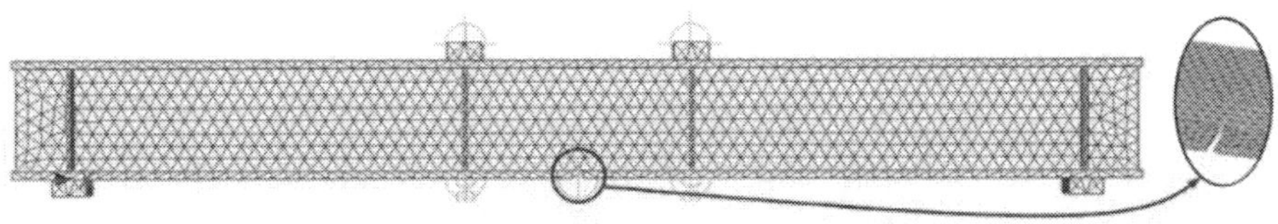

Figure 5. Finite element modeling of the testes W-shaped beams.

3 DISCUSSION OF EXPERIMENTAL AND NUMERICAL RESULTS

Figure 6 presents the experimental load-deflection relations of the three groups of the beams, Q, T, and M separately. The figure shows that group Q specimens had the same stiffness regardless of the notch size. Group T specimens demonstrated a slight decrease in the specimen stiffness as the notch size increased due to notch location. Group M revealed more noticeable decrease in the beam stiffness as the notch location had more impact. However, the ductility of the three group specimens had the same pattern where the largest ductility was for the highest notch size, 60 mm.

Figure 7(a) depicts the experimental load-deflection relations for the tested beams. It can be seen that when the notch was at a quarter of the span, the beam had the highest load carrying capacity and the least ductility. When the notch was at the third span of the span, the beam load carrying capacity decreased while the ductility increased. The beam had the highest ductility when the notch was at the mid-span but the load carrying capacity was the least. This can be attributed to the beam failure mode as a result of the notch location. More specifically, when the notch is at the furthest location from the beam centerline (a quarter of span), the beam the beam prematurely twisted at low deflection value. On the other hand, when the notch was at the mid-span, the beam was able to sustain more deflection due to beam symmetry effect until lateral torsional buckling occurred. The beam behavior was somewhat in between the latter cases when the notch was at third of the beam span. Figure 7(b) shows the load-top and bottom strain relationships of group T specimens.

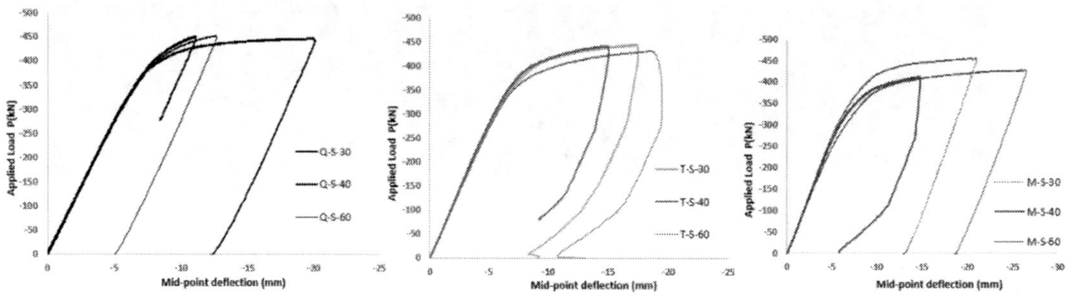

Figure 6. Load-deflection curves.

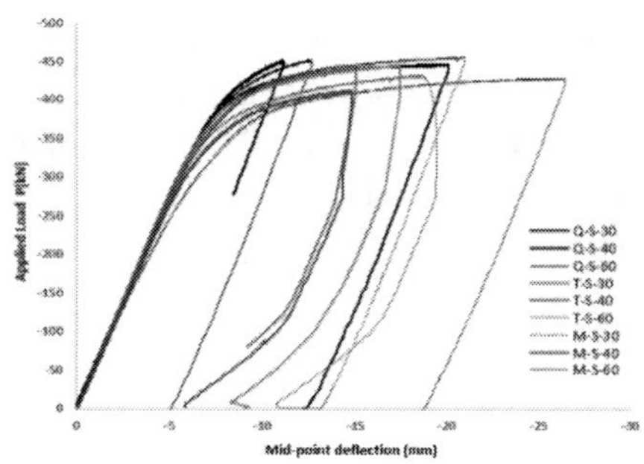

(a)

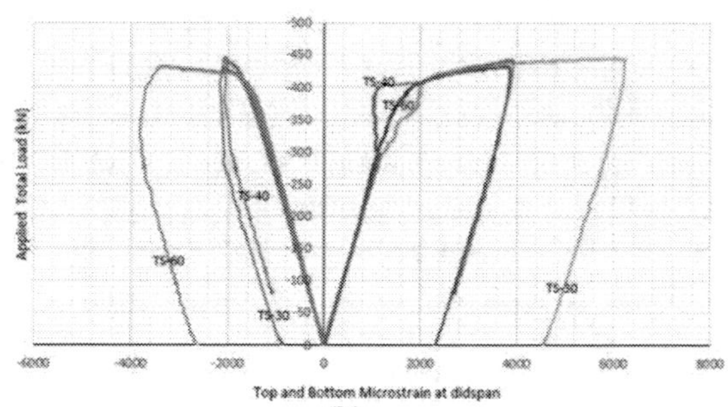

(b)

Figure 7. (a) Load-deflections of all beams, (b) load-strain curve for TS-30, 40, and 60 specimens.

Figures 8 shows the finite element deflection of specimen MS-40. It is clear that the beam numerically predicted behavior matched the beam experimental behavior in terms of buckling failure and location. Both numerical and experimental buckling occurred in the back top flange of the beams simultaneously with plastic deformation arose in the front notch in the bottom flange.

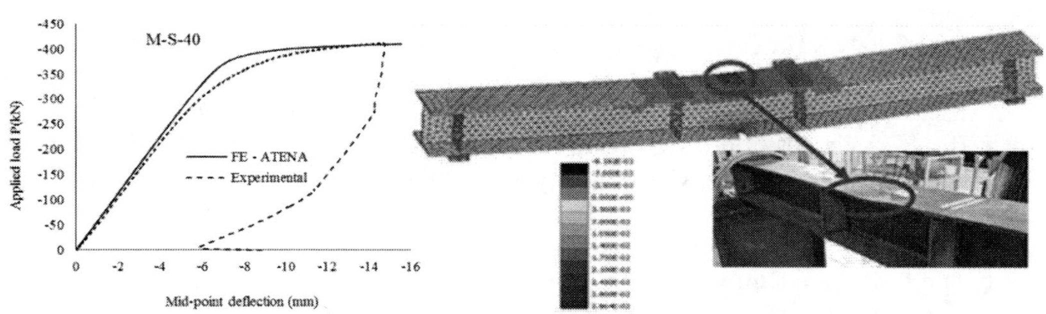

Figure 8. Comparison of FE analysis and experimental results and buckling in specimen M-S-40.

Figure 9 shows the numerical plastic strain in the notch at different loading steps of the test and failure buckling in specimen MS-40. It is obvious from the figure the numerical bucking failure predicted was in agreement with observed experimental. This confirmed the predicted numerical bucking failure. The test results; therefore, verified the accuracy of the numerical simulation of the tested beams generated in this study. Also, both experimental and numerical outputs showed the same pattern of plastic strain occurring in the notch in the front bottom flange buckling arising in the back of the top flange of the beam.

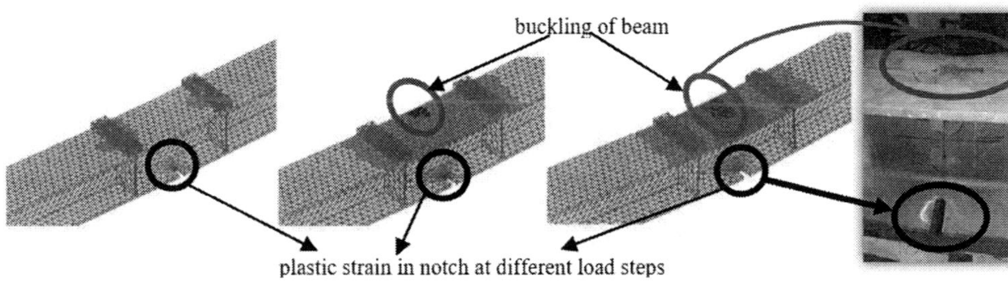

Figure 9. Plastic strain in the notch at different loading step and buckling in specimen MS-40.

4 CONCLUSIONS

Experimental and numerical investigation was performed on ten I-shape steel beams with nine beams were artificially u-shaped notch damaged. The notch location had more effects on the beam behavior, ductility, and failure mode than the notch size effects. The finite element simulations and predictions were in agreement with the experimentally observed results.

Acknowledgments

The authors would like to express their gratitude to the United Arab Emirates University (UAEU) for the financial support of this project under Fund No. 31N296 and No. 31N305. The authors also would like to acknowledge Mr. Zakaria El Karaouf and Mr. Ali Saleh for their contribution to FE simulation.

References

Chen, T., Wang, X., and Qi, M., Fatigue Improvements of Cracked Rectangular Hollow Section Steel Beams Strengthened with CFRP Plates, *Thin-Walled Structures*, 122, 371-377, 2018.

Deng, J., Li, J., Wang, Y., and Xie, W., Numerical Study on Notched Steel Beams Strengthened by CFRP Plates, *Construction and Building Materials*, 163, 622-633, 2018.

Elchalakani, M., Rehabilitation of Corroded Steel CHS Under Combined Bending and Bearing Using CFRP, *J. of Constructional Steel Research*, 125, 26-42, 2016.

Kadhim, M. M. A., Effect of CFRP Plate Length Strengthening Continuous Steel Beam, *Construction and Building Materials*, 28(1), 648-652, 2012.

Kim, Y. J. and Brunell, G., Interaction Between CFRP-Repair and Initial Damage of Wide-Flange Steel Beams Subjected to Three-Point Bending, *Composite Structures*, 93(8), 1986-1996, 2011.

Kim, Y. J., and Harries, K. A., Predictive Response of Notched Steel Beams Repaired with CFRP Strips Including Bond-Slip Behavior, *Intnl. J. of Structural Stability and Dynamics*, 12(01), 1-21, 2012.

Photiou, N., Hollaway, L. C., and Chryssanthopoulos, M. K., Strengthening of an Artificially Degraded Steel Beam Utilizing a Carbon/Glass Composite System, *In Advanced Polymer Composites for Structural Applications in Construction*, 274-283, 2004.

Siwowski, T. W., and Siwowska, P., Experimental Study on CFRP-Strengthened Steel Beams, *Composites Part B: Engineering*, 149, 12-21, 2018.

Tavakkolizadeh, M., and Saadatmanesh, H., Repair of Damaged Steel-Concrete Composite Girders Using Carbon Fiber-Reinforced Polymer Sheets, *J. of Composites for Construction*, 7(4), 311-322, 2003a.

Tavakkolizadeh, M., and Saadatmanesh, H., Strengthening of Steel-Concrete Composite Girders Using Carbon Fiber Reinforced Polymers Sheets, *J. of Structural Engineering*, 129(1), 30-40, 2003b.

Yu, Y., Chiew, S. P., and Lee, C. K., Bond Failure of Steel Beams Strengthened with FRP Laminates–Part 2: Verification, *Composites Part B: Engineering*, 42(5), 1122-1134, 2011.

Interdependence between Structural Engineering and Construction Management
Edited by Ozevin, D., Ataei, H., Modares, M., Gurgun, A., Yazdani, S., and Singh, A.

ISBN: 978-0-9960437-6-2

CRACK PATTERN PREDICTION OF LATERALLY LOADED PANELS WITH OPENINGS BASED ON ANN METHOD

YONGFEI WANG[1,2] and YU ZHANG[1,2]

[1]*Key Lab of Structures Dynamic Behavior and Control of the Ministry of Education, Harbin Institute of Technology, Harbin, China*
[2]*Key Lab of Smart Prevention and Mitigation of Civil Engineering Disasters of the Ministry of Industry and Information Technology, Harbin Institute of Technology, Harbin, China*

In this paper, a Back Propagation Neural Network (BPNN) is used to predict crack pattern for masonry panels with opening subjected to lateral loading. The cellular automata method is used to digitalize the panels, including two steps - dividing a panel into a certain number of cells and calculating cell state values by use of a Von Neumann neighborhood model. These digitalized values are used as input data of NN model, respectively. All the experimental data is collected, including panel configuration, material property, opening ratio and location, state values, and crack pattern. The NN model is trained repeatedly, taking part of the data as a training set, to determine parameters, and the rest of the data is taken to check the model. Well-trained NN models can predict the crack pattern of any other panel. The results show that NN method is suitable for prediction of crack pattern. Comparing the two ways of prediction, the Fragility Coefficient Method gets a more precise pattern. The predicted cracks are distributed successively in some specific areas, especially in high similarity, compared with experimental crack pattern.

Keywords: Cellular automata, Digitalization, Weakness, Fragility coefficient, Back-propagation neural network.

1 INTRODUCTION

For traditional masonry structure, the lateral force is an unneglectable load; for example, great wind pressure could not be ignored for damages on some special buildings. Other lateral forces, like explosions (Varma *et al.* 1997), cyclic lateral force (Davidson and Wang 1985), and out-of-plane shaking force (Tu *et al.* 2010), are also taking effect on the behavior. Golding (1991) and Sinha (2001) did deep research on the design of laterally loaded masonry. One of the key research topics is the failure model. In this paper, an artificial neural network (ANN) model is built to predict the crack pattern of laterally-loaded panels with different openings, and all the original statistics are from Chong (1993)'s experiments. To accomplish the prediction, firstly digitalize the panels by cellular automata (CA) technique. Secondly, the generated data and other aspects which concern the crack distribution are treated as the input data of the ANN model. Crack status (cracked or non-cracking) is used as the output data of the ANN model. Taking one panel as the base panel, it's CA model and crack status are used to train the ANN model. Then, the trained ANN model is used to predict other panels. This method could reflect the main crack distribution, but it still has imperfections such as the disappeared tiny cracks. So, this paper gives an innovative concept: the fragility coefficient.

2 BACK PROPAGATION NEURAL NETWORK ARCHITECTURE

The BPNN contains input layer, hidden layers, and output layer. Input layer contains 11 nodes in total (shown in Figure 1), in which 9 nodes are the state values of one cell (shown in Figure 2), and the other two are the x and y coordinates (these two parameters were used especially for panels with different size, and it's not mentioned in this paper). There are two hidden layers, 18 nodes in the first layer and 10 nodes in the second one. For the direct prediction method (DPM), output layer means the crack pattern of matrix type, only including two separate digits, 0 for non-cracking cells and 1 for cracked ones. Considering the fragility coefficient, output layer is different from DPM's continuous values of 0 to 1. Figure 3 describes the relationship between the DPM and the fragility coefficient method (FCM). Activation function is logsig.

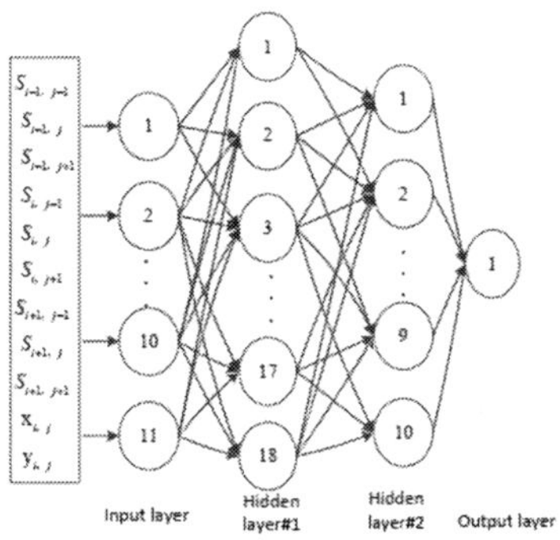

Figure 1. The BPNN model.

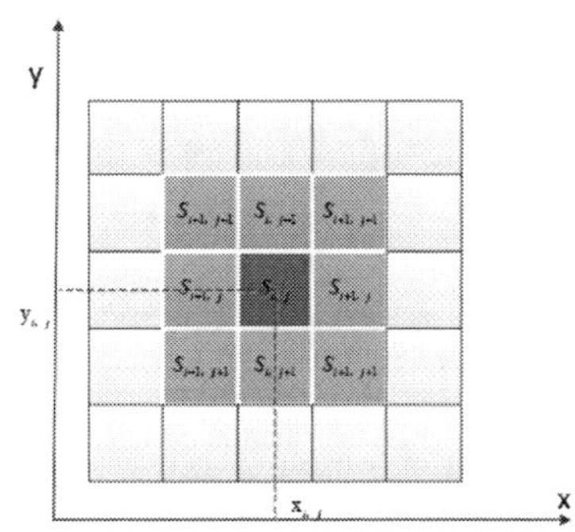

Figure 2. The state value by Moore model of CA.

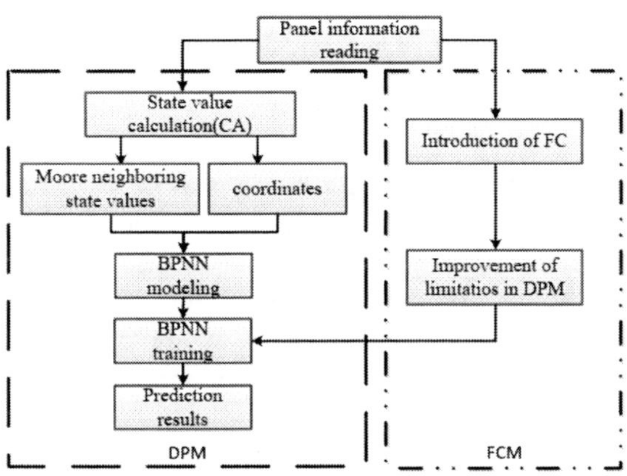

Figure 3. Introduction of two methods.

3 BACK PROPAGATION NEURAL NETWORK MODEL

3.1 Panel Information Digitalization

CA is a localized and dynamic system both in time and space. Cell is a basic unit of a CA system. Every cell has its unique state value, which could reflect the corresponding location's potential information (Maja and Justyna 2010, Morita 2018). The state value of a cell is not only decided by itself, but also these cells around it. These cells are called neighborhoods. There're two main neighborhood models, Von Neumann and Moore neighborhoods showing in Figure 4.

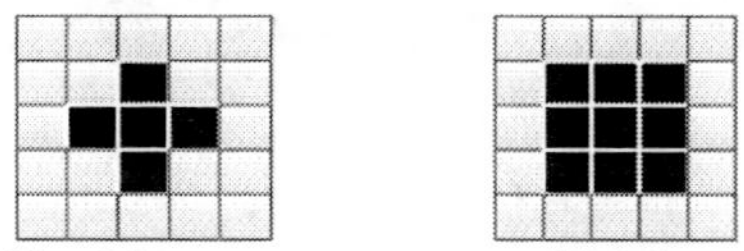

(a) Von Neumann (b) Moore

Figure 4. Two neighborhood models.

3.1.1 *Selection of the state values*

The state value of a panel is calculated in Eq. (1) by Von Neumann neighborhoods as following:

$$\begin{aligned}
&L_{si,j}=1-L_{i,j}, L_{i,j}=L_{i,j-1}+\eta(1-L_{i,j-1}), \ (i=1,2,\dots,M;\ j=1,2,\dots,N)\\
&R_{si,j}=1-R_{i,j}, R_{i,j}=R_{i,j-1}+\eta(1-R_{i,j-1}), (i=1,2,\dots,M; j=N, N-1,\dots,1)\\
&B_{si,j}=1-B_{i,j}, B_{i,j}=B_{i-1,j}+\eta(1-B_{i-1,j}), (i=M,\ M-1;\ j=1,2,\dots,N)\\
&T_{si,j}=1-T_{i,j}, T_{i,j}=T_{i+1,j}+\eta(1-T_{i+1,j}), (i=1,2,\dots,M;\ j=1,2,\dots,N)
\end{aligned} \tag{1}$$

Where $L_{si,j}$, $T_{si,j}$, $B_{si,j}$, $R_{si,j}$ are state values calculated by constraint transiting from left, top, bottom and right side. $L_{i,0}$, $T_{0,j}$, $B_{M+1,j}$, $R_{i,N+1}$ are the initial values of $L_{i,j}$, $T_{i,j}$, $B_{i,j}$, $R_{i,j}$, which reflect the constraint on each border. *M, N* are the numbers of rows and columns of the zoning area. And η is coefficient of transition (Zhang *et al.* 2010 and Huang *et al.* 2013).

The state value $S_{i,j}$ of each cell is defined in Eq. (2) as the average value from its four neighboring cells:

$$S_{i,j}=\frac{\left(L_{i,j}+R_{i,j}+T_{i,j}+B_{i,j}\right)}{4}\left(i=1,2\dots M, j=1,2\dots,N\right) \tag{2}$$

3.1.2 *Division of the panel with opening*

During the digitalization of panel with opening, there is a bit flaw; that is loss of continuity near the opening. Due to its importance to the NN training, division again is needed absolutely. The division method for a representing panel, taking Panel SB02 for example, is shown in Figure 5. According to this division method, all panels can be divided into four parts and the zoning data is shown in Table 1. Taking Area I-VIII-VII for an example, the state value is calculated by transiting the left, top and bottom side constraints. Each parameter is expressed in Figure 6.

Table 1. Sizes of panel division (mm).

size	SB01	SB02	SB03	SB04	SB05	SB06	SB07	SB09
x1	2808	1677	1340	2352	2808	1450	1000	3815
x2	0	2260	2935	910	0	0	900	900
x3	2808	1677	1340	2352	2808	1450	1000	900
y1	1238	900	1500	0	1238	1225	900	900
y2	0	1125	525	2025	0	0	900	900
y3	1238	450	450	450	1238	1225	650	675

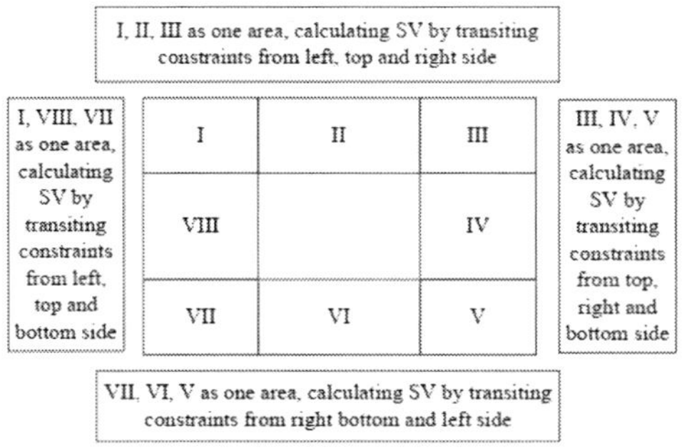

Figure 5. Panel division.

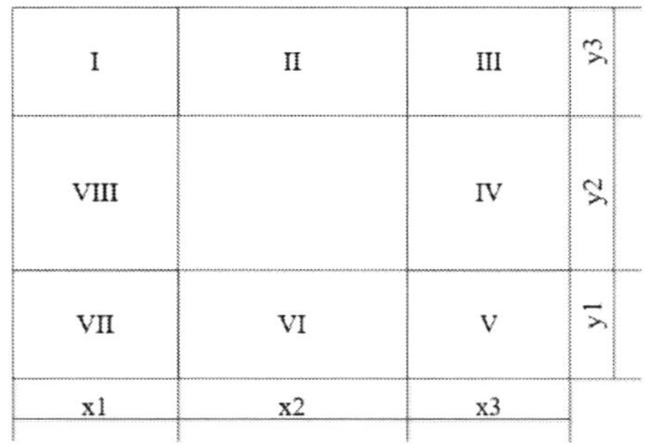

Figure 6. Parameters from division.

3.2 The Fragility Coefficient of The Panel with Opening

Unlike the homogeneous materials like steels, masonry panels are made up of building blocks (like bricks) and bonder (like mortar), which gives them the variability that is another factor influencing the working performance besides material itself and load case.

The data from the previous experiment is not accurate enough to choose a base panel. The information the base panel represents cannot stand for the scientific and general law. On the other hand, predicted crack patterns of other panels may show great differences from the actual ones due to the influence of the masonry's variability. To alleviate the influence, Fragility Coefficient (FC) is proposed to make comparison of the DPM method.

Fragility Coefficient: a parameter that shows whether a unit within the panel is apt to crack or not. To realize it, Eq. (3) is proposed to calculate FC, referring to the feature of inverse proportional function. With an experimental crack pattern, the FC data could be achieved by programming. Figure 7 shows the FC distribution of Panel SB05.

$$\mu = \frac{0.8}{|x_i - x_p| + |y_i - y_p|} \tag{3}$$

Where x_i, y_i are x and y coordinates of the calculating unit, x_p, y_p are x and y coordinates of the reference unit, and 0.8 comes from experience.

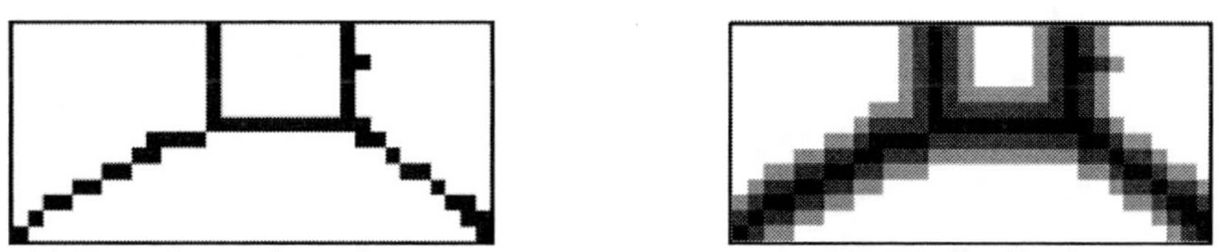

(a) Experimental crack pattern (b) FC distribution

Figure 7. Panel SB05.

4 BPNN TRAINING AND RESULTS

With all data preparation and BPNN model, this study selects different representative base panels in the two methods to compare two prediction results. In each method, a base panel with 512(16×32, from one base panel) groups of data are used to train the BPNN. In DPM, Panel SB02 is used as the base panel. Panel SB05 is the base one in FCM. Training process of the DPM (as an example) is shown in Figure 8. Predicted crack pattern of other panels from SB01 to SB09 are shown in the Table 2.

Table 2. Prediction result.

Panel No.	Experimented crack pattern	Predicted crack patterns	
		DPM	FCM
SB01			
SB02	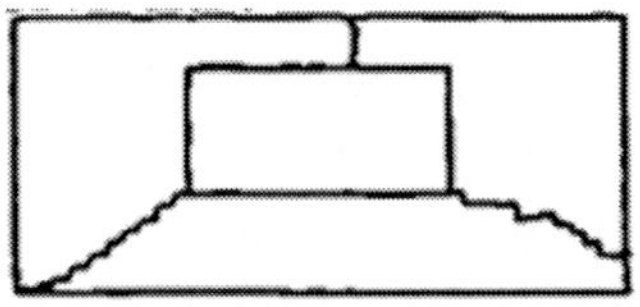	Base Panel	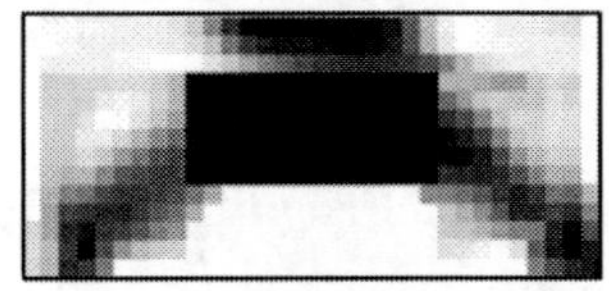
SB03	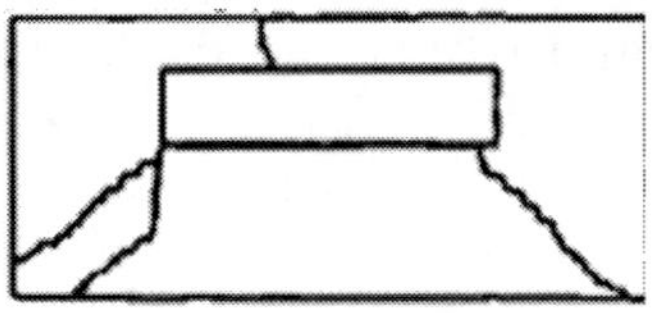	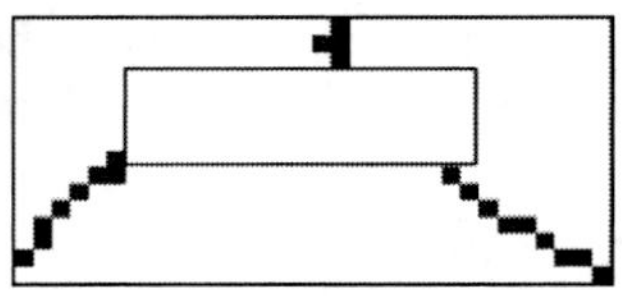	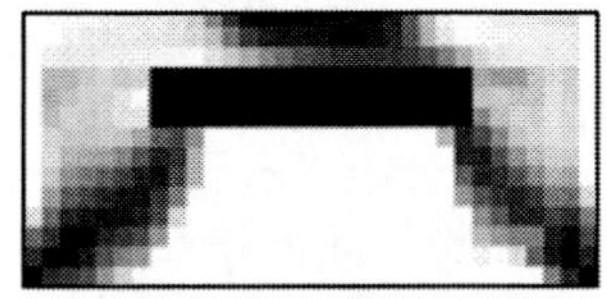
SB04	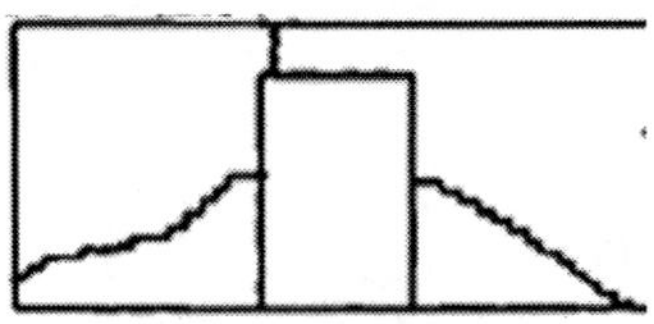	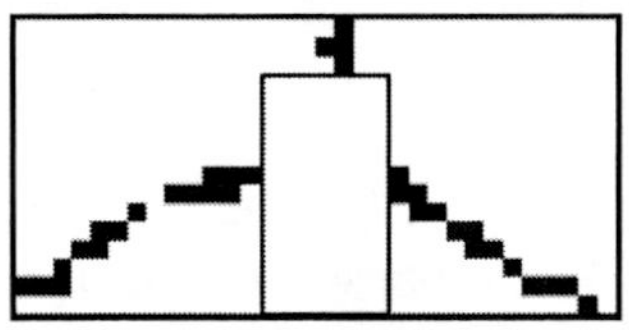	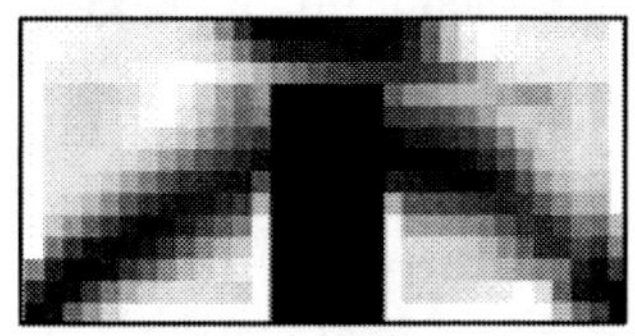
SB05	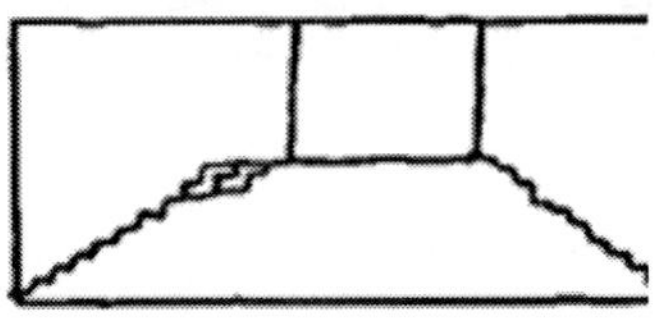	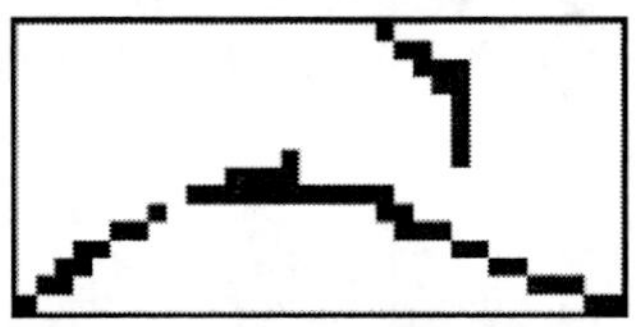	Base Panel
Panel No.	Experimented crack pattern	Predicted crack patterns	
		DPM	FCM
SB09	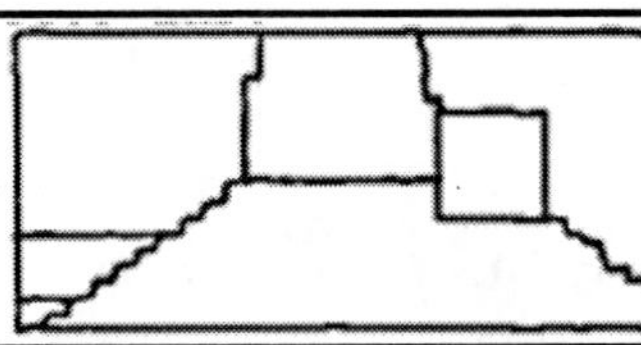	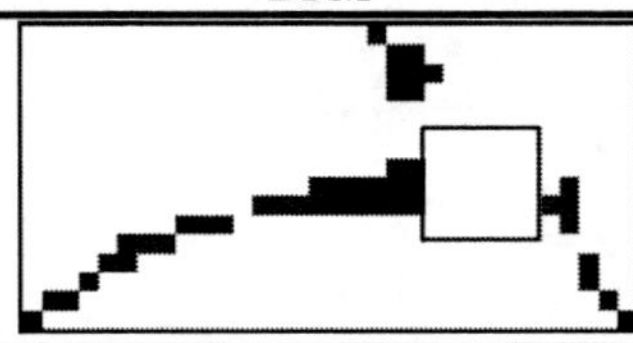	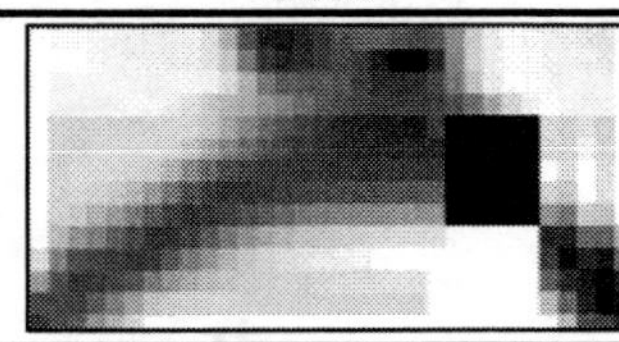

By the comparison of prediction results, it can be found that there are still distinct similarities and differences, although both methods have the same BPNN model. Further analysis shows that:

- Both predicted crack patterns can generally reflect the main location and spreading direction of actual crack.
- In DPM, predicted cracks are located on top of the opening, spreading along the vertical symmetrical axis, such as Panel SB03 and SB04.
- In FCM, a black rectangular area means the opening of panel, and the shadow means location and spreading direction of actual crack. This area contains all potential cracks.

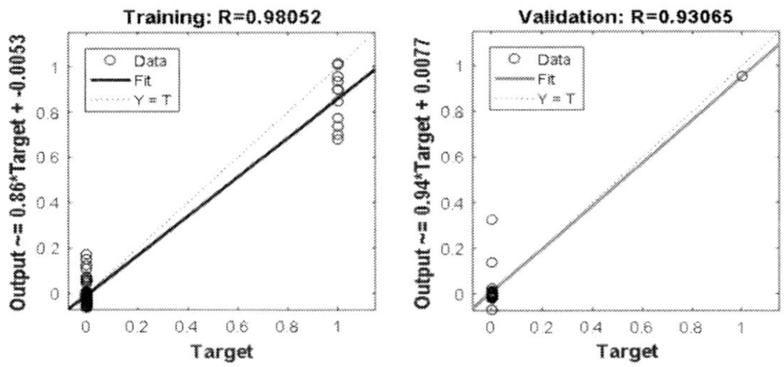

Figure 8. Training of the BPNN.

5 CONCLUSIONS

This paper presents that BPNN predicts the crack pattern of the masonry panels with opening. Comparing two methods of prediction, we have the following conclusions:

- Although only one panel is used for training, with the science of digitalization, different panels have the same feature in state value, giving BPNN the basis to decide a unit is cracked or not. So, it can make the prediction of crack pattern of the masonry panels with opening more accurate.
- FC is presented to describe the location and spreading direction of actual crack. It is better than DPM as it includes abundant information about potential crack pattern.
- Proposed division of panel with opening is a specialized targeted method about digitalization. It is an effective and worthy method to analyze in future.

References

Chong, V. L., *The Behavior of Laterally Loaded Masonry Panels with Openings,* PhD Thesis, University of Plymouth, UK, January, 1993.

Davidson, E. B., and Wang, L. R. L., Study of the Cyclic Lateral Resistance of Low Rise Masonry Wall Panels, *Proceedings of the Third North American Masonry Conference,* 1985.

Golding, J. M., Practical Design of Laterally Loaded Masonry Panels, *Structural Engineer London,* 69(4), 55-65, February, 1991.

Huang, Y. X., Zhang, Y., Zhang, M., and Zhou, G. C., Cellular Automata Method for Mapping Cracking Patterns of Laterally Loaded Wall Panels with Openings, *Engineering Review*, 35(1), 81-88, November, 2013.

Maja, A. S., and Justyna, B., Surface Dynamic Process Simulation with the Use of Cellular Automata, *Acta Physica Polonica B, Proceedings Supplement*, 3(2), 391-398, March, 2010.

Morita, K., Language Recognition by Reversible Partitioned Cellular Automata and Iterative Arrays, *Journal of Cellular Automata*, 13(3), 183-213, 2018.

Sinha, B. P., A Critical Assessment of The Design Method For Laterally Loaded Masonry Panels According to The IS: 1905-1987, *Journal of the Institution of Engineers (India)*: Civil Engineering Division, 14-16, June, 2001.

Tu, Y. H., Chuang, T. H., Liu, P. M., and Yang, Y. S., Out-of-Plane Shaking Table Tests on Unreinforced Masonry Panels in RC Frames, *Engineering Structures*, 32(12), 3925-3935, December, 2010.

Varma, R. K., Tomar, C. P. S., Parkash, S., and Sethi, V. S., Damage to Brick Masonry Panel Walls Under High Explosive Detonations, *American Society of Mechanical Engineers*, Pressure Vessels and Piping Division, 207-216, July, 1997.

Zhang, Y., Zhou, G. C., Xiong, Y., and Rafiq M. Y., Techniques for Predicting Cracking Pattern of Masonry Wallet Using Artificial Neural Networks and Cellular Automata, *Journal of Computing in Civil Engineering,* 24(2), 161-172, March, 2010.

Interdependence between Structural Engineering and Construction Management
Edited by Ozevin, D., Ataei, H., Modares, M., Gurgun, A., Yazdani, S., and Singh, A.

ISBN: 978-0-9960437-6-2

EVALUATION OF THE DYNAMIC CHARACTERISTICS FOR SEISMIC DESIGN OF MULTI-STORY BUILDINGS

AYA ABOELHAMD[1], AMAN MWAFY[1], and SULIMAN GARGOUM[2]

[1]*Dept of Civil and Environmental Engineering, UAE University, Al-Ain, UAE*
[2]*Dept of Civil and Environmental Engineering, University of Alberta, Edmonton, Canada*

The fundamental period of vibration is a critical structural dynamic characteristic in seismic design. Several expressions for the calculation of the fundamental period have been recommended by different building codes and previous studies. However, further studies are still needed to evaluate the design expressions used for the calculation of the fundamental periods and assess the need for further refinement. In this study, comprehensive fundamental period data from two sources is collected and compared with different formulas from building codes and previous studies. The first data set is obtained from 147 instrumented buildings with various lateral force resisting systems (LFRSs). The second set of period data are collected from the dynamic response simulations of selected structures. Different LFRSs are considered, including steel moment resisting frames (SMRFs), reinforced concrete moment resisting frames (RCMRFs), reinforced concrete shear walls (RCSWs), concentrically braced frames (CBFs), eccentrically braced frames (EBFs), masonry structures and pre-cast structures. The correlations between the derived period expressions with those recommended by the design provisions show that the code approach is conservative enough for SMRFs, CBFs, masonry buildings and pre-cast structures. For RCMRFs, EBFs and RCSWs, the design code is slightly unconservative for low-rise buildings. The outcomes of the study help to arrive at more efficient and cost-effective seismic design of buildings with different characteristics.

Keywords: Fundamental period, Dynamic response, Design standards, Instrumented structures, Structural systems.

1 INTRODUCTION

Evaluating the dynamic characteristics of structures is an essential step for the design of buildings when subjected to earthquake loads. As the fundamental period of vibration cannot be precisely calculated before the design of buildings, several empirical formulas for the calculation of this crucial parameter have been recommended by seismic codes and previous studies. The code provisions specify formulas to estimate the fundamental vibration periods according to LFRS (e.g., EN 1998–1 2004, ASCE/SEI-7 2016). Moreover, several previous studies evaluated the fundamental vibration periods of buildings according to the structural LFRS (e.g., Goel and Chopra 1997, 1998, Kwon and Kim 2010). For instance, Goel and Chopra (1997) evaluated the fundamental period formulas of MRFs for the design of buildings. The same authors also proposed a vibration period formula for RCSW buildings (Goel and Chopra 1998). More recently, Kwon and Kim (2010) assessed the fundamental period for different LFRSs and recommended few changes in the code formulas related to certain structural systems.

It is important to note that the data needed to estimate the period of vibration is continuously updated and enriched due to: (i) instrumenting more buildings all over the world, (ii) recording additional data from recent earthquakes for previously instrument structures, and (iii) conducting extensive dynamic response simulation of more realistic building models (Mwafy 2013, Mwafy *et al.* 2015, Ashri and Mwafy 2017). Further assessment studies are still needed for the estimation of the fundamental vibration period of buildings to provide more reliable formulas for the design, considering the diverse vibration period data from both instrumented and simulated structures representing different LFRSs. Therefore, this study aims to collect and classify data representing the dynamic characteristics of instrumented and simulated structures, evaluate the fundamental periods based on different LFRSs, and compare the calculated period database with the current formulas specified by the seismic design code (ASCE/SEI-7 2016).

2 DATA COLLECTION AND PROCESSING

The dynamic characteristics data of this study is collected from two main sources. The first set of data is gathered from the California Geological Survey (CGS) of the California Strong Motion Instrumentation Program (CSMIP 2018). The second data set is obtained from selected simulated structures (Mwafy and Elnashai 2008, Mwafy 2011, Issa and Mwafy 2014, Ashri and Mwafy 2017). In the following sections, the process of collecting and calculating the fundamental periods from each of the above sources is described in more details.

2.1 Period Data from Instrumented Structures

The instrumented period data is collected in this study from CSMIP (2018). A total of 147 instrumented structures with different LFRSs and heights are considered. These structures are classified according to the following LFRSs: (i) SMFRs, (ii) RCMRFs, (iii) CBFs, (iv) EBFs, (v) RCSWs, (vi) masonry structures and (vii) pre-cast structures. For each building, the story layouts are used to specify the locations of the sensors on each floor, as shown in Figure 1 for a sample building. The difference in acceleration between the top and bottom sensors is obtained for each instrumented structure. This difference is then used to derive Fourier and power amplitude spectra. The response spectra of each building are used to calculate the fundamental periods from different earthquake events in both orthogonal directions of the building by reading the period corresponding to the peak value. Figure 2 depicts the adopted process for estimating the fundamental period of vibration from the response spectra.

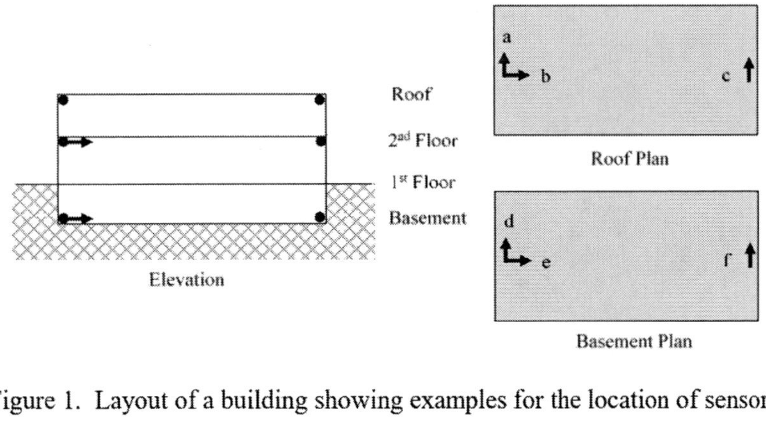

Figure 1. Layout of a building showing examples for the location of sensors.

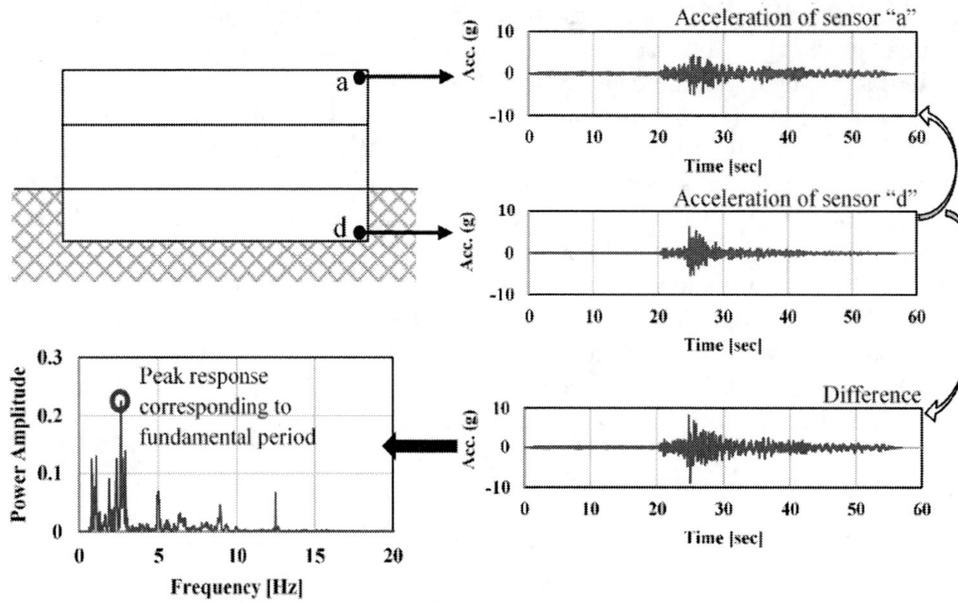

Figure 2. Process of estimating the fundamental period of a sample building.

2.2 Period data from Simulated Structures

A range of simulated structures with different heights and structural systems is also considered in this study. The selected structures are categorized into two main groups, as shown in Table 1. Group A consists of twelve medium-to-high-rise RCSW buildings with different layouts. On the other hand, group B includes low-to-medium-rise RCMRFs. The difference in response between the upper and lower nodes located at the middle interior LFRS of the selected buildings is calculated. The same procedure described previously is then implemented to calculate the fundamental vibration periods of the simulated structures.

Table 1. Characteristics of the simulated structures.

Group	Building reference	No. of stories	Total height (m)	Structural system	Group	Building reference	No. of stories	Total height (m)	Structural system
A	B1	100	321.3	RCSW	B	B13	12	36.0	Special RCMRF
	B2	80	257.3	RCSW		B14	12	36.0	Intermediate RCMRF
	B3	66	212.5	RCSW		B15	12	36.0	Intermediate RCMRF
	B4	60	193.3	RCSW		B16	12	36.0	Ordinary RCMRF
	B5	56	180.5	RCSW		B17	8	25.5	Special RCMRF
	B6	50	161.3	RCSW		B18	8	25.5	Intermediate RCMRF
	B7	40	129.3	RCSW		B19	8	25.5	Intermediate RCMRF
	B8	30	97.3	RCSW		B20	8	25.5	Ordinary RCMRF
	B9	26	84.5	RCSW		B21	2	8	RCMRF
	B10	20	65.3	RCSW		B22	8	28.5	RCMRF
	B11	18	58.9	RCSW		B23	6	24.5	RCMRF
	B12	10	33.3	RCSW		B24	3	13	RCMRF
						B25	2	9.5	RCMRF
						B26	2	9	RCMRF

3 EVALUATION OF DYNAMIC CHARACTERISTICS

Following the calculation of the comprehensive fundamental period data from the CGS database and simulated structures, the results are correlated with the code formulas according to different LFRSs. Two upper bound periods are considered, which represent 1.4xCode and 1.6xCode. These upper bound values were used in previous studies (e.g., Goel and Chopra 1997, Kwon and Kim 2010) and also recommended by design codes for high and medium seismicity regions, respectively (i.e., Sd_1 = 0.3-0.4 and 0.15, respectively, as per ASCE/SEI-7). The calculated periods of MRFs from the instrumented and simulated structures are compared with the code formulas in Figure 3. One building is excluded from the SMRFs database due to its unrealistic period since it was designed for twice the recommended seismic design load, Figure 3(a). Although the code formula is generally conservative for SMRF buildings, it is over-conservative for buildings higher than 60 meters. Hence, the code approach can be refined using the formula presented in Figure 3(a). For RCMRFs, the results of the flat slab-column system are excluded because of their flexible LFRS unlike other instrumented and simulated structures, which consist of efficient beam-column framing systems, Figure 3(b). The envelope of the derived formula is very close to the code formula, which confirm that the design approach is conservative and predicts well the periods of RCMRFs. It is noteworthy that the data used in previous studies included RCMRFs higher than 80 meters, which is uncommon for the RCMRF system. Moreover, two bracing systems are considered in this study, namely CBFs and EBFs. Figure 4(a) shows that the code formula is conservative enough for CBFs. Figure 4(b) also indicates that the design formula is conservative for the EBFs buildings higher than 40 meters, while it slightly overestimates the periods for low-rise structures.

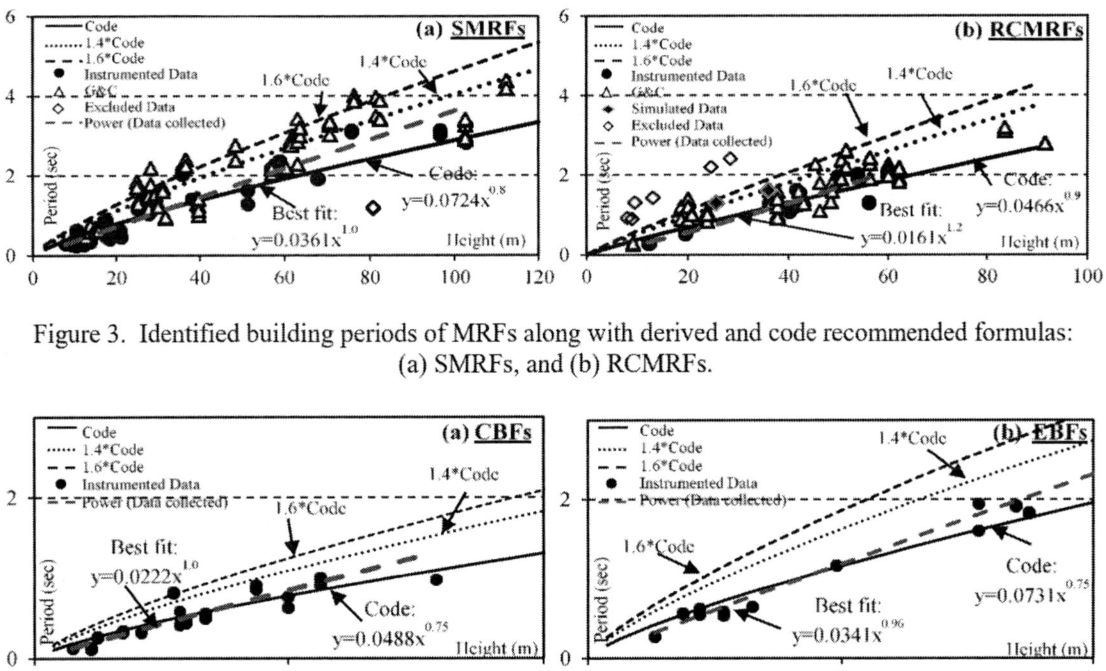

Figure 3. Identified building periods of MRFs along with derived and code recommended formulas: (a) SMRFs, and (b) RCMRFs.

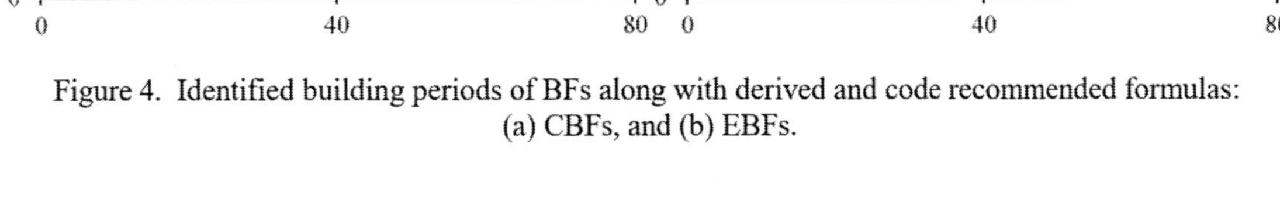

Figure 4. Identified building periods of BFs along with derived and code recommended formulas: (a) CBFs, and (b) EBFs.

Considering the calculated data of the RCSW system, two period points are excluded as the corresponding buildings were designed using the tube-in-tube system (Ashri and Mwafy 2017). The latter LRFS is typically used for super high-rise buildings. In addition to the code approach, the formula suggested by Kwon and Kim (2010) for RCSWs is presented in Figure 5. As the majority of the instrumented building periods is for low-to-mid-rise structures, the corresponding data is magnified in Figure 5(b). It is shown that the code formula is conservative for mid-to-high-rise buildings with heights over 25 meters. For low-rise RCSW buildings, the code approach slightly overestimates the period. It is also shown from the presented results that the suggested formula by Kown and Kim (2010) for RCSWs is over-conservative although it provides a lower bound for this class of structures. It is noteworthy that the simulated data are for buildings designed for medium seismicity regions, while the instrumented data were collected from a high seismic region (CSMIP 2018). Furthermore, the instrumented building data are mainly for RCSW structures lower than 75 meters, which is insufficient data to cover high-rise buildings. Therefore, the results suggest increasing the number of instrumented and simulated high-rise buildings from different regions, which may result in adopting different formulas for RCSWs depending on the seismic regions. Two additional LFRSs are considered in this study, namely masonry structures and pre-cast structures (Figure 6). The period formulas recommended by Kown and Kim (2010) for these two systems are also presented. For both LFRSs, the results show that the code formula provides a conservative lower boundary, while the Kown and Kim (2010) formulas are over-conservative although they may provide a lower bound for these two systems.

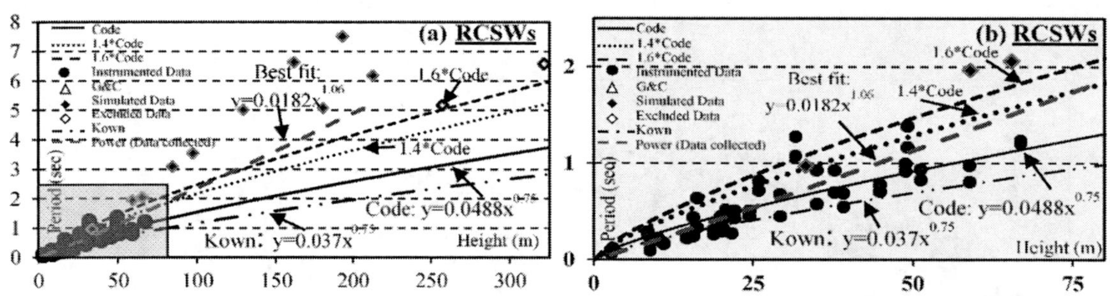

Figure 5. (a) Identified building periods of RCSWs, along with derived and code formulas, and (b) magnified data for the low-mid-rise buildings.

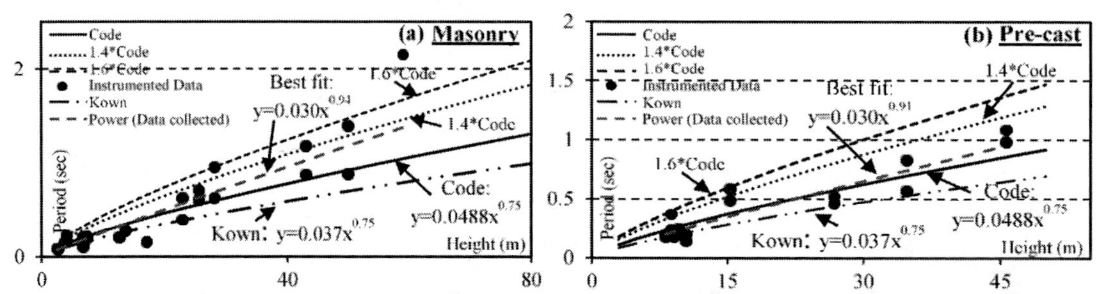

Figure 6. Identified building periods, along with derived and code recommended formulas, for: (a) masonry structures, and (b) pre-cast structures.

4 CONCLUSIONS

This study evaluated the fundamental period data collected and calculated from instrumented and simulated structures by correlating the data with the formulas recommended by design codes and previous studies. Several LFRSs were considered, including SMRFs, RCMRFs, RCSWs, CBFs, EBFs, masonry buildings and pre-cast structures. It is concluded that the design code formulas are conservative in predicting the time periods of SMRFs, CBFs, masonry structures and pre-cast structures. For RCMRFs, EBFs and RCSWs, the code formulas slightly overestimate the fundamental period of low-rise structures while they are conservative enough for mid-rise and high-rise buildings. For RCSW structures, further study is still needed to evaluate the fundamental periods of high-rise buildings located in different seismicity regions. The results of this study, which were obtained from comprehensive database representing instrumented and simulated structures, shed light on the effectiveness of the design codes in estimating the dynamic characteristics of structures, which help to arrive at more efficient and cost-effective seismic design of buildings with different characteristics.

Acknowledgments

This work was supported by the United Arab Emirates University under research grant No. 31N320.

References

ASCE/SEI-7, *Minimum Design Loads for Buildings and Other Structures,* ASCE Standard ASCE/SEI 7-16, American Society of Civil Engineers, Reston, VA, 2016.

Ashri, A. R., and Mwafy, A., *Assessment of Seismic Design Coefficients for Different Structural Systems of Buildings,* Proceedings of the 16th World Conference on Earthquake, 16WCEE 2017, Santiago Chile, 2017.

CSMIP, California Strong Motion Instrumentation Program, Retrieved 1 April 2012, from http://www.conservation.ca.gov/cgs/smip/Pages/Index.aspx, (2018).

EN 1998–1, *Eurocode 8: Design of Structures for Earthquake Resistance. General Rules, Seismic Actions and Rules for Buildings,* CEN, European Committee for Standardization, Brussels, 2004.

Goel, R. K., and Chopra, A. K., Period Formulas for Moment-Resisting Frame Buildings, *Journal of Structural Engineering,* 123(11), 1454-1461, 1997.

Goel, R. K., and Chopra, A. K., Period Formulas for Concrete Shear Wall Buildings, *Journal of Structural Engineering,* 124(4), 426-433, 1998.

Issa, A., and Mwafy, A., *Fragility Assessment of Pre-Seismic Code Buildings and Emergency Facilities in the UAE,* Proceedings of the Second European Conference on Earthquake Engineering and Seismology (2ECEES), Istanbul, Turkey, 2014

Kwon, O. S., and Kim, E. S., Evaluation of Building Period Formulas for Seismic Design, *Earthquake Engineering and Structural Dynamics,* 39(14), 1569-1583, 2010.

Mwafy, A. M., Assessment of Seismic Design Response Factors of Concrete Wall Buildings, *Earthquake Engineering and Engineering Vibration,* 10(1), 115-127, 2011.

Mwafy, A. M., Use of Overstrength and Inelastic Response in Seismic Design, *Structures and Buildings,* 166(6), 282-297, 2013.

Mwafy, A. M., and Elnashai, A. S., Importance of Shear Assessment of Concrete Structures Detailed to Different Capacity Design Requirements, *Engineering Structures,* 30(6), 1590-1604, 2008.

Mwafy, A. M., Hussain, N., and El- Sawy, K., Seismic Performance and Cost- Effectiveness of High- Rise Buildings with Increasing Concrete Strength, *The Structural Design of Tall and Special Buildings,* 24(4), 257-279, 2015.

Interdependence between Structural Engineering and Construction Management
Edited by Ozevin, D., Ataei, H., Modares, M., Gurgun, A., Yazdani, S., and Singh, A.

ISBN: 978-0-9960437-6-2

SEISMIC DESIGN OF FRAME STRUCTURES EQUIPPED WITH INNOVATIVE HYSTERETIC DISSIPATIVE DEVICES

MICHELE PALERMO, VITTORIA LAGHI, STEFANO SILVESTRI, GIADA GASPARINI, and TOMASO TROMBETTI

Dept of Civil, Chemical, Environmental and Materials Engineering, University of Bologna, Bologna, Italy

In the present work, a Performance-Based Seismic Design procedure applied to multi-storey frame structures with innovative hysteretic diagonal steel devices (called Crescent Shaped Braces or CSB) is introduced. CSBs are steel elements of peculiar geometrical shapes that can be adopted in frame buildings as enhanced hysteretic diagonal braces. Based on their "boomerang" configuration and placement inside the frame structure, they are characterized by a lateral stiffness uncoupled from the yield strength and, if properly inserted, by an overall symmetric hysteretic behavior with hardening response at large drifts, thus preventing from global structural instability due to second-order effects. The procedure here presented is intended to guide the structural engineer through all the steps of the design process, from the selection of the performance objectives to the preliminary sizing of the CSB devices, up to the final design configuration. The steps are described in detail through the development of an applicative example.

Keywords: Design procedure, Performance-based seismic design, Viscous dampers, Crescent shaped braces.

1 INTRODUCTION

Most of current force based seismic code design procedures are based on the accomplishment of prescriptive rules leading to a "passive" design, since the designer often obtains a solution without having a clear and complete understanding/control of the structural performances under different earthquake intensities.

A change of paradigm has been postulated since the end of the 20th century with the conceptual framework of Performance Based Seismic Design (PBSD) first proposed by SEAOC (1995), which encompasses the full range of seismic engineering issues toward predictable and controlled seismic performances under established multiple earthquake intensity levels (Bozorgnia and Bertero 2004). PBSD principles would therefore permit to shift from a "passive" to an "active" overarching design approach, in which the designer (i) first selects and identifies multiple performance objectives (the coupling of a building performance level with a given earthquake intensity level) (Bertero and Bertero 2002), (ii) then conceives a conceptual design and a preliminary sizing of structural and non-structural elements (member size, reinforcements) and (iii) finally develops the detailed design.

Among various design approaches proposed in the literature, some of the authors proposed in 2009 a conceptual approach based on the full exploitation of stiffness, strength, ductility, and energy dissipation properties of a structural system (stiffness-strength-ductility design approach, Trombetti *et al.* 2009). It was soon realized that conventional diagonal bracing systems do not permit the required freedom in the design, so that a novel steel hysteretic dissipative brace, called Crescent Shaped Brace (CSB), was introduced (Palermo *et al.* 2015). Up to now, the mechanical properties of CSBs have been through analytical, numerical and experimental investigations (Palermo *et al.* 2017), while design procedures have been developed only for the particular case of CSBs inserted at the ground floor only (Palermo *et al.* 2014) or for shear-type frames with CSBs inserted at all stories (Kammouh *et al.* 2018), so that the structure stiffness matrix can be easily computed. In the present work a general PBSD design procedure for multi-stories frames equipped with CSB devices is presented.

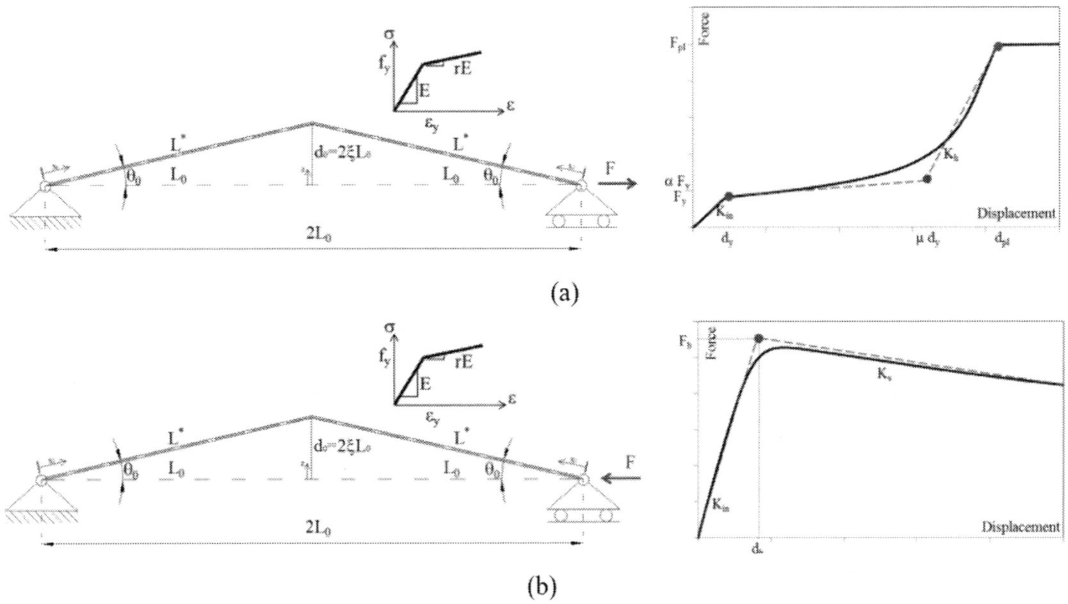

Figure 1. The bilinear symmetric configuration of the CSB and its lateral force-displacement behavior: (a) in tension; (b) in compression (Adapted from Palermo *et al.* 2017).

2 CRESCENT-SHAPED BRACES

A CSB is a hysteretic steel device connecting two points of the structure, either members of two adjacent stories (like diagonal braces) or slabs of the same floor. In its symmetric configuration (Figure 1) the device is composed by a continuous element made by two equal straight segments each one inclined of an angle θ_0 with respect to the horizontal direction. A characteristic dimension of the device is the so-called "initial arm" d_0. When subjected to a lateral force F, the two straight segments react through both internal axial force and bending moment whose signs and relative magnitudes (compression/ tension plus bending) depend on the initial lever arm value and on the direction and sign of the applied force. For instance, the qualitative graphical representation of the lateral force vs lateral displacement behavior of a bilinear symmetric CSB device is represented in Figure 1. Let us first focus the attention on the behavior under lateral forces inducing tensile/bending (Figure 1a). A first linear-elastic behavior is observed until the first yielding of the knee point, followed by a pseudo-plastic range (kind of plateau) and the final hardening range until the device reaches the straight configuration (e.g., the arm reduces to zero). The behavior under lateral forces inducing compression/bending is characterized by a first linear-elastic branch until the yielding of the knee point, followed by a softening branch (Figure 1b). The resulting force-displacement curves can be described in terms of the following parameters: initial stiffness k_{in}, first yielding point (F_y, d_y), ductility capacity (δ), hardening stiffness in tension (k_h), plastic tensile capacity (F_{pl}, d_{pl}), compression capacity (F_b, d_b), softening slope under compression (k_s). These "behavior" parameters can be related to the specific geometrical and mechanical properties of the device (details can be found in Palermo *et al.* 2017).

3 THE DESIGN PROCEDURE

The proposed design procedure allows to obtain a structure, which will follow a so-called objective curve that represents the desirable structural response in a Force-displacement graph, to satisfy multiple performance objectives POs (see Figure 2). It consists of the following phases:

- **PHASE 0. Identification of the POs**. It consists in the identification of the performance objectives. For a frame structure equipped with CSB devices placed as diagonal braces, the following four POs are envisaged:

  PO1: The building has to remain fully operational (no damage) under frequent earthquakes (characterized by minor intensity, EQ1). It is achieved by imposing a target initial stiffness to each CSB device.

  PO2: The building has to remain in operational conditions (limited and repairable damage) under occasional earthquakes (characterized by moderate intensity, EQ2). It is achieved by imposing a target yielding force to each CSB device.

  PO3: The safety of the occupants has to be guaranteed under rare earthquakes (characterized by major intensity, EQ3). It is achieved by ensuring a minimum ductility to each CSB device.

  PO4: The collapse of the building has to be prevented under very rare earthquakes (characterized by extreme intensity, EQ4). It is achieved by ensuring the building stability (no collapse due to P-D effects).
- **PHASE 1. HRS design: linear elastic behavior and yielding point.** It consists in (i) calibrating the initial elastic stiffness of each CSB through a straightforward iterative procedure (based on the one proposed by Lavan and Levy 2006) in order to satisfy inter-storey drifts limitations (as prescribed by most current seismic codes) under occasional earthquakes (EQ1); (ii) calibrating the yielding point of each CSB so that the structure will reach it under frequent earthquakes (EQ2).
- **PHASE 2. HRS design: non-linear behavior.** It consists in: (i) calibrating the ductility of each CSB device, so that they will respond within the "plateau" range under rare earthquakes (EQ3); (ii) evaluation of the final hardening stiffness of each CSB device, so that they will provide additional stiffness under very rare earthquake in order to prevent from P-Delta induced collapse (EQ4).
- **PHASE 3. Final design/verifications through non-linear TH analyses.** It consists in the development of the final design/verifications by means of non-linear time history (TH) analyses considering the actual non-linear characteristics of the CSB devices as obtained from the preliminary design (Phases 1-3).

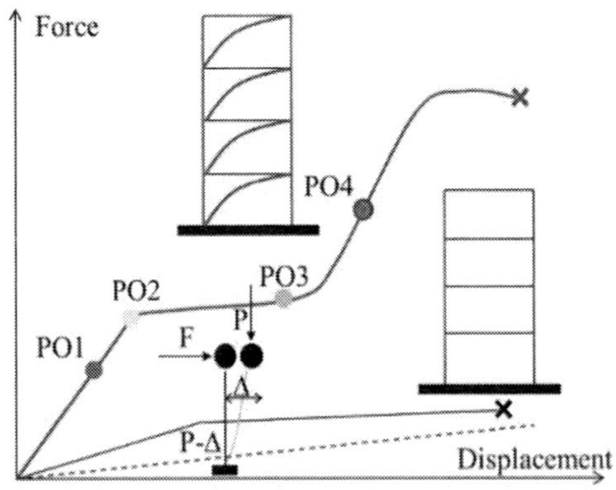

Figure 2. Graphical representation of the "Objective curve" and performance objectives (POs).

4 APPLICATIVE EXAMPLE

The design procedure is here applied to an example case that is a 10-story steel regular MRF located in a high seismic risk region of Southern Italy. The building has a rectangular plan (24 m x 18 m) with equal span width of 6.00m. The building height is 35 m with a constant inter-storey height H= 3.5 m. The moment-resisting frames are composed by columns and beams with European HE and IPE cross section profiles, respectively. Rigid beam-column connections are considered. The seismic design load is equal to 10 kN/mq. Figure 3 provides a plan view and a frame elevation view with the corresponding SAP2000 finite element model. The non-linear behavior of frame elements and CSB devices is modelled through M-θ plastic hinges. The four POs used for the seismic design can be summarized as follows. Under EQ1 intensity level (frequent events) all peak inter-storey drifts are limited to $\overline{ID}$=0.2%. Under the EQ2 intensity level (occasional events) all the CSB devices reach the first yielding. Under EQ3 intensity level (rare events) the ductility demand of all CSB devices is limited to $\lambda \cdot \frac{a_{g,EQ3}}{a_{g,EQ2}}$ (being $a_{g,EQ2}$ and $a_{g,EQ3}$ the design PGA values at EQ2 and EQ3, respectively, the λ value is chosen by the designer). Under EQ4 intensity level all the devices should have enough lateral strength to prevent from P-D collapses. The values of the design PGA are equal to $a_{g,EQ1}$=0.1 g, $a_{g,EQ2}$=0.13 g, $a_{g,EQ3}$= 0.359 g and $a_{g,EQ4}$= 0.469. For the sake of conciseness, the detailed calculations to impose each PO are not here reported (they will be provided in a journal paper). Some selected results are, instead, shown in Figure 4a and b. In particular, Figure 4a graphically displays the iterative procedure to reach the PO1. It can be noted that the algorithm (adapted from Lavan and Levy 2006) converges in few iterations. Figure 4b displays the Force-Displacement structural global response (in terms of base shear vs first storey ID ratio). The colored circles indicate the response under a particular accelerogram (seven artificial accelerograms, generated using the software SIMQKE, and scaled to the four different seismic design intensities have been used to perform the non-linear TH analyses). The squares indicate the average responses from T-H analyses. The graph includes also the pushover curve obtained using a uniform along-the-height distribution of lateral forces. It can be noted that the seismic response of the building is quite close to the pushover curves.

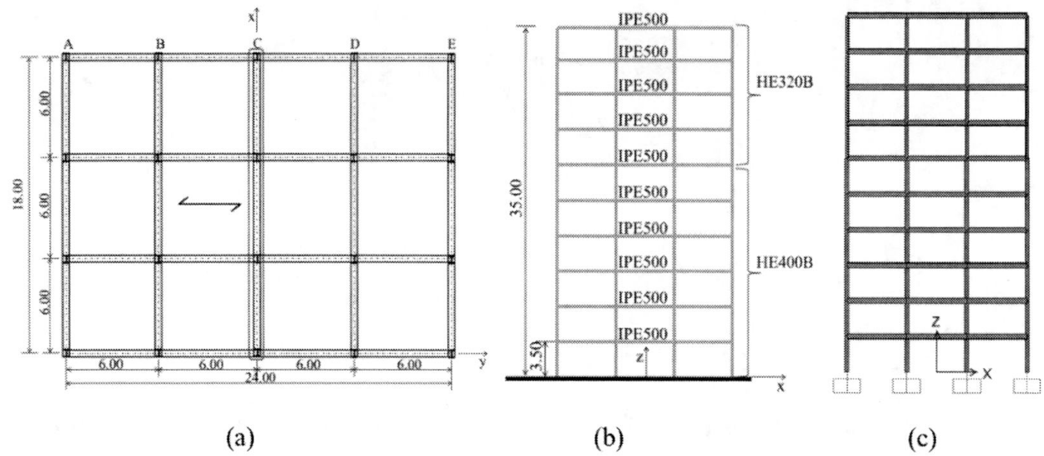

Figure 3. (a) Plan view. (b) Elevation view. (c) SAP2000 bare frame FE model.

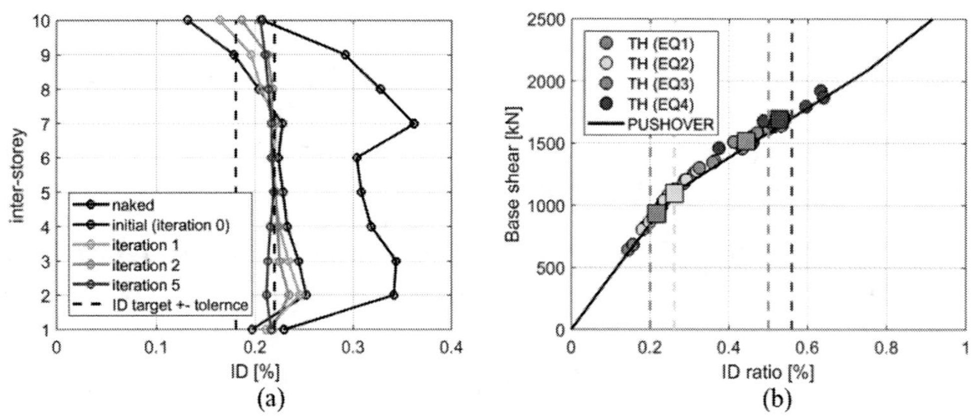

Figure 4. (a) Graphical representation of the iterative procedure to find the CSB initial stiffness. (b) Global response in terms of base shear vs ID ratio.

5 CONCLUSIONS

In the present work a general procedure for the seismic design of frame structures equipped with CSB devises has been presented. The procedure is framed in the context of PBSD. It is composed of four main phases going from the definition of the seismic performance objectives to the definition of the linear and non-linear properties of the CSB devices, up to the final verification of the actual behavior of the building with the added CSB devices through the development of fully non-linear time history analyses. The procedure has been finally applied to a 10-story example frame, considering both cases of moment resisting frame and pendular frame. The results have shown that the procedure efficiently achieved the proposed performance objectives.

Acknowledgments

Financial supports of Department of Civil Protection (DPC-Reluis 2014–2018 Grant—Research line 6: "Seismic isolation and dissipation") is gratefully acknowledged.

References

Bertero, R. D., and Bertero, V. V., Performance-Based Seismic Engineering: The Need for A Reliable Conceptual Comprehensive Approach, *Earthquake Engineering and Structural Dynamics,* 31(3), 627-652, 2002.

Bozorgnia, Y., and Bertero, V. V., *Earthquake Engineering: From Engineering Seismology to Performance-Based Engineering,* CRC Press, 2004.

Kammouh, O., Silvestri, S., Palermo, M., and Cimellaro, G. P., Performance-Based Seismic Design of Multistory Frame Structures Equipped with Crescent-Shaped Brace, *Structural Control and Health Monitoring,* 25(2), e2079, 2018.

Lavan, O., and Levy, R., Optimal Design of Supplemental Viscous Dampers for Linear Framed Structures, *Earthquake Engineering and Structural Dynamics*, 35(3), 337-356, 2006.

Palermo, M., Pieraccini, L., Dib, A., Silvestri, S., and Trombetti, T., Experimental Tests on Crescent Shaped Braces Hysteretic Devices, *Engineering Structures,* 144, 185-200, 2017.

Palermo, M., Ricci, I., Gagliardi, S., Silvestri, S., Trombetti, T., and Gasparini, G., Multi-Performance Seismic Design through an Enhanced First-Storey Isolation System, *Engineering Structures,* 59, 495-506, 2014.

Palermo, M., Silvestri, S., Gasparini, G., and Trombetti, T., Crescent Shaped Braces for The Seismic Design of Building Structures, *Materials and Structures*, 48(5), 1485-1502, 2015.

SEAOC, *Performance-Based Seismic Engineering for Buildings*, Vision 2000 Committee, Structural Engineers Association of California, Sacramento, CA, 1995.

Trombetti, T., Silvestri, S., Gasparini, G., and Ricci, I., Stiffness-Strength-Ductility Design Approaches for Crescent-Shaped Braces, *The Open Construction and Building Technology Journal,* 3, 127-140, 2009.

Interdependence between Structural Engineering and Construction Management
Edited by Ozevin, D., Ataei, H., Modares, M., Gurgun, A., Yazdani, S., and Singh, A.

ISBN: 978-0-9960437-6-2

SEISMIC PERFORMANCE EVALUATION OF PILOTI-TYPE BUILDINGS AFTER AN EARTHQUAKE

HYE-MIN SHIN, KYUNG-JAE SHIN, SU-WOONG LEE, DAE-GEUN KIM, MIN-KI LEE, and JUN-SEOP LEE

Dept of Architecture, Civil, Environmental and Energy Engineering, Kyungpook National University, Daegu, Republic of Korea

On November 15, 2017, the second strongest earthquake occurred in Korea, which was 5.4 in size on the Richter Scale. The duration of the earthquake was short, but the damage was serious. Two recent earthquakes have shown that our country is no longer safe from earthquakes. However, to date, Korean structures are showing a low earthquake resistance, and seismic retrofitting is necessary in preparation for a large-scale earthquake. In this study, reinforcing effect of steel slit damper was analyzed based on the dynamic test results of the previously studied reinforced concrete frame. After that, push over analysis and nonlinear time history analysis using OpenSees were selected for the residential piloti-type building as the target building. In the above Korean earthquake, the damage to the piloti-type building was conspicuous. Through analysis, the vulnerable part of the piloti-type building was identified and the seismic strengthening with the steel slit damper was carried out.

Keywords: Shaking table test, Steel slit damper, OpenSees, Nonlinear analysis.

1 INTRODUCTION

The Gyeongju earthquake that occurred on September 12, 2016 had a short duration of less than 10 seconds, but with a large number of 640 large aftershocks. Vibration was detected not only Gyeongju, Ulsan, Daegu and Busan, but also some parts of Seoul. On November 15, 2017, the second strongest earthquake occurred in Pohang city of 5.4 magnitude. The duration of the earthquake was short, from 4 to 6 seconds, but the damage was incomparable with the 9.12 Gyeongju earthquake. In this study, based on the results of the shaking table tests performed in the previous research (Shin *et al.* 2016), nonlinear static analysis and time history analysis were performed by modeling the test specimens using OpenSees, and the reliability of the steel slit damper modeling was verified by comparing with experimental results. After that, seismic performance evaluation was carried out for the residential piloti-type buildings located in Pohang city through nonlinear analysis of domestic and overseas earthquakes.

2 SHAKING TABLE TEST

2.1 Shaking Table Test Specimens

This study considered a 1-story 1-span RC frame specimen with a non-seismic (NS) design. Its detailed specifications are shown in Figure 1(a) and the specimen includes a column height of

2,750 mm and beam length of 3,700 mm. Its seismic performance was compared with the performances of various seismic reinforcement systems when a 22-ton mass was loaded on its top. Figure 1(b) portrays a specimen reinforced with a steel plate slit damper (NS-D), which was designed so that the stiffness of the damper was greater than of the frame.

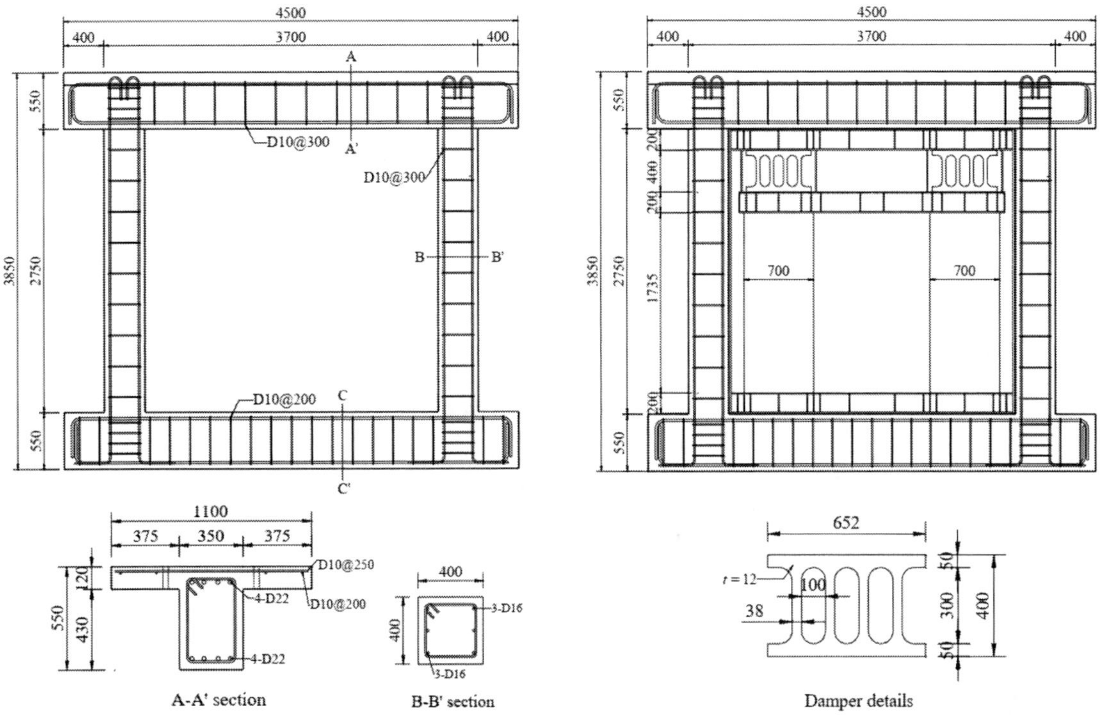

(a) Non-Seismic Frame (NS) (b) Frame with slit damper (NS-D)

Figure 1. Detail of specimens (unit: mm).

2.2 Shaking Table Test Result

The shaking table test was carried out by shaking the specimens stepwise using a 10% to 300% acceleration scale based on the seismic wave data of the El Centro (1940) earthquake, which had a rating of 6.9 on the Richter scale, as listed in Table 1. Here, PGA is the peak ground acceleration, EPA is the effective peak acceleration, and Sa (1.0) is the acceleration in 1 second of the response spectrum. The test results for all the specimens are shown in Figure 2. The NS specimen, which is a non-reinforced frame, ended the experiment at 160% seismic excitation. Figure 2(a) shows the experimental results with 100% and Figure 2(b) shows the results with 160%.

Table 1. Characteristics of El Centro seismic wave.

Seismic wave	PGA (g)	EPA (g)	EPA / PGA	Sa (1.0) (g)	Sa (1.0) / PGA
El Centro	0.319	0.310	0.973	0.454	1.423

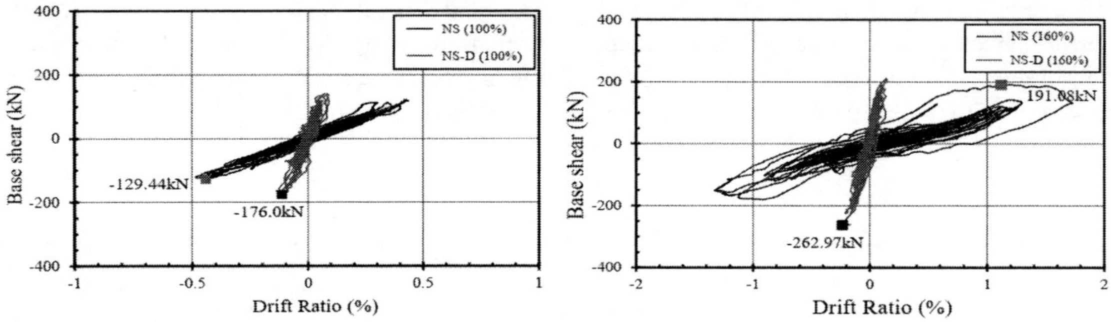

(a) 100% seismic excitation (b) 160% seismic excitation

Figure 2. Base shear force-drift ratio response curve of specimens.

3 OPENSEES ANALYSIS

3.1 Modeling of Specimens

OpenSees analysis modeling was performed with two-dimensional elements, as shown in Figure 3. The column and beam were modeled with nonlinear beam-column elements, and a fiber section was used to model the section (Mazzoni *et al.* 2006). In the case of the beam section, it was modeled not as a T-shaped beam but as a rectangular beam for the sake of convenience of the analysis. Boundary conditions of between column and ground are fixed. In the case of the steel slit damper, it was modeled by replacing each part with steel columns, as shown in Figure 3(b).

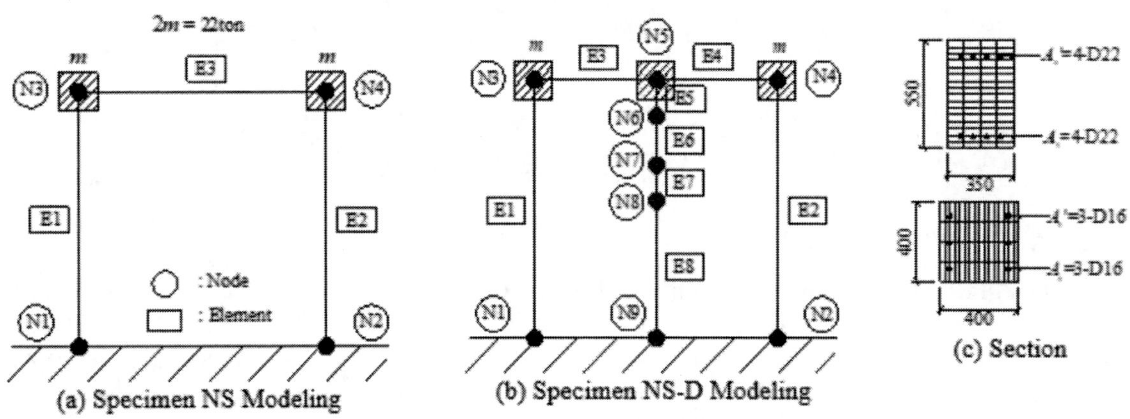

Figure 3. OpenSees modeling.

3.2 Pushover Analysis of Specimens

A pushover analysis was performed with the OpenSees program to determine the approximate behavior and limit condition of the analytic model before performing a nonlinear time history analysis. Figure 4 shows a comparison of the test result of each specimen with the analytic result through OpenSees. The initial stiffness and maximum strength of the analytic model were very close to the test results.

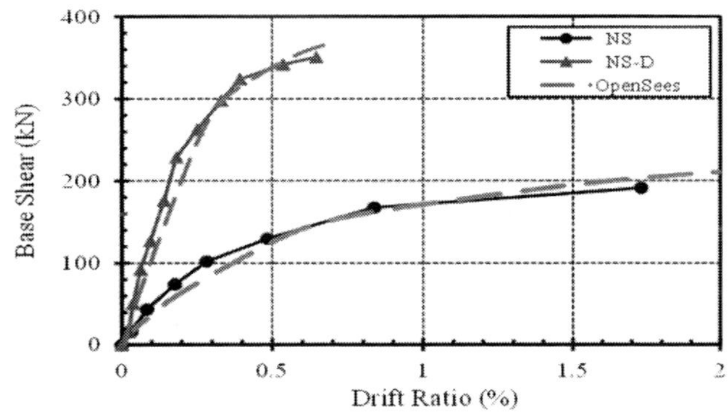

Figure 4. Comparison of experimental result and OpenSees analysis result.

4 SEISMIC PERFORMANSCE EVALUATION OF PILOTI-TYPE BUILDINGS

4.1 Target Building Overview

In order to evaluate the seismic performance of the piloti-type building, multi - family residential building located in Pohang city was selected as the target building. The target building is the 4-story RC ordinary shear wall structure constructed in 2011.

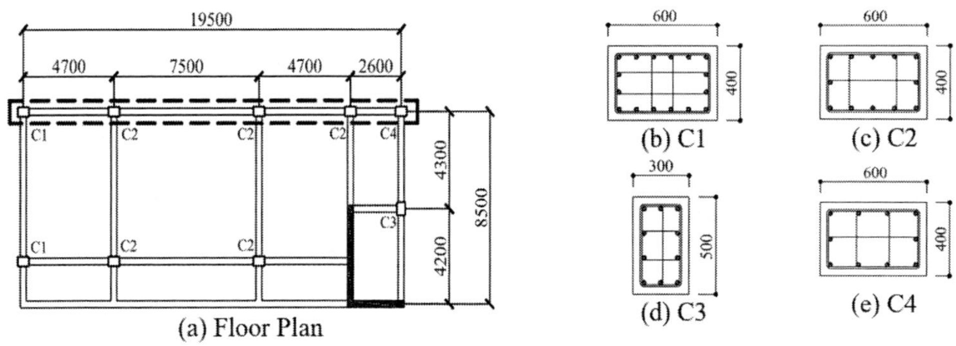

Figure 5. Detail of target building.

4.2 Modeling of Target Building and Reinforcement Overview

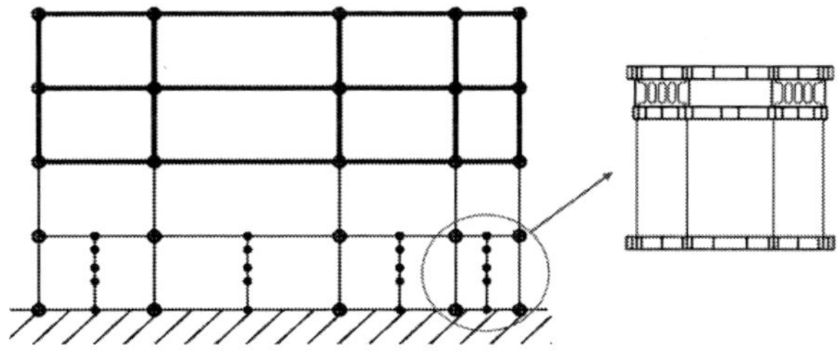

Figure 6. Modeling of piloti-type building and reinforcement overview.

Modeling was carried out to perform computational analysis of piloti-type buildings. In the case of the pillow type building due to the lateral force, the upper wall part behaves almost rigidly and most of the transverse strain is concentrated in the lower part of the frame (Go and Lee 2009). In order to reduce the analysis time and errors, the upper wall part, which is expected to be hardly damaged, is modeled as an Elastic Beam Column member, and the lower frame is modeled in detail with nonlinear beam members. The modeling of the upper wall as an elastic member for the piloti-type building has been proven by previous studies (Lee and Go 2013). Figure 6 is a visualization of the piloti-type modeling.

4.3 Nonlinear Analysis using Seismic Wave

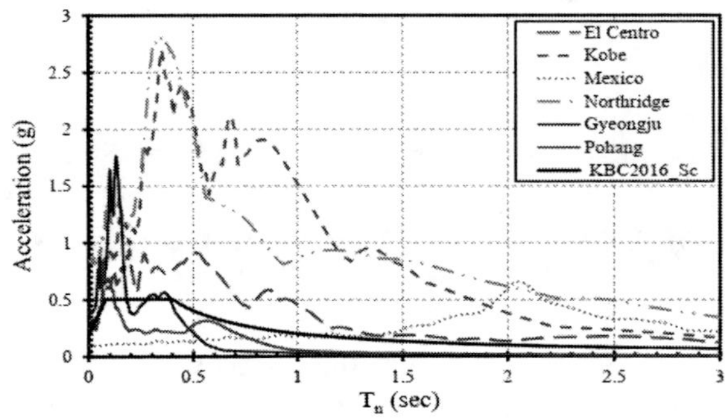

Figure 7. Response spectra of seismic waves with 5% damping ratio in comparison to design response spectrum in Korea (Sc).

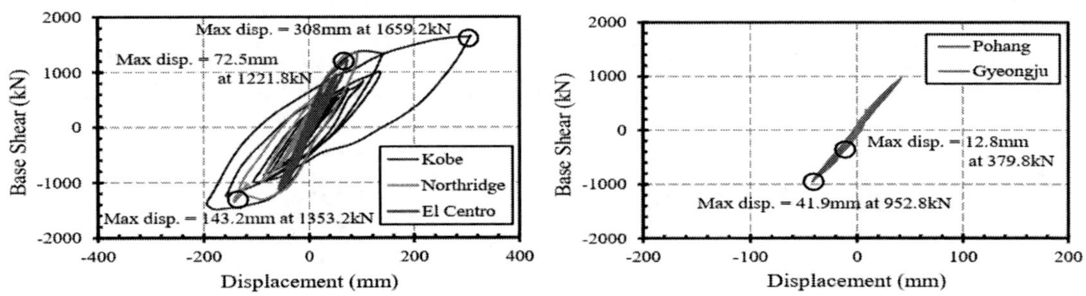

Figure 8. Base shear force- Roof displacement of Piloti-type building.

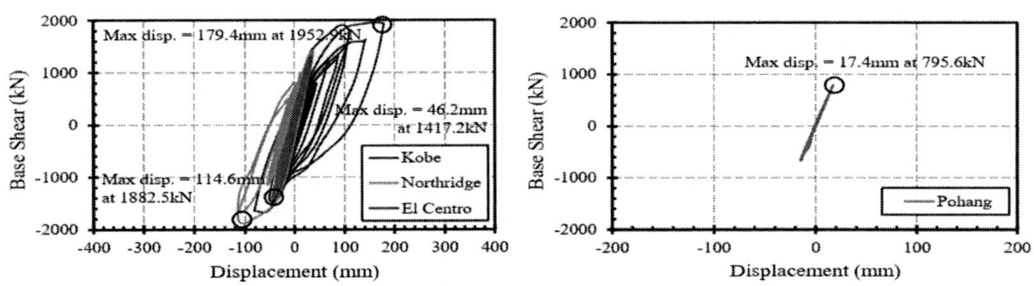

Figure 9. Behavior after reinforcement of piloti-type building.

Recently, the Gyeongju and Pohang earthquakes that occurred in Korea have short - period high frequency characteristics and are different from the earthquake that was used in the study. Figure 7, the response spectra of the Gyeongju and Pohang earthquakes were compared with those of the other earthquake. The nonlinear time history analysis was performed using the Gyeongju earthquake, the Pohang earthquake and the oversea earthquake. Figure 8 is a roof displacement graph for the base shear force of a Piloti-type building with seismic waves. The results of the nonlinear analysis after reinforcement with the steel slit damper are shown in Figure 9. As a result of the reinforcement, the displacement due to the Gyeongju earthquake hardly occurred and the effect of the large displacement was also reduced for the other earthquakes.

5 CONCLUSIONS

1) OpenSees, a structural modeling and analysis program, was used to carry out modeling of the specimen in the same manner as the previous research by Shin *et al.* (2016) and to perform nonlinear static analyses and dynamic analyses.
2) As a result of the response spectrum of Korean seismic waves, it was considered that it will have a short period high frequency characteristic and affect the middle-low story structure.
3) After conducted nonlinear analysis of the piloti-type building, large displacement occurred and reinforcement was performed by using a steel slit damper. As a result of nonlinear time history analysis after reinforcement, the displacement reduction effect was more than 50%.
4) As a result of the analysis, it is judged that proper reinforcement has been done, and additional studies on the modeling method of the slit damper and the seismic strengthening method applicable to the piloti-type building will be need.

Acknowledgments

This work was supported by the National Research Foundation of Korea (NRF) grant funded by the Korea government (MSIT) (No.NRF-2017R1A2B3009984).

References

Mazzoni, S., McKenna, F., Scott, M. H., and Fenves, G. L., *OpenSees Command Language Manual*, USA, 2006.

Go, D. W., and Lee, H. S., Non-linear Time History Analysis of Piloti-Type High-rise RC Buildings, *Journal of the Earthquake Engineering Society of Korea*, 13(1), 35-43, 2009.

Lee, Y.G., Go, D.W., Study on the Seismic Behavior of Piloti-Type RC Buildings using Push-Over Analysis, *Journal of the Architectural Institute of Korea Structure and Construction*, 15(5), 181-188, 2013.

Shin, K.-J., Lee, S.-H., Lee, H.-D., and Kwon, H.-S., Seismic-Strengthening Effect of Steel Shear Wall with Circular Opening Through Shaking Table Test on Reinforced Concrete Frame, *Journal of the Regional Association of Architectural Institute of Korea*, 18(2), 107-114, 2016.

Interdependence between Structural Engineering and Construction Management
Edited by Ozevin, D., Ataei, H., Modares, M., Gurgun, A., Yazdani, S., and Singh, A.

ISBN: 978-0-9960437-6-2

MODEL ANALYSIS OF HIGH-RISE STRUCTURES FOR WIND AND EARTHQUAKE FORCES

VANKUDOTHU BHIKSHMA[1], MOHAMMED HAMRAJ[2], and RUTTAL TEJASWINI[2]

[1]*Dept of Civil Engineering, Osmania University, Hyderabad, India*
[2]*Muffakham Jah College of Engineering and Technology, Hyderabad, India*

The objective of the present work is to study and analyze the effect of wind and earthquake loads on a chosen symmetrical plan of a multi-storied building to assess the behavior of the structure in all the seismic zones based on IS 1893 and subjected to different wind velocities based on IS 875 codes. For the present study, six cases (6 floors, 9 floors and 12 floors each for earthquake and wind) are considered under which 30 structures are modelled in ETABSv16.0 software and static analysis is carried out. Structural parameters like story drifts, story displacements and story shears are studied. Finally, it is found that story drifts are found to be maximum at 2nd floor (i.e., at 6 m) and 40% increase is observed in parameters when wind speed is increased from 33 to 39 m/s and by 21% when wind speed is increased from 50 to 55 m/s. With increase in wind speed on moving to higher seismic zones, with increase in number of stories from 6 to 9 story drifts increased by 109%, story displacements increased by 185% and increased by 15% and 62% on moving to higher seismic zones. Also, wind forces on the structure subjected to variation in increasing wind speeds and increasing heights in zone II and zone III but in no case are they greater than the earthquake forces in zone IV and V.

Keywords: Lateral loads, Wind forces, Story drifts, Story displacements shears.

1 INTRODUCTION

Considering the ever-increasing population as well as limited space, horizontal expansion is no longer a viable solution especially in metropolitan cities. As there has been a considerable increase in the number of tall buildings in recent times, effects of lateral loads like wind loads and earthquake forces are attaining increasing importance (Reddy and Sandip 2014). The Indian sub-continent is highly vulnerable to natural disasters such as earthquakes, draughts, floods, cyclones, landslides, and avalanches. Hence, the response of structures under wind and earthquake effects is a very important area where the researchers should concentrate and bring out effective disaster mitigating techniques, so that the structures remain functional. As height of buildings increases, the effects of both earthquakes and wind increases can be found out by analyzing buildings for earthquake and wind forces (Suchita and Magadum 2014). Architectural re-interpretations of the building type, the high cost of land in urban areas, the need to preserve agricultural production, the concept of skyscrapers, and the influence of cultural significance have all contributed to forcing buildings upward (Rajmani 2015). Dynamic actions are caused on buildings by both wind and earthquakes, but design for wind forces and for earthquake effects are distinctly different (Narla and Vardhan 2017). By the study of past earthquakes occurred in multistoried building shows that if they are not well designed and constructed with adequate strength it leads to the complete collapse of the structure. So, to acquire safety against additional deformations there is need to study of detailed considerations to design earthquake resistance structures. Generally structural engineers traditionally use linear static analysis to compute design forces, moments and displacements of a structure resulting from loads acting on it (Radha Devi and NagaSai 2018). The basic intent of analysis for earthquake resistant structures is that buildings should be able to resist minor earthquakes without damage.

2 NEED FOR THE PRESENT STUDY

Lateral loads can develop high stresses, produce sway moment or cause vibration. Therefore, it is very important for the structure to have sufficient strength against vertical loads together with adequate stiffness to resist lateral forces. Wind and earthquake loads are random in nature and it is difficult to predict them. They are estimated based on a probabilistic approach. Therefore, it is necessary to know which force is governing on the chosen structure for design.

For a design of case when a building is situated in a particular wind and earthquake zone, the major designing loads for such a building can be adopted directly from the analytical conclusions of the present work subjected to parameters chosen for deciding the final design of structure. For the analysis, 30 models were created and subjected to wind and earthquake loads separately.

3 OBJECTIVES

The primary objective of our study was to understand behavior of the chosen structure under the action of seismic loads and wind loads in different seismic zones II, III, IV, and V (IS 1893 Part-1 2002) subjected to different wind speeds 33m/s, 39m/s, 44m/s, 47m/s, 50m/s and 55m/s (IS 875 Part 3 1987) respectively and to find the governing force among wind and earthquake on the structure through model analysis.

4 ANALYTICAL METHODOLOGY

For the present study, a symmetrical plan is considered and modelled for performing the wind analysis and seismic analysis separately on different floors (varying heights) of model at various locations (different zones).

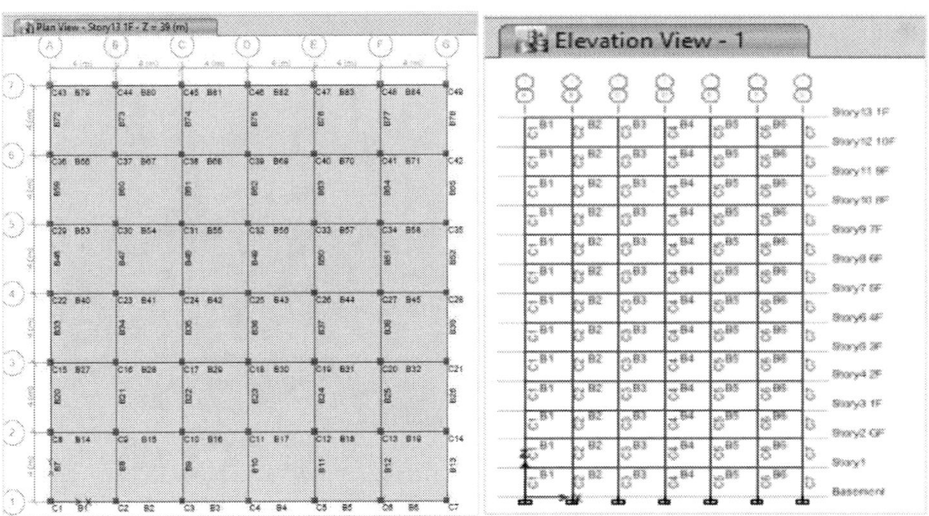

Figure 1. Plan and elevation of the model chosen for analysis with labels.

The selected plan of building consists of 49 columns. Models with different stories (6, 9 and 12) are modelled with floor height of 3.0m each and slab thickness of 0.125m with grade of concrete chosen to be M40 and steel Fe500. Soil under foundation is taken as medium soil. Loads considered for study are dead load, live load (3kN/m^2), seismic load and wind load.

Figure 1 shows the plan and elevation of the 12-floor symmetrical model with labels, which is modelled in ETABS for present study on which different loads considered are applied.

4.1 Seismic Load

Seismic base shear is given by Eq. (1),

$$V_B = A_h x W \tag{1}$$

where $A_h$is the Design horizontal acceleration spectrum value using fundamental natural period in the considered direction of vibration is given by Eq. (2)

$$A_h = \frac{z}{2} * \frac{I}{R} * \frac{S_a}{g} \tag{2}$$

4.2 Wind Load

Design wind pressure is calculated by Eq. (3)

$$P_z = 0.6\ V_z^2,\ Where\ V_z = V_b * K_1 * K_2 * K_3 \tag{3}$$

where $\mathbf{V_z}$: design wind speed.

$\mathbf{V_b}$: basic wind speed.

All the parameters required for calculating wind loads are taken from Indian Standard Code IS 875-Part 3 (1987) and these wind loads and wind pressure coefficients are applied on to model and analyzed using method of exposure from shell objects in ETABS v16.0.

5 ANALYTICAL TEST RESULTS

The numerical models were carried out using ETABS v16.0 software. A total of 30 different structures were modeled in ETABS separately, and then analyzed for all the wind and earthquake zones. The safety of structures was then checked against allowable limits prescribed using inter-story drifts.

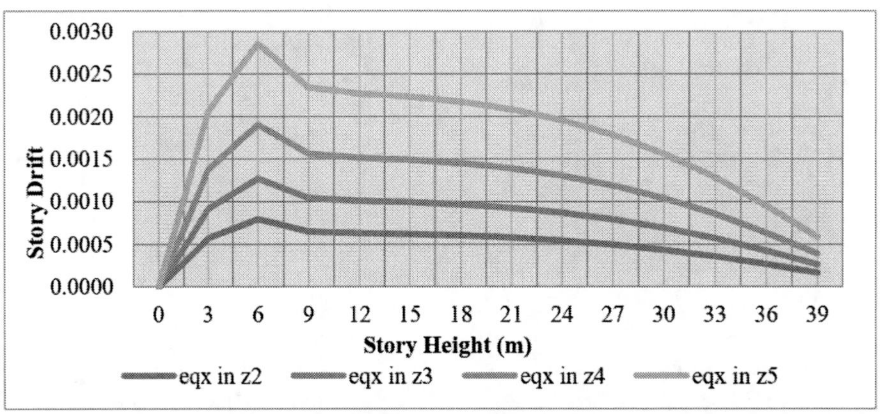

Figure 2. 12th Floor model with earth quake load.

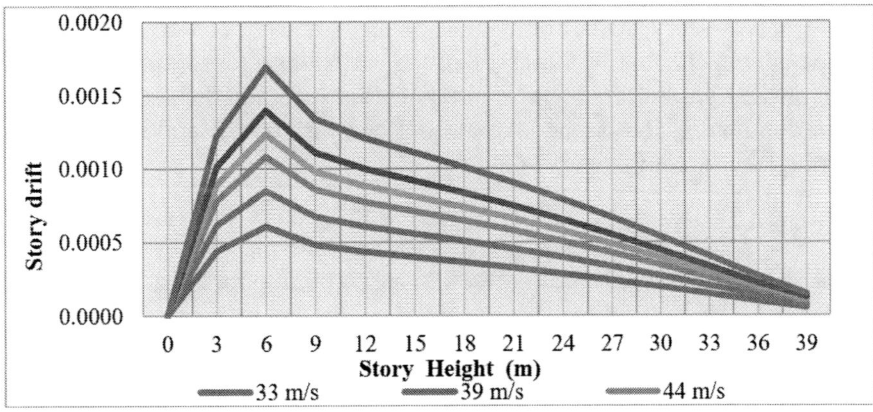

Figure 3. 12th Floor model with wind load.

From the analytical results' obtained parameters considered for study story drifts, displacements and shears are tabulated and compared in aspects of increasing height of the models, increasing seismic zones and wind velocities for which graphs are drawn and conclusions were inferred. Figures 2 and 3 show the behavior of 12-story building subjected to different earthquake zones and wind speeds, respectively. With an increase of wind velocity or seismic zone, the magnitude of story drift increases, and the maximum story drift is observed at the second floor.

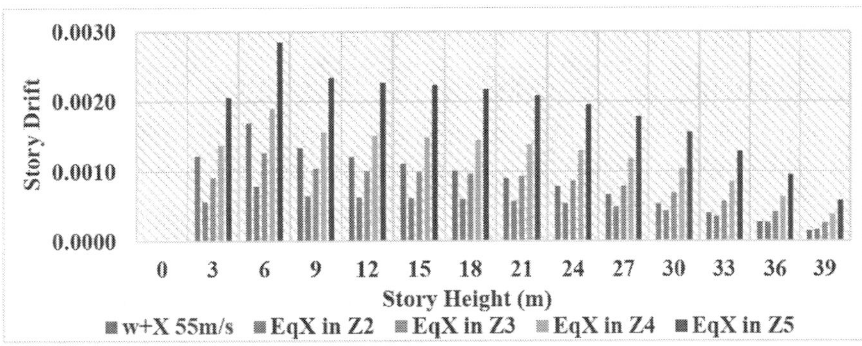

Figure 4. Story drift for earthquake with wind speed of 55m/s on 12 floor model.

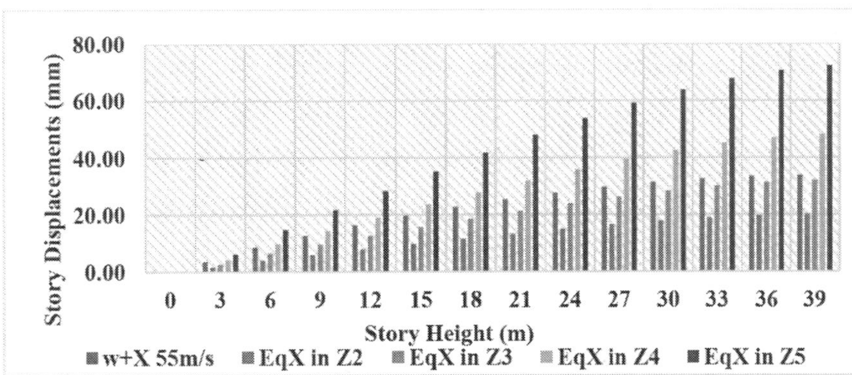

Figure 5. Story displacement for earthquake with wind speed of 55m/s on 12 floor.

In Figures 4 and 5, story drifts and displacements of 12 floor building subjected speed of 55m/s are compared with 12 floor building in all the earthquake zones to find the governing force among the both on 12 floor building and found that story drift and displacement on model due to wind forces are dominating the earthquake at all floors in zone II and zone III but in no case greater than earthquake forces in zone IV and zone V.

6 CONCLUSIONS

The effective parameters for wind forces affecting any building are the area subjected to wind as well as the intensity of wind defined by the code according to location. The effects of both earthquake forces and wind forces on multi-story buildings increase with increase in height of building. In the present study, with the increase of stories from 6 (21 m) to 9 (30 m), story drift, story shear and story displacements increase by 109%, 109% and 185%, respectively. When number of stories increase from 9 (30 m) to 12 (39 m), story drift, story shears, and story displacement increase by 53%, 53%, and 96%, respectively, with the increase in wind speed. Similarly, moving on to higher earthquake zones, story drift, story shears and story displacements increase by 15%, 15% and 62%, and increase by 9%, 9% and 42% respectively with increase in number of stories from 6 (21 m) to 9 (30 m) and 9 (30 m) to 12 (39 m).

On comparing the effect of wind forces with the effect of earthquake forces on performance of 12 story building subjected to wind speed of 55 m/s, we found that wind is the governing force in zone II and III and earthquake is the governing force in zone IV and V. Similarly, the effect of wind and earthquake forces is compared with regards to the performance of 6 story, 9 story and 12 story models subjected to different wind speeds. It was found that for medium height of the buildings subject to higher intensity of wind speeds, wind is the governing force over earthquake in zone II and to considerable height of the building in zone III. Thus, it can be presented that although seismic zone II is not severe for considering the structure to be designed for seismic forces, structures with moderate heights are to be designed for safety against wind loads compared to earthquake loads in zone II and zone III, considering the wind speed the structure is likely to experience.

The percentage increase in the parameters story drift, story displacement and story shear for same height of the building (models chosen 21 m, 30 m, 39 m) is 40% when wind speed is increased from 33 to 39m/s and 27%, 14%, 13%, and 21% when wind speeds are increased from 39 to 44 m/s, 44 to 47 m/s, 47 to 50 m/s and 50 to 55 m/s respectively. Similarly, the parameters increased by 60% when earthquake zone is changed from zone II to zone III and increased by 50% when changed from zone III to zone IV and zone V. Magnitude of the parameters on the same building increased with increase in wind speed and on moving to higher seismic zones (increase in zone factors). All the parameters considered in the study on a building increased by 178% when speed on the building is increased from 33m/s to 55m/s and increased by 260% when moved from zone II to zone V.

References

IS 875-Part 3, *Code of Practice for Design Loads (Other Than Earth Quake) for Buildings and Structures,* Bureau of Indian Standards, New Delhi, 1987.

IS 1893-Part 1, *Criteria for Earth Quake Resistance Design of Structures,* Bureau of Indian Standards, New Delhi, 2002.

NagaSai, A., and Radha Devi, G., Seismic and Wind Effects on High Rise Structure Using ETABS, *International Journal for Research in Applied Science and Engineering Technology*, 6, (II), 2018.

Narla, M., and Vardhan, M., A., Analysis Of G+20 RC Building in Different Zones Using ETABS, *International Journal of Professional Engineering Studies,* VIII (3), 2017.

Rajmani, A., Analysis of Wind and Earthquake Load For Different Shapes of High Rise Building, *International Journal of Civil Engineering and Technology*, 2(2), 2015.

Reddy, K.R.C., and Sandip, A.T., The Effect of Zone Factors on Wind and Earthquake Loads of High-Rise Structures, *IOSR Journal of Mechanical and Civil Engineering*, 53-58, 2014.

Suchita, H., and Magadum, V., Severity of Earthquake Forces Over Wind Forces for Multistoried RCC Building, Earthquake Loads of High-Rise Structures, *IOSR Journal of Mechanical and Civil Engineering*, 71-75, 2014.

Interdependence between Structural Engineering and Construction Management
Edited by Ozevin, D., Ataei, H., Modares, M., Gurgun, A., Yazdani, S., and Singh, A.

ISBN: 978-0-9960437-6-2

A STUDY OF HUMAN GAIT FOR VIBRATION SERVICEABILITY EVALUATION

MEHDI SETAREH and MATT TILLER

Virginia Tech, Blacksburg, USA

The vibration serviceability problem of structures such as building floors, footbridges, large balconies, and monumental stairs due to human movements has become an important issue, which requires careful attention of designers and engineers. Design engineers often create Finite Elements models of structures to predict their dynamic properties and their responses due to human movements. Therefore, one important aspect of the dynamic analysis of structures is the identification and quantification of excitation forces due to human movements. As part of this research study, measurements of ground reaction forces (GRF) from a number of individuals were conducted at the Virginia Tech Vibration Testing Laboratory using an instrumented force platform. The subjects walked at various speeds synchronized using a metronome. The same tests were subsequently repeated while the subjects ascended and descended over an instrumented mock stair. Considering the periodicity of human excitation force, the Fourier Series parameters (Fourier Coefficients and corresponding phase angles) of the measured forcing functions were computed. Correlations between the measured GRF's on a flat surface and an inclined surface (stair) were established. The results of the analysis were compared to the available information in the literature in an attempt to check their accuracy.

Keywords: Ground reaction force, Fourier coefficients, Dynamic load factors, Stair forcing function, Flat-surface forcing function.

1 INTRODUCTION

To predict the dynamic response of a structure subjected to human activities, the designer needs to define the excitation force accurately. In general, the human induced ground reaction force (GRF) on floors or inclined surfaces such as stairs, can be defined using Fourier Series as shown in Eq. (1):

$$F(t) = G\left[1 + \sum_{i=1}^{n} \alpha_i \sin(2\pi i f_p t + \phi_i)\right] \tag{1}$$

where $F(t)$ is the applied force, G is the weight of person, α_i is the dynamic load factor (DLF) or Fourier coefficient for the *i-th* harmonic, f_p is the step frequency, t is the time variable, and ϕ_i is the phase angle associated with the *i-th* harmonic.

A number of studies with the aim of computing the DLF for flat surfaces have been conducted and different values and equations have been recommended (Blanchard *et al.* 1977, Bachmann and Ammann 1987, Rainer *et al.* 1988, Bachmann *et al.* 1995, Young 2001, Yao *et al.* 2006). However, few studies have focused on defining DLFs for stair vibration serviceability evaluation purposes. The following is a brief summary of attempts to define the GRF and the DLF values for stairs:

Bishop *et al.* (1993, 1995) found that, in general, ascent has larger first harmonic dynamic load factor (α_1) than descent; however, the opposite is true for the higher harmonics. Kerr and Bishop (1997, 2001) noted much larger second harmonic DLF for descents in particular for higher footfall rates. Also, the first harmonic DLF of fast descents (4.3 Hz) was significantly larger than for normal descents (1.85 Hz). They found no major differences between the DLFs of higher harmonics for descents and ascents. They measured forces up to 3 times the body weight for fast descents and about twice the body weight for slow descents. For fast ascents the measured forces were up to 2.5 times the subjects' body weight.

Kerr (1998) conducted experimental studies to measure forcing function of individuals on stairs. He found that the most important factor affecting the DLFs of various harmonics was the step frequency. Twenty-five individuals participated in the tests, ascending and descending the test stair. He found that the DLF for the first harmonic of ascent was larger than for descent. However, the second harmonic DLF for very fast descent (greater than 4 Hz) was about three times greater than for ascent. Comparing to the applied forces on flat floors, the magnitude of forces on stairs was found to be larger and higher footfall rates were also possible.

Kasperski and Czwikla (2012) studied the differences between the right versus left foot loading in terms of DLF of different harmonics, and step frequency when one ascends or descends a stair. Using a force plate, the researchers measured the applied forces from eight people walking on floor and over four small wooden steps. They found that the first harmonic of descents to be larger than ascents at the same step frequency, which is not consistent with the findings of Bishop *et al.* (1993, 1995) and Kerr (1998).

This paper presents an experimental study of GRF by measuring the exerted footfall forces by a number of subjects using an instrumented force platform/plate. The measurements were conducted on a flat surface and on a stair with a typical slope. The DLF's for various harmonics for each measured force were then computed. Relationships between the values for the sloped surface (stair) and flat surface were then established. In addition, comparison between the values obtained from the tests and the information available in the literature on the GRF for stairs were made.

2 DESCRIPTION OF THE TESTS

The GRF measurement tests were conducted at the Virginia Tech Vibration Testing Laboratory. A Bertec force platform was used to measure the applied footfall forces on a flat surface. A total of 84 human subjects participated in these tests. A mock stair was used for the stair GRF measurements. A force plate, made of strain gages, was placed within one of the stair treads to enable the force measurements. First, each subject stood still on the force platform/plate to measure his/her weight. Then, using a metronome to synchronize their movements, the subject walked over the force platform/plate measuring the right and left feet GRF separately. The following step frequencies were used: 100, 110, 120, 129, 135, 140, 155, 160, 163, 178, 190, 210 steps/min. These were selected to range from a slow walk to a fast run. The same tests were repeated to measure the GRF for the stair. The mock stair had a very high natural frequency to ensure the accuracy of the force measurements. Figures 1 and 2 show the weight and force measurements on the flat surface and on the stair, respectively.

(a) (b)

Figure 1. GRF measurement on a flat surface: (a) Weight measurement, (b) Gait measurement.

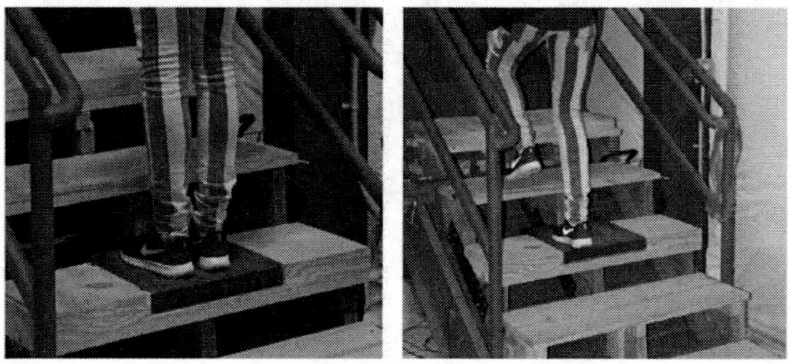

(a) (b)

Figure 2. GRF measurements on the instrumented mock stair: (a) Weight measurement, (b) Gait measurement.

Once the GRF measurements were completed, they were normalized by the individual's weight (G). Figure 3 shows a typical measured GRF on the flat surface and stair.

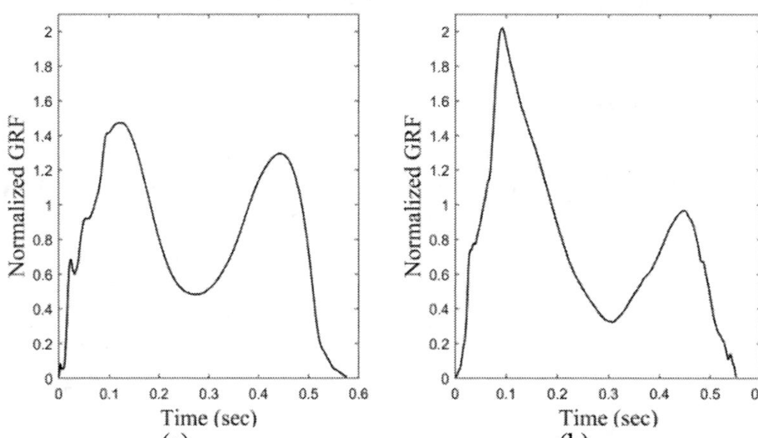

(a) (b)

Figure 3. Typical measured GRF: (a) Flat surface (2 steps/sec), (b) Stair (descending 2 steps/sec).

3 MEASURED DLF AND THE CORRELATION BETWEEN THE FLAT SURFACE AND STAIR

The DLF values were computed by combining the right and left feet forcing functions and conducting a Fourier Series analysis. The effects of half-harmonics were neglected in this study. The above were repeated for both the flat surface and stair. The computed DLF values on the stair were divided by their counterparts for the flat surface for each test subject to compute the DLF ratios. These values were then averaged and plotted versus the step frequency, f_p. Figure 4 shows these plots for the first harmonics of the forcing function.

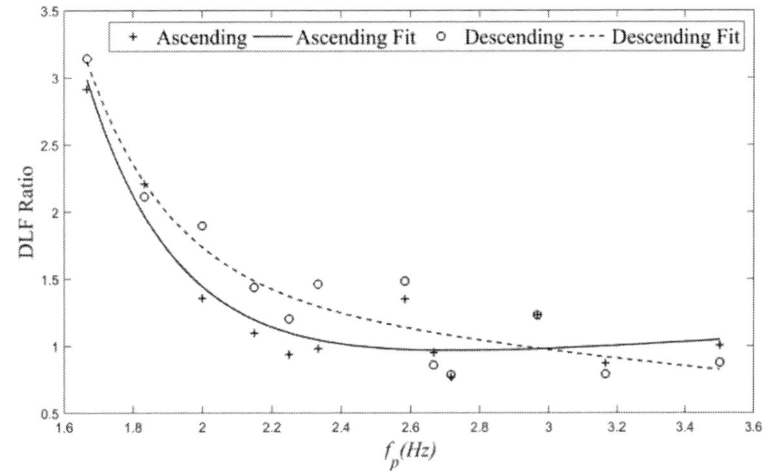

Figure 4. Variation of the first harmonic DLF ratio.

Curve-fitting of the graph showed a consistent trend of the DLF ratios with step frequency. This is represented by approximate exponential distributions shown below:

$$Ascent\ DLF\ Ratio = 1{,}323e^{-3.84f_p} + 0.62e^{0.15f_p} \tag{2}$$

$$Descent\ DLF\ Ratio = 1{,}758e^{-4.19f_p} + 2.51e^{-0.32f_p} \tag{3}$$

The R^2 goodness-of-fit for ascending and descending DLF ratios were 0.91 and 0.92, respectively. DLF ratios for ascending and descending were greater than 1 for majority of tested step frequencies. Descending stair DLF ratios were larger than ascending DLF ratios for most step frequencies less than 2.7 Hz. There were multiple instances where DLF ratios dropped below 1. The lowest DLF ratios for ascending and descending occurred at 2.7 Hz, where DLF ratios were 0.77 and 0.78, respectively.

Ascending DLF ratios for the first harmonic were largest at the lowest tested step frequencies with a maximum value of 2.9 at 1.7 Hz. Descending DLF ratios showed a similar trend to the ascending ratios. They were highest at the lowest step frequencies and had a maximum value of 3.1 at 1.7 Hz. DLF ratios for descent at 2 and 3 Hz were measured to be 1.9 and 1.2 respectively. These values are approximately 25 to 50 percent lower than an average ratio of 2.5 reported by Kerr and Bishop (1997).

Between 2.3 and 3 Hz there is a slight discrepancy in the trend of the overall data. It is possible that in this range subjects switched from a walk to a run. This resulted in a larger coefficient of variation at 2.6 Hz than that of the surrounding step frequencies, and thus potentially less accurate data.

Descending DLF values for the first harmonic were larger than ascending values for all step frequencies below 2.7 Hz with the exception of the 1.8 Hz case. At step frequencies greater than 3 Hz, ascending DLF values were larger than descending. These trends are consistent with the results from Kasperski and Czwikla (2012) at frequencies below 2.7 Hz, and with that of Bishop *et al.* (1993, 1995) and Kerr (1998) at above 3 Hz.

Figure 5 shows second harmonic DLF ratios plotted versus step frequency.

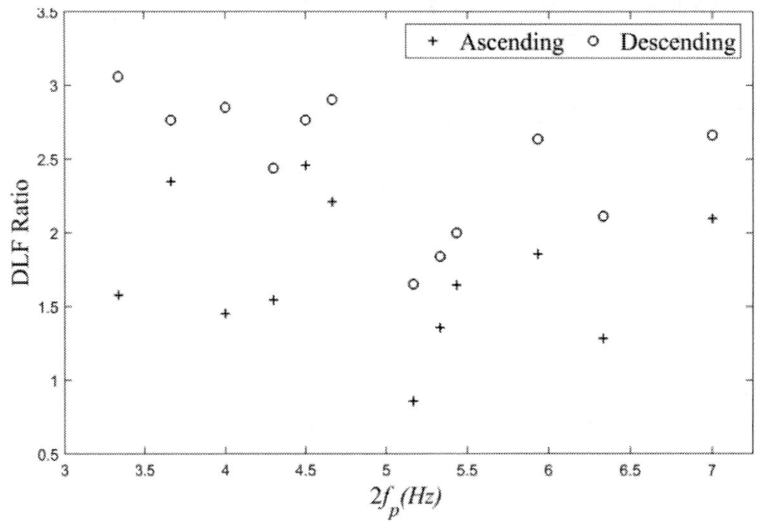

Figure 5. Variation of the second harmonic DLF ratio.

DLF ratios for the second harmonic did not show a trend with step frequencies. Descending ratios remained consistently larger than ascending ratios throughout the range of step frequencies used in this study, which is consistent with the observations of Kerr and Bishop (1997). The average descending second harmonic DLF ratio was 2.5 across the range of tested step frequencies.

4 CONCLUSION

A comparative study of human gait on a flat and inclined surface (stair) was conducted. Dynamic load factors were computed from the measured footfall forces of a number of individuals. It was found that the DLF values for ascending and descending forces were, for majority of cases, larger than that of walking on a flat surface. First harmonic DLF ratios for ascending and descending forces showed a decreasing exponential trend relative to increasing step frequency. First harmonic DLF ratios for descending forces were larger than those of ascending forces for the large majority of step frequencies below 2.7 Hz. Second harmonic DLF ratios for ascending and descending forces showed no trend relative to step frequency. Second harmonic descending DLF ratios were larger across all frequencies tested; ranging from 1 to 2 times larger than ascending DLF ratios.

Acknowledgments

The research presented here was supported by the National Science Foundation under grant number CMMI-1335004. This support is gratefully acknowledged. Any opinions, findings, and conclusions expressed in this paper are those of the writers and do not necessarily reflect the views of the National Science Foundation.

References

Bachmann, H., and Ammann, W., *Vibrations in Structures Induced by Man and Machines,* International Association of Bridge and Structural Engineering, Zurich, Switzerland, 1987.

Bachmann, H., Ammann, W. J., Deischl F., Eisenmann, J., Floegl, I., Hirsch, G. H., Klein, G. K., Lande, G. J., .Mahrenholtz, O, Natke, H. G., Nussbaumer, H., Pretlove, A. J., Rainer, J. H., Saemann, E. U., and Steinbeisser, L., *Vibration Problems in Structures: Practical Guidelines,* Birkhauser Verlag, Basel, Boston, Berlin, 1995.

Bishop, N. W. M., Willford, M., and Pumphrey, R., Multi-Person Excitation of Modern Slender Staircases, *Engineering for Crowd Safety*, Elsevier Science Publishers, Amsterdam, 399-408, 1993.

Bishop, N. W. M., Willford, M., and Pumphrey, R., Human Induced Loading of Flexible Staircases, *Safety Science*, 18, 261-276, 1995.

Blanchard, J., Davies, B. L., and Smith, J. W., *Design Criteria and Analysis for Dynamic Loading of Footbridges*, Proceedings of the DOE and DOT TRRL Symposium on Dynamic Behaviour of Bridges, Crowthorne, UK, 1977.

Kasperski, M., and Czwikla, B., A Refined Model for Human Induced Loads on Stairs, *Topics on the Dynamics of Civil Structures, Vol. 1,* Proceedings of the Society for Experimental Mechanics Conference, 26, 27-39, 2012.

Kerr, S. C., and Bishop, N. W. M., Human Induced Loading of Flexible Staircases, *Innovation in Civil and Structural Engineering*, Topping, B. H. V., and Leeming, M. B., (Eds.), Civil-Comp Press, Edinburgh, Scotland, 311-317, 1997.

Kerr, S.C., *Human Induced Loading on Staircases*, PhD. Thesis, Mechanical Engineering Department, University of London, London, UK., 1998.

Kerr, S.C., and Bishop, N. W. M., Human Induced Loading on Flexible Staircases, *Engineering Structures*, 23, 37-45, 2001.

Rainer, J. H., Pernica, G., and Allen, D. E., Dynamic Loading and Response of Footbridges, *Canadian Journal of Civil Engineering*, 15, 66–71, 1988.

Yao, S., Wright, J. R., Pavic, A., and Reynolds, P., Experimental Study of Human-Induced Dynamic Forces Due to Jumping on a Perceptibly Moving Structure, *Journal of Sound and Vibration*, 296, 150–165, 2006.

Young, P., *Improved Floor Vibration Prediction Methodologies*, Proceedings of Arup Vibration Seminar on Engineering for Structural Vibration - Current Developments in Research and Practice, London, UK, 2001.

Interdependence between Structural Engineering and Construction Management
Edited by Ozevin, D., Ataei, H., Modares, M., Gurgun, A., Yazdani, S., and Singh, A.

ISBN: 978-0-9960437-6-2

EFFECTS OF HUMAN-STRUCTURE INTERACTIONS ON THE DYNAMIC PROPERTIES OF BUILDING FLOORS SUSCEPTIBLE TO VIBRATIONS

MEHDI SETAREH and STEPHANIE RENARD

School of Architecture and Design, Virginia Tech, Blacksburg, USA

Excessive vibrations of building floors due to human movements have become an important vibration serviceability problem for building designers and owners. A series of vibration tests on a full-scale laboratory floor with different numbers of humans in various postures were conducted. Using this data, the dynamic properties of a two-degree-of-freedom (2-DOF) dynamic system representing groups of people in different postures were computed. A 3-DOF model representing the floor and humans was developed and its dynamic properties were defined in terms of non-dimensional parameters. The dynamic properties of the floor were measured when occupied by groups of people in different postures and compared to those predicted using the 3-DOF dynamic model considering the identified human models. The results showed that the predicted properties were within the range of those found from the measurements, which validated the identified human dynamic models. This study also showed how the presence of humans can affect the natural frequency and damping ratio of a floor system.

Keywords: Human models, Simplified models, Human vibrations, Human dynamic parameters, Floor vibrations, Excessive vibrations.

1 INTRODUCTION

Structural engineers generally consider the presence of humans on building floors as added weight (live load). This is a correct assumption as long as the structure is not susceptible to vibrations. When vibration serviceability issues are considered, humans cannot simply be assumed as static weights (masses). Various research studies and design guides have recommended humans to be considered as additional masses resulting in a reduction in floor natural frequencies and/or added damping.

An accurate approach is to consider human as a dynamic system consisting of a mass (or several masses), spring(s) and damping element(s). Various lumped-mass models of human body with various degrees of freedom have been proposed and studied in the literature. A few recent studies are mentioned below:

Pedersen and Hansen (2004) considered the effects of the human-structure interaction (HSI) on floors using SDOF models to represent the humans and a specific mode of a floor to form a 2-DOF dynamic model. The human parameters were assumed to be f_H (natural frequency) = 6 Hz and ξ_H (damping ratio) = 20%. The results of the study showed that the presence of humans have different effects on the low and high frequency floors.

Sim *et al.* (2006) developed a 2-DOF crowd model using the individual human dynamic properties recommended by Wei and Griffin (1998) and Matsumoto and Griffin (2003) for the seated and standing individuals. They conducted an analytical study, which showed that passive crowd add mass and damping to the structure, resulting in a reduction in natural frequency and response compared to the empty structure.

Agu and Kasperski (2011) conducted a probabilistic study to check how random scatter of individual human dynamic properties in a group can affect the natural frequency and damping ratio *(f and ξ)* of the structure with a crowd. They concluded that due to the randomness of dynamic properties of human body, the mean values for individual human dynamic properties cannot be used for all cases to represent the human-structure dynamic interactions.

Zheng (2013) developed a continuous model of a standing human body from the available natural frequency measurements of standing subjects by Matsumoto and Griffin (2003). By incorporating the available information in biomechanics and his interaction human body model, Zheng (2013) identified parameters for a 2-DOF human model to be f_1= 5.78 Hz and ξ_1= 36.9%, and f_2= 13.2 Hz and ξ_2= 44.5%. The identified damping ratio of the human body varied from 2.5% to 38.8%, and the mean damping ratio of the four individuals were from 8.6% to 22.5%.

This paper uses the dynamic parameters for a 2-DOF model to represent groups of people measured through an extensive experimental and analytical program. The measured effects of the presence of human on the resonance frequency and damping ratio of the floor structures were compared to those found by a simplified 3-DOF system representing a particular mode of vibration of the structure and the 2-DOF human model. The results show that the proposed model provide a very good representation of the dynamic properties of groups of people in terms of their interaction with the structure and can provide a good prediction of building floors' dynamic response.

Figure 1. Test structure at the Virginia Tech Vibration Testing Laboratory.

2 DESCRIPTION OF THE TEST STRUCTURE AND DYNAMIC TESTS

The test structure is comprised of a two-story steel structure located at the Virginia Tech Vibration Testing Laboratory. The structure has a single square bay with center to center column dimensions of 30 ft. by 30 ft. as shown in Figure 1. The floor is made of a 2 in. steel deck with 2 in. of concrete topping, supported by 20LH06 steel joists on the first floor and W16x26 steel beams on the second floor as shown in Figure 2. The joists and beams are supported by W18x40 girders. The test floor did not include any non-structural elements.

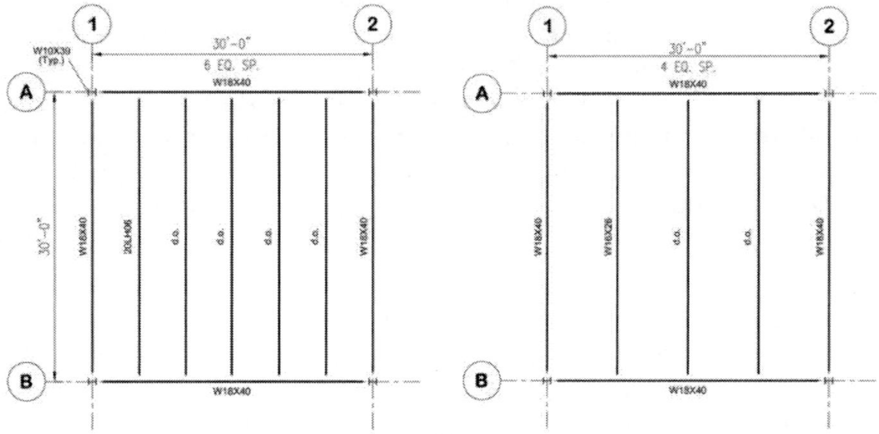

(a)

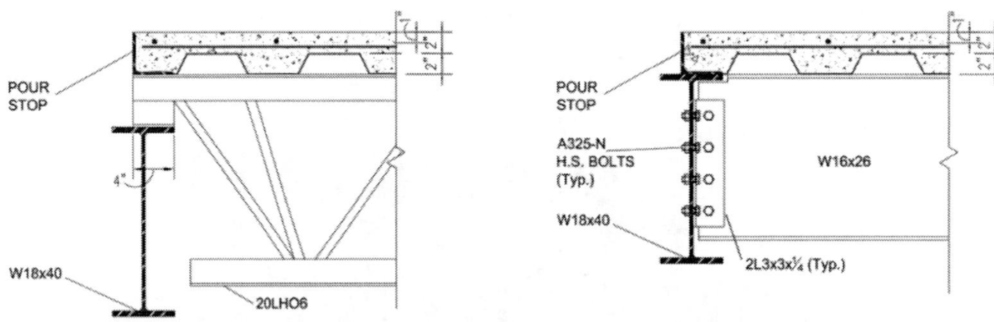

(b)

(c)

Figure 2. Details of the test structure: (a) Plan views of the first and second floors first, respectively, (b) Section through the floors, (c) Views from below the floors.

Full modal tests of the floors were conducted using an APS-400 electrodynamic shaker on each floor. The shaker was placed at the quarter point from a corner column to excite several lower modes of vibration of the floor. The test on the first floor was repeated with two groups of human subjects and also after the floor was covered by 2-inch concrete blocks to reduce the floor natural frequencies. A burst chirp excitation with 30 seconds on and 15 seconds off and a frequency bandwidth of 3-20 Hz was used. The measured records were used in MEscope VES (2013) to estimate the natural frequencies, damping ratios and mode shapes of the structure. Figure 3 shows the shaker and accelerometers during one of the stages of the modal test.

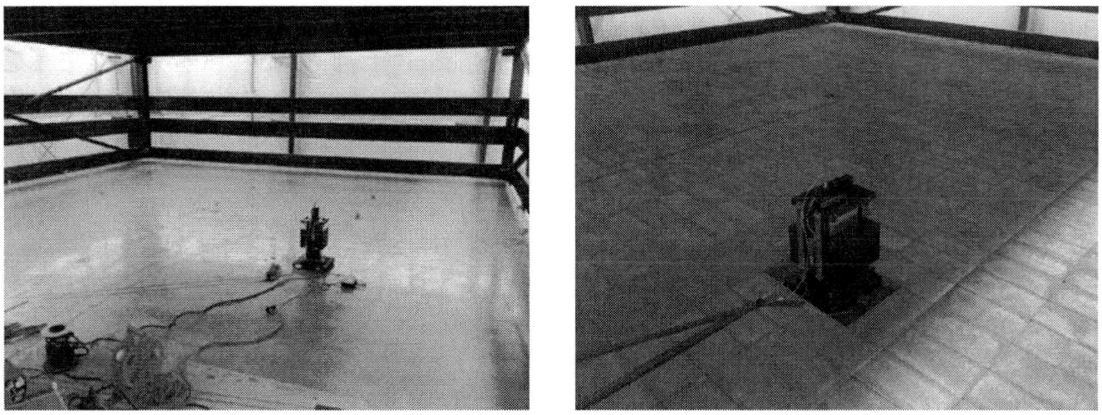

(a) (b)

Figure 3. Modal test setups for the floors: (a) Without concrete blocks, (b) With concrete blocks.

The effects of a group of people on the dynamic properties of the test floors were used to estimate the crowd dynamic properties. Different groups of human subjects with various sizes were placed on the floor in three postures: sitting on chairs, standing erect, and standing with the bent knees (Setareh and Gan 2016).

PCB 393C accelerometers were placed at different locations on the floors and at the drive point and the shaker was placed at the same location as for the modal test. A burst chirp excitation with 30 seconds on and 15 seconds off and a frequency band of 3-15 Hz was used. Low levels of vibration within the range typically resulted from human movements were applied. Their magnitudes were at the lower limits of those used in the biomechanics research.

3 HUMAN DYNAMIC PARAMETERS AND MODELING

Using the results of these tests, the dynamic parameters of the 2-DOF human model were estimated by minimizing errors between the measured resonance frequencies and response of the floor and their analytical counterparts (Gan and Setareh 2015). This resulted in the human dynamic parameters with natural frequencies between 3.4 and 20.4 Hz and damping ratios between 35% to 290% for the three different postures.

A 3-DOF analytical system representing the floor and the 2-DOF human model was developed. The floor response in terms of non-dimensional dynamic parameters was computed. Using the frequency response functions, the resonance frequencies and damping ratios of the floors occupied by different numbers of human subjects were computed. These values were compared to those from the four different test measurements on the first and second floors of the test structure for the first three modes of vibrations of each floor by computing the ratios between the measured resonance frequencies and damping ratios to their counterparts predicted by the analytical models.

Figure 4 shows the variations of the floor resonance frequency ratios and ratios of damping ratio of the second mode of vibration of each tested floor as a function of mass ratio (mass of humans to the effective mass of floor), for the three different postures. The floor natural frequencies (without people) ranged from 6.7 Hz to 9.4 Hz with damping ratios between 0.15% to 0.35%. The results show that the human models and the 3-DOF dynamic representation of the effects of HSI on the floor provide excellent prediction of the variation in the floors' resonance frequencies as the frequency ratios are very close to 1 (maximum discrepancy of 3%). As expected, the variation in the damping ratios are larger. However, on average the predictions of the resonance frequencies were within 2% for all the floors. This value for damping ratios was about 25%. The results show that the human parameters found from the previous studies and the consideration of 3-DOF models to represent the human-structure interactions on building floor are accurate and provide good results for vibration serviceability analysis of building floors.

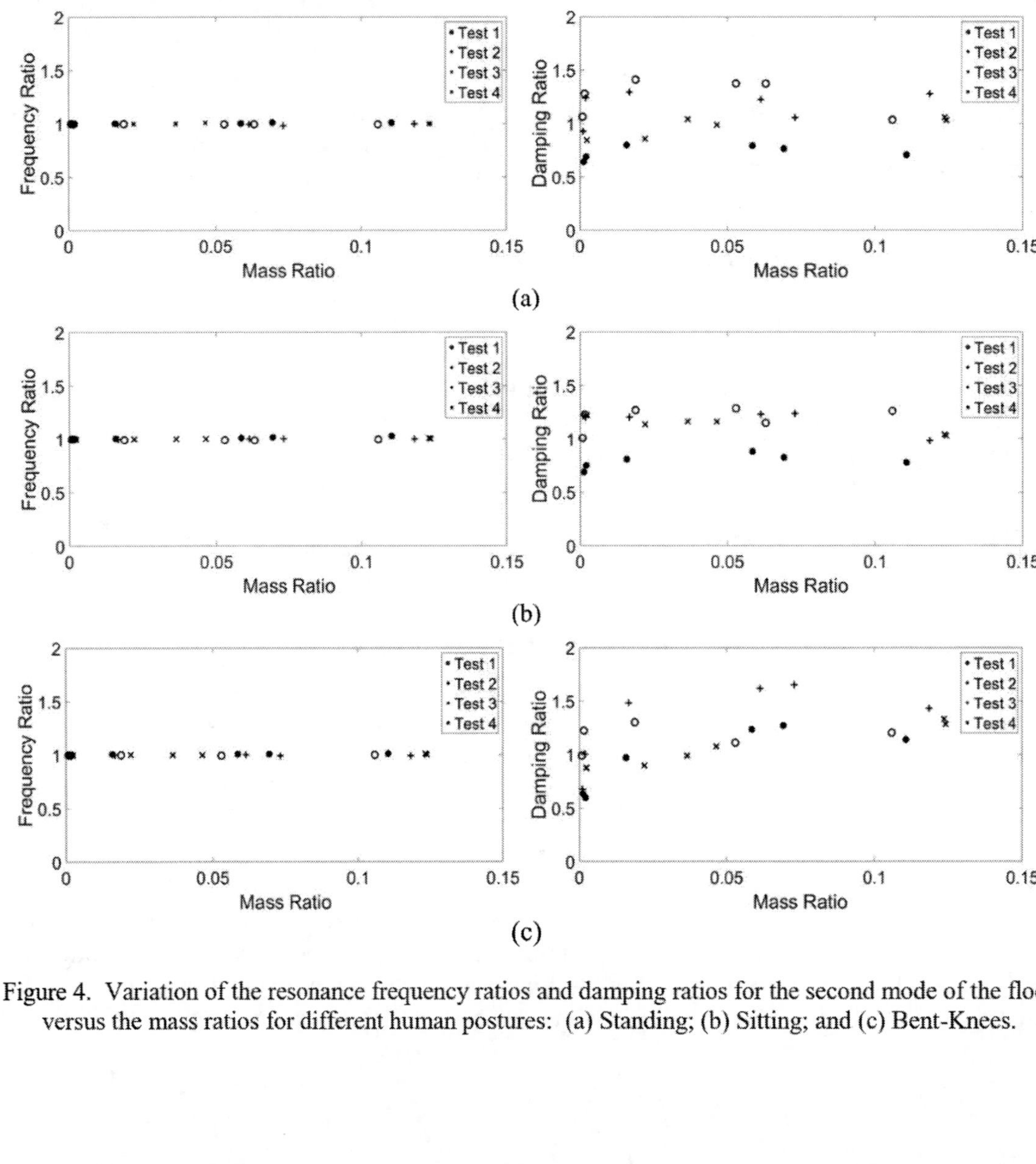

Figure 4. Variation of the resonance frequency ratios and damping ratios for the second mode of the floor versus the mass ratios for different human postures: (a) Standing; (b) Sitting; and (c) Bent-Knees.

4 CONCLUSION

This paper used a simplified 3-DOF model to check the accuracy of the dynamic parameters for the 2-DOF human model that were found through a separate study. It showed the accuracy of the modeling approaches used and the closeness in the prediction of the resonance frequency and damping ratio when human-structure interactions for floor vibration serviceability are considered.

Acknowledgements

The research presented here was supported by the National Science Foundation under grant number CMMI-1335004. This support is gratefully acknowledged. Any opinions, findings, and conclusions expressed in this paper are those of the writers and do not necessarily reflect the views of the National Science Foundation. Mr. Bishal Khadka conducted some of the analysis presented here. His technical assistance is appreciated.

References

Agu, E., and Kasperski, M., Influence of the Random Dynamic Parameters of the Human Body on the Dynamic Characteristics of the Coupled System of Structure-Crowd, *Journal of Sound and Vibration*, 330, 431-444, 2011.

Gan, S., and Setareh, M., *Human-Structure Interaction Studies*, School of Architecture and Design, Virginia Tech, Blacksburg, Virginia, 2015.

Matsumoto, Y., and Griffin, M. J., Mathematical Models for the Apparent Masses of Standing Subjects Exposed to Vertical Whole-Body Vibration, *Journal of Sound and Vibration*, 260(3), 431-451, 2003.

ME'scope VES, ME'scope VES 6.0, *Vibrant Technology*, Inc., Scotts Valley, California, 2013.

Pedersen, L., and Hansen, L., Human Damping and Its Capacity to Control Floor Vibrations, *Smart Structures and Materials 2004: Damping and Isolation,* Proceedings *of SPIE*, 5386, SPIE, Bellingham, WA, 2004.

Setareh, M., and Gan, S., Study of Human-Structure Dynamic Interactions, *Proceedings of the International Modal Analysis Conference, IMAC-XXXIV*, Orlando, Florida, 4, 391-399, 2016.

Sim, J., Blakeborough, A., and Williams, M., *Modelling Effects of Passive Crowds on Grandstand Vibration, Structures and Buildings,* Proceedings of the Institution of Civil Engineers, 159(585), 261-272, 2006.

Wei, L., and Griffin, M. J., Mathematical Models for the Apparent Mass of the Seated Human Body Exposed to Vertical Vibration, *Journal of Sound and Vibration*, 212(5), 855-874, 1998.

Zheng, Q., *Models of a Standing Human Body in Structural Vibration*, PhD Thesis, School of Mechanical, Aerospace and Civil Engineering, University of Manchester, UK, 2013.

Interdependence between Structural Engineering and Construction Management
Edited by Ozevin, D., Ataei, H., Modares, M., Gurgun, A., Yazdani, S., and Singh, A.

ISBN: 978-0-9960437-6-2

SEISMIC EVALUATION OF SQUARE HSS BRACES IN SCBF USING REGRESSION ANALYSIS

MAHMOUD FAYTAROUNI[1], ONUR SEKER[2], BULENT AKBAS[3], and JAY SHEN[1]

[1]*Dept of Civil, Construction and Environmental Engineering, Iowa State University, Ames, USA*

[2]*Dept of Civil Engineering, MEF University, Istanbul, Turkey*

[3]*Dept of Civil Engineering, Gebze Technical University, Kocaeli, Turkey*

Since the 1990s, structural engineering practice geared toward the use of hollow structural sections (HSS), notably square HSS, for their economy, and ease of design and construction. According to the AISC Seismic Provisions, during a severe earthquake, these braces could undergo post-buckling axial deformations 10 to 20 times their yielding deformation. However, recent experimental studies indicate that braces made of square HSS, depending on their size, width-to-thickness, and slenderness ratio, are vulnerable to fracture even prior to 10. Therefore, relying on past experimental studies comprised of a few square HSS specimens to develop seismic requirements for SCBF with square HSS could lead to underestimation of the seismic risk. This paper aims to evaluate the fracture risk of braces in existing SCBFs designed in accordance with AISC 341-05 and AISC 341-16 through incremental dynamic analyses (IDA) along with experimentally developed regression model that estimates fracture.

Keywords: Seismic risk, Incremental dynamic analysis, Ductility capacity and demand, Fracture.

1 INTRODUCTION

The seismic design criteria adopted in the design codes for special concentrically braced frames (SCBFs) is tailored based on the concept of capacity design. In the design procedure of SCBFs, plastic deformations are confined to braces only, leaving other structural members such as beams and columns to remain elastic. According to the commentaries in AISC 341-05 (2005) and AISC 341-16 (2016): During a severe earthquake, braces in SCBF could undergo post-buckling axial deformations 10 to 20 times their yield deformation. This ductility demand range is based on the experimental and analytical work conducted by Goel (1992). The experimental portion of the study included braces varying between large- and small-size sections of various shapes such as wide flanges, angles, and tubes. However, since 1990s engineering practice geared toward the use of hollow structural sections (HSS), notably square HSS, for their ease of design and construction. Numerous experimental studies have been conducted on conventional bracings made of square HSS over the last decades. The results of these tests indicate that braces with square HSS are vulnerable to fracture even prior to 10, the lower bound of the expected ductility demand (Goel 1992).

In this study, experimental cyclic tests on square HSS braces conducted from 1978 to 2013 have been thoroughly surveyed. A total of 79 square HSS braces were collected from 16 experimental programs. For each specimen, the ductility capacity reached fracture was estimated and reported. Collected specimens with sizes of HSS 5x5 and HSS 6x6 reached a mean ductility capacity of 9.0, while with a larger size of HSS 10x10, the possessed ductility capacity at fracture was 8.0. This experimental observation agrees with the analytical findings of Shen *et al.* (2017). Their study concludes that braces in SCBF are often likely to fracture prior to the expected story drift ratio (SDR) demand. Therefore, relying on past experimental results comprised of a few test results to develop seismic requirements for SCBF with square HSS could lead to brace fracture earlier than expected as well as underestimation of the overall seismic risk. The purpose of this paper is to evaluate the fracture risk of braces in existing SCBFs designed following AISC 341-05 (2005) and AISC 341-16 (2016) through incremental dynamic analyses (IDA) along with the experimentally developed regression model that estimates fracture life.

2 CASE STUDY BUILDINGS AND CALIBRATION

2.1 Building Description

The seven-story office buildings adopted in the NEHRP Recommended Seismic Provisions: Design Examples under publication number of FEMA P-751 (NEHRP 2012), and FEMA P-1051 (NEHRP 2016) were investigated in this study. The seven-story office building is located in Los Angeles, California. Figure 1(a) illustrates the plan dimensions of the building. The height of the first story, as shown in Figure 1(b), is 22 feet while all remaining stories are 13 feet in height. In Table 1, and throughout the paper, the SCBF designed in FEMA P-751 (NEHRP 2012) is identified as Frame A, whereas Frame B is given to that designed in FEMA P-1051 (NEHRP 2016). Note that the designs of Frames A and B are in compliance with AISC 341-05 (2005) and AISC 341-16 (2016), respectively. Detailed information on the design process and buildings can be found in (NEHRP 2012 and 2016).

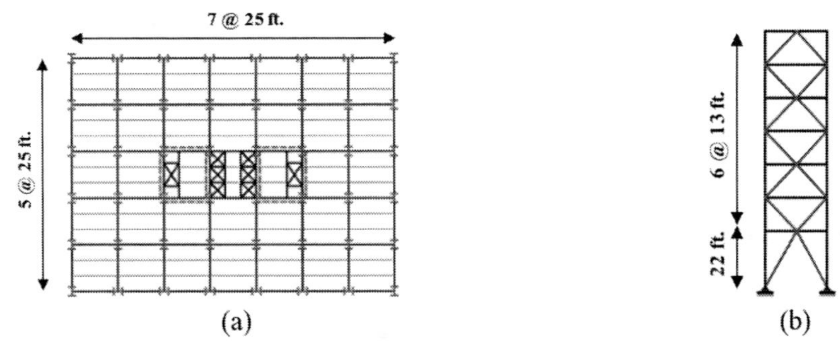

Figure 1. (a) floor plan view, and (b) elevation of the two-story X bracing configuration.

Table 1. Member sizes of the considered seven-story SCBFs.

Story	Frame A			Frame B		
Level	Braces	Columns	Beams	Braces	Columns	Beams
7	HSS 5 ½ ×5 ½×5/16	W 14×53	W 18×35	HSS 5×5×3/8	W 14×68	W 18×40
6	HSS 6×6×1/2	W 14×61	W 18×35	HSS 6×6×1/2	W 14×159	W 18×50
5	HSS 6×6×5/8	W 14×61	W 18×35	HSS 7×7×1/2	W 14×159	W 18×50
4	HSS 6×6×5/8	W 14×132	W 18×35	HSS 7×7×1/2	W 14×342	W 18×60
3	HSS 7×7×1/2	W 14×132	W 18×35	HSS 7×7×1/2	W 14×342	W 18×40
2	HSS 8×8×5/8	W 14×233	W 18×35	HSS 8×8×1/2	W 14×550	W 18×65
1	HSS 9×9×5/8	W 14×233	W 27×102	HSS 9×9×5/8	W 14×550	W 30×132

2.2 Analytical Models of the Frames and Validation Study

Frames A and B were built as two-dimensional (2D) analytical models within *OpenSees* (McKenna *et al.* 2000). The modeling conditions were validated using the shaking table test of the nearly full-scale chevron CBF frame presented in (Okazaki *et al.* 2012). The one-story tested frame shown in Figure 2(a) comprised of braces and columns with square HSS sections, whereas the beam section was a built-up wide-flange. The unit of the sections presented in Figure 2(a) is inches. The frame was subjected to a series of ground shaking using the same ground motion with different intensities. The focus herein is on the experimental results of the overall frame and brace response at 42% motion.

First, a fundamental period of T_1=0.20 seconds was determined from the eigenvalue analysis, which was closely matching to T_1 of the test (Okazaki *et al.* 2012). Accordingly, the frame was subjected to the ground shaking time history measured during the test. Figure 2 compares the analytical and experimental time history response of the frame in terms of lateral displacement and the east side brace's hysteresis. As Figures 2(b) and (c) illustrate, a good agreement was observed from both the displacement time history and brace response under the motion. One can notice that the main characteristics of this dynamic test were captured quite well, and thus the analytical models (i.e., Frames A and B) deemed to be sufficiently reliable for analyzing the frames under several ground motions.

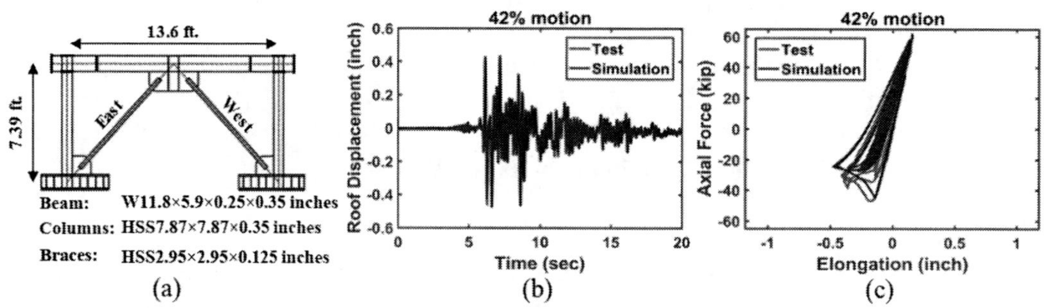

Figure 2. Comparison of the simulated and the experimental responses under 42% motion: (a) tested frame by Okazaki *et al.* (2012), (b) roof displacement time history, (c) brace response.

3 DUCTILITY-BASED REGRESSION MODEL FOR EVALUATING SQUARE HSS

Braces designed in accordance with AISC 341-05 (2005) and AISC 341-16 (2016) are required to satisfy the limits on width-to-thickness (*b/t*) and slenderness (*KL/r*) ratios. These seismic limits are set to achieve adequate ductility on the order of 10 to 20 without experiencing fracture. However, the comprehensive survey of the experimental cyclic tests on square HSS, reported in Figure 3, indicates that fracture for specimens designed to satisfy *b/t* and *KL/r* is more likely to occur within the expected ductility demand range (10-20). The collected specimens included 79 square HSS braces from 16 experimental programs. For each specimen, the ductility capacity at fracture (μ_c) was estimated from the published plots. Note that both ductility capacity (μ_c) and demand (μ_d) are defined as the axial brace deformation in either tension or compression, whichever is larger, normalized by yielding or buckling deformation. Figure 3(a), (b), and (c) report the interaction between μ_c and *b/t*, *KL/r*, and the material parameters (E/F_y), where E is elastic modulus, and F_y is yielding strength. Referring to Figure 3, one might observe the following: (a) majority of the tested specimens that satisfy b/t and KL/r ratios stipulated in AISC 341-05 (2005) and AISC 341-16 (2016) experienced fracture prior to attaining a ductility capacity of 20 and 15, respectively, and (b) although μ_c appeared to be inversely proportional to b/t and directly proportional to the ratio of KL/r and E/Fy to a certain degree, an apparent trend is lacking due to the robust interaction between these parameters.

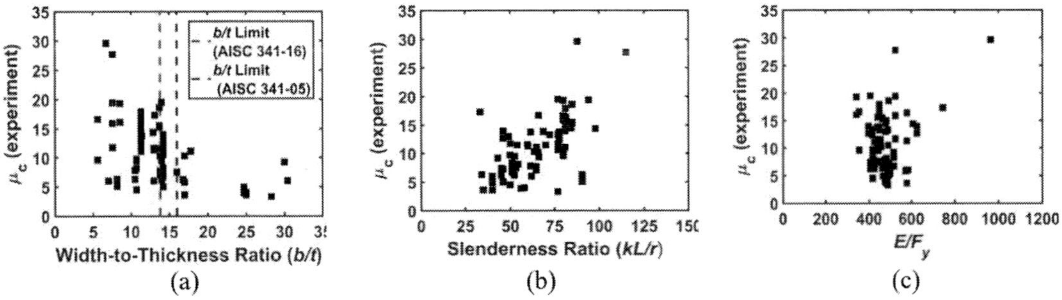

Figure 3. Ductility capacity of the collected 79 specimens at fracture: (a) interaction between ductility capacity at fracture (μ_C) and width-to-thickness ratio (b/t), (b) interaction between μ_C and slenderness ratio (KL/r), and (c) interaction between μ_C with the mechanical properties (E/Fy).

3.1 Regression Analysis

Multiple nonlinear regression analyses were performed on the collected data (Figure 3) to attain a relationship between ductility capacity at fracture (μ_c) and the geometric- material parameters (i.e., b/t, KL/r, and E/F_y). The regression effect summary has shown that the combined interaction between the multiple predictors such as b/t, KL/r, and E/F_y all have a profound impact and play an essential role in determining μ_c at fracture. Consequently, and after several regression iterations, two polynomial equations to the 5th degree were found to be adequate to estimate μ_c at fracture for the specimens satisfying limiting b/t and KL/r ratios specified in AISC 341-05 (2005) and AISC 341-16 (2016). It is noteworthy that the estimation addressed in Eq. (1) is only applicable to the specimens that satisfy AISC 341-05 (2005) limits, whereas Eq. (2) is driven for those satisfying AISC 341-16 (2016). Table 2 summarizes the coefficients that offer the optimal fit. The statistical measure of the predictive model fit adjusted-R^2 of 0.6 and 0.68 for Eqs. (1) and (2), respectively. Accordingly, these equations are utilized to estimate the ductility capacity at fracture for the braces in Frames A and B.

Table 2. Regression coefficients in Eq. (1) and Eq. (2).

	Eq. (1)				Eq. (2)			
C(n)	2	3	4	5	2	3	4	5
C_1	-5.32E-01	-1.40E-01	1.71E-02	4.05E-03	-1.57E+00	-1.08E+00	1.49E-01	5.15E-02
C_2	-1.44E-03	-4.14E-04	-1.91E-06	2.58E-07	3.66E-03	-8.96E-05	-5.78E-07	9.71E-08
C_3	2.15E-05	-1.70E-06	1.16E-08	-1.66E-11	-4.35E-04	-5.65E-06	4.17E-08	-5.85E-11

$$\mu_c = -26.84 + 1.048b/t + 0.241KL/r - 0.023E\Big/F_y + \sum_{n=2}^{5} C_1(n)x(b/t - 11.78)^n + \cdots$$
$$C_2(n)x(KL/r - 67.43)^n + C_3(n)x(E/F_y - 475.85)^n \quad (1)$$

$$\mu_c = -95.62 + 6.484b/t + 0.159KL/r + 0.055E\Big/F_y + \sum_{n=2}^{5} C_1(n)x(b/t - 10.7)^n + \cdots$$
$$C_2(n)x(KL/r - 72.68)^n + C_3(n)x(E/F_v - 474.86)^n \quad (2)$$

4 DUCTILITY DEMAND ON BRACES UNDER SELECTED GROUND MOTIONS

4.1 Selected Ground Motion Records

The seismic demand on the braces in Frames A and B was investigated using five ground motion (GM) records. The GM records were selected from the Pacific Earthquake Engineering Research (PEER) Center database (PEER 2018). The records were scaled, as presented in Figure 4, so that the geometric mean of the 5% damping response spectra of the scaled suite of ground motions is not lower than the target response spectrum over a period range. Note that the analytical first modal periods, T_1, were 1.07 seconds for Frame A and 0.94 seconds for Frame B.

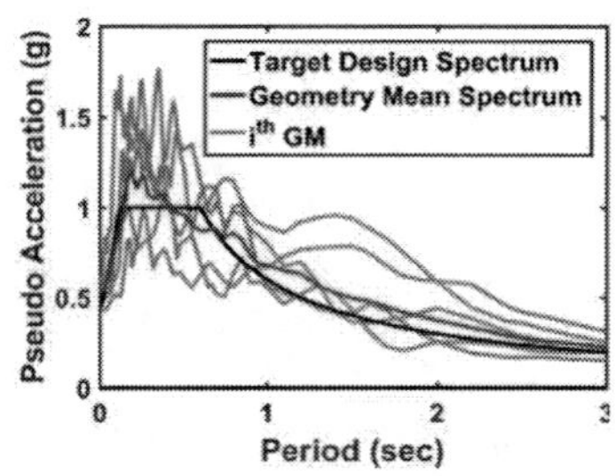

Figure 4. Response spectra of the ground motions used for IDA.

4.2 IDA Results and Conclusions

Incremental dynamic analysis (IDA) was carried out, where frames were subjected to a series of nonlinear dynamic analyses using records of increasing intensities. Once the IDA curves were established, the time history analyses, during which one of the seven stories undergo an SDR of 2% or 4%, were selected first. The ground shaking intensities corresponding to these response quantities are considered to represent the design level and collapse prevention level earthquakes, respectively. Subsequently, for the selected data points on each IDA curve, hysteretic loops of the braces are plotted. Then, the distribution of peak ductility demand (μ_d) on each brace along the height is investigated. In order for the discussions on the seismic risk assessment to be pertinent, the demand is required to be paired with its counterpart. Therefore, fracture life (i.e., ductility capacity) of each brace section used in the design was also estimated by means of the previously derived equations (i.e., Eqs. (1)-(2)) based on the inclusive collection of test data.

Figure 5 presents the estimated ductility capacity (μ_c) established with the *b/t*, *KL/r* and *E/Fy* of the braces along with the peak ductility demand (μ_d) on them. Blue and green solid lines in Figure 5 depict the median ductility demands as reliable response indicators when the frames are subjected to an SDR of 2% or 4%, respectively, while red solid line shows the estimated ductility capacity for each brace section. Note that peak μ_d represents the maximum ductility demand attained by any of the two braces located in each story level at any time step during the selected time history analysis.

One can observe from Figure 5 that peak ductility demand (μ_d) distribution of the two frames over ten GMs indicate that the braces located at the first story level attain the highest ductility demand among other stories. This concentration seems to result in significant demand reduction over higher stories. For example, a noticeable variation can be seen between the first two stories, where the peak μ_d of the braces is abruptly reduced almost by half from the first to second story. Further, the peak μ_d obtained from the first stories in both Frames A and B are within 10 and 20 at SDR of 2%, which is the ductility demand range that braces in SCBFs are expected to achieve without fracture at design level earthquakes according to AISC Seismic (AISC 341 2005, 2016). It is also notable that the ductility-based fracture life (i.e., μ_c) estimation by regression analysis indicates that likelihood of fracture for the braces other than the first story braces is quite low at 2% SDR for both frames. The first story braces, however, might experience fracture even at 2% SDR, since the demand exceeded the estimated capacity under some of the ground motions. As indicated in the green line in Figure 5, the median demand on the first story braces is about 28 and 31 in Frames A and B, respectively, when the frames were pushed further until an SDR of 4%. Comparing the estimated capacity and the demand shows that the first story brace in Frame A and the first three stories in Frame B were susceptible to fracture. Considering the number of GM records adopted in this study, further analysis is required to duly quantify and evaluate square HSS' vulnerability to fracture and its impact on seismic risk.

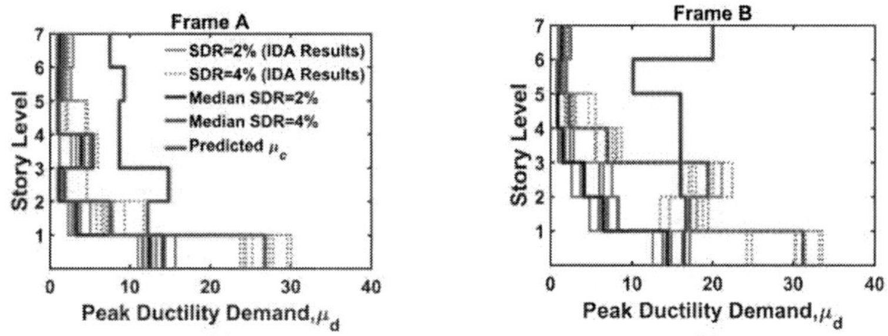

Figure 5. Peak ductility demand (μd) distribution of braces along the building height obtained from IDA of the selected ground motions at story drift ratios (SDR) of 2% and 4%: (a) Frame A, (b) Frame B.

References

AISC 341-05, *Seismic Provisions for Structural Steel Buildings,* ANSI/AISC, Chicago, IL, American Institute of Steel Construction, 2005

AISC 341-16, *Seismic Provisions for Structural Steel Buildings,* ANSI/AISC, Chicago, IL, American Institute of Steel Construction, 2016.

Goel, S. C., Cyclic Post Buckling Behavior of Steel Bracing Members, *Stability and Ductility of Steel Structures Under Cyclic Loading*, Boca Raton, FL: CRC Press, 75-104, 1992.

McKenna, F., Fenves, G. L., and Scott, M. H., Open system for earthquake engineering simulation, *University of California, Berkeley, CA*, 2000.

NEHRP, Recommended Seismic Provisions: Design Examples: FEMA P-751, 2012.

NEHRP, Recommended Seismic Provisions: Design Examples: FEMA P-1051, 2016.

Okazaki, T., Lignos, D. G., Hikino, T., and Kajiwara, K., Dynamic Response of a Chevron Concentrically Braced Frame, *Journal of Structural Engineering*, *139*(4), 515-525, 2012.

Pacific Earthquake Engineering Research Center, 325 Davis Hall, University of California, Berkeley, CA 94720, 2018.

Shen, J., Seker, O., Akbas, B., Seker, P., Momenzadeh, S., and Faytarouni, M., 2017. Seismic performance of concentrically braced frames with and without brace buckling. *Engineering Structures*, *141*, 2017.

Interdependence between Structural Engineering and Construction Management
Edited by Ozevin, D., Ataei, H., Modares, M., Gurgun, A., Yazdani, S., and Singh, A.

ISBN: 978-0-9960437-6-2

EFFECT OF SITE AMPLIFICATION ON SEISMIC SAFETY EVALUATION OF FLYOVER PIER

MOHAMMAD RAIHAN MUKHLIS[1] and MD. ABDUR RAHMAN BHUIYAN[2]

[1]*Institute of Earthquake Engineering Research, Chittagong University of Engineering and Technology, Chittagong, Bangladesh*
[2]*Dept of Civil Engineering, Chittagong University of Engineering and Technology, Chittagong, Bangladesh*

Bangladesh is a developing country in which a lot of multi-span simply/continuous supported flyovers are being constructed in its major cities. Being situated in a seismically active region, seismic safety evaluation of flyovers is essential for seismic risk reduction. Effects of site amplification on seismic safety evaluation of flyover piers are the main concern of this study. In this regard, failure mode, lateral strength and displacement ductility of piers of a typical multi-span simply supported flyover have been evaluated by Japan Road Association (JRA) recommended guidelines, with and without considering site amplification. Ultimate flexural strengths of piers have been computed using the pushover analysis results. Shear capacity of piers have been calculated using the guidelines of JRA. Lateral strengths have been determined depending on the failure modes of the piers. Displacement ductility of piers has been computed using yield and ultimate displacements of the piers obtained from the pushover analysis results. Selected earthquake time history is used in seismic safety evaluation of the flyover piers. Finally, the ductility design method is used to conduct the seismic safety evaluation of the piers with and without considering site amplification. From the numerical results, it has been revealed that the effects of site amplification on seismic safety evaluation of bridge structures should be carefully taken into account.

Keywords: Failure mode, Lateral strength, Shear wave velocity, Ground response, Displacement ductility, Damage state.

1 INTRODUCTION

Bridge, in general, is a structure that crosses over a body of water, traffic or other obstructions permitting smooth and safe passage of vehicles. Flyover, one of different bridge forms, is an elevated structure carrying highway over roads, railways and other features. So, in the subsequent discussion of this paper, the terms 'bridge' and 'flyover' are synonymously used.

It has been well known that Site Amplification (SA) occurs mainly due to local site effects. Local site effects include the characteristics of site soil controlling the shaking of the ground during earthquake (Navidi 2012). If a lightweight flexible structure is built on a very stiff rock foundation, a valid assumption is that the input motion at the base of the structure is the same as the free-field earthquake motion. If the structure is very massive and stiff, and the foundation is relatively soft, the motion at the base of the structure may be significantly different than the free-field surface motion. For designing structures it is important to consider the effect of the site amplification on the response of the structure. Several seismic codes and standards, such as, JRA (2002), have been developed to evaluate seismic safety of bridge structures. The main theme of seismic safety evaluation is that the structures shall resist earthquakes of small to moderate magnitudes without damage while for the large magnitude earthquake excitations the reparability and no collapse condition of the structures shall be ensured (Bhuiyan and Alim 2017).

Based on the above background, the study aims at evaluating the failure mode, lateral strength, displacement ductility demand and capacity to obtain the seismic safety status of a typical pier of flyover, constructed in a major city of Bangladesh, with and without considering site amplification phenomenon. In this regard, the guidelines recommended by JRA (2002) are used. The nonlinear static pushover analysis method has been adopted to obtain lateral strengths and ductility capacity of pier by considering its flexural strength, shear strength and failure mode. Finally, the seismic safety of the pier has been evaluated.

2 MODELING OF THE PIER

2.1 Physical Model

Kadamtali flyover (22.34°N; 91.82°E) has been constructed to reduce traffic jam in Chittagong city of Bangladesh. Figure 1 shows a 3D view of the flyover. The flyover is approximately 1127 m long and 8.54 m wide. It is spanning around 630 m with 22 spans of variable length from 21.3 m to 42.0 m. There are 21 piers with variable heights ranging from 4.66 m to 8.5 m. Figure 2 shows the transverse section of the typical pier of the flyover. The reinforcement with yield strength of 413 N/mm^2 and concrete with compressive strength of 30 N/mm^2 are used in the flyover. The deck of the flyover is comprised of four pre-stressed concrete girders with 200 mm reinforced concrete slab. The girders rest on elastomeric neoprene bearing over concrete bearing pad installed on top of each pier. Each elastomeric neoprene bearing with a cross-section of 500 mm × 350 mm consists of 4 numbers of 3 mm steel plates with a total thickness of 70 mm. For this study, a typical pier (Pier 7 as shown in Figure 1) of the flyover has been selected for analysis purpose. Geometric dimensions of the typical pier are presented in Table 1.

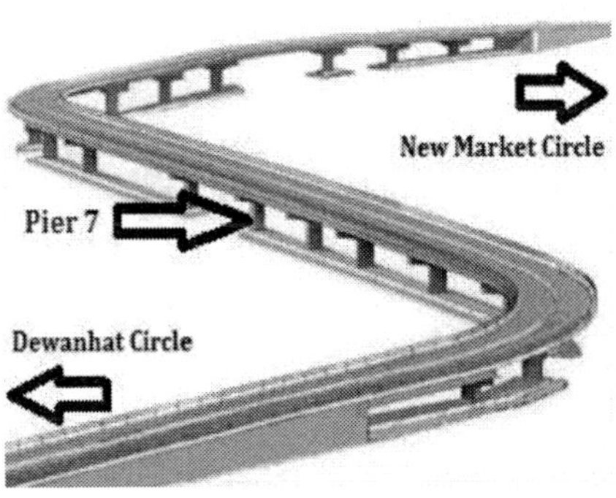

Figure 1. 3-D view of Kadamtali flyover (Mukhlis and Bhuiyan 2017).

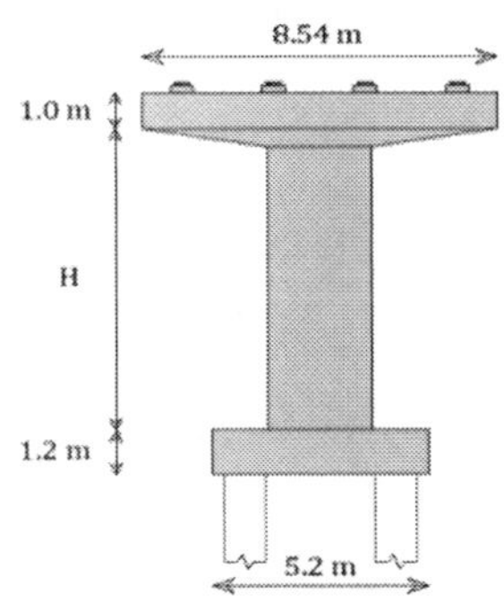

Figure 2. Transverse section of typical Pier.

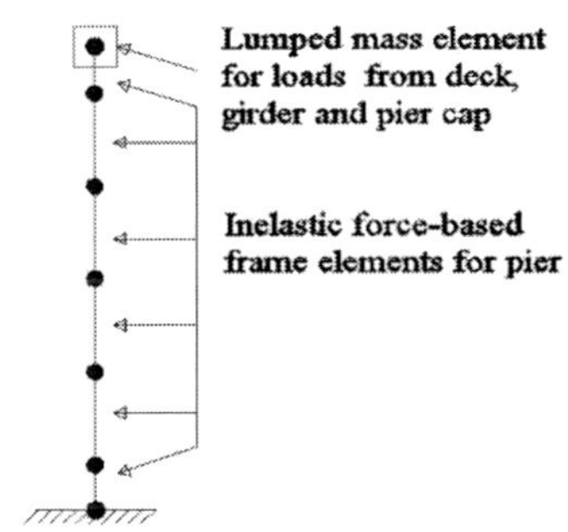

Figure 3. Analytical model of typical pier.

Table 1. Geometric dimensions of a typical pier of Kadamtali flyover.

Pier No.	Pier height, H (m)	Pier Dimension (m x m)	Longitudinal Reinforcement
7	8.27	1.2 x 2.5	66-Y25 bar

2.2 Analytical Model

The superstructure and substructure of the system are modeled as a lumped mass system divided into a number of small discrete segments using nonlinear seismic analysis software SeismoStruct 2016. The mass of each segment is assumed to be distributed between two adjacent nodes. The body of the flyover pier is modeled using fiber elements. The section of bridge pier has been modelled with original geometric dimension as force based inelastic frame element of beam-column element types where, 198 section fibres with 5 integration sections have been used for discretization. The typical piers are then subdivided into 6 elements along its height. The loads from deck, prestressed concrete girders and pier caps are calculated and modeled as lumped mass element on the pier top. The base of the pier is assumed to be fixed neglecting the foundation movement effect. Analytical model of the typical pier is shown in Figure 3.

3 LATERAL STRENGTH AND DUCTILITY CAPACITY OF THE PIER

3.1 Pushover Analysis of the Pier

The sectional analysis has been conducted by Response 2000 to obtain the moment-curvature (M-Φ) relationship of pier cross section which is used to derive the force-displacement relationship of the pier. In addition, SeismoStruct 2016 is used to conduct the pushover analysis in order to derive the force-displacement relationships of the pier along with the bilinear idealization, from which the yield displacement (δ_y), ultimate displacement (δ_u) and ultimate strength (P_u) are found to be 37 mm, 175 mm and 2454 kN respectively.

3.2 Evaluation of Failure Mode, Lateral Strength and Ductility Capacity of the Pier

Failure mode, lateral strength and ductility capacity of the typical pier is analyzed according to the procedure suggested by JRA (2002), depending on the flexural strength (P_u), shear strength (P_s) and shear strength under static loading (P_{s0}) In this study, safety factor depending on importance of bridges and the type of ground motion, α is taken as 3.0 assuming important bridge in near field region. Modification factor on the effects of alternating cyclic loading, c_c is taken as 0.6 for Type I earthquake and 1.0 for calculating P_{s0} in this study. The value of average Shear Stress of Concrete, τ_c is found to be 0.37 N/mm^2 depending on $\mathbf{f_c}'$ value, modification factor c_e is found to be 0.79 depending on effective height, d of pier section and modification factor c_{pt} is found to be 1.5 depending on tensile reinforcement ratio of the pier. Using these values, failure mode, lateral strength and ductility capacity of the typical pier are calculated as shown in Table 2.

Table 2. Failure mode, lateral strength and ductility capacity of the typical pier.

Pier No.	Ultimate Strength, P_u (kN)	Shear Strength, P_s (kN)	Shear Strength, [for c_c=1] P_{s0} (kN)	Failure Criteria	Failure Mode	Lateral Strength, Pa (kN)	Ductility Capacity, μa
7	2454	3002	3509	$P_u \leq P_s$	Flexural Failure	2454	2.26

4 SEISMIC SAFETY EVALUATION OF THE PIER

Seismic safety of bridge components are generally expressed in terms of damage conditions of those components subjected to seismic ground motions. Piers are generally the most critical components of bridge. Different forms of Engineering Demand Parameters (EDP) such as displacement ductility demand (μ_d) is generally used to measure the Damage State (DS) of the bridge piers (Bhuiyan and Alam 2012, Bhuiyan and Alim 2017). Table 3 summarizes the definitions of various damage states and their corresponding damage criteria available in the literature. Displacement ductility demand (μ_d) has been estimated from the results of time history analysis (THA) for specific ground motion history with/without considering site amplification (SA) using Eq. (1), where, d_u = maximum pier displacement from THA, d_y = yield displacement of pier.

$$\mu_d = \frac{d_u}{d_y} \quad (1)$$

Table 3. Damage states of bridge pier.

Bridge component	EDP (Displacement Ductility, μ_d)	Damage states (DS)	Physical Phenomenon
Pier	$\mu_d < 1.0$	No Damage	No Physical phenomenon
	$1.0 < \mu_d < 1.2$	Slight (DS=1)	Cracking and spalling
	$1.2 < \mu_d < 1.76$	Moderate (DS=2)	Moderate cracking and spalling
	$1.76 < \mu_d < 4.76$	Extensive (DS=3)	Degradation without collapse
	$\mu_d > 4.76$	Collapse (DS=4)	Failure leading to collapse
Reference	Hwang *et al.* (2001)	FEMA (2003)	FEMA (2003)

4.1 THA without Considering Site Amplification (SA)

For simplicity in the analysis and unavailability of sufficient seismic data, site specific ground motion data was not developed. Time history analysis is conducted by applying 6.7 magnitude Northridge 1994 earthquake ground motion records with a recorded PGA of 0.45g at 4.06 s as shown in Figure 4 at the fixed base of pier. Rayleigh damping has been chosen for the pier under consideration with 5% damping ratio using two fundamental modes in transverse and longitudinal direction with a period of 0.525 s and 0.252 s respectively found from the eigenvalue analysis. The displacement response at the top of pier is represented with respect to time as shown in Figure 5 with a maximum displacement of 154 mm at 7.28 s without considering SA.

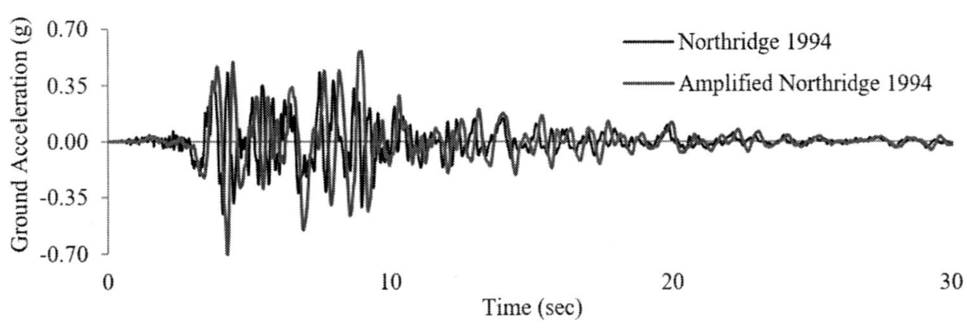

Figure 4. Ground motion accelerations for THA.

4.2 THA Considering Site Amplification (SA)

In this study, one dimensional site response analysis software program "Strata" (SHAKE-type) has been used for the modeling of soil profile. Shear wave velocity (V_s) is estimated using empirical formula proposed by Hossain and Ansary (2015) as shown in Eq. (2) for different soil layer up to 15.5 m with a maximum value of 399 m/s at that depth. The shear wave velocity of half-space bedrock is taken as 800 m/s according to BNBC Final Draft (2015).

$$V_s=169\ N^{0.2638}D^{0.2396} \tag{2}$$

Where, V_s = shear wave velocity (ft/s); D = depth (ft); N = corrected SPT value. Ground response of the borehole site are evaluated by reproducing the ground motion time history at the ground surface level and computing the acceleration transfer function, which has been found to be 2.78, by applying original Northridge 1994 earthquake ground motions at the bedrock layer with equivalent linear analysis. The amplified Northridge 1994 earthquake at the ground surface level of pier is shown in Figure 4 indicating PGA of 0.70g at 4.22 s. Time history analysis is then conducted by applying amplified Northridge 1994 earthquake ground motion records at the fixed base of pier. The displacement response at the top of pier is represented with respect to time as shown in Figure 5 with a maximum displacement of 268 mm at 4.14 s considering SA.

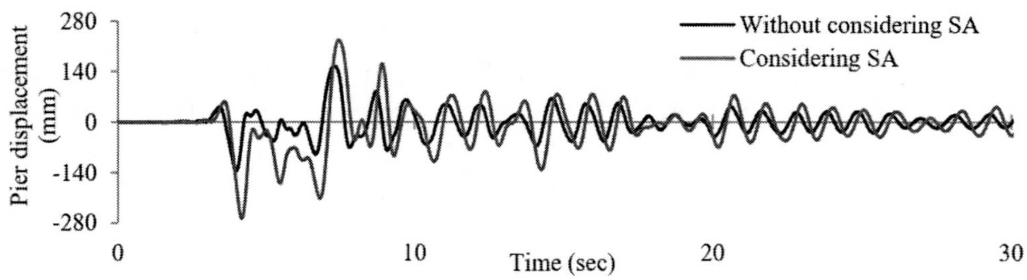

Figure 5. Displacement response at pier top.

4.3 Damage State Evaluation of Typical Pier

Damage state of bridge pier subjected to the selected ground motion intensity is shown in Table 4. Maximum pier displacement and displacement ductility demand of pier for different modeling considerations are shown in Figure 6 and Figure 7 respectively.

Table 4. Damage state evaluation of pier.

Modeling Considerations of Bridge Pier	Earthquake Name	PGA (g)	d_u (mm)	d_y (mm)	$\mu_d = \frac{d_u}{d_y}$	Damage State	Physical Damage Condition
Without Considering SA	Northridge-1994	0.45	154	37	4.16	DS 3	Extensive
Considering SA	Northridge-1994	0.70	268	37	7.24	DS 4	Collapse

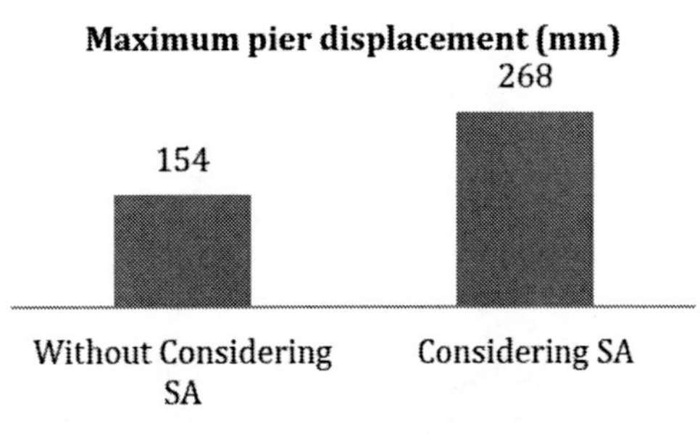

Figure 6. Maximum displacement of pier (mm).

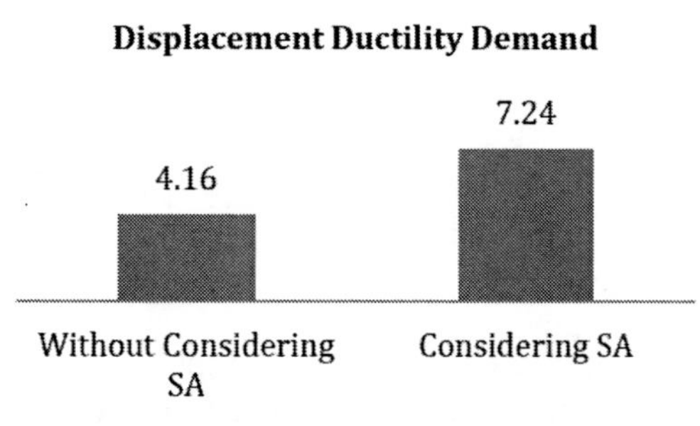

Figure 7. Displacement ductility demand of pier.

It has been observed that maximum pier displacement (d_u) and displacement ductility demand (μ_d) increases 74% if site amplification is considered in the model. Consequently, the physical damage state is changed from extensive damage of the pier to collapse of pier leading to failure in response to the applied ground motion of Northridge 1994 earthquake for considering SA in the analysis procedure.

5 CONCLUSIONS

Effects of site amplification (SA) on seismic safety evaluation of flyover piers have been analytically evaluated. A typical pier has been selected for the detailed evaluation of displacement ductility capacity and displacement ductility demand corresponding to considerations of SA. In both cases, the pier displacement ductility demand of 4.16 and 7.24 corresponding to a ground motion intensity of 0.45g exceeds the pier displacement ductility capacity of 2.26, leading to the unsafe condition. Seismic safety of pier is also evaluated based on the damage state using the displacement ductility demand of pier with and without considering SA. Displacement ductility demand of pier is found to be extreme with a 74% increased value when SA is included in the pier model. Moreover, the physical damage state is drastically changed from extensive damage to collapse of pier leading to ultimate failure. Hence, from the numerical results, it has been revealed that site amplification has great impact on seismic response analysis and safety evaluation of bridge piers and should be carefully considered in the modeling and analysis procedure.

References

Bhuiyan, M. A. R., and Alam, M. S., Seismic Vulnerability Assessment of a Multi-Span Continuous Highway Bridge Fitted with Shape Memory Alloy Bars and Laminated Rubber Bearings, *Earthquake Spectra*, Earthquake Engineering Research Institute, 28(4), 1379-1404, November, 2012.

Bhuiyan, M. A. R., and Alim, H., Seismic Safety Evaluation of Abdul Mannan Overpass in Chittagong, Bangladesh, *Malaysian Journal of Civil Engineering*, Universiti Teknologi Malaysia, 29(3), 250-272, 2017.

BNBC Final Draft, *Bangladesh National Building Code,* Housing and Building Research Institute, Dhaka, Bangladesh, 2015.

FEMA, *HAZUS-MH MR1: Technical Manual, Earthquake Model,* Federal Emergency Management Agency, WA, USA, 2003.

Hossain, A. S. F. M., and Ansary, M. A., Shear Wave Velocity as a Function of Standard Penetration Number and Depth in Dhaka City, Bangladesh, *IOSR Journal of Mechanical and Civil Engineering*, 12(3), 31-39, June, 2015.

Hwang, H., Liu, J. B., and Chiu, Y., *Seismic Fragility Analysis of Highway Bridges,* Technical Report of MAEC RR-4 Project, Mid-America Earthquake Center, USA, 2001.

JRA, *Specifications for Highway Bridges - Part V: Seismic Design*, Japan Road Association, Tokyo, Japan, 2002.

Mukhlis, M. R., and Bhuiyan, M. A. R., Lateral Strength and Safety Evaluation of Piers of Kadamtali Flyover in Chittagong, Bangladesh, *International Journal of Advanced Structures and Geotechnical Engineering*, BRCORP, 6(2), 45-56, April, 2017.

Navidi, S., Development of Site Amplification Model for Use in Ground Motion Prediction Equations, Ph.D. Dissertation, University of Texas at Austin, Texas, USA, 2012.

Interdependence between Structural Engineering and Construction Management
Edited by Ozevin, D., Ataei, H., Modares, M., Gurgun, A., Yazdani, S., and Singh, A.

ISBN: 978-0-9960437-6-2

COMPARATIVE STUDY ON THE BENDING PROPERTIES FOR TWO TYPES OF COMPOSITE BEAMS

XIAN LIANG, SHUJIN DUAN, RUNMING NIU, and FANG LI

Research Institute of Structural Engineering, Shijiazhuang Tiedao University, Shijiazhuang, China

The difference between the steel and concrete composite-laminated action beam and the double steel and concrete composite action beam is in the connection of the top concrete slab and the steel top flange. The bending properties about the composite-laminated beam and the double composite beam in the negative flexural region are investigated in this paper. The relation of the cross section bending moment-curvature for the two types of composite beams under the action of negative bending moment are drawn by the whole process analysis method, and then the corresponding cracking moment, elastic ultimate bending moment and plastic ultimate bending moment are obtained. The analysis results have a good agreement with test data. Although the sectional bending stiffness and bending-carrying capacity of the composite-laminated beam and the double composite beam are comparable in the elastic state, the crack resistance of the composite-laminated beam is much better than that of the double composite beam.

Keywords: Composite-laminated beam (CLB), Double composite beam (DCB), Bending moment-curvature relation, Crack resistance.

1 INTRODUCTION

The double steel and concrete composite action beam (referred to as double composite beam, DCB) is a kind of advantageous structure used in the negative flexural region, it is proposed by Reiner (1996). In the DCB, attaching additional concrete slab to steel bottom flange can improve local buckling strength and increase the sectional stiffness (Duan *et al.* 2010). But a crack may appear in advance because of height descending of sectional neutral axis, which then influences bridge durability.

The experimental results conducted by Nie *et al.* (1991) have shown that the cracking of the top concrete slab could be relieved by reducing shear connectors. Apart from that, changing the arrangement of the shear connectors, i.e., no connection is set in the certain range, the cracking bending moment of the top concrete slab can be improved, and the maximum width of cracks can be reduced (Liu and Chang 2008). Based on the facts, a new type of structure, steel-concrete composite-laminated action beam (referred to as composite-laminated beam, CLB) is proposed by Duan *et al.* (2016). In the CLB, the top concrete slab and the steel top flange are connected by the uplift-restricted and slip-permitted connectors (referred as uplift-restricted connector as follows) (Nie *et al.* 2015), meanwhile the bottom concrete slab and the steel bottom flange are connected by shear connectors. With the uplift-restricted connectors, a laminated interface is formed between the top concrete slab and the steel top flange, which allows the top concrete slab slides freely in horizontal direction (Wang *et al.* 2018), meanwhile a composite interface is formed between the bottom concrete slab and the steel bottom flange in the action of shear connectors. Not only in the CLB but also in the DCB, the structure composited by the bottom concrete slab and the steel profile is referred as bottom composite beam.

In the negative flexural region, applying the uplift-restricted connectors, the tensile stress of the top concrete slab is effectively released; meanwhile the top concrete slab and bottom composite beam in the CLB have the same bending curvature, which is not equivalent to the laminated beam in the general sense. Beyond that, the CLB is different from the classical laminated beam because it is no longer consistent with the assumption of material continuity, result from the cracking of the top concrete slab.

The main cross-sectional characteristics of the DCB and the CLB are shown in Figure 1(a) and (b) respectively. The bending moment-curvature relations (referred to M-Ø curves) in the negative flexural region of the two types of beams are obtained by the whole process analysis. Combined with the test results conducted by Wang (2018, in press), the bending properties including cracking moment, sectional bending stiffness and bending-carrying capacity of the two types of composite beams are discussed.

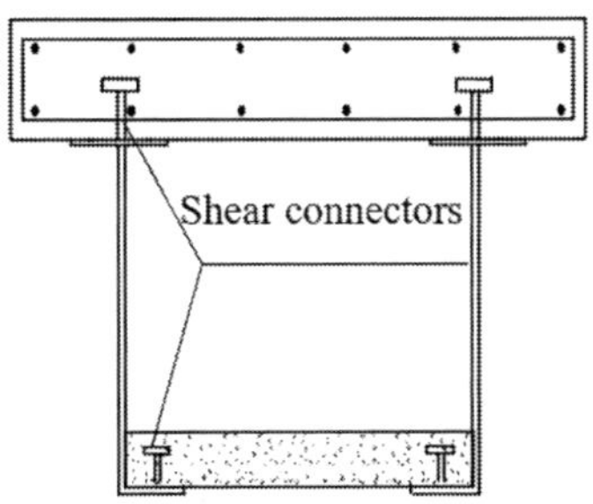

(a) DCB

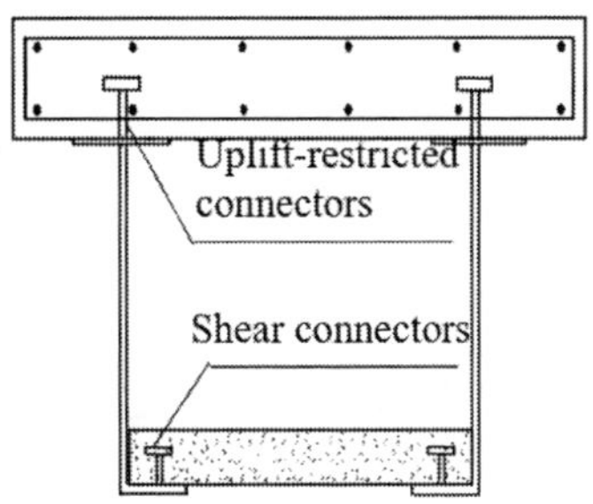

(b) CLB

Figure 1. Cross Section

2 APPROACH OF M-Ø CURVE BY WHOLE PROCESS ANALYSIS

The basic assumptions are: (1) The member strain distribution remains linear from the beginning of stress to the member failure. (2) The tensile strength of concrete is ignored. (3) Does not consider the composite interface slip. (4) The cross section is a compact one, that means the steel profile wouldn't lose the stability under pressure before the cross-sectional strength is reached, and the connectors are not damaged.

The sectional bending stiffness and strength can be reflected from the M-Ø curves, which is the basis to study the bending performance of cross sections. The M-Ø curves will be deduced by the whole process analysis method. At first, the cross section is divided into numerous strips and the stress on each strip is uniform. Then the internal force and moment of the cross section are calculated according to the deformation conditions and the stress-strain relations of materials. At last, based on the equilibrium equations, the M-Ø curves are obtained.

Take the position of sectional neutral axis as the coordinate origin in the DCB. The bending deformations of the top concrete slab and the bottom composite beam are independent when load acts on the CLB, i.e. there are two neutral axes in the CLB. Take the top concrete slab neutral axis as itself coordinate origin, at the same time, take the bottom composite beam neutral axis as itself coordinate origin. In this case, the sectional stress distribution can be schematically depicted as Figure 2(a) and (b).

2.1 Stress-Strain Relations of Materials

The stress-strain relations of materials are selected from the Code for Design of Concrete Structures (GB50010 2010).

(1) The nonlinear stress - strain relation of concrete in compressive zone is shown in Figure 3 (a) and in Eq. (1).

$$\sigma_c = \begin{cases} f_c\left[1-\left(1-\dfrac{\varepsilon_c}{\varepsilon_0}\right)^2\right](\varepsilon_c \le \varepsilon_0) \\ f_c(\varepsilon_0 \le \varepsilon_c \le \varepsilon_{cu}) \end{cases} \tag{1}$$

where σ_c is compressive resistance of concrete; f_c is concrete strength of axial compressive; ε_0 is the concrete compressive strain when σ_c=f_c; ε_{cu} is ultimate compressive strain of concrete, ε_{cu}≤0.033;

(2) The stress-strain relation of steel is shown in Figure 3(b), where, E_s is elasticity modulus of steel; ε_y is yield strain of steel, ε_y=f_y/E_s, f_y is yield strength; $\varepsilon_{s,u}$ is ultimate strain of steel, f_u is ultimate strength of steel, and E_s'=0.01E_s.

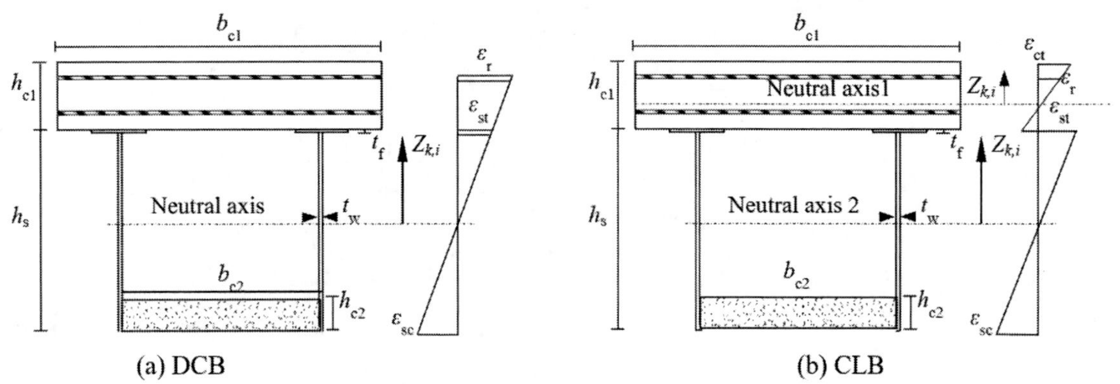

Figure 2. Analysis diagrams.

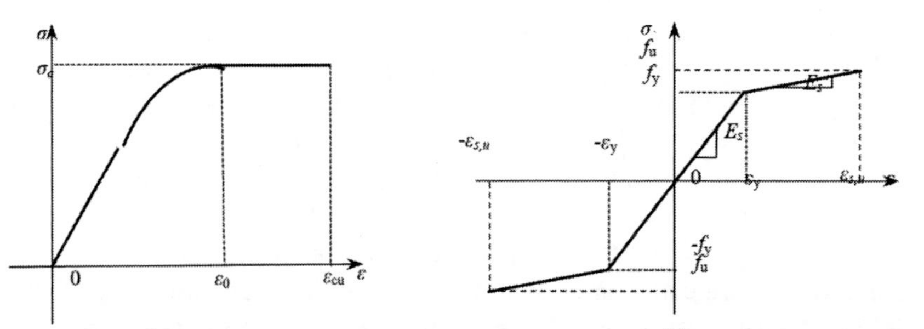

Figure 3. Stress – strain relations.

2.2 The Internal Force and Moment

In this and next sections, we will only take the DCB as the analysis object. At first, divide the cross section into different zones according to materials and section shape, such as steel reinforcement zone, steel profile zone, concrete zone, and so on. Then the different zones are divided into numerous strips. With the plane cross-section assumption and a given φ, the section strain $\varepsilon_{k,i}$ in i strip of k zone can be represented as in Eq. (2):

$$\varepsilon_{k,i} = z_{k,i}\varphi \tag{2}$$

where $z_{k,i}$ is the distance from the i strip of k zone to the neutral axis of cross section. The internal force in the i strip of k zone $N_{k,i}$, as follows in Eq. (3):

$$N_{k,i} = \sigma_{k,i}\Delta A_{k,i} \tag{3}$$

where $\sigma_{k,i}$ is the strain in the i strip of k zone, $\sigma_{k,i}=\varepsilon_{k,i}E_k$, E_k is the elasticity modulus of k zone; $\Delta A_{k,i}$ is the area of this strip. The bending moment in the i strip of k zone $M_{k,i}$, as fin Eq. (4):

$$M_{k,i} = N_{k,i}z_{k,i} \tag{4}$$

The total internal force and moment of the cross section can be written in the form in Eq. (1):

$$M = \sum_{k=1}^{m}\sum_{i=1}^{n} M_{k,i} \tag{5}$$

At last, the equilibrium equations are given according to the mechanical characteristics of composite beam in the negative bending zone as follows in Eq. (6):

$$\sum N_i = 0, \sum M_i = M \tag{6}$$

In general, it is defined as the cross-sectional ultimate strength of the DCB, when the strain of steel reinforcement or steel profile reaches the ultimate strain of the material. Define M_{u1} as the cross sectional bending moment when the strain of steel reinforcement reaches ultimate tensile strain $\varepsilon_{s,u}$, and define M_{u2} as the cross sectional bending moment when the strain of steel profile reaches ultimate compressive strain $\varepsilon_{s,u}$. Then the cross-sectional ultimate strength of the DCB takes the smaller one of M_{u1} and M_{u2}.

In the CLB, the top concrete slab and the bottom composite beam act independently under the action of negative bending moment except the consistence of vertical deformation. Define M_{u1}' as the cross-sectional bending moment when the bending-carrying capacity of the top concrete slab is reached and define M_{u2}' as the cross-sectional bending moment when the bending-carrying capacity of the bottom composite beam is reached. The cross-sectional ultimate strength of the CLB is the smaller one of M_{u1}' and M_{u2}'.

2.3 M-Ø Curves

The M-Ø curves could be obtained by the equations (2)-(6), the specific steps are shown as below:

(1) $\varphi_j=\varphi_{j-1}+\Delta\varphi$, where $\Delta\varphi$ could be set and the $\varphi_1=0$.

(2) Using the formula of force equilibrium equations and internal force formulas, the location of neutral axis can be obtained.

(3) The strain of every stripe can be obtained by the location of neutral axis and φ_j.
(4) Determine whether φ_j has achieved the ultimate curvature of the section, if have not, repeat steps (1)-(3). After numerous cycles the M-Ø curves can be obtained.

3 NUMERICAL EXAMPLE AND DISCUSSION

Cracking bending moment, cross sectional bending stiffness and bending-carrying capacity of the DCB and the CLB are analyzed and compared in this part. Through the whole process analysis, the M-Ø curves of two types of composite beams are obtained, combined with test results (Wang 2018, to appear), the crack resistance and bending-carrying capacity of two types of beams are discussed.

The cross sections dimensional size of test specimens (one DCB and two CLBs) are shown in Figure 4. The main characteristics of the DCB and the CLB are almost same except the difference of the connectors in the interface between top concrete slab and steel top flange, the layout of the connector is shown in Figure 1. Material properties of test specimens are summarized in Table 1.

Table 1. Material parameters.

	Steel profile	Reinforcement		Concrete
Area	9504mm^2	904mm^2	E_c	3.3×10^4 MPa
E_s	2×10^5 MPa	2.06×10^5	f_c	28MPa
f_y	378MPa	455MPa	ε_0	0.00196
ε_y	17×10^{-6}	22×10^{-6}	ε_u	0.0033

Under the action of negative bending moment, the strain of the bottom concrete slab first reaches ultimate compressive strain of concrete, soon the strain of the steel bottom flange reaches ultimate compressive strain. At this time, the ultimate strength of cross section is reached. The M-Ø curves can be derived from φ_j and the corresponding M. The M-Ø curves including analysis results and test results (represented by CLB(E) and DCB(E)) are given in Figure 5. The results of CLB(E) are the average values of the two test CLBs.

According to Figure 5, we can find that the analysis results have a good agreement with the tests results (Wang 2018, to appear). The specific values of the cracking bending moment, the yield bending moment and the ultimate bending moment are given in Table 2.

Table 2. Yield bending moment and ultimate bending moment (kN·m).

	Cracking bending moment		Yield bending moment		Ultimate bending moment	
	DCB	**CLB**	**DCB**	**CLB**	**DCB**	**CLB**
Theoretical analysis values	29.6	54.0	242.6	226.3	297.1	251.1
Experimental values	26.8	52.6	231.4	216.8	295.7	256.1

From Table 2, the yield bending moment of the CLB and the DCB is almost equivalent, and the service state of the bridge in engineering is generally within this range. Although the ultimate bending moment of the CLB is smaller than that of the DCB, it is much higher than that of the classical composite beams.

The cracking bending moment of the CLB is about two times that of the DCB. The top concrete slab in the DCB is almost in the axial tension state, the axial tensile cracks are generated, however the flexural cracks are generated in the CLB. So that, the distance between two cracks and the maximum cracks width in the CLB is much smaller than that in the DCB.

4 CONCLUSION

By analyzing and comparing bending properties of the CLB and DCB, it can be concluded that:

(1) The analyzed results have a good agreement with the test results. The cross-sectional bending stiffness and strength of the CLB is almost equivalent to the ones of DCB in the elastic state.

(2) The CLB has a good crack resistance, that the cracking moment of CLB is about two times that of DCB.

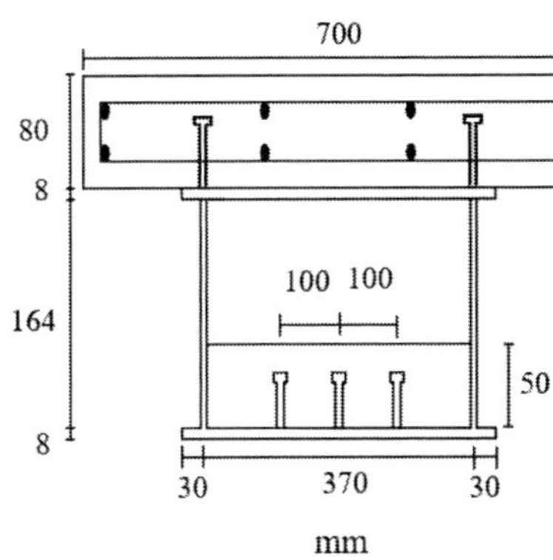

Figure 4. Cross section of test beams.

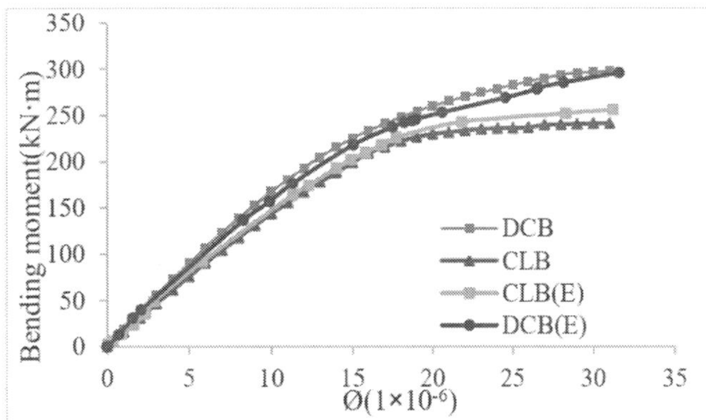

Figure 5. M-Ø curves.

Acknowledgments

The work is supported by key project of Natural Science Foundation of Hebei Education Department, China (Contact No. ZD2018025). and the Shijiazhuang Tiedao University's Graduate Innovation Fund Project, China (No. YC2018001).

References

Duan, S. J., Huo, J. H., and An, R. M., Research on the Ultimate Bearing Capacity of Double Steel-concrete Composite Continuous Beam (in Chinese), *J. of The China Rlwy. Soc.*, 32(05), 82-87, 2010.

Duan, S. J., Niu, R. M., and An, R. M., Steel-Concrete Composite-Laminated Beam, *Chinese patent*; ZL 2016 2 1090805. 0,2017.04.19, 2016.

GB50010, Code for Design of Concrete Structures, *China Arch. and Building Press,* Beijing, China, 2010.

Liu, W. H., and Chang, D. B., Experiment Study on the Crack Resistance of Composite Beams (in Chinese), *Journal of Jilin Architectural and Civil Engineering*, (03), 1-3, 2008.

Nie, J. G., and Wei, J., The Actual Performance of Shear Connector in Steel - Concrete Composite Beam (in Chinese), *Journal of Zhengzhou Institute of Technology*, (04), 43-47.1991.

Nie, J. G., Tao, M. X., and Nie, X., *et al.*, New Technique and Application of Uplift- Restricted and Slip-Permitted Connection (in Chinese), *China Civil Engineering Journal*, (04), 7-14+58, 2015.

Reiner, S., Bridges with Double Composite Action, *Structural Engineering International*, 1, 32-36,1996.

Wang, Y. Y., Duan, S. J., and Niu, R. M., Analysis of Bending Stiffness and Contact Slip of Double Steel-concrete Combination and Lamination Beam (in Chinese), *Engrg. Mechanics*, (S1), 265-269, 2018.

Wang, Y. Y., Experimental Study on the Static Behavior of Steel-Concrete Composite and Lamination Beam (in Chinese), *Shijiazhuang Tiedao University*, 2018 (in press).

Interdependence between Structural Engineering and Construction Management
Edited by Ozevin, D., Ataei, H., Modares, M., Gurgun, A., Yazdani, S., and Singh, A.

ISBN: 978-0-9960437-6-2

EXPERIMENTAL STUDY ON CONTINUOUS BEAM TYPE SQUARE CFST BEAM-TO-COLUMN CONNECTION

TAKASHI FUJINAGA[1] and CHARLES G. CLIFTON[2]

[1]*Research Center for Urban Safety and Security, Kobe University, Kobe, Japan*
[2]*Dept of Civil and Environ. Engineering, University of Auckland, Auckland, New Zealand*

Concrete-filled steel tubular (CFST) columns are used widely as building structure components because of their good performance. It can be complex, however, to connect beams into CFST columns, especially when rigid beam-to-column connections are required. In order to simplify the fabrication requirements and cost of these connections, a continuous beam type connection through the column is proposed. This can develop the full plastic moment capacity of the beam. In this paper, the experiment testing results of two types of CFST beam-to-column connection specimens are presented and the connection performance and load transfer mechanisms are examined. The continuous beam type connection performance is compared to that of traditionally used through-diaphragm type of connection, which is more complex and demanding to fabricate. Both specimens showed good seismic performance, exhibiting stable hysteretic behavior with high energy absorption. The maximum experimental strength exceeded the calculated value. For a continuous beam type specimen, there is the potential for the column flange to be affected by the local deformation from the steel beam under bending induced beam tension. However, such effects were not observed in practice until late in the test after extensive beam plastic action had occurred.

Keywords: Concrete-filled steel tube, Connection performance, Energy absorption, Fillet welding, Full plastic moment capacity, Ultimate strength.

1 INTRODUCTION

Concrete-filled steel tubular (CFST) columns are used widely as building structure components because of their good structural performance under earthquake, fire and gravity loading and their ease of construction. Nevertheless, connecting beams into CFST columns can be especially complex when rigid beam-to-column connections are required. CFST beam-to-column connections are generally designed as thorough-diaphragm type, outer-diaphragm type, and inner-diaphragm type. Design formulas for these three types of beam-to-column connection are given in AIJ CFT Recommendations (Architectural Institute of Japan 2008) and in similar recommendations for New Zealand.

To simplify the fabrication requirements and costs of these connections, a continuous beam type connection has been proposed for CFST beam-to-column connections that, uses I-section beams passing through CFST columns (Azizinamini and Prakash 1993). This continuous beam type connection can easily develop the full plastic moment capacity of the steel beam and considerably simplify the design of CFST beam-to-column connection. For circular CFST columns, load transfer mechanism and design formulas are discussed, but few examples exist for square columns for CFST beam-to-column connections (Azizinamini and Schneider 2004, Schneider and Alostaz 1998).

As described herein, the experiment testing results of two types of square CFST beam-to-column connection specimens are presented and the connection performance and load transfer mechanisms are examined.

2 METHODOLOGY

2.1 Specimens

The dimensions of the specimens are presented in Figure 1. Specimens are cruciform shape and simulating the square CFST beam-to-column connection. Two specimens of 1) Continuous beam type specimen and traditionally used 2) Through-diaphragm type specimen have been planned and designed so that beam yields prior to other parts (See Table 1).

Continuous beam type specimen required an I-section beam slot to be cut in column steel tube by plasma within 1mm tolerance. After putting the I-section beam through the slot, the I-section beam and outer faces of the column were fillet-welded all round.

A through-diaphragm type specimen has been designed according to AIJ CFT Recommendations. Connections of this type are more complex to design and demanding to fabricate.

The standard tensile and compressive tests were conducted for steel and concrete to ascertain the mechanical properties of the materials used. The measured results are presented in Tables 2 and 3. The relation between the concrete material age and strength are portrayed in Figure 3.

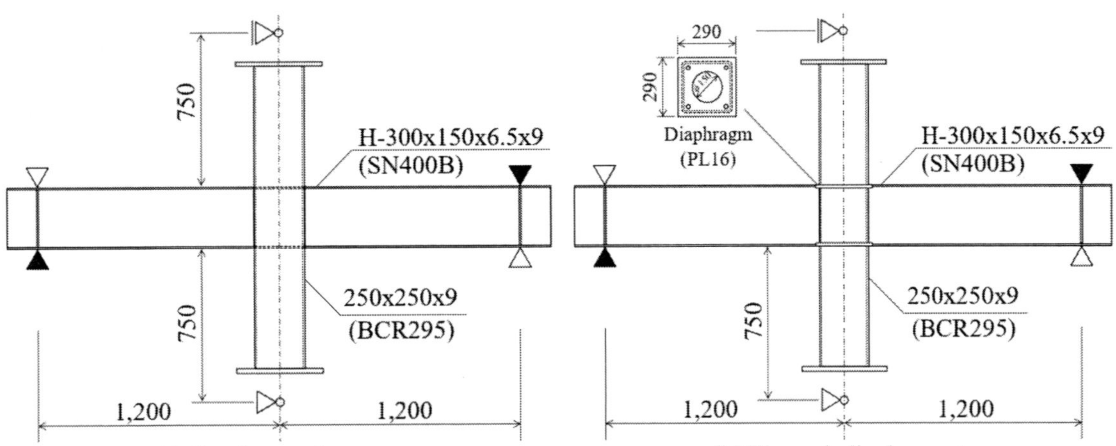

(a) Continuous beam type. (b) Through diaphragm type.

Figure 1. Dimension of specimens.

Table 1. Dimension of specimens.

Specimen	Column	Beam	Diaphragm	Strength of Concrete (MPa)
Continuous Beam Type	250x250x9 (BCR295)	H-300x150x6.5x9 (SN400B)	-	87.1
Through Diaphragm Type			PL16 (SN400B)	84.7

Table 2. Mechanical properties of steel.

		${}_sE$ (GPa)	f_y (MPa)	f_u (MPa)	Elongation (%)
Beam	I-section flange	203	328	474	29.4
	I-section web	203	350	472	23.6
Column	Steel Pipe	188	343	413	27.2
Diaphragm		208	277	437	30.7

${}_sE$: Young's modulus of steel, f_y: yield stress of steel, f_u: tensile strength of steel

Table 3. Material age and strength of concrete.

Age (days)	${}_cE$ (GPa)	f_c' (MPa)	ε_0	f_t (MPa)
7	-	70.7	-	-
14	-	71.3	-	-
21	-	74.6	-	-
28	38.5	83.1	0.00278	-
53	40.5	84.7	0.00276	4.92
74	41.0	89.4	0.00259	4.97
97	39.3	87.1	0.00275	5.18

${}_cE$: Young's modulus of concrete, f_c': compressive strength of concrete, ε_0: strain at compressive strength, f_t: split tensile strength

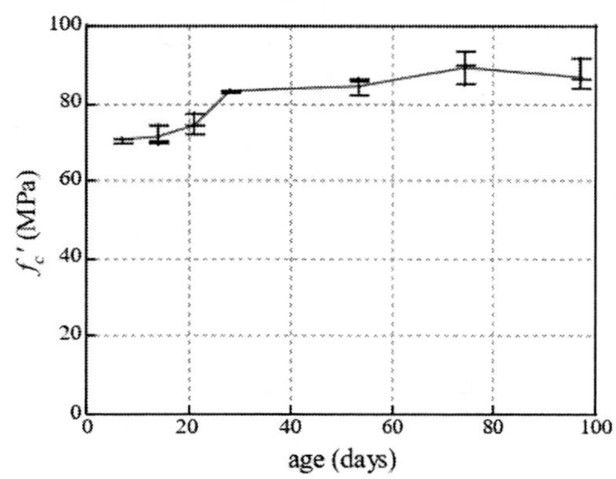

Figure 2. Change of concrete strength.

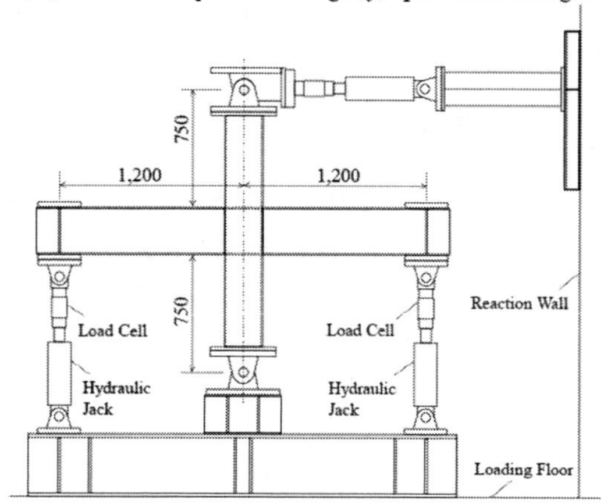

Figure 3. Experimental Setup.

2.2 Loading Method

The top end of the specimens was supported by hinge and roller, the bottom end was supported by hinge. Shear force was added by hydraulic jacks set at both beam ends (See Figure 3). Stiffening devices for the displacement to the out-of-plane direction were not set. Loading was controlled by rotational angle and the targeted rotation angles were 0.005rad, 0.01rad, 0.02rad, and 0.03rad. Rotation angle R is calculated from the relative vertical displacement of both beam ends and the distance between loading points.

3 EXPERIMENTAL RESULTS

3.1 Shear Force and Rotation Angle Relation

Figure 4 presents the measured average shear force - rotation angle relationships. The measured behavior is shown as red solid lines. Black circles show the point at which the beam flange began yielding. Arrows indicate points at which local buckling was observed at the I-section beam flange. Green and blue lines show shear forces calculated using yield flexural strength and ultimate flexural strength. Table 4 presents experimental data and calculated data.

Both specimens showed good seismic performance exhibiting stable hysteretic behavior with high energy-absorbing capacity. The maximum experimental strength was observed at 0.03rad in each specimen. Regarding the continuous beam type specimen, the experimental value exceeded a calculated value by 10%. For the through diaphragm specimen, it exceeded the calculated value by 40%.

Table 4. Test results.

Specimen		Yield Strength (kN)			Ultimate Strength (kN)			Initial Rigidity (kN/rad)		
		Cal	Exp	Exp/Cal	Cal	Exp	Exp/Cal	Cal	Exp	Exp/Cal
Continuous Beam Type	+	147	114	0.78	165	177	1.07	17.3	12.6	0.73
	-		-105	-0.71		-189	-1.14			
Through Diaphragm Type	+		81	0.55		223	1.35	17.4	16.5	0.95
	-		-95	-0.65		-231	-1.39			

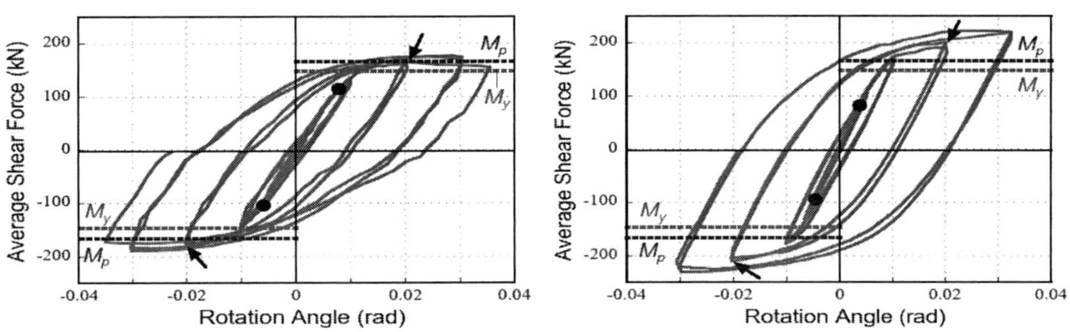

(a) Continuous beam type. (b) Through diaphragm type.

Figure 4. Shear force – rotation angle relation.

3.2 Strain Hysteresis

Figure 5 presents the strain distributions at the column flange above and below the steel beam. They were put in the position of 25mm away from the surface of the steel beam flange. For the continuous beam type specimen, the strain reached the yield strain at 0.005rad rotation angle. It rose thereafter. Parts for which the strain rose were only those just above and below the beam. The strain at the edge remained a small value. However, the strain value of the through diaphragm type specimen remained small. This tendency is the same as the range of minus rotation angle. Therefore, the column flanges of the continuous beam type specimen might be affected by the local deformation from the steel beam under bending induced beam tension.

However, such effects were not observed in practice until late in the test after extensive beam plastic action had occurred. Local deformation of the column flange above or below the I-section beam becomes large in a tensile region but remains a small value in a compression region, which indicates that the filled concrete contributes greatly to the transmission of the compressive load at the connection panel without a diaphragm plate.

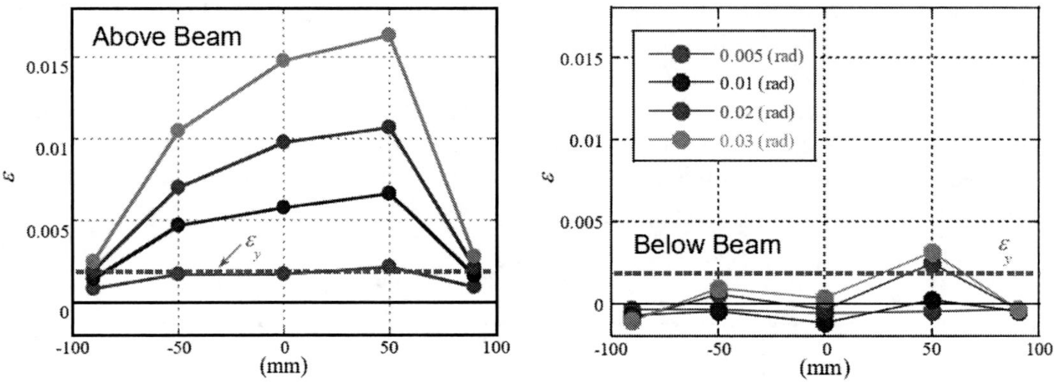

(a) Continuous beam type.

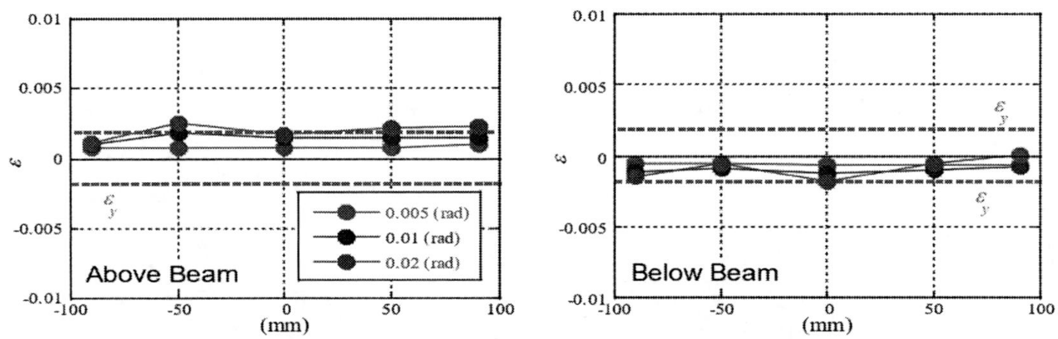

(b) Through diaphragm type.

Figure 5. Strain distribution at steel tube flange.

3.3 Shear Deformation of Panel Zone

Figure 6 shows the relation between the bending moment from the steel beam, ${}_bM$, and the panel shear deformation ${}_p\gamma$. The bending moment from the I-section steel beam was calculated using the strain gauge put in the position of 150mm away from the column surface. The shear deformation of the panel was calculated from the rosette strain gauge value put at the center of the panel web. Shear deformation of the panel of the through diaphragm type specimen was greater than that of the continuous beam type specimen. The rosette gauge value is only the local strain value. Therefore, it is considered that the overall panel deformation could not be measured in the shear deformation of the continuous beam.

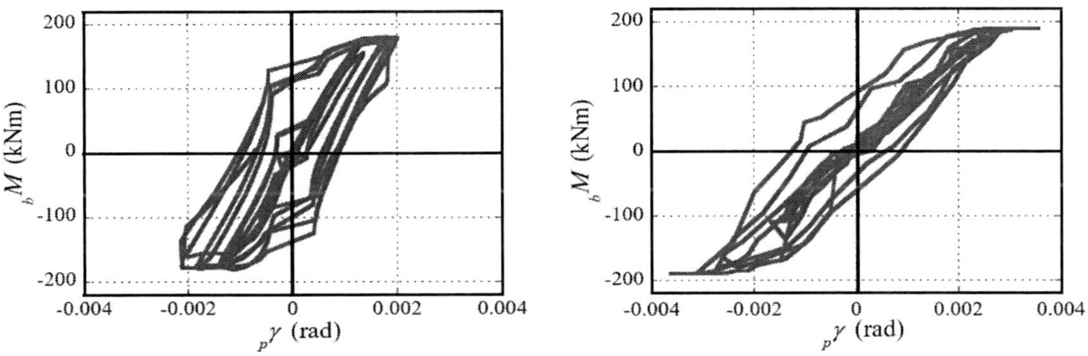

(a) Continuous beam type. (b) Through diaphragm type.

Figure 6. $_bM - \gamma$ relation.

4 CONCLUDING REMARKS

1) Both specimens showed good performance with stable behavior and high energy-absorbing capacity. The maximum experimental strength exceeded the calculated value.

2) For the continuous beam type specimen, the column flange might be affected easily by local deformation from the steel beam under tensile conditions. However, local buckling of the steel column flange was not observed until the end of loading. Filled concrete contributes to transmission of the compressive load at the connection panel.

Acknowledgments

Funding from the Construction Engineering Research Institute is gratefully acknowledged. We would like to thank Ms. Mizuho Murata for the help with experiments.

References

Architectural Institute of Japan, *Recommendations for Design and Construction of Concrete Filled Steel Tubular Structures*, Oct, 2008.

Azizinamini, A., and Prakash, B., A Tentative Design Guideline for a New Steel Beam Connection Detail to Composite Tube Columns, *Engineering Journal*, AISC, 30, 108-115, 1993.

Azizinamini, A., and Schneider, S. P., Moment Connections to Circular Concrete-Filled Steel Tube Columns, *Journal of Structural Engineering*, 130(2), 213-222, 2004.

Schneider, S. P., and Alostaz, Y. M., Experimental Behavior of Connections to Concrete-Filled Steel Tubes, *Journal of Construction Steel Research*, 45(3), 321-352, 1998.

Interdependence between Structural Engineering and Construction Management
Edited by Ozevin, D., Ataei, H., Modares, M., Gurgun, A., Yazdani, S., and Singh, A.

ISBN: 978-0-9960437-6-2

SEMI-COMPOSITE STRUCTURES FOR SEWER RENOVATION

MASAAKI NAKANO[1], ZIHAI SHI[1], YUKARI NAKAMURA[1], YUKITOSHI IWASA[2], and KIMITAKA UJI[3]

[1]*Research and Development Centre, Nippon Koei Co., Ltd., Ibaraki, Japan*
[2]*Tokyo Sewerage Service Co., Ltd., Tokyo, Japan*
[3]*Dept of Civil and Environmental Engineering, Tokyo Metropolitan University, Tokyo, Japan*

In the composite pipe construction method defined in the Japan Sewage Works Association's Design and Construction Guidelines for Sewer Pipe Rehabilitation, a liner pipe is constructed inside an existing sewer and the annular space behind the liner is filled with cementitious grout under pressure to form a highly integrated structure. The Guidelines for sewer renovation in Japan define a composite pipe as a composite structure that requires complete integration between liner materials and an existing pipe. However, in general, owing to the refinement of the small and enclosed working space inside sewers and the significant extension of the pipe to be renovated, mechanical shear connectors are not used to rigidly connect a renovation layer with an existing pipe. To fully explore the advantages of the composite pipe method in sewer renovation while maintaining safe design, a semi-composite structure has been applied in the renovation design of aging sewers using nonlinear FEM analysis for more than 50 sewers (over a total length of 300 km), primarily in big cities in Japan. This paper focuses on the mechanics of the structure.

Keywords: Nonlinear FEM, No-tension interface, Limit state design, Damage indices, SPR method.

1 INTRODUCTION

The general practice of sewer renovation in Japan, from site survey and structural design to renovation construction, is subject to Design and Construction Guidelines for Sewer Pipe Rehabilitation stipulated by the Japan Sewage Works Association (2017). According to the Guidelines, a renovated sewer pipe must be structurally comparable with or superior to a newly constructed sewer pipe in terms of its load-carrying capacity, material and structural endurance, and hydraulic capacity.

In this provisional code, renovation methods are classified into two categories, i.e., the composite pipe method and independent (or stand-alone) pipe method. The concept of the composite pipe method is to construct a composite structure by rigidly attaching an inner lining, i.e., a renovation layer, to an existing pipe. The renovated sewer is expected to bear external loads using the combined resistance of the two structural components. On the contrary, in the independent pipe method, a new pipe is constructed inside an existing pipe, and the new pipe is designed to fully resist external forces without relying on the existing pipe.

The SPR method is one of the typical composite pipe renovation methods put into practical use in Japan (Figure 1). It involves the spiral formations of an inner lining by polyvinyl chloride profiles with steel reinforcement in existing pipes and the filling of the annular gap between a lining and a pipe with cementitious grout. The method can be implemented while water is running through a sewer. A pipe-manufacturing machine is developed to fit the shape of the pipe, and the method is applicable to all types of cross sections. The mortar is mixed with emulsion to increase bond strength, and different specifications are available to produce different levels of strength. The steel reinforcement of a specific shape can be installed in the profile (Figure 2).

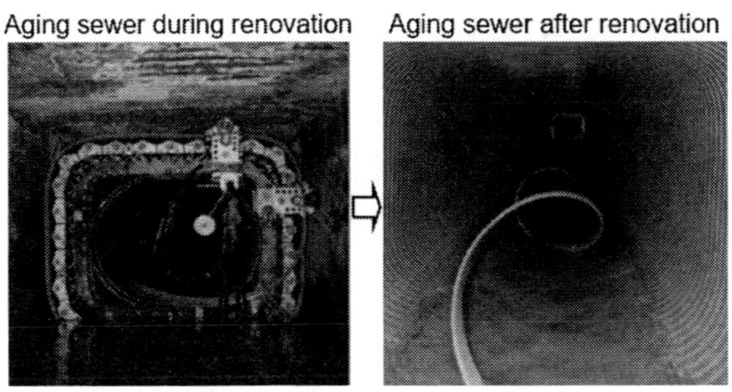

Figure 1. Sewer renovation by the SPR method.

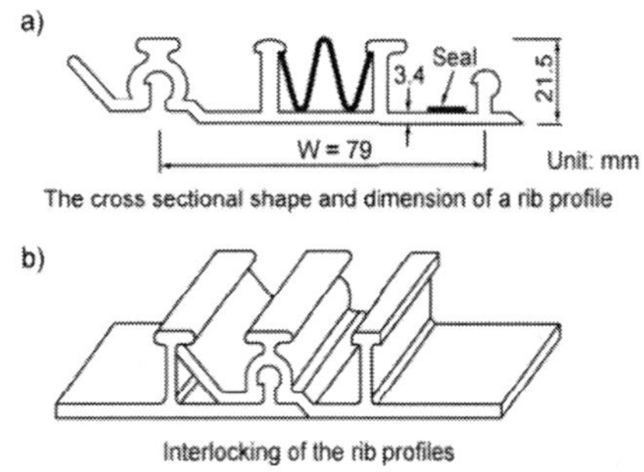

Figure 2. PVC rib profiles used in the SPR renovation method.

2 NO-TENSION INTERFACE MODELING AND THE SEMI-COMPOSITE STRUCTURE

Even though the Guidelines require a direct tension test to be performed on concrete-mortar test specimens to ensure no occurrence of interfacial debonding at the ultimate failure of test specimens, debonding has been reported during fracture tests at the interface of pipe specimens renovated by employing various pipe-reforming methods. Figure 3 shows two typical examples of fractured circular and box-culvert pipe specimens renovated using different pipe-reforming methods. In each case, the localized debonding of the interface between the original pipe and renovation layer has occurred at high loads. This type of interfacial debonding in renovated pipe specimens shows that the requirement for complete integration in the composite pipe method may not always be satisfied.

Numerical studies on these test specimens under the rigid bond condition at the interface showed that the maximum tensile stress could reach approximately 3.0 MPa or higher at peak load. Compared with the typical tensile strength of 2.0 MPa for various types of cementitious grout, debonding at the interface of a composite pipe appears to be inevitable at high loads. To fully explore the advantages of the composite pipe method in sewer renovation while maintaining safe design, a semi-composite structure is proposed for this method based on the following assumptions (Shi *et al.* 2016):

1) Perfect bond after construction: The complete integration of the original pipe and renovation layer is achieved at the completion of renovation construction.
2) No-tension interface: When tensile stress occurs at the interface under external loads, the perfect bond is replaced by free surface modeling to terminate stress transfer.
3) Perfect bond under compression: A perfect bond for the interface is assumed in the compression zones of the renovation layer. This allows for the continuous transfer of shear and compressive stresses between the original pipe and renovation layer.

The perfect integration of the original pipe and renovation layer in the compression zones maximizes the strength of the renovated pipe, and the free interface deformation of the structural components in the no-tension zones compromises the integration effect of a composite structure in these regions. This potential debonding of the interface in the tension zones of the renovation layer in an otherwise fully-integrated composite pipe leads to the unique semi-composite structure of the composite pipe method. No-tension interface modeling in numerical analysis is illustrated in Figure 4, where the interface is modeled using dummy elements and dual nodes with the same coordinates. The spring coefficients connecting the dual nodes are assumed to be infinite for the rigid bond condition. When tensile stress occurs, the spring connections in the vertical and horizontal directions are removed to allow for the free deformation of the interfaces.

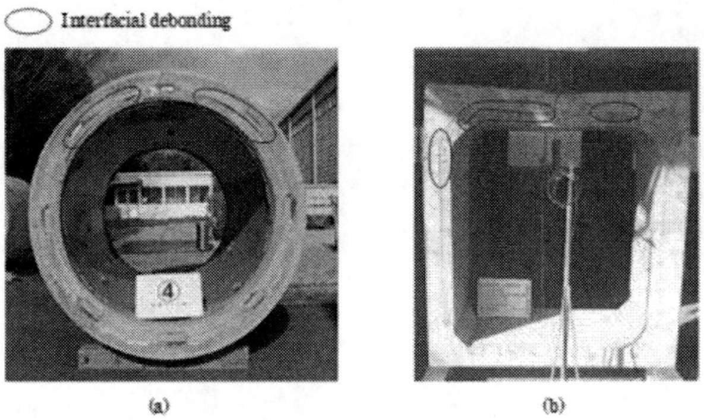

Figure 3. Interfacial debonding of renovated pipe specimens at failure loads.

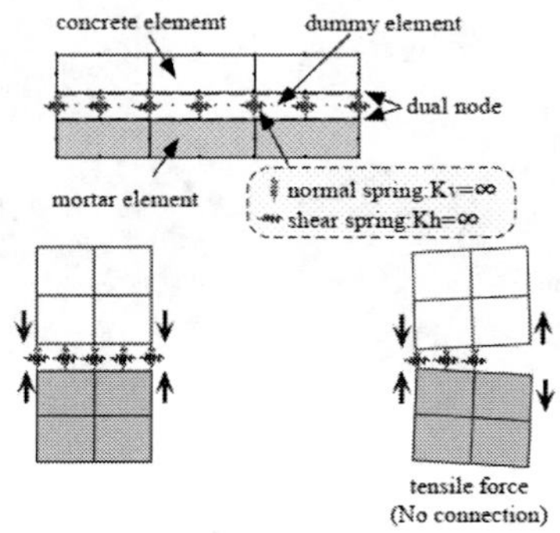

Figure 4. No-tension interface modeling in numerical analysis.

3 NUMERICAL ANALYSES OF FRACTURE IN RENOVATED PIPE SPECIMENS

3.1 Case Studied

For developing the SPR renovation method, a large number of fracture tests have been carried out on renovated pipe specimens to verify the effectiveness of the method in restoring the structural integrity of aging pipes (Nakano *et al.* 2013). Figure 5 shows a 1500 × 1500 mm rectangular pipe, which was selected from the fracture tests performed on renovated pipe specimens. Two cases are examined. One is the standard renovation of an original pipe with a standard doubly-reinforced cross section. The other is a double-layered pipe with no bond strength between the original pipe and renovation layer. This pipe is prepared by placing a thin film on the inner surface of the original pipe before renovation.

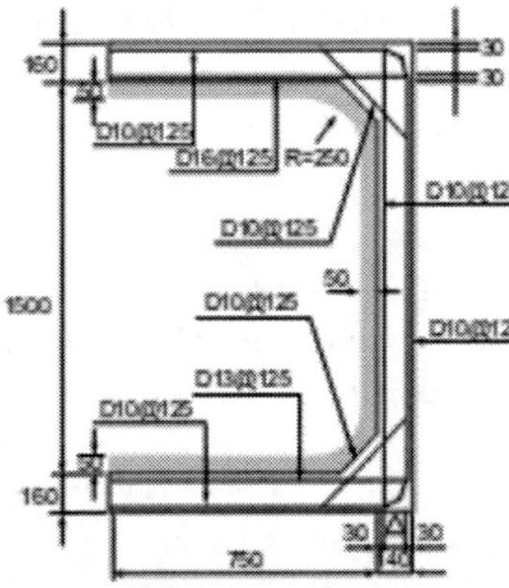

Figure 5. Structural dimensions and rebar arrangement of rectangular pipe.

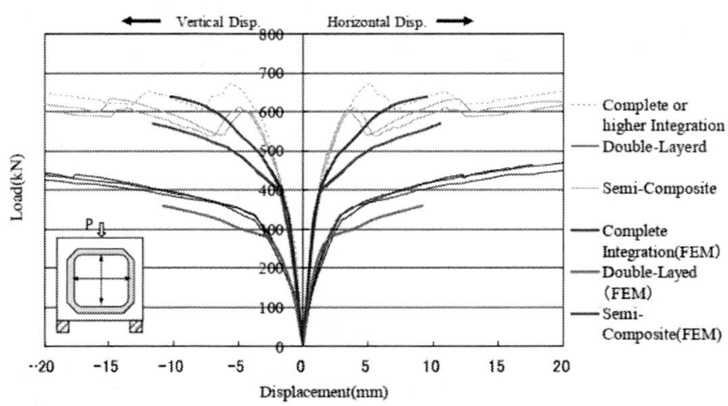

Figure 6. Numerical and test results.

Table 1 and Figure 6 show the results of the fracture tests. The maximum load of the double-layered pipe is approximately 30% smaller than that of the standard renovated pipe. This clearly shows that the bond behavior at the interface significantly affects the load carrying capacity of the renovated pipe. Even though composite pipes should not be modeled in actual design based on the assumption of perfect connection to the end of structural failure, the double-layered no-bond interface model clearly underestimates the load carrying capacity of the renovated pipe. The maximum load of one of the standard renovated pipes is higher than that of the other two specimens because the integration between the renovation materials and original pipe is higher for this pipe.

Table 1. Comparison between test and numerical results on load carrying capacity of rectangular pipe.

Cases		Maaximum Load(kN/m)			FEM/Test
		Test	Test(Ave.)	FEM	
Case1	Complete or higher	673.7	673.7	640	95%
Case2	Double-layered renovated	488.1 473.1 474	478.4	390	82%
Case3	Semi-composite	622.1 634.4	628.3	570	91%

3.2 Numerical Results

Numerical studies on these fracture tests are carried out in three different conditions at the interface using the active crack approach, in which all microscopic mechanical states (cracking, yielding, crack shear, and remaining stiffness of fractured materials) are integrated for macroscopic constitutive modeling (Maekawa *et al.* 2003). Case 1 is a renovated pipe under the rigid bond condition at the interface (complete integration), case 2 is a double-layered renovated pipe in which the transmission of shear stress is neglected from the beginning of loading, and case 3 is a semi-composite renovated pipe modeled with the no-tension interface model.

Figure 6 shows the numerical results of the load–displacement relations for cases 1 to 3. The test results of the load–displacement relations are also shown for comparison. The discrepancy between the numerical and test results is less than 18%, which is considered accurate in predicting the load-carrying capacity of reinforced concrete structures even though the adhesion strength and frictional resistance of the material in the actual test specimens are not considered in cases 2 and 3.

The damage indices of the second strain invariant of the deviatoric strain tensor are calculated based on the local strain at each integration point of the finite elements, as shown in Eq. (1) to Eq. (3). The scalar values are averaged over a certain region and used for studying the mechanics aspects of the semi-composite structure. Here, εx and εy are two-dimensional tensors, D is the averaged damage index, $\bar{D}$ is the local damage index ($\sqrt{J_2'}$), w(x) is the weight function according to the distance from the target integration point, and L is the averaged length (according to previous research, L = 150 mm). The damage indices of the general concrete material of the openings of bending cracks due to the yield of axial rebars or oblique cracks when the averaged region is set as 150 mm are evaluated by exceeding 0.0010 to 0.0015 (Saito *et al.* 2011) and (Tsuchiya *et al.* 2012).

$$\sqrt{J_2'} = \sqrt{\left(\frac{\varepsilon_x-\varepsilon_y}{2}\right)^2 + \left(\frac{\gamma_{xy}}{2}\right)^2} \tag{1}$$

$$\overline{D} = \frac{\int_A D \cdot w(x) dA}{\int_A w(x) dA} \tag{2}$$

$$w(x) = \begin{cases} 1 - x/L & x \leq L \\ 0 & x > L \end{cases} \tag{3}$$

In each case, the damage indices at the loads where vertical displacement sharply increases and at the maximum loads are shown in Figure 7 and Figure 8, respectively. Large deformation occurs when $\sqrt{J_2'}$ exceeds 0.001 at the top plate of the original pipe and outside the side wall at a haunch in the complete integration pipe (case 1) and double-layered pipe (case 2) and at the top plate close to the haunch in the semi-composite structure (case 3). As the shear stress at the interface is not transmitted from the beginning in case 2, the stresses at the top plate of the original pipe are higher and cracks are propagated because of its higher stiffness. In case 1, as the strain continuity at the cross section is maintained, the strains at the top plate in the renovated portion are high. On the contrary, the abrupt increase in the displacement in the semi-composite structure is at the same level as case 1, but the damage condition is different, and debonding occurs close to the haunch. At the maximum load in the complete integration pipe, the damage of the renovated portion is large based on the strain distribution in the midspan of the top plate. In the double-layered pipe, crack propagation in the original pipe is larger than that in the renovated portion. In the semi-composite structure, owing to the influence of the negative bending of the top plate close to the haunch, the damaged area is broader compared to the other two cases.

Load becomes the maximum when the averaged $\sqrt{J_2'}$ in the top plate and the side wall of the original pipe reaches approximately 0.004 and 0.003, respectively, for all cases. This shows that the effect of renovation appears in the suppression of the above two damage stages occurring in the original pipe.

Figure 9 shows the relationship between the vertical displacement and damage index ($\sqrt{J_2'}$) at the midspan of the top plate of the original pipe and the renovated portion for each case. The damage at the top plate of the original pipe is dominant in case 2, which is almost the same as the behaviors of the original pipe before renovation. In case 1, it can be seen that damage indices increase similarly in the renovated portion and original pipe portion, leading to fracture. In contrast, in case 3, the damage index of the original pipe increases in the early stage, but the damage index of the renovated portion increases rapidly when displacement exceeds 5 mm. Cracks occur even in the midspan of the top plate of the renovated portion in addition to the damage at the end of haunch due to negative bending.

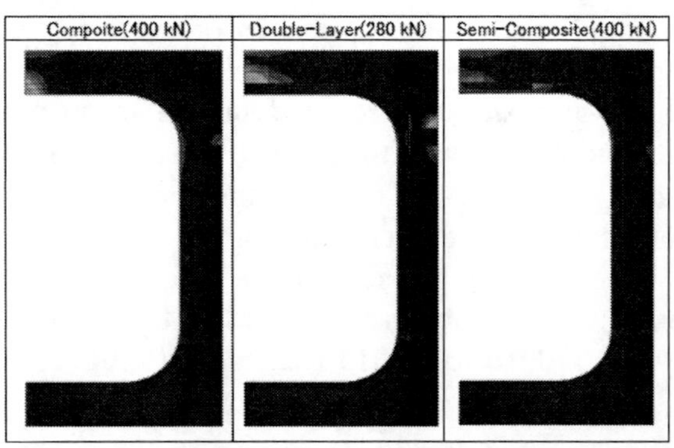

Figure 7. Distribution of damage index ($\sqrt{J_2'}$).

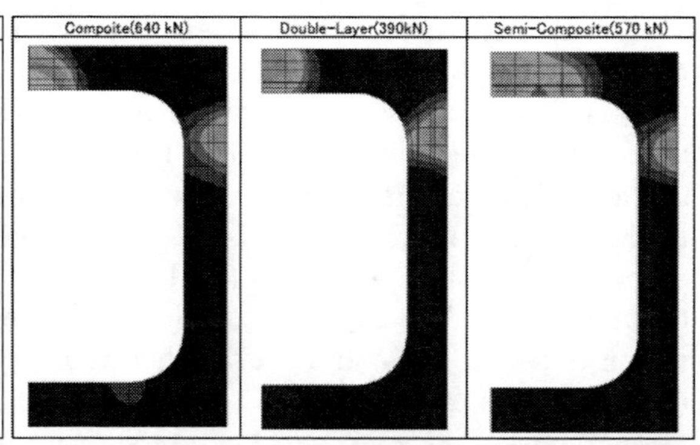

Figure 8. Distribution of damage index (averaged $\sqrt{J_2'}$).

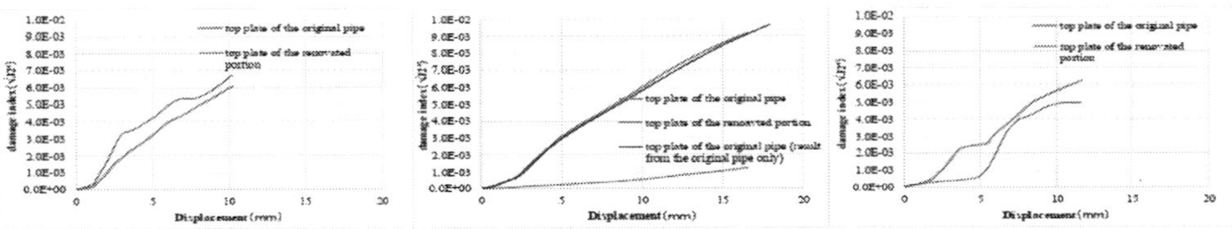

Composite — Double Layered — Semi-Composite

Figure 9. Relation between the vertical displacement and damage index (averaged $\sqrt{J_2'}$).

In case 1, ideal reinforcement is possible where the renovated portion and original pipe portion resist external force. However, in case 2, as the original pipe part resists external force with its normal renovation thickness, the renovation effect is not fully demonstrated. On the contrary, in case 3, the boundary condition changes depending on the stage of deformation or damage because debonding occurs in the portion where the tensile stress occurs at the interface and the degrees of resistance between the original pipe and renovated portion change.

4 CONCLUSIONS

Based on the results of the numerical analysis, it can be concluded that the no-tension interface model and semi-composite structure combined with the fracture-mechanics-based modeling theories accurately predict the maximum loads of composite sewer pipes. By utilizing the damage indices of the second strain invariant of the deviatoric strain tensor, the characteristics and effectiveness of the semi-composite structure are discussed and compared with the renovated pipe under the rigid bond condition and the double-layered pipe at the concrete-mortar interface.

The no-tension interface model is a conservative modeling approach for evaluating the bonding strength of a composite pipe based on safe-design consideration. Even though it may not reflect the actual debonding process in a renovated pipe under loading, the abovementioned accuracy assessment indicates that the final semi-composite structure agreed reasonably well with the actual structural state of a test specimen at its maximum load.

Acknowledgments

The test data were provided by Sekisui Chemical Co., Ltd. and Adachi Construction and Industry Co., Ltd. Their contributions to this study are sincerely acknowledged.

References

Japan Sewage Works Association, Design and Construction Guidelines for Sewer Pipe Rehabilitation, Tokyo, 2017.

Maekawa, K., Pimanmas, A., and Okamura, H., *Nonlinear Mechanics of Reinforced Concrete*, CRC Press, London, 2003.

Nakano, M., Shi, Z., Shi, Z., Nakamura, Y., Nakamura, Y., Takahashi, Y., and Haibara., T., Renovation Design of Aging RC Sewers as Semi-Composite Structures Based on Non-Linear Response Analysis, *VIII International Conference on Fracture Mechanics of Concrete and Concrete Structures,* Van Mier, J. G. M., Ruiz, G., Andrade, C. Yu, R. C., and Zhang, X. X. (eds.), 2013.

Saito, S., Maki, T., Tsuchiya, S., and Watanabe, T., Damage Assessment of Beams by Nonlinear Finite Element Analysis, *Journal of JSCE, Materials and Concrete Structures*, 67(2), 166-180, April, 2011.

Shi, Z., Nakano, M., and Takahashi, Y., *Structural Analysis and Renovation Design of Ageing Sewers: Design Theories and Case Studies*, De Gruyter Open, 2016.

Tsuchiya, S., Maki, T., Saito, S., Watanabe, T., and Maekawa, K., Damage Assessment and Load-Carrying Capacity of RC Planar Members by Nonlinear Finite Element Analysis, *Journal of Materials and Concrete Structures*, 68(3), 209-224, September, 2012.

Interdependence between Structural Engineering and Construction Management
Edited by Ozevin, D., Ataei, H., Modares, M., Gurgun, A., Yazdani, S., and Singh, A.

ISBN: 978-0-9960437-6-2

STUDY OF THE COMPOSITE ACTION OF HOLLOWCORE PANELS USED IN GIRDER-SLAB SYSTEM

NATHALIE ROY[1], SERGE PARENT[2], and MÉLISSA BARRIÈRE[3]

[1]*Dept of Civil and Building Engineering, Université de Sherbrooke, Sherbrooke, Canada*
[2]*SteelSSALG, Sherbrooke, Canada*
[3]*Exp Consultants, Drummondville, Canada*

Floor construction with precast hollowcore panels produced by Lafarge Precast Edmonton results in a commonly used girder-slab system. Continuity between the elements is ensured by bent rebars and shear studs. Once all these elements are installed, a structural concrete is poured between the reinforced concrete panels and over the entire floor. The extent of composite action between the rigid diaphragm and the steel beams is not known. Therefore, its potential benefit is not taken into account in the current design procedures for the steel structure. The main components of this research project are the following: an experimental program consisting of a series of 6 large-scale shear tests were carried out. The outcome of this research shows that there is a potential for a composite action between a hollowcore plank and a standard hot rolled W shape. It was found that there is enough confinement to develop the steel stud strength when the beam is connected to the precast prestressed concrete panels using a 1/2" shear stud embedded between the planks and under two to three inches of concrete topping.

Keywords: Precast prestressed concrete panel, Floor system, Shear transfer device (STD), Nelson stud, Embedment, Push-out tests.

1 INTRODUCTION

This paper describes the tests performed to study the composite action of precast prestressed hollowcore panels used in girder slab system. The tests were carried out in the structural laboratory at the Université de Sherbrooke in collaboration with Lafarge Precast Edmonton, located in Alberta, Canada. These tests were conducted in order to characterize the shear stud behavior between the hollowcore precast floor panels, and the steel frame to which it connects. These results will be instrumental for a subsequent research project aimed at determining the degree of composite action when the panels are connected onto the steel beams.

Lafarge Precast Edmonton provided 12 - 8" thick standard hollowcore precast floor panels. Consultants SteelSSALG supplied six – 4' high W12x40 stubs with 1/2" standard shear studs, and six rigid skids used to support the samples during the loading. The technical staff at Université de Sherbrooke assembled the panel sections, installed the concrete forms around the samples, and poured a 5000 psi concrete topping as shown in Figure 1. Three pairs of identical full-scale samples were tested considering different shear stud layout for each set.

2 EXPERIMENTAL PROGRAM

The objective of this program is to evaluate the shear strength of a standard 1/2" stud at 3 different spacing. The six samples are named in a format that represents the number of stud used and by the sample number. The six samples are described as follows: NB1S1 and NB1S2 have one stud per beam flange, NB2S1 and NB2S2 have two studs equally spaced per beam flange, NB3S1 and NB3S2 have three studs equally spaced per beam flange. For each sample, S1 represents the first specimen and S2 the second.

2.1 Description of Tested Specimens

Each specimen consists of two sets of two hollowcore planks of 3' x 4' x 8" mounted on each flange of a W12x40 stub. Each flange supports an identical stud layout (one, two or three 1/2" stud AWS type A - F_u=61 ksi) embedded within a 5000 psi - 2" to 3" thick concrete topping poured between and over the 3' x 4' hollowcore panels. The concrete topping was poured in two sequences; the first pour for samples NB1S1, NB1S2, NB3S1 took place on November 14th 2017, while the second pour occurred on November 29th 2017 for samples NB2S1, NB2S2 and NB3S2. The 4" long 1/2" stud has a length to diameter ratio equal to 8 which is greater than 4 as required by clause 17.7.2.1 of the Design of Steel Structures Canadian standard (CAN/CSA S16-14 2014). Figure 1 illustrates the test setup.

Figure 1. Test setup.

2.2 Test Setup

The methodology followed for the push-out tests is similar to the one reported in Jayas and Hosain (1988). The tests were performed using a dynamic press MTS with a capacity of 2563 kips. Loading was applied in displacement control at a rate of 7.8x10^{-5} in/s. For all tests, the load and the vertical displacement were respectively measured by the load cell and the displacement sensor inside the MTS machine. Displacement transducers were added to the steel beam into the vicinity of the panel-topping interface to measure the vertical displacement of the W-beam. The load-slip curves, presented in the section 3, shows the slip recorded by a displacement transducer located at the mid-height of the web of the W12x40 beam. More details on the experimental program can be found in Barriere *et al.* (2018).

2.3 Material Characterization

2.3.1 *Compression tests on concrete topping*

In order to characterize the concrete topping, compression tests were carried out on 12 small 6"x12" cylinders from the two different concrete pours. The tests were performed according to ASTM C39/C39M-18 standard, on a MTS 1124 kips capacity press and loading was applied in displacement control at a rate of 4×10^{-4} in/s. The results are presented in Table 1.

Table 1. Compressive strength of concrete topping at 28 days.

Specimen	Strength (psi)
S1 – batch 1	5323
S2 – batch 1	5207
S3 – batch 1	5062
Average batch 1	5192
S1 – batch 2	4960
S2 – batch 2	4844
S3 – batch 2	4888
Average batch 2	4902
Global Average	5047

2.3.2 *Tests on shear stud*

In order to characterize the shear studs, tensile tests were performed on three specimens where the following dimensions were measured: length, section diameter, and reduced notch diameter and notch width. The tests were conducted on a MTS 647 Hydraulic Wedge grip-112 kips capacity press. Loading was applied in displacement control at a rate of 9.8×10^{-4} in/s up to the yield strength, and at a rate of 3.3×10^{-3} in/s up to the ultimate strength. The load and the displacement were respectively measured by the load cell and the displacement sensor of the MTS machine.

The results for tensile capacity are presented in Table 2. These values, obtained with the stress-strain curve, are higher than the values supplied by Nelson Stud Company shown as 49 ksi and 61 ksi for the yield and ultimate strength respectively.

Table 2. Tensile strength of Nelson studs.

Specimen	Fy (ksi)	Fu (ksi)
S1	58.0	73.5
S2	51.3	74.3
S3	51.8	74.7
Average	57.3	74.2

3 RESULTS

In this section, Table 3 presents for each specimen the vertical displacement measured at the maximum load P_{max}, the observed failure mode of the stud, and the ultimate shear load. The load-slip curve for each of the 6 samples is then presented in Figure 2.

As per CAN/CSA S16-14, Clause 17.7.2.2, the factored resistance of a shear connector in a solid slab (q_{rs}) is given by Eq. (1);

$$q_{rs} = 0.50 \varphi_{sc} A_{sc} \sqrt{f'_c E_c} \leq \varphi_{sc} F_u A_{sc} \tag{1}$$

Table 3. Test results.

Specimen	Shear load per stud at failure - P_{max} (kips)	Stud failure mode	Slip at P_{max} (in)
NB1S1	21.56	Sheared	0.0045
NB1S2	14.55	Bent	0.0150
NB2S1	25.77	Sheared	0.0740
NB2S2	29.74	Sheared	0.0820
NB3S1	19.15	Sheared/Bent	0.0590
NB3S2	18.84	Sheared/Bent	0.0230

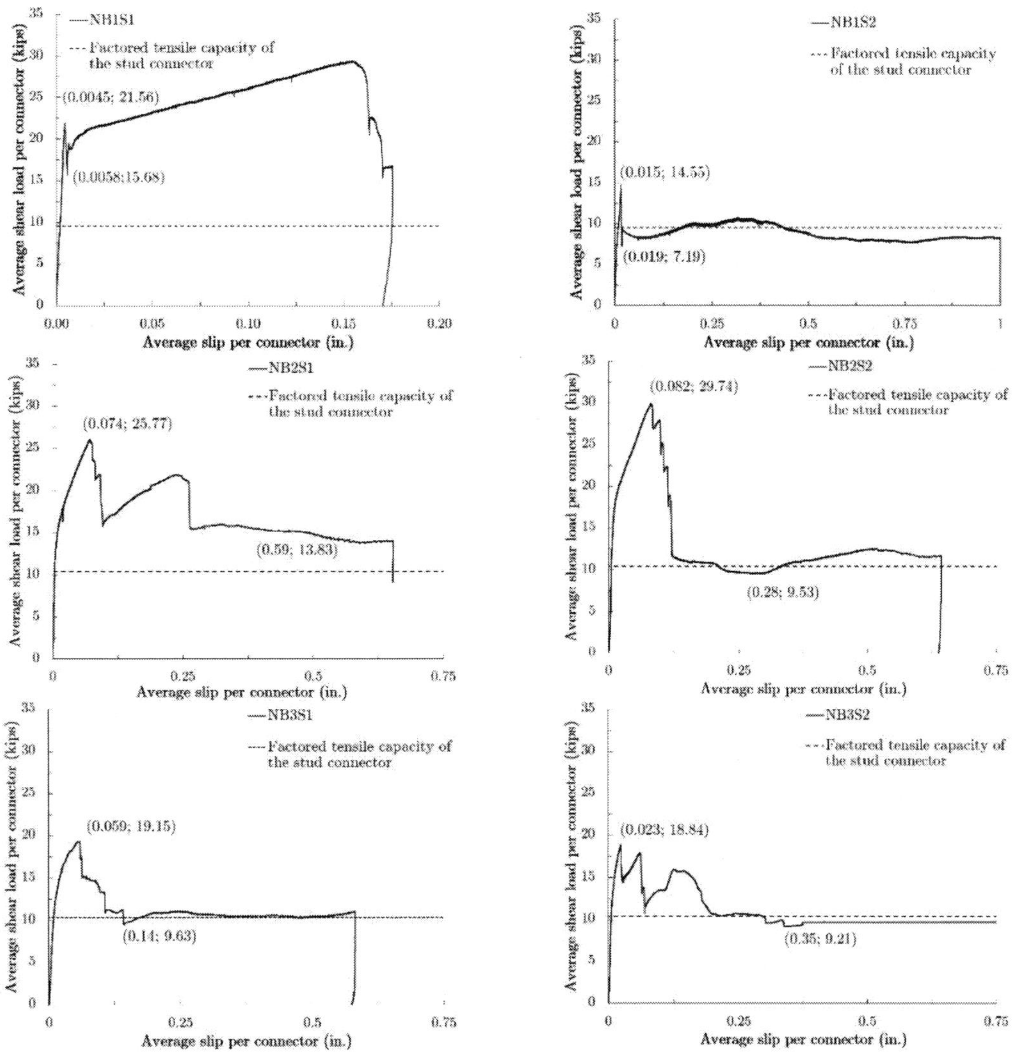

Figure 2. Load-slip curves for the 6 specimens.

Where q_{rs} represents the factored shear strength of a shear connector (N), F_u the tensile strength of the stud (MPa), and A_{sc} the cross-sectional area of the connector (mm^2), $\varphi_{sc} = 0.80$, $f'_c = 35$ MPa and Ec = 26622 MPa. The tensile strength, referenced by the supplier, of a 1/2" diameter steel stud is equal to 420 MPa, which corresponds to a factored shear strength, qrs, of 42.6 kN = 9.6 kips. This value appears on each load-slip curve. For each curve, the factored tensile strength of a standard 1/2" shear stud is shown along with the recording of the load as a function of the vertical displacement. For each curve except one (NB1S1), a peak at the onset of the cracked concrete was captured, followed by a plateau associated to the progressive failure of the shear stud. These results are discussed in detail in section 4, Analysis.

4 ANALYSIS

The results obtained from these six samples are now discussed in this section. From the load/slip curves (Figure 1), it was observed that all samples reached a maximum load higher than the theoretical strength of the studs. This can be seen by the peak in the average shear load per connector followed by a plateau close to the capacity of the stud. It is fair to assume that the peaks correspond to the added contribution of the shear capacity of the concrete before cracking. Therefore, the shear stress inside the concrete was calculated by considering P_{max} minus the strength of the shear studs divided by the sheared concrete plane. An evaluation of the concrete shear surface was made by adding the average concrete topping thickness to the 8" of concrete between the panels. It was observed that all failures occurred in one shear plane between the planks. The end of each plateau corresponds to the complete split failure of the specimen. Finally, the shear studs of all samples were either sheared off or bent over the W12 beam, which confirms this configuration offers an excellent confinement around the studs.

The first sample NB1S1 within the one-stud series reached a failure load of 21.56 kips which is highly in excess of the theoretical strength of 9.6 kips for a ½" stud. This over strength is probably due to an unexpected restraint within the W12 column in contact with the bottom-skid, and shall not be considered as a reliable result. The second sample NB1S2 shows a failure load of 14.55 kips which is slightly higher than the stud capacity. If the theoretical stud shear strength is subtracted from the measured failure load, this leaves a stress of 5.8 psi within the concrete. The only reliable sample within this one-stud series shows a displacement at failure higher than the 2 and 3-studs series, which is due to a lesser amount of stud, and a possible rotation of the hollowcore panel with only one stud at its midspan. For sample NB1S2, the plateau slightly remains under the shear stud strength of 9.8 kips. This observation is most likely due to some voiding in the concrete observed around the failed stud, as detailed in (Barriere *et al.* 2018).

For the two studs series, both specimens NB2S1 and NB2S2 respectively failed at 25.77 and 29.74 kips per stud. Failure was observed at the interface between the stud and W12 flange. The first peak corresponds to a concrete shear stress contribution of 52.2 psi for sample NB2S1, and 72.5 psi for sample NB2S2. All studs within this group were sheared-off from the beam flange. For these two samples, the plateau mostly shifted above the shear stud strength. Sample Both samples definitely exhibit a better contribution of the concrete shear strength between the panels.

The three studs series NB3S1 and NB3S2 respectively shows a maximum load per stud of 19.15 and 18.84 kips. The observed mode of failure was a combination of bent and sheared studs. For these two samples, an almost identical concrete stress of 40.6 psi for NB3S1 and 39.2 psi for NB3S2 are calculated. Both samples of this series shown a loading plateau almost equal to the shear stud strength. These results lead to the conclusions presented next.

5 CONCLUSIONS

Tests on floor construction with precast Hollow panels were performed in the structural laboratory of the University of Sherbrooke in order to describe the shear stud capacity characterizing the composite action between the precast panels and the steel beam supporting it. The tests were carried out at the request of the Lafarge Precast Edmonton at December 2017 and

January 2018. Six full scale specimens were load tested until failure to determine if the confinement around the shear stud was sufficient to develop its tensile strength. Based on the results of five reliable samples presented in this paper, the following conclusions are issued:

1. When the stud is confined between two standard hollowcore planks and under 2" to 3" of concrete topping, the results show that there is potential for a composite action between the planks and the steel beam supporting it;
2. For two- and three- studs set-up, it appears that it would be acceptable to calculate the shear strength of a single stud q_{rs} using Clause 17.7.2.2 of S16 with marginal modification;
3. Since the overall strength of a three stud configuration is comparable to a two stud layout, the results show there would be no need to use more than 2 studs per 4' wide panel.

A series of full scale tests in bending should be performed in order to assess the applicability of Clause 17.9.3 of S16 needed to evaluate the value of Q_r which is the value of the sum of the factored resistances of all shear connectors between points of maximum and zero moment. This will be needed to calculate the factored bending capacity of a composite beam.

Acknowledgments

The authors would like to acknowledge the financial contribution of Lafarge Precast Edmonton, and Natural Sciences and Engineering Research Council of Canada. The authors would also like to thank the technical staff at the Université de Sherbrooke.

References

ASTM C39/C39M-18, *Standard Test Method for Compressive Strength of Cylindrical Concrete Specimens*. ASTM International, West Conshohocken PA, 2018.

Barrière, M., Roy, N, and Parent, S., *Study of the Composite Action of Hollow Core Panels Used in Girder-Slab System*, Prepared for Lafarge Precast Edmonton, Civil Engineering Department, Université de Sherbrooke, 47 p., 2018.

CAN/CSA-S16-14, *Limit States Design of Steel Structures - 11th Edition*, Canadian Standard Association, Mississauga, Canada, 2014.

Jayas, B. S., and Hosain, M. U., Behavior of Headed Studs in Composite Beams: Push-Out Tests, *Canadian Journal of Civil Engineering,* 15th Edition, 240-253, 1988.

Interdependence between Structural Engineering and Construction Management
Edited by Ozevin, D., Ataei, H., Modares, M., Gurgun, A., Yazdani, S., and Singh, A.

ISBN: 978-0-9960437-6-2

SEISMIC PERFORMANCE OF STRUCTURAL FRAMES AND JOINTS IN WOOD TALL BUILDINGS

EMAN SALEH and JAMSHID MOHAMMADI

Dept of Civil, Architectural and Environmental Engineering, Illinois Institute of Technology, Chicago, USA

There has been a growing interest in using wood in tall buildings by employing cross-laminated timber (CLT) in a structure's moment frame system. However, still much needs to be learned about the performance of CLT frame systems subject to lateral loads such as earthquakes. The ductility needed in CLT frames in seismic load environments is primarily provided through connections. Thus, a proper connection detail is required to ensure an adequate seismic response, especially at extreme loads. Current strength specifications of connectors are mainly applicable in cases of pure tension or shear; while in real applications, connections are subjected to a combination of these loads. This paper is part of analytical and simulation studies on seismic performance of wood structural frames being conducted at Illinois Institute of Technology. These studies include simulations to investigate: (1) Moment-resistance capacity of CLT panels; and (2) Resistance capacity of typical joints used in CLT frame systems. Limited experimental data available on a 6-story wood frame system are considered as a means of verification of the aforementioned analytical investigations. Focus of this paper is specifically on the significance of interaction between tension and shear as reported in literature. Several interaction equations as implemented in current analytical models for pushover analysis of CLT frames are explained and their adequacy to predict the nonlinear behavior of CLT panels under lateral loads is discussed.

Keywords: Pushover analysis, CLT, Interaction equations, Failure criteria, Connectors, Nonlinear behavior.

1 INTRODUCTION

Currently, there is a growing interest in using wood as a primary material for tall buildings (Mohammadi and Ling 2017). Tall buildings made solely of timber materials are those in the 8 to the 9-story range. In combination with other materials, taller constructions are also possible and have been considered by structural engineers and architects both in real applications and as concept designs (e.g., Timber Tower Research Project, *Skidmore*, *Owings, and Merrill*, Chicago, IL, 2013). The interest in using wood as an alternative material for tall buildings is due to several unique properties of wood including (1) large strength to density ratio; (2) flexibility in architectural design; and (3) favorable resistance to fire.

Multi-storey timber buildings belong to two main construction types - one is the conventional light-frame constructions that consist of combinations of sheathing and framing elements connected with nails or staples; the second type is the mass timber constructions such as cross-laminated timber (CLT or Xlam). CLT is an engineered wood made of layers of lumber boards that are stacked crosswise (usually at 90-degree angle) and glued together. CLT is mostly used for mid and high rise buildings due to its dimensional stability, stiffness, and rigidity. Moreover, CLT possesses attractive features such as lightweight, low cost and favorable thermal and vibrational performance that make it an alternative material for tall building construction as a stand-alone material or in combination with steel and concrete. Experimental and analytical investigations have shown that CLT structures have a remarkable seismic performance with little or no residual deformations and minor damage. Entire CLT buildings and assemblies under monotonic and cyclic loads have also been tested (Popovski *et al.* 2010, Tomasi and Smith 2014, Gavric *et al.* 2015a). Moreover, uniaxial and 3D shake table tests were performed on CLT structures and substructures to simulate the ground motion of an earthquake (Ceccotti and Follesa 2006, Dujic *et al.* 2010, Hristovski *et al.* 2012, Ceccotti *et al.* 2013). These studies showed that the main factors that affect the seismic performance of CLT panels are: (1) the mechanical properties number and type of connectors (2) aspect ratio of the panels (3) boundary conditions (4) the presence of vertical joints or openings in the substructures (5) vertical load. It is noted that the seismic performance of timber structures is not solely derived from the performance of CLT only; an appropriate design for the structural component and details is required to ensure sufficient performance (Sustersic *et al.* 2015, Casagrande *et al.* 2017, Reynolds *et al.* 2017). In addition, several modeling techniques have been proposed to simulating the behavior of CLT structures and sub-assemblies subject to seismic loads (Filiatrault and Folz 2002, Hummel 2017).

2 MECHANICAL MODELS USED FOR CLT PANELS

CLT panels and walls have a significant in-plane stiffness, thus the predominant kinematic behavior is rocking, sliding or combination of both, while shear and bending deformations are significantly lower (Gavric *et al.* 2015a). Therefore, the energy dissipation is mainly achieved through joints connecting structural members. In general, the behavior of the CLT wall and connection is nonlinear. Consequently, to predict the force-displacement of a CLT wall system, a force-displacement relationship for each connector (angle brackets, hold-downs, screws) must be selected. In our studies, a trilinear force-displacement relationship is adopted for each connector (Figure 1) and calibrated using CLT connections test results based on EN 12512 provisions (Gavric *et al.* 2012, 2015b).

Figure 2 shows the kinematic model adopted for CLT wall. This model neglects any in-plane deformations of the CLT panel and ignores any interaction between rocking and sliding.

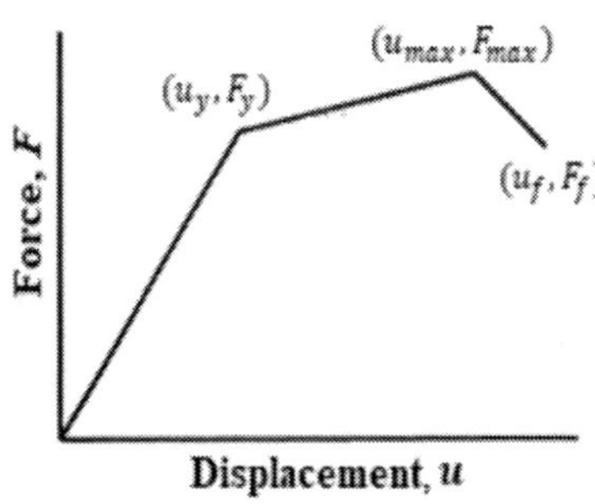

Figure 1. Trilinear force-displacement relationship that is chosen to model the CLT connectors (Gavric *et al.* 2015a).

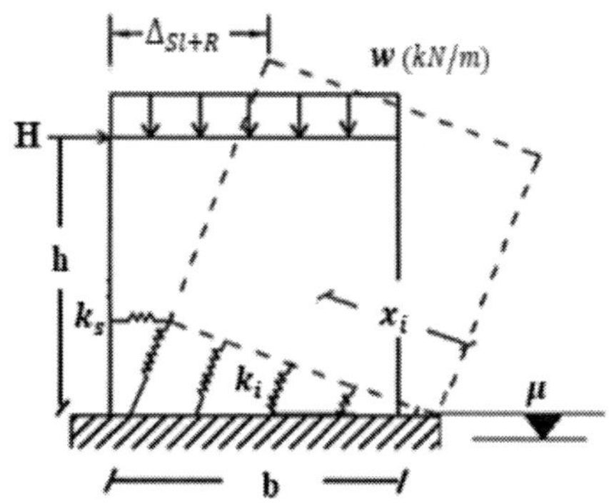

Figure 2. Kinematic model adopted for CLT wall subjected to lateral and vertical loading.

The rocking component of the displacement, Δ_R is a function of the wall rotation and height, $\Delta_R = h\theta$. Considering the force in the spring, i to be linear elastic, it can be obtained from the wall rotation and distance from the rotation point, $F_i = x_i\theta$. Solving the equilibrium at the lower right corner results in:

$$Hh - w\frac{b^2}{2} - F_i x_i = 0 \tag{1}$$

From Eq. (1) the rocking lateral displacement component is equal to (Eq. (2)):

$$\Delta_R = \frac{\left(Hh - \frac{wb^2}{2}\right)h}{\sum_{i=1}^{n} k_i x_i^2} \tag{2}$$

Applying a simple friction model, where the wall starts sliding when the horizontal force exceeds the force of static friction, the friction force equals:

$$F_\mu = \mu\, w\, b \tag{3}$$

The sliding component H is obtained from Eq. (4):

$$H - k_s \Delta_{sl} - F_\mu = 0 \tag{4}$$

From Eq. (4), the sliding lateral displacement component Δ_{sl} equals (Eq. (5)):

$$\Delta_{sl} = \frac{H - F_\mu}{k_s} \tag{5}$$

Considering the connector force in tension, $F_i(d_i)$ and in shear, $F_{si}(d_i)$, the nonlinear wall resistance as a function of displacement, $H(D)$ equals to the minimum of Eq. (6a) and Eq. (6b):

$$H(D) = \sum_{i=1}^{n} \frac{F_i(d_i)x_i}{h} + \frac{wb^2}{2h} \tag{6a}$$

$$H(D) = \sum_{i=1}^{n} F_{si}(d_i) + F_\mu \tag{6b}$$

3 FAILURE CRITERIA

Since the behavior of CLT walls subjected to a lateral load is predominantly consists of rocking and sliding, a two-directional loading will require connection elements such as hold-downs and angle brackets. And as such, criterion frequently used to predict the ultimate capacity and failure is based on maximum normal stress and maximum shear stress (Steeve and Wingate 2012):

$$\sigma_{max} = \frac{\sigma}{2} + \sqrt{\left(\frac{\sigma}{2}\right)^2 + \tau^2} \tag{7}$$

$$\tau_{max} = \sqrt{\left(\frac{\sigma}{2}\right)^2 + \tau^2} \tag{8}$$

Dividing Eq. (7) and Eq. (8) by the maximum normal and shear strength respectively yield to Eq. (9) and Eq. (10):

$$\frac{R_T}{2} + \sqrt{\left(\frac{R_T}{2}\right)^2 + (KR_S)^2} = 1 \tag{9}$$

$$\sqrt{\left(\frac{R_T}{2K}\right)^2 + (R_S)^2} = 1 \tag{10}$$

Where $R_T = {}^{\sigma}/_{\sigma_{max}}$, $R_S = {}^{\tau}/_{\tau_{max}}$ and $K = {}^{\tau_{max}}/_{\sigma_{max}}$

To study the influence of different K values in the analytical model to predict the force-displacement relationship of CLT wall, several models are investigated:

- Model 1: Maximum uncoupled tension and shear stress (no tension-shear interaction).
- Model 2: K=0.5, this value is commonly used in interaction equations for members loaded in combined tension and compression.
- Model 3: Maximum normal stress criterion with K=1/5 as an interaction equation for hold-downs and maximum shear stress criterion with K=1.5 as an interaction equation for the bracket angles. It is reported in Gavric *et al.* (2015b) that the tensile strength of the hold downs is almost five times greater than its shear strength. However, the average uplift resistance of angle brackets was 63% of the shear resistance for wall panel connections and 87% for a wall-foundation connection.
- Model 4: Maximum uncoupled tension stress of hold-downs and maximum shear stress criterion with K= 1.5.

The accuracy of the above models to predict the actual behavior of CLT walls under lateral loads has been verified by comparing the models' predictions of the maximum lateral resistance and the ultimate displacement capacity to experimental data (Gavric *et al.* 2015a). The ratio between experimentally measured data and the analytical predictions are displayed in Table 1.

Table 1. The maximum lateral resistance (F_{max}) and the ultimate displacement capapacity (Δ_f) of tested CLT shear walls to that of the analytical models (Gavric *et al.* 2015a).

Wall panel test configuration	**Test Description**	**Model.1**		**Model.2**		**Model.3**		**Model.4**	
		$F_{max-exp}$ / $_{max-ana}$	Δ_{f-exp} / Δ_{f-ana}	$F_{max-exp}$ / $F_{max-ana}$	Δ_{f-exp} / Δ_{f-ana}	$F_{max-exp}$ / $F_{max-ana}$	Δ_{f-exp} / Δ_{f-ana}	$F_{max-exp}$ / $F_{max-ana}$	Δ_{f-exp} / Δ_{f-ana}
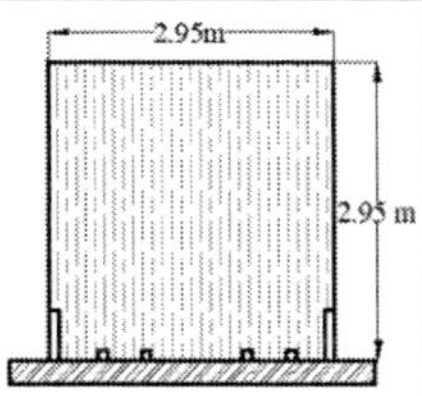	Test.1: n=2, m=4, w= 18.5kN/m	1.00	1.1	1.28	1.15	1.07	1.1	1.01	1.1
	Test.2: n=2, m=4, w= 18.5kN/m	1.02	0.9	1.3	0.94	1.09	0.92	1.03	0.9
HTT22 hold-downs and BMF	Test.3: n=2, m=2, w= 18.5kN/m	0.84	1.07	1.07	0.91	0.88	1.03	0.87	0.87
90×48×3×116 mm angle brackets.	Test.4: n=2, m=4, w= 9.25kN/m	1.08	1.08	1.4	1.1	1.2	1.09	1.08	1.08
Number of hold-downs = n	$\bar{x}$	0.99	1.04	1.26	1.03	1.06	1.04	1.00	0.99
Number of angle brackets = m	COV %	10.4	8.9	11.0	11.5	12.5	8.0	9.0	12.1

Model 1 and Model 4 (Table 1) predicted the CLT maximum resistance and ultimate capacity relatively well. However, in Test 3 both models underestimated the maximum lateral resistance reaching almost a 15% difference. Model 2 significantly underestimated the maximum lateral resistance and showed about 26% lower values than those from experimental results. These underestimated values are rather remarkable and reported to be due to the predicted premature failure of hold-downs in shear as opposed to the real performance where plasticization does not simultaneously apply in tension and shear. Model 3 predictions are relatively good, and calibrating K based on boundary conditions and connections type could lead to better performance of this model.

4 DISCUSSIONS AND CONCLUSIONS

Our study in understanding the seismic behavior of frames made up of CLT panels and walls is in progress. Currently, our research is focusing on: (1) providing an insight into lateral load resistance capability of CLT panels when used in multi-story frame systems; (2) identifying the type of structural panels and joints that offer favorable performance for use in tall buildings; and (3) identifying other areas where one may pursue using wood, as sole material or in combination with other materials, in tall buildings. The interaction models for tension and shear reported in the literature and briefly reviewed in this paper are especially useful in modeling wood frame systems in nonlinear push-over analysis and being pursued in our studies. Limited experimental data available on 6-story wood frame system is also being considered as a means of verification for the results from our analytical investigations (Van de Lindt *et al.* 2010).

The ductility needed for frames can be achieved through a proper design for connections and their components (such as hold-downs and angle brackets). As reported in literature, the traditional interaction equation of shear and tension should not be used for hold-downs since it predicts its premature failure and underestimates the behavior. The literature reports that it is a good practice to include the contribution of angle brackets to resist uplift.

References

Casagrande, D., Doudak, G., Mauro, L., and Polastri, A., Analytical Approach to Establishing the Elastic Behavior of Multipanel CLT Shear Walls Subjected to Lateral Loads, *Journal of Structural Engineering,* 144(2), 04017193, 2017.

Ceccotti, A., and Follesa, M., *Seismic Behavior of Multi-Storey X-Lam Buildings*, International Workshop on Earthquake Engineering on Timber Structures, Coimbra, Portugal, November, 2006.

Ceccotti, A., Sandhaas, C., Okabe, M., Yasumura, M., Minowa, C., and Kawai, N., SOFIE project–3D Shaking Table Test on a Seven Story Full Scale Cross Laminated Timber Building, *Earthquake Engineering and Structural Dynamics,* 42(13), 2003-2021, 2013.

Dujic, B.; Strus, K.; Žarnic, R., and Ceccotti, A., *Prediction of The Dynamic Response of a 7-Storey Massive Xlam Wooden Building Tested on a Shaking Table,* Proceedings of the 11th World Conference on Timber Engineering, Riva del Garda, Italy, 2010.

Filiatrault, A., and Folz, B., Performance-Based Seismic Design of Wood Framed Buildings, *Journal of Structural Engineering,* 128(1), 39–47, 2002.

Gavric, I., Fragiacomo, M., and Ceccotti, A., Cyclic Behavior of CLT Wall Systems Experimental Tests and Analytical Prediction Models, *Journal of Structural Engineering*, 141(11), 04015034, 2015a

Gavric, I., Fragiacomo, M., and Ceccotti, A., Cyclic Behavior of Typical Metal Connectors for Cross-Laminated (CLT) Structures, *Materials and Structu*res, 48(6), 1841-1857, 2015b.

Gavric, I., Rinaldin, G., Fragiacomo, M., Ceccotti, A., and Amadio, C., *Experimental-Numerical Analyses of The Seismic Behavior of Cross-Laminated Wall Systems*, Proc., 15th World Conf. on Earthquake Engineering WCEE 2012, Sociedade Portuguesa de Engenharia, 2012.

Hristovski, V., Dujic, B., Stojmanovska, M., and Mircevska, V., Full-Scale Shaking Table Tests of Xlam Panel Systems and Numerical Verification: Specimen 1, *Journal of Structural Engineering* 139(11), 2010–2018. 2012.

Hummel, J., *Displacement-Based Seismic Design for Multi-Storey Cross-Laminated Timber Buildings* (Vol. 8), Kassel university press GmbH, 2017.

Mohammadi, J., and Ling L., Can Wood Become an Alternative Material for Tall Building Construction?, *Practice Periodical on Structural Design and Construction,* 22(4), 2017.

Popovski, M., Schneider, J., and Schweinsteiger, M., Lateral Load Resistance of Cross-Laminated Wood Panels, *In World Conference on Timber Engineering*, 20-24, June 2010.

Reynolds, T., Foster, R., Bregulla, J., Chang, W. S., Harris, R., and Ramage, M., Lateral-Load Resistance of Cross-Laminated Timber Shear Walls, *Journal of Structural Engineering*, 143(12), 06017006, 2017.

Steeve, B. E., and Wingate, R. J., *Aerospace Threaded Fastener Strength in Combined Shear and Tension Loading,* 2012.

Sustersic, I., Fragiacomo, M., and Dujic, B., Seismic Analysis of Cross-Laminated Multistory Timber Buildings Using Code-Prescribed Methods: Influence of Panel Size, Connection Ductility, and Schematization, *Journal of Structural Engineering*, 142(4), E4015012, 2015.

Tomasi, R., and Smith, I., Experimental Characterization of Monotonic and Cyclic Loading Responses of CLT Panel-To-Foundation Angle Bracket Connections, *Journal of Materials in Civil Engineering*, 27(6), 04014189, 2014.

Van de Lindt, J. W., Pei, S., Pryor, S. E., Shimizu, H., and Isoda, H., Experimental Seismic Response of a Full-Scale Six-Story Light-Frame Wood Building, *Journal of Structural Engineering, ASCE*, 136(10), 1262-1272, 2010.

Interdependence between Structural Engineering and Construction Management
Edited by Ozevin, D., Ataei, H., Modares, M., Gurgun, A., Yazdani, S., and Singh, A.

ISBN: 978-0-9960437-6-2

PERFORMANCE ASSESSMENT OF RC BEAM-COLUMN JOINT REINFORCED WITH DIFFERENT SUPER ELASTIC SHAPE MEMORY ALLOY TYPES

MUMTASIRUN NAHAR[1], A.H.M.M. BILLAH[2], HABIBUR RAHMAN KAMAL[3], and KAMRUL ISLAM[4]

[1]*Dept of Civil Engineering, Military Institute of Science and Technology, Dhaka, Bangladesh*
[2]*Dept of Civil Engineering, Lakehead University, Thunder Bay, Canada*
[3]*Dept of Civil and Environmental Engineering, North South University, Dhaka, Bangladesh*
[4]*Dept of Civil, Geological and Mining Engineering, Ecole Polytechnique de Montreal, Montreal, Canada*

Superelastic Shape Memory Alloys (SE SMAs) are smart materials that have the ability to undergo large inelastic deformation upon stress removal (superelasticity) or heating (shape-memory alloy effect). If such smart materials can be used in the plastic hinge regions of reinforced concrete (RC) beam-column joints as reinforcement, they can undergo large deformations during an earthquake and can return to their undeformed/ original shape after the earthquake. This paper represents the seismic performance of beam-column joint reinforced with three different types of SMA (e.g Ni-Ti, Cu-Al-Mn, Fe based) and compared the result with regular steel-RC beam-column joint. An analytical investigation has been considered to evaluate the seismic performance of smart RC beam-column joint and regular steel-RC joint under reversed cyclic loading. The performance of the beam-column joint is compared in terms of load-story drift ratio and energy dissipation capacity. All SMA-RC beam-column joints show nearly 40-60% higher displacement than steel-RC joint before yielding. Among three types of SMA Ni-Ti dissipate 24% higher energy than any other types of SMA.

Keywords: Super-elasticity, Plastic hinge, Load-storey drift, Energy dissipation.

1 INTRODUCTION

The overall behavior of reinforced concrete moment resisting frame in recent earthquakes all over the world has highlighted the poor performance of the beam-column joint. Beam-column joints are more susceptible to failure than any other members of the building due to severe damage to the joint zone. Steel is being used as reinforcement around the world, but lack of corrosion resistance makes steel vulnerable for long time use. On the other hand, earthquake energy is dissipated through yielding of reinforcement causes permanent deformation of the reinforcement. To address this problem of deteriorating civil infrastructures, researchers are seeking new material, which has enhanced deformation capability and ductility. One of these innovative materials is shaped memory alloy (SMA), which remembers the "Parent" shape.

SMA has introduced a new era for the improvement of the overall performance of concrete frames. An experimental study conducted by Nehdi *et al.* (2011) found that longitudinal SMA bar experienced very negligible residual strain compared to steel. Less residual strain helps material undergo large deformation without permanent damage under cyclic load. In particular,

SMAs have distinct thermomechanical properties, including superelasticity, shape-memory alloy effect, and hysteretic damping. These properties can be effectively utilized in the beam-column joint under reversed cyclic loading. A study by Abdulridha *et al.* (2013) shows that under cyclic loading, the recovery of post-yield deformation of the SMA beam is almost 78% higher than conventionally reinforced beams. Due to the high superelastic property of SMAs, it can be used in the beam-column joint for significant energy dissipation during a seismic event. In this article, it is proposed to use superelastic SMAs in conjunction with steel-RC beam-column joint. Here the beam-column joint is reinforced with three different types of SMAs (Ni-Ti, Fe-based, and Cu-based) in the plastic hinge region and compare the results with steel-RC beam-column joint. In this study, four beam-column joints have been designed and analytically investigated using finite element software SeismoStruct (2015). The prime objective of this study is to investigate the behavior of Steel-RC and SMA-RC beam-column joints under cyclic load in terms of load-story drift ratio, and energy dissipation capacity.

2 DETAILS OF SPECIMENS

In this study, a seven-story RC moment resisting frame building is considered. The exterior beam-column joint is isolated from the building at the points of contra-flexure from the mid-height of the second floor to mid-height of the third floor. The cross-section of the column is 250 mm by 500 mm and reinforced with 12-20 mm diameter longitudinal bar. Cross section of the beam is 250 mm by 400 mm and longitudinal reinforcement is similar for all types of joint except SMA is used only in the plastic hinge region for three joints while the other joint is fully reinforced with steel. The detailing of reinforcement of the joint is shown in Figure 1. The beam-column joints are designated as steel-RC (reinforced with regular steel), SMA-RC-1 (reinforced with SMA-1), SMA-RC-2 (reinforced with SMA-2), SMA-RC-3 (reinforced with SMA-3). SMA-RC-1 is reinforced with 8-20 mm SMA-1 bars, SMA-RC-2 reinforced with 8-16 mm SMA-2 bars, SMA-RC-3 reinforced with 8-28 mm SMA-3 bars. Material properties used for three different types of SMA are shown in Table 1.

Table 1. Properties of different types of SMA (Billah *et al.* 2016).

SMA Types	Alloy	ε_s (%)	E (GPa)	f_Y (MPa)	f_{P1} (MPa)	f_{T1} (MPa)	f_{T2} (MPa)	Reference
SMA-1	$NiTi_{45}$	6	62.5	401.0	510	370	130	Alam *et al.* 2008
SMA-2	FeNCATB	13.5	46.9	750	1,200	300	200	Tanaka *et al.* 2010
SMA-3	CuAlMn	9	28	210	275	200	150	Shrestha *et al.* 2013

f_y (austenite to martensite starting stress); f_{P1} (austenite to martensite finishing stress); f_{T1} (Martensite to austenite starting stress); f_{T2} (martensite to austenite finishing stress); ε_s (superelastic plateau strain length).

In the SMA-RC beam-column joints, SMA is used only in the plastic hinge region of the beam. Rest of the part is detailed with regular steel. The plastic hinge region was calculated using Paulay and Priestley (1992) equation. The plastic hinge length is found to be 320 mm from the face of the column for the RC beam-column joint.

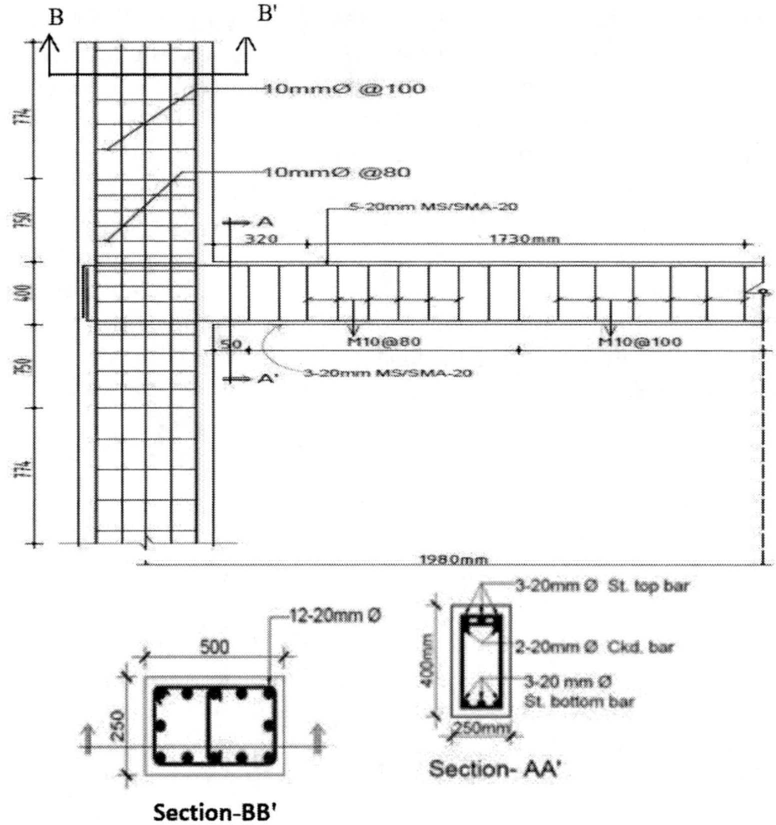

Concrete	
Compressive strength (MPa)	35
Tensile strength (MPa)	3.5
Steel	
Yield strength (MPa)	413.7
Strain hardening parameter	0.005
Modulus of elasticity (GPa)	200

Figure 1. Reinforcement details of beam–column joint and material properties.

3 MOMENT-CURVATURE RELATIONSHIP

The sizes of the rebar are selected in such a way that the axial forces developed in the rebar are almost the same. Figure 2 shows the moment-curvature graph for all types of beam-column joint. From the moment-curvature graph, it is observed that all the beam-column joints have similar stiffness and comparable moment capacity.

4 FINITE ELEMENT MODELING

Nonlinear models of the steel-RC and SMA-RC beam-column joints are developed using SeismoStruct (2015). Fiber modeling technique has been used to represent the distribution of material nonlinearity along the length and cross-section of the beams and columns. Beams and columns are divided into four and two elements, respectively, in the longitudinal direction, and each element, in turn, is subdivided into 200 × 200 fiber elements in the transverse direction. One of the longitudinal elements of the beam represents the plastic hinge region at the beam-column joint. A constant axial load of 1112 kN is applied at the top of the column and reversed cyclic loading is provided at the tip of the beam. A detailed description of finite element modeling approach can be found in Nahar (2018). For validation two beam-column joint one is reinforced with regular steel and another with SMA in the plastic hinge region with a dimension of 250 mm x 400 mm is developed. Reversed cyclic loading is applied at the beam tip and the result is compared in terms of load-story drift ratio and energy dissipation capacity. Ultimate beam tip load is 12% and 6% higher for steel-RC and SMA-RC joint comparing to the result found in the experimental works performed by Alam *et al.* (2008). Energy dissipation capacity is 5.5% higher for both types of the joint while compared with the experimental results by Alam *et al.* (2008). From this validation results, it can be concluded that FE modeling using Seismostruct can predict the experimental results with reasonable accuracy. Moreover, previous studies have also demonstrated the accuracy of the FE program to predict the seismic behavior of SMA-RC bridge pier and SMA-RC frames (Alam *et al.* 2012, Billah and Alam 2012, Billah and Alam 2016).

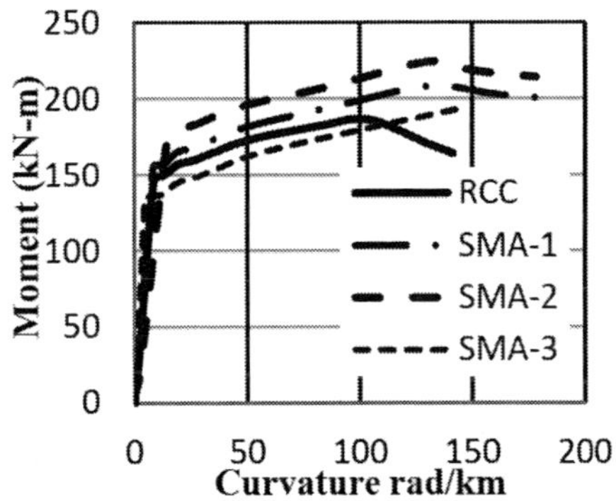

Figure 2. Moment-curvature relationships for steel-RC, SMA-RC-1, SMA-RC-2, SMA-RC-3, beam-column joint.

5 RESULTS AND DISCUSSIONS

5.1 BeamTip Load Versus Story Drift

Figure 3 illustrates the story drift relationship of the beam–column joint with beam tip load. All of the beam-column joints show drifts up to 7%. At 3% story drift, Ni-Ti SMA carry 4 to 8% higher load than SMA-RC-2 and steel-RC joint respectively.

In this study, three performance criteria namely core crushing, yielding and spalling are considered and noted in the figure. All SMA-RC beam-column joints show higher drift before yielding, For Fe-based SMA, i.e., SMA-RC-2, core crushing drift occurs earlier before yielding of reinforcement due to its high yield strength than any other SMAs. At 3% collapse drift defined by Kircil and Polat (2006) SMA-RC-1 carry 5% to 6% higher load than the other two types of SMA.

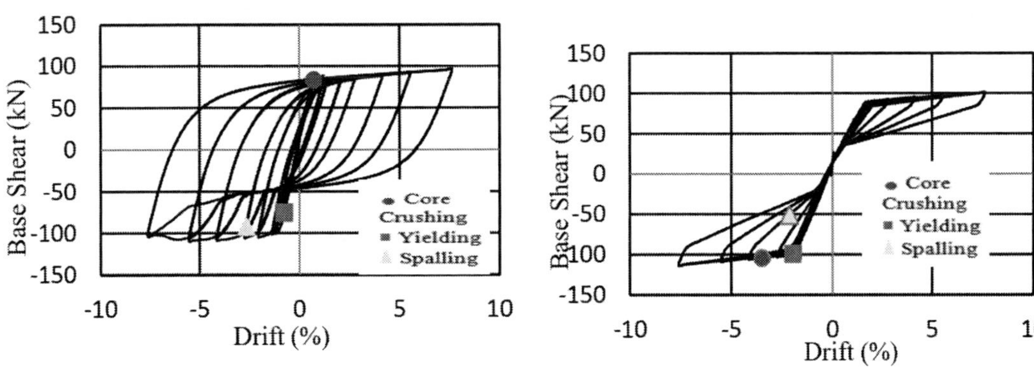

(a) Steel-RC (b) SMA-

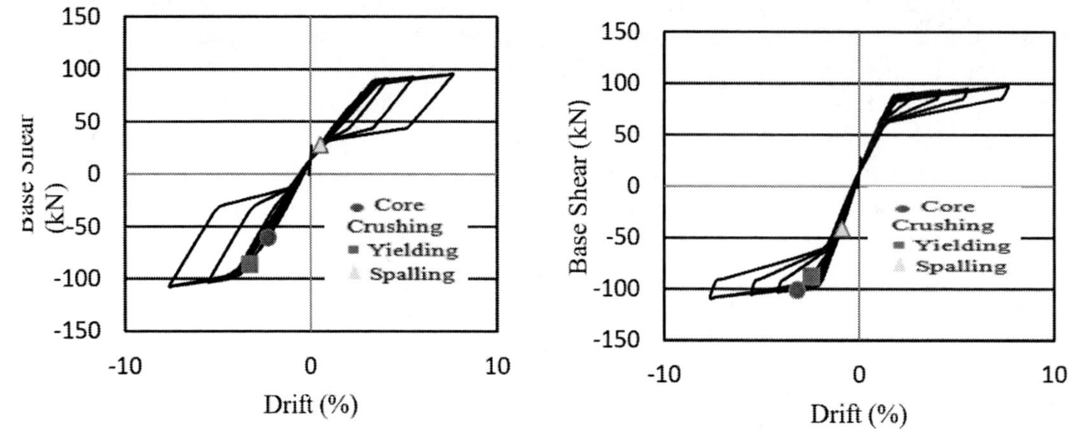

(c) SMA-2 (d) SMA-3

Figure 3. Beam tip load-story drift relationship for (a) steel-RC, (b) SMA-1, (c) SMA-2, (d) SMA-3, beam–column joint.

5.2 Cumulative Energy Dissipation Capacity versus Story Drift

Figure 4 shows the plot of cumulative energy dissipation versus story drift for the four types of the beam–column joints considered in this study. From the figure, it is evident that energy dissipation capacity is higher for the RC-beam–column joint than any other SMA-RC joint due to its larger hysteretic loop. At 4% collapse drift the RC beam-column joint can dissipate 49 kNm of energy, which is almost 65% greater than any other SMA-RC joint.

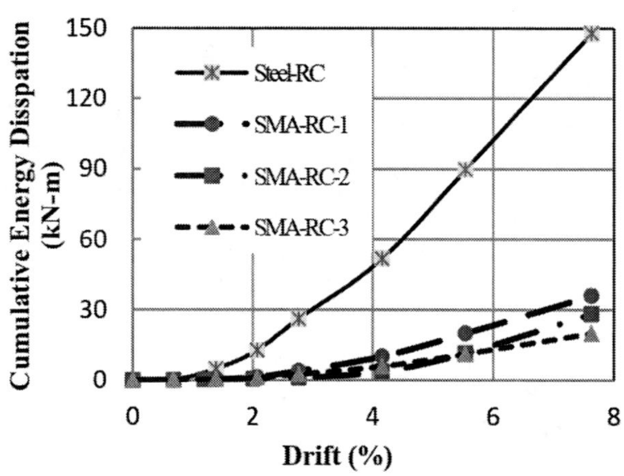

Figure 4. Cumulative energy dissipation-story drift relationships of specimens.

6 CONCLUSIONS

This paper presents a comparative numerical study to investigate the cyclic behavior of Steel-RC and SMA-RC beam-column joints in terms of load-story drift ratio and energy dissipation capacity and shows the potential of using SMA rebars in the plastic hinge region of the RC beam-column joints. The key findings from the current study are summarized below:

- All SMA-RC beam-column joints show similar load carrying capacity compared to steel-RC joint, however, Ni-Ti based SMA carry 7% higher load than any other types of joint at 6% drift ratio.
- The load carrying capacity of SMA-RC beam-column joint is 21-35% higher than steel-RC joint before yielding of reinforcement due to SMAs large superelastic property.
- Although Fe-based SMA dissipates less energy compared to other types of SMA at the lowest drift, however, it can dissipate a significant amount of energy at higher drift.
- Due to the inelastic property of SMA and notably higher load carrying capacity of SMA-RC joint compare to the steel-RC joint before yielding provides a better solution against permanent deformation during seismic events.

References

Alam, M. S., Youssef, M. A., and Nehdi, M., Analytical Prediction of The Seismic Behaviour of Superelastic Shape Memory Alloy Reinforced Concrete Elements, *Engineering Structures,* 30(12), 3399–3411, 2008.

Alam, M. S., Moni, M., and Tesfamariam, S., Seismic Overstrength and Ductility of Concrete Buildings Reinforced with Superelastic Shape Memory Alloy Rebar, *Engineering Structures,* 34, 8-20, 2012.

Abdulridha, A., Palermo, D., Foo, S., and Vecchio, F. J., Behavior and Modeling of Superelastic Shape Memory Alloy Reinforced Concrete Beams, *Engineering Structure,* 49, 893–904, 2013.

Billah, A. H. M. M., and Alam, M. S., Seismic Performance of Concrete Columns Reinforced with Hybrid Shape Memory Alloy (SMA) and Fiber Reinforced Polymer (FRP) Bars, *Construction and Building Materials*, 28(1), 730–742, 2012.

Billah, A. H. M. M., and Alam, M. S., Performance-based Seismic design of Shape Memory Alloy–Reinforced Concrete Bridge Piers. I: Development of Performance-based Damage States, *Journal of Structural Engineering*, 142(12), 04016140, 2016.

Kircil, M. S., and Polat, Z, Fragility Analysis of Mid-Rise R/C Frame Buildings, *Engineering Structures*, 28(9), 1335–1345, 2006.

Nehdi, M., Alam, M. S., and Youssef, M. A., Seismic Behaviour of Repaired Super Elastic Shape Memory Alloy Reinforced Concrete Beam-Column Join, *Smart Structures and Systems*, 7(5), 329-348, 2011.

Nahar, M., *Seismic Collapse Safety Assessment of RC Beam-Column Joint Reinforced with Superelastic Shape Memory Alloy*, MSc Thesis, Department of Civil Engineering, Military Institute of Science and Technology, 2018.

Paulay, T. and Priestley, M. N. J., *Seismic Design of Reinforced Concrete and Masonry Buildings,* New York, John Willey & Sons, Inc, 1992.

SeismoStruct, SeismoSoft, A Computer Program for Static and Dynamic Nonlinear Analysis of Framed Structures, V 7.0.4., 2015.

Interdependence between Structural Engineering and Construction Management
Edited by Ozevin, D., Ataei, H., Modares, M., Gurgun, A., Yazdani, S., and Singh, A.

ISBN: 978-0-9960437-6-2

BOND AND SHEAR-STRENGTHENING PERFORMANCE OF FRCM COMPOSITES

ADEL YOUNIS and USAMA EBEAD

Dept of Civil and Architectural Engineering, Qatar University, Doha, Qatar

This paper is aimed at studying the bond and shear-strengthening performance of fabric reinforced cementitious matrix (FRCM) systems. Three FRCM systems were compared, namely, polyparaphenylene benzobisoxazole (PBO)-FRCM, Carbon-FRCM, and Glass-FRCM. At first, six double-shear specimens were tested to investigate the FRCM/concrete bond, with the test variables including the fabric type and the bond length. After that, seven shear-critical reinforced concrete (RC) beams were tested under three-point loading, considering the fabric type and strengthening configuration (full/intermittent) as the test variables. As for the double-shear test results, the failure observed was fabric/matrix debonding in carbon-FRCM, matrix/concrete debonding in PBO-FRCM, and fabric rapture in glass-FRCM. The FRCM/concrete bond increased with the bonded length, and the PBO-FRCM showed the highest bond to concrete. Regarding the RC beam tests, the FRCM-strengthened beams showed the same failure mode that is debonding at the FRCM/concrete interface. Nonetheless, FRCM had successfully strengthened the beams in shear: an average gain of 57% in the load carrying capacity was achieved as compared to the non-strengthened reference. Indeed, the full-length strengthening resulted in a better structural improvement compared to the intermittent-strengthening configuration. Amongst the three systems, carbon-FRCM systems were the most efficient in shear-strengthening RC beams.

Keywords: Textile-reinforced mortar, Fabric-reinforced cementitious matrix, Bond capacity, Debonding, Rehabilitation.

1 INTRODUCTION

Deterioration of reinforced concrete (RC) structures is inevitable due to the various load and/or environmental exposures encountered during their life cycle (Younis *et al.* 2018), therefore, different strengthening techniques were proposed as a remedial measure such as ferrocement (Ebead 2015) and fiber reinforced polymer (FRP) (Ebead 2011, Ebead and Saeed 2014). Recently, fabric reinforced cementitious matrix (FRCM) has emerged as a viable solution for RC strengthening. FRCM includes dry fibers impregnated in a cementitious matrix, and is surface-applied for strengthening RC and masonry structures (Akbari Hadad *et al.* 2018, Ebead *et al.* 2017, Pino *et al.* 2016, Younis *et al.* 2017a). FRCM was demonstrated as an efficient technique for shear strengthening RC beams (Wakjira and Ebead 2018a, Wakjira and Ebead 2018b, Younis *et al.* 2017b, Younis *et al.* 2017c). Nonetheless, it is agreed that a sufficient understanding of the FRCM/concrete bond is essential to maximize its potential use in RC strengthening: some research efforts have been dedicated in the past few years for such purpose (Raoof *et al.* 2016, Shrestha *et al.* 2017, Younis and Ebead 2018). Given that, the current paper has a two-fold objective: (a) to investigate the bond between different FRCM systems and concrete; (b) to investigate the effectiveness of FRCM systems in shear-strengthening RC beams, with more focus on the failure behavior.

2 EXPERIMENTAL PROGRAM

2.1 Materials

Ready-mixed concrete with a 28-day compression strength of 30 MPa was used to cast the RC beams and the double-shear specimens. Grade 500B (ISE/104 Committee 2005) steel bars were used to reinforce the concrete beams. The yield stress, yield strain, and modulus of elasticity of the reinforcing bars were measured as 595 MPa, 0.266%, and 224 GPa, respectively. Three FRCM systems were adopted, namely, carbon-FRCM (Ruredil 2016a), polyparaphenylene benzobisoxazole (PBO)-FRCM (Ruredil 2016b), and glass-FRCM (SIKA 2016). The FRCM system typically included a single layer of fabric impregnated between an internal (attached to concrete) and external mortar layers. Table 1 lists the mechanical properties of each fabric provided by the manufacturer.

Table 1. Properties of fabric (warp direction) for the FRCM systems.

Fabric type	Area per unit width (mm^2/mm)	Elastic modulus (GPa)	Tensile Strength (GPa)	Ultimate strain (%)
PBO	0.045	270	5.80	2.15
Carbon	0.047	240	4.8	1.8
Glass	0.047	80	2.6	3.25

2.2 Test Specimens

For the double-shear test, six 150-mm concrete cubes were prepared: for each, FRCM system was applied to create a double-shear connection (Figure 1-a). A uniform bond width of 100 mm was considered. Two test parameters were investigated as shown in Table 2: (i) fabric type (glass/carbon/PBO); and (ii) bond length (75 or 100 mm).

For the beam test, seven shear-critical RC beams (of 2,500 mm in length, 150 mm in width, and 330 mm in depth as shown in Figure 1-b) were cast, of which one was kept non-strengthened and used as a reference. The rest of the beam specimens were strengthened considering two test variables as shown in Table 3, namely, (i) fabric type (carbon/PBO/glass); and (ii) strengthening configuration: a single full-length FRCM plate or a set of three intermittent 120-mm wide FRCM strips were applied along the critical shear zone.

2.3 Test Setup

For the double-shear test, FRCM was axially loaded by stretching the fabric from the specimen's opposite sides using a hydraulic jack operated at an approximate rate of 4 mm/min (Figure 1-a). A curved-edge steel plate was placed between the hydraulic jack and the stretched fabric so that the load was uniformly distributed within the fabric mesh and, hence, stress concentrations were mitigated. A cylindrical load cell of 500-kN capacity was placed between the hydraulic jack and the test specimen to monitor the load applied. The FRCM slip was measured using a linear variable displacement transducer (LVDT).

The RC beams were tested under three-point loading: each beam was subjected to a displacement-controlled loading at a rate of 1 mm/min. Vertical displacement was measured by an LVDT at the location of the loading point (Figure 1-b). Data acquisition of the measurements was performed at a frequency of 1 Hz.

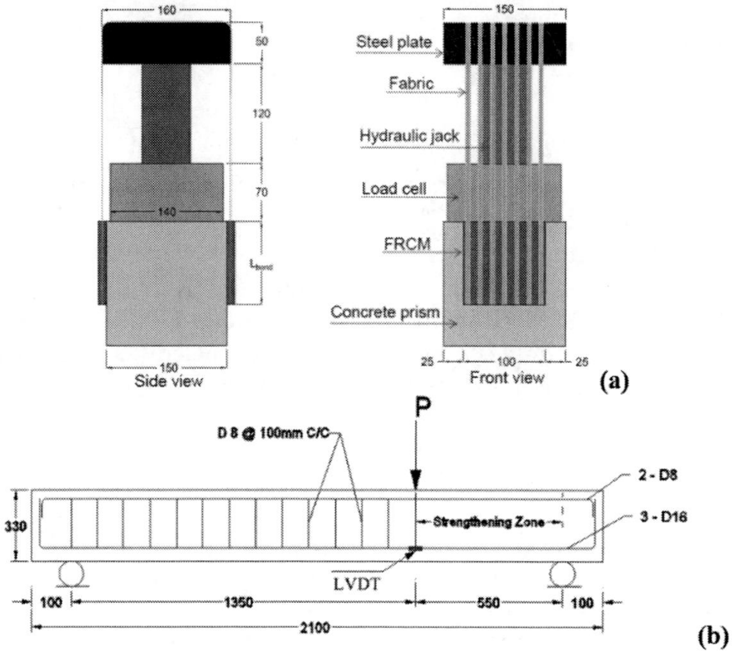

Figure 1. Details for (a) double-shear test specimen, and (b) RC beam specimen.

3 RESULTS AND DISCUSSION

3.1 Bond Performance

Table 2 presents a summary of the double-shear test results. The ultimate shear stress (τ_u) was calculated as the failure load (P_u) divided by the total FRCM bonded area. The average shear-stress capacity of the FRCM/concrete bond was 530, 273, and 265 kPa for PBO-, carbon-, and glass-FRCM systems, respectively; thus, the PBO-FRCM had the highest bond to concrete. As intuitively expected, the FRCM/concrete bond increased with the bond length: the average improvement in the FRCM/concrete bond as a result of increasing the bond length (from 75 to 100 mm) was 94, 37, and 81% for carbon-, PBO-, and glass-FRCM systems, respectively.

Table 2. Summary of the bond test results.

Specimen	Fabric type	Bond length (mm)	P_u (kN)	τ_u (kPa)	δ_u (mm)	Failure mode*
C-L75	Carbon	75	3.34	222	4.2	FD
C-L100	Carbon	100	6.48	324	5.5	FD
P-L75	PBO	75	7.85	523	≈ zero	DB
P-L100	PBO	100	10.75	538	0.03	DB
G-L75	Glass	75	3.38	225	0.03	FR
G-L100	Glass	100	6.11	305	0.09	FR

*FD: Fabric/matrix debonding; DB: Debonding at the concrete/matrix interface; FR: Fabric rapture.

Regarding the failure modes, the carbon-FRCM system showed debonding at the fabric/matrix interface (FD) (Figure 2-a). The PBO-FRCM system exhibited debonding at the concrete/matrix interface (DB) (Figure 2-b). The failure mode for the glass-FRCM system, however, was a premature rupture of the stretched fabric (FR) (Figure 2-c) – mostly attributed to the relatively lower tensile strength of the glass fibers (Table 1). Thus, the bond failure was more brittle in glass- and PBO-FRCM systems; this can be further evidenced by their lower FRCM slip at failure (δ_u) as compared to that of carbon-FRCM (Table 2).

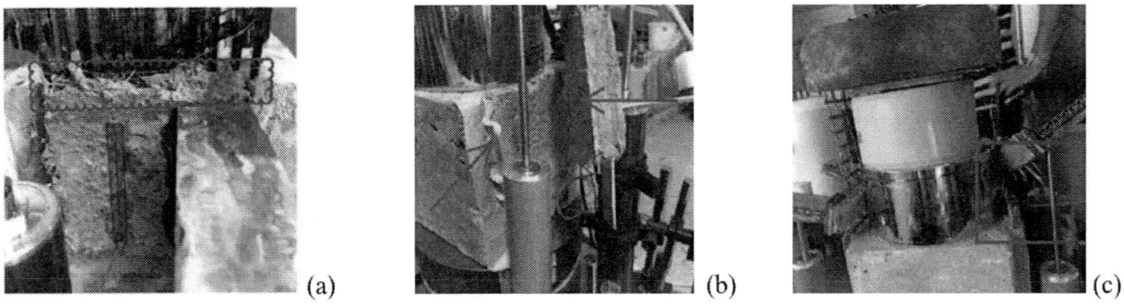

Figure 2. Failure modes – (a) FD in carbon-FRCM; (b) DB in PBO-FRCM; and (c) FR in glass-FRCM.

3.2 Beam Specimens

Table 3 presents a summary of the test results for the RC beam specimens. The load carrying capacity of the FRCM-strengthened beams was generally higher than that of the non-strengthened reference; this demonstrates the successful implementation of FRCM systems in shear-strengthening RC beams. The average improvement in the load carrying capacity was 86.2% for carbon, 39.0% for PBO, and 46.6% for glass FRCM systems. The carbon-FRCM strengthening system showed the most improvement in P_u amongst the three systems, despite it having lower bond capacity than the PBO-FRCM. This can be attributed to the fact that, unlike PBO-FRCM, the failure at the carbon-FRCM/concrete bond occurred not at the concrete/matrix interface, suggesting a better load-transfer mechanism between the RC beam and the FRCM strengthening layer. The effect of the FRCM amount was proved significant, evidently from the fact that the full-strengthened RC beams showed a significantly higher improvement in P_u as compared to those with the intermittent strengthening configuration. Concerning deformational behavior, the deflection at failure (δ_u) for the FRCM-strengthened beams was generally higher than that of the reference. Also, higher values of δ_u were reported for the carbon-FRCM-strengthened beams compared to those with glass- and PBO-FRCM strengthening systems. Finally, the full-length-strengthened beams showed slightly higher values of δ_u compared to those with the intermittent strengthening scheme.

Table 3. Summary of the beam test results.

Specimen	Fabric type	Strengthening pattern	P_u (kN)	Gain in P_u (%)	δ_u (mm)	$\frac{\delta_u}{\delta_{u,ref}}$
R	-	-	104.0	-	3.25	-
C-F	Carbon	Full	209.7	101.6	7.75	2.14
C-I	Carbon	Intermittent	177.6	70.8	7.55	2.13
P-F	PBO	Full	151.3	45.5	5.35	1.65
P-I	PBO	Intermittent	137.7	32.4	4.6	1.42
G-F	Glass	Full	167.9	61.4	6.10	1.88
G-I	Glass	Intermittent	137.0	31.7	4.43	1.36

Figure 3 shows schematic drawings for the crack patterns and failure modes of the tested beams. The reference beam showed a typical shear-failure within the critical span, in which a major crack started at the load point and propagated downwards at approximately 45°. Against this, the FRCM-strengthened beams generally showed an FRCM-debonding mode of failure (mostly observed near the point load). Compared to the intermittent strengthening configuration, the full-length strengthening led to higher ultimate loads because of the large bonded area between the FRCM and concrete substrate. Nonetheless, it was realized that applying the FRCM in an externally-bonded form had not achieved its full strengthening potential: Wakjira and Ebead (2018a, 2018b) reported that near-surface embedding of the FRCM could effectively mitigate the FRCM/concrete deboning and thus maximize the shear-strengthening effectiveness of FRCM.

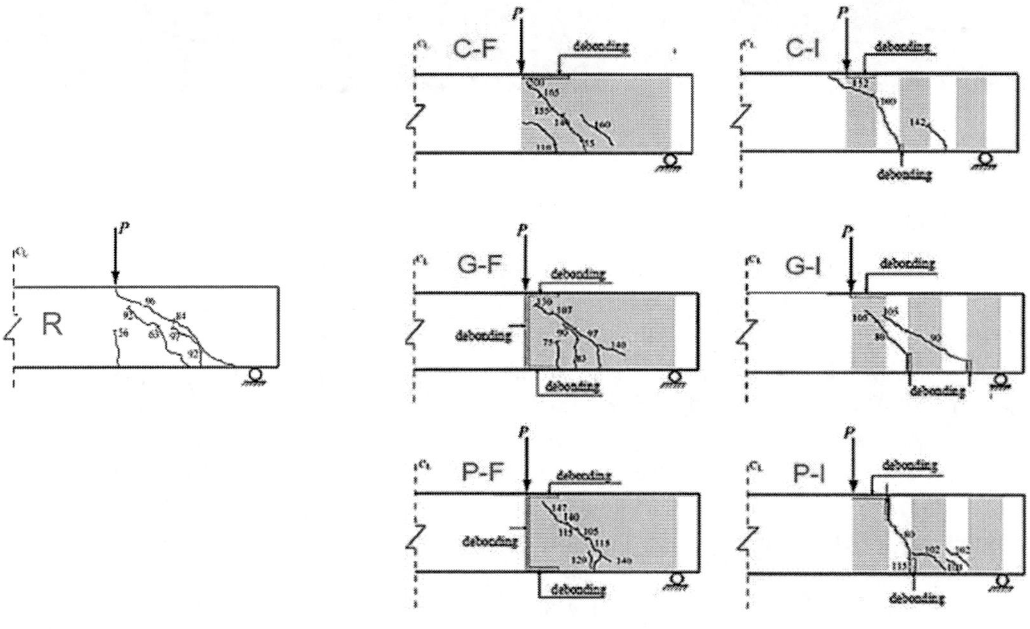

Figure 3. Crack patterns and modes of failure of the tested beams.

4 CONCLUSIONS

This paper investigated the bond and shear-strengthening performance of three FRCM systems. Based on the study results, the following conclusions have been drawn:

- The modes of failure observed in carbon-, PBO-, and glass-FRCM double-shear tests are fabric/matrix debonding, matrix/concrete debonding, and fabric rapture, respectively. Amongst the three systems, the PBO-FRCM showed the highest bond to concrete. The FRCM/concrete bond was observed to improve with an increase in the bond length.
- All strengthened beams were subjected to the same failure mode, namely, debonding at the FRCM/concrete interface. The average gain in the load carrying capacity achieved with FRCM-strengthening was 57%, indicating the successful implementation of FRCM to shear-strengthen RC beams. The full-length strengthening yielded a better strength improvement than that of the intermittent counterpart. Amongst the three systems, carbon-FRCM systems were the most efficient in shear-strengthening RC beams.

Acknowledgments

This paper was made possible by Internal grant # QUST-CENG-SPR-14/15-15 "Textile reinforced mortar for shear strengthening of reinforced concrete beams" from Qatar University. The findings achieved herein are solely the responsibility of the authors.

References

Akbari Hadad, H., Nanni, A., Ebead, U. A., and El Refai, A., Static and Fatigue Performance of FRCM-Strengthened Concrete Beams, *Journal of Composites for Construction*, 22(5), 2018.

Ebead, U., Hybrid Externally Bonded/Mechanically Fastened Fiber-Reinforced Polymer for RC Beam Strengthening, *ACI Structural Journal*, American Concrete Institute, 108(6), 669, 2011.

Ebead, U., Inexpensive Strengthening Technique for Partially Loaded Reinforced Concrete Beams: Experimental Study, *Journal of Materials in Civil Engineering*, American Society of Civil Engineers, 11(10), 1–11, 2015.

Ebead, U. A., and Saeed, H., Flexural and Interfacial Behavior of Externally Bonded/Mechanically Fastened Fiber-Reinforced Polymer Strengthened Reinforced Concrete Beams, *ACI Structural Journal*, American Concrete Institute, 111(4), 741–752, 2014.

Ebead, U., Shrestha, K. C., Afzal, M. S., El Refai, A., and Nanni, A., Effectiveness of Fabric-Reinforced Cementitious Matrix in Strengthening Reinforced Concrete Beams, *Journal of Composites for Construction*, American Society of Civil Engineers, 21(2), 04016084, 2017.

ISE/104 Committee, *BS 4449:2005: Steel for the Reinforcement of Concrete. Weldable Reinforcing Steel. Bar, Coil and Decoiled Product*, BSI, 2005.

Pino, V., Hadad, H. A., De Caso, F., Nanni, A., Ebead, U. A., and El Refai, A., *Performance of FRCM Strengthened RC Beams Subject to Fatigue,* Proceedings of the 4th International Conference in Sustainable Construction Materials and Technologies (SCMT4), University of Nevada, Las Vegas, Las Vegas, Nevada, USA, 2016.

Raoof, S. M., Koutas, L. N., and Bournas, D. A., Bond Between Textile-Reinforced Mortar (TRM) and Concrete Substrates: Experimental Investigation, *Composites Part B: Engineering*, 98, 350–361, 2016.

Ruredil, *Technical Datasheet, Ruredil X Mesh C10 Data Sheet,* 2016a.

Ruredil, *Technical Datasheet, Ruredil X Mesh Gold Data Sheet,* 2016b.

Shrestha, K. C., Ebead, U., and Younis, A., *Effect of Surface Roughening on Concrete/TRM Bond*, Proceedings of the Ninth International Structural Engineering and Construction Conference, Resilient Structures and Sustainable Construction, ISEC Press, Valencia, Spain, St-43, 2017.

SIKA, *Technical Datasheet, Sikawrap-350G Grid Data Sheet,* 2016.

Wakjira, T. G., and Ebead, U., Hybrid NSE/EB Technique for Shear Strengthening of Reinforced Concrete Beams Using FRCM: Experimental Study, *Construction and Building Materials*, 164, 164–177, 2018a.

Wakjira, T. G., and Ebead, U., FRCM/Internal Transverse Shear Reinforcement Interaction in Shear Strengthened RC Beams, *Composite Structures*, 201, 326–339, 2018b.

Younis, A., and Ebead, U., Bond Characteristics of Different FRCM Systems, *Construction and Building Materials*, 175, 610–620, 2018.

Younis, A., Ebead, U. A., and Shrestha, K. C., *Tensile Characterization of Textile Reinforced Mortar,* Proceedings of the Ninth International Structural Engineering and Construction Conference, Resilient Structures and Sustainable Construction, ISEC Press, Valencia, Spain, St-28, 2017a.

Younis, A., Ebead, U. A., and Shrestha, K. C., *FRCM Shear Strengthening for Concrete Beams,* Proceedings of the Ninth International Structural Engineering and Construction Conference, Resilient Structures and Sustainable Construction, ISEC Press, Valencia, Spain, St-27, 2017b.

Younis, A., Ebead, U., and Shrestha, K. C., Different FRCM Systems for Shear-Strengthening of Reinforced Concrete Beams, *Construction and Building Materials*, 153, 514–526, 2017c.

Younis, A., Ebead, U., and Judd, S., Life Cycle Cost Analysis of Structural Concrete Using Seawater, Recycled Concrete Aggregate, and GFRP Reinforcement, *Construction and Building Materials*, Elsevier Ltd, 175, 152–160, 2018.

Interdependence between Structural Engineering and Construction Management
Edited by Ozevin, D., Ataei, H., Modares, M., Gurgun, A., Yazdani, S., and Singh, A.

ISBN: 978-0-9960437-6-2

FRP-FLEXURAL STRENGTHENING OF STEEL BEAMS: A FINITE ELEMENT ANALYSIS STUDY

BIBEK REGMI BAGALE and AZADEH PARVIN

Dept of Civil and Environmental Engineering, The University of Toledo, Toledo, USA

Fiber reinforced polymer (FRP) strengthening has been widely used for concrete structures. However, limited studies have been conducted on its application to steel structures. The use of FRP materials to retrofit steel members is an effective replacement for welded or bolted steel plates and offer high strength-to-weight ratio, resilience to environmental degradation, and a robust fatigue performance. Past studies have shown that the application of FRP to steel is a promising technical solution to help against corrosion and also to increase the load carrying capacity of steel members. In the present study, finite element analysis (FEA) model of a steel beam strengthened with carbon fiber reinforced polymer (CFRP) laminates was developed using ANSYS Mechanical finite element software program. The model was validated by comparing the FEA and experimental results of an existing study in the literature and they were in good agreement. A parametric study was conducted on the validated steel beam model with the aim to investigate the effects of basalt fiber reinforced polymer (BFRP) and two types of adhesives on its performance under static loading. The FRP laminates provided an increase in flexural capacity and pseudo ductility of the beams. The optimum bond length was found to be 2/5 of the span length for the beam in this study.

Keywords: High-strength material, Bond length, Retrofit, Load capacity, Flexure, ANSYS.

1 INTRODUCTION

Large number of steel buildings and bridges in the US are deteriorating due to aging, harsh environmental exposure, creep, fatigue and lack of proper maintenance, (Al-Saidy *et al.* 2004). Most steel structures today are in need of repair and retrofit. Traditional retrofit techniques involve cutting out and replacing steel plates or attaching steel plates by welding or bolting externally which are problematic in terms of scale and workability (Zhaoa *et al.* 2007). Use of fiber reinforced polymers offers a great alternative retrofit scheme which is less labor intensive and has the ability to increase stiffness as well as being corrosion resistant (Linghoff *et al.* 2010). Colombi *et al.* (2006) conducted experimental and analytical study on steel beams reinforced with pultruded CFRP strips and observed 9% to 23% increase in load capacity and also considerable increase in stiffness. In addition to flexural strengthening, the fatigue life of the beams reinforced with CFRP plates was significantly longer than that of the beams repaired only with the welding method (Jiao *et al.* 2012). The strength of external FRP retrofit depends mainly on the bond between FRP and steel. Study performed by Linghoff *et al.* (2009) concluded that surface preperation plays an important role in the bond strength. Limited studies have been performed on FRP appication to steel structures as opposed to concrete structures. Furthermore, most FRP strengthening of steel structures involves carbon fiber reinforced polymer (CFRP). However, to the best of authors knowledge no study has been performed on basalt fiber reinforced polymer (BFRP) stregthened steel members. The focus of the present study is to numerically investigate the effect of CFRP and BFRP plates and various adhesives on the performance of steel beams.

2 FEA MODELING

Finite element analysis modeling of the steel beams was performed using the software program ANSYS Mechanical APDL. In order to reduce calculation time and due to symmetry only half of the beam adopted from experimental study by Lenwari *et al.* (2005) was modeled. The dimensions and material properties of the beams are listed in Table 1. SOLID185 and SOLID65 elements with eight nodes and three degrees of freedom at each node were employed to model steel beams and FRP plates, respectively. Steel pads were modeled using SOLID45 element, and placed at the locations of loading points and supports to distribute the load and to avoid stress concentration. For the validation purpose, beams retrofitted with three CFRP laminate lengths of 500, 650, and 1,200 mm were modeled. Figure 1 shows the steel and FRP materials stress versus strain relationships, where steel was assumed as multi-linear with elastic yield stress of 250 MPa and tangent modulus of 1,450 MPa; the FRP laminate and adhesive were assumed to depict elastic-linear behavior until failure.

Table 1. Material properties.

Description	Dimensions	Elastic modulus (GPa)	Poisson's ratio	Ultimate strain (%)	Density (g/cm^3)
Steel Beam	W100 x 17.2	200	0.26		7.8
SikaCarboDur H514 (C)	50 mm x 1.4 mm	300	0.3	>0.45%	1.6
SikaCarbodur S512 (C')	50 mm x 1.2 mm	165	0.3	>1.7%	1.6
Sikadur30	1 mm thick	12.2	0.3	>0.29%	
Sikadur300	1 mm thick	3.5	0.3	>1.5%	
BFRP	50 mm x 1.4 mm	89	0.3	>3.15%	2.7

Bilinear bond slip model was employed for the interface between the steel beam (substrate) and the FRP laminate (reinforcement) based on the study by (Xia and Teng 2005) (Figure 2). To model the interface the elements TARGET170 and CONTA174 were used. The bond-slip model requires the maximum shear, slip at maximum shear, and slip at debonding. The model proposed by Xia and Teng (2005) helps calculate the maximum elastic slip given by Eq. (1):

$$\delta_1 = \tau_f t_a / G_a \tag{1}$$

δ_1 is the slip at local-bond strength τ_f which is calculated from Eq. (2):

$$\tau_f = 0.8\ \sigma_{max} \tag{2}$$

where σ_{max} is the tensile strength of adhesive; t_a is the thickness of the adhesive layer, G_a is the shear modulus of the adhesive. The value of slip at complete debonding is given by Eq. (3):

$$\delta_f = 62\ (\sigma_{max}/G_a)^{0.56}\ (t_a^{0.27}/\ \tau_f) \tag{3}$$

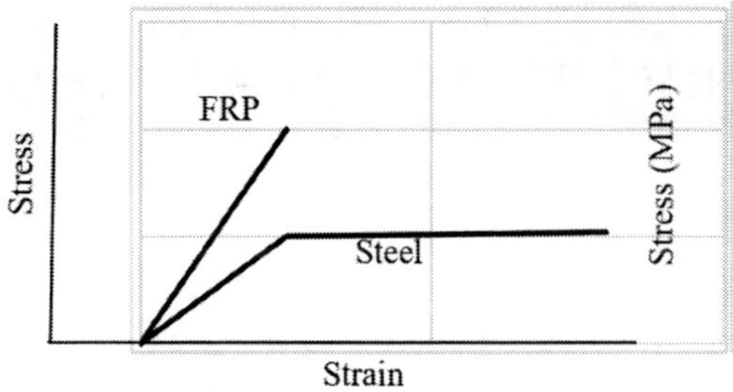

Figure 1. Stress-strain curves.

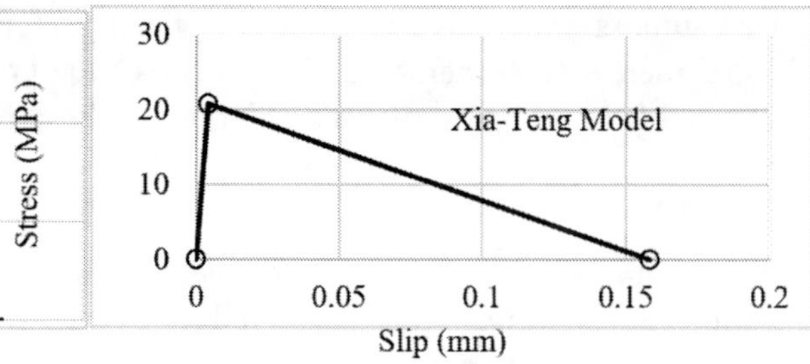

Figure 2. Bond-slip model.

3 VALIDATION OF FEA BEAM MODELS

The experimental study by Lenwari *et al.* (2005) was selected for validation of FEA models in order to perform further parametric study. The beams in the experiment had two types of failures; debonding and rupture of FRP plates. To avoid localized buckling of compression flange and yielding in compression, a steel plate was attached to the compression flange of the beams in the experiment which was also modeled in FEA. The beam models with three different lengths of CFRP plates were validated through comparing the experimental data with FEA results discussed in later section. Figure 3 shows the schematic diagram of the four-point bending test for the simply-supported steel beam.

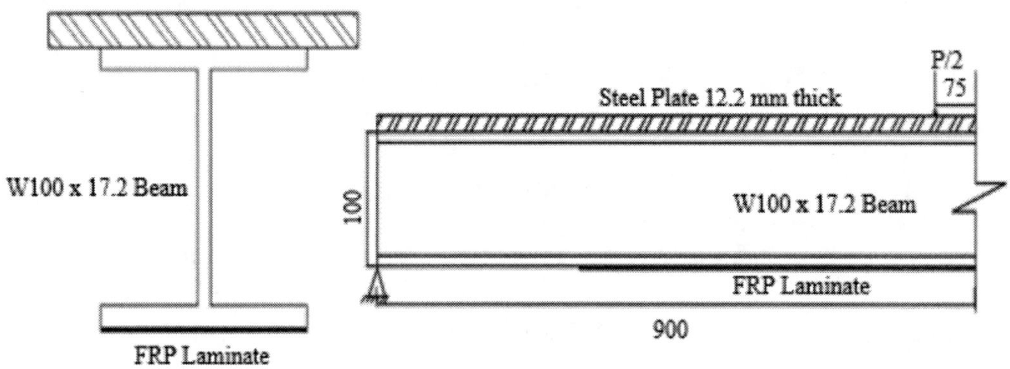

Figure 3. Schematic diagram of beam setup.

4 RESULTS AND DISCUSSIONS

The comparison of experiment and FEA and parametric study results are shown in Table 2. The FEA models were identified as follows: the first letter indicates the member being a beam, first number refers to bond length, second letter indicates the FRP types (CFRP: C for high, and C' for low modulus, and B for BFRP) and the last number indicates adhesive types (Sikadur 30 and Sikadur 300). For example, B750-C'-30 represents the beam with 750 mm bond length, SikaCarbodur S512 FRP plate and SikaDur30 adhesive. The validated FEA beam models, B500-C-30, B650-C-30 and B1200-C-30 correspond to B50-1, B65-1 and B120-1 specimens of the experimental study, respectively. Whereas, B750-C-30, B750-C'-30, B750-B-30 and B750-B-300 represent the beams modeled to investigate the effect of parameters such as bond length, FRP and adhesive types. The failure load results of validated FEA beam models and tested specimens from the literature (Lenwari *et al.* 2005) were in good agreement with less than 4% discrepancy. The modes of failure in FEA steel beam models were also identical to the ones in the experiment.

Table 2. Load and mode of failure.

Beam Identification	Mode of Failure	Failure Load (KN)		Difference between Experimental and FEA Results (%)
		Experiment	FEA	
B500-C-30	Debonding	90.2	94.16	4
B650-C-30	Debonding	105.9	109.24	3
B750-C-30	Plate rupture	-	137.67	-
B750-C'-30	Plate rupture	-	148.1	-
B750-B-30	Plate rupture	-	154.88	-
B750-B-300	Debonding	-	83.6	-
B1200-C-30	Plate rupture	143	144.2	1

As shown in Figure 4 (a), after certain increase in the FRP laminate length, the load deflection curves were almost similar. The FRP laminate length ranged from 500 to 1200 mm and when it changed from 750 mm to 1200 mm the failure load did not change significantly. When BFRP laminate was employed, the strengthened beam observed lower failure load, but had higher displacement capacity making it more ductile as illustrated in Figure 4 (b). For the BFRP strengthened beam, the laminate length was 750 mm and two different types of adhesives were tested, BFRP-SikaDur30 combination failed by rupture whereas BFRP-SikaDur300 combination failed at a much lower load by debonding as shown in Figure 4(c).

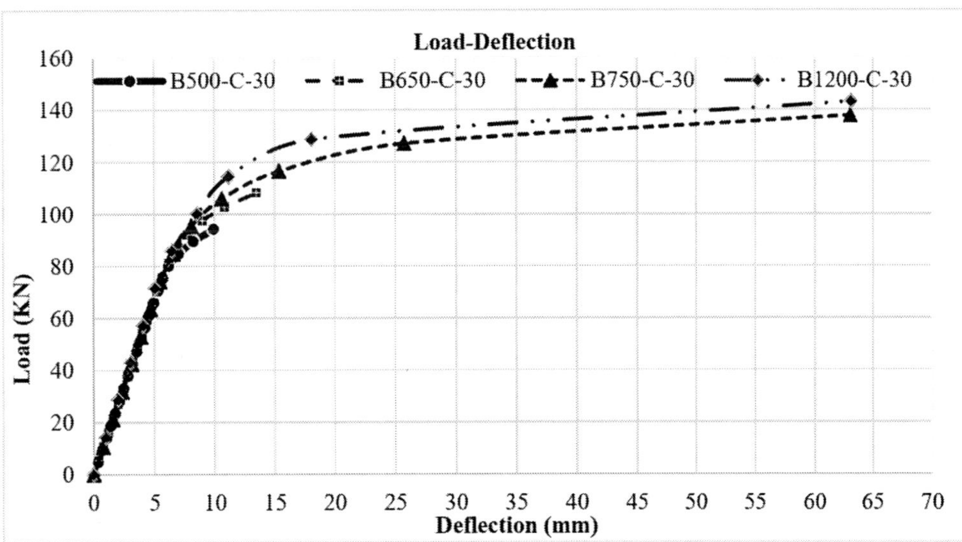

Figure 4 (a). CFRP-strengthened steel beam.

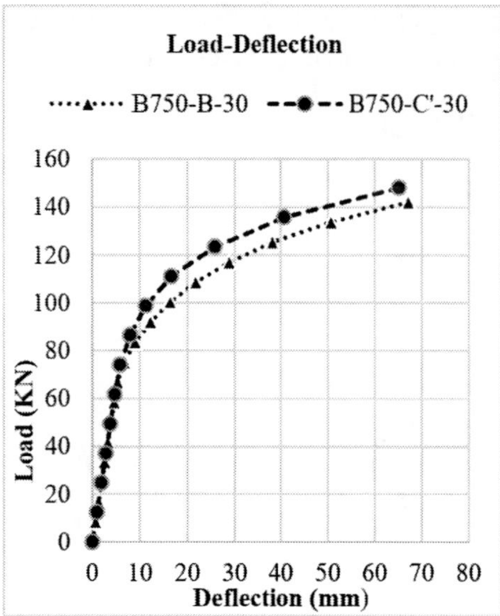

Figure 4 (b). Beam with CFRP and BFRP plates bonded with Sikadur30.

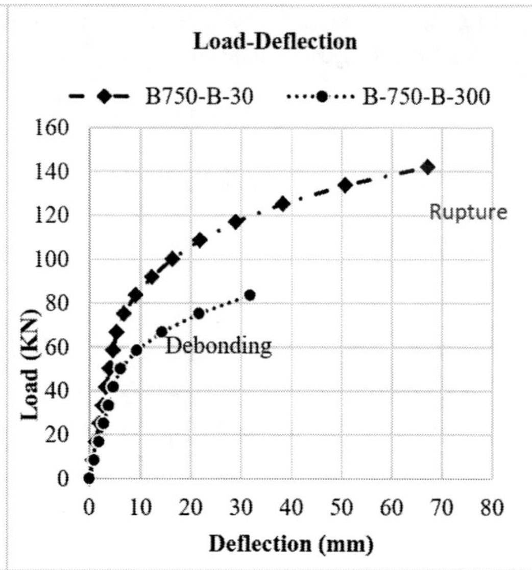

Figure 4 (c). Beams with BFRP plates bonded with different types of adhesives.

5 CONCLUSIONS

In the present study, FEA beam models were developed and validated with the experimental data available in the literature. The effect of bond length, various FRP, and adhesive types were explored. The following conclusions were drawn.

i. Application of CFRP laminates to the tension flange of steel beam increased the flexural capacity and pseudo ductility of the beam.
ii. For partial length lay-ups, after certain increase in the laminate length load capacity did not change significantly (the optimum length was 750 mm equivalent to 2/5 of the beam span) which is useful to know for the design of an equivalent steel beam.
iii. Combination of BFRP laminate with SikaDur30 provided higher load capacity for the beam as compared to bonding the BFRP to the steel substrate using SikaDur300.
iv. In summary, the BFRP laminate application to steel beams, currently in need of maintenance, repair or retrofit, offers an innovative strengthening technique as an alternative to applying steel plates by welding or bolting.

Acknowledgments

Graduate student, Mr. Aman Dhakal, contributed in modeling the interface between the FRP plate and steel beam.

References

Al-Saidy A. H., Klaiber, F. W., and Wipf, T. J., Repair of Steel Composite Beams with Carbon Fiber-Reinforced Polymer Plates, *Journal of Composites for Construction,* ASCE, 8(2), 163-172, April, 2004.

Colombi, P., and Poggi, C., An Experimental, Analytical and Numerical Study of the Static Behavior of Steel Beams Reinforced by Pultruded CFRP Strips, *Composites Part B: Engineering*, 37(1), 64-73, January, 2006.

Jiao, H., Mashiri, F., and Zhao, X. L., A Comparative Study on Fatigue Behaviour of Steel Beams Retrofitted with Welding, Pultruded CFRP Plates and Wet Layup CFRP Sheets, *Thin-Walled Structures*, Elsevier, 59, 144-152, October, 2012.

Lenwari, A., Thaksin, T., and Albrecht, P., Flexural Response of Steel Beams Strengthened with Partial-Length CFRP Plates, *Journal of Composites for Construction,* ASCE, 9(4), 296-303, August, 2005.

Linghoff D., Al-Emrani, M., and Kliger, R., Performance of Steel Beams Strengthened with CFRP Laminate-Part 1: Laboratory Tests, *Composites Part B: Engineering,* Elsevier, 41(7), 509-515, January, 2010.

Linghoff, D., Haghani, R., and Al-Emrani, M., Carbon-Fibre Composites for Strengthening Steel Structures, *Thin-Walled Structures,* Elsevier, 47(10), 1048-1058, May, 2009.

Xia, S. H., and Teng, J. G., Behaviour of FRP-to-Steel Bonded Joints, *Proceedings of the International Symposium on Bond Behaviour of FRP in Structures,* Chen and Teng (eds), 411-418, International Institute for FRP in Construction, 2005.

Zhaoa, X. L., and Zhang, L., State-of-the-Art Review on FRP-Strengthened Steel Structures, *Engineering Structures,* Elsevier, 29(4), 1808-1823, August, 2007.

Interdependence between Structural Engineering and Construction Management
Edited by Ozevin, D., Ataei, H., Modares, M., Gurgun, A., Yazdani, S., and Singh, A.

ISBN: 978-0-9960437-6-2

FINITE STRAIN MEASUREMENT USING IMAGE ANALYSIS UNDER REVERSE SHEAR AFTER FORWARD SHEAR

YASUYUKI KATO

Dept of Mechanical Engineering, Nihon University, Tokyo, Japan

The purpose of this research is to investigate the progress of local deformation under finite deformation by using image analysis based on the Natural Strain theory. Since Natural Strain used in the image analysis can satisfy the addition law of strain on an identical line element and can remove the rigid body rotation from shearing strain component, it is an effective strain for representing stress and strain behavior under large deformation. Therefore, in this research, these features of Natural Strain theory will be incorporated into the method of image analysis. In the present study, using the test pieces made of high purity tough pitch copper, the local deformation occurring under large simple shear is investigated by comparing the strains in each element from the upper position to the middle position with the average strain in the gauge length. In order to investigate the progress of local deformation caused by differences of strain hardening in the material, the experiments are conducted under different deformation histories which are given by applying reverse shear after applying forward shear with different sizes. Consequently, it is revealed that if the value of plastic strain obtained by integrating over the whole deformation path is almost the same, the progress of local deformation is approximately the same even if the deformation path is different.

Keywords: Finite deformation, Elasto-plasticity, Large deformation, Natural strain, Plastic strain, Deformation path, Ductile metal.

1 INTRODUCTION

In general, the electrical measurements such as displacement meter and strain gauge have been widely used for the finite strain measurements. Among these measurements, if one considers the case that the strain gauge is used, the strain gauge is peeled off with an increase of deformation. Hence, there is a limit in the range that it can be measured continuously. On the other hand, in the case of using image analysis, it is possible to measure the strain continuously over the wide range from the infinitesimal deformation to the large deformation. Moreover, the measurement by conventional displacement meter is an average strain in the gauge length of the test piece. On the other hand, in the case of using image analysis, it is possible to measure the strain directly at the position where the picture image is taken. Hence, the measurement by image analysis is very effective for local strain measurements. In this research, using the features of Natural Strain theory effectively into the image analysis, the progress of local deformation under large simple shear is investigated. The Natural Strain theory used in this research is the strain expression which is obtained by integrating the infinitesimal strain increment on an identical line element over the whole process of deformation path (Kato and Moriguchi 2003). Consequently, since the shearing strain component of Natural Strain is defined by pure angular strain, the rigid body rotation can be removed completely from the shearing strain component. Moreover, since the addition law of strain can be satisfied, it is possible to represent subsequent strain by adding the strain from anywhere with reference to the deformed intermediate state.

In our previous study (Kato and Futami 2013), in order to confirm the accuracy of this image analysis, the experiments of three types of different deformation paths, i.e., the proportional loading of tension and shear, the shear after applying tension and the tension after applying shear, were conducted. In those experiments, strain measurements by the image analysis have been compared with measurements by conventional displacement meter, and the validity of this image analysis has been verified. However, in these studies, experiments have been done in the range of uniform deformation fields where the local deformation does not still occur. Hence, as for the larger deformation after local deformation began to occur, detailed measurements have not been done yet in our previous study. Therefore, in the previous report (Kato 2015), the progress state of local deformation with respect to the uniaxial state such as tension or shear has been investigated.

Moreover, in the present study, in order to examine the progress of local deformation in more detail, the experiments are conducted under the different deformation paths which are given by applying the reverse shear after applying the forward shear with different sizes. Then, the relation between the developmental process of local deformation and the deformation history of simple shear is revealed.

2 MEASUREMENT METHOD OF PRINCIPAL STRAIN AND LOCAL DEFORMATION

The Figure 1 (a) shows the three-line elements ℓ_{oa} , ℓ_{ob} , ℓ_{oc} which are located at arbitrary directions $\beta_{oa}, \beta_{ob}, \beta_{oc}$. Among these line elements, if we consider a line element ℓ_{oa}, the relation between original un-deformed state and deformed current state of the line element can be derived as in Eq. (1),

$$\ell_a = \boldsymbol{D}\,\ell_{oa} = \begin{bmatrix} D_{11} & D_{12} \\ 0 & D_{22} \end{bmatrix} \begin{bmatrix} \|\ell_{oa}\| \cos\beta_{oa} \\ \|\ell_{oa}\| \sin\beta_{oa} \end{bmatrix} \tag{1}$$

where, $\boldsymbol{D}$ is a deformation gradient tensor. In this instance, the extensional strain component of the Natural Strain theory is expressed by Eq. (2),

$$\varepsilon_a = \frac{1}{2} ln\left(\frac{\ell_a{}^T \ell_a}{\ell_{oa}{}^T \ell_{oa}}\right) = \frac{1}{2} ln\left(\frac{\ell_{oa}{}^T \boldsymbol{D}^T \boldsymbol{D} \ell_{oa}}{\ell_{oa}{}^T \ell_{oa}}\right) = \frac{1}{2} ln\left(D_{11}{}^2 cos^2\beta_{oa} + 2D_{11}D_{12} sin\beta_{oa} cos\beta_{oa} + D_{12}{}^2 sin^2\beta_{oa} + D_{22}{}^2 sin^2\beta_{oa}\right) \tag{2}$$

Similarly, the extensional strain components for other line elements, i.e., ε_b and ε_c , can also be derived. Hence, the following relation is obtained (Kato and Futami 2013).

$$\begin{bmatrix} D_{11}{}^2 \\ D_{11}D_{12} \\ D_{12}{}^2 + D_{22}{}^2 \end{bmatrix} = \begin{bmatrix} cos^2\beta_{oa} & 2\,sin\beta_{oa} cos\beta_{oa} & sin^2\beta_{oa} \\ cos^2\beta_{ob} & 2\,sin\beta_{ob} cos\beta_{ob} & sin^2\beta_{ob} \\ cos^2\beta_{oc} & 2\,sin\beta_{oc} cos\beta_{oc} & sin^2\beta_{oc} \end{bmatrix}^{-1} \begin{bmatrix} exp(2\varepsilon_a) \\ exp(2\varepsilon_b) \\ exp(2\varepsilon_c) \end{bmatrix} \tag{3}$$

As shown in Figure1 (c), if we consider the special case that the line element ℓ_{oa} is fixed on the base surface in an initial state, i.e., $\beta_{oa} = 0$, $\beta_{ob} = \beta_{oa} + \pi/4$, $\beta_{oc} = \beta_{oa} + \pi/2$, the components of deformation gradient tensor are represented through Eq. (4-6).

$$D_{11} = exp(\varepsilon_a) \tag{4}$$

$$D_{12} = exp(2\varepsilon_b - \varepsilon_a) - \frac{1}{2}\left(exp(\varepsilon_a) + exp(2\varepsilon_c - \varepsilon_a)\right) \tag{5}$$

$$D_{22} = \sqrt{exp(2\varepsilon_c) - \left\{exp(2\varepsilon_b - \varepsilon_a) - \frac{1}{2}\left(exp(\varepsilon_a) + exp(2\varepsilon_c - \varepsilon_a)\right)\right\}^2} \tag{6}$$

Moreover, the principal value of strain e_1 and its direction θ_m can be determined from the polar decomposition of deformation gradient, and these are represented in Eq. (7) and Eq. (8).

$$e_1 = ln\,\lambda_1 = ln\left\{\frac{\sqrt{D_{11}^2 + D_{12}^2 + D_{22}^2 + 2D_{11}D_{22}} + \sqrt{D_{11}^2 + D_{12}^2 + D_{22}^2 - 2D_{11}D_{22}}}{2}\right\} \tag{7}$$

$$\theta_m = \frac{\pi}{4} - \frac{1}{2}\tan^{-1}\left\{\frac{D_{11}^2 + D_{12}^2 - D_{22}^2}{2D_{12}D_{22}}\right\} \tag{8}$$

In the present study, in order to evaluate the local deformation quantitatively, the differences of principal strain Δe_1 between the principal strain in each position in a test piece e_{1i} and the strain measured by displacement meter e_{1d} (see Eq. (9)) are measured.

$$\Delta e_1 = e_{1i} - e_{1d} \tag{9}$$

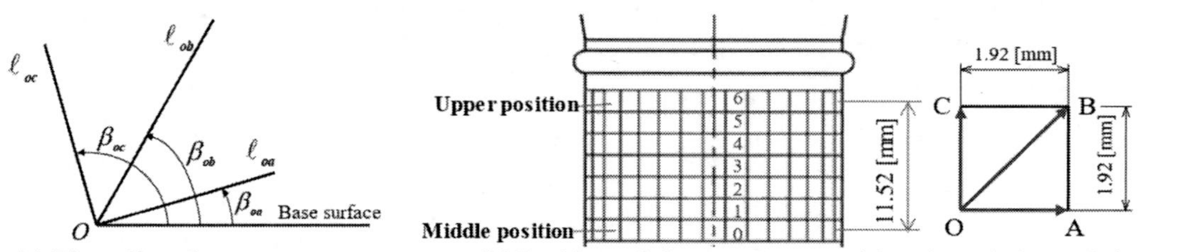

(a) Three line elements. (b) Position of elements. (c) Enlarged view of elements.

Figure 1. Measurement method of local deformation.

3 EXPERIMENTAL METHOD

3.1 Experimental Equipment and Test Specimen

Since it is necessary to keep the axial displacement zero during the process of applying a large simple shear, a multi-axial loading machine, which can apply the axial displacement and the shearing displacement at the same time, was used for conducting this experiment. As for the test piece used in experiments, hollow cylindrical specimens made from annealed high purity tough pitch copper, i.e., outer diameter 22 [mm], inner diameter 16 [mm] and gauge length 30 [mm] , are adopted. As shown in Figure 1 (b), in order to perform image measurements, scribe lines were preliminarily drawn with grid pattern (uniform interval 1.92 [mm]) on the surface of the test piece. Then, after attaching this test piece to the testing machine, the image measurement has been carried out by applying the torque with keeping the axial displacement zero. Here, the image measurements are performed on seven small elements (N=0~6) from the middle position to the upper position on the surface of a specimen (see Figure 1 (b)). The photographic camera used in this study is high pixel camera made from Canon (21.1 megapixel), and in order to measure the deformation on an identical line element, the image photographs of each element, where the resolution is 0.0014 [mm/px], were measured by rotating the camera and moving it up and down.

3.2 Experimental Conditions

In this study, in order to examine the progress of local deformation, the strain measurements in each small element are conducted on the following different deformation paths:

(I) Simple shear in forward direction only: extensional strain e_1 = 1.1 [-] (slip $K = D_{12}$= 2.7 [-], torsional angle φ = 490 [deg.])

(II) Simple shear in reverse direction after forward direction: e_1 = 0.7 [-] (K = 1.52 [-], φ = 275 [deg.]) ⇒ e_1 = - 0.2 [-] (slip K = - 0.34 [-], torsional angle φ = - 62 [deg.])

(III) Simple shear in reverse direction after forward direction: e_1 = 0.5 [-] (K = 1.05 [-], φ = 190 [deg.]) ⇒ e_1 = - 0.6 [-] (slip K = - 1.27 [-], torsional angle φ = - 230 [deg.])

(IV) Simple shear in reverse direction after forward direction: e_1 = 0.3 [-] (K = 0.62 [-], φ = 114 [deg.]) ⇒ e_1 = - 1.0 [-] (slip K = - 2.36 [-], torsional angle φ = - 427 [deg.]).

4 EXPERIMENTAL RESULTS

The progress of local deformation under different deformation paths is firstly examined by comparing the results of image measurement at middle position with upper position. As one example of these experimental results, Figure 2 shows the result for deformation path II. Here, Figure 2 (a) represents the relation of principal deviatoric stress and principal deviatoric strain. On the other hand, Figure 2 (b) shows the comparison of photographs of small elements taken at the place of middle and upper position in the test piece with respect to points a and c in Figure 2 (a). From comparison of photographs at the point c_0 and c_6, it can be confirmed that since the results of both images are different, the local deformation had already begun to occur. Similarly, Figure 3 shows the result of deformation path IV that the shear deformation in the forward direction is the smallest in these experiments. It is revealed from Figure 3 (a) that the local deformation begins to occur under larger reverse shear as compared with the result of deformation path II.

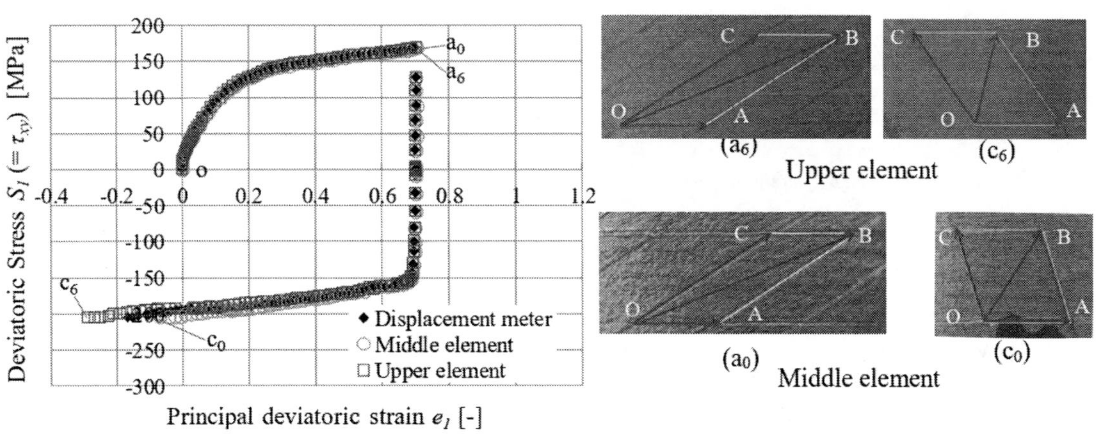

(a) Principal deviatoric stress and strain diagram. (b) Image photographs.

Figure 2. Simple shear in reverse direction after forward direction (deformation path II).

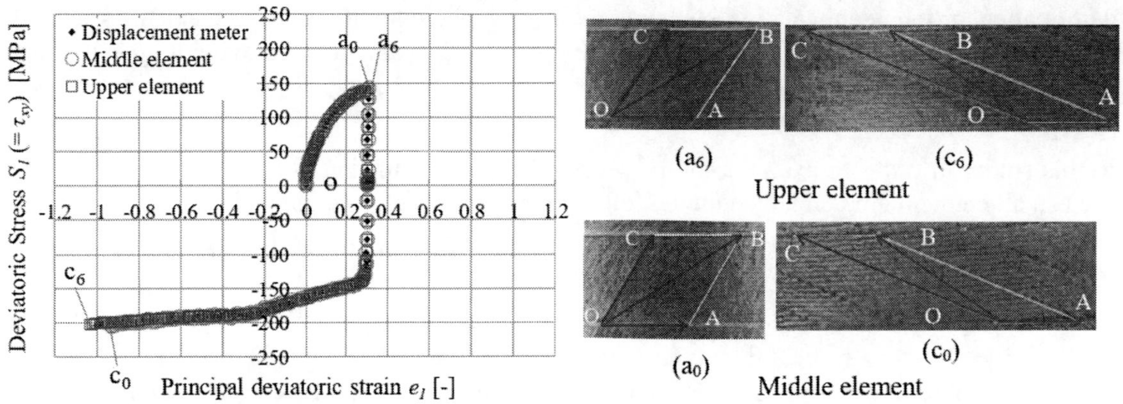

(a) Principal deviatoric stress and strain diagram. (b) Image photographs.

Figure 3. Simple shear in reverse direction after forward direction (deformation path IV).

Next, in order to confirm the progress of local deformation at each position in the test piece, the state of local deformation is investigated at a point b in Figure 4 (a) where the shear deformation measured by using the displacement meter becomes zero. Figure 4 (b) shows the distributions of the differences of principal strain Δe_1 definded by Eq. (9). In this Figure, as for the deformation path II, which is the largest forward shear, it can be comfirmed that since the distribution of Δe_1 changes from a plus through zero to minus, the local deformation has already occurred. On the other hand, as for the case of deformation path III and IV, since the distributions of Δe_1 show little change, the local deformation has not yet occurred.

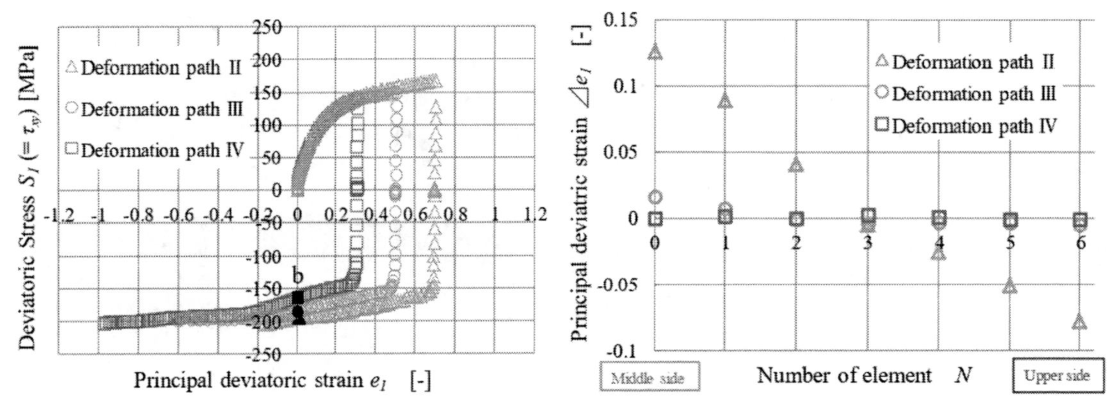

(a) Principal deviatoric stress and strain diagram. (b) Principal steain Δe_1 in each element.

Figure 4. Status of local deformation at point b.

Moreover, in order to see whether there is correlation between the local deformation and the strain history in each deformation path, the progress of local deformation is examined at point c in Figure 5 (a) where the plastic strain integrated along the deformation path becomes the same value, i.e., $\|e_1\|$= 1.6[-]. Figure 5 (b) shows the distributions of the differences of principal strain Δe_1 obtained under the condition at point c. It can be seen that since the differences of principal strain obtained by three different deformation paths are almost the same distribution, the progress of local deformation is almost the same if the total plastic strain obtained by integrating along each deformation path is the same.

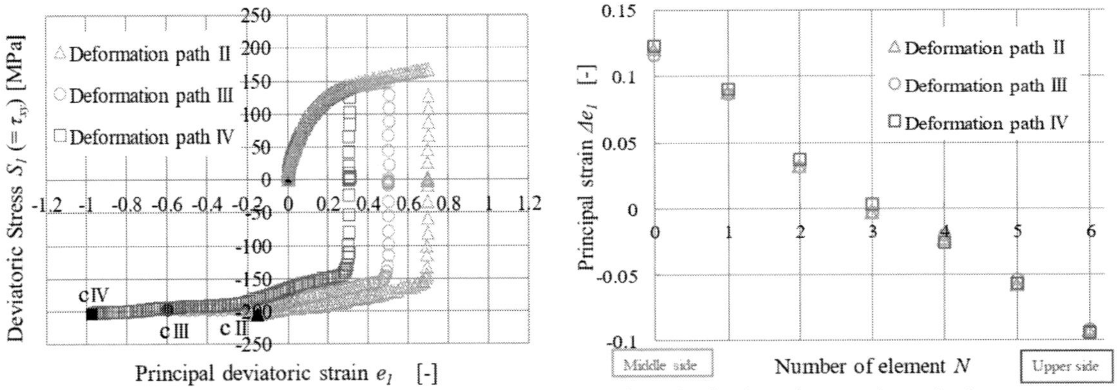

(a) Principal deviatoric stress and strain diagram.

(b) Principal steain Δe_1 in each element.

Figure 5. Status of local deformation at point c.

5 CONCLUDING REMARKS

In this study, focusing on the simple shear that the sectional shape of a test piece does not significantly change, the progress of local deformation was investigated under different deformation paths and the following results were derived.

(1) It is confirmed that since the local deformation does not occur in the final state of forward shear, the local deformation arises during the process of applying the reverse shear.

(2) The local deformation was examined at the place where the shear deformation measured by using the displacement meter becomes zero. As a result, it is revealed that the local deformation is likely to occur as the value of simple shear in the forward direction is larger.

(3) Moreover, it was revealed that the occurrence of local deformation is almost the same when the values of plastic strain obtained by integrating along each deformation path are the same value even if the deformation paths are different.

References

Kato, Y., Image Analysis Based Finite Strain Measurement of Local Deformation under Uniaxial Loading Using Natural Strain, *Implementing Innovative Ideas in Structural Engineering and Project Management*, Yazdani, S., and Singh, A (eds.), 1165-1170, *ISEC* Press, 2015.

Kato, Y., and. Futami, M., Finite Strain Measurement Based on Image Analysis using Natural Strain (Measurement and Evaluation in Uniform Deformation Field), *New Development in Structural Engineering and Construction, Vol 1*, Yazdani, S. and Singh, A. (eds.), 171-176, RPS Publishing, 2013.

Kato, Y., and Moriguchi, Y., Finite Deformation Analysis Using Natural Strain (on Anisotropy in Yield Surface Under Large Uni-Axial Tension and Shear), *System-Based Vision for Strategic and Creative Design*, Bontempi, F. (ed.), 669-675. Rotterdam, Balkema, 2003.

Interdependence between Structural Engineering and Construction Management
Edited by Ozevin, D., Ataei, H., Modares, M., Gurgun, A., Yazdani, S., and Singh, A.

ISBN: 978-0-9960437-6-2

COMPUTATIONAL SIMULATION OF HYGROTHERMAL PERFORMANCE OF PLASTERS WITH ENHANCED MOISTURE ACCUMULATION CAPABILITY

JAN KOČÍ, JAN FOŘT, and JIŘÍ MADĚRA

Dept of Materials Engineering and Chemistry, Czech Technical University in Prague, Prague, Czech Republic

Series of computational simulations are performed within this paper in order to investigate the hygrothermal response of several plasters with enhanced accumulation properties. The newly developed plasters are modified from the reference sample by adding various amounts of super absorbent polymers. Then, the basic physical, thermal and hygric properties are determined in the laboratory conditions and subsequently used as an input parameter in the computational simulations. The simulation output showed that even a thin layer of exterior and interior plaster may significantly affect the hygrothermal performance of the entire building envelope due to increased moisture and thermal buffering of the surface layers. Differences in relative humidity distribution across the studied construction were generally up to 10 % between individual plasters, differences of temperature distribution were mostly negligible, except for the cases when sudden changes of surface temperature were observed. Then, the thermal buffering was evident and the differences of temperature in surface layers were up to 4 °C among studied plasters.

Keywords: Hygrothermal analysis, Advanced plasters, Buffering effect.

1 INTRODUCTION

Moisture level significantly affects the durability of constructions, their thermal performance, and quality of indoor air (Abdul Hamid and Wallentén 2017, Sehizadeh and Ge 2016, Maljaee *et al.* 2016, Lohonayi and Korany 2013). As the building envelopes are subjected to the effect of changing weather conditions from the exterior and numerous sources of water vapor from the interior, the proper design may significantly increase the performance of the construction. Nowadays, there exist many ways of how to control the moisture loads both in the constructions and in the interior of the buildings. As far as material selection is concerned, applying advanced types of renders for both exterior and interior surface of the building envelope might be a prospective solution regarding the building envelope performance. Addition of vermiculite, perlite or super absorbent polymers (SAP) to the mixture increases the absorption capacity of those mortars (Gonzáles *et al.* 2001, Yang *et al.* 2011, Hasegawa *et al.* 2009), which can significantly help to control moisture levels in both construction and interior space.

However, a proper understanding of the behavior of moisture that is retained in the pore space of the building materials is essential for the prediction of hygric or hygrothermal performance. As a result of adding agents to increase moisture accumulation capacity, the moisture buffering of the plasters will be affected, which can imply also changes to thermal buffer properties of the material. Since the plasters on the exterior side are exposed to low temperatures during the year, also phase changes of retained liquid moisture may be observed, which will further affect the hygrothermal performance of whole construction. For that reason, the assessment the effect of applying new types of renders should be done using computational simulation with an advanced computational model that will include the water/ice phase change feature.

In the proposed paper, the newly developed plasters with enhanced moisture accumulation capability are analyzed by the computational simulation. First, the necessary material parameters are determined in the series of laboratory experiments. Those parameters are then used as input data in the computational model. The objective of the computational simulation is to study complex behavior building envelope provided by advanced plasters. Based on the simulation results, recommendations for further application of the interior plasters are given.

2 STUDIED MATERIALS

The investigation of hygrothermal performance was done on the brick wall provided with three types of modified plasters on both exterior and interior surfaces. The scheme of the studied envelope is shown in Figure 1.

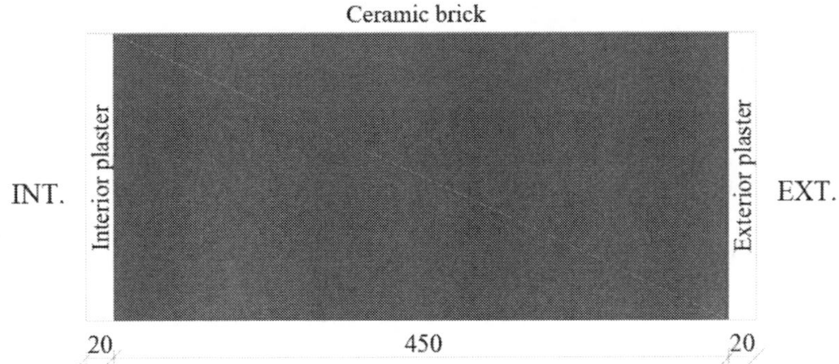

Figure 1. Scheme of the studied building envelope.

The ceramic brick was marked as CB, and the plasters were marked as P-ref, P-2%, and P-3%. The P-ref refers to a reference plaster Knauf MV1 core plaster, while P-2% and P-3% refer to the modified reference plaster by adding 2 wt% or 3 wt% of SAP (Favor PAC 300 by Evonik Ltd.), respectively. The basic physical, thermal and hygric properties are summarized in Table 1, where the following symbols are used: ρ_v is the bulk density, ρ_{mat} is the matrix density, ψ is the total open porosity, λ is the thermal conductivity, c is the specific heat capacity, μ is the water vapor diffusion resistance factor, κ_{app} is the apparent moisture diffusivity, and w/c is the water-cement ratio.

The sorption and desorption isotherms of studied plasters, which are crucial parameters for description of water vapor transport and accumulation capabilities are shown in Figure 2. All the material parameters were obtained in the laboratories of the Department of Materials Engineering and Chemistry, FCE CTU Prague within the frame of this project. The material parameters of ceramic brick were measured in the same laboratories, but the data was adopted from Čáchová *et al.* (2014).

Table 1. Basic material parameters of studied materials.

Parameter/Material	CB	P-ref	P-2%	P-3%
ρ_v (kg·m^{-3})	1389	1593	1225	1203
ρ_{mat} (kg·m^{-3})	2581	2548	2574	2563
ψ (-)	0.279	0.375	0.524	0.531
λ (W·m^{-1}·K^{-1})	0.59	0.53	0.38	0.36
c (J·kg^{-1}·K^{-1})	825	920	1087	1099
μ (-)	22.1	16.8	9.8	7.3
κ_{app} (m^2·s^{-1})	$2.15\cdot10^{-9}$	$9.93\cdot10^{-8}$	$6.80\cdot10^{-7}$	$1.02\cdot10^{-6}$

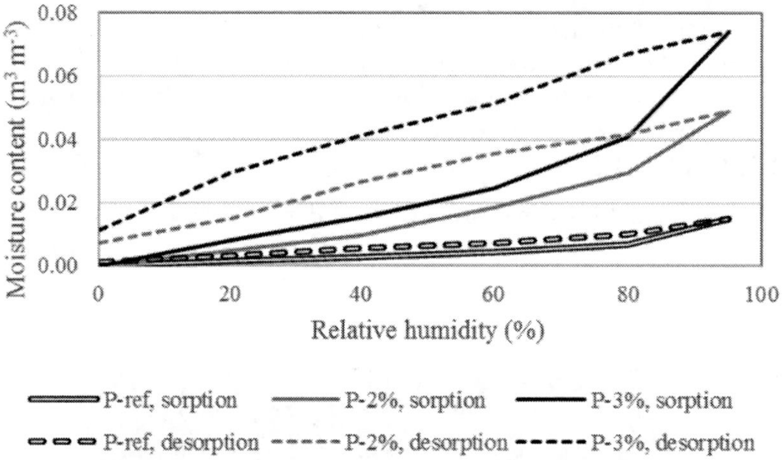

Figure 2. Sorption and desorption isotherms of studied plasters.

3 COMPUTATIONAL MODEL

The simulations were performed under time-dependent boundary conditions using the finite element method. Computer simulations of the hygrothermal performance of studied building envelope were conducted for the time period of three years using the HEMOT simulation tool, which is a pre-processing tool for the general finite element package SIFEL (Kruis *et al.* 2010). In the simulations, a slightly modified version of Künzel's (Künzel 1995) mathematical model of coupled heat and moisture transport was used, which is based on a calculation of temperature and partial pressure of water vapor along the studied domain. The input parameters for the model are bulk density, matrix density, open porosity, thermal conductivity, specific heat capacity, water vapor diffusion resistance factor, sorption isotherm, moisture diffusivity, and water retention function. The input parameters are shown in Table 1. The model includes the feature for a description of water/ice phase change using the effective specific heat capacity c_{eff} defined as in Eq. (1);

$$c_{eff} = \begin{cases} c_s & (T \le T_1) \\ \left[1-\cos\left(\dfrac{2\pi(T-T_1)}{T_2-T_1}\right)\right]^n \left(\dfrac{c_{eff,\max}-c_s}{2^n}\right)+c_s & \left(T_1 < T \le \dfrac{T_1+T_2}{2}\right) \\ \left[1-\cos\left(\dfrac{2\pi(T_2-T)}{T_2-T_1}\right)\right]^n \left(\dfrac{c_{eff,\max}-c_l}{2^n}\right)+c_l & \left(\dfrac{T_1+T_2}{2} < T < T_2\right) \\ c_l & (T \ge T_2) \end{cases}, \quad (1)$$

where T (K) is the temperature, T_1 and T_2 (K) are limiting values of the temperature range defining the phase change, c_{eff} (J·kg^{-1}·K^{-1}) is the effective specific heat capacity, c_s and c_l (J·kg^{-1}·K^{-1}) are the specific heat capacity of the material in frozen and unfrozen state, respectively, $c_{eff, max}$ (J·kg^{-1}·K^{-1}) is the maximum effective specific heat capacity calculated from the heat released/consumed during the phase change. More details on the c_{eff} model and calculation procedure are given by Kočí *et al.* (2018).

Boundary conditions for the interior were kept constant at 21 °C and 55 % of relative humidity during the simulation. The exterior conditions were represented by hourly weather data from the Test Reference Year for the location of Šerák, the Czech Republic that can induce very severe conditions to the constructions.

4 RESULTS AND DISCUSSION

As the objective of the paper was to predict the hygrothermal behavior of the construction, selected outputs are provided in the following figures. Figure 3 shows the distribution of relative humidity in the studied element for the winter period (February 1), while Figure 4 shows the same variable for the summer period (July 1). The presented results clearly show that even a thin layer of exterior and interior plaster may significantly affect the moisture distribution in the entire building envelope.

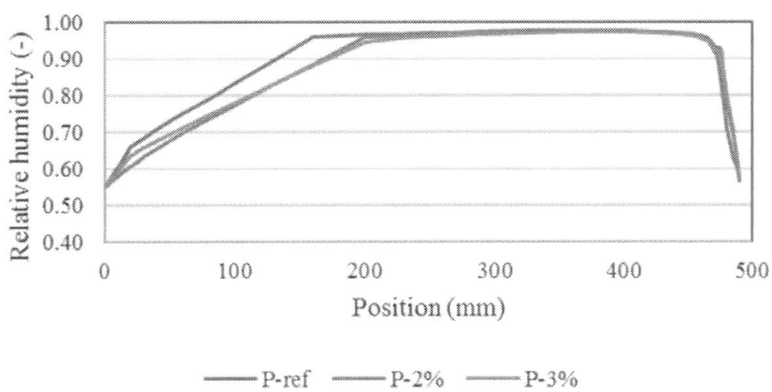

Figure 3. Relative humidity distribution in the studied element on February 1.

Differences of relative humidity in the brick masonry in studied envelopes are generally up to 10 % in the summer period and up to 6 % in the winter period. In extreme cases, the observed differences were more than 20 %. The better performance was observed for plaster P-3%, however, the differences compared to P-2% were practically negligible. Since adding SAP to the plaster decreases the strength of the material, it would be more practical to use P-2% than P-3% in the practice. However, it would be worth of investigation how P-1% would respond when analyzed using the same scenario, i.e. same building envelope loaded with same boundary condition.

As the moisture buffering feature may be in relation to the thermal buffer effect, also the temperature distribution in the studied envelope was investigated. The highest differences were observed on exterior plaster surface and adjacent layers in case of rapid temperature changes of the weather, especially in the winter period. Those changes are mostly induced by solar radiation, which may heat up the surface by several degrees in a short period. In such case, the thermal buffer effect due to the presence of the absorbent was most evident. Example of such sudden heat-up on April 10 is shown in Figure 5. The temperature differences are apparent only in the layers closed to the surface, in the remaining part of the construction differences are almost negligible. However, for the most time of the year, the temperature differences between individual studied envelopes are very low, generally not more than 0.1 °C.

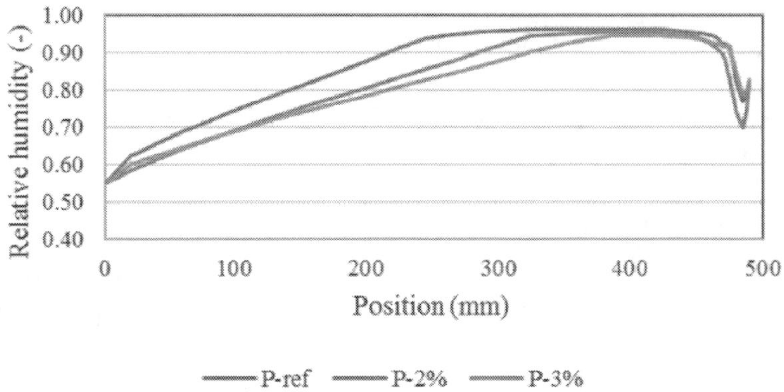

Figure 4. Relative humidity distribution in the studied element on July 1.

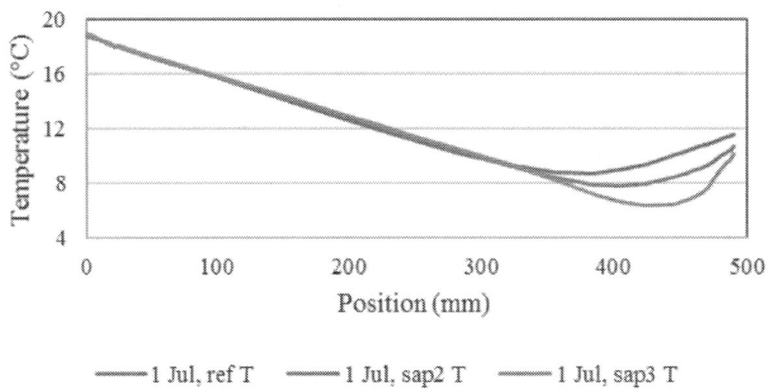

Figure 5. Temperature distribution in the studied element on April 10.

5 CONCLUSIONS

Three different plasters with enhanced moisture accumulation properties were studied in this paper. The plasters were modified by adding super absorbent polymer and their basic physical, hygric and thermal properties were measured. Then the hygrothermal performance was analyzed using computational modeling and several conclusions were drawn:

- Exterior plaster with enhanced accumulation properties may significantly affect the overall hygric performance of the construction,
- Differences in relative humidity distribution were generally up to 10 %, in extreme cases more than 20 %,
- Differences of temperature distribution were mostly negligible, except the cases when sudden changes of surface temperature were observed. Then, the thermal buffering was evident and the differences of temperature in surface layers were up to 4 °C,
- Another mixture with 1-2 wt% of super absorbent polymer should be investigated as well to find a compromise between material's strength and its hygrothermal properties.

Acknowledgments

This research has been supported by the Czech Science Foundation, under project No. 17-01365S.

References

Abdul Hamid, A., and Wallentén, P., Hygrothermal Assessment of Internally Added Thermal Insulation on External Brick Walls in Swedish Multifamily Buildings, *Building and Environment*, Elsevier, 123, 351-362, October 2017.

Čáchová, M., Koňáková, D., Vejmelková, E., Keppert, M., Polozhiy, K., and Černý, R., Pore Structure and Thermal Characteristics of Clay Bricks, *Advanced Materials Research*, Scientific.Net, 982, 104-107, July 2014.

Gonzáles, J. C., Molina-Sabio, M., and Rodríguez-Reinoso, F, Sepiolite-Based Adsorbents as Humidity Controller, *Applied Clay Science*, Elsevier, 20(3), 111-118, November 2001.

Hasegawa, T., Iwasaki, H., Shibutani, Y., and Abe, I., Preparation of Superior Humidity-Control Materials from Kenaf, *Journal of Porous Materials*, Springer US, 16(2), 129-134, 2009.

Kočí, J., Maděra, J., and Černý, R., *Formulation of a Hygrothermal Model for Description of Ice-forming Process in Porous Building Materials*, AIP Conference Proceedings, 2040, 040011, 2018.

Kruis, J., Koudelka, T., and Krejčí, T., Efficient Computer Implementation of Coupled Hydro-Thermo-Mechanical Analysis, *Mathematics and Computers in Simulations*, Elsevier, 80(8), 1578-1588, April 2010.

Künzel, H., *Simultaneous Heat and Moisture Transport in Building Components: One- and Two-Dimensional Calculation Using Simple Parameters*, IRB Verlag, Stuttgart, 1995.

Lohonayi, A. J., and Korany, Y., Comparison of Building Enclosures Designed to the Minimum Requirements of the 2011 NECB, *Energy and Buildings*, Elsevier, 66, 143-153, November 2013.

Maljaee, H., Ghiassi, B., Lourenço, P. B., and Oliveira, D. V., FRP-Brick Masonry Bond Degradation under Hygrothermal Conditions, *Composite Structures*, Elsevier, 147, 143-154, July 2016.

Sehizadeh, A., and Ge, H., Impact of Future Climates on the Durability of Typical Residential Wall Assemblies Retrofitted to the Passivehaus For the Eastern Canada Region, *Building and Environment*, Elsevier, 97, 111-125, February 2016.

Yang, H., Peng, Z., Zhou, Y., Zhao, F., Zhang, J., Cao, X., and Hu, Z., Preparation and Performances of a Novel Intelligent Humidity Control Composite Material, *Energy and Buildings*, 43(2-3), 386-392, February-March, 2011.

Interdependence between Structural Engineering and Construction Management
Edited by Ozevin, D., Ataei, H., Modares, M. Gurgun, A., Yazdani, S., and Singh, A.

ISBN: 978-0-9960437-6-2

ESTIMATION OF SHEAR-WAVE VELOCITY STRUCTURE USING MICROTREMOR MEASUREMENTS

ZORIGT TUMURBAATAR and HIROYUKI MIURA

Department of Architecture, Hiroshima University, Higashi-Hiroshima, Japan

Due to the recent construction boom in Ulaanbaatar city (UB), Mongolia, newly urban areas are rapidly expanded to the surrounded area of the city. According to the previous researches and reports, the thickness of the sedimentary basin in UB reaches 120m at the maximum in the area along the Tuul river. Therefore, the evaluation of site amplification is one of the essential parts of the estimation of earthquake hazard in this area. In this paper, the shear-wave velocity (Vs) structures in UB are estimated from single microtremor observations with the existing microtremor array observations. We carried out the microtremor observations at 50 sites and computed horizontal-to-vertical (H/V) spectral ratios. Inversion analysis is performed to the observed H/V spectral ratios based on the diffuse field approach (DFA) to determine the Vs structures. The site characterizations are evaluated from the amplification factors of the Vs structures.

Keywords: H/V spectral ratio, Inversion technique, Soil model, Site effect.

1 INTRODUCTION

Ulaanbaatar (UB), the capital city of Mongolia, exhibits dynamic urban and industrial development with a rapidly increasing population. Since the population of UB has grown approximately three times larger in the last two decades, the urbanized areas have been expanded to all parts of the city area. Especially in the last six years, moderate- and high-rise buildings have been newly constructed in the Tuul riverside areas located in the southern part of UB. UB is located at the valley of the Tuul river widely covered with alluvial deposits with 30 km length and 4-10 km width. According to existing borehole data and geological map, the thickness of the deposit is around 10-80 meters in most of the area.

According to the previous seismicity studies in Mongolia, the Emeelt fault located approximately 20 km from UB in the west-south-west direction has been one of the active faults that would provide great impact to UB (Adiya 2016). In order to consider countermeasures against the earthquakes, the evaluation and prediction of the strong ground motions due to the earthquakes are indispensable. The evaluation of the site amplification effects is one of the important issues in evaluating the seismic hazard and risk because the earthquake ground motions at the surface strongly depend on the site effects. Recently PS-logging surveys and spectral analysis of surface wave (SASW) method were conducted and the shear-wave velocity (Vs) profiles were estimated at around 40 sites in UB (JICA 2013). However, since the depths of the investigations were much less than 30 m (less than 10 m in most of the sites), the detailed Vs structures including much deeper profiles are not obtained in UB.

The use of microtremors (ambient noise) is a low-cost and handy method for site characterization. Nakamura (1989) has introduced this technique to estimate the fundamental resonant frequency of sediments using horizontal-to-vertical (H/V) spectral ratio at a single station. Microtremor H/V spectral ratio has been used to estimate Vs profiles by assuming that the microtremors are mainly composed of surface waves (Arai and Tokimatsu 2005). Recently, diffuse field theory has been proposed to understand the theoretical background of H/V spectral ratio by assuming a diffuse wavefield containing all types of elastic waves (Sánchez-Sesma *et al.* 2011). Based on the theory, the H/V spectral ratio is computed from the horizontal-to-vertical ratio of imaginary parts of Green's function.

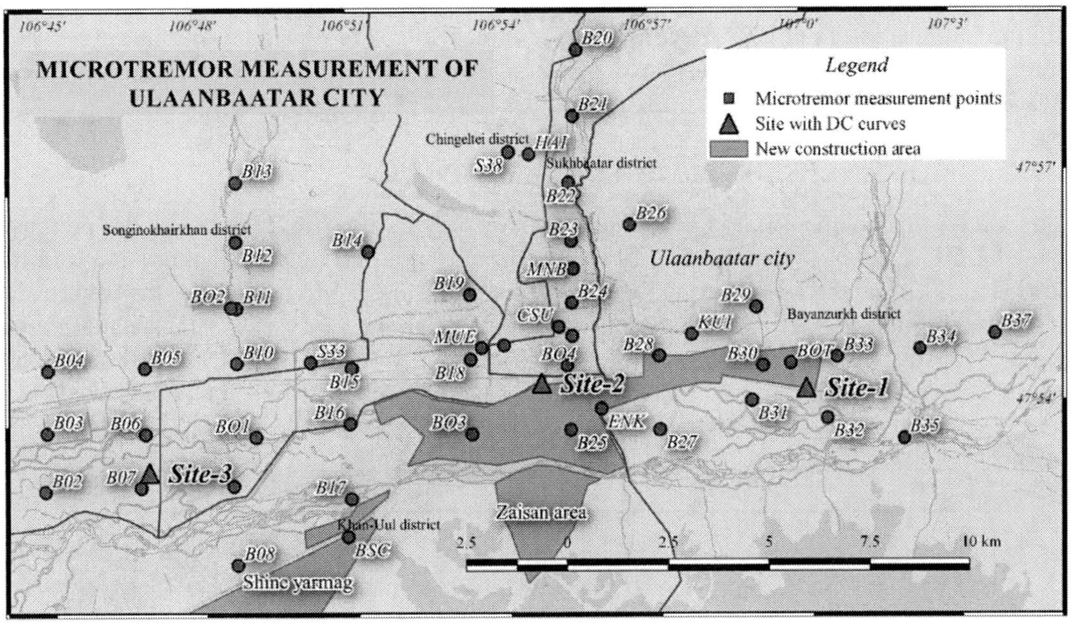

Figure 1. Microtremor observation points at Ulaanbaatar city with reference site-1, 2 and 3.

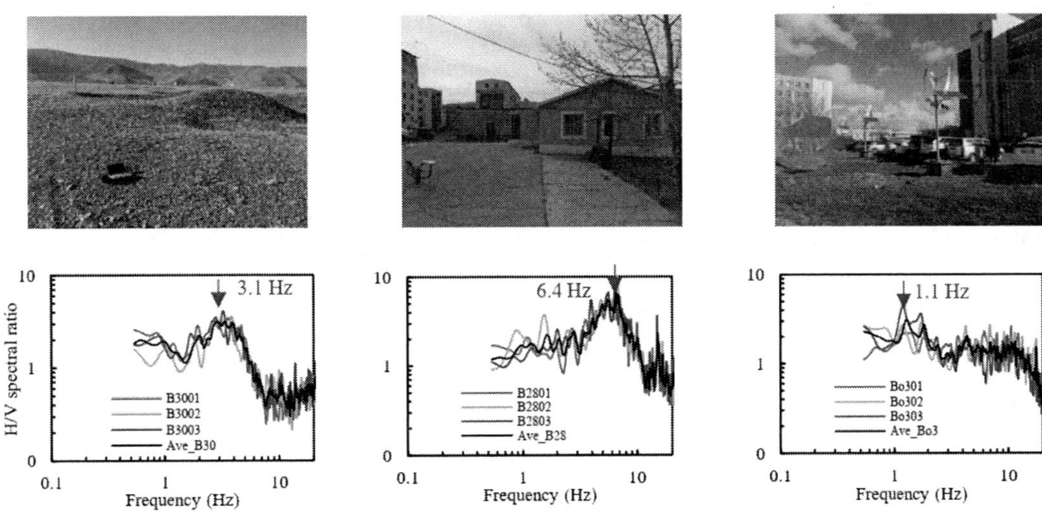

Figure 2. Typical site observation photos and observed H/V spectral ratios at B30, B28 and BO3.

In this paper, microtremor observations at multiple sites in UB are performed, and the inversion analysis based on the diffuse field theory is applied to the observed data in order to identify the Vs structures and to characterize the site amplification in the city.

2 OBSERVATION OF MICROTREMOR IN UB

Figure 1 shows the distribution of the microtremor observation sites in UB. Totally 50 sites are selected over the UB area. The Site-1, 2 and 3 in Fig. 1 are the reference sites where surface wave dispersion curves (DC) were obtained from microtremor array observations (Odonbaatar 2011). Due to the heavy traffic in the central part of UB during the daytime, most of the measurements are conducted in early mornings. Ground photographs for the observation condition and the observed H/V spectral ratios at typical sites are illustrated in Figure 2. The shapes and the predominant frequencies of the spectral ratios largely vary at site-by-site, indicating the variability of the site conditions in UB.

3 INVERSION ANALYSIS OF MICROTREMOR DATA

In this study, first joint inversion analysis is applied at the Site-1 to 3 based on the previously obtained DC curves by Odonbaatar (2011) and the observed H/V spectral ratio to avoid non-uniqueness of Vs and thickness during the inversion process. Second, considering the results of the joint inversion and the existing PS-logging data, single inversion analysis of H/V spectral ratios is applied to the rest of the sites.

Table 1. Initial inversion parameter ranges.

Layer	Vs (m/s)	Vp (m/s)	Depth (m)	Density (kg/m^3)
1	200-600	350-1400	0-15	1600-1800
2	250-1000	450-2400	0-50	1700-1900
3	1200-1500	2600-2950	0-100	1800-2000
4	1800	3280		2000-2200

3.1 Joint Inversion of H/V Spectral Ratio and Dispersion Data

In the joint inversion of the DC curves and H/V spectral ratio, the misfit between the observed values and the theoretical values is minimized by changing Vs and thickness through the iteration based on the Heuristic approach such as Simulated Annealing technique. We performed ten times inversions at each site with different initial sets and selected the best model with the smallest misfit. The Vs of the bedrock in UB is assumed to be 1800 m/s considering the previous study by Odonbaatar (2011). Table 1 shows the parameter ranges for Vs, thickness, Vp and density in the inversion analysis. We assumed five or six-layer model for the initial models. If closer values of the shear-wave velocities in adjacent layers are estimated in the inversion, we reduced the number of layers and applied the inversion calculation again.

Figure 3 shows the observation data of the DC and H/V spectral ratio at the Site-1 to 3. Figure 4 shows the estimated Vs structures by the joint inversion analysis. The theoretical values from the estimated Vs structures are also shown in Fig. 3. Although the theoretical values from the Vs structures by Odonbaatar (2011) could not reproduce the H/V spectral ratio, the theoretical values of this study show good agreement with the observation data for not only the DC but also the H/V ratios. The results show that the depth to the bedrock is estimated at 50 m to 80 m at the sites.

3.2 Single inversion results of the H/V spectral ratio

We apply the single inversion analysis for the rest of the sites using the H/V spectral ratios. The inversion analysis is not applied to the sites where the peak value of the H/V spectral ratio did not reach a factor of 2 because reliable Vs structure could not be estimated by the inversion due to the uncertainties of the observed values.

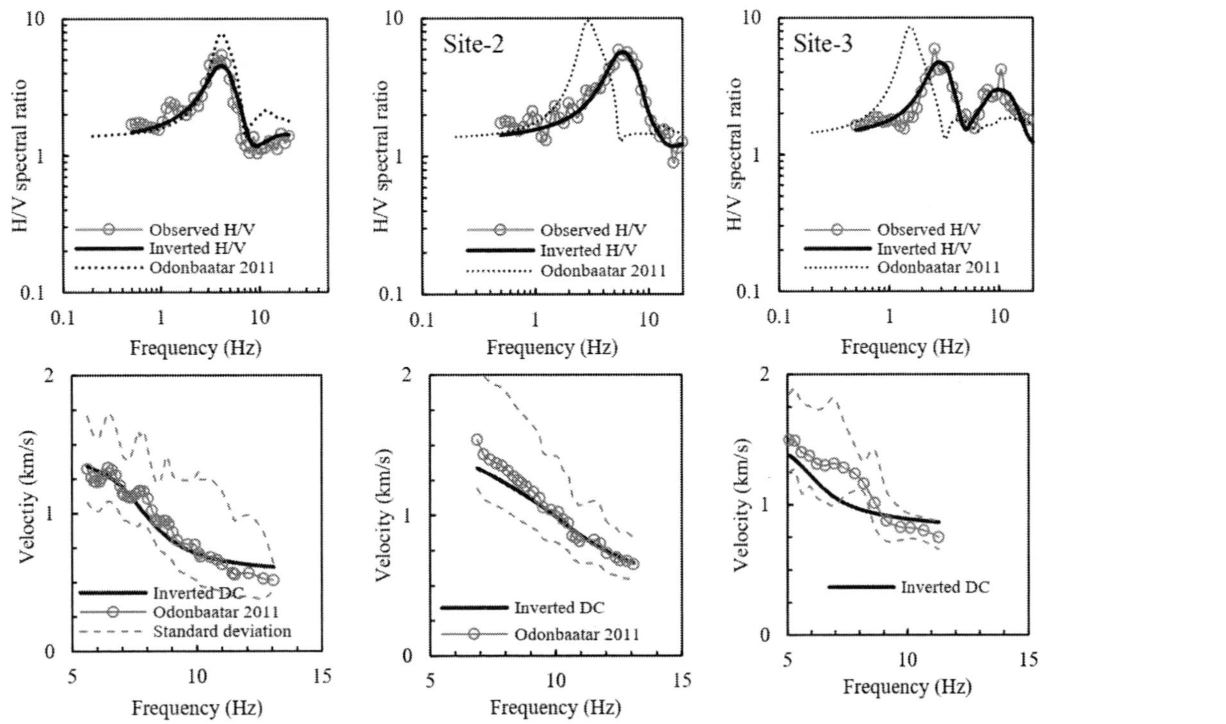

Figure 3. Theoretical and observed H/V spectral ratio and DC curve of joint inversion at Site-1, 2 and 3.

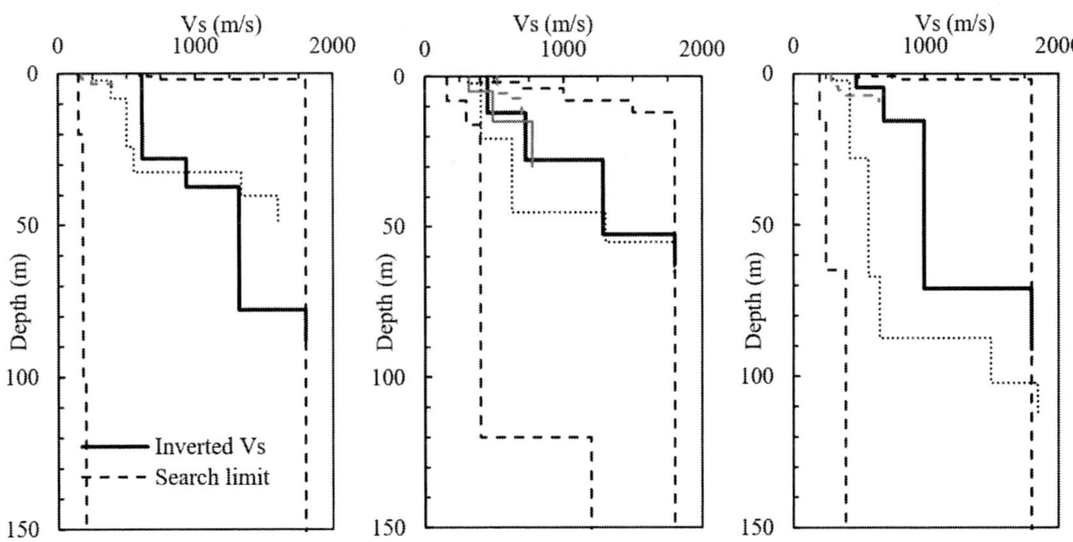

Figure 4. Shear-wave velocity profiles of joint inversion at Site-1, 2 and 3.

The Vs structures at 31 sites are estimated by the analysis. The search limits for Vs and thickness are determined considering the results of the joint inversions and the existing explorations. Figure 5 shows the Vs structures obtained from the inversion analysis. We confirmed that these models well reproduced the observed H/V spectral ratios.

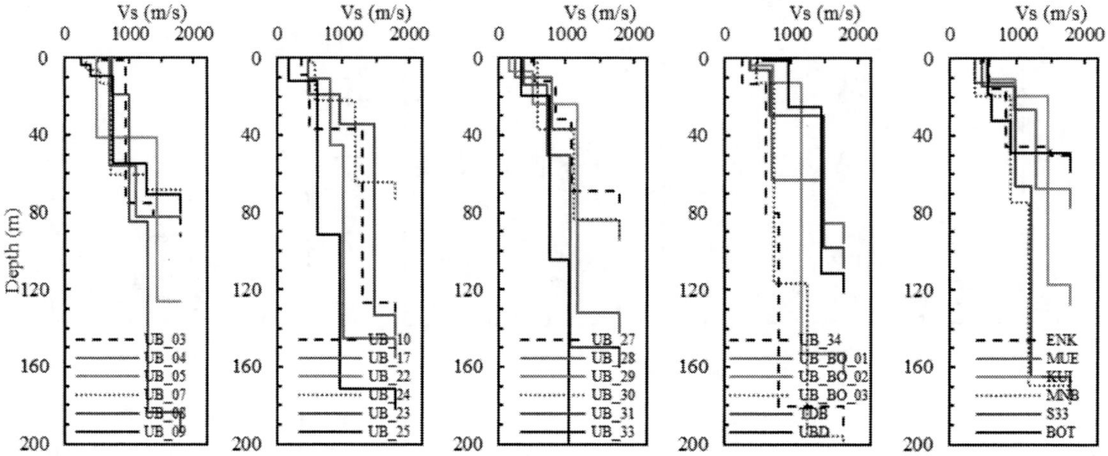

Figure 5. Estimated Shear-wave velocity profiles of all sites from microtremor single inversion.

3.3 Comparison with Borehole Data

In order to validate the obtained Vs structures in this study, the soil profiles are compared with the other data source. The geotechnical map of UB includes the cross-sections of soil profiles. Figure 6 shows the comparison of the soil profiles and the Vs structures obtained in this study. We confirm that the boundary of the Vs structures almost corresponds to the soil layers. The Vs of the gravel with sand or clay near the surface corresponds to 300 to 500 m/s. The Vs of the weak weathered and fissured alevrolit corresponds to 1000 to 1200 m/s.

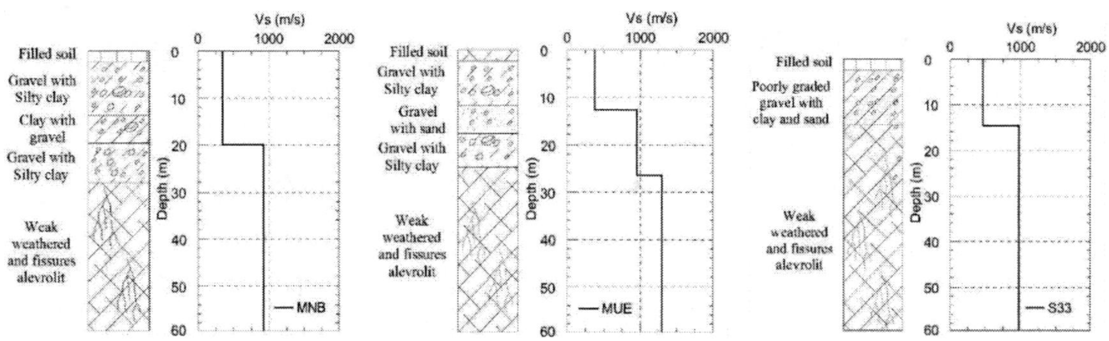

Figure 6. Comparison with estimated Vs profiles and borehole data.

4 SITE AMPLIFICATIONS OF ESTIMATED V_S PROFILES

Site amplifications for SH wave are computed from the estimated Vs structures based on the Haskell's multiple reflection theory. Figure 7 shows the examples of the site amplification at the sites near mountain, central area, and the Tuul river, where the thickest deposits are expected from the inversion analysis and the commercial and residential constructions including mid- and high-rise buildings are newly developed. Most of the old residential apartments are located near the central area. Finally, lowest site amplification is expected near mountain area because of the shallow deposit. The results show that the large amplifications are found at 3 to 10 Hz that would greatly affect the mid-rise buildings.

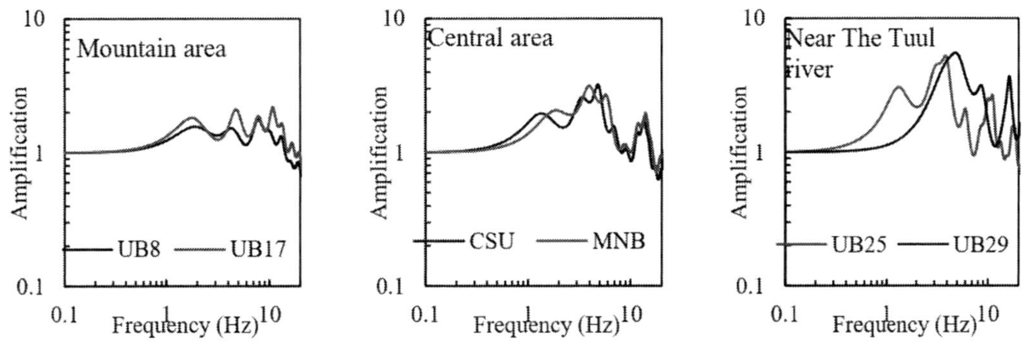

Figure 7. Estimated theoretical site amplification.

5 CONCLUSIONS

In this research, the shear-wave velocity (Vs) profiles of Ulaanbaatar city, Mongolia are estimated by the inversion analysis of the microtremor H/V spectral ratios based on the Diffuse Field Assumption. The joint inversion results with the surface wave dispersion curves are applied to the three sites where the microtremor array observations were conducted in the previous study by Odonbaatar (2011). The single inversion analysis is applied to the rest of the site. From these inversion analyses, the Vs structures up to the bedrock with the Vs of 1800 m/s are estimated, showing good correlation with the shallow borehole data. The site amplifications are evaluated from the estimated Vs structures. The results of this study would be useful for strong ground motion predictions and building damage assessment due to anticipating future earthquakes.

References

Adiya M., *Seismic Activity near Ulaanbaatar: Implication for Seismic Hazard Assessment*, (doctoral dissertation) Geophysics [physics.geo-ph], University of Strasbourg, 2016.

Arai, H., and Tokimatsu, K., S-wave velocity profiling by joint inversion of microtremor dispersion curve and horizontal-to-vertical (H/V) spectrum, *Bulletin of the Seismological Society of America*, *95*(5), 1766–1778. https://doi.org/10.1785/0120040243, 2005.

JICA, *The Project for Strengthening the Capacity of Seismic Disaster Risk Management in Ulaanbaatar City*, Japan International Cooperation Agency and Emergency Management Department of the Capital City, Report of Mongolia, Asian Disaster Reduction Center, Urban Disaster Research Institute Tokyo Electric Power Services Co., Ltd., 3, 2013.

Nakamura, Y., A Method for Dynamic Characteristics Estimation of Subsurface using Microtremor on the Ground Surface, *Quarterly Report of RTRI*, 30(1), 25-33, 1989.

Odonbaatar, C., Site effects characterization in the basin of Ulaanbaatar , Ph.D. Thesis, University of Strasbourg, Strasbourg, 2011.

Sánchez-Sesma, F. J., Rodríguez, M., Iturrarán-Viveros, U., Luzón, F., Campillo, M., Margerin, L., Rodríguez-Castellanos, A., A Theory for Microtremor H/V Spectral Ratio: Application for a Layered Medium, *Geophysical Journal International*, 186(1), 221–225. https://doi.org/10.1111/j.1365-246X.2011.05064.x , 2011.

Interdependence between Structural Engineering and Construction Management
Edited by Ozevin, D., Ataei, H., Modares, M., Gurgun, A., Yazdani, S., and Singh, A.

ISBN: 978-0-9960437-6-2

VERIFICATION OF SIGNAL POLE SHAFT FOUNDATION LRFD DESIGN USING FINITE ELEMENT METHOD

KUANG-YUAN HOU, CHUNG C. FU, YUNCHAO YE, and CHAORAN XU

Dept of Civil and Environmental Engineering, University of Maryland, College Park, USA

This study presents verification methods of current shaft foundation design under linear lateral and torsional loads. Shaft is used as a foundation to support mast arm signal pole structure and transfer loads from superstructures to ground. The capacity of current shaft foundation deployed in the State of Maryland is re-verified due to higher sub-structural strength requirement against super-structural fatigue proposed from the AASHTO LRFD Specification (2015). The shaft foundation verification includes embedment length and torsional capacity. For embedment length check, lateral reactions between soil and shaft are verified by comparing analytical method and finite element method. Wind-induced torque is a design concern for the shaft of a single pole cantilever structure. However, torsional capacity of the shaft foundation of signal pole structures is rarely mentioned in the current design specification. By verifying finite element models with analytical method, torsional effect is further simulated to finite element models to evaluate the adequacy of current shaft foundation design. Results show existing design of shaft foundation could meet requirements to resist lateral force under extreme conditions. For torsional effect, current torsional capacity is less than the torque induced by wind load in the worst soil conditions.

Keywords: Pile foundation, Soil-structure interaction, Lateral capacity, Torsional capacity.

1 INTRODUCTION

In Maryland, drilled shaft is used as the foundation of mast arm traffic signal structures. The drilled shaft can be referred as a bored-cast-in-situ pile and it is one type of deep foundations which facilitate load from superstructures to transfer to deeper firm strata (Murthy 2003). Drilled shafts are generally utilized for bridges and large-scale structures since they may resist high vertical force, lateral force, ground moment and torsional load. The mast arm signal pole structures in the State of Maryland are designed based on 1994 AASHTO. However, fatigue design was not made in the AASHTO Specifications until 2001, and thus fatigue load demand on mast arm signal support structures. To provide more resistance against fatigue, the size of super-structures is required to be modified. Along with size enlargement, reactions from super-structure applying on the shaft also increase. Therefore, the capacity of shaft is supposed to be verified to meet higher requirement. Shaft embedment length is a main factor of checking to provide adequate load capacity and acceptable displacement with various soil properties.

2 LITERATURE REVIEW

2.1 Lateral Soil Reaction

According to the category in Broms (1964), behaviors of shaft can be classified as long pile and short pile for free-head and fixed-head piles by a dimensionless quantity, $\beta = (\frac{kD}{4EI})^{0.25}$, shown in Table 1, where EI is stiffness of pile section, k is coefficient of horizontal subgrade reaction, D is diameter of pile, and L is length of pile.

Table 1. Pile classification under laterally loaded (Broms 1964).

Pile type	Long Pile	Short Pile
Free-head Pile	$\beta L > 2.5$	$\beta L < 2.5$
Fixed-head Pile	$\beta L > 1.5$	$\beta L < 1.5$

Shaft foundation is usually classified as a short pile which behaves rigidly. Failure of laterally loaded pile occurs either when bending moment of pile reaches ultimate or lateral earth pressure reaches ultimate lateral soil resistance. Soil is assumed as a series of linear or non-linear springs connected with pile foundations to represent the interaction between soil and piles. Unlike rigid analyses merely viewing soil reaction as a force, the soil-structure interaction, soil's response to lateral loads and deformation of soil, is considered. Those assumptions are essential to conduct numerical analyses and finite element model establishment. Soil spring is a concept proposed by Winkler which is numerical solutions of soil lateral force termed as soil. Soils surrounded with pile foundation are approximated by a series of independent elastic springs. However, Winkler's hypothesis does not account for soil deformation and limited with surface load applied. The p-y method simulates the soil resistance as nonlinear behavior, where p is the soil pressure per unit length of the pile and y is the pile deflection which is applicable for numerical model by using a two-dimensional finite difference analysis.

2.2 Torsional Capacity

In the AASHTO Standard Specifications for Structural Support for Highway Luminaires, and Traffic Signals (2015), torsional capacity of shaft foundation is not considered in the design procedure. However, torsional load transfer between drilled shaft and soil is relatively critical when wind load is applied. In the analytical method, torsional capacity consists lateral resistance and bottom resistance and could be expressed as in Eq. (1) (Stuedlein *et al.* 2016):

$$T_{total} = T_s + T_b \tag{1}$$

where T_{total} is total torsional resistance, T_s is lateral torsional resistance and T_b bottom torsional resistance. The torsional resistance is caused by either cohesion of soil or friction between soil and shaft. For cohesive soil, the resistance is caused by cohesion of soil related to soil contacting area of foundation. The adhesion factor, α, is the ratio between cohesion and undrained shear strength. For cohesionless soil, the resistance is mainly from the friction between shaft and soil caused by earth lateral force and self-weight of foundation. The β-Method (O'Neill and Hassan 1994) is employed to estimate unit torsional resistance for shaft surrounded with cohesionless soil under axial loading. β factor is the load transfer ratio of effective soil stress and soil resistance. Comparison of different methods is shown and discussed below.

3 ANALYTICAL VERIFICATION

3.1 Embedment Length

For cohesive soil, the shaft embedment length is determined by shaft diameter (D), ultimate shear strength of soil, moment, and shear at groundline. The required embedment length (L) can be defined in Eq. (2) as:

$$L = 1.5D + q[\,1 + \sqrt{2 + \frac{(4H+6D)}{q}}\,] \tag{2}$$

where $H = \frac{M_F}{V_F}$, $q = \frac{V_F}{9cD}$, M_F is factored moment at groundline, V_F is Factored shear at groundline, and c is the ultimate shear strength of cohesive soil. For cohesionless soil, shaft embedment length is determined by shaft diameter, angle of internal friction, effective unit weight, moment, and shear at groundline (Eq. (3) and Eq. (4)),

$$L^3 - \frac{2V_F L}{K_p \gamma D} - \frac{2M_F}{K_p \gamma D} = 0 \tag{3}$$

$$K_p = \tan^2(45 + \frac{\varphi}{2}) \tag{4}$$

where γ is unit weight of cohesionless soil.

3.2 Torsional Capacity

There are four analytical methods including Florida Structure Design office method (FSDM), Florida District 7 (FD7), Colorado DOT (CDOT), and Illinois DOT (IDOT) used for torsional capacity check as Table 2,

Table 2. Torsional capacity methods.

Method	Soil Type	Unit resistance (r_s)	Lateral resistance (T_s)	Bottom resistance (T_b)
FSDM	cohesionless	$K_0\sigma'_{vz}tan\delta$	$0.5\pi D^2 Lr_s$	$DWtan\delta/3$
FD7	cohesionless/cohesive	$\alpha S_u K_0 + \sigma'_{vz}tan\delta$	$0.5\pi D^2 Lr_s$	$4D(W + Q_a)tan\delta/9$
CDOT	cohesionless	2L(1-sinφ) $\sigma'_{vz}tan\delta/3D$	$0.5\pi D^2 Lr_s$	$DWtan\delta/3$
	cohesive	Su	$0.5(\pi D^2)(L\text{-}1.5D)Su$	$(\pi D^3)Su/12$
IDOT	cohesionless	$\beta\sigma'_{vz}$	$0.5\pi D^2 Lr_s$	
	cohesive	αSu	$0.5\pi D^2 Lr_s$	

where K_0 is the at-rest lateral earth pressure coefficient, σ'_{vz} is the effective vertical stress at the midpoint of the layer of interest, and δ is the effective soil-shaft interface friction angle, D is the shaft diameter, and L is the shaft length, W is the shaft weight, the axial load applied on the drilled shaft, and S_u is the average undrained shear strength over the depth of interest. The major differences between those methods are unit resistance estimate and bottom resistance. The β-Method for shafts under axial loading is used by IDOT. β is the load transfer ratio for effective-stress normalized unit shaft resistance and has been correlated to depth and the STP blow count, N. In CDOT Design Method, the unit shaft resistance in cohesive soils is equal to the undrained shear strength over the depth of interest, and the side resistance in cohesive soils for the top 1.5D of the shaft is neglected.

4 ANALYTICAL VERIFICATION FOR GENERAL SOIL IN MARYLAND

In the embedment length check, soil conditions are assumed as general soil properties in Maryland shown in Table 3.

Table 3. Assumed properties of soil for hypothetical cases.

Soil Type	Soil Category	Shear Strength	Unit Weight	Internal Friction Angle
Cohesive	Stiff clay	2.16 (ksf)	N/A	N/A
Cohesionless	Clean gravel-sand	N/A	0.12 (kcf)	30

4.1 Embedment Length Check

The external loads including ground and moment are factored by LRFD Strength I. Both resultant shear force and moment are the root mean square of combined loads in the x and z directions and the load factor is 1.6. The highest required length case is 75 feet arm structure in cohesive soil and its required length is 9.33 feet close to current design length 10 feet.

4.2 Torsional Capacity Check

In cohesive soil, the torsional capacity estimated by all of four methods is enough to resist horizontally rotation factored by LRFD Strength I load combination in worst condition. In cohesionless soil, most of torsional capacity is insufficient when external load reaches the peak value. The torsional check result is listed in Table 4.

Table 4. Torsional capacity check (kip-ft).

		Cohesive		
Method	**Lateral**	**Bottom**	**Total Capacity**	**Maximal Torque**
FD7	290.80	19.34	310.14	> 131.81
CDOT	214.14	36.19	253.33	> 131.81
IDOT	298.57	N/A	298.57	> 131.81
		Cohesionless		
Method	**Lateral**	**Bottom**	**Total Capacity**	**Maximal Torque**
FSDOM	43.53	14.51	58.04	< 131.81
FD7	37.00	19.34	56.34	< 131.81
CDOT	72.55	17.65	90.20	< 131.81
IDOT	161.81	N/A	161.81	> 131.81

5 FINITE ELEMENT MODELING

The shaft foundation dimension and soil properties are given in Table 3. Two-dimensional (2D) model is designed for Broms' simplified analysis mentioned in AASHTO, as shown in Figure 1. The model is built on XY plane and the external force including shear force and moment at groundline. The soil spring is assumed homogeneous and located at the other side of shear force. Three-dimensional (3D) model is designed for all external forces including three translational forces (i.e., moments about the z and x axes, and torsion) as shown in Figure 2.

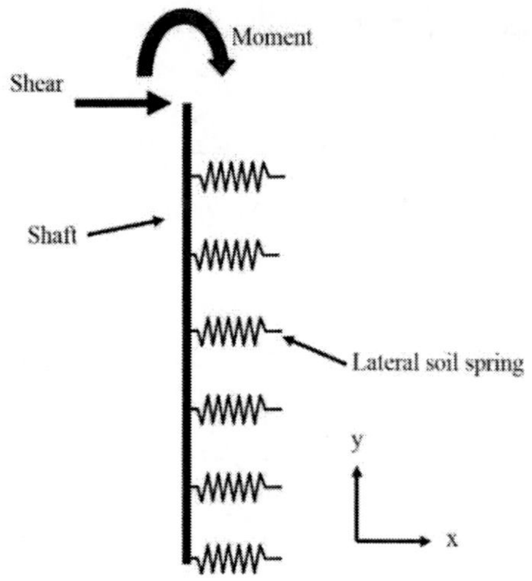

Figure 1. 2D Model demonstration.

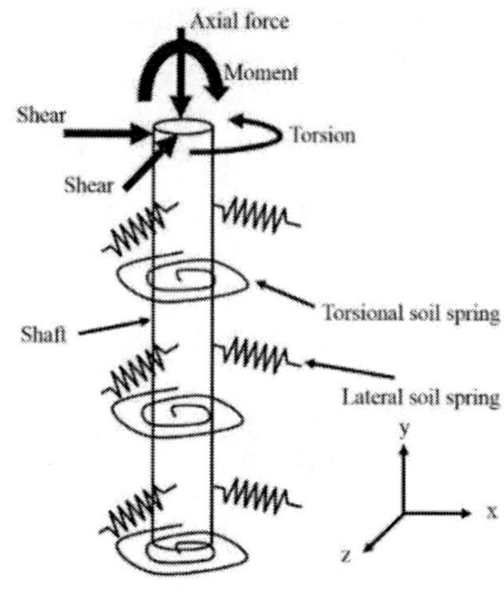

Figure 2. 3D Model demonstration.

The method to estimate soil subgrade modulus is first proposed by Terzaghi in 1955. Modulus of subgrade reaction of soil could be transformed into stiffness of soil spring representing the relationship between soil deformation and reaction from lateral, vertical, and torsional directions. Rigid analyses are used for light weight and short pile loaded laterally such as highway sign (Coduto 2001). Since shaft is defined as short pile and external loading is static and specified, the soil spring is considered under linear range. Based on soil properties mentioned before, lateral and vertical soil springs stiffness could be obtained from the Table 5 (Fu and Wang 2014). For cohesive soil, the stiffness of lateral soil spring is associated with undrained shear. The soil properties are assumed homogeneous; therefore, the stiffness of cohesive lateral soil spring stays constant along depth. For cohesionless soil, the stiffness of lateral soil spring is a function of depth. Thus, the stiffness distribution is triangular. Vertical soil spring is designed for controlling shaft's vertical displacement under acceptable range. The vertical soil spring is defined as a point spring which is either associated with undrained shear strength for cohesive soil or average standard penetration blow count. Torsional spring is currently used in pile foundation analysis program such as FB-MultiPier (2011). The torsion-angle (T-θ) curve is assumed as a hyperbolic curve and torsional stiffness listed in Table 5 is the slope of T-θ curve function at certain angle.

For finite element model, beam element with tension, compression, and bending capabilities is used to represent shaft body and spring-damper element without mass is used as soil spring to restrict shaft's horizontal, vertical and torsional degree of freedom. By assigning external force at shaft top, the maximal reaction moment is used as an indicator to compare the difference between finite element methods and analytical method. The results of 2D model ($\boldsymbol{M_{f2D}}$), 3D model ($\boldsymbol{M_{f3D}}$) and analytical method ($\boldsymbol{M_u}$) are listed in Table 6.

Table 5. Parameters for soil spring.

	Lateral springs		**Vertical springs**		**Torsional springs**	
Soil type	P_u	k_h	$q_{max}(ksf)$	k_q	G	k_t *(kip-ft)*
Clay	$9c_uB$	$67c$	$9c_u$	$8N_{corr}$	$kz/2(1+v)$	$4\pi Gr^2\Delta L$
Sand	$3\gamma Bk_px$	n_px	$10q_{max}/z_c$	$10q_{max}/z_c$	$\beta C_u/2(1+v)$	$4\pi Gr^2\Delta L$

Table 6. Maximal moment from Analytical and FEM methods.

Length (ft)	Soil Type	M_u (kip-ft)	M_{f2D} (kip-ft)	Diff %	M_{f3D} (kip-ft)	Diff %
50	Cohesive	147.07	149.56	1.66	149.43	1.57
	Cohesionless	133.62	139.92	4.5	139.89	4.48
60	Cohesive	166.28	166.14	-0.08	165.82	-0.27
	Cohesionless	145.09	151.78	4.4	151.81	4.42
70	Cohesive	199.06	198.87	-0.09	198.37	-0.34
	Cohesionless	175.58	182.38	3.78	182.29	3.68
75	Cohesive	206.98	206.80	-0.08	206.24	-0.35
	Cohesionless	183.11	189387	3.56	189.75	3.49

6 CONCLUSION

The finite element model has been demonstrated to be an effective approach to simulate shaft foundation of mast arm signal pole structure under linear range in the previous chapters. Results from traditional analytical method should be more conservative but easier to obtain. Based on general soil properties in Maryland, the existing design of shaft foundation can resist lateral force sufficiently. However, along with increasing arm length, required embedment lengths are close but adequate to the design lengths. Torsional capacity is less than the torque induced by wind load in the worst case when the shaft foundation is installed in cohesionless soil.

The outcome from two-dimensional and three-dimensional models is consistent in either cohesive or cohesionless soils. Results from various dimensions of shaft are also consistent in 2D and 3D models. Therefore, 3D model could be simplified by vector summing moments and shears along two horizontal axes as 2D model in lateral reaction analysis. The lateral reaction performance of the 3D model assigned with all external loads seems less conservative than Broms' analytical method. Therefore, foundation design based on the current AASHTO design specifications would cover lateral soil reaction capacity even considering the torsional effect (Hou 2017).

References

AASHTO LRFD Specification for Structural Supports for Highway Signs, Luminaires, and Traffic Signals, (2015).

Broms, M., Lateral Resistance of Piles in Cohesive Soils, *Journal of the Soil Mechanics and Foundations Division © ASCE / March 1964.*

Coduto, D. P., Foundation Design Principles and Practice Second Edition, 2001, ISBN 0-13-589706-8.

FB-MultiPier, Florida Bridge Software Institute, Version 5.1.1, 2011.

Fu, C. C., and Wang, S., *Computational Analysis and Design of Bridge Structures,* CRC Press, USA, 2014.

Hou, K. Y., Verification of Shaft Foundation LRFD Design Using Finite Element Method, University of Maryland, College Park, 2017.

Murthy, V. N. S., Geotechnical Engineering Principles of Soil Mechanics and Foundation Engineering, 2003, ISBN 0-8247-0873-3.

O'Neill, M. W., and Hassan, K. M., *Drilled Shafts: Effects of Construction on Performance and Design Criteria,* Proceedings of the International Conference on Design and Construction of Deep Foundations, December 1994, Vol. 1, 1994, pp. 137-187.

Stuedlein, A. W., Barbosa, A. R., and Li, Q., Evaluation of Torsional Load Transfer for Drilled Shaft Foundations Final Report, SPR 304-701, Oregon Department of Transportation, 2016.

Interdependence between Structural Engineering and Construction Management
Edited by Ozevin, D., Ataei, H., Modares, M. Gurgun, A., Yazdani, S., and Singh, A.

ISBN: 978-0-9960437-6-2

STRUCTURAL OPTIMIZATION USING AN ADAPTIVE RBF APPROACH

ERICA JAROSCH, QIAN WANG, LUCAS SCHMOTZER, and YONGWOOK KIM

Dept of Civil and Environmental Engineering, Manhattan College, Riverdale, USA

This paper presents an adaptive radial basis functions (RBFs) metamodeling method for design optimization of structures. Various numerical techniques have been developed and adopted in structural and multidisciplinary optimization. To evaluate responses of a structural or mechanical engineering system, finite element (FE) analyses are routinely used. An FE code shall be integrated with an optimization algorithm in a nested analysis and design of structures. Therefore, software input/output programming is required. A metamodeling method, on the contrary, expresses structural responses using an approximate function, so that the FE software is not directly coupled in the numerical optimization loop. Any optimization algorithm can be applied to find the optimal design, based on the explicit response functions. In this study, numerical examples were created and FE analyses were first performed at sample points. Subsequently, metamodels were constructed and a gradient-based optimization algorithm was applied. At the optimal point of one adaptive iteration, accuracy of the RBF metamodel was checked, and additional sample points were added to the sample pool to improve the model accuracy. The adaptive iterations continued, until the convergence of the objective function was achieved. The proposed optimization method worked well for a numerical example, and the optimal result was found within a few adaptive iterations.

Keywords: Optimal design, Finite element (FE), Adaptive metamodels, Radial basis functions (RBFs).

1 INTRODUCTION

Traditional structural and mechanical engineering system optimization relies on sensitivity analyses of system responses, since gradient-based optimization methods are widely available (Kirsch and Rozvany 1994, Arora and Wang 2005). Sensitivity-free methods, such as a genetic algorithm (GA), can also be used (Goldberg 1989); however, these involve significant computational resources, when expensive response simulations are needed, i.e., FE analyses. Approximate models have been developed to replace the implicit response functions such that the approximate functions can be directly used in a numerical optimization loop. These models are commonly referred to as a surrogate or metamodel (Jin *et al.* 2001, Fang *et al.* 2005). In literature, different types of metamodeling approaches are available, including response surface method (Bi *et al.* 2010), kriging (Jin *et al.* 2001), support vector machine (Basudhar and Missoum 2008), and RBFs (Fang *et al.* 2005).

When RBF metamodels are used, there are no errors at any sample points. Some studies showed that RBFs could be used to generate very accurate approximation functions (Fang *et al.* 2005, Fang and Wang 2008). There is a need to develop and apply accurate metamodeling methods to the design of practical civil engineering structures. An optimization method based on an adaptive RBFs is the focus of the current study. A traditional gradient-based algorithm can be used as the optimization engine.

In this paper, the structural optimization problem is first introduced. An adaptive metamodeling method based on RBFs is introduced to create explicit approximate functions of implicit structural responses, with the overall optimization steps being outlined. As a numerical example, a roof dome structure is introduced and a global buckling constraint is considered. The optimization results obtained using adaptive and traditional RBFs are compared. Finally, some concluding remarks are presented.

2 STRUCTURAL OPTIMIZATION PROBLEM

In a traditional structural optimization formulation, the objective of a design is to minimize a cost function (Eq. (1)):

$$C(\mathbf{x}) \tag{1}$$

subject to

$$\mathbf{g}(\mathbf{x}) \leq \mathbf{0} \tag{2}$$

$$\mathbf{x}^L \leq \mathbf{x} \leq \mathbf{x}^U \tag{3}$$

where x is a design variable vector, representing the structural size, geometry, or topology variables that need to be determined. Eq. (2) defines response or performance requirements for a given structure, including force, displacement, and buckling constraints. In Eq. (3), x^L and x^U are the lower and upper limits of the design variable vector x.

3 AN ADAPTIVE RBF OPTIMIATION APPROACH

In this section, the proposed adaptive RBF optimization approach and procedure are introduced. Numerical examples are discussed in the next section.

3.1 A Metamodeling Method

A metamodel of a response function $g(\mathbf{x})$ is expressed using RBFs, as (Fang *et al.* 2005, Fang and Wang 2008):

$$\tilde{g}(\mathbf{x}) = \sum_{i=1}^{n} \lambda_i \phi(\|\mathbf{x} - \mathbf{x}_i\|) \tag{4}$$

where ϕ is a basis function, λ_i is coefficient, and n is the sample size, respectively. To improve the accuracy of Eq. (4), an augmented RBF model can be used, when Eq. (4) is augmented with a linear or quadratic function (Fang *et al.* 2005, Fang and Wang 2008).

3.2 An Adaptive Technique

To progressively improve the accuracy of an RBF metamodel, an adaptive scheme is developed in this work. The key idea is to further refine the RBF model in the region of interest, i.e., the neighborhood of the optimal point of the current design iteration. The optimal point is used as an additional sample point that is included in the sample pool. The metamodel function can be re-constructed and used in the optimization in the subsequent iteration. This adaptive refinement will be repeated in each optimization iteration, until the objective function value converges.

3.3 Numerical Optimization Algorithms

Different numerical optimization algorithms are available to solve an optimization problem (Arora 2011). These include gradient-based methods, such as sequential quadratic programming (SQP) and generalized reduced gradient (GRG), and gradient-free methods such as a genetic algorithm (GA) (Goldberg 1989). In this research, the GRG solver in Excel (Microsoft 2013) was used as the optimization engine.

3.4 Overall Optimization Steps

The optimization procedure starts with the generation of initial sample points. FE analyses are conducted and RBF-metamodels are constructed using all the samples. Numerical optimization can be executed using any optimization algorithm to find an optimal design point of the current iteration. If the accuracy of a metamodel needs to be improved, the current optimal point is included in the sample pool. Therefore, an additional FE analysis should be conducted. The overall procedure of the design optimization method is highlighted as follows

(i) Generate initial samples.
(ii) Perform FE analyses at the sample points and evaluate constraint function values.
(iii) Create RBF metamodels of constraint functions using all available sample points.
(iv) Perform design optimization using the metamodel functions and obtain the current optimal design point.
(v) Perform an FE analysis at the current optimal design point to verify the accuracy of the metamodel.
(vi) Check the convergence of the objective function. If convergence is achieved, stop. If not, continue to the next step.
(vii) Add the current optimal design point to the sample pool. Go to Step (iii).

4 A NUMERICAL EXAMPLE

To study the proposed optimization method, a roof dome structure subject to a global buckling constraint was solved.

4.1 A 120-Member Roof Dome Structure

The example was a 120-member roof dome structure. This structure was studied in literature (Keveh and Talatahari 2010). In this study, the design objective of the structure was to minimize the total volume of the structure, subject to a global buckling constraint, as follows in Eq. (5-7):

$$C(\mathbf{x}) = \sum_{i=1}^{5} L_i x_i \tag{5}$$

$$\lambda_1(\mathbf{x}) \geq \lambda_1^{Limit} \tag{6}$$

$$5 \text{ in}^2 \leq x_i \leq 15 \text{ in}^2 \tag{7}$$

Due to the symmetry of the structure, five cross-sectional areas were considered as design variables (x_i, i=1,..,5). The lower and upper bound limits of the variables were 5 in^2 and 15 in^2, respectively. The Young's modulus was 30,450 ksi. In this work, a global buckling constraint was considered and the lower bound limit of the first buckling factor was λ_1^{Limit} =100. Figure 1 shows the three-dimensional (3D) view and top view of the un-deformed roof structure. SAP2000 Software (Computer and Structures 2011) was used to build the FE model and analyze the structure. Linear elastic buckling analyses were performed.

Two optimization methods were applied using two different metamodeling methods. Method 1 was the global RBF created using 51 sample points based on Latin Hypercube sampling. Method 2 was the adaptive RBF started with 21 initial sample points; one additional sample point was added in each subsequent iteration of the design optimization.

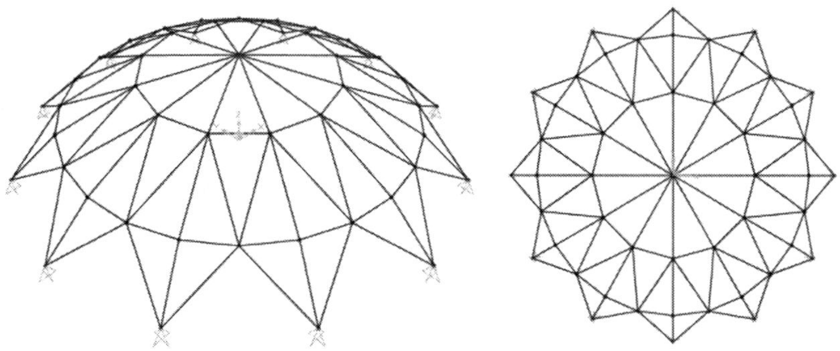

Figure 1. A 120-member roof dome structure.

Table 1. Optimal designs of the roof dome structure.

	RBF (51 samples)	Adaptive RBF (23 samples)	Lower bound	Upper bound
A_1 (in^2)	8.621	8.966		
A_2 (in^2)	5.000	5.000		
A_3 (in^2)	5.000	5.000	5.000	15.000
A_4 (in^2)	5.000	5.000		
A_5 (in^2)	5.000	5.000		
Volume (in^3)	137239	138388		
1st buckling factor (RBF)	100.000	100.000		
1st buckling factor (FE)	98.192	100.000		
1st buckling factor (% error)	1.8%	0.0%		

Table 1 shows the optimal designs of the roof structure using the two methods. For the adaptive RBF method, three iterations and a total of 21+1+1=23 sample points were used to achieve the final optimal design. The optimal design points obtained using the two optimization methods were verified using additional FE analyses, and the results were compared with those estimated using metamodels. Both methods worked well and the optimal designs were achieved.

5 SUMMARY AND CONCLUSIONS

An adaptive metamodeling method was developed and used in structural optimization. The adaptive metamodeling method combined an RBF metamodeling technique and a gradient-based numerical algorithm. The adaptive method was compared with a traditional global metamodeling approach in which the sample size was fixed. The new method improved the accuracy of the RBF metamodel during the optimization iterations, especially in the neighborhood of the optimal design point. To study the performance of the optimization method, a roof truss structure was optimized. The method worked well and accurate optimization results were obtained. The work provides an effective method for design optimization of complex engineering systems that require expensive FE analyses.

References

Arora, J. S., *Introduction to Optimum Design*, Third Edition, Academic Press, 2011.

Arora, J. S., and Wang, Q., Review of Formulations for Structural and Mechanical System Optimization, *Structural and Multidisciplinary Optimization*, Springer, 30(4), 251-272, 2005.

Basudhar A., and Missoum, S., Adaptive Explicit Decision Functions for Probabilistic Design and Optimization using Support Vector Machines, *Computers and Structures*, Elsevier, 86, 1904–17, 2008.

Bi, J., Fang, H., Wang, Q., and Ren, R., Modeling and Optimization of Foam-Filled Thin-Walled Columns for Crashworthiness Designs, *Finite Elements in Analysis and Design*, Elsevier, 46(9), 698-709, 2010.

Computer and Structures, Inc. (CSI). SAP2000 Software, V15. Berkeley, California: Computer and Structures, Inc., 2011.

Fang, H., Rais-Rohani, M., Liu, Z., and Horstemeyer, M. F., A Comparative Study of Metamodeling Methods for Multi-Objective Crashworthiness Optimization, *Computers and Structures*, Elsevier, 83, 25-26, 2121-2136, 2005.

Fang, H., and Wang, Q., On The Effectiveness of Assessing Model Accuracy at Design Points for Radial Basis Functions, *Communications in Numerical Methods in Engineering*, Wiley, 24(3), 219-235, 2008.

Goldberg, D., *Genetic Algorithms in Search, Optimization and Machine Learning*, Addison-Wesley Professional, Reading, MA, 1989.

Jin, R., Chen, W., and Simpson, T. W., Comparative Studies of Metamodeling Techniques under Multiple Modeling Criteria, *Structural and Multidisciplinary Optimization*, Springer, 23, 1-13, 2001.

Keveh, A., and Talatahari, S., Optimum Design of Skeletal Structures using Imperialist Competitive Algorithm, *Computers and Structures*, Elsevier, 88, 1220-1229, 2010.

Kirsch, U., and Rozvany, G. I. N., Alternative Formulations of Structural Optimization, *Structural and Multidisciplinary Optimization* Springer, 7, 32-41, 1994.

Microsoft, *Microsoft Excel 2013*. Redmond, WA: Microsoft Corporation, 2013.

Interdependence between Structural Engineering and Construction Management
Edited by Ozevin, D., Ataei, H., Modares, M., Gurgun, A., Yazdani, S., and Singh, A.

ISBN: 978-0-9960437-6-2

A COMPARISON BETWEEN PROBABILISTIC AND POSSIBILISTIC NUMERICAL SCHEMES FOR STRUCTURAL UNCERTAINTY ANALYSIS

HAMZA SEKKAK and MEHDI MODARES

Dept of Civil, Architectural, and Environmental Engineering, Illinois Institute of Technology, Chicago, USA

Analysis of a structure is a crucial procedure to ensure its reliable design and performance. Because of complexities present in the closed-form structural analysis methods, numerical schemes have become the standard of practice. These schemes are generally performed deterministically. However, the input parameters defining the material and geometric properties may possess uncertainties. They can arise from various sources including modeling, manufacturing, and construction errors. The quantification of uncertainties is generally based on either probability theories (using random variables) or possibility theories (using interval and fuzzy variables). In this work, the finite-element-based probabilistic and possibilistic methods are discussed and compared. A case study of static and dynamic uncertainty analyses of a structure using the aforementioned schemes is performed. The results of those analyses suggest that the incorporation of uncertainty in the analysis provides a higher level of confidence. Moreover, they are compared for both accuracy and computational efficiency. Based on the results, it is observed that the determination of approach must be based on the problem complexity as well as the level of available information.

Keywords: Finite element method, Monte-Carlo simulations, Interval, Fuzzy.

1 INTRODUCTION

In reliable design of a structure, its stability and safety must be guaranteed. This may be achieved by predicting the overall behavior of the structure through analytical schemes over its design lifetime. These schemes construct structural models and consider the applied external static and dynamic loads to obtain the structures' response (deformations) and load effects (internal forces, moments, and stresses). Because of the limitations and complexities present in closed-form analytical methods, numerical and approximate methods such as finite element analysis are widely used (Hughes 1987).

Conventional structural analysis is performed deterministically. However, there exist uncertainties arising from numerous sources that may drastically alter the analytical results. Those uncertainties can stem from: a) physical imperfections (material and geometric properties), and b) modeling inaccuracies (approximate solutions, truncation errors, etc.)

Though conventional methods of structural design use a series of load amplification and strength reduction factors that are based on statistical models of historical data, the effects of uncertainties are not considered in structural analysis schemes used in practice. Therefore, for a reliable design, it is essential to incorporate uncertainties in analytical schemes.

In general, uncertainty is defined as the inability to predict the outcome of an event before it occurs. There exist two categories of uncertainties: 1) aleatoric, and 2) epistemic. Aleatoric uncertainties are inherent in the parameters and are irreducible. On the other hand, epistemic uncertainties stem from lack of knowledge and are reducible. The isomorphic paradigms of uncertainty analysis include probabilistic approaches based on stochasticity (using aleatoric random variables) as well as possibilistic set-based approaches (using epistemic interval and fuzzy variables). In terms of limitations, probabilistic approaches require a well-defined distribution of information, whereas possibilistic approaches require assumption on the bounds. In order to overcome those shortcomings, the polymorphic uncertainty analysis methods are introduced in recent years that combine both probabilistic and possibilistic approaches. The examples of those methods include imprecise probability, Dempster-Shafer functions, and fuzzy random variables.

In this work, the isomorphic-based approaches of structural uncertainty analysis using finite-element-based probabilistic and possibilistic methods are discussed and compared. For this comparison, three distinct methods of uncertainty analysis are utilized that includes: a) Stochastic finite element analysis based on Monte Carlo Simulations, b) Interval Analysis, and c) Fuzzy analysis. As a case study, a finite element model of a structural frame system is constructed and its static response and dynamic natural frequencies are determined. The problem initially assumes known values for material properties leading to obtaining deterministic results. Then, uncertainties in material properties are considered using random, interval, and fuzzy variables respectively. Following that, the corresponding uncertainty analyses for the aforementioned methods are performed. Finally, the results of each method are compared for their level of confidence and computational efficiency.

2 METHODOLOGY

2.1 Deterministic Finite Element Method

2.1.1 *Static problem*

The force equilibrium equation for a static problem is defined as a set of linear equations as in Eq. (1):

$$[K]\{U\}=\{F\} \tag{1}$$

where, $[K]$ is the stiffness matrix, $\{U\}$ is displacement vector (response), and $\{F\}$ is the load vector. The static displacements (response) are determined as in Eq. (2):

$$\{U\}=[K]^{-1}\{F\} \tag{2}$$

2.1.2 *Dynamic problem*

The equation of motion for an undamped, free vibration dynamic problem is defined as a set of linear homogeneous ordinary differential equations as in Eq. (3):

$$[M]\{\ddot{U}\}+[K]\{U\}=\{0\} \tag{3}$$

where, $[M]$ is the mass matrix, and $\{\ddot{U}\}$ is the acceleration vector.

The natural circular frequency is calculated by solving the following eigenvalue problem in Eq. (4):

$$\left([K]-\omega^2[M]\right)\{\varphi\}=\{0\} \tag{4}$$

where, the values of ω are natural circular frequencies (eigenvalues) and the vectors $\{\varphi\}$ are the corresponding mode shapes (eigenvectors).

2.2 Structural Uncertainty Analysis

Analysis of a structure with uncertainty, using an enhanced numerical scheme such as uncertain finite element method, pertains to obtaining the structure's uncertain response and load effects (uncertain output) due to the applied uncertain forces/moments and displacements (uncertain input). Figure 1 shows the uncertain analysis scheme schematically.

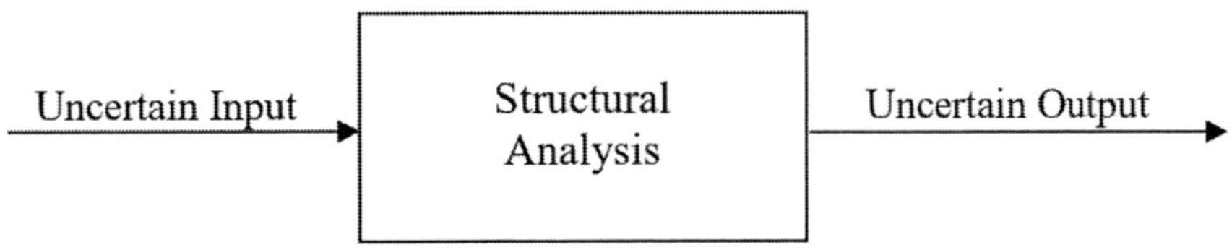

Figure 1. Structural uncertainty analysis scheme.

The following describes three methods of structural uncertainty analysis used in this work.

2.2.1 *Stochastic Finite Element Method (Monte Carlo Simulation)*

In Stochastic Finite Element Method, the input uncertainty is quantified by random variables. A random variable is defined by its prescribed probability function that yields the probability of existence of that variable in a given subset of the real space, Eq. (5). Figure 2 shows the probability density function of a random variable.

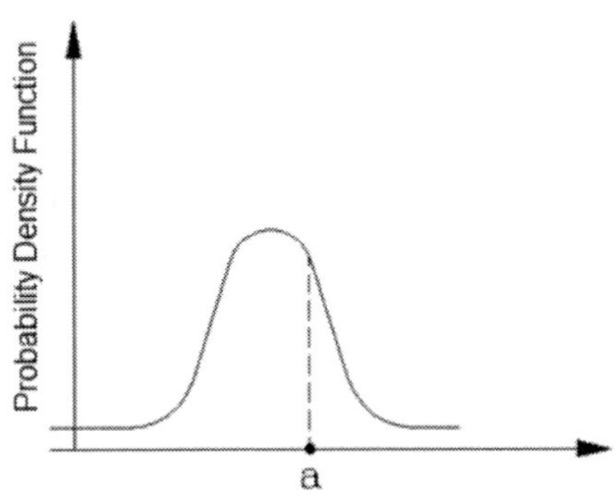

Figure 2. Random variable defined by its probability density function.

$$F_x(a)=P\left([x\le a]\right)=\int_{-\infty}^{a} f(x)dx \tag{5}$$

where, F(x) is the cumulative probability function and f(x) is the probability density function in Eq. (5). In this approach, uncertainty analysis is performed by using Monte Carlo Simulations which are based on random sampling in an iterative process. Using the uncertain input as random variables in the analysis, the output values will possess intrinsic stochastic uncertainties. For structural design purposes, assuming bounded variables for the input, the quantities of interest are the upper and lower bounds of the uncertain output values.

2.2.2 *Interval Finite Element Method*

In Interval Finite Element Method, the input uncertainty is quantified by intervals (Moore and Lodwick 2003). A real interval is defined as a closed set bounded by extreme values, Eq. (6). Figure 3 shows a closed interval schematically.

$$\tilde{x} = [x_l, x_u] = \{x \in \Re \mid x_l \leq x \leq x_u\} \tag{6}$$

Figure 3. A closed interval.

Interval analysis has been used to quantify the uncertainty that arise from various sources including truncation and measurement errors. Using the uncertain input in the analysis as intervals, the output values will possess interval or set-based uncertainties. For structural design purposes, the quantities of interest are the upper and lower bounds of the uncertain output input values.

2.2.3 *Fuzzy Finite Element Method*

In Fuzzy Finite Element Method, the input uncertainty is quantified by fuzzy variables. A fuzzy variable is defined by its membership function, which yields the intervals of confidence for the corresponding levels of presumption (α-cuts) that are bound between 0 and 1 shown in Figure 4.

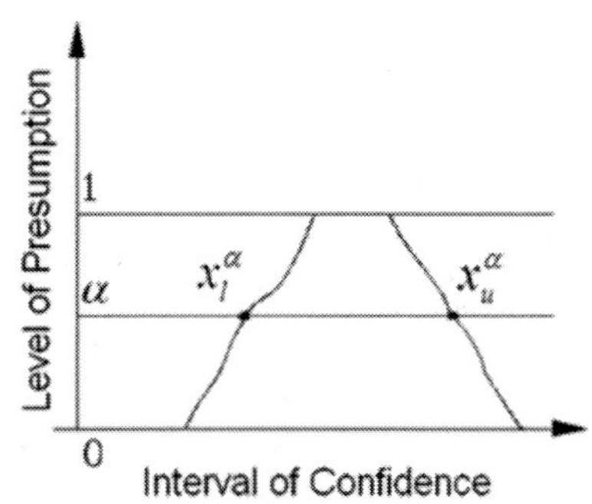

Figure 4. A Fuzzy membership function.

In analysis, defined by their membership functions the uncertain fuzzified input values are used to construct output values as fuzzy membership functions. For structural design purposes, the quantities of interest are the upper and lower bounds of the uncertain output input values for a desired level of presumption.

3 CASE STUDY

3.1 Problem Definition

As a case study, an uncertain 2D moment frame structure, with uncertainty in its material properties, is statically and dynamically analyzed using the aforementioned methods (Figure 5). It consists of three bays and 2 stories, all spaced at distance L. The beams and columns are W-shaped steel members (W18x40) with properties summarized in Table 1.

Table 1. Member properties.

Property	Value
A (in.^{2})	11.8
I (in.^{4})	612
ρ (lb/in.^{3})	0.2839
L (in.)	144

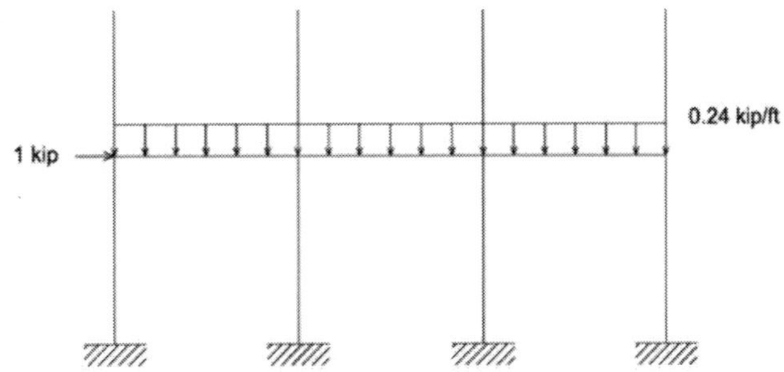

Figure 5. 2D frame with uncertain material properties.

3.2 Quantification of Uncertainty

The uncertainty is considered in the material properties of the structure in the modulus of elasticity. Independent uncertainties are considered for each member. In stochastic finite element method, bounded random variables (μ±σ) are used, in which μ is the mean and σ is the standard deviation. In interval Finite element method, the lower and upper bounds for the modulus of elasticity for each member are considered. In Fuzzy finite element method, the membership function for modulus of elasticity for each member is quantified by six alpha cuts (Table 2).

Table 2. Quantification of material uncertainty.

Approach	Uncertain Variable	Uncertain Modulus of Elasticity (x 29,000) ksi
Bounded Random	(μ±σ)	[0.90,1.10]
Interval	[E_{Lower},E_{Upper}]	[0.90,1.10]
	α=1	[1.00,1.00]
	α=0.8	[0.98,1.02]
Fuzzy	α=0.6	[0.96,1.04]
	α=0.4	[0.94,1.06]
	α=0.2	[0.92,1.08]
	α=0	[0.90,1.10]

3.3 Solution

Uncertain static and dynamic analyses on the frame structure are performed using stochastic, interval, and fuzzy finite element methods. Table 3 summarizes static results for lateral displacement of the roof at the right corner, as well as dynamic results for fundamental natural circular frequency (Modares *et al.* 2006, Vanmarcke and Grigoriu 1983).

Table 3. Summary of results.

Method		Static Problem Horizontal Displacement at Point C	Dynamic problem Fundamental Natural Circular Frequency
Stochastic Finite Element Method (Monte Carlo Simulation)		[0.0289,0.0334]	[4.8131,5.1791]
Interval Finite Element Method		[0.0282,0.0344]	[4.7455,5.2464]
Fuzzy Finite Element Method	α=1	[0.0310,0.0310]	[5.0022,5.0022]
	α=0.8	[0.0304,0.0316]	[4.9520,5.0520]
	α=0.6	[0.0298,0.0323]	[4.9012,5.1013]
	α=0.4	[0.0292,0.0330]	[4.8498,5.1501]
	α=0.2	[0.0287,0.0337]	[4.7980,5.1985]
	α=0	[0.0282,0.0344]	[4.7455,5.2464]

3.4 Observations

It is observed that the results of stochastic finite element method using Monte Carlo simulations are inner-bounds of interval finite element method. Also, interval finite element method, due to its set-based approach, is more computationally feasible than simulation processes. Moreover, fuzzy finite element method, because of discretization of information based on level of presumption, can yield more detailed information of the results.

Comparing probabilistic (stochastic) and possibilistic (interval and fuzzy) analyses shows that in presence of sufficient information, probabilistic methods can yield accurate results. However, in lack of sufficient information, possibilistic method can be used to obtain robust results.

4 CONCLUSIONS

In reliable design of a structure, the uncertainties must be considered in structural analysis schemes through probabilistic and possibilistic methods. Depending on the level of available information, probabilistic (with more information) or possibilistic method (with limited information) can be utilized. Thereby, the combination of probabilistic and possibilistic methods can be used for developing more reliable and robust analytical schemes.

References

Hughes, T. J., *The Finite Element Method: Linear Static and Dynamic Finite Element Analysis*, Englewood Cliffs, NJ: Prentice-Hall, 1987.

Modares, M., Mullen, R. L., and Muhanna, R. L. Natural Frequencies of a Structure with Bounded Uncertainty, *Journal of Engineering Mechanics*, 132(12), 1363-1371, 2006.

Moore, R., and Lodwick, W., Interval Analysis and Fuzzy Set Theory, *Fuzzy Sets and Systems*,135(1), 5-9, 2003.

Vanmarcke, E., and Grigoriu, M., Stochastic Finite Element Analysis of Simple Beams, *Journal of Engineering Mechanics*, 109(5), 1203-1214, 1983.

Interdependence between Structural Engineering and Construction Management
Edited by Ozevin, D., Ataei, H., Modares, M., Gurgun, A., Yazdani, S., and Singh, A.

ISBN: 978-0-9960437-6-2

CONSTRUCTION METHODOLOGY FOR STRAIN GAGE INSTALLATION IN AIRPORT PAVEMENTS

GREGORY KIOA[1] and AMARJIT SINGH[2]

[1]*Hensel Phelps Construction Co., Honolulu, USA*
[2]*Dept of Civil and Environmental Engineering, University of Hawaii, Honolulu, USA*

Installing strain gages in a pavement can give useful information on pavement distress before visible marks appear on the pavement surface. The installation of sensors at existing and busy airports comes at a price, with a very short construction window, because of the need for least disruption to air traffic owing to the expensive concessions paid by airlines almost a year in advance. As such, a proposed project on a runway at Kahului airport, Maui, Hawaii could only be granted three six-hour nights, from 11 pm to 5 am, while Honolulu airport executed a project on a taxiway that has a 7-day window, on which this article will focus. Among the many constraints – both for construction and airport safety – is the requirement that concrete attain 5000-psi strength before the taxiway is reopened to air traffic. This means that the major work of concreting on the taxiway must be completed before the end of the 4th day of the construction window. Thus, scheduling must be done in 15-minute slots. In addition, precision is key, and high productivity is required. This article describes the construction method behind this unique project.

Keywords: Delamination, Neutral axis, Milling, Data Acquisition Cabinet, Slippage, Airplane stress, Asphalt.

1 INTRODUCTION

Pavement distresses at airports are expensive to repair. Moreover, the longer a distress goes unchecked, the worse its effect is. Whereas, design improvements have potential in pavement management, normal distresses cannot be eliminated, perhaps only mitigated. With this in mind, a real-time study is underway to monitor distresses in a taxiway.

Installing early detection sensors to measure delamination is an involved task: the pavement needs to be cordoned off, the surface must be milled, sensors installed, and a trench dug to take the wires from the sensors off the runway to a data acquisition center, resting on a concrete pad. The wires must be pulled, and then the trench and milled area must be concreted and asphalted.

2 LAYOUT AND ORIENTATION OF SENSORS

Airport pavements are designed to withstand the rolling weight of a fully loaded airplane. Over time, due to the impacts that are constantly applied on the pavement, and the heat during summer, pavement distresses appear on the surface, such as slippage, cracking, and surface shoving. The location and orientation of sensors in the pavement are established through studying the major Airbus and Boeing airplanes frequenting the airport. Each different airplane has unique wheel spacing and landing gear, which really determines the location of the sensors and the spacing between them. The nine different types of aircraft analyzed were the B737-300, B747-300, B757-300, B767-300, B767-300 ER, B777-300 Baseline, B777-300 ER, A330-300, A380-800.

The colored dots in Figure 1 indicate the different landing gear location of the distinctive airplanes mentioned above with the sensors positioned in transverse and longitudinal directions at appropriate spacing to enable data capture.

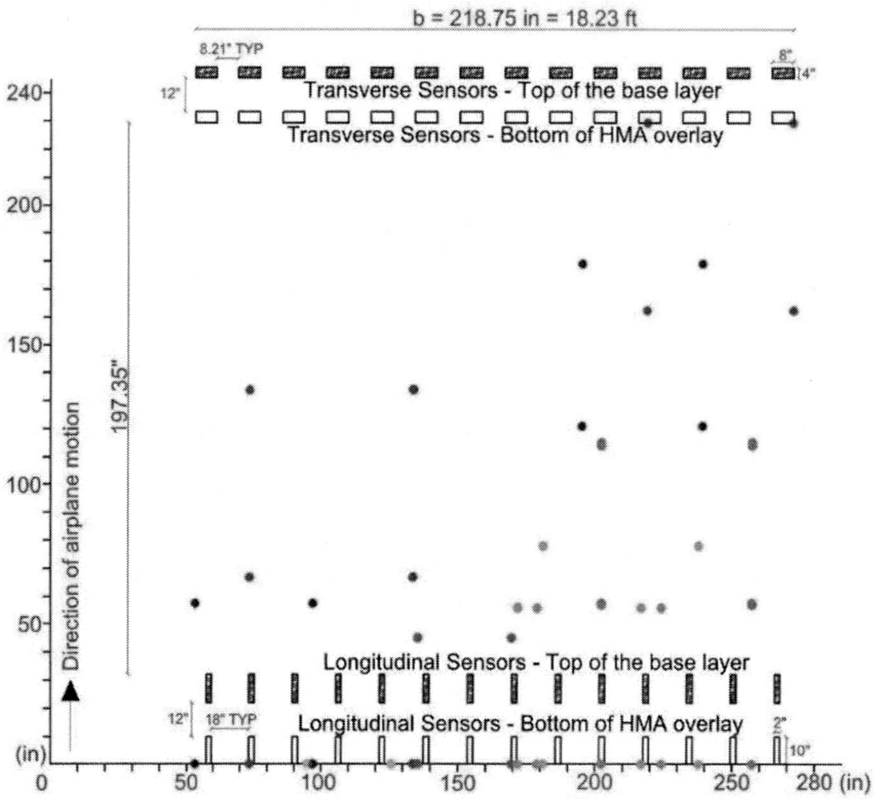

Figure 1. Position of the wheels of nine different aircraft (Singh and Melo 2017).

3 SCIENCE OF THE SYSTEM

The concept behind this new idea of installing strain gage sensors on airport pavements is to detect delamination of the runway surface. In a tightly bonded, homogenous pavement that does not show any form of delamination, compression is registered at the top surface, while tension is registered at the bottom (Figure 2). However, when the surface delaminates, the upper HMA layer acts independently of the lower layer. Thus, we have two independent neutral axes, as in Figure 2. This finding created the logic that if the change in strain pattern can be detected, delamination will also be detected (Singh and Cook 2016). Thus, half the sensors (28) are on the bottom of the top layer, and half are on top of the bottom layer (28).

4 DESIGN AND DRAWINGS OF THE SYSTEM

The surface must have sensors installed at two elevations, and wires must be taken off-pavement as far as the safety regulations require. And then, a platform is required for the DAQ.

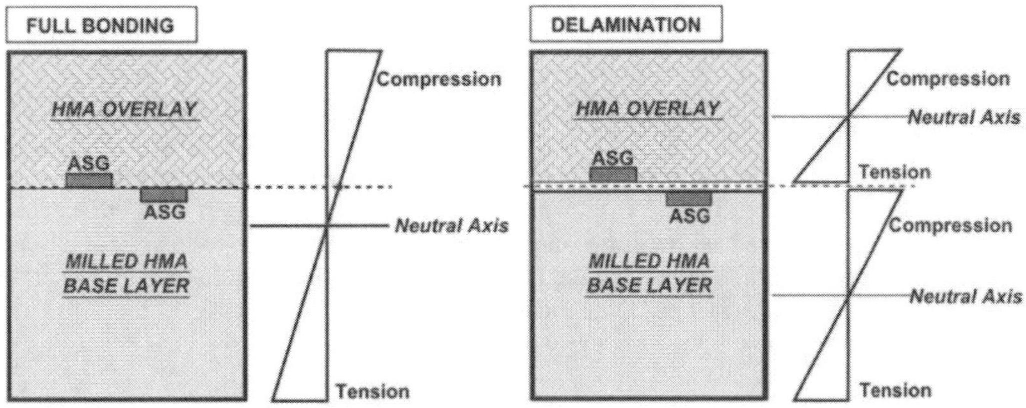

Figure 2. Neutral axis in bonding and delamination (Modified from Cook *et al.* 2016).

5 DESIGN AND DRAWINGS OF THE SYSTEM

There are many different types of elements put together to ensure that the system will achieve its design intent. First, we know that the surface must have sensors installed at two elevations; we know that wires must be taken off-pavement as far as the safety regulations require. And then, a platform is required for the data acquisition cabinet (Singh and Cook 2015).

6 MILLING OF EXISTING ASPHALT

The milling of existing asphalt to a depth of 3 inches is an essential part of the strain gage sensor installation (Figure 3). This is the area where the sensors will be installed to collect the data. Besides, the upper 3 inches is the top course of any pavement that is redone over time. This is what creates the weak interface that causes a disconnect between the upper layer and base layer.

Figure 3. The milled area for the sensor location.

7 TRENCHING AND EXCAVATION

There is a long trench excavated for the conduits from the taxiway to off-runway, while another excavation is made for the pull box. The length is determined by (i) minimum distance off-runway required for airport safety, and (ii) maximum distance before signals from sensors degrade (usually 500 ft.). The trench must be designed to accommodate lighting cans and at least four 2-inch conduits and be at a depth of 18 inches or more to remain safe under vehicular traffic.

8 SENSORS

It is important to ensure that the 56 sensors are buried in the asphalt at a depth of 3". The thickness of the sensor is ½-inch; hence, the recessed sensors must be buried deeper than 3". This thickness, together with the strength of the asphalt concrete, will avoid any instant damages to the sensor caused by the weight of the aircraft. Figure 4 shows the sensors in the pavement.

All the sensors were set in place with asphalt emulsion tack coating to secure the sensors in place. Electricians finalized all wiring connections and routed all 56 wires through the two lighting cans. Wires were then pulled the next day from the lighting cans to the pullbox and up to the DAQ platform. The electricians came back to connect the sensor wires to the DAQ system, connect the camera, and hook up the photovoltaic solar system to power the DAQ equipment.

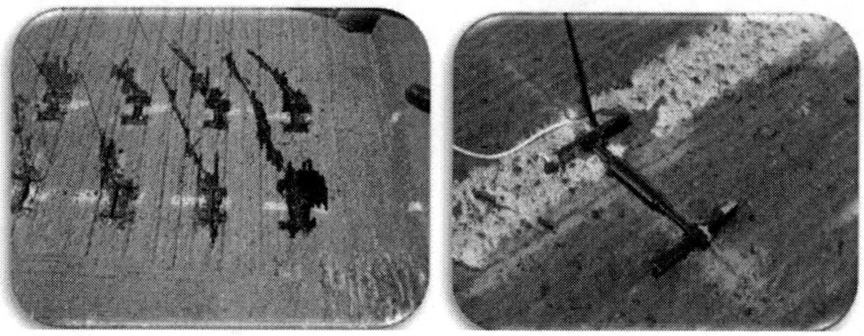

Figure 4. Sensors are secured in place.

9 HIGH STRENGTH CONCRETE (5000-PSI) AND DAQ PLATFORM

The use of 5000-psi concrete in the trench was to ensure that its strength adequately withstands the weight of the aircraft and prevent this distressed spot from settling under aircraft rolling weight. For these reasons, only concrete is allowed in the trench for the base layer and below.

However, the DAQ platform (Figure 5), which is off-taxiway and 10' x 10' x 4", serves only as a rigid platform for the DAQ cabinet and solar panels, and can be sufficient at only 3000-psi. Nevertheless, it is problematic to order two batches of concrete from the concrete supplier for such small quantities. Hence, 5000-psi was used for the platform, as well.

Figure 5. DAQ platform concrete and equipment.

10 ASPHALT PAVING

The depth of the pavement is only 3" which calls for two lifts of asphalt. The first lift is placed, and laborers spread the first layer of asphalt evenly on the paving area. A walk-behind-compactor and a road roller are used to achieve a compaction of 100%. The same process is performed on the second and final layer of asphalt. But the final layer also uses sheep foot rollers. The asphalt paving is also performed on the taxiway trench, while the trenching area off-runway was topped off with dirt. Figure 6 has pictures on the asphalt paving.

Figure 6. Placing hot mix asphalt.

11 DAQ PLATFORM AND EQUIPMENT

Though this is a slab-on-grade for all practical purposes, the weight on the platform of the DAQ cabinet and solar power is not excessive. However, to avoid any damage in the event that vehicular traffic drives over the platform, it is necessary for the platform to be at least 4" thick and be reinforced, for which a wire roll is sufficient.

The equipment installed on the DAQ platform (data logger and multiplexor) are programmed to record and communicate data transmitted from the sensors. The equipment is charged by solar panels positioned above the equipment. To maximize the electricity generated, the solar panels are oriented south. A battery helps to store energy for night operations.

12 PROJECT SCHEDULING

The duration for the sensor installation project was five working days. Some activities are best performed at night such as, trenching, saw cutting, grinding, and chipping. Other activities are best when performed during day, such as concrete placement, forming, wiring, sensor installation. There is one primary reason why night work is best for this project: to avoid the severe jet blast.

In the beginning of the project, the taxiway was barricaded and closed off to aircraft. The contractors milled the asphalt overlay and excavated the trench and pullbox. They recessed the sensors and saw-cut the asphalt base layer for the sensor wires. While the saw cutting on the taxiway took place, the excavation in soil in the off-runway trench also commenced. A line was sprayed on the dirt by the surveyor indicating the path to be excavated. The asphalt on the runway trench was saw cut and busted. The sequence of these activities was setup to avoid any conflict or clash resulting in potential standby time. Figures 7 shows a sample schedule of work.

13 CONCLUSIONS

Strain gauge sensor installation on an operating airport can be challenging due to the limitation of time. Selecting the right contractor to perform the task is very important. They must be well equipped, have many experienced workers, and be familiar with ground, site, and electrical work.

It is important that before a contractor is selected, to meet with them, go through their past work history, and make sure they can meet the work schedule. The success of the strain gage sensor installation at the Honolulu airport depends a lot on good decision making.

SCHEDULE: Night 1 - Excavating & Trenching

Activity ID	Tasks	Dur. (min)	Time (hours)																							
			40.0		80.0		120.0		160.0		200.0		240.0		280.0		320.0		360.0		400		440.0		480.0	
N1-01	Close-off Runway	20	3																							
N1-02	Survey & Layout	40	3	3																						
N1-03	Spray Paint Saw Cut Line	20			1																					
N1-04	Sawcutting	80				2	2	2	2																	
N1-05	Milling	120					3	3	3	3	3	3														
N1-06	Trenching Off-Runway	180							3	3	3	3	3	3	3	3	3									
N1-07	Breaking Existing Ashphalt	60							2	2	2															
N1-08	Trenching On-Runway	180								2	2	2	3	3	3	3	3	3								
N1-09	Set Pullbox	40																2	2							
N1-10	Install Conduit Off-Runway	120										4	4	4	4	4	4									
N1-11	Install Conduit On-Runway	120																4	4	4	4	4	4			
N1-12	Form DAQ	40																2	2							
N1-13	Electrical Stub up on DAQ	40																		2	2					
N1-14	Set Lighting Cans	20																		2						
N1-15	Concrete Placement	120																		4	4	4	4	4	4	
N1-16	Clean-Up & Clear out Equipment	20																								4

SCHEDULE: Night 2 - Sensor and Wire Preparation

Activity ID	Tasks	Dur. (min)	Time (hours)																							
			40.0		80.0		120.0		160.0		200.0		240.0		280.0		320.0		360.0		400		440.0		480.0	
N2-01	Sensor Layout	40	3	3	3	3																				
N2-02	Recessed Sensor Prep	120					3	3	3	3	3	3														
N2-03	Saw Cutting for Sensor Wire	180											2	2	2	2	2	2	2	2	2					
N2-04	Clean-Up & Clear out Equipment	20																				4				

Figure 7. Work tasks and schedule duration.

Acknowledgments

The authors acknowledge grant no. TA-2011R from HDOT Airports for this project. The contents presented in this article are the views of the authors only and do not necessarily represent the views or policies of the project sponsors. This article does not constitute a standard, specification, or regulation.

References

Cook, K., Garg, N., Singh, A., and Flynn, M., Detection of Delamination in the HMA Layer of Runway Pavement Structure Using Asphalt Strain Gauges, *Journal of Transportation Engineering,* ASCE, 142(11), 04016047, 10.1061/(ASCE)TE.1943-5436.0000869, Nov, 2016.

Singh, A., and Cook, K., Procurement Management and Construction Methods for the Installation of Pavement Monitoring Systems in Airport Pavements, *International Conference on Industrial Engineering and Operations Management,* Dubai, UAE, March 3-5, 2015.

Singh, A., and Cook, K., Analysis of Runway Pavement Distresses Using Embedded Instrumentation, *ASCE Construction Congress,*, San Juan, Puerto Rico, June 2-4, 2016.

Singh, A., and Melo, M., Sensor Design and Layout for Airport Asphalt Pavement Instrumentation to Test for Delamination, in *International Conference on Industrial Engineering and Operations Management 2017,* Rabat, Morocco, April 11-13, 2017.

Interdependence between Structural Engineering and Construction Management
Edited by Ozevin, D., Ataei, H., Modares, M., Gurgun, A., Yazdani, S., and Singh, A.

ISBN: 978-0-9960437-6-2

THE FOUNDATIONS OF CONTINUUM DAMAGE MECHANICS

SIAMAK YAZDANI[1], SEVENN BORGERSEN[2], ASLI PELIN GURGUN[3], and HOSSEIN NAZARI[4]

[1]*Dept of Civil Engineering, California State University, Pomona, USA*
[2]*SEB Corp, Eagan, Minnesota, USA*
[3]*Dept of Civil Engineering, Yildiz Technical University, Istanbul, Turkey*
[4]*Dept of Mechanical Engineering, NDSU, Fargo, USA*

Damage Mechanics has become a useful theory in describing the nonlinear behavior of solids driven by the nucleation and growth of cracks and microcracks. This approach, based on the first principles of mechanics and thermodynamics, has also been combined with classical theories of plasticity to address a wide range of loading applications. In spite of the many different damage mechanics models and representations that are proposed, the foundation of damage mechanics is not well understood or at least not thoroughly published giving rise to the many inaccurate definitions and formulations. The intent of this paper is to provide the background of the continuum damage mechanics outlining the fundamentals on which this field theory is set up. The internal variable theory of continuum thermodynamics is reviewed and is shown that with Legendre transformation technique, various potential functions can be developed for damage mechanics formulation in either stress or strain space. The concept of constrained or neighboring equilibrium state is also introduced and is explained. The paper will conclude with the derivation of the general damage potential and a suggestion is given for the isotropic damage formulation with the resulting uniaxial stress-strain relation.

Keywords: Response tensor, Plasticity, Dissipation inequality, Isothermal deformations.

1 INTRODUCTION

The observed nonlinear behavior in brittle solids stems from two main meso-structural changes within the materials. One is the development and propagation of microcracks and micro-voids. The process of cracking destroys material bonds and leads to the reduction in the material stiffness. Failure occurs where these microcracks link up and form a major crack and fault line. This behavioral pattern occurs mostly when zero or low confining pressure exists. Uniaxial tension, uniaxial compression, biaxial compression and tension are typical load paths under which microcracks propagation and coalescence dominate solid behavior.

When the lateral confining pressure is large, the formation of microcracks gets inhibited and/or delayed. The classical experimental work of Hueckel and Maier (1977) on rocks has shown that under sufficiently large confining pressure permanent deformations due to plastic flow and void closures occur and that these meso-structural changes, as the other category of internal changes, lead to no alteration of material stiffness. To model such diverse behavior observed in solid materials, different classes of theories have been proposed.

There are three levels of damage mechanics theories that one may employ to model nonlinear behavior of brittle solids. One is the micro-level modeling incorporating material science approach where the formulation is intended to correlate closely with the physics of micro-structural changes. Although, this approach is fundamental, its usefulness is rather limited in engineering applications where complex loading paths, such as proportional and non-proportional paths, load reversal, etc, must be considered. The next level used by researchers is referred to as the meso-level modeling where microcrack development and kinetics at a smaller scale than a continuum is utilized. This approach is interesting and promising involving discrete field theories such as fracture mechanics to model microcrack initiation and propagation in the medium and at the interfaces. The utility of this approach is restricted to simple load paths and the end an averaging or smearing formalism is used to predict macro behavior; hence losing the fine details used up to that point. For most engineering applications and modeling the third approach, namely the macro-level modeling (Saboori *et al.* 2014, 2015) is overwhelmingly used and appears in numerical simulation codes such as ANSYS and ABAQUS.

This intent of this paper is therefore to present a continuum level class of damage mechanics theories that is based on the first principles of mechanics and thermodynamics. Some fundamental concepts critical to the understanding of the theory is explained in the body of the paper.

2 FORMULATION

Central to the continuum damage mechanics is the concept of neighboring equilibrium state (Ortiz 1985, Yazdani and Karnawat 1996, Yazdani 1993). This can be explained further in a schematic representation of a crack growth in Figure 1. From fracture mechanics, we know that an active and stable crack of length a would reach another equilibrium state of length b once acted upon a sufficiently large load that overcomes local fracture resistance.

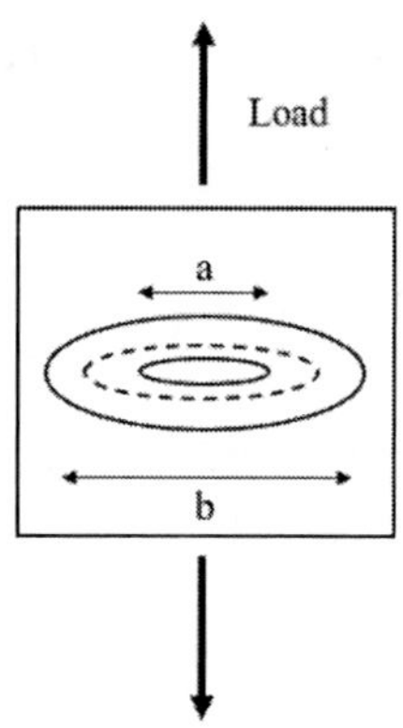

Figure 1. Schematic representation of crack growth.

The concept of the neighboring equilibrium state would allow for the mathematical formulation of crack growths at lengths a and b, but also for any assumed length in between; and hence the term neighboring equilibrium state is used. Assuming that the material internal energy under stress exists, the internal variable theory of thermodynamics (Lubliner 1972) can be used to cast an equilibrium state in terms of the Helmholtz Free Energy (HFE), $A(\epsilon, k)$, or the Gibbs Free

Energy (GFE), $G(\boldsymbol{\sigma}, k)$ where $\boldsymbol{\epsilon}$ is the strain tensor, $\boldsymbol{k}$ represents a cumulative damage parameter, and $\boldsymbol{\sigma}$ denotes the Cauchy stress tensor. The thermodynamics functions HFE and GFE are related to each other through an elegant mathematical transformation technique known as "Legendre" transformation.

When formulation is desired to be developed in terms of the strain components, then the HFE function is used. When stress space formulation is desired, the GFE function is utilized. For isothermal deformations, as assumed here, the internal dissipation inequality as a consequence of the second law of thermodynamics, can be expressed either in terms of the HFE as shown in Eq. (1) and Eq. (2):

$$\dot{A} + \boldsymbol{\sigma}:\dot{\boldsymbol{\epsilon}} \geq 0 \tag{1}$$

or in terms of GFE as:

$$\dot{\boldsymbol{G}} - \boldsymbol{\epsilon} : \dot{\boldsymbol{\sigma}} \geq \boldsymbol{0} \tag{2}$$

where the "super dots" denote the time rate. For the rest of this paper the foundation of the damage mechanics will be presented in the stress space using GFE.

Cracking and microcracking in brittle solids is known to destroy material bonds and thus affecting the compliance of the solid. Denoting the compliance tensor as $\boldsymbol{C}(k)$ and the initial compliance of the undamaged state as $\boldsymbol{C}^{\boldsymbol{0}}$ is then shown in Eq. (3):

$$\boldsymbol{C}(\boldsymbol{k}) = \boldsymbol{C}^{\boldsymbol{0}} + \boldsymbol{C}^{c}(\boldsymbol{k}) \tag{3}$$

where $\boldsymbol{C}^{c}$ is the added compliance tensor reflecting the effects of damage. The total deformation can also be shown in terms of the components as Eq. (4):

$$\begin{aligned} \boldsymbol{\epsilon} &= \boldsymbol{\epsilon}^{\boldsymbol{0}} + \boldsymbol{\epsilon}^{D} + \boldsymbol{\epsilon}^{r}(k) \\ &= \boldsymbol{C}^{\boldsymbol{0}}:\boldsymbol{\sigma} + \boldsymbol{C}^{c}(k):\boldsymbol{\sigma} + \boldsymbol{\epsilon}^{r}(k) \end{aligned} \tag{4}$$

where $\boldsymbol{\epsilon}^{\boldsymbol{0}} = \boldsymbol{C}^{0}:\boldsymbol{\sigma}$ is the elastic component of the deformation, $\boldsymbol{\epsilon}^{D}(k) = \boldsymbol{C}^{c}(k):\boldsymbol{\sigma}$ is referred to as the elastic damage deformation, and $\boldsymbol{\epsilon}^{r}(k)$ is the inelastic or un-recoverable/permanent deformation caused by the misfit of crack surfaces. The schematic representation of these components is shown in Figure 2 for a uniaxial stress path.

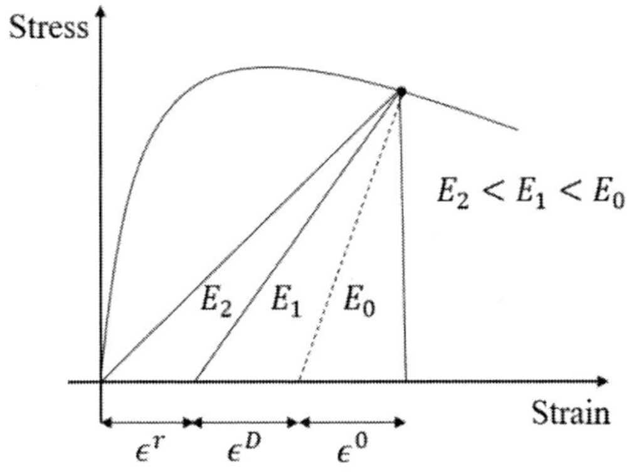

Figure 2. Schematic representation of deformation components.

Eq. (2) leads to two fundamental equations that:

$$\epsilon = \frac{\partial G}{\partial \boldsymbol{\sigma}} \tag{5}$$

and

$$\frac{\partial G}{\partial k}\dot{k} \geq 0 \tag{6}$$

where, Eq. (5) states that the GFE is a potential to the strain tensor and where Eq. (6) is called the dissipation inequality. Integrating Eq. (5) and using the decomposition structure shown in Eq. (4), a general form of the GFE as obtained as:

$$G(\boldsymbol{\sigma}, k) = \frac{\mathbf{1}}{\mathbf{2}}\boldsymbol{\sigma}{:}\,\boldsymbol{C}{:}\,\boldsymbol{\sigma} + \epsilon^r(k){:}\,\boldsymbol{\sigma} - A^i(k) \tag{7}$$

where $A^i(k)$ is the inelastic component of the HFE associated with the surface tension of the crack surfaces and arises as the constant of the integration. Incorporating Eq. (7) into Eq. (6) yields:

$$\frac{\partial G}{\partial k} = \frac{1}{2}{:}\,\boldsymbol{\sigma}{:}\frac{\partial \boldsymbol{C}^c}{\partial k}{:}\,\boldsymbol{\sigma} + \frac{\partial \epsilon^r(k)}{\partial k}{:}\,\boldsymbol{\sigma} + \frac{\partial A^i(k)}{\partial k} \geq 0 \tag{8}$$

where it is assumed that damage is irreversible; i.e. $\dot{k} \geq 0$.

Utilizing Eq. (8), we can now obtain a damage potential $\boldsymbol{\psi}(\boldsymbol{\sigma}, \boldsymbol{k})$ by introducing a positive function $\boldsymbol{g}^2(\boldsymbol{\sigma}, \boldsymbol{k})$ as shown in Eq. (9) such that:

$$\psi(\boldsymbol{\sigma}, k) = \frac{1}{2}\boldsymbol{\sigma}{:}\frac{\partial \boldsymbol{C}^c}{\partial k}{:}\,\boldsymbol{\sigma} + \frac{\partial \epsilon^r}{\partial k}{:}\,\boldsymbol{\sigma} + \frac{\partial A^i(k)}{\partial k} - g^2(\boldsymbol{\sigma}, k)$$

$$= \frac{1}{2}\boldsymbol{\sigma}{:}\frac{\partial \boldsymbol{C}^c}{\partial k}{:}\,\boldsymbol{\sigma} + \frac{\partial \epsilon^r}{\partial k}{:}\,\boldsymbol{\sigma} - \frac{1}{2}t^2(\boldsymbol{\sigma}, k) = 0 \tag{9}$$

where $t^2(\boldsymbol{\sigma}, k) = -2\frac{\partial A^i}{\partial k} + 2g^2(\boldsymbol{\sigma}, k)$ is regarded as the damage function. We note that the identification of either of the functions $\frac{\partial A^i}{\partial k}$ or $g^2(\boldsymbol{\sigma}, k)$ is not necessary as long as the damage function $t(\boldsymbol{\sigma}, k)$ is obtained through experiments. The final general form of the damage surface, $\psi(\boldsymbol{\sigma}, k)$ can then be stated by introducing a damage response tensor $\boldsymbol{R}$ and a flow-rule for $\dot{\boldsymbol{\epsilon}}^r$ (Eq. (10) and Eq. (11)) with **M** being identified as the second order inelastic flow tensor, such that:

$$\dot{\boldsymbol{C}}^c(k) = \dot{k}\boldsymbol{R} \tag{10}$$

$$\dot{\boldsymbol{\epsilon}}^r(k) = \dot{k}\boldsymbol{M} \tag{11}$$

Introducing these into the structure of Eq. (9) yields the general form of the damage surface as Eq. (12):

$$\psi(\boldsymbol{\sigma}, k) = \frac{1}{2}\boldsymbol{\sigma}{:}\,\boldsymbol{R}{:}\,\boldsymbol{\sigma} + \boldsymbol{M}{:}\,\boldsymbol{\sigma} - \frac{1}{2}t^2(\boldsymbol{\sigma}, k) = 0 \tag{12}$$

Different tensors of ***R*** and ***M*** define different models that are published in the literature. If ***R*** is postulated as an isotropic tensor, the predicted results would be considered isotropic. This is not usually observed in brittle solids, but under blast loading some isotropic and uniform distribution of damage has been reported, making isotropic from of $\boldsymbol{R}$ a viable choice in some cases.

3 EXAMPLE – UNIAXIAL LOAD PATH

With $\boldsymbol{R}$ speculated to be proportional to an isotropic tensor, and with $\boldsymbol{M}$ being set as a null tensor for simplicity, the uniaxial stress can be reduced from Eq. (12) to be (Eq. 13)):

$$\sigma_1 = t(k) \tag{13}$$

which implies that the damage function could be experimentally determined from a simple uniaxial test. When the tensor M is set to be a null tensor, no inelastic damage strain, $\epsilon^r(k)$, is predicted and the response is classified as being elastic-perfectly brittle. The associated uniaxial strain, ϵ_1 then obtained by integrating the rate form of the Eq. (4), in general or for the case of simply load path as considered here simply from Eq. (4) and Eq. (13) as Eq. (14):

$$\epsilon_1 = \left(\frac{1}{E_0} + k\right)\sigma_1 \tag{14}$$

A normalized stress-strain curve for using the damage function by Ortiz (1985) is shown in Figure 3, where f_t is the uniaxial tensile strength of solid and ϵ_t is the corresponding strain.

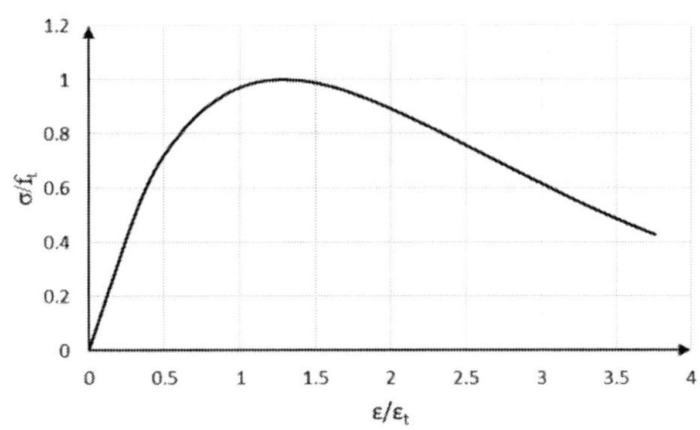

Figure 3. Normalized stress-strain curve of a solid under uniaxial tensile loading.

4 CONCLUSIONS

Damage mechanics is a viable field theory for solids where the development of cracks and microcracks render the material damaged and hence more compliant. A class of damage mechanic theories is outlined in this paper defining all critical assumptions and terms not usually covered in papers on damage mechanics. This approach was shown to be based on the first principles of mechanics and thermodynamics utilizing the second law to obtain dissipation inequality leading to the formulation of a damage potential. In this way, the number of material parameters is usually small, and pertain to the physical attributes of the macro behavior such as elasticity, Poisson's ratio, bi-axial strength ratios, etc. The general form of the damage surface was derived and damage response tensors were defined. An example of stress-strain relation was provided for a special case of isotropic tensor for $\boldsymbol{R}$, making the interpretation of the damage parameter rather straight forward and easy to understand.

References

Hueckel, T. and G. Maier, Incremental Boundary Value Problems in the Presence of Coupling of Elastic and Plastic Deformations, A Rock Mechanics Oriented Theory, *International Journal Solids and Structures,* 13, 1-15, 1977.

Lubliner, J., On the Thermodynamic Foundations of Nonlinear Solid Mechanics, *International Journal Nonlinear Mechanics*, 7, 237-254, 1972.

Ortiz, M., A Constitutive Model for the Behavior of Concrete, *Mechanics of Materials*, 4(1), 67-93, 1985.

Saboori, A., Yazdani, S., Reberg, A., and Tolliver D., Anisotropic Damage Model for Concrete Subjected to Freeze-Thaw Process, *International Journal of Civil and Structural Engineering*, 5(1), 42-51, 2014.

Saboori, A., Yazdani, S., and Tolliver D., Anisotropic Damage Mechanics Modeling for Concrete Under Biaxial Fatigue Loading, *Open Journal of Civil Engineering*, 5(1), 8-16, 2015.

Yazdani, S., On a Class of Continuum Damage Mechanics Theories, *International Journal of Damage Mechanics*, 2(2), 62-176, 1993.

Yazdani, S., and Karnawat, S., A Constitutive Theory for Brittle Solids with Application to Concrete, *International Journal of Damage Mechanics,* 5(1), 93-110, 1996.

Interdependence between Structural Engineering and Construction Management
Edited by Ozevin, D., Ataei, H., Modares, M., Gurgun, A., Yazdani, S., and Singh, A.

ISBN: 978-0-9960437-6-2

THE EFFECT OF AGE ON THE COMPRESSIVE STRENGTH OF CONCRETE BY USING ULTRASONIC PULSE VELOCITY

QADRI AHMED YOUSIF[1], IDRIS BEDIRHANOGLU[2], MEHMET ENVER AYDIN[3], and ZIWAR ZEBARI[4]

[1]*Dept of Physics, Masaka Company, Duhok, Iraq*
[2]*Dept of Civil Engineering, New York University Abu Dhabi, Abu Dhabi, UAE,*
[3]*Dept of Physics, Dicle University, Diyarbakir, Turkey*
[4]*Dept of Physics, Ministry of Education, Duhok, Iraq*

In this study, the effect of age on the strength development of concrete was investigated. For this purpose, specimens casted with ready mix concrete for different grades of concrete were obtained from ready-mix concrete firms available in Diyarbakir city which is the one of the biggest cities of Turkey. A few sets with different grade of concrete cube specimens were used in the experimental work. Each set included over 40 specimens. For measuring concrete strength development, two different tests were used: ultrasonic pulse velocity testing (UPV) and uniaxial compressive strength testing. After UPV and compressive strength tests, strength development for early ages was evaluated and defined. The aim of this work is to define the path of strength development of certain grades of concrete used in the east regions of Turkey. Another aim of the work is to be able to determine concrete strength through non-destructive UPV tests. As a result, it was seen that concrete strength development for early ages is not linearly proportional to age. Further, UPV measurements can give information on the progress of concrete strength.

Keywords: Nondestructive test, Ready-mix concrete, Axial compression test, Samples.

1 INTRODUCTION

Concrete is a composite material consisting of water, cement, sand, and gravel. It is a most widely used material in building construction, because of easily availability of aggregate that forms around 70% of the volume of concrete. Most of the existing buildings in developing countries suffer severe seismic damage due to insufficient concrete quality (Ilki *et al.* 2006, Bedirhanoglu 2009). Concrete quality that is used in any field of construction (i.e., buildings, bridges, roads) must be checked and controlled (Bedirhanoglu 2014). Fortunately, today's technology has more techniques and science branches that can be used in this area.

One of the techniques of physics and engineering sciences widely used to identify concrete quality, and physical or mechanical properties is called non-destructive testing (NDT), (ACI 228.2R 2014). NDT is a technique used to inspect the physical and mechanical properties of a material without causing any damage. In NDT, the major purpose is to determine errors in a piece being tested. One of the most important factors in the development of a measurement and control system, called Non-Destructive Inspection, is the need to minimize or even eliminate the risk that is caused by a destructive testing method, which is the fundamental method of classical quality control. Quality control of the material with NDT is economical due to not damaging the materials. Non-destructive testing methods are used almost as a quality control tool at every step of production, during operation, during periodic maintenance and during the development of new products. Non-destructive testing evaluation is fast and sensitive (Hellier 2003).

The material inspection, which means the examination of the structural and internal properties of the materials, is divided into two sections: non-destructive testing and destructive testing (DT). Non-destructive material inspection can be used for different purposes such as examining of defects that occur in materials because of various reasons. Ultrasonic pulse velocity testing method (UPV) is one of the NDT test methods that works by sending high sound frequency wave through the material for analyzing physical and mechanical properties (Hellier 2003).

UPV method is used to ensure the properties of concrete and natural rocks. In this study, the physical and mechanical properties of concrete are aimed to define through measuring the velocity of ultrasonic pulse velocity. This test is carried out by travelling a pulse of ultrasonic wave energy through concrete and measuring the pass time of the pulse. Higher velocities indicate good quality and continuity of the material, but slower velocities through concrete structures indicate low-quality concrete or may indicate concrete with defects (cracks or voids) (Neville 1996 and Luthi 2013).

The UPV method is used to evaluate the quality of concrete with measurement of velocity propagating through the bulk of the material to find out whether the quality of concrete is good or bad in terms of density, uniformity, homogeneity etc. (IS 1992).

Compressive strength of concrete is evaluated more commonly by destructive methods compared to nondestructive testing. Destructive testing is the most reliable method for measuring properties of the concrete. On the other hand, this method damages the material and costs more labor. So, it is vital to improve new techniques for measuring properties of concrete. At this aspect UPV is a good alternative however more researches are needed to improve the reliability of the technique. For that reason, significant number of experiments are needed to be done with different types of concrete in different regions. This point has been our main source of motivation for carrying this study. For this purpose, a series of experimental and empirical studies were carried out by our research group (Bedirhanoğlu and Şahin 2005, Sahin and Bedirhanoglu 2014, Bedirhanoglu 2014, Zebari *et al.* 2017, Abdullah *et al.* 2017, Yousif 2018). In this study, the effect of the age of concrete on its compressive strength was investigated.

2 EXPERIMENTAL WORK

2.1 Specimens and Tests

Three groups of cube samples were tested in this section. Each group consisted of at least 40 samples. A total of at least 130 samples were prepared. Weights of each sample were measured, and then UPV and axial compression tests were performed (Figure 1). Therefore, three data sets for each sample were generated. These measurements were made on the days that we specified earlier. During tests specimens were kept in water with around 21°C.

All these samples were taken from three different locations. So, we can split these samples into three groups. The first group is the C20 concrete produced by the Emirler Concrete Plant for the text building. The second group is taken from Bes Concrete Plant and has 25 concrete classes. The third group is again from Bes Concrete Plant C30 concrete.

After 24 hours from casting, specimens were removed from the plastic mold with air pressure. The fist measurements were done on three specimens and others were hold in water. In each day at least three specimens were tested. The UPV and compressive strength tests were made after 1, 2, 3, 4, 5, 6, 7, 8, 9, 10, 14, 21 and 28 days.

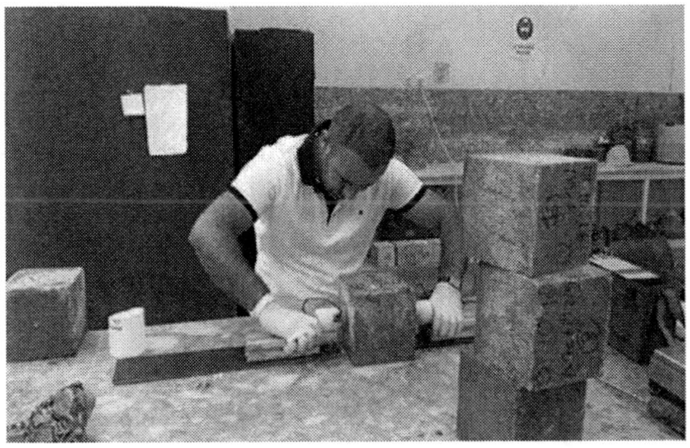

Figure 1. Testing set up of the UPV test.

2.2 Experimental Results

Compressive strength test was conducted by using three specimens at each testing day. Generally, for each day of testing compressive and UPV tests were carried out on three specimens while in order to have an idea about other specimens UPV test were carried on all other specimens, too. At each time of testing minimum three specimens were used and mean values of these three specimens were used to draw graphs given in Figures 2 to 3. Figures 2 to 3 show relationship between age and compressive strength of concrete both with UPV measurement and compressive strength tests for three groups of specimens. All graphs clearly show that strength of concrete suddenly increases with age at very early first few days. After, the increase in compressive strength stops suddenly for next few days.

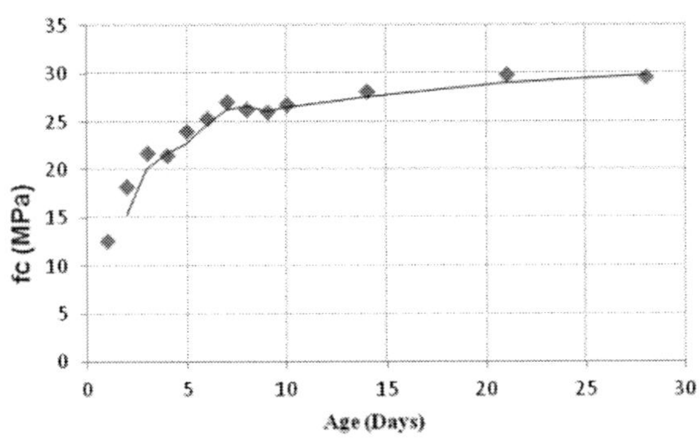

Group 1

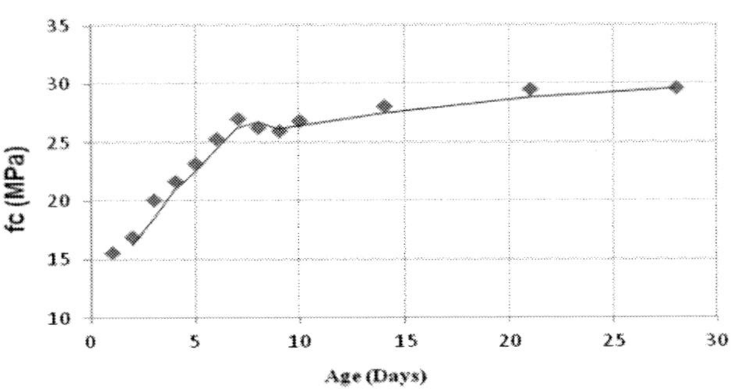

Group 2

Figure 2. Compression strength and age relationships graphs for groups 1 and 2.

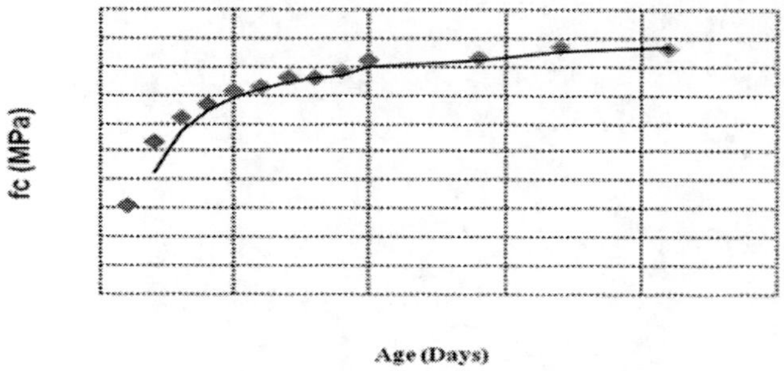

Group 3

Figure 2 (contd). Compression strength and age relationships graphs for Group 3.

After a few days of recession, compressive strength of concrete starts to increase in a very slow manner. Relationships between UPV and compressive tests are given in Figure 4. It can be concluded that there are some reasonable relationships between UPV and compressive strength for different ages of different concrete grades.

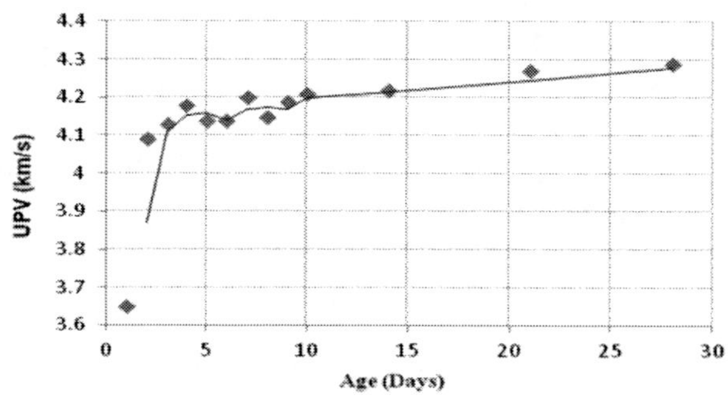

Group 1

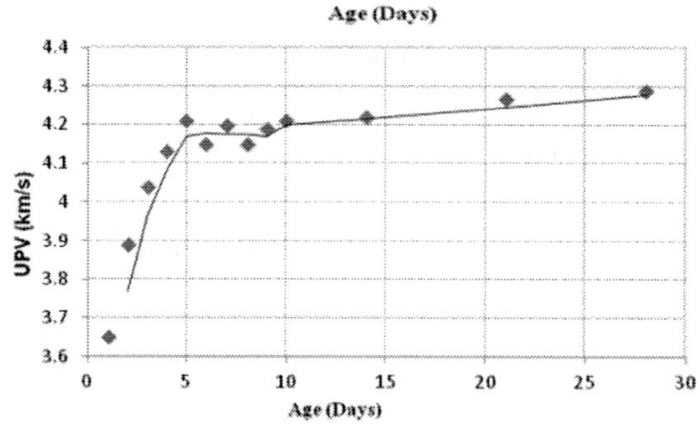

Group 2

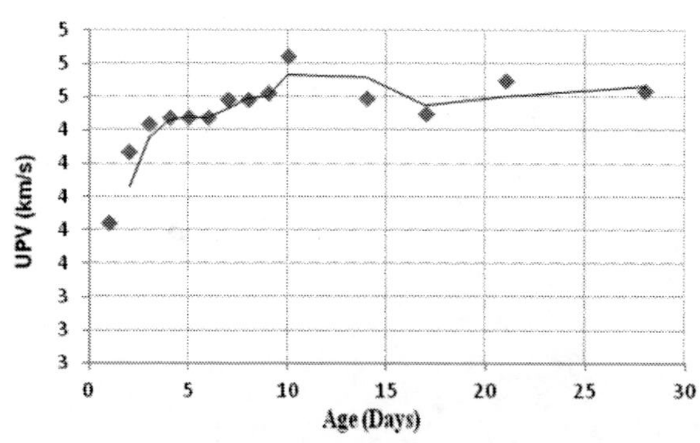

Group 3

Figure 3. UPV and age relationships graphs for all three groups.

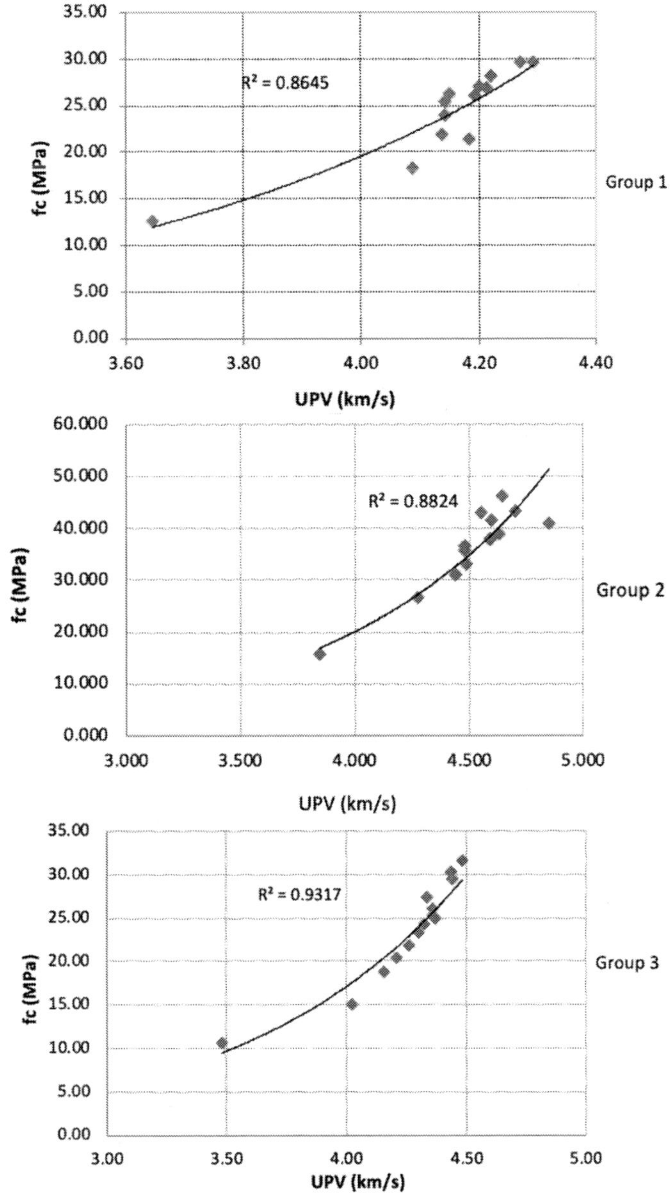

Figure 4. UPV and compressive strength relationships for all three groups.

3 SUMMARY AND CONCLUSIONS

This paper investigated the effect of age of concrete on the strength development through using UPV measurement compared with compression tests. This relation is schematically illustrated as in Figure 5. As a summary, it can be concluded that UPV values increase with age of concrete both in case of UPV and compression tests. However, this increase is not linear. Another important finding is recession in UPV values at certain ages such as between 5-10 days where this was not seen in relation between compressive strength and age of concrete.

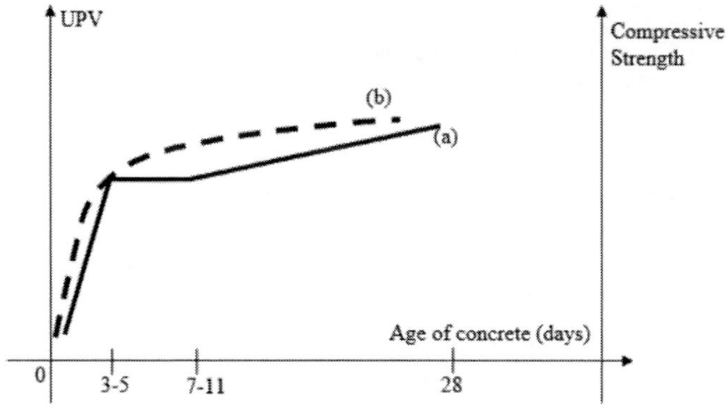

Figure 5. Schematic representation of relationships between a) UPV and age of concrete, b) compressive strength and age of concrete.

Acknowledgments

We thank our friends Vaheel Abdullah, Melik Osal and Emre Güven for their help during experiments. We also thank the Laboratory of Civil Engineering, Department of Dicle University, and Dicle University Department of Physics for the facilities that they provided. We appreciate the financial support provided by MASAKA General Trading – Import and Export LTD Company, KRG, Iraq.

References

Abdullah, V., Aydin M. and Bedirhanoglu, I., Beton Elastisite Modülünün Ultrasonic Ses Dalgası Yayılma Hızı Ile Tahmin Edilmesi (Prediction of Elastic Modulus of Concrete Through Using Ultrasonic Pulse Velocity), *Dicle Üniversitesi Mühendislik Fakültesi Mühendislik Dergisi,* 8(3), 475-484, 2017.

ACI 228.2R, *Non-destructive Test Methods for Evaluation of Concrete in Structures.*

Bedirhanoğlu, I. and Şahin, U. Düşük, Orta ve Yüksek Dayanımlı Betonların Elastisite Modüllerinin Bulanık Mantık İle Belirlenmesi (Defining Elastic Modulus of Concrete with Lowö Normal and High Strength Through Using Fuzzy Logic). Bilimde Modern Yöntemler Sempozyumu -BMYS'2005, 16-18 Kasım 2005. Grand Yükseliş, KOCAELİ. ISBN: 9758047590, 9789758047598

Bedirhanoglu, I., The behavior of reinforced concrete members with low strength concrete under earthquake loads: an investigation and improvement, PhD Thesis, Istanbul Technical University, Turkey, 2009.

Bedirhanoglu, I., A Practical Neuro-Fuzzy Model for Estimating Modulus of Elasticity of Concrete, *Structural Engineering and Mechanics,* 51(2), 249-265, 2014.

Hellier, C., *Handbook of Non-Destructive Testing Evaluation,* McGraw. Hill, Chapter 1, 2003.

Ilki, A., Bedirhanoglu, I. Basegmez, IH., Demir, C. and Kumbasar, N., Shear Retrofit of Low Strength Reinforced Concrete Short Columns with GFRP composites. Third International Conference on FRP Composites in Civil Engineering (CICE 2006), Miami, USA, 2006.

Luthi, T., Non-Destructive Evaluation Method, 2003-2013, Anax Imperator, Micro X-ray Tomographic View, Iwan Jerjen, Empa, 2013.

Neville, A. M., *Properties of Concrete,* 4th edition. John Wiley and Sons, Inc., New York. IS, 1996.

Sahin and Bedirhanoglu, I., A Fuzzy Model Approach to Stress-Strain Relationship of Concrete in Compression, *Arabian Journal for Science and Engineering,* 39(6), 4515-4527, 2014.

Yousif, Q. A., *Beton Yaşının Basınç Dayanımı Üzerine Olan Etkisinin Ultrases Yöntemi İle Belirlenmesi (In Turkish) (Determining Effect of the Age of the Concrete on the Compressive Strength Through with UPV),* MSc Thesis, Institute of Science, Dicle University, Diyarbakir, 2018.

Zebari, Z., Bedirhanoglu, I., and Aydin M., Beton Basınç Dayanımının Ultrasonic Ses Dalgası Yayılma Hızı Ile Tahmin Edilmesi (Prediction of Compressive Strength of Concrete Through Using Ultrasonic Pulse Velocity), *Dicle Üniversitesi Mühendislik Fakültesi Mühendislik Dergisi,* 8(1), 43-52, 2017.

Interdependence between Structural Engineering and Construction Management
Edited by Ozevin, D., Ataei, H., Modares, M., Gurgun, A., Yazdani, S., and Singh, A.

ISBN: 978-0-9960437-6-2

PREDICTION OF UNEVEN DISTRIBUTION OF SNOW LOAD BASED ON GENERALIZED REGRESSION NEURAL NETWORK

ZHICHENG REN and QINGWEN ZHANG

[1]*Key Lab of Structures Dynamic Behavior and Control of the Ministry of Education, Harbin Institute of Technology, Harbin, China*
[2]*Key Lab of Smart Prevention and Mitigation of Civil Engineering Disasters of the Ministry of Industry and Information Technology, Harbin Institute of Technology, Harbin, China*

At present, there are mainly three research ways on the distribution of snow load: field measurement, experimental simulation, and numerical simulation. They all need expensive equipment, specimen model used only once, and slight defective simulation software. Artificial Neural Networks (ANNs) rely on a large number of input functions to estimate unknown or known occurrences. This paper applies ANNs to predict the uneven distribution of snow loads. First, the snow load database is collected and set up, including all the data from field measurement and experimental simulation. Second, part of the data from the $1m^3$ model is used as a training set to train Backpropagation Neural Networks (BPNN) and Generalized Regression Neural Networks (GRNN). The parameters of the two neural networks are determined (such as the smoothing factor in the GRNN), and the rest of the data is used to examine the model; the prediction and measurement results of the two models are compared. Finally, based on different wind direction angles and wind speed, a GRNN model is established to predict the uneven snow load distribution. Through comparative analysis, GRNN is more suitable for predicting the uneven distribution of snow loads, especially in predicting snow thickness. In particular, the prediction effect of the model on the snow load distribution trend is good, and the prediction ability of the neural network model is displayed.

Keywords: Snow height, Backpropagation neural network, Smoothing factor.

1 INTRODUCTION

A neural network is an active and marginal interdisciplinary subject. It is of great theoretical significance to study its background and development process. The prediction problem has been studied and concerned by many domestic and foreign scholars for many years (Fu 2010). Artificial Neural Network (ANN) is a mathematical model that applies information processing similar to the structure of synaptic connections in the brain. It is also often referred to as Neural Networks (NN) or quasi-neural Networks directly in engineering and academia (Li *et al.* 2016). Since the beginning of the 21st century, great progress has been made in the research of ANNs, and research on various aspects of neural networks is increasing. Su *et al.* (2016), in order to expand the application range of typical roof wind tunnel test data, established a wind load prediction model for the long-span spherical roof, which was based on a generalized regression neural network and combined with a typical spherical roof series wind tunnel test.

Nowadays, there are more and more structural damage and collapse accidents caused by snowstorms, causing huge economic losses and casualties. The uneven distribution of snow load, especially for today's more and more buildings with large span, thin shell, and roof as special forms, has a greater impact on structural safety. At present, there are three research methods of snow load: field observation, numerical simulation, and wind tunnel test. Zhou and Gui (2008) summarized and analyzed the research status of these three methods. Li *et al.* (2016) conducted a practical measurement study on snow load on the surface of typical low-rise structures, such as flat roof, arched roof, high and low span roof, and double span and double slope roof. Liu *et al.* (2018) and Zhang *et al.* (2018) measured the distribution of snow around the buildings. Field observation research is indispensable, but field observation also has its disadvantages: heavy workload, high requirements, high cost, and many error-influencing factors. Wang *et al.* (2016) used the fine quartz sand particles to simulate wind-blown snow for a wind tunnel test in order to predict the snow load distribution of roofing. In terms of numerical simulation, the use of CFD technology to analyze wind-induced snow drift has begun to be widely carried out. Xiao (2017) used numerical simulation (CFD) combined with FAE method to study the snow distribution coefficient of the large-span roof structure surface.

In this paper, two neural network models are established, and the prediction results of the two neural networks are compared and analyzed, and the more accurate neural network model is selected.

2 EXPERIMENT

In order to increase the accuracy of the neural network model prediction results and reduce errors, a large amount of data must be provided for the training model, and only high-quality data can guarantee the efficiency, accuracy and other performance of the model. Therefore, before the model is built, a large number of snow load experiments are needed to collect experimental data.

The experimental equipment in this paper includes snowmaker, snow screening machine, 3×2 fan matrix, 4m×3m flat thin wood board, and various wooden models, as shown in Figures 1-4. Among them, the snow screening frequency of the snow screening machine can be adjusted. This experiment considers that as long as the snowfall is sufficient, the snow height can be easily measured, preventing the snow from being too small and the error too large. Snow height is also not one of the experimental variables, so it is fixed to 20Hz. The 3×2 fan matrix can change the wind speed by adjusting the fan frequency, and its wind direction is fixed. In addition, the experimental measuring tools include anemometers, thermometers, hygrometers, and rulers, among others.

For the experiment of uneven distribution of snow load under different wind speeds, this paper adopts changing fan frequency to realize the control of wind speed, so as to achieve accurate control and multi-gradient change of wind speed. Meanwhile, the experiment speed is fast, the efficiency is high, and the accuracy is good. The $1m^3$ cube model was placed in the center of the pad, and the pad was taken as the carrier of snow distribution. When the fan frequency was 15Hz, 20Hz, 25Hz, and 30Hz respectively, the snow cover height on the washer 20 minutes after snow fall was measured and the experimental results were recorded.

Since the pad is 4m long and 3m wide, the pad is separated by 5cm in the length and width directions, and the snow height is measured along the length direction and the width direction respectively, so as to ensure sufficient data can be collected. When establishing a coordinate system for data, the width direction of the pad is taken as the x-axis, the length direction is taken as the y-axis, and the origin is set at the center of the pad.

Figure 1. Snowmaker.

Figure 2. 3×2 Fan matrix.

Figure 3. Snow screening machine.

Figure 4. $1m^3$ Cube model.

3 PREDICTIVE SIMULATION ANALYSIS OF $1M^3$ CUBE MODEL

The structure of the BP neural network model is relatively simple, consisting of one input layer, several hidden layers, and one output layer. The structure of the BP neural network model in this paper is shown in Figure 5. GRNN is composed of four layers in structure, namely input layer, mode layer, summation layer, and output layer. A typical GRNN structure is shown in Figure 6.

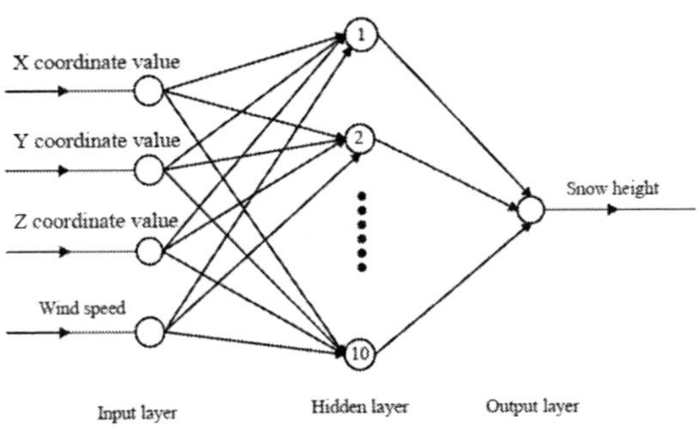

Figure 5. BPNN model.

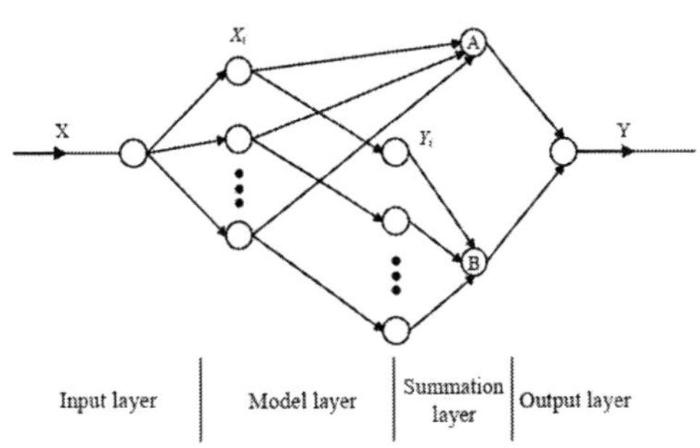

Figure 6. Typical GRNN model.

3.1 Prediction Results with Different Wind Speed

Wind speed is one of the most important factors affecting the uneven distribution of snow load. Therefore, firstly establish a snow load model with only wind speed influence and try to establish a neural network model that can predict the snow load distribution under different wind speeds. Based on the collected snow load experimental data, the snow load distribution prediction model of $1m^3$ cube model can be established. Three of the data sets and results corresponding to the four wind speeds are selected as the training sets, and the other one is used as the detection set.

Considering the distribution of snow cover, the overall distribution trend, the peak and minimum values of snow cover height, and the average value of snow cover height are all important indicators. Therefore, for the evaluation of the prediction results, this paper adopts the overall distribution trend of snow cover, the peak (maximum) of snow cover height, the minimum and average value of snow cover height, as well as the correlation coefficient R between the predicted results and the measured results by comparing the predicted results with the experimental results. The expression is shown in Eq. (1).

$$R=\frac{\sum_{i=1}^{n}(x_i-\overline{x})(y_i-\overline{y})}{\sqrt{\sum_{i=1}^{n}(x_i-\overline{x})^2\cdot(y_i-\overline{y})^2}} \tag{1}$$

Among them, x_i is the predicted value, y_i is the measured value, n is the sample capacity, $\overline{x}$, $\overline{y}$ are the average of the predicted value and the measured value respectively.

The totals of collected data points of the training set and the predicted output are both 4941. For comparison, the data processing results of the snow load distribution experimental data and the prediction results of the neural network model under different wind speed conditions are shown in Table 1. The snow distribution trend is represented by the contour map drawn by the surfer software; the results are shown in Figure 7.

Table 1. Comparison between neural network model prediction results and measured results (1.5m/s).

		The average value of snow height $\overline{h}$ (mm)	The peak values of snow height h_{max} (mm)	The minimum values of snow height h_{min} (mm)	correlation coefficient R
Measured results		2.42	62.00	0.00	--
Prediction results	BP	2.00	50.21	0.91	0.76
	GRNN	3.60	48.00	0.00	0.78

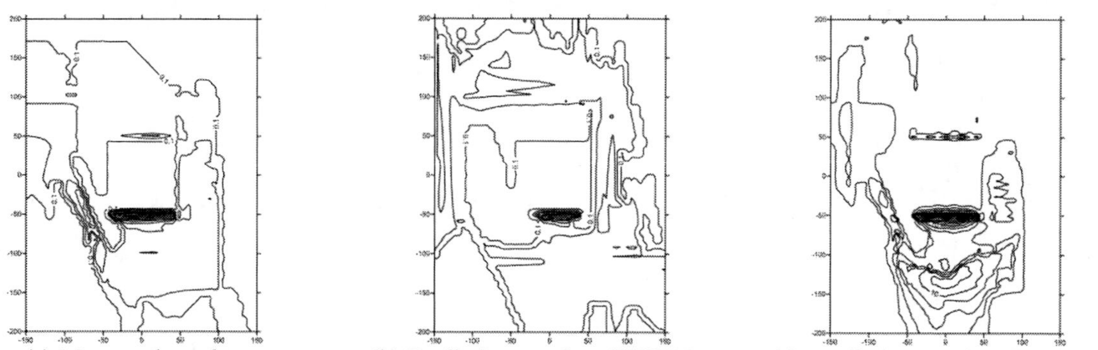

(a) Measured results. (b) Prediction results of BPNN. (c) Prediction results of GRNN.

Figure 7. $1m^3$ model snow distribution trend contour comparison (1.5m/s).

In addition, the cross section where the position where the snow peak appears is found, and the abscissa uses the ratio of the Y coordinate of the measuring point to the span L of the model, and the ordinate uses the ratio of the snow height h of the measuring point to the maximum value h_{max} of the snow depth of the ground area, draw a two-dimensional coordinate map to compare the degree of fitting of the predicted results with the measured results; the results are shown in Figure 8.

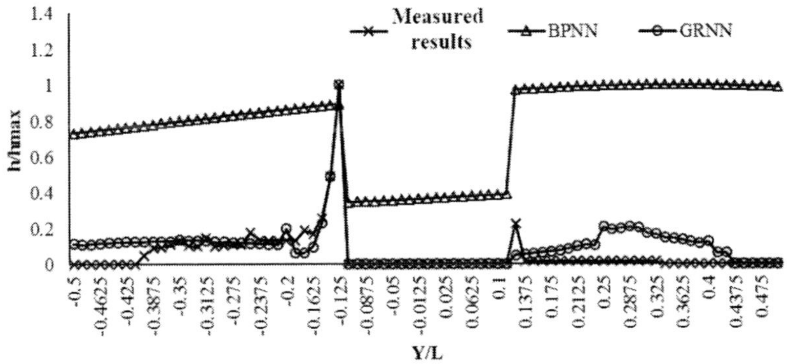

Figure 8. The degree of fitting of the $1m^3$ model prediction results to the measured results (1.5m/s).

3.2 Predictive Analysis of Different Wind Direction Angles Under the Same Wind Speed

The previous predictions of wind speed affecting the uneven distribution of snow load are obtained. Through comparative analysis, it can be seen that GRNN has a better effect on the prediction of snow load distribution. The wind direction angle is also one of the important factors affecting snow load distribution. Therefore, considering the experimental data, a GRNN model for predicting the snow load distribution under different wind direction angles of the same wind speed is established. For the experiment of $1m^3$ cube model, the wind speed is 2m/s, and the wind direction angle has four values, 0°, 15°, 30°, and 45°. Three of the data sets and results corresponding to the four wind direction angles are selected as the training set, and the other one is used as the detection set. Similar to the wind speed prediction model, in this model, the results are shown in Table 2, Figure 9, and Figure 10.

Table 2. Comparison between neural network model prediction results and measured results (0°).

	The average value of snow height $\bar{h}$ (mm)	The peak values of snow height h_{max} (mm)	The minimum values of snow height h_{min} (mm)	The correlation coefficient R
Measured results	1.88	38.00	0.00	--
Prediction results	2.49	34.10	0.00	0.80

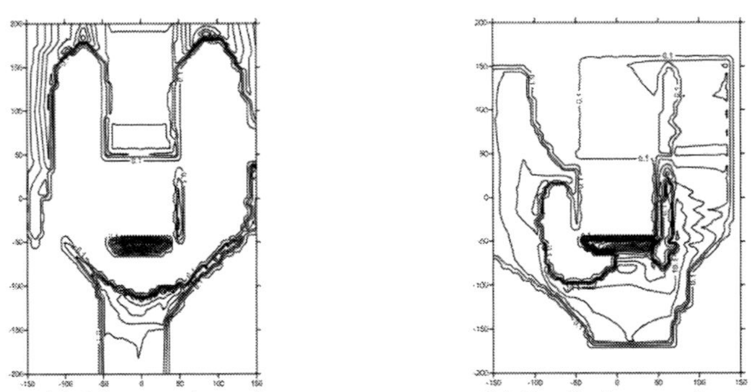

(a) Measured results. (b) Prediction results.

Figure 9. $1m^3$ model snow distribution trend contour comparison (0°).

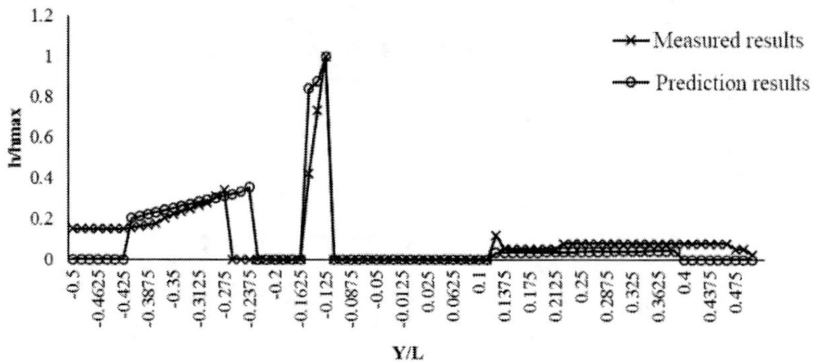

Figure 10. The degree of fitting of the 1m^3 model prediction results to the measured results (0°).

4 CONCLUSIONS

In this paper, the neural network is used to predict the uneven distribution of snow load. The following conclusions can be drawn:

(1) There is a good result in predicting the distribution trend of snow load, especially in the prediction of the position where the peak appears. Although there are some errors in the specific values, the errors are also within a certain range and have certain reference value;
(2) Different neural networks have different accuracy for predicting the results. Compared with BPNN, GRNN has better prediction result;
(3) In the case of predicting the same wind speed with different wind direction angles, the generalized regression neural network has a correlation coefficient of 0.80, indicating a high correlation.

References

Fu, H., *MATLAB Neural Network Application Design*, Mechanical Industry Press, Beijing, 2010.

Li, F. H., Meng, F., and Zhou, X. Y., Field Measurement Research on Distribution of Snow Load on Typical Low-Rise Roofs, *Journal of Natural Disasters*, 25(1), 158-168, 2016.

Liu, M. M., Liu, B. Y., and Zhang, Q. W., Study on Snow Distribution Law of High and Low Cross-Room Area, *Proceedings of the First National Snow Ice Engineering Symposium*, 102-106, Shang Hai, 2018.

Su, N., and Sun, Y., Wind Load Prediction of Large-Span Domes Based on Generalized Regression Neural Network and Its Application, *Journal of Building Structures*, 37(6), 101-106, 2016.

Wang, W. H., and Huang, H. J., Study of Snow-Load Distribution on Roof by Wind Tunnel Test, *Journal of Experiments in Fluid Mechanic*, 30(5), 23-28, 2016.

Xiao, Y., Research Method and Prevention Measures of Wind-Induced Snow Drift in Large-Span Roof, *ShanXi Architecture*, 43(6), 55-56, 2017.

Zhou, X. Y., and Gui, M., Simulation of The Wind-Induced Snowdrift: State of The Art, *Engineering Mechanics*, 25(7), 5-10, 2008.

Zhang, Q. W., Liu, M. M, and Fan, F., Joint Experimental Study on Distributed Wind Distribution Around Buildings, *Proceedings of the First National Snow Ice Engineering Symposium*, 98-101, Shang Hai, 2018.

Interdependence between Structural Engineering and Construction Management
Edited by Ozevin, D., Ataei, H., Modares, M., Gurgun, A., Yazdani, S., and Singh, A.

ISBN: 978-0-9960437-6-2

VISCOELASTIC ANALYSIS OF FLEXIBLE PAVEMENT EMBEDDED WITH CLSM BASES SUBJECTED TO SINUSOIDAL AND IMPULSIVE LOADS

YEN-CHU LIANG[1], HER-YUNG WANG[2], and LI-JENG HUANG[2]

[1]*Dept of Aeronautics and Astronautics, Air Force Academy, Kaohsiung, Taiwan (ROC)*
[2]*Dept of Civil Engineering, National Kaohsiung University of Science and Technology, Kaohsiung, Taiwan (ROC)*

This paper presents finite element analysis of 4-layered flexible pavements embedded visco-elastic asphalt concrete (AC) top layer and controlled low-strength materials (CLSM) sub-layers as well as graded crushed stone subgrades layers subjected to sinusoidal and impulsive surface wheel loads. Two-dimensional deformation assumption is employed. Time-dependent stress-strain relations are treated using Laplace transform based on correspondence principle. Scheme based on fast Fourier transform (FFT) is employed for inversion of Laplace transformed to obtain time-domain results. A typical 4-layered pavement is analyzed in which relaxation modulus of top visco-elastic asphalt layer is expressed in Prony series. Two kinds of CLSMs, i.e., CLSM-B80/30% and CLSM-B130/30% are compared with graded crushed stones and AC used for base materials. Time-domain responses of vertical displacement and stress at bottom of AC layer are compared. The vertical settlements using CLSM bases (B80/30% and B130/30%) are smaller than that using graded crushed stones even considering visco-elastic behavior of asphalt layer and show to be good materials employed as base substitutes for graded crushed stone in flexible pavement design.

Keywords: FEM, Numerical Laplace inversion, FFT, Asphalt concrete.

1 INTRODUCTION

Flexible pavements have been widely employed in highway engineering and effective rapid pavement construction had been achieved by using the controlled low strength materials (CLSM) (Lin *et al.* 2007, Bassani *et al.* 2015). CLSM is a kind of flowable fill defined as self-compacting cementitious material that is in a flowable state at the initial period of placement and has a specified compressive strength of 1200 psi or less at 28 days or is defined as excavatable if the compressive strength is 300 psi or less at 28 days as depicted in ACI-229R (2005). The CLSM is suitable for application to some pavements under repeated excavation and backfill is required such as pipeline construction. The authors had conducted some preliminary studies on engineering properties of CLSM (Huang *et al.* 2014) and the numerical analyses on static and free vibration analysis of CLSM bases in flexible pavements (Huang *et al.* 2017a, Huang *et al.* 2017b).

Asphalt mixtures have special viscoelastic behavior as depicted by Huang (1993). Their mechanical response exhibits time and rate dependency. Numerical schemes combined with finite element models and Laplace transform method or time-incremental skills have been applied for flexible pavement analysis (Zocher *et al.* 1997, Evangelista *et al.* 2005, Malakouti *et al.* 2014). This study is aimed at visco-elastic analysis of 4-layered flexible pavements using finite element method. The top layer is considered as Hot Mix Asphalt (HMA), which is modeled as viscoelastic material; while 4 kinds of base materials are employed, i.e., graded crushed stones, CLSM-B80/30%, CLSM-B130/30% and AC. The numbers B80 and B130 mean the different levels of binder content (i.e. 80 and 130 kg/m^3) and different percentages fly ash (i.e. 30%) substituting to Portland cement were previously chosen for mix design. A comparison study is performed on time responses of vertical displacement and stress located at top of base layer under the sinusoidal and impulsive loadings.

2 FINITE ELEMENT MODELLING OF 4-LAYERED FLEXIBLE PAVEMENTS

Stress-strain relation of a visco-elastic material can be expressed in convolution integrals

$$\sigma(t) = E(t) \otimes \dot{\varepsilon}(t) = \int_0^t E(t-\tau) \bullet \frac{\partial \varepsilon}{\partial \tau} d\tau \tag{1a}$$

$$\varepsilon(t) = D(t) \otimes \dot{\sigma}(t) = \int_0^t D(t-\tau) \bullet \frac{\partial \sigma}{\partial \tau} d\tau \tag{1b}$$

For uni-axial stress state, where *E(t)* is the relaxation modulus, *D(t)* is the creep compliance, *t* is the current time, τ is a temporal variable. Taking Laplace transform of Eq. (1a) we have

$$\hat{\sigma}(s) = s\hat{E}(s) \bullet \hat{\varepsilon}(s) \tag{2}$$

From the correspondence principle in the theory of linear visco-elasticity, we can replace *E* by $s\hat{E}(s)$ in FEM and the calculations of nodal displacements and element stresses in Laplace transform domain are the same as for elastic analysis.

The relaxation modulus for a visco-elastic material can be expressed as the Prony series:

$$E(t) = E_\infty + \sum_{k=1}^{N} E_k \, e^{-t/\rho_k} \tag{3}$$

where E_∞, E_k, ρ_k are evaluated from experiment.

3 FFT FOR NUMERICAL LAPLACE INVERSION

There are various schemes developed for numerical Laplace inversion (Davies and Martin 1979). In this paper we employ the scheme based on FFT using integration of $\hat{F}(s)$ on imaginary axis with trapezoidal rule and expressed by Huddleston and Byrne (1999):

$$f(t_k) = \frac{1}{2\pi i} \int_{C-i\infty}^{C+i\infty} \hat{F}(s) \bullet e^{st_k} ds \approx \frac{e^{Ct_k}}{\pi} \sum_{j=1}^{N_\omega} \frac{1}{2} [\hat{F}(C+i\omega_j) e^{i\omega_j t_k} + \hat{F}(C+i\omega_{j+1}) e^{i\omega_{j+1} t_k}] \Delta\omega_j \tag{4}$$

where $s_j = C + i\omega_j$, C and ω_j denotes the real and imaginary parts of the complex-valued sampling points s_j. From the theory of Laplace inversion C should be located on the right-hand side of all poles of $\hat{F}(s)$.

4 NUMERICAL EXAMPLES

4.1 Validation of Viscoelastic Analysis Using FEM for a Visco-elastic Hollow Cylinder

We consider the thick-walled cylinder, with internal radius $a = 2m$ and external radius $b = 4m$, analyzed by previous researchers (Zocher *et al.* 1997, Evangelista *et al.* 2005), as shown in Figure 1. They analyzed the visco-elastic response for $p(t) = p_0 H(t)$ with $p_0 = 100\ Pa$. Visco-elastic material properties are: $E_\infty = 0.1 kPa,\ \ E_1 = 0.4 kPa,\ \ \rho_1 = 1.0 \sec,\ \ \nu = 0.3$.

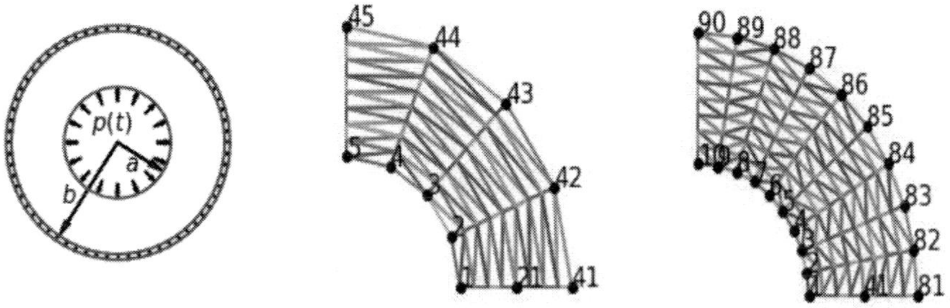

(a) Geometric model. (b) 64 triangular elements. (c) 144 triangular elements.

Figure 1. Geometric model and FEM meshes of the thick-walled visco-elastic cylinder.

4.1.1 *Sinusoidal pressure input*

For sinusoidal loading: $p(t) = p_0 \sin \Omega t$, where $\Omega = 0.5 rad / \sec$. The analytical solution is

$$u_r(t) = p_0 \frac{a^2 b(1+\nu)(1-2\nu)}{a^2 + (1-2\nu)b^2}(\frac{b}{r} - \frac{r}{b})D_S(\Omega, t) \tag{5}$$

where,

$$D_S(\Omega,t) = \Omega[\frac{D_0 + D_1}{\Omega}\sin\Omega t + \frac{D_1}{\Omega^2 + 1/\lambda^2}(\frac{1}{\lambda}e^{-t/\lambda} - \frac{1}{\lambda}\cos\Omega t - \Omega\sin\Omega t\}] \tag{6}$$

with $D_0 = 1/E_0,\ D_1 = (1/E_\infty - 1/E_0),\ \lambda_1 = E_0\rho_1 / E_\infty,\ E_0 = E_\infty + E_1$.

The FEM predictions compared well with the analytical solutions as shown in Figure 2(a).

4.1.2 *Impulsive pressure input*

For impulsive pressure case: $p(t) = p_0\ \delta(t)$ is taken into account. The analytical solution is

$$u_r(t) = p_0 \frac{a^2 b(1+\nu)(1-2\nu)}{a^2 + (1-2\nu)b^2}(\frac{b}{r} - \frac{r}{b})\frac{D_1}{\lambda}e^{-t/\lambda} \tag{7}$$

The comparison between the finite element solutions and the analytical solutions is good as shown in Figure 2(b). Polynomial fitting techniques with 3rd order are employed for impulsive oscillatory responses.

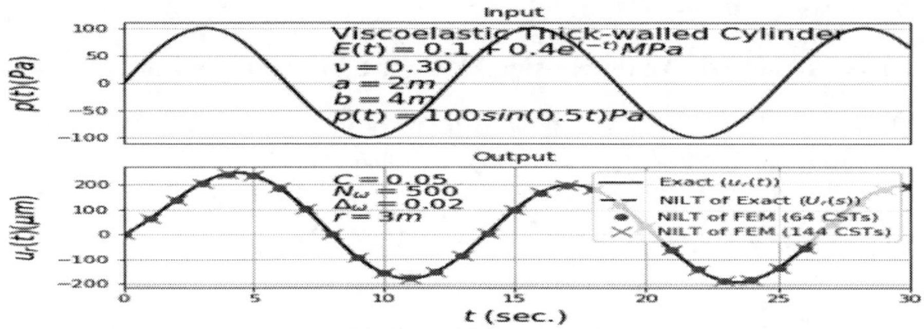

(a) Sinusoidal pressure input.

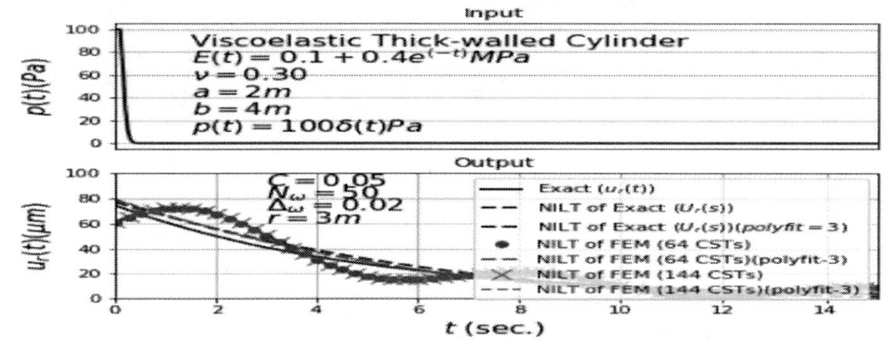

(b) Impulsive pressure input.

Figure 2. Comparison between FEM and analytical solutions for the thick-walled visco-elastic cylinder.

4.2 Visco-elastic Analysis of 4-layered Pavements

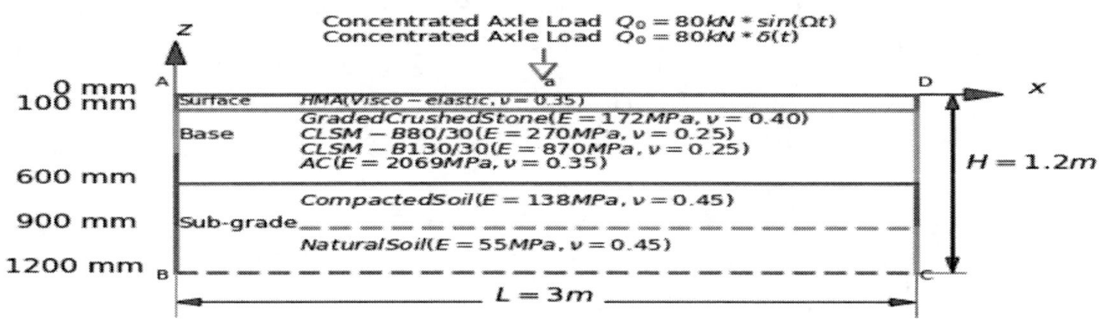

(a) Geometric model.

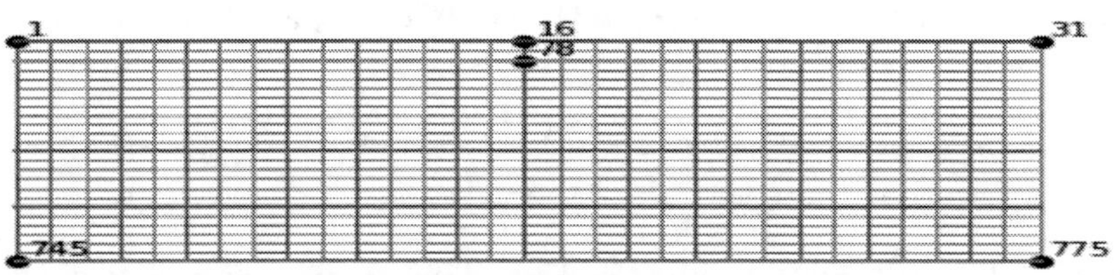

(b) 720 rectangular elements.

Figure 3. Geometric model and FEM meshes of the 4-layered pavements.

A 4-layered flexible pavement with different materials in base layer is illustrated in Figure 3(a). The material constants for AC, graded crushed stone, compacted soil and natural soil are the same as Chang and Chang (1998). The material constants for CLSM-B80/30% and CLSM-B130/30% are obtained from Sheen *et al.* (2014). Visco-elastic material properties in Prony series of the HMA layer are obtained from Lee (1996). Hinge supports along *AB, BC,* and *CD* are prescribed. Totally, $30\times24=720$ *Q4* elements with $31\times25=775$ nodes are employed (Figure 3(b)).

4.2.1 *Sinusoidal vehicle loading*

The time-domain vertical displacements and stresses located at $(x,z)=(1.5m,-0.1m)$ (i.e., the node 78) due to sinusoidal vertical load are shown in Figure 4(a) and 4(b), respectively. The vertical settlements using CLSM bases (B80/30% and B130/30%) are smaller than that using graded crushed stones even considering visco-elastic behavior of top asphalt layer.

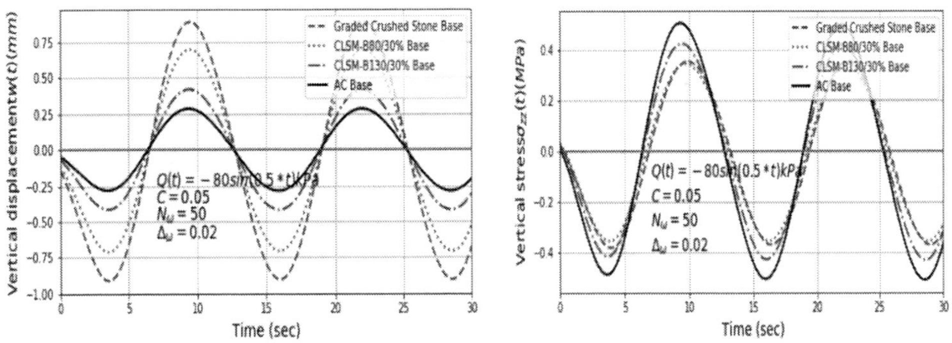

(a) Vertical displacement. (b) Vertical stress.

Figure 4. Displacements and stress of 4-layered flexible pavement for sinusoidal vehicle loading case.

4.2.2 *Impulsive vehicle loading*

The time-domain vertical displacements and stresses of the node 78 due to impulsive vertical load are shown in Figure 5(a) and 5(b), respectively. The vertical settlements using CLSM bases (B80/30% and B130/30%) are also smaller than that using graded crushed stones.

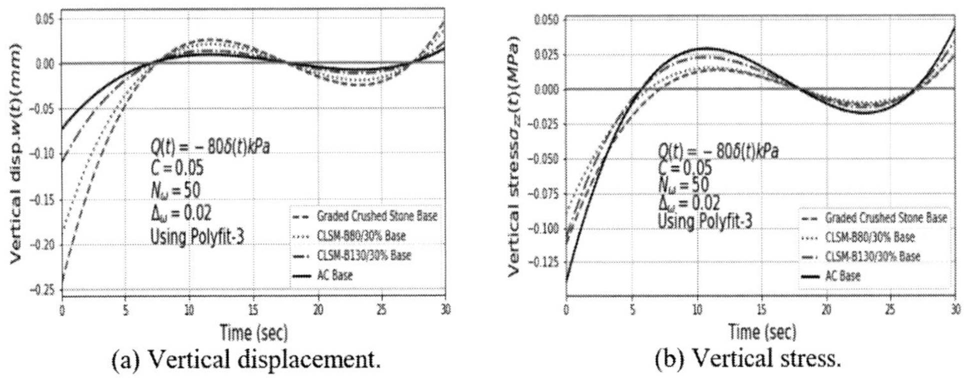

(a) Vertical displacement. (b) Vertical stress.

Figure 5. Displacements and stress of 4-layered flexible pavement for impulsive vehicle loading case.

5 CONCLUDING REMARKS

In this paper the finite element analyses of 4-layered flexible pavements with visco-elastic surface layer embedded with four different base materials subjected to surface sinusoidal and impulsive loads were studied. Correspondence principle and numerical Laplace inversion based on FFT is employed. The vertical displacement responses at top base layer using CLSMs are bounded by those using graded crushed stone base and AC base. The vertical settlements using CLSM bases (B80/30% and B130/30%) are smaller than that using graded crushed stones even considering visco-elastic behavior of asphalt layer.

References

ACI-229R, *Controlled Low-Strength Materials (Reproved 2005),* Farmington Hills (MI) 2005.

Bassani, M., Khosravifar, S., Goulias, D. G., and Schwartz,C. W., Long-term Resilient and Permanent Deformation Behavior of Controlled Low-Strength Materials for Pavement Applications, *Journal of Transportation Geotechnics*, 2, 108-118, 2015.

Chang, D. W., and Chang, C. E., Study on Data Interpretation and Structural Index of Flexible Pavement, *Tamkang Journal of Science and Engineering*, 1(2), 83-96,1998.

Davies, B., and Martin, B., Numerical Inversion of the Laplace Transform: A Survey and Comparison of Methods, *Journal of Computational Physics,* 33, 1-32, 1979.

Evangelista, F., Parente, E., and Soares, J. B., *Visco-elastic and Elastic Structural Analysis of Flexible Pavements,* CILAMCE 2005-ABMEC & AMC, Guarapari, Espirito Santo, Brazil, October 19-21, 2005.

Huang, Y. H., *Pavement Analysis and Design*, Prentice-Hall, Inc., 1993.

Huang, L. J., Sheen, Y. N., and Le, D. H., *On the Multiple Linear Regression and Artificial Neural Networks for Strength Prediction of Soil-Based Controlled Low-Strength Material,* 3rd International Conference for Advanced Materials Design and Mechanics and Workshop on Android Robotics, Singapore, May 23-24, 2014.

Huang, L. J., Wang, H. Y., Huang, W. L., and Lin, M. C., *Finite Element Analysis of Controlled Low-Strength Material Pavement Bases Part I- Static Analysis,* ISEC-9, Resilient Structures and Sustainable Construction, Valencia, Spain, July 24-29, 2017a.

Huang, L. J., Wang, H. Y., Huang, W. L., and Lin, M. C., *Finite Element Analysis of Controlled Low-Strength Material Pavement Bases Part II- Free Vibration Analysis,* ISEC-9, Resilient Structures and Sustainable Construction, Valencia, Spain, July 24-29, 2017b.

Huddleston, T., and Byrne, P., Numerical Inversion of Laplace Transforms, University of South Alabama, April 1999. Retrieved from http://code.activestate.com/recipes/128243-numerical-inversion-of-laplace-transforms-using-the-fft-algorithm/ on January 2019.

Lee, H. J., *Uniaxial Constitutive Modeling of Asphalt Concrete Using Viscoelasticity and Continuum Damage Modeling*, PhD Dissertation, Department of Civil Engineering, North Caroline State University, Raleigh, NC, USA, 1996.

Lin, D. F., Luo, H. L., Wang, H. Y., and Hung, M. J., Successful Application of CLSM on a Weak Pavement Base/Subgrade for Heavy Truck Traffic, *Journal of Performance of Constructed Facilities*, 21(1), 70-77, 2007.

Malakouti, M., Ameri, M., and Malekzadeh, P., Incremental Layerwise Finite Element Formulation for Viscoelastic Response of Multilayered Pavements, *International Journal for Transportation Engineering*, 1(3), 183-198, 2014.

Sheen, Y. N., Huang, L. J. and Le, D. H., *Engineering Properties of Controlled Low-Strength Material Made with Residual Soil and Class F Fly Ash,* 3rd International Conference for Advanced Materials Design and Mechanics and Workshop on Android Robotics. Singapore, May 23-24, 2014.

Zocher, M. A., Groves, S. E., and Allen, D. H., Three-Dimensional Finite Element Formulation for Thermoviscoelastic Orthotropic Media, *International Journal for Numerical Methods in Engineering*, 40, 2267-2288, 1997.

Interdependence between Structural Engineering and Construction Management
Edited by Ozevin, D., Ataei, H., Modares, M., Gurgun, A., Yazdani, S., and Singh, A.

ISBN: 978-0-9960437-6-2

THE MYTH OF SAW-CUT JOINTS IN CONCRETE SLABS AND WALLS

MUANGSANGOP SENIWONGSE

Bermello Ajamil and Partners, New York, USA

This paper is concerned with the world-wide practice of providing shallow saw-cut joints or v-grooves at regular spacing, for structural purposes, in concrete walkway slabs, slabs-on-grade, and parapet walls. The primary purpose of this paper is to raise awareness among engineers that the practice is a myth and has no structural advantages, and for the engineers to *think* of a better and more positive/reliable approach to concrete crack control. The other purpose is to make aware the need for engineers to think and observe actual behavior of existing structures, and use that observation to improve future design, and not repeat unsound or useless practices. Elimination of useless saw-cut joints in future construction would save the whole world billions of dollars. The author demonstrates that the saw cuts have no structural impact by showing the pictures taken in New York of typical cracks seen on walkway slabs and parapet walls, and on garage floor slabs that were built with saw-cut joints. Note that this type of actual field observation is more representative than the crack observation on small concrete models tested in any engineering laboratories. For comparison, the author shows the examples of crack-free, maintenance-free garage floor slabs and parapet walls without any saw cuts that have been in service in Thailand for the past 30 years.

Keywords: Construction, Crack, Creep, Design, Reinforcing, Sequence, Shrinkage.

1 INTRODUCTION

This paper addresses the myth associated with the commonly adopted practice in USA of providing saw-cut joints in concrete slabs, slabs-on-grade and walls at some regular spacing, with the belief that these saw-cut or v-groove joints would prevent concrete cracks at other locations in the slabs or walls, and that cracks if formed would be neatly at and in line with the initiated saw-cuts. This paper presents opposing view based on years of observation of actual behavior of these structural concrete. In fact, cracks can be observed on concrete walkway slabs, slabs-on-grade in many parking garages and driveways, and on bridge parapet walls, all built with saw-cut joints. The occurrence of cracks in the mentioned slabs, some very extensive, is so common that it seems to be an acceptable local practice, and the saw-cut details are even promoted to be standard details by various agencies. The purpose of this paper is to bring to the attention of design engineers of the above observations. Examples of crack-free concrete slabs and walls without any saw cuts in two major projects are shown. This paper shows that saw-cuts in concrete can be eliminated from future design which would result in the saving of billions of dollars worldwide.

2 SAW- CUT JOINTS IN STRUCTURAL CONCRETE SLABS AND WALLS

Saw-cut joints in this paper refer to the shallow saw-cut joints and v-groove or u-groove joints on the faces of structural concrete slabs and walls. These are seen on almost all concrete walkway slabs, concrete driveway and garage concrete floors. The joints are seen at maximum spacing of six feet to ten feet in orthogonal directions as shown diagrammatically in Figure 1. Figure 2 shows concrete slab on ground: a) with saw cut of one-quarter of an inch deep and b) an imitation of saw cut using v-groove or u-groove of one-quarter of an inch deep.

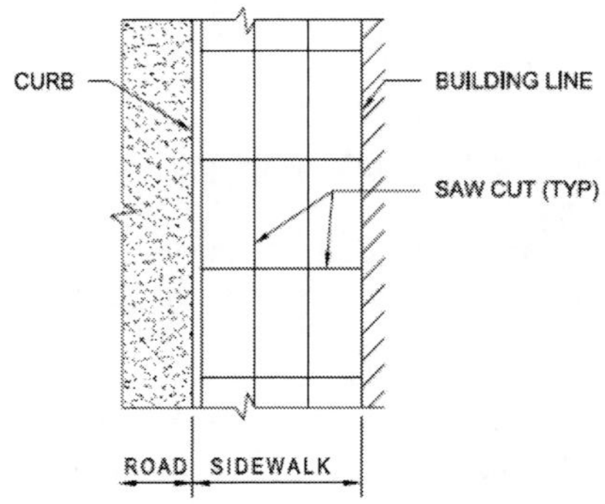

Figure 1. Plan view of example concrete slab on grade walkway with saw cut joints.

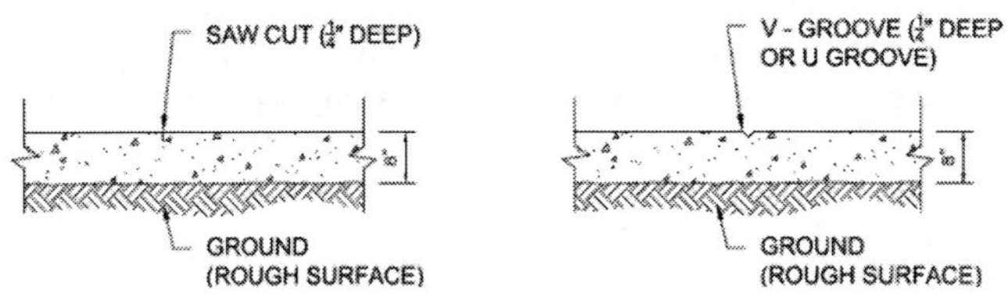

Figure 2. Typical concrete slab on grade with shallow saw cut at top surface.

The saw cuts are found in many sizes and variations. Some are seen to have wider groove (about half an inch) at very fourth or fifth saw cut lines (panels). These wider grooves are sealed with joint sealer.

Note that Figure 2 above indicates the slabs are on a rough ground surface intending to convey the message that the slabs in a) and b) are not free to slide but restrained from sliding by cohesion and friction between the bottom of the concrete slab and the ground surface.

Similar saw-cut joints or v-grooves are seen on structural concrete walls such as retaining walls, and parapets of highway bridges.

The concrete slabs thickness varies from 4 inches to about 15 inches for slab-on-grade or walkway. The amount of reinforcing steel depends on the individual design. For bridge parapet walls, they are either rectangular in cross section or of New Jersey's parapet type, and the reinforcement is designed for large lateral loads according to the AASHTO Bridge Design Specifications. It is noted here that there is no known correlation between the amount of reinforcement in the slabs and walls with the spacing of the saw cuts or v-grooves.

2.1 The Myth of Saw-Cut Joints

Having practiced and observed structural concrete behavior since 1970 in major projects in not less than 16 countries, the author cannot see any merits or structural advantages of providing shallow saw-cut joints on concrete floors or walls. The reason or the myth why saw-cut joints are provided is the belief that concrete cracks would take place at and along the saw-cut joints in the form of deeper and wider joints, but all at the saw-cut locations, and less likely to crack at anywhere else on the slabs or walls. That is the wishful thinking. There is no creditable engineering mechanics that support concrete crack control using shallow saw-cut joints.

2.2 The Effects of Making Saw-Cut Joints

It will be evident in Section 3 that saw-cut joints have no structural advantages on the behavior and performance of the slabs or walls. The only effect - they are seen to be effective in trapping of dirt and cigarette ends on the walkways as shown typically in Figure 3.

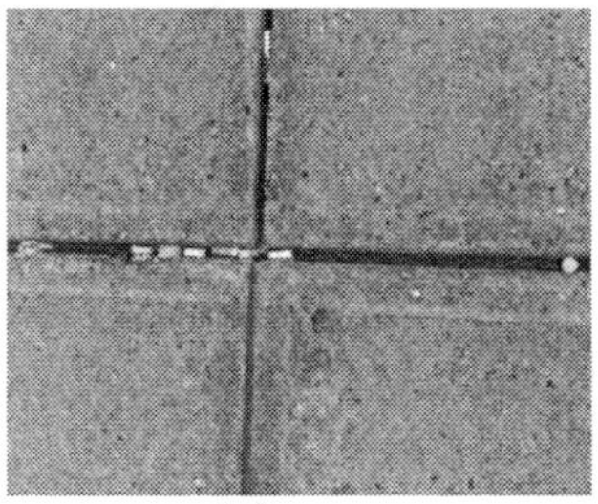

(a)

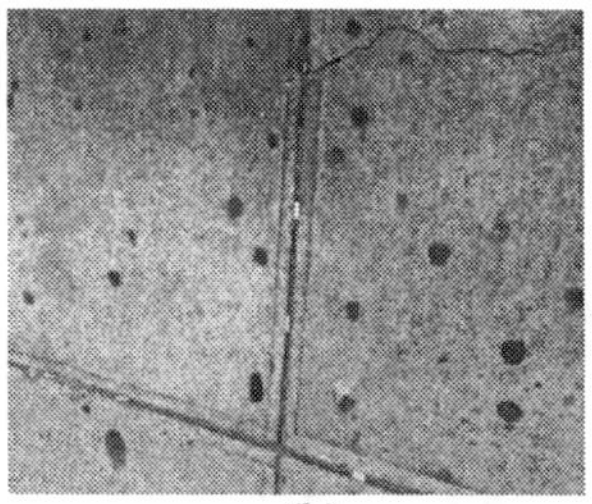

(b)

Figure 3 (a), (b). Cigarette ends trapped in saw-cut grooves in walkway slabs.

3 OBSERVATION OF CRACKS IN CONCRETE WITH SAW-CUT JOINTS

This Section shows typical concrete cracks commonly seen in concrete slabs and walls that incorporated saw cuts. All pictures were taken in 2018. Figure 4 shows concrete cracks in the sidewalk slabs. Figure 5 shows concrete cracks in the slab-on-grade parking garage floor in a six-month old apartment building in New York. Figure 6 shows cracks in concrete parapet wall of a bridge with v-grooves at about five to six feet spacing. The cracks are seen in this case at about one foot away from the groove joints.

(a)

(b)

Figure 4 (a), (b). Cracks in sidewalk concrete slabs on Park Ave. in New York City.

4 REPAIR AND MAINTENANCE OF SAW-CUT JOINTS

All joints require maintenance and repair, sooner or later. All pictures of repairing in this Section were taken on Park Avenue in New York in 2018. Figure 7 shows maintenance work on the grooves. Figure 8 shows the demolition of extensively cracked walkway and recast of new slabs.

(a) (b)

Figure 5. Cracks in slab-on-grade garage floors (a) condominium in NY, (b) shopping mall in Massachusetts.

(a) (b)

Figure 6 (a), (b). Cracks in NY bridge parapet walls with saw-cut joints.

(a) (b)

Figure 7. (a) Saw cutting grooves in walkway slab, (b) Taping saw-cut groove before applying sealant.

(a) (b)

Figure 8. (a) Removal of cracked walkway slab, (b) Recast walkway slab.

5 CRACK-FREE FLOORS AND WALLS WITHOUT SAW-CUT JOINTS

This Section shows examples of concrete garage floors and bridge parapet walls which have been in service since 1990s and are still in excellent conditions without any cracks. The concrete slabs and walls were designed and built conventionally without any saw cuts. The pictures in Figures 9 and 10 were taken in 2008, after some 15-20 years after the completion of construction of both projects. Figure 9 shows the underground bottommost basement garage floor of the United Nations' Conference Building in Bangkok, built in the early 1990s (Seniwongse 1991). Figure 10 shows the pictures of the parapet walls of 40-km long expressway bridges in Bangkok that was built around 1988-1991(Seniwongse 1989). The deck itself is continuous using innovative link slabs between spans (see Jointless Bridge (Seniwongse 1990)).

(a) (b)

Figure 9. Crack-free garage basement floor in UN-ESCAP Conference Building (Seniwongse 1991).

(a) (b)

Figure 10. Crack-free parapet walls on 40-Km viaduct in Bangkok (Seniwongse 1989).

6 DISCUSSION AND CONCLUSION

This paper intends to call upon engineers to exercise field observation and common sense in the design and construction of concrete slabs and walls. The pictures in Sections 3-5 are self-explanatory. Complicated computer programs and mathematical modeling do not always deliver good design and construction, but common sense and engineering intelligent always do. It is obvious that saw cuts in concrete are useless, and money is wasted in making the saw cuts, and later on the maintenance and repair. The cost of making the saw- cut joints (not the cost of slab/wall) is about \$20 - \$30 per linear foot of joints (Rhode Island, 1998). The cost of repairing of existing saw-cut joints is about \$40 - \$60 per linear foot (Appendix A). For an example city having a total 200,000,000 feet length of joints in walkways, the cost of making saw cuts would be 200,000,000 ft x \$25 per foot = \$5,000,000,000. That is the cost in one example city. The potential saving for the whole world would be billions of dollars, and the concrete would be just fine as shown in Section 5.

For crack-free concrete, design proper reinforcing steel, and construct using proper concrete pouring sequence to allow dissipation of concrete creep and shrinkage, using full face construction joints as shown in Figure 11. Fresh concrete is placed in alternate segments of say 10 ft. long. Allow the concrete to cure and shrink for a few days, then place concrete in the closure segments. In fact, the mentioned techniques were used in the two projects in Section 5. Other design techniques and concepts for crack-free concrete can be found in the references by Leonhardt (1988) and Seniwongse (1972, 1973, 1989, 1990, 1991, 2006, 2008, and 2010).

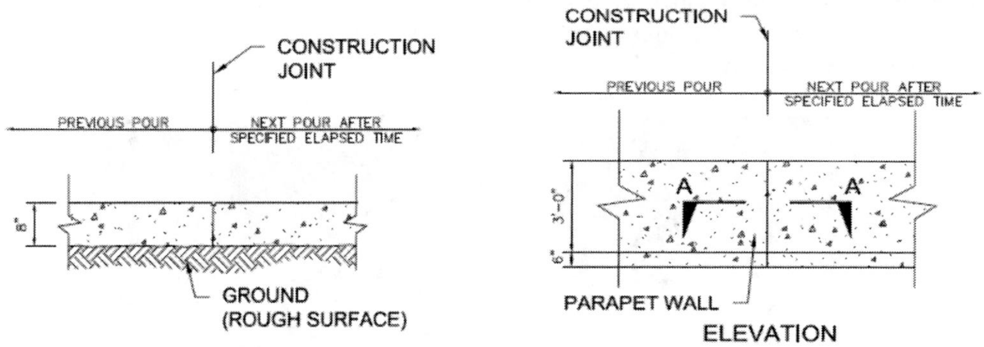

(a) (b)

Figure 11. Full construction joint with proper concrete pour sequencing.

Appendix A. Estimate Cost of Repair of Saw Cut in Concrete

Assume, for example, it is required to repair a total 20 ft long saw cut joint on concrete walkway on Park Avenue in New York City, and the Contractor is in White Plains in New York. Two technicians driving from White Plains to Park Avenue and back after work would take 6 hours, and actual setting up time, saw-cutting concrete and cleaning up would take 3 hours. The cost of the car and concrete saw is \$ 500 per day. The direct cost would be \$(2 x 20 x 9 + 500) = \$ 860. Applying 15 percent insurance and profit, the total cost would be \$(1.15 x 860) = \$989 for 20 ft saw cut. Estimate cost for 1 ft. is \$989/20 = \$49.45, rounded to \$50/ft.

References

Leonhardt Fritz, *Cracks and Crack Control in Concrete Structures*, PCI Journal, Vol 33, July/August 1988.

Rhode Island Department of Transportation, *Engineers Estimate for Louisquisset Bridge No. 748*, Contract 9844, March 24, 1998.

Seniwongse, M., *Limit Design of Reinforced Concrete Floor Slabs*, MSc Thesis, Civil Engineering, University of New South Wales, Sydney, Australia, 1972.

Seniwongse, M., *Creep and Shrinkage Effects in Concrete Structures*, Research Report No. 14, Civil Engineering, Prince of Songkla University, Hadyai, Thailand, 1973.

Seniwongse, M., *The Development of the Design of the Second Stage Expressway in Bangkok,* Second East Asia-Pacific Conference on Structural Engineering and Construction, Chiengmai, Thailand, 1989.

Seniwongse, M., *Jointless Bridge*, 1990 Engineering Technology Symposium, Engineering Institute of Thailand, Bangkok, Thailand, 125-130, 1990.

Seniwongse, M., The Construction of the New Conference Complex for the UN-ESCAP, *1991 Engineering Technology Symposium,* Engineering Institute of Thailand, Bangkok, Thailand, 475-489, 1991.

Seniwongse, M., *Long Term Performance of Cracked Concrete Slab Bridge Decks,* National Concrete Bridge Conference, Reno, Nevada, USA, 2006.

Seniwongse, M., *Slab-on-grade Versus Suspended Slab Alternative, Presented at the 2008* ASCE Architectural Engineering Institute Conference, Denver, Colorado, USA, September 24-27, 2008.

Seniwongse, M., Slab-on-Grade versus Framed Slab, *Journal of Architectural Engineering,* USA, 2010.

Interdependence between Structural Engineering and Construction Management
*Edited by*Ozevin, D., Ataei, H., Modares, M., Gurgun, A., Yazdani, S., and Singh, A.

ISBN: 978-0-9960437-6-2

DEVELOPMENT OF PAVEMENT PERFORMANCE MODEL FOR PROPER REHABILITATION AND MAINTENANCE USING FIRST ORDER MARKOV CHAIN

AMJAD ISSA and SAMEER ABU EISHEH

Dept of Civil Engineering, An-Najah National University, Nablus, Palestine

The development of pavement performance model is an important step in prediction the future condition of pavement section and accordingly identifying the right road rehabilitation and maintenance in the right time for the right section. The first order Markov chain probabilistic model is used to predict the degradation of flexible pavement in Palestine. A pilot study is conducted on part of the road network in Nablus City. Visual road condition assessment is performed, and the Pavement Condition Index (PCI) is used in rating pavement sections by dividing the roads each 100 m length. The prediction of the pavement condition rating for each section in the first five to ten years of section age will enhance the applying of preventive maintenance strategy and consequently urges the Local Governmental Units to use the limited allocated budgets specified for pavement maintenance in a cost-effective manner by applying maintenance actions such as crack sealing, surface patching, micro-surfacing, milling and overlay, etc.

Keywords: Flexible pavement, Degradation model, Pavement condition index, Preventive maintenance, Maintenance proper time.

1 INTRODUCTION

1.1 Pavement Rehabilitation and Maintenance Status in Palestine

The foundation and alignment of some existing roads in Palestine were established during the British mandate of Palestine. A considerable portion of those roads was constructed as Macadam roads following old footpaths with relatively tortured alignment due to the prevailing mountainous terrain and limited resources. During the last three decades most of the roads have been in a deteriorated condition, while some of the urban streets have received some maintenance activities, but not in an organized or prioritized manner. Considering the evaluation of existing pavement evaluation, most of the types are of structural capacity evaluation, functional evaluation, safety evaluation, and distress evaluation. The current road maintenance practices in Palestine are not based on scientific systematic methods. Most of the municipalities use the "worst-first" criterion. Therefore, the current practices are not applying road inventories, road pavement condition surveys, overall evaluation of pavement conditions, identification of the proper maintenance and rehabilitation measures, and setting priorities.

1.2 Importance of Developing Pavement Performance Model

The development of pavement performance model is an important step in predicting the future conditions of pavement sections, and accordingly, in identifying the right road preventive rehabilitation and maintenance measures at the right time for the right section. The lack of systematic maintenance procedures in Palestine requires proposing a new cost-effective system, which will be part of the future Pavement Management System (PMS). Accordingly, it is essential to model the deterioration process of the pavement sections using an applicable and practical model. The first order Markov chain (MC) probabilistic model is used to predict the degradation of flexible pavement in Palestine.

1.3 Literature Review

The future conditions of pavement sections are usually predicted for the purpose of maintaining road networks. Pavement performance has been traditionally presented using a performance curve, which shows how the pavement condition declines over time in the absence of maintenance and rehabilitation (M&R) works (Huang 2004, Shahin 1994). The pavement condition has been historically presented by indicators, such as Pavement Condition Index (PCI), Present Serviceability Index (PSI), and Distress Rating (DR) (AASHTO 1993, ASTM 2007, Shah *et al.* 2013). The emphasis on developing effective models for predicting pavement performance had become of great importance with the emergence of pavement management science over three decades ago. Therefore, an effective pavement prediction model is a significant component of any advanced PMS (Abaza *et al.* 2004, Jorge and Ferreira 2012, Khan *et al.* 2014). Generally, there are two types of models used in predicting future pavement conditions: deterministic and stochastic models (Amin 2015, Li *et al.* 1997). Both types of models can predict future pavement conditions; however, the stochastic-based models have gained wider use in pavement management applications, because they make possible the consideration of risk assessment. This can be attributed to the fact that pavement performance has been identified as probabilistic, which requires assigning different levels of uncertainty (i.e., probability) to different pavement condition outcomes (Abaza 2015). The stochastic model that was used by several researchers over the last three decades had mainly relied on deploying different forms of the discrete-time Markov model (Butt *et al.* 1987, Li *et al.* 1996). The homogenous and non-homogenous chains are the two most popular forms of the deployed discrete-time Markov model (Abaza 2015). In this paper, the homogeneous MC is used, considering the traffic is similar on all roads.

1.4 Research Hypotheses

Homogenous MC is one of several methods devoted to the prediction of future pavement conditions. It is used in several PMSs in many countries around the world. The use of MC requires calculating the transition probability matrix (TPM) from the available historical data of pavement. The indicated research illustrated that the MC theory relies on a discrete probabilistic approach that is widely used in developing probabilistic models for predicting pavement deterioration. MC has been used by many researchers (Butt *et al.* 1999, Abaza and Shahein 2007). Accordingly, there is an agreement that the principles and conditions of the MC theory (the stochastic process is discrete in time and a finite state space) are applicable to pavement deterioration modeling. Additionally, the first order Markov property that future state of the process depends on its current state and not past state holds in pavement deterioration (i.e., future network condition depends on its present condition but not past condition). This means that the present condition contains implicitly a given history (load and resistance) or that whatever the history, a given present distribution turns always to a given future distribution.

2 RESEARCH METHODOLOGY

To establish an appropriate pavement deterioration model, information regarding the road pavement conditions was collected considering visual inspection. After that, the pavement deterioration model using MC is then estimated. All the indicated studies considered using 1st order MC from two consecutive inspection years without considering uncertainty. However, this research deals with the transition probabilities (TP) calculation from age of pavement sections, which leads to providing an uncertainty modeling of the TP. The form of the TPM used in the paper is composed of the diagonal probabilities in addition to extra probabilities concerning the first row of the TP matrixes follows:

$$TPM = \begin{bmatrix} P_{11} & P_{12} & 0 & 0 & 0 \\ 0 & P_{22} & P_{23} & 0 & 0 \\ 0 & 0 & P_{33} & P_{34} & 0 \\ 0 & 0 & 0 & P_{44} & P_{45} \\ 0 & 0 & 0 & 0 & P_{55} \end{bmatrix}$$

Figure 1. TPM matrix.

Where p_{12}, is the probability of the pavement section condition to deteriorate from very good to good condition. The methodology followed in this research can be highlighted to include the interesting issue of using the homogenous first order MC as a stochastic model in the prediction of the performance of flexible pavements through the consideration of the calculated TPM.

3 DATA COLLECTION

The database was categorized by roadway length, sections, uniformity of cross-section, and thickness of pavement structure. The length of road sections was taken to be 100m. The selected pilot roads and related lengths are presented in Table 1. These include three arterial and one collector roads. Figure 1 shows a typical division of an arterial road into sections in each direction. The data collected were based on 2 years of inspection: 2012 and 2016, respectively. Percentage rating of the Road Condition Survey for all the surveyed sections of 138 showed that 60 percent of the sections were in very good and good conditions while 29 percent were in fair condition. The rest of the sections were approximately in poor or very poor condition.

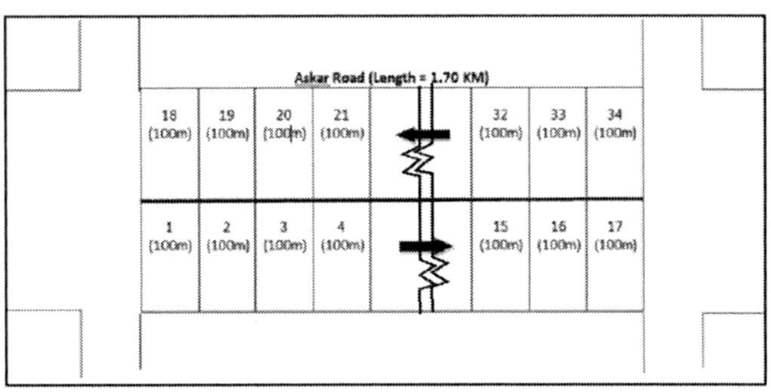

Figure 2. Section identification example.

Table 1. Selected roads in the pilot case study.

Road Name	***Length (KM)***	***Classification***
Yaser Arafat	1.50	Arterial
Tunis	1.20	Arterial
Askar	1.70	Arterial
Mohammad Bin Rashid	2.50	Collector
Total	**6.90**	

4 MODEL FORMATION

The model in this research is developed based on statistical analysis correlating pavement performance as indicated by PCI with pavement age and/or design service life. For each treatment, representative in-service projects were selected and their corresponding needed data were used to develop pavement performance models. The pavement condition rating and the corresponding number of sections are illustrated in Table 2. The resulting Markov model output, considering the collected data for the surveyed selected road sections in this research, is shown in Table 3. After the data was filtered, all the surveyed pavement segments wereused in the development of the TPM. The formations of TM and the TPM are shown in Tables 4 and 5. The calculations required for the Markov model were conducted using Eq. (1):

$$p_{ij} = \frac{n_{ij}}{n_i} \tag{1}$$

Where p_{ij} is the probability of the pavement section to deteriorate from the current condition to the next condition,n_{ij} is the number of the pavement sections deteriorated from i to j condition state, and n_i is the total number of pavement sections in condition state i.

Table 2. Pavement condition rating and corresponding no. of sections.

PCI	**No. of Sections**
Very Good (VG): [5]	58
Good (G): [4]	44
Fair (F): [3]	20
Poor (P): [2]	12
Very Poor (VP): [1]	4
Total	**138**

Table 3. Data on pavement conditions for the selected roads.

PCI	**No. of Sections**
VG to VG	45
VG to G	13
G to G	30
G to F	14
F to F	15
F to P	5
P to P	9
P to VP	3
VP to VP	4
Total	**138**

Table 4. Transition matrix for the selected roads.

Rating Value	**VG**	**G**	**F**	**P**	**VP**	**No. of Sections**
VG	45	13	0	0	0	58
G	0	30	14	0	0	44
F	0	0	15	5	0	20
P	0	0	0	9	3	12
VP	0	0	0	0	4	4
Total						**138**

Table 5. TPM for selected roads.

Rating Value	**VG**	**G**	**F**	**P**	**VP**	**Total**
VG	0.78	0.22	0	0	0	1
G	0	0.68	0.32	0	0	1
F	0	0	0.75	0.25	0	1
P	0	0	0	0.75	0.25	1
VP	0	0	0	0	1	1

5 RESULTS AND DISCUSSION

Table 6 presents the Markov probability distribution for the first six years, which are calculated based on the TPM given in Table 5. The probability distribution within each state vector is the main characteristic that makes the Markov model more attractive than the deterministic models. In Table 6, the state of maximum probability is shown in bold face. This path is the most likely deterioration path that the sections will follow. For example, at year four, there is a probability of 0.370 for the sections to stay at PCI value of 5, but the highest probability value of 0.486 for that year is at PCI value of 4. Figure 2 shows the plot of the Markov model for selected roads pavement sections. In the Markov model, the predicted future determination trend starts at the last rating, which the sections had received. The line represents the average states of the pavement determined from a probability distribution based on the TPM established from the entire data set. This figure also shows that the prediction of performance trend of pavement sections indicates the pavement reaches the minimum accepted performance level of value 2at age of approximately eight years, which is unacceptable value for twenty years pavement design service life, and accordingly, the rate of deterioration is high. The Markov probabilistic model for the selected roads gave approximately the expected deterioration behavior. The pavement reached the minimum acceptable performance level at age of eight years which means that the sections with poor or very poor conditions must have major rehabilitation, while the remaining sections from fair to very good sections may have preventive maintenance or localized repair actions.

Table 6. Probability distribution values for MC for selected roads.

Rating Value	Year 1	Year 2	Year 3	Year 4	Year 5	Year 6
VG (5)	**0.780**	**0.608**	**0.475**	0.370	0.289	0.225
G (4)	0.220	0.361	0.444	**0.486**	**0.500**	**0.493**
F (3)	0.000	0.070	0.168	0.268	0.357	0.428
P(2)	0.000	0.000	0.018	0.055	0.109	0.171
VP (1)	0.000	0.000	0.000	0.004	0.018	0.045

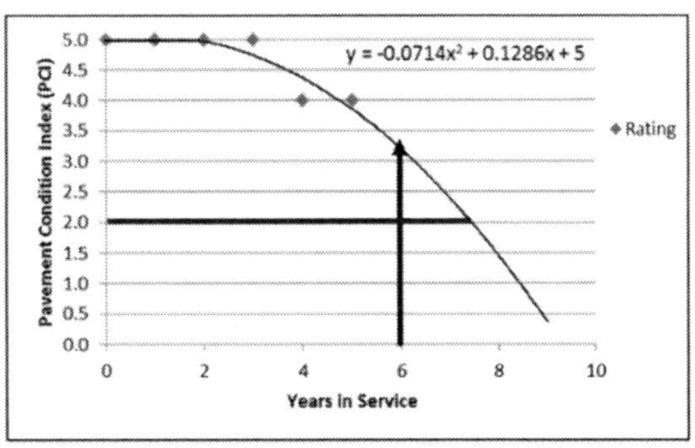

Figure 3. A graphical presentation of the estimated pavement performance model.

6 CONCLUSIONS AND RECOMMENDATIONS

The PMS includes different modules such as inspection and inventory, performance model, and prioritization. The use of first order MC as part of the PMS in Palestine will assist in developing a systematic process of defining the right treatments for the right sections at the right time. The

1[st]order MC is used to estimate the performance of urban arterial and collector roads in Nablus City. The results indicate that it is possible to model the performance of the existing roads using MC up to eight years, which means that this model is appropriate in case of applying preventive maintenance measures such as seal coating, surface patching, milling and overlay, etc. The newly proposed estimate is suitable for applying preventive maintenance treatments, which is usually of limited cost, and do not need much more finance to compare with preventive and other maintenance strategies. The main recommendations are:

- To adopt and apply the new estimated model in other Palestinian municipalities;
- To collect data and perform the inspection process bi-annually;
- To estimate the new model for rural roads using the first order MC.

Acknowledgments

The authors would like to thank the Department of Public Works in Nablus Municipality, and students of GP (CED, An Najah National University) for providing the pavement database used in the paper.

References

Abaza K. A., Simplified Staged-Homogenous Markov Model for Flexible Pavement Performance Prediction, *Journal of Road Materials and Pavement Design,* 17(2), 365-381,2015.

Abaza, K. A., Ashur, S. A., and Al-Khatib, I. A., Integrated Pavement Management System with A Markovian Prediction Model, *Journal of Transportation Engineering,* 130(1), 24–33, 2004.

Abaza, O., and Shahein, L., Developing Pavement Performance Functions for Palestinian Roadway Network, *An-Najah University Journal of Research,* 21, 23-51, 2007.

AASHTO (American Association of State Highway and Transportation Officials), *AASHTO Guide For The Design Of Pavement Structures*, Washington, DC: AASHTO, 1993.

ASTM (American Standard Testing Manual), *Standard Practice for Roads and Parking Lot Pavement Condition Index Surveys*, D6433-07, Philadelphia. Retrieved from http://www.astm.org/Standards/D6433.htm, 2007.

Amin, X. and Reza, XX, *The Pavement Performance Modeling: Deterministic vs. Stochastic Approaches*, S. Kadry & A. El Hami (Eds.), Numerical Methods for Reliability and Safety Assessment, Switzerland: Springer International, 179–196, 2015.

Butt, A., Shahin, M., Feighan, K., and Carpenter, S., Pavement Performance Prediction Model Using The Markov Process, *Transportation Research Record,* 1123, 12–19, 1987.

Huang, Y., *Pavement Analysis and Design*. 2nd ed., Upper Saddle River, NJ: Pearson/Prentice Hall, 2004.

Jorge, D., and Ferreira, A., Road Network Pavement Maintenance Optimization Using The HDM-4 Pavement Performance Prediction Models, *International Journal of Pavement Engineering,* 13(1),39–51, 2012.

Khan, M. U., Mesbah, M., Ferreira, L., and Williams, D. J., Development of Road Deterioration Models Incorporating Flooding for Optimum Maintenance and Rehabilitation Strategies, *Road and Transport Research: A Journal of Australian and New Zealand Research and Practice*, 23(1), 3–24, 2014.

Li, N., Haas, R., and Xie, W.-C., Investigation of The Relationship Between Deterministic and Probabilistic Prediction Models in Pavement Management, *Transportation Research Record,* 1592, 70–79, 1997.

Li, N., Xie, W.-C., and Haas, R., *Reliability-Based Processing of Markov Chains for Modeling Pavement Network Deterioration, Transportation Research Record,* 1524, 203–213, 1996.

Shah, Y., Jain, S., Tiwari, D., and Jain, M., *Modeling The Pavement Serviceability Index for Urban Roads in Noida, International Journal of Pavement Research and Technology,* 6(1), 66–72, 2013.

Shahin, M., *Pavement Management for Airports, Roads, And Parking Lots*, New York, NY: Chapman and Hall, 1994.

Interdependence between Structural Engineering and Construction Management
Edited by Ozevin, D., Ataei, H., Modares, M., Gurgun, A., Yazdani, S., and Singh, A.

ISBN: 978-0-9960437-6-2

IMPACT OF CONCRETE MATERIAL INPUTS ON PERFORMANCE OF CONTINUOUSLY REINFORCED CONCRETE PAVEMENT

GAUHAR SABIH and RAFIQUL A. TAREFDER

Dept of Civil Engineering, The University of New Mexico, Albuquerque, USA

Continuously reinforced concrete pavement (CRCP) is increasingly being used in the United States based on its longer service life with minimal maintenance. CRCP provides excellent performance for heavily loaded traffic volume, if designed and constructed properly. The performance of CRCP is evaluated based on punch-outs and pavement roughness according to the Pavement ME (mechanistic-empirical) Design procedure which is the state-of-the-art pavement design tool. Design of CRCP is dependent on various factors including traffic, climate, material inputs and pavement geometry. This study focuses on evaluating the effects of concrete mechanical and thermal properties and traffic volume on CRCP performance. CRCP design simulations were conducted using Pavement ME Design software version 2.3 to evaluate the effects of material input data on CRCP performance and it was observed that there is significant difference between the performance predictions with time series data in comparison to the default data. The impact of material inputs is more significant on punch-outs distress as compared to the pavement roughness. This necessitates the importance of lab tested material inputs for accurate design of CRCP. The impact of traffic volume on CRCP performance was also evaluated and it was found that with increase in traffic volume, the pavement distresses increases and the performance decreases.

Keywords: Punch outs, Pavement roughness, Mechanistic, Elastic modulus, Coefficient of thermal expansion, Modulus of rupture.

1 INTRODUCTION

Continuously reinforced concrete pavement (CRCP) has the potential to provide a long-term solution of highway needs with zero-maintenance. This type of pavement can be used in heavily trafficked areas and in challenging environmental conditions. As for any other pavement, accurate design and quality construction practices are necessary to ensure desired performance. CRCP has no constructed transverse, contraction or expansion, joints except at bridges or at pavement ends (Plei and Tayabji 2012). Longitudinal steel reinforcement is used to keep the transverse cracks closely spaced and reduce the amount of opening at the cracks. CRCP usually has longitudinal reinforcing steel percentages in the range of 0.65 percent to 0.85 percent. The transverse steel bars are used to support the longitudinal reinforcement and keep them in place. The bond area between the concrete and the bars is also critical and Federal Highway Administration (FHWA) recommends at a minimum of 0.030 square inch per cubic inch of concrete (FHWA 1990). A well-performing CRCP can be identified by a reasonably regular transverse cracking pattern with desirable crack spacing of two feet to eight feet that in turn keeps the cracks tight and provides a high level of load transfer across the cracks. With the advancement in pavement design and performance analysis procedures and the advent of Pavement Mechanistic-Empirical (ME) Design software, the importance of concrete material inputs came to the fore front. Prior research has proved that concrete mechanical and thermal properties including compressive strength, elastic modulus, modulus of rupture (MOR) and coefficient of thermal expansion (CTE) has a direct impact on design and performance of rigid pavements (Sabih and Tarefder 2016). The design methodology of Pavement ME Design is coded into a software program that calculates all the necessary mechanistic responses of the trial design structure required for performance predictions, computes damage and other associated parameters and finally predicts performance in terms of pavement distresses over the design life (Rao 2004). This design procedure takes in to account various design factors including concrete material inputs, traffic, climate and etcetera. In terms of material properties, the design procedure uses 3 input levels with level-1 or time series data being the most accurate one.

Although, there has been numerous studies on effects of material inputs on pavement performance of jointed plain concrete pavements but very few studies are available for CRCP. This study focuses on the significance of concrete material inputs and traffic volume on the design and performance of CRCP and for this purpose simulations are conducted in Pavement ME Design version 2.3 and the results are analyzed to compare the effects of material properties on CRCP design.

2 NOTATIONS

The notations/abbreviations used in this study are presented in Table 1.

Table 1. Notations and abbreviations.

Symbol/Abbreviation	Description
CRCP	Continuously reinforced concrete pavement is a type of rigid pavements
Pavement ME	Latest pavement design tool being used in United States based on mechanistic-empirical principles
CTE	Concrete material property related to thermal expansion
MOR	Concrete strength property related to flexure
Elastic modulus	Concrete mechanical property related to stiffness
CA-ID	Concrete mixture designation based on coarse aggregate type
IRI	International roughness index measured in in/mile
ESAL	Equivalent single axle load based on traffic
AADTT	Average annual daily truck traffic
psi	Pounds per square inch (load per area)

3 METHODOLOGY

Six concrete paving mixes (designated as CA-ID-1 to CA-ID-6) were used for this study and time series data was generated for compressive strength, elastic modulus and modulus of rupture (MOR) by laboratory testing. Coefficient of thermal expansion (CTE) testing was also conducted for all the mixes at the age of 28 days according to AASHTO T-336 (2011) test protocol. The sensitivity analysis of various material and traffic inputs on the performance of CRCP was performed by conducting the simulations in Pavement ME design software version 2.3. The major design inputs are shown in Table 2. The performance parameters including punch-outs and IRI were analyzed with respect to the variability of different inputs.

Table 2. CRCP design parameters for simulation work.

Parameter	Value
Design Life	30 years
Design Thickness	10 in.
Shoulders	Tied PCC Shoulders
Steel Reinforcement (%)	0.72
Steel Bar Diameter	3/4 in. (#6 bar)
Steel Depth	3.5 in.
Initial IRI	63 in/mile
Threshold IRI	172 in/mile
Threshold Punch-outs	10 per mile
Reliability	90%
Modulus of Rupture of Concrete	As per CA-ID
Elastic Modulus of Concrete (28 days)	As per CA-ID
Poisson's Ratio	0.2
Climate Station	Albuquerque, New Mexico
AADTT	4000
Traffic ESALS	$30x10^6$
Base Course Thickness	6 in.
Base Course Resilient Modulus	40,000 psi

4 IMPACT OF INPUT LEVELS ON CRCP PERFORMANCE

Pavement ME design has 3 levels of inputs which the designer can use, based on the available data and accuracy required. Level-1 has the highest level of accuracy with all the tested data for the specific project including CTE, MOR, and elastic modulus. While level-3 has the lowest level of accuracy in the design by using the default values of CTE and using compressive strength parameter at 28 days (AASHTO 2015).

The simulations were conducted in pavement ME design for all the paving mixes (CA-ID-1 to CA-ID-6) to contrast the impact of level 1 and level-3 inputs on CRCP performance. The lab tested data for concrete mechanical properties including MOR and elastic modulus and CTE were used for level-1 design simulations while, for level-3 design, default CTE value and compressive strength at the age of 28 days were used. CRCP thickness was 10 in and other design parameters were kept constant (as per Table 1) to compare the effects of input levels.

4.1 Effects on Pavement Roughness

The comparison for terminal values of transverse cracking for CA-ID-1 to CA-ID-6 is presented in Figure 1, which shows that there is significant variation in IRI values between the results of level 1 and level 3 inputs for all the mixes. The variation in IRI with input levels ranges from 1 to 66.7 in/mile which is highly significant. With these results, it is evident that the CRCP must be designed with the accurately tested level-1 inputs for the paving mix to be used so that the designed pavement can last for the entire service life. The ME default CTE data and level-3 inputs will not produce accurate design for these paving mixes.

4.2 Effects on Punch Outs

The comparative summary for CRCP punch outs for impact of material input levels is presented in Figure 2. It is evident that there is a significant impact on punch outs between the two material input levels. The difference in terminal punch out values ranges between 15.1 to 32.8 per mile.

This shows the importance of using accurate concrete material inputs to be used for CRCP design.

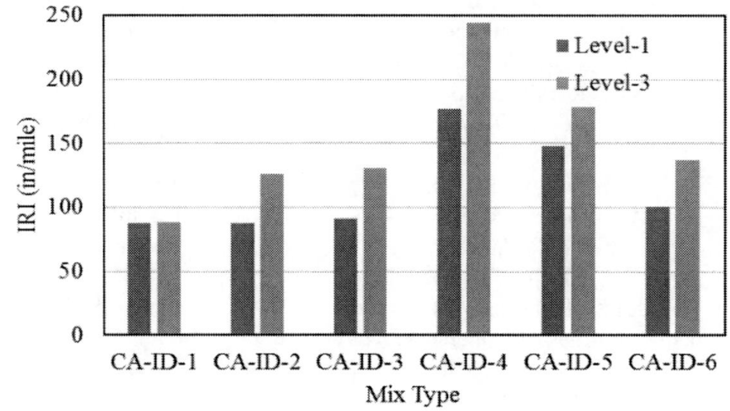

Figure 1. Impact of Input Levels on IRI.

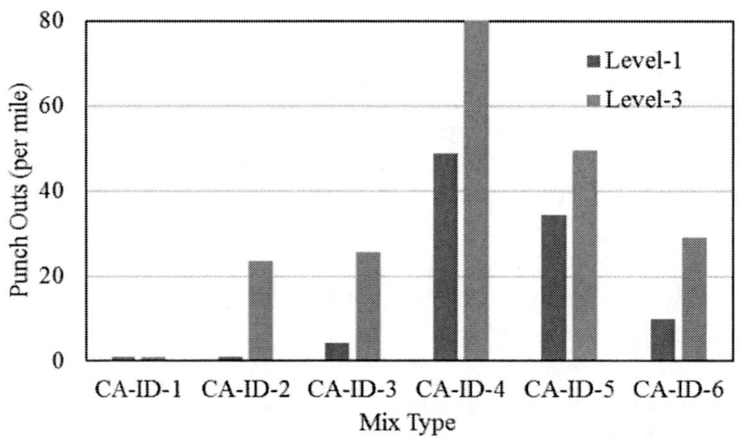

Figure 2. Impact of Input Levels on CRCP Punch Outs.

4.3 Percent Change in CRCP Performance with Level-1 & Level-3 Inputs

The simulation results were analyzed to compare the percentage change in performance parameters including IRI and punch outs with level-1 and level-3 inputs. The results are presented in Figure 3. The results show that the %age change in IRI ranges from 1 to 43.8% while the percentage change in punch outs ranges from 1 to 95.8%. It indicates that the impact of material input levels is more significant on the CRCP punch outs then the pavement roughness. The CRCP designed with level-3 material inputs may be under-designed or over-designed and to obtain accurate CRCP design, level-1 material inputs should be used.

5 EFFECTS OF TRAFFIC VOLUME ON CRCP PERFORMANCE

The effects of traffic volume on CRCP performance were evaluated by conducting simulations with level-1 inputs for all paving mixes with varying traffic volumes ranging from AADTT of 2000 to 9000 (14.9 to 67.2 million ESALs). The slab thickness was kept constant at 10 in. and other design factors were kept constant as shown in Table 1. The analysis of results shows similar trend for all paving mixes and the results for CA-ID-3 are presented in succeeding paras.

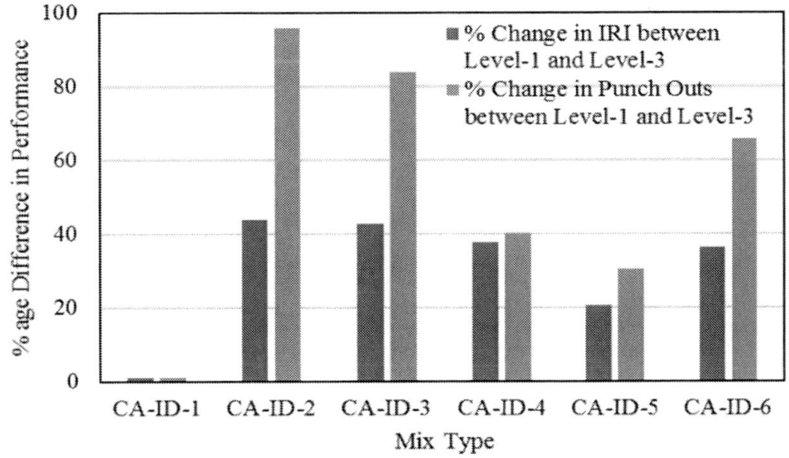

Figure 3. Percent Change in CRCP Performance between Level-1 and Level-3 Inputs.

5.1 Comparison of Effects of Traffic Volume for CA-ID-3

The comparison of the effects of traffic volume on punch outs and IRI are shown in Figures 4 and 5 for CA-ID-3. The results show that the traffic volume has a significant impact on both the performance parameters. As the traffic volume increases, the performance of CRCP decreases with increase in IRI and punch outs. With these results, it is evident that CRCP design be performed with the specific traffic volume for any specific project.

Percentage increase in CRCP distresses was determined for every 1000 increase in AADTT to evaluate the impact of traffic volume on pavement performance. The results show that the traffic volume has a more significant impact on punch outs with a % increase of 8.9 to 88.6% with increased traffic volume whereas the increase in IRI is in the range of 2.2 to 9.6%.

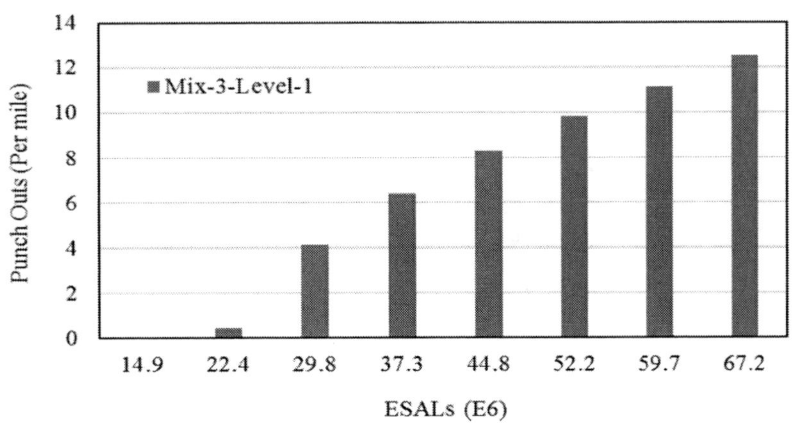

Figure 4. Impact of Traffic Volume on Punch Outs for CA-ID-3.

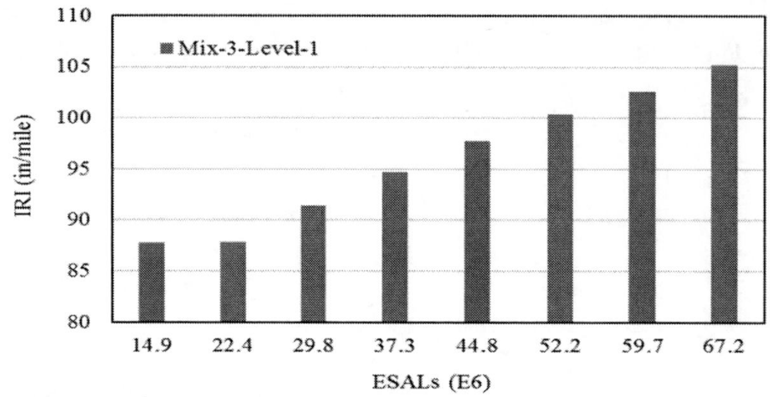

Figure 5. Impact of Traffic Volume on IRI for CA-ID-3.

6 CONCLUSIONS

The analysis of the simulation results of this study dictates that there is significant difference between the predicted performance of CRCP designed with time series data of concrete material inputs and the default material inputs. Both performance parameters including punch-outs and pavement roughness are affected but punch-outs is the most significant with a percent change of up to 95.8% between the two input levels. This necessitates the importance of using time series data of concrete material inputs in design of CRCP. Similarly, the effects of traffic volume on the performance of designed CRCP shows that there is significant impact of traffic volume on pavement performance, thus CRCP should be designed with accurately anticipated traffic volume. Finally, it can be concluded that using time series data for concrete material properties and actual traffic volume will result in accurate CRCP design which can last for the designed service life.

References

AASHTO T336, *Standard Test Method for the Coefficient of Thermal Expansion of Hydraulic Cement Concrete. Standard Specifications for Transportation Materials and Methods of Sampling and Testing,* 31st ed., American Association of State Highway and Transportation Officials, Washington, D.C., 2011.

AASHTO, *Mechanistic – Empirical Pavement Design Guide, A Manual of Practice,* American Association of State Highway and Transportation Officials, Washington, D.C., 2015.

FHWA, *Continuously Reinforced Concrete Pavement,* Technical Advisory T 5080.14, Federal Highway Administration, Washington, DC. http://www.fhwa.dot.gov/pavement, 1990.

Plei, M., and S. Tayabji, *Continuously Reinforced Concrete Pavement Performance and Best Practices,* Report FHWA-HIF-12-039. Springfield, VA, 2012.

Rao, C., Calibration of Mechanistic-Empirical Performance Model for Continuously Reinforced Concrete Pavement Punch-Outs, *Transportation Research Record,* 1896, 15–22, 2004.

Sabih, G., and Tarefder, R. A., Impact of Variability of Mechanical and Thermal Properties of Concrete on Predicted Performance of Jointed Plain Concrete Pavements, *International Journal of Pavement Research and Technology,* 9(6), 436-444, 2016.

Interdependence between Structural Engineering and Construction Management
Edited by Ozevin, D., Ataei, H., Modares, M., Gurgun, A., Yazdani, S., and Singh, A.

ISBN: 978-0-9960437-6-2

DEVELOPMENT OF A DECISION MAKING TOOL FOR PAVEMENT MAINTENANCE MANAGEMENT UNDER SUSTAINABILITY CRITERIA

PEDRO ESPINOZA-VALAREZO, TATIANA GARCÍA-SEGURA, LAURA MONTALBÁN-DOMINGO, and EUGENIO PELLICER

School of Civil Engineering, Universitat Politècnica de València, Valencia, Spain

This paper focuses on solving the problem of budget allocation for the pavement maintenance management of an urban road network. To do this, a decision-making tool is developed. The tool is based on a quantitative valuation of six sustainable criteria to obtain a priority ranking. The relative weights are calibrated with the aim of obtaining a higher benefit-cost ratio to a horizon of 4 years. The tool is applied to a sample of Valencia's urban road network. The work concludes that decision-making based on sustainable criteria increases the ratio benefit-cost in 53%, compared with the plan based only on the condition of the pavement, which is typical in management agencies lacking an assessment tool. The application of the tool to an urban road network that differs in conditions with the sample analyzed must be preceded by a calibration similar to the one raised, in order to define the optimal scenario in relative weights.

Keywords: Infrastructure management, Urban pavements, Budget allocation, Valencia.

1 INTRODUCTION

A nation's road network is one of its greatest patrimonial assets. The road infrastructure provides a fundamental basis for economic performance and development. Likewise, it contributes significantly to environmental damage during its construction, maintenance, and use (Santero and Horvath 2009). Schliessler and Bull (2003) said that a country that allows the deterioration of its road infrastructure may have a surcharge of vehicular operation in a range of between 1 and 3% of its national GDP. On the other hand, Hajj *et al.* (2010) said that the typical cost of maintaining a road network is equal to 15-20% of the cost of its rehabilitation or reconstruction. To get an efficiently pavement manage, it is necessary to use standardized tools and processes such as the implementation of a pavement management system, known as PMS (Pavement Management System). A PMS is a set of tools used and designed to give, assess, and maintain pavements in acceptable service conditions for a period of time (AASHTO 1985). Steger (1978) pointed out that a PMS is due to two fundamental characteristics: the monitoring of pavements and the use of a decision-making tool for maintenance work. Instead, Hudson *et al.* (1979) established that a PMS should be developed in such a way that it can be used by the managers of all or most levels in charge of pavement management. This paper focuses on solving the existing problem with respect to the maintenance management of pavements. This problem has been previously dealt by other authors. Torres-Machi *et al.* (2017), raised an optimization of budget allocation, in which there were ST x N possible solutions, being "N" the number of pavements and "S" the number of possible solutions to be applied in a horizon of analysis of "T" years. The authors focused on the development of an optimization tool to maximize the long term effectiveness of maintenance and minimize the emissions of GHG (global greenhouse gas) with the constraints of a limited budget and a minimum threshold of condition. Meneses and Ferreira (2015), instead, dealt with the problem as an optimization whose objective function was to minimize maintenance costs and maximize the residual value of the pavement network considering a minimum threshold of condition and an annual budget as restrictions. Based on the aforementioned, this work focuses on the development of a decision-making tool for the annual management of pavement maintenance whose main objective is to achieve a greater long-term benefit-cost ratio, guaranteeing the sustainability of the infrastructures. Another of the objectives pursued is the easy adaptability of the tool by the managers and decision makers of a road network.

2 METHODOLOGY

The proposed methodology solves a multi-criteria decision-making problem based on SAW (simple additive weighting) method. Six quantitatively valuation criteria are proposed to establish an annual prioritization of the pavement maintenance that allows accommodating the available budget of a network efficiently. Value functions are used to normalize each criterion to values between zero and one and rate the level of satisfaction of an alternative with respect to certain criteria. The relative weights of each criterion, conditions the prioritization and the long term results. Thus, this study calibrates the weights to obtain the best benefit-cost ratio and define the best scenario. Results are compared to the conventional situation, in which only user's comfort is considered. For the weight's calibration, a long-term economic evaluation (four years) is done.

2.1 Criteria for Decision Making

The proposed valuation criteria are as follows:

Criterion one: User's comfort. This is used to rate the comfort of the road network's users. With the aim of evading a subjective valuation, the valuation of the comfort was made based on the condition of the pavement. For this purpose, a step function is used (Table 1), which score the values between zero and one depending on the PCI (Pavement Condition Index).

Table 1. Rule for the construction of the value function of the criterion 1.

PCI	F (PCI)
0 – 10	1.00
10 – 40	0.75
40 – 70	0.50
70 - 85	0.25
85 - 100	0.00

Criterion two: Accident cost. In order to ensure sustainability and reduce the damage to society and users, it was decided to include this criterion in the decision-making. This aims to prioritize maintenance work on the roads that have a higher accident cost per year. For its evaluation, Eq. (1) proposed by Forkenbrock *et al.* (1997) should be used. With this equation it is possible to obtain the accident cost per mile vehicle travel. PSR assesses the pavement condition, taking values between zero (poor) and five (excellent), PASRES takes value of one or zero in presence or lack of pass restriction, ADTLANE refers to the traffic expressed in miles vehicles per lane, RIGHTSH is the curb width expressed in feet, LANES takes values of one if the road has more than four lanes or zero if not. TOPCURV and TOPGRAD refer to the curvature in horizontal and vertical tracing of the road (Forkenbrock *et al.* 1997).

$$\frac{\text{Cost of Accidents}}{\text{Millions VMT}} = 1.587.580\left(0{,}994^{\text{PSR}}\right)\left(1{,}111^{\text{TOPCURV}}\right)\left(1{,}442^{\text{PASRES}}\right)\left(1{,}741^{\text{ADTLANE}}\right)\left(0{,}952^{\text{RIGHTSH}}\right)\left(0{,}936^{\text{LANES}}\right)\left(1{,}085^{\text{TOPGRAD}}\right) \quad (1)$$

Criterion three: Economic efficiency. In order to achieve obtain the highest benefit-cost ratio, economic efficiency is raised as a criterion. Carnahan *et al.* (1987), quantifies the efficiency of maintenance by increasing the PCI (Δ PCI). For the calculation of the economic efficiency, the Eq. (2) is proposed, in which the product of Δ PCI and the average daily traffic intensity (IMD) are divided, by the total cost of the maintenance's works.

$$\text{Ef Ec}=\frac{\Delta\text{PCI}^{*}\ \text{IMD}}{\text{Cost Actuation}} \quad (2)$$

Criterion four: Proximity to social and tourist infrastructure. This criterion seeks to prioritize those roads close to relevant social and tourist infrastructures. For its valuation, it is proposed to classify the infrastructures in 3 levels of importance (N2, N1 and N0), according to the nature of the infrastructures that are in a circumference of 500 meters in diameter, drawn from the middle point of the road. A value function (Table 2) prioritizes with a convex function the N2, from the N1 with linear function and N0 with concave function.

Table 2. Functions used for the valuation of the criterion 4.

Proximity to infrastructure level	**Value function**
N2	$V=\frac{1}{1-e^{-3,5}}\left[1-e^{\frac{-3,5(100-PCI)}{100}}\right]$
N1	$V=\frac{1}{1-e^{-(10)^{-9}}}\left[1-e^{\frac{-(10)^{-9}(100-PCI)}{100}}\right]$
N0	$V=\frac{1}{1-e^{3,5}}\left[1-e^{\frac{3,5(100-PCI)}{100}}\right]$

Criterion five: Environmental efficiency. This criterion seeks to prioritize efficient maintenances from the environmental point of view. For its assessment, Eq. (3) is used. This equation is similar to (2) but replaces the monetary cost for the environmental cost, quantified as the kilograms of CO_2 generated with the maintenance.

$$\text{Ef Amb}=\frac{\Delta\text{PCI}^{*}\ \text{IMD}}{\text{Kg of CO2}} \quad (3)$$

Criterion six: Inconvenience caused. This criterion seeks to prioritize those alternatives that generate minor discomfort to the society, that is to say, that they take the least amount of time for execution.

Criteria two, three, five and six need to be normalized to values between zero and one. Therefore, a value of one corresponds to the highest value (Eq. (4)), in the case of criteria two, three and five and the lowest value (Eq. (5)), in the case of criterion six.

$$rij = \frac{X-Xmin}{Xmax-Xmin} \quad (4)$$

$$rij = 1-\frac{X-Xmin}{Xmax-Xmin} \quad (5)$$

2.2 Long Term Comparison

The long-term outcome of the decision-making tool depends on the relative weights assigned to each criterion. For the comparison of maintenance strategies, a LCCA (Life Cycle Cost Assessment) is used over a 4-year horizon. The objective of the evaluation is to obtain a greater benefit-cost ratio. For the costs, only those incurred by the management agency in the maintenance works during the 4 years of analysis are considered. On the other hand, the quantification of benefits is carried out by calculating the increase in the residual value of the road network from the beginning of the year zero with respect to the end of 4 year period. Fullana and Puig (2012) defined residual value as the value of life remaining at the end of the analysis. Babashamsi *et al.* (2016) defined as a useful life of a pavement treatment, the time remaining for the pavement to deteriorate to a threshold established. The residual value was calculated under the previous premises using the deterioration curves of the pavements. It was defined a minimum threshold of PCI of 40, because a total reconstruction would be needed for values lower than 40. The residual value was monetized by multiplying the total reconstruction cost (CTR) by the percentage of the area under the corresponding deterioration curve at the time of analysis, obeying Eq. (6) and Figure 1.

$$\mathrm{VR} = \mathrm{CTR} * \frac{\mathrm{A2}}{(\mathrm{A1+A2})} \tag{6}$$

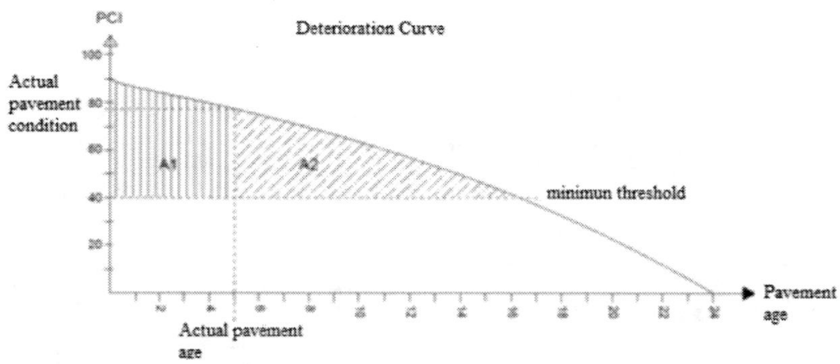

Figure 1. Residual value calculation through deterioration curves.

3 CASE STUDY

The case study focused on a sample of Valencia's urban road network, consisting of 13 asphalt roads. The characterization of the sample was made during the year 2016, including a geometric data extraction, pass restrictions and condition evaluation (PCI) made by images taken with the Google Earth Pro. The volume of traffic was extracted from the database of the Valencia Council. The typical actions carried out in the network were extracted from the research carried out by Guaita (2016). The unitary emissions of CO_2 of maintenance actions were taken from Torres-Machi *et al.* (2015). The annual budget for the maintenance of the network (€865282.32) was obtained proportionally to the area of the sample. Another restriction considered was to make at least a minimum investment of €10000.00, as well as allowing an investment of €100000.00 additional loading, or paying the deficit or super habit to the next year's budget. Due to the lack of performance data for the Valencia pavements, Eq. (7) proposed by (George *et al.* 1989) was used for evaluating the reduction of PCI over time.

$$PCI(t) = 90 - a\left[e^{age^{b}} - 1\right] log\left[\frac{ESAL}{SNC^{c}}\right] \tag{7}$$

3.1 Decision Making

For decision making and comparison purposes, a first scenario was raised (E1) in which it was assigned a weight of 100% to criterion 1, and a second scenario (E2) was raised in which weights are calibrated to obtain the best benefit-cost ratio. E2 calibration obtained weights of 30%, 10%, 30%, 10%, 10% and 10% for the six criteria, respectively. Table 3 shows the decision-making scheme for E2 in 2016. This strategy suggests maintaining Viyarroya and Ayora streets during 2016. After performing the maintenance, the PCI of these roads increased to 100 and deteriorated during the following years using the deterioration curves. The process was repeated until the end of 2019 and the increment in the network's residual value (benefit) was calculated. This amount was divided by the total cost of maintenance executed in the 4 years given the benefit -cost ratio, which was 2.9. From the results, it worth noting that criterion 1 and 3 should take the same weight to avoid being given priority only in total rehabilitation maintenance works with low PCI values (as criterion 1) or preventive maintenance with low-cost maintenance actions (as criterion 3). Criteria 1 and 3 should take greater values than criteria 2, 4, 5, and 6. However, these last criteria cannot be forgotten as environment and social aspects should be taken into account for sustainable purposes. Table 4 compares E1 and E2. As mentioned above, a weight of 100% is allocated to the criterion of comfort in order to simulate the conventional decision making when there is a lack of a decision tool. As Table 4 shows, decision-making under sustainable criteria (E2) generates 1257024.77€ additional value increments, investing 762091.43 € less than E1 condition-based decision making. While the benefit-cost ratio of E1 is 1.9, the benefit-cost ratio of E2 is 2.9.

Table 3. Decision making of E2 at the beginning of 2016.

Weights:	30	10	30	10	10	10			Maintenance
Street	***Crit1***	***Crit2***	***Crit3***	***Crit4***	***Crit5***	***Crit6***	***Val***	***Cost***	2016
VIYARROYA	0.50	0.02	1.000	0.104	0.111	1.00	0.5737	€46,060.00	Yes
AYORA	0.50	0.11	0.365	0.550	1.000	0.94	0.5202	€2,246.40	Yes
GIORGETA	0.50	1.00	0.095	0.400	0.376	0.37	0.3927	€878,421.28	No
BELTRAN	0.75	0.02	0.066	0.399	0.316	0.61	0.3791	€99,865.60	No
URUGUAY	0.50	0.10	0.200	0.182	0.538	0.84	0.3765	€878,421.28	No
PEREZ GALDOS	0.50	0.35	0.094	0.480	0.374	0.37	0.3354	€206,093.26	No
JESUS	0.50	0.07	0.076	0.802	0.301	0.34	0.3244	€4,399.56	No
CARTEROS	0.50	0.08	0.131	0.100	0.347	0.80	0.3217	€403,873.08	No
BRASIL	0.00	0.02	0.436	0.013	0.674	0.97	0.2989	€109,622.80	No
ARCHIDUQUE	0.50	0.04	0.061	0.430	0.241	0.47	0.2865	€649,100.80	No
CUENCA	0.50	0.01	0.020	0.340	0.036	0.64	0.2585	€187,107.00	No
TRES FORQUES	0.50	0.00	0.000	0.300	0.000	0.00	0.1800	€ 9,539.64	No
TOBEÑAS	0.00	0.01	0.088	0.080	0.105	0.94	0.1398	€955,555.38	No
							Total	€ 48,306.40	
							Sup.	€ 816 975.92	

Table 4. E1-E2 Comparison.

Esc	Crit1	Crit2	Crit3	Crit4	Crit5	Crit6	Δ Res Val	Inversion (Cost (€))	B/C
E1	100	0	0	0	0	0	6 573 975.66 €	3 460 811.50 €	1.90
E2	30	10	30	10	10	10	7 831 000.43 €	2 698 720.07 €	2.90

4 CONCLUSIONS

Of the developed work it was concluded that 30%, 10%, 30%, 10%, 10%, 10% are the relative weights of each criterion to obtain the highest cost-benefit ratio. It was concluded that the decision making should not be prioritized according to the condition as a unique criterion. Decision-making based on sustainable criteria allows obtaining a higher long-term benefit-cost ratio. Another important conclusion is that criterion 1 and 3 should have the same weight to compensate for rehabilitation maintenance works and preventive maintenance. Therefore, this strategy suggests that it is beneficial to carry out preventive maintenance but also maintain the infrastructures under worse conditions. As a recommendation, it is emphasized that the application of the tool to an urban road network that differs in conditions with the sample analyzed must be preceded by a calibration similar to the one raised, in order to define the optimal scenario in relative weights. The level of importance of urban infrastructures should be defined when evaluating the criterion 4. It is also recommended to adjust the tool with performance curves based on historical data with the purpose of estimating the maintenance actions more precisely.

References

American Association of State Highway and Transportation Officials (AASHTO), *Guidelines on Pavement Management*, AASHTO, Washington, D.C, 1985.

Babashamsi, P., Yusoff, N.I. M., Ceylan, H., Nor, N.G.M., and Jenatabadi, H.S., Sustainable Development Factors in Pavement Life-Cycle: Highway/Airport Review, *Sustainability*, *8*(3), 2016.

Carnahan, J. V., Davis, W.J., Shahin, M.Y., Keane, P.L., and Wu, M.I., Optimal Maintenance Decisions for Pavement Management, *Journal of Transportation Engineering*, *113*(5), 554–572, 1987.

Forkenbrock, D. J., and Foster, N. S. J., Accident Cost Saving and Highway Attributes, *Transportation*, *24*, 79–100, 1997.

Fullana, P., and Puig, R., *El Análisis del Costo del Ciclo De Vida: Una Herramienta para Evaluar Mejor las Inversiones y Decisiones Técnicas en Pavimentacion*, 2012.

George, K. P., Rajagopal, A. S., and Lim, L. K., Models for Predicting Pavement Deterioration, *Transportation Research Board*, *1215*(1), 25–32, 1989.

Guaita, B., and Serra, J. R., *Propuesta de Implementación de un Sistema de Gestión de Pavimentos en la Ciudad de Valencia*. Master's Thesis, Universitat Politècnica de València, Valencia, 2016.

Hajj, E. Y., Loria. L., and Sebaaly, P. E., Performance Evaluation of Asphalt Pavement Preservation Activities. In *Proceeding of the 89th General Annual Meeting of the Transportation Research Board*. Transportation Research Board. Washington, D.C., doi: 10.3141/2150-05, 2010.

Hudson, W. R., Haas, R., and Pedigo, R. D., *Pavement Management System Development*, Transportation Research Board, Washington D.C., 1979.

Meneses, S., and Ferreira, A., Flexible Pavement Maintenance Programming Considering the Minimisation of Maintenance and Rehabilitation Costs and the Maximisation of the Residual Value of Pavements. *International Journal of Pavement Engineering*, *16*(7), 571–586, 2015.

Santero, N. J., and Horvath, A., Global Warming Potential of Pavements. *Environmental Research Letters*, *4*(3), 034011, 2009.

Schliessler, A., and Bull, A. Roads *A New Approach for Road Network Management and Conservation*, No. LC/L. 693, 2003.

Steger, A. R., *Pavement Management*, Report prepared for AASHTO Region 3 Subcommittee on Design Meeting, Oklahoma City, June 7-9, 1978.

Torres-Machí, C., Chamorro, A., Pellicer, E., Yepes, V., and Videla, C., Sustainable Pavement Management, *Transportation Research Record*, *2523*, 56–63, 2015.

Torres-Machi, C., Pellicer, E., Yepes, V., and Chamorro, A., Towards a Sustainable Optimization of Pavement Maintenance Programs Under Budgetary Restrictions, *Journal of Cleaner Production*, *148*, 90–102, 2017.

Interdependence between Structural Engineering and Construction Management
Edited by Ozevin, D., Ataei, H., Modares, M., Gurgun, A., Yazdani, S., and Singh, A.

ISBN: 978-0-9960437-6-2

TECHNICAL AND ENVIRONMENTAL ASSESSMENT FOR SOIL STABILIZATION USING COAL ASH

DIANA GÓMEZ[1], GLORIA CARVAJAL[1], ALEJANDRA BALAGUERA[1], and YHAN PAUL ARIAS[2]

[1]*Universidad de Medellín, Medellín, Colombia*
[2]*Universidad Nacional de Colombia, Medellín, Colombia*

In most Latin American countries, low-volume roads are composed of unpaved roads; this is considered a problem of economic, social and environmental interest. There are different stabilization alternatives for this type of roads with traditional materials, i.e., Portland cement (OPC) and lime (L), both of which have a high environmental impact due to anthropogenic CO_2 emissions. This paper presents the results of the environmental assessment of an industrial residue Coal Ash (CA) with pozzolanic characteristics. The residue was alkaline activated with $Ca(OH)_2$ from commercial lime (L). The binary system (CA+L) is called (CLM) and forms a material with cementing properties, and when it is mixed with soil, it increases the capacity to support loads. The CLM as a soil stabilizer is proposed along with the modification of some construction processes associated with lime technology and Portland cement. Finally, a technical and environmental comparison is made for conventional stabilizers and the binary system CLM. The results showed that stabilization of a silty soil with CLM can achieve a reduction of 58% and 75% in CO_2 emissions when compared with L and OPC, respectively.

Keywords: Sustainability, Pozzolans, CO_2 emissions, Latin America.

1 INTRODUCTION

Soil characteristics such as resistance and durability can be improved through a stabilization process (Hossain and Mol 2011, Zhang *et al.* 2015). The energy industry, metal manufacturing and other similar industries generate large quantities of industrial by-products such as those obtained from coal combustion (García-Lodeiro *et al.* 2007) blast furnace slag (Provis 2015), biomass ash (Zelaya *et al.* 2017) and construction and demolition wastes that cause a high degree of pollution on landfills (Chowdhury *et al.* 2010). These non-conventional materials may in some cases be combined with highly alkaline materials, such as lime, and create materials with cementitious characteristics "pozzolanic" (Duxson *et al.* 2007) or alkali such as NaOH (Hoyos-Montilla *et al.* 2018) contributing to sustainable development in the construction sector, partially replacing OPC (Zhang 2015) and other calcium-based soil stabilizers.

Currently, there are not universal specifications to address the environmental impacts of the use and application of by-products in road construction. To meet this challenge, the use of methodologies to quantify impacts in different stages of the life cycle is proposed, contributing to the reduction of gas emissions into the atmosphere compared to traditional materials; OPC industry being the second largest producer of greenhouse gases and occupying the third place (after aluminum and steel) in energy consumption (Rashad 2014). According to reports, the production of 1.0 metric ton of OPC releases between 0.73 - 0.99 metric tons of gaseous carbon dioxide $CO_2(g)$ (Hasanbeigi *et al.* 2012). The calcination processes of limestone $CaCO_3$, dolomite Mg_2CO_3 and the impurities thereof can theoretically represent 0.785 $CO_2(g)$ per tons in the production of calcium oxide and 1.092 $CO_2(g)$ in the production of magnesium oxide. Emissions associated with energy use depend on the efficiency of the process, the use of fuel and the emissions that are indirectly produced by electricity generation (IPCC 2014). Colombia has a limited road network when compared to other Latin American developing countries. Its tertiary road network was estimated at more than 142,284 km and, despite information gaps, only 18.74% is considered to be in good conditions, while 40.13% is in poor conditions and the remaining 41.13% is under regular conditions (DNP 2016).

2 METHODOLOGY

2.1 Materials

Fine granular soil of high plasticity from a region of western Colombia was used to improve its bearing capacity, see Table 1. The soil was dosed with two stabilizers of conventional use: Common Purpose Ordinary Portland Cement (OPC) and commercially quicklime (L), and an alternative stabilizer composed of a binary mixture of calcium hydroxide and coal ash (CLM). The stabilizing residue is composed of commercial grade slaked lime and residue from coal combustion in Medellin- Colombia, with 25% and 75% of mass, respectively.

Table 1. Characteristics of natural soil.

ASTM D 4318-10			ASTM D 854-10		ASTM D 1557-10	ASTM D 2487-11	ASTM M 145-91
LL %	PL %	PI %	Specific gravity	Dry-density kN/m3	Opt. Moisture content %	Classification USC	Classification ASSHTO
66	48	17	2.71	14.65	26	MH	A-7-5

The chemical composition was obtained by means of XRF on a Phillips PW 2400 X-ray fluorescence spectrometer. The test was carried out on pressed powder ash and soil. See Table 2.

Table 2. FRX of waste (%) in mass.

Material	SiO_2	Al_2O_3	Fe_2O_3	*CaO*	*MgO*	Na_2O	SO_3	TiO_2	**loss on ignition 110°C to 1000°C**
Coal Ash	41.9	31.1	6.4	7.4	1.5	5.6	1.1	1.3	2.1
Silt	41.2	33.6	9.2	0.1	0.4	0.1	0.1	1.4	13.6
Lime	1.5	1.0	0.1	65.6	0.1	0.1	0.3	1.1	30.3

The mineralogical content of the materials was determined by X-Ray Diffraction (XRD) on a PANalytical X′Pert MPD PRO equipment, between 5° to 60° with a step size of 0.02° and counting time of 56 seconds. A CuKα1 (λ=1.54059 Å) source was used. Figure 1-a shows the diffractograms for the coal ash and soil evaluated. The particle-size distribution (PSD) of ash and soil are shown in

Figure 1 (b) was measured in Master Sizer 3000 equipment. A D80 of 49.1 μm for Coal Ash and 40.89 μm for soil is presented. During the soil mechanical stabilization, every product: OPC, L, and CLM, were incorporated in mass with a dosage ranging from 0% to 24%.

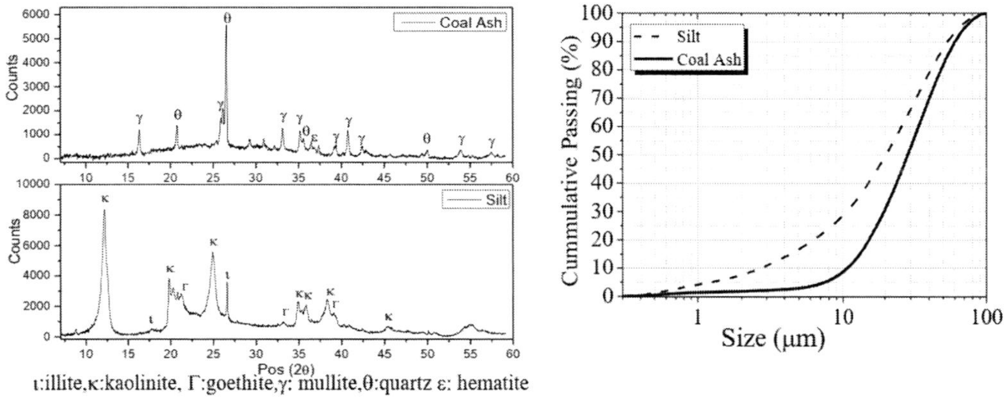

Figure 1. (a) Mineralogical phases of silt and coal ash (b) Particle size distribution of silt and coal ash.

2.2 Methods

2.2.1 *Soil stabilization*

Modified Proctor trials were performed on OPC, L and CLM stabilizers, considering standard Invias 142 de (INVIAS 2012), which is derived from ASTM D 1557-09. The results show that, in all cases, the increase of the stabilizer percentage also increases water demand, being more noticeable for dosages above 14%. See Figure 2.

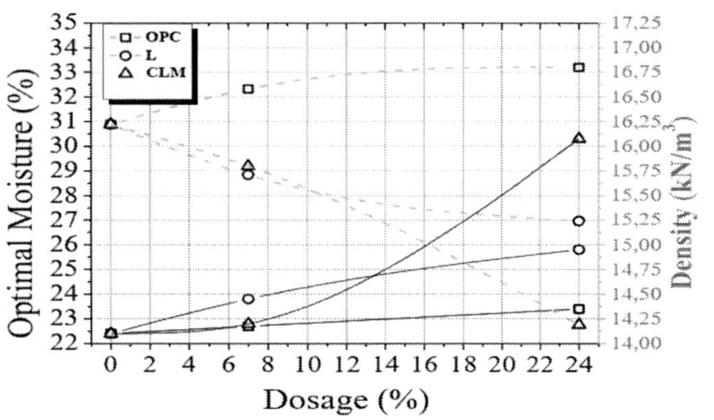

Figure 2. Modified Proctor for stabilized soil.

2.2.2 Emissions quantification

The CO_2 emissions quantification is performed in three stages: stage one, take theoretical values from a stoichiometric balance, see Eq. (1) and Eq. (2), the theoretical $CO_2(g)$ emissions can be determined obtaining one metric ton of Lime (L) and Ordinary Portland Cement (OPC).

$$1C_a\,CO_3 \xrightarrow{\Delta \approx (950°C} 1C_a\,O^{*} + 1CO_2 \tag{1}$$

$$Ca\,CO_3 + 1Al_2O_3\,SiO_2 + Fe_2O_3 \xrightarrow{\Delta \approx (1450°C)} \frac{C3S + C3A + C4AF}{Clinker^{*}} \tag{2}$$

According to Eq. (1), the carbonate contains 56.03% CaO and 43.97% $CO_2(g)$ by weight from which an estimated 0.44 tons of $CO_2(g)$ is obtained. For the OPC case, a ton of clinker was considered, it contains 0.65 tons of CaO coming from $CaCO_3$ (IPCC 2014). The amount of $CaCO_3$ needed to obtain 0.65 tons of CaO is equivalent with 1.1601 tons of $CaCO_3$, therefore, the amount of $CO_2(g)$ released by calcination of $CaCO_3$ in the process of producing a ton of Clinker is 0.5101 tons, see Eq. (2). Stage two includes the values reported in the literature to generate the brute product and established that the CO_2 emissions for obtaining these products can vary as a function of the raw material extraction and their purity, considering a carbon footprint above the theoretical one, this can vary depending on the source. Table 3 based on theoretical sources presents the CO_2 emissions in kilograms per ton of cement and lime respectively, without including the emissions by energy or fossil fuel needed to make the burnt. For the lime, the dissociation of limestone produces up to 0.75 tons of CO_2 per ton of quicklime. In stage three, we used reports like IPCC (2014) when the fossil fuel is included from the cement and lime production processes, and the emissions oscillate between 0.785 and 1.092 CO_2 metric tons for lime (L), and 0.730 to 0.935 CO_2 metric tons for Ordinary Portland Cement (OPC).

Table 3. World cement production and CO_2 emissions in 2015.

Region	North America	CIS	Eur.	Asia+ oceania**	Middle East	Africa	China Korea Japan	Brazil	South America*	Central America	India
Emission kg CO_2/t Gross	1000	910	815	805	800	790	780	780	760	750	715
Emission kg CO_2/t Gross Lime*** (For the all-region 750)											

Notes: specific emission does not include emissions from the use of electricity. *Excl. Brazil, **Excl. CN, India, CIS, and Japan. Source: Adapted from GNR 2015 ***(CSC 2018, European Commission 2001).

3 RESULTS AND DISCUSSIONS

3.1 Effect of the Stabilizer on the Soil

In all cases, with a gradual increase of the stabilizers in the soil, an increase of the mechanical stabilization occurred. See Figure 3. When defining the optimum value of 9.0% for the stabilizer L, it was found that to obtain the same conditions of compression strength with OPC and CLM, additions of approximately 17% and 15% are required.

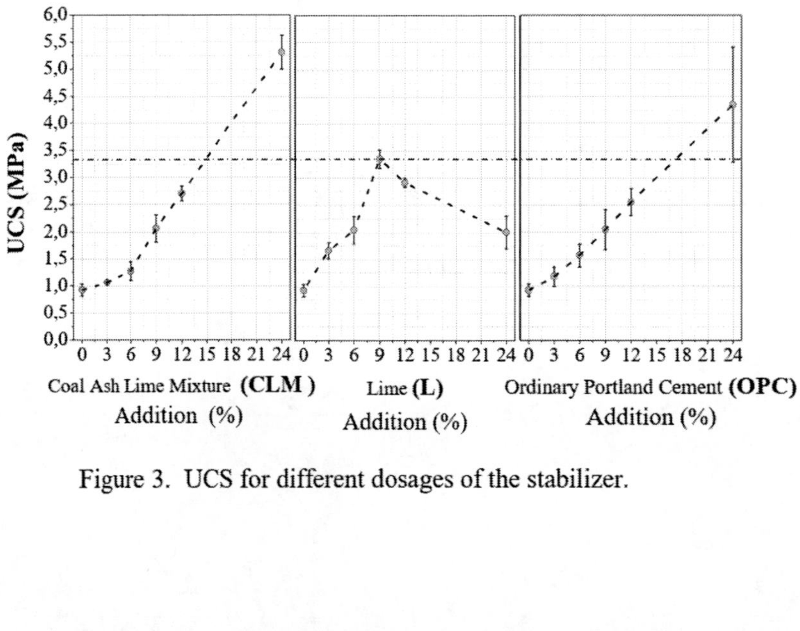

Figure 3. UCS for different dosages of the stabilizer.

3.2 Emissions Quantification

Table 4 presents the average values for $CO_2(g)$ emissions for stabilizers OPC, L and CLM determined. The stabilizers OPC, L, and CLM environmental assessment was addressed using 3.3 MPa in the soil's bearing capacity, corresponding with a UCS value defined from the maximum obtained experimentally for L (see Figure 1). Therefore, the silty soil stabilization A-7-5 requires dosing of 9% L, 17% OPC, and 15% CLM to obtain the same mechanical performance.

Table 4. CO_2 emissions.

	Stoichiometry emissions			Gross emissions (extraction & purity)			Gross + energy & fuel emisions		
	Lime	*Cement*	*Coal Ash/lime*	*Lime*	*Cement*	*Coal Ash/lime*	*Lime*	*Cement*	*Coal Ash/lime*
	100%	100%	75%/25%	100%	100%	75%/25%	100%	100%	75%/25%
Emission Max	-	-	-	-	0.733	-	1.092	0.73	0.273
Emission Min	-	-	-	-	0.549	-	0.785	0.935	0.196
Average	0.44	0.51	0.11	0.75	0.638	0.188	0.939	0.833	0.256
Deviation	-	-	-	-	0.062	-	0.217	0.145	0.054
	Stabilizer emissions for soil CO_2 /tone								
	Lime	*Cement*	*Coal Ash/lime*	*Lime*	*Cement*	*Coal Ash/lime*	*Lime*	*Cement*	*Coal Ash/lime*
	9%	17%	14%	9%	17%	14%	9%	17%	14%
Average	0.04	0.087	0.015	0.07	0.108	0.026	0.084	0.142	0.036
Deviation	-	-	-	-	0.011	-	0.02	0.025	0.008

According to the stoichiometric analysis, the emissions generated by the use of CLM are 61.11% and 82.24% lower than when using L and OPC, respectively. Similarly, the analysis considering industrial production indicated a reduction of 61.11% and 75.80% with respect to L and OPC. When including the energy processes for the use of fuel or energy for burning, the decrease was around the order of 57.53% and 74.66% with respect to L and OPC. The results show that under conditions of productivity in stage three the system with the highest CO_2 emissions is represented by OPC, followed by L and finally the CLM. See Figure 4.

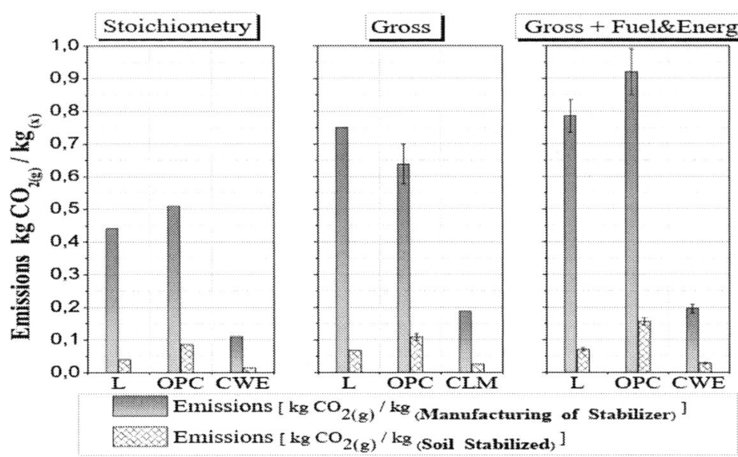

Figure 4. CO_2 emissions for different soil stabilizers.

4 CONCLUSIONS

It is possible to achieve improvements in the bearing capacity of the Lime clayish soil at 7 days of curing by incorporating stabilizers such as OPC, L, and CLM, in mass percentages of 17%, 9%, and 15%, respectively, achieving increases of UCS of an order of 366% with respect to the soil without a stabilizing product. Independently of the three stages evaluated in the environmental aspect the results obtained show how the use of industrial waste with coal ash lime (CLM), when applied to soil stabilization, can decrease the CO_2 emissions in constructive processes based on L and OPC for soils stabilization. Thus, they obtained the closest conditions to the productive chain decreases of 75% and 57% with respect to OPC and L in stage 3. OPC and L are the ones causing more damage, including the extraction process as it entails: fuel consumption, the use of machinery, transportation of the material, among others. This reveals that the use of alternative materials in the construction sector may reduce the impacts on human health, the soil and other ecosystems.

Acknowledgments

Special thanks to Universidad de Medellin, COLCIENCIAS, and Red INNOVIAL.

References

Chowdhury, R., Define, A., and Fry, T., A Life Cycle Based Environmental Impacts Assessment of Construction Materials Used in Road Construction, *Resources, Conservation and Recycling,* 54(4), 250-255, 2010.

CSC, Global cement Database. Retrieved from http://www.wbcsdcement.org on January, 2018.

Duxson, P., Fernández-Jiménez, A., Provis, J. L., Lukey, G. C., Palomo, A., and van Deventer, J. S., Geopolymer Technology: The Current State of The Art, *Materials Science,* 42(6), 2917–2933, 2007.

European Commission, Integrated Pollution Prevention and Control (IPPC). *Document on Best Available Techniques (BREF) in the Cement and Lime Manufacturing Industries,* 2001.

García-Lodeiro I., Palomo, A., and Fernández-Jiménez, A., Alkali-Aggregate Reaction in Activated Fly Ash Systems, *Cement and Concrete Research,* 37(2), 175–183, 2007.

Hasanbeigi, A., Price, L., and Lin, E., Emerging Energy-Efficiency and CO_2 Emission-Reduction Technologies for Cement and Concrete Production: A Technical Review, *Renewable and Sustainable Energy Reviews,* 16(8), 6220-6238, 2012.

Hossain, K.M.A., and Mol, L., Some Engineering Properties of Stabilized Clayey Soils Incorporating Natural Pozzolans and Industrial Wastes, *Construction and Building Materials,* 25 (8), 2011.

Hoyos-Montilla, A. A., Arias-Jaramillo, Y. P., and Tobón, J. I. Evaluation of Cements Obtained by Alkali-Activated Coal, 68 (332), *Materiales de Construcción*, 2018.

IPCC (Intergovernmental Panel on Climate Change), Mitigation of Climate Change: Working Group III Contribution to the Fifth Assessment Report, and Emisiones de la Industria de Los Minerales, 2014, Retrieved from www.ipcc.ch/index.htm on 2018.

INVIAS 2012, Especificaciones Generales de Construcción de Carreteras, Sección 100, norma 142 y 152, Bogotá D.C., Colombia, 2018.

Rashad, A., A Comprehensive Overview about The Influence of Different Admixtures and Additives on The Properties of Alkali-Activated Fly Ash, *Materials and Design,* 53, 1005–1025, 2014.

Provis, J., Advances in Understanding Alkali-Activated Materials, *Cement and Concrete Research,* 78 (part A), 110-125, 2015.

Zelaya, J., Ochoa, J. C., and Arias, Y. P., The Use of Colombian Palm Oil Fuel Ash in Alkali Activated Cement Compressed Stabilized Earth Blocks, *Sustainability Policy and Practice,* 13(2), 2017.

Zhang, M., Zhao, M., Zhang, G., Nowak, P., Coen, A., and Tao, M., Calcium-Free Geopolymer as a Stabilizer for Sulfate-Rich Soils, Applied Clay Science, 108, 199-207, 2015.

Interdependence between Structural Engineering and Construction Management
Edited by Ozevin, D., Ataei, H., Modares, M., Gurgun, A., Yazdani, S., and Singh, A.

ISBN: 978-0-9960437-6-2

TEAMWORK IN SUCCESSFUL CONSTRUCTION MANAGEMENT OF ROAD IMPROVEMENT PROJECTS

SINIT SENIWONGSE

Ardmore Roderick LLC, Chicago, USA

This paper addresses the methodology and the art of construction management of road infrastructure projects from the viewpoint of practicing engineers using teamwork amongst organizations, which include owners, design engineers, consultants, construction managers, contractors, subcontractors, vendors, local residents, publics, and owners of other utilities. The author points out that most construction management text books and research papers have been written by academia without actual construction or management experience, and that successful construction management of infrastructures is not a simple task. To demonstrate this point, the collapse of the roof of the tunnel in the Big Dig project in Massachusetts and the collapse of the pedestrian bridge during installation in Florida in 2017 are mentioned. Both collapses cost the lives of road users and millions of dollars in litigations. The paper gives the factors contributing to successful teamwork as leadership, communication, collaboration, trust, transparency, common goal and mutual benefits. Using these traits, the author describes the construction management using teamwork concept in the successful completion of many road restoration projects in Chicago. It is concluded that the approach can be used equally well in other infrastructure projects of any sizes in USA or any other countries.

Keywords: Collaboration, Communication, Leadership, Role, Transparency, Trust.

1 INTRODUCTION

This paper addresses the concept of teamwork and how to make it work in major infrastructure design and construction from the viewpoint and experience of professional construction management practitioners. There are a countless number of text books and research papers on teamwork in project management, but those are mainly written by academia and research students without real practical experience. Managing inter-organizations that are teamed up to execute construction projects is not a simple task; hence the results are not always successful. Examples of past infrastructure projects of different degrees of success are given to highlight the factors that have impact on the outcome of the project. The factors are listed and briefly described. The author then demonstrates the application of these factors for the successful construction of road restoration and improvement projects in Chicago for the past few years.

2 PREVIOUS PUBLICATIONS ON TEAMWORK

Teamwork and Project Management by Smith (2002) is one of many fundamental classroom text books giving standard information. Fong and Lung (2007) published their research finding on relationship between trust and human perceptions of inter organizational teamwork and the factors that affect the individual's perception. They collected data via questionnaire surveys from a sample of construction practitioners. They used correlation and multiple regression analyses of the collected data to come up with their hypothesis on teamwork.

3 EFFORTS IN SUCCESSFUL CONSTRUCTION MANAGEMENT

Design and construction of major infrastructure projects are complicated and involve many organizations. Some projects were successfully managed, but many projects had problems. The major parties in construction projects are the Owner, the Design Engineers, the Construction Manager, and the Contractor and subcontractors. Each organization has its own stakeholders, skills and objectives. Each organization works for their self-interest and to maximize their profits. Most major construction projects ended with disputes, claims and litigations that took years to settle after completion of projects.

Successful construction management in this paper refers to the management that results in completion of the construction according to the design and specifications within Contract time and within budgets with good safety record and without post-construction major litigations/claims/disputes. Each participating organization is satisfied with the result and ends the project with good memory and experience. This statement is of course easily said, but difficult to achieve in real life construction. The impact of successes and failures in road construction projects are given below from published references.

3.1 The Cost/Impact of Failure in Construction Projects

Two examples are quoted in this Section from published references to remind the importance of construction management in practice. One example is the Big Dig project in Massachusetts. The Associated Press (2008) published that, "The $14.79 billion Big Dig, which had an initial price tag of $2.6 billion, was plagued by problems and cost overruns throughout the two decades it took to design and build". It reported that the consortium that oversaw the design and construction of the nation's costliest and most complex highway project had to pay nearly $450 million to settle tunnel collapse lawsuit.

Another more recent example was the collapse of the pedestrian bridge at Florida International University (FIU) in March 2018 (ENR 2018). The case is still under investigation, but it highlights the importance of good and effective construction management, which would have prevented the collapse.

4 TEAMWORK APPROACH IN CONSTRUCTION MANAGEMENT

In an effort to minimize or eliminate confrontation, conflicts, disputes, claims and litigations, a new construction management approach has been adopted for some time, called Teamwork or Partnering approach. Abudayyeh (1994) described the Partnering approach by California Department of Transportation (Caltrans) implemented to reduce or eliminate claims and litigation.

4.1 Defining Teamwork

Teamwork can be defined as "the process of working collaboratively with a group in order to achieve a goal". It is the gathering together a group of individuals or organizations in hopes that their collective talents, skills, and experiences will provide a more efficient and successful outcome. Successful teamwork benefits the group as a whole.

4.2 Teamwork in Engineering Design and Construction Projects

Teamwork in construction projects is more complex than other non-engineering projects. In construction or engineering projects, teamwork is a must within an organization and between organizations for the projects to be successful. This is because there are many organizations, each with different roles and responsibilities from project conception to the end of the project. Each organization by its nature and type of work consists of individuals of different background, skills, qualifications and professional experiences.

4.3 Basic Factors for Successful Team in Construction Management

Seniwongse (1992) described the approach that he used in the successful construction management of international infrastructure projects in the 1990s. It consists of the following considerations:

- **Selection of key personnel:** Selection of the right persons to the key positions within an organization is most important for the success of any organization. Key positions are the leadership roles in each discipline, such as the project manager, lead design engineers, resident engineer etc.
- **Selection of organizations:** Selection of the right organizations for the inter-organizational projects is crucial for the successful result. In construction, there are many organizations involved: Owners, Designers (Engineers/Architects), Construction Managers, Contractors, Subcontractors, and Vendors. For the projects to be successful, all organizations must fulfill their roles of responsibility, and in sequence as required by the project schedule. The selection is by the prequalification process.
- **Financial Remuneration, Contract Sum and Contract Period:** Private organizations cannot survive if the expenses are more than the earnings. The main objective for any organization is still maximizing the profit for Stakeholders. The financial rewards must be reasonable. In fact, it is one of the main motivations. The time (Contract period) for each organization to complete the assigned tasks must be reasonable in order for them to perform quality work.
- **Transparency:** In order for all organizations involved in the project to work cooperatively as a team, they must trust that the others are doing the same for the common goal. Transparency is crucial for developing trust among the inter-organizations.
- **Win-Win Approach:** No organization will be cooperative, if they lose but the others win. The core of the teamwork is for all partners in the team to achieve their intended benefits upon the successful execution of the construction project.
- **Leadership:** For all the above to happen, someone has to take the lead to organize the teamwork. In construction, it is the role of the Construction Managers (CM). The essential attributes of teamwork are communication, commitment, providing support, and sharing ideas and responsibility. For teamwork to be achieved, there has to be great leadership in command. Teamwork cannot work without someone guiding the groups, and ensuring that every group knows their role, and is on the same page. A leader's role is to also coordinate the activity amongst the groups, as well as encourage communication and collaboration.

5 EXAMPLES OF TEAMWORK IN SUCCESSFUL PROJECTS

This section demonstrates the art of construction management to execute and complete road restoration projects in Chicago, using teamwork approach described in Section 4. The methodology can be applied to any other construction projects of any sizes.

5.1 Road Restoration and Improvement in Chicago

The projects are to restore the roads and walkways that are excavated/damaged by the new water and sewer main installation. The projects are under the jurisdiction of the Water Management Department of the City of Chicago. Existing water and sewer mains in Chicago are more than hundred years old. The City has started since 2009 to remove old mains and replaced with new ones with improved alignment and grade design. New catch basins are installed. In the process, the roads and sidewalks are excavated. Once the new pipes are installed and passed the tests, the road restoration phase begins.

Figure 1 shows the typical cross section of the road restoration design of one of the projects. It is noted that many of these projects are taking places simultaneously in all parts of Chicago: north, south, west and east.

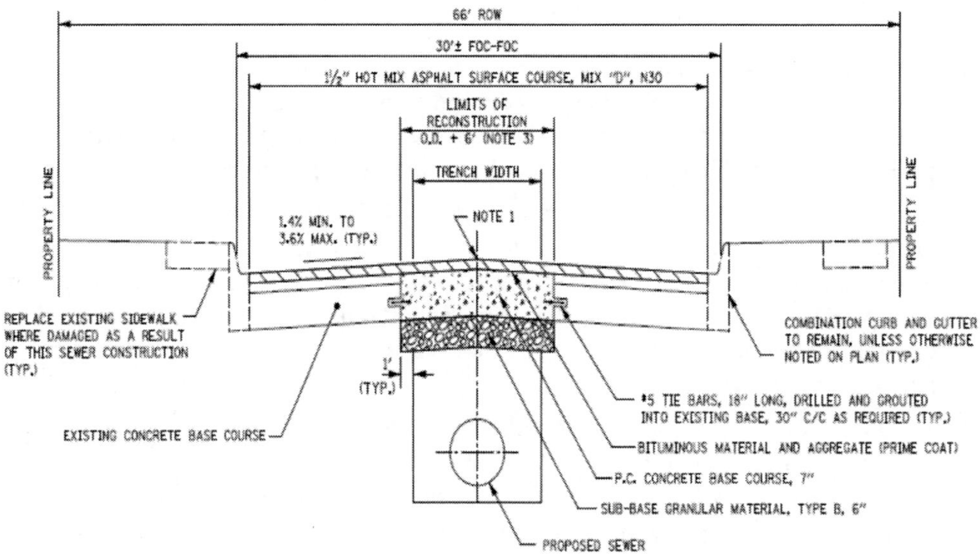

Figure 1. Example of partial pavement restoration cross section (City of Chicago 2014).

5.2 The Team

Many organizations take part in the project. The Project Owner in this case is the City of Chicago Department of Water Management. It acts on behalf of the city for the public. The Engineers are the designers for the Water and Sewer (W/S) Project, and for the Road Restoration (RR) Project. The Consultant is the construction management team (CM). In this case, the author represents the CM for the RR Project. The CM is the center of all operation of the RR Project.

One other group in this Team of RR project that cannot be overlooked is the local residents. In the big city like Chicago, the neighborhood is diversified from north to south, east to west. The importance of achieving cooperation from local residents cannot be overemphasized.

5.3 The Methodology of RR Construction Management

The CM (Consultant) has to coordinate with all parties. The process begins once the city notifies the consultant that the installation of water main is near completion, usually around 2 weeks prior. The project manager (CM) then reaches out to the city engineers to get an update of when the crews and equipment will be demobilized from the site.

Just prior to the demobilization, it is the responsibility of the restoration project manager and consultants to schedule a walkthrough with the city engineer to see what has been done and what remains to be done by the RR Contract. After the completion of the walkthrough, the restoration project manager is responsible to notify the contractor of when work can begin. It is vital that this information is communicated to the contractor in a timely manner. The contractor and the city do not communicate with one another. The contractor has 30 days after the city demobilizes from site to complete restoration. Beyond the 30 days, liquidated damages will be assessed.

During restoration, it is the project manager responsibility to communicate with the General Contractor and utilities companies/ departments to resolve any conflicting site conditions/utilities. This has to be done effectively for the construction to be on schedule. Meetings on construction progress/issues are held every two weeks. All projects have been completed with good success.

6 CONCLUSIONS

This paper has demonstrated the art of construction management and the power of teamwork to bring about the successful road restoration and improvement projects in the City of Chicago. For citywide infrastructure projects of this size, there are many organizations, many steps, and millions of dollars involved. The paper shows how to manage inter-organizations as one team to achieve a common goal to complete the assigned construction project per specifications within schedule and within budgets. The successful management of road restoration projects in Chicago confirms essential attributes of teamwork are leadership, communication, commitment, providing support, and sharing ideas and responsibility. The approach can be adopted for construction management of worldwide projects of any sizes.

References

Abudayyeh, O., Partnering: A Team Building Approach to Quality Construction Management, *Journal of Management in Engineering*, 10(6), 26-29, November 1994.

Associated Press, *Big Dig Contractors Settle Tunnel Collapse Lawsuit for Nearly $450 Million*, January 23, 2008.

City of Chicago, Department of Water Management, *Project No. 6942 Sewer Improvement Program*, Typical Section, Sheet 8 of 11, May 2014.

ENR, Theories Outrun Reliable Facts in Florida Bridge Collapse, *Engineering News-Record*, March 22, 2018.

Fong, P. S., and Lung, B. W., Interorganizational Teamwork in the Construction Industry, *Journal of Construction Engineering and Management*, 133(2), 157-168, February 2007.

Seniwongse, M., *Construction Supervision and Management of International Projects in Thailand,* Third International Conference on Modern Techniques in Construction, Engineering and Project Management, Singapore, 155-160, March 24-26, 1992.

Smith, K. A., *Teamwork and Project Management*, McGraw-Hill, December, 2002.

Interdependence between Structural Engineering and Construction Management
Edited by Ozevin, D., Ataei, H., Modares, M., Gurgun, A., Yazdani, S., and Singh, A.

ISBN: 978-0-9960437-6-2

COMPARATIVE INVESTIGATION OF TWO DESIGN METHODS FOR SLAB-ON-GRADE FOUNDATIONS SUBJECTED TO CONCENTRATED LOADS

MOHAMAD NAZIEH JLILATI, VICKI LEE BROWN, and XIAOCHAO TANG

Dept of Civil Engineering, Widener University, Chester, USA

The concrete slab-on-grade is a type of shallow foundation commonly used in warehouse and other industrial facilities. The slab is constructed in formwork set into the ground with concrete poured directly onto compacted subgrade. Slab-on-grade foundations designed by the traditional linear elastic method may have overly conservative slab thicknesses. As the slab-on-grade foundation can account for up to 15% of construction costs for the projects in which they are employed, reduction in the slab thickness can result in significant cost savings. An alternate design method (Shentu *et al.* 1997) has been proposed but has not been widely accepted due to lack of confidence in its correlation with test results. The objectives of this study were to experimentally define a reasonable factor of safety for the Shentu method and compare with the traditional design method. Test slabs of varying thicknesses, 5 ft by 5 ft in area, were built on top of compacted local soils and aggregates inside a testing box and tested to failure under static loads to simulate storage rack post loads; companion slabs were constructed on top of compacted soils and aggregates and field-tested. Concrete mixes with and without fibers were used in the study. Factors of safety were determined. Test results indicate that the Shentu method is too liberal in its computations while the traditional design method is overly conservative.

Keywords: Experimental, Modulus of subgrade reaction, Factor of safety, Economic factor.

1 INTRODUCTION

A concrete slab-on-grade (SOG) is a common type of shallow foundation that is typically used in industrial buildings. The most popular type of SOG is unreinforced concrete poured directly onto compacted base. The design of concrete SOGs are typically governed by point loads from large storage racking systems. The point loads are transmitted to the foundation through various sized baseplates. The magnitude of the loads can fluctuate depending on the size of the storage rack and what is being stored, but typical loads can be upwards of 20 kips (90kN). The failure of a concrete SOG typically stems from onset cracking caused by moment. SOGs can also fail because of punching from the rack base plate. However, moment failure is more common than the punching failure mode. These two failure modes govern the design criteria. While it is not uncommon for micro cracks to form, because of shrinkage, concrete SOGs are designed to remain uncracked. Design charts from Ringo and Anderson (1996) are currently commonly used to design SOGs, which often results in a slab thickness ranging from 6 in to 10 in (15 cm to 25 cm);

however, Shentu *et al.* 1997 has revealed that the typical industrial slab thickness of 6 in to 10 in may be overly conservative. Shentu used finite element analysis to develop an analytical method for computing the ultimate load-carrying capacity of concrete SOGs. The Shentu design method, which is less conservative than the traditional design method, has not been widely accepted due to lack of confidence in the correlation between the Shentu equations and test results.

The main reason for the conservatism in the traditional design method, resulting in design of thicker slabs, is because traditional design does not take advantage of the elastoplastic properties of concrete. The linear elastic theory works well when loads are small, but as loads increase the accuracy decreases. The Shentu method takes advantage of increased load carrying capacity by accounting for the elastoplastic properties of concrete. The traditional method uses onset cracking as the failure criteria; the Shentu method relies on concrete's ability to carry additional load after onset cracking. If the Shentu method is correct, and concrete can in fact handle additional load after onset cracking, the required slab thickness could be reduced.

As in all shallow foundations, the subgrade is a very important parameter when considering the design of concrete SOGs. In typical reinforced concrete construction, the required tensile strength for flexure is obtained using steel reinforcement. However, most SOGs do not use any additional steel reinforcing. The flexural strength of a concrete SOG relies on ground contact and interaction. The modulus of subgrade reaction (k) is one of the most important parameters when designing SOGs. The modulus of subgrade reaction (k), represents the relationship between the applied load and the deformation of the subgrade. Both the traditional and Shentu methods rely on this subgrade parameter.

The objectives of this study were to experimentally define a reasonable factor of safety for the Shentu Method and another factor of safety for the traditional method based on linear elasticity. Test slabs of varying thicknesses (3-inch or 4-inch thick), 5 ft by 5 ft (1.5 m by 1.5 m) in area, were built on top of compacted local soils and aggregates inside a testing box at Bartholomew Civil Engineering Laboratory at Widener University and tested to failure under static loads to simulate storage rack post loads; companion slabs were constructed on top of compacted soils and aggregates and field-tested. Concrete mixes with and without fibers were used in the study.

2 METHODS AND PROCEDURES

2.1 Analytical Procedures

Equations for the traditional and Shentu methods were adjusted to represent ultimate load capacity; using slab thickness, modulus of subgrade reaction, and 28-day concrete compressive strength as universal parameters. Eq. (1) is the traditional design method equation:

$$P_u = \frac{d^2}{A \log\left(\frac{Bd^3}{C}\right)} \tag{1}$$

Where:
P_u = Ultimate load capacity (lbs)
d = Thickness of Slab (in)
$A = 0.03/\sqrt{f'_c}$
$B = 915{,}000\ \sqrt{f'_c}$
f'_c = 28-Day Concrete Compressive Strength (psi)
$C = kd^4$
k = Modulus of subgrade reaction (pci)

Eq. (2) is the Shentu Method equation:

$$P_u = 1.72\left[\left(\frac{kR_1}{E_c}\right)10^4 + 3.60\right] \tag{2}$$

Where:
R_1 = ½ the width or diameter of the column base plate (in), whichever is smaller
E_c = Modulus of Elasticity of Concrete (psi) = $57{,}000(f'_c)^{0.5}$
f'_t = Tensile Strength in Flexure of Concrete (psi) = $6.5(f'_c)^{0.5}$

Eq. (3) gives the design load capacity for both the traditional and Shentu methods.

$$P_a = \frac{P_u}{F.S.} \tag{3}$$

Where:
P_a = Design or Allowable Load Capacity (lbs)
$F.S.$ = Factor of Safety
The recommended factor of safety for the traditional design method based on linear elasticity is between 1.5 and 2. The Shentu method recommends a factor of safety of 3.

2.2 Experimental Procedures

2.2.1 *Concrete mix designs*

Three concrete mix designs were used for this study: plain concrete, shrinkage fiber-reinforced concrete (MasterFiber Mac 100), and structural fiber-reinforced concrete (SikaFiber Force 650). The plain concrete mix design was used both with and without geogrid reinforcement (Polypropylene triaxial geogrid with 40 mm × 40 mm × 40 mm triangular apertures). The fiber types and geogrid reinforcement were chosen with economic impact in mind. The fiber reinforcement types were relatively cost effective and required little additional labor.

A total of 13 concrete slabs were tested: six plain concrete, three structural fiber concrete, two plain concrete with geogrid reinforcement, and two shrinkage fiber concrete slabs. The slabs were tested either as (1) pre-cast and tested in the lab test box, (2) cast in place and tested in the lab test box, or (3) cast in place and tested in the field. A base mix design, with a target compressive strength of 4500 psi, was provided by Conewago Enterprises, a design-build general contractor with considerable experience in SOG. The plain concrete mix design is shown in Table 1.

Table 1. Plain concrete mix design.

Material	Volume Per Cubic Yard (ft^3)	Weight Per Cubic Yard (lb)
Water	5.45	340
Cement	3.46	680
Coarse Aggregate #67	10.7	1782
Sand	7.03	1150
Air	0.41	1.50%
Total	**27.0**	**3952**

A similar mix design was used for the fiber/geogrid reinforced slabs by adding (4 lbs/yd^3) of the various fibers. Concrete strength was verified by testing 4-inch by 8-inch cylinders in accordance with ASTM C39M (ASTM 2008).

2.2.2 *Subgrade Preparation*

Prior to testing slabs, the subgrade must be prepared. A 36" soil layer was installed in the 6.5-foot by 6.5-foot test box in 6" lifts, with water added to ensure the soil was at its optimum level of (16.8%) to achieve the maximum unit weight of (112 lbs/ft^3). A plate compactor was used to compact the soil to create the subgrade modulus value desired (150~190 pci), tested in accordance with ASTM D1196M-12 (ASTM 2016). Eight inches of #67 coarse aggregate was then added and compacted as shown in Figure 1a.

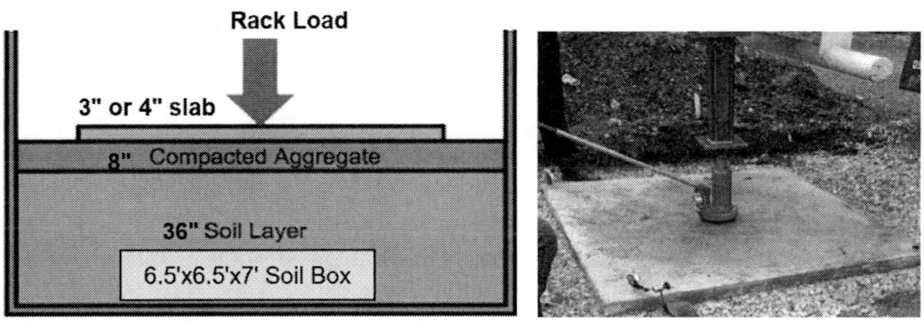

(a) (b)

Figure 1. Testing facilities: (a) soil box layers; (b) in field test arrangements.

The outside procedure for testing was slightly different, as seen in Figure 1b. Top soil was removed, with coarse aggregate filled in the excavation to a height of 12-inch above the compacted subgrade and then compacted using a plate compactor. For both soil box and field tested slabs, standard Proctor Tests from ASTM 698 (ASTM 2014) and Static Plate Load Tests from ASTM D1196M-12 (ASTM 2016) were performed.

2.2.3 *Testing the Slabs*

A bottle jack was used throughout the slab testing. Before the jack could be used, a calibration curve was developed to convert the pressure on the bottle jack gauge into load in pounds. The calibration curve generates an equation, which may be used to determine the load in pounds that has been applied. The calibration curve shown in Figure 2 was developed using a universal testing machine to apply load.

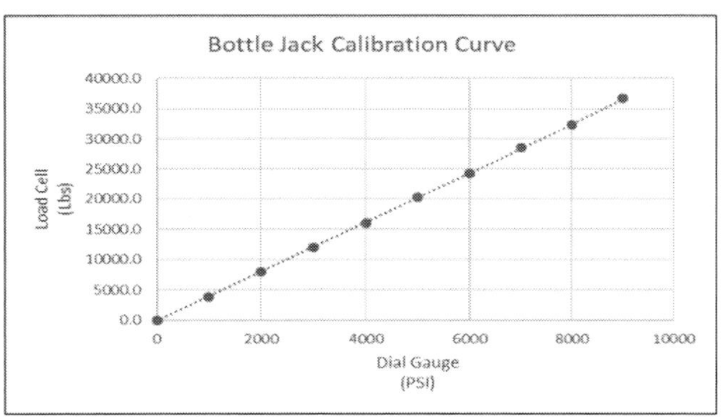

Figure 2. The bottle jack calibration curve.

The bottle jack was positioned above the centroid of the concrete slab, with the contact area a 3-inch by 3-inch steel plate. The slab dimensions were controlled by the soil box dimensions; therefore the slab thickness and the steel plate dimensions were calculated to remove the boundary condition influence on the slab strength, based on the radius of the relative stiffness. The adopted criteria to specify failure of the experimental models was a 15% force drop, where the ultimate load is specified as the highest load reached before failure, as shown in Figure 3.

3 RESULTS AND DISCUSSION

The slab loading test results are summarized in Table 2. The safety factor was examined for the traditional method as well as for the Shentu method. The limiting factor in calculating a valid safety factor is that said factored load must be less than the proportional limit of the slab. Figure 3 shows the load vs. deflection curve for a plain, cast-in-place slab. The elastic zone remains approximately linear until the slab reaches the proportional limit of 25 kips. To design safe and economical slabs, theoretical computations should remain below the proportional limit. It was assumed that design loads should stay below one-half of the ultimate load. This will ensure that design load-carrying capacities will be below one-half of the proportional limit.

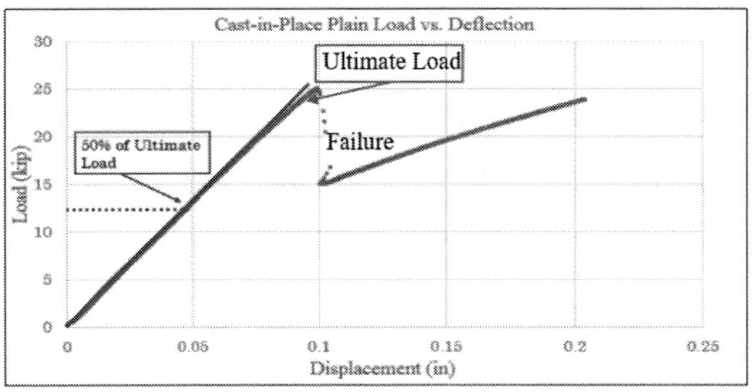

Figure 3. Cast-in-place plain load vs. deflection graph.

The implicit safety factor is calculated by dividing the theoretical traditional or Shentu load by the experimental ultimate load. From this calculation, the traditional method's implicit safety factor ranges from 2.61 to 6.67, so there is no need to further reduce the allowable load. For the Shentu Method, implicit safety factors ranged from 0.36 to 0.91, so the in-use factor of safety is not enough to ensure a safe design. Therefore, adjustment factors should be made for both methods. The proportional limit of the slabs ranged between 4.5 kips and 17.5 kips. Prior to a reduction factor being applied, only the traditional method is under the proportional limit for all slabs tested.

As shown in Table 2, the traditional method multiplied by an adjustment factor of 0.9 still allows for design loads under the proportional limit. The Shentu method with an adjusted factor of safety of 6 brought the design load under the proportional limit. As it might be confusing to have a factor of safety less than one, the authors recommend referring to the 0.9 adjustment to the traditional method as an economic factor which has the effect of increasing the traditional method's design load. However, to reduce the Shentu method allowable design load a factor of safety of 6 was proposed. The proposed adjustment factors of 0.9 for the traditional method and 6 for the Shentu method ensure a safe but not overly conservative design.

Table 2. Modified factor of safety results.

<table>
<tr><th>Slab Type & Thickness</th><th>K (pci)</th><th>Linear-Elastic Method Load (kpis)</th><th>Linear-Elastic Method Load / 0.9 (kpis)</th><th>Shentu Method Load (kips)</th><th>Shentu Method Load / 6 (kips)</th><th>Allowable load (kips) = Ultimate/2</th></tr>
<tr><td>Pre-cast shrinkage fiber 3"</td><td>178</td><td>3.40</td><td>3.78</td><td>24.89</td><td>4.15</td><td>4.50</td></tr>
<tr><td>Pre-cast shrinkage fiber 4"</td><td>178</td><td>5.20</td><td>5.78</td><td>44.30</td><td>7.38</td><td>7.50</td></tr>
<tr><td>Pre-cast plain 3"</td><td>178</td><td>3.40</td><td>3.78</td><td>24.89</td><td>4.15</td><td>5.50</td></tr>
<tr><td>Pre-cast plain 4"</td><td>178</td><td>5.20</td><td>5.78</td><td>44.30</td><td>7.38</td><td>7.50</td></tr>
<tr><td>Pre-cast geogrid 3"</td><td>178</td><td>3.40</td><td>3.78</td><td>24.89</td><td>4.15</td><td>10.50</td></tr>
<tr><td>Pre-cast geogrid 4"</td><td>178</td><td>5.20</td><td>5.78</td><td>44.30</td><td>7.38</td><td>12.00</td></tr>
<tr><td>Pre-cast plain 3"</td><td rowspan="4">164</td><td rowspan="4">3.90</td><td rowspan="4">4.33</td><td rowspan="4">28.60</td><td rowspan="4">4.77</td><td>10.50</td></tr>
<tr><td>pre-cast structural fiber</td><td>12.45</td></tr>
<tr><td>Cast in place plain 3"</td><td>13.00</td></tr>
<tr><td>Cast in place structural fiber 3"</td><td>12.95</td></tr>
<tr><td>Cast in place plain 3" (Outside)</td><td rowspan="3">174</td><td rowspan="3">3.90</td><td rowspan="3">4.33</td><td rowspan="3">28.90</td><td rowspan="3">4.82</td><td>8.43</td></tr>
<tr><td>Cast in place plain 3" (Outside)</td><td>7.90</td></tr>
<tr><td>Cast in place structural fiber 3" (Outside)</td><td>8.05</td></tr>
</table>

4 CONCLUSION

The objectives of this study were to experimentally define a reasonable factor of safety for the Shentu method and more economical adjustment to the linear elastic method. From the test results, it was found that the Shentu method's recommended factor of safety of 3 is very unconservative; rather, a safety factor of 6 should be used with the Shentu Method. On the other hand, the linear elastic method load should be factored by (1.11) as an economic factor that results in a safe and economically viable design. This is because the traditional method proved highly conservative with respect to safety, whereas the Shentu method was very liberal.

Acknowledgments

The authors wish to thank Aaron Schneck, Bradley Smith, Thomas Jones IV, Timothy Sabatino, and William Orangers for conducting experimental work. Financial support from Conewago Enterprises is gratefully acknowledged.

References

ASTM, ASTM C39M: *Standard Test Method for Compressive Strength of Cylindrical Concrete Specimens*, ASTM International, West Conshohocken, PA, 2008.

ASTM, ASTM D1196 / D1196M-12: *Standard Test Method for Nonrepetitive Static Plate Load Tests of Soils and Flexible Pavement Components, for Use in Evaluation and Design of Airport and Highway Pavements*, ASTM International, West Conshohocken, PA, 2016.

ASTM, ASTM D698: *Standard Test Methods for Laboratory Compaction Characteristics of Soil Using Standard Effort (12,400 ft-lbf/ft^3 (600 kN-m/m^3))*, ASTM International, West Conshohocken, PA, 2014.

Ringo, B. C., and Anderson R. B., *Designing Floor Slabs on Grade: Step-by-Step Procedures, Sample Solutions, and Commentary*, Aberdeen Group, Addison, IL, 1996.

Shentu, L., Jiang, D., and Hsu, C., Load-Carrying Capacity for Concrete Slabs on Grade, *Journal of Structural Engineering,* 123(1), 95-103, 1997.

Interdependence between Structural Engineering and Construction Management
Edited by Ozevin, D., Ataei, H., Modares, M., Gurgun, A., Yazdani, S., and Singh, A.

ISBN: 978-0-9960437-6-2

EFFECTS OF AN ENZYME ON THE PERFORMANCE OF SUBGRADES

DANIELA IONESCU and JOE PETROLITO

Dept of Engineering, La Trobe University, Bendigo, Australia

An efficient transportation system requires a good network of roads provided with pavements that can withstand both the heavy and frequent traffic demands and environment variations while requiring minimal maintenance. However, roads are sometimes built where the sub-grade soil does not comply with the specifications and needs to be replaced with an acceptable soil that satisfies these requirements. To minimize the construction cost and the impact on the environment that the road construction would have, stabilizing agents (lime, cement or fly ash), are commonly used to improve the characteristics of in-situ soils. However, this process adds to the cost of road construction and maintenance. Therefore, alternative, environmental-friendly solutions have been investigated to reduce these costs. In recent years, enzymes have been used successfully to stabilize mostly fine-grained soils. This paper reports on an on-going research program on the use of various stabilizing agents to improve the properties of in-situ sub-grades for rural roads. Two locally-sourced soils were used in this investigation. The effects of cement, cement-fly ash mix and Perma-zyme binders on the properties of the selected soils were investigated. A series of tests was performed to establish the strength, stress-strain and deformation characteristics of the supplied soils. The results were compared to establish if Perma-zyme is suitable as stabilizing agent for the supplied soils.

Keywords: Cementing binder, Flexible pavement, Laboratory investigation, Nontraditional additive, Road foundation durability, Stabilizing agent.

1 INTRODUCTION

The economy of modern societies relies heavily on transportation. An efficient transportation system requires roads with pavements that can withstand both the heavy and frequent traffic demands and environment variations while requiring minimal maintenance. In Australia, unsealed roads are constructed to reach remote areas, due to their relative lower construction costs. These roads represent about two-thirds of the country's road network (ARRB 2009a). In Victoria, more than half (> 57%) of the total road length constructed is represented by unsealed roads (ARRB 2009b). However, only approximately 53% of these roads are formed and surfaced with a 50-mm gravel layer placed on the top of the sub-grade. To lower construction costs, local councils may use the existing subgrade soil, which is often not suitable for road construction, and marginal or non-standard materials supplied from local natural deposits (Austroads 2018). Also, the absence of a seal exposes the subgrade to the environmental conditions, leading to increased maintenance costs or premature failures. Hence, the maintenance of Australia's road network constitutes a significant portion of its budgetary spending (Stevenson 2014).

Various mechanical and chemical stabilizing methods are commonly used to improve road performance and minimize maintenance costs. While some of these methods, such as mechanical compaction and granular modification, may be less costly, they are not always efficient (Ionescu 2006). In recent times, enzymes have been used for subgrade stabilization (Agarwal and Kaur 2014, Renjith *et al.* 2017). However, the evidence to support their effectiveness is mainly based on manufacturers' claims. Since enzymes catalyze very specific chemical reactions, it is difficult to discern a general stabilization mechanism for them (Tingle *et al.* 2007). Furthermore, Milburn and Parsons (2004) reported that silty soils treated with enzymes provided no improvement in the performance of soils tested, whereas Tingle *et al.* (2007) reported significant and accelerated improvement in strength. Hence, there is a need for independent assessment of the performance of enzyme-stabilized soils. Consequently, this paper compares the effects of two traditional stabilizers and an enzyme stabilizer on two locally-sourced subgrade soils.

2 PAVEMENT REQUIREMENTS

Flexible pavements must be strong enough to support the traffic load and reduce the stress on the subgrade. Often, the in-situ subgrade does not meet the desired strength (CBR) and stiffness requirements (ARRB 2009a). Hence, chemical stabilization of the in-situ soils is carried out to improve their performance and prolong the pavement life. The type of chemical additive used depends on the percentage of fines in the surface material of the subgrade, climatic conditions, traffic volumes and construction logistics (ARRB 2012).

3 EXPERIMENTAL WORK

The physical characteristics of the materials used in this study were determined in accordance with the relevant Australian Standards (primarily AS1289 2005). The governing factors were the compliance with current specifications and the feasibility of obtaining an acceptable material.

3.1 Soils Classification

Representative soil samples were supplied from two sites in Victoria, Australia. One soil sample (S1) was sourced from Colac East, in the Colac Otway Shire, south-east Victoria. The second soil sample (S2) was obtained from Denyer's Pit, approximately 15 kilometres west of Kerang, in the Gannawarra Shire, north Victoria. Particle size distribution (PSD) plots and grading characteristics for the two materials are presented in Figure 1 and Table 1, respectively. The consistency characteristics of the soil fractions finer than 0.425 mm are summarized in Table 2. Soil S1 can be described as poorly-graded gravelly sand with non-plastic silt (SP-SM), whereas S2 is classified as a low plasticity silty sand-gravel poorly-graded mix (SM). The supplied soils have a silt content of 10-14%, and closely fit the enzyme treatment envelope, rendering them suitable for enzyme stabilization.

Table 1. Grading characteristics of the supplied materials.

Material	Particle shape	D_{10} (mm)	D_{30} (mm)	D_{60} (mm)	D_{max} (mm)	C_u	C_c
S1	sub-angular	0.07	0.75	2.6	19.0	37.1	3.1
S2	sub-rounded	0.06	0.2	2.7	63.0	45.0	0.2

Table 2. Consistency characteristics of supplied soils.

Materials	Atterberg Limits			Linear
	Liquid Limit (%)	Plastic Limit (%)	Plasticity Index	Shrinkage (%)
S1	40	-	NP	0.4
S2	21	24	3	1

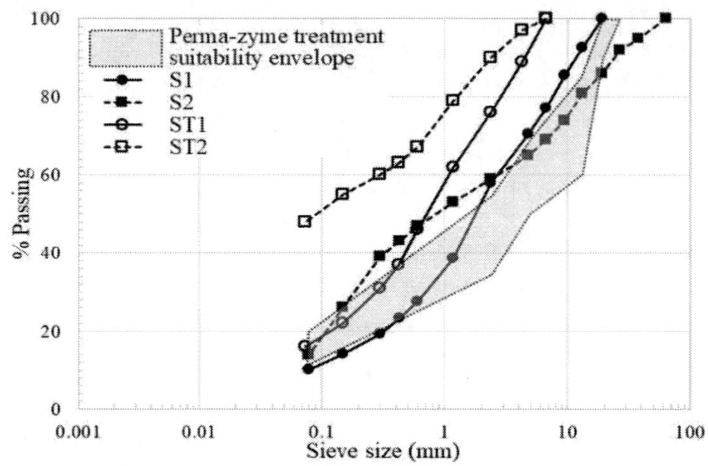

Figure 1. Particle size distributions of soils used in the current study.

3.2 Stabilizing Agents

Cement has been used to stabilize silty soils and provides the most significant improvement of any traditional stabilizer (AustStab 2011). Furthermore, cementitious blends provide satisfactory stabilization of non-plastic silty soils (Austroads 2002). Hence, for the purpose of comparisons, three stabilizing agents were used in this study, namely a cement, a cement-fly ash (50/50) blend and an enzyme (Perma-zyme). The general-purpose cement (GP), the Class F (non-self-cementing) fly ash (FA) and Perma-zyme (PZ) were locally supplied. The enzyme used is a low-energy, sustainable by-product that is non-toxic and non-flammable, and it is produced during the natural fermentation process of cane sugars and other organic compounds. This enzyme is an environmental-friendly and biodegradable natural organic compound that acts as a catalyst. It lowers the surface tension of water, which promotes fast and thorough penetration and dispersal of moisture, resulting in an increased compaction rate, as well as higher density for clays and silts.

3.3 Soil Samples Preparation

The supplied soils were oven dried at 40°C, and then they were subjected to stabilization and left to cure at 25° C for different lengths of time, as per the manufacturer's specification (Vicroads 2012). Table 3 summarizes the soil samples used to determine the dry density-moisture content relationship.

Table 3. Soil samples used during the testing procedure.

Treatment type	Tests performed					
	Compaction test samples		CBR test samples		Triaxial test samples	
Modified PSD	No	No	No	No	Yes	Yes
None	S1	S2	S1	S2	ST1	ST2
3% cement	S1-GP	S2-GP	S1-GP	S2-GP	ST1-GP	ST2-GP
3% cement-fly ash mix	S1-GP-FA	S2-GP-FA	S1-GP-FA	S2-GP-FA	ST1-GP-FA	ST2-GP-FA
1 mL enzyme to 300 mL water	S1-EZ	S2-EZ	S1-EZ	S2-EZ	ST1-EZ	ST2-EZ

A standard triaxial cell was used for the CU tests. The sample size ratio is defined as the diameter of the triaxial specimen (38 mm) divided by the maximum particle dimension. It has been argued that the sample size effects become negligible as the sample size ratio approaches six (Marachi *et al.* 1972, Ionescu 2004). Hence, parallel gradations to the PSD of the supplied soils were used for the triaxial specimens. These gradations, PT1 and PT2, shown in Figure 1, have a have a maximum particle dimension of 6.7 mm and a corresponding sample size ratio of 5.7. Prior to testing, the soil samples were subjected to treatment in the same manner as for the compaction tests. The list of soil samples used during the CU triaxial tests is summarized in Table 3.

4 RESULTS AND DISCUSSION

4.1 Effect of the Additives on Compaction

Standard compaction curves are presented in Figure 2. Soil S2, having a larger number of fractions, attained better compaction at a lower moisture content when compared with the compaction results for soil S1. The untreated soils achieved a maximum dry density (MDD) of 1.73 t/m^3 at an optimum moisture content (OMC) of 12% for S1, whereas S2 achieved a higher MDD (1.86 t/m^3) at a lower OMC (11%). The results of the compaction tests summarized in Table 4 fall in the typical range for non-plastic/low plasticity silty sand-gravel poorly-graded soils. Overall, independent of the soil used, the stabilized soil required more water (9% to 40%) to reach the MDD, although the increase in MMD was insignificant. Cement stabilization produced the best result for soil S1 (MDD = 1.76 t/m^3, OMC = 17%), closely followed by the cement-fly ash mix stabilization. There was no improvement in the MMD for S1 treated with the enzyme. However, a 33% higher moisture content was required to achieve the MDD. Enzyme stabilization resulted in slightly larger compaction for soil S2 at the same OMC, whereas the cement and cement-fly ash blend stabilization produced similar results as for the compaction of the untreated soil. These results contrast to some extent with the expected results, e.g., a lower OMC for enzyme treated soils. Also, the overall improvement of the compaction of the supplied soils due to the additives is insignificant, and more water was required for compaction. These findings agree with Milburn and Parsons (2004), who reached inconclusive results for enzyme-stabilized soils.

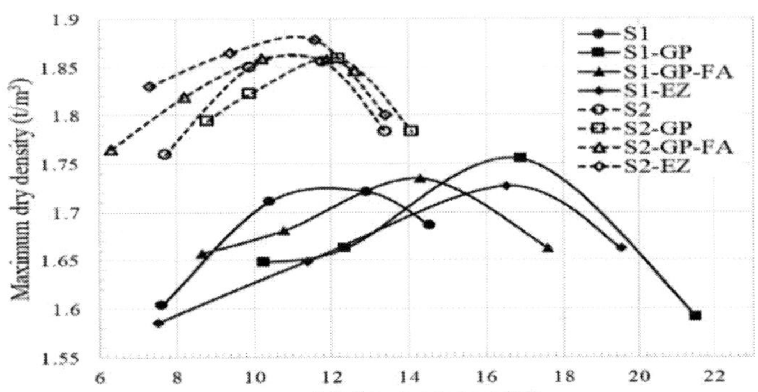

Figure 2. Maximum dry density-moisture content relationship for untreated and treated soils.

4.2 Effect of the Additives on Shear Strength

The shear strength characteristics of the naturally occurring and treated soils were estimated from a series of consolidated undrained (CU) triaxial tests. Constant confining pressures of 100, 200 and 400 kPa were used at a strain rate of 0.02%/s. The triaxial test specimens were prepared at the OMC estimated from the compaction tests. During the preparation of the specimens, a split cylinder was used to support the silicone membrane. The material was placed inside the cylinder in five lifts, with each lift being subjected to a predetermined number of blows from a tamping hammer to achieve the MDD. The relative compaction attained using this method was 95-98% of the MDD. A summary of the physical characteristics of the triaxial test specimens is presented in Table 5. The porosity of the triaxial test specimens was around 33-34% (e_o = 0.5-0.52), and the specimens classified as medium dense (35% < ID < 65%), based on the relative density estimated for the compaction level assumed.

Table 4. Compaction results for the naturally occurring and treated soils.

Parameter	S1	S1-GP	S1-GP-FA	S1-EZ	S2	S2-GP	S2-GP-FA	S2-EZ
MDD (t/m^3)	1.73	1.76	1.74	1.73	1.86	1.86	1.86	1.88
OMC (%)	12	17	14	16	11	12	11	11

The critical state friction angle (ϕ'_{cs}) was back-calculated using Bolton's (1986) dilatancy index and its values are summarized in Table 5. Soil ST1 stabilized with cement-fly ash blend (ST1-GP-FA) displayed the highest values for the critical state friction angle, with about 10% improvement of the shear strength of untreated soil ST1. Stabilization with cement and enzyme of soil ST1 resulted in 5-15% lower values for ϕ'_{cs}. Soil ST2 stabilized with cement (ST1-GP) displayed the highest values for the critical state friction angle, with about 11% improvement in the shear strength of untreated soil ST2, followed by soil stabilized with cement-fly ash blend (ST2-GP-FA) with only a 4% increase in ϕ'_{cs}. Stabilization with the enzyme of soil ST2 produced a 3% decrease in the ϕ'_{cs} value. This correlates well with the degree of packing for the two soils and the Milburn and Parsons (2004) findings from triaxial tests on enzyme-stabilized soils.

Table 5. Triaxial test specimen characteristics and results for the naturally occurring and treated soils.

Parameter	ρ_d (t/m^3)	ρ_{bd} (t/m^3)	S (%)	e_o	I_D (%)	ϕ'_{cs} (°)
ST1	1.73	2.6	82.0	0.51	57.8	34.8
ST1-GP	1.74	2.6	83.7	0.50	64.7	33.2
ST1-GP-FA	1.73	2.6	83.3	0.50	63.4	38.2
ST1-EZ	1.71	2.6	90.0	0.52	50.4	29.8
ST2	1.76	2.65	84.1	0.50	44.1	27.2
ST2-GP	1.75	2.65	83.0	0.51	39.3	30.2
ST2-GP-FA	1.76	2.65	84.2	0.50	44.4	28.5
ST2-EZ	1.77	2.65	84.8	0.50	46.7	26.3

Note: ρ_{bd} = density of oven dried particle

5 CONCLUSIONS

The effects of different stabilizing agents on the performance of two sub-grades were presented in this paper. The effect of stabilizing agents on the compaction performance was minimal, although it seems that the stabilized soils are less sensitive to changes in water content. There was not a definite trend regarding the shear strength of the stabilized soils. It appears that the cementing additives increase the shear strength of the tested soils, although the effect is not the same for each soil. Despite the lower cost for the enzyme, it appears that the enzyme did not improve the performance of the studied soils. These results agree with some past research, although they contradict other results. It is believed that the curing time specified by the manufacturer is far too short, and longer curing time may be required to achieve better performance. Moreover, the change in the PSDs of the soil used in the triaxial tests may not be representative of the gradation of naturally-occurring soils.

Further investigation is required to evaluate the effects of the selected stabilizing agents for longer curing times. In addition, large-scale triaxial cells may need to be used during the testing program to better evaluate the shear strength of the supplied soils.

Acknowledgments

The authors acknowledge with thanks the use of La Trobe University's laboratories for the testing program. Also, they gratefully acknowledge the help provided by Gannawarra Shire Council and Country Vet Wholesaling with the supply of soils and additives. The assistance with preparation and testing provided by Sandra Gadd, Toby Hastie, Mackenzie Stone, Harry Oakley-Warren, David Osborne, Mitchell Wilson, and Kellie White is also recognized.

References

Agarwal, P., and Kaur, S., Effect of Bio-Enzyme Stabilization on Unconfined Compressive Strength of Expansive Soil, *International Journal of Research in Engineering and Technology*, 3(5), 30-33, May, 2014.

ARRB, *Guide to Pavement Technology Part 6: Unsealed Pavements*, Australian Road Research Board, Austroads Ltd, Sydney, 2009a.

ARRB, *Unsealed Roads Manual: Guidelines to Good Practice*, Australian Road Research Board, ARRB Group, Sydney, 2009b.

ARRB, *Guide to Pavement Technology Part 2: Pavement Structural Design*, Australian Road Research Board, Austroads Ltd, Sydney, 2012.

AS1289, *Methods of Testing Soils for Engineering Purposes*, Standards Australia, Sydney, 2005.

AustStab, *Pavement Recycling and Stabilisation Guide*, Australian Stabilisation Industry Association, North Sydney, 2011.

Austroads, *Mix Design for Stabilized Pavement Materials*, AP-T16-028, Austroads Ltd, Sydney, 2002.

Austroads, *Appropriate Use of Marginal and Non-Standard Materials in Road Construction and Maintenance*, AP-T333-18, Austroads Ltd, Sydney, 2018.

Bolton, M., Strength and Dilatancy of Sand, *Geotechnique*, 16(2), 91-128, 1986.

Ionescu, D., *Evaluation of the Engineering Behaviour of Railway Ballast*, Wollongong, 2004.

Ionescu, D., Mechanical Properties of Quarry By-Products, *Progress in Mechanics of Structures and Materials*, Moss, P. J. and Dhakal, R. P. (eds.), 519–526, CRC Press, 2006.

Marachi, N. D., Chan, C. K., and Seed, H. B., Evaluation for Properties of Rockfill Materials, *Journal of Soil Mechanics and Foundation Division*, ASCE, 98(SM 1), 95-114, January, 1972.

Milburn, J. P., and Parsons, R., Report No. K-TRAN: KU-01-8, Kansas Department of Transportation and Kansas University, May 2004.

Renjith, R., Robert, D., Fuller, A., Setunge, S., O'Donnell, B., and Nucifora, R., Enzyme Based Soil Stabilization for Unpaved Road Construction, *MATEC Web of Conferences*, EDP Sciences, 138, 01002, 2017.

Stevenson, A., *Starting Point for the Sustainable Road*, Engineers Australia Civil Edition, 86(1), 46-48, January, 2014.

Tingle, J. S., Newman, J. K., Larson, S. L., Weiss, C. A., and Rushing, L. F., Stabilization Mechanisms of Nontraditional Additives, *Transportation Research Record: Journal of the Transportation Research Board*, Transportation Research Board of the National Academies, 1989-2(1), 59–67, January 2007.

Vicroads, *Preparation of Cement Stabilised Materials to Establish the Dry Density - Moisture Content Relationship*, RC 301.06, Vicroads, Kew, 2012.

Interdependence between Structural Engineering and Construction Management
Edited by Ozevin, D., Ataei, H., Modares, M., Gurgun, A., Yazdani, S., and Singh, A.

ISBN: 978-0-9960437-6-2

CALCULATION OF THE NATURAL VIBRATIONAL FREQUENCY OF THE PULSE-LOADED EMBEDDED FOUNDATION

ALEKSEY KOLESNIKOV[1], VLADIMIR POPOV[2], and TATIANA KOSTIUK[1]

[1]*Novosibirsk State University of Architecture and Civil Engineering (SIBSTRIN), Novosibirsk, Russia*
[2]*Khristianovich Institute of Theoretical and Applied Mechanics SB RAS, Novosibirsk, Russia*

Applicability of formulas obtained within the framework of wave model and Russian design code SP 26.13330.2012 were considered to determine the dynamic stiffness in respect to vertical and horizontal vibrations of shallow and pile embedded foundations. The target is to calculate the system's varying natural vibrational frequencies regarding the deepening. Obtained results are compared with experimental data of a series of pulse dynamic loadings under half-full-scale conditions on foundation models. Values of transversal wave velocities are found directly on test section from test results. It is found that deepening results in natural vibrational frequencies growth, both in case of pile and shallow foundations. Full deepening of more than 1.6 times increases frequencies at horizontal vibrations, and up to 1.4 times at vertical vibrations; the effect is stronger for shallow foundations. Benefit of the results obtained by wave model against the SP 26.13330.2012 method is demonstrated, as well as the good agreement between calculation and experimental results, which permits finding reliable amplitude-frequency characteristics of the foundations. Results obtained in accordance with SP 26.13330.2012 show lower values of natural vibrational frequencies and do not completely regard the deepening effect. Maximal discrepancy with experimental data is 33% for vertical vibrations on shallow and pile foundations. In case of horizontal vibrations, the maximal discrepancy is from 20% for shallow foundation to 27% for shallow and pile foundation.

Keywords: Machine foundation, Wave model, Eigenfrequency of the foundation.

1 GENERAL APPEARANCE

Today, there are special requirements for the accuracy of the results of the foundation vibrations calculation (Zabylin 1983, Koloushek 1965). Thus, the models in use have to give a fuller picture about the dynamic characteristics of the "embedded foundation-base" system (Baranov 1967, Novak and El Sharnouby, 1983, Kolesnikov and Popov 2009, Kolesnikov and Popov 2010). The increased stiffness of the foundation, due to the interaction with the ambient medium of the embedded side edges, results in the increased frequencies of the natural vibrations, which is proven experimentally; the experiments were carried out in different conditions (Koloushek 1965, Kolesnikov and Popov 2010). Hence, the effect of the dense contact of the foundation side surface and ambient soil considered for the calculation and design of the machine foundations permits determining more accurately the values of the amplitude-frequency characteristics of the whole system vibrations (Baranov 1967).

At the same time, when calculating or investigating the characteristics of shallow and pile embedded foundations, there are some discrepancies with the experimental results (Kolesnikov and Popov 2010). It is perhaps necessary to take into account the foundation geometry and other individual characteristics in order to increase the calculation accuracy, but it requires detailed analysis (Nuzhdin and Kolesnikov 2004, Kolesnikov and Popov 2017). The present work addresses to the engineering calculations involving the formulas from Baranov (1967), Nuzhdin and Kolesnikov (2004) to determine the natural frequencies at the vertical and horizontal vibrations of the foundations, with due regard to their deep-grid position and comparison of the results with the data obtained experimentally in a test landfill.

2 INVESTIGATION TECHNIQUE

To evaluate the effect of the deepening on the natural frequencies at the vertical and horizontal vibrations of the foundations, the experimental investigations were carried out in accordance with the schematic in Figure 1.

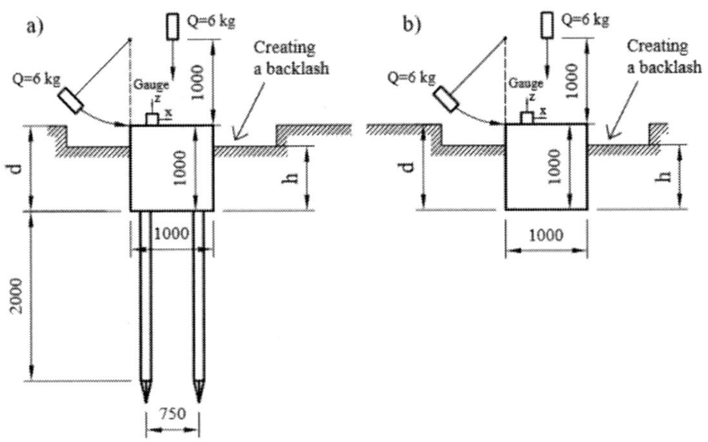

Figure 1. Schematic of tested foundations F-1 (a) and F-2 (b):
h – the height of backfill, *d* – the depth of foundation.

Two specially made foundations were used; one of them is the pile foundation (F-1), and the other is the shallow foundation (F-2). The foundation F-1 was tested as a monolithic reinforced concrete pile cap, its overall sizes are 1.0×1.0×1.0 m; it is based on four rigidly fixed piles with the diameter of 114 mm (the working length 2 m) and is made from metal tubes with the wall thickness of 6 mm. The distance between the pile axes was 6.5 diameters (750 mm), which excluded their mutual influence in the cluster (Kolesnikov and Popov 2017). The monolithic shallow foundation F-2 had the same sizes 1.0×1.0×1.0 m. Thoroughly leveled pit walls served as a concrete form for the test foundations. To the depth of 9.3 m, the soil of the test landfill was the loess slightly wet hard sand clay, its specific weight 17.0 kN/m^3, the deformation modulus E = 14 MPa; the bed layer is semi-solid clay loam. There was no subterranean water in the landfill. The series of dynamic impulse loadings (20 times) were applied for different versions of deepening, the technique was the same. The impulse loading was applied with the aid of a steel block shaped as a parallelepiped of 6 kg. To produce the vertical loading, the block fell freely onto the tested foundation surface from the height of 1.0 m. To produce the horizontal loading, a pendulum was used; it was hung of a string above the foundation side edge at a height of 1.0 m and bumped the upper foundation edge. At the complete foundation deepening, some soil near the impact point was removed to exclude interferences. To determine the vibration natural frequencies, the special automatic vibrometer was used. It permitted registering the vibrations and simultaneously processing the obtained information. The measurement data for each foundation F-1 and F-2 were compared with the engineering calculation results (the analytic relations from Baranov 1967, Nuzhdin and Kolesnikov 2004 and SP 26.13330.2012). The natural frequencies of the foundation vibrations, according to the wave model, λ_z at the vertical and λ_x at the horizontal vibrations of the foundations are defined in Eq. (1):

$$\lambda_z = \sqrt{K_z/m},\ \lambda_x = \sqrt{K_x/m}, \tag{1}$$

where K_z, K_x are the foundation vertical and horizontal stiffness, respectively, m is the foundation mass.

The approach to the determination of the embedded foundation dynamic stiffness based on wave models suggests summing-up the values found separately for the side surface, bottom and piles if there are any (Zabylin 1983, Kolesnikov and Popov 2009). The vertical (z) and horizontal (x) stiffness of the embedded test foundations are found as follows:

For the pile foundations: summing-up the stiffness on the side surface of the pile cap K_1 and stiffnesses K_2 of each pile from N in the base is calculated as in Eq (2):

$$K_z = K_{z1} + \sum_{i=1}^{N} K_{z2_i},\ K_x = K_{x1} + \sum_{i=1}^{N} K_{x2_i}, \tag{2}$$

For the shallow foundations: summing-up the stiffnesses on the side surface of the deep-grid foundation K_1 and under its bottom K_3 can be found in Eq. (3).

$$K_z = K_{z1} + K_{z3},\ K_x = K_{x1} + K_{x3}. \tag{3}$$

The vertical stiffness by the side surface of the pile cap or foundation is found in Eq. (4) as follows:

$$K_{z1} = V_{s0}^2 \rho_0 S_{w1} h, \tag{4}$$

where ρ_0 is the backfill soil density and V_{s0} is the speed of transversal waves in the foundation (pile cap) backfill, h is the backfill height, S_{w1} is the coefficient detected by the dimensionless vibration frequency $a_0 = r\omega/V_{s0}$ and geometrical shape in the embedded foundation (pile cap) plane, ω is the circular vibrational frequency, r is the characteristic size, $r = r_0$ for the circular in plane foundation with the radius of r_0, $r = \sqrt{ab/\pi}$ for the body rectangular in plane, with the sides a and b (Baranov 1967, Zabylin 1983, Nuzhdin and Kolesnikov 2004).

The vertical stiffness of a single pile is detected by the formula shown in Eq. (5).

$$K_{z2} = f_{19.1} E_p A_p / r_p, \tag{5}$$

where E_p is the pile material elasticity modulus, A_p is the pile cross section area, r_p is the circular pile radius, or $r_p = c/\sqrt{\pi}$ in the case of a rectangular pile with the side c, $f_{19.1}$ is the coefficient found in accordance with Novak and Sharnouby (1983).

The vertical stiffness under the foundation bottom is found using the formula in Eq. (6).

$$K_{z3} = V_s^2 \rho r F_{1z}, \tag{6}$$

where ρ is the soil density and V_s is the transversal wave speed in the soil under the foundation bottom, F_{1z} is the coefficient detected in accordance with Zabylin (1983).

Horizontal stiffness by the side surface of the pile cap or foundation is found by the formula in Eq. (7).

$$K_{x1} = V_{s0}^2 \rho_0 S_{u1} h, \tag{7}$$

where S_{u1} is the coefficient in accordance with (Baranov 1967, Zabylin 1983, Nuzhdin and Kolesnikov 2004) related to the dimensionless vibration frequency a_0 and geometrical shape of the embedded foundation (pile cap).

The horizontal stiffness of the single pile is found by the formula in Eq. (8).

$$K_{x2} = f_{11.1} E_p I_p / r_p^3, \tag{8}$$

where I_p is the pile cross section inertia moment, $f_{11.1}$ is the coefficient detected in accordance with Chowdhury and Dasgupta (2008).

The horizontal stiffness under the foundation bottom is found by the formula in Eq. (9).

$$K_{x3} = V_s^2 \rho r F_{1x}. \tag{9}$$

3 INVESTIGATION RESULTS

Tables 1 and 2, respectively, present the measurement and calculation results for the natural vibrational frequencies of the foundations F-1 and F-2. The natural vibrational frequencies were registered for each tested foundation along with the calculations by the formulas in Eq. (2), (3) for different levels of their relative deepening *h*/*d*=1; 0,75; 0,5; 0,25; 0.

Table 1. Natural frequencies of foundation F-1.

Item No	**Deepening, *h/d***	**Measured frequency, Hz**		**Calculated frequency, Hz**					
				SP 26.13330.2012		**Baranov (1967)**		**Nuzhdin and Kolesnikov (2004)**	
		Hor.	Vert.	Hor.	Vert.	Hor.	Vert.	Hor.	Vert.
1.	1.00	42.47	54.54	31.04	36.17	39.97	56.59	41.33	59.01
2.	0.75	41.43	52.25	30.09	35.46	36.96	54.33	39.57	56.23
3.	0.50	38.71	50.00	28.60	34.75	35.92	51.96	37.73	53.29
4.	0.25	33.85	47.61	26.48	33.27	34.85	49.49	35.79	50.19
5.	0.00	30.03	45.40	24.17	31.72	33.75	46.88	33.75	46.88

Table 2. Natural frequencies of the foundation F-2.

Item No	**Deepening, *h/d***	**Measured frequency, Hz**		**Calculated frequency, Hz**					
				SP 26.13330.2012		**Baranov (1967)**		**Nuzhdin and Kolesnikov (2004)**	
		Hor.	Vert.	Hor.	Vert.	Hor.	Vert.	Hor.	Vert.
1.	1.00	36.10	45.76	28.83	30.30	34.68	44.74	38.27	47.79
2.	0.75	32.25	42.30	27.95	29.70	32.08	41.83	35.01	44.29
3.	0.50	28.80	39.50	26.56	29.10	29.25	38.70	31.41	40.49
4.	0.25	25.45	35.25	24.60	27.85	26.11	35.30	27.34	36.29
5.	0.00	22.30	31.15	22.45	26.56	22.55	31.54	22.55	31.54

The calculations were carried out in accordance with Baranov (1967) for the foundation circular in the plane and Nuzhdin and Kolesnikov (2004) for the body rectangular in the plane and in accordance with SP 26.13330.2012. Their mutual influence was not regarded when determining the total pile stiffness in the base (Kolesnikov and Popov 2017).

It should be noted that the calculations involved the values of the transversal wave speeds V_s, detected directly in the test landfill by the method described in Chowdhury and Dasgupta (2008). The speed values for the test landfill soil are detected as V_{s1} = 146 m/s (the vertical loading) and V_{s2} = 105 m/s (the horizontal loading). Different values obtained for different impact direction prove the anisotropy of elastic wave speeds in the soil (Petrashen' 1980).

It follows from the results of the experiments carried out in the test landfill that the increased depth leads to the increased natural vibrational frequencies, both for the pile and shallow foundations. The dependence of the frequency on the depth is almost linear (see Figure 2).

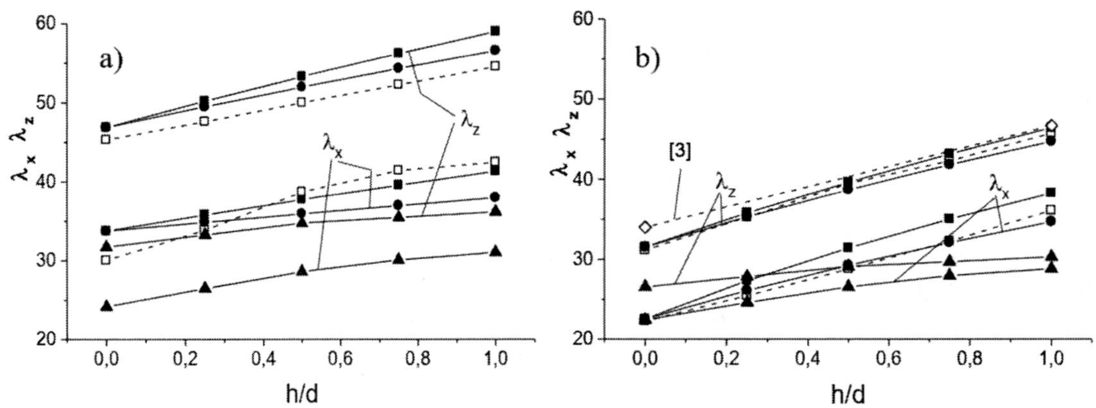

Figure 2. Correlation of the resonant frequencies λ_x, λ_z from the relative depth of foundation h/d: a. F-1, b. F-2. Dashed lines - measurements results of the authors (□) and Koloushek (1965)(◊), solid lines - the calculations results according to Baranov (1967) (•), Nuzhdin and Kolesnikov (2004) (■) and SP 26.13330.2012 (▲).

The complete deepening stimulates the 1.5 times growth of the frequency at the horizontal vibrations for the pile foundation, for the shallow foundation the growth exceeds 1.6 times, at the vertical vibrations for the pile foundation the growth is 1.3 times, and for the shallow foundation, it is 1.4 times. Note that the deepening influences more considerably in the case of the shallow foundations on the variation of the natural vibrational frequencies, than in the pile foundation case. Figure 2b presents the data from Koloushek (1965) as the vindication of the obtained results for the foundation of the same sizes and with similar soil properties. The results presented in Tables 1 and 2, obtained in accordance with SP 26.13330.2012, illustrate much lower values of the calculated natural vibrational frequencies, than the ones detected in the test measurements.

The Tables and Figure 2 show the results obtained in the engineering calculations involving the formulas (2), (3) for the detection of the stiffnesses of the embedded pile and shallow foundations. Then, calculating the natural vibrational frequencies of the foundations, we find the coefficients S_{w1} and S_{u1} in accordance with Baranov (1967) for the body circular in the plane, and in accordance with Nuzhdin and Kolesnikov (2004) for the body rectangular in the plane. The presented frequency values permit stating that the calculations in the framework of the used approximations enable having a good agreement with the experimental data.

Considering the results obtained with S_{w1} and S_{u1}, found by the formulas in accordance with Baranov (1967) for the body circular in the plane, the discrepancy with the experimental data at the horizontal vibrations ranges from 6 % to 12.5 % for the pile foundations, and up to 4 % for the shallow foundations. At the vertical vibrations, the discrepancy with the experimental data is lower: up to 4 % for the pile foundations, about 2 % for the shallow foundations.

Evaluating the results obtained with S_{w1} and S_{u1}, calculated in accordance with Nuzhdin and Kolesnikov (2004) for the body rectangular in the plane, we have that at the horizontal vibrations, the discrepancy with the experimental data ranges from 3 % to 12.5% for the pile foundations, and 6 % for the shallow foundations. Here, the maximal discrepancy is also registered at the depth of h/d = 0 during the calculation of the base pile stiffness. At the vertical vibrations, the discrepancy with the experimental data is below 8 % for the pile foundations and 4 % for the shallow foundations.

4 CONCLUSIONS

It is found that the relations and formulas (Baranov 1967, Nuzhdin and Kolesnikov 2004) obtained within the framework of the wave model and used to calculate the dynamic stiffnesses at the vertical and horizontal vibrations of the embedded shallow and pile foundations permit obtaining more accurate results when detecting the resonance frequencies than the way proposed by SP 26.13330.2012; it is proven by the comparison of the calculation results and experimental data obtained under the half-full-scale conditions. The vibrational frequencies found by the considered formulas show better approximation to the frequencies detected experimentally than the current standard methods. Comparison of the results of the detection of the natural vibrational frequencies did not show any evident advantage of utilization of the formulas for the bodies circular or rectangular in plane, which makes it desirable to carry out further investigation of the embedded foundation under dynamic loadings, as the difference between their sides is essential, as well the analysis of the amplitude characteristics of the system.

References

Baranov, V. A., On The Calculation of Forced Vibrations of the Deepening Foundation (In Russ.), *Voprosy Dinamiki I Prochnosti, Trudyi Rizhskogo Politekhnicheskogo Instituta*, (14), 195-209, 1967.

Chowdhury, I. and Dasgupta, S. P., *Dynamics of Structure and Foundation - A Unified Approach: 1. Fundamentals*, CRC Press, 2008.

Kolesnikov, A. O. and Popov, V. N., Evaluation of Deepening Influence of the Pile-Cap Under the Vibrations of the Pile Foundation (In Russ.), *Stroitel'naya Mekhanika Inzhenernyikh Konstruktsiy I Sooruzheniy*, (2), 55-61, 2009.

Kolesnikov, A. O. and Popov, V. N., Experimental Investigation of Pile-Cap Deepening at Vibrations of the Pile Foundation, *Stroitel'naya Mekhanika Inzhenernyikh Konstruktsiy I Sooruzheniy*, (4), 75-80, 2010.

Kolesnikov, A. O. and Popov, V. N., Dynamic Reactions on Contours of Circular Cuts Taking into Account Their Mutual Arrangement at Plate Vibrations (In Russ.), *Izvestiya VNIIG im. B.Ye. Vedeneeva*. 283, 3-10, 2017.

Koloushek, V., *Dynamics of Building Structures (In Russ.)*, Stroiizdat Publishing, Moscow, 1965.

Novak, M. and El Sharnouby, B., Stiffness and Damping Constants for Single Piles, *Geotechnical Engineering*, 109, 961-974, 1983.

Nuzhdin, L. V. and Kolesnikov, A. O., To an Estimation of Dynamic Reactions on the Contour of a Rectangular Cut in an Infinite Plate (In Russ.), *Izvestiya vuzov. Stroitel'stvo*, (5), 22-31, 2004.

Petrashen', G. I., *Propagation of Waves in Anisotropic Elastic Media (In Russ.)*, Nauka Publishing, Leningrad, 1980.

Zabylin, M. I., *Calculation of Machine Foundations (In Russ.)*, NISI Publishing, Novosibirsk, 1983.

AUTHOR INDEX

AUTHOR INDEX

AUTHOR INDEX

AUTHOR INDEX

AUTHOR INDEX

ISEC PRESS
PO Box 1334
Fargo, North Dakota 58107
USA

ISBN 978-1-5108-8867-8

CURRAN ASSOCIATES INC.
proceedings.com

4th International School and Conference on Optoelectronics, Photonics, Engineering and Nanostructures (Saint Petersburg OPEN 2017)

Journal of Physics: Conference Series Volume 917

Saint-Petersburg, Russia
3 - 6 April 2017

Part 1 of 2

IOP
Institute of Physics